W0018655

SPIN 2002

Previous Proceedings in the Series of Spin Physics Symposia

	Year	Held in	Publisher	ISBN
14th	2000	Osaka, Japan	AIP Conference Proceedings vol. 570	0-7354-0008-3

Other Related Titles from AIP Conference Proceedings

672 Short Distance Behavior of Fundamental Interactions: 31st Coral Gables Conference on High Energy Physics and Cosmology
Edited by Behram N. Kursunoglu, Metin Camcigil, Stephan L. Mintz, and Arnold Perlmutter, June 2003, 0-7354-0139-X

667 Increasing the AGS Polarization
Edited by A. D. Krisch, A. M. T. Lin, and T. Roser, May 2003, 0-7354-0130-6

646 Theoretical Physics: MRST 2002: A Tribute to George Leibbrandt
Edited by V. Elias, R. J. Epp, and R. C. Myers, December 2002, 0-7354-0101-2

601 Theoretical High Energy Physics: MRST 2001: A Tribute to Roger Migneron
Edited by V. Elias, D. G. C. McKeon, and V. A. Miransky, December 2001, 0-7354-0045-8

596 Art and Symmetry in Experimental Physics: Festschrift for Eugene D. Commins
Edited by Dmitry Budker, Philip H. Bucksbaum, and Stuart J. Freedman, November 2001, 0-7354-0040-7

588 Physics with an Electron Polarized Light-Ion Collider: Second Workshop, EPIC 2000
Edited by Richard G. Milner, October 2001, 0-7354-0028-8

549 Intersections of Particle and Nuclear Physics: 7th Conference, CIPANP2000
Edited by Zohreh Parsa and William J. Marciano, December 2000, 1-56396-978-5

To learn more about these titles, or the AIP Conference Proceedings Series, please visit the webpage **http://proceedings.aip.org/proceedings**

SPIN 2002

15th International Spin Physics Symposium

Upton, New York 9–14 September 2002

and

Workshop on Polarized Electron Sources and Polarimeters

Danvers, Massachusetts 4–6 September 2002

***SPONSORING ORGANIZATIONS* (SPIN)**
Brookhaven National Laboratory (BNL)
Brookhaven Science Associates (BSA)
Department of Energy (DOE)
Riken BNL Research Center (RBRC)
National Science Foundation (NSF)
The International Spin Physics Committee

***SPONSORING ORGANIZATIONS* (PESP)**
The International Spin Physics Committee
MIT-Bates Linear Accelerator Center

EDITORS
Yousef I. Makdisi
Alfredo U. Luccio
William W. MacKay
Brookhaven National Laboratory
Upton, New York

Melville, New York, 2003
AIP CONFERENCE PROCEEDINGS ■ VOLUME 675

Editors:

Yousef I. Makdisi
Alfredo U. Luccio
William W. MacKay

Collider-Accelerator Department
Brookhaven National Laboratory
Building 911B
Upton, NY 11973
USA

E-mail: Makdisi@bnl.gov
Luccio@bnl.gov
Mackay@bnl.gov

L.C. Catalog Card No. 2003107891
ISBN 0-7354-0136-5
ISSN 0094-243X
Printed in the United States of America

Contents

Preface...xxiii
Local Organizing Committee xxv
International Committee.......................................xxvi

PART ONE

15TH INTERNATIONAL SPIN PHYSICS SYMPOSIUM

PLENARY TALKS

A Beautiful Spin 3
X. Ji
New Results from the Muon *g-2* Experiment.........................13
E. P. Sichtermann for the Muon *g-2* Collaboration
Semi-inclusive Deep-Inelastic Scattering...........................23
C. A. Miller
Experimental Verification of the GDH Sum Rule: A Survey Including the Extension of the GDH Integral to Virtual Photons..................33
K. Helbing for the GDH-Collaboration
Comments about Vladimir L. Solovianov..............................46
A. D. Krisch
Vladimir Leonidovich Solovianov....................................47
N. Tyurin
Inclusive Asymmetries: A Retrospective View of V. L. Solovianov's Work..48
M. N. Ukhanov
Small Angle Scattering of Polarized Protons 58
B. Z. Kopeliovich
Spin and the Three-Nucleon Force....................................69
B. von Przewoski
Nucleon Electromagnetic Form Factors and Densities 78
J. J. Kelly
Nucleon Spin Structure Functions g_1 and g_2 from Polarized Inclusive Scattering 88
T. D. Averett
Spin Physics at RHIC 98
L. C. Bland
Exploring Nucleon Spin Structure in Longitudinally Polarized Collisions..112
M. Stratmann
Acceleration of Polarized Protons at RHIC 122
H. Huang
Results on DIS Exclusives and Generalized Parton Distributions.....132
M. Vanderhaeghen

Parity Violating Electron Scattering 142
K. S. Kumar
Acceleration of Polarized Protons and Deuterons at COSY 153
A. Lehrach, U. Bechstedt, J. Dietrich, R. Gebel, B. Lorentz, R. Maier, D. Prasuhn, A. Schnase, H. Schneider, R. Stassen, H. Stockhorst, and R. Tölle
Spin on the Lattice 166
K. Orginos for the RBC Collaboration
Single-Spin Asymmetries and Transversity 176
P. G. Ratcliffe
Polarized Jets and Storage Cell Targets for Storage Rings 186
E. Steffens
Parity Violation in $\vec{p}p$ and $\vec{n}p$ Experiments 196
W. D. Ramsay
A Future Linear Collider with Polarised Beams: Searches for New Physics 206
G. Moortgat-Pick
Laser-Polarized Noble Gases for Magnetic Resonance Imaging 217
G. D. Cates, Jr.
Looking into the Future of Spin and QCD 225
J. Soffer
Summary of DUBNA-SPIN01 Workshop 235
A. V. Efremov

PARALLEL SESSIONS

SPIN AND SYMMETRY

Determination of the Polarization of the Decay Positrons in Polarized Muon Decay 241
K.-U. Köhler, K. Bodek, A. Budzanowski, N. Danneberg, W. Fetscher, C. Hilbes, L. Jarczyk, K. Kirch, S. Kistryn, J. Klement, A. Kozela, J. Lang, G. Llosá Llácer, T. Schweizer, J. Smyrski, J. Sromicki, E. Stephan, A. Strzałkowski, and J. Zejma
Neutron Electric Dipole Moment 246
R. E. Mischke for the EDM Collaboration
Measurement of α_Ω in $\Omega^- \rightarrow \Lambda K^-$ Decays 251
L.-C. Lu, A. Chan, Y. C. Chen, C. Ho, P. K. Teng, W. S. Choong, Y. Fu, G. Gidal, P. Gu, T. Jones, K. B. Luk, B. Turko, P. Zyla, C. James, J. Volk, J. Felix, R. A. Burnstein, A. Chakravorty, D. M. Kaplan, L. M. Lederman, W. Luebke, D. Rajaram, H. A. Rubin, N. Solomey, Y. Torun, C. G. White, S. L. White, N. Leros, J.-P. Perroud, H. R. Gustafson, M. J. Longo, F. Lopez, H. K. Park, M. Jenkins, K. Clark, E. C. Dukes, C. Durandet, T. Holmstrom, M. Huang, L. C. Lu, and K. S. Nelson
A Clean Measurement of the Neutron Skin of ^{208}Pb through Parity Violating Electron Scattering 256
R. Suleiman for the Jefferson Lab Hall A Collaboration

Nuclear Anapole Moments and the Parity-Nonconserving Nuclear Interaction 262
C.-P. Liu
HAPPEX Parity Violation Experiments at Jefferson Lab 267
D. S. Armstrong and the HAPPEX and Hall A Collaborations
The G^0 Experiment at Jefferson Lab 272
L. Lee for the G^0 Collaboration

PROBES OF NUCLEON STRUCTURE AT HIGH MOMENTUM TRANSFER

Quark and Gluon Polarization in the Nucleon (I)

New Precision Results on the Spin Structure Function g_1^d 279
P. Lenisa for the HERMES Collaboration
A First Measurement of the Tensor-Polarized Structure Function b_1^d 284
M. Cantalbrigo for the HERMES Collaboration
The Spin Structure Function g_2 289
S. Rock for the Real Photon Collaboration
Recent Measurements of Longitudinal and Transverse Unpolarized Structure Functions, and Their Impact on Spin Asymmetry Measurements 294
C. E. Keppel
An Instability in the Matrix Solution of DGLAP Equations 299
M. Goshtasbpour and A. Shafi'i
Deeply Virtual Compton Scattering on Deuterium and Neon at HERMES 303
F. Ellinghaus for the HERMES Collaboration
Exclusive Electroproduction of Vector and Pseudoscalar Mesons at HERMES 308
C. Schill for the HERMES Collaboration

Quark and Gluon Polarization in the Nucleon (II)

Hard Exclusive Electroproduction of Two Pions off Proton and Deuteron at HERMES 313
P. di Nezza and R. Fabbri for the HERMES Collaboration
Prospects on Constraining ΔG from Inclusive Jet Production in Polarized pp Collisions at RHIC in 2003 318
B. Surrow for the STAR Collaboration
Measuring ΔG in PHENIX Using Electrons to Tag Heavy-Flavor Production 323
K. N. Barish for the PHENIX Collaboration
J/ψ Production in $\sqrt{s}$=200 GeV p+p Collisions with the PHENIX Detector at RHIC 328
H. D. Sato for the PHENIX Collaboration
Charmed Hadron Production in Polarized *pp* Collisions 333
T. Morii and K. Ohkuma

Polarized Hadroproduction of Open Heavy Quarks at NLO QCD at JHF and RHIC 338
I. Bojak

Quark and Gluon Polarization in the Nucleon (III)

Measurement of Polarised Parton Distributions at HERMES 343
A. Bruell for the HERMES Collaboration
Determining Spin-Flavor Dependent Distributions 348
G. P. Ramsey
NLO QCD Corrections to A_{LL}^{π} 353
B. Jäger, M. Stratmann, and W. Vogelsang
On Spin Content of the Proton 358
G. Musulmanbekov
Recent Status of Polarized Parton Distributions 365
M. Hirai for the Asymmetry Analysis Collaboration
***D* Meson Production in Neutrino DIS as a Probe of Polarized Strange Quark Distribution** 370
K. Sudoh
Future Measurements of the Spin Dependent Proton Flavor Structure with the PHENIX Muon Arms 375
N. Bruner for the PHENIX Collaboration
Resummation for Single-Spin Asymmetries in *W*-Boson Production 380
P. M. Nadolsky and C.-P. Yuan
The Quark-Antiquark Asymmetry of the Nucleon Strange Sea 385
M. Wakamatsu

Nuclear Structure Single-Spin Effects (I)

Single Muon Production in Transversely Polarized p+p Collisions at $\sqrt{s}$=200 GeV in the PHENIX Experiment 390
H. Kobayashi and A. Taketani for the PHENIX Collaboration
Single-Spin Transverse Asymmetry in Charged Hadron Production in $\sqrt{s}$=200 GeV p+p Collisions at PHENIX 395
K. Okada for the PHENIX Collaboration
Analyzing Powers for Forward $p^{\uparrow}+p\rightarrow\pi^0+X$ at STAR 400
G. Rakness for the STAR Collaboration
STAR Forward π^0 Detector Upgrade 407
A. Ogawa for the STAR Collaboration
Neutral Pion Measurements from PHENIX in Polarized Proton Collisions at RHIC 412
B. Fox for the PHENIX Collaboration
Midrapidity Spin Asymmetries at STAR 418
J. Balewski for the STAR Collaboration

Relative Luminosity Measurement in STAR and Implications for Spin Asymmetry Determinations 424
J. Kiryluk for the STAR Collaboration

Nuclear Structure Single-Spin Effects (II)

Azimuthal Asymmetries in Meson Electroproduction at HERMES 429
D. Hasch for the HERMES Collaboration
Single-Spin Asymmetries at CLAS 434
H. Avakian
Azimuthal Asymmetries: Access to Novel Structure Functions 439
K. A. Oganessyan, L. S. Asilyan, E. De Sanctis, and V. Muccifora
Perturbative and Nonperturbative Aspects of Azimuthal Asymmetries in Polarized *ep* 445
K. A. Oganessyan
Single Transverse-Spin Asymmetry in $pp^{\uparrow}\rightarrow\pi X$ and $ep^{\uparrow}\rightarrow\pi X$ 449
Y. Koike
Azimuthal Asymmetries in Fragmentation Processes at KEKB 454
K. Hasuko, M. Grosse Perdekamp, A. Ogawa, J. Soeren Lange, and V. Siegle
Collins Analyzing Power and Azimuthal Asymmetries 459
A. V. Efremov, K. Goeke, and P. Schweitzer
Transverse Spin Effects in Proton-Proton scattering and $Q\bar{Q}$ Production 464
S. V. Goloskokov

Nuclear Structure Single-Spin Effects (III)

Single Spin Asymmetries, Unpolarized Cross Sections, and the Role of Partonic Transverse Momentum 469
U. D'Alesio and F. Murgia
Sivers Effect and Transverse Single Spin Asymmetries in Drell-Yan Processes 474
M. Anselmino, U. D'Alesio, and F. Murgia
Handedness inside the Proton 479
D. Boer
Parton Distributions in Light-Cone Gauge: Where Are the Final-State Interactions? 484
F. Yuan and X. Ji
Transversity in Exclusive and Inclusive Processes 489
L. Gamberg, G. R. Goldstein, and K. A. Oganessyan
Role of T-Odd Functions in High-Energy Hadronic Collisions 494
E. Di Salvo

Nuclear Structure "Perspectives on Current Spin Experiments"

Prospects of the Gluon Polarization Measurement at PHENIX 499
Y. Goto for the PHENIX Collaboration

Spin Physics at STAR 504
A. Ogawa for the STAR Collaboration

Status of the COMPASS Experiment 509
F. Tessarotto for the COMPASS Collaboration

The Gluon Spin Structure Function from SLAC E161 516
S. Rock for the Real Photon Collaboration

SPIN AND HADRON DYNAMICS, LOW AND HIGH ENERGY

Spin and Hadron Dynamics (I)

Results from EDDA@COSY: Spin Observables in Proton-Proton Elastic Scattering 523
H. Rohdjeß for the EDDA-Collaboration

Elastic Polarized Proton Scattering at RHIC 528
S. Bültmann for the PP2PP Collaboration

Properties of the Spin-Flip Amplitude of Hadron Elastic Scattering and Possible Polarization Effects at RHIC 533
O. V. Selyugin

SPIN @U-70: An Experiment to Measure the Analyzing Power A_n in Very-High-$P_\perp^2$ *p-p* Elastic Scattering at 70 GeV 538
V. G. Luppov, L. V. Alexeeva, V. A. Anferov, E. D. Courant, Y. S. Derbenev, G. Fidecaro, M. Fidecaro, F. Z. Khiari, S. V. Koutin, A. D. Krisch, M. A. Leonova, A. M. T. Lin, W. Lorenzon, V. S. Morozov, D. C. Peaslee, C. C. Peters, R. S. Raymond, D. W. Sivers, J. A. Stewart, S. M. Varzar, V. K. Wong, K. Yonehara, D. G. Crabb, Y. M. Ado, A. G. Afonin, V. I. Belousov, B. V. Chujko, A. N. Davidenko, N. A. Galyaev, V. I. Garkusha, V. N. Grishin, V. A. Kachanov, Y. V. Kharlov, V. I. Kotov, A. V. Kusnetsov, V. A. Medvedev, Y. M. Melnik, V. V. Mochalov, A. I. Mysnik, S. B. Nurushev, A. F. Prudkoglyad, P. A. Semenov, V. L. Solovianov, V. P. Stepanov, V. A. Teplyakov, S. M. Troshin, A. G. Ufimtsev, M. N. Ukhanov, A. E. Yakutin, V. N. Zapolsky, V. G. Zarucheisky, N. S. Borisov, V. V. Fimushkin, V. A. Nikitin, P. V. Nomokonov, I. A. Rufanov, Y. K. Pilipenko, P. P. J. Delheij, W. T. H. van Oers, and A. N. Zelenski

Spin and Hadron Dynamics (II)

Longitudinal Spin-Transfer in Λ Production at HERMES 543
H. C. Chiang for the HERMES Collaboration

Transverse Polarisation of Λ and $\bar{\Lambda}$ Hyperons in Quasi-Real Photon Nucleon Scattering 548
A. Bruell for the HERMES Collaboration

The Polarized Deuteron Breakup Experiment at COSY 553
F. Rathmann, S. Barsov, S. Dymov, A. Kacharava, A. Khoukaz, V. Komarov, A. Kulikov, A. Kurbatov, N. Lang, I. Lehmann, B. Lorentz, G. Macharashvili, A. Mussgiller, H. Paetz gen. Schieck, R. Schleichert, H. Seyfarth, E. Steffens, H. Ströher, Y. Uzikov, S. Yaschenko, and B. Zalikhanov

New Results on Spin Rotation Parameter A in the πp-Elastic Scattering in the Resonance Region 558
I. G. Alekseev, P. E. Budkovksy, V. P. Kanavets, L. I. Koroleva, B. V. Morozov, V. M. Nesterov, V. V. Ryltsov, D. N. Svirida, A. D. Sulimov, V. V. Zhurkin, Y. A. Beloglazov, A. I. Kovalev, S. P. Kruglov, D. V. Novinsky, V. A. Shchedrov, V. V. Sumachev, V. Y. Trautman, N. A. Bazhanov, and E. I. Bunyatova

Dubna "Delta-Sigma" Experiment: Results of Treatment and Analysis of Statistics Accumulated in 2001 Data Taking Run on Energy Dependence of $\Delta\sigma_L(np)$ 563
V. I. Sharov, N. G. Anischenko, V. G. Antonenko, S. A. Averichev, L. S. Azhgirey, V. D. Bartenev, N. A. Bazhanov, A. A. Belyaev, N. A. Blinov, N. S. Borisov, S. B. Borzakov, Y. T. Borzunov, Y. P. Bushuev, L. P. Chernenko, E. V. Chernykh, V. F. Chumakov, S. A. Dolgii, A. N. Fedorov, V. V. Fimushkin, M. Finger, M. Finger Jr., L. B. Golovanov, G. M. Gurevich, A. Janata, A. D. Kirillov, V. G. Kolomiets, E. V. Komogorov, A. D. Kovalenko, A. I. Kovalev, V. A. Krasnov, P. Krstonoshich, E. S. Kuzmin, V. P. Ladygin, A. B. Lazarev, F. Lehar, A. de Lesquen, M. Y. Liburg, A. N. Livanov, A. A. Lukhanin, P. K. Maniakov, V. N. Matafonov, E. A. Matyushevsky, V. D. Moroz, A. A. Morozov, A. B. Neganov, G. P. Nikolaevsky, A. A. Nomofilov, T. Panteleev, Y. K. Pilipenko, I. L. Pisarev, Y. A. Plis, Y. P. Polunin, A. N. Prokoficv, V. Y. Prytkov, P. A. Rukoyatkin, V. A. Schedrov, O. N. Schevelev, S. N. Shilov, R. A. Shindin, Y. A. Shishov, V. B. Shutov, M. Slunečká, V. Slunečková, A. Y. Starikov, G. D. Stoletov, L. N. Strunov, A. L. Svetov, Y. A. Usov, T. Vasiliev, V. I. Volkov, E. I. Vorobiev, I. P. Yudin, I. V. Zaitsev, A. A. Zhdanov, and V. N. Zhmyrov

Spin and Hadron Dynamics (III)

Diffractive Heavy Pseudoscalar-Meson Productions by Weak Neutral Currents 569
A. Hayashigaki, K. Suzuki, and K. Tanaka

Hyperon Polarization from Unpolarized *pp* and *ep* Collisions 574
Y. Koike

The Challenge of Hyperon Polarization 579
S. Troshin and N. Tyurin

Measurement of Single Transverse-Spin Asymmetry in Forward Production of Photons and Neutrons in pp Collisions at $\sqrt{s}$=200 GeV 584
A. Bazilevsky, L. Bland, A. Bogdanov, G. Bunce, A. Deshpande, H. En'yo, B. Fox, Y. Fukao, Y. Goto, J. Haggerty, K. Imai, W. Lenz, D. von Lintig, M. Liu, Y. Makdisi, R. Muto, S. Nurushev, E. Pascuzzi, M. L. Purschke, N. Saito, F. Sakuma, S. Stoll, K. Tanida, M. Togawa, J. Tojo, Y. Watanabe, and C. Woody

SPIN PHYSICS WITH PHOTONS AND ELECTRONS

Nucleon Structure: GDH Experiments

HERMES Measurements of the Generalized GDH Integral and of Quark-Hadron Duality 591
W.-D. Nowak for the HERMES Collaboration

Inclusive and Exclusive Spin Structure Measurements in the Resonance Region 596
G. E. Dodge for the CLAS Collaboration

Measurement of Spin Structure Functions at Low to Moderate Q^2 Using CLAS 601
K. V. Dharmawardane for the CLAS Collaboration

Proton and Deuteron Spin Structure Function Measurements in the Resonance Region 606
F. R. Wesselmann

Precision Measurement of Neutron Asymmetry A_1^n in the Valence Quark Region 610
X. Zheng for the Jefferson Lab Hall A Collaboration

The Search for Higher Twist Effects in the Spin-Structure Functions of the Neutron 615
K. M. Kramer for the Jefferson Lab E97-103 Collaboration

Nucleon Electromagnetic Form Factors

Measurement of the Electric Form Factor of the Neutron at MAMI 620
M. Seimetz for the A1 Collaboration

The Electric Form Factor of the Neutron via Recoil Polarimetry to Q^2=1.47 (GeV/c)^2 625
B. Plaster, R. Madey, A. Y. Semenov, S. Taylor, A. Aghalaryan, E. Crouse, G. MacLachlan, S. Tajima, W. Tireman, C. Yan, A. Ahmidouch, B. D. Anderson, H. Arenhövel, R. Asaturyan, O. Baker, A. R. Baldwin, H. Breuer, R. Carlini, E. Christy, S. Churchwell, L. Cole, S. Danagoulian, D. Day, M. Elaasar, R. Ent, M. Farkhondeh, H. Fenker, J. M. Finn, L. Gan, K. Garrow, P. Gueye, C. Howell, B. Hu, M. K. Jones, J. J. Kelly, C. Keppel, M. Khandaker, W.-Y. Kim, S. Kowalski, A. Lung, D. Mack, D. M. Manley, P. Markowitz, J. Mitchell, H. Mkrtchyan, A. K. Opper, C. Perdrisat, V. Punjabi, B. Raue, T. Reichelt, J. Reinhold, J. Roche, Y. Sato, I. A. Semenova, W. Seo, N. Simicevic, G. Smith, S. Stepanyan,

V. Tadevosyan, L. Tang, P. Ulmer, W. Vulcan, J. W. Watson, S. Wells, F. Wesselmann, S. Wood, C. Yan, S. Yang, L. Yuan, W.-M. Zhang, H. Zhu, and X. Zhu

Measurement of the Charge Form Factor of the Neutron G_E^n from $\vec{d}(\vec{e}, e'n)p$ at Q^2=0.5 and 1.0 $(GeV/c)^2$ 630
N. Savvinov for the E93-026 Jlab Collaboration

Measuring G_E^n at High Momentum Transfers.......... 634
B. Reitz

Proton Elastic Form Factor Ratio: The JLab Polarization Experiments.......... 639
C. F. Perdrisat, V. Punjabi, and the Jefferson Lab Hall A and $G_{Ep}(III)$ Collaborations

Photon Experiments and Processes

Dispersion Relations in Virtual Compton Scattering.......... 646
B. Pasquini, D. Drechsel, M. Gorchtein, A. Metz, and M. Vanderhaeghen

First Photo-Pion Double Polarization Experiments Using Polarized $\vec{H}\vec{D}$ at LEGS 651
S. Hoblit, K. Ardashev, C. Bade, M. Blecher, C. Cacace, A. Caracappa, A. Cichocki, C. Commeaux, A. d'Angelo, R. Deininger, J.-P. Didelez, C. Gibson, K. Hicks, A. Honig, T. Kageya, M. Khandaker, O. Kistner, A. Lehmann, F. Lincoln, M. Lowry, M. Lucas, J. Mahon, H. Meyer, L. Miceli, D. Morizzianni, B. Norum, B. M. Preedom, T. Saitoh, A. M. Sandorfi, R. di Salvo, C. Schaerf, C. Thorn, K. Wang, X. Wei, and C. S. Whisnant

Spin Physics in Deep-Inelastic Semi-inclusive Reactions with an 11-GeV Electron Beam at Hall A of Jefferson Laboratory.......... 656
X. Jiang

Diagonal Spin Basis (DSB) as a Completely Symmetrized Description of Interacting Fermions.......... 661
S. M. Sikach

About the Processes with Two Fermions and Two Photons in an Intensive Laser Wave (Nonlinear and Spin Effects) 666
S. M. Sikach

SPIN PHYSICS WITH NUCLEI

Pionic 0^- State and Nuclear Structure

Model Independent Spin Parity Determination by the $(d, {}^2He)$ Reaction and Possible Evidence of 0^- state in ^{12}B.......... 671
H. Okamura, T. Uesaka, K. Suda, H. Kumasaka, R. Suzuki, A. Tamii, N. Sakamoto, and H. Sakai

High-Resolution Study of Pionic 0^- State in ^{16}O 676
T. Wakasa, G. P. A. Berg, H. Fujimura, K. Fujita, K. Hatanaka, M. Itoh, J. Kamiya, T. Kawabata, Y. Kitamura, E. Obayashi, H. Sakaguchi, N. Sakamoto, Y. Sakemi, Y. Shimizu, H. Takeda, M. Uchida, Y. Yasuda, H. P. Yoshida, and M. Yosoi

Polarization Transfer in the ^{16}O (p,p') Reaction at Forward Angles and Structure of the Spin-Dipole Resonances 681
T. Kawabata, H. Akimune, G. P. A. Berg, B. A. Brown, H. Fujimura, H. Fujita, Y. Fujita, M. Fujiwara, K. Hara, K. Hatanaka, K. Hosono, T. Ishikawa, M. Itoh, J. Kamiya, M. Nakamura, T. Noro, E. Obayashi, H. Sakaguchi, Y. Shimbara, H. Takeda, T. Taki, A. Tamii, H. Toyokawa, M. Uchida, H. Ueno, T. Wakasa, K. Yamasaki, Y. Yasuda, H. P. Yochida, and M. Yosoi

Spectroscopy of ^{120}Sn Homologous Levels via the ^{123}Sb $(\overrightarrow{p,\alpha})$ ^{120}Sn Reaction 686
P. Guazzoni, L. Zetta, A. Covello, A. Gargano, Y. Eisermann, G. Graw, R. Hertenberger, H.-F. Wirth, M. Jaskola, B. Bayman, and W. E. Ormand

Study of the Spin Dependent ^{3}He-Nucleus Interaction at 450 MeV 691
J. Kamiya, K. Hatanaka, Y. Sakemi, T. Wakasa, H. P. Yoshida, E. Obayashi, K. Hara, Y. Kitamura, Y. Shimizu, K. Fujita, N. Sakamoto, H. Sakaguchi, M. Yosoi, M. Uchida, Y. Yasuda, Y. Shimbara, T. Adachi, T. Noro, and T. Kawabata

Isoscalar Spin Response in the Continuum Studied via the ^{12}C $(\vec{d},\vec{d}')$ Reaction at 270 MeV 696
Y. Satou, S. Ishida, H. Kato, H. Sakai, H. Okamura, N. Sakamoto, T. Uesaka, A. Tamii, T. Wakasa, T. Ohnishi, K. Sekiguchi, K. Yako, K. Suda, M. Hatano, Y. Maeda, and T. Ichihara

Determination of the Gamow-Teller Quenching Factor via the $^{90}Zr(n,p)$ Reaction at 293 MeV 700
K. Yako, H. Sakai, M. B. Greenfield, K. Hatanaka, M. Hatano, J. Kamiya, Y. Kitamura, Y. Maeda, C. L. Morris, H. Okamura, J. Rapaport, T. Saito, Y. Sakemi, K. Sekiguchi, Y. Shimizu, K. Suda, A. Tamii, N. Uchigashima, and T. Wakasa

Few-Body Physics and Nuclear Properties

Experimental Studies on Three-Nucleon Systems at RCNP 705
K. Hatanaka, K. Sagara, Y. Shimizu, T. Yagita, Y. Sakemi, T. Wakasa, H. P. Yoshida, J. Kamiya, M. Yoshimura, H. Sakai, A. Tamii, K. Yako, Y. Maeda, T. Saito, T. Ishida, S. Minami, K. Tsuruta, T. Noro, K. Sekiguchi, H. Akiyoshi, and V. P. Ladygin

Polarization Transfer Measurement for *d-p* Elastic Scattering — A Probe for Three Nucleon Force Properties 711
K. Sekiguchi, H. Sakai, H. Okamura, A. Tamii, T. Uesaka, K. Suda, N. Sakamoto, T. Wakasa, Y. Satou, T. Ohnishi, K. Yakou, S. Sakoda, H. Kato, Y. Maeda, M. Hatano, J. Nishikawa, T. Saito, N. Uchigashima, N. Kalantar-Nayestanaki, and K. Ermisch

Study of the ^{3}He (^{3}H) Spin Structure via $\vec{d}d \to {}^3$He *n* (^{3}H*p*) Reaction 715
T. Saito, V. P. Ladygin, T. Uesaka, M. Hatano, A. Y. Isupov, H. Kato, H. Kumasaka, N. B. Ladygina, Y. Maeda, A. I. Malakhov, J. Nishikawa, T. Ohnishi, H. Okamura, S. G. Reznikov, H. Sakai, N. Sakamoto, S. Sakoda, K. Sekiguchi, K. Suda, R. Suzuki, A. Tamii, N. Uchigashima, and K. Yako

Extraction of Neutron Density Distributions from Proton Elastic Scattering at Intermediate Energies 720
H. Takeda, H. Sakaguchi, S. Terashima, T. Taki, M. Yosoi, M. Itoh, T. Kawabata, T. Ishikawa, M. Uchida, N. Tsukahara, Y. Yasuda, T. Noro, M. Yoshimura, H. Fujimura, H. P. Yoshida, E. Obayashi, A. Tamii, and H. Akimune

Tensor Analysing Powers for ^{7}Li Induced Transfer Breakup Reactions 725
N. J. Davis, R. P. Ward, K. Rusek, N. M. Clarke, G. Tungate, J. A. R. Griffith, S. J. Hall, O. Karban, I. Martel-Bravo, J. M. Nelson, J. Gómez-Camacho, T. Davinson, D. G. Ireland, K. Livingston, E. W. Macdonald, R. D. Page, P. J. Sellin, C. H. Shepherd-Themistocleous, A. C. Shotter, and P. J. Woods

Polarizations For ^{12}C(*p,2p*) Reactions at 1 GeV 730
H. P. Yoshida, T. Noro, O. V. Miklukho, V. A. Andreev, M. N. Andronenko, G. M. Amalsky, S. L. Belostotski, O. A. Domchenkov, O. Y. Fedorov, K. Hatanaka, A. A. Izotov, A. A. Jgoun, J. Kamiya, A. Y. Kisselev, M. A. Kopytin, E. Obayashi, A. N. Prokofiev, D. A. Prokofiev, H. Sakaguchi, V. V. Sulimov, A. V. Shvedchikov, H. Takeda, S. I. Trush, V. V. Vikhrov, T. Wakasa, Y. Yasuda, and A. A. Zhdanov

Momentum Transfer Dependence of Spin Isospin Modes in Quasielastic Region 734
T. Wakasa, H. Sakai, M. Ichimura, K. Hatanaka, M. B. Greenfield, H. Okamura, K. Kawahigashi, A. Tamii, H. Otsu, Y. Nakaoka, T. Ohnishi, K. Yako, K. Sekiguchi, T. Yagita, J. Kamiya, S. Sakoda, K. Suda, H. Kato, M. Hatano, and Y. Maeda

ACCELERATION AND STORAGE OF POLARIZED BEAMS

Acceleration and Storage of Polarized Beams (I)

Spin Tune in the Single Resonance Model with a Pair of Siberian Snakes 741
D. P. Barber, R. Jaganathan, and M. Vogt

The Analysis of Depolarization Factors in the Last RHIC Run 746
V. Ptitsyn, A. U. Luccio, and V. H. Ranjbar

Spin Coupling Resonances and Suppression in the AGS 751
V. H. Ranjbar, S. Y. Lee, L. Ahrens, M. Bai, K. Brown, W. Glenn, H. Huang, A. U. Luccio, W. W. MacKay, V. Ptitsyn, T. Roser, and N. Tsoupas

Acceleration and Storage of Polarized Beams (II)

Exact Solutions for the n-axis and Spin Tune in Model Storage Rings 756
S. R. Mane
Spin-Orbital Function Formalism and ASPIRRIN Code 761
E. A. Perevedentsev, V. Ptitsyn, and Y. M. Shatunov
Spin Flipping and Polarization Lifetimes of a 270 MeV Deuteron Beam 766
V. S. Morozov, M. Q. Crawford, Z. B. Etienne, M. C. Kandes, A. D. Krisch, M. A. Leonova, D. W. Sivers, V. K. Wong, K. Yonehara, V. A. Anferov, H. O. Meyer, P. Schwandt, E. J. Stephenson, and B. von Przewoski
RHIC Spin Flipper Commissioning 771
M. Bai, A. U. Luccio, W. W. MacKay, V. Ranjbar, and T. Roser
99.9 % Spin-Flip Efficiency in the Presence of a Strong Siberian Snake 776
V. S. Morozov, B. B. Blinov, Z. B. Etienne, A. D. Krisch, M. A. Leonova, A. M. T. Lin, W. Lorenzon, C. C. Peters, D. W. Sivers, V. K. Wong, K. Yonehara, V. A. Anferov, P. Schwandt, E. J. Stephenson, B. von Przewoski, and H. Sato

Acceleration and Storage of Polarized Beams (III)

The Relativistic Stern-Gerlach Interaction as a Tool for Attaining the Spin Separation... 781
P. Cameron, M. Conte, A. U. Luccio, W. W. MacKay, M. Palazzi, and M. Pusterla
The Relativistic Stern-Gerlach Interaction and Quantum Mechanics Implications 786
P. Cameron, M. Conte, A. U. Luccio, W. W. MacKay, M. Palazzi, and M. Pusterla
Macroscopic Quantum Processors Based on Stored High-Energy Polarized Beams 791
D. Sivers
Preserving Polarization through an Intrinsic Depolarizing Resonance with a Partial Snake at the AGS 794
H. Huang, L. Ahrens, M. Bai, K. A. Brown, J. W. Glenn, A. U. Luccio, W. W. MacKay, C. Montag, V. Ptitsyn, V. Ranjbar, T. Roser, H. Spinka, N. Tsoupas, D. G. Underwood, and K. Zeno
Matching Quadrupoles for AGS Helical Snake 799
E. D. Courant

POLARIMETRY PROTON AND ELECTRON BEAMS

Polarimetry Proton and Electron Beams (I)

Proton Beam Polarimetry at BNL 807
H. Spinka

RHIC pC CNI Polarimeter: Experimental Setup and Physics Results 812
I. G. Alekseev, A. Bravar, G. Bunce, R. Cadman, A. Deshpande, S. Dhawan, D. E. Fields, H. Huang, V. Hughes, G. Igo, K. Imai, O. Jinnouchi, V. P. Kanavets, J. Kiryluk, K. Kurita, Z. Li, W. Lozowski, W. W. MacKay, Y. Makdisi, S. Rescia, T. Roser, N. Saito, H. Spinka, B. Surrow, D. N. Svirida, J. Tojo, D. Underwood, and J. Wood

RHIC pC CNI Polarimeter: Status and Performance from the First Collider Run 817
O. Jinnouchi, I. G. Alekseev, L. C. Bland, A. Bravar, G. Bunce, R. Cadman, A. Deshpande, S. Dhawan, D. E. Fields, H. Huang, V. Hughes, G. Igo, K. Imai, V. P. Kanavets, J. Kiryluk, K. Kurita, Z. Li, W. Lozowski, W. W. MacKay, Y. Makdisi, A. Ogawa, S. Rescia, T. Roser, N. Saito, H. Spinka, B. Surrow, D. N. Svirida, J. Tojo, D. Underwood, and J. Wood

Measurement of Analyzing Powers for Polarized Proton Scattering on CH_2 Target at Proton Momentum Range from 1.75 to 5.3 GeV/c 826
L. S. Azhgirey, V. A. Arefiev, I. Atanasov, S. N. Basilev, Y. P. Bushuev, V. V. Glagolev, M. K. Jones, D. A. Kirillov, P. P. Korovin, G. F. Kumbartzky, P. K. Manyakov, J. Mušinský, L. Penchev, C. F. Perdrisat, N. M. Piskunov, V. Punjabi, I. M. Sitink, V. M. Slepnev, I. V. Slepnev, and E. Tomasi-Gustafsson

Polarimetry Proton and Electron Beams (II)

The Absolute Polarimeter for RHIC 830
A. Bravar

Absolute Calibration of the RHIC CNI Polarimeters Using 125 GeV/A C Ions 836
G. Igo and I. Tanihata

Spin Asymmetry for Proton Deuteron Collisions at Forward Angles 841
N. H. Buttimore

High-Precision Electron Beam Polarization Measurement with Compton Polarimetry at Jefferson Laboratory 846
F. Marie, E. Burtin, C. Cavata, S. Escoffier, D. Lhuillier, D. Neyret, T. Pussieux, and P. Bertin

POLARIZED ION SOURCES AND TARGETS

Polarized Sources and Targets

Summary of the PST 2001 Workshop 855
V. P. Derenchuk

Workshop on Testing QCD through Spin Observables in Nulear Targets 856
D. G. Crabb

Status of Frozen-Spin Polarized HD Targets for Spin Experiments 857
T. Kageya, C. M. Bade, A. Caracappa, F. C. Lincoln, M. M. Lowry, J. C. Mahon, L. Miceli, A. M. Sandorfi, C. E. Thorn, X. Wei, and C. S. Whisnant

Brute Force with a Gentle Touch: Vibration Isolation Techniques Used to Increase HD Target Polarization 862
C. M. Bade, A. Caracappa, T. Kageya, F. C. Lincoln, M. M. Lowry, J. C. Mahon, L. Miceli, A. M. Sandorfi, C. E. Thorn, X. Wei, and C. S. Whisnant

Target Polarization Measurements with a Crossed-Coil NMR Polarimeter 867
A. Caracappa and C. Thorn

Polarized Atomic Hydrogen Beam Tests in the Michigan Ultra-cold Jet Target 872
K. Yonehara, Z. B. Etienne, M. C. Kandes, K. J. Klein, A. D. Krisch, M. A. Leonova, V. G. Luppov, V. S. Morozov, C. C. Peters, R. S. Raymond, D. E. Saam, D. L. Sisco, D. R. Southworth, N. S. Borisov, V. V. Fimushkin, and A. F. Prudkoglyad

Polarized Ion Sources

Summary Report of PESP-2002 Workshop 876
M. Farkhondeh

Polarization Optimization Studies in the RHIC Optically Pumped Polarized H^- Ion Source 881
A. Zelenski, J. Alessi, B. Briscoe, A. Kponou, S. Kokhanovski, V. Klenov, V. LoDestro, D. Raparia, J. Ritter, and V. Zubets

Polarized D^- Operation and Development of the IUCF Ion Source CIPIOS 887
V. P. Derenchuk and A. S. Belov

D^- Charge Exchange Ionizer for the JINR Polarized Ion Source POLARIS 892
V. P. Ershov, V. V. Fimushkin, G. I. Gai, L. V. Kutuzova, Y. K. Pilipenko, V. P. Vadeev, A. I. Valevich, and A. S. Belov

A Precision Lamb-Shift Polarimeter for the Polarized Gas Target at ANKE/COSY 897
R. Engels, R. Emmerich, J. Ley, M. Mikirtytchiants, H. Paetz gen. Schieck, F. Rathmann, H. Seyfarth, and A. Vassiliev

Polarized Solid Targets

Solid State Polarized Targets for the Study of Nucleon Spin Structure 902
G. R. Court

Thin Scintillating Polarized Targets for Spin Physics 907
B. van den Brandt, E. I. Bunyatova, P. Hautle, and J. A. Konter
Polarized Solid Proton Target for RI Beam Experiments 911
T. Wakui, M. Hatano, H. Sakai, A. Tamii, and T. Uesaka
The Status of the University of Michigan Polarized Proton Target 916
R. S. Raymond, D. G. Crabb, V. V. Fimushkin, A. D. Krisch, A. M. T. Lin, V. G. Luppov, C. C. Peters, A. I. Mysnik, A. F. Prukoglyad, and K. Yonehara
Radiation Damage Effects in Polarized Deuterated Ammonia 919
P. M. McKee

Polarized ABS Internal Targets (I)

The Polarized Internal Gas Target of ANKE at COSY 924
F. Rathmann, R. Brüggemann, R. Engels, S. Geisler, A. Gussen, P. Jansen, H. Kleines, F. Klehr, P. Kravtsov, S. Lemaître, B. Lorentz, S. Lorenz, M. Mikirtytchiants, M. Nekipelov, V. Nelyubin, H. Paetz gen. Schieck, U. Rindfleisch, J. Sarkadi, H. Seyfarth, E. Steffens, H. Ströher, V. Trofimov, A. Vassiliev, and K. Zwoll
The HERMES Polarized Atomic Beam Source 929
A. Nass for the HERMES Target Group
Design of a Polarized Atomic H Source for a Jet Target at RHIC 934
T. Wise, M. A. Chapman, W. Haeberli, H. Kolster, and P. A. Quin

Polarized ABS Internal Targets (II)

Nuclear Polarization of Molecular Hydrogen Recombined on Drifilm 939
P. Lenisa and U. Stösslein
Beam Induced Depolarizing Resonances in the HERMES Hydrogen/ Deuterium Target 944
D. Reggiani for the HERMES Collaboration
Polarized Internal Target Experiments (PINTEX) at the Indiana Cooler 949
B. v.Przewoski
Polarized H^- Jet Polarimeter for Absolute Proton Polarization Measurements in RHIC 954
A. Zelenski, J. Alessi, A. Bravar, G. Bunce, M. A. Chapman, D. Graham, W. Haeberli, H. Hseuh, V. Klenov, H. Kolster, S. Kokhanovski, A. Kponou, V. Lodestro, W. MacKay, G. Mahler, Y. Makdisi, W. Meng, J. Ritter, T. Roser, E. Stephenson, T. Wise, and V. Zoubets

FUTURE FACILITIES AND EXPERIMENTS

Future Experiments

The Physics Programme of HERMES Run-II 963
D. Hasch for the HERMES Collaboration
Future of COMPASS Experiment 968
D. Neyret for the COMPASS Collaboration
Future Spin Experiments at SLAC 973
S. Rock for the Real Photon Collaboration
Research Perspectives at Jefferson Lab: 12 GeV and Beyond 978
K. de Jager
Perspectives for a Next-Generation Electron-Nucleon Scattering Facility in Europe 983
R. Kaiser

Future Collider Facilities and Physics

The Electron-Ion Collider: Status and Plans 988
R. G. Milner
Status of the e-Ring Design for EIC 993
D. E. Berkaev, A. V. Otboev, Y. M. Shatunov, R. Milner, C. Tschalaer, F. Wang, B. Parker, V. Ptitsyn, and D. P. Barber

PART TWO

WORKSHOP ON POLARIZED ELECTRON SOURCES AND POLARIMETERS

Spin Filters as High-Performance Spin Polarimeters 1001
N. Rougemaille, G. Lampel, J. Peretti, H.-J. Drouhin, Y. Lassailly, A. Filipe, T. Wirth, and A. Schuhl
Strained Gaasp Photocathode with GaAs Quantum Well 1006
Y. Yashin, Y. Mamaev, A. Rochansky, and D. Vinokurov
Transmission Polarimetry at MIT Bates 1011
T. Zwart, E. C. Booth, M. Farkhondeh, W. A. Franklin, E. Ihloff, J. L. Matthews, E. Tsentalovich, and W. Turchinetz
A Novel Imaging Spectrometer for Energy-Distribution Measurements of Photoelectrons from GaAs Cathodes 1016
C. D. Schröter, A. Rudenko, A. Dorn, R. Moshammer, and J. Ullrich
High-Power Diode Laser System for SHR 1019
D. Cheever, M. Farkhondeh, W. Franklin, E. Tsentalovich, and T. Zwart
Helicity-Correlated Effects for SAMPLE Experiment 1024
M. Farkhondeh, W. Franklin, E. Tsentalovich, and T. Zwart
Recent Polarized Photocathode R&D at SLAC 1029
D.-A. Luh, A. Brachmann, J. E. Clendenin, T. Desikan, E. L. Garwin, S. Harvey, R. E. Kirby, T. Maruyama, C. Y. Prescott, and R. Prepost

A Zero-Degree Inline Optical Electron Polarimeter **1034**
A. S. Green, M. A. Rosenberry, and T. J. Gay
Polarized Electrons Using the PWT RF Gun **1037**
J. E. Clendenin, R. Kirby, Y. Luo, D. Newsham, and D. Yu
The SLAC Polarized Electron Source **1042**
J. E. Clendenin, A. Brachmann, T. Galetto, D.-A. Luh, T. Maruyama, J. Sodja, and J. L. Turner
Status of the Jefferson Lab Polarized Beam Physics Program and Preparations for Upcoming Parity Experiments **1047**
J. Grames, P. Adderley, M. Baylac, J. Clark, A. Day, J. Hansknecht, M. Poelker, and M. Stutzman
The Polarized Electron Source at ELSA **1053**
W. von Drachenfels, F. Frommberger, M. Gowin, W. Hillert, M. Hoffmann, and B. Neff
The MIT-Bates Compton Polarimeter **1058**
W. A. Franklin, T. Akdogan, D. Dutta, M. Farkhondeh, M. Hurwitz, J. L. Matthews, E. Tsentalovich, W. Turchinetz, T. Zwart, and E. Booth
200 keV Polarized Electron Source at Nagoya University **1063**
K. Wada, M. Yamamoto, T. Nakanishi, S. Okumi, T. Gotoh, C. Suzuki, F. Furuta, T. Nishitani, M. Miyamoto, M. Kuwahara, T. Hirose, R. Mizuno, N. Yamamoto, H. Matsumoto, and M. Yoshioka
Basic R&D Studies for Lower Emittance Polarized Electron Guns **1068**
C. Suzuki, T. Nakanishi, S. Okumi, F. Furuta, K. Wada, T. Nishitani, M. Yamamoto, T. Hirose, M. Kuwahara, R. Mizuno, N. Yamamoto, H. Matsumoto, M. Yoshioka, H. Horinaka, K. Wada, T. Matsuyama, and H. Kobayakawa
Effect of Atomic Hydrogen Exposure on Electron Beam Polarization from Strained GaAs Photocathodes **1073**
M. Baylac, P. Adderley, J. Clark, T. Day, J. Grames, J. Hanskneckt, M. Poelker, P. Rutt, C. Sinclair, and M. Stutzman
Status of Jefferson Lab's Load Locked Polarized Electron Gun **1078**
M. L. Stutzman, P. Adderley, M. Baylac, J. Clark, A. Day, J. Grames, J. Hansknecht, and M. Poelker
Suppression of the Surface Charge Limit in Strained GaAs Photocathodes **1083**
T. Maruyama, A. Brachmann, J. E. Clendenin, T. Desikan, E. L. Garwin, R. E. Kirby, D.-A. Luh, C. Y. Prescott, J. Turner, and R. Prepost
Status of the Polarized Source at MAMI **1088**
K. Aulenbacher, V. Tioukine, M. Wiessner, and K. Winkler
Ultra-stable Flashlamp-Pumped Laser **1093**
A. Brachmann, J. Clendenin, T. Galetto, T. Maruyama, J. Sodja, J. Turner, and M. Woods
MIT-Bates Polarized Source **1098**
M. Farkhondeh, W. Franklin, E. Tsentalovich, T. Zwart, and E. Ihloff

APPENDIX A

PHENIX Collaboration 1103
STAR Collaboration 1107
Muon *g*-2 Collaboration 1110

APPENDIX B

SPIN 2002 Symposium Agenda 1111
SPIN 2002 List of Participants 1115
Photographs 1135

APPENDIX C

PESP 2002 Workshop Agenda 1143
PESP 2002 List of Participants 1147
PESP 2002 Scientific Program 1151

Author Index 1153

Preface

It was our pleasure to host the 15th *International Spin Physics Symposium.* Brookhaven National Laboratory did the same twenty years ago when we were in the planning stages of polarized proton acceleration in the AGS, which would then constitute the highest energy reached in polarized proton acceleration. In the two decades since, BNL is sporting yet another advance in the domain of spin physics. The Relativistic Heavy Ion Collider (RHIC), equipped with the appropriate hardware of Siberian Snakes and polarimeters, thanks to the generous contribution from RIKEN, Japan, has demonstrated its capability to accelerate and collide polarized protons to unprecedented energies of 200 GeV in the center of mass system. The full capability of 500 GeV is slated for the near future. This promises to open new venues for investigation of spin physics in hadronic interactions.

This is the 2nd Symposium in the new format that covers high energy and nuclear physics phenomena. The first was SPIN 2000 in Osaka Japan. The program certainly attests to this strong overlap between spin physics research in both fields.

With the help and acquiescence of the International Committee, the Local Organizing Committee admits to some bias in choosing young speakers, where possible, for the plenary presentations. We were quite lucky in that our choices proved up to the task. This points to a promising and vibrant field of research.

The symposium was preceded with a one-day tutorial/workshop on spin physics issues relating to theoretical, experimental, and acceleration techniques. The intent was to introduce new entrants into the field as well as those who needed a refresher to the terminology and concepts in preparation for the symposium. We are happy to note that a crowd of about 60 people attended, and most stayed for the duration. We extend our thanks to Werner Vogelsang who organized the day's activities. We hope this becomes a feature of future spin physics symposia.

This symposium will break new grounds in that the proceedings will be published in both the traditional book form as well as a searchable compact disk format possibly paving the way to electronic publishing.

Unlike recent symposia, we chose not to have poster sessions. Instead, all contributions were assigned into parallel sessions. While this complicated our task, it provided a better chance for all to be seen and heard. In addition, the parallel sessions' organizers were encouraged to include longer overviews and introductory talks. As such these sessions were run in the spirit of mini workshops to encourage better dialog. The proceedings are arranged to include the plenary talks followed by the parallel sessions presentations in the order they were delivered in the symposium. One plenary session was dedicated to our colleague Vladimir L. Solovianov who passed away this past year.

As this volume was being prepared for publication, we learnt of the untimely death of Vernon Hughes. Professor Hughes was a pioneer, a leader and a visionary in the field of spin physics research. He will be dearly missed.

Unfortunately, and due to the prevailing political and security climate, some of our colleagues could not make it to the Symposium due to visa problems. The Local Organizing Committee Chair requested and was granted approval from the International Committee to extend an invitation to include in the proceedings those contributions that were not presented. We are glad to see that this indeed was the case.

In keeping with the tradition, the proceedings also include those from the Polarized Electron Sources and Polarimeters Workshop, PESP 2002, which was held at MIT-Bates September 4-6, 2002 with M. Farkhondeh as the Chair.

We would like to extend our thanks and gratitude to the agencies whose support was instrumental to the success of this symposium. This includes the Department of Energy, The National Science Foundation, the RIKEN BNL Research Center, Brookhaven Science Associates, and the International Spin Committee. We are indebted to Brookhaven National Laboratory for providing the facilities and infrastructure and to the staff of Berkner Hall and Flik Dining Services for their tremendous help and support, in particular Ruth Comas and Pat Carollo. Appreciation goes to Pat Yalden's artistic talent for the creative design of the Symposium logo.

Our thanks go to the Collider-Accelerator Department, the home of the Chair of the International Committee and the Chair of the Local Organizing Committee for providing the logistical support during the many months spent in preparation for the symposium. We extend our special thanks to Mimi Luccio, Elaine Lowenstein, and Nada Makdisi who looked after the Companion's program, Anna Petway who designed the symposium pocket brochure and coordinated the excursion to the Museum of Natural History and the Rose Center; Penny Lo Presti and Mei Bai for putting together the Book of Abstracts, Melanie Covitz and Jesse Becker who expertly handled all financial matters. The expert support from John Gould, Frank Donato, Frank Nasse and Nick Franco who took care of the web design, registration and computing support is worthy of note.

Finally, the symposium Chair wants to express gratitude and appreciation to Mary Campbell assisted by Shannon Burke without whom this symposium would not have been possible. They put in a heroic effort and spent untold hours leading to and during the symposium.

Alfredo Luccio
Waldo MacKay
Yousef Makdisi

Local Organizing Committee

Yousef Makdisi (Chair)	BNL
M. Bai	BNL
L. Bland	BNL
A. Bravar	BNL
A. Bruell	MIT-Bates
G. M. Bunce	BNL
K. de Jager	JLAB
A. L. Deshpande	RBRC
B. Fox	RBRC
M. Grosse Perdekamp	RBRC
W. Guryn	BNL
T. J. Hallman	BNL
H. Huang	BNL
A. Luccio	BNL
W. Mackay	BNL
N. Saito	RIKEN/ RBRC
M. J. Tannenbaum	BNL
L. Trueman	BNL
W. Vogelsang	RBRC/BNL

International Committee

T. Roser (Chair)	BNL
A.D. Krisch (Past-Chair)	Michigan
D.P. Barber	DESY
J.M. Cameron	IUCF
O. Chamberlain*	Berkeley
E.D. Courant*	BNL
D.G. Crabb	Virginia
A.V. Efremov	JINR
H. Ejiri	RCNP
G. Fidecaro*	CERN
W. Haeberli*	Wisconsin
W. Happer	Princeton
V.W. Hughes*	Yale
K. Imai	Kyoto
R.G. Milner	MIT
C.Y. Prescott (Past-Chair)	SLAC
Y.M. Shatunov	Novosibirsk
L.D. Soloviev	IHEP
Volker Soergel	Heidelberg
E. Steffens	Erlangen
W.T.H. van Oers	Manitoba

* Honorary Member

PART ONE

15th International Spin Physics Symposium

A Beautiful Spin

Xiangdong Ji

Department of Physics, University of Maryland, Colleage Park, MD 20742

Abstract. Spin is a beautiful concept that plays an ever important role in modem physics. In this talk, I start with a discussion of the origin of spin, and then turn to three themes in which spin has been crucial in subatomic physics: a lab to explore physics beyond the standard model, a tool to measure physical observables that are hard to obtain otherwise, a probe to unravel nonperturbative QCD. I conclude with some remarks on a world without spin.

THE ORIGIN OF SPIN

I would like to begin my talk by thanking the organizers for giving me this wonderful opportunity to talk about spin as a beautiful concept in modern physics. RHIC spin collaborations have started taking data and the first results have come in. It is a great occasion to celebrate the fruit of many years of planning and hard work.

What is the origin of spin? As an English word, it existed before the 12th century. As a fundamental observable of subatomic particles, it was introduced by G. Uhlenbeck and S. Goudsmit in 1925 (and also Kronig). Many know the story that as graduate students, the two were not really sure that they did the right thing to introduce the spin degrees of freedom. They, in fact, at one point tried to withdraw their paper, but failed. Upon receiving the paper, their advisor, P. Ehrenfrest, made the following remark, "This is a good idea. Your idea may be wrong, but since both of you are so young without any reputation, you would not lose anything by making a stupid mistake." [1] Because of this potential "stupid mistake", one of the most beautiful concepts in modern physics was born.

It is interesting to note that one of our heros, S. Goudsmit, spent more than two decades here at Brookhaven during his career. Goudsmit was born in the Hague in Netherlands, and was educated at the universities of Amsterdam and Leidan, where he obtained his Ph. D. in 1927. He emigrated to America shortly afterwards, serving as a professor of physics at the University of Michigan and North Western. Since 1948, he moved BNL where he remained until his retirement in 1970.

Speaking of BNL, I would like to remind you some of the major events that happened here which are related to spin physics. In 1957, parity-violation in K-decay ($\theta-\tau$ puzzle) was discovered at the Cosmotron. Spin turned out to be a key element in many parity-violation experiments. In the 1980s, a series of experiments were done at AGS, measuring single-spin asymmetries and hyperon polarizations. More recently, the muon $g-2$ experiment has achieved a record precision of better than one part per million. The RHIC spin is a whole new and exciting program on QCD spin physics. Finally, the electron-ion collider, a planned machine with spin as its key feature, may find its home

CP675, *Spin 2002: 15th Int'l. Spin Physics Symposium and Workshop on Polarized Electron Sources and Polarimeters,* edited by Y. I. Makdisi, A. U. Luccio, and W. W. MacKay

here at BNL.

Back to the question of the orgin of spin. At present, we believe it is due to the spacetime symmetry. The $3+1$ dimensional Minkowski spacetime has symmetries under continuous spacetime translations and rotations. These transformations form the well-known Poincaré group. Mathematically, it is a contracted SO(3,2) group generated by ten generators. Four of them, P_μ, is the energy-momentum vector, and the other six, $J^{\mu\nu}$, are the generators of Lorentz transformations. The group has two Casimir operators, allowing defining two universal physical observables for any physical system. The first one, $P^\mu P_\mu$, introduces the concept of mass. The second one, $W^\mu W_\mu$, originated from the Pauli-Lubanski spin vector $W^\mu \sim \varepsilon^{\mu\nu\alpha\beta} J_{\alpha\beta} P_\gamma$, defines the concept of *spin*.

The above understanding, however, is not the final story. More revolutions in the concept of spacetime await us. According to string or M-theory, the 3+1 spacetime is a result of classical and low-energy approximations. At higher-energy, spacetime is 10 or 11 dimensional. Then an outstanding question is how do we end up seeing only 3+1? Is it due to compactification, or we live just in a 3+1 brane? It has been known for a long time that the strong gravitational fluctuations in the spacetime curvature exist at short distance scales. The meaning of spacetime as we know it ceases to exist there.

The main thesis of my talk is that spin is a beautiful concept that plays a central role in modern physics. In the following, I will elaborate on this in terms of three major themes: Theme 1) spin as a laboratory to explore physics beyond the standard model; Theme 2) spin as a tool to measure physical observables that are hard to obtain otherwise; and Theme 3) spin as a probe to unravel nonperturbative QCD dynamics.

THEME I: SPIN AS A LAB TO EXPLORE PHYSICS BEYOND STANDARD MODEL

The main research frontier in particle physics today is to discover physics beyond the standard model. Spin is playing a unique role in this pursuit. There have been many experiments done, underway, or planned, that exploit the spin degree of freedom to search for nonstandard-model physics. A partial list includes the muon $g-2$ experiment at BNL, the SLAC E158 experiment using Moller scattering to measure the eletron weak charge, the Jefferson Lab Qweak experiment using polarized electron-proton scattering to measure the weak charge of the proton, the newly proposed neutron electric dipole moment (EDM) experiment at LANSCE, the TRIC experiment searching for P-even/T-odd nuclear interactions, the PSI experiment to measure the polarization of positrons from muon decay, the RHIC spin experiment measuring parity-asymmetry at high-energy, and so on. I will highlight a few of these here.

The spin of a particle supports the notion of an intrinsic magnetic moment

$$\vec{\mu} = g\frac{e\hbar}{2mc}\vec{s}\,. \tag{1}$$

For a pointlike Dirac particle without interactions, the Lande g-factor is $g_D = 2$. Hence $a = g-2/2$ is a measure of underlying dynamics. The latest result for muon $g-2$ from

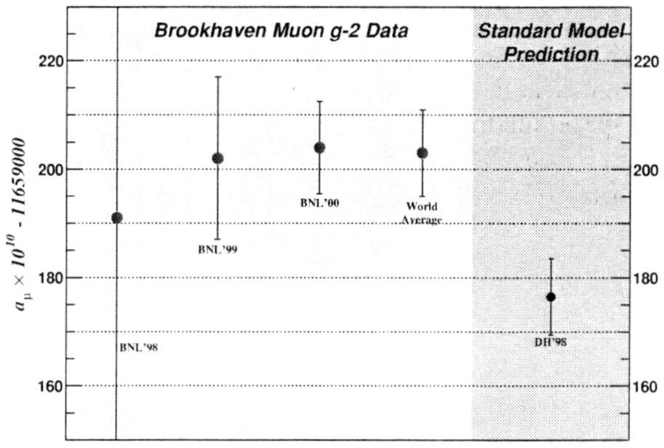

FIGURE 1. New BNL g-2 data.

the experiment here is [2],

$$a_\mu^{BNL} = 11659204(7)(5) \times 10^{-10} \,. \tag{2}$$

This yields a world average, $a_\mu^{world} = 11659208(8) \times 10^{10}$. The lastest theoretical number is $a_\mu^{th} = 11659182.1(7.2) \times 10^{-10}$, after taking into account the infamous sign flip for light-light scattering contribution, and averaging over several latest calculations [3]. The discrepancy is $\delta a_\mu = 21(11) \times 10^{-10}$, about 1.9σ. Taking the above discrepancy seriously, what are the implications for SUSY? A recent analysis indicates that 1) there must be at least two sparticles with mass less than 760 GeV, 2) for models with gaugino mass unification, there must be one sparticle with mass less than 580 GeV, 3) there is no lower bound on $\tan\beta$ [3]. Clearly, pushing for a much better measurement at this point should go hand in hand with theoretical development in understanding the hadronic corrections. In particular, there are still large uncertainties in calculating the vacuum polarization contribution and light-by-light scattering.

For a static particle, the only 3-vector present is the spin. Hence its electric dipole momentum (EDM) must be $\vec{d} = d\vec{s}$. Clearly, d is T-odd, and CP-odd in field theory where the CPT theorem holds. In the standard model, the CP violation in the CKM matrix leads to a neutron EDM of order 10^{-31} ecm, which is several orders of magnitude higher than the current experimental limit, 0.63×10^{-25} ecm. However, the baryon number asymmetry in the Universe requires additional sources of CP violation which can contribute significantly to the neutron EDM. Many weak-scale SUSY models predict the neutron EDM on the order of $10^{-25} \sim 10^{-27}$ ecm. Therefore, one can expect that an improvement on the current limit can reveal a lot about the constraints on the these models. The EDM collaboration at LANSCE has proposed to push the current limit two-order of magnitude lower, i.e., 10^{-27} ecm. This is clearly a very exciting development.

The spin of a particle can lead to simple correlations like $\vec{s} \cdot \vec{p}$ in the decay rate and cross sections. Since it is a pseudo scalar, it is a natural place to look for parity violation. In fact, C. S. Wu used this to find one of the first evidences of parity violation. It has been observed recently that this correlation can be used in high-precision measurement of weak charges. In the standard model, the vector couplings of the electron and proton to the neutral current Z-boson are

$$g_V^e \sim g_V^P \sim 1/4 - \sin^2\theta_W \; . \tag{3}$$

Hence a measurement of the couplings can be translated directly into a measurement of $\sin^2\theta_W$ itself. Like $\alpha_s(Q^2)$, $\sin^2\theta_W$ in the minimal subtraction scheme also runs with the probing scale. Its value has been measured very accurately at the Z-pole, thanks to experiments at LEP. At low-energy, however, the atomic parity experiment and neutrino scattering data seem to be inconsistent with the standard model prediction. A new experiment at SLAC (E158) plans to measure the asymmetry in polarized Moller scattering $\vec{e}e \to ee$, in which the scattering electron is polarized. The experiment will have a 6% measurement on the (vector) weak charge of the electron at $Q^2 = 0.03$ GeV^2. A similar experiment at JLab (Qweak) plans to measure the asymmetry in polarized ep elastic scattering. They have planned a 4% measurement on the vector weak charge of the proton at $Q^2 = 0.03$ GeV^2, which will yields a 0.3% accuracy on $\sin^2\theta_W$.

What new physics can one learn about in the weak charge measurements? The discprepancy from the standard model prediction will be sensitive to a new Z-boson of mass ranging from 0.6 to 1 TeV. It can also detect new contact interactions arising from scalar Higgs of mass ~ 10 GeV. New physics can also modify the gauge boson propagators, resulting in new parameters like X, and the future measurements can constraint it to a level of $\delta X \sim 0.1$.

THEME II: SPIN AS A TOOL TO MEASURE OBSERVABLES THAT ARE HARD TO OBTAIN OTHERWISE

There are many observables that are interesting in physics but hard to measure experimentally. For instance, we would like to know how strange quarks contribute to the electromagnetic form factors of the nucleon, but it is hard to isolate just the strange quarks. The experiments involving spin degrees of freedom can make a significant difference here. Actually, there is a long list of interesting physical observables that can be measured by exploiting the spin degrees of freedom: The strangeness content of the nucleon can be measured in polarized eletron-nucleon scattering; the electromagnetic form factors of the nucleon can be obtained by measuring the recoil polarizations, the spin-dependent nucleon-nucleon scattering can be used to measure the parity-violating nucleon forces. The neutron density in a large nucleus is hard to obtained using electromagnetic interactions but can be measured in parity-violating electron scattering.

The sizable strange effects may have been seen in various nucleon matrix elements: the scalar charge from the $\pi - N$ sigma term and chiral perturbation calculations of the baryon octet mass splitting, the proton spin content measured in polarized deep-inelastic scattering, and the strange quark contribution to the momentum sum rule from neutrino scattering. On the other hand, the strange vector current defines form factors,

$$\langle P'|\bar{s}\gamma^\mu s|P\rangle = \overline{U}(P')\left[F_1^s(Q^2)\gamma^\mu + F_2^s(Q^2)\frac{i\sigma^{\mu\nu}q_\nu}{2M}\right]U(P) \; . \tag{4}$$

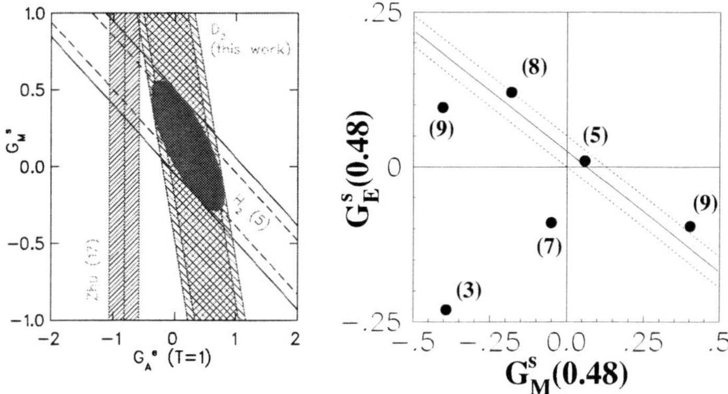

FIGURE 2. The results on the strange quark form factors from the SAMPLE and HAPPEX collaborations [4, 5].

From these, one obtains the strange magnetic moment $\mu_s = F_2(0)$ and strange radius $\langle r_s^2 \rangle = -6dG_E^s(Q^)/dQ^2$. R. McKeown was the first to realize that the strange form factors can be measured in polarized parity-violating eletron-nucleon scattering in which the interference between the Z and γ exchanges is probed [6]. The SAMPLE experiment at MIT-Bates have measured the electron asymmetries on both proton and deuteron targets and the result is shown in the left panel of Fig. 2. The strange magnetic moment seems to be small, and the non-zero axial contribution is inconsistent with the best theoretical calculation [7]. The HAPPEX experiment at Jlab measured the electron asymmetry at a different kinematics and the result is a linear constraint on the strange electric and magnetic form factors [5].

The nucleon electromagnetic form factors tell us detailed information about the charge and current distributions in the nucleon. Despite many years of effort since the mid 1950s, they are still not well determined. Polarization execpriments are helping in a big way in the precision measurements! Recently a JLab Hall-A experiment has measured the polarization transfer in eP-elastic scattering. The ratio of the transverse and longitudinal polarizations yields the ratio of the form factors

$$\frac{G_{Ep}}{G_{Mp}} = -\frac{P_t}{P_l}\frac{(E_e+E_e')}{2M}\tan\frac{\theta_e}{2} \ . \tag{5}$$

The big surprise of the new data is that the ratio dropped steadily as a function of Q^2 [8], differing from popular expectations. The significance of the result is still under debate. The same method has been and will be used again to measure the form factors of the neutron.

The nucleon-nucleon interactions has a parity-violating component when taking into account the underlying weak-boson exchanges. These interactions have been parametrized in terms of effective parity-odd meson-nucleon couplings and optimal values of these couplings have been estimated (DDH) [10]. Many of the couplngs can

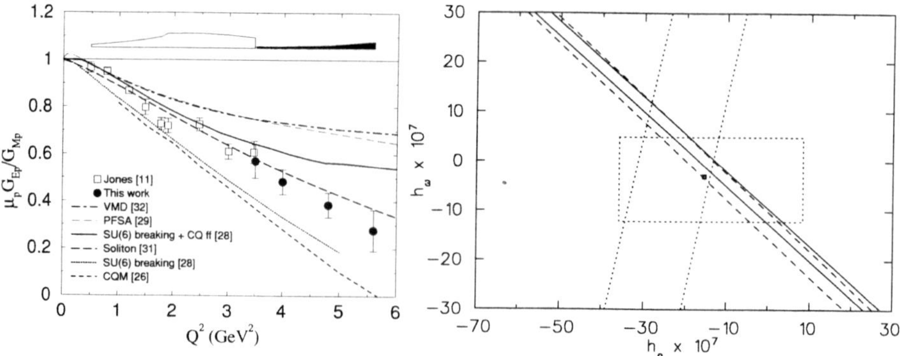

FIGURE 3. Left: The ratio of the electric to magnetic form factors of the proton from the recoil-polarization measurement [8]. Right: the parity-violating nucleon-nucleon interactions probed in polarized nucleon-nucleon scattering [9].

only be measured in spin-dependent processes. For example, the parity-violating pion-nucleon coupling can be measured in spin-dependent pion photoproduction [11, 12]. Other experiments include measuring asymmetries and polarizations of γ-rays in nuclear transitions, asymmetries in polarized proton-proton elastic scattering and thermal neutron-nuclei scattering, and the neutron spin rotation through nuclei matter. A good summary of these experiments can be found in [13]. Shown on the right panel of Fig.3 is the constraint from parity-violating nucleon-nucleon scattering on the ρ and ω meson couplings, together with the latest result from TRIUMF [9].

THEME III: SPIN AS A PROBE TO UNRAVEL NONPERTURBATIVE QCD DYNAMICS

To find how nonperturbative QCD works is one of the greatest challenges in theoretical physics. At present, besides numerically solving the theory on a spacetime lattice, another approach to gain a better understanding is to learn from experimental data. Spin offers a unique way to probe into the internal structure of the nucleon and nonperturbative QCD dynamics. Over the years and in particular after the EMC experiment on polarized deep-inelastic scattering [14], QCD spin physics has evolved into a large subfield. A partial list of important topics would include the spin structure functions of the nucleon—g_1 and g_2, the gluon polarization $\Delta g(x)$, the quark transversity distribution $\delta q(x)$, the single spin asymmetries, the generalized parton distributions, the semi-inclusive deep-inelastic scattering, the GDH sum rule and its generalizations, etc. All of these subjects will be discussed and debated extensively at this conference. Due to time limitation, I cannot go through all of them in this introductory talk.

Let's start with the spin structure of the nucleon, as it has been the major interest in the field for more than a decade. According to QCD, we can write down a decomposition of the nucleon spin as [15]

$$\frac{\hbar}{2} = \frac{1}{2}\Delta\Sigma(\mu^2) + L_q(\mu^2) + J_q(\mu^2) \; . \tag{6}$$

This decomposition is independent of gauge fixing but depends on the resolution scale μ^2. After a decade of careful experimental study following the original EMC experiment, we are now quite certain that the total quark spin $\Delta\Sigma = \delta u + \delta d + \delta s$ contributes only about 30% of the nucleon spin [16], contrary to the prediction of the naive quark model. The total quark contribution $J_q = 1/2\Delta\Sigma + L_q$, spin plus orbital, has been shown to be related to a sum rule of the generalized parton distributions [15]. The total gluon angular momentum J_g contains the gluon helicity contribution Δg which can be measured in high-energy scattering.

It has been speculated that the small size of $\Delta\Sigma = \Delta\ \Sigma_v + \Delta\Sigma_s$ results from a cancellation between the valence and sea contributions. While inclusive DIS cannot separate valence from sea, semi-inclusive processes can [17]. A surprise from the recent SMC and HERMES experiments is that the data do not show the expected large negatively polarized sea [18]! The data are not very accurate yet at this point, and a better precision is needed to make a firm conclusion. The validity of factorization might also be a concern. Let me point out that a measurement of strange sea contribution Δs can be a useful test of SU(3) flavor symmetry. Fortunately, the RHIC spin can help here. Through W-boson production, the sea quark polarization can be measured rather cleanly [19].

Gluon polarization $\Delta g(x)$ is a well-defined quantity theoretically (gauge invariant), just like the quark helicity distribution,

$$\Delta g(x) = \frac{i}{2x}\int \frac{d\lambda}{2\pi} e^{i\lambda x} \langle PS|F^{+\alpha}(0)L(0,\lambda n)\tilde{F}^{+}_{\alpha}(\lambda n)|PS\rangle \; , \tag{7}$$

where L is a light-cone gauge link. Although its role in the spin decomposition is not transparent, $\Delta g(x)$ can be interpreted as the gluon helicity in the light-cone gauge $A^+ = 0$. So far, some information about Δg has been obtained from NLO fit to the $g_1(x)$ structure function data [16], and large p_t hadron production by HERMES [20]. An interesting question here is if Δg is as large as some preliminary results indicate, why do the gluons contribute so much to the spin of the nucleon? After all, the nucleon spin is $\hbar/2$ and the quarks do not seem to contribute much more than$\hbar/2$.

To make an accurate measurement of the gluon distribution is clearly a high priority for the future high-energy spin physics. The gluon polarization can be measured from direct photon production at RHIC, jet and high-P_t hadron production at RHIC and COMPASS, open charm and heavy quark production at SLAC, COMPASS, and RHIC, and Q^2 evolution, for example, at EIC. There are many talks at this conference that are devoted to this topic. One important question one should keep asking is: are the perturbative mechanisms clean? Eventually, are we going to get a good determination of the integral?

Generalized parton distributions (GPDs) have become a very popular subject in theory and experiment for several years now [21, 22, 23] . These studies were motivated by finding the spin structure of the nucleon [15]! It is a new type of parton "distributions" which contains much more information than any other nucleon observable considered

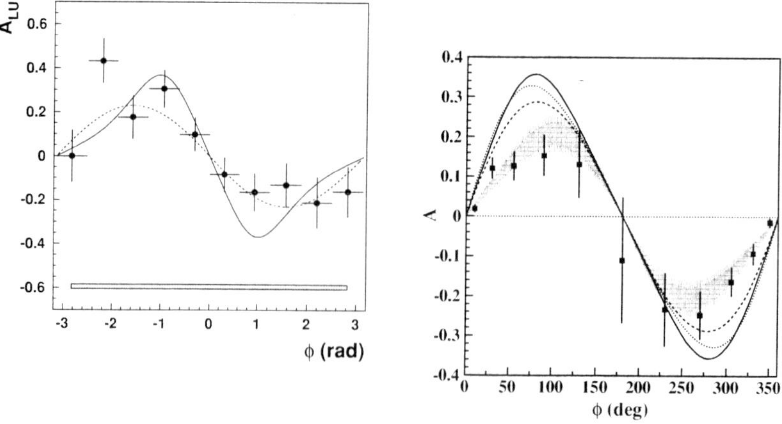

FIGURE 4. The single-spin asymmetry from HERMES (left) and CLAS (right) collaborations.

so far. In various limits, they reduce to elastic form factors and Feynman parton distributions. They depend on three variables: the parton momentum fraction x, the form-factor resolution $t = -q^2$, and the skewness ξ. For $\xi = 0$, the GPDs have a nice interpretation in the impact parameter space [24].

GPDs can be measured in real experiments, such as deeply-virtual Compton [25, 26] scattering and exclusive meson production [27]. Deeply virtual Compton scattering is an exclusive production of high-energy photons in deep-inelastic kinematics in which the single quark scattering dominates—it is a Compton scattering on a single quark! If one compares this to deep-inelastic scattering, DVCS is a noninvasive surgery on the proton in which the structural information is obtained without breaking it into pieces. The Zeus and H1 collaborations at HERA were first to report the DVCS type of events. At low lepton energy, however, it is difficult to measure the DVCS process directly. Rather, one can isolate the interference between the DVCS and Bethe-Heitler amplitudes. The HERMES collaboration at DESY [28] and CLAS at JLab [29] have measured the photon production asymmetry, and they are found to be consistent with the dominance of the DVCS amplitude. Exclusive meson productions are also sensitive to GPDs. Different flavors and spins of mesons help to disintangle the contributions of different GPDs.

The quark transversity distribution is one of the three twist-two quark distributions describing the state of a quark beam in a nucleon beam [30, 31, 32, 33]. Contrary to what's commonly believed, it is not a transverse-spin distribution because a fast-moving quark cannot be in a transverse-spin eigenstate. In a non-relativistic quark model, however, the transversity distribution is the same as the helicity distribution and a quark can be in the transverse-spin eigenstates [34, 35]. The knowledge of the distribution allows access to the tensor charge of the nucleon which is useful, among others, in the neutron EDM calculation.

To measure the transversity distribution is very difficult! It is a chirally-odd quantity, requires another chiral-odd observable present in a hard process. It vanishes asymptotically at a large resolution scale. The most clean process is the Drell-Yan scattering with transversely polarized protons. One can measure it in inclusive jet production and direct γ production. It can also be accessd through chiral-odd quark fragmentation functions. In most cases, however, there are other competing processes that one must isolate. A review of measuring the transversity distribution can be found in [36].

Finally, I would like to briefly mention transverse single-spin asymmetries. These asymmetries have been observed in (single transversely polarized) proton-proton scattering and more recently in electron-proton scattering. The hyperon polarization in unpolarized proton-proton scattering is a closely related phenomenon. A good reference about the subject is a recent review article by Barone and Ratcliffe [37] (see also [38]). The asymmetry is time-reversal-"odd" and arises only when there are initial and final state interaction phases. Moreover, it requires viable mechanisms for helicity flips, such as Collins fragmentation functions[39], Sivers distributions [40, 41, 42, 43], or twist-three correlations[44, 45, 46]. Recently, it was shown by Belitsky, Yuan, and myself that in the light-cone gauge, a gauge link at spatial infinity is crucial to have a nonvanishing Collins and Sivers functions [47, 48].

CLOSING REMARKS

In non-relativistic quantum mechanics, spin is considered an extra degree of freedom that one can put in by hand. In relativistic theories, however, spin is an essential part of a theory, as we have mentioned in the begining of the talk. In particular, the spin-statistics theorem followed from causality has profound implications about the fundamental structure of matter and the stability of many-body systems, even in the non-relativistic limit. For example, the atomic structure which lays the foundation for chemistry and biology critically depends on the operation of the Pauli principle. The supersymmetric theories, which have recently become very popular in particle physics, assume a symmetry between fermions and bosons or different spin states. The importance of spin is neatly summarized in the preface to S. Tomonaga's book "The Story of Spin," by T. Oka who wrote that spin, "is a mysterious beast, and yet its practical effect prevail the whole of science. The existence of spin, and the statistics associated with it, is the most subtle and ingenious design of Nature— without it the whole universe would collapse." [1]

To conclude the talk, I have compiled a top ten list of disasters in a world without spin; for fun:

1. Chemistry would have been messed up.
2. You can't get rid of the glare from the sunlight.
3. You can't have an MRI if you've gotten brain cancer.
4. Quarks aren't compelled to have color.
5. Neutron stars would never be stable.
6. Weak interactions can't be chiral.
7. There wouldn't be any gauge theories.
8. You can't play around with supersymmetry.
9. You would be stuck in 26-dimensions.
10. Spin doctors in Washington, DC would be jobless!

ACKNOWLEDGMENTS

This work is supported partly by the DOE grant DE-FG-02-93ER-40762.

REFERENCES

1. Tomonaga, S.-i., *The Story of Spin*, The University of Chicago Press, Chicago, 1997.
2. Bennett, G. W., et al., *Phys. Rev. Lett.*, **89**, 101804 (2002).
3. Byrne, M., Kolda, C., and Lennon, J. E., *hep-ph/0208067* (2002).
4. Hasty, R., et al., *Science*, **290**, 2117 (2000).
5. Aniol, K. A., et al., *Phys. Lett.*, **B509**, 211–216 (2001).
6. Mckeown, R. D., *Phys. Lett.*, **B219**, 140–142 (1989).
7. Zhu, S.-L., Puglia, S. J., Holstein, B. R., and Ramsey-Musolf, M. J., *Phys. Rev.*, **D62**, 033008 (2000).
8. Gayou, O., et al., *Phys. Rev. Lett.*, **88**, 092301 (2002).
9. Berdoz, A. R., et al., *Phys. Rev. Lett.*, **87**, 272301 (2001).
10. Desplanques, B., Donoghue, J. F., and Holstein, B. R., *Ann. Phys.*, **124**, 449 (1980).
11. Chen, J.-W., and Ji, X.-D., *Phys. Rev. Lett.*, **86**, 4239–4242 (2001).
12. Chen, J.-W., and Ji, X.-D., *Phys. Lett.*, **B501**, 209–215 (2001).
13. Haeberli, W., and Holstein, B. R., *nucl-th/9510062* (1995).
14. Ashman, J., et al., *Nucl. Phys.*, **B328**, 1 (1989).
15. Ji, X.-D., *Phys. Rev. Lett.*, **78**, 610–613 (1997).
16. Filippone, B. W., and Ji, X.-D., *hep-ph/0101224* (2001).
17. Close, F. E., and Milner, R. G., *Phys. Rev.*, **D44**, 3691–3694 (1991).
18. Ackerstaff, K., et al., *Phys. Lett.*, **B464**, 123–134 (1999).
19. Bunce, G., Saito, N., Soffer, J., and Vogelsang, W., *Ann. Rev. Nucl. Part. Sci.*, **50**, 525–575 (2000).
20. Airapetian, A., et al., *Phys. Rev. Lett.*, **84**, 2584–2588 (2000).
21. Ji, X.-D., *J. Phys.*, **G24**, 1181–1205 (1998).
22. Radyushkin, A. V., *hep-ph/0101225* (2000).
23. Goeke, K., Polyakov, M. V., and Vanderhaeghen, M., *Prog. Part. Nucl. Phys.*, **47**, 401–515 (2001).
24. Burkardt, M., *Phys. Rev.*, **D62**, 071503 (2000).
25. Ji, X.-D., *Phys. Rev.*, **D55**, 7114–7125 (1997).
26. Belitsky, A. V., Muller, D., and Kirchner, A., *Nucl. Phys.*, **B629**, 323–392 (2002).
27. Radyushkin, A. V., *Phys. Lett.*, **B385**, 333–342 (1996).
28. Airapetian, A., et al., *Phys. Rev. Lett.*, **87**, 182001 (2001).
29. Stepanyan, S., et al., *Phys. Rev. Lett.*, **87**, 182002 (2001).
30. Ralston, J. P., and Soper, D. E., *Nucl. Phys.*, **B152**, 109 (1979).
31. Cortes, J. L., Pire, B., and Ralston, J. P., *Z. Phys.*, **C55**, 409–416 (1992).
32. Jaffe, R. L., and Ji, X.-D., *Phys. Rev. Lett.*, **67**, 552–555 (1991).
33. Jaffe, R. L., and Ji, X.-D., *Nucl. Phys.*, **B375**, 527–560 (1992).
34. He, H.-X., and Ji, X.-D., *Phys. Rev.*, **D54**, 6897–6902 (1996).
35. He, H.-X., and Ji, X.-D., *Phys. Rev.*, **D52**, 2960–2963 (1995).
36. Jaffe, R. L., *hep-ph/9710465* (1997).
37. Barone, V., Drago, A., and Ratcliffe, P. G., *Phys. Rept.*, **359**, 1–168 (2002).
38. Liang, Z.-T., and Boros, C., *Int. J. Mod. Phys.*, **A15**, 927–982 (2000).
39. Collins, J. C., *Nucl. Phys.*, **B396**, 161–182 (1993).
40. Sivers, D. W., *Phys. Rev.*, **D41**, 83 (1990).
41. Boer, D., and Mulders, P. J., *Phys. Rev.*, **D57**, 5780–5786 (1998).
42. Brodsky, S. J., Hwang, D. S., and Schmidt, I., *Phys. Lett.*, **B530**, 99–107 (2002).
43. Collins, J. C., *Phys. Lett.*, **B536**, 43–48 (2002).
44. Qiu, J.-W., and Sterman, G., *Phys. Rev. Lett.*, **67**, 2264–2267 (1991).
45. Ji, X.-D., *Phys. Lett.*, **B289**, 137–142 (1992).
46. Qiu, J.-W., and Sterman, G., *Phys. Rev.*, **D59**, 014004 (1999).
47. Ji, X.-d., and Yuan, F., *Phys. Lett.*, **B543**, 66–72 (2002).
48. Belitsky, A. V., Ji, X., and Yuan, F., *hep-ph/0208038* (2002).

New Results from the Muon $g-2$ Experiment

E.P. Sichtermann[11], G.W. Bennett[2], B. Bousquet[9], H.N. Brown[2], G. Bunce[2], R.M. Carey[1], P. Cushman[9], G.T. Danby[2], P.T. Debevec[7], M. Deile[11], H. Deng[11], W. Deninger[7], S.K. Dhawan[11], V.P. Druzhinin[3], L. Duong[9], E. Efstathiadis[1], F.J.M. Farley[11], G.V. Fedotovich[3], S. Giron[9], F.E. Gray[7], D. Grigoriev[3], M. Grosse-Perdekamp[11], A. Grossmann[6], M.F. Hare[1], D.W. Hertzog[7], X. Huang[1], V.W. Hughes[11], M. Iwasaki[10], K. Jungmann[5], D. Kawall[11], B.I. Khazin[3], J. Kindem[9], F. Krienen[1], I. Kronkvist[9], A. Lam[1], R. Larsen[2], Y.Y. Lee[2], I. Logashenko[1,3], R. McNabb[9], W. Meng[2], J. Mi[2], J.P. Miller[1], W.M. Morse[2], D. Nikas[2], C.J.G. Onderwater[7], Y. Orlov[4], C.S. Özben[2], J.M. Paley[1], Q. Peng[1], C.C. Polly[7], J. Pretz[11], R. Prigl[2], G. zu Putlitz[6], T. Qian[9], S.I. Redin[3,11], O. Rind[1], B.L. Roberts[1], N. Ryskulov[3], P. Shagin[9], Y.K. Semertzidis[2], Yu.M. Shatunov[3], E. Solodov[3], M. Sossong[7], A. Steinmetz[11], L.R. Sulak[1], A. Trofimov[1], D. Urner[7], P. von Walter[6], D. Warburton[2], and A. Yamamoto[8].
(Muon $g-2$ Collaboration)

[1]*Department of Physics, Boston University, Boston, Massachusetts 02215*
[2]*Brookhaven National Laboratory, Upton, New York 11973*
[3]*Budker Institute of Nuclear Physics, Novosibirsk, Russia*
[4]*Newman Laboratory, Cornell University, Ithaca, New York 14853*
[5] *Kernfysisch Versneller Instituut, Rijksuniversiteit Groningen, NL 9747 AA Groningen, The Netherlands*
[6] *Physikalisches Institut der Universität Heidelberg, 69120 Heidelberg, Germany*
[7] *Department of Physics, University of Illinois at Urbana-Champaign, Illinois 61801*
[8] *KEK, High Energy Accelerator Research Organization, Tsukuba, Ibaraki 305-0801, Japan*
[9]*Department of Physics, University of Minnesota,Minneapolis, Minnesota 55455*
[10] *Tokyo Institute of Technology, Tokyo, Japan*
[11] *Department of Physics, Yale University, New Haven, Connecticut 06520*

Abstract. The Muon $g-2$ collaboration has measured the anomalous magnetic g value, $a = (g-2)/2$, of the positive muon with an unprecedented uncertainty of 0.7 parts per million. The result $a_\mu^+(\text{expt}) = 11\,659\,204(7)(5)\times 10^{-10}$, based on data collected in the year 2000 at Brookhaven National Laboratory, is in good agreement with the preceeding data on a_μ^+ and a_μ^-. The measurement tests standard model theory, which at the level of the current experimental uncertainty involves quantum electrodynamics, quantum chromodynamics, and electroweak interaction in a significant way.

CP675, *Spin 2002: 15th Int'l. Spin Physics Symposium and Workshop on Polarized Electron Sources and Polarimeters,* edited by Y. I. Makdisi, A. U. Luccio, and W. W. MacKay

INTRODUCTION

The magnetic moment $\vec{\mu}$ of a particle with charge e, mass m, and spin $\vec{s}$ is given by

$$\vec{\mu} = \frac{e}{2mc} g \vec{s} \tag{1}$$

in which g is the gyromagnetic ratio. For point particles with spin-$\frac{1}{2}$, Dirac theory predicts $g = 2$.

Precision measurements of the g values of leptons and baryons have historically played an important role in the development of particle theory. The g value of the proton, for example, is found to differ sizeably from 2, which provides evidence for a rich internal proton (spin) structure. The lepton g values deviate from 2 only by about one part in a thousand, consistent with current evidence that leptons are point particles with spin-$\frac{1}{2}$. The anomalous magnetic g value of the electron, $a_e = (g_e - 2)/2$, is among the most accurately measured quantities in physics, and is presently known with an uncertainty of about four parts per billion (ppb) [1]. The value of a_e is described in terms of standard model (SM) field interactions, in which it has a leading order contribution of $\alpha/(2\pi)$ from the so-called Schwinger term. Nearly all of the measured value is contributed by QED processes involving virtual electrons, positrons, and photons. Particles more massive than the electron contribute only at the level of the present experimental uncertainty.

The anomalous magnetic g value of the muon, a_μ, is more sensitive than a_e to processes involving particles more massive than the electron, typically by a factor $(m_\mu/m_e)^2 \sim 4 \cdot 10^4$. A series of three experiments at CERN [2, 3, 4] measured a_μ to a final uncertainty of about 7 parts per million (ppm), which is predominantly of statistical origin. The CERN generation of experiments thus tested electron-muon universality and established the existence of a $\sim$ 59 ppm hadronic contribution to a_μ. Electroweak processes are expected to contribute to a_μ at the level of 1.3 ppm, as are many speculative extensions of the SM.

The present muon $g-2$ experiment at Brookhaven National Laboratory (BNL) has determined a_{μ^+} of the positive muon with an uncertainty of 0.7 ppm from data collected in the year 2000, and aims for a similar uncertainty on a_{μ^-} of the negative muon from measurements in 2001. The continuation of the experiment by a single running period, if funded, should bring the design goal precision of 0.4 ppm within reach.

EXPERIMENT

The concept of the experiment at BNL is the same as that of the last of the CERN experiments [2, 3, 4] and involves the study of the orbital and spin motions of polarized muons in a magnetic storage ring.

The present experiment (Figure 1) is situated at the Alternating Gradient Synchrotron (AGS), which in the year 2000 delivered up to 60×10^{12} protons in twelve 50 ns (FWHM) bunches over its 3 s cycle. The 24 GeV protons from the AGS were directed onto a rotating, water-cooled nickel target. Pions with energies of 3.1 GeV emitted from

FIGURE 1. Top view of the $g-2$ apparatus. The beam of longitudinally polarized muons enters the superferric storage ring magnet through a superconducting inflector magnet located at 9 o'clock and circulates clockwise after being placed onto stored orbit with three pulsed kickers modules in the 12 o'clock region. Twenty-four lead scintillating-fiber calorimeters on the inner, open side of the C-shaped ring magnet are used to measure muon decay positrons. The central platform supports the power supplies for the four electrostatic quadrupoles and the kicker modules.

the target were captured into a 72 m straight section of focusing-defocusing magnetic quadrupoles, which transported the parent beam and naturally polarized muons from forward pion decays. At the end of the straight section, the beam was momentum-selected and injected into the 14.2 m diameter storage ring magnet [5] through a field-free inflector [6] region in the magnet yoke. A pulsed magnetic kicker [7] located at approximately one quarter turn from the inflector region produced a 10 mrad deflection which placed the muons onto stored orbits. Pulsed electrostatic quadrupoles [8] provided vertical focusing. The magnetic dipole field of about 1.45 T was measured with an NMR system [9] relative to the free proton NMR frequency ω_p over most of the 9 cm diameter circular storage aperture. Twenty-four electromagnetic calorimeters [10] read out by 400 MHz custom waveform digitizers (WFD) were used on the open, inner side of the C-shaped ring magnet to measure muon decay positrons. The decay violates parity, which leads to a relation between the muon spin direction and the positron energy spectrum in the laboratory frame. For positrons above an energy threshold E, the muon-decay time-spectrum

$$N(t) = N_0(E)\exp\left(\frac{-t}{\gamma\tau}\right)[1 + A(E)\sin(\omega_a t + \phi(E))], \tag{2}$$

in which N_0 is a normalization, $\gamma\tau \sim 64\,\mu$s is the dilated muon lifetime, A an asymmetry factor, ϕ a phase, and ω_a the angular difference frequency of muon spin precession and momentum rotation. In our measurements, the NMR and WFD clocks were phase-locked to the same LORAN-C [11] frequency signal.

The muon anomalous magnetic g value is evaluated from the ratio of the measured frequencies, $R = \omega_a/\omega_p$, according to:

$$a_\mu = \frac{R}{\lambda - R}, \tag{3}$$

in which $\lambda = \mu_\mu/\mu_p$ is the ratio the muon and proton magnetic moments. The value with smallest stated uncertainty, $\lambda = \mu_\mu/\mu_p = 3.183\,345\,39(10)$ [12], results from measurements of the microwave spectrum of ground state muonium [13] and theory [14, 15].

Important improvements made since our preceeding measurement [16] include: the operation of the AGS with 12 beam bunches, which contributed to a 4-fold increase in the data collected; a new superconducting inflector magnet, which greatly improved the homogeneity of the magnetic field in the muon storage region; a sweeper magnet in the beamline, which reduced AGS background; additional muon loss detectors, which improved the study of time dependence; and further refined analyses, in particular of coherent betatron oscillations.

DATA ANALYSIS

The analysis of a_μ follows, naturally, the separation of the measurement in the frequencies ω_p and ω_a. Both frequencies were analyzed independently by several groups within the collaboration. The magnetic field frequencies measured during the running period were weighted by the distribution of analyzed muons, both in time and over the storage region. The frequency fitted from the positron time spectra was corrected by +0.76(3) ppm for the net contribution to the muon spin precession and momentum rotation caused by vertical beam oscillations and, for muons with $\gamma \neq 29.3$, by horizontal electric fields [17]. The values of $R = \omega_a/\omega_p$ and a_μ were evaluated only after each of the frequency analyses had been finalized; at no earlier stage were the absolute values of both frequencies, ω_p and ω_a, known to any of the collaborators.

The frequency ω_p

The analysis of the magnetic field data starts with the calibration of the 17 NMR probes in the field trolley using dedicated measurements taken during and at the end of the data collection period. In these calibration measurements, the field in the storage region was tuned to very good homogeneity at two specific calibration locations. The field was then measured with the NMR probes mounted in the trolley shell, as well as with a single probe plunged into the storage vacuum and positioned to measure the field values in the corresponding locations. Drifts of the field during the calibration measurements were determined by remeasuring the field with the trolley after the measurements with the plunging probe were completed, and in addition by interpolation of the readings from nearby NMR probes in the outer top and bottom walls of the vacuum chamber. The difference of the trolley and plunging probe readings forms a calibration of the trolley probes with respect to the plunging probe, and hence with respect to each other. The

TABLE 1. Systematic uncertainties for the ω_p analysis. The uncertainty "Others" groups uncertainties caused by higher multipoles, the trolley frequency, temperature, and voltage response, eddy currents from the kickers, and time-varying stray fields.

Source of errors	Size [ppm]
Absolute calibration of standard probe	0.05
Calibration of trolley probe	0.15
Trolley measurements of B_0	0.10
Interpolation with fixed probes	0.10
Uncertainty from muon distribution	0.03
Others	0.10
Total systematic error on ω_p	0.24

plunging probe, as well as a subset of the trolley probes, were calibrated with respect to a standard probe [18] at the end of the running period in a similar sequence of measurements in the storage region, which was opened to air for that purpose. The leading uncertainties in the calibration procedure result from the residual inhomogeneity of the field at the calibration locations, and from position uncertainties in the active volumes of the NMR probes. These uncertainties were evaluated from measurements in which the trolley shell was purposely displaced and known field gradients were applied using the so-called surface and dipole correction coils of the ring magnet. The size of these uncertainties is estimated to be 0.15 ppm, as listed in Table 1. The uncertainty in the absolute calibration of the standard probe amounts to 0.05 ppm [18]. The dependencies of the trolley NMR readings on the supply voltage and on other parameters were measured to be small in the range of operation. An uncertainty of 0.10 ppm ("Others" in Table 1) is assigned, which includes also the measured effects from the transient kicker field caused by eddy currents and from AGS stray fields.

The magnetic field inside the storage region was measured 22 times with the field trolley during the data collection from January to March 2000. Figure 2a shows the field value measured in the storage ring with the center trolley probe versus the azimuthal angle. The field is seen to be uniform to within about ±50 ppm of its average value over the full azimuthal range, including the region near 350° where the inflector magnet is located. Non-linearities in the determination of the trolley position during the measurements — from the measured cable lengths and from perturbations on the readings from fixed probes as the trolley passes — are estimated to affect the azimuthal average of the field at the level of 0.10 ppm. Fig. 2b shows a 2-dimensional multipole expansion of the azimuthally averaged readings from the trolley probes,

$$B_y = \sum_{n=0}^{\infty} C_n r^n \cos(n\phi) - \sum_{n=0}^{\infty} D_n r^n \sin(n\phi), \tag{4}$$

$$B_x = \sum_{n=0}^{\infty} C_n r^n \sin(n\phi) + \sum_{n=0}^{\infty} D_n r^n \cos(n\phi), \tag{5}$$

where the coefficients C_n and D_n are the normal and skew multipoles, and r and ϕ denote the polar coordinates in the storage region. The multipole expansion was truncated in the

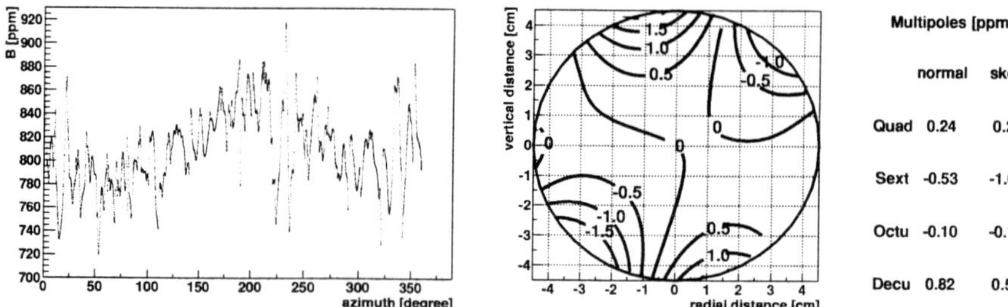

FIGURE 2. The NMR frequency measured with the center trolley probe relative to a 61.74 MHz reference versus the azimuthal position in the storage ring (left), and (right) a 2-dimensional multipole expansion of theazimuthal average of the field measured with 15 trolley probes with respect to the central field value of 1.451 275 T. The multipole amplitudes are given at the storage ring aperture, which has a 4.5 cm radius as indicated by the circle.

analysis after the decupoles ($n = 4$). Measurements with probes extending to larger radii show that the neglect of higher multipoles is at most 0.03 ppm in terms of the average field encountered by the stored muons, in agreement with magnet design calculations. The field averaged over azimuth is seen to be uniform to within 1.5 ppm of its value.

The measurements with the trolley relate the readings of 370 NMR fixed probes in the outer top and bottom walls of the storage vacuum chamber to the field values in the beam region. The fixed NMR probes are used to interpolate the field when the field trolley is 'parked' in the storage vacuum just outside the beam region, and muons circulate in the storage ring. Since the relationship between the field value in the storage region and the fixed probe readings may change during the course of the data collection period, the field mappings with the trolley were repeated typically two to three times per week, and whenever ramping of the magnet or a change in settings required such. The uncertainty associated with the interpolation of the magnetic field between trolley measurements is estimated from the spread of the difference between the dipole moments evaluated from the fixed probe measurements and from the trolley probe measurements in periods of constant magnet settings and powering. It is found to be 0.10 ppm.

Since the field is highly uniform, the field integral encountered by the (analyzed) muons is rather insensitive to the exact location of the beam. As in earlier works [16, 19], the radial equilibrium beam position was determined from the debunching of the beam following injection and the vertical position from the distribution of counts in scintillation counters mounted on the front faces of the positron calorimeters. The position uncertainty amounts to 1 – 2 mm, which contributes 0.03 ppm uncertainty to the field integral.

The result for field frequency ω_p weighted by the muon distribution is found to be,

$$\omega_p/(2\pi) = 61\,791\,595(15)\,\text{Hz} \quad (0.2\,\text{ppm}), \tag{6}$$

where the uncertainty has a leading contribution from the calibration of the trolley probes and is thus predominantly systematic. A second analysis of the field has been performed using additional calibration data, a different selection of fixed NMR probes,

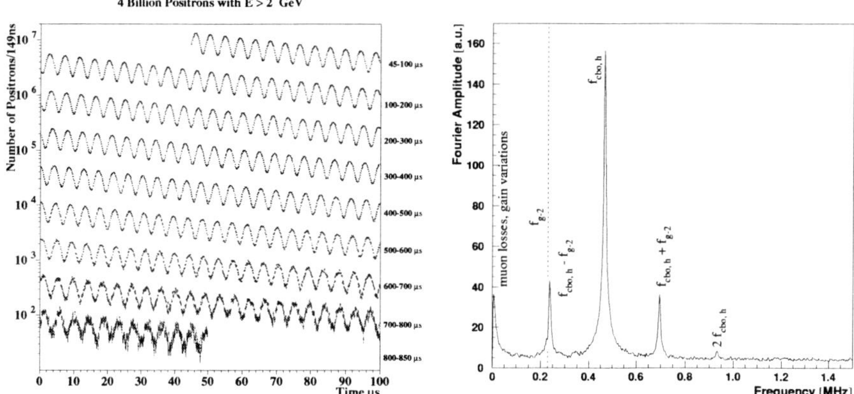

FIGURE 3. The time spectrum for $4 \cdot 10^9$ positrons with energies greater than 2 GeV collected from January to March 2000, after corrections for pile-up and for the bunched time structure of the injected beam (left) were made, and (right) the Fourier transform of the time spectrum, in which muon decay and spin precession (cf. Eq. 2) has been suppressed to emphasize other effects.

and a different method to relate the trolley and fixed probe readings. The results from these analyses are found to agree to within a fraction of the total uncertainty on ω_p.

The frequency ω_a

The event sample available for analysis from data collection in the year 2000 amounts to about $4 \cdot 10^9$ positrons reconstructed with energies greater than 2 GeV and times between 50 μs and 600 μs following the injection of a beam bunch. Figure 3a shows the time spectrum corrected for the the bunched time structure of the beam and for overlapping calorimeter pulses, so called pile-up [16].

The leading characteristics of the time spectrum are those of muon decay and spin precession (cf. Eq. 2). Additional effects exist, as seen from the Fourier spectrum in Figure 3b, and require careful consideration in the analysis. These effects include detector gain and time instability, muon losses, and oscillations of the beam as a whole, so-called coherent betatron oscillations (CBO).

Numerically most relevant to the determination of ω_a are CBO in the horizontal plane. CBO are caused by injecting the beam through the relatively narrow $18(\text{w}) \times 57(\text{h})\,\text{mm}^2$ aperture of the 1.7 m long inflector channel into the 90 mm diameter aperture of the storage region, and have been observed directly with fiber harp monitors plunged into the beam region for this purpose. The CBO frequency is determined by the focusing index of the storage ring, and is numerically close to twice the frequency ω_a for the quadrupole settings employed in most measurements so far. Since the calorimeter acceptances vary with the radial muon decay position in the storage ring and with the momentum of the decay positron, the time and energy spectra of the observed positrions are modulated

TABLE 2. Systematic uncertainties for the ω_a analysis. The uncertainty "Others" groups uncertainties caused by AGS background, timing shifts, vertical oscillations and radial electric fields, and beam debunching/randomization.

Source of errors	Size [ppm]
Coherent betatron oscillations	0.21
Pileup	0.13
Gain changes	0.13
Lost muons	0.10
Binning and fitting procedure	0.06
Others	0.06
Total systematic error on ω_a	0.31

with the CBO frequency. These modulations affect the normalization N_0, the asymmetry A, and the phase ϕ in Eq. 2 at the level of 1%, 0.1%, and 1 mrad at beam injection. When not accounted for in the function fitted to the data, the modulations of the asymmetry and phase with a frequency $\omega_{\mathrm{cbo,h}} \simeq 2 \times \omega_a$ may manifest themselves as artificial shifts of up to 4 ppm in the frequency values ω_a determined from individual calorimeter spectra. The circular symmetry of the experiment design results in a strong cancellation of such shifts in the joined calorimeter spectrum.

Several approaches have been pursued in the analysis of ω_a. In one approach, the time spectra from individual positron calorimeters was fitted in narrow energy intervals using a fit function as in Eq. 2 extended by the number, asymmetry, and phase modulations. Other approaches made use of the cancellation in the joined calorimeter spectra and either fitted for the residual of the leading effects, or accounted for their neglect in a contribution to the systematic uncertainty. The results are found to agree, on ω_a to within the expected 0.5 ppm statistical variation resulting from the slightly different selection and treatment of the data in the respective analyses. The combined result is found to be,

$$\omega_a/(2\pi) = 229\,074\,11(14)(7)\,\mathrm{Hz}\ \ (0.7\,\mathrm{ppm}), \tag{7}$$

in which the first uncertainty is statistical and the second systematic. The combined systematic uncertainty is broken down by source in Table 2.

RESULTS AND DISCUSSION

The value of a_μ was evaluated after the analyses of ω_p and ω_a had been finalized,

$$a_{\mu^+} = 11\,659\,204(7)(5) \times 10^{-10}\ \ (0.7\,\mathrm{ppm}), \tag{8}$$

where the first uncertainty is statistical and the second systematic. This new result is in good agreement with the previous measurements [4, 16, 19, 20] and drives the present world average,

$$a_\mu(\mathrm{exp}) = 11\,659\,203(8) \times 10^{-10}\ \ (0.7\,\mathrm{ppm}), \tag{9}$$

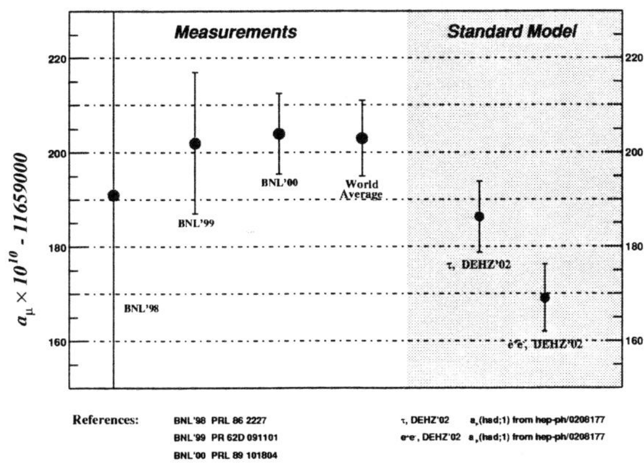

FIGURE 4. Recent measuremens of a_μ and standard model evaluations using the evaluations in Ref. [21] of the lowest order contribution from hadronic vacuum polarization.

in which the uncertainty accounts for known correlations between the systematic uncertainties in the measurements. Figure 4 shows our recent measurements of a_{μ^+}, together with two SM evaluations discussed below.

In the SM, the value of a_μ receives contributions from QED, hadronic, and electroweak processes, $a_\mu(\text{SM}) = a_\mu(\text{QED}) + a_\mu(\text{had}) + a_\mu(\text{weak})$. The QED and weak contributions can, unlike the hadronic contribution, be evaluated perturbatively, $a_\mu(\text{QED}) = 11\,658\,470.57(29) \times 10^{-10}$ [22] and $a_\mu(\text{weak}) = 15.1(4) \times 10^{-10}$ [23]. The hadronic contribution is, in lowest order, related by dispersion theory to the hadron production cross sections measured in e^+e^- collisions and, under additional assumptions, to hadronic τ-decay. Clearly, the hadronic contribution has a long history of values as new data appeared and analyses were refined.

Shortly before the SPIN-2002 conference, Davier and co-workers released a new and detailed evaluation [21], which now incorporates the high precision e^+e^- data [24] in the region of the ρ resonance from CMD-2 at Novosibirsk, more accurate e^+e^- measurements [25, 26] in the 2–5 GeV energy region from BES in Beijing, preliminary results from the final ALEPH analysis [27] of hadronic τ-decay at LEP1, as well as additional CLEO data [28, 29]. The authors note discrepancies between the e^+e^- and τ data at the present levels of precision, and obtain separate predictions for the contribution to $a_\mu(\text{SM})$ from lowest order hadronic vacuum polarization, $a_\mu(\text{had},1) = 685(7) \times 10^{-10}$ from e^+e^- data and $a_\mu(\text{had},1) = 702(6) \times 10^{-10}$ from τ data. Higher order contributions include higher order hadronic vacuum polarization [30, 31] and hadronic light-by-light scattering [32, 33, 34, 35, 36].

Open questions concern the SM value of a_μ, in particular the hadronic contribution, and the experimental value of a_{μ^-} at sub-ppm precision. The former should benefit from further theoretical scrutiny and from radiative-return measurements at e^+e^- factories. We are currently analyzing a sample of about 3×10^9 decay electrons from a_{μ^-} data collected in the year 2001. Stay tuned!

ACKNOWLEDGMENTS

This work was supported in part by the U.S. Department of Energy, the U.S. National Science Foundation, the U.S. National Computational Science Alliance, the German Bundesminister für Bildung und Forschung, the Russian Ministry of Science, and the U.S.-Japan Agreement in High Energy Physics.

REFERENCES

1. Van Dyck, R. S., Schwinberg, P. B., and Dehmelt, H. G., *Phys. Rev. Lett.*, **59**, 26 (1987).
2. Charpak, G., et al., *Phys. Lett*, **1**, 16 (1962).
3. Bailey, J., et al., *Nuovo Cimento*, **A9**, 369 (1972).
4. Bailey, J., et al., *Nucl. Phys.*, **B510**, 1 (1979).
5. Danby, G., et al., *Nucl. Instrum. Meth.*, **A457**, 151 (2001).
6. Yamamoto, A., et al., *Nucl. Instrum. Meth.*, **A491**, 23 (2002).
7. Efstathiadis, E., et al., *Accepted for publication in Nucl. Instrum. Meth.* (2002).
8. Semertzidis, Y., et al., *Submitted to Nucl. Instrum. Meth.* (2002).
9. Prigl, R., et al., *Nucl. Instrum. Meth.*, **A374**, 118 (1996).
10. Sedykh, S., et al., *Nucl. Instrum. Meth.*, **A455** (2000).
11. Superintendent of Documents, *LORAN-C User's Handbook*, U.S. Government Printing Office #050-012-00331-9, 1992.
12. Groom, D., et al., *Eur. Phys. J.*, **C15**, 1 (2000).
13. Liu, W., et al., *Phys. Rev. Lett.*, **82**, 711 (1999).
14. Kinoshita, T., hep-ph/9808351 (1998).
15. Kinoshita, T., and Nio, M., "," in *Frontier Tests of Quantum Electrodynamics and Physics of the Vacuum*, edited by E. Zavattini, D. Bakalov, and C. Rizzo, Heton Press, Sofia, 1998, p. 151.
16. Brown, H. N., et al., *Phys. Rev. Lett.*, **86**, 2227 (2001).
17. Farley, F., and Picasso, E., "," in *Quantum Electrodynamcs*, edited by T. Kinoshita, World Scientific, Singapore, 1990, p. 479.
18. Fei, X., Hughes, V., and Prigl, R., *Nucl. Instrum. Meth.*, **A394**, 349 (1997).
19. Brown, H. N., et al., *Phys. Rev.*, **D62**, 091101 (2000).
20. Carey, R. M., et al., *Phys. Rev. Lett.*, **82**, 1632 (1999).
21. Davier, M., Eidelman, S., Hocker, A., and Zhang, Z., hep-ph/0208177 (2002).
22. Mohr, P. J., and Taylor, B. N., *Rev. Mod. Phys.*, **72**, 351 (2000).
23. Czarnecki, A., and Marciano, W. J., *Phys. Rev.*, **D64**, 013014 (2001).
24. Akhmetshin, R. R., et al., *Phys. Lett.*, **B527**, 161 (2002).
25. Bai, J. Z., et al., *Phys. Rev. Lett.*, **84**, 594 (2000).
26. Bai, J. Z., et al., *Phys. Rev. Lett.*, **88**, 101802 (2002).
27. ALEPH Collaboration, ALEPH 2000-030 CONF 2002-019 (2002).
28. Anderson, S., et al., *Phys. Rev.*, **D61**, 112002 (2000).
29. Edwards, K. W., et al., *Phys. Rev.*, **D61**, 072003 (2000).
30. Krause, B., *Phys. Lett.*, **B390**, 392 (1997).
31. Alemany, R., Davier, M., and Hocker, A., *Eur. Phys. J.*, **C2**, 123–135 (1998).
32. Hayakawa, M., and Kinoshita, T., hep-ph/0112102 (2001).
33. Bijnens, J., Pallante, E., and Prades, J., *Nucl. Phys.*, **B626**, 410–411 (2002).
34. Knecht, M., and Nyffeler, A., *Phys. Rev.*, **D65**, 073034 (2002).
35. Knecht, M., Nyffeler, A., Perrottet, M., and De Rafael, E., *Phys. Rev. Lett.*, **88**, 071802 (2002).
36. Blokland, I., Czarnecki, A., and Melnikov, K., *Phys. Rev. Lett.*, **88**, 071803 (2002).

Semi-Inclusive Deep-Inelastic Scattering

C.A. Miller

TRIUMF

Abstract. Some prominent new semi-inclusive DIS results from JLAB and HERMES that have appeared since SPIN 2000 are presented. Polarized quark distributions have been extracted from the complete HERMES data set on H and D from 1996-2000, based on Monte Carlo "purities". The data quality permits for the first time the separation of the $\bar{u}$, $\bar{d}$ and $s+\bar{s}$ sea quark polarizations. Also, both laboratories have produced new single-spin asymmetries in the azimuthal distributions of semi-inclusive hadrons. New SSAs from longitudinally polarized proton and deuteron targets are presented, as well as beam-helicity SSAs from unpolarized protons.

POLARIZED QUARK DISTRIBUTIONS

The last decade has seen remarkable progress in defining the polarized quark distributions $\Delta q_f(x) \equiv q_f^{\uparrow\uparrow}(x) - q_f^{\uparrow\downarrow}(x)$, where e.g. $q_f^{\uparrow\uparrow}(x)$ represents the probability of finding a quark of any particular flavor f with momentum fraction x of the target nucleon, and with its spin parallel to that of the nucleon. The most precise and clearly interpreted experimental tool has been inclusive deeply inelastic lepton scattering (DIS), applied at the CERN, SLAC and DESY laboratories. In this process, the beam lepton emits a virtual photon with energy ν and invariant mass $-Q^2$, which can be considered to be absorbed by a single quark provided that $Q^2 > 1\,\mathrm{GeV}^2$. However, the information available from this process has inherent limitations, as the cross sections are sensitive to only the *square* of the charge of the quark absorbing the exchanged virtual photon. Hence sea quarks can not be distinguished from valence quarks, and only one particular flavor non-singlet combination of distributions can be directly inferred from a combination of inclusive DIS data on both proton and "neutron" targets: $\Delta q_3(x,Q^2) = \Delta u + \Delta\bar{u} - \Delta d - \Delta\bar{d} = 6(g_1^p - g_1^n)$. Here the $g_1(x)$'s are the spin structure functions extracted from the measured cross section polarization asymmetries that correlate beam and target spins. Further inference from inclusive data requires an additional assumption of SU(3) flavor symmetry relating the spin structure of various hadrons. This allows the use of hyperon beta decay to constrain the first moment of another non-singlet flavor combination involving the strange sea. The celebrated result of this approach is that quark helicities seem to make a small net contribution to the nucleon spin, and the strange sea appears to be somewhat negatively polarized [1, 2, 3].

The questionable assumption of SU(3) flavor symmetry can be avoided at the cost of some complexity in the extraction of more information from the DIS measurements. Leading hadrons produced by fragmentation of the struck quark carry information about that quark's flavor. Hence identification of the type of hadrons detected together with the scattered lepton in these *semi-inclusive* measurements effectively "tags" the quark flavor.

CP675, *Spin 2002: 15th Int'l. Spin Physics Symposium and Workshop on Polarized Electron Sources and Polarimeters*, edited by Y. I. Makdisi, A. U. Luccio, and W. W. MacKay

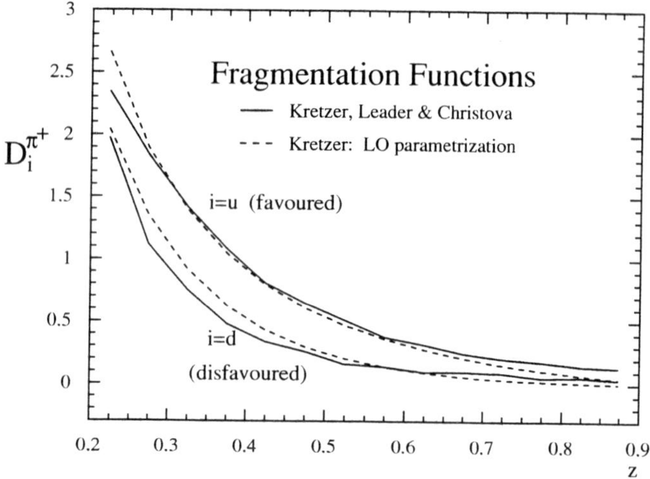

FIGURE 1. A comparison at $\langle Q^2\rangle = 2.5\,\mathrm{GeV}^2$ of the π^+ fragmentation functions from the leading-order parameterization of ref. [4] fitted to e^+e^- data (dashed curves) with those extracted from a combination of HERMES semi-inclusive DIS multiplicity data and the singlet fragmentation function $D_{\Sigma}^{\pi^+}$ from e^+e^- data [5] (solid curves).

This method obviously depends on knowledge of the set of probabilities that each quark flavor f will fragment into each type of hadron h carrying energy fraction $z = E_h/\nu$ of the virtual photon energy, information which is encoded in *fragmentation functions* $D_f^h(z)$. These functions can now be extracted from hadron multiplicity distributions either in flavor-enriched jets at the $e^+e^- \to Z^0$ pole, or in DIS measurements on both proton and "neutron" targets. Fig. 1 shows a comparison of a parameterization [4] based on e^+e^- data with the results of a hybrid analysis of pion multiplicities from DIS on a hydrogen target in combination with the singlet fragmentation function $D_{\Sigma}^{\pi^+}$ from e^+e^- data [5]. The consistency of the results lends support to the universality of the extracted fragmentation functions, particularly in the kinematic conditions of this DIS experiment.

Semi-inclusive DIS measurements require a combination of a polarized lepton beam with good duty factor, polarized targets of both hydrogen and e.g. deuterium, and a spectrometer with substantial acceptance for the produced hadrons. The first attempt to extract polarized quark distributions from semi-inclusive asymmetries was made several years ago by the SMC experiment [6]. As the detected hadrons were unidentified except for charge, and the statistical precision was limited, different flavors of sea quarks could not be distinguished in the analysis. Now the HERMES collaboration has released a new preliminary five-flavor $(u, d, \bar{u}, \bar{d}, s = \bar{s})$ extraction from their entire 1996-2000 data set comprising identified semi-inclusive pions from a hydrogen target and identified pions and kaons from a deuterium target. Important features of this experiment include pure nuclear-polarized atomic gas targets in a polarized electron storage ring, and the capability in the spectrometer for hadron identification. The quality of the resulting deuterium asymmetry data is shown in Fig. 2. They have been integrated over the range

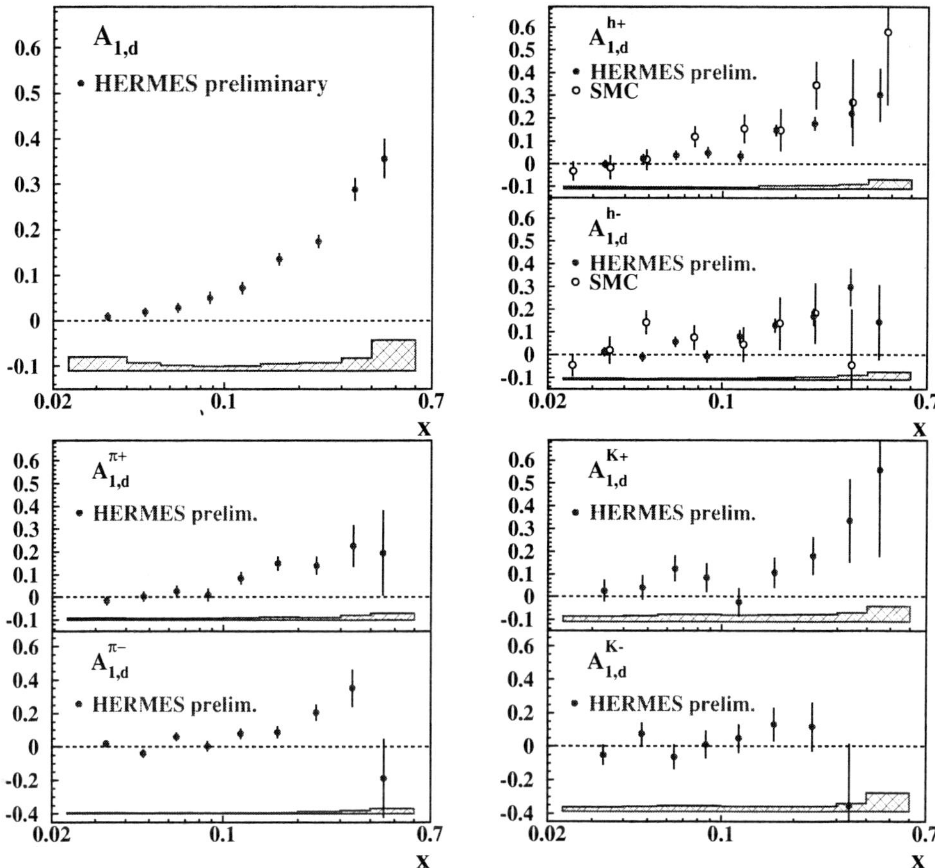

FIGURE 2. Preliminary photo-absorption asymmetries A_1 on the deuterium target from the HERMES collaboration. The upper left panel shows the inclusive asymmetries, the upper right the semi-inclusive asymmetries for all unidentified hadrons for comparison with SMC [6], and the lower panels are for identified pions and kaons as indicated. No corrections have been applied at this stage for the effects of experimental acceptance, which LEPTO simulations suggest are small.

$0.2 < z < 0.8$. The lower limit is chosen to suppress hadrons from target fragmentation. The absence of any z-dependence of the asymmetries in this accepted range is taken as evidence for the dominance of current fragmentation. Kinematic migration due to instrumental and radiative effects has been fully unfolded using information internal to a DIS Monte Carlo event simulation incorporating the significant radiative diagrams and vertex corrections, as well as a detailed simulation of the spectrometer.

The experimental data set thus consists in each x bin of a vector of asymmetries on both hydrogen and deuterium targets, both inclusive and semi-inclusive for various hadrons:

$$\vec{A} = (A_{1,\mathrm{p}}, A_{1,\mathrm{d}}, A^{\pi^\pm}_{1,\mathrm{p}}, A^{\pi^\pm}_{1,\mathrm{d}}, A^{\mathrm{K}^\pm}_{1,\mathrm{d}}). \tag{1}$$

To these data are fit independently in each x bin a vector of five quark polarizations:

$$\vec{Q} = \left(\frac{\Delta u}{u}, \frac{\Delta d}{d}, \frac{\Delta \bar{u}}{\bar{u}}, \frac{\Delta \bar{d}}{\bar{d}}, \frac{\Delta s + \Delta \bar{s}}{s + \bar{s}} \right). \tag{2}$$

The leading-order relationship that is fit is

$$A_1^h(x) \overset{g_2=0}{\simeq} \frac{\sum_f e_f^2 \, \Delta q_f(x) \int dz \, D_f^h(z)}{\sum_{f'} e_{f'}^2 \, q_{f'}(x) \int dz \, D_{f'}^h(z)} \frac{(1+R(x))}{(1+\gamma^2)} \tag{3}$$

$$= \sum_f \underbrace{\frac{(1+R(x))}{(1+\gamma^2)} \frac{e_f^2 \, q_f(x) \int dz D_f^h(z)}{\sum_{f'} e_{f'}^2 \, q_{f'}(x) \int dz D_{q_{f'}}^h(z)}}_{\mathcal{P}_f^h(x) \;\equiv\; \text{Purity}} \frac{\Delta q_f(x)}{q_f(x)}. \tag{4}$$

For simplicity, the weak logarithmic dependences of all functions on Q^2 are suppressed here. The factor involving the kinematic value $\gamma^2 = Q^2/\nu^2$ and R, the ratio of longitudinal to transverse cross sections, accounts for the longitudinal component included in parametrizations of the unpolarized parton distribution functions $q_f(x)$. This set of equations may be represented in matrix form

$$\vec{A}(x) = \mathcal{P}_f^h(x) \cdot \vec{Q}(x), \tag{5}$$

where all of the information that comes from previous unpolarized measurements — both distribution and fragmentation functions — is combined in the *purity* functions $\mathcal{P}_f^h(x)$, each of which represents the conditional probability that an observed hadron of type h was produced by a struck quark of flavor f with momentum fraction x. The purities used in the new HERMES analysis were extracted from the above-mentioned Monte Carlo simulation (but now excluding QED radiative and detector resolution effects) as $\mathcal{P}_f^h(x) = N_f^h(x)/\sum_{f'} N_{f'}^h(x)$, where N_f^h is the number of hadrons of type h, in the geometric experimental acceptance and in the interval $0.2 < z < 0.8$, that were produced from quarks of flavor f. The simulation employs the CTEQ5 Low–Q^2 parametrization [2] for the unpolarized PDF's, and JETSET fragmentation parameters that were tuned to fit hadron multiplicities measured at HERMES [7]. The simulated purities were then augmented by the $1+R$ factor evaluated as per ref. [8], as well as nuclear corrections for the deuteron target based on a d-state probability $\omega_D = 0.05 \pm 0.01$.

The over-constrained linear system Eq. 5 is solved for the unknown $\vec{Q}$ vector by the usual χ^2 minimization. The precision with which this system can be solved clearly depends on the degree of orthogonality of the purities — i.e. how well flavor-tagging works. Positive hadrons alone can distinguish valence u and d quarks. In this case, u-quark dominance of h^+ production from both proton and neutron targets is helpful. Negative hadrons can perform a similar separation for sea quark flavors, as production of h^- from sea quarks is relatively larger than h^+, and again dominated by $\bar{u}$ quarks. The general separation of valence and sea quarks is mainly done by hadron charge. Identified kaons available in the data set from the HERMES deuterium target have a

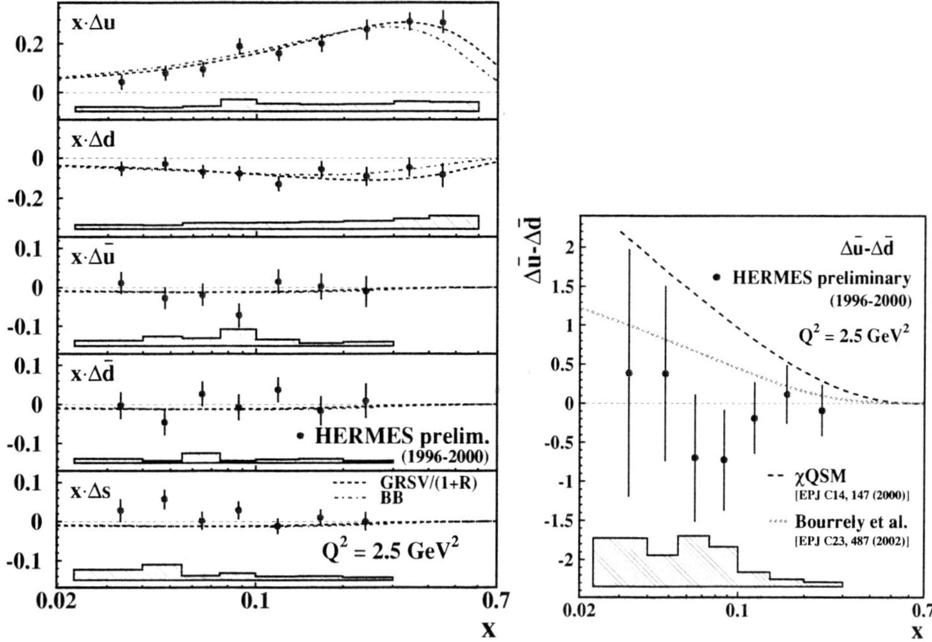

FIGURE 3. Left panel *(a)*: Extracted polarized quark distributions as a function of Bjørken x, compared to two NLO QCD fits to world inclusive data [3, 9]. Right panel *(b)*: The light quark sea flavor asymmetry $\Delta\bar{u} - \Delta\bar{d}$ in the polarized distributions, compared to theoretical predictions based on the chiral quark soliton model [10] and on a statistical model [11]

larger sensitivity to the strange sea, but still at only the 10% level due to u or $\bar{u}$ quark dominance. The systematic uncertainties in kaon purities are presently large due to the uncertainties in both the fragmentation functions and unpolarized parton densities. However, examination of Eq. 4 reveals that the effects of individual unpolarized densities $q_f(x)$ tend to cancel when the polarized densities $\Delta q_f(x)$ are finally derived from the quark polarizations (the solutions of that equation) by multiplying by the unpolarized densities.

The polarized densities $\Delta q_f(x)$ resulting from the analysis are shown in Fig. 3*a)*, compared to two NLO QCD fits to world inclusive data that assume SU(3) flavor symmetry. The systematic uncertainties include contributions from the $A_1^{(h)}$ and from the unpolarized PDF's and purities. Further support for the assumption of pure current fragmentation in the employed z-range is given by the finding that the extracted densities are little affected by omission of the inclusive asymmetries from the extraction, even though the inclusive data tends to have better statistical precision.

The extracted Δu and Δd are consistent with previous semi-inclusive results [6, 12]. The sea distributions, extracted separately here for the first time, are consistent with zero. There is no indication that the strange polarization is negative, in contrast to the findings from the inclusive data that are based on SU(3) flavor symmetry. Another surprise shown

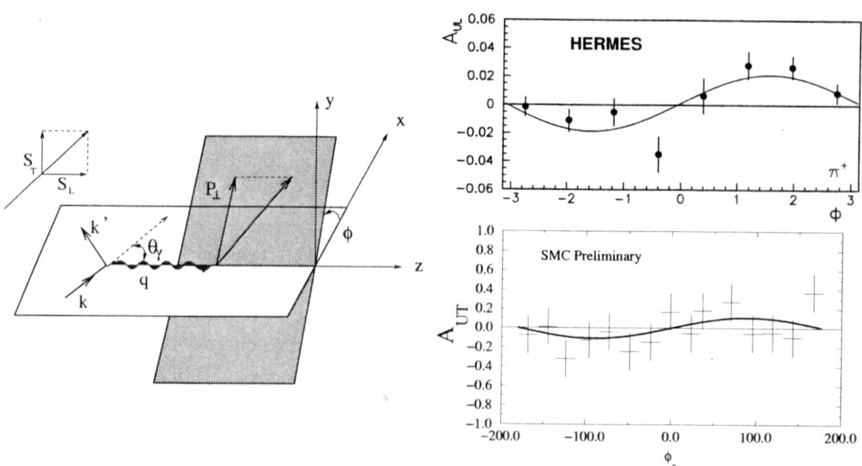

FIGURE 4. *a)* Kinematic geometry for semi-inclusive DIS. *b)* Single-spin asymmetries for longitudinal (HERMES) and transverse (SMC) target polarization.

in Fig. 3*b)* is that there is no evidence of the positive flavor asymmetry $\Delta\bar{u} - \Delta\bar{d}$ in the light quark sea that is predicted by some theoretical models [10].

SINGLE-SPIN AZIMUTHAL ASYMMETRIES

In 1999, HERMES [13] and SMC [14] released observations of semi-inclusive single-spin asymmetries that depend on the orientation of the hadron's $P_\perp$ azimuthally about the direction of the virtual photon (see Fig. 4*a*)). Relative to the lepton beam axis, the target polarization was transverse for SMC, and longitudinal for HERMES. In either case, the linear polarization of the virtual photon in the lepton plane selects transverse polarization of the struck quark. In leading twist, $P_\perp$ can arise from two sources: primordial p_T of the quark in the target, or k_T produced in the fragmentation process. A single-spin asymmetry requires the participation of some time-odd object correlating that p_T or k_T with some spin. Theoretical explanations of these asymmetries have focussed on a T-odd fragmentation process sensitive to the transverse polarization of the fragmenting quark [15]. The associated "Collins fragmentation function", designated $H_1^{\perp(1)}(z)$, is T-odd not in the fundamental sense, but only through the soft interactions in the final state. The wonderful implication of such an interpretation is that this "quark polarimeter" could provide access to the chiral-odd, and hence otherwise-inscrutable, *transversity* distribution. Transversity is the last remaining unmeasured one of the three leading-twist quark distributions that survive integration over intrinsic p_T. It is related to the probability of finding a transversely polarized quark in a nucleon polarized transversely with respect to its infinite momentum.

Unfortunately, almost all of the existing data revealing target-related single-spin asymmetries have been measured with a *longitudinally* polarized target. However, even

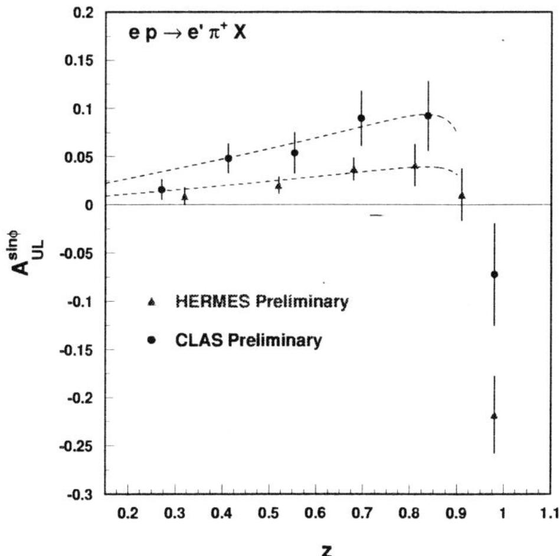

FIGURE 5. Analyzing powers in the $\sin\phi$ moment of single-spin azimuthal asymmetries with a longitudinally polarized hydrogen target, from both HERMES and CLAS. The arbitrarily normalized curves are from ref. [18].

these data can be interpreted [16] in terms of the same Collins fragmentation function, operating together with transversity-related distribution functions — one twist-2 ($h_{1L}^{\perp}$, representing the probability of finding a transversely polarized quark in a longitudinally polarized nucleon) and one twist-3 (h_L). Hence such data from the HERMES collaboration, and now emerging from the CLAS collaboration at JLAB as well, offer important evidence that the Collins function has a substantial magnitude. However, there are presently too may unknowns for an unambiguous interpretation. Some assumptions must be made — e.g. neglecting either the interaction-dependant part of h_L, or neglecting $h_{1L}^{\perp}$ altogether. It has then been found possible to explain most of the existing data [17]. For example, the z-dependence of the asymmetries appear to be explained by new theoretical calculations done at the one-loop level in a chiral- invariant model [18], as shown in Fig. 5. Their model also predicts azimuthal asymmetries for $e^+e^- \to 2$ jets up to of order 5%, consistent with preliminary DELPHI data within their rather large systematic uncertainty [19].

HERMES has now also released results on the same observable for both pions and kaons from a deuterium target. These are shown in Fig. 6, where selected examples are compared in the lower panels to theoretical predictions where the transversity distributions are calculated in the chiral quark soliton model [20]. In these calculations, the interaction-dependent part of h_L was neglected. The Collins function was parameterized to fit the HERMES proton data, now resulting in a magnitude which corresponds to the "optimistic" (large) extreme from DELPHI: $\frac{\langle H_1^{\perp(1)}\rangle}{\langle D_1\rangle} = (12.5 \pm 1.4)\%$. Previous estimates of the Collins function were only half this size, due to an error in the sign of the contribution of that component of the target polarization that is orthogonal to the virtual photon direction. (This component appears even in the HERMES experiment with target polar-

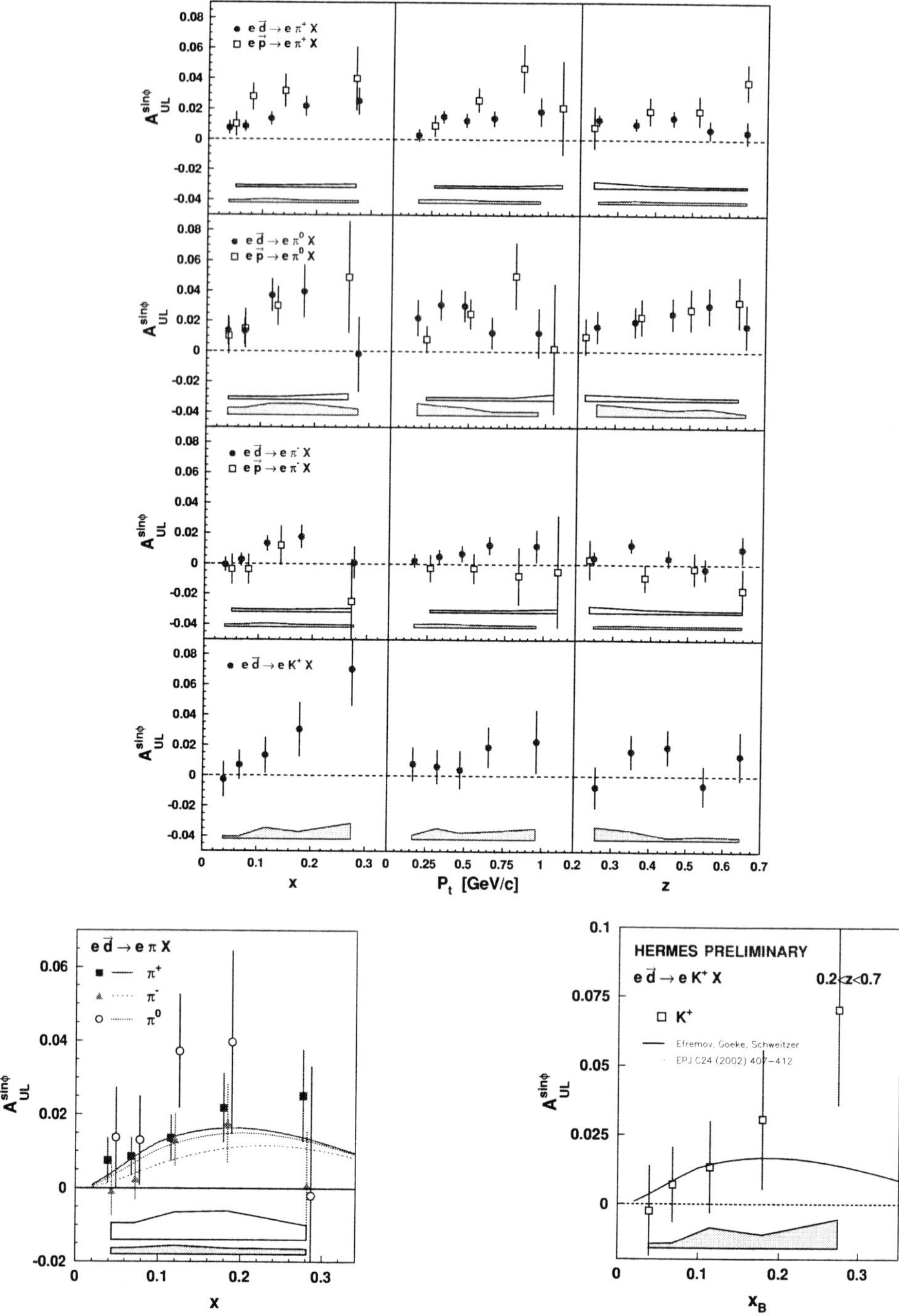

FIGURE 6. Preliminary HERMES analyzing powers in the $\sin\phi$ moment of single-spin asymmetries with a longitudinally polarized deuterium target. The curves are theoretical calculations from ref. [20]

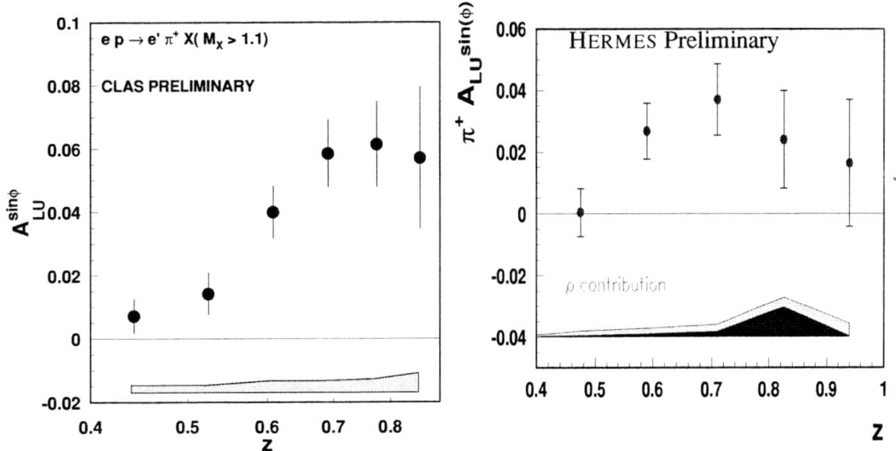

FIGURE 7. Analyzing powers in the $\sin\phi$ moment for single-spin azimuthal asymmetries with a longitudinally polarized positron beam.

ization parallel to the *lepton* beam.) Another result of the corrected sign is that the sign of the SMC preliminary asymmetry with a transversely polarized target no longer appears to be theoretically consistent with that of the HERMES results with a longitudinally polarized target, although those SMC data are not statistically definitive.

Beam-helicity related single-spin azimuthal asymmetries are also understood to be generated by the same Collins fragmentation function, but in conjunction with an unknown twist-3 chiral-odd distribution function designated $e(x)$, whose first moment is related to the pion-nucleon σ term [21]:

$$\int_{-1}^{1} dx[e^u + e^d](x) = \frac{\sigma}{m_{av}} \tag{6}$$

$$m_{av} \equiv \frac{1}{2}(m_u + m_d) \simeq 5\,\text{MeV} \tag{7}$$

Some newly released results for the beam-related single-spin asymmetry from both HERMES and CLAS are shown in Fig. 7. The increased magnitude at larger z is again what is expected of the Collins fragmentation function, while no x-dependence of the asymmetry is observed.

Recently, this arena of single-spin asymmetries has seen some theoretical turmoil, not only because of the detected sign error mentioned above. Even more excitement was precipitated by Brodsky, Hwang and Schmidt [22], who presented a model calculation including a short-distance (hard) "final state interaction" (FSI) between struck quark and spectator. They showed that this mechanism can also generate a single-(transverse)spin azimuthal asymmetry, although no quantitative predictions are yet available based on this model. Contrary to naive appearances, it's a leading twist mechanism. This model inspired John Collins to realize [23] that such a mechanism is the old Sivers Effect [24, 25] in disguise. The model calculation reveals that the T-odd *distribution* function representing the Sivers Effect as the correlation between intrinsic quark p_T and transverse quark polarization in an unpolarized nucleon does not violate fundamental time reversal invariance after all (contrary to Collins' own apparent proof of a few years ago). Furthermore, Ji and Huang [26] explained how parton distributions in the light cone guage can still be interpreted as parton densities, in spite of the claim by Brodsky *et al.* that the nucleon wave function does not contain any information about the FSI.

Thus we are now left with two competing explanations for target-related single-spin asymmetries: the T-odd Collins fragmentation function, and the T-odd distribution function introduced by Sivers. Fortunately they can be distinguished [27] by new data now being recorded by both HERMES and COMPASS with transversely polarized targets, through the different azimuthal dependences of the asymmetry — the Sivers effect depends on the azimuthal angle *difference* between the target spin axis and hadron plane, through $\sin(\theta_h^l - \theta_S^l)$, while the Collins effect depends on the *sum* of these angles relative to the lepton scattering plane via $\sin(\theta_h^l + \theta_S^l)$. If the Collins effect is confirmed to be substantial, the first measurements of transversity can be extracted.

For more details on this rich and rapidly evolving subject of single-spin asymmetries, please find in this volume the theoretical talk by Phillip Ratcliffe, and the experimental talks by Harut Avakian and Delia Hasch.

REFERENCES

1. Ashman, J., et al., *Phys. Lett.*, **B 206**, 364 (1988).
2. Lai, H., et al., *Eur. Phys. J.*, **C 12**, 375 (2000).
3. Glück, M., et al., *Phys. Rev.*, **D 63**, 094005 (2001).
4. Kretzer, S., *Phys. Rev.*, **D 62**, 054001 (2000).
5. Kretzer, S., Leader, E., and Christova, E., *Eur. Phys. J.*, **C 22**, 269 (2001).
6. Adeva, B., et al., *Phys. Lett.*, **B 420**, 180 (1988).
7. Menden, F., *Determination of the Gluon Polarization in the Nucleon*, Ph.D. thesis, Albert-Ludwigs-Universitaet Freiburg, Breisgau (2001), hERMES Note 01-073.
8. Whitlow, L. W., et al., *Phys. Lett.*, **B250**, 193–198 (1990).
9. Blümlein, J., and Böttcher, H., *Nucl. Phys.*, **B636**, 225–263 (2002).
10. Dressler, B., Göke, K., Polyakov, M. V., and Weiss, C., *Eur. Phys. J.*, **C14**, 147–157 (2000).
11. Bourrely, C., Soffer, J., and Buccella, F., *Eur. Phys. J.*, **C23**, 487–501 (2002).
12. Ackerstaff, K., et al., *Phys. Lett.*, **B 464**, 123 (1999).
13. Airapetian, A., *Phys. Rev. Lett.*, **84**, 4047–4051 (2000).
14. Bravar, A., et al., *Nucl. Phys. (Proc. Suppl.)*, **B 79**, 521 (1999).
15. Collins, J. C., *Nucl. Phys.*, **B396**, 161–182 (1993).
16. Boglione, M., and Mulders, P. J., *Phys. Lett.*, **B478**, 114–120 (2000).
17. Ma, B.-Q., Schmidt, I., and Yang, J.-J., *Phys. Rev.*, **D66**, 094001 (2002).
18. Bacchetta, A., Kundu, R., Metz, A., and Mulders, P. J., *Phys. Rev.*, **D65**, 094021 (2002).
19. Efremov, A. V., et al., *Czech. J. Phys.*, **49**, S75 (1999).
20. Efremov, A. V., Goeke, K., and Schweitzer, P., *Eur. Phys. J.*, **C24**, 407–412 (2002).
21. Jaffe, R. L., and Ji, X.-D., *Nucl. Phys.*, **B375**, 527–560 (1992).
22. Brodsky, S. J., Hwang, D. S., and Schmidt, I., *Phys. Lett.*, **B530**, 99–107 (2002).
23. Collins, J. C., *Phys. Lett.*, **B536**, 43–48 (2002).
24. Sivers, D. W., *Phys. Rev.*, **D41**, 83 (1990).
25. Sivers, D. W., *Phys. Rev.*, **D43**, 261–263 (1991).
26. Ji, X.-d., and Yuan, F., *Phys. Lett.*, **B543**, 66–72 (2002).
27. Boglione, M., and Mulders, P. J., *Phys. Rev.*, **D60**, 054007 (1999).

Experimental Verification of the GDH Sum Rule

A survey including the extension of the GDH integral to virtual photons

K. Helbing
on behalf of the GDH-Collaboration

Physikalisches Institut, Universität Erlangen-Nürnberg, 91058 Erlangen, Germany

Abstract. Sum rules involving the spin structure of the nucleon like those due to Bjorken, Ellis and Jaffe and the one due to Gerasimov, Drell and Hearn offer the opportunity to study the structure of strong interactions. At long distance scales in the confinement regime the Gerasimov-Drell-Hearn (GDH) Sum Rule connects static properties of the nucleon like the anomalous magnetic moment κ and the nucleon mass m, with the spin dependent absorption of real photons with total cross sections $\sigma_{3/2}$ and $\sigma_{1/2}$ [1]:

$$\int_0^\infty \frac{d\nu}{\nu}\left(\sigma_{3/2}(\nu)-\sigma_{1/2}(\nu)\right)=\frac{2\pi^2\alpha}{m^2}\cdot\kappa^2 \tag{1}$$

Hence the full spin-dependent excitation spectrum of the nucleon is related to its static properties. The sum rule has not been investigated experimentally until recently. For the first time this fundamental sum rule is verified by the GDH-Collaboration with circularly polarized real photons and longitudinally polarized nucleons at the two accelerators ELSA and MAMI. The investigation of the response of the proton as well as of the neutron allows to perform an isospin decomposition. Data from the resonance region up to the onset of the Regge regime are shown.

The "sum" on the left hand side of the GDH Sum Rule can be generalized to the case of virtual photons. This allows to establish a Q^2 dependency and to study the transition to the perturbative regime of QCD. This is the subject of several experiments e.g. at JLAB for the resonance region and of the HERMES experiment at DESY for higher Q^2.

Moreover, this paper covers the status of theory concerning the GDH Sum Rule, the different experimental approaches and the results for the absorption of real and virtual photons will be reviewed.

1. INTRODUCTION

The GDH Sum Rule has been derived in parallel by several authors in the second half of the 1960ies. Today mostly Gerasimov [1], Drell and Hearn [2] are referenced. Both works are based on a dispersion theoretic derivation. Hosoda and Yamamoto [3] in 1966 used the current algebra formalism to derive the same sum rule. Drell and Hearn rated the sum rule to be of purely academic interest only since it would be impossible to experimentally test it. In contrast, Hosoda and Yamamoto were convinced that it would be straight forward to test it. Moreover, they thought that the sum rule only holds for

[1] Here $3/2$ and $1/2$ identify relative spin orientation of the photon and the nucleon parallel or anti-parallel respectively in the nucleon rest frame; α denotes the fine-structure constant and ν the energy of the photon.

CP675, *Spin 2002: 15th Int'l. Spin Physics Symposium and Workshop on Polarized Electron Sources and Polarimeters,* edited by Y. I. Makdisi, A. U. Luccio, and W. W. MacKay

a Dirac particle and hence a verification could prove that the nucleon complies to this assumption. Iddings [4] in 1965 on the other hand was already all the way there to write down the sum rule for $Q^2 = 0$ but didn't explicitly do so. Nonetheless, his work already contains a version of what is called today a generalization of the GDH sum or integral.

For most of the further discussion we concentrate here on the dispersion theoretic derivation used by Gerasimov, Drell and Hearn. Only fundamental constraints enter this derivation: Lorentz invariance and gauge invariance allow to write the Compton-forward amplitude in a simple form; unitarity provides the Optical Theorem; causality and the so-called *No-Subtraction-Hypothesis* lead to the Kramers-Kronig dispersion relation; the Low-Theorem [5, 6] is based again on Lorentz and gauge invariance.

Among these constraints the *No-Subtraction-Hypothesis* is the only assumption which is open to reasonable questions. To reduce the dispersion relation for the spin-flip Compton forward amplitude $f_2(\nu)$ from a contour integral in the complex plane to an integration along the real axis one has to presume that $f_2(\nu)/\nu \rightarrow 0$ for $\nu \rightarrow \infty$. A violation of this hypothesis would lead to a weird behavior of this spin-flip amplitude and the corresponding differential forward cross sections, namely:

$$\lim_{\nu \to \infty} \frac{1}{d\Omega} \left(d\sigma_{3/2} - d\sigma_{1/2} \right) \Big|_{\theta=0} = \infty \tag{2}$$

On the other hand for the total cross section Regge arguments related to the Froissart bound ensure the following behavior:

$$\lim_{\nu \to \infty} \left(\sigma_{3/2}^{\mathrm{tot}} - \sigma_{1/2}^{\mathrm{tot}} \right) = 0 \tag{3}$$

A possible failure of the GDH Sum Rule would be related to a violation of the No-Subtraction-Hypothesis. There have been several attempts in the past to find reasons for such a failure. Here some of them are reviewed:

Based on the current algebra derivation by Hosoda and Yamamoto the authors Chang, Liang and Workman [8] have argued that an anomaly in the charge density commutator gives rise to a modification of the GDH Sum Rule. Pantförder [9] was able to show that the contribution from this anomaly cancels going to the infinite-momentum limit which ultimately leads to the GDH Sum Rule.

It was questioned if the Low-Theorem holds to all orders of electromagnetic coupling. While Low [5] showed the derivation only in the lowest order Gell-Mann and Goldberger stated in their original paper [6] that their derivation should be "exactly correct in any known theory". Later Roy and Singh [10] established the low theorem to the order α^2.

Haim Goldberg [11] suspected that the photoproduction of gravitons violates the GDH Sum Rule. Contributions at very high energies from photonuclear reactions other than those of strong interactions may not be ignored a priori. On the other hand, the contribution of these effects at high energies to the sum rule will be largely damped due to the weighting with the inverse of the photon energy.

Already in 1968 right after the discovery of the GDH Sum Rule Abarbanel and Goldberger [12] considered a $J = 1$ Regge fixed pole being a possible source for the failure of the No-Subtraction-Hypothesis. Despite such a fixed pole is forbidden by the Froissart theorem for purely hadronic processes and such a behavior has never been observed so far it cannot be ruled out completely for electro-weak processes. Nevertheless, it should be mentioned that it is not quite clear if such a fixed pole in the case of real Compton scattering would not violate the Landau-Yang theorem which forbids two photons to have a total angular momentum of $J = 1$.

Further examples for possible failures of the sum rule that have been considered can be found in Ref. [7]. To summarize, today no stringent indication for a modification of the GDH Sum Rule exists.

2. THE GDH-EXPERIMENT AT ELSA AND MAMI

2.1. Experimental concept

The primary aim of the GDH-Collaboration[2] is obviously to verify the Gerasimov-Drell-Hearn Sum Rule. The central issue of the experimental conception of the GDH-Collaboration is the reduction of systematic uncertainties in order to provide a setup compatible with the fundamental character of the sum rule.

2.1.1. Region of integration

The GDH integrand on the left hand side of the GDH Sum Rule will be determined from the resonance region up to the onset of the Regge regime. This is achieved by the use of two electron accelerators with high duty cycle:

0.14 - 0.8 GeV	MAMI (Mainz)
0.7 - 3.1 GeV	ELSA (Bonn)

At ELSA a completely new experimental area was setup for the GDH measurements while at MAMI the existing tagging facility in the A2-Hall was available. At MAMI two primary electron energy settings are used to cover the energies from pion threshold up to 800 MeV. Five primary electron energy settings at ELSA allow to cover photon energies up to 3.1 GeV. The circular polarization of the photons is given by the helicity transfer of the longitudinal polarization of the electrons.

2.1.2. Beam polarization

At both accelerators the polarization of the electron beam is achieved by high intensity sources with strained super-lattice GaAs-crystals. The typical polarization of the delivered electron beam is $65 - 75\%$ [17, 18].

---

[2] For a list of participants of the GDH-Collaboration be referred to the author list of Ref. [14].

The race-track of the electrons at MAMI is deterministic. Hence, almost no polarization is lost on the way from the source to the experiment. Møller polarimetry is provided simultaneously to the photon tagging by a magnetic tagging spectrometer.

ELSA is a storage type accelerator with depolarizing resonances. The spin of the electrons has to be transported vertically in ELSA and rotated to the longitudinal direction in the external beam line for the experiment. Because of these more delicate circumstances of spin maintenance a dedicated 2-arm Møller spectrometer with large acceptance was built. It enables fast spin diagnostics in all 3 vector components [19, 27].

2.1.3. *Frozen spin target*

A new solid state polarized frozen-spin target has been developed for the GDH measurements [16]. The central part of this new target consists of a ^{3}He/^{4}He dilution refrigerator that is installed horizontally along the beam axis. The refrigerator includes an internal superconducting holding coil to maintain the nucleon polarization in the frozen-spin mode longitudinally to the beam. The design of the dilution refrigerator and the use of an internal holding coil enabled for the first time the measurement of a spin-dependent total cross section in combination with a polarized solid state target. Due to the low fringe field of the holding coil and the horizontal alignment allows the detection of emitted particles with an angular acceptance of almost 4π (see below). Butanol provided polarized protons. In addition, ^{6}LiD was used to obtain polarized neutrons. Typical values of polarization of the protons in the butanol that have been reached during data taking are 70-80%.

2.1.4. *Detector concepts*

Two detector concepts are used to meet the special requirements for the different energy ranges: The DAPHNE detector at MAMI and the GDH-Detector at ELSA.

DAPHNE [20] is well suited for charged particle detection and for the identification of low multiplicity states. It is essentially a charged particle tracking detector having a cylindrical symmetry. In addition it has a useful detection efficiency for neutral pions. In forward direction a silicon microstrip device called MIDAS [22] extends the acceptance for charged particles.

The GDH-Detector [21] has been specifically designed for measurements of total cross sections and is perfectly suited for situations where the contributing channels are not well known and extrapolations due to unobserved final states are not advisable. The concept of the GDH-Detector is to detect at least one reaction product from all possible hadronic processes with almost complete acceptance concerning solid angle and efficiency. This is achieved by an arrangement of scintillators and lead. The over all acceptance for any hadronic process is better than 99 %.

Both detection systems have similar components in forward direction. The electromagnetic background is suppressed by about 5 orders of magnitude by means of a threshold Č detector [21]. The Č detector is followed by the Far-Forward-Wall – a component similar to the central parts of the GDH-Detector – to complete the solid angle coverage [23].

2.2. Results

2.2.1. Systematic studies

Measurements of unpolarized total photoabsorption cross sections were performed [24, 25, 26] to ensure that both detection systems are operational even for measurements of differences of cross sections. An unprecedented data quality has been reached in unpolarized measurements on ^{1}H, ^{2}H and ^{3}He in the photon energy range from pion threshold to 800 MeV as well as on Carbon and Beryllium in the energy range from 250 MeV to 3100 MeV.

Systematic studies with respect to spin have been performed with an unpolarized butanol target in frozen spin mode with all possible holding field configurations. In any case the false asymmetry of $\sigma_{3/2} - \sigma_{1/2}$ turned out to be less than 2 μb [28].

2.2.2. Polarized cross sections in the resonance region

Fig. 1 shows the doubly polarized results for $\sigma_{3/2} - \sigma_{1/2}$ on the proton that have been obtained until now. For comparison also the unpolarized cross section is plotted. The data taken at MAMI have already been published [13] while the analysis of the data collected at ELSA is preliminary [27, 28]. One observes that the data sets for the different energy settings at the two accelerators match each other very well. The three major resonances known from the unpolarized total cross section are present in the difference as well - they are even more pronounced.

Here, it is not possible to summarize the wealth of information obtained for the single resonances especially with respect to partial channels. Much of it is currently in the process of being published. As examples Refs. [14, 15] might serve. A more detailed view of our results for the second resonance region is shown in Fig. 2 on the left. The differences of the total cross sections and the single pion contribution are compared to the corresponding predictions of a unitary isobar model called MAID [29]. One observes that the parameterization fails to describe our data. This might lead to a refined understanding of the involved resonance couplings to the multipoles $M_{2-}^{1/2}, E_{2-}^{1/2}$.

In Fig. 2 on the left are shown the separate helicity states of the total photoabsorption cross sections $\sigma_{3/2}$ and $\sigma_{1/2}$. The separated helicity states are obtained by adding resp. subtracting our polarized cross section difference from the unpolarized data. As already visible in the cross section difference in Fig. 1 the third resonance is a clearly distinct feature of the spectrum. Besides, there is another structure one might call the "forth" resonance which comes more conspicuous in this representation than in the difference. This structure might be due to the F_{35} and the F_{37} contributions.

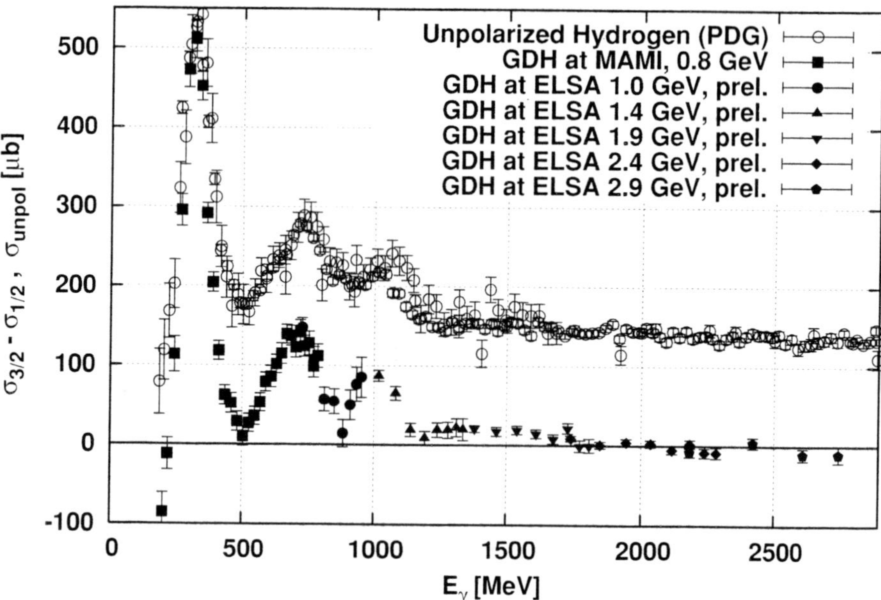

FIGURE 1. Difference of the polarized total photoabsorption of the proton in comparison to the unpolarized cross section

2.2.3. High-energy behavior

Regge fits are able to describe many unpolarized total cross sections simultaneously [30]. All data follow a simple power law, namely $\sigma_T \simeq c_1 \cdot s^{\alpha_R(0)-1} + c_2 \cdot s^{\alpha_P(0)-1}$ with $\alpha_R(0)$ being the ρ, ω trajectory intercept and $\alpha_P(0)$ being that of the Pomeron. For real photo absorption these fits are valid down to photon energies as low as 1.2 GeV. The diagram in Fig. 3 on the left substantiates this statement. Shown is the behavior of the normalized[3] reduced χ^2 under inclusion of data points of lower photon energies. One observes that for the unpolarized hydrogen data [31] the reduced χ^2 remains unchanged as long as data below 1.2 GeV is not included. Hence, it appears that the Regge behavior is established down to very low photon energies starting just above the third resonance.

In the polarized case Regge fits have recently been applied to deep inelastic scattering data [32, 33, 34]. The extrapolation of these fits to $Q^2 = 0$ indicate that the integrand of

[3] In order to have comparable values the χ^2 for the unpolarized data was scaled by a factor of 0.66 which reflects that the error calculation in Ref. [31] was too optimistic. The polarized data from the GDH-Collaboration does not show this problem. However, to have comparable variations of the reduced χ^2 – for displaying purposes only – the polarized values were raised to the 10th power. This in turn reflects the inferior statistics of the polarized data.

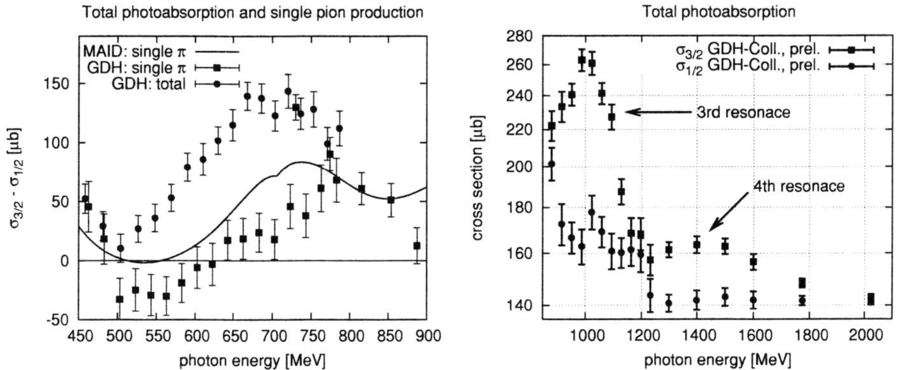

FIGURE 2. Left: Comparison of the differences of the total cross sections and the single pion contribution with the corresponding predictions from MAID ; Right: Separated helicity state total cross sections $\sigma_{3/2}$ and $\sigma_{1/2}$ for the third and a "forth" resonance.

the GDH Sum Rule on the proton could be negative at higher energies. Our polarized data up to 2.8 GeV photon energy disagree with these Regge fits (Fig. 3) but indicate a sign change at the highest energies.

Bass and Brisudová [32] have argued that the polarized cross section difference for the absorption of *virtual* photons can be described by the following Regge behavior:

$$\sigma_{3/2} - \sigma_{1/2} = \left[c_1\, s^{\alpha_{a_1}-1} \cdot I + c_2\, s^{\alpha_{f_1}-1} + c_3 \frac{\ln s}{s} + \frac{c_4}{\ln^2 s} \right] F(s, Q^2)$$

where I denotes the isospin of the nucleon. The logarithmic terms are due to Regge cuts and can be neglected at $Q^2 = 0$ [35]. Also $F(s,Q^2)$ simplifies to a constant at the real photon point and can be incorporated into c_1 and c_2. α_{a_1} and α_{f_1} are the Regge intercepts of the respective trajectories. Hence in the case of real photons the expression for the Regge behavior simplifies considerably to

$$\sigma_{3/2} - \sigma_{1/2} = c_1\, s^{\alpha_{a_1}-1} \cdot I + c_2\, s^{\alpha_{f_1}-1} \tag{4}$$

The intercept of the f_1 trajectory is rather well defined by the deep inelastic scattering data to be about -0.5. The situation is less clear with α_{a_1} where the values from different fits range from about -0.2 to +0.5. For the further calculations here we adopt a value of +0.2.

The right hand diagram of Fig. 3 shows our fit via c_1 and c_2 to the polarized data of the GDH-Collaboration. It seems that the fit works well and it indicates a sign change at about 2 GeV photon energy. The diagram on the left shows that indeed the reduced χ^2 of the fit is about unity if only data above 1.3 GeV is included. In contrast to the unpolarized case at 1.2 GeV the χ^2 already deviates from unity. We consider this as another indication for a 4th resonance as discussed in section 2.2.2.

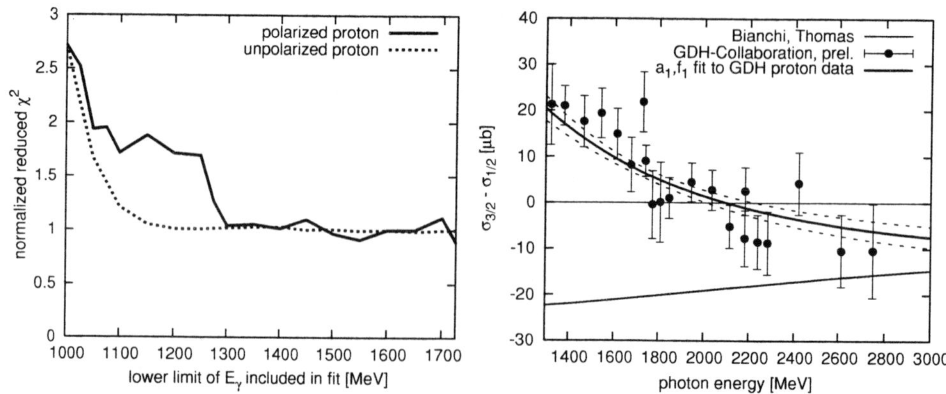

FIGURE 3. Left: Normalized reduced χ^2 as a function of the lowest energy of the data included in the fit; Right: Comparison of the difference of the total cross sections with a Regge fit to DIS data extrapolated to $Q^2 = 0$ and with our fit to the real photon data

2.2.4. Verification of the GDH Sum Rule for the proton

The GDH Sum Rule prediction for the proton amounts to 205 μb. The experimental value of the running GDH integral up to 2.8 GeV clearly overshoots this prediction (Fig. 4). The *preliminary* value of the GDH sum up to 2.8 GeV is $225 \pm 5_{\text{stat}}$ μb. This includes an unmeasured negative contribution at the pion threshold up to 200 MeV taken from Ref. [29]. Since at threshold only a simple E_{0+} amplitude contributes this can be regarded to be a reliable estimate.

The integrand $\sigma_{3/2} - \sigma_{1/2}$ remains positive from about 230 MeV on up to about 2 GeV as seen in Fig. 3 on the right. One has to assume a sign change of the integrand at higher energies as indicated by the data and the Regge fit in order to obtain a better agreement between the measurement and the GDH prediction. Indeed our Regge fit gives a contribution ranging between about -20μb and $-35\mu b$ above 3 GeV. The relatively wide range is a result of the flexibility in choosing the a_1 trajectory intercept as discussed in Sec. 2.2.3. This intercept can be determined much better with the neutron data already taken by the GDH-Collaboration at ELSA (see Sec. 2.3).

To summarize, given the data and taking into account the extrapolation to higher energies with our Regge fit we obtain agreement with the GDH Sum Rule prediction. The level of precision obtained is of the order of 5% [13, 27, 28]. This represents the first verification of the GDH Sum Rule ever.

2.3. Future measurements on the GDH Sum Rule with real photons

The measurements of the GDH-Collaboration on the proton have been extended up to 3.1 GeV photon energy at ELSA. The data are currently under analysis. It will be highly interesting to see if the indications for a sign change of the GDH integrand consolidate

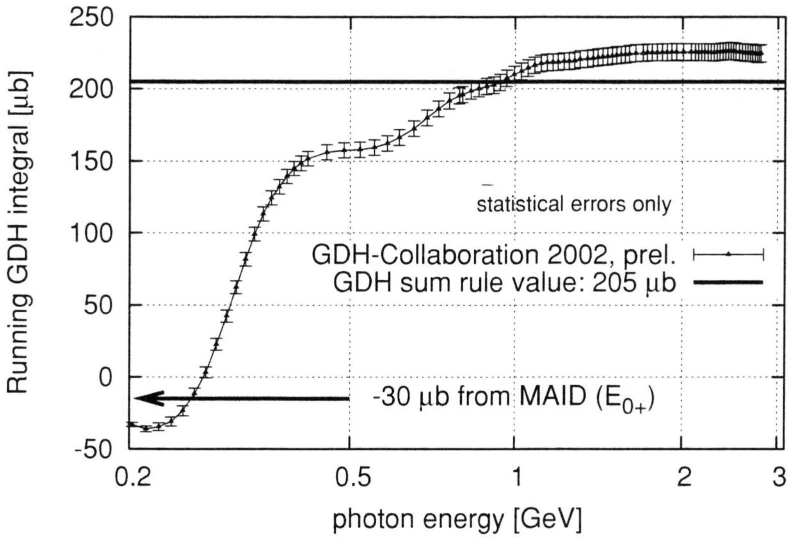

FIGURE 4. Running GDH integral for the proton

and if our Regge fit still describes the data with higher precision. These data will also allow to extrapolate to high energies with a decreased error.

There are several other experiments planed that can verify our findings. These experiments use the laser backscattering technique to obtain polarized photons instead of bremsstrahlung produced by polarized electrons. The principal layout of the detection systems are similar to that developed of the GDH-Experiment at ELSA (see Sec. 2.1.4).

- The LEGS facility at BNL will use a polarized HD-target to cover photon energies up to 470 MeV [36].
- The GRAAL facility also uses the HD-target technique and will cover photon energies up to 550 MeV [37]. The currently envisaged start of the experiment is in 2003.
- At SPRING-8 dynamically polarized PE-foils will be used to cover 1.8 – 2.8 GeV [38]. The anticipated beginning of data taking is at the end of 2003.

Beyond the energy coverage of the GDH-Collaboration there are experiments planed to extend the measurements to higher energies:

- At JLAB a proposal is being prepared to measure at photon energies starting from 2.5 to 6 GeV. A frozen spin target similar to that of the GDH-Collaboration is envisaged to be developed.
- SLAC has an approved proposal to measure total cross section asymmetries in the energy regime from 4 to 40 GeV. The anticipated starting date is in 2005.

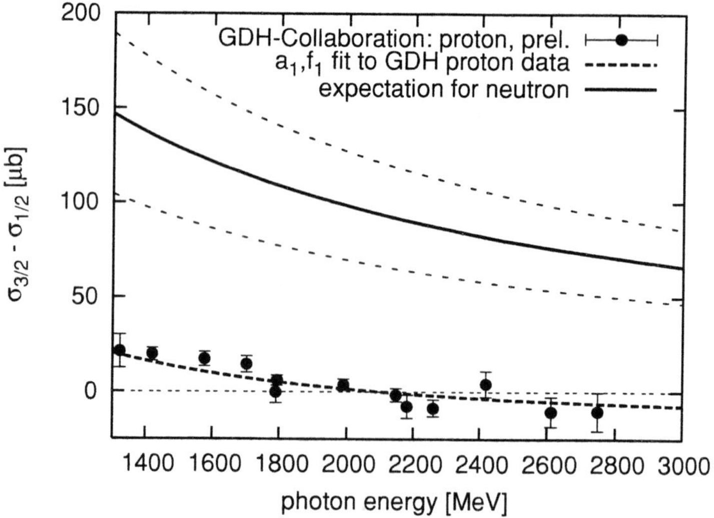

FIGURE 5. Prediction for the neutron obtained from our fit to the proton data.

2.3.1. Neutron prospects

The GDH Sum Rule prediction for the neutron is 233 μb which is almost 30 μb higher that the value for the proton. Moreover, the estimate using MAID [29] for the GDH integral up to 500 MeV is about 25 μb less than in the case of the proton. Relying on this estimate – which already provided a good prediction for the proton in this energy domain – one ends up with a total strength of about 50 μb that the neutron should have in excess over the proton at energies above the delta resonance. There are three possible explanations:

- A much larger difference in the Regge regime for the neutron than was seen for the proton.
- Completely unanticipated behavior of the higher resonances.
- A failure of the isovector GDH Sum Rule.

Here, we will concentrate on the first option. Revisiting Eqn. 4 one observes that the isospin I for the neutron has just the opposite sign and hence the contribution due to the a_1 trajectory simply flips sign. The effect, however, is large taking the coefficients already obtained for our fit to the proton data of the GDH-Collaboration. Fig. 5 shows that the prediction one obtains for the neutron is of the order of almost 100% asymmetry for the neutron whereas that of the proton is close to zero. This would indeed compensate the "missing" strength as argued above.

The GDH-Collaboration has already taken data using a polarized ^{6}LiD target. The data will be analyzed within the next months. As is the case for the proton, there is no other data from other experiments available for the neutron as well.

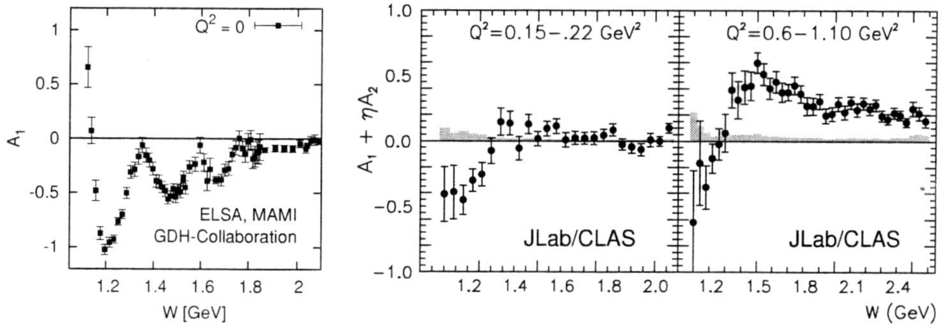

FIGURE 6. Evolution of A_1 with Q^2

3. VIRTUAL PHOTON ABSORPTION

The spin structure of a virtual photon is more involved than that of real photon in that it has a longitudinal polarization component:

$$\frac{d\sigma}{dE'd\Omega} = \Gamma_V \cdot \left[\sigma_T + \varepsilon\sigma_L + P_e P_t \left(\sqrt{1-\varepsilon^2} A_1 \sigma_T \cos\psi + \sqrt{2\varepsilon(1-\varepsilon)} A_2 \sigma_T \sin\psi\right)\right] \tag{5}$$

where $A_1 = \left(\sigma_{1/2} - \sigma_{3/2}\right) / \left(\sigma_{1/2} + \sigma_{3/2}\right) \equiv \sigma_{TT'}/\sigma_T$ and $A_2 = \sigma_{LT'}/\sigma_T$. The observables σ_L, ε and $\sigma_{LT'}$ are due to the longitudinal polarization component only. Fig. 6 shows how the asymmetry A_1 evolves from the real photon point. The data are from the GDH-Collaboration and from the Hall B experiment at JLAB [39]. One observes that apart from the delta resonance no higher resonances are visible in the Hall B data. The delta resonance looses strength when going to finite Q^2 and contributions above $W > 1.5$ GeV change sign.

The GDH Sum Rule is valid only at the real photon point $Q^2 = 0$. There it is a prediction with a high relative precision of 10^{-6} because the anomalous magnetic moment of the nucleon is very well known. There are several ways to generalize the integral on left hand side of the GDH Sum Rule. Amongst these choices experimentalists often favor a version which is a straight forward generalization of the GDH Sum Rule:

$$I^{\text{GDH}}(Q^2) = -\int_0^\infty \frac{d\nu}{\nu} \cdot \left(\sigma_{3/2}(\nu, Q^2) - \sigma_{1/2}(\nu, Q^2)\right) \tag{6}$$

On the other hand theorists prefer the following notation called first moment of g_1:

$$\Gamma_1(Q^2) = \int_0^1 dx\, g_1(x, Q^2) = Q^2/2 \int_0^\infty \frac{d\nu}{\nu} G_1(\nu, Q^2) \tag{7}$$

Going to $Q^2 \to \infty$ in the isovector case one obtains the Bjorken Sum Rule:

$$\Gamma_1^p - \Gamma_1^n = g_A/6 \quad \text{known with a relative precision of } 3 \cdot 10^{-3}. \tag{8}$$

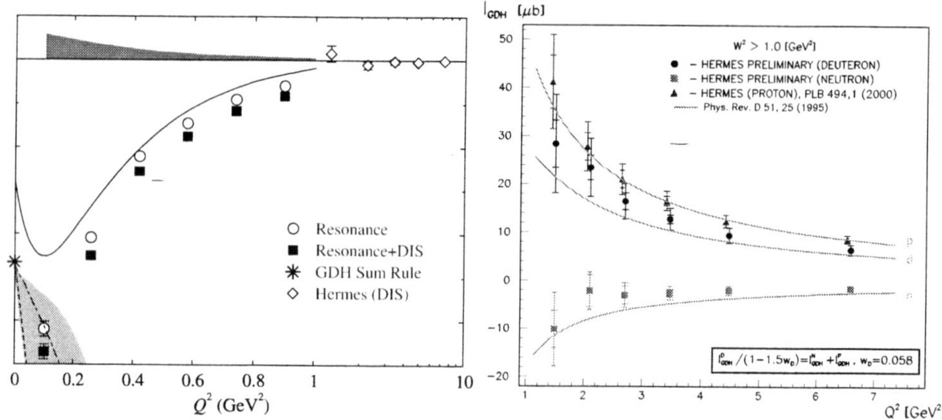

FIGURE 7. Left: Hall A measurements [43] for $I(Q^2)$ vs. Q^2 for the neutron with and without an estimate of the DIS contribution; dotted (dot-dashed) line: ChPT calculations of Ref. [41] (Ref. [42]); solid line: Ref. [29]; open boxes: data from HERMES [44]; for 1 GeV$^2 < Q^2$ a semi-log scale has been adapted. Right: HERMES data from Ref. [45] for the proton, deuteron and the neutron in comparison to a model by Soffer and Teryaev.

On the other end the GDH Sum Rule fixes the slope at $Q^2 = 0$: $\Gamma_1/Q^2 = -\kappa^2/8m^2$.

Recently Ji and Osborne [40] have developed a unified formalism to describe the generalized sum with respect to the virtual Compton forward scattering spin-amplitudes $S_1(\nu,Q^2), S_2(\nu,Q^2)$:

$$\bar{S}_1(0,Q^2) = 4\int_{\nu_0}^{\infty} \frac{d\nu}{\nu} G_1(\nu,Q^2) \tag{9}$$

were $\bar{S}_{1,2}$ are the amplitudes without the elastic intermediate state. The right hand side of Eqn. 9 is measurable while the left hand side is calculable. At $Q^2 > 1$ GeV^2 QCD operator product expansions should yield the value of $\bar{S}_1$ while at $Q^2 < 0.1$ GeV^2 chiral perturbation theory calculations are used. However, it turns out that these calculations have by far not yet converged. A comparison of calculations in the heavy baryon approach by Ji, Kao and Osborne [41] with a recent calculation by Bernard, Hemmert and Meissner [42] shows that both approaches do not agree and moreover, that the chiral expansion has not yet converged. The level of uncertainty for $\Gamma_1(Q^2)$ at $Q^2 = 0.1$ GeV^2 is of the order of 50%.

Despite, there is no stringent sum rule established at finite Q^2, one can study the transition from hadronic degrees of freedom to partonic structure. Fig. 7 shows two examples for experiments that have been performed in this respect. Plotted is in cases $I(Q^2)$ as defined in Eqn. 6 as a function of Q^2. The left hand diagram mainly shows the results obtained at Hall A at JLAB [43]. The right hand diagram shows data for the HERMES collaboration at DESY [45]. The data predominantly show a $1/Q^2$ behavior at high Q^2 while the interpretation at about $Q^2 < 0.3$ is not clear.

REFERENCES

1. S.B. Gerasimov, Sov. J. Nucl. Phys. **2**, 430 (1966)
2. S.D. Drell, A.C. Hearn, Phys. Rev. Lett. **16**, 906 (1966)
3. M. Hosoda and K. Yamamoto, Prog.Theor.Phys. **36**, 425 (1966)
4. C.K. Iddings, Phys.Rev. **138B**, 446 (1965)
5. F.E. Low, Phys. Rev. **96**, 1428 (1954)
6. M. Gell-Mann, M.L. Goldberger, Phys. Rev. **96**, 1433 (1954)
7. R. Pantförder, arXiv:hep-ph/9805434
8. L.N. Chiang, Y. Liang and R.L. Workman, Phys. Lett. B **329**, 514 (1994)
9. R. Pantförder, H. Rollnik and W. Pfeil, Eur. Phys. J. C **1**, 585 (1998)
10. S.M. Roy and V. Singh, Phys.Rev.Lett. **21**, 861 (1968)
11. H. Goldberg, Phys. Lett. B **472**, 280 (2000)
12. H.D.I. Abarbanel, M.L.Goldberger, Phys. Rev. **165**, 1594 (1968)
13. J. Ahrens *et al.* [GDH Collaboration], Phys. Rev. Lett. **87**, 022003 (2001)
14. J. Ahrens *et al.* [GDH Collaboration], Phys. Rev. Lett. **84**, 5950 (2000)
15. J. Ahrens *et al.* [GDH Collaboration], Phys. Rev. Lett. **88**, 232002 (2002)
16. C. Bradtke *et al.*, Nucl. Instr. Meth. A **436**, 430 (1999)
17. W. Hillert, M. Gowin and B. Neff, AIP Conference Proceedings **570**, 961 (2001);
    M. Hoffmann *et al.*, AIP Conference Proceedings **570**, 756 (2001)
18. K. Aulenbacher *et al.*, Nucl. Instrum. Meth. A **391**, 498 (1997)
19. B. Kiel, PhD thesis, Erlangen, 1996
20. G. Audit *et al.*, Nucl. Instr. Meth. A **301**, 473 (1991)
21. K. Helbing *et al.*, Nucl. Instrum. Meth. A **484**, 129 (2002)
    K. Helbing, Diploma thesis, Bonn, 1993
22. S. Altieri *et al.*, Nucl. Instr. Meth. A **452** 185 (2000)
23. T. Speckner, diploma thesis, Erlangen, 1998;
    G. Zeitler, diploma thesis, Erlangen, 1998
24. M. MacCormick *et al.*, Phys. Rev. C **53**, 41 (1996)
25. K. Helbing *et al.*, in preparation for EPJ C;
    K. Helbing, PhD thesis, Bonn, 1997;
    M. Sauer, PhD thesis, Tübingen, 1998
26. T. Michel, PhD thesis, Erlangen, 2001
27. T. Speckner, PhD thesis, Erlangen, 2002
28. G. Zeitler, PhD thesis, Erlangen, 2002
29. D. Drechsel *et al.*, Nucl. Phys. A **645** 145 (1999);
    `http://www.kph.uni-mainz.de/MAID/maidnew/maid2000.html`
30. A. Donnachie and P. V. Landshoff, Phys. Lett. B **296**, 227 (1992)
31. T. A. Armstrong *et al.*, Phys. Rev. D **5**, 1640 (1972).
32. S.D. Bass, M.M. Brisudova, Eur. Phys. J. A **4** 251 (1999)
33. N. Bianchi, E. Thomas, Phys. Lett. B **450** 439 (1999)
34. S. Simula, M. Osipenko, G. Ricco and M. Taiuti, Phys. Rev. D **65**, 034017 (2002)
35. J. Kuti, Personal communication.
36. C. S. Whisnant *et al.* [LEGS SPIN Collaboration], Proceedings of GDH 2000, World Scientific
37. F. Renard [GRAAL Collaboration], Proceedings of GDH 2000, World Scientific
38. T. Iwata, Proceedings of GDH 2000, World Scientific
39. V. D. Burkert, arXiv:hep-ph/0211185.
40. X. D. Ji and J. Osborne, J. Phys. G **27**, 127 (2001)
41. X. D. Ji, C. W. Kao and J. Osborne, Phys. Lett. B **472**, 1 (2000)
42. V. Bernard, T. R. Hemmert and U. G. Meissner, Phys. Lett. B **545**, 105 (2002)
43. M. Amarian *et al.*, arXiv:nucl-ex/0205020.
44. K. Ackerstaff *et al.* [HERMES Collaboration], Phys. Lett. B **444**, 531 (1998)
45. A. Airapetian *et al.* [HERMES Collaboration], Phys. Lett. B **494**, 1 (2000)

Comments about Vladimir L. Solovianov

A.D. Krisch

Spin Physics Center
University of Michigan
Ann Arbor, MI 48109-1120 USA

Vladimir L. Solovianov was my close collaborator and friend for 14 years until his death on 26 June 2001.

His death was very unusual in High Energy Physics, because he died "in the line of duty". He apparently fell 6 meters from the top of the shielding while inspecting the installation of his latest experiment SPIN@U-70. I said apparently because he was first seen by one of his colleagues walking to his office with a broken leg and refusing help. I can recall a few other examples of people killed at High Energy Accelerators:

- a student was killed in the CEA explosion in the 1960's;
- a high voltage engineer was killed around 1970 at the ZGS.

However, I can recall no other leader of a large High Energy Physics group who ever died "in the line of duty".

Since tomorrow is September 11, 2002, we can recall that many firemen's families received a posthumous medal for their heroic death "in the line of duty" just one year ago in nearby New York. Perhaps Vladimir's family should also receive such a medal.

Vladimir Solovianov was a physicist who was unusually capable, dedicated, hardworking, tough, and admirable. His death is a major loss to Spin Physics.

Vladimir L. Solovianov 1940-2001.

CP675, *Spin 2002: 15th Int'l. Spin Physics Symposium and Workshop on Polarized Electron Sources and Polarimeters,* edited by Y. I. Makdisi, A. U. Luccio, and W. W. MacKay

Vladimir Leonidovich Solovianov

Nikolai Tyurin

Institute for High Energy Physics, Protvino, Moscow Region 142281,Russia

Vladimir L. Solovianov was born in 1940 in the town of Krivio Rog in the Ukraine. He was graduated from Moscow State University in 1964 and joined the new Institute for High Energy Physics which was created one year earlier.

The beginning of Solovianoiv's scientific career saw the development of hodoscopes for the experimental studies of elastic scattering at the U-70 accelerator. He already contributed to the early measurements (1969) of the diffractive cone for pp-scattering which showed that the slope is loagrithmically rising.

In the early seventies Vladimir played an essential role in the Soviet-French polarization experiment HERA, the measurement and analysis of the polarization in elastic pp and pbar-p scattering and charge exchange processes. In pp elastic scattering the change of polarization from positive to negative as a function of the momentum transfer was discovered. This particular phenomenon was very important for the development of models of binary reactions.

He contributed to the development of cooperation and joint experiments with Saclay, CERN, and the US labs. The obtained results in these experiments and, in particular E704 at Fermilab continue to be important for the progress of our field. He played a key role in the organization of the two Spin Symposia in Protvino in 1986 and 1998.

Valdimir Solovianov also made key contributions in physics proposals and the preparations of all spin projects at U-70 and UNK. These include PROZA, NEPTUNE, RAMPEX, and the latest SPIN@U-70.

His creative role in all aspects of physics research and its organization, the very scientific and friendly atmosphere he always supported, will be remembered by Valdimir's colleagues.

CP675, *Spin 2002: 15th Int'l. Spin Physics Symposium and Workshop on Polarized Electron Sources and Polarimeters,* edited by Y. I. Makdisi, A. U. Luccio, and W. W. MacKay

Inclusive Asymmetries
(a retrospective view of V. L. Solovianov's work)

Mikhail N. Ukhanov

Institute for High Energy Physics,
Protvino, Moscow region, Russia

Abstract. In this talk I would like to present a retrospective view of the work done by Vladimir L. Solovianov. He began his work in sixties in Dubna. And it happened to be that he participated in many experiments, which were in the main stream of the experimental program over the decades. In the last years he initiated several projects, which were again at the edge of the investigation of the spin puzzle in the High Energy Physics domain.

SEVERAL FACTS OF HIS LIFE

He did his diploma project in the Dubna Joint Institute for Nuclear Research. After graduation from the Moscow State University in 1964 he began his career in the Institute for High Energy Physics in Protvino. In 1975 he defended his PhD in High Energy Physics devoted to the studies of the energy dependence of p-p elastic scattering cross section at 35-55 GeV. In the beginning of 80-s a new big accelerator facility was started next to the existing U-70 ring. Solovianov became the spokesman of NEPTUN - one of the experimental programs at UNK. The NEPTUN experiment goal was the investigation of the polarization phenomena in interaction of the primary UNK beam with internal polarized target. In 2001 together with Professor A.D. Krisch from the University of Michigan he began the SPIN@U-70 project aimed at studies of elastic scattering of the high intensity extracted 70 GeV proton beam with polarized proton target.

SEVERAL EXPERIMENTS OF HIS LIFE

In his first experiment the asymmetry of π^+ production in polarized proton beam interaction with hydrogen target at 612 MeV was measured at Dubna [1]. Part of these studies comprised his diploma project and resulted in several publications. In Protvino he entered a group, which conducted a series of experiments with polarized target aimed at investigation of elastic scattering of various particle species on the polarized protons in the energy range from 35 GeV to 55 GeV [2]. Then he began studies of elastic scattering at the CERN SPS accelerator at 150 GeV [3]. A very significant part of his activities was devoted to the measurements of inclusive asymmetry in $\pi^- p$ interactions at 40 GeV [4]. In the mid nineties he took part in the commissioning and

CP675, *Spin 2002: 15th Int'l. Spin Physics Symposium and Workshop on Polarized Electron Sources and Polarimeters*, edited by Y. I. Makdisi, A. U. Luccio, and W. W. MacKay running of the FNAL E704 experiment where many interesting results were obtained [5].

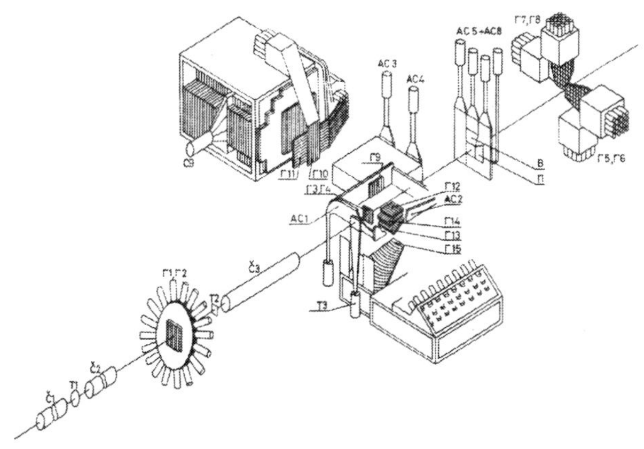

FIGURE 1. Apparatus to study the elastic scattering on polarized target at IHEP U-70 accelerator in Protvino

ELASTIC SCATTERING AT U-70

Setup Layout

This setup [6] was created to study the polarization effects in the elastic scattering of positive and negative particles of various species on a polarized proton target at the U-70 accelerator in IHEP, Protvino, see fig. 1. A propanediol target was embedded in the magnetic field provided by two superconducting Helmholz coils. The field direction was horizontal. The coil design allowed detection of the recoil protons in the horizontal plane. The setup could do the measurements of both the spin rotation parameter in the horizontal plane and the polarization parameter in the vertical plane.

The beam line design provided the beams of positive and negative particles with only a small change in the position of the first five magnetic elements and installation of a magnetic shield near the U-70 ring. The particle fluxes per burst were 3M for negative sign and 1M for positive. Beam contents were determined by a set of threshold Cherenkov counters. For the negative beam the percentage was 97.9/1.8/0.3 for π^-/K^-/anti-proton. For the positive one it was 94/5/1 for $p/\pi^+/K^+$.

Polarization and Spin Rotation Parameter Measurements

The polarization parameter P has been measured at 40 GeV for $\pi^- p$-, $K^- p$-, *p(bar) p*-interactions [7]. The event reconstruction uses information from the scintillation counters in order to determine the kinematics of each event. For quasi-elastic scattering on bound nucleons the Fermi momentum of the target particle results in different kinematics where, in particular, the coplanarity of the incident and outgoing momenta and the exact angular correlation between the outgoing particle are no longer verified. This feature is used to eliminate a large fraction of the background events. The anticoincidence counters around the target provide a good rejection of reactions with more than two charged particles in final state. The normalization of successive runs with opposite target polarization was made using three different monitors:

-Counting of the incident particles.

-Counting events in the plane containing the target polarization.

-Counting the quasi-elastic events outside the elastic peak.

The latter monitor gave the most stable result. All three monitors gave the results, which were consistent within errors.

Subtraction of the background under the "elastic peak" uses the data obtained with a carbon target. The normalization of the carbon data to the polarized data was done in the region outside the "elastic peak" of angular correlation distribution.

A total of 1.5M events were recorded for further analyses. During the exposition two matrices were used to select the events for recording. One, called PHI-matrix, required only two particles in the final state, which should satisfy the coplanarity constraint. Another one put constraint on the scattering angle – the THETA-matrix. The events, without the THETA trigger, come mostly from the quasi-elastic scattering on the bound protons. They were used for the background subtraction. After the final selection about 40K double scattering events retained with incident π^- and only 786 with K^-.

The method of data reduction and analyses in measurement of the spin rotation parameter are essentially the same as for polarization measurement [8].

The results of polarization and spin rotation parameters are presented on fig 2 and 3 respectively. Measured P value in $\pi^- p$ elastic scattering extends up to $|t|$ = 1.8 $(GeV/c)^2$. The characteristic dip is observed at the same value $|t| = 0.6$ $(GeV/c)^2$ as at lower energies. The polarization in $K^- p$ at 40 GeV is consistent with zero up to $|t| \approx 0.5$ $(GeV/c)^2$. Data for *p(bar)-p* scattering exhibits an important structure. In particular, large polarization at $|t| > 0.6$ $(GeV/c)^2$ shows an important contribution of the single helicity flip amplitude.

The final selection of the spin rotation parameter measurements retains about 20K events. The magnitude of R shows no significant variation with energy from 3.8 to 45 GeV. The t-dependence is weak and similar to those obtained in earlier experiments at 6 and 16 GeV.

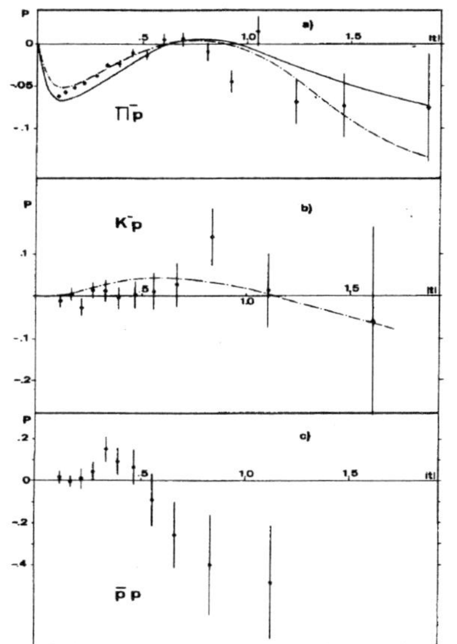

FIGURE 2. Transfer momentum distribution of the polarization parameter in a) $\pi^- p$, b) $K^- p$, c) *p(bar)-p* interactions at 40 GeV. Lines are prediction of various models see[7] for references.

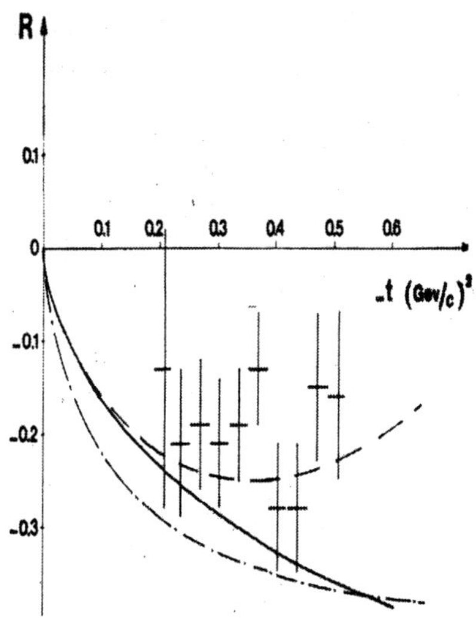

FIGURE 3. Results for R in $\pi^- p$ elastic scattering at 40 GeV/c. The models prediction are represented by dashed and dot-dashed lines respectively see [8] for references. The solid line represents the function $R = -cos\ \theta_p$.

POLARIZATION MEASUREMENT AT 150 GEV

This experiment was done at the CERN SPS accelerator in 1978 [3]. The setup layout is shown on fig. 4. The polarization parameter was measured in the elastic *pp* scattering at 150 GeV. The transfer momentum covers range from $0.2 < |t| < 3$ $(GeV/c)^2$. The measured elastic cross section spans 5 orders of magnitude going down to ~10 nb/$(GeV/c)^2$ at $|t| = 3$ $(GeV/c)^2$.

The incoming beam reached the Polarized Target (PT) through a hole in the yoke of the target magnet. The target consisted of 15 cm long 2 cm diameter propanediol sample kept at 0.55 K in 2.5T magnetic field. The average polarization was 90%. A carbon target was used for ~20% of time to measure the background distribution from scattering on bound protons. It consists of 10 graphite discs 5 mm thick equally spaced in a copper container of the same size as the PT.

In order to have a good background rejection one needs precise measurements and redundancy in determination of the kinematics of reaction. The scattered proton was detected by counter and forward hodoscopes located at 49 meters. The trajectory, bent by two magnets by 13 mrad, was measured by four sets of multi wire proportional chambers (MWPC) with a precision of ±0.07 mrad.

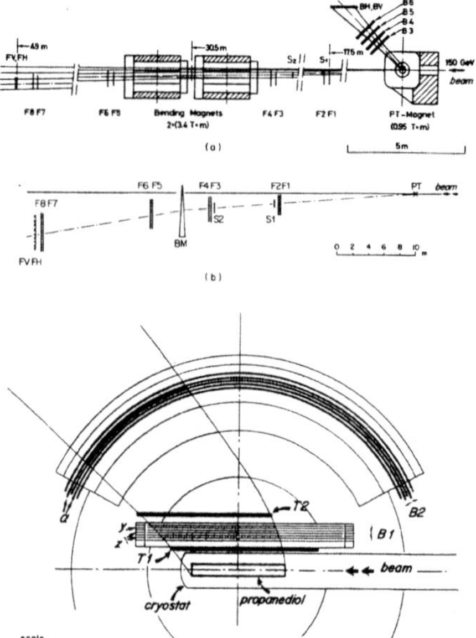

FIGURE 4. Layout of the apparatus used to measure the polarization parameter at 150 GeV. The target region is shown in lager scale at the bottom.

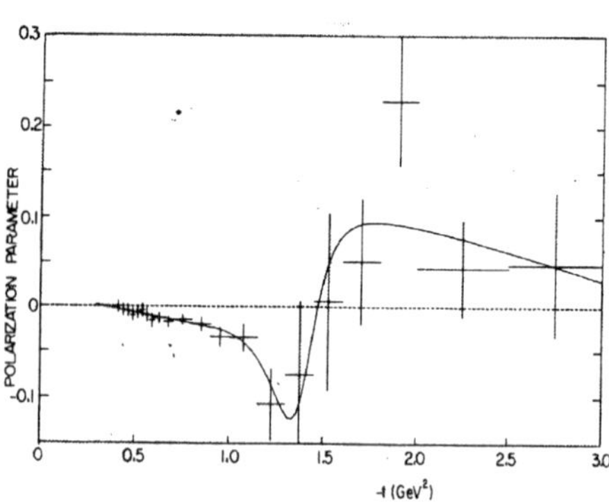

FIGURE 5. The polarization $P(t)$, calculated from the separately fitted functions dN^+/dt and dN^-/dt, is shown (solid line) together with the experimental points.

The recoil proton was detected by the counters T1 and T2 located alongside the target cryostat, and after 3.8 m by two scintillation hodoscopes. Also the time of flight to the hodoscopes was measured. The recoil proton momentum was measured by six MWPCs: two of them inside the magnet and four outside the filed region.

The relative errors in momentum determination were 1% in the forward arm, 2% for low t and ($2\%p$) for $-t > 0.8$ (GeV/c)^2 in the backward arm.

The data analysis were done in three steps: I) at least one track required in both arms; II) the tracks must originate from a vertex, with a vertex resolution of 1 mm; III) after adjustment of the beam parameters, the kinematical 4C-fit should give the positive result for the elastic scattering hypothesis. The selection criteria required the vertex to be within 1 cm from the target axis and a 6% cut on the chi-square kinematical fit probability.

From the shape of the chi-square distribution the signal to background ratio was found to be between 3% at low t values up to 10% at high t. Further checks on coplanarity, angular and momentum correlation supported this conclusion.

The polarization parameter dependence versus transfer momentum is shown on fig. 5. The solid curve represents the polarization calculated from independent fit of the measured differential cross sections for two target polarizations.

The polarization parameter evidently becomes negative at around $|t| = 1$ (GeV/c)^2 and it tends to change the sign in the region where dip is observed in the differential cross section. The t dependence is similar to the data obtained at 150 GeV in Fermilab jet-target experiment, where recoil proton polarization was measured with a polarimeter. At large t-values the polarization is definitely positive despite the errors, agrees with the trend, observed in the Fermilab experiment at 300 GeV.

POLARIZATION IN $\pi^- P \to \pi^0 N$ AT 40 GEV

The experiment was carried out at U-70 accelerator [4, 9]. The beam line design was essentially the same as it was for the elastic scattering measurements [6]. In this reaction the final state particles are all neutral. The gamma-quanta from π^0 are detected by the electromagnetic calorimeter (ECAL). The veto system suppresses production of the charged particles in the ECAL aperture and gamma-quanta outside it. The main part of the veto is made of the interlaced layers of lead-scintillator, which are viewed by photomultipliers (PM). The hermeticity of the veto system is one of the main issues. The overall suppression factor due to the veto system was about 10^{-5} from the initial beam intensity.

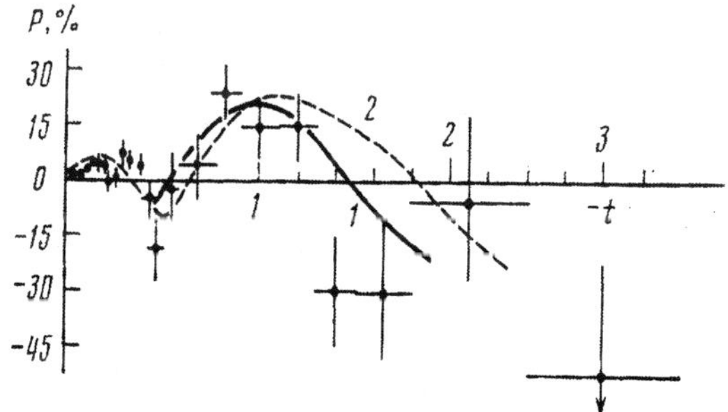

FIGURE 6. Results for P in $\pi^- p$ elastic scattering at 40 GeV/c. Model predictions are also shown.

The neutrons were detected by two blocks, placed symmetrically relative to the beam axis. Each block had acceptance of 30 degree in azimuth and $0.15 < |t| < 1.3$ (Gev/c)^2. Each block consisted of 40 cells (5H x 8V). The cells were made of a scintillator 10x10x50 cm^3 viewed by a PM. The neutron detection efficiency was about 30-35% for neutrons with energy greater than 200 MeV. All cells were calibrated in the muon beam.

The ECAL was calibrated in the electron beams with energy from 10 to 40 GeV. The energy resolution was $15\%/\sqrt{E} + 3\%$ FWHM. Coordinate resolution for the single gamma was 3 mm. The separation of two showers, required to reconstruct the energy with efficiency about 90%, was 35 mm.

The scintillation hodoscope was mechanically attached to the ECAL platform. It was used to monitor the beam position stability and provided the survey data for the ECAL position relative to the beam hodoscopes.

In 1983 an upgrade was done to the electromagnetic calorimeter. As a result the effective mass resolution was improved and available t-range was extended until $|t| = 3$ $(GeV/c)^2$. The result of the polarization measurement in the charge exchange reaction at 40 GeV is presented on the fig. 6. For the first time oscillation of the polarization asymmetry was observed. At the transfer momentum $|t| < 0.4$ $(GeV/c)^2$ the asymmetry is about 4-6% and positive. At the $|t|$ value ~0.25 a minimum is seen with the confidence level of 99%.

SINGLE SPIN ASYMMETRY IN INCLUSIVE π^0 PRODUCTION AT 200 *GEV IN PP* INTERACTIONS

The well known state of the art polarized proton and polarized anti-proton beams facility was designed at Fermilab and commissioned in E704 experiment [5]. The design made use of the parity violating decay of the Lambda hyperons, which provided the polarized beams. The polarization of the beam is determined by tagging the beam particles trajectories. The tagged beam polarization values, which range from 0.65 to –0.65, are divided into three parts with average polarization of +0.45, -0.45 and 0. These values were confirmed by two independent measurements using reactions of known analyzing power. A set of spin-rotating magnet changed the direction of polarization from the transverse-horizontal to the vertical direction at the experimental target and reversed the sign 4 or 5 times per hour. The whole setup consisted of many detectors like proportional chambers, Cherenkov counters, hodoscopes and calorimeters. Photons from the decay of neutral mesons produced in 100 cm hydrogen target were detected in the Central Electromagnetic Calorimeters CEMC-1 and CEMC2, located symmetrically to the left and to the right of the beam axis at 10 m from the target. The photon-pair background under the π^0 peak, due to uncorrelated pairs and π^0s produced outside the target region, varied form 15% to 20%. The effective mass resolution varied from ±9 MeV/c^2 at x_F=0.1 to ±30 MeV/c^2 at x_F=0.7. The asymmetry in the mass region outside the π^0 peak from 250 MeV to 450 MeV was found to be consistent with zero within statistical errors of less then 3% over the entire range of x_F.

In principle, each of the two CEMC detectors can be used to determine the asymmetries either with the two oppositely polarized parts of the beam or with beam polarization reversal by the spin-rotating magnets. The observed consistency among the different ways to calculate the asymmetry provides a check of instrumental errors. The calculated "false asymmetry" for events tagged with an average zero beam polarization was found to be less then 1% for P_T up to 3 GeV/c and less 3% for higher P_T values. The relative systematic error proportional to A_n is about 10% and mainly due to the uncertainty in the beam polarization and a small contribution from uncorrelated photon pairs and neutral mesons produced outside the target volume.

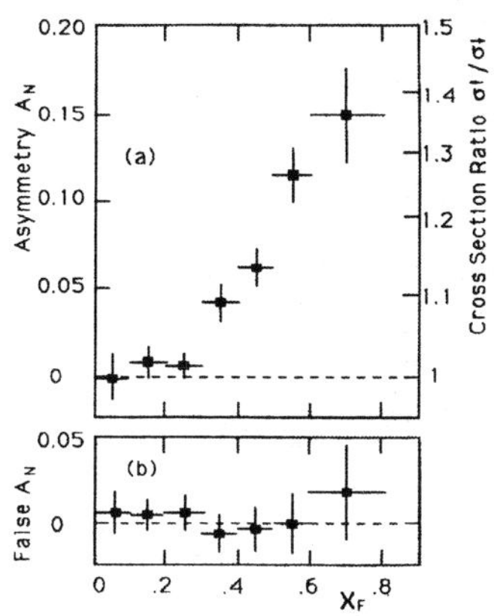

FIGURE 7. a) Plot of the asymmetry parameter A_n as a function of x_F averaged over p_T b) The "false asymmetry calculated for events with beam polarization of zero. Only the statistical errors are shown.

The results on asymmetry measurements versus x_F is shown on the fig. 7. The growth of asymmetry when π^0 momentum gets close to the polarized incident proton momentum allude to the valence quark responsibility.

RAMPEX EXPERIMENT

The experiment is aimed at measurement of the single spin inclusive asymmetries in $\pi^- p\uparrow$ interactions at 40 GeV and $pp\uparrow$ interactions at 70 GeV for various final state particles [12]. The experiment uses beam line #14 at U-70 accelerator in IHEP, Protvino.

Incident beam particles strike the polarized proton propanediol target. The secondary particles are detected in the setup, which consist of the charged particle spectrometer and two calorimeters. The charged particle spectrometer has a traditional composition: five sets of proportional chambers, analyzing magnet and two multi-channel threshold Cherenkov counters for particle identification.

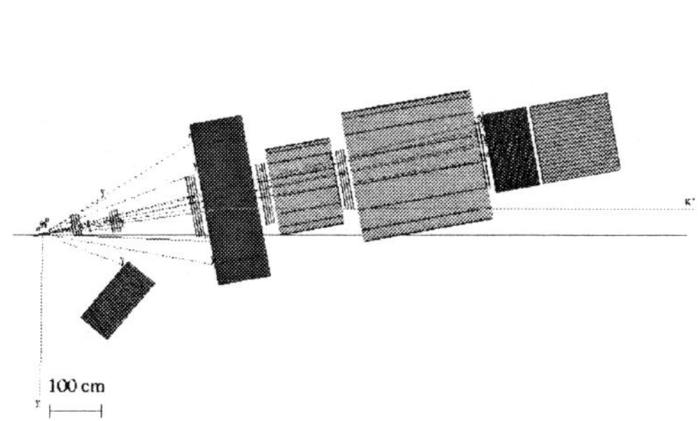

FIGURE 8. Layout of the RAMPEX Experiment

The central axis of the setup is rotated by 10 degrees relative to the beam direction which corresponds to 90 degrees in the center of mass system. The trigger system has two sources - the electromagnetic and hadron calorimeters. In both calorimeters the trigger signal comes from fast analog sum of signals of the calorimeter cells weighted to be proportional to the transverse energy deposition in the calorimeter. The spectrometer acceptance allows detection of the final state particles produced with transverse momentum more than 3 GeV/c and $-0.3 < x_F < +0.6$.

The whole apparatus was commissioned in 2000. Accumulated data is being analyzed and the apparatus tuning has been performed in the series of tune-up runs.

SPIN@U70 SPECTROMETER

The SPIN@U-70 experiment was approved by the IHEP (Protvino) PAC in 2001. The main goal of the experiment is the measurement of spin effects in the elastic scattering of 70 GeV protons with a polarized proton target in the domain of extremely high transverse momentum of the recoil proton.

Movable recoil arm of the spectrometer consist of (starting from the interaction point) a pair of focusing quadrupole magnet (Q1, Q2), swiping dipole magnet (M1), another pair of quadrupoles (Q3, Q4) and bending magnet (M2). The momentum of the recoil proton is measured in the spectrometer magnet (M3), which bends the particle by 12 degrees upward.

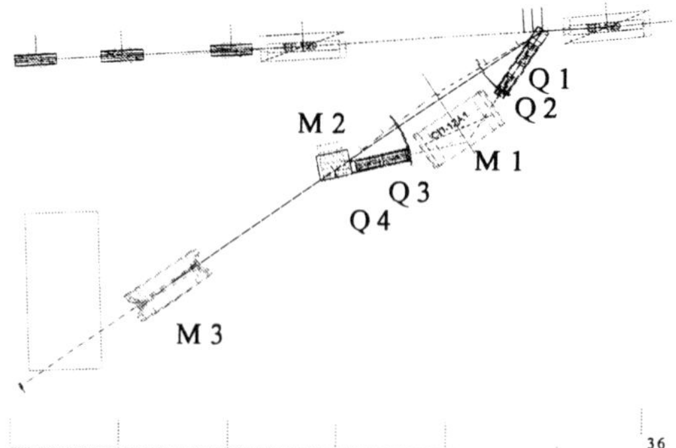

FIGURE 9. Layout of the SPIN @ U70 Spectrometer.

Since elastic production cross section at transverse momentum squared of 12 $(GeV/c)^2$ is about 0.3 pb one needs high intensity beam to reach such values. The experimental setup utilizes the beam line #8 of the U-70 accelerator in IHEP, Protvino. The high intensity proton beam comes from the U-70 ring using the bent crystal slow extraction technique

The beam parameters were measured in the tune up run in April 2002. During the run the design values of the beam size and the stability of the beam position at the interaction point were achieved. The recoil arm magnets setting were also tuned during this run using the polyethylene target equivalent and simplified detector set by detection of elastic events at low transverse momentum. More detailed report on the.results of this run was given at the Prague Spin and Symmetry Symposium in July 2002 and in report given by Luppov V.G. on this conference [11] .

ACKNOWLEDGMENTS

I would like to thank the organizing committee for the hospitality and support they provided me during my stay at BNL. I am also grateful to the IHEP directorate who encouraged and supported my trip to this conference.

REFERENCES

1. Azhgirey L. S. et al. *Sov Journal of Nucl. Phys.* **v. 1**, p. 122, (1965).
2. Gaidot A. et al. *Physics Letter,* **57B**, 389-392, (1975)
3. Fidecaro G. et al. *Nuclear Physics,* **B173**, 513-545, (1980)
4. Avvakumov I. A. et al. *Pribory i Tehnika Eksperimenta (in Russian),* **5**, 46-50, (1987)
5. Grosnick D. P. et al., *Nuclear Instruments and Methods,* **A290**, 269, (1990)
6. Raoul J.C. et al., *Nuclear Instruments and Methods,* **125**, 585, (1975)
7. Gaidot A. et al., *Physics Letters,* **57B,** 389-392, (1975)
8. Pierrard J. et al., *Physics Letters,* **61B,** 107-109, (1976)
9. Apokin V. D. et al. *Sov Journal of Nucl. Phys.* **v. 45**, p. 1355, (1987)
10. Adams D.L. et al., *Z. Phys.* **C56,** 181, (1992)
11. Ufimtsev A. G. *Czechoslovak Journal of Physics,* **53A,** ???, (2003) (in print) also report by Luppov V.G. on this conference.
12. Akimenko S. et al. *IHEP preprint,* **97-58**, (1997)

Small Angle Scattering of Polarized Protons

B.Z. Kopeliovich

Max-Planck Institut für Kernphysik, Postfach 103980, 69029 Heidelberg
Institut für Theoretische Physik der Universität, D-93040 Regensburg,
Joint Institute for Nuclear Research, Dubna, Russia

Abstract.
Experiment E950 at AGS, BNL has provided data with high statistics for the left-right asymmetry of proton-carbon elastic scattering in the Coulomb-nuclear interference region of momentum transfer. It allows to access information about spin properties of the Pomeron and has practical implications for polarimetry at high energies. Relying on Regge factorization the results for the parameter r_5, ratio of spin-flip to non-flip amplitudes, is compared with the same parameter measured earlier in pion-proton elastic and charge exchange scattering. While data for $\mathrm{Im}r_5$ agree (within large systematic errors), there might be a problem for $\mathrm{Re}r_5$. The πN data indicate at a rather small contribution of the f-Reggeon to the spin-flip part of the iso-scalar amplitude which is dominated by the Pomeron. This conclusion is supported by direct analysis of data for elastic and charge exchange pp and pn scattering which also indicate at a vanishing real part of the hadronic spin-flip amplitude at energies 20 GeV and higher. This is a good news for polarimetry, since the E950 results enhanced by forthcoming new measurements at AGS can be safely used for polarimetry at RHIC at higher energies.

INTRODUCTION

It is usually assumed that small angle elastic scattering at high energies is dominated by Pomeron exchange. At the same time, definition for the Pomeron varies depending on a model (a Regge pole, pole plus cuts, two-gluon model, DGLAP, BFKL, two-component Pomeron, etc.) what led to a confusion among the community. In what follows, we do not assume any model, unless otherwise specified. We treat the Pomeron as a shadow of inelastic processes. i.e. the dominant contribution to the elastic amplitude which has vacuum quantum numbers in the crossed channel and is related to the main bulk of inelastic channels via the unitarity relation.

Here we are interested in spin properties of such a shadow, namely, the spin-flip part of the elastic pp amplitude related to the Pomeron. Naively, treating the Pomeron perturbatively, one may expect it to conserve s-channel helicity as the quark-gluon vertex does. However, even perturbatively a quark gains a substantial anomalous color-magnetic moment which leads to a spin-flip, like it happens in QED in $g-2$ experiments. Besides, there are many nonperturbative mechanisms generating a Pomeron spin-flip, which are overviewed in [1].

We will present the results in terms of the parameter r_5 which is defined in [1] and is proportional to ratio of the spin-flip to non-flip forward elastic amplitudes.

$$r_5 = \frac{2m_N \Phi_5}{\sqrt{-t}\,\mathrm{Im}\,(\Phi_1 + \Phi_3)}\,, \tag{1}$$

CP675, *Spin 2002: 15th Int'l. Spin Physics Symposium and Workshop on Polarized Electron Sources and Polarimeters*, edited by Y. I. Makdisi, A. U. Luccio, and W. W. MacKay

where the helicity amplitudes are defined as,

$$\Phi_1 = \langle ++|\hat{M}|++\rangle \; ; \quad \Phi_3 = \langle +-|\hat{M}|+-\rangle \; ; \quad \Phi_5 = \langle ++|\hat{M}|+-\rangle \; . \qquad (2)$$

Parameter r_5 may vary with energy, in particular, it is expected to rise [2].

In this talk I focus on the best of our knowledge of $r_5(s)$ which is still a challenge. Importance of this task is two-fold. First of all, it reflects the underlying dynamics and data for r_5 should be compared with numerous and diverse model predictions. Second of all, the polarization program at RHIC needs reliable and fast polarimetry. The currently available polarimeter, is based on the effect of Coulomb-nuclear interference (CNI) [3, 4] which is fully predicted theoretically provided that r_5 is known.

CNI REVISITED

It is not easy to access the spin-flip part of the Pomeron amplitude since it hardly contributes to single spin asymmetry $A_N(t)$. Indeed, although the Pomeron is not a Regge pole, but if r_5 does not vary steeply with energy, one should not expect a large phase shift between the spin-flip and non-flip parts of the Pomeron amplitude. Of course r_5 can be extracted from spin correlation A_{SL} which is, however, difficult to measure.

A unique source of a spin-flip amplitude with a right phase (i.e. with about 90^0 phase shift) is Coulomb scattering. This real amplitude proportional to the anomalous magnetic momentum of proton, interferes with the imaginary non-flip part of the Pomeron amplitude leading to a sizeable spin asymmetry A_N which is nearly independent of energy. The latter fact, as well as possibility to predict the effect, are crucial for polarimetry at high energy.

If the spin-flip part of the Pomeron amplitude were zero, the CNI contribution to single spin asymmetry would be fully predicted [3],

$$A_N(t) = \frac{4\,(t/t_p)^{3/2}}{3(t/t_p)^2+1}\,A_N(t_p)\;. \qquad (3)$$

Here

$$t_p = -8\sqrt{3}\,\frac{\pi\,\alpha}{\sigma_{tot}^{pp}}\;, \qquad (4)$$

is the position of the maximum of $A_N(t)$ which is equal to

$$A_N(t_p) = \frac{\sqrt{-3t_p}}{4m_N}\,(\mu_p-1)\;, \qquad (5)$$

where $\mu_p - 1 \approx 1.79$ is the anomalous magnetic moment of the proton.

Predictions for $A_N(t)$ [3] shown by thick curve in Fig. 1 (left panel) are compared with data from the experiment E704 at 200 GeV [5]. Apparently, agreement is good, although the error bars are quite large.

One may conclude that CNI provides a perfect absolute polarimeter which can be safely used at high energies. Life, however, is more difficult, but also more exciting. The

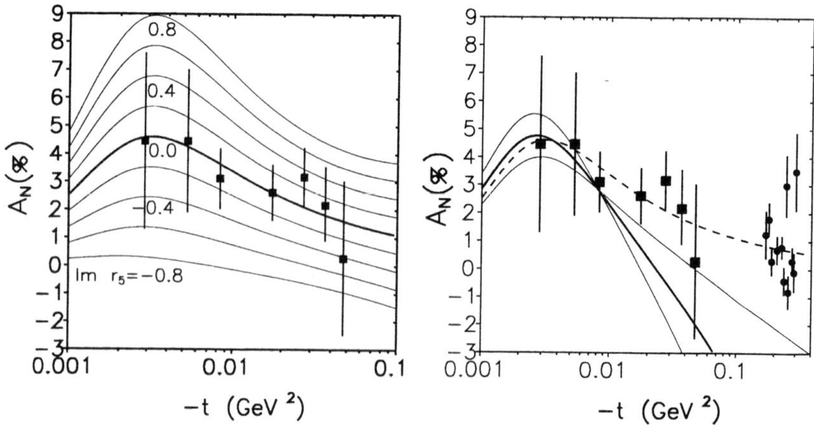

FIGURE 1. Data from the E704 experiment at Fermilab [5] (squares) compared to theoretical calculations. Left panel: calculations with $\rho = 0$, $\delta = 0$ and $\mathrm{Re}\, r_5 = 0$ for different values of $-0.8 < \mathrm{Im}\, r_5 < 0.8$. Thick curve corresponds to $r_5 = 0$. Right panel: r_5 correspond to the results of the E950 experiment at BNL [6] as given by (9). Thin curves show the corridor of uncertainty. Round points show results of other experiments, see in [5]. The dashed curve corresponds to $r_5 = 0$.

Pomeron amplitude may have a nonzero spin-flip part r_5 which affects the spin asymmetry [7, 8]. Fig. 1 (left panel) demonstrates how $A_N(t)$ varies versus $\mathrm{Im}\, r_5$ assuming $\mathrm{Re}\, r_5 = 0$.

Such a sensitivity to r_5 of the CNI effects leads to two-fold consequences:

- CNI polarimetry turns out to be less certain than has been originally expected;
- A_N in the CNI region is an observable maximally sensitive to r_5 and can be used to determine the magnitude of the Pomeron spin-flip.

CNI WITH NUCLEAR TARGETS: THE E950 EXPERIMENT

In order to make use of a CNI polarimeter one should first of all calibrate it, i.e. perform measurements of r_5 with proton beams or target with known polarization. Such beams are available at AGS, but only at energies not much above 20GeV. In this energy range contribution of the sub-leading Reggeons is still important and can substantially contribute to r_5 giving it a steep energy dependence. It would be too risky to rely on a value of r_5 measured at these energies for polarimetry at much higher energies. Especially dangerous are the iso-vector Reggens ρ and a_2 which are spin-flip dominated. To get rid of these unwanted contributions it was suggested in [9] to use CNI on iso-scalar nuclei, in particular carbon. However, two important questions were raised:

- Can one use r_5 measure on nuclear target for polarimetry in pp scattering?
- How should expression for CNI asymmetry be modified in the case of nuclear target?

As for the first question, it has been known since 50s [10] that r_5 remains unchanged, if to treat nuclear effects within the optical model. An updated proof and discussion of possible corrections can be found in [11], as well as the expression for r_5 on a nuclear target,

$$r_5^{pA}(t) = \frac{1 - i\rho_{pA}(t)}{1 - i\rho_{pN}} r_5^{pN} \tag{6}$$

Here ρ_{pN} is the ratio of real to imaginary parts of the forward elastic pN amplitudes. We keep t-dependence of $\rho_{pA}(t)$ since it might be quite steep within the CNI range of t.

The CNI effects for nuclei are substantially modified by nuclear formfactors [9, 11] which are steep functions of t

$$\frac{16\pi}{(\sigma_{tot}^{pA})^2}\frac{d\sigma_{pA}}{dt}A_N^{pA}(t) = \frac{\sqrt{-t}}{m_N}F_A^h(t)\left\{F_A^{em}(t)\frac{t_c}{t}\Big[(\mu_p - 1)[1 - \delta_{pA}(t)\,\rho_{pA}(t)]\right.$$
$$\left. - \; 2\left[\mathrm{Im}\,r_5^{pA}(t) - \delta_{pA}(t)\,\mathrm{Re}\,r_5^{pA}(t)\right]\Big] - 2F_A^h(t)[\mathrm{Re}\,r_5^{pA}(t) - \rho_{pA}(t)\,\mathrm{Im}\,r_5^{pA}(t)]\right\}, \tag{7}$$

where

$$\begin{aligned}\frac{16\pi}{(\sigma_{tot}^{pA})^2}\frac{d\sigma_{pA}}{dt} &= \left(\frac{t_c}{t}\right)^2\left[F_A^{em}(t)\right]^2 - 2\left[\rho_{pA}(t) + \delta_{pA}(t)\right]\frac{t_c}{t}F_A^h(t)F_A^{em}(t) \\ &+ \left[1 + \rho_{pA}^2(t) - \frac{t}{m_p^2}|r_5^{pA}(t)|^2\right]\left[F_A^h(t)\right]^2 .\end{aligned} \tag{8}$$

Here $t_c = -8\pi\alpha/\sigma_{tot}^{pA}$; $\delta_{pA}(t)$ is the Coulomb phase for pA scattering calculated in [12] with high accuracy; the ratio of real to imaginary parts of the pA amplitude, $\rho_{pA}(t)$ and nuclear formfactors, electromagnetic $F_A^{em}(t)$ and hadronic $F_A^h(t)$, are calculated in [12] with a realistic nuclear density.

A first time precise measurement of $A_N^{pA}(t)$ performed by the E950 collaboration for proton-carbon elastic scattering with 22 GeV polarized beam at AGS [6]. The authors fitted the data with expressions (7)-(8) and found

$$\mathrm{Im}\,r_5 = -0.161 \pm 0.226; \quad \mathrm{Re}\,r_5 = 0.088 \pm 0.058\,. \tag{9}$$

The authors added linearly the errors, statistical one and two systematic related to the row asymmetry and the beam polarization. The resulting error seems to be overestimated and may be treated as an upper bound. I have repeated the fit adding quadratically the first two errors, but treating the error in the beam polarization as an overall normalization. I have arrived to similar central values of the parameters, but smaller errors which might be treated as a lower bound,

$$\mathrm{Im}\,r_5 = -0.156 \pm 0.170; \quad \mathrm{Re}\,r_5 = 0.084 \pm 0.042\,. \tag{10}$$

The renormalization factor for the beam polarization was found to be $N = 1.001 \pm 0.120$. The result of the fit and fitted data are depicted in fig. 2

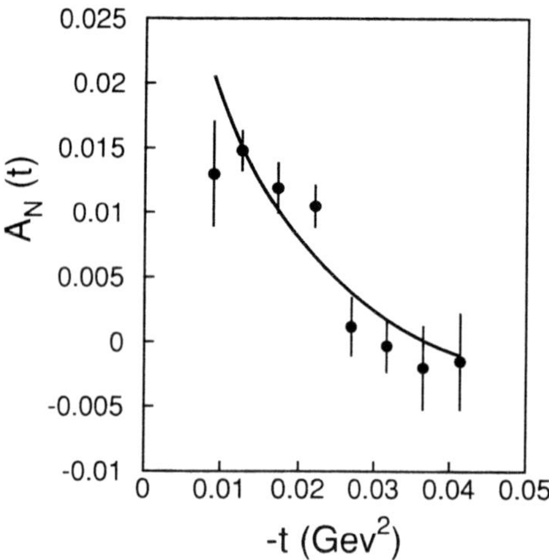

FIGURE 2. The data from [6] with statistical and systematic errors summed quadratically, while the uncertainty in the beam polarization is treated and an overall normalization.

Note that this values of r_5 correspond to the iso-scalar part of elastic pp amplitude. As far as it is known (with a considerable uncertainty) at energy 22 GeV one may consider using it for polarimetry at higher energies. This would be appropriate if energy variation of the Pomeron part of r_5 is small and if the sub-leading iso-scalar Reggeon (ω and f) contribution to r_5 is small at 22 GeV. The latter assumption has been questioned recently and possible corrections for polarimetry are discussed in [8].

Assuming no energy dependence of r_5 one can use (9) to predict A_N^{pp} at energy 200 GeV and compare with the E704 data (including data at larger t, see [5]), as is depicted in Fig. 1 (right panel) by thick solid curve, while the corridor related to the errors in (9) is shown by thin solid curves. In spite of large uncertainties in (9) one may conclude that data do not support such a prediction. Of course this comparison is based of unjustified assumption of no energy dependence of r_5, nevertheless, the observed disagreement should be considered as a warning.

REGGE FACTORIZATION: ANALYSIS OF πN DATA

The iso-scalar part of r_5^{NN} extracted from pA may be compared with πN data. They should be related provided Regge factorization holds. Amplitude analyses of πN elastic and charge exchange scattering up to energy 40 GeV are available [13] and the results contain $r_5(t)$ for iso-scalar part of the scattering amplitude. It turns out that all the analyses demonstrate no t-dependence of $r_5(t)$ within error bars for $|t| < 0.5\,\mathrm{GeV}^2$, what is not surprising since the $\sqrt{-t}$ factor is removed. In order to reduce uncertainties data for iso-scalar $r_5(t)$ for each analysis was fitted by a constant within this t-interval. The results for $\mathrm{Im}\, r_5$ are depicted in Fig. 3 by round points, while the E950 value is shown by a square.

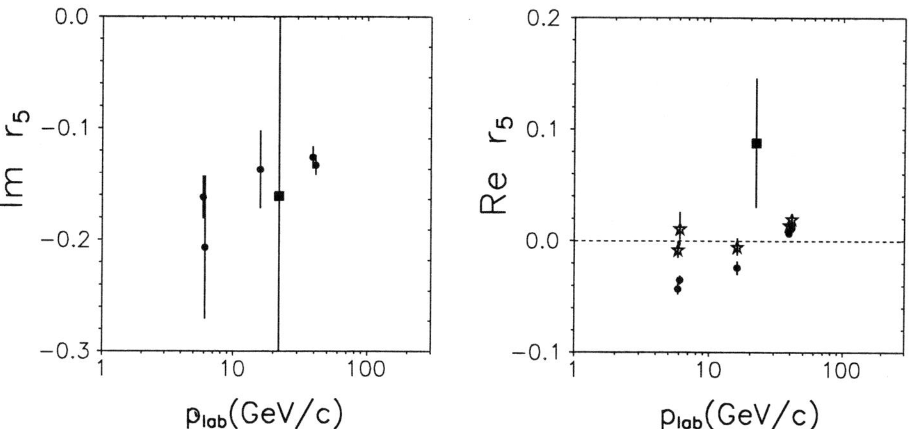

FIGURE 3. Comparison of the results of the E950 experiment at BNL (square points) with the results of amplitude analyses [13] of πN data. Left panel: data for $\mathrm{Im}\, r_5$ are shown by full round dots. Right panel: round dots show the phase uncorrected results of [13] for $\mathrm{Re}\, r_5$, star points are corrected for the phase of the non-flip amplitude.

Apparently, the πN data prefer negative $\mathrm{Im}\, r_5$, however they do not specify energy dependence. Within large error bars they are consistent either with no energy dependence, or with $\mathrm{Im}\, r_5$ rising with energy. The former case would correspond to a net contribution of the Pomeron spin-flip, while the latter possibility would mean that f-Reggeon contribution to $\mathrm{Im}\, r_5$ exists and is negative. Thus, we conclude that $\mathrm{Im}\, r_5^f \leq 0$. Since the phase of the f-amplitude is given by the signature factor, $\eta(t) = i - \cot[\pi\alpha_f(t)/2)]$ we should expect from this consideration that $\mathrm{Re}\, r_5^f \geq 0$.

Data for $\mathrm{Re}\, r_5$ extracted from the same analyses [13] are depicted by round points in Fig. 3 (right panel). The real part of the spin flip amplitude was determined in those analyses relative to the imaginary part of the non-flip amplitude, i.e. assuming it pure imaginary. Thus, one should introduce a correction for a nonzero real part of the non-flip amplitude, $\Delta \mathrm{Re}\, F_{+-} = \rho_{\pi N}\, \mathrm{Im}\, F_{+-}$. Using $\mathrm{Im}\, r_5$ found above, new corrected values for $\mathrm{Re}\, r_5$ were determined and plotted in Fig. 3 (right panel) by star points. These results are in agreement with the above expectation $\mathrm{Re}\, r_5^f \geq 0$, preferring, however, zero and energy independent value. The point from the E950 experiment shown by a square is somewhat higher, but still is compatible with these results.

Thus, available amplitude analyses of πN data at energies $6 - 40\,\mathrm{GeV}$ indicate at the dominance of the Pomeron amplitude with

$$\mathrm{Im}\, r_5 \approx -0.12; \quad \mathrm{Re}\, r_5 \approx 0\,, \tag{11}$$

and vanishing contribution of the f-Reggeon.

COMPARISON WITH THEORETICAL EXPECTATIONS

One can find in the literature a variety model predictions for the spin-flip part of the Pomeron amplitude. Many of them are collected and discuss in [1]. Here we list them briefly mentioning the underlying physical ideas.

- Treating the gluon-quark vertex as an analog to the iso-scalar photon-proton one can relate the anomalous color-magnetic moment of a quark to the iso-scalar part of the anomalous magnetic momentum of the proton [14]. After installation of such a quark-gluon vertex into the two-gluon model for the Pomeron one gets [14], $\mathrm{Im}\, r_5 = 0.13$. Although the order of magnitude is correct, the sign is opposite to data presented in Fig. 3.
- Helicity of the proton is not equal to the sum of quark helicities. Therefore, the proton may flip its helicity even if quarks do not (as the leading order pQCD predicts). A quark-diquark model of the proton leads to nonzero $\mathrm{Im}\, r_5 = -(0.05 - 0.15)$, dependent on the diquark size $(0.5 - 0.2\,\mathrm{fm})$ [7]. Within the uncertainty this prediction agrees with the data.
- Modeling the Pomeron-proton coupling via two pion exchange [15, 16] one arrives at a conclusion that iso-scalar Reggeons ($I\!P$, f, ω) are predominantly spin non-flip, while iso-vectors (ρ, a_2) mostly flip the proton spin. Prediction of [16] for the Pomeron is $\mathrm{Im}\, r_5 = 0.06$, what has incorrect sign. A similar pion cloud model developed [17] with some differences in details predicts $\mathrm{Im}\, r_5 = -0.3$, $\mathrm{Re}\, r_5 = -0.06$, what also disagree with the data.
- The phenomenological model [18] assuming that the spin-flip part can be deduced from the impact parameter distribution of matter in the proton and fitted to data predicts correct sign, $\mathrm{Im}\, r_5 \approx -(0.01 - 0.02)$, but modulo too small value.

One should not treat this comparison as a way to confirm or reject models. None of the models under discussion may pretend to be a dominant mechanism. The dynamics suggested by other models can contribute as well.

Note that analysis of pp elastic data performed in [19] led to parameters $\mathrm{Im}\, r_5 = -0.054$ which is too small, but has the right order of magnitude and correct sign compared to data plotted in Fig. 3. The analysis performed in [19] was based on a specific modeling of the odderon amplitude which introduces a strong sensitivity of polarization to r_5. Besides, the contribution of the sub-leading Reggeons largely contributing to r_5 (ρ, a_2) was neglected, instead this this contribution was attributed to the Pomeron.

πN VS E950 DATA: HOW SHAKY IS THE THEORY BRIDGE?

The results of amplitude analyses of πN data are good news for polarimetry at RHIC. Absence of energy dependent contribution of iso-scalar sub-leading Reggeons to r_5 suggested by the data would allow one to use the result of measurement of A_N by the E950 experiment for polarimetry at higher RHIC energies. However, the central value of $\mathrm{Re}\, r_5$ which follows from the E950 data is different from zero and indicates that the Reggeon contribution might be important. Then, one may expect r_5 to vary with energy and the polarimetry gets an uncertainty.

Moreover, the fitting parameters $\mathrm{Re}\, r_5$ and $\mathrm{Im}\, r_5$ strongly correlate as it is demonstrated in [6]. For example, if to enforce and fix $\mathrm{Re}\, r_5 = 0$, the χ^2 doubles and $\mathrm{Im}\, r_5$ changes sign. Thus, it is difficult to bring together the results of study of different reactions πN and pC.

Facing such a problem one should check how reliable are assumptions done in order to make a link between iso-scalar amplitudes in pp and πN scattering.

1. First of all, how precise is factorization connecting r_5 in pp and πN? In all the models listed above it is provided. Even in the two-gluon model which does not obey Regge factorization, r_5 must be the same for πN and pp. It is known, however that that Regge cuts corresponding to eikonal multi-Pomeron rescatterings violate Regge factorization. However, as is discussed above and proven in [10] these corrections do not alterate r_5.
2. Only sub-leading Reggeon, f, contributes to the iso-scalar amplitude in πN scattering, while both f and ω are present in pp. Moreover, in order to respect duality f and ω should be exchange degenerate, i.e. their contributions are expected to add up in $\mathrm{Re}\, r_5$ and nearly cancel in $\mathrm{Im}\, r_5$. This is different from πN where f-Reggeon should contribute equally to $\mathrm{Re}\, r_5$ and $\mathrm{Im}\, r_5$. However, this difference does not explain the observed difference between πN and pp. If f-Reggeon does not contribute to r_5 in πN, according to factorization and exchange degeneracy both the f and ω contributions to pp must be zero as well.

Although we did not find any good reason to disbelieve the theoretical link between πN and pp, it is still possible that this is the origin of the problem. On the other hand the observed contradiction is not dramatic since the errors of the E950 data are pretty large. In order to progress further, the accuracy of A_N in pC elastic scattering should be improved.

DIRECT INFORMATION FROM *NN* DATA

There is another narrow place in the theoretical bridge between πN and NN reactions: it might be a contribution to NN of sub-leading Reggeons which are forbidden for πN. For instance, besides ω there might be other iso-scalar mesons which are suppressed or forbidden (e.g. have negative G-parity) for πn scattering. This was suggested in [20] as $\varepsilon(0^{++})$ and $\omega'(1^{--})$ exchange degenerate Reggeons. Indeed, analysis [21] of data for pp and np elastic scattering up to 12GeV shown in Fig. 4 demonstrates an iso-scalar spin-flip NN amplitude (left panel) which falls with energy much steeper than iso-vector one (right panel). The iso-scalar Regge trajectory turns out to be displaced by one unit down compared to the ρ-Reggeon trajectory: $\alpha_\varepsilon(t) = \alpha_\rho(t) - 1 = -0.5 + 0.9t$ [21].

It is important to establish whether the large value of $\mathrm{Re}\, r_5$ observed by the E950 experiment is related to the tail of this low-energy mechanism. If so, then $\mathrm{Re}\, r_5$ will steeply vanish at higher energies what should affect the polarimetry. In this case the shape of

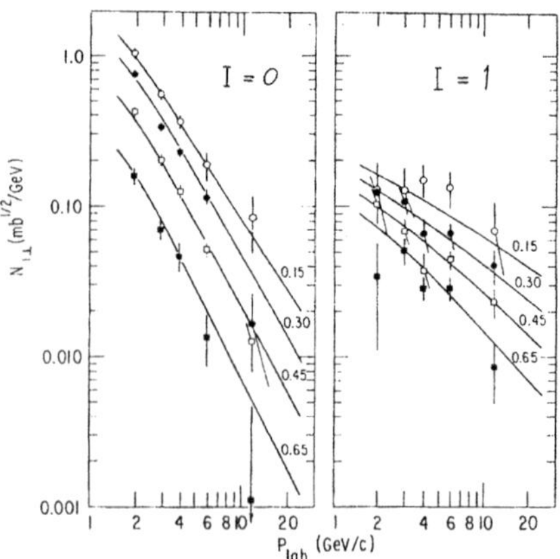

FIGURE 4. Dependence of the spin-flip amplitude on lab momentum for iso-scalar (left panel) and iso-vector (right panel) exchanges. Points are the result of the analysis of data on elastic and charge-exchange pp and pn scattering performed in [21] for different bins in t, and the curves are the results of Regge fit.

$A_N(t)$ in the CNI region would change substantially (not supported by preliminary data at 100 GeV).

One can estimate such a low-energy contribution to r_5 at 20 GeV relying on the extrapolation of the iso-scalar spin-flip NN amplitude done in [21] depicted in Fig. 4. The iso-scalar amplitude is determined by measurement of A_N and cross sections of elastic and charge-exchange pp and pn scattering,

$$N^0_{1\perp} = \left[(A_N\sigma)_{pp} + (A_N\sigma)_{pn} - \frac{1}{2}(A_N\sigma)_{cex}\right] \Big/ (4\,|N^0_0|)\,, \tag{12}$$

where $|N^0_0|^2 = (\sigma_{pp} + \sigma_{pn})/2$. At $t = -0.15$ this amplitude is predicted to be, $N^0_1 \approx 0.03\sqrt{\text{mb}}/\text{GeV}$. The non-flip amplitude equals to $N^0_0 \approx \sigma_{tot}/(4\sqrt{\pi})\exp(5t) \approx 4.2\sqrt{\text{mb}}/\text{GeV}$. Taking into account the factor $\sqrt{-t}/m_N$ in N^0_1, one arrives at the estimate at $t = -0.15\,\text{GeV}^2$,

$$\text{Re}\,r_5(p_{lab} = 22\,\text{GeV}/c) \approx 0.02\,, \tag{13}$$

which is too small to explain the value of $\text{Re}\,r_5$ in (9).

This estimate agrees well with the measurements of single-spin asymmetry in pp and pn performed at 24 GeV at BNL [22]. Neglecting the small charge-exchange contribution (it steeply falls with energy) in (12) one gets at $t = -0.15\,\text{GeV}^2$,

$$\text{Re}\,r_5(p_{lab} = 24\,\text{GeV}/c) = 0.016 \pm 0.010\,, \tag{14}$$

Thus, both extrapolation of Argonne data to higher energies and direct measurements at AGS at 24 GeV confirm that $\text{Re}\,r_5$ is about order of magnitude smaller than what follows from the E950 data.

It is also very improbable that $r_5(t)$ could vary substantially at $0 < -t < 0.15\,\text{GeV}^2$. As it was mentioned above, in πN data r_5 remains unchanged up to $-t = 0.5\,\text{GeV}^2$.

CONCLUSIONS AND OUTLOOK

The E950 experiment has provided first high statistics measurements for CNI asymmetry in proton-carbon elastic scattering. On the one hand, these data bring information about the spin-flip part of the hadronic amplitude which is tempting to associate with the Pomeron. On the other hand, if it true, one can use the found parameters for r_5 to predict $A_N(t)$ at higher energies and use pC scattering as a polarimeter at RHIC.

At the same time, amplitude analysis of data for πN and NN elastic and charge-exchange scattering allow to single out the iso-scalar part of the spin-flip amplitude. The values of $\mathrm{Re}\,r_5$ extracted from these data are sufficiently small to be neglected. This is a great news for the CNI polarimetry which can be safely used at high energies. This value of $\mathrm{Re}\,r_5$ is, however, much smaller than found from the E950 data. To resolve this controversy one needs new and more precise data for CNI spin asymmetry and in a wider energy range.

ACKNOWLEDGMENTS

I am thankful to Larry Trueman for interesting discussions and to Yousef Makdisi for inviting me to speak at the Conference. This research was performed during visit at the BNL Nuclear Theory Group, and I am grateful to Larry McLerran for hospitality. This work has been supported by a grant from the Gesellschaft für Schwerionenforschung Darmstadt (GSI), grant No. GSI-OR-SCH, and also by the grant INTAS-97-OPEN-31696.

REFERENCES

1. N.H. Buttimore, B.Z. Kopeliovich, E. Leader, J. Soffer, T.L. Trueman, Phys. Rev. **D59** (1999) 114010.
2. B.Z. Kopeliovich and B. Povh, Mod. Phys. Lett. **A13** (1998) 3033.
3. B.Z. Kopeliovich and L.I. Lapidus, Sov. J. Nucl. Phys. **19** (1974) 114.
4. N.H. Buttimore, E. Gotsman, E. Leader, Phys. Rev. **D18** 694 (1978).
5. The E704 Collaboration, N. Akchurin et al., Phys. Lett. **B229** (1989) 299.
6. E950 Collaboration, J. Tojo et al., Phys. Rev. Lett. 89 (2002) 052302.
7. B.Z. Kopeliovich and B.G. Zakharov, Phys. Lett. **B226** (1989) 156.
8. T.L. Trueman, *CNI Polarimetry and the Hadronic Spin Dependence of pp Scattering*, hep-ph/9610429.
9. B.Z. Kopeliovich, *High-Energy Polarimetry at RHIC*, hep-ph/9801414.
10. H.S. Köhler, Nucl. Phys. **1**, 433 (1956); I.I. Levintov, Doklady Akad. Nauk U.S.S.R. **107**, 240 (1956), Soviet Phys. JETP **1**, 175 (1956); C. Bourrely, J. Soffer and D. Wray, Nucl. Phys. **87B**, 32 (1975);**91B**, 33 (1975); C. Bourrely, E. Leader and D. Wray, Il Nuovo Cimento **35A**, 559 (1976).
11. B.Z. Kopeliovich and T.L. Trueman, Phys. Rev. **D64** (2001) 034004.
12. B.Z. Kopeliovich and A.V. Tarasov, Phys. Lett. **B497**, 44 (2001).
13. F. Halzen and C. Michael, Phys. Lett. **36B** (1971) 367; G.Cozzika et al., Phys. Lett. **40B** (1972) 281; P. Johnson et al., Phys. Rev. Lett. **30** (1973) 242; J. Pierrard et al., Nucl. Phys. B**107** (1976) 118; V.D. Apokin et al., Sov. J. Nucl. Phys. **38** (1983) 574 [Yad. Fiz. **38** (1983) 956]; Yu.M. Kazarinov et al., Preprint P1-85-426, Dubna 1985 (in Russian).
14. M. G. Ryskin, Yad. Fiz. **46**(1987) 611 [Sov. J. Nucl. Phys. **46** (1987) 337].

15. J. Pumplin and G.L. Kane, Phys. Rev. **D11** (1975) 1183.
16. K.G. Boreskov, A.A. Grigiryan, A.B. Kaidalov and I.I. Levintov, Sov. J. Nucl. Phys. **27** (1978) 432.
17. S.V. Goloskokov, S.P. Kuleshov and O.V. Selyugin, Z. Phys. **C50** (1991) 455.
18. C. Bourrely, J. Soffer and T.T. Wu, Phys. Rev. **D19** (1979) 3249; Nucl. Phys. **B247** (1984) 15.
19. A.F. Martini and E. Predazzi, Phys. Rev. **D66** (2002) 034029.
20. E.L. Berger, A.C. Irving and C. Sorensen, Phys. Rev. **D17** (1978) 2971.
21. S.L. Kramer et al., Phys. Rev. **D17** (1978) 1709.
22. D.G. Crabb et al., Nucl. Phys. **B121** (1977) 231; *ibid* **B201** (1982) 365.

Spin and the Three-Nucleon Force

B. von Przewoski

IUCF, Milo B. Samson Lane, Bloomington, IN 47405, USA

Abstract. Theoretical descriptions of three-nucleon systems, which are based on a summation over pairwise NN interactions, fail to describe existing data. For instance, the binding energy of the triton is underpredicted as well as the unpolarized cross section in pd elastic scattering around $\theta_{cm} = 100°$. It has been demonstrated that the inclusion of a three-nucleon-force, which has been adjusted to reproduce the measured binding energy of the triton, also reduces the discrepancy of the calculation with the measured cross section. In the case of many spin observables - the most notorious being the analyzing power in nd scattering below 10 MeV ("A_y puzzle") - inclusion of any one of the modelled three-nucleon-forces does not explain the data. Several precise data sets of spin observables exist between 100 MeV and the pion production threshold. A theoretical description of the three-nucleon system must be able to reproduce all spin observables simultaneously.

INTRODUCTION

In general, a three-body interaction does not simply consist of the sum of pairwise interactions. Instead, the addition of a third object to an existing pair alters the potential between the two objects. A well known example is the three-body problem in astronomy, which deals with the gravitational interaction between sun, earth and moon. In this case, the three-body force arises from the tidal deformation. In the nuclear case, the excitation of an intermediate state is analogous to the tidal effect.

Nuclear properties are fairly well explained in terms of pairwise interactions. Therefore, the effects of a three-nucleon force are expected to be small. Evidence of the existence of a three-nucleon force comes largely from studies of the bound 3N system. 2N calculations fail to reproduce the measured binding energy of the triton. Inclusion of a three- nucleon force increases the calculated binding energy from 7.6 to 8.5 MeV, in agreement with the measured value. Similarly, the calculated charge radius and D- to S- state ratio of the triton disagree with the measured values by $\sim$10% unless a three-nucleon force is included in the calculation.

The three-nucleon system is described by so-called Faddeev equations [2], which can be solved exactly. Although the original equations were formulated in 1961, only recent advances in computing speed made it practical to solve them at energies higher than a few MeV. The importance of a three-nucleon force is assessed in the following way. First, a Faddeev calculation is performed using a 2N potential alone. The use of different 2N potentials such as CD Bonn or NijmegenI,II results in an error band for the 2N prediction. In the case of the triton binding energy variation of individual phases within a reasonable range does not reproduce the measurement. Then, other Faddeev calculations with the inclusion of different 3N potentials are performed and compared to the data. Sometimes, as in the case of the triton binding energy, the agreement with

CP675, *Spin 2002: 15th Int'l. Spin Physics Symposium and Workshop on Polarized Electron Sources and Polarimeters,* edited by Y. I. Makdisi, A. U. Luccio, and W. W. MacKay

the data is improved.

So far, three 3N force models have been used in the context of Faddeev calculations: the Tucson-Melbourne force (TMF)[3, 4, 5] and its modified, chirally symmetric form (TMF') and the Urbana IX 3NF[6]. The TMF is based on meson exchange and virtual Δ excitation. The Argonne 3NF contains a phenomenological short range interaction. None of the Faddeev calculations at energies above the deuteron threshold contain the Coulomb interaction. Also, the calculations are non-relativistic. There are also attempts to describe the 3N system by models based on chiral perturbation theory and on quark models [7]. The minimum requirement of these calculations is usually that they describe the bound 3N system. For a comprehensive review of both the data base and the theoretical decription of the 3N system see [8]

SPIN EFFECTS AT LOW ENERGIES

Interest in the three-nucleon system first arose in the sixties. The Coulomb interaction is important at low energies, but is not yet calculable above the breakup threshold. Therefore, experimental results in nd scattering are desirable. However, pd scattering is experimentally much easier than nd scattering. Therefore an early experimental goal was to compare pd and nd scattering at energies below 20 MeV [1]. On the theoretical side, Faddeev calculations even at energies below 20 MeV were not feasible until the late eighties when increased computing speed allowed such calculations up to j=2. Concurrently, interest in low energy pd and nd scattering was renewed and several data sets for nd, pd and dp scattering[9, 10] below 10 MeV became available. When precise low-energy analyzing power data in nd scattering were compared with rigorous Faddeev calculations the so-called A_y-puzzle emerged. The maximum of the analyzing power at $\theta_{cm} \sim 120°$ is underpredicted by state-of-the-art Faddeev calculations. Attempts to correct the problem by adjusting parameters of the two-nucleon force within acceptable range are not successful.

ENERGIES UP TO THE PION PRODUCTION THRESHOLD

Since Coulomb effects decrease with energy, but three-nucleon effects, which are dominated by virtual Δ excitation, increase with energy, experimental and theoretical efforts turned towards p+d reactions at energies above $\sim$100 MeV but below the pion production threshold at $\sim$200 MeV. Faddeev calculations at these energies are converging if $j \leq 5$. They do not contain any relativistic effects.

Groups at KVI, RIKEN and IUCF are actively persuing programs to measure spin observables in pd scattering. At KVI angular distributions of the proton analyzing power (A_y^p) have been measured at 108, 120, 135, 150, 170 and 190 MeV[11, 12]. The data are compared to Faddeev calculations with and without three-nucleon force. Calculations using either the AV18, CD-Bonn or Nijmegen I,II 2N potential do not reproduce the measured angular distributions of A_y^p. If the Urbana IX, TM or TM' three nucleon force is included, the calculated angular distributions change, but the agreement with the data does not improve. After including a 3NF in the calculation the diffenence between data and theory is as large as +/-0.15 at $\theta_{cm} \geq 90°$.

At RIKEN a program is under way to measure spin observables with either vector or tensor polarized deuteron beam. Angular distributions of differential cross section, deuteron vector analyzing power (A_y^d) and deuteron tensor analyzing powers (A_{xz}, A_{xx} and A_{zz} have been measured at 140, 200 and 270 MeV incident deuteron energy [13]. Again, the data are compared to Faddeev calculations using different 2N potentials (AV18, CD-Bonn or Nijm I,II,'93) alone and with either the Argonne or the Tucson-Melbourne three-nucleon force included. While the theory reproduces the cross section data quite well after including either of the two three-nucleon forces, it fail to provide an adequate description of the tensor analyzing power data. In fact, the χ^2 of the difference between data and theory increases in the case of A_{xx} and A_{yy} after the three-nucleon force is included. A double scattering experiment to measure $K_{ij}^{y'}$ and $P^{y'}$ was performed at RIKEN [14] as well.

At IUCF a program has been recently completed to measure spin correlation coefficients in pd elastic scattering at 135 and 200 MeV incident proton energy in pd elastic scattering as well as in dp breakup at 270 MeV incident deuteron energy. In conjunction with previous measurements, the data from IUCF at 135 MeV constitute the most comprehensive set of spin observables at any one energy to date.

SPIN-1/2 ON SPIN-1 OBSERVABLES

For a comprehensive discussion of the formalism pertaining to spin correlation measurements and definitions of observables see [15].

Either one or both of beam and target may be polarized along one of three directions x,y,z, where x denotes the horizontal, y the vertical and z the longitudinal axis of a cartesian coordinate system. For vector polarized deuterium, 15 polarization observables are conceivable, just by counting the number of combinations of beam and target polarization. Some of them, like the longitudinal analyzing power, are forbidden in a two-particle final state due to parity conservation while others are redundant, since they can be obtained simply by rotation of the coordinate system about the z-axis. In total, seven spin observables exist for elastic scattering; namely the proton and deuteron analyzing powers A_y^p and A_y^d and the spin correlation coefficients C_{xx}. C_{yy}, C_{zz}, C_{zx} and C_{xz}.

If the deuteron is tensor polarized, one can identify 20 combinations of beam and target polarization, cases where the proton is unpolarized included. Of these combinations seven are related to others by rotation around the z-axis, and three are forbidden by parity conservation for a two-particle final state. The remaining observables are the three tensor analyzing powers A_{xx}, A_{yy} and A_{xz}, and the seven spin correlaton coefficients $C_{xx,y}$, $C_{xy,x}$, $C_{xy,z}$, $C_{xz,y}$, $C_{xz,z}$, $C_{yy,y}$ and $C_{yz,x}$.

In summary, 17 polarization observables exist for $\vec{p}\vec{d}$ elastic scattering. Out of those, 15 have been measured at IUCF.

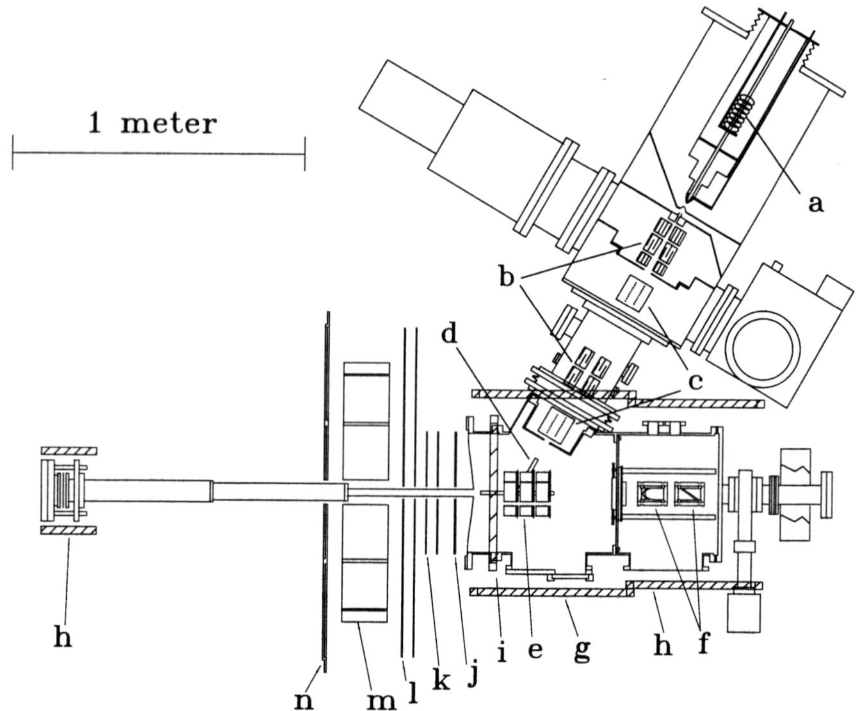

FIGURE 1. The PINTEX (Polarized Internal Target EXperiments) facility at the Indiana Cooler. a:dissociator, b: sextupole system, c:remotely controlled transition units, d:target cell, e:silicon barrel, f: beam position monitors, g:Helmholtz coils, h:compensating Helmholtz coils, i:z-field coil, j:start scintillator, k,l:wire chambers, m:stopping scintillator, n:veto scintillator, h:correction steerer

$\vec{D}(\vec{P},P)D$ AT 135 AND 200 MEV AT IUCF

The experiment was performed at the PINTEX (`http://www.iucf.indiana.edu/Experiments/PINTEX/pintex.html`) facility of the Indiana Cooler, a light ion storage ring with electron cooling. A layout of the PINTEX facility is shown in fig.1.

The proton beam was injected at either 135 or 200 MeV. The beam polarization was on the order of 70% and either vertical or longitudinal. Longitudinal beam polarization at the target location was achieved by two spin precession solenoids in the ring.

The target was a 145 mm long and 12 mm diameter storage cell made from 0.05 mm thick aluminum. Polarized deuterons were produced in an atomic beam source (ABS)[16]. and injected into the target cell. The target thickness was $\sim 10^{13}$ atoms/cm^2. The aluminum was teflon-coated in order to avoid wall depolarization.

The target was either purely vector- or purely tensor polarized. Since the target was operated in a weak magnetic field in order to minimize orbit distortions, the theoretical maximum vector and tensor polarizations were +/-2/3 and +/-1 respectively.

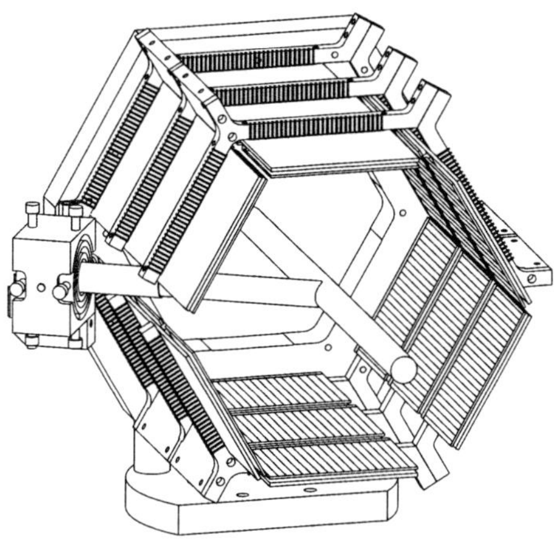

FIGURE 2. Silicon barrel surrounding the target cell.Each detecor measures 4x6cm ^{2} and has 28 strips.

The spin alignment axis of the target was cycled between horizontal, vertical and longitudinal at 2s intervals. When the target was tensor- polarized, two additional orientations of the holding field were added to the sequence. For these orientations, the horizontal or vertical field coils were energized simultaneously in order tilt the spin alignment axis at 45°. This is necessary to generate a target with a t_{21} tensor moment.

The outgoing proton and deuteron from elastic scattering were detected in coincidence. The forward going particle (either p or d) was detected in a stack consisting of a ΔE scintillator, two wire chambers and two stopping scintillators. Laboratory angles up to ~45° were covered. The recoil was detected in the so-called silicon barrel, an array of eighteen silicon strip detectors surrounding the target cell. The silicon detectors were oriented such that they measured the azimuth of the recoil. Fig.2 shows the silicon barrel surrounding the target cell. The alignment fixture which holds the cell at the feedtube is also visible in Fig.2. The detector arrangement provides almost complete coverage of the angular distribution.

Normalization

Both beam and target polarization have to be known in order to determine analyzing powers and spin correlation coefficients.

The target polarization is normalized to known vector and tensor analyzing powers at 135 MeV[13]. At 200 MeV no precise measurements of deuteron analyzing powers exist. The only available data [17] at that energy have large errors. Therefore, the calibration standard at 135 MeV had to be transported to 200 MeV. The target polarization does not change during an energy ramp of the beam and can be determined at 135 MeV prior to the energy ramp. Then, after the beam energy ramp is complete the target asymmetry is measured at 200 MeV with the *same* beam. Thus, the analyzing power at the higher energy is calibrated by this procedure.

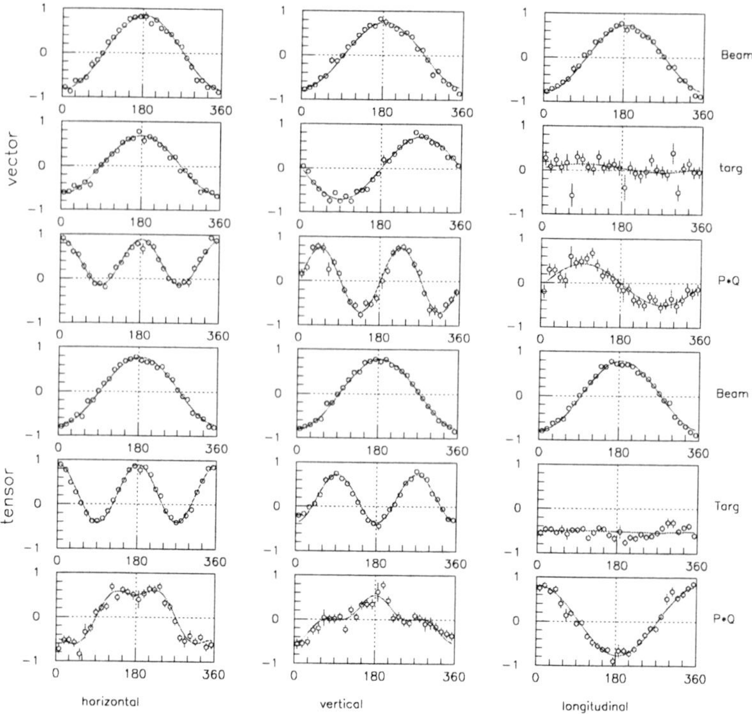

FIGURE 3. Distributions of the azimuthal angle Φ at 135 MeV. The lines are fits of combinations trigonometric functions. See text for details

Calibration of the /it beam analyzing power was achieved by acquiring a data sample with a mixture of unpolarized hydrogen and deuterium in the target cell. This way, asymmetries from pp and pd scattering are measured concurrently. Since the analyzing power for elastic pp scattering is known at 135 and 200 MeV, the beam polarization can be determined at both energies. The analyzing power for pd scattering is then obtained by dividing the concurrently measured pd asymmetry by the beam polarization.

In order to determine the longitudinal beam polarization, the atomic beam source was operated with a mixture of hydrogen and deuterium gas. When the ABS was operated with the gas mixture, the RF transition units were turned off, since efficient transitions could not be made for hydrogen and deuterium at the same time. Without the RF transition units, the sextupole system polarizes hydrogen and deuterium to maximum values of $P_z(h)$=1/2 and ($P_z(d)$=1/3, $P_{zz}(d)$=-1/3) respectively. Since the spin correlation coefficient C_{zz} for elastic $\vec{p}\vec{p}$ scattering is known, the longitudinal beam polarization is determined.

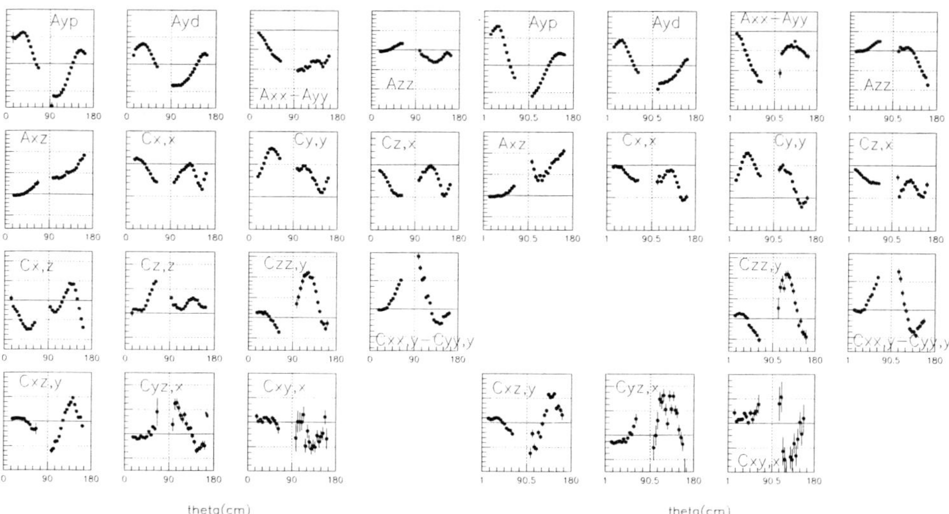

FIGURE 4. Preliminary results at 135 MeV (left) and 200 MeV (right). The vertical scale has been omitted on purpose.

Azimuthal dependencies

If one writes down the general spin-dependent cross section [15], one sees that the spin observables have characteristic azimuthal (ϕ) dependencies. For instance, the tensor analyzing power varies with $\cos(2\phi)$ or the tensor spin correlation coefficient $C_{xy,x}$ varies with $\sin(3\phi)$. By cleverly combining countrates such that certain ϕ dependencies cancel while others are retained, different observables can be isolated. For example, if one averages over all target polarizations one retains only terms that depend on the beam analyzing power. Nearly complete azimuthal coverage of the detector system allows one to then determine A_y^p by fitting a cosine function to the ϕ distribution. Fig.3 shows examples of azimuthal distributions for vector (three upper rows) and tensor (three lower rows) target poalrization at 135 MeV. The columns correspond to horizontal, vertical and longitudinal spin alignment axis of the target. Counting from the top, the first and and fourth row contain terms proportional to the beam polarization only. The second and fifth row contain terms proportional to the target polarization only and the third and sixth row contain terms proportional to the product of beam and target polarization. ϕ distributions like those shown in Fig.3 are fitted for each center-of-mass polar angle bin.

Preliminary results

The amplitudes of the trigonometric functions used to fit the ϕ distributions are used to extract the spin dependent observables as a function of the center-of-mass angle. Figs. 4 and 5 show preliminary angular distributions of analyzing powers and spin correlation coefficients at 135 and 200 MeV respectively. Since the absolute calibration of the observables is still preliminary, the vertical scale has been omitted. The spin correlation coefficients $C_{x,z}$ and $C_{z,z}$ can only be measured using longitudinally polarized beam. No data were taken at 200 MeV with longitudinally polarized beam. One of the spin correlation coefficients ($C_{y,y}$ at 200 MeV) has been measured previously using an optically pumped polarized deuterium target. This experiment was also performed at IUCF[18].

$\vec{D}\ \vec{P}$ BREAKUP AT IUCF

An experiment to measure spin observables in dp breakup at 270 MeV incident deuteron energy was also completed using the PINTEX facility. This experiment was performed in reverse kinematics to maximize the acceptance of the existing forward detector stack ($\sim$50%). For the trigger, we required two charged particles in the forward detector shown in Fig.1. The silicon barrel was not used. The target in this case contained polarized hydrogen and its polarization direction was cycled between the horizontal, vertical and logitudinal direction. The Cooler was filled alternatingly with tensor- or vector-polarized deuterons, or with unpolarized deuterons. Although data were taken with both beam and target polarized, the first goal is to extract the longitudinal analyzing power A_z^p from the data. Unlike in a two-body final state, A_z^p is allowed by parity conservation. In fact, it can be shown[19] that axial observables such as A_z are sensitive to spin operators that occur in 3N potentials but not in 2N potentials. A previous measurement of A_z at 9 MeV [20] showed that A_z^p is very small ($\sim 10^{-3}$) as predicted by a Faddeev calculation. In contrast to that, Faddeev calculations predict that A_z^p is sizable ($\sim$0.1) at 270 MeV incident deuteron energy. At this point the data are under analysis, however, there is already clear evidence of a non-vanishing longitudinal analyzing power in pd breakup.

CONCLUSIONS

Advanced experimental techniques that employ stored beams and internal targets have resulted in precise data for pd elastic scattering and breakup. A large body of high quality data between a few MeV and the pion production threshold is available. Faddeev calculations up to the pion production threshold are now feasible due to increased computing speed. There are discrepancies between the data and Faddeev calculations regardless of whether a three-nucleon force is included or not. More importantly, Faddeev calculations that are based on different two-nucleon forces exhibit differences among each other that are large compared with the experimental uncertainties.

ACKNOWLEDGMENTS

This work was supported by NSF grants PHY-9602872, PHY-9722556, PHY-9901529 and DOE grant DOE-FG02-88ER404308. We are grateful to the operators at IUCF, in particular Gary East and Terry Sloan, for their tireless efforts in providing us with stable, high intensity beam.

REFERENCES

1. Taylor et al.*Phys. Rev. C*, **1**, 808 (1970).
2. Faddeev, L.D., *JETP (Sov. Phys.)*, **12**, 1014 (1961).
3. Friar, J.L. et al. *Phys. Rev. C*, **59**, 53 (1999).
4. Hüber, D. et al. *nucl-th/9910034*.
5. Coon, S.A. et al. *nucl-th/0101003*.
6. Pudliner, B.S et al. *Phys. Rev. C*, **56**, 1720 (1997).
7. Fujiwara, Y., *Phys. Rev. C*, **66**, 021001 (2002).
8. Glöckle, W. et al. *Phys. Rep.*, **274**, 107 (1996).
9. Shimizu, S. et al.*Phys. Rev. C*, **52**, 1193 (1995).
10. Tornow, W. et al.*Phys. Lett. B*, **257**, 273 (1991).
11. Ermisch, K. et al.*Phys. Rev. Lett.*, **86**, 5862 (2001).
12. Ermisch, K. and Kalantar-Nayestanaki, N. *KVI newsletter*, **9**, August (2002).
13. Sekiguchi, K. et al.*Phys. Rev. C*, **65**, 034003-1 (2002).
14. Sakai, H. et al.*Phys. Rev. Lett.*, **84**, 5288 (2000).
15. Ohlsen, G.G. *Rep. Proc. Phys.*, **35**, 717 (1972).
16. Wise, T. et al.*Nucl. Instr. Meth. A*, **336**, 410 (1993).
17. Garcon et al.*Nucl. Phys. A*, **458**, 287 (1986).
18. Cadman, R.V. et al.*Phys. Rev. Lett.*, **86**, 967 (2001).
19. Knutson, L. et al.*Phys. Rev. Lett.*, **73**, 3062 (1993).
20. George, E.A. et al.*Phys. Rev. C*, **54**, 1523 (1996).

Nucleon Electromagnetic Form Factors and Densities

James J. Kelly

Department of Physics, University of Maryland, College Park, MD 20742

Abstract. We review data for nucleon electromagnetic form factors, emphasizing recent measurements of G_E/G_M that use recoil or target polarization to minimize systematic errors and model dependence. The data are parametrized in terms of densities that are consistent with the Lorentz contraction of the Breit frame and with pQCD. The dramatic linear decrease in G_{Ep}/G_{Mp} for $1 \leq Q^2 \leq 6$ (GeV/c)^2 demonstrates that the charge is broader than the magnetization of the proton. High-precision recoil polarization measurements of G_{En} show clearly the positive core and negative surface charge of the neutron. Combining these measurements, we display spatial densities for u and d quarks in nucleons.

INTRODUCTION

The electromagnetic structure of structure of nucleons provides fundamental tests of the QCD confinement mechanism, as calculated on the lattice or interpreted with the aid of models. From elastic electron scattering one obtains the Sachs electric and magnetic form factors, which are closely related to the charge and magnetization densities. Dramatic improvements in the quality of these measurements have recently been achieved by using beams that combine high polarization with high intensity and energy together with either polarized targets or measurements of recoil polarization. In this paper we review the current status of nucleon elastic form factors, emphasizing recent polarization measurements, and analyze these data using a model that permits visualization of the underlying charge and magnetization densities.

Matrix elements of the nucleon electromagnetic current operator J^μ take the form

$$\langle N(p',s')|J^\mu|N(p,s)\rangle = \bar{u}(p',s')e\Gamma^\mu u(p,s) \tag{1}$$

where the vertex function

$$\Gamma^\mu = F_1(Q^2)\gamma^\mu + \kappa F_2(Q^2)\frac{i\sigma^{\mu\nu}q_\nu}{2m} \tag{2}$$

features Dirac and Pauli form factors, F_1 and F_2, whose dependence upon the spacelike invariant four-momentum transfer $Q^2 = q^2 - \omega^2$ probes the nucleon structure. The interpretation is simplest in the nucleon Breit frame in which a nucleon approaches with initial momentum $-\vec{q}_B/2$, receives three-momentum transfer $\vec{q}_B$ without energy transfer, and departs with final momentum $\vec{q}_B/2$ where $q_B^2 = Q^2 = q^2/(1+\tau)$ with $\tau = Q^2/4m^2$. In the Breit frame for a particular value of Q^2, the current separates into electric and

CP675, *Spin 2002: 15th Int'l. Spin Physics Symposium and Workshop on Polarized Electron Sources and Polarimeters*, edited by Y. I. Makdisi, A. U. Luccio, and W. W. MacKay

magnetic contributions [1]

$$\bar{u}(p',s')\Gamma^{\mu}u(p,s) = \chi_{s'}^{\dagger}\left(G_E + \frac{i\vec{\sigma}\times\vec{q}_B}{2m}G_M\right)\chi_s \tag{3}$$

where χ_s is a two-component Pauli spinor and where the Sachs form factors are given by

$$G_E = F_1 - \tau\kappa F_2 \qquad G_M = F_1 + \kappa F_2 \tag{4}$$

Early experiments with modest Q^2 suggested that

$$G_{Ep} \approx \frac{G_{Mp}}{\mu_p} \approx \frac{G_{Mn}}{\mu_n} \approx G_D \tag{5}$$

where $G_D(Q^2) = (1+Q^2/\Lambda^2)^{-2}$ with $\Lambda^2 = 0.71$ (GeV/c)^2 is known as the dipole form factor [2, 3].

FORM FACTORS FROM POLARIZATION MEASUREMENTS

In the one-photon exchange approximation, the differential cross section for elastic scattering of an electron beam from a stationary nucleon target is given by

$$\frac{d\sigma}{d\Omega} = \frac{\sigma_{NS}}{\varepsilon(1+\tau)}\left(\tau G_M^2 + \varepsilon G_E^2\right) \tag{6}$$

where $\varepsilon = (1+(1+\tau)2\tan^2\theta_e/2)^{-1}$ is the transverse polarization of the virtual photon for electron scattering angle θ_e and σ_{NS} is the cross section for a structureless Dirac target. Thus, the traditional Rosenbluth technique separates the electric and magnetic form factors by varying ε, but extraction of G_E becomes extremely difficult at large Q^2 because the magnetic contribution becomes increasingly dominant and because it is difficult to control the kinematic variation and radiative corrections with sufficient accuracy when both the form factors and the kinematic coefficients vary rapidly over the acceptance. Alternatively, the electromagnetic ratio

$$g = \frac{G_E}{G_M} = -\sqrt{\frac{\tau(1+\varepsilon)}{2\varepsilon}}\frac{P_x'}{P_z'} \tag{7}$$

can be obtained by comparing the components of the nucleon recoil polarization along the momentum transfer direction, denoted by $\hat{z}$, and in the $\hat{x}$ direction transverse to $\hat{z}$ in the scattering plane [4, 5]. For the proton, both components can be measured simultaneously using a polarimeter in the focal plane of a magnetic spectrometer, thereby minimizing systematic uncertainties due to beam polarization, analyzing power, and kinematic parameters. The systematic uncertainty due to precession of the proton spin in the magnetic spectrometer is usually much smaller than the uncertainties in comparing the cross sections obtained with different kinematical conditions and acceptances needed for the Rosenbluth method.

Figure 1 compares recent high-precision measurements of the proton electromagnetic ratio performed at Jefferson Laboratory (JLab) [6, 7] with earlier Rosenbluth data obtained at SLAC [8, 9]. I had expected the new JLab experiments to confirm the SLAC findings that $G_{Ep} \approx G_{Mp}/\mu_p$, merely improving their precision, but instead we found the surprisingly strong linear decrease shown in Fig. 1. The systematic uncertainties, primarily due to spin precession in the magnetic spectrometer, are shown by the hatched region and were substantially reduced in the second experiment at larger Q^2 where the deviation from unity is strongest. A recent re-analysis of the SLAC data failed to discover any systematic correction which could account for this disagreement [10] and I am unaware of any plausible mechanism which could produce a failure of the one-photon approximation at this level. Hopefully, a recent *super Rosenbluth* experiment [11] designed to minimize systematic uncertainties will help clarify this discrepancy. Until those results become available, I will rely on the recoil polarization measurements and discard the cross section data for $Q^2 > 1$ (GeV/c)^2 with larger, and possibly seriously underestimated, systematic uncertainties.

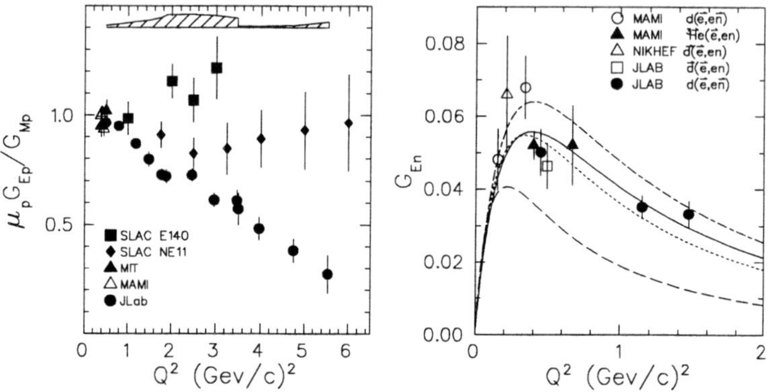

FIGURE 1. Recent G_E/G_M data for the proton (left) and neutron (right). The hatched region indicates the systematic uncertainty in JLab recoil polarization measurements for the proton. The proton data are from: SLAC E140 [8], SLAC NE11 [9], MIT [12], MAMI [13], and JLab [6, 7]. The neutron data are from: MAMI $d(\vec{e},e'\vec{n})$ [14], MAMI $^3\vec{\mathrm{He}}(\vec{e},e'n)$ [15, 16], NIKHEF $\vec{d}(\vec{e},e'n)$ [17], JLab $\vec{d}(\vec{e},e'n)$ [18], and JLab $d(\vec{e},e'\vec{n})$ [19]. See text for explanation of G_{En} curves.

For the neutron one must use quasifree scattering from a neutron in a nuclear target and correct for the effects of Fermi motion, meson-exchange currents, and final-state interactions also. Both recoil and target polarization for quasifree knockout give similar PWIA formulas for the form factor ratio, but the systematic errors and the nuclear physics corrections are appreciably different. Therefore, confident extraction of G_{En}/G_{Mn} benefits from comparison of data for both recoil and target polarization. Data from recent experiments of these types are also shown in Figure 1 and additional data from Mainz on $d(\vec{e},e'\vec{n})$ at $Q^2 = 0.6$ and 0.8 and from JLab on $\vec{d}(\vec{e},e'n)$ at $Q^2 = 1.0$ (GeV/c)^2 are expected soon. Details of the high-precision JLab experiment on $d(\vec{e},e'\vec{n})$ at $Q^2 = 0.45$, 1.15, and 1.47 (GeV/c)^2 may be found in Ref. [19]; corrections for nuclear physics and acceptance averaging have not yet been applied but are expected to be small. Several two-parameter fits based upon the Galster parametrization are shown also. Our fit (solid) to these data and additional data from Refs. [20, 21] remains rather close to the original Galster fit (dotted) to electron-deuteron elastic scattering data for $Q^2 < 0.7$ (GeV/c)^2 [22] despite the rather larger model dependence of the Rosenbluth method. The highest curve shows a fit by Schmieden [23] to a subset of the polarization data for $Q^2 < 0.7$ (GeV/c)^2. The dashed line shows the commonly quoted Platchkov analysis of more recent elastic scattering data (not shown) that used the Paris potential, but other realistic interactions produce variations greater than the spread between the highest and lowest curves in this figure [24]; thus, the model dependence of the Rosenbluth method is at least this large while the model dependence of the recoil polarization method for $Q^2 > 0.5$ (GeV/c)^2 is less than 10% [25]. Therefore, polarization techniques provide much more accurate measurements of G_{En} than the Rosenbluth method and show that G_{En} for $Q^2 < 1.5$ (GeV/c)^2 is substantially larger than the common Platchkov parametrization, although it remains compatible with the Galster parameterization.

Polarization measurements of nucleon electromagnetic ratios and selected cross section data for the form factors relative to G_D are compared in Figs. 2-3 with representative calculations. The chiral soliton model of Holzwarth (dotted lines) predicted the linear behavior of G_{Ep}/G_{Mp} but fails to reproduce neutron form factors [26, 27]. The light-cone diquark model (long dashes) needs only 5 parameters to obtain a reasonable fit for modest Q^2 [28], but except for G_{En} its form factors fall too rapidly for $Q^2 > 1$ (GeV/c)^2. The point-form spectator approximation (PFSA) using pointlike constituent quarks and a Goldstone boson exchange interaction fitted to spectroscopic data successfully describes a wider range of Q^2 (short dashes) without fitting additional parameters to the form factors [29]. Finally, a light-front calculation using one-gluon exchange and constituent-quark form factors fitted to $Q^2 < 1$ (GeV/c)^2 provides a good fit (dash-dot) up to about 4 (GeV/c)^2 [30]. However, none of the available theoretical calculations provides a truly quantitative description for all four form factors over a wide range of Q^2. The differences between these models are largest for G_{En}, which is especially sensitive to small mixed-symmetry and deformed components of the nucleon wave function. Clearly it will be very important to extend the G_{En} data to larger Q^2 — a proposal to measure ${}^3\vec{H}e(\vec{e},e'n)$ up to 3.4 (GeV/c)^2 has been approved at JLab [31] and proposals for higher Q^2 are under development.

Figures 2-3 also show fits made by Lomon [32] using an extension of the Gari-Krümpelmann model [33] that interpolates between vector meson dominance (VMD) at low Q^2 and perturbative QCD (pQCD) at high Q^2. This type of parametrization is very useful for nuclear physics calculations, but offers no insight into the spatial distributions of charge and magnetization within nucleons. In the next section we offer an alternative phenomenology in terms of spatial densities.

FITTED DENSITIES

Relativistic Inversion

Although rigorous comparisons between theory and experiment must be made at the level of form factors, for many it would seem desirable to extract charge and magnetization densities from the corresponding form factors because our intuition is

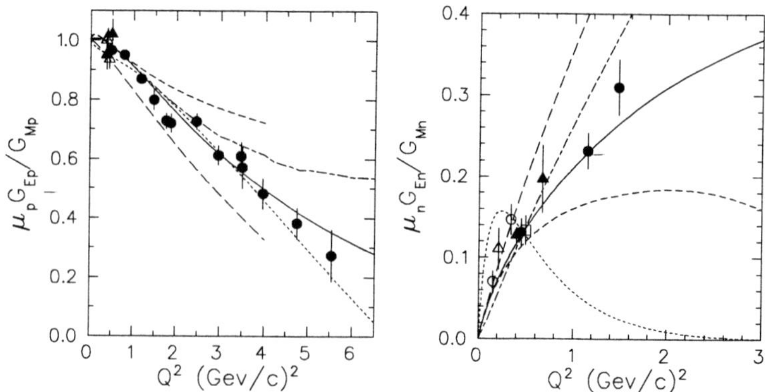

FIGURE 2. Polarization data for electromagnetic ratios are compared with representative calculations: chiral soliton (dotted) [26, 27], light-cone diquark (long dashes) [28], PFSA (short dashes) [29], light-front OGE with constituent form factors (dash-dot) [30]. The data have the same legend as Fig. 1.

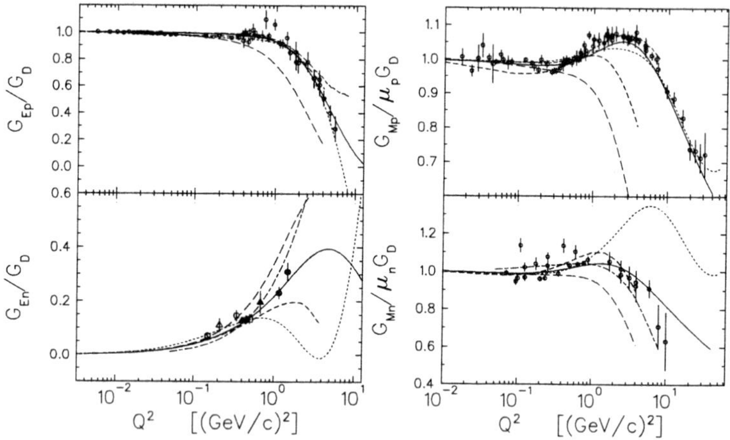

FIGURE 3. Selected form factor data are compared with representative calculations: chiral soliton (dotted) [26, 27], light-cone diquark (long dashes) [28], PFSA (short dashes) [29], light-front OGE with constituent form factors (dash-dot) [30].

usually stronger in space than in momentum transfer. Intrinsic charge and magnetic form factors, $\tilde{\rho}_{ch}(k)$ and $\tilde{\rho}_m(k)$, may be defined in terms of the Sachs form factors by

$$\tilde{\rho}_{ch}(k) = G_E(Q^2)(1+\tau)^{\lambda_E} \qquad \mu\tilde{\rho}_m(k) = G_M(Q^2)(1+\tau)^{\lambda_M} \tag{8}$$

where the intrinsic spatial frequency k is related to the invariant momentum transfer Q by the Breit-frame boost $k^2 = \frac{Q^2}{1+\tau}$ and where the model-dependent exponents, λ_E and λ_M will be discussed shortly. Due to the Lorentz contraction of spatial distributions in the Breit frame, a measurement with Breit-frame momentum transfer $q_B = Q$ probes a reduced spatial frequency k in the rest frame. In fact, the intrinsic frequencies accessible to elastic scattering with spacelike momentum transfer are limited to $k < 2m$ such that the asymptotic Sachs form factors in the limit $Q^2 \to \infty$ are determined by the intrinsic form factors in the immediate vicinity of the limiting frequency $k_m = 2m$. This limitation can be understood as a consequence of relativistic position fluctuations, known as of *zitterbewegung*, that smooth out radial variations on scales smaller than the Compton wavelength.

Using a quark cluster model Licht and Pagnamenta [34] originally proposed to use $\lambda_E = \lambda_M = 1$, but these choices do not conform with pQCD scaling unless one imposes upon both form factors the somewhat artificial constraint $\tilde{\rho}(k_m) = 0$. Mitra and Kumari [35] then demonstrated that a more symmetric version of the quark cluster model that is also applicable to inelastic transitions suggests $\lambda_E = \lambda_M = 2$ and is compatible with pQCD scaling without constraining $\tilde{\rho}(k_m)$. For the present work we employ $\lambda_E = \lambda_M = 2$ and refer to Ref. [36] for a more comprehensive analysis. Although we cannot claim that $\rho_{ch}(r)$ and $\rho_m(r)$ are the true charge and magnetization densities in the nucleon rest frame because the boost operator for a composite system depends upon the interactions among its constituents, this model can be used to fit the form factor data using an intuitively appealing spatial representation that is consistent with relativity and with pQCD.

Fitting Procedures

The model dependence of the fitted form factor can be minimized by expanding the density in a complete set of radial basis functions, such that

$$\rho(r) = \sum_n a_n f_n(r) \Longrightarrow \tilde{\rho}(k) = \sum_n a_n \tilde{f}_n(k) \tag{9}$$

where

$$\tilde{f}_n(k) = \int_0^\infty dr\, r^2 j_0(kr) f_n(r) \tag{10}$$

represents basis functions in momentum space. The expansion coefficients, a_n, are fitted to form factor data subject to several minimally restrictive constraints. An arbitrarily large number of terms can be included by using a penalty function to constrain high-frequency contributions with an envelope of the form

$$k > k_{max} \Longrightarrow |\tilde{\rho}(k)| < |\tilde{\rho}(k_{max})|(k_{max}/k)^4 \tag{11}$$

where k_{max} is the largest frequency for which experimental data are available. This condition ensures that fitted density does not have an unphysical cusp at the origin but does permit the density sufficient flexibility to estimate the uncertainty due to the absence of data for $k > k_{max}$. In addition, one constrains the density for very large radii. Details of these procedures can be found in Ref. [36] and references cited therein.

Analyses of this type are often described as model independent because a complete basis can reproduce any physically reasonable density; if a sufficient number of terms are included in the fitting procedure the dependence of the fitted density upon the assumptions of the model is minimized. By contrast, simple parametrizations like the

Galster model severely constrain the shape of the fitted density. As shown in Ref. [36], virtually identical results are obtained using either the Laguerre-Gaussian expansion (LGE) or Fourier-Bessel expansion (FBE).

Results

We fit all four nucleon electromagnetic form factors using a data selection that emphasizes recent polarization methods where available. For G_{Mp} and G_{Mn} we employ the highest quality cross section data in each range of Q^2. For G_{Ep} we use the recoil polarization data from Refs. [6, 7, 13, 12] and chose cross section data from Refs. [37, 38] for low Q^2 but omitted the higher Q^2 Rosenbluth data from Refs. [8, 9]. For G_{En} we use recoil and target polarization data corrected for nuclear physics effects and use the results of an analysis of the deutron quadrupole form factor by Schiavilla and Sick [20]; Rosenbluth data for elastic or quasielastic scattering from deuterium were omitted. We also include the measurement of $\langle r^2 \rangle_n$ by Kopecky *et al.* [21] using the energy dependence for the transmission of thermal neutrons through liquid ^{208}Pb. A more complete review of these selections and omissions can be found in Ref. [36].

Figure 4 shows fits to the form factor data using the LGE model with $\lambda = 2$. Where data are available the widths of the error bands are governed by the statistical quality of data while for large Q^2 the growth of these uncertainties is limited by the large-k constraint specified by Eq. (11). These fits are generally very good, but in the G_{Mn} data there remain appreciable systematic differences between data sets that probably reflect errors in the efficiency calibration for some of the experiments.

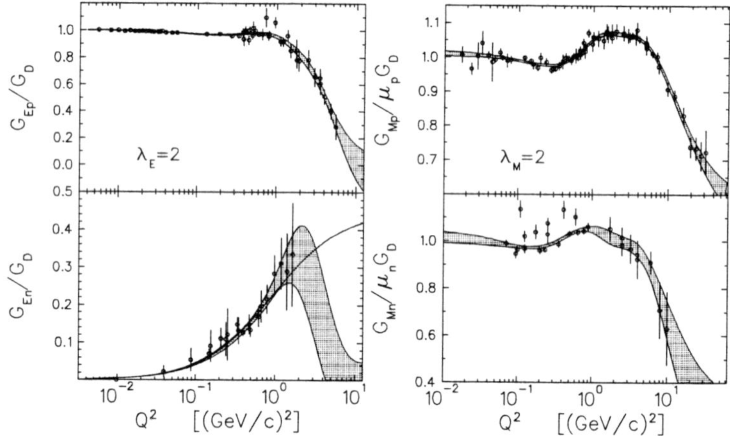

FIGURE 4. LGE fits to selected data for nucleon form factors using $\lambda_E = \lambda_M = 2$. For G_{En} we also show a Galster fit.

Figure 5 compares the four fitted densities. For the proton we find that the charge is distributed over a larger volume than the magnetization. The difference between rms radii is not large, 0.883(14) for charge versus 0.851(26) fm for magnetization, but the large difference in interior densities reflects the strong decrease in G_{Ep}/G_{Mp} for

$Q^2 > 1$ (GeV/c)^2. The magnetization density for the neutron is very similar to that for the proton, but closer examination shows that its distribution is slightly wider. For the purposes of comparing shapes, the neutron charge density is shown scaled to the interior magnetization. Despite the limited range of Q^2 and larger uncertainties in the G_{En} data, the neutron charge density is determined with useful precision.

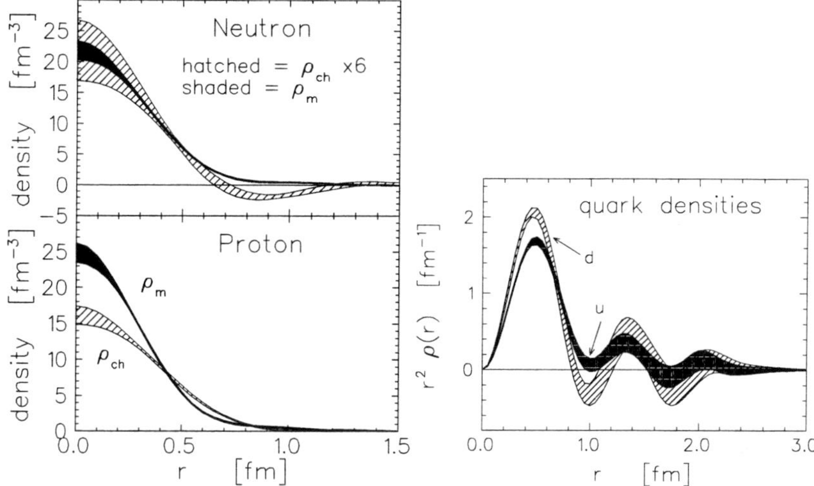

FIGURE 5. Nucleon electromagnetic densities are shown on the left and quark densities on the right using $\lambda_E = \lambda_M = 2$.

The neutron charge density features a positive interior and negative surface. In the meson-baryon picture these characteristics are explained in terms of quantum fluctuations of the type $n \leftrightarrow p\pi^-$ in which the light negative meson is found at larger radius than the heavier positive core. Alternatively, in the quark model these features arise from incomplete cancellation between u and d quark distributions that are similar but not identical in shape. Using a symmetric two-flavor quark model of the nucleon charge densities

$$\rho_p(r) = \frac{4}{3}u(r) - \frac{1}{3}d(r) \qquad \rho_n(r) = -\frac{2}{3}u(r) + \frac{2}{3}d(r) \tag{12}$$

one can obtain the quark densities

$$u(r) = \rho_p(r) + \frac{1}{2}\rho_n(r) \qquad d(r) = \rho_p(r) + 2\rho_n(r) \tag{13}$$

where $u(r)$ is the radial distribution for an up quark in the proton or a down quark in the neutron while $d(r)$ is the distribution for a down quark in the proton or an up quark in the neutron. These quark densities are also displayed in Fig. 5 weighted by r^2 to emphasize the surface region. We find that the u distribution is slightly broader than the d distribution, which is consistent with the repulsive color hyperfine interaction between like quarks needed to explain the $N-\Delta$ mass splitting. The slightly negative $d(r)$ near 1 fm suggests a $\bar{d}$ contribution from the pion cloud. The secondary lobes near 1.4 fm appear to be robust features of the data — elimination of these features seriously degrades fits to data for $Q^2 \sim 1$ (GeV/c)^2 — and might arise from mixed symmetry or $\ell = 2$ admixtures with larger radii than the dominant S-state configuration.

CONCLUSIONS

The advent of highly polarized electron beams with large currents and high energy coupled with advances in recoil polarimetry and polarized targets permit much more precise measurements of the nucleon electromagnetic form factor ratio, G_E/G_M, than was possible with the Rosenbluth technique at large Q^2. Despite the fact that Rosenbluth data suggested $\mu_p G_{Ep}/G_{Mp} \approx 1$ for $Q^2 < 6$ (GeV/c)^2, recoil polarization measurements at Jefferson Laboratory show a strong, nearly linear, decrease for $Q^2 > 1$ (GeV/c)^2 that demonstrates that the charge density is significantly broader than the magnetization density of the proton. Similarly, recoil and target polarization measurements show that G_{En} is substantially larger than the commonly quoted Platchkov analysis of deuteron elastic scattering using the Paris potential yet remains surprisingly close to the original Galster parametrization despite the prohibitively large model dependence of Rosenbluth separations for G_{En}.

We have developed a phenomenological model of these form factors in terms of spatial densities that is consistent with pQCD at large Q^2 and with the Lorentz contraction of the Breit frame relative to the rest frame. The model dependence of the fitted densities is minimized by using an expansion in a complete set of basis functions with minimally restrictive constraints upon the behavior for either large frequency or large radius. The flexibility of the fitted form factor for frequencies beyond the measured momentum transfer provides an estimate of the incompleteness error in the extracted density. The error envelopes for the magnetization densities and the proton charge density are quite narrow, but the uncertainty in the neutron charge density is significantly larger because the data are still limited to $Q^2 < 1.5$ (GeV/c)^2 and are not as precise as for the other form factors. Nevertheless, the precision is already quite useful and will improve when the next generation of $^3\vec{He}(\vec{e},e'n)$ experiments reaches about 3.4 (GeV/c)^2 [31].

We find that the neutron and proton magnetizations densities are quite similar but that the proton charge density is significantly broader. The neutron charge density results from incomplete cancellation between u and d quark densities with slightly different shapes that leaves a positive core surrounded by negative surface charge. By comparing the fitted neutron and proton charge densities, we find that the distribution of like quarks in the nucleon is broader than the distribution of the unlike quark.

ACKNOWLEDGMENTS

I thank D. Quing, S. Simula, M. Radici, and F. Coester for tabulated calculations. The support of the U.S. National Science Foundation under grant PHY-9971819 is gratefully acknowledged.

REFERENCES

1. Sachs, R. G., *Phys. Rev.*, **126**, 2256 (1962).
2. Hughes, E. B., Griffey, T. A., Yearian, M. R., and Hofstadter, R., *Phys. Rev.*, **139**, B458 – B471 (1965).
3. J. R. Dunning, et al., *Phys. Rev.*, **141**, 1286–1297 (1966).
4. Dombey, N., *Rev. Mod. Phys.*, **41**, 236–246 (1969).
5. Arnold, R. G., Carlson, C. E., and Gross, F., *Phys. Rev.* **C**, **23**, 363 (1981).
6. Jones, M. K., et al., *Phys. Rev. Lett.*, **84**, 1398–1402 (2000).
7. Gayou, O., et al., *Phys. Rev. Lett.*, **88**, 092301 (2002).
8. Walker, R. C., et al., *Phys. Rev.* **D**, **49**, 5671–5689 (1994).
9. Andivahis, et al., *Phys. Rev.* **D**, **50**, 5491–5517 (1994).
10. Arrington, J., Are Recoil Polarization Measurements of G_E^p/G_M^p Consistent with Rosenbluth Separation Data (2002), arXiv:nucl-ex/0205019.
11. Arrington, J., et al., Jefferson laboratory proposal e01-001 (2001).
12. Milbrath, B. D., et al., *Phys. Rev. Lett.*, **82**, 2221 (1999).
13. Pospischil, T., et al., *Eur. Phys. J. A*, **12**, 125–127 (2001).
14. Herberg, C., et al., *Eur. Phys. J. A*, **5**, 131–135 (1999).
15. Golak, J., Ziemer, G., Kamada, H., Witala, H., and Glöckle, W., *Phys. Rev.* **C**, **63**, 034006 (2001).
16. Rohe, D., et al., *Phys. Rev. Lett.*, **83**, 4257–4260 (1999).
17. Passchier, I., et al., *Phys. Rev. Lett.*, **82**, 4988–4991 (1999).
18. Zhu, H., et al., *Phys. Rev. Lett.*, **87**, 081801 (2001).
19. Madey, R., et al., Neutron Electric Form Factor up to $Q^2 = 1.47$ (GeV/c)^2 (2002), to be published in Proceedings of Electron Nucleus Scattering VII by Eur. Phys. J. A.
20. Schiavilla, R., and Sick, I., *Phys. Rev.* **C**, **64**, 041002(R) (2001).
21. Kopecky, S., et al. *Phys. Rev.* **C**, **56**, 2220–2237 (1997).
22. Galster, S., et al., *Nucl. Phys.*, **B32**, 221–237 (1971).
23. Schmieden, H., "Form Factors of the Neutron," in *Proceedings of the 8th International Conference on the Structure of Baryons*, edited by D. W. Menze and B. Metsch, World Scientific, Singapore, 1999, pp. 356–367.
24. Platchkov, S., et al., *Nucl. Phys.*, **A510**, 740–758 (1990).
25. Arenhövel, H., *Phys. Lett.*, **B199**, 13 (1987).
26. Holzwarth, G., *Zeit. Phys.* **A**, **356**, 339 (1996).
27. Holzwarth, G., Electromagnetic Form Factors of the Nucleon in the Chiral Solution Model (2002), arXiv:hep-ph/0201138.
28. Ma, B.-Q., Qing, D., and Schmidt, I., *Phys. Rev.* **C**, **65**, 035205 (2002).
29. Wagenbrunn, R., Boffi, S., Klink, W., Plessas, W., and Radici, M., *Phys. Lett.* **B**, **511**, 33–39 (2001).
30. Simula, S., Relativistic quark models (2001), arXiv:nucl-th/0105024.
31. Cates, G., et al., Jefferson laboratory proposal e02-013 (2002).
32. Lomon, E. L., Effect of recent R_p and R_n measurements on extended Gari-Krümpelmann model fits to nucleon electromagnetic form factors (2002), arXiv:nucl-th/0203081.
33. Gari, M. F., and Krümpelmann, W., *Zeit. Phys.* **A**, **322**, 689–693 (1985).
34. Licht, A. L., and Pagnamenta, A., *Phys. Rev.* **D**, **2**, 1156–1160 (1970).
35. Mitra, A. N., and Kumari, I., *Phys. Rev.* **D**, **15**, 261–266 (1977).
36. Kelly, J. J., Nucleon Charge and Magnetization Densities from Sachs Form Factors (2002), arXiv:hep-ph/0204239.
37. Simon, G. G., Schmitt, C., Borkowski, F., and Walther, V. H., *Nucl. Phys.*, **A333**, 381–391 (1980).
38. Price, L. E., Dunning, J. R., Goitein, M., Hanson, K., Kirk, T., and Wilson, R., *Phys. Rev.* **D**, **4**, 45–53 (1971).

Nucleon Spin Structure Functions g_1 and g_2 from Polarized Inclusive Scattering

Todd D. Averett

College of William and Mary, Department of Physics, Williamsburg, VA 23187

Abstract. This paper will present a survey of recent experimental results for the g_1 and g_2 spin structure functions. Over the past decade, these structure functions (and the virtual photon asymmetries A_1 and A_2) have been well-measured in the large-Q^2 scaling region using inclusive polarized deep-inelastic scattering. New precision results from Jefferson Lab are now becoming available which cover kinematic regions that were previously poorly measured or completely unmeasured. Topics covered will include: recent experiments at Jefferson Lab which have made precise measurements of g_1 in the resonance region for both proton and neutron to investigate the Q^2 evolution of the GDH sum rule, a new measurement of A_1 for the neutron in the large-x region where valence quark dynamics dominate, new precision results for g_2 from Jefferson Lab, a review of the g_1 results and NLO analyses in the scaling region for the SLAC, HERMES, and SMC data. Finally, a survey of results expected from several new experiments and longer-term experimental programs are discussed.

INTRODUCTION AND FORMALISM

Polarized inclusive lepton scattering has been extensively used as a tool for probing the internal structure of the nucleon. Early experiments at SLAC and CERN discovered that the spin of the proton could not be accounted for by the contribution from quark spin alone. Since this early work, the idea of using polarized inclusive lepton scattering as a probe of nucleon structure has matured into a mainstay of current nucleon structure research. Experimental results over the last decade from SLAC, CERN, and DESY have confirmed that the quark contribution to the nucleon spin is only $\sim 23\%$ and have also confirmed the prediction of the fundamental Bjorken Sum Rule. New experiments at Jefferson lab continue to explore the spin structure of the nucleon by making precise measurements of the spin structure function g_1 and g_2 as well as the asymmetries A_1 and A_2. The experiments provide new information about nucleon structure through precise measurements in both the deep inelastic and the resonance regions.

The basic inclusive process involves the exchange of a virtual photon between a polarized lepton beam and a polarized proton, deuteron, or ^{3}He target. Only the scattered electron is detected and uniquely determines the kinematics of the exchanged photon. The virtual photon energy, ν, and momentum, $\vec{q}$, are defined by the difference in initial and final electron energy and momentum as $\nu = E - E'$ and $\vec{q} = \vec{k} - \vec{k}'$. It is this photon which probes the electromagnetic structure of the nucleon. The Lorentz invariant $Q^2 = \vec{q} \cdot \vec{q} - \nu^2$ and the Bjorken scaling variable $x = Q^2/(2M\nu)$ are typically used to describe the scattering process.

If the spin of the lepton and target are known before scattering, then the following

CP675, *Spin 2002: 15th Int'l. Spin Physics Symposium and Workshop on Polarized Electron Sources and Polarimeters*, edited by Y. I. Makdisi, A. U. Luccio, and W. W. MacKay

cross section differences can be used to extract the two nucleon spin structure functions $g_1(x,Q^2)$ and $g_2(x,Q^2)$ as follows:

$$\frac{d\sigma}{d\Omega dE'}(\downarrow\Uparrow - \uparrow\Uparrow) = \frac{4\alpha^2 E'}{MQ^2 \nu E}\left[\left(E + E'\cos\theta\right) g_1(x,Q^2) - \frac{Q^2}{\nu} g_2(x,Q^2)\right] \tag{1}$$

$$\frac{d\sigma}{d\Omega dE'}(\downarrow\Rightarrow - \uparrow\Rightarrow) = \frac{4\alpha^2 E'\sin\theta}{MQ^2 \nu E}\left[\nu g_1(x,Q^2) + 2E g_2(x,Q^2)\right] \tag{2}$$

where the arrows refer to the lepton and nucleon spin respectively. Because the virtual photon is the real probe of the nucleon, the spin structure functions g_1 and g_2 can be related to the asymmetries for virtual photon absorption as follows:

$$A_1(x,Q^2) = \frac{\sigma_T^{1/2} - \sigma_T^{3/2}}{\sigma_T^{1/2} + \sigma_T^{3/2}} = \frac{g_1(x,Q^2) - \gamma^2 g_2(x,Q^2)}{F_1(x,Q^2)} \tag{3}$$

$$A_2(x,Q^2) = \frac{2\sigma_{TL}}{\sigma_T^{1/2} + \sigma_T^{3/2}} = \gamma\frac{g_1(x,Q^2) + g_2(x,Q^2)}{F_1(x,Q^2)} \tag{4}$$

where $\gamma^2 = Q^2/\nu$ and the σ_T refer to the absorption of a transverse virtual photon with the total helicity of the photon-nucleon system of $1/2$ or $3/2$. The quantity σ_{TL} is an interference cross section for absorption of a transverse photon in the initial state and longitudinal in the final state.

DEEP INELASTIC SCATTERING AND THE NUCLEON SPIN PUZZLE

At large Q^2, the photon-nucleon interaction is less sensitive to QCD effects and may be treated perturbatively. In the scaling limit, $Q^2 \to \infty$ and $\nu \to \infty$ while Q^2/ν remains finite, the g_1 structure function has simple interpretation in the quark parton model,

$$g_1(x,Q^2) = \frac{1}{2}\sum_i^{N_f} e_i^2\left[\Delta q_i(x,Q^2) + \Delta\bar{q}_i(x,Q^2)\right] \tag{5}$$

where the sum is over quark flavors, $\Delta q_i = q_i^\uparrow - q_i^\downarrow$, and the q_i represent the probability of finding a quark of a given flavor with its spin parallel or anti-parallel to the spin of the nucleon. By integrating g_1 over all x we obtain (for the proton)

$$\Gamma_1^p(Q^2) = \int_0^1 g_1^p(x,Q^2)dx = \frac{4}{18}\Delta u + \frac{1}{18}\Delta d + \frac{1}{18}\Delta s \tag{6}$$

where the Δu, Δd, and Δs are the net contributions of the quark flavors to the overall spin of the nucleon. Using angular momentum conservation, a sum rule exists for the total spin of the nucleon:

$$\frac{1}{2}\Delta\Sigma + \Delta G + L_z = \frac{1}{2} \tag{7}$$

where $\Delta\Sigma$, ΔG, and L_z are the total quark, gluon, and angular momentum contributions to the spin of the nucleon. Taking the difference between proton and neutron yields the famous Bjorken sum rule [1] (in the scaling limit),

$$\Gamma_1^{(p-n)}(Q^2) = \int_0^1 g_1^p(x,Q^2) - g_1^n(x,Q^2)dx = \frac{1}{6}\frac{g_A}{g_V} \tag{8}$$

where g_A/g_V is the ratio of axial to vector coupling constants. This sum rule can also be extended to finite Q^2 using perturbative QCD and is considered a rigorous test of the quark-parton model and QCD.

Experimental and Theoretical Results

Over the past two decades there has been an extensive experimental program at SLAC, DESY, and CERN to accurately measure g_1 for the proton, deuteron, and ^{3}He over the widest kinematic range possible. Published results are summarized in the paper by Blümlein and Böttcher [2] and are reproduced in Table 1. Data from the most recent SLAC measurement [3] on proton and deuteron are shown in Figure 1.

TABLE 1. Summary of experimental programs and their kinematic coverage measuring the spin structure functions g_1 and g_2 in the deep inelastic region.

Experiment	x–range	Q^2–range (GeV^2)	Ref.
Proton			
E143(p)	0.027 – 0.749	1.17 – 9.52	[5]
HERMES(p)	0.028 – 0.660	1.13 – 7.46	[6]
E155(p)	0.015 – 0.750	1.22 – 34.72	[3]
SMC(p)	0.005 – 0.480	1.30 – 58.0	[7]
EMC(p)	0.015 – 0.466	3.50 – 29.5	[8]
Deuteron			
E143(d)	0.027 – 0.749	1.17 – 9.52	[5]
E155(d)	0.015 – 0.750	1.22 – 34.79	[9]
SMC(d)	0.005 – 0.479	1.30 – 54.8	[7]
Neutron			
E142(n)	0.035 – 0.466	1.10 – 5.50	[10]
HERMES(n)	0.033 – 0.464	1.22 – 5.25	[11]
E154(n)	0.017 – 0.564	1.20 – 15.0	[12]/[13]

One of the best tools for interpreting this data is through a next-to-leading-order (NLO) analysis. In this type of analysis, functional forms are chosen to describe the spin dependent parton distributions as a function of x. The world data is then fit and Q^2 dependence is accounted for using the DGLAP evolution method [4]. Recent results from the NLO analysis by Blümlein and Böttcher [2] were presented.

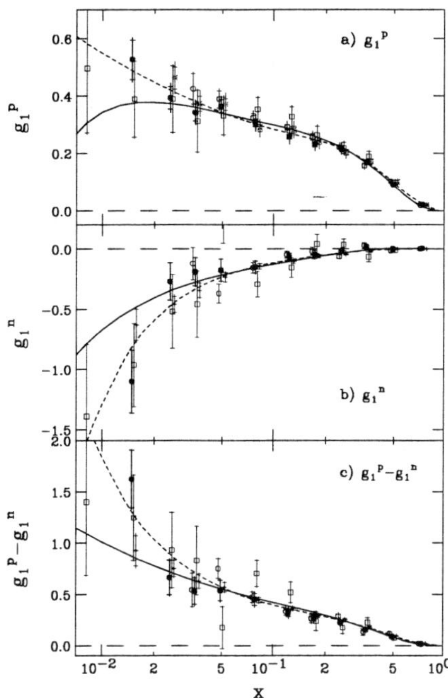

FIGURE 1. SLAC results for g_1^p, g_1^n, and g_1^{p-n} (using proton and deuteron data) at $Q^2 = 5$ GeV^2 are shown as solid circles. Data from E143 (open circles), SMC (squares), and HERMES (stars) are also shown. The solid and dashed curves are NLO QCD fits and a simple parameterization, respectively [3].

A_1 in the valence region

The measurements presented above provide a wealth of data and information about the spin content of the nucleon. However, in the large x region, which is dominated by valence quark dynamics, the data quality remains poor, particularly for the neutron. This is because 1) the large-x region is typically also at large Q^2 where the cross section is small, and 2) previous experiments have focused on the low-x region to test the convergence of the Bjorken sum rule as $x \to 0$. The large-x region provides an important opportunity to test basic models of nucleon structure which are largely free from the effects of sea quark pairs. In this region, it is more interesting to study the asymmetry A_1 which is essentially dominated by the absorption of a virtual photon by a single quark which flips its spin. The simplest model for the nucleon in this region is the SU(6) symmetric wave function which predicts $A_1^n = 0$ for all x. More realistic models of the nucleon will allow SU(6) symmetry breaking by the addition of some sort of interaction between the valence quarks. Popular models include constituent quark models (CQM) [14] where quarks interact by a hyperfine, or spin-spin, interaction. Models based on pQCD analyses, LSS(BBS,HHC) [15] and LSS 2001 [16] use polarized parton distributions and DGLAP evolution to predict the large-x behavior. Both the CQM and pQCD models predict $A_1^n \to 1$ as $x \to 1$. Other models include a statistical models

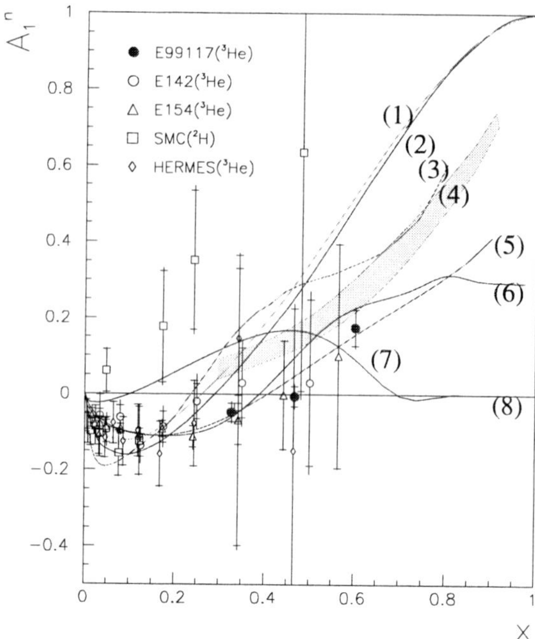

FIGURE 2. A_1^n results compared with theoretical predictions and existing world data. The nuclear target used by each experiment is shown in brackets. Curves: predictions of g_1^n/F_1^n from pQCD HHC based LSS(BBS) parameterization (1) and BBS parameterization (2); g_1^n/F_1^n from chiral soliton model at $Q^2 = 3$ (GeV/c)^2 (3); predictions of A_1^n from constituent quark model (shaded band) (4); g_1^n/F_1^n from LSS 2001 parameterization at $Q^2 = 5$ (GeV/c)^2 (5); predictions of A_1^n from statistical model at $Q^2 = 4$ (GeV/c)^2 (6), meson-cloud bag model at $Q^2 = 3$ (GeV/c)^2 (7), and from basic SU(6) symmetry (8).

based on a NLO fit of world data [17], chiral soliton model [18], and meson cloudy bag model [19]. All models predict A_1^n will be positive for $x > 0.4$, and most predict a dramatic rise as $x \to 1$. The existing world data give no clear indication of an increase as x increases, and instead seem to show a slightly negative trend. Because the structure of the nucleon should become less complicated in the valence region, a precision measurement of the x dependence of A_1^n at large x provides fertile ground for testing basic models of the nucleon.

Experiment E99-117 ran in Hall A at Jefferson Lab in the summer of 2001. Polarized electrons were scattered from a polarized ^{3}He target and were detected in one of two independent high-resolution spectrometers at a scattering angle of 15.5°. From this data, A_1^n was determined at $x = 0.33$, 0.48, and 0.61 with corresponding $Q^2 = 2.7$, 3.6, and 4.9 GeV^2. The preliminary data, shown in Figure 2 are most consistent with the statistical [17] and LSS 2001 pQCD [16] models.

The g_2 structure function

All of the results presented up to this point have been focused on g_1 or A_1 which have relative simple parton model interpretations in the deep-inelastic region. The g_2 spin structure function however, does not. This structure function is sensitive to higher-twist effects, such as quark-gluon correlations, and is best interpreted in the framework of the Operator Product Expansion (OPE) [20, 21]. Here, the unknown hadronic current is expanded in a series of operators with coefficients which are not easily calculated due to their non-perturbative nature. The terms are grouped according to their *twist*, with terms of increasing twist successively suppressed by powers of $1/Q$. The leading term is twist-2 and is the dominant contribution to both g_1 and g_2. It corresponds to scattering from a single, non-interacting, quark. The next term is twist-3, and is only present in g_2 at leading order in α_S. The twist-3 term includes contributions from scattering from a quark which is simultaneously exchanging a gluon with the rest of the nucleon. It is an inherently non-perturbative effect and can be isolated by measuring g_2 and subtracting the twist-2 contribution using the following relationship from Wandzura and Wilczek [22]:

$$g_2^{ww}(x,Q^2) = -g_1(x,Q^2) + \int_x^1 \frac{g_1(y,Q^2)}{y} dy \tag{9}$$

This expression provides a method of calculating the *twist-2* part of g_2 only using measured data from g_1 due to the fact that g_1 and g_2 both contain the same twist-2 matrix element. By making a precise measurement of g_2 and comparing it to the g_2^{ww} calculation, one can extract the higher twist contributions to the nucleon structure.

Jefferson Lab experiment E97-103 also ran during the summer of 2002 in Hall A. A transversely polarized ^{3}He target was used to make a series of precise measurements of g_2 at $x \simeq 0.2$ in the range $0.58 < Q^2 < 1.36$ GeV^2. This measurement made an improvement in statistical errors by a factor of ~ 15 over previous measurements and will allow us, for the first time, to make definitive statements about the size of these higher twist effects. Preliminary data (errors only) are shown in Figure 3 along with calculations of the twist-2 g_2^{ww} contribution using models of g_1^n [3, 23, 2]. Recent g_2 data for proton and deuteron from SLAC experiment E155x [24] is also now available and was presented at the conference. It is further discussed in these proceedings by S. Rock.

RESONANCE REGION

The experiments presented above were all measurements in the deep-inelastic region where QCD effects can be largely treated perturbatively. As one begins to lower Q^2, the nucleon begins to respond more as a whole object. Specifically, at low Q^2, the nucleon resonances begin to become prominent and the scattering can no longer be interpreted using simple quark models and pQCD. At $Q^2 = 0$, the photon which is exchanged between the electron and nucleon is real and the following fundamental sum rule, first derived by Gerasimov, and also by Drell and Hearn [25], allows one to relate the photon-nucleon absorption cross sections to bulk properties of the nucleon such as its mass M

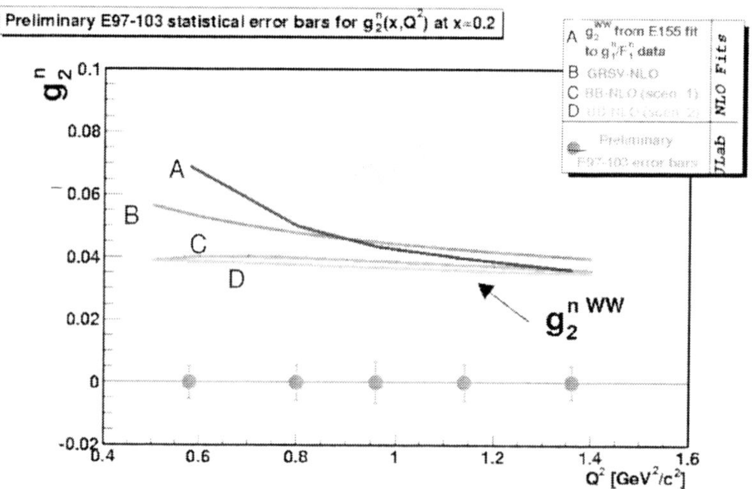

FIGURE 3. Expected errors for g_2^n as a function of Q^2 from Jefferson Lab experiment E97-103. Twist-2 g_2^{ww} predictions using various models for g_1^n are also shown.

and anomalous magnetic moment, κ.

$$I(Q^2=0) = \int_{\nu_0}^{\infty} \left[\sigma_{1/2}(\nu) - \sigma_{3/2}(\nu)\right] \frac{d\nu}{\nu} = -\frac{2\pi^2\alpha}{M^2}\kappa^2 \tag{10}$$

The left side of the GDH sum rule can be generalized to $Q^2 > 0$ by replacing the real photon cross sections by the virtual photon absorption cross sections, $\sigma_T^{1/2}(\nu,Q^2)$ and $\sigma_T^{3/2}(\nu,Q^2)$. To generalize the right hand side, it was pointed out by Ji and Osborne [26] that the *both* the Bjorken *and* the GDH sum rules can be related to the the virtual Compton amplitude $S_1(Q^2)$. This quantity can be calculated at very low Q^2 using chiral perturbation theory and at very large Q^2 using perturbative QCD. This important relation provides the formalism needed to connect the very low Q^2 behavior where the nucleon is described by hadronic degrees of freedom, to the large Q^2 region where the quarks are the relevant degrees of freedom. Clearly, a measurement of the Q^2 dependence of the GDH integral for $Q^2 < 1$ GeV^2 provides a powerful tool to investigate the behavior of the nucleon in the transition region between hadron-like and quark-like behavior.

Experiment E94-010 at Jefferson Lab recently published the first results [27] on the Q^2 evolution of the GDH sum rule for the neutron in the region $0.1 < Q^2 < 0.9$ GeV^2. This measurement was made by inclusively scattering longitudinally polarized electrons from a longitudinal or transversely polarized ^{3}He target, from the quasi-elastic, through resonance, and into the deep-inelastic regions. Figure 4 shows the results of this experiment, which for the first time show the expected decrease towards the real GDH sum rule prediction as $Q^2 \to 0$. This decrease is largely due to the increasing asymmetry from the Δ^{1232} resonance as Q^2 decreases.

This data provides clear motivation to continue this measurement to even lower Q^2 to study the approach to $Q^2 = 0$. Experiment E97-110 will soon measure the neutron GDH sum rule from $Q^2 = 0.3$ GeV^2 down to $Q^2 = 0.02$ GeV^2. Expected results from this experiment are shown in Figure 5.

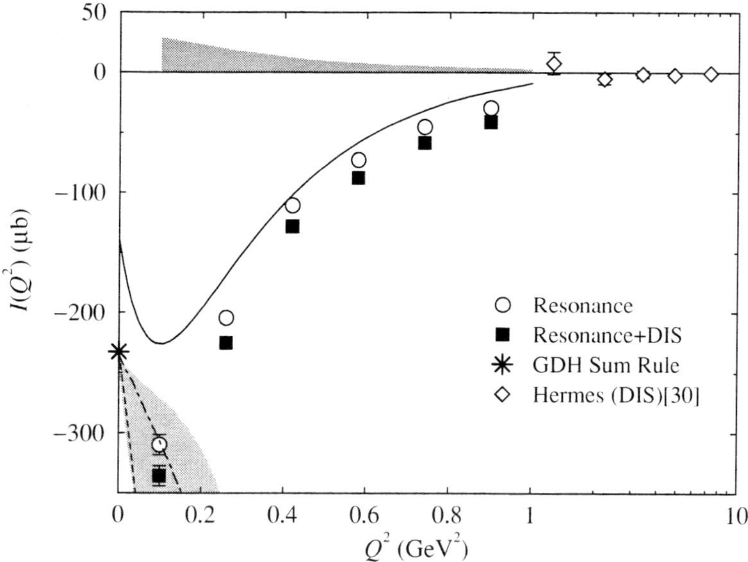

FIGURE 4. Measurements of the neutron GDH integral from JLab experiment E94-010 are shown with open circles. Black squares include a contribution from the deep inelastic region. Also shown are the model predictions of Drechsel [28] (solid line), Ji [26] (dashed), and Bernard [29] (dash-dotted). The HERMES data [30] is shown by open diamonds on a log scale.

Preliminary results from Jefferson Lab Hall B were presented at the conference where the low Q^2 behavior of the GDH sum rule for the proton was measured using a polarized ammonia target. These results are presented by G. Dodge in these proceedings.

Spin Duality

A new experiment is now underway at Jefferson Lab, again using the polarized ^{3}He to look for quark-hadron duality in the neutron spin structure function g_1. Duality has been observed in the unpolarized case [31] where the quark-like behavior of the nucleon ($F_2(x)$ structure function) in the deep inelastic region is an accurate average of the resonance structure seen at low Q^2. Experiment E01-012 is currently measuring $g_1(x)$ in the resonance region up to $Q^2 = 5.4$ GeV^2. This resonance data will be compared to g_1 in the deep-inelastic region where smooth, non-resonant behavior is observed. If duality holds, the curves for g_1 in the resonance region will follow the trend of the smooth deep-inelastic curve. A plot showing an example of expected results is shown in Figure 6.

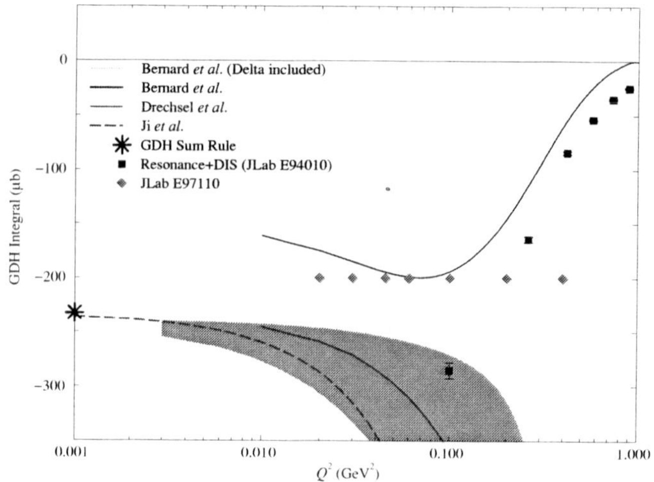

FIGURE 5. Expected errors for the GDH integral from Jefferson Lab experiment E97-110 are shown (diamonds) as a function of Q^2 (log scale) at low Q^2. Published data from the previous JLab measurement (E94-010) at larger Q^2 are shown as blue squares [27]. Model predictions from Bernard [29], Ji [26] and Drechsel [28] are also shown.

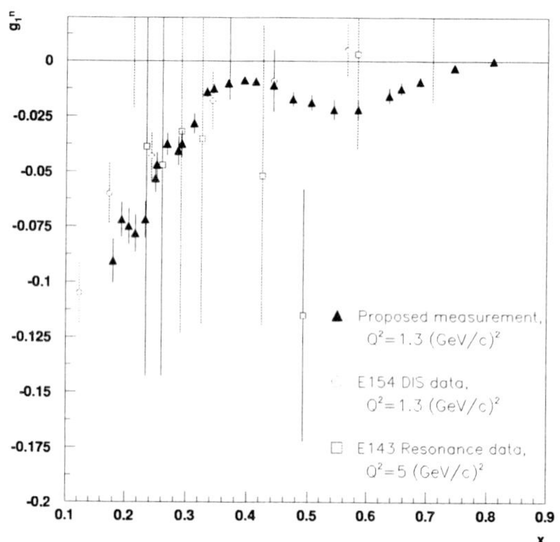

FIGURE 6. Exected errors only for g_1^n from JLab experiment E01-012 at $Q^2 = 1.3$ GeV^2 are shown as black triangles. Also shown is the resonance and deep inelastic data from SLAC experiment E143 [5] and E155 [3].

REFERENCES

1. J.D. Bjorken, Phys. Rev. **148** (1966) 1467; Phys. Rev. **D1** (1970) 1376.
2. J. Blümlein and H. Böttcher, hep-ph/0203155, and Nuc. Phys. **B636** (2002) 225.
3. P.L. Anthony et al., E155, Phys. Lett. **B493** (2000) 19.
4. V.N. Gribov and L.N. Lipatov, Sov. J. Nucl. Phys. **15** (1972) 438, 675; Yu L. Dokshitzer, Sov. Phys. JTEP **46** (1977) 641; G. Altarelli and G. Parisi, Nucl. Phys. **B126** (1977) 298.
5. K. Abe et al., E143, Phys. Rev. **D58** (1998) 120003.
6. A. Airapetian et al., HERMES collaboration, Phys. Lett. **B442** (1998) 484
7. B. Adeva et al., SMC collaboration, Phys. Rev. **D58** (1998) 112001.
8. J. Ashman et al., EMC collaboration, Phys. Lett. **B206** (1988) 364; Nucl. Phys. **B328** (1989) 1.
9. P.L. Anthony et al., E155, Phys. Lett. **B463** (1999) 339.
10. P.L. Anthony et al., E142, Phys. Rev. **D54** (1996) 6620.
11. K. Ackerstaff et al., HERMES collaboration, Phys. Lett. **B404** (1997) 383.
12. K. Abe et al., E154, Phys. Rev. Lett. **79** (1997) 26.
13. K. Abe et al. (E154), Phys. Lett. **B405** (1997) 180.
14. N. Isgur, Phys. Rev. **D59** (1999) 034013.
15. E. Leader, A.V. Sidorov, and D.B. Stamenov, hep-ph/9708335.
16. E. Leader, A.V. Sidorov, and D.B. Stamenov, hep-ph/0111267.
17. C, Bourrely, J. Soffer, and F. Buccella, Eur. Phys. J **C23** (2002) 479; hep-ph/0109160v1.
18. H. Weigel, L. Gamberg, and H. Reinhardt, Phys. Lett. **B399** (1997) 287; Phys. Rev. **D55** (1997) 6910.
19. F.M. Steffens, H. Holtmann, and A.W. Thomas, Phys. Lett. **B358** (1995) 139.
20. E. Shuryak and A. Vainshtein, Nuc. Phys. **B201** (1982) 141.
21. R. Jaffee and X. Ji, Phys. Rev. **D43** (1991) 724.
22. S. Wandzura and F. Wilczek, Phys. Lett. **B72** (1977) 195.
23. M. Gluck, E. Reya, and A. Vogt, Eur. Phys. J. **C5** (1998) 461.
24. P.L. Anthony, et al., Phys. Lett. **B553** (2003) 18.
25. S.B. Gerasimov, Sov. J. of Nucl. Phys. **2** (1966) 430; S. D. Drell and A.C. Hearn, Phys. Rev. Lett. **16** (1966) 908.
26. X. Ji, C. Kao, and J. Osborne, Phys. Lett. **B472** (2000) 1.
27. M. Amarian, et al., Phys. Rev. Lett. **89** (2002) 242301.
28. D. Drechsel, SS. Kamalov, and L. Tiator, Phys. Rev. **D3** (2001) 114010. Note, model I_c is plotted.
29. V. Bernard, T.R. Hemmert, and Ulf.-G. Meissner, hep-ph/0203167.
30. K. Ackerstaff et al., HERMES collaboration, Phys. Lett. **B444** (1998) 531.
31. I. Niculescu *et al.*, Phys. Rev. Lett. **85** (2000) 1186; I. Niculescu *et al.*, Phys. Rev. Lett. **85** (2000) 1182.

Spin Physics at RHIC

L.C.Bland

Brookhaven National Laboratory, Upton, NY USA

Abstract. The physics goals that will be addressed by colliding polarized protons at the Relativistic Heavy Ion Collider (RHIC) are described. The RHIC spin program provides a new generation of experiments that will unfold the quark, anti-quark and gluon contributions to the proton's spin. In addition to these longer term goals, this paper describes what was learned from the first polarized proton collisions at $\sqrt{s}$=200 GeV. These collisions took place in a five-week run during the second year of RHIC operation.

I. INTRODUCTION

The Relativistic Heavy Ion Collider (RHIC) at Brookhaven National Laboratory was built to study the collisions of heavy ions having energies up to 100 GeV/nucleon. The physics goal of these studies is to find a new state of matter where quarks and gluons are not confined in hadrons. PHENIX and STAR are two large experiments and PHOBOS and BRAHMS are two small experiments designed to study relativisitic heavy ion collisions at RHIC. Details about these detectors can be found in Ref. [1].

In addition to having the capability to produce heavy ion collisions at energies up to $\sqrt{s_{NN}}$ = 200 GeV, RHIC is the first high-energy accelerator designed to allow the acceleration, storage and collisions of polarized proton beams with center of mass energies up to $\sqrt{s}$ = 500 GeV. Helical dipole magnets in each ring serve as 'Siberian Snakes' that help preserve the beam's polarization during acceleration and storage [2]. A pair of Siberian Snake magnets in each ring make the vertical direction the stable spin axis. These magnets were commissioned during RHIC run 2 to provide collisions of transversely polarized protons at $\sqrt{s}$=200 GeV. Spin rotator magnets positioned on either side of the STAR and PHENIX interaction regions will permit future studies with longitudinally polarized proton beams. Final preparations of the PHENIX and STAR detectors for the study of high-energy polarized proton collisions will soon be completed. In addition, the PP2PP experiment is being built to study the elastic scattering of polarized proton beams at RHIC.

This paper briefly describes the long-term physics goals of the RHIC spin program. These goals have been described in a recent review [3]. What is new in this area is the first collisions of polarized protons at RHIC. This paper provides a description of what was learned from the first RHIC run producing collisions of polarized protons. Many details about the run were described in parallel session talks at SPIN 2002. This overview serves as a roadmap to these more detailed descriptions.

CP675, *Spin 2002: 15th Int'l. Spin Physics Symposium and Workshop on Polarized Electron Sources and Polarimeters*, edited by Y. I. Makdisi, A. U. Luccio, and W. W. MacKay

II. PHYSICS GOALS OF THE RHIC SPIN PROGRAM

The RHIC spin program aims to unravel the spin structure of the proton. Understanding how the valence quarks, gluons and sea quark-antiquark pairs conspire to produce the spin of the proton is equally as important as understanding how these constituents produce the proton's mass. The quark contribution ($\frac{1}{2}\Delta\Sigma$), the gluon contribution (ΔG) and possible orbital angular momentum of the proton's constituents must sum to the proton spin of $\frac{1}{2}$. To date, spin structure studies have relied on deep inelastic scattering (DIS) of charged leptons as was reviewed at this Symposium [4, 5]. The contribution quarks make to the overall spin of the proton has been found to be substantially smaller than expected ($\Delta\Sigma = 0.23 \pm 0.04 \pm 0.06$). Polarization of the gluons that bind the quarks and/or orbital angular momentum of the proton's constituents must carry the rest. Since gluons carry no *electric* charge, real or virtual photons only probe the gluon through its splitting into $q\overline{q}$ pairs. Use of this photon-gluon fusion process provides a means of establishing the gluon contribution to the proton's spin that is being actively pursued by the COMPASS experiment at CERN [6] and is a planned pursuit at SLAC [7]. COMPASS aims to probe gluon polarization by detecting D mesons from $\gamma^* g \to c\overline{c}$ interactions employing a polarized μ beam and a polarized target. One of the goals of the RHIC spin program is to utilize the *color* charge of the quarks to couple directly to the gluon field within the proton. Existing measurements of quark polarization from DIS experiments enable the use of quarks as an 'analyzer' of gluon polarization in $\vec{p}\vec{p}$ collisions, particularly for final states where a large transverse momentum (p_T) photon is produced. Gluon polarization will be probed at RHIC by measurements of longitudinal double spin asymmetries (A_{LL}).

High energy collisions of protons produce hadronic jets, photons, vector bosons ($W^{\pm}, Z, \gamma^*$) and other particles having large p_T or large mass. From studies at unpolarized hadron colliders, it is known that perturbative QCD can provide quantitative predictions of the cross sections for such processes, particularly when next-to-leading order (NLO) contributions are included. Extending these phenomenlogical analyses to include spin degrees of freedom leads to the expectation that pQCD will provide a quantitative framework to interpret spin observables for jets, photons, and other particles produced in polarized proton collisions.

As previously described [8], the 'golden probe' of gluon polarization at RHIC will be photon production in longitudinally polarized proton collisions. It is expected that the QCD Compton process ($qg \to q\gamma$) will be the dominant source of photons, with physics backgrounds arising from fragmentation photons and the $q\overline{q} \to \gamma g$ process, the latter process corresponding to time-reversed photon-gluon fusion. Experimentally, photon production is challenging because the total cross section is small relative to prolific sources of photons from neutral meson decays. Simulations for both PHENIX and STAR have demonstrated that these backgrounds can be minimized. Coincident detection of the away-side jet with the prompt photon can provide information about the Bjorken x values of the initial-state interacting partons. Recent results from the Tevatron [9] suggest that NLO pQCD can provide a quantitative description of photon production cross sections.

The production of $W^{\pm}$ bosons in the collision of longitudinally polarized protons is expected to result in spectacularly large parity violation. As in charged current weak interactions that gives rise to the flavor sensitivity to neutrino deep inelastic scattering, parity violating spin asymmetries for $W^{\pm}$ production in $\vec{p}p$ collisions will isolate valence quark ($\Delta u/u, \Delta d/d$) and sea anti-quark ($\Delta \overline{u}/\overline{u}, \Delta \overline{d}/\overline{d}$) polarizations [10]. Utilization of the electroweak interaction is expected to be a more precise measure of these polarizations than semi-inclusive DIS experiments that are presently underway. Imprecise knowledge of fragmentation functions is a dominant systematic error for semi-inclusive DIS studies.

The collision of transversely polarized proton beams can be used to study the transversity structure function. Although this is a twist-2 function of equal importance as the longitudinal spin distribution functions ($\Delta q(x)$,$\Delta \overline{q}(x)$ and $\Delta G(x)$) for understanding the proton's spin structure, transversity cannot be probed in inclusive DIS because of helicity conservation. Transverse spin asymmetry measurements at RHIC are expected to shed light on the transversity structure functions and should provide a timely complement to new measurements of transverse spin effects in semi-inclusive DIS [5] and in jet studies from e^+e^- collisions [11].

The development of the RHIC accelerator complex to produce polarized proton collisions with a luminosity of $0.8(2.0) \times 10^{32}$ cm^{-2}s^{-1} at $\sqrt{s}$=200(500) GeV and with beam polarizations of 70% will be required to attain these long term physics goals. Completion and full development of the PHENIX and STAR experiments is also required. Intermediate physics goals can be attained in the first RHIC spin runs, providing an understanding of how to reach the requisite precision to measure small spin effects in the collider environment while providing the first glimpses of the proton's spin structure probed in high-energy polarized proton collisions.

III. SPIN ASYMMETRY MEASUREMENTS IN A COLLIDER

A collider has significant differences, both advantageous and potentially problematic, from traditional fixed target experiments that severely impact the measurement of spin observables. One positive aspect is that in a collider polarization reversals occur at the bunch crossing frequency. For RHIC run 2, the inverse of this frequency was 214 nsec. This is expected to be reduced to 107 nsec in subsequent runs. These time scales are very short compared to most sources of time dependent variations in a detector's response, such as gain drifts, thereby reducing sensitivity to this class of systematic errors. Rapid polarization reversals place stringent demands on timing of detector signals to preserve the correct association of events with the polarization state of the beams. The detector response must also be minimal to backgrounds which are not associated with the passing beams. Out-of-time background events effectively obtain random polarization assignments.

One significant challenge that is presented by collider experiments is the possibility of accidental correlation between polarization directions and luminosity of the colliding beams. This correlation can arise because each ring is injected with individual bunches of protons with particular polarization orientations. An injection pattern is chosen so that in one ring (Yellow) alternating bunches have opposite polarization direction and the other ring (Blue) has bunch pairs with alternating polarization (Fig. 1). Due to the potential correlation between luminosity and polarization, accurate measurement

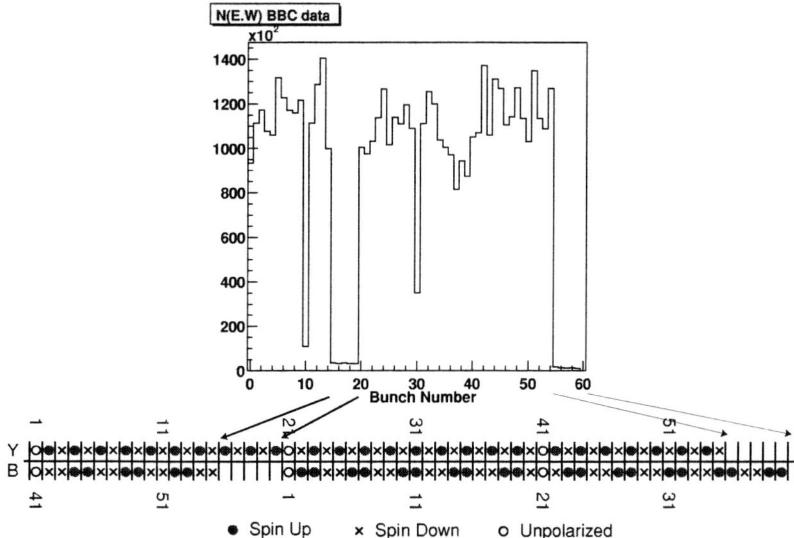

FIGURE 1. STAR data from a bunch crossing scaler system with input from a beam-beam counter coincidence [12] providing a measure of luminosity. There are two groups of five bunch crossings with few counts, corresponding to the 'abort gaps' (RF cycles with no beam) in the Blue and Yellow rings. Counts in the abort gaps arise from single beam backgrounds. The pattern of polarization directions for the Blue and Yellow beams at STAR is also shown. The non-uniformity of the yield versus bunch number produces differences in the relative luminosity for different polarization directions.

of the relative luminosity of bunch crossings with different polarization directions is required. This is accomplished by counting experiments using fast detectors capable of discriminating collisions from beam-related backgrounds. An example from RHIC run 2 is shown in Fig. 1, with more details described in Ref. [12]. Methods are being developed to average out any relationship between intensity and polarization by reversing the spins of the stored beams and by recogging the beams so that different bunches collide.

When measuring transverse single spin asymmetries, an azimuthally symmetric detector configuration can be used to measure spin-dependent cross ratios (spin up/down yields with left/right detectors), thereby making detector acceptance and luminosity asymmetries higher order corrections [13]. For observables involving longitudinal polarization there is no spin-dependent modulation of scattering rates that varies harmonically with the azimuthal angle. Hence, accurate measurements must be made to prevent the relative luminosity determination from being a limiting systematic error of future A_{LL} measurements. The large cross section reactions needed for high rate relative luminosity monitors must themselves have minimal longitudinal double spin asymmetries. Identifying suitable monitoring reactions is one of the primary goals for RHIC run 3 when longitudinally polarized proton collisions are planned for the PHENIX and STAR experiments.

It is planned to develop an AC dipole that will efficiently reverse the polarization of the stored beam [14]. This tool, used in conjunction with changes in which beam bunches overlap (recogging), will be fully developed in subsequent RHIC runs. These developments are expected to further reduce spin dependent relative luminosity as a major source of systematic error for spin asymmetry measurements.

Backgrounds in a collider also differ in important ways from those for fixed target experiments. One difference between the two environments is that there are two beams in a collider, traveling in opposite directions, that can lead to twice the background. As with the spin information, timing plays a critical role in discriminating signals from collisions versus those from backgrounds. Time differences between counters mounted fore and aft of the interaction region can be used to identify collisions relative to beam-related backgrounds.

IV. FIRST POLARIZED PROTON COLLISIONS

The first collisions of transversely polarized protons occurred during RHIC run 2 in the period from early December, 2001 to late January, 2002. The first three weeks of this run were dedicated to accelerator commissioning, with a particular focus on commissioning the 'Siberian Snake' magnet pairs in each ring. Officially, data taking for $\sqrt{s}$=200 GeV $\vec{p}\vec{p}$ collisions began on December 20th, initially at low luminosity. Two weeks into the run, the average luminosity increased to 0.5×10^{30} cm^{-2} s^{-1}, and stayed at that value until the end of the run. Knowledge of the beam polarization is subject to the caveats discussed below.

The RHIC polarimeters detect low-energy recoil carbon ions produced by $p+^{12}$C elastic scattering reactions at small $|t|$, where the Coulomb scattering amplitude is comparable in magnitude to the strong-interaction amplitude. Details about the design [15] and performance [16] of these Coulomb-Nuclear Interference (CNI) polarimeters were described at this conference. At present, the effective analyzing power of the CNI polarimeters is measured only near the injection energy [17]. Ultimately, these polarimeters will be calibrated by determining the beam polarization from proton elastic scattering from a polarized hydrogen gas jet target, effectively transferring knowledge of the target polarization to provide knowledge of the beam polarization. The first opportunity to make these measurements will occur during RHIC run 4. Prior to that, the effective analyzing power of the CNI polarimeter can only be determined by special stores in RHIC involving an acceleration ramp followed by a deceleration ramp. Measurements with the CNI polarimeter at the injection energy, flattop energy and then again at the injection energy can transfer the calibration of the polarimeter effective analyzing power to high energy, if there is found to be minimal polarization loss in both ramps. Deceleration ramps were attempted during RHIC run 2, but were unsuccessful because of beam loss during deceleration. Continuation of this development is planned for run 3.

For RHIC run 2, knowledge of the magnitude of the beam polarization at the collision energy results from the assumption that the effective analyzing power of the CNI polarimeter does not depend on energy. Model calculations suggest only a small increase in the effective analyzing power with increasing energy [18]. Based on the assumption that the effective analyzing power is independent of beam energy, the beam polarization at the collision energy averaged 17% in the Yellow ring and 13% in the Blue ring

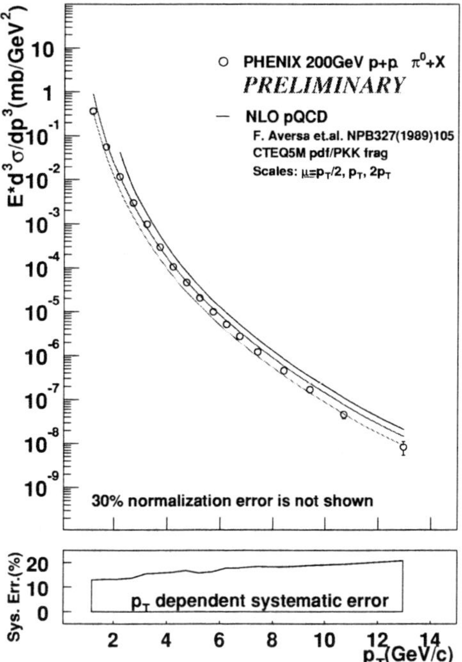

FIGURE 2. PHENIX results for mid-rapidity $p+p \rightarrow \pi^0 + X$ invariant cross sections versus p_T measured at $\sqrt{s} = 200$ GeV in RHIC run 2. The data are compared to NLO pQCD calculations.

over the last two weeks of the run, with the difference reflecting subtle effects in the acceleration ramps of the two rings. The ratio of the measured CNI polarimeter spin asymmetries at the collision energy to those at the injection energy for a given RHIC fill was on average smaller than unity. Hence, the measured asymmetries at 100 GeV set an upper limit on the beam polarization at the collision energy, since acceleration is highly unlikely to increase the beam polarization. The polarization magnitude was limited by the AGS, the last accelerator in the chain used to inject the RHIC rings. Failure of the AGS power source necessitated use of a less powerful backup, resulting in a slower acceleration cycle, thereby reducing the polarization.

The physics goals of the first polarized proton collisions at $\sqrt{s}$=200 GeV included:

- the measurement of unpolarized observables to provide the requisite *pp* reference data for the RHIC heavy-ion program,
- commissioning of the accelerator complex for polarized proton collisions,
- making the first measurements of spin asymmetries in a polarized proton collider,
- identifying reactions that will provide the basis for local polarimeters in subsequent runs which utilize the spin rotator magnets, and

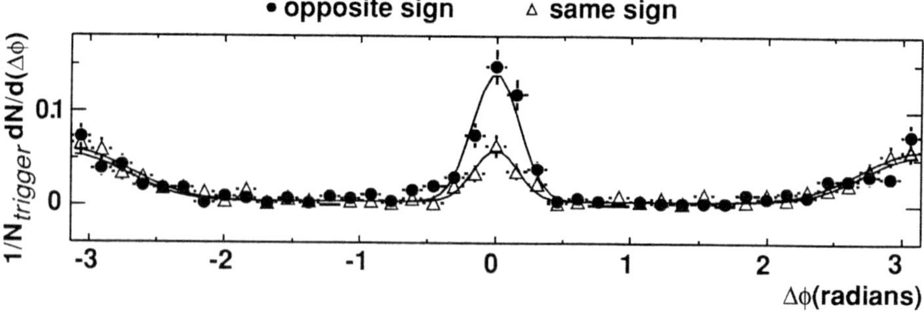

FIGURE 3. STAR results for azimuthal angle correlations between high p_T charged hadrons produced in $p+p$ collisions at $\sqrt{s}$=200 GeV [22]. The correlations are between a trigger particle with $4 < p_T^{trig} <$ 6 GeV/c and associated particles with 2 GeV/c $< p_T < p_T^{trig}$. Both the near side ($\Delta\phi \sim 0$) and away side ($|\Delta\phi| \sim \pi$) correlations are consistent with expectations from jet and dijet events.

- commissioning of the PP2PP Roman pots for the study of proton elastic scattering at small $|t|$.

These goals represent important benchmarks of progress to both the RHIC spin program and the RHIC heavy-ion program. As is evident in the following description, these goals were met in the first polarized proton collision run.

IV.a Unpolarized proton observables

Many of the results for unpolarized pp collision observables were reported at the recent Quark Matter 2002 conference [19]. Minimum bias data samples were recorded at all RHIC experiments. High p_T triggers based on the central arm electromagnetic calorimeters (EMC) enabled PHENIX to obtain the p_T variation of mid-rapidity π^0 production out to 12 GeV/c (Fig. 2). These data are compared with next-to-leading order pQCD calculations, and agree with the calculations over an impressive eight orders of magnitude [20]. This good agreement between data and theory bodes well for using pQCD to interpret spin observables at RHIC energies. PHENIX was also able to measure J/Ψ production cross sections in their central detector arms by detecting e^+e^- coincidences and in their south muon arm by detecting $\mu^+\mu^-$ coincidences [21]. Polarization observables for open charm production in $\vec{p}\vec{p}$ collisions is anticipated to be another important channel for studying gluon polarization. PHENIX plans to commission their north muon arm during RHIC run 3, thereby completing the baseline detector.

STAR obtained a significant pp minimum bias data sample that will provide critical reference data for its heavy-ion program. The relatively low luminosity for the first polarized proton collisions minimizes the influence of 'pile-up' background in the STAR time projection chamber (TPC). In future years, higher luminosity pp collisions will require sophisticated offline analysis algorithms to discriminate charged particle tracks associated with the triggered event from pileup tracks that are produced in the ~400

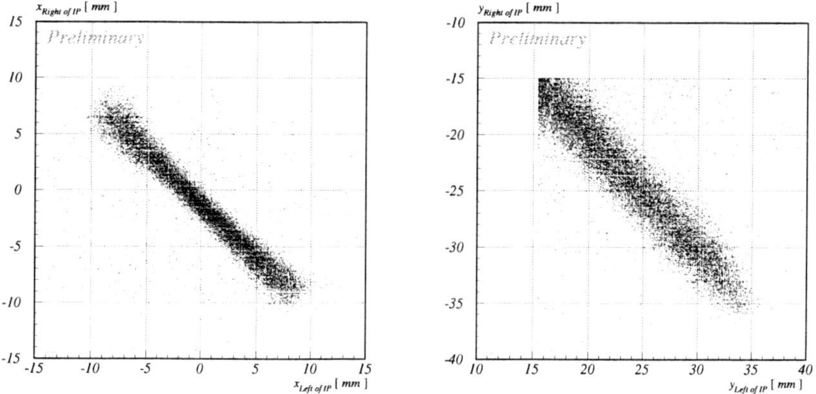

FIGURE 4. Results from PP2PP obtained during a single RHIC fill with reduced emittance and special focusing to permit positioning their Roman pots close to the beam. The left (right) figure shows the horizontal (vertical) position correlation for events detected in silicon detectors mounted in the Blue and Yellow rings either side of the interaction point. Elastic scattering events from pp collisions can be cleanly identified.

bunch crossings before or after the triggered event. Even with only a minimum bias trigger, STAR embarked on its program of jet physics, exploiting the complete azimuthal coverage of the TPC for $|\eta| \leq 1.4$. As shown in Fig. 3, evidence for jet and di-jet events was obtained from di-hadron correlation studies [22]. Cone and k_T jet reconstruction alogrithms are presently being applied to the reconstructed TPC tracks for charged particles. The results from these analyses are qualitatively consistent with expectations. Quantitative results for jet yields and correlations require detailed understanding of the TPC efficiency. With the commissioning of the half barrel EMC (2π azimuthal coverage for $0 \leq \eta \leq 1$) in RHIC run 3, STAR will be able to trigger on and fully reconstruct jets. Complete coverage from the barrel EMC ($-1 \leq \eta \leq +1$), and the coverage from the endcap EMC ($1 \leq \eta \leq 2$), will be added for subsequent runs. Jet physics is an important component of the STAR spin physics program.

RHIC run 2 also saw the commissioning of the PP2PP experiment. Their goal is to measure proton elastic scattering cross sections and spin observables for polarized proton collisions at $\sqrt{s}$=200 and 500 GeV. The experiment utilizes 'Roman pots' to place silicon detectors close to the beams, observing the elastically scattered protons in coincidence. The emphasis is on measuring the $|t|$ dependence for elastic scattering in the vicinity of the maximal interference between the Coulomb and strong interaction amplitudes. Elastic scattering events were observed [23] during the commissioning run in a dedicated store where the beams were scraped to reduce their emittance (Fig. 4). Further runs are planned in RHIC run 3, and beyond, to attain the goals of the physics program.

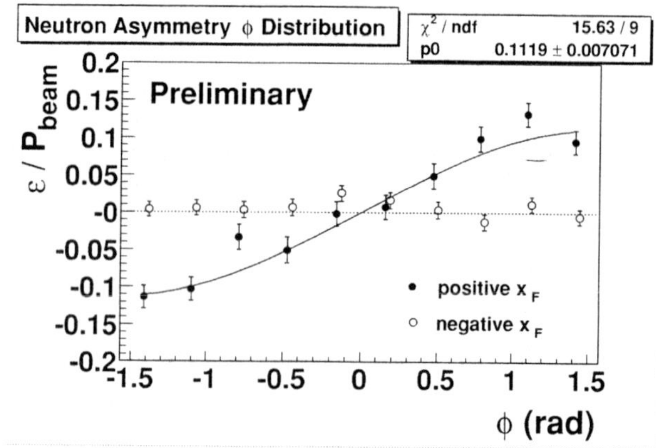

FIGURE 5. Results from the PHENIX spin group for the analyzing power for forward neutron production. Shown in the figure are the measured spin-dependent asymmetries scaled by the Blue ring beam polarization, assuming the effective analyzing power of the CNI polarimeter does not depend on energy.

IV.b First spin asymmetry measurements and local polarimetry

A suitable context for understanding the first polarized proton collisions at RHIC, which used transverse polarization, is the E-704 experiment at Fermi National Laboratory. Prior to RHIC, E-704 studied polarized proton interactions at the highest $\sqrt{s}$. E-704 was a fixed target experiment using a high-energy polarized proton beam produced by hyperon decay, resulting in polarized proton interactions at $\sqrt{s}$=20 GeV, an order of magnitude lower than the first RHIC spin run. Large analyzing powers, increasing in magnitude as Feynman x increases beyond $x_F \sim 0.3$, were found by E-704 for both charged [24] and neutral pion [25] production at moderate p_T. At mid-rapidity, analyzing powers were found to be consistent with zero. Theoretical models that aimed to understand the E-704 analyzing powers generally predicted similar trends at RHIC energies, even though they attributed the large Feynman x analyzing powers to different dynamics.

The E-704 results motivated two separate large x_F measurements during RHIC run 2. The goal of both of these measurements was to observe reactions with large analyzing powers that could provide the basis for local polarimeters at the PHENIX and STAR experiments. Spin rotator magnets will be used during RHIC run 3 to produce nominally longitudinal beam polarization at PHENIX and STAR. Local polarimeters sensitive to either remnant vertical or radial polarization components are required to properly tune the spin rotator magnets. Space constraints at PHENIX and STAR differ, implying that different techniques for local polarimetry are likely to be needed.

One experiment was conducted at interaction point 12 (IP12) in the RHIC ring by members of the PHENIX spin group [26]. They designed and constructed a calorimeter consisting of a 5 (horizontal) × 12 (vertical) matrix of $PbWO_4$ detectors that was positioned 18 meters distant from the interaction point immediately following the DX magnet east of IP12, and hence sensitive to spin effects from collisions associated with the polarization direction of the proton beam in the Blue ring. The objective was to be sensitive to possible analyzing powers associated with large x_F γ or π^0 production. The geometry of their calorimeter implied that 50 GeV energy deposition corresponded to a maximum p_T of 150 MeV/c. The experiment could discriminate incident γ and neutrons by use of counters that were located in front of and behind the calorimeter. Through the course of the run, a portion of a Zero Degree neutron Calorimeter (ZDC) module was installed west of IP12, again 18 meters distant from the interaction point, to confirm results that were obtained from analysis of the data during the run.

The conclusion from this experiment (Fig. 5) is that there are sizable analyzing powers for low p_T neutron production in polarized proton collisions at $\sqrt{s}$=200 GeV. These results will serve as the basis of local polarimetry at the PHENIX experiment in RHIC run 3 by equipping the ZDC's fore and aft of PHENIX with position sensitive shower maximum detectors.

At STAR, electromagnetic calorimeters were mounted upstream of the DX magnet in close proximity to the beam pipe and 7.5 meters distant from the STAR interaction point [27]. A Pb-scintillator sampling calorimeter, equipped with a two orthogonal planes of scintillator strips serving as a position-sensitive shower maximum detector was mounted north of the beam. Simple 4×4 matrices of lead glass detectors were mounted in three locations, directly south of and above and below the beam pipe. Since the calorimeters were mounted east of STAR, they were sensitive to spin effects from collisions associated with the polarization direction of the proton beam in the Yellow ring. The calorimeter mounted north of the beam was able to identify neutral pions with good invariant mass resolution, and was sensitive to transverse momenta in the range, $1 \leq p_T \leq 4$ GeV/c.

The conclusion from the STAR measurement is that forward neutral pion production at RHIC energies has large analyzing powers that increase with increasing Feynman x. The results at $\sqrt{s}$=200 GeV (Fig. 6) bear a strong similarity to the E-704 results [25] and to theoretical predictions based on the lower energy data. Given that the effective analyzing power of the CNI polarimeter is measured only at the RHIC injection energy, these results are *lower limits* on the π^0 analyzing power at $\sqrt{s}$=200 GeV. Attempts are underway to equip STAR with electromagnetic calorimetry mounted at large pseudo-rapidity as a more permanent Forward π^0 Detector (FPD). The new FPD will enable further study of the Feynman x and p_T dependence of analyzing powers for forward π^0 production, and can also serve as a local polarimeter for the tuning of the spin rotator magnets.

In addition to these two experiments aimed at identifying local polarimeters for the future, analyzing powers for particle production at mid-rapidity will result from RHIC run 2. STAR reported the analyzing power for leading charged particles at mid-rapidity out to p_T=5 GeV/c [31]. As expected by naive pQCD, A_N is found to be zero at mid-rapidity. PHENIX will ultimately be able to make quantitative statements to much higher p_T from their mid-rapidity neutral pion and charged hadron data.

In summary, the polarized proton run at $\sqrt{s}$=200 GeV during RHIC run 2 accomplished important goals, including:

- measurement of mid-rapidity particle production cross sections over a broad range of p_T. These measurements are well described by NLO pQCD, which bodes well

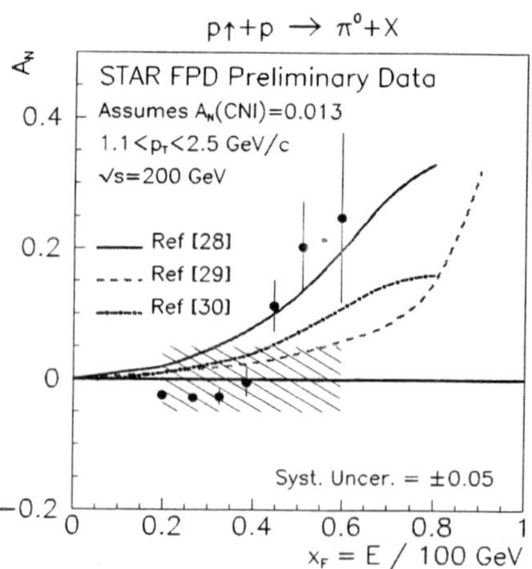

FIGURE 6. STAR results for the analyzing power for events dominated by π^0 production at large pseudorapidity ($3.4 \leq \eta \leq 4.1$). The data are compared to calculations available prior to the measurement. Ref. [28] is based on the Collins effect, corresponding to a spin dependent fragmentation. Ref. [29] is based on a twist-3 quark-gluon correlation responsible for the spin effect evaluated at p_T=1.5 GeV/c. Ref. [30] is based on the Sivers effect, where the spin effects arise from a correlation between the quark spin and its transverse momentum in the distribution function.

for future RHIC studies of the spin structure of the proton,

- the clear observation of jet events in *pp* collisions. Jet production and γ+jet coincidences for longitudinally polarized proton collisions will be a primary means of establishing gluon polarization,
- the observation of *pp* elastic scattering collisions. Only small backgrounds are observed in silicon detectors placed close to the beams, thereby demonstrating that RHIC is a well controlled accelerator providing a clean environment for studing *pp* elastic scattering for colliding beams in the CNI regime (small $|t|$), and
- the observation of large spin asymmetries in the collision of transversely polarized protons. These measurements demonstrate proper handling of the spin labeling of bunches and matching with the RHIC polarimeter results. The reactions that give rise to these large analyzing powers will be used as local polarimeters at PHENIX and STAR in future runs, and themselves represent important transverse spin physics results.

Low polarization from the AGS is an important issue that must be addressed in future RHIC spin runs. New polarimeter and accelerator hardware are expected to increase the polarization of beams stored in RHIC.

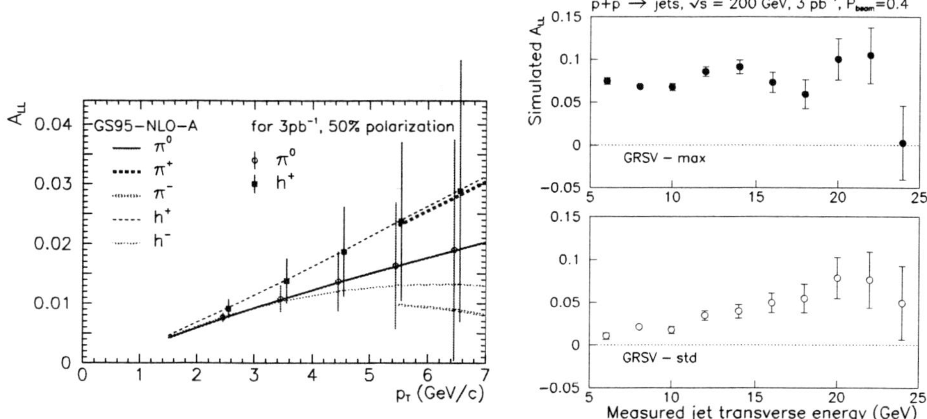

FIGURE 7. Simulations for longitudinal spin correlations (A_{LL}) for (left) inclusive particle production at PHENIX using the polarized parton distributions from Ref. [32] and (right) inclusive jet production at STAR for $0 \leq \eta \leq 1$ using polarized parton distributions from Ref. [33]. All $2 \to 2$ subprocesses are included in the simulations. The STAR simulations account for trigger and reconstruction biases.

V. PLANS FOR FUTURE RUNS

Spin rotator magnets have now been installed at the PHENIX and STAR experiments and will be commissioned during RHIC run 3. These magnets will precess the proton beam's polarization from the vertical stable spin direction set by the two helical dipole Siberian Snakes in each ring, to longitudinal at the interaction points of these two experiments, and then back again. Tuning the spin rotator magnets requires robust local polarimetry to establish that vertical and radial polarization components are small. The analyzing power results from RHIC run 2 should provide the requisite feedback for this important development work. In addition to this development, the luminosity for polarized proton collisions in RHIC run 3 is projected to be 10^{31} cm^{-2}s^{-1} by using a tighter focus (β^*=1 m) and by increasing the number of bunches in the Blue and Yellow ring from 55 to 110. The high-power generator is expected to be repaired and used for the AGS. The projected beam polarization for RHIC run 3 is 40%, with a goal of further improvements on that value. Polarized proton collisions will be at $\sqrt{s}$=200 GeV for run 3.

With longitudinal polarization for both proton beams, and the projected improvement in the polarized beam operation of the AGS injector to RHIC, it should be possible to embark on the first measurements of A_{LL} for mid-rapidity π^0 production at PHENIX and for mid-rapidity ($0 \leq \eta \leq 1$) jet production at STAR (Fig. 7). Projections for the integrated luminosity and the beam polarization for polarized beam operations in RHIC run 3 suggest sufficient statistical sensitivity to discriminate extreme scenarios for the degree to which gluons are polarized within the proton. Systematic errors need to be sufficiently controlled to provide robust measurements of A_{LL}. Careful relative luminosity measurements for bunch crossings with equal and opposite longitudinal polarizations are required as is an absolute calibration of the analyzing power of the RHIC polarimeter for proton beam energy of 100 GeV.

Looking beyond RHIC run 3, there are several accelerator and detector developments on the horizon:

- Polarized beams will be accelerated to 250 GeV to allow collisions at $\sqrt{s}$=500 GeV. The highest collision energies are important for determining gluon polarization at small x and necessary for studying spin asymmetries for $W^{\pm}$ production.
- The polarized hydrogen gas jet target is expected to be first available for RHIC run 4. This will ultimately provide a robust means of determining the beam polarization with a projected accuracy of 5% expected to be attained during RHIC run 5.
- Improvements in the AGS polarization are expected for runs 4 and 5, leading to the goal of 70% polarization for the beams stored in RHIC. A new fast polarimeter, based on the RHIC CNI polarimeter, has been built for the AGS for run 3, and several new hardware improvements are being considered for the AGS for runs 4 and 5. These include a strong partial Siberian Snake, expected to be available for run 5.
- The STAR barrel and endcap electromagnetic calorimeters will be completed and commissioned for RHIC run 5. These calorimeters provide 2π azimuthal coverage for the pseudorapidity interval $-1 \leq \eta \leq 2$, playing a crucial role in spin physics measurements involving jets, photons and high-energy electrons and positrons.
- Upgrades to PHENIX are planned to improve triggering capabilities for the muon arms required to fully exploit their $W^{\pm}$ physics program. There are also plans to improve their mid-rapidity charged particle tracking. The latter will enhance jet reconstruction capabilities at PHENIX.

The success of the first polarized proton collision run upholds the promise that the RHIC spin program will provide important new insight to the spin structure of the proton.

ACKNOWLEDGMENTS

This report summarizes the work of many people who are part of the RHIC spin collaboration (RSC). The RSC has members from the RHIC accelerator groups, the PHENIX collaboration, the STAR collaboration, the PP2PP experiment and the theory community that frequently communicate and exchange important information about polarized proton interactions at RHIC. RIKEN has played a critical role in supporting the development of the hardware for polarized proton collisions at RHIC and the creation of the RIKEN/BNL Research Center, a stimulating environment for exploring RHIC physics. The US Department of Energy and the National Science Foundation are playing central roles in supporting the RHIC spin program.

REFERENCES

1. 'RHIC accelerator and detectors', Nucl. Instr. and Meth., special dedicated volume (to be published).
2. Haixin Huang,'Acceleration of Polarized Protons at RHIC', *these proceedings*.

3. Gerry Bunce, Naohito Saito, Jacques Soffer and Werner Vogelsang, Ann. Rev. Nucl. Part. Sci. **50**, 525 (2000).
4. Todd Averett, 'Structure functions (g1 and g2) from DIS experiments', *these proceedings.*
5. Andrew Miller, 'New results on semi-inclusive DIS measurements', *these proceedings.*
6. F. Tessarotto, 'Status of the COMPASS experiment', *these proceedings.*
7. Steve Rock, 'HERMES + SLAC E161', *these proceedings.*
8. L. C. Bland, 'Plans for spin physics at RHIC', in *Physics with a High Luminosity Polarized Electron Ion Collider*, eds. L. C. Bland, J. T. Londergan and A. P. Szczepaniak, World Scientific (2000).
9. V. M. Abazov, et al. (D0 collaboration), Phys. Rev. Lett. 87, 251805 (2001).
10. N. Saito, Nucl. Phys. A **638**, 575 (1998).
11. K. Hasuko, 'Azimuthal Asymmetries in Fragmentation Processes using e^+e^- Collisions', *these proceedings.*
12. Joanna Kiryluk, 'Relative Luminosity Measurement in STAR and Implications for Spin Asymmetry Determination', *these proceedings.*
13. Hal Spinka, 'Proton beam polarimetry at BNL', *these proceedings.*
14. Mei Bai, 'RHIC Spin Flipper Commissioning', *these proceedings.*
15. Kazu Kurita (for Igor Alekseev), 'RHIC pC CNI Polarimeter: Experimental Setup and Physics Results', *these proceedings.*
16. Osamu Jinnouchi, 'RHIC pC CNI Polarimeter: Status and Performance from the First Collider Run', *these proceedings.*
17. J. Tojo, et al., Phys. Rev. Lett. **89**, 052302 (2002).
18. T. L. Trueman, 'Energy Dependence of CNI Analyzing Power for Proton-Carbon Scattering', RHIC Spin Note 1, hep-ph/0203013 (2002).
19. *16th International Conference on Ultra-Relativistic Nucleus-Nucleus Collisions*, Nantes, 18-24 July 2002 (proceedings to be published).
20. B. Fox, 'Neutral Pion Measurements from PHENIX in Polarized Proton Collisions at RHIC', *these proceedings.*
21. H. Sato, 'J/ψ Production in $\sqrt{s}$ = 200 GeV $p+p$ Collisions with the PHENIX Detector at RHIC', *these proceedings.*
22. C. Adler, et al. (STAR collaboration), 'Disappearance of back-to-back high pT hadron correlations in central Au + Au collisions at sqrt(snn) = 200 GeV', submitted to Phys. Rev. Lett (nucl-ex/0210033).
23. S. Bueltmann, 'Elastic Polarized Proton Scattering at RHIC', *these proceedings.*
24. D. L. Adams, et al., Phys. Lett. **B264**, 462 (1991).
25. D. L. Adams, et al., Phys. Lett. **B261**, 201 (1991).
26. Y. Fukao, 'Measurement of Single Transverse-spin Asymmetries in Forward Production of Photons and Neutrons in pp Collisions at $\sqrt{s}$=200 GeV', *these proceedings.*
27. G. Rakness, 'Analyzing Powers for Forward $p_\uparrow + p \rightarrow \pi^0 + X$ at STAR', *these proceedings.*
28. M. Anselmino, U. D'Alesio and F. Murgia (private communication); M. Anselmino, M. Boglione and F. Muriga, Phys. Rev. D **60**, 054027 (1999).
29. Jianwei Qiu (private communication); Jianwei Qiu and George Sterman, Phys. Rev. D **59**, 014004 (1998).
30. M. Anselmino, U. D'Alesio and F. Murgia (private communication); M. Anselmino, M. Boglione and F. Murgia, Phys. Lett. **B362**, 164 (1995); M. Anselmino and F. Murgia, Phys. Lett. **B442** 470 (1998).
31. J. Balewski, 'Midrapidity Spin Asymmetries at STAR', *these proceedings.*
32. T. Gehrmann and W.J. Stirling, Phys. Rev. D **53** 6100 (1996).
33. M. Gluck, E. Reya, M. Stratmann and W. Vogelsang, Phys. Rev. D **63**, 094005 (2001).

Exploring Nucleon Spin Structure in Longitudinally Polarized Collisions

Marco Stratmann

Inst. for Theor. Physics, Univ. of Regensburg, D-93040 Regensburg, Germany
E-mail: marco.stratmann@physik.uni-regensburg.de

Abstract. We review how RHIC is expected to deepen our understanding of the spin structure of longitudinally polarized nucleons. After briefly outlining the current status of spin-dependent parton densities and pointing out open questions, we focus on theoretical calculations and predictions relevant for the RHIC spin program. Estimates of the expected statistical accuracy for such measurements are presented, taking into account the acceptance of the RHIC detectors.

LESSONS FROM POLARIZED DEEP INELASTIC SCATTERING

Before reviewing the prospects for spin physics at the BNL-RHIC we briefly turn to longitudinally polarized deep-inelastic scattering (DIS) and what we have learned from more than twenty years of beautiful data [1]. To next-to-leading order (NLO) in the strong coupling α_s, the DIS structure function g_1, which parametrizes our ignorance about the nucleon spin structure, can be expressed as

$$g_1(x,Q^2) = \frac{1}{2} \sum_{q=u,d,s} e_q^2 \left[(\Delta q + \Delta\bar{q}) \otimes \left(1 + \frac{\alpha_s}{2\pi}\Delta C_q\right) + \frac{\alpha_s}{2\pi}\Delta g \otimes \Delta C_g \right](x,Q^2) \quad . \tag{1}$$

The symbol $\otimes$ denotes the usual convolution, and $\Delta C_{q,g}$ are the perturbatively calculable coefficient functions, which are known even up to next-to-next-to-leading order [2]. The Δf, $f = (q, \bar{q}, g)$, are the spin-dependent parton distributions, defined as

$$\Delta f(x,\mu) \equiv f^+(x,\mu) - f^-(x,\mu) \quad , \tag{2}$$

where f^+ (f^-) is the number density of a parton type f with helicity "+" ("−") in a proton with positive helicity, carrying a fraction x of the proton's momentum. Once they are known at some initial scale μ_0, their scale μ dependence is governed by a set of evolution equations

$$\mu \frac{d}{d\mu} \begin{pmatrix} \Delta q(x,\mu) \\ \Delta g(x,\mu) \end{pmatrix} = \begin{pmatrix} \Delta\mathcal{P}_{qq} & \Delta\mathcal{P}_{qg} \\ \Delta\mathcal{P}_{gq} & \Delta\mathcal{P}_{gg} \end{pmatrix} \otimes \begin{pmatrix} \Delta q \\ \Delta g \end{pmatrix} (x,\mu) \quad . \tag{3}$$

So far, the spin-dependent $j \to i$ splitting functions entering these evolution equations have been calculated up to NLO accuracy [3].

Figure 1 compares the available information on $g_1(x,Q^2)$ for DIS off a proton target to results of a typical NLO QCD fit. From such types of analyses a pretty good knowledge

CP675, *Spin 2002: 15th Int'l. Spin Physics Symposium and Workshop on Polarized Electron Sources and Polarimeters*, edited by Y. I. Makdisi, A. U. Luccio, and W. W. MacKay

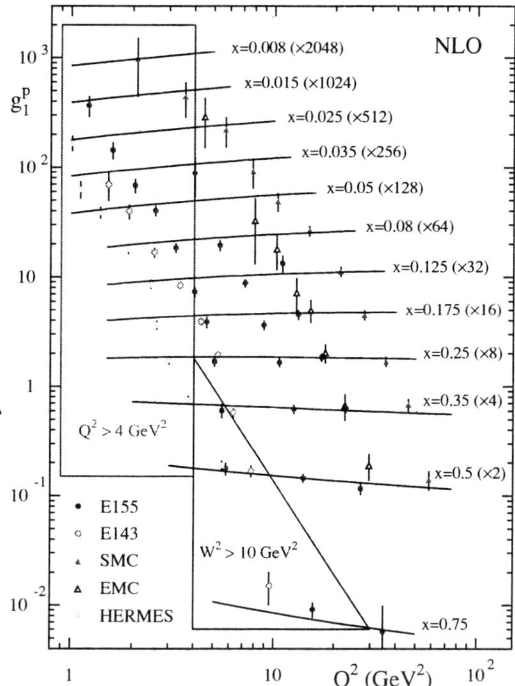

FIGURE 1. Available information on $g_1(x,Q^2)$ as collected by fixed-target experiments [1] compared to results of a typical NLO QCD fit (solid lines). The indicated rectangular and triangular regions contain data which would not pass kinematical cuts of $Q^2 > 4\,\mathrm{GeV}^2$ and $W^2 > 10\,\mathrm{GeV}^2$, respectively, usually imposed in all fits to unpolarized DIS data.

of certain combinations of different quark flavors has emerged, and it became clear that quarks contribute only a small fraction to the proton's spin. However, there is still considerable lack of knowledge regarding the polarized gluon density Δg, which is basically unconstrained by present data, the separation of quark and antiquark densities and of different flavors, and the orbital angular momentum of quarks and gluons inside a nucleon. With the exception of orbital angular momentum RHIC can address all of these questions as will be demonstrated in the following [4].

There is also an important difficulty when analyzing polarized DIS data in terms of spin-dependent parton densities: compared to the unpolarized case the presently available kinematical coverage in x and Q^2 and the statistical precision of polarized DIS data are much more limited [1]. As a consequence, one is forced to include data into the fits from (x,Q^2)-regions where corresponding fits of unpolarized leading-twist parton densities start to break down, see Fig. 1. Data from RHIC, taken at "resolution" scales μ where perturbative QCD and the leading-twist approximation are supposed to work, can shed light on the possible size of unwanted higher-twist contributions in presently available sets of polarized parton distributions.

SPIN PHYSICS AT RHIC

Prerequisites

The QCD-improved parton model has been successfully applied to many high energy scattering processes. The predictive power of perturbative QCD follows from the universality of the parton distribution and fragmentation functions which is based on the factorization theorem. To be specific, let us consider the inclusive production of a hadron H, e.g., a pion, in longitudinally polarized pp collisions at a c.m.s. energy $\sqrt{S}$. The cross section can be written in a factorized form as a convolution of perturbatively calculable partonic cross sections $d\Delta\hat{\sigma}_{ab}^{c}$ describing the hard scattering $ab \to cX$ and appropriate combinations of parton densities $\Delta f_{a,b}$ and fragmentation functions D_c^H embodying the non-perturbative physics:

$$\frac{d\Delta\sigma^H}{d\Gamma} = \sum_{abc} \int dx_a\, dx_b\, dz\, \Delta f_a(x_a,\mu)\, \Delta f_b(x_b,\mu)\, \frac{d\Delta\hat{\sigma}_{ab}^{c}}{d\Gamma}(x_a,x_b,z,S,\Gamma,\mu)\, D_c^H(z,\mu)\,. \quad (4)$$

Here, Γ stands for any appropriate set of kinematical variables like the transverse momentum p_T and/or rapidity y of the observed hadron. The D_c^H are the parton-to-hadron fragmentation functions. Their scale μ-dependence is governed by a set of equations very similar to (3). The factorization scale μ, introduced on the r.h.s. of (4), separates long- and short-distance phenomena. μ is completely arbitrary but usually chosen to be of the order of the scale characterizing the hard interaction, for instance p_T. Since the l.h.s. of (4) has to be independent of μ (and other theoretical conventions), any residual dependence of the r.h.s. on the actual choice of μ gives an indication of how well the theoretical calculation is under control and can be trusted. In particular, leading order (LO) estimates suffer from a strong, uncontrollable scale dependence and hence are not sufficient for comparing theory with data. Figure 2 shows a typical factorization scale dependence for various "high-p_T" processes and experiments as a function of p_T. Clearly, the situation is only acceptable at collider experiments where one can easily reach p_T values in excess of 5 GeV. p_T values of the order of 1-2 GeV, accessible at fixed-target experiments, are not sufficient to provide a large enough scale μ for performing perturbative calculations reliably. For simplicity we have not distinguished between renormalization and initial/final-state factorization scales in (4) which can be chosen differently.

In practice, spin experiments do not measure the polarized cross section $d\Delta\sigma/d\Gamma$ itself, but the longitudinal spin asymmetry, which is given by the ratio of the polarized and unpolarized cross sections, i.e., for our example above, Eq. (4), it reads

$$A_{\mathrm{LL}}^H \equiv \frac{d\Delta\sigma^H/d\Gamma}{d\sigma^H/d\Gamma}\quad . \quad (5)$$

The unpolarized cross section $d\sigma^H/d\Gamma$ is given by Eq. (4) with all polarized quantities replaced by their unpolarized counterparts. At RHIC one can also study doubly transverse spin asymmetries [4] but here we will focus on longitudinal polarization only.

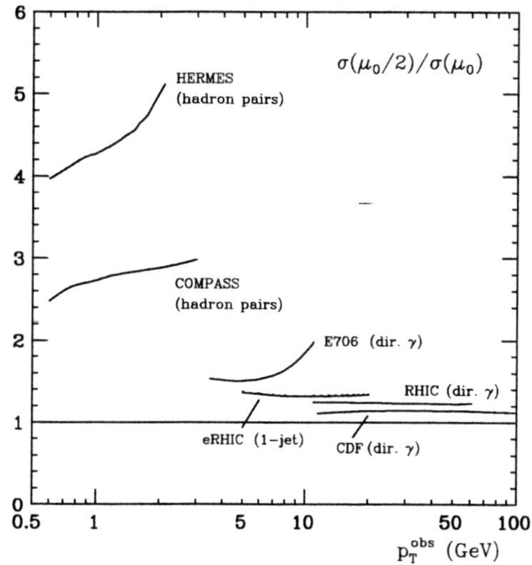

FIGURE 2. Typical factorization scale dependence for various "high-p_T" processes and experiments as a function of p_T. Shown is the cross section ratio for two choices of scale, p_T and $p_T/2$.

Accessing Δg

The main thrust of the RHIC spin program [4] is to pin down the so far elusive gluon helicity distributions $\Delta g(x,\mu)$. The strength of RHIC is the possibility to probe $\Delta g(x,\mu)$ in a variety of hard processes [4], in each case at sufficiently large p_T where perturbative QCD is expected to work. This not only allows to determine the x-shape of $\Delta g(x,\mu)$ for $x \gtrsim 0.01$ but also verifies the universality property of polarized parton densities for the first time. In the following we review the status of theoretical calculations for processes sensitive to Δg, experimental aspects can be found, e.g., in [5].

The "classical" tool for determining the gluon density is high-p_T prompt photon production due to the dominance of the LO Compton process, $qg \to \gamma q$. Exploiting this feature, both RHIC experiments, PHENIX and STAR, intend to use this process for a measurement of Δg. Apart from "direct" mechanisms like $qg \to \gamma q$, the photon can also be produced by a parton, scattered or created in a hard QCD reaction, which fragments into the photon. Such a contribution naturally arises in a QCD calculation from the necessity of factorizing final-state collinear singularities into a photon fragmentation function D_c^γ. However, since photons produced through fragmentation are always accompanied by hadronic debris, an "isolation cut" imposed on the photon signal in experiment, e.g., a "cone", strongly reduces such contributions to the cross section.

The NLO QCD corrections to the direct (non-fragmentation) processes have been calculated in [6] and lead to a much reduced factorization scale dependence as compared to LO estimates. In addition, Monte Carlo codes have been developed [7, 8], which allow to include various isolation criteria and to study also photon-plus-jet observables. The latter are relevant for Δg measurements planned at STAR [4, 5]. Since present comparisons between experiment and theory are not fully satisfactory in the unpolarized

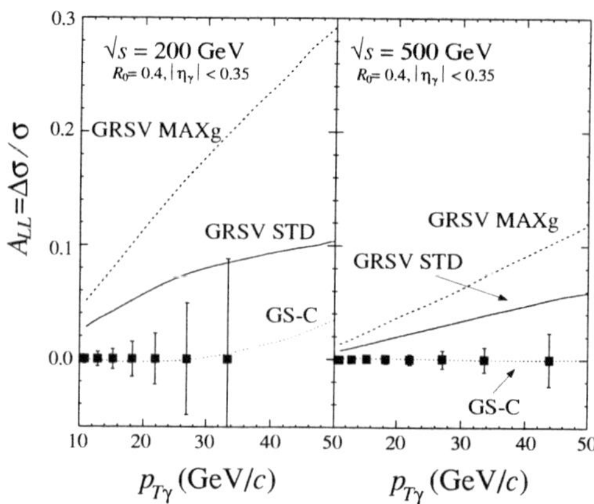

FIGURE 3. $A_{\rm LL}$ for prompt photon production in NLO QCD as a function of p_T for different sets of parton densities. The "error bars" indicate the expected statistical accuracy $\delta A_{\rm LL}$, Eq. (6), for the PHENIX experiment. Figure taken from [8].

case, in particular in the fixed-target regime, considerable efforts have been made to push calculations beyond the NLO of QCD by including resummations of large logarithms [9]. It is hence not unlikely that a better understanding of prompt photon production can be achieved soon. Figure 3 shows $A_{\rm LL}^{\gamma}$ as predicted by a NLO QCD calculation [8] as a function of the photon's transverse momentum p_T. The applied rapidity cut $|\eta| \leq 0.35$ matches the acceptance of the PHENIX detector. The important result is that the expected statistical errors $\delta A_{\rm LL}$ are considerably smaller than the changes in $A_{\rm LL}^{\gamma}$ due to different spin-dependent gluon densities over a wide range of p_T. Hence RHIC should be able to probe Δg in prompt photon production. $\delta A_{\rm LL}$ may be estimated by the formula

$$\delta A_{\rm LL} = \frac{1}{P^2\sqrt{\mathscr{L}\sigma_{\rm bin}}}\,, \qquad (6)$$

where P is the polarization of one beam, $\mathscr{L}$ the integrated luminosity of the pp collisions, and $\sigma_{\rm bin}$ the unpolarized cross section integrated over the p_T-bin for which the error is to be determined. Unless stated otherwise, $P = 0.7$ and $\mathscr{L} = 320\,(800)\ {\rm pb}^{-1}$ is used in Eq. (6) for pp collisions at $\sqrt{S} = 200\,(500)$ GeV [4].

Jets are another key-process to pin down Δg at RHIC: they are copiously produced at $\sqrt{S} = 500\,$GeV, even at high p_T, $15 \lesssim p_T \lesssim 50\,$GeV, and gluon-induced gg and qg processes are expected to dominate in accessible kinematical regimes. Due to limitations in the angular coverage, jet studies will be performed by STAR only. As jet surrogates, PHENIX can look for high-p_T leading hadrons, such as pions, whose production proceeds through the same partonic subprocesses as jet production. Hadrons have the advantage that they can be studied also at $\sqrt{S} = 200\,$GeV and down to lower values in p_T than jets as they do not require the observation of clearly structured "clusters" of particles (jets). Contrary to jet production, a fragmentation function has to be introduced into the theoretical framework, cf. Eq. (4), to take care of final-state collinear singularities.

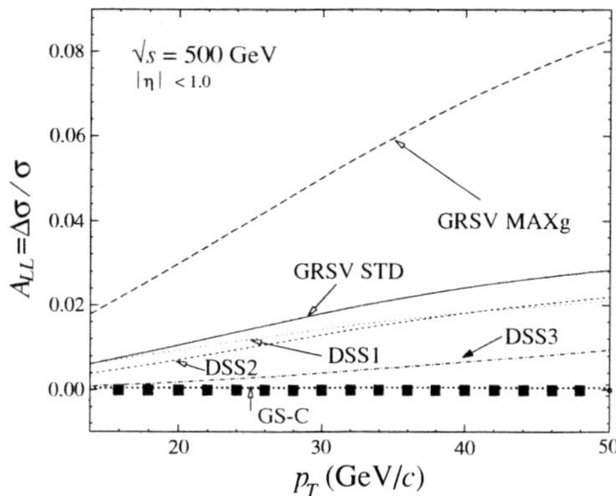

FIGURE 4. As in Fig. 3 but now for high-p_T jet production. The "error bars", Eq. (6), are for the STAR experiment taking into account its acceptance. Figure taken from [11].

In case of pion production, the D_c^π are, however, fairly well constrained by e^+e^- data. It should be also emphasized that in the unpolarized case, the comparison between NLO theory predictions with jet production data from the Tevatron is extremely successful. The same is true for first preliminary data on the p_T-spectrum for pions at $\sqrt{S} = 200\,\text{GeV}$ from PHENIX [10].

The NLO QCD corrections to polarized jet production are available as a Monte Carlo code [11]. Apart from a significant reduction of the scale dependence, they are also mandatory for realistically matching the procedures used in experiment in order to group final-state particles into jets. For single-inclusive high-p_T hadron production the task of computing the NLO corrections has been completed only very recently [12, 13]. Figure 4 shows A_{LL} for single-inclusive jet production at the NLO level as a function of the jet p_T. A cut in rapidity, $|\eta| \leq 1$, has been applied in order to match the acceptance of STAR. The asymmetries turn out to be smaller than for prompt photon production, but thanks to the much higher statistics one can again easily distinguish between different spin-dependent gluon densities. Results for single-inclusive π^0 production are presented in Fig. 5. Note that here the expected statistical accuracy refers to only a very moderate integrated luminosity and beam polarization as targeted for the upcoming run of RHIC. Even under these assumptions a first determination of Δg can be achieved.

The last process which exhibits a strong sensitivity to Δg is heavy flavor production. Here, the LO gluon-gluon fusion mechanism, $gg \to Q\bar{Q}$, dominates unless p_T becomes rather large. Unpolarized calculations have revealed that NLO QCD corrections are mandatory for a meaningful quantitative analysis. In the polarized case they have been computed recently in case of single-inclusive heavy quark production [14]. Again, one observes a strongly reduced scale dependence for charm and bottom production at RHIC energies. It turns out that the major theoretical uncertainty stems from the unknown precise values for the heavy quark masses [14]. Since the heavy quark mass already sets a large scale, one can perform calculations for small transverse momenta or even for total cross sections which, in principle, give access to the gluon density at smaller x-values than relevant for high-p_T jet, hadron or prompt photon production.

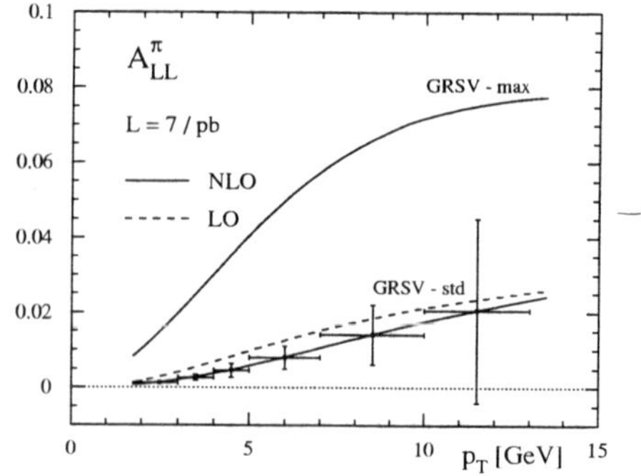

FIGURE 5. As in Fig. 3 but now for high-p_T π^0-production. Note that here the statistical "error bars" δA_{LL} have been estimated by assuming only $P = 0.4$ and $\mathscr{L} = 7\,\mathrm{pb}^{-1}$ in Eq. (6) which is a realistic target for the next RHIC run. Figure taken from [13].

Heavy flavors are not observed directly at RHIC but only through their decay products. Possible signatures for charm/bottom quarks at PHENIX are inclusive-muon or electron tags or μe-coincidences. The latter provide a much better c/b-separation which is an experimental problem. In addition, lepton detection at PHENIX is limited to $|y| \leq 0.35$ and $1.2 \leq |y| \leq 2.4$ for electrons and muons, respectively. Since heavy quark decays to leptons proceed through different channels and have multi-body kinematics, it is a non-trivial task to relate, e.g., experimentally observed p_T-distributions of decay muons to the calculated p_T-spectrum of the produced heavy quark. One possibility is to model the decay with the help of standard event generators like PYTHIA [15] by computing probabilities that a heavy quark with a certain (p_T, y) is actually seen within the PHENIX acceptance for a given decay mode. Figure 6 shows a prediction for the charm production asymmetry A_{LL} at PHENIX in NLO QCD for the inclusive-electron tag. The sensitivity to Δg is less pronounced than for the processes discussed above. It remains to be checked if heavy flavor production at RHIC can be used to extend the measurement of Δg towards smaller x-values.

Further Information on Δq and $\Delta\bar{q}$

Inclusive DIS data only provide information on the sum of quarks and antiquarks for each flavor, i.e., $\Delta q + \Delta\bar{q}$. At RHIC a separation of Δu, $\Delta\bar{u}$, Δd, and $\Delta\bar{d}$ can be achieved by studying $W^{\pm}$-boson production. Exploiting the parity-violating properties of $W^{\pm}$-bosons, it is sufficient to measure a single spin asymmetry, A_L^W, with only one of the colliding protons being longitudinally polarized. The idea is to study A_L^W as a function of the rapidity of the W, y_W, relative to the polarized proton [16]. In LO it is then easy to show [16, 4] that for W^+-production, $u\bar{d} \to W^+$, and large and positive (negative) y_W, A_L^W is sensitive to $\Delta u/u$ ($\Delta\bar{d}/\bar{d}$). Similarly, W^--production probes $\Delta d/d$ and $\Delta\bar{u}/\bar{u}$. The NLO

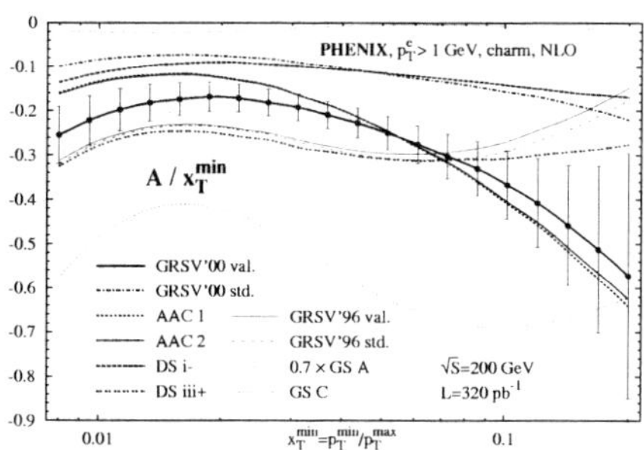

FIGURE 6. NLO single-inclusive charm production asymmetry (rescaled by $1/x_T^{\rm min}$) as a function of $x_T^{\rm min} \equiv p_T^{\rm min}/p_T^{\rm max}$ for different sets of parton densities. The "error bars", Eq. (6), are for the PHENIX experiment and include a detection efficiency for the channel $c \to eX$. Figure taken from [14].

QCD corrections for $A_{\rm L}$ as well as the factorization scale dependence are small [17]. Experimental complications [4] arise, however, from the fact that neither PHENIX nor STAR are hermetic, which considerably complicates the reconstruction of y_W. Therefore it is important to understand $A_{\rm L}$ on the decay-lepton level. Here, fully differential NLO cross sections are available as a MC code [18]. The anticipated sensitivity of PHENIX on the flavor decomposed quark and antiquark densities is illustrated in Fig. 7.

Semi-inclusive DIS measurements, $ep \to HX$, are another probe to separate quark and antiquark densities. HERMES has recently published first preliminary results [19]. The accessible x-range for the Δq and $\Delta\bar{q}$ densities is comparable to that of RHIC, see Fig. 7, but at scales $Q \simeq 1-2\,{\rm GeV}$ rather than M_W. The combination of both measurements can provide an important test of the QCD scale evolution for polarized parton densities and of the possible relevance of higher twist contributions at low scales.

Towards a Global Analysis of Upcoming Data

Having available at some point in the near future data on various different reactions, one needs to tackle the question of how to set up a "global QCD analysis" for spin-dependent parton densities. The strategy is in principle clear from the unpolarized case: an ansatz for the densities, Eq. (2), at some initial scale μ_0, given in terms of some functional form with a set of free parameters, is evolved, Eq. (3), to a scale μ relevant for a certain data point. A χ^2-value is assigned that represents the quality of the comparison of the theoretical calculation to the experimental point. The parameters are varied until eventually a global minimum in χ^2 is reached mutually for all data points. In practice, this approach is not fully viable since the numerical evaluations of the cross sections in NLO QCD are usually time-consuming as they require several tedious integrations. Hence the computing time for a QCD fit easily becomes excessive.

In the unpolarized case, the wealth of DIS data already provides a pretty good knowledge of the parton densities, and reasonable approximations can be made for the most time-consuming processes. For instance, one can absorb all NLO corrections into some pre-calculated "correction factors" K, and simply multiply them in each step of the fit

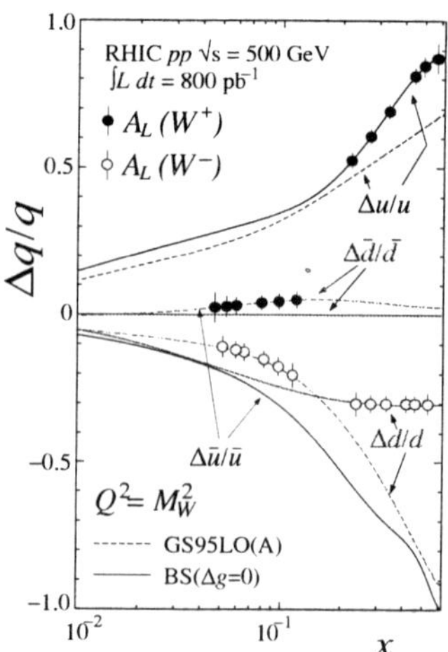

FIGURE 7. Expected statistical accuracy for $\Delta q/q$ from A_L overlayed on two sets of parton densities. The full [open] circles refer to $A_L(W^+)$ [$A_L(W^-)$]. Figure taken from [4].

to the LO approximation for the cross sections which can be evaluated much faster. In the polarized case, it is in general not at all clear whether such a strategy will work. Here, parton densities are known with *much* less accuracy so far. It is therefore not possible to use K-factors reliably. In addition, spin-dependent parton densities as well as partonic cross sections may oscillate, i.e., have zeros, in the kinematical regions of interest such that predictions at LO and the NLO can show marked differences. Clearly, in the polarized case the goal *must* be to find a way of implementing efficiently, and without approximations, the *exact* NLO expressions for all relevant hadronic cross sections. A very simple and straightforward method based on "double Mellin transformations" was proposed in [20]. Recently, its actual practicability and usefulness in a global QCD analysis has been demonstrated [21] in a case study based on fictitious prompt-γ data.

EXPLORING PHYSICS BEYOND THE STANDARD MODEL

Spin observables are also an interesting tool to uncover important new physics. One idea is to study single spin asymmetries A_L for large-p_T jets. In the standard model A_L can be only non-zero for parity-violating interactions, i.e., QCD-electroweak interference contributions, which are fairly small. The existence of new parity-violating interactions could lead to sizable modifications [22] of A_L. Possible candidates are new quark-quark contact interactions, characterized by a compositeness scale Λ. RHIC is surprisingly sensitive to quark substructure at the 2 TeV scale, and is competitive with the Tevatron despite the much lower c.m.s. energy [22]. Other candidates for new physics are possible new gauge bosons, e.g., a leptophobic Z'. Of course, high luminosity and precision as well as a good knowledge of polarized and unpolarized parton densities and of the standard model "background" are mandatory. For details, see [22, 4].

SUMMARY AND OUTLOOK

With first data from RHIC hopefully starting to roll in soon, we can address many open, long-standing questions in spin physics like the longitudinally polarized gluon density. With data from many different processes taken at high energies, where perturbative QCD should be at work, a first global analysis of spin-dependent parton densities will be possible. For a long time to come RHIC will provide the best source of information on polarized parton densities, certainly much improving our knowledge of the spin structure of the nucleon, and, perhaps, the next "spin surprise" is just round the corner. Future projects like the EIC [23], which is currently under scrutiny, would help to further deepen our understanding by probing aspects of spin physics not accessible in hadron-hadron collisions. The structure function g_1 at small x or the spin content of circularly polarized photons are just two examples.

REFERENCES

1. For a review on experimental efforts and results in polarized deep-inelastic scattering, see: E. Hughes and R. Voss, *Annu. Rev. Nucl. Part. Sci.* **49**, 303 (1999).
2. W.L. van Neerven and E.B. Zijlstra, *Nucl. Phys.* **B417**, 61 (1994), (E) **B426**, 245 (1994).
3. R. Mertig and W.L. van Neerven, *Z. Phys.* **C70**, 637 (1996); W. Vogelsang, *Phys. Rev.* **D54**, 2023 (1996); *Nucl. Phys.* **B475**, 47 (1996).
4. For a review on RHIC spin, see: G. Bunce, N. Saito, J. Soffer, and W. Vogelsang, *Annu. Rev. Nucl. Part. Sci.* **50**, 525 (2000).
5. L.C. Bland, these proceedings.
6. L.E. Gordon and W. Vogelsang, *Phys. Rev.* **D48**, 3136 (1993); A.P. Contogouris et al., *Phys. Lett.* **B304**, 329 (1993); *Phys. Rev.* **D48**, 4092 (1993).
7. L.E. Gordon, *Nucl. Phys.* **B501**, 197 (1997); L.E. Gordon and G.P. Ramsey, *Phys. Rev.* **D59**, 074018 (1999); S. Chang et al., *Phys. Rev.* **D58**, 074002 (1998).
8. S. Frixione and W. Vogelsang, *Nucl. Phys.* **B568**, 60 (2000).
9. E. Laenen et al., *Phys. Lett.* **B438**, 173 (1998); *Phys. Rev. Lett.* **84**, 4296 (2000); S. Catani et al., *JHEP* **9807**, 024 (1998); **9903**, 025 (1999).
10. H. Torii, talk presented at "Quark Matter 2002", Nantes, France, 2002.
11. D. de Florian et al., *Nucl. Phys.* **B539**, 455 (1999).
12. D. de Florian, `hep-ph/0210442`.
13. B. Jäger, A. Schäfer, M. Stratmann, and W. Vogelsang, `hep-ph/0211007`.
14. I. Bojak and M. Stratmann, `hep-ph/0112276` and work in progress.
15. T. Sjöstrand et al., `hep-ph/0108264`.
16. P. Chiappetta and J. Soffer, *Phys. Lett.* **B152**, 126 (1985); C. Bourrely and J. Soffer, *Phys. Lett.* **B314**, 132 (1993); *Nucl. Phys.* **B423**, 329 (1994); **B445**, 341 (1995).
17. A. Weber, *Nucl. Phys.* **B403**, 545 (1993); B. Kamal, *Phys. Rev.* **D57**, 6663 (1998); T. Gehrmann, *Nucl. Phys.* **B534**, 21 (1998).
18. P.M. Nadolsky, these proceedings [`hep-ph/0210190`].
19. M. Beckmann, `hep-ex/0210049`; A. Miller, these proceedings.
20. D.A. Kosower, *Nucl. Phys.* **B520**, 263 (1998).
21. M. Stratmann and W. Vogelsang, *Phys. Rev.* **D64**, 114007 (2001).
22. P. Taxil and J.-M. Virey, *Phys. Lett.* **B364**, 181 (1995); **B383**, 355 (1996); **B441**, 376 (1998); *Phys. Rev.* **D55**, 4480 (1997).
23. For information concerning the EIC project, see: `http://www.bnl.gov/eic`.

Acceleration of Polarized Protons at RHIC[1]

Haixin Huang

Collider-Accelerator Department, Brookhaven National Laboratory, Upton, NY 11973, USA

Abstract. Relativistic Heavy Ion Collider (RHIC) ended its second year of operation in January 2002 with five weeks of polarized proton collisions. Polarized protons were successfully injected in both RHIC rings and maintained polarization during acceleration up to 100 GeV per ring using two Siberian snakes in each ring. This is the first time that polarized protons have been accelerated to 100 GeV. The machine performance and accomplishments during the polarized proton run will be reviewed. The plans for the next polarized proton run will be outlined.

INTRODUCTION

The highly successful program of QCD and electroweak tests at the hardron colliders at CERN and FNAL has provided a wealth information on the Standard Model of particle physics. One aspect of our understanding which has not benefited from such experiments at high-energy colliders, however, is the area of spin physics, both spin structure of the proton itself and the spin-dependence of the fundamental interactions.

The RHIC center-of-mass energy range of 200 to 500 GeV is ideal for such studies in the sense that it is high enough for perturbative QCD to be applicable and low enough so that the typical momentum fraction of the valence quarks is about 0.1 or larger. This guarantees significant levels of parton polarizations. RHIC spin program utilizes two Siberian snakes [1] and four spin rotators in each ring to accelerate polarized protons to 250 GeV with 70% polarization. The first polarized proton collider run in RHIC took place from December 2001 to January 2002. Polarized beams were successfully accelerated to 100 GeV. They were stored and collided with a peak luminosity of about 1.5×10^{30} cm^{-2}s^{-1}.

SPIN DYNAMICS

In the presence of magnetic fields, the spin is governed by the Thomas-BMT (Bargmann, Michel, and Telegdi) Equation [2],

$$\frac{d\vec{S}}{dt} = \frac{e}{\gamma m}\vec{S}\times\left(G\gamma\vec{B}_{\perp} + (1+G)\vec{B}_{\parallel}\right), \tag{1}$$

[1] This work was supported by the Department of Energy of United States.

CP675, *Spin 2002: 15th Int'l. Spin Physics Symposium and Workshop on Polarized Electron Sources and Polarimeters*, edited by Y. I. Makdisi, A. U. Luccio, and W. W. MacKay

where $\vec{S}$ is the spin vector of a particle in the frame that moves with the particle's velocity, $\vec{B}_\perp$ and $\vec{B}_\parallel$ are defined in the laboratory rest frame with respect to the particle's velocity. $G = \frac{g-2}{2}$ is the gyromagnetic anomaly of the proton, and γ is the Lorentz factor. It is similar to the equation of motion for a particle moving in magnetic fields,

$$\frac{d\vec{v}}{dt} = \frac{e}{\gamma m}\vec{v} \times \left(\vec{B}_\perp\right), \tag{2}$$

where $\vec{v}$ is the velocity of the particle. By comparing the two equations, one can see that for a pure vertical field, spin rotates $G\gamma$ times faster than orbital motion. Eq. (1) also shows that the spin precession with a vertical field is independent of energy (γ) while the horizontal field has to linearly increase with energy to obtain the same spin precession. For general spin manipulation, longitudinal fields are used at low energy and transverse fields are used at high energy.

In a perfect planar synchrotron with vertically oriented guiding magnetic field, the spin vector of a proton beam precesses around the vertical axis $G\gamma$ times per orbital revolution. The number of precessions per revolution is called the spin tune ν_{sp} and is equal to $G\gamma$ in this case. In a real circular accelerator, the horizontal magnetic field that arises from various sources, such as the vertical closed orbit and the vertical betatron oscillation, kicks the spin vector away from the precessing around the vertical axis. Normally, this perturbation is small. However, when the spin precession frequency coincides with the frequency at which the spin vector gets kicked by the horizontal magnetic field, the spin vector is kicked away coherently and a spin depolarizing resonance is encountered. The spin resonance strength ε_k is defined as the Fourier amplitude of the spin perturbing field.

In general, a spin resonance is located at

$$\nu_{sp} = G\gamma = k \pm l\nu_y \pm m\nu_x \pm n\nu_{syn}, \tag{3}$$

where k, l, m and n are integers, ν_x and ν_y are horizontal and vertical betatron tunes, and ν_{syn} is the tune of the synchrotron oscillation. There are three main types of depolarizing resonances: imperfection resonances at $\nu_{sp} = k$, intrinsic resonances at $\nu_{sp} = l \pm \nu_y$ and coupling resonances at $\nu_{sp} = n \pm \nu_x$. The imperfection resonance is due to the vertical closed orbit error, and its strength is proportional to the size of the closed orbit distortion. The intrinsic resonance comes from the vertical betatron motion and are determined by the size of the vertical betatron oscillation. The larger the betatron oscillation, the stronger the resonance. The coupling resonance is caused by the vertical motion with horizontal betatron frequency due to linear coupling [3]. The strength of coupling resonance is proportional to the coupling coefficient in addition to the beam emittance. The acceleration of polarized proton beam in RHIC from 25 GeV/c to 250 GeV/c crosses numerous spin resonances. Fig. 1 shows the imperfection and intrinsic spin resonance strength as a function of the beam energy in RHIC. Because the RHIC lattice contains 3 superperiods and 27 effective FODO cells in each superperiod, the imperfection spin resonances at $G\gamma = k \times 81$ and intrinsic spin resonances at $G\gamma = k \times 81 \pm (\nu_y - 12)$ with $k = 3,5$ are strongly enhanced. The factor $(\nu_y - 12)$ is because there are total 12 insertion regions in RHIC and the phase advance of each insertion region

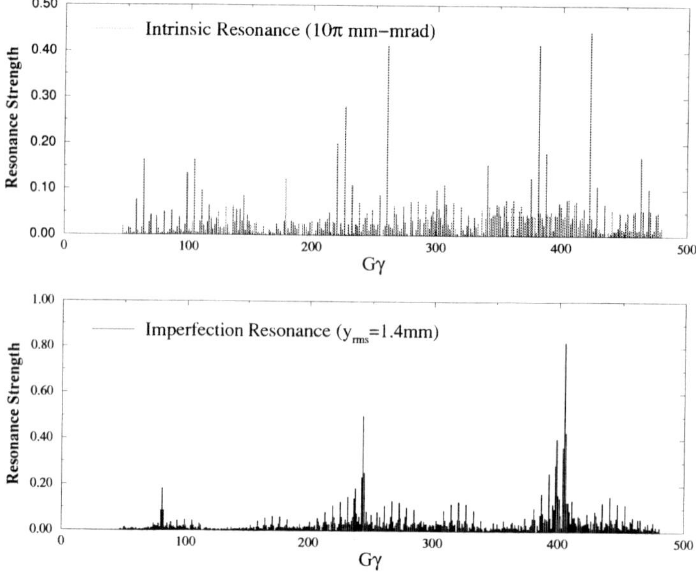

FIGURE 1. RHIC spin resonance spectrum. The top plot is for the intrinsic resonances and the bottom plot is for imperfection resonances. A lattice of RHIC 2000 polarized proton commissioning was used.

is π. The spin resonances with $k =$ even integers are weakened because arrangement of the focusing and defocusing quadrupoles on either side of each interaction point (IP) is antisymmetric.

When a polarized beam is uniformly accelerated through an isolated spin resonance, the final polarization P_f is related to the initial polarization P_i by the Froissart-Stora formula[4]

$$P_f = (2e^{-\pi|\varepsilon_k|^2/2\alpha} - 1)P_i, \tag{4}$$

where α is the rate of change of spin tune per radian of the orbit angle due to acceleration:

$$\alpha = \frac{d(G\gamma)}{d\theta}, \tag{5}$$

and θ is the orbital angle in the synchrotron. In the AGS, a few weak intrinsic resonances were not corrected with any scheme. The final polarization were affected by the value of α.

For a ring with a partial snake with strength s, the spin tune ν_{sp} is given by

$$\cos\pi\nu_{sp} = \cos\frac{s\pi}{2}\cos G\gamma\pi, \tag{6}$$

where $s = 1$ would correspond to a full snake which rotates the spin by 180°. When s=1, the spin tune is always 1/2 and energy independent. Thus, all imperfection, intrin-

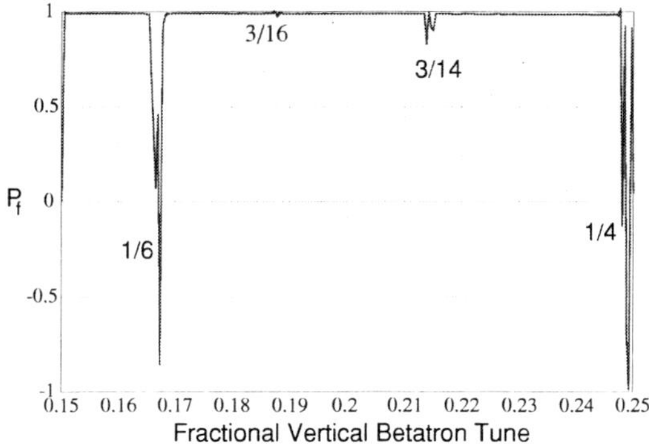

FIGURE 2. Vertical polarization after acceleration through a strong intrinsic resonance and a moderate imperfection resonance shown as a function of the vertical betatron tune.

sic and coupling resonance conditions can be avoided. However, when the spin resonance strength is large, a new class of spin-depolarizing resonance can occur. These resonances, due to coherent higher-order spin-perturbing kicks, are called snake resonances [5] and located at

$$\Delta\nu_y = \frac{k \pm \nu_{sp}}{n}, \tag{7}$$

where $\Delta\nu_y$ is the fractional part of vertical betatron tune, n and k are integers, and n is called the Snake resonance order. Fig. 2 shows the result of a simple spin tracking through an energy region (using RHIC ramp rate) with an intrinsic resonance strength of 0.5 and an imperfection resonance strength of 0.05. There are clearly regions of the betatron tunes that should be avoided. It should be noted that when coupling is present, Eq. (7) also applies to horizontal tune ν_x.

POLARIZED PROTONS IN RHIC INJECTOR

The Brookhaven polarized proton facility complex is shown schematically in Fig.3. The polarized H^- beam from the optically pumped polarized ion source (OPPIS) [6] was accelerated through a radio frequency quadrupole and the 200 MeV LINAC. The OPPIS source produced 10^{12} polarized protons per pulse. The beam polarization at 200 MeV was measured with elastic scattering from a carbon fiber target. During the run, the polarization measured by the 200 MeV polarimeter was about 70%. The beam was then strip-injected and accelerated in the AGS Booster up to 2.5 GeV or $G\gamma = 4.7$. The vertical betatron tune of the AGS Booster was chosen to be 4.9 in order to avoid crossing the intrinsic resonance $G\gamma = 0 + \nu_y$ in the Booster. The imperfection resonances at $G\gamma = 3,4$ were corrected by harmonic orbit correctors.

Only one of the twelve rf buckets in the AGS was filled, and the beam intensity varied between 1.3 -1.7$\times 10^{11}$ protons per fill. At the AGS, a 5% partial Siberian snake [7] that rotates the spin by 9° is sufficient to avoid depolarization from imperfection resonances up to the required RHIC transfer energy [8]. Full spin flip at the four strong intrinsic

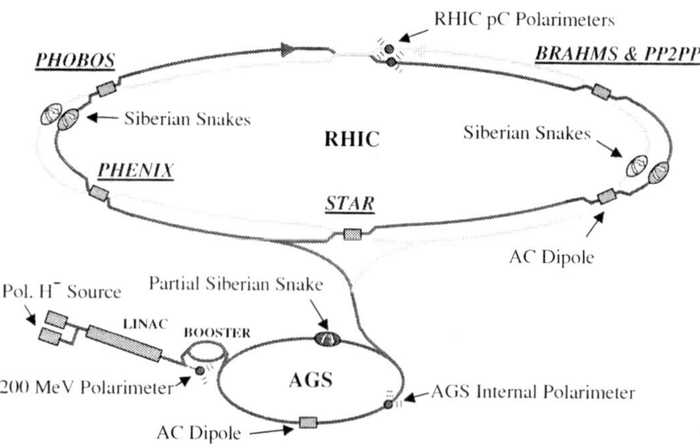

FIGURE 3. The Brookhaven polarized proton facility complex, which includes the OPPIS source, 200 MeV LINAC, the AGS Booster, the AGS, and RHIC. The clockwise ring is called Blue ring and the counter-clockwise ring is called Yellow ring. There are six IPs in RHIC. Two large detectors, STAR and PHENIX resides at six and eight o'clock IP, respectively. Two polarimeters are installed at straight section in 12 o'clock region. Two snakes are installed at three and nine o'clock sections in each ring.

resonances can be achieved with a strong artificial rf spin resonance excited coherently for the whole beam by firing an ac dipole [9]. The remaining polarization loss in the AGS is caused by coupling resonances and weak resonances. The polarized proton beam was accelerated up to $G\gamma = 46.5$ or 24.3 GeV. The normal main magnet ramp rate gives $\alpha = 4.8 \times 10^{-5}$. The ramp rate was much slower in last run which gave $\alpha = 2.4 \times 10^{-5}$, due to the fact that a back-up AGS main magnet power supply had to be used. This reduced the polarization level at the AGS extraction energy from 40% achieved before down to ~25%. Faster acceleration and a future, much stronger partial snake should eliminate depolarization in the AGS [10].

POLARIZED PROTONS IN RHIC

The basic construction unit for RHIC snake is a superconducting helical magnet producing a 4T dipole filed that rotates though 360° in a length of 2.4 meters. These magnets are assembled in group of four to build four Siberian snakes (two for each ring) for RHIC [11] [12]. The 9 cm diameter bore of the helical magnets can accommodate 3 cm orbit excursions at injection. Fig. 4 shows the orbit and spin trajectory through a RHIC snake. The superconducting helical dipoles were constructed at BNL using thin cable placed into helical grooves that have been milled into a thick-walled aluminum cylinder. With two snakes in each ring, the stable spin direction is vertical in RHIC and independent of beam energy.

The first RHIC polarized proton run lasted eight weeks, including commissioning and physics running. 60 bunches pattern was used in each ring (55 filled, 5 empty for an abort gap). Different alternating spin bit pattern were used for two rings to provide

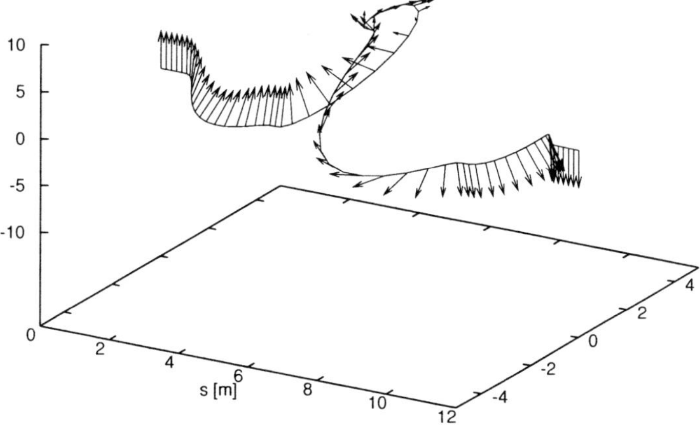

FIGURE 4. Orbit and spin tracking through the RHIC helical snake at $\gamma = 25$. The two axes without label are in the unit of centimeter. The spin tracking shows the reversal of the vertical polarization.

four different spin collision combinations. Specifically, the spin pattern for Blue ring is ↑↓↑↓ ....EEEEE, and the spin pattern for Yellow ring is ↑↑↓↓ ....EEEEE, where E means empty bunch. For the last six days, three unpolarized bunches were put into patterns at every twenty bunches. They were useful for bunch-by-bunch polarization study. The beam was injected into RHIC with 3 meter β^* lattice and accelerated up to 100 GeV without beta-squeeze. The total intensity of 4×10^{12} was reached. The typical beam emittance is 25 π mm-mrad in both transverse planes.

The fractional betatron tune space ranged between 0.20 and 0.225 for the horizontal tune and between 0.225 to 0.25 for the vertical. The vertical betatron tune was chosen to avoid 3/14 snake resonance(see Fig. 2). During the run it was found that when coupling was strong, the coupling snake resonance could also cause polarization loss. One indication of the strong coupling is shown in Fig. 5.

In RHIC the primary source of coupling comes from quadrupole rolls in the triplet quadrupoles at the six interaction regions. In addition, a small longitudinal field is introduced by each helical dipole snake. Much effort was devoted to compensate the rolls in the triplet quadrupoles through a system of local and global corrections outlined in Ref. [13]. These efforts produced some success at injection and flattop. However, problems during the acceleration ramp persisted since a dynamic correction technique has yet to be implemented. The problem of coupling was further enhanced since the fractional tune space during operation left little distance between the horizontal and vertical tunes. Spin tracking results in the Blue ring using the program SPINK [14] and including rolls in the triplet quadrupoles (without correction) show clearly the onset of the coupled snake resonances as the horizontal tune crosses the 3/14 snake resonance location in Fig. 6. In later RHIC fills, the horizontal betatron tune was moved away from the coupling snake resonance and polarization is preserved through the ramp.

Close attention was also paid to the orbit. The imperfection resonance strength is proportional to the rms orbit error. Fig. 7 shows an orbit history in the ramp for fill 2244.

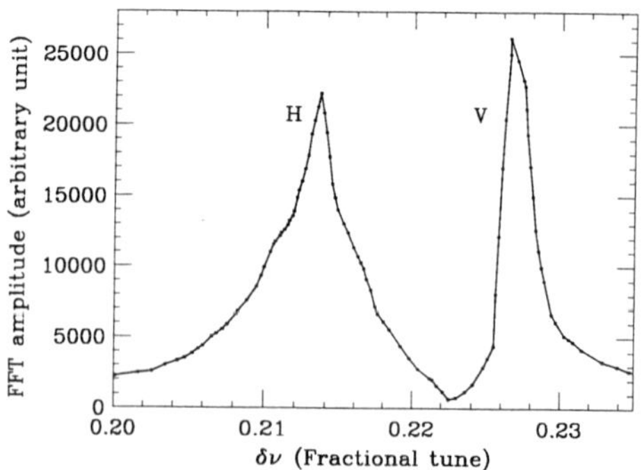

FIGURE 5. Snapshot of the FFT spectrum from the RHIC tune measurement along the ramp in the Yellow ring. For this particular ramp, the polarization preservation efficiency dropped to 20%. This snapshot was taken when crossing one strong resonance location ($G\gamma \sim 100$) along the RHIC ramp with the tunemeter kicker fired only in one plane. The double peak is the clear evidence of strong coupling, where the horizontal tune (lower peak) is clearly overlapping the 3/14 snake resonance.

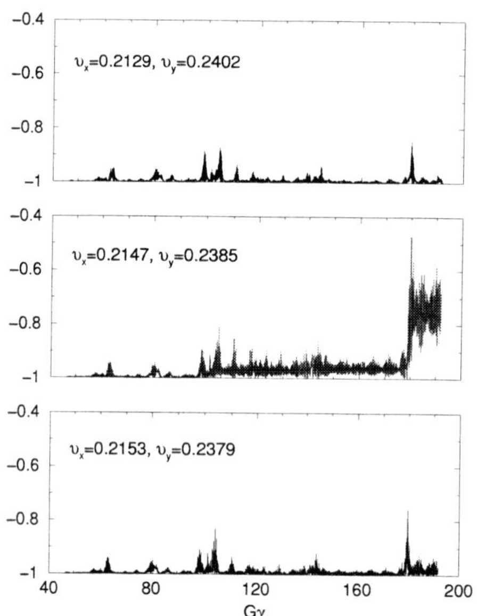

FIGURE 6. Spin tracking results for strongly coupled Blue ring with an emittance of 25 π mm-mrad and y_{rms}=0.6 mm. The three plots show polarization versus $G\gamma$ with the fractional part of horizontal betatron tune near 3/14. The middle plot shows depolarization due to the coupling snake resonance at 3/14.

In this fill, both rms orbit errors for vertical and horizontal orbits were under 1 mm.

Polarization was measured in RHIC at injection and store with RHIC polarimeters (one in each ring). The details of the polarimeters are given in Ref. [15]. Due to the 5% partial snake in the AGS and several vertical bends in the AGS to RHIC transfer

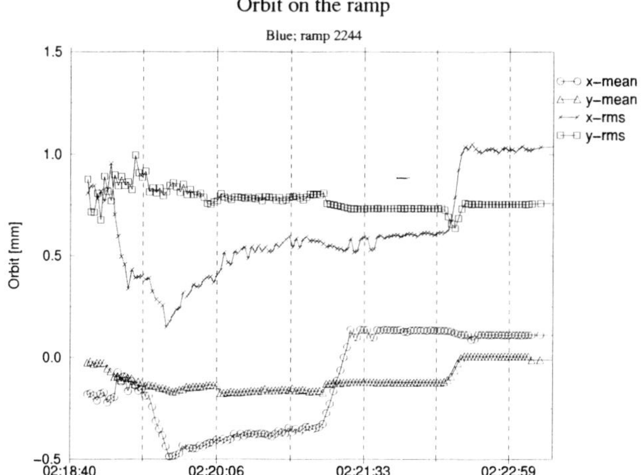

FIGURE 7. RHIC orbit in Blue ring for fill 2244 vs time in the ramp.

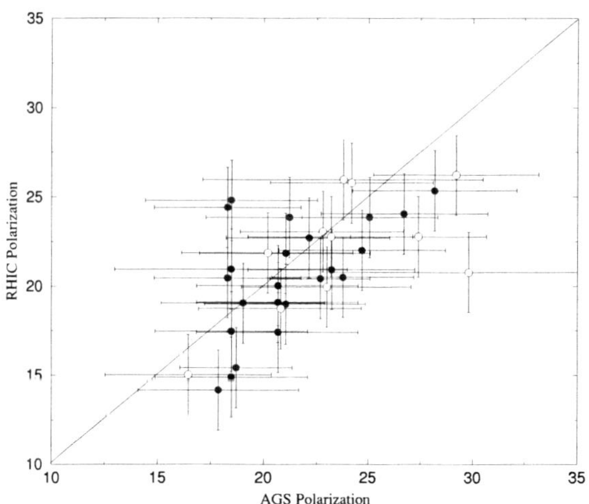

FIGURE 8. Comparison of polarization measured at the AGS polarimeter(at $G\gamma = 46.5$) and the RHIC injection for part of the run. The diagonal line is plotted for a ratio of one. The filled (open) dots are polarization measured at Blue (Yellow) ring injection.

line, the spin transmission efficiency of AGS to RHIC transfer line is less than 100% at $G\gamma = 46.5$ [16]. Nevertheless, the polarization measured at the RHIC injection tracked the AGS polarimeter measurement (at $G\gamma = 46.5$) as shown in Fig. 8. There were about 1000 polarization measurements over the eight weeks running time. With 55 bunches and total intensity of 3.5×10^{12} per ring, the measuring time was about 40 seconds at store to get 20 million events. Since beam intensities were higher at injection, the

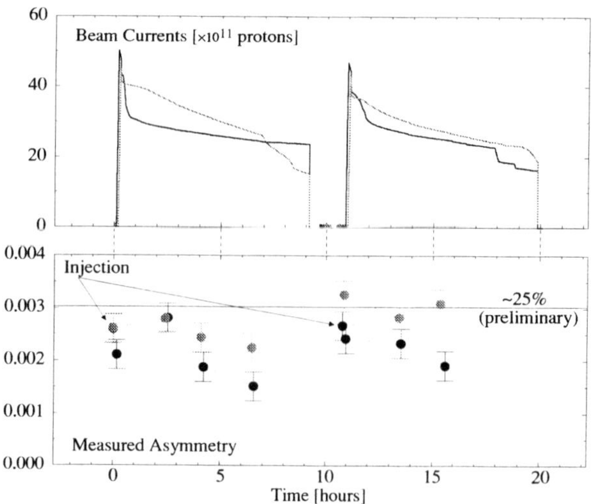

FIGURE 9. Beam intensity and measured asymmetry in the Blue and Yellow rings for two typical stores. The solid curves in the top plot are the beam intensities and the solid dots in the bottom plot are physics asymmetries. Dark (light) lines and symbols are for Blue (Yellow) ring. The asymmetries stayed as constant statistically.

measuring time was about the same. Although the analyzing power at 100 GeV for the RHIC polarimeter is unknown, it is expected to be similar at injection energy. Under this assumption the polarization measured at store was typically about 20-25%. One example is shown in Fig.9.

In summary, to preserve polarization in RHIC, the resonance strengths should be kept lower enough that the two snakes can overcome. A smaller vertical emittance is preferred and vertical orbit error should be as small as possible. The betatron tunes should be carefully set in the good tune window and monitored along the ramp and at store. In addition, attention should be paid to the coupling correction.

OUTLOOK

Spin Rotators are required at the IPs used by PHENIX and STAR to allow measurements of spin effects with longitudinally polarized protons. The spin rotators rotate the polarization from the vertical direction into the horizontal plane on one side of the IP and restore it to the vertical direction on the other side. Eight spin rotators have been installed in the RHIC after last run. They will be commissioned in next run.

Because the same bunches collide for a given experiment, periodically reversing the spin will reduce systematic errors for asymmetry measurements even further. A full spin flip can be induced by slowly sweeping the ac dipole frequency across the spin precession frequency. To achieve this, the spin precession tune needs to be tuned away from its nominal value 0.5. This can be achieved by adjusting the helical snakes' axis angles. The polarization reversal will take about a few seconds. A brief test has been done during last run and the results are very promising [17]. Same technique will also be used to determine how well the two snake axes are set.

A new vertical survey was done after the run and it revealed that the vertical distortion of magnets in the 12 o'clock area is in the order of several milimeters. Such a big misalignment increases the imperfection resonance strength significantly. It is our high priority to restore the alignment of dipole magnets and it will be done before next run.

The real time tune control system has been tested successfully last run for a few RHIC fills along the up ramp [18]. It is essential to have the tune control system operational next run to control betatron tunes along the ramp. Moreover, the system is crucial for down ramp to decelerate beam back to injection energy.

To calibrate the analyzing power of RHIC polarimeters at any energy above injection, the polarized hydrogen jet target will be needed but it will not be ready until 2004 [19]. An alternative way is to ramp down the energy and measure the asymmetry again at injection. If the asymmetry after down ramp is similar to the measurement before the up ramp, polarization is preserved. The analyzing power at storage energy can then be extracted from the asymmetries measured at injection and store.

Recently, an 11.4% partial Siberian snake was used to successfully accelerate polarized protons through a strong intrinsic depolarizing spin resonance at $0+\nu_y$ in the AGS [10]. No noticeable depolarization due to $0+\nu_y$ was observed. This opens up the possibility of using a 20% partial Siberian snake in the AGS to overcome all weak and strong intrinsic spin resonances. With a new strong partial snake in the AGS and higher polarization from the source, the desired 70% polarization could be reached at RHIC.

REFERENCES

1. Ya.S. Derbenev and A.M. Kondratenko, Part. Accel. **8**, 115 (1978).
2. L.H. Thomas, Philos. Mag. **3**, 1 (1927); V. Bargmann, L. Michel, and V.L. Telegdi, Phys. Rev. Lett. **2**. 435 (1959).
3. H. Huang, T. Roser, A. Luccio, Proc. of 1997 IEEE PAC, Vancouver, May, 1997, p.2538.
4. M. Froissart and R. Stora, Nucl. Instrum. Meth. **7**, 297(1960).
5. S.Y. Lee and S. Tepikian, Phys. Rev. Lett. **56**, 1635 (1986).
6. A. Zelenski, *et al.*, in *Proceedings of the 9th International Conference on Ion Sources*, Rev. Sci. Inst., Vol.73, No.2, p.888 (2002).
7. T. Roser, in *Proceedings of the 8th International Symposium on High-Energy Spin Physics*, Minneapolis, 1988, AIP Conf. Proc. No 187 (AIP, New York,1989), p.1442.
8. H. Huang, *et al.*, Phys. Rev. Lett. **73**, 2982 (1994).
9. M. Bai, *et al.*, Phys. Rev. Lett. **80**, 4673 (1998).
10. H. Huang, et al., these proceedings.
11. V.I. Ptitsyn and Yu.M. Shatunov, NIM **A398**, (1997)126.
12. E. Willen et al., Proc. of PAC99, New York, 3161(1999).
13. F. Pilat, et al., Proc. of EPAC2002, Paris, p. 1178.
14. A.U.Luccio *Trends in Collider Spin Physics* World Scientific, p. 235 (1997).
15. O. Jinnouchi, et al., these proceedings; I. Alekseev, et al., these proceedings.
16. W. MacKay, private communication.
17. M. Bai, et al., these proceedings.
18. C. Shuletheiss, et al., Proc. of EPAC2002, Paris, p. 2094.
19. A. Bravar, et al., these proceedings.

Results on DIS Exclusives and Generalized Parton Distributions

M. Vanderhaeghen

Institut für Kernphysik, Johannes Gutenberg-Universität, D-55099 Mainz, Germany

Abstract. We discuss how generalized parton distributions (GPDs) enter in a variety of hard exclusive processes such as deeply virtual Compton scattering (DVCS) and hard meson electroproduction reactions on the nucleon. We discuss the links between GPDs and elastic nucleon form factors as well as the information contained in the second moment of (generalized) parton distributions. We subsequently show some key observables which are sensitive to the various hadron structure aspects of the GPDs, and discuss their experimental status.

INTRODUCTION

Generalized parton distributions (GPDs), are universal non-perturbative objects entering the description of hard exclusive electroproduction processes (see Refs. [1, 2, 3] for reviews and references). In leading twist there are four GPDs for the nucleon, i.e. H, E, $\tilde{H}$ and $\tilde{E}$, which are defined for each quark flavor (u, d, s). These GPDs depend upon the different longitudinal momentum fractions $x+\xi$ ($x-\xi$) of the initial (final) quark and upon the overall momentum transfer $t=\Delta^2$ to the nucleon (see Fig. 1). As the momentum fractions of initial and final quarks are different, one accesses quark momentum correlations in the nucleon. Furthermore, if one of the quark momentum fractions is negative, it represents an antiquark and consequently one may investigate $q\bar{q}$ configurations in the nucleon. Therefore, these functions contain a wealth of new nucleon structure information, generalizing the information obtained in inclusive deep inelastic scattering.

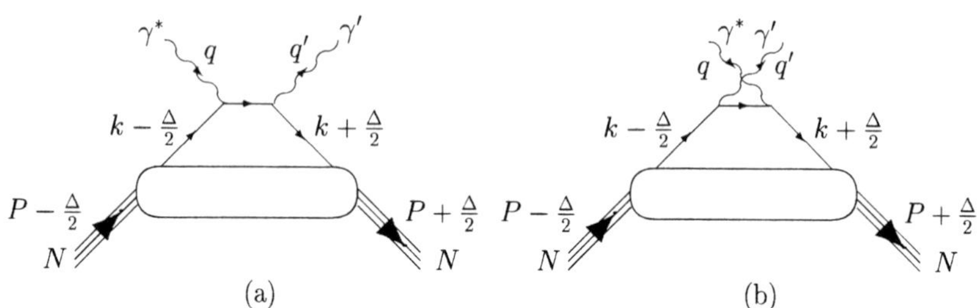

FIGURE 1. "Handbag" diagrams for the DVCS process, containing the GPDs.

In particular, the GPDs allow to access the fraction of the nucleon spin carried by the

CP675, *Spin 2002: 15th Int'l. Spin Physics Symposium and Workshop on Polarized Electron Sources and Polarimeters*, edited by Y. I. Makdisi, A. U. Luccio, and W. W. MacKay

quark total angular momentum (J^u,J^d, etc.), $\bar{q}q$ components of the nucleon wave function (in particular the D-term [4]), the strength of the skewedness effects in the GPDs (encoded in their ξ-dependence), the quark structure of $N \to N^*,\Delta$ transitions, flavor $SU(3)$ breaking effects, and others. Furthermore, it has been shown that by a Fourier transform of the t-dependence of GPDs, it is conceivable to access the distributions of parton in the transverse plane [5, 6].

NUCLEON ELECTROMAGNETIC FORM FACTORS

We start by discussing the t-dependence of the GPDs which is directly related to nucleon elastic form factors (FFs) through sum rules. In particular, the nucleon Dirac and Pauli FFs $F_1(t)$ and $F_2(t)$ can be calculated from the GPDs H and E through the following sum rules for each quark flavor ($q = u,d$)

$$F_1^q(t) = \int_{-1}^{+1} dx\, H^q(x,\xi,t), \quad F_2^q(t) = \int_{-1}^{+1} dx\, E^q(x,\xi,t)\,. \tag{1}$$

We can choose $\xi = 0$ in the previous equations, and model $H(x,0,t)$ and $E(x,0,t)$ subsequently. For the GPD $H(x,0,t)$, a plausible ansatz at low $-t$ is a Regge form as discussed in [3]. This leads to the following integrals to calculate the Dirac FFs for u- and d-quark flavors :

$$F_1^u(t) = \int_0^{+1} dx\, u_v(x)\, \frac{1}{x^{\alpha_1' t}}, \quad F_1^d(t) = \int_0^{+1} dx\, d_v(x)\, \frac{1}{x^{\alpha_1' t}}, \tag{2}$$

where $u_v(x)$ and $d_v(x)$ are the u- and d-quark valence distributions, and where α_1' is the slope of the leading Regge trajectory. The proton and neutron Dirac FFs then follow from

$$F_1^p(t) = e_u F_1^u(t) + e_d F_1^d(t), \quad F_1^n(t) = e_u F_1^d(t) + e_d F_1^u(t), \tag{3}$$

with $e_u = +2/3$ ($e_d = -1/3$) the u (d) quark charges respectively.

In the ansatz of Eq. (2), the Regge slope α_1' is the only free parameter, which can be determined from the proton Dirac radius $< r_1^2 >$, yielding $\alpha_1' \simeq 1.1$ GeV^{-2}. Such value is close to the expectation from Regge slopes for meson trajectories, therefore supporting the ansatz of Eq. (2).

To calculate the electric and magnetic nucleon FFs, one also needs to calculate the Pauli FF F_2. For F_2, we use an ansatz based on a valence quark distribution for the valence part of $E(x,0,t)$ entering in (1) as :

$$F_2^u(t) = \int_0^{+1} dx\, \kappa_u \frac{1}{2} u_v(x)\, \frac{1}{x^{\alpha_2' t}}, \quad F_2^d(t) = \int_0^{+1} dx\, \kappa_d d_v(x)\, \frac{1}{x^{\alpha_2' t}}, \tag{4}$$

where κ_u and κ_d are given by $\kappa_u = 2\,\kappa_p + \kappa_n$, and $\kappa_d = \kappa_p + 2\,\kappa_n$.

In Fig. 2, the predictions of the above Regge ansatz are shown for the nucleon FFs (taking $\alpha_1^{'} = \alpha_2^{'}$, supported by the universality of meson Regge slopes). For both proton and neutron magnetic FFs, the Regge parametrization catches the basic features of the data for $-t < 0.5$ GeV^2. It also reproduces the decreasing trend with $-t$ for the ratio of electric to magnetic proton FFs, and yields a remarkable good description for the neutron electric FF up to $-t \simeq 1$ GeV^2. At larger values of $-t$, the Regge form expectedly falls short of the data as one expects a transition to the perturbative behavior. For $-t > 1$ GeV^2, an overlap representation linking the nucleon Dirac FF to GPDs has been given [10, 11]. A topic for further study is to incorporate both small-t and large-t regimes in a unified parametrization, needed to perform the Fourier transform for the t-dependence of GPDs in order to map out the distribution of partons in the transverse plane.

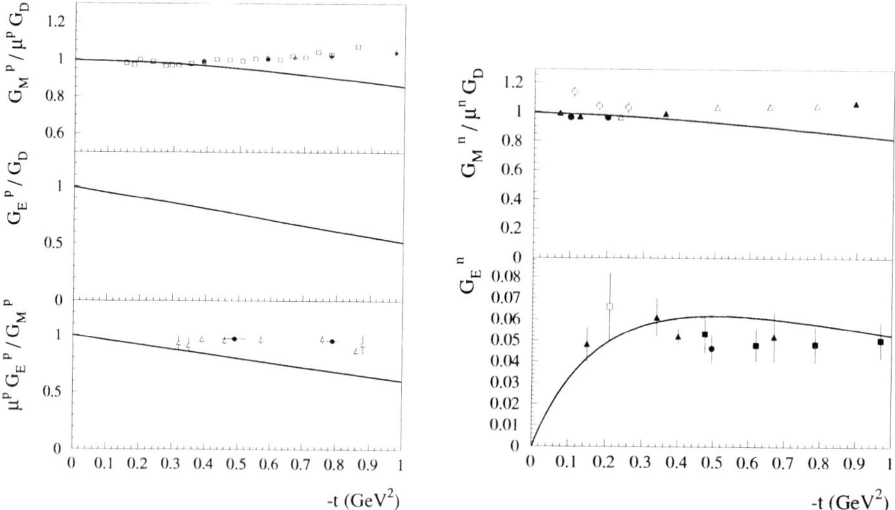

FIGURE 2. Left side : proton magnetic (upper panel) and electric (middle panel) form factors compared to the dipole form $G_D(t) = 1/(1-t/0.71)^2$, as well as the ratio of both form factors (lower panel). Right side : neutron magnetic (upper panel) and electric (lower panel) form factors. The curves correspond to the Regge ansatz of Eqs.(2) and (4) , with $\alpha_1^{'} = 1.1$ GeV^{-2}, $\alpha_2^{'} = 1.1$ GeV^{-2}.

ANGULAR MOMENTUM SUM RULE AND PROTON MOMENTUM FRACTIONS CARRIED BY VALENCE QUARKS

The second moments of the quark helicity independent GPDs are given by the nucleon form factors of the symmetric energy momentum tensor. At zero momentum transfer ($t = 0$), this leads to a sum rule [12] for the fraction of the nucleon angular momentum

(J^q) carried by a quark of the flavor q :

$$J^q = \frac{1}{2}\int_{-1}^{1} dx\, x\, \{\, H^q(x,\xi,0) + E^q(x,\xi,0)\,\} = \frac{1}{2}\Delta\Sigma + L^q\,. \tag{5}$$

As the quark spin part $\Delta\Sigma$ is measured through polarized DIS experiments as $\Delta\Sigma \approx 30\%$, the knowledge of the second moment of the GPDs H and E provides an access to the quark orbital angular momentum (L^q).

A relation involving second moments of quark distributions has been proposed [3] for the ratio of proton to neutron anomalous magnetic moments :

$$\frac{\kappa^p}{\kappa^n} = -\frac{1}{2}\frac{4M_2^{d_{val}} + M_2^{u_{val}}}{M_2^{d_{val}} + M_2^{u_{val}}}\,, \tag{6}$$

in terms of the momentum fractions carried by valence u- and d-quarks :

$$M_2^{q_{val}} = \int_0^1 dx x q_{val}(x)\,. \tag{7}$$

In Fig. 3, the *rhs* of Eq. (6), is seen to be scale *independent* and the relation (6) is numerically verified, for all parton distributions, to an accuracy at the one percent level! It may be interesting to investigate this further within different nucleon structure models (see e.g. [15]).

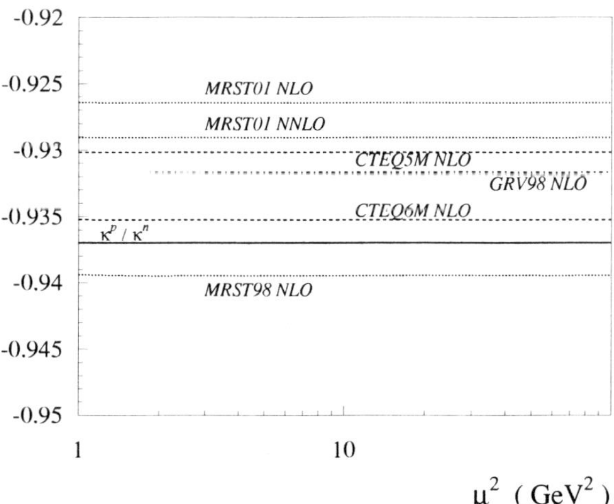

FIGURE 3. Scale dependence of the *rhs* of (6) for various parton distributions. Dotted curves : MRST98 NLO , MRST01 NLO , MRST01 NNLO [7]. Dashed curves : CTEQ5M NLO , CTEQ6M NLO [13]. Dashed-dotted curve : GRV98 NLO($\overline{\text{MS}}$) [14]. The constant solid curve shows the experimental value for κ^p/κ^n.

DVCS BEAM-HELICITY ASYMMETRY

We next turn to the DVCS observables and their dependence on the GPDs. At intermediate lepton beam energies, one can extract the imaginary part of the DVCS amplitude through the $\vec{e}p \to ep\gamma$ reaction with a polarized lepton beam, by measuring the out-of-plane angular dependence (in the angle ϕ) of the produced photon [16]. It was found in Refs. [17, 18] that the resulting electron single spin asymmetry (SSA)

$$\mathscr{A}^{\rm SSA} = \frac{\sigma_{e,h=+1/2} - \sigma_{e,h=-1/2}}{\sigma_{e,h=+1/2} + \sigma_{e,h=-1/2}}, \tag{8}$$

with $\sigma_{e,h}$ the cross section for an electron of helicity h, can be sizeable for HERMES (E_e = 27 GeV) and JLab (E_e = 4 - 11 GeV) beam energies. The SSA for the $\vec{e}p \to ep\gamma$ reaction has recently been measured in pioneering experiments at HERMES [19] and JLab/CLAS [20], as shown in Fig. 4. They display already at the accessible values of $Q^2 \simeq 1 - 2.5$ GeV^2 predominantly a $\sin\phi$ dependence, indicating a dominance of the twist-2 DVCS amplitude. Furthermore, the observed magnitude is in good agreement with the theoretical calculations [21, 22] in terms of GPDs. Once the leading order mechanism is confirmed by experiment, the measured helicity difference is directly proportional to the GPDs along the line $x = \xi$.

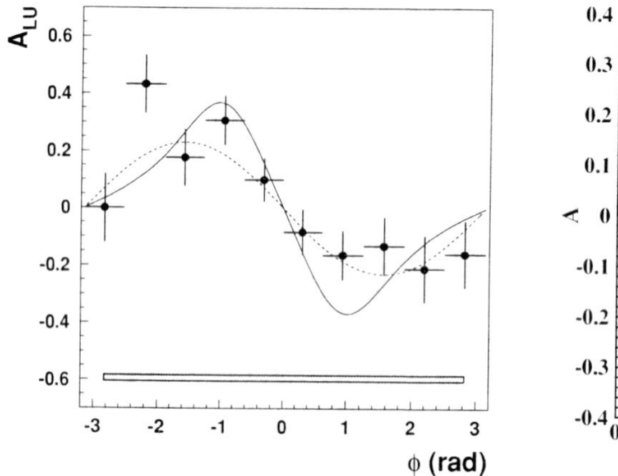

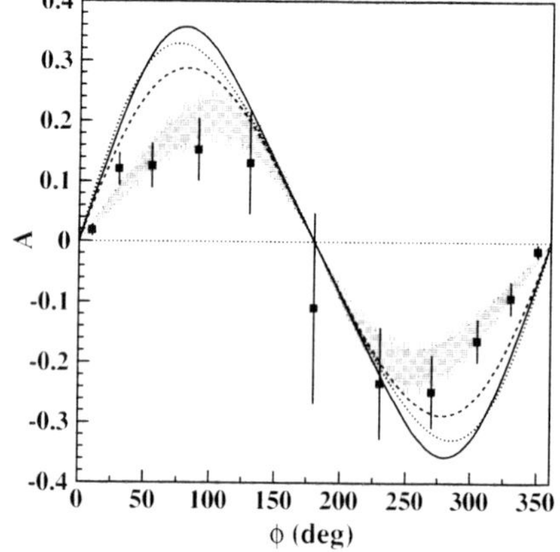

FIGURE 4. The DVCS beam helicity asymmetry as measured at HERMES [19] (left) and JLab/CLAS [20] (right). Full curves : twist-2 + twist-3 predictions of Ref. [21].

Dedicated experiments to measure the SSA with improved accuracy in a large kinematic range are already planned and underway both at JLab and HERMES.

DVCS BEAM-CHARGE ASYMMETRY

Besides the beam-helicity asymmetry for the $\vec{e}p \to ep\gamma$ reaction, which accesses the imaginary part of the DVCS amplitude, one gets access to the real part of the DVCS amplitude by measuring both $e^+p \to e^+p\gamma$ and $e^-p \to e^-p\gamma$ processes. In those reactions, besides the mechanism where the photon originates from a quark (handbag diagrams of Fig. 1), the photon can also be emitted by the lepton lines, in the so-called Bethe-Heitler (BH) process. However, in the difference $\sigma_{e^+} - \sigma_{e^-}$, the BH drops out, and one measures the real part of the BH-DVCS interference [23]

$$\sigma_{e^+} - \sigma_{e^-} \sim \Re\left[T^{BH} T^{DVCS*}\right] , \tag{9}$$

which is sensitive to the GPDs away from the line $x = \xi$.

It has been shown in [21] that this beam-charge asymmetry (BCA) gets a sizeable contribution from the D-term. The latter encodes $q\bar{q}$ scalar-isoscalar correlations in the nucleon as shown in Fig. 5, and has been estimated in the chiral quark soliton model [24].

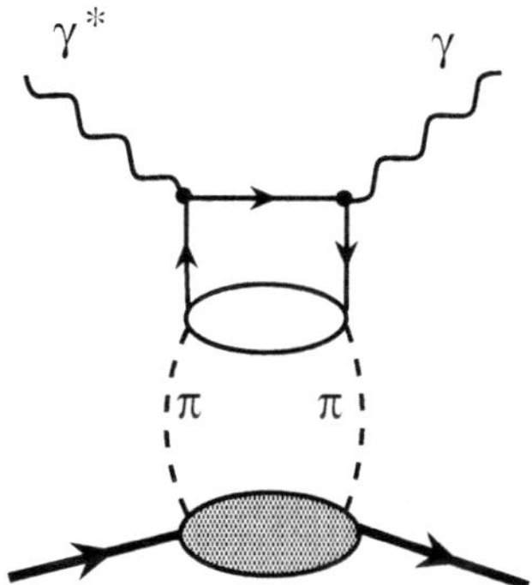

FIGURE 5. Model contribution to the D-term entering the GPDs H and E.

The DVCS BCA has been accessed experimentally at HERMES [25], and the preliminary data are shown in Fig. 6, together with the theoretical predictions. The measured asymmetry shows a $\cos\phi$ dependence with magnitude $\sim 0.10 - 0.15$, and favors the calculations which include the D-term. This opens up the perspective to study systematically (mesonic) $q\bar{q}$ components in the nucleon. Further measurements, with improved statistics, of the DVCS BCA are planned at HERMES.

DOUBLE DEEPLY VIRTUAL COMPTON SCATTERING

In the DVCS observables, as discussed above, the GPDs enter in general in convolution integrals over the average quark momentum fraction x, and only ξ can be accessed experimentally. A particular exception is when one measures an observable proportional

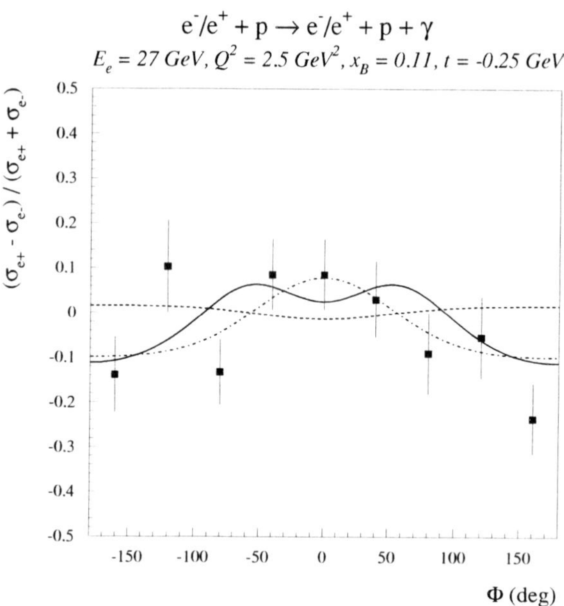

FIGURE 6. The DVCS beam-charge asymmetry with preliminary HERMES data [25]. Theoretical predictions from [21]. Dashed-dotted (dashed) curves : twist-2 DVCS with (without) D-term. Full curve : twist-3 effects in addition to the D-term.

to the imaginary part of the amplitude, such as discussed for the beam helicity asymmetry in DVCS. Then, one actually measures directly the GPDs at some specific point, $x = \xi$, which is certainly an important gain of information but clearly not sufficient to map out the GPDs independently in both quark momentum fractions. In absence of any model-independent "deconvolution" procedure at this moment, existing analyses of DVCS experiments have to rely on some global model fitting procedure.

The double DVCS (DDVCS) process, i.e. the scattering of a spacelike virtual photon from the nucleon with the production of a virtual photon in the final state, i.e. the $lp \to lpe^+e^-$ reaction, provides a way around this problem of principle. Compared to the DVCS process with a real photon in the final state, the virtuality of the final photon in DDVCS yields an additional lever arm, which allows to vary both quark momenta x and ξ independently [26]. In particular, the imaginary part of the DDVCS amplitude maps out the GPD where its first argument lies in the range $x < \xi$. Although one does not access the whole range in x, clearly, the gain of information on the GPDs is tremendous as no deconvolution is involved to access this region of the GPDs. Furthermore, $x < \xi$ is just the range where the GPDs contain wholly new information on mesonic ($q\bar{q}$) components of the nucleon, which is absent in the forward limit (where $\xi = 0$). However to construct sum rules, one also needs information in the region $x > \xi$. To access the range $x > \xi$ one would need two spacelike virtual photons, necessitating to select the two-photon exchange process in elastic electron nucleon scattering.

In Fig. 7,the dependence of the estimated cross section and SSA for the $ep \to epe^+e^-$ process is shown [26] as function of the virtuality $q'^2$of the outgoing lepton pair, in kinematics accessible at JLab. As the twist-2 SSA basically displays a $\sin\Phi$ structure, we show its value at $\Phi = 90^o$. As is seen from Fig. 7, the $ep \to epe^+e^-$ cross section scaled with the factor $N^{-1}q'^2$, with $N = (\alpha_{em}/4\pi) \cdot 4/3$ reduces to the $ep \to ep\gamma$ cross section when approaching the real photon point. Similarly, the SSA for the $ep \to epe^+e^-$ process reduces to the corresponding SSA for the $ep \to ep\gamma$ process. When going to larger virtualities q'^2, the DDVCS shows a growing deviation from the $1/q'^2$ behavior and the magnitude of the SSA decreases. The strong sensitivity of the SSA on q'^2, as seen from Fig. 7, should therefore allow to map out the GPDs in the range $x < \xi$. Although the cross sections arc small, their measurement seems feasible with a dedicated experiment at JLab and at a future high-energy, high-luminosity lepton facility.

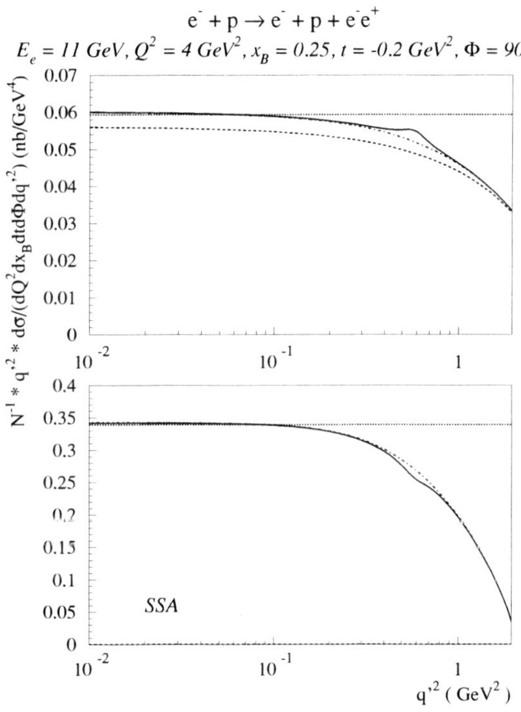

FIGURE 7. Cross section (upper panel) and SSA (lower panel) of the $ep \to epe^+e^-$ process as function of the e^+e^- virtuality q'^2. Dashed curves : BH; dashed-dotted curves : BH + DDVCS, full curves : BH + DDVCS + ρ_L^0, the latter being a background process. The dotted curves are the corresponding results for $ep \to ep\gamma$. The $ep \to epe^+e^-$ cross section is scaled with $N^{-1} \cdot q'^2$, in order to yield the $ep \to ep\gamma$ cross section in the limit $q'^2 \to 0$. Calculations from Ref. [26].

HARD MESON ELECTROPRODUCTION

The GPDs reflect the structure of the nucleon independently of the reaction which probes the nucleon. In this sense, they are universal quantities and can also be accessed, in different flavor combinations, through the hard exclusive electroproduction of mesons - $\pi^{0,\pm}, \eta, ..., \rho^{0,\pm}, \omega, \phi, ...$ - for which a QCD factorization proof was given [27]. This factorization theorem applies when the virtual photon is longitudinally polarized, which corresponds to a small size configuration compared to a transversely polarized photon.

For the longitudinal vector meson (V_L) electroproduction processes $\gamma_L^* + N \to V_L + N$ at large Q^2, the GPDs enter in different isospin combinations for $V_L = \rho_L^0, \rho_L^+, \omega_L$, allowing for a flavor decompostion of GDPs [17, 28].

An $\gamma_L^* + N \to V_L + N$ observable of particular interest is the transverse spin asymmetry (TSA) for a nucleon polarized perpendicular to the reaction plane [3]. The TSA is proportional to the imaginary part of the *interference* of the amplitudes which contain the GPDs H and E respectively. Therefore, the TSA provides a unique observable to extract the GPD E. Besides, one may expect that the theoretical uncertainties for the meson electroproduction cross sections largely disappear for the TSA, as it involves a ratio of cross sections, suggesting that the leading order expression is already accurate at accessible values of Q^2 (of a few GeV^2). Due to its linear dependence on the GPD E, the TSA for longitudinally polarized vector mesons opens up the perspective to extract the total angular momentum contributions J^u and J^d of the $u-$ and d-quarks to the proton spin. Due to the different u- and d-quark content of the vector mesons, the asymmetries for the ρ_L^0, ω_L and ρ_L^+ channels are sensitive to different combinations of J^u and J^d, with ρ_L^0 production sensitive to $2J^u + J^d$, ω_L to $2J^u - J^d$, and ρ_L^+ to $J^u - J^d$.

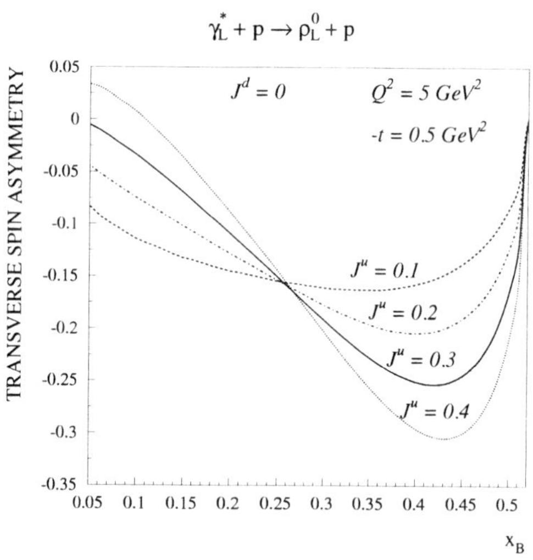

FIGURE 8. x_B dependence of the transverse spin asymmetry for the $\gamma_L^* \vec{p} \to \rho_L^0 p$ reaction. The estimates are given using the model of Ref. [3] for the GPDs E^u and E^d. The sensitivity is shown to different values of J^u (for a value $J^d = 0$).

In Fig. 8, the TSA for ρ_L^0 production is shown. One observes that it displays a pronounced sensitivity to J^u. It will therefore be very interesting to provide a first measurement of this asymmetry in the near future, for a transversely polarized target, such as it currently available at HERMES.

OUTLOOK

We have seen some very promising first glimpses of GPDs entering hard exclusive reactions at existing facilities. A dedicated program aiming at the extraction of the full physics potential contained in the GPDs will require a dedicated facility combining high luminosity and a good resolution.

REFERENCES

1. X. Ji, J.Phys.G **24**, 1181 (1998).
2. A.V. Radyushkin, in the Boris Ioffe Festschrift 'At the Frontier of Particle Physics / Handbook of QCD', edited by M. Shifman (World Scientific, Singapore, 2001).
3. K. Goeke, M.V. Polyakov, M. Vanderhaeghen, Prog.Part.Nucl.Phys. **47**, 401 (2001).
4. M.V. Polyakov, and C. Weiss, Phys.Rev.D **60**, 114017 (1999).
5. M. Burkardt, Phys.Rev.D **62**, 071503 (R) (2000); hep-ph/0207047.
6. M. Diehl, Eur.Phys.J. **C25**, 223 (2002).
7. A.D. Martin, R.G. Roberts, W.J. Stirling, R.S. Thorne, Phys.Lett.B **531**, 216 (2002).
8. M.K. Jones *et al.*, Phys.Rev.Lett. **84**, 1398 (2000).
9. O. Gayou *et al.*, Phys.Rev.Lett. **88**, 092301 (2002).
10. A.V. Radyushkin, Phys.Rev.D **58**, 114008 (1998).
11. M. Diehl, T. Feldmann, R. Jakob, and P. Kroll, Eur.Phys.J. **C8**, 409 (1999).
12. X. Ji, Phys.Rev.Lett. **78**, 610 (1997).
13. J. Pumplin *et al.*, JHEP **0207**, 012 (2002).
14. M. Glück, E. Reya, and A. Vogt, Eur.Phys.J. **C5**, 461 (1998).
15. M.M. Giannini, E. Santopinto, A. Vassallo, M. Vanderhaeghen, Phys. Lett. B **552**, 149 (2003).
16. P. Kroll, M. Schürmann and P.A.M. Guichon, Nucl.Phys. **A598**, 435 (1996).
17. M. Vanderhaeghen, P.A.M. Guichon, and M. Guidal, Phys.Rev.Lett. **80**, 5064 (1998); Phys.Rev.D **60**, 094017 (1999).
18. P.A.M. Guichon, M. Vanderhaeghen, Prog.Part.Nucl.Phys. **41**, 125 (1998).
19. A. Airapetian *et al.* (HERMES Collaboration), Phys.Rev.Lett. **87**, 182001 (2001).
20. S. Stepanyan *et al.* (CLAS Collaboration), Phys.Rev.Lett. **87**, 182002 (2001).
21. N.Kivel, M.V.Polyakov, M.Vanderhaeghen, Phys.Rev.D **63**, 114014 (2001).
22. A.V. Belitsky, D. Müller, and A. Kirchner, Nucl.Phys. **B629**, 323 (2002).
23. S.J. Brodsky, F.E. Close and J.F. Gunion, Phys.Rev.D **6**, 177 (1972).
24. V. Petrov *et al.*, Phys.Rev.D **57**, 4325 (1998).
25. F. Ellinghaus (on behalf of the HERMES Collaboration), hep-ex/0207029.
26. M. Guidal and M. Vanderhaeghen, Phys.Rev.Lett. **90**, 012001 (2003).
27. J.C. Collins, L.L. Frankfurt, and M. Strikman, Phys.Rev.D **56**, 2982 (1997).
28. L. Mankiewicz, G. Piller, and T. Weigl, Eur.Phys.J. **C5**, 119 (1998).

Parity Violating Electron Scattering

Krishna S. Kumar

Department of Physics, University of Massachusetts, Amherst, MA 01002

Abstract. We report on a mature experimental program to measure the parity violating asymmetry in the elastic scattering of longitudinally polarized electrons from unpolarized ^{1}H, ^{2}H, ^{4}He and ^{208}Pb targets. One focus is the measurement of the nucleon neutral weak form factors at intermediate four-momentum transfer $(0.1 < Q^2 < 1)$ (GeV/c)^2 which provide information about the impact of virtual strange quarks on the charge and current distributions inside nucleons. Another focus is the neutral current elastic amplitude at very low Q^2, which can provide stringent tests of the standard model and possess unique sensitivity to new physics at the TeV scale. Finally, the elastic neutral weak amplitude from scattering off a heavy spinless nucleus is very sensitive to the presence of a neutron skin. We report on recent technical progress in the design and scope of the experimental techniques. The physics implications of the published measurements are discussed and the current status and anticipated results experiments under construction are summarized.

INTRODUCTION

More than 40 years ago, soon after the discovery of parity violation in beta decay, Zel'dovich speculated that there might be an analogous parity violating neutral current interaction[1]. He noted that if such an interaction existed, then parity violation would be manifested in a scattering reaction due to the interference between the weak and electromagnetic amplitudes. He predicted that if one scatters longitudinally polarized electrons off unpolarized protons and flipped the sign of the beam polarization, the fractional difference in the cross-section would be:

$$A_{\rm LR} \equiv \frac{\sigma_R - \sigma_L}{\sigma_R + \sigma_L} \simeq \frac{|A_Z|}{|A_\gamma|} \simeq \frac{4\pi\alpha}{Q^2} \simeq 10^{-4}Q^2 \quad (1)$$

In the mid-seventies, parity violation in deep inelastic electron nucleon scattering was first observed at SLAC[2], from which the electron-quark weak neutral current coupling could be extracted. The measurement was an important test of the SU(2)$\times$U(1) gauge theory of electroweak interactions, and the extracted value of the electroweak mixing angle $\sin^2\theta_W$ matched the corresponding value obtained from neutral current neutrino scattering experiments.

Over the past 20 years, the experimental techniques employed to measure these tiny left-right asymmetries have been steadily refined such that statistical errors approaching 0.01 parts per million (ppm) and systematic errors of a few parts per billion (ppb) are possible[3]. This has spawned an important series of experiments that is the focus of the review.

CP675, *Spin 2002: 15th Int'l. Spin Physics Symposium and Workshop on Polarized Electron Sources and Polarimeters,* edited by Y. I. Makdisi, A. U. Luccio, and W. W. MacKay

PHYSICS MOTIVATION

Depending on the choice of target and kinematic variables, the measurement of the weak neutral current amplitude can probe very different physics. We elaborate on the main thrusts below.

Strangeness in Nucleons

Over the past decade, several experimental programs have focused on probing for the manifestations of strangeness in nucleon properties, such as mass, spin, momentum, magnetic moment and charge radius. A clean measurement of the contribution of strange quarks to any of these properties would be a dramatic proof of non-trivial dynamics of sea quarks inside nucleons, providing a new window into nonperturbative QCD.

There are some indications that the strange quarks might contribute to the mass (via $\pi-$proton scattering measurements) and the spin (via spin-dependent deep inelastic scattering measurements). While these experiments are sensitive to the strange scalar and axial vector matrix elements, parity violating elastic electron scattering can access vector strangeness matrix elements sensitive to the the contribution of strange quarks to nucleon charge and magnetic moment distributions[4, 5].

Elastic electron nucleon electromagnetic scattering is well described by the Dirac and Pauli (or equivalently the Sachs electric and magnetic) form factors. One can introduce equivalent neutral weak form factors that would be accessible in parity violating elastic electron scattering. If one assumes the validity of the standard model (weak isospin symmetry) and charge symmetry and that only three flavors are active, then one needs three elastic electroweak electron nucleon amplitudes to achieve flavor separation[6]. Thus, for a given value of Q^2, if the proton and neutron electromagnetic form factors are well measured, the measurement of the neutral weak form factors at the same value of Q^2 allows the extraction of the strange form factors.

The exact calculation of the strange form factors from QCD is currently difficult since it involves nonperturbative dynamics of sea quarks. Various model approaches are used, such as chiral perturbation theory, quark models, lattice gauge theory, Skyrme models and dispersion relations[7, 8]. Most models attempt to model the low Q^2 behaviour of the electric form factor G_E^s and the magnetic form factor G_M^s:

$$\mu_s \equiv G_{Ms}(0); \rho_s \equiv \frac{dG_{Es}}{d\tau}; \tau \equiv \frac{Q^2}{4M_p^2} \tag{2}$$

known as the strange magnetic moment μ_s and the strange radius ρ_s.

These models agree on neither the signs or magnitude of the strange form factors. Nevertheless, some guidence can be obtained from the size of the magnetic moment of the Λ baryon ($0.6\mu_N$) and the charge radius of the neutron ($\rho_n \sim 3$). Experiments must therefore achieve sensitivity significantly smaller than these values for the corresponding leading moments of the strange form factors. The full exploration further requires measurements over the range $0.1 < Q^2 < 1$ (GeV/c)^2, as well as forward and backward angle measurements off ^{1}H, ^{2}H and ^{4}He targets.

Neutron Skin of a Heavy Nucleus

In a heavy nucleus such as ^{208}Pb, the difference between neutron radius R_n and proton radius R_p is believed to be several percent. Analogous to the classic measurement of R_p via elastic electron electromagnetic scattering, R_n can-be measured via elastic electroweak scattering[9]. Experimentally, there is some controversy as to how well R_n is known[10]; the best guess is $\sim 5\%$.

The parity violating asymmetry in elastic scattering off a heavy spinless nucleus is proportional to the ratio of the neutron to proton form factors since the weak neutral current coupling of protons is much smaller than that of neutrons. A precise measurement of R_n can have impact on nuclear theory, atomic parity violation and neutron star structure. Relativistic mean field models tend to favor larger neutron skins than nonrelativistic models because of a larger symmetry energy. The asymmetry measurement at an optimal value of Q^2 can pin down R_n and help eliminate an entire class of models. Knowledge of R_n at the 1% level can reduce the uncertainty in atomic structure of heavy isotopes that can cloud the interpretation of atomic parity violation measurements that test the Standard Model at the level of radiative corrections[11]. Finally, a precision R_n measurement improves our understanding of the equation of state of neutron rich matter, which are important for constraining the structure of neutron stars[12].

Physics Beyond the Standard Model

The electroweak interaction has been tested to high precision at high energy colliders, especially by measurements of the properties of the W and Z bosons and their couplings to leptons and quarks. While all data are consistent with the standard model, experiments continue to probe for the indirect effects of new physics at the TeV scale by making more and more precise measurements of electroweak parameters[13].

Weak neutral current interactions at $Q^2 \ll M_Z^2$ can probe for heavy Z' bosons or leptoquarks whose effects might be highly suppressed in measurements on the Z pole. Since Z pole measurements are imaginary, there are no interference terms with new, real amplitudes. At low Q^2 on the other hand, interference effects might be measurable if sufficient accuracy is achieved[14]. The goal of low energy neutral current measurements are to reach a sensitivity on contact interactions at the 10 TeV level, similar sensitivity to that of the highest energy e^+e^- collider in operation, LEP200.

The NuTeV collaboration at Fermilab has reported a 3σ discrepancey with the standard model in deep inelastic neutrino-nucleus scattering[15]. A 2σ descrepancy has been reported in the measurement of the parity-violating amplitude for the 6S-7S transition in atomic Cesium[16, 17]. In parity violating electron scattering, two experiments are being pursued, one studying electron-electron scattering and one studying electron-proton scattering,, each potentially more sensitive to new physics than the above two published measurements.

TABLE 1. Summary of Recent and Planned Experiments

Experiment	Target	$\theta_{electron}$	Q^2 (GeV/c)^2	A_{LR} (ppm)	Physics Sensitivity	Status
SAMPLE	^{1}H	150°	0.1	8.0	$\mu_s + 0.4G_A^Z$	Published
HAPPEX	^{1}H	12.5°	0.47	15.0	$G_E^s + 0.39G_M^s$	Published
SAMPLEII	^{2}H	150°	0.1	3.0	$\mu_s + 2.0G_A^Z$	Published
SAMPLEIII	^{2}H	150°	0.04	1.0	$\mu_s + 3.0G_A^Z$	Completed
A4	^{1}H	35°	0.1 − 0.25	1.5 − 6.0	$G_E^s + \alpha G_M^s$	running
E158	^{1}H	5 mrad	0.02	0.2	$\sin^2\theta_W$	running
HAPPEX-H	^{1}H	6°	0.11	1.5	$\rho_s + \mu_p\mu_s$	2003
HAPPEX-He	^{4}He	6°	0.11	10.0	ρ_s	2003
G0	^{1}H	10°	0.3 − 0.8	2.0 − 10.0	$G_E^s + \alpha G_M^s$	2003
Lead	^{208}Pb	6°	0.01	0.5	$R_p - R_n$	2004
G0	^{1}H	110°	0.3 − 0.8	3.0 − 10.0	G_M^s	2004-7
G0	^{2}H	110°	0.3 − 0.8	3.0 − 10.0	G_A^Z	2004-7
A4	^{1}H	145°	0.23 − 0.5	2.0 − 6.0	G_M^s	future
A4	^{2}H	145°	0.23 − 0.5	2.0 − 6.0	G_A^Z	future
Helium-4	^{4}He	30°	0.6	10.0	G_E^s	future
Qweak	^{1}H	9°	0.02	0.2	$\sin^2\theta_W$	future

The Experimental Program

The physics thrusts described above have spawned a rich program of experiments laboratories around the world, including the MIT-Bates Linear Accelerator (Bates), the Thomas Jefferson National Accelerator Facility (Jlab), the Stanford Linear Accelerator Center (SLAC) and the Mainz Microtron (MAMI). The salient experimental parameters of these experiments are summarized in Table 1. As can be seen from the status column, very interesting experimental results are expected in the near future.

PUBLISHED MEASUREMENTS

SAMPLE

The SAMPLE experiment was carried out at Bates, with the principal goal of measuring μ_s. This was accomplished by measuring the parity violating amplitude in elastic electron proton scattering at backward angles at $Q^2 \sim 0.1$ (GeV/c)^2. The neutral current amplitude has two primary contributions, one from G_M^s and one from the axial form factor G_A. In order to separate these contributions, three separate runs were carried out: $E_{beam} = 200$ MeV/c^2 on hydrogen deuterium targets and $E_{beam} = 125$ MeV/c^2 on a deuterium target.

Elastic events at backward angles were detected by the Čerenkov light produced as they pass through air. This detection technique can provide a very large solid angle on very thick targets. At 200 MeV/c^2, scattered electrons at backward angles from inelastic scattering did not have very high efficiency for producing Čerenkov light, thus avoiding spurious asymmetry contributions from unknown inelastic amplitudes. The Čerenkov light was collected into 10 shielded phototubes via concave mirrors from a total solid angle of 0.7 steradian. The electrons impinged on a 40cm long liquid hydrogen target with an average beam current ranging from 40μA and 60μA.

For the hydrogen target, at an incident beam energy of 200 MeV/c^2, the measured asymmetry is[18] $A_p = -4.92 \pm 0.61 \pm 0.73$ ppm, while for the deuterium target with the same beam energy, the measured asymmetry is[19] $A_d = -6.79 \pm 0.64 \pm 0.55$ ppm, from which the following parameters are extracted at $Q^2 = 0.1$ (GeV/c)^2:

$$G_M^s = 0.14 \pm 0.29 \pm 0.31; G_A^e(T=1) = 0.22 \pm 0.45 \pm 0.39. \tag{3}$$

The final physics run with a beam energy of 125 MeV/c^2, in order to improve the accuracy on the measurement of G_A, has recently been completed and the final result is expected soon.

HAPPEX

The HAPPEX experiment was carried out in Hall A at Jlab, where the emphasis was on the measurement of the electron-proton weak neutral current amplitude at forward angles, where both electric and magnetic form factors contribute. A CW beam of 3.3 GeV/c^2 struck a 15 cm liquid hydrogen target. Scattered electrons at $\theta_{\text{lab}} \sim 12.5°$ were detected by a pair of 5.5 msr precision spectrometers, whose optics provides a mass focus that spatially separate the elastic events from all inelastic events. This allowed the use of integrating flux counters, thus making high rates feasible.

The integrating detector in each spectrometer consisted of a sandwich of five lead and lucite layers viewed by a single phototube. Separate low rate runs which tracked individual particles established that the backgrounds in the integrated signal were negligible. The polarized beam at Jlab is exceptionally quiet, ideal for parity violation experiments. The beam jitter in intensity, position, angle and energy are small enough so that the impact of false asymmetries due to helicity-correlated beam motion can be studied to a sensitivity of 10 ppb within a few days.

The beam polarization was 39% during the first half of data taking, while a strained GaAs photocathode provided 70% polarization during the second half of the experiment.The beam polarization was measured both with a conventional Moller polarimeter as well as a newly commissioned Compton polarimeter, which provided a non-invasive, continuous monitor of the beam polarization. The final combined result of all the data taking runs is[20] $A_{\text{LR}} = -14.60 \pm 0.94(\text{stat}) \pm 0.54(\text{syst})$ ppm. Combining the above result with data on nucleon electromagnetic form factors allows the extraction of a linear combination of electric and magnetic strange form factors:

$$(G_E^s + 0.392 G_M^s)/(G_M^{p\gamma}/\mu_p) = 0.091 \pm 0.054 \pm 0.039. \tag{4}$$

The result is shown normalized the proton magnetic form factor to underscore the sensitivity of the measurement. It implies that the linear combination of strange form factors that was probed is less than 10% of the proton electromagnetic form factor.

EXPERIMENTS IN PROGRESS

A4

The A4 experiment is taking place at MAMI and is measuring the weak neutral current amplitude in the elastic electron proton scattering at a scattering angle of $\theta_{\rm lab} \sim 35 \pm 5°$. Their first measurement is with an incident energy of 855 MeV/c^2, giving a Q^2 of 0.23 (GeV/c)^2. The unique feature of the experiment is a large acceptance lead fluoride (PbF_2) calorimeter.

A 20μA 70% longitudinally polarized beam scatters off a 20 cm liquid hydrogen target. Data are obtained in 20ms time windows locked to the 50 Hz power line freqnuency. A water Čerenkov luminosity monitor detects charged particles at forward angle and is used to normalize the data and study the effects of spurious asymmetries. The beam polarization is measured by a Compton backscattered laser polarimeter.

The calorimeter is designed to detect elastic electrons at a rate of 10 MHz amongst a background rate of 100 MHz made up of inelastic electrons, soft pions and photons. The calorimeter is made up of an array of 15 radiation length long PbF_2 crystals with a front surface of 25 × 25 mm. Light from showers in the crystal array are summed, integrated over 20 ns, digitized and histogrammed in real time by custom-built fast electronics. the system was carefully monitored to ensure that there was minimal dead time and non-linearity.

The experiment has accumulated more than 600 hours of data in 2001 and 2002. The error on the physics asymmetry is approaching 1 ppm, while the standard model expectation is -5.7 ppm. More work on systamatic corrections, estimation of the neutron electromagnetic form factors as well as standard model radiative corrections are required before any information on strange form form factors can be extracted.

The A4 collaboration plans to double their accumulated statistics at this kinematic point and then measure the neutral current amplitude at $Q^2 \sim 0.1$ (GeV/c)^2, which is achieved in the same detector configuration with a reduced incident beam energy. They then plan to turn the calorimeter apparatus around by 180° to look at electrons scattered at backward angles and measure the asymmetry at $Q^2 \sim 0.23$ and 0.5 (GeV/c)^2. This would complement the forward angle measurements made by A4 and HAPPEX.

E158

The E158 experiment is taking place at SLAC, using the longitudinally polarized 50 GeV/c^2 electron beam incident on a 1.5 m long liquid hydrogen target. The goal of the experiment is to measure the weak neutral current amplitude for electron-electron (Møller) scattering to a precision of about 8%, from which the most precise determination of the weak mixing angle $\sin\theta_{\rm W}$ can be extracted at $Q^2 \ll M_Z^2$.

In order to select charged particles at very forward angles, E158 employs a novel geometry where the primary and scattered beam travel through a foward angle spectrometer consisting of dipoles and quadrupoles. After the particles travel through a dipole "chicane", a fiducial collimator accepts particles with $4.5 < \theta_{\rm lab} < 8$ mrad. A series of four quadrupoles then exploits the energy angle correlation of Møller electrons to bring them to a ring focus downstream (accepting the full range of the azimuth), while separating them from electrons that scatter from target protons.

The Møller electrons are absorbed by a radiation hard calorimeter made up of layers of copper and fused silica fibers. The light from the fibers is collected by 60 shielded photomultiplier tubes and integrated over the duration of each beam pulse to measure the relative scattered flux. The raw asymmetry is small, of the order of 130 ppb and the goal is to measure it to an accuracy of 10 ppb.

The electron beam was produced by a novel gradient-doped GaAs photocathode that produced a charge 6×10^{11} electrons per pulse, at a pulse length of 200 ns, repetition rate of 120 Hz and 80% polarization. The experiment required unprecedented control and monitoring of the electron beam parameters. Each beam pulse was measured with a relative accuracy of 50 ppm in intensity and 2 μm in position using newly developed rf electronics[21]. This accuracy, coupled with novel integrating electronics, allowed the relative flux to be measured with an accuracy of 200 ppm for each pair of beam pulses.

The charge asymmetry was kept well below 1 ppm and was verified by independent devices to within 5 ppb. Much effort was expended in understanding the detailed properties of the laser beam that impinged on the photocathode and feedbacks were implemented to control the helicity-correlated differences in the intensity and position of the laser beam[22]. The electron beam position was controlled to be within 20nm on target and was verified to within 2 nm by redundant measurements. The high power liquid hydrogen target was shown to keep density fluctuations in the scattered flux below 50 ppm per pulse[23].

The experiment has completed two data taking runs, each accumulating about one-quarter of the total approved statistics, for a cumulative error on the asymmetry of the order of 15 ppb. This would provide an error on the weak mixing angle $\sin^2\theta_W \sim 0.0017$. It is anticipated that the remaining statistics will be collected in a final run in late 2003.

EXPERIMENTS UNDER CONSTRUCTION

G0

The G0 experiment at Jlab plans to make a complete set of forward and backward angle asymmetry measurements on hydrogen and deuterium in the range $0.16 < Q^2 < 0.95\ (\text{GeV/c})^2$. It is based on a novel toroidal spectrometer which detects scattered events in the range $62^\circ < \theta < 78^\circ$. When the spectrometer is oriented in the forward direction, recoil protons are detected in this range, which corresponds to electrons scattering in the range $15^\circ < \theta < 5^\circ$. The entire range of Q^2 is thus obtained simultaneously, with an incident beam energy of 3 GeV/c^2.

The spectrometer would then be rotated by 180° and backward angle electrons would be detected. In this configuration, data would be taken at three different beam energies with both hydrogen and deuterium targets. In this way, the electric and magnetic strange form factors as well as the axial form factor would be each independently obtained at $Q^2 \sim 0.3, 0.5, 0.8\ (\text{GeV/c})^2$.

The spectrometer is made up of eight superconducting coils in a common cryostat, with a diameter of 4m and an operating current of 5kA. The geometry provides line-of-sight sheilding of the detectors from the target. The total solid angle of 0.9 sr is accepted. Recoil protons and electrons are momentum-analyzed and detected by plastic scintillators that are placed on the focal surface to accept specific Q^2 values from the entire 20 cm target length.

For forward angle events, time of flight is used to reject backgrounds. The high degree of segmentation keeps instantaneous rates below 1 MHz. The beam microstructure is reduced from 497 MHz by a factor of 16 to provide $\sim$ 32ns timing windows. For backward angle measurements, additional detectors are added at the entrance of the spectrometer to separate electrons from pions and protons.

Custom electronics have been developed to count particles at very high rates in each beam helicity window while rejecting backgrounds, with minimal dead time. Data are recorded by shift registers feeding scalers and by time-to-digital converters. Kinematics corresponding to elastic scattering as well as the $N-\Delta$ transition will be employed.

Installation of the apparatus is nearing completion and the commissioning will take place in Fall 2002. It is anticipated that the first physics run will begin in late 2003. Backward angle measurements will take place between 2005-7.

HAPPEX-II

The HAPPEX collaboration in Hall A at Jlab is preparing for two new experiments, scheduled to begin in summer 2003. These experiments will detect scattered electrons at $\theta_{\text{lab}} \sim 6^\circ$, which is achieved by the HRS high resolution spectrometers in conjunction with septum magnets. One experiment will measure the weak neutral current amplitude off hydrogen, while the other will make the measurement of ^{4}He, both employing an incident beam energy of 3.2 GeV/c^2.

For the hydrogen measurement, the physics asymmetry is 1.7 ppm and the goal is to measure it with a statistical error of 5% and a systematic error of 3%. The total rate into the two spectrometers is 125 MHz. The combination of small statistical error and high rate requires significant improvements in the monitoring and control of electron beam fluctuations and an improved detector technology capable of absorbing high radiation.

For the helium measurements, a dense gas target capable of holding 10 atmospheres of Helium at a temperature of 20 K is being developed. The spectrometer system in Hall A easily separates elastic events from the inelastic events that have lost more than 20 MeV. The physics asymmetry is about 15 ppm and will be measured to an accuracy of 3%. Significant improvement in reducing normalization errors from those of the HAPPEX measurement are required, most notably in the beam polarization measurement.

The hydrogen and helium measurements combine, along with published SAMPLE data at the same Q^2, to cleanly separate G_E^s and G_M^s with little additional uncertainty on unknown form factors. If one assumes a reasonably smooth behaviour of the form factors in the range $0 < Q^2 < 0.1$ (GeV/c)^2, then the two measurements alone form a powerful constraint on ρ_s and μ_s. One measures the combination $\rho_s + \mu_p \mu_s$ to an accuracy of ± 0.31 while the other measures ρ_s to an accuracy of ± 0.5.

THE FUTURE

^{208}Pb

This experiment to measure the neutron skin will use the same experimental configuration as the HAPPEX-II experiment. The electron beam with an energy of 850 MeV/c^2 and an intensity of 50μA would impinge on a high power ^{208}Pb target consisting of a 0.5mm foil of lead sandwiched between two 0.15mm sheets of diamond. The sandwich would be in hard thermal contact with a copper block that is cooled by liquid helium. The scattered electrons are momentum analyzed by the septum-HRS combination of the Hall A spectrometers, and the flux up to an energy loss of 4 MeV would be integrated by radiation-hard copper quartz sandwich detectors.

The physics asymmetry is 0.51 ppm and the required statistical accuracy is 3%. This experiment requires not only stringent control of the electron beam fluctuations, but also the best possible control of normalization errors such as the beam polarization and the knowledge of the absolute value of Q^2. These aspects are under study and most of the technical challenges will be addressed in the HAPPEX-II measurements that will be carried out before this experiment is launched.

Qweak

Qweak at Jlab is a newly approved experiment to measure the weak neutral current elastic electron-proton amplitude at $Q^2 = 0.03$ (GeV/c)^2. A 1.165 GeV/c^2 beam with an intensity of 180 μA would impinge on a 35cm liquid hydrogen target. A toroidal magnet would accept elastically scattered electrons over the full range of the azimuth at a rate of 7.3 GHz. The signal would be intercepted by quartz bars whose response would be measured by photomultiplier tubes and integrated in 33ms beam windows.

The estimated left-right asymmetry is 0.3 ppm and the goal is to measure it to a total accuracy of 4%. This would measure $\sin^2\theta_W$ to a precision of 0.0007 and probe for new physics beyond the standard model in the lepton quark sector at the 10 TeV level. The asymmetry has a 30% contribution from nucleon form factors that must be subtracted; this would be accomplished by exploiting the precision strange form factor measurements that would be completed before the experiment begins collecting data. It is planned to install the experiment in Hall C at Jlab and a research and development plan is underway to complete construction in four years.

Experiments at 12 GeV at Jefferson Laboratory

The upgrade of Jefferson Laboratory to 12 GeV provides new opportunities for deep inelastic scattering measurements, providing unprecedented high luminosity to explore quarks at large x_{Bjorken}, thus allowing new precision studies of valence quark distributions. One measurement that is particularly attractive is precision measurement of the ratio of the down quark to the up quark distribution $d(x)/u(x)$ as $x \to 1$. The basic idea is that the parity violating left-right asymmetry due to deep inelastic scattering at high x off hydrogen has very little uncertainty due to unknown parameters of the standard model and is very sensitive to the value of $d(x)/u(x)$.

On the other hand, the opposite is true for a measurement of the left-right asymmetry at high x from an isoscalar target such as deuterium. Here, a precision measurement of the asymmetry provides a measurement of the weak mixing angle $\sin^2\theta_W$ in a manner complementary to measurements using Møller scattering or elastic electron proton scattering. If new physics is discovered at Fermilab or the Large Hardron Collider, such measurements might prove to be very important in pinning down the nature of the new physics.

Fixed Target Møller Scattering at a Linear Collider

One of the goals of precision electroweak physics measurements is to probe the scalar sector of the electroweak theory via ultra-precise measurements of fundamental electroweak parameters. For example, a measurement of the weak mixing angle $\sin^2\theta_W$ to a precision better than 0.00005 would measure the mass of the standard model Higgs boson to within 10% of itself[24]. Such a measurement is extremely challenging but several methods are being explored in future high energy collider experiments.

Measurements of such precision are very complementary if they are done at different energy scales. One promising method to reach such sensitivity at $Q^2 \ll M_Z^2$ is with fixed target Møller scattering with the electron beam at a future linear collider[25]. The figure of merit of the measurement rises linearly with incident beam energy E_{beam}, a unique feature of Møller scattering due to the fact that the cross-section drops only linearly with E_{beam}.

One idea is to develop a test beam and fixed target facility at a linear collider that would use the "exhaust" electron beam after the primary interaction at the collision point with the positron beam[26]. Thus, it would be possible to run fixed target experiments simultaneously with the electron-positron annihilation experiments. It is estimated the $\sin^2\theta_W$ would be measured with a precision of 0.00007 with physics runs of 10^7s at 250 GeV/c^2 and 500 GeV/c^2.

CONCLUSION

The technique of parity-violating electron scattering has made giant strides over the past two decades. The HAPPEX experiment has set important limits on the size of strange form factors and the SAMPLE experiment has suggested that the proton axial current may be important. New experiments that are about to come on line will clarify the picture. If axial current effects turn out to be small, the new measurements should set stringent limits on the size of the strange form factors or better still, establish their existence.

Asymmetry measurements with accuracy better than 10 ppb are now possible, and this allows the measurement of elastic weak neutral current amplitudes at very forward angle. This will allow unique tests of the standard model and searches for physics at the TeV scale. It also allows a precision measurement of the neutron skin in a heavy nucleus, a measurement that was impossible using electron scattering until the advent of parity violating electron scattering.

ACKNOWLEDGMENTS

It is a pleasure to thank the organizers for a stimulating meeting. The contributions of all the collaborations of the experiments discussed in this review are acknowledged.

REFERENCES

1. Zel'dovich,Ya.B., *J.Exptl.Theoret.Phys.* (U.S.S.R.), **36**, 1959, pp. 964-966.
2. Prescott,C.Y., et.al., *Phys.Lett.*, **B84**, 1979, 524.
3. Kumar,K.S. and Souder,P.A., *Prog.Part.Nucl.Phys.*, **45**, 2000, pp. S333-S395.
4. Kaplan,D.B. and Manohar,A., *Nucl.Phys.*, **B310**, 1988, 527.
5. McKeown,R.D., *Phys.Lett.*, **B219**, 1989, 140.
6. Beck,D.H., *Phys. Rev.*, **D39**, 1989, 3248.
7. Musolf,M.J., et.al., *Phys. Rep.*, **239**, 1994, 1, and references therein.
8. Beck,D.H. and Holstein,B.R., *Int.J.Mod.Phys.*,**E10**, 2001, pp. 1-41.
9. Donnelly,T.W., Dubach,J. and Sick,I., *Nucl.Phys.*, **A503**, 1989, 589.
10. Horowitz,C.J., et.al., *Phys.Rev.*, **C63**, 2001, 025501.
11. Pollock,S.J., Fortson,E.N. and Wilets,L., *Phys.Rev.*, **C46**, 1992, 2587.
12. Horowitz,C.J. and Piekarweicz,J., *Phys.Rev.Lett.*, **86**, 2001, 5647.
13. Ramsey-Musolf,M.J., *Phys.Rev.*, **C60**, 1999, 015501.
14. Kumar,K.S., et.al., *Mod.Phys.Lett.*, **A10**, 1995, 2979.
15. Zeller,G.P., et.al., *Phys.Rev.Lett.*, **88**, 2002, 091802.
16. Wood,C.S., et.al., *Science*, **275**, 1997, 1759.
17. Dzuba,D.A., Flambaum,V.V. and Ginges,J.S., *Phys.Rev.*, **D66**, 2002, 076013.
18. Spayde,D., *et.al.*, *Phys.Rev.Lett.*, **84**, 2000, 1106.
19. Hasty,R., et.al., *Science*, **290**, 2000, 2117.
20. Aniol,K.A., et.al., *Phys.Lett.*, **B509**, 2001, pp. 211-216.
21. Whittum,D.H. and Kolomensky,Y. *Rev.Sci.Instrum.*, **70**, 1999, pp. 2300-2313.
22. Humensky,T.B., et.al., SLAC-PUB-9381, submitted to *Nucl.Instrum.Meth.A*.
23. Gao,J., Gustafsson,K.K., et.al., SLAC-PUB-9565, submitted to *Nucl.Instrum.Meth.A*.
24. Fisher,P., Becker,U. and Kirkby,J., *Phys.Lett.*, **B356**, 1995, pp. 404-408.
25. Kumar,K.S., "Fixed Target Møller Scattering at the NLC", in *Snowmass 1996, New Directions for High-Energy Physics*, 1996, pp. 1030-1034.
26. Keller,L., *et.al.*, SLAC-PUB-8725 (2001).

Acceleration of Polarized Protons and Deuterons at COSY

A. Lehrach, U. Bechstedt, J. Dietrich, R. Gebel, B. Lorentz, R. Maier, D. Prasuhn, A. Schnase, H. Schneider, R. Stassen, H. Stockhorst, R. Tölle

Forschungszentrum Jülich GmbH, Postfach 1913, 52425 Jülich, Germany

Abstract. At the cooler synchrotron COSY, protons and deuterons are accelerated up to 3.65 GeV/c. Vertically polarized proton beams with more than 70% polarization have been delivered in recent years to internal as well as to external experiment areas at different momenta up to the maximum momentum of COSY. In a strong-focusing synchrotron like COSY, imperfection and intrinsic resonances cause polarization losses during acceleration. The existing magnet system of COSY allows to overcome all imperfection resonances by exciting adiabatic spin flips without polarization losses. A tune-jump system consisting of two fast quadrupoles has been developed to handle intrinsic resonances in COSY. This magnet system is being successfully utilized at all intrinsic resonances in the momentum range of COSY. For the acceleration of vertically polarized deuterons, additional correction provisions are not necessary to preserve polarization during acceleration because depolarizing resonances are not crossed in the momentum range of COSY at ordinary transversal betatron tunes.

In future upgrades we are planning to install spin rotators to provide longitudinally polarized beam to internal as well as to external experiments. Such devices can also be used as Siberian snake to overcome depolarizing resonances. Since strong first order depolarizing resonances are crossed in COSY, it will be possible to study snake resonances in detail. In particular the behavior of the spin field and the mechanism of polarization loss during snake resonance crossing are of major interest for high energy polarized beam acceleration. In this paper the status of the polarized beam acceleration at COSY is presented and the upgrade plans are discussed.

INTRODUCTION

The COoler SYnchrotron and storage ring COSY at the Forschungszentrum Jülich accelerates protons and deuterons to momenta between 600 MeV/c and 3.65 GeV/c [1, 2]. At present four internal and four external experiments are in operation. In addition, polarized beams are produced and accelerated at COSY. A polarized ion source originally developed by a collaboration of the universities of Bonn, Erlangen, and Cologne [3] provides vector polarized proton beam and all possible combinations of vector and tensor polarized deuteron beams [4]. The polarized H^- or D^- ion beam delivered by this source is pre-accelerated in the cyclotron JULIC and injected by charge exchange into the COSY ring. To increase the intensity of polarized beams in COSY, a superconducting linear accelerator (COSY-SCL) is being designed and developed to replace the existing cyclotron [5]. The new COSY injector linac is planned to deliver polarized H^- and polarized D^- ions for injection into COSY with an energy of at least 50 MeV. The number of particles accepted in COSY is to be risen up to its space charge limit, which is a significant improvement especially for polarized ions. The improved capabilities will

CP675, *Spin 2002: 15th Int'l. Spin Physics Symposium and Workshop on Polarized Electron Sources and Polarimeters,* edited by Y. I. Makdisi, A. U. Luccio, and W. W. MacKay

enable us to fully exploit the unique experimental opportunities of the COSY facility.

The main diagnostic tool to develop polarized beams in COSY is the EDDA detector [6], primarily designed to measure the pp-scattering excitation function during synchrotron acceleration. The polarization is determined by measuring the asymmetry of scattering between the circulating COSY beam and carbon or CH_2-fiber targets. Additional polarimeters are installed in the injection beamline to COSY, in the COSY ring and in the extraction beamline of COSY. The intensity of the polarized beam in COSY can be increased by stacking injection with an electron-cooler and the beam quality can be further improved with an stochastic cooling systems [7, 8]. The layout of the accelerator complex COSY is shown in Fig. 1.

POLARIZED PROTON AND DEUTERON BEAM ACCELERATION

For an ideal planar circular accelerator with a vertical guide field, the particle spin vector precesses around the vertical axis. Thus the vertical beam polarization is preserved. The spin motion in an external electro-magnetic field is governed by the so called Thomas-BMT equation [9], leading to a spin tune of $\nu_{sp} = \gamma G$, which describes the number of spin precessions of the central beam per revolution in the ring. G is the anomalous magnetic moment of the particle (e.g. G = 1.7928 for protons, -0.1423 for deuterons), and $\gamma = E/m$ the Lorentz factor. During acceleration of a polarized beam, depolarizing resonances are crossed if the precession frequency of the spin γG is equal to the frequency of the encountered spin-perturbing magnetic fields. In a strong-focusing synchrotron like COSY two different types of strong depolarizing resonances are excited, namely imperfection resonances caused by magnetic field errors and misalignments of the magnets, and intrinsic resonances excited by horizontal fields due to the vertical focusing. For deuterons the spin tune is about 25 times lower than for protons at the same energy. Depolarizing resonances for deuterons are therefore 25 times further apart compared to those for protons. Depolarizing resonances for deuterons are also about 13 times weaker at low energies and 25 times weaker for high energies. Another important difference is, that deuterons are spin-1 particles. Hence they appear in three spin states (1,0,-1) relative to an arbitrary quantization axis, compared to two spin states (1,-1) of a spin-$\frac{1}{2}$ particle like protons. Three vector components and five components of a second-rank tensor are required to describe spin-1 polarization. The spin dynamics in circular accelerators of spin-1 particles have been discussed in [10].

Polarized protons have been accelerated in many accelerators like the ZGS, KEK, SATURNE II, AGS and RHIC. Polarized deuterons have first been accelerated to energies in the GeV range at KEK and very recently at IUCF [11, 12].

ACCELERATION OF POLARIZED PROTONS AT COSY

In the momentum range of COSY, five imperfection resonances have to be crossed. The existing correction dipoles of COSY are utilized to overcome all imperfection

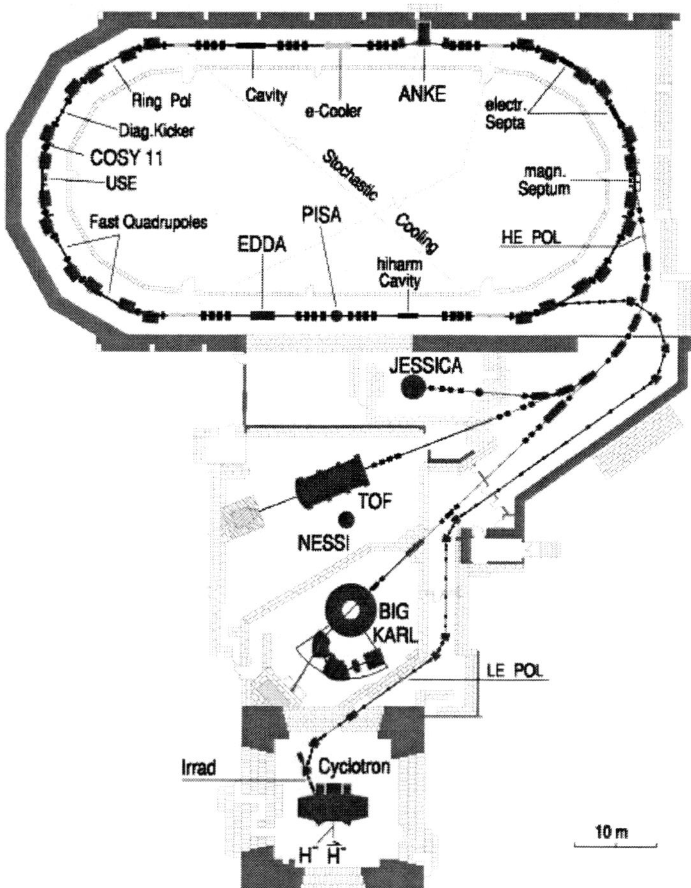

FIGURE 1. The layout of the existing accelerator complex COSY in Jülich, which includes polarized and unpolarized sources, the Cyclotron JULIC and the Cooler Synchrotron COSY. The position of the three different polarimeters (Low Energy Polarimeter, High Energy Polarimeter, Ring Polarimeter) and the four internal experiments (ANKE, COSY 11, EDDA, PISA) are indicated. The beam is also delivered to four external experiment areas (Big Karl, JESSICA, NESSI, TOF).

resonances by exciting adiabatic spin flips without polarization losses. The number of intrinsic resonances depends on the superperiodicity of the lattice. In principle the magnetic structure in the arcs of COSY allows to adjust a superperiodicity of P=2 or 6. A tune-jump system consisting of two fast quadrupoles has especially been developed to handle intrinsic resonances at COSY.

Imperfection Resonances

The imperfection resonances in the momentum range of COSY are listed in Table 1. They are crossed during acceleration, if the number of spin precessions per revolution of the particles in the ring is an integer ($\gamma G = k$, k: integer). The resonance strength depends on the vertical closed orbit deviation.

TABLE 1. Resonance strength ε_r and the ratio of preserved polarization P_f/P_i at imperfection resonances for a typical vertical orbit deviation y_{co}^{rms}, without considering synchrotron oscillation.

γG	E_{kin} **MeV**	P **MeV/c**	y_{co}^{rms} **mm**	ε_r 10^{-3}	P_f/P_i
2	108.4	463.8	2.3	0.95	-1.00
3	631.8	1258.7	1.8	0.61	-0.88
4	1155.1	1871.2	1.6	0.96	-1.00
5	1678.5	2442.6	1.6	0.90	-1.00
6	2201.8	2996.4	1.4	0.46	-0.58

A spin flip occurs at all resonances without considering synchrotron oscillation. However, the influence of synchrotron oscillation during resonance crossing cannot be neglected (Fig. 2). At the first imperfection resonance, the calculated polarization with a momentum spread of $\Delta p/p = 1 \cdot 10^{-3}$ and a synchrotron frequency of $f_{syn} = 450Hz$ is about P_f/P_i[1] $\approx$ -0.85. The resonance strength of the first imperfection resonance has to be enhanced to $\varepsilon_r = 1.6 \cdot 10^{-3}$ for a beam with momentum spread of $\Delta p/p = 1 \cdot 10^{-3}$ to excite spin flips with polarization losses of less than 1%. At the other imperfection resonances the effect of the synchrotron oscillation is smaller, due to the lower momentum spread at higher energies. Vertical correction dipoles or a partial snake can be used to preserve polarization at imperfection resonances by exciting adiabatic spin flips. Simulations indicate that an excitation of the vertical orbit with existing correction dipoles by 1 mrad is sufficient to adiabatically flip the spin at all imperfection resonances. In addition, the solenoids of the electron-cooler system inside COSY are available for use as a partial snake. They are able to rotate the spin around the longitudinal axis by about 8° at the maximum momentum of COSY. A rotation angle of less than 1° of the spin around the longitudinal axis already leads to a spin flip without polarization losses at all five imperfection resonances [13].

Intrinsic Resonances

The number of intrinsic resonances depends on the superperiodicity P of the lattice, which is given by the number of identical periods in the accelerator. COSY is a synchrotron with a racetrack design consisting of two 180° arc sections connected by

[1] Ratio of beam polarization before (i) and after (f) crossing a depolarizing resonance.

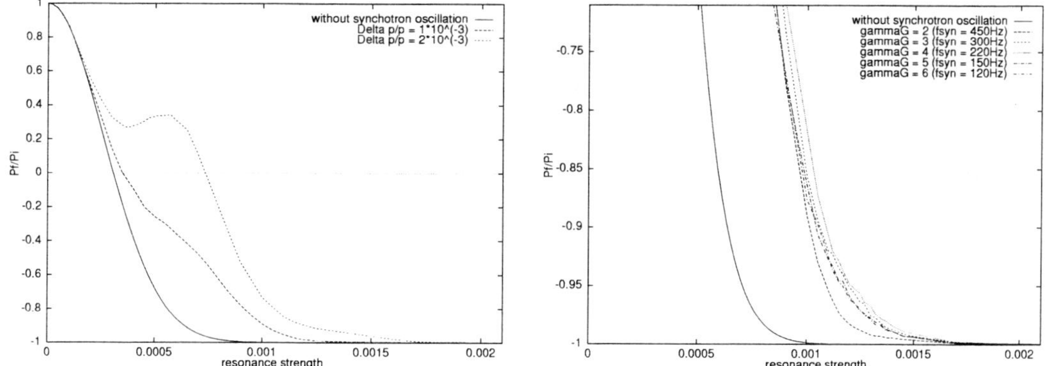

FIGURE 2. Effect of synchrotron oscillation during crossing imperfection resonances in COSY. Ratio of preserved beam polarization P_f/P_i after crossing the first imperfection resonance for two different momentum spreads of $\Delta p/p = 1 \cdot 10^{-3}$ and $\Delta p/p = 2 \cdot 10^{-3}$ with a synchrotron frequency of $f_{syn} = 450Hz$ (left), and ratio of preserved beam polarization (cutaway in the spin flipping region) after crossing different imperfection resonances for a momentum spread of $\Delta p/p = 1 \cdot 10^{-3}$ taking the synchrotron frequencies at the various resonance energies into account (right). The corresponding synchrotron tunes are in the range between $\nu_{syn} = 6 \cdot 10^{-4}$ and $8 \cdot 10^{-5}$.

straight sections. The straight sections can be tuned as telescopes with 1:1 imaging, giving a 2π betatron phase advance. In this case the straight sections are optically transparent and only the arcs contribute to the strength of intrinsic resonances. One then obtains for the resonance condition $\gamma G = k \cdot P \pm (Q_y - 2)$, where k is an integer and Q_y is the vertical betatron tune. The magnetic structure in the arcs allows adjustment of the superperiodicity to P= 2 or 6. The corresponding intrinsic resonances in the momentum range of COSY are listed in Table 2.

TABLE 2. Resonance strength ε_r of intrinsic resonances for a normalized emittance of 1π mm mrad and a vertical betatron tune of Q_y= 3.61 for different superperiodicities P.

P	γG	E_{kin} **MeV**	P **MeV/c**	ε_r 10^{-3}
2	$6 - Q_y$	312.4	826.9	0.26
2	$0 + Q_y$	950.7	1639.3	0.21
2,6	$8 - Q_y$	1358.8	2096.5	1.57
2	$2 + Q_y$	1997.1	2781.2	0.53
2	$10 - Q_y$	2405.2	3208.9	0.25

Tune-Jump System

A tune-jump system was developed to preserve polarization at intrinsic resonances by increasing the crossing speed significantly. This is accomplished by abruptly changing the vertical betatron tune during resonance crossing in the range of microseconds. The magnet system consists of two pulsed air core quadrupoles and is designed to achieve polarization losses of less than 5% at the strongest intrinsic resonance, and of less than 1% at all other intrinsic resonances in COSY [14]. To meet this goal, a vertical tune jump of more than $\Delta Q_y = 0.06$ in $10\mu s$ is needed. The existing stainless steel vacuum chamber at the location of the tune-jumping quadrupoles had to be replaced by a ceramic vacuum chamber. A layer of $10\mu m$ titanium was sputtered on the inside surface of the ceramic chamber. To avoid double crossing of resonances, the fall time of the tune jump can be adjusted for different jump widths and acceleration rates. The maximum fall time is $40\,ms$. Fig. 3 shows the polarization of the COSY beam measured during acceleration around the strongest intrinsic resonance $\gamma G = 8 - Q_y$. This resonance excites a natural

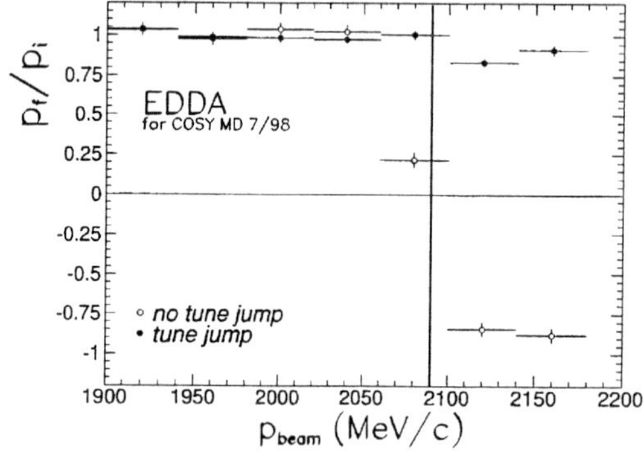

FIGURE 3. Ratio of preserved beam polarization P_f/P_i after crossing the strongest intrinsic resonance at 2090 MeV/c with and without tune jump measured during acceleration with the EDDA detector.

spin flip. The polarization loss depends on the vertical emittance of the beam. With a tune jump, the polarization was almost preserved. Particle losses during tune jumping due to emittance increase can be kept low by adjusting to beam orbit carefully at the position of the tune-jump quadrupoles. The tune jump method can be extended to all other intrinsic resonances because they are at least a factor three weaker than the strongest resonance.

Optimized Optics

To optimize the optics for a polarized beam, phase advances and betatron amplitudes have been determined along the ring. The measurements were done by exciting continuous betatron oscillations and observing the beam response with a network analyzer between a pair of beam position monitors. With the phase advance of the straight sections matched to 2π, the superperiodicity of the COSY lattice is determined by the arcs. Both arcs are composed of three unit cells that are each mirror-symmetrical. A half-cell has a

QD-bend-QF-bend structure (Fig. 1). The superperiodicity equals six if all unit cells operate with the same quadrupole settings. In this case only one intrinsic resonance occurs ($\gamma G = 8 - Q_y$) but the transition crossing takes place at about 1600 MeV/c. To accelerate the beam to maximum momentum, the strength of the horizontally focusing quadrupoles of the inner unit cells in the arcs is enhanced by about 40% to shift the transition energy above the maximum momentum. At the same time, the strength of the horizontally focusing quadrupoles in the outer unit cells is decreased by 20% to keep the betatron tunes constant. The superperiodicity of the beam optics is then $P = 2$. Consequently, four additional intrinsic resonances are introduced (Table 2), which can be suppressed if the harmonics of the corresponding spin-perturbing fields are corrected. Theoretical studies of the COSY lattice revealed the possibility of suppressing the strength of intrinsic resonances using the vertically focusing quadrupoles of the inner unit cells in the arcs, leading to a modified P=2-optics [15]. This new method avoids the drawbacks associ-

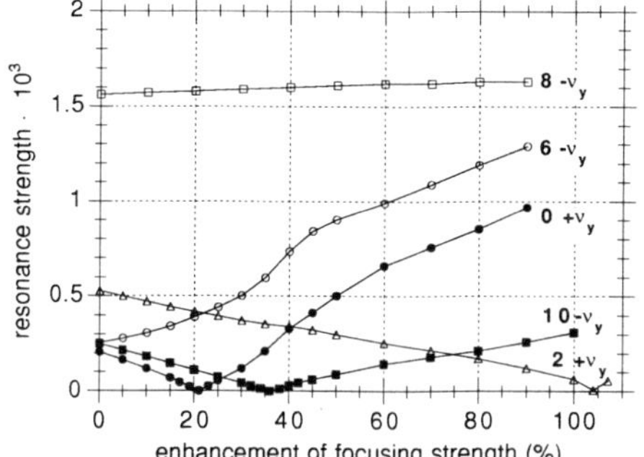

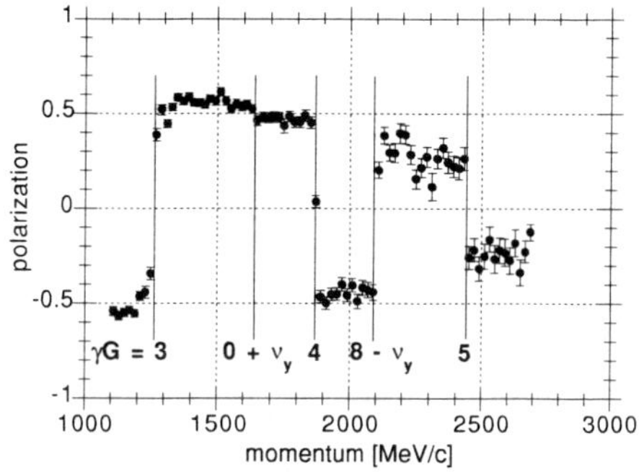

FIGURE 4. The graph on the left side shows the resonance strength of depolarizing resonances in case of a modified P=2-optics versus the enhancement of focusing strength of the vertically focusing quadrupoles in the inner unit cells. The betatron tune is fixed by reducing the strength of the vertically focusing quadrupoles in the outer unit cells. In this calculation the focusing strength of the horizontally focusing quadrupoles in the inner unit cells is enhanced by about 40%. The graph on the right side shows the polarization during acceleration measured with the EDDA detector in the momentum range between 1.1 GeV/c and 2.7 GeV/c. The spin was flipped at the imperfection resonances $\gamma G = 3$, 4 and 5. At the second intrinsic resonance $\gamma G = 0 + Q_y$ the polarization was almost preserved by adjusting a modified P=2-optics. The third intrinsic resonance $\gamma G = 8 - Q_y$ excites a natural spin flip with some polarization losses.

ated with the non-adiabatic nature of tune jumps, which otherwise would be necessary to preserve polarization at all intrinsic resonances of COSY. The method is called suppression of intrinsic spin harmonics, and can also be used at other accelerators like the Brookhaven AGS [16].

During a running period in the year 1998, the new method to overcome intrinsic resonances has been confirmed by measurements with polarized beam (Fig. 4). The polarization was preserved at the two intrinsic resonances, $\gamma G = 6 - Q_y$ and $\gamma G = 0 + Q_y$, by modifying the optics during acceleration. To avoid polarization losses at the first intrinsic resonance, $\gamma G = 6 - Q_y$ at 827 MeV/c, the acceleration of the beam started with P=6 optics. The ratio of preserved polarization was $P_f/P_i = 0.97 \pm 0.05$. At about 900 MeV/c, the COSY beam optics was then switched to superperiodicity P=2 to shift the transition energy. As expected, crossing $\gamma G = 0 + Q_y$ at 1640 MeV/c led to polarization losses (P_f/P_i = 0.13±0.05) in this mode. After suppressing the strength of intrinsic resonances using the vertically focusing quadrupoles in the inner unit cells, the ratio of the preserved polarization at this intrinsic resonance could be significantly increased to P_f/P_i = 0.88± 0.05 [15].

However, due to symmetry-breaking installations in the COSY ring (e.g. ANKE spectrometer and electron-cooler magnets) the superperiod of the accelerator lattice in COSY is reduced to $P = 1$, leading to five additional intrinsic resonances in the energy range of COSY: ($\gamma G = -1 + Q_y, 7 - Q_y, 1 + Q_y, 9 - Q_y, 3 + Q_y$). To preserve polarization up to maximum momentum of COSY tune jumps are utilized at all ten intrinsic resonances.

Beam Set-up for Polarized Beams

The beam is usually set-up with unpolarized beam, due to better accuracy of COSY diagnostics with about a factor of ten higher beam intensity [17]. Polarization optimization becomes much more efficient and particle losses due to emittance growth can be kept low if the beam position is aligned carefully during the acceleration ramp, especially at the location of the tune-jump quadrupoles. Dynamic tune measurements are carried out to adjust the transversal betatron tunes during acceleration [18]. The vertical betatron tune is fixed close to 3.62 during acceleration in order to optimize the distance between intrinsic resonances for consecutive tune jumps. The horizontal tune is set at around 3.60 during acceleration. After closed orbit and betatron tune correction, the beam manipulations for polarized beam are applied. The magnet currents and trigger times for the tune-jump quadrupoles and vertical correction dipoles are set to values used at previous polarized beam times, as can be seen in Fig. 5. The tuning of magnet

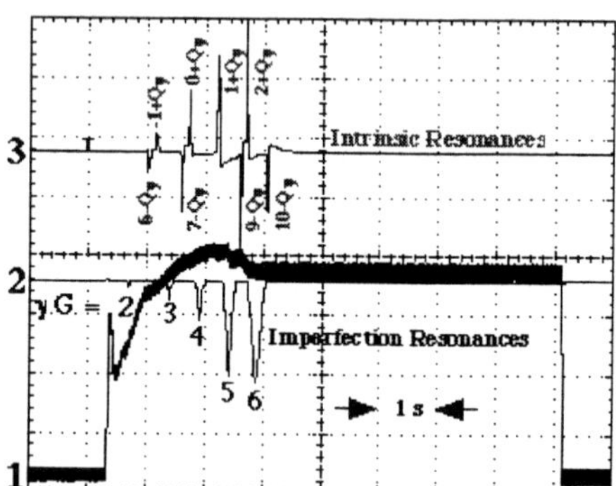

FIGURE 5. Trace 1 shows the beam current, trace 2 the current of vertical correction dipoles, and trace 3 the current of the tune-jump system versus time, applied at various depolarizing resonances.

currents and trigger times is done after switching to polarized beam by utilizing polarization measurements. For any applied correction to preserve polarization, the number of particle are observed. Particle losses due to correction dipole and tune-jump quadrupole fields during depolarizing resonance crossing are kept below 10% in total. For extracted polarized beams the momentum has to be chosen carefully to avoid polarization losses. The beam is extracted via a third-order betatron resonance. Corresponding intrinsic resonances lead to significant polarization losses. Momenta near imperfection resonances can also not be provided. Since the beam is stored for relatively long times at extraction energy and because the momentum spread increase during the extraction process, higher order depolarizing resonances can also lead to polarization losses. After excluding these momenta, one still has to carefully adjust the tunes to prepare the stored beam for extraction.

Polarized Beam Acceleration at COSY

During a running period in the year 2000, the polarized beam was accelerated to 3300 MeV/c [17]. The spin was flipped at the imperfection resonances $\gamma G = 2, 3, 4, 5$ and 6 using correction dipoles. To avoid polarization losses at all intrinsic resonance tune jumps were applied. The measured polarization after the optimization for polarized beam is shown in Fig. 6.

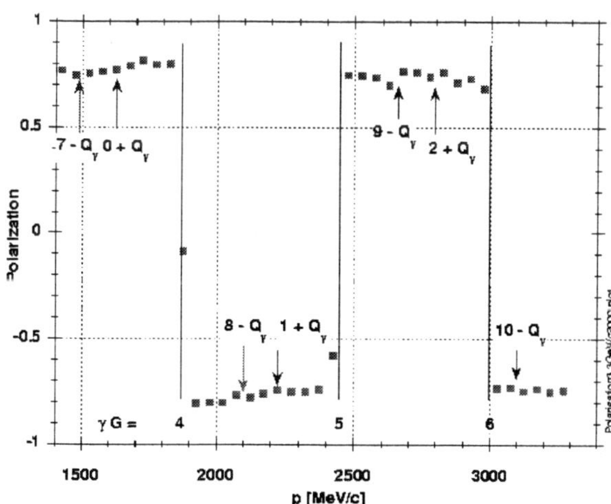

FIGURE 6. Vertical beam polarization during acceleration measured with the EDDA detector in the momentum range between 1100 MeV/c and 3300 MeV/c.

The polarization losses up to final momentum were rather small, only in the order of a few percent. Between $2 \cdot 10^9$ to $5 \cdot 10^9$ polarized protons have been stored at final momentum. Below the strongest intrinsic resonances the beam has been extracted with a polarization of about 80%. Above this resonance we recently reached about 60% polarization of the extracted beam.

ACCELERATION OF POLARIZED DEUTERONS AT COSY

In January 2002, $2 \cdot 10^{11}$ unpolarized deuterons have been accelerated to maximum momentum of COSY. The first injection and acceleration of polarized deuterons is scheduled for February 2003. The polarized source is capable to produce any kind of possible vector and tensor polarization [4]. In Fig. 7 the depolarizing resonance momentum is plotted versus fractional betatron tune.

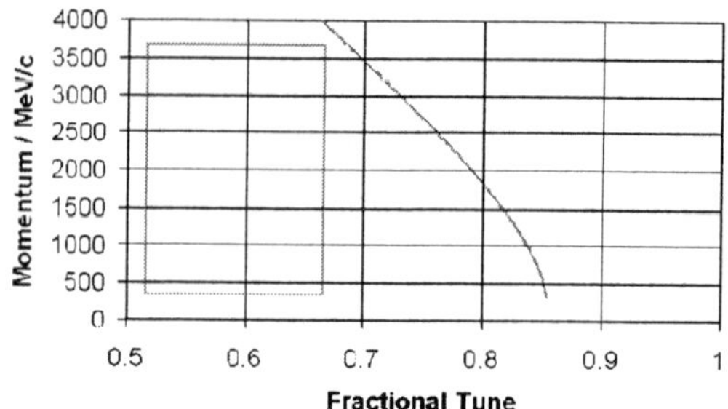

FIGURE 7. Depolarizing first order resonances for deuterons in the momentum range of COSY. The resonance momentum is plotted versus fractional tune. The rectangular box indicates the ordinary operation range of COSY.

No first order depolarizing resonance is crossed in the momentum range of COSY at an ordinary transversal betatron tune. However, one intrinsic resonance is crossed ($\gamma G = -4 + Q_y$) if the vertical betraron tune in COSY is pushed up to unusually high values in the range between 3.7 and 3.85. This could be an interesting option to study depolarizing resonance crossing of polarized deuterons.

SIBERIAN SNAKE

Siberian snakes are used to eliminate depolarizing resonances in circular accelerators. The spin is rotated by 180° in the snake, forcing the spin tune to be a half integer, independent of the beam energy. This concept has been proposed by Ya.S. Derbenev and A.M. Kondratenko [19]. If only one Siberian snake is used, the invariant spin axis[2] is in the horizontal plane. This is an interesting feature to deliver longitudinally polarized beam to internal and external experiments at COSY.

To preserve the polarization using a Siberian snake with a solenoid field, the direction of the spin vector has to be longitudinal at the symmetry point of the snake, which is point in the ring opposite to the position of the snake. This can be achieved if the spin is prepared in the injection beamline by one or two additional solenoid magnets. Another possibility is to inject the beam vertically polarized with the snake turned off, and switch on the snake after injection. During ramping the snake, the spin direction changes from vertical to longitudinal at the symmetry point, and the spin tune of the

[2] Invariant of spin motion for the central beam, called invariant spin field for a particle in the six dimensional phase space [20, 21].

central beam from $\nu_{sp} = \gamma G$ (without snake) to the nearest half-integer tune (180° spin rotation in the snake). To avoid crossing depolarizing resonances during this process, the snake can be turned on at half-integer spin tune. Then the spin tune for the central beam stays half integer for any snake strength. This condition is satisfied whenever the kinetic energy E_{kin} is given by: $E_{kin} = 370MeV + k \cdot 523MeV$, where k is an integer. One can also use tune jumps to avoid crossing depolarizing resonances during turning on the snake, if the snake is ramped fast enough, because the maximum fall time of the tune-jump system in COSY is limited to $40\,ms$. To realize a Siberian snake in the momentum range of COSY, only solenoid magnets are suitable. A solenoid field not only rotates to spin, but also the transversal phase space. For a spin rotation of 180° the transversal phase space is rotated by 32.2°. This can be compensated with two skewed quadrupole doublets. Different snake schemes for COSY have been investigated [13]. One possible magnet arrangement consists of four skewed quadrupoles, with maximum field gradients of 34.2 T/m and -32.2 T/m rotated by 21.5 and 15.2° in each doublet, and one solenoid located in between (Fig. 8). The required integral field of the solenoid is

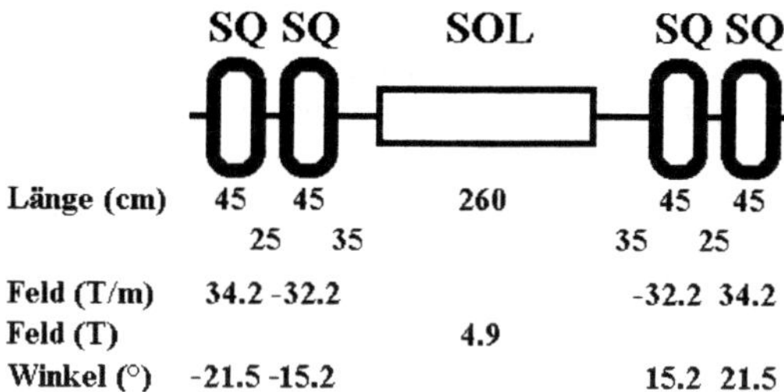

FIGURE 8. Magnet arrangement for a Siberian snake consisting of four skewed quadrupoles (SQ) and one solenoid magnet (SOL).

12.4 Tm at 3.3 GeV/c. Superconducting magnet technology has to be used to achieve an acceptable length of the snake. The total length of such a magnet system is 5.6 m, and it would fit into one of the straight sections of COSY. The required gradient of the skewed quadrupoles to compensate coupling is more then a factor of three higher compared to the maximum gradient of the focusing quadrupoles in COSY. Calculation of the beam optics indicated that the betatron amplitude in the snake is rather small, leading to large betatron amplitudes at other places in the ring. Another option would be to run COSY fully coupled without skew quadrupoles. Further investigations of the beam optics with Siberian snake magnets in the COSY lattice are needed.

If such a magnet system is utilized as Siberian snake, the magnets have to be ramped during acceleration in a few seconds to final field, which is a real challenge for superconducting magnets. Another option is to use this system as a spin rotator. The spin of the vertically polarized beam is rotated after acceleration. Suitable ramp times in the range of a few minutes can be applied. This method provides longitudinally polarized beam at all stored energies at the symmetry point of the snake in the ring. At five different energies in intervals of 523 MeV, longitudinally polarized beam can also be provided to external experiments. These energies depend on the bending angles of the beam in the different extraction beamlines.

For a polarized deuteron beam the rotation angel of the spin in a longitudinal field is about a factor three lower the one for protons. Hence longitudinally polarized deuteron beams with the proposed magnet system can be delivered up to about 1 GeV/c.

CONCLUSION

The solenoids of the electron-cooler acting as a partial snake and vertical correction dipoles were successfully used in COSY to preserve the polarization by exciting adiabatic spin flips. Both methods are available for all five imperfection resonances in the momentum range of COSY. Since solenoids introduce transversal coupling, which excites depolarizing coupling resonances, the vertical correction dipoles are preferred to overcome imperfection resonances in COSY. With the standard optics of COSY, five intrinsic resonances are excited. Measurements confirm, that three of these resonances can be suppressed by changing the optics during acceleration. Due to symmetry-breaking installations, the superperiodicity of the COSY lattice is reduced to one, leading to ten intrinsic resonances. It has been shown that the tune-jump system can handle all ten intrinsic resonances in the momentum range of COSY. Polarization measurements during acceleration confirm that the developed concept allows the acceleration of a vertically polarized proton beam with polarization losses of only a few percent up to the maximum momentum of COSY. The polarization losses at individual depolarizing resonances are within the accuracy of the polarization measurement. Highly polarized proton beams are routinely delivered to internal and external experiments at different momenta. The installation of a Siberian snake in COSY could also provide a longitudinally polarized beam to internal as well as to external experiments at certain momenta. The first acceleration of polarized deuterons is scheduled for next year. Since depolarizing resonances are not crossed with a deuteron beam at ordinary transversal betatron tunes in the momentum range of COSY, no additional corrections have to be applied. To increase the intensity of the polarized beams in COSY by typically one order of magnitude, a superconducting injector linac is presently being developed and built.

ACKNOWLEDGMENTS

We are indebted to all members of the COSY team and the collaboration of the polarized source for their support. We are especially grateful to the EDDA collaboration for their sophisticated measurement of the beam polarization during acceleration.

REFERENCES

1. Maier R.,*Nucl. Inst. and Meth.* **A 390**, 1 (1997).
2. Stockhorst H. et al., Progress and Developments at The Cooler Synchrotron COSY, Proc. European Particle Accelerator Conference EPAC 2002, Paris (2002).
3. Eversheim P.D. et al., The Polarized Ion-Source for COSY, Proc. International Symposium on High-Energy Spin Physics SPIN 1996, Amsterdam (1996), (published in World Scientific Singapore (1997)).
4. Gebel R., private communication.

5. Maier R. et al., The Superconducting Injector Linac for the Cooler Synchrotron COSY, Internal Report Forschungszentrum Jülich, October 2001 (edited by Jungwirth H.); Maier R. et al., The proposed Superconducting Injector Linac for the Cooler Synchrotron COSY at the FZ-Jülich, Proc. International Linac Conference LINAC 2002, Korea (2002), (to be published).
6. Schwarz V. et al., EDDA As Internal High-Energy Polarimeter, Proc. International Spin Physics Symposium SPIN 1998, Protvino, (published in World Scientific Singapore (1999)).
7. Prasuhn D., Dietrich J., Maier R., Stassen R., Stein H.J., Stockhorst H., *Nucl. Inst. and Meth.* **A 441**, 167 (2000).
8. Prasuhn D. et al., Cooling at COSY, Workshop on Beam Cooling and Related Topics, Bad Honnef 2001, Forschungszentrum Jülich, Matter and Material, Volume **13**, ISBN 3-89336-316-5, ISSN 1433-5506.
9. Thomas L.H. *Phil. Mag.* **3**, 1 (1927); Bargman V., Michel L., Telegdi V.L., *Phys. Rev. Letters* **2**, 43 (1959).
10. Bell J.S., *CERN yellow reports* **75-11**; Huang H., Lee S.Y., Ratner L., Proc. Particle Accelerator Conference PAC 1992, Washington (1992), (published in IEEE (1993)).
11. Sato, H. et al., *Nucl. Inst. and Meth.* **A 385**, 391 (1997).
12. Morozov V.S. et al., Proc. International Spin Physics Symposium SPIN 2002, Brookhaven National Laboratory (2002), (to be published in in AIP Conference Proc., edited by Makdisi Y.).
13. Lehrach A. et al., Status of the polarized beam at COSY, Proc. International Symposium on High-Energy Spin Physics SPIN 1996, Amsterdam (1996), (published in World Scientific Singapore(1997)).
14. Lehrach A., PhD-thesis Universität Bonn (1997), Jülich Report Juel-3501, ISSN 0944-2952 (1998).
15. Lehrach A., Gebel R., Maier R., Prasuhn D., Stockhorst H., *Nucl. Inst. and Meth.* **A 439**, 26 (2000).
16. Lehrach A., AGS Lattice Changes to eliminate Weak Intrinsic Resonances, Workshop on Increasing the Polarization at the AGS, Michigan (2002), (to be published in AIP Conference Proc., edited by Krisch A.D., Roser T.).
17. Stockhorst H. et al., The Medium Energy Proton Synchrotron COSY, Proc. European Particle Accelerator Conference EPAC 2000, Vienna (2000).
18. Dietrich J., Mohos I., Broadband FFT Method for Betatron Tune Measurements in the Acceleration Ramp at COSY-Jülich, 8th Beam Instrumentation Workshop, Stanford, AIP Conference Proc. No. 451, 454 (1998).
19. Derbenev Ya.S., Kondratenko A.M., *Part. Accel.* **8** 115 (1978).
20. Derbenev Ya.S., Kondratenko A.M., *Sov. Phys. JETP* **37(6)**, 968 (1973).
21. Heinemann K., Hoffstätter G.H., *Phys. Rev. E* **54**, 4240 (1996).

Spin on the Lattice

Konstantinos Orginos[1]

RIKEN-BNL Research Center, Brookhaven National Laboratory, Upton, NY 11973, USA

Abstract. I review the current status of hadronic structure computations on the lattice. I describe the basic lattice techniques and difficulties and present some of the latest lattice results; in particular recent results of the RBC group using domain wall fermions are also discussed.

Understanding the basic properties of matter requires the understanding of the nucleon structure. Quantum Chromodynamics (QCD) is the theory describing strong interactions and hence is responsible for the properties of the nuclear matter. Although QCD has been around for more than twenty years, its non-perturbative nature is an obstacle to the direct connection of low energy physics to quarks and gluons, the fundamental degrees of freedom of the theory. Unlike QED, non-perturbative techniques had to be developed in order to understand the QCD predictions at low energies. The lattice formulation of QCD is both a non-perturbative way to define the theory and a very powerful tool in understanding its properties.

Deep inelastic scattering of leptons on nucleons has been an important tool in understanding the structure of hadrons. Over the last few decades experiments at SLAC, Fermilab, CERN, DESY, and more recently at RHIC and JLAB, have measured the quark and gluon light cone distribution functions of the nucleon. These experiments have substantially advanced our knowledge of the properties of hadrons. However, we would also like to study how this observed rich phenomenology arises form first principles, i.e. QCD. With todays advances in computer technology, algorithms, and recent developments in lattice regularization of fermions, lattice calculations can complement the experimental effort and promote our understanding of the non-perturbative nature of QCD.

The Lattice Formulation

The continuum Euclidean path integral can be defined using the lattice regulator [1]. In order to preserve gauge invariance the lattice gauge fields are link variables

$$U_\mu(x) = e^{i\int_x^{x+\hat{\mu}} d\tau A_\mu(\tau)}, \tag{1}$$

where A_μ are the continuum gauge fields. The fermion fields live on the sites of the lattice. Naive discretization of the continuum fermionic action leads to the so-called fermion doubling problem. This problem can be avoided by either reinterpreting the

[1] For the RBC collaboration. The current members of the RBC collaboration are: Y. Aoki, T. Blum, N. Christ, M. Creutz, C. Dawson, T. Izubuchi, L. Levkova, X. Liao, G. Liu, R. Mawhinney, Y. Nemoto, J. Noaki, S. Ohta, K. Orginos, S. Prelovsek, S. Sasaki and A. Soni. Plenary talk presented at SPIN2002

CP675, *Spin 2002: 15th Int'l. Spin Physics Symposium and Workshop on Polarized Electron Sources and Polarimeters*, edited by Y. I. Makdisi, A. U. Luccio, and W. W. MacKay

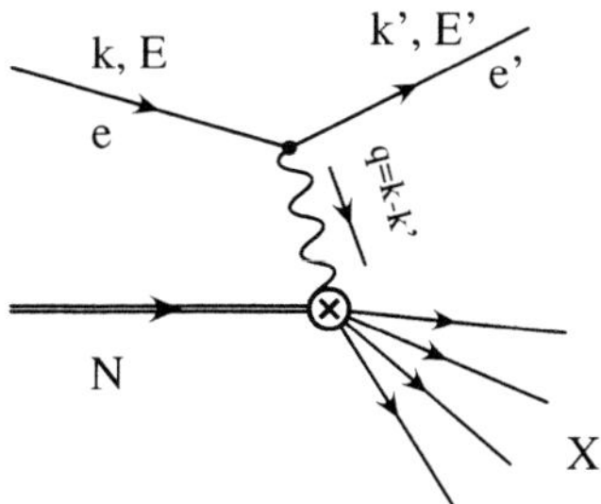

FIGURE 1. Deep inelastic scattering.

additional light fermions as extra flavors (the Kogut-Susskind approach) or by introducing an irrelevant dimension 5 operator that breaks chiral symmetry on the lattice and gives mass proportional to the inverse lattice cutoff to the fermion doublers (the Wilson approach). Recently, new lattice fermionic actions that both preserve chiral symmetry on the lattice and do not suffer from the fermion doubling problem have been introduced. Such fermionic actions are the domain wall fermions [2, 3, 4, 5], the overlap fermions [6], and the fixed point fermions [7, 8, 9]. Having defined the lattice theory, correlation functions can be evaluated using Monte-Carlo integration in Euclidean space.

However, parton distribution functions are defined in the Minkowski space, and hence cannot be directly computed in lattice QCD. Using the operator product expansion we can relate moments of the structure functions to forward matrix elements of gauge invariant local operators (for a pedagogical review see [10]). These matrix elements can then be computed using lattice QCD.

In a deep inelastic process (see Fig. 1) the cross section is given by

$$\frac{d^2\sigma}{d\Omega dE'} = \frac{1}{2m_N}\frac{\alpha^2}{q^4}\frac{E'}{E} l^{\mu\nu} W_{\mu\nu} \tag{2}$$

where $l^{\mu\nu}$ is the lepton tensor, $W_{\mu\nu}$ is the hadronic tensor, q is the momentum transfer, m_N is the nucleon mass. The initial and final energy and momentum of the lepton are (E,k) and (E',k') respectively.

The hadronic tensor can be decomposed in the symmetric $W^{\{\mu\nu\}}$ and anti-symmetric $W^{[\mu\nu]}$ pieces:

$$W^{\mu\nu} = W^{[\mu\nu]} + W^{\{\mu\nu\}} \tag{3}$$

The symmetric piece defines the unpolarized structure functions F_1 and F_2 (F_3 also for neutrino scattering).

$$W^{\{\mu\nu\}}(x,Q^2) = \left(-g^{\mu\nu} + \frac{q^\mu q^\nu}{q^2}\right) F_1(x,Q^2) + \left(p^\mu - \frac{\nu}{q^2}q^\mu\right)\left(p^\nu - \frac{\nu}{q^2}q^\nu\right)\frac{F_2(x,Q^2)}{\nu}, \tag{4}$$

while the anti-symmetric defines the polarized structure functions g_1 and g_2

$$W^{[\mu\nu]}(x,Q^2) = i\varepsilon^{\mu\nu\rho\sigma} q_\rho \left(\frac{s_\sigma}{\nu}(g_1(x,Q^2) + g_2(x,Q^2)) - \frac{q\cdot s p_\sigma}{\nu^2} g_2(x,Q^2)\right). \tag{5}$$

where p_μ and s_μ are the nucleon momentum and spin vectors, $\nu = q \cdot p$, $s^2 = -m_N^2$, $x = Q^2/2\nu$ and $Q^2 = -q^2$.

At the leading twist in the operator product expansion the moments of the structure functions can be factorized at a scale μ in hard perturbative contributions (the Wilson coefficients) and low energy matrix elements of local gauge invariant operators:

$$
\begin{aligned}
2\int_0^{1^-} dx x^{n-1} F_1(x,Q^2) &= \sum_{q=u,d} c_{1,n}^{(q)}(\mu^2/Q^2, g(\mu))\, v_n^{(q)}(\mu), \\
\int_0^1 dx x^{n-2} F_2(x,Q^2) &= \sum_{q=u,d} c_{2,n}^{(q)}(\mu^2/Q^2, g(\mu))\, v_n^{(q)}(\mu), \\
2\int_0^1 dx x^{n} g_1(x,Q^2) &= \frac{1}{2}\sum_{q=u,d} e_{1,n}^{(q)}(\mu^2/Q^2, g(\mu))\, a_n^{(q)}(\mu), \\
2\int_0^1 dx x^{n} g_2(x,Q^2) &= \frac{1}{2}\frac{n}{n+1}\sum_{q=u,d} [e_{2,n}^{(q)}(\mu^2/Q^2, g(\mu))\, d_n^{(q)}(\mu) - \\
&- e_{1,n}^{(q)}(\mu^2/Q^2, g(\mu))\, a_n^{(q)}(\mu)]
\end{aligned} \tag{6}
$$

where $c_{i,n}^{(q)}, e_{i,n}^{(q)}$ are the Wilson coefficients and $v_n^{(q)}(\mu), a_n^{(q)}, d_n^{(q)}(\mu)$ are the non-perturbative matrix elements. At the leading twist $v_n^{(q)}(\mu)$ and $a_n^{(q)}$ are related to the parton model distribution functions $\langle x^n \rangle_q$ and $\langle x^n \rangle_{\Delta q}$:

$$
\langle x^{n-1} \rangle_q = v_n^{(q)} \qquad \langle x^n \rangle_{\Delta q} = \frac{1}{2} a_n^{(q)} \tag{7}
$$

In order to extract $v_n^{(q)}(\mu), a_n^{(q)}$, and $d_n^{(q)}(\mu)$ we need to compute non-perturbatively the following matrix elements:

$$
\begin{aligned}
\frac{1}{2}\sum_s \langle p,s|\mathcal{O}^q_{\{\mu_1\mu_2\cdots\mu_n\}}|p,s\rangle &= 2v_n^{(q)}(\mu) \times [p_{\mu_1}p_{\mu_2}\cdots p_{\mu_n} + \cdots - tr] \\
-\langle p,s|\mathcal{O}^{5q}_{\{\sigma\mu_1\mu_2\cdots\mu_n\}}|p,s\rangle &= \frac{1}{n+1} a_n^{(q)}(\mu) \times [s_\sigma p_{\mu_1}p_{\mu_2}\cdots p_{\mu_n} + \cdots - tr] \\
\langle p,s|\mathcal{O}^{[5]q}_{[\sigma\{\mu_1]\mu_2\cdots\mu_n\}}|p,s\rangle &= \frac{1}{n+1} d_n^q(\mu) \times [(s_\sigma p_{\mu_1} - s_{\mu_1}p_\sigma)p_{\mu_2}\cdots p_{\mu_n} + \cdots - tr]
\end{aligned} \tag{8}
$$

$\{\}$ implies symmetrization and $[]$ anti-symmetrization of indices. The nucleon states $|p,s\rangle$ are normalized so that $\langle p,s|p',s'\rangle = (2\pi)^3 2E(p)\delta(p-p')\delta_{s,s'}$ and $s^2 = -m_N^2$. The operators $\mathcal{O}$ are

$$
\begin{aligned}
\mathcal{O}^q_{\mu_1\mu_2\cdots\mu_n} &= \left(\frac{i}{2}\right)^{n-1} \bar{q}\gamma_{\mu_1} \overleftrightarrow{D}_{\mu_2} \cdots \overleftrightarrow{D}_{\mu_n} q - trace \\
\mathcal{O}^{5q}_{\sigma\mu_1\mu_2\cdots\mu_n} &= \left(\frac{i}{2}\right)^{n} \bar{q}\gamma_\sigma\gamma_5 \overleftrightarrow{D}_{\mu_2} \cdots \overleftrightarrow{D}_{\mu_n} q - trace
\end{aligned} \tag{9}
$$

where $\overleftrightarrow{D} = \overrightarrow{D} - \overleftarrow{D}$ and $\overrightarrow{D}$, $\overleftarrow{D}$ are covariant derivatives acting on the right and the left respectively.

In Drell-Yan processes the transversity distribution $\langle x \rangle_{\delta q}$ can be measured (for details see [11, 12, 13]). The relevant matrix element is

$$\langle p,s|\mathcal{O}^{\sigma q}_{\rho\nu\{\mu_1\mu_2\cdots\mu_n\}}|p,s\rangle = \frac{2}{m_N}\langle x^n\rangle_{\delta q}(\mu)\times[(s_\rho p_\nu - s_\nu p_\rho)p_{\mu_1}p_{\mu_2}\cdots p_{\mu_n}+\cdots - tr] \quad (10)$$

and the operators $\mathcal{O}^{\sigma q}$ are

$$\mathcal{O}^{\sigma q}_{\rho\nu\mu_1\mu_2\cdots\mu_n} = \left(\frac{i}{2}\right)^n \bar{q}\gamma_5\sigma_{\rho\nu}\overleftrightarrow{D}_{\mu_1}\cdots\overleftrightarrow{D}_{\mu_n}q - trace. \quad (11)$$

Lattice matrix elements

In order to calculate on the lattice the needed matrix elements we have to compute nucleon three point functions

$$C^{\Gamma}_{3pt}(\vec{p},t,\tau) = \sum_{\alpha,\beta}\Gamma^{\alpha,\beta}\langle J_\beta(\vec{p},t)\mathcal{O}(\tau)\bar{J}_\alpha(\vec{p},0)\rangle \quad (12)$$

and nucleon two point functions

$$C_{2pt}(\vec{p},t) = \sum_{\alpha,\beta}\left.\frac{1+\gamma_4}{2}\right|_{\alpha,\beta}\langle J_\beta(\vec{p},t)\bar{J}_\alpha(\vec{p},0)\rangle \quad (13)$$

where $\bar{J}(\vec{p},0)$ and $J(\vec{p},t)$ are creation and annihilation operators of states with the quantum numbers of the nucleon. For unpolarized matrix elements $\Gamma = \frac{1+\gamma_4}{2}$ while for the polarized $\Gamma = \frac{1+\gamma_4}{2}i\gamma_5\gamma_k$ $(k \neq 4)$. The $\frac{1+\gamma_4}{2}$ factor is for projecting out the positive parity part of the baryon propagator i.e. the nucleon. For the proton a typical choice is

$$J_\alpha(\vec{p},t) = \sum_{\vec{x},a,b,c} e^{-i\vec{p}\cdot\vec{x}}\epsilon^{abc}\left[u^a(x,t)C\gamma_5 d^b(x,t)\right]u^c_\alpha(x,t) \quad (14)$$

where $C = \gamma_4\gamma_2$ the charge conjugation matrix, α is a spinor index and a,b,c are color indices. When $t \gg \tau \gg 0$

$$\begin{aligned} C_{2pt}(\vec{p},t) &= Z_N\frac{E_N(\vec{p})+m_N}{E(\vec{p})}e^{-E_N(\vec{p})t}+\cdots \\ C^{\Gamma}_{3pt}(\vec{p},t,\tau) &= Z_N\sum_{\alpha,\beta,s}\Gamma_{\alpha\beta}U^{\alpha}_N(p,s)\langle p,s|\mathcal{O}|p,s\rangle\bar{U}^{\beta}_N(p,s)e^{-E_N(\vec{p})t}+\cdots \end{aligned} \quad (15)$$

where $U(p,s)$ is the nucleon spinor which satisfies the Dirac equation and $\langle 0|J_\alpha(\vec{p},t)|p,s\rangle = \sqrt{Z_N}U^\alpha(p,s)$. From Eq. 15 and Eq. 8 (or Eq. 10) the required matrix elements can be extracted from the ratio of three point functions over two point functions. In practice we would like to achieve the asymptotic behavior of Eq. 15 with as small as possible t and τ. For that reason the interpolating operator J is tuned so that the overlap with the exited nucleon states would be as small as possible. For more details on the technical aspects of the lattice calculation the reader may refer to [14, 15, 16, 17].

In order to reduce the computational cost of calculating the above correlation functions some times the so-called quenched approximation is used. In this approximation the quark loop contributions to the path integral are ignored. Quenching reduces the computational cost by several orders of magnitude, while for certain quantities it introduces a systematic error $\sim 10\%$.[2] In addition, lattice computations are typically performed with heavier quark masses than the physical up and down quarks. Hence we have to perform extrapolations to the chiral limit. If the quark masses are light enough, chiral perturbation theory [18, 19, 20] can be used to calculate the dependence of the matrix elements on the quark mass.[3] Finally the lattice matrix elements have to be renormalized, typically to $\overline{\text{MS}}$, and extrapolated to the continuum limit.

Renormalization

The renormalized operators at scale μ are obtained from the lattice operators calculated at lattice spacing a from

$$\mathcal{O}^{ren}(\mu) = Z(\mu;a)\mathcal{O}^{lat}(a) \tag{16}$$

in the case of multiplicatively renormalized operators. In general, there is operator mixing and as a result the above relation becomes

$$\mathcal{O}_i^{ren}(\mu) = Z(\mu;a)\left[\mathcal{O}_i^{lat}(a) + \sum_{j\neq i} a^{d_j - d_i} Z_{ij}(\mu;a)\mathcal{O}_j^{lat}(a)\right], \tag{17}$$

where $\mathcal{O}_j$ are a set of operators allowed by symmetries to mix, and d_j is the dimension of each operator. It is evident that if mixing with lower dimensional operators occur, the mixing coefficients are power divergent as we approach the continuum limit. Hence we have to compute these terms non-perturbatively in order to accurately renormalize the operators. Higher dimensional operators are typically ignored since their effects vanish in the continuum limit. In certain cases we may want to compute these coefficients in order to remove part of the systematic error introduced by the finite cutoff.

The mixing of lattice operators is more complicated than that of the continuum operators, since on the lattice we do not have all the continuum symmetries. In particular, $O(4)$ rotational symmetry in Euclidean space is broken down to the hypercubic group $H(4)$. As a result, an irreducible representation of $O(4)$ is reducible under $H(4)$ and hence mixing of operators that would not occur in the continuum can occur on the lattice. For a detailed analysis of the $H(4)$ group representations see [21, 22] and references therein. In lattice calculations we have to select carefully the lattice operators so that

---

[2] Note that there are quantities for which the quenched approximation introduces uncontrollable errors.

[3] In the case of the quenched approximation the so-called quenched chiral perturbation theory is used.

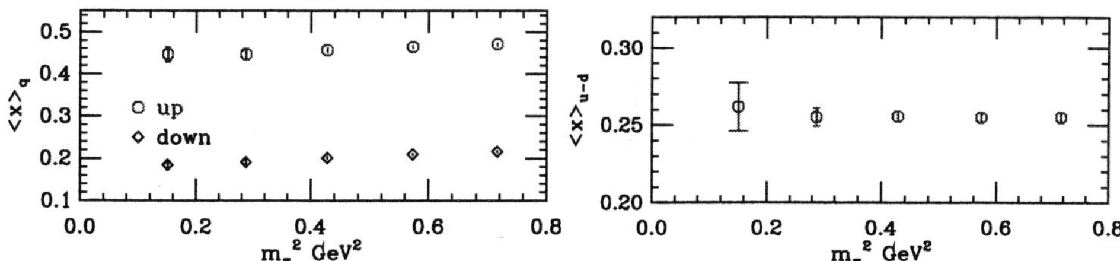

FIGURE 2. Quark density $\langle x \rangle_q$ vs. the pion mass squared. **[left]** The connected up (octagons) and down (diamonds) quark contributions. **[right]** The flavor non-singlet $\langle x \rangle_{u-d}$.

mixing with lower dimensional operators does not occur and hence no power divergent coefficients in Eq. 17 are encountered. This turns out to be a significant constraint on how many moments can be practically computed on the lattice.

Another symmetry that is broken on the lattice for Wilson fermions is chiral symmetry. This results in mixings with lower dimensional operators for the d_n matrix elements. Fortunately, in this case we can use lattice fermions, such as domain wall or overlap and fixed point fermions, that respect chiral symmetry on the lattice. For Wilson fermions, the renormalization of d_2 has been done non-perturbatively as described in [16].

The renormalization constants for all the operators relevant to structure function calculations have been computed perturbatively for Wilson fermions, improved and unimproved [23, 24, 25]. Moreover, the RI-MOM scheme has been used to renormalize non-perturbatively both local [26] and derivative operators [27, 28]. In the Schroedinger functional scheme (developed by the ALPHA collaboration), all local operator renormalizations and the renormalization of v_2 have been computed [29, 30]. In addition, work is underway for computing the constants for flavor singlet operators [31]. For domain wall fermions, all local operators have been renormalized non-perturbatively [32] using the RI-MOM scheme, and also perturbatively [33].

LATTICE RESULTS

In the last several years, the lattice community (QCDSF/UKQCD and LHP/SESAM collaborations) has made a substantial effort to compute the first few moments of the nucleon structure functions. Apart from the constraints imposed by the renormalization of the operators mentioned above, the requirement of having nucleon states with non-zero momentum[4] limits the number of moments we can compute. These are the first three moments of the unpolarized structure functions, the first two moments of the polarized structure functions, and the first two moments of the transversity. These computations have been performed both in quenched and in full QCD with improved and unimproved Wilson fermions [15, 34, 16, 17].

The RBC group has recently begun quenched computations with domain wall fermions [35]. Our current results are restricted only to those matrix elements that can

[4] operators with more than one derivative need non-zero momentum nucleon states see Eq. 8 and Eq. 10

be computed with zero momentum nucleon states. We use the DBW2 gauge action which is known to improve the domain wall fermion chiral properties [36, 37]. We have 416 lattices of size $16^3 \times 32$ at $\beta = 0.870$ with lattice spacing $a^{-1} = 1.3$GeV, providing us with a physical volume ($\sim (2.4fm)^3$) large enough to reduce finite size effects known to affect some nucleon matrix elements, such as g_A [38, 39]. Using fifth dimension length $L_s = 16$ we achieve a residual mass $m_{\text{res}} \sim 0.8$MeV [36, 37]. The input quark masses ranged from 0.02 to 0.10, providing pion masses ranging from 390MeV to 850MeV. Further technical details of our calculation can be found in [35].

Unpolarized Structure Functions

The first three moments of the unpolarized structure functions have been computed by QCDSF in the quenched approximation. The needed chiral and continuum extrapolations have also been performed. A summary of recent results can be found in [40]. In comparison with MRS phenomenological results, the lattice results are typically higher. Also, v_3 is smaller than v_4, while $v_3 > v_4$ is phenomenologically expected. The same computations have been performed by LHP/SESAM in full QCD [17]. Their results indicate that dynamical fermions have only a small effect on the matrix elements they studied.

It has been argued that the main reason for such discrepancies is the fact that lattice computations are performed at rather heavy quark masses and then extrapolated linearly to the chiral limit [41, 42, 43]. For that reason, we need computations at much lighter quark masses in order to see whether there is a disagreement with phenomenological expectations. In quenched QCD, a study with very light quark masses has been done [44] indicating that the linear behavior persists down to 300MeV pion masses.

In Fig. 2 we present our results for the quark density distribution $\langle x \rangle_q$ (v_2). We plot the unrenormalized result for $\langle x \rangle_u$, $\langle x \rangle_d$ and the flavor non-singlet $\langle x \rangle_{u-d}$. Down to 380MeV pion mass no significant curvature within our statistical errors can be seen.[5] The ratio $\langle x \rangle_u / \langle x \rangle_d$ is 2.41(4), linearly extrapolated to the chiral limit, is in agreement with the quenched Wilson fermion results [15, 17].

Polarized Structure Functions

The nucleon axial charge g_A is related to the first moment of the polarized structure function g_1. The current experimental value for g_A/g_V measured from neutron beta decays is 1.2670(30) [45]. Lattice calculations, quenched and dynamical, have been underestimating this quantity typically by 10% to 20% [46, 15, 47, 16, 17, 40]. For earlier calculations see also [48, 49].

One of the systematic errors believed to affect these calculations is the finite volume. In order to study this effect we performed two calculations. One with spatial volume $2.4^3 fm^3$ and another with spatial volume $1.2^3 fm^3$. Our results are shown in Fig. 3[right]. Between these two volumes it is clear that there is a finite volume effect of about 20% at

[5] In [36] we had an indication of some curvature but this effect went away as we doubled the statistics.

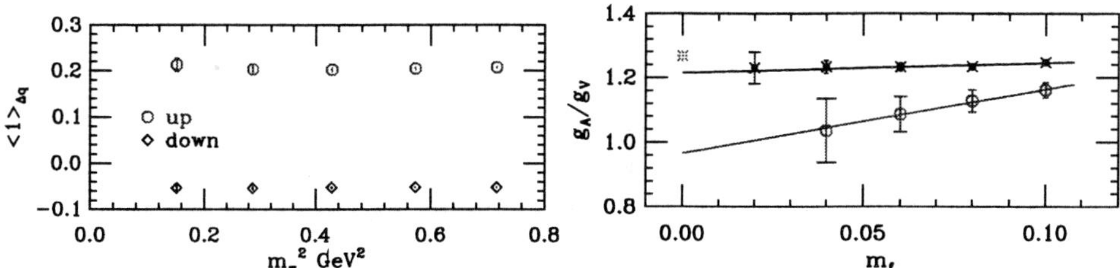

FIGURE 3. Helicity $\langle 1\rangle_{\Delta q}$ vs. the pion mass squared. **[left]** The connected up (octagons) and down (diamonds) quark contributions. **[right]** The nucleon axial charge g_A i.e. flavor non-singlet $\langle 1\rangle_{\Delta u-\Delta d}$.

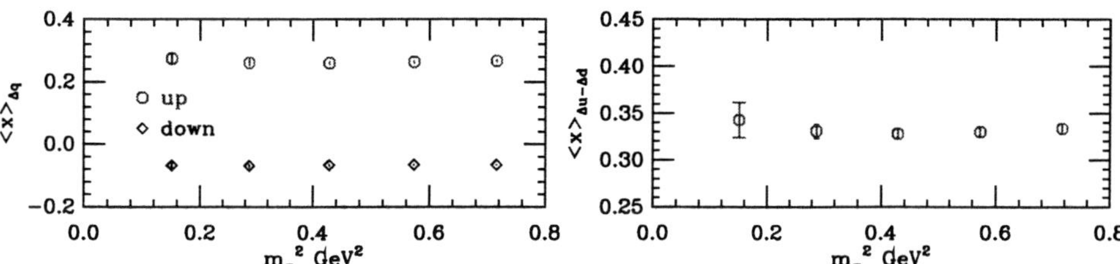

FIGURE 4. Helicity $\langle x\rangle_{\Delta q}$ vs. the pion mass squared. **[left]** The connected up (octagons) and down (diamonds) quark contributions. **[right]** The flavor non-singlet $\langle x\rangle_{\Delta u-\Delta d}$.

the chiral limit. In addition, the linearly extrapolated to the chiral limit value for g_A/g_V is 1.21(2). For a detailed analysis of this computation see [39]. Note that for domain wall fermions g_A/g_V does not require renormalization, since the finite renormalization constants of the axial and the vector currents Z_A, Z_V are equal [32, 50]. In Fig. 3[left] we present the up and down quark contributions of $\langle 1\rangle_{\Delta q}$ for the proton renormalized using $Z_A = .77759(45)$ [37]. In Fig. 4 we present our unrenormalized data for $\langle x\rangle_{\Delta q}$. The ratio $\langle x\rangle_{\Delta u}/\langle x\rangle_{\Delta d}$ linearly extrapolated to the chiral limit is roughly -4, consistent with other lattice results [16, 17]. The lowest moment of the transversity $\langle 1\rangle_{\delta q}$ is also measured. In Fig. 5 we plot the unrenormalized contributions for both the up and down quark, and the flavor non-singlet combination $\langle 1\rangle_{\delta u-\delta d}$. Again the quark mass dependence is very mild and there is no sign of a chiral log behavior. The ratio $\langle 1\rangle_{\delta u}/\langle 1\rangle_{\delta d}$ linearly extrapolated to the chiral limit is also roughly -4.

For computing moments of g_2 we need to calculate the twist 3 matrix elements d_n. We computed the d_1 matrix element which contributes to the first moment of g_2. If chiral symmetry is broken the operator $\mathcal{O}_{34}^{[5]q} = \frac{1}{4}\bar{q}\gamma_5 \left[\gamma_3 \overleftrightarrow{D}_4 - \gamma_4 \overleftrightarrow{D}_3\right] q$ which is used to compute d_1 mixes with the lower dimensional operator $\mathcal{O}_{34}^{\sigma q} = \bar{q}\gamma_5\sigma_{34}q$. Hence in Wilson fermion calculations a non perturbative subtraction has to be performed. This has been done for d_2 by QCDSF [16, 40]. With domain wall fermions this kind of mixing is proportional to the residual mass ($\sim m_{\rm res}/a$), which in our case is negligible. Thus we expect that a straightforward computation of d_1 with domain wall fermions provides directly the physically interesting result. In Fig. 6 we present our unrenormalized results

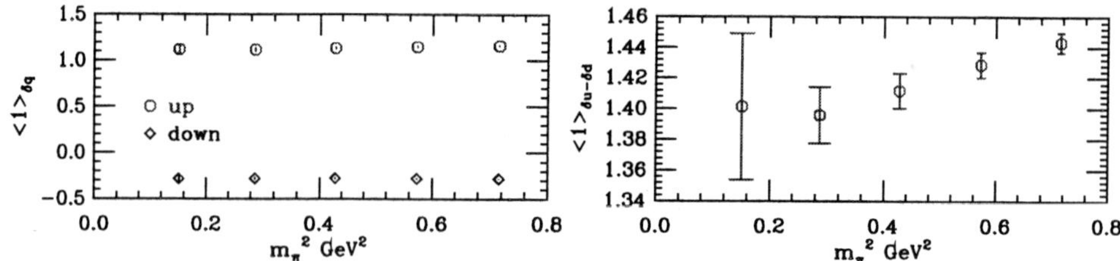

FIGURE 5. Transversity $\langle 1 \rangle_{\delta q}$ vs. the pion mass squared. **[left]** The connected up (octagons) and down (diamonds) quark contributions. **[right]** The flavor non-singlet $\langle 1 \rangle_{\delta u - \delta d}$.

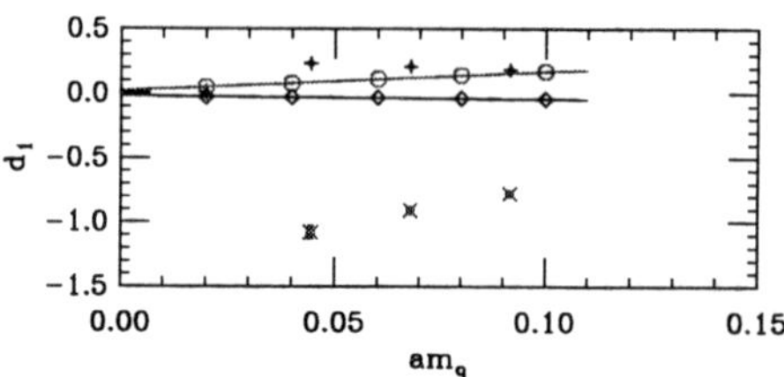

FIGURE 6. The connected d_1 matrix element vs. quark mass for the up (octagons) and down (diamonds) quarks. The up (fancy squares) and down (fancy diamonds) quark for Wilson fermions [17].

for d_1 as a function of the quark mass. For comparison we also plot the unsubtracted quenched Wilson results for $\beta = 6.0$ from [17]. The fact that our result almost vanishes at the chiral limit is an indication that the power divergent mixing is absent for domain wall fermions. The behavior we find for the d_1 matrix element is consistent with that of the subtracted d_2 computed by QCDSF [16, 40] with Wilson fermions.

CONCLUSIONS

In conclusion, lattice computations can play an important role in understanding the hadronic structure and the fundamental properties of QCD. Although some difficulties still exist, several significant steps have been made. Advances in computer technology are expected to play a significant role in pushing these computations closer to the chiral limit and in including dynamical fermions. RBC has already begun preliminary dynamical domain wall fermion computations [51] which we expect to be pushed forward with the arrival of QCDOC [52]. In the near future, we also expect to complete the non-perturbative renormalization of the relevant derivative operators in quenched QCD.

ACKNOWLEDGMENTS

I wish to thank Tom Blum, Chulwoo Jung, Shigemi Ohta, and Shoichi Sasaki for helpful discussions. I also wish to thank the RIKEN BNL research center, BNL, and the U.S. DOE for providing the facilities essential for the completion of this work.

REFERENCES

1. Wilson, K. G., *Phys. Rev.*, **D10**, 2445–2459 (1974).
2. Kaplan, D. B., *Phys. Lett.*, **B288**, 342–347 (1992).
3. Kaplan, D. B., *Nucl. Phys. Proc. Suppl.*, **30**, 597–600 (1993).
4. Shamir, Y., *Nucl. Phys.*, **B406**, 90–106 (1993).
5. Furman, V., and Shamir, Y., *Nucl. Phys.*, **B439**, 54–78 (1995).
6. Narayanan, R., and Neuberger, H., *Nucl. Phys.*, **B412**, 574–606 (1994).
7. DeGrand, T., Hasenfratz, A., Hasenfratz, P., and Niedermayer, F., *Nucl. Phys.*, **B454** (1995).
8. Bietenholz, W., and Wiese, U. J., *Nucl. Phys.*, **B464**, 319–352 (1996).
9. Hasenfratz, P., et al., *Int. J. Mod. Phys.*, **C12**, 691–708 (2001).
10. Manohar, A. V., *Proceedings of the Seventh Lake Louis Winter Institute (World Scientific)* (1992).
11. Jaffe, R. L., and Ji, X.-D., *Phys. Rev. Lett.*, **67**, 552–555 (1991).
12. Jaffe, R. L., and Ji, X.-D., *Nucl. Phys.*, **B375**, 527–560 (1992).
13. Barone, V., Drago, A., and Ratcliffe, P. G., *Phys. Rept.*, **359**, 1–168 (2002).
14. Martinelli, G., and Sachrajda, C. T., *Nucl. Phys.*, **B316**, 355 (1989).
15. Gockeler, M., et al., *Phys. Rev.*, **D53**, 2317–2325 (1996).
16. Gockeler, M., et al., *Phys. Rev.*, **D63**, 074506 (2001).
17. Dolgov, D., et al., *Phys. Rev.*, **D66**, 034506 (2002).
18. Arndt, D., and Savage, M. J., *Nucl. Phys.*, **A697**, 429–439 (2002).
19. Chen, J.-W., and Ji, X.-d., *Phys. Lett.*, **B523**, 107–110 (2001).
20. Chen, J.-W., and Savage, M. J., *Nucl. Phys.*, **A707**, 452–468 (2002).
21. Mandula, J. E., Zweig, G., and Govaerts, J., *Nucl. Phys.*, **B228**, 109 (1983).
22. Gockeler, M., et al., *Phys. Rev.*, **D54**, 5705–5714 (1996).
23. Capitani, S., and Rossi, G., *Nucl. Phys.*, **B433**, 351–389 (1995).
24. Beccarini, G., Bianchi, M., Capitani, S., and Rossi, G., *Nucl. Phys.*, **B456**, 271–295 (1995).
25. Capitani, S., et al., *Nucl. Phys.*, **B593**, 183–228 (2001).
26. Martinelli, G., Pittori, C., Sachrajda, C. T., Testa, M., and Vladikas, A., *Nucl. Phys.*, **B445** (1995).
27. Gockeler, M., et al., *Nucl. Phys.*, **B544**, 699–733 (1999).
28. Capitani, S., et al., *Nucl. Phys. Proc. Suppl.*, **106**, 299–301 (2002).
29. Guagnelli, M., Jansen, K., and Petronzio, R., *Phys. Lett.*, **B459**, 594–598 (1999).
30. Guagnelli, M., Jansen, K., and Petronzio, R., *Phys. Lett.*, **B493**, 77–81 (2000).
31. Palombi, F., Petronzio, R., and Shindler, A., *Nucl. Phys.*, **B637**, 243–271 (2002).
32. Blum, T., et al., *Phys. Rev.*, **D66**, 014504 (2002).
33. Aoki, S., Izubuchi, T., Kuramashi, Y., and Taniguchi, Y., *Phys. Rev.*, **D60**, 114504 (1999).
34. Gockeler, M., et al., *Phys. Lett.*, **B414**, 340–346 (1997).
35. Orginos, K., *Nucl. Phys. Proc. Suppl. (Lattice 2002)* (2002).
36. Orginos, K., *Nucl. Phys. Proc. Suppl.*, **106**, 721–723 (2002).
37. Aoki, Y., *Nucl. Phys. Proc. Suppl.*, **106**, 245–247 (2002).
38. Sasaki, S., Blum, T., Ohta, S., and Orginos, K., *Nucl. Phys. Proc. Suppl.*, **106**, 302–304 (2002).
39. Ohta, S., *Nucl. Phys. Proc. Suppl. (Lattice 2002)* (2002).
40. Gockeler, M., et al., *Nucl. Phys. Proc. Suppl. (Lattice 2002)* (2002).
41. Detmold, W., Melnitchouk, W., Negele, J. W., Renner, D. B., and Thomas, A. W., *Phys. Rev. Lett.*, **87**, 172001 (2001).
42. Detmold, W., Melnitchouk, W., and Thomas, A. W., *Phys. Rev.*, **D66**, 054501 (2002).
43. Thomas, A. W., *Nucl. Phys. Proc. Suppl. (Lattice 2002)* (2002).
44. Gockeler, M., Horsley, R., Pleiter, D., Rakow, P. E. L., and Schierholz, G., *Nucl. Phys. Proc. Suppl. (Lattice 2002)* (2002).
45. Hagiwara, K., et al., *Phys. Rev.*, **D66**, 010001 (2002).
46. Fukugita, M., Kuramashi, Y., Okawa, M., and Ukawa, A., *Phys. Rev. Lett.*, **75**, 2092–2095 (1995).
47. Gusken, S., et al., *Phys. Rev.*, **D59**, 114502 (1999).
48. Liu, K. F., Dong, S. J., Draper, T., Wu, J. M., and Wilcox, W., *Phys. Rev.*, **D49**, 4755–4761 (1994).
49. Dong, S. J., Lagae, J. F., and Liu, K. F., *Phys. Rev. Lett.*, **75**, 2096–2099 (1995).
50. Dawson, C., *Nucl. Phys. Proc. Suppl. (Lattice 2002)* (2002).
51. Izubuchi, T., *Nucl. Phys. Proc. Suppl. (Lattice 2002)* (2002).
52. Boyle, P. A., et al., *Nucl. Phys. Proc. Suppl. (Lattice 2002)* (2002).

Single-Spin Asymmetries and Transversity

Philip G. Ratcliffe

Dip.to di Scienze CC.FF.MM., Univ. degli Studi dell'Insubria—sede di Como, Italy and INFN—sezione di Milano, Italy

Abstract. A pedagogical introduction to single-spin asymmetries (SSA's) and transversity is presented. Discussion in some detail is made of certain aspects of SSA's in lepton–nucleon and in hadron–hadron scattering and the role of pQCD and evolution in the context of transversity.

1. PREAMBLE

Single-spin asymmetries are one of the oldest forms of high-energy spin measurement, the reason being accessibility: the only requirement is *either* beam *or* target polarised, for Λ^0 production neither is necessary! However, after early interest (due to large experimental effects), a theoretical "dark age" descended: pQCD had apparently nothing to say, save that such asymmetries are zero! We now know that the rich phenomenology is matched by a richness of the theoretical framework: the main topic of my talk.

One might argue the inapplicability of pQCD to existing SSA data owing to the low Q^2 accessed while there are several non-pQCD models that can explain some (though not all) of the data. Examples may be found in [1–4]. However, I shall examine SSA's purely from within the pQCD framework.

Transversity too has a long history: the concept (though not the term) was introduced in 1979 by Ralston and Soper via the Drell–Yan process. The leading order (LO) anomalous dimensions were first calculated by Baldracchini *et al.* [6] but forgotten. They were recalculated by Artru and Mekhfi [7] and it turns out that they had also been obtained by various groups as part of the g_2 evolution [8–11]. A complete classification of chirally-odd densities including transversity, is due to Jaffe and Ji [12]. However, as yet there are no experimental data on transversity. This is owing to the inaccessibility (discussed later) of transversity in inclusive deeply-inelastic scattering (DIS).

After introducing single-spin asymmetries and transversity, I shall discuss SSA's in lepton-nucleon and hadron-hadron scattering in some detail and close with a few brief comments and conclusions. A large part of what follows is taken from the *Physics Reports* by Barone, Drago and Ratcliffe [13] and from a forthcoming book by Barone and Ratcliffe [14]. Thus, much credit is due to my two collaborators.

2. INTRODUCTION

Generically, SSA's reflect correlations of the form $\mathbf{s}\cdot(\mathbf{p}\wedge\mathbf{k})$, where $\mathbf{s}$ is a polarisation vector while $\mathbf{p}$ and $\mathbf{k}$ are particle/jet momenta. An example is $\mathbf{s}$ a (transverse) target

CP675, *Spin 2002: 15th Int'l. Spin Physics Symposium and Workshop on Polarized Electron Sources and Polarimeters*, edited by Y. I. Makdisi, A. U. Luccio, and W. W. MacKay

polarisation, **p** a beam direction, and **k** that of a final-state particle. Thus, polarisations in SSA's will typically be transverse (but see later). Transforming the basis from transverse spin to the more familiar helicity, $|\uparrow / \downarrow\rangle = \frac{1}{\sqrt{2}}[|+\rangle \pm \mathrm{i}\,|-\rangle]$, such an asymmetry takes on the (schematic) form

$$A \sim \frac{\langle\uparrow|\uparrow\rangle - \langle\downarrow|\downarrow\rangle}{\langle\uparrow|\uparrow\rangle + \langle\downarrow|\downarrow\rangle} \sim \frac{2\,\mathrm{Im}\,\langle+|-\rangle}{\langle+|+\rangle + \langle-|-\rangle}\,. \tag{1}$$

The presence of both $|+\rangle$ and $|-\rangle$ in the numerator implies a spin-flip amplitude while the precise form indicates interference between spin-flip and non-flip amplitudes, with a non-trivial relative phase difference.

It was soon realised [15] that in the Born approximation and massless limit a gauge theory, such as quantum chromodynamics (QCD), cannot furnish either requirement since fermion helicity is conserved and tree diagrams are real. Quoting from [15]: "*... observation of significant polarizations in the above reactions would contradict either QCD or its applicability.*" Later, however, examining the three-parton correlators related to g_2, Efremov and Teryaev [16] found a way out: the mass scale relevant to spin flip is not that of a current quark but the hadron and the two-loop nature of the diagrams can give rise to an imaginary part. Nonetheless, it was a while before the complexity of the new structures was fully exploited (see, *e.g.*, [17, 18]).

Transversity is the third twist-two partonic density. At this point it is important to make the distinction between partonic *densities*—$q(x)$, $\Delta q(x)$, $\Delta_T q(x)$, ... and DIS *structure functions*—F_1, F_2, g_1, g_2, ... In the leading-twist unpolarised and helicity-dependent cases there is a simple connection between the two: DIS structure functions are weighted sums of partonic densities; in contrast, there is no DIS transversity structure function and g_2 does not correspond to a partonic density. The absence of transversity in DIS is illustrated in Fig. 1. Note that chirality flip is not a problem if the quarks connect

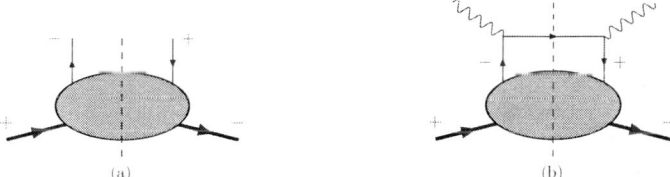

FIGURE 1. (a) Chirally-odd hadron–quark amplitude, (b) chirality-flip forbidden DIS diagram.

to different hadrons, as in Drell–Yan processes.

The three twist-two structures are then:

$$f(x) = \int \frac{\mathrm{d}\xi^-}{4\pi}\, \mathrm{e}^{\mathrm{i}xP^+\xi^-} \langle PS|\overline{\psi}(0)\gamma^+\psi(0,\xi^-,\mathbf{0}_\perp)|PS\rangle\,, \tag{2a}$$

$$\Delta f(x) = \int \frac{\mathrm{d}\xi^-}{4\pi}\, \mathrm{e}^{\mathrm{i}xP^+\xi^-} \langle PS|\overline{\psi}(0)\gamma^+\gamma_5\psi(0,\xi^-,\mathbf{0}_\perp)|PS\rangle\,, \tag{2b}$$

$$\Delta_T f(x) = \int \frac{\mathrm{d}\xi^-}{4\pi}\, \mathrm{e}^{\mathrm{i}xP^+\xi^-} \langle PS|\overline{\psi}(0)\gamma^+\gamma^1\gamma_5\psi(0,\xi^-,\mathbf{0}_\perp)|PS\rangle\,. \tag{2c}$$

The γ_5 matrix signals spin dependence while the extra γ^1 matrix in $\Delta_T f(x)$ signals the chirality-flip that precludes transversity contributions in DIS.

For somewhat similar reasons the LO QCD evolution of transversity is non-singlet like: quark–gluon mixing would require a chirality-flip in a quark propagator—see

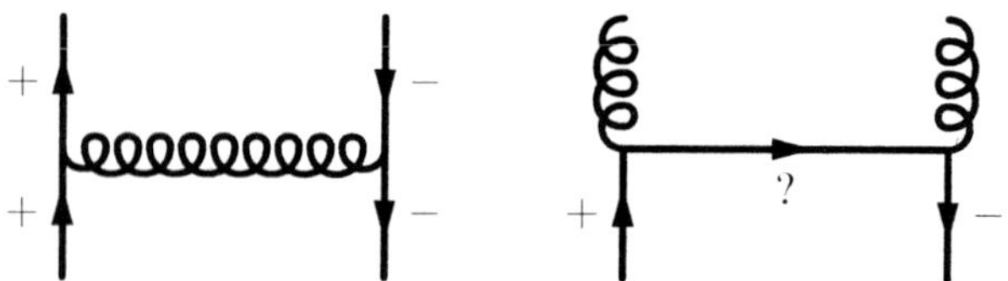

FIGURE 2. Left, transversity evolution kernel; right, disallowed gluon–fermion mixing.

Fig. 2. The LO quark–quark splitting functions are then:

$$P_{qq}^{(0)} = C_{\mathrm{F}}\left(\frac{1+x^2}{1-x}\right)_+, \tag{3a}$$

$$\Delta P_{qq}^{(0)} = P_{qq}^{(0)} \quad \text{(by helicity conservation)}, \tag{3b}$$

$$\Delta_T P_{qq}^{(0)} = C_{\mathrm{F}}\left[\left(\frac{1+x^2}{1-x}\right)_+ - 1 + x\right] = P_{qq}^{(0)}(x) - C_{\mathrm{F}}(1-x). \tag{3c}$$

Note that the first moments of both $P_{qq}^{(0)}$ and $\Delta P_{qq}^{(0)}$ vanish (leading to conservation laws and sum rules), but not so of $\Delta_T P_{qq}^{(0)}$. The effects of evolution are shown in Fig. 3.

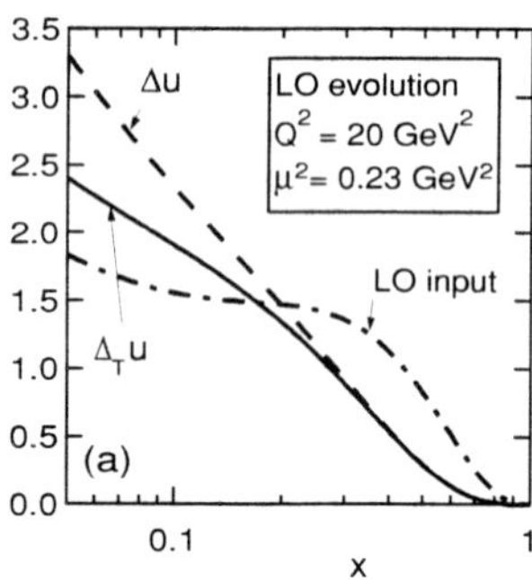

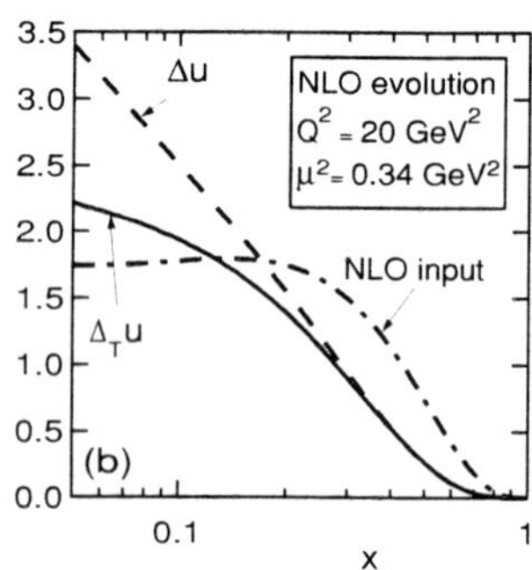

FIGURE 3. The Q^2-evolution of $\Delta_T u(x,Q^2)$ and $\Delta u(x,Q^2)$ compared at (a) LO and (b) NLO; from [19].

By considering hadron–parton helicity amplitudes (see the figure alongside), Soffer [20] constructed an interesting bound involving transversity. Taking into account all relevant symmetries there are only two independent amplitudes, in terms of which all three densities are expressed:

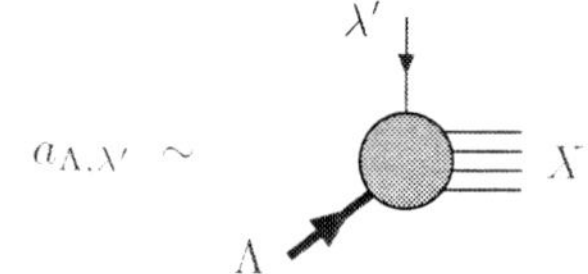

$$f(x) \propto \mathrm{Im}(A_{++,++} + A_{+-,+-}) \propto \Sigma_X (a^*_{++}a_{++} + a^*_{+-}a_{+-}), \tag{4a}$$

$$\Delta f(x) \propto \mathrm{Im}(A_{++,++} - A_{+-,+-}) \propto \Sigma_X (a^*_{++}a_{++} - a^*_{+-}a_{+-}), \tag{4b}$$

$$\Delta_T f(x) \propto \mathrm{Im}A_{+-,-+} \propto \Sigma_X\, a^*_{--}a_{++}. \tag{4c}$$

A straight-forward Schwartz-type inequality: $\sum_X |a_{++} \pm a_{--}|^2 \geq 0$ then translates into $f_+(x) \geq |\Delta_T f(x)|$ or $f(x) + \Delta f(x) \geq 2|\Delta_T f(x)|$, which is precisely the Soffer bound. Notice that it involves all three leading-twist structures.

3. A DIS DEFINITION FOR TRANSVERSITY

The other twist-two densities are naturally defined via DIS, where the parton picture is formulated and many model calculations performed. When translated to Drell–Yan (DY) processes, large K factors appear $\sim O(\pi\alpha_s)$. At RHIC energies this corresponds to a $\sim 30\%$ correction, at EMC/SMC nearly 100%. Pure DY coefficient functions are known, but are scheme dependent. Moreover, a $\ln^2 x/(1-x)$ term appears that is not found for spin-averaged or helicity-dependent DY. Together with recent problems arising in connection with vector–scalar current products, this suggests an interesting check.

One might invoke a Higgs–photon interference mechanism, which, though experimentally hardly viable, does provide a DIS-type definition for transversity since the presence of a scalar vertex forces chirality flip. Care must be taken over the extra renormalisation contribution from the scalar vertex, which factorises into the running mass (or Higgs coupling). One then needs to calculate diagrams of the form of Fig. 4 (and correspondents for DY) in order to obtain the relevant higher-order Wilson coefficients.

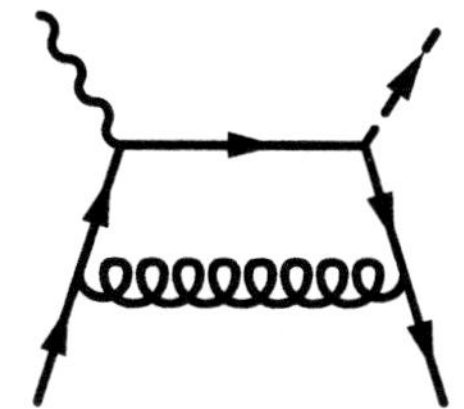

FIGURE 4. Hypothetical Higgs–photon interference mechanism.

Fig. 5 illustrates the effect of the next-to-leading order (NLO) Wilson coefficient [21]. In contrast to the helicity asymmetry [22], where the difference between the LO and NLO is small (the large coefficient of the δ-function is identical in the numerator and denominator), here there are important differences between the spin-averaged denominator and the transversity-dependent numerator. The principal culprits are the δ-function coefficient and the new $\frac{\ln^2 x}{1-x}$ term. The DIS–DY "transformation" coefficients for the unpolarised and transversity cases are:

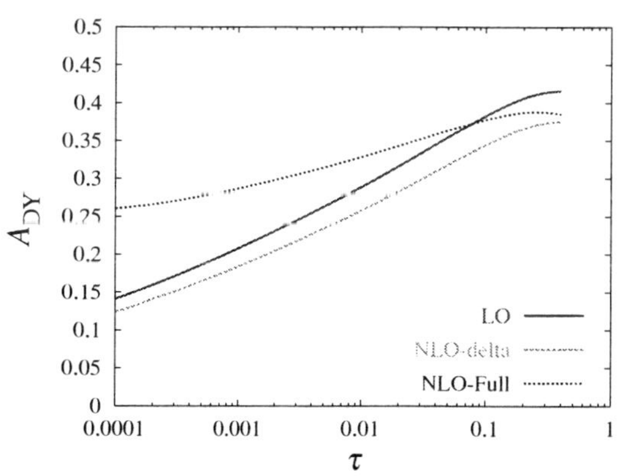

FIGURE 5. Transversity asymmetry (valence only) for DY; $\tau = Q^2/s$, $s = 4{\cdot}10^4\,\mathrm{GeV}^2$.

$$C^f_{q,DY} - 2C^f_{q,DIS} = \frac{\alpha_s}{2\pi}\frac{4}{3}\left[\frac{3}{(1-z)_+} + 2(1+z^2)\left(\frac{\ln(1-z)}{1-z}\right)_+ - 6 - 4z \right.$$
$$\left. + \left(\frac{4}{3}\pi^2 + 1\right)\delta(1-z)\right], \quad \text{(5a)}$$

$$C^h_{q,DY} - 2C^h_{q,DIS} = \frac{\alpha_s}{2\pi}\frac{4}{3}\left[\frac{3z}{(1-z)_+} + 4z\left(\frac{\ln(1-z)}{1-z}\right)_+ - 6z\frac{\ln^2 z}{1-z} + 4(1-z) + \left(\frac{4}{3}\pi^2 - 1\right)\delta(1-z)\right]. \quad (5b)$$

4. T-ODD STRUCTURES

We now wish to generalise the $\mathbf{k}_\perp$-integrated density functions to include all possible correlations between the quark and parent-hadron spins, later on we shall find we also need $\mathbf{k}_\perp$-dependent generalisations. Thus, some extra notation will be required (see [13]). In this way objects like $\Delta^T_L f$ have a simple interpretation:

- *subscripts* $0, L$ and T in density and fragmentation functions denote the *quark* polarisation state,
- *superscripts* $0, L$ and T denote the parent or off-spring *hadron* polarisation state.

The superscript is dropped when equal to the subscript.

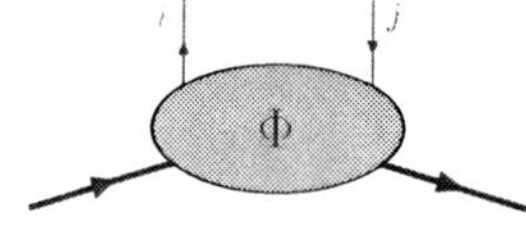

FIGURE 6. Quark–quark correlation matrix.

The aim then is to parametrise the quark–quark correlation matrix (see Fig. 6) in the most general manner, while respecting the natural properties of hermiticity, parity, and time-reversal invariance, though, as we shall see later, this last may be relaxed. The most general decomposition of Φ over a complete basis of Dirac matrices is

$$\Phi(k,P,S) = \tfrac{1}{2}\left\{S\,\mathbb{1} + V_\mu\,\gamma^\mu + A_\mu\gamma_5\gamma^\mu + \mathrm{i}P_5\gamma_5 + \mathrm{i}\,T_{\mu\nu}\,\sigma^{\mu\nu}\gamma_5\right\}, \quad (6)$$

where the quantities S, V^μ, A^μ, P_5 and $T^{\mu\nu}$ are to be constructed from the vectors k^μ, P^μ and the pseudovector S^μ.

Relaxing T invariance allows two new twist-two structures:

$$V^\mu = \cdots + \frac{1}{M}A'_1\,\varepsilon^{\mu\nu\rho\sigma}P_\nu k_{\perp\rho}S_{\perp\sigma} \quad \text{and} \quad T^{\mu\nu} = \cdots + \frac{1}{M}A'_2\,\varepsilon^{\mu\nu\rho\sigma}P_\rho k_{\perp\sigma}. \quad (7)$$

These give rise to two $\mathbf{k}_\perp$-dependent T-odd density functions, $f^\perp_{1T}$ and $h^\perp_1$ [23]:

$$\Phi^{[\gamma^+]} = \cdots - \frac{\varepsilon^{ij}_\perp k_{\perp i}S_{\perp j}}{M} f^\perp_{1T}(x,\mathbf{k}^2_\perp) \quad \text{and} \quad \Phi^{[\mathrm{i}\sigma^{i+}\gamma_5]} = \cdots - \frac{\varepsilon^{ij}_\perp k_{\perp j}}{M} h^\perp_1(x,\mathbf{k}^2_\perp). \quad (8)$$

The partonic interpretation is as follows. The density $f^\perp_{1T}$ relates to the number density of unpolarised quarks in a transversely polarised nucleon:

$$\begin{aligned} P_{q/N\uparrow}(x,\mathbf{k}_\perp) - P_{q/N\downarrow}(x,\mathbf{k}_\perp) &= P_{q/N\uparrow}(x,\mathbf{k}_\perp) - P_{q/N\uparrow}(x,-\mathbf{k}_\perp) \\ &= -2\frac{|\mathbf{k}_\perp|}{M}\sin(\phi_k - \phi_S)\,f^\perp_{1T}(x,\mathbf{k}^2_\perp). \end{aligned} \quad (9)$$

The T-odd density $h^\perp_1$ measures quark transverse polarisation in an unpolarised hadron:

$$P_{q\uparrow/N}(x,\mathbf{k}_\perp) - P_{q\downarrow/N}(x,\mathbf{k}_\perp) = -\frac{|\mathbf{k}_\perp|}{M}\sin(\phi_k - \phi_s)\,h^\perp_1(x,\mathbf{k}^2_\perp). \quad (10)$$

It is convenient to define two quantities $\Delta_0^T f$ and $\Delta_T^0 f$ (related to $f_{1T}^\perp$ and $h_1^\perp$ respectively) by absorbing the factors $|\mathbf{k}_\perp|/M$:

$$\Delta_0^T f(x,\mathbf{k}_\perp^2) \equiv -2\frac{|\mathbf{k}_\perp|}{M} f_{1T}^\perp(x,\mathbf{k}_\perp^2) \quad \text{and} \quad \Delta_T^0 f(x,\mathbf{k}_\perp^2) \equiv -\frac{|\mathbf{k}_\perp|}{M} h_1^\perp(x,\mathbf{k}_\perp^2). \tag{11}$$

The question now arises as to why we should entertain such T-odd quantities at all. There are various attitudes: Anselmino *et al.* [24] (among others) advocate initial-state effects, which prevent implementation of naïve time-reversal invariance. The suggestion is that the colliding hadrons interact strongly with non-trivial relative phases, akin to those arising from final-state effects. An alternative has been proposed by Anselmino *et al.* [25]: they apply a general argument on time reversal for particle multiplets suggested by Weinberg [26]. If the internal structure of hadrons is described at some low momentum scale by a chiral lagrangian, time reversal might be realised in a "non-standard" manner that could mix the multiplet components. According to this approach, the *u* (*d*) density transforms into the *d* (*u*) density, and time-reversal invariance simply establishes a relation between the *u* and *d* sectors.

Finally, Collins [27] has recently reconsidered his proof of the vanishing of $f_{1T}^\perp$ and $h_1^\perp$, based on the field-theoretical expressions of the two densities. He noticed that on reinstating the link operators into quark–quark bi-locals the densities do not simply change sign under T; a future-pointing Wilson line becomes past-pointing. Consequently, time-reversal invariance, does not constrain $f_{1T}^\perp$ and $h_1^\perp$ to be zero, but relates processes probing Wilson lines in opposite directions. Collins thus predicts the Sivers asymmetry, *e.g.*, to have opposite signs in DIS and in DY.

5. LEPTON-NUCLEON SCATTERING

One might hope to access transversity through exclusive leptoproduction of vector mesons (see Fig. 7 alongside). However, Mankiewicz *et al.* [28] showed that the chirally-odd contribution to vector-meson production is actually zero at LO in α_s. Diehl *et al.* [29] and Collins *et al.* [30] later extended this, observing that the chirally-odd contribution vanishes due to angular-momentum and chirality conservation in the hard scattering and so holds at leading twist to all orders in α_s. Thus, exclusive vector-meson leptoproduction cannot probe (off-diagonal) transversity densities.

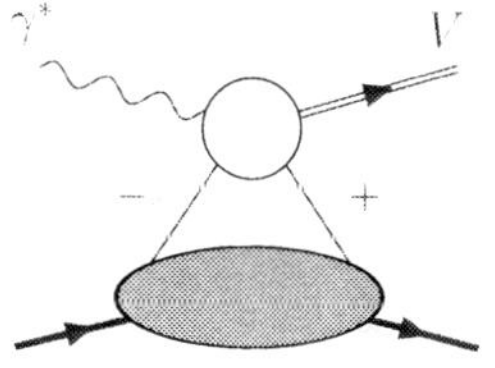

FIGURE 7. Exclusive production of vector-mesons.

The cross-section for production off a longitudinally polarised target is [31]:

$$\frac{\mathrm{d}^5\sigma(\lambda_N)}{\mathrm{d}x\,\mathrm{d}y\,\mathrm{d}z\,\mathrm{d}^2\mathbf{P}_{h\perp}} = -\frac{4\pi\alpha_{\text{em}}^2 s}{Q^4}\lambda_N \sum_a e_a^2 x(1-y)\sin(2\phi_h)$$
$$\times I\left[\frac{2(\hat{\mathbf{h}}\cdot\kappa_\perp)(\hat{\mathbf{h}}\cdot\mathbf{k}_\perp) - \kappa_\perp\cdot\mathbf{k}_\perp}{MM_h} h_{1La}^\perp(x,\mathbf{k}_\perp) H_{1a}^\perp(z,\kappa_\perp)\right]. \tag{12}$$

Transversity is not present here, but the asymmetry does depend on the Collins function $H_1^\perp \propto \sin(2\phi_h)$, also on a $\mathbf{k}_\perp$-dependent density function $h_{1L}^\perp$.

Summarising, in the context of semi-inclusive DIS there are four candidate leading-twist reactions to determine $\Delta_T f$: namely, inclusive leptoproduction of

1. a transversely polarised hadron from a transversely polarised target;
2. an unpolarised hadron from a transversely polarised target;
3. two hadrons from a transversely polarised target;
4. a spin-one polarised or unpolarised hadron from a transversely polarised target.

6. HADRON-HADRON SCATTERING

Let us now examine single-hadron production with a transversely polarised target. The process is exemplified in Fig. 8 alongside: A is transversely polarised and the unpolarised (or spinless) hadron h is produced at large transverse momentum $\mathbf{P}_{hT}$, therefore pQCD is applicable. In typical experiments A and B are protons while h is a pion. According to the factorisation theorem, the differential cross-section for the reaction may be written formally as

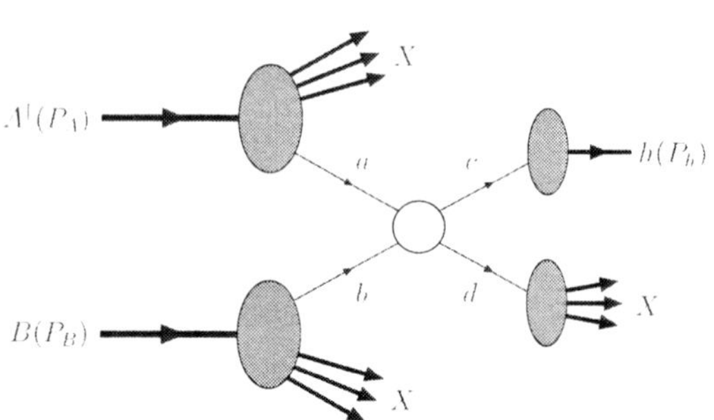

FIGURE 8. Single-hadron production with a transversely polarised target.

$$d\sigma = \sum_{abc} \sum_{\alpha\alpha'\gamma\gamma'} \rho^a_{\alpha'\alpha} f_a(x_a) \otimes f_b(x_b) \otimes d\hat{\sigma}_{\alpha\alpha'\gamma\gamma'} \otimes D^{\gamma'\gamma}_{h/c}(z), \tag{13}$$

f_a (f_b) is the density of parton a (b) inside hadron A (B), $\rho^a_{\alpha\alpha'}$ is the parton a spin density matrix, $D^{\gamma\gamma'}_{h/c}$ is the fragmentation matrix of parton c into hadron h and $d\hat{\sigma}/d\hat{t}$ is the elementary cross-section:

$$\left(\frac{d\hat{\sigma}}{d\hat{t}}\right)_{\alpha\alpha'\gamma\gamma'} = \frac{1}{16\pi\hat{s}^2}\frac{1}{2}\sum_{\beta\delta} M_{\alpha\beta\gamma\delta} M^*_{\alpha'\beta\gamma'\delta}. \tag{14}$$

Here $M_{\alpha\beta\gamma\delta}$ is the amplitude for the hard partonic process, displayed in Fig. 9. For an unpolarised produced hadron, the off-diagonal elements of $D^{\gamma\gamma'}_{h/c}$ vanish, *i.e.*, $D^{\gamma\gamma'}_{h/c} \propto \delta_{\gamma\gamma'}$. Helicity conservation then implies $\alpha = \alpha'$ and thus there can be no dependence on the spin of hadron A. Consequently, all SSA's must vanish.

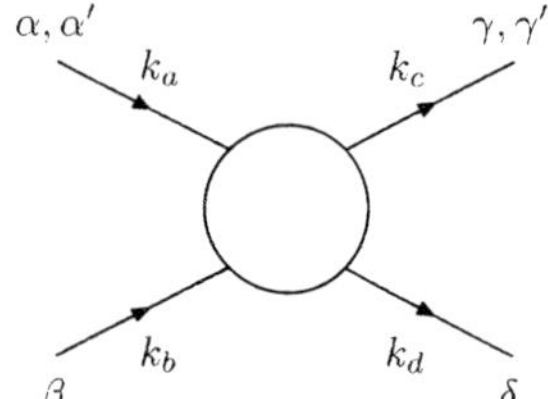

FIGURE 9. Partonic hard scattering amplitude.

To avoid this conclusion, intrinsic quark transverse motion or higher-twist effects must be invoked; this can be done in three different ways:

1. κ_T in hadron h allows $D^{\gamma\gamma'}_{h/c}$ to be non-diagonal (a fragmentation T-odd effect), the Collins effect [32];

2. $\mathbf{k}_T$ in hadron A implies that $f_a(x_a)$ should be replaced by the $P_a(x_a, \mathbf{k}_T)$, which may depend on the spin of hadron A (a density T-odd effect), the Sivers effect [33];
3. $\mathbf{k}'_T$ in hadron B implies that $f_b(x_b)$ should be replaced by $P_b(x_b, \mathbf{k}'_T)$; a transverse spin of parton b in the unpolarised hadron B may then couple to the transverse spin of parton a in A (a density T-odd effect), see [34].

It should be stressed that all these intrinsic-κ_T, -$\mathbf{k}_T$, or -$\mathbf{k}'_T$ effects are T-odd. Note too that when intrinsic quark transverse motion is taken into account, the QCD factorisation theorem is not proven.

Assuming, for discussion purposes, factorisation to be valid, the cross-section is

$$E_h \frac{d^3\sigma}{d^3\mathbf{P}_h} = \sum_{abc} \sum_{\alpha\alpha'\beta\beta'\gamma\gamma'} \int dx_a \int dx_b \int d^2\mathbf{k}_T \int d^2\mathbf{k}'_T \int d^2\kappa_T \frac{1}{\pi z} \times P_a(x_a, \mathbf{k}_T)\, \rho^a_{\alpha'\alpha}\, P_b(x_b, \mathbf{k}'_T)\, \rho^b_{\beta'\beta} \left(\frac{d\hat{\sigma}}{d\hat{t}}\right)_{\alpha\alpha'\beta\beta'\gamma\gamma'} D^{\gamma'\gamma}_{h/c}(z, \kappa_T), \quad (15)$$

where

$$\left(\frac{d\hat{\sigma}}{d\hat{t}}\right)_{\alpha\alpha'\beta\beta'\gamma\gamma'} = \frac{1}{16\pi\hat{s}^2} \sum_\delta M_{\alpha\beta\gamma\delta} M^*_{\alpha'\beta'\gamma'\delta}\,. \quad (16)$$

The Collins mechanism requires intrinsic quark transverse motion inside the produced hadron h while neglecting all other quark transverse momenta (the spin of A is along y):

$$E_h \frac{d^3\sigma(\mathbf{S}_T)}{d^3\mathbf{P}_h} - E_h \frac{d^3\sigma(-\mathbf{S}_T)}{d^3\mathbf{P}_h} = -2\,|\mathbf{S}_T| \sum_{abc} \int dx_a \int \frac{dx_b}{\pi z} \int d^2\kappa_T \times \Delta_T f_a(x_a)\, f_b(x_b)\, \Delta_{TT}\hat{\sigma}(x_a, x_b, \kappa_T)\, \Delta^0_T D_{h/c}(z, \kappa^2_T), \quad (17)$$

where $\Delta_{TT}\hat{\sigma}$ is a partonic spin-transfer asymmetry. The Sivers effect relies on T-odd density functions and predicts a form

$$E_h \frac{d^3\sigma(\mathbf{S}_T)}{d^3\mathbf{P}_h} - E_h \frac{d^3\sigma(-\mathbf{S}_T)}{d^3\mathbf{P}_h} = |\mathbf{S}_T| \sum_{abc} \int dx_a \int \frac{dx_b}{\pi z} \int d^2\mathbf{k}_T \times \Delta^T_0 f_a(x_a, \mathbf{k}^2_T)\, f_b(x_b) \frac{d\hat{\sigma}(x_a, x_b, \mathbf{k}_T)}{d\hat{t}} D_{h/c}(z), \quad (18)$$

where $\Delta^T_0 f$ (related to $f^\perp_{1T}$) is a T-odd density. Finally, the effect studied by Boer gives rise to an asymmetry involving the other T-odd density $\Delta^0_T f$ (related to $h^\perp_1$):

$$E_h \frac{d^3\sigma(\mathbf{S}_T)}{d^3\mathbf{P}_h} - E_h \frac{d^3\sigma(-\mathbf{S}_T)}{d^3\mathbf{P}_h} = -2|\mathbf{S}_T| \sum_{abc} \int dx_a \int \frac{dx_b}{\pi z} \int d^2\mathbf{k}'_T \times \Delta_T f_a(x_a)\, \Delta^0_T f_b(x_b, \mathbf{k}'^2_T)\, \Delta_{TT}\hat{\sigma}'(x_a, x_b, \mathbf{k}'_T)\, D_{h/c}(z), \quad (19)$$

where $\Delta_{TT}\hat{\sigma}'$ is a partonic initial-spin correlation.

As already mentioned, Efremov *et al.* [35] pointed out that SSA's can arise in pQCD at higher twist via gluonic poles in diagrams involving qqg correlators. Such asymmetries were evaluated by Qiu and Sterman, who studied direct photon production [17, 18] and hadron production [36]. The extension to chirally-odd contributions was made by Kanazawa *et al.* [37, 38]. The results may be summarised in

$$\begin{aligned} d\sigma = \sum_{abc} \Big\{ & G_F^a(x_a, y_a) \otimes f_b(x_b) \otimes d\hat{\sigma} \otimes D_{h/c}(z) \\ & + \Delta_T f_a(x_a) \otimes E_F^b(x_b, y_b) \otimes d\hat{\sigma}' \otimes D_{h/c}(z) \\ & + \Delta_T f_a(x_a) \otimes f_b(x_b) \otimes d\hat{\sigma}'' \otimes D_{h/c}^{(3)}(z) \Big\}. \end{aligned} \tag{20}$$

The first term (not containing transversity) is a chirally-even mechanism studied by Qiu and Sterman, the second term is the chirally-odd contribution analysed by Kanazawa and Koike, and the third contains a twist-three fragmentation function $D_{h/c}^{(3)}$.

Admitting twist-three contributions, the SSA in DY is [39]

$$\begin{aligned} A_T^{\mathrm{DY}} = & |\mathbf{S}_{1\perp}| \frac{2\sin 2\theta}{1+\cos^2\theta} \sin(\phi - \phi_{S_1}) \frac{M}{Q} \\ & \times \frac{\sum_a e_a^2 \left[x_1 \widetilde{f}_T^a(x_1) \bar{f}_a(x_2) + x_2 \Delta_T f_a(x_1) \bar{h}_a(x_2) \right]}{\sum_a e_a^2 f_a(x_1) \bar{f}_a(x_2)}, \end{aligned} \tag{21}$$

where $\widetilde{f}_T(x)$ and $\bar{h}(x)$ are twist-three T-odd density functions. The existence of such T-odd density functions has been advocated by Boer [34] to explain an anomalously large $\cos 2\phi$ term seen in unpolarised DY data. As presented, such contributions would require initial-state interactions—this may be considered unlikely. Hammon *et al.* [40] have shown that SSA's may arise from gluonic poles in twist-three multiparton correlation functions (see Fig. 10 alongside). The corresponding SSA is then

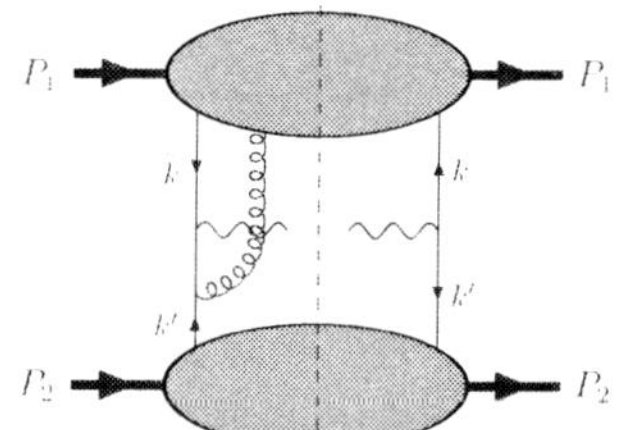

FIGURE 10. A twist-three gluon-pole contribution to DY.

$$\begin{aligned} A_T^{\mathrm{DY}} \propto & |\mathbf{S}_{1\perp}| \frac{2\sin 2\theta}{1+\cos^2\theta} \sin(\phi - \phi_{S_1}) \frac{M}{Q} \\ & \times \frac{\sum_a e_a^2 [G_F^a(x_1, x_1) \bar{f}_a(x_2) + \Delta_T f_a(x_1) E_F^a(x_2, x_2)]}{\sum_a e_a^2 f_a(x_1) \bar{f}_a(x_2)}. \end{aligned} \tag{22}$$

Comparing this with the previous expression we may identify

$$f_T^{\mathrm{eff}}(x) \sim G_F(x,x) \sim \int dy \, \mathrm{Im}\, G_A^{\mathrm{eff}}(x,y), \tag{23a}$$

$$h^{\mathrm{eff}}(x) \sim E_F(x,x) \sim \int dy \, \mathrm{Im}\, E_A^{\mathrm{eff}}(x,y). \tag{23b}$$

Thus, T-odd functions at twist three, can explain A_T^{DY} via quark–gluon interactions, without initial-state effects.

7. CONCLUSIONS

The study of single-spin asymmetries has become a very complex area of high-energy spin physics. A plethora of new structure and fragmentation functions has opened the way to explaining much existing phenomenology. However, in order to distinguish and separate out the various mechanisms proposed, a large amount of diverse high-energy data will be necessary and it is difficult (if not indeed irrelevant and even misleading) to single out at a few key experiments. In other words, all new data will be very welcome.

REFERENCES

1. B. Andersson, G. Gustafson and G. Ingelman, *Phys. Lett.* **B85** (1979) 417.
2. T.A. DeGrand and H.I. Miettinen, *Phys. Rev.* **D24** (1981) 2419; *erratum, ibid.* **D31** (1985) 661.
3. R. Barni, G. Preparata and P.G. Ratcliffe, *Phys. Lett.* **B296** (1992) 251.
4. J. Soffer and N.A. Tornqvist, *Phys. Rev. Lett.* **68** (1992) 907.
5. J. Ralston and D.E. Soper, *Nucl. Phys.* **B152** (1979) 109.
6. F. Baldracchini, N.S. Craigie, V. Roberto and M. Socolovsky, *Fortschr. Phys.* **30** (1981) 505.
7. X. Artru and M. Mekhfi, *Z. Phys.* **C45** (1990) 669.
8. J. Kodaira, S. Matsuda, K. Sasaki and T. Uematsu, *Nucl. Phys.* **B159** (1979) 99.
9. I. Antoniadis and C. Kounnas, *Phys. Rev.* **D24** (1981) 505.
10. A.P. Bukhvostov, É.A. Kuraev, L.N. Lipatov and G.V. Frolov, *Nucl. Phys.* **B258** (1985) 601.
11. P.G. Ratcliffe, *Nucl. Phys.* **B264** (1986) 493.
12. R.L. Jaffe and X.-D. Ji, *Nucl. Phys.* **B375** (1992) 527.
13. V. Barone, A. Drago and P.G. Ratcliffe, *Phys. Rep.* **359** (2002) 1; hep-ph/0104283.
14. V. Barone and P.G. Ratcliffe, *Transverse Polarisation of Quarks in Hadrons* (World Sci., in press).
15. G.L. Kane, J. Pumplin and W. Repko, *Phys. Rev. Lett.* **41** (1978) 1689.
16. A.V. Efremov and O.V. Teryaev, *Phys. Lett.* **B150** (1985) 383.
17. J.-W. Qiu and G. Sterman, *Phys. Rev. Lett.* **67** (1991) 2264.
18. J.-W. Qiu and G. Sterman, *Nucl. Phys.* **B378** (1992) 52.
19. A. Hayashigaki, Y. Kanazawa and Y. Koike, *Phys. Rev.* **D56** (1997) 7350; hep-ph/9707208.
20. J. Soffer, *Phys. Rev. Lett.* **74** (1995) 1292; hep-ph/9409254.
21. P.G. Ratcliffe, work in progress.
22. P.G. Ratcliffe, *Nucl. Phys.* **B223** (1983) 45.
23. D. Boer and P.J. Mulders, *Phys. Rev.* **D57** (1998) 5780; hep-ph/9711485.
24. M. Anselmino and F. Murgia, *Phys. Lett.* **B442** (1998) 470; hep-ph/9808426.
25. M. Anselmino, V. Barone, A. Drago and F. Murgia, Non-standard time reversal and transverse single-spin asymmetries, hep-ph/0209073.
26. S. Weinberg, *The Quantum Theory of Fields. Vol. 1: Foundations* (Cambridge U. Press, 1995).
27. J.C. Collins, *Phys. Lett.* **B536** (2002) 43; hep-ph/0204004.
28. L. Mankiewicz, G. Piller and T. Weigl, *Eur. Phys. J.* **C5** (1998) 119; hep-ph/9711227.
29. M. Diehl, T. Gousset and B. Pire, *Phys. Rev.* **D59** (1999) 034023; hep-ph/9808479.
30. J.C. Collins and M. Diehl, *Phys. Rev.* **D61** (2000) 114015; hep-ph/9907498.
31. A.M. Kotzinian and P.J. Mulders, *Phys. Lett.* **B406** (1997) 373; hep-ph/9701330.
32. J.C. Collins, *Nucl. Phys.* **B396** (1993) 161; hep-ph/9208213.
33. D. Sivers, *Phys. Rev.* **D41** (1990) 83.
34. D. Boer, *Phys. Rev.* **D60** (1999) 014012; hep-ph/9902255.
35. A.V. Efremov and O.V. Teryaev, *Yad. Fiz.* **36** (1982) 242; *transl., Sov. J. Nucl. Phys.* **36** (1982) 140.
36. J.-W. Qiu and G. Sterman, *Phys. Rev.* **D59** (1999) 014004; hep-ph/9806356.
37. Y. Kanazawa and Y. Koike, *Phys. Lett.* **B478** (2000) 121; hep-ph/0001021.
38. Y. Kanazawa and Y. Koike, *Phys. Lett.* **B490** (2000) 99; hep-ph/0007272.
39. D. Boer, P.J. Mulders and O.V. Teryaev, *Phys. Rev.* **D57** (1997) 3057; hep-ph/9710223.
40. N. Hammon, O. Teryaev and A. Schäfer, *Phys. Lett.* **B390** (1997) 409; hep-ph/9611359.

Polarized Jets And Storage Cell Targets For Storage Rings

Erhard Steffens

Physikalisches Institut, University of Erlangen-Nürnberg
steffens@physik.uni-erlangen.de

Abstract. The state of the art of polarized gas targets for storage rings is reviewed. The storage cell technique is inevitable for producing densities with sufficient luminosity for electron scattering experiments. Different sources of polarized atoms for jets and storage cell targets are compared. Remarkable progress has been achieved over the last 15 years, which has led to a vast amount of experimental results with high statistics and good systematic precision. For hydrogen or deuterium gas targets based on atomic beam sources and cold storage cells an areal density of 10^{14} atoms/cm^2 with high polarization has been achieved. For a next generation experiment this figure could be considerably enhanced.

I. INTRODUCTION

Targets discussed in this review are employed in spin-dependent scattering experiments on nucleon structure and on the interaction between nucleons. The most direct approach to *nucleon* targets is hydrogen for the proton, and deuterium as weakly bound proton-neutron system. Both hydrogen isotopes can be polarized by Stern-Gerlach separation in inhomogeneous magnetic fields and subsequent rf transitions in a very flexible manner. Due to its strong binding hydrogen atoms recombine at elevated densities unlike ^{3}He which can be compressed as polarized gas to high pressure. Therefore hydrogen targets are ideally suited for storage rings with their high stored current of several mA up to more than a hundred mA resulting in luminosities of 10^{29} to several 10^{31}/cm^{2}s. Polarized gas targets in storage rings offer the following distinct advantages:

- No dilution of polarization due to their isotopic purity unlike solid polarized target material where a high contamination of background material is present, e.g. about 6/7 of nitrogen nucleons and 1/7 of polarizeable protons in NH_3.
- Detection of low-energy recoils enabled.
- High systematic precision due to rapid switching of the sign of P_z (and P_{zz}) on a ms time scale if required.
- No background of unwanted reactions which might have a much higher cross section, e.g. pion production in pp scattering at threshold (PINTEX) which would be completely dominated by pions produced on heavier nuclei; the same holds for deeply virtual Compton scattering (DVCS) on the nucleon as observed at HERMES for the first time.

CP675, *Spin 2002: 15th Int'l. Spin Physics Symposium and Workshop on Polarized Electron Sources and Polarimeters*, edited by Y. I. Makdisi, A. U. Luccio, and W. W. MacKay

II. METHODS FOR POLARIZED GAS TARGETS

Beam Of Polarized Atoms: Jet Target

After pioneering work at the Stanford tandem accelerator in the 1970s [1], the first polarized Jet target experiment in an electron ring was performed at the 2 GeV VEPP-3 storage ring in Novosibirsk [2]. The target thickness achieved in 1985 was $2 \cdot 10^{11}$D-atoms/cm^2. A Jet target of similar density is presently employed at the EDDA experiment at COSY (Jülich) [3]. A hydrogen Jet of well known polarization will be used at RHIC (BNL) in order to establish a polarization standard up to proton energies of 250GeV [4].

Whereas these targets are based on atomic beam sources (ABS) a new approach studied at Michigan [5] makes use of hydrogen atoms cooled to < 1K by means of walls covered with superfluid He. In a strong magnetic field of several Tesla the thermal energy E_{th} becomes small compared with the magnetic energy $E_{magn} = \pm\mu_B \cdot B$. Then one electron spin state is trapped in the high field region, while the other state with $m_J = +1/2$ is extracted from the solenoid field and accelerated by the fringe field to a fairly monochromatic atomic beam. After focusing by a parabolic mirror and a large-bore sextupole magnet a dense target spot is produced. Here the dependence of the target areal density t on the inverse velocity helps to enhance the density.

Polarized Atoms Injected Into A Storage Cell

There are two ways of injecting polarized atoms into a cell:

- *Ballistic Flow*, i.e. from an atomic beam source for H or D atoms;
- *Flow driven by pressure gradient*, i.e. from a laser-driven source for H, D or ^{3}He working with gas at a stagnation pressure above the one present in storage cells.

The storage cell has been proposed by W. Haeberli in 1965 [6], probably inspired by the storage bulb of the hydrogen maser [7]. Applications as storage ring targets were proposed by Schüler [8], and Haeberli [9], in the early 80s. The basic idea is sketched in Figure 1. A particle current I_t of polarized atoms is fed into the center of a T-shaped storage cell, e.g. ballistically via its feed tube. Beam particles with current I_b circulate along the axis of the straight beam tube. Target atoms diffuse from the center outwards through all three tubular structures, resulting in a triangular density profile with central *volume* density ρ_0 which is given by $\rho_0 = I_t/C_{tot}$, with C_{tot} being the total conductance from the center outwards. The target *areal* density t is then given by

$$t = L \cdot \rho_o = L \cdot I_t / C_{tot} \quad (1)$$

The first test of a storage cell has been performed in 1980 by the Wisconsin group [10]. H atoms from an atomic beam source were injected into a Drifilm-coated Pyrex vessel. The estimated average number of wall bounces per atom was about 900. The polarization was determined by the left-right asymmetry in elastic (α,p)-scattering. In a weak magnetic field, the proton polarization of the upper two substates is ½. After subtracting a 50% background a proton polarization of $P_z = 0.43 \pm 0.07$ was deduced

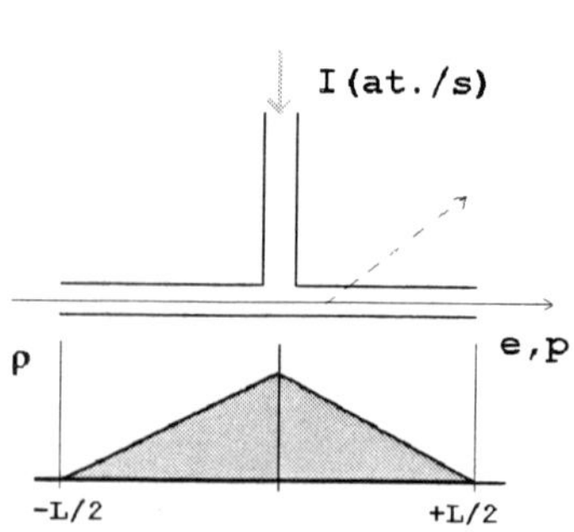

FIGURE 1. Schematic of a storage cell target located in a vacuum chamber on axis of the stored beam. The energetic beam of electrons or protons pass the straight beam tube. Scattered particles may leave the cell via its thin walls and exit from the vacuum chamber by means of an exit window into the detector. A beam of atoms from the source is injected into the cell center via a feed tube. The atoms form a triangular density distribution

which is compatible with no depolarization. The areal density viewed by the Si detector system was $1.1 \cdot 10^{12}/cm^2$, which was more than a factor of 5 improvement over free jets.

Design Of Storage Cells

Requirements: The cell walls must not limit the acceptance of the ring which depends on the local machine functions. As t goes with the inverse total conductance and for a round beam tube roughly with $1/r^3$, r being the inner tube radius, it is important to provide small beam size at the target position. If σ_i are the rms beam radii (i = x, y) then depending on the machine type the following condition for the half axis of an elliptical beam tube might be adequate:

$$a_i = 15\,\sigma_i + 1\text{mm} \qquad (i = x, y) \tag{2}$$

The cell length determines strongly the available target density t (see equation 1). For vanishing conductance of the feed tube t scales with L^2. On the other hand, L must be chosen in accordance with the detector requirements which may limit the maximum length to 250mm (PINTEX, [11, 12]) or 400mm (HERMES, [13, 14]).

The cell walls should be as thin as possible for the passage of scattered particles. In case of an electron storage ring they have to be of metal in order to shield the bunch field from the detector and to avoid rf heating of the target chamber. A suitable coating like Teflon or Drifilm is applied for minimum depolarization and recombination. By cell cooling the density is enhanced with $(T_{cell})^{-1/2}$. The minimum temperature is determined by the onset of recombination below 60K.

Example: The PINTEX Cell. This experiment [11] runs at the IUCF cooler ring with proton beams of up to about 450MeV. It is optimized for the detection of elastic pp and pd scattering and pion production near threshold by Si strip detectors close to the cell and tracking and scintillating detectors downstream the exit window[12]. Very thin cell walls at room temperature are provided by suspending $450\mu g/cm^2$ Teflon foils by means of a system of fins, as shown in Figure 2. The detector is designed such that the symmetry of the cell is taken into account. A weak guide field in all three directions x,y,z is applied by a system of orthogonal coils and can be switched rapidly

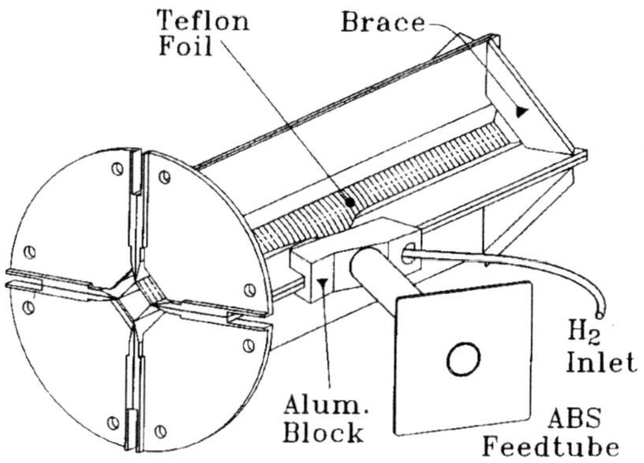

FIGURE 2. Schematic picture of the PINTEX target cell [11], consisting of four quadrants. The 250mm long beam tube of quadratic cross section $10\times10mm^2$ is formed by Teflon foils suspended by four pairs of fins fixed by braces. A polarized H or D atomic beam is injected via the feed tube. Unpolarized gas can be admitted to the cell center as well by means of a capillary. Scattered particles may exit under all azimuthal directions except the horizontal and vertical plane.

in order to change the polarization components $P_{x,y,z}$ in the reaction system.

Example: The HERMES Cell: The cell [15] is designed for operation in the 27.5GeV HERA electron ring with very short bunches and peak currents of more than 100A. It consists of 75µm thick pure Al walls coated with Drifilm and has an elliptical cross section of $21.0\times8.9mm^2$. The cell is suspended by cooling rails cooled with cold He gas to temperatures of 100K and below. Throughout the experiment the electron beam is surrounded by conducting walls of slowly varying cross section for suppressing the excitation of wake fields. A system of tungsten collimators upstream of the cell serves to protect the cell walls from Synchrotron radiation. With the help of the involved target diagnostics even small changes of the cell wall properties can be detected. After operating with beam clear signs of detoriation of the Drifilm coating are visible. It has turned out that under normal operating conditions a layer of frozen water is formed which serves as *renewable* surface with excellent properties [14, 15]. The water is supplied via the atomic beam by the ABS dissociator [16] which is run with H_2 or D_2 and a small oxygen admixture for optimum dissociation fraction. O_2 is converted in the discharge to water.

III. SOURCES OF POLARIZED GAS

An overview of operating targets and targets under design or construction is given in Table 1. Target types are labeled "Source-Target" with ABS = atomic beam source,

TABLE 1. Comparison of operating targets and targets under design or construction.

Experiment	Target Gas	Type	Beam Part.	Status
PINTEX / IUCF	H, D	ABS - SC	p	terminated (2002)
EDDA / COSY	H	ABS - J	p	operational
DEUTERON / VEPP-3	D	ABS – SC	e	operational
HERMES / HERA	H, D	ABS - SC	e	operational
ANKE / COSY	D, H	ABS - SC	p	under construction
Polarized Jet for RHIC	H	ABS - J	p	under design
BLAST / Bates	H, D	ABS - SC	e	commissioning
" "	H	LDS – SC	e	under development
" "	^{3}He	OP – SC	e	proposed

LDS = laser-driven source, OP = optically pumped source, and J = jet, SC = storage cell. In the following, three source types are discussed.

1. Atomic Beam Source: A modern ABS starts with the formation of a cold hydrogen atomic beam from a dissociator with cold Al nozzle and powerful differential pumping system, based on turbomolecular and cryogenic pumps of total pumping speed of about 10^4l/s. Spin-dependent focusing is provided by high-field permanent sextupole magnets with pole tip fields of up to 1.6T [17]. Nuclear polarization P_z (P_{zz}) is produced by a system of rf-transitions with nearly 100% efficiency. For H and D in two substates beams with polarization close to the maximum possible values can be produced and switched rapidly.

The production of cold atomic beams of high brilliance has been studied over the last 15 years at various labs incl. ANL, Bonn, ETH Zürich, CERN, Dubna, Novosibirsk, and by a Heidelberg-Marburg-Munich-Wisconsin collaboration. The particle current accepted by a standard compression tube of 10mm diameter and 100mm length which can be measured to a few % error, went up by a factor of 3. The sources exhibit very low contamination in the atomic beam important for cryogenic cells, and they are extremely reliable as proven by the HERMES ABS which has been operated over the last 6 years for most of the calendar time with very high availability. The particle flow rate of H beams in two substates seems to saturate at values below 10^{17}/s. Systematic studies of beam formation are performed at DESY by the HERMES target group. The recent work on the target test bench was described at this symposium by Nass [16]. The measured velocity distributions and intensities as function of gas flow and nozzle temperature are reproduced by DSMC simulations. A recent attempt to increase the beam intensity by applying a hollow dense *carrier jet* failed due to the strong admixture of carrier gas to the central beam which can be reproduced by recent simulations [16].

2. Ultracold Source: The status of the Michigan Jet was reviewed at PST01 by Luppov [5]. The improved version comprises a 12T solenoid, 100mW of cooling power at 300mK for the separation cell with parabolic mirror, a large-aperture rf transition and sextupole magnet, and a powerful cryoabsorber as beam catcher. By means of a slit-type compression tube of $11.0 \cdot 1.4$mm^2 a maximum intensity of $2.2 \cdot 10^{15}$H/s in two substates has been detected, corresponding to an areal density of $1.1 \cdot 10^{12}$H/cm^2.

3. Laser-Driven Source: The basic idea is to employ the high number of polarized photons from a laser to polarize hydrogen. As there are no UV-lasers available for direct optical pumping of ground-state hydrogen atoms an indirect approach is to optically pump a small admixture of K atoms and to transfer their polarization by spin exchange collisions to the hydrogen atoms [18]. The arrangement is shown schematically in Figure 3. The glass cell with Drifilm coating is kept between 150 and 200C in order to avoid condensation of potassium. The storage cell must be operated at a similar temperature, which is about five times higher than cryogenic cells used with AB sources. The anticipated flow rates and target densities are well beyond ABS figures but the feasibility of such targets has not been demonstrated yet.

The main difficulty is to maintain a high dissociation fraction α at the exit of the pumping cell and in the target cell, in particular its long-term performance. By the

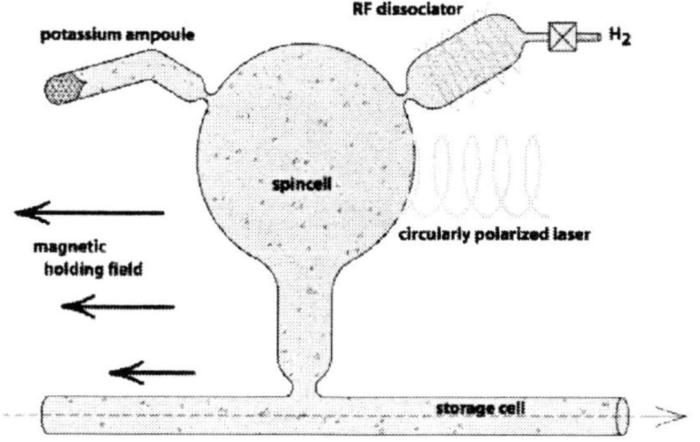

FIGURE 3. Schematic picture of the Laser-Driven Source [19]. Atomic hydrogen is injected into the heated spin-exchange cell where the interaction with optically pumped K atoms takes place. Circular polarized photons at 770nm from a Ti-Sapphire laser are shined in parallel to the holding field. The gas then flows into the storage cell target which is kept at a similar temperature as the cell in order to avoid condensation of potassium.

presence of molecules the polarization of atoms is diluted. Assuming vanishing nuclear polarization of molecules the quality factor QF is given by

$$QF = t \cdot P_{av}^2 = t \cdot P_{at}^2 \cdot \boldsymbol{\alpha}^2 \tag{3}$$

t being the areal target density, P_{av} the average nuclear polarization of target particles, and P_{at} the polarization of atoms.
The LDS development started at Argonne in 1984 [20] and was continued by work at Erlangen, Illinois, and recently MIT [19]. Probably the most systematic approach has been chosen at Erlangen [21] by studying the basic processes using an involved diagnostic system, including

- a Breit-Rabi polarimeter to measure electron and nuclear polarization of atoms;
- a Faraday monitor to measure K density and polarization;
- an optical monitor to measure α in the glass tube of the dissociator by comparing intensities from Balmer and molecule spectra;
- the measurement of α at the source exit via intensity modulations in the BRP.

The results are used to get a better understanding of the various processes taking place in this rather aggressive environment. An important finding was that spin-temperature equilibrium is attained in such sources [22] which enables to generate in addition to electron polarization nuclear polarization as well.

A test experiment on the LDS target at the IUCF cooler ring has been performed by the CE66 Argonne-Erlangen-Illinois-Colorado collaboration in 1996-99. The nuclear polarization of the hydrogen and deuterium targets was determined by elastic scattering of polarized protons. The results were presented at PST99 [23]. A common experience of all targets to date is that the atomic fraction α decreases considerably when the K ampoule is heated to the operating temperature. Clearly the surface detoriation in the presence of the alkali needed for optical pumping seems to be the most serious limitation for a LDS target [23]. The α values measured in the course of the test experiment were around 0.4 and the electron polarization around 0.5. The average nuclear polarization was up to about 0.15 for hydrogen, and about 0.10 for deuterium. –This result may be compared with typical parameters of the HERMES target by calculating the quality factor QF according to equation 3.

The resulting QF's are:

$$QF\ (CE66) \quad = 0.12^2 \times 4\cdot 10^{14}/cm^2 = \mathbf{6}\cdot 10^{12}/cm^2 \qquad (4)$$
$$QF\ (HERMES) = 0.85^2 \times 1\cdot 10^{14}/cm^2 = \mathbf{75}\cdot 10^{12}/cm^2$$

Despite the much higher flow rate and four times higher density the LDS target turns out to be more than an order of magnitude lower in quality factor. Further progress is ultimately linked to improved wall properties with a considerably higher α.

Target Polarimetry

There are two scenarios of determining the target polarization required for evaluating spin-dependent quantities:

- Measurement of absolute target polarization by scattering of beam particles with known analyzing power and sufficient scattering rate. In this case monitoring of the electron polarization P_e for setting up and control is sufficient.
- In case of no useful reaction with beam particles due to insufficient rate like in deep-inelastic electron scattering an independent absolute target polarimeter is required.

The 2nd approach is realized in case of the HERMES target for which the so-called Breit-Rabi polarimeter (BRP) has been developed [24]. The diagnostic system is shown in Figure 4.

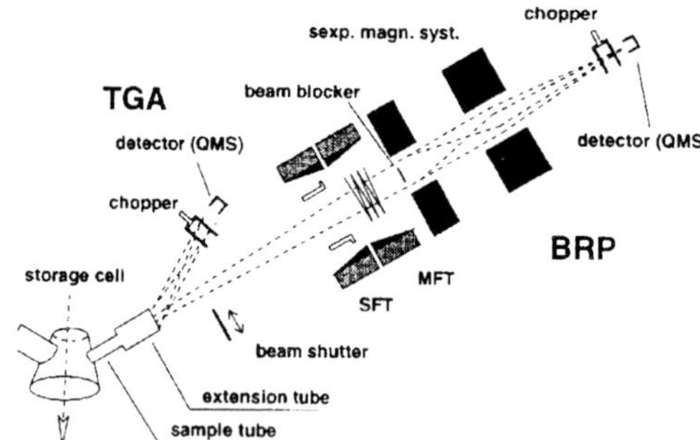

Fig. 4: The HERMES target diagnostics [24]. The sample beam from the storage cell formed by the sample tube enters the Breit-Rabi polarimeter (BRP) where its atomic substate population is determined to 0.01 in precision within about 1min. By means of the extension tube a flow of target gas is directed to the target gas analyzer (TGA) under an angle of 7° to the BRP axis where the atomic fraction α is measured.

The key problem is to relate the target parameters as seen by the high energy beam to those of the sample of target gas extracted in the cell center. These so-called Sampling Corrections are obtained from simulations which are constrained by the result of measurements. In this way total errors $\delta P/P$ of 3-4% have been achieved [14]. By measuring substate populations as function of the guide field one is – in particular at low field – sensitive to spin exchange between nuclear and electron polarization. Their rate is density dependent. Knowing the spin exchange cross section the density can be extracted.

IV. PERFORMANCE OF OPERATIONAL TARGETS

In Table 1 the operational targets including the recently terminated PINTEX target [12] are listed. The EDDA target at COSY (Jülich) is designed as Jet target in a single substate [25] and $t = 2\cdot10^{11}$ H/cm^2 are obtained. For the test of time-reversal invariance (TRIC) the addition of a cell is in progress in order to increase t to about 10^{13}/cm^2. – The PINTEX target [11,12] has been operated from 1993 to summer 2002 with protons of energy up to 450MeV and has enabled the collection of a vaste amount of pp and pd scattering data. The target cell with extremely thin Teflon walls has already been described in section II (see Figure 2). In 2000, the target was upgraded to deuterium and successfully employed for experiments. As a surprise a considerable reduction of P_{zz} to about half of the expected value was observed [26] which could be traced back to spin exchange at weak guide field. – The DEUTERON target at Novosibirsk reaches back to the first deuterium jet target in an electron ring operated in the mid-80s [2] which was complemented by the first storage cell in a storage ring in 1988. Attempts of an Argonne-Novosibirsk collaboration to implement a LDS deuterium target were unsuccessful and terminated. The target now comprises a cryogenic ABS with superconducting sextupoles and a storage cell at nitrogen temperature [27]. A record ABS intensity of $8\cdot10^{16}$ D/s has been achieved. Future plans include to employ ee scattering on polarized target electrons for the measurement of beam polarization. – The HERMES target [14] has been operated with longitudinal polarized hydrogen in 1996-97 and with longitudinal polarized deuterium in 1998-2000. It is operated more than 8 months per year plus setting up and testing. Since 2001 the target runs with transverse polarized hydrogen for which a new guide field magnet with high uniformity has been developed [28] required to inhibit a new class of densely-spaced depolarizing resonances.

V. STATUS OF FUTURE FACILITIES

As shown in Table 1, there are three new facilities where polarized gas targets will be applied, two at proton rings and one at an electron ring.

1. The ANKE target [29, 30] at COSY: The state of the art ABS for polarized H and D atoms is operational and has produced up to $7\cdot10^{16}$ H/s into the compression tube. A Lamb shift polarimeter is used to optimize the system of rf transitions. The target chamber with storage cell and Si strip detector is under construction and will be installed at the beginning of 2003.
2. The Jet polarimeter [31, 32] for RHIC: A polarimeter for RHIC with high statistics based on CNI is relative only. The idea is to relate elastic scattering of unpolarized RHIC protons on a hydrogen Jet of well known polarization to the same reaction between the polarized RHIC beam and an unpolarized (spin averaged) Jet target. The Jet will be produced by an ABS and the polarization measured by a Breit-Rabi-type polarimeter. For the required precision of 3% of the target polarization the molecular background at the interaction point has to be measured.
3. The BLAST target [33] at Bates: The BLAST detector is based on a 4π toroid spectrometer with the target cell at the center. Apart from the planned OP source for ^{3}He and a LDS target source [19], an ABS target is foreseen. This target utilizes the former AmPS ABS which is modified in order to meet the requirements imposed by the spectrometer, e.g. by placing the turbopumps outside the coil system. Other

sensitive components are heavily shielded against magnetic fields. At present the target is in the commissioning phase [33].

VI. CONCLUSIONS AND OUTLOOK

The status of gas targets for storage rings may be summarized as follows.

- These targets represent a mature technology which is well understood and they have enabled new Spin experiments, as ***single users*** in medium-energy storage rings like the IUCF Cooler, VEPP-3 and COSY, and in future at the Bates south hall ring, and as *parallel user* in a high-energy storage ring like HERA.
- A characteristic feature is the extremely low background which enables to study reactions with low cross sections, e.g. pion production at threshold, and to utilize open spectrometers as for PINTEX, HERMES, or BLAST where the full angular range is measured at once.
- New experiments based on state of the art gas targets should lead to a much improved performance. BLAST is a first example. A similar proposal exists for COSY, the so-called TOROID experiment to study heavy meson production at threshold [34].

In the outlook we might speculate about possible improvements. The presently available technology for H & D targets is the atomic beam source with (cold) storage cell. The ABS intensity seems to be limited to about 10^{17} atoms/s in 2 substates. By increasing the cell length considerably the target densities could be strongly improved. This is particularly important for electron rings where luminosity is a crucial factor. For a long cell the total conductance (see equation 1) is dominated by the feed tube and might be as low as 5l/s for a cold cell which results in a central density of $\rho_o \cong 2\cdot10^{13}/\text{cm}^3$. With a 1m long cell areal densities in the order of $t \cong 10^{15}/\text{cm}^2$ are conceivable. This would lead for an improved HERMES experiment as a single user of the HERA electron ring to integrated luminosities of up to 3fb^{-1} per year, which is about 20× the luminosity of the "Golden Year" 2000 of HERMES! We should not neglect that such a long, high density target puts stringent requirements on the design of the tracker system. Also there will be a much higher spin exchange rate which might inhibit a precise measurement of the target polarization by means of a sampling BRP. On the other hand at such luminosities it appears feasible to normalize the product of beam and target polarization to the inclusive DIS asymmetry which is well known.

ACKNOWLEDGMENTS

I would like to express my sincere gratitude to W. Haeberli for the pleasant and fruitful collaboration on polarized gas targets over the last 15 years. Valuable help in the preparation of this talk by D. Eversheim, M. Henoch, H. Kolster, P. Lenisa, A. Nass, F. Rathmann, D. Toporkov, J. Wilbert, T. Wise and A. Zelenski is gratefully acknowledged. My thanks are due to the Organizers, in particular to Th. Roser and Y. Makdisi, for inviting me to give this talk, and to the Bundesministerium für Bildung und Forschung, Germany, for financial support.

REFERENCES

1. Hanna, S.S., in *Proc. 4th Symp. on Pol. Phen. in Nucl. Reactions*, Zürich 1975, edited by W. Grüebler and V. König, Basel: Birkhäuser, 1976, p. 407.
2. Dmitriev, V. F., et al, *Phys. Lett.* **157 B**, 143 (1985).
3. Felden, O., et al, in *Proc. 9th Int. Workshop on Pol. Sources and Target, PST01*, Nashville 2001, edited by V.P. Derenchuk and B. v. Przewoski, Singapore: World Scientific, 2002, p. 73.
4. Makdisi, Y.I., in *Proc. PST01 (see ref. 3)*, p. 253.
5. Luppov, V.G., et al, in *Proc. PST01 (see ref. 3)*, p. 32.
6. Haeberli, W., in *Proc. 2nd Int. Symp. on Pol. Phen. in Nucl. Reactions*, Karlsruhe 1965, edited by P. Huber and H. Schopper, Basel: Birkhäuser 1966, p. 64.
7. Kleppner, D., et al, Phys. Rev. **A 138**, 972 (1965).
8. Schüler, K.P., in *Proc. High Energy Physics with Polarized Beams and Targets*, Marseille 1980, edited by C. Joseph and J. Soffer, Experientia Supplement **38**, Basel: Birkhäuser, 1981, p. 460.
9. Haeberli, W., in *Proc. Int. Workshop on Nuclear Physics with Stored, Cooled Beams*, Indiana 1984, edited by P. Schwandt and H.O. Meyer, AIP Conf. Proc. No. **128**, New York: AIP, 1985, p. 251.
10. Barker, M.D., et al, in *Proc. 4th Symp. on Pol. Phen. in Nucl. Reactions*, Santa Fe 1980, edited by G.G Ohlsen et al, AIP Conf. Proc. No. **69**, New York: AIP, 1981, p. 931.
11. Dezarn, D.A., et al, *Nucl. Instr. Methods* **A 362,** 36 (1995).
12. v. Przewoski, B.:" Spin and the Three-Nucleon Force Effekt", invited talk, these proceedings.
13. Ackerstaff, K., et al, *Nucl. Instr. Methods* **A 417**, 230 (1998).
14. Lenisa, P.: "The HERMES Internal Polarized H/D Target", these proceedings.
15. Baumgarten, C., et al: "The Polarized H/D Internal Target Storage Cell for the HERMES Experiment at HERA", to appear in *Nucl. Instr. Methods* **A**.
16. Nass, A.: "The HERMES Polarized Atomic Beam Source", these proceedings.
17. Vassiliev, A., et al, *Rev. Sci. Instr.* **71**, 3331 (2000).
18. Coulter, K.P., et al, *Phys. Rev. Lett.* **68,** 174 (1992).
19. Crawford, C., et al, in *Proc. PST01 (see ref. 3)*, p. 78.
20. Green, M.C., in *Proc. of the Workshop on Polarized Targets in Storage Rings*, Argonne 1984, edited by R.J. Holt, Argonne National Lab. ANL-84-50, p. 307.
21. Schmidt, F., et al, in *Proc. Int. Workshop on Polarized Sources and Targets, PST99*, Erlangen 1999, edited by A. Gute, S. Lorenz and E. Steffens, Univ. of Erlangen-N., ISBN 3-00-005510-X, p. 212.
22. Stenger, J., et al, Phys. Rev. Lett. **78**, 4177 (1997).
23. Jones, C., in *Proc. PST99 (see ref. 21)*, p. 204, and Cadman, R.V., et al, *Phys. Rev. Lett.* **86**, 967 (2001)
24. Baumgarten, C., et al, *Nucl. Instr. Methods* **A 482**, 606 (2002).
25. Eversheim, D.: "The TRIC Experiment: A P-even Time Reversal Invariance Test at COSY", these proceedings.
26. v.Przewoski, B., et al, in *Proc. PST01 (see ref. 3)*, p. 57.
27. Dyug, M.V., et al, in *Proc. PST01 (see ref. 3)*, p. 62.
28. Reggiani, D.: "Beam Induced Depolarizing Resonances in the HERMES Hydrogen and Deuterium Target", these proceedings.
29. Rathmann, F., et al: "Atomic Beam Source for the Polarized Internal Gas Target of ANKE at COSY", these proceedings.
30. Engels, R., et al: " A Precision Lamb-shift Polarimeter for the Polarized Gas Target at ANKE", these proceedings.
31. Zelenski, A.: "Status of the Polarized Atomic H-Jet Development...", these proceedings.
32. Wise, T.: "Polarized Jet for Calibration of RHIC Proton Beam Polarization", these proceedings.
33. Kolster, H.: "The BLAST Internal Gas Target at the MIT-Bates Linear Accelerator Center", these proceedings.
34. Rathmann, F., et al, *Czech. J. Phys.* **52**, C319 (2002)

Parity Violation in $\vec{p}p$ and $\vec{n}p$ Experiments

W.D. Ramsay

Department of Physics and Astronomy, University of Manitoba, Winnipeg, MB, R3T 2N2, Canada

Abstract.
Parity violation experiments involving only two nucleons provide a way to study the non-leptonic, strangeness-conserving part of the weak interaction in a clean measurement free of nuclear structure uncertainties. Such measurements are particularly appropriate for discussion at this conference as their success depends critically on the ability to accurately control and measure spin. Although simple in principle, the experiments are technically very demanding and great pains must be taken both in the preparation of the incident polarized beams and the measurement of the resultant parity violating asymmetries, which may be masked by a multitude of systematic effects. At low and intermediate energies, $\vec{p}p$ experiments are sensitive to the medium range part of the parity violating nucleon-nucleon force, usually parameterized in terms of rho and omega meson exchange. The pion does not contribute to parity violation in the pp experiments, as the π^0 is its own antiparticle and parity violation would also imply CP violation. I review existing pp measurements with particular emphasis on the recent 221 MeV $\vec{p}p$ measurement at TRIUMF which permitted the weak meson-nucleon coupling constants h_ρ^{pp} and h_ω^{pp} to be determined separately for the first time. The $\vec{n}p$ experiments, on the other hand, are used to extract the weak pion nucleon coupling, f_π, describing the longest range part of the parity violating nucleon-nucleon force. The np system is the only two nucleon system that is sensitive to f_π. I also review these experiments, with specific details of the $\vec{n}p \rightarrow d\gamma$ experiment now under preparation at Los Alamos National Laboratory.

INTRODUCTION

Sometimes is is necessary to repeat what we know. All mapmakers should place the Mississippi at the same location, and avoid originality. – Saul Bellow

In preparing a review talk, one becomes acutely aware of essentially "repeating what we know". In this review I will, however, do just that, concentrating in particular on what we know about pp and np parity violation experiments in which we control the spin of the incident nucleon, in other words experiments that use the nucleon spin as a tool, rather than experiments concerned with the nature of the nucleon spin itself. I will give a historical overview of $\vec{p}p$ and $\vec{n}p$ parity violation experiments, with technical details of the experiments I am most familiar with – the TRIUMF 221 MeV $\vec{p}p$ experiment and the Los Alamos $\vec{n}p \rightarrow d\gamma$ experiment now being installed at LANSCE. What I cover is a biased personal selection, and I refer readers interested in more background to two fine reviews of the field by Adelberger and Haxton [1] and Haeberli and Holstein [2].

CP675, *Spin 2002: 15th Int'l. Spin Physics Symposium and Workshop on Polarized Electron Sources and Polarimeters*, edited by Y. I. Makdisi, A. U. Luccio, and W. W. MacKay

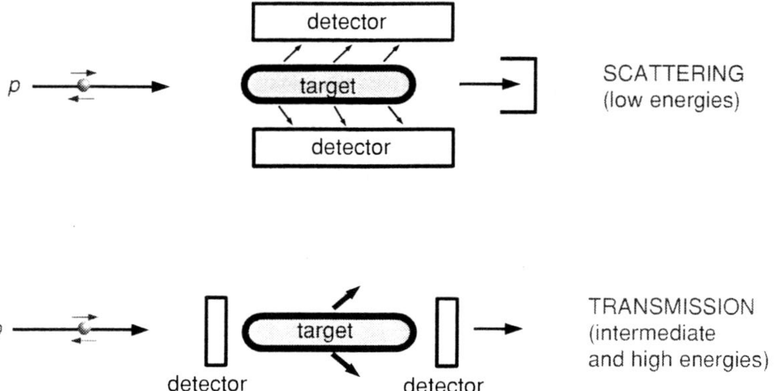

FIGURE 1. Types of $\vec{p}p$ experiments. The low-energy experiments use scattering geometry, while the intermediate and high-energy experiments use transmission geometry.

Parity and the Weak Interaction

The parity operation reflects all space coordinates through the origin ($\vec{r} \to -\vec{r}$), which is equivalent to a mirror reflection (which reverses only one space coordinate)[1] and a 180^0 rotation (which reverses the other two). Since we can assume rotational invariance, the parity operation is often thought of as simply a mirror reflection. If a process is not identical to its mirror image, it is said to be parity violating (PV) or parity nonconserving (PNC). In physics, parity violation is exclusive to processes involving the weak interaction, such as:

1. $\mu \to e^- + \nu_\mu \overline{\nu}_e$
2. $n \to p + e^- + \overline{\nu}_e$; $\Lambda \to p + e^- + \overline{\nu}_e$
3. $K^+ \to \pi^+ \pi^-$
4. $pp \to pp$

The first three examples, in the leptonic, semi-leptonic and hadronic $\Delta s = 1$, sectors, clearly involve the weak interaction; the decays shown would disappear without it. The last example, however, in the purely hadronic $\Delta s = 0$ sector, is so dominated by the strong interaction that, were the weak interaction to disappear, the process would be virtually unchanged. The only way to see the effects of the weak interaction in such processes is to look for parity violation, and the experimental signal is very small. In pp scattering, for example, the parity violating part of the scattering cross section is typically of order 10^{-7} of the parity conserving part.

[1] The answer to your kids' question, "Why does a mirror reverse left-right but not up-down?" is that it doesn't; it reverses fore and aft. You do the left-right reversal when you turn the paper around to look at it in the mirror.

$\vec{P}P$ EXPERIMENTS

Figure 1 shows typical $\vec{p}p$ parity violation experiments. They scatter a longitudinally polarized beam of protons from a hydrogen target and measure the difference in cross section for right-handed and left-handed proton helicities. The low energy experiments use scattering geometry, in which the detectors measure the scattered protons directly. The intermediate and high energy experiments use transmission geometry in which the change in scattering cross section is deduced from the change in transmission through the target. Transmission geometry uses a simpler detector arrangement, but can't be used

TABLE 1. Summary of $\vec{p}p$ parity violation experiments. The long times taken to achieve small uncertainties reflects the time taken to understand and correct for systematic errors. In cases where authors reported both statistical and systematic uncertainties, this table shows the quadrature sum of the two.

Lab/Energy	Technical Details	A_z (10^{-7})	Where Reported
Los Alamos 15 MeV	scattering 3 atm x 38cm hydrogen gas 4 liquid scintillators	$+1 \pm 4$	1974 Phys. Rev. Lett. [3]
	scattering 6.9 atm hydrogen gas 4 plastic scintillators	-1.7 ± 0.8	1978 Argonne Conference [4]
Texas A&M 47 MeV	scattering 39 atm x 42cm hydrogen gas 4 plastic scintillators	-4.6 ± 2.6	1983 Florence Conference [5]
Berkeley 46 MeV	scattering 80 atm hydrogen gas target He ion chamber around target	-1.3 ± 1.1 -1.63 ± 1.03	1980 Santa Fe Conference [6] 1985 Osaka Conference [7]
SIN (PSI) 45 MeV	scattering 100 atm hydrogen gas annular ion chamber	-3.2 ± 1.1 -2.32 ± 0.89 -1.50 ± 0.22	1980 Phys. Rev. Lett. [8] 1984 Phys. Rev. D. [9] 1987 Phys. Rev. Lett. [10]
Los Alamos 800 MeV	transmission 1 m liquid hydrogen gas ion chambers	$+2.4 \pm 1.1$	1986 Phys. Rev. Lett. [11]
Bonn 13.6 MeV	scattering 15 atm hydrogen gas hydrogen ion chambers	-1.5 ± 1.1 -0.93 ± 0.21	1991 Phys. Lett. B [12] 1994 private communication [13]
TRIUMF 221 MeV	transmission 40 cm liquid hydrogen hydrogen ion chambers	$+0.84 \pm 0.34$	2001 Phys. Rev. Lett. [14]
Argonne ZGS 5130 MeV	transmission 81 cm water target ion chambers and scintillators	$+26.5 \pm 7.0$	1986 Phys. Rev. Lett. [15]

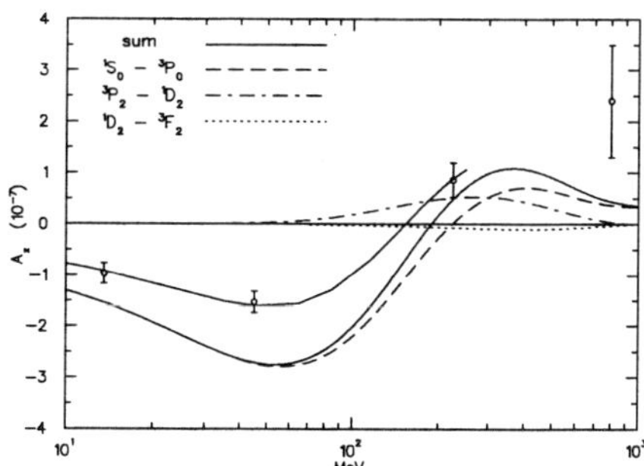

FIGURE 2. Partial wave decomposition of A_z. The broken curves and lower solid sum curve are calculated by Driscoll and Miller [23] using the DDH best guess couplings [16]. The upper solid sum curve is calculated by Carson *et al.* [24] with adjusted couplings.

at low energies because the energy loss in the target is too high to permit a sufficiently thick target. The quantity reported by both types of experiments is the parity violating longitudinal analyzing power, $A_z = \frac{\sigma^+ - \sigma^-}{\sigma^+ + \sigma^-}$, where σ^+ and σ^- are the scattering cross sections for positive and negative helicity. Because the statistical precision required on A_z is typically $\pm 10^{-8}$, it would take too long to count the requisite 10^{16} scattered particles, and all $\vec{p}p$ experiments so far have used current mode detection (as opposed to counting individual scattered particles).

Historical Summary

A roughly historical summary of $\vec{p}p$ parity violation experiments is given in Table 1. The long time taken to acquire measurements at a reasonable selection of energies and with small experimental uncertainties reflects the technical difficulty of these measurements. Running time is dominated by the time required to understand, and correct for, the various sources of systematic error. The time required to get the desired statistical precision is normally small by comparison.

Interpretation of the Results

Although it is known that the weak force is carried by the W and Z bosons, it is also known that their range is very small ($\sim 0.002\,fm$), and most authors assume that at low and intermediate energies the protons never get close enough for direct W and Z exchange. The interaction is normally treated in a meson exchange model with one strong, parity conserving vertex and one weak, parity non-conserving vertex. The weak vertex is parameterized in terms of a set of weak meson-nucleon coupling constants, $f_\pi, h_\rho^{0,1,2}, h_\omega^{0,1}$, where the subscript denotes the exchanged meson and the superscript gives the isospin change [16]. Mesons heavier than the ω-meson are not included because of the hard core of the nucleon-nucleon force. Further, for the pp interaction

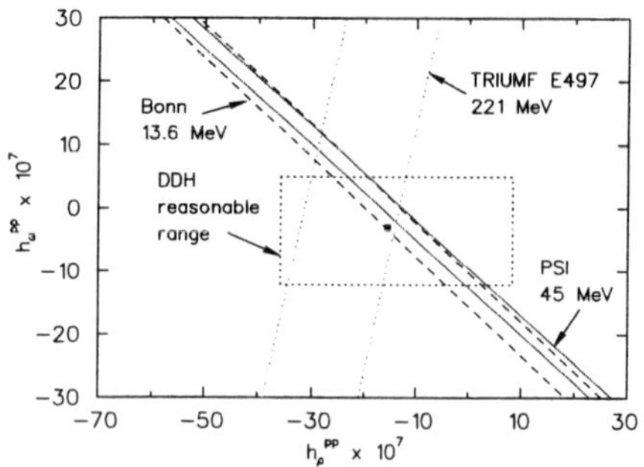

FIGURE 3. Constraints on the weak meson-nucleon couplings imposed by experiments in the energy range where the meson exchange model is normally used. The bands are based on calculations by Carlson *et al.* [24] using the AV18 potential [25] and CD-Bonn strong couplings [26]. (Figure modified from [14])

there is no π exchange because the π^0 is its own antiparticle and parity violation would also imply CP violation, another factor of 10^3 suppression. CP invariance also excludes other neutral scalar and pseudoscalar mesons such as the η, η', S, and δ^0 from consideration (Barton's theorem [17]).

The weak couplings were calculated by Desplanques, Donoghue and Holstein [16], and subsequently by a number of other authors [18, 19, 20, 21, 22]. The range of calculated values is large, and an experimental determination is needed.

Figure 2 shows A_z as a function of energy, broken down into contributions from various partial wave mixings [23]. Since the parity is $(-1)^\ell$, where ℓ is the orbital angular momentum, one would not normally expect partial waves differing by one unit in ℓ (S-P, P-D, etc.) [2] to mix, but due to the weak force, some small mixing occurs. At low energy, A_z is dominated by the contribution of S-P mixing. At the energy of the TRIUMF experiment, A_z arises almost exclusively from P-D mixing. The *shape* of the curves is set by the strong interaction, while the multiplying factor is set by the weak interaction. By adjusting the multiplying factors to fit the data, the weak meson-nucleon couplings can be extracted. The lower solid line is the total A_z calculated by Driscoll and Miller [23] using the DDH [16] weak couplings. The upper solid line is from a calculation by Carlson *et al.* [24] using adjusted values of the weak couplings.

pp experiments are sensitive to the combinations $h_\rho^{pp} = h_\rho^{(0)} + h_\rho^{(1)} + \frac{1}{\sqrt{6}} h_\rho^{(2)}$ and $h_\omega^{pp} = h_\omega^{(0)} + h_\omega^{(1)}$. Using the AV18 strong potential [25] and CD-Bonn values for the strong couplings [26], Carlson *et al.* [24] calculate that

$$\begin{aligned} A_z(13.6MeV) &= 0.059h_\rho^{pp} + 0.075h_\omega^{pp} \\ A_z(45MeV) &= 0.10h_\rho^{pp} + 0.14h_\omega^{pp} \\ A_z(225MeV) &= -0.038h_\rho^{pp} + 0.010h_\omega^{pp} \end{aligned}$$

[2] The notation in Fig. 2 is ${}^{(2S+1)}\ell_J$, where S is the total spin (0 or 1), ℓ = S,P,D,F is orbital angular momentum of 0,1,2,3, and J is the total angular momentum.

TABLE 2. Overall corrections for systematic errors. The table shows the average value of each coherent modulation, the net correction made for this modulation, and the uncertainty resulting from applying the correction.

Property	Average Value	$10^7 \Delta A_z$
$A_z^{uncorrected}(10^{-7})$	$1.68 \pm 0.29(stat.)$	
$y * P_x(\mu m)$	-0.1 ± 0.0	-0.01 ± 0.01
$x * P_y(\mu m)$	-0.1 ± 0.0	0.01 ± 0.03
$\langle yP_x \rangle (\mu m)$	1.1 ± 0.4	0.11 ± 0.01
$\langle xP_y \rangle (\mu m)$	-2.1 ± 0.4	0.54 ± 0.06
$\Delta I/I(ppm)$	15 ± 1	0.19 ± 0.02
$position + size$		0 ± 0.10
$\Delta E(meV at OPPIS)$	7–15	0.0 ± 0.12
electronic crosstalk		0.0 ± 0.04
Total		$0.84 \pm 0.17(syst.)$
$A_z^{corr}(10^{-7})$	$0.84 \pm 0.29(stat.) \pm 0.17(syst.)$	
$\chi_\nu^2(23 sets)$	1.08	

These constraints are shown graphically in Fig. 3. Notice that the precision results from Bonn and PSI determine essentially the same combination of couplings.

The TRIUMF *pp* experiment [14] is a transmission experiment, as depicted in the bottom panel of Fig. 1. A 221 MeV longitudinally polarized proton beam was passed through a 40 cm long liquid hydrogen target, which scattered about 4% of the beam. Hydrogen filled ion chambers located upstream and downstream of the target measured the change in transmission when the spin of the incident protons was flipped from right-handed to left-handed. Although a very good optically pumped polarized ion source [27, 28, 29] was used that minimized the changes in beam properties other than helicity, other beam properties still change very slightly. These helicity-correlated beam property changes cause a systematic shift in the A_z distribution, and corrections must be made. To do this, the TRIUMF group continuously measured the helicity correlated changes in beam properties and made corrections based on the sensitivities determined in separate control measurements. The data before and after correction are shown in Fig. 4. The main effect visible to the eye is from *first moments of transverse polarization* resulting from the distribution of transverse polarization components across the beam. All the corrections are summarized in Table 2. The measured A_z actually came half from true parity violation and half from false effects.

$\vec{N}P \rightarrow D\gamma$ EXPERIMENTS

Unlike the $\vec{p}p$ experiments just discussed, which are sensitive to ρ and ω exchange, $\vec{n}p \rightarrow d\gamma$ experiments are sensitive almost exclusively to pion exchange, and measure the

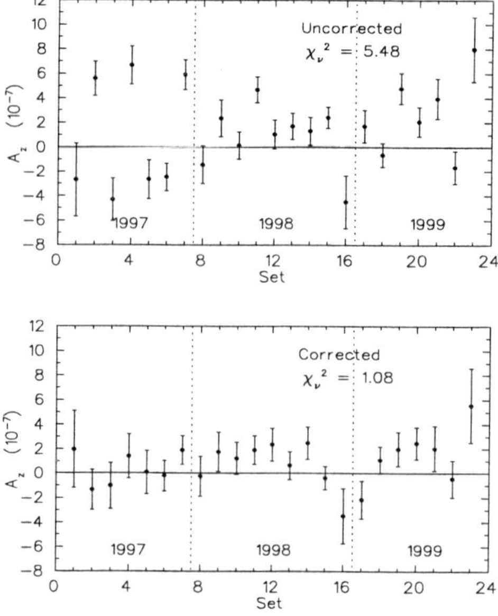

FIGURE 4. Effect of corrections to the E497 data. The main effect visible to the eye is for first moments of transverse polarization.

weak pion-nucleon coupling, f_π.[3] Measurements such as the circular γ-ray polarization from ^{21}Ne [30], or the longitudinal analyzing power in $p\alpha$ scattering [31], provide constraints on a combination of π, ρ, and ω couplings. The most precise limit on f_π alone is believed to be from measurements of circularly polarized gamma rays from a parity mixed doublet in ^{18}F [32, 33]. These results, however, are only about 10% of theoretical predictions [16, 34, 35], which give $f_\pi \sim 4 \times 10^{-7}$, and are also at odds with the large value of f_π deduced from measurements of the anapole moment in ^{133}Cs [36, 37], although the ^{133}Cs experiment is also sensitive to $h_\rho^{(0)}$, and, as pointed out by Wilburn and Bowman, [38] the disagreement may not be too significant.

$np \to d\gamma$ radiative capture measurements can be made with unpolarized neutrons, as was done in the Leningrad experiment [39], but the circular polarization, P_γ, of the capture gammas must be measured and the analyzing power of the polarimeters is very low. In an $\vec{n}p \to d\gamma$ experiment, the incident cold neutrons are polarized vertically and the gamma rays produced by neutron capture in the hydrogen target are expected to be emitted slightly more in the direction opposite to the neutron spin. The up-down asymmetry $A_\gamma \approx -0.11 f_\pi$ provides a clean measure of f_π free of nuclear structure uncertainties [40]. Previous measurements at ILL Grenoble gave $A_\gamma = (6 \pm 21) \times 10^{-8}$ [41] and $A_\gamma = (-1.5 \pm 4.8) \times 10^{-8}$ [42], but neither result was accurate enough to impose a significant constraint.

---

[3] Some authors quote $H_\pi = f_\pi \frac{g_\pi}{\sqrt{32}}$, where g_π is the strong pion-nucleon coupling.

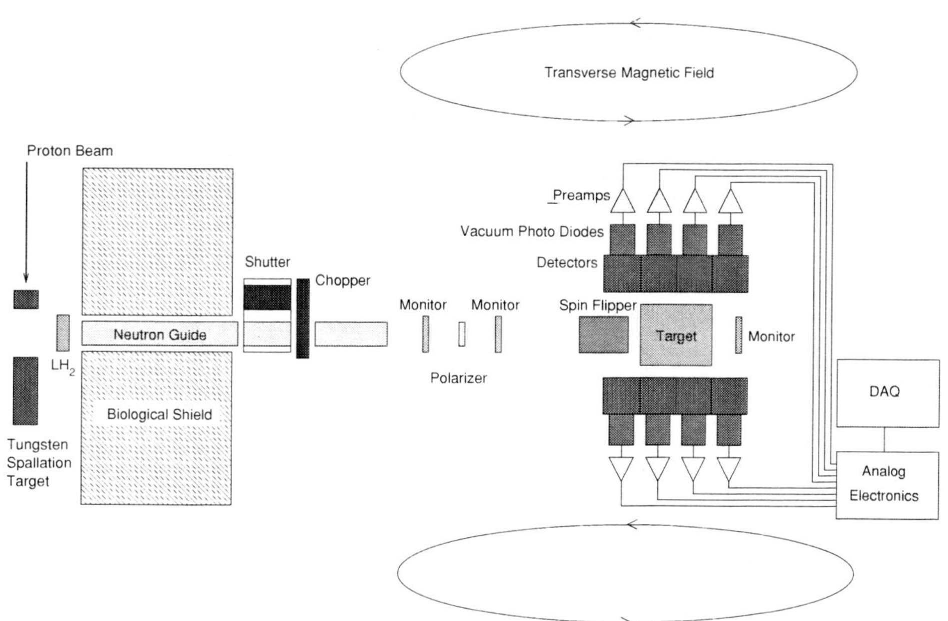

FIGURE 5. Layout of apparatus for the $\vec{n}p \rightarrow d\gamma$ experiment at LANSCE.

An experiment is now being prepared at Los Alamos to measure the gamma ray asymmetry in $\vec{n}p \rightarrow d\gamma$ with an uncertainty of $\pm 0.5 \times 10^{-8}$ [40]. The expected asymmetry is $A_\gamma \approx -5 \times 10^{-8}$.

The apparatus is shown schematically in Fig. 5. A pulsed, 800 MeV, 100-150 μA proton beam impinges on a tungsten spallation target. Neutrons from the spallation target are cooled in a liquid hydrogen moderator and transported to the experiment in a neutron guide. The neutron guide prevents the $1/r^2$ intensity fall-off that would otherwise occur, and also enhances the fraction of low energy neutrons. The peak neutron flux through the 9.5 cm x 9.5 cm guide is 6×10^7 n/ms at 8 meV.

The neutrons are polarized by passing through a polarized ^3He spin filter. Neutrons with spin parallel to the ^3He spin pass through the filter, while neutrons with spin anti-parallel to the ^3He spin are captured. The neutron polarization is determined by the front monitors located before and after the spin filter. If half the neutrons are captured, the polarization of the downstream neutron beam is 100%. A test ^3He spin filter was demonstrated in a fall 2000 test run and the final version will be tested at the University of Michigan in fall 2002.

The polarized neutrons are captured in a 30 cm long, 20 L liquid para-hydrogen target operating at 17 K. The low temperature is important to keep a small (0.03%) equilibrium ortho-hydrogen fraction, as the spin 1 ortho-hydrogen molecule allows spin-flip scattering and destroys the asymmetry. The cross section for scattering from ortho-hydrogen is approximately 20 times that for scattering from para-hydrogen, so the neutron detector downstream of the target will indicate any change in the ortho-hydrogen fraction.

Approximately 60% of the neutrons are captured in the para-hydrogen, producing a deuteron and a 2.2 MeV gamma ray. The gamma rays are measured by an array of 48 15 x 15 cm^2 CsI(Tl) detectors surrounding the target.

The neutron spin is flipped 20 times per second by a 30 kHz RF spin filter. The RF itself has a very small (< 0.1 mm) skin depth in, and is well contained by, the conducting shell surrounding the spin filter. In the fall 2000 test runs at LANSCE, the on-axis spin flipper efficiency was >95%.

The neutron beam is pulsed at 20 Hz, so the data acquisition operates in 50 ms "frames". The pulsed beam makes it possible to identify neutrons by their time of flight over the 22 m flight path. The fast neutrons arrive at the target first, and for the first 9 ms the neutron energy is above the 15 meV required to excite a para-to-ortho transition in the para-hydrogen target. For this reason, the first 9 ms of a frame will have no physics asymmetry and can be used to measure background. From 9 ms to 30 ms, the neutron energy falls from 15 meV to 1.5 meV. The amplitude of the RF spin filter is synchronized with this fall to ensure that fast and slow neutrons are all rotated by 180^0. After 40 ms, a "frame overlap chopper" blocks the neutrons so that slow neutrons from one frame are not still arriving when fast neutrons from the next pulse arrive. During the 40 ms to 50 ms dead interval, electronic noise can be checked. Different systematic errors have a different dependence on neutron energy and their time-of-flight signatures can be used to identify them. In addition, the ^3He spin filter direction and the overall holding field can be reversed for further cancellation of systematic errors.

The beamline, FP12, is now almost complete at LANSCE. The experimental cave is scheduled for installation in early 2003 and commissioning runs are planned for summer and fall of 2003.

SUMMARY

pp and np parity violation experiments provide a means to study the hadronic, $\Delta s = 0$ part of the weak interaction. $\vec{n}p \to d\gamma$ experiments are sensitive to the long range part of the nucleon-nucleon interaction and constrain the weak pion-nucleon coupling constant, f_π. Despite decades of experimental and theoretical work, the strength of this coupling is still very uncertain. The new, precision experiment now under construction at the LANSCE pulsed neutron source at Los Alamos should finally lay this question to rest. $\vec{p}p$ parity violation experiments are sensitive to the shorter range part of the nucleon-nucleon force and constrain the combinations $h_\rho^{pp} = h_\rho^{(0)} + h_\rho^{(1)} + \frac{1}{\sqrt{6}} h_\rho^{(2)}$ and $h_\omega^{pp} = h_\omega^{(0)} + h_\omega^{(1)}$. Prior to 2001, low energy experiments had constrained only a linear combination of approximately equal parts h_ρ^{pp} and h_ω^{pp}. With the addition of the TRIUMF 221 MeV result in 2001, h_ρ^{pp} and h_ω^{pp} are now separately constrained. The data so far are not sufficient to determine all 6 couplings, $f_\pi, h_\rho^{0,1,2}, h_\omega^{0,1}$, and much careful experimental work remains to be done. Nonetheless, one should remember that, in the words of Charles Babbage, "errors using inadequate data are much less than those using no data at all".

REFERENCES

1. E.G. Adelberger and W.C. Haxton, Ann. Rev. Nucl. Part. Sci. **35**, 501 (1985).
2. W. Haeberli and Barry R. Holstein, in *Symmetries and Fundamental Interactions in Nuclei*, edited by W.C. Haxton and E.M. Henley, (World Scientific, 1995) p. 17.
3. J.M. Potter *et al.*, Phys. Rev. Lett. **33**, 1307 (1974).
4. D.E. Nagle *et al.*, in *Proceedings of the 3rd International Conference on High Energy Beams and Polarized Targets* (Argonne, 1978), edited by L.H. Thomas, AIP Conference Proceedings 51, New York 1979, p. 224.
5. D.M. Tanner *et al.*, in *Proceedings of the International Conference on Nuclear Physics* (Florence, 1983), (Typographia, Bologna, 1983), p. 697.
6. P. von Rossen *et al.*, in *Proceedings of the 5th International Symposium on Polarization Phenomena in Nuclear Physics* (Santa Fe, 1980), edited by G.G. Ohlsen *et al.*, AIP Conference Proceedings 69, New York, 1981, p. 1442.
7. P. von Rossen *et al.*, in *Proceedings of the 6th International Symposium on Polarization Phenomena in Nuclear Physics* (Osaka, 1985), J. Phys. Soc. Japan **55**, Suppl. p. 1016 (1986).
8. R. Balzer *et al.*, Phys. Rev. Lett. **44**, 699 (1980).
9. R. Balzer *et al.*, Phys. Rev. C **30**, 1409 (1984).
10. S. Kistryn *et al.*, Phys. Rev. Lett. **58**, 1616 (1987).
11. V. Yuan *et al.*, Phys. Rev. Lett. **57**, 1680 (1986).
12. P.D. Eversheim *et al.*, Phys. Lett. B **256**, 11 (1991)
13. P.D. Eversheim, private communication (1994).
14. A.R. Berdoz *et al.*, Phys. Rev. Lett. **87**, 272301 (2001).
15. N. Lockyer *et al.*, Phys. Rev. D **30**, 860 (1984).
16. B. Desplanques, J.F. Donoghue, and B.R. Holstein, Ann. Phys.(N.Y.) **124**, 449 (1980).
17. G. Barton, Nuovo Cimento **19**, 512 (1961).
18. V.M. Dubovik and S.V. Zenkin, Ann. Phys.(N.Y.) **172**, 100 (1986).
19. G.B. Feldman, G.A. Crawford, J. Dubach, and B.R. Holstein, Phys. Rev. C **43**, 863 (1991).
20. N. Kaiser and U-G. Meissner, Nucl. Phys. A **499**, 699 (1989).
21. U-G. Meissner and N. Kaiser, Nucl. Phys. A **510**, 759 (1990).
22. U-G. Meissner and H. Wiegel, Phys. Lett. B **447**, 1 (1999).
23. D.E. Driscoll and G.A. Miller, Phys. Rev. C **39**, 1951 (1989); *ibid.*, **40**, 2159 (1989).
24. J.A. Carlson, R. Schiavilla, V.R. Brown, and B.F. Gibson, Phys. Rev. C **65**, 035502, (2002); R. Schiavilla, private communication (2001).
25. R.B. Wiringa, *et al.*, Phys. Rev. C **51**, 38 (1995).
26. R. Machleidt, Phys. Rev. C **63**, 24001 (2001).
27. A.N. Zelenski *et al.*, in *Proceedings of the 12th International Symposium on High Energy Spin Physics (SPIN96)*, edited by C.W. de Jager *et al.*, (World Scientific, Amsterdam, 1997), p. 637.
28. A.N. Zelenski *et al.*, in *Proceedings of the 6th Conference on Intersections Between Particle and Nuclear Physics*, edited by T.W. Donnelly, AIP Conference Proceedings 412, New York, 1997, p.328.
29. C.D.P. Levy *et al.*, in *Proceedings of the International Workshop on Polarized Beams and Polarized Gas Targets* (Cologne, 1995), edited by H.P. gen. Schieck and L. Sydow (World Scientific, Singapore, 1996), p. 120; A.N. Zelenski, *ibid.*, p. 111.
30. E.D. Earle *et al.*, Nucl. Phys. A **396**, 221 (1983).
31. J. Lang *et al.*, Phys. Rev. C **34**, 1545 (1986).
32. S.A. Page *et al.*, Phys.Rev. **C35**, 1119 (1987).
33. M. Bini *et al.*, Phys. Rev. Lett. **55**, 795 (1985).
34. D.B. Kaplan and M.J. Savage, Nucl. Phys. A **556**, 653 (1993).
35. E.M. Henley *et al.*, Phys. Lett. B **440**, 449 (1998).
36. C.S. Wood *et al.*, Science **275**, 1759 (1997).
37. V.V. Flambaum and D.W. Murray, Phys. Rev. C **56**, 1641 (1997).
38. W.S. Wilburn and J.D. Bowman, Phys. Rev. C **57**, 3425 (1998).
39. V.A. Knyazkov *et al.*, Nucl. Phys. A **417**, 209 (1984).
40. W.M. Snow *et al.*, Nucl. Inst. Meth. A **440**, 729 (2000).
41. J.F. Caviagnac *et al.*, Phys. Lett. B **67**, 148 (1977).
42. J. Alberi *et al.*, Can. J. Phys. **66**, 542 (1988).

A Future Linear Collider with Polarised Beams: Searches for New Physics

Gudrid Moortgat-Pick

DESY, Deutsches Elektronen–Synchrotron, D–22603 Hamburg, Germany[1]

Abstract. There exists a world–wide consensus for a future e^+e^- Linear Collider in the energy range between $\sqrt{s}$ =500–1000 GeV as the next large facility in HEP. The Linear Collider has a large physics potential for the discovery of new physics beyond the Standard Model and for precision studies of the Standard Model itself. It is well suited to complement and extend the physics program of the LHC. The use of polarised beams at a Linear Collider will be a powerful tool. In this paper we will summarize some highlights of high precision tests of the electroweak theory and of searches for physics beyond the Standard Model at a future Linear Collider with polarised e^- and e^+ beams.

BEAM POLARISATION AT A LINEAR COLLIDER

The next large experiment in high energy physics after the start of the Large Hadron Collider (LHC) will most probably be a future Linear Collider (LC) in the energy range between LEP and O(1 TeV). The existing world–wide proposals are designed with high luminosity of about $\mathscr{L} = 3.4 \cdot 10^{34}\mathrm{cm}^{-2}\mathrm{s}^{-1}$ at $\sqrt{s} = 500$ GeV and $\mathscr{L} = 5.8 \cdot 10^{34}\mathrm{cm}^{-2}\mathrm{s}^{-1}$ at $\sqrt{s} = 800$ GeV, see e.g. [1]. A LC will not only be well suited to complement but also to extend the physics program of the Hadron Colliders, Tevatron and LHC[2].

A LC has a large potential for the discovery of new particles and is – due to its clear experimental signatures – very well suited for the precise analysis of new physics (NP) as well as of the Standard Model (SM) [1]. Providing precision studies in this energy range, in particular in the electroweak sector, also small traces of physics beyond the SM might be found, even if high scale particles of NP might not be directly produced at the LHC or the first phase of a LC. For these studies the GigaZ option of the LC, i.e. running with very high luminosity at the Z and WW threshold, is crucial.

An important tool at a LC is the use of polarised beams. In the following we will summarize some highlights of searches and analyses of new physics with the help of polarised beams [2, 1, 3].

Already in the base line design of a LC it is foreseen to use electron beams polarised to around 80% via a strained photocathode technology [1] similar to those at the SLC where in the last year of running 1994/95 $P_{e^-} = (77.34 \pm 0.61)\%$ ([2] and references therein) was reached. In order to generate also polarised positrons the use of a helical undulator

---

[1] gudrid@mail.desy.de

[2] LHC/LC working group, see http://www.ippp.dur.ac.uk/~georg/lhclc/

CP675, *Spin 2002: 15th Int'l. Spin Physics Symposium and Workshop on Polarized Electron Sources and Polarimeters*, edited by Y. I. Makdisi, A. U. Luccio, and W. W. MacKay

is favoured. It allows to produce polarised photons which generate positrons via pair production with a designed polarisation degree of about 40% up to 60% [4]. There already exists a world–wide collaboration supporting the activities to get a prototype for a polarised positron source at the 50 GeV Final Focus Target Beam at SLAC [5].

It is foreseen to measure the polarisation with Compton polarimetry and it is assumed that one could reach even an accuracy better than $\Delta(P_{e^\pm} < 0.5\%)$ [6]. The simultaneous use of Møller polarimetry will also be studied at the LC [7]. However, the reachable accuracy with Compton and Møller polarimetry will not be sufficient for the high precision tests at GigaZ. For this purpose one uses a modified Blondel Scheme [8], see next section, where one expresses the polarisation via polarised cross sections. Therefore one can avoid absolute measurements of polarisation and uses polarimetry only for relative measurements.

After a short introduction into the physics of beam polarisation we begin our summary with high precision studies of the SM as a motivation for New Physics searches.

Introductory remarks

Within the Standard Model (SM) only $(V-A)$ couplings happen in the s–channel and therefore the configurations LR and RL are possible for the e^-e^+ helicities. That means that once the e^- polarisation is chosen also the e^+ polarisation is fixed. For these processes an additional simultaneous positron polarisation leads to an enhancement (or suppression) of the fraction of the colliding particles, which is expressed by the effective luminosity

$$\mathscr{L}_{eff}/\mathscr{L} := \tfrac{1}{2}[1 - P_{e^-}P_{e^+}], \tag{1}$$

and of the effective polarisation

$$P_{eff} := [P_{e^-} - P_{e^+}]/[1 - P_{e^-}P_{e^+}]. \tag{2}$$

In Table 1 we list $\mathscr{L}_{eff}$ and P_{eff} for some characteristic values of P_{e^-} and P_{e^+}. One can see that even with a completely polarised e^- beam the fraction of colliding particles is not enhanced. With simultaneously polarised positrons, however, this fraction will be enhanced by about 50%.

It is well–known that with suitably polarised beams one can suppress background processes, e.g. the main SM backgrounds $e^+e^- \to W^+W^-$ and ZZ. Some scaling factors $\sigma^{pol}/\sigma^{unpol}$ for these processes are given in Table 2.

Moreover, beyond the SM there are also coupling structures in the s-channel possible where also the configurations LL and RR could lead to strong signals. Simultaneous polarisation of both beams would therefore allow fast and easy diagnostics, as we will explain for an example below.

These given 'rules' are not generally valid for t–channel exchanges since in that case the helicity of the incoming e^- is only coupled to the outgoing particle at the vertex and not to the incoming e^+. This can be easily seen when studying the well–known Bhabha background. For small energies the s–channel with its LR and RL coupling characteristics dominates. However, for higher energies, also LL and RR coupling is

TABLE 1. Fraction of colliding particles ($\mathscr{L}_{eff}/\mathscr{L}$) and the effective polarisation (P_{eff}) for different beam polarisation configurations, which are characteristic for (V-A) processes in the s–channel [9].

	RL	LR	RR	LL	P_{eff}	$\mathscr{L}_{eff}/\mathscr{L}$
$P_{e^-} = 0$, $P_{e^+} = 0$	0.25	0.25	0.25	0.25	0.	0.5
$P_{e^-} = -1$, $P_{e^+} = 0$	0	0.5	0	0.5	-1	0.5
$P_{e^-} = -0.8$, $P_{e^+} = 0$	0.05	0.45	0.05	0.45	-0.8	0.5
$P_{e^-} = -0.8$, $P_{e^+} = +0.6$	0.02	0.72	0.08	0.18	-0.95	0.74

TABLE 2. Scaling factors $\sigma^{pol}/\sigma^{unpol}$ for the dominating SM background processes $e^+e^- \to W^+W^-$ and ZZ for different configurations of beam polarisation .

($P_{e^-} = \mp 80\%$, $P_{e^+} = 0, \pm 60\%$)	$e^+e^- \to W^+W^-$	$e^+e^- \to ZZ$
$(+0)$	0.2	0.76
(-0)	1.8	1.25
$(+-)$	0.1	1.05
$(-+)$	2.85	1.91

possible via the t–channel contributions. Another example for the importance of having both beams polarised, is the single W background since with e^- polarisation only the W^- signal can be suppressed. For the corresponding signal from W^+ the polarisation of e^+ is needed.

ELECTROWEAK HIGH PRECISION ANALYSES OF THE SM

Electroweak precision tests with an unprecedented accuracy – at high energies as well as at the Z resonance and the WW threshold – would provide a high sensitivity to effects of NP, even if new particles are not directly produced. In the following section we list some examples for these high precision measurements at a LC, e.g. the measurement of triple gauge couplings. After that we will have a look at the physics prospects of GigaZ.

Anomalous couplings in $e^+e^- \to W^+W^-$

In order to test the SM with high precision one can carefully study triple gauge boson couplings, which are generally parametrized in an effective Lagrangian e.g. by the C– and P–conserving couplings g_1^V, κ_V, λ_V, with $V = \gamma, Z$. In the SM at tree level the

TABLE 3. Sensitivity for anomalous triple gauge couplings with different configurations of beam polarisation [1].

error $[10^{-4}]$:	Δg_Z^1	$\Delta\kappa_\gamma$	λ_γ	$\Delta\kappa_Z$	λ_Z
unpolarised beams					
$\sqrt{s} = 500$ GeV	38.1	4.8	12.1	8.7	11.5
$\sqrt{s} = 800$ GeV	39.0	2.6	5.2	4.9	5.1
only electron beam polarised, $\lvert P_{e^-}\rvert = 80\%$					
$\sqrt{s} = 500$ GeV	24.8	4.1	8.2	5.0	8.9
$\sqrt{s} = 800$ GeV	21.9	2.2	5.0	2.9	4.7
both beams polarised, $\lvert P_{e^-}\rvert = 80\%$, $\lvert P_{e^+}\rvert = 60\%$					
$\sqrt{s} = 500$ GeV	15.5	3.3	5.9	3.2	6.7
$\sqrt{s} = 800$ GeV	12.6	1.9	3.3	1.9	3.0

couplings have to be $g_1^V = 1 = \kappa_V$, while the λ_V are identical to zero.

These couplings can be determined by measuring the angular distribution and polarisation of the $W^\pm$'s. A simultaneous fit of all couplings results in a strong correlation between the $\gamma-$ and $Z-$couplings. It turns out that the polarisation of the beams is very powerful for separating these couplings: e.g. the polarisation of $P_{e^-} = \pm 80\%$ (together with $P_{e^+} = \mp 60\%$) improves the sensitivity up to a factor 1.8 (2.5), see Table 3 [10, 1].

Transversely polarised beams in $e^+e^- \to W^+W^-$

Another promising possibility to study the origin of electroweak symmetry breaking is the use of transversely polarised e^+e^- beams in the process $e^+e^- \to W^+W^-$. This beam polarisation projects out $W_L^+W_L^-$ [11]. The asymmetry with respect to the azimuthal angle of this process focusses on the LL mode. This asymmetry is very pronounced at high energies reaching about 10%. The advantage of this observable is that at high energies this asymmetry peaks at larger angles and not in beam direction where the analysis might be difficult. One has to note, however, that for the use of transverse beams the polarisation of both beams is needed. The effect does not occur if only one beam is polarised since the cross section is given by:

$$\sigma = (1 - P_{e^-}^L P_{e^+}^L)\sigma_{unp} + (P_{e^-}^L - P_{e^+}^L)\sigma_{pol}^L + P_{e^-}^T P_{e^+}^T \sigma_{pol}^T. \quad (3)$$

GigaZ

With the GigaZ option of the LC the process $e^+e^- \to Z \to f\bar{f}$ is studied and the effective electroweak leptonic mixing angle at the Z resonance can be measured via the left–right asymmetry

$$A_{LR} = \frac{2(1 - 4\sin^2\Theta_{eff}^\ell)}{1 + (1 - 4\sin^2\Theta_{eff}^\ell)^2} \quad (4)$$

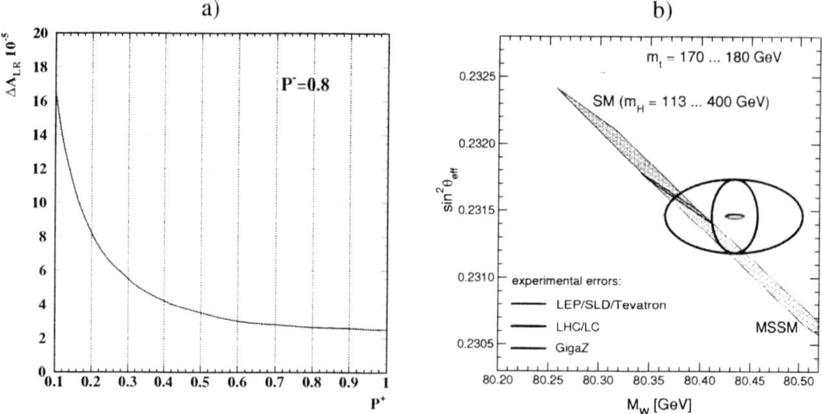

FIGURE 1. Test of the electroweak theory: a) The statistical error on the left–right asymmetry A_{LR} of $e^+e^- \to Z \to \ell\bar{\ell}$ at GigaZ as a function of the positron polarisation P_{e^+} for fixed electron polarisation $P_{e^-} = \pm 80\%$. In b) the allowed parameter space of the SM and the MSSM in the $\sin^2\theta^{\ell}_{eff}$-$M_W$ plane is shown together with the experimental accuracy reachable at GigaZ.

of this process. Since one gets only a gain in statistical power if the error due to the polarisation measurement $\Delta A_{LR}(pol)$ is smaller than the statistical error $\Delta A_{LR}(stat)$ one has to know $P_{e^\pm}$ extremely accurately. Using only polarimetry even $\Delta P_{e^\pm} < 0.5\%$ would not be sufficient. Therefore one uses a modified Blondel Scheme [10, 1] and expresses A_{LR} via polarised rates:

$$A_{LR} = \sqrt{\frac{(\sigma^{++}+\sigma^{+-}-\sigma^{-+}-\sigma^{--})(-\sigma^{++}+\sigma^{+-}-\sigma^{-+}+\sigma^{--})}{(\sigma^{++}+\sigma^{+-}+\sigma^{-+}+\sigma^{--})(-\sigma^{++}+\sigma^{+-}+\sigma^{-+}-\sigma^{--})}}. \tag{5}$$

With this method, polarimetry has to be used only for calibration and one can reach a spectacular accuracy for the effective leptonic weak mixing angle of about $\Delta(\sin^2\theta^{\ell}_{eff}) \sim 1\times 10^{-5}$ [1]. The polarisation of the positron beam is absolutely needed but already a polarisation of about $P_{e^+} = |40\%|$ would be sufficient, see Fig. 1a, to measure this observable with an unprecedented accuracy.

As an example of the potential of the GigaZ $sin^2\theta^{\ell}_{eff}$ measurement, Fig. 1b compares the present experimental accuracy on $sin^2\theta^{\ell}_{eff}$ and M_W from LEP/SLD/Tevatron and the prospective accuracy from the LHC and from a LC without GigaZ option with the predictions of the SM and the MSSM [12]. With the measurement of $\sin^2\theta^{\ell}_{eff}$ and M_W at GigaZ a very sensitive test of the theory will be possible.

It should also be mentioned that with the help of polarised beams a LC could be sensitive to electroweak dipole form factors. They have been analysed in [13] with regard to CP violation of the τ lepton via CP–odd triple product correlations. Sensitivity bounds for the real and imaginary parts of these form factors have been set up to $O(10^{-19})$ ecm.

In the same context one should not forget that also for searches of heavy gauge bosons, as e.g. for the Z', the use of polarised beams enhances the discovery range. Also for contact interations the sensitivity can be enhanced significantly [14, 2].

BEYOND THE STANDARD MODEL

Supersymmetry

Supersymmetry is widely regarded as one of the best motivated extensions of the SM. However, since the SM particles and their SUSY partners are not mass degenerate, SUSY has to be broken, which leads even for its minimal version, the Minimal Supersymmetric Standard Model (MSSM), to about 105 free new parameters. In specific scenarios of SUSY breaking one has only a few parameters: 5 in mSUGRA, 4 in AMSB and 5 in GMSB. However, one should note that one of the most favoured motivations for SUSY – the unification of the gauge couplings – does not require a specific SUSY breaking scheme.

In order to exactly pin down the structure of the underlying model it is unavoidable to extract the parameters without assuming a particular breaking scheme. Since the LC with its very clear signatures provides a measurement of the particle masses of the level $O(100)$ MeV, and of cross sections and branching ratios at the % level, the LC is well suited for revealing the underlying structure of the model. Different step–by–step procedures have been worked out to determine the general MSSM parameters and to test fundamental SUSY assumptions, as e.g. the equality of quantum numbers or of couplings of the particles and their SUSY partners, as model independently as possible. It turns out that the use of polarised beams plays a decisive role in this context.

Stop mixing angle in $e^+e^- \to \tilde{t}_1\tilde{t}_1$

As demonstrated in [1] the mass and the mixing angle in the $\tilde{t}$ sector can be extracted with high precision via the study of polarised cross sections for light stop production. At a high luminosity LC and with $P(e^-) = 80\%$ and $P(e^+) = 60\%$ an accuracy of $\delta(m_{\tilde{t}_1}) \approx 0.8$ GeV and $\delta(\cos\theta_{\tilde{t}}) \approx 0.008$ could be reachable, see Fig. 2a. Similar studies have been done for the $\tilde{\tau}$ sector [15].

Quantum numbers in $e^+e^- \to \tilde{e}^+_{L,R}\tilde{e}^-_{L,R}$

SUSY transformations associate chiral leptons to their scalar SUSY partners: $e^-_{L,R} \leftrightarrow \tilde{e}^-_{L,R}$ and the antiparticles $e^+_{L,R} \leftrightarrow \tilde{e}^+_{R,L}$. In order to prove this association between scalar particles and chiral quantum numbers the use of polarised beams is necessary [16]. The process occurs via γ and Z exchange in the s–channel and via $\tilde{\chi}^0_i$ exchange in the t–channel. As already mentioned in the general introduction one has direct coupling between the SM particle and its scalar partner only in the t–channel. Therefore one has to project out the t–channel exchange in order to test the association of chiral quantum numbers to the scalar SUSY partners.

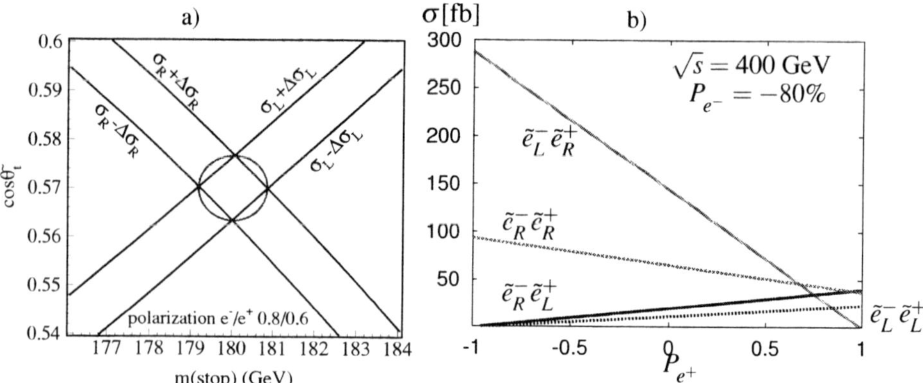

FIGURE 2. a) Contours of $\sigma_R(\tilde{t}_1\tilde{t}_1)$ and $\sigma_L(\tilde{t}_1\tilde{t}_1)$, $\tilde{t}_1 \to c\tilde{\chi}_1^0$ as function of $m_{\tilde{t}_1}$ and $\cos\theta_{\tilde{t}}$ for $\sqrt{s} =$ 500 GeV, $\mathscr{L} = 1$ ab^{-1}, $|P_{e^-}| = 80\%$, $|P_{e^+}| = 60\%$; b)Test of selectron quantum numbers in $e^+e^- \to \tilde{e}^+_{L,R}\tilde{e}^-_{L,R}$ with fixed electron polarisation $P(e^-) = -80\%$ and variable positron polarisation $P(e^+)$. For $P(e^-) = -80\%$ and $P(e^+) < 0$ both pairs $\tilde{e}_L^-\tilde{e}_R^+$ and $\tilde{e}_R^-\tilde{e}_R^+$ still contribute. For $P(e^+) = -60\%$ the pair $\tilde{e}_L^-\tilde{e}_R^+$ dominates by more than a factor 3 [16, 1, 2].

With completely polarised $e_L^-e_L^+$ beams only the pair $\tilde{e}_L^-\tilde{e}_R^+$ would contribute. Due to their L,R coupling character $\tilde{e}_L$, $\tilde{e}_R$ can be discriminated via their decay characteristics and can be identified via their charge. One has to note that a polarised e^+ beam is necessary. Even completely polarised e^- would not be sufficient, since the s-channel exchange could not be switched off completely. However, partially polarised beams of maximal $P_{e^-} = -80\%$ and $P_{e^+} = -60\%$ would be sufficient to probe this association, since in this case the pair $\tilde{e}_L^-\tilde{e}_R^+$ dominates by a factor of 3 in our example, Fig. 2b.

Gaugino/higgsino sector

The SUSY partners of the charged and neutral gauge bosons are the charginos $\tilde{\chi}^\pm_{1,2}$ and neutralinos $\tilde{\chi}^0_{1,\ldots,4}$. Since SUSY is broken the electroweak eigenstates mix. Strategies have been worked out to determine the mixing angles via polarised rates in $e^+e^- \to \tilde{\chi}_i^\pm\tilde{\chi}_j^\mp$ and $e^+e^- \to \tilde{\chi}_i^0\tilde{\chi}_j^0$ and to derive the underlying MSSM parameters ([17, 1] and references therein). Even if only the lightest particles $\tilde{\chi}_1^+\tilde{\chi}_1^-$, $\tilde{\chi}_1^0\tilde{\chi}_2^0$ were accessible, it would be sufficient for determining the fundamental MSSM parameters M_1, Φ_{M_1}, M_2 and μ, Φ_μ, i.e. the U(1), the SU(2), and the higgsino mass parameters with their CP–violating phases. The ratio of the two Higgs vev's $\tan\beta = v_2/v_1$ can only be derived via this sector if $\tan\beta$ is not too large [17]. In Fig. 3a, it is demonstrated how to determine $|M_1|$ and Φ_{M_1} with polarised cross sections $\sigma(e^+e^- \to \tilde{\chi}_1^0\tilde{\chi}_2^0)$ and the light masses $m_{\tilde{\chi}^0_{1,2}}$. In this context the beam polarisation is needed in order to resolve ambiguities and to improve the statistics.

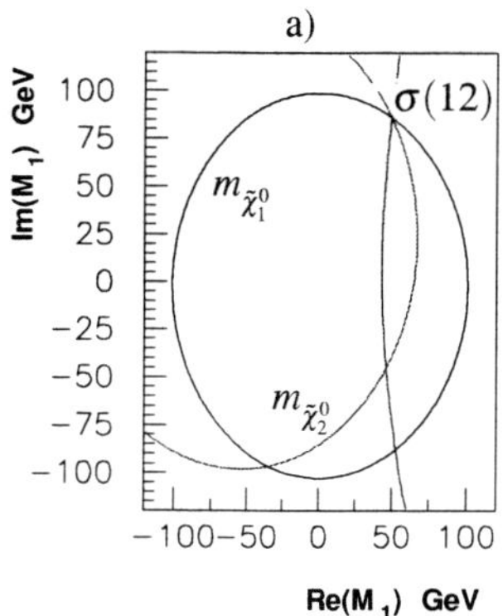

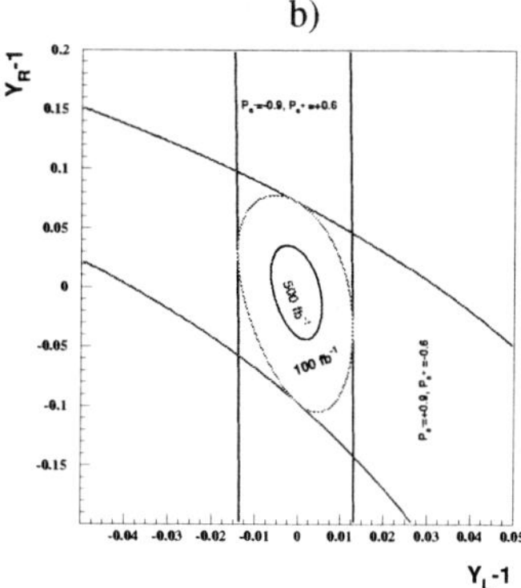

FIGURE 3. a) The contours of two neutralino masses (1,2) and one neutralino production cross section $\sigma(e^+e^- \to \tilde{\chi}_1^0\tilde{\chi}_2^0)$ in the $Re(M_1)$, $Im(M_1)$ plane; b) Contours of the cross sections $\sigma(e_R^+e_L^- \to \tilde{\chi}_1^0\tilde{\chi}_2^0)$ and $\sigma(e_L^+e_R^- \to \tilde{\chi}_1^0\tilde{\chi}_2^0)$ in the plane of the Yukawa couplings $g_{\tilde{W}}$ and $g_{\tilde{B}}$ normalized to the SU(2) and U(1) gauge couplings g and g' $\{Y_L = g_{\tilde{W}e\tilde{e}}/g, Y_R = g_{\tilde{B}e\tilde{e}}/g'\}$.

Once the parameters are determined one can efficiently test whether the gauge couplings g_{Bee} and g_{Wee} are identical to the Yukawa couplings $g_{\tilde{B}e\tilde{e}}$ and $g_{\tilde{W}e\tilde{e}}$, respectively, by studying the polarised cross sections with a variable ratio of $g_{\tilde{B}e\tilde{e}}/g_{Bee}$ and $g_{\tilde{W}e\tilde{e}}/g_{Wee}$ and comparing them with the experimental values [17], see Fig. 3b.

The case of high $\tan\beta$*:* τ *polarisation*

In case of high $\tan\beta$, $\tan\beta > 10$, the chargino and neutralino sector is insensitive to this parameter. But even in the Higgs sector the case $\tan\beta > 10$ will lead to rather large uncertainties [18]. However, in many scenarios one could then determine $\tan\beta$ from another sector whose particles are relatively light: the $\tilde{\tau}$ sector.

The polarisation of τ's from $\tilde{\tau}_i \to \tau\tilde{\chi}_j^0$ is sensitive to $\tan\beta$ [19] and the τ polarisation can be rather accurately measured at a LC via e.g. the τ decays into π's, see Fig. 4a. It has been shown in [20] that in case of a sufficient higgsino admixture in the $\tilde{\chi}_j^0$ it is even possible to determine high $\tan\beta$ as well as A_τ, without any assumptions on the SUSY breaking mechanism: after determining the $\tilde{\tau}$ mixing angle via polarised rates, preferably in the configuration σ_{RL} due to WW background suppression, one can determine $\tan\beta$ from the τ polarisation in the decay $\tilde{\tau}_1 \to \tau\tilde{\chi}_1^0$, see Fig. 4b. Even for high $\tan\beta \geq 20$ one can reach an accuracy of about 10%.

Extended SUSY models

In case of R–parity violating SUSY non–standard couplings could occur which produce a scalar particle in the s–channel: $e^+e^- \to \tilde{\nu} \to e^+e^-$. The process gives a signif-

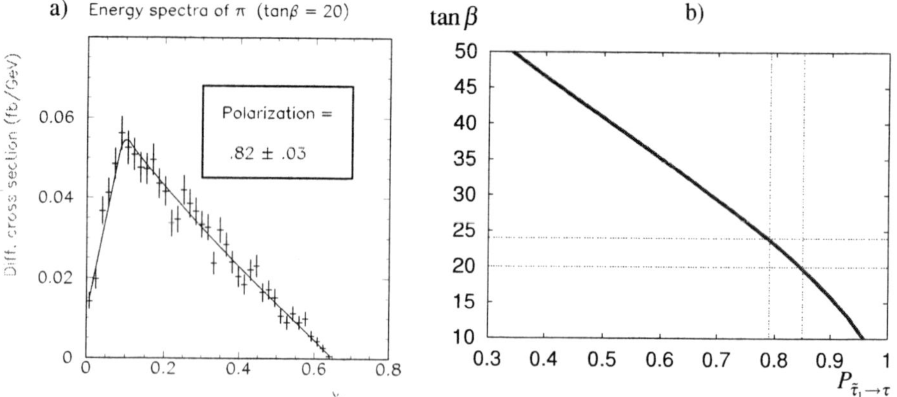

FIGURE 4. a) Determining the τ polarisation $P_{\tilde{\tau}_1 \to \tau}$ from the pion energy distribution in the decay $\tilde{\tau}_1 \to \tau \tilde{\chi}_1^0 \to \nu_\tau \pi \tilde{\chi}_1^0$. It leads to an accurate determination of high $\tan\beta$: e.g. $\tan\beta = 20 \pm 2$, see b) [20].

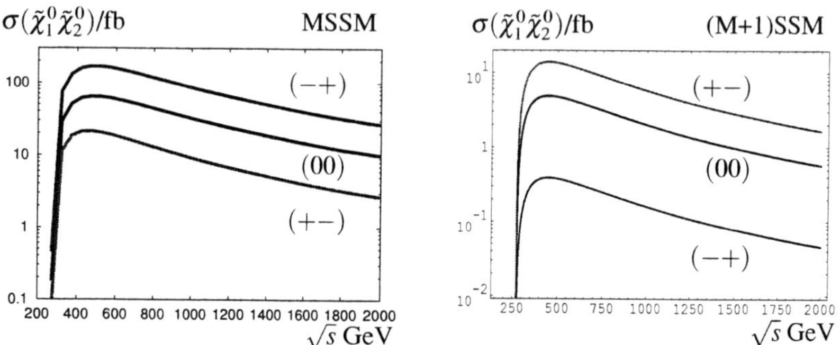

FIGURE 5. Cross sections for the process $\sigma(e^+e^- \to \tilde{\chi}_1^0 \tilde{\chi}_2^0)$ with polarised beams ($P_{e^-} = \pm 80\%, P_{e^+} = \mp 60\%$) for an example in the MSSM and the (M+1)SSM, where the mass spectra of the light neutralinos are similar [22].

icant signal over the background. Since it requires both left–handed e^- and e^+ beams – the *LL* configuration– it can be easily analysed and identified by the use of beam polarisation ([21, 2]): here simultaneously polarised beams enhance the signal by about a factor of 10, see Table 4.

One could also extend the MSSM without changing the gauge group, by introducing an additional Higgs singlet: it leads to the (M+1)SSM with one additional neutralino. Since the mass spectra of the four light neutralinos could be similar to those in the MSSM in some parts of the (M+1)SSM/MSSM parameter space, a distinction between these models might be difficult via spectra and rates alone. However, polarisation effects might then indicate the different coupling structure in the (M+1)SSM [22] and help disentangling the models, Fig. 5.

TABLE 4. Sneutrino production in R–parity violating SUSY: Cross sections of $e^+e^- \to \tilde{\nu} \to e^+e^-$ for unpolarised beams, $P_{e^-} = -80\%$ and unpolarised positrons, and for $P_{e^-} = -80\%$, $P_{e^+} = -60\%$. The study was made for $m_{\tilde{\nu}} = 650$ GeV, $\Gamma_{\tilde{\nu}} = 1$ GeV, an angle cut of $45^0 \le \theta \le 135^0$ and the R–parity violating coupling $\lambda_{131} = 0.05$

	$\sigma(e^+e^- \to e^+e^-)$ including $\sigma(e^+e^- \to \tilde{\nu} \to e^+e^-)$	Bhabha–background
unpolarised	7.17 pb	4.50 pb
$P_{e^-} = -80\%$	7.32 pb	4.63 pb
$P_{e^-} = -80\%, P_{e^+} = -60\%$	8.66 pb	4.69 pb
$P_{e^-} = -80\%, P_{e^+} = +60\%$	5.97 pb	4.58 pb

Large extra dimensions

Another approach for physics beyond the SM, which could also resolve the hierachy problem, is the introduction of large extra dimensions. At a LC the process $e^+e^- \to \gamma G$ is promising. Running at two different $\sqrt{s}$ could provide a determination of the number of the extra dimensions [23, 1]. The use of polarised beams in this context enlarges on the one hand the sensitivity to the new scale M_* and suppresses on the other hand the main background $e^+e^- \to \nu\nu\gamma$ significantly. The ratio $S/\sqrt{B}$ is enhanced by a factor of about 2.1 (4.4) if $P_{e^-} = +80\%$ (and $P_{e^+} = -60\%$) is used.

SUMMARY

A LC in the TeV range with its clean initial state of e^+e^- collisions is ideally suited for the search for new physics, for the determination of both Standard Model and New Physics couplings with high precision and for revealing the structure of the underlying model. The use of polarised beams plays a decisive role in this context. We have shown that simultaneous polarisation of both beams can significantly expand the accessible physics opportunities[2]. The use of simultaneously polarised e^-, e^+ beams has several advantages: determining quantum numbers of new particles, providing higher sensitivity to non–standard couplings, increasing rates and background suppression, raising the effective luminosity and expanding the range of measurable experimental observables e.g. with the help of transversely polarised beams.

ACKNOWLEDGMENTS

The author would like to thank Yousef Makdisi and his friendly and helpful organizing team for a very interesting conference! G.M.–P. was partially supported by BNL.

---

[2] For updates see POWER group (Polarisation at Work in Energetic Reactions): http://www.desy.de/~gudrid/POWER/.

REFERENCES

1. J. A. Aguilar-Saavedra *et al.*, ECFA/DESY LC Physics Working Group Collaboration, [hep-ph/0106315].
2. G. Moortgat-Pick and H. M. Steiner, Eur. Phys. J. directC **6** (2001) 1 [hep-ph/0106155]; J. Erler *et al.*, in *Proc. of the APS/DPF/DPB Summer Study on the Future of Particle Physics (Snowmass 2001)* ed. N. Graf, arXiv:hep-ph/0112070; G. Moortgat-Pick *et al.*, arXiv:hep-ph/0210212.
3. T. Abe *et al.* [American Linear Collider Working Group Collaboration], in *Proc. of the APS/DPF/DPB Summer Study on the Future of Particle Physics (Snowmass 2001)* ed. N. Graf,SLAC-R-570 *Resource book for Snowmass 2001, 30 Jun - 21 Jul 2001, Snowmass, Colorado.*
4. V.E. Balakin and A.A. Mikhailichenko, INP 79–85 (1979); K. Flöttmann, DESY-93-161, 1993; K. Flöttmann, DESY-95-064, 1995.
5. see Talk by John Sheppard on the POWER meeting, June 8-9, 2002, Durham: http://www.desy.de/˜gudrid/power/schedule_power_fin.html and also http://www.slac.stanford.edu/˜achim/positrons/.
6. V. Gharibyan, N. Meyners, K.P. Schüler, LC–DET–2001–047.
7. G. Alexander and I. Cohen, Nucl. Instrum. Meth. A **486** (2002) 552 [arXiv:hep-ex/0006007].
8. A. Blondel, Phys. Lett. **B202** (1988) 145.
9. see also talk by Klaus Desch, International Workshop on Linear Colliders August 26-30, 2002 Jeju Island, Korea: http://lcws2002.korea.ac.kr/.
10. R. Hawkings and K. Moenig, EPJ*direct* C 8 (1999) 1; K. Mönig, LC–PHSM–2000–059; W. Menges, LC–PHSM–2001–022.
11. J. Fleischer, K. Kolodziej and F. Jegerlehner, Phys. Rev. D **49** (1994) 2174.
12. J. Erler, S. Heinemeyer, W. Hollik, G. Weiglein and P. M. Zerwas, Phys. Lett. B **486** (2000) 125.
13. B. Ananthanarayan, S. D. Rindani and A. Stahl, LC-PHSM-2002-006, hep-ph/0204233.
14. A. Leike, S. Riemann, Z. Phys.C75 (1997) 341; S. Riemann, hep-ph/9610513, Proceedings of the 1996 DPF / DPB Summer Study on New Directions for High-energy Physics; R. Casalbuoni, S. De Curtis, D. Dominici, R. Gatto, S. Riemann, hep-ph/0001215, LC-TH-2000-006; S. Riemann, Proceedings of 4th International Workshop on Linear Colliders (LCWS 99), Sitges, Barcelona, Spain; S. Riemann, LC-TH-2001-007; A. A. Pankov and N. Paver, arXiv:hep-ph/0209058.
15. A. Bartl, H. Eberl, S. Kraml, W. Majerotto, W. Porod and A. Sopczak, Z. Phys. C **76** (1997) 549 [arXiv:hep-ph/9701336]; A. Bartl, H. Eberl, S. Kraml, W. Majerotto, W. Porod and A. Sopczak, arXiv:hep-ph/9604221; A. Bartl, H. Eberl, S. Kraml, W. Majerotto and W. Porod, Z. Phys. C **73** (1997) 469 [arXiv:hep-ph/9603410]; A. Bartl, K. Hidaka, T. Kernreiter and W. Porod, hep-ph/0207186 and Phys. Lett. B **538** (2002) 137.
16. C. Blöchinger, H. Fraas, G. Moortgat-Pick and W. Porod, Eur. Phys. J. C **24** (2002) 297.
17. S. Y. Choi, J. Kalinowski, G. Moortgat-Pick and P. M. Zerwas, Eur. Phys. J. C **22** (2001) 563 S. Y. Choi, J. Kalinowski, G. Moortgat-Pick and P. M. Zerwas, Eur. Phys. J. C **23** (2002) 769 [hep-ph/0202039].
18. J. F. Gunion, T. Han, J. Jiang, S. Mrenna and A. Sopczak, in *Proc. of the APS/DPF/DPB Summer Study on the Future of Particle Physics (Snowmass 2001)* ed. N. Graf, arXiv:hep-ph/0112334; V. D. Barger, T. Han and J. Jiang, Phys. Rev. D **63** (2001) 075002 [arXiv:hep-ph/0006223]; J. L. Feng and T. Moroi, Phys. Rev. D **56** (1997) 5962 [arXiv:hep-ph/9612333].
19. M. M. Nojiri, Phys. Rev. D **51** (1995) 6281 [arXiv:hep-ph/9412374]; M. M. Nojiri, K. Fujii and T. Tsukamoto, Phys. Rev. D **54** (1996) 6756 [arXiv:hep-ph/9606370]; M. Guchait and D. P. Roy, Phys. Lett. B **535** (2002) 243 [arXiv:hep-ph/0205015].
20. E. Boos, U. Martyn, G. Moortgat–Pick, M. Sachwitz, S. Vologdin, hep-ph/0211040.
21. M. Heyssler, R. Rückl, H. Spiesberger, Proceedings of the LC–Workshop, Sitges 1999; Private communication with H. Spiesberger.
22. S. Hesselbach, F. Franke, H. Fraas, hep-ph/0003272; F. Franke and S. Hesselbach, Phys. Lett. B **526** (2002) 370.
23. A. Vest, LC–TH–2000–058. G. Wilson, LC-PHSM-2001-010.

Laser-polarized Noble Gases for Magnetic Resonance Imaging

Gordon D. Cates, Jr.

Departments of Physics and Radiology, University of Virginia, Charlottesville, VA 22904

Abstract. This paper describes a technique in which noble gases such as ^{3}He and ^{129}Xe are polarized using optical pumping techniques, and used as a source of signal for magnetic resonance imaging. Techniques for polarizing the gas and a few representative examples of clinical applications are presented. Also emphasized is the connection between noble-gas imaging and the use of polarized gaseous targets in nuclear and particle physics experiments such as E142, an experiment at SLAC to study the spin structure of the neutron.

INTRODUCTION

Noble-gas imaging is a new type of magnetic resonance imaging (MRI) that utilizes laser-polarized noble gases. The noble gas, typically ^{3}He or ^{129}Xe, is polarized using optical pumping techniques. It is subsequently inhaled by the subject, and an MRI scanner, tuned to the Larmor frequency of the noble gas nucleus, is used to make an image. The result is unprecedented resolution of the gas space of the lungs. In the case of ^{129}Xe, the gas is also absorbed into the blood and transported throughout the body. Imaging of the brain and other parts of the body have also been demonstrated.

There are now two established techniques for polarizing large quantities of noble gases. One approach, known as spin-exchange optical pumping, involves the optical pumping of alkali-metal atoms such as rubidium (Rb) and potassium (K) and subsequent spin-exchange collisions with noble-gas atoms [1]. Another approach, known as metastability exchange, involves the direct optical pumping of helium (He) in a metastable 2^3S_1 state, and subsequent collisions in which polarization is transferred to ground state atoms [2, 3]. While both polarization techniques were demonstrated in the 1960's, a great deal of basic research[4] and the evolution of laser technology have played a key role in making them practical.

Noble-gas imaging was first proposed by G. Cates and W. Happer at Princeton and M. Albert at Stony Brook in 1993. The first images, of the gas space of the excised lungs of a mouse, were published in Nature in 1994 [5]. Images of the gas space of human lungs followed shortly thereafter, and were made by both a Princeton/Duke collaboration in the US [6] and a Mainz/Heidelberg/Ecole Normale Superieure collaboration in Europe [7]. Both collaborations drew on experience gained in building large polarized ^{3}He targets. In the US, in 1992, a polarized ^{3}He target, based on spin-exchange optical pumping was used at SLAC (E142) for the study of the spin structure functions of the neutron [8, 9]. Shortly thereafter another polarized ^{3}He target, based on metastability exchange, was used at Mainz to measure the electric form factor of the neutron [10]. In both cases the

CP675, *Spin 2002: 15th Int'l. Spin Physics Symposium and Workshop on Polarized Electron Sources and Polarimeters,* edited by Y. I. Makdisi, A. U. Luccio, and W. W. MacKay

amount of gas involved was on the order of a quantity corresponding to a liter at STP, thus establishing a necessary requirement for noble-gas imaging, and setting the stage for medical applications.

Noble gas imaging is an excellent example of an unexpected technological spin-off resulting from otherwise basic research. From, its beginnings in 1960, a great deal of atomic physics research has been done on spin-exchange optical pumping. The same is also true of metastability exchange. The nuclear and particle physics communities have also badly needed targets rich in polarized neutrons, stimulating the development of techniques for producing large quantities of polarized ^{3}He. While none of this activity was conducted in anticipation of an application in medical imaging, it is clearly the case that this new type of MRI would not otherwise have come into being.

POLARIZING GAS AND POLARIZED TARGETS

Spin-exchange optical pumping

Spin-exchange optical pumping is a two step process in which 1) an alkali metal such as Rb or K is optically pumped and 2) the nuclei of a noble gas such as ^{3}He or ^{129}Xe is polarized in subsequent spin-exchange collisions. For reasons that will be explained more below, the approach is often quite different depending on whether it is ^{3}He or ^{129}Xe that is being polarized.

For the case of ^{3}He, the gas is typically polarized in a glass cell containing up to 10 atmospheres of ^{3}He, around 70 Torr of nitrogen (N_2), and on the order of 10-100 milligrams of rubidium metal. The density of Rb vapor is controlled by heating the cell. The cell is irradiated with circularly polarized light with a wavelength of 795 nm, which corresponds to the D_1 transition between the $5^2S_{1/2}$ ground state and the $5^2P_{1/2}$ first excited state. The N_2 is included in the cell to induce "radiationless quenching" of the excited Rb atoms. Once the Rb atoms are polarized, polarization is transferred to the ^{3}He through spin-exchange collisions. The interaction responsible is a hyperfine interaction between the valence electron of the Rb atoms and the spin of the ^{3}He nuclei.

In order to maintain high Rb polarization, it is essential that the photon absorption rate per Rb atom be greatly in excess of the electronic relaxation rate. In equilibrium, the photon absorption rate is given by

$$\text{Photon absorption rate} = \gamma_{sd}\,[\text{Rb}]\,\text{V}\,P_{\text{Rb}} \tag{1}$$

where γ_{sd} is the electronic relaxation rate, [Rb] is the number density of Rb, V is the volume of Rb vapor, and P_{Rb} is the Rb polarization. If we assume that the Rb number density is set by other factors, eqn. (1) sets a requirement on the required laser power. Conceptually, enough laser power is required to replace the angular momentum lost to electronic spin relaxation.

Another critical factor in achieving high ^{3}He polarization is ensuring that the Rb-^{3}He spin-exchange rate is greatly in excess of ^{3}He spin-relaxation rates unrelated to spin exchange. The time evolution of a sample of ^{3}He is given by

$$P_{\text{He}}(t) = P_{\text{Rb}}\left(1 - e^{(\gamma_{se}+\Gamma)t}\right)\frac{\gamma_{se}}{\gamma_{se}+\Gamma} \tag{2}$$

where $P_{\text{He}}(t)$ is the time dependent nuclear polarization of the ^{3}He, γ_{se} is the Rb-^{3}He spin-exchange rate, and Γ is the spin-relaxation rate due to all processes other than spin exchange. Under most circumstances, γ_{se}^{-1} is somewhere between 5 and 20 hours. The value of Γ^{-1}, which is a number characteristic of each cell and a quantity often referred to as a cell's lifetime, must be long compared to γ_{se}^{-1} to achieve high polarization. A great deal of effort has gone into establishing techniques for achieving long values of Γ^{-1} on a consistent basis.

Polarizing ^{129}Xe has associated with it very different challenges than is the case when polarizing ^{3}He. In particular, the electronic spin relaxation rate associated with 5-10 atmospheres of ^{129}Xe is on the order of several thousand times faster than what one typically encounters when polarizing ^{3}He. This makes it difficult to maintain high Rb polarization. One solution to this is to use a very dilute mixture, on the order of 1%, of Xe, with the remainder of the gas being mostly ^{4}He or N_2 [11]. Such a mixture can still be used at relatively high pressures while maintaining a manageable electronic spin relaxation rate. The high pressures are desirable because they broaden the absorption line of the Rb, better matching the broad spectral output of the high-power diode lasers. The dilute mixture can be flowed through a polarization chamber quickly because Rb-^{129}Xe spin-exchange rates are much higher than is the case for Rb and ^{3}He. The Xe can then be frozen out of the gas mixture and accumulated. There are certainly other approaches to polarizing ^{129}Xe, and the options can be expected to become more numerous as better lasers are developed.

The E142 polarized ^{3}He target

The development of target technology for experiments such as SLAC E142 played a critical role in both in inspiring and enabling noble-gas imaging. When E142 was proposed, spin-exchange experiments were typically performed in cells not unlike the small spherical cell shown in Fig. 1, with volumes of 10 cm^3 or less. Cells with volumes as large as 35 cm^3 were being developed for use at labs such as TRIUMF and Bates, but had not yet been demonstrated. The SLAC target cells, also shown in Fig. 1, had volumes of around 150 cm^3 in volume. The SLAC target, at the time it was proposed, represented more than an order of magnitude increase in the quantity being polarized than what had previously been achieved. It marked the first time that (at STP) liter-type quantities of ^{3}He had been polarized, and presented the possibility for considering other applications that would require significant quantities of ^{3}He. The SLAC target cells were comprised of two chambers, an upper "polarization chamber" in which spin exchange took place and a lower "target chamber" through which the electron beam passed.

As emphasized earlier, laser power is a limiting factor in determining the quantity of noble gas that can be polarized. When E142 took data, five titanium::sapphire lasers were used, each pumped by an argon ion laser. Collectively they produced 20-25 W.

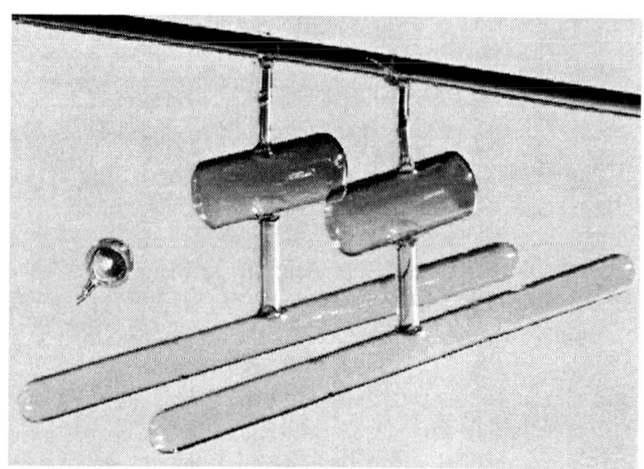

FIGURE 1. Shown are two glass cells that were used for the SLAC E142 polarized ^{3}He target together with another glass cell that was more typical at that time of experiments involving spin-exchange optical pumping.

Each of the five systems cost on the order of \$100K and drew close to 50 kW out of the wall. Shortly after E142 took data, however, a solid-state technology became available known as fiber-coupled high-power diode laser arrays. In these devices each of a dozen or more emitters on a single chip are coupled to fiber optics that can be bundled to produce a bright source of light. The diode array systems have a much broader linewidth than Ti::sapphire systems, around 1000 GHz as opposed to roughly 30 GHz, but they are still quite effective for optical pumping of Rb. A system producing 20-30 W costs around \$20-25K, and only draws a few hundred Watts from the wall. Diode laser systems played an important role in E154 (the follow-on experiment to E142) and are critical to medical imaging where large power consumptive lasers are completely impractical.

Metastability exchange

For the purposes of this paper I have focused on spin-exchange optical pumping, but it is also important to describe metastability exchange. The technique relies on producing helium atoms in the metastable 2^3S_1 state using an rf discharge, and optically pumping them using a transition to the 2^3P_0 ($\lambda = 1.08\mu$). The polarized metastables subsequently collide with other helium atoms in metastability exchange collisions, during which a significant amount of angular momentum is transferred to ground state atoms.

There are both advantages and disadvantages to metastability exchange over spin-exchange optical pumping. On the positive side, metastability exchange is more efficient than spin-exchange optical pumping in terms of the number of photons required to produce a polarized ^{3}He nucleus. Also, the cross section for the metastability exchange collisions is quite large compared to the cross section for alkali-metal—noble-gas spin exchange, so the process proceeds much more rapidly. On the negative side, metastabil-

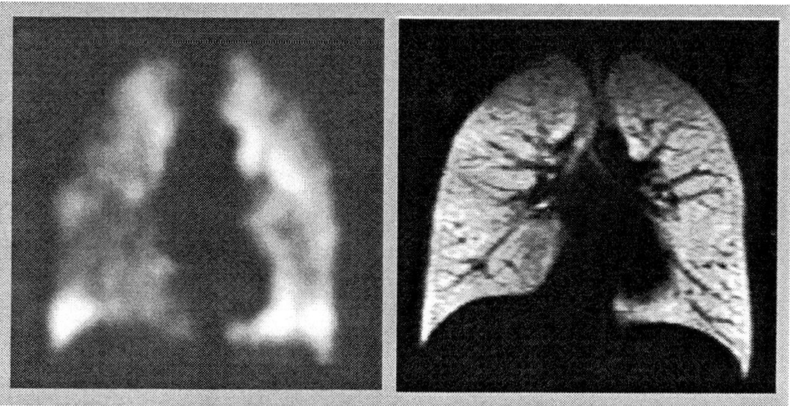

FIGURE 2. Shown are two images off the gas space of human lungs (from different subjects). At left is a traditional ventilation scan in which the subject inhales radioactive gas and an image is made using a gamma camera. At right is an MRI in which the signal source is inhaled nuclear-polarized ^{3}He. Both images were made at UVa.

ity exchange only works for pressures on the order of one Torr. Thus, once the ^{3}He is polarized, it must be compressed in order to reach pressures over an atmosphere. The apparatus for metastability exchange ends up being significantly more complex than is the case for spin-exchange optical pumping, but the performance of a well designed system can be impressive. At the time of this writing most noble-gas imaging in the United States is performed using spin-exchange optical pumping, and most noble-gas imaging in Europe is performed using metastability exchange.

NOBLE-GAS IMAGING

The rapid growth of noble-gas imaging is due in part to existing difficulties in lung imaging. The current state-of-the-art method for imaging the gas space of the lungs is a nuclear medicine technique known as a ventilation scan. The subject inhales a radioactive gas, and a "gamma camera" is used to image the gamma rays that emanate from the subject's chest. An example of a ventilation scan is shown in Fig. 2 on the left. While the image certainly reveals larger features, the practical resolution is on the order of a centimeter. In contrast, the image on the right in Fig. 2 is an example of noble-gas imaging using ^{3}He. The improvement in resolution is striking, with details on the order of a millimeter becoming visible.

Polarizing gas in a clinical environment

In the early experiments with noble-gas imaging, the apparatus used to polarize the gas was very much the sort of equipment that one would expect to find in a physics

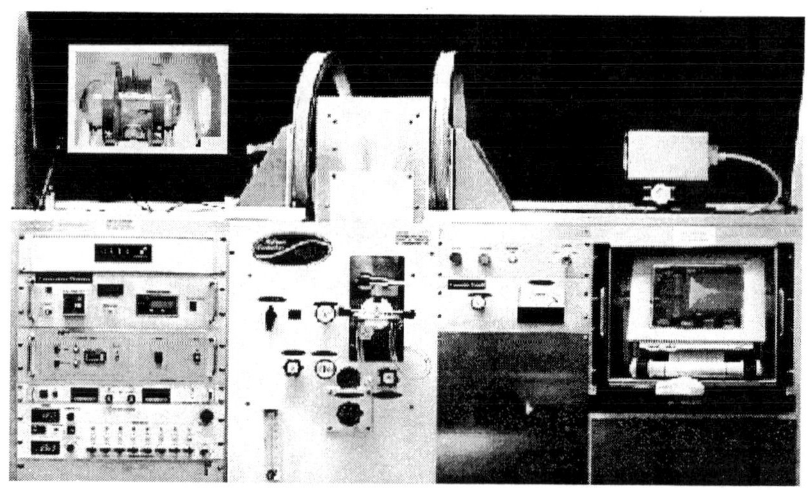

FIGURE 3. Shown is the first prototype commercial noble-gas polarizer. Shown in the inset is the "polarization chamber", used for polarizing ^{3}He by spin exchange. The polarization chamber can be seen to be similar to the polarization chamber of the SLAC target cells shown in Fig. 1.

laboratory. Among the implications of this was that physicists were required to operate the equipment. While this was appropriate for early experiments, it is impractical for extensive clinical studies. Medical research generally involves substantial numbers of trials. Only by establishing repeatability can patterns be identified. This places demands, however, on the equipment that is being used to polarize the noble gases. Ideally one would like the apparatus to be easy to operate so a well-trained technician could handle the job. It also needs to be very reliable. Often the patients being imaged are in ill health, and cannot tolerate or would be unwilling to reschedule a procedure should technical difficulties prevent the noble gas from being ready for an image.

The apparatus pictured in Fig. 3 is a prototype of the first commercial noble-gas polarizer. It is built to produce batches of ^{3}He on a repeated basis. It is self contained, the controls are accessible from the front of the machine, and it is relatively easy to operate. Subsequent versions of the device are even more compact and user friendly.

The connections between the prototype polarizer shown in Fig. 3 and the technology developed for SLAC E142 are many. As an example, the inset in Fig. 3 shows the polarization chamber in which ^{3}He is polarized. It can be seen that the polarization chamber is nearly identical in structure to the polarization chamber used in the polarized ^{3}He target for E142 as depicted in Fig. 1.

Finding a clinical niche

While noble-gas imaging clearly provides images of the gas space of the lungs that are dramatically improved over what was previously available, a great deal of clinical research is needed before it can be established as a new clinical tool. With a reliable

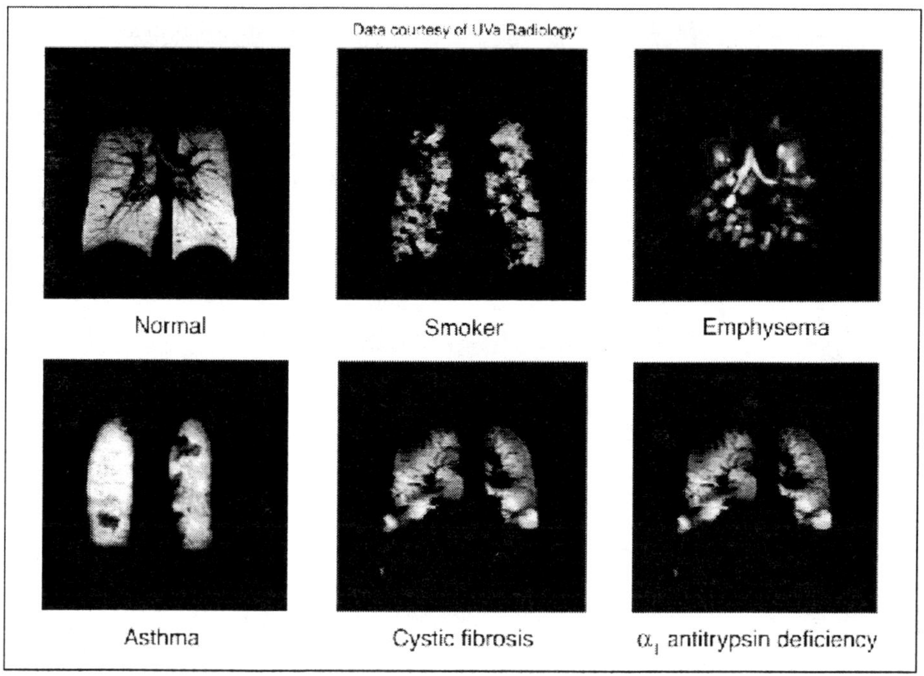

FIGURE 4. Shown are several examples of pulmonary pathologies in contrast to a healthy volunteer.

source of polarized gas, however, it becomes possible for physicians to conduct studies on a regular basis on a wide variety of pathologies. At UVa, over 300 studies have been performed on both healthy volunteers as well as patients suffering from a variety of conditions. A few representative examples are shown in Fig. 4.

A nice example of a new insight that has come from noble gas imaging pertains to asthma. In Fig. 4, on the image labeled asthma, small dark areas are visible that are not present on a normal lung. Researchers at UVa, seeing such features, came to refer to them as ventilation defects. They were first seen in images of volunteers that believed themselves to be completely healthy, but turned out to have mild asthma. Noble-gas imaging had accidentally detected mild asthma in volunteers that were otherwise asymptomatic. Establishing a means to detect asthma even in its mildest form may have important consequences. There are studies that suggest that treating asthma when it is quite mild is a good strategy to prevent more serious disease later. Research at UVa has now established noble-gas imaging as a useful tool for studying asthma [12], and additional research is continuing with NIH funding.

Another application of noble-gas imaging is the visualization of the dynamic process of moving gas into and out of the lungs. Because the signal from polarized ^{3}He is so large, it is possible to obtain images very quickly, making it possible to construct real-time movies of the breathing process. Dynamic imaging appears to be useful for patients who have received a single lung transplant. In patients in which no rejection is taking place, dynamic imaging reveals that gas moves quickly and in larger quantities into the grafted lung. The native lung, being in poor condition, receives less gas at a slower rate. In patients suffering from bronchiolitis obliterans (rejection of the transplanted lung) the native and transplanted lung move gas at similar rates and in similar quantities. To a physician examining the patient using conventional techniques such as spirometry (which measures volumes of exhaled gas), the condition is often not apparent until it is too late to treat. Using anti-rejection drugs early on can greatly improve the patient's chances of recovery.

CONCLUDING REMARKS

It is not possible within the scope of this paper to do a thorough job reviewing all the medical applications that are being explored using noble-gas imaging. There are numerous groups in United States, in Europe, and Japan that are actively pursuing research in noble-gas imaging, and I have essentially restricted myself to describing a few of the activities at UVa. Still, I hope it is clear that particularly where pulmonary function is concerned, noble-gas imaging is already having an important impact. More exotic applications, such as dissolved-phase imaging using ^{129}Xe have yet to prove themselves, but may expand the range of applications even further.

I hope I have also made the point that noble-gas imaging grew in an unexpected manner out of basic research in atomic and nuclear physics. This lesson, that basic research can have important and unanticipated results, is a message worth carrying beyond our immediate community so that public support of all research can remain generous.

REFERENCES

1. Bouchiat, M. A., Carver, T. R., and Varnum, C. M., *Phys. Rev. Lett.*, **5**, 373 (1960).
2. Walters, G. K., Colegrove, F. D., and Schearer, L. D., *Phys. Rev. Lett.*, **8**, 439 (1962).
3. Colegrove, F. D., Schearer, L. D., and Walters, G. K., *Phys. Rev.*, **132**, 2561 (1963).
4. T. G. Walker, T. G., and Happer, W., *Rev. Mod. Phys.*, **69**, 629 (1997).
5. Albert, M. S., Cates, G. D., Driehuys, B., Happer, W., Samm, B., Springer Jr., C., and Wishnia, A., *Nature*, **370**, 199 (1994).
6. MacFall, J. R., Charles, H. C., Black, R. D., Middleton, H., Swartz, J., Saam, B., Driehuys, B., Erickson, C., Happer, W., Cates, G. D., Johnson, G. A., and Ravin, C. E., *Radiology*, **200**, 553 (1996).
7. Ebert, M., Grossman, T., and Heil, W. *et al.*, *Lancet*, **347**, 1297–1299 (1996).
8. Anthony, P. L. *et al.* (the SLAC E-142 collaboration), *Phys. Rev. Lett.*, **71**, 959 (1993).
9. Anthony, P. L. *et al.* (the SLAC E-142 Collaboration), *Phys. Rev. D*, **54**, 6620 (1996).
10. Meyerhoff, M. *et al.* (the A3 collaboration), *Phys. Lett. B*, **327**, 959 (1994).
11. Driehuys, B., Cates, G. D., Miron, E., Sauer, K., Walter, D., and Happer, W., *Appl. Phys. Lett.*, **69**, 1668 (1996).
12. Altes, T. A., Powers, P. L., Knight-Scott, J., Rakes, G., Platts-Mills, T. A., de Lange, E. E., Alford, B. A., Mugler III, J. P., and Brookeman, J. R., *J. of Magn. Reson. Imaging*, **13**, 378–384 (2001).

Looking into the Future of Spin and QCD

Jacques Soffer

Centre de Physique Théorique
CNRS Luminy Case 907
13288 Marseille Cedex 09 France

Abstract. Several stimulating open questions in high energy spin physics will be described together with the striking progress recently achieved in this field. In view of the new experimental facilities and the new tools, soon available, I will try to anticipate what we will learn next. The prospects for the future are excellent and clearly, some exciting times are ahead of us.

INTRODUCTION

Although this is the concluding lecture of SPIN2002, I was asked NOT to give a summary talk, but rather to " look into the future ", in other words to foresee, for example, what will be the highlights of SPIN2010! Soon I realized it is not an easy task, but it was too late because I had already accepted it. Let me first focus on one of the key words and quote twice a great man, A. Einstein who said:

- *I never think about the FUTURE, it comes too soon*
- *To make predictions is very difficult, mainly when it concerns the FUTURE*

In preparing this lecture I was thinking "I need to learn how to read off the crystal ball" and I came across the existence of the *Institute for the Future, 2744 Sand Hill Rd. Menlo Park, CA 94025*, which didn't help.

Turning now to physics questions, I want to stress the fact that *spin* is a powerful and elegant tool which will continue, as in the past, to play a crucial role in physics. One of the main goals of high energy particle physics is to understand the fundamental structure of hadrons. In particular concerning the nucleon, which is composed of quarks, antiquarks and gluons, it is essential to determine the role of each parton in making the nucleon spin. It is widely accepted that hadron dynamics is accurately described by perturbative QCD, which has been tested so far to a certain level of accuracy. However it is urgent to improve it and to confirm its validity, by means of new spin experiments which will test, for the first time, the spin sector of QCD.

The future of high energy physics lies mainly in discovering physics beyond the Standard Model and this is the essential motivation for building a multi-TeV proton collider, like LHC, or new e^+e^- linear colliders. However, spin can be also very relevant for searching new physics, which might exist at the TeV energy scale. One way is to detect indirect effects at low energy, by measuring spin-dependent observables with a very high precision and we will present a few specific cases. Another way is to use several hundreds GeV polarized proton beams, like at RHIC, to carry out a general spin physics programme [1] and to uncover some direct effects in pp collisions, as we

CP675, *Spin 2002: 15th Int'l. Spin Physics Symposium and Workshop on Polarized Electron Sources and Polarimeters*, edited by Y. I. Makdisi, A. U. Luccio, and W. W. MacKay

will see. One should not forget the prospects we also have with polarized e^+e^- linear colliders [2].

Before I go on, I would like to apologize for leaving out so many interesting topics because of my ignorance or by lack of time.

WINDOWS FOR PHYSICS BEYOND THE STANDARD MODEL

Neutron Electric Dipole Moment

The existence of a neutron electric dipole moment (EDM) d_n requieres parity P and time reversal T violation and the latter one implies CP violation, due to the CPT theorem which has very strong physics grounds. So a natural way to improve our understanding of the origin of CP violation is to search for non-zero d_n. In the Standard Model, predictions from the Cabibbo-Kobayashi-Maskawa mixing matrix give the upper limit $d_n < 10^{-31}e$cm, which is around six orders of magnitude from the present experimental limits $d_n < (6-10) \cdot 10^{-26}e$cm. However beyond the Standard Model there are many sources of CP violation (Susy, Strong CP, electroweak baryogenesis, etc...) which lead to $d_n < 10^{-26}e$cm. A new experiment is planned at Los Alamos Nat.Lab. [3], whose goal is to improve the sensitivity, in order to push the limit to $10^{-27}e$cm by 2004 and to $10^{-28}e$cm, ten years from now.

Parity Violating Asymmetries

The most accurate determination of the electroweak mixing angle $sin^2\theta_w$, at the energy scale of the Z pole, comes from the SLC and LEP experiments. This fundamental physical constant is running with Q^2, according to the radiative corrections, so its value at lower Q^2 is well predicted by the Standard Model. This can be tested, for example, by measuring parity violation in Møller scattering, as planned by the E158 experiment at SLAC [4]. They are using a 50GeV polarized electron beam which is scattered off unpolarized atomic electrons and measure the asymmetry $A_{PV} = (\sigma_R - \sigma_L)/(\sigma_R + \sigma_L)$. It is expected to be very small, of the order of 10^{-7}, since it goes like $(1/4 - sin^2\theta_w)$. The goal of E158 is to reach a precision $\delta sin^2\theta_w = \pm 0.0008$ by the end of 2003. With this achievement, the sensitivity to new physics is up to $\Lambda = 15$TeV for compositeness, for a new boson Z' of mass $M_{Z'} = 1$TeV and also for lepton flavor violation, to some extend. We look forward to the impact of the results of this new challenging experiment.

The Anomalous Magnetic Moment of the Muon

Another way to probe possible extensions of the Standard Model is to make a high precision measurement of the anomalous magnetic moment of the muon. It is characterized by the value of $a_\mu = (g-2)/2$, which vanishes for the case of a pointlike elementary Dirac particle since $g = 2$. For the muon the theoretical value of a_μ in the Standard Model is obtained by adding up QED, hadronic and weak contributions and the current most recent estimate is $a_\mu(SM) = 11\ 659\ 177(7) \cdot 10^{-10}$, predicted with an accuracy of 0.6ppm. A twenty years old CERN experiment on positive muon had obtained an accuracy of only 10ppm and this has motivated a new experiment at BNL (E821), which has recently reported [5]$a_{\mu^+} = 11\ 659\ 204(7)(5) \cdot 10^{-10}$, a value with an accuracy of 0.7ppm. This experiment plans to reduce the uncertainty to about 0.35ppm and the data obtained so far for negative muon, is also expected to be released soon. On the theoretical side one should try to improve the understanding of QCD corrections and the use of better data (for example on e^+e^- collisions) in order to reduce the still rather large uncertainty on hadronic corrections. We note that $a^{exp}_{\mu^+} > a^{th}_{\mu^+}$, so may be Susy is round the corner!

SPIN IN ELASTIC SCATTERING

Spin Properties of the Pomeron

In the framework of Regge phenomenology, two-body hadronic reactions at very high energy s and small momentum transfer t, are described by the Pomeron exchange. This important trajectory, whose fundamental nature is not yet fully elucidated, couples to hadrons and the forward nonflip amplitude F^P_{nf} is directly related to the total cross section, by the optical theorem. For simplicity, one usually assumes that the flip coupling of the Pomeron vanishes, so for the single flip amplitude, one has $F^P_{sf} = 0$. This is theoretically groundless and essentially based on s-channel helicity conservation, in a perturbative treatment of the Pomeron. It should be checked, for example in the forward direction, by measuring total cross sections in pure spin states and for $t \neq 0$ by measuring spin dependent observables, like the analyzing power (or left-right asymmetry) A_N, in pp elastic scattering. This is by all means an interesting physics programme for the pp2pp experiment at RHIC. A good knowledge of the spin properties of the Pomeron has also relevant implications for polarimetry at high energy, a crucial issue for spin measurements at RHIC. One possible method is to measure A_N in proton-proton or proton-nucleus elastic scattering [6], at very small momentum transfer $10^{-3} \leq -t \leq 10^{-2}$, the so-called Coulomb-nuclear interference (CNI) region. It originates from the interference between the hadronic non-flip amplitude F^{had}_{nf} and the electromagnetic (Coulomb) spin-flip amplitude, which are out of phase. The theoretical prediction depends strongly on the existence of a spin-flip hadronic amplitude F^{had}_{sf}. This contribution is characterized by the ratio $r_5 = (m_p/\sqrt{-t})F^{had}_{sf}/ImF^{had}_{nf}$ and a new BNL-AGS experiment (E950), which has measured A_N in proton-carbon elastic scattering in the CNI region at 21.7GeV/c, has obtained [7] $Rer_5 = 0.088 \pm 0.058$ and $Imr_5 = -0.161 \pm 0.226$. Although it is tempting to associate this with a spin-flip contribution of the Pomeron, one needs to wait for further results at higher RHIC energies with better accuracy.

Nucleon Electromagnetic Form Factors

The electromagnetic form factors of the proton contain important informations on its internal structure and they can be measured in electron-proton elastic scattering. However for the region of large four-momemtum transfer squared, say $Q^2 > 1GeV^2$, from the unpolarized cross section, it is not possible to separate the electric form factor $G_{E_p}(Q^2)$ from the magnetic form factor $G_{M_p}(Q^2)$, which gives the dominant contribution. $G_{Mp}(Q^2)$ is known to behave approximately as a dipole $[1+Q^2/0.71]^{-2}$ and the same behavior was assumed for $G_{Ep}(Q^2)$, such that $\mu_p G_{Ep}(Q^2)/G_{Mp}(Q^2) = 1$, where μ_p is the proton magnetic moment. In fact this ratio of form factors can be measured using the recoil polarization technique, because it is proportional to the ratio of the transverse to longitudinal components of the recoil proton in the elastic reaction $\vec{e}p \to e\vec{p}$. According to a recent JLab experiment [8], one finds that the above naive assumption is not true and one has, instead, $\mu_p G_{Ep}(Q^2)/G_{Mp}(Q^2) = 1 - 0.13\ (Q^2 - 0.04)$, up to $Q^2 = 5.6 GeV^2$. If one extrapolates this linear trend to higher Q^2, the electric form factor should exhibit a zero near $Q^2 = 8 GeV^2$ and this measurement will be done in the near future. The theoretical implications of this result in the framework of various models were discussed by J.J. Kelly [9], who also presented some new results for the neutron electromagnetic form factors.

IS THE RIDDLE OF THE PROTON SPIN STRUCTURE SOLVED ?

This basic question is with us for nearly 15 years and although hudge progress have been made, both on experimental and theoretical sides, we don't have yet the final answer. Let us recall the fundamental proton spin sum rule

$$1/2 = 1/2\Delta\Sigma + \Delta g + L_q + L_g \tag{1}$$

where $\Delta\Sigma = \sum_i(\Delta q_i + \Delta\bar{q}_i)$ is the fraction of the proton spin carried by quarks and antiquarks. The summation runs over all the different flavors, $u, d, s, ...$, and all these quantities are first moments of the corresponding x-dependent distributions, Δg is the spin carried by the gluon and $L_{q,g}$ are the quark and gluon orbital angular momentum contributions. In the naive parton model $\Delta g = 0$ and one ignores $L_{q,g}$, so in order to fulfill the above sum rule one has $\Delta\Sigma = 1$. However in 1988 the EMC data [10] led to a small $\Delta\Sigma$, a big surprise which was interpreted as the fact that the sea quarks carry a fraction of the proton spin, negative and large, so to make effectively $\Delta\Sigma \sim 0$. This inclusive polarized Deep Inelastic Scattering (DIS) data has been greatly improved since then, not only for the proton but also for the neutron, but the earlier trend remains, since we have now $\Delta\Sigma = 0.23 \pm 0.04 \pm 0.06$ at $Q^2 = 5 GeV^2$ [11]. One should keep in mind that in this type of experiment one measures the asymmetry $A_1(x, Q^2)$, from which one extracts the spin-dependent structure function $g_1(x, Q^2) = 1/2\sum_i[\Delta q_i(x, Q^2) + \Delta\bar{q}_i(x, Q^2)]$. Clearly this does not allow the flavor separation, or to disentangle quarks from antiquarks, and some attempts to achieve it were done by measuring semi-inclusive polarized DIS [12]. The HERMES Collaboration has produced some interesting new results showing that, at least in the limited kinematic region they have investigated, the sea quarks contribution is not large and negative. Within a rather low accuracy, there is perhaps an indication for $\Delta s(x) > 0$ and the data are consistent with flavor symmetry, *i.e.* $\Delta\bar{u}(x) \sim \Delta\bar{d}(x)$. This can be better checked in the future at RHIC, by measuring parity-violating helicity asymmetries A_L^{PV} in $W^{\pm}, Z$ production. This idea was first proposed nearly 10 years ago [13] and according to the Drell-Yan production mechanism, one gets simple expressions for these asymmetries. The results of recent calculations [14] for $A_L^{PV}(W^{\pm})$ are shown in Fig. 1

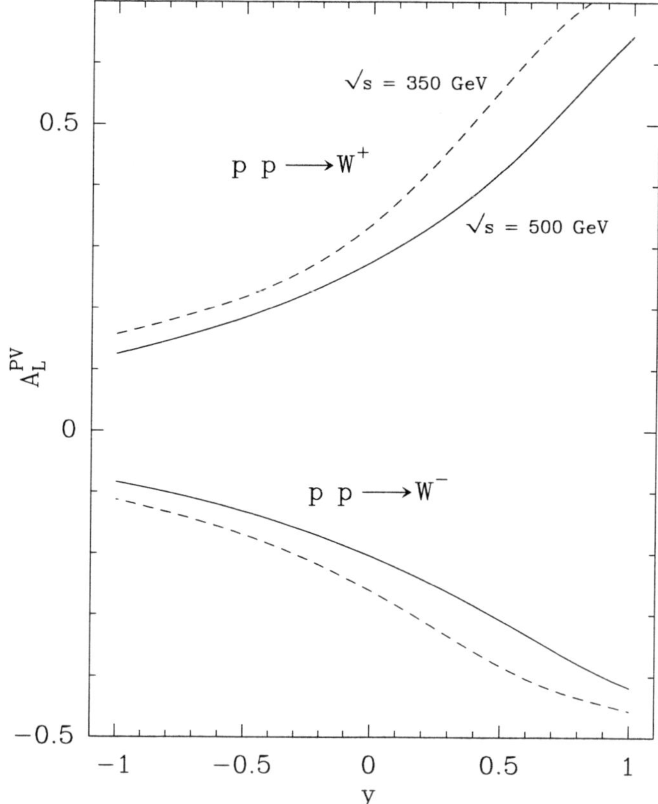

FIGURE 1. The parity violating asymmetry A_L^{PV} for $pp \to W^{\pm}$ production versus the W rapidity at $\sqrt{s} = 350$GeV (dashed curve) and $\sqrt{s} = 500$GeV (solid curve) (Taken from Ref. [14]).

Their general trend can be easily understood and, for example at $\sqrt{s} = 500$GeV near $y = +1$, $A_L^{PV}(W^+) \sim \Delta u/u$ and $A_L^{PV}(W^-) \sim \Delta d/d$, evaluated at $x = 0.435$. Similarly for near $y = -1$, $A_L^{PV}(W^+) \sim -\Delta\bar{d}/\bar{d}$ and $A_L^{PV}(W^-) \sim -\Delta\bar{u}/\bar{u}$, evaluated at $x = 0.059$. Given the expected rates for $W^{\pm}$ production at RHIC-BNL and the high degree of the proton beam polarization [1], it will be possible to check these predictions to a high accuracy.

Although, since 1988 a major effort has been made to explore the small and medium x-regions, for the purpose of checking the fundamental Bjorken sum rule [15], it is also important to improve the existing high x data on the spin dependent structure functions. The preliminary results from a JLab experiment (E99-117) were presented [16] and they clearly show that $A_1^n(x)$ changes sign for $x \sim 0.45$ and becomes positive and large for $x \sim 0.6$. This important feature will have to be confirmed in the future, up to $x \sim 0.75$ by JLab upgrade to 12GeV.

Finally, we still have a very poor knowledge on the contribution of the gluon to the proton spin, which is very badly constrained by the QCD fits to the available polarized

DIS data, whose Q^2 coverage is rather limited. There are various direct ways to get access to $\Delta g(x)$ and, for example the COMPASS experiment at CERN will do it by means of charm particles production, through the photon-gluon process $\gamma g \to c\bar{c}$, with c ($\bar{c}$) hadronizing into D^0 ($\bar{D}^0$). The final uncertainty on the measurement of $\Delta g(x)/g(x)$ from open charm is estimated to be $\delta(\Delta g(x)/g(x)) = 0.11$, in the x-range $0.02 < x < 0.4$ at $Q^2 = 10\mathrm{GeV}^2$. The HERMES experiment at DESY is also expected to collect some data with a lower accuracy, in a smaller x-range and at lower Q^2.

There are also several very promising methods for extracting $\Delta g(x)$ in polarized pp collisions at RHIC, by measuring the double helicity asymmetry A_{LL}. Some of the key processes which will be investigated, both at $\sqrt{s} = 200GeV$ and $500GeV$, are high-p_T prompt photon production $pp \to \gamma + X$, jet production $pp \to jets + X$ and heavy flavor production $pp \to c\bar{c} + X$, $pp \to b\bar{b} + X$, $pp \to J/\psi + X$. Carefull studies about the expected accuracy have been completed recently [17] and it was also proposed to get very soon, some relevant determination of $\Delta g(x)$ by looking at the production of leading high-p_T pions, which are copiously produced even with a moderate luminosity. In all these reactions the subprocesses are either $gq \to \gamma q$, $gq \to gq$, $gg \to gg$, etc.., for all of which, the double helicity asymmetry a_{LL}^{ij} is positive. Therefore the sign of the predicted A_{LL} is always positive because there is a strong prejudice to anticipate $\Delta g(x) > 0$. However one can speculate that Susy particles are not too heavy [2], so that they could be produced in the above processes. In this case, one should remember that the subprocesses $gg \to \tilde{g}\tilde{g}$, $gq \to \tilde{g}\tilde{q}$, etc.., all have their $\tilde{a}_{LL}^{ij}$ negative and large, -100% or so [18], and this could reduce the effect, leading to a misinterpretation of the experimental results.

Clearly when a precise determination of $\Delta\Sigma$ and Δg will be obtained, by using the spin sum rule Eq. (1), it will be possible to estimate the orbital angular momentum contributions, which are unknown so far. According to recent theoretical developments, they are related also to generalized (off-forward) parton distributions (OFPD), which might be measured, for example, in deeply virtual Compton scattering (DVCS), where a real photon is produced and in diffraction meson production [19].

TRANSVERSE SPIN PHYSICS

Quark Transversity in a Nucleon

In addition to the unpolarized and polarized quark distributions $q_i(x,Q^2)$ and $\Delta q_i(x,Q^2)$, discussed so far, there exists a third quark distribution, also at leading-twist order, called transversity and denoted $h_1^{q_i}(x,Q^2)$ or $\delta q_i(x,Q^2)$. It is not accessible in DIS because it is chiral-odd and there is no corresponding transversity for gluons, due to helicity conservation [20]. The quark transversity can be measured, in conjonction with a chiral-odd fragmentation function, in the single-spin azimuthal asymmetry for inclusive pion production $ep^\uparrow \to e\pi X$, with a transversely polarized target. There is a first indication of the effect in the HERMES data [21] on $A_{UL}(\phi)$, with a longitudinally polarized target. Another possibility is to measure the double-transverse asymmetry A_{TT} in Drell-Yan dimuon production in pp collisions at RHIC. To this process $p^\uparrow p^\uparrow \to \mu^+\mu^- X$, corresponds an asymmetry which involves the product $h_1^{q_i}(x,Q^2) \cdot h_1^{\bar{q}_i}(x,Q^2)$. The calculation of A_{TT} has been done at $\sqrt{s} = 200GeV$ in next-to-leading order [22] and using the positivity bound on transversity distributions [23], but it leads to an effect of a few precents, which will not be easy to detect by lack of statistics. Finally, it is important to notice the small size of double-transverse asymmetries in other reactions like jet production or prompt photon production. We anticipate a strong dilution of A_{TT} in the gluon induced processes and such an example is displayed in Fig. 2. So, in future measurements, we should always observe $|A_{TT}| << |A_{LL}|$, a definite test of QCD.

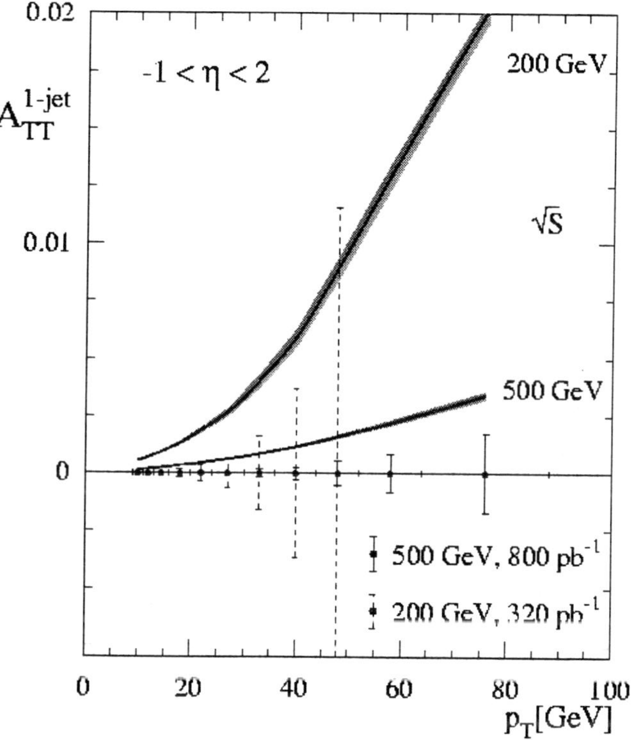

FIGURE 2. Upper bounds for A_{TT} for single-inclusive jet production at RHIC versus p_T for $\sqrt{s} = 200GeV$ and $500GeV$ (Taken from Ref. [24]).

Single Transverse Spin Asymmetries

The study of single transverse spin asymmetries, usually denoted A_N as above, in one-particle inclusive hadron collisions $a + b \to c + X$ (with either a, b or c, transversely polarized) is strongly motivated by large effects observed in a vast collection of intriguing data, for example, in hyperon production [25] or in single pion production [26]. According to naive theoretical arguments, they were expected to be zero for several years. They are non-zero *only* if there is an interference between a non-flip and a single-flip amplitudes, out of phase. Several possible mechanisms have been proposed recently [20] and, in this respect, I would like to give a warning: make sure that the proposed mechanism to explain A_N is also appropriate to describe, *both in shape and magnitude*, the corresponding unpolarized cross section in the *same kinematic region*, because it contains the bulk of the underlying dynamics. We show in Fig. 3 some preliminary results obtained at $\sqrt{s} = 200GeV$ at RHIC [27], which confirm the trend of the FNAL E704 data [26] at $\sqrt{s} = 19.4GeV$. We look forward to the future and to some results at higher p_T, which will strongly test the different theoretical models, provide the authors dare to make predictions in this new kinematic region, prior to the release of the data.

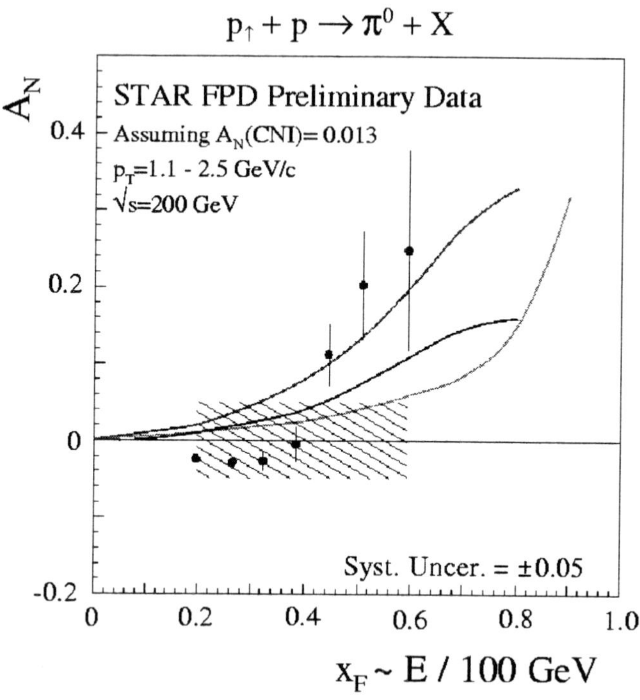

FIGURE 3. Preliminary data from STAR at RHIC and theory predictions at $p_T = 1.5GeV/c$ (All three are private communications). Upper curve, Collins effect [28]-Middle curve, Sivers effect [29]-Lower curve, Twist-3 effect [30].

HUNTING FOR NEW PHYSICS AT RHIC

Like we said earlier, the use of high energy polarized proton beams at RHIC is a new tool to try to uncover new physics. As a specific example, let us consider the parity violating asymmetry A_L^{PV}, in single-jet production. A Standard Model asymmetry is generated from a QCD electroweak interference and this contribution, at $\sqrt{s} = 500GeV$, gives a positive asymmetry which rises, with the transverse momentum of the jet, up to 2-3 %

for $p_T \sim 150 GeV$, with a small uncertainty related to that of the parton distributions. However, instead of this small effect, one could observe sizable effects from contact interactions with a new energy scale $\Lambda > 1TeV$ or a new vector boson Z' with a mass up to $1TeV$ [31]. It was shown that RHIC can be competitive with Tevatron run II, because high luminosity and polarization give a larger sensitivity despite the lower energy. RHIC is also a unique place to probe the chiral structure of new interactions.

Another example is the possibility to detect a large CP-violation in lepton pair production in polarized pp collisions at RHIC. Let us consider the single-transverse asymmetry A_T in the reaction $p^{\uparrow}p \to l^{\pm}\nu X$, where $l^{\pm}$ is a charged lepton. It was shown that some mechanisms beyond the Standard Model can generate a non-zero asymmetry, which was estimated to be rather large in phenomenological tensor interactions [32]. This interesting speculation should be checked in future experiments at RHIC.

To complete the exciting talk we heard this morning on *Applications of Spin in Other Fields* [33], let me mention one recent piece of work, where a *Soft Spin Model* is used to study the fluctuations and market friction in financial trading [34], a subject of great interest! There are also some topics which may belong to "Science Fiction and the Future of Spin", like Quantum information, Quantum teleportation or Quantum computation. No doubt that a successful quantum computer would be exponentially faster than a classical computer.

Finally, let me refer to a nice lecture "History of Spin and Statistics" by A. Martin [35] who stressed the fact that the stability of the world is due to the electrons having a spin-1/2. In agreement with a statement made in the opening talk of SPIN2002 [36], I also claim:

A world without spin would collapse: so keep it UP (or DOWN as you wish)!

ACKNOWLEDGMENTS

I would like to thank the organizers, in particular Yousef Makdisi, for the opportunity to deliver this closing lecture at the very successful conference "SPIN2002". I also thank many participants of this meeting for helpful discussions.

REFERENCES

1. Bunce G., Saito N., Soffer J. and Vogelsang W., *Annu. Rev. Nucl. Part. Sci.* **50**, 525 (2000).
2. Moortgat-Pick G., these proceedings.
3. Mischke R.E., these proceedings.
4. Relyea D., these proceedings.
5. Bennett G.W. *et al.*, *Phys. Rev. Lett.* **89**, 101804 (2002).
6. Buttimore N.H., Kopeliovich B., Leader E., Soffer J. and Trueman T.L., *Phys. Rev.* **D59**, 114010 (1999) and references therein; Kopeliovich B., these proceedings.
7. Tojo J. *et al.*, *Phys. Rev. Lett.* **89**, 052302 (2002).
8. Gayou O. *et al.*, *Phys. Rev. Lett.* **88**, 092301 (2002).
9. Kelly J.J., these proceedings.
10. Ashman J. *et al.*, *Phys. Lett.* **B206**, 364 (1988).
11. Averett T., these proceedings.
12. Miller A., these proceedings.
13. Bourrely C., and Soffer J., *Phys. Lett.* **B314**, 132 (1993).
14. Bourrely C., Buccella F. and Soffer J. *Euro. Phys. J.* **C23**, 487 (2002).
15. Bjorken J.D.,*Phys. Rev.* **D1**, 1376 (1970).
16. Zheng X., these proceedings.
17. Stratmann M., these proceedings.
18. Bourrely C., Renard F., Soffer J. and Taxil P., *Phys. Rep.* **177**, 319 (1989) and references therein.
19. Vanderhaegen M., these proceedings.
20. Ratcliffe P.G., these proceedings.
21. Airapetian A., *et al.*, *Phys. Rev. Lett.* **84**, 4047 (2000).
22. Martin O., Schäfer A., Stratmann M., and Vogelsang W., *Phys. Rev.* **D57**, 3084 (1998) and *Phys. Rev.* **D60**, 117502 (1999).
23. Soffer J., *Phys. Rev. Lett.* **74**, 1292 (1995).
24. Soffer J., Stratmann M., and Vogelsang W., *Phys. Rev.* **D65**, 114024 (2002).
25. Pondrom L., *Phys. Rep.* **122**, 57 (1985).
26. Adams D.L.,*et al.*, *Phys. Lett.* **B264**, 462 (1991).
27. Rakness G., these proceedings.
28. Anselmino M. *et al.*, *Phys. Rev.* **D60**, 054027 (1999).
29. Anselmino M. *et al.*, *Phys. Lett.* **B442**, 470 (1998).
30. Qiu J. and Sterman G., *Phys. Rev.* **D59**, 014004 (1998).
31. Taxil P. and Virey J.-M.,*Phys. Lett.* **B441**, 376 (1998) and references therein.
32. Kovalenko S., Schmidt I. and Soffer J., *Phys. Lett.* **B503**, 313 (2001).
33. Cates G., these proceedings.
34. Rosenow B., arXiv:cond-mat/0107018.
35. Martin A., Proceedings X Séminaire Rhodanien de Physique "Spin in Physics", Torino 3-8/10/02, Frontier Group (Eds. M. Anselmino, F. Mila and J. Soffer) p.17.
36. Ji X., these proceedings.

Summary of DUBNA-SPIN01 Workshop

A. V. Efremov [1]

Joint Institute for Nuclear Research, Dubna, 141980 Russia

Abstract. The main results of IX International Workshop on High Energy Spin Physics, Dubna, August 2–9 are outlined.

This series of workshops began exactly twenty years ago at Dubna under the chairmanship of Lev Iosifovich Lapidus, the well-known scientist, remarkable person and the great enthusiast of the spin physics, who made a noticeable contribution to its development. Next year he would be of 75 and therefore this Workshop was devoted to his memory. The recollections about L.I.Lapidus, his scientific and organizational activity and also the basic reports at the first Workshop were the subject of S.B. Nurushev's (IHEP) talk.

During subsequent years, the Workshop was held in IHEP (Protvino) in the intermediate years between the large Symposia on spin physics, but in 1997 it again returned to Dubna. The present Workshop assembled about 90 scientists from the former Soviet Union countries, Germany, Poland, USA, Japan, etc., including 40 physicists from JINR. Traditionally at the workshop, hot theoretical and experimental problems of high and intermediate energy spin physics were discussed. Wide programs of spin effect investigations, developed and realized at JINR laboratories, were presented in the program of the Workshop. These are the study of spin phenomena on unique polarized beams of deuterons, neutrons and protons in the Laboratory of High Energies, wide spectrum of theoretical works at the Bogoliubov Laboratory of Theoretical Physics. A significant place in the workshop program was given to talks on the current and planned studies of spin effects in the largest experiments: SMC, HERMES and NOMAD and on the program of COMPASS and RHIC. In total, the workshop program included 25 invited and more than 40 original talks.

L.I.Lapidus, as nobody else, clearly understood, how subtle and sensitive tools are the spin investigations of the particle and nuclei properties. They touch the most hidden and intimate features of their interaction. That is why the results of such investigations not once changed the "fashion" in particle physics, forcing theorists to reexamine our ideas of the mechanism of particle interactions. In the fifties it was the discovery of P- and C-parity violation; in 1976 it was the observation of large transverse polarization of Λ-hyperons, produced by a nonpolarized beam which sharply contradicted to the naive parton model, dominant at that time, and led to the necessity to modify it. (By the way, although 26 years passed since then, there is no yet unique understanding of the hyperons

[1] Supported by grants RFBR-00-02-16696, INTAS-00/587.

CP675, *Spin 2002: 15th Int'l. Spin Physics Symposium and Workshop on Polarized Electron Sources and Polarimeters,* edited by Y. I. Makdisi, A. U. Luccio, and W. W. MacKay

polarization nature. Why, for example, $\tilde{\Lambda}$-hyperon is unpolarized, but $\tilde{\Sigma}^-$-hyperon is strongly polarized?) And finally notorious "Spin Crisis" of 1987, the impossibility to explain the nucleon spin by spins of its quarks only, which caused the stormy activity of theorists over the world and dedicated experiments (COMPASS, RHIC, HERMES) in order to check the hypotheses proposed for its resolution. In particular, the hypothesis of a large gluon spin contribution, the most popular now.

The selection of a best method in the future measurements of the gluon spin contribution was also discussed at the Workshop (J.Pretz–COMPASS, A.Bravar–RHIC, A.Tkabladze–HERMES). Unfortunately, one has to state that by all means there is no such a best method. Some of them (e.g. charm detection) provide a good statistical error but have a large systematic one, and the other (e.g. large p_T hadrons) – vice versa. It was the reason of great interest caused by talk of K.Kowalik (Warsaw) about the development of a new approach to the selection of necessary events based on the application of neuron networks. Meanwhile the theorists waiting for the results of new experiments try to estimate the contribution of gluons from the analysis of world data using the QCD evolution equations. The result is that this contribution has the necessary sign and is sufficiently large ($\Delta G/G \approx 0.2$ in the region of $x \approx 0.1$, about twice as small as the HERMES result) in order to explain the "Spin Crisis" (D.Stamenov, Sofia).

A new approach to the Q^2-dependence of the asymmetry A_1 taken into account in polarized DIS was presented by A.Kotikov (Dubna). And a new strategy for systematic extraction of the parton distribution and fragmentation functions based on interplay of semi-inclusive DIS and e^+e^- annihilation was presented by K.Christova (Sofia).

The question about the contribution of "sea" quarks to the nucleon spin was raised in the talk of M.Polyakov (Bochum) who noted that the usually assumed equality of the contributions of different flavors contradicts the Pauli principle that forbids two identical quarks to be in the same state. However, the fraction of different flavors depends on the model of the spin formation. For example, for chiral models in a large N_C limit, more natural is $\Delta\tilde{u} = -\Delta\tilde{d}$. Therefore, the measurement of this fraction in future experiments offers the possibility of the models check.

The role of the parton orbital angular momentum contribution remains also unclear yet. Its measurement requires the study of special processes, the so-called Deeply Virtual Compton scattering (DVCS) or a meson electroproduction (π or ρ), whose first probe at HERMES were presented by A.Borissov, J.Ely (USA), and by E.Thomas (Frascati). Unfortunately, the data are not yet sufficient to answer the question on the orbital angular momentum contribution. A new development of the theory of these processes and connected with them Generalized Distribution Functions which unify the usual parton distribution functions and light-front wave functions were the subject of talks by A.Schaefer (Regensburg), B.Postler (Wuppertal), O.Teryaev, and I.Anikin (Dubna).

One of the most important characteristics of the nucleon spin structure is the transversity distribution. It is one of the three main characteristics of the quark spin density matrix of nucleons presenting the helicity flip element. Due to this, its determination requires the measurement of special spin azimuthal asymmetries of secondary hadrons. The first experiments on the measurements of such asymmetries at HERMES were presented by K.Oganesyan (Yerevan), and the extraction of the proton transversity from the asymmetries was presented by A.Efremov (Dubna).

An important step in understanding the nucleon spin structure is the check of the Gerasimov-Drell-Hearn sum rule[2]. The matter is that it could answer to the question how the parton picture of nucleon with its infinite number of partons converts into the classical quark picture of the nucleon consisting of three quarks. The data of a new experiment of JLab (USA) for the ^3He target were presented giving the possibility to check the sum rule for the neutron (P.Zolnerczuk's talk).

New experiments on checking this sum rule are done at ELSA in Bonn and MAMI in Mainz (K.Helbing, Erlangen and I.Preobrazhenski, Petersburg) and also at HERMES (A.Nagaytsev, Dubna).

An additional puzzle was added to the problem of the transverse polarization of the Λ-hyperon, mentioned above. It turned out that the Λ-hyperon from the electroproduction process at HERMES is also transversally polarized strongly enough, but with *opposite* sign with respect to hadron processes (O.Grebenyuk–PNPI, V.Aleksakhin–Dubna)! Why? There is no idea thus far.

The spin transfer from polarized positron to Λ in the current fragmentation region ($x_F > 0$) at HERMES was the subject of the talk by S.Belostotski (PNPI). It was found that it is compatible with zero, while it is negative in the case of target fragmentation ($x_F < 0$). It is clear that much better statistics is still needed in order to try to distinguish between various models of the Λ spin structure, in particular, in the limit of $z \to 1$. One should also remember that the Λ are often produced via resonance decays like $\Sigma^0 \to \Lambda + \gamma$ and $\Sigma(1385) \to \Lambda + \pi$. Those Λ are difficult experimentally to sort out from the Λ produced directly from a string. This complication of the analysis results inevitably in the necessity of Monte-Carlo simulations with its intrinsic uncertainties.

New data are available also on the left-right asymmetry of pions produced by the transversely polarized beam with the energy 22 GeV (AGS) on the carbon target (S.Nurushev, IHEP). Comparison with previously obtained data confirmed that this asymmetry is practically independent either of the energy or of the type of a target, which makes process very convenient for the polarimetry aim (A.Bogdanov–Moscow). As for the theory, at least two possibilities for explaining this phenomenon are known, one of which was developed in Dubna yet in the beginning of the 80'es and connected with the asymmetry in the parton subprocess, and the second with the "Collins effect", i.e. with the left-right asymmetry of the transversely polarized quark fragmentation process due to a new T-odd fragmentation function. Which of them will prove to be correct, future will show.

Traditionally, many reports on spin physics of intermediate energies were presented and discussed at the workshop with different polarized beams and targets at the accelerators of Gatchina (V.Sumachev), Dubna (L.Azhgirey, N.Piskunov, V.Ladygin, V.Sharov), Novosibirsk (D.Toporkov), and RIKEN in Japan (K.Hatanaka, T.Uesaka, H.Sakai). As a rule, the obtained spin characteristics of processes on nuclei (spin asymmetries, alignments, the polarization transfer, and so forth) at short internuclear distances do not follow the predictions of standard approach to the nucleus as to the system of nucleons

---

[2] By the way, as it follows from the Nurushev's memorial talk, all necessary formulas for this sum rule were contained already in the work by Lapidus and Chou Kuan-chao (1961). However, for the first time they were written in the modern form by S.Gerasimov.

bounded by two-body forces and require the introduction of some new elements (three-body forces, multi-quark configurations and so forth).

Processes with the participation of polarized particles were always among difficult both for theorists and for the experimentalists. First, working with the polarized targets, an experimentalist has to fight with thermal "chaos", that tends to destroy the alignment of spins. Liquid helium temperatures are necessary for this aim. Other difficulties, like the depolarizing resonances, arise in the acceleration of polarized particles and in controlling of polarized beam. Second, spin phenomena are very insidious. As a rule, they are most noticeable in those kinematic regions where the process itself is least probable. Therefore, at the workshop, a significant place was given to the spin processes technique (target, sources, polarimeters, etc.). Among the presented talks, I would noted works on the polarization of electrons and positrons by a laser beam (M.Galinski, Minsk and A.Potylitsyn, Tomsk) and, especially, works on producing and accelerating the polarized deuterons at the Nuclotron of JINR (Yu.Pilipenko, Dubna). The latter allows us to hope that the successful completion of these works will allow JINR to preserve its noticeable place in the spin physics.

Finally, I have to stress that the organization of a workshop like this and participation in it of many scientists from Russia and FSU countries would be impossible without the financial support of the Russian Foundation for Basic Research, International organizing committee for spin symposia, UNESCO and JINR. And we are grateful to them for this support.

The IX Workshop Proceedings (JINR-E1, 2-2002-103, ed. A.Efremov and O.Teryaev) are available from JINR Publishing Dept. The next X Workshop on High Energy Physics is planned for September 2003 in Dubna.

Spin and Symmetry

Determination of the Polarization of the Decay Positrons in Polarized Muon Decay

K.-U. Köhler*, K. Bodek†, A. Budzanowski**, N. Danneberg*, W. Fetscher*, C. Hilbes*, L. Jarczyk†, K. Kirch‡*, S. Kistryn†, J. Klement*, A. Kozela***, J. Lang*, G. Llosá Llácer*, T. Schweizer*, J. Smyrski†, J. Sromicki*, E. Stephan§, A. Strzałkowski† and J. Zejma†

*Institute for Particle Physics, ETH Zürich, Switzerland
†Institute of Physics, Jagellonian University, Kraków, Poland
**H. Niewodniczanski Institute of Nuclear Physics, Kraków, Poland
‡Paul Scherrer Institut, Villigen, Switzerland
§Institute of Physics, University of Silesia, Katowice, Poland

Abstract. The standard model of electroweak interactions predicts that the positrons from the decay of polarized positive muons are mainly longitudinally polarized. The measurement of the two transverse polarization components of the positron P_{T_1} and P_{T_2} is a sensitive tool to look for contributions from additional, exotic interactions and for the violation of time reversal invariance in this purely leptonic decay.
The μ_{P_T} - experiment at the Paul Scherrer Institute determines the three positron polarization components simultaneously with the same apparatus by making use of three different effects. By examining the temporal dependence of annihilation-in-flight of the decay positrons with polarized electrons a possible non-zero value of the transverse polarization is determined. The phase of this transverse polarization can be measured by making use of the decay asymmetry. Using the dependence of annihilation-in-flight on the angle between electron polarization and positron momentum leads to the determination of the longitudinal polarization which not only completes the measurement of the entire polarization vector but also serves as a sensitivity check.
In order to deduce the polarization at the time and location of the muon decay a method based on Monte-Carlo simulations with full spin-dependence implemented is being developed.

INTRODUCTION

Precision measurements of muon decay provide low energy tests of the standard model. Only a few years ago it has been shown that $V-A$, as one of the basic assumptions of the standard model, follows from the results of a selected set of muon decay experiments (including inverse muon decay) [1].
However, the experimental limits obtained up to now still allow for substantial contributions from non-standard couplings in addition to the $V-A$ interaction. The limits on these couplings can be efficiently improved by performing experiments with polarized muons and their decay positrons.
The measurement of the positrons' transverse polarization component P_{T_1} (as defined below) as a function of the positron energy, in particular, offers the possibility to obtain the low energy Michel parameter η without the suppression factor m_e/m_μ, which makes

CP675, *Spin 2002: 15th Int'l. Spin Physics Symposium and Workshop on Polarized Electron Sources and Polarimeters*, edited by Y. I. Makdisi, A. U. Luccio, and W. W. MacKay

the determination of η from the electron energy spectrum extremely difficult. The simultaneous determination of the polarization component P_{T_2} allows one to test time reversal invariance in a purely leptonic decay.

OBSERVABLES IN THE DECAY OF POLARIZED MUONS

The kinematic variables for the decay of polarized positive muons are illustrated in figure 1.

While the e^+ from μ^+ decay is mainly longitudinally polarized (polarization P_L), there also is a transverse polarization component P_{T_1} lying in the plane of muon polarization $\mathbf{P}_\mu$ and positron momentum $\mathbf{k}_e$.

Within the standard model, P_{T_1} is negligibly small at large positron energies, but substantial at lower energies and reaches the value $-1/3$ in the limiting case of a positron at rest (see figure 1, $\eta = 0$). Due to the low rate at small positron energies the energy averaged transverse polarization predicted by the standard model is $< P_{T_1} >= -0.003$ which at present cannot be detected.

The second transverse polarization component P_{T_2}, which is perpendicular to the plane spanned by muon polarization and positron momentum (see figure 1), is exactly equal to zero according to the standard model.

In contrast to the determination of η from the energy spectrum of the decay positrons, the energy dependence of P_{T_1} yields η without the suppression factor m_e/m_μ. With the experimental knowledge that $V-A$ is dominant [1], and neglecting exotic contributions in second order, one obtains

$$\eta = \frac{1}{2} Re\{g^S_{RR}\} \qquad (1)$$

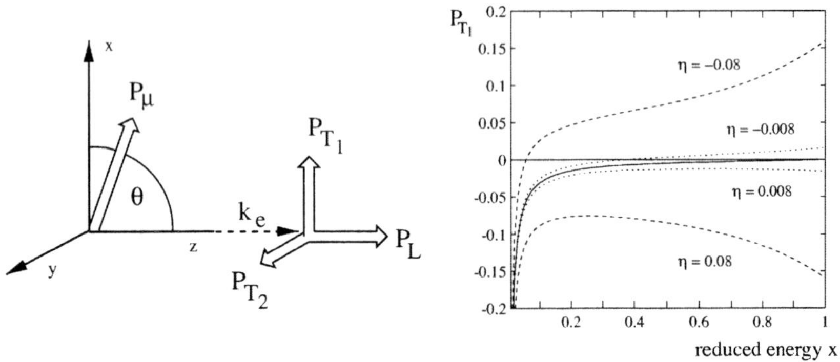

FIGURE 1. *On the left:* Polarization components of the decay positrons. *On the right:* Transverse positron polarization P_{T_1} as a function of the reduced positron energy x. The standard model predicts $\eta = 0$ (solid curve). The present experimental limit is $\eta =$

Here, g^{S}_{RR} represents a scalar, charge-changing interaction with right-handed charged leptons [1]. In the general case there will be a phase between $V-A$ and an additional interaction which leads to a transverse component P_{T_2}. Correspondingly one derives a value for $Im\{g^{S}_{RR}\}$ from the energy dependence of P_{T_2}. A non-zero value for this polarization component violates time reversal invariance.

The transverse polarization has been measured previously with a precision of $\Delta P_{T_1} = \Delta P_{T_2} = 23 \times 10^{-3}$ [2]. A more precise value of P_{T_1} and thus of η is needed for a model-independent determination of the Fermi coupling constant G_F: The influence of the uncertainty in the experimental value of η on the value of G_F is at present 20 times larger than the one of the more precisely known muon life time [3].

EXPERIMENTAL SETUP AND METHODS

The setup [4] of the μ_{P_T} - experiment at the Paul Scherrer Institute in Villigen, Switzerland is shown in figure 2.

A beam of highly polarized muons ($P_\mu \approx 91\%$) enters the beryllium stop target in bunches every 20 ns. The polarization of the stopped muons precesses in a homogeneous magnetic field with the same frequency as the accelerator high frequency. Thus the polarization of new muons entering the beryllium target is added coherently to the polarization of the muons that are already in the target.

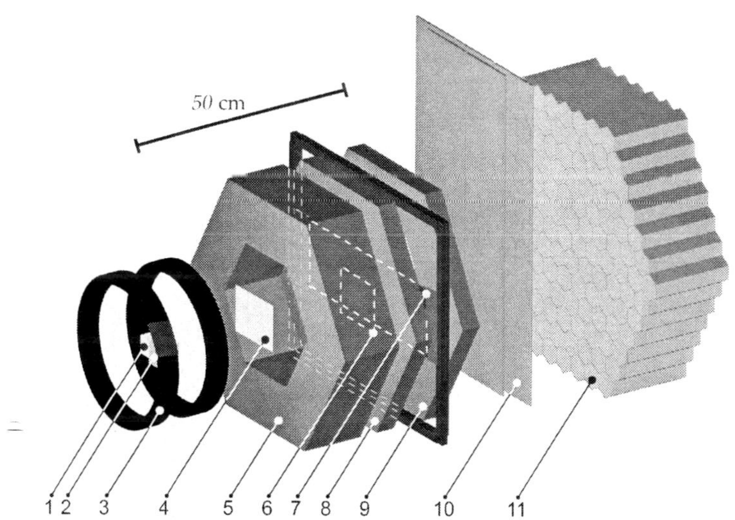

FIGURE 2. Experimental setup: 1 - Be target; 2 - C moderator; 3 - spin precession magnet; 4 and 6 - trigger scintillation counters; 5, 8 and 9 - drift chambers; 7 - magnetized foil within iron return yoke; 10 - veto scintillation counters; 11 - BGO calorimeter. A telescope of scintillation counters above the target region and cosmic trigger scintillation counters on top and below the BGO wall are not shown.

Decay positrons emitted approximately parallel to the B-field are tracked by means of a set of drift chambers and can annihilate with polarized electrons in a magnetized foil. The direction of the magnetic field applied to that foil is reversed approximately every hour.
The two annihilation quanta are then detected by a hexagonal calorimeter consisting of 127 BGO crystals.
A valid annihilation event requires a coincidence of two plastic scintillator counters before the magnetized foil with two separated clusters of BGO detectors and anticoincidences with the last drift chamber and the scintillation counter array in front of the BGO calorimeter (9 and 11 in figure 2).

DATA ANALYSIS AND PRELIMINARY RESULTS

The orientation of the plane in which the two annihilation quanta are emitted depends on the relative orientation of the transverse positron polarization and the polarization of the electrons in the magnetized foil. As the muon polarization precesses due to the magnetic field in the beryllium target, the axis of the transverse positron polarization also precesses with the same frequency. Therefore, a transverse polarization $\vec{P}_T \neq \vec{0}$ would be detected as a harmonic time dependence of the annihilation rate for a given pair of BGO detectors.
Since the accepted decay positrons are emitted into a cone whose axis coincides with the symmetry axis of the apparatus and is perpendicular to the precession plane of the muon polarization, there is a small remnant μSR effect (i.e., a time-dependent rate variation due to the decay asymmetry with respect to the precessing muon polarization). This effect depends on the azimuthal emission angle of the positron and yields time zero, i.e. the position of the precessing muon polarization vector $\vec{P}_\mu$ at a given time. This information allows one to distinguish the two perpendicular components of the transverse polarization of the positrons.
In addition one can make use of the fact that the positrons usually hit the magnetized foil off the symmetry axis and therefore "see" a component of the electron polarization which is either parallel or anti-parallel to the positrons' longitudinal polarization. For positrons hitting a given spatial area of the foil, a difference in annihilation rates for the two different directions of the magnetic field in the magnetized foil can be observed. From this asymmetry the longitudinal polarization of the positrons can directly be deduced.

In fall of 1999 the first data taking run yielded a data sample containing approximately 11×10^6 annihilation events that could be analyzed.

The preliminary result for the longitudinal polarization is $P_L = 1.09 \pm 0.15$. This is in good agreement with the standard model expectation and with the current experimental limit $P_L = 1.00 \pm 0.04$ [5] and proves the sensitivity of this experiment to polarization.

From the analysis of the measured data the absolute value of the transverse polarization at the time of annihilation in the magnetized foil is determined. The energy dependence of two orthogonal transverse polarization components P_1 and P_2 is given in figure 3. The preliminary results for the energy averaged values are $P_1 = 0.006 \pm 0.016$

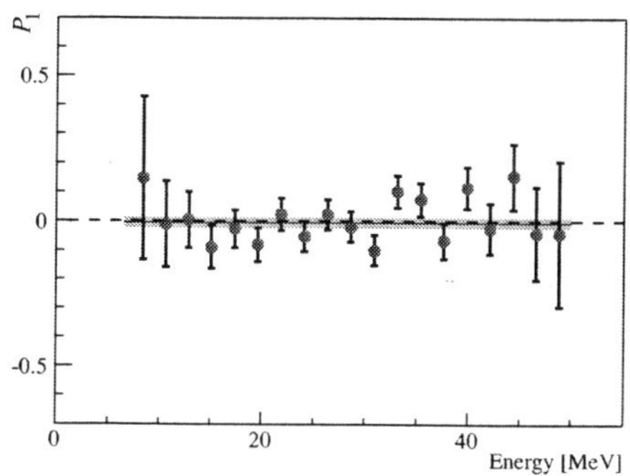

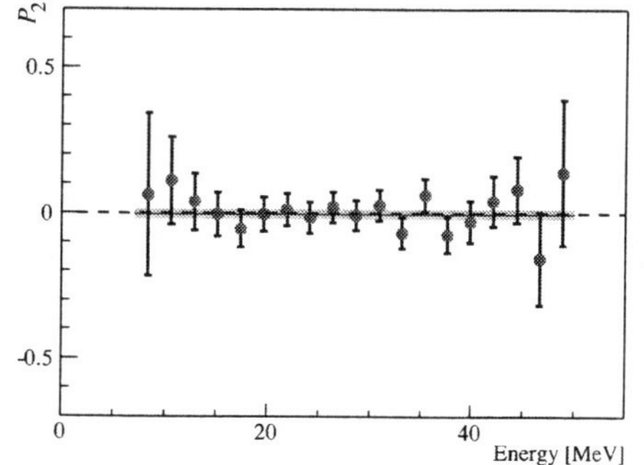

FIGURE 3. Energy dependence of two orthogonal transverse polarization components of the positrons from polarized muon decay at the time of annihilation (preliminary result). The grey bars indicate the one sigma standard deviations of the average values given in the text.

and $P_2 = 0.004 \pm 0.016$.

To obtain the transverse polarization components P_{T_1} and P_{T_2} at the time of muon decay a method based on Monte-Carlo simulations is being used. Our simulation package is based on GEANT 3.21 and features full spin transport for positrons with different initial polarization distributions. After tracking from the beryllium target to the magnetized foil, the resulting polarization distributions at the time of annihilation can then be compared to the experimental distribution.

REFERENCES

1. W. Fetscher, H.-J. Gerber and K.F. Johnson, Phys. Lett. **173B** (1986) 102.
2. H. Burkard *et al.*, Phys. Lett. **160 B** (1985) 343.
3. W. Fetscher and H.-J. Gerber, in *Precision Tests of the Standard Electroweak Model, ed. P. Langacker* (World Scientific, Singapore, 1995).
4. I. Barnett *et al.*, Nucl. Instr. Meth. **A 455** (2000) 329.
5. J. Bartels, D. Haidt and A. Zichichi *(editors)*, The European Physical Journal C - Review of Particle Physics **15** (Springer, Hamburg, 2000).

Neutron Electric Dipole Moment

R. E. Mischke (for the EDM collaboration)

Physics Division, Los Alamos National Laboratory, Los Alamos, NM 87545

Abstract. The status of experiments to measure the electric dipole moment of the neutron is presented and the planned experiment at Los Alamos is described. The goal of this experiment is an improvement in sensitivity of a factor of 50 to 100 over the current limit. It has the potential to reveal new sources of T and CP violation and to challenge calculations that propose extensions to the Standard Model. The experiment employs several advances in technique to reach its goals and the feasibility of meeting these technical challenges is currently under study.

INTRODUCTION

A permanent electric dipole moment (EDM) of the neutron requires an interaction that violates both P and T invariance. The weak interaction violates P maximally. Because of widespread acceptance of the CPT theorem, CP violation implies T violation, so most electroweak models of CP violation give predictions for non-zero values for the neutron EDM. The present limit for the neutron EDM is $< 0.63\times10^{-26}$ e-cm [1]. The standard model prediction is very small ($\sim10^{-31}$ e-cm), which opens a wide window for new physics and several models predict values near the present limit [2]. New experiments are underway that will reach well into this window and will probe for physics beyond the Standard Model.

PREVIOUS AND PROPOSED EXPERIMENTS

The most recent published results for the neutron EDM and projected results from the experiments discussed below are summarized in Table 1. The ILL result [1] includes the data from an older version of the experiment [3]; the validity of the quoted result has been questioned because a statistics limited value has been combined with an earlier more precise, but systematics limited result [4].

TABLE 1. Previous and projected results for the neutron EDM.

Limit (e-cm)	CL%	Lab (Ref.)	year
$< 1.2 \times 10^{-25}$	95	ILL [3]	1990
$< 0.97 \times 10^{-25}$	90	PNPI [5]	1996
$< 0.63 \times 10^{-25}$	90	ILL [1]	1999
projected:			
$< 1 \times 10^{-26}$	90	ILL [6]	2004
$< 7 \times 10^{-28}$	90	PSI [7]	2005(?)
$< 2 \times 10^{-28}$	95	LANL [8]	2010

CP675, *Spin 2002: 15th Int'l. Spin Physics Symposium and Workshop on Polarized Electron Sources and Polarimeters*, edited by Y. I. Makdisi, A. U. Luccio, and W. W. MacKay

A schematic drawing of the ILL experiment is shown in Fig. 1. This experiment stores ultracold neutrons (UCN) in the main storage cell with parallel E and B fields. A polarizing foil is used to polarize the incident UCN and to analyze the outgoing UCN. The density of UCN was 0.6 UCN/cm^3 and the measurement time was 130 s. The Ramsey separated field technique is used to look for a change in the Larmor precession frequency upon reversal of the E field with respect to the static B field. The average B field is monitored by an Hg comagnetometer, which is crucial for accounting for magnetic field fluctuations.

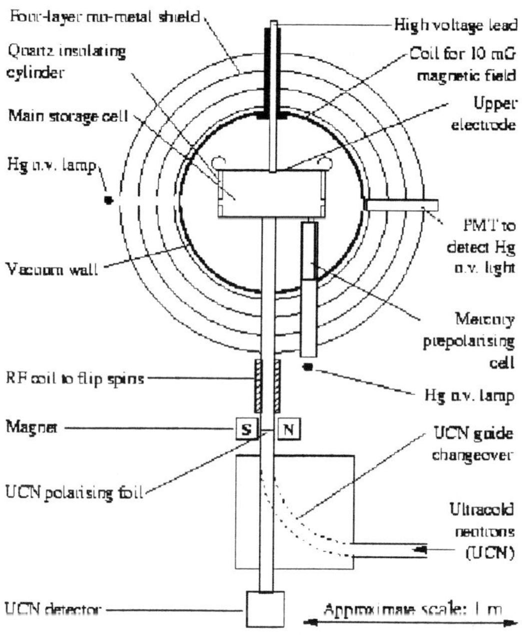

FIGURE 1. Schematic of the current ILL experiment from Ref. 1.

The collaboration is working on a series of incremental improvements that should result in a projected limit of $< 1 \times 10^{-26}$ e-cm by 2004 [6]. The improvements include an increase in the UCN flux, longer UCN storage time with better wall coatings for the storage cell, and an increased E field with a shorter cell.

A schematic of the planned PSI experiment [7] is shown in Fig. 2. The technique is similar to that used in the experiment at ILL. A solid D_2 source is being developed to produce an unprecedented flux of UCN with 10^3 UCN/cm^3. The experiment features multiple pairs of cells to store the UCN. Having adjacent cells with opposite directions for the E field will help to reduce systematic errors. However, the large volume places severe demands on the magnetic shielding, and active shielding is planned to keep the spatial uniformity to 10^{-12} T/cm and the temporal stability to 10^{-14} T at mHz frequencies. Cs magnetometers will be located adjacent to the cells. This experiment expects to reach a sensitivity of $< 5 \times 10^{-28}$ e-cm.

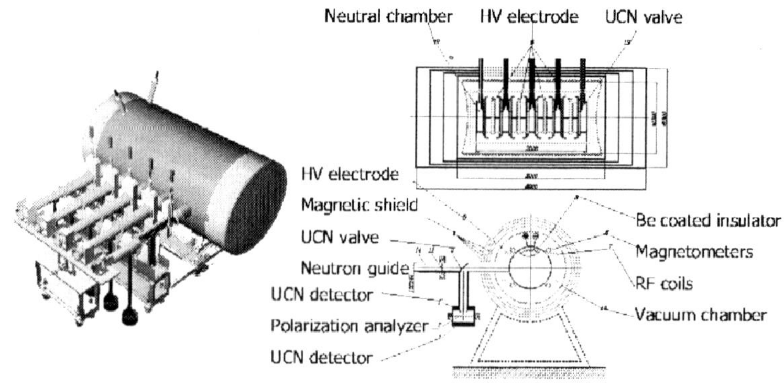

FIGURE 2. Schematic of the proposed PSI experiment from Ref. 4.

LANL EXPERIMENT

A schematic of the design concept for the LANL EDM experiment [8] is shown in Fig. 3. The electric field is formed by a central HV electrode with two ground electrodes. Cells to contain the ultracold neutrons are between the plates and are not visible in the figure. The volume is filled with superfluid ^{4}He surrounded by a superconducting shield. A beam of cold polarized neutrons enters the apparatus from the right. Those neutrons that downscatter in liquid ^{4}He [9] and are trapped in the target cells are used in the measurement. The expected density is ~500 UCN/cm^3 at LANL and a storage time of 500 s is assumed. The spins of the UCN will be rotated by 90° by a $\pi/2$ pulse and then they will precess about a static magnetic field that is parallel to the strong E field. The volume-averaged magnetic field will be monitored by a dilute mixture of polarized ^{3}He in the cells. SQUIDS will be used to measure the spin precession frequency of the ^{3}He. The UCN and ^{3}He precess at slightly different rates and the relative alignment of the spins is detected by neutron capture on ^{3}He because the capture cross section is spin dependent. Scintillation light in ^{4}He from energy deposited by the capture products can reveal a frequency shift when the E field is reversed. The (2σ) limit of the LANL experiment is expected to reach 9×10^{-28} e-cm and will be at least 2×10^{-28} e-cm after the experiment is moved to the SNS, which is under construction at ORNL.

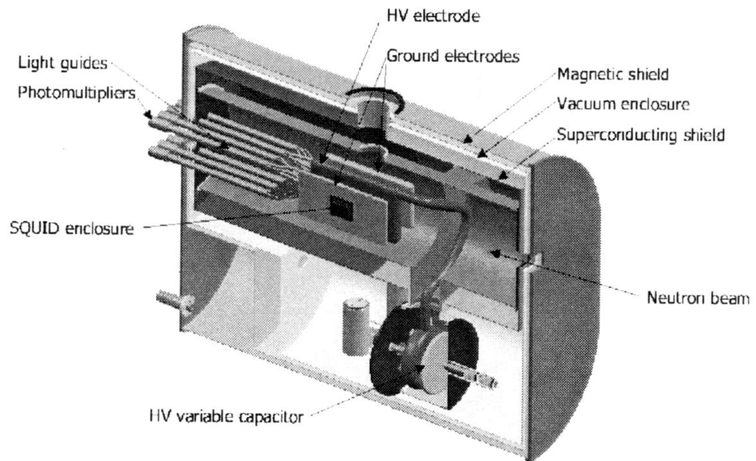

FIGURE 3. Schematic of the proposed LANL experiment.

Because the experiment will employ several advances in technique to reach its goals, the feasibility of meeting these technical challenges is currently under study. The challenges include requirements on the uniformity of the distribution of ^{3}He in the cell, on the uniformity of the E and B fields, and on the noise level for the SQUIDS, etc. The distribution of ^{3}He in liquid ^{4}He was the subject of a test run at LANSCE and a uniform distribution was measured. A byproduct of this test is a measurement of the diffusion coefficient of ^{3}He in ^{4}He for temperatures below 1 K [10]. The electric and magnetic fields have been modeled using a finite element code and it has been determined that the design requirements of 1% uniformity for the E field and 0.1% uniformity for the B field can be met with the apparatus shown in Fig. 3. The SQUID noise measurements have been reported in Ref. 11 and show that the design requirement of less than 5 $\mu\Phi_0/\text{Hz}^{1/2}$ can be achieved.

An apparatus to produce a beam of polarized ^{3}He is being constructed and studies are being planned to determine how to transport and store the ^{3}He without losing polarization. An apparatus to purify ^{4}He has been assembled and will be put into operation soon. A number of topics related to the high voltage are being addressed.

HIGH VOLTAGE TEST

A test apparatus is being constructed as shown in Fig. 4. In addition to validating the variable capacitor concept for this application, the apparatus will serve as a test bed for several engineering issues and for materials studies. The central volume containing the HV and ground electrodes will be filled with liquid ^{4}He. The HV electrode will be charged to 50 kV while the ground electrode is close to the HV plate. Then the HV contact actuator will be retracted and the plates will be separated, lowering the capacitance and raising the voltage. The electric field in the gap will be measured using the Kerr effect, i.e. the medium polarization induced by the electric field.

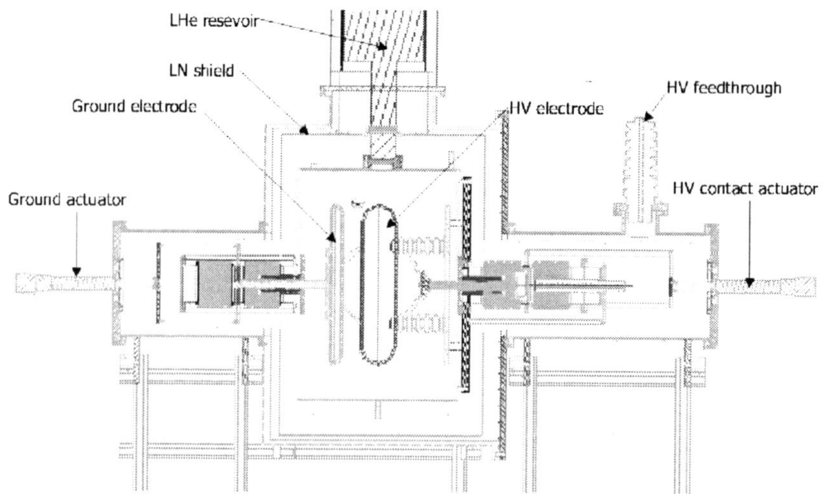

FIGURE 4. Schematic of the high voltage test apparatus.

ACKNOWLEDGMENTS

This work was supported by Laboratory Directed Research and Development (LDRD) funds from Los Alamos National Laboratory.

REFERENCES

1. Harris, P.G. *et al.*, *Phys. Rev. Letters* **82**, 904-907 (1999).
2. Pendelbury, J.M. and Hinds, E.A., *Nucl. Instr. and Meth.*, **A440**, 471-478 (2000).
3. Smith, K.F. *et al.*, *Phys. Lett. B* **234**, 191 (1990).
4. Lamoreaux, S.K. and Golub, R., *Phys. Rev. D* **61,** 051301-3 (2000).
5. Altarev, I.S. *et al.*, *Phys. At. Nucl.* **59**, 1152 (1996).
6. Harris, P.G. *et al.*, *Nucl. Instr. and Meth.*, **A440**, 479-482 (2000).
7. Alexandrov, E. *et al.*, PSI Annual Report, http://ucn.web.psi.ch/publications_ucn.htm
8. Budker, D., *et al.*, (EDM collaboration), A New Search for the Neutron Electric Dipole Moment, http://p25ext.lanl.gov/edm/edm.html
9. Golub, R. and Pendelbury, J.M., *Phys. Lett.* **53A**, 133 (1975).
10. Lamoreaux, S.K. *et al.*, *Europhys. Lett.*, **58** 718-724 (2002).
11. Espy, M.A. *et al.*, *IEEE Trans. Appl. Supercond.* **9**, 3696-3699 (1999).

Measurement of α_Ω in $\Omega^- \rightarrow \Lambda K^-$ Decays

Lan-Chun Lu (Representing the *HyperCP* Collaboration)[*],
A. Chan[†], Y.C. Chen[†], C. Ho[†], P.K. Teng[†], W.S. Choong[**], Y. Fu[**],
G. Gidal[**], P. Gu[**], T. Jones[**], K.B. Luk[**], B. Turko[**], P. Zyla[**], C. James[‡],
J. Volk[‡], J. Felix[§], R.A. Burnstein[¶], A. Chakravorty[¶], D.M. Kaplan[¶],
L.M. Lederman[¶], W. Luebke[¶], D. Rajaram[¶], H.A. Rubin[¶], N. Solomey[¶],
Y. Torun[¶], C.G. White[¶], S.L. White[¶], N. Leros[||], J.-P. Perroud[||],
H.R. Gustafson[††], M.J. Longo[††], F. Lopez[††], H.K. Park[††], M. Jenkins[‡‡],
K. Clark[‡‡], E.C. Dukes[§§], C. Durandet[§§], T. Holmstrom[§§], M. Huang[§§],
L.C. Lu[§§] and K.S. Nelson[§§]

[*]*Physics Department, University of Virginia, Charlottesville, VA 22901, U.S.A.*
[†]*Academia Sinica, Nankang, Taipei 11529, Taiwan, Republic of China*
[**]*Lawrence Berkeley Laboratory and University of California, Berkeley, CA 94720, USA*
[‡]*Fermilab, Batavia, IL 60510, USA*
[§]*Universidad de Guanajuato, León, Mexico*
[¶]*Illinois Institute of Technology, Chicago, IL 60616, USA*
[||]*Université de Lausanne, Lausanne, Switzerland*
[††]*University of Michigan, Ann Arbor, MI 48109, USA*
[‡‡]*University of South Alabama, Mobile, AL 36688, USA*
[§§]*University of Virginia, Charlottesville, VA 22901, USA*

Abstract. The *HyperCP* experiment (E871) at Fermilab has collected the largest sample of hyperon decays in the world. With a data set of over a million $\Omega^- \rightarrow \Lambda K^-$ decays we have measured the product of $\alpha_\Omega \alpha_\Lambda$ from which we have extracted α_Ω. This preliminary result indicates that α_Ω is small, but non-zero. Prospects for a test of *CP* symmetry by comparing the α parameters in Ω^- and $\overline{\Omega}^+$ decays will be discussed.

INTRODUCTION

In the quark model, the Ω baryon is predicted to have spin $J = 3/2$. The spin of it has not yet been determined experimentally, but measurements made by Deutschmann *et al.* [1] and Baubillier *et al.* [2] have ruled out $J = 1/2$ and found consistency with $J = 3/2$. Angular momentum conservation allows the ΛK system in the decay $\Omega \rightarrow \Lambda K$ to be P and D, corresponding to parity conserving and parity violating amplitudes respectively. Parity violation is characterized by the parameter α_Ω defined as $\alpha_\Omega = 2Re(P^*D)/(|P|^2 + |D|^2)$ [3]. A non-zero α_Ω is the signature of parity violation in this decay.

Although the main goal of the *HyperCP* experiment at Fermilab is to search for *CP* violation in $\Xi \rightarrow \Lambda\pi \rightarrow p\pi\pi$ decays with a precision at the 10^{-4} level, the topological similarity of $\Omega \rightarrow \Lambda K \rightarrow p\pi K$ decays to $\Xi \rightarrow \Lambda\pi \rightarrow p\pi\pi$ decays has enabled us to

CP675, *Spin 2002: 15th Int'l. Spin Physics Symposium and Workshop on Polarized Electron Sources and Polarimeters,* edited by Y. I. Makdisi, A. U. Luccio, and W. W. MacKay

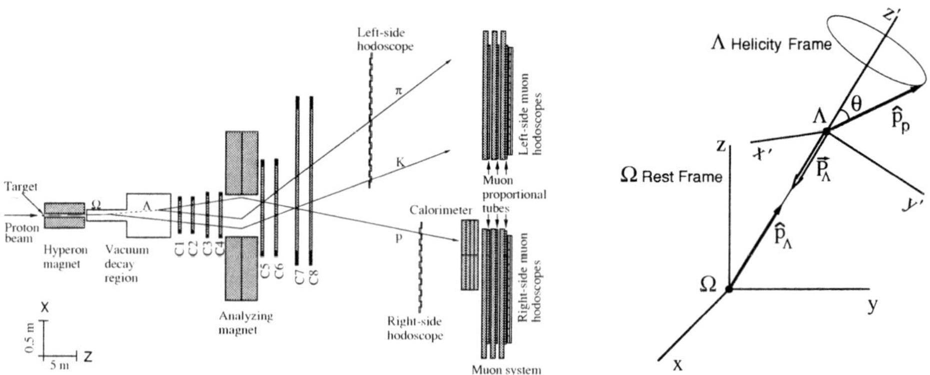

FIGURE 1. (a) Plan view of the *HyperCP* spectrometer. (b) Proton direction in the Λ helicity frame.

collect a large sample of Ω events [4]. Nineteen million Ω^- and $\overline{\Omega}^+$ were acquired during RUN-I (1997) and RUN-II (1999), allowing us to measure α_Ω for both Ω^- and $\overline{\Omega}^+$ decays at the 10^{-3} level. A difference between $|\alpha_\Omega|$ and $|\alpha_{\overline{\Omega}}|$ would be evidence of *CP* violation in $\Omega \to \Lambda K \to p\pi K$ decays.

THE *HYPERCP* SPECTROMETER

Figure 1 (a) shows the spectrometer used in the *HyperCP* experiment. Hyperons are produced by 800-GeV protons from the Tevatron striking a target. Omegas and other charged particles travel through a curved magnetic channel (collimator) followed by a vacuum decay pipe. The trajectories of the K from the Ω decay and the proton and π from the Λ decay are measured by four proportional wire chambers upstream of the analyzing magnet. The K and the π are deflected to the left side of the spectrometer, and the proton is deflected to the right side in the field of the analyzing magnet. After four downstream proportional wire chambers the K and the π strike the Left-side Hodoscope, and the proton strikes the Right-side Hodoscope before depositing energy in the calorimeter. Two muon stations are located downstream of the calorimeter. These are used to identify muons from rare kaon and hyperon decays. The Left-side Hodoscope, Right-side Hodoscope, and the calorimeter were used to form the triggers, which had a rate of $50 \sim 80$ KHz. The samples of Ω and $\overline{\Omega}^+$ were taken alternatively by switching the sign of the Hyperon Magnet and the Analyzing Magnet.

ANALYSIS METHOD

For unpolarized Ω, the angular distribution of the proton in $\Omega \to \Lambda K \to p\pi K$ decays is expressed as

$$\frac{dN}{d\cos\theta} = \frac{N_0}{2}(1 + \alpha_\Omega \alpha_\Lambda \cdot \cos\theta), \quad \alpha_\Lambda \equiv \frac{2Re(S^*P)}{|S|^2 + |P|^2}, \tag{1}$$

where θ is the polar angle of proton in the Λ helicity frame, and α_Λ is the decay parameter for $\Lambda \to p\pi$ decays. As illustrated in Figure 1 (b), $\vec{P}_\Lambda$ represents the Λ polarization, $\hat{p}_\Lambda$ is the Λ momentum, and $\hat{p}_p$ is the unit momentum of the proton in the Λ helicity frame. In reality, the proton $\cos\theta$ distribution is distorted by the spectrometer acceptance. To correct for the acceptance we use a Hybrid Monte-Carlo method (HMC) [5] in our data analysis. We take all variables from each real event except $\cos\theta$, generate Monte-Carlo events (to distinguish them from normal Monte-Carlo events, we call them HMC fake events) with uniform $\cos\theta$, and then let all the HMC fake events go through the software spectrometer, triggers, event selection cuts, etc. to simulate the behavior of real events in the experiment. Assuming the Monte-Carlo code describes the spectrometer perfectly, the distortion by the acceptance of the proton angular distribution of fake events should be exactly the same as for real events.Matching the fake event $\cos\theta$ distribution to the real event $\cos\theta$ distribution by minimizing the χ^2 in Eq. (2), without explicitly computing the acceptance correction, gives us the unknown $\alpha_\Omega\alpha_\Lambda$,

$$\chi^2(X) = \sum_{k=1}^{20} \frac{[N_r(k) - N_f(X,k)]^2}{\sigma_k^2}, \tag{2}$$

here $X \equiv \alpha_\Omega\alpha_\Lambda$, $\sigma_k^2 = N_r(k) + N_f(X,k)$, and $N_r(k)$ and $N_f(X,k)$ are numbers of real events and fake events in bin k respectively.

RESULTS

To select good $\Omega \to \Lambda K \to p\pi K$ decays, we require events to meet the topology of three tracks and two vertices. The initial filtering process, which is a geometric fit, gets rid of most events that have the wrong topology. All three-track combinations in an event are required to have a Λ vertex and a Ω vertex under the hypothesis of $\Omega \to \Lambda K \to p\pi K$ decay. Those three tracks that best match the $\Omega \to \Lambda K \to p\pi K$ hypothesis are kept for the further study. If the $p\pi$ invariant mass and ΛK invariant mass are ± 50 MeV of the Λ and Ω PDG masses, we consider this event as a Ω decay candidate. Additional cuts are required to get a clean Ω sample including: 1) cuts on z positions of Λ and Ω vertices, 2) Λ vertex downstream of the Ω vertex, 3) x-projection and y-projection of the Ω track at the target within the xy-dimension of the target, 4) a cut on $K \to 3\pi$ mass, and 5) a cut on the geometric fit χ^2.

We have obtained a preliminary result for the Ω^- from the RUN-I data. Figure 2 (a) shows the ΛK invariant mass of 1.2 million events after all event selection cuts. Under the signal region (marked by arrows), the ratio of background to signal is 0.7%. The raw $\alpha_\Omega\alpha_\Lambda$ before background correction, defined as S_m, is measured to be $(1.32 \pm 0.29) \times 10^{-2}$. The comparison of proton angular distributions of real and fake events after the matching is shown in Figure 2 (b). The bias from the background was investigated using the side bands of the ΛK invariant mass. A careful study shows that the contributions from the side bands at low mass region and high mass region are essentially the same, and the mean value is $S_b = (21.77 \pm 1.80) \times 10^{-2}$. Assuming the background under the mass peak has the same shape as side bands, we use the formula $\alpha_\Omega\alpha_\Lambda = N_m S_m / N_s -$

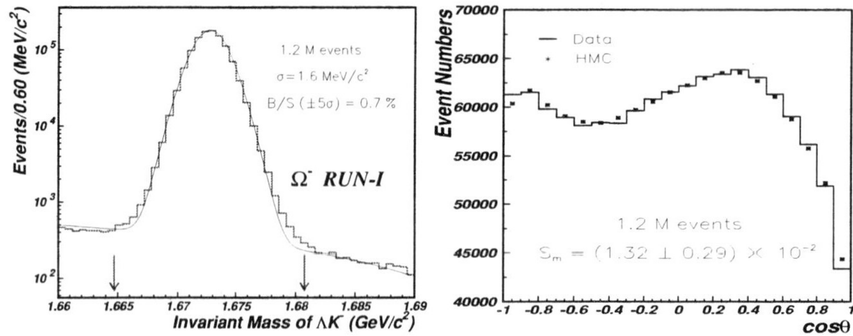

FIGURE 2. (a) ΛK invariant mass. (b) The $\cos\theta$ distributions after matching.

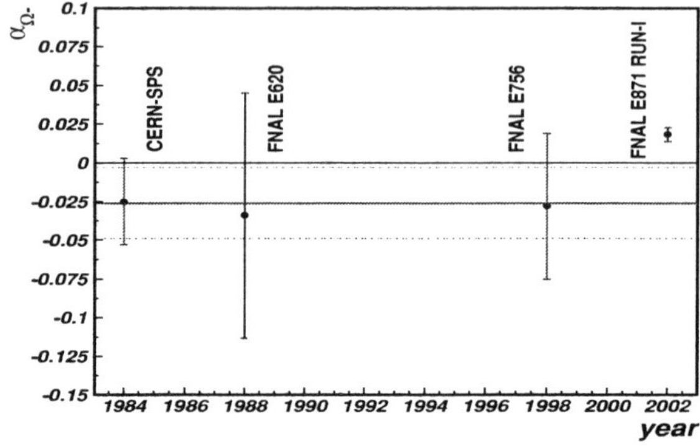

FIGURE 3. Different measurements of α_Ω for Ω^-. The lower solid line is for PDG value and the error is marked by the two dashed lines. The upper solid line marks the position of zero.

$N_b S_b / N_s$ to obtain the $\alpha_\Omega \alpha_\Lambda$, where N_m, N_s, and N_b are numbers of measured events, signal events, and background events respectively. With this background correction, our preliminary result is: $\alpha_\Omega \alpha_\Lambda = [1.18 \pm 0.29 \text{ (stat)}] \times 10^{-2}$. Using PDG value for α_Λ (0.642 ± 0.013 [6]), α_Ω is extracted:

$$\alpha_\Omega = [1.84 \pm 0.46 \text{ (stat)} \pm 0.04 \text{ (sys PDG)}] \times 10^{-2}, \tag{3}$$

where 0.04×10^{-2} is the error propagated from the error of α_Λ. The stability of our result with different cuts and different data samples indicates that the systematic error is expected to be smaller than the statistical error. However, the systematic errors are still under investigation. The small but non-zero of α_Ω value indicates parity violation in $\Omega^- \to \Lambda K^-$ decays. Figure 3 shows the comparison of our preliminary measurement from RUN-I data with previous experimental results which are all consistent with zero within the error bars.

Analysis of Ω^- and $\overline{\Omega}^+$ data samples from RUN-II have just begun. With similar event selection cuts, we get about 4.6 million Ω^- and 1.9 million $\overline{\Omega}^+$. The measured raw $\alpha_\Omega \alpha_\Lambda$ before background correction are $S_m = (1.41 \pm 0.11) \times 10^{-2}$ and $S_m =$

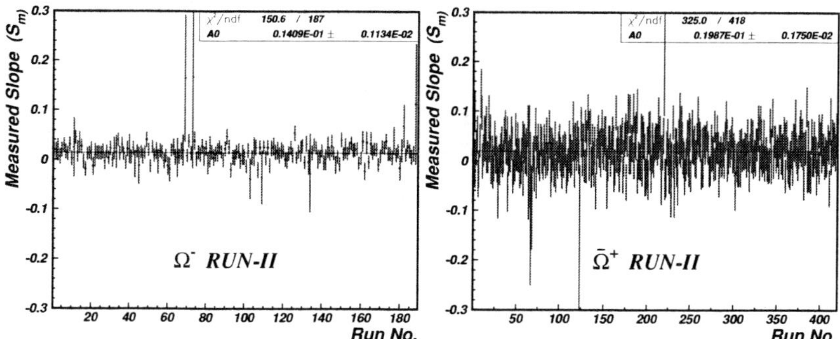

FIGURE 4. Raw $\alpha_\Omega \alpha_\Lambda$ versus run number (RUN-II data).

$(1.99 \pm 0.18) \times 10^{-2}$ for Ω^- and $\overline{\Omega}^+$ respectively. The Ω^- result is consistent with RUN-I data with over two times smaller statistical error. Figure 4 shows the raw $\alpha_\Omega \alpha_\Lambda$ versus different run number of RUN-II data for both Ω^- and $\overline{\Omega}^+$.

By measuring the difference of $\alpha_\Omega \alpha_\Lambda$ between Ω^- and $\overline{\Omega}^+$, which is defined as

$$\Delta(\alpha_\Omega \alpha_\Lambda) \equiv \alpha_\Omega \alpha_\Lambda - \alpha_{\overline{\Omega}} \alpha_{\overline{\Lambda}} \quad \text{or} \quad A_{\Omega\Lambda} \equiv \frac{\alpha_\Omega \alpha_\Lambda - \alpha_{\overline{\Omega}} \alpha_{\overline{\Lambda}}}{\alpha_\Omega \alpha_\Lambda + \alpha_{\overline{\Omega}} \alpha_{\overline{\Lambda}}} \approx A_\Omega + A_\Lambda, \tag{4}$$

we can test CP-violation in $\Omega \to \Lambda K \to p\pi K$ decays. The statistical precisions of $\Delta(\alpha_\Omega \alpha_\Lambda)$ and $A_{\Omega\Lambda}$ are estimated to be 2×10^{-3} and 9×10^{-2} respectively for RUN-II data. Systematic errors are expected to be very similar between Ω^- and $\overline{\Omega}^+$ and should almost cancel in this comparison.

ACKNOWLEDGMENT

We thank the staffs of Fermilab and the participating institutions for their important contributions. We are grateful to the Fermilab Computing Division for their support of the PC-farm. This work is supported by the U.S. Department of Energy and the National Science Council of Taiwan.

REFERENCES

1. M. Deutschmann *et al.*, Phys. Lett. 73B, (1978) 96.
2. M. Baubillier *et al.*, Phys. Lett. 78B, (1978) 342.
3. T. D. Lee and C. N. Yang, Phys. Rev. 108, (1957) 1645.
4. E.C. Dukes, K.B. Luk, *et al.*, Fermilab Proposal P-871 (1994).
5. G. Bunce, Nucl. Instr. Meth. 172, (1980) 553.
6. Particle Data Group, Phys. Rev. D66, (2002) 010001-821.

A Clean Measurement of the Neutron Skin of ^{208}Pb Through Parity Violating Electron Scattering

Riad Suleiman[1]

Massachusetts Institute of Technology, Cambridge, MA 02139

Abstract. The difference between the neutron radius R_n of a heavy nucleus and the proton radius R_p is believed to be on the order of several percent. This qualitative feature of nuclei, which is essentially a neutron skin, has proven to be elusive to pin down experimentally in a rigorous fashion. A new Jefferson Lab experiment will measure the parity-violating electroweak asymmetry in the elastic scattering of polarized electrons from ^{208}Pb. Since the Z-boson couples mainly to neutrons, this asymmetry provides a measure of the size of R_n that can be interpreted with as much confidence as the traditional electron scattering data. The projected experimental precision corresponds to a 1% determination of R_n, which will have a big impact on nuclear theory and its application to neutron rich matter such as neutron stars.

INTRODUCTION

The size of a heavy nucleus is one of its most basic properties. However, because of a neutron skin of uncertain thickness, the size does not follow from measured charge radii and is relatively poorly known. For example, the root mean square neutron radius in ^{208}Pb, R_n, is thought to be about 0.25 fm larger than the proton radius $R_p \approx 5.45$ fm. An accurate measurement of R_n would provide the first clean observation of the neutron skin. This is thought to be an important feature of all heavy nuclei.

Donnelly, Dubach and Sick [1] suggested that parity-violating electron scattering can measure neutron densities. This is because the Z-boson couples primarily to the neutron at low Q^2. Therefore, one can deduce the weak-charge density and the closely related neutron density from measurements of the parity-violating asymmetry in polarized elastic scattering $A = (\sigma_R - \sigma_L)/(\sigma_R + \sigma_L)$. In PWIA, the relationship between the asymmetry and the neutron form factor is given by:

$$A = -\frac{G_F Q^2}{4\pi\alpha\sqrt{2}}\left[1 - 4\sin^2\theta_W - \frac{F_n(Q^2)}{F_p(Q^2)}\right], \qquad (1)$$

where G_F is the Fermi constant, $\alpha = \frac{1}{137}$ is the fine structure constant, $\sin^2\theta_W$ is the Weinberg angle, and $F_n(Q^2)$ and $F_p(Q^2)$ are the neutron and proton form factor of

[1] For the Jefferson Lab Hall A Collaboration.

CP675, *Spin 2002: 15th Int'l. Spin Physics Symposium and Workshop on Polarized Electron Sources and Polarimeters*, edited by Y. I. Makdisi, A. U. Luccio, and W. W. MacKay

the nucleus. This is similar to how the charge and proton densities are deduced from unpolarized cross sections.

PHYSICS IMPACT AND INTERPRETATION

A single measurement of R_n with 1% accuracy can have an impact on several areas of physics, including nuclear theory, atomic parity violation, and neutron star structure.

In a recent analysis of neutron radii in nuclear mean field models, Furnstahl [2] showed that the variable range of R_n allowed by a large set of viable nuclear models was associated primarily with the density dependence of the nuclear symmetry energy. This is an energy cost associated with having unequal numbers of neutrons and protons. A measurement of R_n would pin down this one parameter and could potentially demonstrate that an entire class of models is less likely than another. For example, relativistic mean field models tend to favor larger neutron skins than non-relativistic models because of a larger symmetry energy.

The impact of an accurate R_n measurement on atomic parity violation experiments has been analyzed in [3]. Knowledge of R_n at the 1% level is needed for interpreting atomic physics measurements of the Weinberg angle at the level of the Standard Model weak radiative corrections.

Measurements of the equation of state of neutron rich matter are important for calculating the structure of neutron stars [4]. The radius of a neutron star is deduced from optical and X-ray observations. To find out possible exotic phases of dense matter one needs to combine the high density measurements of neutron stars with low density precision measurements of R_n in nuclei. As a second example of application, the proton fraction of neutron rich matter in beta equilibrium depends on the symmetry energy, which is calibrated by R_n. A large symmetry energy favors more protons, and if the proton fraction is high enough then the following "URCA" process can cool neutron stars $n \rightarrow p+e+\bar{\nu}$; $p+e \rightarrow n+\nu$ where the $\nu, \bar{\nu}$ carry off energy. URCA cooling might explain recent Chandra observations of the neutron star 3C58, a remnant of the supernova seen in the year 1186 that appears to be unexpectedly cold [5]. A neutron skin larger than about 0.2 fm may imply that URCA cooling is possible, while a smaller skin implies it is probably not possible.

The physics interpretation of the experiment can be summarized as follows. From the measured asymmetry one may deduce the weak form factor, which is the Fourier transform of the weak-charge density at the momentum transfer of the experiment. One must correct for Coulomb distortions, which has been done accurately by Horowitz [6]. The weak-charge density can be compared directly to theoretical calculations and this will constrain the density dependence of the symmetry energy. The weak density can be directly applied to atomic PNC because the observables have approximately the same dependence on nuclear shape. From the weak-charge density one can also deduce a neutron density at one Q^2 by making small corrections for known nucleon form factors. The uncertainty in these corrections for a realistic experiment have been estimated and are small [3]. The corrections considered were Coulomb distortions (which was by far the biggest), strangeness and the neutron electric form factor, parity admixtures, dis-

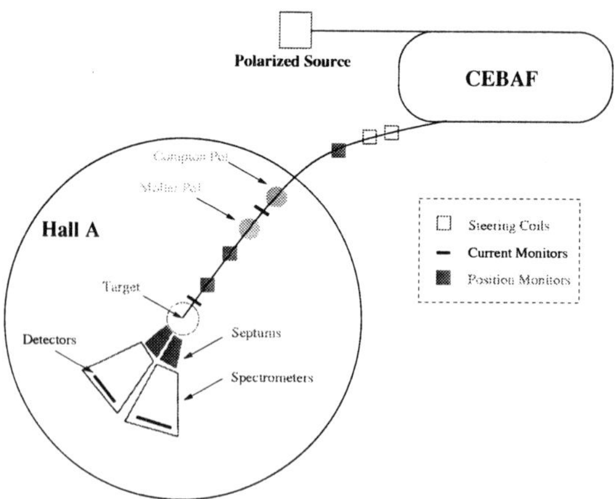

FIGURE 1. Layout of Hall A at Jefferson Lab.

persion corrections, meson exchange currents, isospin admixtures, radiative corrections, and possible contamination from excited states and target impurities.

Finally from a low Q^2 measurement of the point neutron density one can deduce R_n. This requires knowledge of the surface thickness to about 25% to extract R_n to 1%. The spread in surface thickness among successful mean field models is much less than 25%, hence we can extract R_n with the desired accuracy. In summary, the physics results of the experiment are the weak-charge density, the point neutron density, and R_n.

EXPERIMENT

Experimental Overview

The ^{208}Pb experiment [7] will take place in Hall A at Jefferson Lab (see Fig. 1). The two Hall A 3.7 msr spectrometer systems supplemented by septum magnets focus elastically scattered electrons onto total-absorption detectors in their focal planes. Separate studies at lower rates are required to measure backgrounds, acceptance, and Q^2. The experimental conditions are listed in Tab. 1.

Choice of Nucleus and Kinematics

The target nucleus was chosen to be ^{208}Pb. The advantages of Pb are that it is very well known and has a simple doubly magic shell structure, and most importantly that it has the largest separation to the first excited state (2.6 MeV) of any heavy nucleus, thus permitting flux integration of the elastically scattered electrons.

The choice of kinematics (incident energy and scattering angle) is guided by the objective of minimizing the running time required for a 1% accuracy in R_n. The three

TABLE 1. Summary of the experimental conditions.

Measured Asymmetry ($p_e A$)	0.51 ppm
Beam Energy	850 MeV
Beam Current	50 μA
Beam Polarization	80%
Target	10% *r.l.* Pb
Scattering Angle	6°
Required Statistical Accuracy	3%
Energy Cut (due to detector)	4 MeV
Detected Rate (each Spectrometer)	860 MHz
Running Time	30 days

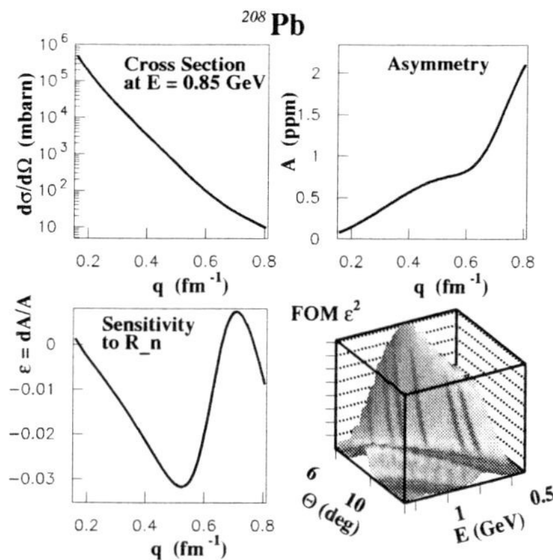

FIGURE 2. Cross section, parity violating asymmetry, and sensitivity to R_n for ^{208}Pb elastic scattering at 0.85 GeV. The fourth plot shows the variation of FOM$\times\varepsilon^2$ with energy and angle, showing an optimum at 0.85 GeV for a 6° scattering angle which corresponds to $q = 0.45$ fm^{-1}

ingredients which enter into this optimization are: the cross section $d\sigma/d\Omega$, the parity violating asymmetry A, and the sensitivity to the neutron radius $\varepsilon = dA/A = (A1-A)/A$ where A is the asymmetry computed from a mean field theory (MFT) calculation [6] and $A1$ is the asymmetry for the same MFT in which the neutron radius is increased by 1%. These three ingredients, which each vary with energy and angle, are plotted in Fig. 2 for a beam energy of 0.85 GeV which turns out to be optimum energy. The minimum running time is equivalent to maximum product in FOM $\times\ \varepsilon^2 = R \times A^2 \times \varepsilon^2$, where R is the detected rate and is proportional to $d\sigma/d\Omega$ and "FOM" is the conventionally defined figure of merit for parity experiments, FOM $= R \times A^2$. Note that rather than only maximizing the conventional FOM, parity violating neutron density measurements take into account the sensitivity (ε) to R_n which varies with kinematics. Figure 2 shows the product FOM $\times\ \varepsilon^2$ for ^{208}Pb. The running time needed to reach a 1% accuracy in R_n is 30 days.

A novel feature of the ^{208}Pb experiment is a high powered lead target which will withstand 40 Watt for a 50 μA beam. Improving the thermal properties of the target is necessary since lead has a low melting temperature. A 0.5 mm foil of lead will be sandwiched between two 0.15 mm sheets of diamond, which is pure ^{12}C. This assembly is clamped in a spring loaded copper block assembly which is cooled by liquid helium. The copper block has a hole to allow the beam to pass through; the beam only sees ^{208}Pb and ^{12}C. The diamond has an extremely high thermal conductivity, and calculations show this target should be stable up to 100 μA.

The target thickness required to maximize the rate in the momentum bite defined by the detector is 0.5 mm (10% *r.l.*). A thicker target suffers more radiative loss and hence less rate. By integrating the rate up to a cutoff of 4 MeV, we reduce the running time by 25% compared to a cutoff that would exclude the first excited state. The resulting contamination from inelastic scattering constitutes a fraction 0.5% of our signal. The systematic errors from inelastics and from ^{12}C background are tolerable [3].

Systematics

Measuring a tiny asymmetry of 0.5 ppm to 3% absolute accuracy is a major challenge involving the following considerations:

1. The experimental systematic error must be much smaller than the statistical goal (1.5×10^{-8}), hence a goal of $\leq 10^{-9}$. The main issues here are with the control of false asymmetries associated with helicity correlated beam parameters such as intensity, energy, and position. This will require good setup and feedback loops in the source, as well as betatron matching in the accelerator. The betatron matching will ensure maximal dampening of helicity correlated beam positions on target. To measure the beam parameters accurately we are installing microwave cavity beam position and current monitors. The position monitors supplement the existing stripline monitors, which provides a complementary method with presumably different systematics to help unfold beam fluctuation noise from instrumentation noise. This experiment requires beam position differences to be less than 1 nm with an accuracy of 0.1 nm averaged over the whole experiment. The charge asymmetry must be maintained to less than 100 ppb with an accuracy of 10 ppb.
2. Because of the high rates (860 MHz per spectrometer), the statistical error in each 30 msec window will be 140 ppm. All other noises, e.g. instrumental noises, must be kept well below this.
3. The normalization of the asymmetry must be better than 3%. There are two main issues: the Q^2 measurement and the beam polarization. We expect to be able to measure Q^2 to 0.3%. The polarization must be measured to 1% preferably, or at least 2%. With a polarization accuracy of 1% (2%) we can extract R_n to 1% (1.2%) respectively. To achieve this the Hall A Compton Polarimeter must be upgraded to use a green laser and new photon and electron detectors must be installed.
4. Since we must integrate our detected signal, the backgrounds must be measured separately; in addition, pedestals and nonlinearities need to be controlled at the few tenths of percent level.

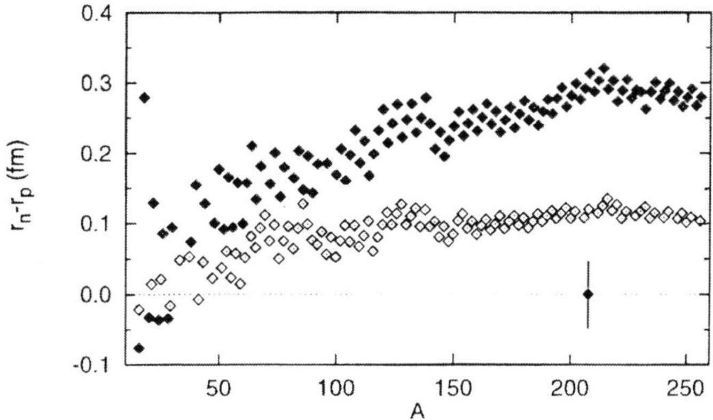

FIGURE 3. The difference between neutron radii $R_n = r_n$ and proton radii $R_p = r_p$ for several nuclei of different mass number A. The filled symbols are for the relativistic mean field NL1 interaction while the open symbols are for the non-relativistic zero range Skyrme skiii interaction. This figure is taken from calculations of Ring *et. al.* [8]. A possible 1% measurement in ^{208}Pb is indicated by the error bar which has been arbitrarily placed at $R_n - R_p = 0$.

CONCLUSION

The neutron skin is a qualitative feature of heavy nuclei which has never been cleanly observed. Measurement of R_n will have a fundamental impact on nuclear physics. This experiment requires state-of-the-art control of systematic errors and very accurate polarimetry. Figure 3 shows the projected error bar compared to two different theoretical calculations.

ACKNOWLEDGMENTS

I am specially grateful to P. A. Souder, R. W. Michaels, and G. M. Urciuoli for the many discussions about this experiment.

REFERENCES

1. T. W. Donnelly, J. Dubach, and I. Sick, Nucl. Phys. A **503**, 589 (1989).
2. R. J. Furnstahl, nucl-th/0112085.
3. C. J. Horowitz, S. J. Pollock, P. A. Souder, and R .W. Michaels, Phys. Rev. C **63**, 025501, (2001).
4. C. J. Horowitz and J. Piekarweicz, Phys. Rev. Lett. **86**, 5647 (2001); C. J. Horowitz and J. Piekarweicz, Phys. Rev. C **64**, 062802 (2001).
5. New York Times, April 11 edition (2002). See also P. Slane, D. Helfand, and S. Murray, astro-ph/0204151; D. G. Yakovlev *et. al.*, astro-ph/0204233.
6. C. J. Horowitz, Phys. Rev. C **57**, 3430 (1998).
7. P. A. Souder, R. W. Michaels, and G. M. Urciuoli, Jefferson Lab Proposal E-00-003, (2000).
8. P. Ring *et. al.*, Nucl. Phys. A **624**, 349 (1997).

Nuclear Anapole Moments and the Parity-nonconserving Nuclear Interaction

Cheng-Pang Liu

TRIUMF Research Facility, 4004 Wesbrook Mall, Vancouver, BC, Canada V6T 2A3

Abstract.
The anapole moment is a parity-odd and time-reversal-even electromagnetic moment. Although it was conjectured shortly after the discovery of parity nonconservation, its existence has not been confirmed until recently in heavy nuclear systems, which are known to be the suitable laboratories because of the many-body enhancement. By carefully identifying the nuclear-spin-dependent atomic parity nonconserving effect, the first clear evidence was found in cesium. In this talk, I will discuss how nuclear anapole moments are used to constrain the parity-nonconserving nuclear force, a still less well-known channel among weak interactions.

INTRODUCTION

Tests of the unified electroweak theory have long been an important subject in physics. Compared with successes gained in the leptonic, semileptonic, and flavor-changing hadronic sectors, it is fair to say that the flavor-conserving hadronic weak interaction is not well-constrained. Experimentally, this sector could only be studied in nuclear systems, therefore, the major challenge comes from the sensitivity required to separate the parity-nonconserving (PNC) observables from much larger strong and electromagnetic (EM) backgrounds. Despite a number of difficulties, these studies are of fundamental importance because the hadronic neutral weak interaction only appears in the flavor-conserving sector. One also hope that these studies can reveal more information about the hadronic dynamics which can not be probed by parity-conserving (PC) observables.

Several precise and interpretable nuclear PNC measurements have already been made and put constraints on the PNC nucleon-nucleon (NN) interaction. These results along with new developments using polarized proton or neutron beam will be reviewed in the plenary talk by W. D. Ramsay. The focus of this short presentation is the nuclear anapole moment, which provides another window to examine the PNC NN interaction and have to be measured in atomic PNC experiments.

It was first noted independently by Vaks and Zel'dovich [1] that the PNC mechanism allows a parity-odd EM coupling to an elementary particle (actually to a composite system as well) by inducing an exotic electromagnetic moment, called the "anapole moment" (AM). Later on, Flambaum, Khriplovich, and Sushkov [2] pointed out that the AM of a nucleus grows roughly as the nucleon number to the power of two-thirds, and this suggested it might be possible to measure this nuclear AM in heavy atoms.

These theoretical conjectures finally realized when Colorado group [3] announced the first clear evidence of nuclear AM using the polarized atomic cesium beam. By carefully

CP675, *Spin 2002: 15th Int'l. Spin Physics Symposium and Workshop on Polarized Electron Sources and Polarimeters,* edited by Y. I. Makdisi, A. U. Luccio, and W. W. MacKay

identifying the hyperfine dependence of atomic PNC effects, a 7σ determination was reported. And the error bar is so small that a very good constraint on the PNC NN interaction could be deduced (if one does the calculation right). In the context of this symposium, it is more than adequate to recognize this discovery by atomic physicists which contributes to the nuclear spin physics at the lowest energy end.

NUCLEAR AMs AND THE PNC *NN* INTERACTION

According to the multipole expansion, the electromagnetic moments are classified as charge C_J, transverse electric E_J, and transverse magnetic M_J, where J denote the angular momentum. Normally, parity (P) and time-reversal (T) invariance only allow charge moments of even order (C_0, total charge; C_2, charge quadrupole; ...etc.) and transverse magnetic moments of odd order (M_1, magnetic dipole; M_3, magnetic octupole; ...etc.). The vector moment C_1, which is P- and T-odd, is the charge dipole moment. The vector moment E_1, which is P-odd but T-even, is exactly the "anapole moment". Often in the literature, the anapole operator $\vec{a}$, generated by the current density operator $\vec{j}(\vec{r})$, is defined as

$$\vec{a} = -\pi \int d^3r\, r^2 \vec{j}(\vec{r}) \equiv \frac{G_F}{\sqrt{2}} \kappa_{am} \vec{I}.^{1}$$

It is clear that this operator gives vanishing expectation values unless the wave function is not a parity eigenstate or the current is axial-vector—both are linked to weak interactions at the fundamental level. In the last part of the equation above, a dimensionless quantity κ_{am}, which characterizes the strength of a nuclear AM, is defined through the Fermi constant (the typical scale of weak interaction) and the nuclear spin vector $\vec{I}$ (the only intrinsic vector of an elementary or composite particle).

An illustrative picture of the AM is the toroidal current winding. Because the r^2 weighting factor, the currents on the outer part of the torus give larger moments than the inner part, and this leads to a net AM. Suppose in a system where parity is a good symmetry, the left- and right-handed current windings should be equally probable, therefore no net AM occurs. However, any non-equal mixture of these two by some PNC mechanism will results in a chiral current and thus an AM. Also noteworthy is that the magnetic field generated by the toroidal current winding is confined, therefore, unless a particle is inside the torus, there in no interaction. This contact character of interactions with AMs is the same as the low energy neutral weak interaction, a result anticipated by the unified electroweak theory.

Although one believes the nuclear AMs have their origin in the couplings of quarks and weak bosons, $W^{\pm}$ and Z^0, a hadronic theory from the first principle is still unavailable. Instead, various models are designed to describe the nonperturbative dynamics of hadrons. For the PNC NN interaction, the widely-used framework is a one meson exchange model including π, ρ, and ω, with one of the meson couplings is PC and the

[1] The current conservation plays a role in defining the form of E_1 operator, and the definition for κ is different from what Khriplovich *et al.* adopted. For more details, see Ref. [10].

other PNC. The six PNC meson coupling constants in this model, h_π, h^0_ρ, h^1_ρ, h^2_ρ, h^0_ω, and h^1_ω, as defined by Desplanques, Donoghue, and Holstein (DDH) [4], undermine the physics of how the fundamental couplings of quark and weak bosons are modified by the strong interaction. [2] The theoretical benchmark is given by the so-called DDH "best values" along with some reasonable ranges. It is the hope that experiments could constrain these couplings well and justify the hadronic theories.

Given this PNC NN potential, nuclear AMs arise in three ways: i) one-body contribution, where weak radiative corrections are induced in the form of single nucleon loop or pole diagrams, often called as the nucleonic AM, ii) two-body contribution, where mesons induce extra EM currents by coupling photons to nucleon-antinucleon pairs and mesons in-flight, iii) polarization mixing, where a parity eigenstate state is mixed by opposite-parity states, thus the normally forbidden EM couplings are allowed. While the one-body contribution is incoherent, many-body effects would possibly enhance the two-body and polarization mixing terms.

EXPERIMENTAL RESULTS AND DEDUCED CONSTRAINTS

With the increasing accuracy, atomic PNC experiments have been an important part of the low-energy precision tests of the standard model. The dominant PNC effect comes from the tree-level Z^0 exchange between atomic electrons and the nucleus with an axial-vector coupling at electrons and a vector coupling at the nucleus, A(e)-V(N). This is a nuclear-spin-independent (NSI) effect in which every nucleon contributes coherently. On the other hand, The V(e)-A(N) exchange gives a nuclear-spin-dependent (NSD) effect, but is much suppressed because nucleons contribute incoherently and electrons have a weaker vector coupling to Z^0.

Although the interaction of electrons with the nuclear AM comes at a higher order, i.e., $G_F\alpha$, it actually dominates the NSD effect in heavy nuclei because the electron coupling is not suppressed by (1 - 4 $\mathrm{Sin}^2\theta_W$) and the nuclear many-body enhancement grows as $A^{2/3}$. Therefore, the extraction of nuclear AM involves: i) an atomic many-body calculation relating the experiment result to the PNC electron-nucleus interaction, ii) the identification of NSD PNC effect by comparing results from different hyperfine levels, and iii) the subtraction of contributions due to Z^0 exchange and hyperfine interaction.

So far, nuclear AMs in cesium and thallium have been reported. The cesium experiment by Colorado group showed a clear evidence, however, the thallium experiments by Seattle [5] and Oxford [6] groups had large error bars so the results are consistent with zero. The extracted AMs in terms of κ_{am} are: κ_{am}(Cs) = 0.090 ± 0.016 [7] and κ_{am}(Tl) = 0.376 ± 0.400 (Seattle's only). [3]

In order to constrain the PNC meson couplings using these results, one has to perform a model calculation of the nuclear AM and then express κ_{am} in terms of these couplings.

---

[2] h_π was named ad f_π originally by DDH, however, it is changed in order not confuse with pion decay constant sometimes.

[3] The Oxford result is not quoted here, see Ref. [10] for discussion.

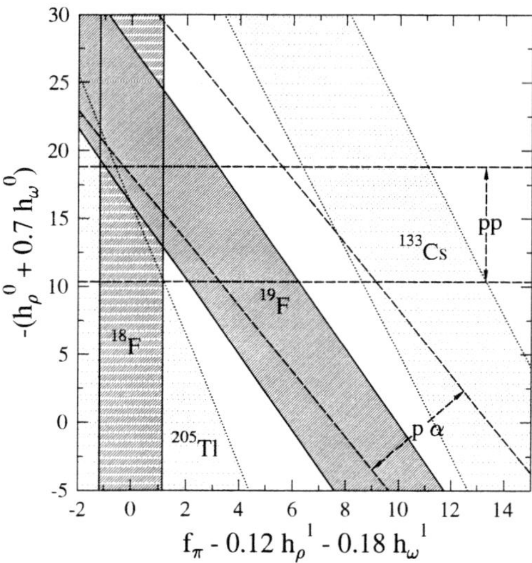

FIGURE 1. Constraints on the PNC meson couplings ($\times 10^7$). The error bands are one standard deviation.

Because both Cs and Tl are heavy nuclei, the nuclear structure is the most important issue. There have been quite a few calculations with various treatments, a brief summary and survey could be found in Ref. [8]. Roughly speaking, the calculations based on the single particle approximation, which treat the Cs as a $1g_{7/2}$ proton plus the closed core and Tl as a $3s_{1/2}$ proton hole plus the closed core, tend to predict larger AMs than calculations which consider many-body effects.

The constraints on the PNC meson couplings is presented in Fig. 1. The Cs and Tl bands are plotted based on the shell model results of Ref. [9, 10], a full two-body calculation in which all the exchange currents are included and the polarization mixing is handled by the closure approximation with the aid of nuclear systematics found in light nuclei.

Apparently, the anapole constraints are not in agreement with the existing nuclear PNC results, and also with each other (only a small part of the Tl band is shown here, and the central line of this band has a negative x-intercept). The result for the AM of Tl is rather confusing because the experiment gave a positive value, but all the calculations predict negative. Therefore, it is very possible that the tension between Cs and Tl bands is due to this sign problem. One can also observe that the Cs result tests a similar combination of PNC couplings as $p\alpha$ and ^{19}F, but favors lager values. By combining Cs and pp bands, the allowed region does fall into the DDH reasonable ranges, with $h_\pi \sim 9$. However, the stringent limit set by the ^{18}F result, $|h_\pi| \leq 1$, which has been performed by five groups, definitely rules out this possibility.

The big discrepancy between the anapole constraints and existing nuclear PNC results is certainly a puzzle to be sorted out. The first criticism of the theory would be on our still limited knowledge of the structure of heavy nuclei. By the way, the atomic many-body theory, which is the key to the interpretation of experiments, should also be examined.

With only one certain result in Cs, obviously we need more experimental inputs to clarify the current situation. There are several new measurements being in progress or proposal. For example, a new Cs measurement will double-check the existing result, an improved Tl experiment hopefully can solve the sign problem, results of odd-neutron nuclei like Dy, Yb, and Ba would produce constraints roughly perpendicular to what odd-proton nuclei do, and the study of a chain of Fr isotopes should reduce some of the theoretical uncertainties.

However, it ought to be emphasized that, if any of these results, when available, is going be to used for constraining PNC meson couplings reliably, a good nuclear structure calculation is still the top necessity.

SUMMARY

The nuclear anapole moment, a manifestation of nuclear parity-nonconservation which has been conjectured for a long time, is clearly discovered in the atomic PNC experiment of cesium. The precision of this result makes it sensible to constrain the PNC meson couplings. However, a big discrepancy is found by comparing this new constraint with existing nuclear PNC results, most possibly due to the nuclear structure uncertainties. In order to constrain the hadronic theory reliably, this puzzle should be further addressed.

ACKNOWLEDGMENTS

The author would like to thank Profs. W. van Oers and M. J. Ramsey-Musolf for encouraging this presentation at SPIN 2002 symposium.

REFERENCES

1. Zel'dovich, Ya.B., *Sov. Phys. JETP* **6**, 1184 (1958), and citation therein.
2. Flambaum, V.V., and Khriplovich, I.B., *Sov. Phys. JETP* **52**, 835 (1980); Flambaum, V.V., Khriplovich, I.B., and Sushkov, O.P., *Phys. Lett.* **B146**, 367 (1984).
3. Wood, C.S., Bennett, S.C., Cho, D., Masterson, B.P., Roberts, J.L., Tanner, C.E., and Wieman, C.E., *Science* **275**, 1759 (1997).
4. Desplanques, B., Donoghue, J., and Holstein, B.R., *Ann. Phys. (N.Y.)* **124**, 449 (1980).
5. Vetter, P., Meekhof, D.M., Majumder, P.K., Lamoreaux, S.K., and Forstan, E.N., *Phys. Rev. Lett.* **74**, 2658 (1995).
6. Edwards, N.H., Phipp, S.J., Baird, P.E.G., and Nakayama, S., *Phys. Rev. Lett.* **74**. 2654 (1995).
7. Flambaum, V.V., and Murray, D.W., *Phys. Rev.* **C56**, 1641 (1997).
8. Dmitriev, V.F., and Khriplovich, I.B., arXiv: nucl-th/0201041.
9. Haxton, W.C., Liu, C.P., Ramsey-Musolf, M.J., *Phys. Rev. Lett.* **86**, 5247 (2001).
10. Haxton, W.C., Liu, C.P., Ramsey-Musolf, M.J., *Phys. Rev.* **C65**, 045502 (2002).

HAPPEX Parity Violation Experiments at Jefferson Lab

D.S. Armstrong*† and the HAPPEX and Hall A Collaborations †

*Dept. of Physics, College of William & Mary, Williamsburg VA 23187, USA
†Thomas Jefferson National Accelerator Facility. Newport News, VA 23606, USA

Abstract. The HAPPEX program of measurements of parity-violation in elastic electron scattering, in Hall A of Jefferson Lab, is presented. The results of the recently completed measurement on the proton at $Q^2 = 0.48$ GeV^2 are briefly reviewed. The plans are presented for the upcoming HAPPEX II measurement on the proton at $Q^2 = 0.1$ GeV^2, as well as the companion measurement with a ^{4}He target at the same momentum transfer. These experiments are sensitive to strange quark contributions to the vector structure of the nucleon. The two new experiments will provide a precision measurement of the strangeness radius parameter, and the combination of the two experiments will also determine the strange contribution to the proton's magnetic moment.

INTRODUCTION

Our knowledge of the electromagnetic form factors of the nucleon has become increasingly refined in recent years [1]. However, a basic question remains open: how does the quark-antiquark sea contribute to the distributions of charge and magnetization represented by these form factors? As the nucleon possesses no net strangeness, strange quarks contributions represent a promising window onto the effects arising from the sea.

A significant role for strange quarks in certain nucleon properties is suggested by results on a number of observables. Deep inelastic neutrino scattering experiments [2, 3] indicate that the quark structure functions $s(x)$ and $\bar{s}(x)$ are significant at low x. Analyses of spin-dependent deep inelastic scattering [4, 5, 6, 7] suggest that strange quarks may contribute a sizable fraction to the nucleon's spin. The pion-nucleon sigma term, accessible through low-energy π scattering, gives information on scalar strange matrix elements. The comparison of the πN scattering results with hyperon mass systematics indicates sizable scalar strange matrix elements for the nucleon [8, 9]. Low-energy antiproton annihilation measurements [10] find large enhancements of ϕ meson production over predictions based on the OZI rule, which have been interpreted as evidence of large polarized $s\bar{s}$ components in the nucleon [11, 12].

Parity violation in electron scattering provides another avenue by which to investigate strange quarks in the nucleon. In contrast to the techniques discussed above, the primary sensitivity here is to the *vector* matrix elements, $< N|\bar{s}\gamma^{\mu}s|N >$, about which we have little information.

In electron scattering, the interference of the γ exchange and Z^0 exchange diagrams leads to parity violation in the scattering rate, which can be used as a probe of hadron structure. In essence, while the photon's coupling to the quarks is proportional to their

CP675, *Spin 2002: 15th Int'l. Spin Physics Symposium and Workshop on Polarized Electron Sources and Polarimeters,* edited by Y. I. Makdisi, A. U. Luccio, and W. W. MacKay

electric charges, the Z^0 coupling is proportional to their weak charges. Thus a measurement of parity violation allows a quark flavor decomposition of the vector structure of the nucleon.

The vector matrix elements of the nucleon can be represented in terms of the Sachs form factors, $G_E(Q^2)$ and $G_M(Q^2)$, (Q^2 is the momentum transfer) which parameterize the electric and magnetic response, respectively. Similarly, one defines form factors for each flavor i, G_E^i and G_M^i; of particular interest here are the strange quark versions, G_E^s and G_M^s.

The measured asymmetry is $A = \frac{\sigma_R - \sigma_L}{\sigma_R + \sigma_L}$ where σ_R and σ_L are the cross sections for scattering with right- and left-handed helicity electrons respectively. The dominant parity-conserving term from the square of the γ-exchange diagram drops out from the difference, and one is left with a term proportional to the γ-Z interference.

For elastic scattering from the proton, the asymmetry is given by

$$A = -A_0 \tau \left(2 - 4\sin^2\theta_W - \frac{\varepsilon G_E^0 + \tau G_M^0}{\varepsilon G_E^{\gamma p} + \tau G_M^{\gamma p}} \right) - A_A$$

where $G_{E,M}^0(\tau) = \left(G_{E,M}^u + G_{E,M}^d + G_{E,M}^s \right)/3$ are the SU(3) flavor singlet form factors, ε and τ are kinematic factors, θ is the lab scattering angle, and $A_0 = \frac{G_F M_p^2}{\sqrt{2}\pi\alpha}$. A_A is the axial vector form factor, which is small at the forward scattering angles used here.

The asymmetry involves the term $(\varepsilon G_E^0 + \tau G_M^0)/(\varepsilon G_E^{\gamma p} + \tau G_M^{\gamma p})$ which allows one to extract $G_{E,M}^0$ if the electromagnetic form factors are known. Assuming charge symmetry one can extract the strange quark form factors from the G^0's. Combining with the nucleon electromagnetic form factors allows the extraction of some combination of G_E^s and G_M^s; the exact combination depends on the kinematics of the measurement.

At low Q^2, it is convenient to consider the leading moments of the form factors, given by the strangeness magnetic moment $\mu_s = G_M^s(0)$ and radius parameter $\rho_s = \frac{dG_E^s(\tau)}{d\tau}|_{\tau=0}$. Many theoretical predictions for the strange quark form factors are available, using a variety of models of non-perturbative QCD, and they range for μ_s from -0.6 to +0.4, and from -3.0 to 3.2. for ρ_s. Recent reviews of the field are available elsewhere [13, 14, 15].

HAPPEX - I

The Hall A Proton Parity Experiment (HAPPEX) [16] measured the asymmetry in scattering of longitudinally-polarized 3.3 GeV electrons from a liquid hydrogen target at $Q^2 = 0.48$ GeV^2. Two identical high-resolution magnetic spectrometers, both located at 12.3° to the beam, were used to detect the scattered electrons. Elastic scattering events were detected in the focal plane of each spectrometer using total absorption counters made from a lead-lucite sandwich.

The data-taking took place in two runs. In the 1998 run, the electron beam current was typically 100 μA, and the polarization was about 39%. The polarized electrons were produced using a circularly-polarized laser shining on a bulk GaAs photocathode. In the

1999 run, the photocathode was replaced with a "strained crystal" GaAs photocathode, which afforded higher polarization ($\sim 73\%$) with reduced intensity (35 μA) [1]. The beam polarization was monitored periodically using a Mott polarimeter at the accelerator injector, a Møller polarimeter, and (for the 1999 run) a Compton polarimeter, which enabled continuous non-invasive polarization measurements [17].

The helicity of the beam was reversed at 30 Hz (to average over 60 Hz noise), in a pseudorandom manner. The detector signals were integrated in customized 16-bit ADCs during each 33 ms helicity window; by integrating the signal, data acquisition dead-time was eliminated.

Helicity-correlated changes in the beam energy were of the order of 10^{-8}, and typical helicity-correlated beam position differences, integrated over the runs, were a few nm. The total effect of these helicity-correlated variations in beam properties on the experimental asymmetry was small ($<< 1$ ppm).

The focal-plane detector signals are normalized by the beam current, as measured by two independent rf cavity monitors. Non-linearity in the detector response or the electronics can produce false asymmetries if the intensity exhibits significant helicity-correlated fluctuations. These were kept at the 1 ppm level using a slow-feedback system at the electron source, based on the voltage applied to the helicity-reversal Pockels cell.

Backgrounds from inelastic scattering events and scattering from the target windows were measured with runs at reduced beam intensity, taken with the full instrumentation package of the spectrometers. The background processes contributed 0.2% and 1.5%, respectively, to the measured signal.

The two data-taking runs yielded consistent results, and the combined asymmetry is $A = (15.0 \pm 1.1) \times 10^{-6}$ [16], leading to the combination of strange quark form factors $G_E^s + 0.392 G_M^s = 0.025 \pm 0.020 \pm 0.014$, at $Q^2 = 0.48$ GeV^2. The first error is the combined statistical and experimental systematic errors, and the second arises from uncertainties in the electromagnetic form factors.

The result is consistent with the absence of strange quark effects, however, note that in several models G_E^s and G_M^s are predicted to have opposite sign. Thus it may be that the results reflect an 'accidental' cancellation of the form factors. Also, some models with substantial strange quark contributions predict form factors that cross zero at this Q^2, but that are large at other momentum transfers. Several model predictions are ruled out, however the data are consistent with other models which include substantial strange quark contributions. Measurements at different kinematic points are therefore of interest.

HAPPEX-II AND ^{4}HE

In 2003, the HAPPEX experiment will be extended down to $Q^2 = 0.1$ GeV^2 [18]. To achieve this lower Q^2, a pair of septum magnets will be installed, which will allow the spectrometers to view a smaller scattering angle (6°). The expected asymmetry at this kinematics is small ($A \sim 1.6$ ppm), however the scattering rate will be 65 MHz per

[1] More recently currents in excess of 100 μA have been achieved with the high-polarization source.

spectrometer, so a statistical precision of 4.6% will be obtained in the approved 700 hr run. New radiation-hard focal plane detectors, consisting of 5-layer sandwiches of brass and fused silica, have been constructed and tested for this experiment.

The experiment demands increased attention to helicity-correlated fluctuations in beam parameters, since the statistical precision corresponds to an ambitious 75 ppb. These fluctuations will be reduced by using simultaneous feedback at the electron source on the beam position and beam current asymmetries.

Another challenging requirement is the liquid hydrogen target. Beam-induced density fluctuations on the 30-Hz helicity-flip time scale, such as due to local boiling, will spoil the statistical precision. We require such fluctuations to be below 100 ppm in order not to degrade the statistics. Two different designs of thin-walled high-power target cells are being developed.

Systematic errors will include the polarization and Q^2 measurements (2% and 0.6%), background contributions (0.75%), the electromagnetic form factors (1.4%) and radiative corrections (1%). The measurement will yield $\rho_s + \mu_p \mu_s$ to a precision of $\pm$ 0.3 which is compared to model predictions in Fig. 1 (see [13] for theory references).

An approved companion measurement to HAPPEX-II will be a measurement of the elastic scattering from ^{4}He, at the same Q^2 [19]. Essentially the same instrumentation will be adopted. Scattering from a spin-zero target has the advantage that there are no magnetic or axial-vector contributions - only charge scattering contributes. Thus the asymmetry can be interpreted directly in terms of the strangeness radius ρ_s.

The detected elastic scattering rate will be 12 MHz in each spectrometer, and the predicted asymmetry is $A = 8.4$ ppm. The 700 hr run will provide a statistical error of 2.2% on A, yielding a measurement of ρ_s to a precision (including systematics) of 0.5. This will considerably reduce the allowed model space for the strange form factors (see Figure 1), and, in combination with HAPPEX-II, will allow us to disentangle the electric from the magnetic contributions. The combined experiments will provide a measurement of μ_s to ± 0.22.

CONCLUSIONS

The strange quark contributions to the vector structure of the nucleon remain elusive, and are, to date, consistent with zero, despite a variety of models which suggest substantial effects. This could, however, be due to accidental cancellations between the magnetic and electric form factors, or it may be that they happen to have Q^2 dependences which make the effects small at the measured kinematics. The next-generation HAPPEX measurements on hydrogen and helium will provide a precise separation of the strange electric and magnetic form factors at $Q^2 = 0.1$. If, after these measurements, the strange contributions remain consistent with zero, this will beg the interesting question: what mechanism suppresses the effect of the $q\bar{q}$ sea on the electromagnetic structure of the nucleon?

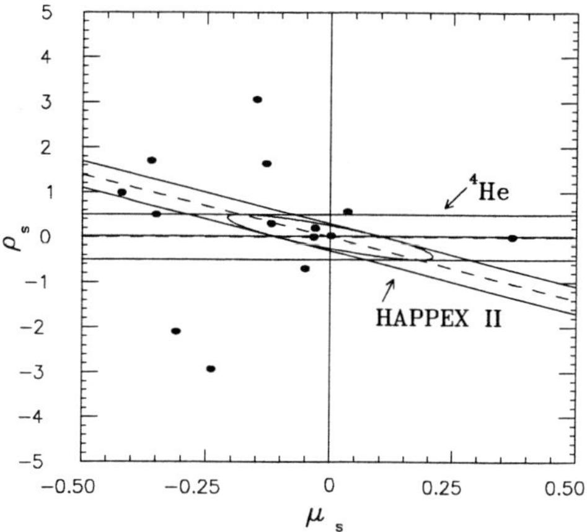

FIGURE 1. Various model predictions (points) for the leading moments of the strange form factors, ρ_s *vs.* μ_s, compared with the expected precision of the HAPPEX-II and ^{4}He parity experiments (bands). The location of the bands shown is arbitrary, only the slopes and widths are relevant.

ACKNOWLEDGMENTS

This work was supported by the National Science Foundation (grants PHY-9602901, PHY-0099557) and by DOE contract DE-AC05-84ER40150 under which the Southeastern Universities Research Association (SURA) operates The Thomas Jefferson National Accelerator Facility.

REFERENCES

1. See contributions to these proceedings from J. Kelly, M. Seimetz, B. Plaster, N. Savvinov, B. Reitz, and C.F. Perdrisat.
2. T. Adams, *et al.*, *Phys. Rev.* D **64**, 112006 (2001).
3. A. O. Bazarko, *et al.*, *Z. Phys.* C **65**, 189 (1995).
4. K. Abe, *et al.*, *Phys. Rev. Lett.* **79**, 26 (1997).
5. B. Adeva, *et al.*, *Phys. Rev. D* **58**, 112002 (1998).
6. H. Lipkin and M. Karliner, *Phys. Lett.* B **461**, 280 (1999).
7. A. Miller, these proceedings.
8. M. Pavan, *πN Newsletter* **19**, 118 (1999).
9. J. Gasser, H. Leutwyler, and M.E. Saino, *Phys. Lett.* B **253**, 260 (1991).
10. C. Amsler, *et al.*, *Phys. Lett.* B **346**, 363 (1995).
11. J. Ellis, *et al.*, *Phys. Lett.* B **353**, 319 (1995).
12. M.G. Sapozhnikov, *Nucl. Phys.* **A692**, 63 (2001).
13. K.S. Kumar and P.A. Souder, *Prog. Part. Nucl. Phys.* **45**, S333 (2000).
14. D.H. Beck and B.R. Holstein, *Int. J. Mod. Phys.* E **10**, 1 (2001).
15. D.H. Beck and R.D. McKeown, *Ann. Rev. Nucl. Part. Sci.* **51**, 189 (2001).
16. K.A. Aniol, *et al.*, *Phys. Rev. Lett.* **82**, 1096 (1999); K.A. Aniol, *et al.*, *Phys. Lett.* B **509**, 211 (2001).
17. M. Baylac, *et al.*, *Phys. Lett.* B **539**, 8 (2002).
18. Jefferson Lab experiment 99-115, K.S. Kumar and D. Lhuillier, spokespersons.
19. Jefferson Lab experiment 00-114, D.S. Armstrong and R. Michaels, spokespersons.

The G^0 Experiment At Jefferson Lab

L. Lee *(for the G^0 Collaboration)*

Dept. of Physics and Astronomy, Univ. of Manitoba, Winnipeg, MB, Canada, R3T 2N2

Abstract. The electron-proton parity-violation G^0 experiment at Jefferson Lab aims to make a determination of the 'strange' quark currents in the proton. Two *new* proton ground state matrix elements will be measured which are sensitive to point-like 'strange' quarks and hence to the quark-antiquark sea in the proton. The matrix elements of interest are the elastic-scattering vector weak neutral-current 'charge' and 'magnetic' form factors, G^Z_E and G^Z_M, respectively. By measuring the very small parity-violating asymmetries in elastic electron-proton scattering at momentum transfers between 0.1 and 1.0 GeV^2, and combining these asymmetries with previously measured electromagnetic form factors, *new* information about the proton weak form factors can be obtained. This new high precision experiment is presently in the installation and commissioning phase.

INTRODUCTION

The detailed structure of the nucleon at low energies is not well understood within the framework of quark and gluon degrees of freedom. For example, relatively little is known about the importance of the quark-antiquark sea at these energies. However, since strange quarks contribute only to the sea, direct information about the role of the quark-antiquark sea can be gleaned through determinations of the strangeness content of the nucleon. In particular, the strange quark contributions to the 'electric' and 'magnetic' properties of the nucleon can be determined through measurements of the nucleon weak vector form factors [1,2].

Parity-violating (PV) electron-nucleon elastic scattering offers a unique opportunity to study the electroweak structure of the nucleon [1]. The electroweak interaction takes place at first order through two processes, involving the exchange of a virtual photon (γ) and the exchange of a Z^0, respectively, with 4-momentum transferred $-Q^2$. Parity violation arises through the interference of the γ and the Z^0 exchange amplitudes and can be characterized by the ratio of the helicity dependent to helicity independent cross sections, or the PV asymmetry:

$$A = \left[\frac{1}{P}\right]\frac{\sigma_R - \sigma_L}{\sigma_R + \sigma_L} \tag{1}$$

where σ_R and σ_L are the cross sections for right- and left-handed electrons, respectively. For elastic electron-proton scattering, the PV asymmetry can be expressed as:

CP675, *Spin 2002: 15th Int'l. Spin Physics Symposium and Workshop on Polarized Electron Sources and Polarimeters,* edited by Y. I. Makdisi, A. U. Luccio, and W. W. MacKay

$$A = \left[\frac{-G_F Q^2}{4\sqrt{2}\pi\alpha} \right] \frac{\varepsilon G_E^\gamma G_E^Z + \tau G_M^\gamma G_M^Z - (1 - 4\sin^2\theta_W)\varepsilon' G_M^\gamma G_A^e}{\varepsilon (G_E^\gamma)^2 + \tau (G_M^\gamma)^2} \qquad (2)$$

where G^γ_E and G^γ_M are the usual proton electromagnetic form factors, G^Z_E and G^Z_M are the proton weak vector form factors, and G^e_A is the axial form factor of the proton. At a given Q^2, measurement of the PV asymmetry at forward and backward angles allow for the determination of the proton weak vector form factors via a Rosenbluth type separation. When combined with knowledge of the usual proton and neutron electromagnetic form factors, the strange quark contributions (G^S_E and G^S_M) to the nucleon's structure can be extracted. This extraction relies only on the SU(3) and charge symmetry of the nucleon.

THE G^0 EXPERIMENT

In pursuit of some of the goals described above, a number of dedicated PV asymmetry experiments [3,4,5] have been proposed, developed and several have been completed over the last decade. These experiments have either measured a linear combination of form factors (e.g. $G^S_E + 0.39G^S_M$) [4] or have utilized kinematical suppression to isolate one of the form factors (G^S_M) [5]. In contrast, the G^0 experiment [6] proposes to perform the full separation of the strange electric G^S_E, magnetic G^S_M and axial G^e_A form factors and to characterize the Q^2 evolution of these observables by carrying out the measurements at three different momentum transfers 0.3, 0.5 and 0.8 $(\text{GeV}/c)^2$. To accomplish this, a first set of asymmetry measurements will be performed at forward electron scattering angles between 7° to 15°, accessing a range of Q^2 between 0.1 and 1.0 $(\text{GeV}/c)^2$. This first measurement is carried out at a single electron beam energy and makes use of a hydrogen target. Following this, a second set of asymmetry measurements will be performed at backward electron scattering angles centered at 110° and will require measurements on both hydrogen and deuterium targets. Due to the limited Q^2 acceptance at these backward angles, three separate measurements are planned at incident electron beam energies of 424, 525 and 799 MeV, which correspond to Q^2 = 0.3, 0.5 and 0.8 $(\text{GeV}/c)^2$, respectively. Figure 1

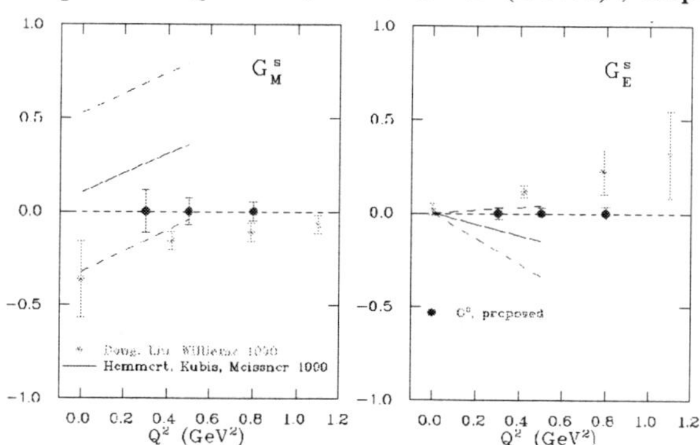

FIGURE 1. The projected errors for the G^0 measurement compared with two different models.

shows the projected total errors for the G^0 measurements along with two theoretical predictions, one based on a Chiral Perturbation Theory calculation [7] and the other based on a Lattice QCD calculation [8]. The projected errors in figure 1 include contributions from statistical uncertainties ($\Delta A/A = 5\%$), uncertainties from systematic effects, as well as uncertainties associated with the 'known' values of the proton and neutron electromagnetic form factors. The overall uncertainties are dominated by the statistical errors.

An important consideration for these types of parity-violation measurements is false asymmetries associated with helicity-correlated modulations in the beam properties. The G^0 requirements for the beam properties are summarized in Table 1.

TABLE 1. Beam Requirements for G^0

Beam parameter	Nominal value	Helic. Corr. Asym (in 30 days)
Current	40 μA	< 1 ppm
Energy	3 GeV	$< 2.5 \times 10^{-8}$
Position	-	< 20 nm
Angle	-	< 2 nrad

The G^0 Apparatus

The G^0 measurement will be carried out at Jefferson Lab, in the Hall C experimental area. The experiment will require a source capable of delivering a 40 μA electron beam with 70% polarization to a 20 cm long cryogenic target. A specially designed superconducting toroidal spectrometer, with azimuthally symmetric angular acceptance has been constructed. In the forward angle mode (shown in figure 2), individually scattered particles will be detected and counted by a set of 16 scintillator pairs, located at the focal surface of each spectrometer octant. The bend angle through the magnet is ~35° and eight sets of collimators (not shown in figure 2) located in the magnet gaps shield the detectors from a direct view of the beam. Shown in figure 3 are: one octant of scintillator pairs; and eight complete sets of scintillator arrays, mounted within eight light-tight modules, and installed on the detector support structure.

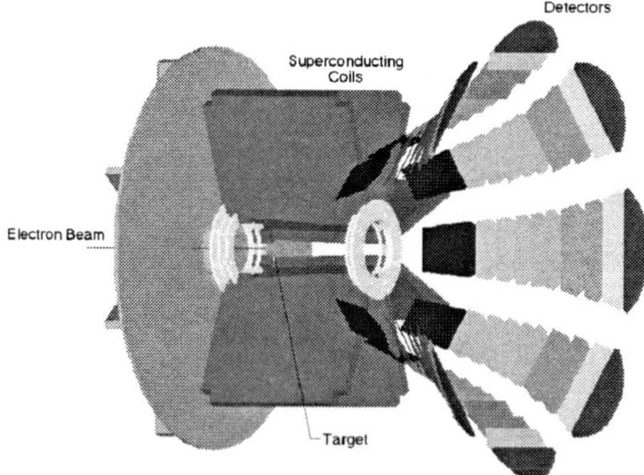

FIGURE 2. Schematic layout of the G^0 spectrometer and the forward angle mode of the experiment.

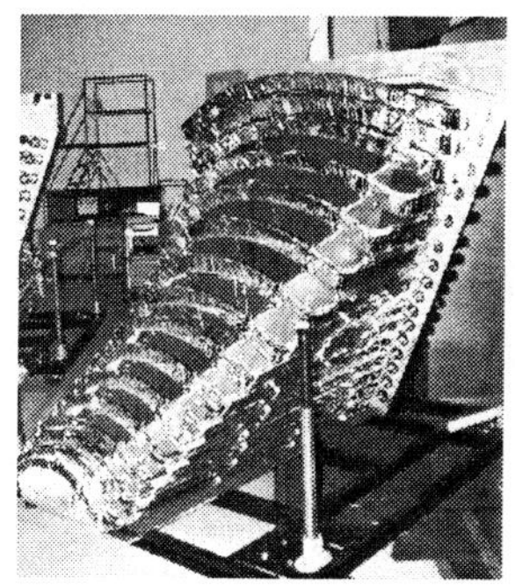

FIGURE 3. One set of Focal Plane scintillators and all eight sets loaded onto the main Detector support structure.

In this first phase of the experiment, the beam energy is fixed at 3 GeV and recoil protons from the e-p scattering process are detected at $\theta_p = 70° \pm 10°$ (or $\theta_e = 11° \pm 4°$). Time-of-flight measurements over a 32 ns time-window will be used to supplement momentum selection by the spectrometer and to discriminate between elastic and inelastic scattering processes. Custom time-encoding electronics allow readout at high detector rates, of the order of 2 MHz per scintillator pair. To enable proper operation of the time-encoding electronics, the electron beam will be pulsed at 31.25 MHz (the 16th subharmonic of the usual 499 MHz pulse structure).

In the second phase of the experiment, the detector and spectrometer system will be rotated back-to-front to detect the back-scattered electrons at $\theta_e = 110° \pm 10°$, allowing a reasonable lever arm for a Rosenbluth separation. For this backward angle configuration, each incident beam energy will correspond to a different Q^2 measurement, and time-of-flight techniques will not adequately discriminate between elastic and inelastic scattering processes. Instead, an additional array of scintillation detectors located near the spectrometer cryostat exit of each octant, will enable kinematic separation of the elastic and inelastic electrons.

Results from the SAMPLE experiment [5] at the MIT-Bates laboratory have shown the importance of measuring the axial form factor complete with radiative corrections. In order to perform the separation of the proton axial form factor from the weak vector form factors, further asymmetry measurements must be carried out using a deuterium target. However, for the deuterium measurements, pion backgrounds are a significant issue and special low-index aerogel Cerenkov counters will be required to provide π/e separation. The schematic layout of the G^0 backward angle configuration is shown in figure 4.

SUMMARY

The G^0 experiment proposes to measure and fully separate the strange electric (G^S_E), magnetic (G^S_M), and axial (G^e_A) form factors of the nucleon and to characterize the Q^2 evolution of these observables by carrying out the measurements at different

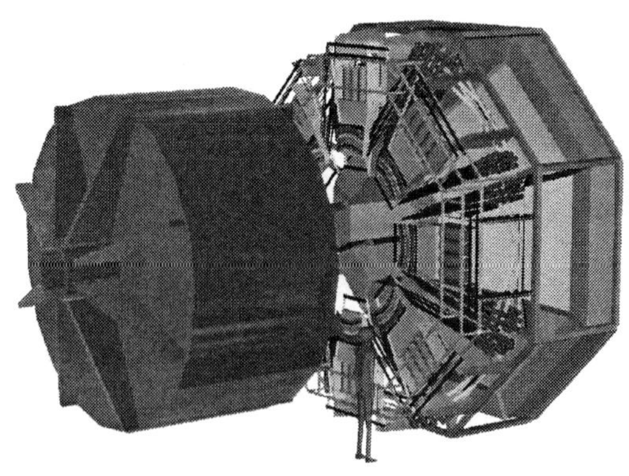

FIGURE 4. Schematic layout of the G^0 backward angle configuration.

momentum transfers. Determinations of these observables may shed light on the role, at low energies, of strange quarks and the quark-antiquark sea in the nucleon. As well, G^e_A may provide valuable information on the proton anapole moment. Presently, this experiment is in the installation and commissioning phase at Jefferson Lab in Hall C. All subsystems for the first phase, forward angle measurements have been installed. Commissioning of the G^0 apparatus will take place between October 2002 to January 2003, followed by the actual 'physics' run some time later (possibly in late 2003). Upon completion of the forward angle measurements, the 'turn-around' of the G^0 apparatus will be carried out, and data-taking for the backward angle program could start in 2005.

ACKNOWLEDGMENTS

The G^0 experiment is supported by grants from NSERC (Canada), CNRS/IN2P3 (France), DOE (USA) and NSF (USA).

REFERENCES

1. Kaplan, D., and Manohar, A., *Nucl. Phys.* **B310**, 527-547 (1988); McKeown, R.D., *Phys. Lett.* **B219**, 140-142 (1989); Beck, D.H., *Phys. Rev.* **D39**, 3248-3256 (1989).
2. Musolf, M. et al., *Phys. Rep.* **239**, 1-178 (1994).
3. Beck, D.H. and McKeown, R.D., *Ann. Rev. Part. Sci.* **51**, 189-217 (2001).
4. Armstrong, D., *HAPPEX Parity-violation Experiments at Jefferson Lab*, these proceedings.
5. Hasty, R. et al., *Science* **290**, 2117-2119 (2000); Hasty, R., *The SAMPLE Experiment at 125 MeV*, these proceedings.
6. http://www.npl.uiuc.edu/exp/G0/G0Main.html; Jefferson Lab proposals E00-006 and E01-116, D.H.Beck spokesperson.
7. Hemmert, T. et al., *Phys. Rev.* **C60**, 45501 (1999).
8. Dong, S.J. et al., *Phys. Rev.* **D58**, 74504 (1998).

Probes of Nucleon Structure at High Momentum Transfer

Spin 2002

New Precision Results on the Spin Structure Function g_1^d

P. Lenisa
on behalf of the HERMES Collaboration

Università di Ferrara and INFN - Sez. Ferrara, 44100 Ferrara, ITALY

Abstract. The Hermes experiment studies the spin structure of the nucleon using the 27.6 GeV longitudinally polarized positron beam of HERA and an internal target of pure gases. Recently, HERMES presented preliminary results on the deuteron spin structure function g_1^d in the kinematic range $0.002 < x < 0.85$ and $0.1 < Q^2 < 20$ GeV^2 based on a restricted data set. Here, new, precise results are presented using the superior statistics of the 2000 data taking period. A reduction of the systematic uncertainty which could be achieved in this preliminary analysis relies in particular on the excellent performance of HERA and of the HERMES target.

INTRODUCTION

Deep-inelastic lepton-nucleon scattering is well established as a powerful tool for the investigation of the nucleon structure. The introduction of polarization variables like in the scattering of polarized leptons off polarized nucleons provides informations on the spin composition of the nucleons. This is the case of the polarized structure function g_1 which, in the parton model, is interpreted as a measure of the number of quarks with same (q^+) or opposite (q^-) helicity with respect to the nucleon:

$$g_1(x) - \frac{1}{2}\sum_f e_f^2(q_f^+ - q_f^-) - \frac{1}{2}\sum_f e_f^2 \Delta q_f(x) \tag{1}$$

Recently, HERMES presented preliminary results on the deuteron spin structure function g_1^d in the kinematic range $0.002 < x < 0.85$ and $0.1 < Q^2 < 20$ GeV^2 based on a restricted data set. In this contribution, new, precise results are presented using the superior statistics of the 2000 data taking period.

THE HERMES EXPERIMENT

The HERMES experiment is located in the East straight section of the HERA storage ring at the DESY laboratory in Hamburg. In 2000 the positron beam of 27.6 GeV with injected beam currents of about 45 mA has been used. The positrons become transversely polarized by the emission of synchrotron radiation. Longitudinal polarization at the interaction point is achieved by spin rotators situated upstream and downstream the HERMES experiment. The beam elicity is usually reversed every $2 \div 3$ months to

CP675, *Spin 2002: 15*[th] *Int'l. Spin Physics Symposium and Workshop on Polarized Electron Sources and Polarimeters*, edited by Y. I. Makdisi, A. U. Luccio, and W. W. MacKay account for possible systematic effects. Two Compton-back scattering polarimeters measure the beam polarization [1]. An average polarization value of 0.55 with an uncertainty of 0.02 has been measured during the year.

Target

A beam of Deuterium atoms is generated in a micro-wave dissociator which forms part of the atomic beam source (ABS). The beam of nuclear polarized atoms is injected into the center of a thin-walled storage cell via a side tube and the atoms then diffuse to the open ends of the cell where they are removed by a high speed pumping system. A longitudinal magnetic holding field provides a quantization axis for the spins. The beam emerging from a second side tube is analysed by a Breit-Rabi polarimeter (BRP) to measure its atom polarization and a target gas analyser (TGA) to determine its atomic fraction [2]. During the atom diffusion process, relaxation by wall and spin exchange collisions and wall recombination changes the polarization and the atomic fraction of the target gas. The atom polarization and atomic fraction values measured by the BRP and TGA must be corrected for these effects to obtain the average target polarization as seen by the positron beam which is described by the following expression:

$$P_T = \alpha_0 [\alpha_r + (1 - \alpha_r)\beta] P_a, \tag{2}$$

where α_0 is the atomic fraction accounting for unpolarized molecules, α_r is the relative atomic fraction surviving recombination, P_a is the nuclear polarization of the atoms and $\beta = P_m/P_a$ the relative polarization of the recombined molecules respect to the atomic polarization ($0 \leq \beta \leq 1$). In the 2000 data taking period, the target density and thus the luminosity could be increased by a factor of about 2 respect to 1999, by reducing the working temperature from 100 K down to 60 K and using a cell with a smaller cross section. The average polarization was 85 % with a total uncertainty of $\pm 3\%$.

In contrast to the solid targets, the Hermes gaseous target does not suffer from radiation damage (flowed gas), its polarization can be continuously measured and rapidly reversed, and is not diluted by non-polarizable material. In 2000 the extremely good and stable performances of both HERA beam and HERMES target allowed the collection of about 10 million positron DIS candidates, a statistics 5 times higher than in any of the previous data-taking years.

Spectrometer

The HERMES detector [3] is a forward spectrometer with a dipole magnet providing a field integral of 1.3 Tm. A horizontal iron plate shields the HERA beam lines against the field thus dividing the spectrometer into two identical halves with a minimum vertical acceptance of ± 40 mrad. The acceptance extends to ± 140 mrad vertically and to ± 170 mrad horizontally. Tracking in each detector half is accomplished by 42 drift chamber planes. Positron identification is accomplished using a probability method based on signals of three sub-systems: the lead-glass block calorimeter, a transition-radiation detector and a preshower hodoscope. For positrons in the momentum range of 2.5 to 27

GeV, the identification efficiency exceeds 98 % with a negligible hadron contamination, the average polar angle resolution is 0.6 mrad, and the average momentum resolution is $1 \div 2$ %.

MEASUREMENT

Extraction Formalism

The kinematic range of this measurement accesses intervals in the Bjorken variables of $0.002 < x < 0.85$ and $0.1 < y < 0.91$. Inclusive events are selected by the requirements: $Q^2 > 0.1$ GeV and $W^2 > 3.24$ GeV (where $-Q^2$ and W^2 are squared 4-momenta of the photon and of the hadronic final state). For each of the 2-dimensional $[x, y]$ bins the kinematic range is divided in, the asymmetry between cross-sections of parallel or antiparallel helicities of the beam and the target is calculated:

$$A_{\parallel}^{meas} = \frac{\sigma^{\rightleftarrows} - \sigma^{\rightrightarrows}}{\sigma^{\rightleftarrows} + \sigma^{\rightrightarrows}} = \frac{1}{P_b P_t} \frac{(N/L)^{\rightleftarrows} - (N/L)^{\rightrightarrows}}{(N/L)^{\rightleftarrows} + (N/L)^{\rightrightarrows}}$$

The number of events selected (N) are corrected per spin state for the background arising from charge symmetric processes. The corresponding luminosities (L) used for normalization are measured with Bhabha scattering and are corrected for a small cross section asymmetry caused by residual electron polarization of the target. P_b (P_t) is the beam (target) polarization.

The Born asymmetry has been derived from the measured asymmetry by subtraction of the radiative background according to the formula:

$$A_{\parallel}^{Born} = \frac{1}{\lambda} \left[A_{\parallel}^{meas} \cdot \left(\lambda + \frac{\sigma_{bg}^{unpol}}{\sigma_{DIS}^{unpol}} \right) - \frac{\sigma_{bg}^{pol}}{\sigma_{DIS}^{unpol}} \right]$$

where $0.9 < \lambda < 1.12$ is a spin independent parameter, $\sigma_{\rm bg}^{\rm (un)pol}$ are the (un)polarized radiative tails, $\sigma_{\rm DIS}^{\rm unpol}$ is the unpolarized DIS cross section.

The radiative corrections have been calculated using the POLRAD program [4]. The statistical error had to be enlarged to account for radiative background by about a factor 2 at low x. The structure function ratio g_1^d/F_1^d and the spin structure function g_1^d have been extracted according to

$$\frac{g_1^d}{F_1^d} = \frac{1}{1+\gamma^2} \left[\frac{A_{\parallel}^{Born}}{D} + (\gamma - \eta) A_2^d \right] \tag{3}$$

and

$$g_1^d = \left(\frac{g_1^d}{F_1^d} \right) \cdot (F_1^d)_{par} = \left(\frac{g_1^d}{F_1^d} \right) \cdot \frac{(1+\gamma^2) F_2^d}{2x(1+R)} \tag{4}$$

The first term in Eq. 3 is the virtual-photon asymmetry, where D is the effective polarization of the virtual-photon, γ and η are kinematic factors, A_2^d is parameterized

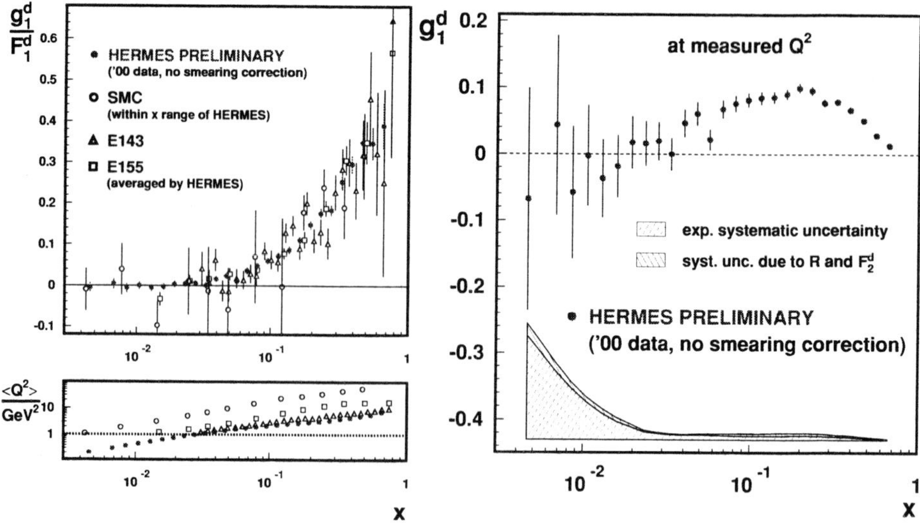

FIGURE 1. Comparison between different g_1^d/F_1^d measurements [9] (left): data are shown at the measured Q^2 and the plotted errors are the quadratic sum of the statistical and systematic uncertainties. HERMES result on g_1^d (right): the error bars are statistical only and the shaded histograms show the systematic uncertainty estimation.

using the Wandzura-Wilczek relation [5]. F_1^d in Eq. 4 has been parametrized in terms of the ratio $R = \sigma_L/\sigma_T$ [6] and the structure function F_2^d, which has been expressed by $F_2^d = F_2^p/2\left(1 + F_2^n/F_2^p\right)$, using the parameterizations for F_2^p [7] and F_2^n/F_2^p [8].

Results

The HERMES result on g_1^d is presented in Fig. 1. The contribution to the systematic uncertainty from the experiment is dominated by the accuracy of the beam and target polarization measurements. The extraction formalism adds a significant contribution at lowest x due to the insufficient knowledge of the radiative corrections and at most 1.8 % at highest x due to the knowledge of A_2^d. The uncertainty arising for the unpolarized structure function F_2^d is estimated to be at most 3 %, whereas the uncertainty due to R is as large as 14 % at lowest x and 2.7 % at highest x. Furthermore, the stability of the results under cut variations were studied and the data have been investigated for non-statistical fluctuations by division into sub-samples defined by relevant parameters like dead-time, beam current and time periods. No additional systematic effects could be detected beyond the statistical accuracy of the data.

The HERMES result is the most precise available measurement of g_1^d in the $0.002 < x < 0.85$ and $0.1 < Q^2 < 20$ GeV^2 kinematic range (Fig. 2): it will allow the extraction of g_1^n and an accurate test of the Bjorken sum rule.

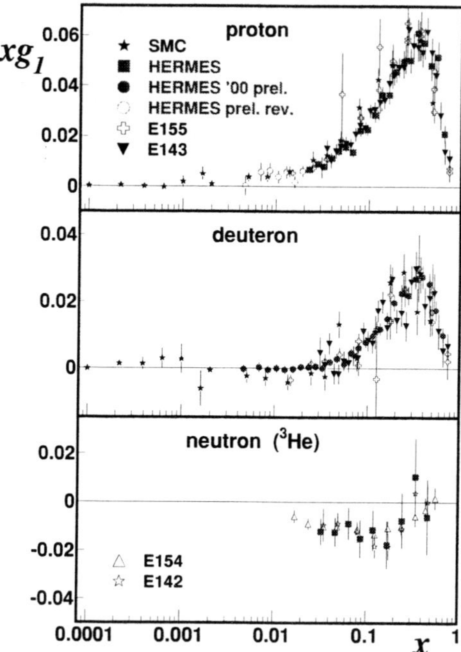

FIGURE 2. World data on xg_1 [9, 10]. Data are shown at the measured Q^2 and the plotted errors are the quadratic sum of systematic and statistical uncertainties.

REFERENCES

1. D. P. Barber *et al*, PLB, 343, 436, 1995
   M. Beckmann *et al*, NIMA, 479, 334, 2001
2. a schematic rapresentation of the HERMES target is given in P. Lenisa *The HERMES Polarized Internal Target*, Proceedings of PST2001, International Workshop on Polarized Beam and Targets, Nashville-Indiana, (2001).
3. K. Ackerstaff *et al*, NIMA, 417, 230, 1998
4. I. V. Akushevich *et al*, JPG, 20, 513, 1994
5. S. Wandzura *et al*, PLB, 72, 195, 1977
6. L. W. Whitlow *et al*, PLB, 250, 193, 1990
7. ALL97 parameterization: H. Abramowicz *et al*, hep-ph 9712415
8. NMC Coll., P. Amaudruz *et al*, NPB, 371, 3, 1992
9. SMC Coll., B. Adeva *et al*, PRD, 60, 072004, 1999
   E143 Coll., K. Abe *et al*, PRD, 58, 112003, 1998
   E155 Coll., P. L. Antony *et al*, PLB, 463, 339, 1999
10. Hermes Coll., K. Ackerstaff *et al*, PLB, 404, 383, 1997
    Hermes Coll., A. Airapetian *et al*, PLB, 442, 484, 1998

A First Measurement of the Tensor-Polarized Structure Function b_1^d

Marco Contalbrigo

(on behalf of the HERMES Collaboration)

INFN - Sezione di Ferrara e Dipartimento di Fisica dell'Università di Ferrara
Via del Paradiso 12, 44100 Ferrara, ITALIA

Abstract. The HERMES experiment studies the spin structure of the nucleon using the 27.6 GeV longitudinally polarized positron beam of HERA and an internal target of pure gases. In addition to the well-known spin structure function g_1, measured precisely with longitudinally polarized proton and deuteron targets, the use of a tensor-polarized deuteron target provides access to the tensor polarized structure function b_1^d. The latter, measured with an unpolarized beam, quantifies the dependence of the parton momentum distribution on the nucleon spin. HERMES had a 1-month dedicated run with a tensor polarized deuterium target during the 2000 data taking period. Here preliminary results on the tensor-polarized structure function b_1^d are presented for the kinematic range $0.002 < x < 0.85$ and $0.1\ \mathrm{GeV}^2 < Q^2 < 20\ \mathrm{GeV}^2$.

INTRODUCTION

The HERMES experiment [1] has been designed to measure the nucleon spin structure functions from deep inelastic scattering (DIS) of polarized positrons and electrons from polarized gaseous targets (H, D, ^{3}He). Three of the main leading-twist structure functions are listed in the table below, along with their interpretation in the Quark-Parton Model. The sums are over quark and antiquark flavours q and the dependences on Q^2 and Bjorken x are omitted for simplicity:

	Proton	Deuteron
F_1	$\frac{1}{2}\sum_q e_q^2[q^+ + q^-]$	$\frac{1}{3}\sum_q e_q^2[q^+ + q^- + q^0]$
g_1	$\frac{1}{2}\sum_q e_q^2[q^+ - q^-]$	$\frac{1}{2}\sum_q e_q^2[q^+ - q^-]$
b_1	$--$	$\frac{1}{2}\sum_q e_q^2\left[2q^0 - (q^- + q^+)\right]$

The unpolarized structure function F_1 measures the quark momentum distribution summed over all the possible helicity states. The polarized structure function g_1 is sensitive to the spin structure of the nucleon, and measures the imbalance of quarks with the same (q^+) or opposite (q^-) helicity with respect to the nucleon they belong to. For targets of spin 1 such as the deuteron, the tensor structure function b_1 compares the quark momentum distribution between the zero-helicity state of the hadron (q^0) and the average of the helicity-1 states ($q^+ + q^-$). As the deuteron is a weakly-bound state of spin-half nucleons, b_1^d was initially predicted to be small [2]. More recently, coherent double scattering models have predicted a sizable b_1^d at low x [3, 4, 5], violating the sum rule which suggests a vanishing first moment of b_1^d [6]. Although b_1^d describes basic

CP675, *Spin 2002: 15th Int'l. Spin Physics Symposium and Workshop on Polarized Electron Sources and Polarimeters*, edited by Y. I. Makdisi, A. U. Luccio, and W. W. MacKay

properties of the spin-1 deuterium nucleus, and may affect the experimental determination of g_1^d, it has not yet been measured. In 2000, HERMES collected a dedicated data set with a tensor polarized deuterium target for the purpose of making a first measurement of b_1^d. Preliminary results from these data are presented in this paper.

HERMES SETUP

The HERMES experiment is installed in the HERA ring where the beam positrons self-polarize by emission of synchrotron radiation (Sokolov-Ternov effect) along the direction of the bending magnetic field. Longitudinal beam polarization, needed for the g_1^d measurement, is obtained with two spin rotators placed immediately upstream and downstream of the HERMES apparatus. As b_1^d appears in the symmetric part of the hadronic tensor [2], its measurement neither depends on the beam polarization nor is diluted by the depolarization of the virtual photon. At HERMES , an "unpolarized beam" is achieved by grouping together data with opposite beam polarization.

A unique feature of the HERMES experiment is its tensor-polarizable gaseous target [7]. Deuterium has a total of six hyperfine states depending on the relative orientation of the electron and nuclear spins. An atomic beam source (ABS) generates a Deuterium atomic beam and selects the two hyperfine states with the desired nuclear polarization. A cylindrical 40 cm long, 75 μm thick Al tube (target cell) confines the polarized gas along the positron beam line, where a longitudinal magnetic field provides the quantization axis for the nuclear spin. Every 90 seconds, a diagnostic system measures the atomic and molecular abundances and the atomic polarization inside the cell, then the polarization of the injected gas is changed. The vector V and tensor T atomic polarizations are

$$V = \frac{n^+ - n^-}{n^+ + n^- + n^0} \qquad T = \frac{n^+ + n^- - 2n^0}{n^+ + n^- + n^0}, \tag{1}$$

where n^+, n^-, n^0 are the atomic populations inside the cell with positive, negative and zero spin projection onto the beam axis. In 2000 the running conditions were very stable, the recombination on the cell surface was negligible, and no correction was needed for the residual polarization of the recombined atoms. An average tensor polarization greater than 80 % was obtained with a residual vector polarization of only 1 %. In contrast to the solid targets used by other experiments, the HERMES target does not suffer from radiation damage thanks to the continuous flow of the target gas. Its polarization can be continuously measured, rapidly reversed, and is not diluted by non-polarizable material.

The HERMES detector is a forward spectrometer with a dipole magnet providing a field integral of 1.3 T·m. A horizontal iron plate shields the HERA beam lines from the field, thus dividing the spectrometer into two identical halves with a minimum vertical acceptance of ± 40 mrad. The acceptance extends to ± 140 mrad vertically and to ± 170 mrad horizontally. In this analysis 36 drift chamber planes in each detector half were used for tracking. Positron identification is accomplished using a probability method based on signals of three subsystems: a lead-glass block calorimeter, a transition-radiation detector, and a preshower hodoscope. For positrons in the momentum range of 2.5 to 27 GeV, the identification efficiency exceeds 98 % with a negligible hadron contamination, the average polar angle resolution is 0.6 mrad, and the average momentum resolution is 1 – 2 %.

MEASUREMENT

Depending on the beam (P_B) and target (V and T) polarizations, the lepton-nucleon DIS cross section measured by the experiment is sensitive to the vector A_1 (A_2 is here neglected) and tensor A_{zz} asymmetries of the virtual photon nucleon cross section

$$\sigma_{\rm meas} = \sigma_U \left[1 + P_B D_\gamma V A_1 + \frac{1}{2} T A_{zz}\right] \qquad \begin{array}{cccc} & P_B & V & T \\ \sigma^+ & 1 & 1 & 1 \\ \sigma^- & 1 & -1 & 1 \\ \sigma^0 & 0 & 0 & -2 \end{array} \tag{2}$$

σ_U is the unpolarized cross section $(\sigma^+ + \sigma^- + \sigma^0)/3$ and D_γ is polarization transfer from the lepton beam to the virtual photon. Here and in the following no higher-twist contribution (g_2, b_3, b_4) is considered. The vector asymmetry A_1 is related to g_1^d; it is measured with a polarized beam ($P_B = 1$) by comparing the cross sections for a target with the same σ^+ and opposite σ^- helicity as the beam:

$$A_1 = \frac{\sigma^+ - \sigma^-}{2\sigma_U D_\gamma} = \frac{g_1}{F_1} = A_1^{\rm meas}\left[1 + \frac{1}{2} T A_{zz}\right]. \tag{3}$$

Here $A_1^{\rm meas}$ is the asymmetry typically measured by experiments, where $(\sigma^+ + \sigma^-)/2$ is used rather than σ_U. The tensor term, until now neglected in g_1^d measurements, is needed to ensure that the denominator is proportional to σ_U and so to F_1. The tensor asymmetry A_{zz}, and the related b_1^d, provide the missing information on the difference between the cross section for the zero-helicity target state and the spin-averaged states of helicity 1:

$$A_{zz} = \frac{(\sigma^+ + \sigma^-) - 2\sigma^0}{3\sigma_U} = -\frac{2}{3}\frac{b_1}{F_1}. \tag{4}$$

At HERMES , the target residual vector polarization in the σ^0 case and in the case of the sum $\sigma^+ + \sigma^-$ is very small and has a negligible effect on the A_{zz} measurement. The vector contribution is further reduced by combining together data with opposite beam helicities. The kinematic range accessible for the b_1^d measurement is limited to the intervals $0.002 < x < 0.85$ and $0.1 < y < 0.91$ due to detector acceptance, resolution, and trigger requirements. Inclusive deep-inelastic events are selected by the requirements: $Q^2 > 0.1$ GeV^2 and $W^2 > 3.24$ GeV^2 . The total x-range is divided into 6 bins. For each x-bin the following yield asymmetry is calculated:

$$A_{zz} = \frac{1}{T} \cdot \left[\frac{(N/L)^+ + (N/L)^- - 2\cdot(N/L)^0}{(N/L)^+ + (N/L)^- + (N/L)^0}\right]. \tag{5}$$

The numbers of events selected (N) are corrected per spin state for the background arising from charge symmetric processes. The corresponding luminosities (L) used for normalization are measured from Bhabha scattering off the target gas electrons. The radiative corrections on A_{zz} are calculated using POLRAD [8]. For this preliminary result the DIS radiative tail is neglected due to the small size of A_{zz} and the elastic radiative tail is estimated from a not fully updated parameterization of the deuteron quadrupole

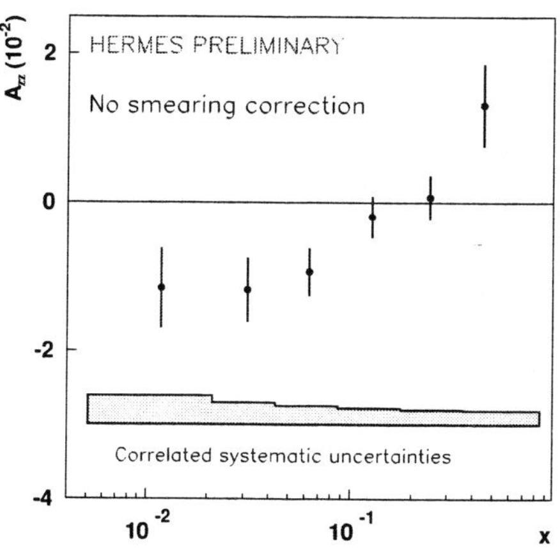

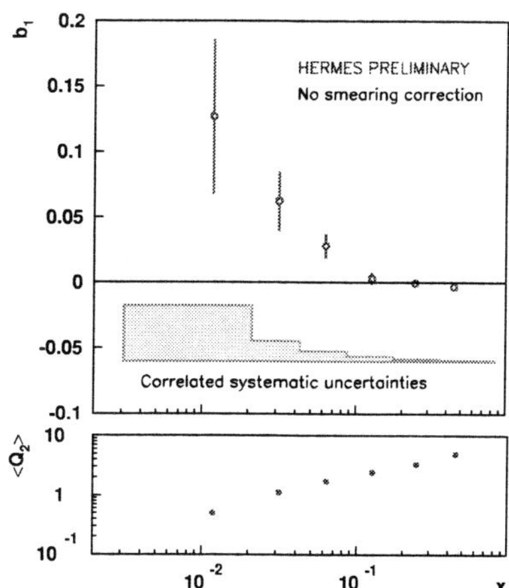

FIGURE 1. The tensor asymmetry A_{zz} (left) and the tensor-polarized structure function b_1^d (right) as measured by HERMES . The error bars are statistical only and the shaded bands show the estimated systematic uncertainties. The bottom panel on the right plot shows the average Q^2 of the measurements.

form factor [9]. The statistical error was enlarged by almost a factor of 2 at low x to account for radiative background. The measured A_{zz} is presented in Fig. 1. The systematic uncertainties are correlated over the kinematical bins, being dominated by the target density normalization between different injection modes of the ABS. The target polarization measurement, the misalignment of the spectrometer, and the hadron contamination give negligible systematic effects. Data with different beam polarizations were checked to give compatible A_{zz} results. No systematic error has been estimated from the still incomplete radiative correction. A_{zz} is found to be less than 2 %. From this result, the influence of the tensor asymmetry bias on a g_1^d measurement is estimated to be less than 0.5 – 1.0 %.

The spin structure function b_1^d is extracted from the tensor asymmetry via the relation $b_1^d = -\frac{3}{2}A_{zz}\frac{(1+\gamma^2)F_2^d}{2x(1+R)}$, where the structure function F_1^d has been expressed in terms of the ratio $R = \sigma_L/\sigma_T$ [10] and the structure function F_2^d (γ is a kinematic factor). $F_2^d = F_2^p(1+F_2^n/F_2^p)$ is calculated using parameterizations for F_2^p [11] and F_2^n/F_2^p [12]. Fig. 1 displays the result for b_1^d, which is small but different from zero. The structure function b_2^d has also been extracted, using the Callan-Gross relation $b_2^d = \frac{2x(1+R)}{(1+\gamma^2)}b_1^d$. The data indicate a rise at low x as predicted by the most recent theoretical models [3, 4, 5]. A comparison of the measured b_2^d with the prediction of one of the models [5] is given in Fig. 2.

In conclusion HERMES has provided the first direct measurement of the structure function b_1^d in the kinematic range $0.002 < x < 0.85$ and $0.1\ \mathrm{GeV}^2 < Q^2 < 20\ \mathrm{GeV}^2$. The preliminary result for the tensor asymmetry is sufficiently small to produce an effect of

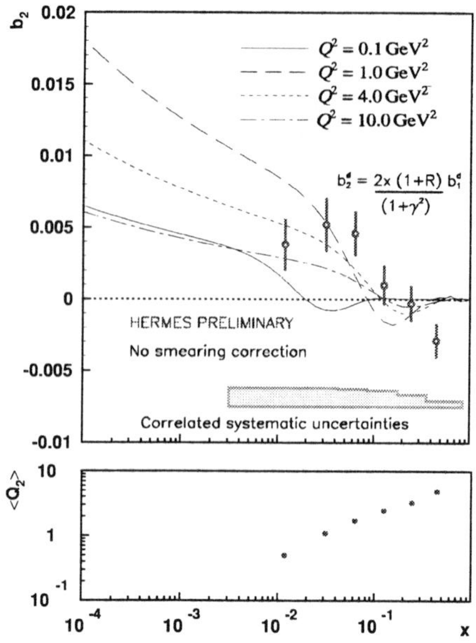

FIGURE 2. The tensor-polarized structure function b_2^d. The error bars are statistical only and the shaded bands show the estimated systematic uncertainties. The bottom panel shows the average Q^2 of the measurements. The curves are from calculations within the re-scattering model of Ref. [5] for Q^2 values in the range of the HERMES measurement.

less than 1 % on the measurement of g_1^d. The dependence of b_1^d on Bjorken x is in qualitative agreement with expectations based on coherent double scattering models [3, 4, 5] and favors a sizeable b_1^d at low-x. This suggests a significant tensor polarization of the sea-quarks, violating the Close-Kumano sum rule [6]. This observation is analogous to the well-established $\bar{u} - \bar{d}$ asymmetry of the sea and the consequent violation of the Gottfried sum rule [13].

REFERENCES

1. HERMES Coll., K. Ackerstaff *et al*, *Nucl. Instrum. Methods* A **417**, 230 (1998)
2. P. Hoodbhoy *et al*, *Nucl. Phys.* B **312**, 571 (1989), H. Khan *et al*, *Phys. Rev.* C **44**, 1219 (1991)
3. N. N. Nikolaev *et al*, *Phys. Lett.* B **398**, 245 (1997)
4. J. Edelmann *et al*, *Phys. Rev.* C **57**, 254 (1998)
5. K. Bora *et al*, *Phys. Rev.* D **57**, 6906 (1998)
6. F. E. Close *et al*, *Phys. Rev.* D **42**, 2377 (1990)
7. C. Baumgarten *et al*, *Nucl. Instrum. Methods* A **482**, 606 (2002)
8. I. V. Akushevich *et al*, *J. Phys.* G **20**, 513 (1994)
9. Kobushkin *et al*, *Phys. At. Nucl.* G **58**, 1477 (1995)
10. L. W. Whitlow *et al*, *Phys. Lett.* B **250**, 193 (1990)
11. ALLM97 parameterization: H. Abramowicz *et al*, hep-ph 9712415
12. NMC Coll., P. Amaudruz *et al*, *Nucl. Phys.* B **371**, 3 (1992)
13. S. Kumano, *Phys. Rept.* **303**, 183 (1998)

The Spin Structure Function g_2

Stephen Rock for the Real Photon Collaboration

University of Mass, Amherst MA 01003

Abstract. We have measured the spin structure functions g_2^p and g_2^d over the kinematic range $0.02 \leq x \leq 0.8$ and $0.7 \leq Q^2 \leq 20$ GeV^2 by scattering 29.1 and 32.3 GeV longitudinally polarized electrons from transversely polarized NH_3 and ^{6}LiD targets. Our measured g_2 approximately follows the twist-2 Wandzura-Wilczek calculation. The twist-3 reduced matrix elements d_2^p and d_2^n are less than two standard deviations from zero. The data are inconsistent with the Burkhardt-Cottingham sum rule. The Efremov-Leader-Teryaev integral is consistent with zero within our measured kinematic range.

The deep-inelastic spin structure functions of the nucleons, $g_1(x,Q^2)$ and $g_2(x,Q^2)$, depend on the spin distribution of the partons and their correlations. The function g_1 can be primarily understood in terms of the quark parton model (QPM) and perturbative QCD with higher twist terms at low Q^2. The function g_2 is of particular interest since it has contributions from quark-gluon correlations and other higher twist terms at leading order in Q^2 which cannot be described perturbatively. By interpreting g_2 using the operator product expansion (OPE) [1, 2], it is possible to study contributions to the nucleon spin structure beyond the simple QPM.

The structure function g_2 can be written [3]:

$$g_2(x,Q^2) = g_2^{WW}(x,Q^2) + \overline{g_2}(x,Q^2)$$

where

$$g_2^{WW}(x,Q^2) = -g_1(x,Q^2) + \int_x^1 \frac{g_1(y,Q^2)}{y} dy,$$

$$\overline{g_2}(x,Q^2) = -\int_x^1 \frac{\partial}{\partial y}\left(\frac{m}{M} h_T(y,Q^2) + \xi(y,Q^2)\right)\frac{dy}{y},$$

x is the Bjorken scaling variable and Q^2 is the absolute value of the virtual photon four-momentum squared. The twist-2 term g_2^{WW} was derived by Wandzura and Wilczek [4] and depends only on g_1. The function $h_T(x,Q^2)$ is an additional twist-2 contribution [3, 5] that depends on the transverse polarization density. The h_T contribution to $\overline{g_2}$ is suppressed by the ratio of the quark to nucleon masses m/M [5] and its effect is thus small for up and down quarks. The twist-3 part (ξ) comes from quark-gluon correlations. Low-precision measurements of g_2 exist for the proton and deuteron [6, 7, 8], as well as for the neutron [9, 10]. Here, we report new, precise measurements of g_2 for the proton and deuteron.

CP675, *Spin 2002: 15th Int'l. Spin Physics Symposium and Workshop on Polarized Electron Sources and Polarimeters,* edited by Y. I. Makdisi, A. U. Luccio, and W. W. MacKay

Electron beams with energies of 29.1 and 32.3 GeV and longitudinal polarization $P_b = (83.2 \pm 3.0)\%$ struck approximately transversely polarized NH_3 [12] (average polarization $< P_t >= 0.70$) or ^{6}LiD ($< P_t >= 0.22$) targets. The beam helicity was randomly chosen pulse by pulse. Scattered electrons were detected in three independent spectrometers centered at 2.75°, 5.5°, and 10.5°. The two small-angle spectrometers were the same as in SLAC E155 [11], while the large-angle spectrometer had additional hodoscopes and a more efficient pre-radiator shower counter. Further information on the experimental apparatus can be found in references [11, 12, 13]. The approximately equal amounts of data taken with the two beam energies and opposites signs of target polarization gave consistent results.

The measured asymmetry, $\tilde{A_\perp}$, differs from $A_\perp$ because the target polarizations were not exactly perpendicular to the beam line. We determined $\tilde{A_\perp}$ using:

$$\tilde{A_\perp} = \frac{1}{f_{RC}}\left[\frac{C_1}{fP_t}\left(\frac{A_{raw}}{P_b} - A_{EW}\right) + C_2\frac{\sigma_p}{\sigma_d}\tilde{A^p_\perp}\right] + A_{RC}$$

where A_{raw} is the measured counting rate asymmetry from the two beam helicities, including small corrections for pion and charge symmetric backgrounds, dead-time and tracking efficiency, and A_{EW} is the electroweak asymmetry. The target dilution factor, f, is the fraction of free polarizable protons (≈ 0.13) or deuterons (≈ 0.18). C_1 and C_2 are nuclear corrections. The quantities f_{RC} and A_{RC} are radiative corrections determined using a method similar to E143 [12]. The detailed results for $\tilde{A_\perp}$ are shown in Ref. [14]. The multiplicative uncertainties due to target and beam polarization and dilution factor combined are 5.1% (proton) and 6.2% (deuteron). are small compared to the statistical errors. We determined $g_2(x,Q^2)$ from $\tilde{A_\perp}$ (dominant contribution) and the previously measured g_1.

The data cover the kinematic range $0.02 \leq x \leq 0.8$ and $0.7 \leq Q^2 \leq 20$ GeV^2 with an average Q^2 of 5 GeV^2. Tables of the complete results are in Ref. [14]. Figure 1 (left) shows the values of xg_2 as a function of Q^2 for several values of x along with results from E143 [12] and E155 [8]. The systematic error on xg_2 is much smaller than the statistical error. The former includes the systematic errors on $\tilde{A_\perp}$, the 5% normalization uncertainty of g_1, the 2% uncertainty of F_2, and the systematic errors of R. The data approximately follow the Q^2 dependence of g_2^{WW} (solid curve), although for the proton, the data points are slightly lower than g_2^{WW} at low and intermediate x, and higher at high x. The predictions of Stratmann [15] are closer to the data.

We obtained values at the average Q^2 for each x bin by using the Q^2-dependence of g_2^{WW}. Figure 1 (right) show the averaged xg_2 of this experiment. The figure also has xg_2^{WW} calculated using our parameterization of g_1. The combined new data for p disagree with g_2^{WW} with a χ^2/dof of 3.1 for 10 degrees of freedom. For d the new data agree with g_2^{WW} with a χ^2/dof of 1.2 for 10 dof. The data for g_2^p are inconsistent with zero (χ^2/dof=15.5) while g_2^d differs from zero only at $x \sim 0.4$. Also shown in Fig. 1 (right) is the bag model calculation of Stratmann [15] which is in good agreement with the data, chiral soliton models calculations [16, 17] which are too negative at $x \sim 0.4$, and the bag model calculation of Song [5] which is in clear disagreement with the data.

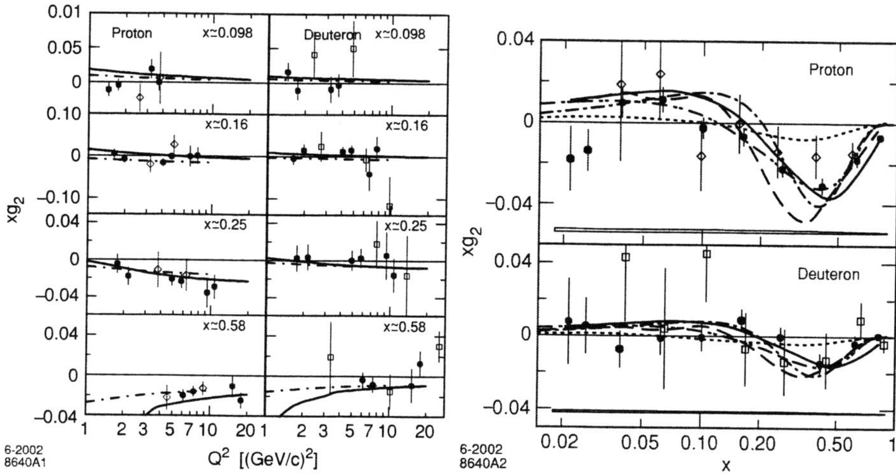

FIGURE 1. LEFT) xg_2^p and xg_2^d as a function of Q^2 for selected values of x from this experiment (solid), E143 [12] (open diamond) and E155 [8] (open square). Errors are statistical, the systematic errors are small. The curves show xg_2^{WW} (solid) and the bag model of Stratmann [15] (dash-dot).
RIGHT) The Q^2-averaged structure function xg_2 from this experiment (solid circle), E143 [7] (open diamond) and E155 [8] (open square). The errors are statistical; systematic errors are shown as the width of the bar at the bottom. Also shown is our twist-2 g_2^{WW} at the average Q^2 of this experiment at each value of x (solid line), the bag model calculations of Stratmann [15] (dash-dot-dot) and Song [5] (dot) and the chiral soliton models of Weigel and Gamberg [16] (dash dot) and Wakamatsu [17] (dash)

The OPE allows us to write the hadronic matrix element in deep-inelastic scattering in terms of a series of renormalized operators of increasing twist [1, 2]. The moments of g_1 and g_2 for even $n \geq 2$ at fixed Q^2 can be related to twist-3 reduced matrix elements, d_n, and higher-twist terms which are suppressed by powers of $1/Q$. Neglecting quark mass terms:

$$d_n = 2\frac{n+1}{n}\int_0^1 dx\, x^n \overline{g_2}(x, Q^2).$$

The matrix element d_n measures deviations of g_2 from the twist-2 g_2^{WW} term. Note that some authors [2, 18] define d_n with an additional factor of two. We calculated d_2 with the assumption that $\overline{g_2}$ is independent of Q^2 in the measured region. This is not unreasonable since d_2 depends only logarithmically on Q^2 [1]. The part of the integral for x below the measured region was assumed to be zero because of the x^2 suppression. For $x \geq 0.8$ we used $\overline{g_2} \propto (1-x)^m$ where m=2 or 3, normalized to the data for $x \geq 0.5$. Because $\overline{g_2}$ is small at high x, the contribution was negligible for both cases. We obtained values of d_2^p =0.0025±0.0016±0.0010 and d_2^d =0.0054±0.0023±0.0005 at an average Q^2 of 5 GeV^2. We combined these results with those from SLAC experiments on the neutron (E142 [9] and E154 [10]) and proton and deuteron (E143 [12] and E155 [8])

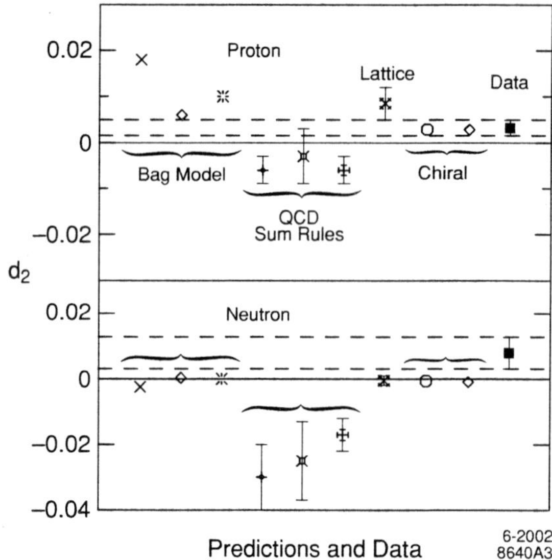

FIGURE 2. The twist-3 matrix element d_2 for the proton and neutron from the combined data from this and other SLAC experiments (E142 [9], E143 [12], E154 [10] and E155 [8] (DATA). The region between the dashed lines indicates the experimental errors. Also shown are theoretical model values from left to right: bag models [5, 15, 19], QCD Sum Rules [20, 21, 22], Lattice QCD [18] and chiral soliton models [16, 17].

to obtained average values d_2^p =0.0032±0.0017 and d_2^n =0.0079±0.0048. These are consistent with zero (no twist-3) to within two standard deviations. The values of the 2nd moments alone are: $\int_0^1 dx\, x^2 g_2(x,Q^2)$=−0.0072± 0.0005±0.0003 (p) and −0.0019± 0.0007±0.0001 (d).

Figure 2 shows the experimental values of d_2^p and d_2^n plotted along with theoretical models from left to right: bag models (Song [5], Stratmann [15], and Ji [19]); sum rules (Stein [20], BBK [21], Ehrnsperger [22]); chiral soliton models [16, 17]; and lattice QCD calculations ($Q^2 = 5$ GeV^2, $\beta = 6.4$) [18]. The lattice and chiral calculations are in good agreement with the proton data and two standard deviations below the neutron data. The sum rule calculations are significantly lower than the data. The Non Singlet combination, $3 \cdot (d_2^p - d_2^n) = -0.0141 \pm 0.0170$ is consistent with an instanton vacuum calculation of ~ 0.001 [23].

The Burkhardt-Cottingham (BC) sum rule [24] for g_2 at large Q^2, $\int_0^1 g_2(x)dx = 0$, was derived from virtual Compton scattering dispersion relations. It does not follow from the OPE since $n = 0$. Its validity depends on the lack of singularities for g_2 at $x = 0$, and a dramatic rise of g_2 at low x could invalidate the sum rule. We evaluated the BC integral in the measured region of $0.02 \leq x \leq 0.8$ at $Q^2 = 5$ GeV^2. The results for the proton and deuteron are −0.044±0.008±0.003 and −0.008±0.012±0.002 respectively. Averaging with the E143 and E155 results which cover a slightly more restrictive x range gives −0.042±0.008 and −0.006±0.011. This does not represent a conclusive test of the sum rule because the behavior of g_2 as $x \to 0$ is not known. However, if we assume that $g_2 = g_2^{WW}$ for $x < 0.02$, and use the relation $\int_0^x g_2^{WW}(y)dy = x\left[g_2^{WW}(x) + g_1(x)\right]$, there is an additional contribution of 0.020 (p) and 0.004 (d). This leaves a $\sim 2.8\sigma$ deviation from zero for the proton.

The Efremov-Leader-Teryaev (ELT) sum rule [25] involves the valence quark contributions to g_1 and g_2: $\int_0^1 x[g_1^V(x) + 2g_2^V(x)]dx = 0$. If the sea quarks are the same in protons and neutron this becomes $\int_0^1 x[g_1^p(x) + 2g_2^p(x) - g_1^n(x) - 2g_2^n(x)]dx = 0$. We evaluated this ELT integral in the measured region using the fit to g_1. The result at $Q^2 = 5$ GeV^2 is $-0.013\pm 0.008\pm0.002$, which is consistent with the expected value of zero. Including the data of E143 [12] and E155 [8] leads to -0.011 ± 0.008. The extrapolation to x=0 is not known, but is suppressed by a factor of x. The values of the 1st moments at $Q^2 = 5$ GeV^2 are: $\int_0^1 dx\, xg_2(x,Q^2)$=$-0.0157 \pm 0.0012 \pm 0.0005$ (p) and $-0.0037 \pm 0.0016 \pm 0.0002$ (d).

In summary, our results for g_2 follow approximately the twist-2 g_2^{WW} shape, but deviate significantly at some values of x. The twist-3 matrix elements d_2 are less than two standard deviations from zero. The data over the measured range are inconsistent with the BC sum rule and consistent with the ELT integral.

REFERENCES

1. E. Shuryak and A. Vainshtein, Nuc. Phys. B **201**, 141 (1982).
2. R. Jaffe and X. Ji, Phys. Rev. D **43**, 724 (1991).
3. J. L. Cortes, B. Pire and J. P. Ralston, Z. Phys. C **55**, 409 (1992).
4. S. Wandzura and F. Wilczek, Phys. Lett. B **72**, 195 (1977).
5. X. Song, Phys. Rev. D **54**, 1955 (1996).
6. SMC: D. Adams *et al.*, Phys. Lett. B **336**, 125 (1994); **396**, 338 (1997).
7. E143: K. Abe *et al.*, Phys. Rev. Lett. **76**, 587 (1996).
8. E155: P. Anthony *et al.*, Phys. Lett. B **458**, 529 (1999).
9. E142: P. Anthony *et al.* Phys. Rev. D **54**, 6620 (1996).
10. E154: K. Abe *et al.*, Phys. Lett. B **404**, 377 (1997).
11. E155: P. Anthony *et al.*, Phys. Lett. B **463**, 339 (1999); B **493**, 19 (2000).
12. E143: K. Abe *et al.*, Phys. Rev. D **58**, 112003 (1998).
13. E154: K. Abe *et al.*, Phys. Rev. Lett. **79**, 26 (1997).
14. P. Anthony *et al.*, SLAC-PUB-8813 (hep-ex/0204028).
15. M. Stratmann, Z. Phys. C **60**, 763 (1993).
16. H. Weigel and L. Gamberg, Nucl. Phys. A **680**, 48 (2000).
17. M. Wakamatsu, Phys. Lett. B **487**, 118 (2000).
18. M. Göckeler *et al.*, Phys. Rev. D **63**, 074506 (2001).
19. X. Ji and P. Unrau, Phys. Lett. B **333**, 228 (1994).
20. E. Stein *et al.*, Phys. Lett. B **343**, 369 (1995).
21. I. Balitsky, V. Braun and A. Kolesnichenko, Phys. Lett. B **242**, 245 (1990); **318**, 648 (1993) (Erratum).
22. B. Ehrnsperger and A. Schafer, Phys. Rev. D **52**, 2709 (1995).
23. J. Balla, M.V. Polyakov, and C. Weiss, Nucl. Phys. B **510**, 327 (1998).
24. H. Burkhardt and W. N. Cottingham, Ann. Phys. **56**, 453 (1970).
25. A. V. Efremov, O. V. Teryaev and E. Leader, Phys. Rev. D**55**, 4307 (1997).

Recent Measurements of Longitudinal and Transverse Unpolarized Structure Functions, and Their Impact on Spin Asymmetry Measurements [1]

C. E. Keppel

Hampton University / Jefferson Lab

Abstract. New measurements are presented of the separated longitudinal and transverse proton structure functions in the nucleon resonance region ($1 < W^2 < 4$ GeV^2), spanning the four-momentum transfer range $0.2 < Q^2 < 4.0$ (GeV/c)^2. The results are from Jefferson Lab experiment E94-110, which measured unpolarized inclusive electron-proton cross sections in Hall C for the purpose of performing Rosenbluth-type separations. Results of the analysis of the data are presented, as well as a discussion of their non-trivial relevance to spin asymmetry measurements.

MOTIVATION

The determination of the nucleon spin structure functions, g_1 and g_2, from electron spin asymmetry measurements requires knowledge of the longitudinal/transverse (L/T) separated unpolarized structure functions. At lower values of Q^2, the region of the nucleon resonances covers larger fractions of the Bjorken x range. Determining the small Q^2 behavior of the structure functions at large x, therefore, requires precision L/T measurements in the resonance region. High precision measurements of $R(x,Q^2) = \sigma_L/\sigma_T$, the ratio of longitudinal to transverse cross sections, have been available for over a decade in the deep inelastic scattering (DIS) regime. However, in the region of the resonances there is very little data on $R(x,Q^2)$ and, therefore, the longitudinal and transverse structure functions, $F_L(x,Q^2)$ and $F_1(x,Q^2)$. The small number of measurements that exist vary in the range ($-0.1 < R < 0.4$) and have typical errors of 100% or more. The world's R data prior to the recent E94-110 experiment at Jefferson Lab are plotted in figure 1, for all $Q^2 < 9$ $(GeV/c)^2$. This data set can provide neither knowledge of the resonance structure nor information on the Q^2 dependence of $R(x,Q^2)$ in this regime.

UNPOLARIZED STRUCTURE FUNCTIONS AND SPIN ASYMMETRIES

Both the virtual photon spin asymmetries, A_1 and A_2, and the spin structure functions can be determined from measurements of the parallel and perpendicular electron spin

[1] Work supported in part by a grant from the National Science Foundation.

CP675, *Spin 2002: 15th Int'l. Spin Physics Symposium and Workshop on Polarized Electron Sources and Polarimeters*, edited by Y. I. Makdisi, A. U. Luccio, and W. W. MacKay

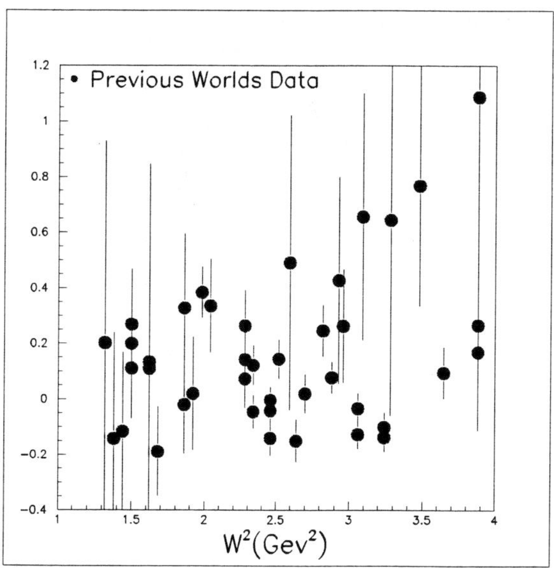

FIGURE 1. World's data on $R(W^2, Q^2)$ in the resonance region for all $Q^2 < 9\ (GeV/c)^2$ prior to E94-110.

asymmetries, $A_{||}$ and $A_{\perp}$. For the transverse asymmetry,

$$A_1 = \frac{C}{D}(A_{||} - dA_{\perp}). \tag{1}$$

The factor C is kinematic only, while the photon depolarization factor,

$$D = \frac{1-\varepsilon E'/E)}{1+\varepsilon R}, \tag{2}$$

is a function of the unpolarized structure function ratio R. ε is defined in the following section. For $\varepsilon R << 1$, the fractional uncertainty in A_1 coming from an uncertainty in R of δR is

$$\frac{\Delta A_1(\delta R)}{A_1} = \frac{\varepsilon}{1+\varepsilon R}\delta R \approx \varepsilon \delta R. \tag{3}$$

Using either the average of the previous world's resonance region data, $R = 0.06$, or an extrapolation of SLAC DIS fits, $R \approx 0.21$, both of which are used in the literature, results in a difference of $\delta R = 0.15$. At $\varepsilon = 0.5$ this leads to an uncertainty in A_1 of 9%.

The spin structure function g_1 can be extracted via

$$g_1 = \frac{F_1(A_1 - \gamma A_2)}{(1+\gamma^2)} \propto F_1(1+\varepsilon R) \propto \frac{F_2(1+\varepsilon R)}{1+R}. \tag{4}$$

Therefore, at $\varepsilon \approx 1$, only knowledge of F_2 is needed for the extraction of g_1. However, F_2 can not be measured independently from R except at $\varepsilon = 1$, and all previous extractions of F_2 in the resonance region have required an assumption for the value of R. The percent difference in F_2 extracted from cross section measurements assuming a value for R of either 0.2 or 0.0 (both choices, again, are used in the literature) varies from $\approx 2\%$ to $\approx 16\%$ at typical JLab resonance region kinematics. [10]

EXPERIMENT

In the one photon exchange approximation, the cross section for unpolarized inclusive electron-proton scattering can be expressed in terms of the helicity coupling between the photon and proton as

$$\frac{d\sigma}{d\Omega dE'} = \Gamma\left[\sigma_T(x,Q^2) + \varepsilon\sigma_L(x,Q^2)\right], \tag{5}$$

where σ_T and σ_L are the photo-absorption cross sections for pure transversely and longitudinally polarized photons, respectively, Γ is the flux of transverse virtual photons and ε is the photon relative longitudinal polarization. In terms of the structure functions $F_1(x,Q^2)$ and $F_L(x,Q^2)$, the double differential cross section can be written as

$$\frac{d\sigma}{d\Omega dE'} = \Gamma\frac{4\pi^2\alpha}{x(W^2 - M_p^2)}\left[2xF_1(x,Q^2) + \varepsilon F_L(x,Q^2)\right]. \tag{6}$$

Direct correspondence between equations 5 and 6 shows that $F_1(x,Q^2)$ is purely transverse, while the combination

$$F_L(x,Q^2) = \frac{1 + 4M_p^2x^2}{Q^2}F_2(x,Q^2) - 2xF_1(x,Q^2) \tag{7}$$

is purely longitudinal.

For E94-110, the separation of the measured differential cross section into longitudinal and transverse strengths was accomplished via the Rosenbluth technique [1]. Measurements were made over a range in ε at fixed x, Q^2 and $d\sigma/\Gamma$ was fit linearly in ε. The intercept of the fit gave σ_T (and therefore $F_1(x,Q^2)$), while the slope gave the structure function ratio $R(x,Q^2) = \sigma_L/\sigma_T = F_L(x,Q^2)/2xF_1(x,Q^2)$. These separations were performed at all Q^2, W^2, where enough range in ε existed to allow a good fits. A single Rosenbluth fit and the extracted value of R is shown in figure 2. In total, over 180 Rosenbluth fits were performed and the separated $F_1(x,Q^2)$ and $F_L(x,Q^2)$ structure functions were fit globally as a function of x and Q^2. This data allows, for the first time, both the resonance structure and the Q^2 dependence of the separated structure functions to be studied.

RESULTS AND CONCLUSIONS

The Rosenbluth extracted values (triangles) of $2xF_1$ and F_L are plotted in figure 3 as a function of x for various Q^2 bins and include $\approx 80\%$ of the data. Also plotted are the results of a two-dimensional fitting of the data in Q^2 and W^2 (solid line). A fit to DIS data

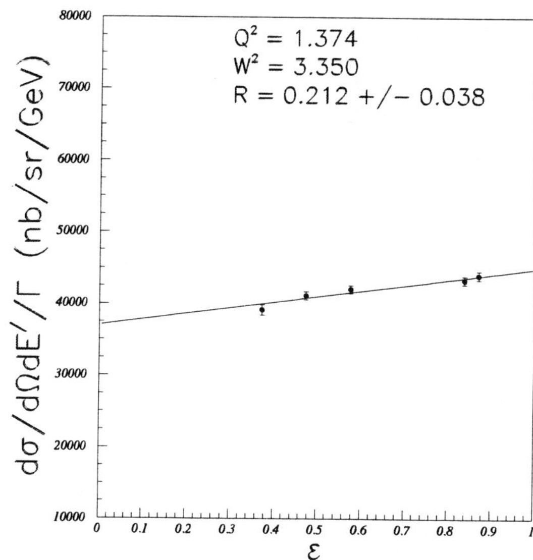

FIGURE 2. A example Rosenbluth separation, with the W^2, Q^2 and extracted value for the $R(W^2,Q^2)$ on the plot.

from SLAC [9] (dashed line) is seen to extend smoothly to the current data. Combining the DIS and resonance regimes will allow for the lower Q^2 moments of the separated structure functions to be extracted for the first time. Resonant structure can be seen for the first time in $R(W^2,Q^2)$, which is plotted versus W^2 in figure 4.

The data represent the first high precision measurement of the resonance region L/T separated unpolarized structure functions for the proton. The large kinematic range measured allows for a determination of the Q^2 dependence of individual resonance regions. The new data will allow for the systematic uncertainties in asymmetry measurements associated with uncertainties in the unpolarized structure functions to be reduced significantly.

REFERENCES

1. M. N. Rosenbluth, Phys. Rev. 79, 615 (1956)
2. F.W. Brasse *et al.*, Nucl. Phys. B110, 413 (1976)
3. L. W. Whitlow *et al.*, Phys. Lett B250, 193 (1990)
4. L.H. Tao, Ph.D. Thesis, The American University (1994)
5. C.E. Keppel, Ph.D. Thesis, The American University (1994)
6. I. Niculescu, R. Ent, C.E. Keppel, Phys. Rev. Lett. 85, 1186 (2000)
7. C.E. Carlson and N.C. Mukhopadhyay, Phys. Rev. D41, 2343 (1990)
8. C.E. Carlson and N.C. Mukhopadhyay, Phys. Rev. D47, 1737 (1993)
9. K. Abe *et al.*, Phys. Lett. B452, 194 (1999)
10. I. Niculescu. Ph.D. Thesis, Hampton University (1999)

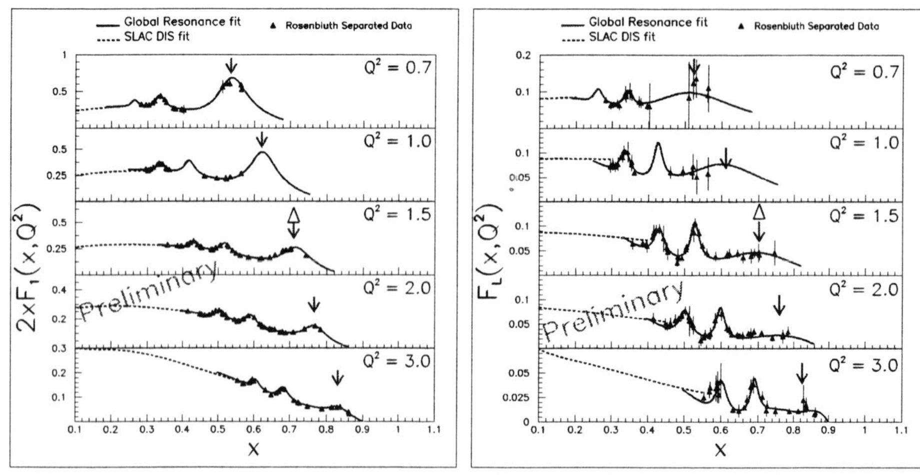

FIGURE 3. Extracted $2xF_1$ (Left), and F_L (Right), as a function of Bjorken x for various ranges in Q^2. The Rosenbluth separated data (blue triangles) are plotted with the full uncertainties (statistical + systematic). Also plotted are the results of a two-dimensional fit in Q^2 and W^2 to the data (solid curve). The position of the $\Delta P_{33}(1232)$ resonance is indicated by the red arrow.

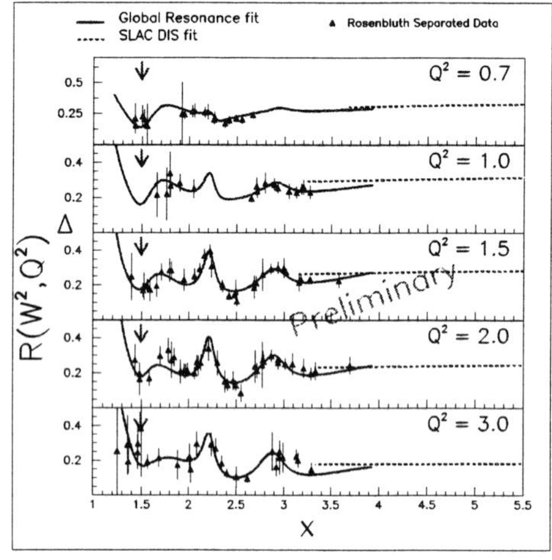

FIGURE 4. R as a function of W^2 for various ranges in Q^2. The Rosenbluth separated data (triangles) are plotted with the full uncertainties (statistical + systematic).

An Instability in the Matrix Solution of DGLAP Equations

Mehrdad Goshtasbpour and Ali Shafi'i

Dept. of Physics, Shahid Beheshti University, Evin 19834, Tehran, Iran

Abstract. Following, the matrix solution outlined by Ratcliffe [1], and using data points ($g_1^{p,n,d}$ or) $F_2^{p,n,d}(x_i, Q_{ij}^2)$ as the only parameters -thus doing away with the need for a set of free parameters used in a phenomenological fit to (polarized) parton distributions usually adopted (GRV, MRS, CTEQ, ...)- we arrive at a system of linear equations. Taking the unpolarized case, where there is plenty of data, and a possible combination of unknowns q_8, Σ, and g, where $q_3(x_i, Q_{i1}^2) = 3F_2^{(p-n)}(x_i, Q_{i1}^2)$ can be known, there remains three linear equations with three unknowns at every x_i, each decomposing one of the data points $F_2^p(x_i, Q_{ij}^2), j = 1,2,3$. The unknowns would be at (x_i, Q_{i1}^2), since the matrix solution of the nonsinglet and singlet DGLAP evolution equations can be used to determine the unknowns at $(x_i, Q_{ij}^2), j = 2,3$ in terms of those at (x_i, Q_{i1}^2).

The determinant of coefficients of these equations is too small to permit a stable solution within the range of the errors of F_2^p values. The significance of the problem and possible solutions for it are discussed.

INTRODUCTION

The simpler nonsinglet evolution equation:

$$\frac{dq(x,Q^2)}{d(\ln Q^2)} = \frac{\alpha_s(Q^2)}{2\pi} \int_x^1 \frac{dy}{y} P\left(\frac{x}{y}\right) q(y,Q^2), \tag{1}$$

where q represents the usual q_3 or q_8 the triplet and the octet quark parton distributions and P the respective splitting function, and the more complex ones for singlet distribution Σ and for g are familiar. The mentioned method by matrix solution consists primarily of discretizing the Bjorken x variable, via x-bins, and solving the resulting matrix equation exactly as a function of Q^2. Thus, parameterization can use available data points in x-bins as the only parameters, doing away with free parameters used in other phenomenological fits to the data.

A numeric code for the matrix solution was prepared and satisfactorily tested, separately for the nonsinglet and the singlet, in comparison to the available parameterized parton distributions, in particular GRV, as e.g, may be seen in the figures which show the LO evolution for the nonsinglet.

CP675, *Spin 2002: 15th Int'l. Spin Physics Symposium and Workshop on Polarized Electron Sources and Polarimeters,* edited by Y. I. Makdisi, A. U. Luccio, and W. W. MacKay

PROBLEM

Once the program code for evolution of the nonsinglet and singlet, q_3, q_8, Σ, and g, at LO or NLO, is well solved, one is ready to turn to data points ($g_1^{p,n,d}$ or) $F_2^{p,n,d}(x_i, Q_{ij}^2)$. Now, the problem to solve is the decomposition of each, e.g, $F_2^p(x_i, Q_{ij}^2)$ in terms of its nonsinglet and singlet components, via a system of equations, which turn out to be linear. After solving these equations, evolution and recomposition into evolved data points is not a difficult matter.

Using

$$q_3(x_i, Q_{i1}^2) = (3/x_i)F_2^{(p-n)}(x_i, Q_{i1}^2) = (6/x_i)(F_2^p(x_i, Q_{i1}^2) - F_2^d(x_i, Q_{i1}^2)/(1-w_D)), \quad (2)$$

where w_D is the probability of D-state for deutron (absorbing its coefficient), q_3 can be calculated. Thus, eventually, a decomposition can be made to leave us with three linear equations, each decomposing one of the data points $F_2^p(x_i, Q_{ij}^2)$, $j = 1,2,3$. At this stage, the matrix solution of nonsinglet and singlet DGLAP evolution equations is already used to allow us to have only three unknowns q_8, Σ, and g at (x_i, Q_{i1}^2).

The determinant of coefficients of these equations is too small, to permit a stable solution within the range of the errors of F_2^p values. A closer look indicates that part of the problem arises due to similarity of two columns of the determinant as q_8 and Σ have very close values in the large and middle x (or small $s(x)$) range where the evolution begins.

As the difference of q_8 and Σ depends on a very small strange distribution s, one would be tempted to try a change of variables to s and v, where $q_8 = v - 2s$ and $\Sigma = v + s$, while fixing or giving the small s values. Thus, dropping one equation and one unknown. We note that this procedure fixes the problem of divergence; as it can be clearly checked numerically, when, e.g, s values are provided from known GRV distributions, while working with real data otherwise. Lets see if there is a more fundamental way of rendering known the value for one of the three variables, namely s. In other words, reducing the number of variables and equations.

Using the data points $F_2^d(x_i, Q_{i1}^2)$, we have:

$$q_8 = 6v - 18(1/x_i)F_2^d/(1-w_D) + 2(\alpha_s/3\pi)c_g * g, \quad (3)$$

and

$$\Sigma = -3v/2 + 9(1/x_i)F_2^d/(1-w_D) - (\alpha_s/3\pi)c_g * g, \quad (4)$$

where * means: convolution and c_g are defined in the usual way; furthermore, both equations are meant to be at (x_i, Q_{i1}^2).

Thus, using these replacements, we reduce the number of unknowns to two, v and g at (x_i, Q_{i1}^2). Note that now we have used an overall of three data points, $F_2^p(x_i, Q_{ij}^2)$, $j = 1,2$ and $F_2^d(x_i, Q_{i1}^2)$, which can not determine all of the four independent original unknowns q_3, q_8, Σ, and g, all at (x_i, Q_{i1}^2). Indeed, the two linear equations of decomposition of $F_2^p(x_i, Q_{ij}^2)$, $j = 1,2$ are of the form:

$$(1/x_i)F_2^p(x_i, Q_{i1}^2) = (1/6)q_3 + (1/18)q_8 + (2/9)\Sigma + (\alpha_s/3\pi)C_g * g, \quad (5)$$

$$(1/x_i)F_2^p(x_i,Q_{i2}^2) = (A_3/6)q_3 + (A_8/18)q_8 + (2A_s/9)\Sigma + ((A_g)\alpha_s/3\pi)C_g * g, \quad (6)$$

where

$$f_k(x_i,Q_{i2}^2) = (A_k)f_k(x_i,Q_{i1}^2), f_k = q_3, q_8, \Sigma, g, \quad (7)$$

and A_k, where $k = 3,8,s,g$, are known coefficients determined via evolution equations. Thus, all four variables are kept at (x_i,Q_{i1}^2). Now, using (2), (3), (4), which incidently are not independent equations, there remains only one independent equation (6) in the two unknowns v and g at (x_i,Q_{i1}^2), and a second equation via another set of data points is needed.

It may be seen that the points $F_2^d(x_i,Q_{i2}^2)$ are not in a sense independent of a relatively simple nonsinglet evolution, and thus provide an exellent check for the solution to the nonsinglet evolution equation, as the evolution of q_3 defined through (2) together with $F_2^p(x_i,Q_{i2}^2)$ data points correspond to them.

To create the second equation, what remains are $F_2^p(x_i,Q_{i3}^2)$. Then the two independent equations are (6) and:

$$(1/x_i)F_2^p(x_i,Q_{i3}^2) = (B_3/6)q_3 + (B_8/18)q_8 + (2B_s/9)\Sigma + ((B_g)\alpha_s/3\pi)C_g * g, \quad (8)$$

These two together with (2) are exactly the equations that can be made for $F_2^d(x_i,Q_{ij}^2), j = 2,3$. Whether the problem can be solved in this manner is not yet clear.

REFERENCES

1. P. G. Ratcliffe, HEP-PH/0012376

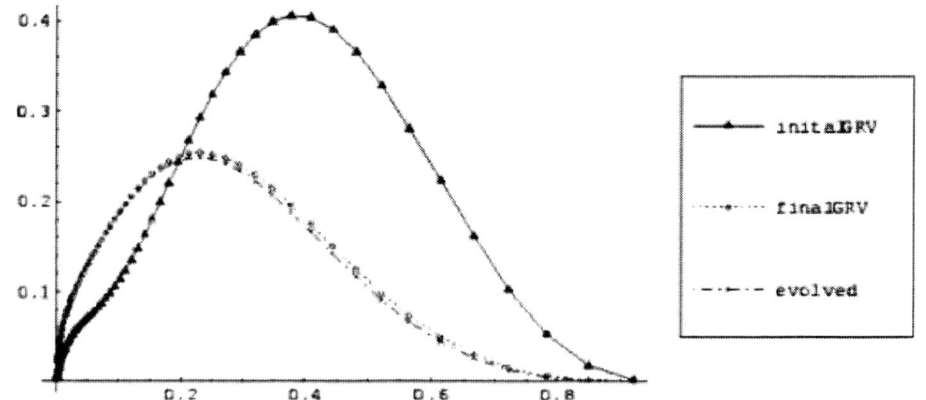

FIGURE 1. Our LO Evolution of q_3 of GRV from $Q^2 = .75GEV^2$ to $Q^2 = 5000GEV^2$ as compared to GRV itself

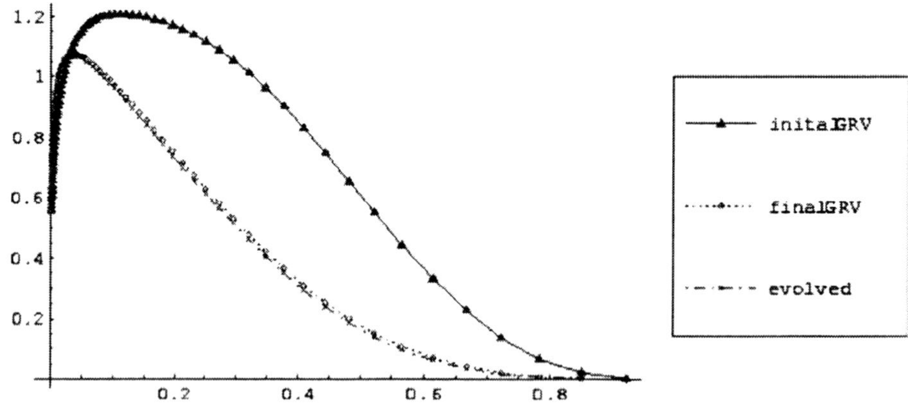

FIGURE 2. Evolution of q_8, similar to previous figure

Deeply–Virtual Compton Scattering on Deuterium and Neon at HERMES

F. Ellinghaus*[1]

(On behalf of the HERMES Collaboration)

*DESY Zeuthen, Platanenallee 6, 15738 Zeuthen, Germany

Abstract. We report the first observation of azimuthal beam–spin asymmetries in hard electroproduction of real photons off nuclei. Attributed to the interference between the Bethe–Heitler process and the deeply–virtual Compton scattering process, the asymmetry gives access to the latter at the amplitude level. This process appears to be the theoretically cleanest way to access generalized parton distributions. The data have been accumulated by the HERMES experiment at DESY, scattering the HERA 27.6 GeV positron beam off deuterium and neon gas targets.

INTRODUCTION

Hard scattering processes, such as inclusive deeply–inelastic scattering (DIS), semi–inclusive DIS and hard exclusive scattering have an important property in common, namely the possibility to separate the exactly calculable perturbative parts of the reaction from the non–perturbative parts. This factorization property is well established in the case of inclusive and semi–inclusive DIS and has been extensively used to investigate the structure of the nucleon. Only a few years ago, factorization theorems have been established for some hard exclusive reactions, where the produced particle is e.g. a photon [1, 2, 3]. Their description in the theoretical framework of generalized parton distributions (GPDs) [4, 5, 1] takes into account the dynamical correlations between partons of different momenta in the nucleon. The ordinary parton distribution functions and form factors turn out to be the limiting cases and moments of GPDs, respectively. Of particular interest is the second moment of two unpolarized quark GPDs, which for the first time offers a possibility to determine the total angular momentum carried by the quarks in the nucleon [5]. Recent theoretical ideas indicate that GDPs are able to describe correlations between the longitudinal and transverse structure of the nucleon [6, 7]. For the case of coherent hard exclusive processes on nuclei it was pointed out very recently [8] that information about the energy, pressure, and shear forces distributions inside nucleons and nuclei become accessible.

[1] E-mail: Frank.Ellinghaus@desy.de

CP675, *Spin 2002: 15th Int'l. Spin Physics Symposium and Workshop on Polarized Electron Sources and Polarimeters,* edited by Y. I. Makdisi, A. U. Luccio, and W. W. MacKay

0-7354-0136-5/03/$20.00

DEEPLY–VIRTUAL COMPTON SCATTERING

In deeply–virtual Compton scattering (DVCS) a high energetic virtual photon is absorbed by a parton inside the nucleon and a real photon is produced. This process is considered to be the theoretically cleanest way to access GPDs. The DVCS cross section can be obtained through a measurement of the exclusive photon production cross section after subtracting the background from the Bethe–Heitler (BH) process which has an identical final state and is calculable exactly in QED. First results on the DVCS cross section at high energies have been published recently by H1 [9] and ZEUS [10] at DESY. At the lower energies of HERMES at DESY and CLAS at Jlab, the DVCS cross section is expected to be much smaller than the BH cross section and thus a measurement with sufficient precision is not yet feasible. However, the DVCS amplitudes are directly accessible through the interference between the DVCS and BH processes. The leading–order and leading–twist interference term [11]

$$I = \pm \frac{4\sqrt{2}}{tQx_B} \frac{m_p e^6}{\sqrt{1-x_B}} \times \left[\cos\phi \frac{1}{\sqrt{\varepsilon(\varepsilon-1)}} \mathrm{Re}\tilde{M}^{1,1} - P_l \sin\phi \sqrt{\frac{1+\varepsilon}{\varepsilon}} \mathrm{Im}\tilde{M}^{1,1} \right] \quad (1)$$

depends on the charge and the polarization of the incident lepton, where +(-) in front of the expression denotes a negatively (positively) charged lepton with polarization P_l. Here m_p represents the proton mass, t the square of the four–momentum transfer to the target, $-Q^2$ the virtual–photon four–momentum squared, x_B the momentum fraction of the nucleon carried by the struck quark, and ε is the polarization parameter of the virtual photon. The azimuthal angle ϕ is defined as the angle between the lepton scattering plane, i.e. the plane defined by the incoming and the outgoing lepton trajectories, and the photon production plane made up by the virtual and real photons. The linear combination of DVCS amplitudes $\tilde{M}^{1,1}$ can be expressed as a linear combination of GPDs convoluted with hard scattering amplitudes.

Appropriate cross section asymmetries allow the separate access to the real and imaginary part of $\tilde{M}^{1,1}$. The beam–spin asymmetry (BSA)

$$A_{LU}(\phi) = \frac{1}{<|P_l|>} \frac{\overrightarrow{N}(\phi) - \overleftarrow{N}(\phi)}{\overrightarrow{N}(\phi) + \overleftarrow{N}(\phi)} \sim \sin\phi \,\mathrm{Im}\tilde{M}^{1,1} \quad (2)$$

is proportional to the imaginary part of $\tilde{M}^{1,1}$, where the average polarization of the beam is given by $<|P_l|>$ and $\overrightarrow{N}$ ($\overleftarrow{N}$) represents the normalized yield for positive (negative) beam helicity. The subscripts L and U denote a longitudinally polarized beam and an unpolarized target. Measurements of the BSA on the proton have already been carried out by HERMES [12] and CLAS [13]. The new preliminary result on the proton based on HERMES data collected in 2000 is shown in the left panel of figure 1. The asymmetry indeed shows the expected $\sin\phi$ modulation. Note that although the average kinematic values are slightly different compared to those from the already published BSA from the 1996/97 running period [12], the results are consistent with each other.

Recently, the first measurement of the beam–charge asymmetry, accessing the real part of the same combination of DVCS amplitudes, has been carried out at HERMES [14] via the scattering of positron and electron beams off an unpolarized hydrogen target.

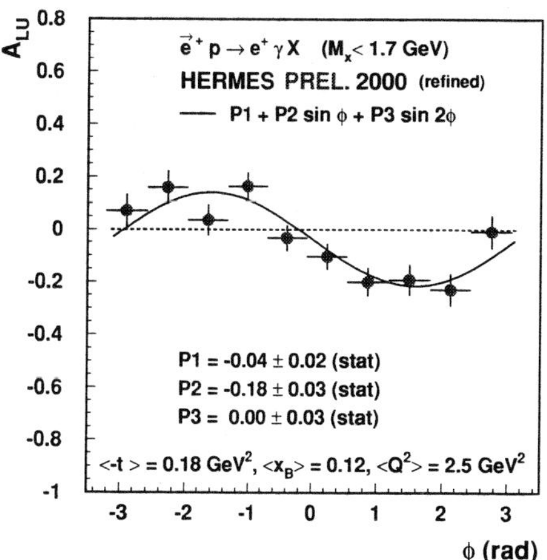

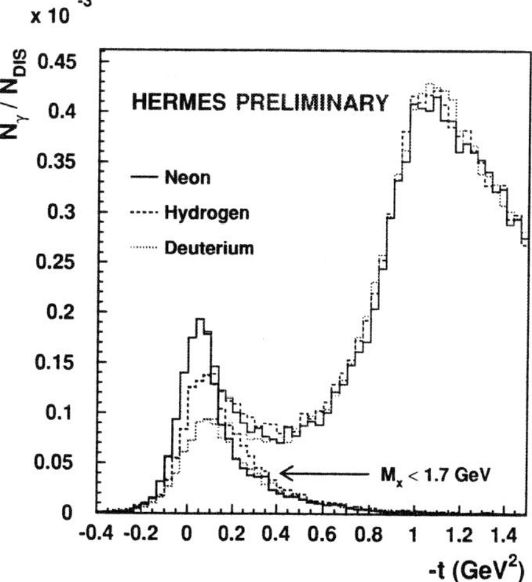

FIGURE 1. Left panel: Beam–spin asymmetry $A_{LU}(\phi)$ for the hard exclusive electroproduction of photons off the proton. Exclusive events are defined through the missing mass constrained $M_x < 1.7$ GeV. Right panel: $-t$ distribution of single-photon yields for neon, deuterium and hydrogen normalized to the number of DIS events. The exclusive events ($M_x < 1.7$ GeV) are shown separately.

DVCS ON NUCLEI

The data presented in the following have been accumulated by the HERMES experiment [15] at DESY during the 2000 running period. The HERA 27.6 GeV positron beam was scattered off polarized and unpolarized deuterium and unpolarized neon gas targets. Both the unpolarized and the spin–averaged polarized–target data have been used in this analysis. Selected events contained exactly one photon and one charged track identified as the scattered positron. The kinematical requirements were $Q^2 > 1$ GeV^2, $W^2 >$ 4 GeV^2 and $\nu < 23$ GeV. Here W denotes the photon–nucleon invariant mass and ν is the virtual–photon energy. The angle between the real and the virtual photon was required to be within 2 and 70 mrad. In the right panel of figure 1 the normalized yield of single photon events is shown versus $-t$ for neon and deuterium in comparison to hydrogen. Note that negative values of $-t$ appear due to the finite resolution of the spectrometer. Since the recoiling nucleus is not detected in the HERMES spectrometer, the missing mass $M_x = \sqrt{(q+P-k)^2}$ is calculated from q, P and k, the four–momenta of virtual photon, target nucleus and real photon, respectively. Note that for this analysis the target mass is set to the proton mass in order to keep the same $M_x < 1.7$ GeV definition for exclusive events regardless of the target. The cross section for the exclusive events is dominated by the BH contribution, i.e. when going from nucleon to nuclei it increases with the square of the charge diminished by the form factor squared. This explains the differences in the single–photon yield for the different targets at small values of $-t$, i.e. for exclusive events as shown in the right panel of figure 1.

In figure 2 the azimuthal dependences of the BSAs on deuterium and neon are shown

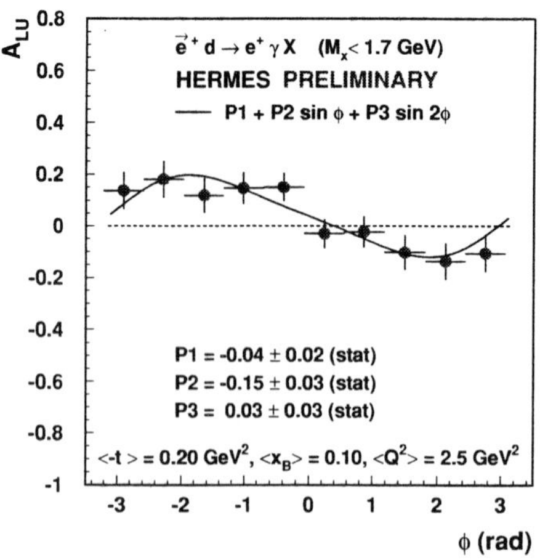

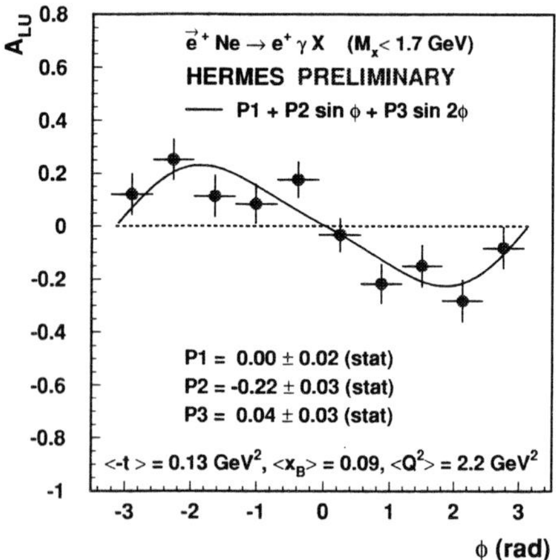

FIGURE 2. Beam–spin asymmetries $A_{LU}(\phi)$ for the hard electroproduction of photons off deuterium (left panel) and neon (right panel) for events with a missing mass $M_x < 1.7$ GeV.

for events with a missing mass M_x below 1.7 GeV. At the given average kinematics as indicated in the plots, the data exhibit the expected $\sin\phi$ behavior represented by the fit to the function $P_1 + P_2\sin(\phi) + P_3\sin(2\phi)$. Note that x_B is calculated using the proton mass, i.e. $x_B = Q^2/2m_p\nu$. As an independent method the $\sin\phi$–weighted moments

$$A_{LU}^{\sin\phi} = \frac{2}{\overrightarrow{N} + \overleftarrow{N}} \sum_{i=1}^{\overrightarrow{N}+\overleftarrow{N}} \frac{\sin\phi_i}{(P_l)_i} \tag{3}$$

are shown in figure 3 versus the missing mass M_x. The moments are non–zero only in the exclusive region and integrating $A_{LU}^{\sin\phi}$ up to $M_x < 1.7$ GeV yield the same results as the fits for the parameter P_2. Note, that negative values of the missing mass are again a consequence of the finite momentum resolution of the spectrometer, in that case $M_x = -\sqrt{-M_x^2}$ was defined. In contrast to the case of DVCS on the proton, DVCS on the deuteron has received little consideration in the literature [16, 17, 18]. The BSA on the deuteron is expected to be slightly smaller [17] than the one on the proton. This is in agreement with our results when comparing the left panels in figure 1 and figure 2, where the BSAs on the proton and the deuteron, achieved in a similar average kinematic region, amount to -0.18 ± 0.03 (stat) ± 0.03 (sys) and -0.15 ± 0.03 (stat) ± 0.03 (sys), respectively. However, present theoretical predictions assume $Q^2 \geq 4$ GeV^2 in order to avoid possible large target–mass corrections which have not yet been calculated for spin–1 targets. In addition, since at HERMES the recoiling nucleus is presently not detected, the ratio of coherent to incoherent production can not be inferred from the measurement. For nuclei heavier than the deuteron no predictions are available yet.

In summary, beam–spin asymmetries in the hard electroproduction of real photons off nuclei have been measured for the first time. Sizeable asymmetries of $-0.15 \pm$

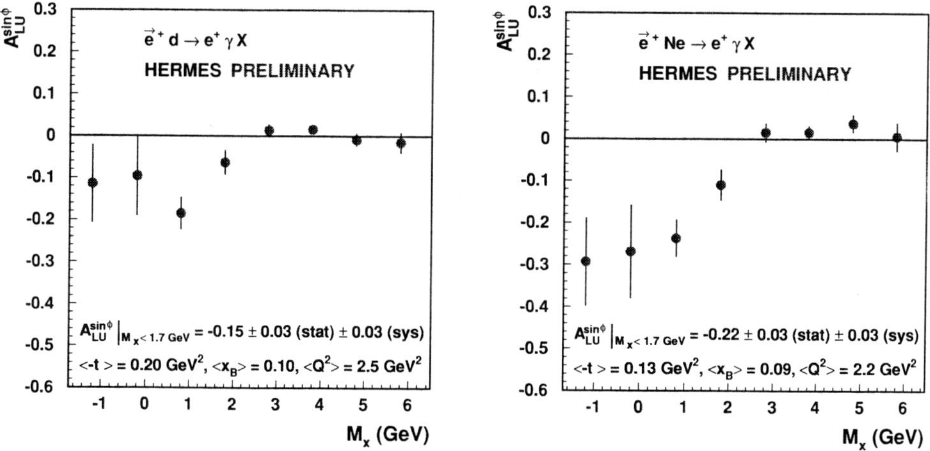

FIGURE 3. $\sin\phi$–weighted moments $A_{LU}^{\sin\phi}$ for the hard electroproduction of photons off deuterium (left panel) and neon (right panel) versus the missing mass M_x.

0.03 (stat) ± 0.03 (sys) and -0.22 ± 0.03 (stat) ± 0.03 (sys) have been found in the exclusive region for deuterium and neon, respectively. The corresponding asymmetry on the proton amounts to -0.18 ± 0.03 (stat) ± 0.03 (sys). This value is in agreement with the already published data from an earlier running period.

REFERENCES

1. A.V. Radyushkin, Phys. Rev. **D 56** (1997) 5524
2. X. Ji and J. Osborne, Phys. Rev. **D58** (1998)
3. J.C. Collins and A. Freund, Phys. Rev. **D 59** (1999) 074009
4. D. Müller et al., Fortschr. Phys. **42** (1994) 101
5. X. Ji, Phys. Rev. Lett. **78** (1997) 610
6. M. Burkardt, Phys. Rev. **D 62** (2000) 071503
7. M. Diehl, Eur. Phys. J. **C 25** (2002) 223
8. M.V. Polyakov, hep-ph/0210165
9. H1 collaboration, C. Adloff et al., Phys. Lett. **B 517** (2001) 47
10. P.R.B. Saull [for the ZEUS collaboration], Proc. of the International Europhysics Conference on High Energy Physics, Budapest, 2001
11. M. Diehl et al., Phys. Lett. **B 411** (1997) 193
12. HERMES collaboration, A. Airapetian et al., Phys. Rev. Lett. **87** (2001) 182001
13. CLAS collaboration, S. Stepanyan et al., Phys. Rev. Lett. **87** (2001) 182002
14. F. Ellinghaus [for the HERMES collaboration], Nucl. Phys. **A 711** (2002) 171
15. HERMES collaboration, K. Ackerstaff et al., Nucl. Instr. and Meth. **A 417** (1998) 230
16. E.R. Berger et al., Phys. Rev. Lett. **87** (2001) 142302
17. A. Kirchner and D. Müller, hep-ph/0202279
18. F. Cano and B. Pire, Nucl. Phys. **A 711** (2002) 133

Exclusive Electroproduction of Vector and Pseudoscalar Mesons at HERMES

Christian Schill
(for the HERMES collaboration)

INFN-Laboratori Nazionali di Frascati, 00044 Frascati, Italy

Abstract. The exclusive production of vector and pseudoscalar mesons in deep-inelastic lepton scattering gives a mean to access the recently introduced generalized parton distributions (GPDs). The GPDs provide a unified description of hadronic structure, which can be investigated with many different reactions. Recent results are presented on the exclusive electroproduction of π^+ mesons and on diffractive ρ^0 and ϕ production on the proton, collected by the HERMES collaboration at DESY. The experimental results for the ρ^0 and ϕ cross section are compared to calculations based on a description of the reactions in terms of GPDs. It is shown that quark exchange is the dominant production mechanism for ρ^0 production, while for ϕ production only gluon exchanges contributes. For exclusively produced π^+, an asymmetry in the azimuthal distribution around the virtual photon direction has been observed which depends on the orientation of the target spin.

INTRODUCTION

Recently, a growing interest has appeared for the generalized parton distributions (GPDs) of the nucleon, which provide a unified formalism for the description of inclusive deep-inelastic scattering, deeply-virtual compton scattering, exclusive meson production and electromagnetic form factor measurements [1]. In contrast to the usual parton distribution functions, which can be interpreted as probabilities to find a parton of a certain momentum fraction x and with a certain spin orientation in the nucleon, generalized parton distributions describe the correlation between two partons with different momenta $x+\xi$ and $x-\xi$.

There exist four chirally-even GPDs for each quark flavor q. The functions H^q and $\tilde{H}^q$ conserve the nucleon helicity, E^q and $\tilde{E}^q$ do not conserve it. In the forward limit, $\xi=0$ and $t=0$, the GPDs are equivalent to the ordinary parton densities. Here, t is the squared four-momentum transfer to the nucleon. However, the generalized parton distributions contain more information. It was shown that the second moment of the GPDs can be related to the total angular momentum of quarks and gluons in the nucleon [2]. Through their dependence on t, GPDs describe also transverse degrees of freedom in the nucleon.

The exclusive production of a vector or pseudoscalar meson M in deep-inelastic electron nucleon scattering is a way to experimentally gain information about GPDs:

$$e+p \longrightarrow e+p+M \tag{1}$$

CP675, *Spin 2002: 15th Int'l. Spin Physics Symposium and Workshop on Polarized Electron Sources and Polarimeters*, edited by Y. I. Makdisi, A. U. Luccio, and W. W. MacKay

The total cross section for this reaction can be factorized into a hard lepton scattering coefficient, which can be calculated in QCD, the wave function of the produced meson, and GPDs in quadratic combinations [3]. It turned out that the produced meson acts as a helicity filter. For the production of vector mesons the cross-section contains the unpolarized GPDs H^q and E^q, for pseudoscalar mesons the polarized ones $\tilde{H}^q$ and $\tilde{E}^q$.

So far, very few experimental data exist for exclusive reactions. The reasons are the small cross-sections involved and the high missing mass resolution required to ensure exclusivity.

In this contribution, experimental data are presented for the exclusive production of ρ^0, ϕ and π^+ mesons. These data have been collected by the HERMES experiment at DESY in Hamburg. A 27.5 GeV polarized electron or positron beam in the HERA storage ring at DESY is scattered off a longitudinally polarized or unpolarized hydrogen gas target. The HERMES forward spectrometer [4] features excellent particle identification capabilities. It can detect the scattered lepton and identifies pions, kaons and protons in a wide momentum range.

VECTOR MESON PRODUCTION

In this section data are presented on the above mentioned exclusive ρ^0 and ϕ production off the proton. The factorization of the cross-section has only been proven for scattering longitudinally polarized virtual photons [3]. In order to obtain information about the longitudinal part of the total cross section, the spin density matrix elements of the produced meson have been extracted from its decay angle distribution [5]. Assuming that the helicity of the virtual photon is entirely transferred to the produced meson (the so called *s-channel helicity conservation*), the longitudinal cross-section can be derived from the measured total one by using these matrix elements. The ratio of the longitudinal to the transverse part of the cross-section is given by: $R = \sigma_L/\sigma_T = r^{04}_{00}/\varepsilon(1-r^{04}_{00})$, where ε is the virtual-photon polarization parameter and r^{04}_{00} is one of the spin density matrix elements.

The results on the longitudinal cross section for ρ^0 and ϕ production [6] are shown in figure 1, compared with a model calculation in the framework of GPDs [7, 8]. The HERMES kinematic region covers an intermediate range of W between 4 and 6 GeV, where W is the photon-nucleon center-of-mass energy. Besides the dominant quark exchange, also gluon exchange needs to be considered in the calculation. In the case of ρ^0 production the gluon exchange contribution becomes relevant only at large Q^2 and W. As can be seen in figure 1 (left panel), the prediction for the ρ^0 cross section is in agreement with the data.

For the ϕ production cross section shown in figure 1 (right panel), only gluon exchange should contribute, since the proton contains only a small amount of s quarks. This is confirmed by the good agreement between data and the model calculation.

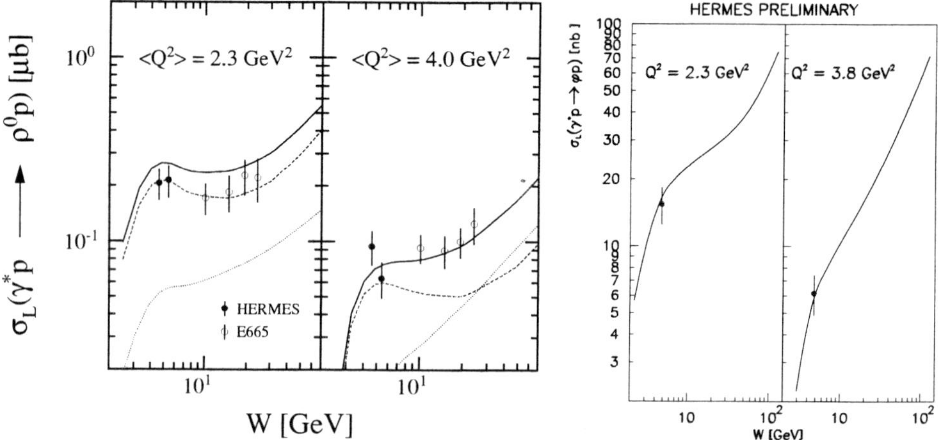

FIGURE 1. In the left panel the longitudinal cross section for exclusive ρ^0 production is compared to the results of a GPD-based calculation [7, 8]. The dotted curves represent the gluon-exchange contribution, the dashed curves the quark-exchange and the solid curves their sum. In the right panel, the longitudinal cross section for exclusive ϕ production is shown together with the results of a GPD-based calculation [7, 8], involving only gluon exchange.

PSEUDOSCALAR MESON PRODUCTION

Only a quadratic combination of GPDs appears in the unpolarized cross section for exclusive meson leptoproduction. Considering in addition the polarization degree of freedom helps to disentangle the various contributions by measuring further quantities. It has been shown that for exclusive π^+ production from a transversely polarized nucleon the interference between the two polarized GPDs $\tilde{H}^q$ and $\tilde{E}^q$ can lead to a large asymmetry in the distribution of the azimuthal angle ϕ [9, 10, 11]. Here, ϕ is the angle of the outgoing meson around the direction of the virtual photon with respect to the lepton scattering plane, as shown in figure 2.

At HERMES, exclusive meson production has been studied using a longitudinally polarized hydrogen target [12]. With respect to the virtual photon, however, the target polarization vector has a certain transverse component as well.

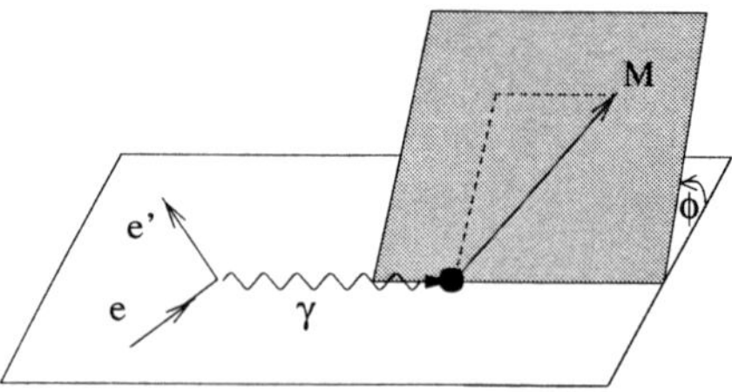

FIGURE 2. Definition of the azimuthal angle ϕ for exclusive leptoproduction of the meson M.

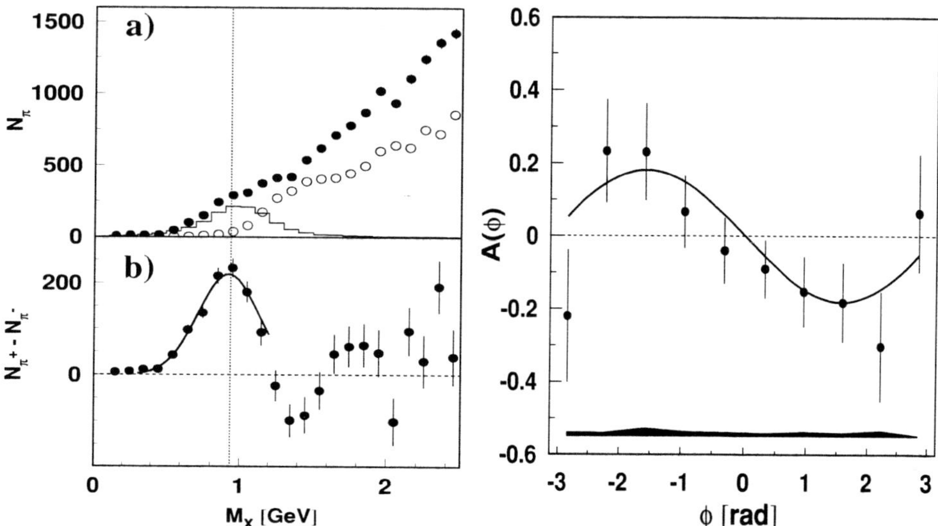

FIGURE 3. Left panel: a) Missing mass spectra for π^+ *(filled circles)* and π^- *(open circles)* electro-production on the proton [12]. The histogram is a Monte Carlo prediction for exclusive π^+ production. b) Difference between the π^+ and the normalized π^- distribution. The curve is a Gaussian fit to the data, the dotted line indicates the nucleon mass. **Right panel:** Longitudinal target-spin asymmetry for exclusive π^+ production. The curve is a fit to the data with the function $A(\phi) = A_{\mathrm{UL}}^{\sin\phi} \cdot \sin\phi$.

In the present setup, the identification of exclusive events is not possible on an event-by-event basis since the missing mass resolution of the spectrometer is limited to about 230 MeV for this process. However, the background of non-exclusive events can be subtracted from the data. As an estimate for the non-exclusive background for π^+ production, the π^- yield was used, since exclusive production of π^- is forbidden on the proton due to charge conservation. In the upper part of the left panel of figure 3, the π^+ and π^- missing mass spectra are displayed. In the lower part the difference of the π^+ and the normalized π^- spectrum is shown. A clear peak of exclusive π^+ production can be seen at the mass of the proton.

The ϕ-dependence of the cross section can be described by a cross section asymmetry defined as follows:

$$A(\phi) = \frac{1}{|P_t|} \frac{N^{\leftarrow}(\phi) - N^{\rightarrow}(\phi) - |P_t| A_{\mathrm{bg}}(\phi) N_{\mathrm{bg}}}{N^{\leftarrow}(\phi) + N^{\rightarrow}(\phi) - N_{\mathrm{bg}}}, \tag{2}$$

where $N^{\rightarrow}$ ($N^{\leftarrow}$) is the number of events with the direction of the target spin parallel (anti-parallel) to the electron beam momentum and P_t the target polarization. N_{bg} is the number of events of the non-exclusive background estimated from π^- production and A_{bg} is the non-exclusive background asymmetry of π^+, estimated for $1.3 < M_x < 2.0$ GeV. The HERMES data are displayed in the right panel of figure 3. The measured asymmetry shows a clear ϕ-dependence and can be parameterized by the function

$$A(\phi) = A_{\rm UL}^{\sin\phi} \cdot \sin\phi \tag{3}$$

with $A_{\rm UL}^{\sin\phi} = -0.18 \pm 0.05(\text{stat.}) \pm 0.01(\text{syst.})$.

For this experimental result there exist no theoretical predictions yet, due to the above mentioned two components of the target polarization vector in the center-of-mass frame. For a longitudinally polarized target the polarized cross section σ_S has the form

$$\sigma_S \propto [S_T \sigma_L + S_L \sigma_{LT}] A_{\rm UL}^{\sin\phi} \sin\phi \tag{4}$$

with contributions from the longitudinal (L) virtual photon amplitude and from the interference (LT) of the longitudinal and the transverse photon amplitude. The longitudinal S_L and transverse S_T components of the target polarization vector are determined in the photon-nucleon center-of-mass frame. For a quantitative model calculation of the measured asymmetry, an evaluation of the term $\sigma_{\rm LT}$ is necessary, which requires a next-to-leading twist calculation [11], that is not yet available. For a entirely transversely polarized target the second term in equation (4) vanishes and only leading twist amplitudes contribute [9, 10]. New measurements at HERMES with a transversely polarized hydrogen target, which have started in 2002, will provide a more direct access to GPDs.

SUMMARY AND OUTLOOK

In this contribution results from the HERMES experiment for the exclusive electroproduction of ρ^0, ϕ and π^+ mesons on the proton have been presented. Some of these results show a good agreement with model calculations performed in the framework of generalized parton distributions. To obtain more information on these functions data on many different exclusive reactions will be needed. In 2004, HERMES will continue measuring exclusive reactions with the help of a new recoil detector, which will allow better identification of exclusive reactions [13]. In addition, measurements with a transversely polarized hydrogen target have started in 2002. This will result in the first experimental information about transverse azimuthal target spin asymmetries, which have been predicted to be large [9].

REFERENCES

1. M. Diehl, T. Feldmann, R. Jakob and P. Kroll, *Nucl. Phys. B*, **596**, 33 (2001) (and references therein).
2. X. Ji, *Phys. Rev. D*, **55**, 7114 (1997).
3. J. Collins, L. Frankfurt and M. Strikmann, *Phys. Rev. D*, **56**, 2982 (1997).
4. HERMES Collaboration, *Nucl. Instr. Meth. A*, **417**, 230 (1998).
5. HERMES Collaboration, *Eur. Phys. J. C*, **18**, 303 (2000).
6. HERMES Collaboration, *Eur. Phys. J. C*, **17**, 389 (2000).
7. M. Vanderhaegen, P. Guichon, and M. Guidal, *Phys. Rev. Lett.*, **80**, 5064 (1998).
8. M. Vanderhaegen, P. Guichon, and M. Guidal, *Phys. Rev. D*, **60**, 094017 (1999).
9. L.L. Frankfurt, P.V. Pobylitsa, M.V. Polyakov, and M. Strikman, *Phys. Rev. D*, **60**, 14010 (1999).
10. L.L. Frankfurt, M.V. Polyakov, M. Strikman, and M. Vanderhaegen, *Phys. Rev. Lett.*, **84**, 2589 (2000).
11. A.V. Belitzky and D. Müller, *Phys. Lett. B*, **513**, 349 (2001).
12. HERMES Collaboration, *Phys. Lett. B*, **535**, 85 (2002).
13. HERMES Collaboration, *DESY PRC Proposal 02-01, HERMES Internal Note 02-003* (2001).

Hard Exclusive Electroproduction of Two Pions off Proton and Deuteron at HERMES

Pasquale di Nezza* and Riccardo Fabbri (*for the HERMES Collaboration*)†

**INFN, Laboratori Nazionali di Frascati, Via Enrico Fermi 40, 00044 Frascati, Italy*
E_mail: Pasquale.DiNezza@lnf.infn.it
†Department of Physics, Ferrara University, Via delle Scienze 12, 44100 Ferrara, Italy
E_mail: rfabbri@fe.infn.it

Abstract. Exclusive electroproduction of $\pi^+\pi^-$ pairs off hydrogen and deuterium targets has been studied with the HERMES experiment. The angular distribution of the π^+ in the $\pi^+\pi^-$ rest system has been studied in the invariant mass range $0.3 < m_{\pi\pi} < 1.5$ GeV. Theoretical models derived in the framework of the Generalized Parton Distributions show that this angular distribution receives only contributions from the interference between the isoscalar channel $I = 0$ and the isovector channel $I = 1$.

1. INTRODUCTION

The analysis of hard exclusive production of $\pi^+\pi^-$ pairs off unpolarized targets of hydrogen and deuterium at HERMES is presented. Recent theoretical studies [1, 2] have shown that the exclusive process $e^+p \longrightarrow e^+p\ \pi^+\pi^-$ can be described in the framework of Generalized Parton Distributions (*GPD*s) [3, 4, 5]. The diagrams relevant for this reaction at leading twist are shown in Fig. 1. The pion pairs may be produced through gluon or quark exchange with the target, either from quark (Fig. 1a,b,c) or from gluon fragmentation (Fig. 1d). In the range of the considered $\pi^+\pi^-$ invariant mass ($m_{\pi\pi}$), both resonant and non-resonant contributions are present. The particular state describing the $\pi^+\pi^-$ pair, with the angular momentum quantum number l being odd or even, also defines unequivocally the quantum numbers C-Parity (C) and Strong Isospin (I) of this state. At small values of the Bjorken variable x_{Bj}, pions are produced mostly in the isovector state, because the dominant mechanism is two-gluon exchange with positive C-parity (Fig. 1a) [6]. At large x_{Bj} the production of pion pairs is dominated by $q\bar{q}$ exchange (Fig. 1b,c,d) [7], which leads to a sizable admixture of pion pairs with isospin zero.

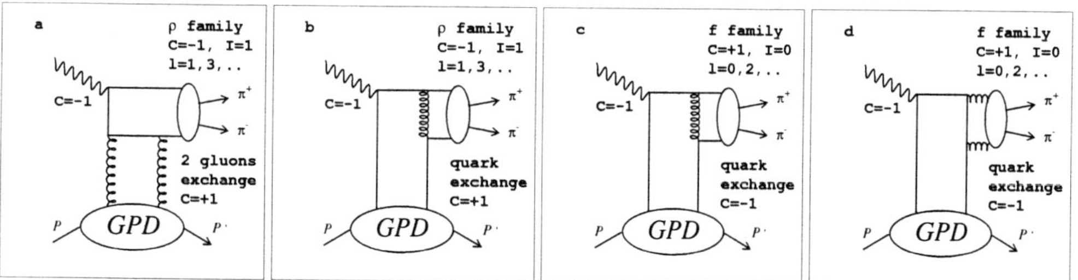

FIGURE 1. Leading twist diagrams for the hard exclusive reaction $e^+p \rightarrow e^+p\ \pi^+\pi^-$. The gluon exchange (*a*) gives rise to pions in the isovector state only, while the quark exchange (*b,c,d*) gives rise to pions both in the isoscalar and in the isovector state.

CP675, *Spin 2002: 15th Int'l. Spin Physics Symposium and Workshop on Polarized Electron Sources and Polarimeters*, edited by Y. I. Makdisi, A. U. Luccio, and W. W. MacKay

In the HERMES kinematics the $\pi^+\pi^-$ cross section is dominated by the isovector channel, i.e. ρ^0 production. The less copious isoscalar $\pi^+\pi^-$ production can be investigated by studying the interference between the odd and even l wave production. An observable suitable to probe this interference appears to be the *intensity density* [1, 2], defined as the lth Legendre Polynomial $P_l(cos\theta)$ moment

$$\langle P_l(cos\theta)\rangle^{\pi^+\pi^-} = \frac{\int_{-1}^{1} dcos\theta\, P_l(cos\theta)\, \frac{d\sigma^{\pi^+\pi^-}}{dcos\theta}}{\int_{-1}^{1} dcos\theta\, \frac{d\sigma^{\pi^+\pi^-}}{dcos\theta}}, \tag{1}$$

where θ is the scattering angle of π^+ meson in the $\pi^+\pi^-$ rest frame with respect to the direction of the recoiled system [7]. Some of the above defined intensities are non-zero only if the interference between the isoscalar and the isovector channels is present. Experimentally they are obtained as average values of $P_l(cos\theta)$. In this paper we describe the measurement of the intensity density for $l = 1$ which is calculated by the average $\langle cos\theta\rangle$.

2. DATA ANALYSIS

The data have been accumulated with the HERMES forward spectrometer [8] during the running period 1996-2000 of HERA. The 27.6 GeV positron beam was scattered off hydrogen and deuterium targets, respectively. Events have been selected with one positron track and two oppositely charged hadron tracks ($E_h > 1.0$ GeV) without additional neutral clusters in the calorimeter. The exclusivity of these events was ensured by imposing a further cut on the inelasticity $\Delta E = \frac{M_X^2 - M_P^2}{2M_P}$, where M_X is the invariant mass of the undetected system and M_p is the proton mass. As explained above, to enhance the isoscalar production and consequently the interference between the isovector and isoscalar channel, it was also required that $x_{Bj} > 0.1$. Moreover, in order to enter the *hard* regime of the process, constraints on Q^2 ($Q^2 > 1$ GeV^2) and W ($W > 2$ GeV) were imposed, where W is the invariant mass of the virtual-photon nucleon system.

The most important source of background to the exclusive channel comes from fragmentation of partons in semi-inclusive deep inelastic scattering (*DIS*) at low ΔE. Due to the instrumental resolution and smearing, those events may contaminate the exclusive sample. The shape of the *DIS* background is well reproduced by a *MonteCarlo* simulation (*MC*). It is based on the *LEPTO* generator using the *LUND* model and the detector response was simulated by a *GEANT*-based *MonteCarlo* code. *MC* data were normalized to experimental data at $\Delta E > 2$ GeV and then subtracted. In order to illustrate our procedure, in Fig. 2 both the unsubtracted (left panel) and full subtracted (right panel) $\pi^+\pi^-$ data are shown as a function of the inelasticity ΔE. Here only data within the $\pi^+\pi^-$ invariant mass window $0.6 < m_{\pi\pi} < 1.0$ GeV, around the ρ^0 mass, are shown. Exclusive events were then selected for $\Delta E < 0.625$ GeV to maximize the ratio of the exclusive signal over the background (Sg/Bg) and minimize its relative statistical error. For these events the intensity density was calculated in 10 bins of $m_{\pi\pi}$. The intensity density of the background, $\langle cos\theta\rangle_{Bg}$, was evaluated using data for $\Delta E > 2$ GeV, assuming its ΔE independence. This assumption was tested checking the stability of $\langle cos\theta\rangle_{Bg}$ in different ΔE bins. In every bin of $m_{\pi\pi}$, the value of $\langle cos\theta\rangle_{Bg}$, weighted by the ratio Sg/Bg for the chosen ΔE cut, was subtracted. The background corrections range between $10-70\%$ in the various bins.

3. RESULTS

In Fig. 3 the preliminary HERMES results on the $m_{\pi\pi}$-dependence of the intensity density $\langle cos\theta\rangle$ are shown, for the proton in the left panel and for the deuteron in the right one.

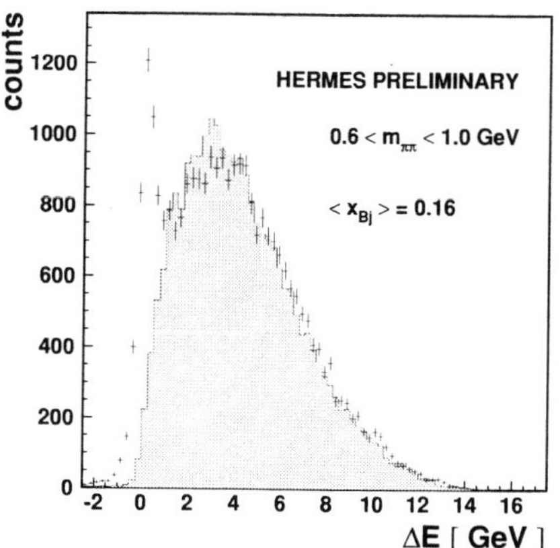

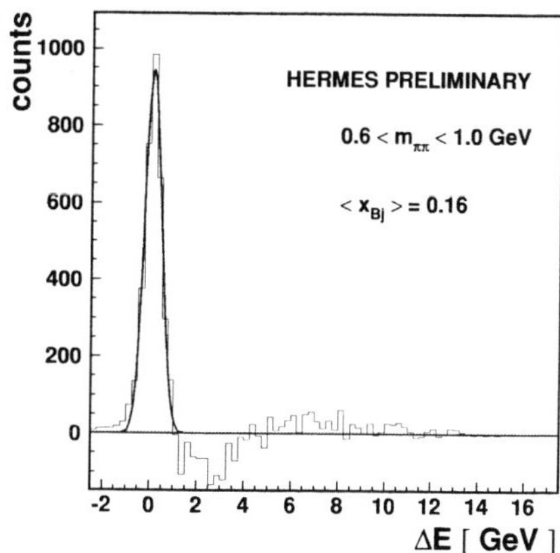

FIGURE 2. Left panel: *DIS MC* (shaded) normalized to HERMES data (crosses) at $\Delta E > 2$ GeV. Right panel: the exclusive $\pi^+\pi^-$ channel obtained by subtracting the *DIS*-MC spectrum from the experimental one. In both panels only the ρ^0 invariant mass window is considered.

The distributions show a clear angular asymmetry whose size changes with $m_{\pi\pi}$. At low $m_{\pi\pi}$ ($m_{\pi\pi} < 0.6$ GeV) the asymmetry may be due to an interference between the lower tail of the ρ^0 meson and the non-resonant $\pi^+\pi^-$ S-wave production ($I = 1$ and $I = 0$ interference). This interference is present over the entire invariant mass region considered. At large $m_{\pi\pi}$ ($m_{\pi\pi} > 1.0$ GeV), additionally, an interference between the upper tail of the ρ^0 meson and the f-type mesons arises and is superimposed. In particular, the possible change of sign of the asymmetry at $m_{\pi\pi} \approx 1.3$ GeV may be understood as being caused by the interference of the broad ρ^0 tail and the f_2 resonance (1.270 GeV). Note that no similar behavior is seen at the narrow f_0 resonance (0.980 GeV), possibly due to the experimental resolution.

In Fig. 4 the x_{Bj}-dependence of the intensity density $\langle cos\theta \rangle$ is shown for the proton within

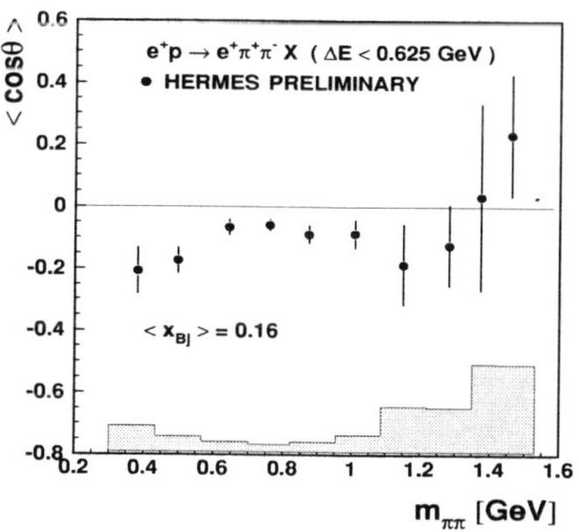

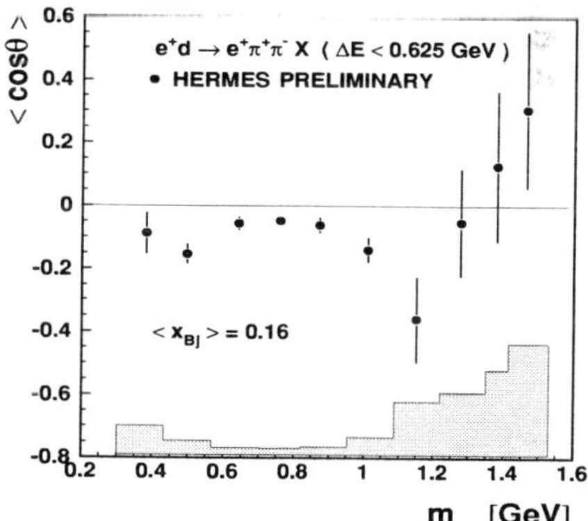

FIGURE 3. $m_{\pi\pi}$-dependence of the intensity density $\langle cos\theta \rangle$ for the proton (left panel) and the deuteron (right panel). Shaded areas in both panels represent the systematic uncertainty.

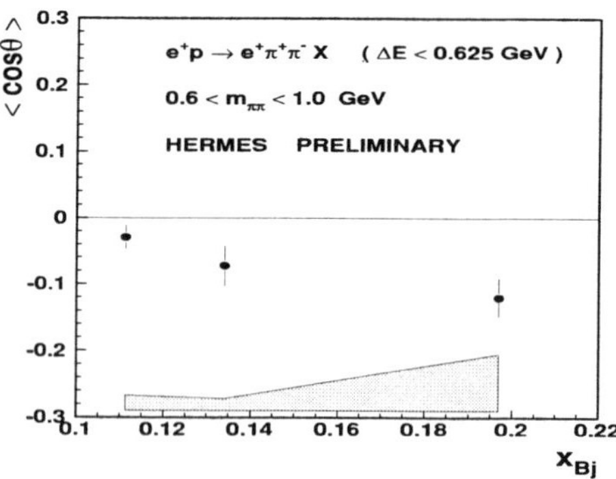

FIGURE 4. x_{Bj}-dependence of the intensity density $\langle cos\theta \rangle$ in the ρ^0 invariant mass window for the proton. The shaded area represents the systematic uncertainty.

the above defined ρ^0 window. The size of the $\langle cos\theta \rangle$ asymmetry increases with x_{Bj}. This experimental finding is in agreement with theoretical expectations according to which at increasing x_{Bj} the $\pi^+\pi^-$ production becomes increasingly dominated by $q\bar{q}$ exchange, leading to a sizable admixture of isoscalar and isovector pion pairs. As explained above this leads to an enhancement of the interference term.

The systematic uncertainties have been evaluated considering all hadrons as pions, using different exclusive cuts and applying various procedures for the normalization of the *MC* generated background to the data. The main contribution has been found to originate from using different exclusive cuts. Using the diffractive *DIPSI MC* generator [9], effects due to the acceptance were found to be negligible. Radiative corrections were shown in [10] to be smaller than 1% in the H1 and ZEUS kinematics. At larger x_{Bj}, where the HERMES analysis is performed, they are even smaller, and for that reason have been neglected.

The comparison with the theoretical predictions available so far is promising. In the left (right) panel of Fig. 5 the experimental dependence of the intensity density $\langle cos\theta \rangle$ on $m_{\pi\pi}(x_{Bj})$ for

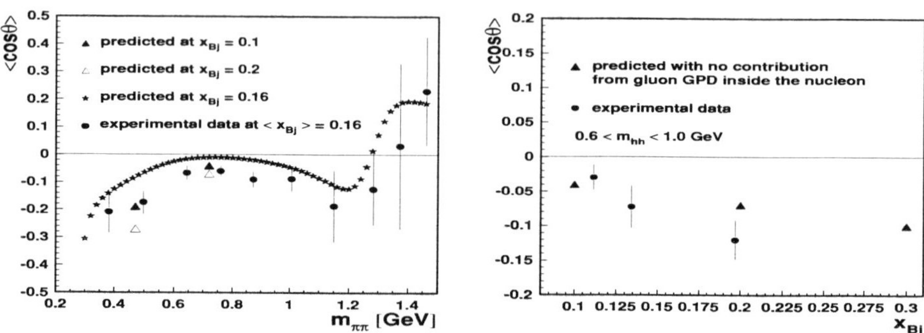

FIGURE 5. The experimental $m_{\pi\pi}$-dependence (left panel) and x_{Bj}-dependence (right panel) of the intensity density $\langle cos\theta \rangle$ for the proton are compared with theoretical predictions in [1, 2]. *Triangles* (*stars*) show the predictions with the gluon *GPD* neglected [1] (included [2]). In the left panel, *Triangles* have been slightly shifted for a better visibility.

the proton is compared with predictions performed at leading twist [1, 2]. The shape of the theoretical distribution is nicely reproducing the data, in particular at the f_2 meson mass where the asymmetry changes sign. The reasonable agreement of the leading twist predictions with data may be understood as to arise from the cancellation of higher twist effects in this kind of asymmetry [11].

4. CONCLUSIONS

The $l = 1$ intensity density in $\pi^+\pi^-$ hard exclusive electroproduction was measured for the first time, for both proton and deuteron at $x_{Bj} > 0.1$. The quantity $\langle cos\theta \rangle$ is sensitive to the interference between the isoscalar channel ($I = 0$) and the isovector channel ($I = 1$). The absolute value of the asymmetry measured in the $l = 1$ intensity density $\langle cos\theta \rangle$ shows a minimum at $m_{\pi\pi} = m_{\rho^0}$. At smaller invariant mass the asymmetry may be interpreted as originating from the interference of the ρ^0 lower tail with the non-resonant $\pi^+\pi^-$ production (S-wave), and at larger invariant mass from the interference of the ρ^0 upper tail (P-wave) with $\pi^+\pi^-$ production from the f_2 (D-wave). The interference signal in the ρ^0 invariant mass window was shown to increase in size with x_{Bj}. This behavior may be understood as to arise from the increased dominance of $q\bar{q}$ exchange at larger values of x_{Bj}, which leads to a sizable admixture of isoscalar and isovector pion pairs.

This is the first evidence of an $I = 0$ admixture in $\pi^+\pi^-$ exclusive electroproduction. This fact by itself is interesting in view of the importance of the scalar meson sector.

In the *GPD* framework, theoretical predictions are available with and without inclusion of the *GPD* describing the two-gluon exchange. Both predictions appear to be consistent with the data. Therefore these results on exclusive $\pi^+\pi^-$ production, together with results from different exclusive channels (e.g. Deeply Virtual Compton Scattering [12], exclusive π^+ production [13]) may lead to a better modeling of *GPD*s.

ACKNOWLEDGMENTS

We are deeply grateful to N. Bianchi, A. Borissov, M. Diehl, W.-D. Nowak, B. Pire and M.V. Polyakov for useful discussions and suggestions.

REFERENCES

1. B. Lehmann-Dronke et al., Phys. Lett. B **475**, (2000) 147.
2. B. Lehmann-Dronke et al., Phys. Rev. D **63**, (2001) 114001.
3. A.V. Radyushkin, JLAB-THY-00-33.
4. X. Ji, J. Phys. G **24**, (1998) 1181.
5. X. Ji, Phys. Rev. D **55**, (1997) 7114.
6. M.V. Polyakov, Nucl.Phys. B **555** (1999) 231.
7. M. Diehl, T. Gousset, B. Pire, hep-ph/9909445, Talk given at 6th INT/JLAB Workshop on Exclusive and Semiexclusive Processes at High Momentm Trasfer, NewPort News, VA, 19-23 May 1999.
8. K. Ackerstaff et al., (HERMES Collaboration), N.I.M. A **417**, 230 (1998).
9. M. Arneodo, L. Lamberti, M. Ryskin, Comput. Phys. Commun. **100** (1997) 195.
10. I. Akushevich and P. Kuzhir, Phys. Lett. B **474** (2000) 411.
11. M. Garcon, hep-ph/0210068.
12. A. Airapetian et al. (HERMES Collaboration), Phys. Rev. Lett. **87** (2001) 182001.
13. A. Airapetian et al. (HERMES Collaboration), Phys. Lett. B **535** (2002) 85.
14. Ph. Haegler, B. Pire, L. Szymanowski and O.V. Teryaev, Phys. Lett.B **535** (2002), 117.

Prospects on Constraining ΔG from Inclusive Jet Production in Polarized pp Collisions at RHIC in 2003

B. Surrow for the STAR Collaboration[1]

Brookhaven National Laboratory
Department of Physics
Upton, NY 11973-5000
USA

Abstract. The anticipated increase in luminosity and polarization for the RHIC spin run in 2003 together with the installation of spin rotators at the STAR interaction region will allow the first measurement of A_{LL} in inclusive jet production at $\sqrt{s} = 200\,\text{GeV}$. This data should provide hints of the gluon polarization, $\Delta G/G$, of the proton. In the long-term, the determination of the gluon polarization of the proton will be made through prompt-photon production and photon-jet coincidences. Other possibilities include di-jet production and heavy flavor production.

The measurement of A_{LL}, the expected event rates, and simulation results based on the anticipated RHIC performance in 2003 will be described together with a discussion of various systematic error sources.

INTRODUCTION

The spin of elementary particles is as fundamental to their nature as the mass. The proton is a fermion of $J = 1/2$. The proton spin is understood to be made up of contributions arising from the quark spin, the gluon spin, and orbital angular momentum. The fundamental question in this regard is how the proton spin is distributed among those contributions. It was found in polarized lepton-nucleon experiments that only about $1/3$ of the proton spin is carried by quarks and anti-quarks, contrary to the expectation of the constituent quark model that the proton spin would be carried predominantly by its three valence quarks. A significant fraction of the proton spin must therefore be carried by gluons and orbital angular momentum. The role of the gluons in making up the missing proton spin is currently only very poorly constrained from scaling violations in deep-inelastic scattering fixed-target experiments. A need for a new generation of experiments to explore the spin structure of the proton is clearly apparent. The current spin physics effort at RHIC at BNL focuses on the collision of polarized protons to gain a deeper understanding of the spin structure of the proton in a new, previously unexplored territory. The first polarized proton run from December 2001 until January 2002 is the beginning of a multi-year experimental program which aims to address a

[1] For the full author list and acknowledgments, see appendix to the proceedings.

CP675, *Spin 2002: 15th Int'l. Spin Physics Symposium and Workshop on Polarized Electron Sources and Polarimeters,* edited by Y. I. Makdisi, A. U. Luccio, and W. W. MacKay

variety of topics related to the spin structure of the proton. A recent review of the RHIC spin program can be found in [1].

The measurement of the double longitudinal spin asymmetry, A_{LL}, for photon production allows the extraction of the gluon polarization, $\Delta G/G$. In LO QCD employing factorization of the underlying hard process, the asymmetry measured for $\vec{p}+\vec{p} \rightarrow \gamma + \text{jet} + X$ is represented as: $A_{LL} = \frac{\Delta G(x_g)}{G(x_g)} \cdot A_1^p(x_q) \cdot \hat{a}(g+q \rightarrow \gamma + q)$. The ratio of the polarized and unpolarized structure functions, $A_1^p(x_q)$, is measured in polarized deep-inelastic scattering and $\hat{a}(g+q \rightarrow \gamma+q)$ is calculated in pQCD. Hence a measurement of A_{LL} for prompt photons detected in coincidence with the away-side quark-jet allows an extraction of the gluon polarization $\Delta G(x_g)/G(x_g)$.

THE POLARIZED PROTON COLLIDER RHIC

The first collisions of polarized protons occurred in December 2001, ushering in a new era to complement the ongoing relativistic heavy-ion program. RHIC is the first accelerator to accelerate and collide polarized protons, ultimately at high luminosity, and at a center-of-mass energy of up to 500 GeV. An overview of the polarized proton collider RHIC can be found in [4].

The first polarized proton run at RHIC was carried out at $\sqrt{s} = 200$ GeV. A transverse polarization of about 0.2 was achieved at the injection energy of 24.6 GeV and was approximately maintained when the proton beams were accelerated to 100 GeV.

The collision of longitudinal polarized protons at $\sqrt{s} = 200$ GeV is foreseen for the first time at RHIC in 2003 with the installation of spin rotators at the STAR and PHENIX interaction regions. The anticipated polarization is about 0.4 and the instantaneous luminosity of $1 \cdot 10^{31}\ \text{s}^{-1}\text{cm}^{-2}$ is an order of magnitude larger compared to the first polarized proton run in 2002. The expected integrated luminosity at the STAR experiment for longitudinal polarized protons amounts to approximately $3\,\text{pb}^{-1}$. Those RHIC performance expectations on polarization and luminosity serve as the basis for simulations of A_{LL} in inclusive jet production.

THE STAR EXPERIMENT

A detailed description of the STAR experiment can be found in [5]. The STAR experiment was upgraded for the polarized proton program with the installation of a beam-beam counter (BBC) [6] and a forward-pion detector (FPD) [7] prior to the first polarized proton run of transverse polarized protons. Both components will play a crucial role in measuring A_{LL}. The BBC provides the principal relative luminosity measurement. An upgraded FPD detector system will be used to tune the spin rotators at the STAR interaction region [8].

Installation of the STAR electromagnetic calorimeters [5] is partially complete. The STAR barrel electromagnetic calorimeter has been completed for $0 < \eta < 1$ and $\Delta\phi = 2\pi$ for the RHIC run in 2003. Part of the STAR endcap electromagnetic calorimeter has been installed. In subsequent years, it is expected that the barrel electromagnetic calorimeter will have complete azimuthal coverage for the interval $-1 < \eta < 1$. The endcap electromagnetic calorimeter will provide complete azimuthal coverage for $1.09 < \eta < 2$. Both systems will be crucial to study inclusive jet production and ultimately prompt-photon production and photon-jet coincidences.

The simulations below are based on the partial calorimeter coverage in 2003 of $0 < \eta < 1$ and $\Delta\phi = 2\pi$.

INCLUSIVE JET PRODUCTION

Besides the golden channel ($\vec{p}+\vec{p} \to \gamma+\text{jet}+X$) to measure ΔG, inclusive jet production provides another way of constraining ΔG accessible in the 2003 run. The sensitivity to ΔG arises from gluon-initiated processes such as: $gg \to gg$ and $qg \to qg$. Jet production also includes the $qq^{(\prime)} \to qq^{(\prime)}$ processes.

The yield for the anticipated condition in 2003 for 10 days of running at an efficiency of 33% has been estimated from a PYTHIA simulation. For $p_T = 5-10\,\text{GeV}$, $1 \cdot 10^6$ jets are expected to be recorded. This number reduces to $1 \cdot 10^3$ for $p_T = 30-35\,\text{GeV}$.

Figure 1 shows the result of a PYTHIA-based simulation for A_{LL} in inclusive jet production including trigger and jet reconstruction efficiency effects.

Besides the PYTHIA event generation, the response of electromagnetic (e.g. photons) and hadronic energy deposition in the barrel electromagnetic calorimeter has been separately accounted for. In addition, polarization effects are added after the PYTHIA event generation using separate polarized and unpolarized structure functions and the partonic, process-dependent LO pQCD results for $\hat{a}_{LL}$.

A jet trigger requiring $E_T > 5\,\text{GeV}$ for a $(\Delta\eta = 1)$ x $(\Delta\phi = 1)$ patch of the barrel electromagnetic calorimeter has been applied. Jets were reconstructed using a cone algorithm with a seed of 1 GeV and a radius of $R = 0.7$. The result of two simulations are shown using two different input parameterizations for ΔG, GRSV-std and GRSV-max [2]. The latter reflects the limiting case of ΔG [3].

The reconstructed values for A_{LL} for the two different ΔG parameterizations reveal a different dependence on the measured jet transverse energy between approximately $5-15\,\text{GeV}$. It is the region of small values in the measured jet transverse energy where the sensitivity to ΔG is expected to be dominant.

Beyond a measured jet transverse energy of 15 GeV, the expected uncertainty does not permit drawing any further conclusions on discriminating different ΔG parameterizations under the conditions expected for 2003.

A clear sensitivity in A_{LL} for ΔG has been also established from a NLO calculation using two different input parameterizations (GRSV-std and GRSV-max) based on the anticipated RHIC performance [9]. No detector effects have been included in those studies.

Besides discussing a possible sensitivity to ΔG for inclusive jet production during the RHIC run in 2003, a major emphasis with the first measurement of A_{LL} will be the understanding of various systematic error sources which contribute to the measurement of A_{LL} and thus ultimately to the extraction of ΔG.

Systematic errors in the measurement of A_{LL} arise from various uncertainties. The

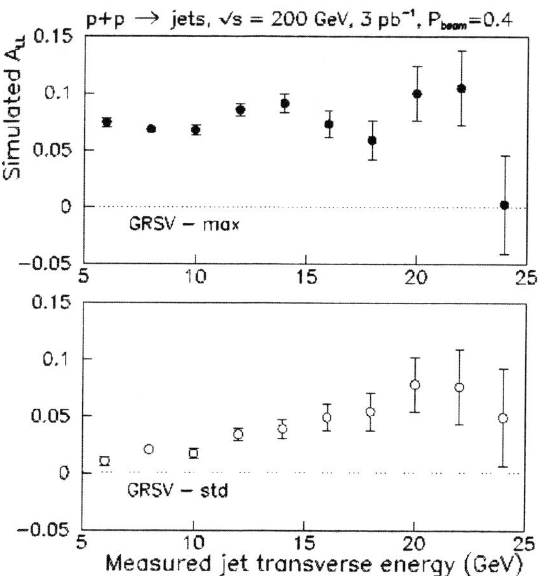

FIGURE 1. A_{LL} from a PYTHIA-based simulation for inclusive jet production including trigger and jet reconstruction efficiency effects. Only statistical errors are shown.

absolute polarization uncertainty at injection is known at present to about 20%. The analyzing power of the RHIC CNI polarimeter [10] at full beam energy has not been measured yet. Ultimately, the RHIC beam polarization will be determined from a polarized gas jet target experiment to about 5%. This will not be available for the first RHIC run of longitudinal polarized protons [11]. Therefore other means have to be explored such as the RHIC down-ramping development [12] to estimate the beam polarization at full beam energy by constraining the analyzing power of the RHIC CNI polarimeter at full energy.

The measurement of the relative luminosity will be crucial in the measurement of A_{LL}. More details can be found in [6].

Differences in the fragmentation for quark and gluon jets could lead to a possible trigger bias. It is therefore important to employ different jet triggers to see if those changes on the trigger level can be accounted for by simulations.

The collision of longitudinal polarized protons at $\sqrt{s} = 200\,\text{GeV}$ is foreseen for the first time at RHIC in 2003 with the installation of spin rotators at the STAR interaction region. It is planned to determine the operating point of the spin rotator magnets by minimizing left/right and up/down spin-dependent asymmetries with the upgraded FPD detector system [8].

Figure 2 shows preliminary results of uncorrected distributions for jet production obtained during the first run of transverse polarized protons in 2002 at $\sqrt{s} = 200\,\text{GeV}$. A cone jet finder for charged particles employing the STAR TPC only has been used with: $R = 0.7$, seed $> 1\,\text{GeV}$, $E_T > 5\,\text{GeV}$ and $|\eta^{jet}| < 0.7$. The uncorrected transverse energy distribution and the difference in azimuth angle for a di-jet sample is shown. This first

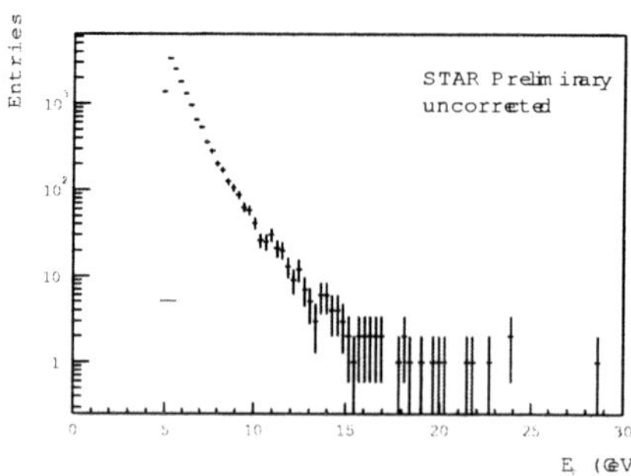

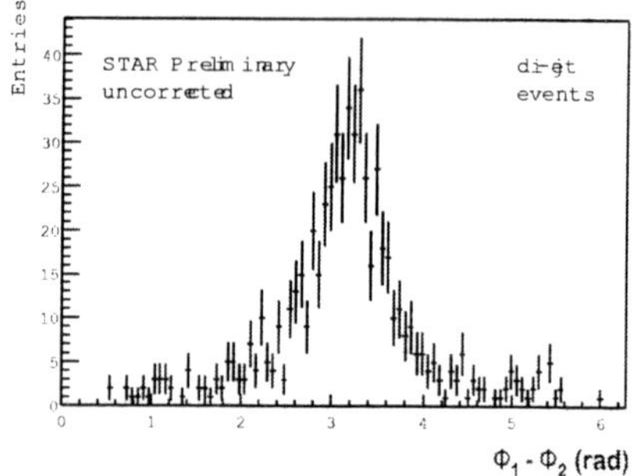

FIGURE 2. Preliminary results of uncorrected distributions for jet production obtained during the first run of transverse polarized protons in 2002 at $\sqrt{s} = 200\,\text{GeV}$.

look at jet production is quite encouraging for the STAR spin program in 2003.

SUMMARY AND OUTLOOK

The collision of longitudinal polarized protons at $\sqrt{s} = 200\,\text{GeV}$ is foreseen for the first time at RHIC in 2003 with the installation of spin rotators at the STAR interaction region.

The result of a PYTHIA-based simulation including detector effects and anticipated 2003 operating conditions indicated that STAR would be able to measure A_{LL} as a function of the jet transverse energy for the first time. Various systematic errors sources will have to be carefully understood in order that, at a later stage, extractions of ΔG can be reliably performed.

REFERENCES

1. G. Bunce et al., *Ann. Rev. Nucl. Part. Sci.* 50 (2000) 525.
2. M. Glück, E. Reya, M. Stratmann and W. Vogelsang, *Phys. Rev.* D63 (2001) 094005.
3. M. Glück, E. Reya, and A. Vogt, *Eur.Phys.J.* C5 (1998) 461.
4. H. Huang, *these proceedings.*
5. STAR Collaboration, *Nucl.Instrum.Meth.* to be published (Special volume edition).
6. J. Kiryluk, *these proceedings.*
7. G. Rakness, *these proceedings.*
8. A. Ogawa, *these proceedings.*
9. W. Vogelsang, *private communications.*
10. O. Jinnouchi, *these proceedings.*
11. A. Bravar, *these proceedings.*
12. H. Spinka, *these proceedings.*

Measuring ΔG in PHENIX Using Electrons to Tag Heavy-flavor Production [1]

Kenneth N. Barish for the PHENIX Collaboration[2]

University of California at Riverside, Riverside, CA 92506, USA
E-mail: Kenneth.Barish@ucr.edu

Abstract. Heavy flavor production can be used as a probe of the gluons contribution to the protons spin, ΔG. In this paper [1] we discuss the prospects for PHENIX to tag heavy flavor production with single electron and muon/electron coincidences. We have estimated our sensitivity to ΔG using a full detector simulation which includes the effects of the trigger and dilutions due to conversions in the inner chambers and π^0 Dalitz decays.

INTRODUCTION

Heavy flavor production, $c\bar{c}$ and $b\bar{b}$, is dominated by gluon-gluon interactions and gives rise to a double spin asymmetry

$$A_{LL} \sim \frac{\Delta G(x_A)}{G(x_A)} \otimes \frac{\Delta G(x_B)}{G(x_B)} \otimes \hat{a}_{LL}^{gg \to Q\bar{Q}} \quad , \tag{1}$$

from which ΔG can be extracted. A_{LL} is the measured double longitudinally polarized asymmetry and $\hat{a}_{LL}$ is the partonic level asymmetry, or analyzing power, which is calculable within the framework of pQCD.

Below we explore tagging heavy flavor production in PHENIX using single electrons and μ-e-coincidences. This is made possible by an electron trigger which utilizes the PHENIX electromagnetic calorimeter and ring imaging cerenkov counter. The following simulations are based on the event generator Pythia and PHENIX acceptances [2]. We have simulated the full response and reconstruction of the PHENIX Multiplicity and Vertex detector (MVD) and have used the parameterizations of the gluon polarization provided by Gehrmann and Sterling [3] and leading order calculations for the analyzing power [4]. Recently, next-to-leading order calculations have been performed [5], solidifying the theoretical framework for this measurement.

We find that a measurement using heavy flavor production extends the accessible x_g-range for PHENIX, and even more importantly it provides an alternative way to access the gluon polarization with different systematic and theoretical uncertainties. This will permit a cross check of the results obtained from direct photon production.

---

[1] This work is supported by the United States DOE Grant DOE-FG03-01ER41171.

[2] For the full PHENIX Collaboration author list and acknowledgements, see Appendex "Collaborations" of this volume.

CP675, *Spin 2002: 15th Int'l. Spin Physics Symposium and Workshop on Polarized Electron Sources and Polarimeters,* edited by Y. I. Makdisi, A. U. Luccio, and W. W. MacKay

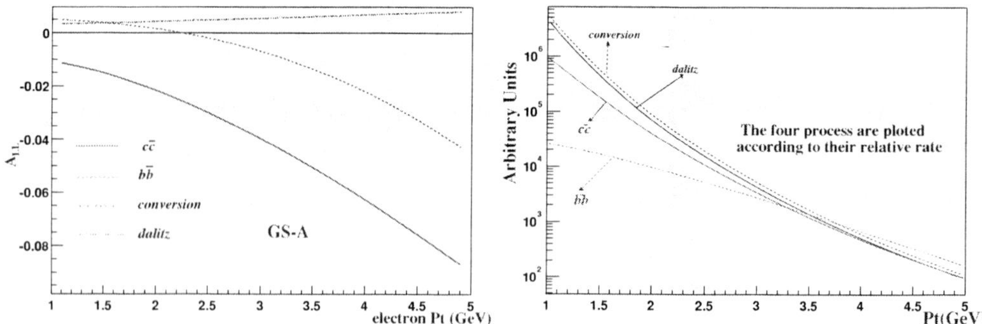

FIGURE 1. (left) Input asymmetries for charm and bottom production, and Dalitz and conversion decays. (right) Relative rates of single electrons from $b\bar{b}$ and $c\bar{c}$ QCD jet events and conversion and Dalitz processes from minimum bias Pythia events versus transverse momentum in GeV at $\sqrt{s} = 200\,\text{GeV}$.

SINGLE ELECTRON MEASUREMENTS

Single electron samples provide large numbers of charmed events with significant backgrounds from π^0-Dalitz decays and gamma conversions. The charm production cross section is not well known at RHIC center of mass energies, but will be measured soon by PHENIX and has been estimated to be between $200 < \sigma^{c\bar{c}} < 350\mu\text{b}$ at $\sqrt{s} = 200\,\text{GeV}$ [6]; the branching ratio for leptonic charm decays is about 10%. The power of PHENIX's electron identification was demonstrated in heavy-ion collisions [7]. Single electron yields come from different sources including charm decays, bottom decays, π^0 Dalitz decays, and conversion electron each of which can have different asymmetries, see Fig. 1. The difference between the charm and bottom asymmetries come about because of the mass dependence of the analyzing power and the different decay kinematics. The Dalitz and conversion ($\pi^0 \rightarrow e^+e^-\gamma$) asymmetries is the same as for π^0's. We use Pythia coupled with a simulation of the PHENIX detector to determine the relative input rates, see Fig. 1.

PHENIX's multiplicity vertex detector (MVD) can be used to help identify electrons which have come from conversions in the beam pipe or Dalitz decay electrons. We find that a pulse height in association with a separation cut between charged particle tracks (10 degrees) rejects 68% of the Dalitz decay electrons and 75% of the beam pipe conversion electrons, while keeping 78% of the signal electrons, see Fig. 2. The values at an opening angle of 0° represents only a pulse height cut. We have estimated our sensitivity to A_{LL} for heavy flavor production taking into account the diluting effect of the conversion and Dalitz electrons which are not rejected by the MVD, see Fig. 3.

The MVD cuts can be inverted to produce a sample of events which contain mostly electrons from conversions and Dalitz decays. These come from QCD jet events with π^0's. Again, we can estimate the asymmetry from this sample, see Fig. 4. The asymmetry at low transverse momentum has flipped sign, giving us a handle on false asymmetries caused by acceptance effects. Further, the asymmetry can be used in conjunction with the direct π^0 measurement in a global analysis that will give us a handle on our systematic errors.

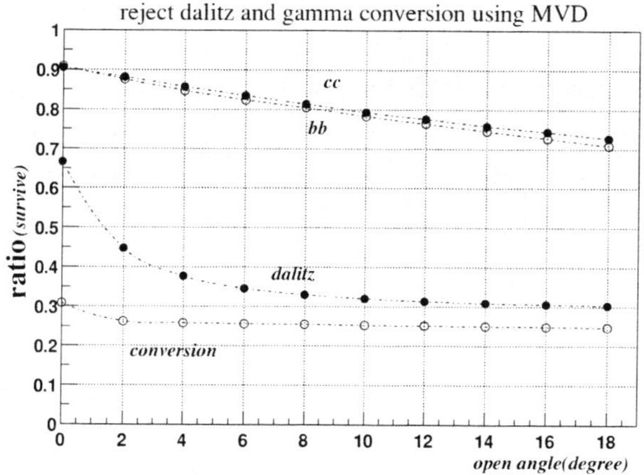

FIGURE 2. Rejection of Dalitz decay electrons using a pulse-height and angular minimum separation cut between charged particle tracks in the PHENIX multiplicity vertex detector. A pulse height and 10 degree separation cut rejects 68% of the Dalitz decay electrons and 75% of the beam pipe conversion electrons, while keeping 78% of the signal electrons.

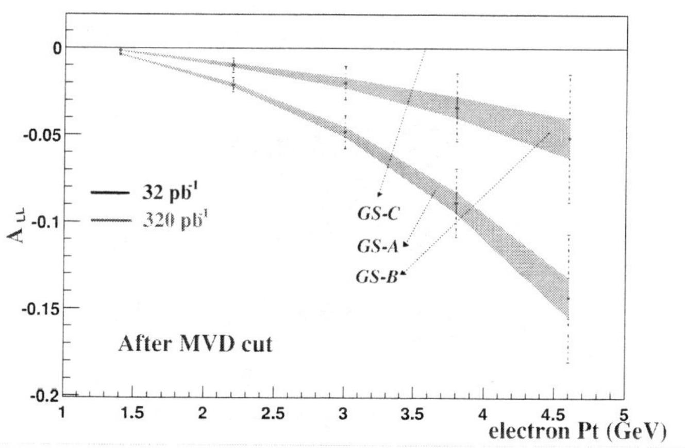

FIGURE 3. Projected double spin asymmetry A_{LL} and statistical and background subtraction errors based on a 32 pb^{-1} and 320 $^{-1}$. Events have been tagged online by an electron with $p_T > 1$ GeV in the central arm, and an offline MVD cut which rejects Dalitz and conversion electrons has been applied.

HEAVY FLAVOR PRODUCTION TAGGED IN μ e COINCIDENCES

In addition to the electron in the central detector it is possible to require a muon detected in one of the forward muon arms in coincidence. This requirement removes all background from conversions and Dalitz decays and enhances the $b\bar{b}$ yield in the event sample. The x_g-distributions are shown in Fig. 5. In the μ e channel the kinematic range

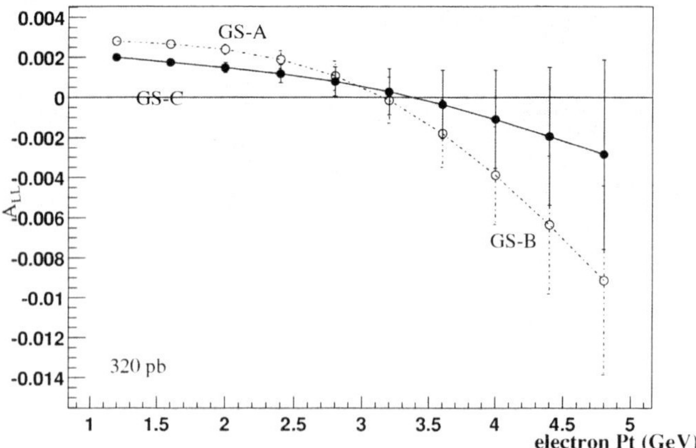

FIGURE 4. Projected double spin asymmetry A_{LL} and statistical and background subtraction errors based on a 320 $^{-1}$. Events have been tagged online by an electron with $p_T > 1$ GeV in the central arm, and an offline MVD cut enhancing conversion and Dalitz electrons which have come from π^0 QCD jet events.

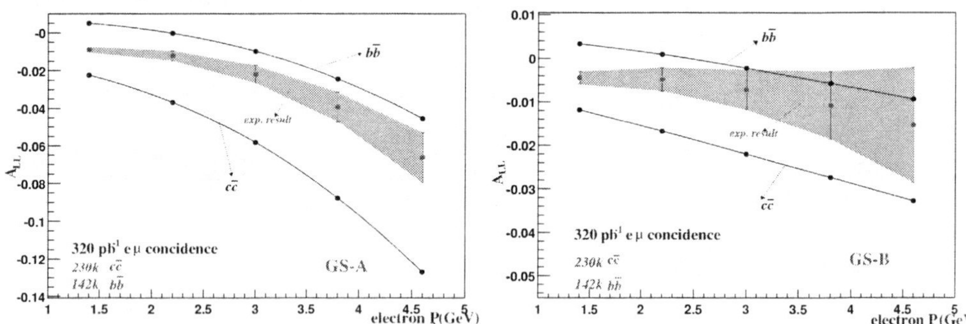

FIGURE 5. Projected double spin asymmetry A_{LL} and statistical errors based on a 10 weeks at design luminosity. The asymmetry corresponds to parameterizations "A" and "B" of the gluon polarization from Gehrmann and Sterling. Events have been tagged online by an electron in the central arm and by an additional (offline) muon in one of the forward arms.

reaches down to $x_g \approx 0.02$, and, unlike PHENIX's other measurements, can be roughly reconstructed.

Fig. 5 shows the expected experimental asymmetries for 10 weeks of data taking at design luminosity based on Gehrmann Stirling A and B. The pure charm and bottom asymmetries are also shown in the plots. At high transverse momentum, bottom begins to dominate. In 320pb^{-1} of eμ coincidences we expect approximately 230K charm events and 142K bottom events if we require the electron to have $p_t > 1$ GeV and the muon to have a momentum > 2 GeV into the muon arm acceptance. The statistics allows us to differentiate between Gehrmann Sterling A and B. Further, the e-μ channel will allow us to distinguish between charm and bottom using the asymmetry at high p_t and

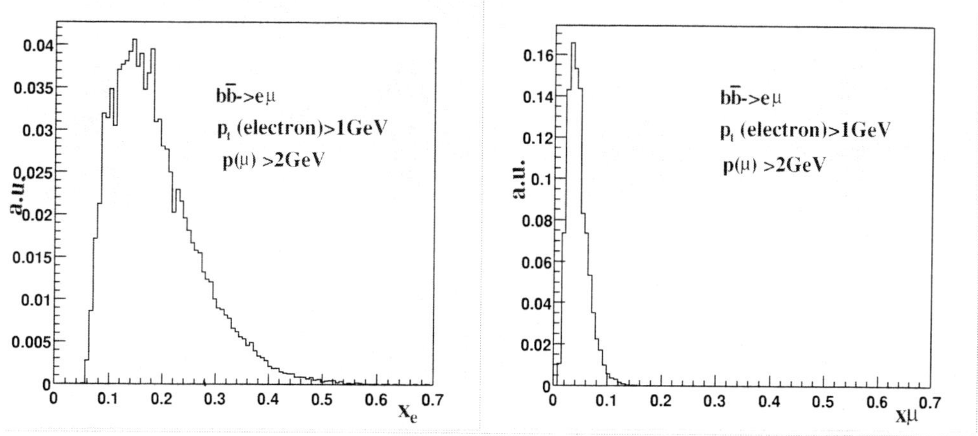

FIGURE 6. Parton kinematics in μ-e-events. The difference in kinematics between x_e and x_μ stem from the different acceptance for electrons and muons in the central arms and muon arms respectively.

comparisons between like and unlike sign electron muon pairs.

SUMMARY AND OUTLOOK

ΔG can be measured in PHENIX using single electrons and the quality of the measurement is improved if the background from Dalitz decays and photon conversions can be identified using an inner tracker. The additional requirement of a muon allows for an additional measurement that helps separate the charm and bottom contributions. The heavy flavor channels provide more independent measurements in PHENIX, helping to control experimental and theoretical systematic errors and the different channels cover different kinematic regions. Both of these measurements require a central arm trigger. The EMCal trigger worked in this past p+p run, and the EMCal/RICH trigger will be ready for the next run.

REFERENCES

1. A copy of the transparencies for this talk can be found at http://www.phenix.bnl.gov/phenix/WWW/publish/barish/talks/spin2002
2. "Sensitivity of ΔG Through Open Heavy Quark Production using Electron Decay Channels at PHENIX," DNP 2000.
3. T. Gehrmann, W.J. Stirling, Z.Phys. **C65** (1995) 461.
4. M. Karliner and R. Robinett, Phys.Lett.**B324** 1994.
5. I. Bojak and M. Stratmann, hep-ph/0112276; I. Bojak these proceedings.
6. P. L. McGaughey et. al., Int. J. Mod. Phy. A10 2999 (1995); Y. Akiba, private communication.
7. K. Adcox et al. [PHENIX] Phys.Rev.Lett.**88**:192303, 2002.

J/ψ Production in $\sqrt{s}$ = 200 GeV p+p Collisions with the PHENIX Detector at RHIC

H.D. Sato[1]
for the PHENIX Collaboration

University of Kyoto, Sakyo-ku, Kyoto 606-8502, Japan

Abstract. The PHENIX experiment at the Relativistic Heavy Ion Collider (RHIC) detects J/ψ particles in the rapidity range $|y| < 0.35$ using the electron-pair decay channel and in $1.2 < y < 2.2$ using the muon-pair decay channel. Cross sections for the J/ψ production in those rapidity ranges have been measured in p+p collisions at $\sqrt{s} = 200$ GeV in the RHIC Run 2001-2002 period (Run-2). As a result, $Br(J/\psi \rightarrow \mu^+\mu^-)d\sigma_{J/\psi}/dy|_{y=1.7}$ = 37 ± 7 (stat.) ± 11 (syst.) nb and $Br(J/\psi \rightarrow e^+e^-)d\sigma_{J/\psi}/dy|_{y=0}$ = 52 ± 13 (stat.) ± 18 (syst.) nb are obtained. Total cross section has been extracted by fitting these results to be $\sigma_{J/\psi}$ = 3.8 ± 0.6 (stat.) ± 1.3 (syst.) μb which is consistent with the perturbative QCD prediction.

INTRODUCTION

Heavy quarkonium production has been playing a very important role in high energy hadron physics. Its production in high-energy heavy-ion collisions is considered to be one of the best probes for the earliest stages of the new state of matter, called "Quark-Gluon Plasma" [1, 2]. Also its production asymmetries in longitudinally-polarized p+p collisions are expected to contain information on the polarized gluon density. The elucidation of the production mechanism of the heavy quarkonium is a key to this measurement.

The importance of the measurement of the production cross-section for J/ψ in p+p collisions is twofold: (1) testing theoretical models for the production mechanism and (2) providing the reference point for measurements in heavy ion collisions. In this paper, the first results of the production cross section for J/ψ in p+p collisions at $\sqrt{s}$ = 200 GeV are presented.

THE PHENIX EXPERIMENT

The PHENIX experiment consists of two independent spectrometers which cover different pseudo-rapidity (η) regions. Two Central Arms, West and East Arms, cover $|\eta| < 0.35$, $\Delta\phi$(azimuthal coverage) $= \pi$ and measure hadrons, electrons and photons.

---

[1] a JSPS Research Fellow

CP675, *Spin 2002: 15th Int'l. Spin Physics Symposium and Workshop on Polarized Electron Sources and Polarimeters,* edited by Y. I. Makdisi, A. U. Luccio, and W. W. MacKay

Two Muon Arms, North and South Arms, cover $1.2 < \eta < 2.4$ and $-2.2 < \eta < -1.2$ respectively with a full azimuth and measure muons. J/ψ particles are identified with the invariant mass of e^+e^- pairs measured in the Central Arms and $\mu^+\mu^-$ pairs measured in the Muon Arms. In Run-2, both Central Arms and the South Muon Arm were operational. In addition to these Arms, three kinds of interaction trigger counters have been used during the run. They are Beam-Beam Counters (BBC), Normalization Trigger Counters (NTC) and Zero-Degree Calorimeters (ZDC), which cover different pseudo-rapidity ranges. For the J/ψ analysis, only BBC-trigger events are used. The coverage of the BBC is $3.0 < |\eta| < 3.9$.

An approximately 100 nb^{-1} integrated-luminosity with a good vertex cut is useful for physics analyses out of 150 nb^{-1} recorded during the Run-2. For the J/ψ analysis described in this paper, 81 nb^{-1} is used for the muon channel and 48 nb^{-1} for the electron channel with appropriate run selections.

MUON CHANNEL MEASUREMENT

The South Muon Arm in PHENIX measures muons in the pseudo-rapidity range $-2.2 < \eta < -1.2$ and with a momentum $p > 2$ GeV/c. It consists of the Muon Tracking (MuTr) chambers inside the conical Muon Magnet to measure muon momenta and the Muon Identifier (MuID) chambers interleaved between the steel absorber to discriminate muons from charged hadrons and provide triggers.

A simple NIM-logic with a memory look-up is used to trigger events with muons in p+p collisions. Each MuID plane is divided into four quadrants by both vertical and horizontal lines at its center. If a muon passes through a certain quadrant, a coincidence of fired planes gives a "quadrant trigger". The number of the fired quadrants is counted and if it is one, a single-muon trigger is issued while a dimuon trigger is issued if it is more than one. Dimuon-trigger events are used for the J/ψ analysis. Inefficiency due to hardware dead time is measured to be small (1-2%).

Figure 1 shows invariant mass spectra for both opposite-sign muon pairs and same-sign pairs. There is a significant enhancement for the opposite sign pairs in the J/ψ mass region. Assuming the same spectra for opposite-sign and same-sign muon pairs from background, which is confirmed with simulation and real data, the number of J/ψ is obtained to be 36 ± 7 (stat.) ± 4 (syst.).

Acceptance times reconstruction efficiency for $J/\psi \to \mu^+\mu^-$ is obtained using a GEANT [3] simulation with the same reconstruction software as for the real data to be (1.63 ± 0.31)%. Dominant systematic uncertainties come from (1) MuID chamber efficiencies (11%), (2) MuTr chamber efficiencies (10%) and (3) Unknown spin-alignment (λ) of J/ψ (10%). For (3), $|\lambda| < 0.3$ is assumed which is consistent with both the lower energy and higher energy experiments [4, 5].

The integrated luminosity $\mathscr{L}$ used in this analysis is obtained as $\mathscr{L} = N_{BBC}/\varepsilon_{BBC}^{inela}\sigma_{inela}$ where N_{BBC} (1.73×10^9) is the number of BBC triggers with a small error, $\varepsilon_{BBC}^{inela}$ (0.51) is BBC efficiency for p+p inelastic events obtained using the PYTHIA [6] event generator and GEANT simulation and σ_{inela} is the p+p inelastic cross section which is obtained using the fit in [7] (42 mb). The value $\varepsilon_{BBC}^{inela}\sigma_{inela}$

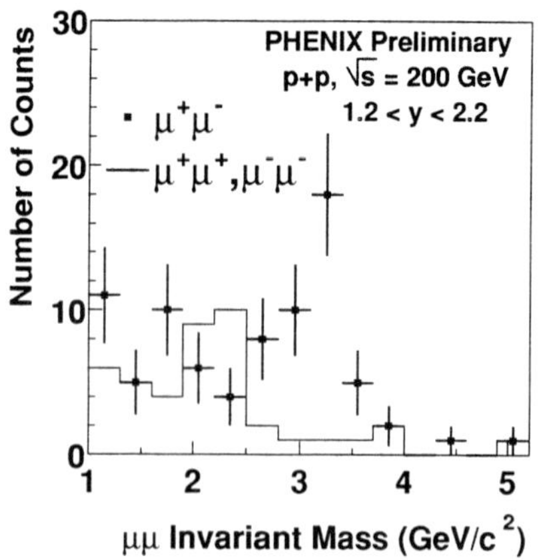

FIGURE 1. Invariant mass spectra for unlike-sign and like-sign muon pairs. The error bars include statistical errors only.

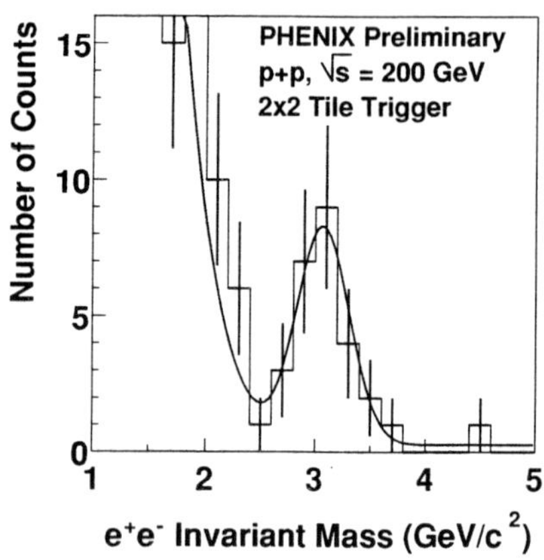

FIGURE 2. Invariant mass spectrum for e^+e^- pairs with a Gaussian fit to the J/ψ peak. The error bars include statistical errors only.

is compared with the one obtained with the van der Meer scan results and they are consistent within 20%. BBC efficiency for p+p $\rightarrow J/\psi \rightarrow \mu^+\mu^-$ events is also needed for the normalization and obtained to be 0.74 with simulation.

Consequently, the branching fraction for the decay $J/\psi \rightarrow \mu^+\mu^-$ times the cross section for the J/ψ production in the Muon Arm acceptance $Br(J/\psi \rightarrow \mu^+\mu^-)d\sigma_{J/\psi}/dy|_{y=1.7}$ is obtained to be 37 ± 7 (stat.) $\pm$ 11 (syst.) nb.

The average transverse momentum ($\langle p_T \rangle$) of J/ψ is also obtained to be 1.66 ± 0.18 (stat.) $\pm$ 0.09 (syst.) GeV/c in the limit of our measurement ($0 < p_T < 5$ GeV/c). A correction on $\langle p_T \rangle$ due to missing high p_T events is expected to be small (3%) assuming the function form which describes the p_T spectra of the lower energy measurements well [8].

ELECTRON CHANNEL MEASUREMENT

Electrons are tracked by the Central Tracking Chambers (Drift Chambers and Pad Chambers) and identified by the Ring-Imaging Čerenkov detectors and Electro-Magnetic Calorimeters (EMCal) in the Central Arms [9].

The EMCal was used also for triggering electrons. A trigger is fired when at least one energy sum of EMCal 2×2 towers exceeds the threshold, which is 0.8 GeV/c. The efficiency of this trigger for $J/\psi \rightarrow e^+e^-$ is obtained to be $0.90^{+0.06}_{-0.07}$ with a Monte-Carlo simulation tuned to reproduce single-photon efficiencies in the real data well.

Figure 2 shows an invariant mass spectrum for e^+e^- pairs. The peak of J/ψ is clearly seen with a small background. The number of J/ψ is obtained to be 24 ± 6 (stat.) $\pm$ 4 (syst.).

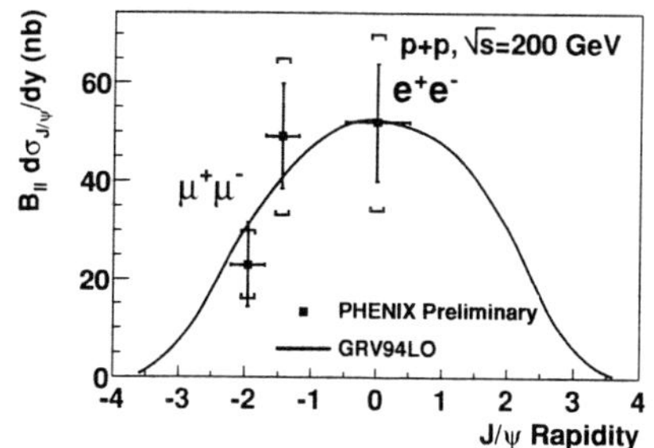

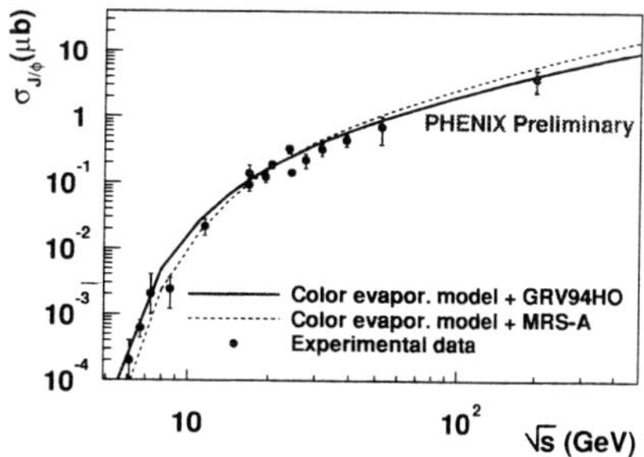

FIGURE 3. J/ψ rapidity differential cross section including both the e^+e^- and $\mu^+\mu^-$ measurements. The error bars show statistical errors while systematic errors are shown with the brackets. The curve is for the PYTHIA prediction with the GRV94-LO PDFs.

FIGURE 4. Center-of-mass energy dependence of the total production cross section for J/ψ in nucleon-nucleon collisions. The error bars include both statistical and systematic errors added in quadrature. The two curves show the color-evaporation model predictions described in [10].

Acceptance times reconstruction efficiency for $J/\psi \to e^+e^-$ is obtained to be (1.63 $\pm$ 0.20)%. The azimuthal coverage of the Central Arm detectors gives p_T dependence of the acceptance. An systematic error on the acceptance due to the uncertainty on the p_T distribution is estimated to be 7%.

The integrated luminosity used for the electron channel analysis and BBC efficiency for p+p $\to J/\psi \to e^+e^-$ events are obtained in the same way as the muon analysis.

As a result, the branching fraction times the cross section for the J/ψ production at mid rapidity $Br(J/\psi \to e^+e^-)d\sigma_{J/\psi}/dy|_{y=0} = 52 \pm 13(\text{stat.}) \pm 18(\text{syst.})$ nb is obtained.

RAPIDITY DISTRIBUTION AND TOTAL CROSS SECTION

Figure 3 shows the rapidity differential cross section for the J/ψ production including both the electron and muon channel measurements, which is consistent with the PYTHIA distribution with the GRV94-LO parton distribution functions (PDFs). The choice of PDF affects the rapidity distribution slightly, thus changes the acceptance estimation. This effect, however, is found to be small (3%). Total cross section is extracted to be $\sigma(p+p \to J/\psi X) = 3.8 \pm 0.6(\text{stat.}) \pm 1.3(\text{syst.})\mu\text{b}$ using $Br(J/\psi \to l^+l^-)$ [2] = $(5.9 \pm 0.1)\%$ [7].

Figure 4 shows the center-of-mass energy dependence of the total cross section for the J/ψ production in nucleon-nucleon collisions including the results of the lower energy experiments. The solid and dotted lines show the color-evaporation model predictions with two different sets of PDFs, QCD scales and charm quark masses described in [10]. The color-octet model can also reproduce these experimental results including the color-

[2] The average value of $Br(J/\psi \to e^+e^-)$ and $Br(J/\psi \to \mu^+\mu^-)$

octet matrix elements described in [11]. Our result is consistent with the perturbative QCD prediction with appropriate parameters and gluon density.

CONCLUSION

We have successfully identified J/ψ particles in $\sqrt{s}$ = 200 GeV p+p collisions both at forward rapidity ($1.2 < \eta < 2.2$) using the muon decay channel and at central rapidity ($|\eta| < 0.35$) using the electron decay channel with the PHENIX detector at RHIC. Cross sections for the J/ψ production in those rapidity ranges are obtained to be $Br(J/\psi \to \mu^+\mu^-)d\sigma_{J/\psi}/dy|_{y=1.7} = 37 \pm 7(\text{stat.}) \pm 11(\text{syst.})$ nb and $Br(J/\psi \to e^+e^-)d\sigma_{J/\psi}/dy|_{y=0} = 52 \pm 13(\text{stat.}) \pm 18(\text{syst.})$ nb respectively. Total production cross section is extracted from these measurements to be $\sigma_{J/\psi} = 3.8 \pm 0.6(\text{stat.}) \pm 1.3(\text{syst.})\mu\text{b}$ which is consistent with the perturbative QCD calculation.

REFERENCES

1. T. Matsui and H. Satz, Phys. Lett. **B178**, 416 (1986).
2. R.L. Thews *et al.*, Phys. Rev. **C63**, 054905 (2001).
3. GEANT User's Guide, 3.15, CERN Program Library.
4. C. Akerlof *et al.*, Phys. Rev. **D48**, 5067 (1993), A. Gribushin *et al.*, Phys. Rev. **D53**, 4723 (1996), T. Alexopoulos *et al.*, Phys. Rev. **D55**, 3927 (1997), C. Biino *et al.*, Phys. Rev. Lett. **58**, 2523 (1987).
5. T. Affolder *et al.*, Phys. Rev. Lett. **85**, 2886 (2000).
6. T. Sjöstrand, Comp. Phys. Comm. **82**, 74 (1994).
7. K. Hagiwara *et al.*, Phys. Rev. **D66**, 010001 (2002).
8. M. H. Schub *et al.*, Phys. Rev. **D52**, 1307 (1995), **D53**, 570 (1996), A. Gribushin *et al.*, Phys. Rev. **D62**, 012001 (2000).
9. J. T. Mitchell *et al.*, Nucl. Instr. Meth. **A482**,491 (2002).
10. J. F. Amundson *et al.*, Phys. Lett. **B390**, 323 (1997).
11. M. Beneke and I. Z. Rothstein, Phys. Rev. **D54** 2005 (1996).

Charmed Hadron Production in Polarized pp Collisions

Toshiyuki Morii* and Kazumasa Ohkuma†

*Division of Sciences for Natural Environment, Faculty of Human Development, Kobe University, Nada, Kobe 657-8501, JAPAN
Electronic address: morii@kobe-u.ac.jp
†Department of Physics, Faculty of Engineering, Yokohama National University, Hodogaya, Yokohama 240-8501, JAPAN
Electronic address: ohkuma@phys.ynu.ac.jp

Abstract. To extract information about polarized gluons in the proton, production of charmed hadrons, in particular, Λ_c^+ baryon in pp collisions was studied. We calculated the transverse momentum distribution and the pseudo-rapidity distribution of the spin correlation asymmetry A_{LL} between the initial proton and the produced Λ_c^+. Those statistical sensitivities were also calculated under the condition of RHIC experiment. We found that the pseudo-rapidity distribution of A_{LL} is promising for testing the model of polarized gluons in the proton and also the spin-dependent fragmentation model of a charm quark decaying into Λ_c^+ baryon.

INTRODUCTION

The Relativistic Heavy Ion Collider (RHIC) at Brookhaven National Laboratory has just started to explore the internal structure of proton. One of the important purposes of those RHIC experiments is to study the behavior of polarized gluons in the proton. As is well known, the proton spin is given by the sum of the spin carried by quarks $\Delta\Sigma$ and gluons ΔG, and their orbital angular momenta $\langle L_Z \rangle$. In these years, a great deal of efforts have been made for extracting those components from polarized structure functions of nucleons[1]. Based on the next–to–leading order QCD analyses on the polarized structure functions $g_1(x)$, the contribution of quarks to the proton spin is well known. However, knowledge on polarized gluons in the proton is still poor. To understand the origin of the nucleon spin, it is very important to know how gluons polarize in the nucleon. So far, several interesting processes have been proposed for extracting ΔG. Here we also propose a different process to obtain more detailed information on polarized gluons, expecting the forthcoming RHIC experiment. The processes which we propose here are the polarized charmed hadron production, i.e. $p\vec{p} \rightarrow \vec{\Lambda}_c^+ X$ and $p\vec{p} \rightarrow \vec{D}^* X$, in the polarized proton–unpolarized proton collision,[1] which will be observed at the RHIC experiment. We study which observables are useful for extracting information about polarized gluons in the proton and also discuss its sensitivity.

[1] Though we have calculated even for D^* production, we focus only on the Λ_c^+ production in this report, because the main point of the result remain unchanged.

CP675, *Spin 2002: 15th Int'l. Spin Physics Symposium and Workshop on Polarized Electron Sources and Polarimeters,* edited by Y. I. Makdisi, A. U. Luccio, and W. W. MacKay

Λ_C^+ BARYON PRODUCTION IN PROTON–PROTON COLLISION

In the process on which we focus here, the Λ_c^+ baryon is expected to have some advantageous properties for probing behavior of polarized gluons in the proton. Those properties are as follows;

1. A charm quark which is one of constituents of the Λ_c^+ baryon is dominantly produced via gluon-gluon fusion in pp reaction, because charm quarks are tiny contents in the proton. Thus, the cross section of this process is directly proportional to the gluon distribution in the proton.
2. Since the Λ_c^+ baryon is composed of a charm quark and antisymmetrically combined light up and down quarks, the spin of Λ_c^+ baryon is expected to be almost equal to the spin of its constituent charm quark.
3. Since a charm quark is heavy, it must be very rare for the charm quark to change its spin arrangement during its fragmentation into a Λ_c^+ baryon. In other words, the spin direction of the charm quark produced in the subprocess is expected to be kept in the Λ_c^+ baryon produced in the final state.

After all, the spin of the Λ_c^+ is in strong correlation to the polarization of gluons in the proton. Therefore, by observing the spin correlation between the polarized proton in the initial state and the polarized Λ_c^+ in the final state, we can get, rather clearly, information on the polarized gluon in the proton.

SPIN CORRELATION ASYMMETRY AND ITS STATISTICAL SENSITIVITY

To study the polarized gluon distribution in the proton, we introduced the spin correlation asymmetry of the target polarized-proton and produced Λ_c^+ baryon [2];

$$A_{LL} = \frac{d\sigma_{++} - d\sigma_{+-} + d\sigma_{--} - d\sigma_{-+}}{d\sigma_{++} + d\sigma_{+-} + d\sigma_{--} + d\sigma_{-+}} \equiv \frac{d\Delta\sigma/dX}{d\sigma/dX}, \quad (X = p_T \text{ or } \eta), \tag{1}$$

where $d\sigma_{+-}$, for example, denotes the spin-dependent differential cross section with the positive helicity of the target proton and the negative helicity of the produced Λ_c^+ baryon. p_T and η, which are represented by X in Eq.(1), are transverse momentum and pseudo-rapidity of produced Λ_c^+, respectively.

According to the quark-parton model, $d\Delta\sigma/dX$ can be expressed as

$$\frac{d\Delta\sigma}{dX} = \int_{Y^{\min}}^{Y^{\max}} \int_{x_a^{\min}}^{1} \int_{x_b^{\min}}^{1} G_{p_A \to g_a}(x_a, Q^2) \Delta G_{\vec{p}_B \to \vec{g}_b}(x_b, Q^2) \Delta D_{\vec{c} \to \vec{\Lambda}_c^+}(z) \times \frac{d\Delta\hat{\sigma}}{d\hat{t}} J dx_a dx_b dY, \quad (X, Y = \eta \text{ or } p_T \ (X \neq Y)), \tag{2}$$

with

$$J \equiv \frac{2s\beta p_T^2 \cosh\eta}{z\hat{s}\sqrt{m_c^2 + p_T^2\cosh^2\eta}}, \quad \beta \equiv \sqrt{1 - \frac{4m_p^2}{s}}$$

where $G_{p_A \to g_a}(x_a, Q^2)$, $\Delta G_{\vec{p}_B \to \vec{g}_b}(x_b, Q^2)$ and $\Delta D_{\vec{c} \to \vec{\Lambda}_c^+}(z)$ represent the unpolarized gluon distribution function, the polarized gluon distribution function and the spin-dependent fragmentation function of the outgoing charm quark decaying into a polarized $\vec{\Lambda}_c^+$, respectively. $d\Delta\hat{\sigma}/d\hat{t}$ is the spin-dependent differential cross section of the subprocess and J is the Jacobian which transforms the variables z and $\hat{t}$ into p_T and η. In the expression of Eq.(2), p_T and η are described as X or Y.

Statistical sensitivities of A_{LL} for the p_T and η distribution are estimated by using the following formula;

$$\delta A_{LL} \simeq \frac{1}{P} \frac{1}{\sqrt{b_{\Lambda_c^+} \, \varepsilon \, L \, T \, \sigma}}. \tag{3}$$

To numerically estimate the value of δA_{LL}, here we use following parameters: operating time; T =100-day, the beam polarization; P =70%, a luminosity; $L = 8 \times 10^{31}$ (2×10^{32}) cm^{-2} sec^{-1} for $\sqrt{s} = 200$ (500) GeV, the trigger efficiency; $\varepsilon = 10\%$ for detecting produced Λ_c^+ events and a branching ratio; $b_{\Lambda_c^+} \equiv \mathrm{Br}(\Lambda_c^+ \to pK^-\pi^+) \simeq 5\%$ [3]. The purely charged decay mode is needed to measure the polarization of produced Λ_c^+. σ denotes the unpolarized cross section integrated over suitable p_T or η region.

NUMERICAL CALCULATIONS

To carry out the numerical calculation of A_{LL}, we used, as input parameters, $m_c = 1.20$ GeV, $m_p = 0.938$ GeV and $m_{\Lambda_c^+} = 2.28$ GeV[3]. We limited the integration region of η and p_T of produced Λ_c^+ as $-1.3 \lesssim \eta \lesssim 1.3$ and 3 GeV $\lesssim p_T \lesssim$ 15(40) GeV, respectively, for $\sqrt{s} = 200(500)$ GeV. The range of η and the lower limit of p_T were selected in order to get rid of the contribution from the diffractive $\Lambda_c^!$ production. As for the upper limit of p_T, we took it as described above, for simplicity, though the kinematical maximum of p_T of produced Λ_c^+ is slightly larger than 15 GeV and 40 Gev for $\sqrt{s}$ =200 GeV and 500 GeV, respectively. In addition, we took the AAC[4] and GRSV01 [5] parameterization models for the polarized gluon distribution function and the GRV98 [6] model for the unpolarized one. Though both of AAC and GRSV01 models excellently reproduce the experimental data on the polarized structure function of nucleons $g_1(x)$, the polarized gluon distributions for those models are quite different. In other words, the data on polarized structure function of nucleons $g_1(x)$ alone are not enough to distinguish the model of gluon distributions. Since the process is semi-inclusive, the fragmentation function of a charm quark to Λ_c^+ is necessary to carry out numerical calculations. For the unpolarized fragmentation function, we used Peterson fragmentation function, $D_{c \to \Lambda_c^+}(z)$ [3, 7]. However, since we have no data, at present, about polarized fragmentation functions for the polarized Λ_c^+ production, we took the

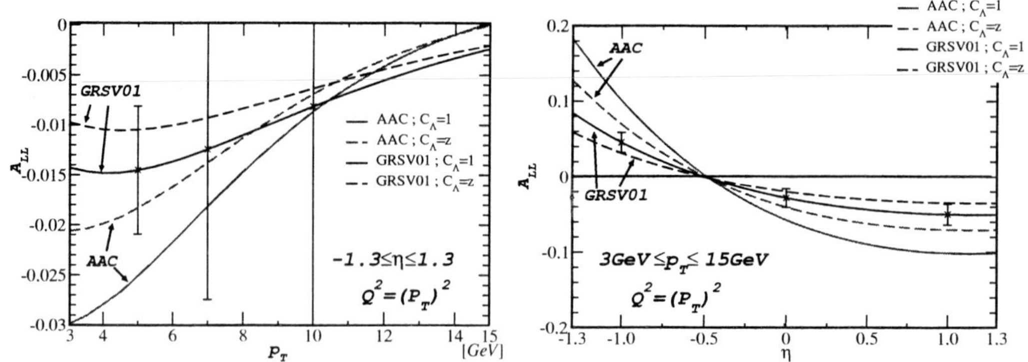

FIGURE 1. A_{LL} as a function of p_T (left panel) and η (right panel) at $\sqrt{s}$ =200 GeV

following ansatz for the polarized fragmentation function $\Delta D_{\vec{c}\to\vec{\Lambda}_c^+}(x)$,

$$\Delta D_{\vec{c}\to\vec{\Lambda}_c^+}(z) = C_{c\to\Lambda_c^+} D_{c\to\Lambda_c^+}, \tag{4}$$

where $C_{c\to\Lambda_c^+}$ is a scale-independent spin transfer coefficient. In this analysis, we studied two cases: (A) $C_{c\to\Lambda_c^+} = 1$ (non-relativistic quark model) and (B) $C_{c\to\Lambda_c^+} = z$ (Jet fragmentation model [8]). As we discussed before, if the spin of Λ_c^+ is same as the spin of charm quark produced in subprocess, the model (A) might be a reasonable scenario.

Numerical results of A_{LL} are shown in Fig. 1 and Fig. 2. In those figures, statistical sensitivities, δA_{LL}, are also attached to the solid line of A_{LL} which is calculated for the case of the GRSV01 parametrization model of polarized gluon and the non-relativistic fragmentation model. [2] From these results, we see that the η distributions of A_{LL} are more effective observables than the p_T distributions at $\sqrt{s}$ =200 GeV and 500 GeV. As shown in the right panel of Fig. 2 given at $\sqrt{s}$ =500 GeV, we could distinguish the parametrization models of polarized gluon as well as the models of the spin-dependent fragmentation function though the magnitude of A_{LL} is rather small. At $\sqrt{s}$ =200 GeV, the magnitude of A_{LL} for η distribution becomes larger, though statistical sensitivities are not so small. If the integrated luminosity at $\sqrt{s}$ =200 GeV could be large and the detection efficiency, ε, is improved, this observable could be promising to distinguish not only the models of $\Delta G(x)$ but also the models of $\Delta D(z)$. For the p_T distribution of A_{LL}, δA_{LL} become rapidly larger with increasing p_T and we cannot say anything from those region. However, if we confine the kinematical region in rather small p_T range like $p_T = 3 \sim 5(10)$ GeV at $\sqrt{s} = 200(500)$ GeV, it might be still effective.

---

[2] Note that as shown from Eq.(4), δA_{LL} does not depend on both of the model of polarized gluons and the model of fragmentation functions.

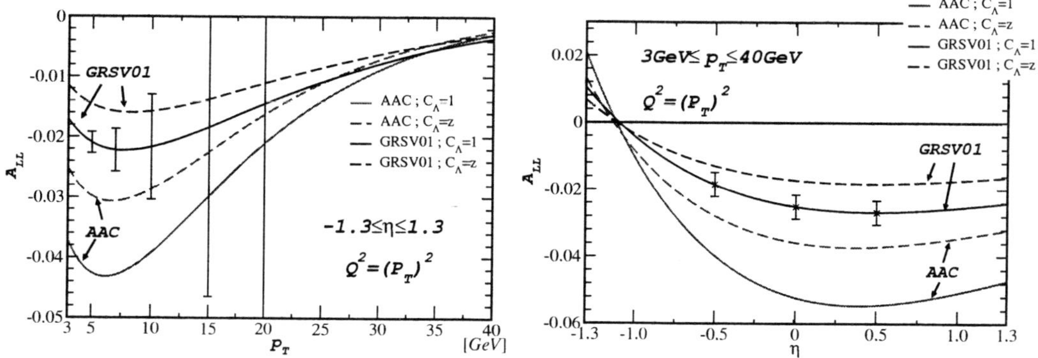

FIGURE 2. The same as in Fig. 1, but for $\sqrt{s}=500$ GeV

CONCLUDING REMARK

To extract information on the polarized gluon distribution in the proton, the charmed hadron production processes at RHIC experiments have been proposed. (Actually, only Λ_c^+ process was discussed in this report.) The spin correlation asymmetry A_{LL} between the initial proton and the produced Λ_c^+ was calculated for p_T and η distributions with statistical sensitivities which were estimated using RHIC parameters. We found that A_{LL} is rather sensitive to the model of $\Delta G(x)$ and $\Delta D(z)$. The η distribution of A_{LL} could be promising for distinguishing the parametrization model of polarized gluons as well as the model of spin-dependent fragmentation of a charm quark into Λ_c^+.

ACKNOWLEDGMENTS

One of the authors, (K.O), would like to thank the organizers and fellowship committee of SPIN2002 for giving a chance to present this contribution at the symposium.

REFERENCES

1. For a review see: H. Y. Cheng, *Int. J. Mod. Phys.* A**11**, 5109, (1996); B. Lampe and E. Reya, *Phys. Rep.* **332**, 1,(2000); H. Y. Cheng, *Chin. J. Phys.* **38**, 753, (2000) [hep-ph/0002157]; B. W. Filippone and X. Ji, hep-ph/0101224.
2. K. Ohkuma, K. Sudoh and T. Morii, *Phys. Lett.* B**491**,117, (2000) [*Erratum-ibid.* B**543**, 323, (2002)]; K. Ohkuma, T. Morii and S. Oyama, hep-ph/0201144.
3. K. Hagiwara *et al*, *Phys. Rev.* D**66**, 010001,(2002).
4. Y. Goto *et al.* [Asymmetry Analysis Collaboration], *Phys. Rev.* D **62**, 034017, (2001).
5. M Glück, E. Reya, M. Stratmann and W. Vogelsang, *Phys. Rev.* D **63**, 094005, (2001).
6. M. Glück, E. Reya and A. Vogt, *Eur. Phys. J.* C **5**, 461, (1998)
7. C. Peterson, D. Schlatter, I. Schmitt and P. M. Zerwas, *Phys. Rev.* D **27**, 105, (1983).
8. A. Bartl, H. Fraas and W. Majerotto, *Z. Phys.* C **6**, 335, (1980).

Polarized Hadroproduction of Open Heavy Quarks in NLO QCD at JHF and RHIC

Ingo Bojak

CSSM, The University of Adelaide, SA 5005, Australia

Abstract. We present the complete next-to-leading order QCD corrections to the polarized hadroproduction of heavy flavors. This reaction can be studied experimentally in polarized *pp* collisions at the JHF and at the BNL RHIC in order to constrain the polarized gluon density. It is demonstrated that the dependence on the unphysical renormalization and factorization scales is strongly reduced beyond the leading order. We also discuss how the high luminosity at the JHF can be used to control remaining theoretical uncertainties. An effective method for bridging the gap between theoretical predictions for heavy quarks and experimental measurements of heavy meson decay products is introduced briefly.

INTRODUCTION

The gluon helicity density Δg remains weakly constrained [1, 2]. Current data are compatible with $\Delta g \equiv 0$ at a low input scale $\mu_{f0}^2 = 0.4$ GeV^2, but even full saturation cannot be excluded [1]. Hence the gluonic contribution to the nucleon spin is unknown: $-0.8 \lesssim \Delta g_{n=1}(5\text{ GeV}^2) \lesssim 1.7$. Polarized DIS data has been the only source of information so far. But the severely restricted kinematical range, in which approximate "scaling" holds, thwarts attempts to pin down the gluon. Furthermore, the helicity sum rule is useless for constraining Δg without independent angular momentum measurements. Data from exclusive processes may ameliorate the polarized parton fits.

Two experiments will be or are collecting data for the hadroproduction of open heavy quarks: the JHF and the BNL RHIC [3]. We provide here the first corresponding complete calculation in NLO QCD. Note that our calculation is also necessary for obtaining the *resolved* photon contributions in NLO QCD for our older NLO photoproduction analysis [4, 5].

RESULTS FOR THE JHF

In Fig. 1 *left* we show the ratio of $\Delta\sigma_{ij}/\Delta\sigma_{\text{tot}}$ at JHF energies of $\sqrt{S} = 10$ GeV, with the subprocesses $ij = gg,\ q\bar{q},\ gq + g\bar{q}$ and $\Delta\sigma_{\text{tot}} = \sum \Delta\sigma_{ij}$. The ratio for the gluon-gluon fusion is shown in thick lines and the ratio for the quark-antiquark annihilation is shown by thin lines. The gluon-(anti)quark subprocess contributes little and is omitted. Note that given the total asymmetry $A \equiv \Delta\sigma_{\text{tot}}/\sigma_{\text{tot}}$, if we require $A \equiv \sum A_{ij}$ of the subprocesses, then the shown ratio corresponds to the subprocess asymmetry contribution $A_{ij} = \Delta\sigma_{ij}/\Delta\sigma_{\text{tot}}$.

CP675, *Spin 2002: 15th Int'l. Spin Physics Symposium and Workshop on Polarized Electron Sources and Polarimeters*, edited by Y. I. Makdisi, A. U. Luccio, and W. W. MacKay

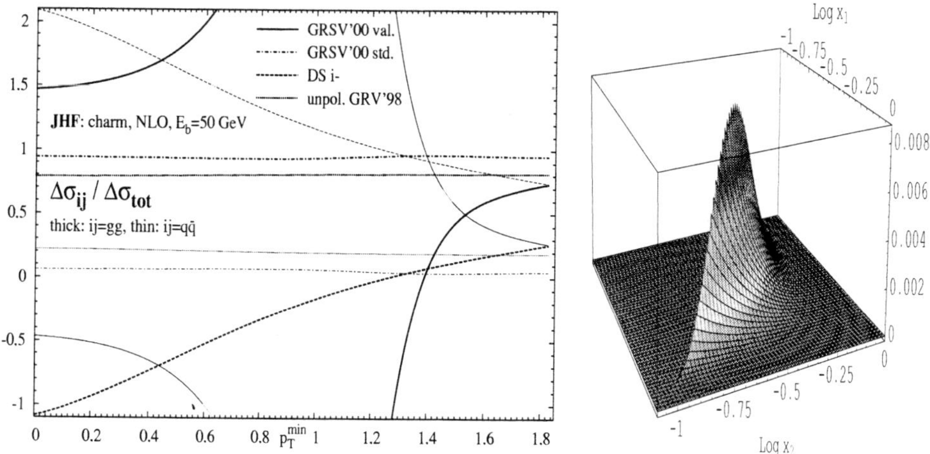

FIGURE 1. *Left:* Subprocess importance depending on $p_T \geq p_T^{\min}$ for different helicity densities at the JHF. *Right:* Gluon-gluon contribution to $\Delta\sigma$ depending on the x_1 and x_2 of the gluons. See text for details.

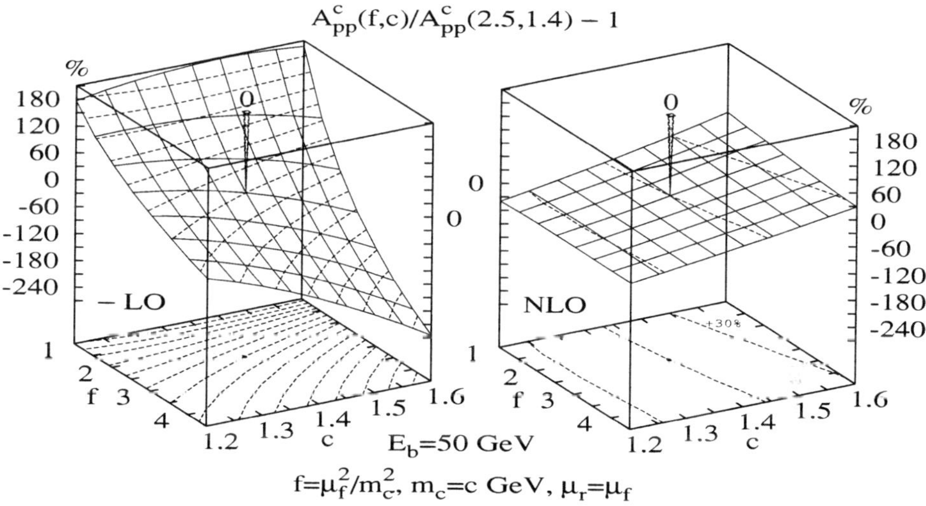

FIGURE 2. Deviation in % of the asymmetry from the marked value as $\mu_f^2 = \mu_r^2$ and m_c are varied. The GRSV'00 std. helicity densities [1] are used. LO is multiplied by -1 and 30% contour steps are shown.

Several helicity density sets have been used [1, 2], and $\sigma_{ij}/\sigma_{\text{tot}}$ with the unpolarized GRV'98 distributions [6] is shown for comparison. The curves depend on a cut in transverse momentum $p_T \geq p_T^{\min}$. The polarized GRSV'00 std. and the unpolarized GRV'98 curves show optimal behavior: the gluon-gluon subprocess dominates and there is almost no dependence on the p_T-cut. For the smaller GRSV'00 val. set Δg the quark-antiquark subprocess contributes significantly. Also there is now a strong dependence on the cut (although the "pole" is due to $\Delta\sigma_{\mathrm{tot}} = 0$). For the very small DS i- Δg, quark-antiquark annihilation dominates. Hence at the small JHF energy, one cannot simply assume gluon-gluon fusion dominance.

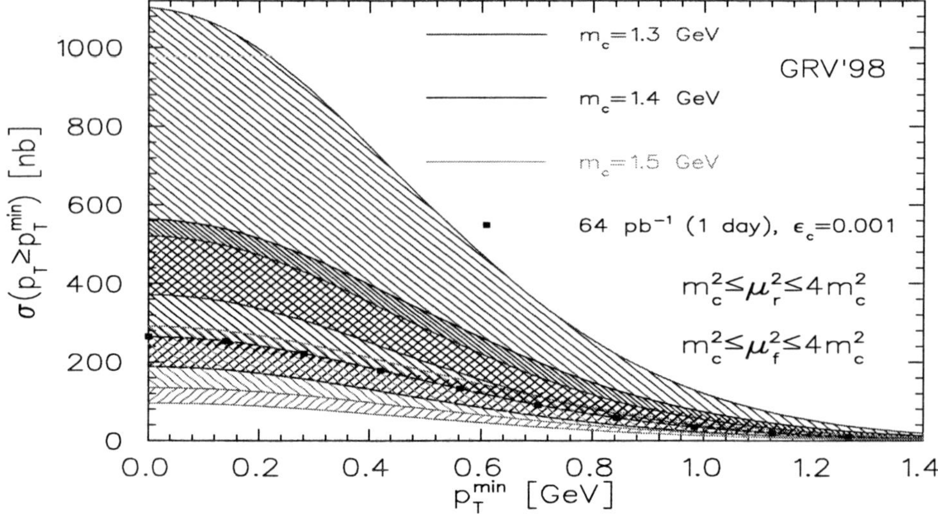

FIGURE 3. The NLO cross section at JHF depending on a cut $p_T \geq p_T^{\mathrm{min}}$ as μ_f^2, μ_r^2 and m_c are varied. The GRV'98 distributions [6] are used. The statistical error after one day using $\mathcal{L} = 64\ \mathrm{pb}^{-1}$ is shown.

Given a large enough Δg, which regions of x does gluon-gluon fusion probe? From kinematics we have $x_1 x_2 \geq 4m^2/S$ for the momentum fractions of the gluons. In Fig. 1 *right* we show

$$\frac{x_1 x_2}{\alpha_s^2/m^2}\, \Delta g(x_1, \mu_f^2) \Delta g(x_2, \mu_f^2) \Delta\hat{\sigma}_{gg}(x_1 x_2)\,, \tag{1}$$

with $\mu_f^2 = 2.5 \cdot (1.4\ \mathrm{GeV})^2$, Δg of the GRSV'00 std. set, and the partonic cross section $\Delta\hat{\sigma}_{gg}$. The $x_1 x_2$ is multiplied to give the appropriate volume impression with logarithmic axes and a general α_s^2/m^2 dependence has been divided out. Apart from that the integrals over x_1 and x_2 of (1) gives the hadronic $\Delta\sigma_{gg}$. The main contribution comes from the kinematic edge $x_1 x_2 \simeq 4m^2/S$. Furthermore it is peaked at $x_1 \simeq x_2$. Hence $\hat{x} \simeq \sqrt{4m^2/S}$ is mainly probed. At the JHF for charm $\hat{x}_c \simeq 0.3$ and at the BNL RHIC $\hat{x}_c \simeq 0.01$, 0.006 for $\sqrt{S} = 200$, 500 GeV. For photoproduction with $x_2 \equiv 1$ one can similarly show $\hat{x} \simeq 4m^2/S$. Hence the COMPASS experiment [7], at the same center-of-mass energy $\sqrt{S} = 10$ GeV as the JHF, further complements the probed range with $\hat{x}_c \simeq 0.08$.

In Fig. 2 we show the deviation in % from a central prediction of the asymmetry ($\mu_r^2 = \mu_f^2 = 2.5\ m_c^2$ and $m_c = 1.4$ GeV) as renormalization and factorization scales, which are set equal, and the charm mass are varied. We see a massive reduction in the the theoretical uncertainty in NLO as compared to LO, basically only the dependence on m_c of $\pm 30\%$ is left. However, Fig. 3 shows that this amazing improvement is partly due to strong cancellations in $\Delta\sigma/\sigma$. The underlying uncertainty of σ is massive. However,

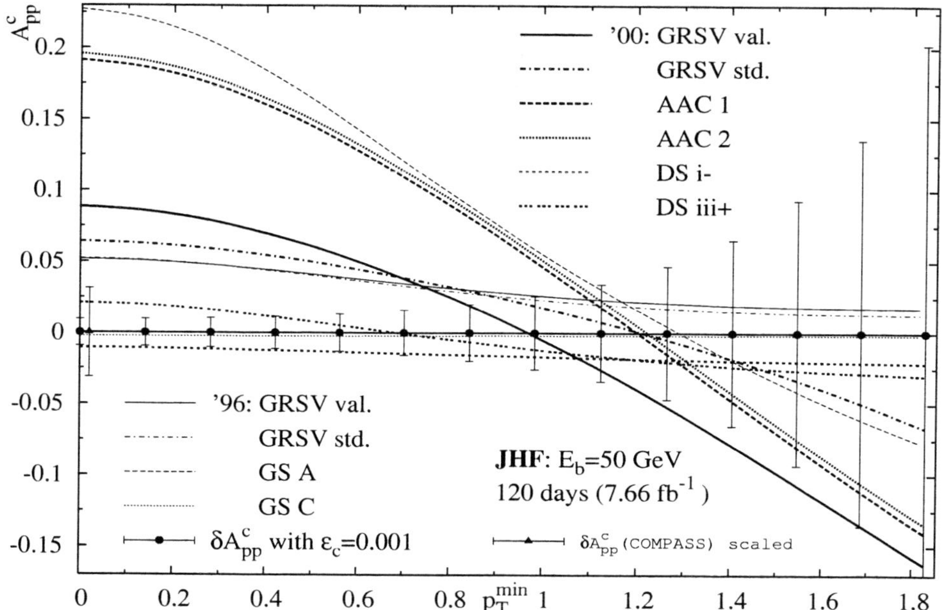

FIGURE 4. The NLO charm asymmetry A at $\sqrt{S} = 10\,\text{GeV}$ for JHF depending on a cut $p_T \geq p_T^{\text{min}}$. An estimate for the statistical error after 120 days using $\mathscr{L} = 7.66\,\text{fb}^{-1}$ is shown.

the error bars of the shown statistical accuracy of *one day* of measurements at the JHF are too small to be seen. Hence there is hope that the scales can be pinned down first in precise unpolarized measurements.

Finally, we present in Fig. 4 predictions for the JHF with a range of older and newer helicity distributions [1, 2]. When compared to the expected statistical uncertainty at the JHF with a detection efficiency of $\varepsilon_c = 0.001$, we see that one will be able to see any but the smallest gluons and that one can even clearly distinguish between several groups of sets. The situation is similar at the BNL RHIC, as we will see below.

RESULTS FOR THE BNL RHIC

Obtaining similar predictions for the PHENIX experiment at the BNL RHIC is more involved, because its detector cannot reasonably be approximated by an uniform detection efficiency ε_c. We use the software employed by PHENIX to generate a ε_{eff} depending on p_T and the pseudo-rapidity η, which takes into account hadronization, decay product cuts, and the detector acceptance. For Fig. 5, see [8] for more details, ε_{eff} is then convoluted with our double differential partonic results

$$\tilde{\sigma}_{\text{eff}}(p_T^e > 1\,\text{GeV}) = \int_0^{p_T^{\text{max}}} dp_T \int_{-\eta^{\text{max}}}^{\eta^{\text{max}}} d\eta\; \varepsilon_{\text{eff}}(p_T, \eta; p_T^e > 1\,\text{GeV})\, \frac{d^2\tilde{\sigma}}{dp_T d\eta}\,. \quad (2)$$

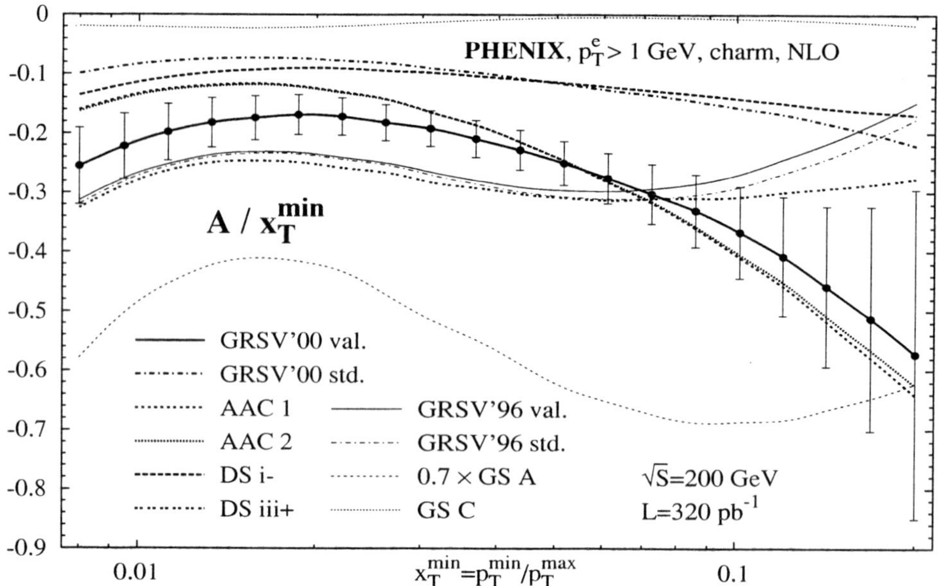

FIGURE 5. The NLO charm asymmetry A at $\sqrt{S} = 200$ GeV for PHENIX depending on $x_T^{\min} = p_T^{\min}/p_T^{\max}$. A is rescaled by $1/x_T^{\min}$. An estimate for the statistical error using $\mathscr{L} = 320\,\mathrm{pb}^{-1}$ is shown.

CONCLUSIONS

We have shown that the NLO predictions for the JHF and the BNL RHIC show great promise for pinning down Δg at x values of about 0.3 and 0.01 (200 GeV), 0.006 (500 GeV), respectively. The theoretical uncertainties for the asymmetry are much improved in NLO, but the large uncertainty of the unpolarized cross section should be reduced first by employing the fantastic luminosity of the JHF. We have also shown a method of taking into account the complicated experimental setup of PHENIX at the BNL RHIC.

REFERENCES

1. M. Glück *et al.*, *Phys. Rev.*, **D63**, 094005 (2001).
2. M. Glück *et al.*, *Phys. Rev.*, **D53**, 4775 (1996); T. Gehrmann and W.J. Stirling, *Phys. Rev.*, **D53**, 6100 (1996); D. de Florian *et al.*, *Phys. Rev.*, **D57**, 5803 (1998); Y. Goto *et al.*, AA Collab., *Phys. Rev.*, **D62**, 034017 (2000); D. de Florian and R. Sassot, *Phys. Rev.*, **D62**, 094025 (2000).
3. F. Bradamante, *Prog. Part. Nucl. Phys.*, **44**, 339 (2000); G. Bunce *et al.*, *Ann. Rev. Nucl. Part. Sci.*, **50**, 525 (2000).
4. I. Bojak and M. Stratmann, *Phys. Lett.*, **B433**, 411 (1998), *Nucl. Phys.*, **B540**, 345 (1999) and Erratum *Nucl. Phys.*, **B569**, 694 (2000).
5. I. Bojak, Ph.D. Thesis, Universität Dortmund, 2000, `hep-ph/0005120`.
6. M. Glück *et al.*, *Eur. Phys. J.*, **C5**, 461 (1998).
7. COMPASS Collaboration, G. Baum *et al.*, reports CERN/SPSLC 96-14, -30.
8. I. Bojak and M. Stratmann, `hep-ph/0112276`, submitted to *Phys. Rev.*, **D**.

Measurement of Polarised Parton Distributions at HERMES

Antje Bruell
(on behalf of the HERMES collaboration)

Massachusetts Institute of Technology, 77 Massachusetts Avenue, Cambridge, MA 02139, USA

Abstract. During the last years the HERMES collaboration has measured inclusive and semi-inclusive double-spin asymmetries in deep inelastic scattering on polarised hydrogen and deuterium targets in the kinematic range $0.02 < x < 0.6$ and $1\,\text{GeV}^2 < Q^2 < 10\,\text{GeV}^2$. With the installation of a Ring Imaging Čerenkov detector in 1998, charged pions and kaons could be identified and for the first time allowed to separately determine the polarisations of all valence and sea quark distributions. Of special interest are the difference between the polarisation of the u- and d- sea quarks where various phenomenological models predict a significant flavour asymmetry and the polarisation of the strange quarks which have been suggested to have a negative contribution to the nucleon spin.

INTRODUCTION

The study of the spin structure of the nucleon has attracted great interest since the EMC experiment [1] reported that only a small fraction of the spin of the proton is carried by the spins of the quarks. Several experiments at CERN, SLAC and DESY have since confirmed this result. To further investigate the spin structure of the nucleon, both the SMC experiment and the HERMES experiment have measured not only inclusive deep-inelastic scattering but also semi-inclusive channels, in which a final state hadron is detected in coincidence with the scattered lepton. As the identity of fast hadrons from the fragmentation process is correlated with the flavour of the struck quark this "flavour-tagging" can be used to determine the polarised parton distribution functions for each quark flavour.

Assuming factorisation of the hard scattering and fragmentation processes, the photo-absorption double spin asymmetry A_1^h for the production of a hadron of type h can be written in LO QCD as

$$A_1^h(x,z,Q^2) = \frac{1+R(x,Q^2)}{1+\gamma^2} \frac{\sum_f e_f^2 \Delta q_f(x,Q^2) \int_{z_{min}}^{z_{max}} D_f^h(z,Q^2)dz}{\sum_f e_f^2 q_f(x,Q^2) \int_{z_{min}}^{z_{max}} D_f^h(z,Q^2)dz}. \quad (1)$$

Here $q_f(x,Q^2)$ denote the unpolarised parton densities, $\Delta q_f(x,Q^2) = q_f^{\uparrow}(x,Q^2) - q_f^{\downarrow}(x,Q^2)$ are the helicity dependent quark distributions, and $R = \sigma_L/\sigma_T$ is the ratio of the longitudinal and transverse virtual photon absorption cross sections. The fragmentation functions $D_f^h(z,Q^2)$ represent the probability that a struck quark of flavour f fragments into a hadron of type h with energy E_h and fractional energy $z = E_h/\nu$.

CP675, *Spin 2002: 15th Int'l. Spin Physics Symposium and Workshop on Polarized Electron Sources and Polarimeters*, edited by Y. I. Makdisi, A. U. Luccio, and W. W. MacKay

SEMI-INCLUSIVE ASYMMETRIES

At HERMES, longitudinally polarised positrons of 27.5 GeV are scattered deep-inelastically on polarised pure atomic hydrogen and deuterium gas targets. A large acceptance forward spectrometer with excellent lepton/hadron separation detects the scattered positrons in coincidence with final state hadrons. For the HERMES data on a longitudinally polarised hydrogen target a threshold Cerenkov detector provided pion identification for momenta above 4 GeV. Since the installation of the RICH detector in 1998, data have been collected on a longitudinally polarised deuterium target and pions, kaons and protons are identified over the entire momentum range.

In Fig. 1 the semi-inclusive asymmetries for charged hadrons, pions and kaons are presented as a function of x. Deep-inelastic scattering events were selected by imposing the kinematical constraints $Q^2 > 1$ GeV^2, $W^2 > 10$ GeV^2 and $y < 0.85$. To enhance the contribution of hadrons from the current fragmentation region and to supress contributions from exclusive events, the selection criteria $0.2 < z < 0.8$ and $x_F > 0.1$ were applied. A Monte Carlo simulation was used to correct the asymmetries for kinematic smearing in the variable x due to instrumental resolution and QED radiative effects.

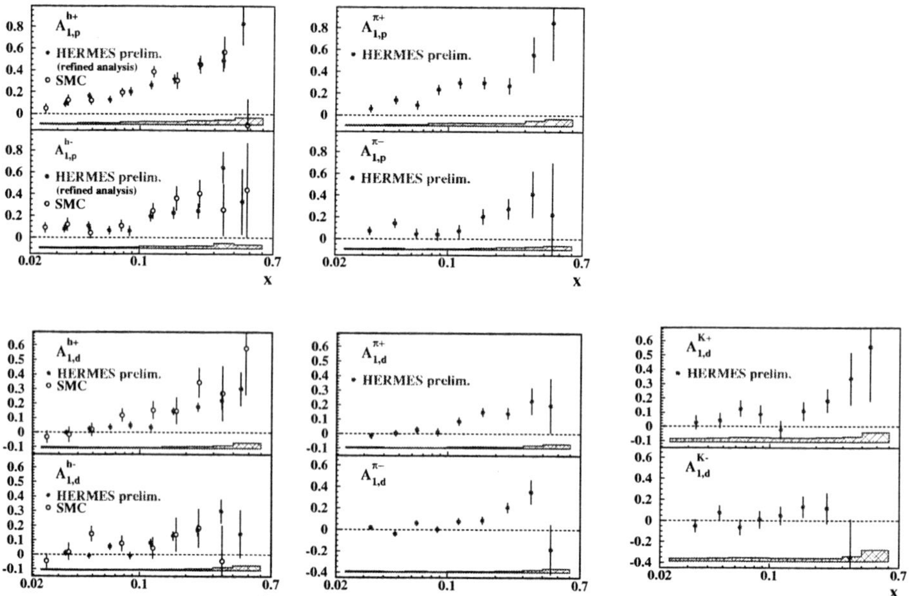

FIGURE 1. Semi-inclusive asymmetries for charged hadrons, pions and kaons on hydrogen (top panels) and deuterium (bottom panels) targets.

The most striking feature of the new preliminary HERMES results is the difference between the negative kaon asymmetry which is found to be consistent with zero over the entire x range and all other asymmetries which are positive for $x > 0.1$.

HELICITY DEPENDENT QUARK DISTRIBUTIONS

After integration over Q^2, Eq. 1 can be rewritten for any set of measured asymmetries

$$\vec{A}(x) = P(x) \cdot \vec{Q}(x) . \tag{2}$$

The vectors $\vec{A}(x)$ and $\vec{Q}(x)$ contain the measured inclusive and semi-inclusive asymmetries from the different targets and the polarisations $\frac{\Delta q}{q}$ of the different quark flavours to be extracted, respectively. The elements of the matrix $P(x)$ depend on the fragmentation functions, the unpolarised parton densities and the cross section ratio $R(x,Q^2)$. The matrix $P(x)$ also contains the influence of the limited acceptance of the spectrometer and was determined from a Monte Carlo simulation of the HERMES experiment where the fragmentation process was modelled in the LUND string model with parameters tuned to the HERMES hadron multiplicities [2]. The CTEQ5L parametrisation [3] was used for the unpolarised quark distributions.

Equation (2) has been solved for the vector $\vec{Q}(x)$ by a least squares minimisation technique. To avoid large uncertainties from correlations between the elements of the $\vec{Q}(x)$, the polarisation of the sea quarks was set to zero for $x > 0.3$.

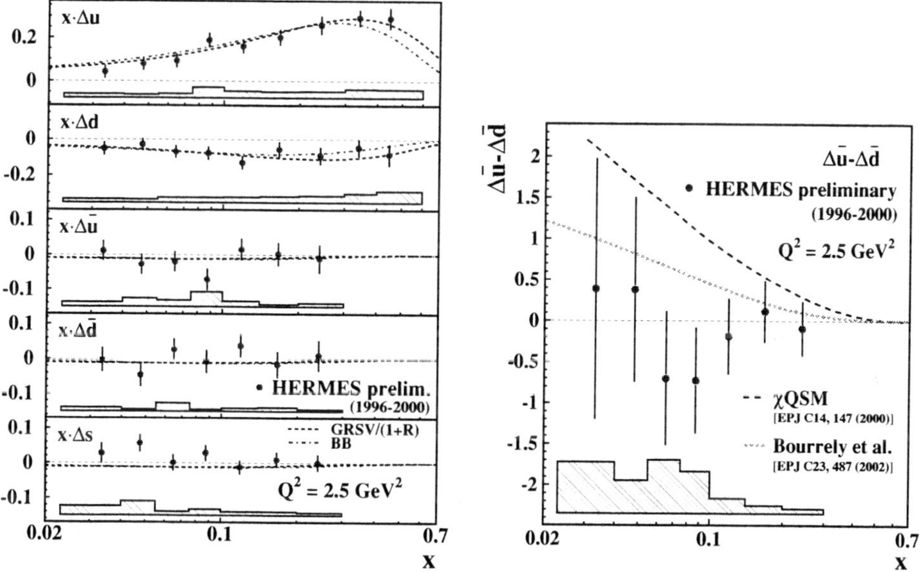

FIGURE 2. Left: the x-weighted polarized quark distributions $x\Delta q(x)$ as a function of x, as extracted from HERMES inclusive and semi-inclusive asymmetries on polarized hydrogen and deuterium targets. All data are evolved to $Q^2 = 2.5$ GeV^2. The dashed curves represent the GRSV00 parametrisation [4], the dot-dashed curves correspond to the results of a LO QCD analysis by Blümlein and Böttcher [5]. **Right:** the difference $\Delta\bar{u} - \Delta\bar{d}$ at $Q^2 = 2.5$ GeV^2 as a function of x in comparison to the predictions of two recent phenomenological models [6, 7].

The resulting x-weighted polarised parton densities, evolved to $Q^2 = 2.5$ GeV^2, are shown in Fig. 2. Also shown are two recent parametrisations from LO analyses of only inclusive polarised deep-inelastic scattering data. Good agreement between the data and the parametrisations is observed for the up and down quark polarisations. In contrast to the parametrisations, the polarisation of the strange sea quarks extracted from the HERMES asymmetries tends to be positive. Using the same fitting formalism, the value of $\Delta\bar{u} - \Delta\bar{d}$ was extracted (right panel of Fig. 2). No breaking of the flavor symmetry in the light sea quark polarisation is observed at the present level of experimental accuracy.

FRAGMENTATION AT HERMES

To apply the concept of flavour-tagging in the determination of parton distributions the fragmentation process has to be understood. Especially at the relatively low center-of-mass energy of the HERMES experiment ($\sqrt{s} \sim 7$ GeV), two important questions have to be addressed:

- do the hard and the soft processes factorise ?
- can one clearly separate the current and the target fragmentation region ?

At present, the assumption of factorisation is supported by two measurements:

1. The measurement of the pion multiplicity [8]
   The differential multiplicity, i.e. the number of pions produced in deep-inelastic scattering normalised to the total number of inclusive deep-inelastic scattering events has been determined for both neutral and charged pions. As expected from isospin symmetry, the agreement between neutral and charged pions is excellent, at least up $z \sim 0.7$, where a possible contribution from exclusive channels might become important. In the left panel of Fig. 3 the neutral pion multiplicities as measured at HERMES are compared to the EMC measurements [9] as a function of z. As a significant Q^2 dependence of the fragmentation process is expected by perturbative QCD, the HERMES results have been evolved from the average measured Q^2 of 2.5 GeV^2 to the average Q^2 of the EMC experiment ($Q^2 = 25$ GeV^2). The excellent agreement between the two experiments strongly supports the fact that the fragmentation process at HERMES is essentially the same as for the EMC experiment at a much higher center-of-mass energy.
2. The measurement of the flavour asymmetry in the light quark sea [10]
   At HERMES the flavour asymmetry of the light quark sea has been determined from the ratio of the differences between charged pion yields for proton and neutron targets. Using the factorised ansatz for semi-inclusive deep-inelastic scattering one can derive an expression which factorises into two independent functions of x and z and thus can be rearranged to extract the ratio $(\bar{d}(x) - \bar{u}(x))/(u(x) - d(x))$. Plotting this quantity for fixed values of x as a function of z provides a test of the assumed form of factorisation. As shown in the right panel of Fig. 3 no z dependence is observed, strongly supporting the assumption of factorisation between the hard scattering process (depending on the parton distributions $q_i(x)$) and the hadronisation of the struck quarks (described by the fragmentation functions $D_q^{\pi^\pm}(z)$). It

should be noted, however, that the statistical precision of the data presented here does not allow to exclude a z dependence of the order of 10-20%.

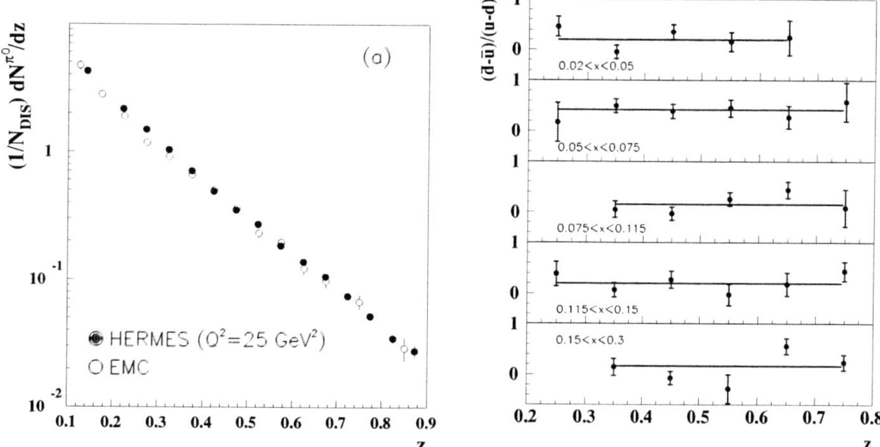

FIGURE 3. **Left:** multiplicity of neutral pions as a function of z, as measaured by HERMES and EMC. The HERMES data have been evolved to $Q^2 = 25$ GeV^2, the average Q^2 of the EMC muon data. **Right:** the flavour asymmetry of the light quark sea as extracted from HERMES semi-inclusive data as function of z for different values of x.

The current and the target fragmentation region are expected to be reasonably separated if the rapidity difference is larger than about 4 [11]. For pions in the HERMES kinematics this corresponds to the requirement of a minimum z value of about 0.2-0.3. As all HERMES analyses of semi-inclusive events used for the extraction of parton distributions have imposed a minimum z value of 0.2, contributions from the target fragmentation region are expected to be small.

REFERENCES

1. EMC, J. Ashman *et al.*, Phys. Lett. **B206** (1988) 364 and Nucl. Phys. **B328** (1989) 1.
2. F. Menden, Ph.D. thesis, University of Freiburg, 2000.
3. H.L. Lai *et al.*, Eur. Phys. J. **C12** (2000) 375.
4. M. Glück *et al.*, Phys. Rev. **D63** (2001) 094005.
5. J. Blümlein and H. Böttcher, hep-ph/0203155 (2002).
6. B. Dressler *et al.*, Eur. Phys. J. **C14** (2000) 147.
7. C. Bourrely *et al.*, Eur. Phys. J. **C23** (2002) 487.
8. HERMES, A. Airapetian *et al.*, Eur. Phys. Jour. **C21** (2001) 599.
9. EMC, J.J. Aubert *et al.*, Zeit. Phys. **C18** (1983) 189.
10. HERMES, K. Ackerstaff *et al.*, Phys. Rev. Lett. **81** (1998) 5519.
11. P.J. Mulders, Proc. of the workshop on Physics with an Electron Polarized Light-Ion Collider, R.G. Milner ed., AIP Proc. 588 (2001) 75, hep-ph/0010199.

Determining Spin-Flavor Dependent Distributions

Gordon P. Ramsey

Loyola University Chicago and Argonne National Laboratory

Abstract. Many of the present and planned polarization experiments are focusing on determination of the polarized glue. There is a comparable set of spin experiments which can help to extract information on the separate flavor-dependent polarized distributions. This talk will discuss possible sets of experiments, some of which are planned at BNL, CERN, DESY and JHF, which can be used to determine these distributions. Comments will include the estimated degree to which these distributions can be accurately found and the possible effects, if any, that the unpolarized distributions may have on this analysis.

INTRODUCTION

During the past 20 years, considerable progress has been made in understanding the nature of polarized distributions within nucleons. Various theoretical models, coupled with data from polarized deep-inelastic scattering (PDIS) have allowed extraction of polarized quark distributions. The net result is that valence disctibutions and the up and down sea flavors are relatively well determined, but the polarized strange sea, gluons and their corresponding orbital angular momenta are unknown. The most recent efforts have generated theoretical calculations and experiments at RHIC, HERA, and CERN, designed to determine ΔG. Many of these experiments are in progress.

We now have the theoretical and experimental techniques to pursue more detail into the flavor dependence of quark spin. Future efforts should include calculations and design of experiments to determine the spin contributions of all quark (and antiquark) flavors. This paper will discuss existing theoretical models, suggest a possible set of experiments and comment on the feasiblity of determining these distributions.

THEORETICAL MODELS

Models for the spin contributions of the valence quarks are based mostly upon modifications to the constituent quark model (CQM). [1, 2] These have the basic form:

$$\begin{aligned}\Delta u_v(x) &= M(x)[u_v(x) - 2d_v(x)/3] \\ \Delta d_v(x) &= M(x)(-d_v(x)/3)\end{aligned}$$

where $M(x)$ is a modification factor to the CQM. In the Carlitz-Kaur model, $M(x)$ is a "dilution" factor due to creation of gluons and the sea from valence quarks at small-

CP675, *Spin 2002: 15th Int'l. Spin Physics Symposium and Workshop on Polarized Electron Sources and Polarimeters,* edited by Y. I. Makdisi, A. U. Luccio, and W. W. MacKay

x. In the Relativistic Constituent Quark Model (Isgur), it represents a possible range of hyperfine interactions of the valence quarks with the other constituents. A statistical model, based upon the Pauli exclusion principle, [3] generates a valence distribution in a similar form,

$$\begin{aligned}\Delta u_v &= u_v - d_v \\ \Delta d_v &= -d_v/3.\end{aligned}$$

These three models predict valence distributions that are qualitatively similar, but give a range of possible extrema for Δu_v and Δd_v in the valence region, which can be tested with suitable polarization measurements.

The chiral quark model (χQM) [4] predicts integrals of the valence distribution over x, with free parameters that can be fit with data. For appropriate ranges of these parameters, this model is consistent with the integral predictions of others. Lattice calculations of the moments of up and down valence distributions are consistent with the χQM for Δu_v, but considerably less negative than the χQM prediction for Δd_v. The exist a number of NLO fits of quark distributions to data, with assumed parametrizations of Δu_v and Δd_v. [5, 6] These make certain assumptions about the symmetry of the polarized sea and could change with more experimental information. All of these valence models are consistent with the Bjorken Sum Rule and are similar in form. However, the differences are large enough to be distinguished by experimental measurements.

There is considerably more variance in the models for sea quarks, depending upon the assumptions made about how the polarized sea is generated. Sea models can be split into two categories: those based entirely on theoretical assumptions and the models that are a phenomenological combination of theory and experimental data. We will consider models providing a completely broken SU(3) polarized sea, where each flavor of quark/antiquark is determined separately. There is considerable theoretical evidence for broken SU(3). [7, 8, 9] Lattice calculations of the moments of the up and down sea have also indicated that this asymmetry could exist.

The statistical model mentioned above [3] combines the Pauli exclusion principle, F_2 data and axial-vector couplings, F and D to represent the polarized up quarks in terms of the unpolarized antiquarks. All other flavors are assumed to be unpolarized. This places a tight restriction on the size of the polarized sea. A light-cone model of meson-baryon fluctuations puts the intrinsic $q\bar{q}$ pairs with the valence quarks in an energetically favored state. [10] In this model, coupling to virtual $K^+\Lambda$ hyperons is the source of intrinsic $s\bar{s}$ pairs. Thus, the antiquarks are unpolarized and the light flavored quarks are polarized opposite to that of the proton.

In the chiral quark model, [4] the polarized sea determined by chiral fluctuations of the valence quarks, creating Goldstone bosons, which result in the prediction that $\Delta\bar{q} = 0$ all flavors. As with the valence quarks, ranges for the integrals of the polarized sea quarks are predicted. The meson cloud model is similar, with pseudo-scalar mesons replacing the Goldstone bosons. [9] In contrast to the χQM, the result is that $\Delta\bar{d} = \Delta u$ and the remaining quarks are unpolarized. In a chiral quark-soliton model, [7] quark fields interact with massless pions, yielding an asymmetry for the polarized up and down quarks, related to the unpolarized up and down antiquarks. This model has been phenomenologically tested in polarized semi-inclusive processes. [8]

TABLE 1. Sea flavor contributions by type of model

Model	Δu	$\Delta\bar{u}$	Δd	$\Delta\bar{d}$	Δs	$\Delta\bar{s}$	Δc	$\Delta\bar{c}$
Statistical	$\bar{u}-\bar{d}$	$\bar{u}-\bar{d}$	0	0	0	0	0	0
L-Cone	–	–	< 0	0	< 0	0	0	0
χQM	0.83	0	-0.39	0	-0.07	0	-0.003	0
M-cloud	Δu	0	0	Δu	–	–	0	0
CQSM	–	$\Delta\bar{d}-Cx^{\alpha}(\bar{d}-\bar{u})$	–	see $\Delta\bar{u}$	0	0	0	0

Most predictions of heavy quark contributions to proton spin indicate that they are likely small. The χQM prediction gives $\Delta c \approx -0.003$ and $\Delta\bar{c} = 0$. Similarly, an analysis using the operator product expansion and the axial anomaly predicts that $\Delta c = -0.0024 \pm 0.0035$, consistent with the χQM.[11] Instanton models tend to predict somewhat larger contributions from the heavier quarks. [12] These range from $\Delta c = -0.012 \pm 0.002 \rightarrow -0.020 \pm 0.005$. Thus, Δc is at most a very small fraction of the total polarized sea and will likely prove quite difficult to measure.

Table 1 contains key results from some of the models described above for comparison. The most significant differences are in the predictions for the polarization of the antiquarks. This distinction can also be carried over to the theoretically motivated phenomenological models.

Phenomenological models range from those grounded in theoretical constraints and use data to fit parameters to the ones which are primarily parametrizations determined by fits to data. Models in which the polarized sea is created by gluons, that pass polarization "information" to the quarks by the splitting process, are in the former category. [5, 13] In the GGR model, [13] the flavor asymmetry of the polarized sea is caused by the asymmetry in the unpolarized distributions. Specific forms for the parametrization of the separate distributions come from axial-vector constraints and data.

Most direct data fits [5, 6] assume minimal SU(3) breaking of the polarized sea. Similarly, LO/NLO moment fits to data [14] result in only a small amount of SU(3) breaking, but a stronger asymmetry of the sea is possible within the cited error analysis. These theoretical and phenomenological models provide a sufficient variance for experiments to be able to distinguish between their fundamental assumptions.

EXPERIMENTS

Polarized valence distributions can be fine-tuned by measuring asymmetries in pion production. By taking differences of these asymmetries for π^+ and π^- production, the valence contributions can be extracted. This results in:

$$\Delta A^{\pi} \equiv A_p^{\pi^+} - A_p^{\pi^-} = \frac{4\Delta u_v - \Delta d_v}{4u_v - d_v}. \tag{1}$$

In the valence models previously discussed, these asymmetries differ by 0.2 for $x < 0.5$ and by 0.1 for $0.5 \leq x \leq 0.9$. Similarly, differences in π^0 production for p and $\bar{p}$ yield large asymmetries for $0.1 \leq p_T/\sqrt(s) \leq 0.3$, but high energy $\bar{p}$ beams with sufficient luminosity for good statistics are difficult to achieve.

Present measurements of $\Delta(q+\bar{q})/(q+\bar{q})$ for the light quark flavors at HERA are providing a good start at finding the contributions of these flavors to the spin of the proton. [15] We would like to determine the individual spin contributions of each quark and antiquark flavor. For this, a combination of polarization experiments will be necessary. Charged current interactions are a useful tool in investigating kinematic dependences of both the polarized valence and sea quark distributions. The single spin asymmetries in parity-violating W production ($A_L^{W^\pm}$) can yield valuable information about the polarization of light quark flavors. [16]

$$A_L^{W^+}(y) = \frac{\Delta u(x_a)\bar{d}(x_b) - \Delta\bar{d}(x_a)u(x_b)}{u(x_a)\bar{d}(x_b) + \bar{d}(x_a)u(x_b)} \tag{2}$$

$$A_L^{W^-}(y) = \frac{\Delta d(x_a)\bar{u}(x_b) - \Delta\bar{u}(x_a)d(x_b)}{d(x_a)\bar{u}(x_b) + \bar{u}(x_a)d(x_b)} \tag{3}$$

For example, at $y = 0$, $x \approx M_W/\sqrt{s}$ and the asymmetry measures combinations of u and $\bar{d}$ or d and $\bar{u}$. For $y = -1$, x is small and the second terms in each numerator and denominator dominate, so we can separately probe $\bar{u}$ and $\bar{d}$. At $y = +1$, x is of moderate value, the first terms in each numerator and denominator dominate so that both u and d polarizations can be measured. This would provide a more complete picture of the light quark polarizations. However, a limited kinematic range will be probed at RHIC. Therefore, this should be combined with other experiments to probe the sea polarization.

Combinations of polarized sea flavors can be investigated in a number of different experiments. For $W^\pm$ production in the HERA kinematic range, g_5/F_1 is extracted from the measured asymmetry. This yields the following combinations: $g_5^{W^-} = \Delta u + \Delta\bar{d} + \Delta\bar{s} + \Delta c$ and $g_5^{W^+} = \Delta\bar{u} + \Delta d + \Delta s + \Delta\bar{c}$. However, the uncertainties in the hadronic energy scale of the calorimeter are of comparable size to the asymmetries. Measurements may be difficult at RHIC as well, since these asymmetries are generally small at its kinematic range. Measurement of g_1 in polarized $e^\pm p \to \nu(\bar{\nu})X$ scattering at HERA could yield a similar combination of polarized flavors.

Parity-violating ν scattering (p and n) measurement of g_3 at the proposed Japan Hadron Facility would give: $\frac{1}{2}[g_3^{\nu(p+n)} - g_3^{\bar{\nu}(p+n)}] \sim \Delta s + \Delta\bar{s} - \Delta c - \Delta\bar{c}$. However, since g_3 comes from $W_{\mu\nu}^{\perp}$, which is small, this may be difficult to distingish.

Parity conserving double spin asymmetries in Z production ($A_{LL}^{Z^0}$) provide a valuable tool in investigating the polarization of the antiquarks. This asymmetry is given by:

$$A_{LL}^{Z^0}(y) \sim \Sigma_i \frac{\Delta q_i(x_a)\Delta\bar{q}_i(x_b) + \Delta\bar{q}_i(x_a)\Delta q_i(x_b)}{q_i(x_a)\bar{q}_i(x_b) + \bar{q}_i(x_a)q_i(x_b)} \tag{4}$$

Predicted asymmetries of ~ 0.10 for $\sqrt(s) = 500$ GeV could be distiguishable from zero with 400-500 events at RHIC. This would provide an excellent test of the light-cone, χQM and instanton models that predict zero polarization for antiquarks.

Polarized Drell-Yan experiments at both RHIC (at 50-100 GeV) or the proposed Japan Hadron Facility (JHF) at 50 GeV provide promising ways to extract more precise information about the polarization of the sea. At these energies, the cross sections are larger and the asymmetries are moderately sized. This makes the competing predictions easy to distinguish. The cross sections decrease rapidly with energy, so experiments at larger $\sqrt{s}$ are not good candidates for this set of measurements. The RHIC luminosity is low at 50 GeV, (the injection energy) but probably suitable at 100 GeV. Polarized beams at the JHF are quite appropriate for lepton pair production experiments and would be excellent for determining the relative size of the polarized sea. [17] These measurements in principle could distinguish the flavor dependence of the polarized sea. Combined with the experiments described above, they would give a complete picture of the sea polarization.

CONCLUSION

There has been considerable progress in narrowing the polarizations of the lighter quark flavors, Δq_v, Δu_{tot} and Δd_{tot}. There exist many theoretical predictions for polarizations of the valence, sea quark and antiquark flavors. The experiments described here include most possibilities for determining the spin contributions of four quark and antiquark flavors. Many of the suggested measurements are feasible and should be done, since a combination of experiments would give the best range of information about quark spin. This opportunity opens up numerous possibilities for polarization experiments at RHIC, HERA, COMPASS and the JHF.

REFERENCES

1. Carlitz and Kaur, Phys. Rev. Lett. 38, 673 (1977).
2. Isgur, N. Phys. Rev. D59, 034013 (1999) and hep-ph/9809255.
3. Bourrely and J. Soffer, Nucl. Phys. B445, 341 (1995) and Eur. Phys. J. C23, 487 (2002).
4. X. Song, Phys. Rev. D57, 4114 (1998).
5. Bartelski and Tatur, Phys. Rev. D65, 034002 (2002) and hep-ph/0107202.
6. E. Leader, Siderov, Stamenmov, Eur. Phys. J. C23, 479 (2002) and hep-ph/0111267.
7. Wakamatsu and Watabe, Phys. Rev. D62, 017506 (2000).
8. T. Morii and Yamanishi, Phys. Rev. D61, 057501 (2000), erratum: Phys. Rev. D62, 059901 (2000) and D. de Florian and R. Sassot, Phys. Rev. D62, 094025 (2000).
9. S. Kumano and Miyama, Phys. Rev. D65, 034012 (2002).
10. S. Brodsky and B-Q. Ma, Phys. Lett. B381, 317 (1996).
11. A. Manohar, Phys. Lett. B242, 94 (1990).
12. Blotz and Shuryak, Phys. Lett. B439, 415 (1998).
13. L. Gordon, M. Goshtasbpour and G. Ramsey, Phys. Rev. D58, 094017 (1998).
14. J. Blümlein and Böttcher, Nucl. Phys. B636, 225 (2002).
15. See talk by A. Bruell in this proceedings and `http://www.desy.de`.
16. Bourrely and Soffer, Phys. Rev. D51, 2108 (1995).
17. J. C. Peng, *et. al.*, hep-ph/0007341 and S. Kumano, hep-ph/0207151.

NLO QCD Corrections to A_{LL}^{π}

Barbara Jäger*, Marco Stratmann* and Werner Vogelsang†

*Inst. for Theor. Physics, Univ. of Regensburg, D-93040 Regensburg, Germany
†RBRC and Physics Department, Brookhaven National Laboratory, Upton, NY 11973, U.S.A.

Abstract. We present a calculation for single-inclusive large-p_T pion production in longitudinally polarized pp collisions in next-to-leading order QCD. The corresponding double-spin asymmetry A_{LL}^{π} for this process will soon be used at BNL-RHIC to measure Δg.

THEORETICAL FRAMEWORK

Very inelastic pp collisions with longitudinally polarized beams at the BNL-RHIC will open up unequaled possibilities to measure the so far elusive polarized gluon density Δg. RHIC has the advantage of operating at high energies ($\sqrt{S} = 200$ and 500 GeV), where the underlying theoretical framework, i.e., perturbative QCD, is expected to be under good control. In addition, it offers various different channels in which Δg can be studied, such as prompt-γ, heavy flavor, jet or inclusive-hadron production [1, 2]. In this way, RHIC will provide the best source of information on Δg for a long time to come.

The basic concept that underlies most of spin physics at RHIC is the factorization theorem. It states that large momentum-transfer reactions may be factorized at a scale μ_F into long-distance pieces that contain the desired information on the spin structure of the nucleon in terms of its *universal* parton densities, such as Δg, and parts that are short-distance and describe the hard interactions of the partons. The latter can be evaluated using perturbative QCD. The factorization scale μ_F is not further specified by the theory but usually chosen to be of the order of the hard scale in the reaction.

In the following, we consider the spin-dependent cross section

$$d\Delta\sigma \equiv \frac{1}{2}\left[d\sigma^{++} - d\sigma^{+-}\right] \ , \tag{1}$$

where the superscripts denote the helicities of the protons in the scattering, for the reaction $pp \to \pi X$, where the pion is at high transverse momentum p_T, ensuring large momentum transfer. The statement of the factorization theorem is then

$$\begin{aligned} d\Delta\sigma &= \sum_{a,b,c} \int dx_a \int dx_b \int dz_c \, \Delta f_a(x_a,\mu_F)\,\Delta f_b(x_b,\mu_F)\,D_c^{\pi}(z_c,\mu_F') \\ &\quad \times d\Delta\hat{\sigma}_{ab}^{c}(x_a P_A, x_b P_B, P_\pi/z_c, \mu_R, \mu_F, \mu_F') \ , \end{aligned} \tag{2}$$

where the sum is over all contributing partonic channels $a+b \to c+X$, with $d\Delta\hat{\sigma}_{ab}^{c}$ the associated partonic cross section, defined in complete analogy with Eq. (1). Besides

CP675, *Spin 2002: 15th Int'l. Spin Physics Symposium and Workshop on Polarized Electron Sources and Polarimeters*, edited by Y. I. Makdisi, A. U. Luccio, and W. W. MacKay

the factorization scale μ_F for the initial-state partons $\Delta f_{a,b}$, there is also a factorization scale μ_F' for the absorption of long-distance effects into the parton-to-pion fragmentation functions D_c^π. The renormalization scale μ_R in (2) is associated with the running of α_s.

It is planned for the coming RHIC run (early 2003) to attempt a first measurement of

$$A_{LL}^{\pi} = \frac{d\Delta\sigma}{d\sigma} = \frac{d\sigma^{++} - d\sigma^{+-}}{d\sigma^{++} + d\sigma^{+-}} \tag{3}$$

for high-p_T pion production. The main underlying idea here is that the spin asymmetry A_{LL}^{π} is very sensitive to Δg through the contributions from polarized quark-gluon and gluon-gluon scatterings. In general, a leading-order (LO) estimate of (2) or (3) merely captures the main features, but does not usually provide a quantitative understanding. For instance, the dependence on the unphysical scales μ_F, μ_F', and μ_R is expected to be much reduced when going to higher orders in the perturbative expansion. Hence, only with knowledge of the next-to-leading order (NLO) QCD corrections can one reliably extract information on the parton distribution functions from the reaction. A NLO calculation of A_{LL}^{π} has been completed very recently [3], and here we briefly sketch the results; for details, see [3]. We note that the PHENIX collaboration has recently presented first, still preliminary, results for the *un*polarized cross section for $pp \to \pi^0 X$ at $\sqrt{S} = 200$ GeV, which are well described by a NLO QCD calculation [4].

The partonic cross sections $d\Delta\hat{\sigma}_{ab}^{c}$ in (2) have to be summed over all final states (excluding c which fragments) and integrated over the entire phase space of X. The LO results, which have been known for a long time [5], are obtained from evaluating all tree-level $2 \to 2$ QCD scattering diagrams. At NLO, we have $\mathcal{O}(\alpha_s)$ corrections to the LO reactions, and also additional new processes, giving rise to 16 different channels in total, like $qq \to qX$, $qg \to gX$, etc. At intermediate stages the NLO calculation will necessarily show singularities that represent the long-distance sensitivity. In addition, for those processes that are already present at LO, real $2 \to 3$ and virtual one-loop $2 \to 2$ contributions will individually have infrared (IR) singularities that only cancel in their sum. Virtual diagrams will also produce ultraviolet (UV) poles that need to be removed by the renormalization of the strong coupling constant at a scale μ_R. We choose $n = 4 - 2\varepsilon$ dimensional regularization to make these singularities manifest. Subtractions of poles will generally be made in the $\overline{\text{MS}}$ scheme. We use the HVBM prescription [6] to describe polarizations of particles in n dimensions.

At $\mathcal{O}(\alpha_s^3)$, virtual corrections, which we have calculated adopting two different methods, only contribute through their interference with the Born diagrams. Firstly, one could make use of known $\overline{\text{MS}}$-renormalized one-loop vertex and self-energy insertions as given in [7]. Only the UV-finite box diagrams have to be calculated from scratch. The second approach makes use of the fact that helicity amplitudes for all one-loop $2 \to 2$ QCD scattering diagrams were presented in [8]. These results will not immediately yield the answer for the HVBM prescription but the transformation is straightforward.

In the $2 \to 3$ contributions, the two unobserved partons need to be integrated over their entire phase space which we perform *analytically*. In this way the final answer is much more amenable to a numerical evaluation, giving stable results in a short time. This may become important when experimental data will become available, and one is aiming to extract Δg from them within a "global analysis" [9]. Phase space integrations are organized best in the rest frame of the two unobserved partons. Extensive partial fractioning of the matrix elements then always leads to a "master integral" which can be done analytically. Singularities when the invariant mass of the unobserved partons vanishes are made manifest with help of the usual "+"-distributions.

All genuine IR singularities cancel in the sum of all contributions. However, the limit $\varepsilon \to 0$ still cannot be taken as a result of collinear divergencies. These remaining poles need to be factored into the bare parton distribution and fragmentation functions, depending on whether their origin was in the initial or final state. This standard procedure introduces the factorization scales μ_F and μ_F' in Eq. (2). We note that we have simultaneously computed also the NLO corrections for the unpolarized case, where we fully agree at an *analytical level* with results available in the literature [10]. This provides an extremely powerful check on the correctness of all our calculations.

Finally, we note that the same NLO calculation was presented in [11] based on MC phase space integration techniques. Such an approach has the advantage of being very flexible as it may be used for any IR-safe observable, with any experimental cut. However, the numerical integrations are delicate and time-consuming. Early comparisons show very good agreement of the numerical results.

NUMERICAL RESULTS

For our numerical calculations we assume the same kinematic coverage as in the unpolarized PHENIX measurement mentioned above [4]: $\sqrt{S} = 200\,\text{GeV}$, pion transverse momenta in the range $2 \leq p_T \leq 13$ GeV, and pseudorapidities integrated over $|\eta| \leq 0.38$. We also always take into account that the pion measurement is at present possible only over half the azimuthal angle. To calculate the NLO/LO polarized cross section (2) we use the spin-dependent GRSV parton densities ("GRSV-std") and the pion fragmentation functions of [13]. To investigate the sensitivity of A_{LL}^{π} to Δg, we also use a set, for which Δg is assumed to be particularly large ("GRSV-max"). For the NLO (LO) unpolarized cross section, we use the CTEQ5M (CTEQ5L) [14] densities.

Figure 1 shows our results for the unpolarized and polarized cross sections at NLO and LO, where we have chosen the scales $\mu_R = \mu_F = \mu_F' = p_T$. The lower part of the figure displays the "K-factor", $K = d(\Delta)\sigma^{\text{NLO}}/d(\Delta)\sigma^{\text{LO}}$. One can see that in the unpolarized case the corrections are roughly constant and about 50% over the p_T-region considered. In the polarized case, we find generally smaller corrections which become of similar size as those for the unpolarized case only at the high-p_T end. The cross section for p_T-values smaller than about 2 GeV is outside the domain of perturbative calculations as indicated by rapidly increasing NLO corrections and, therefore, is not considered here.

Figure 2 shows the improvement in scale dependence of the spin-dependent cross section when going from LO to NLO. In each case the shaded bands indicate the uncertainties from varying the unphysical scales in the range $p_T/2 \leq \mu_R = \mu_F = \mu_F' \leq 2p_T$. The solid lines are for the choice where all scales are set to p_T. One can see that the scale dependence indeed becomes much smaller at NLO.

Results for A_{LL}^{π} are given in Fig. 3. We have again chosen all scales to be p_T. As expected from the larger K-factor for the unpolarized cross section shown in Fig. 1, the

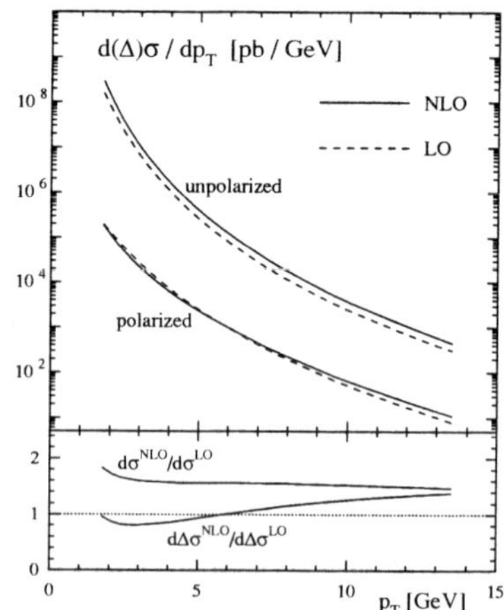

FIGURE 1. Unpolarized and polarized π^0 production cross sections in NLO (solid) and LO (dashed) at $\sqrt{S} = 200$ GeV. The lower panel shows the K-factor in each case. Figure taken from [3].

asymmetry is somewhat smaller at NLO than at LO, showing that inclusion of NLO QCD corrections is rather important for the analysis of the data in terms of Δg.

We also conclude from the figure that there are excellent prospects for determining $\Delta g(x)$ from A_{LL}^{π} measurements at RHIC: the asymmetries found for the two different sets of polarized parton densities, which mainly differ in the gluon density, show marked differences, much larger than the expected statistical errors in the experiment, indicated in the figure. The latter may be estimated by the formula $\delta A_{LL}^{\pi} = 1/(P^2\sqrt{\mathscr{L}\sigma_{\rm bin}})$, where P is the polarization of one beam, $\mathscr{L}$ the integrated luminosity, and $\sigma_{\rm bin}$ the unpolarized

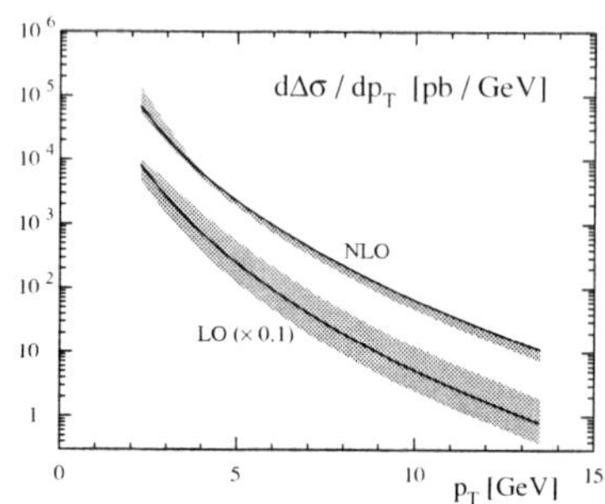

FIGURE 2. Scale dependence of the polarized cross section for π^0 production at LO and NLO [3] in the range $p_T/2 \leq \mu_R = \mu_F = \mu_F' \leq 2p_T$. We have rescaled the LO results by 0.1 to separate them better from the NLO ones. In each case the solid line corresponds to the choice where all scales are set to p_T.

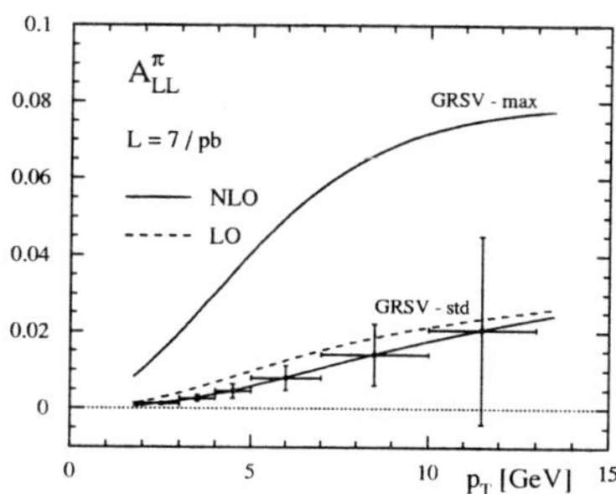

FIGURE 3. Spin asymmetry for π^0 production in NLO (solid lines). The dashed line shows the asymmetry at LO for the GRSV "standard" set. The "error bars" indicate the expected statistical accuracy targeted for the upcoming run of RHIC (see text). Figure taken from [3].

cross section integrated over the p_T-bin for which the error is to be determined. We have used very moderate values $P = 0.4$ and $\mathscr{L} = 7$/pb, which are targets for the coming run.

To conclude, we have presented the results of a largely analytical computation of the NLO partonic hard-scattering cross sections relevant for the spin asymmetry A^{π}_{LL} for high-p_T pion production in longitudinally polarized hadron-hadron collisions. The asymmetry turns out to be a promising tool to provide first information on Δg even for the rather moderate luminosities targeted for the coming run with polarized protons at RHIC.

B.J. is supported by the European Commission IHP program under contract HPRN-CT-2000-00130. W.V. is grateful to RIKEN, Brookhaven National Laboratory and the U.S. Department of Energy (contract number DE-AC02-98CH10886) for providing the facilities essential for the completion of this work.

REFERENCES

1. For a review on RHIC spin, see: G. Bunce et al., *Annu. Rev. Nucl. Part. Sci.* **50**, 525 (2000).
2. M. Stratmann, these proceedings.
3. B. Jäger, A. Schäfer, M. Stratmann, and W. Vogelsang, hep-ph/0211007.
4. H. Torii, talk presented at *Quark Matter 2002*, Nantes, France, 2002.
5. See, for example: J. Babcock et al., *Phys. Rev. Lett.* **40**, 1161 (1978); *Phys. Rev.* **D19**, 1483 (1979); R. Gastmans and T.T. Wu, *The ubiquitous photon*, Clarendon Press, Oxford, 1990.
6. G. 't Hooft and M. Veltman, *Nucl. Phys.* **B44**, 189 (1972); P. Breitenlohner and D. Maison, *Commun. Math. Phys.* **52**, 11 (1977).
7. M.A. Nowak, M. Praszalowicz, and W. Slominski, *Annals Phys.* **166**, 443 (1986).
8. Z. Kunszt, A. Signer, and Z. Trocsanyi, *Nucl. Phys.* **B411**, 397 (1994).
9. M. Stratmann and W. Vogelsang, *Phys. Rev.* **D64**, 114007 (2001).
10. R.K. Ellis and J.C. Sexton, *Nucl. Phys.* **B269**, 445 (1986); F. Aversa et al., *ibid.* **B327**, 105 (1989).
11. D. de Florian, hep-ph/0210442.
12. M. Glück, E. Reya, M. Stratmann, and W. Vogelsang, *Phys. Rev.* **D63**, 094005 (2001).
13. B.A. Kniehl, G. Kramer, and B. Pötter, *Nucl. Phys.* **B582**, 514 (2000).
14. CTEQ Collaboration, H.-L. Lai et al., *Eur. Phys. J.* **C12**, 375 (2000).

On Spin Content of the Proton

G. Musulmanbekov

Join Institute for Nuclear Research, Dubna

Abstract. It is shown, in the frame of the model proposed by author, that spin of the proton could be described by the orbital momentum of quark and qluon condensate circulating around valence quarks.

In this paper we describe the spin of a nucleon as arising from the orbital momentum of quark and gluon condensate circulating around valence quarks. Considerations are performed in the frame of so–called Strongly Correlated Quark Model (SCQM) [1]. The ingredients of the model are the following. Single quark of definite color embedded in vacuum begins to polarize its surrounding that results in formation of quark and gluon condensate. At the same time it experiences the pressure of the vacuum because of zero point radiation field or vacuum fluctuations which act the quark tending to destroy the ordering of the condensate. Suppose that we place the corresponding antiquark in the vicinity of the first one. Owing to their opposite signs color polarization fields of quark and antiquark interfere destructively in the overlapped space regions eliminating each other at most in space around the middle–point between the quarks. This effect leads to decreasing of condensates density in the same space region and overbalancing of the vacuum pressure acting on quark and antiquark from outer space regions. As a result the attractive force between quark and antiquark emerges and quark and antiquark start to move towards each other. The density of the remaining condensate around quark (antiquark) is identified with hadronic matter distribution. At maximum displacement in $\bar{q}q-$ system, that corresponds to small overlapping of polarization fields, hadronic matter distributions have maximum extent and values. The closer they to each other, the larger destructive interference effect and the smaller hadronic matter distributions are around quarks and the larger their kinetic energies. In that way quark and antiquark start to oscillate around their middle–point. For such interacting $\bar{q}q-$ pair located on X axis at the distance $2x$ from each other the total Hamiltonian is

$$H = \frac{m_{\bar{q}}}{(1-\beta^2)^{1/2}} + \frac{m_q}{(1-\beta^2)^{1/2}} + V_{\bar{q}q}(2x), \tag{1}$$

were $m_{\bar{q}}$, m_q- current masses of valence antiquark and quark, $\beta = \beta(x)-$ their velocity depending on displacement x and $V_{\bar{q}q}-$ quark–antiquark potential energy with separation $2x$. It can be rewritten as

$$H = \left[\frac{m_{\bar{q}}}{(1-\beta^2)^{1/2}} + U(x)\right] + \left[\frac{m_q}{(1-\beta^2)^{1/2}} + U(x)\right] = H_{\bar{q}} + H_q, \tag{2}$$

CP675, *Spin 2002: 15th Int'l. Spin Physics Symposium and Workshop on Polarized Electron Sources and Polarimeters,* edited by Y. I. Makdisi, A. U. Luccio, and W. W. MacKay

were $U(x) = \frac{1}{2}V_{\overline{q}q}(2x)$ is potential energy of quark or antiquark. Quark (antiquark) with the surrounding cloud (condensate) of quark – antiquark pairs and gluons, or hadronic matter distribution, forms the constituent quark. It is natural to assume that the potential energy of quark (antiquark), $U(x)$, corresponds to the mass M_Q of constituent quark:

$$2U(x) = C_1 \int_{-\infty}^{\infty} dz' \int_{-\infty}^{\infty} dy' \int_{-\infty}^{\infty} dx' \rho(x, \mathbf{r}') \approx 2M_Q(x) \tag{3}$$

where C_1 is dimensional constant and hadronic matter density distribution, $\rho(x, \mathbf{r}')$, is defined as

$$\rho(x, \mathbf{r}') = C_2 \left|\varphi(x, \mathbf{r}')\right| = C_2 \left|\varphi_Q(x' + x, y', z') - \varphi_{\overline{Q}}(x' - x, y', z')\right|. \tag{4}$$

Here C_2 is a constant, φ_Q and $\varphi_{\overline{Q}}$ are density profiles of the condensates around quark and antiquark located at distance $2x$ from each other. We consider by convention the condensates around quark and antiquark having opposite color charges. They look like compressive stress and tensile stress (around defects) in solids. Generalization to three–quark system in baryons is performed according to $SU(3)_{color}$ symmetry: in general, pair of quarks have coupled representations

$$3 \otimes 3 = 6 \oplus \overline{3}$$

in $SU(3)_{color}$ and for quarks within the same baryon only the $\overline{3}$ (antisymmetric) representation occurs. Hence, an antiquark can be replaced by two correspondingly colored quarks to get color singlet baryon and destructive interference takes place between color fields of three valence quarks (VQs). Putting aside the mass and charge differences of valence quarks we may say that inside baryon three quarks oscillate along the bisectors of equilateral triangle. Therefore, keeping in mind that quark and antiquark in mesons and three quarks in baryons are strongly correlated, we can consider each of them separately as undergoing oscillatory motion under the potential (3) in 1+1 dimension. Hereinafter we consider VQ oscillating along $X-$ axis and $Z-$ axis is perpendicular to the plane of oscillation XY. Density profiles of condensates around VQs are taken in gaussian form. It has been shown in papers [2] that the wave packet solutions of time dependent Schrodinger equation for harmonic oscillator move in exactly the same way as corresponding classical oscillators. These solutions are called "coherent states". This relationship justifies (partly) our semiclassical treatment of quantum objects.

We define the mass of constituent quark at maximum displacement as

$$M_{Q(\overline{Q})}(x_{\max}) = \frac{1}{3}\left(\frac{m_\Delta + m_N}{2}\right) \approx 360\, MeV,$$

where m_Δ and m_N are masses delta–isobar and nucleon correspondingly. The parameters of the model, namely, maximum displacement, $x_{\max}$, and parameters of 3–D gaussian function, $\sigma_{x,y,z}$, for hadronic matter distribution around VQ are chosen to be

$$x_{\max} = 0.64\ fm,\ \sigma_{x,y} = 0.24\ fm,\ \sigma_z = 0.12\ fm. \tag{5}$$

They are adjusted by comparison of calculated and experimental values of inelastic cross sections, $\sigma_{in}(s)$, and inelastic overlap function $G_{in}(s,b)$ for pp and $\overline{p}p-$ collisions [3]. The current mass of valence quark is taken to be 5 MeV. The behavior of potential (3) evidently demonstrates the relationship between constituent and current quark states inside a hadron (Fig. 1). At maximum displacement quark is nonrelativistic, constituent one (VQ surrounded by condensate), since the influence of polarization fields of other quarks becomes minimal and the VQ possesses the maximal potential energy corresponding to the mass of constituent quark. At the origin of oscillation, $x=0$, antiquark and quark in mesons and three quarks in baryons, being close to each other, have maximum kinetic energy and correspondingly minimum potential energy and mass: they are relativistic, current quarks (bare VQs). This configuration corresponds to so called "asymptotic freedom". In the intermediate region there is increasing (decreasing) of constituent quark mass by dressing (undressing) of VQs due to decreasing (increasing) of the destructive interference effect. This mechanism meets local gauge invariance principle. Indeed, destructive interference of color fields of quark and antiquark in mesons and three quarks in baryons depending on their displacements can be treated as phase rotation of wave function of single VQ in color space ψ_c on angle θ depending on displacement x of the VQ in coordinate space

$$\psi_c(x) \rightarrow e^{ig\theta(x)}\psi_c(x). \tag{6}$$

Phase rotation,in turn, leads VQ dressing (undressing) by quark and gluon condensate that corresponds to the transformation of gauge field $A_\mu = (\varphi,0,0,0)$

$$A_\mu(x) \rightarrow A_\mu(x) + \partial_\mu\theta(x). \tag{7}$$

Here we drop color indices of $A_\mu(x)$ and consider each quark of specific color separately as changing its effective color charge, $g\theta(x)$, in color fields of other quarks (antiquark) due to destructive interference. Thus gauge transformations (6, 7) map internal (isotopic) space of colored quark onto coordinate space. On the other hand this dynamical picture of VQ dressing (undressing) corresponds to chiral symmetry breaking (restoration). Due to this mechanism of VQs oscillations nucleon runs over the states corresponding to the specific terms of the infinite series of Fock space

$$|B\rangle = a_1\,|q_1q_2q_3\rangle + a_2\,|q_1q_2q_3\overline{q}q\rangle + ...$$

The proposed model has some important consequences. Inside hadrons quarks and accompanying them gluons, as well, are strongly correlated. Nucleons are nonspherical object: they are flattened along the axis perpendicular to the plane of quarks oscillations. There are two reasons of that: first, because VQs undergo plane oscillations and second, owing to the flatness of hadronic matter distributions around VQs (according to (5)). From the form of the quark potential (Fig.1) one can conclude that dynamics of VQ corresponds to nonlinear oscillator and VQ with its surrounding can be treated as nonlinear wave packet. Moreover, our quark – antiquark system turned out to be identical to, so called, "breather" solution of (nonlinear) sine–Gordon (SG) equation [4].

So far we dealt with scalar polarization field around VQ. How can one include spin in the frame of these classical considerations? According to the prevailing belief, the spin

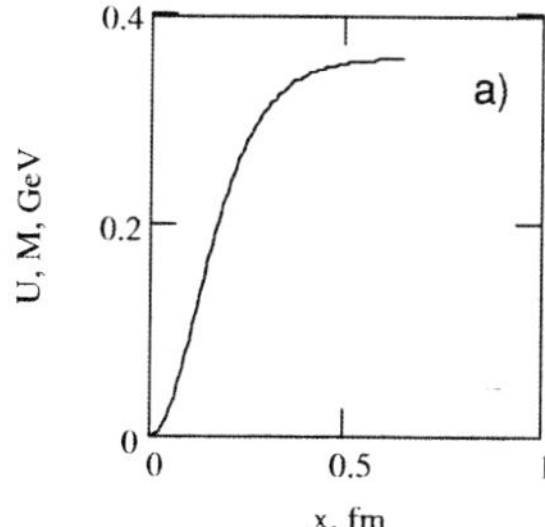

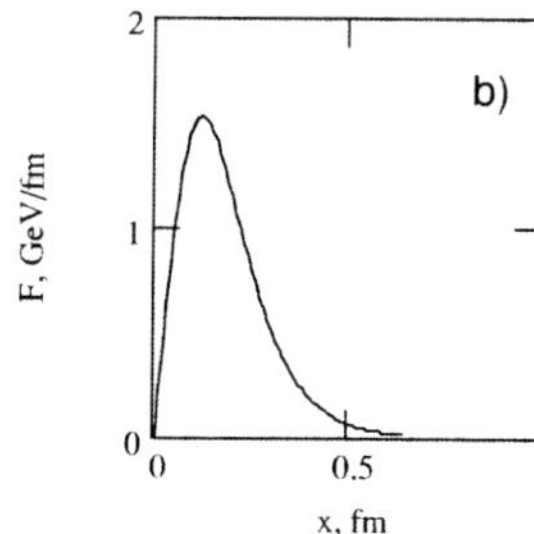

FIGURE 1. a) Potential energy of valence quark and mass of constituent quark; b) "Confinement" force.

is quantum feature of microparticle and has no classical analog. However, Belinfante [5] as early as in 1939 showed that the spin of electron may be regarded as an angular momentum generated by a circulating flow of energy, or a momentum density, in the wave field of electron. Furthermore, a comparison between calculations of angular momentum in the Dirac and electromagnetic fields, performed by Ohanian [6], shows that the spin of the electron is entirely analogous to the angular momentum carried by classical circularly polarized wave.

We follow the classical picture of electron spin considered by R. Feynman, and later, Hiqbie [7, 8]. Let us imagine point electric charge, e, and magnetic bar, μ, settled close nearby [8]. In surrounding space electric, **E,** and magnetic, **B,** fields arise. Although vectors **E** and **B** are constant the circulating flow of energy is created around such a system which is described by Poynting's vector

$$\mathbf{S} = \varepsilon_0 c^2 \mathbf{E} \times \mathbf{B}. \tag{8}$$

The volume density of angular momentum is given by

$$\mathbf{L} = \mathbf{r} \times \mathbf{P}, \tag{9}$$

where **P** is the volume density of linear momentum defined as

$$\mathbf{P} = \mathbf{S}/c^2 = \mathbf{D} \times \mathbf{B} == \frac{q(\mathbf{r} \times \mu)}{(4\pi c)^2 \varepsilon_0 r^6} = \frac{r_0(\mathbf{r} \times \mu)}{(q/m) 4\pi r^6}. \tag{10}$$

Here $r_0 = q_e^2/4\pi\varepsilon_0 m_e c^2-$ classical radius of electron. Substitution of (12) into (11) gives

$$\mathbf{L} = \mathbf{r} \times \mathbf{P} = (...)\mathbf{r} \times (\mathbf{r} \times \mu) = (...) \left[\mathbf{r}(\mathbf{r} \cdot \mu) - \mu r^2\right] \tag{11}$$

When we align μ with $Z-$ axis and use spherical polar coordinates, $z-$ component of **L** is

$$L_z = -(...)\mu r^2 \sin^2\theta. \tag{12}$$

The total angular momentum of the field extending from $r = a$ to infinity is given by the integral

$$s = \int_a^\infty L_z dV = -(2/3)\frac{r_0\mu}{a(q_e/m_e)}. \tag{13}$$

For $a = (2/3)r_0$ we have

$$s = -\mu/(q_e/m_e), \tag{14}$$

which just happens to be the entire spin angular momentum of the electron. Moreover, Feynman showed that if we extend the classical coulomb field all the way down to 2/3 of the classical electron radius, r_0, the entire mass of the electron is contained in its field.

By analogy we suppose that intersecting chromoelectric, $\mathbf{E}_{ch}$, and chromomagnetic, $\mathbf{B}_{ch}$, fields create around quark (antiquark) circulating flow of energy, the color analog of Poynting's vector

$$\mathbf{S}_{ch} = c^2\mathbf{E}_{ch} \times \mathbf{B}_{ch}. \tag{15}$$

Spin of quark (antiquark) then is defined by the integral

$$\mathbf{s} = \int_a^\infty \mathbf{r} \times (\mathbf{E}_{ch} \times \mathbf{B}_{ch}) d^3\mathbf{r}. \tag{16}$$

Obviously, lower limit of integration, a, is connected with the size of quark (antiquark). Now we must take into account correlated oscillations of quarks inside hadrons which lead to periodic changes in values of $\mathbf{E}_{ch}$ and $\mathbf{B}_{ch}$. At maximal displacement of quark from the origin of oscillation $\mathbf{E}_{ch}$ and $\mathbf{B}_{ch}$ around it are expanded to the maximum (size of constituent quark) and so is for circulating flow $\mathbf{S}_{ch}$. At minimum displacement $\mathbf{E}_{ch}$ and $\mathbf{B}_{ch}$, owing to destructive interference described above, are confined in small region around the quark and circulating flow of energy $\mathbf{S}_{ch}$ is concentrated in a narrow shell. For consistency of the model we must demand that spins of quarks are perpendicular to the plane of oscillation. Recalling that entire spin of quark is due total angular momentum of circulating flow we compare it with circular flow of ideal liquid and apply hydrodynamic considerations. Equations of hydrodynamics for ideal liquid flow read

$$\frac{\partial \xi}{\partial t} + \nabla \times (\xi \times \mathbf{v}) = 0, \tag{17}$$

$$\xi = \nabla \times \mathbf{v}, \tag{18}$$

where $\mathbf{v}$ is vector of liquid velocity and ξ is its vorticity defining circulation around unit area. Equation (17) implies angular momentum conservation law applied to liquid flow. In other words, if σ is the area of vortex field in ideal liquid then

$$\xi \cdot \sigma = \oint_\sigma \mathbf{v} \cdot d\mathbf{r} = const, \tag{19}$$

that means the larger area of circulation the less the vorticity and vice versa. In our case $\mathbf{v}$ is velocity of circulation flow depending on the distance from the vortex center. Following these considerations we assume that the spin of quark is conserved quantity during oscillation and

$$\mathbf{s} \propto \xi \cdot \sigma. \tag{20}$$

Rigorously, the case at hand is the conservation of quark spin module because quarks inside hadronic system can interact and what is conserved is the total angular momentum of the system. Thus, according to our approach quark spin is given by circulating flow of polarized vacuum (sea quarks and gluons) and recalling that entire mass of constituent quark (antiquark) is contained in it's color field one can consider quarks as to be vortices in vacuum. Now gauge field $A_\mu(x)$ in (7) contains along with the scalar part, φ, vector components, $\mathbf{A}$, as well. Whereas vortex is singularity (hole) in vacuum the quantization of spin follows from nonsimple-connectedness of vacuum structure. The representation of quark spin as created by circulating flow of quark and gluon condensate gives alternative solution to "spin crisis" ,[9], that implies the essential deviation of quantity

$$\Delta\Sigma = \Delta u + \Delta d + \Delta s, \tag{21}$$

defined as the relative amount of the nucleon spin carried by intrinsic spins of the quarks, from one. The effect gets clear explanation in proposed picture. In DIS the transverse size of virtual photon d emitted by incident lepton is given by $d^2 \propto 1/Q^2$. Because of potential (irrotational) character of velocity field in (19) contribution to $\Delta\Sigma$ in specific interaction at random impact parameter depends on if the center of quark vortex (singularity) falls into the area covered by transverse size of virtual photon. If it is the circled integral (19) gets nonzero value proportional to spin of valence quark and this occurrence gives contribution to $\Delta\Sigma$. Otherwise the integral equals zero and there is no contribution to (21). This consideration reminds the well known Aharonov – Bohm effect. Preliminary estimation gives $\Delta\Sigma \simeq 0.02$ at $Q^2 = 5\ (GeV/c)^2$. If $Q^2 \to 0$, then $\Delta\Sigma \to 1$. Therefore, according to our approach proton spin is defined by spins of valence quarks, as naive quark model gives , and the (intrinsic) spin of valence quark, in turn, is defined by orbital momentum of circulating flow of quark and gluon condensate around it. Irrotational character of the condensate velocity field suggests the zero polarization of sea quarks. Experimental nonzero value of $\Delta\Sigma$, (0.2–0.3), however, could be the manifestation of nonzero polarization of sea quarks. To study the contribution of sea quarks to the spin of nucleons one needs to use microscopic treatment of the mechanism of condensation (e. g., instanton mechanism).

As noted above, in our scheme nucleons are nonspherical, flattened objects. This geometrical feature of extended nucleons can lead to observable effects. Since inside a proton spins of quarks are perpendicular to the plane of their oscillations there should be the difference between longitudinal $\sigma_{tot}^L = \frac{1}{2}(\overset{\leftarrow}{\vec{\sigma}} + \overset{\rightarrow}{\vec{\sigma}})$ and transversal $\sigma_{tot}^{\perp} = \frac{1}{2}(\sigma^{\uparrow\downarrow} + \sigma^{\uparrow\uparrow})$ total cross section in polarized (anti)proton–proton collisions at high energies. Calculated values of $\sigma_{tot}, \sigma_{tot}^L$ and $\sigma_{tot}^{\perp}$ with the data on σ_{tot} for unpolarized proton–proton and antiproton–proton collisions in the energy interval $\sqrt{s} = 10 - 10^3\ GeV$ are shown in Fig 2. Details of calculations are given in paper [10]. One can see that

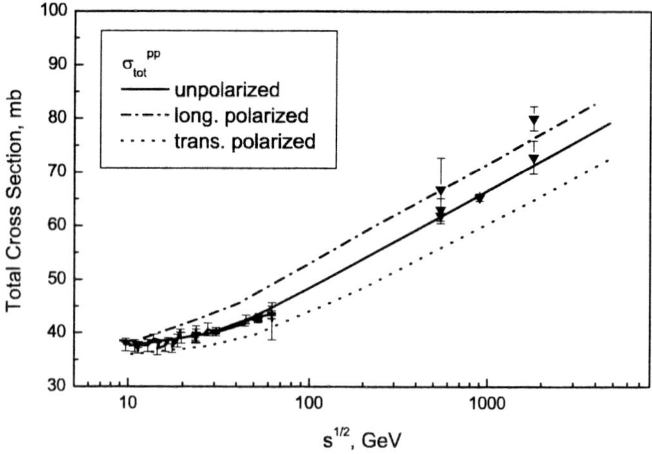

FIGURE 2. Total cross sections: dot–dashed curve is for longitudinally polarized protons; solid curve – for unpolarized protons; dotted curve – for transversely polarized protons; data are taken from Review of Part. Prop.

the differences between these three cross section are significant. The measurements of these cross sections could be performed at RHIC with longitudinally and transversally polarized proton–proton beams.

This research was partly supported by the Russian Foundation of Basics Research, grant 01-07-90144.

REFERENCES

1. G. Musulmanbekov, *Nucl. Phys. Proc. Suppl.* **B71,** 117–120 (1999).
2. E. Schrodinger, *Naturwissenschaften*, **14** 664 (1926); C.C. Yan, *Am. J. Phys.*, **62** 147 (1994).
3. G. Musulmanbekov in *Proc. of the VIIIth Blois Workshop*, Ed. V. A. Petrov, World Scientific, Singapore, 2000, p. 341–351.
4. G. Musulmanbekov in *Frontiers of Fundamental Physics*, Kluwer Acad./Plenum Pub., New York, 2001, p. 109–120.
5. F.J. Belinfante, *Physica* **6,** 887 (1939).
6. H.C. Ohanian, *Am. J. Phys.*, **54,** 500 (1986).
7. J. Hiqbie, *Am. J. Phys.*, **56,** 378–379 (1988).
8. R.P. Feynman, R. B. Leighton and M. Sands, *The Feynman Lectures on Physics*, Addison–Wesley Pub. Co. Inc., London, 1963, Vol.2,; Execises.
9. J. Ashman, et al., *Nucl. Phys.,* 1–18 **B328** (1988).
10. G. Musulmanbekov, in *Proc. IXth Blois Workshop*, Ed. V. Kundrat, Inst. of Physics AS CR, Prague, 2002, p. 339–346.

Recent Status of Polarized Parton Distributions

M. Hirai
(Asymmetry Analysis Collaboration) [1]

Radiation Laboratory, RIKEN (The institute of Physics and Chemical Research), Wako, Saitama 351-0198, Japan

Abstract. We study an influence of precise data on uncertainty of polarized parton distribution functions. This analysis includes the SLAC-E155 proton target data which are precise measurements. Polarized PDF uncertainties are estimated by using the Hessian matrix. We examine correlation effect between the antiquark and gluon uncertainties. It suggests that reducing the gluon uncertainty is needed to determine the polarized antiquark distribution clearly.

INTRODUCTION

Polarized parton distribution functions (polarized PDF's) have so far been optimized from polarized deep inelastic scattering (polarized DIS) world data [1, 2]. We could obtain only a slight piece of information about polarized antiquark and gluon distributions. At this stage, the antiquark $SU(3)_f$ flavor symmetry is assumed in most of the polarized PDF analyses. The $SU(3)_f$ symmetry breaking is already known as the Gottfried sum rule violation in the unpolarized case. In principle, the polarized PDF analysis should take account of the symmetry breaking. However, we must not only determine a shape but also a sign of each polarized antiquark distribution. It needs more precise data to improve the current status. Semi-inclusive DIS experiments [3] are also expected to separate antiquark flavor distributions. However, the separated distributions may not be credible due to ambiguity of the fragmentation functions. Then, antiquark flavor distributions cannot be decomposed clearly. The current knowledge of the polarized gluon distribution is still poor. The polarized gluon distribution is suggested as the positive distribution; however, there is large difference between various parameterization results.

We would like to know ambiguity of polarized PDF's quantitatively. PDF uncertainty plays an important role in illustrating the ambiguity. Furthermore, it is important to show the phenomenological uncertainty of predicted physical quantities (e.g., scattering cross-sections and spin asymmetries) with parameterized PDF's and their uncertainties in our work. A purpose of this analysis is to clarify the current knowledge about the polarized PDF's from the polarized DIS world data by using their PDF uncertainty. In this analysis, the polarized PDF's are optimized including precise SLAC-E155 proton target data [4]. Then, we examine an influence of the precise data on the polarized PDF uncertainty, which is estimated by the Hessian method.

[1] A fortran program of the AAC PDF library is available from http://spin.riken.bnl.gov/aac/.

CP675, *Spin 2002: 15th Int'l. Spin Physics Symposium and Workshop on Polarized Electron Sources and Polarimeters,* edited by Y. I. Makdisi, A. U. Luccio, and W. W. MacKay

PARAMETERIZATION OF THE POLARIZED PDF'S

The polarized PDF's are determined by using spin asymmetry A_1 of the polarized DIS experiments from the EMC, SMC, SLAC-E130, E142, E143, E154, E155, and HERMES:

$$A_1(x,Q^2) = \frac{2x[1+R(x,Q^2)]}{F_2(x,Q^2)} g_1(x,Q^2), \tag{1}$$

where F_2 is the unpolarized structure function. The function $R(x,Q^2) = \sigma_L/\sigma_T$ is the ratio of absorption cross sections for longitudinal and transverse virtual photons, and it is determined from experimental data in reasonably wide Q^2 and x ranges in the SLAC experiments [5]. The polarized structure function g_1 is expressed with polarized PDF's:

$$g_1(x,Q^2) = \frac{1}{2}\sum_{i=1}^{n_f} e_i^2 \Big\{\Delta C_q(x,\alpha_s)\otimes[\Delta q_i(x,Q^2)+\Delta\bar{q}_i(x,Q^2)] + \Delta C_g(x,\alpha_s)\otimes\Delta g(x,Q^2)\Big\}, \tag{2}$$

where e_i is the electric charge of quarks, and ΔC_q, ΔC_g are Wilson's coefficient functions. The convolution $\otimes$ is defined by $f(x)\otimes g(x) = \int_x^1 dy/y\, f(x/y)g(y)$. The polarized PDF's $\Delta f(\equiv f^{\uparrow} - f^{\downarrow})$ are defined as helicity distributions in the nucleon. In the AAC analysis, the polarized PDF $\Delta f(x)$ is defined at initial Q^2 by the weight function form:

$$\Delta f(x) = Ax^{\alpha}(1+\lambda x^{\gamma})f(x)\ , \tag{3}$$

where $f(x)$ is the unpolarized PDF, and A, α, λ, and γ are free parameters. Optimized PDF's are four distributions; $\Delta u_v(x)$, $\Delta d_v(x)$, $\Delta\bar{q}(x)$, and $\Delta g(x)$, and these are evolved from the initial Q^2(=1 GeV^2) to the same Q^2 of experimental data by the DGLAP equation [7]. In particular, the gluon distribution $\Delta g(x)$ contributes to the structure function with the non-zero coefficient function ΔC_g in the NLO case.

This analysis uses two constraint conditions. First, the positivity condition is used to restrict large-x behavior of the polarized PDF's. This condition corresponds to the probabilistic interpretation of the parton distributions in the LO: $|\Delta f(x)| \leq f(x)$. It needs not to be satisfied strictly in the NLO analysis. However, the polarized antiquark and gluon distributions tend to badly break the positivity limit: $|\Delta f(x)| \gg f(x)$. Such excessive behavior is due to the large experimental errors in the large-x region. Hence, this behavior should be limited by this condition.

Next, the SU(3)$_f$ flavor symmetry is assumed: $\Delta\bar{u}(x) = \Delta\bar{d}(x) = \Delta\bar{s}(x) = \Delta s(x)$. Using this assumption, one can fix the first moments of the valence quarks with hyperon decay constants, then $\Delta u_v = 0.926$ and $\Delta d_v = -0.341$ are obtained. Note that the Bjorken sum rule is satisfied automatically by fixing first moments. Furthermore, the spin content $\Delta\Sigma$ is obtained by $\Delta\Sigma_{N_f=3} = \Delta u_v + \Delta d_v + 6\Delta\bar{q}$. Since, the antiquark contribution is emphasized, then the spin content determination is susceptible to the antiquark behavior.

In the analysis, we choose the modified minimal subtraction ($\overline{\text{MS}}$) scheme, and the GRV parameterization for the unpolarized PDF's at the NLO analysis [8]. The total χ^2 is minimized by the CERN subroutine `MINUIT`.

UNCERTAINTY ESTIMATION

Fortunately PDF uncertainty estimation method has been developed in the last several years (see a brief review [6]). The polarized PDF uncertainty comes from several error sources, e.g., experimental errors, unpolarized PDF, Λ_{QCD}, and so on. However, it is difficult to incorporate these errors into uncertainty estimation simultaneously. In the present analysis, the polarized PDF uncertainty is estimated from experimental errors by using the Hessian matrix H_{ij} which is defined as a second order derivative matrix in the expanded $\chi^2(a_i)$ function around its minimum point. The PDF uncertainty $\delta\Delta f(x)$ can be obtained easily by the inverse matrix of the Hessian and linear error propagation:

$$[\delta\Delta f(x)]^2 = \Delta\chi^2 \sum_{i,j} \frac{\partial\Delta f(x)}{\partial a_i} H_{ij}^{-1} \frac{\partial\Delta f(x)}{\partial a_j}, \tag{4}$$

where $\Delta\chi^2 (= \chi^2(a_i) - \chi^2_{min})$ is defined as the difference from the minimum χ^2. It determines a confidence level of the PDF uncertainty, and it depends on the χ^2 distribution $K(s)$ with N degrees of freedom. Here, N is the number of optimized parameters. In our estimation, the value of $\Delta\chi^2$ is obtained by the following equation: $\int_0^{\Delta\chi^2} K(s)\, ds = \sigma$, where $\sigma (= 0.683)$ corresponds to 1 σ error of a standard distribution in order to compare with general experimental errors. The statistical and systematic errors are added in quadrature, so that it could be overestimation. The proper estimation exists between the overestimated uncertainty and the uncertainty from only the statistical error.

RESULTS AND DISCUSSIONS

The best fitting result is $\chi^2(/d.o.f.) = 346.33(0.90)$. The first moments of new results and the AAC pervious results (AAC00, NLO set2) [1] are shown in Table. 1. A correlation coefficient $\rho_{\bar{q}g}$ between the first moment of the antiquark and gluon distributions is $\rho_{\bar{q}g} = -0.836$, and there is strong correlation between two distributions. The uncertainties of the new results become smaller than those of the previous results. The gluon first moment and spin content $\Delta\Sigma$ still have large uncertainty. The fixed first moments Δu_v and Δd_v do not have uncertainty, then the $\Delta\Sigma$ uncertainty is six times as large as the antiquark uncertainty. Thus, the spin content is subject to the uncertainty of the antiquark distribution. Figure 1 shows the uncertainty of the new antiquark distribution. The antiquark uncertainty becomes rather large in the region $x < 0.01$, however the experimental data scarcely exist. The polarized DIS spin asymmetries $A_1^{p,d}(x)$ approaches rapidly to zero in the rang $x < 0.004$. It is insufficient to clarify small-x behavior of the antiquark distribution. Therefore, the antiquark determination has extrapolating ambiguity in small-x region. It is needed tight constraint condition or other experiment.

In addition, Figure 1 shows comparison between the PDF uncertainties of new results and the previous results. There are no significant improvements of the valence quark uncertainties. On the SU(3)$_f$ symmetry assumption, the fixing first moments strongly restricts the behavior of valence quark distributions. In contrast, the antiquark and gluon uncertainties are reduced in the range $0.01 < x < 0.5$, where the E155 proton data exist.

TABLE 1. First moments of the polarized antiquark, gluon, and spin content $\Delta\Sigma$ with their uncertainties at $Q^2 = 1$ GeV^2.

	$\Delta\bar{q}$	Δg	$\Delta\Sigma$
New	-0.062 ± 0.023,	0.499 ± 1.268,	0.213 ± 0.138
AAC00	-0.057 ± 0.038,	0.532 ± 1.949,	0.241 ± 0.228

The precise polarized DIS data can reduce the antiquark uncertainty mainly. On the other hand, the gluon uncertainty changes in response to antiquark uncertainty reduction due to a strong correlation between two distributions. Since the gluon contribution to the structure function $g_1(x)$ is smaller than the quark and antiquark contributions, we can extract only a little information of the gluon distribution in spite of the NLO analysis. Actually, the gluon uncertainty is still large. It indicates the difficulty of determining the gluon distributions from the polarized DIS data. Therefore, the uncertainty reduction of the gluon distribution is due to the strong correlation rather than the NLO contribution.

In order to examine the correlation effect on the parameterization, we re-analyzed the $\Delta g(x) = 0$ case in which the fixed gluon distribution does not have uncertainty. The polarized PDF uncertainties of the $\Delta g(x) = 0$ case are compared to those of the $\Delta g(x) \neq 0$ case in Figure 2. The gluon distribution slightly exists at high-Q^2 due to Q^2 evolution of the singlet type DGLAP equation. The valence quark uncertainties scarcely change. Drastic improvement of the antiquark uncertainty is due to vanished the large gluon uncertainty. the obscure gluon distribution brings about the larger antiquark uncertainty by the complementary relation.

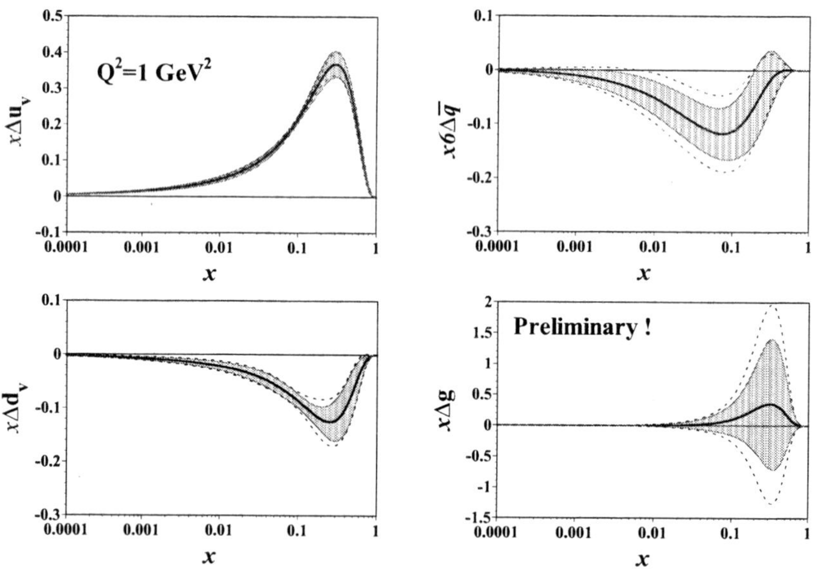

FIGURE 1. Polarized PDF's with their uncertainties at $Q^2 = 1$ GeV^2. Dashed curves are the uncertainties of previous results (AAC NLO-2)

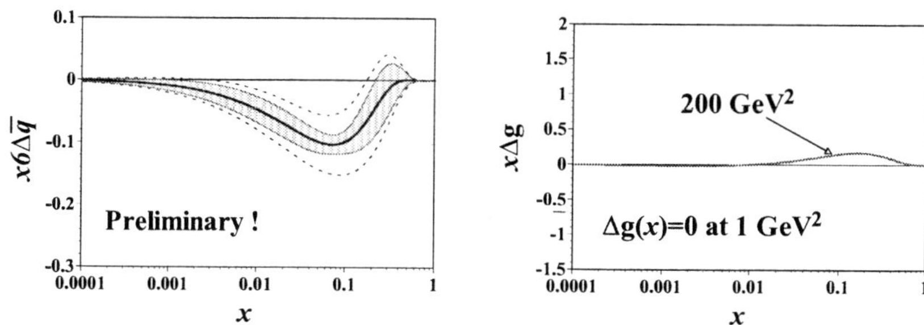

FIGURE 2. Polarized antiquark and gluon distributions with their uncertainties at $Q^2 = 1$ GeV^2. The shaded portion shows the uncertainty of $\Delta g(x) = 0$ results, and the dashed curves are the uncertainties of new results ($\Delta g(x) \neq 0$).

SUMMARY

By this analysis, the polarized PDF's were optimized from the polarized DIS world data which included the SLAC-E155 proton target data. The polarized PDF uncertainties were estimated by the Hessian method. The E155 precise measurements scarcely improve the valence quark uncertainties, but they can reduce the antiquark and gluon uncertainties. These, however, are still wrapped in large uncertainty. The SU$(3)_f$ symmetry, which we are obliged to assume, restricts strongly the valence quark behavior by fixing first moments, and the spin content determination depends on the antiquark behavior. Additionally, there is the strong correlation between the antiquark and gluon distributions. If the gluon distribution is clarified by RHIC-Spin at BNL, the uncertainty of the antiquark distribution can be reduced to some extent. Similarly, the complementary relation can reduce the large uncertainty of the spin content which comes from the extrapolating issue of the antiquark behavior. Then, we will be able to investigate the antiquark flavor dependence in detail.

REFERENCES

1. AAC, Y. Goto *et al.*, Phys. Rev. **D62** (2000) 034017.
2. De Florian and R. Sassot, Phys. Rev. D62 (2000) 094025; M. Glück, E. Reya, M. Stratmann, and W. Vogelsang, Phsy. Rev. **D63** (2001) 09400; E. Leader, A.V. Sidorov, and D.B. Stamenov, Eur.Phys.J. C23 (2002) 479-485; J. Blümlein and H. Böttcher, Nucl. Phys. B **B636** (2002) 225-263. Fortran program librarys of polarized PDF's are available from http://www-spires.dur.ac.uk/hepdata/pdf.html.
3. SMC, B. Adeva *et al.*, Phys. Lett. **B420**, 180 (1998); HERMES, K. Ackerstaff *et al.*, Phys. Lett. **B464** 123 (1999).
4. SLAC-E155, P. L. Anthony *et al.*, Phys. Lett. **B493** (2000) 19.
5. L. W. Whitlow, S. Rock, A. Bodek, S. Dasu and E. M. Riordan, Phys. Lett. **B250**, 193 (1990); SLAC-E143, K. Abe *et al.*, Nucl. Phys **B452** (1999) 194
6. M. Botje , J. Phys. **G** 28 (2002) 779-790.
7. V. N. Gribov and L. N. Lipatov, Sov. J. Nucl. Phys. 15 (1972) 438 and 675; G. Altarelli and G. Parisi, Nucl. Phys. B 126 (1977) 298; Yu. L. Dokshitzer, Sov. Phys. JETP 46 (1977) 641.
8. M. Glück, E. Reya, and A. Vogt, Eur. Phys. J. **C5** (1998) 461.

D Meson Production in Neutrino DIS as a Probe of Polarized Strange Quark Distribution

Kazutaka Sudoh

Radiation Laboratory, RIKEN (The Institute of Physical and Chemical Research), Wako, Saitama 351-0198, JAPAN
E-mail: `sudou@rarfaxp.riken.go.jp`

Abstract. Semi-inclusive $D/\bar{D}$ productions in neutrino deep inelastic scattering are studied including $\mathcal{O}(\alpha_s)$ corrections. Supposing a future neutrino factory, cross sections and spin asymmetries in polarized processes are calculated by using various parametrization models of polarized parton distribution functions. It is found that $\bar{D}$ production is promising to directly extract the strange quark distribution.

INTRODUCTION

Flavor structure of sea quark distributions has been actively studied in recent years. Most of parametrization models are assumed the flavor $SU(3)_f$ symmetry for the sea quark distributions. However, there is an attempt to include the violation of the $SU(3)_f$ symmetrty [1]. Knowledge about the polarized sea quark distributions remains still poor.

A study of heavy flavor production in deep inelastic scattering (DIS) is one of the most promising ways to access the parton density in the nucleon. Since the heavy quark mass scale is quite larger than Λ_{QCD}, it is considered that we can treat heavy quarks purely perturbatively. Heavy quarks are produced only at the short distance scale within the framework of a fixed flavor number scheme (FFNS), where only light quarks (u, d, and s) and gluons are considered as active partons, and any heavy quarks (c, b, ...) contribution is calculated in fixed order α_s perturbation theory.

Charged current (CC) DIS is effective to extract the flavor decomposed polarized parton distribution function (PDF), since $W^{\pm}$ boson changes the flavor of parton. Since there is no intrinsic heavy flavor component in the FFNS, we can extract information about the parton flavor in the nucleon from the study of heavy flavor production in CC DIS. Actually, the NuTeV collaboration reported a measurement of unpolarized s and $\bar{s}$ quark distributions by measuring dimuon cross sections in neutrino-DIS [2].

In this work, to extract information about the polarized PDFs we investigated $D/\bar{D}$ meson production in CC DIS including $\mathcal{O}(\alpha_s)$ corrections in neutrino and polarized proton scattering; $\nu+\vec{p}\to l^-+D+X$, $\bar{\nu}+\vec{p}\to l^++\bar{D}+X$. The leading order process is due to W boson exchange $W^+s(d)\to c$. In addition, several processes are taken account of $\mathcal{O}(\alpha_s)$ next-to-leading order (NLO) calculations, in which gluon radiation processes $W^+s(d)\to cg$, virtual corrections to remove singularity coming from soft gluon radiation, and boson-gluon fusion processes $W^+g\to c\bar{s}(\bar{d})$ are considered. These processes might be observed in the forthcoming neutrino experiments.

CP675, *Spin 2002: 15th Int'l. Spin Physics Symposium and Workshop on Polarized Electron Sources and Polarimeters,* edited by Y. I. Makdisi, A. U. Luccio, and W. W. MacKay

CHARM PRODUCTION IN CC DIS

We have numerically calculated the spin-independent and -dependent cross sections, and the spin asymmetry A^D which is defined by

$$A^D \equiv \frac{d\sigma(+)/dx - d\sigma(-)/dx}{d\sigma(+)/dx + d\sigma(-)/dx} = \frac{d\Delta\sigma/dx}{d\sigma/dx}, \tag{1}$$

where $+$ and $-$ denote the helicity of the target proton. The spin-dependent cross section can be written in terms of the polarized structure functions g_i as follows:

$$\frac{d^3\Delta\sigma^{\nu p}}{dxdydz} = \frac{G_F^2 s}{2\pi(1+Q^2/M_W^2)^2}\left[(1-y)g_4^{W^\mp} + y^2 x g_3^{W^\mp} \pm y(1-\frac{y}{2})x g_1^{W^\mp}\right], \tag{2}$$

where $Q^2 = -q^2$ and G_F, s, and M_W denote the Fermi coupling, center of mass energy squared, and $W^\pm$ boson mass, respectively. Note that the $+$ and $-$ in front of the 3rd term correspond to when initial beam is anti-neutrino and neutrino, respectively. Kinematical variables x and y are Bjorken scaling variable and inelasticity defined according to the standard DIS kinematics, and z is defined by $z = P_p \cdot P_D / P_p \cdot q$ with P_p, P_D and q being the momentum of proton, D meson, and $W^\pm$ boson, respectively. The polarized structure functions g_i in $\nu\vec{p}$ scattering are obtained by the following convolutions:

$$\begin{aligned}\mathcal{G}_i^c(x,z,Q^2) &= \Delta s'(\xi,\mu_F^2)D_c(z) \\ &+ \frac{\alpha_s(\mu_R^2)}{2\pi}\int_\xi^1 \frac{d\xi'}{\xi'}\int_{\max(z,\zeta_{\min})}^1 \frac{d\zeta}{\zeta}\left\{\Delta H_i^q(\xi',\zeta,\mu_F^2,\lambda)\Delta s'(\frac{\xi}{\xi'},\mu_F^2)\right. \\ &\left.+ \Delta H_i^g(\xi',\zeta,\mu_F^2,\lambda)\Delta g(\frac{\xi}{\xi'},\mu_F^2)\right\} D_c(\frac{z}{\zeta}),\end{aligned} \tag{3}$$

where $\Delta s'$ means $\Delta s' \equiv |V_{cs}|^2\Delta s + |V_{cd}|^2\Delta d$ with CKM parameters. $\Delta H_i^{q,g}$ are coefficient functions of quarks and gluons, which can be calculated by using perturbative QCD. The argument ξ is the slow rescaling parameter, and ξ' and ζ are the partonic scaling variables which are defined for the parton momentum p_i as

$$\xi = \frac{Q^2}{2P_p\cdot q}\left(1+\frac{m_c^2}{Q^2}\right),\quad \xi' = \frac{Q^2}{2p_{s,g}\cdot q}\left(1+\frac{m_c^2}{Q^2}\right),\quad \zeta = \frac{p_{s,g}\cdot p_c}{p_{s,g}\cdot q}. \tag{4}$$

$D_c(z)$ represents the fragmentation function of an outgoing charm quark decaying to D meson. For the fragmentation function, we adopted the parametrization proposed by Peterson *et al.* [3] and recently developed by Kretzer *et al.* [4]. $\mathcal{G}_i$ is related to the polarized structure functions through $\mathcal{G}_1 \equiv g_1/2$, $\mathcal{G}_3 \equiv g_3$, and $\mathcal{G}_4 \equiv g_4/2\xi$. Similar analyses have been done by Kretzer *et al.*, in which charged current charm production at NLO in ep and νp scattering is discussed [5].

NUMERICAL RESULTS

In numerical calculations, we set a charm quark mass $m_c = 1.4$ GeV, an initial neutrino energy $E_\nu = 200$ GeV, and the factorization scale μ_F which is equal to the renor-

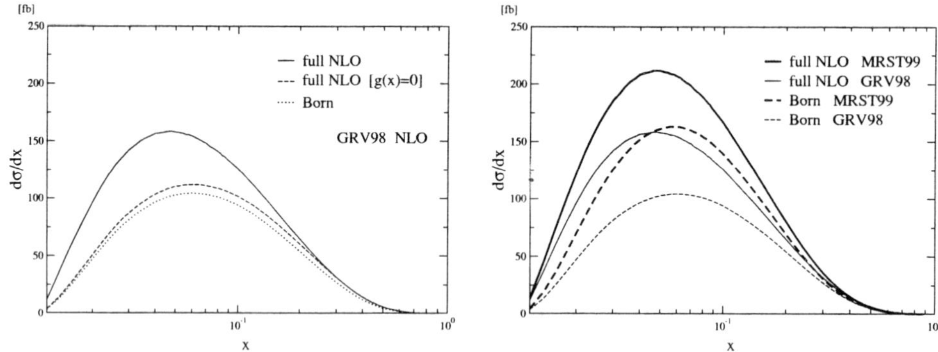

FIGURE 1. The spin-independent differential cross sections for the process $\nu\vec{p} \to l^- DX$ at $E_\nu = 200$ GeV as a function of x. We show the contribution from each diagram (left panel) and parametrization dependence of the unpolarized PDFs (right panel). Solid, dashed, and dotted lines in the left panel show the cross section in full NLO, full NLO with $g(x) = 0$, and LO calculation, respectively. Solid and dashed lines in the right panel represent the case of MRST99 and GRV98 parametrizations, and bold and normal lines are full NLO and LO cross sections, respectively.

malization scale μ_R as $\mu_F^2 = \mu_R^2 = Q^2 + m_c^2$. We used the GRV98[6] and MRST99[7] parametrizations as the unpolarized PDFs. As for the polarized PDFs, we adopted the AAC00[8], BB02[9], GRSV01[10], and LSS02[11] parametrizations which are now widely used.

We show the spin-independent differential cross section for D meson production $\nu\vec{p} \to l^- DX$ in Fig. 1. The left panel in Fig. 1 represents the contribution to the cross section from each diagram. Solid, dashed, and dotted lines indicate the contribution form full NLO, full NLO with $g(x) = 0$, and LO diagrams to the x differential cross section, respectively. Hence, the difference between the dashed and dotted lines comes from the gluon radiation process $W^+ s(d) \to cg$ and virtual corrections, while the difference between the solid and dashed lines comes from the boson-gluon fusion process ($W^+ g \to c\bar{s}(\bar{d})$). As shown in Fig. 1, the contribution from the NLO boson-gluon fusion process is considerably large. This is because the gluon distribution is sufficiently larger than the strange quark distribution, though the short distance matrix element in NLO is suppressed by the strong coupling constant α_s.

Comparison of the cross section using MRST99 and GRV98 parametrization for the unpolarized PDFs is represented in the right panel in Fig. 1. The gap between two dashed lines stems from the difference of behavior of the unpolarized strange quark distribution, since only the strange quark distribution contributes to the cross section at the LO level. We found that the parametrization dependence is quite large. It is indicated that even the unpolarized PDFs still has large ambiguity, in spite of the analyses of unpolarized processes are investigated for a long time. Therefore, this reaction is effective to determine the unpolarized strange quark distribution, because the charm quark production in CC DIS using neutrino beams is sensitive to the strange density in the nucleon.

The spin-dependent cross section for D meson production is presented in Fig. 2. We

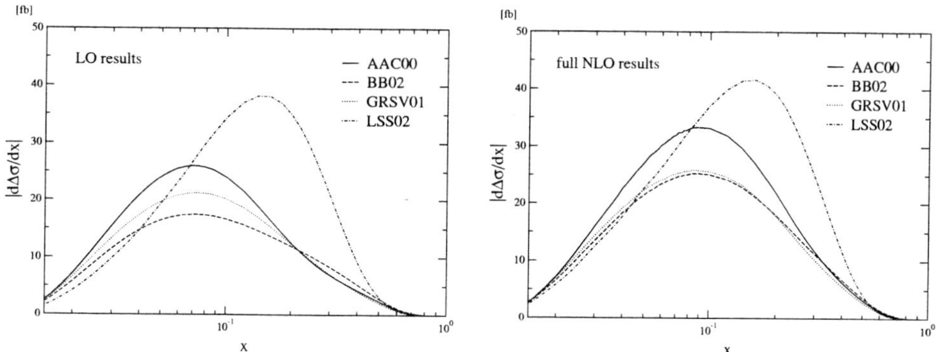

FIGURE 2. The x distribution of spin-dependent cross sections for D meson production $\nu\vec{p} \to l^- DX$ in LO (left panel) and full NLO (right panel). Solid, dashed, dotted, and dot-dashed lines show the case of AAC00 set-2, BB02 scenario-2, GRSV01 standard set, and LSS02 $\overline{\mathrm{MS}}$ parametrizations, respectively.

show the comparison between LO (left panel) and full NLO (right panel) results with various parametrization models of the polarized PDFs. We see large contribution from NLO corrections and the parametrization model dependence of the polarized PDFs. As well as the unpolarized case, the cross sections are dominated by the LO process and boson-gluon fusion process at NLO. Contribution from the gluon radiation and virtual corrections are not significant in the cross sections.

Both the LO and full NLO results by the LSS parametrization are quite large compared with other parametrization. In the LSS parametrization, the polarized strange quark distribution has the peak at $x \sim 0.2$, whereas the polarized gluon distribution is not significant in this x region. Therefore, the LO cross section becomes large, and the difference between the LO and full NLO cross sections is consequently small for the LSS parametrization. On the other hand, the NLO cross sections by GRSV and BB parametrizations are quite similar in whole x regions, though we see some difference in the LO cross section.

We show the spin asymmetry A^D in Fig. 3 as a function of x. Left panel and right panel in Fig. 3 represent asymmetries for D production and $\bar{D}$ production, respectively. For D production, s, d quarks and gluon distribution contribute to the asymmetry A^D. A^D is dominated by valence d_v quark at large x ($x > 0.3$), though the d quark component is quite highly suppressed by CKM. On the contrary, for $\bar{D}$ production, $\bar{s}$, $\bar{d}$ quarks and gluon component contribute to the asymmetry A^D. The $\bar{d}$ quark contribution is almost negligible. Therefore, the asymmetry is directly affected by the shape of the $\bar{s}$ quark distribution.

As shown in both figures, spin asymmetries strongly depend on parametrization models. We see that the case of the LSS parametrization is quite different from the ones of other parametrizations. In particular, the asymmetry by the LSS parametrization in $\bar{D}$ production goes over 1 at $x \sim 0.3$, though the asymmetry should be less than 1. This is because the polarized s quark distribution in their parametrization extremely violates the positivity condition at $x \sim 0.3$. Measurement of $\bar{D}$ production in this reaction is effective to test the parametrization models of the polarized PDFs. In semi-inclusive DIS, we

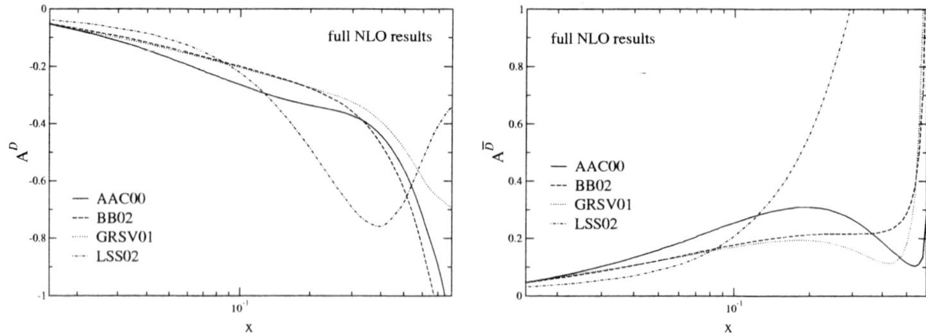

FIGURE 3. Comparison of the spin asymmetries in NLO for D production $\nu\vec{p} \to l^- DX$ (left panel) and $\bar{D}$ production $\bar{\nu}\vec{p} \to l^+ \bar{D}X$ (right panel) with various parametrization models of the polarized PDFs. Several lines are the same as Fig. 2.

have an additional ambiguity coming from the fragmentation function. However, the ambiguity can be neglected in the x distribution of A^D, since the kinematical variable related to fragmentation is integrated out in this distribution.

SUMMARY

In summary, semi-inclusive $D/\bar{D}$ meson productions in CC DIS in neutrino-polarized proton scattering are discussed. The cross sections and the spin asymmetries are calculated including $\mathcal{O}(\alpha_s)$ corrections with various parametrization models of the polarized PDFs. The $\bar{D}$ production is promising to extract the sea quark density. This is not the case for the D production because of the large d_v contribution over the sea quark contribution. If the gluon polarization $\Delta g(x,Q^2)$ is fixed by RHIC experiments with high accuracy, we can directly extract the strange sea $\Delta s(x,Q^2)$.

REFERENCES

1. E. Leader, A. V. Sidorov, and D. B. Stamenov, Phys. Lett. B **462**, 189 (1999).
2. NuTeV Collaboration, M. Goncharov *et al.*, Phys. Rev. D **64**, 112006 (2001); G. P. Zeller *et al.*, Phys. Rev. D **65**, 111103 (2002).
3. C. Peterson, D. Schlatter, I. Schmitt, and P. M. Zerwas, Phys. Rev. D **27**, 105 (1983).
4. S. Kretzer and I. Schienbein, Phys. Rev. D **59**, 054004 (1999).
5. M. Glück, S. Kretzer, and E. Reya, Phys. Lett. B **398**, 381 (1997); S. Kretzer and M. Stratmann, Eur. Phys. J. C **10**, 107 (1999).
6. M. Glück, E. Reya, and A. Vogt, Eur. Phys. J. C **5**, 461 (1998).
7. A. D. Martin, R. G. Roberts, W. J. Stirling, and R. S. Thorne, Eur. Phys. J. C **14**, 133 (2000).
8. Asymmetry Analysis Collaboration, Y. Goto *et al.*, Phys. Rev. D **62**, 034017 (2000).
9. J. Blümlein and H. Böttcher, Nucl. Phys. B **636**, 225 (2002).
10. M. Glück, E. Reya, M. Stratmann, and W. Vogelsang, Phys. Rev. D **63**, 094005 (2001).
11. E. Leader, A. V. Sidorov, and D. B. Stamenov, Eur. Phys. J. C **23**, 479 (2002).

Future Measurements of Spin Dependent Proton Flavor Structure with the PHENIX Muon Arms

N. Bruner for the PHENIX Collaboration[1]

University of New Mexico, Albuquerque, NM 87131, USA

Abstract. The Relativistic Heavy Ion Collider (RHIC) debuted as the first and only polarized p+p collider during the 2001–2002 physics run. The beams were transversely polarized and collided at a center of mass energy of 200 GeV. From this run, the PHENIX experiment measured the absolute cross section and single spin asymmetries in neutral and charged hadronic channels.

In light of these successes, prospects for future measurements of flavor decomposed quark and anti-quark spin distributions in a polarized proton using the PHENIX Muon Arms are discussed. The first of two Muon Arms was commissioned in the RHIC 2001–2002 run.

INTRODUCTION

Recent measurements have demonstrated the complexity of the nucleon flavor structure, in particular, a large asymmetry in sea quark densities [1, 2] and non-zero quark spin-asymmetries [3, 4]. Future measurements that more precisely characterize the quark and anti-quark spin-dependent distributions in the nucleon will increase our understanding of this complex structure.

To resolve the proton's $\bar{d}$ and $\bar{u}$ spin structure, we can take advantage of the parity-violating production of $W^{\pm}$ bosons, in which the spin direction of the q and $\bar{q}$ that form the W can be precisely determined. The differential cross-section for $pp \to W^+$ is [5]

$$\frac{d\sigma^{W+}}{dy} - G_F \pi \frac{\sqrt{2}}{3} \frac{M_W^2}{s} \left[u(x_1, M_W^2)\bar{d}(x_2, M_W^2) + u(x_2, M_W^2)\bar{d}(x_1, M_W^2) \right], \tag{1}$$

with quark densities, $q(x, Q^2)$, in terms of x_i, the Bjorken-x of the parton from the ith proton, and $Q^2 = M_W^2$. The W^- cross section is given by interchanging the quark flavors. The s-quark contribution is ignored. For an unpolarized measurement, the ratio $(d\sigma(W^-)/dy)/(d\sigma(W^+)/dy)$ is sensitive to $\bar{d}/\bar{u}$, already shown not to be unity over a range of x [1]. By extension of Eq. 1 to polarized cross sections, the single-spin, longitudinal asymmetry, $A_L = \frac{\sigma_- - \sigma_+}{\sigma_- + \sigma_+}$, can be expressed as

$$A_L^{W+} = \frac{\Delta u(x_1)\bar{d}(x_2) - \Delta \bar{d}(x_1) u(x_2)}{u(x_1)\bar{d}(x_2) + \bar{d}(x_1) u(x_2)} \tag{2}$$

[1] For the full PHENIX Collaboration author list and acknowledgments, see Appendix "Collaborations" of this volume.

CP675, *Spin 2002: 15th Int'l. Spin Physics Symposium and Workshop on Polarized Electron Sources and Polarimeters*, edited by Y. I. Makdisi, A. U. Luccio, and W. W. MacKay

and

$$A_L^{W-} = \frac{\Delta d(x_1)\bar{u}(x_2) - \Delta\bar{u}(x_1)d(x_2)}{d(x_1)\bar{u}(x_2) + \bar{u}(x_1)d(x_2)}, \quad (3)$$

where Δu, Δd, $\Delta\bar{u}$, and $\Delta\bar{d}$ are the parton spin densities for u, d, $\bar{u}$, and $\bar{d}$ quarks, respectively, and x_1 refers to the x for the parton from the polarized proton. It has been shown that these asymmetries approach the individual parton spin asymmetries for large values of x_1 or x_2 [10]. Namely, for $x_1 \gg x_2$ (corresponding to larger $|y_W|$),

$$A_L^{W+} \approx \frac{\Delta u(x_1)}{u(x_1)} \quad \text{and} \quad A_L^{W-} \approx \frac{\Delta d(x_1)}{d(x_1)}. \quad (4)$$

For $x_2 \gg x_1$,

$$A_L^{W+} \approx \frac{-\Delta\bar{d}(x_1)}{\bar{d}(x_1)} \quad \text{and} \quad A_L^{W-} \approx \frac{-\Delta\bar{u}(x_1)}{\bar{u}(x_1)}. \quad (5)$$

These approximations make forward rapidity the regions of interest. Work is underway (see P. Nadolsky, these proceedings) to express Δq in terms of $W \to l\nu$ kinematics.

RHIC SPIN AND THE PHENIX MUON ARMS

The Relativistic Heavy Ion Collider (RHIC) was designed to collide beams of polarized protons as well as heavy ions. During the 2001–2002 physics run, RHIC debuted as the first and only polarized proton–proton collider, with transversely polarized beams colliding at $\sqrt{s}$ = 200 GeV with a maximum beam polarization of 25%. These values are expected to increase to 70% polarization at $\sqrt{s}$ = 500 GeV. The RHIC facility layout, including hardware relevant for polarized proton collisions, is shown in Fig. 1. The Siberian Snakes and pC polarimeters were commissioned in the 2001–2002 run. The polarized jet target in Fig. 1 is a future enhancement to enable an absolute polarization measurement. Spin rotator magnets will be installed for the 2002–2003 run to deliver longitudinally polarized beams to the STAR and PHENIX experiments.

Details of the PHENIX detector design and performance can be found in Ref. [6]. The Muon Arms, shown in Fig. 2, are located in the forward rapidity regions, $-1.1 > \eta > -2.2$ and $1.2 < \eta < 2.4$, with full azimuthal coverage. Each arm has two main components, a muon identifier and a muon tracker. The muon identifier is composed of 5 layers of transversely-oriented plastic proportional tubes interleaved with steel absorbers, providing a coarse $x-y$ track position while providing excellent hadron rejection. The muon tracker consists of three stations of tracking chambers, each with 4 to 6 radially segmented readout planes, inside a radial magnetic field. For the 2001–2002 run, both muon identifiers and the South muon tracker were installed. The North muon tracker has been installed for the 2002–2003 run.

In order to determine the charge of a muon with p = 50 GeV/c, each plane of the tracking chambers must have 100μm resolution. Although work continues on track reconstruction software, South Arm alignment studies from the 2001-2002 run indicate that this goal is achievable. Plot *a* of Fig. 3 shows the difference between the fitted

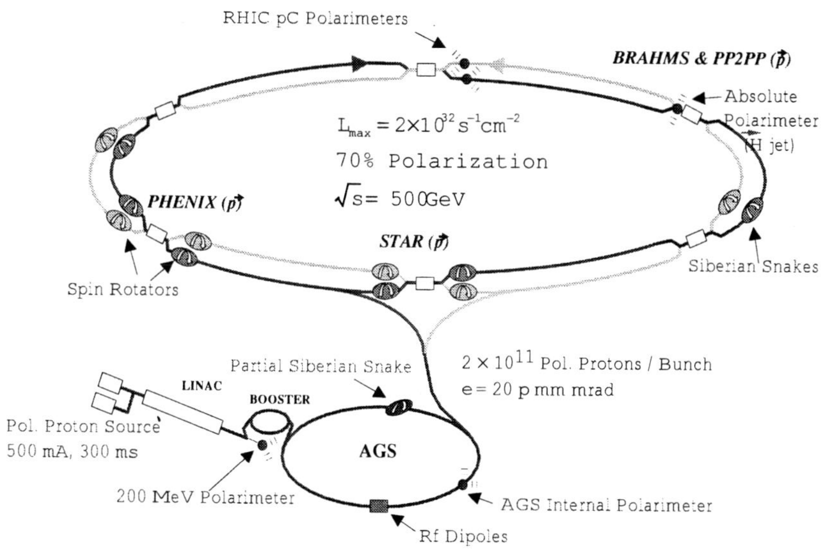

FIGURE 1. RHIC-Spin facility layout.

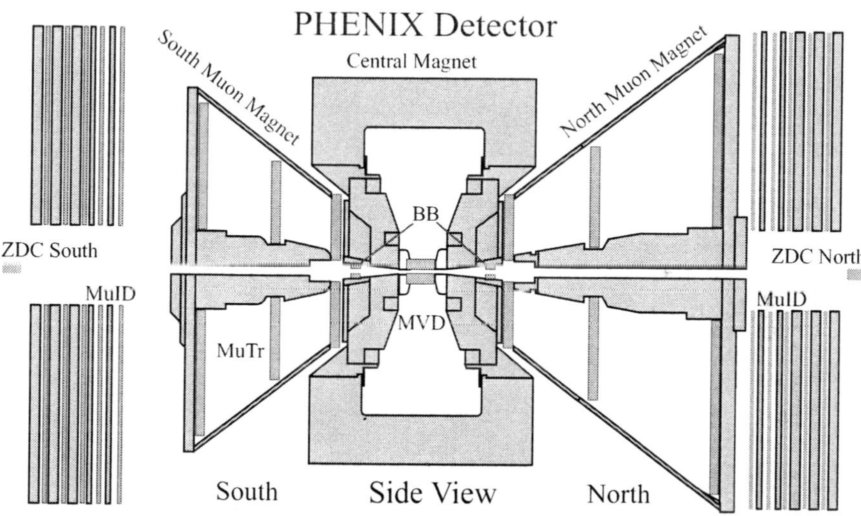

FIGURE 2. Side view of the PHENIX detector.

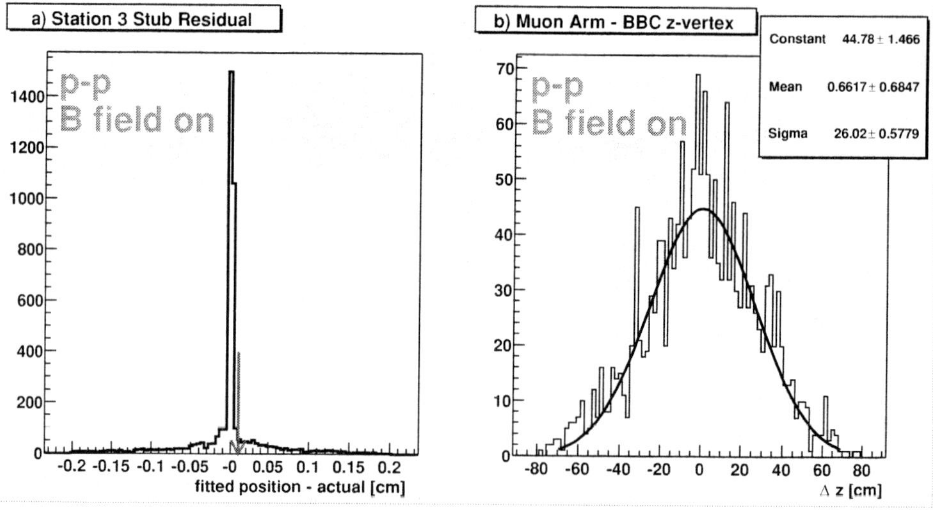

FIGURE 3. South Muon Arm alignment results from the 2001-2002 run. Plot *a* is the difference between track fitted position - actual hit for a plane in the third tracking station. Plot *b* is the difference between the z-position of the vertex determined using dimuons and using the Beam-Beam Counter fitted to a Gaussian.

position of a track in a given plane and the actual hit position after alignment. Plot *b* of Fig. 3 shows the difference between the z-position of the vertex determined using dimuons and using the Beam-Beam Counter. The PHENIX Muon Arms therefore have the capability and are located in the correct region for the $A_L^{W^\pm}$ measurement.

PREDICTED PHENIX SENSITIVITIES

The unpolarized cross sections calculated by RESBOS [7] at $\sqrt{s} = 500$ GeV are 1.2 nb for W^+, 380 pb for W^-, 300pb for Z. The yield estimated by PYTHIA [8] for 800 pb^{-1} and $p_T \geq 20$ GeV/c is ~ 8000 (~ 8000) W^+s (W^-s) in the muon arms [9]. At $p_T \geq 20$ GeV/c, the backgrounds from Drell-Yan and heavy flavor are expected to be negligible and leptons from W's will dominate leptons from Z production. The expected PHENIX Muon Arm sensitivities to quark spin-asymmetries are shown in Fig. 4 as a function of Bjorken-x for each flavor.

OUTLOOK

$W^\pm$ production at RHIC provides a unique probe of spin-dependent proton flavor structure because of the ability to resolve the individual quark helicities and their spin asymmetries at large $|y_W|$. The PHENIX Muon Arms are positioned to measure $W^\pm \to \mu^\pm \nu$ in the regions of high y_μ which is easily related to y_W. Work continues on track re-

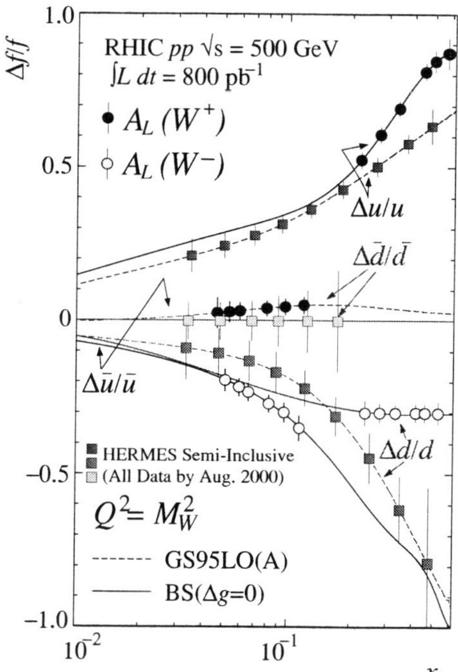

FIGURE 4. Predicted sensitivities to quark spin asymmetries, $\Delta f/f$, as a function of x for RHIC running at $\sqrt{s}$ = 500 GeV with $\int \mathcal{L} dt$ = 800 pb^{-1}. Data are overlaid on the BS [10] and GS95LO(A) [11] parton densities. The precision of the expected HERMES results from their 1995–2000 dataset are shown for comparison [4].

construction, $W^{\pm}$ efficiencies and backgrounds, and trigger rate reduction. Both Muon Arms will be fully characterized for a future RHIC spin run at $\sqrt{s}$ = 500 GeV.

REFERENCES

1. R. S. Towell, *et al.*, Phys. Rev. D**64**, 052002 (2001).
2. K. Ackerstaff, *et al.*, Phys. Rev. Lett. **81**, 5519 (1998).
3. B. Adeva, *et al.*, Phys. Lett. B**420**, 180 (1998).
4. K. Ackerstaff, *et al.*, Phys. Lett. B**464**, 123 (1999).
5. This equation is from the following reference with the mass eigenstates replaced with the weak. Particle Data Group, *Physical Review D: Review of Particle Physics*, The American Physical Society, Ridge, NY (2002).
6. PHENIX Collaboration, Accepted for publication in a Special Issue of Nucl. Inst. Meth. **A**.
7. C. Balazs and C.-P. Yuan, Phys. Rev. D**56**, 5558 (1997).
8. T. Sjostrand, Comput. Phys. Commun. **82**, 74 (1994).
9. G. Bunce, *et al.*, *Prospects for Spin Physics at RHIC*, hep-ph/0007218 (2000).
10. C. Bourrely and J. Soffer, Nucl. Phys. **B**445, 341 (1995).
11. T. Gehrmann and W. J. Stirling, Phys. Rev. D**53**, 6100 (1996).

Resummation for Single-spin Asymmetries in W-boson Production

Pavel M. Nadolsky* and C.-P. Yuan†

*Department of Physics, Southern Methodist University, Dallas, TX 75275-0175, U.S.A.
†Department of Physics & Astronomy, Michigan State University, East Lansing, MI 48824-2320, U.S.A.

Abstract. To measure spin-dependent parton distribution functions in the production of $W^{\pm}$-bosons at the Relativistic Heavy Ion Collider, an accurate model for distributions of charged leptons from the W-boson decay is needed. We present the single- and double-spin lepton-level cross sections of order $\mathcal{O}(\alpha_S)$, as well as resummed cross sections, which include effects of multiple parton radiation effects. We also present a program RHICBOS for the numerical analysis of such cross sections in $\gamma^*, W^{\pm}$, and Z^0 boson production.

The measurement of longitudinal spin asymmetries in the production of $W^{\pm}$ bosons at the Relativistic Heavy Ion Collider (RHIC) will provide an essential probe of spin-dependent quark distributions at high scales Q^2 [1, 2]. At pp center-of-mass energy $\sqrt{s} = 500$ GeV, about 1.3×10^6 W^+- and W^--bosons will be produced by the time the integrated luminosity reaches 800 pb^{-1}. Due to the parity violation in the $Wq\bar{q}$ coupling, this process permits nonvanishing single-spin asymmetries $A_L(\xi)$, defined here as

$$A_L(\xi) \equiv \frac{\frac{d\sigma(p^{\rightarrow}p\rightarrow WX)}{d\xi} - \frac{d\sigma(p^{\leftarrow}p\rightarrow WX)}{d\xi}}{\frac{d\sigma(p^{\rightarrow}p\rightarrow WX)}{d\xi} + \frac{d\sigma(p^{\leftarrow}p\rightarrow WX)}{d\xi}}, \text{ where } \xi = y_W, y_l, p_{Tl}, \ldots. \quad (1)$$

Here y_W is the rapidity of the W-boson; y_l and p_{Tl} are the rapidity and transverse momentum of the charged lepton from the W-boson decay in the lab frame, respectively. The *lowest-order* expression for the asymmetry $A_L(y_W)$ with respect to the rapidity y_W of the W-boson is particularly simple if the absolute value of y_W is large. In that case, $A_L(y_W)$ reduces to the ratio $\Delta q(x)/q(x)$ of the polarized and unpolarized parton distribution functions [3]. Furthermore, $A_L(\xi)$ tests the flavor dependence of quark polarizations.

The original method for the measurement of the W-boson production at RHIC is based on the direct reconstruction of the asymmetry $A_L(y_W)$ [1]. Unfortunately, such reconstruction is obstructed by specifics of the detection of $W^{\pm}$-bosons at RHIC. First, RHIC detectors do not monitor energy balance in particle reactions. Due to the lack of information about the missing momentum carried by the neutrino, the determination of y_W is in general ambiguous and depends on assumptions about the dynamics of the process. Second, due to the correlation between the spins of the initial-state quarks and final-state leptons, the measured value of $A_L(y_W)$ is strongly sensitive to experimental cuts imposed on the observed charged lepton. This feature is illustrated in Fig. 1, which

CP675, *Spin 2002: 15th Int'l. Spin Physics Symposium and Workshop on Polarized Electron Sources and Polarimeters,* edited by Y. I. Makdisi, A. U. Luccio, and W. W. MacKay

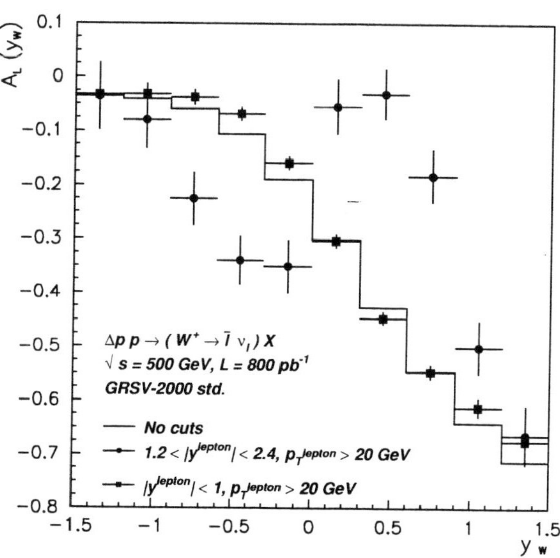

Figure 1. Dependence of the asymmetry $A_L(y_W)$ on the cuts imposed on the momentum of the observed antilepton in the process $p^{\rightarrow}p \to (W^+ \to \bar{l}\nu_l)X$. The asymmetry is calculated using the resummation method described in the paper. The GRSV standard set [4] of the polarized PDF's was used. The error bars are calculated according to Eq. (13) in Ref. [1] assuming the integrated luminosity $\mathscr{L} = 800$ pb^{-1} and beam polarization 70%.

shows the asymmetry in the W^+-boson production calculated without constraints on y_l and p_{Tl} (solid line), and with constraints $1.2 < |y_l| < 2.4$, $p_{Tl} > 20$ GeV (circles) and $|y_l| < 1$, $p_{Tl} > 20$ GeV (boxes). According to the Figure, there is a substantial difference between asymmetries calculated with and without selection cuts. This difference arises due to the different dependence of the unpolarized and polarized cross sections on angular distributions of leptons in the W-boson decay. For instance, at the lowest order in W^+production

$$A_L(y_W, y_l) = \frac{-\Delta u(x_a)\bar{d}(x_b)(1-\cos\theta^*)^2 + \Delta\bar{d}(x_a)u(x_b)(1+\cos\theta^*)^2}{u(x_a)\bar{d}(x_b)(1-\cos\theta^*)^2 + \bar{d}(x_a)u(x_b)(1+\cos\theta^*)^2}, \tag{2}$$

where $x_{a,b} \equiv (Q/\sqrt{s})e^{\pm y_W}$, and θ^* is the polar angle of the antilepton in the rest frame of the W-boson. Since y_l is related to $\cos\theta^*$, as

$$y_l = y_W + \frac{1}{2}\ln\frac{1+\cos\theta^*}{1-\cos\theta^*},$$

it is clear that restrictions on the range of integration of y_l strongly affect $A_L(y_W)$.

Since the only straightforward signature of the W-bosons at RHIC is the observation of secondary charged leptons, it is important to understand differential cross sections of spin-dependent W-boson production *at the lepton level*. Given that the radiative corrections are sizeable ($\sim 30\%$) both in the numerator and denominator of Eq. (1), and that the measurement results will be used in the next-to-leading order (NLO) PDF analysis, it is necessary to derive these cross sections at NLO accuracy (*i.e.*, at order $\mathcal{O}(\alpha_S)$).

Furthermore, most of the W-bosons are produced with small, but non-zero transverse momenta. Such non-zero q_T is acquired through radiation of soft and collinear partons, which cannot be approximated by finite-order perturbative calculations. In order to obtain reliable predictions for differential cross sections, dominant logarithmic terms $\alpha_S^n \ln^m \left(q_T^2/Q^2\right)$ (where $0 \leq m \leq 2n-1$) associated with such radiation should be summed through all orders of the perturbative series. In our work [5], we performed a complete lepton-level study for the production of $W^{\pm}$, γ^*, and Z^0 bosons for arbitrary longitudinal polarizations of incident protons. This study combined the $\mathcal{O}(\alpha_S)$ contributions with the all-order sum of small-q_T logarithmic corrections. The resummation of the logarithms $\ln^m(q_T^2/Q^2)$ was performed with the help of the impact parameter space (b-space) resummation technique [6]. It extended the methodology developed for the unpolarized vector boson production [7] to the spin-dependent case.

Our study goes beyond those in the previous publications [8, 9, 10] in several aspects. It presents the fully differential NLO cross section at the lepton level, which was not available before. The resummed single- and double-polarized cross sections for the production of *on-shell* vector bosons were presented earlier in Ref. [10]. We have derived a more complete resummed cross section, which also accounts for the decays of vector bosons, *i.e.*, for spin correlations in the final state. The lepton-level cross section includes several additional angular structure functions, which do not contribute at the level of on-shell vector bosons. Moreover, resummation is needed not only for the parity-conserving angular function $1+\cos^2\theta^*$, which contributes to the on-shell cross section, but also for the parity-violating angular function $2\cos\theta^*$, which affects angular distributions of the decay products. The estimate of parity-violating contributions is more complicated since it involves γ_5-matrices and Levi-Civita tensors both from the electroweak Lagrangian and spin-projection operators. As a result, special care is needed to treat finite terms that arise in the factorization of collinear poles in $d \neq 4$ dimensions. We perform the calculation using the dimensional reduction and find that the resummed cross sections satisfy helicity conservation conditions for the incoming quarks. In addition, the coefficient functions in Ref. [10] were obtained in a non-conventional factorization scheme and cannot be used with the existing PDFs. In contrast, our results are fully consistent with the $\overline{MS}$ factorization scheme.

The resummed cross sections are incorporated in a numerical program RHICBOS for Monte-Carlo integration of the differential cross sections [11]. We are not able to discuss all aspects of our numerical study in this short report. However, as an example we show lepton-level asymmetries $A_L(y_l)$ for various cuts on p_{Tl} (Fig. 2). We find that these asymmetries can be accurately measured for both W^+- and W^--bosons. These directly observed asymmetries can effectively discriminate between different PDF sets; hence, they provide a viable alternative to the less accessible asymmetry $A_L(y_W)$.

It is also useful to study distributions with respect to the transverse momentum p_{Tl} of the charged lepton, not only because they are sensitive to the PDF's, but also because they probe in detail dynamics of the QCD radiation. As was discussed above, the transverse momentum distributions for vector bosons are affected by the multiple parton radiation, which can be described only by means of all-order resummation. In addition,

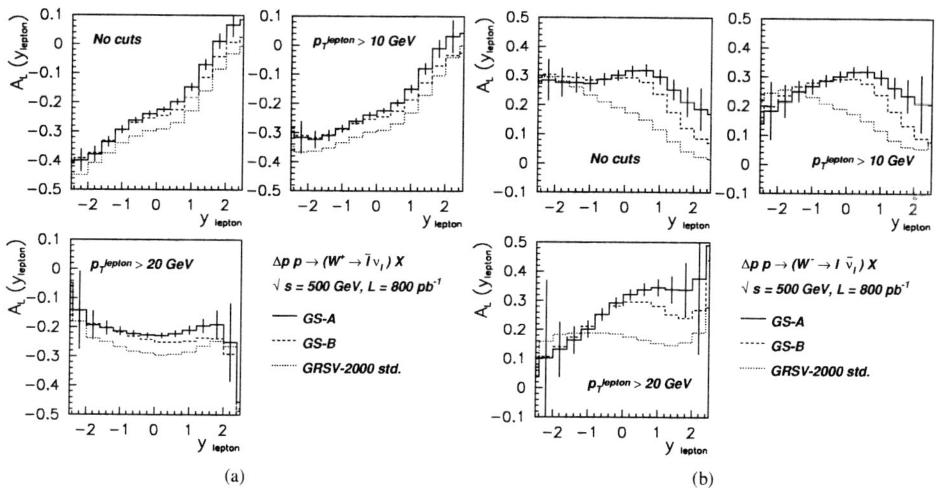

Figure 2. Asymmetries $A_L(y_l)$ for various selection cuts on p_{Tl} in (a) W^+-boson production and (b) W^--boson production. The asymmetries are derived using the Gehrmann-Stirling PDF sets A and B [12] and GRSV-2000 standard PDF set [4]. Statistical errors are estimated as in Fig. 1.

the distributions at very small q_T are sensitive to nonperturbative contributions characterized by large impact parameters $b \gtrsim 1\ \text{GeV}^{-1}$. As a result, the shape of the lepton-level distribution $d\sigma/dp_{Tl}$ around its peak at about $p_{Tl} = M_W/2$ is affected by both perturbative and nonperturbative QCD radiation. Remarkably, the shape of the Jacobian peak can be predicted by the theory, even though it cannot be calculated at any finite order of α_S. The prediction is possible because the perturbative soft and collinear contributions are systematically approximated in the resummation formalism. The nonperturbative contributions currently cannot be derived in a systematical way; but there is substantial indirect evidence (spin independence of the perturbative soft radiation, quark helicity conservation) that such contributions do not depend on the proton spin and type of the vector boson. The impact of the nonperturbative contributions is illustrated in Fig. 3, which shows the number of events for the difference $d\sigma(p^{\rightarrow}p)/dp_{Tl} - d\sigma(p^{\leftarrow}p)/dp_{Tl}$ of single-spin cross sections at $\mathscr{L} = 800\ \text{pb}^{-1}$. This rate was calculated using two parameterizations [13, 14] of the nonperturbative part, which were found in the unpolarized vector boson production. It can be seen that the sensitivity to the nonperturbative input is small, but, nonetheless, visible near the Jacobian peak. For comparison, we also included the $\mathscr{O}(\alpha_S)$ finite-order cross section calculated using the phase space slicing method. The finite-order curve substantially deviates from the resummed curves, and, moreover, its shape can be drastically modified by varying the phase space slicing parameter q_T^{sep}. In contrast, the resummed curve is determined unambiguously once a parameterization of the nonperturbative input is obtained from the double-spin γ^* production, single-spin or double-spin Z^0 production, or even the unpolarized W-production. Needless to say, the resummation predictions for the shape of the Jacobian peak, which directly follow from fundamental principles of QCD, must be tested at RHIC.

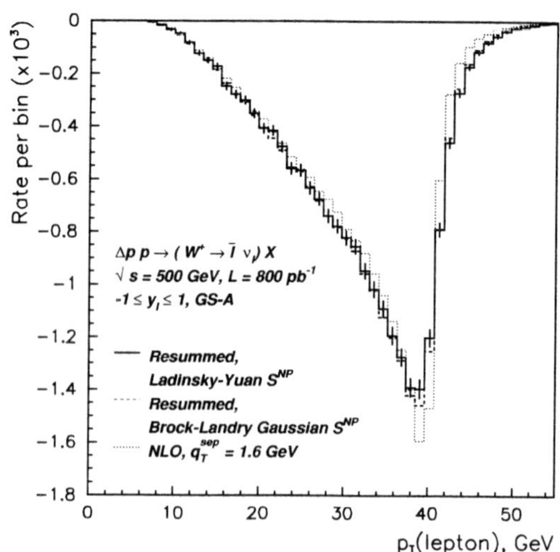

Figure 3. The single-spin charged lepton transverse momentum distribution for W^+-boson production discussed in the main text. The nonperturbative parts of the resummed cross sections were calculated using the Ladinsky-Yuan parameterization [13] (solid) and the most recent Brock-Landry Gaussian 2 parameterization [14] (dashed).The $\mathcal{O}(\alpha_S)$ finite-order cross section is shown as a dotted line.

ACKNOWLEDGMENTS

The work of P. M. N. was supported by the U.S. Department of Energy, National Science Foundation, and Lightner-Sams Foundation. The research of C.-P. Y. has been supported by the National Science Foundation under grant PHY-0100677.

REFERENCES

1. Bunce, G., Saito, N., Soffer, J., and Vogelsang, W., *Ann. Rev. Nucl. Part. Sci.*, **50**, 525–575 (2000).
2. Bourrely, C., Soffer, J., and Leader, E., *Phys. Rept.*, **59**, 95–297 (1980); Craigie, N. S., Hidaka, K., Jacob, M., and Renard, F. M., *Phys. Rept.*, **99**, 69–236 (1983).
3. Bourrely, C., and Soffer, J., *Nucl. Phys.*, **B423**, 329–348 (1994); **B445**, 341–379 (1995); Nadolsky, P. M., hep-ph/9503419 (1995).
4. Gluck, M., Reya, E., Stratmann, M., and Vogelsang, W., *Phys. Rev.*, **D63**, 094005 (2001).
5. Nadolsky, P. M., and Yuan, C.-P., in preparation.
6. Collins, J. C., Soper, D. E., and Sterman, G., *Nucl. Phys.*, **B250**, 199 (1985).
7. Balazs, C., and Yuan, C.-P., *Phys. Rev.*, **D56**, 5558–5583 (1997).
8. Kamal, B., *Phys. Rev.*, **D53**, 1142–1152 (1996); **D57**, 6663–6691 (1998).
9. Gehrmann, T., *Nucl. Phys.*, **B498**, 245–266 (1997); **B534**, 21–39 (1998).
10. Weber, A., *Nucl. Phys.*, **B382**, 63–96 (1992); **B403**, 545–571 (1993).
11. The Fortran code and input grids for RhicBos can be downloaded at http://schwinger.physics.smu.edu/~nadolsky/RhicBos .
12. Gehrmann, T., and Stirling, W. J., *Phys. Rev.*, **D53**, 6100 (1996).
13. Ladinsky, G. A., and Yuan, C.-P., *Phys. Rev.*, **D50**, 4239 (1994).
14. Landry, F. J., Ph. D. Thesis, Michigan State University, E. Lansing, MI 48824 (2001), also available as UMI-30-09135-mc (microfiche).

The Quark-Antiquark Asymmetry of the Nucleon Strange Sea

M. Wakamatsu

Department of Physics, Osaka University, Toyonaka, Osaka 560-0043, JAPAN
wakamatsu@miho.rcnp.osaka-u.ac.jp

Abstract. Theoretical predictions are given for the light-flavor sea-quark distributions in the nucleon including the strange quark ones on the basis of the flavor SU(3) version of the chiral quark soliton model. Careful account is taken of the SU(3) symmetry breaking effects due to the mass difference between the strange and nonstrange quarks, which is the only one parameter necessary for the flavor SU(3) generalization of the model. A particular emphasis of study is put on the *light-flavor sea-quark asymmetry* as well as the *particle-antiparticle asymmetry* of the strange quark distributions in the nucleon.

INTRODUCTION

The key observation that motivates our investigation here is that, in their semi-phenomenological fit, Glück, Reya and Vogt prepared the initial PDF at pretty low energy scale around 600 MeV, in contrast to the common consensus of perturbative QCD [1], and concluded that sea-quark (or antiquark) components are absolutely necessary even at this low energy scale. Moreover, even the isospin asymmetry of the sea-quark distributions are established by the celebrated NMC measurement. The origin of this sea-quark asymmetry is definitely nonperturbative, and cannot be radiatively generated through the perturbative QCD evolution processes. We certainly need some low energy (nonperturbative) mechanism which generates sea-quark excitations.

In our opinion, the Chiral Quark Soliton Model (CQSM) is the simplest and most powerful effective model of QCD, which fulfills the above requirement. Although it is still a toy model in the sense that the gluon degrees of freedom are only implicitly treated, it has several nice features that are not possessed by many other effective models of QCD like the MIT bag model [2, 3]. Among others, most important in the above-explained context would be its field theoretical nature, i.e. the proper account of the polarization of Dirac sea quarks, which enables us to make reasonable estimation not only of quark distributions but also of *antiquark* distributions [4, 5]. We have already shown that, *without introducing any adjustable parameter*, it reproduces almost all qualitatively noticeable features of the recent DIS observables including the NMC measurement as well as the famous EMC finding [6]. What was lacking for the flavor SU(2) CQSM is the neglect of hidden strange quark degrees of freedom in the nucleon. Here, we attack this problem by using the flavor SU(3) generalization of the CQSM, which is constructed on the basis of the SU(2) model with some additional dynamical assumptions [7, 8].

CP675, *Spin 2002: 15th Int'l. Spin Physics Symposium and Workshop on Polarized Electron Sources and Polarimeters*, edited by Y. I. Makdisi, A. U. Luccio, and W. W. MacKay

FLAVOR SU(3) CHIRAL QUARK SOLITON MODEL

The model lagrangian is a straightforward generalization of the SU(2) model with flavor octet collective meson fields, except for one important new feature, i.e., the existence of the sizably large SU(3) symmetry breaking term due to the mass difference between the strange and nonstrange quarks. This mass difference Δm_s is the only one additional parameter of our effective model.

The basic dynamical assumption of the flavor SU(3) CQSM is as follows. First, the lowest energy classical solution is obtained by the embedding of the SU(2) self-consistent mean-field into the SU(3) matrix, analogous to the flavor SU(3) Skyrme model. The next is the collective quantization of the symmetry restoring rotational motion of the soliton in SU(3) collective coordinate space. Finally, we assume that the SU(3) symmetry breaking effects can be treated perturbatively. Actually, we have taken account of several possible SU(3) breaking corrections consistently, which are all first order in Δm_s. The detail may be found in our recent article [8].

COMPARISON WITH HIGH ENERGY DATA

The model predictions are compared with the available high-energy data in the following way. First, after some trial, only one parameter of the SU(3) CQSM, i.e. Δm_s, is fixed to be 100 MeV. Then, we use the predictions of the model as initial distributions given at the low energy model scale, simply assuming the smallness of the gluon distributions at this energy scale. The scale dependence of the PDF is then taken into account by solving the standard DGLAP equation at the NLO. The intial energy scale of this Q^2-evolution is fixed to be $Q^2_{ini} = 0.30\,\mathrm{GeV}^2$ throughout the study.

Skipping the details, we show in Fig.1 the final predictions of the SU(3) CQSM for the unpolarized s- and $\bar{s}$-quark distributions at the model energy scale. The left figure shows the result obtained in the chiral limit, i.e. by neglecting the SU(3) symmetry breaking effects, while the right figure is obtained after introducing Δm_s correction. One sees that the $s - \bar{s}$ asymmetry of the unpolarized distribution functions certainly exists. The difference function $s(x) - \bar{s}(x)$ has some oscillatory behavior with several zeros as a function of x. This is of course due to the two general constraints of the PDF, i.e. the *positivity constraint* for the unpolarized distributions and the *strangeness quantum number conservations*. Comparing the two figures, one observes that $s(x) - \bar{s}(x)$ is extremely sensitive to the SU(3) breaking effects.

Fig.2 shows the theoretical predictions for the longitudinally polarized strange quark distributions. In the chiral limit case, the s and $\bar{s}$ are both negatively polarized. After introducing Δm_s correction, $\Delta s(x)$ remains large and negative, while $\Delta\bar{s}(x)$ becomes very small. As a consequence, the s-$\bar{s}$ asymmetry of the longitudinally polarized distributions is much more profound than that of the unpolarized distributions. This is reasonable because, for the spin-dependent distributions, there is no conservation laws which prevents the generation of asymmetry.

Now we make some preliminary comparison with the existing high energy data. In Fig.3, we compare the theoretical strange-quark distributions evolved to $Q^2 = 4\,\mathrm{GeV}^2$

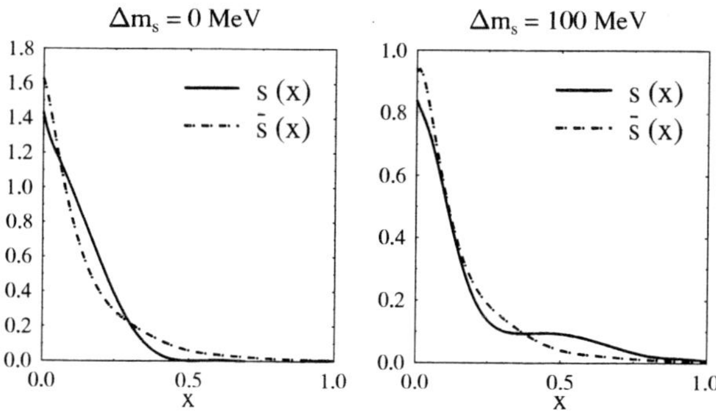

FIGURE 1. Theoretical unpolarized s- and $\bar{s}$- distributions at the model energy scale.

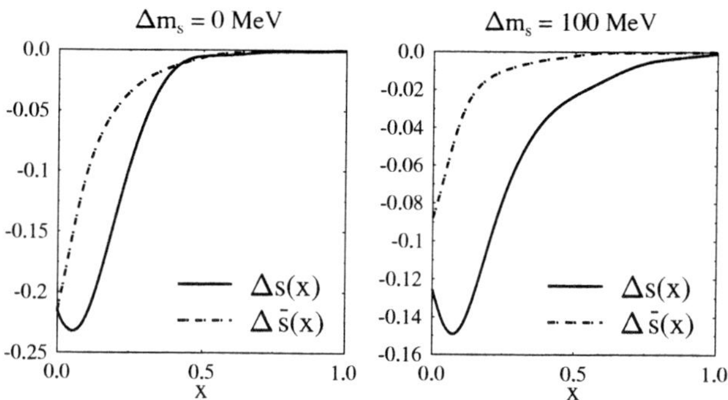

FIGURE 2. Theoretical longitudinally polarized s- and $\bar{s}$- distributions at the model energy scale.

with the corresponding CCFR (NLO) fit of the neutrino-induced charm production, which was carried out under the constraint $\bar{s}(x) = s(x)$ [9]. One can say that, after inclusion of the SU(3) symmetry breaking corrections, the theory reproduces the qualitative feature of CCFR fit.

Recently, Barone et al. carried out quite elaborate global analysis of the DIS data, especially by using all the available neutrino data [10]. This enables them to obtain some interesting information even for the asymmetry of the s and $\bar{s}$ distributions. The thick and thin shaded areas in Fig. 4 are the allowed regions, respectively obtained by Barone et al. and by CCFR analysis. One clearly sees that the theory reproduces the qualitative tendency of the data only after including the SU(3) breaking effects.

Although we have no space to show them, we also find that the predictions of the $SU(3)$ CQSM for the longitudinally polarized distributions including the strange quarks are qualitatively consistent with the recent elaborate analyses carried out by Leader, Sidorov and Stamenov [11].

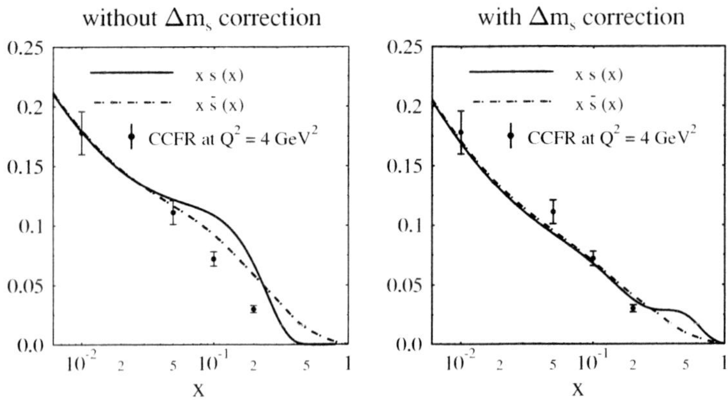

FIGURE 3. The longitudinally polarized s- and $\bar{s}$- distributions at the model energy scale.

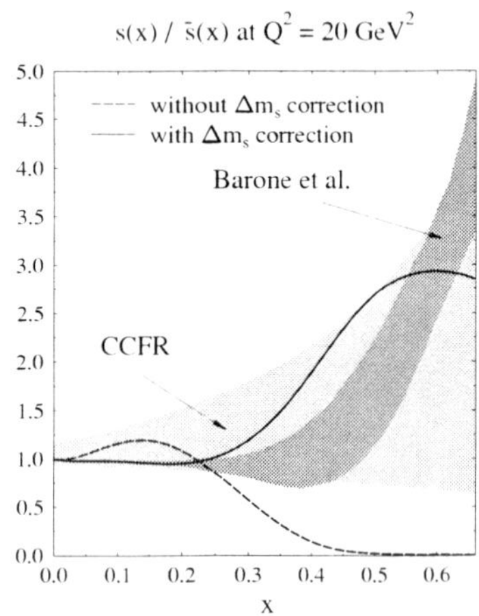

FIGURE 4. The longitudinally polarized s- and $\bar{s}$- distributions at the model energy scale.

So far, we have shown that the SU(3) CQSM can give reasonable and unique predictions for the hidden strange quark distributions in the nucleon. A natural question is whether it does not destroy the success of the SU(2) CQSM already obtained for u,d-quark dominated observables. To verify it, we have compared the predictions of the both versions of the CQSM with the corresponding EMC and SMC data for the longitudinally polarized structure functions for the proton, the neutron and the deuteron to find that both models reproduces equally well the general tendency of the experimental data.

Next, turning to the problem of isospin asymmetry of sea quark distribution, we recall that the SU(2) CQSM predicts that $\bar{u}(x) - \bar{d}(x) < 0$, while $\Delta\bar{u}(x) - \Delta\bar{d}(x) > 0$. Now the question is what the predictions of the SU(3) CQSM is like. We have compared the predictions of both versions of the CQSM. We find that, for the unpolarized sea quark distributions, both models give nearly the same answer, which we already know is consistent with the NMC observation. In contrast to the unpolarized distributions, the magnitude of $\Delta\bar{u}(x) - \Delta\bar{d}(x)$ turns out to become sizably smaller when going from the SU(3) model to the SU(2) one. Still, the positive polarization of $\Delta\bar{u}$ and negative polarization of $\Delta\bar{d}$ in the proton is a definite prediction of the both version of the CQSM, which should be contrasted with the prediction of other models like the naive meson cloud convolution model.

CONCLUSION

To summarize, an incomparable feature of the CQSM as compared with many other effective models like the MIT bag model is that it can give reasonable predictions also for the *antiquark distribution functions*. We emphasize that this feature is essential for giving any reliable predictions for *strange distributions* in the nucleon, which totally have *non-valence character*.

With a single parameter, the SU(3) CQSM predicts that the $s(x) - \bar{s}(x)$ difference function has some oscillatory x-dependence, due to the positivity constraint for the spin-averaged distributions and the strangeness quantum number conservation. We have also shown that, after introducing the SU(3) breaking effects, the x dependence of $s(x) - \bar{s}(x)$ and $s(x)/\bar{s}(x)$ are qualitatively consistent with the global analysis of Barone et al. The $s - \bar{s}$ asymmetry of longitudinally polarized sea is more profound than that of unpolarized sea. The model predicts that the polarization of s-quark is large and negative, while the polarization of $\bar{s}$-quark is very small. The model also predicts quite large isospin asymmetry of the sea-quark distributions not only for the unpolarized distributions but also for the longitudinally polarized ones.

An important lesson learned from our investigation is that the nonperturbative QCD dynamics due to the spontaneous chiral-symmetry breaking manifest most clearly in the *spin* and *isospin* dependence of *antiquark* distributions in the nucleon. What is absolutely required for future experiments is therefore the flavor and valence $\oplus$ sea-quark decomposition of PDF.

REFERENCES

1. M. Glück, E. Reya, and A. Vogt, Z. Phys. **C67**, 433 (1995).
2. D.I. Diakonov, V.Yu. Petrov, and P.V. Pobylitsa, Nucl. Phys. **B306**, 809 (1988).
3. M. Wakamatsu and H. Yoshiki, Nucl. Phys. **A524**, 561 (1991).
4. D.I. Diakonov et al., Phys. Rev. **D56**, 4069 (1997).
5. M. Wakamatsu and T. Kubota, Phys. Rev. **D60**, 034020 (1999).
6. M. Wakamatsu and T. Watabe, Phys. Rev. **D62**, 054009 (2000).
7. M. Wakamatsu, Prog. Theor. Phys. **107**, 1037 (2002).
8. M. Wakamatsu, hep-ph / 0209011.
9. CCFR Collaboration, A. Bazarko et al., Z. Phys. **C65**, 189 (1995).
10. V. Barone, C. Pascaud, and F. Zomer, Eur. Phys. J. **C12**, 243 (2000).
11. E. Leader, A. Sidorov, and D. Stamenov, Phys. Lett. **B488**, 283 (2000).

Single Muon Production in Transversely Polarized p+p Collisions at $\sqrt{s}=200$ GeV in the PHENIX experiment

H. Kobayashi* and A. Taketani[1†] for the PHENIX Collaboration[2†]

*RIKEN BNL Research Center, Building 510A, Upton, NY 11973-5000, U. S. A.
†RIKEN, 2-1 Hirosawa Wako Saitama 351-0198, Japan

Abstract. During the operation of the Relativistic Heavy Ion Collider (RHIC) in 2001-2002, the PHENIX experiment accumulated data on single muon production from collisions of transversely polarized protons at $\sqrt{s}=200$ GeV. It was observed that the single muon sample is dominated by muons from pion and kaon decays. Subsequent transverse spin asymmetry analysis is discussed.

INTRODUCTION

The detection of single muons in proton-proton interactions using the Relativistic Heavy Ion Collider (RHIC) provide a means to study spin physics. For collisions at a $\sqrt{s}$ of 200 GeV, it is predicted that, at a modest muon transverse momentum p_T ($1 < p_T <$ 3 GeV$/c$), single muon production is dominated by muons from pion or kaon decays; whereas, at high p_T ($3 < p_T < 10$ GeV$/c$), it is dominated by muons from open heavy flavor production. The heavy flavor production is dominantly created through the gluon-gluon fusion, so it allows one to measure the gluon polarization in the proton directly if the proton beams are longitudinally polarized.

During the RHIC running period in 2001-2002 (Run2), the beams were transversely polarized and a large number of low p_T single muons and a modest number of high p_T single muons were detected. Even though the sample is dominated by muons from pion or kaon decays, it is still interesting to look at the single-spin transverse asymmetry (A_N) for the single muon production because the Fermilab E704 experiment found large single transverse spin asymmetries in the reactions using polarized proton beams, ($p_{\uparrow}p \to \pi^+X$ and $p_{\uparrow}p \to \pi^-X$) at $\sqrt{s} = 19.4$ GeV [1]. Specifically, they reported that the A_N for π^+ varied from 0 to 0.3 as a function of Feynman x (x_F) over the range $-0.2 < x_F < 0.9$. The A_N for π^- varied similarly in this range of x_F, but with an opposite sign. Similar large spin effects had been observed at lower energies in the BNL E925 experiment at 22 GeV$/c$ incident proton beam energy ($\sqrt{s} = 6.56$ GeV) [2] and earlier in Argonne at 11.75 GeV$/c$ ($\sqrt{s} = 5.00$ GeV) [3]. It is interesting to look at the

[1] Presenting author.
[2] For the full PHENIX Collaboration author list and acknowledgments, see Appendix "Collaborations" of this volume.

CP675, *Spin 2002: 15th Int'l. Spin Physics Symposium and Workshop on Polarized Electron Sources and Polarimeters,* edited by Y. I. Makdisi, A. U. Luccio, and W. W. MacKay

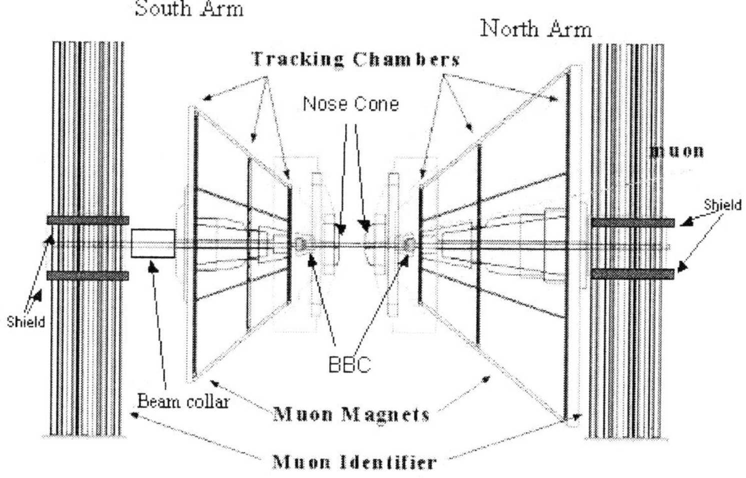

FIGURE 1. Apparatus of the PHENIX muon arms looking from side.

transverse spin asymmetry A_N for pion and kaon production through the inclusive muon production because it has never been measured at such a higher center of mass energy $\sqrt{s} = 200$ GeV. The $\sqrt{s}$ variation may help to isolate the model because it is expected that the perturbative QCD is applicable with less ambiguity at a higher energy.

Currently, there is no rigorous model that enables one to interpret the properties of these spin effects. Various theoretical models have been proposed to explain the analyzing powers observed in pion production. (1) higher twist effects [4], (2) correlation of $k_\perp$ and spin in the structure function [5, 6], or in fragmentation function [7, 8, 9], (3) orbital angular momentum of valence quarks inside a polarized hadron [10, 11, 12], and (4) a quark recombination model with a relativistic description for the parton-parton interaction [13].

EXPERIMENTAL SETUP

The RHIC is a collider accelerator which consists of two rings [14]. In Run2, transversely polarized beams colliding at $\sqrt{s}=200$ GeV was achieved. Siberian snakes were employed [15] to accelerate the beams without loosing their polarization. Polarimeters using Coulomb nuclear interaction were successfully used to measure the beam polarization [16]. The operation of these devices was commissioned as required for physics measurements.

The PHENIX detector consists of a central arm and two muon arms. The z-axis was defined on the beam line from the south to north direction. As shown in the figure 1, the muon arms are located in the forward regions covering the mid-rapidity range of $1.2 < |\eta| < 2.4$ over the full azimuthal range. The south muon arm, one of the two PHENIX muon arms, was operational during Run2 for the first time. The north muon arm has been installed for the coming 2002-2003 run (Run3). Each muon arm consists of a muon identifier (MuID) and a muon tracker (MuTr). The MuID consists of 5 sandwiches of plastic proportional tubes and steel absorber with 10 interaction length. It is used as a trigger counter as well as a "road" (rough track) finder to help locate tracks in the MuTr. In order to get a good efficiency for identification of muons, minimum momentum of 2 GeV$/c$ is required for the muon tracks. The MuTr consists of 3 layers of cathode strip readout wire chambers situated in a magnetic field. The chamber resolution is designed as 100μm at each cathode plane to achieve $\Delta p/p \sim 3\%$ at $p = 3 \sim 10$ GeV$/c$. Upstream of each muon arm, there is an absorber made of copper which covers all of the geometrical acceptance of the muon arm in order to reduce hadron multiplicity in the muon arms. Upstream of the muon arms, closer to the beam, there is a pair of Beam Beam Counter (BBC). The counters can determine z position of collision point by measuring timing difference between north and south BBC detectors.

ANALYSIS

In Run2, the PHENIX accumulated the integrated luminosity of 0.15 pb^{-1}. The polarization was 14% on average for one beam, 17% for the other beam.

The minimum bias event was defined as the BBC north-south coincidence or 1 layer penetration (shallow) of tracks in the MuID then 188 million events were obtained. To select muon track candidates, 4 layer penetration (deep) of tracks in the MuID was required then total number of tracks was reduced to 981k. After more detailed track quality cuts were applied, 203k tracks were obtained as a clean muon sample.

In the Figure 2, the vertex distribution is shown for all minimum-bias events (top right) and events with a muon track (top left). By dividing these two distribution, the muon yield was determined as a function of vertex position in z (bottom left). This distribution has a clear slope that indicates that there are more muons as the distance between the vertex position and the south muon arm increases. This feature is explained by the fact that the produced muons are dominated by pions or kaons for which the decay length before going into the absorber depends on the collision vertex position. The kink from the decay is negligible since it is small (kink angle for pion is $\theta_\pi < 0.01$ rad and that for kaon is $\theta_K < 0.1$ rad).

The number of muons originating from pion decay (N_π) can be expressed as:

$$N_\pi(L) = N_\pi^0 \exp\left[\frac{-L}{c\tau_\pi\gamma_\pi}\right] \tag{1}$$

where, as shown in the figure 3, L is the distance between the absorber and the collision point corrected by track polar angle θ. Together with muons coming from kaon decay (N_K), the total number of observed muons can be expressed as:

$$N_\mu(L) = N_\pi^0\left(1 - \exp\left[\frac{-L}{c\tau_\pi\gamma_\pi}\right]\right) + N_K^0\left(1 - \exp\left[\frac{-L}{c\tau_K\gamma_K}\right]\right) \tag{2}$$

$$\simeq L\left(\frac{N_\pi^0}{c\tau_\pi\gamma_\pi} + \frac{N_K^0}{c\tau_K\gamma_K}\right) \qquad (L \ll c\tau\gamma) \tag{3}$$

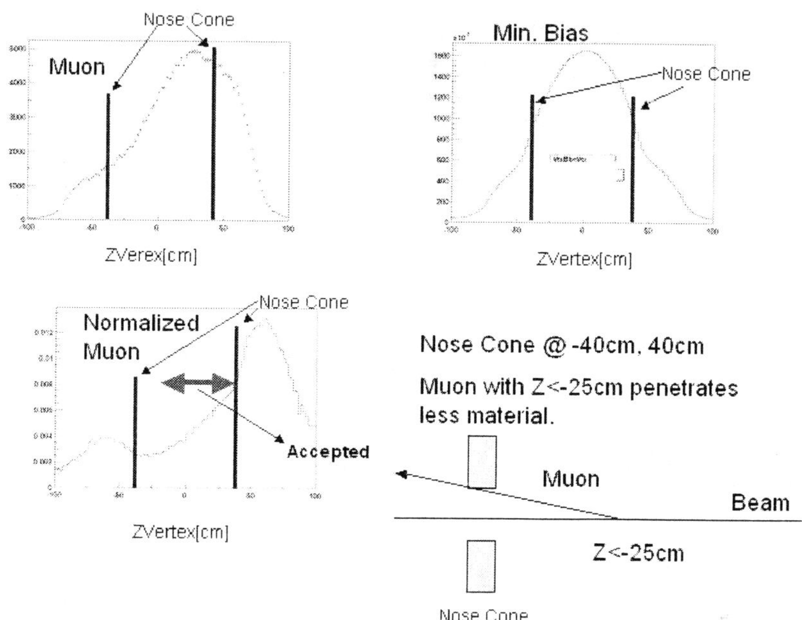

FIGURE 2. Vertex distribution of collisions that contains muon tracks (top left), minimum bias with BBC north-south coincidence (top right) and the muon track vertex divided by the minimum bias BBC vertex distribution (bottom left).

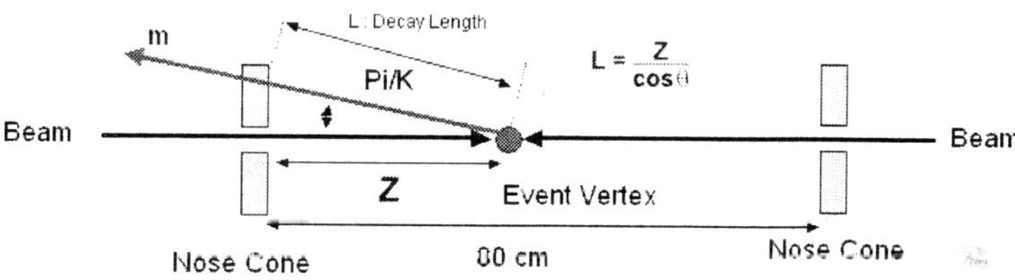

FIGURE 3. Schematic view of a collision vertex and the absorber (nose cone).

where the expression was simplified using the approximation $L \ll c\tau\gamma$.

The other possible contributions to the muon yield are muons from heavy flavor production and hadrons which punch-through the absorbers. These contributions do not depend upon L and thus result in an offset to the histogram at the bottom left of Figure 2. In order to determine the flat distribution, the function form of (3) plus offset was fitted. The vertex distribution of events with muon tracks can be reproduced by a simulation which taking into account of the in-flight decay of π and K. Figure 4 shows the simulated vertex distribution which reproduces the decay length dependence well.

Based on the vertex study, we can say that we are principally detecting muons from π and K decays. A study of the single spin transverse asymmetry is underway. Significance

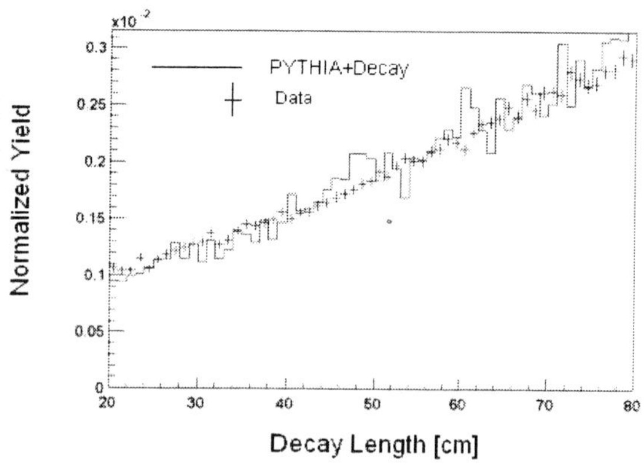

FIGURE 4. Comparison of the measured vertex distribution with a simulation.

of this measurement can be estimated from the statistics that we obtained from the vertex study. Statistics of $N_+ + N_- = 100$k and beam polarization $P_b = 0.2$ gives the statistical precision $\delta A_N = 0.015$ at average values $\langle x_F \rangle = 0.04$ and $\langle p_T \rangle = 1.5$ GeV$/c$.

SUMMARY

Single muons were observed in the transversely polarized proton-proton collision at $\sqrt{s} = 200$ GeV. Those are dominated by decays of π and K. Single spin asymmetry A_N measurement using single muon production is feasible with current data.

REFERENCES

1. D. L. Adams *et al.*, Phys. Lett. **B264**, 462-466 (1991).
2. C. E. Allgower *et al.*, Phys. Rev. **D65**, 092008 (2002).
3. W. H. Dragoset, *et al.*, Phys. Rev. **D18**, 3939-3954 (1978).
4. J. W. Qiu and G. Sterman, Phys. Rev. **D59**, 014004 (1999).
5. D. W. Sivers, Phys. Rev. **D41**, 83 (1990).
6. D. W. Sivers, Phys. Rev. **D43**, 261-263 (1991).
7. J. C. Collins, Nucl. Phys. **B396**, 161-182 (1993).
8. J. C. Collins, S. F. Hepplemann and G. A. Ladinsky, Nucl. Phys. **B420**, 565-582 (1994).
9. M. Anselmino, M. Boglione and F. Murgia, Phys. Rev. **D60**, 054027 (1999).
10. C. Boros,and Z. T. Liang and T. C. Meng, Phys. Rev. Lett. **70**, 1751-1754 (1993).
11. Z. T. Liang and T. C. Meng, Z. Phys. **A344**, 171-180 (1992).
12. S. M. Troshin and N. E. Tyurin, Phys. Rev. **D52**, 3862-3871 (1995).
13. K. Suzuki and N. Nakajima, H. Toki and K. I. Kubo, Mod. Phys. Lett. **A14**, 1403-1412, (1999).
14. H. Huang, These proceedings.
15. H. Huang, *et al.*, Phys. Rev. Lett. **73**, 2982-2985 (1994).
16. O. Jinnouchi, These proceedings.

Single-spin Transverse Asymmetry in Charged Hadron Production in $\sqrt{s} = 200$ GeV p+p Collisions at PHENIX

K. Okada for the PHENIX Collaboration[1]

RIKEN (The Institute of Physical and Chemical Research), Wako, Saitama 351-0198, Japan

Abstract. We report on the status of the single-spin transverse asymmetry (A_N) measurements for charged hadrons with the PHENIX Central-arm detector. Data from transverse polarized proton-proton collision were collected during the last run (2001-2002) at the Relativistic Heavy Ion Collider at Brookhaven National Laboratory. The energy ($\sqrt{s} = 200$ GeV$/c$) is ten times higher than the previous experiments for A_N. Based on the charged hadron yield and the polarizations, an A_N measurement at the mid-rapidity region will be achieved 6-7% statistical error at the high transverse momentum range (6-8 GeV$/c$). The prospect for future runs is also discussed.

INTRODUCTION

Presently, one of the most important topics in the spin physics community is the investigation of the carrier of the proton spin. Because in the 1980's, DIS experiments revealed that the spin of quarks contribute only a part of proton spin. The next target is the gluon component (ΔG) which can be accessed by proton-proton collisions. The Relativistic Heavy Ion Collider (RHIC) at Brookhaven National Laboratory (BNL) provides the polarized proton-proton collisions at the highest energy in the world. ΔG signal appears in the double-spin longitudinal asymmetries (A_{LL}). A_{LL} in hadron production is expected to have different features for neutral pions and charged hadrons reflecting the different contributions of u, d quark and their polarizations. These differences grow with the transverse momentum (p_T) of the particles. The PHENIX detector is one of the multi-purpose detectors at RHIC. Its central arm detector covers the mid-rapidity region and, with the electromagnetic calorimeter trigger, is optimized to collect the high p_T particles. So, PHENIX can measure ΔG via these particles [1].

During the last RHIC run (RHIC RUN2), data were collected for transversely polarized proton-proton collision at $\sqrt{s} = 200$ GeV. From these data, perturbative QCD (pQCD) calculations can be confronted with new measurements of the production cross section and the left-right single-spin transverse asymmetry (A_N). In the latter case, the E704 experiment at FNAL measured A_N for neutral pions at mid-rapidity with a p_T range from 1 to 3 GeV$/c$ but at a smaller $\sqrt{s}$ of 19.4 GeV [2]. The data were consistent with zero, confirming the expectations of lowest order (twist-2) pQCD. The PHENIX

[1] For the full PHENIX Collaboration author list and acknowledgements, see Appendix "Collaborations" of this volume.

CP675, *Spin 2002: 15th Int'l. Spin Physics Symposium and Workshop on Polarized Electron Sources and Polarimeters,* edited by Y. I. Makdisi, A. U. Luccio, and W. W. MacKay

data can reach to higher p_T region than E704 and thus possibly be sensitive to higher order effects.

EXPERIMENTAL SETUP

During RHIC RUN2, proton-proton data were collected from December 2001 to January 2002. The total luminosity acquired by PHENIX was 0.15 pb^{-1} (January 8th to January 22nd). The averaged polarization was 14% in the blue beam and 17% in the yellow beam, as measured with the RHIC CNI polarimeters[3]. Each bunch in the beam had alternate polarization direction to reduce time-dependent variations both in the detector and in the beam.

The PHENIX Central Arm Detectors[4][5] consist of a west and an east spectrometer, each of which covers 90 degrees in azimuthal angle (ϕ) and ± 0.35 in pseudorapidity (η) where z-axis is aligned with the beam direction. A particle from an interaction, after traversing through a magnetic field, passes through Drift Chambers (DC), Pad Chambers (PC), Ring Image Cherenkov Counters (RICH) and Electromagnetic Calorimeters (EMCal). The azimuthal symmetric magnetic field exists inside of DC radius (~2m). A charged particle is kicked by the field in ϕ direction ($\int \mathrm{B} \cdot \mathrm{dl} = 0.78[\mathrm{T} \cdot \mathrm{m}]$). Its trajectory is reconstructed from the DC hits and, via the field map, used to determine the momentum. In addition, we used Beam-Beam Counters (BBC) situated at the beam forward region ($3.0 < |\eta| < 3.9$) to give a measure of the vertex position.

The simple interaction trigger (called the minimum-bias trigger) is formed from the BBC hit data and is used to collect for the low p_T sample Due to the rate, a factor 10 to 80 of prescale was needed. The trigger system is very important to maximize the yield of the high p_T sample. For this purpose, an EMCal level-1 trigger was installed just prior to the run. An energy threshold was set for a trigger unit which was 4 PMTs (covers about $10 \times 10[\mathrm{cm}^2]$ perpendicular to the particle direction). Other trigger units made from 16 PMTs were also functional, but were not used in this analysis. Fig. 1 shows the trigger turn-on curve as a function of the energy deposited in a trigger unit. For neutral pions, the trigger performance was studied and is well understood. Fig. 2 shows that the energy turn-on curve for photons is well reconstructed in the Monte-Carlo simulation. For charged hadrons, the situation differs from the case of neutral pions. Since the EMCal has less than one nuclear interaction length, some of the hadrons traverse the EMCal as minimum ionized particles. The total energy deposit distribution for $\pi^{\pm}$ had been obtained from the test beam measurements done at AGS [6] and CERN. By fitting empirical function, the efficiency of the charged pions as a function of the momenta were calculated and shown in Fig. 3. There are uncertainties in the fraction of energy deposit in the trigger unit to the total energy deposit.

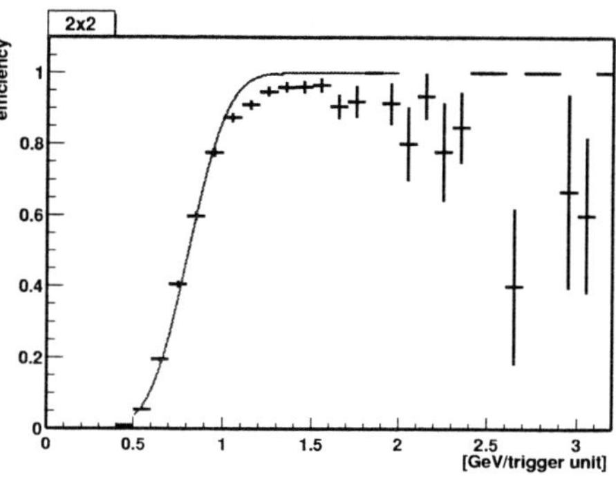

FIGURE 1. The EMCal trigger turn-on curve as a function of the energy deposit in the trigger unit.

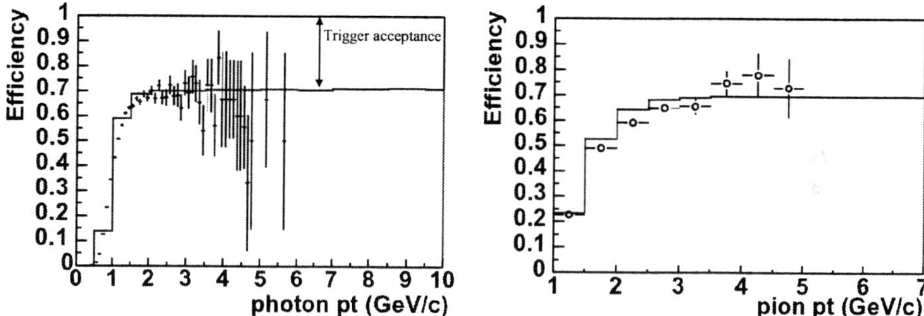

FIGURE 2. (left) The EMCal trigger photon efficiency as a function of transverse momentum (p_T). The points are from data; the line shows the prediction determined using each EMCal tile response. The plateau is consistent with the trigger acceptance. (right) Here the photon clusters are selected from π^0 daughters and plotted as a function of p_T of the π^0.

ANALYSIS

A charged track is defined as a track in the DC which projects to hits in the PC and back to the vertex point obtained from BBCs. Backgrounds such as low momentum electron not coming from the vertex point are eliminated with this combination. But in the case that a photon converts to electrons in front of the DC or a particle decays in flight, the kick offsets the magnetic field kick and such daughters will be assigned an incorrect momentum. The number of these daughters were estimated to be less than 10% in the p_T range p_T below 4 GeV$/c$. For the events taken with the EMCal trigger, this background is absent because it deposits too little energy in the EMCal to fire the trigger.
Fig. 4 shows the yield of the charged particles from 32 million minimum-bias triggered events and 19 million EMCal triggered events where, for the latter sample, an additional off-line EMCal deposit energy cut (E_{dep}[GeV] > 0.3312p[GeV$/c$] $+ 0.1636$) was

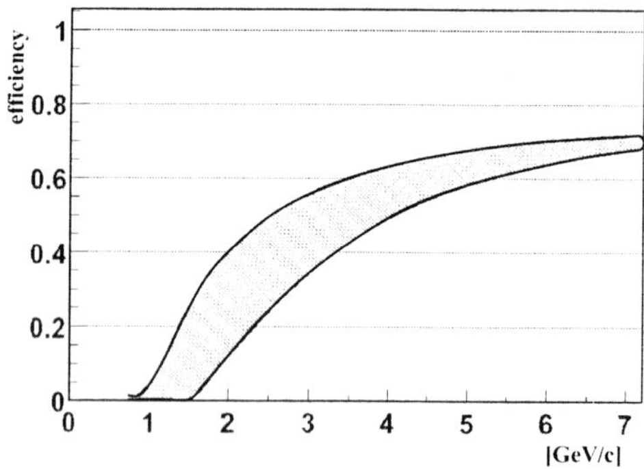

FIGURE 3. The EMCal $\pi^{\pm}$ efficiency computed with an empirical fit of the energy response of charged pions which was extracted from test beam data. The filled band represents to the width of the trigger turn-on curve.

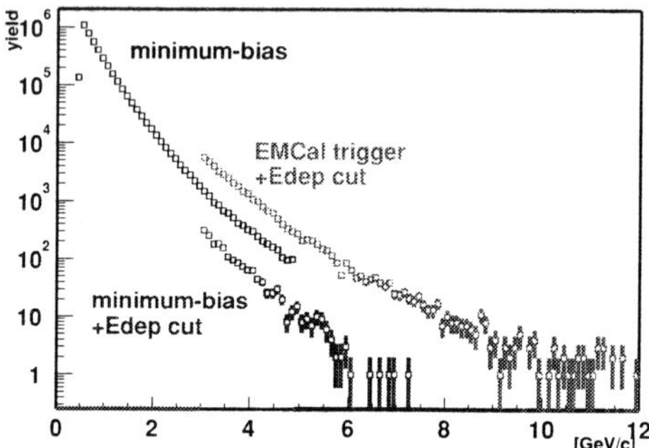

FIGURE 4. The charged hadron ($h^{\pm}$) yield for the RUN2 data set. The minimum-bias and the EMCal trigger events cover the low p_T and the high p_T regions, respectively.

imposed to minimize the p_T-dependence of the efficiency. The EMCal trigger not only increases the yield of high p_T particles, but also it significantly rejects the background contribution. From the comparison with the yield from minimum-bias+EMCal deposit cut sample, its gain was factor 20 ($p_T > 4$ GeV$/c$). Anti-protons might introduce another bias on the negative hadron component of the EMCal trigger sample, because they were favored due to the larger EMCal energy deposit of the annihilation process. The charged hadron production cross section will be obtained by the minimum-bias events. The A_N analysis at high p_T region will be performed using the EMCal triggered events.

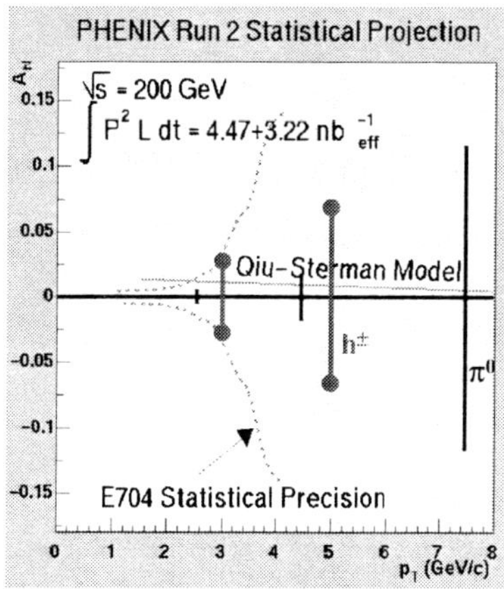

FIGURE 5. The estimated statistical error on the A_N compared with the expected PHENIX RUN2 π^0 precision and the E704 π^0 results.

SUMMARY AND OUTLOOK

Based on the yield of charged hadrons, the expected statistical error on the A_N measurement is shown in Fig. 5. It is calculated by $\delta A_N = \frac{1}{P} \cdot \frac{1}{\sqrt{N}}$, where P is polarization and N is the number of tracks, and we took 15% for P as a typical value. We can test A_N of charged hadrons at 7% level for the p_T range from 4 to 6 GeV/c. E704 couldn't reach this high in p_T. Fig. 5 also shows the estimation for neutral pions in RUN2 and E704 result on neutral pions. There are more statistics of neutral pions, because they have higher EMCal trigger efficiency and are less affected by reconstruction inefficiencies. The analysis is still on going to check the tracking resolution and the quality for the cross section measurement. The systematic error estimation for A_N measurement has yet to be undertaken.

During RHIC RUN3 and RUN4, we are planning to take data with longitudinally polarized proton-proton collisions[1]. The analysis procedure of the charged hadron sample described here is expected to be the one used in the future gluon polarization measurement made via this channel.

REFERENCES

1. Y. Goto, these proceedings.
2. D. L. Adams *et al.*, Phys. Rev. **D53**, 4747-4755 (1996).
3. O. Jinnouchi, these proceedings.
4. D. P. Morrison *et al.*, Nucl. Phys. **A638**, 565-569 (1998).
5. J. T. Mitchell *et al.*, Nucl. Instrum. Meth. **A482**, 491-512 (2002).
6. A. V. Bazilevskii *et al.*, Instruments and Experimental Techniques, **42-2**, 167-173 (1999).

Analyzing Powers for Forward $p_\uparrow + p \to \pi^0 + X$ at STAR

G. Rakness for the STAR Collaboration[1]

Indiana University Cyclotron Facility
Bloomington, IN 47408 USA

Abstract. Preliminary results of the analyzing power for the production of forward, high-energy π^0 mesons from collisions of transversely polarized protons at $\sqrt{s} = 200\,\text{GeV}$ from STAR are presented. The kinematic ranges covered by the data are $x_F \approx 0.2 - 0.6$ and $p_T \approx 1 - 3\,\text{GeV/c}$. The analyzing power at $\sqrt{s} = 200\,\text{GeV}$ is found to be comparable to that observed at $\sqrt{s} = 20\,\text{GeV}$.

INTRODUCTION

Perturbative QCD makes a qualitative prediction that the single-spin transverse asymmetry, known as the analyzing power (A_N), for $2 \to 2$ parton scattering at large transverse momentum should be zero based on helicity conservation. In the late 1980's, the experiment E704 at Fermi National Laboratory measured A_N for the production of charged and neutral pions in $p_\uparrow + p$ collisions at $\sqrt{s} = 20\,\text{GeV}$ and $p_T = 1 - 3\,\text{GeV/c}$ over the Feynman-x range of $0 - 0.8$ [1, 2]. The analyzing power for π^+ mesons was found to increase as a function of Feynman-x from $A_N = 0$ at $x_F = 0$ to $A_N \approx 0.4$ at $x_F = 0.8$. For π^- mesons, the analyzing power was found to be approximately equal in magnitude and opposite in sign to the π^+ results, while for π^0 mesons, the analyzing power was found to be approximately half the size observed for π^+ mesons.

This result has inspired several theory groups to develop models to account for the observed large analyzing power. Most models attribute forward pion production to collisions between a quark in one proton and a gluon in the other. There are many different plausible mechanisms by which one might expect transverse spin effects, all of which could contribute to some degree. One perturbative QCD approach attributes transverse spin effects to twist-3 gluon correlations before or after the primary quark-gluon coupling [3, 4]. A second approach attributes A_N to the transversity distribution function and a T-odd Heppelmann-Collins fragmentation function [5]. Another approach is to include initial state interactions to introduce transverse spin effects before the primary quark-gluon coupling [6, 7]. All of these models predict the large analyzing power observed at E704 should persist to collision energies an order of magnitude greater [8, 9]. Here we present the measurement of the analyzing power for the production of π^0 mesons from the STAR experiment at the Relativistic Heavy Ion Collider (RHIC) at Brookhaven Na-

[1] For the full author list and acknowledgements, see the appendix to these proceedings.

CP675, *Spin 2002: 15th Int'l. Spin Physics Symposium and Workshop on Polarized Electron Sources and Polarimeters*, edited by Y. I. Makdisi, A. U. Luccio, and W. W. MacKay

tional Laboratory, studying $p_\uparrow + p$ collisions with total energy $\sqrt{s} = 200\,\text{GeV}$ available to the system.

EXPERIMENTAL CONDITIONS

The analyzing power for a reaction with a transversely polarized beam interacting with an unpolarized target is determined from a spin-dependent asymmetry,

$$\cos\phi\, P_{beam} A_N = \frac{N_+ - RN_-}{N_+ + RN_-}, \qquad (1)$$

and requires the concurrent measurements of three independent quantities. The magnitude of the transverse polarization of the beam is P_{beam}. The number of measured π^0 events observed when the polarization direction is up(down) is $N_{+(-)}$. The relative luminosity is given by $R = \mathscr{L}_+ LT_+ / \mathscr{L}_- LT_-$, where $\mathscr{L}_{+(-)}$ is the luminosity and $LT_{+(-)}$ is the livetime of the detector for different polarization states. The spin-dependent asymmetry corresponds to the right-hand side of equation 1. The azimuthal angle between the polarization vector and the normal to the reaction plane is ϕ. Parity constrains the asymmetry to be zero when the π^0 is emitted along the direction of the polarization vector.

Data were collected during the polarized proton run at RHIC in January 2002, and, as such, resulted from the first observations of polarized protons in a collider environment. A typical RHIC fill lasted $6-8$ hours with collision luminosities on the order of $10^{30}\,\text{cm}^{-2}\text{sec}^{-1}$ at the STAR interaction region. The so-called "yellow" proton beam rotated counterclockwise around RHIC when viewed from above, while the "blue" proton beam went clockwise. Each beam contained 55 filled bunches and 5 empty bunches which collided every 213 nsec at the STAR interaction region, resulting in 50 chances for collisions per revolution. The beam polarization vector was oriented vertically, perpendicular to the proton momentum direction. In the polarization pattern for the yellow beam, the polarization direction alternated every bunch, while the blue pattern alternated every second bunch. The data presented here refer to spin asymmetries with respect to the direction of the yellow polarization vector, averaging over the blue polarization.

The average beam polarization for each fill was given by the Coulomb-Nuclear Interference (CNI) polarimeter located at 12 o'clock in RHIC [10, 11]. This detector measured the asymmetry of elastic proton-carbon collisions by observing recoil carbon atoms at approximately 90 degrees. At the RHIC injection energy ($\approx 25\,\text{GeV}$ per beam), the analyzing power of the CNI reaction has been measured [12] and can be used to deduce the absolute polarization of the proton beam. At RHIC collision energies ($\approx 100\,\text{GeV}$ per beam), the analyzing power for this reaction has not yet been measured. Therefore, for these proceedings the beam polarization is determined by using the average value of the measured CNI asymmetry at collision energy within each fill, divided by the measured analyzing power of the CNI reaction at injection energy. With this assumption, the average luminosity-weighted beam polarization for the data presented here is $P_{beam} \approx 16\%$. As the asymmetry measured at collision energy was consistent with the asymmetry measured at injection energy for these runs, the P_{beam} quoted here represents

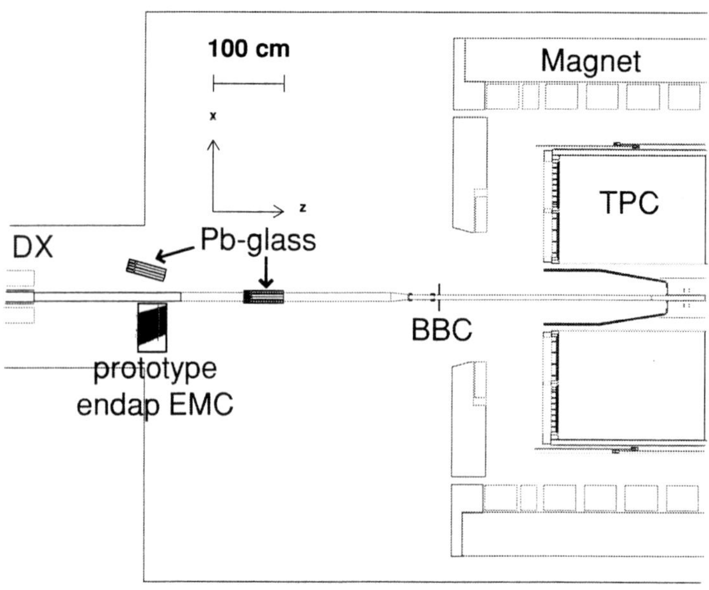

FIGURE 1. Schematic of the FPD and the east half STAR, looking from above. In this figure, the yellow beam moves from right to left, and collisions occur at the right-hand edge.

a likely upper limit at the collision energy, since the beam acceleration is not expected to enhance P_{beam}.

A forward π^0 detector (FPD) comprising four arms was installed at STAR approximately 750 cm from the interaction region and very close to the beam pipe (Fig. 1). Its location was such that significant energy deposition corresponded to positive x_F particle production with respect to the yellow beam. An electromagnetic Pb-scintillator sampling calorimeter of ≈ 21 radiation lengths subdivided into 12 towers was placed to the left of the oncoming yellow beam. This detector is a prototype of 1/60 of the endcap electromagnetic calorimeter (pEEMC), currently being installed at STAR. The pEEMC has two layers of preshower readout and a shower-maximum detector (SMD) made of orthogonal layers of finely segmented scintillator strips to measure the longitudinal and transverse profiles of photon showers. A 4×4 array of $3.8 \times 3.8 \times 45\,\text{cm}^3$ Pb-glass detectors was placed to the right of the oncoming beam as well as above and below the beam. Readout of all FPD calorimeters was triggered for events that deposited ≈ 20 GeV electron-equivalent energy in any one calorimeter. The kinematic ranges covered by the FPD were $1 < p_T < 3$ GeV/c and $0.2 < x_F < 0.6$. A valid coincidence from scintillator annuli mounted around the beam on both sides of the STAR magnet was required in the offline analysis of the data. These scintillator annuli are called the STAR beam-beam counters (BBC) [13]. A dedicated beam study in which the beams were steered out of collision at the STAR interaction region determined that approximately 98% of the observed FPD triggers accompanied by a BBC coincidence came as a result of $p+p$ collisions.

DATA ANALYSIS

Neutral π mesons were reconstructed with the pEEMC from two cluster events in the SMD according to the formula,

$$M_{\gamma\gamma} = E_\pi \sqrt{1 - z_\gamma^2} \sin(\frac{\phi_{\gamma\gamma}}{2}) \approx E_{tot} \sqrt{1 - z_\gamma^2} \frac{d_{\gamma\gamma}}{2 z_{vtx}}. \quad (2)$$

The energy of the π^0, E_π, was taken to be the total energy deposited in all of the towers in the calorimeter, E_{tot}. The opening angle between the photons, $\phi_{\gamma\gamma}$, was determined by the measurement of two values: the vertex position, z_{vtx}, given by the time difference measured by the east and west STAR BBC's, and the distance between the two photons at the calorimeter, $d_{\gamma\gamma}$. Both $d_{\gamma\gamma}$ and the di-photon energy sharing parameter, $z_\gamma = |E_1 - E_2|/(E_1 + E_2)$, were measured by an analysis of the energy deposited in the strips of the two orthogonal SMD planes. Typical events had these SMD distributions fit with two peaks used to model the transverse profile of the electromagnetic shower from the incident photons. The value of $d_{\gamma\gamma}$ was determined from the fitted centroids of the peaks, while z_γ was determined from the fitted area under each peak. Background at low invariant mass was reduced by constraining z_γ as indicated in Figure 2, to ensure that both photons deposit significant energy in the SMD. This algorithm resulted in a mass resolution of 20 MeV/c^2 for π^0 energies from 20 – 80 GeV, limited by the measurement of $\phi_{\gamma\gamma}$ (Fig. 2). Due to the finite size of the collision diamond, making an assumption of a fixed value for $z_{vtx,fixed}$ would result in a mismeasurement of $\phi_{\gamma\gamma}$. The peak in the invariant mass distribution, reconstructed using $z_{vtx,fixed}$, was found to be linearly correlated with z_{vtx} as determined from charged tracks reconstructed with the STAR time projection chamber, with which a subset of the FPD data was accumulated [14]. This provides evidence that the observed π^0 mesons were produced in $p+p$ collisions.

The absolute energy scale for each tower was determined from the π^0 peak in the invariant mass distribution. The invariant mass was sorted according to the calorimeter tower with the greatest energy deposition in each event, and then the gain for each tower was adjusted to match the known mass of the π^0 meson. Since typical events involve multiple calorimeter towers, the gain matching procedure was iterative. After approximately five iterations, this procedure converged to a stable set of values for each fill with an absolute uncertainty better than 1%. Small drifts of the gain on the order of a few percent were observed for many towers. It has been checked that the position of the π^0 peak had negligible dependence on the spin-state of the yellow beam and was independent of π^0 energy, as shown in Figure 2. The energy calibration of the Pb-glass arrays were also performed with π^0 mesons, although the mass resolution was significantly worse since the positions of the photons at the detector were not as well measured.

The FPD data were compared with a simulation of $p+p$ collisions using PYTHIA together with a full GEANT simulation of the pEEMC response. Minimum-bias PYTHIA events with more than 25 GeV of energy within a box of size comparable to the pEEMC were run through GEANT. The simulated detector responses were processed through the analysis algorithm as if they were data. The simulation was found to compare well with the data for an over-determined set of kinematic variables, bolstering the evidence

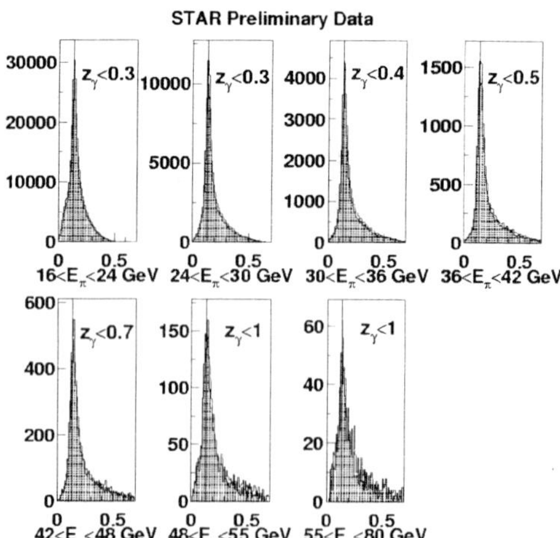

FIGURE 2. Preliminary distributions of the diphoton invariant mass spectra sorted into the energy bins used in the asymmetry analysis. The distributions are not corrected for acceptance or efficiency effects. The vertical line is drawn at 135 MeV/c^2. The absolute gain calibration of the pEEMC has been determined by the position of the mass peak in spin-summed distributions. Bin-by-bin constraints are applied to the energy sharing parameter to reduce background at small invariant mass. The filled area represents events collected when the yellow spin polarization direction is up, while unfilled represents spin down. There is negligible dependence of the peak position on either spin or energy.

that the FPD was measuring π^0 mesons resulting from $p-p$ collisions. A comparison of the data and the simulation for the p_T and the energy spectra can be seen in Figure 3.

The orientation of the yellow beam polarization for FPD triggered events was determined by measuring the time difference between the FPD trigger and spin direction bits provided by RHIC. The relative luminosity of collisions with polarization direction of the yellow beam oriented up or down was measured by counting the coincidences of charged particles fore and aft of the collision vertex by the STAR BBC sorted by the yellow beam spin direction bits [13]. The live time of the FPD data acquisition system was measured by counting the number of events acquired divided by the number of events which satisified the trigger condition. No appreciable spin-dependence of the live time was observed. Correcting the π^0 yield by the luminosity and live time resulted in a normalized π^0 yield with fill-to-fill stability on the order of 15%. The value of the relative luminosity correction (R in Equation 1) was typically on the order of 1.15, and is understood to come from variations in the beam intensity from bunch-to-bunch [13].

RESULTS

The measured analyzing power is not strongly affected by cuts used to identify π^0 mesons. The A_N for the π^0 candidate events in the mass range $70 < M_{\gamma\gamma} < 300\,\mathrm{MeV/c^2}$ shown in Figure 2 is consistent with A_N for the energy spectra observed with the

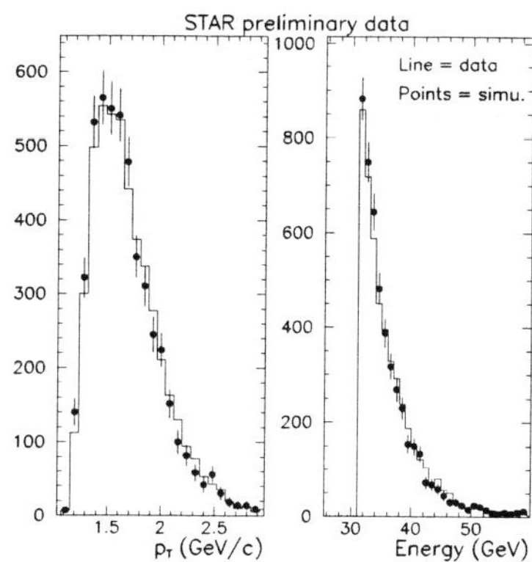

FIGURE 3. The spin-summed transverse momentum and energy distributions for events seen with the Pb-scintillator sampling calorimeter (pEEMC), uncorrected for efficiency or acceptance effects. The histogram is data from a single fill subjected to the data analysis algorithm described in the text. The points are a Monte-Carlo simulation using events generated from a PYTHIA minimum-bias sample together with a GEANT model of the pEEMC, subjected to identical analysis constraints as the data. The simulation agrees well with the data for virtually all observables, indicating minimal contributions from background sources other than $p+p$ collisions.

pEEMC. The value of Feynman-x is approximately $E_{tot}/100\,\text{GeV}$. A preliminary analysis of the simulation in Figure 3 indicates that events which trigger the FPD are composed of 95% photons, 95% of which are daughters from π^0 decay. Non-photon triggers predominantly come from hadron showers, while other photon triggers mostly come from other meson decays, such as the ω, η, and η'.

The analyzing power for the energy spectra in the pEEMC is shown in the upper-left plot in Figure 4. Also displayed in Figure 4 is A_N for the energy spectra measured with the Pb-glass arrays. The analyzing power observed on the left side with the Pb-scintillator sampling calorimeter is consistent with A_N measured with the Pb-glass array on the right side, even though systematic effects arising from hadronic contributions in these two detector arms are different. The analyzing powers observed above and below the beam pipe with Pb-glass arrays are consistent with zero.

The preliminary systematic uncertainty is taken to be constant throughout the energy range covered by the detectors, its value being $\delta A_N = 0.05$. This estimate has three primary, approximately equal, components: the average difference between the left and right detectors, the difference between the energy spectra asymmetry and the asymmetry for identified π^0 mesons, and the time dependence of the spin-dependent asymmetry seen with the pEEMC. The systematic uncertainty does not include the asymmetric normalization uncertainty from the beam polarization.

In summary, we have observed that the analyzing power for π^0 mesons at $\sqrt{s} = 200\,\text{GeV}$ is similar in magnitude and x_F dependence to that measured at collision energies an order of magnitude smaller.

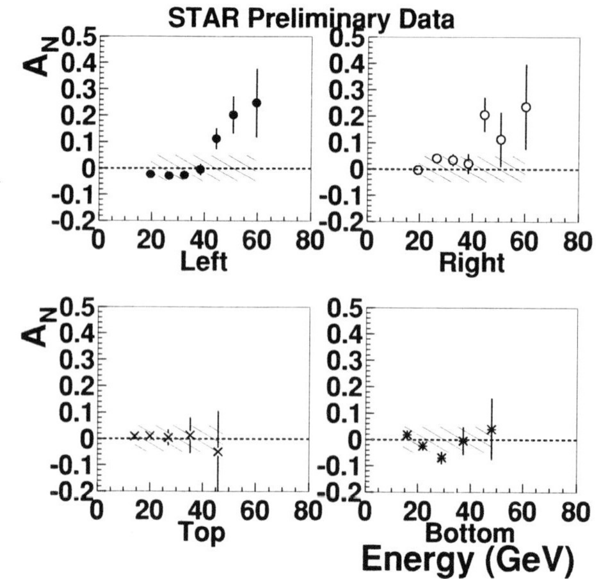

FIGURE 4. Preliminary results of the analyzing power for the energy spectra in the four arms of the FPD detector measured with a vertically polarized proton beam. The lines on the data points represent the statistical uncertainty, while the hatched area represents an estimate of the systematic uncertainty. The results from the Pb-scintillator sampling calorimeter to beam-left are consistent with the Pb-glass array to beam-right, while the Pb-glass arrays above and below the beam are consistent with zero. The analyzing power for identified π^0 mesons with the Pb-scintillator sampling calorimeter is consistent with these data, but with significantly larger statistical uncertainties. The size and shape of the analyzing power is similar to that seen for π^0 mesons at E704 [1, 2].

REFERENCES

1. D. L. Adams, *et al.*, Phys. Lett. B **261**, 201 (1991).
2. D. L. Adams, *et al.*, Phys. Lett. B **264**, 462 (1991).
3. J. Qiu and G. Sterman, Phys. Rev. D **59**, 014004 (1998).
4. Y. Koike, 'Single Transverse Spin Asymmetry in $p_\uparrow + p \to \pi + X$ and $e_\uparrow + p \to \pi + X$,' *these proceedings.*
5. M. Anselmino, M. Boglione and F. Murgia, Phys. Rev. D **60**, 054027 (1999).
6. M. Anselmino, M. Boglione and F. Murgia, Phys. Lett. B **442**, 470 (1998).
7. M. Anselmino, M. Boglione and F. Murgia, Phys. Lett. B **362**, 164 (1995).
8. V. Barone, A. Drago and P. G. Ratcliffe, Phys. Rep. **359**, 1 (2002).
9. U. D'Alesio, 'Single Spin Asymmetries, Unpolarized Cross Sections, and the Role of Partonic Intrinsic Transverse Momentum,' *these proceedings.*
10. K. Kurita, 'RHIC *pC* CNI Polarimeter: Experimental Setup and Physics Results,' *these proceedings.*
11. O. Jinnouchi, 'RHIC *pC* CNI Polarimeter: Status and Performance from the First Collider Run,' *these proceedings.*
12. C. A. Allgower, *et al.*, Phys. Rev. D **65**, 092008 (2002).
13. J. Kiryluk, 'Relative Luminosity Measurement in STAR and Implications for Spin Asymmetry Determinations,' *these proceedings.*
14. A. Ogawa, 'STAR Forward π^0 Detector Upgrade,' *these proceedings.*

STAR Forward π^0 Detector Upgrade

Akio Ogawa
for the STAR collaboration [1]

Brookhaven National Laboratory, Department of Physics
Upton, NY, 11973-5000, USA

Abstract.
Forward rapidity in hadron-hadron collisions is an interesting place to look because one can access high x quarks, which may be highly polarized, and also low x gluons. The STAR experiment at RHIC had a prototype Forward π^0 Detector (FPD) composed of electromagnetic calorimeters in the forward region during the January 2002 run with the first polarized proton-proton collisions. In this paper, I will discuss some aspects of the data, such as the correlation measurement between the FPD and mid-rapidity detectors and measurements in the negative x_F region. These data may provide clues to the origin of the single spin asymmetries A_N which were measured at FNAL-E704 [1]. At the end, I will discuss the plan for the FPD upgrade for next RHIC run.

1. PHYSICS OF THE FORWARD π^0 DETECTOR

The STAR Forward Pion Detector (FPD) covers forward pseudo-rapidity ($3 < \eta < 4$), high energy (20 GeV $< E$), high Feynman x_F ($0.2 < x_F$) and moderate transverse momentum ($1 < p_T < 4$ GeV). The FPD measures primarily electromagnetic energy and is capable of π^0 reconstruction up to an energy of more than 60 GeV. By supplementing the capabilities of the STAR detector to identify particles at large rapidity and moderate transverse momentum, the FPD allows quantitative measurements related to several interesting phenomena both in polarized proton-proton collisions and in deuteron-gold collisions.

1.1. Physics of Polarized Proton-Proton Collisions

In polarized proton-proton collisions, the transverse single spin asymmetry (A_N) for the $p^{\uparrow} + p \rightarrow \pi^0$+$X$ reaction was measured at FNAL-E704 [1] at $\sqrt{s} = 20$ GeV and found to be large. After more than a decade, this asymmetry still remains a mystery.

In perturbative QCD at leading twist with co-linear factorization, this A_N has to be zero. If one takes a step beyond this simple scheme and allows particles to have k_T, there are 3 terms which can contribute to the asymmetry. The first is the Sivers effect [2, 3], which is an initial state correlation between k_T, the internal transverse momentum of the quark, and the transverse spin of the nucleon. The second is nucleon transversity

[1] For the full author list and acknowledgements,see the appendix to the proceedings.

CP675, *Spin 2002: 15th Int'l. Spin Physics Symposium and Workshop on Polarized Electron Sources and Polarimeters,* edited by Y. I. Makdisi, A. U. Luccio, and W. W. MacKay

and the Collins effect [4, 5] in the fragmentation process. The third involves a correlation between k_T and quark spin within the unpolarized nucleon. This is believed to be very small [6]. There are also calculations of twist-3 contributions [8, 7] which have structures similar to models in the k_T picture but are not identical.

There are experiments either underway or planned that will challenge this mystery. Semi-inclusive polarized DIS experiments have reported [9, 10, 11] azimuthal asymmetries which may have the same origin as the E704 asymmetry. Azimuthal correlations in di-jet events at e^+e^- colliders will be sensitive to the Collins-Heppelmann fragmentation function [13, 12, 15]. The RHIC spin program will measure azimuthal asymmetries within a jet at mid-rapidity. Such measurements are expected to be sensitive to transversity, if e^+e^- collider experiments find these chiral-odd fragmentation functions to be non-zero [12, 14].

STAR has just reported [16] A_N for π^0 production at forward rapidity at an order of magnitude higher $\sqrt{s}$ compared to the E704 experiment. In the coming years, it is important for STAR to continue these measurements to further explore this phenomenon.

The polarization in the next RHIC run is expected to be increased by a factor of ~ 2 (from 0.2 to 0.4) and the luminosity will be increased by a factor of ~ 10 (from 10^{30} to $10^{31}/cm^2/sec$). Thus the figure of merit (P^2L) should increase by a factor of 40. A few hours (or a single RHIC fill) will have the same statistical significance as last year's entire data set. There may also be a measurement at $\sqrt{s} = 500$ GeV. These improvements allow us to map the asymmetry as a function of $\sqrt{s}$, x_F and transverse momentum p_T. The models described above predict the p_T and $\sqrt{s}$ dependence of A_N. Measurements at STAR will provide important tests of these models.

The Sivers effect is in the initial state, and is based on the incident quark having k_T relative to the beam axis. On the other hand, the Collins effect is in the fragmentation process, and depends on hadronic k_T relative to the final state quark or the jet. One reason why these different effects are presently indistinguishable is because there are only inclusive measurements. With the addition of hadron calorimetry at forward rapidity, we should be able to distinguish which of these effects are contributing.

At RHIC, the colliding beams are both polarized, with the Yellow beam heading toward the prototype FPD and the Blue beam heading away from it. Just by sorting the data in a different way, *i.e.* averaging over the Yellow beam polarization and calculating the spin asymmetry based on the Blue beam polarization, we should be able to determine A_N at negative Feynman x. All three models described above predict that A_N has to be zero at negative x_F. There are some speculations [17] that we may see a non-zero value arising from three-gluon correlations. Unfortunately, only a limited amount of data from the prototype FPD can be sorted on the Blue beam polarization. Results at negative x_F from the next RHIC run may give new insights to the A_N puzzle.

All of the above will provide more hints about the physics behind the E704 mystery. If the Sivers picture turns out to be most important, then A_N measures $f_{1T}^{\perp}$, from which we may learn something about orbital angular momentum of the quarks in the nucleon. If the Collins effect is the dominant contribution, then A_N measures the transversity distribution which is the last unmeasured quark distribution at leading twist. If A_N is described by either of these perturbative QCD models based on two-to-two parton hard scattering, then measuring the second final state parton will constrain the kinematics and x region of the spin-dependent distribution function. STAR has measured the azimuthal

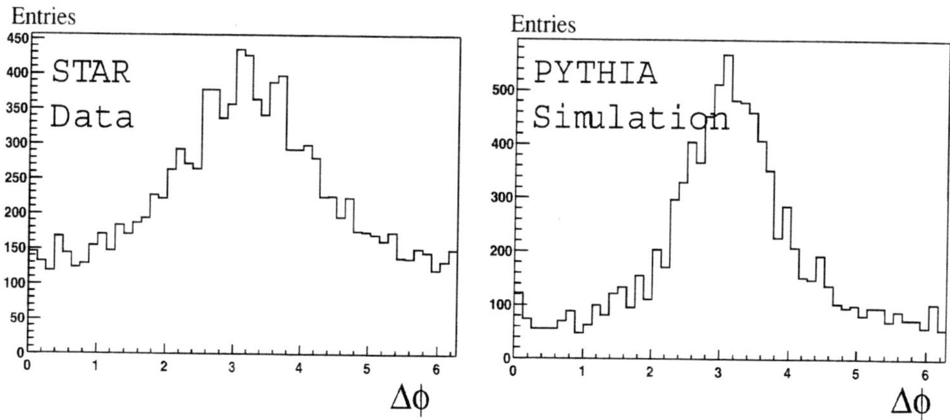

FIGURE 1. The left figure shows STAR data for the azimuthal angle difference between the π^0 detected with the FPD and the leading charged particle detected at mid rapidity. The right figure shows the same distribution from a PYTHIA simulation.

angle correlation between the π^0 at the FPD and the leading charged particle at mid-rapidity. PYTHIA simulations show fair agreement with the data (Fig. 1) and indicate that the FPD is sensitive to high Bjorken-x quarks (0.3 and above) and low x gluons (down to 0.001). Over the next 3 years, the STAR barrel and endcap electromagnetic calorimeters will increase their acceptance to be azimuthally complete for $-1 < \eta < 2$. Correlations between the FPD and midrapidity detectors will be studied.

1.2. Tuning Spin Rotators

One of the main goals for the STAR spin physics program is to measure longitudinal double-spin asymmetries to determine the gluon polarization in a nucleon. In the RHIC ring, the stable polarization direction is vertical. To get longitudinal polarization, spin rotators are being installed before and after the STAR and PHENIX interaction points (IP). Due to steering magnets between the spin rotators and the STAR IP, the horizontal polarization precession is more than a full turn when the beam energy is 250 GeV. Therefore an *in situ* polarimeter is required to tune the STAR spin rotators.

A left-right and up-down symmetric FPD will serve as a local polarimeter at STAR, providing a continuous monitor of unwanted vertical and radial polarization components during runs that aim to measure longitudinal two-spin asymmetries. Given the measured large A_N for forward rapidity π^0 production and with expected luminosity and polarization improvements in RHIC, the FPD will be able to give feedback within a typical RHIC fill of a few hours length.

1.3. Physics in Deuteron-Gold Collisions

To reliably interpret data from relativistic heavy ion collisions at RHIC, it is critical to know the initial state gluon density in the colliding heavy ions. Recent models suggest that the color fields of the colliding ions are so dense as to reach saturation [18]. d-Au collisions at RHIC may provide a key test of the gluon saturation model [19]. At forward rapidity in the deuteron direction, high x quarks will be scattered by low x gluons in gold nuclei. The FPD can measure the ratio of energetic π^0's produced in the incident deuteron direction to the corresponding yield in the direction of the incident gold ion beam. That ratio is expected to be sensitive to the gluon density in a heavy nucleus.

2. UPGRADE PLAN

With these motivations, we are proposing to construct electromagnetic calorimeters similar to what is shown in Fig. 2, with the left and right calorimeters each consisting of:

- a 7×7 matrix of lead-glass counters which were built by the IHEP, Protvino group for the E-704 experiment at FNAL.
- two finely segmented scintillator-strip shower maximum detectors.
- a 7-element lead-glass preshower detectors with a lead radiator in front.

For top and bottom, we will use a simpler 5×5 matrix of lead-glass counters due to the limitation in the available space.

These detectors will be read out with other STAR detectors for full event reconstruction, as well as without the slow detectors to accumulate greater statistics. The data from the upgraded FPDs will be included in bunch-sorted scalers operating at the 9 MHz RHIC bunch crossing frequency. This readout mode will provide STAR with an *in situ* polarimeter.

REFERENCES

1. A. Bravar *et al.*, Phys. Rev. Lett.77: 2626, 1996; D. L. Adams *et al.*, Phys. Rev. D53 :4747, 1996
2. D. W. Sivers, Phys. Rev. D41: 83, 1990
3. M. Anselmino, M. Boglione and F.Murgia, Phys. Rev. D60: 054027, 1999 [arXiv:hep-ph/9901442].
4. J. C. Collins, S. F. Heppelmann, G. A. Ladinsky, Nucl. Phys. B396 :161, 1993
5. M. Anselmino, M. Boglione and F. Murgia, Phys. Lett. B442: 470, 1998
6. D. Boer, Proceedings of Spin2002
7. J. Qiu and G. Sterman, Phys. Rev. D59, 014004, 1998
8. Y. Koike, Proceedings of Spin2002
9. HERMES collaboration, Phys. Rev. Lett.84:4047, 2000
10. SMC collaboration, Nucl. Phys. Proc.Suppl.79:520,1999
11. A. V. Efremov, K.Goeke, P.Schweitzer, Aug 2002. 9pp. e-Print Archive: hep-ph/0208124
12. M. Grosse Perdekamp, A. Ogawa, K. Hasuko, S. Lange, V. Siegle, Nucl. Phys. A711:69, 2002
13. D. Boer, R. Jakob, P. J. Mulders, Phys. Lett. B424 :143, 1998; X. Artru, J. Collins, Z. Phys. C69 :277, 1996

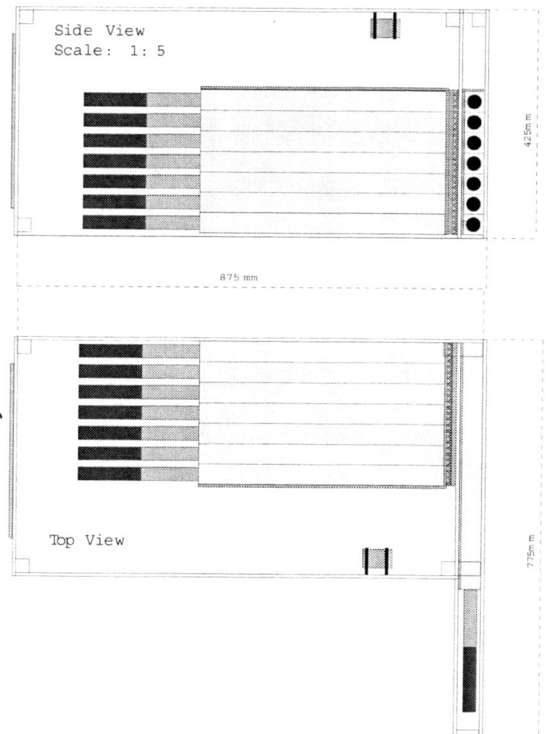

FIGURE 2. Schematic of a calorimeter module proposed for the upgrade of the STAR Forward π^0 Detector. The module consists of a main calorimeter built from an 7×7 matrix of Pb-glass detectors, two 48-strip scintillator shower maximum detectors, and a 7-element Pb-glass preshower detector. A simple lead radiator must be mounted upstream of the preshower detector to optimize the efficiency of the calorimeters as a π^0 detector.

14. R. L. Jaffe *et al.*, Phys. Rev. Let. 80 :1166, 1998; Phys. Rev. D57 :5920, 1998; Boffi, R. Jakob, M. Radici hep-ph/9907374
15. K. Hasuko, Proceedings of Spin2002
16. G. Rakness, Proceedings of Spin2002
17. X. Ji, Phys. Lett. B289: 137, 1992
18. L. McLerran, Lectures given at 40th Internationale Universitatswochen fuer Theoretishche Physik: Dense Matter (IUKT 40), Scladming, Asutria, 3-10 Mar 2001, hep-ph/0104285.
19. A. Dumitru and J. Jalilian-Marian, Phys. Rev. Lett. 89: 022301, 2002

Neutral Pion Measurements from PHENIX in Polarized Proton Collisions at RHIC

B. Fox for the PHENIX Collaboration[1]

RIKEN-BNL Research Center, Brookhaven National Laboratory, Upton, NY 11973-5000, USA

Abstract. This report presents the preliminary result for the absolute neutral pion (π°) production cross section at $\sqrt{s} = 200$ GeV for $|\eta| < 0.50$ which was obtained from the proton-proton data collected by the PHENIX experiment in January, 2002 as a part of Run02 at Relativistic Heavy Ion Collider (RHIC). We compare this result to the prediction of a next-to-leading order perturbative QCD calculation and find good agreement. Since the beams were transversely polarized, these data will be used to measure the single-spin transverse asymmetry (A_N). So, we also discuss the statistical precision for such a measurement. In future runs, the gluon polarization ($\Delta G/G$) will be probed by measuring double-spin longitudinal asymmetries (A_{LL}). We present an estimate for the statistical precision which is expected for the neutral pion measurement in the upcoming run and compare it to predictions from pQCD calculations using different polarized gluon densities.

INTRODUCTION

A detailed understanding of hadron production in proton-proton collisions is essential to both the heavy-ion and spin physics programs at the Relativistic Heavy Ion Collider (RHIC). For the heavy-ion program, proton-proton data provide the reference to which hadron production in heavy-ion collisions can be compared so that novel phenomena, such as jet energy loss or suppression, can be delineated from more prosaic effects. In the spin physics program, hadron production is a key probe of transverse and longitudinal spin structure functions and thus an understanding of the unpolarized cross section with next-to-leading order (NLO) perturbative QCD (pQCD) calculations provides the theoretical underpinnings for the physics interpretation of the polarized data.

During the 2001/02 run, RHIC was successfully operated for the first time as a proton collider at a center of mass energy ($\sqrt{s}$) of 200 GeV with transversely polarized beams.[1] From the data collected by the PHENIX experiment [2], we report the spin-averaged neutral pion (π°) cross section measurement at mid-rapidity and compare it with a next-to-leading order (NLO) perturbative QCD calculation [3, 4].

From this data sample, we also anticipate extracting a measurement of the single spin transverse asymmetry (A_N) for neutral pions produced at $x_F \sim 0$ with p_t up to $\sim$8 GeV/c. Interest in this measurement arises from the observation of large ($\sim 30\%$) asymmetries in $pp_\uparrow \rightarrow \pi X$ at forward angles by the Fermilab E704 experiment [5, 6] at $\sqrt{s} = 19.4$ GeV and single-spin azimuthal asymmetries in semi-inclusive deep-inelastic

---

[1] For the full PHENIX Collaboration author list and acknowledgments, see Appendix "Collaborations" of this volume.

CP675, *Spin 2002: 15th Int'l. Spin Physics Symposium and Workshop on Polarized Electron Sources and Polarimeters,* edited by Y. I. Makdisi, A. U. Luccio, and W. W. MacKay

scattering by the HERMES experiment at DESY [7]. Such large asymmetries were surprising because, at leading order, pQCD predicted only small effects. Presently, it is recognized that it is possible to have large asymmetries due to for example, the Sivers's effect [8], the Collin's effect [9], twist-three contributions [10] or combinations of the three. We report on the projected statistical precision of the measurement at PHENIX and compare it to that of the E704 measurement and a prediction of the Qiu-Sterman twist-three calculation.

During the upcoming run in 2003, the protons will be longitudinally polarized so that the gluon polarization (ΔG) can be probed by measuring double-spin, longitudinal asymmetries (A_{LL}). Since the acceptance of the PHENIX detector is limited, we intend to measure the asymmetry for neutral pions as an alternative to jets. We report on the anticipated statistical precision of the forthcoming dataset and compare it with pQCD predictions for the asymmetry performed with different gluon polarization densities.

EXPERIMENTAL SETUP

In Run-02, the PHENIX experiment operated with two central arm spectrometers, one muon arm spectrometer, and other detectors for triggering. This work utilized the electromagnetic calorimeters (EMCal) in the central arms, each of which has an azimuthal coverage of 90° and a pseudo-rapidity coverage of ± 0.35. This detector consists of six lead scintillator sampling calorimeter (PbSc) sectors and two lead glass (PbGl) sectors. In this paper, we will report only the measurement done with five of the six PbSc sectors. These sectors have nominal energy and position resolutions of $8.2\%/\sqrt{E\ (\mathrm{GeV})} \oplus 1.9\%$ and $5.7\ \mathrm{mm}/\sqrt{E\ (\mathrm{GeV})} \oplus 1.6$ mm, respectively.

During the proton-proton run in 2001-2002, PHENIX recorded an integrated luminosity of $0.15\ \mathrm{pb}^{-1}$. For the central arm, these data were collected by using two triggers: the minimum bias (MB) trigger and the newly installed electromagnetic calorimeter (EMCal) triggers. The MB trigger was formed by applying a loose (± 75 cm) cut to the interaction position reconstructed with timing information from the beam-beam counters which detect charged particles in the pseudo-rapidity range of 3.0 to 3.9. Events collected with this trigger covered the p_t range up to ~ 5 GeV/c. The sample at high p_t was collected with a coincidence between the MB trigger and one of the EMCal triggers. The latter consisted of two types: a 2x2 non-overlapping tower sum trigger with an 0.8 GeV threshold and several 4x4 overlapping tower sum triggers with thresholds ranging from 2 to 3 GeV. For this work, the 2x2 trigger, which had a rejection factor of 90, provided the data at higher p_t. Data collected with 4x4 triggers were used for systematic studies. In the analysis, a more stringent cut of ± 30 cm was imposed on the interaction vertex position.

ANALYSIS PROCEDURE

The cross section is the ratio of the yield of neutral pions after being corrected for efficiency, acceptance, and smearing ($\mathscr{N}_{\pi^0}$) to the effective integrated luminosity ($\mathscr{L}$)

where, for the MB sample, these two quantities are computed according to:

$$\mathcal{N}_{\pi^\circ} = \frac{N_{\pi^\circ}(p_t)\cdot C^{reco}_{\pi^\circ}(p_t)}{\varepsilon_{\pi^\circ}(p_t)} \qquad \frac{1}{\mathcal{L}} = \sigma^{pp} \times \frac{\varepsilon^{MB}_{trig}}{N^{MB}_{trig}}$$

with: $N_{\pi^\circ}(p_t)$ is the raw yield of π°'s, $C^{reco}_{\pi^\circ}(p_t)$ is the reconstruction efficiency, smearing, and acceptance correction, $\varepsilon_{\pi^\circ}(p_t)$ is the correction for the bias imposed on the event geometry by the MB trigger requirement, σ^{pp} is the proton-proton inelastic cross section, ε^{MB}_{trig} is the efficiency of the MB trigger, and N^{MB}_{trig} is the number of observed MB triggers. For the 2x2 trigger sample, the π° yield was also corrected for the trigger efficiency and the luminosity was corrected for the prescale factors.

The raw yield of neutral pions in each p_t bin $[N_{\pi^\circ}(p_t)]$ was determined from the invariant mass distribution for two photon clusters. To extract these quantities, the spectrum for each bin was fit over a variety of ranges with a gaussian for the π° peak and several functions – including a gaussian and some order of polynomial – for the combinatorial background. The systematic error was estimated from the variation in these fits and the run-to-run stability of the yield.

The acceptance,[2] efficiency, and smearing correction $[C^{reco}_{\pi^\circ}(p_t)]$ was computed from a Monte Carlo simulation of the calorimeter which had been tuned using results from the test beam measurements[11] and the analyzed data itself. The p_t dependences of the mass and the width of the π° peak are very sensitive to the calorimeter calibration and performance parameters. So, the systematic error in the correction factor was estimated from the change in it as the parameters in the Monte Carlo were varied over the range for which the peak's mass and width in the simulation was consistent with that in data up to $p_t \sim 9$ GeV/c, the maximum value for which these quantities could be determined with precision.

TABLE 1. Summary of the p_t dependent systematic error; there is also a normalization error of 30% not noted here since it is independent of p_t.

Correction Term	Source	Estimate
N_{π°	Background subtraction	5%
	Hot/Warm towers	2-3%
	Run dependence	10%(MB) 6%(2x2)
$C^{reco}_{\pi^\circ}(p_t)$	Fast MC statistical error	1%
	Edge towers	5%
	Position resolution	0-1%
	Energy absolute calibration	3-8%
	Energy non-linearity	0-10%
	Energy resolution	3%
$\varepsilon^{2\times2}_{\pi^\circ}(p_t)$	2x2 high p_t trigger threshold	10%

[2] The result is corrected from $|\phi|<\pi/2$ and $|\eta|<0.35$ of the detector to $|\phi|<\pi$ and $|\eta|<0.5$ by assuming the cross section has no ϕ and η dependences at mid-rapidity.

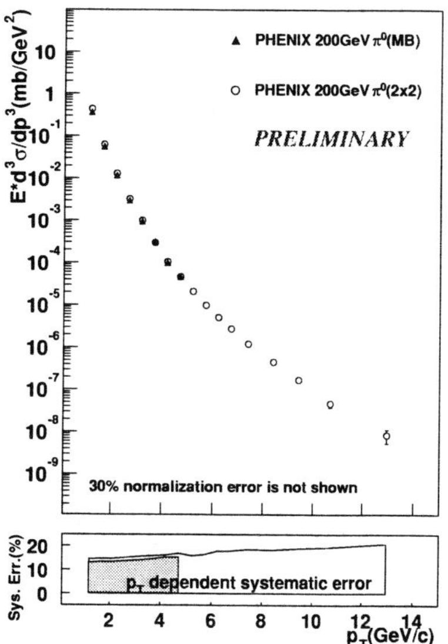

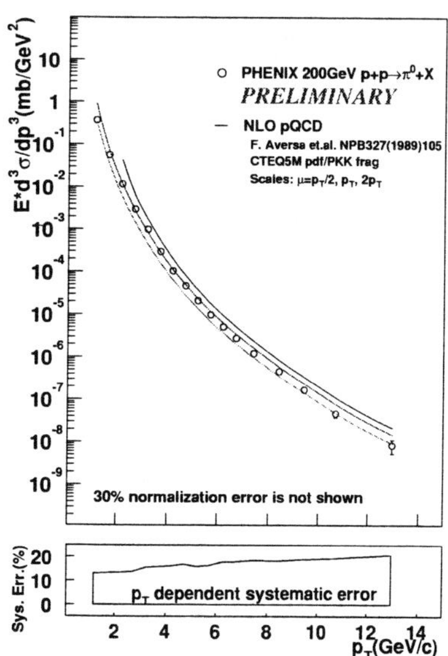

FIGURE 1. [left] The inclusive neutral pion cross section for the MB (open circle) and the 2x2 (solid triangles) trigger samples. [right] a comparison of the combined measurement with an NLO pQCD calculation using $p_t/2$ (top line), p_t (middle line), and $2p_t$ (bottom line) renormalization and factorization scales. The p_t dependent systematic error of the data is shown in the lower box of each panel.

Using the MB dataset, the π° threshold curve in p_t determined for the 2x2 trigger and found to plateau at 80% above a p_t of $\sim$3 GeV/c. A systematic error of 10% was assigned to this quantity based upon a comparison between this threshold curve and that determined with the Monte Carlo simulation to which the trigger performance had been included by using the measured efficiencies for the tiles that formed the trigger.

The bias for π° detection arising from the MB trigger condition $[\varepsilon_{\pi^\circ}^{MB}(p_t)]$ was measured to be 75%, independent of p_t up to $\sim$5 GeV/c, by using the data sample collected with a 4x4 trigger which, unlike the 2x2 trigger, did not impose the MB requirement. This value was consistent with an estimate from a PYTHIA+GEANT simulation of the experiment and thus also used to correct the data at higher p_t.

The MB trigger efficiency $[\varepsilon_{trig}^{MB}]$ of 51% was obtained from a PYTHIA+GEANT simulation of the experiment. Presently, we have assigned a normalization error of 30% based on the difference between the cross section measurement from a van der Meer/vernier scan and the total (elastic+inelastic) $p+p$ cross section. We anticipate that this error will be reduced to $\sim$15% with further analysis.

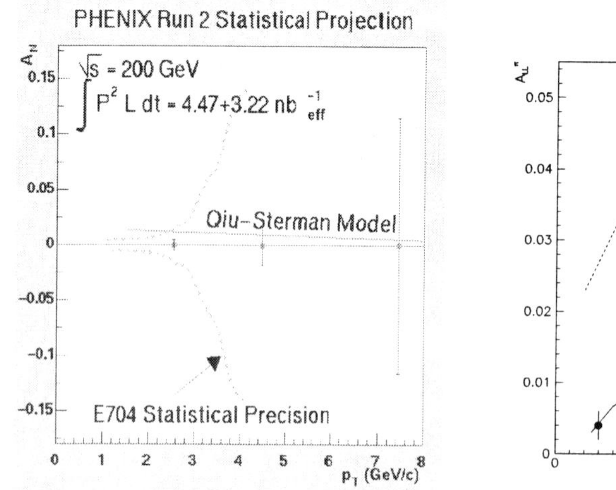

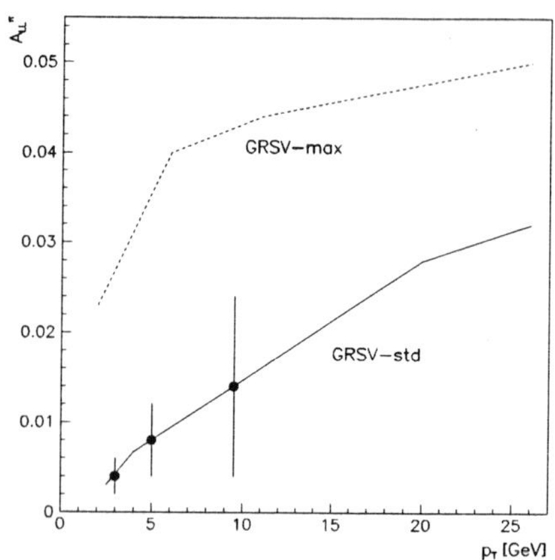

FIGURE 2. [left] The expected precision for a measurement of the single-spin transverse asymmetry (A_N) (points) in comparison to the precision of the E704 measurement (dashed line) and a prediction from the Qiu-Sterman model (solid line). [right] The anticipated precision for a measurement of the double-spin longitudinal asymmetry (A_{LL}) (points) from Run-3 data in comparison with predictions from NLO pQCD calculations using two polarized gluon densities (dashed and solid lines).

RESULTS AND DISCUSSION

The left panel of Figure 1 shows the measured cross sections for the MB and the 2x2 trigger samples along with the p_t dependent systematic error which are separately tabulated in Table 1. The results from the two samples are consistent within the error. In the right panel of this figure, this result is compared with an NLO pQCD calculation[3] using the formalism of F. Adversa *et al.*[4] with the CTEQ5M parton distribution functions[14] and the PKK fragmentation functions[15]. The data for the lower p_t range is shown from the MB trigger samples to avoid the larger systematic error of the 2x2 trigger samples. Over the full p_t range, this calculation is consistent with our measurement within the systematic errors. This measurement provides a baseline for high p_t heavy-ion physics[17].

FUTURE MEASUREMENTS

Since the protons were transversely polarized with, on average, 14% polarization in one beam and 17% in the other (as measured by the RHIC CNI polarimeters [16]), we anticipate making a measurement of A_N. The left panel of Figure 2 shows the statistical precision of such a measurement in comparison to the precision of the E704 result and a prediction based upon the Qiu-Sterman twist-three calculation.[3][10]

In the upcoming run, we expect higher polarization (> 40%) and an integrated luminosity of at least 3 pb^{-1}. In the right panel of Figure 2, the anticipated precision for the measurement of the double-spin, longitudinal asymmetry from this dataset assuming that we achieve the minimal of the expected performance in the collider. This precision is compared with predictions from NLO pQCD calculations using two polarized gluon densities.

REFERENCES

1. H. Huang, these proceedings.
2. J. T. Mitchell *et al.*, Nucl. Instrum. Meth. **A482** 491 (2002).
3. *Private communication with W. Vogelsang.*
4. F. Aversa, P. Chiappetta, M. Greco and J.-Ph. Guillet, Nucl. Phys. **B327**, 105 (1989).
5. D. L. Adams, *et al.*, Phys. Lett. **B264**, 462-466 (1991).
6. D. L. Adams *et al.*, Phys. Rev. **D53**, 4747-4755 (1996).
7. A. Airapetian *et al*, Phys. Rev. Lett. **84** 4047-4051 (2000).
8. D. W. Sivers, Phys. Rev. **D41** 83 (1990).
9. J. C. Collins, Nucl. Phys. **B396** 161-182 (1993).
10. J. W. Qiu and G. Sterman, Phys. Rev. **D59** 014004 (1999).
11. G. David, *et al.*, IEEE Trans. Nucl. Sci. **42** (1995) 306; G. David, *et al.*, IEEE Trans. Nucl. Sci.**45** (1998) 705; G. David, *et al.*, IEEE Trans. Nucl. Sci. **47** (2000) 1982; references therein.
12. C. Albajar *et al.*, Nucl. Phys. **B335** 261 (1990).
13. B. Alper *et al.*, Nucl. Phys. **B100** 237 (1975).
14. H. L. Lai *et al.* (CTEQ5), Eur. Phys. Jour. **C12** 375(2000).
15. B. A. Kniehl, G. Kramer and B. Potter, Nucl. Phys. **B582** 514 (2000).
16. O. Jinnouchi, these proceedings.
17. D. G. d'Enterria *et al.* (PHENIX), Proceedings for the XVI International Conference on Ultrarelativistic Nucleus-Nucleus Collisions, in press.

---

[3] It should be noted that the formalism of Qiu and Sterman is applicable to the high x_F (> 0.5) regime whereas, in our measurement, we are probing the $x_F \sim 0$ regime.

Midrapidity Spin Asymmetries at STAR

J. Balewski[a] for the STAR Collaboration [1]

[a]*Indiana University Cyclotron Facility, Bloomington, IN 47408,USA*

Abstract. Single- and double-spin asymmetries have been measured for the leading charged particle at mid-rapidity in collisions of polarized protons at $\sqrt{s}$ of 200 GeV. Raw asymmetries consistent with zero were measured at the sensitivity level of 10^{-3} over the p_T range from 0.2 to several GeV/c. Instrumental asymmetries are also zero with similar precision.

INTRODUCTION

This year, the STAR detector joined with other experiments at RHIC recording the first collisions of polarized protons at $\sqrt{s}$ = 200 GeV. These measurements represent a leap by an order of magnitude in the energy of studies with polarized proton beams. Both beams were polarized vertically with alternating spin directions, allowing for extraction of possible single- and double-spin transverse asymmetries for a number of processes. The analysis reported here focused on the determination of asymmetries for the detected charged particle of highest transverse momentum (p_T) within the pseudorapidity range $|\eta| \leq 1.4$ in each event recorded with a minimum-bias trigger. For p_T well in excess of 1 GeV/c, it is expected that this leading charged particle arises from jet production, where leading-twist perturbative QCD (pQCD) predicts vanishingly small single-spin effects at mid-rapidity [1].

We have developed an analysis procedure taking advantage of the full azimuthal symmetry of the STAR detector to permit simultaneous and independent extraction of single- and double-spin physics asymmetries and a number of possible instrumental asymmetries [2]. The multiple null tests incorporated in this procedure provide a stringent assessment of the robustness of transverse spin asymmetry measurements.

EXPERIMENTAL SETUP AND EVENT SELECTION

The Time Projection Chamber (TPC) positioned in the uniform magnetic field of the STAR detector (see Fig. 1) enabled reconstruction of the event vertex and of the momentum for charged particles emerging from that vertex, with p_T up to several GeV/c and $\eta \equiv -\ln(\tan\theta/2)$ up to 1.4. Beam-Beam Counters (BBC) situated at $3.4 < |\eta| < 5.0$ on either end of STAR served to provide both the minimum bias trigger and a relative luminosity ($\mathscr{L}_\tau$) monitor for the different beam spin combinations [3].

---

[1] For the full author list and acknowledgments, see the appendix to the proceedings.

CP675, *Spin 2002: 15th Int'l. Spin Physics Symposium and Workshop on Polarized Electron Sources and Polarimeters,* edited by Y. I. Makdisi, A. U. Luccio, and W. W. MacKay

The leading charged particle (LCP) was chosen as the charged track with the largest p_T among reconstructed primary tracks with at least 20 TPC hits, out of a possible 45 hits. The azimuthal direction (ϕ) of the LCP – measured from the detector-fixed positive x-axis, directed into the plane of Fig. 1 – defined the quantization axes x', y' for each event. ϕ distributions of LCP's were accumulated for several bins in η and p_T, for both charge signs, and for the four spin combinations (τ) of the colliding beams. Typical characteristics of the measured LCP's are shown in Fig. 2.

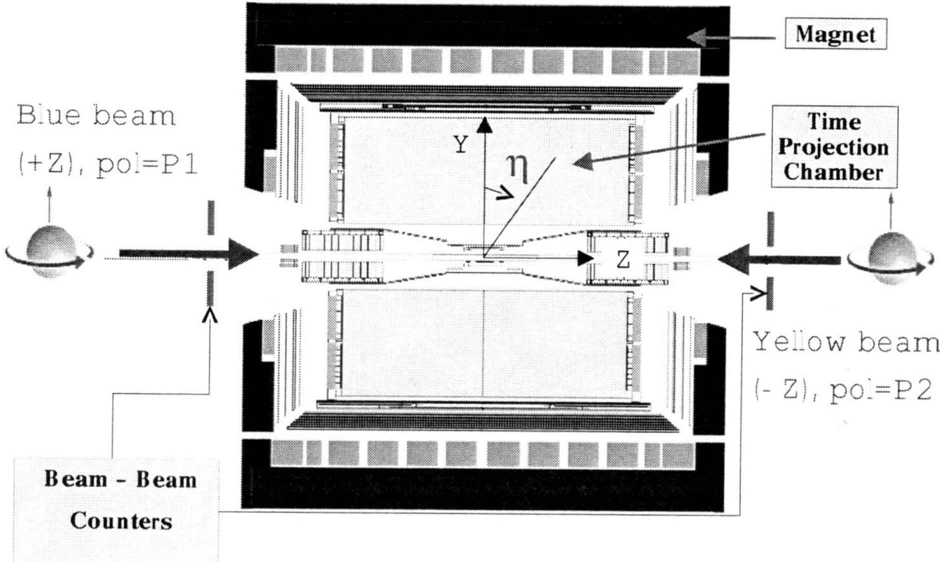

FIGURE 1. View of the STAR detector and of the detector-fixed reference frame used in this analysis.

EXTRACTION OF SPIN OBSERVABLES

For a given reaction process of choice – here the production of LCP's at some η ,p_T, and charge, the yields ($N_{\pm\pm}$) may depend on the polarizations of the beams ($\pm P_1, \pm P_2$, each assumed here to have equal magnitudes for the two spin orientations):

$$N_\tau(\phi,\eta) = \mathscr{L}_\tau \cdot \sigma(\eta) \cdot eff(\phi,\eta) \cdot [\ 1 \pm A_{y'}(\eta) P_1 \cos(\phi)\ \mp A_{y'}(-\eta) P_2 \cos(\phi) \\ \pm \frac{A_{y'y'} + A_{x'x'}}{2}(|\eta|) P_1 P_2\ \pm \frac{A_{y'y'} - A_{x'x'}}{2}(|\eta|) P_1 P_2 \cos(2\phi)\ ] \qquad (1)$$

where $\tau = \{++,+-,-+,--\}$ denotes the polarization state (with the first coefficient indicating the spin orientation of beam 1), σ the unpolarized cross section, *eff* the detection efficiency, $A_{y'}$ the analyzing power, and A_{jj} the relevant double-spin correlation coefficients. Beam 1 (2) is heading in the $+z$ $(-z)$ direction in Fig. 1. The dependences of the factors in Eq. (1) on other variables such as p_T or charge sign have been suppressed for simplicity.

Equation (1) includes all the polarization observables allowed by symmetry principles for a detector with full azimuthal coverage when the two beam polarizations are purely transverse. The indistinguishability of the two proton beams is reflected in the anti-symmetry of $A_{y'}$, and the symmetry of $A_{x'x'}$ and $A_{y'y'}$, about $\eta = 0$, assumed in Eq. (1). In pQCD treatments of polarized proton-proton scattering [1], the analyzing power $A_{y'}$ is commonly referred to as A_T, while the two-spin correlation combination $(A_{y'y'} - A_{x'x'})/2$ is called A_{TT}. The other allowed combination $(A_{x'x'} + A_{y'y'})/2$, which we refer to below as A_Σ, is normally neglected in pQCD treatments, but can be quite large in low energy pp interactions [4].

Raw experimental yields were normalized to the relative luminosities $\mathscr{L}_\tau$ measured by the BBC, and these normalized yields ($H_\tau = N_\tau/\mathscr{L}_\tau$) were then combined to form 3 independent ratios:

$$
\begin{aligned}
R_1(\phi,\eta) &= \frac{H_{++}(\phi,\eta)+H_{+-}(\phi,\eta)-H_{-+}(\phi,\eta)-H_{--}(\phi,\eta)}{\Sigma_\tau H_\tau(\phi,\eta)} , \\
R_2(\phi,\eta) &= \frac{H_{++}(\phi,\eta)-H_{+-}(\phi,\eta)+H_{-+}(\phi,\eta)-H_{--}(\phi,\eta)}{\Sigma_\tau H_\tau(\phi,\eta)} , \\
R_3(\phi,\eta) &= \frac{H_{++}(\phi,\eta)-H_{+-}(\phi,\eta)-H_{-+}(\phi,\eta)+H_{--}(\phi,\eta)}{\Sigma_\tau H_\tau(\phi,\eta)} . \qquad (2)
\end{aligned}
$$

In the absence of instrumental asymmetries, these ratios extract the physics asymmetries as follows:

$$R_1(\phi,\eta) = A_T(\eta)P_1 \cdot \cos(\phi) \ , \qquad (3)$$

$$R_2(\phi,\eta) = -A_T(-\eta)P_2 \cdot \cos(\phi) \ , \qquad (4)$$

$$R_3(\phi,\eta) = A_\Sigma(|\eta|)P_1P_2 + A_{TT}(|\eta|)P_1P_2 \cdot \cos(2\phi) \ . \qquad (5)$$

Next, the ϕ-dependence of the ratios R_1 and R_2 were fitted by the trigonometric series

$$F(\phi) = a_0 + a_1 \cdot \cos(\phi) + a_2 \cdot \cos(2\phi) \ . \qquad (6)$$

The same formula (6) was fitted to the ratio R_3, but we named the coefficients b_0, b_1, b_2 in this case. The orthogonality of the three terms in $F(\phi)$ ensures that the three coefficients can be determined without correlations among them.

We thus extract a total of 9 coefficients, each a function of p_T, η and charge sign, four of which give the physics asymmetries in Eqs. (3,4,5), and five of which should vanish in the absence of instrumental asymmetries. In particular, the coefficients a_0 and b_0 are sensitive to errors in relative luminosity monitoring. In this sense, extraction of A_Σ (from b_0) is subject to the same problems we will face in subsequent years in measuring A_{LL} with longitudinally polarized beams in an attempt to extract information on the

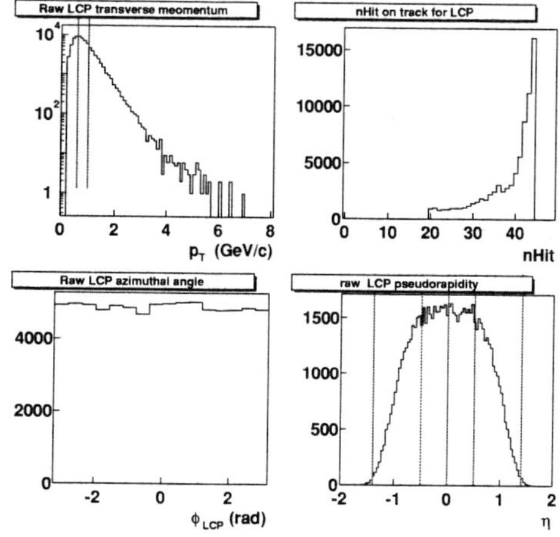

FIGURE 2. Characteristics of selected LCPs: p_T, number of points on track, ϕ and η distributions. Dashed vertical lines mark bins in p_T and η . Spectra are not corrected for the detection efficiency. The 'beam-gas' backgrounds and the false reconstructed vertex contributions are also not subtracted.

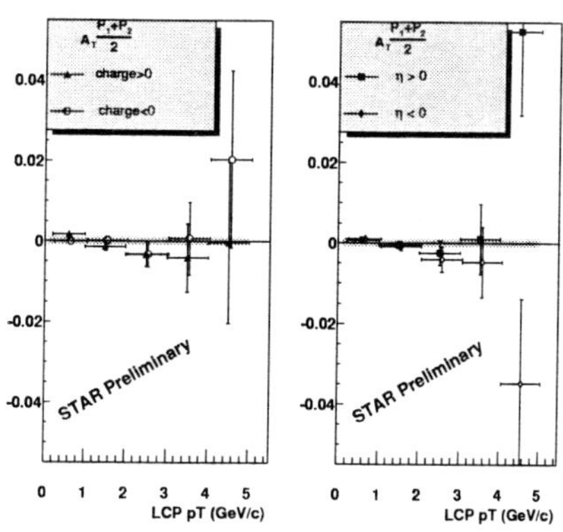

FIGURE 3. Product of transverse spin analyzing power A_T and beam polarization for the choices of opposite charges (left) and pseudorapidities (right). Total of 6.1 million events with three or more reconstructed primary tracks were used. The spin-dependent luminosity corrections contribute to a_1 in the second order and were not applied here. The average $(P_1 + P_2)/2 \simeq 0.11$ for the runs included.

gluon polarization in a polarized proton [5]. It is difficult to distinguish among a real ϕ-independent two-spin asymmetry in a reaction of interest, a similar asymmetry in the luminosity-monitoring reaction, and a luminosity monitoring error that happens to be different when the beam spins are parallel *vs*. antiparallel.

RESULTS

A total of 6.1 million minimum bias events were analyzed to extract the coefficients a_1 and the analyzing power A_T. Since the luminosity monitor scalers were not fully commissioned for all of these runs, a smaller sample of 2.8M events was subjected to the full analysis to extract all nine coefficients. The beam polarizations were monitored in RHIC with Coulomb-nuclear interference polarimeters [6], whose effective analyzing power at 100 GeV is not yet known. Under the assumption that the polarimeters have the same analyzing power as their measured values at the RHIC injection energy of 24.3 GeV, the time-averaged beam polarizations were $P_1 = 0.12$, $P_2 = 0.17$ for the 2.8M events, and $(P_1 + P_2)/2 = 0.11$ for the 6.1M events. The statistical error of these average polarizations is below 0.01, but they are subjected to large normalization uncertainties.

By averaging the a_1 values obtained from the fits to R_1 and R_2, we obtain the single-spin asymmetry $A_T(P_1 + P_2)/2$. The results are shown as a function of p_T in Fig. 3, for the two charge signs (integrated over all η) in the left frame and separately for positive and negative η (summed over both charge signs) on the right. The values of A_T are consistent with zero within statistical limits of 0.01 for p_T of 1.5 GeV/c and of 0.16 at p_T of 4.5 GeV/c. The gray band in Fig. 3 depicts the estimated systematic error for a_1 of 10^{-3}. This estimated systematic error is based on the observed values of the instrumental asymmetries, the consistency of a_1 values extracted for different choices of cuts (e.g. on the minimal number of the primary tracks in the event or on the minimum number of TPC hits required for the LCP track) and for different event samples with large or small 'beam-gas' backgrounds.

For the sample of 2.8M events, we have extracted each of the nine coefficients in each of 12 kinematic bins (2 charge signs, 2 η bins, 3 p_T bins). All the measurements are consistent with zero within 3σ. The total χ^2 value for the deviation of these 108 measurements from zero is 107. We show a sample of the results in Fig. 4.

The unphysical coefficients a_0 and a_2 extracted from eqs. (3,4) contain important cross-check information about possible systematic errors. For example, a non-zero a_0 might arise from a non-zero analyzing power of a process measured by the luminosity monitor (BBC) folded with a possible azimuthal non-uniformity of that detector. Fig. 4a shows the p_T-dependence of a_0, also consistent with zero within statistical limits of 10^{-3}.

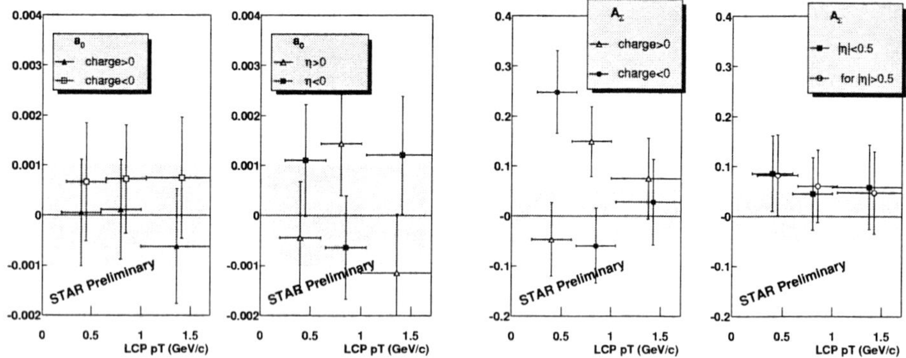

FIGURE 4. a) Instrumental asymmetry a_0 is consistent with zero. b) Physically allowed ϕ independent double-spin correlation coefficient $A_\Sigma = b_0/P_1P_2$. The last p_T bin in both figures contains any p_T above 1 GeV/c. The η range was chosen as $|\eta| > 0.5$ or $|\eta| < 0.5$ to accommodate the symmetry of eq. (5). A sample of 2.8M events with one or more reconstructed primary tracks was used. The spin-dependent luminosity corrections were applied to the measured yields. The gray band depicts the estimated systematic error of 10^{-3}.

The values of the physically allowed double-spin correlation coefficient $A_\Sigma = b_0/P_1P_2$ are shown in fig. 4b. Its average is marginally consistent with zero, deviating by 2.3σ. This deviation most likely represents a purely statistical fluctuation but could be a hint of a spin-dependent A_Σ contribution in LCP production, or in a process measured by the luminosity monitor. Note that the sensitivity level to the spin correlations is only of order 0.1 even at low p_T, because the product of beam polarizations was only of order 0.02.

SUMMARY

The first measurements of the single- and double-spin asymmetries for the leading charged particle at mid-rapidity in pp collisions at $\sqrt{s} = 200$ GeV yield values statistically indistinguishable from zero. The results are consistent with theoretical predictions for charged pions [7], [8] extrapolated to mid-rapidity.

All together, there were three coefficients extracted from each equation in formula (2), leading to simultaneous measurement of 4 physics raw asymmetries (eq. 3 - 5), plus 5 terms sensitive to instrumental asymmetries. The instrumental asymmetries consistent with zero confirm that STAR is a valuable detector for measurements of colliding polarized protons.

In the same runs, STAR has observed a significant non-zero transverse spin asymmetry for π^0 measured at forward rapidity, presented at this conference by G.Rakness [9]. The simultaneous measurement of zero and non-zero asymmetries with the same apparatus builds our confidence that STAR is ready to pursue the more challenging measurement of A_{LL} in the near future.

REFERENCES

1. Leader, E., *Spin in Particle Physics (Cambridge University Press)* (2001).
2. Meyer, H., *Phys. Rev.*, **C59**, 2074 (1997).
3. Kiryluk, J., *these proceedings* (2002).
4. Meyer, H., *Phys. Rev. Lett.*, **81**, 3096 (1998).
5. Bland, L., *RIKEN Rev.*, **28**, 8 (2000), URL `hep-ex/0002061`.
6. Jinnouchi, O., *these proceedings* (2002).
7. Qui, J., *et al.*, *Phys. Rev.*, **D59**, 014004 (1998).
8. Anselmino, M., *et al.*, *Phys. Rev.*, **D60** (1999), URL `hep-ph/9901442`.
9. Rakness, G., *these proceedings* (2002).

Relative Luminosity Measurement in STAR and Implications for Spin Asymmetry Determinations

Joanna Kiryluk for the STAR Collaboration[1]

University of California Los Angeles, CA 90095-1547 USA

Abstract. The Relativistic Heavy Ion Collider (RHIC) has collided, for the first time, transversely polarized proton beams at $\sqrt{s} = 200$ GeV. STAR has measured the relative luminosities R for different spin orientations with Beam-Beam Counters to a statistical accuracy of 10^{-3} to 10^{-4} per run. Systematic effects on R, of particular importance in preparing for future measurements of double longitudinal spin asymmetries, have been studied in detail and found to be of order 10^{-3}.

INTRODUCTION

The long-term goal of the spin program of the STAR collaboration is to study the spin structure of the nucleon. In particular, we aim to determine directly and precisely the gluon polarization by measuring the double longitudinal spin asymmetries A_{LL} in polarized proton-proton collisions. The double spin asymmetry, for example from inclusive jet production $\vec{p}+\vec{p} \to \text{jet}+X$, is defined as:

$$A_{\text{LL}} = \frac{1}{P_1 P_2} \times \frac{(N^{\uparrow\uparrow}+N^{\downarrow\downarrow}R_1) - (N^{\downarrow\uparrow}R_2+N^{\uparrow\downarrow}R_3)}{(N^{\uparrow\uparrow}+N^{\downarrow\downarrow}R_1) + (N^{\downarrow\uparrow}R_2+N^{\uparrow\downarrow}R_3)}, \tag{1}$$

where N^{ij} are the spin dependent yields for different spin orientation of the beams, $ij \equiv \uparrow\uparrow, \uparrow\downarrow, \downarrow\uparrow, \downarrow\downarrow$. The beam polarizations $P_{1(2)}$ from the RHIC CNI polarimeters [1] and the relative luminosities $R_1 = \mathscr{L}^{\uparrow\uparrow}/\mathscr{L}^{\downarrow\downarrow}$, $R_2 = \mathscr{L}^{\uparrow\uparrow}/\mathscr{L}^{\downarrow\uparrow}$ and $R_3 = \mathscr{L}^{\uparrow\uparrow}/\mathscr{L}^{\uparrow\downarrow}$ at STAR require separate measurements. The anticipated spin asymmetries A_{LL} are of the order of a few per cent, and statistical significance is achieved if $\delta A_{LL} \sim 10^{-3}$. The beam polarization must be measured to an accuracy of 10^{-1}, and the relative luminosities to 10^{-3} or better. The process/detector used for the luminosity measurements is required to have high rates, small background, and relatively small ($A_{LL}^{\text{Lumi}} < 10^{-3}$) spin dependence.

In December 2001 and January 2002, the Relativistic Heavy Ion Collider (RHIC) at Brookhaven National Laboratory collided transversely polarized proton beams at $\sqrt{s} = 200$ GeV. STAR measured the relative luminosities R for different spin orientations with the newly installed Beam-Beam Counters (BBC). Systematic effects on R were studied in detail. Finally the transverse single-spin asymmetries in $p+p^{\uparrow} \to A+X$,

[1] For the full author list and acknowledgements, see the appendix to the proceedings.

CP675, *Spin 2002: 15th Int'l. Spin Physics Symposium and Workshop on Polarized Electron Sources and Polarimeters,* edited by Y. I. Makdisi, A. U. Luccio, and W. W. MacKay

where A denotes charged particle(s) detected in the BBC, were determined using the azimuthal segmentation of the BBC.

BEAM BEAM COUNTERS

The BBC consists of scintillator annuli of small and large hexagonal tiles as shown in Fig. 1. The counters are mounted around the beam pipe beyond the east and west poletips of the STAR magnet at 3.7 m from the interaction point. The 2×18 arrays of small hexagonal tiles cover a full ring of 9.6 cm inner diameter and 48 cm outer diameter, corresponding to pseudorapidities in the region of $3.4 < |\eta| < 5.0$. The acceptance of the small hexagons is about 40% of the total proton-proton cross section. The 2×18 arrays of large hexagonal tiles span 38 cm to 193 cm in diameter, corresponding to pseudorapidities of $2.1 < |\eta| < 3.6$. During the 2002 proton-proton running period, all small tiles and one third of the large tiles were installed. We are planning to install the remaining outer tiles for the 2003 run.

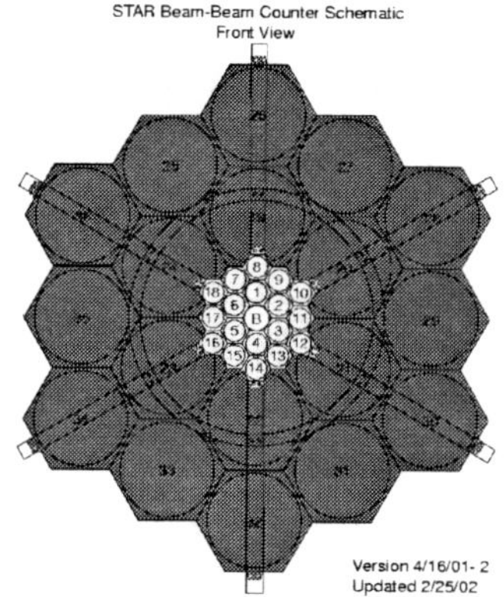

FIGURE 1. Schematic front-view of the Beam-Beam Counters.

RELATIVE LUMINOSITY MEASUREMENT

A signal from any of the 18 tiles on the east side and any of the 18 tiles on the west side of the interaction region constitutes a BBC coincidence. The number of BBC coincidences is a measure of the luminosity. The BBC coincidences were used to reject beam-gas events in physics triggers, provide a minimum bias trigger for full readout of STAR, to monitor the absolute beam luminosity $\mathcal{L}$ (15% accuracy achieved), and to measure the relative luminosities R with high precision. The BBC coincidences were summed for each bunch crossing number for each run of data taking. An example of the luminosity versus bunch crossing number is shown in Fig. 2a. During the 2002 proton-proton running period RHIC collided 55 transversely $(\uparrow,\downarrow)$ polarized proton bunches using a fixed fill pattern. Each bunch crossing can be uniquely related to the beam spin orientations. The relative luminosities were determined for each run, from

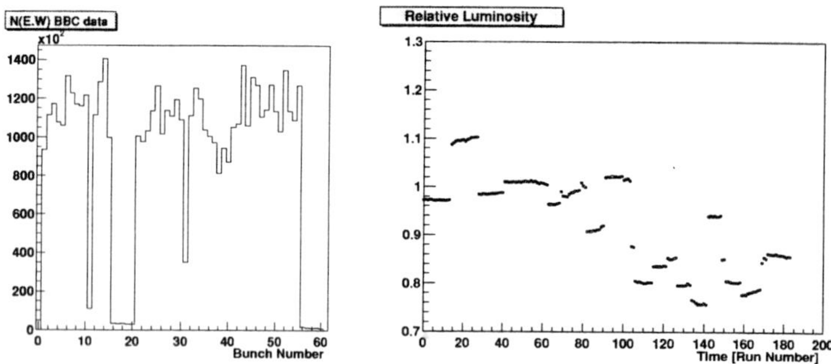

FIGURE 2. (a) The number of coincidences between the BBC on either side of the interaction region versus bunch crossing for one data run (b) The relative luminosity $R_3 = \mathscr{L}^{\uparrow\uparrow}/\mathscr{L}^{\uparrow\downarrow}$ for the 184 data runs collected from January 11 to 23, 2002. The R_3 is approximately constant during a single RHIC store, but can vary substantially from store to store.

distributions such as the one shown in Fig. 2a. Figure 2b shows the measured values for $R_3 = \mathscr{L}^{\uparrow\uparrow}/\mathscr{L}^{\uparrow\downarrow}$ during most of the 2002 proton-proton running period. The statistical uncertainties in R_3 are in the range of 10^{-4} to 10^{-3} for each point. Relative luminosities R_1 and R_2 were determined with similar statistical accuracy. Two groups of five bunch crossings: 16-20 and 56-60 at STAR, cf. Fig. 2a, are the so-called abort gaps. During these crossings, empty bunches interact with filled ones. The number of counts in the abort gap is a measure of the beam-gas background. Beam-gas background was reduced during the running period, from about 10% at the beginning to about 2% at the end. The difference between the background-corrected and background-uncorrected relative luminosity determinations was found to be a few times 10^{-3}. The future aim is to further reduce beam-gas background to the level of 10^{-4}.

SINGLE SPIN ASYMMETRIES

The BBC data allow classification of the counted occurrences by azimuth. Of the 18 inner tiles, the 4 tiles labeled 1, 7, 8, and 9 in Fig. 1 are referred to as Top, the tiles 4, 13, 14, and 15 are called Bottom. The remaining tiles are labeled Left and Right for the groups on the left and right, respectively. A hit coincidence with any of the BBC tiles on the other side of the interaction region is implied in all cases below. In an experiment with a transversely polarized beam and a left-right symmetric detector, the single spin asymmetry A_N can be determined by measuring beam polarization and the asymmetry of yields,

$$\varepsilon^{\text{Left-Right}} = \frac{\sqrt{\left(N_L^{\uparrow\uparrow}+N_L^{\uparrow\downarrow}\right)\left(N_R^{\downarrow\downarrow}+N_R^{\downarrow\uparrow}\right)} - \sqrt{\left(N_L^{\downarrow\downarrow}+N_L^{\downarrow\uparrow}\right)\left(N_R^{\uparrow\uparrow}+N_R^{\uparrow\downarrow}\right)}}{\sqrt{\left(N_L^{\uparrow\uparrow}+N_L^{\uparrow\downarrow}\right)\left(N_R^{\downarrow\downarrow}+N_R^{\downarrow\uparrow}\right)} + \sqrt{\left(N_L^{\downarrow\downarrow}+N_L^{\downarrow\uparrow}\right)\left(N_R^{\uparrow\uparrow}+N_R^{\uparrow\downarrow}\right)}} \simeq A_N \times P, \tag{2}$$

in which N_L^{ij} and N_R^{ij}, $ij \equiv \uparrow\uparrow, \uparrow\downarrow, \downarrow\downarrow, \downarrow\uparrow$, are the spin dependent yields from the detector on the left (N_L) and right (N_R) side of the plane spanned by the momentum and spin vectors of the polarized beam under study (Yellow or Blue at RHIC). In Eq. (2), false asymmetries due to differences in luminosity or acceptance cancel to lowest order.

At RHIC both beams are polarized, thus to measure single spin asymmetry one needs to sum over the yields for both spin orientations of one of the beams. At STAR the Yellow (Blue) beam heads towards the BBC East (West) detectors. Hence, one expects that $\varepsilon^{\mathrm{Left-Right}}_{\mathrm{BBCEast}}/P_{\mathrm{Yellow}} = \varepsilon^{\mathrm{Left-Right}}_{\mathrm{BBCWest}}/P_{\mathrm{Blue}}$. The single spin asymmetry A_N can alternatively be obtained using a method which requires the knowledge of the relative luminosities for different beam spin orientations. This method will be used for determining A_{LL}. The measured single spin asymmetry, using the Left (Right) part of the detector and the luminosity ratios is defined as follows:

$$\varepsilon^{\mathrm{Left(Right)}} = (-)\frac{\left(N^{\uparrow\uparrow}_{\mathrm{L(R)}} + N^{\uparrow\downarrow}_{\mathrm{L(R)}} R_3\right) - \left(N^{\downarrow\downarrow}_{\mathrm{L(R)}} R_1 + N^{\downarrow\uparrow}_{\mathrm{L(R)}} R_2\right)}{\left(N^{\uparrow\uparrow}_{\mathrm{L(R)}} + N^{\uparrow\downarrow}_{\mathrm{L(R)}} R_3\right) + \left(N^{\downarrow\downarrow}_{\mathrm{L(R)}} R_1 + N^{\downarrow\uparrow}_{\mathrm{L(R)}} R_2\right)} \simeq A_N \times P. \tag{3}$$

where $R_3 = \mathscr{L}^{\uparrow\uparrow}/\mathscr{L}^{\uparrow\downarrow}$, $R_1 = \mathscr{L}^{\uparrow\uparrow}/\mathscr{L}^{\downarrow\downarrow}$ and $R_2 = \mathscr{L}^{\uparrow\uparrow}/\mathscr{L}^{\downarrow\uparrow}$, as are defined previously. A detailed comparison between the two methods tests the consistency of the results as well as systematic differences associated with the methods. The scatter of the differences: $\Delta_1 = \varepsilon^{\mathrm{Left-Right}} - \varepsilon^{\mathrm{Left}}$ and $\Delta_2 = \varepsilon^{\mathrm{Left-Right}} - \varepsilon^{\mathrm{Right}}$, by run was found to be statistical in nature, while the mean $\Delta_{1(2)} = \pm 0.00013(4)$. The mean is sensitive to the systematic uncertainty δR in the determination of the relative luminosity R: $R_{\mathrm{meas}} = R_{\mathrm{true}} + \delta R$ and thus $|\Delta_{1(2)}| \sim \delta R/2R$. We conclude that systematic effects in relative luminosity measurement associated with this method are $\delta R < 10^{-3}$.

The single spin asymmetries were calculated, run by run using Eq. (2). They were determined with (i) Left and Right (ii) Top and Bottom groups of tiles in the East and West BBC. Because of the larger acceptance asymmetries (about 50%) observed for the West BBC than for the East BBC (about 10%), the West BBC were excluded from the present analysis. In the following, only results obtained with the East BBC are presented. The average of the measured asymmetries $\varepsilon^{\mathrm{Left-Right}}$ by run was evaluated for each RHIC fill for which the beam polarization was measured [1]. Internally self-consistent results were obtained for each RHIC fill, showing no evidence for polarization loss to within the statistical accuracy. The data shown in Fig. 3a,b are the asymmetries from the BBC East: Left-Right (filled circles) and Top-Bottom (open squares) yields summed over the polarization states of (a) the Blue and (b) the Yellow beam. The asymmetries are shown as a function of the single spin asymmetries measured by the CNI polarimeter in the Yellow and Blue rings, respectively. The indicated uncertainties on the CNI and BBC asymmetries are statistical only. The point to point systematic uncertainty is about 3 $\times 10^{-4}$ and the overall systematic uncertainties, for example from beam-gas background, are under study. The asymmetries measured with the Top-Bottom groups are consistent with zero, as expected for transverse beam polarizations along the vertical axis. A weak correlation is observed between the physics asymmetry determined with the Left-Right BBC tiles and with the CNI polarimeter in the Yellow ring, as presented in Fig. 3a. The asymmetries measured with the BBC are of order 10^{-3} in this case, comparable in size to those in the CNI polarimeter. The left-right asymmetries observed in the backward direction are consistent with zero, cf. Fig. 3b, from the Left-Right yields in the East BBC when the data are summed over the Yellow beam polarization states.

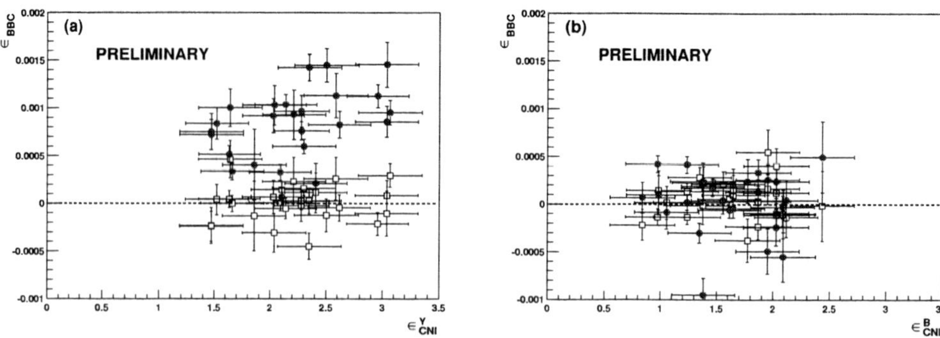

FIGURE 3. The single spin asymmetry determined with the Left-Right (filled circles) and Top-Bottom (open squares) tiles of the East BBC versus the measured CNI asymmetry $\varepsilon_{CNI} = P \times A_N^{CNI}$ (in parts per thousand) for the (a) Yellow (b) Blue RHIC beam. The indicated uncertainties on the CNI and BBC asymmetries are statistical only.

SUMMARY

The spin program with STAR at RHIC began December 2001, when transversely polarized protons were collided at $\sqrt{s} = 200$ GeV. A crucial component of the STAR detector was the Beam-Beam Counters. The BBC was used to (i) suppress beam-gas background, (ii) monitor the overall luminosity and (iii) measure the relative luminosities. Precise measurement of the relative luminosities is critical for the longitudinal double-spin asymmetry measurement to determine the gluon polarization. From the first proton-proton run at RHIC, the statistical uncertainty for the luminosity (run by run) is $10^{-3} - 10^{-4}$. The systematic uncertainties on R are found to be of order 10^{-3}. Both uncertainties are expected to be smaller in future, higher-luminosity running periods, but additional tests will have to be carried out to gauge systematic errors from possible spin-dependence in the luminosity-monitoring reaction with longitudinal beam polarizations. The azimuthal segmentation in the BBC allowed the determination of the transverse single-spin asymmetries in $p + p^{\uparrow} \rightarrow A + X$, where A denotes charged particle(s) detected in the BBC, for $3.4 < \eta < 5.0$. The BBC East Left-Right asymmetries were found to be of order 10^{-3} in the forward direction, and consistent with zero in the backward direction. The Top-Bottom asymmetries were found to be consistent with zero as expected.

REFERENCES

1. Jinnouchi, O., *these proceedings* (2002).

Azimuthal Asymmetries in Meson Electroproduction at HERMES

Delia Hasch [1]
(On behalf of the HERMES Collaboration)

INFN - Laboratori Nazionali di Frascati, 00044 Frascati, Italy

Abstract. The measurement of single-spin azimuthal asymmetries for pseudoscalar meson production in semi-inclusive deep-inelastic scattering of 27.6 GeV electrons off a longitudinally polarised hydrogen and deuterium target is reported by the HERMES experiment. A significant target-spin asymmetry amplitude in the azimuthal distribution of charged and neutral pions and positively charged kaons relative to the lepton scattering plane has been observed. The dependence on the relevant kinematic variables which are the Bjorken variable x, the meson fractional energy z and the meson transverse momentum $P_{\perp}$ has been investigated as well. The results are compared to predictions of model calculations which are base on a fragmentation function that varies with the transverse polarisation of the struck quark. In addition, data from the measurement of a single beam-spin azimuthal asymmetry in the electroproduction of positive pions in semi-inclusive and semi-exclusive deep-inelastic scattering will be presented.

INTRODUCTION

Polarized deep-inelastic lepton scattering has been the main experimental basis for our understanding of the spin structure of the nucleon for decades. In addition to inclusive and semi-inclusive measurements, recently single-spin azimuthal asymmetries in semi-inclusive scattering have been recognized as a powerful tool to access further information about the spin structure of the nucleon. A complete tree-level analysis of the semi-inclusive deep inelastic scattering (SIDIS) cross section [1] identifies a series of 8 distribution functions of the nucleon at leading twist along with an analogous set of 8 fragmentation functions. Each of these function describes qualitatively different information about the hadronic structure and formation. The functions $f_1(x)$, $g_1(x)$ and $h_1(x)$ have a special property: they are the only distribution functions which survive at leading twist integration over the intrinsic quark transverse momentum. The new leading twist distribution function $h_1(x)$, the so-called transversity or helicity-*flip* distribution, represents the transverse spin distribution of quarks in a nucleon polarized transversely to the virtual photon in the infinite momentum frame [2]. The afromentioned decomposition of the full polarized SIDIS cross section reveals how experiments may access the new distribution and fragmentation functions: by measuring moments of the azimuthal asymmetries in SIDIS.

---

[1] E-mail: delia.hasch@lnf.infn.it

CP675, *Spin 2002: 15th Int'l. Spin Physics Symposium and Workshop on Polarized Electron Sources and Polarimeters,* edited by Y. I. Makdisi, A. U. Luccio, and W. W. MacKay

TARGET-SPIN AZIMUTHAL ASYMMETRIES

HERMES has measured the azimuthal distribution of charged and neutral pions and of K^+ in the scattering of unpolarized (spin averaged) positrons from longitudinally polarized hydrogen [3] and deuterium targets. The target-spin cross section asymmetry $A_{\rm UL}$ for an unpolarized beam (U) and a longitudinally polarized target (L) was evaluated as

$$A_{\rm UL}(\phi) = \frac{1}{|P_{\rm L}|} \cdot \frac{N^{\rightarrow}(\phi)/L^{\rightarrow} - N^{\leftarrow}(\phi)/L^{\leftarrow}}{N^{\rightarrow}(\phi)/L^{\rightarrow} + N^{\leftarrow}(\phi)/L^{\leftarrow}}, \tag{1}$$

were $N^{\rightarrow(\leftarrow)}$ is the number of pions or kaons detected at angle ϕ for the target-spin antiparallel (parallel) to the beam momentum, $L^{\rightarrow(\leftarrow)}$ the respective dead-time corrected luminosities and $P_{\rm L}$ the mean longitudinal target polarization. Here ϕ is the azimuthal angle of the meson around the virtual photon direction, with respect to the lepton scattering plane. For the hydrogen target, the measured asymmetries show a significant $\sin\phi$ moment in the case of π^+ and π^0 production, while no ϕ-dependence is seen in π^- production. The $\sin 2\phi$ moments of the asymmetries were found to be consistent with zero in all cases. The new data from a deuterium target are shown in Fig. 1, integrated over the experimental acceptance in the kinematic variables x, $P_\perp$, z, y and Q^2. In Fig. 1, fits of the functions $f_1(\phi) = P_0 + P_1 \sin\phi$ and $f_2(\phi) = P_0 + P_1 \sin\phi + P_2 \sin 2\phi$ are shown, where all coefficients P_0 are compatible with zero. The $\sin\phi$ and $\sin 2\phi$ amplitudes P_1 and P_2 represent the analyzing powers $A_{\rm UL}^{\sin\phi}$ and $A_{\rm UL}^{\sin 2\phi}$ of the cross section asymmetry. The analyzing powers $A_{\rm UL}^{\sin\phi}$ have been studied as a function of x, $P_\perp$ and z and are shown in Fig. 2 together with the asymmetries from the proton target. The results for the two different targets show a similar behaviour in their kinematic dependences on x and z, but not $P_\perp$. The monotonic increase of $A_{\rm UL}^{\sin\phi}$ with increasing x for all mesons suggests that single-spin asymmetries are associated with valence quark contributions. Two mechanisms have been proposed to explain the measured single-spin asymmetries. One is the combination of chiral-odd transversity-related dis-

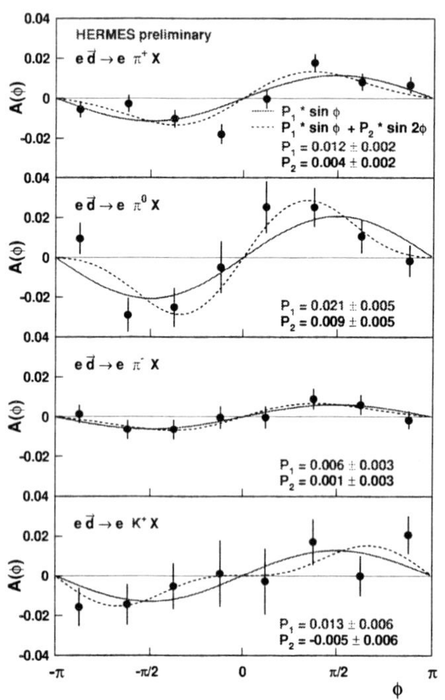

1: Cross section asymmetries $A_{\rm UL}(\phi)$ for π^+, π^-, π^0 and K^+. The error bars give the statistical uncertainty.

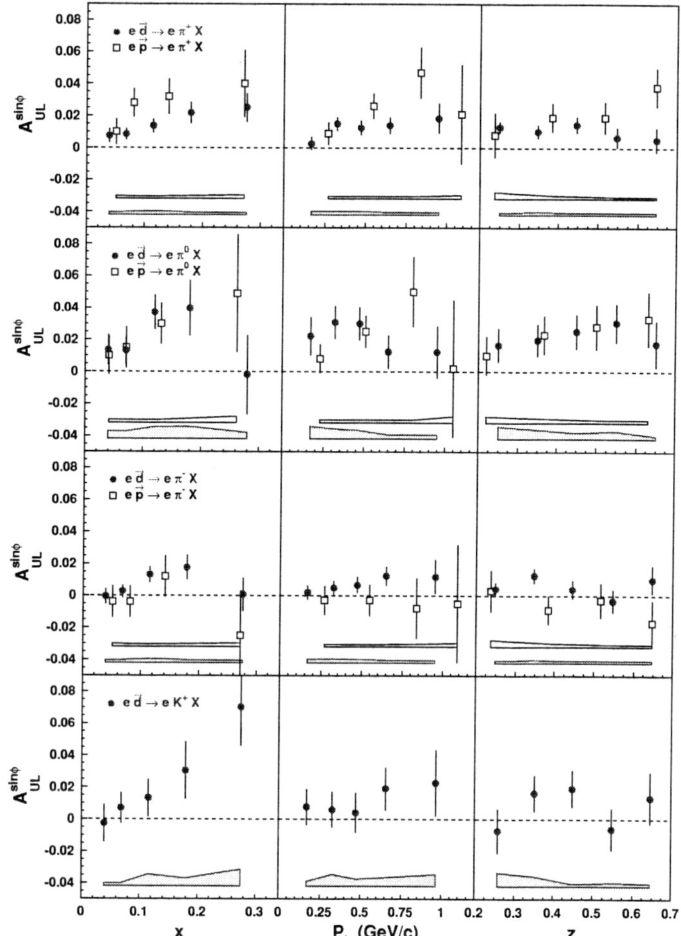

FIGURE 2. $A_{UL}^{\sin\phi}$ for semi-inclusive π^+, π^0, π^- and K^+ production on a deuterium *(filled circles)* and a proton *(open squares)* target as function of x, $P_\perp$ and z. The error bars give the statistical uncertainty, the bands give the systematic uncertainties for the deuteron *(filled band)* and for the proton measurement *(white band)*.

tribution functions and chiral-odd fragmentation functions such as the Collins fragmentation function $H_1^\perp$ [4]. The other one is a final-state interaction of the outgoing meson with the target remnant (Sivers Effect) [5, 6]. No calculations are available for the latter scenario to compare with data. Recent publications [7, 8] have reported transversity-related model calculations of $A_{UL}^{\sin\phi}$ for scattering of positrons off a longitudinally polarized deuterium target within the kinematic range of the HERMES experiment. As an input, the transversity distributions calculated in the chiral quark soliton model (χQSM) [8], in the SU(6) quark spectator diquark model [7] and in a perturbative QCD model [7] have been used.

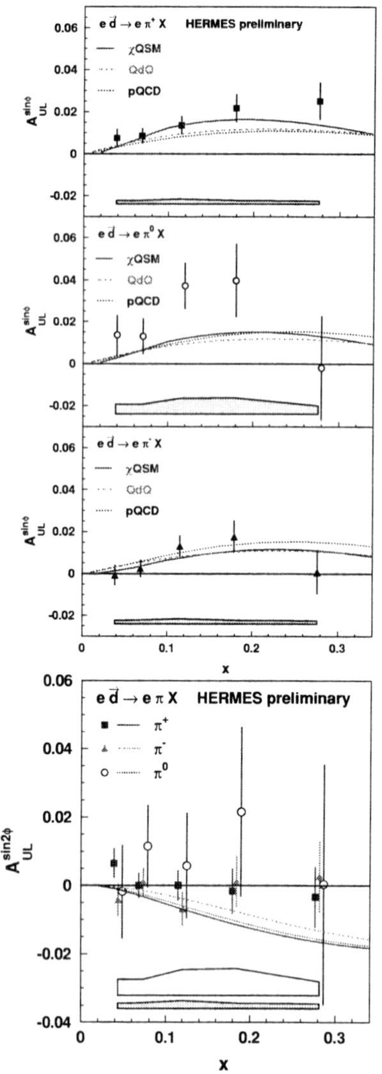

3: Comparison of $A_{UL}^{\sin\phi}$ and $A_{UL}^{\sin 2\phi}$ with predictions from model calculations (see text).

The results of some model calculations for pions are displayed in Fig. 3 together with the experimental data. The model predictions that are shown here describe the data fairly well at least for charged pions. In Fig. 3 also the dependence of $A_{UL}^{\sin 2\phi}$ on x is shown. Integrated over the measured x-range, it is compatible with zero for all mesons. Also shown are results from χQSM calculations [8]. According to this model, a deviation of the analyzing power from zero originates from effects of twist-3 distribution functions calculated by Wandzura-Wilczek type relations from the twist-2 transversity distribution function. However, as can be seen from the lower panel of Fig. 3, the data do not follow the prediction of negative analyzing power.

For a more sophisticated interpretation of the data one must consider in detail which terms in the SIDIS cross-section of ref. [1] contribute to the $A_{UL}(\phi)$ asymmetry. In the theoretical decomposition, the longitudinal and transverse components of the target polarization are measured with respect to the virtual photon direction, not to the lepton beam axis. The analyzing power $A_{UL}^{\sin\phi}$ thus contains a mixture of the cross section moments $\langle\sin\phi\rangle_{UL}$ and $\langle\sin\phi\rangle_{UT}$ (theoretical convention). The $\langle\sin\phi\rangle_{UT}$ moment is directly proportional to the product of transversity and Collins fragmentation function. However, the present HERMES measurements are most directly related to $\langle\sin\phi\rangle_{UL}$, which is more complex as it is subleading in Q and contains interaction-dependent twist-3 functions. All theoretical calculations adressing the HERMES measurements of $A_{UL}(\phi)$ agree that $A_{UL}^{\sin\phi}$ is dominated by higher-twist effects due to the longitudinal target polarization.

HERMES begun already measurements with a transversely polarized target. A precise measurement of $A_{UT}(\phi)$, sensitive at leading twist to the product of $h_1(x)$ and $H_1^{\perp}(z)$, will form a cornerstone of the new HERMES Run-II [9] that started in 2002. Moreover, by scattering on a transversely polarized target, it will be possible to distinguish the Collins and Sivers mechanism through their different dependence on the angle ϕ_S between the transverse target polarization vector and the lepton scattering plane.

BEAM-SPIN AZIMUTHAL ASYMMETRIES

At higher twist, in addition to $f_1(x)$, $g_1(x)$ and $h_1(x)$, other distribution functions survive integration over the intrinsic quark transverse momentum in the tree-level analysis of the SIDIS cross section of Ref. [1]. At twist-3 level these additional functions are the chiral-even polarized distribution function $g_T(x)$ and two chiral-odd distribution functions, the polarized one $h_L(x)$ and the unpolarised distribution function $e(x)$. The combination of the latter distribution function and the Collins fragmentation function gives rise to a beam-spin azimuthal asymmetry

$$A_{\mathrm{LU}}(\phi) = \frac{1}{|P_{\mathrm{B}}|} \cdot \frac{N^{\rightarrow}(\phi)/L^{\rightarrow} - N^{\leftarrow}(\phi)/L^{\leftarrow}}{N^{\rightarrow}(\phi)/L^{\rightarrow} + N^{\leftarrow}(\phi)/L^{\leftarrow}}, \tag{2}$$

where now $N^{\rightarrow(\leftarrow)}(\phi)$ is the number of pions at azimuthal angle ϕ for opposite beam helicity states, $L^{\rightarrow(\leftarrow)}$ the respective dead-time corrected luminosities and P_{B} the mean beam polarization. A significant beam-spin azimuthal asymmetry was reported recently by Jefferson Lab's CLAS collaboration [10]. HERMES has measured the analyzing power $A_{\mathrm{LU}}^{\sin\phi}$ for π^+ production from scattering longitudinal polarized positrons from an unpolarized (spin averaged) hydrogen target. The asymmetry shows a significant dependence on z, as displayed in Fig. 4. The beam-spin asymmetry provides a complementary tool to access information about the Collins fragmentation.

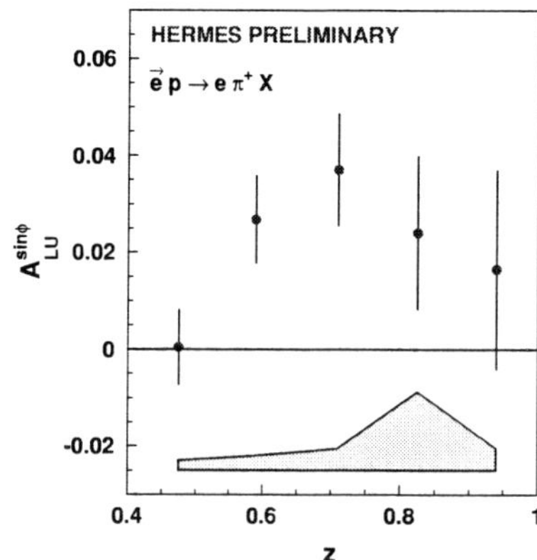

4: $A_{\mathrm{LU}}^{\sin\phi}$ for π^+ electroproduction on a proton target.

REFERENCES

1. P.J. Mulders and R.D. Tangermann, *Nucl. Phys.* B **461**, 197 (1996).
2. For a recent review see V. Barone, A. Drago, and P. G. Ratcliffe, *Phys. Rept.* **359**, 1-168 (2001).
3. A. Airapetian et al., HERMES collaboration, *Phys. Rev. Lett.* **84**, 4047 (2000), *Phys. Rev.* D **64**, 097101 (2001).
4. J. Collins, *Nucl. Phys.* B **396**, 161 (1993).
5. S. Brodsky, D. Hwang and I. Schmidt, *Phys. Lett.* B **530**, 99 (2002).
6. X. Ji and F. Yuan, hep-ph/0206057.
7. B. Ma, I. Schmidt, and J. Yang, hep-ph/0209114, accepted by Phys. Rev. D.
8. A.V. Efremov, K. Goeke, P. Schweitzer, hep-ph/0112166, Eur. Phys. J. C **in press**, (2002).
9. D. Hasch, Contribution to this Symposium.
10. H. Avakian, Contribution to this Symposium.

Single-Spin Asymmetries at CLAS

H.Avakian

Thomas Jefferson National Accelerator Facility, Newport News, Virginia 23606

Abstract. Significant single-spin azimuthal asymmetries have been observed in semi-inclusive pion production in the deep-inelastic scattering (DIS) of longitudinally polarized electrons off unpolarized hydrogen and polarized NH_3 targets at CLAS detector at Thomas Jefferson National Accelerator Facility (JLab). Issues related to the separation of current fragmentation, and the factorization of dependencies on x-Bjorken and the fractional energy z, at lowest beam energies where the DIS region is accessible, are also discussed.

Single-spin asymmetries (SSA) in hadronic reactions have been among the most difficult phenomena to understand from first principles in QCD. Large SSAs have been observed in hadronic reactions for decades [1, 2]. Recently, significant SSAs were reported in pion production in semi-inclusive DIS (SIDIS) by the HERMES collaboration at HERA [3, 4] for a longitudinally polarized target, by the SMC collaboration at CERN for a transversely polarized target [5], and by the CLAS collaboration at the Thomas Jefferson National Accelerator Facility (JLab)[6] with a polarized beam.

In general, such single-spin asymmetries require a correlation of a particle spin direction and the orientation of the production or scattering plane and they are related to the orbital angular momentum of partons in the nucleon. The interference of wavefunctions with different orbital angular momentum responsible for single-spin asymmetries [7, 8, 9, 10, 11, 12], also yields the helicity-flip Generalized Parton Distribution (GPD) E [13] entering Deeply Virtual Compton Scattering[14, 15] and the Pauli form factor F_2. The connection of SSAs and GPDs also has been discussed in terms of transverse distribution of quarks in the nucleon [16].

It is also argued that in both semi-inclusive [17] and hard exclusive [18, 19] pion production, scaling sets in for cross section ratios and spin asymmetries at lower squared four-momentum transfer, Q^2, than it does for the absolute cross section. This makes it possible for the measurement of spin-asymmetries to be a major tool for the study of parton distributions in the Q^2 domain of a few GeV^2.

The complete tree-level description of single pion electroproduction up to order $1/Q$ containing contributions from twist-2 and twist-3 distribution and fragmentation functions has been given in Ref. [20]. Terms contributing to the $\sin\phi$ moment are given by:

$$\sigma_{UT}^{\sin\phi} \propto S_T(1-y)sin(\phi+\phi_S)\sum_{a,\bar{a}} e_a^2 x h_1^a(x) H_1^{\perp a}(z) \quad (1)$$

$$\sigma_{UT}^{\sin\phi} \propto S_T(1-y)sin(\phi-\phi_S)\sum_{a,\bar{a}} e_a^2 x f_{1T}^{\perp}(x) D_1^a(z) \quad (2)$$

CP675, *Spin 2002: 15th Int'l. Spin Physics Symposium and Workshop on Polarized Electron Sources and Polarimeters*, edited by Y. I. Makdisi, A. U. Luccio, and W. W. MacKay

$$\sigma_{UL}^{\sin\phi} \propto S_L sin\phi\,(2-y)\sqrt{1-y}\frac{M}{Q}\sum_{a,\bar{a}} e_a^2 x^2 h_L^a(x)[H_1^{\perp a}(z)+\tilde{H}/2z] \quad (3)$$

$$\sigma_{LU}^{\sin\phi} \propto \lambda_e sin\phi\, y\sqrt{1-y}\frac{M}{Q}\sum_{a,\bar{a}} e_a^2 x^2 e^a(x) H_1^{\perp a}(z). \quad (4)$$

Here the first and second subscripts specify the beam and target polarizations with U,L,T denoting unpolarized, longitudinally polarized and transversely polarized cases, respectively. The azimuthal angle ϕ is defined by a triple product:

$$\sin\phi = \frac{[\vec{k}_1 \times \vec{k}_2]\cdot\vec{P}_\perp}{|\vec{k}_1 \times \vec{k}_2||\vec{P}_\perp|}$$

where $\vec{k}_1$ and $\vec{k}_2$ are the initial and final electron momenta, and $\vec{P}_\perp$ is the transverse momentum of the observed hadron with respect to the virtual photon $\vec{q}$. The λ_e is the helicity of the electron and S_L and S_T are proton longitudinal and transverse polarizations, $\sum_{a,\bar{a}} \rightarrow$ sum over quarks and anti-quarks. The x is the Bjorken variable, y and z are fractions of the incoming electron energy carried by the virtual photon and virtual photon energy carried by the final pion, respectively.

Eq. 1,2 correspond to leading twist contributions from Collins [21, 20, 23, 17] and Sivers [7, 8, 9, 11, 12] effects, and give access to chiral-odd transversity, $h_1(x)$[22], and *time-reversal odd* (T-odd) Sivers function, $f_{1T}^{\perp}$, respectively. Collins ($H_1^{\perp}$) and Sivers T-odd functions in Eq. 1-4 are first moments of corresponding transverse momentum dependent functions [20, 23]. Eq. 3,4[20, 24, 25] correspond to twist-3 contributions, and may be related to observed SSA at HERMES [3] with polarized target and at CLAS [6] with polarized beam.

Distribution and fragmentation functions responsible for non zero beam SSA in SIDIS (Eq.4) identified by Levelt and Mulders [25], include the twist-3 unpolarized distribution function $e(x)$ introduced by Jaffe and Ji [22] and the Collins function [21]. Contributions to beam SSA were also predicted to arise from the absorbtive part of the one-loop corrections to the $\gamma^* q \rightarrow qg$ and $\gamma^* g \rightarrow q\bar{q}$ [26].

Important issues at low beam energies are the separation of current and target fragmentation regions and the presence of factorization when the fragmentation functions depend only on the fractional energy, z. At low beam energies in DIS the current fragmentation region (CFR) is contaminated with events coming from the target fragmentation region (TFR). At low beam energies and low z a significant overlap is expected of current and target fragmentation regions [27]. A quantitative estimate is available from LUND Monte-Carlo (MC)[28]. The LUND model was successfully used by different experimental groups for different processes and in a wide energy range. A very good agreement of kinematic distributions measured at CLAS and LUND simulation was shown to take place at energies as low as 4.3 GeV. All this is making the LUND-MC a major tool in SIDIS studies. The relative yield of current fragmentation events extracted from LUND-MC for different beam energies (see. Fig. 1) shows no significant dependence of the CFR fraction on the beam energy in range $0.5 < z < 0.8$. In addition, we find that in that z-range kinematic distributions of final state electrons and pions at CLAS

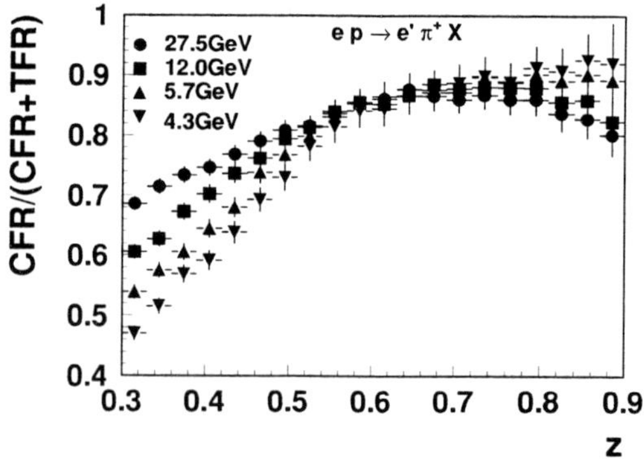

FIGURE 1. The separation of CFR with $0.5 < z < 0.8$ is not changing significantly with beam energy.

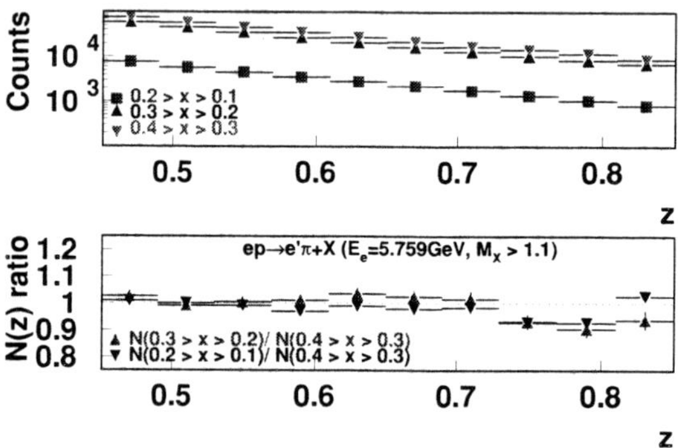

FIGURE 2. z-distributions of pions in $ep \to e'\pi^+ + X$ for different bins in x (top plot) and their ratios (bottom plot) for CLAS E1 data at 6 GeV.

are in good agreement with the LUND-MC even though the LUND MC was developed and tuned at much higher beam energies.

Factorization in the z-distributions of pions was studied using the 6 GeV data from the CLAS E1 experiment. Distributions of pions in z, for different x bins, are shown in Fig. 2. No significant dependence (<10%) within statistical uncertainties was observed in the z-distributions for different values of x (bottom plot in Fig.2).

SSA were measured in single pion production scattering of longitudinally polarized electrons off unpolarized liquid-hydrogen and polarized NH_3 targets. The total number of π^+ events in the DIS range ($Q^2 > 1$ GeV^2,$W^2 > 4$ GeV^2) selected by quality, vertex, acceptance, fiducial, and kinematic cuts was $\approx 4 \times 10^5$ and 2.7×10^6 respectively. The ϕ-dependent spin asymmetries are formed by extracting moments of the cross section for the two helicity states weighted by the corresponding ϕ-dependent functions. The $\sin\phi$ moment is thus given by:

$$A_{LU[UL]}^{\sin\phi} = \frac{2}{P^{\pm}N^{\pm}} \sum_{i=1}^{N^{\pm}} \sin\phi_i, \quad (5)$$

where $N^{\pm}$ and $P^{\pm}$ are the number of events and the luminosity weighted polarization for positive/negative helicities of the electron and proton, respectively. The azimuthal moment $A_{LU[UL]}^{\sin\phi}$ can be computed for each polarization state, and the comparison of two results provides a strong test of systematics. The $\sin\phi$ azimuthal moment of the cross section , unlike the $\cos\phi$ azimuthal moment, typically is not affected significantly by the detector acceptance and is practically an "acceptance free" observable. The sum of all contributions to the systematic uncertainties including the acceptance, beam polarization, particle mis-identification and radiative corrections is limited to 20% of the value of the measured asymmetry in all kinematic ranges (Fig. 3).

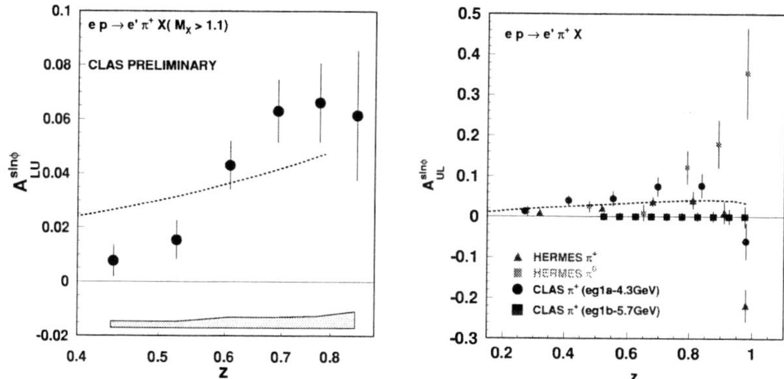

FIGURE 3. The left panel shows the beam-spin azimuthal asymmetry ($\sin\phi$ moment of the cross section) extracted from hydrogen data at 4.3GeV as a function of z in a range $0.15 < x < 0.4$. The band represents the systematic uncertainties. The right panel shows the comparison of target SSA from CLAS 4.3 GeV and HERMES data. The squares show the projected error bars expected from the CLAS 5.7GeV running with NH_3 target. The dashed line shows the z-shape of the fit to Collins function $H_1^{u,\pi^+}(z)$ from HERMES A_{LU}.

While the target SSA ($A_{UL}^{\sin\phi}$) analyzed in terms of the fragmentation effect (Eq. 3), in addition to a contribution from the Collins function, contains other contributions [29, 30], the beam SSA, depends only on the convolution of $e(x)$ and the Collins fragmentation function [20, 25], making it a very attractive observable for extraction of the z-shape of the Collins fragmentation function at large z, where the analyzing power is large.

The z-dependence of the beam SSA for positive pions measured at CLAS (Fig. 3) differs significantly from the shape of the z-dependence of the target SSA measured by HERMES [3, 4]. Preliminary results on beam SSA presented by the HERMES collaboration [31] and target SSA from CLAS (see Fig.3) confirm, that there is a difference between z-shapes of beam and target SSA, rather than a difference between measurements performed at different beam energies.

The x-dependence of beam SSA in SIDIS, analyzed in terms of the fragmentation effect, is defined by the ratio of the twist-3 unpolarized distribution function $e(x)$ and the leading twist distribution function $f_1(x)$. The first extraction of the twist-3 distribution function from CLAS beam SSA data, assuming factorization, was reported recently by Efremov et al.[32].

In conclusion, we have presented measurements of beam and target single-spin asymmetries at CLAS in single-pion electro-production off the unpolarized and longitudinally polarized protons. Beam and target SSA measured at CLAS and HERMES are in good agreement, suggesting there is no significant scale dependence for SSA observables. Higher statistics in the DIS range of ongoing experiments with CLAS at 6 GeV using unpolarized hydrogen and deuteron targets, as well as polarized NH_3 and ND_3 targets (3 billion triggers accumulated so far) will enable the extraction of the important Q^2 dependence at fixed x for different final state particles.

REFERENCES

1. K. Heller *et al.*, 'Proceedings of Spin 96',Amsterdam,Sep.1996,p23
2. Fermilab E704 collaboration (A. Bravar et al.), Phys.Rev.Lett. **77**, 2626 (1996).
3. HERMES Collaboration, A. Airapetyan *et al.*, Phys. Rev. Lett. **84**, 4047 (2000).
4. HERMES Collaboration, A. Airapetyan *et al.*, Phys. Rev. D **64**, 097101 (2001).
5. A. Bravar, Nucl. Phys. (Proc. Suppl.) **B 79**, 521 (1999).
6. CLAS Collaboration, H. Avakian *et al.*, in preparation.
7. D. Sivers, Phys.Rev. **D43**, 261 (1991).
8. M. Anselmino and F. Murgia, Phys. Lett. B **442**, 470 (1998).
9. S. Brodsky *et al.*, Phys. Lett. B **530**, 99(2002).
10. J. Collins, Phys. Lett. B **536**, 43 (2002).
11. X. Ji, F. Yuan e-Print Archive: hep-ph/0206057.
12. A. Belitsky,X. Ji and F. Yuan hep-ph/0208038.
13. S. Brodsky *et al.*, Nucl. Phys. **B642**, 344 (2002).
14. X. Ji, Phys. Rev. Lett. **78**, 610 (1997); Phys. Rev. D55, 7114 (1997).
15. A.V. Radyushkin, Phys. Lett. **B380**, 417 (1996); Phys. Rev. D **56**, 5524 (1997).
16. M. Burkardt, hep-ph/02091179.
17. A. Bacchetta *et al.* Phys.Rev. D **65**, 094021 (1999).
18. L.L. Frankfurt *et al.*, Phys.Rev. D **60**, 014010 (1999) Phys. Rev. Lett. **84**, 2589 (2000).
19. A. Belitsky and D. Muller, Phys. Lett., **B 513**, 349 (2001).
20. P.J. Mulders and R.D. Tangerman, Nucl. Phys.
21. J. Collins, Nucl. Phys. **B396**, 161 (1993).
22. R.L.Jaffe and X.Ji, Nucl. Phys. **B 375**, 527 (1992).
23. A. Kotzinian, Nucl. Phys. **B 441**, 234 (1995). **B 461**, 197 (1996).
24. A. M. Kotzinian *et al.*, Nucl.Phys. **A666**, 290-295 (2000).
25. J. Levelt and P. J. Mulders, Phys. Lett. B **338**, 357 (1994).
26. K. Hagiwara, K. Hikasa and N. Kai, Phys. Rev. D **27**, 84 (1983);
27. P.J. Mulders, hep-ph/0010199.
28. L. Mankiewicz, A. Schafer, and M. Veltri, Comput. Phys. Commun. **71**, 305 (1992).
29. P.J. Mulders and M. Boglione, Phys.Lett B **478**, 114; (2000).
30. H. Avakian, Proceedings of DIS-2000, Liverpool University 2000.
31. A. Miller, this proceedings .
32. A. Efremov *et al.*, hep-ph/0208124.

Azimuthal Asymmetries: Access to Novel Structure Functions

K. A. Oganessyan*†, L. S. Asilyan*, E. De Sanctis* and V. Muccifora*

*INFN-LNF, I-00044 Frascati, via Enrico Fermi 40, Italy
†DESY, Notkestrasse 85, 22603 Hamburg, Germany

Abstract. One of the most interesting consequence of non-zero intrinsic transverse momentum of partons in the nucleon is the nontrivial azimuthal dependence of the cross section of hard scattering processes. Many of the observable asymmetries contain unknown functions which provide essential information on the quark and gluon structure. Several of them have been studied in the last few years; we discuss their qualitative and quantitative features in semi-inclusive DIS.

INTRODUCTION

The study of the structure of hadrons, bound states of quarks and gluons, in the context of quantum chromodynamics (QCD), is one of the challenges of elementary particle physics requiring new nonperturbative approaches in the field theory.

This talk focuses on a discussion of correlations between the spins of hadron or quark and the momentum of the quark with respect of that of the hadron in polarized hard scattering processes. Signatures of these correlations appear in high-energy scattering processes as correlations between the azimuthal angles of the (transverse) spin and the (transverse) momentum vectors.

In particular, we discuss single-spin (transversely polarized target or longitudinally polarized beam) and double-spin (longitudinally polarized beam and transversely polarized target) azimuthal asymmetries in single hadron electroproduction in DIS.

The general form of the factorized cross sections of hard scattering process is written as [1]

$$d\sigma = H^0 \otimes f_2 \otimes f_2' + \frac{1}{Q^n} H^1 \otimes f_2 \otimes f_{2+n}' + \mathscr{O}\left(\frac{1}{Q^{n+1}}\right), \tag{1}$$

with $n = 1$ and 2 for the polarized and unpolarized case, respectively. The perturbatively calculable coefficient functions are denoted H^0, H^1 and are convoluted with the non-perturbative soft parts f_t, f_t', where t denotes the twist. The cross section is related to helicity-dependent amplitudes $\mathscr{A}_{\Lambda\lambda,\Lambda'\lambda'}$ [2], which describes a process where a target of helicity Λ emits a parton of helicity λ, and the scattered parton with helicity λ' is reabsorbed by a hadron of helicity Λ'. In the spin-1/2 case, due to helicity conservation, parity and time invariance, for each twist assignment there are three independent helicity amplitudes. The familiar distribution functions $f_1 \Leftrightarrow (\mathscr{A}_{++,++} + \mathscr{A}_{+-,+-})$,

CP675, *Spin 2002: 15th Int'l. Spin Physics Symposium and Workshop on Polarized Electron Sources and Polarimeters*, edited by Y. I. Makdisi, A. U. Luccio, and W. W. MacKay

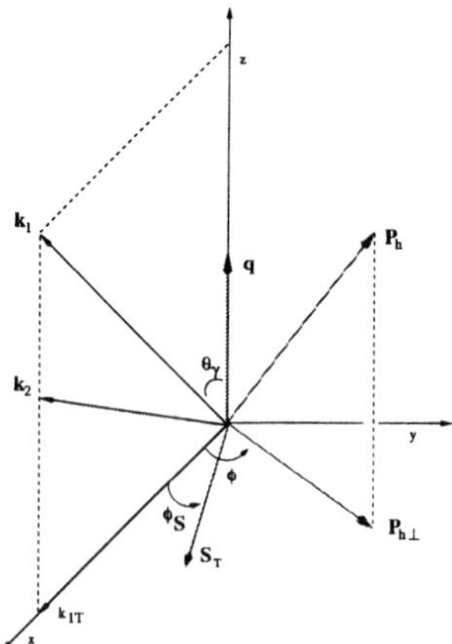

FIGURE 1. The kinematics of semi-inclusive DIS: k_1 (k_2) is the 4-momentum of the incoming (outgoing) charged lepton, $Q^2 = -q^2$, where $q = k_1 - k_2$, is the 4-momentum of the virtual photon. The momentum P (P_h) is the momentum of the target (observed) hadron. The scaling variables are $x = Q^2/2(P \cdot q)$, $y = (P \cdot q)/(P \cdot k_1)$, and $z = (P \cdot P_h)/(P \cdot q)$. The momentum k_{1T} ($P_{h\perp}$) is the incoming lepton (observed hadron) momentum component perpendicular to the virtual photon momentum direction, and ϕ is the azimuthal angle between $P_{h\perp}$ and k_{1T}, S_T is the target spin vector and ϕ_S is its azimuthal angle.

$g_1 \Leftrightarrow (\mathscr{A}_{++,++} - \mathscr{A}_{+-,+-})$, and $g_T \Leftrightarrow (\mathscr{A}_{++^*,+-})$[1] can be measured in inclusive DIS. The chiral-odd nature of $h_1 \Leftrightarrow (\mathscr{A}_{+-,-+})$, $h_L \Leftrightarrow (\mathscr{A}_{++^*,++} - \mathscr{A}_{+-,+-})$, and $e \Leftrightarrow (\mathscr{A}_{++^*,++} + \mathscr{A}_{+-,+-})$ makes their experimental determination difficult. Since electroweak and strong interactions conserve chirality, these functions cannot occur alone, but have to be accompanied by a second chiral odd quantity. For this reason, up to now no experimental information on these functions is available.

TRANSVERSE TARGET SINGLE-SPIN AZIMUTHAL ASYMMETRY

To access the transversity distribution function, h_1 (also commonly denoted δq), in semi-inclusive DIS off transversely polarized nucleons (see the relevant kinematics in Fig.1), one can measure the azimuthal angular dependences in the production of leading spin-0 or (on average) unpolarized hadrons from transversely polarized quarks with nonzero transverse momentum. This production is described by the intrinsic transverse momentum dependent fragmentation function $H_1^{\perp}(z)$, which is chiral-odd and also T-odd, i.e., non-vanishing only due to final state interactions [3]. Due to its chiral-odd structure it is a natural partner to isolate chiral-odd distribution functions, such as h_1, h_L, and e. It is worthy to note that a clean separation of current and target fragmentation

[1] The asterisk indicates bad component of quark/gluon field.

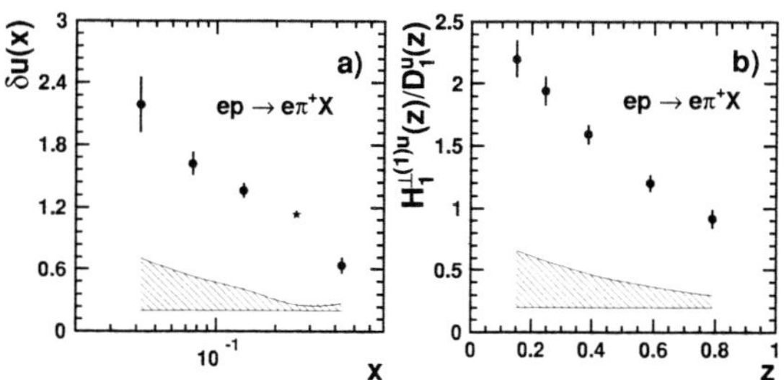

FIGURE 2. **a)** The transversity distribution $\delta u(x) \equiv h_1^u(x)$ and **b)** the ratio of the fragmentation functions $H_1^{\perp(1)u}(z)$ over $D_1(z)$ as they would be measured at HERMES [13, 14]. The hatched bands show projected systematic uncertainties.

effects in the data is required [4, 5].

The observable moment is defined as the appropriately weighted integral over ϕ of the cross section asymmetry [2]:

$$A_{UT}^{\sin(\phi+\phi_S)} \equiv \frac{\int d\phi \sin(\phi+\phi_S)\left[d\sigma^{\Uparrow} - d\sigma^{\Downarrow}\right]}{\frac{1}{2}\int d\phi \left[d\sigma^{\Uparrow} + d\sigma^{\Downarrow}\right]}, \tag{2}$$

where $\Uparrow$ ($\Downarrow$) denotes the up (down) transverse polarization of the target in the virtual-photon frame. This asymmetry is given by [3, 6, 7]:

$$A_{UT}^{\sin(\phi+\phi_S)} \propto \frac{h_1^u(x)}{f_1^u(x)} \cdot \frac{H_1^{\perp(1)u}(z)}{D_1^u(z)}. \tag{3}$$

An indication of a non-zero $H_1^{\perp(1)}(z)$ comes from the single spin asymmetry measured for pions produced in semi-inclusive DIS of leptons off a longitudinally polarized target at HERMES [8].

A larger asymmetry with a transversely polarized target is expected [9, 10]. However, the existence of the competing mechanism, which, due to asymmetric distribution of quarks transverse momenta in a hadron, also gives a transverse spin asymmetry at leading twist [11, 12], turns the transversity measurement challenging. For distinguishing the different mechanisms and for its complete description (x-dependence at large x, Q^2-evolution, etc.) results from different scattering processes and different kinematics are required. New data from HERMES and COMPASS measurements on transversely polarized targets will give possibility to measure the transversity [13]. Fig.2 show the expected accuracies for the reconstruction of $h_1^u(x)$ and $H_1^{\perp(1)u}(z)/D_1(z)$ at HERMES with transversely polarized proton target [13, 14]. In addition, analysis of data from $e^+e^- \to \pi^+\pi^-X$ process, which expected to show a similar azimuthal correlations is under way at the BELLE B-factory at KEK [15].

---

[2] The first and second subscripts indicate the polarizations of beam and target, respectively. We use U for unpolarized, L for longitudinally polarized and T for transversely polarized particles.

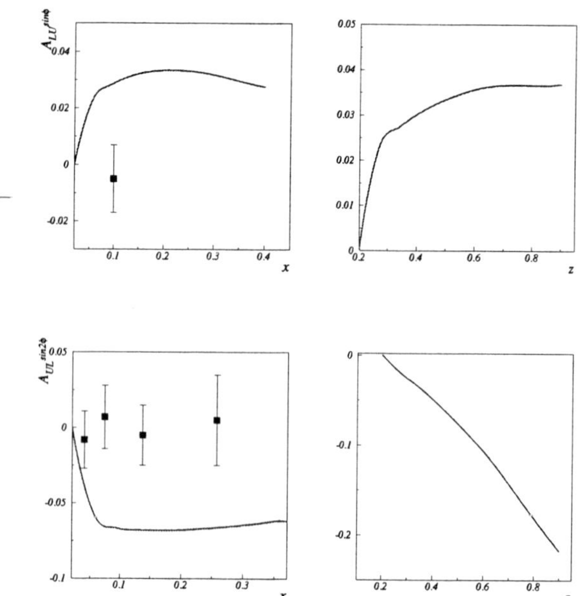

FIGURE 3. The single beam spin $A_{LU}^{sin\phi}$ and the single target spin $A_{UL}^{sin2\phi}$ asymmetries for π^+ production as a function of x and z. Data are from HERMES experiment [8]. Error bars show the statistical and the systematical uncertainties.

BEAM SINGLE-SPIN AZIMUTHAL ASYMMETRY

At order $1/Q$ a $\sin\phi$ asymmetry was predicted [16] for longitudinally polarized beam and unpolarized target. It probes the interaction-dependent distribution function $\tilde{e}(x)$ ($\tilde{e}(x) = e(x) - \frac{m}{M}\frac{f_1(x)}{x}$) [7, 16] in combination with the above mentioned fragmentation function $H_1^{\perp}$. It is important to notice that this asymmetry is related to the left-right asymmetry in the hadron momentum distribution with respect to the electron scattering plane,

$$A_{LU} = \frac{\int_0^{\pi} d\phi d\sigma - \int_{\pi}^{2\pi} d\phi d\sigma}{\int_0^{\pi} d\phi d\sigma + \int_{\pi}^{2\pi} d\phi d\sigma}, \tag{4}$$

which is $2/\pi$ times $A_{LU}^{\sin\phi}$,

$$A_{LU}^{\sin\phi} = \frac{\int d\phi \sin\phi [d\sigma^{\rightarrow} - d\sigma^{\leftarrow}]}{\frac{1}{2}\int d\phi [d\sigma^{\rightarrow} + d\sigma^{\leftarrow}]} \quad \propto \frac{\sum_a e_a^2 \tilde{e}^a(x) H_1^{\perp(1)a}(z)}{\sum_a e_a^2 f_1^a(x) D_1^a(z)}. \tag{5}$$

To evaluate this asymmetry as well as the single target-spin $\sin 2\phi$ asymmetry, $A_{UL}^{sin2\phi}$ [7, 17], we use the MIT bag model [18] as input.

For the weighted T-odd fragmentation function, $H_1^{\perp(1)}(z)$, we use the Collins ansatz [3] for the analyzing power with the factor $\eta = 1.6$ [9]. The x- and z-dependences

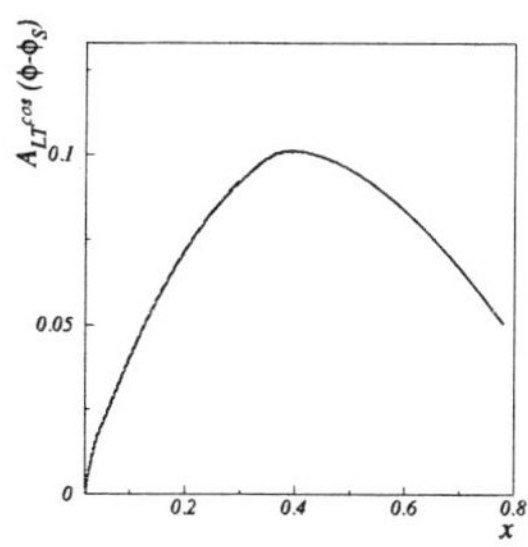

FIGURE 4. The double-spin asymmetry $A_{LT}^{cos(\phi-\phi_S)}$ for π^+ production as a function of x.

of the $A_{LU}^{sin\phi}$ and $A_{UL}^{sin2\phi}$ for π^+ are shown in Fig.3. The asymmetry $A_{LU}^{sin\phi}$ amounts to about 3%, while $A_{UL}^{sin2\phi}$ is larger (about 6%). This is in disagreement with the asymmetries measured by HERMES [8], which, as shown in the figure, are consistent with zero within the errors. This clearly indicate the necessity of a serious improvement of MIT bag model.

DOUBLE-SPIN AZIMUTHAL ASYMMETRY

The double spin azimuthal asymmetry is related to the $g_T(x)(g_2(x))$ distribution function.

Accounting for transverse momenta of the quarks, a longitudinal quark spin asymmetry exists in a transversely polarized nucleon target. The relevant leading twist distribution $g_{1T}(x, p_T^2)$ can be determined from the measurement of this asymmetry in semi-inclusive DIS [7, 19, 20] in the case of longitudinally polarized beam and transversely polarized target.

The observable $\cos(\phi - \phi_S)$ moment in the cross section is defined in the following way

$$A_{LT}^{\cos(\phi-\phi_S)} = \frac{\int d\phi \cos(\phi - \phi_S) \cdot [\sigma^{\leftarrow\Uparrow} + \sigma^{\rightarrow\Downarrow} - \sigma^{\leftarrow\Downarrow} - \sigma^{\rightarrow\Uparrow}]}{\int d\phi \cdot [\sigma^{\leftarrow\Uparrow} + \sigma^{\rightarrow\Downarrow} + \sigma^{\leftarrow\Downarrow} + \sigma^{\rightarrow\Uparrow}]} \tag{6}$$

$$A_{LT}^{\cos(\phi-\phi_S)} \propto \frac{\sum_a e_a^2 g_{1T}^{(1)}(x) z D_1(z)}{\sum_a e_a^2 f_1(x) D_1(z)}. \tag{7}$$

The x-dependence of $A_{LT}^{cos(\phi-\phi_S)}$ calculated for π^+ production is shown in Fig.4 [19, 20]: as seen it is a sizable asymmetry that may provide an alternative way to measure $g_2(x)$.

CONCLUSION

We have discussed the qualitative and quantitative features of nontrivial azimuthal dependence of the cross section of the semi-inclusive DIS which provided essential information on nucleon spin structure.

REFERENCES

1. J. Qiu and G. Sterman, Nucl. Phys. B **353** (1991) 137; Phys. Rev. Lett. 67 (1991) 2264.
2. R.L. Jaffe, hep-ph/9602236.
3. J.C. Collins, Nucl. Phys. B **396**, 161 (1993).
4. Ed.L. Berger, Preprint ANL-HEP-CP-87-45 (Argonne, IL, 1987).
5. P.J. Mulders, hep-ph/0010199.
6. A. Kotzinian, Nucl. Phys. B **441**, 234 (1995).
7. P.J. Mulders and R.D. Tangerman, Nucl. Phys. B **461** (1996) 197 and Nucl. Phys. B **484** (1997) 538.
8. HERMES Collaboration, A. Airapetian et al., Phys. Rev. Lett. **84**, 4047 (2000); Phys. Rev. D **64** (2001) 097101.
9. K.A. Oganessyan, et. al., Nucl. Phys. A **689** (2001) 784.
10. A. Bacchetta, et.al., Phys. Rev. D **65** (2002) 094021.
11. S.J. Brodsky, D.S. Hwang, I. Schmidt, Phys. Lett. B **530** (2002) 99.
12. J.C. Collins, Phys. Lett. B bf 536 (2002) 43.
13. V.A. Korotkov, W.-D. Nowak, K.A. Oganessyan, Eur. Phys. J. C **18** (2001) 639.
14. HERMES Collaboration, HERMES Internal Note 00-003.
15. K. Hasuko, et. al., for the Belle collaboration, these proceedings.
16. J. Levelt, P.J. Mulders, Phys. Lett. B **338** (1994) 357.
17. E. De Sanctis, W.-D. Nowak, K.A. Oganessyan, Phys. Lett. B **483** (2000) 69.
18. R.L. Jaffe, X. Ji, Nucl. Phys. B **375** (1992) 527.
19. A.M. Kotzinian, P.J. Mulders, Phys. Lett. B **406**, 373 (1997).
20. K.A. Oganessyan, P.J. Mulders, E. De Sanctis, Phys.Lett. B **532** (2002) 87.

Perturbative and Nonperturbative Aspects of Azimuthal Asymmetries in Polarized *ep*

K. A. Oganessyan

LNF-INFN, I-00040, Enrico Fermi 40, Frascati, Italy
DESY, Notkestrasse 85, 22603 Hamburg, Germany

Abstract. We discuss the possibilities of testing of perturbative quantum chromodynamics through azimuthal asymmetries in polarized *ep* in the context of future collider and fixed target facilities, such as Electron Ion Collider and TESLA-N.

It is widely recognized that the study of the distributions in the azimuthal angle ϕ of the detected hadron in hard scattering processes provide interesting variables to study in both non-perturbative and perturbative regimes. They are of great interest since they test perturbative quantum chromodynamics (QCD) predictions for the short-distance part of strong interactions and yield an important information on the long-distance internal structure of hadrons which computed in QCD by non-perturbative methods.

We discuss here the perturbative aspects of spin-independent, single- and double-spin azimuthal symmetries in leptoproduction processes. The non-perturbative aspects of some of them have been discussed in these proceedings (see Refs. [1-5]).

SPIN-INDEPENDENT $\cos\phi$, $\cos 2\phi$ ASYMMETRIES

Different mechanisms to generate spin-independent azimuthal asymmetries in semi-inclusive processes have been discussed in the literature. Georgi and Politzer [6] found a negative contribution to $\langle\cos\phi\rangle$ in the first order in α_S perturbative theory and proposed the measurement of this quantity as a clean test of QCD. However, as Cahn [7] showed, there is a contribution to $\langle\cos\phi\rangle$ from the lowest-order processes due to this intrinsic transverse momentum. The measurements indicate large negative $\cos\phi$ asymmetry [8, 9] and the simple QCD-improved parton model rather well describes the data, where essential contribution comes from non-perturbative intrinsic transverse momentum effects.

Recently, the expectation values of $\cos\phi$ and $\cos 2\phi$ in the deep inelastic electroproduction of single particles in the high Q^2 region have been observed by ZEUS detector at HERA [10] for the first time. For hadrons produced at large transverse momenta the results are in agreement with QCD predictions [11, 12]. Since the non-perturbative contributions to $\langle\cos 2\phi\rangle$ are of $1/Q^2$, its precise measurement may provide clear evidence for a pQCD and an alternative possibility to measure $F_L(x,Q^2)$ [13].

CP675, *Spin 2002: 15th Int'l. Spin Physics Symposium and Workshop on Polarized Electron Sources and Polarimeters,* edited by Y. I. Makdisi, A. U. Luccio, and W. W. MacKay

DOUBLE-SPIN $cos\phi$, $cos2\phi$ ASYMMETRIES

The $\cos\phi$ and $\cos 2\phi$ asymmetries arise also in semi-inclusive DIS of longitudinally polarized electrons off longitudinally polarized protons. Recently, the non-perturbative phenomena of these asymmetries has been considered [14] and a sizable negative $\cos\phi$ asymmetry for π^+ electroproduction was predicted. Perturbative contributions proportional to $\alpha_s(Q^2)\dots g_1\dots$ will likely appear at the same point where the twist-3 function $(M/Q)\dots g_L^\perp$ appears [14, 15, 16], like contributions proportional to $\alpha_s(Q^2)\dots f_1\dots$ appear at the same point where the function $(M/Q)\dots f_1^\perp$ appears [6, 15, 14]. Then the $\cos\phi$ dependence of the double longitudinal spin asymmetry,

$$A_{LL} = \frac{d\sigma^{++} + d\sigma^{--} - d\sigma^{+-} - d\sigma^{-+}}{d\sigma^{++} + d\sigma^{--} + d\sigma^{+-} + d\sigma^{-+}}, \tag{1}$$

for charged pion electroproduction have been examined [17]. The subscript LL denotes the longitudinal polarization of the beam and target, and the superscripts $++, --$ $(+-,-+)$ denote the helicity states of the beam and target respectively, corresponding to antiparallel (parallel) polarization.

In general, due to parity conservation in the electromagnetic and strong interactions, Eq.(1) can be written in terms of the spin-independent (σ) and double-spin ($\Delta\sigma$) cross sections of semi-inclusive DIS

$$A_{LL} = \frac{2\sum_{m=1}\Delta\sigma_{LL}^m \cos([m-1]\cdot\phi)}{4\sum_{m=1}\sigma_{UU}^m \cos([m-1]\cdot\phi)}, \tag{2}$$

Up to sub-leading order $1/Q$ the Eq.(2) can be rewritten as

$$A_{LL} = \frac{\Delta\sigma_{LL}^1/2\sigma_{UU}^1 + \langle\cos\phi\rangle_{LL}\cdot\cos\phi}{1 + 2\langle\cos\phi\rangle_{UU}\cdot\cos\phi}, \tag{3}$$

where $\langle\cos\phi\rangle_{UU}$ and $\langle\cos\phi\rangle_{LL}$ are unpolarized and double polarized $\cos\phi$ moments, respectively. The A_{LL} depends on and allows the simultaneous determination of the spin-independent and double-spin $\cos\phi$-moments of the cross section. A $\cos 2\phi$ asymmetry only appears at order $1/Q^2$, unless one allows for T-odd structure functions [2, 4].

BEAM/TARGET SINGLE-SPIN AZIMUTHAL ASYMMETRIES

The interest in the single beam asymmetry in the pion electroproduction in semi-inclusive deep inelastic scattering of longitudinally polarized electrons off unpolarized nucleon resides in the fact that they probe the antisymmetric part of the hadron tensor, which entirely due and particularly sensitive to final state interactions (FSI). In longitudinally polarized electron scattering it shows up as a $< \sin\phi >$ asymmetry for the produced hadron and expressed as

$$\langle\sin\phi\rangle = \pm\langle\frac{\vec{s}\times\vec{k}'\cdot\vec{P}_{h\perp}}{|\vec{s}\times\vec{k}'||\vec{P}_{h\perp}|}\rangle, \tag{4}$$

where $\vec{s}$ denoted the spin vector of the electron (the upper (lower) sign for right (left) handed electrons), $\vec{k}$ ($\vec{k}'$) and $\vec{P}_{h\perp}$ are three vectors of incoming (outgoing) electron and the produced hadrons transverse momentum about virtual photon direction. It is important that this asymmetry is related to the left-right asymmetry in the hadron momentum distribution with respect to the electron scattering plane,

$$A = \frac{\int_0^{\pi} d\phi d\sigma - \int_{\pi}^{2\pi} d\phi d\sigma}{\int_0^{\pi} d\phi d\sigma + \int_{\pi}^{2\pi} d\phi d\sigma}, \tag{5}$$

which is $4/\pi$ times $< \sin\phi >$.

Among various proposed tests to measure the gluon self-coupling, the tests based on observation of time-reversal-odd (T-odd) processes, which are highly sensitive to FSI, are particularly promising. In the one-loop order, which is the leading order for T-odd quantities in pQCD [18], the left-right asymmetry is quite sensitive to the gluon self-coupling [19], and even the qualitative observation of such effects is sufficient to establish its existence. Furthermore, they are fairly insensitive to the poorly known gluon distribution and fragmentation functions [19]. The magnitude of this asymmetry for kinematical conditions (assuming $\sqrt{s} = 300$GeV) at HERA collider have been evaluated using parton model expressions at leading order [20]. The T-odd single beam spin $\sin\phi$ asymmetry does hardly exceed 10^{-3} in neutral current interactions and 10^{-2} in charged current processes. Although the asymmetries appear small at first sight, a measurement may still be possible. An experimental determination of it is therefore a challenging task. In particular by including only hadrons above a minimal transverse momentum in the measurement, the asymmetries can be suppressed to $\mathcal{O}\ (\alpha_S^{(1)})$. It is worthy to not that in the electromagnetic scattering the difference in magnitude between the opposite helicity measurements and the asymmetry $< \sin 2\phi >$ is proportional the parity-violating effects.

Within the last few years, after appearance of experimental results from HERMES [21, 22] and SMC [23] collaborations on single target-spin azimuthal asymmetries in semi-inclusive pion electroproduction the subject obtained much more attention. It is considered as one of the ways to access and measure third, completely unknown distribution function, so-called, transversity. This twist-2 function is equally important for the description of quarks in nucleons as the more familiar function $g_1(x)$; their information is complementary (for details see Refs. [24, 25]).

In the perturbative regime the single target-spin asymmetries also should be different from zero in the one-loop order. They are remaining out of attention, while they may give a valuable information about the behavior of QCD interactions under discrete symmetry operations and may test our understanding of FSI at the perturbative level.

The future collider and fixed target facilities, such as Electron Ion Collider (EIC) [26] and TESLA-N [27] will be very important for deeper study of considered azimuthal asymmetries. In particular, the wide and tunable range of collision energies at EIC will allow to study the nonperturbative and perturbative effects of azimuthal asymmetries consistently. Polarization of electron and proton spins will allow to measure all discussed asymmetries as well as their different combinations simultaneously. Some of the asymmetries are expected to be small, thus the facilities with high luminosity, such as

EIC, TESLA-N are required.

From the theoretical site, a realistic estimates of these azimuthal asymmetries for the future collider and fixed target facility kinematics are needed.

I am grateful to Abhay Deshpande and Werner Vogelsang for many valuable discussions and the Physics Department at Brookhaven National Laboratory for hospitality.

REFERENCES

1. K.A. Oganessyan, these proceedings.
2. D. Boer, these proceedings.
3. Y. Koike, these proceedings.
4. L. Gamberg, these proceedings.
5. U. D'Alesio, these proceedings.
6. H. Georgi, H.D. Politzer, Phys. Rev. Lett. **40** (1978) 3.
7. R.N. Cahn, Phys. Lett. B **78** (1978) 269; Phys. Rev. D **40** (1989).
8. M. Arneodo, et al., (EMC Collaboration), Z. Phys. C **34** (1987) 277.
9. M. R. Adams, et al., (Fermilab E665 Collaboration), Phys.Rev. D**48** (1993) 5057.
10. ZEUS Collaboration, Phys.Lett. **B481** (2000) 199.
11. M. Ahmed, T. Gehrmann, Phys. Lett. B **465** (1999) 297.
12. K.A. Oganessyan, et.al, Eur.Phys.J. C **5** (1998) 681.
13. T. Gehrmann, hep-ph/0003156.
14. K.A. Oganessyan, P.J. Mulders, E. De Sanctis, Phys. Lett. **B532** (2002) 87.
15. P.J. Mulders and R.D. Tangerman, Nucl Phys. B **461** (1996) 197.
16. A. Brandenburg, D. Muller, K.A. Oganessyan, in preparation.
17. K.A. Oganessyan, L.S. Asilyan, M. Anselmino, E. De Sanctis, hep-ph/0208208.
18. A. De Rujula, R. Petronzio, B. Lautrup, Nucl. Phys. **B146** (1978) 50.
19. K. Hagiwara, K. Hikasa, N. Kai, Phys. Rev. **D27** (1983) 84.
20. M. Ahmed, T. Gehrmann, Phys. Lett. **B465** (1999) 297.
21. HERMES Collaboration, A. Airapetian, et al., Phys. Rev. Lett. **84** (2000) 4047.
22. HERMES Collaboration, A. Airapetian, et al., Phys. Rev. D **64** (2001) 097101.
23. A. Bravar, Nucl. Phys. Proc. Suppl. **79** (1999) 520.
24. V. Barone, A. Drago, P.G. Ratcliffe, Phys. Rep. **359** (2002) 1.
25. P.G. Ratcliffe, these proceedings.
26. http://www.phenix.bnl.gov/WWW/publish/abhay/Home_of_EIC/Whitepaper/Final/
27. M. Anselmino, et. al., hep-ph/0011299, DESY 00-160, TPR 00-20.

Single Transverse-Spin Asymmetry in $pp^{\uparrow} \to \pi X$ and $ep^{\uparrow} \to \pi X$

Yuji Koike

Department of Physics, Niigata University, Ikarashi, Niigata 950–2181, Japan

Abstract. Cross section formulas for the single spin asymmetry in $p^{\uparrow}p \to \pi(\ell_T)X$ and $ep^{\uparrow} \to \pi(\ell_T)X$ are derived and its characteristic features are discussed.

In this report we discuss the single transverse-spin asymmetry for the pion production with large transverse momentum A_N in $pp^{\uparrow} \to \pi(\ell)X$ and $p^{\uparrow}e \to \pi(\ell)X$ relevant for RHIC-SPIN, HERMES and COMPASS experiments. According to the QCD factorization theorem, the polarized cross section for $pp^{\uparrow} \to \pi X$ consists of three twist-3 contributions:

$$(A) \qquad G_a(x_1,x_2) \otimes q_b(x') \otimes D_c(z) \otimes \hat{\sigma}_{ab\to c},$$

$$(B) \qquad \delta q_a(x) \otimes E_b(x'_1,x'_2) \otimes D_c(z) \otimes \hat{\sigma}'_{ab\to c},$$

$$(C) \qquad \delta q_a(x) \otimes q_b(x') \otimes \widehat{E}_c(z_1,z_2) \otimes \hat{\sigma}''_{ab\to c},$$

where the functions $G_a(x_1,x_2)$, $E_b(x'_1,x'_2)$ and $\widehat{E}_c(z_1,z_2)$ are the twist-3 quantities representing, respectively, the transversely polarized distribution, the unpolarized distribution, and the fragmentation function for the pion. $\delta q_a(x)$ is the transversity distribution in $p^{\uparrow}$. a, b and c stand for the parton's species, sum over which is implied. δq_a, E_b and $\widehat{E}_c$ are chiral-odd. Corresponding to the above (A) and (C), the polarized cross section for $ep^{\uparrow} \to \pi X$ (final electron is not detected) receives two twist-3 contributions:

$$(A') \qquad G_a(x_1,x_2) \otimes D_a(z) \otimes \hat{\sigma}_{ea\to a},$$

$$(C') \qquad \delta q_a(x) \otimes \widehat{E}_a(z_1,z_2) \otimes \hat{\sigma}'_{ea\to a}.$$

The (A) and (B) contributions for $pp^{\uparrow} \to \pi X$ have been analyzed in [1] and [2], respectively, and it has been shown that (A) gives rise to large A_N at large x_F as observed in E704, and (B) is negligible in all kinematic region. Here we extend the study to the (C) term (see also [3]) at RHIC energy and also the asymmetry in ep collision.

CP675, *Spin 2002: 15th Int'l. Spin Physics Symposium and Workshop on Polarized Electron Sources and Polarimeters*, edited by Y. I. Makdisi, A. U. Luccio, and W. W. MacKay

The transversely polarized twist-3 distributions $G_F(x_1,x_2)$ relevant to the (A) term is given in [2]. Likewise the twist-3 fragmentation function for the pion (with momentum ℓ) is defined as thelightcone correlation function as ($w^2=0$, $\ell\cdot w=1$)

$$\frac{1}{N_c}\sum_X\int\frac{d\lambda}{2\pi}\int\frac{d\mu}{2\pi}e^{-i\frac{\lambda}{z_1}}e^{-i\mu(\frac{1}{z_2}-\frac{1}{z_1})}\langle 0|\psi_i(0)|\pi X\rangle\langle\pi X|gF^{\alpha\beta}(\mu w)w_\beta\bar{\psi}_j(\lambda w)|0\rangle$$

$$=\frac{M_N}{2z_2}\left(\gamma_5\not{\ell}\gamma_\nu\right)_{ij}\varepsilon^{\nu\alpha\beta\lambda}w_\beta\ell_\lambda\widehat{E}_F(z_1,z_2)+\cdots. \tag{1}$$

Note that we use the nucleon mass M_N to normalize the twist-3 pion fragmentation function. There is another twist-3 fragmentation function which is obtained from (1) by shifting the gluon-field strength from the left to the right of the cut. The defined function $\widehat{E}_{FR}(z_1,z_2)$ is connected to $\widehat{E}_F(z_1,z_2)$ by the relation $\widehat{E}_F(z_1,z_2)=\widehat{E}_{FR}(z_2,z_1)$, which follows from hermiticity and time reversal invariance. Unlike the twist-3 distributions, the twist-3 fragmentation function does not have definite symmetry property.

Following [1]-[3], we analyze the asymmetries focussing on the soft-gluon pole contributions with $G_F(x,x)$ and $\widehat{E}_F(z,z)$. In the large x_F region, i.e. production of pion in the forward direction of the polarized nucleon, the main contribution comes from the large-x and large-z region of distribution and fragmentation functions, respectively. Since G_F and $\widehat{E}_F$ behaves as $G_F(x,x)\sim(1-x)^\beta$ and $\widehat{E}_F(z,z)\sim(1-z)^{\beta'}$ with β, $\beta'>0$, $|(d/dx)G_F(x,x)|\gg|G_F(x,x)|$, $|(d/dz)\widehat{E}_F(z,z)|\gg|\widehat{E}_F(z,z)|$ at large x and z. In particular, the valence component of G_F and $\widehat{E}_F$ dominate in this region. We thus keep only the valence quark contribution for the derivative of these soft-gluon pole functions ("valence-quark soft-gluon approximation") for the pp collision. For the ep case, we include all the soft-gluon pole contribution.

In general A_N is a function of $S=(P+P')^2\simeq 2P\cdot P'$, $T=(P-\ell)^2\simeq -2P\cdot\ell$ and $U=(P'-\ell)^2\simeq -2P'\cdot\ell$ where P, P' and ℓ are the momenta of $p^\uparrow$, unpolarized p (or e), and the pion respectively. In the following we use S, $x_F=\frac{2\ell_\parallel}{\sqrt{S}}=\frac{T-U}{S}$ and $x_T=\frac{2\ell_T}{\sqrt{S}}$ as independent variables. The polarized cross section for the (C) term reads

$$E_\pi\frac{d^3\Delta\sigma(S_\perp)}{d\ell^3}=\frac{2\pi M_N\alpha_s^2}{S}\varepsilon^{\alpha\ell wS_\perp}\sum_a\int_{z_{min}}^1\frac{dz}{z^2}\int_{x_{min}}^1\frac{dx}{x}\frac{1}{xS+U/z}\int_0^1\frac{dx'}{x'}$$

$$\times\delta\left(x'+\frac{xT/z}{xS+U/z}\right)$$

$$\left\{\sum_b\delta q^a(x)q^b(x')\left[-z_1^2\frac{\partial}{\partial z_1}\widehat{E}_F^a(z_1,z)\right]_{z_1=z}\left(\frac{-2p_\alpha}{T}\widehat{\sigma}^I_{ab\to a}+\frac{-2p'_\alpha}{U}\widehat{\sigma}^{II}_{ab\to a}\right)\right.$$

$$+\sum_b\delta q^a(x)q^b(x')\left[-z^2\frac{d}{dz}\widehat{E}_F^a(z,z)\right]\frac{xp_\alpha+x'p'_\alpha}{|xT+x'U|}\left(\widehat{\sigma}^I_{ab\to a}+\widehat{\sigma}^{II}_{ab\to a}\right)$$

$$+\delta q^a(x)G(x')\left[-z_1^2\frac{\partial}{\partial z_1}\widehat{E}_F^a(z_1,z)\right]_{z_1=z}\left(\frac{-2p_\alpha}{T}\widehat{\sigma}^I_{ag\to a}+\frac{-2p'_\alpha}{U}\widehat{\sigma}^{II}_{ag\to a}\right)$$

$$+\delta q^a(x)G(x')\left[-z^2\frac{d}{dz}\widehat{E}_F^a(z,z)\right]\frac{xp_\alpha+x'p'_\alpha}{|xT+x'U|}\left(\widehat{\sigma}^I_{ag\to a}+\widehat{\sigma}^{II}_{ag\to a}\right)\Bigg\}, \tag{2}$$

where the lower limits for the integration variables are $z_{min}=-(T+U)/S=\sqrt{x_F^2+x_T^2}$ and $x_{min}=-U/z(S+T/z)$. Using the invariants in the parton level, $\hat{s}=(xp+x'p')^2=xx'S$, $\hat{t}=(xp-\ell/z)^2=xT/z$ and $\hat{u}=(x'p'-\ell/z)^2=x'U/z$, the partonic hard cross sections read

$$\widehat{\sigma}^I_{qq'\to q}=\frac{\hat{s}\hat{u}}{18\hat{t}^2}-\frac{\hat{s}}{54\hat{t}}\delta_{qq'},\quad \widehat{\sigma}^I_{q\bar{q}'\to q}=\frac{\hat{s}\hat{u}}{18\hat{t}^2},\quad \widehat{\sigma}^I_{\bar{q}q'\to\bar{q}}=-\frac{\hat{s}\hat{u}}{18\hat{t}^2},$$

$$\widehat{\sigma}^I_{\bar{q}\bar{q}'\to\bar{q}}=-\frac{\hat{s}\hat{u}}{18\hat{t}^2}+\frac{\hat{s}}{54\hat{t}}\delta_{qq'},\quad \widehat{\sigma}^{II}_{qq'\to q}=-\frac{7\hat{s}\hat{u}}{18\hat{t}^2}-\frac{\hat{s}}{54\hat{t}}\delta_{qq'},\quad \widehat{\sigma}^{II}_{q\bar{q}'\to q}=-\frac{\hat{s}\hat{u}}{9\hat{t}^2},$$

$$\widehat{\sigma}^{II}_{\bar{q}q'\to\bar{q}}=\frac{\hat{s}\hat{u}}{9\hat{t}^2},\quad \widehat{\sigma}^{II}_{\bar{q}\bar{q}'\to\bar{q}}=\frac{7\hat{s}\hat{u}}{18\hat{t}^2}+\frac{\hat{s}}{54\hat{t}}\delta_{qq'},$$

$$\widehat{\sigma}^I_{qg\to q}=\frac{\hat{s}\hat{u}}{8\hat{t}^2}+\frac{\hat{s}+\hat{u}}{16\hat{t}}+\frac{5}{72},\qquad \widehat{\sigma}^{II}_{qg\to q}=-\frac{9\hat{s}\hat{u}}{16\hat{t}^2}-\frac{9\hat{u}}{16\hat{t}}-\frac{1}{16}. \tag{3}$$

At large x_F ($-U\gg -T$), σ^I becomes more important because of $1/T$ factor in (2).

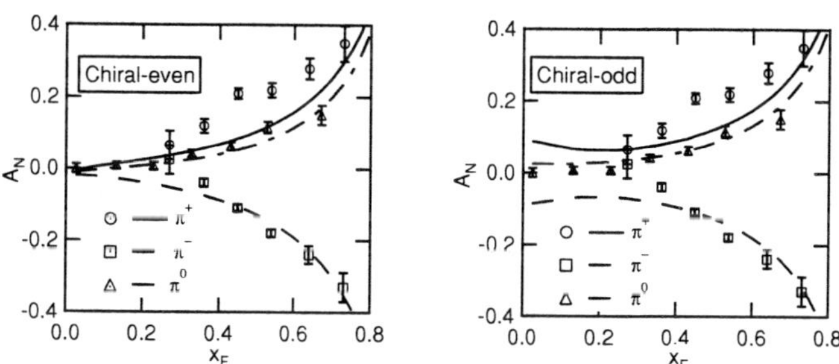

FIGURE 1. A_N^{pp} at $\sqrt{S}=20$ GeV and $\ell_T=1.5$ GeV.

To estimate the above contribution, we introduce a simple model ansatz as $\widehat{E}_F^a(z,z)=K_aD_a(z)$ with a flavor dependent factor K_a. K_a's are determined to be $K_u=-0.11$ and $K_d=-0.19$ so that (2) approximately gives rise to A_N^{pp} observed in E704 data at $\sqrt{S}=20$ GeV and $\ell_T=1.5$ GeV. As noted before, $\widehat{E}_F(z_1,z_2)$ does not have definite symmetry property unlike the twist-3 distribution $G_F(x_1,x_2)$. Nevertheless we assume $\left[(\partial/\partial z_1)\widehat{E}_F(z_1,z)\right]_{z_1=z}=(1/2)(d/dz)\widehat{E}_F(z,z)$. We refer the readers to [2] for the adopted distribution and fragmentation functions. The result for A_N^{pp} from the (C) (chiral-odd) term is shown in Fig. 1 in comparison with the (A) (chiral-even) contribution. (See

[2] for the detail.) Both effects give rise to A_N^{pp} similar to the E704 data. The origin of the growing A_N at large x_F is (i) large partonic cross sections in (3) ($\sim 1/\hat{t}^2$ term) and (ii) the derivative of the soft-gluon pole functions. With the parameters K_a fixed, A_N^{pp} at RHIC energy ($\sqrt{S} = 200$ GeV) is shown in Fig.2 at $l_T = 1.5$ GeV. Both (A) and (C) contributions give slightly smaller A_N^{pp} than the STAR data reported at this conference [4]. Fig.3 shows the A_N^{pp} as a function of ℓ_T at $\sqrt{S} = 200$ GeV and $x_F = 0.6$, indicating quite large l_T dependence in both (A) and (C) contributions at $1 < l_T < 4$ GeV region.

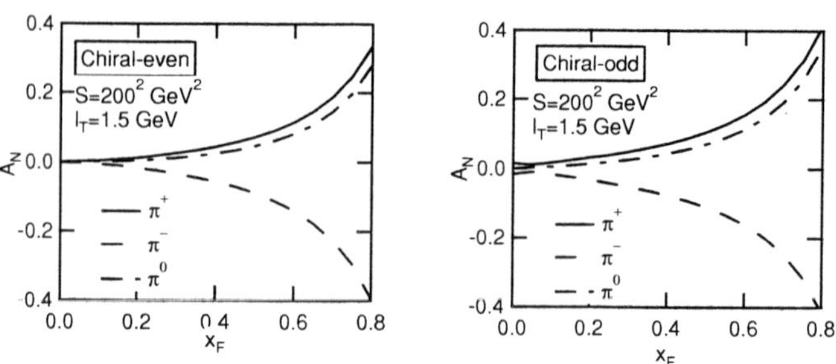

FIGURE 2. A_N^{pp} at $\sqrt{S} = 200$ GeV and $\ell_T = 1.5$ GeV.

We next discuss the asymmetry A_N^{ep} for $p^{\uparrow}e \to \pi(\ell)X$ where the final electron is not observed. In our $O(\alpha_s^0)$ calculation, the exchanged photon remains highly virtual as far as the observed π has a large trasverse momentum ℓ_T with respect to the ep axis. Therefore experimentally one needs to integrates only over those virtual photon events to compare with our formula.

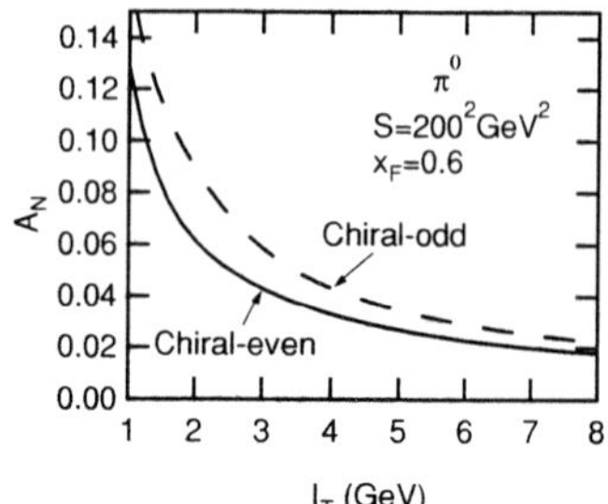

FIGURE 3. ℓ_T dependence of A_N^{pp} at $\sqrt{S} = 200$ GeV and $x_F = 0.6$.

Using the twist-3 distribution and fragmentation functions used to describe pp data, we show in Fig. 4 A_N^{ep} corresponding to (A')(chiral-even) and (C')(chiral-odd) contributions. Remarkable feature of Fig. 4 is that in both chiral-even and chiral-odd contributions (i) the sign of A_N^{ep} is opposite to the sign of A_N^{pp} and (ii) the magnitude of A_N^{ep} is much larger than that of A_N^{pp}, in particular, at large x_F, and it even overshoots one. (In our convention, $x_F > 0$ corresponds to the production of π in the forward hemisphere of the initial polarized proton both in $p^{\uparrow}p$ and $p^{\uparrow}e$ case.) The origin of these features can be traced back to the color factor in the dominant diagrams for the *twist-3 polarized* cross sections in ep and pp collisions. Of course, the asymmetry can not exceeds one, and thus our model estimate needs to be modified. First, the applied kinematic range of our formula should be reconsidered: Application of the twist-3 cross section at such small ℓ_T may not be justified. Second, our model ansatz of $G_F^a(x,x) \sim q^a(x)$ and $\widehat{E}_F^a(z,z) \sim D^a(z)$ is not appropriate at $x \to 1$ and $z \to 1$, respectively. The derivative of these functions, which is important for the growing A_N^{pp} at large x_F, eventually leads to divergence of A_N^{pp} at $x_F \to 1$ as $\sim 1/(1-x_F)$.

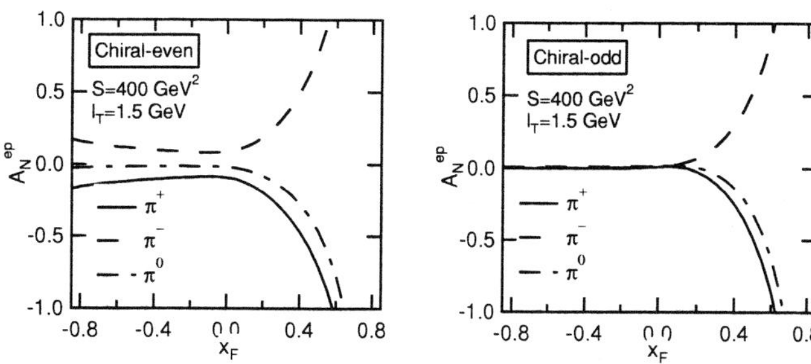

FIGURE 4. A_N^{ep} at $\sqrt{S} = 20$ GeV.

As a possible remedy to this pathology we tried the following. For the (A) (chiral-even) contribution we have a model $G_F^a(x,x) \sim q_a(x) \sim_{x\to 1} (1-x)^{\beta_a}$ where $\beta_u = 3.027$ and $\beta_d = 3.774$ in the GRV distribution we adopted. Tentatively we shifted $\beta_{u,d}$ as $\beta_a \to \beta_a(x) = \beta_a + x^3$, which suppresses the divergence of A_N at $x_F \to 1$ but still causes rising behavior of A_N at large x_F. This avoids overshooting of one in A_N^{ep} but reduces A_N^{pp}, typically by factor 2. So the twist-3 contribution to A_N^{pp} shown in Fig. 1 and 2 is further reduced, making the deviation from E704 and STAR data bigger.

To summarize we have studied the A_N for pion production in pp and ep collisions. Although our approach provides a systematic framework for the large ℓ_T production, applicability of the formula to the currently available low ℓ_T data still needs to be tested, in particular, comparison of ℓ_T dependence with data is needed.

ACKNOWLEDGMENTS

I would like to thank Daniel Boer, Jianwei Qiu and George Sterman for useful discussions. This work is supported in part by the Grant-in-Aid for Scientific Research of Monbusho.

REFERENCES

1. J. Qiu and G. Sterman, Phys. Rev. **D59** (1999) 014004.
2. Y. Kanazawa and Y. Koike, Phys. Lett. **B478** (2000) 121; **B490** (2000) 99.
3. Y. Koike, hep-ph/0106260 (Proceedings of DIS2001, Bologna, Italy, April, 2001.)
4. G. Rakness, these proceedings.

Azimuthal Asymmetries in Fragmentation Processes at KEKB

Kazumi Hasuko*†, Matthias Grosse Perdekamp*, Akio Ogawa***, Jens Soeren Lange*‡ and Viktor Siegle*

*RIKEN BNL Research Center, Upton, NY 11973-5000, USA
†RIKEN, Wako, Saitama 351-0198, Japan
**Brookhaven National Laboratory, Upton, NY 11973-5000, USA
‡University of Frankfurt, Frankfurt 60486, Germany

Abstract. In unpolarized electron-positron annihilation, there may exist interesting and possibly non-zero azimuthal asymmetries, which measure novel chiral-odd fragmentation functions, such as the Collins-Heppelmann function, $H_1^\perp$, and the two-pion interference fragmentation function, $\delta\hat{q}^h$. We will present the experimental method to extract these functions using e^+e^- collision data from the Belle experiment at KEK B-factory (KEKB). In addition to the considerable interest in the properties of these new fragmentation functions, they are expected to be a powerful tool in accessing proton quark transversity distributions.

INTRODUCTION

The study of spin effects in high-energy interactions provides sensitive tests for models of strong interaction dynamics. Recently much attention has been paid to transverse spin phenomena. In this paper we discuss future measurements of two of the relevant objects, the chiral-odd fragmentation functions describing the fragmentation of transversely polarized quarks into one or two charged pions. Chiral-odd fragmentation functions may arise from the non-perturbative dynamic in the fragmentation process and have been the subject of extensive theoretical discussion. However, experimentally these functions are presently unknown.

Collins *et al.* pointed out that there exists a non-trivial azimuthal angle asymmetry in the fragmentation of transversely polarized quarks into single pions [1]. In unpolarized electron-positron annihilations these fragmentation functions can be accessed through the observation of azimuthal asymmetries that are proportional to the product of chiral-odd fragmentation functions. In the following we discuss the experimental method employed to extract the Collins-Heppelmann and interference fragmentation functions from e^+e^- collisions at Belle.

While there is substantial interest in the symmetry properties of these new fragmentation functions, they are also a useful tool to access proton quark transversity distributions, δq, in semi inclusive deep inelastic lepton scattering (SIDIS) experiments (HERMES at DESY and COMPASS at CERN) and polarized proton-proton scattering experiments (STAR and PHENIX at BNL, RAMPEX at Protvino). At leading twist transversity distributions remain the last unknown quark distribution functions and their

CP675, *Spin 2002: 15th Int'l. Spin Physics Symposium and Workshop on Polarized Electron Sources and Polarimeters,* edited by Y. I. Makdisi, A. U. Luccio, and W. W. MacKay

knowledge is essential for a complete understanding of nucleon structure. Presently it is thought that transverse single spin asymmetries A_T in SIDIS and pp scattering offer the most practical way to measure transversity distributions. This possibility relies on the presence of the spin-dependent quark fragmentation functions (FF) which we intend to extract from Belle data. The experimental asymmetries A_T are proportional to $\sum_q \delta q \times a_i^f \times FF$, where a_i^f are the transversity dependent partonic initial-final-state asymmetries which can be calculated from pQCD.

The analyzing power in this process arises from the spin dependence of the partonic cross section as well as from the spin dependence of the fragmentation process: Collins suggested that the quark spin direction might be reflected in the azimuthal distribution of a final state pion [1]. Collins further demonstrated that the symmetry properties of the process do not require the proposed FF to be identical to zero.

Recent result from HERMES [2] and SMC [3] in fact seem to suggest that these FF and δq are different from 0.

SPIN-DEPENDENT FRAGMENTATION FUNCTIONS

We intend to study the following spin-dependent quark fragmentations in e^+e^- annihilation:

- The Collins-Heppelmann function $H_1^\perp$ describes the fragmentation of a transversely polarized quark into a charged pion and the azimuthal distribution of the final state pion with respect to the initial quark momentum (jet-axis).
- Interference fragmentation functions $\delta\hat{q}^{h_1,h_2}$ parameterize the fragmentation of transversely polarized quarks into pairs of hadrons including interference between different partial wave amplitudes; e.g. $\pi^+\pi^-$ pairs in the ρ-σ invariant mass region.

BELLE EXPERIMENT

The spin-dependent fragmentation functions discussed above can be extracted from the data taken by the Belle detector at KEKB. KEKB is an asymmetric storage ring that collides 8 GeV electrons against 3.5 GeV positrons [4]. The experimental data are recorded at the $\Upsilon(4S)$ resonance and in the continuum 60 MeV below the resonance, corresponding to integrated luminosities of more than $100 fb^{-1}$ on resonance and $\approx 10\%$ of this off-resonance.

The Belle detector is a general purpose, spectrometer based on a 1.5 T superconducting solenoid magnet. Charged particles are reconstructed with a three-layer double-sided silicon vertex detector (SVD) and a central drift chamber (CDC) that consists of 50-layer segmented into 6 axial and 5 stereo super-layers. The CDC covers the polar angle range between 17° and 150° in the laboratory frame, which corresponds to 92% of the full solid angle in the center of mass frame. Together with the SVD, a transverse momentum resolution of $(\sigma_{p_t}/p_t)^2 = (0.0019 p_t)^2 + (0.0030)^2$ is achieved, where p_t is in GeV/c.

Particle identification (PID) for charged hadrons is provided by a combination of three sub-system devices: a sub-system of 1188 aerogel Čerenkov counters (ACC)

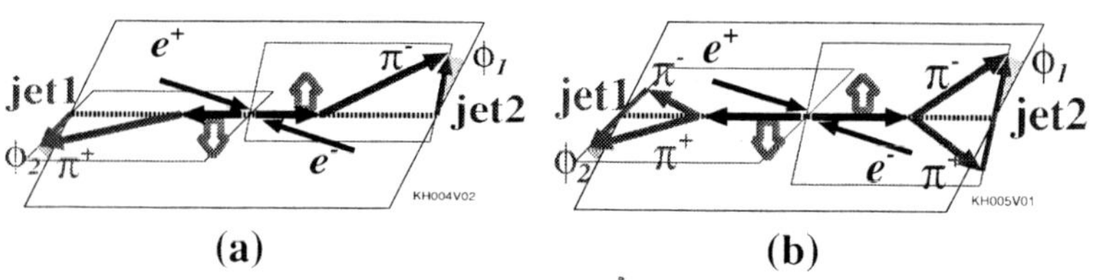

FIGURE 1. Kinematics of $e^+e^- \longrightarrow \pi^+_{jet1}\pi^-_{jet2}X$ (a) and $e^+e^- \longrightarrow (\pi^+\pi^-)_{jet1}(\pi^+\pi^-)_{jet2}X$ (b)

covering the momentum range 1-3.5 GeV/c, a time-of-flight scintillation counter subsystem (TOF) for track momenta below 1.5 GeV/c, and dE/dx information from the CDC for particles with very low or high momenta. Information from these three devices is combined to give the likelihood of a particle being a kaon, L_K, or pion, L_π. Kaon-pion separation is then accomplished based on the likelihood ratio $L_\pi/(L_\pi + L_K)$. The pion identification efficiencies are measured using a high momentum D^{*+} data samples, where $D^{*+} \longrightarrow D^0\pi^+$ and $D^0 \longrightarrow K^-\pi^+$. With this pion selection criterion, the typical efficiency for identifying pions in the momentum region 0.5 GeV/c $< p <$ 4 GeV/c is (88.5±0.1)%. By comparing the D^{*+} data sample with a Monte Carlo sample, the systematic error in the PID is estimated to be 1.4% for the mode with three charged tracks and 0.9% for the modes with two [5].

Surrounding the charged PID devices, a Cs(Tl) electromagnetic calorimeter and muon/K_L detector using iron plates interleaved with resistive plate counters are equipped. A detailed description of the Belle detector can be found elsewhere [6].

ANALYSIS METHOD AND SENSITIVITY

The Collins-Heppelmann function $H_1^\perp$ can be obtained from e^+e^- collisions using the recipe introduced in reference [7]: First identify a sample of two-jet events from light quark production. In each event, two unlike-sign pion tracks are selected, one in each event hemisphere. The fundamental observable is the angle between the two "pion-planes" formed by the pion momentum vectors and the jet axis. The pion planes include the angles ϕ_1 and ϕ_2 with the event plane defined by the beam axis and jet axis. The definition of the angles is shown in Figure 1-(a). The measured angular dependence in the angle $\phi_1 + \phi_2$ is then proportional to the product of the Collins-Heppelmann fragmentation function on each side taken at their respective fractional pion energy z_1 and z_2: $A(\phi_1 + \phi_2) = H_1^\perp(z_1)H_1^\perp(z_2)$.

We use a model by Jaffe et al. [8] in order to discuss some of the properties of the interference fragmentation function $\delta\hat{q}$· This fragmentation function describes quark fragmentation into two pions in a state which is a linear superposition of s-wave and p-wave states. These two partial waves are active in the ρ-region. The effect is an s-p interference and the fragmentation function peaks just above and below the ρ resonance and is changing it's sign across the ρ. This sign change should help to identify the fragmentation function and discriminate against possible systematic effects.

In extracting the interference fragmentation function we follow the recipe provided

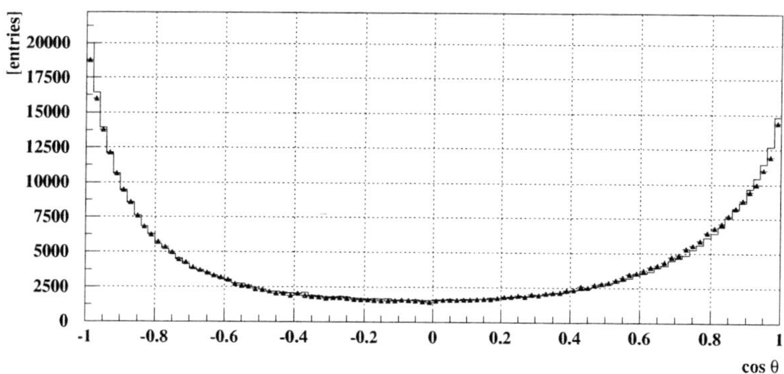

FIGURE 2. The distribution of the angle between two unlike-sign charged hadron tracks for data (triangle) and MC (histogram)

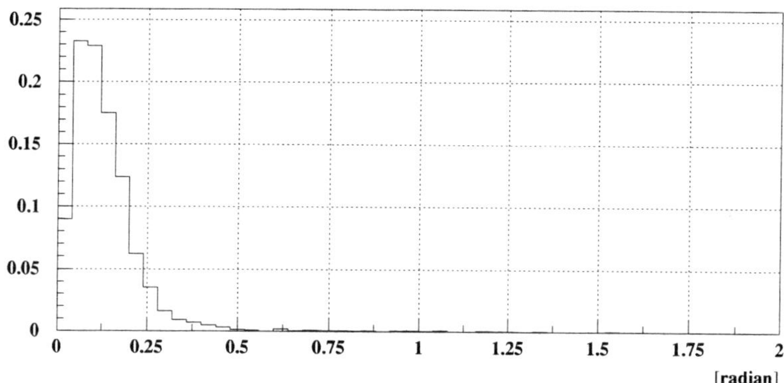

FIGURE 3. The distribution of the angle between thrust axis and q ($\bar{q}$) direction

by Atru and Collins [9]: In a sample of two-jet events, for each event $\pi^+\pi^-$ pairs are identified in each hemisphere of the event and the angle between the planes formed by the two pion pairs is measured. The distribution in this angle is the product of twice the interference fragmentation function evaluated at the invariant masses of the two pion pairs m_{12}, m_{34} and $z_{12,34}$, the longitudinal momentum fractions of the pairs. The kinematics is shown in Figure 1-(b). Schematically, the angular distribution is proportional to $f(z_{12}, m_{12}, Q^2) \times f(z_{34}, m_{34}, Q^2)$.

Figure 2 shows the distribution of the angle between two unlike-sign hadron tracks for Belle continuum data and Monte Carlo generated data [10]. The distribution clearly displays jet-like correlations with peaks for the near side and away side jets. The jet axis can be measured by using the thrust axis. The difference between the reconstructed thrust axis and the initial q ($\bar{q}$) direction in MC is shown in Figure 3. Three- or more-jet events are rejected by applying a thrust cut of $T > 0.8$.

The essential experimental requirements for fragmentation function measurements with Belle lie in the ability to identify and to precisely measure the momenta and charge sign of pions. The momentum resolution must be high enough to reconstruct the $\pi\pi$ mass to about 100MeV. This will permit a verification of the invariant mass dependence of the fragmentation function in the ρ mass region. These requirements are easily within the reach of the Belle detector [6, 11] which has been designed to meet the challenging demands of B-physics.

The high luminosities at KEKB and Belle's superior particle identification will allow fragmentation function measurements over a large range in $z = E_h/E_q$ and for different combinations of final state hadrons. The relevant fragmentation functions scale as $\log Q^2$. Subsequently analyzing powers at Belle are expected to be four times larger than for measurements using LEP data. Taking into account the larger luminosity at KEKB, the sensitivity $\Delta H_1^\perp / H_1^\perp$ will be higher by a factor 20 at Belle compared to LEP (this assumes that off-resonance and resonance data can be used).

CONCLUSION

The existence of spin-dependence in fragmentation processes would be interesting and a powerful tool for the study of transverse nucleon quark spin structure. A recent discussion of the prospects of future programs to access nucleon transversity using fragmentation function information from Belle can be found in reference [12].

The high luminosities at the KEK B-factory and the excellent momentum resolution and particle identification capabilities of the detector make Belle an ideal place for measurements of spin-dependent fragmentation functions.

REFERENCES

1. Collins, J.C., *Nucl. Phys.* **B396**, 161 (1993); Collins, J.C., Heppelmann, S.F., and Ladinsky, G., *Nucl. Phys.* **B420**, 563 (1994).
2. Airapetian, A. *et al.*, *Phys. Rev. Lett.* **84**, 4047 (2000).
3. Bravar, A., *Nucl. Phys. (Proc. Suppl.)* **B79**, 520 (1999).
4. KEK B Factory Design Report No. 95-7, 1995 (unpublished).
5. Iijima, T. *et al.*, *Nucl. Instrum. Meth.* **A379**, 457-459 (1996).
6. Belle Collaboration, Mori, S. *et al.*, *Nucl. Instrum. Meth.* **A479**, 117-232 (2002).
7. Boer, D., Jakob, R., and Mulders, P.J., *Phys. Lett.* **B424**, 143-151 (1998).
8. Jaffe, R.L., Jin, X., Tang, J. *et al.*, *Phys. Rev. Lett.* **80**, 1166 (1998);
9. Artru, X., and Collins, J., *Z. Phys.* **C69**, 277(1996).
10. These MC events are generated with the CLEO group's QQ program; see `http://www.lns.cornell.edu/public/CLEO/soft/QQ`; The detector response is simulated using GEANT, R. Brun *et al.*, GEANT 3.21, CERN Report DD/EE/84-1 (1984).
11. Belle Collaboration, Gordon, A. *et al.*, *Phys. Lett.* **B542**, 183-192 (2002).
12. Boer, D., "Transversity Single Spin Asymmetries" in *9th International Workshop on Deep Inelastic Scattering (DIS 2001)*, edited by G. Bruni *et al.*, World Scientific, 2001.

Collins Analyzing Power and Azimuthal Asymmetries

A. V. Efremov[a] [1], K. Goeke[b], and P. Schweitzer[c] [2]

[a] *Joint Institute for Nuclear Research, Dubna, 141980 Russia*
[b] *Institute for Theoretical Physics II, Ruhr University Bochum, Germany*
[c] *Dipartimento di Fisica Nucleare e Teorica, Università di Pavia, Italy*

Abstract. Spin azimuthal asymmetries in pion electro-production in deep inelastic scattering off longitudinally polarized protons, measured by HERMES, are well reproduced theoretically with no adjustable parameters. Predictions for azimuthal asymmetries for a longitudinally polarized deuteron target are given. Using this the z-dependence of the Collins fragmentation function is extracted for the first time. The first information on $e(x)$ is extracted from CLAS A_{LU} asymmetry.

INTRODUCTION

Recently azimuthal asymmetries have been observed in pion electro-production in semi inclusive deep-inelastic scattering off longitudinally (with respect to the beam) [1, 2] and transversely polarized protons [3]. These asymmetries contain information on the T-odd "Collins" fragmentation function $H_1^{\perp a}(z)$ and on the transversity distribution $h_1^a(x)$ [4][3]. $H_1^{\perp a}(z)$ describes the left-right asymmetry in fragmentation of transversely polarized quarks into a hadron [5, 6, 7] (the "Collins asymmetry"), and $h_1^a(x)$ describes the distribution of transversely polarized quarks in nucleon [4]. Both $H_1^{\perp a}(z)$ and $h_1^a(x)$ are twist-2, chirally odd, and not known experimentally. Only some years ago experimental indications to $H_1^{\perp}$ in e^+e^--annihilation have appeared [8], while the HERMES and SMC data [1, 2, 3] provide first experimental indications to $h_1^a(x)$.

Here we explain the observed azimuthal asymmetries [1, 2] and predict pion and kaon asymmetries from a deuteron target for HERMES by using information on $H_1^{\perp}$ from DELPHI [8] and the predictions for the transversity distribution $h_1^a(x)$ from the chiral quark-soliton model (χQSM) [9]. Our analysis is free of any adjustable parameters. Moreover, we use the model prediction for $h_1^a(x)$ to extract $H_1^{\perp}(z)$ from the z-dependence of HERMES data. For more details and complete references see Ref.[10, 11, 12]. Finally, using the new information on $H_1^{\perp}(z)$, we extract the twist-3 distribution $e^a(x)$ from very recent CLAS data [13].

---

[1] Supported by grants RFBR-00-02-16696, INTAS-00/587 and Heisenberg-Landau Programm.

[2] Supported by the contract HPRN-CT-2000-00130 of the European Commission.

[3] We use the notation of Ref.[5, 6] with $H_1^{\perp}(z)$ normalized to $\langle P_{h\perp}\rangle$ instead of M_h.

CP675, *Spin 2002: 15th Int'l. Spin Physics Symposium and Workshop on Polarized Electron Sources and Polarimeters,* edited by Y. I. Makdisi, A. U. Luccio, and W. W. MacKay

TRANSVERSITY AND COLLINS PFF

The χQSM is a quantum field-theoretical relativistic model with explicit quark and antiquark degrees of freedom. This allows an unambiguous identification of quark *and* antiquark distributions in the nucleon, which satisfy all general QCD requirements due to the field-theoretical nature of the model [14]. The results of the parameter-free calculations for unpolarized and helicity distributions agree within $(10-20)\%$ with parameterizations, suggesting a similar reliability of the model prediction for $h_1^a(x)$ [9].

$H_1^\perp$ is responsible in e^+e^- annihilation for a specific azimuthal asymmetry of a hadron in a jet around the axis in direction of the second hadron in the opposite jet [5]. This asymmetry was probed using the DELPHI data collection [8]. For the leading particles in each jet of two-jet events, averaged over quark flavors, the most reliable value of the analyzing power is given by $(6.3\pm2.0)\%$. However, the larger "optimistic" value is not excluded

$$\left|\frac{\langle H_1^\perp\rangle}{\langle D_1\rangle}\right| = (12.5\pm1.4)\% \tag{1}$$

with unestimated but presumably large systematic errors.

THE AZIMUTHAL ASYMMETRY

In [1, 2] the cross section for $l\vec{p}\to l'\pi X$ was measured in dependence of the azimuthal angle ϕ, i.e. the angle between lepton scattering plane and the plane defined by momentum of virtual photon $\mathbf{q}$ and momentum $\mathbf{P}_h$ of produced pion. The twist-2 and twist-3 azimuthal asymmetries read [6][4]

$$A_{UL}^{\sin2\phi}(x) \propto \sum_a e_a^2 h_{1L}^{\perp(1)a}(x)\langle H_1^{\perp a/\pi}\rangle \Big/ \sum_a e_a^2 f_1^a(x)\langle D_1^{a/\pi}\rangle\,, \tag{2}$$

$$A_{UL(1)}^{\sin\phi}(x) \propto \frac{M}{Q}\sum_a e_a^2 x h_L^a(x)\langle H_1^{\perp a/\pi}\rangle \Big/ \sum_a e_a^2 f_1^a(x)\langle D_1^{a/\pi}\rangle\,, \tag{3}$$

$$A_{UL(2)}^{\sin\phi}(x) \propto -\sin\theta_\gamma\cdot\sum_a e_a^2 h_1^a(x)\langle H_1^{\perp a/\pi}\rangle \Big/ \sum_a e_a^2 f_1^a(x)\langle D_1^{a/\pi}\rangle\,, \tag{4}$$

with $\sin\theta_\gamma\approx 2x\sqrt{1-y}(M/Q)$ and $A_{UL}^{\sin\phi}=A_{UL(1)}^{\sin\phi}+A_{UL(2)}^{\sin\phi}$. In Eqs.(2-4) the pure twist-3 $\tilde{h}_L$ terms are neglected. The results of Ref.[16] justify to use this WW-type approximation in which $xh_L=-2h_{1L}^{\perp(1)}=2x^2\int_x^1 d\xi\, h_1(\xi)/\xi^2$.

We assume isospin symmetry and favoured fragmentation for D_1^a and $H_1^{\perp a}$, i.e. $D_1^\pi\equiv D_1^{u/\pi^+}=D_1^{d/\pi^-}=2D_1^{\bar{u}/\pi^0}$ etc. and $D_1^{\bar{u}/\pi^+}=D_1^{u/\pi^-}\simeq 0$ etc.

[4] Note a sign-misprint in Eq.(115) of [6] for the $\sin\phi$-term Eq.(3). It was corrected in Eq.(2) of [15]. The conventions in Eqs.(2–4) agree with [1, 2]: Target polarization opposite to beam is positive, and z axis is parallel to $\mathbf{q}$ (in [6] it is anti-parallel).

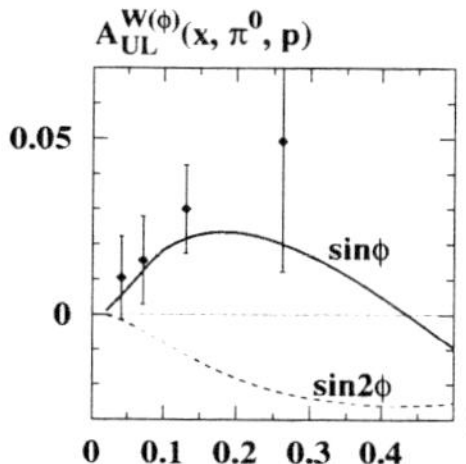

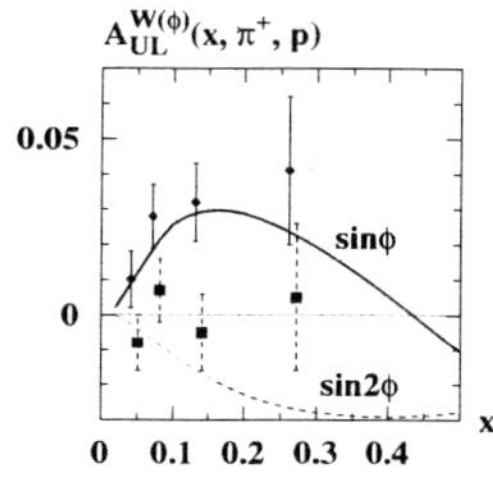

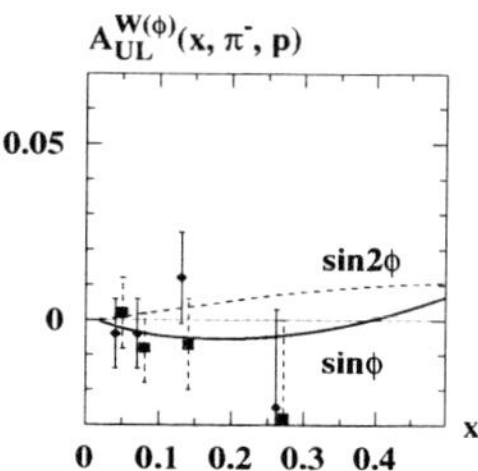

FIGURE 1. Azimuthal asymmetries $A_{UL}^{W(\phi)}$ weighted by $W(\phi) = \sin\phi$, $\sin 2\phi$ for pions as function of x. Rhombus (squares) denote data for $A_{UL}^{\sin\phi}$ ($A_{UL}^{\sin 2\phi}$).

HERMES ASYMMETRIES

When using Eq.(1) to explain HERMES data, we assume a weak scale dependence of the analyzing power. We take $h_1^a(x)$ from the χQSM [9] and $f_1^a(x)$ from Ref.[17], both LO-evolved to the average scale $Q_{\rm av}^2 = 4\,{\rm GeV}^2$.

In Fig.1 HERMES data for $A_{UL}^{\sin\phi}(x)$, $A_{UL}^{\sin 2\phi}(x)$ [1, 2] are compared with the results of our analysis. We conclude that the azimuthal asymmetries obtained with $h_1^a(x)$ from the χQSM [9] combined with the "optimistic" DELPHI result Eq.(1) for the analyzing power are consistent with data.

We exploit the z-dependence of HERMES data for π^0, π^+ azimuthal asymmetries to extract $H_1^\perp(z)/D_1(z)$. For that we use the χQSM prediction for $h_1^a(x)$, which introduces a model dependence of order $(10-20)\%$. The result is shown in Fig.2a. The data can be described by a linear fit $H_1^\perp(z) = (0.33 \pm 0.06) z D_1(z)$. The average $\langle H_1^\perp\rangle/\langle D_1\rangle = (13.8 \pm 2.8)\%$ is in good agreement with DELPHI result Eq.(1)[5]. The errors are the statistical errors of the HERMES data. Numerically it is close to those calculated from chiral perturbation theory [18].

The approach can be applied to predict azimuthal asymmetries in pion and kaon production off *a longitudinally polarized deuterium* target, which are under current study at HERMES. The additional assumption used is that $\langle H_1^{\perp K}\rangle/\langle D_1^K\rangle \simeq \langle H_1^{\perp\pi}\rangle/\langle D_1^\pi\rangle$. The predictions are shown in Fig.2b. The "data points" estimate the expected error bars. Asymmetries for $\bar{K}^0$ and K^- are close to zero in our approach.

Interestingly all $\sin\phi$ asymmetries change sign at $x \sim 0.5$ (unfortunately the HERMES cut is $x < 0.4$). This is due to the negative sign in Eq.(4) and the harder behaviour of $h_1(x)$ with respect to $h_L(x)$. This prediction however is sensitive to the favoured fragmentation approximation.

We learn that transversity could be measured also with a *longitudinally* polarized target, e.g. at COMPASS, simultaneously with ΔG.

---

[5] SMC data [3] yield an opposite sign, $\frac{\langle H_1^\perp\rangle}{\langle D_1\rangle} = -(10 \pm 5)\%$, however, seem less reliable.

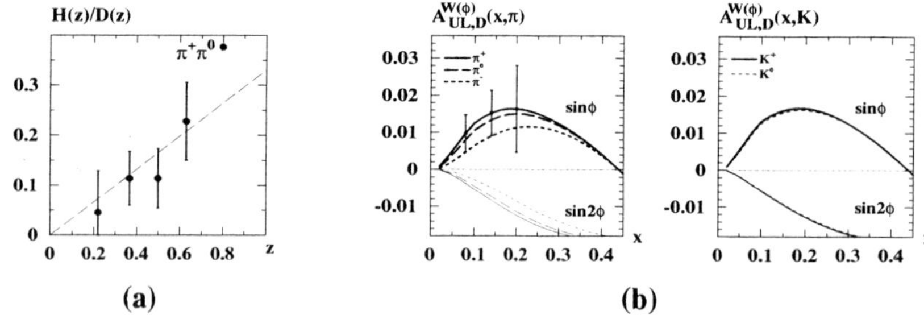

FIGURE 2. **(a).** $H_1^{\perp}/D_1$ vs. z, as extracted from HERMES data for π^+ and π^0 production. **(b).** Predictions for $A_{UL}^{\sin\phi}$, $A_{UL}^{\sin 2\phi}$ from a deuteron target for HERMES. Asymmetries for $\bar{K}^0$, K^- are close to zero in our approach.

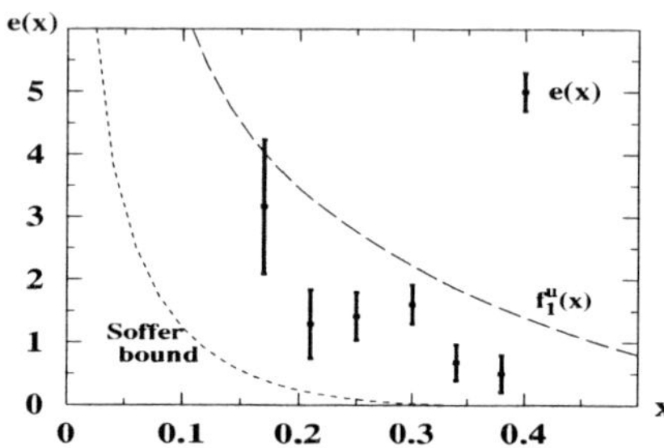

FIGURE 3. The flavour combination $e(x) = (e^u + \frac{1}{4}e^{\bar{d}})(x)$, with errorbars due to statistical error of CLAS data, vs. x at $\langle Q^2\rangle = 1.5\,\mathrm{GeV}^2$. For comparison $f_1^u(x)$ and the twist-3 Soffer bound are shown.

EXTRACTION OF E(X) FROM CLAS DATA

Very recently the $\sin\phi$ asymmetry of π^+ produced by scattering of polarized electrons off unpolarised protons was reported by CLAS collaboration [13]. This asymmetry is interesting since it allows to access the unknown twist-3 structure functions $e^a(x)$ which are connected with nucleon σ-term:

$$\int_0^1 \mathrm{d}x \sum_a e^a(x) = \frac{2\sigma}{m_u + m_d} \approx 10\,. \tag{5}$$

The asymmetry is given by [6]

$$A_{LU}^{\sin\phi}(x) \propto \frac{M}{Q}\,\frac{\sum_a e_a^2 e^a(x)\langle H_1^{\perp a/\pi}\rangle}{\sum_a e_a^2 f_1^a(x)\langle D_1^{a/\pi}\rangle}\,. \tag{6}$$

Disregarding unfavored fragmentation and using the Collins analysing power extracted from HERMES in Sect., which yields for z-cuts of CLAS $\langle H_1^{\perp\pi}\rangle/\langle D_1^{\pi}\rangle =$

0.20 ± 0.04, we can extract $e^u(x) + \frac{1}{4}e^{\bar{d}}(x)$. The result is presented in Fig.3. For comparison the Soffer lower bound [19] from twist-3 density matrix positivity, $e^a(x) \geq 2|g_T^a(x)| - h_L^a(x)$,[6] and the unpolarized distribution function $f_1^u(x)$ are plotted. The obtained points for $e(x)$ are close to calculations as in bag model [21] as in χQSM. One can guess that the large number in the sum rule Eq.(5) might be due to, either a strong rise of $e(x)$ in the small x region, or a δ-function at $x = 0$ [22].

REFERENCES

1. Airapetian A. *et. al., Phys. Rev. Lett.* **84**, 4047 (2000).
2. Airapetian A. *et al., Phys. Rev.* **D64**, 097101 (2001).
3. Bravar A., *Nucl. Phys. (Proc. Suppl.)* **B79**, 521 (1999).
4. Ralston J. and Soper D.E.,*Nucl.Phys.* **B152**, 109 (1979).
   Jaffe R.L. and Ji X., *Phys. Rev. Lett.* **67**, 552 (1991); *Nucl. Phys.* **B375**, 527 (1992). Cortes J.L., Pire B. and Ralston J.P., *Z. Phys.* **C55**, 409 (1992).
5. Boer D., Jakob R. and Mulders P.J., *Phys. Lett.* **B424**, 143 (1998).
6. Mulders P. and Tangerman R., *Nucl. Phys.* **B461**, 197 (1996). [Erratum: ibid. **B484**, 538 (1996)].
7. Collins J., *Nucl. Phys.* **B396**, 161 (1993).
   Artru X. and Collins J.C., *Z. Phys.* **C69** 277 (1996).
8. Efremov A.V., Smirnova O.G. and Tkatchev L.G., *Nucl. Phys. (Proc. Suppl.)* **74**, 49 (1999) and **79**, 554 (1999).
9. Pobylitsa P.V. and Polyakov M.V., *Phys. Lett.* **B389**, 350 (1996).
   P. Schweitzer *et al., Phys. Rev.* **D64**, 034013 (2001).
10. Efremov A.V. *et al., Phys. Lett.* **B478**, 94 (2000).
11. Efremov A.V. *et al., Phys. Lett.* **B522**, 37 (2001) [Erratum-ibid. **B544**, 389 (2001)]
12. Efremov A.V. *et al., Eur. Phys. J.* **C24**, 407 (2002), hep-ph/0112166.
13. Avakian H. [CLAS-Coll.], Talk at "BARYONS-2002".
14. Diakonov D.I. *et al., Nucl. Phys.* **B480**, 341 (1996).
15. Boglione M. and Mulders P., *Phys. Lett.* **B478**, 114 (2000).
16. Dressler B. and Polyakov M.V., *Phys. Rev.* **D61**, 097501 (2000).
17. Glück M., Reya E. and Vogt A., *Z. Phys.* **C 67**, 433 (1995).
18. Bacchetta A., Kundu R., Metz A. and Mulders P.J., hep-ph/0206309.
19. Soffer J., *Phys. Rev. Lett.* **74**, 1292 (1995).
20. Balla J., Polyakov M.V. and Weiss C., *Nucl. Phys.* **B510**, 327 (1998).
21. Signal A.I., *Nucl. Phys.* **B497**, 415 (1997), hep-ph/9610480.
22. Burkardt M. and Koike Y., *Nucl. Phys.* **B632**, 311 (2002).

[6] For $g_T^a(x)$ we use the Wandzura-Wilczek approximation $g_T^a(x) = \int_x^1 d\xi \, g_1^a(\xi)/\xi$ and neglect consistently $\widetilde{g}_2^a(x)$ which is strongly suppressed in the instanton vacuum [20]. For $h_L^a(x)$ we use the analogous approximation, as described in Sect.III.

Transverse Spin Effects in Proton-proton Scattering and $Q\bar{Q}$ Production

S.V. Goloskokov

Bogoliubov Laboratory of Theoretical Physics, Joint Institute for Nuclear Research, Dubna 141980, Moscow region, Russia

Abstract. We discuss transverse spin effects caused by the spin-flip part of the Pomeron coupling with the proton. The predicted spin asymmetries in proton-proton scattering and QQ production in proton-proton and lepton-proton reactions are not small and can be studied in future polarized experiments.

In this report, we discuss what future facilities can be used to ascertain the existence of the spin spin-flip part of the Pomeron coupling. The spin structure of the Pomeron was analysed by different authors (see [1, 2] and reference therein). Its manifestation can be investigated in the elastic pp scattering at low t [3]. We shall analyze here the single spin asymmetries in the elastic pp scattering near the diffraction minimum and of the $Q\bar{Q}$ production in the pp reaction. The double spin asymmetries of $Q\bar{Q}$ production in lepton-proton reactions for a longitudinally polarized lepton and a transversely polarized proton will be studied too. It will be shown that these asymmetries are sensitive to the spin-flip part of the Pomeron coupling and predicted asymmetries are not small, about 10%. They can be used to obtain information about the spin structure of the Pomeron coupling. To study spin effects in diffractive reactions, we use the two- gluon exchange model , which is directly connected with the Pomeron. On the other hand, diffractive processes can be expressed in terms of the generalized or skewed parton distribution (GPD) in the nucleon $F_\zeta(x)$, $K_\zeta(x)$ [4]. The connection of the two-gluon model with GPD will be shown.

The two-gluon coupling with the proton can be parametrized in the form [5]

$$\begin{aligned} V_{pgg}^{\alpha\beta}(p,t,x_P,l_\perp) &= B(t,x_P,l_\perp)(\gamma^\alpha p^\beta + \gamma^\beta p^\alpha) \\ &+ \frac{iK(t,x_P,l_\perp)}{2m}(p^\alpha \sigma^{\beta\gamma} r_\gamma + p^\beta \sigma^{\alpha\gamma} r_\gamma) + ... \end{aligned} \tag{1}$$

Here m is the proton mass. In the matrix structure (1) we wrote only the terms with the maximal powers of a large proton momentum p which are symmetric in the gluon indices α,β. The structure proportional to $B(t,...)$ determines the spin-non-flip contribution. The term $\propto K(t,...)$ leads to the transverse spin-flip at the vertex.

The spin-dependent cross section of diffractive processes are expressed in terms of the soft gluon coupling (1), which in the case of hadron production is convoluted with the hard hadron production amplitude. The spin asymmetries are expressed in terms of the functions $B(t,...)$ and $K(t,...)$ integrated over the gluon transverse momentum $l_\perp$.

CP675, *Spin 2002: 15th Int'l. Spin Physics Symposium and Workshop on Polarized Electron Sources and Polarimeters*, edited by Y. I. Makdisi, A. U. Luccio, and W. W. MacKay

The helicity-non-flip and helicity-flip amplitudes of the polarized proton off the spinless particle (a meson or unpolarized proton) can be written in terms of the invariant functions $\tilde{B}$ and $\tilde{K}$

$$F_{++}(s,t) = is[\tilde{B}(t)]f(t); \;\; F_{+-}(s,t) = is\frac{\sqrt{|t|}}{m}\tilde{K}(t)f(t), \tag{2}$$

where f(t) is determined by the Pomeron coupling with the other hadron. The functions $\tilde{B}$ and $\tilde{K}$ are defined by the integrated over $l_\perp$ structures from (1).

There are some models that provide spin-flip effects which do not vanish at high energies. In the model [1], the amplitudes K and B have a phase shift caused by the soft Pomeron rescattering effect. The vector diquarks in the diquark model [2] generate the K amplitude which is out of phase with the Pomeron contribution to the amplitude B. The value $|\tilde{K}|/|\tilde{B}| \sim 0.1$ found in [1, 2] will be used in our estimations of the asymmetry in diffractive hadron production.

The single spin asymmetry is determined by

$$A_N \sim 2\frac{\sqrt{|t|}}{m}\frac{\mathrm{Im}(BK^*)}{|B|^2}. \tag{3}$$

The models [1, 2] describe the experimental data on single spin transverse asymmetry A_N [6] quite well. Thus, the weak energy dependence of spin asymmetries in exclusive reactions is now not in contradiction with the experiment [1, 7].

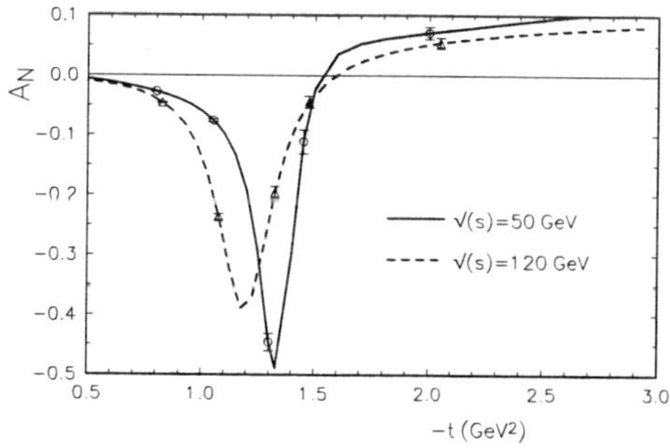

FIGURE 1. Predictions of the model [1] for single-spin transverse asymmetry of the pp scattering at RHIC energies [7]. Error bar indicates expected statistical errors for the PP2PP experiment at RHIC.

In the diffraction minimum the imaginary part of the amplitude B is equal to zero and the asymmetry is determined by the product ($\mathrm{Re}B\,\mathrm{Im}K$). Thus, the asymmetry near the diffraction minimum is sensitive to the imaginary part of the spin–flip K, amplitude and study of the A_N asymmetry in the PP2PP experiment at RHIC can give a direct information about the energy dependence of the amplitude K. There are model predictions for A_N in this region. The model [1] predicts a large negative value of A_N asymmetry near the diffraction minimum which weakly depends on s at the RHIC energy range [7] and A_N is of about 10% for $|t| \sim 3\text{GeV}^2$ (Fig.1). Other model predictions for single-spin asymmetry at small momentum transfer was discussed in [8].

The single spin asymmetry of diffractive $Q\bar{Q}$ production in polarized pp reaction was estimated in [9]. It was found that A_N is determined by Eq. (3) which is modified by the amplitude of hard $Q\bar{Q}$ production. The expected value of the single spin asymmetry in this case is about 5% in the RHIC energy range.

Let us study now diffractive double spin asymmetries of $Q\bar{Q}$ production in lepton-proton reactions for a longitudinally polarized lepton and a transversely polarized proton within the two-gluon model. This model should describe the cross sections of hard and light quark production at small $x < 0.1$ as well. The spin-dependent cross section can be written in the form

$$\frac{d^5\sigma(\pm)}{dQ^2 dy dx_p dt dk_\perp^2} = \left(\frac{(2-2y+y^2)}{(2-y)}\right) \frac{C(x_P,Q^2)\,N(\pm)}{\sqrt{1-4(k_\perp^2+m_q^2)/M_X^2}}. \tag{4}$$

Here $C(x_P,Q^2)$ is a normalization function which is common for the spin average and spin dependent cross section; $N(\pm)$ is determined by a sum of graphs integrated over the gluon momenta l and l'. The function $N(+)$ determines the spin-average cross section. It has the form

$$N(+) = \left(|\tilde{B}|^2 + |t|/m^2 |\tilde{K}|^2\right) \Pi^{(+)}(t,k_\perp^2,Q^2), \tag{5}$$

with

$$\tilde{B} \sim \int_0^{l_\perp^2<k_0^2} \frac{d^2l_\perp (l_\perp^2 + \vec{l}_\perp \vec{r}_\perp)}{(l_\perp^2+\lambda^2)((\vec{l}_\perp+\vec{r}_\perp)^2+\lambda^2)} B(t,l_\perp^2,x_P,\ldots) = F_{x_P}^g(x_P,t,k_0^2)$$

$$\tilde{K} \sim \int_0^{l_\perp^2<k_0^2} \frac{d^2l_\perp (l_\perp^2 + \vec{l}_\perp \vec{r}_\perp)}{(l_\perp^2+\lambda^2)((\vec{l}_\perp+\vec{r}_\perp)^2+\lambda^2)} K(t,l_\perp^2,x_P,\ldots) = K_{x_P}^g(x_P,t,k_0^2), \tag{6}$$

where $k_0^2 \sim (k_\perp^2+m_q^2)/(1-\beta)$. The connection of the two-gluon structure functions from (1) with GPD $F_{x_P}^g(x_P,t,k_0^2)$ and $K_{x_P}^g(x_P,t,k_0^2)$ is written. This connection is general and is shown for a vector meson and $Q\bar{Q}$ production in [5].

The function $N(-)$ which determines the spin-dependent cross sections looks like

$$N(-) = \sqrt{\frac{|t|}{m^2}}\,(\tilde{B}\tilde{K}^* + \tilde{B}^*\tilde{K})\,\Big[\frac{(\vec{Q}\vec{S}_\perp)}{m}\Pi_Q^{(-)}(t,k_\perp^2,Q^2) + \frac{(\vec{k}_\perp\vec{S}_\perp)}{m}\Pi_k^{(-)}(t,k_\perp^2,Q^2)\Big]. \tag{7}$$

The other form of interference between B and K with respect to (3) appears in (7) because we consider here the double spin effects. The large value of asymmetry will appear for a small phase shift between the amplitudes.

The asymmetry is approximately proportional to the ratio of polarized and spin–average gluon distribution functions

$$A_{LT}^{Q\bar{Q}} \sim C^{Q\bar{Q}} \frac{K_\zeta^g(\zeta)}{F_\zeta^g(\zeta)} \quad \text{with } \zeta = x_P \text{ and } |\tilde{K}|/|\tilde{B}| \sim 0.1 \tag{8}$$

The spin-dependent contribution to the asymmetry which is proportional to $\vec{k}_\perp \vec{S}_\perp$ in (7) will be analyzed for the case when the transverse jet momentum $\vec{k}_\perp$ is parallel to the target polarization $\vec{S}_\perp$. The asymmetry is maximal in this case. To observe this contribution to asymmetry, it is necessary to distinguish experimentally the quark and antiquark jets. If we do not separate events with $\vec{k}_\perp$ for the quark jet, e.g., the resulting asymmetry will be equal to zero because the transverse momentum of the quark and antiquark are equal and opposite in sign.

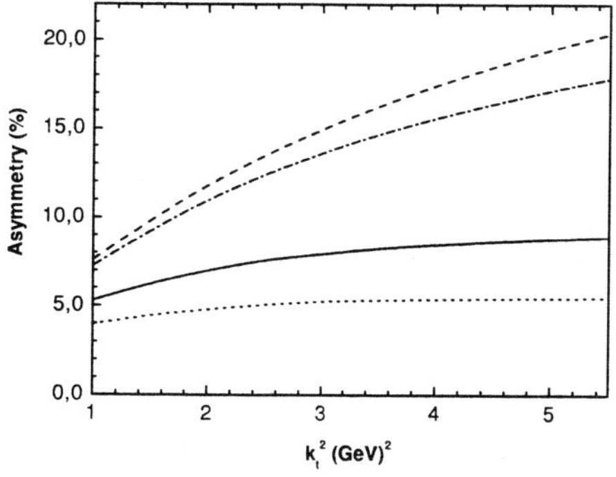

FIGURE 2. The A^k_{lT} asymmetry in diffractive heavy $Q\bar{Q}$ production at $\sqrt{s}=20$GeV for $x_P=0.1$, $y=0.5$, $|t|=0.3$GeV^2: dotted line-for $Q^2=0.5$GeV^2; solid line-for $Q^2=1$GeV^2; dot-dashed line-for $Q^2=5$GeV^2; dashed line-for $Q^2=10$GeV^2.

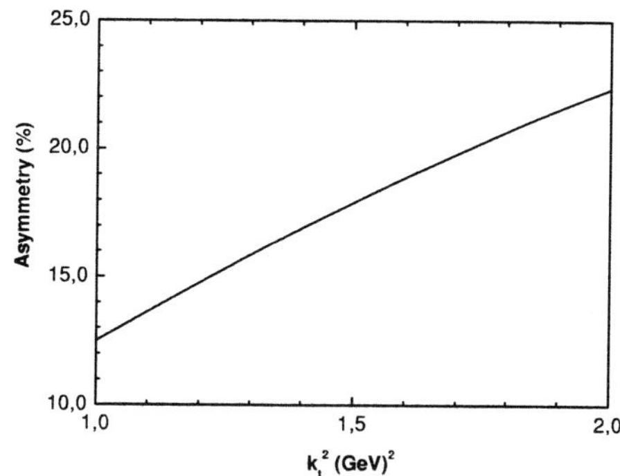

FIGURE 3. The A^k_{lT} asymmetry in diffractive light $Q\bar{Q}$ production for $Q^2=5$GeV^2, $x_P=0.1$, $y=0.5$, $|t|=0.3(GeV)^2$ at $\sqrt{s}=7$GeV.

The predicted asymmetry for heavy $c\bar{c}$ production at the COMPASS energies is shown in Fig.2. The asymmetry for light quark production is approximately of the same order of magnitude. The expected A_{lT} asymmetry for light quark production at HERMES is shown in Fig.3. The function $C_k^{Q\bar{Q}}$ in (8) is quite large, about 1.5 at the HERMES energy for $k_\perp^2=1.3$GeV^2, $Q^2=5$GeV^2, $x_P=0.1$, $y=0.5$, and $|t|=0.3$GeV^2. The spin–dependent cross section vanishes for $Q^2 \to 0$, while the spin–average cross section is constant in this limit. As a result, the asymmetry can be estimated as $A_{lT} \propto Q^2/(Q^2+Q_0^2)$ with $Q_0^2 \sim 1$GeV^2. This shows a possibility of studying the polarized gluon distribution $K_\zeta^g(x)$ in the COMPASS and HERMES experiment at $Q^2 \geq 0.5$GeV^2.

The contribution to asymmetry $\propto \vec{Q}\vec{S}_\perp$ in (7) is simpler to study experimentally. The expected asymmetry in this case is not small too, about 5% or a little bit smaller. The predicted $C_Q^{Q\bar{Q}}$ in (8) in this case is about 0.3. In contrast to the A^k_{lT} term, the A^Q_{lT} asymmetry has a strong mass dependence [5].

Similar analyses have been carried out for diffractive J/Ψ production which is described by the same two-gluon exchange. Unfortunately, the double spin A_{lT} asymmetry in this case is proportional to x_P which is fixed here by $x_P \sim (m_V^2+Q^2+|t|)/(sy)$. As a result the expected asymmetry is small and $C_g(J/\Psi) \sim 0.007$ for vector meson production in the HERMES energy range. It is difficult to expect experimental study of such small asymmetry.

I this report we have studied the transverse asymmetries caused by the spin structure of the Pomeron coupling with the proton. Similar to the low energy experiments [6] we predict the large negative asymmetry in the vicinity of the diffractive minimum in elastic pp scattering at the RHIC energies. If the weak energy dependence of the asymmetry in this region is found in the PP2PP experiment at the RHIC, it will give a definite indication of the spin-dependent Pomeron coupling. Another possibility is to study the single-spin asymmetry of diffractive $Q\bar{Q}$ production at RHIC which is predicted to be about 5%.

The diffractive hadron leptoproduction for a longitudinally polarized lepton and a transversely polarized proton at high energies has been studied. The A_{lT} asymmetry is found to be proportional to the ratio of structure functions $A_{lT} = CK^g/F^g$. If the Pomeron has a spin structure, the ratio K^g/F^g should have a weak x dependence at low $x < 0.1$. The A_{lT} asymmetry can be used to get information on the transverse distribution $K^g_{x_P}(x_P,t)$ from experiment if the function C in (8) is not small. We predict that $C^{Q\bar{Q}}_k \sim 1$ for the term $\propto \vec{k}_\perp \vec{S}_\perp$ and $C^{Q\bar{Q}}_Q \sim 0.3$ for the contribution $\propto \vec{Q}\vec{S}_\perp$ in the asymmetry (7). We can see that the expected values of C are quite large and such asymmetries might be excellent objects to study transverse spin effects in the proton– gluon coupling.

The results presented here should be applicable to the reactions with heavy quarks. For processes with light quarks, our predictions can be used in the small x region ($x \leq 0.1$ e.g.) where the contribution of quark GPD is expected to be small. In the case of light quark production the polarized u and d quark GPD might be studied together with the gluon distribution in the region of not small $x \geq 0.1$. Such experiments can be conducted at the HERMES and COMPASS spectrometers for a transversely polarized target and future eRHIC. We conclude that important information on the spin–dependent GPD $K_\zeta(x)$ at small x can be obtained from the asymmetries in diffractive pp and lp reactions for longitudinally polarized lepton and transversely polarized hadron targets.

The author is grateful to the Organizing Committee of SPIN2002 for the local financial support. These report was supported in part by the Russian Foundation for Basic Research, Grants 00-02-16696 and 02-02-27409.

REFERENCES

1. Goloskokov S.V., Kuleshov S.P., Selyugin O.V., Z. Phys. **C50** 455-464 (1991).
2. Goloskokov S.V., Kroll P., Phys. Rev. **D60** 014019 1-8 (1999).
3. Buttimore N.H. et al., Phys.Rev. **D59** 114010 1-18 (1999); Kopeliovich B.Z., "PP elastic scattering at low t". This Proceedings.
4. Radyushkin A.V., Phys. Rev, **D56** 5524-5557 (1997). Ji X., Phys.Rev. **D55** 7114-7125 (1997).
5. Goloskokov S.V., Euro. Phys. J. **C24** 413-424 (2002).
6. Fidecaro G. et al., Phys. Lett. **B76** 369-373 (1978), **B105** 309-314 (1981).
7. Akchurin N., Goloskokov S.V., Selyugin O.V., Int.J.Mod.Phys. **A14**, 253-269 (1999).
8. Martini A.F., Predazzi E., Diffractive effects in spin-flip pp amplitudes and predictions for relativistic energies. E-print: hep-ph/0209027.
9. Goloskokov S.V., Phys.Rev. **D53** 5995-5999 (1996).

Single Spin Asymmetries, Unpolarized Cross Sections and the Role of Partonic Transverse Momentum

U. D'Alesio and F. Murgia

Istituto Nazionale di Fisica Nucleare, Sezione di Cagliari, and Dipartimento di Fisica, Università di Cagliari C.P. 170, 09042 Monserrato (CA), Italy

Abstract. Partonic intrinsic transverse momentum can be essential for the explanation of large single spin asymmetries in hadronic reactions in the framework of perturbative QCD. The status of an ongoing program investigating in a consistent way the role of intrinsic transverse momentum both in unpolarized and polarized processes is discussed. We compute inclusive cross sections for hadron and photon production in hadronic collisions and for Drell-Yan processes; the results are compared with available experimental data in several different kinematical situations.

In the last years a lot of experimental and theoretical activity has been devoted to the study of transverse single spin asymmetries (SSA) in hadronic collisions and in semi-inclusive DIS. In fact, perturbative QCD (pQCD) with ordinary collinear partonic kinematics leads to negligible values for these asymmetries, as soon as the relevant scale of the process under consideration becomes large. There are however several experimental results which seem to contradict this expectation; two well known examples are: the large transverse Λ polarization measured in unpolarized hadronic collisions; the large SSA observed in the process $p^{\uparrow}p \to \pi X$. A possible way out from this situation comes from extending the collinear pQCD formalism with the inclusion of spin and partonic intrinsic transverse momentum, $\mathbf{k}_{\perp}$, effects. This leads to the introduction of a new class of spin and $\mathbf{k}_{\perp}$ dependent partonic distribution (PDF) and fragmentation (FF) functions, describing fundamental properties of hadron structure [1].

The role of $\mathbf{k}_{\perp}$ effects in inclusive hadronic reactions has been extensively studied also in the calculation of unpolarized cross sections. It has been shown that, particularly at moderately large p_T (which is the region where SSA are measured to be large) these effects can be relevant and may help in improving the agreement between experimental results and pQCD (at LO and NLO) calculations, which often underestimate the data [2].

Based on these considerations, in this contribution we present a preliminary account of an ongoing program which aims to describe consistently both polarized and unpolarized cross sections (and SSA) for inclusive particle production in hadronic collisions at large energies and moderately large p_T, using LO pQCD with the inclusion of intrinsic transverse momentum effects. Our main goal is not to fit the cross sections as well as possible (including NLO contributions, etc.), but rather to show that in our LO approach they are reproduced up to an overall factor of 2-3, compatible with expected NLO K-

CP675, *Spin 2002: 15th Int'l. Spin Physics Symposium and Workshop on Polarized Electron Sources and Polarimeters,* edited by Y. I. Makdisi, A. U. Luccio, and W. W. MacKay

factors and scale dependences, which reasonably cancel out in SSA and are then out of our present interest.

In a pQCD approach at LO and leading twist with inclusion of spin and $\mathbf{k}_\perp$ effects, the unpolarized cross section for the inclusive process $AB \to CX$ reads

$$d\sigma \propto \sum_{a,b,c} \hat{f}_{a/A}(x_a,\mathbf{k}_{\perp a}) \otimes \hat{f}_{b/B}(x_b,\mathbf{k}_{\perp b}) \otimes d\hat{\sigma}^{ab\to c\dots}(x_a,x_b,\mathbf{k}_{\perp a},\mathbf{k}_{\perp b}) \otimes \hat{D}_{C/c}(z,\mathbf{k}_{\perp C}), \tag{1}$$

with obvious notations. A similar expression holds for the numerator of a transverse SSA ($\propto d\Delta^N\sigma/d\sigma$), substituting for the polarized particle involved the corresponding unpolarized PDF (or FF) with the appropriate polarized one, $\Delta^N f$ or $\Delta^N D$. At leading twist there are four new spin and $\mathbf{k}_\perp$ dependent functions to take into account:

$$\Delta^N f_{q/p^\uparrow} \equiv \hat{f}_{q/p^\uparrow}(x,\mathbf{k}_\perp) - \hat{f}_{q/p^\downarrow}(x,\mathbf{k}_\perp)\,;\; \Delta^N f_{q^\uparrow/p} \equiv \hat{f}_{q^\uparrow/p}(x,\mathbf{k}_\perp) - \hat{f}_{q^\downarrow/p}(x,\mathbf{k}_\perp)\,; \tag{2}$$

$$\Delta^N D_{h/q^\uparrow} \equiv \hat{D}_{h/q^\uparrow}(z,\mathbf{k}_\perp) - \hat{D}_{h/q^\downarrow}(z,\mathbf{k}_\perp)\,;\; \Delta^N D_{h^\uparrow/q} \equiv \hat{D}_{h^\uparrow/q}(z,\mathbf{k}_\perp) - \hat{D}_{h^\downarrow/q}(z,\mathbf{k}_\perp), \tag{3}$$

two in the PDF, Eq. (2), and two in the FF, Eq. (3); the first functions in Eq.s (2),(3) are respectively the Sivers and the Collins function. The second ones are respectively the function introduced by Boer [3] and the so-called “polarizing” FF [4].

The unpolarized PDF and FF are given in a simple factorized form, and the $\mathbf{k}_\perp$ dependent part is usually taken to have a Gaussian shape:

$$\hat{f}_{a/A}(x,\mathbf{k}_{\perp a}) = f_{a/A}(x)\,\frac{\beta^2}{\pi}\,e^{-\beta^2 k_{\perp a}^2}\,; \qquad \hat{D}_q^h(z,\mathbf{k}_{\perp h}) = D_q^h(z)\,\frac{\beta'^2}{\pi}\,e^{-\beta'^2 k_{\perp h}^2}, \tag{4}$$

where the parameter β (β') is related to the average partonic (hadronic) $k_\perp$ by the simple relation $1/\beta(\beta') = \langle k^2_{\perp a(h)}\rangle^{1/2}$. Similar expressions are adopted for the polarized PDF and FF of Eq.s (2),(3).

Using this approach, we have studied the unpolarized cross section for several hadronic processes, analyzing a large sample of available data in different kinematical situations: *i)* The Drell-Yan process $pp \to \ell^+\ell^- X$; *ii)* Prompt photon production in $pp \to \gamma X$; *iii)* Inclusive pion production in $pp \to \pi X$. We find that an overall good reproduction of the corresponding unpolarized cross sections is possible (within the limits indicated above) by choosing, depending on the kinematical situation considered, $\beta = 1.0 - 1.25$ (GeV/c)$^{-1}$ (that is, $\langle k_\perp^2\rangle^{1/2} = 0.8 - 1.0$ GeV/c). The choice of β is related to the set of x dependent PDF utilized; throughout this paper we use the GRV94 set [5]. The optimal choice of β' in case *iii)* (pion production) is commented in the following. We limit ourself to present few indicative results and comments regarding the processes considered and the SSA in pion production. A full account of this analysis will be presented elsewhere [6].

i) At LO and within collinear partonic configuration the final lepton pair produced in Drell-Yan processes cannot have any transverse momentum, q_T, with respect to the colliding beams. Experimental data show however that the lepton pair has a well defined q_T spectrum. As an example, in Fig. 1a we show estimates of the invariant cross section at $E = 400$ GeV as a function of q_T, for several different invariant mass bins (in GeV) at

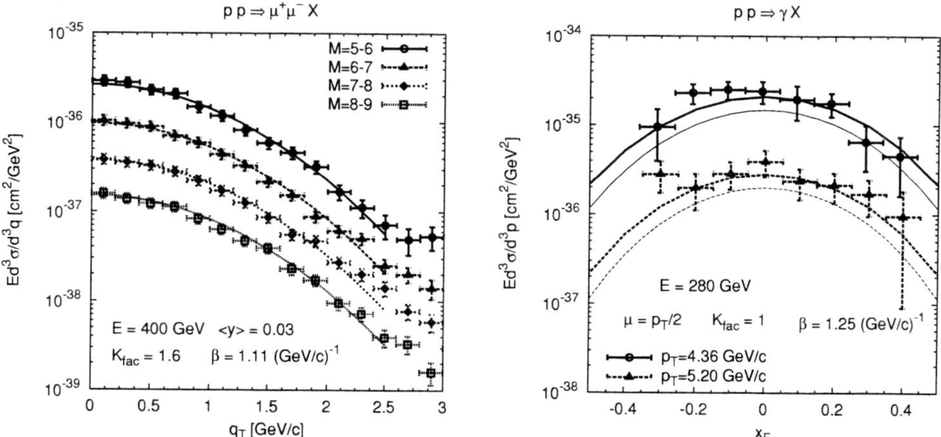

FIGURE 1. The invariant cross section for (a) $pp \to \mu^+\mu^- X$ vs. q_T and (b) $pp \to \gamma X$ vs. x_F; see plots and text for more details.

fixed rapidity $y = 0.03$, and using $\beta = 1.11\ (\mathrm{GeV}/c)^{-1}$; data are from [7]. Theoretical curves are arbitrarily raised by a factor $K_{\rm fac} = 1.6$, which could be well accommodated by NLO K-factors and scale dependences, an issue that as said above we do not address here. Notice how data are well reproduced by a Gaussian dependence up to $q_T = 2 - 2.5$ GeV/c; larger q_T data show a power-law decrease well explained by pQCD corrections. *ii)* Several data for the unpolarized cross section are available in the case of prompt photon production, mainly at central rapidities and moderately large p_T. As an example, in Fig. 1b we show estimates of the invariant cross section for the process $pp \to \gamma X$ at $E = 280$ GeV for two different values of p_T vs. x_F, with $\mathbf{k}_\perp$ effects (thick lines) and without them (thin lines), using $K_{\rm fac} = 1$ and $\beta = 1.25\ (\mathrm{GeV}/c)^{-1}$; data are from [8]. *iii)* For inclusive pion production, $pp \to \pi X$, some experimental results for SSA are also available, and we can see how our approach works for SSA and unpolarized cross sections at the same time. This case is however more intricate, since we can have $\mathbf{k}_\perp$ effects in the fragmentation process also. The z and $\mathbf{k}_\perp$ dependences in the FF are chosen according to Eq. (4); a direct z dependence of the β' parameter seems to be favored, $1/\beta'(z) = \langle k_{\perp\pi}^2(z)\rangle^{1/2} = 1.4\,z^{1.3}(1-z)^{0.2}$ GeV/c.

Unpolarized FF are presently known with much less accuracy than nucleon PDF. In particular, all available sets of parameterizations for the pion FF are for neutral pions (or for the sum of charged pions), since e^+e^- data do not allow to separate among π^+ and π^- case; this can be made under further assumptions, which remain to be tested. In Fig. 2a we present estimates of the invariant cross section for the process $pp \to \pi X$ at E= 200 GeV vs. x_F for different p_T values. We use two sets of FF from Kretzer (K, thin lines) [9] and Kniehl, Kramer, and Pötter (KKP, thick lines) [10], $K_{\rm fac} = 2.4$(K), 1.9(KKP), $\beta = 1.25\ (\mathrm{GeV}/c)^{-1}$. Data are from [11].

Let us now consider the SSA in $p^\uparrow p \to \pi X$, within the same approach and assuming it is generated by the Sivers effect alone, that is from a spin and $\mathbf{k}_\perp$ effect in the PDF inside

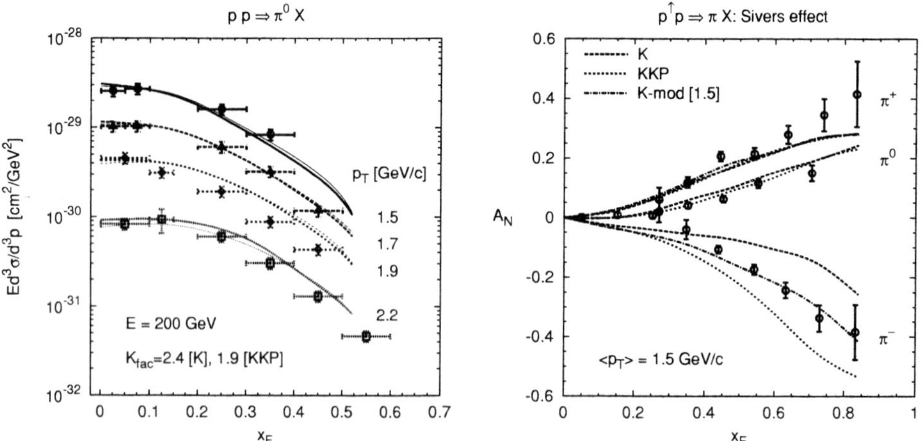

FIGURE 2. The invariant unpol. cross section (a) and the SSA (b) for $p^\uparrow p \to \pi X$ vs. x_F; see plots and text for more details.

the initial polarized proton, described by the Sivers function $\Delta^N f_{q/p^\uparrow}(x, \mathbf{k}_\perp)$. There are other possible sources for SSA, and notably the so-called Collins effect, concerning the fragmentation of a polarized parton into the final observed pion. These effects are not considered here. Analogous studies have already been performed [12], using an effective averaging on $\mathbf{k}_\perp$ and a simplified partonic kinematics. Here we show the first results with full treatment of $\mathbf{k}_\perp$ effects and partonic kinematics. These results are in good qualitative agreement with previous work.

The numerator of the SSA, $d\sigma^\uparrow - d\sigma^\downarrow$ can be expressed in the form of Eq. (1), with the substitution $\hat{f}_{a/A}(x, \mathbf{k}_\perp) \to \Delta^N f_{q/p^\uparrow}(x, \mathbf{k}_\perp)$. For the Sivers function we choose an expression similar to that of the unpolarized distribution, Eq. (4)

$$\Delta^N f_{q/p^\uparrow}(x, \mathbf{k}_\perp) = \Delta^N f_{q/p^\uparrow}(x)\, h(k_\perp) \sin\phi_{k_\perp}\,, \tag{5}$$

where $\phi_{k_\perp}$ is the angle between $\mathbf{k}_\perp$ and the polarization vector of the proton; $\Delta^N f_{q/p^\uparrow}(x)$ and $h(k_\perp)$ are such to fulfill the general positivity bound $|\Delta^N f_{q/p^\uparrow}(x, k_\perp)|/2\hat{f}_{q/p}(x, k_\perp) \leq 1$:

$$\Delta^N f_{q/p^\uparrow}(x) = N_q x^{a_q}(1-x)^{b_q} \frac{(a_q+b_q)^{(a_q+b_q)}}{a_q^{a_q}\, b_q^{b_q}} 2 f_{q/p}(x)\,, \quad |N_q| \leq 1 \tag{6}$$

$$h(k_\perp) = \left(2e\frac{1-r}{r}\right)^{1/2} \frac{\beta^3}{\pi} k_\perp \exp\left[-\beta^2 k_\perp^2/r\right]\,, \quad 0 < r < 1\,. \tag{7}$$

A choice of the parameters in Eq.s (6),(7) which allow to reasonably reproduce the experimental results for the pion SSA is the following (only valence quark contributions to the Sivers function are considered):

$$N_u \;=\; +0.5 \quad a_u = 2.0 \quad b_u = 0.3 \tag{8}$$

$$N_d = -1.0 \quad a_d = 1.5 \quad b_d = 0.2, \qquad r \simeq 0.7.$$

In Fig. 2b we show our preliminary estimates of A_N with Sivers effect at E = 200 GeV and p_T = 1.5 GeV/c, vs. x_F, for three different choices of the pion FF: K, KKP and a modified version of K. Data are from [13]. The SSA for π^+ and π^0 is well reproduced independently of the FF set. Interestingly, the π^- case shows a stronger sensitivity to the relation between the leading and non-leading contributions to the fragmentation process, which cannot be extracted from present experimental information on unpolarized pion cross sections. In fact, our results with the K(KKP) FF sets underestimate (overestimate) in magnitude the π^- asymmetry, while a good agreement is recovered using a somehow fictitious set (K-mod) with an intermediate behavior.

In conclusion, we have presented here preliminary results of an ongoing program dedicated to the study of partonic transverse momentum effects both in unpolarized and polarized cross sections (and SSA) for inclusive particle production in hadronic collisions. These results show that it seems possible to reproduce reasonably well, within pQCD at LO and leading twist and up to a factor of 2-3, unpolarized cross sections for Drell-Yan processes, prompt photon and inclusive pion production in hadronic collisions, in several different kinematical situations. Within the same approach, we have reanalyzed the SSA for $p^\uparrow p \to \pi X$ taking into account Sivers effect alone; we have found reasonable agreement with data and with previous theoretical results obtained with a simplified treatment of $\mathbf{k}_\perp$ effects and partonic kinematics, whose main results are therefore confirmed by our analysis. The next steps of this program are the study of the pion SSA with Collins effect, of the SSA in photon production, and of the unpolarized cross section and the transverse polarization for Λ production in unpolarized hadronic collisions. The extension of our analysis to RHIC kinematics, where a thorough program on SSA measurements is in progress, is of great interest. First estimates of the SSA in $p^\uparrow p \to \pi X$ seem to be in reasonable agreement with preliminary results from RHIC [14].

REFERENCES

1. For review papers on the subject, see, e.g., Liang, Z.-T., and Boros, C., *Int. J. Mod. Phys.* **A15**, 92 (2000); Anselmino, M., e-Print Archive: hep-ph/0201150.
2. Wang, X.-N., *Phys. Rev.* **C61**, 064910 (2000); Wong, C.-Y., and Wang, H., *Phys. Rev.* **C58**, 376 (1998); Zhang, Y., Fai, G., Papp, G., Barnaföldi, G., and Lévai, P., *Phys. Rev.* **C65**, 034903 (2002);
3. Boer, D., *Phys. Rev.* **D60**, 014012 (1999).
4. Anselmino, M., Boer, D., D'Alesio, U., and Murgia, F., *Phys. Rev.* **D63**, 054029 (2001); *Phys. Rev.* **D65**, 114014 (2002).
5. Gluck, M., Reya, E., and Vogt, A., *Z. Phys.* **C67**, 433 (1995).
6. D'Alesio, U., and Murgia, F., work in preparation.
7. Ito, A.S., *et al.*, *Phys. Rev.* **D23**, 604 (1981).
8. Bonesini, M., *et al.*, *Z. Phys.* **C38**, 371 (1988).
9. Kretzer, S., *Phys. Rev.* **D62**, 054001 (2000).
10. Kniehl, B.A., Kramer, G., and Pötter, B., *Nucl. Phys.* **B582**, 514 (2000).
11. Donaldson, G., *et al.*, *Phys. Lett.* **B73**, 375 (1978).
12. Anselmino, M., Boglione, M., and Murgia, F., *Phys. Lett.* **B362**, 164 (1995); Anselmino, M., and Murgia, F., *Phys. Lett.* **B442**, 470 (1998).
13. Adams, D.L. *et al.* (E704 Collab.), *Phys. Lett.* **B261**, 197 (1991); *Phys. Lett.* **B264**, 462 (1991).
14. Rakness, G., these proceedings; Bland, L.C., these proceedings.

Sivers Effect and Transverse Single Spin Asymmetries in Drell-Yan Processes

M. Anselmino*, U. D'Alesio† and F. Murgia†

*Dipartimento di Fisica Teorica, Università di Torino and INFN, Sezione di Torino, Via P. Giuria 1, I-10125 Torino, Italy
†INFN, Sezione di Cagliari and Dipartimento di Fisica, Università di Cagliari, C.P. 170, I-09042 Monserrato (CA), Italy

Abstract. Sivers asymmetry, adopted to explain transverse single spin asymmetries (SSA) observed in inclusive pion production, $p^\uparrow p \to \pi X$ and $\bar{p}^\uparrow p \to \pi X$, is used here to compute SSA in Drell-Yan processes; in this case, by considering the differential cross section in the lepton-pair invariant mass, rapidity and transverse momentum, other mechanisms which may originate SSA cannot contribute. Estimates for RHIC experiments are given.

Single spin asymmetries in high energy inclusive processes are a unique testing ground for QCD; they cannot originate from the simple spin pQCD dynamics – dominated by helicity conservation – but need some non perturbative chiral-symmetry breaking in the large distance physics.

Among the best known transverse single spin asymmetries (SSA) let us mention: *i*) the large polarization of Λ's and other hyperons produced in $pN \to \Lambda^\uparrow X$; *ii*) the large asymmetry $A_N = \frac{d\sigma^\uparrow - d\sigma^\downarrow}{d\sigma^\uparrow + d\sigma^\downarrow}$ observed in $p^\uparrow p \to \pi X$ and $\bar{p}^\uparrow p \to \pi X$ processes; *iii*) the similar azimuthal asymmetry observed in $\ell p^\uparrow \to \ell \pi X$.

Several models [1] to explain the data within QCD dynamics can be found in the literature; here we focus on a phenomenological approach based on the generalization of the factorization theorem with the inclusion of parton intrinsic motion $\mathbf{k}_\perp$. The cross section for a generic process $AB \to CX$ then reads:

$$d\sigma = \sum_{a,b,c} \hat{f}_{a/A}(x_a, \mathbf{k}_{\perp a}) \otimes \hat{f}_{b/B}(x_b, \mathbf{k}_{\perp b}) \otimes d\hat{\sigma}^{ab \to c \ldots}(x_a, x_b, \mathbf{k}_{\perp a}, \mathbf{k}_{\perp b}) \otimes \hat{D}_{C/c}(z, \mathbf{k}_{\perp C}), \tag{1}$$

where the $\hat{f}$'s ($\hat{D}$'s) are the $\mathbf{k}_\perp$ dependent parton distributions (fragmentation functions).

Even if Eq. (1) is not formally proven in general, it has been shown that intrinsic $\mathbf{k}_\perp$'s are indeed necessary in order to be able to explain, within pQCD and the factorization scheme, data on (moderately) large p_T production of pions and photons [2].

When dealing with polarized processes the introduction of $\mathbf{k}_\perp$ dependences opens up the way to many possible spin effects; these can be summarized, at leading twist, by new polarized distribution functions and fragmentation functions,

$$\Delta^N f_{q/p^\uparrow} \equiv \hat{f}_{q/p^\uparrow}(x, \mathbf{k}_\perp) - \hat{f}_{q/p^\downarrow}(x, \mathbf{k}_\perp) = \hat{f}_{q/p^\uparrow}(x, \mathbf{k}_\perp) - \hat{f}_{q/p^\uparrow}(x, -\mathbf{k}_\perp), \tag{2}$$

$$\Delta^N f_{q^\uparrow/p} \equiv \hat{f}_{q^\uparrow/p}(x, \mathbf{k}_\perp) - \hat{f}_{q^\downarrow/p}(x, \mathbf{k}_\perp) = \hat{f}_{q^\uparrow/p}(x, \mathbf{k}_\perp) - \hat{f}_{q^\uparrow/p}(x, -\mathbf{k}_\perp), \tag{3}$$

CP675, *Spin 2002: 15th Int'l. Spin Physics Symposium and Workshop on Polarized Electron Sources and Polarimeters*, edited by Y. I. Makdisi, A. U. Luccio, and W. W. MacKay

$$\Delta^N D_{h/q^\uparrow} \equiv \hat{D}_{h/q^\uparrow}(z,\mathbf{k}_\perp) - \hat{D}_{h/q^\downarrow}(z,\mathbf{k}_\perp) = \hat{D}_{h/q^\uparrow}(z,\mathbf{k}_\perp) - \hat{D}_{h/q^\uparrow}(z,-\mathbf{k}_\perp)\,, \quad (4)$$

$$\Delta^N D_{h^\uparrow/q} \equiv \hat{D}_{h^\uparrow/q}(z,\mathbf{k}_\perp) - \hat{D}_{h^\downarrow/q}(z,\mathbf{k}_\perp) = \hat{D}_{h^\uparrow/q}(z,\mathbf{k}_\perp) - \hat{D}_{h^\uparrow/q}(z,-\mathbf{k}_\perp)\,, \quad (5)$$

which have a clear meaning if one pays attention to the arrows denoting the polarized particles. All the above functions vanish when $k_\perp = 0$ and are naïvely T-odd. The ones in Eqs. (3) and (4) are chiral-odd, while the other two are chiral-even. The fragmentation in Eq. (4) is the Collins function [3], while the distribution in Eq. (2) was first introduced by Sivers [4]. Some of the above functions have been widely used for a phenomenological description of the observed SSA [5].

Despite its successful phenomenology, the Sivers function was always a matter of discussions and its very existence rather controversial; in fact in Ref. [3] a proof of its vanishing was given, based on time-reversal invariance. Ways out based on initial state interactions or non standard time-reversal properties [6] were discussed. Very recently a series of papers [7] have resurrected Sivers asymmetry in its full rights: a quark-diquark model calculation has given an explanation of the HERMES azimuthal asymmetry different from the Collins effect and has shown that initial state interactions can give rise to SSA in Drell-Yan processes. Moreover Collins recognized that *i*) such a new mechanism is compatible with factorization and is due to the Sivers asymmetry (2), *ii*) his original proof of the vanishing of $\Delta^N f_{q/p^\uparrow}$ is incorrect.

Some issues concerning factorizability and universality of these effects are still open to debate; however, we feel now confident to use Sivers effects – and equally all functions in Eqs. (2)-(5) – in SSA phenomenology. The natural process to test the Sivers asymmetry is Drell-Yan where there cannot be any effect in fragmentation processes and, by suitably integrating over some final configurations, other possible effects vanish. SSA in Drell-Yan processes are particularly important now, as ongoing or imminent experiments at RHIC will be able to measure them.

Let us consider a Drell-Yan process, that is the production of $\ell^+\ell^-$ pairs in the collision of two hadrons A and B: the difference between the single transverse spin dependent cross sections $d\sigma^\uparrow$ for $A^\uparrow B \to \ell^+\ell^- X$ and $d\sigma^\downarrow$ for $A^\downarrow B \to \ell^+\ell^- X$ from the Sivers asymmetry of Eq. (2), is

$$d\sigma^\uparrow - d\sigma^\downarrow = \sum_{ab} \int \left[dx_a\, d^2\mathbf{k}_{\perp a}\, dx_b\, d^2\mathbf{k}_{\perp b}\right] \Delta^N f_{a/A^\uparrow}(x_a,\mathbf{k}_{\perp a})\, \hat{f}_{b/B}(x_b,\mathbf{k}_{\perp b})\, d\hat{\sigma}^{ab\to\ell^+\ell^-}. \quad (6)$$

We consider the differential cross section in the variables $M^2 = (p_a + p_b)^2$, y and $\mathbf{q}_T$, that is the squared invariant mass, the rapidity and the transverse momentum of the lepton pair. Notice that we do not look at the angular distribution of the lepton pair production plane, which is integrated over.

We take the hadron A as moving along the positive z-axis, in the A-B c.m. frame and measure the transverse polarization of hadron A, $\mathbf{P}_A$, along the y-axis.

In the kinematical regions such that: $q_T^2 \ll M^2 \ll M_Z^2$ and $k_{\perp a,b}^2 \simeq q_T^2$, the asymmetry becomes

$$A_N = \frac{\sum_q e_q^2 \int d^2\mathbf{k}_{\perp q}\, d^2\mathbf{k}_{\perp \bar{q}}\, \delta^2(\mathbf{k}_{\perp q} + \mathbf{k}_{\perp \bar{q}} - \mathbf{q}_T)\, \Delta^N f_{q/A^\uparrow}(x_q,\mathbf{k}_{\perp q})\, \hat{f}_{\bar{q}/B}(x_{\bar{q}},\mathbf{k}_{\perp \bar{q}})}{2\sum_q e_q^2 \int d^2\mathbf{k}_{\perp q}\, d^2\mathbf{k}_{\perp \bar{q}}\, \delta^2(\mathbf{k}_{\perp q} + \mathbf{k}_{\perp \bar{q}} - \mathbf{q}_T)\, \hat{f}_{q/A}(x_q,\mathbf{k}_{\perp q})\, \hat{f}_{\bar{q}/B}(x_{\bar{q}},\mathbf{k}_{\perp \bar{q}})}\,, \quad (7)$$

where $x_q \simeq \frac{M}{\sqrt{s}} e^y$ and $x_{\bar{q}} \simeq \frac{M}{\sqrt{s}} e^{-y}$, with $a,b = q,\bar{q}$ and $q = u,\bar{u},d,\bar{d},s,\bar{s}$.

On the other hand the SSA generated by the distribution function in Eq. (3) would lead to a contribution of the kind [8]

$$\sum_q h_{1q}(x_q, \mathbf{k}_{\perp q}) \otimes \Delta^N f_{\bar{q}^\uparrow/B}(x_{\bar{q}}, \mathbf{k}_{\perp \bar{q}}) \otimes d\Delta\hat{\sigma}^{q\bar{q}\to\ell^+\ell^-}\,, \tag{8}$$

where h_{1q} is the transversity of quark q (inside hadron A) and $d\Delta\hat{\sigma}$ is the double transverse spin asymmetry $d\hat{\sigma}^{\uparrow\uparrow} - d\hat{\sigma}^{\uparrow\downarrow}$. Such an elementary asymmetry has a $\cos 2\phi$ dependence [8], where ϕ is the angle between the transverse polarization direction and the normal to the $\ell^+\ell^-$ plane; when integrating over all final angular distributions of the $\ell^+\ell^-$ pair – as we do – the contribution of Eq. (8) vanishes.

Analogously, other mechanisms [9], based on higher twist quark-gluon correlation functions, lead to expressions of A_N vanishing upon integration over the leptonic angles.

In order to give numerical estimates, we introduce here a simple model for the Sivers asymmetry (2), and for the unpolarized distributions, which is similar to the one introduced for the polarizing fragmentation function (see Eq. 5) in Ref. [10].

Let us start from the most general expression for the number density of unpolarized quarks q, inside a proton with transverse polarization $\mathbf{P}$ and three-momentum $\mathbf{p}$. One has

$$\hat{f}_{q/p^\uparrow}(x, \mathbf{k}_\perp) = \hat{f}_{q/p}(x, k_\perp) + \frac{1}{2}\Delta^N f_{q/p^\uparrow}(x, k_\perp)\,\hat{\mathbf{P}}\cdot\hat{\mathbf{p}}\times\hat{\mathbf{k}}_\perp\,. \tag{9}$$

In our configuration one simply has $\hat{\mathbf{P}}\cdot\hat{\mathbf{p}}\times\hat{\mathbf{k}}_\perp = (\hat{\mathbf{k}}_\perp)_x = \cos\phi_{k_\perp}$.

We consider simple factorized and Gaussian forms (see also [11]):

$$\hat{f}_{q/p}(x, k_\perp) = f_{q/p}(x)\, g(k_\perp) = f_{q/p}(x)\,\frac{\beta^2}{\pi}\, e^{-\beta^2 k_\perp^2}\,; \tag{10}$$

by imposing the positivity bound $|\Delta^N f_{q/p^\uparrow}(x, k_\perp)| \leq 2\hat{f}_{q/p}(x, k_\perp)$, we can write

$$\Delta^N f_{q/p^\uparrow}(x, k_\perp) = 2\mathcal{N}_q(x)\, f_{q/p}(x)\,\frac{\beta^2}{\pi}\,\sqrt{2e\,(\alpha^2-\beta^2)}\,k_\perp\, e^{-\alpha^2 k_\perp^2}\,, \tag{11}$$

with

$$\mathcal{N}_q(x) = N_q x^{a_q}(1-x)^{b_q}\,\frac{(a_q+b_q)^{(a_q+b_q)}}{a_q^{a_q}\, b_q^{b_q}}\,, \qquad |N_q| \leq 1\,. \tag{12}$$

Inserting the above choice of $\Delta^N f(x, \mathbf{k}_\perp)$ and $\hat{f}(x, \mathbf{k}_\perp)$ into Eq. (7) one can perform analytical integrations; assuming β independent of x (see below) one gets

$$\begin{aligned} A_N(M, y, \mathbf{q}_T) &= \mathcal{Q}(q_T, \phi_{q_T})\,\mathcal{A}(M, y) \\ &= 2\frac{r^2}{(1+r)^2}\left(2e\frac{1-r}{r}\right)^{1/2}\beta\, q_T\cos\phi_{q_T}\,\exp\left[-\frac{1}{2}\frac{1-r}{1+r}\beta^2 q_T^2\right] \\ &\times\ \frac{1}{2}\,\frac{\sum_q e_q^2\,\Delta^N f_{q/p^\uparrow}(x_q)\, f_{\bar{q}/p}(x_{\bar{q}})}{\sum_q e_q^2\, f_{q/p}(x_q)\, f_{\bar{q}/p}(x_{\bar{q}})}\,. \end{aligned} \tag{13}$$

where ϕ_{q_T} is the azimuthal angle of $\mathbf{q}_T$ and $r \equiv \beta^2/\alpha^2 < 1$. $\mathscr{Q}(q_T)$ has a maximum when $q_T = q_T^M = \sqrt{(1+r)/(1-r)}/\beta$, where its value is $\mathscr{Q}(q_T^M) \equiv \mathscr{Q}_M = [2r/(1+r)]^{3/2}$. Notice that only the position of the maximum depends on the parameter β.

All the parameters of the model have been fixed in a complementary analysis of unpolarized inclusive particle production and pion SSA [13]. As a result of this study we have: $\beta = 1.25\,(\mathrm{GeV}/c)^{-1}$ ($\langle k_\perp^2 \rangle^{1/2} = 0.8$ GeV/c, independent of x), and

$$\begin{array}{lllll} N_u = 0.5 & a_u = 2.0 & b_u = 0.3\,, & & \\ N_d = -1.0 & a_d = 1.5 & b_d = 0.2\,, & r \simeq 0.7\,. & (14) \end{array}$$

For the unpolarized partonic distributions, $f_{q/p}(x)$, we adopt the GRV94 set [12].

One further uncertainty concerns the sign of the asymmetry: as noticed by Collins and checked by Brodsky [7], the Sivers asymmetry has opposite signs in Drell-Yan and SIDIS, respectively related to s-channel and t-channel elementary reactions. As in $p-p$ interactions we expect that large x_F regions are dominated by t-channel quark processes, we think that the Sivers function extracted from $p-p$ data should be opposite to that contributing to D-Y processes. Our numerical estimates will then be given with the same parameters as in Eq. (14), *changing the signs of* N_u and N_d. Given these considerations, even a simple comparison of the sign of our estimates with data might be significant.

In Fig. 1 we show A_N at $\sqrt{s} = 200$ GeV as a function of y averaged over two kinematical ranges $6 \le M \le 10$ GeV and $10 \le M \le 20$ GeV (on the left) and as a function of M averaged over the ranges $|y| < 2$ and $0 < y < 2$ (on the right). We have fixed $q_T = q_T^M$ ($\simeq 1.9$ GeV/c), and $\phi_{q_T} = 0$, which maximizes the $\mathbf{q}_T$-dependent part of the asymmetry; on the other hand A_N is reduced by a factor of 50% at $q_T \simeq 0.6$ GeV/c.

We can also consider the asymmetry averaged over $\mathbf{q}_T$ up to a value of $q_T = q_{T1}$ (integrating over ϕ_{q_T} in the range $[0, \pi/2]$ only, otherwise one would get zero). In our simple model (for a full account of this study see [11]), for $q_{T1} \ge 1.7$ GeV/c we would get $\langle A_N \rangle \simeq 0.4 A_N(q_T^M)$ (for $q_{T1} = 0.6$ GeV/c $\langle A_N \rangle \simeq 0.2 A_N(q_T^M)$).

Our numerical estimates show that A_N can be well measurable within RHIC expected statistical accuracy. The actual values depend on the assumed functional form of the Sivers function and its role with valence quarks only.

Transverse single spin phenomenology, within QCD dynamics and the factorization scheme, is a rich and interesting subject. It combines simple pQCD spin dynamics with new long distance properties of quark distribution and fragmentation; the experimental measurements are relatively easy and clear, many have been and many more will be performed in the near future, both at nucleon-nucleon and lepton-nucleon facilities.

Very recently a large transverse SSA (contrary to naive expectations) has been observed at $\sqrt{s} = 200$ GeV (at RHIC) in $p^\uparrow p \to \pi X$ processes [14]; a reasonable agreement with these preliminary data has been found in our approach.

We have presented here the explicit formalism for computing single transverse spin asymmetries in Drell-Yan processes, within a generalized QCD factorization theorem formulated with $\mathbf{k}_\perp$ dependent distribution functions. Simple Gaussian forms have been assumed and available data from other processes have been exploited, in order to give estimates for single spin effects in D-Y production at RHIC, which should be of interest for the incoming measurements. Again, sizable and measurable values have been found.

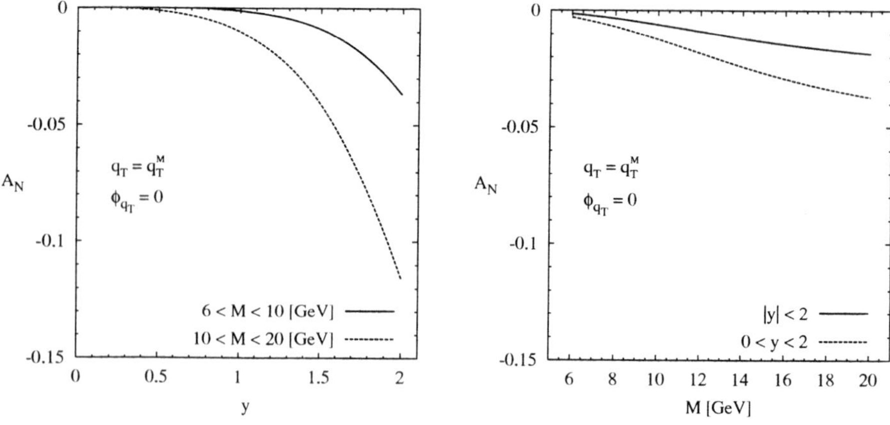

FIGURE 1. Single spin asymmetry A_N for the Drell-Yan process, at RHIC energies, $\sqrt{s} = 200$ GeV, as a function of y and averaged over M (left) and as a function of M and averaged over y (right), (see text).

ACKNOWLEDGMENTS

One of us (U.D.) would like to thank the organizers for their kind invitation to a fruitful and interesting Symposium. U.D. and F.M. thank COFINANZIAMENTO MURST-PRIN for partial support.

REFERENCES

1. For review papers on the subject, see Liang, Z.-T., and Boros, C., *Int. J. Mod. Phys.* **A15**, 92 (2000); Anselmino, M., e-Print Archive: hep-ph/0201150.
2. Wang, X.-N., *Phys. Rev.* **C61**, 064910 (2000); Wong, C.-Y., and Wang, H., *Phys. Rev.* **C58**, 376 (1998); Zhang, Y., Fai, G., Papp, G., Barnaföldi, G., and Lévai, P., *Phys. Rev.* **C65**, 034903 (2002).
3. Collins, J.C., *Nucl. Phys.* **B396**, 16 (1993).
4. Sivers, D., *Phys. Rev.* **D41**, 83 (1990); *Phys. Rev.* **D43**, 261 (1991).
5. Anselmino, M., Boglione, M., and Murgia, F., *Phys. Lett.* **B362**, 164 (1995); *Phys. Rev.* **D60**, 054027 (1999); Anselmino, M., and Murgia, F., *Phys. Lett.* **B442**, 470 (1998); Boglione, M., and Leader, E., *Phys. Rev.* **D61**, 114001 (2000); Anselmino, M., Boer, D., D'Alesio, U., and Murgia, F., *Phys. Rev.* **D63**, 054029 (2001).
6. Anselmino, M., Barone, V., Drago, A., and Murgia, F., e-Print Archive: hep-ph/0209073.
7. Brodsky, S.J., Hwang, D.S., and Schmidt, I., *Phys. Lett.* **B530**, 99 (2002); *Nucl. Phys.* **B642**, 344 (2002); Collins, J.C., *Phys. Lett.* **B536**, 43 (2002).
8. Boer, D., *Phys. Rev.* **D60**, 014012 (1999).
9. Hammon, N., Teryaev, O., and Schäfer, A., *Phys. Lett.* **B390**, 409 (1997); Boer, D., Mulders, P.J., and Teryaev, O., *Phys. Rev.* **D57**, 3057 (1998); Boer, D., and Mulders, P.J., *Nucl. Phys.* **B569**, 505 (2000); Boer, D., and Qiu, J., *Phys. Rev.* **D65**, 034008 (2002).
10. Anselmino, M., Boer, D., D'Alesio, U., and Murgia, F., *Phys. Rev.* **D65**, 114014 (2002).
11. Anselmino, M., D'Alesio, U., and Murgia, F., e-Print Archive: hep-ph/0210371.
12. Gluck, M., Reya, E., and Vogt, A., *Z. Phys.* **C67**, 433 (1995).
13. D'Alesio, U., and Murgia, F., work in preparation; these proceedings.
14. Rakness, G., these proceedings; Bland, L.C., these proceedings.

Handedness Inside the Proton

Daniël Boer

Department of Physics and Astronomy, Vrije Universiteit Amsterdam
De Boelelaan 1081, NL-1081 HV Amsterdam, The Netherlands

Abstract. The transversity of quarks inside unpolarized hadrons and its phenomenology are discussed. Several experimental suggestions are proposed that would allow further study of this intrinsic handedness.

INTRODUCTION

As pointed out in Ref. [1] there exists an experimental indication –a $\cos 2\phi$ azimuthal asymmetry in the Drell-Yan process– for nonzero transversity of quarks inside *unpolarized* hadrons. The idea is that transverse polarization of a noncollinear quark inside an unpolarized hadron in principle can have a preferred direction and therefore, does not need to average to zero. This preferred direction signals an intrinsic handedness. For example expressed in the infinite momentum frame, the transverse quark polarization is orthogonal to the directions of the proton and the (noncollinear) quark momentum:

$$S_T^q \propto P_{\text{hadron}} \times p_{\text{quark}}. \tag{1}$$

Clearly, this must be related to orbital angular momentum, but how exactly is still an open question.

Whether there is indeed nonzero intrinsic handedness remains to be tested and here we will discuss ways of how one would be able to pursue this issue. Some theoretical aspects of this quark-transversity distribution function (usually denoted by $h_1^\perp$) [2] will be reviewed and its main experimental signatures will be pointed out. In particular, unpolarized and single spin asymmetries will be discussed for the Drell-Yan (DY) process and semi-inclusive DIS. Important in the latter case are polarized Λ production observables. Special emphasis will be put on how to distinguish the various asymmetries compared to those arising from other mechanisms, like the Sivers effect [3].

"T-ODD" DISTRIBUTION FUNCTIONS

The quark-transversity function $h_1^\perp$ is a function of the lightcone momentum fraction x and transverse momentum p_T of a quark inside an unpolarized hadron. At first sight, this intrinsic handedness function appears to violate time reversal invariance, when the incoming hadron is treated as a plane-wave state. However, already in a simple gluon-exchange model calculation [4] such so-called "T-odd" distribution functions turn out to

CP675, *Spin 2002: 15th Int'l. Spin Physics Symposium and Workshop on Polarized Electron Sources and Polarimeters,* edited by Y. I. Makdisi, A. U. Luccio, and W. W. MacKay

be nonzero. In recent work [5, 6, 7] it has been demonstrated that the proper gauge invariant definition of transverse momentum dependent functions does indeed allow for such seemingly time reversal symmetry violating functions[1]. Therefore, here we will simply assume that there exists no symmetry argument that forces the intrinsic handedness to be absent.

A large part of the $h_1^\perp$ phenomenology was already presented in Refs. [2, 1]. The function $h_1^\perp$ enters the asymmetries discussed below without suppression by inverse powers of the hard scale in the process (but the operator associated to the function is not twist-2 in the OPE sense). The other unsuppressed transverse momentum dependent "T-odd" function is the Sivers effect function, denoted by $f_{1T}^\perp$. It parameterizes the probability of finding an unpolarized quark (with x and p_T) inside a transversely polarized hadron.

UNPOLARIZED DRELL-YAN

As mentioned, there exists data that is compatible with nonzero $h_1^\perp$. A large $\cos 2\phi$ angular dependence in the unpolarized DY process $\pi^- N \to \mu^+\mu^- X$ was observed, for deuterium and tungsten and with π-beam energies ranging between 140 and 286 GeV [8, 9, 10]. Conventionally, the differential cross section is written as

$$\frac{1}{\sigma}\frac{d\sigma}{d\Omega} \propto \left(1+\lambda\cos^2\theta+\mu\sin^2\theta\,\cos\phi+\frac{\nu}{2}\sin^2\theta\cos 2\phi\right), \tag{2}$$

where ϕ is the angle between the lepton and hadron scattering planes in the lepton center of mass frame (see Fig. 3 of Ref. [1]). The perturbative QCD (pQCD) prediction for very small transverse momentum (Q_T) of the muon pair is $\lambda \approx 1, \mu \approx 0, \nu \approx 0$. More generally, i.e. also for larger Q_T values, one expects the Lam-Tung relation $1-\lambda-2\nu = 0$ to hold (at order α_s). However, the data (with invariant mass Q of the lepton pair in the range $Q \sim 4-12$ GeV) is incompatible with this pQCD relation (and with its $\mathcal{O}(\alpha_s^2)$ modification as well [11]). Several explanations have been put forward in the literature, but these will not be reviewed here.

In Ref. [1] we have observed that within the framework of transverse momentum dependent distribution functions, the $\cos 2\phi$ asymmetry can only be accounted for by the function $h_1^\perp$ or else will be $1/Q^2$ suppressed. We obtained $\nu \propto h_1^{\perp\pi}\, h_1^{\perp N}$ and this expression was used to fit the function $h_1^\perp$ from the data. This approach has several aspects in common with earlier work by Brandenburg, Nachtmann and Mirkes [11], where the large values of ν were generated from a nonperturbative, nonfactorizing mechanism that correlates the transverse momenta and spins of the quark and anti-quark that annihilate into the virtual photon (or Z, but not W, boson). But the description of ν as a product of two $h_1^\perp$ functions implies that these correlations are not necessarily factorization breaking and this type of effects will then not be specific to hadron-hadron scattering. We also note that since the function $h_1^\perp$ is a quark helicity-flip matrix element, it offers a natural explanation for $\mu \approx 0$.

---

[1] The name "T-odd" is thus a misnomer, since it seems to suggest a violation of time reversal invariance, which turns out not to be the case. Other names, like naive or artificial T-odd, have been suggested.

POLARIZED DRELL-YAN

Instead of colliding two unpolarized hadrons, one can also use a polarized hadron to become sensitive to the polarization of quarks inside an unpolarized hadron. In principle, this provides a new way to measure the transversity distribution function h_1. In this case the transverse hadron spin (S_T) dependent differential cross section may be parameterized by (choosing $\mu = 0$ and $\lambda = 1$)

$$\frac{d\sigma(pp^{\uparrow} \to \ell\bar{\ell}X)}{d\Omega\, d\phi_S} \propto 1 + \cos^2\theta + \sin^2\theta \left[\frac{\nu}{2}\cos 2\phi - \rho\,|S_T|\sin(\phi + \phi_S)\right] + \ldots, \quad (3)$$

where ϕ_S is the angle of the transverse spin compared to the lepton plane, cf. Ref. [1].

The analyzing power ρ is (within this framework) proportional to the product $h_1^{\perp} h_1$ [2, 1]. Hence, the measurement of $\langle \cos 2\phi \rangle$ (e.g. at RHIC in $pp \to \mu^+\mu^- X$ or at Fermilab in $p\bar{p} \to \mu^+\mu^- X$) combined with a measurement of the single spin azimuthal asymmetry $\langle \sin(\phi + \phi_S) \rangle$ (also possible at RHIC) could provide information on h_1. In other words, a nonzero function $h_1^{\perp}$ will imply a relation between ν and ρ, which in case of one (dominant) flavor (usually called u-quark dominance) and Gaussian transverse momentum dependences, is approximately given by

$$\rho \approx \frac{1}{2}\sqrt{\frac{\nu}{\nu_{\max}}}\,\frac{h_1}{f_1}, \quad (4)$$

where $\nu_{\max}$ is the maximum value attained by $\nu(Q_T)$. This relation depends on the magnitude of h_1 compared to f_1 and since h_1 is not known experimentally, in Fig. 1 we display two options for ρ, using the function ν which was fitted from the 194 GeV data [1] (which has $\langle Q_T \rangle \lesssim 3$ GeV) and is extrapolated to larger Q_T values (the theoretically expected turn-over of ν has not yet been seen in experiments).

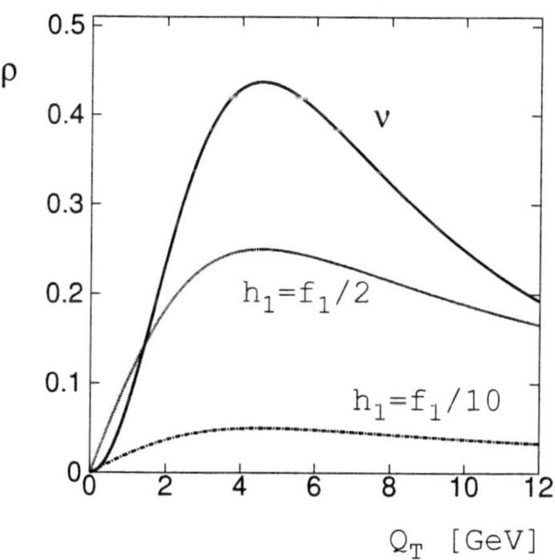

FIGURE 1. Results for ρ using Eq. (4), for h_1 equal to $f_1/2$ and $f_1/10$.

We note that the Sivers function $f_{1T}^{\perp}$ will generate a different angular single spin asymmetry, namely proportional to $(1 + \cos^2\theta)\,|S_T|\sin(\phi - \phi_S)\, f_{1T}^{\perp}\, f_1$.

HADRON PRODUCTION SINGLE SPIN ASYMMETRIES

Large single transverse spin asymmetries have been observed in the process $p\,p^{\uparrow} \to \pi X$ [12]. It has been suggested that these asymmetries can arise from various "T-odd" functions with transverse momentum dependence. There are three options:

$$h_1^{\perp} \otimes h_1 \otimes D_1; \quad f_{1T}^{\perp} \otimes f_1 \otimes \dot{D}_1; \quad h_1 \otimes f_1 \otimes H_1^{\perp}.$$

The first two options are similar to those described in the previous section, now accompanied by the unpolarized fragmentation function D_1. The third option contains the Collins effect function $H_1^{\perp}$ [13], which is the fragmentation function analogue of $h_1^{\perp}$, but in principle is unrelated in magnitude. The last two options were investigated in [14, 15].

We note that the first two options also occur in jet production asymmetries: $p + p^{\uparrow} \to$ jet $+X$ (with only neutral current contributions for the first option). However, as Koike has pointed out, the first option is a double transverse spin asymmetry on the parton level and is thus expected to be small, like the example of Ref. [16] or the double transverse spin asymmetry in DY. It is therefore more likely to obtain information on $h_1^{\perp}$ from hadron production asymmetries in semi-inclusive DIS (SIDIS).

SEMI-INCLUSIVE DIS

First some comments on unpolarized asymmetries in SIDIS. The $\langle \cos 2\phi \rangle$ in SIDIS at values of Q^2 similar to those of the unpolarized DY data, has been measured by the EMC collaboration [17, 18]. No significant asymmetry was observed due to the large errors. But in the present picture of "T-odd" functions the $\langle \cos 2\phi \rangle$ in SIDIS would be $\propto h_1^{\perp} H_1^{\perp}$, which thus can be quite different in magnitude. A consistent picture should emerge by also comparing to $\langle \cos 2\phi \rangle \propto H_1^{\perp} H_1^{\perp}$ in e^+e^- annihilation (e.g. doable at BELLE). However, one should keep in mind that there is another source of a $\cos 2\phi$ asymmetry, namely one that stems from double gluon radiation in the hard scattering subprocess. Fortunately, this forms a calculable background which only dominates in the large Q_T region (close to Q).

Another test would be to look at $\langle \cos 2\phi \rangle$ for a jet instead of a hadron: $e\,p \to e'\,\mathrm{jet}X$. The contribution from $h_1^{\perp}$ will then be absent.

Apart from these unpolarized asymmetries, one can also consider polarized hadron production asymmetries in SIDIS, most notably polarized Λ production. In that case the intrinsic handedness can lead to the following asymmetries:

- $\sin(\phi_{\Lambda}^e + \phi_{S_T^{\Lambda}}^e)$ and $\sin(3\phi_{\Lambda}^e - \phi_{S_T^{\Lambda}}^e)$ in $e\,p \to e'\,\Lambda^{\uparrow} X$ (transverse Λ polarization)
- $\sin(2\phi_{\Lambda}^e)$ in $e\,p \to e'\,\vec{\Lambda} X$ (longitudinal Λ polarization)

These particular angular dependences should be absent for charged current exchange processes, like $\nu\,p \to e\,\Lambda^{\uparrow} X$ or $\nu\,p \to e\,\vec{\Lambda} X$ [19].

These asymmetries are distinguishable from other mechanisms via the y and ϕ^e dependences. For instance, the first asymmetry for transversely polarized Λ production, can be distinguished from the asymmetry due to the so-called polarizing fragmentation functions [20, 21] (also called the Sivers fragmentation function, although it is in principle unrelated in magnitude to the Sivers distribution function). Moreover, the asymmetries should vanish after integration over Q_T, leaving only possibly a $\sin(\phi^e_{S^\Lambda_T})$ asymmetry (which is a twist-3, and hence suppressed, asymmetry) [22, 23].

CONCLUSIONS

The chiral-odd, "T-odd" distribution function $h_1^\perp$ offers an explanation for the large unpolarized $\cos 2\phi$ asymmetry in the $\pi^- N \to \mu^+\mu^- X$ data. Nonzero $h_1^\perp$ would relate unpolarized and polarized observables in a distinct way and thus in principle offers a new way to access h_1 in $p\,p^\uparrow \to \mu^+\mu^- X$.

There are several ways of differentiating $h_1^\perp$ dependent asymmetries from those due to other mechanisms and the suggestion of nonzero $h_1^\perp$ can be explored using a host of existing (Fermilab, BELLE) and near-future data (RHIC, COMPASS, HERMES).

ACKNOWLEDGMENTS

I thank Arnd Brandenburg, Stan Brodsky, Dae Sung Hwang, Yuji Koike and Piet Mulders for fruitful discussions on this topic. The research of D.B. has been made possible by financial support from the Royal Netherlands Academy of Arts and Sciences.

REFERENCES

1. Boer, D., *Phys. Rev.*, **D60**, 014012 (1999).
2. Boer, D., and Mulders, P. J., *Phys. Rev.*, **D57**, 5780–5786 (1998).
3. Sivers, D. W., *Phys. Rev.*, **D41**, 83 (1990).
4. Brodsky, S. J., Hwang, D. S., and Schmidt, I., *Phys. Lett.*, **B530**, 99–107 (2002).
5. Collins, J. C., *Phys. Lett.*, **B536**, 43–48 (2002).
6. Ji, X.-d., and Yuan, F., *Phys. Lett.*, **B543**, 66–72 (2002).
7. Belitsky, A. V., Ji, X., and Yuan, F., hep-ph/0208038 (2002).
8. Falciano, S., et al., *Z. Phys.*, **C31**, 513 (1986).
9. Guanziroli, M., et al., *Z. Phys.*, **C37**, 545 (1988).
10. Conway, J. S., et al., *Phys. Rev.*, **D39**, 92–122 (1989).
11. Brandenburg, A., Nachtmann, O., and Mirkes, E., *Z. Phys.*, **C60**, 697–710 (1993).
12. Adams, D. L., et al., *Phys. Lett.*, **B264**, 462–466 (1991).
13. Collins, J. C., *Nucl. Phys.*, **B396**, 161–182 (1993).
14. Anselmino, M., Boglione, M., and Murgia, F., *Phys. Lett.*, **B362**, 164–172 (1995).
15. Anselmino, M., Boglione, M., and Murgia, F., *Phys. Rev.*, **D60**, 054027 (1999).
16. Kanazawa, Y., and Koike, Y., *Phys. Lett.*, **B490**, 99–105 (2000).
17. Aubert, J. J., et al., *Phys. Lett.*, **B130**, 118 (1983).
18. Arneodo, M., et al., *Z. Phys.*, **C34**, 277 (1987).
19. Boer, D., Jakob, R., and Mulders, P. J., *Nucl. Phys.*, **B564**, 471–485 (2000).
20. Mulders, P. J., and Tangerman, R. D., *Nucl. Phys.*, **B461**, 197–237 (1996).
21. Anselmino, M., Boer, D., D'Alesio, U., and Murgia, F., *Phys. Rev.*, **D63**, 054029 (2001).
22. Kanazawa, Y., and Koike, Y., *Phys. Rev.*, **D64**, 034019 (2001).
23. Koike, Y., *these proceedings* (2002).

Parton Distributions in Light-Cone Gauge: Where Are the Final-State Interactions? [1]

Feng Yuan and Xiangdong Ji

Department of Physics, University of Maryland, College Park, Maryland 20742

Abstract. We show that the final-state interaction effects in the single target spin asymmetry discovered by Brodsky et al. can be reproduced by either a standard light-cone gauge definition of the parton distributions with a prescription of the light-cone singularities consistent with the definition with a gauge link involving the gauge potential at the spatial infinity.

INTRODUCTION

Recently, Brodsky and collaborators have re-examined the significance of the parton distributions measurable in deep-inelastic and other high-energy scattering. They found that the final state interactions (FSI) between the struck quark and target spectators yield distinct physical effects such as shadowing and single-spin asymmetry [1, 2]. These effects are of course contained in the light-cone gauge-link explicitly present in the definition of the parton distributions in the non-singular gauges, in which the gauge potential vanishes at the spacetime infinity. In the light-cone gauge, however, the gauge-link vanishes by choice, and the parton distributions in the conventional definition become parton *densities* which are entirely determined by the ground state light-cone wave functions.

In this paper, we argue that the standard definition of the parton distributions in the light-cone gauge requires a unique prescription for the light-cone singularities -the one that is constrained to reproduce the light-cone gauge link in the covariant gauge. Alternatively, if one demands the initial state wave function be real, then the usual light-cone gauge link in the definition of the parton distributions is incomplete. It ought to be supplemented with an additional contribution. In the non-singular gauges, this new eikonal factor does not contribute. But in the gauges such as the light-cone gauge where the gauge potential does not vanish asymptotically, the additional gauge link is responsible for the final state interactions. We use the example of the single spin asymmetry discussed in Ref. [2] to show that the FSI physics is faithfully reproduced in this approach.

---

[1] Supported by the US Department of Energy DE-FG02-93ER-40762.

CP675, *Spin 2002: 15th Int'l. Spin Physics Symposium and Workshop on Polarized Electron Sources and Polarimeters,* edited by Y. I. Makdisi, A. U. Luccio, and W. W. MacKay

SIVERS FUNCTION IN COVARIANT GAUGE

The transverse-momentum parton distribution in Covariant gauge is defined,

$$\begin{aligned} f(x,k_\perp) &= \frac{1}{2}\int \frac{d\xi^- d^2\xi_\perp}{(2\pi)^3} e^{-i(\xi^- k^+ - \vec{\xi}_\perp\cdot\vec{k}_\perp)} \\ &\times \langle P|\overline{\psi}(\xi^-,\xi_\perp)L^\dagger_{\xi_\perp}(\infty,\xi^-)\gamma^+ L_0(\infty,0)\psi(0)|P\rangle \ , \end{aligned} \quad (1)$$

where the path-ordered light-cone gauge link[3, 4, 5, 6]

$$L_{\xi_\perp}(\infty,\xi^-) = P\exp\left(-ig\int_{\xi^-}^{\infty} A^+(\xi^-,\xi_\perp)d\xi^-\right) \ . \quad (2)$$

In hard scattering, the gauge link $L(\infty,0)$ arises from the final state interactions between the struck quark and the gluon field in the target spectators[3, 6].

In the non-singular gauges, the above definition yields the correct gauge-invariant parton distributions. As an example, let us first calculate the asymmetrical part of the transverse momentum distribution in a nucleon due to its transverse polarization, the so-called Sivers function [7]. Since we are only interested in the matter of principle, we use the simple model introduced in Ref. [2]to study the polarization asymmetry discussed there.

Expanding Eq. (1) to the first order in g and dropping the leading term which does not yield any asymmetry in the transverse momentum distribution, we have

$$\begin{aligned} f_{1T}^{\perp}(x,k_\perp) &= \frac{1}{2}\sum_n\int \frac{d\xi^- d^2\xi_\perp}{(2\pi)^3} e^{-i(\xi^- k^+ - \vec{\xi}_\perp \vec{k}_\perp)}\langle P|\overline{\psi}(\xi^-,\xi_\perp)|n\rangle \\ &\times \langle n|\left(-ie_1\int_0^\infty A^+(\xi^-,0)d\xi^-\right)\gamma^+\psi(0)|P\rangle + \text{h.c.} \ , \end{aligned} \quad (3)$$

where e_1 is the charge of the struck quark and n represents the intermediate di-quark states. The notation $f_{1T}^{\perp}$ follows Ref. [8] except the kinematic factor is included here. At one-loop order, we have contributions from Fig. 1,

$$\begin{aligned} f_{1T}^{\perp}(x,k_\perp) &= \frac{-ig^2 e_1 e_2}{4(2\pi)^3\Lambda(k_\perp^2)}\int \frac{d^4q}{(2\pi)^4}\overline{U}(PS)(\not{k}+m)\gamma^+(\not{k}+\not{q}+m)U(PS)\frac{1}{q^2+i\varepsilon} \\ &\times \frac{2(1-x)-q^+}{q^+ + i\varepsilon}\frac{1}{(k+q)^2-m^2+i\varepsilon}\frac{1}{(P-k-q)^2-\lambda^2+i\varepsilon} + \text{h.c.}, \end{aligned} \quad (4)$$

where q^μ is the gluon momentum. M, m and λ are the masses of the nucleon, quark and diquark, respectively. $U(PS)$ is the on-shell spinor for the nucleon with momentum P and polarization S. $\Lambda(k_\perp^2)$ denotes

$$\Lambda(k_\perp^2) = k_\perp^2 + x(1-x)\left(-M^2 + \frac{m^2}{x} + \frac{\lambda^2}{1-x}\right) \ . \quad (5)$$

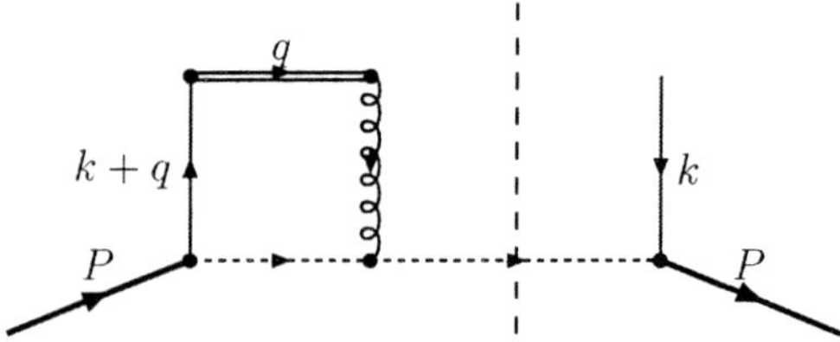

FIGURE 1. One-loop contribution to the spin-dependent transverse momentum distribution in the nucleon

The final result for the Sivers function is

$$f_{1T}^{\perp}(x,k_{\perp}) = \frac{g^2 e_1 e_2}{(2\pi)^4} \frac{(1-x)(m+xM)}{4\Lambda(k_{\perp}^2)} \varepsilon^{+\alpha\beta\gamma} k_{\perp\alpha} P_{\beta} S_{\gamma} \frac{1}{k_{\perp}^2} \ln \frac{\Lambda(k_{\perp}^2)}{\Lambda(0)} \,. \tag{6}$$

With this Sivers function, the SSA found by Brodsky et al.[2] is reproduced, and so this calculation demonstrates that the standard definition of the parton distribution in the non-singular gauge does take into account properly the effects of the final-state interactions [4].

SIVERS FUNCTION IN LIGHT-CONE GAUGE

In the light-cone gauge $A^+ = 0$, however, the light-cone gauge link L vanishes. Where are the final state interactions? To find the answer, we consider all contributions to $f(x,k_{\perp})$ at one-loop order in both Feynman and light-cone gauges[5]. In the light-cone gauge, the gluon propagator has singularity at $q \cdot n = 0$ and requires a regularization[6]. The parton distribution in Eq. (1) is valid for the light-cone gauge only when the following light-cone gauge propagator is used[5]

$$D^{\mu\nu}(q) = \frac{-i}{q^2}\left(g^{\mu\nu} - \frac{q^{\mu}n^{\nu} + q^{\nu}n^{\mu}}{q \cdot n + i\varepsilon}\right) , \tag{7}$$

where the direction of q^{μ} is toward the struck quark in its initial state. In other words, one now does not really has a freedom to choose the regularization for the light-cone singularity, contrary to the popular belief.

In general, if the gauge potential does not vanish at large ξ^-, it has a non-vanishing contribution to a gauge link at $\xi^- = \infty$. Therefore, the definition of the parton distribution in Eq. (1) is no longer gauge invariant because the two light-cone links generated by $\psi(0)$ and $\psi(\xi^-, \xi_{\perp})$ are not connected at $\xi^- = \infty$. If one makes a gauge transformation $U(\xi)$ which does not vanish at $\xi^- = \infty$, an SU(3) matrix $U^{\dagger}(\xi^- = \infty, \xi_{\perp})U(\xi^- = \infty, 0)$ pops up in the distribution after the transformation. Therefore, Eq. (1) must be modified to a form that is invariant under a singular gauge transformation. Motivated by the above consideration, we modify the eikonal phase in Eq. (1) to,

$$L_0(\infty,0) \to \Delta L = P\exp\left(-ig\int_0^{\infty} d\xi_{\perp} \cdot A_{\perp}(\xi^- = \infty, \xi_{\perp})\right) , \tag{8}$$

where the path in the transverse direction is largely arbitrary.

Consider the parton distribution in Eq. (1) with the gauge link ΔL,

$$f_{1T}^{\perp}(x,k_\perp) = \frac{1}{2}\sum_n \int \frac{d\xi^- d^2\xi_\perp}{(2\pi)^3} e^{-i(\xi^- k^+ - \vec{\xi}_\perp \vec{k}_\perp)} \langle P|\overline{\psi}(\xi^-,\xi_\perp)|n\rangle \times \langle n| \left(-ie_1 \int_{\xi_\perp}^{\infty} d\xi'_\perp \cdot A_\perp(\infty,\xi'_\perp) \right) \gamma^+ \psi(0)|P\rangle + \text{h.c.}\,. \quad (9)$$

Going to the momentum space, we have

$$f_{1T}^{\perp}(x,k_\perp) = \frac{ig^2 e_1 e_2}{4(2\pi)^3 \Lambda(k_\perp)} \int \frac{d^4q}{(2\pi)^4} \overline{U}(PS)(\not{k}+m)\gamma^+(\not{k}+\not{q}+m)U(PS) \times \frac{e^{iq^+\infty}}{q^+} \frac{(2(1-x)-q^+)}{(k+q)^2-m^2+i\varepsilon} \frac{1}{(P-k-q)^2-\lambda^2+i\varepsilon} \frac{1}{q^2+i\varepsilon} + \text{h.c.} \quad (10)$$

where $1/q^+$ comes from the $n^- q^\perp/q^+$ term in the light-cone propagator for the gluon. Using

$$\lim_{L\to\infty} \frac{e^{iq^+L}}{q^+} = i\pi\delta(q^+)\,, \quad (11)$$

which is true in the sense of principal-valued distribution, we recover the result in Eq. (6).

A consistency check follows when replacing q^+ by $q^+ + i\varepsilon$ in Eq. (11). The exponential factor becomes $\exp(-\varepsilon\infty) = 0$. Therefore, if Eq. (7) is used from the light-cone gauge propagator, the new gauge link does not contribute. However, *the parton distributions defined with the extra gauge link free one from choosing a specific prescription for* $1/q^+$[6]. Any prescriptions in fact will yield the same result.

SUMMARY AND DISCUSSIONS

To summarize, we have shown that the final state interactions can be taken into account in the light-cone gauge by either a gauge propagator chosen according to the physics of light-cone gauge link in the usual parton distribution, or an extra gauge link at $\xi^- = \infty$ in the parton distribution.

In the Drell-Yan porcess, the gauge link in the parton distributions arises from the initial state interactions rather than from the final state. Correspondingly, the gauge link in Eq.(1) for the parton distributions will end up with $\xi^- = -\infty$, and the extra gauge link in light-cone gauge in Eq. (8) will take integral at $\xi^- = -\infty$ as well. As a consequence, the Sivers function for DY process will have overall sign difference from the DIS process[2, 4, 6], and the naive universality of the parton distributions will not be valid any more.

Several interested transverse momentum dependent parton distribution functions are sensitive to the quark orbital angular momentum of the proton, which is very important to understand the proton spin sum rule[9]. There are many observables which are potentially sensitive to, although they do not directly *measure* the orbital angular momentum itself. For example, the Pauli form factor $F_2(Q^2)$ of the proton, the generalized parton distributions, higher-twist structure functions, and the $P_\perp$-dependent parton distributions. All of these observables have been recently correlated in the framework of light-cone wave functions for three-quark Fock state of the proton[10].

ACKNOWLEDGMENTS

The authors thank A. Belitsky for collaboration, and S. Brodsky for a number of useful discussions.

REFERENCES

1. S. J. Brodsky, P. Hoyer, N. Marchal, S. Peigne, and F. Sannino, hel-ph/0104291.
2. S. J. Brodsky, D. S. Hwang, and I. Schmidt, Phys. Lett. B**530**, 99 (2002); Nucl. Phys. **B642**, 344 (2002).
3. J. C. Collins and D. E. Soper, Nucl. Phys. B **194**, 445 (1982). J. C. Collins and D. E. Soper, Nucl. Phys. B **193**, 381 (1981). [Erratum-ibid. B **213**, 545 (1983).]
4. J. C. Collins, hep-ph/0204004; also J. C. Collins, Phys. Rev. D **57**, 3051 (1998).
5. X. Ji, F. Yuan, Phys. Lett. **B543**, 66 (2002).
6. A. Belitsky, X. Ji, and F. Yuan, hep-ph/0208038.
7. D. W. Sivers, Phys. Rev. D **41**, 83 (1990); Phys. Rev. D **43**, 261 (1991).
8. P. J. Mulders and R. D. Tangerman, Nucl. Phys. B **461**, 197 (1996); Erratum-ibid. **484**. 538 (1997); D. Boer and P. J. Mulders, Phys. Rev. D **57**, 5780 (1998).
9. X. Ji, Phys. Rev. Lett. **78**, 610 (1997).
10. X. Ji, J.P. Ma, F. Yuan, hep-ph/0210430.

Transversity in Exclusive and Inclusive Processes

Leonard Gamberg*, Gary R. Goldstein† and Karo A. Oganessyan**‡

*Division of Science, Penn State-Berks Lehigh Valley College, Reading, PA 19610, USA [1]
†Department of Physics and Astronomy, Tufts University, Medford, MA 02155, USA [2]
**INFN-Laboratori Nazionali di Frascati, Enrico Fermi 40, I-00044 Frascati, Italy
‡DESY, Notkestrasse 85, 22603 Hamburg, Germany

Abstract. Both meson photoproduction and semi-inclusive deep inelastic scattering can potentially probe transversity properties of the nucleon. We explore how that potential can be realized dynamically. The role of rescattering in both exclusive and inclusive meson production as a source for single spin asymmetries is examined. Using a dynamical model, we evaluate the spin independent $\cos 2\phi$ asymmetry associated with transversity of quarks inside unpolarized hadrons, at HERMES kinematics. We also explore the effects of rescattering on the transversity distribution of the nucleon.

Transversity

It is well known that the leading twist transversity distribution $h_1(x)$ [1] and its first moment, the tensor charge, being chiral odd cannot be accessed in deep inelastic scattering. However, $h_1(x)$ can be probed when at least two hadrons are present, *e.g.* Drell Yan [2] or semi-inclusive deep inelastic scattering (SIDIS). In the latter process at leading twist, the property of quark transversity can be measured via the azimuthal asymmetry in the fragmenting hadron's momentum and spin distributions. For example, spinless hadrons produced in the so-called Collins asymmetry [3] depend on the transverse momentum of quarks in the target, k_T, and fragmentation functions, p_T [4]. Including transverse momentum leads to an increase in the number of leading twist distribution and fragmentation functions (*e.g.* $h_{1T}(x,k_T), h_{1T}^{\prime}(x,k_T), H_1^{\perp}(z,p_T)$). Allowing time reversal odd (T-odd) quark distribution functions [5, 6, 7], $h_1^{\perp}(x,k_T)$, $f_{1T}^{\perp}(x,k_T)$ suggests they enter the semi-inclusive unpolarized momentum, and polarized spin, asymmetries. That is, the distribution $f_{1T}^{\perp}(x,k_T)$ representing the number density of *unpolarized* quarks in transversely polarized nucleons, maybe entering the recent measurements of SSAs at HERMES and SMC in semi-inclusive pion electroproduction [8]. Alternatively, $h_1^{\perp}(x,k_T)$ which describes the transfer of transversity to quarks inside unpolarized hadrons may enter transverse momentum dependent asymmetries. Beyond the T-odd properties, the existence of these distributions are a signal of the *essential* role played by the intrinsic transverse quark momentum and the corresponding angular momentum of quarks inside the target and fragmenting hadrons in these hard scattering processes.

Further insight into transversity has come from analyzing quark-target helicity flip

---

[1] Supported by a Research Development Grant, Penn State Berks.
[2] Supported by the US Department of Energy DE-FG02-29ER40702.

CP675, *Spin 2002: 15th Int'l. Spin Physics Symposium and Workshop on Polarized Electron Sources and Polarimeters,* edited by Y. I. Makdisi, A. U. Luccio, and W. W. MacKay

amplitudes in deeply virtual Compton scattering (DVCS)[9]. Angular momentum conservation requires that helicity changes are accompanied by transferring 1 or 2 units of orbital angular momentum; *again highlighting* the essential role of intrinsic k_T and orbital angular momentum in determining transversity.

The Exchange Picture

The interdependence of transversity on quark *orbital* angular momentum and k_T is more general than suggested in the above discussion on SSAs and the GPD analysis of transversity. This behavior arises in ref. [10] where we study the vertex function associated with the tensor charge in exclusive meson production. Again, angular momentum conservation results in the transfer of orbital angular momentum $\ell = 1$ carried by the dominant $J^{PC} = 1^{+-}$ mesons to compensate for the non-conservation of helicity across the vertex. Transverse momentum dependence arises from the axial vector mesons that *dominate* the tensor coupling (They are the C-odd – $h_1(1170)$, $h_1(1380)$, $b_1(1235)$). These mesons are in the $(35 \otimes \ell = 1)$ multiplet of the $SU(6) \otimes O(3)$ symmetry group that best represents the mass symmetry among the low lying mesons. Along with axial vector dominance this symmetry results in the isoscalar and isovector contribution to the tensor charge

$$\delta u(\mu^2) - \delta d(\mu^2) = \frac{5}{6}\frac{g_A}{g_V}\frac{M_{a_1}^2}{M_{b_1}^2}\frac{\langle k_T^2 \rangle}{M_N M_{b_1}}, \quad \delta u(\mu^2) + \delta d(\mu^2) = \frac{3}{5}\frac{M_{b_1}^2}{M_{h_1}^2}\delta q^v. \tag{1}$$

Each depends on two powers of the average intrinsic quark momentum $\langle k_T^2 \rangle$, because the tensor couplings involve helicty flips associated with kinematic factors of 3-momentum transfer. as required by angular momentum conservation.

The k_T dependence can be understood on fairly general grounds from the kinematics of the exchange picture in exclusive pseudoscalar meson photoproduction. For large s and relatively small momentum transfer t simple combinations of the four helicity amplitudes involve definite parity exchanges. The four independent helicity amplitudes can have the minimum kinematically allowed powers,

$$f_1 = f_{1+,0+} \propto k_T^1, \quad f_2 = f_{1+,0-} \propto k_T^0, \quad f_3 = f_{1-,0+} \propto k_T^2, \quad f_4 = f_{1-,0-} \propto k_T^1.$$

However, in single hadron exchange (or Regge pole exchange) parity conservation requires $f_1 = \pm f_4$ and $f_2 = \mp f_3$ for even/odd parity exchanges. These pair relations, along with a single hadron exchange model, force f_2 to behave like f_3 for small t. This introduces the k_T^2 factor into f_2. However for a non-zero polarized target asymmetry to arise there must be interference between single helicity flip and non-flip and/or double flip amplitudes. Thus this asymmetry must arise from rescattering corrections (or Regge cuts or eikonalization or loop corrections) to single hadron exchanges. That is, one of the amplitudes in

$$P_y = \frac{2Im(f_1^* f_3 - f_4^* f_2)}{\sum_{j=1...4} |f_j|^2} \tag{2}$$

must acquire a different phase. In fact rescattering reinstates $f_2 \propto k_T^0$ by integrating over loop k_T, which effectively introduces a $\langle k_T^2 \rangle$ factor [11]. This is true for the *inclusive process* as well, where only one final hadron is measured; a relative phase in a helicity flip three body amplitude is required.

Rescattering and SIDIS

Recently a rescattering approach was applied to the calculation of SSA in pion electroproduction, using a QCD motivated quark-diquark model of the nucleon [12] (BHS). In Ref. [13, 6] the rescattering effect is interpreted as giving rise to the T-odd Sivers [14] $f_{1T}^{\perp}$ function; the number density of unpolarized quarks in a transversely polarized target. Being T-odd, this asymmetry vanishes at tree level. The important lesson beyond the model calculation, is that, theoretically, final state interactions are essential for producing non-zero SSAs. Furthermore, the phenomenological determination of quark spin distributions can be disentangled from measurements of SSAs.

We have investigated the rescattering contributions to the transversity distribution $h_1(x)$ and the T-odd function $h_1^{\perp}(x)$ and corresponding asymmetries in SIDIS. Collins [3] considered one such process, the production of pions from transversely polarized quarks in a transversely polarized target. The corresponding SSA involves the convolution of the transversity distribution function and the T-odd fragmentation function, $h_1(x) \star H_1^{\perp}(z)$ [15, 16]. The transversity distribution function is defined through the light-cone quark distribution with gauge link indicated,

$$\begin{aligned} s_T^i \Delta f_T(x,k_T) &= \frac{1}{2}\sum_n \int \frac{d\xi^- d^2\xi_\perp}{(2\pi)^3} e^{-i(\xi^- k^+ - \vec{\xi}_\perp \vec{k}_\perp)} \langle P|\overline{\psi}(\xi^-,\xi_\perp)|n\rangle \\ & \langle n| \left(-ie_1 \int_0^\infty A^+(\xi^-,0)d\xi^-\right) \gamma^+ \gamma^i \gamma^5 \psi(0)|P\rangle + \text{h.c.}, \end{aligned} \tag{3}$$

where e_1 is the charge of the struck quark and n represents intermediate diquark states. Performing the loop integration over q^{μ}, we obtain

$$\begin{aligned} s_T^i \Delta f_T(x,k_T) = \frac{e_1 e_2 g^2}{2(2\pi)^4} \frac{1-x}{\Lambda(k_T^2)} \Bigg\{ \left(S_T^i \left[\left(m + xM \right)^2 + k_T^2 \right] + 2k_T^i \mathbf{S}_T \cdot \mathbf{k}_T \right) \\ \times \frac{1}{k_T^2 + \Lambda(0)^2 + \lambda_g^2} \left(\ln \frac{\Lambda(k_T^2)}{\Lambda(0)} + \ln \frac{k_T^2 + \lambda_g^2}{\lambda_g^2} \right) - \left(S_T^i k_T^2 + 2k_T^i \mathbf{S}_T \cdot \mathbf{k}_T \right) \frac{1}{k_T^2} \ln \frac{\Lambda(k_T^2)}{\Lambda(0)} \Bigg\}, \end{aligned} \tag{4}$$

where $\Lambda(k_T^2) = k_T^2 + x(1-x)\left(-M^2 + \frac{m^2}{x} + \frac{\lambda^2}{1-x}\right)$. The (Abelian) gluon mass (usually chosen at $\lambda_g \approx 1\ GeV$) is indicative of χSB scale and appears here to regulate the IR divergence. This one loop contribution constitutes the next order term in an eikonalization. The first part has the same nucleon spin dependent structure as a tree level model calculation (modified by the log terms) - it is leading twist and a combination of $h_{1T}(x,k_T)$ and $h_{1T}^{\perp}(x,k_T)$. The second part has a different structure than tree level - it appears as a rescattering effect only. It is IR finite and, in this model, is proportional to the one loop result for $f_{1T}^{\perp}$ [13] and $\mathscr{P}_y$ in BHS. The ratio of $h_{1T}(x,k_T)$ to $h_{1T}^{\perp}(x,k_T)$ will differ from the tree level. When combined with a measure of transversely polarized quarks, the fragmentation function $H_1^{\perp}(z)$, the integrated $h_1(x)$ will contribute to the observable weighted meson azimuthal asymmetry from a transversely polarized nucleon [4, 16].

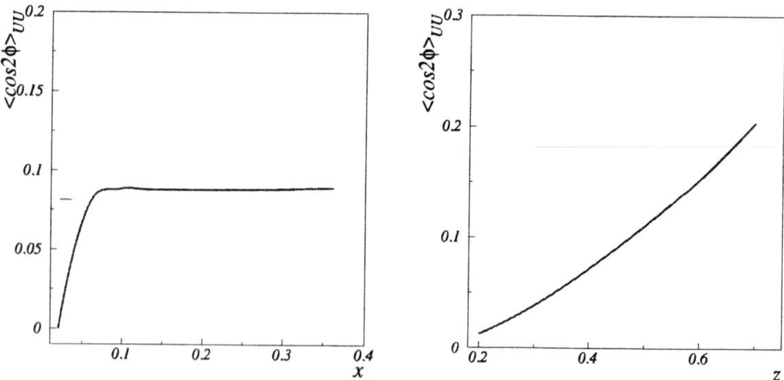

FIGURE 1. Left Panel: The $\langle\cos 2\phi\rangle_{UU}$ asymmetry for π^+ production as a function of x. Right Panel: The $\langle\cos 2\phi\rangle_{UU}$ asymmetry for π^+ production as a function of z.

As mentioned in the introduction, the T-odd structure function $h_1^\perp(x,k_T)$ is of great interest theoretically, since it vanishes at tree level, and experimentally, since its determination does not involve polarized nucleons. Repeating the calculation above *without nucleon polarization* leads to the result [17]

$$\frac{\varepsilon_{+-\perp j}k_{Tj}}{M}h_1^\perp(x,k_T) \quad = \quad \frac{e_1 e_2 g^2}{2(2\pi)^4}\frac{(m+xM)(1-x)}{\Lambda(k_T^2)}\varepsilon_{+-\perp j}k_{Tj}\frac{1}{k_T^2}\ln\frac{\Lambda(k_T^2)}{\Lambda(0)}. \tag{5}$$

It is leading twist and IR finite. Being T-odd it will appear in SIDIS observables along with T-odd fragmentation functions. In particular, the following weighted semi-inclusive DIS cross section projects out a leading $\cos 2\phi$ asymmetry [16],

$$\langle\frac{|P_{h\perp}^2|}{MM_h}\cos 2\phi\rangle_{UU} = \frac{\int d^2P_{h\perp}\frac{|P_{h\perp}^2|}{MM_h}\cos 2\phi d\sigma}{\int d^2P_{h\perp}d\sigma} = \frac{8(1-y)\sum_q e_q^2 h_1^{\perp(1)}(x)z^2H_1^{\perp(1)}(z)}{(1+(1-y)^2)\sum_q e_q^2 f_1(x)D_1(z)} \tag{6}$$

where the subscript UU indicates unpolarized beam and target(Note: The non-vanishing $\cos 2\phi$ asymmetry originating from kinematical and dynamical effects only appears at order $1/Q^2$ [18, 4, 19]). The functions $h_1^{\perp(1)}(x)$ and $H_1^{\perp(1)}(z)$ are the weighted moments of the distribution and fragmentation functions

$$h_1^{\perp(1)}(x) \equiv \int d^2k_T\frac{k_T^2}{2M^2}h_1^\perp(x,k_T), \quad H_1^{\perp(1)}(z) \equiv z^2\int d^2p_T\frac{k_T^2}{2M_h^2}H_1^\perp(z,-p_T). \tag{7}$$

k_T, p_T, are the transverse momentum of the quark in the target proton, and fragmenting quark respectively and M, M_h are the mass of the target proton and produced hadron. We evaluate the $\langle\cos 2\phi\rangle_{UU}$ asymmetry obtained from the approximation,

$$\langle\cos 2\phi\rangle_{UU} \approx \frac{MM_h}{\langle P_{h\perp}^2\rangle}\langle\frac{|P_{h\perp}^2|}{MM_h}\cos 2\phi\rangle_{UU} \tag{8}$$

in the HERMES kinematic range corresponding to 1 GeV^2 $\leq Q^2 \leq$ 15 GeV^2, 4.5 GeV $\leq E_\pi \leq$ 13.5 GeV, $0.2 \leq z \leq 0.7$, $0.2 \leq y \leq 0.8$, and taking $\langle P_{h\perp}^2\rangle = 0.25$ GeV^2

as input. The Collins ansatz [3, 4] for the analyzing power of transversely polarized quark fragmentation function $H_1^{\perp(1)}(z)$, has been adopted [20]. For $D_1(z)$, the simple parameterization from Ref. [21] was used. In Fig1 the $\langle\cos 2\phi\rangle_{UU}$ of Eq.(6) for π^+ production on a proton target is presented as a function of x and z, respectively. Using $\Lambda_{QCD} = 0.2\,GeV$ and $\mu = 0.8\,GeV$, Fig.1 indicates that the $\cos 2\phi$ asymmetry related to is large enough (about 8%) to be measured [22].

Conclusions

The interdependence of intrinsic transverse quark momentum and angular momentum conservation are intimately tied with the studies of transversity. This is demonstrated from analyses of the tensor charge in the context of the vector dominance approach to exclusive meson photo-production, to SSAs in SIDIS[10]. In the study of the tensor charge we find $\langle k_T^2\rangle$ factor that appears in rescattering models in meson photoproduction where interference phenomena are non-zero due to rescattering. In the case of unpolarized beam and target we have predicted at HERMES energies the sizable $\cos 2\phi$ asymmetry associated with the asymmetric distributions of transversely polarized quarks inside of unpolarized hadrons.

Acknowledgements

L.G. thanks the organizers of SPIN 2002 for the invitation to present this work. I also thank Daniel Boer, Dennis Sivers and Feng Yuan for useful discussions.

REFERENCES

1. X. Artu and M. Mekhfi, Z. Phys. C45 (1990) 669; R. L. Jaffe and X. Ji, Phys. Rev. Lett. 67 (1991) 552; Nucl. Phys. B375 (1992) 527.
2. J. Ralston and D. E. Soper, Nucl. Phys. B152 (1979) 109.
3. J.C. Collins, Nucl. Phys. B396 (1993) 161.
4. A. M. Kotzinian, Nucl. Phys B441 (1995) 234.
5. M. Anselmino and F. Murgia, Phys. Lett B442 (1998) 470; M. Anselmino, V. Barone, A. Drago and F. Murgia, hep-ph/0209073.
6. J. C. Collins, Phys. Lett. B536 (2002) 43.
7. R. D. Tangerman and P. J. Mulders, Phys. Lett. B352 (1995) 129; Phys. Rev. D51 (1995) 3357; Nucl. Phys. B461 (1996) 197.
8. A. Airapetian *et al.*, Phys. Rev. Lett. 84 (2000) 4047; A. Bravar, Nucl. Phys. Proc. Suppl. 79 (1999) 520.
9. P. Hoodbhoy and X. Ji, Phys. Rev. D58 (1998) 054006; M. Diehl, Eur. Phys. J. C19 (2001) 485.
10. L. Gamberg and G. R. Goldstein, Phys. Rev. Lett. 87 (2001) 242001.
11. G. R. Goldstein and J. F. Owens, Phys. Rev. D7 (1973) 865; Nucl. Phys. B71 (1974) 461.
12. S. Brodsky, D.S. Hwang and I. Schmidt, Phys. Lett. B530 (2002) 99.
13. X. Ji and F. Yuan, Phys. Lett. B543 (2002) 66; A.V. Belitsky, *et al.*, hep-ph/0208038.
14. D. Sivers, Phys. Rev D 41 (1990) 83 ; Phys. Rev. D 43 (1991) 261.
15. A. M. Kotzinian and P. J. Mulders, Phys. Lett. B406 (1997) 373.
16. D. Boer and P. J. Mulders, Phys. Rev. D57 (1998) 5780.
17. G. R. Goldstein and L. Gamberg, hep-ph/0209085.
18. R.N. Cahn, Phys. Lett. B78 (1978) 269; Phys. Rev. D40 (1989) 3107.
19. K.A. Oganessyan, et.al., Eur. Phys. J. C5 (1998) 681.
20. K.A. Oganessyan, N. Bianchi, E. De Sanctis, and W.D. Nowak, Nucl. Phys. A689 (2001) 784.
21. E. Reya, Phys. Rep. 69 (1981) 195.
22. L. Gamberg, G. R. Goldstein and K.A. Oganessyan, hep-ph/0301018 .

Role of T-odd Functions in High Energy Hadronic Collisions

E. Di Salvo

Dipartimento di Fisica ad INFN- sez. Genova
Via Dodecaneso 33 - 16146 Genova - Italy

Abstract. I propose a simple model for predicting the enegy behavior of T-odd, chiral odd function $h_1^\perp$. Furthermore I illustrate a method for extracting $h_1^\perp$ and the transversity function from Drell-Yan. The method may be applied also to other reactions.

INTRODUCTION

The T-odd functions[1-3] have become important in the last ten years, since when high energy physicists realized that such functions could be used as *polarimeters* for extracting chiral odd functions, especially transversity[4-6]. Here I consider the T-odd, chiral odd function $h_1^\perp$[7] of a quark inside the proton. In particular I propose a simple model, which allows to predict the behavior of this function at varying proton momentum. Moreover I am concerned with asymmetries relative to unpolarized and singly polarized Drell-Yan (DY), *i. e.*,

$$pp \to \mu^+\mu^- X. \tag{1}$$

I show how to extract $h_1^\perp$ from this reaction and I suggest an alternative method for determining transversity.

GENERAL FORMULAE

The single transverse spin asymmetry for reaction (1) is defined as

$$A = \frac{d\sigma_\uparrow - d\sigma_\downarrow}{d\sigma_\uparrow + d\sigma_\downarrow}, \tag{2}$$

where $d\sigma_{\uparrow(\downarrow)}$ refer to cross sections with opposite polarizations of one of the proton beams. In one-photon approximation,

$$d\sigma_\uparrow - d\sigma_\downarrow \propto d\Gamma L^{\mu\nu} H^a_{\mu\nu}, \tag{3}$$

$$L_{\mu\nu} = k_\mu k'_\nu + k'_\mu k_\nu - g_{\mu\nu} k\cdot k', \tag{4}$$

$$H^a_{\mu\nu} = \int d^2p_{1\perp} Tr\left[\gamma_\mu \Phi_{\chi.o.}(x_1, \mathbf{p}_{1\perp})\gamma_\nu \bar{\Phi}_{\chi.o.}(x_2, \mathbf{p}_{2\perp}) + (1 \leftrightarrow 2)\right]. \tag{5}$$

CP675, *Spin 2002: 15th Int'l. Spin Physics Symposium and Workshop on Polarized Electron Sources and Polarimeters,* edited by Y. I. Makdisi, A. U. Luccio, and W. W. MacKay

Here $d\Gamma$ is the phase space element. k and k' are the four-momenta of the muons. x_1 and x_2 are the longitudinal fractional momenta of the annihilating quark and antiquark, $\mathbf{p}_{1\perp}$ and $\mathbf{p}_{2\perp}$ their transverse momenta with respect to the initial beams and $\Phi_{\chi.o.}$ and $\bar{\Phi}_{\chi.o.}$ the chiral odd components of their correlation matrices. The index 1 in x and $\mathbf{p}_\perp$ refers to the transversely polarized proton, the index 2 to the unpolarized one. $\mathbf{p}_{2\perp}$ is chosen in such a way that the transverse momentum of the muon pair with respect to the proton beam in the laboratory frame, *i. e.*,

$$\mathbf{Q}_\perp = \mathbf{p}_{1\perp} + \mathbf{p}_{2\perp}, \tag{6}$$

is kept fixed. Lastly the sum over flavors has been omitted.

PARAMETRIZATION OF THE T-ODD CORRELATION MATRIX

In the laboratory frame, at sufficiently high energies, the chiral odd component of the correlation matrix of the transversely polarized proton can be parametrized as[7]

$$\Phi_{\chi.o.} = \frac{1}{4} x_1 \mathscr{P} \gamma_5 \left\{ [\not{S}, \not{n}_+] h_{1T} + \frac{1}{\mu} [\not{r}_\perp, \not{n}_+] h_1^\perp \right\}. \tag{7}$$

Here h_{1T} is the transverse momentum dependent transversity distribution, while $h_1^\perp$ will be illustrated in a moment. Moreover

$$\begin{aligned} r_\perp &= p_{1a} S - p_{1b} n_a \equiv (0, -p_{1b}, p_{1a}, 0), & (8)\\ p_{1a} &= \mathbf{p}_{1\perp} \cdot \mathbf{S} \times \mathbf{n}, \qquad p_{1b} = \mathbf{p}_{1\perp} \cdot \mathbf{S}, & (9)\\ n_+ &\equiv (1, \mathbf{n}), \qquad n_a \equiv (0, \mathbf{S} \times \mathbf{n}). & (10) \end{aligned}$$

$\mathscr{P}\mathbf{n}$ and $S \equiv (0, \mathbf{S})$ are respectively the momentum and the Pauli-Lubanski four-vector of proton 1, $\mathbf{S}$ and $\mathbf{n}$ being unit vectors such that $\mathbf{S} \cdot \mathbf{n} = 0$. Lastly μ is an undetermined mass scale, which was set equal to the proton mass by various authors[7-9]; as I shall show, this is not the most suitable choice.

The second term of parametrization (7) is T-odd and gives a nonvanishing contribution also when the proton is unpolarized. In this case, given a unit vector $\mathbf{s}$ not parallel to $\mathbf{n}$, the density of quarks whose spin component along $\mathbf{s}$ is positive, minus the density of quarks for which this spin component is negative, amounts to

$$\delta q_\perp = -\frac{r_\perp \cdot s_0}{\mu} h_1^\perp, \tag{11}$$

where $s_0 \equiv (0, \mathbf{s})$. Eq. (11) is a consequence of eq. (7) for an unpolarized proton. The two equations exhibit the meaning of the function $h_1^\perp$: in an unpolarized proton, a quark with nonzero transverse momentum is polarized perpendicularly to its momentum and to the proton momentum, in agreement with parity conservation.

A MODEL FOR T-ODD FUNCTIONS

A proton may be viewed as a bound state of the active quark with a set X of spectator partons. In order to take into account coherence effects, I project the bound state onto scattering states with a fixed third component of the total angular momentum with respect to the proton momentum, J_z, and with a spin component $s = \pm 1/2$ of the quark along the unit vector **s** introduced in the previous section. For the sake of simplicity, I assume X to have spin zero, moreover I choose a state with $J_z = 1/2$. Then

$$|J_z = 1/2; s; X\rangle = \alpha| \rightarrow, L_z = 0; s; X\rangle + \beta| \leftarrow, L_z = 1; s; X\rangle. \tag{12}$$

Here $\rightarrow$ ($\leftarrow$) and L_z denote the components along **n**, respectively, of the quark spin and orbital angular momentum, while α and β are Clebsch-Gordan coefficients. Then the probability of finding a quark with $J_z = 1/2$ and spin component s along **s**, in a longitudinally polarized proton with a positive helicity, is

$$\begin{aligned} |\langle P,\Lambda = 1/2|J_z = 1/2; s; X\rangle|^2 &= \alpha^2|\langle P,\Lambda = 1/2| \rightarrow, L_z = 0; s; X\rangle|^2 \\ &+ \beta^2|\langle P,\Lambda = 1/2| \leftarrow, L_z = 1; s; X\rangle|^2 + I, \end{aligned} \tag{13}$$

$$I = 2\alpha\beta Re[\langle P,\Lambda = 1/2| \rightarrow, L_z = 0; s; X\rangle\langle(\leftarrow, L_z = 1; s; X)|P,\Lambda = 1/2\rangle]. \tag{14}$$

Expanding the amplitudes in partial waves yields

$$I = 2\sum_{l,l'=0}^{\infty} Re\left[ie^{-i\phi}A_l B^*_{l'}\right] P_l(cos\theta)P^1_{l'}(cos\theta). \tag{15}$$

Here A_l and B_l are related to partial wave amplitudes; moreover θ and ϕ are respectively the polar and the azimuthal angle of the quark momentum, assuming **n** as the polar axis and, as the azimuthal plane, the one through **n** and **s**. In the Breit frame one has

$$P_l(cos\theta) \sim 1, \qquad P^1_l(cos\theta) \sim \frac{|\mathbf{p}_{1\perp}|}{x\mathscr{P}}. \tag{16}$$

Then eq. (15) yields

$$I \sim \frac{|\mathbf{p}_{1\perp}|}{x\mathscr{P}}(Acos\phi + Bsin\phi), \tag{17}$$

where A and B are real functions made up with A_l and B_l. Since **s** is an axial vector, parity conservation implies $A = 0$. Therefore eqs. (13) and (17) imply that the interference term I is T-odd and that the final quark is polarized perpendicularly to the proton momentum and to the quark momentum, independent of the proton polarization. Comparing eq. (17) with eq. (11) yields

$$\mu = x\mathscr{P}. \tag{18}$$

Eqs. (18) predicts that the quark tranverse polarization in an unpolarized (or spinless) hadron decreases as $\mathscr{P}^{-1}$.

EXTRACTING CHIRAL ODD FUNCTIONS FROM DY

The transversity function

Eqs. (3), (5) and (7) imply that the numerator of the DY asymmetry (2) is of the form

$$d\sigma_\uparrow - d\sigma_\downarrow \propto \int d^2 p_{1\perp} \left[\left(\frac{p_{2a}}{x_2 \mathscr{P}} h_{1T} + \frac{\mathbf{p}_{1\perp} \cdot \mathbf{p}_{2\perp}}{x_1 x_2 \mathscr{P}^2} h_1^\perp \right) \bar{h}_1^\perp + (1 \leftrightarrow 2) \right], \tag{19}$$

assuming the constraint (6). Here

$$p_{2a} = \mathbf{p}_{2\perp} \cdot \mathbf{S} \times \mathbf{n}. \tag{20}$$

In order to extract the transversity, *i. e.*, $h_1 = \int d^2 p_\perp h_{1T}$, from DY, I define the following weighted asymmetry[10,11]:

$$\langle A_1 \rangle = \frac{\Sigma_n d\sigma^{(n)} Q_a^{(n)}}{M_P \Sigma_n d\sigma^{(n)}}, \qquad Q_a = p_{1a} + p_{2a}. \tag{21}$$

Here M_P is the proton rest mass and $d\sigma^{(n)}$ the differential cross section at a fixed transverse momentum, the sum running over the data. Eq. (19) implies

$$\sum_n d\sigma^{(n)} Q_a^{(n)} \propto h_1(x_1) \bar{h}_{1(1)}^\perp (x_2) + \bar{h}_1(x_1) h_{1(1)}^\perp (x_2), \tag{22}$$

$$h_{1(1)}^\perp (x_2) = \int d^2 p_\perp p_{2a}^2 h_1^\perp. \tag{23}$$

This allows to extract h_1 and $\bar{h}_1$, provided $h_{1(1)}^\perp$ and $\bar{h}_{1(1)}^\perp$ are known. These functions have to be inferred from an independent analysis, for example with the method I exhibit in the next subsection. According to my model, one has $\langle A_1 \rangle \propto \mathscr{P}^{-1}$.

The function $h_1^\perp$

$h_1^\perp$, which is washed out by the weighted asymmetry (21), can be singled out by using an alternative weight function. Indeed, defining

$$\langle A_1' \rangle = \frac{\Sigma_n (d\sigma_\uparrow^{(n)} - d\sigma_\downarrow^{(n)}) (Q_a^{(n)})^2}{M_P^2 \Sigma_n d\sigma^{(n)}}, \tag{24}$$

formula (19) yields

$$\langle A_1' \rangle \propto h_{1(1)}^\perp (x_1) \bar{h}_{1(1)}^\perp (x_2) + \bar{h}_{1(1)}^\perp (x_1) h_{1(1)}^\perp (x_2), \tag{25}$$

where $h_{1(1)}^\perp (x_1)$ is defined analogously to eq. (23). Formula (19) implies that the weight functions $\mathbf{Q}_\perp^2$ and $(\mathbf{Q}_\perp \cdot \mathbf{S})^2$ could be used instead of Q_a^2. Moreover $h_1^\perp$ may be extracted also from unpolarized DY; in this case the weighted asymmetry is defined as

$$\langle A_1''\rangle = \frac{\Sigma_{n_+} d\sigma^{(n_+)}\left[\mathbf{Q}_\perp^{(n+)}\right]^2 - \Sigma_{n_-} d\sigma^{(n_-)}\left[\mathbf{Q}_\perp^{(n_-)}\right]^2}{M_P^2 \Sigma_n d\sigma^{(n)}}. \tag{26}$$

Here the symbols $\pm$ refer to events whose transverse momenta $\mathbf{Q}_\perp$ are at the right (+) or at the left (-) of the plane through **n** and the unit vector **s**. The model I have elaborated predicts that both $\langle A_1'\rangle$ and $\langle A_1''\rangle$ decrease as $\mathcal{P}^{-2}$.

DISCUSSION

I have suggested how to extract h_1 and $h_1^\perp$ from DY. In particular, the T-odd function $h_1^\perp$ can be inferred from unpolarized proton-proton collisions and it can be used as a quark polarimeter in order to get h_1 from single transverse spin asymmetry. Similar methods could be elaborated for semi-inclusive deep inelastic scattering(SIDIS)[12], for $e^+e^- \to \pi X$ and, with some assumptions, also for $pp \to \pi X$. In these reactions, in order to infer the tranversity, the Collins function[13] is needed, which can be deduced, for example, from e^+e^- collisions. Moreover in such kinds of experiments, as well as in DY, one is faced with the problem of disentangling the contributions of quarks and antiquarks of any flavor, as pointed out by Boglione and Leader[14]. In this connection, a comparison between SIDIS and DY results is particularly helpful, since the two cross sections depend on the quark and antiquark functions according to different combinations.

REFERENCES

1. J. Collins: Nucl. Phys. B **396** (1993) 161
2. R.L. Jaffe: Phil. Trans. Roy. Soc. Lond. A **359** (2001) 391
3. A. V. Efremov: hep-ph/0001214 and hep-ph/0101057
4. X. Ji: Phys Lett. B **284** (1992) 137
5. R.L. Jaffe, X. Jin and J. Tang: Phys. Rev. Lett. **80** (1998) 1166; Phys. Rev. D **57** (1998) 5920
6. R.L Jaffe: "2nd Topical Workshop on Deep Inelastic Scattering off polarized targets: Theory meets Experiment (Spin 97)", Proceedings eds. J. Blümlein, A. De Roeck, T. Gehrmann and W.-D. Nowak. Zeuten, DESY, (1997) p. 167
7. D. Boer and P.J. Mulders: Phys. Rev. D **57** (1998) 5780
8. R.J. Ralston and D.E. Soper: Nucl Phys. B **152** (1979) 109
9. P.J. Mulders and R.D. Tangerman: Nucl Phys. B **461** (1996) 197
10. A.M. Kotzinian and P.J. Mulders: Phys. Rev. D **54** (1996) 1229
11. A.M. Kotzinian and P.J. Mulders: Phys. Lett. B **406** (1997) 373
12. HERMES coll., Airapetian et al.: Phys. Rev. Lett. **84** (2000) 4047
13. J. Collins: Nucl Phys. B **396** (1993) 161
14. M. Boglione and E. Leader: Phys. Rev. D **61** (2000) 114001

Prospects of the Gluon Polarization Measurement at PHENIX

Y. Goto for the PHENIX Collaboration [1]

RIKEN (The Institute of Physical and Chemical Research), Wako, Saitama 351-0198, Japan
and
RIKEN BNL Research Center, Brookhaven National Laboratory, Upton, NY 11973, USA

Abstract. The Relativistic Heavy Ion Collider (RHIC) started operation as a polarized proton collider in December, 2001 with transverse-spin beams. From the data collected by the PHENIX experiment, we will report measurements of single transverse-spin asymmetries. From 2003, we will start the gluon polarization measurement with longitudinal-spin collisions. In this article, we report on the systematic studies performed in the previous run and then discuss the processes by which PHENIX intends to measure the gluon polarization during the next several years.

INTRODUCTION

Since deep inelastic scattering (DIS) experiments of polarized leptons from polarized nucleons showed that only 10–30% of the proton spin is carried by the quarks and anti-quarks, we have been pursuing origin of the missing spin. In the PHENIX experiment [1] at the Relativistic Heavy Ion Collider (RHIC), the gluon polarization can be measured over a large range of gluon momentum fraction (x_{gluon}) by using many processes.

The RHIC started operation as a polarized proton collider in December, 2001 with

TABLE 1. Summary of the physics processes to be measured at PHENIX as probes of the gluon polarization. The full luminosity corresponds to 320 pb $^{-1}$ at $\sqrt{s}$ = 200 GeV and 800 pb^{-1} at $\sqrt{s}$ = 500 GeV for 10-week operation.

	luminosity	channels	x_{gluon} coverage
run-3 (2003)	3 pb^{-1} @ $\sqrt{s}$ = 200 GeV	π^0 / $\pi^{\pm}$	0.05–0.2
		single-electron	0.005–0.1
baseline plan (2004–)	10% – full luminosity @ $\sqrt{s}$ = 200 GeV and 500 GeV	prompt-γ	0.04–0.3
		J/ψ e-μ coincidence	0.005–0.2
upgrade plan (2005–)	full luminosity @ $\sqrt{s}$ = 200 GeV and 500 GeV	γ–jet heavy flavor with displaced vertex	wider

[1] For the full PHENIX Collaboration author list and acknowledgements, see Appendix "Collaborations" of this volume.

CP675, *Spin 2002: 15th Int'l. Spin Physics Symposium and Workshop on Polarized Electron Sources and Polarimeters*, edited by Y. I. Makdisi, A. U. Luccio, and W. W. MacKay

transverse-spin collisions. In this run (run-2), we measured single transverse-spin asymmetries (A_N) of neutral pions, charged hadrons, J/ψ's, single-muons, *etc*. In addition, we performed many systematic studies for the future measurement of the gluon polarization. One of the important systematic studies was the measurement of the relative luminosity, and another was the development of a device to monitor polarization of beams at the PHENIX collision point, called the local polarimeter.

Beginning in run-3, we will start to probe the gluon polarization with measurements in many channels. Table 1 summarizes the physics processes which will be measured and are discussed in this article.

SYNOPSIS OF THE RUN-2 (2001–2002)

In run-2, PHENIX was operated with both Central Arms and the South Muon Arm. First level trigger systems were developed and operational for the polarized proton run. One of them was an EM calorimeter trigger which selected high-p_T particles in the Central Arms [2], and another one was a muon-identification trigger in the Muon Arm [3]. The DAQ system was upgraded to deal with high event rate in the proton run and was able to handle 1 kHz of triggers with a 70 MB/sec bandwidth.

The maximum value of the luminosity reached up to 1.5×10^{30} cm^{-2} sec^{-1} and the integrated luminosity recorded at PHENIX was 0.15 pb^{-1}. The beam polarization was 14% on average in the Blue ring, 17% on average in the Yellow ring, and the maximum value achieved was 25%. This low polarization was a result of the slow ramp rate of the backup motor generator in the AGS, the use of which was necessitated by the breakdown of the primary system during run-2. Despite this, we have many A_N measurements. In the Central Arms which cover mid-rapidity region ($x_F \sim 0$), we measured neutral pions [4], charged hadrons [2], J/ψ's [3], *etc*. In the South Muon Arm which covers forward-rapidity region ($1.2 < \eta < 2.2$), we also measured J/ψ's [3] and single-muons [5]. We have already reported cross sections of π^0 and J/ψ, which are vital for the systematic study to understand the detector performance for asymmetry measurements.

In addition, we performed many systematic studies for the future double longitudinal-spin asymmetry (A_{LL}):

$$A_{LL} = \frac{1}{P_B \cdot P_Y} \cdot \frac{N_{++} - R \cdot N_{+-}}{N_{++} + R \cdot N_{+-}}$$

where P_B and P_Y are the polarization of colliding beams in the RHIC rings, N_{++} (N_{+-}) are the number of events or yields from collisions with parallel (antiparallel) beam helicity, and R is ratio of the luminosities for collisions with parallel (L_{++}) and antiparallel (L_{+-}) beam helicity (L_{++}/L_{+-}).

In the A_{LL} measurement, the relative luminosity (R) measurement is very important to normalize the parallel-helicity yield and the antiparallel-helicity yield. In the measurement of neutral pions and charged hadrons which will be discussed later, our goal is to measure a 0.3% level asymmetry, so we require a sub-0.1% level measurement of the relative luminosity. To meet this challenge, we made crossing-by-crossing scalers to measure the counts in four trigger detectors for each of the 120 bunch crossing at PHENIX. For this luminosity-scaler counters, we used the Beam-Beam Counter (BBC) which covers $3.0 < |\eta| < 3.9$, the Normalization Trigger Counter (NTC) which extends the rapidity coverage of the BBC, and the Zero-Degree Calorimeter (ZDC) [6] which detects neutrons at the most forward region. For the other scaler, we used the minimum-bias trigger made by the BBC and the NTC. By using these crossing-by-crossing scalers, we found the relative luminosity measurement had a 0.3% systematic error in good RHIC fills. By investigating the vertex distribution and other data for each crossing, we found that bunch-by-bunch characteristics of the RHIC beam gave rise to this systematic uncertainty. We are still investigating the relationship between this uncertainty and the accelerator parameters. To improve on this uncertainty in run-3, we plan to install an additional luminosity telescope as a fourth luminosity-scaler counter. We will also have recogging and spin-flip of the beam [7]. With this procedure, we average the bunch-by-bunch characteristics of the RHIC beam and thus anticipate a 10-times better relative luminosity measurement.

In run-2, we also developed a local polarimeter to be used at PHENIX. For the operation with the spin rotators in run-3, we need to confirm that we have longitudinal polarized protons at the PHENIX collision point. Since spin dynamics between the spin rotators is completely transparent to the rest of the accelerator, we need a local polarimeter.

To develop this polarimeter, we installed two calorimeters (one hadron calorimeter and one EM calorimeter) at a previously uninstrumented collision point (IP12) and measured A_N for neutrons, photons and neutral pions at the most forward region [8]. The kinematic region, $p_T < 0.3$ GeV/c and $x_F > 0.2$, was covered. Both the hadron calorimeter and the EM calorimeter showed $\sim 10\%$ asymmetry for neutrons over a wide x_F region. We presently don't understand the physics process that results in this unexpectedly large asymmetry. In any case, this asymmetry gives a basis for the local polarimeter at PHENIX. We are implementing position-sensitive counters in the ZDC [6] to measure the neutron asymmetry at PHENIX. The position-sensitive counters will be comprised by 7-channel hodoscopes for both X- and Y-directions at the shower maximum position of the ZDC.

BASELINE PLAN (2003–)

In run-3 which will start in 2003, we anticipate an integrated luminosity of about 3 pb^{-1} at $\sqrt{s}$ = 200 GeV with 50% beam polarization. In this run, our main goal is to measure the double longitudinal-spin asymmetries (A_{LL}) for neutral pions and for charged hadrons. For the gluon polarization measurement, these channels serve as an alternative to the jet measurement in the limited acceptance. Expectation of the non-zero A_{LL} measurement of neutral pions and charged hadrons is shown in Fig.1, which is compared with the GS95 NLO polarized PDF model-A [9]. When extracting the gluon polarization, all channels will be utilized.

In run-4 and beyond, we expect 10% to 100% of the design luminosity (320 pb^{-1} at $\sqrt{s}$ = 200 GeV; 800 pb^{-1} at $\sqrt{s}$ = 500 GeV). With the much improved statistics, we will be able to distinguish the differences between the asymmetries for neutral pions and charged hadrons. This difference of the asymmetries is caused by different species

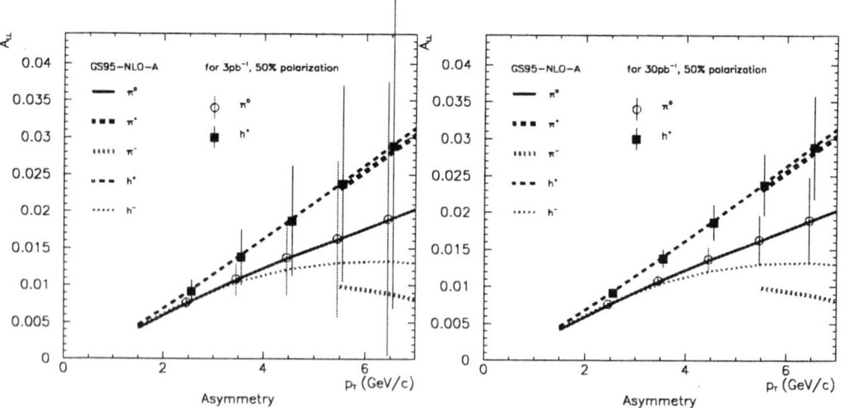

FIGURE 1. Expectation for the neutral pions and charged hadron A_{LL} measurement in run-3 (left) and run-4 (right).

of initial quarks because the origin of π^+ is mainly u-quark, and those of π^0 and π^- are both u-quark and d-quark. In these runs, we will start the measurement of A_{LL} with prompt photons and heavy flavor (channels like single-electrons, J/ψ's and the electron–muon coincidence channel [10]).

The prompt photon process is a clean channel dominated by the gluon Compton process ($gq \rightarrow \gamma q$). By measuring A_{LL}, we can directly factor out the gluon polarization in the leading-order. With the high performance EM calorimeters at PHENIX, we can remove backgrounds from π^0 decays effectively. Before the background reduction, we have about 10 times larger background photons in the low momentum region. After the reduction by reconstructing π^0 and applying the isolation cut, background level can be well lower than 1 [11]. This measurement will cover the x_{gluon} region of 0.04 to 0.3.

By measuring heavy flavor production, the x_{gluon} coverage will be extended down to 0.005. The heavy flavor is produced by the gluon fusion process ($gg \rightarrow Q\bar{Q}$). Wide region is covered by many channels complementarily. In these measurements, one of the most clean channel is the electron–muon coincidence channel. In the invariant mass plots of unlike-sign pairs and like-sign pairs, bottom pairs can be identified as high invariant mass pairs. Background is evaluated by the like-sign pairs. By this measurement, x_{gluon} region 0.02 to 0.2 will be covered [10].

UPGRADE PLAN (2005–)

We also plan to pursue the gluon polarization with upgraded detectors at full luminosity. In the full luminosity runs, we will measure A_{LL} of prompt photon and heavy flavor production with higher statistics and cleaner channels. The major upgrade plan for the gluon polarization is a silicon detector upgrade. We also plan to have the Time Projection Chamber (TPC) surrounding the silicon detector. In the current strawman design, the silicon detector system consists of four layers of barrel detectors which cover $|\eta| < 1$ and almost full azimuthal angle, and four layers of endcap detectors to match with the Muon Arms.

With this detector, by identifying displaced vertices, we expect much better heavy flavor production measurement. One channel we will see is a B-meson identification with a J/ψ decay which will be detected as a displaced muon-pair vertex in the endcap part and an electron-pair vertex in the barrel part. In the barrel part, we will have good displaced vertex resolution evaluated with distance of the closest approach (DCA) value smaller than 50 μm at $p_T > 1$ GeV/c. Another channel is a bottom pair measurement. This is an extension of the electron–muon coincidence measurement. By detecting a muon and a displaced vertex of bottom decay, we expect about thirty times larger statistics than the electron–muon coincidence measurement with the baseline detector.

The silicon detector will also serve as a tracker, which will show much better performance in combination with the TPC upgrade. By using this and the EM calorimeter, we will measure the prompt photon + jet production. This will enable us to reconstruct the kinematics of the event, pin down the x_{gluon}, and make the gluon polarization measurement much more sensitive and clean.

SUMMARY

The polarized proton collision run in 2001-2002 at PHENIX was very successful. Highly selective trigger systems were developed and operational for the proton run. From these data, we have reported cross sections of π^0 and J/ψ, and will report the A_N measurements in the near future. Studies of the relative luminosity measurement and the local polarimeter were performed as needed for the gluon polarization measurement in the next runs. We plan to measure A_{LL} to probe the gluon polarization using many channels (neutral pions, charged hadrons, and single-electrons in run-3; prompt photons and heavy flavor production channels in run-4 and beyond) with both the baseline detector and the upgraded detectors.

REFERENCES

1. D. P. Morrison, Nucl. Phys. A **638**, 565c (1998).
2. K. Okada, these proceedings.
3. H. D. Sato, these proceedings.
4. B. Fox, these proceedings.
5. A. Taketani, these proceedings.
6. C. Adler, et al., Nucl. Instrum. Meth. A **470**, 488 (2001).
7. M. Bai, these proceedings.
8. Y. Fukao, these proceedings.
9. T. Gehrmann and W. J. Stirling, Phys. Rev. D **53**, 6100 (1996).
10. K. N. Barish, these proceedings.
11. A. Bazilevsky, RIKEN Rev. **28**, 15 (2000).

Spin Physics at STAR

Akio Ogawa
for the STAR collaboration [1]

*Brookhaven National Laboratory, Department of Physics
Upton, NY, 11973-5000, USA*

Abstract. The question of how the spin degrees of freedom in the nucleon are organized has still not been fully answered even after recent polarized deep inelastic scattering experiments. Studying polarized proton-proton collisions will add new and unique information to improve our understanding of the spin structure of the nucleon.

The Relativistic Heavy Ion Collider (RHIC) successfully accelerated and collided polarized proton beams in the beginning of 2002. STAR is one of the two large detectors at RHIC. STAR has been taking heavy ion collision data since 2000 and will have excellent capability for spin physics as well.

In this paper, an overview of the STAR spin program is given, covering a wide range of physics topics including determination of gluon polarization, flavor separation of quark polarizations, and quark transversity. Some details about the STAR detector, including future upgrade plans, are presented. Results from the 2002 run with transversely polarized protons are summarized.

INTRODUCTION

The spin of the nucleon is known to be 1/2 [1, 2, 3, 4]. QCD describes the dynamics of the nucleon's constituent partons: spin-1/2 quarks and spin-1 gluons. Exactly how the quarks are confined in the nucleon is not fully understood. One of the central mysteries in QCD is the spin structure of the nucleon. Polarized charged lepton deep inelastic scattering (DIS) experiments [5, 6, 7, 8] found that the spins of the quarks and anti-quarks carry only about $1/4$ of the nucleon spin, contrary to the expectation of 2/3 from the constituent quark model or the Ellis-Jaffe Sum Rule [9]. The spin of the nucleon is built up from the spin and angular momentum of quarks and gluons:

$$\frac{1}{2} = \frac{1}{2}\Delta\Sigma + \Delta G + L_z^q + L_z^G. \tag{1}$$

Recent experiments are trying to find the missing contributions to the nucleon spin in either gluon polarization or angular momentum of quarks and gluons.

[1] For the full author list and acknowledgements, see the appendix to the proceedings.

CP675, *Spin 2002: 15th Int'l. Spin Physics Symposium and Workshop on Polarized Electron Sources and Polarimeters*, edited by Y. I. Makdisi, A. U. Luccio, and W. W. MacKay

STAR SPIN PHYSICS PROGRAM

The STAR spin physics program will study nucleon spin structure in experiments that are sensitive to gluon polarization, the flavor decomposition of quark and anti-quark spins, and quark transversity. One of the advantages of the RHIC spin program is that it extends spin structure studies to higher Q^2 and lower x_{Bj} than fixed target polarized DIS experiments. Measurements at large Q^2 are important to minimize uncertainties associated with pQCD theory. Data from a wide Q^2 range is needed for a reliable global pQCD analysis.

Gluon Polarization

Measurements of the longitudinal double spin asymmetry A_{LL} in polarized proton collisions provide sensitivity to gluon polarization. The cleanest way to determine the gluon polarization is to study A_{LL} for direct photon and jet production. In leading-order pQCD, $\sim 90\%$ of the direct photon production cross section is from the QCD Compton sub-process, $qg \to q\gamma$. When the direct photon is detected in coincidence with the away side jet, A_{LL} can be approximated as:

$$A_{LL}^{pp\to\gamma+Jet} = \frac{\Delta G}{G} \cdot A_1^p \cdot \hat{a}_{LL}(qg \to q\gamma). \tag{2}$$

The quark polarization weighted by the squared electric charge, A_1^p, is measured well in polarized DIS experiments. The double spin asymmetry for the QCD Compton sub-process, $\hat{a}_{LL}$, is calculable in pQCD. The wide acceptance of the STAR detector is ideal for containing the jets, which is important to determine the initial state kinematics.

When the direct photon is detected at forward rapidity in the acceptance of the STAR endcap electromagnetic calorimeter (EMC), presently under construction, the quark polarization is larger and the QCD Compton sub-process double spin asymmetry increases, providing greater sensitivity to gluon polarization. It also enables us to access lower Bjorken x gluons, which is important to determine the integral of ΔG.

Determination of A_{LL} for direct photons detected in coincidence with away side jets is a long term goal of the STAR spin program, requiring that RHIC delivers the design luminosity ($\sim 10^{32}\text{cm}^{-2}\text{s}^{-1}$) and beam polarizations ($P_{beam} \sim 70\%$) and that the STAR barrel and endcap EMC are completed.

Sensitivity to gluon polarization is also provided by measurements of A_{LL} for inclusive jet and di-jet production. These processes have much higher cross sections. STAR should be able to study these processes starting with the next RHIC run, as discussed in B. Surrow's proceedings [10].

Quark and Anti-Quark Polarization with Flavor Decomposition

Understanding the polarization of quarks and anti-quarks of specific flavors is important for a complete description of how the nucleon spin is built from its constituents.

Flavor dependence of unpolarized sea anti-quark distributions has been observed [11]. Models that aim to understand this flavor asymmetry make different predictions for $\bar{u}$ and $\bar{d}$ polarization.

Anti-quark polarization measurements from semi-inclusive DIS are still limited [12]. $W^{\pm}$ bosons produced in pp collisions select spin and flavor and are therefore a unique and ideal tool for studying flavor decomposition of the quark spin at RHIC. $W^{+(-)}$ is produced in $u+\bar{d}$ $(d+\bar{u})$ collisions and is detected by its decay to $e^{+(-)}$. When the electron(positron) is detected with the endcap EMC ($1.0 < \eta < 2.0$) from a polarized proton propagating away from (toward) the endcap, the purity of $W^{-(+)}$ coming from a $\bar{u}$ quark ($\bar{d}$ quark) in the polarized nucleon is $\sim 98\%$ ($\sim 75\%$).

Transverse Spin Physics

Physics with transverse spin has received a lot of interest recently. Transversity is the last unmeasured quark distribution function at leading twist, and studying it will give another hint about the missing spin. The RHIC spin program will measure azimuthal asymmetries within a jet at mid-rapidity. Such measurements are expected to be sensitive to transversity, if e^+e^- collider experiments find these chiral-odd fragmentation functions to be non-zero [13, 14, 15, 16].

The transverse single spin asymmetry (A_N) for the $p^{\uparrow}+p \rightarrow \pi^0 + X$ reaction was measured at FNAL-E704 [17] at $\sqrt{s} = 20$ GeV and found to be large. After more than a decade, this asymmetry still remains a mystery. STAR will measure A_N for particle production over a range of x_F and p_T comparable to that explored by E704. This asymmetry may be sensitive to transversity or may give hints about the orbital angular momentum of quarks in the nucleon.

THE STAR DETECTOR

STAR (Fig. 1) is one of the two large detectors at RHIC. The STAR detector is designed to cover a relatively large acceptance at mid-rapidity (2π in ϕ, $-1.4 < \eta < 1.4$) with tracking detectors to measure thousands of charged particles coming from Au-Au collisions. The STAR barrel and endcap EMCs will increase their acceptance to be azimuthally complete for $-1 < \eta < 2$ in the next 3 years. At forward rapidity, a Beam-Beam Counter (BBC) and a Forward π^0 Detector (FPD) were installed prior to the first polarized proton run. A new scaler system counts 131,072 channels corresponding to the 2^{17} input patterns from 17 different physics signals for every bunch crossing, or every 107 nsec. This scaler system was used to monitor bunch-by-bunch relative luminosity and to perform spin dependent counting experiments. A detailed description of the STAR experiment can be found in [18].

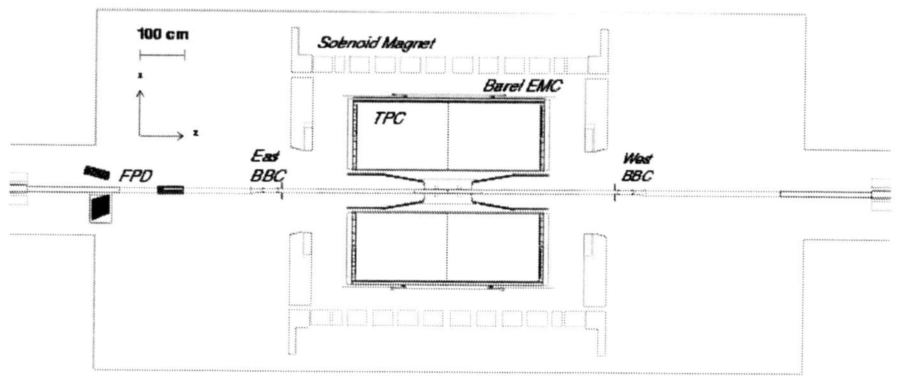

FIGURE 1. The top view of the STAR detector as of the January the 2002 proton-proton run

HIGHLIGHTS OF PHYSICS FROM 2002 RUN

In December 2001, the first polarized proton-proton collider began operation. RHIC accelerated beams to provide polarized proton collisions at $\sqrt{s} = 200$ GeV. The average polarization, measured by the Coulomb Nuclear Interference (CNI) polarimeters [19], was about 0.2 at the injection. The top luminosity was around $2 \times 10^{30}\mathrm{cm}^{-2}\mathrm{s}^{-1}$ and STAR recorded ~ 300 nb^{-1} in 33 days of operation. During this run, STAR collected 16 million minimum bias events, 3.5 million FPD events with full readout of the mid-rapidity detectors, 11 million FPD standalone events, 0.8 million EMC triggered events, and 8 billion scaler events with BBC coincidences.

At this symposium, G. Rakness reported [20] on A_N for inclusive π^0 production at forward rapidity. J. Balewski reported [21] on A_N for leading charged particles at mid rapidity. A preliminary measurement indicates that at large x_F, A_N is as large at $\sqrt{s} = 200$ GeV as at $\sqrt{s} = 20$ GeV and that A_N is consistent with zero at small x_F. J. Kiryluk reported [22] on A_N measurements with the BBC and showed that STAR can measure relative luminosity with systematic errors of less than 10^{-3} in the collider environment.

PLANS FOR THE NEXT AND FUTURE RUNS

The polarization in the next RHIC run is expected to be increased by a factor of 2 (from 0.2 to 0.4) and the luminosity will be increased by a factor of 10 (from 10^{30} to $10^{31}\mathrm{cm}^{-2}\mathrm{s}^{-1}$). STAR will have an upgraded FPD and BBC, and the barrel EMC will increase its acceptance to 2π in ϕ and $0 < \eta < 1$. These will enable us to do precise measurements of A_N for processes that were measured in the last run and to use these asymmetries for tuning the spin rotators which are being installed to produce longitudinal polarization at the STAR interaction point [23]. Once the spin rotators are tuned, we should be able to start studying double spin asymmetries for single jet and dijet production [10].

Beyond the next RHIC run, STAR will be completing the barrel and endcap EMC. STAR will have an exciting spin physics program including direct photon, W, and transversity measurements.

REFERENCES

1. F. Hund, Zeitschr. Z. Phys. 41: 239, 1927
2. T. Hori, Zeitschr. Z. Phys. 44: 834, 1927
3. D.M. Dennison, Proc. of the Roy. Soc. of London 115: 483, 1927
4. G.M. Murphy and H. Johnston, Phys. Rev. 46: 95, 1933
5. EMC, J. Ashman *et al.*, Phys. Lett. B206: 364, 1988; EMC, J. Ashman *et al.*, Nucl. Phys. B328: 1, 1989
6. SMC, B. Adeva *et al.*, Phys. Lett. B302: 533, 1993
7. E-143, K. Abe *et al.*, Phys. Rev. Lett. 74:346, 1995
8. HERMES, K. Ackerstaff *et al.*, Phys. Lett. B442:484, 1998; HERMES, K. Ackerstaff *et al.*, Phys. Lett. B404:3, 1997
9. J. Ellis and R.L. Jaffe, Phys. Rev. D9: 1444, 1974; Phys. Rev. D10: 1669 1974
10. B. Surrow, Proceedings of Spin2002
11. NMC, P. Amaudruz *et al.*, Phys. Rev. Lett. 66:2712, 1991; NuSea, R.S. Towell *et al.*, Phys. Rev. D64:052002, 2001
12. SMC, B. Adeva *et al.*, Phys. Lett. B369:93, 1996; HERMES, K. Ackerstaff *et al.*, Phys. Lett. B464:123, 1999
13. M. Grosse Perdekamp, A. Ogawa, K. Hasuko, S. Lange, V. Siegle, Nucl. Phys. A711:69, 2002
14. D. Boer, R. Jakob, P. J. Mulders, Phys. Lett. B424 :143, 1998; X. Artru, J. Collins, Z. Phys. C69 :277, 1996
15. R. L. Jaffe *et al.*, Phys. Rev. Let. 80 :1166, 1998; Phys. Rev. D57 :5920, 1998 Boffi, R. Jakob, M. Radici hep-ph/9907374
16. K. Hasuko, Proceedings of Spin2002
17. A. Bravar *et al.*, Phys. Rev. Lett.77: 2626, 1996; D. L. Adams *et al.*, Phys. Rev. D53 : 4747, 1996
18. STAR Collaboration, Nucl.Instrum.Meth. to be published (Special volume edition).
19. O. Jinnouchi, Proceedings of Spin2002
20. G. Rakness, Proceedings of Spin2002
21. J. Balewski, Proceedings of Spin2002
22. J. Kiryluk, Proceedings of Spin2002
23. A. Ogawa, Proceedings of Spin2002

Status of the COMPASS Experiment

F. Tessarotto

CERN (CH) and INFN-Trieste (I)

on behalf of the COMPASS Collaboration

Abstract. The COMPASS Experiment at CERN has a broad physics program aimed at the study of nucleon spin structure and hadron spectroscopy. It has an outstanding fixed-target apparatus, mostly commissioned in 2001, presently consisting of a solid ^{6}LiD polarised target and a two stage spectrometer with high resolution tracking, particle identification and calorimetry, capable of standing high event rates.

This paper describes the apparatus and its performances during the run of 2002, when 260 TB of polarised muon nucleon scattering data have been collected. First physics signals from the analysis and projections for the expected accuracy of the measurement of the gluon polarisation $\Delta G/G$ from photon gluon fusion are presented too.

THE COLLABORATION

In 1996 two communities which had presented the CHEOPS [1] and HMC [2] Letters of Intent for fixed target experiments at CERN merged in the COMPASS (COmmon Muon and Proton Apparatus for Structure and Spectroscopy) Collaboration and submitted a Proposal [3] which obtained approval in 1997.

COMPASS has a broad physics program with different beams and targets, extending over a decade; the Collaboration consists of about 220 physicists from 27 Institutes.

Its spin program with a polarised muon beam and a polarised target aims to provide a direct measurement of the gluon polarisation $\Delta G/G$ via the asymmetry of the photon-gluon fusion process, accessed by detecting open charm and high-p_T correlated hadron pairs [4]; it also aims to determine the transversity structure function h_1, to perform accurate flavour decomposition of the quark helicity distributions and to measure polarised fragmentation functions.

Using hadronic beams COMPASS will study π and K polarisabilities from Primakoff reactions and test predictions from chiral perturbation theory; perform extensive meson spectroscopy to investigate the presence of exotic states; collect large samples of semi-leptonic decays of charmed mesons and baryons, determine form factors and probe predictions from Heavy Quark Effective Theory; perform a systematic study of charm hadroproduction cross sections and investigate doubly charmed baryon states, among which the Ξ_{cc}^+, recently claimed [5] by SELEX.

CP675, *Spin 2002: 15th Int'l. Spin Physics Symposium and Workshop on Polarized Electron Sources and Polarimeters,* edited by Y. I. Makdisi, A. U. Luccio, and W. W. MacKay

THE COMPASS APPARATUS

In 2002 COMPASS used the CERN μ^+ beam with an energy of 160 ($\pm$ 5) GeV. Muons were generated by decay of secondary π (and K) mesons and had a polarisation $P_B \approx -80\%$. Typical beam intensity was $2.1 \cdot 10^8$ μ^+ per spill, with 4.8 s long spills of 16.8 s period. The track of each muon was measured upstream of the target and its momentum determined from the track measurement.

The polarised target

The target consisted of two cylindrical cells (60 cm long, 3 cm diameter, separated by 10 cm) filled with solid ^{6}LiD, which has a dilution factor f $\approx$ 0.5.

The luminosity was $5 \cdot 10^{32}$ $cm^{-2}s^{-1}$.

A ^{3}He-^{4}He dilution refrigerator cooled the target at T < 100 mK, while a solenoid provided a highly homogeneous 2.5 T magnetic field along the beam axis. Since the COMPASS solenoid (600 mm internal diam.) is not yet available, the solenoid of the SMC experiment (255 mm internal diam.) was used instead: the resulting loss in angular acceptance has a moderate impact on the measurement of $\Delta G/G$, but it significantly affects other physics items.

A dipole magnet (0.5 T) was used to reverse the target spin orientations every 8 hours and to provide transverse polarisation: in this case the target was kept in frozen spin mode (with a relaxation time measureed to be > 1000 h). The two target cells were dynamically polarised (and kept) in opposite directions by microwave irradiation, with frequency modulation. The needed paramagnetic centres were produced by previous intense irradiation of the material by e^- beam.

Maximum polarisation values as high as $P_T = (-0.49, +0.57)$ have been obtained, while typical values during the run were around -0.45 and $+0.52$. About one day was needed to reach $P_T \sim 0.40$. The polarisations were constantly measured in longitudinal mode via 10 NMR coils with an accuracy around 3%.

The spectrometer

To achieve high-resolution tracking over a wide angular and dynamical range the COMPASS spectrometer comprised two magnetic stages: the first (Large Acceptance Spectrometer: LAS) had an acceptance of $\pm$180 mrad and a wide aperture dipole magnet with 1 Tm bending power; the second (Small Acceptance Spectrometer: SAS), covered $\pm$30 mrad and used a 4.4 Tm dipole to analyse high momentum particles.

Tracking

Tracking in the beam region was provided by 9 stations of scintillating fibres hodoscopes, with 0.5 mm (1.0 mm) diameter fibres read by multi-anode PMs capable of withstanding rates above 5 MHz/channel. They provided a total of 21 coordinates with efficiency $\geq$ 99%, better than 500 ps time and 130 μm (250 μm) space resolution. Silicon microstrip detectors covering 50×70 mm^2 area provided four coordinates before

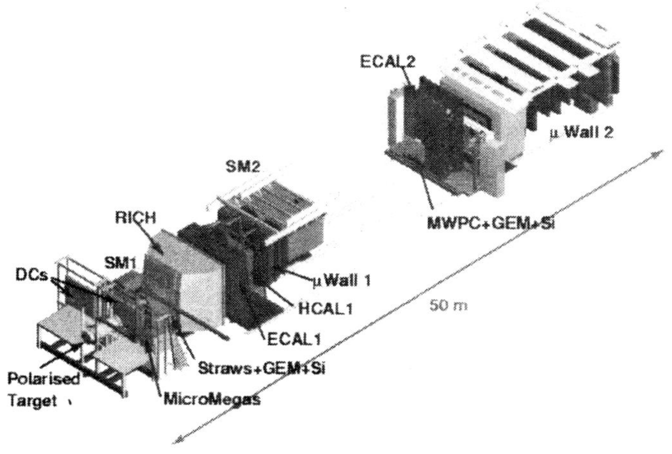

FIGURE 1. Schematic view of the COMPASS setup for the run of 2002.

the target, with analog readout, resolutions of 14 μm and 2.5 ns and $\sim$99% efficiency.

The high flux area around the beam region was covered by micro-pattern detectors: Micromegas and GEMs [6] . The COMPASS Micromegas [7] contained a thin micromesh foil separating an ionization volume from a high field (40 KV/cm, 100 μm gap) amplification region; they had typical efficiencies of 98% and a space resolution of 70 μm. 12 planes of 40 $\times$ 40 cm^2 active area were used.

The COMPASS GEMs (Gas Electron Multiplier) [8] were made of kapton foils having Cu layers on both sides and a large $(10^4/cm^2)$ number of 60 μm diameter holes with the high field inside the holes providing electrons amplification. Each GEM-chamber used three cascaded GEM-foils and provided two-dimensional projective readout, for a total of 40 coordinates. They had 50 μm space resolution, typical efficiencies around 96%, time resolution in the range of 15 ns and covered $31 \times 31 cm^2$

Both Micromegas and GEMs did operate very smoothly during the 2002 run.

Tracking in the lower flux regions of the LAS was performed by 3 Drift Chambers (with 8 coord. each, 1.2 $\times$ 1.2 m^2 active area, 170 μm res.) and by 9 large Straw Tube detectors [9] covering 3.2 $\times$ 2.4 m^2 with double (staggered) layers of 6 mm diameter tubes (10 mm in the outer region), having $\approx$ 300 μm res.

In the SAS a set of old MWPC's refurbished and equipped with fast electronics provided 34 coordinates with 600 μm resolution, while the very external region was covered by drift chambers with 2.6 $\times$ 5.2 m^2 active area.

Calorimetry and Particle Identification

Hadron calorimeters were present on both spectrometers and used at the trigger level too. Resolutions were similar for the two calorimeters HCAL1 and HCAL2, both made of Fe scintillator sandwiches with planar WLS: $\sigma/E \approx 6\% \oplus 60\%/\sqrt{E}$ for pions and $\sigma/E \approx 0.6\% \oplus 24\%/\sqrt{E}$ for electrons. Electro-magnetic calorimetry was only marginally implemented in the SAS, with LG blocks having $\sigma/E = 2.3\% \oplus 5.8\%/\sqrt{E}$.

Muons were identified with high efficiency by large sets of streamer detector planes before and after a 60 cm thick iron absorber for the LAS and by drift tube planes for the SAS.

Hadron identification was provided in the first spectrometer only, by a Ring Imaging Cherenkov Detector (RICH1) [10], designed to perform π-K separation up to 60 GeV/c over the entire acceptance.

RICH1 consisted of a 3 m long C_4F_{10} radiator at atmospheric pressure, a wall of 116 spherical mirrors (3.3 focal length) covering an area of >20 m^2 and two sets of far UV photon detectors placed above and below the acceptance region. Cherenkov photons in the range between 160 and 210 nm were detected using MWPCs equipped with CsI photocathodes [11] and covering 5.3 m^2. The photocathodes were segmented in pads of 8×8 mm^2 from where the induced signals were readout by a system of front-end boards [12] with local intelligence. A total of 83000 channels were digitized by 10 bit ADCs and compared to their individual threshod values at each event.

Despite non-optimal conditions of some photon detectors and sometimes of the radiator gas, RICH1 has been operational during most of the running time.

Trigger and data acquisition

Coincidences between elements of hodoscope planes at different positions along the beam selected scattered muons for triggering purposes: 2 ns wide coincidence matrices filtered signals from 500 trigger channels accepting target pointing "tracks" in the kinematical region of interest. Presence of hadron shower signal was required to provide more selective triggering.

Typical trigger rate was 5 kHz and the overall dead time was $\approx 7\%$.

COMPASS used custom designed, parallel, readout electronics with local pre event building for a total of $\sim$ 190 k channels. A pipelined acquisition system transferred the data via S-Links to 16 PCs used for buffering during spills; network swithches received the data through Gigabit Ethernet and distributed them to 12 Event Builder PCs.

With an event size of $\sim$ 44 kB the typical data flow was 220 MB/s during spill and the average data recording rate was almost 3 TB/day.

A Detector Control System based on PVSS monitored and archived all values of voltage, temperature and other parameters, and handled a complex alarms system. On-line data monitoring was provided by COOOL, a C++ program built on ROOT libraries which sampled events from the DAQ farm and shared the decoding with the off-line software. A flexible Electronic Log-book system was used: edited by the shift crew and accessible via www it also contained many automatically transferred informations, including COOOL plots for each run.

After a setting-up period COMPASS took physics data during about 80 days, 19 of which in transverse polarisation mode. The average combined efficiency of SPS, polarised target, spectrometer and DAQ was almost 70 % and a total of 260 TB of data, corresponding to $\geq$ 5 G events have been collected.

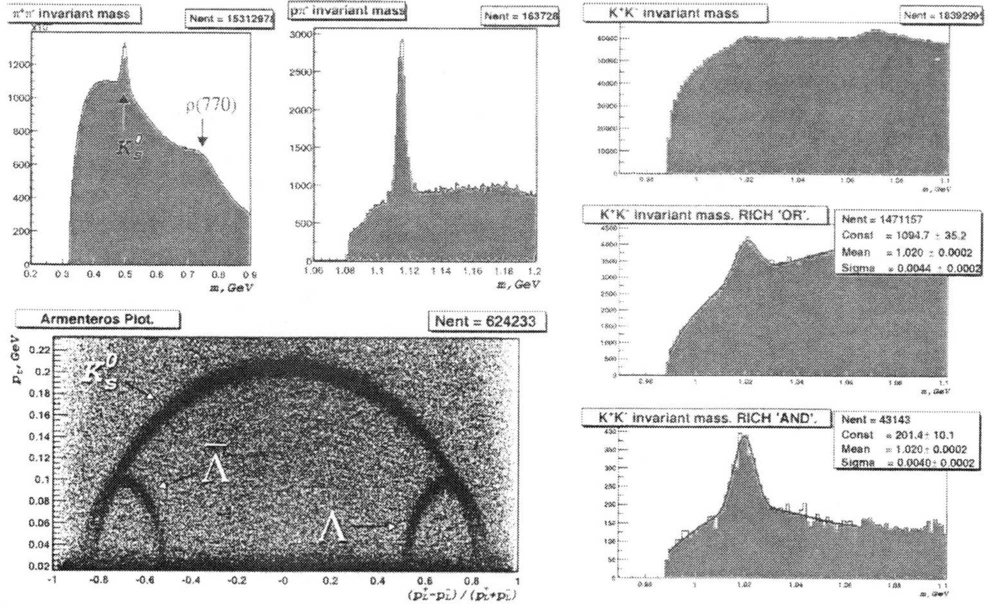

FIGURE 2. Invariant mass and Armenteros plots from a very preliminary analysis of asubsample of 2002 data: K^0_s, Λ ($\bar{\Lambda}$) and $\phi(1020)$ signals are clearly seen.

THE ANALYSIS

For both data recording and reconstruction COMPASS used a farm of 100 dual processor PCs at CERN and the CASTOR file system, on which the 260 TB of data were stored in about 260000 files.

The analysis is performed through a program called CORAL, fully object oriented, with a modular architecture, written in C++, from scratch, by members of the Collaboration.

In a very preliminary analysis on a subsample of the 2002 data, secondary vertices have been fitted to the reconstructed tracks; the distribution of invariant masses for opposite charged pairs with the $\pi^+\pi^-$ hypothesis is shown on the left top of fig. 2: the K^0 peak appears clearly and has a width similar to the one expected from Monte Carlo simulations; a shoulder corresponding to the ρ meson is present too. For vertices reconstructed behind the target the invariant mass for the $\pi^- p$ hypothesis is shown on the top central plot of fig. 2, the Armenteros plot in the left bottom: here too the resolution for Λ and $\bar{\Lambda}$ is compatible with expectations.

A first look at the RICH information provided promising hints: the right part of fig. 2 shows the invariant mass of opposite charged tracks with the K^+K^- hypothesis when none (top), at least one (central) and both (bottom) tracks are asked to have been identified as kaons by the RICH. The peak corresponding to the $\phi(1020)$ meson appears unambiguously when the particle identification information is used.

THE MEASUREMENT OF ΔG

The gluon helicity distribution $\Delta G(\eta)$ is poorly known: NLO QCD analyses of polarised structure functions suggest large positive values, and so does a comparison between high-p_T data and Monte Carlo simulations [13] performed by HERMES.

A major step forward is expected in the incoming years: COMPASS at CERN, STAR and PHENIX at RHIC are planning to provide accurate data for ΔG in the next runs, and Experiment E161 has proposed a high statistics measurement at SLAC.

The gluon polarisation $\Delta G/G$ is measured in COMPASS from the asymmetry of two processes: open charm production and correlated high p_T hadron pair production.

Open charm is predominantly produced by photon gluon fusion: $\gamma^* g \rightarrow c\bar{c}$, at a scale $s > 4m_c^2$; the useful cross section for a muon beam of 160 GeV/c is about 3 nb.

The identification is done by reconstructing D^0 and $\overline{D^0}$: on average 1.2 neutral D mesons are produced per open charm event; they are primarily detected through their golden decay channel: $D^0 \rightarrow K^-\pi^+$, $\overline{D^0} \rightarrow K^+\pi^-$ with a branching ratio of $\approx 4\%$; cuts on the K direction in the D^0 rest frame: $|cos(\theta_K^*)| < 0.5$ and on the D^0 energy fraction $z_D = E_D/E_{\gamma^*} > 0.25$ are needed to reduce background contamination.

A much cleaner charm sample is expected from identifying the $D^{*\pm} \rightarrow D^0\pi^\pm$ decay chain from its unique kinematics. Also, the D^0 statistics will be increased by including $D^0 \rightarrow K^+\pi^-\pi^0$ decays.

The gluon polarization will be determined from the measured charm production asymmetry: $A^{meas.} = (N_{c\bar{c}}^{\uparrow\downarrow} - N_{c\bar{c}}^{\uparrow\uparrow})/(N_{c\bar{c}}^{\uparrow\downarrow} + N_{c\bar{c}}^{\uparrow\uparrow}) = P_B P_T f D A_{c\bar{c}}^{\gamma^* N}$ where $N_{c\bar{c}}^{\uparrow\uparrow}(N_{c\bar{c}}^{\uparrow\downarrow})$ represents the number of charm events with target spin parallel (anti-parallel) to the μ helicity and D is the γ^* depolarisation factor, with $< D > \approx 0.66$ for COMPASS. Using the equation:

$$\int_{4m_c^2}^{2MEy} \Delta\sigma(s)^{\gamma g \rightarrow c\bar{c}} \Delta G(\eta, s) ds = A_{c\bar{c}}^{\gamma^* N} \cdot \int_{4m_c^2}^{2MEy} \sigma(s)^{\gamma g \rightarrow c\bar{c}} G(\eta, s) ds$$

one averaged value for $\Delta G(\eta)$ will be derived, for $\eta \approx 0.1$. All other terms in the above equation are known, in particular $\Delta\sigma(s)^{\gamma g \rightarrow c\bar{c}}$ has been calculated in QCD at NLO [14].

An alternative methode uses the large analysing power of the photon gluon fusion production of light quarks: in order to discriminate this process from the leading order one ($\gamma^* q \rightarrow q$), events with hadron pairs having correlated high transverse momentum are selected. The kinematics of these events allows the reconstruction of η, probing $\Delta G(\eta)$ in different η bins in the range $0.04 < \eta < 0.2$. The abundant production rate provides high statistical accuracy but QCD compton scattering ($\gamma^* q \rightarrow qg$) contributes to these events: the selection of K^+K^- pairs will reduce its contribution but the interpretation of the measurement will remain to some extent model dependent.

The expected statistical errors for 80 days of run for the two methods are presented in fig. 3 together with three parametrisations of $\Delta G/G$ from Gehrmann and Stirling [15].

CONCLUSIONS

The COMPASS Experiment at CERN has a state-of-the-art fixed target apparatus with high resolution tracking, particle identification and calorimetry, capable of standing high event rates; it collected 260 TB of polarised μ-N scattering data in 2002.

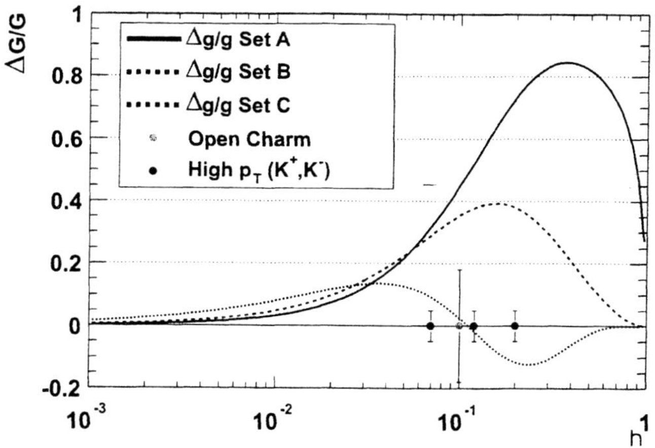

FIGURE 3. Expected statistical errors for $\Delta G(\eta)/G(\eta)$ from 80 days of COMPASS data: the open charm asymmetry measurement will provide one point with low systematic error, high p_T kaon pairs will access different η regions with high statistical accuracy. Not shown are systematical uncertainties

The gluon polarisation $\Delta G/G$ will probably be known with an accuracy around 10% in few years, from direct measurements by COMPASS, STAR, PHENIX and E161.

REFERENCES

1. Alexandrov, Y., et al., CHEOPS *Letter of Intent* (1995), CERN/SPSLC 95-22.
2. Nappi, E., et al., HMC *Letter of Intent* (1995), CERN/SPSLC 95-27.
3. The COMPASS Collaboration, COMPASS *Proposal* (1996), CERN/SPSLC 96-14.
4. Bravar, A., et al., *Phys. Lett.* **B**, **421**, 349 (1998).
5. The SELEX Collaboration (2002), FERMILAB Pub 02/183 E.
6. Sauli, F., *Nucl. Instrum. and Meth.* **A**, **386**, 531 (1997).
7. Thers, D., et al., *Nucl. Instrum. and Meth.* **A**, **469**, 133 (2001).
8. Altunbas, C., et al., *Nucl. Instrum. and Meth.* **A**, **490**, 177 (2002).
9. Bychkov, V. N., et al., *Particles and Nuclei, Letters No.* **2**, **111**, 64 (2002).
10. Baum, G., et al., *Nucl. Instrum. and Meth.* **A**, **333**, 207 (1999).
11. The RD26 Collaboration, RD26 *Status Report* (1996), CERN/DRDC 96-20.
12. Baum, G., et al., *Nucl. Instrum. and Meth.* **A**, **333**, 426 (1999).
13. Airapetian, A., et al., *Z. Phys. Rev. Lett.*, **6584**, 2584 (2000).
14. Bojak, I., and M.Stratmann, *Nucl Phys.* **B**, **540**, 345 (1999).
15. Gehrmann, T., and Sterling, W. J., *Z. Physics* **C**, **65**, 461 (1994).

The Gluon Spin Structure Function From SLAC E161

Stephen Rock for the Real Photon Collaboration

University of Mass, Amherst MA 01003

Abstract. We will determine the gluon spin density $\Delta g(x)$ in the nucleon by measuring the asymmetry for polarized photo-production of charmed quarks from polarized targets in End Station A at SLAC. The quasi-mono-chromatic circularly polarized photon beam will be produced from an oriented diamond crystal. The target will be longitudinally polarized LiD at a temperature of 0.3 K, centered in a 6.5 T magnetic field to obtain high polarization. Photo-production of open charm will be tagged by decays of D mesons into high transverse momentum muons. The asymmetry for single muons will be measured as a function of muon momentum, muon transverse momentum, and photon beam energies with sufficient precision to discriminate among models of $\Delta g(x)$ that differ from each other by as little as 10% in the range $0.1 < x < 0.2$.

MOTIVATION

The spin degree of freedom has, in recent years, opened up a new window into our understanding of nucleon structure. Much of this improvement is due to measurements of the structure functions g_1 using the SLAC high energy, high current, and high polarization electron beam combined with recent major advances in polarized target technology. One of the goals of these experiments has been the determination of the spin distributions of the quarks and gluons using the evolution equations We will continue this program by directly measuring the gluon spin distribution function.

A major goal of the g_1 experiments has been the determination of spin distributions of quarks and gluons in the nucleon ("what carries the spin?"). A "spin crisis", resulting in numerous theoretical papers, arose when early determinations [1] of $\Delta\Sigma = \Delta u + \Delta d + \Delta s$ were found to be much smaller than expected [where $\Delta f = \int_0^1 \Delta f(x,Q^2)dx$, and $\Delta f(x,Q^2)$ are the individual polarized quark spin distribution functions]. At present, the world average for $\Delta\Sigma$ is approximately 0.23 ± 0.07 [2], much smaller than the relativistic quark model prediction of 0.58. One explanation for at least part of this discrepancy is that the strange sea may be highly polarized, but this depends rather strongly on assumptions of SU(3) symmetry between the beta decays in the baryon octet. In any case, the low value of $\Delta\Sigma$ implies strong gluon contributions via loop and radiative corrections in QCD, expected to be stronger in the polarized case than in the unpolarized case due to the triangle diagram axial anomaly.

Many authors have made fits to the existing data on g_1 to parameterize the spin distribution function (SDF's) $\Delta u(x)$, $\Delta d(x)$, $\Delta s(x)$, and $\Delta g(x)$ using the GLAP evolution equations. Due to the imprecision of present data, different theoretical constraints and parameterizations of the x-dependence are used. The result is that the gluon distributions

CP675, *Spin 2002: 15th Int'l. Spin Physics Symposium and Workshop on Polarized Electron Sources and Polarimeters*, edited by Y. I. Makdisi, A. U. Luccio, and W. W. MacKay

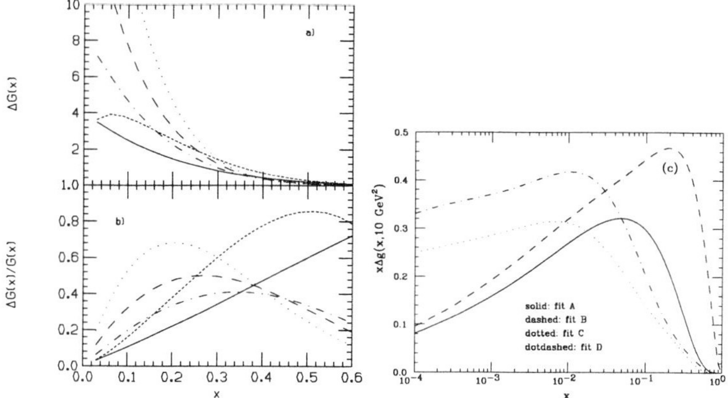

FIGURE 1. Various predictions for the gluon spin density from References [3](Left) and [4](Right).

differ quite significantly, as can be seen in Fig. 1. Some of the fits have maximum values of $\Delta g(x)/g(x)$ close to unity. There are corresponding large range in values in the total spin from the gluon, ΔG, from near zero to over 2. According to GRSV [3], inevitably the large uncertainty in $\Delta g(x)$ implies that the small x behavior of g_1 is completely uncertain and not reliably predictable, with values ranging from less than 5 to greater than 10 at $x = 0.0001$.

METHOD

Our approach for studying polarized gluons in the nucleon is photo- or electro-production of either open charm or inelastic J/ψ's through the hard process of photon-gluon fusion $\gamma + g \to c\bar{c}$. Higher order diagrams have been calculated for both the polarized and unpolarized cases. The relatively large charm quark mass ensures that the relevant momentum transfer scale μ^2 is of order $\mu^2 = 4m_c^2 \approx 9$ GeV^2. where perturbative QCD is expected to work well. Other mechanisms for producing charm are suppressed because of the large ($\sim$ 1.5 GeV) mass of the charm quark. To leading oder for polarized photons and longitudinally polarized nucleons, the polarized experimental cross section difference, $\Delta\sigma_{\gamma p}(k)$, is a convolution of Δg and hard scattering $\Delta\sigma$ [5]

$$\Delta\sigma_{\gamma p}(k) = \int_{x_{min}}^{1} \Delta g(x,Q^2)dx \int_{-1}^{1} \Delta\sigma(\hat{s},\cos(\theta^*))\ \varepsilon(\hat{s},\cos(\theta^*))\ \beta\ d\cos(\theta^*) \quad (1)$$

where k is the photon energy, $x_{min} = 4m_c^2/2Mk$, m_c is the charmed quark mass of about 1.5 GeV, M is the nucleon mass, $s = 2Mk + M^2$ is the square of the c.m. energy in the photon-nucleon system, $\beta = \sqrt{1 - 4m_c^2/\hat{s}}$ is the c.m. velocity of the charmed quark, $\hat{s} = xs$ is the square of the c.m. energy in the photon-gluon system (or the $c\bar{c}$ system), Q^2 is the squared momentum transfer to the gluon (taken in our calculations to be $Q^2 = \hat{s}$),

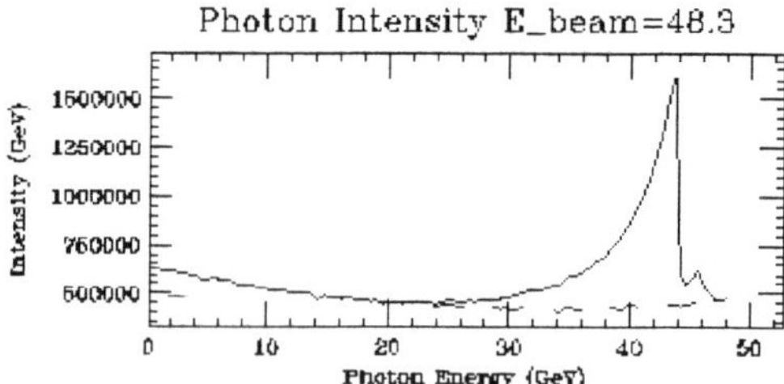

FIGURE 2. Calculated intensity (flux times energy) for collimated coherent bremsstrahlung at the highest energy for this experiment. The dashed lines are incoherent radiation only, while the solid lines include coherent contributions.

θ^* is the c.m. angle of the charmed quarks,

$$\Delta\sigma(\hat{s},\cos(\theta^*)) = \frac{4}{9}\frac{2\pi\alpha\alpha_s(\hat{s})}{\hat{s}}\left[\frac{4m_c^4(\hat{t}^3+\hat{u}^3)}{\hat{t}^2\hat{u}^2} + 2\frac{\hat{t}^2+\hat{u}^2-2m_c^2\hat{s}}{\hat{t}\hat{u}}\right], \tag{2}$$

$\hat{t} = \frac{\hat{s}}{2}[1+\beta\cos(\theta^*)]$, and $\hat{u} = \frac{\hat{s}}{2}[1-\beta\cos(\theta^*)]$. The produced $c\bar{c}$ pair will be detected by their fragmentation into B and $\bar{B}$ mesons and subsequent decay into high p_T muons. The method requires a high energy longitudinally polarized photon beam, a polarized target and a spectrometer to detect the muons.

The Photon Beam

The beam will be produced using collimated coherent bremsstrahlung of a polarized electron beam hitting an oriented diamond crystal. The electron beams will have energies from 9.9 to 48.5 GeV; an intensity of up to 5×10^{10} electrons per pulse; repetition rate of 120 Hz; pulse length of 500 nsec; and polarization of about 83%. A typical photon beam energy spectrum is shown in Fig. 2. The coherent peak is produced by constructive interference of the photons produced at different planes of the diamond crystal.

The Target

The target will be a 5 cm long, 1 cm diameter cylinder filled with ^{6}LiD , which is polarized using the technique of Dynamic Nuclear Polarization (DNP). Using a dilution refrigerator at 300mK, in conjunction with a magnetic field of up to 6.5 T, we expect 70% polarization [6] for the deuteron and ^{6}Li. The setup will be similar to that used in SLAC experiments E143 and E155.

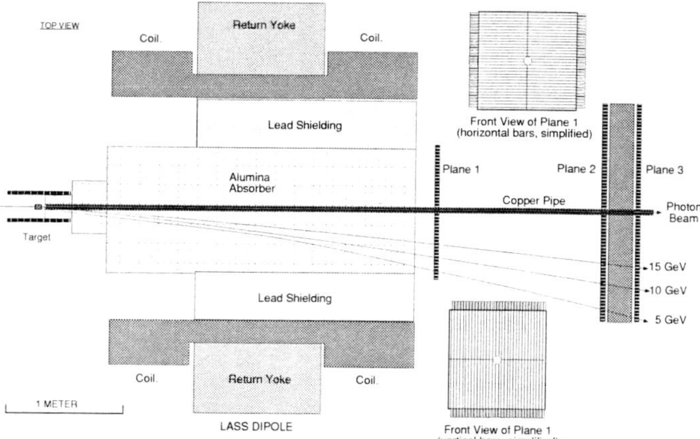

FIGURE 3. Overall plan view of the main components of the spectrometer. The absorber fills most of the gap of the LASS dipole, and also extends into the warm bore of the target magnet. A thick evacuated copper beam pipe contains the photon beam. The three detector planes are made from scintillator hodoscopes. Two simplified front views of the front plane are shown. The dashed curves are typical trajectories for muons with $p_t = 0.7$ and $P = 5$, 10, and 15 GeV.

The Spectrometer

We will identifying the high p_t prompt muons from open charm decay amongst the approximately 1000 times higher rate of long lifetime pions and kaons. The detector shown in Fig. 3 has absorbers are placed very close to the target with enough interaction lengths and radiation lengths to almost completely contain all hadronic and electromagnetic showers. The amount of absorber needed is reduced somewhat by placing the material in a magnetic field, which is also used to determine the momentum of the muons. Tracking of the muons will be done using planes of scintillator hodoscopes. The fine granularity and good timing of this type of system are needed to match hits from muon tracks that will occur on average every 15 to 25 nsec. We plan to use three planes of hodoscopes as seen in Fig. 3. The first two planes have fine resolution for measuring momentum and angles, while the last plane, shielded behind additional lead, has less granularity and will be used to establish a point in space and time on the muon track at a location where the singles rates are much lower. The planes are segmented into top and bottom halves for the horizontal measuring readout (vertical fingers), and left and right halves for the vertical measuring readout (horizontal fingers). This is done to reduce the random hit rate per finger, and to reduce the spread in signal height from attenuation of the light in the scintillator.

RESULTS

The anticipated statistical precision of the experiment is shown in Fig. 4 for the three peak energies of the beam and two different muon momentum ranges. The data points

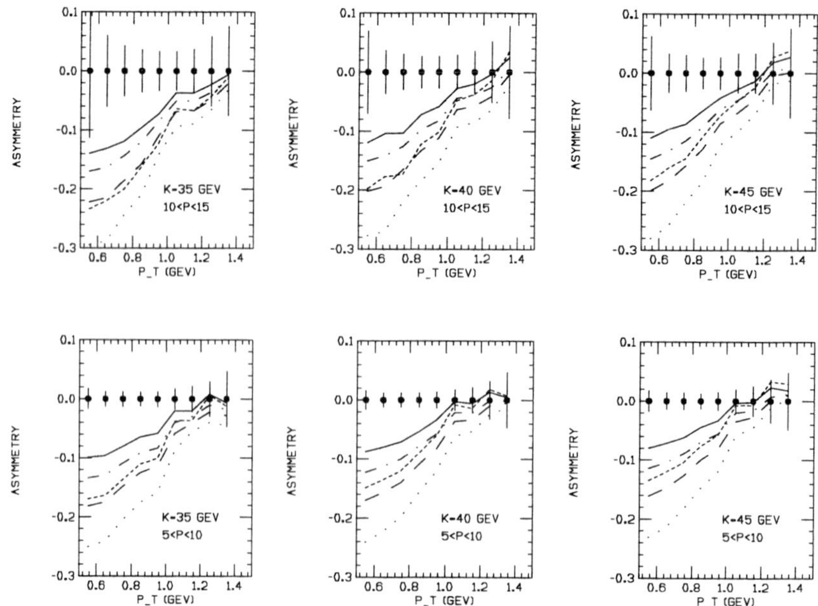

FIGURE 4. Asymmetries for five fits to the gluon polarization as a function of p_T^μ of the detected muon for six kinematic conditions. The points indicate the projected statistical errors. Experimental systematic errors will be highly correlated from point-to-point, and will be approximately 0.10 of the measured asymmetries.

are shown at the arbitrary value of zero, with the errors expected from the experiment. The error estimates include the corrections due to pion and kaon decay and those due to associated production. The errors are as small as 0.012 on the asymmetry over many data points at different energies and p_T^μ. Predicted asymmetry using representative fits are shown here. Our experiment can easily distinguish between them. Averaging over the points will give an error of about 0.006 at each of the energies.

REFERENCES

1. EMC, J. Ashman *et al.*, Phys. Lett. **B206** (1988), 364; Nucl. Phys. **B328** (1989), 1.
2. SLAC E155, P. L. Anthony *et al.*, Phys. Rev. D **54** (1996) 6620.
3. M. Glück, E. Reya, M. Stratmann, and W. Vogelsang, HEP-PH/9910318; Phys. Rev. D **53** (1996) 4775.
4. G. Altarelli,R. D. Ball, S. Forte, and G. Ridolfi, hep-ph/9803237.
5. M. Gluck and E, Reya, Z. Phys. **C39**, 569 (1988).
6. A. Abragam *et al.*, J. Physique - Letts., **41**, L-309 (1980).

Spin and Hadron Dynamics, Low and High Energy

Spin 2002

Results from EDDA@COSY: Spin Observables in Proton-Proton Elastic Scattering

Heiko Rohdjeß for the EDDA-Collaboration [2]

Helmholtz-Institut für Strahlen- und Kernphysik
Universität Bonn, 53115 Bonn, Germany

Abstract. Elastic proton-proton scattering as one of the fundamental hadronic reactions has been studied with the internal target experiment EDDA at the Cooler-Synchrotron COSY/Jülich. A precise measurement of differential cross section, analyzing power and three spin-correlation parameters over a large angular ($\theta_{\text{c.m.}} \approx 35° - 90°$) and energy ($T_p \approx 0.5 - 2.5$ GeV) range has been carried out in the past years. By taking scattering data during the acceleration of the COSY beam, excitation functions were measured in small energy steps and consistent normalization with respect to luminosity and polarization. The experiment uses internal fiber targets and a polarized hydrogen atomic-beam target in conjunction with a double-layered, cylindrical scintillator hodoscope for particle detection. The results on differential cross sections and analyzing powers have been published and helped to improve phase shift solutions. Recently data taking with polarized beam and target has been completed. Preliminary results for the spin-correlation parameters A_{NN}, A_{SS}, and A_{SL} are presented. The observable A_{SS} has been measured the first time above 800 MeV and our results are in sharp contrast to phase-shift predictions at higher energies. Our analysis shows that some of the ambiguities in the direct reconstruction of scattering amplitudes which also show up as differences between available phase-shift solutions, will be reduced by these new measurements.

INTRODUCTION

The nucleon-nucleon (NN) interaction as one of the fundamental processes in nuclear physics has been studied over a broad energy range and its contribution to our understanding of the strong interaction cannot be overstated. NN elastic scattering data, parameterized by energy-dependent phase-shifts, are used as an important ingredient in theoretical calculations of inelastic processes, nucleon-nucleus and heavy-ion reactions. Below the pion production threshold at about 300 MeV elastic scattering is described to a high level of precision by a number of models [1], e.g. phenomenological and meson exchange. More recently, chiral perturbation theory [2] has also made significant progress in this energy domain. Meson exchange models have been pushed further and

---

[1] supported by the BMBF and FZ Jülich

[2] F. Bauer[b], J. Bisplinghoff[a], K. Büßer[b], M. Busch[a], T. Colberg[b], L. Demirörs[b], C. Dahl[a], P.D. Eversheim[a], O. Eyser[b], O. Felden[c], R. Gebel[c], J. Greiff[b], F. Hinterberger[a], E. Jonas[b], H. Krause[b], C. Lehmann[b], J. Lindlein[b], R. Maier[c], A. Meinerzhagen[a], C. Pauli[b], D. Prasuhn[c], H. Rohdjeß[a], D. Rosendaal[a], P. von Rossen[c], N. Schirm[b], W. Scobel[b], K. Ulbrich[a], E. Weise[a], T. Wolf[b], and R. Ziegler[a] (a) Helmholtz-Institut für Strahlen- und Kernphysik, Universität Bonn, Germany, (b) Institut für Experimentalphysik, Universität Hamburg, Germany, (c) Institut für Kernphysik, Forschungszentrum Jülich, Germany

CP675, *Spin 2002: 15th Int'l. Spin Physics Symposium and Workshop on Polarized Electron Sources and Polarimeters*, edited by Y. I. Makdisi, A. U. Luccio, and W. W. MacKay

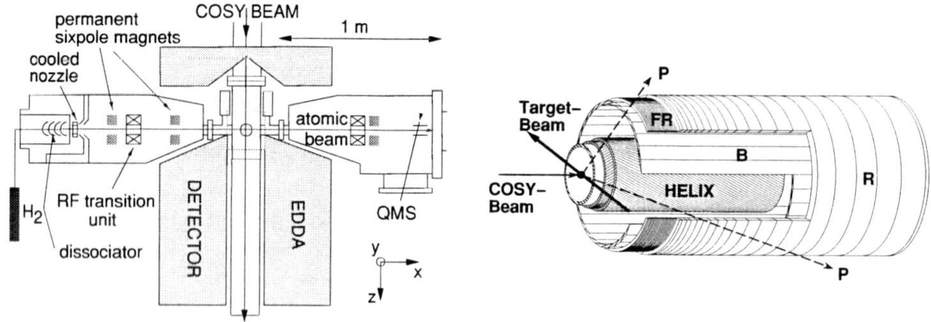

FIGURE 1. Schematic view of the atomic beam target (left) and the EDDA-detector (right).

roughly reproduce experimental data up to 1 GeV. However, at even higher energies, where details of the short-range interaction may become important, the limits of these models remains to be explored. An unambiguous determination of phase shift parameters has been achieved up to about 1 GeV[3, 4, 5, 6, 7]. With increasing energy, however, the number of partial waves to be determined grows, but the quality and density of the experimental data base diminishes. Recently, it has been pointed out that above about 1.2 GeV serious discrepancies between Phase-Shift Analysis (PSA) of different groups [7, 8] exist. An unambiguous, model-independent direct reconstruction of the 5 complex scattering Amplitudes, which could solve this puzzle, is as yet not possible, as it was shown by the Saclay-Geneva group [8]. Only new data, preferably on observables where little or no data exists, will pin down phase shift parameters and remove discrete ambiguities in amplitude reconstruction.

EXPERIMENT

The EDDA-experiment was conceived to provide high-precision elastic-scattering data in the COSY energy range (0.5-2.5 GeV). Unpolarized differential cross [9] and analyzing power excitation functions [10] have been measured in the first phases of the experiment and helped to extend PSA-analysis up to 2.5 GeV [7, 11]. These data have been used to impose strong upper bounds [12] on possible resonant contributions to pp-elastic scattering, as they might arise from coupling to isovector, strangeness zero dibaryonic resonances (e.g. [13, 14]).

Recently, data taking with a polarized beam on the polarized atomic beam target [15] (Fig. 1) has been completed. This allows to access spin correlation parameters A_{BT}, which describe the dependence of the cross section on the relative spin-orientation of the beam (B) and target (T) protons. By orienting the target-spin in different directions, normal (N) to or sideways (S) in the scattering plane, or longitudinal (L) to the beam, three such spin correlation parameters A_{NN}, A_{SS}, and A_{SL} can be measured with a detector of full azimuthal coverage. Here, a storage ring provides a unique experimental environment by combining pure polarized hydrogen targets with fast and easy spin-manipulation in order to minimize systematic errors, a technique pioneered by the

PINTEX [16, 17] collaboration at IUCF at lower energies.

The EDDA-detector [18, 19] (cf. Fig.1) is comprised of two cylindrical double-layered scintillator hodoscopes surrounding the COSY-beampipe downstream of the internal target. Protons from pp elastic scattering are detected in coincidence for scattering angles ranging from 30° to 90° in the center-of-mass. A beam of polarized hydrogen atoms crossing the COSY beam at right angle serves as an internal target with a typical area density of $2 \cdot 10^{11}$ atoms/cm^2 and polarization above 90%. The density of unpolarized hydrogen which builds up in the beam pipe when operating the target reduces the effective polarization to about 70% for accepted scattering events. To this end, the scattering vertex, reconstructed offline with 1 mm resolution, is used to select events originating in the overlap region of the atomic beam target with the COSY beam. The target polarization is aligned by applying a weak (1 mT) magnetic guiding field in either one of 6 possible directions ±x,±y, and ±z in the interaction region.

Data is acquired during acceleration of the COSY beam and for about 5s in the flattop at the desired beam momentum. Beam intensities ranged from $3 \cdot 10^9$ to $1.5 \cdot 10^{10}$ protons stored and accelerated in the ring and provided luminosities in the $1 - 5 \cdot 10^{27}$ 1/(cm^{2}s) range with beam polarizations between 50 and 75%. Due to the limited beam intensity, nine different flattop momenta were chosen to cover the energies above 2100 MeV/c with sufficient statistics in view of dropping cross-sections. Data in the lower energy range is obtained from data recorded during the ramp. For each COSY machine cycle the target and beam polarizations were held constant and then alternated between the twelve different combinations of the beam (±y) and target (±x,±y,±z) spin orientations.

ANALYSIS

Data analysis proceeds in two steps: First the elastic scattering rate for 5° wide bins in the c.m. polar angle $\theta_{c.m.}$ is determined as a function of the azimuthal angle ϕ, by selecting events well within the detector acceptance, which originated at the desired target location and obey elastic scattering kinematics. Due to the analyzing power and the non-vanishing spin correlation coefficients, the scattering rate for each spin combination exhibits characteristic modulations with the azimuthal angle. Secondly, the spin correlation parameters as well as beam and target polarizations are extracted either by calculating certain asymmetries [20, 21] which cancel the influence of detector efficiencies to first order, or by standard χ_2-minimization techniques. Both methods yield consistent results. The overall normalization of the target and beam polarizations is fixed with reference to the EDDA analyzing power data [10] with an uncertainty ranging from 1.5-3% from lower to higher energies.

RESULTS

The results for the three spin correlation parameters were extracted for 12 angle bins between 30° and 90° in $\theta_{c.m.}$ both for the data measured during the flattop and beam acceleration. An example of an excitation function and an angular distribution is shown

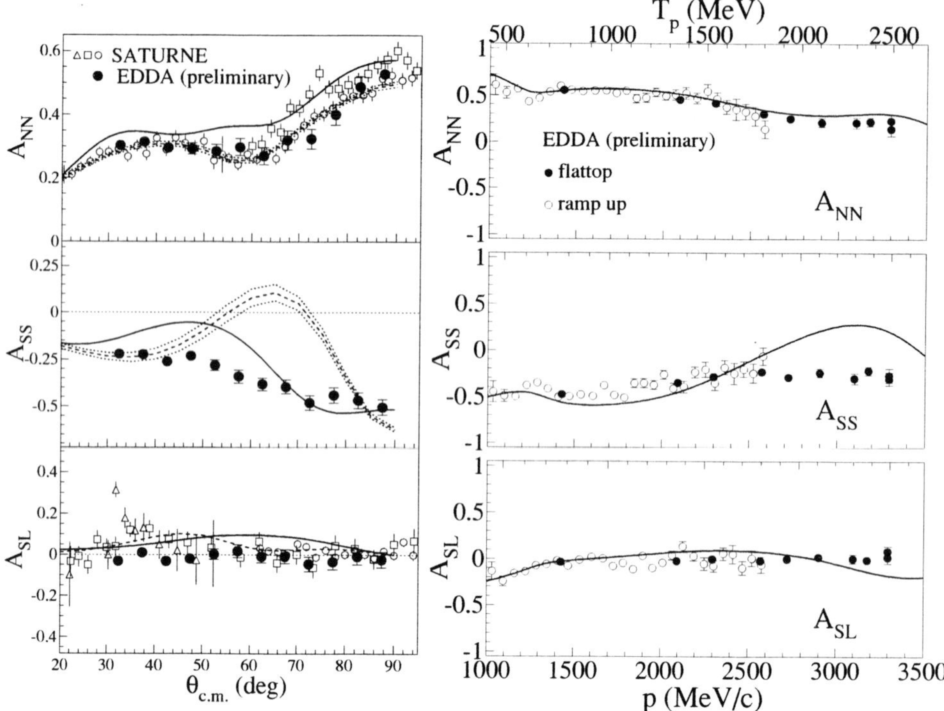

FIGURE 2. Angular distributions at 2572 MeV/c (left) and excitation functions at $\theta_{c.m.} = 47.5°$ (right) of spin correlation parameters A_{NN}, A_{SS}, and A_{SL} in comparison to phase shift predictions of SAID [7] (SM00, solid) and the Saclay-Geneva [8] (dashed line) analysis. On the left, data from SATURNE [22] are shown as open symbols. On the right, closed (open) symbols distinguish data points measured at the flattop (during acceleration) of the COSY machine cycle. All EDDA data are preliminary.

in Fig 2. The results obtained during beam acceleration nicely match those obtained at fixed momenta, and provide reasonably accurate data below 2100 MeV/c.

Previous measurements of spin correlation parameters at these energies were done mainly at SATURNE [22] on A_{NN}, A_{LL}, and A_{SL}In comparison, our new data on A_{NN} is consistent at all energies, however, we find values for A_{SL} more or less compatible with zero at all angles and do not confirm excursions in the angular distributions (cf. Fig. 2) as evinced in the SATURNE data. In contrast the observable A_{SS} has not been measured before above 800 MeV. These data thus put the predictive power of existing parameterization by scattering phase shifts to a true test. In the figures our new data are compared to PSA solutions of the Virginia (SAID, solution SM00) and Saclay-Geneva groups. They fit our data on A_{NN} and A_{SL} reasonably well but are in striking disagreement with the new data on A_{SS} - and each other - in particular at higher energies. PSA solutions at these energies should therefore be used with caution.

This highlights that the world data base on proton-proton elastic scattering to date does not allow to unambiguously determine the scattering phase shifts or amplitudes at energies well above 1GeV. It will be interesting to see to what extent the new spin correlation data on A_{SS} will be a remedy. Since A_{SS} is sensitive to the interference of non-spinflip and double-spinflip helicity amplitudes [23], it provides important information on the spin-dependence of the NN-interaction. To explore this, we carried out a direct reconstruction of the scattering amplitudes along the lines of [8] and found that the addition of our data to the world database removes some – but not all – of the discrete ambiguities. Here, further, significant improvement, can only be obtained by accurate triple polarization observables, i.e. with detection of the polarization of one ejectile. To what extent the PSA variants on the market will converge towards a unique solution remains to be seen until the new EDDA data is included in the data base. First steps in this direction indicate sizable modification of the phase shifts in the central partial waves of the SAID solutions.

ACKNOWLEDGMENTS

The excellent beam support by the COSY accelerator team during all phases of the now completed experimental program of EDDA is warmly acknowledged.

REFERENCES

1. Machleidt, R., and Slaus, I., *J. Phys.*, **G27**, R69 (2001), and references herein.
2. Bedaque, P. F., and van Kolck, U., *nucl-th/0203055, to appear in Ann. Rev. Nucl. Part. Sci.*, **53** (2002), and references herein.
3. Stoks, V. G. J., Klomp, R. A. M., Rentmeester, M. C. M., and de Swart, J. J., *Phys. Rev.*, **C48**, 792–815 (1993).
4. Bystricky, J., Lechanoine-LeLuc, C., and Lehar, F., *J. Phys. (Paris)*, **48**, 199–226 (1987).
5. Bystricky, J., Lechanoine-LeLuc, C., and Lehar, F., *J. Phys. (Paris)*, **51**, 2747 (1990).
6. Nagata, J., Yoshino, H., and Matsuda, M., *Prog. Theor. Phys.*, **95**, 691 (1996).
7. Arndt, R. A., Strakovsky, I. I., and Workman, R. L., *Phys. Rev. C*, **62**, 34005 (2000).
8. Bystricky, J., Lehar, F., and Lechanoine LeLuc, C., *Eur. Phys. J.*, **C4**, 607 (1998).
9. Albers, D., et al., *Phys. Rev. Lett.*, **78**, 1652–1655 (1997).
10. Altmeier, M., et al., *Phys. Rev. Lett.*, **85**, 1819 (2000).
11. Arndt, R. A., Oh, C. H., Strakovsky, I. I., Workman, R. L., and Dohrmann, F., *Phys. Rev.*, **C56**, 3005 (1997).
12. Rohdjess, H., Habilitationsschrift, Universität Bonn (2000).
13. Mulders, P. J., Aerts, A. T., and Swart, J. J. D., *Phys. Rev.*, **D21**, 2653 (1980).
14. Gonzalez, P., LaFrance, P., and Lomon, E. L., *Phys. Rev.*, **D35**, 2142 (1987).
15. Eversheim, P., et al., *Nucl. Phys.*, **A626**, 117c (1997).
16. von Przewoski, B., et al., *Phys. Rev.*, **C58**, 1897 (1998).
17. Rathmann, F., et al., *Phys. Rev.*, **C58**, 658 (1998).
18. Bisplinghoff, J., et al., *Nucl. Instr. and Meth.*, **A329**, 151 (1993).
19. Altmeier, M., et al., *Nucl. Instr. Meth.*, **A431**, 428 (1999).
20. Ohlsen, G. G., *Nucl. Instr. and Meth.*, **109**, 41–59 (1973).
21. Bauer, F., Ph.d. thesis, Universität Hamburg (2001).
22. Ball, J., et al., *CTU Reports*, **4**, 3 (2000).
23. Bourrely, C., Soffer, J., and Leader, E., *Phys. Rept.*, **59**, 95 (1980).

Elastic Polarized Proton Scattering at RHIC

S. Bültmann (for the PP2PP Collaboration [*])

Brookhaven National Laboratory, Physics Department, Upton, NY 11973-5000

Abstract. The PP2PP Collaboration is investigating the elastic scattering process of polarized protons at the Relativistic Heavy Ion Collider (RHIC) at the Brookhaven National Laboratory. The center of mass energy, $\sqrt{s}$, and squared four-momentum transfer, $-t$, accessible to the experiment are $50\,\mathrm{GeV} < \sqrt{s} < 500\,\mathrm{GeV}$ and $4 \cdot 10^{-4}\,\mathrm{GeV}^2/c^2 < |t| < 1.3\,\mathrm{GeV}^2/c^2$. During the 2002 polarized proton run about $3 \cdot 10^5$ elastic events were collected in a 14 hour engineering run at a center of mass energy of $\sqrt{s} = 200\,\mathrm{GeV}$, covering $0.005\,\mathrm{GeV}^2/c^2 < |t| < 0.030\,\mathrm{GeV}^2/c^2$. The experiment and its motivation is outlined and first results from the engineering run are presented.

INTRODUCTION AND MOTIVATION

The capability of RHIC to accelerate protons while maintaining their polarization allows the measurement of unpolarized and polarized scattering processes. The PP2PP experiment is measuring these processes in elastic and diffractive scattering.

Elastic proton-proton scattering has so far been measured up to an energy of $\sqrt{s} = 62$ GeV at the ISR at CERN. The proton-antiproton measurements reached an energy of $\sqrt{s} = 1.8$ TeV at Fermilab. The energy range accessible at RHIC will provide for the first time results in an unexplored kinematic region, with some overlap at energies below $\sqrt{s} = 62$ GeV.

High energy elastic scattering is described phenomenologically by Regge theory, which at higher squared four-momentum transfer, $-t$, greater than a few GeV^2/c^2, lead to the region where pQCD calculations are applicable. Pomeron exchange models have been successfully used to describe the present data for pp and $p\overline{p}$ scattering [1, 2]. In Regge theory the Pomeron is a singularity in the complex angular momentum plane, having isospin $I = 0$ and charge conjugation $C = +1$. A second mediator with the same quantum numbers, except for $C = -1$, called the Odderon, has been proposed [3]. In the pQCD region the Odderon is described as a three-gluon exchange with $C = -1$.

Elastic proton-proton scattering at small $-t$ can be described by two amplitudes, Coulomb interaction, $A_C = -8\pi\alpha_{em}s\frac{G^2(t)}{|t|}$, dominant at $-t$ below 10^{-3} GeV^2/c^2, and hadronic interaction, $A_h = s(i+\rho)\sigma_{tot}e^{\frac{1}{2}Bt}$, dominant above $-t$ =0.01 GeV^2/c^2, plus a Coulomb-Nuclear-Interference (CNI) term of the two in the $-t$-region between the two exchanges. The elastic differential cross section is then given by

$$\frac{\mathrm{d}\sigma_{el}}{\mathrm{d}t} = \frac{1}{16\pi s^2}\left|A_h + A_C \cdot e^{i\alpha_{em}\phi}\right|^2,$$

with σ_{tot} the total scattering cross section, ρ the ratio of real to imaginary part of the forward ($t = 0$) scattering amplitude, B the nuclear slope parameter, $G(t)$ the proton

CP675, *Spin 2002: 15th Int'l. Spin Physics Symposium and Workshop on Polarized Electron Sources and Polarimeters*, edited by Y. I. Makdisi, A. U. Luccio, and W. W. MacKay

electromagnetic form factor and $\alpha_{em}\phi$ the relative Coulomb-hadronic phase.

Because the Coulomb scattering contribution can be calculated, a measurement of the scattering yield per t-interval, dN/dt, in the region below $5 \cdot 10^{-4}$ GeV$^2/c^2$ yields a normalization, the luminosity $\mathscr{L}$, necessary to obtain the total cross section and absolute differential elastic cross section. In the nuclear region, the slope parameter, B, can be measured and in the CNI-region also the ρ-parameter becomes accessible.

At higher $-t$, around 1 GeV$^2/c^2$, the differential cross section features a $\sqrt{s}$-dependent diffractive minimum (dip structure). For a complete description of the scattering process over the entire t-range double Pomeron exchange and triple gluon exchange need to be included [4]. In case of an Odderon contribution, a difference of the differential cross sections for pp and $p\overline{p}$ is expected.

Polarization effects will also be studied in the kinematic region of RHIC. In the CNI-region the interference between the one-photon exchange contribution to the spin-flip amplitude with the hadronic non-flip amplitude leads to the single transverse spin asymmetry, A_N [5]. The maximum of the asymmetry is located at $-t = 3 \cdot 10^{-3}$ GeV$^2/c^2$ with a value of about $A_N \approx 0.04$. A contribution to A_N due to an interference between the Coulomb non-flip and hadronic spin-flip amplitude is also possible, but presently not accurately known and adds to the theoretical uncertainty of A_N.

The double transverse spin asymmetry, A_{NN}, is predicted to show a different t-dependence for a pure Pomeron contribution compared to an admixture of an Odderon contribution to the scattering amplitude [6]. The measurement would enable us to probe the size of the Odderon contribution.

In earlier fixed target measurements, up to $\sqrt{s} = 24$ GeV, the analyzing power, A_N, showed a sign change in the dip region. For RHIC energies predictions for a Pomeron contribution to the spin-flip amplitude have been made [7].

EXPERIMENTAL SETUP

At RHIC energies the scattering angle for elastically scattered protons is of the order of a few milliradians or less. To be able to detect these protons and to separate them from beam protons, suitable locations along the beam line and an appropriate tune for the accelerator have to be found. The PP2PP experiment detects the scattered protons at locations about 60 meters to the left and right of an interaction point (IP). At this location, the outgoing beam and scattered protons are in a common beam pipe, separated from the incoming proton beam. Because of the close vicinity of scattered protons and the beam, both are guided by the same group of magnets. The focusing magnets are determining the parameters relating the position of the scattered protons to the scattering angle at the IP. The parameters describing these magnets have to be known accurately. Neglecting mixing parameters between the two cartesian coordinates perpendicular to the proton beam momentum, horizontally x and vertically y, the equations relating the position, x_D, and the angle, Θ_x^D, at the detection point to the scattering angle, Θ_x^*, are given by

$$\begin{aligned} x_D &= a_{11} \cdot x^* + a_{12} \cdot \Theta_x^* \\ \Theta_x^D &= a_{21} \cdot x^* + a_{22} \cdot \Theta_x^* \end{aligned}$$

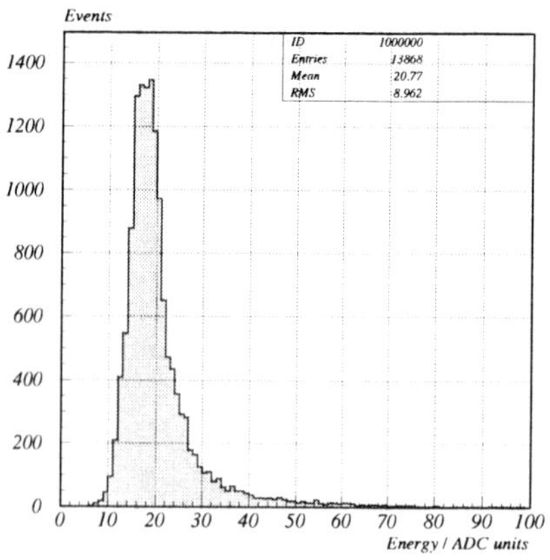

FIGURE 1. Energy distribution of protons. The mean value of 20.8 ADC units corresponds to 200 keV.

with x^* the position at the IP, and the a-parameters determined by the accelerator optics. A second set of equations holds for the y-coordinate, perpendicular to x, with different a-parameters. As apparent from the set of equations, a_{12}, relevant for the magnification of the scattering angle, needs to be maximized, and either a_{11}, providing a scale for x^*, needs to be minimized, or Θ_x^D needs to be measured to eliminate the dependence on x^*. These conditions, together with the requirement to minimize the beam size at the detection location, lead to an accelerator tune that is different from the settings used for other experiments at RHIC, which require a small beam size at the IP.

Because of the small scattering angle, the scattered protons are not leaving the beam pipe, and the detectors need to be moved as close to the beam as possible. This is achieved by mounting the detector packages inside Roman Pots, vacuum vessels with thin windows allowing the detector packages at ambient pressure to penetrate into the evacuated beam pipe. During the 2002 engineering run one detector package above and below the beam was installed on either side of the IP. For a measurement of Θ_x^D a second set of detector packages would have been needed, but was not yet available. Because of that, our measurement is depending on the position of the scattering protons at the IP, which is not well known and adds to the overall uncertainty.

A detector package consisted of four single sided, 400 μm thick silicon microstrip detectors with 70 μm wide strips and 100 μm strip pitch, covering an active detection area of 45 mm by 75 mm. It is complemented by a 8 mm thick scintillator plane of equal area, read out by two photomultiplier tubes for trigger purposes. Two of the silicon detectors were measuring the x-coordinate, the other two the y-coordinate. Almost all of the 16 detectors except two had an efficiency of above 0.95, resulting in a detection efficiency of above 0.99 for scattered protons. The silicon detectors were read out by a chip (SVXIIe [8]) featuring 128 input channels with preamplification, 32 event pipeline, and Wilkenson type ADC. A distribution of the energy deposited by 100 GeV/c protons is shown for one silicon detector in Figure 1. Scintillator counters, covering a pseudo-rapidity range of $2.4 < |\eta| < 5.3$, are installed to veto inelastic events and to provide a

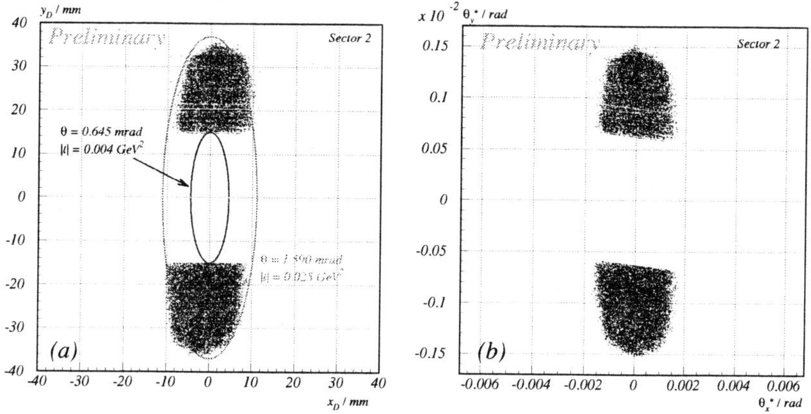

FIGURE 2. Measured coordinates of elastically scattered protons as recorded by the detector package above and below the beam pipe center. In part (a) the coordinates are plotted for y_D versus x_D, while in part (b) the transformation into the coordinate system of scattering angles, Θ_x^* versus Θ_y^*, has been applied.

relative luminosity measurement.

FIRST RESULTS FROM ENGINEERING RUN

During the 2001/2002 RHIC run with 100 GeV/c polarized protons a dedicated accelerator tune on the last day of the run enabled us to record 14 hours of data. The total intensity of each beam was $5 \cdot 10^{11}$ protons, distributed among 55 bunches. This was reached by scraping of the stored beams to an emittance of $\varepsilon = (12\pi) \cdot 10^{-6}$ m. Under these conditions the detector packages were slowly moved into the beam pipe. The closest distance between a top- and bottom-mounted detector package was about 30 mm. The minimum $|t|$-value accessible with the detector packages is determined by half of that distance, giving a lower limit for $-t$ of 0.004 GeV$^2/c^2$ for the scattered protons.

Out of the one million events recorded during this run, about one third could be reconstructed as due to elastically scattered protons. The remaining events were mostly inelastic triggers or triggers belonging to bunch crossings that were not read out properly.

In Figure 2.a the coordinates, x_D and y_D, of reconstructed elastic events are shown as they are measured by the two detector packages above and below the beam. The a_{12} values for the two coordinates x and y were different by a ratio of $a_{12}(y)/a_{12}(x) \approx 3.5$, and also slightly different for the two sides of the IP. As a consequence, t-isolines have an elliptic shape. The horizontal shift of the two detector measurements with respect to each other is due to a magnet misalignment, leading to an interdependence of x- and y-coordinates. Applying the calculated a-parameters to the data yields a set of scattering angles with the two coordinates, Θ_x^* and Θ_y^*. The result is shown in Figure 2.b. In Figure 3 a correlation plot of the measured x_D and y_D coordinates on either side of the IP is shown. The bands of elastically scattered protons is very distinct from the background.

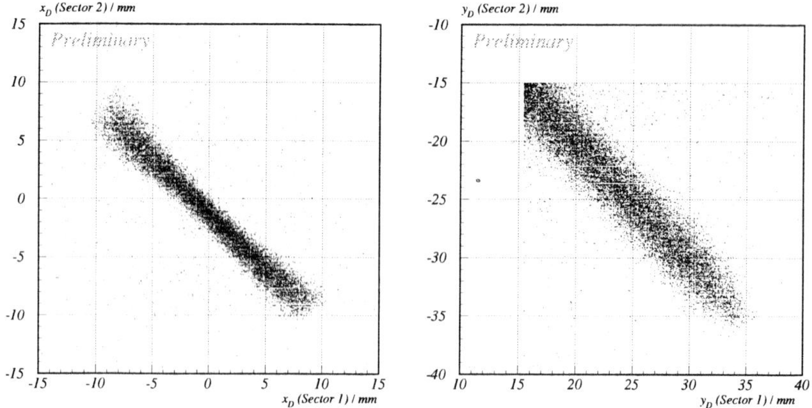

FIGURE 3. Correlation plots of the horizontal, x_D, and the vertical, y_D, coordinate as measured on either side of the IP, referred to as sector 1 and 2.

In conclusion, the first PP2PP engineering run successfully recorded for the first time elastically scattered polarized protons at an energy of $\sqrt{s} = 200$ GeV / c. The continuing analysis of the limited data sample taken might provide a measurement of the slope-parameter, B, in the t-range from -0.01 GeV^2/c^2 to -0.02 GeV^2/c^2, and a possible first look at the analyzing power A_N.

For the next run, the experiment plans to install the second set of detectors to eliminate some of the dependence of the measurement on accelerator parameters, and to take a data sample allowing to determine A_N accurately.

REFERENCES

*. S. Bültmann, I. H. Chiang, B. Chrien, A. Drees, R. Gill, W. Guryn, D. Lynn, P. Pile, A. Rusek, M. Sakitt, S. Tepikian, Brookhaven National Laboratory, USA; J. Chwastowski, B. Pawlik, Institute of Nuclear Physics, Cracow, Poland; M. Haguenauer, Ecole Polytechnique/IN2P3-CNRS, Palaiseau, France; A. A. Bogdanov, S. B. Nurushev, M. F. Runtzo, MEPHI, Moscow, Russia; I. G. Alekseev, V. P. Kanavets, B. V. Morozov, D. N. Svirida, ITEP, Moscow, Russia; M. Rijssenbeek, C. Tang, S. Yeung, SUNY, Stony Brook, USA; K. De, N. Guler, J. Li, University of Texas at Arlington, USA; A. Sandacz, Institute for Nuclear Studies, Warsaw, Poland.

1. Landshoff, P. V., *Pomeron physics: an update*, hep-ph/0010315 (2000).
2. Barone, V., and Predazzi, E., *High-Energy Particle Diffraction*, Berlin (2002).
3. Gauron, P., Leader, E., and Nicolescu, B., Nucl. Phys. B 299, 189 (1984).
4. Donnachie, A., and Landshoff, P. V., Nucl. Phys. B231, 189 (1984).
5. Buttimore, N. H., et al., Phys. Rev. D 59, 114010 (1999).
6. Leader, E., and Trueman, T. L., Phys. Rev. D 61, 077504 (2000).
7. Martini, A. F., and Predazzi, E., Phys. Rev. D 66, 034029 (2002).
8. Lipton, R., Nucl. Instr. and Methods A 418, 85 (1998).

Properties of the Spin-flip Amplitude of Hadron Elastic Scattering and Possible Polarization Effects at RHIC

O.V. Selyugin

BLTPh, JINR, Dubna, Russia

Abstract. In the framework of the RHIC spin program we investigated the polarization effects in elastic proton-proton scattering at small momentum transfer and in the diffraction dip region. The calculations take into account the Coulomb-hadron interference effects including the additional Coulomb-hadron phase. In particular, we show the impact of the form of the hadron potential at large distances on the behavior of the hadron spin-flip amplitude at small angles. The t-dependence of the spin-flip amplitude of high energy hadron elastic scattering is analyzed under different assumptions on the hadron interaction.

INTRODUCTION

Several attempts to extract the spin-flip amplitude from the experimental data show that the ratio of spin-flip to spin-nonflip amplitudes can be non-negligible and may be only slightly dependent on energy [1, 2].

For the definition of new effects at small angles and especially in the region of the diffraction minimum one must know the effects of the Coulomb-hadron interference with sufficiently high accuracy. The Coulomb-hadron phase was calculated in the entire diffraction domain taking into account the form factors of the nucleons [3]. Some polarization effects connected with the Coulomb hadron interference, including some possible odderon contribution, were also calculated [4].

The model-dependent analysis based on all the existing experimental data of the spin-correlation parameters above $p_L \geq 6\ GeV$ allows us to determine the structure of the hadron spin-flip amplitude at high energies and to predict its behavior at superhigh energies [6]. This analysis shows that the ratios

$$Re\ \phi_5^h(s,t)/(\sqrt{|t|}\ Re\ \phi_1^h(s,t)), \quad Im\ \phi_5^h(s,t)/(\sqrt{|t|}\ Im\ \phi_1^h(s,t)) \qquad (1)$$

depend on s and t (see Fig.1 a,b). At small momentum transfers, it was found that the slope of the "residual" spin-flip amplitudes is approximately twice the slope of the spin-non flip amplitude. The obtained spin-flip amplitude leads to the additional contribution to the pure CNI effect at small t (Fig. 1 c).

The dependence of the hadron spin-flip amplitude on t at small angles is closely related with the basic structure of hadrons at large distances. We show that the slope of the so-called "reduced" hadron spin-flip amplitude (the hadron spin-flip amplitude

CP675, *Spin 2002: 15th Int'l. Spin Physics Symposium and Workshop on Polarized Electron Sources and Polarimeters,* edited by Y. I. Makdisi, A. U. Luccio, and W. W. MacKay

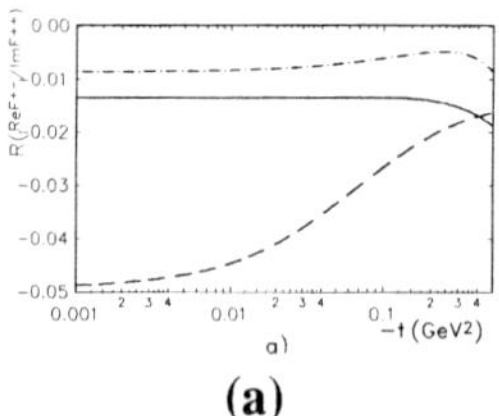

(a)

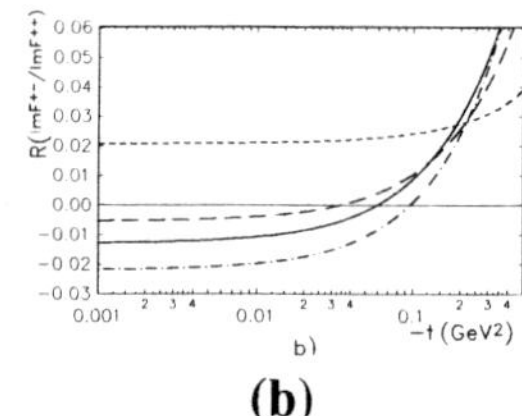

(b)

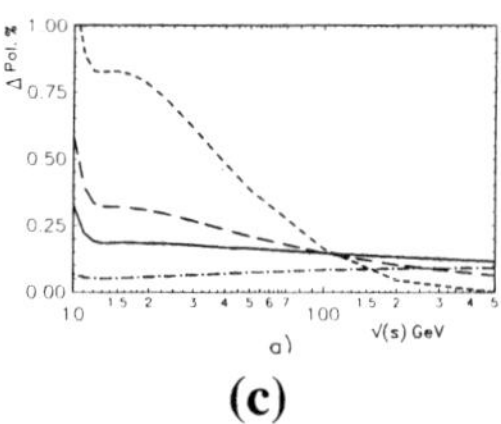

(c)

FIGURE 1. (**a**) and (**b**) Ratio of the imaginary (a) and real (b) part of the "residual" F_h^{+-} to the imaginary F_h^{++}; (**c**) Contribution to the pure CNI effect from the model F_h^{+-} (our calculations at t_{max}, $|t| = 0.001\ GeV^2$ $|t| = 0.01\ GeV^2$,$|t| = 0.1\ GeV^2$ are shown by the full line, the long dashed line, the dashed line and the dotted line respectively.

without the kinematic factor $\sqrt{|t|}$) can be larger than the slope of the hadron spin-non-flip amplitude, as was observed long ago [5].

SLOPE OF THE HADRON AMPLITUDES

For an exponential form of the amplitudes this coincides with the usual slope of the differential cross sections divided by 2. At small t ($\sim 0 \div 0.1\ GeV^2$), practically all semiphenomenological analyses assume:

$$B_1^+ \approx B_2^+ \approx B_1^- \approx B_2^-.$$

If the potentials V_{++} and V_{+-} are assumed to have a Gaussian form in the first Born approximation ϕ_1^h and $\hat{\phi}_h^5$ will have the same form

$$\phi_1^h(s,t) \sim exp(-B\Delta^2), \quad \phi_5^h(s,t) \sim= q\ B\ exp(-B\Delta^2). \tag{2}$$

Therefore, in this special case the slopes of the spin-flip and "residual"spin-non-flip amplitudes are indeed the same. A Gaussian form of the potential is adequate to represent the central part of the hadronic interaction. The form cuts off the Bessel function and the contributions at large distances. If, however, the potential (or the corresponding eikonal) has a long tail (exponential or power) in the impact parameter, the Bessel functions can not be taken in the approximation form and the full integration leads to different results.

If we take

$$\chi_i(b,s) \sim H\ e^{-a\ \rho}, \tag{3}$$

we obtain

$$F_{nf}(s,t) = a/[(a^2+q^2)^{3/2}] \approx 1/[a\sqrt{a^2+q^2}]\ exp(-Bq^2), \tag{4}$$

with $B = 1/a^2$. For the "residual" spin-flip amplitude, on the other hand, we obtain [8]

$$\sqrt{|t|}\tilde{F_{sf}}(s,t) = (3\ a\ q)/[(a^2+q^2)^{5/2}] \approx (3\ aq\ B^2)/(\sqrt{a^2+q^2})\ exp(-2\ Bq^2). \tag{5}$$

In this case, therefore, the slope of the "residual" spin-flip amplitude exceeds the slope of the spin-non-flip amplitudes by a factor of two. A similar behaviour can be obtained with the standard dipole form factor [8].

DETERMINATION OF THE STRUCTURE OF THE HADRON SPIN-FLIP AMPLITUDE

Note that if the "reduced" spin-flip amplitude is not small, the impact of a large B^- will reflect in the behavior of the differential cross section at small angles [7]. The method gives only the absolute value of the coefficient of the spin-flip amplitude. The imaginary and real parts of the spin-flip amplitude can be found only from the measurements of the spin correlation coefficient.

Let us take the spin nonflip amplitude in the standard exponential form with definite parameters: slope B^+, σ_{tot} and ρ^+. For the "residual" spin-flip amplitude, on the other hand, we consider two possibilities: equal slopes $B^- = B^+$ and $B^- = 2B^+$. The results of these two different calculations are shown in Fig.2. It is clear that around the maximum of the Coulomb-hadron interference, the difference between the two variants is very small. But when $|t| > 0.01\ GeV^2$, this difference grows. So, if we try to find the contribution of the pomeron spin-flip, we should take into account this effect. As the value of A_N depends on the determination of the beam polarization, let us calculate the derivative of A_N with respect to t, for example, at $\sqrt{s} = 500\ GeV$.

If we know the parameters of the hadron spin non-flip amplitude, the measurement of the analyzing power at small transfer momenta helps us to find the structure of the hadron spin-flip amplitude. There is a specific point of the differential cross sections and of A_N on the axis of the momentum transfer, - t_{re}, where $|ReF_c^{++}| = |ReF_h^{++}|$. This point t_{re} can be found from the measurement of the differential cross sections [9]. At high energies at the point t_{re} [8] we obtain for pp-scattering

$$ReF^h_{sf}(s,t) - \frac{-1}{2(ImF^h_{nf}(s,t) + ImF^c_{nf}(t))} A_N(s,t)\,\frac{d\sigma}{dt} - ReF^c_{sf}(t). \qquad (6)$$

We can again take the hadron spin-nonflip and spin-flip amplitudes with definite parameters and calculate the magnitude of A_N by the usual complete form while the real part of the hadron spin-flip amplitude is given by (6). Our calculation by this formula and the input real part of the spin-flip amplitude are shown in Fig. 2 c. At the point t_{re} both curves coincide. So if we obtain from the accurate measurement of the differential cross sections the value of t_{re}, we can find from A_N the value of the real part of the hadron spin-flip amplitude at the same point of momentum transfer.

MODEL PREDICTIONS

The model [10] takes into account the contribution of the hadron interaction at large distances and leads to the high-energy spin-flip amplitude. The model gives the large spin effects in the hh-elastic scattering and predicts non-small effects for the $PP2PP$

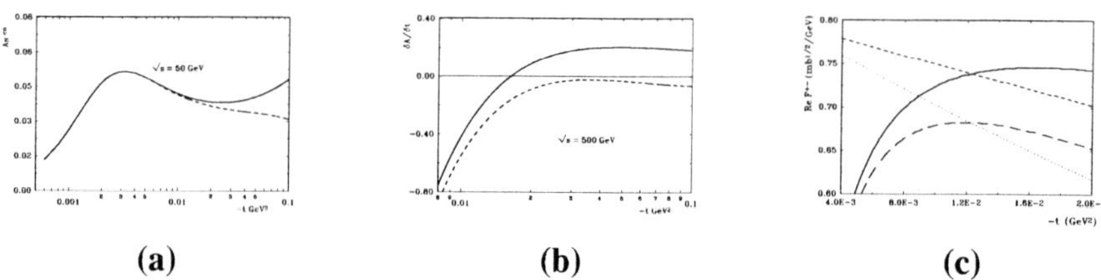

(a) **(b)** **(c)**

FIGURE 2. (**a**) A_N at $\sqrt{s} = 50\ GeV$ (**b**) $-\delta A_N / \delta t$ at $\sqrt{s} = 500\ GeV$ (the solid line is with the slope B_1^- of $\tilde{F}_{sf}$ equal to the slope B_1^+ of F_{nf}; the dashed line is with the $B_1^- \ = \ 2\ B_1^+$. (**c**) The form of $Re(F^{sf})$: solid and long-dashed lines are calculations by (3); short-dashed and dotted lines are model amplitudes with $B_1^- = B_1^+$ and $B_1^- = 2\ B_1^+$, respectively.

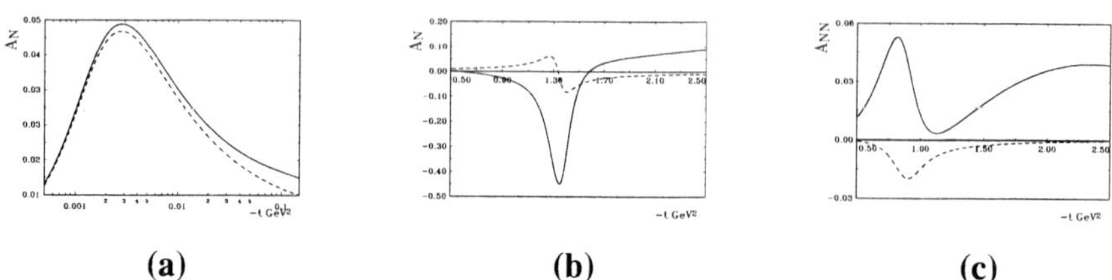

(a) **(b)** **(c)**

FIGURE 3. (**a**) and (**b**) A_N at $\sqrt{s} = 50\ GeV$ (a) the full line is the total A_N; the dashed line is the A_N^{CN}. (**c**). A_{NN} at $\sqrt{s} = 500\ GeV$ in the domain of the dip ; the full line is the total A_{NN}; the dashed line is the A_{NN}^{CN}.

experiment at RHIC especially in the diffraction dip domain [11]. The additional pure CNI effects can be calculated using the Coulomb-nuclear phase [3]. These polarization effects will be present at RHIC energy, even though $F_h^{+-} \to 0$ at high energy. Our model calculations show on Fig.3 for both cases.

The model gives the standard t-dependence of $ReF_h{++}$ and $ImF_h{++}$. Instead of it, in a convenient parameterization of both the modulus and the phase one can obtain the alternative case, in which $ImF_h(s,t)$ has the zero at small t (for details, see [12]). Such an approach enables one to specify the elastic hadron scattering amplitude $F_h(s,t)$ directly from the elastic scattering data. The difference between the phases leads either to central or peripheral distributions of elastic hadron scattering in the impact parameter space. The obtained form of A_N^{CN} at small momentum transfers differs for the two variants beginning at $|t| > 0.05$ GeV^2 (Fig.4 a). The difference reaches 2% at $-t = 0.15\ GeV^2$ and, in principle, can be measured in an accurate experiment. Now let us calculate the Coulomb-hadron interference effect - A_N^{CN} in the two alternatives for higher $|t|$: (i) the diffraction dip is created by the "zero" of the $ImF_h(s,t)$ part of the scattering amplitude and $ReF_h(s,t)$ fills it; (ii) the diffraction dip is created by the "zero" of the $ReF_h(s,t)$ part of the scattering amplitude and $ImF_h(s,t)$ fills it. The results are shown in Fig. 4 (b) for $\sqrt{s} = 50$ GeV and in Fig. 4 (c) at $\sqrt{s} = 500$ GeV.

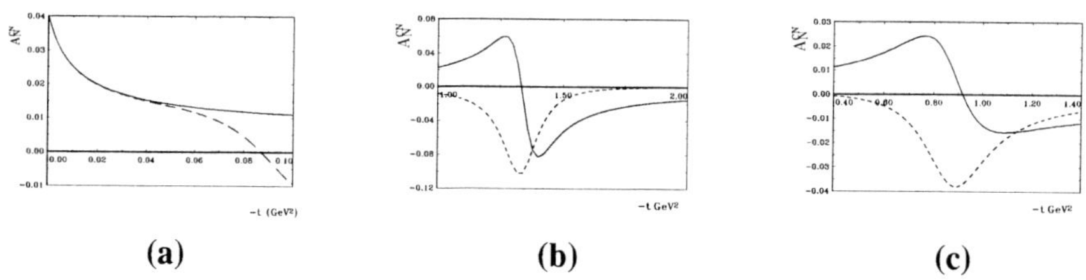

(a) (b) (c)

FIGURE 4. (**a**) A_N^{CN} at $\sqrt{s} = 50\ GeV$ and small t for two models. (**b**) and (**c**) A_N^{CN} at $\sqrt{s} = 50$ GeV and $\sqrt{s} = 500\ GeV$ in the region of the dip (the solid line corresponds to the model I with zero of $\Im F_h$ at dip; the dashed line shows the variant II, with the zero of the ReF_h at the dip).

ACKNOWLEDGMENTS

I would like to express my sincere thanks to the organizer for the kind invitation and financial support; and to S.B. Gerasimov, A.V. Efremov and E. Predazzi for fruitful discussions.

REFERENCES

1. Akchurin N., Buttimore N.H. and Penzo A., *Phys. Rev.* **D 51**, 3944 (1995).
2. Selyugin O.V., *Phys. Lett.* **B333**, 245 (1993).
3. Selyugin O.V., *Phys.Rev.* **D 60** 074028 (1999).
4. Selyugin O.V., *in Proc. "New Trends in High Energy Physics", Crimea, 2000, ed. P. Bogolyubov, L. Jenkovszky,* Kiev, 2000.
5. Predazzi E., Soliani G., *Nuovo Cim.* **A 2** 427 (1967); Hinotani K., Neal H.A., Predazzi E. and Walters G., *Nuovo Cim.*, **A 52** 363 (1979).
6. Selyugin O.V., *Phys. of Atomic Nuclei* **62** 333 (1999).
7. Selyugin O.V., *Mod. Phys. Lett.*, **A 9** 1207 (1994).
8. Predazzzi E. and Selyugin O.V., *Eur. Phys. J.* **A 13**, 471 (2002).
9. O.V. Selyugin, Nucl.Phys. B (Proc.Suppl.) 99A (2001) 60.
10. Goloskokov S.V., Kuleshov S.P., Selyugin O.V., Z. Phys. C **50**, 455 (1991).
11. N.Akchurin, S.V.Goloskokov, O.V.Selyugin, Int.J. of Mod.Phys. **A 14** (1999) 253.
12. V. Kundrát and M. Lokajiček, Z. Phys. C **63**, 619 (1994).

SPIN@U-70: An Experiment to Measure the Analyzing Power A_n in Very-high-$P_\perp^2$ p-p Elastic Scattering at 70 GeV[*]

V.G. Luppov[1**], L.V. Alexeeva[1a], V.A. Anferov[1b], E.D. Courant[1], Ya.S. Derbenev[1], G. Fidecaro[1c], M. Fidecaro[1c], F.Z. Khiari[1d], S.V. Koutin[1a], A.D. Krisch[1#], M.A. Leonova[1a], A.M.T. Lin[1], W. Lorenzon[1], V.S. Morozov[1a], D.C. Peaslee[1e], C.C. Peters[1], R.S. Raymond[1], D.W. Sivers[1f], J.A. Stewart[1g], S.M. Varzar[1a], V.K. Wong[1], K. Yonehara[1], D.G. Crabb[2], Yu.M. Ado[3], A.G. Afonin[3], V.I. Belousov[3], B.V. Chujko[3], A.N. Davidenko[3], N.A. Galyaev[3], V.I. Garkusha[3], V.N. Grishin[3], V.A. Kachanov[3], Yu.V. Kharlov[3], V.I. Kotov[3], A.V. Kusnetsov[3], V.A. Medvedev[3], Yu.M. Melnik[3], V.V. Mochalov[3], A.I. Mysnik[3], S.B. Nurushev[3], A.F. Prudkoglyad[3], P.A. Semenov[3], V.L. Solovianov[3†], V.P. Stepanov[3], V.A. Teplyakov[3], S.M. Troshin[3], A.G. Ufimtsev[3], M.N. Ukhanov[3], A.E. Yakutin[3], V.N. Zapolsky[3], V.G. Zarucheisky[3], N.S. Borisov[4], V.V. Fimushkin[4], V.A. Nikitin[4], P.V. Nomokonov[4], I.A. Rufanov[4], Yu.K. Pilipenko[4], P.P.J. Delheij[5], W.T.H. van Oers[5], and A.N. Zelenski[5h]

[1] *University of Michigan, Ann Arbor, USA*
[2] *University of Virginia, Charlottesville, USA*
[3] *IHEP, Protvino, Russia*
[4] *JINR, Dubna, Russia*
[5] *TRIUMF, Vancouver, Canada*

Abstract. The SPIN@U-70 experiment plans to measure the one-spin analyzing power A_n for 70 GeV proton-proton elastic scattering at large $P_\perp^2$ values of 1 to 12 $(GeV/c)^2$. The Michigan frozen NH_3 polarized proton target (Solid PPT) should later be installed in the Channel 8 extracted beam-line of the 70 GeV U-70 accelerator in IHEP, Protvino. The forward-scattered protons are detected by small scintillation counters placed at about 9 m from the PPT, while the

[*] Supported by a U.S. Department of Energy Research Grant
[**] E-mail: vluppov@umich.edu
[#] Spokesperson for SPIN@U-70 Collaboration, e-mail: krisch@umich.edu
[†] Deceased
[a] Also at: Moscow State University
[b] Also at: IUCF
[c] Also at: CERN
[d] Also at: UCAPS/RI, King Fahd University
[e] Also at: University of Maryland
[f] Also at: Portland Physics Institute
[g] Also at: DESY-Zeuthen
[h] Also at: Brookhaven/INR-Moscow

CP675, *Spin 2002: 15th Int'l. Spin Physics Symposium and Workshop on Polarized Electron Sources and Polarimeters*, edited by Y. I. Makdisi, A. U. Luccio, and W. W. MacKay

recoil-scattered protons are detected by a 35-m-long focusing magnetic spectrometer, with a 12 degree vertical bend, placed at 30 degrees to the beam. A tune-up run for testing the beam and the spectrometer, using a polyethylene target, was carried out in April 2002 at IHEP. The layout and the results of the test run are presented.

The SPIN@U-70 experiment plans to use a high intensity 70 GeV unpolarized extracted proton beam from the U-70 accelerator at IHEP-Protvino in Russia to measure the analyzing power A_n in $p + p_{\uparrow} \rightarrow p + p$ at large-$P_{\perp}^2$. We would scatter the high intensity beam from a polarized proton target and measure the quantity,

$$A_n = \frac{A_{mea}}{P_T} = \frac{1}{P_T}\left[\frac{N(\uparrow) - N(\downarrow)}{N(\uparrow) + N(\downarrow)}\right],$$

where A_{mea} is the measured asymmetry, P_T is the target polarization, and $N(\uparrow)$ and $N(\downarrow)$ are the normalized elastic event rates with the spin up and spin down, respectively.

Our main goal is to determine if the unexpected large value of A_n, discovered in large-$P_{\perp}^2$ proton-proton elastic scattering at the AGS, persists to higher energy and larger $P_{\perp}^2$ [1,2]. At 24 GeV the one-spin analyzing power A_n was found to be 20.4% ± 3.9% near $P_{\perp}^2$ of 7 $(\text{GeV/c})^2$, as shown in Fig. 1. This large and unexpected spin effect has been difficult to reconcile with conventional models of strong interactions such as perturbative Quantum Chromodynamics (PQCD). The validity of PQCD is predicted to improve with increasing energy and increasing $P_{\perp}^2$. This 70 GeV experiment would increase the maximum $P_{\perp}^2$ by a factor of about 1.7; it would also increase the maximum energy for A_n data at high-$P_{\perp}^2$ by a factor of about 2.5.

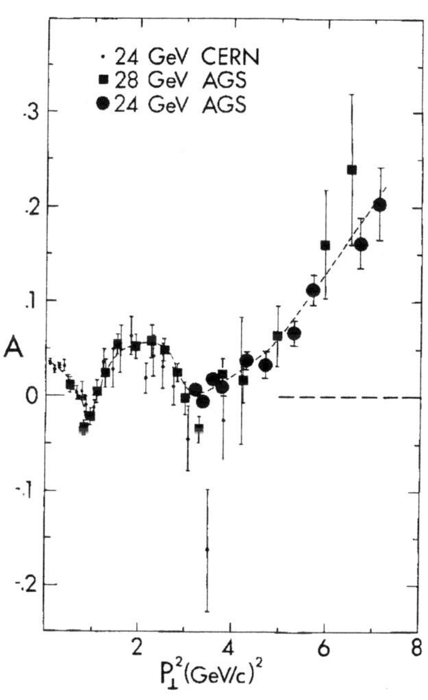

FIGURE 1. The analyzing power A_n is plotted against $P_{\perp}^2$ for spin polarized proton-proton elastic scattering at 24 and 28 GeV.

The experiment would use the Michigan 1-watt-cooling-power solid polarized proton target containing radiation-doped frozen ammonia (NH_3) beads. This target [3] successfully operated with an average polarizaton of 85% during a 3-month run at the AGS with an average beam intensity of 10^{11} protons per sec; this allowed the precise large $P_{\perp}^2$ measurements [1] of A_n shown in Fig. 1.

The SPIN@U-70 spectrometer is shown in Fig. 2. The forward-scattered protons are detected by small scintillation counters placed at about 9 m from the PPT, while the recoil-scattered protons are detected by a 35-m-long focusing magnetic spectrometer, with a 12 degree vertical bend, placed at 30 degrees to the beam. Table 1 lists the angles and momenta for both the forward and recoil protons, as well as the $\int B \cdot dl$ of the recoil-spectrometer magnets for various $P_\perp^2$. The dipole fields for the spectrometer's angles were calculated from the recoil protons' kinematics. The beam optics program TRANSPORT calculated the quadrupoles' gradients needed to focus the recoil protons to fit through the spectrometer's apertures. Most focusing is done by the vertically focusing Q_1 and the horizontally focusing Q_2; this gives the spectrometer a larger vertical acceptance angle $\Delta\varphi'_{lab} (= \Delta\varphi_{lab} \sin\theta_R)$ than its horizontal acceptance angle $\Delta\theta_{lab}$. For each $P_\perp^2$, the horizontal angle θ_R is correlated with the elastic recoil momentum P_R.

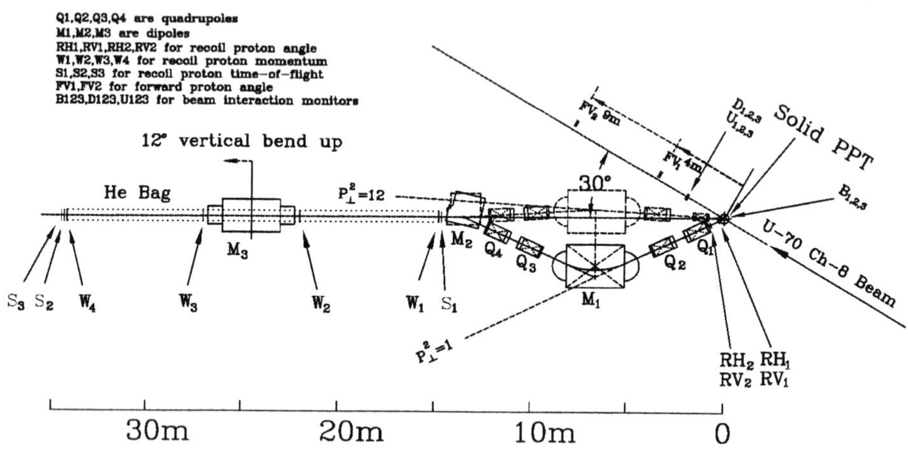

FIGURE 2. SPIN@U-70 spectrometer.

We estimated the event rates and errors in A_n for large-$P_\perp^2$ proton-proton elastic scattering at U-70 using the Michigan solid PPT and the SPIN@U-70 recoil spectrometer in U-70's Channel 8 extracted 70 GeV proton beam-line. Table 2 lists the estimated event rate and error in A_n for each $P_\perp^2$ point. We may run with a lower beam intensity at $P_\perp^2 = 1$ (GeV/c)^2 to reduce accidentals since the statistical precision is around 0.03%. Note that a superconducting quadrupole magnet Q_1 is required in the very-large $P_\perp^2$ region. The measurements of A_n should be rather precise, with an error of less than 1%, for $P_\perp^2$ up to 6.0 (GeV/c)^2, and less than 5% at $P_\perp^2 = 12.0$ (GeV/c)^2.

In spring 2001, installation of SPIN@U-70 started in the Channel 8 extracted beam-line of the 70 GeV U-70 accelerator in IHEP. The 35-m-long recoil Spectrometer's M_1, M_2, and M_3 dipoles and Q_1, Q_2, Q_3, and Q_4 quadrupole magnets, their movable stands, electric and water systems were installed. On 5-10 November 2001, a beam tune-up run showed that the Channel-8 extracted beam at the PPT

TABLE 1. Angles and momenta of elastic protons and magnet strengths. Positive Bl^{eff} corresponds to bending to the right for PPT, M_1 and M_2 and bending up for M_3. The recoil angle after the PPT magnet is θ'_R; it differs from θ_R by $\approx eBl^{eff}_{PPT} / P_R$.

$P_\perp^2$ (GeV/c)^2	θ_F degrees	P_F GeV/c	θ_R degrees	P_R GeV/c	Bl^{eff}_{PPT} kG-m	θ'_R degrees	Bl^{eff}_{M1} kG-m	Bl^{eff}_{M2} kG-m	Bl^{eff}_{M3} kG-m
1	0.825°	69.5	61.44°	1.139	4.42	54.94°	31.5	-15.8	7.94
2	1.176°	68.9	52.19°	1.790	4.46	48.08°	36.3	-18.0	12.48
3	1.452°	68.4	46.22°	2.399	4.50	43.17°	35.7	-17.6	16.73
4	1.690°	67.8	41.87°	2.997	4.54	39.44°	32.1	-15.7	20.90
5	1.906°	67.2	38.48°	3.594	4.57	36.46°	26.4	-12.9	25.06
6	2.107°	66.6	35.72°	4.196	4.61	34.00°	19.1	-9.4	29.26
7	2.296°	66.0	33.41°	4.804	-4.65	34.91°	26.8	-13.1	33.50
8	2.477°	65.4	31.44°	5.423	-4.69	32.76°	17.0	-8.3	37.82
9	2.653°	64.8	29.72°	6.051	-4.73	30.89°	6.2	-3.0	42.20
10	2.824°	64.2	28.20°	6.692	-4.77	29.26°	-5.7	2.8	46.67
12	3.159°	62.9	25.61°	8.015	-4.85	26.48°	-32.1	15.7	55.89

TABLE 2. Event rates and errors in A_n for p-p elastic scattering at U-70.

$P_\perp^2$ (GeV/c)^2	Δt (GeV/c)^2	$\Delta\phi$ mr	$d\sigma/dt$ $\frac{nb}{(GeV/c)^2}$	Events per hour	Hours	Events (N)	ΔA_n $[.85\sqrt{N}]^{-1}$	
1.0	0.06	159	4000	230000	100	2.3 10^7	0.03%	
2.0	0.09	177	90	8600	100	8.6 10^5	0.1%	
3.0	0.25	194	19	5500	100	5.5 10^5	0.2%	
4.0	0.35	210	4.0	1800	100	1.8 10^5	0.3%	
5.0	0.45	225	0.9	550	100	5.5 10^4	0.5%	
6.0	0.56	240	0.22	180	200	3.6 10^4	0.6%	
....	....	...	....	....	.....	....	....	....
7.0	0.67	254	0.055	56	200	1.1 10^4	1.1%	Super Q_1
8.0	0.79	268	0.016	20	300	6.0 10^3	1.5%	
9.0	0.92	282	0.0047	7.3	400	2.9 10^3	2.2%	
10.0	1.06	296	0.0017	3.2	600	1.9 10^3	2.7%	
12.0	1.25	324	0.0003	0.73	800	5.8 10^2	4.9%	

location can be as small as 2.5×2.5 mm FWHM and can stay centered within 0.5 mm. A typical beam intensity was 1 – 4 · 10^{11} protons/pulse. During the run a vertical beam rastering system was successfully tested. During U-70's January-February 2002 shut-down, IHEP finished installing SPIN@U-70. On 11 March 2002, Michigan's shipment of detectors and electronics, needed for the April 2002 run, arrived at Moscow's SVO Airport. This shipment was then impounded by Russian Customs. [Note added: In early November 2002 it was returned to Michigan.]

Although the shipment of equipment was impounded, a short 19-26 April 2002 test run still occurred, using a polyethylene target. Some simple IHEP detectors and electronics were prepared in only a few weeks and were successfully used for the run. The resulting almost 100 to 1 signal to background rate in the $S_1 \cdot S_{1.5}$ coincidence data, shown in Fig. 3, indicates that the 35-m-long Elastic Recoil Spectrometer is quite effective in discriminating against inelastic and other background events.

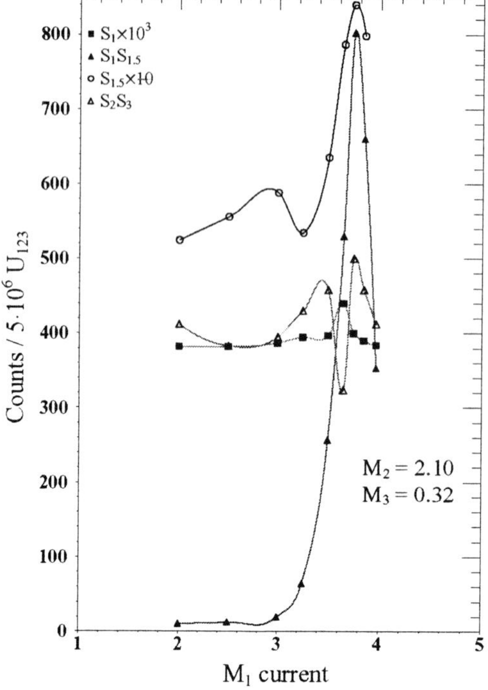

FIGURE 3. Magnet Curve: The Elastic Recoil Spectrometer event rate ($S_1 \cdot S_{1.5}$ coincidences) is plotted against the current in the 3-m-long 68-ton M_1 bending dipole magnet. Note that the ratio of the signal to background is almost 100 to 1 in the elastic peak. The single detector data rates are also shown along with the coincidences between the close-together S_2 and S_3 detectors.

Unfortunately, due to the problems with Russian Customs, the SPIN@U-70 experiment was recently suspended until the problems can be fully resolved.

1. D. G. Crabb *et al.*, Phys. Rev. Lett. **65**, 3241 (1990).
2. P. R. Cameron *et al.*, Phys. Rev. Rap. Comm. **D32**, 3070 (1985);
   D. C. Peaslee *et al.*, Phys. Rev. Lett. **51**, 2359 (1983);
   P. H. Hansen *et al.*, *ibid.* **50**, 802 (1983).
3. D. G. Crabb *et al.*, Phys. Rev. Lett. **64**, 2627 (1990).

Longitudinal Spin-Transfer in Λ Production at HERMES

H. C. Chiang, for the HERMES collaboration

University of Illinois, 1110 W Green St, Urbana, IL, USA 61801-3080

Abstract. Spin transfer in deep-inelastic Λ electroproduction has been studied with the HERMES detector using the 27.6-GeV polarized positron beam in the HERA storage ring. The longitudinal spin transfer $D_{LL'}$ from the virtual photon to the Λ has been extracted as a function of z, the fraction of the virtual photon energy carried by the Λ. The observable $D_{LL'}$ is sensitive to both helicity conservation in the fragmentation process and to hyperon spin structure. Including all data taken at HERMES during the years 1996–2000, a preliminary average value of $D_{LL'} = 0.04 \pm 0.09$ is obtained in the current fragmentation region $x_F > 0$. These results are explained by a Monte Carlo simulation based on the Lund string model. The principal conclusion of these studies is that even in the forward-production region $x_F > 0$, Λ production in medium-energy deep-inelastic scattering is complicated by the influence of the target remnant.

BACKGROUND AND MOTIVATION

It has been proposed that longitudinal spin-transfer through the fragmentation process may be studied by examining Λ hyperons produced in the current fragmentation region. It is of interest to determine the degree to which Λ hyperons "remember" the spin of their parent quarks, since the Λ is potentially useful as a polarimeter for probing the spin structure of the nucleon. Lambda hyperon production was studied at HERMES using deep-inelastic scattering (DIS) of polarized 27.6-GeV positrons off unpolarized gas (H, D, He, Ne, Kr) targets. The longitudinal spin-transfer coefficient $D_{LL'}$, defined as

$$P_q^{\Lambda} = P^{beam} \cdot D(y) \cdot D_{LL'}(z), \tag{1}$$

was measured as a function of z, the fraction of the available energy carried by the Λ. Here P_q^{Λ} is the polarization of a Λ containing struck quark q; this quantity is accessible via the self-analyzing weak decay $\Lambda \rightarrow p + \pi^-$. The variable P^{beam} represents the beam polarization, and $D(y)$ is the photon depolarization factor, calculable from the relative energy transfer $y = \nu/E$. The combination $P^{beam} \cdot D(y)$ provides a measure of the polarization of the struck quark, thus $D_{LL'}$ may be interpreted as the fraction of the struck quark polarization retained by the Λ.

MODEL PREDICTIONS OF $D_{LL'}$

The spin-transfer coefficient $D_{LL'}$ is primarily sensitive to Λ spin structure and the degree of helicity conservation in the fragmentation process, and may be predicted by several

CP675, *Spin 2002: 15th Int'l. Spin Physics Symposium and Workshop on Polarized Electron Sources and Polarimeters,* edited by Y. I. Makdisi, A. U. Luccio, and W. W. MacKay

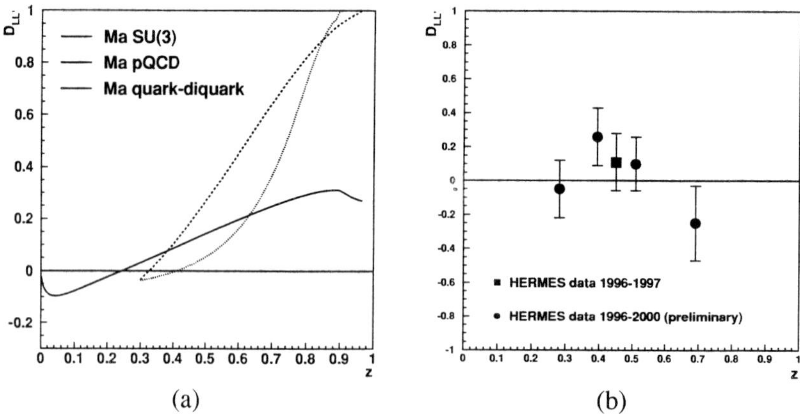

FIGURE 1. Model predictions [2, 3] for the spin-transfer coefficient $D_{LL'}$ as a function of z (a), and HERMES spin-transfer data (b). The models agree that $D_{LL'}$ rises with z, but the data, with an average of 0.04 ± 0.09, hover around zero and appear to decrease slightly at high z.

models. The models presented here differ in their predictions of the Λ spin structure but assume the same ideal conditions: (1) all Λ hyperons are produced directly from the struck quark, and (2) helicity is perfectly conserved throughout fragmentation. The most basic model is the naïve constituent quark model (NCQM), which predicts that $\Delta u = \Delta d = 0$ and $\Delta s = 1$ in the Λ. Because Λ production in DIS is dominated by scattering off up quarks, we expect to see essentially zero spin transfer. The NCQM does not predict $D_{LL'}$ as a function of z; $D_{LL'}$ is simply a constant. A second, more sophisticated way of predicting Λ spin structure is to obtain it from an SU(3) rotation of the experimentally determined proton spin structure [1]. This calculation yields a spin-transfer coefficient of about -0.2 and, like the NCQM, gives no information about z-dependence. Finally, there are several phenomenological models that attempt to describe the z-dependence of $D_{LL'}$; a few of the predictions [2, 3] are shown in Fig. 1(a). The behavior of the curves at high z arises from the high-x behavior of the quark polarization $\Delta q/q$ in the Λ in the various models. All of the phenomenological models agree that $D_{LL'}$ rises as a function of z.

HERMES SPIN-TRANSFER DATA

Lambda candidates were identified by examining events containing at least three tracks: a positron track, and two oppositely-charged hadron tracks. Several kinematic requirements were imposed to ensure that the events were in the DIS region: $Q^2 > 1$ GeV^2, $W > 2$ GeV, and $y < 0.85$. A positive value of x_F was also required, to restrict the data sample to hyperons produced in the forward direction. Details of the analysis may be found in Ref. [4], which presents the $D_{LL'}$ result from the 1996–1997 HERMES data.

Fig. 1(b) shows the new, preliminary results for $D_{LL'}$ as a function of z obtained from all HERMES data collected in the years 1996–2000. The spin-transfer coefficient is small—average value $D_{LL'} = 0.04 \pm 0.09$—and appears to decrease at high z. This trend contradicts the predictions of the phenomenological models previously described. On average the data appear to be consistent with the small values of $D_{LL'}$ predicted by the NCQM and the SU(3) model; however, the shape of the data is poorly described by these models.

There are several possible reasons for the discrepancy between the data and the models. As previously described, the models assume ideal conditions for Λ production. In reality, not all Λ hyperons are produced directly from the struck quark—Λ's may be produced in the decays of heavier hyperons (Σ^*, Σ^0, Ξ^0, and Ξ^-), or may not contain the struck quark at all. The total spin-transfer coefficient $D_{LL'}$ thus depends on the fractional contribution of each subprocess to Λ production and the degree of spin transfer within each of those subprocesses. Spin transfer to Λ hyperons has previously been studied by the OPAL [5] and ALEPH [6] collaborations, both of which observed large Λ polarizations. Using several basic assumptions and a Monte Carlo model accounting for the different Λ production mechanisms, they were able to successfully explain their data. A similar Monte Carlo model, using the same assumptions, has been developed for the HERMES $D_{LL'}$ data.

MONTE CARLO MODELS FOR $D_{LL'}$

Lambda production mechanisms may be classified in the following way: (1) direct production from the struck quark, (2) hyperon parent (Σ^*, Σ^0, Ξ^0, Ξ^-) that contains the struck quark, and (3) Λ or hyperon parent that does not contain the struck quark. The total spin-transfer coefficient $D_{LL'}$ is calculated using

$$D_{LL'} = \sum_{Y} f_q^Y C_q^Y, \tag{2}$$

where f_q^Y is a subprocess's fractional contribution to Λ production, C_q^Y is the individual spin-transfer coefficient for a particular subprocess, and the sum is over all Λ production mechanisms. The subprocess fractions f_q^Y are obtained from a Monte Carlo simulation, and the subprocess spin-transfer coefficients C_q^Y are calculated using several different spin structure models.

We make two assumptions about quarks in the fragmentation process when calculating C_q^Y: first, the helicity of the struck quark is perfectly conserved, and second, quarks created during fragmentation have random spin directions [7]. Thus hyperons that do not contain the struck quark have no net polarization, and $C_q^Y = 0$ for subprocess (3) described above. For the other subprocesses, the average Λ polarization is determined by the probability of a polarized quark fragmenting into a hyperon Y with spin $|S_Y, M_Y\rangle$, and the probability for that hyperon to decay into a Λ with spin $|1/2, M_\Lambda\rangle$. The decay probabilities can be calculated using simple angular momentum conservation arguments and Clebsch-Gordan coefficients [8]. The production probabilities, on the other hand, are model dependent.

It can be shown that the average polarization of a hyperon $\langle P_q^Y \rangle$ produced directly from a polarized struck quark is $\Delta q_Y / q_Y$ and $(5/9) \cdot \Delta q_Y / q_Y$ for spin-1/2 and spin-3/2

TABLE 1. Subprocess spin-transfer coefficients C_q^Y in the NCQM, Burkardt–Jaffe SU(3) model, and Ashery-Lipkin SU(3) model

	NCQM				B-J SU(3)				A-L SU(3)			
	u	d	s	n	u	d	s	n	u	d	s	n
Λ	0	0	1	0	-0.17	-0.17	0.63	0	-0.07	-0.07	0.73	0
Σ^*	5/9	5/9	5/9	0	0.38	0.38	0.55	0	0.38	0.38	0.55	0
Σ^0	-2/9	-2/9	1/9	0	-0.12	-0.12	0.14	0	-0.16	-0.16	0.11	0
Ξ^0	-1/3	0	2/3	0	-0.43	-0.10	0.42	0	-0.33	0.00	0.47	0
Ξ^-	0	-1/3	2/3	0	-0.10	-0.43	0.42	0	0.00	-0.33	0.47	0

hyperons, respectively [8]. Three different models were used to calculate Δq_Y and q_Y: the NCQM, Burkardt–Jaffe SU(3) model (in which the proton spin structure is used to deduce the spin structure of other members of the baryon octet) [1], and Ashery–Lipkin SU(3) model [8], a modified version of the Burkardt–Jaffe SU(3) model in which Δq from only the valence quarks is considered. The spin-transfer coefficients C_q^Y, as calculated by these three models, for Λ, Σ^*, Σ^0, Ξ^0, and Ξ^- hyperons either containing a struck quark (u,d,s) or not containing a struck quark (n) are listed in Table 1.

MONTE CARLO RESULTS

Fig. 2(a) shows the fractions of Λ's produced from various direct sources as a function of z: direct Λ production from string fragmentation, and Λ's resulting from the decays of Σ^*, Σ^0, and Ξ. The plot shows that, on average, 40%-60% of Λ's observed at HERMES are decay products of heavier hyperons, thus it is imperative that $D_{LL'}$ models consider the effects of these intermediate decays. Fig. 2(b) shows the fractions of Λ hyperons containing various flavors of the struck quark as a function of z; on average, about 90% of Λ's do not contain the struck quark. Of the remaining $\sim$10%, most contain struck u quarks, as expected from the semi-inclusive DIS cross section. Strange quark dominance appears when $z > 0.8$, and can likely be explained by the fact that at high z, there is more available energy for the creation of s quarks.

Given the extremely small fraction of Λ hyperons that contain the struck quark, it is not surprising that the observed $D_{LL'}$ is close to zero. Fig. 2(c) shows the Monte Carlo predictions for $D_{LL'}$ in comparison with HERMES data. The solid, dashed, and dotted lines represent results from the NCQM, Burkardt–Jaffe SU(3), and Ashery–Lipkin SU(3) models, respectively. These different methods used to calculate C_q^Y produce very similar predictions for $D_{LL'}$. We therefore conclude that $D_{LL'}$ is sensitive primarily to the subprocess fractions f_q^Y and that HERMES spin-transfer data cannot be used to determine which model most accurately describes hyperon spin structure.

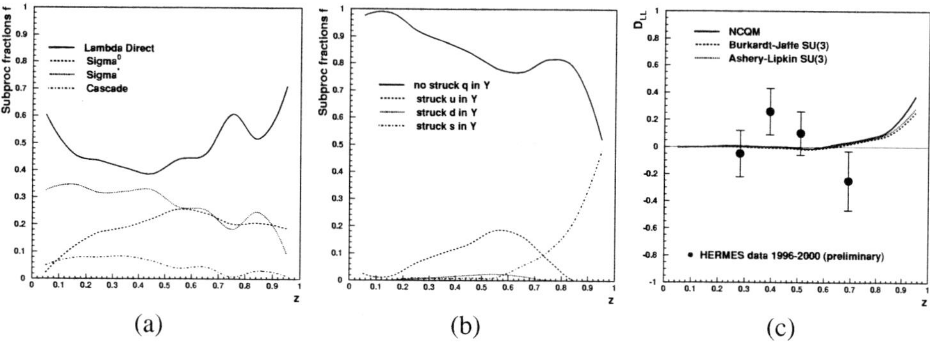

(a) (b) (c)

FIGURE 2. Fractions of Λ's originating from various sources as a function of z (a), fractions of hyperons containing different flavors of the struck quark (b), and Monte Carlo predictions for $D_{LL'}$ (c) (where the solid, dashed, and dotted lines indicate the NCQM, Burkardt–Jaffe SU(3), and Ashery–Lipkin SU(3) models, respectively).

FUTURE WORK

The Monte Carlo models for $D_{LL'}$ are successful in that they predict essentially zero spin transfer to Λ hyperons up to about $z = 0.7$; this is qualitatively consistent with HERMES data. However, the shape of the data is rather poorly matched by the Monte Carlo, suggesting that the spin-transfer models used in this study are too simple. Since the Monte Carlo indicates that many of the Λ hyperons observed at HERMES are produced in the target fragmentation region, it may be of interest to reconsider the polarization of Λ's originating from the target remnant instead of assuming the polarization is simply zero. This study primarily provides a rough explanation for HERMES spin-transfer data; with more sophisticated models, it may be possible to achieve a more detailed understanding in the future.

REFERENCES

1. Burkardt, M., and Jaffe, R. L., *Phys. Rev. Lett.*, **70**, 2537 (1993).
2. Ma, B.-Q., Schmidt, I., and Yang, J.-J., *Phys. Rev.*, **D61**, 034017 (2000).
3. Ma, B.-Q., Schmidt, I., Soffer, J., and Yang, J.-J., *Phys. Rev.*, **D65**, 034004 (2002).
4. Airapetian, A., et al., *Phys. Rev.*, **D64**, 112005 (2001).
5. Ackerstaff, K., et al., *Eur. Phys. J.*, **C2**, 49 (1998).
6. Buskulic, D., et al., *Phys. Lett.*, **B374**, 319 (1996).
7. Gustafson, G., and Häkkinen, J., *Phys. Lett.*, **B303**, 350 (1993).
8. Ashery, D., and Lipkin, H. J., *Phys. Lett.*, **B469**, 263 (1999).

Transverse Polarisation of Λ and $\bar{\Lambda}$ Hyperons in Quasi-Real Photon Nucleon Scattering

Antje Bruell
(on behalf of the HERMES collaboration)

Massachusetts Institute of Technology, 77 Massachusetts Avenue, Cambridge, MA 02139, USA

Abstract. The transverse polarisation in inclusively produced Λ and $\bar{\Lambda}$ hyperons has been studied at HERMES using the 27.5 GeV positron beam of the HERA accelerator at DESY. The average transverse polarisations were found to be $P_n^{\Lambda} = 0.055 \pm 0.006\,(\mathrm{stat}) \pm 0.016\,(\mathrm{syst})$ and $P_n^{\bar{\Lambda}} = -0.043 \pm 0.013\,(\mathrm{stat}) \pm 0.012\,(\mathrm{syst})$ for Λ and $\bar{\Lambda}$, respectively. The dependence of P_n^{Λ} and $P_n^{\bar{\Lambda}}$ on the transverse momentum p_T and on the hyperons' light-cone momentum fraction ζ has been investigated. The measured polarisations for Λ and $\bar{\Lambda}$ exhibit different behavior in the current and target fragmentation regions.

INTRODUCTION

At the beginning of the Fermilab hyperon program [1] it was believed that spin effects in hadronic reactions should be of little importance at high energies. As helicity is conserved in the limit of massless quarks, helicity-flip amplitudes, when calculated in perturbative QCD, are greatly suppressed at high energies. It was thus surprising that a significant polarisation was measured for Λ's produced by 300 GeV unpolarised protons scattered from an unpolarised beryllium target [2]. The measured Λ polarisation was transverse, negative and directed opposite to the normal ($\hat{n} = \hat{p}_{\mathrm{beam}} \times \hat{p}_{\Lambda}$) of the production plane. This transverse "self-polarisation" of Λ's and other hyperons has now been investigated in many scattering experiments with a wide variety of hadronic beams [3]. The polarisation of Λ's is almost always found to be negative (as in the original pN experiment) and to increase linearly with the transverse momentum p_T up to $p_T \simeq 1$ GeV/c where a plateau is reached. Also, a linear rise with x_F is observed: the polarisation is most pronounced when the hyperons are produced in the forward region, where current-quark fragmentation is dominant. One notable exception to this rule is the positive polarisation measured in $K^- p$ scattering, where a similar rise with p_T but the opposite sign of the polarisation is observed.

While hyperon polarisation has been studied extensively in hadron-hadron reactions, very little experimental information exists about the effect in photo- and electroproduction. Transverse polarisation in the inclusive photoproduction of neutral strange particles was investigated 20 years ago at CERN [4] and SLAC [5]. However, the statistical quality of these data is rather poor. No significant p_T dependence was observed in either experiment. The SLAC experiment, however, reported a tendency for negative polarisation values in the forward region.

CP675, *Spin 2002: 15th Int'l. Spin Physics Symposium and Workshop on Polarized Electron Sources and Polarimeters*, edited by Y. I. Makdisi, A. U. Luccio, and W. W. MacKay

TRANSVERSE Λ AND $\bar{\Lambda}$ POLARISATION AT HERMES

The HERMES experiment offers an excellent opportunity to measure transverse Λ polarisation in the reaction $\gamma^* N \rightarrow \vec{\Lambda} X$. The yields in this inclusive reaction, where the beam positrons are scattered at very small angles and are not detected by the HERMES spectrometer, originate mainly from the photoproduction peak in the cross-section at $Q^2 \approx 0$ and are much higher than in the semi-inclusive case. The analysis presented here combines all data collected at HERMES in the years 96-00. The sample includes data taken with both longitudinally polarised hydrogen and deuterium targets and a variety of unpolarised targets. The polarisation of the positron beam was always longitudinal but was frequently reversed, resulting into a negligible value of the product $\langle P_B P_T \rangle$.

Events containing Λ and $\bar{\Lambda}$ hyperons were identified through the reconstruction of secondary vertices from oppositely charged hadron tracks. The secondary vertex was required to be at least 10 cm away from the upstream end of the target and the Λ decay products were identified using a Cerenkov detector. For the polarisation analysis, Λ and $\bar{\Lambda}$ events within a $\pm 2.5\sigma$ invariant mass window of the fitted peak were chosen and the background was estimated using a side-band subtraction method.

Because of the parity-violating nature of the weak decay $\Lambda \rightarrow p\pi^-$, protons are preferentially emitted along the spin direction of their parent Λ. The angular distribution of the decay products of the Λ may thus be used to measure its polarisation:

$$\frac{dN}{d\Omega_p} = \frac{dN_0}{d\Omega_p}(1 + \alpha P_n^{\Lambda} \cos\theta_p), \tag{1}$$

where the symbol $dN_0/d\Omega_p$ denotes the distribution for the (isotropic) decay of an unpolarised Λ sample, θ_p is the angle of proton emission relative to the $\hat{n}$ axis in the Λ rest frame and $\alpha = 0.642 \pm 0.013$ [6] is the analysing power of the parity-violating weak decay. Using the up/down mirror symmetry of the HERMES acceptance, the polarisation of the Λ hyperons is determined from the measured $\langle \cos\theta_p \rangle$ and $\langle \cos\theta_p^2 \rangle$ moments [7]:

$$P_n^{\Lambda} = \frac{1}{\alpha} \frac{\langle \cos\theta_p \rangle}{\langle \cos^2\theta_p \rangle}. \tag{2}$$

Averaged over the full kinematic range of the data, the transverse polarisation of the Λ and $\bar{\Lambda}$ hyperons was measured to be

$$\begin{aligned} P_n^{\Lambda} &= +0.055 \pm 0.006\,(\text{stat}) \pm 0.016\,(\text{syst}), \\ P_n^{\bar{\Lambda}} &= -0.043 \pm 0.013\,(\text{stat}) \pm 0.012\,(\text{syst}). \end{aligned} \tag{3}$$

The systematic uncertainty was estimated by measurements of the false “transverse polarisation” of K_s^0 mesons, and of hadron-hadron pairs which did not originate from $\Lambda(\bar{\Lambda})$ decay. The average virtual photon energy was determined from a Monte Carlo simulation based on the PYTHIA 6.2 event generator to be $\langle \nu \rangle = 16$ GeV.

The good statistical quality of the HERMES data set allows to study the dependence of the Λ and $\bar{\Lambda}$ polarisation on certain kinematic variables. An approximate measure of whether the hyperons were produced in the forward or backward region in the center-of-mass frame of the $\gamma^* N$ reaction is the light cone momentum fraction of the beam positron carried by the outgoing Λ or $\bar{\Lambda}$. In a Monte Carlo simulation this variable $\zeta \equiv (E_\Lambda + p_{z\Lambda})/(E_B + p_B)$ has been found to be correlated to x_F (Fig. 1a). In particular, all events at $\zeta \geq 0.25$ are produced in the forward hemisphere ($x_F > 0$). This correlation between ζ and x_F is also reflected in the ratio of Λ to $\bar{\Lambda}$ yields observed in the data after background subtraction (Fig. 1b): above $\zeta = 0.25$, the ratio of Λ to $\bar{\Lambda}$ yields is constant indicating similar production mechanisms for the two hyperons. At lower values of ζ the ratio increases dramatically, most likely due to differences between the Λ and $\bar{\Lambda}$ formation mechanisms in the target- and current-quark fragmentation regions.

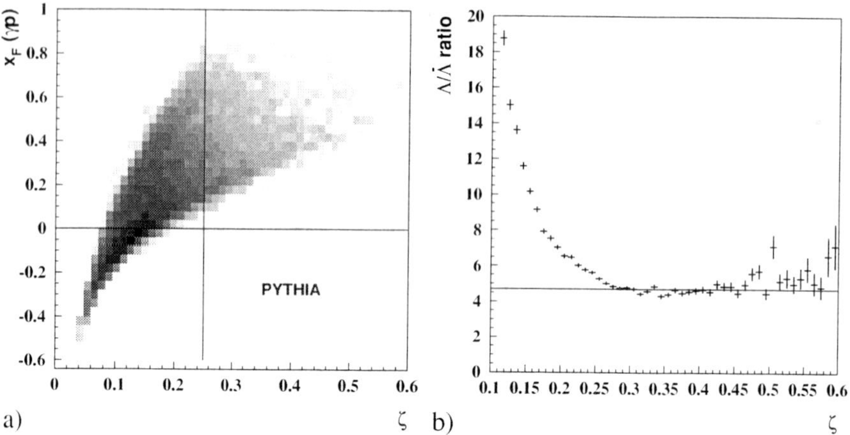

FIGURE 1. (a) Correlation between x_F, evaluated in the $\gamma^* N$ center-of-mass frame, and the light-cone fraction ζ determined in the eN frame, as determined from a PYTHIA Monte Carlo simulation. (b) Ratio of Λ to $\bar{\Lambda}$ yields versus the light cone fraction ζ.

In Fig. 2 the Λ and $\bar{\Lambda}$ polarisations are shown as a function of ζ. A clear difference between the Λ polarisation at low and high ζ is observed. While the Λ and $\bar{\Lambda}$ polarisations are similar at high ζ, at low values of ζ they are of opposite sign.

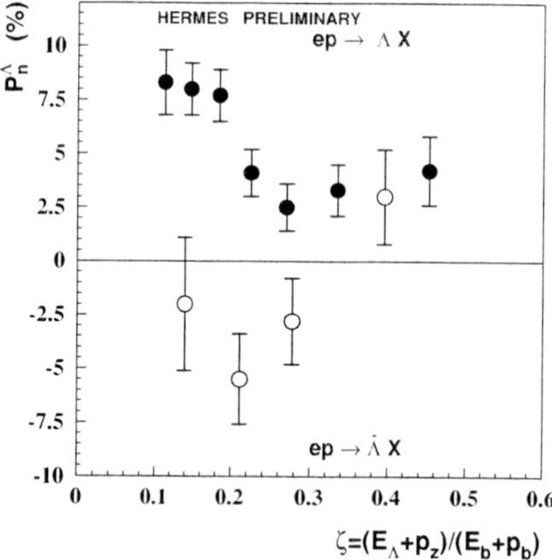

FIGURE 2. (a) Transverse polarisation of Λ and $\bar{\Lambda}$ hyperons as a function of the light cone momentum fraction $\zeta = (E_\Lambda + p_{z\Lambda})/(E_B + p_B)$.

The dependence of P_n^Λ on the transverse momentum p_T of the Λ (defined with respect to the eN system rather than the unavailable γ^*N system) has also been explored. This dependence is shown in Fig. 3 for the two intervals $\zeta < 0.25$ and $\zeta > 0.25$. In both regimes the Λ polarisation rises with p_T, particularly for the forward-going hyperons where this behavior is more pronounced. This kinematic dependence is reminiscent of the linear rise of hyperon polarisation with p_T, up to $p_T \approx 1$ GeV, that has been consistently observed in the forward production of hyperons in hadronic reactions. In the forward region the $\bar{\Lambda}$ polarisation is consistent with zero, also in agreement with hadronic reactions. In the backward region, however, the measured $\bar{\Lambda}$ polarisation favours a negative value.

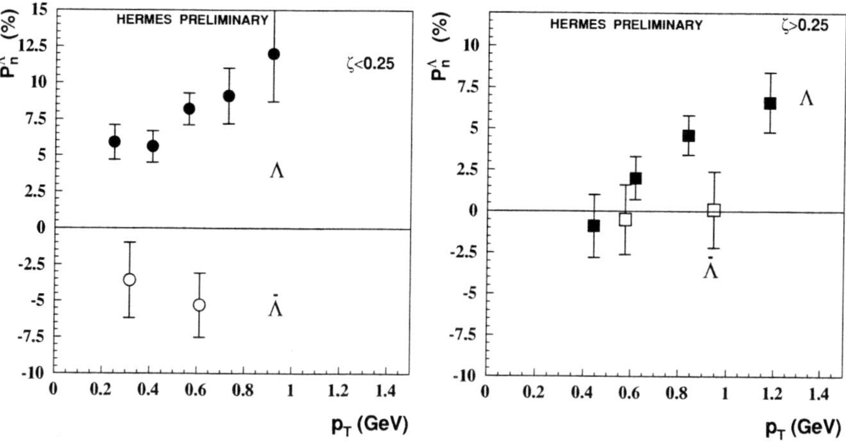

FIGURE 3. Transverse polarisation of Λ and $\bar{\Lambda}$ hyperons as a function of p_T for the intervals $\zeta < 0.25$ (left panel) and $\zeta > 0.25$ (right panel).

DISCUSSION

In contrast to the negative values observed in almost all other reactions, the transverse Λ polarisation measured at HERMES is found to be positive. Furthermore, a negative $\bar{\Lambda}$ polarisation was measured, in contrast to the zero values measured in other reactions. This negative value, however, appears to arise principally from the data in the "backward-production" region $\zeta < 0.25$. It may thus be a consequence of the complex and poorly understood mechanism of the target fragmentation.

The positive Λ polarisation measured at HERMES can possibly be related to the structure of the photon: at the energy of the HERMES experiment, the hadronic component of the photon contributes about 80% of the total γp cross section. Thus a substantial fraction of the Λ and $\bar{\Lambda}$ hyperons might be produced in the interaction of $q\bar{q}$ pairs (or of resonant ρ^0, ω or Φ mesons) with the target. Consequently, the transverse Λ polarisation measured in the forward region might be related to the one measured in the reactions $\pi^{\pm}p \to \Lambda X$ or $K^{\pm}p \to \Lambda X$. Among these reactions, only Λ production from a K^- beam shows a clear polarisation signal, which also is the only hadronic measurement showing a positive value of the polarisation. Fitting the p_T dependence of the Λ polarisation observed in the HERMES forward region with the same functional form as observed in the K^-p reaction results into a good description (i.e. both experiments found an almost linear rise with p_T) if a relative contribution of the K^- meson to the photon of about 7% is assumed. In this approach, the quarks and anti-quarks in the photon beam are on an equal footing, and the Λ and $\bar{\Lambda}$ polarisations should thus become similar at large ζ. This trend is qualitatively supported by the data (Fig. 2). Conversely, Λ and $\bar{\Lambda}$ polarisation are seen to differ in the target fragmentation region, where quarks and anti-quarks are distinguished by the target.

No theoretical calculation is yet able to explain all existing data on Λ and $\bar{\Lambda}$ polarisation. The model of DeGrand and Miettinen [8], however, can at least account for the relative signs and magnitudes of the polarisations in numerous hadron-to-hyperon transitions. In this model of proton-proton scattering, the forward-going Λ's are formed from the recombination of a high-momentum valence $(ud)_0$ diquark from the beam with a strange sea quark from the target. The negative Λ polarisation then arises from the acceleration of the strange quark, via the Thomas precession effect. The positive Λ polarisation observed with K^- beams is conversely indicative of the deceleration of valence strange quarks from the beam. The positive polarisation observed in the HERMES photoproduction data might therefore indicate that the $\gamma \rightarrow s\bar{s}$ hadronic component of the beam is the dominant source of inclusive Λ production.

REFERENCES

1. J. Lach, Nucl. Phys. (Proc. Suppl.) **B50**, (1996) 216.
2. G. Bunce *et al.* , Phys. Rev. Lett **36**, (1976) 1113 .
3. A.D. Panagiotou, Int. J. Mod. Phys. **A5**, (1990) 1197.
4. CERN-WA-004 Collaboration, D.Aston *et al.* , Nucl. Phys. **B195**, (1982) 189 .
5. SLAC-BC-072 Collaboration, K.Abe *et al.* , Phys. Rev. D **29**, (1984) 1877 .
6. Particle Data Group, D.E. Groom *et al.* , Eur. Phys. J. C **15**, (2000) 1 .
7. S. Belostotski, HERMES Internal Note 98-091.
8. T.A. DeGrand and H.I. Miettinen, Phys. Rev. D **24**, (1981) 2419 .

The Polarized Deuteron Breakup Experiment at COSY[1]

F. Rathmann*, S. Barsov*, S. Dymov†, A. Kacharava**, A. Khoukaz‡, V. Komarov†, A. Kulikov†, A. Kurbatov†, N. Lang‡, I. Lehmann*, B. Lorentz*, G. Macharashvili†, A. Mussgiller*, H. Paetz gen. Schieck§, R. Schleichert*, H. Seyfarth*, E. Steffens**, H. Ströher*, Yu. Uzikov†¶, S. Yaschenko** and B. Zalikhanov†

*Institut für Kernphysik, Forschungszentrum Jülich, Germany
†Laboratory of Nuclear Problems, Joint Institute for Nuclear Research, Dubna, Russia
**Physikalisches Institut II, Universität Erlangen-Nürnberg, Germany
‡Institut für Kernphysik, Universität Münster, Germany
§Institut für Kernphysik, Universität zu Köln, Germany
¶Kazakh State University, Almaty, Kazakhstan

Abstract. A study of the deuteron breakup reaction $pd \to (pp)n$ with forward emission of a fast proton pair with small excitation energy $E_{pp} < 3$ MeV has been performed using the ANKE spectrometer at COSY Jülich. The differential cross section of the breakup reaction, averaged up to 8° over the cm polar angle of the total momentum of the pp pairs, has been obtained at six proton beam energies $T_p = 0.6, 0.7, 0.8, 0.95, 1.35$, and 1.9 GeV. A first measurement of the vector analyzing power A_y^p has been carried out, using a polarization normalization obtained with the EDDA detector. In addition, for the first time asymmetries of $\vec{p}d$ elastic scattering at $T_p = 500$ MeV have been recorded with the spectator setup at ANKE.

1. MEASUREMENT OF THE SPIN–AVERAGED CROSS SECTION

The breakup process $p+d \to (pp)+n$ with forward emission of a fast proton pair with small excitation energy $E_{pp} < 3$ MeV was studied at ANKE [1] at six beam energies from 0.6 to 1.9 GeV. The process has never been explored under these conditions. The experiment allows for a complete kinematical reconstruction of the events, whereby the five–fold differential cross sections can be determined.

The study of the deuteron breakup $pd \to (pp)n$ discussed here comprises two phases, *a)* measurements with unpolarized deuterium target to obtain spin–averaged differential cross sections, angular distributions, and using a polarized proton beam the vector analyzing power A_y^p, and *b)* measurements with polarized deuterium target to obtain the other polarization observables[2] (tensor analyzing power, spin–spin and spin–tensor cor-

[1] This work has been supported by the BMBF (contracts RUS 667-97, RUS 99/684, RUS 99/685, RUS 01/691, KAZ 99/001, 06 ER 831, and 06 ER 930), by the European Community (contract INTAS 93-3661), by the Forschungszentrum Jülich (FFE contract 41419786 [COSY–55]).

[2] Details about the polarized target are described elsewhere in these proceedings[2, 3].

CP675, *Spin 2002: 15th Int'l. Spin Physics Symposium and Workshop on Polarized Electron Sources and Polarimeters*, edited by Y. I. Makdisi, A. U. Luccio, and W. W. MacKay

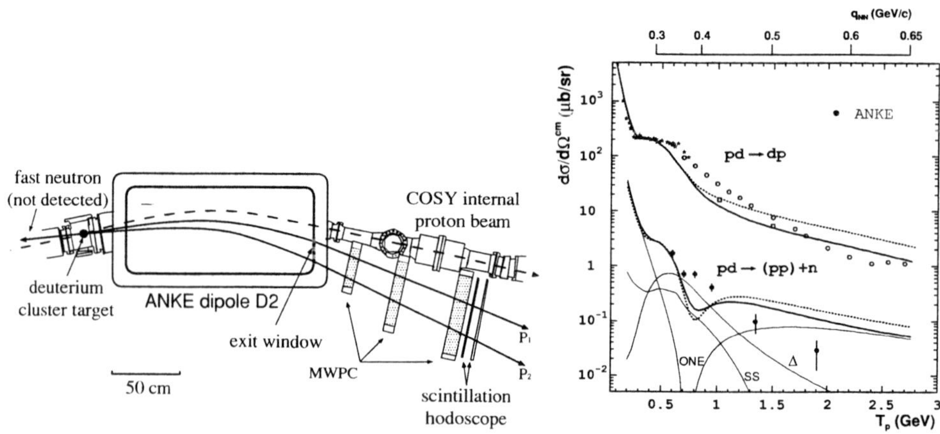

FIGURE 1. Left panel: Top view of the experimental setup with the forward detection system of the ANKE spectrometer. Right Panel: Measured cross section of the process under study in the interval $E_{pp} < 3$ MeV. The calculations with ONE+Δ+SS model are performed using the NN potentials RSC (dotted line) and Paris (solid). The individual contributions with the Paris potential are shown by thin full lines. (Figure from ref. [5].)

relation parameters). The final goal of both phases is the measurement of a complete set of observables. This goal cannot be achieved in one step. The well–defined event vertex of the point–like unpolarized cluster–jet target in the first phase allows precise determination of the angular dependence of the differential cross section. In the second phase, the storage cell target provides an extended vertex distribution along the proton beam direction. Therefore, in order to extract angular distributions of polarization observables, the angular distribution of the differential cross section must be obtained first.

The experimental setup is shown in Fig. 1 (left panel). The protons stored in the COSY ring ($\approx 3\cdot 10^{10}$) impinged on a deuterium cluster jet target [4], which provided a target thickness of about $1.3\cdot 10^{13}$ atoms/cm^2. The produced charged particles, after passing the magnetic field of the dipole D2, were registered by a set of three multiwire proportional chambers (MWPC) and a scintillation counter hodoscope. Each wire chamber contains a horizontal and a vertical anode wire plane (1 mm wire spacing), and two planes of inclined strips, that allowed us to obtain the required resolution of ≈ 0.8 to 1.2% (rms) in the momentum range 0.6 to 2.7 GeV/c. The hodoscope consists of two layers, containing 8 and 9 vertically oriented scintillators (4 to 8 cm width, 1.5 to 2 cm thickness). It provided a trigger signal, an energy loss measurement, and allowed for the determination of the differences in arrival times for particle pairs hitting different counters.

The statistics taken at $T_p > 1$ GeV is insufficient even to determine the three–fold differential cross sections. Therefore, in order to compare the energy dependence of the cross section to the theoretical prediction (ONE+Δ+SS model [5]), the cross section was integrated over the interval $0 < E_{pp} < 3$ MeV and averaged over the pair emission angle in the range $0 < \theta_{pp}^{\rm cm} < 8°$. The results are shown in Fig. 1 (right panel).

2. BEAM POLARIMETRY AT ANKE AND FIRST GLANCE AT A_Y^p

A first short run with polarized beam was mainly utilized to develop a method to determine the beam polarization at ANKE. The EDDA detector [6] was employed to obtain an unambiguous reference value for the beam polarization. A measurement of the beam polarization at EDDA is possible only at energies above ≈ 0.7 GeV (1343 MeV/c). In addition, the employed polyethylene targets at EDDA do not tolerate beam intensities exceeding about $5 \cdot 10^8$ stored protons. According to the previous beam time request, we wanted to perform a measurement of A_y^p at 0.5 and at 1.0 GeV, therefore a macro cycle was realized, which consisted of two flattops at 0.5 and 1.0 GeV (Fig. 2, left panel). The beam polarization was measured in separate cycles with appropriately reduced intensity (micro–pulsing), but otherwise identical to the data–taking cycles. At 0.5 GeV data were taken for 10 min, subsequently the beam was ramped to a short 1 GeV flattop. The beam spin was alternated from cycle to cycle. Polarization data with EDDA were recorded once or twice per day, *a)* during flattop at 1 GeV to monitor the polarization and *b)* during the ramp at 0.7 GeV, with modified flattop durations. Since between 0.5 and 0.7 GeV the depolarizing resonances in COSY are understood and compensated, this procedure provides the beam polarization at 0.5 and at 1.0 GeV. Since the polarization data recorded with EDDA did not provide a continuous monitoring of the beam polarization, the up/down ($\uparrow, \downarrow$) beam spin asymmetry of protons scattered off the deuterium target was measured with ANKE. Elastically and quasi-elastically scattered protons between $\theta_{\text{lab}} = 5°$ to $11°$ were registered in the forward detection system (FD). At the same time proton pairs from the breakup process under study were recorded. The relative luminosity $\mathscr{L}_\uparrow / \mathscr{L}_\downarrow$ was determined from inelastically scattered protons near zero degree, simultaneously detected in the FD. Online track reconstruction provided information (angle and momentum) to select the scattered protons. Both proton count rates and observed asymmetry were sufficiently large to monitor the beam polarization during the run. The resulting polarizations from EDDA and asymmetries from ANKE $\varepsilon(\uparrow,\downarrow) = \frac{N_\uparrow/\mathscr{L}_\uparrow - N_\downarrow/\mathscr{L}_\downarrow}{N_\uparrow/\mathscr{L}_\uparrow + N_\downarrow/\mathscr{L}_\downarrow}$ are shown in Fig. 2 (right panel). The asymmetries from

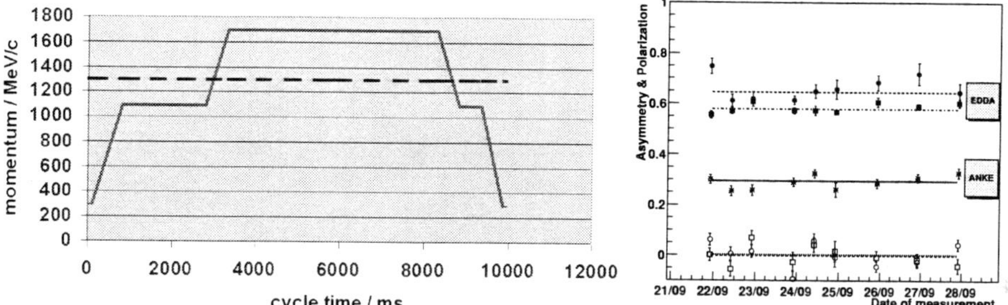

FIGURE 2. Left panel: Schematic picture of the cycle with two flattops, at $T_p = 0.5/1.0$ GeV ($1.090/1.696$ GeV/c). Right panel: EDDA beam polarization and asymmetry measured at ANKE. Filled circles denote the measured beam polarization at 0.7 GeV, filled squares the 1 GeV flattop polarization measured at EDDA. Stars indicate the asymmetry of small–angle scattered protons off the ANKE deuteron target, and the open symbols correspond to the measured false asymmetries.

ANKE were obtained for 2 h runs carried out right after the EDDA polarization measurements. The data are quite stable in time, thus averaging was justified. At $T_p = 0.7$ GeV, $P_{beam}^{\rm EDDA} = 0.645 \pm 0.009$, while at $T_p = 1.0$ GeV, $P_{beam}^{\rm EDDA} = 0.577 \pm 0.001$. The measured asymmetries at ANKE at $T_p = 0.5$ GeV correspond to $\varepsilon(\uparrow,\downarrow) = 0.294 \pm 0.006$. The false asymmetries $\varepsilon(\uparrow,\uparrow)$ and $\varepsilon(\downarrow,\downarrow)$ were obtained by analyzing cycles with the same polarization direction, yielding $[\varepsilon(\uparrow,\uparrow)+\varepsilon(\downarrow,\downarrow)]/2 = -0.002 \pm 0.009$. The effective analyzing power of ANKE, given by the ratio $\varepsilon(\uparrow,\downarrow)/P_{beam}^{\rm EDDA}$, is thus $A_y^{\rm eff} = 0.456 \pm 0.011$.

A clean separation of the *pd* elastic scattering events at small angles was achieved by detection of the scattered deuterons using the vertex spectator detector [11] in coincidence with a fast proton detected in the FD (see Fig. 3 (panel a). A comparison of

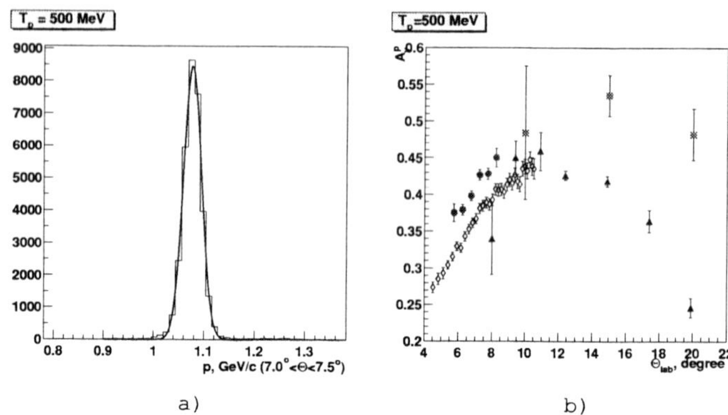

FIGURE 3. Beam polarimetry utilizing *pd* elastic scattering: a) Momentum of protons registered in the FD in coincidence with deuterons stopped in the second layer of spectator detector. b) Comparison of the angular dependence of the proton vector analyzing power measured at ANKE at 500 MeV (circles) and the data obtained at 544 MeV [7] (squares), 796 MeV [8] (rhombuses) and 800 MeV [9] (triangles).

the obtained ANKE data at 500 MeV with other data available from literature is shown in Fig. 3 (panel b). Detection of *pd* elastic scattering events with the spectator counter can be directly applied to measure the beam polarization at ANKE at 796 MeV, utilizing the existing precise data from Irom et al. [8]. It is possible to export a calibrated measurement of the beam polarization from 796 MeV to other energies, higher or lower, as shown in Ref. [10].

A first measurement of the vector analyzing power A_y^p for the breakup process could be obtained. The analyzing power A_y^p is expected to follow a linear function of the neutron emission angle θ_n in the range from 180° to 165°: $A_y^p = \alpha_y^p \cdot (180° - \theta_n)$ (Fig. 4, left panel). In Fig. 4 (right panel) the obtained angular dependence of the observed asymmetry for the breakup process is shown. The dependence exhibits a linear behaviour and the obtained value of the slope parameter $\alpha_y^p = (0.041 \pm 0.011)$ deg^{-1}. Thus, the measured preliminary value differs by about two standard deviations from the theoretical value of 0.020 deg^{-1}, given by the solid curve in Fig. 4 (left panel) for $T_p = 0.5$ GeV.

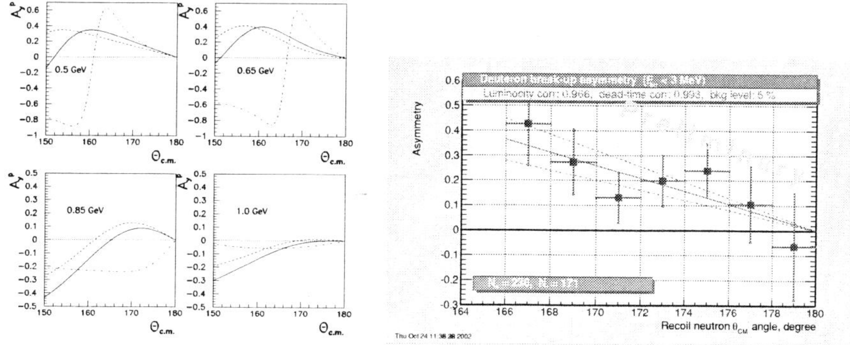

FIGURE 4. Left panel: Vector analyzing power A_y^p of the proton versus the neutron scattering angle θ_{cm} in the $\vec{p}+d \to (pp)_s + n$ reaction at $E_{pp} = 3$ MeV for the different mechanisms at kinetic energies $T = 0.5$, 0.65, 0.85, and 1 GeV: ONE (DWBA) (dashed–dotted), Δ (dotted), ONE $+\Delta+$ SS (dashed) and ONE (DWBA) $+\Delta+$ SS (full) [12]. Right panel: Measured asymmetry of the neutron emission at backward $\theta_{\rm cm}$ angle in the $\vec{p}+d \to (pp)+n$ reaction at $T_p = 0.5$ GeV with excitation energy $E_{pp} < 3$ MeV. The solid line shows a linear fit to the data.

ACKNOWLEDGMENTS

We would like to thank the EDDA collaboration for their support during the polarized measurements, in particular the help by Heiko Rohdjeß and Dieter Prasuhn is gratefully acknowledged.

REFERENCES

1. S. Barsov *et al.*, Nucl. Instr. Meth. **A 462**, 364 (2001).
2. F. Rathmann *et al.*, *The Atomic Beam Source for the Polarized Internal Gas Target of ANKE at COSY*, contribution to these proceedings.
3. R. Engels *et al.*, *A precision Lamb–shift polarimeter for the polarized gas target at ANKE*, contribution to these proceedings.
4. A. Khoukaz *et al.*, Eur. Phys. J. **D 5**, 275 (1999).
5. V.I. Komarov *et al.*, nucl–ex/0210017, accepted for publication in Phys. Lett. B.
6. H. Rohdjeß *et al.*, Proc. 7$^{\rm th}$ Int. Workshop on Polarized Gas Targets and Polarized Beams, Urbana, IL, USA, 1997. R.J. Holt and M.A. Miller (Eds.), AIP Conf. Proc. **421**, 99 (New York, 1998).
7. E.T. Boschitz *et al.*, Phys. Rev. **C 6**, 547 (1972).
8. F. Irom *et al.*, Phys. ReV. **C 28**, 2380 (1983).
9. E. Winkelmann *et al.*, Phys. ReV. **C 21**, 2535(1980).
10. R. Pollock *et al.*, Phys. Rev. **E 55**, 7606 (1997) .
11. Details about the spectator setup of ANKE can be found at www.fz-juelich/ikp/anke/vertex.
12. Yu.N. Uzikov, J. Phys. G.: Nucl. Part. Phys. **28**, B13 (2002).

New Results on Spin Rotation Parameter A in the πp-elastic Scattering in the Resonance Region

I.G. Alekseev*, P.E. Budkovsky*, V.P. Kanavets*, L.I. Koroleva*, B.V. Morozov*, V.M. Nesterov*, V.V. Ryltsov*, D.N. Svirida*†, A.D. Sulimov*, V.V. Zhurkin*, Yu.A. Beloglazov**, A.I. Kovalev**, S.P. Kruglov**, D.V. Novinsky**, V.A. Shchedrov**, V.V. Sumachev**, V.Yu. Trautman**, N.A. Bazhanov‡ and E.I. Bunyatova‡

*Institute for Theoretical and Experimental Physics,
25 B. Cheremushkinskaya, Moscow, 117259, Russia
†E-mail: Dmitry.Svirida@itep.ru
**Petersburg Nuclear Physics Institute, Gatchina, Leningrad district, 188350, Russia
‡Joint Institute for Nuclear Research, Dubna, Moscow district, 141980, Russia

Abstract. The paper presents new experimental data on the spin rotation parameter A obtained recently by ITEP-PNPI collaboration at the ITEP accelerator. The set of measurements was performed in carefully chosen critical points with precision sufficient for choosing the correct branches of partial wave analyses. The data for both π^+ and π^--scattering at 1.0, 1.43 and 1.62 GeV/c is included.

INTRODUCTION

Partial wave analyses (PWA) of the pion-nucleon scattering are the main source of the information about the spectrum and properties of non-strange baryon resonances. Yet in the absence of the spin rotation parameter measurements they possess a principal ambiguity, though of the discrete type (Barrelet). Before the series of measurements presented in this paper there were no experimental data on spin rotation parameters above 0.75 GeV/c in the resonance region.

Current state of the baryon spectroscopy is mainly based on the results of the two partial wave analyses KH80 [1] and CMB [2], carried out in early eighties. Later solutions by former VPI group SM90–SM99–FA02 [3] are believed to be missing too many resonant states. This experiment definitely shows that in the area of the measurements wrong solution branch was chosen by KH80 and CMB analyses.

MOTIVATION OF THE KINEMATIC REGION

A-parameter measurements require proton spin analysis in the final state, thus secondary (analyzing) scattering is necessary, leading to much smaller event rates compared to single spin experiments. This means that such measurements cannot be fulfilled on a

CP675, *Spin 2002: 15th Int'l. Spin Physics Symposium and Workshop on Polarized Electron Sources and Polarimeters,* edited by Y. I. Makdisi, A. U. Luccio, and W. W. MacKay

regular basis in a large number of kinematic points, making the choice of the kinematic region extremely important.

Based on the careful analysis of the PWA ambiguities and discrepancies, the following areas were selected for the A-measurements:

- π^+p at 1.43 GeV/c (120°–140°) and π^-p at 1.00 GeV/c (157°-171°) and at 1.43 GeV/c (155°–172°) — to resolve ambiguities of the PWA solutions and choose correct solution branch;
- π^+p at 1.62 GeV/c (118°–140°) to repeat and confirm our first measurement at 1.43 GeV/c out of the resonance region with a new and completely different polarimeter;
- π^-p at 1.62 GeV/c (118°–140°) and π^+p at 1.00 GeV/c (157°-171°) to test PWA predictions and provide data for the direct amplitude reconstruction.

EXPERIMENTAL SETUP

The SPIN-LM experimental setup is located at the secondary pion beam of the ITEP proton synchrotron and is a joint effort of the PNPI and ITEP groups. It is based on the evaporation type cryo polarized proton target with super-conductive solenoid, several sets of wire chambers for tracking of all particles involved and a thick filter carbon polarimeter. More detailed description can be found in [4]

DATA PROCESSING

For each event the complete kinematic reconstruction was performed based on the tracking of all particles in the magnetic field of the polarized target. Unified χ^2 criterion was used for the elastic event selection and background (mainly quasielastic) determination:

$$\chi^2 = (\Delta\varphi/\sigma_\varphi)^2 + (\Delta\theta/\sigma_\theta)^2$$

where $\Delta\varphi$ and $\Delta\theta$ are the deviations from the elastic kinematics in the azimuthal and polar angles, while σ_φ and σ_θ are the RMS of the corresponding distributions from Monte-Carlo simulations. Typical result is presented in fig. 1 for π^+p at 1.62 GeV/c. The selection χ^2-criterion was chosen to take 6–8% of the background and 85–95% of good events (see fig. 1c) for various momenta and pion sign.

For every event a 3×3 matrix was calculated, describing recoiled proton spin rotation in the magnetic field of the setup along its trajectory from the vertex of the first scattering to the point of the rescattering on the carbon nucleus. Single track events in the polarimeter were selected with the polar angle of the second scattering $> 3^o$. Of them only those were taken for which all the azimuthal angles are allowed by the chambers geometry.

Several thousand events ($(4-16)\cdot 10^3$) in various kinematic ranges and pion signs were selected for the treatment with the method of maximum likelihood to get the polarization parameters. The probability density was built only as a function of the

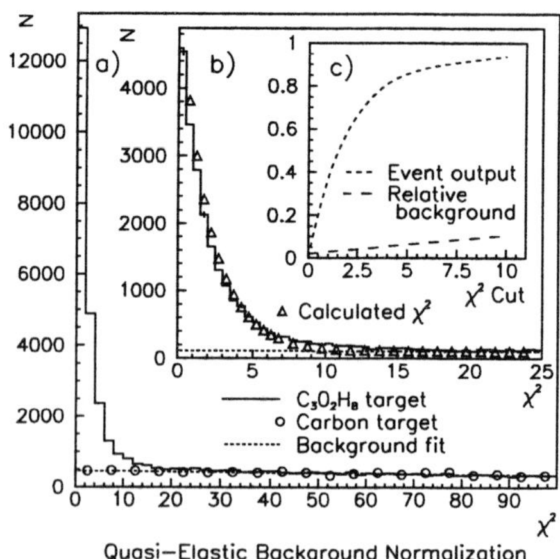

FIGURE 1. a) χ^2-distribution of the events from the polarized target (solid line) and from the carbon target (open dots). b) Real (solid line) and MC expected (triangles) distributions. c) Elastic event output and relative background vs χ^2 cut value.

parameters A and P, while the absolute value of R was calculated using the equation $P^2 + A^2 + R^2 = 1$.

RESULTS

The results for the spin rotation parameter A are presented in fig. 2 compared to the predictions of several partial wave analyses. Only statistical errors are given, all the systematic errors such as false setup asymmetry, uncertainties in the target polarization, pC analyzing power, amount and polarization of the background are negligible compared to the statistical errors.

- The results for π^+p reaction at 1.43 GeV/c and 1.62 GeV/c does not contradict to the predictions given by the analyses SM90 and SM99 and is in strong disagreement with the predictions of KH80 and CMB. This remains true in a wide momentum range, confirming the conclusion that the difference between various PWA comes from the discrete ambiguity of Barrelet type [5] and from the choice of the branch of the transverse amplitude zero trajectory.
- At 1.43 GeV/c in π^-p the new data definitely chooses SM99 solution and has strong discrepancy with CMB, suggesting some correction to KH80 and SM90.
- In π^-p scattering at 1.62 GeV/c the parameter A from this experiment does not deviate much from PWA predictions, but looks to be more close to SM90 and SM99.
- The A result for π^-p at 1.00 GeV/c confirms the PWA's KH80 and SM99 and is in contradiction with the predictions of CMB and SM90.
- In π^+p at this momentum slight correction to KH80 may be suggested.

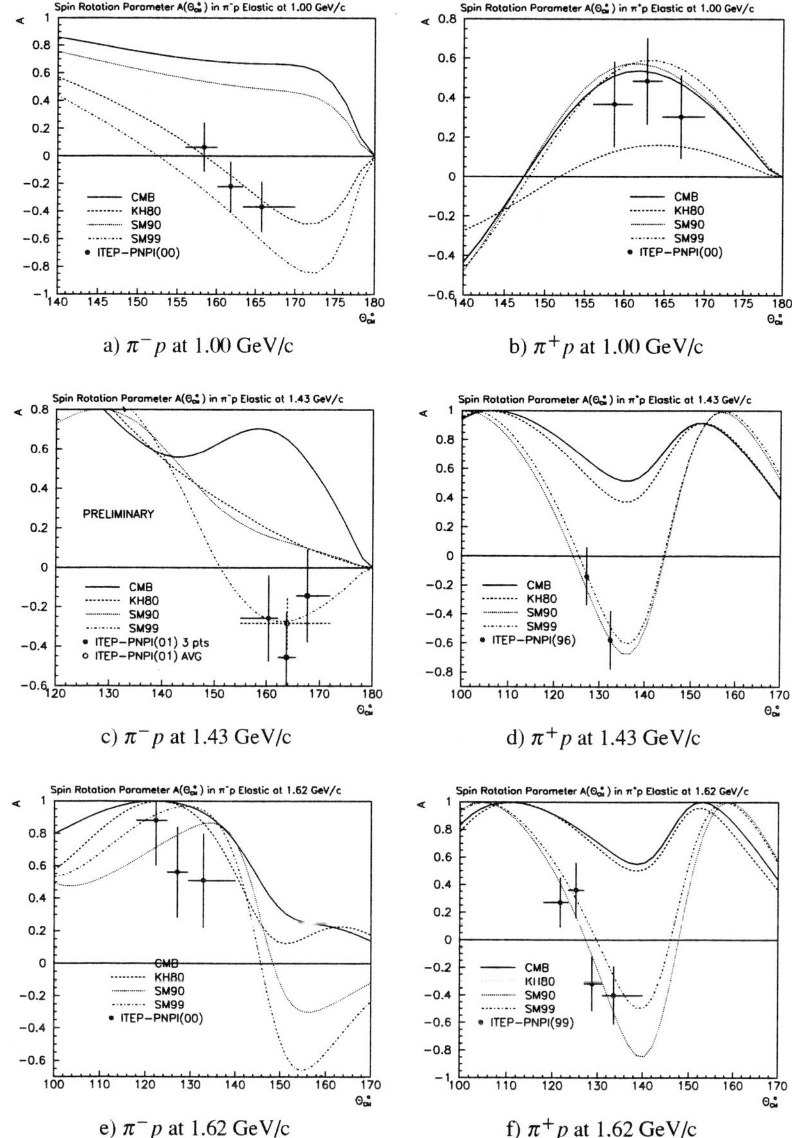

a) $\pi^- p$ at 1.00 GeV/c
b) $\pi^+ p$ at 1.00 GeV/c
c) $\pi^- p$ at 1.43 GeV/c
d) $\pi^+ p$ at 1.43 GeV/c
e) $\pi^- p$ at 1.62 GeV/c
f) $\pi^+ p$ at 1.62 GeV/c

FIGURE 2. Spin rotation parameter A in elastic πp scattering (θ_{CM}-dependence).

Selected results for the normal polarization P are presented in fig. 3 in comparison with other experimental and PWA data. No contradiction can be seen within the errors to the results of other works and PWA predictions.

Since a) the P-parameter is determined from the same statistical material as A and b) the outgoing normal proton spin component does not depend on the target polarization

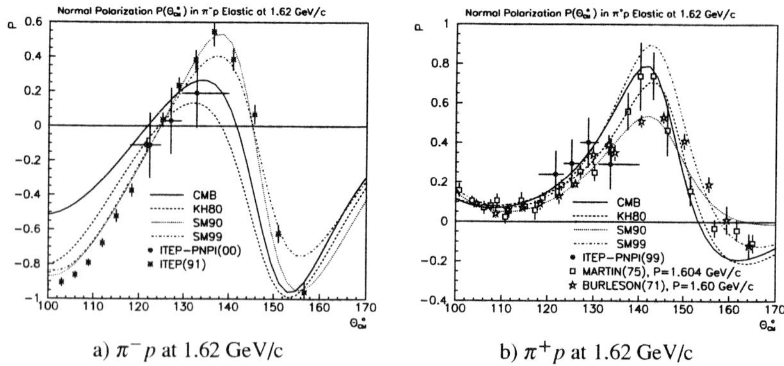

a) $\pi^- p$ at 1.62 GeV/c b) $\pi^+ p$ at 1.62 GeV/c

FIGURE 3. Normal polarization P in elastic πp scattering (θ_{CM}-dependence).

sign, the discrepancies in this parameter is a good measure of the false asymmetries in the setup. The systematic error in A caused by the false asymmetries is at least an order of magnitude smaller than that in P due to the regular target polarization sign reverse. At the same time, A and P measurements have the same scale uncertainty due to the pC analyzing power, and the reasonable agreement of our P result with the world data is an evidence of the good quality of the pC data used in the analysis.

CONCLUSIONS

The direct and one of the most important consequences of the A-parameter measurements is the observation of the discrete ambiguities in the PWA solutions in $\pi^+ p$ scattering at 1.43 GeV/c and 1.62 GeV/c as well as in $\pi^- p$ at 1.00 GeV/c and 1.43 GeV/c. The obtained results allow to make the distinct choice of the solution branch.

From the other hand the satisfactory agreement of the results on the spin rotation parameters with the predictions of the partial wave analyses (except for the discrete ambiguities) gives an evidence of the relatively high accuracy of the amplitude reconstruction by the modern PWA's, and this is in spite of the fact that they are carried out without A and R measurements in large momentum range.

REFERENCES

1. G. Höller, Handbook of Pion-Nucleon Scattering., Physics Data. No 12-1, Fachinformationzentrum, Karlsruhe, 1979.
2. R.E. Cutcosky, et al., Phys. Rev. **D20** (1979) 2839.
3. R.A. Arndt et al., Phys. Rev. **C52** (1995) 2120;
   R.A. Arndt et al., nucl-th/9807087;
   `http://gwdac.phys.gwu.edu/analysis/pin_analysis.html`
4. I.G. Alekseev et al., Phys. Atom. Nucl. **65**, 220 (2002) [Yad. Fiz. **65**, 244 (2002)].
5. E. Barrelet, Nuovo Cim. **A8** (1972) 331.

Dubna "Delta-Sigma" Experiment: Results Of Treatment And Analysis Of Statistics Accumulated In 2001 Data Taking Run On Energy Dependence Of $\Delta\sigma_L(np)$

V.I. Sharov[1], N.G. Anischenko[1], V.G. Antonenko[2], S.A.Averichev[1], L.S. Azhgirey[3], V.D. Bartenev[1], N.A. Bazhanov[4], A.A. Belyaev[5], N.A. Blinov[1], N.S. Borisov[3], S.B. Borzakov[6], Yu.T. Borzunov[1], Yu.P. Bushuev[1], L.P. Chernenko[6], E.V. Chernykh[1], V.F. Chumakov[1], S.A. Dolgii[1], A.N. Fedorov[3], V.V. Fimushkin[1], M. Finger[3,7], M. Finger Jr.[3], L.B. Golovanov[1], G.M. Gurevich[8], A. Janata[9], A.D. Kirillov[1], V.G. Kolomiets[3], E.V. Komogorov[1], A.D. Kovalenko[1], A.I. Kovalev[4], V.A. Krasnov[1], P. Krstonoshich[3], E.S. Kuzmin[3], V.P. Ladygin[1], A.B. Lazarev[3], F. Lehar[10], A. de Lesquen[10], M.Yu. Liburg[1], A.N. Livanov[1], A.A. Lukhanin[5], P.K. Maniakov[1], V.N. Matafonov[3], E.A. Matyushevsky[1], V.D. Moroz[1], A.A. Morozov[1], A.B. Neganov[3], G.P. Nikolaevsky[1], A.A. Nomofilov[1], Tz. Panteleev[6,11], Yu.K. Pilipenko[1], I.L. Pisarev[3], Yu.A. Plis[3], Yu.P. Polunin[2], A.N. Prokofiev[4], V.Yu. Prytkov[1], P.A. Rukoyatkin[1], V.A. Schedrov[4], O.N. Schevelev[3], S.N. Shilov[3], R.A. Shindin, Yu.A. Shishov[1], V.B. Shutov[1], M. Slunečka[3], V. Slunečková[3], A.Yu. Starikov[1], G.D. Stoletov[3], L.N. Strunov[1], A.L. Svetov[1], Yu.A. Usov[3], T. Vasiliev[1], V.I. Volkov[1], E.I. Vorobiev[1], I.P. Yudin[12], I.V. Zaitsev[1], A.A. Zhdanov[4], V.N. Zhmyrov[3]

[1] *Joint Institute for Nuclear Research, Laboratory of High Energies, 141980 Dubna, Russia*
[2] *Russian Scientific Center "Kurchatov Institute", 123182 Moscow, Russia*
[3] *Joint Institute for Nuclear Research, Dzhelepov Laboratory of Nuclear Problems, 141980 Dubna, Russia*
[4] *Peterburg Nuclear Physics Institute, High Energy Physics Division, 188350 Gatchina, Russia*
[5] *Kharkov Institute of Physics and Technology, 310108 Kharkov, Ukraine*
[6] *Joint Institute for Nuclear Research, Frank Laboratory of Neutron Physics, 141980 Dubna, Russia*
[7] *Charles University, Faculty of Mathematics and Physics, V Holešovičkách 2, 180 00 Praha 8, Czech Republic*
[8] *Russian Academy of Sciences, Institute for Nuclear Research, 117312 Moscow, Russia*
[9] *Nuclear Research Institute, 25068 Řež, Czech Republic*
[10] *DAPNIA, CEA/Saclay, 91191 Gif-sur-Yvette Cedex, France*
[11] *Bulgarian Academy of Sciences, Institute for Nuclear Research and Nuclear Energy, Tsarigradsko shaussee boulevard 72, 1784 Sofia, Bulgaria*
[12] *Joint Institute for Nuclear Research, Laboratory of Particle Physics, 141980 Dubna, Russia*

CP675, *Spin 2002: 15th Int'l. Spin Physics Symposium and Workshop on Polarized Electron Sources and Polarimeters*, edited by Y. I. Makdisi, A. U. Luccio, and W. W. MacKay

Abstract. New results on energy dependence of the $\Delta\sigma_L(np)$ over a GeV energy region are presented. Measurements of the *np* spin-dependent total cross section difference $\Delta\sigma_L(np)$ were carried out at the Synchrophasotron of the Laboratory of High Energies of the Joint Institute for Nuclear Research in Dubna. A quasi-monochromatic neutron beam was produced by break-up of accelerated and extracted polarized deuterons. The neutrons were transmitted through a large proton polarized target. The values of $\Delta\sigma_L$ were measured as a difference between the *np* total cross sections for parallel and antiparallel beam and target polarizations, both oriented along the beam momentum. In 2001 data taking run the $\Delta\sigma_L(np)$ value were measured at 1.4, 1.7, 1.9 and 2.0 GeV. A fast decrease of $\Delta\sigma_L(np)$ with increasing energy above 1.1 GeV, as it was first seen from our previous data, was confirmed. The obtained results are also compared with model predictions and with the phase shift analysis fits. The investigations are carrying out under a program of the "DELTA-SIGMA experiment" project. The aims of these studies are to obtain the values of imaginary and real parts of the spin-dependent forward *np*-scattering amplitudes over the energy range of 1.2-3.7 GeV for the first time.

INTRODUCTION

This contribution presents new results for the spin-dependent neutron-proton total cross section difference $\Delta\sigma_L(np)$ obtained in 2001 with a quasi-monochromatic polarized neutron beam and a polarized proton target (PPT). The values of $\Delta\sigma_L(np)$ were measured at neutron beam kinetic energies of 1.4, 1.7, 1.9 and 2.0 GeV. The obtained results are presented together with our previous results and compared with the dynamic model predictions and with the phase shift analysis fits.

Two spin-dependent observables $\Delta\sigma_L$ and $\Delta\sigma_T$ are defined as a difference in the NN total cross sections for antiparallel and parallel beam and target polarizations, oriented longitudinally L and transverse T to the beam direction. The NN total cross section differences $\Delta\sigma_L$ and $\Delta\sigma_T$ together with the spin-independent total cross section σ_{0tot} are linearly related to three non-vanishing imaginary parts of the NN forward scattering amplitudes via optical theorems. The $\Delta\sigma_{L,T}$ data allow a direct determination of the imaginary parts of the spin dependent NN forward scattering amplitudes and check predictions of available dynamic models of strong interactions and provide an important contribution to a database of phase-shift analyses (PSA). It is also possible to deduce the $\Delta\sigma_{L,T}$ nucleon-nucleon isosinglet ($I=0$) parts using the measured *np* quantities and the existing *pp* (isitriplet $I=1$) results.

A large amount of results for *np* elastic scattering and transmission experiments at energies up to 1.1 GeV was accumulated by the end of the 80-th. This data set allowed one the direct determination of the spin dependent *np* forward scattering amplitudes and to perform phase-shift analysis fits for *np* interactions up to this energy boundary. The possibility to extend measurements of the *np* spin-dependent observables to higher energies exists now at the Dubna Synchrophasotron (JINR LHE) only.

The measurements of energy dependences of differences of *np* total cross sections $\Delta\sigma_L$ and $\Delta\sigma_T$ for parallel and antiparallel particle spins oriented longitudinally (L) or transverse (T) were proposed [1,2] at the beginning of the 90-th and started [3-7] in Dubna. To implement the proposed $\Delta\sigma_{L,T}(np)$ experimental program, a large Argonne-Saclay polarized proton target (PPT) was reconstructed in Dubna [8,9], and a new polarized neutron beam line with suitable parameters [10,11] was constructed and tested . A set of necessary neutron detectors with corresponding electronics, rather modern data acquisition system and other needed equipment were also prepared, tuned and tested. Two successful data taking runs were carried out in 1995 and 1997. The energy dependence of $\Delta\sigma_L(np)$ was measured at 1.19, 1.59, 1.79, 2.2, 2.49, and 3.66 GeV [3-7]. The results of these runs are also presented.

In the last few years the measurements of spin-correlation parameters $A_{00kk}(np)$ and $A_{00nn}(np)$ from $np \to np$ elastic charge exchange at 0°(Lab.) were also prepared in frame of the "Delta-Sigma experiment" project. These spin-correlation *np* observables can be simultaneously (and independently) with the $\Delta\sigma_{L,T}(np)$ measurements. A magnetic spectrometer with multiwire proportional chambers for detection of protons from $np \to np$ elastic charge exchange at 0°(Lab.) was installed and tested at the polarized neutron beam line. The data set of *np* polarization observables to be obtained will allow to extract the values of imaginary and real parts of spin dependent forward *np* scattering amplitudes first over this energy range.

DETERMINATION OF THE $\Delta\sigma_{L,T}(np)$ OBSERVABLES

In this paper, we use NN formalism and the notations for elastic nucleon-nucleon scattering observables from [12].

The general expression of the total cross section for a polarized nucleon beam transmitted through a polarized proton target, with arbitrary directions of beam and target polarizations is

$$\sigma_{tot} = \sigma_{0tot} + \sigma_{1tot}(\vec{P}_B, \vec{P}_T) + \sigma_{2tot}(\vec{P}_B, \vec{k})(\vec{P}_T, \vec{k}), \tag{1}$$

where $\vec{P}_B$ and $\vec{P}_T$ are the beam and target polarizations, and $\vec{k}$ is a unit vector in the beam momentum direction. The term σ_{0tot} is the total cross section for unpolarized particles, and the σ_{1tot}, σ_{2tot} are the spin-dependent contributions which connect with the measurable observables $\Delta\sigma_T$ and $\Delta\sigma_L$ by :

$$-\Delta\sigma_T = 2\sigma_{1tot} = [\sigma(\uparrow\uparrow) - \sigma(\downarrow\uparrow)]/(P_B P_T), \tag{2}$$

$$-\Delta\sigma_L = 2(\sigma_{1tot} + \sigma_{2tot}) = [\sigma(\rightrightarrows) - \sigma(\leftrightarrows)]/(P_B P_T). \tag{3}$$

The values of σ_{0tot}, $\Delta\sigma_T$ and $\Delta\sigma_L$ are linearly connected with the imaginary parts of the three independent forward scattering invariant amplitudes $a + b$, c and d via optical theorems :

$$\sigma_{0tot} = (2\pi / K)\,\mathrm{Im}[a(0) + b(0)], \tag{4}$$

$$-\Delta\sigma_T = (4\pi / K)\,\mathrm{Im}[c(0) + d(0)], \tag{5}$$

$$-\Delta\sigma_L = (4\pi / K)\,\mathrm{Im}[c(0) - d(0)], \tag{6}$$

where K is the CM momentum of the incident nucleon. Relations (5) and (6) allow one to extract the imaginary parts of the spin-dependent invariant amplitudes $c(0)$ and $d(0)$ at an angle of 0° from the measured values of $\Delta\sigma_T$ and $\Delta\sigma_L$.

Using the measured values of $\Delta\sigma(np)$ and the existing $\Delta\sigma(pp)$ data at the same energy, one can deduce $\Delta\sigma_{L,T}(I=0)$ as

$$\Delta\sigma_{L,T}(I=0) = 2\Delta\sigma_{L,T}(np) - \Delta\sigma_{L,T}(pp). \tag{7}$$

EXPERIMENTAL SET-UP

A quasi-monochromatic neutron beam was produced by break-up of accelerated and extracted polarized deuterons. Both polarized deuteron and polarized free neutron beam lines [10,11], the two polarimeters, the neutron production target BT, collimators Cl-C4, the spin rotation magnet SRM, the polarized proton target PPT [8,9], neutron beam monitors Ml, M2, transmission detectors T1-T3 and the apparatus for monitoring the neutron beam profiles NP are shown in Fig.l. The associated electronics and the data acquisition system are also described in [3-7].

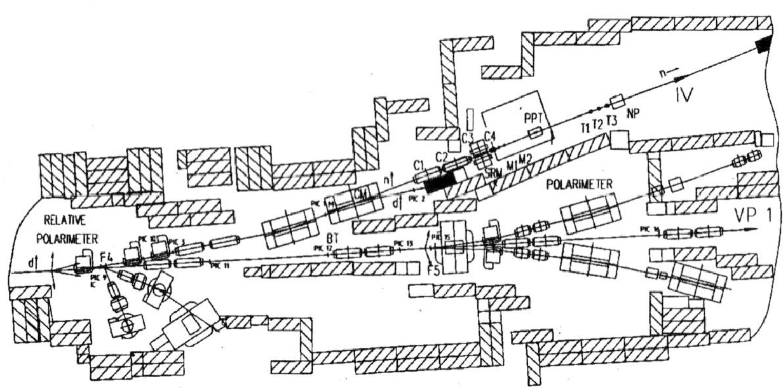

FIGURE 1. Layout of the Set-up in the Experimental Hall.

Deuterons were extracted at energies of 1.4, 1.7, 1.9 and 2.0 GeV for the $\Delta\sigma_L(np)$ measurements. The beam momenta p_d were known with a relative accuracy of $\sim \pm 1$ %. The intensity of the primary polarized deuteron beam was $\sim 2 \cdot 10^9$d/cycle. It was continuously monitored using two calibrated ionization chambers placed in the focal points F3 and F4 of the deuteron beam line before the neutron production target BT. Different characteristics of the accelerated and extracted deuteron beam and the status of the used beam lines were in part available on the Laboratory Ethernet (cycle by cycle).

The beam of free quasi-monochromatic polarized neutrons was obtained by break-up of vector polarized deuterons at 0° in BT. Neglecting the BT thickness, the laboratory momentum of neutrons $p_n = p_d/2$ with a momentum spread of FWHM ~5%. The BT contained 20 cm Be with a cross section of 8 x 8 cm^2. During the data acquisition, the position and X,Y-profiles of the neutron beam were continuously monitored by a neutron profilometer NP with multiwire proportional chambers closely placed downstream the last transmission detector.

The value and direction of neutron beam polarization $\vec{P}_B(n)$ after deuteron break-up at 0° and for $p_n = p_d/2$ are the same as the vector polarization $\vec{P}_B(d)$ of the incident deuteron beam. During the run, the polarization $P_B(d)$ and hence the neutron beam polarization $P_B(n)$ were reversed every cycle, as requested.

The prepared neutron beam has the vertical orientation of $P_B(n)$ as the accelerated and extracted deuteron beam. For a purpose of the $\Delta\sigma_L(np)$ measurements, we have to turn the neutron spins from the vertical to the longitudinal direction. This was done by a spin-rotating magnet SRM in the neutron beam line. The SRM magnetic field was continuously monitored by a Hall probe.

The value of $P_B(d)$ was continuously monitored by relative polarimeter during the data acquisition. The deuteron beam, considered as a beam of quasi-free protons and neutrons, collided with a CH_2 target, and quasifree protons, scattered at 14° lab., were detected in coincidence with the protons detected by recoil arms. The weighted average value of deuteron beam polarization was 0.528 ± 0.004(stat) ± 0.008(syst).

The frozen-spin polarized proton target, reconstructed to a movable device [8,9], was used. The target material was pentanol $C_5H_{12}O$. The pentanol beads were loaded into the thin wall teflon container 200 mm long and 30 mm in diameter placed inside the dilution refrigerator. The number of hydrogen atoms per cm^2 for the target was n_H=(9.138 ±0.10) 10^{23}/cm^2. The P_T measurements were carried out using a computer-controlled NMR system. The average value of proton polarization was ~ 0.6 with uncertainty of ±5 %.

RESULTS AND DISCUSSION

The measured $-\Delta\sigma_L(np)$ values are presented in Fig.2. The errors are statistical only. The results from refs.[3-7] together with the existing $\Delta\sigma_L(np)$ data (see for example [13]), obtained with free polarized neutrons at lower energies, are also shown in Fig.2. We can see that the new results are smoothly connected with the lower energy data and confirm a fast decrease to zero within a 1.2-2.0 GeV energy region, observed previously [3-7].

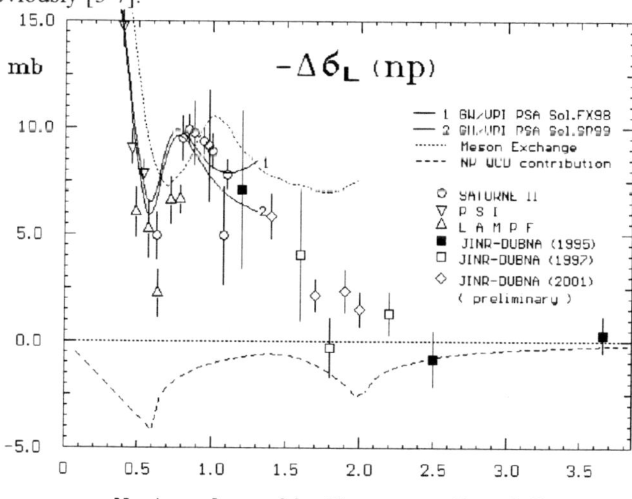

FIGURE 2. Energy dependence of the $-\Delta\sigma_L(np)$ observable obtained with free neutron polarized beams.

The solid curves show the last energy-dependent GW/VPI-PSA [14] fits (SAID FX98 and SP99 solutions) of this observable over the interval from 0.1 to 1.3 GeV. Above 1.1 GeV (SATURNE II), the *np* database is insufficient and a high energy part of the $\Delta\sigma_L(np)$ predictions [14] still disagrees with the measured data.

Below 2.0 GeV, a usual meson exchange theory of NN scattering [15] gives the $\Delta\sigma_L(np)$ energy dependence as shown by the dotted curve in Fig.2. It can be seen that this model provides a qualitative description only.

ACKNOWLEDGEMENTS

The authors thank the JINR, JINR LHE and LNP Directorates for their support of these investigations.

REFERENCES

1. J. Ball, N.S. Borisov, J. Bystricky, A.N. Chernikov et al. In:Proc.Int. Workshop "Dubna Deuteron-91", JINR E2-92-25, p.12, Dubna, 1992.
2. E. Cherhykh, L. Golovanov, A. Kirillov, Yu. Kiselev et al. In: Proc. Int. Workshop "Dubna Deuteron-93", JINR E2-94-95, p.185, Dubna, 1994; Proc. "V Workshop on High Energy Spin Physics", Protvino, 20-24 September 1993, Protvino, 1994, p.478.
3. B.P. Adiasevich, V.G. Antonenko, S.A. Averichev, L.S. Azhgirey et al. Zeitschrift fur Physik **C71** (1996) 65.
4. V.I. Sharov, S.A. Zaporozhets, B.P. Adiasevich, V.G. Antonenko et al. JINR Rapid Communications **3[77]-96** (1996) 13.
5. V.I. Sharov, S.A. Zaporozhets, B.P. Adiasevich, N.G. Anischenko et al. JINR Rapid Communications **4[96]-99** (1999) 5.
6. V.I. Sharov, S.A. Zaporozhets, B.P. Adiasevich, N.G. Anischenko et al. Eur.Phys.J. C **13** (2000) 255.
7. V.I. Sharov, S.A. Zaporozhets, N.G. Anischenko, V.D.Bartenev et al. Czech. Jour.Phys. **Vol.50** (2000) Suppl. SI, 255; Czech. Jour.Phys. **Vol.51** (2001) Suppl. A, A87.
8. F. Lehar, B. Adiasevich, V.P. Androsov, N. Angelov et al. Nucl.Instrum.Methods **A356** (1995) 58.
9. N.A. Bazhanov, B. Benda, N.S. Borisov, A.P. Dzyubak et al. Nucl.Instrum.Methods **A372** (1996) 349.
10. IB. Issinsky, A.D. Kirillov, A.D. Kovalenko and P.A. Rukoyatkin. Acta Physica Polonica **B25** (1994) 673
11. A. Kirillov, L. Komolov, A. Kovalenko, E. Matyushevsky, A.A.Nomofilov, P. Rukoyatkin, V. Sharov, A. Starikov, L. Strunov, A. Svetov. "Relativistic Polarized Neutrons at the Laboratory of High Energy Physics, JINR". Preprint JINR E13-96-210, Dubna, 1996.
12. J.Bystricky, F.Lehar and P.Winternitz. J.Physique (Paris) **39** (1978) 1.
13. C.Lechanoine-Leluc, F.Lehar. Rev. Mod. Phys. **65** (1993) 47.
14. R.A. Arndt, C.H. Oh, I.I. Strakovsky, R.L. Workman and F. Dohrmann: Phys.Rev. C **56** (1997) 3005.
15. T.-S.H. Lee: Phys. Rev. C **29** (1984) 195.

Diffractive Heavy Pseudoscalar-meson Productions by Weak Neutral Currents

A. Hayashigaki*, K. Suzuki † and K. Tanaka **

* *Department of Physics, University of Tokyo, Tokyo 113-0033, Japan*
† *Division of Liberal Arts, Numazu College of Technology, Shizuoka 410-8501, Japan*
** *Department of Physics, Juntendo University, Inba-gun, Chiba 270-1695, Japan*

Abstract. A first theoretical study for neutrino-induced diffractive productions of heavy pseudoscalar-mesons, η_c and η_b, off a nucleon is performed based on factorization formalism in QCD. We evaluate the forward diffractive production cross section in perturbative QCD in terms of the light-cone wave functions of Z boson and $\eta_{c,b}$ mesons, and the gluon distribution of the nucleon. The diffractive production of η_c is governed by the axial vector coupling of the longitudinally polarized Z boson to $Q\bar{Q}$ pair, and the resulting η_c production cross section is larger than the J/ψ one by one order of magnitude. The bottomonium η_b production, which shows up for higher beam energy, is also discussed.

Exclusive diffractive leptoproductions of neutral vector mesons provide unique insight into an interplay between nonperturbative and perturbative effects in QCD. The diffractive processes are mediated by the exchange of a Pomeron with the vacuum quantum numbers, whose QCD description is directly related to the gluon distributions inside the nucleons for small Bjorken-x [1, 2, 3]. The processes also allow us to probe the light-cone wave functions (WFs) of the vector mesons. Relating to the latter point, however, the applicability is apparently limited to probing the neutral vector mesons due to the vector nature of the electromagnetic current.

In this talk, we propose the exclusive diffractive productions of mesons in terms of the neutrino beam. The weak currents allow us to observe both neutral and charged mesons by Z and W boson exchanges, and these mesons can be not only vector but also other types of mesons including pseudoscalar mesons. Thus, such processes may reveal structure of various kinds of mesons, the coupling of the QCD Pomeron to quark-antiquark pair with various spin-flavor quantum numbers, and also information on the CKM matrix elements. There already exist some experimental data for π, ρ, $D_s^{\pm}$, D_s^* [4, 5], D_s^{*+} [6], and J/ψ production [4], although the amount of the data is not enough. On the other hand, there are only a few theoretical calculations, e.g., for the J/ψ production in a vector meson dominance model [7] and for D_s^- production with the generalized parton density [8]. Here, our interest will be directed to diffractive productions of heavy pseudoscalar mesons, η_c and η_b. So far η_c has been observed via the decays of J/ψ or B mesons produced by $p\bar{p}$ and e^+e^- reactions, while η_b has not been observed. The diffractive productions via the weak neutral current will give a direct access to η_c as well as a new experimental method to identify η_b by e.g. measuring the two photon decay.

CP675, *Spin 2002: 15th Int'l. Spin Physics Symposium and Workshop on Polarized Electron Sources and Polarimeters,* edited by Y. I. Makdisi, A. U. Luccio, and W. W. MacKay

We treat the η_c and η_b productions by generalizing the approach in the leading logarithmic order of perturbative QCD, which has been developed successfully for the vector meson electroproductions [1, 2, 3]. We consider the near-forward diffractive productions $Z^*(q)+N(P)\to\eta_Q(q+\Delta)+N'(P-\Delta)$, where $Q=c,b$, and each momentum is labeled in Fig.1. Here the total center-of-mass energy $W=\sqrt{(P+q)^2}$ is much larger than any other mass scales involved, i.e., $W^2\gg Q^2(=-q^2),-t(=-\Delta^2),m_Q^2,\Lambda_{\rm QCD}^2,\cdots$ with m_Q the heavy-quark mass. $-t\ll m_Q^2$ and $m_Q^2\gg\Lambda_{\rm QCD}^2$ are also assumed. The crucial point is that at such high W the scattering of the $Q\bar{Q}$ pair on the nucleon occurs over a much shorter timescale than the $Z^*\to Q\bar{Q}$ fluctuation or the η_Q formation times (see Fig.1). As a result, the production amplitudes obey factorization in terms of the Z and η_Q light-cone WFs. Also, a hard scale m_Q ensures the application of perturbative QCD for the Z WFs even for $Q^2=0$. The $Q\bar{Q}$-N elastic scattering amplitude, sandwiched between the Z and η_Q WFs, is further factorized into the $Q\bar{Q}$-gluon hard scattering amplitude and the nucleon matrix element corresponding to the (unintegrated) gluon density distribution [1, 2, 3]. The participation of the "new players" Z and η_Q requires an extension of the previous works [1, 2, 3] by introducing the corresponding light-cone WFs.

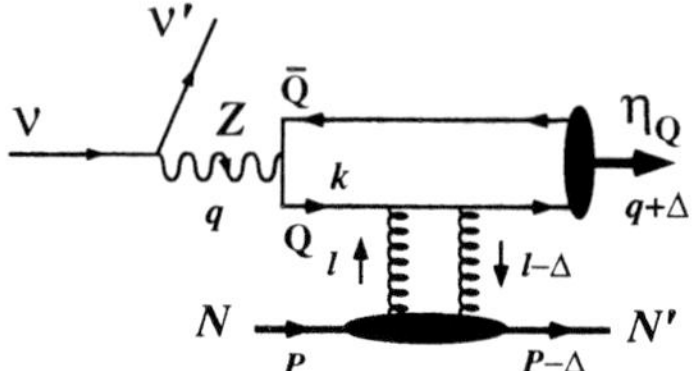

FIGURE 1. A typical diagram for the exclusive diffractive η_Q ($Q=c,b$) productions induced by neutrino (ν) through the Z boson exchange. There are other diagrams by interchanging the vertices on the heavy-quark lines.

First of all, we discuss the extension due to the participation of the Z boson. The $ZQ\bar{Q}$ weak vertex of Fig. 1 is given by $(g_W/2\cos\theta_W)\gamma_\mu(c_V-c_A\gamma_5)$, where $(g_W/2\cos\theta_W)^2=\sqrt{2}G_F M_Z^2$ with G_F the Fermi constant and M_Z the Z mass. $c_V=1/2-(4/3)\sin^2\theta_W, c_A=1/2$ for the c-quark and similarly for the b-quark. As usual, we introduce the two light-like vectors q' and p' by the relations $q=q'-(Q^2/s)p'$, $P=p'+(M_N^2/s)q'$, $s=2q'\cdot p'$ with M_N the nucleon mass, and the Sudakov decomposition of all momenta, e.g., $k=\alpha q'+\beta p'+k_\perp$ ($k_\perp^2=-\vec{k}_\perp^2$). We also introduce the polarization vectors $\varepsilon^{(\xi)}$ ($\xi=0,\pm1$) of the virtual Z boson to satisfy $\sum_\xi(-1)^{\xi+1}\varepsilon_\mu^{(\xi)*}\varepsilon_\nu^{(\xi)}=-g_{\mu\nu}+q_\mu q_\nu/M_Z^2$, which is the numerator of the propagator for the massive vector boson. Because the $q_\mu q_\nu/M_Z^2$ term vanishes when contracted with the neutral current, we conveniently choose as $\varepsilon^{(0)}=q'/Q+p'Q/s$ and $\varepsilon^{(\pm1)}=\varepsilon_\perp^{(\pm1)}=(0,1,\pm i,0)/\sqrt{2}$ for the longitudinal ($\xi=0$) and transverse ($\xi=\pm1$) polarizations respectively. The light-cone WFs for the virtual Z boson can be obtained in analogy with the photon light-cone WF used in the J/ψ electroproductions [1, 2] as

$$\Psi_{\lambda\lambda'}^{Z(\xi)}(\alpha,\vec{k}_\perp)=-\frac{g_W}{2\cos\theta_W}\frac{\sqrt{\alpha(1-\alpha)}\bar{u}_\lambda(k)\not{\varepsilon}^{(\xi)}(c_V-c_A\gamma_5)v_{\lambda'}(q-k)}{\alpha(1-\alpha)Q^2+\vec{k}_\perp^2+m_Q^2}, \tag{1}$$

where $u_\lambda(k)$ ($v_{\lambda'}(q-k)$) denote the on-shell spinor for the (anti)quark with helicity $\lambda^{(\prime)}$.

Next we proceed to the light-cone WFs for the η_Q meson. It is convenient to exploit the correspondence with the case of the J/ψ electroproductions. The light-cone WFs for the heavy vector mesons $V = J/\psi, \Upsilon$ have been discussed in many works, but are still controversial in the treatment of subleading effects like the Fermi motion corrections [2, 3, 9], corrections to ensure the pure S-wave $Q\bar{Q}$ state [10, 11], etc. Here we employ the vector-meson light-cone WFs given by

$$\Psi_{\lambda\lambda'}^{V(\varpi)^*}(\alpha, k_\perp) = \frac{\bar{v}_{\lambda'}(q+\Delta-k)}{\sqrt{1-\alpha}}\gamma^\mu e_\mu^{(\varpi)^*}\mathcal{R}\frac{u_\lambda(k)}{\sqrt{\alpha}}\frac{\phi^*(\alpha,\vec{k}_\perp)}{M_V}, \tag{2}$$

where M_V and $e_\mu^{(\varpi)}$ ($\varpi = 0, \pm 1$) are the mass and the polarization vector of the vector meson with $(q+\Delta)^2 = M_V^2$, $e^{(\varpi)}\cdot(q+\Delta) = 0$, and $e^{(\varpi)*}\cdot e^{(\varpi')} = -\delta_{\varpi\varpi'}$. $\mathcal{R} \equiv [1+(\not{q}+\Delta)/M_V]/2$ denotes the projection operator, $\mathcal{R}^2 = \mathcal{R}$, to ensure the S-wave $Q\bar{Q}$ state in the heavy-quark limit [9, 12]. (This projection operator coincides with that discussed in Ref. [10] up to the binding-energy effects of the quarkonia.) Note that eq. (2) reduces to the vector-meson WFs of Ref. [1] by the replacement $\mathcal{R} \to 1$. Essential difference of eq. (2) from the "perturbative" WFs (1) is that the scalar function $\phi(\alpha, \vec{k}_\perp)$ contains nonperturbative dynamics between Q and $\bar{Q}$. Now the light-cone WFs of the η_Q meson can be derived from eq. (2) utilizing spin symmetry, which is exact in the heavy-quark limit. This symmetry relates the S-wave states, η_Q and the three spin states of the vector meson. Namely, $M_{\eta_Q} = M_V$, and the pseudoscalar state is related to the vector state with longitudinal polarization as $|\eta_Q\rangle = 2\hat{S}_Q^3|V(\varpi=0)\rangle$, where $\hat{S}_Q^3$ is the third component of the hermitean spin operator $\hat{S}_Q^i$ which acts on the spin of the heavy quark Q but does not act on $\bar{Q}$. This implies that the η_Q WFs are given by the replacement $u_\lambda \to 2S^3 u_\lambda$ in eq. (2), where S^3 is a matrix representation of $\hat{S}_Q^3$ as $S^3 = \gamma_5(\not{q}+\Delta)\not{e}^{(0)}/(2M_V)$ which is related to a spin matrix $\sigma^{12}/2 = \gamma_5\gamma^0\gamma^3/2$ in the meson rest frame by a Lorentz boost in the third direction:

$$\Psi_{\lambda\lambda'}^{\eta_Q *}(\alpha, k_\perp) = -\frac{v_{\lambda'}(q+\Delta-k)}{\sqrt{1-\alpha}}\gamma_5\mathcal{R}\frac{u_\lambda(k)}{\sqrt{\alpha}}\frac{\phi^*(\alpha,\vec{k}_\perp)}{M_{\eta_Q}}. \tag{3}$$

This result shows that the η_Q is described by the same nonperturbative WF $\phi(\alpha, \vec{k}_\perp)$ as the vector meson. We also note that, due to the presence of $\mathcal{R}$, the "$Q\bar{Q}\eta_Q$ vertex" involves pseudovector as well as pseudoscalar coupling.

Combining our Z and η_Q WFs with the $Q\bar{Q}$-N elastic amplitude [1], we get the total amplitude $\mathcal{M}^{(\xi)}$ for the polarization $\varepsilon^{(\xi)}$ of the virtual Z boson. We find $\mathcal{M}^{(\pm 1)} = 0$, which reflects conservation of helicity in the high energy limit, and

$$\begin{aligned} i\mathcal{M}^{(0)} &= \frac{-\sqrt{2}\pi^2 W^2}{\sqrt{N_c}}\frac{g_W m_Q c_A}{M_{\eta_Q} Q\cos\theta_W}\alpha_s(Q_{\rm eff}^2)\left[1+i\frac{\pi}{2}\frac{\partial}{\partial \ln x}\right] \\ &\times\ xG(x, Q_{\rm eff}^2)\int_0^1\frac{d\alpha\tilde{Q}}{\alpha(1-\alpha)}\int_0^\infty dbb^2\phi^*(\alpha,b)K_1\left(b\tilde{Q}\right), \end{aligned} \tag{4}$$

where $x = (Q^2 + M_{\eta_Q}^2)/s$, $Q_{\rm eff}^2 = (Q^2 + M_{\eta_Q}^2)/4$, $\tilde{Q} = [\alpha(1-\alpha)Q^2 + m_Q^2]^{1/2}$, K_1 is a modified Bessel function, and $G(x, Q_{\rm eff}^2)$ is the conventional gluon distribution. Eq. (4) is written in the "$\vec{b}$-space" conjugate to the $\vec{k}_\perp$-space via the Fourier transformation; $\vec{b}$ denotes the transverse separation ($b \equiv |\vec{b}|$) between Q and $\bar{Q}$. In derivation of eq. (4) we retain only the leading $\ln(Q_{\rm eff}{}^2/\Lambda_{\rm QCD}^2)$ contribution, which corresponds to the "color-dipole picture" [2, 13]. As expected, the result (4) is proportional to c_A so that the η_Q meson is generated by the axial-vector part of the weak current. (For comparison, we also calculate the diffractive vector meson production via the weak neutral current. Using eq. (2), we find that $\mathcal{M}^{(0)}$ and $\mathcal{M}^{(\pm 1)}$ give the production of the longitudinally and transversely polarized vector mesons, respectively, and that all these amplitudes are proportional to the vector coupling c_V.)

Combining eq. (4) with the Z boson propagator and the weak neutral current by a neutrino, the forward differential cross section for the η_Q production is given by

$$\left.\frac{d^3\sigma(\nu N \to \nu' N' \eta_Q)}{ds\, dQ^2 dt}\right|_{t=0} = \frac{1}{4(8\pi)^3 E_\nu^2 M_N^2 s} \frac{g_W^2}{\cos\theta_W^2} \frac{Q^2}{(Q^2 + M_Z^2)^2} \frac{\varepsilon}{1-\varepsilon} \left|\mathcal{M}^{(0)}\right|^2, \quad (5)$$

where E_ν is the neutrino beam energy in the lab system, and $\varepsilon = [4(1-y) - Q^2/E_\nu^2]/[2\{1 + (1-y)^2\} + Q^2/E_\nu^2]$ with $y = s/(2M_N E_\nu)$. In order to evaluate the corresponding elastic η_Q production rate, we assume the t-dependence as $d^3\sigma/dsdQ^2dt = d^3\sigma/dsdQ^2dt|_{t=0} \exp(B_{\eta_Q} t)$ with a constant diffractive slope, as in the case of the vector meson production. Integrating over t, Q^2 and s, we get the elastic production rate $\sigma(\nu N \to \nu' N' \eta_Q)$ as a function of E_ν.

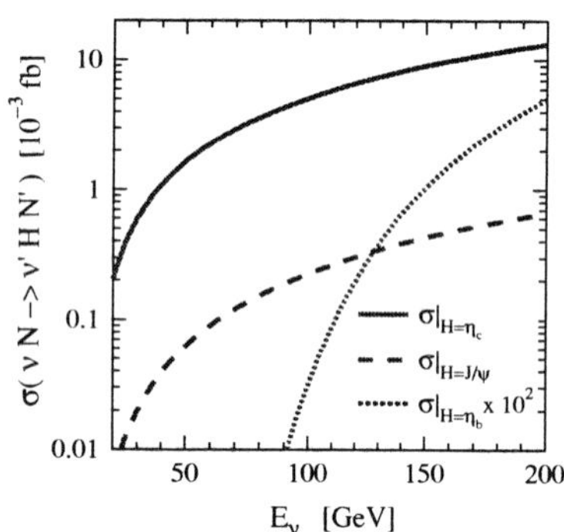

FIGURE 2. The elastic production rates as functions of E_ν. Solid, dashed and dotted curves are for η_c, J/ψ and η_b, respectively. Note that the dotted curve shows the rate multiplied by 10^2.

For numerical computation of $\sigma(\nu N \to \nu' N' \eta_Q)$, we need explicit form of the non-perturbative WF $\phi(\alpha, \vec{b})$ of eq. (4). As mentioned above, we can use the corresponding nonperturbative part of the vector-meson WF and we use the one which was constructed in Ref. [13] based on non-relativistic Cornell potential model for heavy quarkonia with $m_{c(b)} = 1.5(4.9)$ GeV. Also, we use the empirical values for the masses $M_{\eta_c} = 2.98$ GeV, $M_N = 0.94$ GeV and $M_Z = 91.2$ GeV. For η_b, we use an estimate $M_{\eta_b} = 9.45$

GeV [14]. Because the slope B_{η_Q} introduced above is unknown, we assume that B_{η_Q} has the same value as that for the corresponding vector meson: $B_{\eta_c} = 4.5$ GeV^{-2} and $B_{\eta_b} = 3.9$ GeV^{-2} (See Ref. [3]). For the gluon distribution function $G(x, Q^2_{\text{eff}})$ of eq. (4), we employ GRV95 NLO parameterization [15].

We show the elastic η_c production rate, $\sigma(\nu N \to \nu' N' \eta_c)$, by the solid curve in Fig. 2. The result monotonically increases as a function of the beam energy E_ν. Such behavior is similar with that observed in the π-production data by the neutral and charged currents [5]. For comparison, we show the elastic J/ψ production rate $\sigma(\nu N \to \nu' N' J/\psi)$ by the dashed curve. The rate for η_c production is much larger than that for J/ψ by a factor ~ 20. This is mainly due to the relevant weak couplings, c_A for η_c and c_V for J/ψ, as $(c_A/c_V)^2 \cong 7$. Also, most of the remaining factor arises from the different behavior of the Z light-cone WFs (1) between the axial-vector and vector channels, resulting in a few times difference in the overlap integrals with the corresponding meson WFs. In Fig. 2, we also show the η_b production rate $\sigma(\nu N \to \nu' N' \eta_b)$ by the dotted curve. Although the rate for η_b is generally much smaller than that for η_c, the former increases more rapidly than the latter for increasing E_ν. Therefore, the η_b production rate could become comparable with the J/ψ or η_c productions for higher beam energy. It suggests a possibility to observe η_b through the diffractive productions by high intensity neutrino beams available at ongoing or forthcoming neutrino factories.

In conclusion, we have computed the diffractive production cross sections of η_c and η_b mesons via the weak neutral current. Using the new results of the light-cone WFs for Z and $\eta_{c,b}$, the production rates are obtained based on the factorization formalism in QCD. Our results demonstrate that neutrino-induced productions will open a new window to measure $\eta_{c,b}$.

ACKNOWLEDGMENTS

A.H. was supported by JSPS Research Fellowship for Young Scientists.

REFERENCES

1. S.J. Brodsky *et al.*, Phys. Rev. **D50**, 3134 (1994).
2. L. Frankfurt *et al.*, Phys. Rev. **D57**, 512 (1998).
3. M.G.Ryskin *et al.*, Z. Phys. **C76**, 231 (1997).
4. E815 Collaboration, T. Adams *et al.*, Phys. Rev. **D61**, 092001 (2000).
5. E632 Collaboration, S. Willocq *et al.*, Phys. Rev. **D47**, 2661 (1993).
6. CHORUS Collaboration, P. Annis *et al.*, Phys. Lett. **B435**, 458 (1998).
7. J.H. Kühn and R. Rückl, Phys. Lett. **B95**, 431 (1980).
8. B. Lehmann-Dronke and A. Schäfer, Phys. Lett. **B521**, 55 (2001).
9. P. Hoodbhoy, Phys. Rev. **D56**, 388 (1997).
10. I. P. Ivanov and N. N. Nikolaev, JETP Lett. **69**, 294 (1999).
11. J. Hüfner *et al.*, Phys. Rev. **D62**, 094022 (2000).
12. E. L. Berger and D. L. Jones, Phys. Rev. **D23**, 1521 (1981).
13. K. Suzuki *et al.*, Phys. Rev. **D62**, 031501(R) (2000).
14. D.S. Hwang and G.-H. Kim, Z. Phys. **C76**, 107 (1997).
15. M. Glück *et al.*, Z. Phys. **C67**, 433 (1995).

Hyperon Polarization from Unpolarized pp and ep Collisions

Yuji Koike

Department of Physics, Niigata University, Ikarashi, Niigata 950–2181, Japan

Abstract. Cross section formulas for the Λ polarization in $pp \to \Lambda^{\uparrow}(\ell_T)X$ and $ep \to \Lambda^{\uparrow}(\ell_T)X$ are derived and its characteristic features are discussed.

In this report we discuss the polarization of Λ hyperon produced in unpolarized pp and ep collisions relevant for the ongoing RHIC-SPIN, HERMES and COMPASS experiments. According to the QCD factorization theorem, the polarized cross section for $pp \to \Lambda^{\uparrow}X$ consists of two twist-3 contributions:

$$(A) \qquad E_a(x_1,x_2)\otimes q_b(x')\otimes \delta\widehat{q}_c(z)\otimes \hat{\sigma}_{ab\to c},$$

$$(B) \qquad q_a(x)\otimes q_b(x')\otimes \widehat{G}_c(z_1,z_2)\otimes \hat{\sigma}'_{ab\to c},$$

where the functions $E_a(x_1,x_2)$ and $\widehat{G}_c(z_1,z_2)$ are the twist-3 quantities representing, respectively, the unpolarized distribution in the nucleon and the polarized fragmentation function for $\Lambda^{\uparrow}$. $\delta\widehat{q}_c(x)$ is the transversity fragmentation function for $\Lambda^{\uparrow}$. a, b and c stand for the parton's species, sum over which is implied. E_a and $\delta\widehat{q}_c$ are chiral-odd. Corresponding to the above (A) and (B), the polarized cross section for $ep \to \Lambda^{\uparrow}X$ (final electron is not detected) receives two twist-3 contributions:

$$(A') \qquad E_a(x_1,x_2)\otimes \delta\widehat{q}_a(z)\otimes \hat{\sigma}_{ea\to a},$$

$$(B') \qquad q_a(x)\otimes \widehat{G}_a(z_1,z_2)\otimes \hat{\sigma}'_{ea\to a}.$$

The (A) contribution for $pp \to \Lambda^{\uparrow}X$ has been analyzed in [1], where it was shown that (A) gives rise to growing P_{Λ} at large x_F as observed experimentally. Here we extend the study to the (B) term (see also [2]) at RHIC energy and also for the ep collision.

The unpolarized twist-3 distribution $E_{F,D}(x_1,x_2)$ is defined in [1]. Likewise the twist-3 fragmentation function for a polarized Λ (with momentum ℓ) is defined as the lightcone correlation function as ($w^2=0$, $\ell\cdot w=1$)

$$\frac{1}{N_c}\sum_X \int\frac{d\lambda}{2\pi}\int\frac{d\mu}{2\pi}e^{-i\frac{\lambda}{z_1}}e^{-i\mu(\frac{1}{z_2}-\frac{1}{z_1})}\langle 0|\psi_i(0)|\pi X\rangle\langle \pi X|gF^{\alpha\beta}(\mu w)w_\beta\bar{\psi}_j(\lambda w)|0\rangle$$

$$=\frac{M_N}{2z_2}(\ell\!\!\!/)_{ij}\varepsilon^{\alpha\ell w S_\perp}\widehat{G}_F(z_1,z_2)+i\frac{M_N}{2z_2}(\gamma_5\ell\!\!\!/)_{ij}S_\perp^{\alpha}\widehat{G}_F^5(z_1,z_2)+\cdots. \qquad (1)$$

CP675, *Spin 2002: 15th Int'l. Spin Physics Symposium and Workshop on Polarized Electron Sources and Polarimeters,* edited by Y. I. Makdisi, A. U. Luccio, and W. W. MacKay

Note that we use the nucleon mass M_N to normalize the twist-3 fragmentation function for Λ. There is another twist-3 fragmentation functions which are obtained from (1) by shifting the gluon-field strength from the left to the right of the cut. The defined functions $\widehat{G}_{FR}(z_1,z_2)$ and $\widehat{G}^5_{FR}(z_1,z_2)$ are connected to $\widehat{G}_F(z_1,z_2)$ by the relation $\widehat{G}_F(z_1,z_2)=\widehat{G}_{FR}(z_2,z_1)$ and $\widehat{G}^5_F(z_1,z_2)=-\widehat{G}^5_{FR}(z_2,z_1)$, which follows from hermiticity and time reversal invariance. Unlike the twist-3 distributions, the twist-3 fragmentation function does not have definite symmetry property. Another class of twist-3 fragmentation functions $\widehat{G}^{(5)}_D(z_1,z_2)$ is also defined from (1) by replacing $gF^{\alpha\beta}(\mu w)w_\beta$ by $D^\alpha(\mu w)=\partial^\alpha - igA(\mu w)$. Note, however, this is not independent from the above (1).

Following the method of [3] we present the analysis of the (C) term. The detailed analysis shows $\widehat{G}_F(z,z)$ appears as soft-gluon-pole contribution ($z_1=z_2=z$), while $\widehat{G}_D(z_1,z_2)$ appears as a soft fermion pole ($z_1=0$ or $z_2=2$). Physically, the latter is expected to be suppressed, and we include only the former contribution. This observation also applies to $E_{F,D}(x_1,x_2)$ relevant for the (A) term. In the large x_F region, the main contribution comes from large-x and large-z (and small x') region. Since E_F and $\widehat{G}_F$ behaves as $E_F(x,x)\sim(1-x)^\beta$ and $\widehat{G}_F(z,z)\sim(1-z)^{\beta'}$ with $\beta,\ \beta'>0$, $|(d/dx)E_F(x,x)|\gg|E_F(x,x)|$, $|(d/dz)\widehat{G}_F(z,z)|\gg|\widehat{G}_F(z,z)|$ at large x and z. In particular, the valence component of E_F and $\widehat{G}_F$ dominates in this region. We thus keep only the valence quark contribution for the derivative of these soft-gluon pole function ("valence-quark soft-gluon approximation") for the *pp* collision. For the *ep* case, we include all the soft-gluon pole contribution, since the calculation is relatively simple compared to the *pp* case.

In general P_Λ is a function of $S=(P+P')^2\simeq 2P\cdot P'$, $T=(P-\ell)^2\simeq -2P\cdot\ell$ and $U=(P'-\ell)^2\simeq -2P'\cdot\ell$ where P and P' are the momenta of the two nucleons, and ℓ is the momentum of Λ. In the following we use S, $x_F=\frac{2\ell_\parallel}{\sqrt{S}}=\frac{T-U}{S}$ and $x_T=\frac{2\ell_T}{\sqrt{S}}$ as independent variables. The polarized cross section for the (B) term reads

$$
\begin{aligned}
E_\Lambda\frac{d^3\Delta\sigma(S_\perp)}{d\ell^3} &= \frac{2\pi M_N\alpha_S^2}{S}\varepsilon^{\ell w S_\perp}\sum_a\int_{z_{min}}^1\frac{dz}{z^2}\int_{x_{min}}^1\frac{dx}{x}\frac{1}{xS+U/z}\int_0^1\frac{dx'}{x'}\\
&\times\delta\left(x'+\frac{xT/z}{xS+U/z}\right)\\
&\times\left\{\sum_{b,c}q^a(x)q^b(x')\left[-z_1^2\frac{\partial}{\partial z_1}\widehat{G}^a_F(z_1,z)\right]_{z_1=z}\left(\frac{-2p_\alpha}{T}\widehat{\sigma}^I_{ab\to c}+\frac{-2p'_\alpha}{U}\widehat{\sigma}^{II}_{ab\to c}\right)\right.\\
&+\sum_{b,c}q^a(x)q^b(x')\left[-z^2\frac{d}{dz}\widehat{G}^a_F(z,z)\right]\frac{xp_\alpha+x'p'_\alpha}{|xT+x'U|}\left(\widehat{\sigma}^I_{ab\to c}+\widehat{\sigma}^{II}_{ab\to c}\right)\\
&+q^a(x)G(x')\left[-z_1^2\frac{\partial}{\partial z_1}\widehat{G}^a_F(z_1,z)\right]_{z_1=z}\left(\frac{-2p_\alpha}{T}\widehat{\sigma}^I_{ag\to a}+\frac{-2p'_\alpha}{U}\widehat{\sigma}^{II}_{ag\to a}\right)
\end{aligned}
$$

$$+q^a(x)G(x')\left[-z^2\frac{d}{dz}\widehat{G}_F^a(z,z)\right]\frac{xp_\alpha+x'p'_\alpha}{|xT+x'U|}\left(\widehat{\sigma}^I_{ag\to a}+\widehat{\sigma}^{II}_{ag\to a}\right)\Bigg\}, \tag{2}$$

where the lower limits for the integration variables are $z_{min}=-(T+U)/S=\sqrt{x_F^2+x_T^2}$ and $x_{min}=-U/z(S+T/z)$. The partonic hard cross sections are written in terms of the invariants in the parton level, $\hat{s}=(xp+x'p')^2=xx'S$, $\hat{t}=(xp-\ell/z)^2=xT/z$ and $\hat{u}=(x'p'-\ell/z)^2=x'U/z$. They read

$$\widehat{\sigma}^I_{ab\to c}=-\frac{1}{36}\frac{\hat{s}^2+\hat{u}^2}{\hat{t}^2}\delta_{ac}+\frac{7}{36}\frac{\hat{s}^2+\hat{t}^2}{\hat{u}^2}\delta_{bc}+\frac{1}{54}\frac{\hat{s}^2}{\hat{t}\hat{u}}\delta_{ab}\delta_{ac},$$

$$\widehat{\sigma}^{II}_{ab\to c}=\frac{7}{36}\frac{\hat{s}^2+\hat{u}^2}{\hat{t}^2}\delta_{ac}-\frac{1}{36}\frac{\hat{s}^2+\hat{t}^2}{\hat{u}^2}\delta_{bc}+\frac{1}{54}\frac{\hat{s}^2}{\hat{t}\hat{u}}\delta_{ab}\delta_{ac},$$

$$\widehat{\sigma}^I_{a\bar{b}\to c}=-\frac{1}{36}\frac{\hat{s}^2+\hat{u}^2}{\hat{t}^2}\delta_{ac}+\frac{7}{36}\frac{\hat{u}^2+\hat{t}^2}{\hat{s}^2}\delta_{ab},\quad \widehat{\sigma}^{II}_{a\bar{b}\to c}=\frac{1}{18}\frac{\hat{s}^2+\hat{u}^2}{\hat{t}^2}\delta_{ac}+\frac{1}{18}\frac{\hat{u}^2+\hat{t}^2}{\hat{s}^2}\delta_{ab},$$

$$\widehat{\sigma}^I_{qg\to q}=\frac{-1}{8}\left(1-\frac{\hat{s}\hat{u}}{\hat{t}^2}\right)+\frac{1}{288}\left(\frac{-\hat{u}}{\hat{s}}+\frac{\hat{s}}{-\hat{u}}\right)-\frac{\hat{s}}{16\hat{t}}-\frac{\hat{u}}{16\hat{t}},$$

$$\widehat{\sigma}^{II}_{qg\to q}=\frac{9}{16}\left(1-\frac{\hat{s}\hat{u}}{\hat{t}^2}\right)+\frac{\hat{u}}{32\hat{s}}-\frac{\hat{s}}{4\hat{u}}+\frac{9\hat{u}}{16\hat{t}}. \tag{3}$$

Among these partonic cross sections, $\widehat{\sigma}^I$ becomes more important at large x_F because of the $1/T$ factor in (2).

To estimate the above contribution, we introduce a model ansatz as $\widehat{G}_F^a(z,z)=K_a\widehat{q}_a(z)$ with twist-2 unpolarized fragmentation function $\widehat{q}^a(z)$, noting that the Dirac structure of $\widehat{G}_F^a(z,z)$ and $\widehat{q}_a(z)$ is the same [3]. K_a's are taken to be $K_u=-K_d=0.07$ which are the same values used in the relation $G_F(x,x)=K_aq^a(x)$ to reproduce A_N in $p^\uparrow p\to\pi X$ observed at E704 [4]. As noted before, $\widehat{G}_F(z_1,z_2)$ does not have definite symmetry property unlike the twist-3 distribution $E_F(x_1,x_2)$. Nevertheless we assume $\left[(\partial/\partial z_1)\widehat{E}_F(z_1,z)\right]_{z_1=z}=(1/2)(d/dz)\widehat{E}_F(z,z)$. The result for the Λ polarization P_Λ^{pp} at $\sqrt{S}=62$ GeV is shown in Fig. 1(a) together with the R608 data.There (A) (chiral-odd) contribution studied in [1] is also shown for comparison. (For the adopted distribution and fragmentation functions, see [1].) One sees that the tendency of P_Λ^{pp} from the (B)(chiral-even) contribution is quite similar to the R608 data. Rising behavior of P_Λ^{pp} at large x_F comes from (i) the large partonic cross sections in (3) ($\sim 1/\hat{t}^2$ term) and (ii) the derivative of the soft-gluon pole functions. With these parameters K_a, P_Λ^{pp} at RHIC energy ($\sqrt{S}=200$ GeV) is shown in Fig. 1(b) at $l_T=1.5$ GeV.Fig. 2(a) shows the l_T dependence of P_Λ^{pp} of the (B) term,indicating large ℓ_T dependence at $1\le\ell_T\le 3$ GeV. Experimentally, P_Λ^{pp} grows up as ℓ_T increases up to $\ell_T\sim 1$ GeV and stays constant at $1\le\ell_T\le 3$ GeV. So the P_Λ^{pp} observed at R608 can not be wholly ascribed to the twist-3 effect studied here which is designed to describe large ℓ_T polarization.

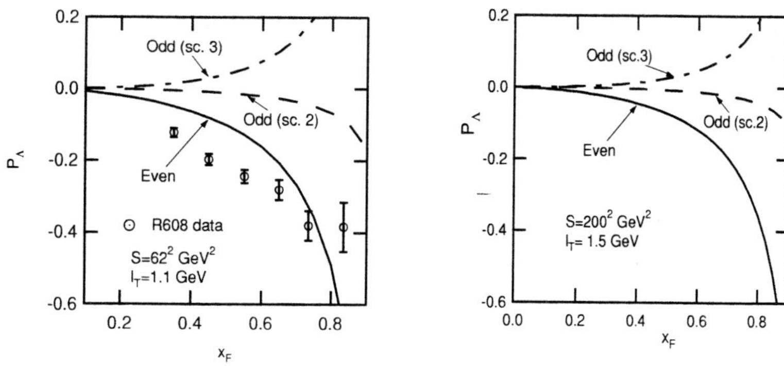

FIGURE 1. (a) P_Λ^{pp} at $\sqrt{S} = 62$ GeV. (b) P_Λ^{pp} at $\sqrt{S} = 200$ GeV.

We next discuss the polarization P_Λ^{ep} in $pe \to \Lambda^\uparrow(\ell)X$ where the final electron is not observed. In our $O(\alpha_s^0)$ calculation, the exchanged photon remains highly virtual as far as the observed Λ has a large transverse momentum ℓ_T with respect to the ep axis. Therefore experimentally one needs to integrates only over those virtual photon events to compare with our formula.

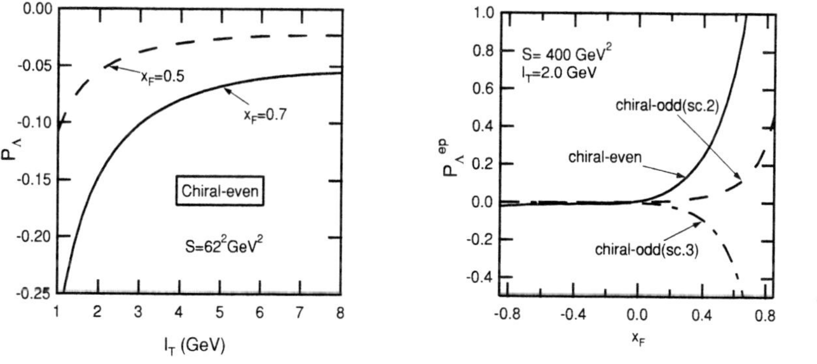

FIGURE 2. (a) ℓ_T dependence of P_Λ^{pp}. (b)P_Λ^{ep} at $\sqrt{S} = 20$ GeV.

Using the twist-3 distribution and fragmentation functions used to describe P_Λ^{pp}, we show in Fig. 2(b) the obtainedP_Λ^{ep} corresponding to (A')(chiral-odd) and (B')(chiral-even) contributions. Remarkable feature of Fig. 2(b) is that in both chiral-even and chiral-odd contributions (i) the sign of P_Λ^{ep} is opposite to the sign of P_Λ^{pp} and (ii) the magnitude of P_Λ^{ep} is much larger than that of P_Λ^{pp}, in particular, at large x_F, and it even overshoots one. (In our convention, $x_F > 0$ corresponds to the production of Λ in the forward hemisphere of the initial proton in the ep case.) The origin of these features can be traced back to the color factor in the dominant diagrams for the *twist-3 polarized* cross sections in ep and pp collisions. Of course, the P_Λ can not exceeds one, and thus our model estimate needs to be modified. First, the applied kinematic range of our formula should be reconsidered: Application of the twist-3 cross section at such small ℓ_T may not be justified. Second, our simple model ansatz of $E_F^a(x,x) \sim \delta q^a(x)$ (in (A) term) and $\widehat{G}_F^a(z,z) \sim \widehat{q}^a(z)$ should be modified at $x \to 1$ and $z \to 1$, respectively. The derivative of these functions, which is important for the growing P_Λ^{pp} at large x_F, eventually leads to divergence of P_Λ at $x_F \to 1$ as $\sim 1/(1-x_F)$.

As a possible remedy for this pathology we tried the following: As an example for the (B) (chiral-even) contribution we have a model $\widehat{G}_F^a(z,z) \sim \widehat{q}_a(z) \sim_{z\to 1} (1-z)^\beta$ where $\beta = 1.83$ in the fragmentation function we adopted. Tentatively we shifted β as $\beta \to \beta(z) = \beta + z^8$, which suppresses the divergence of P_Λ at $x_F \to 1$ but still keeps rising behavior of P_Λ at large x_F. This avoids overshooting of one in P_Λ^{ep} but reduces P_Λ^{pp} seriously. The result obtained by this modification is shown in Figs. 3(a) and (b)

To summarize we have studied the Λ polarization in pp and ep collisions in the framework of collinear factorization. Our approach includes all effects for the large ℓ_T production. One needs to be causious in interpreting the available pp data at relatively low ℓ_T in terms of the derived formula. Determination of the participating twist-3 functions requires global analysis of future pp and ep data.

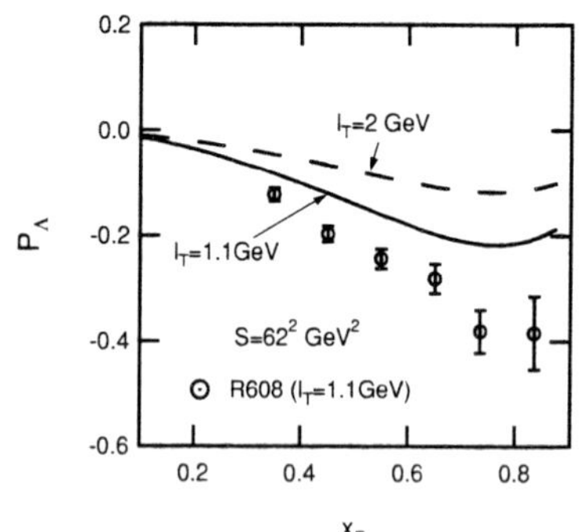

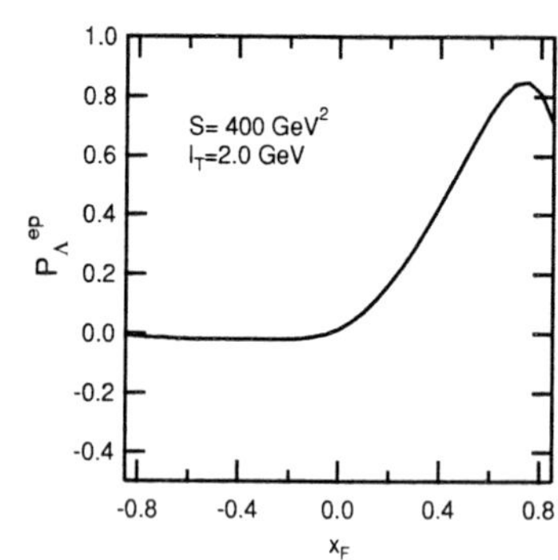

FIGURE 3. (a) P_Λ^{pp} with modified $\widehat{G}_F$. (b) P_Λ^{ep} with modified $\widehat{G}_F$.

ACKNOWLEDGMENTS

This work is supported in part by the Grant-in-Aid for Scientific Research of Monbusho.

REFERENCES

1. Y. Kanazawa and Y. Koike, Phys. Rev. **D64** (2001) 034019.
2. Y. Koike, hep-ph/0106260 (Proceedings of DIS2001, Bologna, Italy, April, 2001.)
3. J. Qiu and G. Sterman, Phys. Rev. **D59** (1999) 014004.
4. Y. Kanazawa and Y. Koike, Phys. Lett. **B478** (2000) 121; **B490** (2000) 99.

The Challenge of Hyperon Polarization

Sergey Troshin and Nikolai Tyurin

Institute for High Energy Physics, Protvino, Moscow Region, 142281, Russia

Abstract. A nonperturbative mechanism for hyperon polarization in inclusive production is considered. The main role belongs to the orbital angular momentum and the polarization of the strange $\bar{s}s$–pairs in the internal structure of the constituent quarks.

We concentrate on a particular problem of Λ–polarization. Experimentally, the situation is stable and clear. Λ–polarization is negative and energy independent. It grows linearly with x_F for $p_\perp > 0.8$ GeV/c, and for large values of the momentum transfer ($0.8 < p_\perp < 3.5$ GeV/c) it is $p_\perp$–independent [1, 2]. It is remarkable that both parameters A_N and D_{NN} show $p_\perp$–dependence similar to polarization [3].

On the theoretical side, Λ–polarization is not understood in perturbative QCD. Indeed, a straightforward collinear factorization leads to very small values of P_Λ [4, 5]. pQCD modifications and in particular account for higher twists result in the dependence $P_\Lambda \sim 1/p_\perp$ [6, 7, 8]. This behavior still does not correspond to the experimental trends. Account for $k_\perp$–effects when the source of polarization is shifted to the polarizing fragmentation functions also leads to falling $P_\Lambda \sim k_\perp/p_\perp$ at large $p_\perp$ values [9]. The overall conclusion is that the dynamics of Λ polarization (as well as other single spin asymmetries) within pQCD is far from being settled. The essential point here is that the vacuum at short distances is perturbative. Potentially Λ–polarization could be even a more serious problem that the nucleon spin problem. And in any case the both problems are interrelated.

It is generally believed that polarization dynamics has its roots hidden in the genuine nonperturbative sector of QCD. The models exploiting confinement and the chiral symmetry breaking have been proposed. Our model considerations [10] are based on the effective Lagrangian approach which in addition to the four–fermion interactions of the original NJL model includes the six–fermion $U(1)_A$–breaking term.

Chiral symmetry breaking generates quark masses:

$$m_U = m_u - 2g_4\langle 0|\bar{u}u|0\rangle - 2g_6\langle 0|\bar{d}d|0\rangle\langle 0|\bar{s}s|0\rangle.$$

In this approach massive quarks appear as quasiparticles, i.e. current quarks surrounded by a cloud of quark–antiquark pairs of different flavors. For example, for the U–quark the ratio

$$\langle U|\bar{s}s|U\rangle / \langle U|\bar{u}u + \bar{d}d + \bar{s}s|U\rangle$$

is estimated as $0.1 - 0.5$. The scale of spontaneous chiral symmetry breaking is

$$\Lambda \simeq 4\pi f_\pi \simeq 1 \quad \text{GeV/c}$$

CP675, *Spin 2002: 15th Int'l. Spin Physics Symposium and Workshop on Polarized Electron Sources and Polarimeters*, edited by Y. I. Makdisi, A. U. Luccio, and W. W. MacKay

and provides the momentum cutoff which determines a transition to the partonic picture. We consider nonperturbative hadron as consisting of constituent quarks located in the central part of the hadron and embedded into a quark condensate.

Respectively, spin of the constituent quark is given by the following "spin balance equation":

$$J_U = 1/2 = S_{u_v} + S_{\{\bar{q}q\}} + L_{\{\bar{q}q\}} = 1/2 + S_{\{\bar{q}q\}} + L_{\{\bar{q}q\}}, \quad (1)$$

where $L_{\{\bar{q}q\}}$ is the orbital angular momentum of quark–antiquark pairs in the structure of the constituent quark. Its value can be estimated [10] with the use of the polarized DIS data:

$$L_{\{\bar{q}q\}} \simeq 0.4 \quad (2)$$

This means that the cloud quarks rotate coherently and significant part of the constituent quark spin is to be associated with the orbital angular momentum. In the model just this orbital motion of quark matter is the origin of asymmetries in inclusive processes. It is to be noted that the only effective degrees of freedom here are quasiparticles. The gluon degrees of freedom are overintegrated, and the six-fermion operator in the NJL Lagrangian simulates the effect of the gluon operator $(\alpha_s/2\pi)G^a_{\mu\nu}\tilde{G}^{\mu\nu}_a$ in QCD. It is also important to note the exact compensation between the spins of $\bar{q}q$-pairs and their orbital momenta:

$$L_{\{\bar{q}q\}} = -S_{\{\bar{q}q\}}, \quad (3)$$

which follows from Eq. (1).

Assumed picture of hadrons implies that overlapping and interaction of peripheral clouds and condensate excitation occur at the first stage of the collision. As a result massive virtual quarks appear in the overlapping region and some mean field is generated. Inclusive production of hyperon results from the two mechanisms: recombination of constituent quark with virtual massive strange quark (soft interactions) or from the constituent quark scattering in the mean field, its excitation and appearance of a strange quark as a result of decay of the parent constituent quark. The second mechanism is determined by interactions at the distances smaller than the constituent quark radius ($r < R_Q \sim 1/\Lambda_\chi$) and is associated with hard interactions. Thus, we adopt a two-component picture of hadron (hyperon) production which incorporates interactions at long and short distances and it is the short distance dynamics which leads to production of polarized Λ's.

Polarization of a strange quark results from the multiple scattering of parent constituent quark Q in the mean field where it gets polarized

$$\mathscr{P}_Q \propto -I\frac{m_Q g^2}{\sqrt{s}} \quad (4)$$

and the polarization is nearly constant in the model since $m_Q \sim m_h/3$ and $I \sim \sqrt{s}$. The second crucial point is correlation between s–quark polarization and polarization of the parent quark Q. Indeed, the total orbital momentum of $\bar{q}q$–pairs in the constituent quark which has polarization $\mathscr{P}_Q(x)$ is

$$L^{\mathscr{P}_Q(x)}_{\{\bar{q}q\}} = \mathscr{P}_Q(x) L_{\{\bar{q}q\}}, \quad (5)$$

where the value $L_{\{\bar{q}q\}}$ on the right hand side enters Eq. (1) written for the constituent quark with polarization $+1$. On the basis of Eq. (3) we suppose that there is a compensation between spin and orbital momentum of strange quarks inside the constituent quark

$$L_{s/Q} = -J_{s/Q} = \alpha \mathcal{P}_Q(x) L_{\{\bar{q}q\}}, \tag{6}$$

where the parameter α determines the fraction of orbital momentum due to the strange quarks. Eq. (6) is quite similar to the conclusion made in the framework of the Lund model but has a different dynamical origin rooted in the mechanism of the spontaneous chiral symmetry breaking.

Final expression for the polarization is

$$P(s,x,p_\perp) = \sin[\alpha \mathcal{P}_Q(x) L_{\{\bar{q}q\}}] \frac{R(s,x,p_\perp)}{1+R(s,x,p_\perp)}. \tag{7}$$

The function R is the cross–section ratio of hard and soft processes. At $p_\perp > \Lambda_\chi$ the function $R(s,x,p_\perp) \gg 1$ and the polarization saturates

$$P(s,x,p_\perp) = \sin[\alpha \mathcal{P}_Q(x) L_{\{\bar{q}q\}}]. \tag{8}$$

Characteristic $p_\perp$–dependence of Λ–polarization follows from Eqs. (7) and (8): polarization is vanishing for $p_\perp < \Lambda_\chi$, it gets an increase in the region of $p_\perp \simeq \Lambda_\chi$ and polarization saturates and becomes $p_\perp$–independent (flat) for $p_\perp > \Lambda_\chi$. The respective scale is the scale of the spontaneous chiral symmetry breaking $\Lambda_\chi \simeq 1$ GeV/c. Such a behavior of polarization follows from the fact that constituent quarks themselves have slow (if at all) orbital motion and are in the S–state, but interactions with $p_\perp > \Lambda_\chi$ resolve the internal structure of constituent quark and feel the presence of internal orbital momenta inside this constituent quark.

To describe the data quantitatively we have to introduce some parameterization. The function

$$R(s,x,p_\perp) = C(x)\exp(p_\perp/m)/(p_\perp^2 + \Lambda_\chi^2)^2 \tag{9}$$

implies typical behavior of cross–sections for hard and soft processes. The form

$$\mathcal{P}_Q(x) = \mathcal{P}_Q^{max} x$$

is suggested by the ALEPH data [11]. The value of Eq. (2) alongside with $m = 0.2$ GeV and $\alpha = 0.8$ provides a rather good fit to the experimental data (Fig. 1 and 2).

In the model the spin transfer parameter D_{NN} is positive since P_Λ has the same sign as $\mathcal{P}_Q$. The model also predicts similarity of $p_\perp$ dependencies for the different spin observables. The respective features were clearly seen in E-704 experiment [3].

Conclusion and the experimental prospects

In this approach the short distance interaction with $p_\perp \Lambda_\chi$ observes a coherent rotation of correlated $\bar{q}q$–pairs inside the constituent quark and not a gas of the free partons. The nonzero internal orbital momenta in the constituent quark means that there are

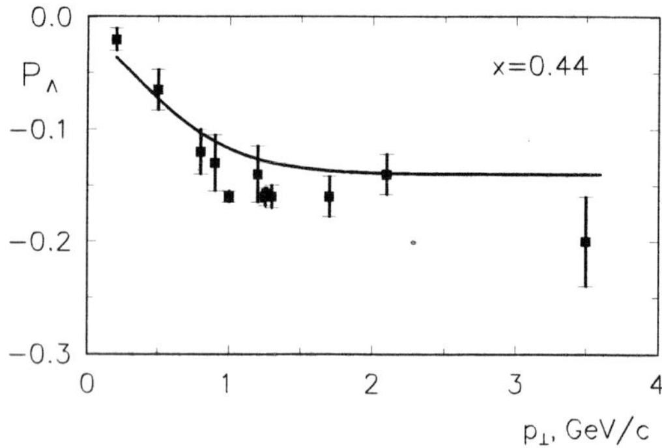

FIGURE 1. Transverse momentum dependence of P_Λ.

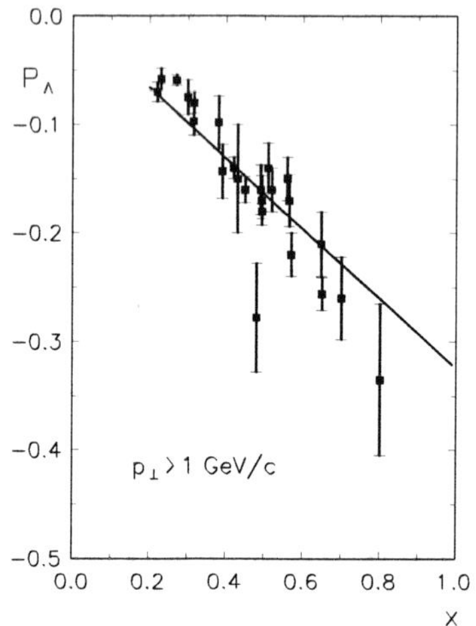

FIGURE 2. Feynman x dependence of P_Λ.

significant multiparton correlations. The important point is what the origin of this orbital angular momentum is. The analogy with an anisotropic extension of the theory of superconductivity seems match well with the adopted picture for a constituent quark. An axis of anisotropy can be associated with the polarization vector of the valence quark located at the origin of the constituent quark.

It seems interesting to perform Λ–polarization measurements at RHIC. When two polarized nucleons are available one could measure three–spin correlation parameters $(n,n,n,0)$ and $(l,l,l,0)$ in the processes

$$p_{\uparrow,\rightarrow} + p_{\uparrow,\rightarrow} = \Lambda_{\uparrow,\rightarrow} + X. \tag{10}$$

It would provide important data to study mechanisms of hyperon polarization.

On the basis of the above model we expect significant P_Λ at RHIC energies. But what seems to be most interesting a diminishing Λ–polarization could serve as a signal for

QGP formation in heavy ion collisions. We do not expect a strong diminishing of P_Λ due to the nuclear effects since the available data show a weak A–dependence and are not sensitive to the type of the target. Respectively the behavior of $P_\Lambda \to 0$ with centrality increase is expected due to chiral symmetry restoration.

REFERENCES

1. G. Bunce at al., Phys. Rev. Lett. **36** , 1113 (1976);
   For a history of the hyperon polarization discovery, see: T. Devlin, in SPIN 94, Proceedings of the 11th International Symposium on High-energy Spin Physics and the 8th International Symposium on Polarization Phenomena in Nuclear Physics, Bloomington, Indiana, 15-22 September 1994, AIP Conf. Proc. Woobury, New York, 1995, eds. K. Heller and S. Smith, p. 354.
2. L. Pondrom, Phys. Rep. **122**, 57 (1985);
   K. Heller, in Proceedings of the 7th International Symposium on High-Energy Spin Physics, Protvino, Russia, 1987, p. 81;
   J. Duryea et al., Phys. Rev. Lett. **67** , 1193 (1991).
   A. Morelos et al., Phys. Rev. Lett. **71** , 2172 (1993);
   K. A. Johns et al., in SPIN 94, Proceedings of the 11th International Symposium on High-energy Spin Physics and the 8th International Symposium on Polarization Phenomena in Nuclear Physics, Bloomington, Indiana, 15-22 September 1994, AIP Conf. Proc. Woobury, New York, 1995, eds. K. Heller and S. Smith, p. 417;
   L. Pondrom, ibid., p. 365;
   A. D. Panagiotou, Int. J. Mod. Phys A **5**, 1197 (1990).
3. A. Bravar, in SPIN 98 Proceedings of the 13th International Symposium on High-Energy Spin Physics, Protvino, Russia, 8-12 September 1998, N. E. Tyurin, V. L. Solovianov, S. M. Troshin, and A. G. Ufimtsev, eds. (World Scientific, Singapore, 1999) p. 167.
4. G. L. Kane, J. Pumplin, and W. Repko, Phys. Rev. Lett. **41**, 1689 (1978).
5. W. G. D. Dharmaratna and G. R. Goldstein, Phys. Rev. D **53**, 1073 (1996)
6. A. V. Efremov and O. V. Teryaev, Sov. J. Nucl. Phys. **36**, 140 (1982).
7. J. Qiu and G. Sterman, Phys. Rev. D**59**, 014004 (1999).
8. Y. Kanazawa and Y. Koike, Phys. Rev. D,**64**, 034019 (2001).
9. M. Anselmino, D. Boer, U. D'Alesio, and F. Murgia, Phys. Rev. D.**63**, 054029 (2001).
10. S. M. Troshin and N. E. Tyurin, Phys. Rev. D**55**, 1265 (1997).
11. D. Buskulic et al. (ALEPH Collaboration), Phys. Lett. B**374**, 319 (1996).

Measurement of Single Transverse-spin Asymmetry in Forward Production of Photons and Neutrons in pp Collisions at $\sqrt{s} = 200$ GeV

A. Bazilevsky*, L. Bland†, A. Bogdanov**, G. Bunce*†, A. Deshpande*, H. En'yo*‡, B. Fox*, Y. Fukao*§, Y. Goto*‡, J. Haggerty†, K. Imai§, W. Lenz†, D. von Lintig†, M. Liu¶, Y. Makdisi†, R. Muto‡§, S. Nurushev||, E. Pascuzzi*, M. L. Purschke†, N. Saito§*, F. Sakuma‡§, S. Stoll†, K. Tanida‡, M. Togawa‡§, J. Tojo§‡, Y. Watanabe*‡ and C. Woody†

*RIKEN BNL Research Center, Brookhaven National Laboratory, Upton, NY 119730-5000, USA
†Brookhaven National Laboratory, Upton, NY 11973-5000, USA
**Moscow Engineering Physics Institute, State University Russia
‡RIKEN, Wako, Saitama 351-0198, Japan
§Kyoto University, Kyoto 606-8502, Japan
¶Los Alamos National Laboratory, Los Alamos, NM 87545, USA
||Institute for High Energy Physics Protovino, Russia

Abstract. The Relativistic Heavy Ion Collider (RHIC) at the Brookhaven National Laboratory (BNL) was commissioned for polarized proton-proton collisions at the center of mass energy $\sqrt{s} = 200$ GeV during the run in 2001-2002. We have measured the single transverse-spin asymmetry $\mathscr{A}_N$ for production of photons, neutral pions, and neutrons at the very forward angle. The asymmetries for the photon and neutral pion sample were consistent with zero within the experimental uncertainties. In contrast, the neutron sample exhibited an unexpectedly large asymmetry. This large asymmetry will be used for the non-destructive polarimeter for polarized proton beams at the collision points in the RHIC interaction region.

INTRODUCTION

There have been many interesting results of the single transverse-spin asymmetries, $\mathscr{A}_N$ in the fixed-target energy region. One of the examples is the E704 results from Fermilab, which were obtained using the polarized proton and anti-proton beams of 200 GeV with the liquid hydrogen target [1]~[3]. Large asymmetries were observed especially in foward region; π^+ and π^- showed mirror asymmetries as functions of x_F and π^0 asymmetry was about a half of the π^+ asymmetry in pp collision case. Since the asymmetry $\mathscr{A}_N$ was expected to be small ($\mathscr{O}(\alpha_s m_q/\sqrt{s})$) in high-energy hadron reaction, the results stimulated many theoretical works [4]~[8].

One of the motivation of our experiment is to see if such effects persist at higher energies, which was only achievable with the completion of the RHIC as the first polarized proton collider. In addition, we have been interested in neutral particle production in particular as a development of "Local Polarimeter", which measures the transverse

CP675, *Spin 2002: 15th Int'l. Spin Physics Symposium and Workshop on Polarized Electron Sources and Polarimeters,* edited by Y. I. Makdisi, A. U. Luccio, and W. W. MacKay

component of the beam polarization at the collision points at RHIC. Such equipment is extremely useful in understanding the spin dynamics in the RHIC accelerator system especially in obtaining the longitudinal beam polarizations by using spin rotators, which is a set of helical dipole magnets, located at RHIC. Since such polarimeter is non-destructive, it can co-exist with the experiments e.g. PHENIX and STAR. For the development of the "Local Polarimeter" at RHIC, STAR experiment has equipped with the Forward Pion Detector (FPD) to cover higher p_T but still forward region to search for any sizable $\mathscr{A}_N$ in neutral pion production[9]. In the context of the "Local Polarimeter" development, our measurement and STAR FPD measurements are complimentary.

EXPERIMENTAL SETUP

In this experiment, two detector systems were located 1800 cm upstream and downstream the collision point called IP12 (see Figure 1). Those were located right after the dipole magnets, which separate incoming and outgoing beams, and placed between two beam pipes. One of the detector systems is Electromagnetic Calorimeter based system (EM-Cal Polarimeter) facing the Blue Beam with the emphasis on photon and π^0 detection, and the other is Hadron Calorimeter based system (H-Cal Polarimeter) facing the Yellow Beam for neutron detection. The geometry of 11 cm-horizontal gap of two beam pipes and rf filter inside the dipole magnet limit the angular coverage within $\sim$3 mrad. Charged particles are swept away by the dipole magnet in front of these detector systems and only neutral particles are expected to be detected by these systems.

In order to separate the beam-gas interaction, which is the beam interaction with the residual gas in the beam pipe, or collision events outside appropriate vertex distribution, two sets of hodoscopes are located at $z = \pm 200$ cm covering pseudorapidity range of $2.3 < |\eta| < 4.0$. Since root-mean-square of vertex distribution is measured to be 54 ± 6.3 cm, and the time resolution of hodoscopes is equivalent to 23 cm, beam collisions within the hodoscopes are well separated from the beam gas events or the events outside the hodoscope. In the high energy region, where we are interested in, contamination from the beam-gas/off-vertex events into the vertex cut is estimated to be $\lesssim 1.1\%$

The beam polarization has been monitored by RHIC CNI Polarimeter [10] throughout the beam time and $17 \pm 0.3 \pm 2.8\%$ for the Yellow Beam and $11 \pm 0.3 \pm 2.5\%$ for the Blue Beam where the first and the second errors represent statistical and systematic ones, respectively.

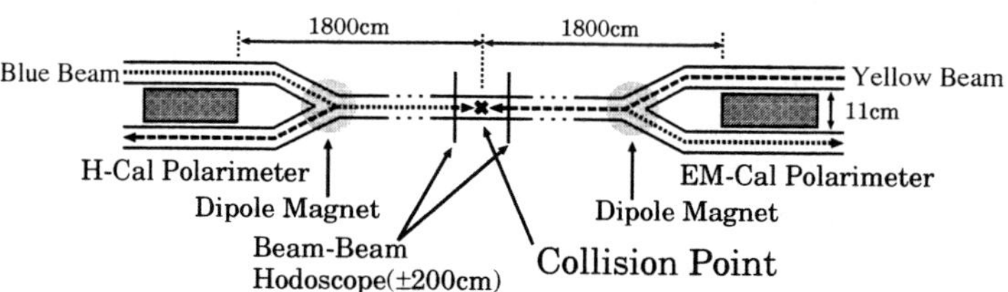

FIGURE 1. Top view of the experimental setup (not to scale) where the beams are indicated as dotted (Blue Beam) and dashed (Yellow Beam) lines with the arrows showing the circulation direction.

EM Calorimeter based system

EM Calorimeter consists of sixty Lead-Tungstate ($PbWO_4$) crystals, which have dimensions of $2.0 \times 2.0 \times 20.0$ cm^3, with read-out by photo-tube. The crystals are arranged in an array of five by twelve so that total volume of the EM Calorimeter is 10 cm wide, 24 cm high, and 20 cm deep, which corresponds to 22 radiation length and $\sim$1 interaction length. This calorimeter was calibrated using electron beams at SLAC and an energy resolution of $\Delta E/E = 10\%/\sqrt{E}$ and position resolution of 0.1 cm were achieved for electron beam. For hadronic showers, position resolution is estimated to be 0.5 cm using GEANT[11].

In front of the EM Calorimeter, pre-shower counter was located. This counter comprises five $PbWO_4$ towers and these towers have same dimension as that of EM Calorimeter. Three scintillation counters were located for particle identification. One was located right after the dipole magnet and in front of the pre-shower counter to make sure neutral particles and called Charge Veto Counter (CVC). And two followed the EM Calorimeter with 1.1 inch-thick iron block in between for neutral hadron identification and referred to as Neutron Counter 1 and 2 (NC1 and NC2, respectively).

The logic of particle identification for photon and neutron is as follows: photons are identified by $\overline{\mathrm{CVC}} \otimes \overline{\mathrm{NC}}$ and neutrons are identified as $\overline{\mathrm{CVC}} \otimes \mathrm{NC}$. We adopted this logic basing on the simulation and the real data by controlling the photon purity using the two photon invariant mass spectrum, in which we can identify $\pi^0 \to 2\gamma$ easily. Our simulation studies show the purity of photon and neutron identification to be 98% and 89%, respectively with the systematic uncertainty of 16%.

Hadron Calorimeter based system

Hadron Calorimeter has sandwich-like structure of 0.5 cm-thick tungsten plates and layer of optical fibers[12]. Its dimensions are 10 cm in width, 10 cm in height and 23 cm in length, which corresponds to $\sim$2 interaction lengths. Its energy response for neutron is calibrated using cosmic ray test and simulation and its resolution is estimated to be $\sim$50%.

The hadron calorimeter is followed by post-shower counter comprising five $PbWO_4$ crystals to measure horizontal position of particle hits. The post-shower detector is identical to the pre-shower detector in the EM-Cal Polarimeter. By using the center-of-gravity of the energy deposits in the post-shower, position resolution is estimated to be 3 to 4 cm from GEANT study.

Photon samples are eliminated by introducing a two-inch lead block followed a plastic scintillation counter (Photon Veto Counter, or PVC). With this PVC, the sample becomes practically purely long-lived neutral hadron. Remaining question is the contaminations from K_L^0s. In the ISR energies, its contribution in the similar kinematical region is estimated to be 3-4% from observing charged kaon samples [13]. In our experiment, a lack of experimental data on the kaon production in this extremely forward region represents a major uncertainty in neutron identification.

ASYMMETRY RESULTS

Energy, and position of the produced photon, π^0, and neutron were measered with the experimental apparatus described above. We have defined "Left" and "Right" and calculated the single transverse-spin asymmmetry $\mathscr{A}_N$ with the Square Root Formula[1]. For the Blue Beam polarization, we can calculate the asymmetry for positive-x_F using the EM-Cal Polarimeter and the one for negative-x_F using the H-Cal Polarimeter. Similarly the asymmetries for positive and negative x_F can be obtained for the Yellow Beam polarization. The asymmetries for photon, π^0, neutron in the EM-Cal Polarimeter, and neutron in the H-Cal Polarimeter are plotted as functions of energy, except neutron samples in the EM-Cal, where observed energy is used instead. These plots show only statistical error for error bar, while there is additional systematic error. Two vertical scales of raw asymmetries uncorrected for beam polarization for each Blue and Yellow Beam are shown on the right side of the plots.

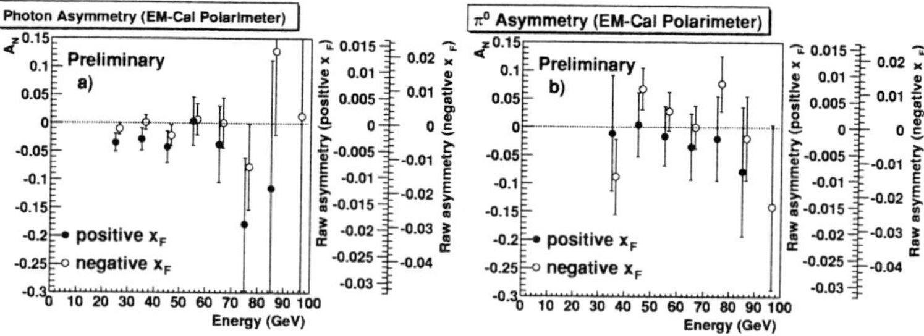

FIGURE 2. a) Photon and b) π^0 asymmetry measured by EM-Cal Polarimeter

Figure 2.a) shows inclusive photon asymmetry as a function of energy. Closed and open circles are for positive and negative x_F, respectively. The analyzing power is small and consistent with zero within systematic errors in both positive and negative x_F. Figure 2.b) shows π^0 asymmetry, and it is consistent with zero within statistical errors.

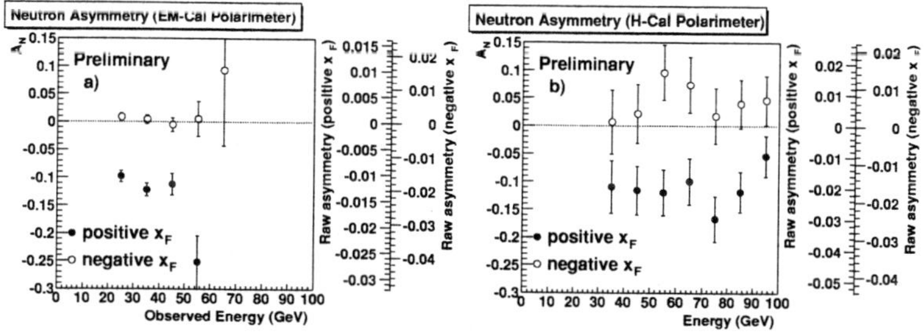

FIGURE 3. Neutron asymmetries measured by a) EM-Cal Polarimeter and b) H-Cal Polarimeter

---

[1] The formula is typically represented as $A_N = \frac{1}{P_B}\frac{\sqrt{N_{\uparrow L}N_{\downarrow R}}-\sqrt{N_{\uparrow R}N_{\downarrow L}}}{\sqrt{N_{\uparrow L}N_{\downarrow R}}+\sqrt{N_{\uparrow R}N_{\downarrow L}}}$ where $N_{\uparrow L}$ means number of particle detected in "Left" by the collision of upward polarized beam for example. Using this formula, we can eliminate luminosity and detector asymmetry and obtain physics asymmetry precisely.

Figure 3.a) and b) show neutron asymmeties which were measured using EM-Cal Polarimeter and H-Cal Polarimeter, respectively. Positive-x_F asymmetries are significantly large in both EM-Cal and H-Cal Polarimeters, while negative-x_F asymmetries are practically zero. Averaged values of $\mathscr{A}_N$ for positive-x_F measured in EM-Cal and H-Cal Polarimeter are -0.109 ± 0.007 and -0.110 ± 0.015, respectively, which are consistent each other within statistical uncertainties.

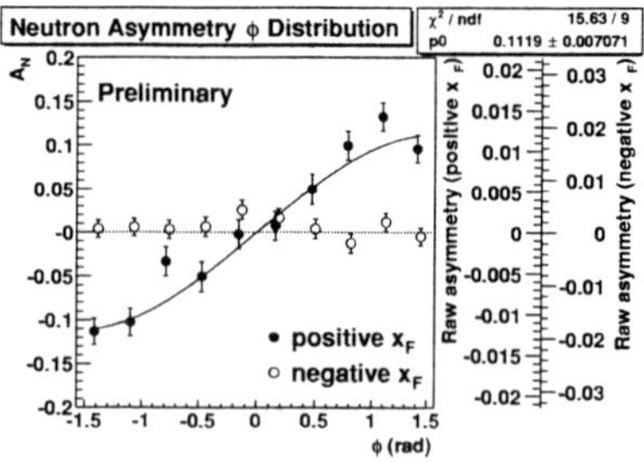

FIGURE 4. Neutron asymmetry versus ϕ angle

Figure 4 shows another neutron asymmetry from the EM-Cal Polarimeter averaged over observed energy as a function of azimuthal angle (ϕ) of produced neutron with respect to the beam polarization. The asymmetries are fitted to a sine curve, which is expected for $\mathscr{A}_N$. The results is -0.112 ± 0.007 with a reduced χ^2 of 1.7 showing a reasonable agreement with the expected $\sin(\phi)$ dependence.

CONCLUSION

We have measured the single transverse-spin asymmetry $\mathscr{A}_N$ for photon, π^0, and neutron production in polarized proton-proton collisions at $\sqrt{s}$=200 GeV for the first time. The asymmetry for photon and π^0 is consistent with zero within error, while significant asymmetry has been observed in forward production of neutron. This makes it possible to develop a non-destructive polarimeter and modified H-Cal Polarimeter is planned to be installed at PHENIX-experiment collision point for spin rotator commissioning.

REFERENCES

1. Adams, D.L. et al., *Phys. Lett.* **B261**, 201-206 (1991)
2. Adams, D.L. et al., *Phys. Lett.* **B264**, 462-466 (1991)
3. Bravar, A. et al., *Phys. Rev. Lett.* **77**, 2626-2629 (1996)
4. Qiu, J., and Sterman. G., *Phys. Rev. Lett.* **67**, 2264-2267 (1991)
5. Collins, J.C., *Nucl. Phys.* **B396**, 161-182 (1993)
6. Efremov, A., Korotkiian, V., and Teryaev., O., *Phys. Lett.* **B348**, 577-581 (1995)
7. Anselmino, M., Boglione, M., Murgia., F., *Phys. Lett.* **B362**, 164-172 (1995)
8. Artru, X., Czyzewski, J., Yabuki. H., *Z. Phys.* **C73**, 527-534 (1997)
9. Rakness, G., *These proceedings*
10. Jinnouchi, O., *These proceedings*
11. *GEANT Detector description and simulation tool* CERN Program Library Long Writeup W5013, CERN Genova (1993)
12. Adler, C. et al., *Nucl. Instrum. Meth.* **A470**, 488-499 (2001)
13. Flauger, W., and Mönnig, F., *Nucl. Phys.* **B109**, 347-356 (1976)

Spin Physics with Photons and Electrons

2002 Spin

HERMES Measurements of the Generalized GDH Integral and of Quark-Hadron Duality

W.-D. Nowak
on behalf of the HERMES Collaboration

DESY Zeuthen, D-15738 Zeuthen, Germany

Abstract. The physics impact of measuring the generalized Drell-Hearn-Gerasimov integral in inclusive deep inelastic scattering (DIS) is discussed. Studies of the transition region from the real-photon point into the DIS region have been performed simultaneously in and beyond the nucleon-resonance region, thereby testing the duality concept. HERMES measurements on the proton and deuteron show consistently that the hard regime holds down to photon virtualities of about 1.5 GeV ^{2}.

PHYSICS MOTIVATION

The Gerasimov-Drell-Hearn sum rule

$$\int_{\nu_0}^{\infty} [\sigma^{\overset{\rightarrow}{\Leftarrow}}(\nu) - \sigma^{\overset{\rightarrow}{\Rightarrow}}(\nu)] \frac{d\nu}{\nu} = -\frac{4\pi^2 I \alpha}{M_t^2} \kappa^2, \tag{1}$$

establishes a connection between the helicity-dependent dynamics and ground state properties of a nucleus [1] (or nucleon [1, 2]). The dynamics is expressed through the difference of the photoabsorption cross-sections for anti-parallel and parallel beam ($\rightarrow$) and target ($\Rightarrow$) spin orientation, $\sigma^{\overset{\rightarrow}{\Leftarrow}}(\nu)$ and $\sigma^{\overset{\rightarrow}{\Rightarrow}}(\nu)$, respectively, while the ground state is described in terms of the anomalous contribution $\kappa = \frac{\mu M_t}{eI} \quad Z$ to the magnetic moment μ of the target with atomic number Z. Here I is the spin of the nucleus, ν the photon energy in the target rest frame, ν_0 the photoabsorption threshold, M_t the nucleus mass, α the electromagnetic fine–structure constant and e the elementary charge.

The GDH integral on the l.h.s. of Eq.(1) can be generalized to the case of non–zero photon virtuality Q^2. In terms of the helicity–dependent virtual–photon absorption cross-sections $\sigma^{\overset{\rightarrow}{\Leftarrow}}(\nu, Q^2)$ and $\sigma^{\overset{\rightarrow}{\Rightarrow}}(\nu, Q^2)$ [3, 4] it reads:

$$I_{GDH}(Q^2) = \int_{\nu_0}^{\infty} [\sigma^{\overset{\rightarrow}{\Leftarrow}}(\nu, Q^2) - \sigma^{\overset{\rightarrow}{\Rightarrow}}(\nu, Q^2)] \frac{d\nu}{\nu}. \tag{2}$$

This integral is well defined over the entire Q^2-range and so allows the study of the transition of the helicity-dependent dynamics from the photoabsorption point ($Q^2 = 0$) into the region of Deep Inelastic Scattering (DIS). Above $Q^2 \approx 1$ GeV^2 perturbative Quantum Chromodynamics (pQCD) can be applied taking twist-2 and twist-3 contributions into account. Below this value there exists no field-theoretical description yet.

CP675, *Spin 2002: 15th Int'l. Spin Physics Symposium and Workshop on Polarized Electron Sources and Polarimeters*, edited by Y. I. Makdisi, A. U. Luccio, and W. W. MacKay

For Q^2-values below 0.3 GeV^2 the QCD-inspired chiral quark-soliton model [5] delivers promising descriptions. In between, one has to resort to phenomenological models [6].

The generalized GDH integral can be described in pQCD by the helicity-dependent nucleon structure functions $g_1(x,Q^2)$ and $g_2(x,Q^2)$. For a spin-$\frac{1}{2}$ target it reads:

$$I_{GDH}(Q^2) = \frac{8\pi^2\alpha}{M}\int_0^{x_0} \frac{g_1(x,Q^2) - \gamma^2 g_2(x,Q^2)}{K}\frac{dx}{x}, \tag{3}$$

where $\gamma^2 = Q^2/\nu^2$, $x = Q^2/2M\nu$, and $x_0 = Q^2/2M\nu_0$. The virtual–photon flux factor $K = \nu\sqrt{1+\gamma^2}$ was chosen in the Gilman notation [7]. In the Q^2-region under consideration, the Burkhardt-Cottingham sum rule $\int_0^1 g_2(x,Q^2)dx = 0$ is expected to hold so that $I_{GDH}(Q^2)$ can be entirely described through $\Gamma_1(Q^2)$, the first moment of g_1:

$$I_{GDH}(Q^2) = \frac{16\pi^2\alpha}{Q^2}\Gamma_1(Q^2). \tag{4}$$

At large Q^2, the first moment $\Gamma_1(Q^2) = \int_0^1 g_1(x,Q^2)dx$ exhibits only the logarithmic Q^2-dependence from QCD evolution.

Two regions in the invariant mass W^2 of the virtual-photon nucleon system are to be distinguished: the DIS region above, and the nucleon-resonance region below $W^2 \approx 4$ GeV^2. While $I_{GDH}(Q^2)$ is defined over the entire W^2-range, nucleon structure functions were originally introduced only in the DIS region; their interpretation in the nucleon-resonance region requires additional assumptions, as e.g. quark-hadron duality (see below). The GDH integral may thus be considered to describe the 'transition' of $\Gamma_1(Q^2)$ down to low virtualities *including* the nucleon-resonance region. To this end it allows to determine at which virtuality the hard $1/Q^2$ regime breaks down.

The difference between the generalized GDH integrals for proton and neutron is of special interest. At large virtualities it is directly related to the Bjorken sum rule:

$$I^p_{GDH}(Q^2) - I^n_{GDH}(Q^2) \xrightarrow{Q^2 large} \frac{16\pi^2\alpha}{Q^2}[\Gamma_1^p(Q^2) - \Gamma_1^n(Q^2)]. \tag{5}$$

This fundamental sum rule is derived from current algebra and isospin symmetry, $\Gamma_1^p(Q^2) - \Gamma_1^n(Q^2) = \frac{1}{6}\cdot g_a \cdot C_{ns}(\alpha_s(Q^2))$, where $g_a = |g_A/g_V| = 1.2670 \pm 0.0035$ is the neutron beta–decay coupling constant. Its Q^2-dependence is given by DGLAP evolution. Measurements of the generalized GDH integral on both proton and neutron thus connect two important quantities characterizing the 'static' nucleon, g_A and κ.

The concept of quark-hadron duality, in the interpretation of Bloom and Gilman [8], connects experimental results obtained in two different regimes. It conjectures a structure function (or a ratio) measured in the DIS region to represent an average of the same function (or ratio) measured in the resonance region. In the pQCD re-formulation of this concept [9] the leading terms describe the non-interacting quarks, i.e. scaling, and the non-leading ones the quark-gluon interactions, characterized by a $1/Q^2$-behaviour. Duality is expected to break down at low virtuality as does the whole pQCD picture. Note that duality for *polarized* structure functions is non-trivial because of the different helicity structure of polarized DIS and nucleon resonances.

DISCUSSION OF EXPERIMENTAL RESULTS

The HERMES results discussed in this contribution were derived from measurements with a longitudinally polarized positron beam of 27.57 GeV incident on a gas target internal to the HERA storage ring at DESY. The gas target was filled with longitudinally polarized atomic Hydrogen (Deuterium) in 1997 (1998-2000). The cross-section asymmetry A_1 for the absorption of virtual photons was calculated from the measured cross-section asymmetry $A_{||}$. The latter was obtained from differences of luminosity-weighted numbers of events for parallel and anti-parallel orientation of virtual-photon and nucleus(on) spin, respectively, normalized by their sum. For details on the experiment as well as a complete description of the analysis procedure necessary to calculate GDH integrals see Ref. [10] and references therein. The large kinematic coverage of the HERMES experiment made it possible to measure for the first time the generalized GDH integrals for deuteron, proton and neutron in the nucleon-resonance region and in the DIS region, *simultaneously*.

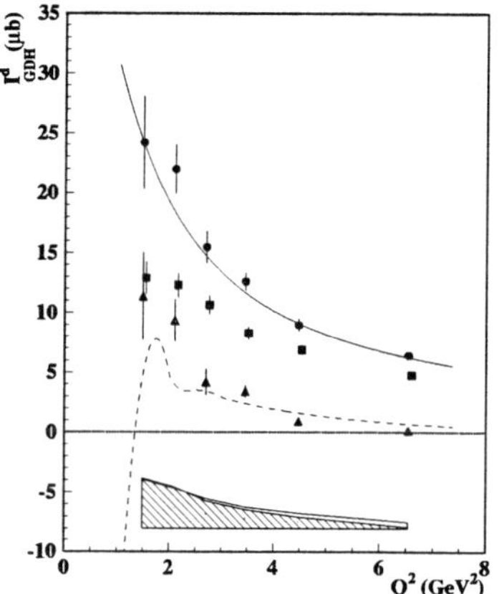

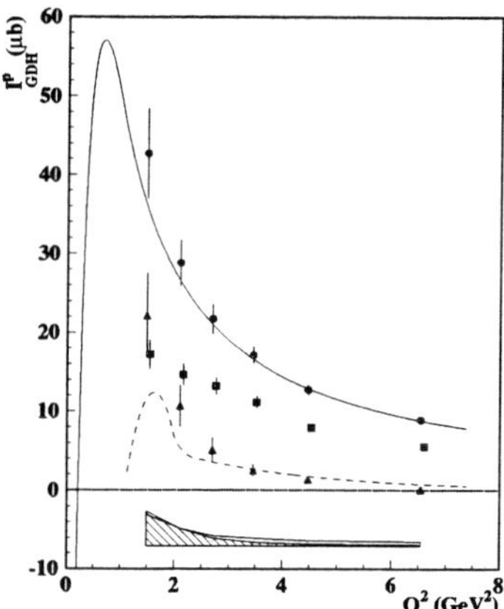

FIGURE 1. Q^2-dependence of the generalized GDH integrals [10] for the deuteron nucleus (left) and the proton (right), shown for the nucleon-resonance region (triangles), DIS region (squares), and full region (circles) including the extrapolation of the unmeasured part above $W^2 = 45$ GeV^2. Solid and dashed curves are predictions for the full integral [6] and the integral in the nucleon-resonance region [11], respectively. The error bars represent the statistical uncertainties. The systematic uncertainties of the full integral are given as a band; the hatched area inside represents the contribution of the nucleon-resonance region alone.

From the HERMES data shown in Fig. 1 it can be seen that the resonance contribution decreases rapidly with increasing Q^2 while the DIS contribution is sizeable over the entire range measured, even down to the lowest measured value of $Q^2 = 1.5$ GeV^2. This holds for both proton and deuteron, as well as for the neutron (not shown here). From Fig. 2 (left panel), which shows HERMES results on the *full* integral for all three targets, the agreement with previous measurements of the first moment of g_1 can be seen. Good agreement exists also with the model of Ref. [6] which is based only on a leading-twist Q^2-evolution of the first moments of g_1 and g_2 without any

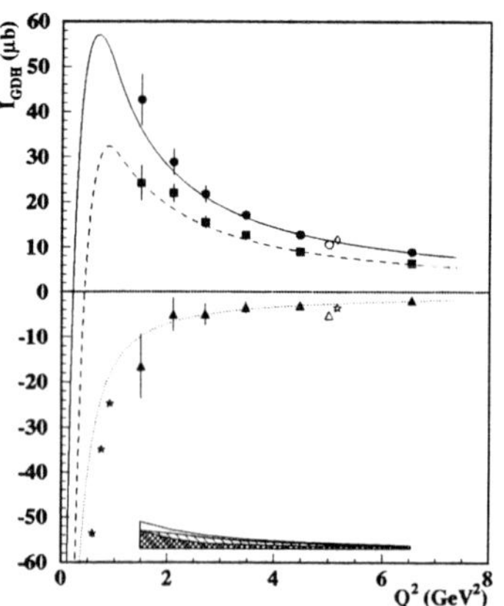

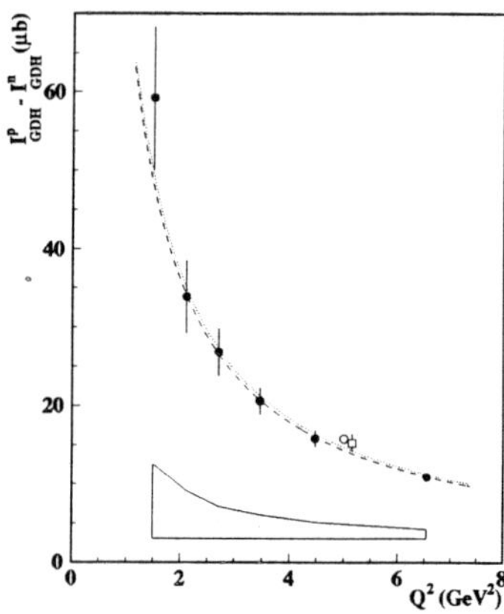

FIGURE 2. Q^2-dependence of the full generalized GDH integral [10]. Left: deuteron (squares), proton (circles), and neutron (triangles) obtained from deuteron and proton. Right: proton-neutron difference. The curves are predictions from Refs. [6] (left) and [12] (right). The error bars represent statistical uncertainties. The bands describe systematic uncertainties; in the left panel open for neutron, lined for deuteron, and cross-hatched for proton. The open symbols at $Q^2 = 5$ GeV^2 show measurements from Refs. [13, 14]. The stars represent the three highest Q^2 bins of the neutron measurement from Ref. [15] including an extrapolation of the unmeasured DIS region.

resonance contribution. The proton-neutron difference (right panel of Fig. 2) is equally well described by the prediction of Ref. [12] and a $1/Q^2$ fit which at $Q^2 = 5$ GeV^2 shows agreement of the HERMES data ($14.3 \pm 0.9_{stat.} \pm 1.3_{syst.}$ μb) with the Bjorken sum rule prediction (16.33 ± 0.45 μb) and with earlier measurements at SLAC, E-143 [13] and E155 [14]. The main result from HERMES with regard to the four measured *full* integrals (proton, deuteron, neutron and proton-neutron difference) is that a clear $1/Q^2$-behaviour is observed, i.e. no effects from higher twists or resonance form factors are seen down to the lowest measured value of $Q^2 = 1.5$ GeV^2. This demonstrates that the hard regime is intact down to a surprisingly low virtuality. Note that at the real-photon point ($Q^2 = 0$) the values for the four full integrals are given by the GDH sum rule to be -204, -0.65, -233 and + 29 μb. Hence at a rather low value of Q^2 a turn-over must occur for proton and deuteron which is not seen in the HERMES kinematics.

Duality in *polarized* lepton-nucleon scattering was studied for the first time at HERMES [16]. In the left panel of Fig. 3 the spin asymmetry A_1, measured in the resonance region, is shown to well describe the world DIS data on average. The average ratio of the measured A_1^{res} to the DIS fit shown is $1.11 \pm 0.16_{stat.} \pm 0.18_{syst.}$. The effect of target-mass corrections was found to be smaller than 5%. In the right panel is shown the Q^2-dependence of the ratio of the first moments of g_1, measured in the nucleon-resonance and in the DIS region, respectively. There is clear evidence that quark-hadron duality holds down to Q^2-values of 1.6 GeV^2.

Acknowledgements. I am indebted to A. Fantoni, H. Jackson, and B. Seitz for valuable comments and discussions.

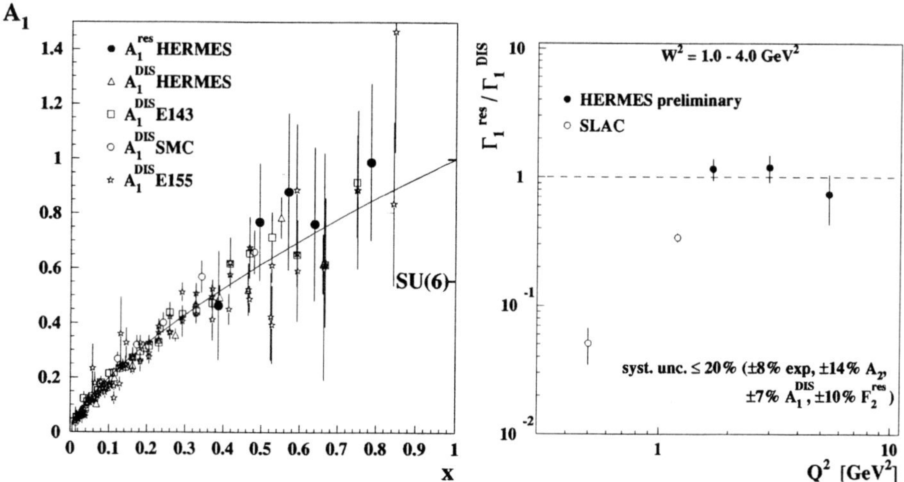

FIGURE 3. Left panel: Spin asymmetry A_1 as a function of x measured at HERMES in the resonance region (full circles) [16]. The error bars represent the statistical uncertainties; the systematic uncertainty for the data in the resonance region is about 16 %. Data obtained in the DIS region (open symbols [17, 13, 18, 14]) are described by a power law fit for $x > 0.3$ (curve). Right panel: Preliminary HERMES data (full circles) on the Q^2-dependence of the ratio of the first moments of g_1, calculated in the nucleon-resonance region ($W^2 = 1 \div 4$ GeV^2) and in the DIS region (using the NMC parameterization [19]), respectively. For comparison SLAC data (open circles) at lower Q^2 [13] are shown, as well.

REFERENCES

1. Gerasimov, S.B., Sov. J. Nucl. Phys. **2** (1966) 430.
2. Drell, S.D. and Hearn, A.C., Phys. Rev. Lett. **16** (1966) 908.
3. Pantförder, R., PhD Thesis, Universität Bonn (1998), BONN-IR-98-06, arXiv:hep-ph/9805434 and references therein.
4. Drechsel,D., Kamalov, S.S. and Tiator, L., Phys. Rev. **D 63** (2001), 114010.
5. Diakonov, D.I., Petrov, V.Yu. and Pobylitsa, P.V., Nucl. Phys. **B 306** (1988) 809.
6. Soffer, J. and Teryaev, O.V., Phys. Rev. **D 51** (1995) 25; Soffer, J. and Teryaev, O.V., Phys. Rev. Lett. **70** (1993) 3373.
7. Gilman, F.J., Phys. Rev **167** (1968) 1365.
8. Bloom, E.D. and Gilman, F.J., Phys. Rev. Lett. **25** (1970) 1140; Phys. Rev. **D 4** (1971) 2901.
9. De Rujula, A., Georgi, H. and Politzer, H.D., Phys. Lett. **B 64** (1976) 428.
10. Airapetian, A. *et al.*, arXiv:hep-ex/0210047, subm. to Eur. Phys. J. C.
11. Aznauryan,I. G., Phys. of At. Nucl. **58** (1995) 1014.
12. Soffer, J. and Teryaev, O.V., Phys.Rev. **D 56** (1997) 7458; Soffer, J. and Teryaev, O.V., arXiv:hep-ph/0207252.
13. E–143 Collaboration, Abe, K., *et al.*, Phys. Rev. **D 58** (1998) 112003.
14. E–155 Collaboration, Anthony, P. L., *et al.*, Phys. Lett. **B 493** (2000) 19; E–155 Collaboration, Anthony, P. L., *et al.*, Phys. Lett. **B 463** (1999) 339.
15. JLab 94010 Collaboration, Amarian, M., *et al.*, arXiv:nucl-ex/0205020.
16. HERMES Collaboration, Airapetian, A. *et al.*, arXiv:hep-ex/0209018, subm. to Phys. Rev. Lett.
17. HERMES Collaboration, Airapetian, A. *et al.*, Phys. Lett. **B 442** (1998) 484.
18. SMC Collaboration, Adeva, B. *et al.*, Phys. Rev. **D 58** (1998) 112001
19. NMC Collaboration, Amaudruz, P., *et al.*, Phys. Lett. **B 364** (1995) 107.

Inclusive and Exclusive Spin Structure Measurements in the Resonance Region

G.E. Dodge for the CLAS Collaboration

Department of Physics, Old Dominion University, Norfolk, VA 23529

Abstract. A program of spin structure function measurements is underway in Jefferson Lab's Hall B. We use polarized electrons incident on polarized NH_3 and ND_3 targets to study proton and deuteron spin observables in and above the resonance region. Results from the first set of data taken in 1998 will be presented for the first moment of g_1 and for the double polarization asymmetry for exclusive pion production, which is sensitive to the spin structure of the nucleon resonances.

MOTIVATION

The spin structure functions g_1 and g_2 of the nucleon have been a topic of great interest for more than two decades. These structure functions have been extensively measured in deep inelastic scattering (DIS) experiments over a wide kinematic range in Bjorken x and momentum transfer Q^2. Particularly interesting is the first moment of g_1, $\Gamma_1 = \int_0^1 g_1(x,Q^2)dx$, which is related to the spin carried by the quarks at large Q^2. Comparisons between these data and theoretical predictions, especially the fundamental Bjorken Sum Rule [1], have yielded reasonable agreement.

At the real photon point, $Q^2 = 0$, the Gerasimov Drell Hearn (GDH) Sum Rule [2] relates the difference of the cross section for photo absorption of a linearly polarized photon antiparallel and parallel to the spin of a nucleon to the anomalous magnetic moment of the nucleon, κ_N. The slope of Γ_1 at low Q^2 is constrained by the value of the GDH Sum Rule, $\Gamma_1(Q^2) = -\frac{Q^2}{8M^2}\kappa_N^2$. Since Γ_1 is negative at low Q^2 and positive in the DIS regime, it must have an interesting behavior as a function of Q^2, including a sign change. An intense experimental effort is now underway at DESY and Jefferson Lab to investigate the low to moderate Q^2 regime, where resonances are expected to play an important role.

EXPERIMENT

At the Thomas Jefferson Laboratory in Newport News, Virginia, an extensive investigation of spin structure functions at low to moderate Q^2 is underway in all three experimental halls. In Hall B, our experimental program encompasses measurements of the double polarization asymmetry for both inclusive and exclusive processes on the proton and the deuteron. We scatter longitudinally polarized electrons from longitudinally polarized NH_3 and ND_3 targets and detect the reaction products in the CEBAF Large

CP675, *Spin 2002: 15th Int'l. Spin Physics Symposium and Workshop on Polarized Electron Sources and Polarimeters*, edited by Y. I. Makdisi, A. U. Luccio, and W. W. MacKay

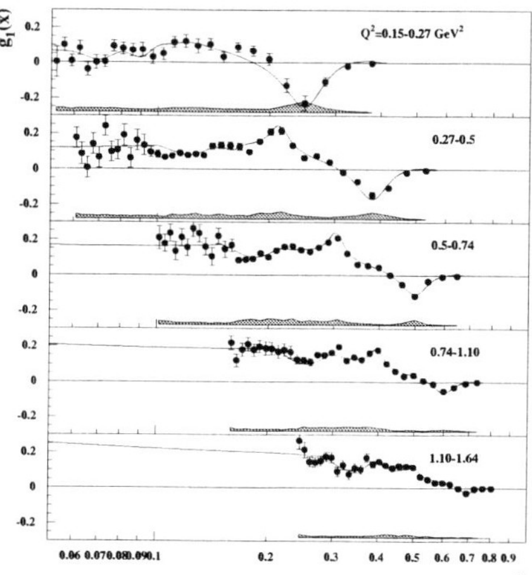

FIGURE 1. g_1 for the proton in five Q^2 bins. The solid line is our parameterization of world data and the shaded band indicates the systematic error.

Acceptance Spectrometer (CLAS). The electron polarization is measured with a Moller polarimeter and the target polarization is monitored with an NMR system. In the fall of 1998 we recorded three billion triggers with beam energies of 2.5 and 4.2 GeV. For seven months in 2000 and 2001 we took an additional 23 billion triggers at beam energies of 1.6, 2.5, 4.2 and 5.7 GeV. Results from the 1998 data are presented here.

INCLUSIVE ANALYSIS AND RESULTS

The raw asymmetry for longitudinally polarized beam and target is given by $A_{raw} = (N^{\uparrow\downarrow} - N^{\uparrow\uparrow})/(N^{\uparrow\downarrow} + N^{\uparrow\uparrow})$ where $N^{\uparrow\downarrow}$ ($N^{\uparrow\uparrow}$) is the yield normalized to the accumulated electron charge for beam and target antiparallel (parallel). The raw asymmetry must be divided by the dilution factor f, which corrects the denominator for target nuclei which are not polarized. After applying radiative corrections, the asymmetry

$$A_1 + \eta A_2 = \frac{1}{DP_eP_tf}A_{raw}, \quad \text{where } D = \frac{1-\varepsilon E'/E}{1+\varepsilon R} \text{ and } \eta = \frac{\varepsilon\sqrt{Q^2}/E}{1-\varepsilon E'/E}, \tag{1}$$

can be extracted. A_1 is the transverse photo absorption asymmetry and A_2 is the ratio of the longitudinal/transverse interference cross section to the total transverse cross section. The product of the electron beam P_e and target P_t polarization is determined by comparing our measured asymmetry in the elastic peak to known values. Finally, we extract g_1

$$g_1(x,Q^2) = \frac{\tau}{1+\tau}[A_1(x,Q^2) + \frac{1}{\sqrt{\tau}}A_2(x,Q^2)]F_1(x,Q^2), \tag{2}$$

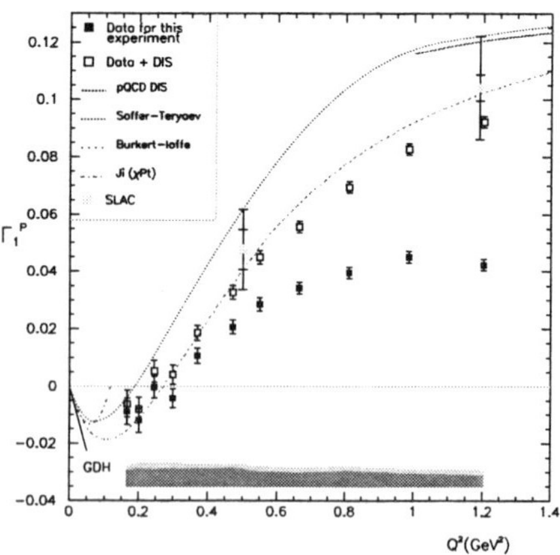

FIGURE 2. Γ_1^p as a function of Q^2 integrated over our measured range in x (solid squares) and including an extrapolation to $x = 0$ (open squares). The systematic error on our measured integral is shown by the dark grey band and the additional contribution from the extrapolation is indicated by the light grey band. Models by Burkert and Ioffe [3] and Soffer and Teryaev [4] are shown as dashed and dotted lines. SLAC data [5] are plotted as the shaded squares. The Chiral Perturbation calculation of Ji [6] is indicated by the dot dashed line.

where $\tau = \nu^2/Q^2$, using a model which includes a parameterization of world data for A_2 and the unpolarized structure function F_1.

Figure 1 shows $g_1(x, Q^2)$ for the proton for several bins in Q^2. The resonant structure is clearly visible, including the negative delta resonance at large x, which decreases as Q^2 increases. The solid line indicates our model, which has been tweaked slightly to accomodate our data. This same model was used to extrapolate to the unmeasured DIS region at low x in order to calculate the integral Γ_1^p shown as the open squares in Figure 2. The solid squares indicate the contribution to the integral from the range in x that we measure, which includes the resonance region up to at least $W = 2$ GeV for all Q^2 bins. The data change sign at $Q^2 \sim 0.25$ GeV^2 and seem to favor the phenomenological model of Burkert and Ioffe [3] which explicitly includes resonance contributions on top of a smooth curve connecting the real photon point to the high Q^2 regime.

From the ND_3 data we have extracted Γ_1^d, shown in Figure 3. The solid triangles indicate the contribution to the integral that we measure and the open triangles include the extrapolation to the DIS regime. In addition to the predictions by Burkert [3] and Soffer [4], the contribution to the integral from resonances alone, as calculated by the AO model [7], is shown as the short dashed line. The sign of Γ_1^d changes at roughly 0.6 GeV^2 and follows the trend of the various models. The statistical precision of the 1998 deuteron data is not as good as for the proton because the target polarization was poor and consequently less data were taken on the deuteron target. Analysis of the much larger 2000-2001 EG1 data set is underway; preliminary results are presented elsewhere in these proceedings [8].

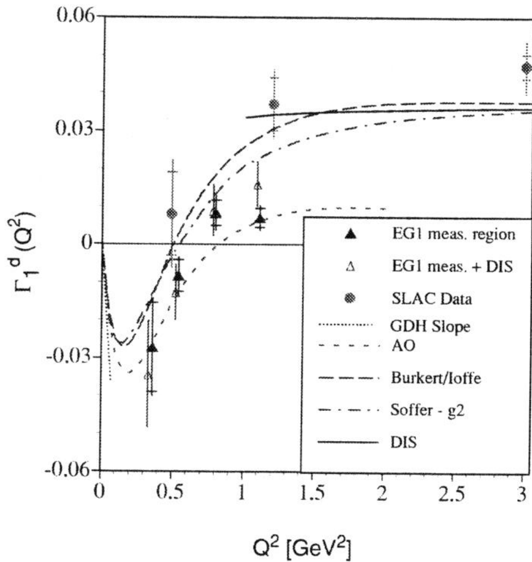

FIGURE 3. Γ_1^d as a function of Q^2 integrated over our measured range in x (solid triangles) and including an extrapolation to $x = 0$ (open triangles). Models by Burkert and Ioffe [3] and Soffer and Teryaev [4] are shown as long-dashed and dash-dotted lines. SLAC data [5] are shown as the filled circles. The resonance contribution shown by the short dashed line is from the AO code [7].

EXCLUSIVE ANALYSIS AND RESULTS

There is also a great deal of interest in the study of polarization observables in exclusive reactions. The cross section σ for an exclusive reaction can be written as

$$\sigma = \sigma_0 + P_e\sigma_e + P_t\sigma_t + P_eP_t\sigma_{et} \tag{3}$$

where σ_0 is the unpolarized cross section and σ_e, σ_t and σ_{et} all require polarized beam and/or target. The nucleon excitations are of fundamental interest and have been fairly well investigated (at least for the proton) in unpolarized measurements. However, of the 18 response functions upon which the cross section depends, only 4 can be measured in unpolarized experiments. With polarized beam and target one can, in principle, gain access to all 18 response functions (which are combinations of 11 independent helicity amplitudes) and study the spin structure of the nucleon resonances.

The large acceptance of the CLAS detector enabled us to measure multi-particle final states. We studied the single pion production reactions $\gamma p \to p\pi^0$ and $\gamma p \to n\pi^+$ with the NH_3 target and $\gamma n \to p\pi^-$ with the ND_3 target. In each case, the exclusive channel was identified using missing mass techniques. After correcting for dilution factor and beam charge asymmetry, we extract $A_{et} = \sigma_{et}/\sigma_0$ and divide by the kinematic factor $\sqrt{1-\varepsilon^2}\cos\theta_\gamma$ to obtain the asymmetry $(A_1 + \eta A_2)/(1+\varepsilon R)$. Figure 4 shows results from the π^+ channel [9] as a function of Q^2 and the polar angle of the emitted pion for various regions in invariant mass W. Predictions of the phenomenological models A0 [7] and MAID2000 [10] are shown as the solid and dashed lines. An AO calculation of the asymmetry due to resonances only is shown as the dotted line. Non-resonant contributions are clearly important at low W. As an example of the sensitivity of this

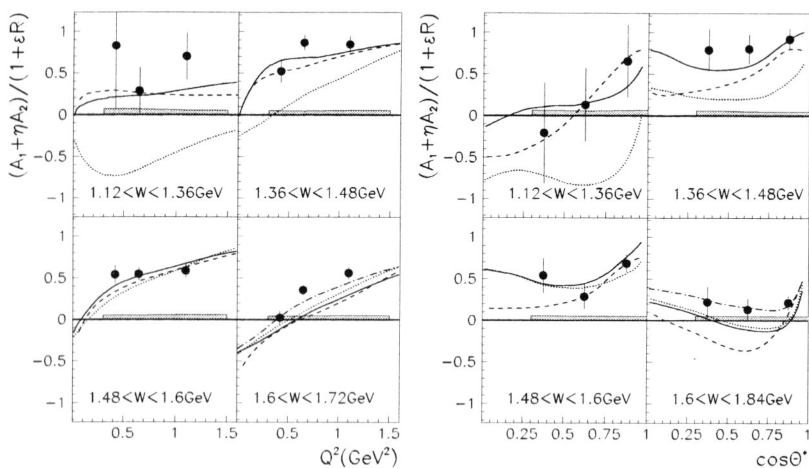

FIGURE 4. The asymmetry $(A_1 + \eta A_2)/(1 + \varepsilon R)$ as a function of Q^2 (left) and $\cos\theta_\pi$ (right) for four different bins in W [9]. See the text for an explanation of the curves.

asymmetry to the details of individual resonance contributions, the asymmetry A_1 for the F_{15} (1680) resonance was increased by 0.4 using AO and plotted as the dot dashed curve in the fourth W panel, resulting in improved agreement with the data. AO also includes a new parametrization for the S_{11}(1535) in which $A_{1/2}$ has been increased, which may explain its better agreement with the data at forward angles. With the vastly improved statistics of the full EG1 data set, we plan to study the resonance region with other exclusive channels (*e.g.*, ρ, ω) in addition to single pion production.

ACKNOWLEDGMENTS

This research is supported by the US Department of Energy under grant DE-FG02-96ER40960.

REFERENCES

1. J.D. Bjorken *et al.*, *Phys. Rev.* **148**, 1467 (1966).
2. S. Gerasimov, *Sov. J. Nucl. Phys.* **2**, 430 (1966); S.D. Drell and A.C. Hearn, *Phys. Rev. Lett.* **16**, 908 (1966).
3. V.D. Burkert and B.L. Ioffe, *Phys. Lett. B* **296**, 223 (1992).
4. J. Soffer and O.V. Teryaev, *Phys. Rev. D* **51**, 25 (1995).
5. K. Abe *et al.*, *Phys. Rev. D* **58**, 112003 (1998).
6. X. Ji and J. Osborne, *J. Phys. G* **27**, 127 (2001).
7. V. Burkert and Z. Li, *Phys. Rev. D* **47**, 46 (1993).
8. K.V. Dharmawardane, these proceedings.
9. R. De Vita *et al.*, *Phys. Rev. Lett.* **88**, 082001 (2002).
10. D. Drechsel *et al.*, *Nucl. Phys.* **A645**, 145 (1999).

Measurement of Spin Structure Functions at Low to Moderate Q^2 using CLAS

K.V. Dharmawardane for the CLAS Collaboration

Department of physics, Old Dominion University, Norfolk, VA 23529.

Abstract. Spin structure functions of the nucleon in the region of large x and small to moderate Q^2 continue to be of high current interest. A large experimental program to measure the spin structure function g_1 and its first moment Γ_1 has been concluded at Jefferson Lab. An overview of the experiment and its kinematic coverage will be discussed. We will also show preliminary results from the 5.7 GeV and the 1.6 GeV data sets.

INTRODUCTION

The inclusive doubly polarized electron-nucleon cross section for longitudinally polarized target and beam can be written as:

$$\frac{d\sigma}{d\Omega dE'} = \Gamma_T[\sigma_T + \varepsilon\sigma_L + P_e P_t(\sqrt{1-\varepsilon^2}A_1\sigma_T cos\psi + \sqrt{2\varepsilon(1+\varepsilon)}A_2\sigma_T sin\psi)], \quad (1)$$

where Γ_T is the transverese flux factor, A_1 and A_2 are the virtual photon asymmetries, ψ is the angle between the target spin and the virtual photon direction, σ_T and σ_L are the total absorption cross sections for transverse and longitudinal virtual photons and ε is the polarization parameter of the virtual photon.

The asymmetry $A_{||}$ for longitudinally polarized beam and target is given by:

$$A_{||} = \frac{\frac{d\sigma^{\uparrow\downarrow}}{d\Omega dE'} - \frac{d\sigma^{\uparrow\uparrow}}{d\Omega dE'}}{\frac{d\sigma^{\uparrow\downarrow}}{d\Omega dE'} + \frac{d\sigma^{\uparrow\uparrow}}{d\Omega dE'}} = D(A_1 + \eta A_2), \quad (2)$$

where $\frac{d\sigma^{\uparrow\downarrow}}{d\Omega dE'}(\frac{d\sigma^{\uparrow\uparrow}}{d\Omega dE'})$ is the differential cross section for the target spin antiparallel (parallel) to the beam helicity and D and η are:

$$D = \frac{1-\varepsilon E'/E}{1+\varepsilon R}, \qquad \eta = \frac{\varepsilon\sqrt{Q^2}/E}{1-\varepsilon E'/E}. \quad (3)$$

The spin structure function g_1 is related to the virtual photon asymmetries A_1 and A_2 by:

$$g_1(x,Q^2) = \frac{\tau}{1+\tau}[A_1 + \frac{1}{\sqrt{\tau}}A_2]F_1(x,Q^2), \quad (4)$$

where F_1 is the well known unpolarized structure function and $\tau = \nu^2/Q^2$.

CP675, *Spin 2002: 15th Int'l. Spin Physics Symposium and Workshop on Polarized Electron Sources and Polarimeters*, edited by Y. I. Makdisi, A. U. Luccio, and W. W. MacKay

The first moment of the spin structure function g_1, $\Gamma_1(Q^2) = \int g_1(x,Q^2)dx$, is a quantity of great interest. At $Q^2 \longrightarrow \infty$ the proton neutron difference of Γ_1 is given by the celebrated Bjorken [1] sum rule, $\Gamma_1^p(Q^2) - \Gamma_1^n(Q^2) = \frac{1}{6}g_A$, where g_A is the weak axial coupling constant. The Bjorken [1] sum rule is based on quark current algebra and has been verified at the level of 5% accuracy. It can be evolved to finite values of Q^2 using pQCD and the Operater-Product-Expansion (OPE).

At the real photon point the Gerasimov-Drell-Hearn (GDH) [2, 3] sum rule predicts:

$$I_{GDH} = \frac{M^2}{8\alpha\pi^2}\int_{\nu_{thr}}^{\infty} \frac{\sigma_{1/2} - \sigma_{3/2}}{\nu} d\nu = -\frac{1}{4}\kappa^2, \tag{5}$$

where κ is the target anomalous magnetic moment. As $Q^2 \longrightarrow 0$ the GDH [2, 3] sum rule implies a negative slope for Γ_1, $\Gamma_1(Q^2) \longrightarrow \frac{Q^2}{2M}I_{GDH}$, which goes through a rapid transition to the deep inelastic limit where it is sensitive to the nucleon spin fraction carried by quarks. The interesting behavior in the transition region is dominated by baryon resonance excitations [4]. Recently Ji and Osborn [5] have shown that chiral perturbation theory can be used to calculate the GDH integral up to $Q^2 \sim 0.1$ GeV^2. Further they point out the possibility of using lattice QCD to calculate the integral in the region $Q^2 \sim 0.1 - 0.5$ GeV^2. This is the domain where Jefferson Lab can play an important role.

The focus of the EG1 experiment, which was carried out in Hall B at Jefferson Laboratory, is on measuring Γ_1 for the proton and deuteron (neutron) in the transition region. These measurements complement the data at the photoabsorption point and in the deep inelastic scattering region and cover a Q^2 range of $0.05 < Q^2 < 4.5$ GeV^2. This opens up the possibility of studying the transition from hadronic to quark degrees of freedom over a wide range of Q^2.

DATA ANALYSIS

A highly polarized electron beam, dynamically polarized $^{15}NH_3$ and $^{15}ND_3$ targets and the CEBAF Large Acceptance Spectrometer (CLAS) in Hall B were used to accumulate over 23 billion triggers with beam energies of 1.6, 2.5, 4.2 and 5.7 GeV. These data are in addition to the three billion triggers recorded in 1998, the results of which are presented elsewhere [6]. Calibration and analysis of the data are still under way. Here we present the asymmetry analysis for the 1.6 and 5.7 GeV data.

In the asymmetry given in equation 2 many factors such as luminosity and acceptance will cancel out. Therefore the experimental raw counting asymmetry A_{raw} can be converted to the longitudinal electron asymmetry $A_{\|}$ by :

$$A_{\|} = \frac{A_{raw}}{P_b P_t f}; \qquad A_{raw} = \frac{N^{\uparrow\downarrow} - N^{\uparrow\uparrow}}{N^{\uparrow\downarrow} + N^{\uparrow\uparrow}}, \tag{6}$$

where $N^{\uparrow\downarrow}(N^{\uparrow\uparrow})$ are the raw counting rates normalized to accumulated beam charge for beam helicity antiparallel (parallel) to the target spin, f is the target dilution factor and P_b and P_t are the beam and target polarizations.

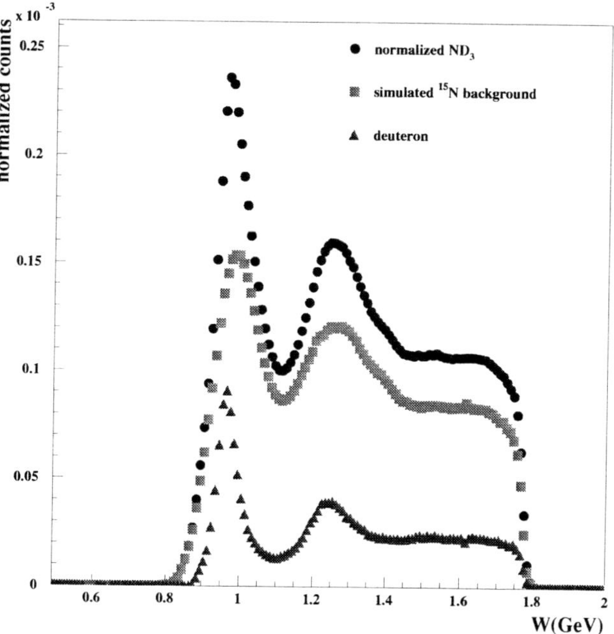

FIGURE 1. The figure above compares the $^{15}ND_3$, simulated ^{15}N background and deuteron invariant mass spectra for 1.6 GeV data. The ^{15}N background has been simulated by the method described in the text.

To remove the contribution from the ^{15}N in ammonia as well as the helium and foils in the target we took extensive data on ^{12}C and ^4He at several different beam energies. In addition we were able to take data on a solid ^{15}N target at some beam energies. The ^{12}C data were used to simulate the ^{15}N spectrum parameterizing the ^{15}N cross section as a function of the ^{12}C cross section:

$$\sigma_{15_N} = (a + b\frac{\sigma_n}{\sigma_D})\sigma_{12_C}. \tag{7}$$

Here σ_n and σ_D are the neutron and deuteron cross sections. The parameters a and b were determined by fitting the limited statistics ^{15}N data with high statistics ^{12}C data. Then a full background spectrum was simulated using the known thicknesses of foils and the ^4He data. The resulting background spectrum was subtracted from the NH_3 and ND_3 spectra (Figure 1) to obtain the target dilution factor.

The product of beam and target polarization was extracted by dividing the measured elastic asymmetry by the product of dilution factor and the theoretical value of the elastic asymmetry. This greatly reduces the systematic uncertainty of the beam and target polarization. The product of beam and target polarization for the 1.6 GeV deuteron data is 0.19 ± 0.004 which gives a target polarization of about 27%. For the 5.7 GeV deuteron

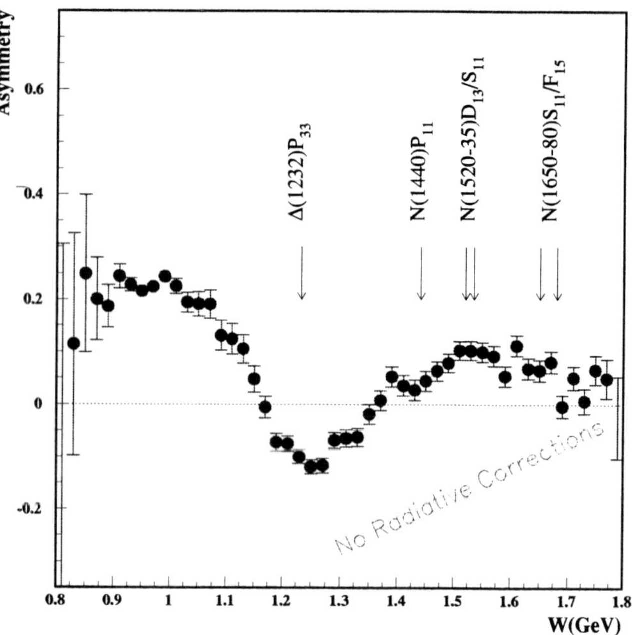

FIGURE 2. Preliminary asymmetry in $\vec{d}(\vec{e},e')$ at E = 1.6 GeV for the Q^2 range 0.187-0.707 GeV^2 as a function of invariant mass.

data elastic electron-proton coincidences were used to determine the product of beam and target polarization to be 0.209 ± 0.026.

The measured asymmetry for the deuteron at a beam energy of 1.6 GeV is shown in Figure 2. The asymmetry is negative in the $\Delta(1232)$ region as expected and is positive for higher resonances. The asymmetry for the deuteron and proton at a beam energy of 5.7 GeV is shown in Figure 3. The asymmetry is nearly zero in the $\Delta(1232)$ region because of the high Q^2, but becomes positive and large for higher resonances.

SUMMARY AND OUTLOOK

The EG1 experiment is dedicated to study the spin structure of the proton, neutron and their excited states. The analysis of the data is still in progress but preliminary asymmetries at beam energies of 1.6 and 5.7 GeV are shown in Figures 2 and 3. Radiative corrections must still be applied and then g_1 and Γ_1 will be extracted in the near future. This enormous data set is expected to provide detailed information on the spin structure of the proton and deuteron (neutron) over a large kinematic range in and above the resonance region.

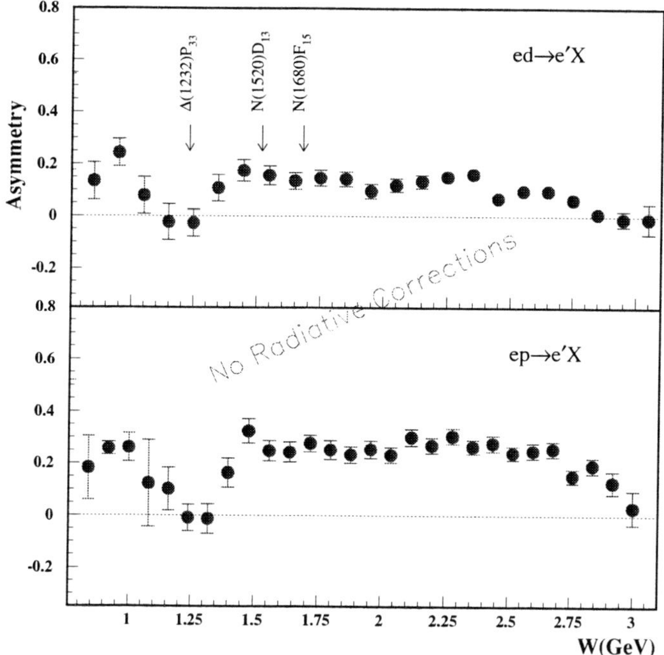

FIGURE 3. Preliminary asymmetry in $\vec{d}(\vec{e},e')$ (top) and $\vec{p}(\vec{e},e')$ (bottom) at E = 5.7 GeV for the Q^2 range 1.31-4.16 GeV^2 as a function of invariant mass. The plot represent only 50% of the 5.7 GeV data set.

ACKNOWLEDGEMENTS

This research is supported by the US Department of Energy under grant DE-FG02-96ER40960

REFERENCES

1. J.D. Bjorken, *Phys. Rev.* **179**, 1547 (1969).
2. S.B. Gerasimov, *Sov. J. Nucl. Phys.* **2**, 430 (1966).
3. S.D. Drell and A.C. Hearn, *Phys. Rev. Lett.* **16**, 908 (1966).
4. V. Burkert and Z. Li, *Phys. Rev.* **D47**, 46 (1993).
5. X. Ji and J. Osborne, *J. of Phys.* **G57**, 127 (2001).
6. G.E. Dodge, *these proceedings*.

Proton and Deuteron Spin Structure Function Measurements in the Resonance Region

Frank R. Wesselmann

University of Virginia, Charlottesville, Virginia 22904
For the RSS Collaboration

Abstract. The RSS collaboration has measured the spin structure functions of the proton and the deuteron at Jefferson Lab using the Hall C HMS spectrometer, a polarized electron beam and a polarized solid target. The asymmetries $A_{\parallel}$ and $A_{\perp}$ were measured in the region of the nucleon resonances ($0.82\,GeV < W < 1.98\,GeV$) at an average four momentum transfer of $Q^2 = 1.3\,GeV^2$. The extracted spin structure functions and their kinematic dependence will make a significant contribution in the study of higher-twist effects and polarized duality tests. A description of the experiment and the latest findings of the analysis will be presented.

INTRODUCTION

Spin structure functions of the nucleon have been measured since the late 1970's, for example at SLAC [1], CERN [2], and DESY [3]. Most of the experimental studies have concentrated on the high Q^2 region, where perturbative QCD works well with electrons scattering from an essentially free constituent. The low and intermediate energy range (up to 6 GeV) polarized electron beam at the Thomas Jefferson National Accelerator Facility (TJNAF) gives us a unique opportunity to study the nucleon spin structure in the lower energy regime where quarks are not asymptotically free and pQCD does not apply. Measuring the spin structure in this Q^2 region will allow us to study the contribution of individual nucleon resonances to the spin structure function.

One of the measurable quantities in polarized electron–nucleon scattering is the cross section asymmetry, simply a normalized difference between the scattering cross sections with the electron and nucleon spins aligned parallel and anti-parallel (or $\pm 90°$ in the transverse case). Experiment E01-006 at Jefferson Laboratory has measured the asymmetries $A_{\parallel}$ and $A_{\perp}$ at $Q^2 \approx 1.3\,GeV^2$ from proton and deuteron targets. We define these asymmetries as:

$$A_{\parallel} = \frac{d\sigma^{\downarrow\uparrow} - d\sigma^{\uparrow\uparrow}}{d\sigma^{\downarrow\uparrow} + d\sigma^{\uparrow\uparrow}} \tag{1}$$

and

$$A_{\perp} = \frac{d\sigma^{\downarrow\rightarrow} - d\sigma^{\uparrow\rightarrow}}{d\sigma^{\downarrow\rightarrow} + d\sigma^{\uparrow\rightarrow}} \tag{2}$$

This high precision measurement of both $A_{\parallel}$ and $A_{\perp}$ will allow us to extract the virtual photon asymmetries $A_1(\nu, Q^2)$ and $A_2(\nu, Q^2)$ without using any assumption or models. A_1 and A_2 provide a description of the quark contribution to the nucleon spin.

CP675, *Spin 2002: 15th Int'l. Spin Physics Symposium and Workshop on Polarized Electron Sources and Polarimeters,* edited by Y. I. Makdisi, A. U. Luccio, and W. W. MacKay

The nucleon spin structure in the resonance region as well as the connection to the DIS measurements can be studied. Also, the effects of quark–gluon interactions, twist-3 terms, and local polarized duality can be explored from this measurement.

EXPERIMENT

The experiment was conducted at Jefferson Lab in Hall C during two months in early 2002 using a continuous, polarized electron beam with energy of 5.7 GeV. The beam polarization was measured by a Moller polarimeter installed upstream of the target; the average beam polarization was about 70%. The beam helicity was flipped at 30 Hz on a pseudo-random basis. To minimize any false asymmetry or bias, we determined the beam charge asymmetry over five minutes intervals and the measured value was fed back to the helicity control device at the injector.

Frozen $^{15}NH_3$ and $^{15}ND_3$ were used as material for proton and deuteron targets, respectively. The target polarization was achieved via a Dynamic Nuclear Polarization technique and measured by an NMR system using pickup coils embedded in the target material. The average target polarization was around 80% for $^{15}NH_3$, and for $^{15}ND_3$ around 20%. For the $A_\perp$ measurement, the whole target was rotated by 90° from its parallel position.

Scattered electrons were detected using the High Momentum Spectrometer (HMS), positioned at a scattering angle of 13.15°. Two different HMS momentum setting were used to cover wider kinematic range, as shown in figure 1, resulting in an average Q^2 of ≈ 1.3 GeV^2. A detector package consisting of hodoscope planes, wire chambers, a Cerenkov counter and a lead glass calorimeter allowed for particle identification and measurement of the event kinematics.

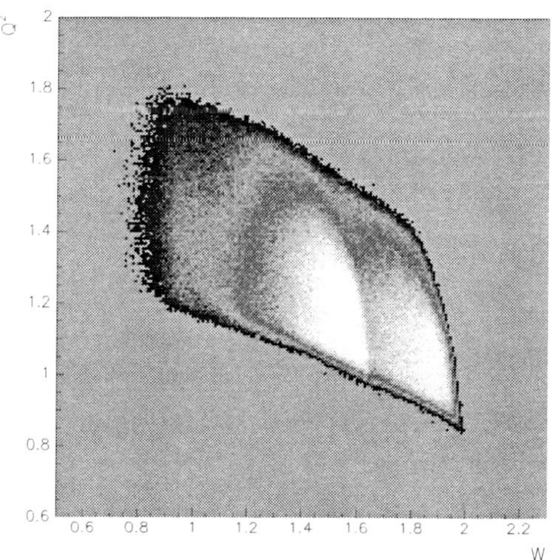

FIGURE 1. Statistical and Kinematic Coverage of RSS Experiment.

RESULTS

Approximately 160 million scattering events were recorded on the proton target, and 350 million on deuteron. Data analysis is currently underway, including offline calibration of the spectrometer optics, tuning of tracking and particle identification and extracting beam and target polarization. The results from the first pass analysis of the count asymmetries, shown in figure 2, are very promising. These preliminary asymmetries have been extracted with online values for the beam and target polarizations, which will later be replaced with offline numbers. The target dilution factor and radiative corrections were not applied to the plotted data.

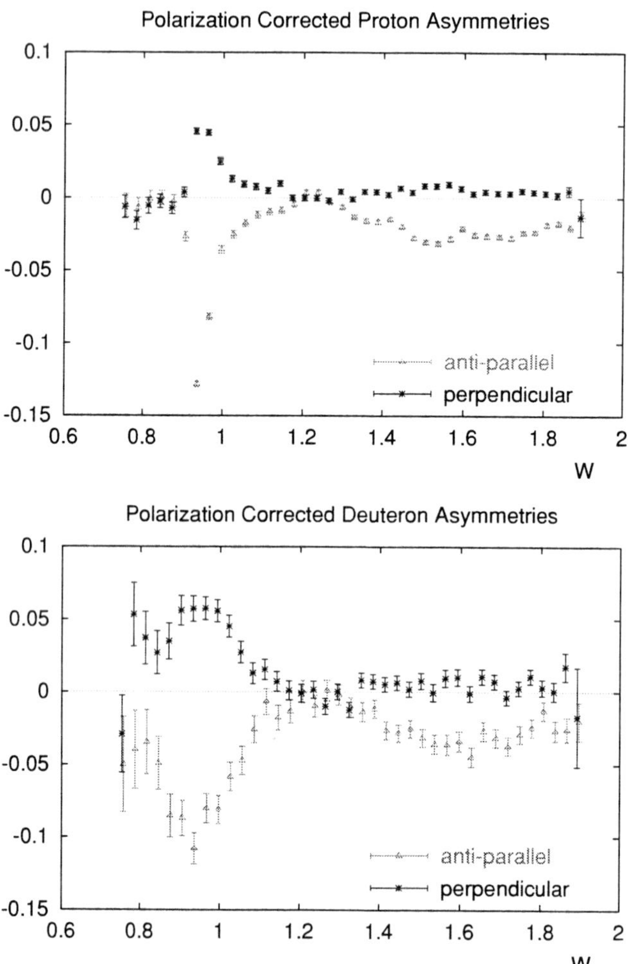

FIGURE 2. Preliminary result for NH_3 and ND_3 asymmetries, with beam–target polarization parallel and perpendicular. The experiment's anti-parallel alignment of target polarization and beam direction results in the negative sign for $A_{\parallel}$.

OUTLOOK

One of main physics goals of the program is to study the W dependence of the transverse and longitudinal polarized structure functions of the proton and deuteron in the resonance region. This measurement will provide the first precision measurement of the transverse asymmetry for the proton and deuteron. It will give us an opportunity to study the effects of twist-3 contributions to the structure functions. The kinematic region was chosen so that the data would connect closely with polarized DIS experiments, permitting direct comparison of extrapolated DIS data with our data and thus the study local duality for the polarized structure functions. Also, these data can be used to test the extended Gerasimov-Drell-Hearn [4] sum rule with a minimum of interpolations or use of fits to the world data on the structure function g_1. Along with other programs at Jefferson Lab [5] [6] [7], this measurement will contribute significantly to the world data on the spin structure functions.

ACKNOWLEDGMENTS

This work was supported by Department of Energy contract DE-FG02-96ER40950, and by the Institute of Nuclear and Particle Physics of the University of Virginia. The Southern Universities Research Association (SURA) operates the Thomas Jefferson National Accelerator Facility for the United States Department of Energy under contract DE-AC05-84ER40140.

REFERENCES

1. Abe, K., et al., *Phys. Rev. Lett.*, **78**, 815–819 (1997).
2. Adams, D., et al., *Phys. Lett.*, **B396**, 338–348 (1997).
3. Ackerstaff, K., et al., *Phys. Lett.*, **B404**, 383–389 (1997).
4. Burkert, V. D., and Ioffe, B. L., *Phys. Lett.*, **B296**, 223–226 (1992).
5. Cates, G., and Meziani, Z.-E., Jefferson lab experiment 94-010.
6. Burkert, V., Crabb, D., and Minehart, R., Jefferson lab experiment 91-023.
7. Kuhn, S. E., Jefferson lab experiment 93-009.

Precision Measurement of Neutron Asymmetry A_1^n in the Valence Quark Region

Xiaochao Zheng* for The Jefferson Lab Hall A Collaboration

*Massachusetts Institute of Technology, Cambridge, MA 02139
Address: Jefferson Laboratory, 12000 Jefferson Avenue, MS 16B, Newport News, VA 23606
Email: xiaochao@jlab.org

Abstract. We have measured the neutron virtual photon asymmetry A_1^n over the kinematic range $0.33 \leq x \leq 0.61$ and $2.7 \leq Q^2 \leq 4.9$ (GeV/c)^2. To extract A_1^n, longitudinal and transverse spin asymmetries have been measured for inclusive $^3\vec{\mathrm{He}}(\vec{e},e')$ scattering, using a 5.7 GeV longitudinally polarized electron beam at Jefferson Lab and a high-density polarized ^{3}He target in Hall A. Preliminary results of A_1^n are presented and compared to existing data and various models, including the predictions of SU(6), broken SU(6) constituent quark models, perturbative QCD based models and chiral soliton model.

A_1^N AT LARGE X

In the scattering of polarized electrons on a polarized target, the virtual photon asymmetry A_1 is defined as

$$A_1 = \frac{\sigma_{1/2} - \sigma_{3/2}}{\sigma_{1/2} + \sigma_{3/2}} \tag{1}$$

where $\sigma_{1/2(3/2)}$ is the total virtual photoabsorption cross section for the nucleon with a projection of $1/2$ $(3/2)$ for the total spin along the direction of photon momentum. A_1 can be expressed as a ratio of the polarized structure functions g_1, g_2 and the unpolarized structure function F_1 as

$$A_1 = \frac{g_1(x,Q^2) - \gamma^2 g_2(x,Q^2)}{F_1(x,Q^2)} \tag{2}$$

with $\gamma^2 = 4M^2x^2/Q^2$ a kinematic factor, M the nucleon mass, $x = x_{Bj}$ the Bjorken variable, and Q^2 the four momentum transfer squared.

To first approximation, the constituent quarks in the nucleon are described by SU(6) wavefunctions as

$$\begin{aligned} |n\uparrow\rangle = & \frac{1}{2}\left|d\uparrow(ud)_{S=0}\right\rangle + \frac{1}{\sqrt{18}}\left|d\uparrow(ud)_{S=1}\right\rangle - \frac{1}{3}\left|d\downarrow(ud)_{S=1}\right\rangle \\ & -\frac{1}{3}\left|u\uparrow(dd)_{S=1}\right\rangle - \frac{\sqrt{2}}{3}\left|u\downarrow(dd)_{S=1}\right\rangle \end{aligned} \tag{3}$$

CP675, *Spin 2002: 15th Int'l. Spin Physics Symposium and Workshop on Polarized Electron Sources and Polarimeters*, edited by Y. I. Makdisi, A. U. Luccio, and W. W. MacKay

where the subscript S denotes the total spin of the diquark state. In this limit where SU(6) is an exact symmetry, both diquark spin states $S = 1$ and $S = 0$ contribute equally to the observables of interest, leading to the predictions $A_1^p = \frac{5}{9}$, $A_1^n = 0$.

However, the SU(6) symmetry is known to be broken. A natural SU(6) symmetry breaking mechanism based on phenomenological arguments is the hyperfine interaction among the quarks, described as $\vec{S}_i \cdot \vec{S}_j\ \delta^3(\vec{r}_{ij})$, where $\vec{S}_i$ is the spin of ith quark. The effect of this perturbation on the wavefunction is to lower the energy of the $S = 0$ 'diquark'. This allows the d-quark in the first term of Eq.(3), which has its spin parallel to that of the neutron, to dominate the high energy tail of the quark momentum distribution that is probed as $x \to 1$. The dominance of this term as $x \to 1$ leads to $A_1^p \to 1$, $A_1^n \to 1$. Hyperfine interaction has been incorporated in the constituent quark model (CQM) in which A_1^p and A_1^n have been calculated in the large x region [1].

Another approach focuses directly on relativistic quarks. Farrar and Jackson [2] in the early 1970's, noted that at $x \to 1$, the scattering is from a high energy quark, and the process can be treated perturbatively. They proceeded to show that a quark carrying nearly all the momentum of the nucleon (i.e. $x \to 1$) must have the same helicity as the nucleon. This is known as hadron helicity conservation. Quark-gluon interactions cause only the $S = 1$, $S_z = 1$ diquark spin projection component, rather than the full S=1 diquark system to be suppressed as $x \to 1$. This gives $d^{\downarrow} = u^{\downarrow} = 0$, $\frac{d^{\uparrow}}{u^{\uparrow}} \to \frac{1}{5}$ as $x \to 1$. Consequently, they obtained the previous limiting value for both the proton and the neutron, namely $A_1^{n,p} \to 1$ for $x \to 1$. This is one of few places where QCD can make an absolute prediction for the x dependence of the structure functions (here a ratio of structure functions). How low in x and Q^2 this picture will work is uncertain. Using this perturbative QCD (pQCD) prediction, A_1^n can be calculated from polarized and unpolarized parton distributions, for example, LSS(BBS) parameterization [3]. A_1^n have also been calculated using LSS parameterization without pQCD constraint [4].

In addition to SU(6), constituent quark models and pQCD based models, there are a few other models which can give a prediction for A_1^n at large x, including statistical model [5] and local duality method [6]. In contrast to other theories, the chiral soliton model [7] and the instanton model [8] predict the possibility that A_1^n can be negative at large x. All world data for A_1^n at $x > 0.4$ have poor statistics and even cannot determine the sign of A_1^n. Therefore, high precision data on A_1^n at large x are greatly needed.

THE EXPERIMENT E99-117

The experiment E99-117 [9] was carried out in Hall A at JLAB in the summer of 2001 to measure A_1^n in the x region $0.33 < x < 0.61$. The kinematics are shown in Table 1. To

TABLE 1. E99-117 kinematics

x_{Bj}	0.331	0.474	0.609
Q^2 (GeV/c)^2	2.738	3.567	4.887
W^2 (GeV/c)^2	6.426	4.846	4.023

measure A_1^n, the asymmetries $A_{\|}$ and $A_{\perp}$ of polarized e^- scattering off a polarized ^{3}He target have been measured in the deep inelastic region. They are defined as

$$A_{\parallel} = \frac{\sigma^{\downarrow\Uparrow} - \sigma^{\uparrow\Uparrow}}{\sigma^{\downarrow\Uparrow} + \sigma^{\uparrow\Uparrow}}, \quad A_{\perp} = \frac{\sigma^{\downarrow\Rightarrow} - \sigma^{\uparrow\Rightarrow}}{\sigma^{\downarrow\Rightarrow} + \sigma^{\uparrow\Rightarrow}} \tag{4}$$

with $\sigma^{\downarrow\Uparrow}$, $\sigma^{\uparrow\Uparrow}$, $\sigma^{\downarrow\Rightarrow}$ and $\sigma^{\uparrow\Rightarrow}$ the electron scattering cross sections with electron spin anti-parallel, parallel, anti-perpendicular and perpendicular to target spin, respectively. A_1 can be extracted from $A_{\parallel}$ and $A_{\perp}$ as

$$A_1 = \frac{A_{\parallel}}{D(1+\eta\xi)} - \frac{\eta A_{\perp}}{d(1+\eta\xi)} \tag{5}$$

where D, d, η, ξ depend on kinematics and the ratio $R \equiv \sigma_L/\sigma_T$.

The experiment used the JLAB longitudinally polarized electron beam at its highest available energy 5.7 GeV and with a 81% polarization. The polarized ^{3}He target in Hall A was a 25 cm long gas target operated at a density of above 12 atmosphere at 0°C. During the experiment the average in-beam polarization with an average beam current of 12 μA was 40%. The scattered electrons were detected by the two standard Hall A High Resolution Spectrometers (HRS) [10] at symmetric positions. Particle identification is achieved by using a CO_2 gas Cherenkov detector and a double-layered lead-glass shower counter. The combined pion rejection factor is found to be better than 10^4 for both HRSs, with a 99% identification efficiency for electrons. This is sufficient regarding pion rates in this experiment.

The raw and physics asymmetries are extracted from data as

$$A_{raw} = \frac{N^+/Q^+ - N^-/Q^-}{N^+/Q^+ + N^-/Q^-}, \quad A_{\parallel,\perp} = \frac{A_{raw}}{fP_bP_t} \tag{6}$$

with $N^{\pm}$, $Q^{\pm}$ the yield and accumulated beam charge for each beam helicity state, f the target dilution factor, typically 0.92~0.94, P_b and P_t the beam and target polarizations. False asymmetries have been checked by measuring asymmetries of polarized e^- beam scattering off unpolarized ^{12}C target. They are found to be negligible compared to the physics asymmetry being measured. $A_1^{^3\mathrm{He}}$ is calculated from physics asymmetries using Eq. (5). Radiative corrections have been made to the ^{3}He asymmetries $A_{\parallel}^{^3\mathrm{He}}$ and $A_{\perp}^{^3\mathrm{He}}$ directly. A ^{3}He model [11] which includes S, S', D states and pre-existing $\Delta(1232)$ components in the ^{3}He ground-state wavefunction has been used to extract A_1^n from $A_1^{^3\mathrm{He}}$.

PRELIMINARY RESULTS

Preliminary results of A_1^n are shown in Fig. 1, along with world data from HERMES and SLAC. The error bars in Fig. 1 only include statistical errors. However, a detailed study has been done for systematic errors showing that the total error is dominated by statistics.

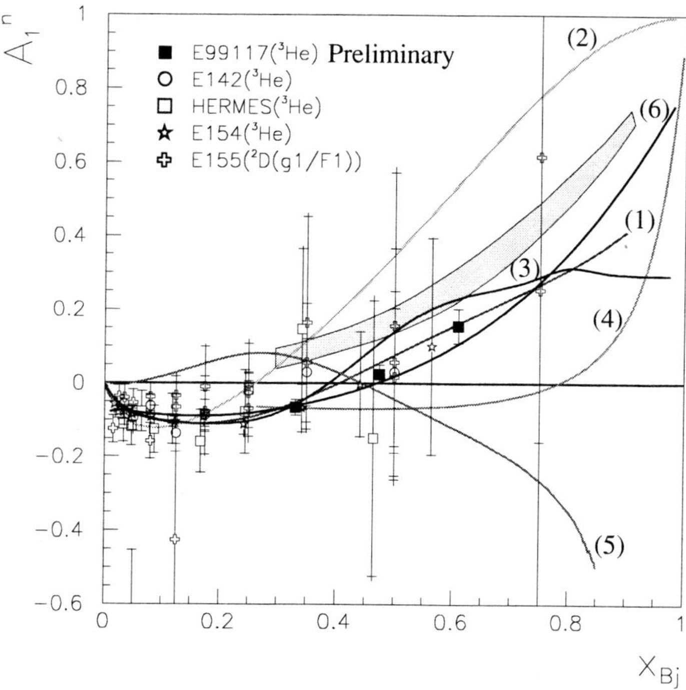

FIGURE 1. Preliminary results of A_1^n compared with world data and theoretical predictions. Curves: predictions of $\frac{g_1}{F_1}$ from pQCD based model using LSS 2001 parametrization at $Q^2 = 5$ (GeV/c)^2 (1), and A_1^n from constituent quark model (light shaded band), pQCD based model using BBS parameterization (2), statistical model at $Q^2 = 4$ (GeV/c)^2 (3), local duality (4), chiral soliton model at $Q^2 = 3$ (GeV/c)^2 (5), and E155 fit at $Q^2 = 4$ (GeV/c)^2 (6).

The $x = 0.33$ datum is in good agreement with existing world data. In the region of $x > 0.4$, the statistical errors of A_1^n have been improved by about one order of magnitude. Also, the data show a clear trend that A_1^n turns to positive values at large x. Compared with theory curves, it is intriguing to note that the constituent quark model gives the correct sign and trend of A_1^n at large x. Two other models which use the world data as input - the LSS parameterization and Soffer's statistical model can be refined if the data from this experiment are included in the inputs; Chiral soliton model calculations are not in agreement with world data at the moment.

Besides A_1^n and g_1^n, the asymmetry A_2^n and structure function g_2^n can also be extracted from our data. The statistical uncertainties of A_2^n and g_2^n are comparable to the latest data from SLAC E155x [12].

Assuming valence quark dominance, one can extract the polarized quark distributions $\Delta u/u$ and $\Delta d/d$ from A_1^n data, d/u ratio [13] and a world fit [14] of A_1^p data. Results show that $\Delta d/d$ is negative at all three x points which agrees with CQM prediction but contradicts the prediction from pQCD based hadron helicity conservation.

SUMMARY AND OUTLOOK

Experiment E99-117 provided precise data on the neutron spin asymmetry A_1^n. Data on the structure functions $g_1^n(x,Q^2)$, $g_2^n(x,Q^2)$and asymmetry A_2^n are also available. The results of this experiment will provide valuable constraints to theoretical calculations. A_1^n in the valence quark region is of great interest in the understanding of the valence quark structure and the constituent quark concept. The measurement of A_1^n in the valence quark region is also an important part of the JLab 12 GeV upgrade [15], in which A_1^n will be measured up to $x = 0.8$ and within a larger Q^2 range of $2 < Q^2 < 10$ (GeV/c)^2.

ACKNOWLEDGMENTS

The work presented was supported in part by funds provided to the Laboratory for Nuclear Science at the Massachusetts Institute of Technology by the U. S. Department of Energy(DOE) under contract number DE-FC02-94ER40818. The Southeastern Universities Research Association operates the Thomas Jefferson National Accelerator Facility for the DOE under contract DE-AC05-84ER40150.

REFERENCES

1. N. Isgur, *Phys. Rev.* **D59**, 034013 (1999); e-Print: hep-ph/9809255
2. G. R. Farrar, D. R. Jackson, *Phys. Rev. Lett.* **35**, 1416(1975).
3. E. Leader, A. V. Sidorov, D. B. Stamenov, *Int. J. Mod. Phys.* **A13**, 5573 (1998); e-Print: hep-ph/9708335
4. E. Leader, A. V. Sidorov, D. B. Stamenov, *Eur. Phys. J.* **C23**, 479 (2002); e-Print: hep-ph/0111267
5. C. Bourrely, J. Soffer, F. Buccella, *Eur. Phys. J.* **C23**, 487 (2002); e-Print: hep-ph/0109160v1.
6. W. Melnitchouk, *Phys. Rev. Lett.* **86**, 35 (2001); e-Print: hep-ph/0106073
7. H. Weigel, L. Gamberg and H. Reinhardt, *Phys. Lett.* **B399**, 287 (1997); *Phys. Rev.* **D55**, 6910 (1997).
8. N. I. Kochelev, talk presented at the Workshop "Deep Inelastic Scattering off Polarized Targets: Theory Meets Experiment", Sept. 1997, DESY-Zeuthen; e-Print: hep-ph/9711226v1.
9. Jlab E99-117, J. P. Chen, Z. -E. Meziani, P. Souder *et al.*, http://hallaweb.jlab.org/physics/experiments/he3/A1n/
10. B. D. Anderson *et al.*, http://hallaweb.jlab.org/equipment/NIM.ps
11. F. Bissey *et al.*; e-Print: hep-ph/0109069.
12. P. L. Anthony *et al.*, submitted to *Phys. Rev. Lett.*; e-Print: hep-ex/0204028
13. W. Melnitchouk, A. W. Thomas, *Phys.Lett.* **B377**, 11 (1996); e-Print: nucl-th/9602038.
14. P. L. Anthony *et al.*, *Phys. Lett.* **B493**, 19 (2000); e-Print: hep-ph/0007248
15. Jefferson Lab, *The Science Driving the 12 GeV Upgrade of CEBAF*, (2001); http://www.jlab.org/div_dept/physics_division/GeV/WhitePaper_V11.ps

The Search for Higher Twist Effects in the Spin-Structure Functions of the Neutron

K. M. Kramer for the Jefferson Lab E97-103 Collaboration

College of William and Mary,Dept. of Physics,P.O. Box 8795,Williamsburg,VA 23185-8795, USA

Abstract. Jefferson Lab experiment E97-103 measured the spin structure function g_2^n from a Q^2 of 0.58 to 1.36 with a nearly constant x of 0.2. Combining this data with a fit to the world g_1^n data, the size of higher twist contributions to g_2^n can be extracted using the Wandzura-Wilczek relation. These higher twist contributions result from quark-gluon correlations and are expected to be larger as Q^2 decreases. This experiment was performed in Hall A with a longitudinally polarized electron beam and a high density polarized ^{3}He target. The physics motivation and an overview of the experiment will be presented.

INTRODUCTION

The Spin Dependent Structure Functions

A virtual photon exchange process in inclusive deep inelastic scattering can be characterized by the differential cross-section:

$$\frac{d^2\sigma}{d\Omega dE'} = \frac{\alpha^2}{Q^4}\frac{E'}{E}L_{\mu\nu}W^{\mu\nu} \tag{1}$$

where $\frac{d^2\sigma}{d\Omega dE'}$ is the differential cross-section, α is the fine structure constant, $L_{\mu\nu}$ is the lepton vertex and $W_{\mu\nu}$ is the nucleon vertex. While the lepton vertex can be calculated exactly using QED, the nucleon vertex uses structure functions to summarize the QCD structure of the nucleon.

The nucleon vertex can be seperated into spin-independent and spin-dependent parts. The spin-dependent parts, being the ones of interest to this experiment, can be written as follows:

$$W^A_{\mu\nu} = \frac{\varepsilon_{\mu\nu\lambda\sigma}q^\lambda S^\sigma}{M\nu}g_1(x,Q^2) + \frac{\varepsilon_{\mu\nu\lambda\sigma}q^\lambda\left(M\nu S^\sigma - q\cdot Sp^\sigma\right)}{M^2\nu^2}g_2(x,Q^2) \tag{2}$$

where M is the nucleon mass, S^μ is the nucleon polarization vector, p_μ is the ingoing nucleon momentum, q_μ is the ingoing electron momentum, ν is the energy transfer and Q^2 is the four-momentum transfer squared. $g_1(x,Q^2)$ and $g_2(x,Q^2)$ are the spin dependent structure functions of the nucleon, each depending only on x and Q^2.

CP675, *Spin 2002: 15th Int'l. Spin Physics Symposium and Workshop on Polarized Electron Sources and Polarimeters*, edited by Y. I. Makdisi, A. U. Luccio, and W. W. MacKay

These structure functions can be isolated by different combinations of the cross-section as shown in these equations:

$$\frac{d^2\sigma^{\downarrow\Uparrow}}{d\Omega dE'} - \frac{d^2\sigma^{\uparrow\Uparrow}}{d\Omega dE'} = \frac{4\alpha^2 E'}{Q^2 E M \nu}\left[(E+E'\cos\theta)g_1(x,Q^2) - 2xMg_2(x,Q^2)\right] \tag{3}$$

$$\frac{d^2\sigma^{\downarrow\Leftarrow}}{d\Omega dE'} - \frac{d^2\sigma^{\uparrow\Leftarrow}}{d\Omega dE'} = \frac{4\alpha^2 E'}{Q^2 E M \nu}E'\sin\theta\left[g_1(x,Q^2) - \frac{2xEM}{Q^2}g_2(x,Q^2)\right] \tag{4}$$

where the single line arrows, $\uparrow$ and $\downarrow$, refer to postive and negative beam helicity and the double line arrows, $\Uparrow$ and $\Downarrow$, refer to the target polarization being either pointing the direction of positive beam helicity or opposing it. The $\Leftarrow$ symbol refers to a target polarization perpendicular to the beam.

The Operator Product Expansion

The nucleon vertex can also be written in terms of vector currents as shown here:

$$W_{\mu\nu}(q,p,S) = \frac{1}{4\pi}\int d^4\xi e^{iq\cdot\xi}\langle pS|\left[J_\mu(\xi),J_\nu(0)\right]|pS\rangle \tag{5}$$

In pQCD, the vector current terms $J_\mu(z)J_\nu(0)$ can be expanded in a power series in $\frac{1}{Q}$ (known as the operator product expansion). The moments of hadronic structure functions can be expressed as:

$$\int_1^{-1} x^{n-1}W(Q^2,x)dx = \frac{1}{4}\sum_{\tau=2}^{\infty} C_n^\tau(Q^2,\mu^2)\bar{\mathcal{O}}_n^\tau(\mu^2)\left(\frac{1}{Q^2}\right)^{\frac{\tau}{2}-1} \tag{6}$$

where τ is the order of the twist.

If one applies this expansion to $g_1(x,Q^2)$ and $g_2(x,Q^2)$ and limits the expansion to the twist-2 and twist-3 operators then we get:

$$\int_0^1 x^n g_1(x,Q^2)dx = \frac{1}{4}\bar{\mathcal{O}}_n^{\{2\}} = \frac{1}{4}a_n \tag{7}$$

$$\int_0^1 x^n g_2(x,Q^2)dx = \frac{1}{4}\frac{n}{n+1}(\bar{\mathcal{O}}_n^{\{3\}} - \bar{\mathcal{O}}_n^{\{2\}}) = \frac{1}{4}\frac{n}{n+1}(d_n - a_n) \tag{8}$$

where n is the set of even integers starting at 2. Notice that g_1 does not contain the twist-3 operator and that the twist-2 term is the same in both g_1 and g_2. This is what allows E97-103 to extract the higher twist effects.

The Wandzura-Wilczek Relation

In their 1977 paper [5], Wandzura and Wilczek postulated if the twist-3 term is small ($\bar{\mathcal{O}}_n^{\{3\}} = 0$) then g_2 can be written :

$$g_2^{WW}(x,Q^2) = -g_1(x,Q^2) + \int_x^1 \frac{dx'}{x'} g_1(x',Q^2) \quad (9)$$

where g_2^{WW}, the leading twist part of $g_2(x,Q^2)$, is completely determined by the twist-2 part of $g_1(x,Q^2)$. Therefore, g_2 can be seperated into two functions:

$$g_2(x,Q^2) = g_2^{WW}(x,Q^2) + \bar{g}_2(x,Q^2) \quad (10)$$

where $\bar{g}_2(x,Q^2)$ is expected to be dominated by the twist-3 term [1] . The Wandzura-Wilczek relation, as it is known, presents a method of isolating higher twist contributions to $g_2(x,Q^2)$: measure $g_2(x,Q^2)$ precisely, use the world data on g_1 to calculate $g_2^{WW}(x,Q^2)$ and the difference will be from $\bar{g}_2(x,Q^2)$.

MEASURING SPIN STRUCTURE FUNCTIONS

Extracting spin structure functions

To extract g_2 one measures the cross-section asymmetry with change in beam helicity or target polarization direction. There are two methods of analyzing the data taken in this method. The first is to measure the cross-section difference divided by the unpolarized cross-section, known as the raw asymmetry, which reduces to the following formulas:

$$A_{\parallel} = \frac{1}{fP_bP_t} \frac{(N/Q)^{\downarrow\Uparrow} - (N/Q)^{\uparrow\Uparrow}}{(N/Q)^{\downarrow\Uparrow} + (N/Q)^{\uparrow\Uparrow}} \quad (11)$$

$$A_{\perp} = \frac{1}{fP_bP_t} \frac{(N/Q)^{\downarrow\Leftarrow} - (N/Q)^{\uparrow\Leftarrow}}{(N/Q)^{\downarrow\Leftarrow} + (N/Q)^{\uparrow\Leftarrow}} \quad (12)$$

where $A_{\parallel}$ is the raw parallel asymmetry, $A_{\perp}$ is the raw perpendicular asymmetry, N is the number of electrons detected within the selected kinematic range, Q is the amount of charge delivered to the target with a particular helicity, f is the dilution factor of the target, P_b is the beam polarization and P_t is the target polarization. These raw asymmetries can then be used to calculate g_1 and g_2 by the following formulas:

$$g_1(x,Q^2) = \frac{F_1(x,Q^2)}{D'} \left[A_{\parallel} + A_{\perp} \tan\theta/2\right] \quad (13)$$

$$g_2(x,Q^2) = \frac{F_1(x,Q^2)}{D'} \frac{y}{2\sin\theta} \left[A_{\perp} \frac{E + E'\cos\theta}{E'} - A_{\parallel} \sin\theta\right] \quad (14)$$

$$D' = \frac{(1-\varepsilon)(2-y)}{y(1+\varepsilon R(x,Q^2))} \quad (15)$$

$$\varepsilon = 1/[1+2(1+\nu^2/Q^2)\tan^2\theta/2] \quad (16)$$

$$y = \nu/E \quad (17)$$

Calculating $g_2(x,Q^2)$ in this manner requires knowledge of $R(x,Q^2)$ and $F_1(x,Q^2)$, both of which can be found in the literature.

The second method simply calculates the unnormalized cross-section featured in equations (3) and (4). This method requires detailed knowledge of the acceptance and detector inefficiencies, but does not depend on knowing $F_1(x,Q^2)$ or $R(x,Q^2)$. The two methods provide a good cross-check of systematic uncertainties.

RUNNING THE EXPERIMENT

E97-103 ran from August 1 to September 17th, 2001. In this time, we measured 5 kinematic points from $Q^2 = 0.58$ to $Q^2 = 1.36$ with $x = 0.2$. The experiment used 10-12 microAmps of CW beam, at energies from 1.1 to 5.7 GeV and with polarizations of 70-80%. Two high-resolution spectrometers with standard electron detector packages measured the scattered electrons. The table below lists the specific kinematic settings:

TABLE 1. Kinematics for experiment E97-103

Q^2 (GeV/c)^2	x	W^2 (GeV/c)^2
0.579	0.17	3.82
0.956	0.20	4.43
0.796	0.18	4.83
1.138	0.20	5.57
1.358	0.21	6.03

The experiment uses a polarized ^{3}He gas target. The ^{3}He gas is polarized by spin-exchange with optically polarized Rb. The target system consists of a glass cell which sits between two Helmholtz coils that provide a 25 Gauss field (another set of Helmholtz coils allows the target to be polarized in the perpendicular position). This glass cell has two chambers: one that is spherical and contains the rubidium vapor and the other which is cylindrical and is where the beam passes through. The cell is filled with roughly 8-10 atmospheres of ^{3}He at room temperature. The spherical chamber of the cell is placed in an oven 170 degrees and exposed to 90 watts of 795 nm circularly polarized laser light. It reached a maximum polarization of above 49% and an average polarization of 41%. Two target cells were used and both performed well.

RESULTS

The statistical errors for each kinematic point are shown in Figure 1. The data is compared to a calculation of $g_2^{n\ WW}$ from the E155 fit to g_1^n data [4] and two calculations of $g_2^{n\ WW}$ using parton models evolved to our kinematics.

There have been two bag model calculations of g_2^n by X. Song [6] and M. Stratmann [7]. Both predict that the higher twist effects are large and that $g_2^n(x,Q^2)$ will be

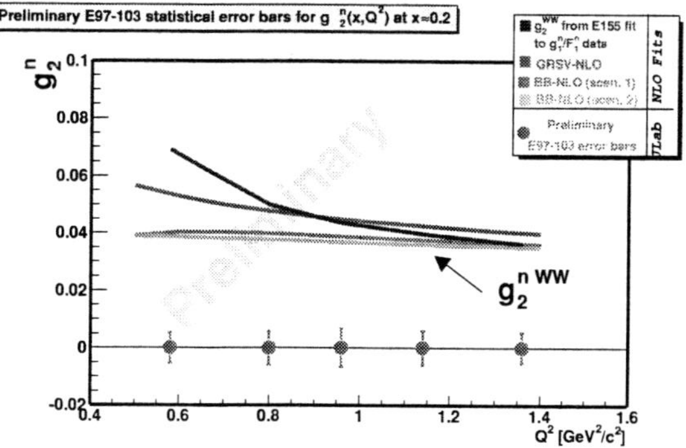

FIGURE 1. Statistical errors of E97-103. Presented with $g_2^{n\ WW}$ calculated using world data on g_1^n and two calculations of $g_2^{n\ WW}$ using two different parton models that are evolved down to our kinematics.

near zero or negative in our kinematic range. The achieved error bars will have the precision to make statements about g_2^n's relation to $g_2^{n\ WW}$ and to the bag model predictions. The significance of the data will also depend on a meaningful value of $g_2^{n\ WW}$ in this kinematic region, as the world g_1^n data is at higher Q^2. There are standard procedures for doing this that produce reasonable error bars for interpreting our data.

SUMMARY

While the spin structure function g_1^n has been the focus of extensive experimental programs over the last decade there have been few dedicated experiments on g_2^n. While g_2^n doesn't have a simple quark-parton model interpretation, it contains important information about quark-gluon correlations in the nucleon. E97-103 measured g_2^n precisely at low Q^2 in a narrow x range. This data will provide information about the magnitude of higher twist effects in the spin structure functions and will be able to guide theorists in understanding the physics of the higher twist terms.

REFERENCES

1. R. L. Jaffe and X. D. Ji, Phys. Rev. D **43**, 724 (1991).
2. F. Bissey, V. Guzey, M. Strikman and A. W. Thomas, Phys. Rev. C **65**, 064317 (2002) [arXiv:hep-ph/0109069].
3. C. Ciofi degli Atti, S. Scopetta, E. Pace and G. Salme, Phys. Rev. C **48**, 968 (1993) [arXiv:nucl-th/9303016].
4. P. L. Anthony *et al.* [E155 Collaboration], Phys. Lett. B **493**, 19 (2000) [arXiv:hep-ph/0007248].
5. S. Wandzura and F. Wilczek, Phys. Lett. B **72**, 195 (1977).
6. X. Song, Phys. Rev. D **54**, 1955 (1996) [arXiv:hep-ph/9604264].
7. M. Stratmann, Z. Phys. C **60**, 763 (1993).
8. A. Nogga, Ph.D. thesis, (2001).
9. P. L. Anthony *et al.* [E155 Collaboration], arXiv:hep-ex/0204028.

Measurement of the Electric Form Factor of the Neutron at MAMI

Michael Seimetz, for the A1 collaboration

Johannes Gutenberg–Universität, Institut für Kernphysik, Becherweg 45, 55099 Mainz, Germany

Abstract.
A new $D(\vec{e},e'\vec{n})p$ experiment has been performed at Mainz University to measure the electric form factor of the neutron, $G_{E,n}$. To this purpose a neutron polarimeter, optimized to withstand high electromagnetic background rates, was built for the Three Spectrometer Hall of the A1 collaboration. We present the experimental setup and the status of our data analysis.

MOTIVATION

The elastic form factors parametrize the nucleon's ability to absorb momentum transfer in a scattering reaction without excitation or production of further particles. They are related to the charge and magnetization distributions, and thereby reflect the inner structure of the nucleon. Precise measurements of G_E and G_M offer insight into the basic constituents of matter, being a crucial test for every nucleon model. Furthermore, precise values of the nucleon electromagnetic form factors are needed as input for many other experiments in nuclear and hadron physics.

Among the four Sachs form factors, $G_{E,n}$ is the least precisely known, for various experimental reasons. First of all, light nuclei (D, ^{3}He) are commonly used due to the lack of a free neutron target, which in general implies model dependencies of the $G_{E,n}$ results on nuclear binding effects. In $G_{E,n}$ measurements based on elastic $e-$D scattering [1] these uncertainties amount to more than 50%. Secondly, results on $G_{E,n}$ obtained from experiments based on the Rosenbluth separation technique are compatible with zero because in the cross section, $G_{E,n}$ is dominated by the much larger magnetic form factor, $G_{M,n}$. Thirdly, absolute neutron cross sections are hard to measure because they require a calibration of the neutron detection efficiency. An alternative approach, developed during the last decade, is offered by double polarization experiments, which provide results with small systematic and model errors for the ratio $G_{E,n}/G_{M,n}$.

The published data points from $D(\vec{e},e'\vec{n})p$, $\vec{D}(\vec{e},e'n)p$, and $^3\vec{He}(\vec{e},e'n)pp$ experiments are shown in Figure 1. The first neutron recoil polarization experiment on Deuterium was performed at MIT-Bates [2] (open circle), albeit with large systematic errors. Two data points using the same reaction were obtained at Mainz (A3) [3] (full circles). In this experiment a neutron spin precession technique [4] was used for the first time with which the systematic errors were reduced significantly. Furthermore, the model uncertainties could be minimized by including calculations of nuclear binding effects, such as final state interactions (FSI) and meson exchange currents (MEC), as published by Arenhövel

CP675, *Spin 2002: 15th Int'l. Spin Physics Symposium and Workshop on Polarized Electron Sources and Polarimeters*, edited by Y. I. Makdisi, A. U. Luccio, and W. W. MacKay

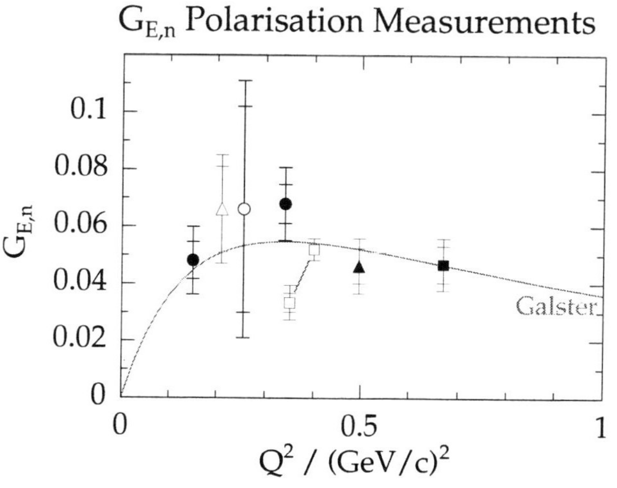

FIGURE 1. $G_{E,n}$ data points from double polarisation experiments. References are given in the text.

and coworkers [5, 6].

Two further data points were obtained at MAMI using polarized ^{3}He targets [7, 8] (squares). For the lower-lying one preliminary calculations of FSI effects [9] exist, introducing a significant shift of $G_{E,n}$ towards the value expected from the experiments with deuteron target. Another two points were acquired using polarized deuterium targets at NIKHEF [10] (open triangle) and JLab [11] (full triangle). All these experimental results agree (within their error range) with the $G_{E,n}$ parametrization originally given by Galster [12], a fact which, from a theoretical point of view, has to be regarded as fortuitous.

In the new A1 experiment, three additional data points at central four-momentum transfers $Q^2 = 0.3, 0.6$ and $0.8\,(\mathrm{GeV}/c)^2$ are acquired, allowing an independent check of the former results and an extension to higher Q^2 values.

EXPERIMENTAL METHOD

In the A1 experiment $G_{E,n}$ is determined from the quasielastic $\mathrm{D}(\vec{e}, e'\vec{n})p$ reaction. It is based on the spin transfer in the elastic scattering of polarized electrons on unpolarized nucleons [13]. The recoil nucleon carries nonvanishing polarization components, P_x and P_z, which are proportional to the form factor products $G_E G_M$ and G_M^2, respectively. The coordinate system is spanned by $\hat{z}$, the direction of momentum transfer, and $\hat{x}$ lying in the electron scattering plane. Then the polarization ratio

$$R_P = \frac{P_x}{P_z} = -\frac{1}{\sqrt{\tau + \tau(1+\tau)\tan^2\frac{\vartheta_e}{2}}} \cdot \frac{G_{E,n}}{G_{M,n}}, \qquad (1)$$

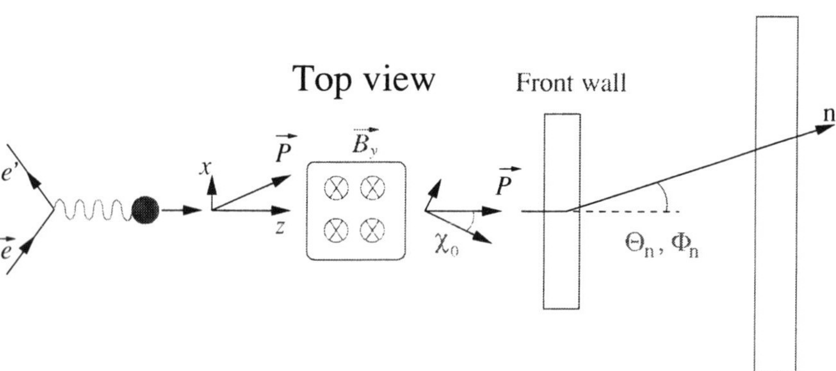

FIGURE 2. Sketch of the A1 neutron polarimeter setup (top view).

with $\tau = Q^2/4M_n^2$ determined by the kinematics of the scattering reaction. Together with the known value of $G_{M,n}$ ([14] and references therein), R_p gives the desired value of $G_{E,n}$. The measurement of R_p requires the use of a neutron polarimeter, which will be illustrated in the following, together with the other components of the experimental setup.

The $D(\vec{e},e'\vec{n})p$ experiment is performed in the A1 Three Spectrometer Hall [15] at Mainz. A beam of polarized electrons, provided by the MAMI accelerator with $P_e \simeq 80\%$, hits a 5 cm long liquid deuterium target cell. The momentum of the scattered electrons is analyzed with high resolution in a magnetic spectrometer. The outgoing neutrons are detected in the neutron polarimeter (Figure 2) at forward angles between 27° and 47°. Their scattering angles and time of flight are measured in a highly segmented wall of plastic scintillators.

At the same time, the neutron polarization is determined using the scintillators as active scattering target: The strong $n-p$ and $n-\mathrm{C}$ scattering reactions in the first scintillator wall carry analyzing power. A transverse polarization, P_t, of the outgoing neutron results in an asymmetry in the azimuthal angle Φ_n,

$$A = P_e A_{\mathrm{eff}} P_t \sin(\Phi_n) , \tag{2}$$

which can be measured as up-down asymmetry in the neutron distribution on a second wall of scintillators. P_e is the polarization of the incident electron beam and A_{eff} the effective analyzing power.

This method is sensitive only to the transverse polarization component, which initially is P_x. However, by precessing the neutron spin in a magnetic dipole field perpendicular to the (x,z) plane both P_x and P_z become accessible. At a certain precession angle χ_0 P_t, and with it the asymmetry (2), vanish. At this angle the effective analyzing power and the absolute electron beam polarization cancel out. Thereby systematic uncertainties are minimized.

Massive concrete shielding protects the large-area scintillators from electromagnetic background. Furthermore, a 5 cm thick lead brick wall is built up inside the magnet gap in front of the 1st wall, which in turn is highly segmented in order to keep the single paddle rates as low as possible. Systematic errors due to charge exchange effects in the lead shielding were investigated during the data taking using a LH_2 target. The rear wall is divided into two parts above and below the beam without direct target view, but in a way that neutrons scattered in the front wall can be detected in the rear wall with angles corresponding to high analyzing power. By these means luminosities of up to 5 cm $LD_2 \times 15\,\mu$A beam current could be used during data taking.

We chose three central four-momentum transfers, $Q^2 = 0.3, 0.6$, and $0.8\ (\mathrm{GeV}/c)^2$. The energy of the electron beam was 698, 855 and 883 MeV, respectively. A Møller polarimeter was used to monitor the electron beam polarization.

DATA ANALYSIS

The reconstruction of the kinematics of the quasielastic $\mathrm{D}(\vec{e}, e'\vec{n})p$ reaction is based on five independent observables. The energy, E'_e, and scattering angles, (ϑ_e, φ_e), of the outgoing electron are determined in a magnetic spectrometer. The neutron scattering angles, (ϑ_n, φ_n), are measured in the neutron polarimeter. In addition, the neutron Time of Flight is obtained allowing cross checks. This requires a good calibration of all scintillators, which is performed in one-arm measurements using minimally ionizing particles, and in the elastic $\mathrm{H}(e, e'p)$ reaction. Thereby a timing resolution around 300 ps is achieved for the first wall scintillators.

The identification of quasielastically scattered neutrons and their separation from charged and uncharged background is a major task of the data analysis. In the A1 experiment various methods are applied. Veto paddles on both scintillator walls of the neutron polarimeter, partly implemented in the hardware trigger, filter out charged particles. Additional veto conditions between each neutron detector and its neighbours refine this separation in the offline analysis. Protons can be discriminated against neutrons since they lose part of their kinetic energy in the lead shielding. Even if they are not stopped, their time of flight is significantly larger than the one of the neutrons. In addition, the energy deposition of the neutrons in elastic $p(n, n'p)$ and quasielastic ${}^{12}C(n, n'p)^{11}B$ reactions is kinematically related to the scattering angles in the front wall, a fact which can be used to enrich the analyzing power.

The azimuthal scattering angles Φ_n in the front wall are combined to the asymmetry

$$A(\Phi_n) = \frac{\sqrt{N^+(\Phi_n)N^-(\Phi_n+\pi)} - \sqrt{N^+(\Phi_n+\pi)N^-(\Phi_n)}}{\sqrt{N^+(\Phi_n)N^-(\Phi_n+\pi)} + \sqrt{N^+(\Phi_n+\pi)N^-(\Phi_n)}}. \tag{3}$$

It shows the sine-like dependence on Φ_n given in Equation (2). The amplitude of the sine depends on the transverse neutron polarization, P_t, and thus on the spin precession angle in the dipole magnet. Data were taken for seven different values of χ, permitting a precise determination of the precession angle χ_0 at vanishing P_t. An example for such a fit is shown in Figure 3.

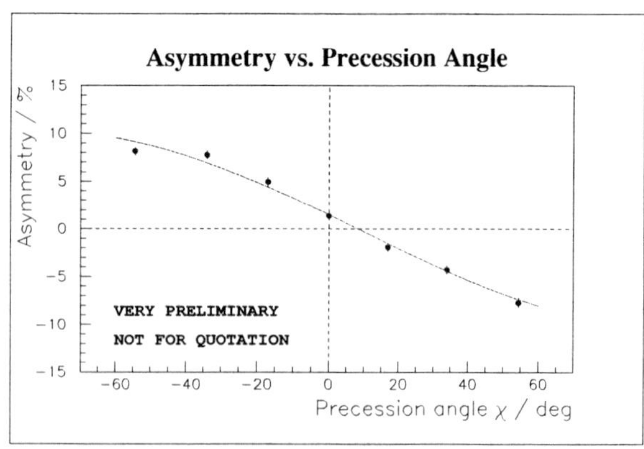

FIGURE 3. Variation of the asymmetry with the spin precession angle χ (example).

The $G_{E,n}$ data taking was completed in August 2002. The analysis is in progress, and first results seem to confirm that $G_{E,n}$ is significantly larger than assumed from former unpolarized experiments. We aim to achieve a relative error of our data points of $\Delta G_{E,n}/G_{E,n} = 10\%$.

ACKNOWLEDGMENTS

This work has been supported by SFB 443 of the Deutsche Forschungsgemeinschaft (DFG) and the Federal State of Rhineland-Palatinate.

REFERENCES

1. Platchkov, S. et al., *Nucl. Phys.* **A510**, 740-758 (1990).
2. Eden, T. et al., *Phys. Rev. C* **50**, 1749-1753 (1994).
3. Herberg, C. et al., *Eur. Phys. J.* **A5**, 131-135 (1999).
4. Ostrick, M. et al., *Phys. Rev. Lett.* **83**, 276-279 (1999).
5. Arenhövel, H., Leidemann, W., and Tomusiak, E.L., *Z. Phys. A* **331**, 123-138 (1988).
6. Arenhövel, H., Leidemann, W., and Tomusiak, E.L., *Few-Body Systems* **15**, 109-127 (1993).
7. Becker, J. et al., *Eur. Phys. J. A* **6**, 329-344 (1999).
8. Bermuth, J., Ph.D. thesis, Mainz (2001).
9. Golak, J. et al., *Phys. Rev. C* **63**, 034006 (2001).
10. Passchier, I. et al., *Phys. Rev. Lett.* **82**, 4988-4991 (1999).
11. Zhu, H. et al., *Phys. Rev. Lett.* **87**, 081801 (2001).
12. Galster, S. et al., *Nucl. Phys.* **B32**, 221-237 (1971).
13. Arnold, R.G., Carlson, C.E., and Gross, F., *Phys. Rev. C* **23**,363-374 (1981).
14. Kubon, G. et al., *Phys. Lett. B* **524**, 26-32 (2002).
15. Blomqvist, K.I. et al., *Nucl. Inst. Meth. A* **403**,263-301 (1998).

The Electric Form Factor of the Neutron via Recoil Polarimetry to $Q^2 = 1.47$ (GeV/c)^2

B. Plaster[*], R. Madey[†**], A. Yu. Semenov[†], S. Taylor[*], A. Aghalaryan[‡], E. Crouse[§], G. MacLachlan[¶], S. Tajima[||], W. Tireman[†], Chenyu Yan[†], A. Ahmidouch[††], B. D. Anderson[†], H. Arenhövel[‡‡], R. Asaturyan[‡], O. Baker[§§], A. R. Baldwin[†], H. Breuer[¶¶], R. Carlini[**], E. Christy[§§], S. Churchwell[||], L. Cole[§§], S. Danagoulian[**††], D. Day[***], M. Elaasar[†††], R. Ent[**], M. Farkhondeh[*], H. Fenker[**], J. M. Finn[§], L. Gan[§§], K. Garrow[**], P. Gueye[§§], C. Howell[||], B. Hu[§§], M. K. Jones[**], J. J. Kelly[¶¶], C. Keppel[§§], M. Khandaker[‡‡‡], W.-Y. Kim[§§§], S. Kowalski[*], A. Lung[**], D. Mack[**], D. M. Manley[†], P. Markowitz[¶¶¶], J. Mitchell[**], H. Mkrtchyan[‡], A. K. Opper[¶], C. Perdrisat[§], V. Punjabi[‡‡‡], B. Raue[¶¶¶], T. Reichelt[****], J. Reinhold[¶¶¶], J. Roche[§], Y. Sato[§§], I. A. Semenova[†], W. Seo[§§§], N. Simicevic[††††], G. Smith[**], S. Stepanyan[‡], V. Tadevosyan[‡], L. Tang[§§], P. Ulmer[‡‡‡‡], W. Vulcan[**], J. W. Watson[†], S. Wells[††††], F. Wesselmann[***], S. Wood[**], Chen Yan[**], S. Yang[§§§], L. Yuan[§§], W.-M. Zhang[†], H. Zhu[***] and X. Zhu[§§]

[*]Massachusetts Institute of Technology, Cambridge, Massachusetts 02139
[†]Kent State University, Kent, Ohio 44242
[**]Thomas Jefferson National Accelerator Facility, Newport News, Virginia 23606
[‡]Yerevan Physics Institute, Yerevan 375036, Armenia
[§]The College of William and Mary, Williamsburg, Virginia 23187
[¶]Ohio University, Athens, Ohio 45701
[||]Duke University, Durham, North Carolina 27708
[††]North Carolina A&T State University, Greensboro, North Carolina 27411
[‡‡]Johannes Gutenberg-Universität, D-55099 Mainz, Germany
[§§]Hampton University, Hampton, Virginia, 23668
[¶¶]University of Maryland, College Park, Maryland 20742
[***]University of Virginia, Charlottesville, Virginia 22904
[†††]Southern University at New Orleans, New Orleans, Louisiana 70126
[‡‡‡]Norfolk State University, Norfolk, Virginia 23504
[§§§]Kyungpook National University, Taegu 702-701, Korea
[¶¶¶]Florida International University, Miami, Florida 33199
[****]Rheinische Friedrich-Wilhelms-Universität, D-53115 Bonn, Germany
[††††]Louisiana Tech University, Ruston, Louisiana 71272
[‡‡‡‡]Old Dominion University, Norfolk, Virginia 23508

CP675, *Spin 2002: 15th Int'l. Spin Physics Symposium and Workshop on Polarized Electron Sources and Polarimeters*, edited by Y. I. Makdisi, A. U. Luccio, and W. W. MacKay

Abstract. The Jefferson Laboratory E93-038 collaboration conducted measurements of the ratio of the electric form factor to the magnetic form factor of the neutron, G_E^n/G_M^n, via recoil polarimetry from the quasielastic $^2\mathrm{H}(\vec{e},e'\vec{n})^1\mathrm{H}$ reaction at three values of Q^2 [viz., 0.45, 1.15, and 1.47 (GeV/c)^2] in Hall C of the Thomas Jefferson National Accelerator Facility. The preliminary results for G_E^n at $Q^2 = 0.45$ and 1.15 (GeV/c)^2 are consistent with the Galster parameterization; however, the preliminary result for G_E^n at $Q^2 = 1.47$ (GeV/c)^2 lies slightly above the Galster parameterization.

INTRODUCTION

The electric form factor of the neutron, G_E^n, is a fundamental quantity needed for an accurate description of both nucleon and nuclear structure. The Jefferson Laboratory E93-038 collaboration conducted measurements of the ratio of the electric form factor to the magnetic form factor of the neutron, $g \equiv G_E^n/G_M^n$, at three values of Q^2 [viz., 0.45, 1.15, and 1.47 (GeV/c)^2] via recoil polarimetry from the quasielastic $^2\mathrm{H}(\vec{e},e'\vec{n})^1\mathrm{H}$ reaction. Data were taken in Hall C of the Thomas Jefferson National Accelerator Facility from September 2000 to April 2001.

EXPERIMENTAL TECHNIQUE

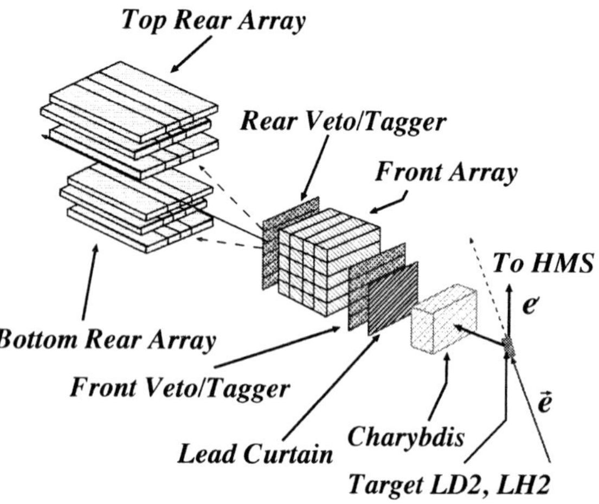

FIGURE 1. A schematic diagram of the polarimeter.

The experimental arrangement is shown in Fig. 1. A beam of longitudinally polarized electrons scattered quasielastically from neutrons in a 15-cm liquid deuterium target. In the plane-wave approximation, the polarization vector of the recoil neutron lies in the scattering plane [1] and consists of two components: The longitudinal component, $P_{L'}$, and the sideways component, $P_{S'}$, are parallel and perpendicular, respectively, to the recoil neutron's momentum vector. The scattered electron was detected and momentum analyzed by the Hall C High Momentum Spectrometer (HMS) in coincidence with the recoil neutron. A neutron polarimeter (NPOL) designed specifically for E93-038 by Madey [2] measured the up-down scattering asymmetry from a transverse projection of the recoil neutron's polarization vector. A dipole magnet (Charybdis) located ahead of the polarimeter precessed the polarization vector through an angle χ which permitted asymmetry measurements from different transverse projections of the polarization vector.

The polarimeter consisted of 20 detectors in the front array and 12 detectors in each of the two (upper and lower) rear arrays for a total of 44 plastic scintillation detectors. A double layer of thin scintillators ("veto/tagger" detectors) located directly ahead of and behind the front array detected incoming and scattered charged particles. The 100 cm $\times$ 10 cm $\times$ 10 cm dimensions of each detector in the front array permitted high luminosity; in addition, a collimator shielded each detector in the rear array from the direct flux of particles from the target. A 10-cm lead curtain located at the entrance of the collimator attenuated the flux of electromagnetic radiation and low-energy charged particles incident on the polarimeter. For the duration of the experiment, the polarimeter was fixed at an angle of 46° relative to the incoming beam, and the mean flight path from the target to the front array was 7 m.

The ratio of the electric form factor to the magnetic form factor of the neutron, $g \equiv G_E^n/G_M^n$, is

$$g = -K \tan \delta \,, \tag{1}$$

where K is a function of kinematic variables and $\tan \delta \equiv P_{S'}/P_{L'}$ [3]. After precession through an angle χ, the transverse projection, and hence the scattering asymmetry, is proportional to $\sin(\chi+\delta)$; therefore g can be obtained by extracting δ from a fit of the scattering asymmetries as a function of χ. A significant advantage of this experimental technique is that the analyzing power of the polarimeter cancels in the ratio of $P_{S'}$ to $P_{L'}$; the beam polarization also cancels provided the polarization is stable.

At each Q^2 point, scattering asymmetry measurements were conducted with precession angles of $\chi = \pm 40°$; in addition, scattering asymmetry measurements with precession angles of $\chi = 0°$ and $\pm 90°$ were conducted at $Q^2 = 1.15$ and 1.47 (GeV/c)^2. The measurements at $Q^2 = 0.45$ and 1.47 (GeV/c)^2 were associated with beam energies of 0.884 and 3.395 GeV, respectively. The measurement at $Q^2 = 1.15$ (GeV/c)^2 is the weighted average of a measurement at $Q^2 = 1.14$ (GeV/c)^2 associated with a beam energy of 2.33 GeV and a measurement at $Q^2 = 1.17$ (GeV/c)^2 associated with a beam energy of 2.42 GeV.

ANALYSIS AND PRELIMINARY RESULTS

Data were collected with an event defined to be a triple coincidence between an electron in the HMS, a neutral particle in the front array of the polarimeter, and a neutral or charged particle in the rear array of the polarimeter. Inelastic events were rejected via the application of a 100 MeV/c missing momentum cut and a relative -3% to $+5\%$ bite on the momentum of the scattered electron.

Typical time-of-flight spectra from a representative run at $Q^2 = 1.17$ (GeV/c)^2 are shown in Fig. 2. The left panel (cTOF) is a histogram of the difference between the measured time-of-flight from the target to the front array of the polarimeter and the time-of-flight calculated for quasielastic scattering. The right panel (ΔTOF) is a histogram of the difference between the measured time-of-flight between a neutral event in the front array and a neutral or charged event in either the upper or lower rear array and the time-of-flight calculated for elastic np scattering. The secondary peak centered at approximately -2.5 ns is the result of π^0 production in a scintillator in the front array. The ΔTOF spectrum shown in Fig. 2 can be decomposed into four ΔTOF spectra for scattering events to either the upper (U) or lower (D) rear array for the R (+) or L ($-$) helicity state of the incoming beam. From the yields in the four ΔTOF spectra, the cross ratio, r, can be calculated; the cross ratio is defined to be the ratio of two geometric means, $(N_U^+ N_D^-)^{1/2}$ and $(N_U^- N_D^+)^{1/2}$, where $N_U^+ (N_D^-)$ is the yield in the ΔTOF peak for neutrons scattered up(down) when the beam helicity was postive(negative). The physical scattering asymmetry is then given by $(r-1)/(r+1)$.

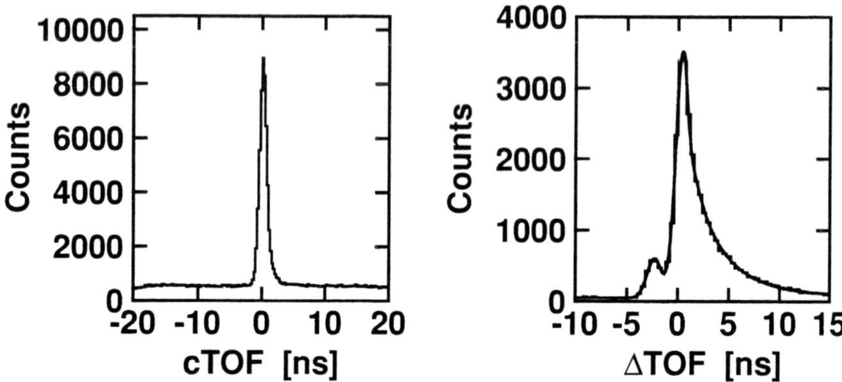

FIGURE 2. Typical cTOF and ΔTOF spectra from a representative run at $Q^2 = 1.17$ (GeV/c)^2.

Our preliminary results for G_E^n are plotted as the filled squares in Fig. 3 together with the current world data on G_E^n obtained via polarization measurements [4-12]. In order to extract G_E^n from our measurements of g, we used the dipole parameterization for G_M^n with a 5% relative uncertainty. Our preliminary results for G_E^n at $Q^2 = 0.45$ and 1.15 (GeV/c)^2 are consistent with the Galster [13] parameterization; however, our preliminary result for G_E^n at $Q^2 = 1.47$ (GeV/c)^2 lies slightly above the Galster parameterization.

The error bars that are plotted for our preliminary results reflect statistical errors only; however, the systematic errors are small compared to the statistical errors. Corrections resulting from the finite acceptance of the polarimeter and final state interactions have not yet been applied to our preliminary results for G_E^n; these effects are currently under investigation.

ACKNOWLEDGMENTS

We thank the TJNAF Hall C scientific and engineering staff for their outstanding support. This work was supported in part by the National Science Foundation, the Depart-

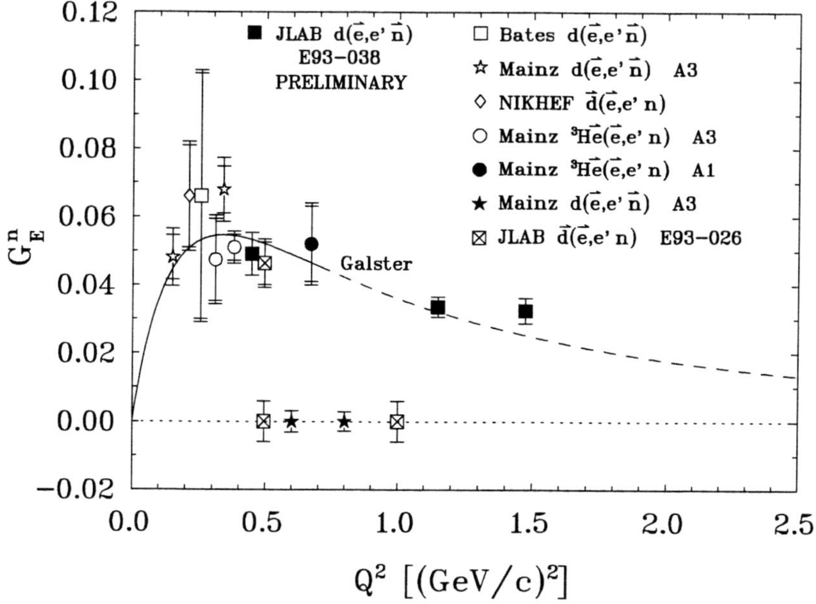

FIGURE 3. The current world data on G_E^n versus Q^2 obtained via polarization measurements. Our preliminary data are represented by filled squares. The Galster parameterization is shown as the solid line for $Q^2 < 0.7$ (GeV/c)^2, and its extension to higher Q^2 is shown as the dashed line. The points on the abscissa are projections.

ment of Energy, and the Deutsche Forschungsgemeinschaft. The Southeastern Universities Research Association (SURA) operates the Thomas Jefferson National Accelerator Facility under the U.S. Department of Energy contract DE-AC05-84ER40150.

REFERENCES

1. N. Dombey, Rev. Mod. Phys. **41**, 236 (1969). J. Scofield, Phys. Rev. **113**, 1599 (1959) and **141**, 1352 (1966). A. I. Akhiezer and M. P. Rekalo, Sov. J. Part. Nucl. **4**, 277 (1974).
2. R. Madey, A. Lai, and T. Eden, *Polarization Phenomena in Nuclear Physics*, edited by E. J. Stephenson and S. E. Vigdor, A.I.P. Proceedings No. 339, 47-54 (1995).
3. R. G. Arnold, C. E. Carlson, and F. Gross, Phys. Rev. C **23**, 363 (1981).
4. T. Eden *et al.*, Phys. Rev. C **50**, R1749 (1994).
5. M. Meyerhoff *et al.*, Phys. Lett. B **327**, 201 (1994).
6. J. Becker *et al.*, Eur. Phys. J. A **6**, 329 (1999).
7. J. Golak, G. Ziemer, H. Kamada, and W. Glöckle, Phys. Rev. C **63**, 034006 (2001).
8. M. Ostrick *et al.*, Phys. Rev. Lett. **83**, 276 (1999).
9. C. Herberg *et al.*, Eur. Phys. J. A **5**, 131 (1999).
10. I. Passchier *et al.*, Phys. Rev. Lett. **82**, 4988 (1999).
11. D. Rohe *et al.*, Phys. Rev. Lett. **83**, 4257 (1999).
12. H. Zhu *et al.*, Phys. Rev. Lett. **87**, 081801 (2001).
13. S. Galster, H. Klein, K. H. Schmidt, D. Wegener, and J. Blechwenn, Nucl. Phys. B **32**, 221 (1971).

Measurement of the Charge Form Factor of the Neutron G_E^n from $\vec{d}(\vec{e},e'n)p$ at $Q^2 = 0.5$ and $1.0\ (GeV/c)^2$

N. Savvinov, for the E93-026 JLab collaboration

Department of Physics, University of Maryland, College Park, USA

Abstract. We determined the electric form factor of the neutron G_E^n via the reaction $\vec{d}(\vec{e},e'n)p$ using a longitudinally polarized electron beam and a frozen polarized $^{15}ND_3$ target at Jefferson Lab. The knocked out neutrons were detected in a segmented plastic scintillator in coincidence with the quasi-elastically scattered electrons which were tracked in Hall C's High Momentum Spectrometer. To extract G_E^n, we compared the experimental beam–target asymmetry with theoretical calculations based on different G_E^n models. We report the preliminary results of the fall 2001 run at $Q^2 = 0.5$ and $1.0\ (GeV/c)^2$.

INTRODUCTION

In a non-relativistic picture, the charge form-factor of the neutron G_E^n is related to the charge distribution in the neutron and thus is important for our understanding of electromagnetic structure of nucleons. Despite a great effort focused on its determination, G_E^n remains poorly known. Two major difficulties faced by experimenters in their studies of G_E^n are its small magnitude and the lack of a free neutron target. Unpolarized cross-section measurements, from which G_E^n was extracted until 1990's, were incapable of overcoming these difficulties and yielded inconsistent or model-dependent results.

Recent technological advances in high duty factor accelerators, polarized sources, polarized targets and recoil polarimetry made possible double polarization methods of determining G_E^n. These methods use asymmetries rather than cross-sections and thus reduce sensitivity to systematic errors.

In the experiment described here a longitudinally polarized electron beam was scattered off a polarized deuterated ammonia target. The polarized electron-neutron scattering cross section consists of helicity-independent and helicity-flip terms. The helicity induced asymmetry, given by the ratio of these two terms, depends on G_E^n . For a general orientation of the target polarization this dependence is complex. In order to minimize the influence of the magnetic form factor G_M^n and maximize the sensitivity to G_E^n , the direction of the target polarization was chosen to be perpendicular to the three-momentum transfer $\vec{q}$ and to lie in the scattering plane. For this case the expression for the electron-neutron asymmetry A_{en} simplifies to the following:

$$A_{en} = \frac{-2\sqrt{\tau(1+\tau)}G_M^n G_E^n}{(G_E^n)^2 + \tau(1+2(1+\tau)\tan^2(\theta_e/2))(G_M^n)^2},$$

CP675, *Spin 2002: 15th Int'l. Spin Physics Symposium and Workshop on Polarized Electron Sources and Polarimeters,* edited by Y. I. Makdisi, A. U. Luccio, and W. W. MacKay

where $\tau = \frac{Q^2}{4M^2}$, M is the neutron mass, and θ_e is the electron scattering angle.

Since neutrons in deuterium are not free, the measured electron-deuteron asymmetry A^V_{ed} differs from A_{en} due to reaction mechanisms such as final state interactions and meson exchange currents. These reaction mechanisms were taken into account in theoretical calculations used for extraction of G^n_E [1].

EXPERIMENTAL SETUP

The experiment E93-026 was conducted in Hall C of Thomas Jefferson Accelerator Facility (Jefferson Lab) in 1998 and 2001. The measurements were taken at two points, $Q^2 = 0.5$ and 1.0 $(GeV/c)^2$.

The beam polarization was measured using a Moeller polarimeter. The average beam polarization during the experiment was 75%. The beam current was limited to 100 nA to avoid excessive thermal and radiation damage to the target polarization. A system of raster magnets was used to distribute these stresses uniformly over the full target cell. The readings of the raster magnets were also used by the reconstruction algorithm to determine the horizontal and vertical position of the interaction point.

The solid polarized target [2] was developed by University of Virginia and was successfully used in two experiments prior to being used in E93-026. The basic components of the polarized target include a superconducting magnet operated at 5 Tesla, a ^4He evaporation refrigerator, a pumping system, a high power microwave tube operating at frequencies around 140 GHz and an NMR system for measuring the target polarization. The target material was polarized using the principle of dynamic nuclear polarization. Target polarization was determined by measuring the impedance change of the series resonant LCR circuit due to the nuclear magnetic moment. The conversion constants between the area of the NMR signal and the target polarization were obtained by a series of thermal equilibrium measurements. The target polarization typically varied between 15% and 35% and averaged to 22%.

After interaction in the target material, the scattered electrons were detected in the High Momentum Spectrometer of Hall C. Recoil nucleons were detected in the neutron detector which consisted of six planes of thick scintillators and two planes of thin scintillators. The latter were used for particle identification. The detector was set along the direction of the three-momentum transfer and was enclosed in a concrete hut open towards the target. The neutron vertical position was determined by the segmentation of the detector while the horizontal position was determined from the time difference of the phototubes.

ANALYSIS AND RESULTS

The electrons in the HMS were reconstructed using the standard HMS reconstruction code extended for the effects of beam raster and target magnetic field. On the neutron detector side a custom tracking algorithm was developed for proper particle identification. Neutrons were defined as events with no hits in the paddles along the track to

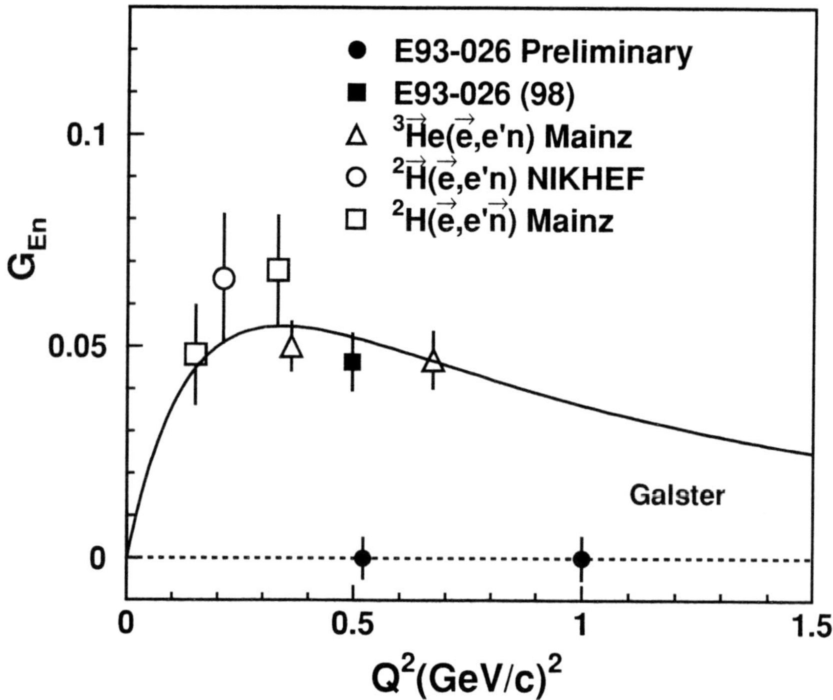

FIGURE 1. Double polarization world data on G_E^n. Filled circle: preliminary E93-026 2001 data (error bars only), filled square: final E93-026 result for 1998 data [3], open triangles: polarized helium data [4], open squares: recoil polarimetry data [5, 6], open circle: NIKHEF polarized deuterium data [7].

the target, within a narrow time interval and within a 100 MeV range of invariant mass around the nucleon mass. A number of other cuts was applied to optimize the dilution factor and limit the recoil momentum to values where nuclear corrections are small.

After event reconstruction and event selection, charge and dead-time normalized yields were produced for each beam helicity state, N^+ and N^-. From these yields the raw asymmetry ε was calculated:

$$\varepsilon = \frac{N^+ - N^-}{N^+ + N^-} = P_B P_T f A_{ed}^V.$$

In order to obtain A_{ed}^V from the raw asymmetry ε, one needs the knowledge of dilution factor f (due to scattering from unpolarized materials) in addition to beam and target polarizations P_B and P_t. Dilution factor was calculated using montecarlo simulations. Finally, A_{ed}^V was corrected for accidental background and radiative effects.

The G_E^n was extracted by comparing the corrected experimental asymmetry to the theoretical asymmetry averaged over the experimental acceptance under different assumptions about the size of the G_E^n. Preliminary results are consistent with the Galster parametrization. The systematic error is expected to be dominated by uncertainty in target polarization (3-5%) and dilution factor ($\sim 3\%$).

REFERENCES

1. Arenhövel, H., et al., *Z. Phys.*, **A331**, 123 (1988).
2. Crabb, D., et al., *Nucl. Instr. Meth.*, **A356**, 9 (1995).
3. Zhu, H., et al., *Phys. Rev. Lett.*, **87**, 081801 (2001).
4. Golak, J., et al., *Phys. Rev.*, **C63**, 034006 (2001).
5. Herberg, C., et al., *Eur. Phys. J.*, **A5**, 131 (1999).
6. Ostrick, M., et al., *Phys. Rev. Lett.*, **83**, 276 (1999).
7. Passchier, I., et al., *Phys. Rev. Lett.*, **82**, 4988 (1999).

Measuring G_E^n at High Momentum Transfers

Bodo Reitz

Thomas Jefferson National Accelerator Facility, 12000 Jefferson Avenue, Newport News, VA 23606, USA

Abstract. Experiment E02-013 at Thomas Jefferson National Accelerator Facility will extend the measured range of the neutron electric form factor G_E^n to Q^2=3.4 (GeV/c)^2 through a measurement of the cross section asymmetry in the reaction $^3\vec{He}(\vec{e},e'n)$. Recent theoretical investigations, motivated by the results on the ratio of the proton electric and magnetic form factor, predict higher values of G_E^n compared to older predictions. The experiment utilizes a polarized ^{3}He target and the polarized CEBAF electron beam. Scattered electrons will be detected in the BigBite spectrometer, recoiling neutrons in an array of scintillators. The experimental and theoretical developments needed to perform the measurement and to extract G_E^n from ^3He will be described. Concepts of extending the measurement of G_E^n to even higher momentum transfers will be discussed.

INTRODUCTION

Elastic electron scattering off the nucleon provides important ingredients to our knowledge of nucleon structure. In the one-photon approximation the interaction is fully characterized by only two independent form factors. There are well founded predictions for the Q^2 dependence of the form factors and their ratio in the limit of high momentum transfers in pQCD [1], indicating that the ratio becomes constant. However recent results [2, 3] on the electric form factor of the proton G_E^p show that the ratio G_E^p / G_M^p declines as Q^2 increases, and therefore pQCD is not applicable up to 10 (GeV/c)^2. According to these measurements, G_M^p and G_E^p behave differently, starting at Q^2 = 1 (GeV/c)^2. The same mechanism causing this deviation should also be present in the neutron, and therefore it is an important question, how G_E^n develops in the Q^2 regime of several (GeV/c)^2.

DATA ON G_E^N

Although experiments to measure G_E^n have been performed at all electromagnetic labs for the last two decades, our knowledge of this quantity at higher momentum transfers is still rather poor compared to the data available on G_E^p, G_M^p, and G_M^n. The reason is manyfold. First of all, there are no targets of free neutrons with sufficient density to perform electron scattering experiments on. Therefore experiments have to be performed on light nuclei, and the contribution from the proton has to be taken into account. Secondly due to the zero charge of the neutron, G_E^n is small at low momentum transfers, whereas at higher momentum transfers, the cross section is dominated by the magnetic form factor. All these factors make the standard method of the Rosenbluth separation very

CP675, *Spin 2002: 15th Int'l. Spin Physics Symposium and Workshop on Polarized Electron Sources and Polarimeters,* edited by Y. I. Makdisi, A. U. Luccio, and W. W. MacKay

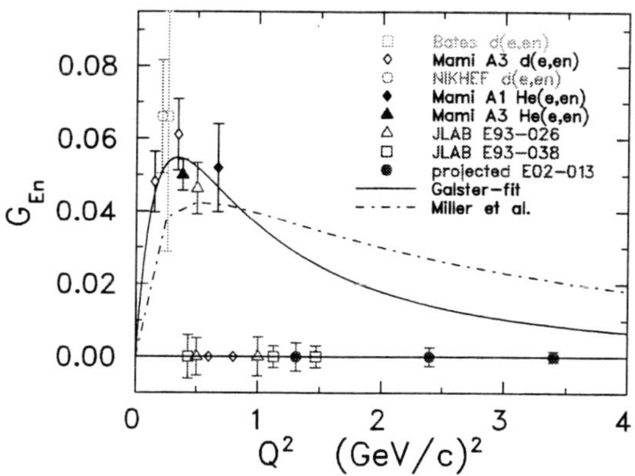

FIGURE 1. Data on G_E^n from double polarization experiments. For experiments which have not published their results yet, only the expected uncertainty is plotted. Also shown are the projected error bars of E02-013. The solid line represents the predictions from [5], the dashed dotted line the ones from [16]

demanding. At SLAC G_E^n was extracted up to $Q^2 = 4$ (GeV/c)^2 from quasi elastic e – d scattering data [4], however the uncertainties are rather large and the result is compatible with $G_E^n = 0$ as well as with the Galster "parameterization", an empirical fit to data on G_E^n obtained at lower values of Q^2 [5].

Double polarization experiments are another tool to study G_E^n. By investigating spin observables, the sensitivity of these reactions is enhanced due to the interference between G_E^n and G_M^n. Figure 1 shows the published results on G_E^n obtained with this method [6] – [12] together with the projected error bars for experiments which have already collected data, but not yet published [13, 14]. Also the projected error bars for the future Jlab Hall A experiment E02-013 are shown [15]. These experimental data are compared to the Galster approximation and to the calculations from [16]. The latter theoretical calculations are also able to reproduce the data on G_E^p from [2, 3].

The double polarization technique was already introduced 20 years ago [17] – [19]. In all of them a polarized electron beam was utilized, together with either a polarized ND_3 or polarized ^{3}He target, or with an unpolarized deuterium target (and a neutron polarimeter).

EXPERIMENT E02–013

The previously described experiments provide an accurate determination of G_E^n up to 1.47 (GeV/c)^2. For an further increase in Q^2 an experimental approach with a much higher figure-of-merit (FOM) is required. Compared to the previous experiments the following items have been optimized for Jlab Hall A Experiment E02-013:

- the solid angle of the electron detector,
- the neutron detector efficiency,
- the type of the (polarized) target.

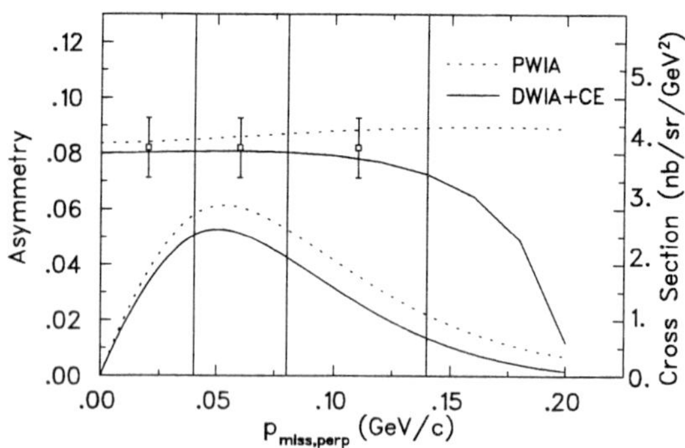

FIGURE 2. The GEA predictions for the cross section and the asymmetry in E02-013. For each Q^2 the asymmetry will be measured in three bins in $p_{m,\perp}$. The error bars show the projected statistical accuracy.

The BigBite spectrometer, which was originally developed and used at NIKHEF [20], is a recent addition to the Jlab Hall A spectrometers. It has a large solid angle of 76 msr for a 40 cm long target. For this experiment the detector package will consist of three drift chambers to obtain the tracking information, a segmented trigger scintillator plane and a lead glass calorimeter for particle identification. The momentum resolution with BigBite for electrons with momenta up to 1.5 GeV/c is 1%, and therefore sufficient to identify quasi elastic scattering events and separate them from other reactions. The luminosity available with the polarized ^{3}He target is about 10^{36} Hz/cm^2, about a factor of 10 higher than the one in a recent Jlab experiment utilizing a polarized ND_3 target [13]. According to our simulations BigBite can still be used at this luminosity, despite of the direct view of the target by the detectors.

The recoiling neutrons will have kinetic energies above 1 GeV at the proposed kinematics. Therefore they can be efficiently detected in an array of neutron detectors, and at the same time relatively high detector thresholds can be utilized to suppress background. The detector will have five layers of 10 cm thick plastic scintillators, with each layer separated by iron converters to further increase the neutron detection efficiency. Two layers of thin scintillators in front of the detector will be used to veto charged particles. The neutron detector will be large enough to match the solid angle of BigBite.

To extract G_E^n from the experimentally measured asymmetry, nuclear effects have to be taken into account. The Generalized Eikonal Approximation (GEA) [21] provides the appropriate framework for this extraction in the proposed range of Q^2. The GEA prediction for the asymmetry as a function of the missing transverse momenta $p_{m,\perp}$ is shown in Fig. 2. The GEA calculations as well as experimental data from Jlab Hall B for the unpolarized reaction ^{3}He(e,e$'$ p) have demonstrated the dominance of quasi-elastic scattering at $p_{m,\perp}$ below 0.15 GeV/c, when a modest cut on $p_{m,\|}$ is applied.

The error budget for the highest Q^2 point is shown in Tab. 1. For each Q^2 we will achieve a statistical accuracy of 14% within each of the three bins in $p_{m,\perp}$. The statistical and systematic uncertainties will contribute equally to the total uncertainty in G_E^n.

TABLE 1. The contributions to the error in G_E^n at Q^2 = 3.4 (GeV/c)^2.

quantity	expected value	rel. uncertainty in G_E^n
raw asymmetry A_{exp}	-0.0233	
⇒ statistical error in G_{En}		14.2%
beam polarization P_e	0.75	3%
target polarization P_{He}	0.40	4%
neutron polarization P_n	$0.86 \cdot P_{He}$	2%
dilution factor D (nitrogen)	0.94	3%
dilution factor V (background)	0.91	4%
correction factor for $A_{\|\|}$ components	0.94	<3%
G_{Mn}	0.057	5%
nuclear correction factor	1.0 – 0.85	5%
⇒ systematic error in G_{En}		10.4%

FUTURE DEVELOPMENTS

The highest Q^2 point at which G_E^n will be extracted in experiment E02-013 is at Q^2 = 3.4 (GeV/c)^2. This limit is based on the achievable luminosity and maximum polarization of the polarized ^{3}He target and the parameters of the BigBite spectrometer. However, this Q^2 is still small compared to the available data on G_E^p, G_M^p, and G_M^n. For testing recent theoretical calculations it is desirable to go beyond that limit. Because the cross section as well as the asymmetry is getting smaller, the FOM has to be further increased.

Improvements necessary to measure G_E^n at 5 (GeV/c)^2 are expected in two areas. An increase of luminosity with the polarized ^{3}He target seems feasible. In its present configuration, the target has its highest FOM at a beam current of 12-15 μA, when the beam induced depolarization time is on the order of 30 hours. To further increase the beam current a higher rate of polarization and a faster delivery of the polarized atoms to the target cell becomes mandatory. Because of the steady advances in solid-state laser technology, 100-200 W lasers suitable for polarizing Rb atoms are becoming available. Changes in the cell design, like utilizing schemes with more than one tube between the pumping cell and the target cell, will provide improved flow of the polarized ^{3}He [22]. To withstand higher beam currents modifications of the end windows of the glass cells are considered.

The second limitation is the maximum momentum accepted in the BigBite spectrometer. The FOM of the experiment is approximately proportional to $E_f^2/E_i^2 = (E_i - Q^2/2M)^2/E_i^2$, where E_i (E_f) is the initial (final) energy of the electron. This illustrates, that by using higher beam energies, and using a medium to large acceptance spectrometer capable of detecting electrons at higher momenta than BigBite will further increase the FOM. With the proposed MAD spectrometer for Jlab Hall A together with the proposed energy upgrade to 12 GeV of Jlab experiments to measure G_E^n up to 5 (GeV/c)^2 will become feasible. A specially designed electron spectrometer, with an even larger angular acceptance than MAD, could further boost the FOM and therefore the possibilities to measure G_E^n.

CONCLUSIONS

After the good progress in the experimental determination of the electromagnetic form factors in the recent years, Jlab experiment E02-013 will push the knowledge of G_E^n even further. It will measure G_E^n up to momentum transfers of 3.4 (GeV/c)^2. There are possibilities that future experiments can go even beyond that limit, by improving the luminosity and polarization of the polarized ^{3}He target, and utilizing the possibilities of the proposed Jlab energy upgrade together with new spectrometers.

ACKNOWLEDGMENTS

Thanks to my collaborators B. Wojtsekhowski, K. McCormick, and G. Cates for the many discussions about E02-013 and sharing their ideas about further experiments on G_E^n. The Southeastern Universities Research Association (SURA) operates the Thomas Jefferson National Accelerator Facility for the United States Department of Energy under contract DE-AC05-84ER40150.

REFERENCES

1. S. J. Brodsky and G.P. Lepage, *Phys. Rev.* **D 22**, 2157 (1981).
2. M. Jones *et al.*, *Phys. Rev. Lett.* **84** , 1398 (2000).
3. O. Gayou *et al.*, *Phys. Rev. Lett.* **88** , 092301 (2002).
4. A. Lung *et al.*, *Phys. Rev. Lett.* **70**, 718 (1993).
5. S. Galster *et al.*, *Nucl. Phys.* **B32**, 221 (1971).
6. T. Eden *et al.*, *Phys. Rev.* **C 50**, R1749 (1994).
7. M. Ostrick *et al.*, *Phys. Rev. Lett.* **83**, 276 (1999).
8. C. Herberg *et al.*, *Eur. Phys. J.* **A5**, 131 1999).
9. I. Passchier *et al.*, *Phys. Rev. Lett.* **82**, 4988 (1999).
10. D. Rohe *et al.*, *Phys. Rev. Lett.* **83**, 4257 (1999).
11. H. Zhu *et al.*, *Phys. Rev. Lett.* **87**, 081801 (2001).
12. J. Golak *et al.*, *Phys. Rev.* **C63**, 034006 (2001).
13. D. Day, J. Mitchell, and G. Warren, JLab proposal E93-026.
14. R. Madey and S. Kowalski, spokespeople, Jlab proposal E93-038.
15. G. Cates, K. McCormick, B. Reitz, and B. Wojtsekhowski, spokespeople, Jlab proposal E02-013.
16. G. Miller, *Phys. Rev.* **C66**, 032201 (2002).
17. N. Dombey, *Rev. Mod. Phys.* **41**, 236 (1969).
18. A. I. Akhiezer and M. P. Rekalo, *Sov. J. Part. Nucl.* **3**, 277 (1974).
19. R. Arnold, C. Carlson, and F. Gross, *Phys. Rev.* **C 23**, 363 (1981).
20. D. J. J. de Lange *et al.*, *Nucl. Instr. and Meth.* **A 406**, 182 (1998).
21. M.M. Sargsian, *Int. J. Mod. Phys* **E10** 405 (2001).
22. B. Wojtsekhowski, *Proc. of the Int. Conf. on Exclusive Processes at High Momentum Transfer*, Newport News (2002), ed. by P. Stoler and A. Radyushkin, World Scientific, to appear.

Proton Elastic Form Factor Ratio: the JLab Polarization Experiments

C.F. Perdrisat*, V. Punjabi† and the Jefferson Lab Hall A and $G_{Ep}(III)$ Collaborations**

*College of William and Mary, Williamsburg, VA 23187
†Norfolk State University,Norfolk, VA 23504
**12000 Jefferson Avenue, Newport News, VA 23666

Abstract. The ratio of the electric and magnetic proton form factors, G_{Ep}/G_{Mp}, has been obtained in two Hall A experiments, from measurements of the longitudinal and transverse polarization of the recoil proton, P_ℓ and P_t, respectively, in the elastic scattering of polarized electrons, $\vec{e}p \to e\vec{p}$. Together these experiments cover the Q^2- range 0.5 to 5.6 GeV^2. A new experiment is currently being prepared, to extend the Q^2-range to 9 GeV^2 in Hall C.

INTRODUCTION

The nucleon elastic form factors describe the internal structure of the nucleon; in the non-relativistic limit, for small four-momentum transfer squared, Q^2, they are Fourier transforms of the charge and magnetization distributions in the nucleon. At high Q^2 values, the nucleon must be treated as a system of three valence quarks; perturbative QCD predicts the Q^2 dependence [1] of the form factors. At Q^2 between 1 and 10 GeV^2, relativistic constituent quark models [2, 3] currently give the best understanding of the nucleon form factors, with the strongest dynamical input; Vector Meson Dominance (VMD) (see e.g. Refs. [4, 5]) also describes the form factors well.

The unpolarized elastic ep cross section is given by:

$$\frac{d\sigma}{d\Omega} = \frac{\alpha^2 E_e^{'} \cos^2\frac{\theta_e}{2}}{4E_e^3 \sin^4\frac{\theta_e}{2}} \left[G_{Ep}^2 + \frac{\tau}{\varepsilon} G_{Mp}^2\right] \left(\frac{1}{1+\tau}\right), \tag{1}$$

where G_{Ep} and G_{Mp} are the electric and magnetic form factors, $\varepsilon = [1+2(1+\tau)\tan^2(\frac{\theta_e}{2})]^{-1}$, θ_e is the scattering angle of the electron in the laboratory and $\tau = Q^2/4M_p^2$, with M_p the proton mass; E_e and $E_e^{'}$ are the energies of the in- and outgoing electrons, respectively. For a given Q^2, G_{Ep} and G_{Mp} can be extracted from cross section measurements made at fixed Q^2 over a range of ε values with the Rosenbluth method. At Q^2 below 1 GeV^2, G_{Ep} and G_{Mp} have been determined by this method and $\mu_p G_{Ep}/G_{Mp}$ has been found to be ≈ 1. At larger Q^2, the cross section becomes dominated by the G_{Mp} contribution; G_{Mp} is known up to Q^2 = 31 GeV^2 [6]. In Fig. 1a), the

CP675, *Spin 2002: 15th Int'l. Spin Physics Symposium and Workshop on Polarized Electron Sources and Polarimeters*, edited by Y. I. Makdisi, A. U. Luccio, and W. W. MacKay

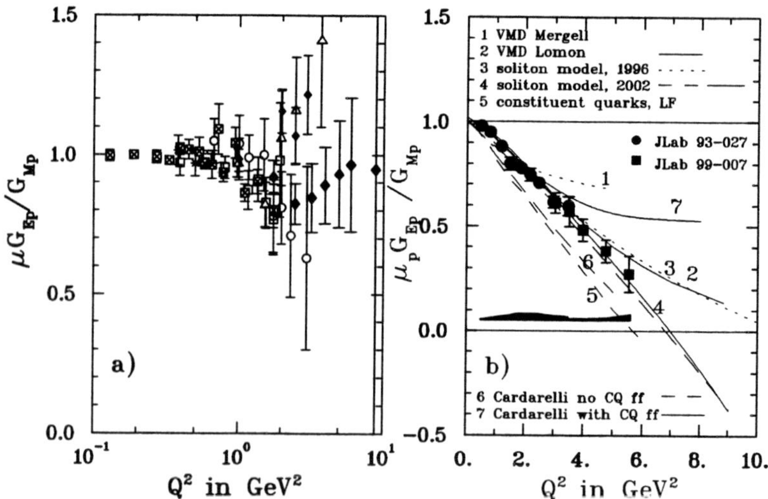

FIGURE 1. a) World data for $\mu_p G_{Ep}/G_{Mp}$ versus Q^2, not including the JLab polarization data. b) the JLab data compared with several theoretical predictions. The systematic uncertainties are show as the black polygon.

error bars on $\mu_p G_{Ep}/G_{Mp}$ from the world cross section data (refs. [7, 8, 9, 10, 11, 12, 13]) are seen to grow with Q^2. Above $Q^2 \approx 1$ GeV^2, systematic differences between different experiments are evident.

The JLab results have been obtained by measuring the recoil proton polarization in $\vec{e}p \rightarrow e\vec{p}$ [14, 15]. In one-photon exchange, the scattering of longitudinally polarized electrons on unpolarized hydrogen results in a transfer of polarization to the recoil proton with two components, P_t perpendicular to, and P_ℓ parallel, to the proton momentum in the scattering plane [16]:

$$I_0 P_t = -2\sqrt{\tau(1+\tau)} G_{E_p} G_{M_p} \tan\frac{\theta_e}{2} \tag{2}$$

$$I_0 P_\ell = \frac{1}{M_p}(E_e + E_{e'})\sqrt{\tau(1+\tau)} G_{M_p}^2 \tan^2\frac{\theta_e}{2} \tag{3}$$

where $I_0 \propto G_{E_p}^2 + \frac{\tau}{\varepsilon} G_{M_p}^2$. Measuring simultaneously these two components and taking their ratio gives the ratio of the form factors:

$$\frac{G_{Ep}}{G_{Mp}} = -\frac{P_t}{P_\ell}\frac{(E_e + E_{e'})}{2M_p}\tan(\frac{\theta_e}{2}). \tag{4}$$

Neither the beam polarization nor the analyzing power of the polarimeter, used to measure P_t and P_ℓ, appear in Eqn. 4.

EXPERIMENTS

In 1998 G_{Ep}/G_{Mp} was measured for Q^2 from 0.5 to 3.5 GeV^2 [14]. Protons and electrons were detected in coincidence in the two high-resolution spectrometers (HRS) of Hall A. The polarization of the recoiling proton was measured in a graphite analyzer focal plane polarimeter (FPP). The data shown here differ from the previously published ones [14]; a reanalysis has been recently completed for all data points obtained in 1998, taking into account the information from optical studies of both HRSs [17]. One outcome of the reanalysis is the much smaller systematic uncertainty; another is the disappearance of the irregularities observed previously around Q^2=2 GeV^2. A new method to analyze the data has been developed, which eliminates the difficulty associated with the vanishing of the normal polarization component at the focal plane, when the precession angle in the dispersive plane is 180^0 (or an integer multiple there of). Details of this reanalysis will appear in the archival paper in preparation [18].

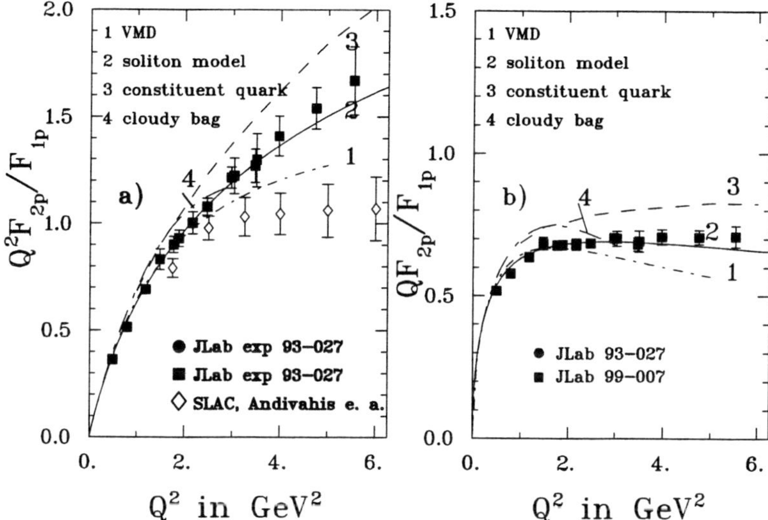

FIGURE 2. a) The ratio Q^2F_{2p}/F_{1p} from the JLab experiments, compared with the data of ref. [12]. b) the ratio QF_{2p}/F_{1p} discussed in the text.

In 2000 new measurements were made at Q^2 = 4.0, 4.8 and 5.6 GeV^2 with overlap points at Q^2 = 3.0 and 3.5 GeV^2 [15]. To extend the measurement to these higher Q^2, two changes were made. First, to increase the figure-of-merit of the FPP, a CH_2 analyzer was used; the thickness was increased from 50 cm of graphite to 100 cm of CH_2 (60 cm for Q^2 = 3.5 GeV^2). Second, the electrons were detected in a lead-glass calorimeter with 9 columns and 17 rows of $15 \times 15 \times 35$ cm^3 blocks placed so as to achieve complete solid angle matching with the HRS detecting the proton. At the largest Q^2 the solid angle of the calorimeter was 6 times that of the HRS. These data were analyzed with the new method mentioned above.

The combined results from both experiments are plotted in Fig. 1b) as the ratio $\mu_p G_{Ep}/G_{Mp}$. If the $\mu_p G_{Ep}/G_{Mp}$-ratio continues its linear decrease with the same slope, it will cross zero at $Q^2 \approx 7.5$ GeV^2. In Fig. 1b), calculations based on VMD [4], a relativistic constituent quark (CQ) [2], and a soliton model[19] are shown. Also shown are results with another relativistic CQ model (rCQM) [3], with and without CQ form factors. Lomon [5] has reworked the Gari-Krumpelmann VMD model [20] and obtains good agreement with the data for reasonable parameters for the vector-meson masses and coupling constants.

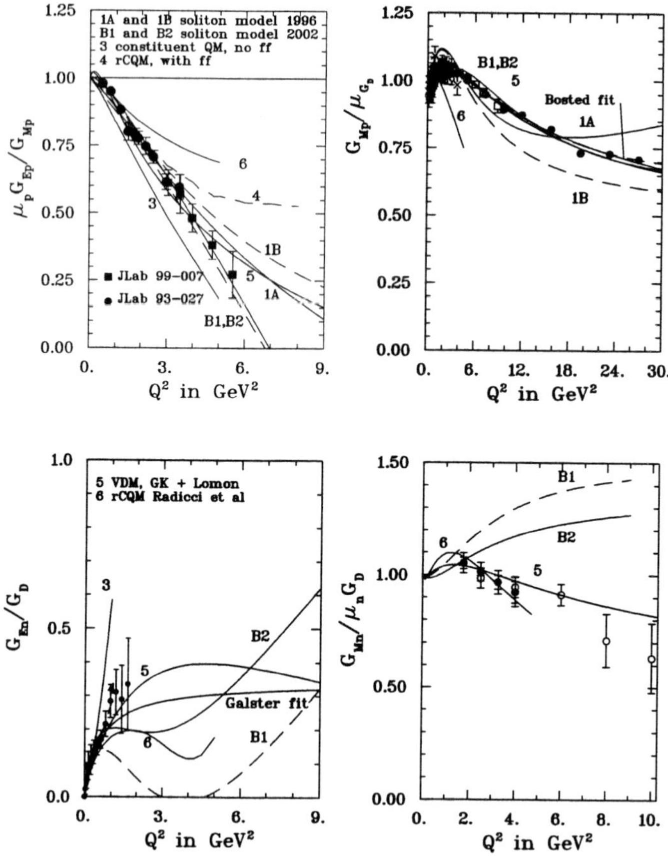

FIGURE 3. Theoretical predictions for G_E and G_M of the proton and neutron, along with selected data. For G_{En} only the results of a recent analysis of elastic ed data from ref. [21] are shown; for G_{Mn} only the larger Q^2 data of refs. [22] and [23] are shown. Curves labeled "Bosted fit" and "Galster fit" are from refs. [24] and [25], respectively.

RESULTS AND DISCUSSION

In Fig. 2a) the JLab data are shown as Q^2 times F_2/F_1. pQCD predicts quenching of the spin flip form factor F_2, or equivalently helicity conservation; higher order contributions should make Q^2F_2/F_1 asymptotically constant. Previous SLAC data supported the pQCD prediction, but the present data are in strong disagreement with it.

Shown in Fig. 2b) is Q times F_2/F_1, which reaches a constant value at $Q^2 \sim 2$ GeV^2. Ralston et al. [26] have proposed that this scaling is due to the non-zero orbital angular momentum part of the proton quark wave function. Miller and Frank [27] have shown that imposing Poincare invariance leads to violation of the helicity conservation rule, and reproduces the QF_2/F_1 behavior. More demanding for models are predictions for

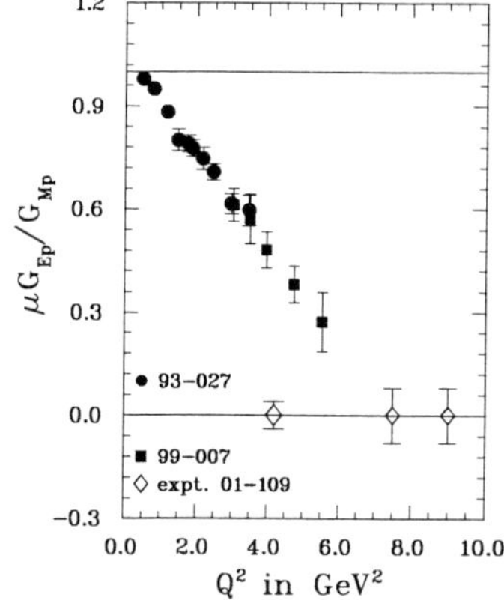

FIGURE 4. The expected statistical error bars for the two new data points at Q^2=7.5 and 9 GeV^2, as well as a control point at 4.3 GeV^2; the systematcis uncertainties should be of the same size as shown in Fig. 1b).

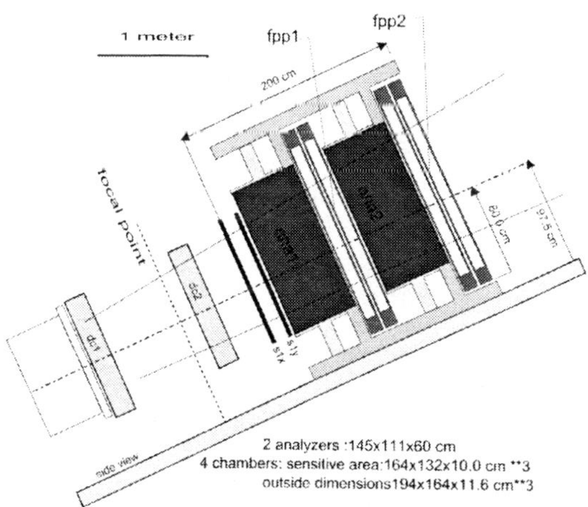

FIGURE 5. The new double focal plane polarimeter for the HMS in Hall C

all four form factors of the nucleon. The VMD fits are done in terms of the isoscalar and isovector form factors and thus naturally include all four form factors. In Fig. 3

predictions from the rCQM with SU(6) symmetry breaking [3], the soliton model [19], the point form model [28], and the VMD model of Ref. [5] are shown. The soliton model does well only for the proton. The recent VMD analysis [5] reproduces G_{Ep}, G_{Mp} and G_{Mn} well, and predicts larger values for G_{En} than the fit of Ref. [25], in agreement with the preliminary data of Ref. [29].

CONTINUATION TO Q^2=9 GEV^2

The regular, quasi-linear decrease of the $\mu_p G_{Ep}/G_{Mp}$ ratio with increasing Q^2 seen in Fig. 1b) suggests that the ratio may become zero. then negative at $Q^2 < 10$ GeV^2. Several theoretical prediction support this expectation. Accordingly, a proposal to continue these measurements to Q^2=9 GeV^2 was submitted to the JLab PAC in 2001. Following approval of this proposal, work has started on the construction of a new calorimeter as well as a new focal plane polarimeter to be installed in the Hall C high momentum spectrometer (HMS). The projected statistical errors for two news points as well as a control point at 4.3 GeV^2 are shown in Fig. 4. The electrons will be detected in a Cerenkov calorimeter with 1744 lead-glass bars of 45 cm length, which is currently being built. The new polarimeter will consists of two polarimeters in series, each with a CH_2 analyzer of 50-60 cm thickness to maximize the efficiency, as illustrated in Fig. 5. The analyzing power for CH_2 has been recently measured in a calibration experiment at the JINR in Dubna, Russia. This calibration showed that the analyzing power for 5.3 GeV/c protons was essentially constant up to a thickness of 80 cm of CH_2 [30]; the proton momentum for 9 GeV^2 is 5.66 GeV/c. However the efficiency stops increasing significantly with a CH_2 thickness of about 60 cm. The new experiment will be ready by the end of 2004.

CONCLUSION

The precise new JLab data on $\mu_p G_{Ep}/G_{Mp}$ show that this ratio continues to drop off linearly with increasing Q^2 up to 5.6 GeV^2. The ratio F_2/F_1 does not follow the $1/Q^2$ behavior predicted by pQCD, and is a distinct signature of the non-perturbative regime dominating the Q^2 range of the two JLab experiments described here. This behavior must be compared with the scaling of $Q^4 G_{Mp}$ seen in Ref. [6], which has been interpreted as indicative of pQCD for the magnetic form factor of the proton. Comparison of model calculations to the JLab data provides a stringent test of models of the nucleon.

ACKNOWLEDGMENTS

We thank our colleagues M. Jones, E. Brash, L. Pentchev and O. Gayou for their essential roles in the completion of these experiments. The Southeastern Universities Research Association manages the Thomas Jefferson National Accelerator Facility under DOE contract DE-AC05-84ER40150. U.S. National Science Foundation grant PHY 99

01182 (CFP) asnd Department of Energy grant DE-FG05-89ER40525 (VP) support our research.

REFERENCES

1. S.J. Brodsky and G.P. Lepage, Phys. Rev. D **22**, (1981) 2157.
2. M.R. Frank, B.K. Jennings and G.A. Miller, Phys. Rev. C **54**, 920 (1996).
3. E. Pace, G. Salme, F. Cardarelli and S. Simula, Nucl. Phys. A **666&667**, 33c (2000). F. Cardarelli and S. Simula, Phys. Rev. C **62**, 65201 (2000).
4. P. Mergell, U.G. Meissner, D. Drechsler Nucl. Phys. B **A596**, 367 (1996) ; and A.W. Hammer, U.G. Meissner and D. Drechsel, Phys. Lett. B **385**, 343 (1996).
5. E. Lomon, Phys. Rev. C **64** 035204 (2001) and nucl-th/0203081.
6. A.F. Sill *et al.*, Phys. Rev. D **48**, 29 (1993).
7. J. Litt *et al.*, Phys. Lett. B **31**, 40 (1970).
8. Ch. Berger *et al.*, Phys. Lett. B **35**, 87 (1971).
9. L.E. Price *et al.*, Phys. Rev. D **4**, 45 (1971).
10. W. Bartel *et al.*, Nucl. Phys. B **58**, 429 (1973).
11. R.C. Walker *et al*, Phys. Rev. **49**,5671 (1994).
12. L. Andivahis *et al.*, Phys. Rev. D **50**, 5491 (1994).
13. B. Milbrath *et al*, Phys. Rev. Lett. **80**, 452 (1998); erratum Phys. Rev. Lett. **82**, 221 (1999).
14. M. K. Jones *et al.*, Phys. Rev. Lett. **84**, 1398 (2000).
15. O. Gayou *et al.*, Phys. Rev. Lett. **88** 092301 (2002).
16. A.I. Akhiezer and M.P. Rekalo, Sov. J. Part. Nucl. **3**, 277 (1974); R. Arnold, C. Carlson and F. Gross, Phys. Rev. C **23**, 363 (1981).
17. L. Pentchev, JLab Technical Note No/ TN-01-052 (2001).
18. V. Punjabi et al., to be submitted to Physical Review C (2002).
19. G. Holzwarth, Z. Phys. A **356**, 339 (1996) and private communication (2002).
20. M.F. Gari and W. Krumpelmann, Phys. Lett. B **274**, 159 (1992).
21. R. Schiavilla and I. Sick, nucl-ex/0107004 (2001).
22. A. Lung et al., Phys. Rev.Lett. **70**, 718 (1993).
23. S. Rock et al. Phys. Rev. Lett. **49**, 1139 (1982).
24. P. E. Bosted, Phys. Rev. **51**, 409 (1995).
25. S. Galster *et al.* Nucl. Phys. **B32**, 221 (1971).
26. J. Ralston *et al.*, in Proc. of 7th International Conference on Intersection of Particle and Nuclear Physics, Quebec City (2000), p. 302, and private communication (2001).
27. G.A. Miller and M.R. Frank, nucl th 0201021 (2002).
28. R.F. Wagenbrunn, S. Boffi, W.H. Klink, W. Plessas and M. Radici, Phys. Lett. B, **511**:33 (2001).
29. see T. Madey *et al.*, to be published in the proceedings of the Elba Workshop "Electron-Nucleus Scattering VII", and contribution to this conference (B. Plaster *et al.*.
30. L.S. Azghirey *et al*, report to this conference and to be published (2002).

Dispersion Relations in Virtual Compton Scattering

B. Pasquini*, D. Drechsel†, M. Gorchtein**, A. Metz‡ and M. Vanderhaeghen†

ECT, Villazzano (Trento), Italy, and Universitá degli Studi di Trento, Povo (Trento), Italy
†Institut für Kernphysik, J. Gutenberg-Universität, D-55099 Mainz, Germany
**Dipartimento di Fisica dell'Universitá degli Studi di Genova, Italy
‡Institut für Theoretische Physik II, Ruhr Universität Bochum, D-44780 Bochum, Germany

Abstract. We describe the application of a dispersion-relation formalism to the virtual Compton scattering (VCS) reaction with the aim to extract information on generalized polarizabilities from VCS observables over a large energy range.

In the Virtual Compton scattering (VCS) process off the proton, denoted as $\gamma^* + p \to \gamma + p$, a spacelike virtual photon interacts with a proton and a real photon is produced. At low energies, this real photon plays the role of an applied quasi-static electromagnetic field, and the VCS process measures the response of the nucleon to this applied field. In particular, the virtuality of the initial photon can be dialed so as to map out the spatial distribution of the polarization effects induced by the real photon, giving access to so-called generalized polarizabilities (GPs).

The VCS process on the proton can be achieved through the $ep \to ep\gamma$ reaction. In this process, the final photon can be emitted either by the proton, which is referred to as the fully virtual Compton scattering (FVCS) process, or by the lepton, which is referred to as the Bethe-Heitler (BH) process. The BH amplitude is exactly calculable from QED if one knows the nucleon electromagnetic form factors. The FVCS amplitude contains, in the one-photon exchange approximation, the VCS subprocess. The latter one can be further separated in a Born term and a non-Born contribution. The Born (B) amplitude refers to the process where the virtual photon is absorbed on a nucleon and the intermediate state remains a nucleon, i.e., contains only properties of the proton in its ground state. The residual non-Born (NB) contribution contains the information of the excitation spectrum and the meson-loop contributions. These nucleon-structure information can be parametrized in terms of 12 non-Born invariant amplitudes, $F_i^{NB}(Q^2, \nu, t)$, $i = 1, ..., 12$, which are functions of 3 invariants: Q^2, $\nu = (s-u)/(4M)$, and t (s, t and u are the Mandelstam variables for VCS, and M is the proton mass). These amplitudes fulfill unsubtracted dispersion relations (DRs) at fixed t and fixed virtuality Q^2,

$$\begin{aligned} \mathrm{Re} F_i^{NB}(Q^2, \nu, t) &= F_i^{pole}(Q^2, \nu, t) - F_i^B(Q^2, \nu, t) \\ &+ \frac{2}{\pi} \mathscr{P} \int_{\nu_0}^{+\infty} d\nu' \, \frac{\nu' \, \mathrm{Im}_s F_i(Q^2, \nu', t)}{\nu'^2 - \nu^2} , \end{aligned} \tag{1}$$

CP675, *Spin 2002: 15th Int'l. Spin Physics Symposium and Workshop on Polarized Electron Sources and Polarimeters*, edited by Y. I. Makdisi, A. U. Luccio, and W. W. MacKay

where F_i^B is the Born contribution defined as in Ref. [1], whereas F_i^{pole} represents the nucleon-pole contribution [2]. In Eq. (1), $\mathrm{Im}_s F_i$ are the discontinuities across the s-channel cuts of the VCS process, starting at the pion-production threshold ν_0, which is the first inelastic channel. However, as can be inferred from Regge theory [3, 4], the high-energy behaviour ($\nu \to \infty$ at fixed t and fixed Q^2) of the $\mathrm{Im}\, F_1$ and $\mathrm{Im}\, F_5$ amplitudes does not guarantee the convergence of the unsubtracted dispersion integrals. In order to avoid this convergence problem, we perform the unsubtracted dispersion integrals of Eq. (1) for F_1 and F_5 along the real ν-axis in the range $-\nu_{max} \leq \nu \leq +\nu_{max}$, and we close the contour by a semi-circle with radius ν_{max} in the upper half of the complex ν-plane, with the result

$$\mathrm{Re}\, F_i^{NB}(Q^2,\nu,t) = F_i^{pole}(Q^2,\nu,t) - F_i^B(Q^2,\nu,t) + F_i^{int}(Q^2,\nu,t) + F_i^{as}(Q^2,\nu,t), \quad (2)$$

for $i = 1,5$. The dispersion integral contributions are typically evaluated up to a maximum photon energy $E_{max} \simeq 1.5$ GeV. The s-channel discontinuities can be expressed, through unitarity equation, in terms of the contribution from all possible intermediate states which can couple to the initial and final photon-nucleon system. Since we are mainly interested in VCS through the $\Delta(1232)$-resonance region, we only consider the main contribution coming from πN intermediate states, using the pion photo- and electroproduction multipoles of the phenomenological MAID analysis [5]. All contributions from higher energies are then absorbed in the asymptotic term. In particular, the asymptotic contribution to the amplitudes F_5 results from t-channel π^0-exchange, while the asymptotic contribution to the amplitude F_1 originates predominantly from t-channel $\pi\pi$ intermediate states. In addition, higher-energy dispersive contributions beyond the one-pion channel mainly affect the F_1 and F_2 amplitudes. Since the present data for the production of those intermediate states are too scarce to evaluate the imaginary parts of the amplitudes directly, we estimate these contributions by an energy-independent constant, fixed at arbitrary Q^2, $\nu = 0$ and $t = -Q^2$, i.e. at the point where the GPs are defined. In this way we introduce two free parameters which can be expressed in terms of the electric, $\alpha(Q^2)$, and magnetic, $\beta(Q^2)$, GPs, which have to be fitted to experimental VCS data at each fixed value of Q^2. However, in order to provide predictions for VCS observables at different values of Q^2, we take the following parametrization for the Q^2 dependence of the scalar GPs

$$\alpha(Q^2) - \alpha^{\pi N}(Q^2) = \frac{(\alpha - \alpha^{\pi N})}{(1+Q^2/\Lambda_\alpha^2)^2}, \qquad \beta(Q^2) - \beta^{\pi N}(Q^2) = \frac{(\beta - \beta^{\pi N})}{(1+Q^2/\Lambda_\beta^2)^2}, \quad (3)$$

where the values at $Q^2 = 0$, $(\alpha - \alpha^{\pi N})$ and $(\beta - \beta^{\pi N})$, are fitted to real Compton scattering (RCS) data [6].

First information on unpolarized VCS observables have been obtained from experiments at the MAMI accelerator [7] at a virtuality $Q^2 = 0.33$ GeV^2, and recently at JLab [8] at $1 < Q^2 < 2$ GeV^2. VCS data in the Q^2 region around 0.05 - 0.1 GeV^2 are under analysis at MIT-Bates [9], and further experimental programs are underway at MAMI [10], and JLab [11] to measure both unpolarized and polarized VCS observables. VCS experiments at low outgoing photon energies can be analyzed in terms of low-energy expansions (LEXs), proposed in Ref. [1]. In the LEX, only the leading term (in the energy of

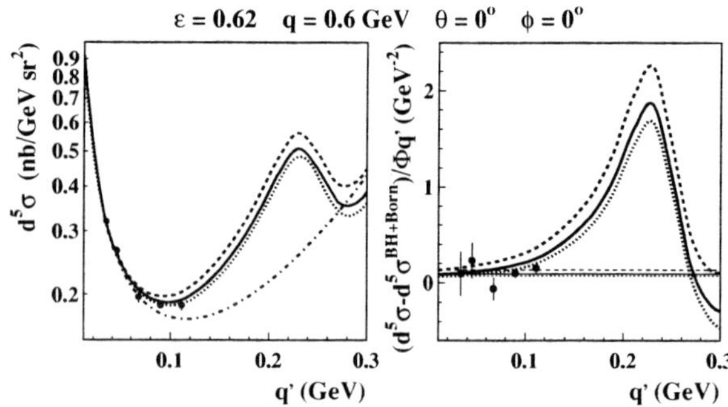

FIGURE 1. Left panel: differential cross section for the reaction $ep \to ep\gamma$ as function of the outgoing-photon energy q′ in MAMI kinematics. The BH + B contribution is given by the dashed-dotted curve. Right panel: Results for $(d^5\sigma - d^5\sigma^{\text{BH+Born}})/\Phi q'$ as function of q′ . The total DR results are obtained with the asymptotic parts calculated according to Eq. (3) with $\Lambda_\alpha = 1$ GeV and three different values of Λ_β : 0.7 GeV (dotted curve), 0.6 GeV (solid curve), and 0.4 GeV (dashed curve). In the right panel, the DR calculations with the full energy dependence of the non-Born contribution (thick curves) are compared to the corresponding results within the LEX formalism (thin horizontal curves). The data are from Ref. [7].

the produced real photon) of the response to the quasi-constant electromagnetic field, due to the internal structure of the system, is taken into account. This leading term depends linearly on the GPs. As the sensitivity of the VCS cross sections to the GPs grows with the photon energy, it is however advantageous to go to higher photon energies, outside the range of validity of the LEX, provided one can keep the theoretical uncertainties under control when approaching and crossing the pion threshold. This situation can be compared to RCS , where it was shown that one uses DR formalism to extract the polarizabilities at energies above threshold, with generally larger effects on the observables. From the results shown in Fig. 1., we note indeed that the region between pion threshold and the $\Delta(1232)$-resonance peak displays an enhanced sensitivity to the GPs through the interference with the rising Compton amplitude due to Δ-resonance excitation and seems quite promising to measure VCS observables. Such an experiment has been proposed at MAMI and is underway [10].

Going to higher Q^2, the VCS process has also been measured at JLab and data have been obtained both below pion threshold at Q^2 = 1 GeV^2, at Q^2 = 1.9 GeV^2, as well as in the resonance region around Q^2 = 1 GeV^2 (see Ref. [8] for a short review of these JLab data). In Fig. 2., we show the results for the $ep \to ep\gamma$ reaction in the resonance region at $Q^2 = 1$ GeV^2 and at a backward angle. These are the first VCS measurements ever performed in the resonance region. The DR calculations reproduce well the Δ region. However, due to scarce information for the dispersive input above the Δ resonance, DRs cannot be extended at present into the second and third resonance regions. Between pion threshold and the Δ resonance, the calculations show a sizable sensitivity to the GPs, in particular to $\alpha(Q^2)$ in this backward angle kinematics, and seem very promising to extract information on the electric polarizability. Besides the measurement in the resonance region, from the JLab data below pion threshold two unpolarized structure functions $P_{LL} - P_{TT}/\varepsilon$ and P_{LT} have been extracted at Q^2 = 1 GeV^2 and at Q^2 =

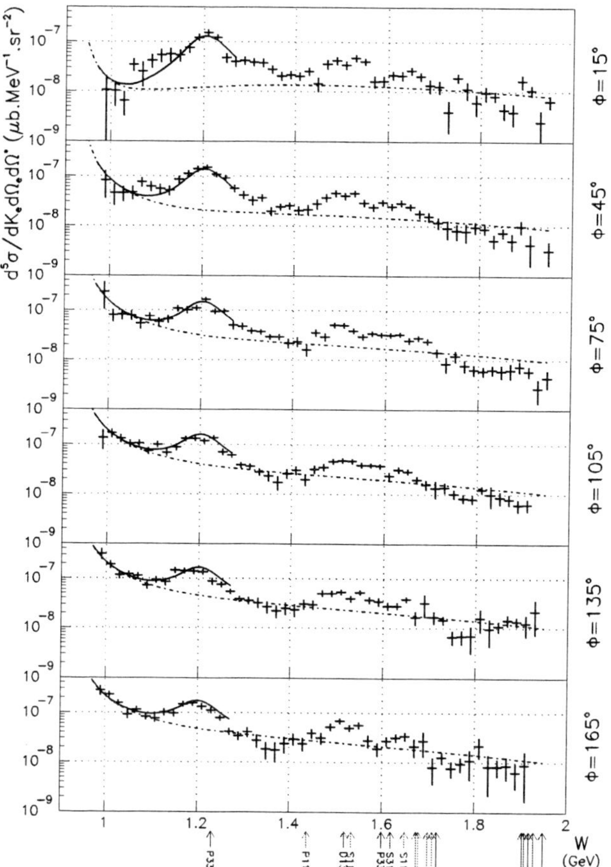

FIGURE 2. Differential cross sections for the $ep \to ep\gamma$ reaction as function of the *c.m.* energy W in JLab kinematics : $Q^2 = 1.0$ GeV^2, electron energy $E_e = 4.032$ GeV, scattering angle $\theta_{cm}^{\gamma*\gamma} = -167.2^o$, and different out-of-plane angles ϕ. BH + B contribution: dashed curve. Total DR result (with the asymptotic parts calculated from Eq. (3) with $\Lambda_\alpha = 1$ GeV and $\Lambda_\beta = 0.45$ GeV): solid curve. The data are from Ref. [8].

1.9 GeV^2 [8]. For this extraction below pion threshold, both the LEX and the DR formalisms can be used. A nice agreement between the results of both methods for the structure functions was found in Ref. [8]. These preliminary results for P_{LL} and P_{LT} are displayed in Fig. 3., alongside the RCS point and the results obtained at MAMI at $Q^2 = 0.33$ GeV^2. By dividing out the form factor G_E, P_{LL} is proportional to the electric GP $\alpha(Q^2)$, whereas P_{LT} is proportional to the magnetic GP $\beta(Q^2)$ plus some correction due to spin-flip GPs which results small in the DR formalism. One sees from Fig. 3. that the best fit value for $\Lambda_\alpha \simeq 0.92$ GeV yields an electric polarizability which is dominated by the asymptotic contribution and has a similar Q^2 behavior as the dipole form factor. However, the best fit value for $\Lambda_\beta \simeq 0.66$ GeV is substantially lower, indicating that

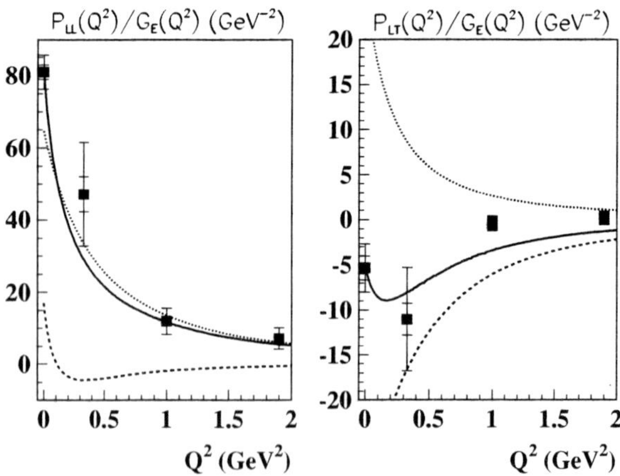

FIGURE 3. Results for the unpolarized VCS structure functions P_{LL} (left panel) and P_{LT} (right panel) divided by the proton electric form factor. Dashed lines: dispersive πN contributions. Dotted lines: asymptotic contributions calculated according to Eq. (3) with $\Lambda_\alpha = 0.92$ GeV (left panel) and $\Lambda_\beta = 0.66$ GeV (right panel). Solid curves: total results, sum of the dispersive and asymptotic contributions. The RCS data are from Ref. [6], the VCS MAMI data at $Q^2 = 0.33$ GeV^2 are from Ref. [7], and the preliminary VCS JLab data at $Q^2 = 1$ GeV^2 and $Q^2 = 1.9$ GeV^2 from Ref. [8] (inner error bars are statistical errors only, outer error bars include systematical errors). The values for P_{LL} at $Q^2 > 0$ were extracted by use of a dispersive estimate for the not yet separated P_{TT} contribution.

the diamagnetism, which is related to pion-clouds effects, drops faster with Q^2. On the other hand, the structure function P_{LT} receives also a large dispersive πN (paramagnetic) contribution, dominated by the Δ-resonance excitation. Due to a the large cancellation between the competing para- and dia- magnetic effects, the net result is relatively small, and gives rise to an interesting structure, in particular in the Q^2 region around $0.05 - 0.1$ GeV^2, where forthcoming data are expected from the MIT-Bates experiment [9].

REFERENCES

1. Guichon, P.A.M., Liu, G.Q., and Thomas, A.W., Nucl. Phys. **A591**, 606 (1995).
2. Drechsel, D., Pasquini, B., Vanderhaeghen, M., to appear in *Phys. Rep.*
3. Pasquini, B., Gorchtein, M., Drechsel, D., Metz, A., Vanderhaeghen, M., *Eur. Phys. J.* **A11**, 185 (2001).
4. Pasquini, B., Drechsel, D., Gorchtein, M., Metz, A., Vanderhaeghen, M., *Phys. Rev.* **C62**, 052201 (R) (2000).
5. Drechsel, D. O. Hanstein, O., S. Kamalov, S., and Tiator, L., *Nucl. Phys.* **A645**, 145 (1999).
6. Olmos de León, V., et al., *Eur. Phys. J.* **A10**, 207 (2001).
7. Roche, J., *et al.*, *Phys. Rev. Lett.* **85**, 708 (2000).
8. Fonvieille, H., in Proceedings of *the 9th International Conference on the Structure of Baryons (Baryons 2002)*; Eds. C. Carlson and B. Mecking, World Scientific, Singapore (2003); hep-ex/0206035.
9. Miskimen, R., spokespersons MIT-Bates experiment, 97-03; and *these proceedings*
10. d'Hose,N., and Merkel, H., spokespersons MAMI experiment, (2001).
11. Hyde-Wright, C., Laveissière, G., private communication.

First Photo-Pion Double Polarization Experiments Using Polarized $\vec{H}\vec{D}$ at LEGS

S. Hoblit*, K. Ardashev†*, C. Bade†*, M. Blecher**, C. Cacace*, A. Caracappa*, A. Cichocki‡, C. Commeaux§, A. d'Angelo¶, R. Deininger†, J.-P. Didelez§, C. Gibson||, K. Hicks†, A. Honig††, T. Kageya***, M. Khandaker‡‡, O. Kistner*, A. Lehmann||*, F. Lincoln*, M. Lowry*, M. Lucas†, J. Mahon†, H. Meyer***, L. Miceli*, D. Morizzianni¶, B. Norum‡, B.M. Preedom||, T. Saitoh***, A.M. Sandorfi*, R. di Salvo¶, C. Schaerf¶, C. Thorn*, K. Wang‡, X. Wei* and C.S. Whisnant§§

*Brookhaven National Laboratory, Upton, NY 11973, USA
†Ohio University, Athens, OH 45701, USA
**Virginia Polytechnic Inst. & State Univ., Blacksburg, VA 24061, USA
‡University of Virginia, Charlottesville, VA 22904, USA
§Université de Paris-sud / IN2P3, Orsay, France
¶Università di Roma-"Tor Vergata" and INFN-Sezione di Roma2, Rome, Italy
||University of South Carolina, Columbia, SC 29208, USA
††Syracuse University, Syracuse, NY 13244, USA
‡‡Norfolk State University, Norfolk, VA 23504, USA
§§James Madison University, Harrisonburg, VA 22807, USA

Abstract. We report preliminary results of π photo-production using polarized γ beams and a polarized HD target. Four observables can be extracted simultaneously from the data, the cross section, the beam asymmetry Σ, and the double-polarization observables G and E. The latter determines the GDH sum rule integral.

In this paper we described the first photo pion double polarization measurements using polarized HD at the Laser-Electron-Gamma-Source (LEGS) facility at Brookhaven National Laboratory. The LEGS facilities produces polarized gamma beams ranging from π threshhold to 470 MeV with either linear or circular polarization, by backscattering laser light from relativistic electrons circulating in a 2.8 GeV storage ring. The main focus of our current experimental program is the measurement of polarized asymmetries and cross sections with the aim of constraining π-production amplitudes, particularly from the neutron for which there is relatively little available data. One of the observables measured at LEGS, with circularly polarized photons on longitudinally polarized nucleons, is the asymmetry that enters the Gerasimov-Drell-Hearn (GDH) sum rule integral. The LEGS energy range covers the first $\sim 65\%$ of the sum rule integral.

Measurements of the GDH sum rule require integrals of hadron cross sections and for this we have assembled a large solid angular detector. In the central angular range, from $\sim 50° - 130°$ lab, 432 sodium iodide detectors surround the target. Forward angles,

CP675, *Spin 2002: 15th Int'l. Spin Physics Symposium and Workshop on Polarized Electron Sources and Polarimeters,* edited by Y. I. Makdisi, A. U. Luccio, and W. W. MacKay

from $\sim 10^\circ - 35^\circ$ lab, are covered by walls of segmented plastic scintilators and lead glass arrays. This system has nearly complete coverage for π° detection and about 3π coverage for charged pions.

The most novel feature of this experiment is the use of a new class of polarized target consisting of molecular HD in the solid phase. At the low temperatures required for target polarization the lowest state of the two identical protons in the H_2 molecule is the *para* configuration with the proton spins opposed, which cannot be polarized. The HD molecule, consisting of non- identical particles, does not have this symmetry restriction and in fact its lowest state has H and D spins parallel with relative orbital angular momentum $L = 0$. There are no phonons that couple an S-wave HD molecule to its crystal lattice and so, once polarized, HD has an extremely long relaxation time. While this is of course very desirable another mechanism must be found to polarize it. For this we introduced a small (a few parts in 10^4) concentration of H_2 which initially has 3/4 of its molecules in the orbital configuration with $S = 1$ and $L = 1$ (*ortho*-). These readily polarize in high magnetic fields and spin-spin coupling between an H in H_2 and an H in HD transfers polarization to HD. This process occurs very rapidly (on the order of hours). Once cooled, the *ortho*-H_2 decays to the magnetically inert *para*-H_2 state with a half life of about six days. Thus, maintaining the target at high field and low temperature for several *ortho* $\rightarrow$ *para* time constants results in a frozen spin target.

The targets we have used so far are two and a half centimeters diameter by five centimeters long. They were made in a dilution refrigerator and the spins were frozen out at 18 mK and 15 tesla over a period of three months. They were then extracted from the dilution refrigerator and placed into a in-beam cryostat at 1.25 K and 0.7 tesla. Apart from the frozen HD the only other material seen by the beam are the cell walls made of CTFE (Chlorotrifluoroethylene) and 2050 μm aluminum wires which represent 20% of the target by weight. These are needed to take away the heat from *ortho*-H_2 to *para*-H_2 conversion. Nonetheless, the yield from this background is easily measured by pumping out the HD material and collecting data from an empty cell.

The target used in these measurements was condensed July 27, 2001 and taken to high fields and low temperatures. On September 17 the field was lowered and polarization measured to be 70% for H and 17% for D. The targets then went through a long period of tests, including five transfers between the dilution refrigerator and the in-beam cryostat. The experimental run did not start until mid November, at which point the polarization had decreased to 30% for H and 6% for D. The measured in-beam relaxation times were 13 days for H and 36 days for D. Unfortunately, an accelerator shutdown occurred 3 1/2 days after the start of the data taking, but the results from this short period are nonetheless impressive.

Missing energy spectra for $\gamma + p \rightarrow \pi^+ n$ and $\gamma + n \rightarrow \pi^\circ n$ are shown in Fig. 1 for the two helicity states $h = 1/2$ (parallel beam+target spins), $h = 3/2$ (antiparallel spins), and their difference. The spectra in both helicity states for $\pi^+ n$ has two components, a narrow spike on top of a broad peak coming from bound protons in deuterium. The latter is almost absent in the helicity difference, since the deuterium polarization was low. The $\pi^\circ n$ spectra are considerably broader since they only come from bound neutrons in deuterium.

In these measurements we flipped between six gamma ray polarization states with randomly chosen intervals: left circular, right circular, 0° linear, -45° linear, $+45^\circ$ linear,

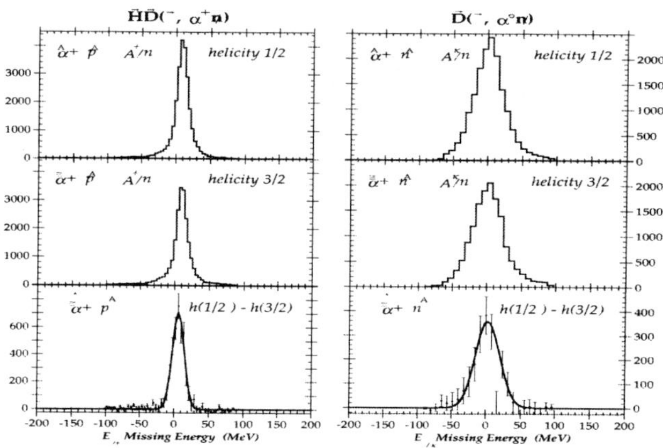

FIGURE 1. Missing energy spectra for π^+n from polarized HD (left panels), and for π°n from polarized D (right panels), for the indicated initial helicity configurations.

and +90° linear. The small unpolarized component in the beam due to bremsstrahlung of the electron in the residual gas of the storage ring vacuum chamber ($\sim 1\%$) was also monitored regularly by blocking the laser from entering the straight section. Since the γ-ray production process is simply Klein-Nishina scattering of light from free electrons in the vacuum, the γ-ray polarization is very precisely determined by measuring the laser Stokes vector (Q, U, V). Here Q is the intensity of linear polarization at 0°, U is the linear intensity at 45° and V is the intensity of right-circular polarization. With a longitudinally polarized target of polarization P_z, the yield at a given angle and energy is determined by four independent quantities: the cross section, the beam asymmetry Σ (independent of target polarization) and two double-polarization quantities, E and G, shown schematically in Fig. 2. The G asymmetry is obtained in reactions on longitudinally polarized nucleons by flipping the beam polarization between linear states oriented at +45° and −45° to the reaction plane. E is the asymmetry entering GDH sum rule, as measured with longitudinal target polarization and circular beam polarization. The dependence of the cross sections on these observables is shown in Eq. (1), where the subscript i refers to the incident polarization state characterized by polarization (Q_i, U_i, V_i).

$$\begin{aligned}\frac{d\sigma_i}{d\Omega}(\theta,\phi,E_\gamma) = \frac{d\sigma}{d\Omega}(\theta,E_\gamma)\cdot\{1+ &\left[Q_i\cdot(E_\gamma)\Sigma(\theta,E_\gamma) - P_z\cdot U_i\cdot(E_\gamma)G(\theta,E_\gamma)\right]\cdot\cos 2\phi \\ + &\left[P_z\cdot Q_i(E_\gamma)\cdot G(\theta,E_\gamma) + U_i(E_\gamma)\cdot\Sigma(\theta,E_\gamma)\right]\cdot\sin 2\phi \\ - &\; P_z\cdot V_i(E_\gamma)\cdot E(\theta,E_\gamma)\}\end{aligned} \tag{1}$$

From the six polarization states we form three asymmetries, the left/right circular, −45°/+45° linear and 0°/90° linear asymmetry. These three asymmetries are then simultaneously fit using the measured stokes parameters for each state to yield Σ, E, and

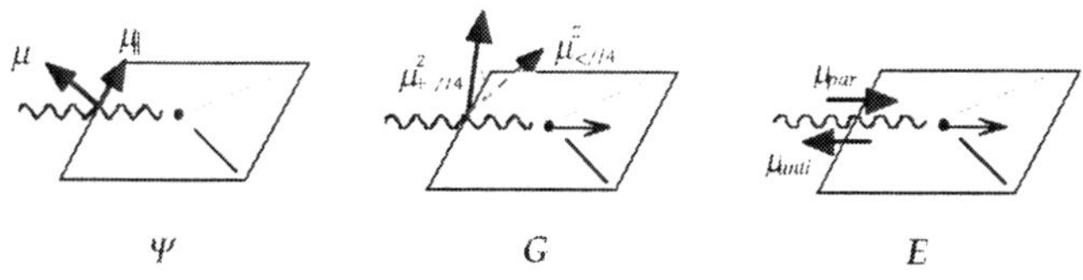

FIGURE 2. Polarization states entering the three measured asymmetries.

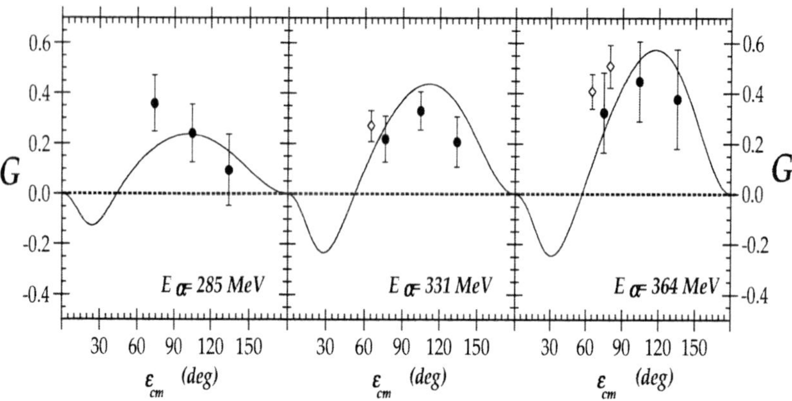

FIGURE 3. Preliminary asymmetries for the $\gamma + p \rightarrow \pi^+ n$ (solid circles), compared with data from Ref. [1], open symbols. The curves are the SM02k multipole solution from SAID.

G. The resulting G asymmetries for π^+n are shown in Fig. 3. Also shown are three of the four points from Khar'kov which comprises the world's published data.

The E asymmetry is a ratio formed from the cross sections with photon and target-nucleon spins parallel (helicity = $\vec{S} \cdot \vec{P}/|P| = 1/2$) and anti-parallel (helicity = 3/2). In terms of the helicity cross sections, $E = (\sigma_{3/2} - \sigma_{1/2})/(\sigma_{3/2} + \sigma_{1/2})$, or $\hat{E}/\sigma^{unp}$ where σ^{unp} is the unpolarized cross section and we denote the numerator by $\hat{E} = (\sigma_{3/2} - \sigma_{1/2})/2$. Examples of these asymmetries are shown for two energy bins in Fig. 4 for the exclusive reactions $\vec{H}\vec{D}(\vec{\gamma}, \pi^+ n)$, left panels, and $\vec{H}\vec{D}(\vec{\gamma}, \pi^\circ p)$, right panels. (At present, only events with pions detected in the central NaI calorimeter are included in the plots of Fig. 4. The analysis of forward going pions is underway.) The π^+n asymmetries are very close to multipole predictions from either SAID or MAID. In contrast, the π°p E-asymmetries near 90° are always larger. This arises from the bound protons in deuterium. Although the D polarization is small and the free H dominates the numerator $\hat{E}$, the contribution of bound protons in D does not cancel in σ^{unp}, the denominator of E. The exclusive cross sections for $D(\gamma, \pi^\circ p)$ have been measured at LEGS and in the central angular range they are about half of the corresponding cross sections on the free proton.[2] It is this large reduction in σ^{unp} that produces the larger asymmetries in

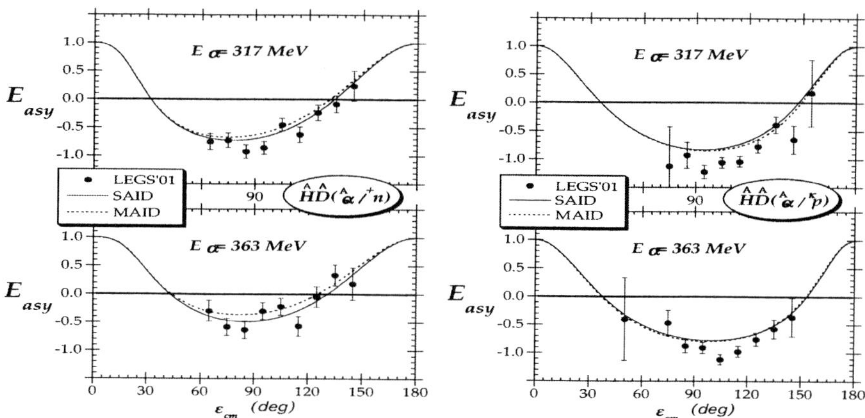

FIGURE 4. Preliminary E asymmetries for exclusive π^+n (left panels) and π°n (right panels) at energies near the Δ, compared with predictions from recent multipole solutions.

Fig. 4.

Charged π production from the neutron requires separating the $D(\gamma,\pi^- p)$ and $D(\gamma,\pi^+ n)$ channels. The recoil proton can serve to tag the reaction of interest, provided that it has sufficient energy to emerge from the target cryostat. Unfortunately, below about 280 MeV this is not the case for more than half of the angular distribution. The only direct method of isolating charged π reactions on the neutron is to track π's in a magnetic field and measure the pion charge. For this we are constructing a *Time Projection Chamber* (TPC) which will be housed in a large-bore 1.8 Tesla field. We expect this to yield the most accurate determination of the neutron multipoles at low energy.

ACKNOWLEDGMENTS

This work is supported by the U. S. Dept. of Energy under contract No. DE- AC02-98CH10886, the U.S. National Science Foundation, and the Istituto Nazionale di Fisica Nucleare - Italy.

REFERENCES

1. A.A. Belyaev *et al.*, Sov. *J. Nucl. Phys.* **40**, 83 (1984).
2. LEGS *collaboration*: K.H. Hicks *et al.*, Proc. Int. Sym. on Electromagnetic Interactions in Nucl. and Hadron Phy., Osaka (ed. M. Fujiwara and T. Shima) page 144, World Scientific River Edge, NJ (2002).

Spin Physics in Deep-Inelastic Semi-Inclusive Reactions with an 11-GeV Electron Beam at Hall A of Jefferson Laboratory

Xiaodong Jiang

Department of Physics and Astronomy, Rutgers University, Piscataway, New Jersey.

Abstract. We outline the physics oppertunities of semi-inclusive deep inelastic measurements with a polarized NH_3 and a polarized ^{3}He target in Jefferson Lab Hall A after the planned 12 GeV CEBAF machine upgrade. In this paper, we estimate statistical uncertainties associated with double-spin and single-spin asymmetries in $(e,e'\pi)$ type measurements.

INTRODUCTION

With the planned 12 GeV upgrade at the Thomas Jefferson National Accelerator Facility in Newport News, Virginia, the combination of a high current CW polarized electron beam and the use of high density polarized targets presents many new physics opportunities, especially in the measurements of spin observables of deep inelastic semi-inclusive scattering (SIDIS) reactions. If factorization between quark scattering and quark fragmentation can be clearly demonstrated, SIDIS can provide direct accesses to quark and quark polarization distributions. The unique feature of quark-flavor tagging capability allows us to study the flavor decomposition of the nucleon spin structure and to access new distribution functions such as the quark transversity distributions.

At the experimental Hall A, after the 12 GeV CEBAF machine upgrade, 85% polarized electron beam with a current of 40 μA can be delivered up to a beam energy of 11 GeV. Luminosities of 10^{38} $cm^{-2}s^{-1}$ can be achieved for unpolarized hydrogen or deuterium targets, a factor of 2×10^5 improvement over the typical HERMES luminosity of unpolarized targets. Assuming no improvement on the polarized target technology, a polarized proton luminosity of 10^{35} $cm^{-2}s^{-1}$ with 80% polarization can be achieved on a polarized NH_3 target, and a ^{3}He luminosity of 10^{36} $cm^{-2}s^{-1}$ with 45% polarization can be achieved with a polarized ^{3}He gas target, an improvement of four orders of magnitudes over the HERMES luminosity in each case.

A new moderate resolution ($\delta p/p = 10^{-3}$) magnetic spectrometer, the Medium Acceptance Device (MAD), is under consideration for Hall A. The MAD spectrometer will operate up to 6 GeV/c in central momentum with a momentum acceptance of $\pm15\%$. The geometrical acceptance of MAD will be ranging from 6 msr at $13°$ to 30 msr at $35°$. In a typical SIDIS measurement, the MAD spectrometer can be used as the electron arm with the the existing Hall A HRS spectrometer serves as the hadron arm. The HRS spectrometer, operates up to 4.3 GeV/c in central momentum, has a momentum acceptance of $\pm4.5\%$ and a solid angle of 6 msr. With an additional SEPTUM magnet, the HRS

CP675, *Spin 2002: 15th Int'l. Spin Physics Symposium and Workshop on Polarized Electron Sources and Polarimeters,* edited by Y. I. Makdisi, A. U. Luccio, and W. W. MacKay

spectrometer is able to access 6° in scattering angle.

AVAILABLE KINEMATICS

At a beam energy of 11 GeV, the accessible kinematics region in (x, Q^2) and (x, W^2) plane are shown in Fig. 1. Constant electron scattering angle (θ_e) lines are plotted from 10° to 60° in addition to the constant $\vec{q}$ angle (θ_q) lines. In order to stay in the deep inelastic region at the highest possible W, the center of the fragmentation cone is limited to a forward angle of $\theta_q > 15°$.

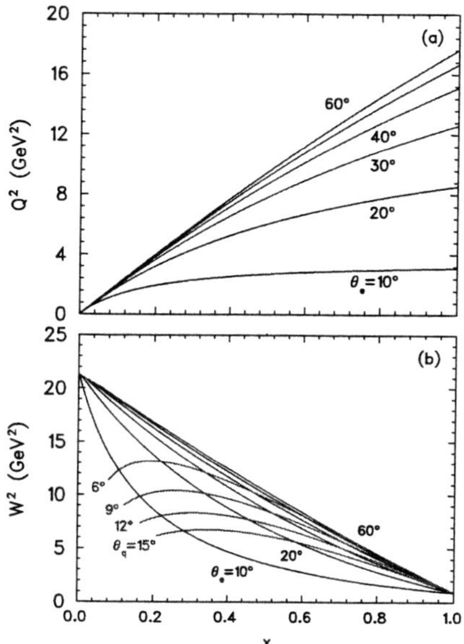

FIGURE 1. Accessible kinematics region for an $(e,e'h)$ measurement at E_0=11 GeV.

In a double-spin asymmetry measurement, the fragmented hadron will be detected along the momentum transfer direction $\vec{q}$, and the target spin will be aligned along the same direction, as illustrated in Fig. 2(left). In a single-spin asymmetry measurement, the fragmented hadron will be detected on the side of $\vec{q}$, within the electron scattering plane as illustrated in Fig. 2(right). The target spin will be aligned in a plane perpendicular to $\vec{q}$, and the Collins angle ϕ is defined as the angle between the target spin and the $(\vec{q}, \vec{p}_\pi)$ plane.

DOUBLE SPIN ASYMMETRIES IN SIDIS

While unpolarized semi-inclusive meson production provides means of extracting spin-averaged quark and antiquark distributions in the nucleon, semi-inclusive production with a *polarized* beam on a *polarized* target offers the prospect of determining the spin-dependence of the individual quark species. Furthermore, by comparing semi-inclusive

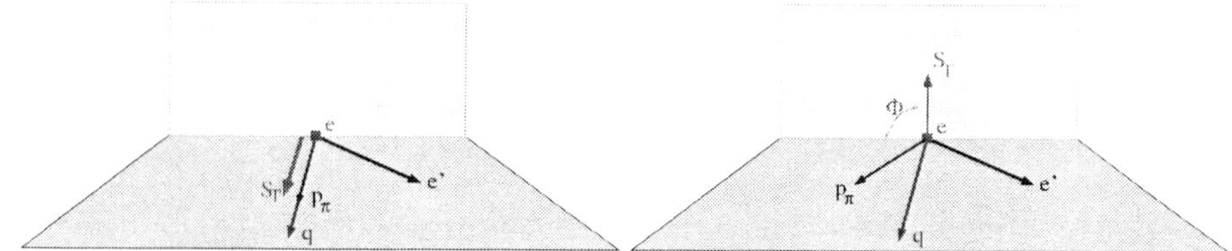

FIGURE 2. Target spin configuration in a double-spin asymmetry measurement (left), and in a single-spin asymmetry measurement (right), viewed toward the electron beam.

data with inclusive DIS measurements, one can directly test the degree to which flavor SU(3) symmetry holds in DIS processes. At large Q^2, the spin asymmetry A_1^h for the production of a hadron h by a polarized virtual photon on a polarized nucleon can be written:

$$A_1^h(x,z) = P_e \cdot P_T \cdot \frac{y(1-\frac{1}{2}y)}{1-y+\frac{1}{2}y^2} \cdot \frac{\sum_q e_q^2 \, \Delta q(x) D_q^h(z)}{\sum_q e_q^2 \, q(x) D_q^h(z)} . \tag{1}$$

where P_e and P_T are beam and target polarization. Measurement of π^+ and π^- (or K^+ and K^-) mesons from proton or neutron targets, together with knowledge of the unpolarized distributions $q(x)$, allows one to extract from Eq. (1) information on the spin-dependent distributions $\Delta q(x)$ and $\Delta\bar{q}(x)$.

Assuming the use of a polarized $^{15}NH_3$ target and a polarized ^3He target in their standard configurations, a total of 1000 hour measurements on each target at $z = E_\pi/\nu = 0.40 \sim 0.5$ in each setting will yield high statistical accuracies on $A_1^{\pi^+}$ and $A_1^{\pi^-}$, as shown in Fig. 3. The measurement time is arranged such that similar statistical accuracies can be achieved for $A_1^{\pi^+}$ and $A_1^{\pi^-}$. Assuming factorization has been clearly demonstrated, the "purity" method used by the HERMES collaboration can be adopted to extract the quark polarization distributions from the measured semi-inclusive asymmetries [1]. The corresponding statistical accuracies are shown in Fig. 4 together with the HERMES published results [1] for comparison.

SINGLE-SPIN ASYMMETRIES IN SIDIS

Following Ref. [2], we calculated single-spin asymmetry A_T as a function of x, y and z for 11 GeV electron beam energy. Typical results of a 1000 hour measurement are shown in Fig. 5, where the z-dependence of A_T are plotted for $x = 0.2$, $Q^2 = 2.5$ GeV^2 and $x = 0.3$, $Q^2 = 3.0$ GeV^2 kinematics. We assume $\delta q(x) = \Delta q(x)$ in this calculation. The AAC parameterization of the polarized nucleon structure functions were used for $\Delta q(x)$, and the CTEQ5M parameterization were used for the unpolarized structure functions. For the fragmentation functions, the parameterization of Aubert et al. [3] was adopted. We assume a typical polarized $^{15}NH_3$ target with 80% polarization, and a ^3He target of 45% polarization. The dilution factors due to the unpolarized nucleons in the target material have been taken into account. It is interesting to note that the π^- production on

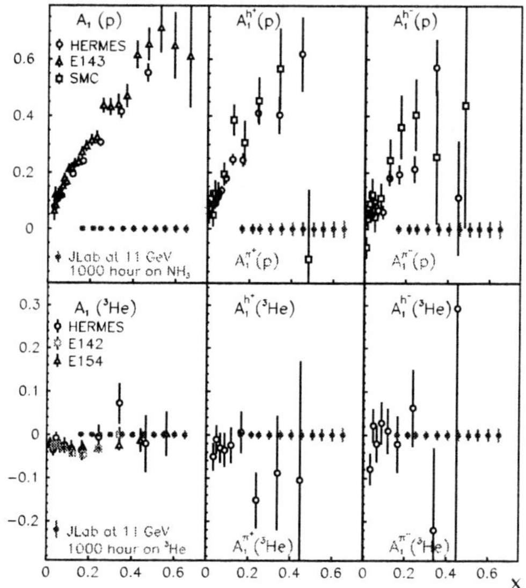

FIGURE 3. Expected semi-inclusive asymmetry measurements with polarized NH_3 and ^{3}He targets. 1000 hours of beam time is assumed for each target. Error bars are statistical only.

a polarized ^{3}He target has a much larger asymmetry with an opposite sign compare to that of the π^+ asymmetry. This trend is very different from the situation of a polarized proton target due to the contribution of d-quark transversity $\delta d(x)$ in the neutron.

ACKNOWLEDGMENTS

This work represents a collective effort by the Jefferson Lab Hall A collaboration. The author thanks H. Avagyan, T. Averett, J.-P. Chen, H. Gao, R. Gilman, D. Higinbotham, X. Ji, J. LeRose, W. Melnitchouk, Z.-E. Meziani, J.-C. Peng, R. Ransome for many discussions. This work is supported by the US Department of Energy and the National Science Foundation.

REFERENCES

1. K. Ackerstaff *et al.*, Phys. Lett. B **464**, 123 (1999).
2. V.A. Korotkov, W.-D. Nowak and K.A. Oganessian, Eur. Phys. J. C **18**, 639 (2001).
3. J.J. Aubert *et al.*, Phys. Lett. B **160**, 417 (1985).

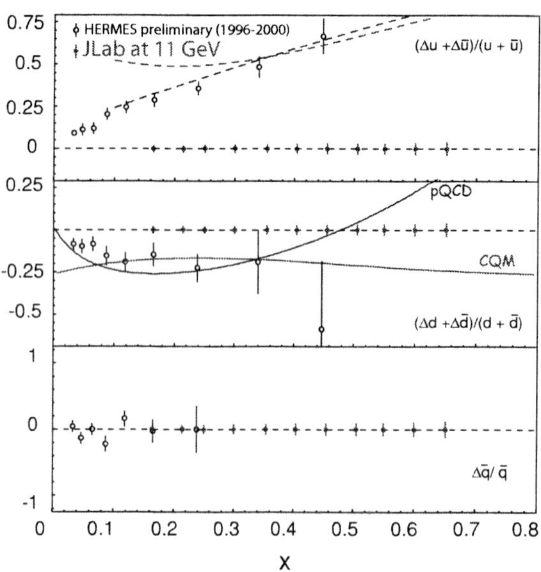

FIGURE 4. Expected results on $\Delta\bar{q}/\bar{q}$ from semi-inclusive asymmetry measurements with polarized NH_3 and ^3He targets. 1000 hours of beam time is assumed for each target. Error bars are statistical only.

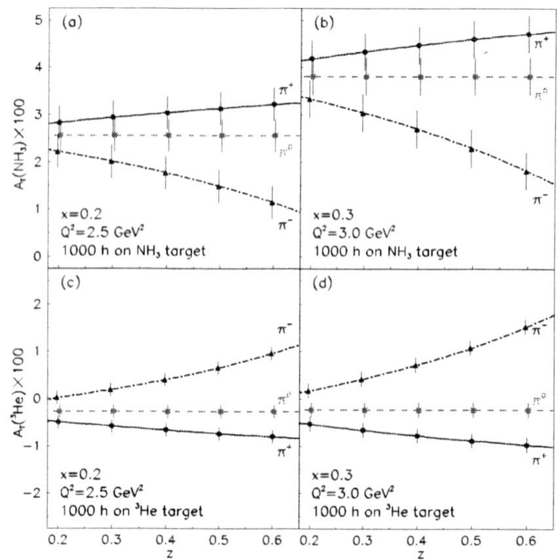

FIGURE 5. The expected precisions in transverse single-spin asymmetry measurements on a polarized $^{15}NH_3$ and a polarized ^3He target, for the kinematics of $x = 0.2$ and $x = 0.3$. 1000 hours of beam time is assumed for each target.

Diagonal Spin Basis (DSB) as a Completely Symmetrized Description of Interacting Fermions

Sergey M. Sikach

Institute of Physics, National Academy of Sciences of Belarus, e-mail: sikach@dragon.bas-net.by

Any given reaction includes participation of an even number of fermions and spin computations in this case are computations of combinations of fermion " sandwiches"

$$\bar{u}^{\sigma'}(p',s')Qu^{\sigma}(p,s) = TrQu^{\sigma}(p,s)\bar{u}^{\sigma'}(p',s'). \tag{1}$$

For the last 40 years one or the other kind of transition operators

$$u^{\sigma}(p,s)\bar{u}^{\sigma'}(p',s') \tag{2}$$

in arbitrary or helicity bases has been offered. Computations in arbitrary basis may be compared in difficulty to the standard method using the quadrating procedure. Helicity , on the other hand, for mass particles is a "bad" (non-covariant) quantum number, and in addition, the operator $u^{\lambda}(p)\bar{u}^{\lambda'}(p')$ does not reflect the dynamics of spin-flip and non-flip interactions. These problems, however, are irrelevant in the introduction in [1] DSB. The essential features of this method are described in [2] (see also [3]).

In the DSB all the fermions of reactions are considered in pairs. Each pair corresponds one fermion line on the Feynman diagrams and forms a transition operator in the matrix element.

In DSB the vectors s and s' are chosen such that they belong to the 2-plane (p,p') or (v,v'); $v = \frac{p}{m}$, $v' = \frac{p'}{m'}$. Satisfying this requirement and $vs = v's' = 0$, we obtain

$$s = \frac{(vv')v - v'}{\sqrt{(vv')^2-1}}, \quad s' = -\frac{(vv')v' - v}{\sqrt{(vv')^2-1}}. \tag{3}$$

Thus, the support vectors (see [2]) for the *in* and *out* states of the fermion line are vectors $q = v'$, $q' = -v$, or $q = q' = v' - v$. With such a choice of signs in picked reference systems vectors $\vec{s}$ and $\vec{s'}$ coincide with the direction of the 3-momentum of the initial particle and are opposite to the 3-momentum of the final particle. By picked reference systems is meant the Breit system for fermion or antifermion lines (t-lines) and s.c.m. for the pair being annihilated or created (s-lines).

From (3) it can easily be seen that in the derived reference systems diagonality gets the meaning of helicity, with $\delta = \lambda$, $\delta' = -\lambda'$. Thus, the DSB is covariant description of

CP675, *Spin 2002: 15th Int'l. Spin Physics Symposium and Workshop on Polarized Electron Sources and Polarimeters*, edited by Y. I. Makdisi, A. U. Luccio, and W. W. MacKay

the helicity in the picked reference systems. Exactly in these systems helicity the pairs of particles as well as such notions as non-flip and flip amplitudes have a clear physical meaning. Indeed, in helicity basis in arbitrary reference system the spin of the initial particle is projected on the 3-momentum $\vec{p}$ and that of the final particle on $\vec{p}'$, than what can be said about the non-flip or flip process? These notions in helicity basis have a only marking meaning.

In our opinion, neglect of this fact is the main reason why the process of constructing operators (2) convenient for calculation has been extended to decades. Attempts to construct covariant operators (2) in helicity basis look unresonable. The helicity of massive particle is a "bad" quantum number, since it is not invariant at a Lorentz transformation. Any declaration of covariancy of operators (2) in the helicity basis usually is a disguised transition to the picked reference system. It should be noted that other fermion pair "droop", since each pair has its own picked reference system. The only exception is $e^+e^- \to \mu^+\mu^-$ type reactions are chosen, as a rule, for examples of calculations of processes.

Before to go to the construction of tetrads for the initial and final states, we introduce into the 2-plane (v, v') two v- and v'- symmetrized, orthonormalized vectors

$$n_0 = \frac{v+v'}{2V_+}\,,\; n_3 = \frac{v-v'}{2V_-}\,; \tag{4}$$

$$V_\pm = \sqrt{\frac{vv' \pm 1}{2}}\,. \tag{5}$$

From (3)–(5) it follows that

$$\begin{aligned} g_{\|}^{\mu\nu} &= v^\mu v^\nu - s^\mu s^\nu = v'^\mu v'^\nu - s'^\mu s'^\nu = n_0^\mu n_0^\nu - n_3^\mu n_3^\nu\,, \\ \tilde{\varepsilon}_{\|}^{\mu\nu} &= \varepsilon^{\mu\nu\rho\sigma} v_\rho s_\sigma = \varepsilon^{\mu\nu\rho\sigma} v'_\rho s'_\sigma = \varepsilon^{\mu\nu\rho\sigma} n_{0\rho} n_{3\sigma} = \frac{1}{2V_+V_-}\varepsilon^{\mu\nu\rho\sigma} v_\rho v'_\sigma\,. \end{aligned} \tag{6}$$

From (6), it follows that the choice of common phase vector $r' = r$ (r is arbitrary vector) [2] leads to the coincidence for both vectors of the tetrads lying in the orthogonal 2-plane, i.e.

$$n_1^\mu = s_1^\mu = s'^\mu_1 = (g_{\|}^{\mu\nu} - g^{\mu\nu})\frac{r_\nu}{r_\perp}\,,\;\; n_2^\mu = s_2^\mu = s'^\mu_2 = -\tilde{\varepsilon}_{\|}^{\mu\nu}\frac{r_\nu}{r_\perp};\; r_\perp = \sqrt{r(g_\| - g)r}\,. \tag{7}$$

Next, let us coincidence the plane Lorentz transformation that transforms v to v'. In the representation of γ– matrix space it is of the form

$$\Lambda(v \to v') = \frac{1+\hat{v}'\hat{v}}{2V_+}\,, \tag{8}$$

and $\Lambda\hat{v}\bar{\Lambda} = \hat{v}'$. From (3) it also follows that $\Lambda\hat{s}\bar{\Lambda} = \hat{s}'$. Transformation (8) does not change vectors lying in the orthogonal 2-plane. Thus, the Lorentz transformation (8) converts the tetrad of the initial particle into the tetrad of final particle:

$$\Lambda(v \to v')\left\{\hat{v}, \hat{s}, \hat{n}, \hat{n}_2\right\}\bar{\Lambda}(v \to v') = \left\{\hat{v}', \hat{s}', \hat{n}', \hat{n}'_2\right\}\,, \tag{9}$$

and this in turn means that the relation between the bispinors of the initial and final states in DSB is of the form

$$u^{\delta}(p',s') = \Lambda(v \to v')u^{\delta}(p,s)\,. \tag{10}$$

In the DSB we choose the normalization

$$\bar{u}^{\delta}(p)u^{\delta}(p) = \bar{u}^{\delta'}(p')u^{\delta'}_{-}(p') = 1\,. \tag{11}$$

Then the relation (10) describes the cases $m' \neq m$.

To restore the generally accepted normalization, it is necessary to multiply the amplitudes calculated in the DSB by factor

$$\prod_{i=1}^{n} \sqrt{2m_i 2m'_i}\,, \tag{12}$$

where n is the number of open fermion lines.

Using the Dirac equation and eq.(10), we can write relation (10) in different representations

$$u^{\delta}(p',s') = \frac{\hat{v}'+1}{2V_+}u^{\delta}(p,s) = \hat{n}_0 u^{\delta}(p,s) = (V_+ - \delta\gamma_5 V_-)u^{\delta}(p,s)\,. \tag{13}$$

Formula (13) makes it possible to express in the DSB the explicit form of the projection operators of the initial (or final) state

$$u^{\delta}(p,s)\bar{u}^{\delta}(p,s) = \frac{1}{4}(\hat{v}+1)(1+\delta\gamma_5\hat{s})\,. \tag{14}$$

As relation (13), the transition operators (2) can be given in different form. Some of them are shown below.

$$\begin{aligned} 4u^{\delta}(p,s)\bar{u}^{\delta}(p',s') &= (\hat{v}+1)\left(\frac{1}{2V_+} - \frac{\delta\gamma_5}{2V_-}\right)(\hat{v}'+1) = \\ &= (V_+ + \delta\gamma_5 V_-)(1 - \delta\gamma_5\hat{n}_0\hat{n}_3) + \hat{n}_0 + \delta\gamma_5\hat{n}_3 = \\ &= \left(1 + \frac{1}{2}(V_+ + \delta\gamma_5 V_-)(\hat{n}_0 + \delta\gamma_5\hat{n}_3)\right)(\hat{n}_0 + \delta\gamma_5\hat{n}_3)\,, \end{aligned} \tag{15}$$

$$\begin{aligned} &4u^{\delta}(p,s)\bar{u}^{-\delta}(p',s') = \\ &= \frac{\delta}{r_\perp}(\hat{v}+1)\left(\frac{1}{2V_-}(\hat{r} - \frac{r(v+v')}{vv'+1}) - \frac{\delta\gamma_5}{2V_+}(\hat{r} - \frac{r(v-v')}{vv'-1})\right)(\hat{v}'+1) = \\ &= \gamma_5(V_+ + \delta\gamma_5 V_- - \hat{n}_0)(\hat{n}_1 + i\delta\hat{n}_2) = \\ &= \gamma_5\left(V_+ + \delta\gamma_5 V_- - \frac{1}{2}(\hat{n}_0 + \delta\gamma_5\hat{n}_3)\right)(\hat{n}_1 + i\delta\hat{n}_2)\,. \end{aligned} \tag{16}$$

Since any interaction operator can be expanded by the complete set of Dirac matrices, it is tempting to calculate the matrix elements of this set in DSB. From (1), (15), (16) it follows that

$$\begin{aligned} &\bar{u}^{\delta}(p',s')\left\{1;\gamma_5;\gamma^{\mu};\gamma_5\gamma^{\mu};\sigma^{\mu\nu}\right\}u^{\delta}(p,s) = \\ &= \left\{V_+\,;\ \delta V_-\,;\ n_0^{\mu}\,;\ -\delta n_3^{\mu}\,;\ V_-[n_0\cdot n_3]^{\mu\nu} - i\delta V_+\widetilde{[n_0\cdot n_3]}^{\mu\nu}\right\}, \end{aligned} \tag{17}$$

$$\bar{u}^{-\delta}(p',s')\left\{1;\gamma_5;\gamma^\mu;\gamma_5\gamma^\mu;\sigma^{\mu\nu}\right\}u^\delta(p,s) = \\ =\left\{0\,;\,0\,;\,\delta V_-(\hat{n}_1+i\delta\hat{n}_2)^\mu\,;\,-V_+(\hat{n}_1+i\delta\hat{n}_2)^\mu\,;\,\delta\left[n_3\cdot(\hat{n}_1+i\delta\hat{n}_2)\right]^{\mu\nu}\right\}, \tag{18}$$

where $[a\cdot b]^{\mu\nu}=a^\mu b^\nu - b^\mu a^\nu$.

The matrix elements (17), (18) can be interpreted as spin characteristics of exchange particles under the scalar, pseudoscalar, vector, axial and tensor interaction, respectively.

Formulas (15), (16) described the fermion t-line. To describe the antifermion line as well as the s-line for the annihilating and creating pairs, one should make use by the relation

$$v^\delta(p,s) = -\delta\gamma_5 u^{-\delta}(p,s)\,, \tag{19}$$

which relates the particle and antiparticle bispinors.

If in (15), (16) we restore by the recipe (12) the normalization $\bar{u}^\delta(p)u^\delta(p)=2m$ and perform the limiting transition $m\to 0$ or/and $m'\to 0$, we will obtain the transition operators for processes involving massless fermions found in [4].

For calculating concrete processes, it may be convenient to utilize a formalism, in which the basis spinor $u^\delta(n_0,n_3;n_1,n_2)$ common for initial and final bispinors and satisfying the conditions

$$\hat{n}_0 u^\delta(n_0,n_3) = u^\delta(n_0,n_3)\,,\ \gamma_5\hat{n}_3 u^\delta(n_0,n_3) = \delta u^\delta(n_0,n_3) \tag{20}$$

is used.

It is easy to see[1] that plane Lorentz transformations $\Lambda(n_0\to v)=\dfrac{1+\hat{v}\hat{n}_0}{\sqrt{2(V_++1)}}$ and $\Lambda(n_0\to v')=\dfrac{1+\hat{v}'\hat{n}_0}{\sqrt{2(V_++1)}}$ to change the basis spinor tetrad to tetrads of the initial and final states respectively. Therefore, the transition operators (15), (16) can be given in the form

$$u^\delta(p,s)\bar{u}^{\delta'}(p',s') = \frac{2}{V_++1}(\hat{v}+1)u^\delta(n_0,n_3)\bar{u}^{\delta'}(n_0,n_3)(\hat{v}'+1) = \\ = \frac{1}{2(V_++1)}(\hat{v}+1)(\hat{n}_0+1)(\delta_{\delta'\delta}+\gamma_5\hat{n}^i\sigma^i_{\delta'\delta})(\hat{v}'+1)\,. \tag{21}$$

In this equality the Bouchiat and Michel relation [5] is used.

In (21), vectors v and v' can be expanded in vectors n_0 and n_3 with the aid of relations (4).

In conclusion of this section, we give the recipe for calculating in DSB exchange diagrams. As an example, we consider the electron-electron scattering. For certainty, let particles 1, 3 and 2, 4 are paired. Then the exchange diagram has the structure

$$\bar{u}^{\delta_4}(p_4)\gamma_\mu u^{\delta_1}(p_1)\bar{u}^{\delta_3}(p_3)\gamma^\mu u^{\delta_2}(p_2) = Tr\gamma_\mu u^{\delta_1}(p_1)\bar{u}^{\delta_3}(p_3)\gamma^\mu u^{\delta_2}(p_2)\bar{u}^{\delta_4}(p_4)\,, \tag{22}$$

i.e. it is expressed in terms of the transition operators entering into direct diagram.

---

[1] For this the relation $\dfrac{V_-}{V_++1}=\dfrac{V_+-1}{V_-}$ is used.

REFERENCES

1. S.M. Sikach// *Covariant Methods in Theoretical Physics*, Institute of Physics, Academy of Sciences of Belarus. Minsk.1981.P.91. (in Russian)
2. S.M. Sikach// Actual Problems of Particle Physics, *Proceedins of 7th International School-Seminar*. Dubna.2002.Vol.2.P.62. [hep-ph/0203139]
3. M.V. Galynskii, S.M. Sikach// Physics of Particles and Nuclei. 1998.Vol.29.P.446. [hep-ph/9910284]
4. CALCUL collab.:
   P. de Gausmaehecker at al.// Nucl.Phys.1982.Vol.B206.P.53.
   F.A. Berends at al.// Nucl.Phys.1982.Vol.B206.P.61.
   F.A. Berends at al.// Nucl.Phys.1984.Vol.B239.P.382.
5. C. Bouchiat, L. Michel// Nucl.Phys.1958.Vol.5.P.416.

About the Processes with two Fermions and two Photons in an Intensive Laser wave (Nonlinear and Spin Effects)

Sergey M. Sikach

Institute of Physics National Academy of Sciences of Belarus, e-mail: sikach@dragon.bas-net.by

The pattern of calculation of amplitudes of a series of processes in the field of an intensive laser wave, in which two fermions $(p;p')$ and two real photons $(k_1;k_2)$ participate, is considered. In relation to one-photon processes, these processes are of the second order on α if the wave intensity $\xi \ll 1$ (i.e., actually absorption from the wave only one quantum). Otherwise, they are competing and essentially nonlinear. One-photon processes have a number of the important physical applications. For example, γe and $\gamma\gamma$ colliders work on their basis. Some of these processes calculated in [1] in Diagonal Spin Basis (see [2] in this issue). Two-photons back Compton-effect in intensive laser wave occurs together with one-photon process and thus influences the formation of spectr and polarisation of photon beam. The amplitudes of two-photon annihilation of a fermion pair in an intensive laser wave are applicable for calculating of induced decay of orthopositronium into two photons. And in this approach there is no necessity to assume the relative momentum equal to zero.

In DSB the calculation is conducted at the level of reaction amplitudes. It essentially simplifies both the calculation and the form of obtained results; those combinations of amplitudes which describe the spin effects are easy to calculate. And these effects are especially essential in nonlinear processes. The calculations are conducted in covariant form. Besides compactness, this provides independence of the frames of reference and energy modes. The masses of fermions are taken into account precisely and are not assumed equal each other (heavy leptons, modes of various decays).

The process under consideration is described by 2^5 amplitudes. Because of parity conservation, we need to calculate 16 amplitudes (in our approach, they are described by 8 formulas). Therefore, here we can present only the pattern of calculation. The necessary formulas are cited in accordance with monograph [3] with the indication paragraph number,where §40 "An electron in the field of the plane electromagnetic wave"; §89 "An annihilation of a positronium"; §101 "Radiation of a photon by an electron in the field of an intensive electromagnetic wave". Unlike [3], we assume $\gamma_5 = i\gamma^0\gamma^1\gamma^2\gamma^3$ (opposite sign) and elementary charge $e > 0$.

Process $Laser + e^+ + e^- \to \gamma_1 + \gamma_2$ described by amplitude

$$\mathbf{M} = ie \int \frac{M\, d^4x}{\sqrt{2\omega_1 2\omega_2 2q_0 2q'_0}} e^{i(k_1+k_2-q-q')+i(\alpha_1 \sin\phi - \alpha_2 \cos\phi)}, \qquad (1)$$

CP675, *Spin 2002: 15th Int'l. Spin Physics Symposium and Workshop on Polarized Electron Sources and Polarimeters*, edited by Y. I. Makdisi, A. U. Luccio, and W. W. MacKay

$$M = \bar{v}^{\delta'}(p')[1+\frac{e}{2pk'}\hat{A}\hat{k}]Q[1-\frac{e}{2pk}\hat{k}\hat{A}]\bar{u}(p), \tag{2}$$

$$Q = \hat{e}_{\lambda_2}\frac{\hat{q}-\hat{k}_1+m_*}{-2qk_1}\hat{e}_{\lambda_1}+\hat{e}_{\lambda_1}\frac{\hat{q}-\hat{k}_2+m_*}{-2qk_2}\hat{e}_{\lambda_2}. \tag{3}$$

The expressions in square brackets are obtained from the Volkov solutions for an electron (positron) in the wave field.

$A = a(l_1\cos\varphi + \mu l_2 \sin\varphi)$ is the 4-potential of the wave (101.2), μ is the polarization, $\varphi = kx$, $\xi = ea/m$ is the wave intensity, $q^\mu = p^\mu - \frac{\xi^2 m^2}{2kp}k^\mu$ is the quasimomentum, $m_*^2 = m^2(1+\xi^2)$ (101.4), $\alpha_1 = -ea(\frac{l_1 p}{kp}+\frac{l_1 p'}{kp'})$, $\alpha_2 = -ea(\frac{l_2 p}{kp}+\frac{l_2 p'}{kp'})$ (101.6).

If in tetrad n_A (4), (6) from [2] for fermion line we make the replacement $p\to k_1$, $p'\to k_2$, then we obtain tetrad h_A common for both photons. In this case, the vector $e_\lambda = \frac{1}{\sqrt{2}}(h_1 - i\lambda h_2)$ simultaneously describes both the emitted photons: k_1 with helicity $\lambda_1 = \lambda$ and k_2 with $\lambda_2 = -\lambda$. Thus, if $\lambda_1+\lambda_2 = 0$, then only one operator $\hat{e}_\lambda$ enters into the amplitude, if $\lambda_1+\lambda_2 = \pm 2$, then $\hat{e}_\lambda$ and $\hat{e}_\lambda^*$ enter into the amplitude. This fact also extremely simplifies both the structure and the calculation of the reaction amplitudes. Thus,

$$\begin{aligned}
&\hat{e}_\lambda^2 = 0\,,\ \hat{e}_\lambda\hat{e}_\lambda^* = -1+\frac{\gamma_5}{k_1k_2}k_1^\mu\sigma_{\mu\nu}k_2^\nu\,;\\
&\hat{e}_\lambda\hat{f}\hat{e}_\lambda = 2e_\lambda f\hat{e}_\lambda\,;\ \hat{e}_\lambda\hat{f}\hat{e}_\lambda^* = \frac{2}{k_1k_2}(fk_2\omega_{-\lambda}\hat{k}_1 + fk_1\omega_\lambda\hat{k}_2)\,,
\end{aligned} \tag{4}$$

where f is arbitrary vector, $\omega_\lambda = \frac{1}{2}(1+\lambda\gamma_5)$.

Let us consider in more detail the case of $\lambda_1+\lambda_2 = 0$. Using Volkov's solution [3] for the electron in the field of the plane wave (40.7), (101.3) and above relations, we obtain the spin structure of the amplitude

$$M = c_\lambda q\left(\frac{1}{k_1 q}+\frac{1}{k_2 q}\right)\bar{v}^{\delta'}(p')\left\{\hat{e}_\lambda + \frac{\xi^2 m^2 e_\lambda k}{2(kp)(kp')}\hat{k} - e\left(\frac{\hat{e}_\lambda\hat{k}\hat{A}}{2kp}-\frac{\hat{A}\hat{k}\hat{e}_\lambda}{2kp'}\right)\right\}u^\delta(p)\,. \tag{5}$$

Making series expansion in Bessel functions $J_n(z)$ (101.7) and integrating in 4-space, we make sure that amplitude (5) of the reaction $sk+q+q' = k_1+k_2$ completely coincides, to an accuracy of the factor preceding the "sandwich", with the s-channel of single-photon reaction $sk+q = q'+k'$ (101.10). sk denotes coherent absorption from the wave of photons (nonlinearity). The factor itself just reflects that exactly two-photon annihilation (or a double Compton back- scattering) takes place.

If $\lambda_2 = \lambda_1 = \lambda$, $e_{\lambda_2} = e_{\lambda_1}^* = e_{-\lambda}$ in the selected gauge. In this case formula simplifying the matrix structure is (B is an arbitrary matrix operator)

$$\hat{e}_{-\lambda}B\hat{e}_\lambda = \frac{1}{2}\gamma_\mu B\gamma_\nu T_{-\lambda}^{\mu\nu}(k_1,k_2) = \frac{1}{2k_1k_2}\gamma_\mu B\gamma_\nu\,(k_1^\mu k_2^\nu + k_2^\mu k_1^\nu - k_1k_2\,g^{\mu\nu}n + i\lambda\varepsilon^{\mu\nu\rho\sigma}k_{1\rho}k_{2\sigma})\,. \tag{6}$$

From (6) one gets (4) and more complicate structure in (2) is

$$\hat{A}\hat{k}\hat{e}_{-\lambda}\hat{f}\hat{e}_{\lambda}\hat{k}\hat{A} = \frac{2a^2}{k_1k_2}[(fk_2)(k_1k)\omega_{\lambda} + (fk_1)(k_2k)\omega_{-\lambda}]\hat{k}. \tag{7}$$

For the laser wave we choose

$$l_1 = \frac{[(kv')v - (kv)v']}{\sqrt{(vv')^2 - 1}k_{\perp}}, \; l_2 = n_2 . \tag{8}$$

Than in (101.6) $\alpha_2 = 0$ and $z = |\alpha_1| = \dfrac{\xi m|kv'/kv - kv/kv'|}{\sqrt{(vv')^2 - 1}k_{\perp}}$ is the Bessel function argument. With such a choice in the DSB we have

$$e\frac{\hat{e}_{\lambda}\hat{k}\hat{A}}{2kp}u^{\delta}(p) = \mu\delta\frac{\xi\hat{e}_{\lambda}\hat{k}}{2m^2k_{\perp}}\sum_{s=-\infty}^{\infty} e^{-is\varphi}\left\{\left(1+\mu\delta\frac{s_3k}{vk}\right)J_{s-\mu}(z)\omega_{-\mu} - \left(1-\mu\delta\frac{s_3k}{vk}\right)J_{s+\mu}(z)\omega_{\mu}\right\}u^{\delta}(p) . \tag{9}$$

Besides,

$$\hat{e}_{\lambda}\hat{k}\omega_{-\lambda} = \sqrt{\frac{2k_2k}{(k_1k)(k_1k_2)}}\hat{k}_1\hat{k}\omega_{-\lambda} , \; \hat{e}_{\lambda}\hat{k}\omega_{\lambda} = \sqrt{\frac{2k_1k}{(k_2k)(k_1k_2)}}\hat{k}_2\hat{k}\omega_{\lambda} .$$

A similar expression is obtained for the structure $v^{\delta'}(p')\hat{A}\hat{k}\hat{e}_{\lambda}$. Than rather trivial calculations by formulas (15),(16) or (17), (18) from [2] remain.

The obtained formulas permit investigating the spin and nonlinear effects of a number of interesting processes: in the region of conversion (linear colliders); induced decay of ortopositronium with coherent absorption (radiation) of an odd number of photons.

In the t-channel (when $e^{\pm} \to e^{\pm}$), this is a double Compton back-scattering (the Compton back-scattering is a basic process in the operation of γ-colliders [5]); 2γ-creation of pairs in the region of conversion (collider testing), etc. Examples of comprehensive studies of single-photon processes are given in [1].

REFERENCES

1. M.V. Galynskii, S.M. Sikach// Physics of Particles and Nuclei.1998.Vol.29.P.496. [hep-ph/9910284].
2. S.M. Sikach// In this issue.
3. L.D. Landau, E.M. Lifshitz// *Quantum Electrodinamics. Pergamon*, New York.1980.
4. S.M. Sikach// Proceedings of 10 Annual Seminar *Nonlinear Phenomena in Complex Systems*, Minsk.2001.P.297. [hep-ph/0103323].
5. I.F. Ginzburg, G.L. Kotkin, V.G. Serbo, V.I. Telnov// Nucl.Instr.Meth.1983.Vol.205.P.47.
   I.F. Ginzburg, G.L. Kotkin, SL. Panfil, V.G. Serbo, V.I. Telnov// Nucl.Instr.Meth.1984.Vol.219.P.5.

Spin Physics with Nuclei

Model Independent Spin Parity Determination by the $(d, {}^{2}\mathrm{He})$ Reaction and Possible Evidence for a 0^- State in ^{12}B

H. Okamura*, T. Uesaka†, K. Suda*, H. Kumasaka*, R. Suzuki*, A. Tamii**, N. Sakamoto‡ and H. Sakai**‡

*Department of Physics, Saitama University, 255 Shimo-Ohkubo, Saitama 338-8570, Japan
†Center for Nuclear Study, University of Tokyo, 7-3-1 Hongo, Bunkyo, Tokyo 113-0033, Japan
**Department of Physics, University of Tokyo, 7-3-1 Hongo, Bunkyo, Tokyo 113-0033, Japan
‡The Institute of Physical and Chemical Research (RIKEN), Wako, Saitama 351-0198, Japan

Abstract. A method of model-independent spin-parity determination is proposed for the $(d, {}^{2}\mathrm{He})$ reaction by using the tensor analyzing power A_{zz} at $\theta = 0°$, which shows extreme values for 0^- and natural-parity states solely by parity-conservation. It is applied to the $^{12}\mathrm{C}(d,{}^{2}\mathrm{He})^{12}\mathrm{B}$ reaction at $E_d = 270$ MeV and a possible indication of 0^- state is found at $E_x = 9.3$ MeV in ^{12}B. The bump at $E_x = 7.5$ MeV appears to have more 2^- strength than 1^- strength, consistent with our earlier work.

INTRODUCTION

In spite of extensive studies particularly by using charge-exchange reactions, the understanding of spin-flip dipole states is still rather limited. The $(d, {}^{2}\mathrm{He})$ reaction can be a powerful tool for pursuing this subject further. Here the two-proton system in the $^{1}S_0$ state is denoted by ^{2}He, though it is unbound. Besides its unique feature as an (n,p)-type reaction with exclusive excitation of spin-flip states, the tensor analyzing powers of the $(d, {}^{2}\mathrm{He})$ reaction, A_{xx} and A_{yy}, show characteristic behavior depending on the spin-parity of the final states. Their usefulness for identifying the spin-parity, particularly 2^-, 1^-, and 0^- in spin-dipoles, was first demonstrated at $E_d = 270$ MeV on the ^{12}C target in our previous work [1]. There, the bump at $E_x = 7.5$ MeV in residual ^{12}B, which had been believed to be dominated by 1^- states, was unexpectedly found to have larger contributions from 2^- states. This finding is supported by the heavy-ion charge-exchange reaction [2] and also theoretically by calculations including the tensor correlation [3] and the deformation effect [4], where the quenching of 1^- states at low excitation energies due to fragmentation is reported. It must be said, however, that the above conclusion is not widely accepted. The aim of this work is to present a method for model-independent spin-parity determination of states excited by the $(d, {}^{2}\mathrm{He})$ reaction and to report the results of its application to the ^{12}C target.

For reactions having a spin-parity structure of $1^+ + 0^+ \to 0^+ + I^\pi$, such as (d,α) and $(d, {}^{2}\mathrm{He})$ reactions on even-even targets, the tensor analyzing power A_{zz} shows extreme values at $\theta = 0°$ and $180°$ for some I^π residual states solely by the requirement of parity-

CP675, *Spin 2002: 15th Int'l. Spin Physics Symposium and Workshop on Polarized Electron Sources and Polarimeters*, edited by Y. I. Makdisi, A. U. Luccio, and W. W. MacKay

conservation:

$$A_{zz}(0^\circ, 180^\circ) = \begin{cases} -2, & \text{if } I^\pi = 0^-, \\ +1, & \text{if } \pi = (-)^I. \end{cases} \tag{1}$$

These relations allow one to determine spin-parity states unambiguously. While the (d, α) cross section rapidly decreases with increasing incident energy, due to momentum mismatch, the importance of the $(d, {}^2\text{He})$ reaction increases at several hundred MeV because of the relative enhancement of the spin-dependent part of the two-body interaction and of the simple reaction mechanism. Furthermore the spin-dipole states, 2^-, 1^-, and 0^-, are fairly strongly excited even at 0°, and two out of three are identified by Eq. (1) without ambiguity. It is a matter of interest to know the distribution of 0^- states, which is expected to reflect pion correlations in nuclei.

EXPERIMENT

The experiment was performed at RIKEN Accelerator Research Facility. The deuteron beam produced by the polarized ion source [5] was accelerated up to 270 MeV by the AVF and ring cyclotrons, to bombard the target. Two states of tensor polarization having theoretical maxima of $p_{ZZ} = -2$ and $+1$ were used, as well as the unpolarized state ($p_{ZZ} = 0$), changed every 5 s to minimize systematic uncertainties. Here the rotation symmetry axis at the ion source defines the polarization axis Z, as distinct from the beam axis z. The beam polarization was monitored continuously throughout the experiment at the beam transport line using $d + p$ elastic scattering [6, 7]. The averaged value of each state was $p_{ZZ} = -1.16$ and $+0.79$. The best way to measure A_{zz} is to align the polarization axis parallel to the beam. The cross section is then related to A_{zz} simply by $d\sigma/d\Omega = (d\sigma/d\Omega)_0[1 + \frac{1}{2}p_{ZZ}A_{zz}]$, where $(d\sigma/d\Omega)_0$ is the cross section for the unpolarized beam. However, this technique, which is common at Tandem accelerators, is rarely used at higher energies, because of difficulties in rotating the spin direction due to the small anomalous gyromagnetic ratio of the deuteron. In this respect RIKEN is a unique facility, that allows one to freely control the direction of the polarization axis [8]. The polarization axis is readily rotated by using the Wien filter downstream of the ion source at energies as low as 14 keV. Although the spin precesses during acceleration by the cyclotrons, the magnitude of polarization is not reduced, owing to the single-turn extraction available at RIKEN. The direction of the polarization axis after acceleration is determined by the beam-line polarimeter. The field strength and the rotation angle of the Wien filter were tuned so that $Z \parallel z$.

^{2}He was measured by the coincidence detection of two protons in close geometries. Protons emitted from the target were momentum-analyzed by the magnetic spectrometer SMART [9] and detected at the first focal-plane by using a pair of multi-wire drift chambers and a plastic scintillator hodoscope. Data acquisition was triggered by requiring the multiplicity of the hodoscope to be greater than or equal to two. Details of the detector system and the analysis procedure can be found in [10]. The whole system was the same as the one used in the previous publication [1] except for the data acquisition system, which has been upgraded to increase speed by a factor of 10 [11]. After correction for the detection efficiency of two protons based on the Monte-Carlo simulation,

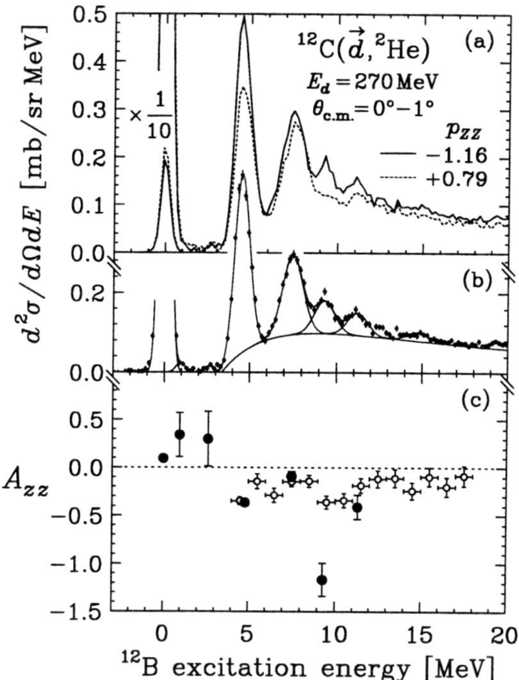

FIGURE 1. Double differential cross sections at $\theta_{c.m.} = 0.67°$ (averaged over 0°–1°) plotted as a function of ^{12}B excitation energy (a), a result of peak-fitting for the spectrum with $p_{ZZ} = -1.16$ (b), and the corresponding A_{zz} spectra (c). A_{zz} for each peak obtained by the fitting is shown by the closed circle, while A_{zz} for the continuum binned in 1 MeV shown by open circles.

the $(d, {}^2\text{He})$ cross section is obtained by integrating the (d, pp) triple-differential cross section over two-proton relative energies. The integration limit was set to be less than 1 MeV to minimize contributions from *p*-*p* partial waves higher than $^1S_0$ [1, 10].

RESULTS AND DISCUSSION

The cross sections at $\theta_{c.m.} = 0.67°$ with $p_{ZZ} = -1.16$ and $+0.79$ are presented in Fig. 1(a). Concerning the 1^+ ground state and the bump at $E_x = 7.5$ MeV, as well as the continuum at $E_x \geq 12$ MeV, only a small difference is observed between the spectra. The large difference at $E_x = 9.3$ MeV is striking; a clear peak appears in the $p_{ZZ} = -1.16$ spectrum but vanishes for $p_{ZZ} = +0.79$, indicating quite a large negative value for A_{zz}.

The contribution from continuum background, however, makes A_{zz} less prominent if it is directly derived from the raw spectra [open circles in Fig. 1(c)]. The dominant contribution of the continuum can be ascribed to quasifree scattering. Following the commonly employed procedure, as in ref. [12] for example, the spectra are fitted by using the least-squares method. Figure 1(b) shows the result of such fitting for the

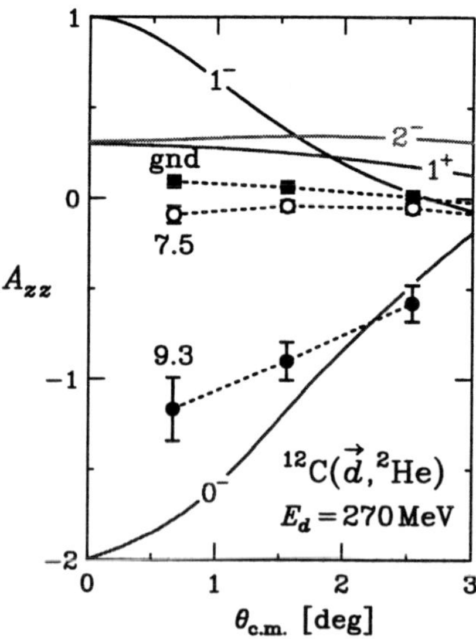

FIGURE 2. Angular distributions of A_{zz} for the ground (closed square), $E_x = 7.5$ MeV (open circle), and $E_x = 9.3$ MeV (closed circle) states. Results of calculation [13] for Gamow-Teller (1_1^+) and spin-dipole (0_2^-, 1_1^-, and 2_2^-) states are also shown by solid curves.

$p_{ZZ} = -1.16$ spectrum. A_{zz} for each state thus deduced from the peak-fitting is presented by the closed circle in Fig. 1(c). Obviously A_{zz} at $E_x \geq 4$ MeV is subject to ambiguities in estimation of the continuum. For various sets of fitting parameters, however, A_{zz} at $E_x = 9.3$ MeV takes large negative values stably less than or equal to -1, suggesting large contributions from 0^- states.

One must remember that Eq. (1) holds only at exact $\theta = 0°$. Indeed very steep angular distributions of A_{zz} are predicted generally by calculations, without strong dependence on the parameters employed, but depending only on spin-parity. Figure 2 shows the angular distributions for the bumps at $E_x = 9.3$ and 7.5 MeV and for the 1^+ ground state together with the results of calculations for Gamow-Teller and spin-dipole states using the adiabatic coupled-channels Born approximation [13]. At this very forward angle, transitions with higher orbital angular-momentum transfers are unlikely to be observed. Since there is no other state predicted to have a negative A_{zz}, and also from the steep angular dependence of the experimental data, it is natural to conclude that the bump at 9.3 MeV is dominated by 0^- states. The target form factors have been obtained by using the effective interactions by Millener and Kurath [14] (other parameters employed in the calculation can be found in ref. [13]) but A_{zz} does not reflect details of the wave functions. Calculations using the interactions by Warburton and Brown [15], for example, give similar results. It is worth noting that both shell-model calculations predict the existence of a 0^- state at $E_x \simeq 9$ MeV and support the above conclusion.

Concerning the 1_1^+ ground state, the monotonous distribution with a small magnitude of A_{zz} is reasonably well described by the calculation. The prediction for 2_2^- state is very similar to that for the 1_1^+ state. Likewise the bump at 7.5 MeV shows a distribution similar to that for the ground state, substantially deviating from the expected behavior for 1^- states. Although the prediction for 2^- is model-dependent, it appears that the bump at 7.5 MeV is dominated not by 1^-, but most likely by 2^-. Ambiguities in estimation of the continuum background do not seriously influence the discussion because A_{zz} of the continuum has a similar value to that of the bump at 7.5 MeV.

SUMMARY AND CONCLUSIONS

The $^{12}C(d,{}^2He)^{12}B$ reaction has been measured at 270 MeV as an application of the method proposed for model-independent spin-parity determination, which uses the tensor analyzing power A_{zz} at $\theta = 0°$ and the relations required by parity-conservation, Eq. (1). Owing to an enhanced sensitivity to A_{zz} achieved by aligning the polarization axis parallel to the beam, a possible indication of a 0^- state has been found at $E_x = 9.3$ MeV in ^{12}B for the first time. The bump at $E_x = 7.5$ MeV appears to have more 2^- strength than 1^- strength at $\theta \sim 0°$, supporting our results in a previous study [1]. These findings, and further studies applying this novel method to other nuclei, will provide valuable information in studies of nuclear structure, e.g., tensor correlations in nuclear spin-excitation modes.

REFERENCES

1. Okamura, H. et al., *Phys. Lett. B*, **345**, 1 (1995).
2. Ichihara, T. et al., *Nucl. Phys.*, **A577**, 93c (1994).
3. Sagawa, H., and Suzuki, T., "Spin-dipole states in ^{12}C and ^{16}O," in *Int. Symposium on New Facet of Spin Giant Resonances in Nuclei*, edited by H. Sakai, H. Okamura, and T. Wakasa, World Scientific, 1998, p. 191.
4. Kurasawa, H., and Suzuki, T., "Spin dipole states of deformed nuclei," in *Int. Symposium on New Facet of Spin Giant Resonances in Nuclei*, edited by H. Sakai, H. Okamura, and T. Wakasa, World Scientific, 1998, p. 183.
5. Okamura, H. et al., "Development of the RIKEN polarized ion source," in *Polarized Ion Sources and Polarized Gas Targets*, edited by L. W. Anderson and W. Haeberli, AIP Conf. Proc. No. 293, AIP, New York, 1994, p. 84.
6. Sakamoto, N. et al., *Phys. Lett. B*, **367**, 60 (1996).
7. Suda, K. et al., "Absolute Calibration of the Deuteron Beam Polarization at Intermediate Energies via the $^{12}C(\vec{d},\alpha)^{10}B^*(2^+)$ Reaction," in *14th International Spin Physics Symposium*, edited by K. Hatanaka, T. Nakano, K. Imai, and H. Ejiri, AIP Conf. Proc. No. 570, AIP, New York, 2001, p. 806.
8. Okamura, H. et al., "Technique for Rotating the Spin Direction at RIKEN," in *High Energy Spin Physics*, edited by K. J. Heller and S. L. Smith, AIP Conf. Proc. No. 343, AIP, New York, 1995, p. 123.
9. Ichihara, T. et al., *Nucl. Phys.*, **A569**, 287c (1994).
10. Okamura, H. et al., *Nucl. Instrum. Methods Phys. Res. A*, **406**, 78 (1998).
11. Okamura, H., *Nucl. Instrum. Methods Phys. Res. A*, **443**, 194 (2000).
12. Raywood, K. J. et al., *Phys. Rev. C*, **41**, 2836 (1990).
13. Okamura, H., *Phys. Rev. C*, **60**, 064602 (1999).
14. Millener, D. J., and Kurath, D., *Nucl. Phys.*, **A255**, 315 (1975).
15. Warburton, E. K., and Brown, B. A., *Phys. Rev. C*, **46**, 923 (1992).

High Resolution Study of Pionic 0^- State in ^{16}O

T. Wakasa*, G. P. A. Berg*, H. Fujimura*, K. Fujita*, K. Hatanaka*, M. Itoh*, J. Kamiya*, T. Kawabata†, Y. Kitamura*, E. Obayashi*, H. Sakaguchi**, N. Sakamoto*, Y. Sakemi*, Y. Shimizu*, H. Takeda**, M. Uchida**, Y. Yasuda**, H. P. Yoshida* and M. Yosoi**

**Research Center for Nuclear Physics (RCNP), Ibaraki, Osaka 567-0047, Japan*
†Center for Nuclear Study (CNS), University of Tokyo, Tokyo 113-0033, Japan
***Department of Physics, Kyoto University, Kyoto 606-8502, Japan*

Abstract. The cross sections and analyzing powers of the $^{16}\mathrm{O}(p,p')^{16}\mathrm{O}(0^-,T=1)$ scattering were measured at a bombarding energy of 295 MeV and an angular range of $14^\circ < \theta_{\mathrm{lab}} < 30^\circ$. The isovector 0^- state at E_x = 12.80 MeV is clearly separated from the neighboring states with an energy resolution of $\Delta E \simeq 30$ keV. The data have been compared with distorted wave impulse approximation (DWIA) calculations. The analyzing powers are sensitve to the effective nucleon-nucleon (*NN*) interaction used in DWIA calculations, and our data support the medium modification of the *NN* interaction in nuclei. The DWIA calculation employing a random phase approximation (RPA) response function predicts an enhancement of the cross sections around a momentum transfer of $q \simeq 1.7$ fm^{-1}, and it gives a reasonable agreement with the data.

INTRODUCTION

Isovector $J^\pi = 0^-$, $0^\pm \to 0^\mp$ excitations are of particular interest since they carry the simplest pion-like quantum number. At low momentum transfers, they have been investigated in beta decay and muon capture experiments [1, 2, 3]. Axial-vector and pseudoscalar currents are responsible for these first-forbidden transitions in nuclear weak processes. Gagliardi *et al.* [1] reported an enhancement of the decay rate by more than a factor of 3 for the first-forbidden beta decay of the 120 keV, 0^- state in ^{16}N. This enhancement can be explained by considering meson-exchange effects [4].

The (p,n) and (p,p') reactions are suited to study these transitions for a wide range of momentum transfer [5]. Orihara *et al.* [6] measured the angular distribution for the $^{16}\mathrm{O}(p,n)^{16}\mathrm{N}(0^-$, 0.12 MeV) reaction at T_p = 35 MeV. They reported discrepancies between distorted wave Born approximation (DWBA) calculations and their data in the large momentum transfer region of q = 1.4–2.0 fm^{-1} that might be due to an enhancement of the pion probability in the nucleus [7, 8, 9, 10, 11]. However, in the proton inelastic scattering to the 0^-, $T=1$ state in ^{16}O at T_p = 65 MeV, such an enhancement was not observed [12]. The differences between (p,n) and (p,p') results might indicate contributions from complicated reaction mechanisms as these low incident energies.

At intermediate energies of $T_p > 100$ MeV, where reaction mechanisms are expected to be simple, there are data only for the 0^-, T=0 transition at T_p = 135 [13, 14], 180 [14], 200 MeV [15], 318 MeV [16], and 400 MeV [17]. Most of these measurements

CP675, *Spin 2002: 15th Int'l. Spin Physics Symposium and Workshop on Polarized Electron Sources and Polarimeters*, edited by Y. I. Makdisi, A. U. Luccio, and W. W. MacKay

were not performed with sufficient energy resolution to separate the 0^-, $T = 0$ state at $E_x = 10.96$ MeV from its strong neighboring doublet (3^+ and 4^+) which is only about 140 keV away. It should be noted that there are no published experimental data for the 0^-, $T = 1$ state at $E_x = 12.80$ MeV in this energy region.

In this article, we present the measurement of cross sections and analyzing powers for the excitation of the 0^-, T=1 (12.80 MeV) unnatural-parity state in ^{16}O using 295 MeV inelastic proton scattering. The results will be compared with distorted wave impulse approximation (DWIA) calculations with shell-model wave functions. This provides information on tensor and spin-spin components of effective nucleon-nucleon (*NN*) interactions. Furthermore, the data will be compared with DWIA calculations employing random phase approximation (RPA) response functions in order to assess the pionic enhancement in a large momentum-transfer region.

EXPERIMENTAL METHODS

The measurement was carried out by using the West-South (WS) beam line and Grand Raiden (GR) spectrometer at the Research Center for Nuclear Physics (RCNP), Osaka University. The WS beam line and GR spectrometer are described in detail in Refs. [19, 20]. Here we only present a brief description of the experimental apparatus and discuss details relevant to the present experiment.

The high resolution WS beam line [20] has been designed and constructed to accomplish complete matching including both lateral and angular dispersion and focus matching with the high-resolution Grand Raiden spectrometer at RCNP. The WS beam line consists of six dipole magnets with a total bending angle of 270°. This beam line is divided into five sections. The beam is focused horizontally and vertically at the end of each section. Beam line polarimeter systems positioned at the ends of the first and second sections allow the measurement of all polarization components of the beam. They are separated by a bending angle of 115° for the determination of horizontal components of the beam polarization. In dispersive mode, lateral and angular dispersions of the WS beam line are $b_{16} = 37.1$ m and $b_{26} = -20.0$ rad, necessary to satisfy dispersion matching conditions for Grand Raiden. The magnifications of the beam line are $(M_x, M_y) = (-0.98, 0.89)$ and $(-1.00, -0.99)$ for dispersive and achromatic modes, respectively.

A windowless and self-supporting ice target [21] was used as an oxygen target. The thin ice target with a thickness of 14.1 mg/cm^2 was mounted on a thin aluminum frame attached to a copper frame that was cooled down to 77K using liquid nitrogen.

Scattered particles were momentum-analyzed by the GR spectrometer. The spectrometer consists of two dipole (D1 and D2) magnets, two quadrupoles (Q1 and Q2), a sextupole (SX), and a multipole (MP). The spectrometer is characterized by a high resolving power of $R = 37{,}000$.

DATA REDUCTION

The elastic scattering data on ^{16}O are shown in Fig. 1. Differential cross sections were normalized to the known $p+p$ cross section at $\theta_{\rm lab}$ = 14° by utilizing the hydrogen present in the ice target. The beam energy was determined to be 295 ± 1 MeV, based on the kinematic energy shift between elastic scattering from ^{1}H and ^{16}O. The beam polarization was continuously monitored with the hydrogen polarimeter in the WS beam line. Its typical value was 0.70 ± 0.01. The hydrogen in the ice target limited the useful scattering angles for inelastic scattering on ^{16}O to larger than 14°. At smaller angles, the $p+p$ events overlap the ^{16}O excited states of interest in this measurement.

The elastic scattering data were analyzed using optical model potentials generated phenomenologically. The solid curves in Fig. 1 are the results with the global optical potential optimized for ^{16}O [22]. The gray bands represent the results by using several optical potentials parameterized for nuclei from ^{12}C to ^{208}Pb with a smooth mass number dependence [22]. The global optical potential for ^{16}O shown by the solid curves reproduces the experimental data fairly well not only for cross sections but also for analyzing powers. Thus, in the following, we will use this optical potential in DWIA calculations for inelastic scattering.

RESULTS AND DISCUSSION

Figure 2 shows the excitation energy spectrum of the $^{16}\mathrm{O}(p,p')$ scattering at T_p = 295 MeV and $\theta_{\rm lab}$ = 30°. The isovector 0^- state at E_x = 12.80 MeV is clearly separated from the neighboring states with an energy resolution of ΔE = 29–34 keV depending on the reaction angle.

We have performed DWIA calculations by using the computer code DWBA98 [23] in which the knock-on exchange amplitude is treated exactly. The one-body density matrix elements (OBDME) for the isovector 0^- transitions of the $^{16}\mathrm{O}(p,p')$ scattering were obtained from Ref. [24]. This shell-model calculation was performed in the $0s$-$0p$-$1s0d$-$0f1p$ configurations by using phenomenological effective interactions. The single particle radial wave functions were generated by using a Woods-Saxon potential, the depth of which was adjusted to reproduce the binding energy. The effective *NN* interaction was taken from the t-matrix parameterization of the free *NN* interaction by Franey and Love [25] at 325 MeV.

Figure 3 compares the preliminary result of the angular distribution for the isovector 0^- state with the DWIA calculation. The calculation reproduces the cross sections around the 2nd maximum at 14° without a normalization factor, while it underestimates and slightly misses the 3rd maximum. Furthermore the analyzing power data are not reproduced by this calculation completely by giving the opposite sign. We have also performed a DWIA calculation by using the density- and energy-dependent in-medium t-matrix evaluated from the G-matrices [26]. The G-matrix calculations were performed by using the Paris *NN* potential. The results are shown in Fig. 3 as the dashed curves. The calculated cross sections give a similar angular distribution compared with those with the Franey and Love free t-matrix, but they are larger by a factor of 2. On the contrary,

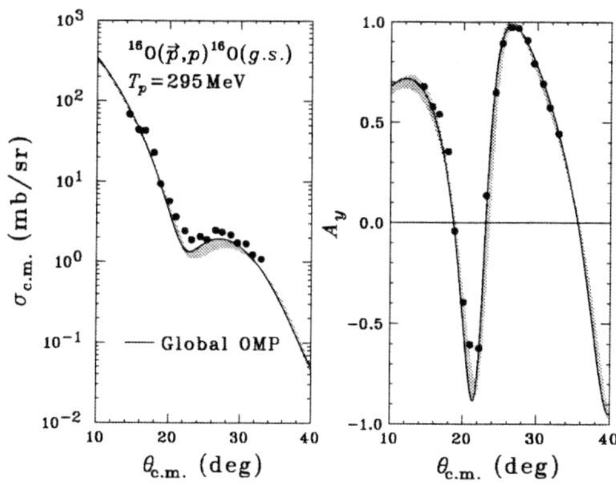

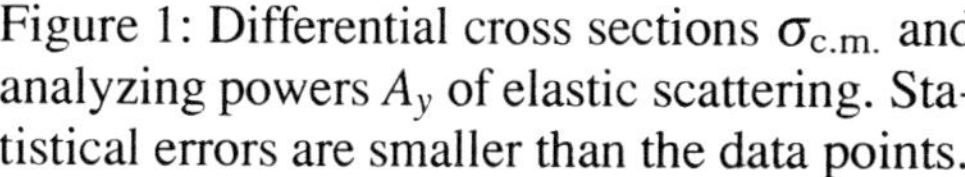
Figure 1: Differential cross sections $\sigma_{c.m.}$ and analyzing powers A_y of elastic scattering. Statistical errors are smaller than the data points.

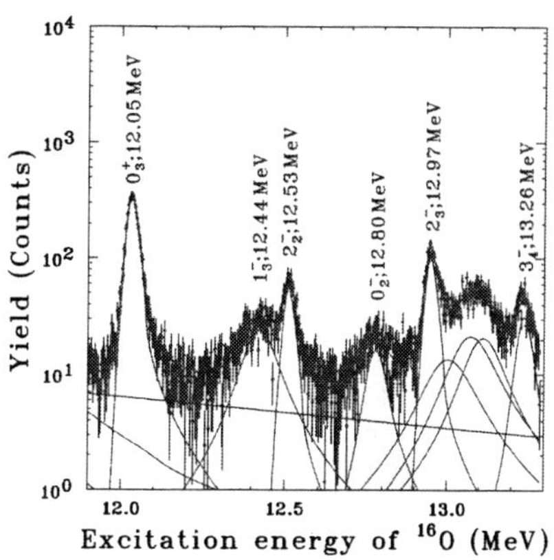

Figure 2: A typical energy spectrum of the $^{16}O(p,p')$ scattering at T_p = 295 MeV and $\theta_{\text{lab}} = 30°$. Results of Hyper-Gaussian peak-fitting are also shown.

the results of DWIA calculations for the analyzing powers depend on the choice of the t-matrix. The calculated analyzing powers with the in-medium t-matrix give the correct sign, but they are smaller compared with the experimental data.

Finally we compared our experimental data with the DWIA+RPA calculation. The RPA calculations are performed without the commonly used universality ansatz ($g'_{NN} = g'_{N\Delta} = g'_{\Delta\Delta}$), namely all of the g's are treated independently [27]. The nonlocality of the mean field is treated by an effective mass m^*. These parameters in the present RPA calculation are (g'_{NN}, $g'_{N\Delta}$, $g'_{\Delta\Delta}$) = (0.6, 0.4, 0.5) and $m^*(0)$=0.7m_N. The formalism of DWIA calculations is described in Ref. [28].

The results of DWIA+RPA and DWIA+free response calculations are shown in Fig. 4 as solid and dashes curves, respectively. The DWIA+RPA calculation predicts an enhancement of the 3rd maximum of the cross sections compared with the 2nd maximum, and it reproduces the 2nd and 3rd maxima simultaneously with a normalization factor of 0.5. Thus the experimental data supports the enhancement of the pionic 0^- mode in nuclei as is predicted in the RPA calculation.

ACKNOWLEDGMENTS

We are grateful to M. Ichimura and H. Sakai for their helpful correspondence. This work is supported in part by the Grants-in-Aid for Scientific Research Nos. 12740151, and 14702005 of the Ministry of Education, Science, Sports and Culture of Japan.

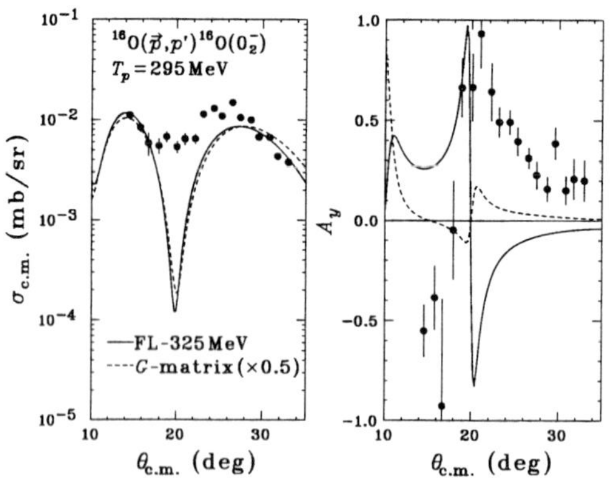

Figure 3: Angular distribution for the isovector 0^- state via $^{16}O(p,p')$ scattering at T_p = 295 MeV. The solid and dashed curves are the results of DWIA calculations. See text for details.

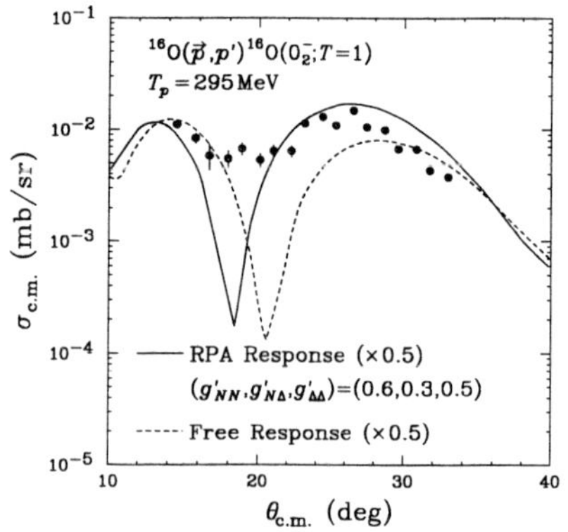

Figure 4: The results of DWIA+RPA and DWIA+free calculations.

REFERENCES

1. G. A. Gagliardi *et al.*, Phys. Rev. Lett. **48**, 914 (1982).
2. P. Guichon *et al.*, Phys. Rev. C **19**, 987 (1979).
3. E. G. Adelberger *et al,*, Phys. Rev. Lett. **46**, 695 (1981).
4. K. Kubodera *et al.*, Phys. Rev. Lett. **40**, 755 (1978).
5. W. G. Love *et al.*, in *Proceedings of International Conference on Spin Excitations, Telluride, Colorado, 1982*, edited by F. Petrovich *et al.*, (1982).
6. H. Orihara *et al.*, Phys. Rev. Lett. **49**, 1318 (1982).
7. C. H. Llewellyn Smith, Phys. Lett. **128B**, 107 (1983).
8. M. Ericson and A. W. Thomas, Phys. Lett. **128B**, 112 (1983).
9. B. L. Friman, V. R. Pandharipande, and R. B. Wiringa, Phys. Rev. Lett. **51**, 763 (1983).
10. E. L. Berger, F. Coester, and R. B. Wiringa, Phys. Rev. D **29**, 398 (1984).
11. D. Stump, G. F. Bertsch, and J. Pumlin, AIP Conf. Proc. **110**, 339 (1984).
12. K. Hosono *et al.*, Phys. Rev. C **30**, 746 (1984).
13. J. J. Kelly *et al.*, Phys. Rev. C **39**, 1222 (1989).
14. J. J. Kelly *et al.*, Phys. Rev. C **41**, 2504 (1991).
15. R. Sawafta *et al.* IUCF Scientific and Technical Report, May 1988–April 1989, p.19.
16. J. J. Kelly *et al.*, Phys. Rev. C **43**, 1272 (1991).
17. J. D. King *et al.*, Phys. Rev. C **44**, 1077 (1991).
18. W. M. Alberico, M. Ericson, and A. Molinari, Nucl. Phys. **A379**, 429 (1982).
19. M. Fujiwara *et al.*, Nucl. Instrum. Methods Phys. Res. A **422**, 484 (1999).
20. T. Wakasa *et al.*, Nucl. Instrum. Methods Phys. Res. A **482**, 79 (2002).
21. T. Kawabata *et al.*, Nucl. Instrum. Methods Phys. Res. A **459**, 171 (2001).
22. S. Hama *et al.*, Phys. Rev. C **41**, 2737 (1990).
23. J. Raynal, Program DWBA98, NEA 1209/05, 1999.
24. T. Kawabata *et al.*, Phys. Rev. C **65**, 064316 (2002), and references therein.
25. M. A. Franey and W. G. Love, Phys. Rev. C **31**, 488 (1985).
26. H. V. von Geramb, AIP Conf. Proc. **97**, 44 (1982).
27. K. Nishida and M. Ichimura, Phys. Rev. C **51**, 269 (1995).
28. K. Kawahigashi, K. Nishida, A. Itabashi, and M. Ichimura, Phys. Rev. C **63**, 044609 (2001).

Polarization Transfer in the ^{16}O(p,p') Reaction at Forward Angles and Structure of the Spin-dipole Resonances

T. Kawabata*, H. Akimune†, G.P.A. Berg**, B. A. Brown‡, H. Fujimura**, H. Fujita**, Y. Fujita§, M. Fujiwara**, K. Hara**, K. Hatanaka**, K. Hosono¶, T. Ishikawa||, M. Itoh**, J. Kamiya**, M. Nakamura||, T. Noro**, E. Obayashi**, H. Sakaguchi||, Y. Shimbara§, H. Takeda||, T. Taki||, A. Tamii††, H. Toyokawa‡‡, M. Uchida||, H. Ueno§, T. Wakasa**, K. Yamasaki†, Y. Yasuda||, H. P. Yoshida** and M. Yosoi||

** Center for Nuclear Study, University of Tokyo, Hongo 7-3-1, Bunkyo-ku, Tokyo 113-0033, Japan*
†Department of Physics, Konan University, Kobe, Hyogo 658-8501, Japan
***Research Center for Nuclear Physics, Osaka University, Ibaraki, Osaka 567-0047, Japan*
‡Department of Physics and Astronomy and National Superconducting Cyclotron Laboratory, Michigan State University, East Lansing, Michigan 48824-1321, USA
§Department of Physics, Osaka University, Toyonaka, Osaka 560-0043, Japan
¶Department of Engineering, Himeji Institute of Technology, Hyogo 678-1297, Japan
||Department of Physics, Kyoto University, Kyoto 606-8502, Japan
††Department of Physics, University of Tokyo, Hongo, Tokyo 113-0033, Japan
‡‡Japan Synchrotron Radiation Research Institute, Hyogo 679-5198, Japan

Abstract. Cross sections and polarization transfer observables in the ^{16}O(p,p') reactions at 392 MeV were measured at forward angles including 0°. The non-spin-flip ($\Delta S = 0$) and spin-flip ($\Delta S = 1$) strengths in transitions to several discrete states and broad resonances in ^{16}O were extracted using a model-independent method. The giant resonances in the energy region of E_x =19–27 MeV were found to be predominantly excited by $\Delta L = 1$ transitions. The strength distribution of spin-dipole transitions with $\Delta S = 1$ and $\Delta L = 1$ were deduced. The obtained distribution was compared with a recent shell model calculation. Experimental results are reasonably explained by distorted-wave impulse-approximation calculations with the shell model wave functions.

INTRODUCTION

Spin-isospin excitation modes in nuclei have been studied intensively, not only because they are of interest in nuclear structure, but also because the relevant operators mediate β-decay and neutrino capture processes. The cross sections of hadronic reactions provide a good measure for the weak interaction response, which is a key ingredient in studies of nucleosynthesis. Gamow-Teller resonances ($\Delta T = 1$, $\Delta S = 1$, $\Delta L = 0$) mediated by the $\vec{\sigma}\vec{\tau}$ operator have been systematically investigated by charge exchange reactions like (p,n) and $(^3\mathrm{He},t)$ reactions with a selectivity for spin-flip transitions at intermediate energies. On the other hand, spin-dipole resonances (SDR; $\Delta T = 1, \Delta S = 1$, $\Delta L = 1$) mediated by the $\vec{\sigma}\vec{\tau}rY_1$ operator have not been studied in any detail although the excitations have recently received attention from the view point of detection of su-

CP675, *Spin 2002: 15th Int'l. Spin Physics Symposium and Workshop on Polarized Electron Sources and Polarimeters*, edited by Y. I. Makdisi, A. U. Luccio, and W. W. MacKay

pernova neutrinos. The detailed structure of the SDR remains unclear with respect to the three different spin states of $J^{\pi} = 2^-$, 1^-, and 0^-. Transitions to the 1^- states can be induced by a probe with spin through the spin-flip and non-spin-flip processes with the $\vec{\sigma}\vec{\tau}rY_1$ and $\vec{\tau}rY_1$ operators. The $\vec{\tau}rY_1$ operator mediates the isovector giant dipole resonance (IVGDR), which has the spin parity of $J^{\pi} = 1^-$, the same as the SDR. Theoretically, the SDR and IVGDR are observed together in (p,p') reactions because they have the same $J^{\pi} = 1^-$ and are located in the same excitation energy region.

The ^{16}O nucleus consists of eight protons and eight neutrons in the $1s_{1/2}$, $1p_{3/2}$, and $1p_{1/2}$ shell orbitals in a simple independent particle model. Since the SDR excitation in p-shell nuclei is described as a coherent sum of $1p$-$1h$ transitions from the p- to the sd-shell orbitals, the SDR excitations in ^{16}O are expected to be strong. Djalali *et al.* identified several 2^- and 1^- states at E_x =19–27 MeV in ^{16}O by comparing a (p,p') spectrum at $E_p = 201$ MeV with a (γ,n) spectrum [1]. They pointed out that the gross structures of the 1^- resonances observed in (p,p') and (γ,n) reactions are similar. This suggests that the IVGDR, which is excited through the Coulomb interaction, is dominant in the proton inelastic scattering at intermediate energies, especially at forward angles. Therefore, spin-flip 1^- states were not identified in the (p,p') measurement of Ref. [1].

Recently, all the polarization transfer (PT) observables for (p,p') reactions were successfully measured at 0° and were found to be a useful spectroscopic tool to study nuclear structure [2], because the total spin transfer $\Sigma \equiv [3-(D_{SS}+D_{NN}+D_{LL})]/4$ provides a clear means to clarify spin-flip or non-spin-flip transitions. The Σ value is unity for spin-flip transitions and zero for non-spin-flip transitions at forward scattering angles where the spin-orbit interaction is negligible [3]. Thus, measurements of PT observables at forward angles enable us to extract spin-flip transitions by introducing spin-flip cross section $(\Sigma d\sigma/d\Omega)$ and non-spin-flip cross section $[(1-\Sigma)d\sigma/d\Omega]$.

In this report, we will present information on the structure of the SDR in ^{16}O, which is obtained from measurements of PT observables in proton inelastic scattering at very forward angles including 0°.

RESULTS AND DISCUSSION

The experiment was performed at the Research Center for Nuclear Physics (RCNP), Osaka University by using a 392-MeV polarized proton beam. The newly developed self-supporting ice target system was used to obtain clean ^{16}O(p,p') spectra [4]. The double differential cross sections measured at $\theta_{lab} = 0°$ and 4° are shown in Fig. 1. All low-lying discrete peaks observed between 6.05 MeV and 13.09 MeV have been identified as those of known transitions.

Four broad resonance states at 20.9, 22.1, 23.0, and 24.0 MeV are observed in the non-spin-flip spectrum at 0° [Fig. 1(c)]. The cross sections of these non-spin-flip states are forward peaked, which could be characterized as IVGDR or $\Delta L = 0$ transitions. The IVGDR in ^{16}O has been already well studied with electromagnetic probes. The total photo-absorption cross sections from Ref. [5] are shown in Fig. 2(a) as a function of photon energy. The reduced $E1$ transition matrix element $B(E1)$ can be easily obtained from the photo-absorption cross sections [6]. By using the $B(E1)$ values, we can estimate

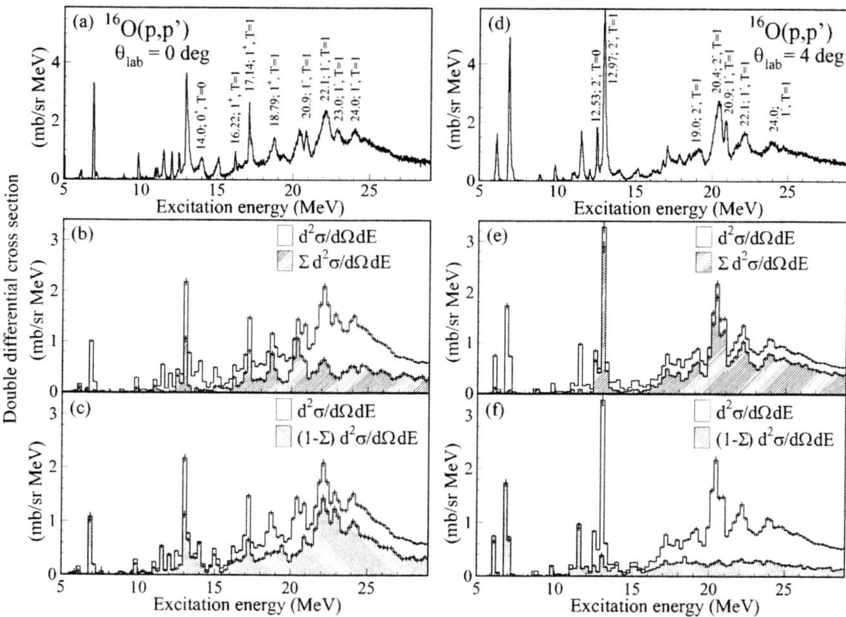

FIGURE 1. Double differential cross sections for the $^{16}\mathrm{O}(p,p')$ reaction at 0° [(a)-(c)] and 4° [(d)-(f)].

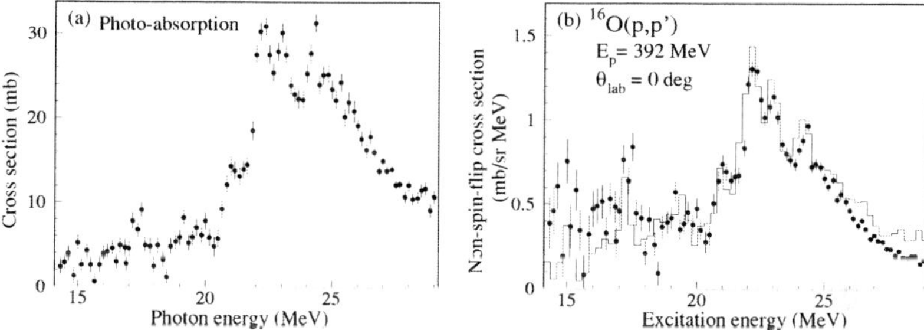

FIGURE 2. (a) The photo-absorption spectrum from Ref. [5]. (b) Comparison of the non-spin-flip $^{16}\mathrm{O}(p,p')$ spectrum at 0° (solid lines) and the corresponding converted photo-absorption spectrum (solid circles). Measurement errors for the non-spin-flip spectrum are not shown in this figure for simplicity.

(p,p') cross sections for the IVGDR. The conversion factors from $B(E1)$ to non-spin-flip cross sections were determined by the DWIA calculation at each excitation energy. After multiplying the converted photon absorption spectrum by a factor of 1.3, the converted spectrum agrees well with the non-spin-flip spectrum from the (p,p') experiment as seen in Fig. 2(b). This means that the IVGDR exhausts most of the non-spin-flip transition strengths at 0°.

In the spin-flip spectra shown in Fig. 1(b) and (e), broad bumps with $\Delta L = 1$ were observed at $E_x = 19.0$, 20.4, 20.9, 22.1, and 24.0 MeV. Since the bumps at 19.0 and 20.4

MeV are not seen in the non-spin-flip spectra [see Fig. 1(c) and (f)], they are inferred to be excited by unnatural parity transitions, and correspond to the SDR (2^-) reported in electron scattering [7, 8, 9]. The other resonances at 20.9, 22.1, and 24.0 MeV, which are seen in both the spin-flip and non-spin-flip spectra, could be due to 1^- excitations with a mixture of spin-flip and non-spin-flip characters. The resonance at 20.9 MeV was assigned as 2^- in Ref. [1], but our result favors the conclusion reported by Ref. [8] that the 20.9-MeV state is 1^-. A bump due to the excitation of a 1^- resonance at $E_x = 23.0$ MeV is clearly seen in Fig. 1(c). However, the corresponding bump is not observed in the spin-flip spectrum at 4° [see Fig. 1(e)]. Thus, we conclude that the 1^- transition to the 23.0-MeV resonance is dominated by a non-spin-flip component.

In Fig. 3(a) and (c), the spin-flip spectrum at $\theta_{lab} = 4°$ is compared with the results of electron scattering experiments of Ref. [9]. The spin parity of the state at $E_x = 23.5$ MeV was tentatively assigned as $J^\pi = 2^-$ in Ref. [7]. Our assignment for the SDR (2^-) is consistent with the result from the electron scattering experiments, but it is rather difficult to get a clear one-to-one correspondence for the 2^- states at 16.82, 17.78, 18.50, and 23.5 MeV in the present experiment. The ratio of the strength of the 20.4 MeV and 19.0 MeV resonance is quite different from the result obtained in electron scattering. This difference might be due to the contribution of the orbital part in the electromagnetic interaction, which does not give a sizable effect in (p,p') scattering at small momentum transfer.

The spin-flip spectrum is compared with the DWIA calculation with the recent shell model wave functions [10] and Franey-Love effective interaction [11] in Fig. 3(b). It is assumed that each shell model state has a Lorentzian shape with a width of 1.0 MeV. The calculation predicts a concentration of discrete spin-flip strengths at $E_x \approx 13$ MeV, which is consistent with the experimental result. In the region of giant resonances, the calculation reproduces the experimental result that the 2^- strength concentrates at an excitation energy below the 1^- strength. This ΔJ splitting is expected due to the spin-orbit interaction, supporting the validity of the present calculations. The strong resonance at 20.4 MeV is predominantly due to a 2^- transition, while the 22.1 MeV is due to both 2^- and 1^- transitions according to the calculation. In addition, the shell model calculation predicts a considerable 0^- strength at higher excitation energies. However, such a 0^- strength could not be separated reliably from the quasifree background. It is noteworthy to mention that a simple $1p$-$1h$ shell model calculation by Picklesimer and Walker [12] has predicted the gross structure of the SDR similar to the recent sophisticated calculation [10], although the quenching problem for spin excitations was not seriously discussed before the 1980s.

SUMMARY

In the present $^{16}O(p,p')$ experiment, spin-flip and non-spin-flip transitions were separated by measuring the polarization transfer (PT) observables.

Non-spin-flip transitions observed at excitation energies between 20 MeV and 27 MeV are well reproduced by a calculation in which the excitation strengths are converted from the photo-absorption cross sections with a normalization factor of 1.3. Therefore,

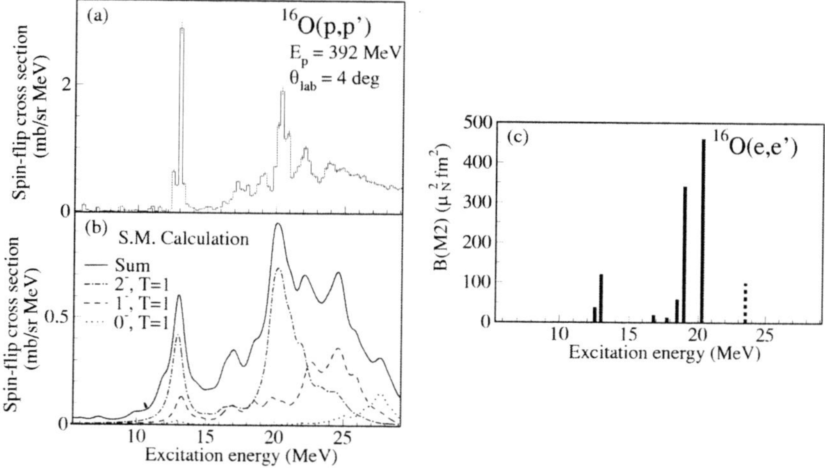

FIGURE 3. (a) Measured spin-flip spectrum in the $^{16}O(p,p')$ reaction at $\theta_{lab} = 4°$. (b) Calculated spin-flip spectra obtained from the shell calculation calculation [10]. The solid line shows the sum of all the transitions up to $\Delta J = 4$. (c) Measured $M2$ strength distribution in ^{16}O from Ref. [9]. The state at $E_x = 23.5$ MeV (dashed line) was tentatively assigned as 2^- in Ref. [7].

we conclude that the major part of the IVGDR strength is exhausted in the resonance region of E_x =20−27 MeV.

Spin-flip strengths observed in the same energy region with the IVGDR are found to be excited with $\Delta L = 1$ angular momentum transfer. The resonances observed at $E_x =$ 20.9, 22.1, and 24.0 MeV carry both IVGDR and SDR (1^-) strengths. The resonances at $E_x = 19.0$ and 20.4 MeV are observed only in the spin-flip spectra and are, therefore, assigned to be 2^- states. The energies of strong 2^- states observed in the present (p,p') experiment agree well with those of the 2^- states reported in electron scattering studies [7, 8, 9]. The recent shell model calculation [10] reproduces well the distribution of the 2^- and 1^- spin flip strengths in ^{16}O measured in this study.

REFERENCES

1. C. Djalali, *et al.*, Phys. Rev. C **35**, 1201 (1987).
2. A. Tamii, *et al.*, Phys. Lett. B **459**, 61 (1999); T. Ishikawa, *et al.*, Nucl. Phys. **A687**, 58c (2001).
3. T. Suzuki, Prog. Theor. Phys. 103, 859 (2000), and references therein.
4. T. Kawabata, *et al.*, Nucl. Instrum. Methods Phys. Res. A **459**, 171 (2001).
5. J. Ahrens, *et al.*, Nucl. Phys. **A251**, 479 (1975).
6. A. Bohr and B.R. Mottelson, *Nuclear Structure* (Benjamin, New York, 1975), Vol. 2, p. 478.
7. A. Goldmann and M. Stroetzel, Phys. Lett. **31B**, 287 (1970); Z. Phys. **239**, 235 (1970).
8. M. Stroetzel and A. Goldmann, Z. Phys. **233**, 245 (1970).
9. G. Küchler, A. Richter, E. Spamer, W. Steffen, and W. Knüpfer, Nucl. Phys. **A406**, 473 (1983).
10. N. Auerbach and B.A. Brown, Phys. Rev. C **65**, 024322 (2002).
11. M.A. Franey and W.G. Love, Phys. Rev. C **31**, 488 (1985).
12. A. Picklesimer and G.E. Walker, Phys. Rev. C **17**, 237 (1978).

Spectroscopy of ^{120}Sn Homologous Levels via the $^{123}Sb(\vec{p},\alpha)^{120}Sn$ Reaction

P. Guazzoni (1), L. Zetta (1), A. Covello (2), A. Gargano (2), Y. Eisermann (3), G. Graw (3), R. Hertenberger (3), H.-F. Wirth (3), M. Jaskola (4), B. Bayman (5), and W.E. Ormand(6)

(1) Dipartimento di Fisica dell'Università and I.N.F.N, I-20133 Milano, Italy
(2) Dipartimento di Fisica dell'Università and I.N.F.N,I-80126 Napoli, Italy
(3) Sektion Physik der Universitaet Muenchen, D-85748 Garching, Germany
(4) Soltan Institute for Nuclear Studies, Warsaw, Poland
(5) Physics and Astronomy Department, University of Minnesota, Minneapolis
(6) Lawrence Livermore National Laboratory, CA-94551 Livermore, USA

Abstract. In order to investigate the spectator role of the $1g_{7/2}$ unpaired proton outside the Z=50 closed shell in the ^{123}Sb nucleus and to test in this region the validity of the concept of homology, which we already tested in the Z=40 and Z=82 regions, the reaction $^{122}Sn(\vec{p},\alpha)^{119}In$ was measured and the reaction $^{123}Sb(\vec{p},\alpha)^{120}Sn$ is currently being studied. In the present contribution the multiplet of states of ^{120}Sn, homologous to the $9/2^+$ ^{119}In G.S. is described in details.

THE EXPERIMENT

The $(\vec{p},\alpha)$ reactions on nuclei around closed or semiclosed shells display several properties that make it a useful spectroscopic tool for supplementing level structure information obtained by other charged-particle reactions. In our previous work concerning Z=40 and Z=82 regions [1-4] we have shown that an interesting behavior can be observed for a number of transitions induced by $(\vec{p},\alpha)$ reactions on near magic target nuclei having one nucleon outside a completely filled magic shell. In this case the unpaired nucleon, slightly bound, may act as spectator in the process. Some distinctive features are displayed: a) weak population of residual nucleus levels below an excitation energy strictly related to the energy gap in the nucleon state spacing at the filling of the magic shell, b) excitation of homologous states (i.e. of states with a close structural relationship) of residual nuclei from (p,α) on adjacent target nuclei one *magic* (leading to parent state transition) and the other *near-magic* (leading to a multiplet of corresponding daughter states) with one more nucleon outside the magic shell.

In this framework, in order to complete spectroscopic study of ^{120}Sn, a high resolution experiment was carried out with the 24 MeV polarized proton beam of the

CP675, *Spin 2002: 15th Int'l. Spin Physics Symposium and Workshop on Polarized Electron Sources and Polarimeters*, edited by Y. I. Makdisi, A. U. Luccio, and W. W. MacKay

Munich MP Tandem accelerator, using the new Stern-Gerlach source [5], the Q3D magnetic spectrograph and the new light ion focal plane detector [6] to study $^{123}Sb(\vec{p},\alpha)^{120}Sn$, to be compared with $^{122}Sn(\vec{p},\alpha)^{119}In$ [7]. The beam current intensity was up to 1.5 μA and the beam polarization 60%.

Angular distributions for cross section and asymmetry have been measured from 6° to 52.5°. High resolution and very low background allowed study of 61 transitions to final states of ^{120}Sn up to an excitation energy of ≈ 4700 keV, with a precision of ± 3 keV. This has allowed a remarkable increase of the knowledge of ^{120}Sn nucleus, because 19 new levels have been identified, with the attribution of energy, spin and parity. To 16 levels spin and parity have been attributed, while the only parity has been attributed to 11 levels. For the attribution of spin and/or parity the methodology introduced by our group for homologous states has been applied. In such a way multiplets of states of ^{120}Sn homologous to the low lying states of ^{119}In have been identified. In particular we found the octet of states [1^+, 2^+, 3^+, 4^+, 5^+, (6^+ fragmented in two levels), 7^+, 8^+] homologous to the G.S. ($9/2^+$) of ^{119}In, the doublet of states [3^- (fragmented in three levels), 4^-] homologous to the level 0.311 $1/2^-$ MeV of ^{119}In , the quadruplets of states [2^-, 3^-, 4^-, (5^- fragmented in four levels)] homologous to the level 0.604 MeV $3/2^-$ of ^{119}In and the sestet of states [1, 2^-, 3^-, 4^-, 5^-, (6^- fragmented in four levels)] homologous to the level 1.044 MeV $5/2^-$ of ^{119}In.

In fig.1 the α-spectrum measured at θ_{lab}=10° is shown. It is possible to recognize two different regions: the homologous region (E_{exc} > 3.6 MeV) intensively populated and the low excitation energy region poorly populated by the pickup of the spectator proton $1g_{7/2}$ together with a neutron pair. The configuration of the homologous states results from the coupling of the spectator proton (not involved in the process) with the one-proton-hole-two-neutron-hole states of the ^{119}In core.

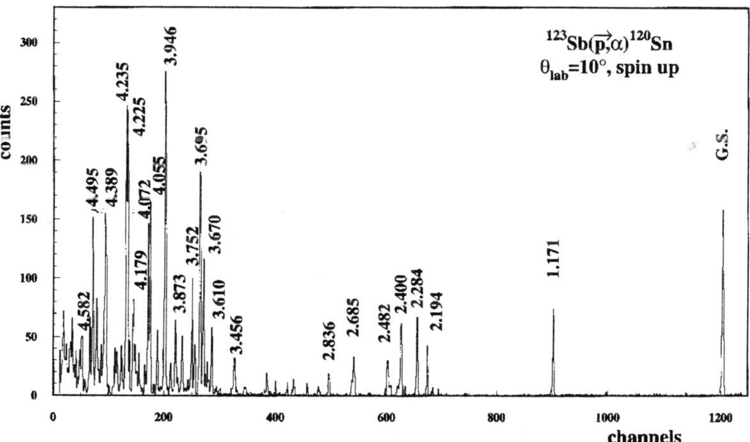

FIGURE 1. The α-spectrum measured at θ_{lab}=10°. It is possible to recognize two different regions: the homologous region (E_{exc}>3.6 MeV) intensively populated and the low excitation energy region, poorly populated.

THE ANALYSIS

Spin and Parity Assignment Via The Homologous State Methodology

In fig.2 the comparison between the measured $\sigma(\theta)$ and $A_y(\theta)$ for the transition to the multiplet of ^{120}Sn states (dots) homologous to the G.S. ($9/2^+$) of ^{119}In and the measured $\sigma(\theta)$ and $A_y(\theta)$ for the transition to the G.S. ($9/2^+$) of ^{119}In is shown. The last cross section is scaled for each level by the proper factor $(2J_i+1)/\Sigma_i(2J_i+1)$ (solid line). In the same figure the cumulative cross section and asymmetry for the octet of levels (dots) homologous to the G.S. ($9/2^+$) of ^{119}In are compared with cross section and asymmetry for the transition to the G.S. ($9/2^+$) of ^{119}In. It is possible to observe the good agreement both in shape and absolute value.

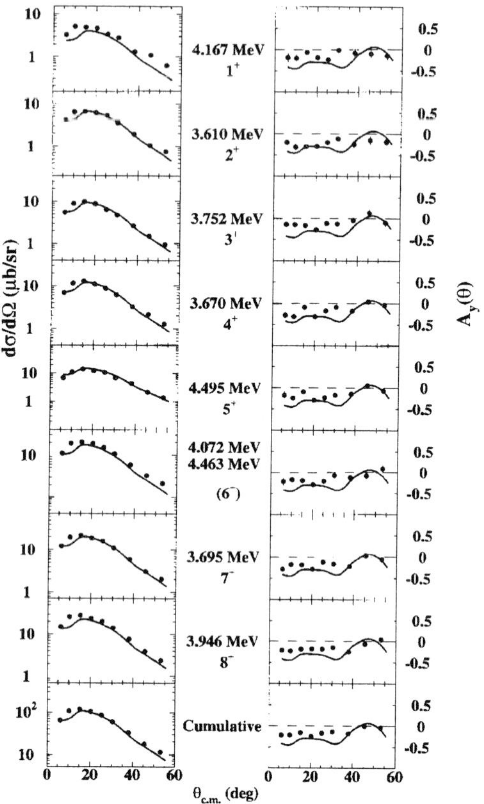

FIGURE 2. Comparison between the measured $\sigma(\theta)$ and $A_y(\theta)$ for the transition to the multiplet of ^{120}Sn states (dots) homologous to the G.S. ($9/2^+$) of ^{119}In and the measured $\sigma(\theta)$ and $A_y(\theta)$ for the transition to the G.S. ($9/2^+$) of ^{119}In. The last one cross section is scaled for each level by the proper factor $(2J_i+1)/\Sigma_i(2J_i+1)$ (solid line).

In case of weak coupling the cross section of a homologous state with spin J_i in a given multiplet can be related to that of the corresponding parent state by the following expression:

$$\sigma_{son}(^{120}Sn,J_i) = \sigma_{parent}(^{119}In) * (2J_i+1)/\Sigma(2J_i+1).$$

In fig.3 the quantities $[(2J_i+1) * \sigma_{parent}(^{119}In)]/\Sigma(2J_i+1)$ are reported for each member of the multiplet vs. J_i, together with the straight $(2J_i+1)$ line.

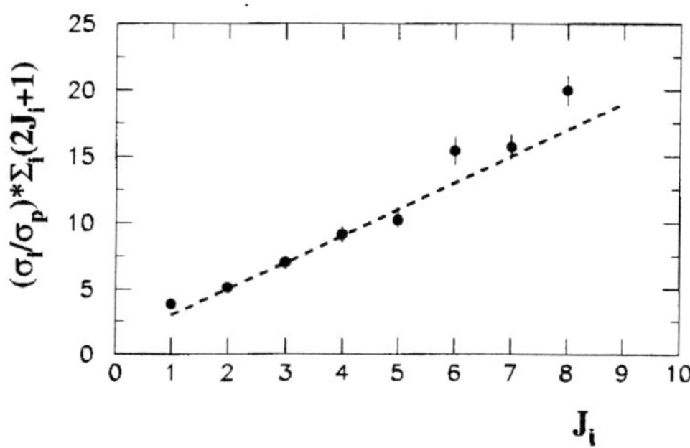

FIGURE 3. Plot of the quantities $[(2J_i+1) * \sigma_{parent}(^{119}In)]/\Sigma(2J_i+1)$ for each member of the multiplet vs. J_i (dots), together with the straight $(2J_i+1)$ line

Microscopic DWBA Calculations

The ^{123}Sb ground state is taken to be a $1g_{7/2}$ proton outside filled proton shells, whereas the 22 valence neutrons move in the $1g_{7/2}$, $2d_{5/2}$, $2d_{3/2}$, $3s_{1/2}$ and $1h_{11/2}$ shells. These neutrons interact via a neutron-neutron pairing force, which spreads the neutrons over the valence shells, with a total neutron angular momentum of zero. In this same picture, the ^{120}Sn ground state consists of filled proton shells, with 20 valence neutrons. Thus the pickup reaction involves the transfer of a $1g_{7/2}$ proton, and $1g_{7/2}$, $2d_{5/2}$, $2d_{3/2}$, $3s_{1/2}$ and $1h_{11/2}$ neutron pairs. Since the assumed interaction is between the incident proton and the mass-center of the three transferred nucleons, it cannot change the relative motion of the three transferred nucleons. Thus we must project from shell-model states such as $|1g_{7/2}(2d_{5/2},2d_{5/2})|$, the part in which the three nucleons have the same relative motion as they will have in the outgoing alpha particle. This is done for each of the possible triples of shell-model states that can connect the initial and final states, and then a coherent sum must be taken to obtain the total transfer form factor.

For the population of the homologous states, one removes a $1g_{9/2}$ proton, as well as a neutron pair. In this way one can reach states with total angular momenta J=1,2,...,8, which consist of a $1g_{7/2}$ proton coupled to a $1g_{9/2}$ proton hole. The form factor is the same for each value of J, except for a factor of $(2J+1)^{1/2}$. Thus apart from Q-value effects, the calculated angular distribution for each of the J=1,2,...,8 final states will be the same, with a 2J+1 factor of proportionality. Figure 4 shows the comparison between experimental and microscopically calculated angular distributions of cross section and asymmetry for the transition to ^{120}Sn 0^+ G.S. together with the cumulative angular distributions for the ^{119}In G.S. homologous multiplet. The calculations were done in finite range approximation, with the previously described microscopic configurations. The overall multiplicative factor, used to give a reasonable fit to the experimental data, is the same for the cumulative and G.S. cross sections.

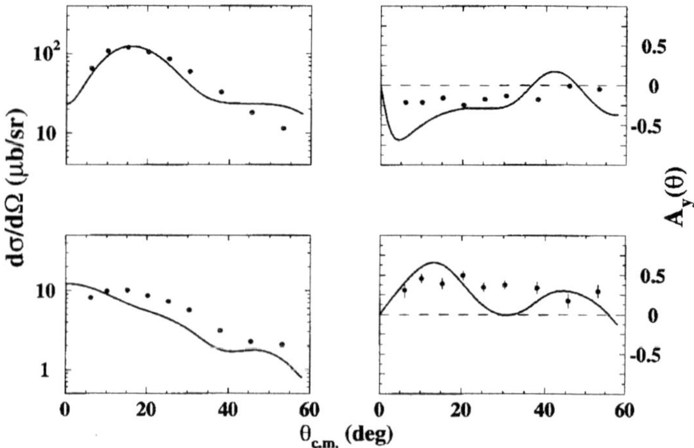

Figure 4. Comparison between experimental and microscopically calculated angular distributions of cross section and asymmetry for the transition to ^{120}Sn 0^+ G.S. (bottom) together with the comparison for cumulative angular distributions of the ^{119}In G.S. homologous multiplet (top), obtained in finite range approximation with the previous microscopic configurations (dots represent experimental values, solid lines microscopic calculations).

ACKNOWLEDGMENTS

This work was performed in part under the auspices of the Italian Ministry for University and Research, under contract CRUI-Progetto Vigoni, the DFG under Grant No. C4-Gr894/2, and the U.S. Department of Energy by the University of California, Lawrence Livermore National Laboratory under contract No. W-7405-Eng-48.

REFERENCES

1. Guazzoni, P., et al., *Z. Phys. A* **356**, 381-391 (1997).
2. Guazzoni, P., et al., *Eur. Phys. J. A* **1**, 365-378 (1998).
3. Guazzoni, P., et al., *Phys. Rev. C* **49**, 2784-2787 (1994).
4. Gu, J.N., et al., *Phys. Rev. C* **55**, 2395-2406 (1997).
5. Hertenberger, R., et al., AIP Conference Proceedings 570, New York: American Institute of Physics, 2001, pp. 825-829.
6. Wirth, H.-F., Ph.D. Thesis, Technische Universität, München, 2001.
7. Guazzoni, P., et al., AIP Conference Proceedings 570, New York: American Institute of Physics, 2001, pp. 664-668.

Study of the Spin Dependent ^{3}He-Nucleus Interaction at 450 MeV

J. Kamiya*, K. Hatanaka*, Y. Sakemi*, T. Wakasa*, H.P. Yoshida*, E. Obayashi*, K. Hara*, Y. Kitamura*, Y. Shimizu*, K. Fujita*, N. Sakamoto*, H. Sakaguchi†, M. Yosoi†, M. Uchida†, Y. Yasuda†, Y. Shimbara**, T. Adachi**, T. Noro‡ and T. Kawabata§

*Research Center for Nuclear Physics (RCNP) Ibaraki, Osaka 567-0047, Japan
†Department of Physics, Kyoto University, Kyoto 606-8502, Japan
**Department of Physics, Osaka University, Toyonaka, Osaka 560-0043, Japan
‡Department of Physics, Kyushu University, Hakozaki, Fukuoka 812-8581, Japan
§RIKEN-BNL Research Center, Brookhaven National Laboratory, Upton, NY 11973, USA

Abstract. The cross sections and analyzing powers were measured for ^{3}He+^{12}C, ^{58}Ni, and ^{90}Zr elastic scattering at $T_{^3\mathrm{He}}$=450 MeV. The incident energy dependence of the volume integral per nucleon for ^{3}He shows the similar behavior to that of protons if the spin-orbit potential is taken into account. This result supports the assumption that interactions between ^{3}He and nucleus are dominated by the interactions between constituent nucleons of ^{3}He and target nucleus in the intermediate energy region. The results of the single folding calculation suggest the interaction between point nucleon and target nucleus is modified in the ^{3}He+nucleus elastic scattering.

INTRODUCTION

The interaction between complex nuclei is one of the most fundamental subjects in nuclear physics and many studies have been performed both experimentally and theoretically. The origin of the nucleus-nucleus spin-orbit interaction is of special interest because it is closely related to the nuclear structure and reaction mechanism. Unlike the case of an electron moving in a Coulomb field, where the force arises from electro magnetic interaction, the difficulties in the nuclear case arise from the complex meson exchange nature of the nucleon-nucleon interaction. ^{3}He elastic scattering is good tool to study the interaction between the complex nuclei because the reaction mechanism is simple compare to more fragile projectiles like deuteron or ^{6}Li. It is also important to determine the optical potential parameters including the spin-orbit potentials to study inelastic scatterings and charge exchange reactions like (^{3}He,t) reactions. Spin dependent part of the interaction between ^{3}He and nucleus, however, has not been studied well because of the lack of the experimental data for polarization observables. The experimental data are limited to cross sections in the intermediate energies, where the phenomenological optical model calculations show that the spin-orbit terms do not affect the cross section data. Therefore, the spin-orbit interaction has been considered to be negligible at intermediate energies. Recently, however, theoretical investigations using microscopic optical models have been reported and the results show the folding model calculations

CP675, *Spin 2002: 15th Int'l. Spin Physics Symposium and Workshop on Polarized Electron Sources and Polarimeters*, edited by Y. I. Makdisi, A. U. Luccio, and W. W. MacKay

predict the large effects of the spin-orbit component on the cross sections and analysing powers for ^{3}He-nucleus elastic scattering at intermediate energies [1]. Therefore it is very effective to perform the experiment at intermediate energy to study the spin dependent ^{3}He-nucleus interactions precisely.

EXPERIMENT

The experiment was performed at RCNP, Osaka University. We measured the cross sections and analyzing powers for ^{3}He+^{12}C, ^{58}Ni, ^{90}Zr elastic scatterings at a bombarding energy of 450 MeV. The analyzing powers A_y, which has the same values as the polarizations for spin $\frac{1}{2}+0 \rightarrow \frac{1}{2}+0$ reactions, were measured by using the double scattering method. The calorimeter for ^{3}He was specially designed and constructed behined the normal focal plane polarimeter system of the Grand Raiden spectrometer. The calorimeter consisted of plastic scintillators and measured the energies of the secondarily scattered ^{3}He particles from the analyzer target. Preceding the A_y measurement, the focal plane polarimeter system was calibrated. At first we measured the absolute value of the polarization for ^{3}He+^{12}C elastic scattering at $\theta_{lab} = 7°$, where the double folding model predicts the large value of the polarization. The absolute value of the polarization was obtained as

$$p_y = 0.547 \pm 0.018. \tag{1}$$

The effective analyzing power A_y^{eff} was calibrated using a plastic scintillator as an analyzer target as

$$A_y^{eff} = \frac{1}{p_y} Asym = 0.232 \pm 0.010, \tag{2}$$

where the p_y is the obtained value in Eq.(1), and $Asym$ is the left-right asymmetry in the second scattering.

RESULTS AND DISCUSSIONS

Fig. 1 shows the results of the angular distribution of the cross sections and analyzing powers for the ^{3}He elastic scattering off ^{12}C, ^{58}Ni, ^{90}Zr nuclei. The experimental results show the large values of the analyzing powers even in the middle angular range. Solid curves shows the results of the phenomenological optical model calculations with central and spin-orbit potentials.

The Fig. 2 shows the incident energy dependence of the volume integral per nucleon for ^{90}Zr target. The upper and lower panels shows the real J_R and imaginary J_I terms, respectively. Solid curves show the volume integrals for protons obtained by fitting the experimental data by Arnold *et al.* [2], while the open circles the experimental results for ^{3}He. The dotted curve is the guide for eyes. The closed circles shows the present results given by the central optical potentials fitted to the cross sections. The closed squares also shows the present results but by using both central and spin-orbit optical potentials fitted to cross sections and analyzing powers. The energy dependence

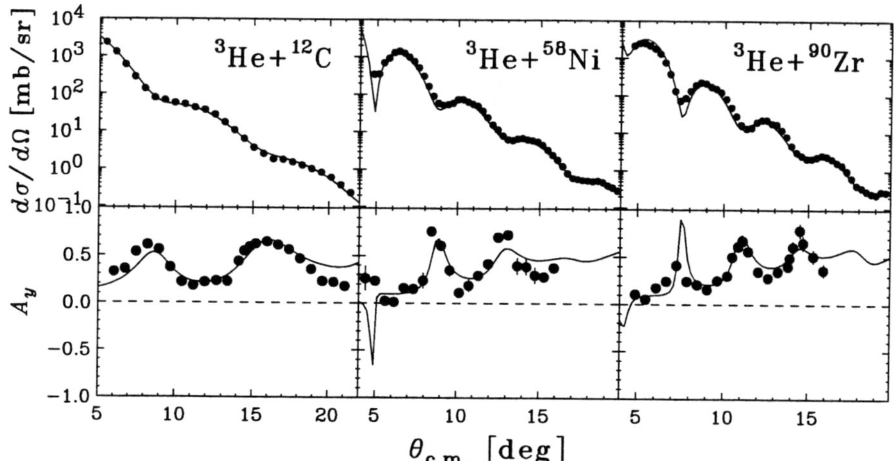

FIGURE 1. The differential cross sections and analyzing powers for ^{3}He+^{12}C, ^{58}Ni, and ^{90}Zr elastic scattering. The experimental data are shown by the closed circles. The solid lines show the results of the optical model calculation which give the minimum χ^2 to the data.

of J_R show the similar behavior to protons in the intermediate energy region, where it decreases monotonically. The behavior of the J_I of ^{3}He was not understandable without spin-orbit effect because it decrease monotonically with increasing the incident energy, while J_I of the protons exhibit a gradual rise above 100 MeV/nucleon reflecting a pion-production effect. However, by including the spin-orbit interaction, the values of J_I of ^{3}He at 150 MeV/nucleon increase and the energy dependence shows the very similar behavior to the protons. This results support the assumption that interactions between ^{3}He and target nucleus are dominated by the interactions between constituent nucleons in ^{3}He and target nucleus in the intermediate energy region. Therefore, the single folding model is possible to be the good model to reproduce the experimental data.

The central component of the single folding model is calculated as

$$V_{SF}^{C}(\mathbf{R}) = \int \rho_{^3\mathrm{He}} v^C(\mathbf{R}-\mathbf{r}')d\mathbf{r}', W_{SF}^{C}(\mathbf{R}) = \int \rho_{^3\mathrm{He}} w^C(\mathbf{R}-\mathbf{r}')d\mathbf{r}', \tag{3}$$

where $\rho_{^3\mathrm{He}}$ represents the nucleon density distribution of ^{3}He, while $v(w)^C$ is the real (imaginary) central part of a proton-target optical potential at the incident energy of $T_p = T_{^3\mathrm{He}}/A$. Assuming a simple harmonic oscillator as a wave function of ^{3}He, the density distribution reduces to a single Gaussian form, allowing its width to be consistent with the ^{3}He charge radius evaluated from electron scattering data. The spin-orbit component is written as

$$U_{SF}^{SO}(\mathbf{R}) = \langle \phi_{^3\mathrm{He}}(\mathbf{r},\rho)|u^{SO}(\mathbf{R}+2/3\rho)|\phi_{^3\mathrm{He}}(\mathbf{r},\rho)\rangle l_n \cdot \sigma_n = F^{SO}(R)L\cdot\sigma, \tag{4}$$

where only neutron in the s state in ^{3}He contributes to the spin-orbit component of the single folding potential. $\rho_{^3\mathrm{He}}(\mathbf{r},\rho)$ represents the internal wave function of ^{3}He, where

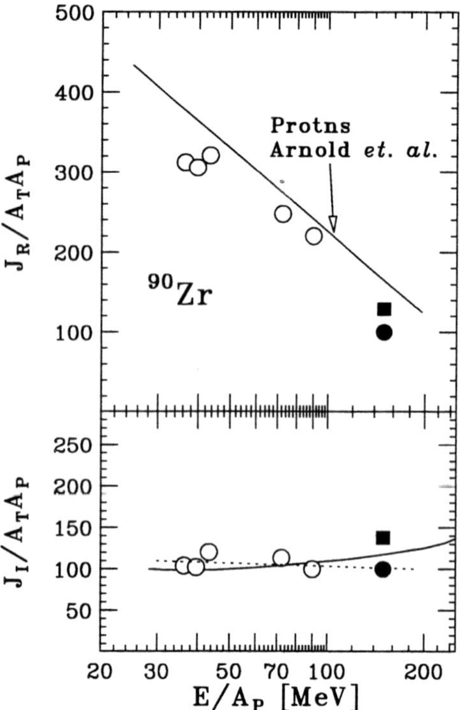

FIGURE 2. The incident energy dependence of the volume integral per nucleon for ^{90}Zr target. The closed circles and squares show the present results given by the only central optical potentials and both central and spin-orbit potentials, respectively. The solid curves shows volume integrals for protons. The dotted curves are guide for eyes.

$\mathbf{r}$ and ρ is the relative coordinate between the two protons and that between the neutron and the center of mass of the two protons in ^{3}He, respectively. l_n and σ_n represent the orbital angular momentum operator and the Pauli spin operator of the neutron in ^{3}He. L and σ are the corresponding operators for ^{3}He-target system. The total single folding potential is written as

$$U_{\mathrm{SF}}(R) = N_R V_{SF}^C(R) + iN_I W_{SF}^C + N_{\mathrm{SO}} F^{\mathrm{SO}}(R) L \cdot \sigma, \tag{5}$$

where renomarization factors (N_R, N_I, N_{SO}) are introduced to examine modification of the strength of the each potential.

Fig. 3 shows the results for the ^{90}Zr target. The dotted curves represent the results of the folding potential without renormalization factors, while the dashed curves are best-fit calculations with renormalization factors (N_R, N_I, N_{SO})=(0.80, 1.08, 0.50). By using the renormalization factors, experimental data can be reproduced except for the cross sections in the backward angles. This results suggest the interaction between point nucleon and target nucleus is modified in the ^{3}He+nucleus elastic scattering. Therefore a double-folding model using density dependent effective nucleon-nucleon interaction

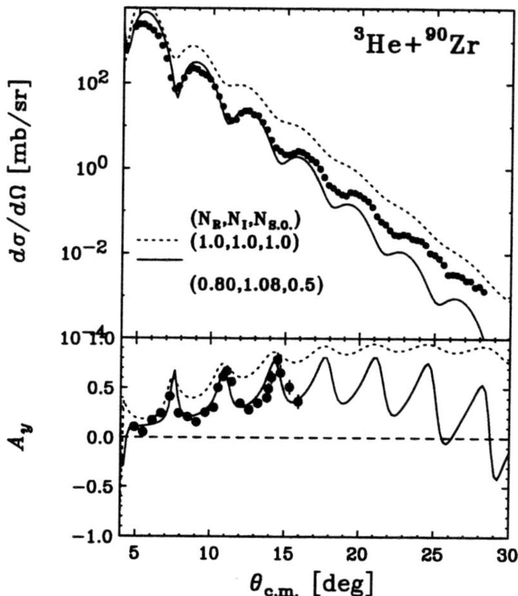

FIGURE 3. The differential cross sections and analyzing powers for ^{3}He+^{90}Zr with the results of the single folding model calculations. The dotted curves are the results of the folding potential without renormalization factors. The dashed curves are best-fit calculations with renormalization factors (N_R, N_I, N_{SO})=(0.80, 1.08, 0.50).

is expected to be the good approach. The calculation is been performed now. The results can be shown near the future.

ACKNOWLEDGEMENTS

We thank Professor Y. Sakuragi for helpful discussions and Dr. M. Katsuma for the nice advise for the folding calculations and many helpful comments. This experiment was performed under the Program No. E157 and E182 at RCNP.

REFERENCES

1. Y. Sakuragi and M. Katsuma, Nucl. Instr. Meth. A **402**, 347 (1998).
2. L. G. Arnold *et al.*, Phys. Rev. C **25**, 936 (1982).

Isoscalar Spin Response in the Continuum Studied via the $^{12}C(\vec{d},\vec{d}')$ Reaction at 270 MeV

Y. Satou*, S. Ishida†, H. Kato**, H. Sakai***, H. Okamura‡, N. Sakamoto†, T. Uesaka‡, A. Tamii**, T. Wakasa§, T. Ohnishi†, K. Sekiguchi†, K. Yako*, K. Suda‡, M. Hatano**, Y. Maeda** and T. Ichihara†

*Center for Nuclear Study, University of Tokyo, Bunkyo, Tokyo 113-0033, Japan
†The Institute of Physical and Chemical Research (RIKEN), Wako, Saitama 351-0198, Japan
**Department of Physics, University of Tokyo, Bunkyo, Tokyo 113-0033, Japan
‡Department of Physics, Saitama University, Urawa, Saitama 338-8570, Japan
§Research Center for Nuclear Physics (RCNP), Ibaraki, Osaka 567-0047, Japan

Abstract. Single and double spin-flip probabilities in inelastic deuteron scattering on ^{12}C have been measured at 270 MeV up to 50 MeV in excitation energy using a focal plane deuteron polarimeter capable of measuring both vector and tensor components of the deuteron polarization. The obtained S_1 values are enhanced relative to the Fermi gas model prediction in highly excited continuum region. The S_2 values are close to zero over the measured excitation energy range.

INTRODUCTION

Studies of polarization phenomena in nucleon induced inelastic scattering at intermediate energies have been one of the active fields of research in nuclear physics. Not only giving an insight into the reaction mechanisms, measurements of spin-flip probability, S_{nn} in particular, provided useful information on spin dependent modes of nuclear excitation. Clear signatures of such specific excitations as the giant Gamow-Teller (GT) resonance and the spin-flip dipole resonance were obtained, and the relative spin response was extracted in highly excited continuum region [1].

The study of spin-flip processes in inelastic deuteron scattering is an important extension to corresponding studies using the nucleon projectile. A new aspect is the selective excitation of isoscalar transitions. This will make the reaction an ideal tool for the study of isoscalar spin-flip modes. Much less is known about this mode due to the lack of efficient probes as well as to the weakness of the effective interaction in this channel. Information on the isoscalar spin response should be useful in elucidating "the quenching mechanisms of spin transitions" and "the problem of the enhancement of the relative spin response in the continuum found in the (p,p') reactions".

Furthermore, spin-1 nature of the deuteron may offer a unique capability to probe double spin-flip transitions, such as the proposed double GT state [2]. Up to now no double GT transition has been identified except for the double beta decay, which represents the ground state to ground state double GT transition exhausting only a minor portion of the double GT sum rule [3]. Experimental determination of the double GT strength distribution should lead to a better understanding of spin-isospin properties of

CP675, *Spin 2002: 15th Int'l. Spin Physics Symposium and Workshop on Polarized Electron Sources and Polarimeters*, edited by Y. I. Makdisi, A. U. Luccio, and W. W. MacKay

nuclei. It will also provide an excellent calibration of structure calculations of the double beta decay nuclear matrix elements, which are necessary ingredients in extracting the neutrino mass from the double beta decay life time measurements.

In the inelastic deuteron scattering two kinds of spin-flip probabilities S_1 and S_2 can be defined as fractions of deuterons undergoing spin-flip by 1 and 2 units along an axis normal to the reaction plane. The single spin-flip probability S_1 is expected to be a good signature of spin excitations, as S_{nn} is in (p,p'), and the double spin-flip probability S_2 a possible probe of double spin-flip excitations. They are given using polarization observables in the following relations:

$$S_1 = \frac{1}{9}(4 - P^{y'y'} - A_{yy} - 2K_{yy}^{y'y'}), \quad (1)$$

$$S_2 = \frac{1}{18}(4 + 2P^{y'y'} + 2A_{yy} - 9K_y^{y'} + K_{yy}^{y'y'}), \quad (2)$$

where the quantities A, P and K refer to the analyzing power, polarizing power and polarization transfer coefficient, one and two indices stand for the vector and tensor polarizations, and lower and upper indices stand for the incident and outgoing beams. Note that the determination of S_1 and S_2 requires vector and tensor polarized beams and a vector and tensor polarimeter.

Recently we have succeeded in extracting S_1 and S_2 in inelastic deuteron scattering on ^{12}C at 270 MeV for an excitation energy range between 4 and 24 MeV [4]; the feasibility of measuring the deuteron spin-flip probabilities over a wide excitation energy range has been demonstrated. As an continuation to the previous work, we have extended the measurement up to 50 MeV in excitation energy of ^{12}C in order to investigate the isoscalar single and double spin-flip strengths in the continuum. The results of the experiment are presented.

EXPERIMENT

The experiment was performed at RIKEN accelerator research facility using the 270 MeV polarized deuteron beams from the Ring Cyclotron. The measured quantities are the differential cross sections and eight polarization observables with respect to the y-axis. The beam polarizations were measured using the beam line polarimeter based on the $\vec{d}+p$ elastic scattering at 270 MeV, and the obtained polarization magnitudes were 60 to 70% of the ideal values. The scattered deuterons were analyzed in momentum with the magnetic spectrometer SMART, and their polarizations were determined with a polarimeter DPOL placed at the focal plane of the spectrometer.

Figure 1 shows the detector arrangement of the polarimeter. It was comprised of three parts: the multiwire drift chamber for track reconstruction, the secondary CH_2 target, and the counter hodoscope to detect charged particles. Measurements of both vector and tensor polarizations of outgoing deuterons are crucial in extracting S_1 and S_2. This was realized by utilizing, as the analyzing reactions, the $\vec{d}+C$ elastic scattering and the change exchange $(\vec{d},2p)$ reaction on hydrogen, which show large angular asymmetries depending, respectively, on the vector and tensor components of the deuteron polariza-

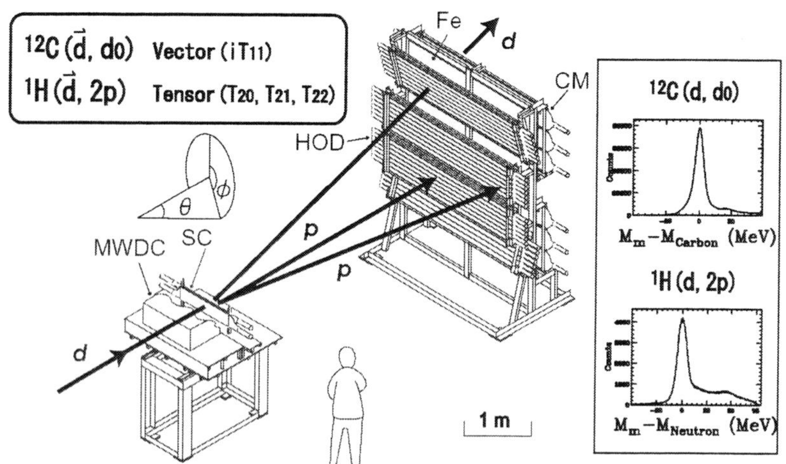

FIGURE 1. Layout of the polarimeter DPOL.

tion. Shown in the inset of Fig. 1 are missing-mass spectra for the $\vec{d}+\mathrm{C}$ elastic scattering and the change exchange $^1\mathrm{H}(\vec{d},2p)$ reaction. We see clear peaks due to the desired reactions. In order to eliminate contributions from parasitic components, such as those arising from the $d+p$ and $^{12}\mathrm{C}(d,2p)$ reactions in the CH_2 scatterer, cuts were applied on the missing-mass spectra in off-line analysis.

RESULTS

Figure 2 (a) shows an excitation energy spectrum of the differential cross section integrated over the laboratory scattering angles between 2.5° and 7.5°. Up to 30 MeV in excitation energy several discrete levels are excited, which include the spin-flip 1^+ (12.71 MeV) and 2^- (18.3 MeV) states and the non-spin-flip 0^+ (7.65 MeV) and 3^- (9.64 MeV) states. All these are isoscalar states. Above 30 MeV in excitation energy the spectrum exhibits a continuum structure.

Figure 2 (b) shows an excitation energy spectrum of the spin-flip probability S_1. The error bars are only statistical ones. We see that the S_1 value in a bin at 20.5 MeV shows a slight enhancement, similarly to the already established spin-flip states such as the 1^+ and 2^- states. In a (d,d') experiment performed at SATURNE, a spin-flip resonance state has been reported at 20.5 MeV, which was tentatively assigned as $J^\pi = 1^+$ [5]. The present result shows consistency with its identification as an isoscalar spin-flip transition. At higher excitation energy region S_1 takes values between 0.1 and 0.3. These values are, in fact, much larger than the prediction of the noninteracting Fermi gas model (dotted curve). Such a large deviation of the S_1 values from the Fermi gas values in the highly excited continuum region has not been identified in previous deuteron scattering experiments. It would be explained either in terms of "the enhancement of the spin

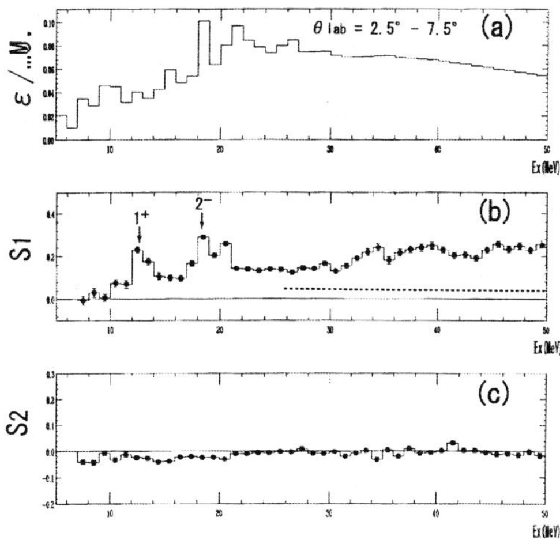

FIGURE 2. Excitation energy spectra of (a) the yield, (b) the spin-flip probability S_1 and (c) the double spin-flip probability S_2. The dotted curve in (b) represents the Fermi gas model prediction.

response in the isoscalar channel" or "the quenching of the effective interaction in the scalar isoscalar channel". Further studies would need to be done to solve this problem. The S_2 values, shown in Fig. 2 (c), are close to zero over the measured excitation energy range up to 50 MeV, and the presence of double spin-flip states could not be indicated in the present experiment.

SUMMARY

We have measured single and double spin-flip probabilities S_1 and S_2 in inelastic deuteron scattering on ^{12}C at 270 MeV up to 50 MeV in excitation energy. An isoscalar spin-flip nature of the state at 20.5 MeV in ^{12}C has been confirmed. The values of S_1 much larger than those of the Fermi gas prediction were observed in the continuum region. The origin of such large S_1 values has not been understood yet; it would need to be addressed in forthcoming studies. The values of S_2 were close to zero; no indication of double spin-flip states could be obtained from the present experiment.

REFERENCES

1. Baker, F. T. *et al.*, *Phys. Rep.*, **289**, 235 (1997).
2. Auerbach, N., Zamick, L., and Zheng, D. C., *Ann. Phys.*, **192**, 77 (1989).
3. Vogel, P., Ericson, M., and Vergados, J. D., *Phys. Lett. B*, **212**, 259 (1988).
4. Satou, Y. *et al.*, *Phys. Lett. B*, **521**, 153 (2001).
5. Johnson, B. N. *et al.*, *Phys. Rev. C*, **51**, 1726 (1995).

Determination of the Gamow-Teller Quenching Factor via the $^{90}Zr(n,p)$ Reaction at 293 MeV

K. Yako*, H. Sakai*, M.B. Greenfield†, K. Hatanaka**, M. Hatano*, J. Kamiya**, Y. Kitamura**, Y. Maeda*, C.L. Morris‡, H. Okamura§, J. Rapaport¶, T. Saito*, Y. Sakemi**, K. Sekiguchi||, Y. Shimizu**, K. Suda††, A. Tamii*, N. Uchigashima* and T. Wakasa**

*Department of Physics, University of Tokyo, Bunkyo, Tokyo 113-0033, Japan
†International Christian University, Mitaka, Tokyo 181-8585, Japan
**Research Center for Nuclear Physics, Osaka University, Ibaraki, Osaka 567-0047, Japan
‡Los Alamos National Laboratory, Los Alamos, NM 87545, USA
§Department of Physics, Saitama University, Saitama, Saitama 338-8570, Japan
¶Department of Physics, Ohio University, Athens, Ohio 45701, USA
||The Institute of Physical and Chemical Research (RIKEN), Wako, Saitama 351-0198, Japan
††Department of Physics, Saitama University, Urawa, Saitama 338-8570, Japan

Abstract. The double differential cross sections at 0°–12° were measured for the $^{90}Zr(n,p)$ reaction at 293 MeV in a wide excitation energy region of 0–70 MeV. The experiment was performed by using the (n,p) facility at the Research Center for Nuclear Physics. The multipole decomposition (MD) technique was applied to the measured cross sections to extract the GT component in the continuum. After subtracting the contribution of the isovector spin-monopole excitation we obtained the GT strength of $S_{\beta+} = 3.0 \pm 0.3 \pm 0.8 \pm 0.5$ up to 31.4 MeV excitation. The quenching factor Q was deduced by using the present result and the $S_{\beta-}$ value obtained from the MD analysis of the $^{90}Zr(p,n)$ spectra. The result is $Q = 0.83 \pm 0.06$ in regards to Ikeda's sum rule value of $3(N-Z) = 30$.

INTRODUCTION

Gamow-Teller (GT) resonances have been extensively studied since its discovery in 1975 [1]. The GT transition involves the operator $\sigma\tau$ and is characterized as spin-flip ($\Delta S = 1$), isospin-flip ($\Delta T = 1$) and no transfer of orbital angular momentum ($\Delta L = 0$). There exists a model-independent sum rule, $S_{\beta-} - S_{\beta+} = 3(N-Z)$, where $S_{\beta-}$ and $S_{\beta+}$ are the GT strength of β^- and β^+ types, respectively [2]. Surprisingly, however, only a half of the GT sum rule value was identified from the (p,n) measurement on targets throughout the periodic table [3]. This problem, so-called the quenching of the GT strengths, has been one of the most interesting phenomena in nuclear physics because it is related to non-nucleonic (Δ-isobar) degrees of freedom in nuclei; the quenching factor sets a strong constraint on the Landau-Migdal parameters, g'_{NN} and $g'_{N\Delta}$, in the π+ρ+g' model [4].

Recently, Wakasa *et al.* have measured the angular distribution of the double differential cross sections for the $^{90}Zr(p,n)$ reaction at 295 MeV [5]. By performing multipole

CP675, *Spin 2002: 15th Int'l. Spin Physics Symposium and Workshop on Polarized Electron Sources and Polarimeters,* edited by Y. I. Makdisi, A. U. Luccio, and W. W. MacKay

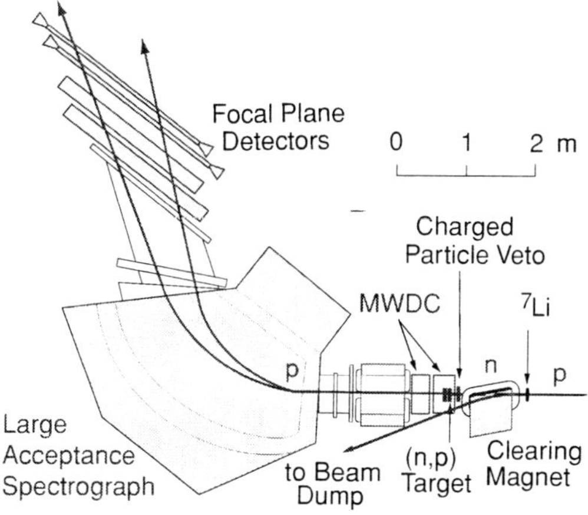

FIGURE 1. A schematic drawing of the RCNP (n,p) facility.

decomposition (MD) analysis, the GT strengths of $S_{\beta^-} = 28.0 \pm 1.6$ has been obtained in the continuum up to 50 MeV excitation in ^{90}Nb [5]. Determination of the Δ-isobar contribution to the GT sum rule, however, requires precise (n,p) cross section data at the same energy. For this purpose we have constructed an (n,p) facility at Research Center for Nuclear Physics (RCNP) and measured the double differential cross sections for the ^{90}Zr$(n,p)^{90}$Y reaction at 293 MeV.

EXPERIMENT

Figure 1 shows a schematic layout of the RCNP (n,p) facility in the WS beam course. A nearly mono-energetic neutron beam was produced by the ^{7}Li(p,n) reaction at 295 MeV. The primary proton beam, after going through the ^{7}Li target, was bent away by 23° by the clearing magnet [6] to a beam dump in the floor. The typical intensity of the beam was 450 nA and the thickness of the ^{7}Li target was 320 mg/cm^2. About 2×10^6/sec neutrons bombarded the target area of $30^W \times 20^H$ mm^2 downstream by 95 cm from the ^{7}Li target. Three ^{90}Zr targets with thicknesses of 200–400 mg/cm^2 and a polyethylene (CH_2) target with a thickness of 46 mg/cm^2 were mounted in a multiwire drift chamber (target MWDC). Wire planes placed between the targets detected outgoing protons and enabled one to determine the target in which a reaction occurred. Charged particles coming from the beam line were rejected by the veto scintillator with a thickness of 1 mm. The ^{1}H(n,p) events from the CH_2 target were used for normalization of the neutron beam flux. The position of outgoing protons were detected by six wire planes installed just behind the targets in the target MWDC. Another MWDC, front end MWDC, was installed at the entrance of the Large Acceptance Spectrometer (LAS). The scattering angle of the (n,p) reaction was determined by the information from the two

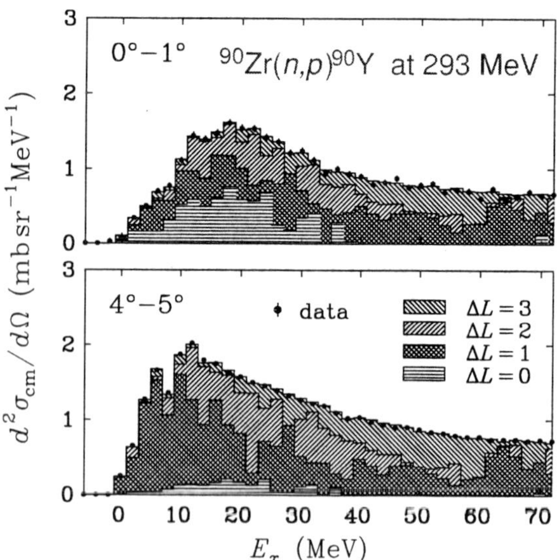

FIGURE 2. The result of MD analysis on the double differential cross sections for the ^{90}Zr$(n,p)^{90}$Y reaction at 293 MeV. The upper and lower panels show the results at angular region close to the maximum of GT and dipole angular distributions.

MWDCs. The outgoing protons were momentum analyzed by LAS and were detected by the focal plane detectors. Blank target data were also taken for background subtraction. The ^{1}H(n,p) cross sections given by the program SAID [7] was used to normalize the ^{90}Zr(n,p) cross sections.

We have obtained the differential cross sections up to 70 MeV excitation energy over an angular range of 0°–12° with a statistical accuracy of 1.7%/2 MeV·1° at 1°–2°. The overall energy resolution expected from the target thicknesses and the energy spread of the beam is 1.5 MeV. The angular resolution is 10 mr which is dominated by the the effect of multi scattering in the ^{90}Zr targets.

ANALYSIS

The MD analysis has been performed on the excitation energy spectra to extract the GT strengths.

First of all, the cross section data was binned in 2-MeV energy intervals to reduce the statistical fluctuation. For each excitation energy bin from 0 MeV to 70 MeV, the experimentally obtained angular distribution $\sigma^{\exp}(\theta_{\rm cm}, E_x)$ has been fitted by means of the least-squares method with the linear combination of calculated distributions $\sigma^{\rm calc}(\theta_{\rm lab}, E_x)$ defined by

$$\sigma^{\rm calc}(\theta_{\rm cm}, E_x) = \sum_{\Delta J^\pi} a_{\Delta J^\pi} \sigma^{\rm calc}_{ph;\Delta J^\pi}(\theta_{\rm cm}, E_x), \tag{1}$$

where the variables $a_{\Delta J\pi}$ are the fitting coefficients with positive values.

The angular distributions for various J^π states have been obtained by DWIA calculations by using the computer code DW81 [8]. To calculate the distortions in the incident and the outgoing channels, the energy-dependent global optical model potentials (OMPs) were used. The OMP for incident neutrons towards ^{90}Zr is taken from Ref. [9]. The OMP for outgoing protons is taken from Ref. [10] and varied as a function of the kinetic energy of the outgoing protons. The effective *NN* interaction is taken from the t-matrix parameterization of the free *NN* interaction by Franey and Love at 325 MeV [11]. Two kinds of the radial wave functions have been tried. One is the harmonic oscillator (HO) shape with a range parameter of $b = 2.12$ fm [12], and the other is the radial wave functions generated from a Woods-Saxon (WS) potential.

The one-body transition densities (OBTDs) are calculated from pure $1p1h$ configurations. The angular distributions for the following final J^π states have been calculated: $1^+(\Delta L = 0), 0^-, 1^-, 2^-(\Delta L = 1), 3^+(\Delta L = 2)$, and $4^-(\Delta L = 3)$. According to the independent-particle model, where the protons fill up to the $2p$ orbital and the neutrons to $1g_{9/2}$ orbital, the $\Delta L = 0$ transitions with $0\hbar\omega$ are completely blocked. Thus, in order to take into account the GT strength due to ground state correlation, the $(\nu 1g_{7/2}, \pi 1g_{9/2}^{-1})$ and $(\nu 1g_{9/2}, \pi 1g_{9/2}^{-1})$ configurations are activated for the $\Delta L = 0$ states. For the transition with $\Delta L \geq 1$, the active neutron particles are restricted to the $1g_{7/2}, 2d_{5/2}, 2d_{3/2}, 1h_{11/2}$, or $3s_{1/2}$ shells, which covers from $N = 51$ to $N = 82$ while the active proton holes are restricted to the $1g_{9/2}, 2p_{1/2}, 2p_{3/2}, 1f_{5/2}$, or $1f_{7/2}$ shells, which covers from $Z = 21$ to $Z = 40$ assuming ^{40}Ca to be a core. All the $0\hbar\omega$ and $1\hbar\omega$ excitations are included in this choice of configuration.

The minimizing procedure was performed for all the possible 58080 combinations. The combination of the *ph* configurations at each energy window was chosen so that the χ^2 value was minimized.

Figure 2 shows the result of the MD analysis. The $\Delta L = 0$ component has a broad (~ 10 MeV in FWHM) bump at $E_x \simeq 20$ MeV mainly due to the isovector spin monopole (IVSM) resonance [13], which is excited through the $r^2\sigma\tau$ operator. The $\Delta L = 0$ component of the cross section, $\sigma_{\Delta L=0}(q,\omega)$, is related to the GT strengths through the proportionality relation, *i.e.* $\sigma_{\Delta L=0}(q,\omega) = \hat{\sigma}_{\rm GT} F(q,\omega) B(\rm GT)$, where $\hat{\sigma}_{\rm GT}$ is the GT unit cross section [5] and $F(q,\omega)$ is the kinematical correction factor [14]. The upper limit energy of integration, $E_x^{\rm max}$, is determined so that it corresponds to $E_x^{\rm max} = 50$ MeV in the (p,n) work [5]. Considering the difference in the Coulomb energy between the ^{90}Y and the ^{90}Nb nuclei and the difference in the reaction Q value, we use the $E_x^{\rm max}$ value of 31.4 MeV and have obtained a total GT strength of $S_{\beta+} = 5.4 \pm 0.3 \pm 0.9$, where the errors are uncertainties of the MD analysis and the GT unit cross section.

The contribution of IVSM is estimated by the DWIA calculations in which all the IVSM strengths are assumed to lie below 31 MeV excitation [5] . After subtracting the IVSM contribution of 2.4 ± 0.8 GT units, we have obtained a total GT strength of $S_{\beta+} = 3.0 \pm 0.3(\rm MD) \pm 0.8(\rm IVSM) \pm 0.5(\hat{\sigma}_{\rm GT})$ up to 31.4 MeV excitation.

By using the $S_{\beta-}$ value by Wakasa *et al.*[5] the quenching factor Q, which is defined by $Q \equiv \frac{S_{\beta^-} - S_{\beta^+}}{3(N-Z)}$, has been deduced to be $Q = 0.83 \pm 0.06$ in regard to Ikeda's sum rule value of $3(N-Z) = 30$. Therefore the quenching of the GT strength due to the ΔN^{-1} admixture into the $1p1h$ GT state is significantly smaller than the quenching of $\sim 50\%$, observed in the previous studies [3] where the GT strengths in the continuum are not taken into account. Then the Landau-Migdal parameters, $g'_{N\Delta}$ and g'_{NN}, have been determined from the quenching factor. The deduction by Suzuki and Sakai [4] in Chew-Low model leads to $g'_{NN} \approx 0.6$ and $0.16 < g'_{N\Delta} < 0.35$ for $g'_{\Delta\Delta} = 0.6$. Therefore the universality ansatz of the Landau-Migdal parameters, *i.e.* $g'_{NN} = g'_{N\Delta} = g'_{\Delta\Delta} (= 0.6 \sim 0.8)$, does not hold.

ACKNOWLEDGMENTS

The authors acknowledge H.P. Yoshida for the support on the trigger system. This project is supported by the Ministry of Education, Science, Sports and Culture of Japan with the Grant-in-Aid for Scientific Research No. 10304018 and the Japan Society for the Promotion of Science.

REFERENCES

1. Doering, R., Galonsky, A., Patterson, D., and Bertsch, G., *Phys. Rev. Lett.*, **35**, 1691 (1975).
2. Ikeda, K., Fujii, S., and Fujita, J. I., *Phys. Rev. Lett.*, **3**, 271 (1963).
3. Gaarde, C. *et al.*, *Nucl. Phys.*, **A369**, 258 (1981).
4. Suzuki, T., and Sakai, H., *Phys. Lett.*, **B455**, 25 (1999).
5. Wakasa, T. *et al.*, *Phys. Rev. C*, **55**, 2909 (1997).
6. Kamiya, J. *et al.*, *RCNP annual report 1998*, p. 113.
7. Arndt, R.A. and Roper, L.D., *Scattering Analysis Interactive Dial-in (SAID) program*, phase shift solution SP02, Virginia Polytechnic Institute and State University, unpublished.
8. Schaeffer, M. A., and Raynal, J., *Program* DW81, unpublished.
9. Quing-biao, S., Da-chun, F., and Yi-zhong, Z., *Phys. Rev. C*, **43**, 2773 (1991).
10. Cooper, E. D., Hama, S., Clark, B. C., and Mercer, R. L., *Phys. Rev. C*, **47**, 297 (1993).
11. Franey, M. A., and Love, W. G., *Phys. Rev. C*, **31**, 488 (1985).
12. Love, W. G., and Franey, M. A., *Phys. Rev. C*, **24**, 1073 (1981).
13. Raywood, K. J. *et al.*, *Phys. Rev. C*, **41**, 2836 (1990).
14. Taddeucci, T. N. *et al.*, *Nucl. Phys.*, **A469**, 125 (1987).

Experimental Studies on Three-Nucleon Systems at RCNP

K. Hatanaka[*], K. Sagara[†], Y. Shimizu[*], T. Yagita[†], Y. Sakemi[*], T. Wakasa[*], H.P. Yoshida[*], J. Kamiya[*], M. Yoshimura[*], H. Sakai[**], A. Tamii[**], K. Yako[**], Y. Maeda[**], T. Saito[**], T. Ishida[†], S. Minami[†], K. Tsuruta[†], T. Noro[†], K. Sekiguchi[‡], H. Akiyoshi[‡] and V.P. Ladygin[§]

[*]RCNP, Osaka University, Mihogaoka, Ibaraki, Osaka 560-0047, Japan
[†]Department of Physics, Kyushu University, Hakozaki, Fukuoka 812-8581, Japan
[**] Department of Physics, University of Tokyo, Bunkyo, Tokyo 113-0033, Japan
[‡]RIKEN, Hirosawa, Wako, Saitama 351-0198, Japan
[§]Joint Institute for Nuclear Research, 141980 Dubna, Russia

Abstract. The results of experimental studies of three-nucleon systems will be presented, i.e. the $\vec{p}d$ elastic scattering at E_p = 250 MeV and the $\vec{d}p \rightarrow {}^3$He+$\gamma$ capture reaction at E_d = 200 MeV. The angular distributions of the cross section and all the proton spin observables were measured for elastic scattering, and the cross section and analyzing powers, A_y, A_{xx} and A_{yy} for the capture reaction. The results are compared with theoretical predictions based on exact solutions of the three-nucleon Faddeev equations and modern realistic nucleon-nucleon potentials combined with three-nucleon forces.

INTRODUCTION

One of the fundamental interests in nuclear physics is to establish the nature of nuclear forces and understand nuclear phenomena based on the fundamental Hamiltonian. Studies of few-nucleon systems offer a good opportunity to investigate these forces. Realistic two-nucleon forces (2NF) [1] fail to reproduce experimental binding energies for light nuclei, clearly showing underbinding. Correct three-nucleon (3N) and four-nucleon (4N) binding energies can be achieved by including the Tucson-Melbourne (TM) [2] or Urbana IX [3] three-nucleon forces (3NF) which are refined versions of the Fujita-Miyazawa force [4], a 2π-exchange between three nucleons with an intermediate Δ excitation. In recent years, it became possible to perform rigorous numerical Faddeev-type calculations for the 3N scattering processes by the tremendous advances in computational capabilities. In addition to the first signal on 3NF effects resulting from discrete states, strong 3NF effects were observed in a study of the minima of the *Nd* elastic scattering cross section at incoming nucleon energies above 60 MeV. The discrepancy between the data and predictions based exclusively on NN forces could be largely removed by including the TM 3NF [5]. For spin observables, however, a recent study at RIKEN [6] showed that the inclusion of the 3NF does not always improve the description of precise data taken at intermediate deuteron energies. Proton vector analyzing power

CP675, *Spin 2002: 15th Int'l. Spin Physics Symposium and Workshop on Polarized Electron Sources and Polarimeters*, edited by Y. I. Makdisi, A. U. Luccio, and W. W. MacKay data at 70–200 MeV have revealed the deficiency of the 3NF [7], which produces large but wrong effects. These results may be caused by a wrong spin structure of present-day 3NF. Clearly the present situation is only the very beginning of the investigation of the spin structure of the 3NF. Precise data at intermediate energies including higher-rank spin observables are needed to provide constraints on theoretical 3NF models.

In the present paper, we present two experimental studies on 3N systems recently performed at the Research Center for Nuclear Physics (RCNP), Osaka University; the $\vec{p}d$ elastic scattering at 250 MeV [8] and the $\vec{d}p \rightarrow {}^3$He+$\gamma$ reaction at E_d = 200 MeV. The experimental results are compared with the theoretical predictions.

EXPERIMENTAL RESULTS AND DISCUSSION

Measurements were performed at the RCNP cyclotron facility. Polarized protons or deuterons were produced in an atomic beam polarized ion source [9], injected into and accelerated by the K = 120 MeV AVF (azimuthally varying field) cyclotron. Subsequently the beam was injected into the K = 400 MeV Ring cyclotron and accelerated to the final energy.

$\vec{p}d$ elastic scattering at 250 MeV

Differential cross sections, analyzing powers, and a complete set of polarization transfer (PT) coefficients were measured for $\vec{p}d$ elastic scattering using self-supporting 99% isotopically enriched deuterated polyethylene foils (CD_2) [10] with total thicknesses of 21 and 44 mg/cm^2. In a later measurement, a gaseous target was used to normalize cross sections taken with the solid CD_2 target. Scattered protons or recoil deuterons in the *pd* scattering were momentum analyzed by the Grand Raiden spectrometer [11]. The horizontal and vertical acceptance of the Grand Raiden was limited by a slit system to $\pm$20 and $\pm$30 msr, respectively. The polarization of elastically scattered protons from CD_2 targets was measured at center of mass scattering angles from 10° to 95° by the focal plane polarimeter (FPP) [12]. The experimental results for the differential cross section, the vector analyzing power and the PT coefficients are shown in Figs. 1 and 2. Only statistical errors, mostly smaller than the size of the data point are shown. The overall uncertainty in the absolute normalization calibrated by the gaseous target measurements is estimated to be 3%. There is also the relative uncertainty of 2.5% attributed to the inhomogeneity of the CD_2 foils. The analyzing power has an uncertainty of only 1% in the absolute normalization owing to the precise calibration of the beamline polarimeter [8]. The the normalization of the PT coefficients have an uncertainty of 2.5% [12].

In the left panel of Fig. 1, the measured differential cross sections are compared with theoretical predictions [13]. The various 2NF predictions are very similar and are depicted by a narrow band (light shaded). The inclusion of the TM 3NF (dark shaded band) leads to a much better description at angles larger than 70°. This supports the claim of the clear evidence [5, 6, 14, 15] of the 3NF from the systematic analysis of the energy dependence of the cross section data. The inclusion of the TM′ (dashed curve) and the Urbana IX (solid curve) 3NF also leads to a good agreement with the data.

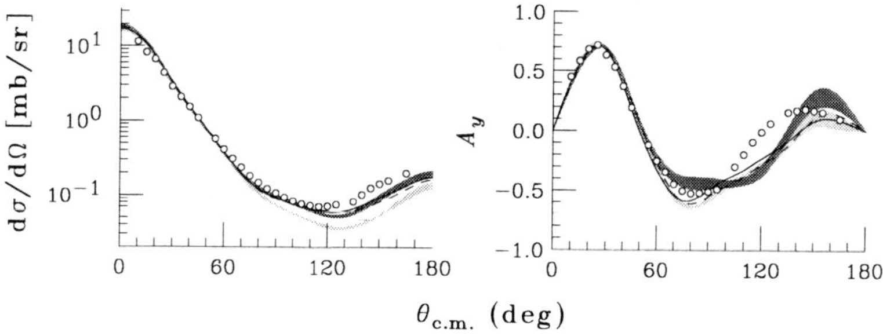

FIGURE 1. The differential cross section $d\sigma/d\Omega$ (left) and proton analyzing powers (right) of elastic $\vec{p}d$ scattering at E_p = 250 MeV. The light shaded bands contain several NN force predictions (AV18, CD-Bonn, Nijm I, II, and 93), the dark shaded bands contain the NN + TM 3NF predictions. The solid and dashed lines are the AV18 + Urbana IX and CD-Bonn + TM′ predictions, respectively.

However, discrepancies remain at angles larger than 120°. In the right panel of Fig. 1, we compare the experimental analyzing power A_y with different nuclear-force predictions. The differences (narrow light shaded band) between the 2NF predictions are rather small at forward angles and become larger at backward angles. These predictions are in good agreement with the experimental data at forward angles, but deviate dramatically at backward angles larger than 60°. By including the TM 3NF (dark shaded band) the agreement with the data becomes better in the minimum around $\theta_{c.m.}$ = 60°–100° but the discrepancies at more backward angles remain. The discrepancy between data and theoretical predictions, which increases with increasing energy [7, 16], may be due to relativistic effects not accounted for in the present nonrelativistic calculations.

The measured PT data are shown in Fig. 2 together with theoretical predictions. The PT coefficients in the horizontal plane ($K_x^{x'}, K_x^{z'}, K_z^{x'}$, and $K_z^{z'}$) are reasonably well described by calculations with 2NF only (light shaded bands). The inclusion of the TM 3NF (dark shaded bands) rather deteriorates the agreement with the experimental data.

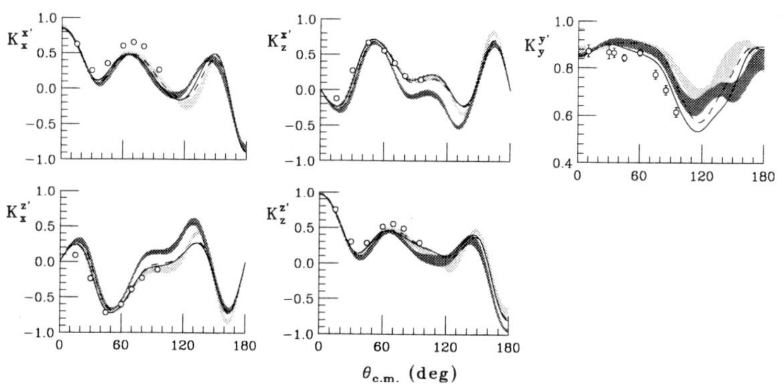

FIGURE 2. Polarization transfer coefficients ($K_x^{x'}$, $K_x^{z'}$, $K_z^{x'}$, $K_z^{z'}$, and $K_y^{y'}$) of elastic $\vec{p}d$ scattering at E_p = 250 MeV. For the description of bands and lines see legend of Fig. 1

The TM′ (dashed curves) and the Urbana IX (solid curves) 3NF do not have a large effect on these PT coefficients and give a reasonably good agreement with the data. In the case of the PT coefficient in the vertical plane ($K_y^{y'}$), the inclusion of the TM 3NF (dark shaded band) and especially the Urbana IX 3NF (solid curve) give results in better agreement with the measurements. This is similar to the case of the analyzing power which is also a polarization observable in the vertical plane. These results clearly indicate that the spin-dependent parts of 3NF are not well described in present-day models.

$p\vec{d}$ radiative capture at E_d = 200 MeV

A liquid hydrogen target was used instead of a solid CH_2 foil to cope with the small cross section of the $p\vec{d}$ capture reaction and large $(d,{}^3\mathrm{He})$ cross sections on carbon and other materials. The target thickness was about 1.5 mm (11 mg/cm^2). The liquid hydrogen was obtained by cooling hydrogen gas with a cryogenic refrigerator. Recoil ^{3}He particles from the capture reaction were emitted into a come out at forward angle cone of $\pm 5°$ in the laboratory frame with energies in the range from 105 to 145 MeV. They were detected by the Large Acceptance Spectrometer (LAS) which has an angular acceptance of ± 60 mr and ± 100 mr in the horizontal and vertical plane, respectively, and a momentum acceptance of ± 15 %. The angular distributions of the cross section and analyzing powers, A_y, A_{xx} and A_{yy}, were measured from 20° to 160° in the center of the mass frame. The absolute value of the cross section was calibrated by a separate measurement with CH_2 foil target around θ_{cm}=90°.

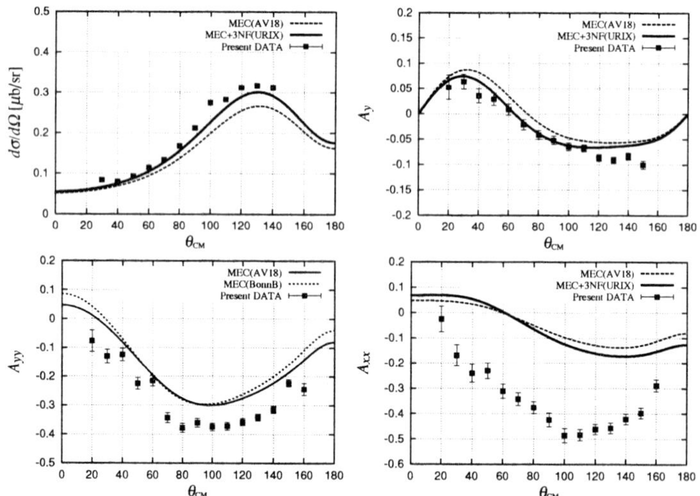

FIGURE 3. Cross section and analyzing powers, A_y, A_{yy}, and A_{xx}, of the $p\vec{d}$ radiative capture at E_d = 200 MeV. Curves represent MEC calculations[13] based on the AV18 NN potential with (solid curves) or without (dashed ones) Urbana IX 3NF.

The experimental results are shown in Fig. 3 where only statistical errors are show. The uncertainties in the absolute normalization are estimated to be about 5 %, 2 % and 3 % for the cross section, the tensor and vector analyzing powers, respectively. These data are compared with Faddeev calculations [13] in which the meson (π and ρ) exchange current (MEC) is explicitly taken into accounts [17]. In the calculations, AV18 NN interactions up to j = 3 are used with or without Urbana IX 3NF which is adjusted to reproduce 3N binding energies. The cross section and A_y disagree with predictions without 3NF, and are well reproduced by including 3NF, as seen in Fig. 3. This fact confirms the existence and necessity of 3NF. Since A_y of the pd elastic scattering is reproduced by calculations with 3NF at some energies and disagrees with the calculations at other energies, reproduction of A_y of the $p\vec{d}$ capture reaction, shown in this paper at E_d = 200 MeV, has to be examined in a wide energy range. Contrary to the good discription of the cross section and A_y, calculations cannot reproduce tensor analyzing powers, A_{yy} and A_{xx}, with or without 3NF. The disagreement of A_{yy} is moderate and similar to the disagreement of tensor analyzing powers of the elastic scattering in the same energy range [6]. However, A_{xx} differs completely from the calculations. The experimental values of A_{xx} and A_{yy} are nearly the same, while the calculated values are quite different.

The energy dependence of A_{yy} and A_{xx} at θ_{cm} = 90° is shown in Figure 4. There are several measurements of A_{yy} below E_d = 200 MeV. However, A_{xx} have been measured only at 17.5 MeV and 200 MeV. Measured A_{yy} are fairly well reproduced by calculations in whole the energy range below 200 MeV. Measured A_{xx} agrees with calculations at 17.5 MeV, however, remarkably disagrees at 200 MeV. Measured A_{xx} have nearly the same values as measured A_{yy} at 17.5 and 200 MeV. Therefore, A_{xx} and A_{yy} are expected to have nearly the same values below 200 MeV. Calculated A_{xx} and A_{yy} agree to each other at low energy and disagree above about 50 MeV. The difference increases with the energy up to 200 MeV. It is expected, therefore, that discrepancies in A_{xx} between the experiment and calculation begin at about 50 MeV and increase with energy up to 200 MeV.

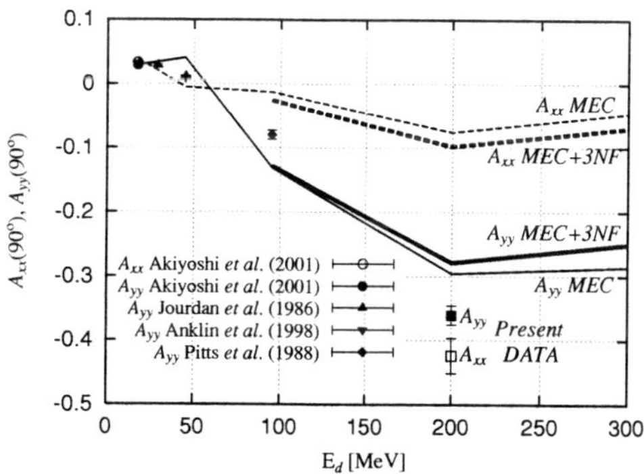

FIGURE 4. Energy dependence of A_{yy} and A_{xx} at θ_{cm} = 90°. Experimental data at 17.5 MeV[18], 29.2MeV[19], 45 MeV[20] and 200 MeV are compared with MEC calculations[13] based on AV18 NN potential with or without Urbana IX 3NF.

SUMMARY

Recent results of experimental studies on three-nucleon systems were presented, i.e. the $\vec{p}d$ elastic scattering at E_p = 250 MeV and the $\vec{d}p \rightarrow {}^3$He+$\gamma$ capture reaction at E_d = 200 MeV. The experimental results are compared with theoretical predictions based on exact solutions of the three-nucleon Faddeev equations and modern realistic nucleon-nucleon potentials combined with three-nucleon forces. For the elastic scattering, the differential cross sections and the vector analyzing powers are reasonably well explained by calculations including 3NF around the cross section minima, but the discrepancies at more backward angles remain. PT data are not always better described by calculations with 3NF. For the capture reaction, calculations with 3NF improve the description of the cross section and the vector analyzing power A_y. There are large discrepancies between tensor analyzing power data and theoretical predictions. These results clearly indicate that the spin-dependent parts of 3NF's are not well described in present-day models. More theoretical studies are needed including relativistic treatments and chiral parturbation theories. From the experimental point of view, a rich spectrum of spin observables will be measured not only for elastic scattering but also for the Nd breakup and capture processes over wide energy range in order to offer further valuable information.

ACKNOWLEDGMENTS

We thank the RCNP staff for their support during the experiment. We also wish to thank Professor H. Toki for his encouragements throughout the work. We are grateful to Dr. G.P.A. Berg for his critical reading of the manuscript.

REFERENCES

1. R.B. Wiringa *et al.*, Phys. Rev. C **51**, 38 (1995); R. Machleidt *et al.*, Phys. Rev. C **53**, R1483 (1996); V.G.J. Stoks *et al.*, Phys. Rev. C **49**, 2950 (1994).
2. S.A. Coon and M.T. Peña, Phys. Rev. C **48**, 2559 (1993).
3. B.S. Pudliner *et al.*, Phys. Rev. C **56**, 1720 (1997).
4. J. Fujita and H. Miyazawa, Prog. Theor. Phys. **17**, 360 (1957).
5. H. Witała *et al.*, Phys. Rev. Lett. **81**, 1183 (1998).
6. K. Sekiguchi *et al.*, Phys. Rev. C **65**, 034003 (2002).
7. E.J. Stephenson *et al.*, Phys. Rev. C **60**, 061001(R) (1999).
8. K. Hatanaka *et al.*, Phys. Rev. C **66**, 044002 (2002).
9. K. Hatanaka *et al.*, Nucl. Instrum. Methods Phys. Res. A **384**, 575 (1997).
10. Y. Maeda *et al.*, Nucl. Instrum. Methods Phys. Res. A **490**, 518 (2002).
11. M. Fujiwara *et al.*, Nucl. Instrum. Methods Phys. Res. A **422**, 484 (1999).
12. M. Yosoi *et al.*, AIP Conf. Proc. **343** (AIP, New York, 1995), p.157.
13. H. Kamada, private communication.
14. N. Sakamoto *et al.*, Phys. Lett. B **367**, 60 (1996).
15. H. Sakai *et al.*, Phys. Rev. Lett. **84**, 5288 (2000).
16. H. Rohdjeß*et al.*, Phys. Rev. C **57**, 2111 (1998).
17. J. Golak et al., Phys. Rev. C **62** 054005 (2000)
18. H. Akiyoshi *et al.*, Phys. Rev. C **64** 034001 (2001).
19. J. Jourdan *et al.*, Nucl. Phys. **A453** 220 (1986).
20. H. Anklin *et al.*, Nucl. Phys. **A636** 189 (1998).

Polarization Transfer Measurement for d–p Elastic Scattering – a Probe for Three Nucleon Force Properties –

K. Sekiguchi*, H. Sakai†, H. Okamura**, A. Tamii†, T. Uesaka**, K. Suda**, N. Sakamoto†, T. Wakasa‡, Y. Satou*, T. Ohnishi*, K. Yakou†, S. Sakoda†, H. Kato†, Y. Maeda†, M. Hatano†, J. Nishikawa**, T. Saito†, N. Uchigashima†, N. Kalantar-Nayestanaki§ and K. Ermisch§

*RIKEN, the Institute of Physical and Chemical Research, Saitama 351-0198, Japan
†Department of Physics, University of Tokyo, Tokyo 113-0033, Japan
**Department of Physics, Saitama University, Saitama 338-8570, Japan
‡Research Center for Nuclear Physics, Osaka University, Osaka 567-0047, Japan
§Kernfysisch Versneller Instituut (KVI), NL-9747 AAGroningen, The Netherlands

Abstract. Precise measurements of the deuteron to proton polarization transfer coefficients for the d–p elastic scattering has been made at 135 MeV/u at RIKEN Accelerator Research Facility. The obtained results are compared with the Faddeev calculations based on modern nucleon–nucleon forces together with Tucson-Melbourne, Tucson-Melbourne′ and Urbana–Argonne type of three nucleon forces.

INTRODUCTION

Recent advance in computational resources has made it possible to obtain rigorous numerical Faddeev–type calculations for the three–nucleon scattering processes by using two–nucleon(2N) and three–nucleon forces (3NF). It has also allowed us to search for 3NF effects by direct comparison between such theoretical predictions and precisely measured data.

In Refs. [1,2] we have reported the precise measurement of the cross section and the deuteron analyzing powers for d–p elastic scattering at incoming deuteron energies of 70, 100, and 135 MeV/u. The data have been compared with the Faddeev calculations with or w/o 3NFs. For the cross section, the large discrepancy between the data and the calculations w/o 3NFs has been found in the cross section minimum and it is essentially removed by taking into account 3NFs. The vector analyzing power A_y^d is also explained by the predictions incorporating 3NFs. However the tensor analyzing power data are not reproduced by any theoretical prediction and these results indicate that the present day 3NF models have deficiencies in the spin parts. In order to assess further the study of 3NF effects, we have measured the deuteron-to-proton polarization transfer coefficients for d-p elastic scattering, which are expected theoretically to have strong sensitivities to the spin dependent parts of 3NF.

CP675, *Spin 2002: 15th Int'l. Spin Physics Symposium and Workshop on Polarized Electron Sources and Polarimeters*, edited by Y. I. Makdisi, A. U. Luccio, and W. W. MacKay

EXPERIMENT

The experiment was performed at the RIKEN Accelerator Research Facility using tensor and vector deuteron beams of 135 MeV/u [3]. A liquid hydrogen (19.8 mg/cm^2) or CH_2 (93.4 mg/cm^2) target was bombarded and scattered protons were momentum analyzed by the magnetic spectrograph SMART [4]. The polarization of the scattered protons were measured with the focal-plane polarimeter DPOL [5]. The measured observables were the deuteron to proton polarization transfer coefficients ($K_y^{y'}$, $K_{xx}^{y'} - K_{yy}^{y'}$, and $K_{xz}^{y'}$) in the angular range of $\theta_{\text{c.m.}} = 90° - 180°$. This measurement also yielded an induced polarization ($P^{y'}$) of the outgoing protons. The relation between the polarizations and the observables is expressed as

$$p_{y'}\left(\frac{d\sigma}{d\Omega}\right) = \left(\frac{d\sigma_o}{d\Omega}\right)\Big(P^{y'} + \frac{3}{2}K_y^{y'}p_y + \frac{2}{3}K_{xz}^{y'}p_{xz} + \frac{1}{3}(K_{xx}^{y'} - K_{zz}^{y'})p_{xx} + \frac{1}{3}(K_{yy}^{y'} - K_{zz}^{y'})p_{yy}\Big),$$

with

$$K_{xx}^{y'} + K_{yy}^{y'} + K_{zz}^{y'} = 0,$$

where x, y, and z are the coordinates of the incident deuterons; x', y', and z' are those of the emitted protons; and $\left(\frac{d\sigma_o}{d\Omega}\right)$ denotes the cross section with unpolarized beams.

RESULTS AND DISCUSSIONS

Figure 1 shows a part of the experimental data $K_{xx}^{y'} - K_{yy}^{y'}$ with open squares. The statistical errors are only shown. The statistical errors are smaller than 0.03 for all the polarization transfer coefficients, and 0.01 for the induced polarization $P^{y'}$. The systematic uncertainties for the polarization transfer coefficients are estimated to be 3% at most.

In Fig. 1, four theoretical predictions in terms of Faddeev theory are shown together with the experimental results. The dark (light) shaded band in the figure is the Faddeev calculations with (w/o) Tucson-Melbourne (TM) 3NF [6] based on the modern nucleon–nucleon(NN) potentials, namely CDBonn [7], AV18 [8], Nijimegen I, II and 93 [9]. The solid line is the calculation with including Urbana IX 3NF [10] based on AV18 potential. The dotted line is the predictions in which TM′ 3NF is taken into account and CDBonn potential is considered as the NN potential. The TM′ 3NF is a modified version of the TM 3NF closer to chiral symmetry.

Comparing the theoretical predictions with the observed values, for $K_{xx}^{y'} - K_{yy}^{y'}$ the clear discrepancies exist between the data and the 2N force predictions and these deviations are explained well by inclusion of 3NFs. All 3NF potentials considered here (TM, TM′, Urbana IX) provide almost the same 3NF effects (magnitude and direction). However for the other polarization transfer coefficients $K_y^{y'}$ and $K_{xz}^{y'}$ which are not shown here,

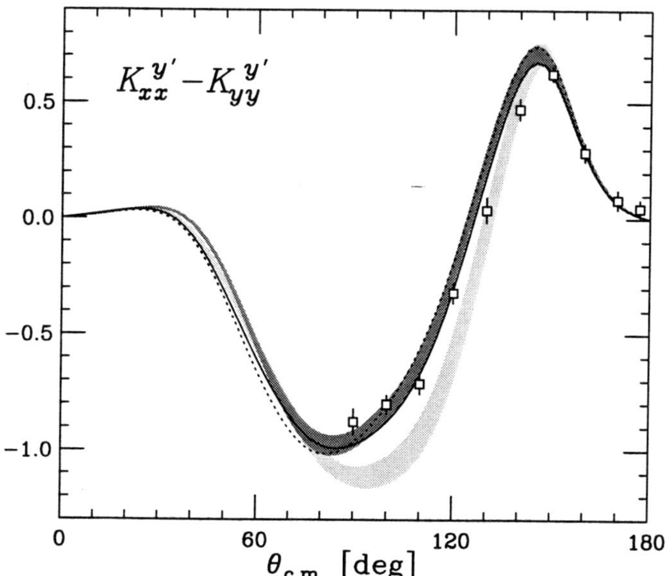

FIGURE 1. Deuteron to proton polarization transfer coefficients $K_{xx}^{y'} - K_{yy}^{y'}$ for *d*-*p* elastic scattering at 270 MeV.

large differences between the data and the 2N force predictions are not reproduced by including the 3NF models.

The results of the comparison for the polarization transfer coefficients reveal reveal that the present 3NF models have deficiencies in its spin parts and that these observables are useful to clarify the spin dependence of 3NF effects.

SUMMARY

In order to study of the properties of the three nucleon forces, we have measured the deuteron to proton polarization transfer coefficients for *d*–*p* elastic scattering at 135 MeV/u which cover the angular range of $\theta_{c.m.} = 90^\circ - 180^\circ$. Highly accurate data have been obtained. These results are compared with the Faddeev calculations with and without the Tucson-Melbourne 3NF, or a modification thereof closer to chiral symmetry TM′, or the Urbana IX 3NF. The large difference are obtained between the data and the 2N force predictions. However not all spin observables are reproduced by incorporating the present three nucleon force models and the results clearly show the deficiency of these models in spin parts.

ACKNOWLEDGMENTS

We would like to thank H. Witała, W. Glöckle and H. Kamada for their strong theoretical support. We would also like to thank S. Nemoto and P. U. Sauer for their useful comments on theoretical issues. We would also like to express our appreciation to the continuous help of the staff of RIKEN Accelerator Research Facility.

REFERENCES

[1] H. Sakai *et al.*, Phy. Rev. Lett. **84** (2000) 5288.

[2] K. Sekiguchi *et al.*, Phy. Rev. C **65** (2002) 034003.

[3] H. Okamura *et al.*, AIP Conf. Proc. **293**, 84 (1994),
H. Okamura *et al.*, *ibid.* **343**, 123 (1995).

[4] T. Ichihara *et al.*, *Nucl. Phys.* **A569**, 287c (1994).

[5] S. Ishida *et al.*, AIP Conf. Proc. **343**, 182 (1995).

[6] S. A. Coon, and M. T. Peña, Phys. Rev. C **48**, 2559 (1993).

[7] R. Machleidt, Phys. Rev. C **63**, 024001 (2001).

[8] R. B. Wiringa, *et al.*, Phys. Rev. C **51**, 38 (1995).

[9] V. G. J. Stoks, *et al.*, Phys. Rev. C **49**, 2950 (1994).

[10] B. S. Pudliner, *et al.*, Phys. Rev. C **56**, 1720 (1997).

Study of ^{3}He (^{3}H) Spin Structure via $\vec{d}d \rightarrow {}^3\mathrm{He}\,n\,({}^3\mathrm{H}\,p)$ Reaction

T. Saito*, V.P. Ladygin†, T. Uesaka**, M. Hatano*, A.Yu. Isupov†, H. Kato*, H. Kumasaka‡, N.B. Ladygina†, Y. Maeda*, A.I. Malakhov*, J. Nishikawa‡, T. Ohnishi§, H. Okamura‡, S.G. Reznikov*, H. Sakai*, N. Sakamoto§, S. Sakoda*, K. Sekiguchi§, K. Suda‡, R. Suzuki‡, A. Tamii*, N. Uchigashima* and K. Yako*

*Department of Physics, University of Tokyo, 7-3-1 Hongo, Bunkyo, Tokyo 113-0033, Japan
†LHE-JINR, 141980, Dubna, Moscow region, Russia
**Center for Nuclear Study (CNS), University of Tokyo, 7-3-1 Hongo, Bunkyo, Tokyo 113-0033, Japan
‡Department of Physics, Saitama University, 255 Shimo-okubo, Saitama 338-8570, Japan
§The Institute of Physical and Chemical Research (RIKEN), 2-1 Hirosawa, Wako, Saitama 351-0198

Abstract. Measurements of the tensor and vector analyzing powers A_{yy}, A_{xx}, A_{xz}, and A_y for the $\vec{d}d \rightarrow {}^3\mathrm{He}\,n$ and $\vec{d}d \rightarrow {}^3\mathrm{H}\,p$ reactions were performed at $E_d = 270$ and 200 MeV over wide angular range. T_{20} at $\theta_{\mathrm{cm}} = 0°$ and 180° were also measured at $E_d = 270$, 200, and 140 MeV. Obtained data were compared with predictions based on one nucleon exchange approximation.

INTRODUCTION

It has been predicted from non-relativistic Faddeev calculations of the three nucleon bound state that the main components of the ^{3}He ground state wave function are a spatially symmetric S-state and a small contribution of a D-state [1].

In the last two decades, the structure of a ^{3}He nucleus has been investigated using reactions of quasielastic knockout of the ^{3}He constituent nucleons. The momentum distribution of the constituent nucleons was extracted by plane wave impulse approximation (PWIA) analyses of $^3\mathrm{He}(e,ep)$ reaction [2] and $^3\mathrm{He}(p,2p)d$ and $^3\mathrm{He}(p,pd)p$ reactions [3]. It was found that the theoretical calculations using modern realistic ^{3}He wave functions did not reproduce the experimentally obtained momentum distribution functions in the region of the internal nucleon momentum $q > 250$ MeV/c. To investigate the spin structure of ^{3}He, spin correlation for the quasi elastic $^3\vec{\mathrm{He}}(\vec{p},pN)$ reactions was measured up to the internal nucleon momentum of $q \sim 400$ MeV$/c$, and the distribution function of the nucleon polarization in a ^{3}He nucleus was extracted by a PWIA analysis [4]. The distribution function by Faddeev calculations, however, did not reproduce the experimental data in the region of $q > 300$ MeV/c.

These deviations indicate that the structure of ^{3}He in the high-momentum region has not been clearly understood. Since various kinds of mesons contribute to the nu-

CP675, *Spin 2002: 15th Int'l. Spin Physics Symposium and Workshop on Polarized Electron Sources and Polarimeters*, edited by Y. I. Makdisi, A. U. Luccio, and W. W. MacKay

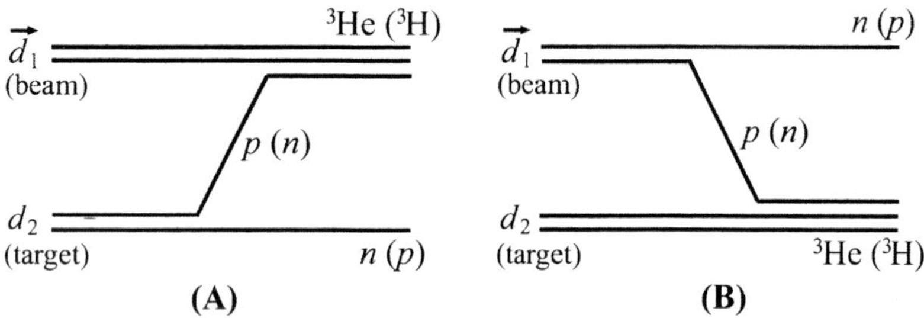

FIGURE 1. ONE processes of the $\vec{d}d \to {}^3\mathrm{He}\,n\,({}^3\mathrm{H}\,p)$ reaction.

clear interaction in the high-momentum region, investigation of high-momentum ^{3}He structure may reveal new physics which has not been observed in the low-momentum region. Since the contribution from the D-state component becomes large in the high-momentum region, measurements of polarization observables sensitive to the D-state is necessary to study the high-momentum ^{3}He structure.

ONE APPROXIMATION

In the framework of One Nucleon Exchange (ONE) approximation, tensor analyzing powers for the $\vec{d}d \to {}^3\mathrm{He}n$ and $\vec{d}d \to {}^3\mathrm{H}p$ reactions at intermediate energies are sensitive to the D-state component of ^{3}He or ^{3}H [5, 6]. The ONE processes of these reactions are shown in Fig. 1. Let $u_d(k_d)$ and $w_d(k_d)$ be the S- and D-state radial wave functions of a deuteron in the momentum space, respectively. Similarly, let $u_h(k_h)$ and $w_h(k_h)$ respectively be the S- and D-state radial functions of a ^{3}He or ^{3}H in the $d+N$ cluster configuration. The tensor analyzing powers for the $\vec{d}d \to {}^3\mathrm{He}n$ and $\vec{d}d \to {}^3\mathrm{H}p$ reactions in the framework of ONE approximation have following characteristics:

1. If only the diagram (A) (see Fig. 1) is considered, they are determined by the ratio of the ^{3}He(^{3}H) wave function components $w_h(k_h)/u_h(k_h)$.
2. Conversely, If only the diagram (B) is considered, they are determined by the ratio of the deuteron wave function components $w_d(k_d)/u_d(k_d)$.
3. If ^{3}He(^{3}H) is scattered at forward angles, the corresponding deuteron internal momentum k_d is very large, hence the contribution from the diagram (B) becomes negligible. Consequently, they are determined by the ^{3}He(^{3}H) structure.
4. Conversely, if ^{3}He(^{3}H) is scattered at backward angles, corresponding internal momentum of ^{3}He(^{3}H) k_h is very large, hence the contribution from the diagram (A) becomes negligible. Consequently, they are determined by the deuteron structure.

Particularly, the tensor analyzing power T_{20} at $\theta_{\mathrm{cm}}({}^3\mathrm{He},{}^3\mathrm{H}) = 0°$ or $180°$ is simply given by [5, 6]

$$T_{20} = \frac{1}{\sqrt{2}} \frac{2\sqrt{2}u(k)w(k) - w(k)^2}{u(k)^2 + w(k)^2}. \tag{1}$$

Here, $u(k)$ and $w(k)$ are respectively replaced by $u_h(k_h)$ and $w_h(h_h)$ if $\theta_{\rm cm} = 0°$, or, by $u_d(k_d)$ and $w_d(k_d)$ if $\theta_{\rm cm} = 180°$. Thus, the tensor analyzing powers for the $\vec{d}d \to {}^3{\rm He}\,n$ and $\vec{d}d \to {}^3{\rm H}\,p$ reactions at forward angles are directly related to the D/S ratio of ^{3}He(^{3}H). With a 270 MeV deuteron beam, the ^{3}He(^{3}H) structure can be investigated up to a relative momentum of the $d+N$ pair of $\sim$ 600 MeV/c in principle.

EXPERIMENT

The experiment was performed at RIKEN Accelerator Research Facility. A polarized deuteron beam extracted from a polarized ion source was accelerated with AVF and Ring Cyclotrons up to the energy of 270, 200, or 140 MeV. The accelerated beam was transported to a spectrometer SMART [7] and were injected onto a target placed in the scattering chamber. Scattered particles (^{3}He,^{3}H, or protons) were momentum analyzed with three quadrupole and two dipole magnets (Q-Q-D-Q-D configuration) and then detected with a multi-wire drift chamber and three plastic scintillators at the focal plane. The direction of the symmetry axis of the beam polarization was controlled with a Wien filter located at the exit of the ion source. The magnitude of the beam polarization was measured with beam-line polarimeters based on the dp elastic scattering. We used a deuterated polyethylene (CD_2) sheet [8] as the deuteron target. Measurement with a carbon foil target was also performed to subtract the contribution from the carbon nuclei in the CD_2 target. We detected ^{3}He for the ${}^3{\rm He}+n$ channel. In the case of ${}^3{\rm H}+p$ channel, we detected ^{3}H (protons) if ^{3}H were scattered in the forward (backward) angles in the center-of-mass frame.

RESULTS AND DISCUSSION

The experimental results of the tensor analyzing power T_{20} at $\theta_{\rm cm} = 0°$ and $180°$ at $E_d = 270, 200,$ and 140 MeV are presented in Fig. 2. The results for the ${}^3{\rm He}+n$ (${}^3{\rm H}+p$) channel are presented by filled (open) symbols. The curves are predictions by ONE approximation [5, 6]. The upper fives symbols (two filled ones are hidden behind the open ones) and the three curves are T_{20} at $\theta_{\rm cm} = 0°$. The lower three symbols and a curve are T_{20} at $\theta_{\rm cm} = 180°$. The solid, dashed, and dot-dashed curves are respectively calculated using ^{3}He wave functions of Urbana, Paris, and Reid soft core potentials. Paris deuteron wave function was used for these calculations. The ONE predictions reproduced the incident energy dependence and the signs of the experimental data. Since T_{20} at 0° and 180° is directly connected with the D/S ratio of ^{3}He (^{3}H) or deuteron by Eq. (1), the difference in the signs of T_{20} at 0° and 180° reflects the difference in the relative sign of $u(k)$ and $w(k)$ for ^{3}He (^{3}H) and deuteron. Angular distributions of the analyzing powers at $E_d = 270$ and 200 MeV are presented in Fig. 3. The meanings of the symbols and curves are same as those in Fig. 2. The ONE predictions [6] reproduced the global features of the experimental data at backward angles, where the tensor analyzing powers depend mainly on the deuteron structure. At forward angles, however, significant discrepancies can be found. Since the tensor analyzing powers at forward angles are mainly determined by the ^{3}He or ^{3}H structure, these discrepancies might be naively ascribed to some problems of the wave function of ^{3}He or ^{3}H. However, since the ONE approximation is very crude, calculation with more detail reaction mechanism is needed to extract information of the ^{3}He or ^{3}H structure in the high-momentum region. Further development in theoretical calculations of four body systems is expected.

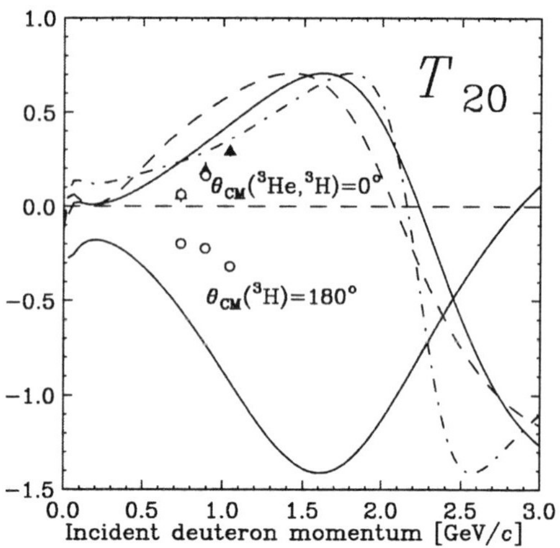

FIGURE 2. The experimental results of T_{20} at $\theta_{\mathrm{cm}} = 0°$ and 180°. The curves are ONE predistions. Explanations are written in the text.

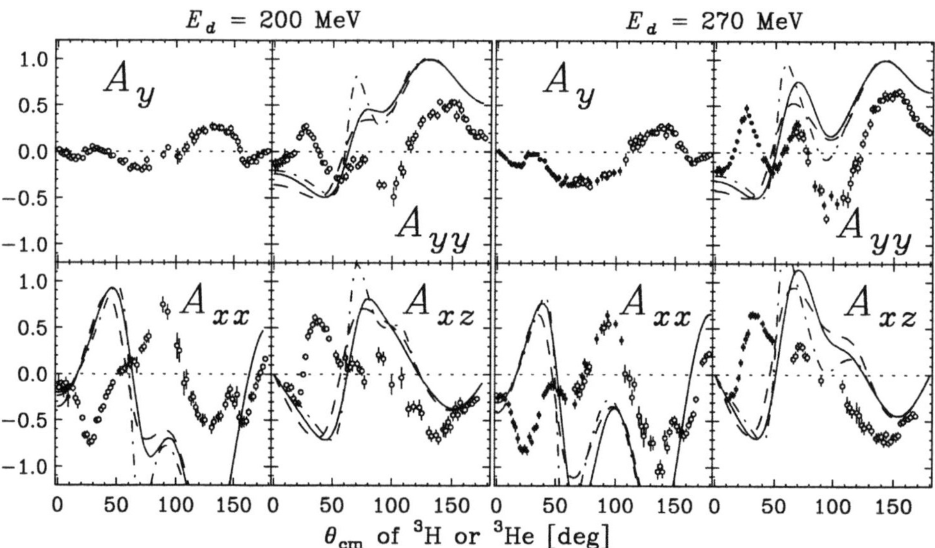

FIGURE 3. Angular distributions of the analyzing powers for the $\vec{d}d \to {}^3\mathrm{He}\,n\,({}^3\mathrm{H}\,p)$ reactions. See text for the explanations.

REFERENCES

1. B. Blankleider and R. M. Woloshyn, *Phys. Rev.* **C29**, 538 (1984).
2. E. Jans *et al.*, *Nucl. Phys.* **A475**, 687 (1987).
3. M. B. Epstein *et al.*, *Phys. Rev.* **C32**, 967 (1985).
4. R. G. Milner *et al.*, *Phys. Lett.* **B379**, 67 (1996).
5. V. P. Ladygin and N. B. Ladygina, *Phys. Atom. Nucl.* **59**, 789 (1996).
6. V. P. Ladygin *et al.*, *Part. Nucl. Lett.* **3[100]-2000**, 74 (2000).
7. T. Ichihara *et al.*, *Nucl. Phys.* **A569**, 287c (1994).
8. Y. Maeda *et al.*, *Nucl. Inst. Meth. Phys. Res.* **A490**, 518 (2002).

Extraction of Neutron Density Distributions from Proton Elastic Scattering at Intermediate Energies

H. Takeda*, H. Sakaguchi*, S. Terashima*, T. Taki*, M. Yosoi*, M. Itoh*, T. Kawabata*, T. Ishikawa*, M. Uchida*, N. Tsukahara*, Y. Yasuda*, T. Noro†, M. Yoshimura†, H. Fujimura†, H.P. Yoshida†, E. Obayashi†, A. Tamii** and H. Akimune‡

*Department of Physics, Kyoto University, Kyoto 606-8502, Japan
†Research Center for Nuclear Physics, Osaka University, Osaka 567-0047, Japan
**Department of Physics, University of Tokyo, Tokyo 113-0033, Japan
‡Department of Physics, Konan University, Kobe 658-8501, Japan

Abstract. Cross sections, analyzing powers and spin rotation parameters of proton elastic scattering from ^{58}Ni and ^{120}Sn have been measured at intermediate energies. Obtained data have been analyzed in the framework of relativistic impulse approximations. In order to explain the ^{58}Ni data, it was necessary to modify *NN* interactions in the nuclear medium by changing coupling constants and masses of σ and ω mesons. For ^{120}Sn, by assuming the same modification of *NN* interactions and by using proton densities deduced from charge densities, the neutron density distribution was searched so as to reproduce ^{120}Sn data at 300 MeV.

INTRODUCTION

Research fields in nuclear study are remarkably spreading due to the recent developments of radioactive isotope beam facilities all over the world. Nuclei far from β stability line are expected to be different not only quantitatively but also qualitatively. For instance neutron rich unstable nuclei are expected to have anomalous structures such as neutron skin and halo. Neutron distributions in nuclei will provide fundamental information for nuclear structure study. Thus it is indispensable to establish procedures to extract neutron density distributions from experimental information.

Protons at intermediate energies are considered to be suitable to extract information inside the nucleus because of the large mean free path in the nuclear medium. Ambiguities due to the target nuclear structure are relatively small in the elastic scattering since the ground state wave functions used for elastic scattering are restricted by charge distributions measured by electron scattering. Although it is hard to obtain neutron distributions from charge distributions, they can be assumed to have the same shapes as with protons for $N \simeq Z$ nuclei. Thus the proton elastic scattering at intermediate energies has been used to discern various microscopic approaches for nuclear interactions. Applying these models to $N \neq Z$ nuclei, neutron density distributions can be extracted from proton elastic scattering data.

CP675, *Spin 2002: 15th Int'l. Spin Physics Symposium and Workshop on Polarized Electron Sources and Polarimeters*, edited by Y. I. Makdisi, A. U. Luccio, and W. W. MacKay

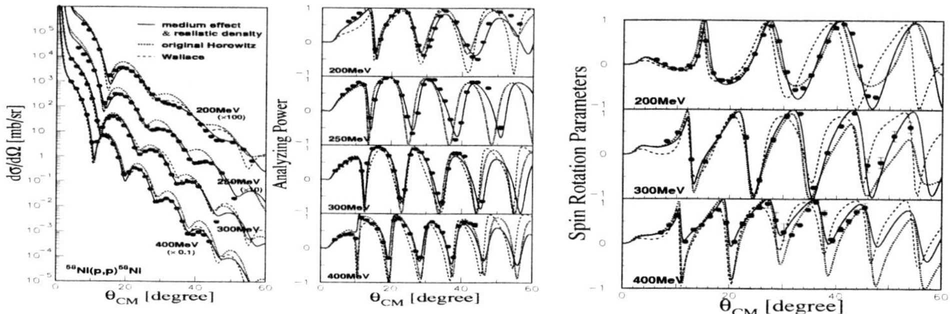

FIGURE 1. Cross sections, analyzing powers and spin rotation parameters of p-^{58}Ni elastic scattering at 200 – 400 MeV. See text for details.

EXPERIMENT

We measured cross sections, analyzing powers and spin rotation parameters of proton elastic scattering from ^{58}Ni and ^{120}Sn at $E_p = 200-400$ MeV. Scattering angles were up to 60° for measurements of cross sections and analyzing powers, and up to 45° for spin rotation parameters. We can determine the scattering amplitudes completely from these measurements since there are only three independent observables in proton elastic scattering from a spin 0 nucleus.

The experiment was performed at the Research Center for Nuclear Physics (RCNP). Polarized protons were injected into the pre-accelerator AVF cyclotron, transported to the six sector ring cyclotron, accelerated to a final energy of 200 – 400 MeV and directed against the target. Scattered protons were momentum analyzed with a high momentum resolution magnetic spectrometer, 'Grand Raiden', and detected by counters at the focal plane. Momenta of scattered protons were measured by detecting their position with vertical drift chambers. The polarization of scattered protons was determined using the focal plane polarimeter (FPP), which measured scattering asymmetries in a carbon analyzer block with multi-wire proportional chambers. Scintillators and hodoscopes were used to trigger the data acquisition and for particle identifications.

NN INTERACTIONS IN MEDIUM

Obtained data were analyzed in the framework of the relativistic impulse approximation (RIA). It has been pointed out[1] that the RIA model with density dependent coupling constants and masses of exchanged σ and ω mesons has been able to explain cross sections and analyzing powers of proton elastic scattering from ^{58}Ni at intermediate energies. In that analysis the neutron distribution has been assumed to be same as protons deduced from the charge distribution except for the normalization factor N/Z. Figure 1 shows the experimental results and some model calculations of cross sections, analyzing powers and spin rotation parameters of ^{58}Ni. Solid circles are our data. Dotted and dashed curves are original RIA calculations by Horowitz *et al.*[2] and Wallace *et al.*[3],

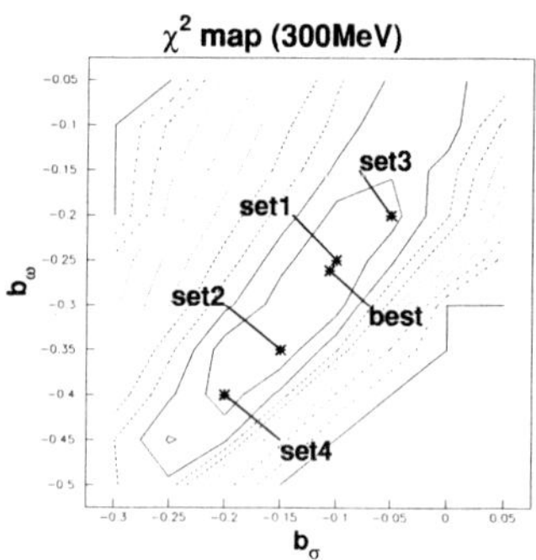

FIGURE 2. χ^2-map in (b_σ, b_ω) parameter space for 300 MeV.

respectively. Solid curves indicate the medium modified RIA model described above. Modifications of coupling constants and masses are parameterized as

$$g_j^2,\ \bar{g}_j^2 \ \rightarrow\ \frac{g_j^2}{1+a_j\rho(r)/\rho_0},\ \frac{\bar{g}_j^2}{1+\bar{a}_j\rho(r)/\rho_0}, \tag{1}$$

$$m_j,\ \bar{m}_j \ \rightarrow\ m_j\left(1+b_j\rho(r)/\rho_0\right),\ \bar{m}_j\left(1+\bar{b}_j\rho(r)/\rho_0\right), \tag{2}$$

where j refers to the σ or ω mesons and ρ_0 stands for the normal density. Newly measured spin rotation parameters are also well explained by the medium modified RIA model. Figure 2 shows χ^2-map in (b_σ, b_ω) parameter space for 300 MeV. Other six parameters were optimized at each grid point. A very strong correlation between b_σ and b_ω is indicated by a narrow valley. The same correlation can be found between a_σ and a_ω and in other energies also.

NEUTRON DISTRIBUTION SEARCH

For $N \neq Z$ nuclei such as ^{120}Sn it can not be expected that the neutron distribution has the same shape as with protons. However, since the elastic scattering is sensitive to both NN interactions in nuclear medium and density distributions of the target nucleus, the neutron density distribution can be extracted from elastic scattering, assuming the same medium modifications fixed by the ^{58}Ni data.

In order to search the neutron distribution we used a sum of Gaussians (SOG) type distribution

$$\rho_n(r) = \frac{N}{2\pi^{3/2}\gamma^3}\sum_i \frac{Q_i}{1+2R_i^2/\gamma^2}\left(e^{-(r-R_i)^2/\gamma^2} + e^{-(r+R_i)^2/\gamma^2}\right). \tag{3}$$

The normalization condition $\int \rho_n(r)\, d^3\vec{r} = N$ results in the constraint for Q_i ($\sum Q_i = 1$). Q_i are searched so as to reproduce ^{120}Sn data at 300 MeV, whereas width γ and position R_i of each Gaussian are fixed with the values listed in a reference[4]. All resulting SOG

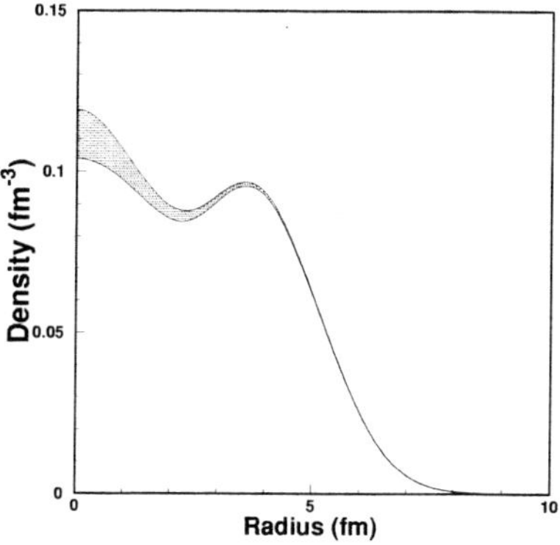

FIGURE 3. Obtained neutron distribution in ^{120}Sn.

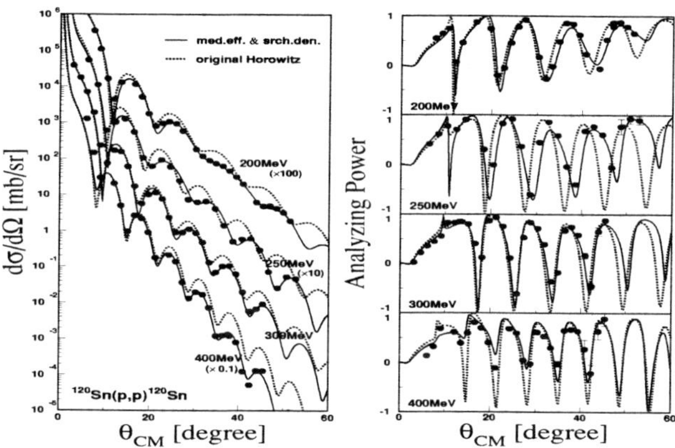

FIGURE 4. Cross sections and analyzing powers of proton elastic scattering from ^{120}Sn at $E_p = 200-400$ MeV.

distributions with good reduced $\chi^2_\nu\,(\equiv \chi^2/\nu)$

$$\chi^2_\nu \leq \chi^2_{\nu\mathrm{min}} + 1, \tag{4}$$

where ν is the number of degrees of freedom and $\chi^2_{\nu\mathrm{min}}$ is the minimum value of χ^2_ν, are possible densities and their superposition determines an error band, which is displayed in Fig. 3 as shaded area. The obtained neutron distribution has a bump structure at the nuclear center. This result is consistent with the wave function of neutrons in $3s_{1/2}$ orbit that is expected to be occupied in ^{120}Sn nuclei. Solid curves in Fig. 4 are the calculations using the best fit neutron density. Original unmodified RIA calculations with relativistic Hartree densities are also displayed in Fig. 4 by dotted curves. It is notable that our data

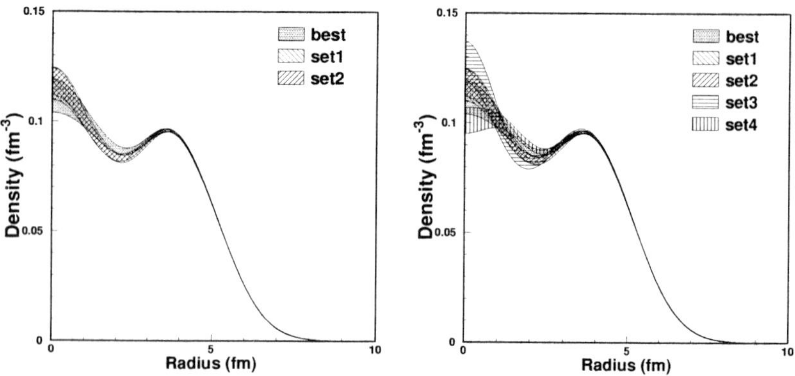

FIGURE 5. The left part shows the distributions deduced with 'set1' and 'set2' parameters. Deduced distributions with all sets in Fig. 2 are superposed in the right part.

indicated by solid circles are well explained by the deduced density not only at 300 MeV but also at other energies although the density search was performed using 300 MeV data only. The difference of the neutron and proton root mean square radii can be evaluated as $\Delta r_{np} = 0.116 \pm 0.015$ fm, which agrees with a result deduced from the SDR sum rule in Ref. [5].

In order to estimate the uncertainty in the deduced distribution due to the ambiguities in our medium modification parameters, we also searched the distributions using various parameter sets indicated in Fig. 2 as 'set1' to 'set4'. Reduced χ^2_ν of the 'set1' and 'set2' parameters are about $\chi^2_{\nu\mathrm{min}} + 1$, while it is $\chi^2_\nu \sim \chi^2_{\nu\mathrm{min}} + 2$ for the 'set3' and 'set4' parameters. The left part of Fig. 5 shows the distribution deduced with the 'set1' and 'set2' parameters. Changes of the distributions are relatively small compared to the error band. However changes become larger if we use the 'set3' and 'set4' parameters as displayed in the right part of Fig. 5. In other words, we can obtain the neutron distribution with small errors if the *NN* interactions in medium are well determined.

ACKNOWLEDGMENTS

We would like to thank Prof. Hatanaka and the staff members of the RCNP for their support and tuning to obtain a clean and high intensity beam during the experiment.

REFERENCES

1. Sakaguchi, H., *et al.*, Phys. Rev. **C57**, 1749 (1998) and references therein.
2. Murdock, D.P., and Horowitz, C.J., Phys. Rev. **C35**, 1442 (1987); Horowitz, C.J.,*et al.*, *Computational Nuclear Physics 1*, Springer-Verlag, Berlin, 1991, Chap. 7.
3. Tjon, J.A., and Wallace, S.J., Phys. Rev. **C32**, 1667 (1985); Phys. Rev. **C36**, 1085 (1987).
4. de Vries, H., *et al.*, Atomic Data and Nuclear Data Tables **36**, 495 (1987).
5. Krasznahorkay, A., *et al.*, Phys. Rev. Lett. **82**, 3216 (1999).

Tensor Analysing Powers for ^{7}Li Induced Transfer Breakup Reactions

N. J. Davis*, R. P. Ward*, K. Rusek†, N. M. Clarke**, G. Tungate**, J. A. R. Griffith**, S. J. Hall**, O. Karban**, I. Martel-Bravo**, J. M. Nelson**, J. Gómez-Camacho‡, T. Davinson§, D. G. Ireland§, K. Livingston§, E. W. Macdonald§, R. D. Page§, P. J. Sellin§, C. H. Shepherd-Themistocleous§, A. C. Shotter§ and P. J. Woods§

*School of Chemistry and Physics, Keele University, Keele, Staffordshire ST5 5BG, England
†Department of Nuclear Reactions, The Andrzej Sołtan Institute for Nuclear Studies, Hoża 69, 00-681 Warsaw, Poland
**School of Physics and Astronomy, University of Birmingham, Edgbaston, Birmingham B15 2TT, England
‡Departmento de Física Atómica, Molecular y Nuclear, Facultad de Física, Universidad de Sevilla, Aptdo. 1065, 41080 Sevilla, Spain
§Department of Physics and Astronomy, University of Edinburgh, Mayfield Road, Edinburgh EH9 3JZ, Scotland

Abstract. The tensor analysing power T_{20} has been measured for the ^{120}Sn(^{7}Li,^{8}Be $\rightarrow 2\alpha)^{119}$In and ^{120}Sn(^{7}Li,^{6}Li$^* \rightarrow \alpha + d)^{121}$Sn transfer breakup reactions at 70 MeV bombarding energy. Coupled channels and continuum discretized coupled channels calculations, incorporating a detector phase space correction, were found to give good agreement with the data.

INTRODUCTION

Transfer breakup reactions are of particular interest because they combine two processes, nucleon transfer and breakup, which are now relatively well understood in isolation. However there has been little published on such transfer breakup reactions. Lithium-7 induced transfer breakup reactions are of relevance to studies of nuclei having cluster structures and to radioactive beam studies where fragmentation is likely. Coupled channels (CC) calculations have been found to describe transfer reactions very well and more recently continuum discretized coupled channels (CDCC) calculations have been very successful in describing breakup reactions [1]. The challenge is to thoroughly test these calculations with the more complex transfer breakup reactions.

EXPERIMENT

The analysing power T_{20} was measured for transfer breakup reactions resulting from heavy ion collisions induced by a 70 MeV $^7\vec{\text{Li}}$ beam on a ^{120}Sn target. Data were obtained for the ^{120}Sn(^{7}Li,^{8}Be $\rightarrow 2\alpha)^{119}$In proton pickup and ^{120}Sn(^{7}Li,^{6}Li$^* \rightarrow \alpha +$

CP675, *Spin 2002: 15th Int'l. Spin Physics Symposium and Workshop on Polarized Electron Sources and Polarimeters,* edited by Y. I. Makdisi, A. U. Luccio, and W. W. MacKay

d)^{121}Sn neutron stripping reactions. The experimental procedure was described earlier by Davis et al. [1].

RESULTS

Data

For the ^{120}Sn(^{7}Li,^{8}Be $\rightarrow 2\alpha$)^{119}In reaction spectra were reconstructed corresponding to breakup via the 0^+ ground state of ^{8}Be. The unresolved ground $9/2^+$ and 0.31 MeV $1/2^-$ first excited states of ^{119}In were found to be populated strongly in these spectra. For the ^{120}Sn(^{7}Li,^{6}Li* $\rightarrow \alpha +$d)^{121}Sn reaction spectra were reconstructed corresponding to breakup via the 2.18 MeV 3^+ state of ^{6}Li. These latter spectra were observed to contain three strong structures corresponding to the unresolved ground $3/2^+$, 0.006 MeV $11/2^-$ first excited and 0.06 MeV $1/2^+$ second excited states and many unresolved states around 1.2 MeV and 2.7 MeV in ^{121}Sn. Many of the states contributing to the two latter structures have uncertain or unknown spin-parities, rendering the analysis of that data difficult or impossible. Thus, data for those two structures will not be presented here but will be reported elsewhere [2].

Calculations

CC and CDCC calculations were performed, using version FRXP.18 of the code FRESCO [3], for the ^{120}Sn(^{7}Li,^{8}Be $\rightarrow 2\alpha$)^{119}In and ^{120}Sn(^{7}Li,^{6}Li* $\rightarrow \alpha +$d)^{121}Sn reactions, respectively. For the ^{120}Sn(^{7}Li,^{8}Be$\rightarrow 2\alpha$)^{119}In reaction the entrance channel optical potential was that of Cook [4] and the exit channel optical potential was determined from single folding using an empirical α+^{120}Sn optical potential [5] and ^{8}Be $= \alpha + \alpha$ cluster wavefunctions with a Gaussian shaped binding potential. The ^{8}Be ground state was assumed to be weakly bound, by just 0.01 MeV, and the potential depth was adjusted to reproduce this. For the ^{120}Sn(^{7}Li,^{6}Li* $\rightarrow \alpha$+d)^{121}Sn reaction the entrance channel optical potential was that of Cook [4] and the exit channel optical potential was determined from single folding using empirical α+^{120}Sn [5] and d+^{120}Sn [6] optical potentials and CDCC ^{6}Li $= \alpha +$d wavefunctions [7] calculated using the $\alpha +$d binding potential proposed by Kubo and Hirata [8]. Spectroscopic amplitudes were obtained from Cohen and Kurath [9] and Turkiewicz et al. [10]. The coupling schemes used are shown in Figures 1 and 2.

To make a reasonable comparison of data with prediction, the detector configuration used for the breakup fragments needs to be considered. The ^{8}Be case is the simplest because the ^{8}Be is in its ground state so $L = 0$ for the breakup. The $L = 0$ breakup gives an isotropic distribution of α fragment directions in the centre of mass of the ^{8}Be. A direct measurement of T_{20} for the transfer breakup reaction is consequently made because it does not matter where the detectors are placed relative to the reaction plane. The ^{6}Li* case is somewhat more complicated because the ^{6}Li* is in the 2.19 MeV 3^+ excited state. This is known to be a pure $L = 2$ $\alpha +$d cluster state, so $L = 2$ breakup

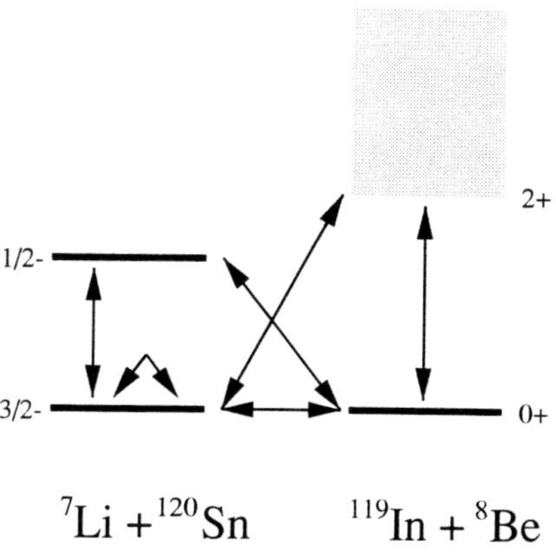

FIGURE 1. Coupling scheme for ^{120}Sn(^{7}Li,^{8}Be $\rightarrow 2\alpha$)^{119}In CC calculations. The spin-parities refer to the projectile/ejectile.

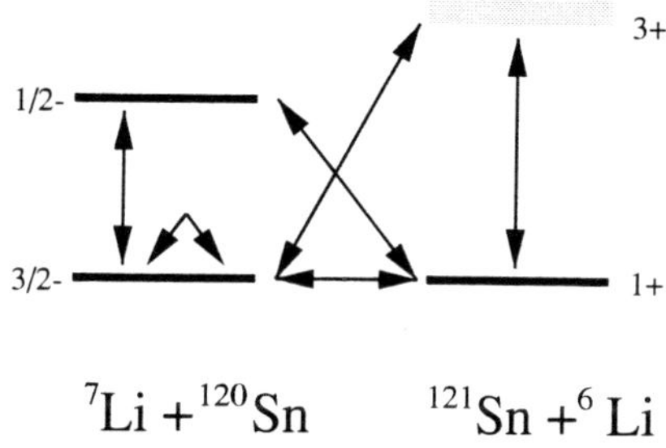

FIGURE 2. Coupling scheme for ^{120}Sn(^{7}Li,^{6}Li$^* \rightarrow \alpha + $d)^{121}Sn CDCC calculations. The spin-parities refer to the projectile/ejectile.

of the state with no $L = 4$ admixture can be assumed to a very good approximation. The $L = 2$ breakup will result in an anisotropic fragment distribution and a consequent phase space effect due to detector positions. A technique was therefore developed to take this into account, as described previously by Davis et al. [1], by which the measured analysing powers T_{kq} are modelled by a combination of the calculated polarization transfer coefficients $X_{kq,k'q'}$ with tensors $I_{k'q'}$ which are related to the detector geometry.

DISCUSSION AND CONCLUSIONS

Calculations for the ^{120}Sn(^{7}Li,^{8}Be $\rightarrow 2\alpha$)^{119}In reaction are compared with the data in Figure 3. The data agree very well with the calculation assuming population of the ^{119}In ground state. The calculation assuming population of the ^{119}In first excited state is very different and does not reproduce the data. This is perhaps an indication that only the

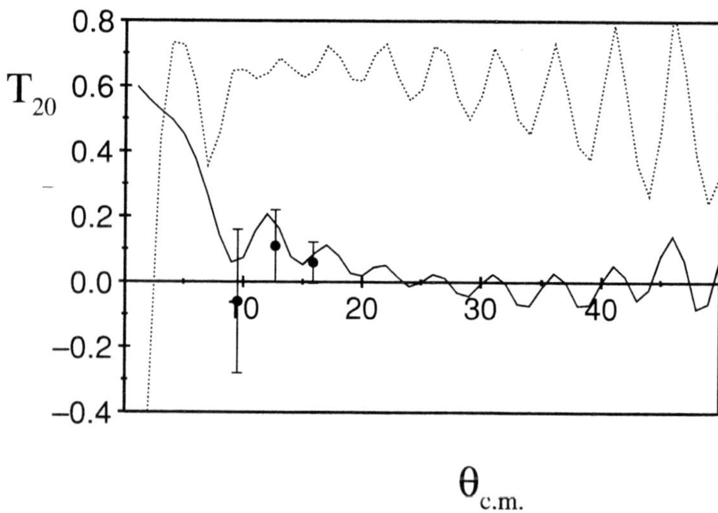

FIGURE 3. Results of CC calculations for the ^{120}Sn(^{7}Li,^{8}Be $\rightarrow 2\alpha$)^{119}In reaction compared with data. The solid and dotted lines assume population of the ^{119}In $9/2^+$ ground state and 0.31 MeV $1/2^-$ first excited state, respectively.

ground state is significantly populated by the reaction.

Calculations without the detector phase space correction for the ^{120}Sn(^{7}Li,^{6}Li$^* \rightarrow \alpha + d$)^{121}Sn reaction assuming population of the ground $3/2^+$, 0.006 MeV $11/2^-$ and 0.06 MeV $1/2^+$ states are shown with the unresolved data in Figure 4. Good agreement is not achieved, although it could be argued on the basis of these calculations alone that the calculation for the $1/2^+$ state, being predominantly negative, represents the data better than the calculations for the other two states, which are predominantly positive.

Calculations with the detector phase space correction included are shown in Figure 5. These illustrate the importance of the correction which leads to far better agreement between the calculations and the data. They also lead to a different conclusion than that arrived at from the uncorrected calculations alone, because once the correction is included all three calculations are very similar and agree with the data equally well. This means the relative contributions from the three states to the data are not important in assessing the success of the calculations.

In conclusion, the results show that the calculations do well in reproducing T_{20} analysing power data in one of the first tests of CC and CDCC calculations for transfer breakup reactions. A more detailed description of this study is in preparation [2].

ACKNOWLEDGMENTS

This work was supported in part by the Engineering and Physical Sciences Research Council of the United Kingdom and by the State Committee for Scientific Research (KBN) of Poland.

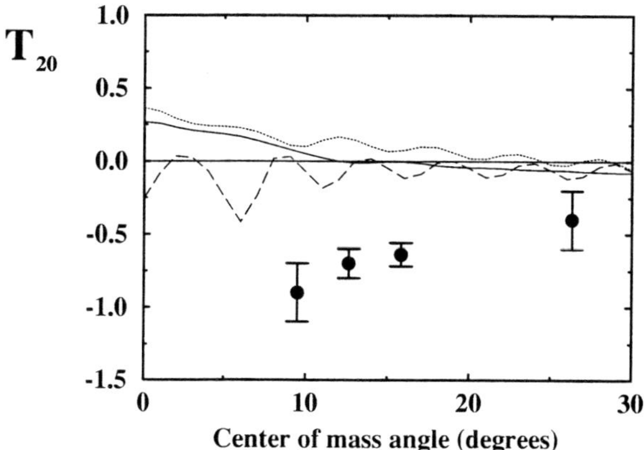

FIGURE 4. Results of CDCC calculations without detector phase space correction for the ^{120}Sn(^{7}Li,^{6}Li$^{*} \rightarrow \alpha + \mathrm{d}$)^{121}Sn reaction. The dotted, solid and dashed lines assume population of the ^{121}Sn $3/2^{+}$ ground state, the $11/2^{-}$ state at 0.006 MeV and the $1/2^{+}$ state at 0.06 MeV, respectively.

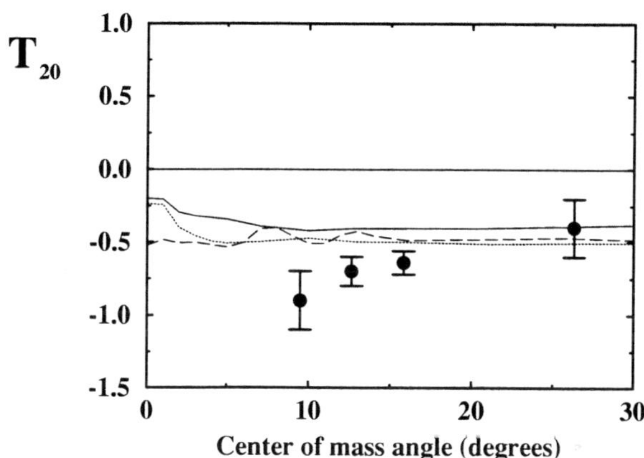

FIGURE 5. Results of CDCC calculations including detector phase space correction for the ^{120}Sn(^{7}Li,^{6}Li$^{*} \rightarrow \alpha + \mathrm{d}$)^{121}Sn reaction. The dotted, solid and dashed lines assume population of the ^{121}Sn $3/2^{+}$ ground state, the $11/2^{-}$ state at 0.006 MeV and the $1/2^{+}$ state at 0.06 MeV in ^{121}Sn, respectively.

REFERENCES

1. Davis, N. J., et al., Phys. Rev. **C52**, 3201 (1995).
2. Davis, N. J., et al., to be submitted for publication in Phys. Rev. C.
3. Thompson, I. J., Comput. Phys. Rep. **7**, 167 (1988).
4. Cook, J., Nucl. Phys. **A388**, 153 (1982).
5. Baron, N., Leonard, R. F., and Stewart, W. M., Phys. Rev. **C4**, 1159 (1971).
6. Childs, J. D., Daehnick, W. W., and Spisak, M. J., Phys. Rev. **C10**, 217 (1974).
7. Keeley, N., and Rusek, K., Phys. Lett. **B427**, 1 (1998).
8. Kubo, K. I., and Hirata, M., Nucl. Phys. **A187**, 186 (1972).
9. Cohen, S., and Kurath, D., Nucl. Phys. **A101**, 1 (1967).
10. Turkiewicz, I. M., et al., Nucl. Phys. **A486**, 152 (1988).

Polarizations for $^{12}C(p,2p)$ Reactions at 1 GeV

H. P. Yoshida*, T. Noro†, O. V. Miklukho**, V. A. Andreev**, M. N. Andronenko**, G. M. Amalsky**, S. L. Belostotski**, O. A. Domchenkov**, O. Ya. Fedorov**, K. Hatanaka*, A. A. Izotov**, A. A. Jgoun**, J. Kamiya*, A. Yu. Kisselev**, M. A. Kopytin**, E. Obayashi*, A. N. Prokofiev**, D. A. Prokofiev**, H. Sakaguchi‡, V. V. Sulimov**, A. V. Shvedchikov**, H. Takeda‡, S. I. Trush**, V. V. Vikhrov**, T. Wakasa*, Y. Yasuda‡ and A. A. Zhdanov**

*Research Center for Nuclear Physics, Osaka University, Ibaraki 567-0047, Japan
†Department of Physics, Kyushu University, Fukuoka 812-8581, Japan
**Petersburg Nuclear Physics Institute, , Gatchina 188350, Russia
‡Department of Physics, Kyoto University, Kyoto 606-8502, Japan

Abstract.
We have measured polarizations P of outgoing protons in $(p,2p)$ reactions at an incident energy of 1 GeV for three kinds of targets. The experimental result shows a distinct reduction from IA calculation values using NN interaction in free space and the reduction is found to be monotonic to the effective mean density estimated with DWIA. This is consistent with the previous result obtained at 392 MeV, though the incident energy is quite different. We have also measured an angular distribution of the polarization for ^{12}C target. All of the data, including both for forward nd backward outgoing protons, show similar reduction from the IA calculation, though, again, outgoing energies are quite different from 130 MeV to 890 MeV. From these results, it is concluded that these reductions are nuclear structure or interaction originated and not caused by the reaction mechanism such as multi-step processes or distortions.

INTRODUCTION

It is a long standing problem that the analyzing power A_y for proton quasifree scattering is reduced from values predicted with NN interactions in free space. Since theoretical calculations based on the Schrödinger equation have failed to reproduce this phenomenon, it has been taken interest in as a phenomenon which shows appearance of a relativistic effect or a medium-modification effect in a hadron level.

The $(p,2p)$ reaction, where both of two outgoing nucleons are detected, this reaction is regarded as an NN scattering in nuclear field. Thus we expect to extract information on modification of the NN interaction in nuclear field from direct comparison of this reaction with the NN scattering in free space.

By a TRIUMF group, it is found that the analyzing power A_y for this reaction leading to a $1s_{1/2}$-hole state shows a distinct reduction from the free p–p scattering values.[1, 2] At RCNP, A_y of this reaction was measured for several target nuclei at 392 MeV and it was pointed out that this reduction is monotonically depends on the *averaged density*, which was estimated by using the DWIA and the local density approximation.[3]

CP675, *Spin 2002: 15th Int'l. Spin Physics Symposium and Workshop on Polarized Electron Sources and Polarimeters*, edited by Y. I. Makdisi, A. U. Luccio, and W. W. MacKay

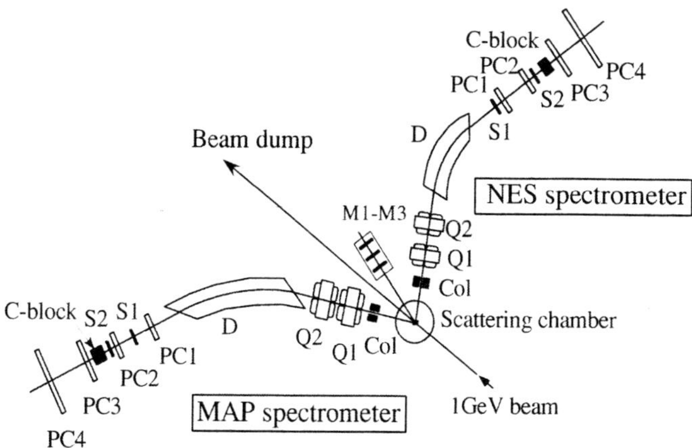

FIGURE 1. Schematic view of the two arm spectrometer system at PNPI. Four sets of MWPC's (PC1–PC4), two trigger scintillators (S1,S2) and a carbon analyzer block form a focal-plane polarimeter system on each spectrometer. Each of PC1–PC4 consists of two MWPC's for measurements of horizontal and vertical positions. Collimeters are used in front of two spectrometers and luminosity is monitored by using the beam-monitor which consists of three scintillators (M1–M3).

In this article, we show our new measurement of polarizations P for the $(p,2p)$ reactions at 1 GeV, which is a significantly different energy from previous measurements at TRIUMF and RCNP. In addition, we have also measured an angular distribution of polarizations for ^{12}C target, where outgoing energies changes significantly, from 740 MeV to 890 MeV for forward outgoing protons and from 130 MeV to 265 MeV for backward outgoing protons.

A discussion is given on the reaction mechanism of this reaction, which is helpful in investigating by measuring polarizations at this higher energy and in such a wide energy range.

EXPERIMENT AND RESULT

Experimental detail

The experiment has been performed by using a 1 GeV unpolarized proton beam from the synchrocyclotron at Petersburg Nuclear Physics Institute(PNPI) in Gatchina. The beam line we used equipped with the two-arm spectrometer system which consists of two QQD-type magnetic spectrometers, called MAP and NES. Each spectrometer has a polarimeter system using a carbon block analyzer on the focal plane and polarizations of both the forward and backward outgoing protons were measured. MAP analyses The forward scattered protons with higher energies were analyzed with MAP, while the backward ones with lower energies were analyzed with NES. The schematic view of the detection system is shown in Fig. 1.

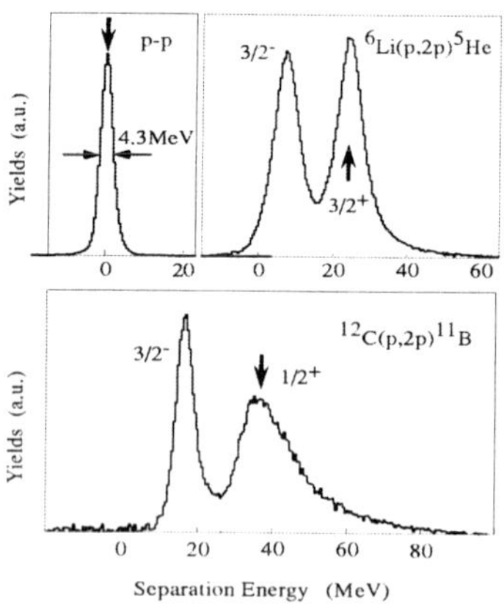

FIGURE 2. Separation energy spectra for p–p scattering and $(p,2p)$ reactions. Each vertical axis is linear scale in arbitrary unit. The states with arrows are used for data analysis.

Throughout the $(p,2p)$ measurement, setting angles and field strengths of spectrometers are kept to those corresponding to the zero-recoil condition for the s-shell knockout from each target nucleus. This is the condition where the cross section for the s-shell knockout gives a maximum and the reaction mechanism is expected to be the simplest. Figure 2 shows typical spectra. The overall energy-resolution for the p–p measurement is 4.3 MeV, as shown in the figure.

Experimental Result

The experimental result is shown in Fig. 3. The polarizations of forward and backward outgoing protons are denoted as P_1 and P_2, respectively.

The data in the left figure are plotted as a function of the averaged density, which is defined in [3]. The solid line is a PWIA calculation by using the free NN interaction. The calculation increases with the averaged density, caused by changes of the kinematics in fact, while the experimental data, both of P_1 and P_2, decrease. And the reduction, the difference between the theoretical and experimental values, is monotonic to the averaged density. This result is consistent with results obtained at 392MeV.[4]

The data in the right figure shows the angular distribution for the ^{12}C$(p,2p)$ reaction. The data corresponding to the NES angle of 64.0, 59.7, and 53.3 degrees are plotted as a function of transferred momentum q. The solid and dash lines are PWIA and DWIA calculations by using the free NN interaction. It shows that the distortion effect, the difference between DWIA and PWIA, is negligibly small for this kinematics. Though the outgoing energies is quite different from 130 MeV to 890 MeV, the data show essentially the same amount of reduction from IA calcurations, independent of energies.

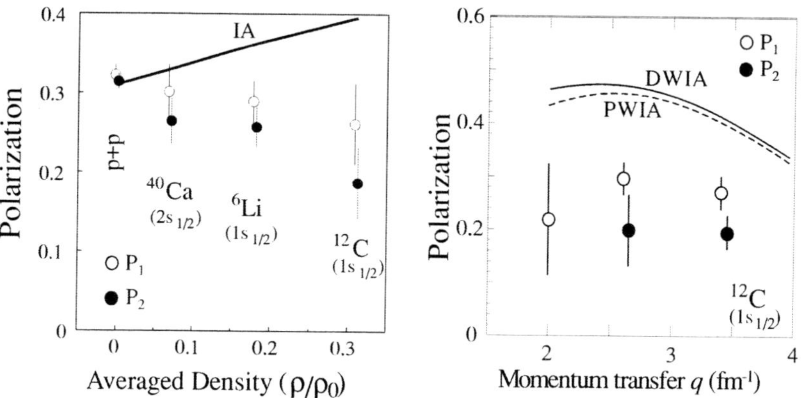

FIGURE 3. *The left panel:* Target dependence of polarization data P for p–p scattering and $(p,2p)$ reactions. The data are plotted as a funcion of the averaged density. *The right panel:* Angular dependence of P for $^{12}C(p,2p)$ reaction. The data are plotted as a funcion of momentum transfer values. In both panels, open circles are polarizations P_1 for forward outgoing protons and black circles are polarizations P_2 for backward ones.

Reaction Mechanism

In the case of nuclear reactions by using hadron projectile, contribution of multi-step processes should be considered carefully. In this measurement at 1 GeV, the incident energies are quite different from previous experiment at 392 MeV and the energies of two emitted protons are also quite different as well. But the result of present data at 1GeV is consistent with that at 392 MeV. This results show that contributions by multi-step processes are not the main reason of the reaction.

REFERENCES

1. Miller, C. A., *et al.*, in *Proc. of the 7th Int. Conf. on Polarization Phenomena in Nuclear Physics*, (Paris, 1990) C6-595–598.
2. Miller, C. A., *et al.*, *Phys. Rev.* **C57**, 1756–1765 (1998).
3. Hatanaka,K., *et al.*, *Phys. Lev. Lett.* **78**, 1014–1017 (1997).
4. Noro, T. *et al.*, *Nucl. Phys.* **A629**, 324c–333c (1998).

Momentum Transfer Dependence of Spin Isospin Modes in the Quasielastic Region

T. Wakasa*, H. Sakai†, M. Ichimura**, K. Hatanaka*, M. B. Greenfield‡, H. Okamura§, K. Kawahigashi¶, A. Tamii†, H. Otsu‖, Y. Nakaoka†, T. Ohnishi††, K. Yako‡‡, K. Sekiguchi††, T. Yagita§§, J. Kamiya*, S. Sakoda†, K. Suda§, H. Kato†, M. Hatano† and Y. Maeda†

*Research Center for Nuclear Physics (RCNP), Ibaraki, Osaka 567-0047, Japan
†Department of Physics, University of Tokyo, Tokyo 113-0033, Japan
**Faculty of Computer and Information Sciences, Hosei University, Tokyo 184-8584, Japan
‡ Division of Natural Sciences, International Christian University, Tokyo 181-8585, Japan
§Department of Physics, Saitama University, Saitama 338-8570, Japan
¶Department of Information Science, Kanagawa University, Kanagawa 259-1293, Japan
‖Department of Physics, Tohoku University, Miyagi 980-8578, Japan
††Institute of Chemical and Physical Research (RIKEN), Saitama 351-0198, Japan
‡‡Center for Nuclear Study (CNS), University of Tokyo, Tokyo 113-0033, Japan
§§Department of Physics, Kyushu University, Fukuoka 812-8581, Japan

Abstract. A complete set of polarization transfer coefficients has been measured for the quasielastic $^{12}C(\vec{p},\vec{n})$ reaction at a bombarding energy of 345 MeV and laboratory scattering angles of 16°, 22°, and 27°. The spin-longitudinal ID_q and spin-transverse ID_p polarized cross sections are deduced. The theoretically expected enhancement in the spin-longitudinal mode is observed. The observed ID_q is consistent with the pionic enhanced ID_q evaluated in a distorted wave impulse approximation (DWIA) calculation employing a random phase approximation (RPA) response function. On the contrary, the theoretically predicted quenching in the spin-transverse mode is not observed. The observed ID_p is not quenched, but rather enhanced in comparison with the DWIA+RPA calculation. Two-step contributions are responsible in part for the enhancement of ID_p.

INTRODUCTION

Isovector spin responses to nuclear mesonic fields are expected to show an enhanced ratio of the spin-longitudinal ($\sigma \cdot \hat{\mathbf{q}}$) to spin-transverse ($\sigma \times \hat{\mathbf{q}}$) response functions for momentum transfer $q > 1\ \mathrm{fm}^{-1}$, as predicted by $\pi + \rho + g'$ meson exchange models of the nuclear mean field [1]. Recent experimental and theoretical investigations of isovector $(\vec{p},\vec{n})$ reactions at $T_p = 346$ [2, 3] and 494 MeV [3, 4, 5, 6] show a signature of the enhancement of the spin-longitudinal cross section ID_q relevant to the spin-longitudinal response function R_L around the quasielastic peak. The enhancement of R_L is attributed to the collectivity induced by the one-pion exchange interaction, and thereby has attracted much interest in connection with both the precursor phenomena of the pion condensation [1] and the enhancement of the pion probability in the nucleus [7, 8, 9, 10, 11]. However the theoretical calculation in a distorted wave impulse approximation (DWIA) employing random phase approximation (RPA) response functions underpredicts ID_q for energy

CP675, *Spin 2002: 15th Int'l. Spin Physics Symposium and Workshop on Polarized Electron Sources and Polarimeters,* edited by Y. I. Makdisi, A. U. Luccio, and W. W. MacKay

transfer greater than that for the quasielastic peak. Furthermore the DWIA+RPA calculation underestimates the spin-transverse cross section ID_p in the quasielastic region. These disagreements between experimental and theoretical results would suggest the importance of the two-step (and possibly higher step) contribution in this region [18].

In this article, we present the measurements of a complete set of polarization transfer coefficients for the quasielastic $(\vec{p},\vec{n})$ reaction on ^{12}C at T_p = 345 MeV and laboratory scattering angles of $\theta_{\rm lab}$ = 16°, 22°, and 27° which correspond to $q_{\rm lab} \simeq 1.2$, 1.7 and 2.0 fm^{-1} at the quasielastic peak [2]. The measured polarization transfer coefficients and cross sections are used to separate the cross sections into non-spin, spin-longitudinal, and spin-transverse polarized cross sections. The experimental polarized cross sections will be compared with theoretical calculations in frameworks of DWIA and RPA. The data will also be compared with a calculation including the two-step contribution.

EXPERIMENTAL METHODS

The measurement was carried out at the Neutron Time-Of-Flight (NTOF) facility [12] at the Research Center for Nuclear Physics (RCNP), Osaka University. The NTOF facility and the neutron detector/polarimeter NPOL2 system are described in detail in Refs. [12, 13, 14]. In the following, therefore, we present a brief description of the detector system and discuss experimental details relevant to the present experiment.

Natural carbon (98.9% ^{12}C) targets used for the cross section and polarization observable measurements with thicknesses of 338 mg/cm^2 and 682 mg/cm^2, respectively, were placed in a beam swinger dipole magnet. Neutrons from the (p,n) reaction traversed various distances within a 100 m time-of-flight (TOF) tunnel, and were detected with NPOL2. Protons downstream of the target were swept by the beam swinger magnet into an aluminum beam stop (Faraday cup). The integrated beam current stopped in the Faraday cup was measured. Typical beam currents were 10 and 50 nA for the cross section and polarization observable measurements, respectively.

The NPOL2 system [14] consists of six planes of two dimensionally position sensitive scintillation detectors: four detectors of liquid scintillator BC519 and two detectors of plastic scintillator BC408. The liquid scintillator BC519 has a high hydrogen-to-carbon ratio of 1.7, which is useful to analyze the neutron polarization using the $\vec{n}+p$ scattering in the scintillator. Each of the six neutron detectors has an effective detection area of approximately 1 m^2 with a thickness of 0.1 m. Thin plastic scintillation detectors are placed in front of each neutron detector in order to distinguish charged particles from neutrons.

The incident neutron energies were determined from the TOF between the target and a given neutron detector. Flight times are measured relative to the cyclotron rf signal. Prominent γ-rays from π^0-decays and prompt de-excitation of inelastically excited states in the target provide a time reference for the absolute timing calibration. Then the transitions to discrete states with known reaction Q values were used to determine the incident beam energy. The beam energy was thus determined to be T_p = 345$\pm$1 MeV. The total full width at half maximum energy resolutions are about 2 and 3 MeV for the cross section and polarization observable measurements, respectively.

DATA REDUCTION

The spin-longitudinal ID_q and spin-transverse ID_p polarized cross sections are related to the unpolarized cross section I and the laboratory-frame polarization transfer coefficients D_{ij} according to [16]

$$ID_q = \frac{I}{4}[1 - D_{NN} + (D_{S'S} - D_{L'L})\cos\alpha_2 - (D_{L'S} + D_{S'L})\sin\alpha_2] , \quad (1)$$

$$ID_p = \frac{I}{4}[1 - D_{NN} - (D_{S'S} - D_{L'L})\cos\alpha_2 + (D_{L'S} + D_{S'L})\sin\alpha_2] , \quad (2)$$

where $\alpha_2 \equiv 2\theta_p - \theta_{\rm lab} - \Omega$. The angle θ_p represents the angle between the incident beam direction and the transverse direction $\hat{\mathbf{p}}$, and the relativistic spin rotation angle Ω is given by [15]

$$\tan(\theta_{\rm cm} - \theta_{\rm lab} - \Omega) = \frac{\sin\theta_{\rm cm}}{\gamma(\cos\theta_{\rm cm} + \beta/\beta_{\rm cm})} , \quad (3)$$

where $\beta_{\rm cm}$ is the velocity of the center-of-mass (cm) frame relative to that of the laboratory frame, β is the velocity of the outgoing nucleon in the cm frame, and $\gamma \equiv 1/\sqrt{1-\beta^2}$.

RESULTS AND DISCUSSIONS

In Fig. 1 the experimental polarized cross sections ID_q and ID_p for ^{12}C are compared with DWIA+RPA calculations. The RPA calculations are performed without the commonly used universality ansatz ($g'_{NN} = g'_{N\Delta} = g'_{\Delta\Delta}$), namely all of the g's are treated independently [17]. The nonlocality of the mean field is treated by an effective mass m^* with radial dependence of

$$m^*(r) = m_N - \frac{f_{\rm WS}(r)}{f_{\rm WS}(0)}[m_N - m^*(0)] , \quad (4)$$

where $f_{\rm WS}$ is the Woods-Saxon radial form factor. The formalism of DWIA calculations is described in Ref. [3].

The dashed curves are the results of DWIA calculations with the RPA response functions employing (g'_{NN}, $g'_{N\Delta}$, $g'_{\Delta\Delta}$) = (0.6, 0.4, 0.5) and $m^*(0)$=0.7m_N. The dotted curves are the DWIA results with the free response functions employing $m^*(0) = m_N$. The calculations reasonably reproduce the observed ID_q at the larger angles of 22° and 27°. This result is consistent with the predicted enhancement of R_L in this momentum-transfer region. It should be noted that the present calculations prefer the smaller $g'_{N\Delta}$ ($\simeq 0.4$) compared with g'_{NN} ($\simeq 0.6$) and the smaller effective mass at the center ($m^* \simeq 0.7m_N$). The calculation at 16° is slightly larger than the observed data. This might mean that the effective interaction in this momentum-transfer region is not so attractive as is expected in the $\pi + \rho + g'$ model.

In the spin-transverse mode, the calculations underestimate ID_p in the quasielastic region by a factor of approximately 2 at all three angles. The RPA correlation quenches

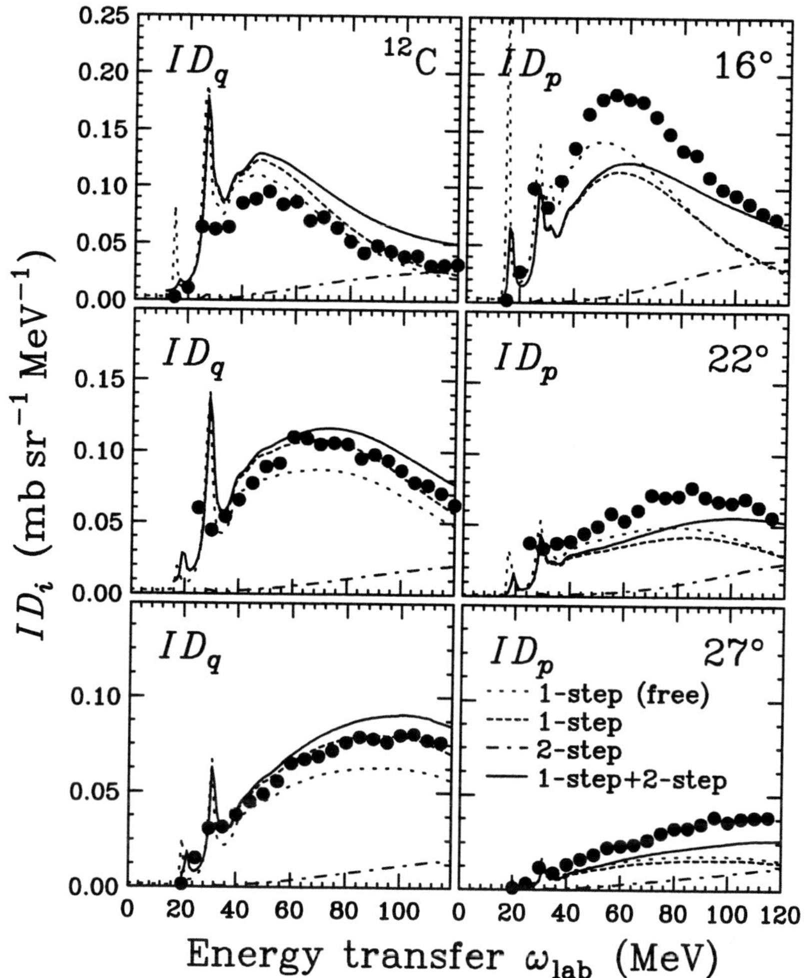

FIGURE 1. The spin-longitudinal ID_q (left panels) and spin-transverse ID_p (right panels) polarized cross sections for the $^{12}C(p,n)$ reaction at T_p = 345 MeV and θ_{lab} = 16°, 22°, and 27°. The dotted and dashed curves represent the results of DWIA calculations with RPA and free response functions, respectively. The dotted-dashed curve are the two-step cross sections. The solid curves are the sum of one- and two-step contributions employing RPA correlations.

ID_p as is predicted, while the experimental results are significantly enhanced. Recently, Nakaoka and Ichimura [18] have pointed out that the two-step contribution for ID_p would be significantly larger than that for ID_q in the present momentum-transfer region. They showed that the 1st- and 2nd-step contributions for ID_q are partly destructive, while those for ID_p are wholly constructive. As a result, the two-step contribution for ID_p is more important than that for ID_q.

The dotted-dashed curves in Fig. 1 are the two-step cross sections calculated by Nakaoka [19]. The solid curves are the sum of one- and two-step contributions employing RPA correlations. Relatively small two-step contributions for ID_q at large angles of 22° and 27° do not affect the agreement between the experimental and theoretical results. On the contrary, the two-step contribution at 16° is fairly large, and the inclusion of this overestimates the experimental ID_q at large energy transfers.

In the spin-transverse mode, at all three angles, two-step contributions relative to one-step ones are significantly large compared with those for ID_q. Two-step contributions account for the underestimation of ID_p in DWIA+RPA calculations at large energy transfers beyond the quasielastic peak. However they are insufficient to explain the underestimation of ID_p around the quasielastic peak. This discrepancy in ID_p might be due to the effects of the higher order (such as $2p2h$) configuration mixing which are not included in the present RPA calculations.

ACKNOWLEDGMENTS

We are grateful to K. Nishida and A. Itabashi for their helpful correspondence. This work is supported in part by the Grants-in-Aid for Scientific Research Nos. 6342007, 12740151, and 14702005 of the Ministry of Education, Science, Sports and Culture of Japan.

REFERENCES

1. W. M. Alberico, M. Ericson, and A. Molinari, Nucl. Phys. **A379**, 429 (1982).
2. T. Wakasa *et al.*, Phys. Rev. C **59**, 3177 (1999).
3. K. Kawahigashi, K. Nishida, A. Itabashi, and M. Ichimura, Phys. Rev. C **63**, 044609 (2001).
4. J. B. McClelland *et al.*, Phys. Rev. Lett. **69**, 582 (1992).
5. X. Y. Chen *et al.*, Phys. Rev. C **47**, 2159 (1993).
6. T. N. Taddeucci *et al.*, Phys. Rev. Lett. **73**, 3516 (1994).
7. C. H. Llewellyn Smith, Phys. Lett. **128B**, 107 (1983).
8. M. Ericson and A. W. Thomas, Phys. Lett. **128B**, 112 (1983).
9. B. L. Friman, V. R. Pandharipande, and R. B. Wiringa, Phys. Rev. Lett. **51**, 763 (1983).
10. E. L. Berger, F. Coester, and R. B. Wiringa, Phys. Rev. D **29**, 398 (1984).
11. D. Stump, G. F. Bertsch, and J. Pumlin, AIP Conf. Proc. **110**, 339 (1984).
12. H. Sakai *et al.*, Nucl. Instrum. Methods Phys. Res. A **369**, 120 (1996).
13. H. Sakai *et al.*, Nucl. Instrum. Methods Phys. Res. A **320**, 479 (1992).
14. T. Wakasa *et al.*, Nucl. Instrum. Methods Phys. Res. A **404**, 355 (1998).
15. N. Hoshizaki, Suppl. Prog. Theor. Phys. **42**, 107 (1968).
16. M. Ichimura and K. Kawahigashi, Phys. Rev. C **45**, 1822 (1992).
17. K. Nishida and M. Ichimura, Phys. Rev. C **51**, 269 (1995).
18. Y. Nakaoka and M. Ichimura, Prog. Theor. Phys. **102**, 599 (1999).
19. Y. Nakaoka, Phys. Rev. C **65**, 064616 (2002).

Acceleration and Storage of Polarized Beams

Spin Tune in the Single Resonance Model with a Pair of Siberian Snakes

D. P. Barber*, R. Jaganathan† and M. Vogt*

*Deutsches Elektronen–Synchrotron, DESY, 22603 Hamburg, Germany.
†The Institute of Mathematical Sciences, Chennai. Tamilnadu 600113, India.

Abstract. Snake "resonances" are classified in terms of the invariant spin field and the amplitude dependent spin tune. Exactly at snake "resonance" there is no continuous invariant spin field.

INTRODUCTION

Spin motion in storage rings and circular accelerators is most elegantly systematised in terms of the invariant spin field (ISF) and the amplitude dependent spin tune (ADST). Here we apply them in the context of snake "resonances". We begin by briefly recapitulating some necessary basic ideas. For more details see [1, 2, 3, 4].

Spin motion in electric and magnetic fields at the 6–dimensional phase space point $\vec{z}$ and position s around the ring, is described by the T–BMT precession equation $d\vec{S}/ds = \vec{\Omega}(\vec{z};s) \times \vec{S}$ [1] where $\vec{S}$ is a spin expectation value ("the spin") and $\vec{\Omega}(\vec{z};s)$ contains the electric and magnetic fields. The ISF, denoted by $\hat{n}(\vec{z};s)$, is a 3–vector *field* of unit length obeying the T–BMT equation along particle orbits $(\vec{z}(s);s)$ and fulfilling the periodicity condition $\hat{n}(\vec{z};s+C) = \hat{n}(\vec{z};s)$ where C is the circumference. Thus $\hat{n}(\vec{M}(\vec{z};s);s+C) = \hat{n}(\vec{M}(\vec{z};s);s) = R_{3\times3}(\vec{z};s)\hat{n}(\vec{z};s)$ where $\vec{M}(\vec{z};s)$ is the new phase space vector after one turn starting at $\vec{z}$ and s and $R_{3\times3}(\vec{z};s)$ is the corresponding spin transfer matrix. The scalar product $J_s = \vec{S}\cdot\hat{n}$ is invariant along an orbit, since both vectors obey the T–BMT equation. Thus with respect to the local $\hat{n}$ the motion of $\vec{S}$ is simply a precession around $\hat{n}$. The field $\hat{n}$ can be constructed at each reference energy where it exists without reference to individual spins.

The chief aspects of the ISF are that: 1) For a turn–to–turn invariant particle distribution in phase space, a distribution of spins initially aligned along the ISF remains invariant (in equilibrium) from turn–to–turn, 2) the ISF determines the maximum attainable time averaged polarisation, $P_{\rm lim} = |\langle\hat{n}(\vec{z};s)\rangle|$, at each s where $\langle\rangle$ denotes the average over phase space, 3) under appropriate conditions J_s is an adiabatic invariant while system parameters such as the reference energy are slowly varied, 4) it provides the main axis for orthonormal coordinate systems at each point in phase space which serve to define the ADST which in turn is used to define the concept of spin–orbit resonance.

These coordinate systems are constructed by attaching two other unit vectors $\hat{u}_1(\vec{z};s)$ and $\hat{u}_2(\vec{z};s)$ to all $(\vec{z},s)$ such that the sets $(\hat{u}_1,\hat{u}_2,\hat{n})$ are orthonormal. Like $\hat{n}$, the *fields* $\hat{u}_1$ and $\hat{u}_2$ are 1–turn periodic in s: $\hat{u}_i(\vec{z};s+C) = \hat{u}_i(\vec{z};s)$ for $i\in\{1,2\}$. With the basis vectors

CP675, *Spin 2002: 15th Int'l. Spin Physics Symposium and Workshop on Polarized Electron Sources and Polarimeters*, edited by Y. I. Makdisi, A. U. Luccio, and W. W. MacKay

$\hat{u}_1$ and $\hat{u}_2$ we can quantify the rate of the above mentioned spin precession around $\hat{n}$: it is the rate of rotation of the projection of $\vec{S}$ onto the $\hat{u}_1, \hat{u}_2$ plane. Except on or close to orbital resonance, the fields $\hat{u}_1(\vec{z};s)$ and $\hat{u}_2(\vec{z};s)$ can be chosen so that the rate of precession is constant and independent of the orbital phases [5, 2, 3, 4]. The number of precessions per turn "measured" in this way is called the spin tune. The spin tune, $\nu_s(\vec{J})$, depends only on the orbital amplitudes (actions) $\vec{J}$, hence the name ADST. The choice of some $\hat{u}_1(\vec{z};s)$ and $\hat{u}_2(\vec{z};s)$ satisfying the condition $\hat{u}_i(\vec{z};s+C) = \hat{u}_i(\vec{z};s)$ for $i \in \{1,2\}$ is not unique. An infinity of others can be chosen by suitable rotations of the $\hat{u}_i$ around $\hat{n}$. These lead to the *equivalence class* of spin tunes obtained by the transformation: $\nu_s(\vec{J}) \Rightarrow \nu_s(\vec{J}) + l_0 + l_1 Q_1 + l_2 Q_2 + l_3 Q_3$ for any integers l. The ADST provides a way to quantify the degree of coherence between the spin and orbital motion and thereby predict how strongly the electric and magnetic fields along particle orbits disturb spins. In particular, the spin motion can become very erratic close to the *spin–orbit resonance* condition $\nu_s(\vec{J}) = m_0 + m_1 Q_1 + m_2 Q_2 + m_3 Q_3$ where the m's are integers and the Q's are the tunes on a torus of integrable orbital motion. At these resonances the ISF can spread out so that P_{lim} is very small. Examples of the behaviour of P_{lim} near spin–orbit resonance and the application of a generalised Froissart–Stora description of the breaking of the adiabatic invariance of J_s while crossing resonances during variation of system parameters can be found in [3, 4, 6]. Note that: 1) the resonance condition is *not* expressed in terms of the spin tune $\nu(\vec{0})$ on the closed orbit, 2) a "tune" describing spin motion but depending on orbital phases could not be meaningful in the spin–orbit resonance condition, 3) if the system is on spin–orbit resonance for one spin tune of the equivalence class, it is on resonance for all others. In general $\hat{u}_1$ and $\hat{u}_2$ do not obey the T–BMT equation along an orbit $(\vec{z}(s);s)$. But at spin–orbit resonance, they can be chosen so that a spin $\vec{S}$ is at rest in its local $(\hat{u}_1, \hat{u}_2, \hat{n})$ system. Then $\hat{u}_1(\vec{z};s)$ and $\hat{u}_2(\vec{z};s)$ do obey the T–BMT equation so that the ISF $\hat{n}(\vec{z};s)$ is not unique.

Nowadays we emphasise the utility of the ISF for defining equilibrium spin distributions. However, it was originally introduced for bringing the combined semiclassical Hamiltonian of spin–orbit motion into action–angle form for calculating the effects of synchrotron radiation [7]. The initial Hamiltonian is written as $H_{\mathrm{s-o}} = \frac{2\pi}{C}(Q_1 J_1 + Q_2 J_2 + Q_3 J_3) + \vec{\Omega} \cdot \vec{S}$. By viewing the spin motion in the $(\hat{u}_1, \hat{u}_2, \hat{n})$ systems, a new Hamiltonian in full action–angle form $H^{\mathrm{aa}}_{\mathrm{s-o}} = \frac{2\pi}{C}(Q'_1 J'_1 + Q'_2 J'_2 + Q'_3 J'_3) + \frac{2\pi}{C}\nu_s(\vec{J}')J_s$ is obtained which is valid at first order in $\hbar$ [5]. This emphasises again that, as with all action–angle formulations, the spin frequency cannot depend on orbital phases. Moreover, it is easy to show that at orbital resonance, (i.e. $k_0 + k_1 Q_1 + k_2 Q_2 + k_3 Q_3 = 0$ for suitable integers k) the "diagonalisation" of the Hamiltonian (i.e. finding the $\hat{u}_1, \hat{u}_2$) might not be possible [5]. Thus at orbital resonance the ADST may not exist. On the other hand, one avoids running a machine on such resonances. The spin tune on the closed orbit $\nu_0 = \nu_s(\vec{0})$ always exists and so does $\hat{n}_0(s) = \hat{n}(\vec{0};s)$.

For our present purposes there are two kinds of orbital resonances: resonances where at least one of the Q's is irrational and those where all are rational. We write the rational tunes as $Q_i = a_i/b_i$ $(i = 1,2,3)$ where the a_i and b_i are integers. Then for the second type, the orbit is periodic over c turns where c is the lowest common multiple of the b_i. In this case the ISF at each $(\vec{z},s)$ can be obtained as the unit length real eigenvector of the c–turn spin map (c.f. the calculation of $\hat{n}_0$ from the 1–turn spin map on the closed orbit.). However, the corresponding eigentune extracted from the complex eigenvalues and which we write as $c\nu_c$ depends in general on the orbital phases at the starting $\vec{z}$. Thus in general ν_c is *not* a spin tune and should not be so named [8]. Nevertheless if c is very large the dependence of ν_c on the phases can be very weak so that it can approximate well the ADST of nearby irrational tunes. This is expected heuristically since the influence of the starting phase can be diluted on forming the spin map for a large number of turns. At non–zero amplitudes, both for irrational or rational Q's the "fake spin tune" obtained as the eigentune of the 1–turn spin map has no physical significance. Of course, it normally depends on the orbital phases.

For non–resonant orbital tunes, the spin tune can be obtained using the SODOM–II algorithm [9] whereby spin motion is written in terms of two component spinors and SU(2) spin transfer matrices. The *functional equation* $\hat{n}(\vec{M}(\vec{z};s);s) = R_{3\times3}(\vec{z};s)\hat{n}(\vec{z};s)$ is then expressed in terms of a Fourier representation, w.r.t. the orbital phases, of the spinors and of the 1–turn SU(2) matrices. The spin tune appears as the set of eigentunes of an *eigen problem for Fourier components* and $\hat{n}$ is reconstructed from the Fourier eigenvectors. SODOM delivers the whole spin field on the torus $\vec{J}$ at the chosen s.

SNAKE "RESONANCES"

In perfectly aligned flat rings with no solenoids, $\hat{n}_0$ is vertical and $\nu_0 = a\gamma_0$ where γ_0 is the Lorentz factor on the closed orbit and a is the particle's gyromagnetic anomaly. In the absence of skew quadrupoles, the primary disturbance to spin is then from the radial magnetic fields along vertical betatron trajectories. The disturbance can be very strong and the polarisation can fall if the particles are accelerated through the condition $a\gamma_0 = \kappa \equiv k_0 \pm Q_2$ where mode 2 is vertical motion. This can be understood in terms of the "single resonance model" (SRM) whereby a rotating wave approximation is made in which the contribution to $\vec{\Omega}$ from the radial fields along the orbit is dominated by the Fourier harmonic at κ with strength $\varepsilon(J_2)$. The SRM can be solved exactly and the ISF is given by [10] $\hat{n}(\phi_2) = \mathrm{sgn}(\delta)\left(\delta\hat{e}_2 + \varepsilon(\hat{e}_1\cos\phi_2 + \hat{e}_3\sin\phi_2)\right)/\sqrt{\delta^2+\varepsilon^2}$ where $\delta = a\gamma_0 - \kappa$, ϕ_2 is the orbital phase, $(\hat{e}_1, \hat{e}_2, \hat{e}_3)$ are horizontal, vertical and longitudinal unit vectors and the convention $\hat{n}\cdot\hat{e}_2 \geq 0$ is used. The tilt of $\hat{n}$ away from the vertical $\hat{n}_0$ is $|\arcsin(\varepsilon/\sqrt{\delta^2+\varepsilon^2})|$ so that it is 90° at $\delta = 0$ for non–zero ε. At large $|\delta|$, the equilibrium polarisation directions $\hat{n}(J_2, \phi_2; s)$, are almost parallel to $\hat{n}_0(s)$ but during acceleration through $\delta = 0$, $\hat{n}$ varies strongly and the polarisation will change if the adiabatic invariance of J_s violated. The change in J_s for acceleration through $\delta = 0$ is given by the Froissart–Stora formula. The ADST which reduces to $a\gamma_0$ on the closed orbit is $\nu_s = \mathrm{sgn}(\delta)\sqrt{\delta^2+\varepsilon^2} + \kappa$. Note that the condition $\delta = 0$ is *not* the spin–orbit resonance condition. On the contrary, as δ passes through zero ν_s jumps by 2ε with our convention for $\hat{n}$ and avoids fulfilling the true resonance condition: for particles with non–zero ε, $a\gamma_0$ is just a parameter. In this simple model ν_s exists and is well defined near spin–orbit resonances for all Q_2. This is also true in more general cases if orbital resonance is avoided.

Polarisation loss while accelerating through $\delta = 0$ can be reduced by installing pairs of Siberian Snakes, devices which rotate spins by π independently of $\vec{z}$ around a "snake axis" in the machine plane. For example, one puts two snakes at diametrically opposite points on the ring. Then $\hat{n}_0 \cdot \hat{e}_2 = +1$ in one half ring and -1 in the other. With the snake axes relatively at 90°, $\nu_0 = 1/2$ for all γ_0. For calculations one often represents the snakes as elements of zero length ("pointlike snakes"). Then if, in addition, the effect of vertical betatron motion is described by the SRM, calculations with SODOM, perturbation theory [11] and the treatment in [8] suggest that $\nu_s(J_2) = 1/2$ too, independently of γ_0 but also of J_2. Thus for Q_2 sufficiently away from 1/2 no spin–orbit resonances $\nu_s(J_2) = k_0 \pm Q_2$ are crossed during acceleration through $\delta = 0$ and the polarisation can be preserved. This is confirmed by tracking calculations. However, such calculations and analytical work show that the polarisation can still be lost if the fractional part of Q_2, $[Q_2]$, is $\tilde{a}_2/2\tilde{b}_2$ where here, and later, $\tilde{a}_2$ and $\tilde{b}_2$ are odd integers with $\tilde{a}_2 < 2\tilde{b}_2$. This is the so called "snake resonance phenomenon" and it also has practical consequences [12, 13], especially for small $\tilde{b}_2$. Such a $[Q_2]$ fits the condition $1/2 = (1-\tilde{a}_2)/2 + \tilde{b}_2[Q_2]$. But calculations (see below) show that exactly at $[Q_2] = \tilde{a}_2/2\tilde{b}_2$ the ADST may not exist. If it doesn't, it isn't in the equivalence class for 1/2. Then we are not dealing with a conventional resonance $\nu_s(J_2) = (1-\tilde{a}_2)/2 + \tilde{b}_2[Q_2]$ and the term *resonance* is misleading! Depolarisation in this model has also been attributed to the fact that for non–zero J_2 the fake spin tune, which depends on ϕ_2, is 1/2 at some values of ϕ_2 [12]. However, the fake spin tune does not describe spin–orbit coherence. Snake "resonances" are usually associated with acceleration but it has been helpful in other circumstances [3, 4, 6] to begin by studying the *static* properties of the system, namely with the ISF. We now do that for the SRM with two snakes for representative, parameters.

Figure 1 shows $P_{\lim}$ (just before a snake) and ν_s for 25000 equally spaced $[Q_2]$'s between 0 and 0.5 for $\varepsilon = 0.4$ and $\delta = 0$. At each $[Q_2]$, $\hat{n}$ is calculated by stroboscopic averaging [1] ($\leq 25\ 10^6$ turns) at 500 equally spaced ϕ_2 in the range $0 - 2\pi$ and $P_{\lim}$ is obtained by averaging over these ϕ_2. The ADST is obtained from SODOM. If the ADST exists SODOM delivers a part of the equivalence class, namely the spectrum $[\pm 0.5 + l_2 Q_2]$ for a range of contiguous even l_2 restricted by the necessarily finite size of the matrix of Fourier coefficients. Only even l_2 are allowed by the algorithm. For irrational Q_2 the range of l_2 is large. For rational Q_2 the spectrum can include ± 0.5 but is otherwise highly degenerate or contains none or just a very few of the required members $[\pm 0.5 + l_2 Q_2]$. Thus the existence of an ADST is easily checked. The central horizontal row of points in figure 1 shows the common member $+0.5$ of the equivalence class of the ADST at the values of $[Q_2]$ where the ADST exists. There is an ADST for most $[Q_2]$'s used. The first row of dots up from the bottom marks $[Q_2]$ values where there is no ADST. As expected, these are all at rational $[Q_2]$'s such as 1/5, 1/4, 2/5 $\cdots$ or $\tilde{a}_2/2\tilde{b}_2 = 1/6, 3/14, 3/10 \cdots$ and the $c\nu_c$ computed for these $[Q_2]$ show ϕ_2 dependence. The curved line shows $P_{\lim}$ and the second row of dots from the bottom marks $[Q_2]$ values where the ISF from the stroboscopic average did not converge for all phases. These coincide with sharp dips in $P_{\lim}$ and are at or near $[Q_2] = \tilde{a}_2/2\tilde{b}_2$, i.e in the snake "resonance" subset of the $[Q_2]$'s in the first row. Thus snake "resonances" are already a static phenomenon. Near such $[Q_2]$'s, the ISF, which for just one orbital mode is a closed curve in three dimensions, becomes extremely complicated as $\hat{n}$ strives to satisfy its defining conditions. Right at $[Q_2] = \tilde{a}_2/2\tilde{b}_2$ the nonconvergence occurs at

$[\phi_2/2\pi] = j/2\tilde{b}_2$ for integers $j = 1,...,2\tilde{b}_2$ and, moreover, *the ISF is discontinuous at these phases*. For $[Q_2] = \tilde{a}_2/4\tilde{b}_2$, $P_{\rm lim}$ and the ISF show no special behaviour. These observations are consistent with the perturbative result [4] that for mid–plane symmetric systems, $\hat{n}$ should be well behaved near even m_2 but may show exotic behaviour close to odd $m_2 = \tilde{b}_2$. Some snake "resonances" such as that at $[Q_2] = 1/30$ are narrower than 0.00002 in $[Q_2]$ and are missed in this scan. $P_{\rm lim}$ also has several dips at values of $[Q_2]$ (e.g. at 0.341) which appear to have no special significance, but which should still be avoided at storage. The results for $0.5 \leq [Q_2] \leq 1.0$ are the reflection in 0.5 of the curves and points shown. Qualitatively similar results are obtained with equally distributed odd pairs of snakes set to give $\nu_0 = 1/2$. The ISF and $P_{\rm lim}$ usually vary significantly with s.

In summary: a snake "resonance" is at root a *static* phenomenon characterised by an invariant spin field which is irreducibly discontinuous in ϕ_2. Moreover, on and near snake "resonance", there is no amplitude dependent spin tune so that the snake "resonances" of this model are not simple spin–orbit resonances. The mechanism, in terms of J_s, for polarisation loss during acceleration through $\delta = 0$ at and near such $[Q_2]$'s is under study.

We thank K. Heinemann, G. H. Hoffstaetter and J.A. Ellison for useful discussions.

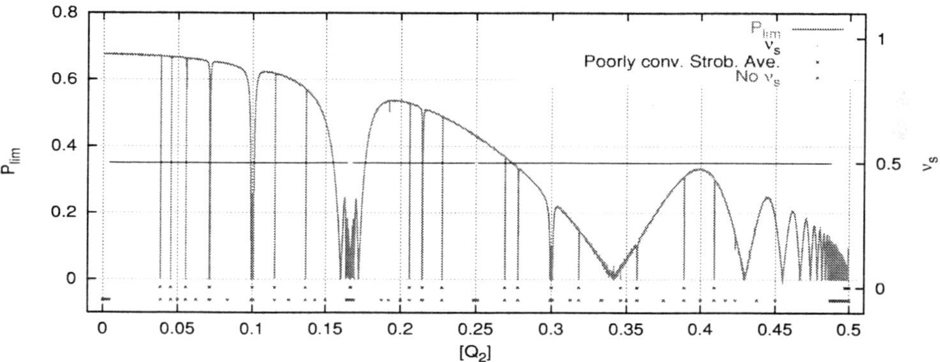

FIGURE 1. $P_{\rm lim}$ (left axis) and a component of the ADST (right axis) for the SRM with $\delta = 0$, $\varepsilon = 0.4$ and with 2 Siberian Snakes with axes at 90° and 0°.

REFERENCES

1. K. Heinemann and G.H. Hoffstaetter, Phys.Rev. **E 54**(4) 4240 (1996).
2. G.H. Hoffstaetter, M. Vogt and D.P. Barber, Phys. Rev. ST Accel. Beams **11**(2) 114001 (1999).
3. G.H. Hoffstaetter, *"A modern view of high energy polarised proton beams"*, Springer, in preparation.
4. M. Vogt, Ph.D. Thesis, University of Hamburg, DESY-THESIS-2000-054 (2000).
5. K. Yokoya, DESY report 86-57 (1986).
6. D.P. Barber, G.H. Hoffstaetter and M. Vogt, SPIN 2000, AIP proceedings 570 (2001).
7. D.P. Barber et al., in *"Quantum Aspects of Beam Physics"*, Monterey, 1998, World Scientific (1999).
8. S.R. Mane, Nucl. Instr. Meth. **A480**, p.328 and **A485**, p.277 (2002).
9. K. Yokoya, DESY report 99-006 (1999), Los Alamos archive: physics/9902068.
10. S.R. Mane, Fermilab technical report TM-1515 (1988).
11. K. Yokoya, SSC CDG report SSC-189 (1988).
12. S.Y. Lee, *"Spin Dynamics and Snakes in Synchrotrons"*, World Scientific (1997).
13. A.Luccio, V.Ptitsyn and V. Ranjbar, a talk on polarisation in RHIC at this Symposium.

The Analysis of Depolarization Factors in the Last RHIC Run

V. Ptitsyn*, A. U. Luccio* and V. H. Ranjbar†

*Brookhaven National Laboratory, Upton NY 11973
†Indiana University, Bloomington IN 47405

Abstract. Polarized proton beams were accelerated succesfully at RHIC up to 100Gev with the use of Siberian Snakes. Although the snakes were designed to preserve polarization, the succesful acceleration and storage of polarized beams was dependent also on beam characteristics, like closed orbit, betatron tunes and even betatron coupling. The high-order spin resonances were observed and evaluated. The paper summarizes depolarizing effects observed during the run.

INTRODUCTION

At the last run polarized protons were accelerated in both RHIC rings up to 100 GeV energy ($G\gamma = 192$) where they were used for colliding beam experiments [1]. The injection energy for polarized beams was at 24.3 Gev ($G\gamma = 46.5$). The polarization preservation during the beam acceleration from the injection to the top energy was provided by means of Siberian Snakes installed into the RHIC rings (two full snakes in each ring).

Although the Snakes succesfully accomplished their task, the polarization preservation was not always perfect. On some acceleration ramps the significant depolarization was observed. It was noted that the polarization preservation efficiency was sensitive to the beam conditions like closed orbit, betatron tune and betatron coupling.

Although, the Siberian Snakes help to avoid first-order imperfection and intrinsic spin resonances, higher-order spin resonances can occur at the so-called snake resonance conditions [2]:

$$\delta\nu_z = \frac{\nu_{sp} \pm k}{l}, \tag{1}$$

where $\delta\nu_z$ is fractional part of vertical betatron tune, and with the Snake setup at RHIC $\nu_{sp} = 1/2$.

Additional spin resonances might be caused by the betatron coupling effect. Coupling originates from quadrupole roll errors or solenoidal fields in the accelerator. This betatron coupling introduces the frequency of the horizontal betatron motion into vertical oscillations which leads to "coupled" snake resonances with conditions:

$$\delta\nu_x = \frac{\nu_{sp} \pm k}{l}, \tag{2}$$

Vertical closed orbit errors are responsible for the generation of even order snake resonances (with l even in (1) and (2)). It has been shown [2] that in the presence of

CP675, *Spin 2002: 15th Int'l. Spin Physics Symposium and Workshop on Polarized Electron Sources and Polarimeters,* edited by Y. I. Makdisi, A. U. Luccio, and W. W. MacKay

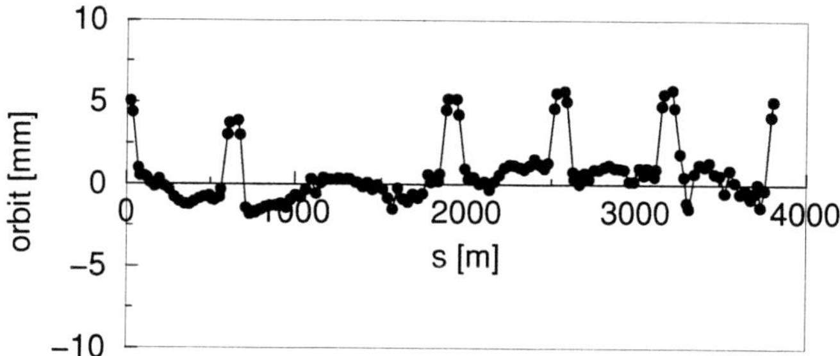

FIGURE 1. Ideal vertical orbit for polarized protons in Blue ring, as seen at Beam Position Monitors. The 5mm orbit bumps were introduced in interaction regions to prevent beam collisions during the acceleration. Their depolarization effect is small.

closed orbit errors each snake resonance will split into two, separated by

$$\delta \nu_z \leq |\frac{1}{\pi l} \arcsin[\sin^2 \frac{\pi \varepsilon_{imp}}{N_s}]|. \tag{3}$$

where ε_{imp} is the imperfection resonance strength.

MACHINE SETUP

The major factors determining the beam depolarization were closed orbit errors, betatron tune values and betatron coupling.

The depolarization effect of the vertical closed orbit errors is proportional to z'' which in fact describes radial fields on the beam orbit. The orbit should be perfectly flat, $z'' = 0$, in order to eliminate the imperfection resonances. Ideal orbit correction, using ideally aligned system of Beam Position Monitors (BPM), can realize this task. But in a real accelerator there are two kinds of errors that prevent a perfect orbit correction. One of them is BPM misalignments relative to quadrupole centers, which in case of the RHIC collider has rms value of about 150 μm. Another possible error is quadrupole vertical misalignments from some ideal plane, which are caused by ground motion as well as limited alignment precision. In the case of RHIC the quadrupole misalignment errors are dominant, with about 0.9mm rms value of quad misalignments over the ring azimuth. In order to compensate for the quadrupole alignment errors a special closed orbit through the BPMs was created based on quadrupole misalignment data. Fig. 1 shows the orbit used as ideal orbit for polarized beam acceleration. Although the orbit does not look flat as measured by BPMs, it should be flat, in reality, assuming that the quadrupole misalignment data are correct. The use of this ideal closed orbit decreases imperfection resonance strength and simplifies the task for the Siberian Snakes.

The precision of orbit correction was limited by triplet gradient errors, but typically was kept below 0.8mm as shown at the left plot in Fig. 2.

The fractional betatron tune space, in which RHIC was operated ranged from 0.20 to 0.25. The vertical tune was placed at 0.23, between snake resonances $3/14$ and $1/4$.

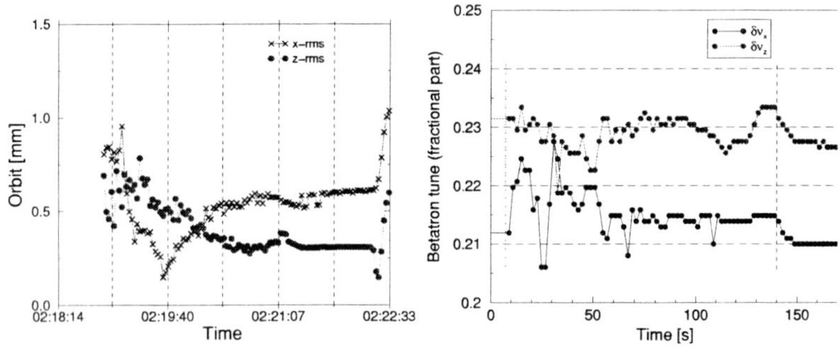

FIGURE 2. The examples for closed orbit and betatron tunes on the ramp. Left plot demonstrates the closed orbit horizontal and vertical rms value during the acceleration. Right plot shows horizontal and vertical betatron tunes during the acceleration.

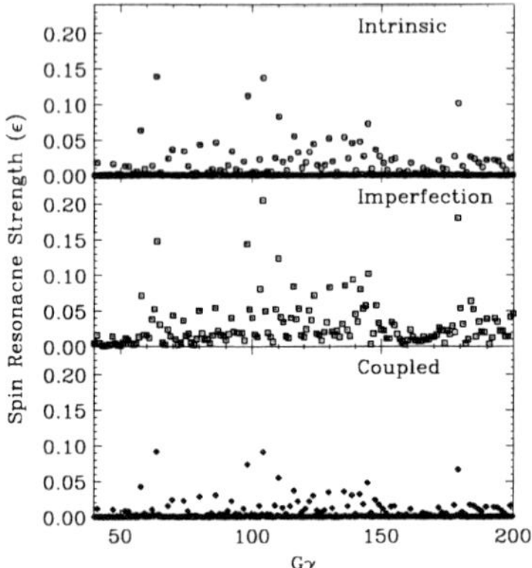

FIGURE 3. Calculated imperfection, intrinsic, and coupled spin resonance strength for RHIC $z_{rms} = 2.2$ mm, using vertical and horizontal emittance of 10π mm-mrad.

During the ramp the tunes did not stay constant. An example of tune excursion along the ramp is shown at the right plot in Fig. 2. In some regions, mainly at the first half of the ramp, the tunes approached too close together (≤ 0.008) so that the betatron coupling effect was clearly seen. Betatron coupling was well corrected at the injection and collision energies [3], but was not very well controlled during the acceleration.

The intrinsic, coupled and imperfection resonance strength is presented in Fig. 3 as calculated by new version of DEPOL code [4]. The maximum imperfection resonance strength $\varepsilon_{imp} < 0.2$ determines the maximum tune splitting using the Eq. (3). Thus the maximal incursion of $1/4$ snake resonances into the operating tune box might be as large as 0.242.

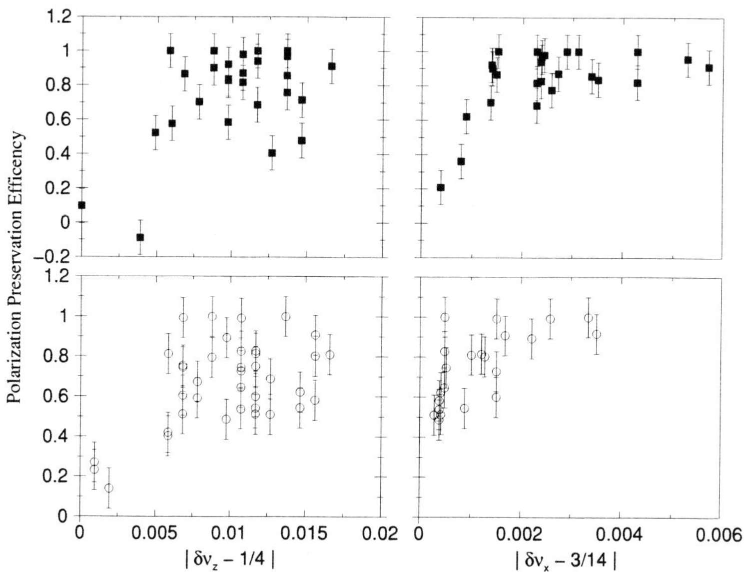

FIGURE 4. $|\delta v_z - 1/4|$ and $|\delta v_x - 3/14|$ versus polarization preservation efficiency (P_f/P_i) shown for the Yellow (square symbol) and Blue (open circles) rings. In the case of $|\delta v_z - 1/4|$, polarization preservation efficiency was scattered between 1.0 and 0.4. This is because other parameters fluctuated each fill.

OBSERVED RESONANCES

The depolarization caused by two high-order spin resonances, $\delta v_x = 3/14$ and $\delta v_z = 1/4$, was observed during the run and clearly identified from the analysis of the polarization transfer efficiency data. The results of the analysis are shown in Fig. 4.

The right graphs of Fig. 4 show the polarization preservation efficiency for the $3/14$ coupled snake resonance versus the resonance proximity parameter, $|\delta v_x - 3/14|$ for both Blue and Yellow rings. The observed resonance width is defined by the value of the betatron coupling and by resonance splitting effect described by Eq. (3). A larger resonance width was observed in Yellow ring which is consistent with the fact that the Yellow ring had usually larger values of both betatron coupling and closed orbit errors.

On several occasions the vertical tune did approach $1/4$ which allowed us to see the depolarization from the $1/4$ snake resonance. The left graphs of Fig. 4 clearly show the onset of the $1/4$ snake resonance. The width of the resonance is defined by the closed orbit errors. From these graphs the estimated imperfection resonance strength is about 0.16 for both rings. This agrees well with DEPOL calculations.

The depolarization from coupled snake resonance $\delta v_x = 3/14$ was confirmed by spin tracking studies. The tracking results in the Blue ring using the program SPINK [5] with betatron coupling effect included show clearly the depolarization as the horizontal tune crosses the $3/14$ snake resonance location in Fig. 5.

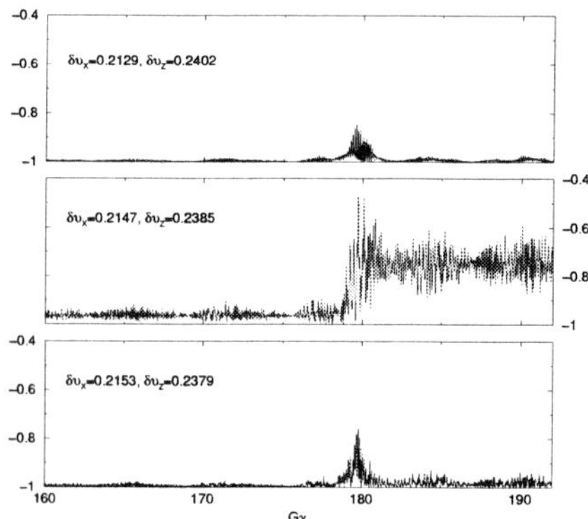

FIGURE 5. Spin Tracking results for strongly coupled Blue ring with an emittance of 25 mm-mrad and $z_{rms} = 0.6$ mm. The three graphs show polarization versus $G\gamma$ with the fractional part of the horizontal betatron tunes near $3/14$.

CONCLUSION

The depolarization effects from high-order spin resonances were observed during the polarized proton run at RHIC. The resonance, $\delta v_x = 3/14$, was caused by the betatron coupling. Another resonance, $\delta v_z = 1/4$, was due to the closed orbit errors.

For the next polarized proton run several improvements should help to avoid the depolarization. For better tune control the PLL feedback system will be applied. The new quadrupole misalignment data should provide a more accurate ideal orbit. With the elimination of gradient errors in the IR triplets, the quality of the closed orbit correction is expected to improve below 0.5 mm rms level. The techniques to measure and correct the betatron coupling on the acceleration ramp are under discussions.

ACKNOWLEDGEMENTS

The authors would like to thank H. Huang, S.Y. Lee, W.W. MacKay, T. Roser, and S. Tepikian for discussions of various aspects of depolarization analysis. We are also grateful to the RHIC polarimetry team.

REFERENCES

1. H.Huang, "Acceleration of Polarized Protons at RHIC", *these proceedings.*
2. S. Y. Lee and S. Tepikian, Phys. Rev. Lett. **56**, 1635 (1986); S. Tepikian, Ph. D. thesis, State University of New York at Stony Brook, (1986) (unpublished).
3. F. Pilat, et. al, Proc. of EPAC2002, Paris, p. 1178.
4. E. D. Courant and R.D. Ruth, BNL 51270 (1980), (unpublished).; V. Ranjbar et. al., Proc. of EPAC2002, Paris, p. 359.
5. A.U.Luccio *Trends in Collider Spin Physics* World Scientific, p. 235 (1997).

Spin Coupling Resonance and Suppression in the AGS

V. H. Ranjbar, S. Y. Lee* and L. Ahrens, M. Bai, K. Brown, W. Glenn, H. Huang, A. U. Luccio, W. W. MacKay, V. Ptitsyn, T. Roser, N. Tsoupas†

*Indiana University Bloomington IN 47405
†BNL, Upton NY 11973

Abstract. Spin depolarizing resonances due to coupling may account for as much as a 30 percent loss in polarization in the AGS. The major source of coupling in the AGS is the solenoidal snake. In the past some preliminary work was done to understand this phenomena [1], and a method to overcome these resonances was attempted [2]. However, in the polarized proton run of 2002, the response of these coupled spin resonances to the strength of the solenoidal snake, skew quadrupoles and vertical and horizontal betatron tune separation was studied to provided a benchmark for a modified DEPOL program [3]. Then using the new DEPOL program, a method method to cure the coupled spin resonances in the AGS via spin matching rather than global or local decoupling was explored.

UPDATE ON MODIFICATION TO DEPOL PROGRAM

In previous papers [10] we reported on the modifications to the well established DEPOL code [3] to include the effects of coupling. We present now some additional modifications which have significantly improved the speed of this code. The central algorithm presented in [10] is created to evaluate the following Fourier integral,

$$\varepsilon_{K_m} = -\frac{1}{2\pi}\int_{s_1}^{s_2}\left[(1+K)(z''+\frac{iz'}{\rho})-i(1+G)(\frac{z}{\rho})'\right]e^{iK\theta(s)}ds. \tag{1}$$

Here ε_K is the spin resonance amplitude and K is the spin resonance tune. The solution, following the original DEPOL paper, was to break up the integral into a sum over all the lattice elements denoted with subscript m. The final closed solution for each element is:

$$\varepsilon_m = \frac{1}{2\pi}\left[\frac{(1+K)(\xi_1+i)}{\rho}z_1e^{iK\theta_1}+\frac{(1+K)(\xi_2-i)}{\rho}z_2e^{iK\theta_2}\right.$$
$$-(1+K)\left((z_2'-\frac{iK}{\rho}z_2)e^{iK\theta_2}-(z_1'-\frac{iK}{\rho}z_1)e^{iK\theta_1}\right)+\left(\frac{K(K^2+G)}{\rho^2}\right)$$

[1] Work performed under the auspices of the US Department of Energy

CP675, *Spin 2002: 15th Int'l. Spin Physics Symposium and Workshop on Polarized Electron Sources and Polarimeters*, edited by Y. I. Makdisi, A. U. Luccio, and W. W. MacKay

$$\times \left[\frac{1}{\sqrt{1+|r_e|}} \left((\frac{iK}{\rho} r_{e_{1,2}} - r_{e_{1,1}}) \left(\frac{(a_2' - \frac{iK}{\rho} a_2) e^{iK\theta_2} - (a_1' - \frac{iK}{\rho} a_1) e^{iK\theta_1}}{K_a - K^2/\rho^2} \right) \right. \right.$$

$$\left. \left. - \left(\frac{(b_2' - \frac{iK}{\rho} b_2) e^{iK\theta_2} - (b_1' - \frac{iK}{\rho} b_1) e^{iK\theta_1}}{K_b - K^2/\rho^2} \right) + r_{e_{1,2}} (a_2 e^{iK\theta_2} - a_1 e^{iK\theta_1}) \right) \right] . \quad (2)$$

Here r_e is the rotation matrix which transforms from the x, x', z, z' coupled basis to the a, a', b, b' locally uncoupled basis (uncoupling each lattice element only). Since for intrinsic resonances K is not an integer Eq. ?? becomes an integral around the lattice an infinite number of times. Previously, a solution was derived by evaluating an appropriately large number of passes over the lattice.

However if we look closely at the behavior of the elements which make up the integral to be evaluated in Eq. 2 it appears that we can factor out the phase element which changes with each period around the lattice. The remaining elements in the sum remain constant for each pass. The factored phase elements can be evaluated analytically using the properties of a geometric series. The results are four separate enhancement functions,

$$\begin{aligned} E_u(N)_{\pm} &= \sum_{n=0}^{N} e^{i2\pi n(K \pm \nu_u)} = \pm e^{iN\pi(K \pm \nu_u)} \frac{\sin(\pi(N+1)(K \pm \nu_u))}{\sin(\pi(K \pm \nu_u))} \\ E_v(N)_{\pm} &= \sum_{n=0}^{N} e^{i2\pi n(K \pm \nu_v)} = \pm e^{iN\pi(K \pm \nu_v)} \frac{\sin(\pi(N+1)(K \pm \nu_v))}{\sin(\pi(K \pm \nu_v))} \end{aligned} \quad (3)$$

Here ν_u and ν_v the betatron tunes in the uncoupled u and v basis. N the number of passes around the lattice. The function once evaluated can then be multiplied by the appropriate terms in the sum over one pass in the lattice.

STUDY OF COUPLING SPIN RESONANCES

The primary source of coupling in the AGS is the partial solenoidal snake. In addition there exists a family of six skew quadrupoles. It has been observed that the bare AGS machine has a net skew quadrupole moment. Coupling studies in the past estimated the average roll to be 0.13 mrad [9]. Additionally closed orbit errors can contribute to coupling via feed down from the sextupole fields present in the AGS combined function magnets and sextupole magnets.

During the 2002 polarized proton run, particular attention was paid to studying the behavior of the $0+\nu_x$ resonance crossing since the analyzing power of the AGS polarimeter was sufficiently large at low energy to generate accurate measurements and the strength of the $0+\nu_x$ coupled spin resonance was large. Initial DEPOL calculations without rolls generated curves which were too broad. It was only by including either a large single roll or selectively placed rolls that a good fit to the measured data was achieved. For all DEPOL calculations shown here selectively distributed rolls were applied to the CF magnets [2] (0.05 mrad per magnet) and applied to the BD magnets [3] (0.25 mrad per magnet). This is not unreasonable considering previous estimates. However it should be emphasized that this configuration is by no means unique. While it was essential to include a net skew quadrupole moment in the bare AGS the distribution and the direction of these rolls is still unclear since our data could fit many different configurations.

In Figs. 1 - 3 one can see the results of our tune scans, snake scans and skew quadrupole scans. All calculations assume a 70% polarization at injection into the AGS. In all cases we were able to achieve a good agreement between DEPOL polarization and measured.

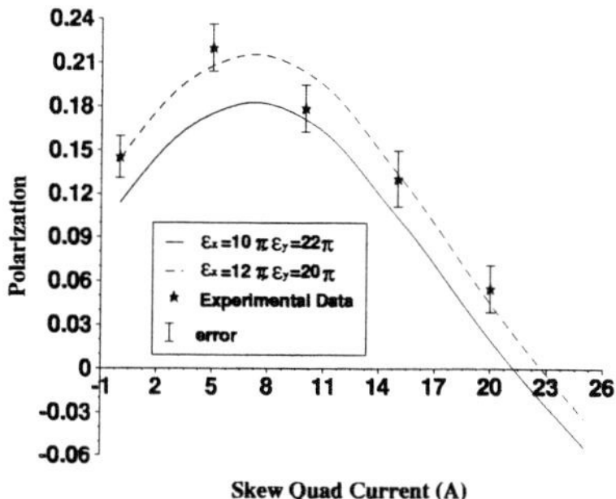

FIGURE 1. Polarization after crossing the $0+\nu_x$ and $0+\nu_z$ resonances with set vertical tune and horizontal tunes ($\nu_z = 8.8$, $\nu_x = 8.78$). Scanning through skew quadrupole input currents from 0 to 25 A. The vertical and horizontal emittances were measured as $(11 \pm 1)\pi$ and $(21 \pm 1)\pi$ mm-mrad.

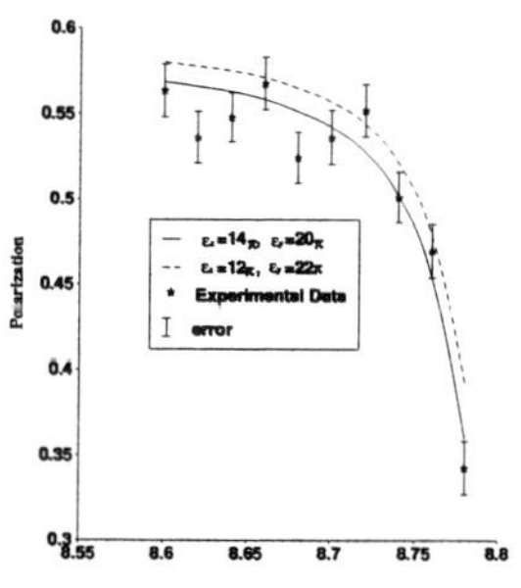

FIGURE 2. Polarization after crossing the $0+\nu_x$ and $0+\nu_z$ resonances with set vertical tune ($\nu_z =$ 8.8) scanning horizontal set tunes. Vertical and horizontal emittances were measured as $(13 \pm 1)\pi$ and $(21 \pm 1)\pi$ mm-mmrad respectively.

---

[2] CF is the label for a family of combined function focusing magnets located at 13,14,17 and 18 positions in each super-period.

[3] BD is the label for a family of combined function focusing magnets located at 11,12,19 and 20 positions in each super-period.

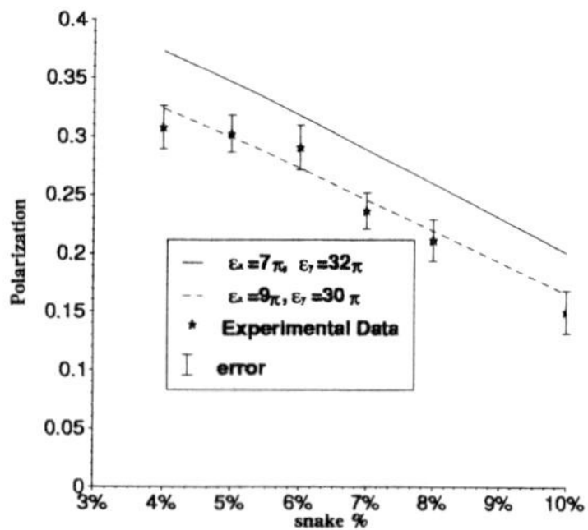

FIGURE 3. Polarization after crossing the $0+\nu_x$ and $0+\nu_z$ resonances with fixed vertical tune and horizontal tune ($\nu_z = 8.8$, $\nu_x = 8.7$) scanning from 4 to 10% partial snake strength. Vertical and horizontal emittances were measured to be $(8 \pm 1)\pi$ and $(30 \pm 1)\pi$ mm-mrad.

SUPPRESSION OF THE COUPLED SPIN RESONANCES

In the AGS, six skew quadrupoles are located in the 17th location at every other super period. To globally decouple the AGS, ideally one should pick a location with as large a phase difference from the existing skew quadrupoles. Unfortunately, we are limited in the number of free locations. The 15th location which has been suggested for the future normal quadrupole could also accommodate a skew quadrupole.

Unfortunately, the field strength required to approach a situation of global decoupling cause a large tune shifts in the AGS which makes identifying the proper strengths necessary to decouple the machine difficult if not impossible. However, since we are concerned with eliminating the coupled spin resonances and not necessarily decoupling the AGS, a spin matching condition may still exist. Using spin matching it may be possible to either partly or totally cancel the coupled spin resonance with the perturbation introduced by the skew quadrupoles. In the plots in Fig. 4 such a spin matching condition appears achievable. For these figures we have fixed our vertical and horizontal tunes ($\nu_z = 8.8, \nu_x = 8.7$) and scanned through various current strengths for the 15th and 17th skew quadrupoles. For these calculations the 15th skew quadrupoles were assumed to have the same size and current to field transfer function as the existing skew quadrupoles in the 17th lattice postion. For all four resonances a solution appears possible. However, overcoming the $36+\nu_x$ requires a current in excess of 1200 A. Since this high current needs to be maintained only during the brief milliseconds of the resonance crossing it should be possible to achieve.

Actually since the calculations where all done using the slower acceleration rate of $\alpha = 2.4 \times 10^{-5}$ generated by the backup Westinghouse power generator and not the usual $\alpha = 4.8 \times 10^{-5}$ which is normally achieved by the Siemens power generator, depolarization could effectively be overcome with a stronger residual resonance.

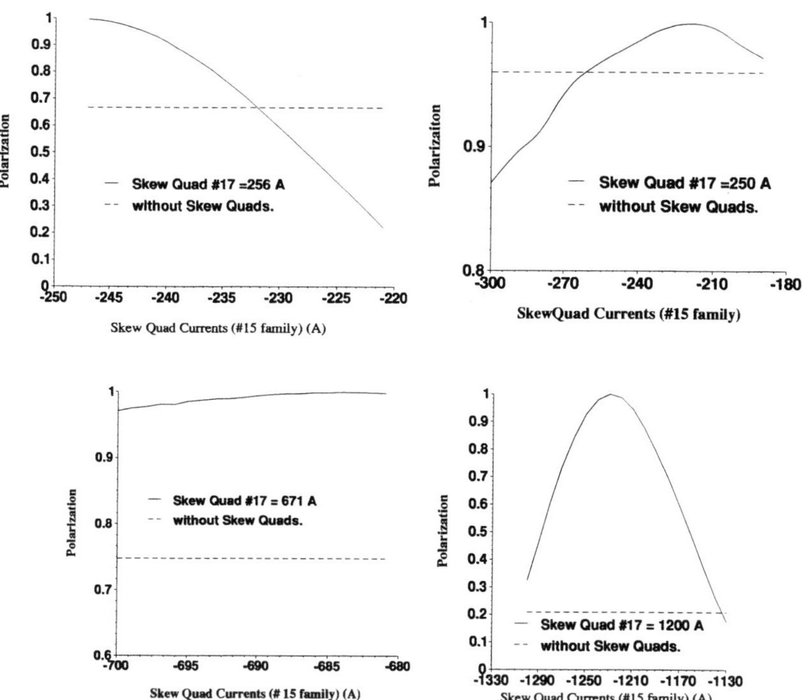

FIGURE 4. Polarization after crossing (clockwise) $0+\nu_x$,$12+\nu_x$, $36-\nu_x$, and $36+\nu_x$ resonance with fixed vertical tune and horizontal tune ($\nu_z = 8.8$, $\nu_x = 8.7$). Scanning currents for a hypothetical skew quadrupole in the 15th lattice position and the 17th skew quadrupole family at fixed.

REFERENCES

1. H. Huang, Ph. D Thesis, Indiana University (1995)
2. M. Bai and T. Roser. C-A/AP/37 (2001).
3. E. D. Courant and R. D. Ruth, BNL Report 51270 (1980).
4. H.Huang et al., Phys. Rev. Lett. **73**, 2982 (1994).
5. M.Bai et al, Phys. Rev **E56**, 6002 (1997).
6. L. C. Teng, FN 229, FNAL Report, (1971).
7. H.Grote and F.C.Iselin, *Methodical Accelerator Design Program Version 8.23*, CERN/SL/90-13(AP) (1990).
8. A. Zelenski, et al., Proceedings of the 9th International Conference on Ion Sources, Rev. Sci.Inst., Vol. **73**, No.2, p.888 (2002).
9. C.J.Gardner, et. al, AGS Studies Report, N224, (1987).
10. V.Ranjbar et al., Mapping out the full spin resonance structure of RHIC, PAC 2001 Proc.
11. J. Y. Lin, Phys. Rev. E **49**, 2347 (1994).

Exact Solutions for the n-axis and Spin Tune in Model Storage Rings

S. R. Mane[1]

Convergent Computing Inc., P.O. Box 561, Shoreham, NY 11786

Abstract. We present a new nonperturbative formalism MILES to calculate the n-axis in storage rings. We employ MILES to obtain the exact solution for the n-axis in several model storage rings. In particular, we display the exact analytical solution for the single resonance model with one Snake. Our solution depends on new types of mathematical function, which we call "sine-factorial" and "sine-Bessel" functions. We confirm the spectrum of Snake resonances found by Lee and Tepikian. Also, under suitable circumstances, we show that the spin tune depends explicitly on the orbital phase. We term such trajectories "exceptional orbits."

INTRODUCTION

We present a new, nonperturbative formalism "MILES" to calculate the $\hat{n}$ axis in storage rings. The name is simply an anagram of the author's old algorithm and program SMILE [1]. We employ MILES to obtain the exact analytical solution for the $\hat{n}$ axis for several storage ring models. The models contain Siberian Snakes. In particular, we solve the model of a planar ring with a single resonance driving term (the "single resonance model") and one Snake (also two, four, etc. Snakes). This model is in some sense the "classic" problem in the field. We also solve the same model using Yokoya's formalism SODOM2 [2]. Our solution contains new types of mathematical functions, which we call "sine-factorial" and "sine-Bessel" functions. We confirm a longstanding conjecture by Yokoya [3] that the spin tune is 1/2 even off-axis for a single resonance model with multiple Snakes. We verify the spectrum of the Snake resonances found by Lee and Tepikian [4]. We show they are higher order depolarizing resonances, following from the standard spin resonance condition. Finally, we show that under suitable circumstances the spin tune can depend explicitly on the orbital phase. We term such a trajectories "exceptional orbits."

MILES ALGORITHM

For brevity, denote a point in the orbital phase space by z. By definition, the $\hat{n}$ axis transforms as a vector field over the orbital phase space, i.e.

[1] Email: srmane@optonline.net

CP675, *Spin 2002: 15th Int'l. Spin Physics Symposium and Workshop on Polarized Electron Sources and Polarimeters,* edited by Y. I. Makdisi, A. U. Luccio, and W. W. MacKay

$$\vec{\sigma}\cdot\hat{n}(z_f) = M\,\vec{\sigma}\cdot\hat{n}(z_i)M^{-1}. \qquad (1)$$

Here M is the spin-orbit map from the initial to the final azimuth. It is simplest to choose M to be the one-turn map. Then we can write, at a fixed azimuth θ_*,

$$\vec{\sigma}\cdot\hat{n}(\vec{\phi}_* + \vec{\mu}) = M\,\vec{\sigma}\cdot\hat{n}(\vec{\phi}_*)M^{-1}. \qquad (2)$$

We denote the orbital action-angle variables by $(\vec{I},\vec{\phi})$ and the orbital tunes by Q_j (j=1,2,3). The one-turn orbital phase advances are $\mu_j = 2\pi Q_j$. We parameterize the map M via

$$M = \begin{pmatrix} f & -g^* \\ g & f^* \end{pmatrix}. \qquad (3)$$

We employ a basis of right-handed orthonormal vectors $(\hat{e}_1,\hat{e}_2,\hat{e}_3)$, which are radial, longitudinal and vertical, respectively. The components of $\hat{n}$ in this basis are (n_1,n_2,n_3) and we define $n_\pm = n_1 \pm in_2$ (so $n_- = n_+^*$). This yields the equations

$$\begin{aligned} n_3(\vec{\phi}_* + \vec{\mu}) &= (ff^* - gg^*)n_3(\vec{\phi}_*) - f^*g^*n_+(\vec{\phi}_*) - fgn_-(\vec{\phi}_*) \\ n_+(\vec{\phi}_* + \vec{\mu}) &= 2f^*gn_3(\vec{\phi}_*) + f^{*2}n_+(\vec{\phi}_*) - g^2n_-(\vec{\phi}_*). \end{aligned} \qquad (4)$$

We now expand in Fourier harmonics

$$\begin{aligned} n_3(\vec{\phi}_*) &= \sum_{\vec{m}} n_{3\vec{m}}\, e^{i\vec{m}\cdot\vec{\phi}_*} \\ n_+(\vec{\phi}_*) &= \sum_{\vec{m}} n_{+\vec{m}}\, e^{i\vec{m}\cdot\vec{\phi}_*}. \end{aligned} \qquad (5)$$

We must also expand f and g in Fourier harmonics. We then solve for the Fourier coefficients $n_{3\vec{m}}$ and $n_{+\vec{m}}$. This is the MILES algorithm. We obtain the solution at other azimuths by tracking.

We have employed MILES to obtain analytical solutions for the following models. In all cases the resonance driving term contains only one Fourier harmonic.

- Single resonance model (SRM)
- SRM with partial Type 3 Snake
- Planar ring with *vertical* resonance driving term and one or two (or more) Snakes
- SRM with one Snake
- SRM with two or more Snakes

We shall present the explicit solution for the SRM with one Snake below.

SRM WITH ONE SNAKE

We consider a model of a planar ring with a single resonance driving term, of strength ε, in the horizontal plane and a single Snake. The Snake is located at $\theta_s = 0$ and its axis points at an angle ξ relative to $\hat{e}_1$. The spin precession vector is

$$\vec{W} = \nu_0\hat{e}_3 + \varepsilon(\cos\phi\,\hat{e}_1 + \sin\phi\,\hat{e}_2) + \pi\delta_p(\theta)(\cos\xi\,\hat{e}_1 + \sin\xi\,\hat{e}_2). \qquad (6)$$

Here ν_0 is a constant and $\delta_p(\theta)$ is the periodic delta function:

$$\delta_p(\theta) = \sum_{j=-\infty}^{\infty} \delta(\theta - 2\pi j). \tag{7}$$

We define $\Omega = \sqrt{(\nu_0 - Q)^2 + \varepsilon^2}$ and $\eta = (\varepsilon / \Omega)\sin(\pi\Omega)$, so $-1 \le \eta \le 1$. We place the origin just *after* the Snake. Then the one-turn map is

$$M = \begin{pmatrix} -\eta e^{-i(\phi_* - \xi + \mu/2)} & -i\sqrt{1-\eta^2}\, e^{-i(\kappa + \xi + \mu/2)} \\ -i\sqrt{1-\eta^2}\, e^{i(\kappa + \xi + \mu/2)} & -\eta e^{i(\phi_* - \xi + \mu/2)} \end{pmatrix}. \tag{8}$$

where

$$\cos(\pi\Omega) \pm i\frac{\nu_0 - Q}{\Omega}\sin(\pi\Omega) = \sqrt{1-\eta^2}\, e^{\pm i\kappa} \tag{9}$$

We also set

$$\begin{aligned} n_3 &= (1-\eta^2)a \\ n_+ &= igb \end{aligned} \tag{10}$$

We expand in Fourier harmonics

$$\begin{aligned} a(\phi_*) &= 2\sum_{m=odd} a_m \sin(m(\phi_* - \xi)) \\ b(\phi_*) &= b_0 + 2\sum_{m=even} b_m \cos(m(\phi_* - \xi)) \end{aligned} \tag{11}$$

Notice that the expression for *a* contains only odd harmonics, all sines, and *b* contains only even harmonics, all cosines. This is a nice pattern. To solve for the Fourier coefficients, we define $\delta = Q - 1/2$ and introduce "sine-factorial" functions

$$\begin{aligned} S_m(\delta) &= \sin(\pi\delta)\sin(2\pi\delta)\cdots\sin(m\pi\delta) \\ C_m(\delta) &= \cos(\pi\delta)\cos(2\pi\delta)\cdots\cos(m\pi\delta) \end{aligned} \tag{12}$$

We also adopt the convention $S_0(\delta) = C_0(\delta) = 1$. Then the solution is

$$\begin{aligned} a_m &= \frac{1}{\cos(m\pi\delta/2)} \sum_{k=0}^{\infty} \frac{C^2_{m/2+k}(2\delta)}{S_k(2\delta) S_{m+k}(2\delta)} (-1)^k \eta^{m+2k} \\ b_m &= \sum_{k=0}^{\infty} \frac{C^2_{(m-1)/2+k}(2\delta)}{S_k(2\delta) S_{m+k}(2\delta)} (-1)^k (\eta e^{i\pi\delta})^{m+2k} \end{aligned} \tag{13}$$

These solutions bear a close resemblance to the power series expansion of a Bessel function

$$J_m = \sum_{k=0}^{\infty} \frac{1}{k!(m+k)!} (-1)^k \left(\frac{\eta}{2}\right)^{m+2k} \tag{14}$$

We also display the solution using the SODOM2 [2] algorithm. This algorithm actually solves for a spinor. We write $\hat{n} = \Psi^\dagger \vec{\sigma} \Psi$ and parameterize

$$\Psi = \begin{pmatrix} A \\ igB \end{pmatrix} \tag{15}$$

We set

$$A(\phi_*) = A_0 + 2\sum_{m=even} A_m \cos(m(\phi_* - \xi)) + 2\sum_{m=odd} A_m \sin(m(\phi_* - \xi))$$
$$B(\phi_*) = B_0 + 2\sum_{m=even} B_m \cos(m(\phi_* - \xi)) - 2\sum_{m=odd} B_m \sin(m(\phi_* - \xi)) \quad (16)$$

The minus sign in the expression for B is significant. It is not a misprint. The answer is

$$A_m = \sum_{k=0}^{\infty} \frac{e^{ik(m+k)\pi\delta}}{S_k(\delta)S_{m+k}(\delta)}(-1)^k \left(\frac{\eta}{2}e^{-i\pi\delta}\right)^{m+2k}$$
$$B_m = \sum_{k=0}^{\infty} \frac{e^{ik(m+k)\pi\delta}}{S_k(\delta)S_{m+k}(\delta)}(-1)^k \left(\frac{\eta}{2}e^{i\pi\delta}\right)^{m+2k} \quad (17)$$

These solutions have a nicer symmetry than the solutions for the vector $\hat{n}$. We call the above functions "sine-Bessel" functions. However, the elegant pattern of the Fourier harmonics in $\hat{n}$ is not so obvious from the spinor solution. Along with sine-factorials, the sine-Bessel functions are *new* mathematical functions. They do not appear to be in the mathematical literature.

The SODOM2 formalism also calculates the spin tune. The solution is $\nu = 1/2$, not only on the closed orbit, but also off-axis, for *all nonresonant orbits* with $|\eta| < 1$. We also find that $\nu = 1/2$ off-axis for a model with 2 or more Snakes (provided $|\eta| < 1$). This confirms Yokoya's [3] conjecture, that $\nu = 1/2$ off-axis for nonresonant orbits in a planar ring with multiple Snakes and a single resonance driving term.

SNAKE RESONANCES

Depolarizing spin resonances occur whenever there are zero denominators in the solution for $\hat{n}$. For a ring with *two* Snakes, such zeroes occur whenever the orbital tune has the form

$$Q = \frac{2k-1}{2(2m-1)}. \quad (18)$$

Here m and k are integers. This is precisely the spectrum of the so-called "Snake resonances" found by Lee and Tepikian [4]. We therefore verify that they obtained the correct resonance spectrum. We also see that Snake resonances are higher-order spin resonances and follow from the standard formula

$$\nu + m'Q = k. \quad (19)$$

We substitute $\nu = 1/2$ (which we have shown is the value off-axis) to obtain

$$Q = \frac{2k-1}{2m'}. \quad (20)$$

A more detailed analysis of the Fourier structure of the one-turn map is required to show that m' must be an odd integer $m' = 2m - 1$, which completes the proof.

EXCEPTIONAL ORBITS

The remaining case to consider is $|\eta|=1$. For $\eta=-1$, the one-turn map is

$$M=\begin{pmatrix} e^{-i(\phi_*-\xi+\mu/2)} & 0 \\ 0 & e^{i(\phi_*-\xi+\mu/2)} \end{pmatrix}. \tag{21}$$

The solution is clearly $\hat{n}=\hat{e}_3$. However the spin tune is not 1/2 but is

$$\nu=\frac{\phi_*-\xi+\mu/2}{\pi}. \tag{22}$$

The spin tune depends explicitly on the orbital phase. We call such a trajectory an "exceptional orbit." On an exceptional orbit, the Stern-Gerlach force *cannot* be neglected. It leads to a secular growth in the amplitudes of the orbital motion. This invalidates the semiclassical approximation normally employed in accelerator physics.

CONCLUSION

We have presented a new nonperturbative formalism "MILES" to calculate the $\hat{n}$ axis in storage rings. We employed it to solve several models analytically. We displayed the analytical solution for a planar ring with a single resonance driving term and one Snake. The solution contained some new mathematical functions, which we call "sine-factorial" and "sine-Bessel" functions. These functions do not appear to be in the mathematical literature. We showed that the spin tune is 1/2 even off-axis, and we also derived the spectrum of the Snake resonances. Finally, we showed that under some circumstances the spin tune depends explicitly on the orbital phase. We termed such trajectories "exceptional orbits." The semiclassical approximation of accelerator dynamics breaks down on such orbits.

REFERENCES

1. S. R. Mane, *Phys. Rev. A* **36**, 120 (1987).
2. K. Yokoya, DESY 99-006 (1999), also at LANL E-Print Physics/9902068 (1999).
3. K. Yokoya, SSC-189 (1988).
4. S. Y. Lee and S. Tepikian, *Phys. Rev. Letters* **56**, 1635 (1986).

Spin-Orbital Function Formalism and ASPIRRIN Code

E.A.Perevedentsev*, V.Ptitsyn† and Yu.M.Shatunov*

*Budker Institute for Nuclear Physics, Novosibirsk, Russia
†C-A Department, Brookhaven National Laboratory, Upton, NY, USA

Abstract. A spin-orbital function formalism is described. The formalism was realized in ASPIRRIN code that does beam polarization calculations at the first order. The code has been used for calculating equlibrium polarization and polarization time in electron rings with the complex geometry of applied magnetic fields as well for resonance strength analysis for proton accelerators.

INTRODUCTION

The ASPIRRIN (*Analysis of SPIn Resonances in RINgs*) is a code designed to calculate beam polarization related quantities for a charged particle beam circulating into an accelerator or storage ring. The code was written some years ago and have been used for polarization calculations for different accelerators. The paper desribes the underlying formalism used by the code, called the spin-orbital function formalism.

The code, and underlying formalism, involves polarization calculation at the first order, thus revealing effect of first-order spin resonances. Therefore results produced by the code have to agree with results obtained by other first order codes, like SLIM and SLICK [1]. The various calculations done for different accelerator lattices have demonstrated a good agreement between the ASPIRRIN and SLIM codes.

SPIN-ORBITAL FUNCTIONS

The ASPIRRIN algorithm is based on the calculation of a special set of functions, named spin-orbital functions. These functions are complex functions and connect the motion of individual particle in a circular accelerator with the motion of particle spin.

For description of the transverse and longitudinal orbital motion one can use the set of canonical variables $(x, p_x, z, p_z, \sigma, p_\sigma)$, which forms the orbital vector X. In linear approximation, the orbital 6-D dynamic is governed by the equation:

$$X' = SHX + Q \qquad (1)$$

where H is symmetrical hamiltonian matrix and S is the fundamental symplectic matrix. The vector Q includes effects from bending field errors.

The description of spin motion in circular accelerators are based on two important quantities , the periodical spin field $\hat{\mathbf{n}}$ and the spin tune ν_s.

CP675, *Spin 2002: 15th Int'l. Spin Physics Symposium and Workshop on Polarized Electron Sources and Polarimeters*, edited by Y. I. Makdisi, A. U. Luccio, and W. W. MacKay

The periodicity of the $\hat{\mathbf{n}}$ can be expressed as:

$$\hat{\mathbf{n}}(x, p_x, z, p_z, \sigma, p_\sigma, \theta + 2\pi) = \hat{\mathbf{n}}(x, p_x, z, p_z, \sigma, p_\sigma, \theta) \quad (2)$$

which means that at the given azimuth θ the $\hat{\mathbf{n}}$ is function of orbital phase space point. The large variations of $\hat{\mathbf{n}}$ over orbital motion phase space in spin resonance zones lead to possible beam depolarization during the resonance crossing. The spin-orbital function formalism aims at the spin field $\hat{\mathbf{n}}$ calculation, taking first-order approximation both in orbital and spin motion. Having it calculated, the analysis can be done for depolarization effects.

The suitable reference system to do the spin calculations is the system formed by $\hat{\mathbf{n}}_0$,η_1, η_2 unit vectors [2]. $\hat{\mathbf{n}}_0$ is the periodical spin solution on the reference beam orbit in an accelerator (without magnet errors and mislaignments). $\hat{\eta}_1$ and $\hat{\eta}_2$ vectors are spin solutions on the reference orbit, that are orthogonal to $\hat{\mathbf{n}}_0$ and to each other. In this reference frame any spin solution, including the vector $\hat{\mathbf{n}}$, can be described by a complex variable C as:

$$\hat{\mathbf{n}} = \sqrt{1 - |C|^2}\hat{\mathbf{n}}_0 + \mathrm{Re}(iC\hat{\eta}^*) \quad (3)$$

where $\hat{\eta} = \hat{\eta}_1 - i\hat{\eta}_2$. The differential equation for C variable can be derived from the equation of spin rotation in an external field. In the first-order approximation the equation for C simplifies to:

$$C' = w_\perp \quad (4)$$

with $w_\perp = \mathbf{w}\hat{\eta}$, where the spin precession vector $\mathbf{w}$ describes the precession due to betatron and synchrotron oscillations and field errors:

$$\begin{aligned}
w_x &= (1+\nu_0)z'' + (\nu_0 + \frac{a}{\gamma_0})K_x p_\sigma + (1+a)K_y x' \\
w_y &= (1+a)\left(K_x' x + K_z' z + \Delta K_y - K_y p_\sigma\right) - (\nu_0 - a)(K_x p_x + K_z p_z) \\
w_z &= -(1+\nu_0)x'' + (\nu_0 + \frac{a}{\gamma_0})K_z p_\sigma + (1+a)K_y z'
\end{aligned}$$

Since, according to (2), $\hat{\mathbf{n}}$ is the function of the orbital phase space point, it can be expanded in a power set of orbit variables. In the first-order approximation, one should leave just the linear part of the expansion, which we present in the form:

$$C = (1+\nu_0)(z'\eta_x - x'\eta_z) + f_0 + F^T S X \quad (5)$$

where F is 6-D vector, $\nu_0 = \gamma a$ and a is the anomalous magnetic moment of the particle. The components of this vector, F_i $(i = 1..6)$ are functions of the ring azimuth θ. Together with the scalar function $f_0(\theta)$ they form the set of functions which we call spin-orbital functions.

To derive the equation which describes the spin-orbital function evolution along θ one can substitute the expression (5) into the spin equation (4). After some transformations it leads to following equations:

$$\begin{aligned}
F' &= SHF + P \quad (6) \\
f_0' &= \Delta K_z F_1 - \Delta K_x F_3 + (1+a)\Delta K_y \eta_y \quad (7)
\end{aligned}$$

The function f_0 is generated by field errors ΔK_i on the reference orbit.

The equation for the vector F is quite similar to the equation (1) governing the orbital motion. The linear parts of the equations (6) and (1) are exactly the same. The components of vector P are present only in dipole or solenoidal magnets:

$$
\begin{aligned}
P_1 &= -(\nu_0 - a)K_y\eta_x + (\nu_0^2 + a)K_x\eta_y; & P_2 &= -(1+a)K_x'\eta_y - (\nu_0 - a)\frac{1}{2}K_y^2\eta_z \\
P_3 &= (\nu_0^2 + a)K_z\eta_y - (\nu_0 - a)K_y\eta_z; & P_4 &= (\nu_0 - a)\frac{1}{2}K_y^2\eta_x - (1+a)K_z'\eta_y \\
P_5 &= (\nu_0 + \tfrac{a}{\gamma_0})(K_x\eta_x + K_z\eta_z) - 1 + a)K_y\eta_y; & P_6 &= 0
\end{aligned}
$$

where $K_i = B_i / < B_z >$ describes normalized magnetic field.

The similarity of linear terms of the equations of (6) and (1) leads to the important conclusion. The well-known transfer matrices for accelerator elements, routinely used for calculation of the linear orbital motion, can be applied for the calculation of spin-orbital functions too. The transformation of vector F trough an accelerator element can be written as:

$$F_{out} = MF_{in} + Y \tag{8}$$

where M is the transfer matrix for given element and the transfer vector Y, generated by the vector P, exists at the bending and solenoidal magnets. Making element-by-element transformation one comes to the one turn transformation in the form:

$$F(2\pi) = M_{rev}F(0) + Y_{rev} \tag{9}$$

The periodicity conditions for the vector $\hat{\mathbf{n}}$ lead to the following conditions for spin-orbital functions:

$$F_i(\theta + 2\pi) = e^{i2\pi\nu}F_i(\theta), \quad f_0(\theta + 2\pi) = e^{i2\pi\nu}f_0(\theta) \tag{10}$$

Then, from expressions (9) and (10) one can find the solution for vector F at $\theta = 0$

$$F(0) = (I \cdot \mathrm{e}^{i2\pi\nu} - M_{rev})^{-1}Y_{rev} \tag{11}$$

The value of vector F at any other element of a ring can be found then by doing again element-by-element transformation (8). Because eigen values of M_{rev} matrix are $\mathrm{e}^{i2\pi\nu_i}$, where ν_i are tunes of the orbital motion, the resonance denominator in (11) shows the first order spin resonances when $\nu = m \pm \nu_i$.

Another consequence of the linear parts identity of (6) and (1) is that the spin-orbital functions can be related through integral transformations to characteristic functions of the orbital motion (β-functions and orbital motion phases).

In an accelerator with only vertical guiding field on the particle reference orbit and without field errors the spin motion is coupled only with vertical orbital motion, thus only F_3 and F_4 functions have not zero values. An introduction of horizontal or solenoidal fields, for example as a part of a spin rotator insertion, leads to exciting a whole set of the spin-orbital functions.

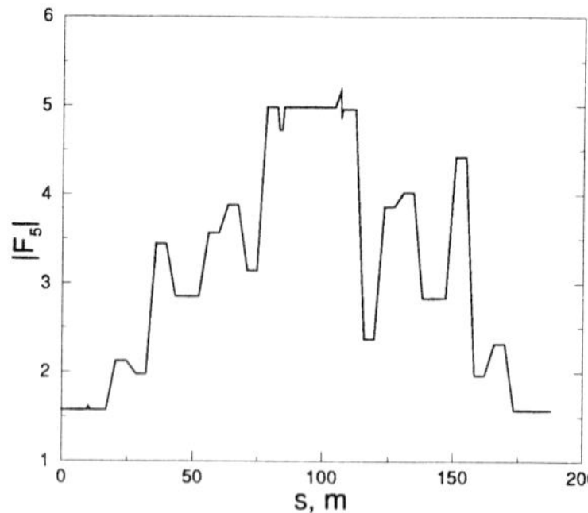

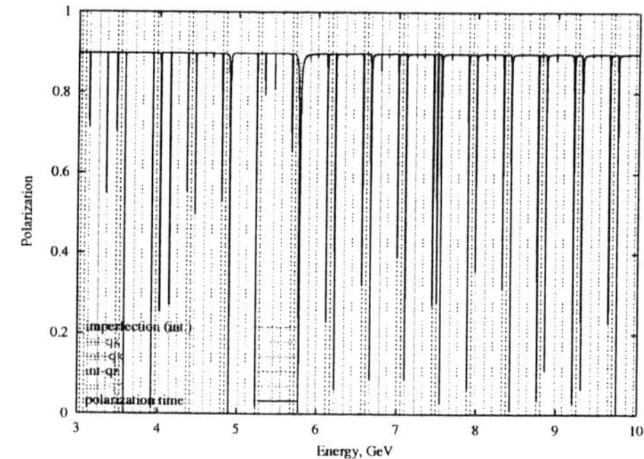

FIGURE 1. The examples of the ASPIRRIN calculations. On the left plot: $|F_5|$ for the Bates SHR with Siberian snake and wigglers. On the right plot: equilibrium polarization at proposed EIC electron ring.

APPLICATION OF THE SPIN-ORBITAL FUNCTIONS

The set of functions F can be used for calculating of various polarization characteristics of polarized beams in accelerators.

Calculation of equilibrium polarization and depolarization time

For electron rings, very important quantity is the partial derivative of $\hat{\mathbf{n}}$ over longitudinal momentum variable, $\mathbf{d} = \partial\hat{\mathbf{n}}/\partial\mathbf{p}_\sigma$, which defines the equilibrium polarization and depolarization time due to synchrotron radiation process. In the spin-orbital function formalism the vector $\mathbf{d}$ is connected with F_5 component of the vector F: $\mathbf{d} = \mathrm{Re}\,(iF_5\hat{\eta}^*)$

Let us note that: $|\mathbf{d}| = |F_5|$ and that $|F_5|$ is periodical function of θ. Thus, with F_5 calculated one can calculate also the equilibrium polarization and depolarization time according to Derbenev-Kondratenko formula [2].

Some examples of the polarization calculation by ASPIRRIN code are shown in Figure 1. Left figure demonstrates the calculated $|F_5|$ for South Hall Ring at MIT-Bates with the solenoidal Siberian snake and polarizing wigglers. On the right plot the polarization degree is shown for the electron ring of electron-ion collider project [3], with the spin depolarizing resonance pattern present.

Like F_5 function (or $|\mathbf{d}|$ vector) is used to calculate depolarizing effect of spin diffusion caused by sudden particle energy changes due to synchrotron radiation, the F_3 and F_1 spin-orbital function can be applied to calculate the spin diffusion caused by particle transverse momentum changes in scattering processes.

Calculation of resonance strength

Another possible application of spin-orbital functions is for the spin resonance strength analysis. We show this on the example of calculation of imperfection resonance strength.

As seen from (7) the f_0 function is generated by the magnetic field errors. Therefore this function should describe imperfection resonances. Indeed, the solution for f_0 is:

$$f_0 = \frac{1}{e^{i2\pi\nu} - 1} \int_{\theta}^{\theta+2\pi} \left(\Delta K_z F_1 - \Delta K_x F_3 + (1+a)\Delta K_y \eta_y\right) d\theta$$

And for an imperfection resonance strength one can get:

$$|w_k| = \frac{1}{2\pi} \left| \int_0^{2\pi} f'_{0(\nu=m)} d\theta \right| = \frac{1}{2\pi} \left| \int_0^{2\pi} \left(\Delta K_z F_1 - \Delta K_x F_3 + (1+a)\Delta K_y \eta_y\right) d\theta \right| \qquad (12)$$

where F_1, F_3 and η_y are calculated at the resonance condition $\nu = m$. Thus, the contributions to the resonance strength from horizontal, vertical and solenoidal field errors are described by the functions F_3, F_1 and η_y respectively. Let us note that in this approach it is not required to calculate the contribution to the resonance strength coming from closed orbit distortions produced by the field errors. It is taken into account automatically.

The similar approach can be used to calculate the resonance strength generated by coupling or gradient errors in an accelerator [4].

CONCLUSION

The set of spin-orbital function F was introduced and was shown to be useful for polarization calculation for particle beams in circular accelerators with complex magnet field configuration. The spin-orbital function formalism was put in the base of the `ASPIRRIN` code, which calculates the equilibrium polarization and depolarization time for electron rings as well as strength of spin resonances.

ACKNOWLEDGMENTS

We would like to thank D. Barber (DESY), I. Koop, A. Otboev (BINP) and F. Wang (MIT) for interesting discussions and valuable comments concerning the ASPIRRIN code.

REFERENCES

1. A.W.Chao, *NIM* **180**, p.29 (1981).
   D.P. Barber, Private notes, (1982).
2. Ya. S. Derbenev, A. M. Kondratenko, *Sov. Phys. JETP*, **37**, p.968 (1973).
3. Yu.M. Shatunov, et al, "Status of the e-ring design for EIC", *these proceedings*.
4. V. Ptitsyn, Ph. D. Theses, p.26 (1997).

Spin Flipping and Polarization Lifetimes of a 270 MeV Deuteron Beam[1]

V.S. Morozov[2*], M.Q. Crawford[*], Z.B. Etienne[3*], M.C. Kandes[*], A.D. Krisch[*], M.A. Leonova[*], D.W. Sivers[4*], V.K. Wong[*], K. Yonehara[*], V.A. Anferov[†], H.O. Meyer[†], P. Schwandt[†], E.J. Stephenson[†] and B. von Przewoski[†]

[*]*Spin Physics Center, University of Michigan, Ann Arbor, MI 48109-1120*
[†]*Indiana University Cyclotron Facility, Bloomington, IN 47408-0768*

Abstract. We recently studied the spin flipping of a 270 MeV vertically polarized deuteron beam stored in the IUCF Cooler Ring. We swept an rf solenoid's frequency through an rf-induced spin resonance and observed the effect on the beam's vector and tensor polarizations. After optimizing the resonance crossing rate and setting the solenoid's voltage to its maximum value, we obtained a spin-flip efficiency of about $94 \pm 1\%$ for the vector polarization; we also observed a partial spin-flip of the tensor polarization. We then used the rf-induced resonance to measure the vector and tensor polarizations' lifetimes at different distances from the resonance; the polarization lifetime ratio $\tau_{\text{vector}}/\tau_{\text{tensor}}$ was about 1.9 ± 0.4.

INTRODUCTION

Polarized beam experiments are now an important part of the programs in many high-energy polarized storage rings. Frequent reversals of the beam polarization direction can greatly reduce the detectors' systematic errors in spin asymmetry measurements. Artificial rf-induced spin resonances can be used to do such reversals, or spin-flips, in a very-well-controlled way. We have successfully spin-flipped beams of spin-$\frac{1}{2}$ particles: a horizontally polarized electron beam at the MIT-Bates Storage Ring [1], and horizontally and vertically polarized proton beams at the IUCF Cooler Ring [2, 3]. We recently studied the more complex spin-flipping of a polarized deuteron beam stored in the IUCF Cooler Ring.

In any flat circular accelerator or storage ring, each deuteron's spin precesses around the Stable Spin Direction, which is defined by the Ring's magnetic structure. With no horizontal magnetic fields present in the Ring, the Stable Spin Direction points along the vertical field direction of the Ring's dipole magnets. Moreover, the spin tune ν_s, which is the number of spin precessions during one turn around the Ring, is proportional to the

[1] Supported by research grants from the U.S. Department of Energy and the U.S. National Science Foundation

[2] E-mail: morozov@umich.edu

[3] Also at: Indiana University Cyclotron Facility, Bloomington, IN 47408-0768

[4] Also at: Portland Physics Institute, Portland, OR 97201, USA

CP675, *Spin 2002: 15th Int'l. Spin Physics Symposium and Workshop on Polarized Electron Sources and Polarimeters*, edited by Y. I. Makdisi, A. U. Luccio, and W. W. MacKay

deuteron's energy

$$\nu_s = G\gamma\,, \tag{1}$$

where $G = (g-2)/2 = -0.1426$ is the deuteron's gyromagnetic anomaly and γ is its Lorentz energy factor.

The deuteron's vertical polarization can be perturbed by a horizontal rf magnetic field from either an rf-solenoid or an rf-dipole magnet. This perturbation can induce an rf spin resonance, which can be used to flip the spin of the stored polarized deuterons [2, 3]. The frequency f_r, at which an rf-induced spin resonance occurs, is given by

$$f_r = f_c(k \pm \nu_s), \tag{2}$$

where f_c is the deuterons's circulation frequency and k is an integer. When an rf solenoid operates at the resonance frequency, its spin rotations accumulate coherently after each turn around the Ring; this can cause large deviations of the polarization from the Stable Spin Direction, which depolarize the beam.

However, sweeping the rf magnet's frequency through f_r can rotate the Ring's Stable Spin Direction by 180° with no depolarization. If the rotation is adiabatic, then each deuteron's vertical spin follows the Stable Spin Direction as it rotates by 180° resulting in a polarization flip. We use a modified [1] Froissart-Stora formula [4] to relate the beam's initial vector polarization P_i, to its final vector polarization P_f, after crossing the resonance,

$$\frac{P_f}{P_i} = (1+\eta)\; exp\left[\frac{-(\pi\varepsilon f_c)^2}{\Delta f/\Delta t}\right] - \eta, \tag{3}$$

where η is the spin-flip efficiency, ε is the resonance strength, and $\Delta f/\Delta t$ is the resonance crossing rate, while Δf is the frequency range during the ramp time Δt.

The polarization of a beam of spin-1 particles is more complex than that of spin-$\frac{1}{2}$ particles; the component of the spin along the vertical axis can have three values: $m_z = +1, 0, -1$; moreover, a spin-1 particle is also usually parameterized by a quantity called the tensor polarization [5]

$$p_{zz} \equiv 1 - 3(N_0/N_T) \tag{4}$$

where N_0 is the number of particles in the $m_z = 0$ state and N_T is the total number of particles. Eq. (4) is defined so that an unpolarized beam with all m_z states equally populated, has $p_{zz} = 0$. To describe a change in the tensor polarization after a frequency ramp we developed Eq. (5),

$$\frac{P_{zz\,f}}{P_{zz\,i}} = \frac{3}{2}\left\{(1+\eta)\; \exp\left[\frac{-(\pi\varepsilon f_c)^2}{\Delta f/\Delta t}\right] - \eta\right\}^2 - \frac{1}{2}, \tag{5}$$

which is an extension of the Froissart-Stora formula [4] to tensor polarization using Eq. (4).

Experimental Apparatus

The apparatus used for this experiment, including the rf-solenoid, the IUCF Cooler Ring and the polarimeter are shown in Fig. 1. For this experiment an unpolarized hydrogen gas cell was used as polarimeter's target. The thickness of the gas target was optimized to about $5 \cdot 10^{13}$ atoms/cm^2 to maximize the statistics while minimizing the background due to both the cell walls and deuteron breakup.

The 270 MeV polarized deuteron beam stored in the Cooler Ring was obtained using the new Cooler Injector Polarized IOn Source (CIPIOS) and the new Cooler Injection Synchrotron (CIS). To reduce the systematic errors in our polarization measurements, we normally cycled the deuteron beam through the four vertical polarization states available in CIPIOS: $|p_z\, p_{zz}\rangle$ of $|1\,1\rangle$, $|-1\,1\rangle$, $|0\,1\rangle$ and $|0\,-2\rangle$, and occasionally through the totally unpolarized state $|0\,0\rangle$ to check systematic errors.

The vector polarization's magnitude in the Cooler Ring, was about 0.6 for the p_z: ± 1 states; the tensor polarization magnitudes, for the p_{zz}: $+1$ and p_{zz}: -2 states, were about 0.8 and -1.6, respectively. Each data point required about an hour to obtain statistical errors of about $\pm$ 2% for the vector polarization and about $\pm$5% for the tensor polarization.

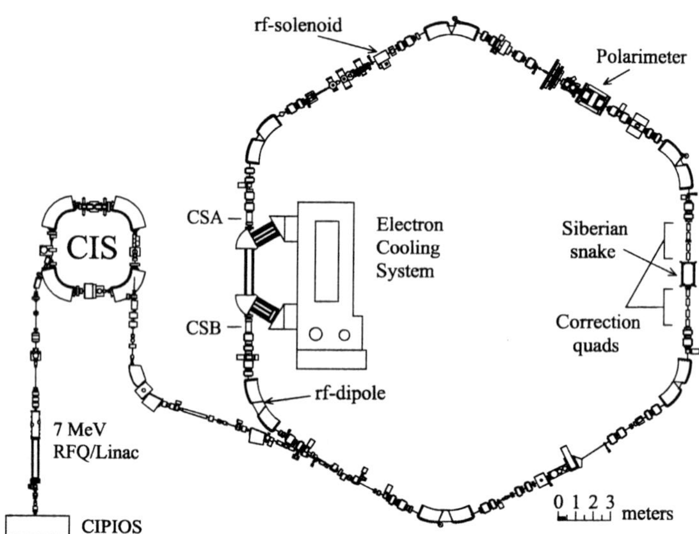

FIGURE 1. Layout of the IUCF Cooler Ring with its new Cooler Injector Synchrotron (CIS) and its new CIPIOS polarized ion source. Also shown inthe Cooler Ring are the rf-dipole, the rf-solenoid, the polarimeter, and the Siberian snake.

Experimental Results (February 2002)

We first found the approximate position of the $f_r = f_c(1-|\nu_s|)$ spin depolarizing resonance by sweeping the rf solenoid's frequency first over 64 kHz and then narrowing the frequency range into those ranges, which caused spin flip. We then more precisely determined f_r by measuring the vertical polarization after running the rf solenoid at different fixed frequencies. From these data we found that the resonance frequency was $f_r = 1\,403\,002 \pm 14$ Hz for the vector plarization and $f_r = 1\,402\,999 \pm 17$ Hz for the tensor polarization.

We next optimized the rf-solenoid's frequency ramp time Δt and frequency range Δf to maximize the spin-flip efficiency for the deuteron's vector polarization. We first linearly ramped the rf-solenoid's frequency from $f_r - 2$ to $f_r + 2$ kHz, with various ramp times Δt, while measuring the polarization after each frequency ramp. The measured vector and tensor polarizations are plotted against the ramp time in Fig. 2. The vector polarization data are fit to Eq. (3) by ignoring the $\Delta t = 500$ ms point, which seems anomalous. The tensor polarization data in Fig. 2 are fit to Eq. (5) as shown by the curves. Next, using the optimum Δt of 2 s, we spin-flipped the deuterons while varying the rf solenoid's frequency range Δf. These studies gave an optimum Δf of ± 0.75 kHz.

After setting Δt, Δf and the voltage to maximize the vector polarization spin-flip efficiency η, we more precisely determined η by measuring the vector polarization while varying the number of frequency sweeps. This measured vector and tensor polarization after n frequency sweeps are each plotted against n in Fig. 3. We fit these data using

$$P_n = P_i \cdot \eta^n, \tag{6}$$

where P_n is the measured radial beam polarization after n spin flips. The fits gave a vector polarization spin-flip efficiency of $94 \pm 1\%$ and a tensor depolarization of $18 \pm 3\%$ per frequency ramp.

We also studied the polarizations' lifetimes near the resonance. We turned the rf solenoid on at a fixed frequency near the depolarizing resonance and observed the polarization's decay with time. From these data we extracted lifetimes of the vector and tensor polarization at three distances from the resonance. Then we calculated the polarization lifetime ratio $\tau_{\text{vector}}/\tau_{\text{tensor}}$ to be about 1.9 ± 0.4.

We would like to thank the entire Indiana University Cyclotron Facility staff for the successful operation of the Cooler Ring

REFERENCES

1. V.S. Morozov *et al.* Phys. Rev. ST-AB **4**, 104002 (2001).
2. D.D. Caussyn *et al.*, Phys. Rev. Lett. **73**, 2857 (1994).
3. B.B. Blinov *et al.*, Phys. Rev. Lett. **88**, 014801 (2002).
4. M. Froissart and R. Stora, Nucl. Instrum. and Methods **7**, 297 (1960).
5. The Madison Convention, *Proc. of 3rd Intl. Symp. on Polarization Phenomena in Nuclear Physics*, Madison, 1970, the University of Wisconsin Press, p. xxv (1971).

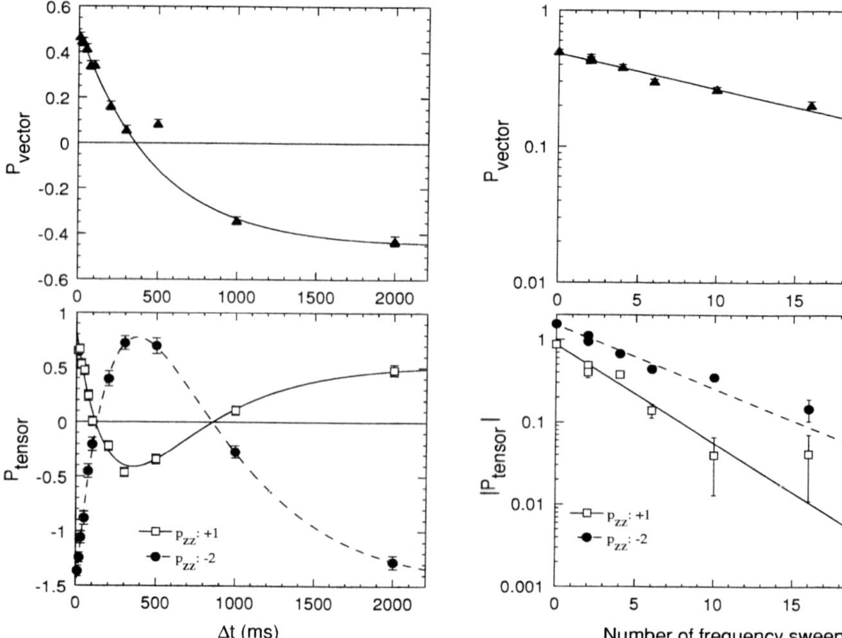

FIGURE2. The measured vector and two tensor deuteron polarizations at 270 MeV are plotted against the rf solenoid ramp time Δt. The curve in the P_{vector} plot is a fit using Eq. (3). The solid and dashed lines in the bottom plot are fits to the P_{tensor} data using Eq. (5).

FIGURE3. The measured vector and two tensor deuteron polarizations at 270 MeV are plotted against the number of frequency sweeps. The solid and dashed curves are fits to Eq. (6).

RHIC Spin Flipper Commissioning

M. Bai*, A.U. Luccio*, W.W. MacKay*, V. Ranjbar† and T. Roser*

*BNL, Upton, NY 11973, USA
†Indiana University, Bloomington, IN 47408, USA

Abstract. An ac dipole with horizontally oriented oscillating magnetic field (spin flipper) was installed in RHIC to reverse the spin direction in the presence of two full Siberian snakes, thereby reducing the systematic errors for the spin physics experiments in RHIC. With two full snakes, the spin vector completes one full precession around the vertical direction in two revolutions, and the spin depolarization resonances due to the machine imperfections and betatron oscillations are eliminated. Since the spin flipper provides an oscillating horizontal dipole field, a "spin resonance" can occur if the spin flipper frequency is placed in the neighborhood of the spin precession frequency [1, 2, 3]. By slowly sweeping the spin flipper frequency across the spin precession frequency, a full spin flip can be achieved. This paper reports the results of the RHIC spin flipper commissioned during the RHIC 2002 polarized proton run. By running the spin flipper at a slightly different configuration, one can also measure the spin precession tune.

INTRODUCTION

Like any other magnets, the spin motion through an ac dipole also obeys the Thomas-BMT equation

$$\frac{d\vec{S}}{dt} = \frac{e}{\gamma m}(1+G\gamma)\vec{S}\times\vec{B}(t), \tag{1}$$

where $\vec{S}$ is the spin vector in the particle rest frame, G is the anomalous gyromagnetic g-factor and γmc^2 is the moving particle energy. $\vec{B}(t)$ is the magnetic field of the ac dipole

$$B(t) = B_o \cos(2\pi f_m t + \chi), \tag{2}$$

where B_o is the amplitude of the oscillating magnetic field, f_m is the oscillation frequency and χ is the arbitrary phase of the ac dipole oscillating magnetic field.

In a perfect planar circular accelerator, the beam's spin vector precesses around vertical direction. With the ac dipole in the machine, the spin vector gets kicked away from the vertical direction every time it passes through the ac dipole. In the frame which rotates at the same oscillation frequency of f_m, the two-component spinor equation then becomes

$$\frac{d\psi_K}{d\theta} = -\frac{i}{2}(\vec{\sigma}\cdot\vec{n})\psi_K, \tag{3}$$

where

$$\vec{n} = G\gamma\hat{e}_3 - B_o\hat{e}_1. \tag{4}$$

Here, $\hat{e}_1$ is the unit vector pointing radially outward, and the unit vector $\hat{e}_3$ points up. Compare this with the spinor equation of an intrinsic spin resonance [4]; the effect of

CP675, *Spin 2002: 15th Int'l. Spin Physics Symposium and Workshop on Polarized Electron Sources and Polarimeters,* edited by Y. I. Makdisi, A. U. Luccio, and W. W. MacKay

the ac dipole on the spin motion is equivalent to a spin resonance located at

$$\nu_s = \nu_m = \frac{f_m}{f_{rev}}, \tag{5}$$

where ν_s is the spin tune, ν_m is the ac dipole oscillation tune and f_{rev} is the beam revolution frequency. The strength of this ac-dipole-induced spin resonance is given by

$$\varepsilon_K = \frac{1+G\gamma}{4\pi}\frac{B_o L}{B\rho}, \tag{6}$$

where L is the length of the ac dipole and $B\rho$ is the magnetic rigidity of the beam.

According to the Froissart-Stora formula, the beam polarization, after crossing through a spin resonance, is

$$P_f = \left(2e^{-\pi|\varepsilon_K|^2/2\alpha} - 1\right) P_0, \tag{7}$$

where P_f and P_0 are the beam polarization after and before the resonance. $\alpha = \frac{d\Delta\nu_m}{d\theta}$ is the resonance crossing rate, where $\Delta\nu_m$ is the width of the ac dipole oscillation tune sweep. To achieve a 99.999% spin flip, the following condition has to be fulfilled

$$\alpha \leq 0.13|\varepsilon_K|^2. \tag{8}$$

SPIN FLIPPING IN RHIC

RHIC spin setup

To eliminate the spin depolarization resonances along the acceleration, two pairs of Siberian snakes were installed in RHIC [5]. In each ring, the two snakes are placed on opposite sides of ring. Each snake consists of four super-conducting helical dipole magnets and rotates the spin vector by 180° around an axis which lies in the horizontal plane and is called the snake axis. The spin precession tune is given by

$$\nu_s = \frac{1}{\pi}|\psi_1 - \psi_2|, \tag{9}$$

where ψ_i is the angle between the axis of the i_{th} snake and the longitudinal direction. In general, the axes of the two snakes are configured to be ±45° away from the longitudinal direction for the RHIC polarized proton run and the nominal spin precession tune is $\frac{1}{2}$. In order to rotate the spin vector by 180°, the two outer helical magnets are powered with the same current with opposite polarities. Similarly the two inner helical magnets are powered with equal but opposite currents.

The RHIC beam polarization was measured with the CNI (Compton Nuclear Interference) polarimeter [7, 8, 9] located at the IP12 region [6]. It measures the left-right asymmetry of the recoiling carbon nucleus. A typical measurement requires 2 million events to reach a 2% statistical error.

RHIC Spin Flip

To induce a full spin flip in RHIC, one can sweep the artificial spin resonance across the spin precession frequency. In order to cross the resonance, one needs to detune the snake axis to move the spin precession tune away from $\frac{1}{2}$.

The RHIC spin flipper is located at the interaction region of IP4 and is common to both beams. Since each Siberian snake in RHIC is energized with two independent power supplies, the control of the spin precession tune in the two rings is independent of each other. Thus, one should in principle be able to achieve spin flipping in one ring without impacting the beam polarization in the other ring.

Two spin flips were tried in the Blue ring at the end of a typical polarized proton store. We first ramped both snakes in the Blue ring to the settings which correspond to $\nu_s = 0.48$. Table 1 lists the snake inner and outer currents set values and read-backs. We

TABLE 1. Blue Siberian Snake Current Setting

	$\nu_s = 0.48$		$\nu_s = 0.5$	
magnet	**set point[A]**	**readback[A]**	**set point[A]**	**readback[A]**
bo3 snake outer	106.11	105.75	99.95	99.72
bo3 snake inner	325.06	324.07	325.06	324.33
bi9 snake outer	106.11	106.40	99.95	100.26
bi9 snake inner	325.06	324.88	325.94	326.09

then ramped the ac dipole magnetic field amplitude from zero to 100 gauss-m in 6000 revolution turns with the resonant frequency fixed at $0.47 f_{rev}$. Here, f_{rev} is the particle's revolution frequency around the ring. The ac dipole frequency was then swept from $0.47 f_{rev}$ to $0.49 f_{rev}$ in 200,000 revolution turns. The ac dipole was gradually turned off at a fixed frequency of $0.49 f_{rev}$ during 6000 turns. With the ac dipole maximum field of 100 gauss-m and the resonance crossing rate

$$\alpha = \frac{d\Delta\nu_m}{d\theta} = \frac{0.02}{200000 \times 2\pi} = 1.6 \times 10^{-8}, \tag{10}$$

the expected spin flip is

$$\frac{P_f}{P_0} = -1.0. \tag{11}$$

Fig. 1 shows the measured asymmetry before and after the spin flipping. The average spin flipping efficiency η is defined as

$$P_i = P_0 \eta^i \tag{12}$$

where P_0 is the beam polarization before the spin flipping and P_i is the beam polarization after the ith spin flipping. After two spin flips, we measured an efficiency of

$$\eta = 0.66. \tag{13}$$

The fact that we did not reach full spin flip can be either due to the spin tune not being 0.48 as we expected or due to the spread of spin tune among the different particles in

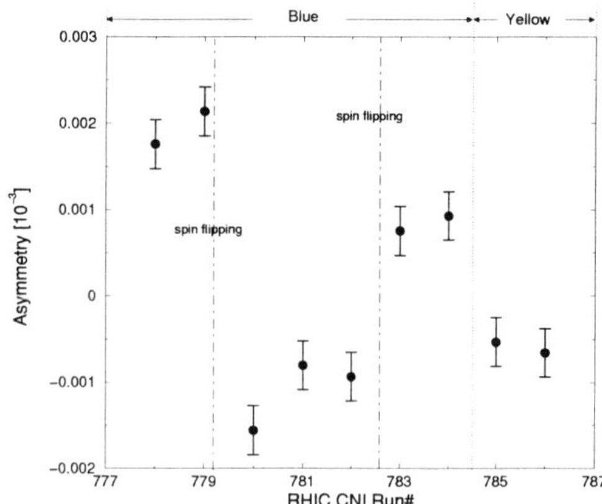

FIGURE 1. This plot shows the measured beam asymmetry before and after the spin flipping in RHIC. The vertical averaged asymmetry in Blue was measured as 0.00195 ± 0.0002 After the 1st spin flipping, the asymmetry was measured as 0.0011 ± 0.000164. The asymmetry was 0.00084 ± 0.0002 after the 2nd spin flipping. The data point of RHIC CNI run 785 and 786 correspond to the measured asymmetry in the Yellow ring after the 2nd spin flipping in Blue.

the bunch. Particles with different betatron oscillation amplitudes experience different focusing forces; hence they have different spin precession frequencies. In general, the spin tune is a function of the betatron oscillation amplitude. Fig. 2 shows the numerical simulation of spin flipping of a beam with a normalized 95% emittance of 25πmm-mrad emittance in both planes at $G\gamma = 191.5$.

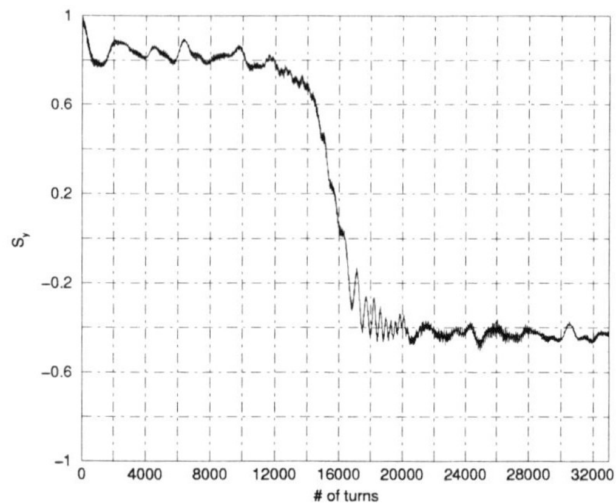

FIGURE 2. This plot shows the spin tracking results of spin flipping of 100 particles at $G\gamma = 191.5$. They are gaussian distributed with 25πmm-mrad in both horizontal and vertical planes.The final polarization after the flipping Only 40% of the initial polarization. With the same machine parameters and spin flipper setup, the single particle tracking yields a full spin flip.

The asymmetry in Yellow was also measured as before the spin flippings in the Blue as 0.0015 ± 0.0003 However, the fact that we measured the Yellow asymmetry as

-0.00075 ± 0.0002 after the 2nd spin flipping in the Blue as shown in Fig. 1 indicated that the spin tune in the Yellow ring was not $\frac{1}{2}$. Otherwise, the spin flipping in the Blue ring should not have done any harm to the beam polarization even though the ac dipole affects both beams. This is also consistent with the two spin flipping attempts in the Yellow ring in which no spin flipping except depolarization was observed even though the same procedure as for the Blue ring was followed.

CONCLUSION

A short study of using the RHIC vertical ac dipole as a spin flipper was performed at the end of the RHIC 2002 polarized proton run. A partial spin flip was obtained in the RHIC Blue ring at the storage energy. However, due to the time limitation, we did not get chance to investigate the cause of depolarization and improve the spin flip efficiency. No spin flipping was obtained in the Yellow ring.

ACKNOWLEDGMENTS

The authors would like to thank Dr. G. Bunce, Dr. Y. Makdisi, Dr. L. Ahrens, Dr. H. Huang and Dr. N. Sato for the fruitful discussions and support. The authors would also like to thank A. Zaltsman, P. Oddo, C. Pai, T. Russo, R. Sanders, L. Hoff, J. Piacentino and B. Oerter for their technical support. This work is performed under the auspices of Department of Energy of U.S.

REFERENCES

1. D.D.Caussyn et al., 'Spin Flipping a Stored Polarized Proton Beam', Phys. Rev. Lett. 73, 2857 (1994).
2. B.B.Blinov et al., 'Spin Flipping in the Presence of a Full Siberian Snake', Phy. Rev. Lett. 81, 2906 (1998).
3. T. Roser, *Handbook of Accelerator Physics and Engineering*, P. 150, edited by A. Chao and M. Tigner.
4. S. Y. Lee, *Spin Dynamics and Snakes in Synchrotrons*, World Scientific, 1997
5. T. Roser et al. 'Accelerating and Colliding Polarized Protons in RHIC with Siberian Snakes', EPAC2002, June. 2002.
6. H. Hunag et al. 'Commissioning CNI Proton Polarimeters in RHIC', EPAC2002, June 2002.
7. O. Jinnouchi et al. 'RHIC pC CNI Polarimeter: Status and Performance from the First Collider Run', Proceedings of 15th International Spin Physics Symposium, BNL, 2002.
8. K. Kurita et al. 'RHIC pC CNI Polarimeter: Experimental Setup and Physics Results', Proceedings of 15th International Spin Physics Symposium, BNL, 2002.
9. G. Igo et al. 'Absolute Calibration of the CNI Polarimeters at RHIC Using 125 GeV/A C Ions',Proceedings of 15th International Spin Physics Symposium, BNL, 2002.

99.9% Spin-Flip Efficiency in the Presence of a Strong Siberian Snake [1]

V.S. Morozov[2*], B.B. Blinov*, Z.B. Etienne[3*], A.D. Krisch*, M.A. Leonova*, A.M.T. Lin*, W. Lorenzon*, C.C. Peters*, D.W. Sivers[4*], V.K. Wong*, K. Yonehara*, V.A. Anferov†, P. Schwandt†, E.J. Stephenson†, B. von Przewoski† and H. Sato**

*Spin Physics Center, University of Michigan, Ann Arbor, MI 48109-1120
†Indiana University Cyclotron Facility, Bloomington, IN 47408-0768
**KEK, High Energy Accelerator Research Organization, Tsukuba, Ibaraki 305-0801, Japan

Abstract. We recently studied the spin-flipping efficiency of an rf-dipole magnet using a 120-MeV horizontally polarized proton beam stored in the Indiana University Cyclotron Facility Cooler Ring, which contained a full Siberian snake. We flipped the spin by ramping the rf dipole's frequency through an rf-induced depolarizing resonance. By adiabatically turning on the rf dipole, we minimized the beam loss, while preserving almost all of the beam's polarization. After optimizing the frequency ramp parameters, we used up to 400 multiple spin flips to measure a spin-flip efficiency of $99.93 \pm 0.02\%$. This result indicates that spin flipping should be possible in very-high-energy polarized storage rings, where Siberian snakes are certainly needed and only dipole rf-flipper magnets are practical.

INTRODUCTION

Polarized beam scattering experiments require frequent reversals in the beam polarization direction in order to reduce systematic errors in the measuredasymmetry. It is best to be able to spin-flip a stored polarized beam while it is circulating. Earlier we succesfully used an rf-solenoid to spin-flip a horizontally polarized proton beam stored in the Cooler Ring containing a Siberian snake with $97 \pm 1\%$ spin-flip efficiency [1]. However, a solenoid's spin rotation decreases linearly with energy because of the Lorentz contraction of its $\int Bdl$; thus, a solenoid is impractical for spin-flipping in high energy rings. Fortunately, a dipole's spin rotation is energy independent. Therefore, we recently studied an rf-dipole's ability to spin-flip a 120 MeV horizontally polarized proton beam stored in the IUCF Cooler Ring operating with a full Siberian snake.

---

[1] Supported by research grants from the U.S. Department of Energy and the U.S. National Science Foundation

[2] E-mail: morozov@umich.edu

[3] Also at: Indiana University Cyclotron Facility, Bloomington, IN 47408-0768

[4] Also at: Portland Physics Institute, Portland, OR 97201, USA

CP675, *Spin 2002: 15th Int'l. Spin Physics Symposium and Workshop on Polarized Electron Sources and Polarimeters,* edited by Y. I. Makdisi, A. U. Luccio, and W. W. MacKay

SPIN MOTION AND SPIN FLIPPING

In a storage ring, the beam's polarization vector precesses around the Stable Spin Direction with a frequency of

$$f_s = f_c\, \nu_s \tag{1}$$

where f_c is the circulation frequency, and ν_s is the spin tune, which is the number of spin precessions during one turn around the ring. With a Siberian snake in the ring, the spin tune is given by

$$\nu_s = \frac{1}{\pi} cos^{-1}\left[cos(\pi G\gamma)cos(\pi s/2)\right], \tag{2}$$

where $G = (g-2)/2 = 1.792847$ is the proton's anomalous magnetic moment, γ is its Lorentz energy factor, and s is the snake strength. For a full (100%) snake, $s = 1$, and then $\nu_s = \frac{1}{2}$, independent of energy.

This spin motion can be perturbed by a horizontal rf magnetic field from either an rf-solenoid or an rf-dipole. Such perturbations can induce an rf depolarizing resonance, which can be used to flip the spin direction of the ring's stored polarized protons [1, 2]. The resonant frequency f_r, at which an rf-induced depolarizing resonance occurs, is given by

$$f_r = f_c(k \pm \nu_s), \tag{3}$$

where k is an integer. Sweeping the rf magnet's frequency through f_r can flip the beam's spin direction. The Froissart-Stora equation [3] relates the beam's polarization after crossing the resonance P_f, to its initial polarization P_i,

$$P_f = P_i\left\{2\, exp\left[\frac{-(\pi\varepsilon f_c)^2}{\Delta f/\Delta t}\right] - 1\right\}, \tag{4}$$

where ε is the resonance strength, and $\Delta f/\Delta t$ is the resonance crossing rate, while Δf is the frequency ramp's range during the ramp time Δt. The resonance strength ε is given by

$$\varepsilon = \frac{\theta_s}{\pi} = \frac{Ge\int Bdl}{2\pi m_p v}, \tag{5}$$

where θ_s is the rf-dipole's spin rotation angle, m_p is the proton's mass and v is its velocity. For a sufficiently strong resonance and low crossing rate, Eq. (4) indicates that P_f is about equal to $-P_i$. The spin-flip efficiency η is defined as

$$\eta = \frac{-P_f}{P_i}. \tag{6}$$

EXPERIMENTAL RESULTS

The apparatus used for this experiment, including the rf-dipole, the IUCF Cooler Ring and the polarimeter were discussed earlier [1, 2]. The 120 MeV horizontally polarized proton beam in the Cooler Ring was obtained using the new Cooler Injector Polarized

IOn Source (CIPIOS) and the new Cooler Injection Synchrotron (CIS). The beam polarization was typically 77% after the 7 MeV Linac and was practically the same at 120 MeV injection into the Cooler Ring.

The circulation frequency in the Cooler Ring was f_c = 1.59784 MHz at 120 MeV. With a nearly-full Siberian snake in the Ring, the spin tune ν_s is very near, but not exactly equal, to $\frac{1}{2}$. Thus, at 120 MeV,Eq. (3) implies that there should be two closely spaced rf spin resonances centered around $\frac{1}{2}f_c = 0.79892$ MHz, with their frequencies at$f_r^- = f_c(1-\nu_s)$ and $f_r^+ = f_c(0+\nu_s)$. Since our snake strength was about 1.02, the spin tune ν_s was about 0.510; thus, the f_r^- resonance should have a frequency slightly below $0.5f_c$.

We first determined that the f_r^- resonance's frequency was near 0.777 MHz by using our new resonance search technique:

- We first measured the beam polarization after sweeping the rf-dipole through some frequency range Δf, which might flip the spin.
- Then we cut the previous Δf range into two equal halves and measured the polarization after sweeping the frequency through each half.
- Then we chose the Δf range which caused spin-flip and repeated the process.

Fig. 1. shows the measured radial polarization plotted against each sweep's central frequency; the horizontal bars show the frequency range Δf for each sweep. We then more precisely determined f_r^- and f_r^+ by measuring the radial polarization at different fixed rf-dipole frequencies as shown in Fig. 2. Notice the very interesting behavior of the vertical polarization data in Fig. 2(b).

Next, we spin-flipped the beam by linearly ramping the rf-dipole's frequency from $f_r^- - 5$ to $f_r^- + 5$ kHz, with various ramp times Δt, while measuring the beam polarization after each frequency ramp. The measured radial polarization is plotted against the ramp time in Fig. 3. This measured polarization is a good fit to the Froissart-Stora formula (Eq. (4)), which is shown by the curve.

To further study the spin-flip efficiency η, we next performed 10 spin-flips while varying the rf-dipole's frequency range Δf, its ramp time Δt and its voltage V. The beam polarization after 10 spin-flips is plotted in Fig. 4 against the rf-dipole's voltage, with its Δf and Δt set to their optimum values. The dashed curve is a fit to the formula obtained by taking the 10th power of $\eta \equiv -P_f/P_i$ in Eq. (4):

$$P_{10} = P_i \cdot \eta^{10} = P_i \left\{ 2\, exp \left[\frac{-(\pi \varepsilon f_c)^2}{\Delta f/\Delta t} \right] - 1 \right\}^{10}. \tag{7}$$

After maximizing the spin-flip efficiency by setting $\Delta t = 27$ ms, $\Delta f = \pm 2$ kHz and $V = 250$ V rms ($\int Bdl = 0.43$ T·mm rms), we more precisely determined this efficiency η by varying the number ofspin-flips; we measured the radial polarization after 0, 100, 200, 300 and 400 spin-flips while keeping Δt, Δf and V all fixed. This measured radial polarization is plotted against the number of spin-flips in Fig. 5 [5]. We then fit these data using

$$P_n = P_i \cdot \eta^n, \tag{8}$$

[5] These data were obtained in a later run than the data in Figs. 1-4.

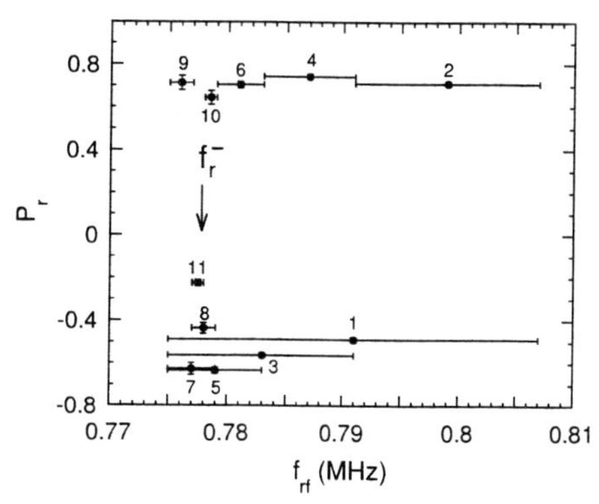

FIGURE 1. The measured radial proton polarization at 120 MeV is plotted against the range of each frequency ramp; each frequency ramp's Δf range is shown by a horizontal bar. The arrow shows f_r^-.

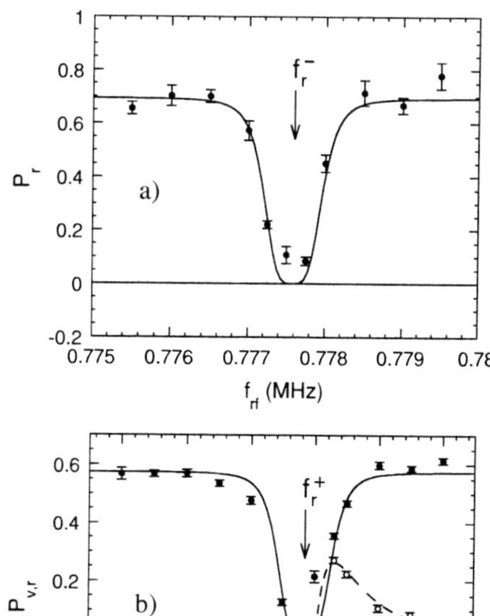

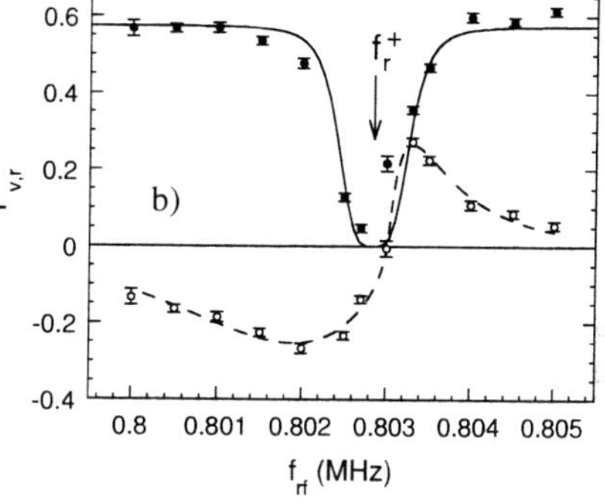

FIGURE 2. The measured radial (black) and vertical (white) proton polarizations at 120 MeV are plotted against the rf-dipole's fixed frequency for the f_r^- (*a*) and f_r^+ (*b*) resonances. The solid curves are fits using a second-order Lorentzian. The arrows show f_r^- and f_r^+.

where P_n is the measured radial beam polarization after n spin-flips and η is the spin-flip efficiency. The best fit gave a spin-flip efficiency of $99.93 \pm 0.02\%$.

ACKNOWLEDGMENTS

We would like to thank the entire Indiana University Cyclotron Facility staff for the successful operation of the Cooler Ring.

REFERENCES

1. B.B. Blinov *et al.*, Phys. Rev. Lett. **81**, 2906 (1998);
   V.A. Anferov *et al.*, *Proc. of 13th Intl. Symp. on High Energy Spin Physics*, eds. N.E. Tyurin *et al.*, World Scientific, 503 (1999).
2. B.B. Blinov *et al.*, Phys. Rev. Lett. **88**, 014801 (2002).
3. M. Froissart and R. Stora, Nucl. Instrum. and Methods **7**, 297 (1960).

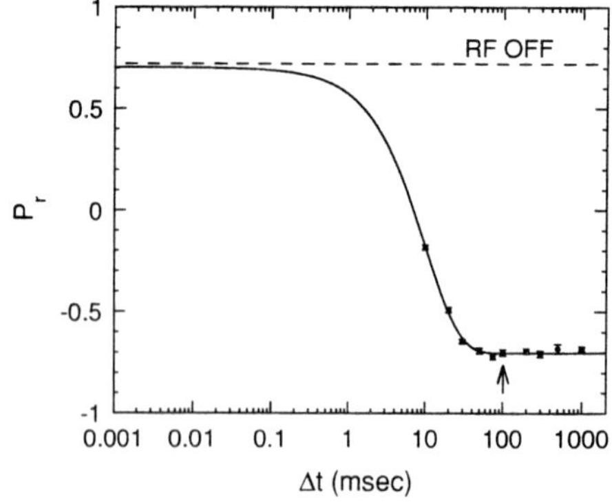

FIGURE 3. The measured radial proton polarization at 120 MeV is plotted against the rf-dipole ramp time Δt. The curve is a fit using Eq. (4). The dashed line shows the polarization with the rf-dipole off; the arrow shows the 100 ms setting used for later studies.

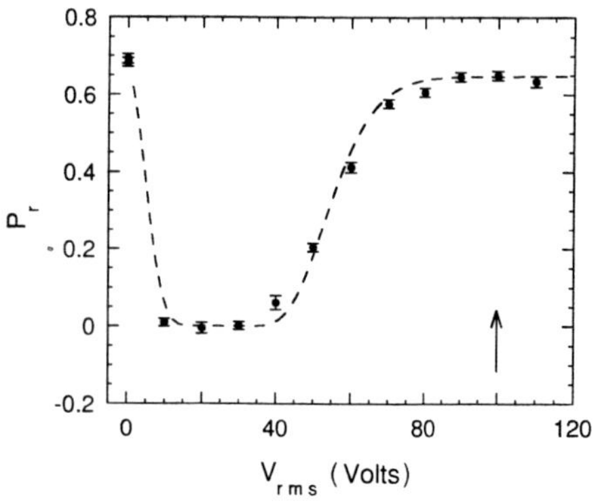

FIGURE 4. The measured radial proton polarization at 120 MeV after 10 spin-flips is plotted against the rms rf-dipole voltage V_{rms}. The dashed curve is a fit using Eq. (7); the arrow shows the 100 V setting used for later studies, which gives $\int Bdl = 0.17$ T·mm rms.

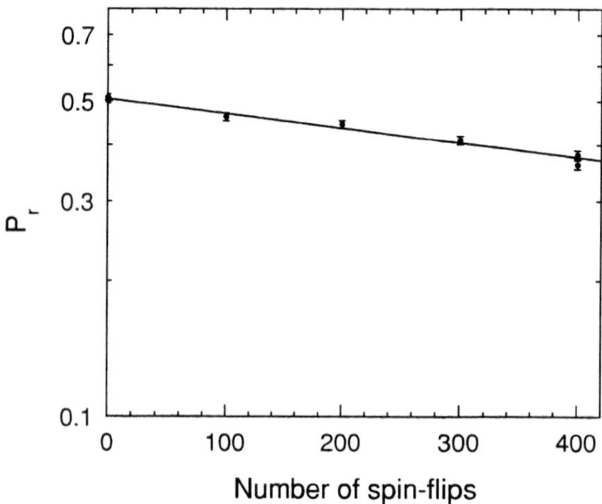

FIGURE 5. The measured radial proton polarization at 120 MeV is plotted against the number of spin-flips. The curve is a fit to the data using Eq. (8).

The Relativistic Stern-Gerlach Interaction as a Tool for Attaining the Spin Separation

P. Cameron*, M. Conte†, A. U. Luccio*, W. W. MacKay*, M. Palazzi† and M. Pusterla**

*Brookhaven National Laboratory, Upton, NY 11973, USA.
†Dipartimento di Fisica dell'Università di Genova, INFN Sezione di Genova, Via Dodecaneso 33, 16146 Genova, Italy.
**Dipartimento di Fisica dell'Università di Padova, INFN Sezione di Padova, Via Marzolo 8, 35131 Padova, Italy.

Abstract. The relativistic Stern-Gerlach interaction is here considered as a tool for obtaining the spin state separation of an unpolarized (anti)proton beam circulating in a ring. Drawbacks, such as spin precessions within the TE rf cavity, spurious kicks due to the transverse electric field and, worst of all, filamentation in the longitudinal phase plane are analyzed. Possible remedies are proposed and their feasibility is discussed.

INTRODUCTION

We have exhaustively demonstrated [1] that the relativistic Stern-Gerlach interaction can play a decisive role in accomplishing the spin states separation of a high energy unpolarized beam of protons and, possibly, of antiprotons, since the single cavity crossing energy kick is

$$dU \simeq 2\gamma^2 B_0 \mu^* \tag{1}$$

where γ is the Lorentz factor, B_0 is the peak magnetic field in the cavity and $\mu^* = |\vec{\mu}^*|$ is the particle magnetic moment: 1.41×10^{-26} JT^{-1} for (anti)protons and 9.28×10^{-24} JT^{-1} for electrons and positrons.

After having crossed $N_{\rm cav}$ cavities and completed $N_{\rm turns}$ revolutions, particles with opposite spin states should be gathered in couples of bunches exhibiting an energy separation

$$\Delta U \simeq 4 N_{\rm turns} N_{\rm cav} \gamma^2 B_0 \mu^* \tag{2}$$

and a momentum spread

$$\frac{\Delta p}{p} = \frac{1}{\beta^2}\frac{\Delta U}{U} \simeq 4 N_{\rm turns} N_{\rm cav} \frac{B_0}{B_\infty}\gamma \tag{3}$$

with $B_\infty = \frac{mc^2}{\mu^*} \simeq 10^{16}\ T$ for (anti)protons. The number of revolutions and the time interval required for reaching the value of the design momentum spread are

CP675, *Spin 2002: 15th Int'l. Spin Physics Symposium and Workshop on Polarized Electron Sources and Polarimeters*, edited by Y. I. Makdisi, A. U. Luccio, and W. W. MacKay

TABLE 1. RHIC, HERA and LHC Parameters

	RHIC	HERA	LHC
E (GeV)	250	820	7000
γ	266.5	874.2	7462.7
$\tau_{\rm rev}(\mu s)$	12.8	21.1	88.9
Δp/p	4.1×10^{-3}	5×10^{-5}	1.05×10^{-4}
N_{SS}	6.67×10^{9}	2.48×10^{7}	1.76×10^{6}
Δt	23.7 hr	523 s	156 s

$$N_{SS}=\frac{B_\infty}{4N_{\rm cav}B_0\gamma}\left(\frac{\Delta p}{p}\right)_{\rm ring}, \quad \text{and} \quad \Delta t=N_{SS}\tau_{\rm rev}. \tag{4}$$

By applying Eq. (4) to three rings, either operating or under development, we find the data gathered in Table 1. From the last row, we can ascertain how the LHC [2] splitting time is rather short making us quite optimistic. Nevertheless, it is wise to analyze all the drawbacks which can haunt the proposed procedure.

SPURIOUS EFFECTS

An effect to be considered is the one regarding the influence on the spin precession of the rf fields.

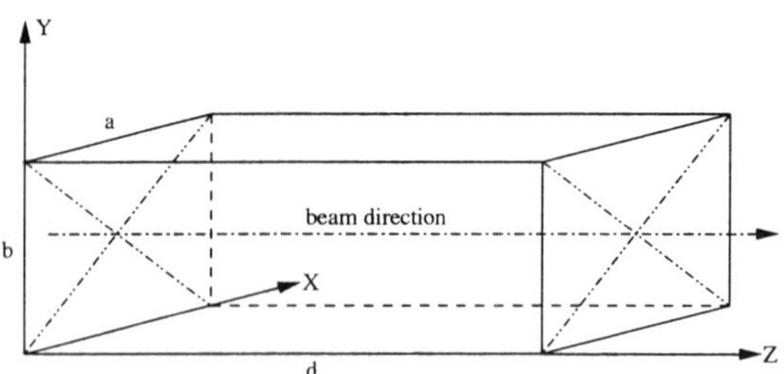

FIGURE 1. Rectangular cavity.

We recall the Thomas-BMT equation

$$\frac{d\vec{S}}{dt}=\vec{\Omega}_s\times\vec{S}, \qquad \text{or} \qquad \frac{d\vec{P}}{dt}=\vec{\Omega}_s\times\vec{P}, \tag{5}$$

with

$$\vec{\Omega}_s=-\frac{e}{\gamma m}\left\{(1+a\gamma)\vec{B}-(\gamma-1)a\frac{(\vec{v}\cdot\vec{B})}{v^2}\vec{v}+\gamma\left[\left(a+\frac{1}{\gamma+1}\right)\frac{\vec{E}\times\vec{v}}{c^2}\right]\right\}, \tag{6}$$

where $\vec{P}=(P_x,P_y,P_z)$ is the beam polarization built up by the Stern-Gerlach energy kicks. The proposed 3 GHz TE_{011} mode, inside a rectangular cavity (see Fig. 1), is characterized by the fields

$$\vec{B} = \begin{pmatrix} 0 \\ -B_0 \frac{b}{d} \sin\left(\frac{\pi y}{b}\right) \cos\left(\frac{\pi z}{d}\right) \cos(\omega t) \\ B_0 \cos\left(\frac{\pi y}{b}\right) \sin\left(\frac{\pi z}{d}\right) \cos(\omega t) \end{pmatrix}, \tag{7}$$

$$\vec{E} = \begin{pmatrix} -\omega B_0 \frac{b}{\pi} \sin\left(\frac{\pi y}{b}\right) \sin\left(\frac{\pi z}{d}\right) \sin(\omega t) \\ 0 \\ 0 \end{pmatrix}. \tag{8}$$

The effects of these fields, which are expected to be negligible, will be analyzed by means of computer simulation, starting with a polarization $\vec{P}_0 = (0, P_{0y}, 0)$ at the cavity entrance and looking for the the polarization state at the cavity exit.

Another effect to be considered is the interaction between the cavity's electric field and the particle's electric charge. We have already [1] demonstrated that, after a single cavity crossing, the energy exchange is

$$\Delta U_E = \left[e\omega B_0 \frac{bd}{\pi^2} \frac{\beta^2}{\beta_{\rm ph}^2 - \beta^2} \sin\left(\frac{\beta_{\rm ph}}{\beta}\pi\right)\right] x', \tag{9}$$

where $\beta_{\rm ph} = \sqrt{1+(d/b)^2}$ is a function of the cavity dimensions b and d (see Fig. 1). For ultrarelativistic particles $\beta_{\rm ph}$ equal to an integer, Eq. (9) reduces to

$$\Delta U_E = \pm \left[e\omega B_0 \frac{bd}{2\pi} \frac{\beta_{\rm ph}}{\beta_{\rm ph}^2 - \beta^2} \left(\frac{\varepsilon}{\beta_{\rm ph}} + \frac{1}{\gamma^2}\right)\right] x', \tag{10}$$

having accounted for the error ε in $\beta_{\rm ph}$. The quantity within square brackets is very small; besides the trajectory slope x' averages continuously to zero every few turns, due to required incoherence of betatron oscillations. However, a rather pleonastic computer simulation confirms the insubstantiality of such an effect. In fact, by defining

$$\delta x = (x)_{\rm rf} - (x)_{\rm norf}, \tag{11}$$

where $(x)_{\rm rf}$ is the path run by the particle after having crossed the cavity with the radio frequency on, and $(x)_{\rm norf}$ is the same path with the rf switched off, we may assess the displacement through the cavity for the four initial conditions at the entrance: $(\pm x_0, 0)$ and $(0, \pm x'_0)$, where the quantities x_0 and x'_0 are compatible with the LHC normalized emittance $\varepsilon^* = 3.75\mu$m. The plot in Fig. 2a (the same for all four cases) exhibits a displacement of 5 nm, i.e. an actually negligible effect.

FILAMENTATION IN THE LONGITUDINAL PHASE PLANE

We have already discussed [3] how the plots in the synchrotron oscillations phase plane are distorted due to the typical non linearity of the phase oscillations equation

$$\frac{d^2\phi}{dt^2} + \Omega_s^2 \sin\phi = 0 \quad \text{(stationary bucket case)}, \tag{12}$$

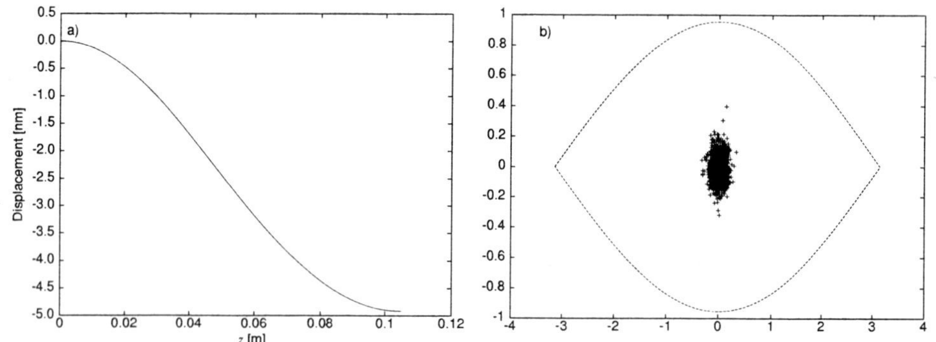

FIGURE 2. a) Particle trajectory inside the cavity. b) Longitudinal phase space of initial bunch.

TABLE 2. LHC Parameters at Collision

	Values	Unit
Revolution frequency	11.2455	kHz
rf frequency	400.7	MHz
Harmonic number h	35640	
rf voltage $V_{\rm rf}$	16	MV
Synchrotron period τ_s	0.042	s
Transition parameter $\eta_{\rm tr}$	3.47×10^{-4}	
Bunch duration	0.28	ns
Bunch length	8.39	cm

with

$$\Omega_s = \omega_s \sqrt{\frac{h|\eta_{\rm tr}|}{2\pi\beta^2\gamma} \frac{qV_{\rm rf}}{mc^2}} \simeq \omega_s \sqrt{\frac{h|\eta_{\rm tr}|}{2\pi\gamma} \frac{qV_{\rm rf}}{mc^2}} \qquad \text{(ultrarelativistic)} \tag{13}$$

where

$$\eta_{\rm tr} = \gamma^{-2} - \alpha_p = \gamma^{-2} - \gamma_{\rm tr}^{-2} \tag{14}$$

is the phase-slip factor (α_p = momentum compaction factor), h is the harmonic number and $V_{\rm rf}$ is the peak rf voltage. The synchrotron period is

$$\tau_s = \frac{2\pi}{\Omega_s} = \tau_{\rm rev} \sqrt{\frac{2\pi\beta^2\gamma}{h|\eta_{\rm tr}|} \frac{V_p}{V_{\rm rf}}} \simeq \tau_{\rm rev} \sqrt{\frac{2\pi\gamma}{h|\eta_{\rm tr}|} \frac{V_p}{V_{\rm rf}}} \qquad \text{(ultrarelativistic)} \tag{15}$$

with $V_p = 938$ MV for (anti)protons.

Concentrating our attention on LHC, we gather in Table 2 a few parameters of interest which will be used together with the data in the third column of Table 1. Starting with a bunch like the one illustrated in Fig. 2b, the simulation program shows (see Fig. 3a) that the filamentation begins scarcely after 10 synchrotron periods, i.e. after about 0.42 s: a time much smaller than $(\Delta t)_{\rm LHC} = 156$ s shown in Table 1. However, the cavity's magnetic field $B_0 = 0.1$ T could be increased by a factor of 10, since the associated electric field would be 300 MV/m, a value already realized in TM cavities. Besides, it would not be impossible to lower the momentum dispersion down to 10^{-5}, perhaps at expense of beam intensity. With these values, a new $(\Delta t)_{\rm LHC} = 1.56$ s will result. After

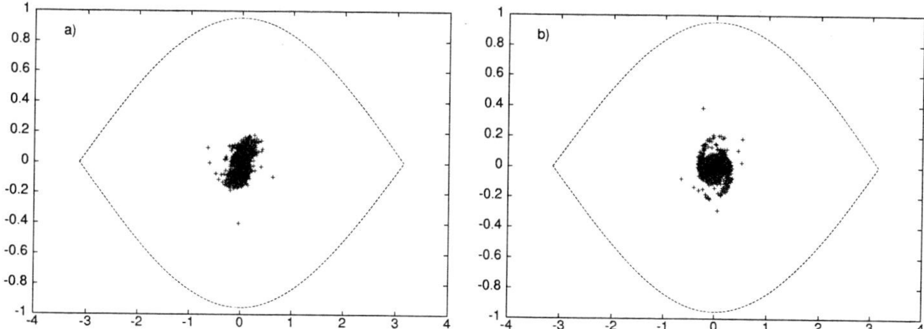

FIGURE 3. a) Filamentation after 10 synchrotron periods. b) Filamentation after 40 synchrotron periods with the more stringent requirements discussed in the text.

40 synchrotron periods, i.e. after 1.68 s, the filamentation is not too bad, as illustrated in Figs. 3b. This means that the desired spin-state separation could occur, although with an efficiency less than 100% due to the "tails" generated by the filamentation phenomenon. Notwithstanding, it is worthwhile to note that there are not so many particles in the tails.

CONCLUSIONS

We have demonstrated that the self polarization of the LHC high energy protons might be attained by making use of the time varying Stern-Gerlach interaction. The bunch length of 8.39 cm (See Table 2.) fits very well the TE wavelength $\lambda = 10$ cm. This of course assumes that the LHC lattice would be capable of maintaining the polarization of a stored beam; however without the addition of several snakes, this is perhaps illusory. As should be clear, what found here is specific of this particular machine. For other rings, e.g. such as Tevatron [4], things have to be reconsidered, perhaps exploiting other physical properties of particle accelerators.

Since most high energy colliders are not designed with polarization in mind, it becomes problematic to refit them later for polarized beams. Perhaps the right approach would be to design a conceptual collider optimized for polarized beams; then the ascertained concepts could be implemented up front in designs for future machines.

REFERENCES

1. M. Conte et al., INFN/TC-00/03 (2000), (http:xxx.lanl.gov/listphysics/0003, preprint 0003069); M. Conte et al., *ICFA Beam Dynamics Newsletter*, **24**, 66 (2001), (http://wwwslap.cern.ch/icfa/);P. Cameron et al., *Proc. of the SPIN2000 Symposium*, AIP Conf. Proc. **570**, 785 (2001).
2. See, for instance, http://lhc-new-homepage.web.cern.ch/lhc-new-homepage/
3. M. Conte, W.W. MacKay and R. Parodi, BNL-52541, (1997).
4. http://www-bdnew.fnal.gov/tevatron/

The Relativistic Stern-Gerlach Interaction and Quantum Mechanics Implications

P. Cameron*, M. Conte†, A. U. Luccio*, W. W. MacKay*, M. Palazzi† and M. Pusterla**

*Brookhaven National Laboratory, Upton, NY 11973, USA.
†Dipartimento di Fisica dell'Università di Genova, INFN Sezione di Genova, Via Dodecaneso 33, 16146 Genova, Italy.
**Dipartimento di Fisica dell'Università di Padova, INFN Sezione di Padova, Via Marzolo 8, 35131 Padova, Italy.

Abstract. The time varying relativistic Stern-Gerlach force, which acts over a charged particle endowed with a magnetic moment, is deduced from the Dirac Hamiltonian finding its coincidence with the classical expression. Possible drawbacks related to the Heisenberg uncertainty principle are discussed.

THE FORMALISM OF THE STERN-GERLACH INTERACTION

The effect of the Stern-Gerlach (SG) force over a charged particle endowed with a magnetic moment $\vec{\mu}^*$ has been evaluated mainly in the perspective of achieving the separation of the two polarization states of fermion beams, either by acting on the trajectory slopes[1][2] while crossing constant magnetic gradient, or by exploiting the energy differences [3][4] related to the interaction between magnetic moments and electromagnetic fields of a suitable rf cavity. The latter system has also been proposed [5] as a polarimeter. In the particle rest frame (PRF), where all the quantities are labelled with a prime, this force takes the expression

$$\vec{f}'_{SG} = \nabla'(\vec{\mu}^* \cdot \vec{B}') = \frac{\partial}{\partial x'}(\vec{\mu}^* \cdot \vec{B}')\hat{x} + \frac{\partial}{\partial y'}(\vec{\mu}^* \cdot \vec{B}')\hat{y} + \frac{\partial}{\partial z'}(\vec{\mu}^* \cdot \vec{B}')\hat{z}, \tag{1}$$

which when boosted back to the laboratory frame (LAB) becomes

$$\vec{f}_{SG} = \frac{1}{\gamma}\frac{\partial}{\partial x}(\vec{\mu}^* \cdot \vec{B}')\hat{x} + \frac{1}{\gamma}\frac{\partial}{\partial y}(\vec{\mu}^* \cdot \vec{B}')\hat{y} + \frac{\partial}{\partial z'}(\vec{\mu}^* \cdot \vec{B}')\hat{z}, \tag{2}$$

where β and γ are the usual relativistic parameters, and the z-axis has been chosen parallel to the particle velocity $\vec{v}$. Bearing in mind the relation between $\vec{B}'$ and the fields $\vec{B}, \vec{E}$ in the laboratory frame, the longitudinal component of the SG force (2) becomes

$$f_z = \mu_x^* C_{zx} + \mu_y^* C_{zy} + \mu_z^* C_{zz}, \quad \text{with} \tag{3}$$

$$C_{zx} = \gamma^2\left[\left(\frac{\partial B_x}{\partial z} + \frac{\beta}{c}\frac{\partial B_x}{\partial t}\right) + \frac{\beta}{c}\left(\frac{\partial E_y}{\partial z} + \frac{\beta}{c}\frac{\partial E_y}{\partial t}\right)\right], \tag{4}$$

$$C_{zy} = \gamma^2\left[\left(\frac{\partial B_y}{\partial z} + \frac{\beta}{c}\frac{\partial B_y}{\partial t}\right) - \frac{\beta}{c}\left(\frac{\partial E_x}{\partial z} + \frac{\beta}{c}\frac{\partial E_x}{\partial t}\right)\right], \tag{5}$$

CP675, *Spin 2002: 15th Int'l. Spin Physics Symposium and Workshop on Polarized Electron Sources and Polarimeters*, edited by Y. I. Makdisi, A. U. Luccio, and W. W. MacKay

$$C_{zz} = \gamma\left(\frac{\partial B_z}{\partial z} + \frac{\beta}{c}\frac{\partial B_z}{\partial t}\right). \tag{6}$$

It is relevant to point out that Eq. (3) can be deduced from the quantum relativistic theory of the spin-$\frac{1}{2}$ charged particle (Dirac). Indeed, if we start from the Hamiltonian

$$H = e\phi + c\vec{\alpha}\cdot(\vec{p} - e\vec{A}) + \gamma_0 mc^2, \tag{7}$$

with the Dirac's matrices

$$\vec{\gamma} = \begin{pmatrix} 0 & \vec{\sigma} \\ -\vec{\sigma} & 0 \end{pmatrix}, \quad \gamma_0 = \begin{pmatrix} I_2 & 0 \\ 0 & -I_2 \end{pmatrix}, \quad \vec{\alpha} = \gamma_0\vec{\gamma} = \begin{pmatrix} 0 & \vec{\sigma} \\ \vec{\sigma} & 0 \end{pmatrix}, \tag{8}$$

where $\vec{\sigma}$ is a vector whose components are the Pauli's matrices

$$\sigma_x = \begin{pmatrix} 0 & -i \\ i & 0 \end{pmatrix}, \quad \sigma_y = \begin{pmatrix} 1 & 0 \\ 0 & -1 \end{pmatrix}, \quad \sigma_z = \begin{pmatrix} 0 & 1 \\ 1 & 0 \end{pmatrix}, \quad \text{and} \quad I_2 = \begin{pmatrix} 1 & 0 \\ 0 & 1 \end{pmatrix}. \tag{9}$$

Here we have chosen the y-axis as the one parallel to the main magnetic field. By means of standard steps it is straightforward to derive the nonrelativistic expression of the Hamiltonian exhibiting the SG interaction with the "normal" magnetic moment

$$\tilde{H} = e\phi + \frac{1}{2m}(\vec{p} - e\vec{A})^2 - \frac{e\hbar}{2m}(\vec{\sigma}\cdot\vec{B}) \tag{10}$$

which coincides with the Pauli equation. In particular this equation is valid in the PRF.

At this stage we must add to the Stern-Gerlach energy the contribution from the anomalous magnetic moment, which gives rise to a factor $1 + a = g/2$, thus obtaining

$$-\frac{g}{2}\frac{e\hbar}{2m}\,\vec{\sigma}\cdot\vec{B} = -\vec{\mu}^*\cdot\vec{B} \qquad \text{with} \qquad \vec{\mu}^* = g\frac{e\hbar}{4m}\vec{\sigma}. \tag{11}$$

In order to obtain the z-component of the SG force in the Laboratory frame, we must boost the whole Pauli term of Eq. (10) by using the unitary operator U [6] in the Hilbert space, which expresses the Lorentz transformation:

$$U^{-1}\left[g\frac{e\hbar}{4m}(\gamma_0\vec{\sigma}\cdot\vec{B}')\right]U = g\frac{eh}{4m}S^{-1}(\gamma_0\sigma_x)SB_x + S^{-1}(\gamma_0\sigma_y)SB_y + S^{-1}(\gamma_0\sigma_z)SB_z \tag{12}$$

where

$$S = \exp\left[\gamma_0(\vec{\gamma}\cdot\hat{v})\tfrac{u}{2}\right] = I_4\cosh\tfrac{u}{2} + \begin{pmatrix} 0 & \sigma_z \\ \sigma_z & 0 \end{pmatrix}\sinh\tfrac{u}{2}, \tag{13}$$

$$\hat{v} = \frac{\vec{v}}{|\vec{v}|}, \quad \cosh u = \frac{1}{\sqrt{1-\beta^2}} = \gamma \ \ \text{(Lorentz factor)}, \quad \text{and} \quad \beta = \frac{v}{c}. \tag{14}$$

Here I_4 is the 4×4 identity matrix. From the algebraic properties of the γ and σ matrices, we obtain (extending the $\vec{\sigma}$ to 4×4 matrices with the 2×2 Pauli matrices repeated in the diagonal blocks)

$$S^{-1}(\gamma_0\sigma_x)S = \gamma_0\sigma_x, \tag{15}$$

$$S^{-1}(\gamma_0\sigma_y)S = \gamma_0\sigma_y, \quad \text{and} \tag{16}$$

$$S^{-1}(\gamma_0\sigma_z)S = \gamma(\gamma_0\sigma_z) + i\sqrt{\gamma^2-1}\gamma_0\gamma_5, \qquad \text{with} \tag{17}$$

$$\gamma_5 = \gamma_x\gamma_y\gamma_z\gamma_0 = i\begin{pmatrix} 0 & I_2 \\ I_2 & 0 \end{pmatrix}. \tag{18}$$

At this stage it is straightforward to deduce the expectation values of the SG force in the LAB system with a defined spin component (along the y-axis in our case) via the expectation values of the Pauli interaction term and Lorentz transformation of the proper force:

$$f_z = \gamma^2\mu^*\left[\left(\frac{\partial B_y}{\partial z} + \frac{\beta}{c}\frac{\partial B_y}{\partial t}\right) - \frac{\beta}{c}\left(\frac{\partial E_x}{\partial z} + \frac{\beta}{c}\frac{\partial E_x}{\partial t}\right)\right]. \tag{19}$$

In fact, only Eq. (16) gives a non null result, while both Eqs. (15) and (17) yield a null contribution to the expectation values quoted above, because of orthogonality of the two spin states $s = \pm\frac{1}{2}$ and the properties of the σ matrices.

It should be noted that the electric field $\vec{E}$ and the time derivative appearing in Eq. (19) are due to the well known Lorentz transformations

$$\vec{E}' = \gamma(\vec{E} + c\vec{\beta}\times\vec{B}) - \frac{\gamma^2}{\gamma+1},\vec{\beta}(\vec{\beta}\cdot\vec{E}) \tag{20}$$

$$\vec{B}' = \gamma\left(\vec{B} - \frac{\vec{\beta}}{c}\times\vec{E}\right) - \frac{\gamma^2}{\gamma+1}\vec{\beta}(\vec{\beta}\cdot\vec{B}), \tag{21}$$

$$\frac{\partial}{\partial x'} = \frac{\partial}{\partial x}, \quad \frac{\partial}{\partial y'} = \frac{\partial}{\partial y}, \quad \text{and} \quad \frac{\partial}{\partial z'} = \gamma\left(\frac{\partial}{\partial z} + \frac{\beta}{c}\frac{\partial}{\partial t}\right). \tag{22}$$

It is interesting to underline that Eq. (19) coincides with the Eq. (3) if we keep in mind that only a single spin component can be measured.

QUANTUM LIMITATIONS

Let us now consider the effect of a static quadrupole field $\vec{B} = (Gy, Gx, 0)$ on the trajectory slopes. Particles with spins parallel to the y-axis undergo a horizontal impulse, or kick, $\delta p_x = \mu^* G\ell_Q/\beta c$ (ℓ_Q = quadrupole length) which generates two slope variations

$$\eta_\pm = \frac{\delta p_x}{p} = \pm\frac{\mu^* G\ell_Q}{\beta^2\gamma mc^2} \quad (p = \text{particle momentum}) \tag{23}$$

Hence particles with opposite spin states will have, on the long run, separate trajectories. Eq. (23) shows the method is not viable at high energy since $|\eta|$ fades off with the growth of γ.

On the contrary, in the case of the energy exchange between particles and the electromagnetic field of a rf cavity, both Eqs. (4) and (5) exhibit a factor γ^2 which is also present in the expression of the energy gained (or lost) during a cavity crossing. Let us recall the typical example [4] dealing with a rectangular cavity excited in its TE_{011} mode:

$$\Delta U_\pm = \pm 2\gamma^2\mu^* B_0 \tag{24}$$

where B_0 is the cavity's peak magnetic field. The energy variation (24) makes the beam's momentum spread vary by the amount

$$\left(\frac{\Delta p}{p}\right)_{\pm} = \pm\frac{2\gamma}{\beta^2}\frac{\mu^* B_0}{mc^2} \simeq \pm 2\gamma\frac{\mu^* B_0}{mc^2} \qquad \text{(ultrarelativistic)} \tag{25}$$

which implies that the energy separation between particles with opposite spin states is more and more effective as the beam's energy increases.

We return to the case of the nonrelativistic transverse kick [7], considering the classical betatron oscillation of a particle, circulating in a ring, as the motion of the center of its quantum wave packet which represents [8],[9] the wave-function of the "corresponding state". Moreover, when the particle crosses a focusing quadrupole, we can treat its motion with the quantum harmonic oscillator thus obtaining the following expressions for the momentum and position uncertainties:

$$\Delta x_0 = \sqrt{\frac{\hbar}{\sqrt{mveG}}} \qquad \text{and} \qquad \Delta p_0 = \sqrt{\hbar\sqrt{mveG}} \tag{26}$$

As soon as the particle leaves the focusing region and travels along either a free space or a defocusing quadrupole, no discrete energy levels exist anymore and a pure continuous spectrum will appear; this implies a dilation of the wave-packet size, according to the relation

$$\Delta x_{wp} = \Delta x_0 \sqrt{1 + \frac{\lambda^2}{(\Delta x_0)^4}\frac{z^2}{4\beta^2}} \simeq \frac{\lambda}{2\beta\Delta x_0} z, \tag{27}$$

where z is the space covered by the particle outside the focusing quadrupole and $\lambda = \hbar/mc$ is the Compton wave length of the particle.

On the other hand, the momentum uncertainty (26) generates an angular deflection $x'_q = \frac{\Delta p_0}{mv}$ which gives rise to a spatial increment $\Delta x_q = \frac{\Delta p_0}{mv} z$ over a length z. Comparing this growth to the size (27) of the swollen wave-packet, we obtain:

$$\frac{\Delta x_q}{\Delta x_{wp}} = \frac{\Delta p_0}{mv} z \frac{2mv\Delta x_0}{\hbar z} = \frac{2\Delta p_0 \Delta x_0}{\hbar} = 2 \tag{28}$$

If we set, for instance, $G = 10$ Tm^{-1} and $v = 3.095 \times 10^7$ ms^{-1} or $\beta = 0.103$ (5 MeV protons), from Eq. (26) we obtain $\Delta x_0 = 1.93 \times 10^{-8}$, while for $z = 10$ m Eq. (27) yields $\Delta x_{wp} = 5.37 \times 10^{-7}$ m. With all these in mind, we may find, for $\ell = 1$ m, $\delta p_x = 4.56 \times 10^{-33}$ kgms^{-1} or $\delta p_x \Delta x_q \simeq 10^{-39}$ Js $\ll \hbar \simeq 10^{-34}$ Js. However, the random nature of the quantum uncertainty does not allow its growth to macroscopic size; on the contrary, the coherence of our classical kicks permits overcoming the uncertainty limit and realizing a macroscopic effect by addition. In our example, the small spatial increments $\delta x = (\delta p_x/p)z = 8.80 \times 10^{-13}$ m will sum up till reaching the value of Δx_q after about 10^6 revolutions.

As far as the energy gain/loss is concerned, the situation is simpler and more straightforward. In fact in the relativistic case, with ΔE from Eq. (24), Δt being of size $\frac{1}{2}\tau_{\rm rf} \simeq 10^{-10}$ s (i.e. a half period of 3 GHz) and $B_0 = 0.1$ T, we obtain:

$$\Delta E\ \Delta t \simeq \begin{cases} \gamma^2 1.85 \times 10^{-34}\ \text{Js} & \text{for electrons,} \\ \gamma^2 2.82 \times 10^{-37}\ \text{Js} & \text{for protons.} \end{cases} \tag{29}$$

From Eq. (29) it is easy to infer that, in the case of protons, the uncertainty principle constraints are overwhelmed with $\gamma \geq 19$ or $\gamma \geq 48.5$ depending whether we have chosen $\hbar$ or h as limiting quantity. As far as electrons are concerned, even nonrelativistic particles skip these constraints.

We have demonstrated the possibility of detecting sizeable effects of the SG force over a free fermion. We have made it clear that, even in the example of the transverse impulse hidden in the uncertainty blur, this kick can repeat turn after turn giving thus rise to a measurable effect, provided that reasonable operating conditions are settled. Energy exchanges between the particle's kinetic energy and the cavity's electromagnetic fields can be detected with an efficiency proportional to γ.

CONCLUDING REMARKS

Under the hypothesis that the PRF is inertial we proved the equivalence of the classical and quantum mechanical relativistic SG force. We are aware that such an issue is an approximation because the actual motion of the test particle is neither straight nor uniform in any real accelerator. However, the time spent inside the cavities is much shorter than the revolution period and this fully justifies our approximation.

Another relevant application to consider is the proposal of constructing an absolute polarimeter. To be more specific, this might consist of a passive rf cavity, placed along the beam axis, which should detect a total energy transfer of the order of

$$\Delta U \simeq 2NP\mu^* B_0 \gamma^2 \qquad (30)$$

where N is the number of particles belonging to a bunch train, P is the beam polarization and B_0 is now the self-field created by the crossing particles.

Finally, it should be noted that the continuous Stern-Gerlach effect has been recently observed for an electron in an atomic ion[10].

REFERENCES

1. T. Niinikoski and R. Rossmanith, Nucl. Instrum. Methods, A255, 460 (1987).
2. Y. Onel, A. Penzo and R. Rossmanith, AIP Conf. Proc., Vol. 150, 1229 (1986),
3. M. Conte, A. Penzo and M. Pusterla, Il Nuovo Cimento, **A108** (1995) 127.
4. M. Conte et al., INFN/TC-00/03, (http:xxx.lanl.gov/listphysics/0003, preprint 0003069); M. Conte et al., *ICFA Beam Dynamics Newsletter*, **24**, 66 (2001) (http://wwwslap.cern.ch/icfa/); P. Cameron et al., *Proc. of the SPIN2000 Symposium*,Osaka, AIP Conf. Proc. 570, 785 (2001).
5. P. Cameron et al., PAC2001, 2403 (2001).
6. R.P. Feynman, Quantum Electrodynamics, W.A. Benjamin Inc., New York 1961.
7. M. Conte and M. Pusterla, Il Nuovo Cimento, **103A**, 1087 (1990).
8. L.I. Schiff, Quantum Mechanics, McGraw-Hill Co. Inc., New York 1949.
9. P. Caldirola, R. Cirelli e G.M. Prosperi, Introduzione alla Fisica Teorica, UTET, Torino 1982. (In Italian)
10. N. Hermanspahn et al., Phys. Rev. Lett. **84**, 427 (2000).

Macroscopic Quantum Processors Based on Stored High-Energy Polarized Beams

Dennis Sivers

Portland Physics Institute, U. Michigan Spin Physics Center

Abtract. The coherent spin rotation of stored particle beams produces macroscopic states with familiar quantum properties. Because massive quantum systems with spin greater than or equal to one display instructive interference effects not accessible with spin one-half particles, we discuss the sequential rotation and decoherence of a stored polarized deuteron beam. Experimental measurements on such systems should produce unique quantum effects.

The purity and lifetime of polarized beams stored in high-energy rings have improved dramatically in recent decades as progress in polarized sources and refinements in accelerator design have been implemented. In a series of experiments based at the IUCF facility at the University of Indiana [1] a group led by Alan Krisch has systematically studied the spin manipulation of stored beams using a Siberian snake as well as various induced-resonance techniques. These experiments demonstrate conclusively that coherent spin rotation using the Froissart-Stora [2] technique of adiabatic resonance crossing can be reliably controlled with existing radio-frequency magnet technology. The resulting large-scale particle beam systems display carefully-designed quantum complexity. An explicit example of such a process involves the coherent rotation of a stored polarized deuteron beam.

The concept of intrinsic angular momentum or "spin" constitutes an intrinsically quantum-mechanical feature of matter that has no classical counterpart. The quantization of spin is built into the fabric of space-time [3]. For example, the spin-statistics theorem [4] inexorably slots particles as fermions or bosons according to the spin quantum number. While we are familiar with quantum systems at the particle or "atomic" level, most quantum effects are lost in macroscopic systems due to averaging over a very large number of possibilities. However, a beam of particles moving with spins aligned – a "polarized beam"—retains the inherently quantum-mechanical nature of the spin degree of freedom in spite of multiple random interparticle effects. The spatial rotation of the spin degree of freedom using electromagnetic fields necessarily involves highly nontrivial coherent effects. The coherence allows the quantum superposition principle to apply to large-scale quantum spin states. In a storage ring, these states return to the same location at periods controlled by the ring circulation frequency and can be studied in controlled experiments. The idea that particle beams can constitute definite quantum states is not unfamiliar [5] but it does seem to be under appreciated. At a time when the quantum nature of the spin degree of freedom is

CP675, *Spin 2002: 15th Int'l. Spin Physics Symposium and Workshop on Polarized Electron Sources and Polarimeters,* edited by Y. I. Makdisi, A. U. Luccio, and W. W. MacKay

under study for field-effect transistors and other examples of "spintronics" [6], it seems worthwhile for accelerator physicists to consider the application of quantum ideas to the manipulation of stored polarized beams.

The range of possible quantum effects involving polarized beams is quite large. The opportunity for interferometry with multiple beams is worth consideration. We would like to illustrate the basic concept with a discussion of the coherent rotation of a polarized deuteron beam. With a beam stored in the x-y plane and the spin quantization axis in the z direction, the most general form of a massive spin-1 density matrix is

$$\rho = \begin{pmatrix} N_+ & 0 & a(N_+N_-)^{1/2}e^{i\eta} \\ 0 & N_o & 0 \\ a(N_+N_-)^{1/2}e^{-i\eta} & 0 & N_- \end{pmatrix} \qquad 1$$

The activation of a coherent Froissart-Stora spin rotation using a dipole (or a solenoid) rf magnet directly breaks the symmetry in the x-y plane. The breaking of the symmetry produces an explicit example of the Higgs mechanism [7] in which the symmetry breaking couples the [0> states to the [+> and [-> quantum states. The effect of the rotation can be illustrated by a simple example. Let c=cos(θ) and s=sin(θ), the coherent rotation of the pure quantum state represented by the idempotent density matrix

$$\rho_+(0) = \begin{pmatrix} 1 & 0 & 0 \\ 0 & 0 & 0 \\ 0 & 0 & 0 \end{pmatrix} \qquad 2$$

produces the density matrix

$$\rho_+(\theta) = \frac{1}{4}\begin{pmatrix} (1+c)^2 & s(1+c)\sqrt{2} & s^2 \\ s(1+c)\sqrt{2} & 2s^2 & s(1-c)\sqrt{2} \\ s^2 & s(1-c)\sqrt{2} & (1-c)^2 \end{pmatrix} \qquad 3$$

The density matrix also represents a pure quantum state. If, however, the rotator magnet is turned off after a rotation through an angle less than 180 degrees, the coherence between the [0> states and the [+> and [-> states which had been maintained because of the broken symmetry during the magnet operation begins to be lost as the original symmetry is restored. The subsequent decoherence occurs with a time scale which depends of the full set of operating parameters of the storage ring. The decoherence may happen within a few cycles or the rotated "pure" state may persist as a quasistable state for thousands of cycles. The diagonal elements of the density matrix are preserved as the phase information is lost. However, a subsequent coherent rotation can detect the amount of "depolarization." Therefore, a systematic set of measurements of the deuteron polarizations, P_z and P_{zz}, after a set of timed sequences involving a coherent rotation, decoherence interval, and further rotation can explore the dynamics of this simple, but interesting, quantum system in some detail. Because of the full quantum nature of the spin degree of freedom, a polarized deuteron beam can provide a quantum workbench to study the Higgs mechanism in action.

While we have chosen a simple example from a spin-1 system to illustrate a nontrivial quantum effect, it should be kept in mind that numerous examples involving spin-1/2 systems can also be formulated. This presentation has not focused on the accelerator physics techniques that allow the storage and manipulation of polarized beams. Many practical aspects are discussed in the papers cited in reference 1. Further discussion concerning the application of Froissart-Stora rotations can be found in the presentations of Vassili Morozov at this conference (7). This talk is intended to point out some of the "sophomore quantum mechanics" of polarized beams. Because these aspects of polarized beams are simple, they can often be ignored. Stored polarized beams do, however, involve some profound information about the nature of space and time so that the manipulation and measurement of the spin degree of freedom in these systems can prove to be very informative.

It is instructive to compare the properties of a stored polarized beam with the properties of other macroscopic quantum systems. Lasers provide macroscopic quantum collection of photons. The range of applications that has been found to date for lasers is enormous. However, the photons in a laser lack one aspect of control that stored particle beams possess, the charged particles in a storage ring can be manipulated by electromagnetic fields in a very controlled manner. Multiple beams or multiple bunches within the same ring can be controlled from can be controlled from a few simple instructions. The phase information contained in the polarized beam can be used to monitor these instructions. In this sense, a polarized beam in a storage ring can be considered a full-fledged quantum processor.

REFERENCES

1. A.M.T. Lin, et al., *SPIN 2000, 14th International Spin Physics Symposium*, Edited by K. Hatanaka, et al. (American Inst. Of Physics, NY, 2001) p.236 and references therein.
2. M. Froissart and R. Stora, *Nucl. Ins. Meth.* 7, 297 1960
3. See, for example, the discussion of the Dirac monopole quantization condition in D. Sivers, *High Energy Spin Physics-1982*, (American Inst. Of Physics, NY, 1983) p. 13
4. R.F. Streater and A.S. Wightman, *PCT, Spin and Statistics, and All That*, (WA Benjamin, NY, 1964)
5. D.P. Barber, et al, *Proc. ICGA Workshop "Quantum Aspects of Beam Physics"*, Monterey 1998, (World Scientific, NY 1999)
6. G. Zorpette, *IEEE Spectrum*, Dec. 2001, p. 30
7. V. Morozov, these proceedings.

Preserving Polarization through an Intrinsic Depolarizing Resonance with a Partial Snake at the AGS[1]

H.Huang*, L. Ahrens*, M. Bai*, K.A. Brown*, J.W. Glenn*, A.U. Luccio*, W.W. MacKay*, C. Montag*, V. Ptitsyn*, V. Ranjbar†, T. Roser*, H. Spinka**, N. Tsoupas*, D.G. Underwood** and K. Zeno*

*C-A Department, Brookhaven National Laboratory, Upton, NY 11973, USA
†Physics Department, Indiana University, Bloomington, IN 47505, USA
**Argonne National Laboratory, Argonne, IL 60439, USA

Abstract. An 11.4% partial Siberian snake was used to successfully accelerate polarized protons through a strong intrinsic depolarizing spin resonance in the AGS. No noticeable depolarization was observed. This opens up the possibility of using a 20% partial Siberian snake in the AGS to overcome all weak and strong spin resonances.

INTRODUCTION

Acceleration of polarized proton beams to high energy in circular accelerators is difficult due to numerous depolarizing resonances. It is particularly difficult in the medium energy range since the limited available straight sections in the existing synchrotrons make it very hard to install a full Siberian snake [1] to correct all types of depolarizing spin resonances. During acceleration, a depolarizing resonance is crossed whenever the spin precession frequency equals the frequency with which spin-perturbing magnetic fields are encountered. In the presence of the vertical dipole guide field in an accelerator, the spin precesses $G\gamma$ times per orbit revolution [2], where $G = (g-2)/2 = 1.7928$ is the gyromagnetic anomaly of the proton, and γ is the Lorentz factor. The number of precessions per revolution is called the spin tune ν_{sp} and is equal to $G\gamma$ in this case.

There are three main types of depolarizing resonances: imperfection resonances, which are driven by magnet misalignments; intrinsic resonances, driven by the vertical betatron motion through quadrupoles; and coupling resonances, caused by the vertical motion with horizontal betatron frequency due to linear coupling [3]. The resonance condition for an imperfection resonance is $\nu_{sp} = n$, where n is an integer. The resonance condition for an intrinsic resonance is $\nu_{sp} = nP \pm \nu_y$, where n is an integer, $P = 12$ is the super-periodicity of the AGS, and ν_y is the vertical betatron tune. The resonance condition for a coupling spin resonance is $\nu_{sp} = n \pm \nu_x$, where ν_x is the horizontal betatron tune; it is only important in the vicinity of a strong intrinsic resonance.

[1] This work was supported by the DOE of USA.

CP675, *Spin 2002: 15th Int'l. Spin Physics Symposium and Workshop on Polarized Electron Sources and Polarimeters*, edited by Y. I. Makdisi, A. U. Luccio, and W. W. MacKay

At the AGS, a 5% partial Siberian snake [5] has been used to overcome the imperfection resonances [6], and an ac dipole has been used to overcome strong intrinsic resonances [7]. For a ring with a partial snake with strength s, the spin tune ν_{sp} is given by

$$\cos\pi\nu_{sp} = \cos\frac{s\pi}{2}\cos G\gamma\pi, \tag{1}$$

where $s = 1$ would correspond to a full snake which rotates the spin by 180°. When s is small, the spin tune is nearly equal to $G\gamma$ except when $G\gamma$ equals an integer n, where the spin tune ν_{sp} is shifted away from the integer by $\pm s/2$. Thus, the partial snake creates a gap in the spin tune at all integers. Since the spin tune never equals an integer, the imperfection resonance condition is never satisfied. Thus the partial snake can overcome all imperfection resonances, provided that the resonance strengths are much smaller than the spin-tune gap created by the partial snake.

By adiabatically exciting a vertical coherent betatron oscillation using a single ac dipole magnet, an artificial spin resonance is excited. If the resonance location is chosen near an intrinsic spin resonance, the spin motion will be dominated by the ac dipole resonance, and full spin flip can be achieved [7]. However, the ac dipole technique only works for strong intrinsic resonances, since it relies on the strength of the intrinsic resonances to induce a strong enough artificial resonance. With a strong enough partial snake, the spin tune gap can be increased to allow placing the betatron tune inside the gap so the intrinsic resonance conditions can also be avoided. Simulations showed that for the first intrinsic resonance at $0+\nu_y$, a 10% partial snake would be strong enough. In a recent experiment at the AGS this was successfully demonstrated. This paper will present the results and discussions.

EXPERIMENT SETUP

The polarized H^- beam from the optically pumped polarized ion source (OPPIS) [8] was accelerated through a radio frequency quadrupole and the 200 MeV LINAC. The beam polarization at 200 MeV was measured with elastic scattering from a carbon fiber target. During the study, the polarization measured by the 200 MeV polarimeter was (66±0.5)%. The beam was then strip-injected and accelerated in the AGS Booster up to 1.5 GeV kinetic energy or $G\gamma = 4.7$. The vertical betatron tune of the AGS Booster was chosen to be 4.9 in order to avoid crossing the intrinsic resonance $G\gamma = 0+\nu_y$ in the Booster. The imperfection resonances at $G\gamma = 3,4$ were corrected by harmonic orbit correctors. Only one bunch of the twelve rf buckets in the AGS was filled, and the beam intensity varied between 1.3 -1.7×10^{11} protons per fill. The polarized proton beam was accelerated up to $G\gamma = 12.5$ or 5.6 GeV kinetic energy passing through just one intrinsic resonance located at $G\gamma = 0+\nu_y$. The resonance crossing speed $\alpha = \frac{d(G\gamma)}{d\theta}$ was 2.4×10^{-5}. Polarization was measured at $G\gamma = 12.5$ during an approximately one second flattop after the partial snake was ramped to zero. At $G\gamma = 0+\nu_y$, the solenoid can generate a 25% partial snake. At the same time, the solenoidal field will generate significant coupling, which will cause sizeable depolarization. In addition, such

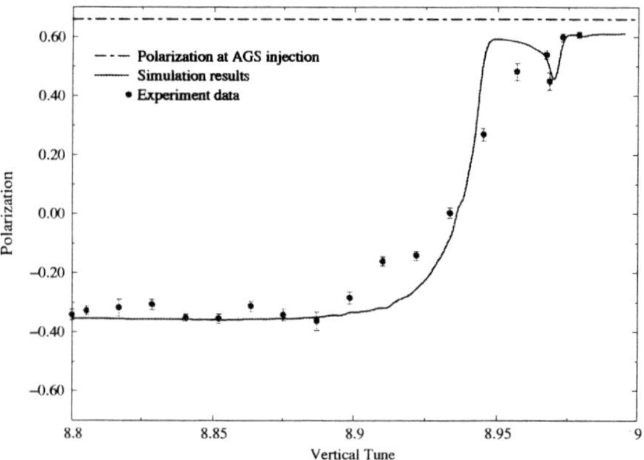

FIGURE 1. The measured vertical polarization as a function of the vertical betatron tune for an 11.4% partial snake. The dots are measured polarization, and the error bars only represent the statistical errors. The solid straight line indicates the polarization level measured at the end of the LINAC. Since the two imperfection resonances in the Booster have been corrected by harmonic orbit correctors, this is also the polarization at AGS injection. The solid curve shows the simulation results.

a strong snake will tilt the stable spin direction away from vertical by 12.5°, reducing the measurable vertical polarization component. An 11.4% partial snake was chosen as a compromise between obtaining a large enough spin tune gap and minimizing the coupling effects. The AGS partial snake was turned on before injection at 6% and ramped up to 11.4% for the first intrinsic resonance at $0+\nu_y$. The orbit was carefully corrected to maintain stable beam as the vertical betatron tune was moved as high as 8.98. During the experiment, the horizontal tune was kept at 8.54, while the beam polarization was measured as a function of the vertical betatron tune.

RESULTS AND DISCUSSION

The experimental data and simulation results are plotted in Fig.1. The simulation is a combination of a DEPOL [9] calculation and a tracking model with two overlapping resonances: one located at $G\gamma = 9$ generated by the partial snake and the intrinsic resonance at $G\gamma = 0+\nu_y$, which changes its location when the vertical betatron tune changes. The strength of the intrinsic resonance was determined from beam size measurements using the AGS ionization profile monitor(IPM). A ± 0.004 vertical tune spread was included in the simulation. The vertical tunes were not measured in the tune window $\nu_y = 8.9$ to 8.96. A fitting of measured tunes vs. set tunes outside the window was done. Inside the window vertical tunes were derived from the set values based on the fitting. The strengths of the coupling resonances located at $G\gamma = 17-\nu_x, 0+\nu_x, 18-\nu_x$, and $1+\nu_x$ were calculated from the extended DEPOL [9] program. Since the coupling resonances are separated from the other two resonances, they can be treated independently. When the intrinsic and imperfection resonances do not overlap ($\nu_y \sim 8.85$), the resonance at $G\gamma = 9$ will flip spin completely while the intrinsic resonance at $G\gamma = 0+\nu_y$

will flip spin partially and cause depolarization. Note the polarization is getting slightly lower when the vertical betatron tune moves closer to the horizontal betatron tune; this is due to stronger coupling resonances. When the two resonances are very close, such as $\nu_y = 8.98$, the intrinsic resonance is overpowered by the resonance at $G\gamma = 9$. Beam essentially just experiences one resonance at $G\gamma = 9$, and full spin flip is observed. When the two resonances are a modest distance apart, such as for $\nu_y > 8.90$, the two resonances interfere with each other, and $G\gamma = 9$ resonance gradually dominates when the vertical betatron tune goes higher. The turning point is $\nu_y \sim 8.94$ because an 11.4% partial snake generates a 0.057 gap on each side of the integer. In general, most data points agree well with the simulation. The remaining discrepancies for data points between $\nu_y = 8.9$ and 8.96 could be explained with different beam size or different vertical betatron tune, but there were no beam size and vertical tune measurements for these data points.

Both experiment and simulation show a polarization dip close to ν_y=8.97. This is caused by a snake resonance [10]. Even when the intrinsic resonance condition can not be met for $\nu_y > 8.95$, depolarization can occur from resonance conditions extended over many turns if the intrinsic resonance is very strong. This happens when the following condition is met

$$\Delta\nu_y = \frac{k \pm \nu_{sp}}{n}, \tag{2}$$

where $\Delta\nu_y$ is the fractional part of vertical betatron tune, n and k are integers, and n is called the snake resonance order. With an 11.4% partial snake, the spin tune is close to 0.057 for $G\gamma \sim 9$. The polarization dip then corresponds to the second order snake resonance ($n = 2$). With the given acceleration rate and intrinsic resonance strength, snake resonances higher than second order do not show a significant effect. The existence of the snake resonance reduced the usable tune space and the betatron tune had to be carefully chosen to avoid depolarization.

At the vertical betatron tune of 8.98, the difference between the polarization at injection and polarization measured after $0 + \nu_y$ is due to the spin mismatch at injection and some additional depolarization from coupling resonances. If spin matching were achieved at AGS injection and the linear coupling were eliminated, the scheme could provide full spin flip through the intrinsic resonance. It also would work for weak intrinsic resonances, such as $G\gamma = 24 \pm \nu_y, 48 - \nu_y$.

OUTLOOK

In conclusion, we have demonstrated for the first time that an 11.4% partial snake can effectively overcome an intrinsic depolarizing resonance when the vertical betatron tune is put close to an integer. The critical element of this operation is to maintain beam stability with betatron tune close to an integer. For AGS, this method has several advantages. First, it works for both strong and weak intrinsic resonances. Currently, there is no effective way to overcome the weak intrinsic resonances in the AGS. Second, if the coupling of a new snake could be reduced, the strength of coupling resonances could also be reduced. Or, if both horizontal and vertical tunes could be put into the gap, both intrinsic and coupling resonances could be avoided. Simulation shows that a 20%

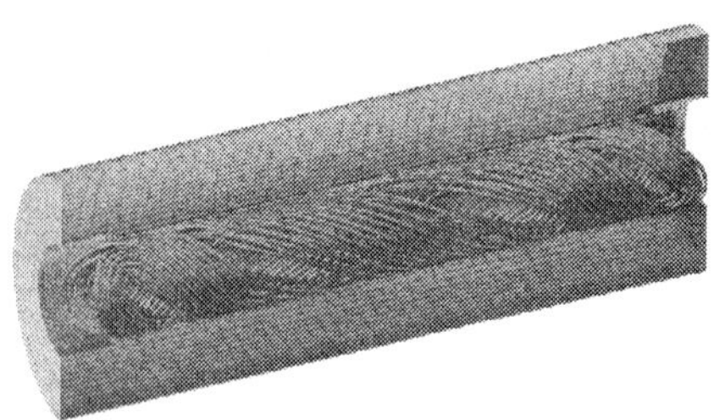

FIGURE 2. The design of superconducting snake magnet. It is 2.6 meters long and the beam pipe is 15 cm by 15 cm.

partial snake is needed for the strongest intrinsic resonance at $36+\nu_y$. This is beyond the capability of the existing solenoidal partial snake. Furthermore, the solenoidal field is the main source of coupling, which causes coupling resonances in the vicinity of strong intrinsic resonances. The better choice would be a helical dipole magnet as has been used in RHIC polarized proton operation. With the constraint of 10-foot AGS straight section, the required field can only be achieved by a super-conducting magnet. With compensating coils, the coupling from the new snake can be greatly reduced. The design of a super-conducting helical AGS snake with strength on the order of 20% to replace the current solenoidal AGS partial snake has already begun. A sketch of the design is shown in Fig. 2. With the given field, the snake strength at AGS injection would be about 24%. At injection energy, the tune shifts are of the order of 0.2 units, and the beta functions fluctuate throughout the ring up to values around 100 m instead of the matched maximum of about 22 m. A solution has been found by E. Courant [11] to eliminate the coupling and beta function mismatch caused by the snake. In addition, such a strong snake in the AGS will tilt the stable spin direction away from vertical significantly. Study of the spin matching at injection and extraction from the AGS is also underway to ensure the best polarization transfer efficiency at AGS injection and extraction.

REFERENCES

1. Ya.S. Derbenev and A.M. Kondratenko, Part. Accel. **8**, 115 (1978).
2. L.H. Thomas, Philos. Mag. **3**, 1 (1927); V. Bargmann, L. Michel, and V.L. Telegdi, Phys. Rev. Lett. **2**. 435 (1959).
3. H. Huang, T. Roser, A. Luccio, Proc. of 1997 IEEE PAC, Vancouver, May, 1997, p.2538.
4. M. Froissart and R. Stora, Nucl. Instrum. Meth. **7**, 297(1960).
5. T. Roser, in *Proceedings of the 8th International Symposium on High-Energy Spin Physics*, Minneapolis, 1988, AIP Conf. Proc. No 187 (AIP, New York,1989), p.1442.
6. H. Huang, *et al.*, Phys. Rev. Lett. **73**, 2982 (1994).
7. M. Bai, *et al.*, Phys. Rev. Lett. **80**, 4673 (1998).
8. A. Zelenski, *et al.*, in *Proceedings of the 9th International Conference on Ion Sources*, Rev. Sci. Inst., Vol.73, No.2, p.888 (2002).
9. V. Ranjbar, *et al.*, Proc. of 2001 IEEE PAC, Chicago, June, 2001, p. 3177; E.D. Courant and R.D. Ruth, BNL report 51270, (1980).
10. S.Y. Lee and S. Tepikian, Phys. Rev. Lett. **56**, 1635 (1986).
11. E. Courant, these proceedings.

Matching Quadrupoles for AGS Helical Snake

E. D. Courant

Brookhaven National Laboratory

Assume: a helical snake inserted in AGS section I-20 with Parameters:
Total length 2.1 m, field strength 3.0 Tesla.
First section: 44.5 cm, twist 180 degrees
Second section: 1.21 m, twist 245 degrees
Third section: 44.5 cm, twist 180 degrees.

Assume (for the sake of manageable analytic computation): abrupt transitions from zero field to full field at ends of helix; abrupt change of pitch between end and center sections.
At injection energy (Ggamma = 4.7) the snake strength is 34.9%, and the trajectory is as shown

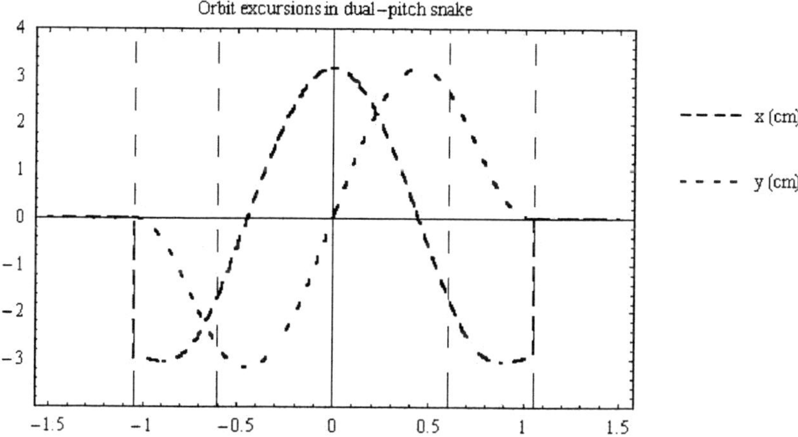

Because of Maxwell's equations, a helical field automatically include nonlinearities, local field gradients and solenoidal fields, varying over the trajectory. Averaging these over the trajectory, we can compute the effective transfer matrix for the helix, which unfortunately is nothing like that for a drift space of the same length. For the whole 10-foot (3.048 m) straight section including the snake, this matrix comes out to be

CP675, *Spin 2002: 15th Int'l. Spin Physics Symposium and Workshop on Polarized Electron Sources and Polarimeters,* edited by Y. I. Makdisi, A. U. Luccio, and W. W. MacKay

0.664572	2.646738	-0.03787	-0.1274
-0.21013	0.664572	0.007435	-0.03248
0.032485	0.127396	0.845606	2.809608
-0.00743	0.037874	-0.10064	0.845606

which includes significant coupling terms (upper right and lower left corners).
The fields on the trajectory (again calculated with the hard-edge approximations) are

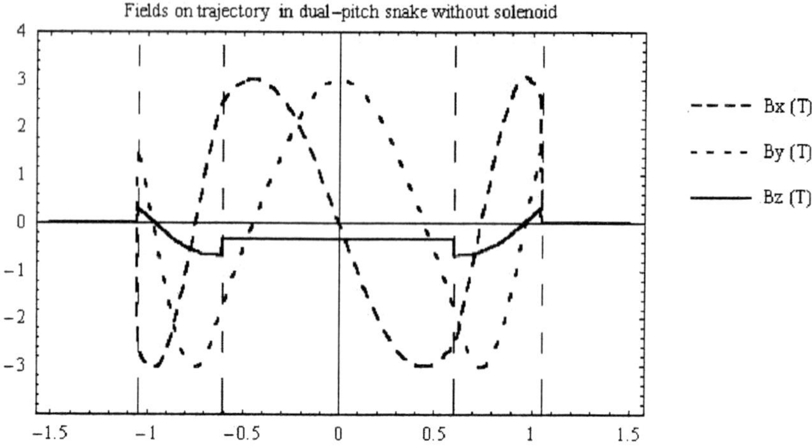

By superimposing a suitable solenoidal field we can eliminate the coupling terms. With a solenoidal field of 0.3371 T the fields are as shown below:

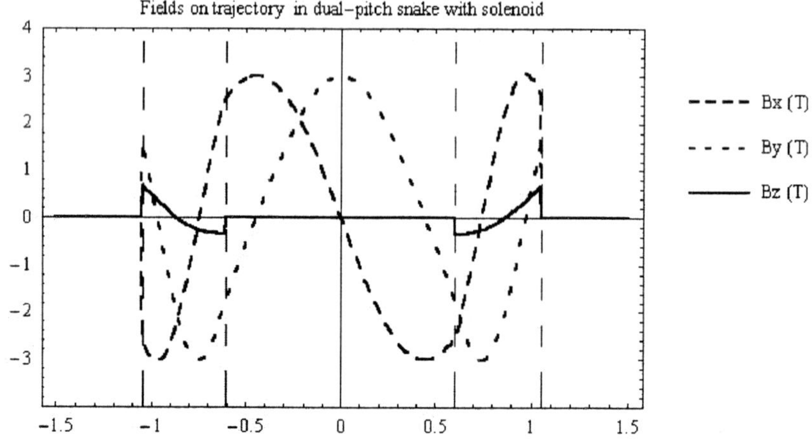

the snake is now 27.7% of a full snake, and . the matrix is altered to

0.666718	2.651499	-7.21E-06	-2.9E-05
-0.2095	0.666718	1.59E-06	-9.05E-06
9.05E-06	2.92E-05	0.848034	2.814629
-1.59E-06	7.21E-06	-0.09978	0.848034

i.e. the coupling is eliminated. But the matrix is a long way from a simple drift matrix; this causes a huge mismatch of the orbit stability. With the tunes of the plain AGS set at (8.1,8.75) the tunes come out as (8.29,8.88), i.e. a tune change of around 0.2 units, and the orbit functions are as shown:

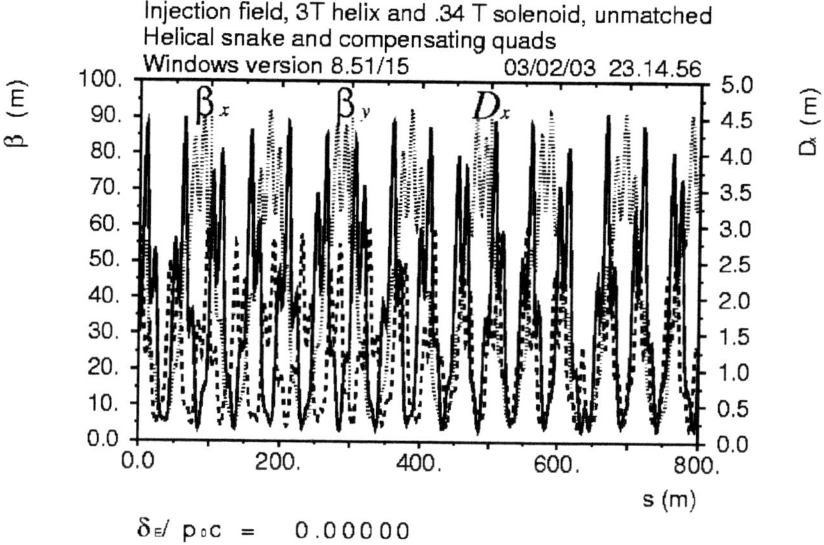

(Here the snake is located in the last superperiod, i.e. at s = 790 m). Recall that in the bare AGS the maximum beta functions are about 22 m, and the dispersion function 1.8 m.)

Using the MAD program we find that with quadrupoles in sections I17, I19, J1, J3 (the sections nearest to the snake at I20) we can match the beta functions perfectly (except, of course, in the section containing the snake).
The quadrupoles are assumed to have effective length of 35 cm, and their strengths are calculated as

I17 = 1.82 T/m,
I19= 1.63 T/m,
J1 = -0.92 T/m,
J3 = 1.92 T/m.

These are quite modest, and should be easy to put in. With these quadrupoles the orbit functions become:

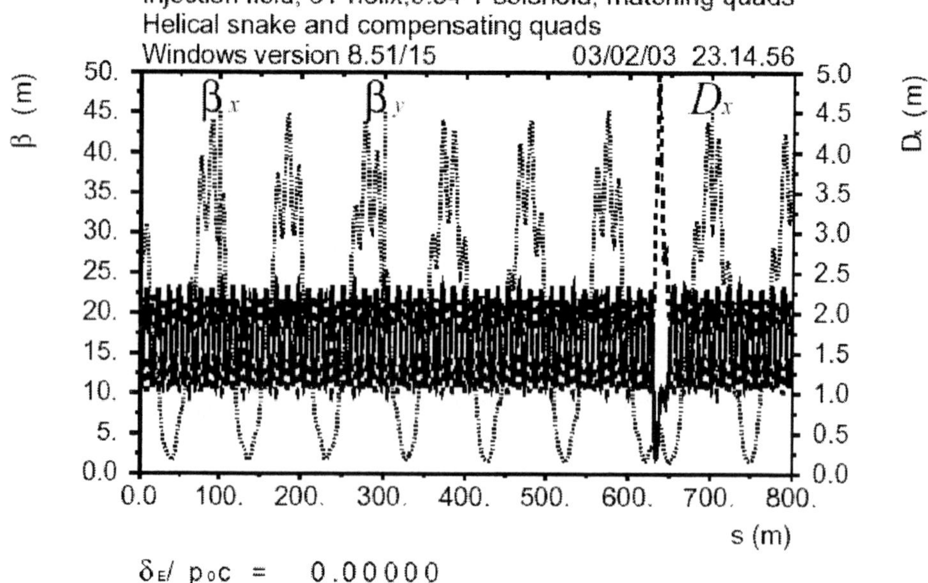

Note that the beta functions are perfectly matched, but the dispersion function is not (no attempt was made to match it – undoubtedly that could be done too). The tunes are 8.65 and 8.97; this could easily be adjusted.

At higher energies the effect of the snake on the orbit is weaker. At Ggamma = 8.9 (the 0 + nu resonance) the unmatched beta functions have a maximum of 39 meters as against 90 at injection, and the solenoid needed to counteract coupling is down to 0.17 Tesla. The compensating quadrupoles are, for one set of tunes,

I17 = 3.46 T/m,
I19= -0.65 T/m,
J1 = 2.12 T/m,
J3 = 0.65 T/m.

Finally, at high energy no compensation is needed. In all cases the tunes in my examples are about 8.65 and 8.95. Another cost of the compensating quads is that they break the 12-fold periodicity of the AGS lattice; therefore in addition to the intrinsic resonances at $G\gamma = 12k \pm \nu$ there are now resonances at

$$G\gamma = k \pm \nu$$

with all integers k. Thus with the corrections at injection energy the spectrum of resonances is now

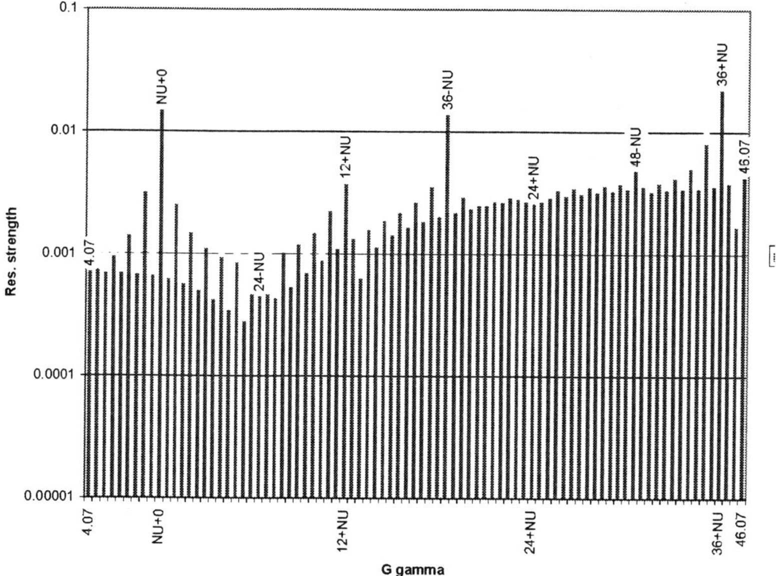

Thus new resonances have now appeared. But the ones near injection energy are weak enough so that, with the partial snake, they will not cause any trouble.

At higher energies the compensating quadrupoles are relatively weaker, and the resonances they cause are correspondingly weaker. With the compensating quadrupoles for $G\gamma = 8.9$ the resonance spectrum is

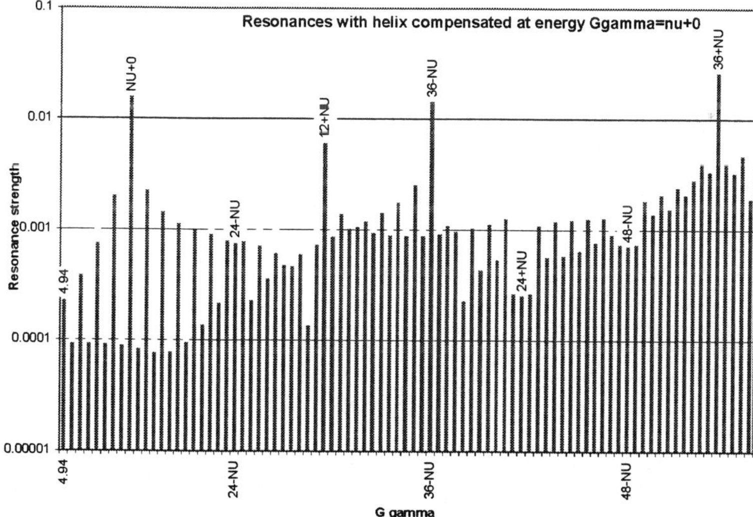

and the symmetry-breaking resonances are weak compared to the main 0+ν resonance, so clearly the will not affect the partial snake's ability to cope with resonances.

Polarimetry Proton and Electron Beams

Proton Beam Polarimetry at BNL

Harold Spinka

HEP Division, Argonne National Laboratory, Argonne, IL 60439 USA

Abstract. A brief overview is presented of the beam polarimeters in the LINAC – Booster – AGS – RHIC complex at Brookhaven National Laboratory. Absolute calibrations, performance, and outstanding issues for these polarimeters, as well as their interactions with the operations of the accelerators, are discussed. The role of systematic effects in the polarimeter data is emphasized.

A new optically pumped polarized ion source was installed at Brookhaven National Laboratory (BNL) in 1999-2000. This source produces more than an order of magnitude higher beam intensity. However, the 200 MeV polarimeter [1] located after the LINAC and the polarimeter [2] internal to the Alternating Gradient Synchrotron (AGS) ring remained relatively unchanged until this past run. As might be expected, the increased beam intensity led to rate problems in both polarimeters. This talk describes recent changes and operating experience with these instruments.

Coulomb-Nuclear Interference (CNI) polarimeters in the blue and yellow RHIC rings were used extensively during the recent polarized proton run period. Additional details will be presented in the following talks by O. Jinnouchi and K. Kurita. Some observations about the performance of these polarimeters and of RHIC will also be provided in this talk.

THE 200 MeV POLARIMETER

The 200 MeV polarimeter [1] consists of pairs of scintillation counters at laboratory angles of ±12° and ±16° to the incident polarized proton beam and a thin carbon strip target. During late 2001, changes were made to handle the higher rates from the new ion source. The ±12° counters were moved to a larger distance from the target, and lead shielding and collimators were added to restrict the origin of the detected charged particles to near the target. Upgrades to the data acquisition system were also made that allowed more events to be recorded in a single LINAC beam pulse.

The absolute calibration of the modified 200 MeV polarimeter was checked using p+d elastic scattering on a CD_2 target. The ±16° scintillation counters were moved to θ_{lab} = ±64°, and deuteron telescopes of three scintillation counters each were added at θ_{lab} = ±42.6°. The analyzing power was taken from measurements at IUCF [3]. The values of beam polarization measured with p+d elastic scattering and that measured using p+C scattering with the old calibration are consistent:

CP675, *Spin 2002: 15th Int'l. Spin Physics Symposium and Workshop on Polarized Electron Sources and Polarimeters,* edited by Y. I. Makdisi, A. U. Luccio, and W. W. MacKay

$$p_d / p_C = 0.987 \pm 0.014 . \quad (1)$$

However, a reduction in photomultiplier gain with time was observed during the LINAC pulse in the calibration experiments. The problems were more severe in the p+d than the p+C measurements. For example, the p+C asymmetry was observed to drop with increasing intensity. The variation in intensity was provided by adjusting slits or a LINAC quadrupole, neither of which is expected to change the beam polarization. These rate problems suggest that systematic effects in the polarimeter may exceed the statistical uncertainty in the calibration (Eq. 1). It will be important to further upgrade the 200 MeV polarimeter to become more rate tolerant.

THE AGS POLARIMETER

The AGS polarimeter [2] was used in efforts to minimize polarization losses through depolarizing resonances with the AGS partial solenoidal snake and ac-dipoles over the past decade. It consists of two telescopes of three scintillation counters and an adjustable degrader to the left and right of the beam at $\theta_{lab} \cong \pm 77°$. Thin carbon fiber targets of diameter ~ 8 μm are used to avoid undesirable rate effects in the photomultipliers due to the high beam currents from the new ion source. The analyzing power drops with momentum as approximately $A \propto P_{lab}^{-1}$. At the extraction momentum of 24 GeV/c, A ~ 0.006, which leads to long measurement times (20 – 30 min.) in order to obtain statistically significant beam polarization results.

Observation of changes in asymmetries with time measured with the AGS polarimeter has led to suggestions of systematic errors comparable to statistical uncertainties. However, these changes could be caused by variations in accelerator conditions as well as problems with the polarimeter. For example, there is evidence for changes in the relative position of the beam and target from spill to spill, possibly causing a systematic error in the measured asymmetries. The origin could be non-reproducible target position as it is moved into the beam each spill, or a variable beam position in the AGS. New diagnostics to display the polarimeter asymmetry as a function of time were implemented for the recent polarized proton run period; no obvious problems were observed during the operation with carbon fiber targets.

A recent absolute calibration of the AGS polarimeter was performed at P_{lab} = 3.8 GeV/c (Gγ = 7.5) for nylon and carbon targets. Scintillation counters at small angles to the beam were added to select pp elastic scattering events (where the analyzing power is well known) from the nylon target. Beam polarizations were then measured to be consistent with those seen at the 200 MeV polarimeter. During these studies it was shown that some previous results at Gγ = 7.5 had contamination of carbon target data by nylon events and vice versa. This was due to the ~ 2.5 cm separation of carbon and nylon targets on the target holder and the large beam size at this momentum. Detailed analyses [4] suggest consistency of all Gγ = 7.5 data from 1996 – 2002, except for those from September 2000.

RHIC AND THE CNI POLARIMETERS

RHIC operated as the first polarized proton beam collider from November 2001 – January 2002. Any new type of accelerator will provide new challenges, and RHIC is no exception. During that period, RHIC operated with nominally 60 equally spaced beam bunches separated by ~ 210 ns or 6° around each ring. These bunches were filled one at a time from a single AGS spill each. The beam polarization direction was vertical. Each ring had 26 bunches with polarization up, 26 with polarization down, and 3 unpolarized bunches in an alternating pattern. In addition, there were 5 adjacent bunches that were unfilled to allow the beam to be dumped safely – this was called the "abort gap." A total of 10 filled bunches in the yellow ring "saw" the abort gap of the blue ring in two of the six interaction regions, and vice versa. The other 45 filled bunches always saw bunches with beam in each of the intersection regions.

For accurate polarimetry, beam conditions must be stable from bunch to bunch and vary slowly with time. Such conditions include the distributions of beam particle angles and transverse and longitudinal positions relative to the nominal trajectory, the spread of momenta of the particles in each bunch, and the polarization of each bunch. Some of these are well constrained by the accelerator design and operations, while others are not. The polarimeter design may also limit sensitivity to certain beam conditions. For example, the small CNI polarimeter ribbon target reduces the sensitivity to transverse shifts of the beam from bunch to bunch.

There are several known mechanisms that could produce different beam phase space among the bunches, in addition to the obvious possible changes in the AGS or the AGS-RHIC transfer line operating conditions. These include: a) One bunch was perturbed to measure the beam tune. Thus, it was affected differently than the other bunches. b) Beam leakage out of filled bunches into the abort gap was observed at times. The phase space of the leakage beam was probably different from the remaining beam. Different bunch intensities and such leakage could also cause different phase space among the bunches and a possible decrease in polarization with time (see Fig. 1). c) Some bunches experienced significantly fewer interactions than others due to the abort gap (see the talk by J. Kiryluk).

These various mechanisms should be studied experimentally and perhaps with simulations to estimate the impact on the polarimeter measurements. Unpolarized beam can probably be used to study some of these effects (see Ref. [5] for a way to search for phase space differences using luminosities measured in various intersection regions).

The CNI polarimeters in each RHIC beam observed carbon recoil nuclei with silicon strip detectors. Very small angle p+C elastic scattering was measured, and the forward proton was not detected. Timing and pulse height information were determined for each strip, which helped to reduce backgrounds. Each polarimeter consisted of three "left-right" pairs of silicon detectors. One pair was horizontal, in the plane of the beam. The other two were at ±45°, and allowed both transverse components of the beam spin to be measured (see the talks of O. Jinnouchi and K. Kurita).

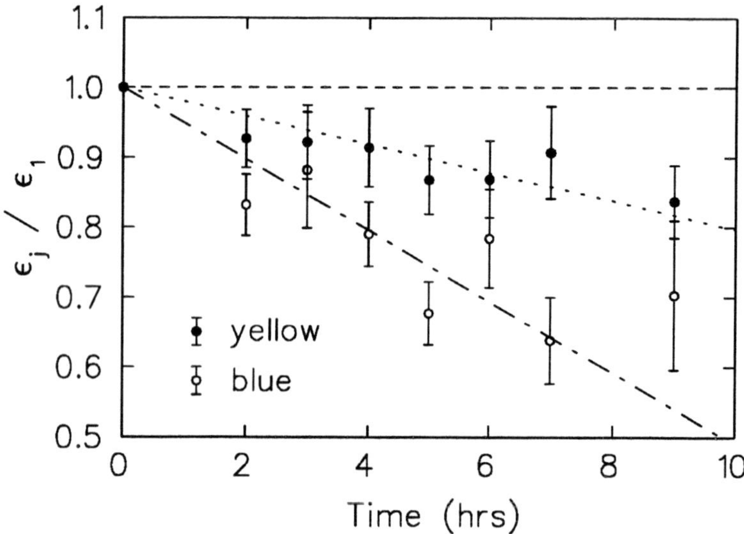

FIGURE 1. The average of the ratios of the asymmetry measured at least 1.5 hours into flattop to the asymmetry at the beginning of flattop is plotted as a function of time since the start of flattop for the blue and yellow rings. At least three polarimeter measurements were required on flattop in the same beam and fill, with the magnitude of the initial asymmetry exceeding 1.0×10^{-3}. Fills with special conditions or with known hardware problems were excluded. It had been expected that these ratios would be consistent with 1.00.

The absolute calibrations of the CNI polarimeters are known only at the RHIC injection momentum (P_{lab} = 24 GeV/c, A ~ 0.013) [2,6]. At the operating energy of 100 GeV, it is possible to demonstrate that A $\geq$ 0.013 based on observed asymmetries at the two energies and the requirement that the beam polarization does not increase during acceleration. It is planned to obtain a better calibration if the beam can be successfully decelerated to 24 GeV/c. Measuring the asymmetries at injection, 100 GeV, and at injection again would allow the calibration to be found if the two asymmetries at injection are (nearly) equal.

Data from the CNI polarimeters are being studied for systematic effects. For example, it was discovered that the "earliest" bunch to be injected into RHIC from the AGS often had anomalous specific luminosity. The cause is not presently understood. However, there may also be systematic effects caused by the non-ideal operation of the polarimeters. Especially important are possible rate effects, and degradation of the silicon detectors from radiation damage, causing an increase in noise and leading to a gain change in the electronics. Of lesser importance are the slow degradation of the ribbon targets with integrated beam intensity, and the reproducibility of the target position from measurement to measurement, which could affect the analyzing power.

Two observations suggest systematic errors in the CNI polarimeter data. The nominal beam polarization was vertical. However, statistically significant up-down asymmetries were observed for a sizable number of runs. These asymmetries were computed from the two pairs of detectors at ±45° to the horizontal plane. Secondly, the three left-right pairs of detectors all viewed the same carbon ribbon target. Thus, the three luminosity asymmetries should be the same within statistics, but they disagreed much more than expected. The luminosity asymmetries were computed from:

$$\varepsilon_{LUM} = \frac{\sqrt{N_{L\uparrow}N_{R\uparrow}} - \sqrt{N_{L\downarrow}N_{R\downarrow}}}{\sqrt{N_{L\uparrow}N_{R\uparrow}} + \sqrt{N_{L\downarrow}N_{R\downarrow}}}, \qquad (2)$$

where $N_{L\uparrow}$ is the sum of counts in the left detector for all bunches with spin up, etc. This asymmetry is dominated by the asymmetry in integrated luminosity for spin up and down when systematic effects are negligible. The data suggest that the systematic errors may be comparable to the statistical uncertainties. The origin of these effects is presently under study.

SUMMARY

The 200 MeV, AGS, and CNI polarimeters are key instruments for the polarized beam program at RHIC. Hardware improvements, calibrations, and studies of the performance of each continue, but additional work remains.

Modifications to the 200 MeV polarimeter were made to better handle the higher beam intensities from the new polarized ion source. Afterwards, the calibration was rechecked using p+d elastic scattering, and the calibration appears unchanged. However, it is desirable to have polarization measurements that are less sensitive to beam intensity for typical currents from the LINAC. Thus, it will be necessary to further upgrade both the p+C and p+d systems to be more rate tolerant.

An absolute calibration of the AGS polarimeter was recently performed at P_{lab} = 3.8 GeV/c and new diagnostics added. Thin carbon fiber targets are required to prevent rate problems in the polarimeter photomultipliers.

Continued studies with the CNI polarimeter data to understand systematic effects from the beam and the polarimeter hardware are ongoing and required. Close consultation of polarimeter physicists with the accelerator physicists may permit the beam related effects to be minimized. Dedicated tests with beam may be required to fully understand the origin of some of the instrumental effects. These studies are essential in order to achieve the stated goal of knowledge of the beam polarization at RHIC with a ±5% absolute calibration.

REFERENCES

1. Khiari, F.Z. et al., *Phys. Rev. D*39, 45-85 (1989).
2. Allgower, C.E. et al., *Phys. Rev. D*65, 092008 (2002).
3. Wells, S.P. et al., *Nucl. Instrum. Meth. In Phys. Res.* **A325**, 205-215 (1993).
4. Cadman, R. et al., BNL Collider-Accelerator note C-A/AP/56, unpublished (2001).
5. Spinka, H., BNL RHIC Spin note #79, unpublished (1999).
6. Tojo, J. et al., *Phys. Rev. Lett.* **89**, 052302 (2002).

RHIC pC CNI Polarimeter: Experimental Setup and Physics Results

I.G. Alekseev[*†], A. Bravar[**], G. Bunce[‡**], R. Cadman[§], A. Deshpande[‡], S. Dhawan[¶], D.E. Fields[||‡], H. Huang[**], V. Hughes[¶], G. Igo[††], K. Imai[‡‡], O. Jinnouchi[§§], V.P. Kanavets[*], J. Kiryluk[††], K. Kurita[‡¶¶], Z. Li[**], W. Lozowski[***], W.W. MacKay[**], Y. Makdisi[**], S. Rescia[**], T. Roser[**], N. Saito[‡‡‡], H. Spinka[§], B. Surrow[**], D.N. Svirida[*], J. Tojo[‡‡§§], D. Underwood[§] and J. Wood[††]

[*]*Institute for Theoretical and Experimental Physics, B. Cheremushkinskaya 25, Moscow, 117259, Russia*
[†]*E-mail: Igor.Alekseev@itep.ru*
[**]*Brookhaven National Laboratory, Upton, NY 11973, USA*
[‡]*RIKEN BNL Research Center, Upton, NY 11973, USA*
[§]*Argonne National Laboratory, Argonne, IL 60439, USA*
[¶]*Yale University, New Haven, CT 06511, USA*
[||]*University of New Mexico, Albuquerque, NM 87131, USA*
[††]*UCLA, Los Angeles, CA 90095, USA*
[‡‡]*Kyoto University, Kyoto 606-8502, Japan*
[§§]*RIKEN (The Institute of Physical and Chemical Research), Wako, Saitama 351-0198, Japan*
[¶¶]*Rikkyo University, Toshima-ku, Tokyo 171-8501, Japan*
[***]*Indiana University Cyclotron Facility, Bloomington, IN 47405, USA*

Abstract.
Acceleration of polarized proton beams and experiments with them at RHIC require fast and reliable measurements of the polarization.

The polarimeter presented here uses very high figure of merit of the elastic pC scattering at very low momenta transfer since the cross section is large. Small (a few percent) analysing power of the reaction makes it necessary to collect about 10^7 events per measurement. A deadtimeless DAQ system for the polarimeter is discussed in this paper. It is based on the waveform digitizer modules with "on-board" event analysis, resulting in typical polarization measurement times of several tens of seconds.

During winter 2001/2002 RHIC polarized run several dedicated data runs were taken by the polarimeter to extract the form of the analyzing power dependence as a function of the momentum transferred at beam energies 24 and 100 GeV. This dependence is extremely important for the theoretical understanding of the CNI process including the contribution of the spin-flip hadronic amplitude. The new data may become an input to some theoretical models predicting the energy dependence of the analyzing power.

INTRODUCTION

A description of the RHIC pC CNI polarimeter is already given in this volume [1]. This paper will focus on the polarimeter DAQ in which main role is played by the wave form digitizers (WFD) and on the results of the preliminary analysis of the dedicated

CP675, *Spin 2002: 15th Int'l. Spin Physics Symposium and Workshop on Polarized Electron Sources and Polarimeters,* edited by Y. I. Makdisi, A. U. Luccio, and W. W. MacKay

polarimeter data runs.

DAQ WITH WAVE FORM DIGITIZERS

The polarimeter operation is based on a small (about a percent) analysing power of the elastic scattering of protons on a carbon nuclei with a very small momentum transfer. Nevertheless figure of merit of this process is fairly good due to a very large crossection. This results in a requirement to collect about 10^7 events in several tens of seconds. We don't like to miss events, because it will lead to unnecessary beam emittance blow up. Actually crossection is so large that we could see one good event in setup per bunch crossing. The DAQ solution to this problem was found in on-line hardware analysis of every pulse individually in each silicon strip and counting just good events.

The block diagram of the WFD module is shown in fig. 1. It contains 4 equal channels each consisting of a 140 MHz RGB ADC, connected to Xilinx [2] FPGA, and some auxiliary logic for CAMAC interface, accelerator clock signals and FPGA power on configuration. RGB-channels of the ADC are used to triple digitization frequency, accepting the input signal with 3 different delays: 0, 2.4 and 4.7 ns.

The FPGA layout which we used in the run is shown in fig. 2. Each 70 MHz clock 6 waveform points come to the FPGA. During this run RHIC operated in 60-bunch mode and the interval between bunch crossings corresponded to exactly 15 periods of our 70 MHz frequency, which was derived from the RHIC clock. So each waveform consists of 90 points. There were 4 possible FPGA operating modes: raw mode, subtract mode, at-mode and read-all mode. All the modes but at-mode were dedicated for debugging purposes. Here we will describe at-mode, which was used for polarimeter running. For each ADC reading pedestal subtraction was made and then the points were stored in a waveform FIFO. Simultaneously the waveform was scanned for the maximum value, which was considered as an event amplitude, and integral over the waveform was calculated. While the next waveform was moving into the FIFO the previous one was looked through for the first point above 1/4 of the maximum. The number of this point gives the event time. FPGA also keeps track of the bunch number. Each bunch crossing am-

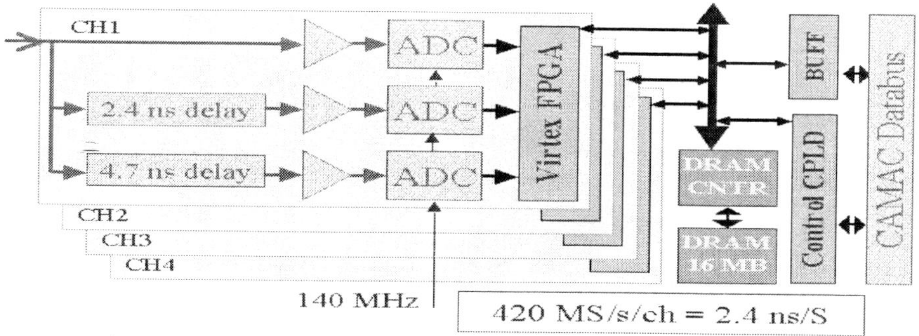

FIGURE 1. Wave form digitizer block diagram

plitude, integral, time and bunch number formed an event. Values of time and amplitude are used as an input to the lookup table, which selected or rejected the event using time of flight to energy correlation for carbon events. Amplitude to integral lookup table was also implemented, but we didn't use it in the run. The events which passed the lookup were put into an event FIFO, which could be read through CAMAC backplane. Simultaneously these events were used to fill a set of internal histograms. This set consisted of 5 histograms: distribution of events over bunch number, energy distribution for positive bunches, energy distribution for negative bunches, energy distribution for not polarized bunches and time versus energy distribution. During polarization measurements no readout of the event FIFO was performed and only distributions were read to PC. Then the distribution of events over bunch number was used to calculate asymmetries. During the dedicated polarimeter run event data (amplitude, integral, time and bunch number) was also read to PC. The percent of events read was determined by CAMAC speed.

DAQ with WFD has no dead time and can accept events in each strip independently as often as one per bunch crossing. During the run up to 600 k events per second were seen in the 48 active strips of the polarimeter. No data transfer was performed during the data taking in regular polarimeter runs. This results in one minute measurement time which includes 30–60 s of data collection and less than 10 s for readout and asymmetry calculation. During the RHIC polarized run polarization was measured about 1000 times.

In future we plan to adopt the FPGA design to 120-bunch mode and to make use of the onboard SDRAM memory for storage of the events.

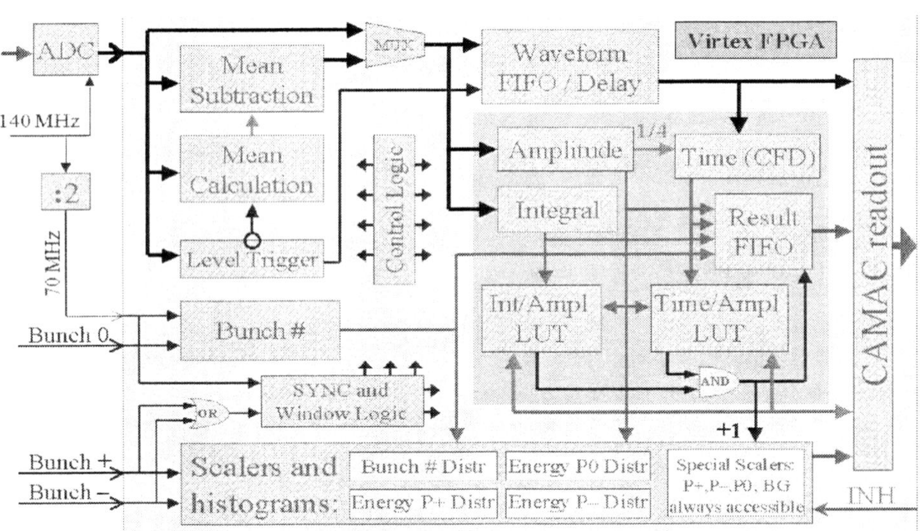

FIGURE 2. Virtex FPGA firmware block diagram

DEDICATED RUNS DATA

During winter 2001/2002 RHIC run we took about 8 hours of running time to get several physics results. Our goals were to measure $-t$ dependence of the asymmetry, to compare it for 24 and 100 GeV and to achieve better understanding of the carbon signal in the silicon and its processing in the readout system.

We have 8 dedicated runs at the energy 24 and 100 GeV. In 5 of them the signal was attenuated before the shaping amplifier to expand $-t$ scale. Parameters of 250 M events were collected in addition to all distributions. These events provide us with better understanding of the selection criteria and energy distribution.

The selection of carbon events is shown in fig. 3a by outer lines in the $time(1/\sqrt{amplitude})$ event distribution. A very careful treatment of the dead layer effect in the silicon was performed, which results in good understanding of each strip energy calibration. Some consistency check is given by extraction of the diffraction cone slope from our data. Resulting slope is shown in fig. 3b. The value $B \approx 60$ (GeV/c)$^{-2}$ agrees with the expectation for pC scattering [3, 4].

Event numbers collected inside WFD were corrected by the event distributions obtained from the at-mode data with applied cuts and then used to calculate asymmetries by the square root formula [5]:

$$A_{phys} = -\frac{\sqrt{LU \cdot RD} - \sqrt{LD \cdot RU}}{\sqrt{LU \cdot RD} + \sqrt{LD \cdot RU}},$$

where LU, LD, RU and RD counts in Left and Right detectors, when bunch polarization is Up and Down correspondingly.

Asymmetry values from the new data are shown in fig. 3 for 24 GeV (c) and 100 GeV (d), normalized to E950 data [5]. Data from the experiment E950 at 22 GeV is also shown in both figures. The data is fit by the formula for analysing power [6].

Here are some conclusions:

- Behavior of the asymmetry at 24 GeV and 100 GeV is similar but not identical.
- The data at 24 GeV qualitatively agree with the E950 results at 22 GeV, although a statistical comparison using only the statistical errors gives a poor χ^2. Systematic errors are being evaluated.
- The asymmetry looks to be zero at $-t > 0.028$ (GeV/c)^2.
- Our data are not consistent with the hypothesis of the absence of the spin-flip amplitude ($r_5 = 0$) and significantly disagree with the hypothesis of purely imaginary spin-flip ($\mathrm{R}er_5 = 0$).
- Resulting $\mathrm{I}mr_5$ and $\mathrm{R}er_5$ are in a very strong correlation so can't be extracted separately from this measurement.
- Maybe energy dependent fit to our data at 24 and 100 GeV could provide better estimation of r_5.
- It would be very interesting to expand the kinematic region to smaller $-t$, but it is difficult with current detector, because of noise and relatively thick dead layer.

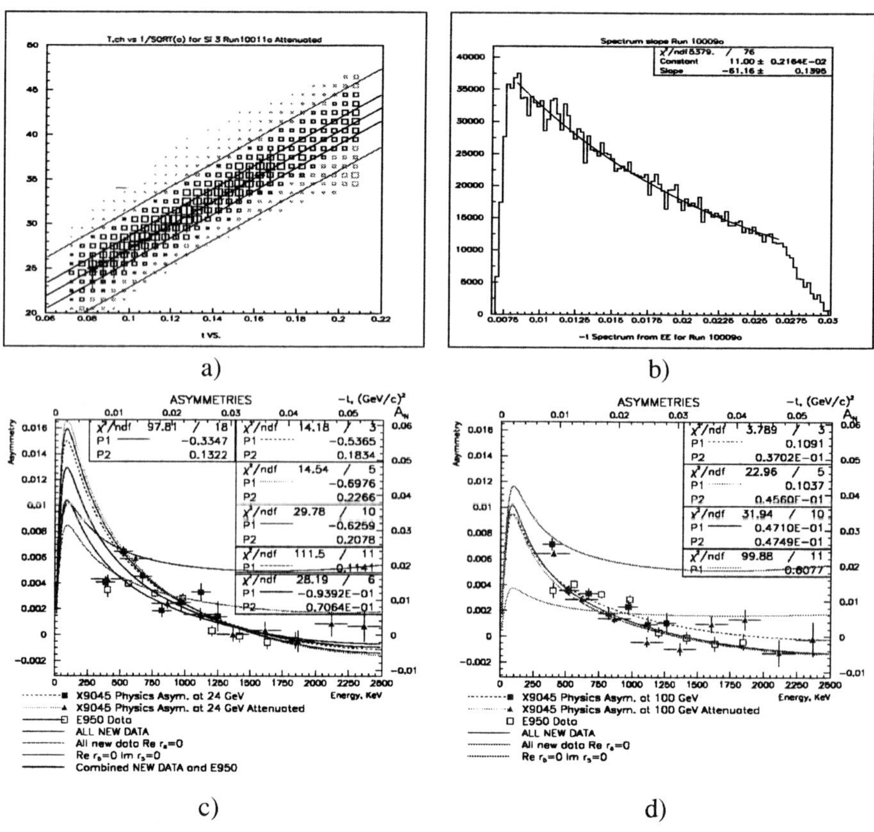

FIGURE 3. a) Distribution of the events in the *time* versus $\frac{1}{\sqrt{amplitude}}$ plane. Outer solid lines show the cut on carbon events. b) Distribution of carbon events over 4-momentum transferred. c) Analysing power of pC elastic scattering at 24 GeV. d) At 100 GeV.

ACKNOWLEDGMENTS

We would like to thank Boris Kopeliovich and Larry Trueman for very useful discussions and help in fitting our data. We are also thankful to BNL instrumentation for preparing silicon strip detectors.

REFERENCES

1. O. Jinnouchi et al. in these proceedings.
2. `http://www.xilinx.com`
3. G. Bellettini et al., *Nucl. Phys.*, **P79** (1966) p. 609.
4. R.J. Glauber et al., *Nucl. Phys.*, **B21** (1970) p. 135.
5. J. Tojo et al., *Phys. Rev. Lett.*, **89** (2002) p. 052302.
6. B.Z. Kopeliovich and T.L. Trueman, *Phys. Rev.*, **D64** (2001) p. 034004.

RHIC pC CNI Polarimeter: Status and Performance from the First Collider Run

O. Jinnouchi*, I.G. Alekseev†, L.C. Bland**, A. Bravar**, G. Bunce**‡, R. Cadman§, A. Deshpande‡, S. Dhawan¶, D.E. Fields||‡, H. Huang**, V. Hughes¶, G. Igo††, K. Imai‡‡, V.P. Kanavets†, J. Kiryluk††, K. Kurita§§‡, Z. Li**, W. Lozowski¶¶, W.W. MacKay**, Y. Makdisi**, A. Ogawa**, S. Rescia**, T. Roser**, N. Saito‡‡‡, H. Spinka§, B. Surrow**, D.N. Svirida†, J. Tojo*, D. Underwood§ and J. Wood††

*RIKEN (The Institute of Physical and Chemical Research), Wako, Saitama 351-0198, Japan
†Institute for Theoretical and Experimental Physics, B. Cheremushkinskaya 25, Moscow, 117259, Russia
**Brookhaven National Laboratory, Upton, NY 11973, USA
‡RIKEN BNL Research Center, Upton, NY 11973, USA
§Argonne National Laboratory, Argonne, IL 60439, USA
¶Yale University, New Haven, CT 06511, USA
||University of New Mexico, Albuquerque, NM 87131, USA
††UCLA, Los Angeles, CA 90095, USA
‡‡Kyoto University, Kyoto 606-8502, Japan
§§Rikkyo University, Toshima-ku, Tokyo 171-8501, Japan
¶¶Indiana University Cyclotron Facility, Bloomington, IN 47405, USA

Abstract. Polarimeters using the proton carbon elastic scattering process in Coulomb Nuclear Interference (CNI) region were installed in two RHIC rings. Polarization measurements were successfully carried out with the high energy polarized proton beams for the first polarized pp collision run. The physics principles, performance, and polarization measurements are presented.

INTRODUCTION

The RHIC polarimeters play a key role in the RHIC spin program, providing fast feed-back to the accelerator, and providing measurements of the beam polarization to the experiments. In planning for the RHIC spin program, it was recognized that the traditional method of proton polarimetry at intermediate energy (at the AGS, for example) would not be sufficiently sensitive at RHIC energies. This method uses the empirical analyzing power of proton-proton elastic scattering at -t=0.15 $(GeV/c)^2$, where there is an observed maximum independent of energy. This analyzing power falls as $1/p_{beam}$, and is about 1% at the RHIC injection energy of 24 GeV/c.

For RHIC polarimetry, we have selected the reaction proton-carbon elastic scattering at very small momentum transfer, -t=0.006 to 0.03 $(GeV/c)^2$. This t-range is in the Coulomb nuclear interference region (CNI), where quantum electrodynamics predicts a significant analyzing power at the peak (0.04), which is essentially constant over the entire RHIC energy range. The figure of merit for proton-carbon CNI scattering, cross-

CP675, *Spin 2002: 15th Int'l. Spin Physics Symposium and Workshop on Polarized Electron Sources and Polarimeters*, edited by Y. I. Makdisi, A. U. Luccio, and W. W. MacKay

section × (analyzing power)^2, becomes greater than the traditional proton-proton elastic scattering polarimeter for energies greater than about 5 GeV/c.

The QED prediction for the CNI analyzing power is based on the interference of the electromagnetic spin flip amplitude, which generates the proton's anomalous magnetic moment, and the hadronic non-flip amplitude. This QED prediction is modified by a potential contribution from an hadronic spin flip amplitude which is not presently calculable. Indeed, from our initial studies of this method of polarimetry near the RHIC injection energy, we have found a significantly lower analyzing power than that predicted from QED alone. The hadronic spin-flip amplitude is expected to diminish with increasing energy, so we expect to have a reasonable sensitivity for polarization measurement over the RHIC energy range, possibly increasing with energy.

The theoretical uncertainty from the unknown hadronic spin flip amplitude leads to the requirement that the proton-carbon CNI polarimeters must be experimentally calibrated over the RHIC energy range. At 22 GeV/c this has been done to ±31%, by making simultaneous measurements of the CNI asymmetry, and the beam polarization using a proton-proton elastic polarimeter. This calibration is ultimately tied to proton-proton elastic measurements with a polarized proton target where the target polarization was measured independently. It is planned for the near future to calibrate the CNI polarimeters at RHIC at higher energy by measuring the beam polarization at injection (24 GeV/c), accelerating to a new energy, measuring the CNI asymmetry, then decelerating back to 24 GeV/c. If the polarization is found to be the same at 24 GeV/c, before and after the down-ramp, we can use this polarization value at the new energy to calculate the analyzing power. If we measure a lower polarization after the down-ramp, we will develop bounds on the analyzing power at the new energy.

For the year 2004 we plan to install a polarized hydrogen jet target in RHIC. The target polarization will be measured to about 2% using a Breit-Rabi polarimeter. The knowledge of the target polarization will be transfered to the beam polarization using proton-proton elastic scattering in the CNI region. For p-p elastic scattering, the analyzing power is the same, whether the beam or target is polarized. The beam polarization is then used to calibrate the RHIC proton-carbon polarimeters. Our plan is to reach a ±5% calibration of the p-C CNI analyzing power, by 2005.

We chose proton-carbon CNI over proton-proton CNI for the RHIC polarimetry because an ultra-thin carbon ribbon target was available [1], which would allow the low energy carbon recoils to escape the target. The target would survive heating from the RHIC beam, provide sufficient luminosity for a quick precision polarization measurement, and be sufficiently thin to avoid pile-up of events in the detector at the same time. For a thicker target, only surface scattering would produce observable recoils in the CNI region, and the target would fail on the other points as well. A hydro-carbon target, for p-p CNI, would not survive the RHIC beam. A polarized hydrogen jet target, which will be used for calibration, is technically difficult, and is very thin. The luminosity available for a polarized jet is orders of magnitude less, and not practical to use as a fast polarimeter. A polarimeter based on an unpolarized hydrogen jet was considered, but is also technically complicated, without providing a major advantage over proton-carbon CNI.

The choice of CNI scattering offers several very clean experimental advantages. The recoil (carbon for the RHIC polarimeters, proton for the polarized jet calibration

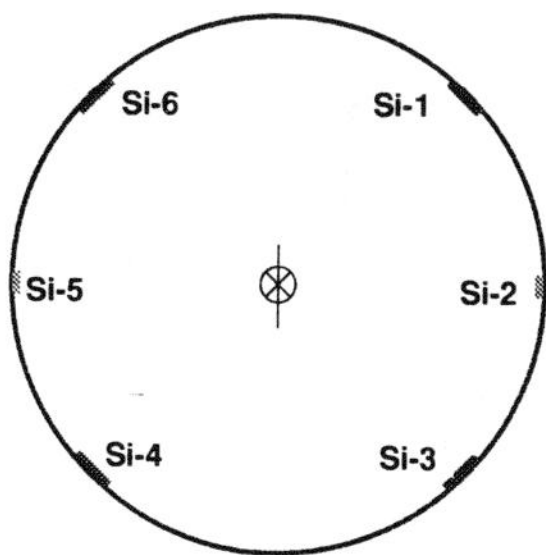

FIGURE 1. The schematic geometry layout of the silicon detectors inside the 15cm radius RHIC beam vacuum pipe. The polarized proton beam direction is into the paper, and the carbon target is represented by the vertical line at the center of the vacuum pipe.

experiment) exits at nearly 90° to the beam, an angle which changes only slowly with energy. Therefore, the same recoil detectors cover the entire RHIC energy range. Furthermore, the recoils are at very low energy so that a time of flight measurement can be made, relative to the rf-bunched beam. The relationship between the velocity and energy of the recoil identifies the mass of the recoil. The CNI reaction is two orders of magnitude larger in cross section than nearby inelastic reactions. Finally, the slow recoil arrives at the detectors well after prompt background, which is timed with the rf-bunched beam.

In the following sections we describe the RHIC polarimeters and the measurements during the 2001-2002 RHIC run, the first run of the first polarized proton collider.

EXPERIMENTAL SETUP

The RHIC polarimeters are located near the 12 o'clock intersection region, with separate polarimeters in each beam, referred to as the Blue and Yellow beams. A schematic of the polarimeters is shown in Fig. 1. The RHIC polarized proton beam passes through an ultra-thin carbon ribbon target, and carbon recoils from CNI scattering are observed in six silicon strip detectors placed as shown.

Very thin carbon ribbon targets have been developed at IUCF [1]. A typical target is 2.5 *cm* long, 3.5-$\mu g/cm^2$ thick (150 Å) and 5-μm wide. The target is mounted on a mechanism which rotates into the beam, with a choice of 3 vertical and 3 horizontal targets. The silicon detectors have twelve 12 mm × 2 mm strips, for a 24 mm total width. The six detectors are mounted inside of the vacuum chamber with readout pre-amplifier boards directly attached to the chamber detector ports through vacuum feed-through connectors. Figure 2 shows a scatter plot of time of flight versus energy for one silicon strip in the polarimeter. The silicon detectors are 15 *cm* from the target, and the RHIC bunch length was about 3 *ns* for this run, which is from the commissioning run in 2000 [2]. The inset in the figure shows the mass distribution derived from velocity and energy (note that the Time-of-flight in the figure includes an offset of 40 *ns*). The carbon and α peaks are clear, with little background under the carbon peak.

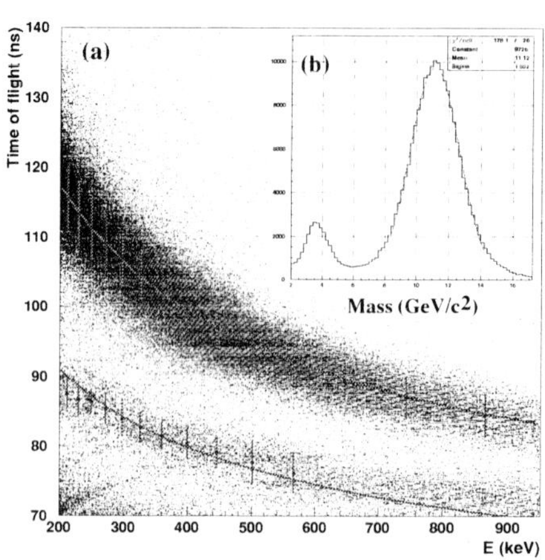

FIGURE 2. (a) The time of flight is plotted as a function of kinematic energy of the detected particle. (b) Sub-figure shows the projected mass distribution. A Carbon mass peak (11.18 GeV/c^2) is clearly separated from an alpha mass peak (3.7 GeV/c^2).

The beam polarization is measured by counting the number of events in the carbon band in each strip versus the azimuthal angle of the strip around the beam (Fig. 1). A vertical polarization generates a left-right asymmetry in the detectors and a radial polarization generates an up-down asymmetry.

The rates are very high, so we chose a readout system without dead time based on wave form digitizers (WFDs) [3]. The WFDs consist of a high frequency video ADC chip (used for laptop screens) and a Xilinx FPGA. The waveform from each strip was digitized every 2.4 *ns*, and pulse height and time of flight, compared to the RHIC rf clock, was determined in real time, and compared to a look up table which accepted the carbon band (as in Fig. 2). On-board scalers kept the number of events for each strip, and for each beam bunch. The 55 beam bunches of polarized protons in RHIC for the 2001/2 run, spaced 212 *ns* apart, alternated in polarization sign. Therefore, the on-board scalers collected data for both signs, and for bunches set up with zero polarization, for each strip. The WFDs were introduced for the 2001/2 run and 48 strips were read out, 8 for each detector (Fig. 1). Also, the orientation of the strips for the left and right detectors (Fig. 1) were set up with the strips perpendicular to the beam direction, to measure scattering angle. The 45° detectors were oriented along the beam direction to reduce the azimuthal acceptance for each strip, reducing the rate compared to the 90° central strips. For the 2001/2 run, we typically had 4×10^{12} protons in each ring, and 2×10^7 carbon elastic events were collected in about 20 seconds, with the target then rotated out of the beam. The data were then transfered to a PC, the asymmetry and various monitor asymmetries were calculated, and the result was sent automatically to the accelerator and experiments in minutes. A detailed description is given in [3], including results from a dedicated polarimeter run.

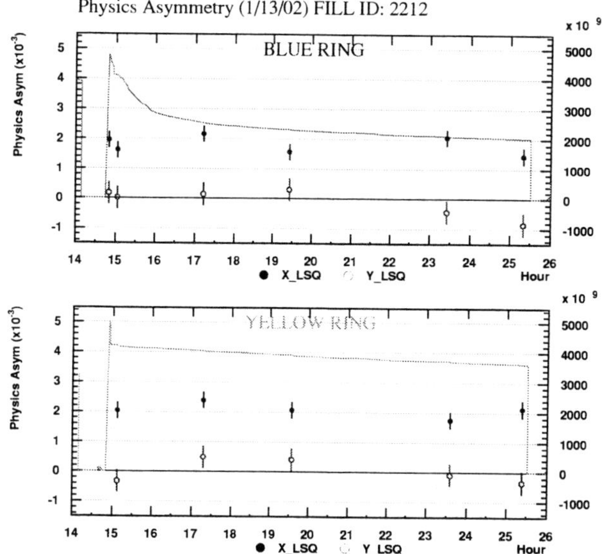

FIGURE 3. Measured physics asymmetries along with the polarized proton intensity as a function of the beam lifetime for a typical fill (Jan. 13th, 2002 fill 2212). Upper (Lower) figure shows Blue (Yellow) ring. Closed points represent the vertical asymmetry ε_X and the open points show the radial components ε_Y; the scale is found at left axis. All the measurements are taken at 100 GeV/c store except for the first Blue measurement, i.e. run 669 was taken at injection energy. The solid curves represent the number of total protons in the ring with the right-handed axis displaying its intensity scale.

POLARIMETER PERFORMANCE

Figure 3 shows polarization measurements for a typical fill. We plot the least square fit to the data for the six detectors (48 strips), with both + and − polarized bunches. The plot shows the raw physics asymmetry, referred to later as ε, uncorrected for analyzing power. Either a vertical polarization is assumed, with a ($sin\ \phi$) dependence, or a radial polarization with a ($cos\ \phi$) dependence. Acceptance and luminosity asymmetries generally cancel in this fit. Ordinarily the measurements were taken at injection energy, right after the acceleration to 100 GeV/c, then every two hours during the store at flat-top. In some cases the injection energy measurements for the Yellow ring were omitted in order to expedite the acceleration. Although the intensity dropped during the store, the vertical polarization asymmetries (closed points) were non-zero and stable. The radial polarization asymmetries (open points), which should be zero, fluctuated around zero. Each measurement point contains 2×10^7 carbon events corresponding to approximately 10% relative error for a 20% polarization.

Many systematic error studies have been carried out. Referring to Fig. 1, the 90° detectors are sensitive to vertical polarization, and the 45° detectors can be used to measure vertical polarization (left-right asymmetry) and radial polarization (up-down asymmetry).

The results for left-right asymmetry between the 90° and 45° detectors were compared

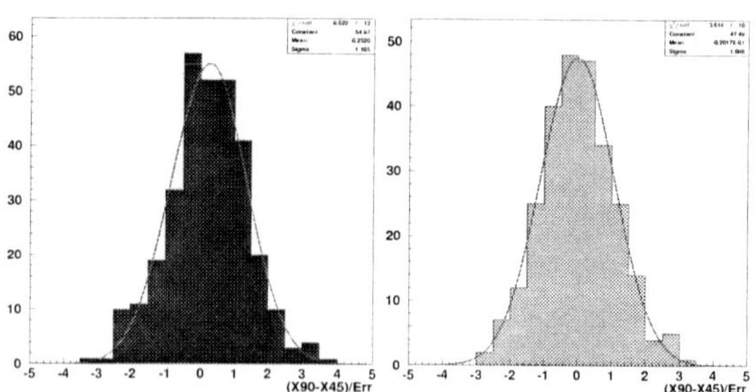

FIGURE 4. Distributions of the difference of asymmetries measured by the 90° and the 45° detectors. Left (Right) figure shows a Blue (Yellow) ring.

(after correction by $\sqrt{2}$ for the smaller analyzing power of the 45° detectors). These were calculated using a square root formula which cancels luminosity asymmetries for + and − polarized bunches, and cancels left-right or up-down detector asymmetries, to third order [4].

The distributions for the measurements for the two beams of this difference of asymmetries, divided by the overall statistical error of the difference, is shown in Fig. 4. A Gaussian fit of the distribution gives $\overline{\varepsilon_{90-45}/\sigma_{stat.}} = 0.25$, $\sigma_{dist.} = 1.11$ for the Blue beam measurements, and $\overline{\varepsilon_{90-45}/\sigma_{stat.}} = -0.02$, $\sigma_{dist.} = 1.10$ for the Yellow beam. We also measured up-down asymmetries, using the 45° detectors. Since no radial polarization was expected (with two Siberian Snakes per ring, the stable polarization direction is vertical), the up-down asymmetry directly measures a false asymmetry. A Gaussian fit to the up-down asymmetries gives $\overline{\varepsilon_{45vert}/\sigma_{stat.}} = 0.16$, $\sigma_{dist.} = 1.11$ for Blue and $\overline{\varepsilon_{45vert}/\sigma_{stat.}} = -0.30$, $\sigma_{dist.} = 1.43$ for Yellow. A cross asymmetry was also formed from the 45° detectors, which must be a false asymmetry. This distribution gave $\overline{\varepsilon_{cross}/\sigma_{stat.}} = -0.17$, $\sigma_{dist.} = 1.30$ for Blue and $\overline{\varepsilon_{cross}/\sigma_{stat.}} = -0.32$, $\sigma_{dist.} = 1.26$ for Yellow.

In another approach to study systematic errors, the results for different bunches were compared by normalizing the number of events in each detector for each bunch by the number of events for each detector observed for a standard or "good bunch". This normalized distribution for each bunch was fit with a ($constant + \varepsilon \times \sin\phi$) distribution where ϕ is the azimuthal position of each detector, and $\phi = 0$ is vertical (Thereby allowing an asymmetry from vertical polarization). The χ^2 for this fit showed that a small number of bunches had anomalous behavior and these bunches were removed from the polarimeter analysis. The remaining χ^2 distribution was broader than the standard χ^2, indicating an additional average systematic error of about $0.5 \times \sigma_{statistical}$. Anomalies for bunches were also studied by calculating specific luminosities for each bunch at the polarimeter (N_{total}/I_{bunch} where I_{bunch} is the bunch current from the wall current monitor), and also at the experiments ($N_{total}/I_{Blue}I_{Yellow}$). N_{total} refers to polarimeter total counts, and to experiment counts in a luminosity monitor respectively. The anomalous bunches from these analyses matched well (for STAR, see [5]) and they were removed from the polarization and the asymmetry analyses.

Finally, the systematic error from bunch dependent effects was explored by mixing

(+) and (−) bunches randomly into two groups, "+" and "−", each with zero (or small) polarization. These groups were used in place of the real (+) and (−) groups to calculate asymmetry using the square root formula. This was done 1000 times for each run, and these false asymmetries from mixed bunches are compared to the real asymmetries in Fig. 5. The false asymmetry distributions are characterized by a Gaussian with $\overline{\sigma_{false}} = 1.12$ for the blue beam and $\overline{\sigma_{false}} = 1.11$ for the yellow beam. The statistical error for these asymmetry measurements (both false and real) is $\sigma_{stat.}(\varepsilon) = 2.8 \times 10^{-4}$.

When we compare the different studies of systematic errors in the polarimeter measurement, which are not independent, we find that each indicates a systematic error of $(0.5 \text{ to } 1) \times \sigma_{stat.}$, for each measurement. The false asymmetries appear to fluctuate around zero, with average offsets as large as 1/3 of the statistical error. We have not identified the origin of these systematic errors, and this is a work in progress.

We have decided to use, as the systematic error for the asymmetry measurement, $\sigma_{systematic}(\varepsilon) = 2.8 \times 10^{-4}$. This is equal to the statistical error for a single measurement. This estimate is larger than that given in the studies above, but it reflects our uncertainty on the origin of the observed systematic errors. This systematic error is roughly $\pm 10\%$ of the measured asymmetry. As we will discuss, the uncertainty of the analyzing power is larger than this.

We have also studied backgrounds, pile-up, and variations in the $-t$ definition from energy loss in the dead layer of the silicon. Backgrounds from inelastic reactions appear to be small ($<$ a few %), from observing the apparent signal-to-background in the carbon mass peak of Figure 2(b). Target empty data were not collected for this run, but were a small fraction in the first AGS pC CNI experiment [6], 3%. Another background under the carbon peak, which could change with the fill or time, is de-bunched beam. In this case, the time-of-flight vs. the beam rf is spoiled. We studied five fills with unusually large de-bunching ($\sim 20\%$) and observed the effect, but also that there was little background under the carbon peak, $<$ few %. Pile-up, two pulses in the same strip from the same crossing, can also confuse the time-of-flight vs. energy correlation. We estimated pile-up at $< 1\%$. Finally, a stable $-t$ acceptance is necessary because the analyzing power is a sharply varying function of $-t$. We estimated, based on fits for the dead-layer thickness for each silicon strip, a variation in A_N of $< \pm 4\%$.

We can estimate the stability of the polarimeters against varying backgrounds, pile-up, or $-t$ acceptance by comparing results for the same beam for different polarimeters. This is done when we compare the 45° and 90° asymmetries in Fig. 4. Pile-up is also tested because the rates for the central strips for the 90° detectors are about twice that of each strip for the 45° detectors. Our estimated systematic error therefore includes these effects.

The analyzing power at injection energy is determined from Ref. [6], for the $-t$ ranges of the RHIC polarimeters. These were $-t = (0.7 \text{ to } 3.0) \times 10^{-2}$ $(\mathrm{GeV/c})^2$ for the blue beam polarimeter, and $-t = (0.6 \text{ to } 2.7) \times 10^{-2}$ $(\mathrm{GeV/c})^2$ for the yellow beam polarimeter. The difference are due to the silicon dead layer thicknesses. The values for A_N are, $A_N(blue) = (1.27 \pm 0.40) \times 10^{-2}$ and $A_N(yellow) = (1.33 \pm 0.41) \times 10^{-2}$, where the errors are the linear additions of the statistical, raw asymmetry systematic, and beam polarization errors from Ref. [6].

Referring to Fig. 5, the measured beam polarization at injection (24 GeV/c) was,

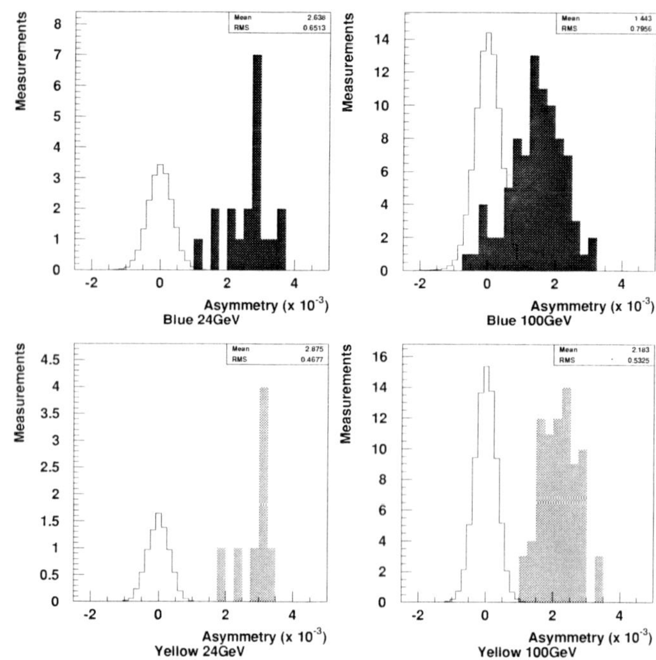

FIGURE 5. Physics asymmetries (solid histograms) are compared to the artificial asymmetries from mixed bunches (open histograms). The measurements are selected from long store (> 4 hours) fills after Jan. 5th. The above (below) left plot shows the asymmetries at 24 GeV/c, while the right plot shows for 100 GeV/c in blue (yellow) beam.

$\overline{P_{blue}(24\text{GeV}/c)} = \overline{\varepsilon}/A_N(blue) = 0.21 \pm 0.005 \pm 0.02 \pm 0.07$ and $\overline{P_{yellow}(24\text{GeV}/c)} = \overline{\varepsilon}/A_N(yellow) = 0.22 \pm 0.007 \pm 0.02 \pm 0.07$. The first and the second errors are the statistical and systematic errors on the raw asymmetry respectively, and the third error is a scale error from the fractional error on A_N. These results use the most stringent bunch selection, based on the polarimeters, STAR, and PHENIX data analyses of beam monitors.

We were not able to calibrate the polarimeters at 100 GeV/c for this run. There are theoretical arguments that A_N should change only slowly with energy over the RHIC range [7]. However, an experimental calibration is required, and is planned for future runs. *If we assume the same analyzing power at 100 GeV/c as for 22 GeV/c,* we find from Fig. 5, $\overline{P_{blue}(100\text{GeV}/c)} = 0.11 \pm 0.002 \pm 0.02 \pm 0.03$ and $\overline{P_{yellow}(100\text{GeV}/c)} = 0.16 \pm 0.002 \pm 0.02 \pm 0.05$.

ISSUES AND PLANS

There are several concerns that must be resolved before next year. One is the systematic errors that some bunches contribute to create the false asymmetry. From data, some bunches were observed to have unphysical ratio between six detectors. The understanding of their mechanism and criteria to discard them is needed. Another issue is the bunch by bunch polarization. Our data for the bunch by bunch analysis is limited by statistics from the previous runs. Further study on this issue is still in progress. High statistics measurements are expected for the coming run. The last concern is the serious gain drops of the silicon strip detectors. There were significantly large leakage current due to silicon radiation damage which effectively reduces the bias voltage on silicon that affects the signal shape. The replacement of the silicons and hardware improvements are planned.

For the future run, determination of the absolute analyzing power at 100 GeV/c is a crucial thing to be done, since our knowledge on the analyzing power of CNI polarimeter is limited to near the injection energy (22 GeV/c). Absolute calibration using the polarized hydrogen jet target is planned and is in preparation towards the first operation in 2004. For the moment, a calibration using a down ramp is adopted as the second best way. As a procedure, the usual polarization measurements are performed at injection, after ramp to 100 GeV/c, then after down ramp again. If the polarization is found to be the same at 24 GeV/c, after the down ramp vs. at injection, we can use this polarization value at 100 GeV/c to calculate the analyzing power at 100 GeV/c. If we measure lower polarization after the down ramp, we will develop bounds on the analyzing power at 100 GeV/c.

Finally, in the next run we hope to improve the pC CNI calibration by using a new pC CNI polarimeter in the AGS, calibrated to p-p elastics in the AGS internal polarimeter.

SUMMARY

The RHIC pC CNI polarimeter proved itself in the successful first polarized proton collision run 2001/2. It worked beautifully throughout the run period. Reliable high statistics (20×10^6 events) measurements were carried out in 1 minute measuring periods, owing to the successful operation of newly adopted wave form digitizer modules. The polarimeter results were broadcast to the experiments immediately after the measurements. Two Siberian snakes per ring worked well, and stable proton polarizations at 100 GeV/c were measured with little or no loss in magnitude over the store. Further detailed off-line analysis is in progress and interesting challenges are expected for the up coming run.

REFERENCES

1. W.R. Lozowski and J.D. Hudson, *NIM in Physics Research*, **A334**, 173 (1993).
2. H. Huang et al., *Proc. of 2001 IEEE PAC, Chicago*, p. 2443 (2001).
3. I.G. Alekseev et al. (2002), detail description is found in this proceedings.
4. G.G. Ohlsen and P.W. Keaton, *NIM in Physics Research*, **109**, 41 (1973).
5. J. Kiryluk et al. (2002), this proceedings.
6. J. Tojo et al., *Phys. Rev. Lett.*, **89**, 052302 (2002).
7. L. Trueman (2002), RHIC Spin Note 1 hep-ph/0203013; B.Z. Kopeliovich, these proceedings.

Measurement of Analyzing Powers for Polarized Proton Scattering on CH_2 Target at Proton Momentum Range from 1.75 to 5.3 GeV/c

L.S. Azhgirey*, V.A. Arefiev*, I. Atanasov†, S.N. Basilev*, Yu.P. Bushuev*, V.V. Glagolev*, M.K. Jones**, D.A. Kirillov*, P.P. Korovin*, G.F. Kumbartzky‡, P.K. Manyakov*, J. Mušinský*, L. Penchev§, C.F. Perdrisat§, N.M. Piskunov*, V. Punjabi¶, I.M. Sitink*, V.M. Slepnev*, I.V. Slepnev* and E. Tomasi-Gustafsson||

*Joint Institute for Nuclear Research, Dubna, Moscow region, 141980 Russia
†Institute of Nuclear Research and Nuclear Energy BAS, Sofia, Bulgaria
**Thomas Jefferson National Accelerator Facility, Newport News, VA 23606 USA
‡Rutgers, The State University of New Jersey, Piscataway, NJ 08855 USA
§College of William and Mary, Williamsburg, VA 23187 USA
¶Norfolk State University, Norfolk, VA 23504 USA
||DAPNIA/SPhN CEA/Saclay, 91191 Gif-sur-Yvette Cedex, France

Abstract.
We report a new measurement of analyzing powers for the reaction $\vec{p} + CH_2 \rightarrow$ one charged particle $+X$, at proton momenta of 1.75, 3.8, 4.5 and 5.3 GeV/c. These results extend the existing data basis, necessary for proton polarimetry at intermediate energy, and confirm the feasibility of an extended polarimeter based on this process. The experiment is performed at the accelerator complex of the JINR-LHE (Dubna).

As all fundamental interactions are spin-dependent, the knowledge of polarization observables is essential for the understanding of the structure of hadrons and for disentangling the reaction mechanism in nuclear reactions. The polarization of intermediate energy protons (i.e. in the range from a few hundreds MeV to a few GeV) is generally measured with full azimuthal acceptance focal plane polarimeters. They consist in large detectors which measure the angular distribution of a charged particle issued from an inclusive reaction, usually the scattering on a carbon target [1].

The availability of high luminosity polarized beams opens the possibility to develop the experimental study of spin degrees of freedom in hadron physics, at intermediate energy. In particular a recent experiment at the Jefferson Laboratory [2], through the measurement of the recoil proton polarization in the elastic scattering of longitudinally polarized electrons on a proton target, showed that the ratio of the two electromagnetic form factors of the proton, electric and magnetic, G_{Ep}/G_{Mp}, decreases monotonically with increasing four momentum squared, Q^2, starting at about 0.8 and up to 6 GeV^2, which corresponds to proton momenta up to 3.7 GeV/c. The extension of this measurement to larger momenta [3] requires the construction of a new polarimeter that will measure the polarization of recoil protons at momenta up to 5.7 GeV/c.

CP675, *Spin 2002: 15th Int'l. Spin Physics Symposium and Workshop on Polarized Electron Sources and Polarimeters*, edited by Y. I. Makdisi, A. U. Luccio, and W. W. MacKay

The experiment was carried out at a slowly extracted beam of vector polarized deuterons produced by the POLARIS ion source and accelerated by the Synchrophasotron at the Laboratory of High Energies, JINR, Dubna. The intensity of the primary beam was up to $2 \cdot 10^9$ particles per spill. The deuteron vector polarization was flipped at each beam spill, one spill over three been unpolarized. The deuteron vector polarization, P_d, was different for the two beam polarization states: $P^{(+)} = 0.568 \pm 0.037$ and $P^{(-)} = -0.612 \pm 0.037$.

The polarized protons were produced by fragmentation of the polarized deuteron beam on an 8 cm thick Be target, installed 60 m ahead of the polarimeter. Two dipoles of the beam transport line separated the break up protons at zero angles from the deuteron beam.

For the proton beam momentum of 5.3 GeV/c the deuteron beam momentum was 9 GeV/c. At this ratio of proton to deuteron momenta the polarization transfer from deuteron to proton is still equal to 1 [4, 7]. For the other proton momenta the corresponding deuteron beam momenta were twice as large as the proton momentum; hence, for the 4 proton momenta of this experiment the proton polarization was equal to the deuteron polarization.

A schematic view of the detection is shown in Fig.1. The incident protons were

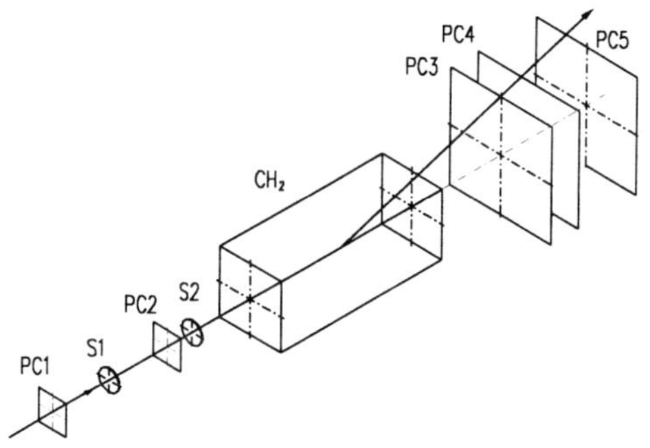

FIGURE 1. A schematic view of the setup.

detected in the proportional chambers PC1, PC2. The polyethylene target thickness was varied between 37.5 and 79.8 g/cm^2. Outgoing particles were detected by PC3-PC5 (POMME [1]) with sensitive area 48×48 cm^2. The trigger was realized by coincidence of signals from scintillation counters S1 and S2 of a diameter of 5 cm. The data acquisition system was capable to record up to 4800 events per beam spill.

In the data analysis the scattered events were classified in bins of the transverse momentum transfer, $p_t = p \sin\theta$, and ϕ, the azimuthal scattering angle.

In Fig.2 a) the analyzing powers for all target thicknesses at p_p=3.8 GeV/c are presented. To evaluate dependence of analyzing power on target thickness, the data were fitted by a polynomial in p_t (see the curves). The fit shows that within error bars there is no dependence of analyzing power on target thickness. Analyzing powers averaged over the various target thickness at the 4 energies are in Fig.2 b).

The calculations of the figure of merit, $\mathcal{F}$, assuming 100% detection efficiency and using fitted values of the analyzer efficiency and of the analyzing power are presented in

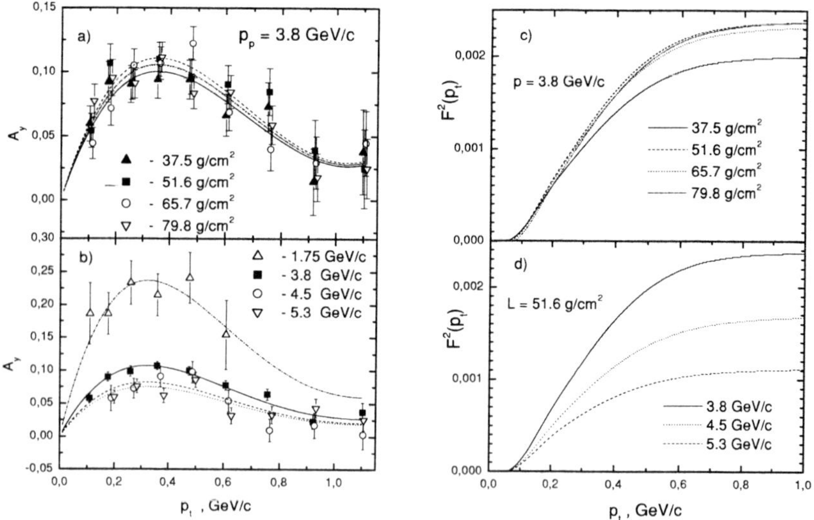

FIGURE 2. Dependence of: a) analyzing power on target thickness; b) analyzing power on energy; c) figure of merit on target thickness; c) figure of merit on energy.

Figs.2 c) and d): the coefficient of merit is defined as:

$$\mathcal{F}^2(p_t) = \int_{a(L)}^{p_t} A_y^2(p_t') \frac{d\varepsilon}{dp_t'} dp_t',$$

where $a(L)$ is the lowest p_t for each target thickness L, $d\varepsilon/dp_t = N(p_t)/(N_{inc}\Delta p_t)$, N_{inc} and $N(p_t)$ are the numbers of particles incident on the target and emitted into interval Δp_t, respectively.

Values of the maximum analyzing powers at the different energies are good observables to estimate the energy dependence. To compare the analyzing powers of CH_2 and C, a fit to existing data for carbon was made using the same approach as for CH_2. We observe that both CH_2- and C analyzing power are linearized when plotted versus the inverse of the proton momentum, as seen in Fig.3. Comparing values of A_y^{max} one can see that the ratio of the analyzing powers on C and CH_2 is approximately 0.89.

Summarizing the results presented here we emphasize the following main features:

- For protons of 3.8 GeV/c, the analyzing power is fairly independent from the amount of material in the analyzer, from 37 to 80 g/cm^2.
- A target thickness larger than the nuclear collision length in the material of the analyzer, and a polarimeter acceptance larger than $p_t \simeq 0.7$ GeV/c, do not improve the figure of merit.

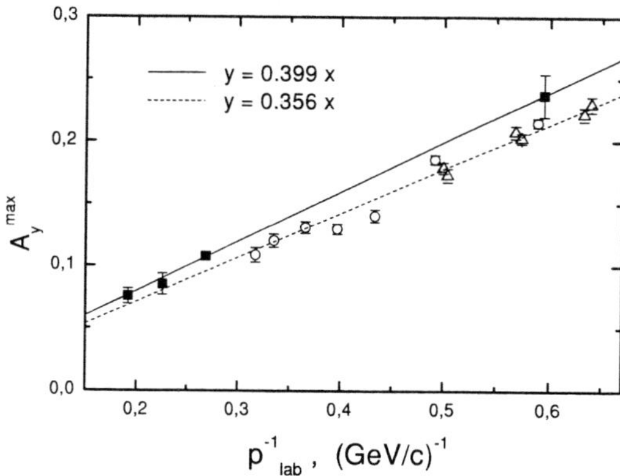

FIGURE 3. Momentum dependence of CH_2 and C data. Solid squares – current data, open circles – [5], open triangles – [6].

- The analyzing power decreases with increasing incident momentum, but it is still sizeable at a proton momentum of 5.3 GeV/c.
- The CH_2 shows a larger analyzing power than the carbon.
- In a wide region of momentum transfer, the analyzing power of both carbon and CH_2 targets has a maximum around $p_t = p \sin\theta \simeq 0.3$ GeV/c.
- High angular resolution of the polarimeter is important to maximize the figure of merit.

This work is supported in part by the Russian Foundation for Basic Research (grants 00-02-016189 and 98-02-16915) and the Grant Agency for Science at the Ministry of Education of the Slovak Republic (grant 1/8041/01). The transportation of POMME to Dubna has been made possible with the help of the Donation Program of UNESCO. The US participants are supported by the US National Science Foundation grants PHY 9901182 (CFP), PHY 0098642 (GFK) and US Department of Energy grant DE-FG05-89ER40525 (VP).

REFERENCES

1. B. Bonin *et al.*, Nucl. Instr. and Meth. A288, 379 (1991); E. Tomasi-Gustafsson, POMME Manual, LNS/Ph/92-07.
2. M. K. Jones *et al.*, Phys. Rev. Lett. **84**, 1398 (2000); O.Gayou *et al.*, Phys. Rev. Lett. **88**, 092301 (2002).
3. Proposal to JLab PAC18: 'Measurement of G_{Ep}/G_{Mp} to Q^2=9 GeV^2 via Recoil Polarization', (Spokepersons: C.F. Perdrisat, V. Punjabi, M.K. Jones and E. Brash), JLab, July 2001.
4. V. Punjabi *et al.*, Phys. Lett. B **350**, 178 (1995).
5. N.E. Cheung *et al.*, Nucl. Instr. & Meth., A **363**, 561 (1995).
6. I.G. Alekseev *et al.*, Nucl. Instr. & Meth., A **434**, 254 (1999).
7. B. Kuehn *et al.*, Phys. Lett. B **334**, 298 (1994); L.S. Azhgirey *et al.*, JINR Rapid Com. No. 3[77]-96, 23 (1996).

The Absolute Polarimeter for RHIC

A. Bravar [1]

Brookhaven National Laboratory
Upton, NY 11973 USA

Abstract. A polarimeter to measure the absolute polarization of the RHIC proton beams using an internal polarized hydrogen gas jet target is being built at BNL. The chosen polarimetric process is elastic pp scattering at very low momentum transfer in the Coulomb Nuclear Interference region. In this talk I'll discuss the beam polarization measurement and the attainable precisions.

INTRODUCTION AND METHOD

The RHIC Spin program [1] aims to determine the spin asymmetries with one or both beams polarized for a variety of processes with high precision, such to allow significant comparison with theoretical predictions and possibly unveil new physics. A crucial requirement is the knowledge of the absolute polarization of the RHIC polarized proton beams to 5% of its value or better and its continous monitoring.

For this purpose an absolute polarimeter using an internal polarized hydrogen gas jet target is being built. The current plan is to install the polarimeter for the year 2004 run with the initial goal of determining the beam polarization to about 10%. In 2005 our plan is to reach the desired precision of 5% on the beam polarization.

The chosen polarimetric process is elastic pp scattering at very small momentum transfer t in the Coulomb Nuclear Interference (CNI) region of $0.001 < |t| < 0.02$ $(\text{GeV}/c)^2$, where the analyzing power A_N^{pp} reaches a maximum value of about 0.045 [2]. The present knowledge of A_N^{pp} from previous experiments and theory, however, is not sufficient to obtain the desired precision. Therefore, a new measurement of A_N^{pp} to about ± 0.001 is required.

The luminosity of the polarized jet target, however, is too low to be practically used as a *fast* polarimeter and to continously monitor the beam polarization. About 24 hours of data taking will be required to accumulate enough elastic pp events to achieve the required statistical precision. While the absolute beam polarization will be determined with the jet target, the beam polarization will be routinely measured and monitored with the *fast* proton-Carbon (pC) polarimeters installed in each RHIC ring [3]. The pC polarimeters also operate in the CNI region. These polarimeters must be calibrated with a beam of known polarization to an accuracy better than 5%. The calibration will proceed in two steps. First, the analyzing power for these *fast* polarimeters, A_N^{pC} will

[1] for the jet target collaboration: I. Alekseev, A. Bravar, G. Bunce, S. Dhawn, W. Heaberli, Z. Li, Y. Makdisi, W. Meng, S. Rescia, E. Stephenson, D. Svirida, T. Wise, A. Zelenski

CP675, *Spin 2002: 15th Int'l. Spin Physics Symposium and Workshop on Polarized Electron Sources and Polarimeters*, edited by Y. I. Makdisi, A. U. Luccio, and W. W. MacKay

be determined using the polarized proton beam with a well measured polarization. This procedure will require the simultaneous measurements of the absolute beam polarization with the polarized gas jet target. Then, this measured A_N^{pC} will be used in successive measurements of the beam polarization.

The polarized target is a free atomic beam jet. The polarized gas jet will deliver polarized protons with a polarization near 90% and a density of about 5×10^{11} p/cm^2. The target polarization will be determined with a Breit-Rabi polarimeter to better than 3%. The jet target has been discussed in different presentations at this symposium [4, 5].

The left-right asymmetry A_N^{pp} in the CNI region arises from the interference of the electromagnetic spin-flip amplitude, due to the anomalous magnetic moment of the proton, with the hadronic spin-non-flip amplitude, which is proportional to the square root of the total hadronic cross section. This effect is fully calculable within QED. However, this asymmetry receives also a contribution from the hadronic spin-flip amplitude [2]. This second QCD effect at present is not predictable and potentially generates a large uncertainty in A_N^{pp}. The analyzing power A_N of pp elastic scattering in the CNI t region has been measured by the E-704 experiment [6], albeit with large statistical errors, confirming the long-standing prediction of a structure in A_N produced by the interference of the electromagnetic and hadronic amplitudes. Using the E-704 result, however, a precision of only 20% could be obtained.

With the use of an internal gas jet target the low energy recoil protons, originating from elastic pp scattering in the CNI region ($0.5 < T_{kin} < 10$ MeV), can be detected with a recoil spectrometer based on silicon detectors. Previous measurements have shown that by detecting the recoil particle only, the elastic scattering events can be selected above a background that is smaller than 3% [7].

This measurement can be performed *in situ* at RHIC, with the same apparatus measuring the beam polarization, at any energy of interest, independent of theoretical assumptions, using the so called *self-calibration* method. The transverse spin asymmetry in elastic pp scattering of a polarized beam on an unpolarized target is identical in magnitude but with opposite sign to the unpolarized beam – polarized target one in the same kinematical region: $A_N^{p\uparrow p} = -A_N^{pp\uparrow}$. This symmetry relation, which holds for elastic scattering only, permits the direct transfer of the target polarization P_{target} to the beam polarization P_{beam}, i.e. P_{beam} can be expressed in terms of P_{target}.

An interesting alternative to calibrate the pC polarimeters is offered by scattering a carbon beam off the polarized gas jet target [8]. This will allow a direct measurement of A_N for Cp elastic scattering in the CNI region. For elastic scattering, scattering a polarized proton beam off a carbon target is equivalent to scattering a carbon beam of the appropriate energy off a polarized proton target in the same kinematical region and $A_N^{pC} = -A_N^{Cp}$. This method will allow to directly measure A_N for pC, which can be then used in the pC polarimeters.

RATE ESTIMATES

To estimate the event yields we assume each RHIC beam is operated at its highest intensity of 2×10^{11} protons per bunch with 112 bunches in the ring at a revolution

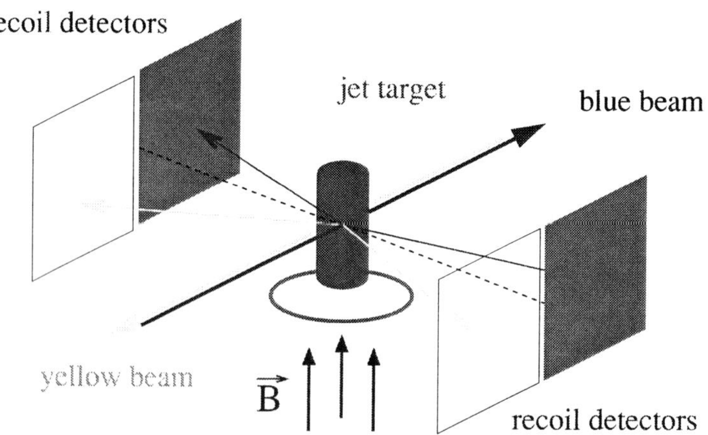

FIGURE 1. Schematic view of the polarimeter set-up showing the jet target, the recoil detectors and the target holding magnetic field.

frequency of 78 kHz. We also assume a jet target density of 5×10^{11} protons per cm^2. With these inputs the luminosity L is given by:

$$\begin{aligned} L &= 2 \times 10^{11} \cdot 112 \cdot 78 \times 10^3 \cdot 5 \times 10^{11} \text{ cm}^{-2}\text{s}^{-1} \\ &\simeq 1 \times 10^{30} \text{ cm}^{-2}\text{s}^{-1} \quad (1\mu\text{b}^{-1}s^{-1}) \end{aligned} \tag{1}$$

The cross section σ^{pp} for elastic pp scattering in the CNI region of $0.001 < |t| < 0.02\ (\text{GeV}/c)^2$ around 250 GeV is 3 mb. In the same t region $\langle A_N^{pp} \rangle \geq 0.03$.

Assuming further a coverage in azimuth of the recoil spectrometer of $\Delta\varphi = 30°$ the acceptance will be $acc = \Delta\varphi/360° = 0.0833$.

The expected instantaneous event rate in the CNI t region follows from these inputs:

$$N = \sigma^{pp} \cdot L \cdot acc = 250 \text{ events}/\text{s}\,. \tag{2}$$

To obtain the desired statistical precision on A_N^{pp} of 3×10^{-4}, about 10^7 pp events are required. This statistics can be accumulated in about 24 hours, where an overall efficiency of 50% has been assumed for the polarimeter. In a few days, therefore, sufficient data can be collected under different conditions (polarized and unpolarized beam and/or target) to measure the absolute beam polarization to better than 5% and calibrate the *fast pC* polarimeters.

KINEMATICS

For an elastic pp scattering process there is essentially one free parameter, the momentum transfer $t = (p_B - p_S)^2 = (p_T - p_R)^2$ between the incident beam proton and the scattered proton or between the target proton and the recoil proton. The angle between the normal to the incident proton direction and the recoil proton ϑ_R is correlated to the energy T_R of the recoil proton by

$$\sin\vartheta_R \simeq \sqrt{T_R/2M_p} \quad . \tag{3}$$

The relationship between the recoil angle and the recoil energy also determines the mass of the scattered system, known as missing mass. For an elastic *pp* process, the missing mass must correspond to the proton mass. This is a very strong correlation to identify elastic *pp* scattering events and to separate them from inelastic ones, when the beam proton dissociates into a higher mass state. The resolution on the missing mass, which will determine the amount of inelastic background below the elastic peak, depends on the angular and energy resolution of the recoil spectrometer. The first is essentially determined by the size of the jet along the beam and the distance of the detectors from the jet. The angular resolution is estimated to be around 4 milliradians. The expected energy resolution of the silicon recoil detectors will be around 100 keV. In addition, the relationship between the time of flight and the energy of the recoil particle identifies the mass of this particle, thus identifiying the recoiling protons.

SETUP

Figure 1 shows the schematic setup of the polarimeter. More details can be found in [4, 5]. The jet target will cross the RHIC beams in the vertical direction with its polarization also directed vertically (i.e. normal to the horizontal plane).

The recoil protons from elastic *pp* scattering will emerge close to 90° with respect to the beam direction and will be detected in the horizontal plane with silicon detectors. These detectors will provide good energy, position and time measurements of the recoil particle. Because of the very low energy of the recoil protons ($0.5 < T_{rec} < 10$ MeV) the detectors must be positioned inside the RHIC vacuum. They will be located in two recoil-arm cylindrical chambers, attached to the central jet target vaccum chamber, on the left and on the right of the beams at about 80 cm from the jet axis.

The recoil protons angle ϑ_R varies between 1° and 5° in the considered t range and the recoil protons emerge in the same hemisphere as the scattered proton. Therefore, two separate sets of silicon detectors will be used for each of the two RHIC beams as illustrated in Figure 1. In order to cover a 15° angle in azimuth on both sides of the beam, each set of detectors will consist of a vertical array of three silicon detectors. The silicon detectors will have a double sided readout in order to measure the recoil angle ϑ_R as well as the azimuthal angle of the recoil particles. To fully stop 10 MeV recoil protons about 800 μ of silicon is required. Rather than build a single thick detector, a stack of two silicon detectors will be used. The second detector will be also used as a veto for punch through particles.

A set of Helmholtz coils centered around the jet axis will generate the necessary magnetic field of about 1.0 – 1.5 kGauss to hold the target polarization. This magnet will give a strong momentum kick to low momentum recoil protons, making the reconstruction of the recoil angle ϑ_R difficult. A second set of Helmholtz coils, coaxial to the first set, with the current circulating in the opposite direction, has been introduced to compensate the momentum kick from the inner magnet and to bring the recoil protons back to the original trajectory. The total displacement from the ideal trajectory of the lowest momentum recoil protons will be less than 3 mm at 80 cm from the target. This magnet configuration will guarantee almost equal left – right acceptances for the recoil protons, thus cancelling acceptance induced errors in the measurement.

The polarimeter will be installed at the Intersection Point (IP) at 12 o'clock in the RHIC ring. At the intersection point the two RHIC beams will be displaced by about 10 mm. The jet target will be centered around one of the two beams. Thanks to the small gas jet diameter of less than 10 mm only one beam at the time will interact with the target gas. Because of the relatively low density of the target and its location away from the experiments, the polarimeter can be operated continously without any impact on the beam and the experiments. By locating the target at the interaction point, the time between two consecutive bunch crossings of the target region is maximized. The bunch wake-field induced pick up signal and pick up signals from capacitive couplings (beam return currents) come at the same time for both beams in the IP and will be less of a problem for the measurement of the recoil protons.

The detection of the forward scattered proton will be difficult because of the proximity of any detector to the beam required to detect the forward scattered proton (very small scattering angle). The problem of operating the forward detectors close to the high intensity beam will make it necessary to operate the polarimeter with the recoil detectors alone. The discriminative power of the recoil technique is powerful enough to suppress almost all the inelastic backgrounds in the CNI t region.

ATTAINABLE PRECISIONS

The beam polarization measurement and calibration of the *fast pC* polarimeters will require the measurement of several spin asymmetries. At each step some measurement errors will accumulate.

The uncertainty on the absolute beam polarization measurement will be dominated by the uncertainty on the target polarization, estimated at about 3%, and the background below the elastic peak, which will contribute an uncertainty around 1%. Statistics will also contribute an error of about 1% to the measurement.

In the presence of background, the measured asymmetry A_N^{mes} differs from the physics asymmetry A_N^{pp} by $|A_N^{pp} - A_N^{BG}| \cdot R$, where $R = N^{BG}/N^S + N^{BG}$ is the ratio of background events to the total and A_N^{BG} is the background asymmetry. To achieve the desired precision, this term must be smaller than 10^{-3}. The two major sources of background are:

1. beam proton dissociation $pp \to Xp$, which represents a physics background that can be correctly modeled;
2. beam – gas and beam – residual target gas interactions, which can be suppressed with the use of appropriate collimators.

The first source of background can be almost fully suppressed with sufficient precision on the measurement of the recoil proton. We estimated this background to be below 1%. The second source will generate an almost uniform background below the elastic peak (these might well be elastic pp scattering events, originating however outside of the target region). This background will be estimated from the reconstructed missing mass distribution by studying the wings of this distribution. Then the polarization of this background will be estimated from the spin-sorted amount of background below the elastic peak and, if necessary, it will be subtracted.

With frequent reversals of the target polarization, bunches of opposite polarization in the RHIC rings and reversal of the beam polarization with the spin flippers several systematic effects in the extraction of A_N^{pp} and P_{beam} can be controlled and minimized. Further, the reversal of the target holding magnetic field will allow the supression of most left – right acceptance effects. In particular, comparing A_N^{pp} for the two RHIC beams, will be a strong indicator for systematic effects.

SUMMARY

The measurement of the absolute beam polarization and the calibration of the *fast pC* polarimeters will require several measurements of spin asymmetries, most of them performed simultaneously, as illustrated below:

$$P_{target} \to A_N^{pp} \to P_{beam}^{pp} \to A_N^{pC} \to P_{beam}^{pC} \quad . \tag{4}$$

The overall anticipated error on the beam polarization based on the expected performance of the pp polarimeter and on the performance of the *fast pC* polarimeters is about 6% of the beam polarization value (i.e. relative error):

$$\frac{\Delta P_{beam}}{P_{beam}} = \frac{\Delta P_{target}}{P_{target}} \oplus \frac{\Delta A_N^{pp}}{A_N^{pp}} \oplus \frac{\Delta P_{beam}^{pp}}{P_{beam}^{pp}} \oplus \frac{\Delta A_N^{pC}}{A_N^{pC}} \oplus \frac{\Delta P_{beam}^{pC}}{P_{beam}^{pC}} \tag{5}$$

$$\frac{\Delta P_{beam}}{P_{beam}} = 3\% \oplus 2\% \oplus 2\% \oplus 4\% \oplus 2\% = 6\% \quad . \tag{6}$$

There are different possible scenarios to improve on that value. For instance, with the jet target continously running, the beam polarization could be determined over longer running periods to a precision of about 4%, while the polarization could be monitored with the *fast pC* polarimeters.

This polarimeter will provide a reliable and efficient tool to measure the absolute beam polarization and to calibrate the *fast pC* polarimeters in a relatively short time. Also, it will allow studying various aspects of the spin dependence in pp elastic scattering.

REFERENCES

1. G. Bunce et al., *Annu. Rev. Nucl. Part. Sci.*, **50**, 525 (2000).
2. N.H. Buttimore et al., *Phys. Rev.*, **D59**, 114010 (1999).
3. O. Jinnouchi et al., *these proceedings* (2002).
4. T. Wise et al., *these proceedings* (2002).
5. A. Zelenski et al., *these proceedings* (2002).
6. N. Akchurin et al., *Phys. Rev.*, **D48**, 3026 (1993).
7. R.E. Breedon et al., *Phys. Lett.*, **B216**, 459 (1989).
8. G. Igo and I. Tanihata, *these proceedings* (2002).

Absolute Calibration of the RHIC CNI Polarimeters Using 125 GeV/A C Ions

George Igo[*] and Isao Tanihata[†]

*Department of Physics and Astronomy, University of California, Los Angeles, Los Angeles, CA 90095
† RIKEN, 2-1 Hirosawa, Wako, Saitama 351-01, Japan

ABSTRACT. A polarized hydrogen jet target will be installed at an intersection point in RHIC in the next few years to make an absolute measurement of the proton beam polarizations possible. We discuss here a procedure to measure the p-C analyzing power $A_{p\text{-}C}$ in the Coulomb-nuclear interference region at beam energies between injection energy and 125 GeV. A carbon ion beam (23 - 125 GeV/A) and the polarized hydrogen jet target would be used in the measurements.

INTRODUCTION

The experimental program at RHIC involving the collision of polarized proton beams began in 2001. It has already begun to provide new and exciting physics. As an example, preliminary measurements[1] of the asymmetry of forward pi-zero mesons produced in the collisions of transversally polarized protons at the highest CM energy ever measured, $\sqrt{s}$ = 200 GeV, show unexpectedly large asymmetries. Unique opportunities will soon be available to study spin effects in hard processes[2] at high luminosities including the measurement of the gluon spin structure function and quark and anti-quark spin structure functions by flavor, the latter in parity violation processes in W and Z production[3]. These measurements require knowledge of the absolute beam polarization at the 5% level. To implement these measurements a polarized jet target is being prepared which will be located at an intersection point in the RHIC rings. Measurements of the p-p asymmetry in the Coulomb-nuclear interference (CNI) angular region will be then possible with unpolarized (polarized) proton beams in both the blue and yellow rings on the polarized (unpolarized) jet target. These measurements taken together can determine the absolute polarizations of the blue and yellow proton beams. Simultaneous measurements of the asymmetries made with the RHIC CNI polarimeters will determine their analyzing power, $A_{p\text{-}C}$.

p-p and p-C Methods

Figures 1 and 2 illustrate the absolute calibration procedures using the accepted p-p method referred to in the Introduction and the p- C method which is being proposed as an alternate to the p-p method in this paper. In the p-p method, (see Figure 1) first

CP675, *Spin 2002: 15th Int'l. Spin Physics Symposium and Workshop on Polarized Electron Sources and Polarimeters*, edited by Y. I. Makdisi, A. U. Luccio, and W. W. MacKay

an unpolarized proton beam is incident on the above mentioned polarized jet target in order to measure the left - right (L-R) asymmetry at a particular energy E and four momentum transfer -t in the CNI range $\varepsilon^{T}_{p-p}(E,-t) = A_{p-p}(E,-t)\,|\vec{P}_T|$ where $A_{p\text{-}p}$ is the p-p analyzing power. In a second set of measurements, polarized proton beams circulating in the blue and yellow rings are incident on the jet target run in an unpolarized mode to measure the asymmetry ε^{P}_{p-p}.

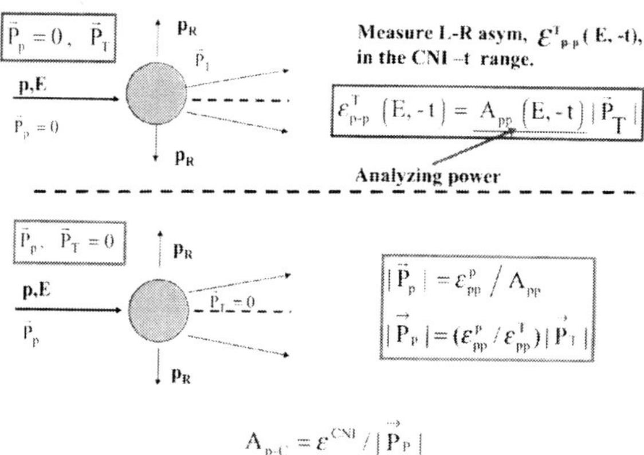

FIGURE 1. Absolute Calibration of the RHIC CNI Polarimeters by the p-p Method

These measurements allow the extraction of the proton beam polarization $|\vec{P}_P| = \varepsilon^{P}_{p-p} / A_{pp}$ also expressible in terms of the asymmetries as $|\vec{P}_P| = (\varepsilon^{P}_{p-p} / \varepsilon^{T}_{p-p})\,|\vec{P}_T|$. During this latter set of measurements, the RHIC CNI polarimeters must be inserted into the blue and yellow beams sufficiently often to monitor the behavior of the asymmetry ε^{CNI} with time as a check on the constancy of the beam polarization during the fill, and finally to determine the analyzing power $A_{p-C} = \varepsilon^{CNI} / |\vec{P}_T|$ where ε^{CNI} is the asymmetry measured with the RHIC CNI polarimeter.

Absolute calibration by the p-C method is illustrated in Figure 2. A carbon ion (C) beam is stored in one of the rings and the asymmetry ε_{p-C} is measured with the jet target polarized. The analyzing power $A_{p\text{-}C}$ can then be extracted directly from the expression $A_{p-C} = \varepsilon_{p-C} / |\vec{P}_T|$.

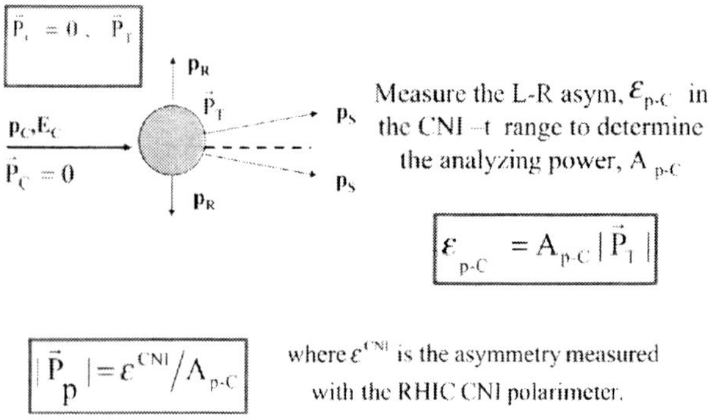
$\vec{P}_t = 0,\ \vec{P}_T$
p_R
$\vec{P}_T$
p_C, E_C
p_S
$\vec{P}_C = 0$
p_S
p_R
Measure the L-R asym, $\varepsilon_{p\text{-}C}$ in the CNI -t range to determine the analyzing power, $A_{p\text{-}C}$
$\varepsilon_{p\text{-}C} = A_{p\text{-}C} \mid \vec{P}_T \mid$
$\mid \vec{P}_p \mid = \varepsilon^{CNI} / A_{p\text{-}C}$
where ε^{CNI} is the asymmetry measured with the RHIC CNI polarimeter.

FIGURE 2. Absolute Calibration of the RHIC CNI Polarimeters by the p-C Method

The p-C method, although more straightforward, is limited by the maximum energy C beam that RHIC can produce. By scaling the ratio of beam momentum to the ion charge, the maximum value of P_C/A_C for a C beam can be seen to be approximately 125 GeV/A. Figure 3 illustrates that a carbon beam, with a certain P_C/A_C, incident on the polarized jet target, will allow the measurement of $A_{p\text{-}C}$ for a proton beam momentum $P_p = P_C$. The p-C method therefore will be useful for beam energies from injection energy up to 125 GeV.

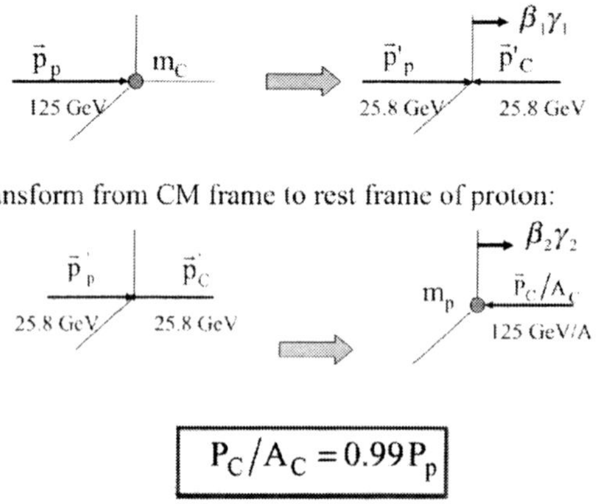
$\vec{p}_p$
m_C
125 GeV
$\vec{p}'_p$
$\beta_1\gamma_1$
$\vec{p}'_C$
25.8 GeV
25.8 GeV
Transform from CM frame to rest frame of proton:
$\vec{p}'_p$
$\vec{p}'_C$
25.8 GeV
25.8 GeV
$\beta_2\gamma_2$
m_p
$\vec{P}_C/A_C$
125 GeV/A
$P_C/A_C = 0.99 P_p$

FIGURE 3. The relation between the momentum of a proton beam [P_p + C_{tgt} , *CNI pol*] and a C beam [P_C / A_C + *pol jet tgt*] at the same $\sqrt{s}$.

A limitation, inherent in the p-C method, is that there is <u>no possibility</u> from kinematics to separate elastic events in the CNI region from events at the same -t when the ^{12}C ion (projectile) is left in an excited state since the recoil protons from the polarized jet target cannot be distinguished in the two cases. In principle, inelastic events could contaminate the desired measurement of $A_{p\text{-}C}$ in the CNI region. Fortunately the elastic cross section is expected to be very large compared to inelastic scattering in the CNI region. Further, it is possible to learn about the relative magnitude of inelastic events from the existing RHIC CNI polarimeter data since elastic events can be distinguished kinematically. No evidence[4,5] of events where the recoil C nucleus was produced in its first excited state at 4.44 MeV appears in the RHIC CNI polarimeter spectra. It should be noted in passing that the extra broadening of the distribution when the carbon nucleus is left in this state due to recoil from the emission of the 4.44 MeV gamma ray is comparable to broadening caused by multiple Coulomb scattering in the thin carbon target of the CNI polarimeter. Most higher lying states in the carbon inelastic spectrum have broad alpha particle decay widths and will decay therefore mainly by alpha particle emission (which will be followed by breakup of the residual Be nuclei into two alpha particles).

Considerable care will be necessary to relate the (E, -t) coverage of the RHIC CNI polarimeters to the coverage obtained from measurements using the p-C method. This has not been studied carefully by us at the present time.

SUMMARY

The p-C method described in this paper is a procedure which allows a measurement of the absolute polarization $A_{p\text{-}C}$ directly. The beam energy range is from injection energy to 125 GeV. The p-C method requires, in addition to the polarized jet target, the production of C beams in RHIC (a costly use of beam time). Inelastic contributions, i.e. excitation of states in C in the CNI region may be a 'show stopper'. However events in the kinematical region of the first excited state (4.44 MeV, 2^-) are not visible in CNI polarimeter spectra at 23 GeV and 100 GeV beam energies. Studies of excitation of higher lying states in carbon-12 may be investigated by moving the RHIC CNI target with respect to the detector array to change the kinematic region of the detected recoils.

Finally, the virtue of having two essentially independent methods to measure the absolute beam polarization cannot be underestimated.

ACKNOWLEDGMENTS

The authors would like to thank G. Bunce, A. Bravar, O. Jinnouchi and N. H. Buttimore for helpful discussions about the p-C method. Thanks are also due to J. Kiryluk for help in preparation of the figures in this paper.. N. H. Buttimore has independently proposed a similar scheme using a carbon analyzer in about the same time frame.

REFERENCES

1. G. Rackness *et al.*, "Forward $\vec{p}+\vec{p}\rightarrow\pi^0+X$ Production at STAR," in *Abstracts, Spin 2002*, Brookhaven National Laboratory, September 9-14, 2002.
2. D. Underwood *et al.*, Part. World 3, 1-15 (1992).
3. C. Bourrely and J. Soffer, Phys. Lett. B314, 132 (1993).
4. J. Tojo *et al.*, Phys. Rev. Lett. 89, 52302-52305 (2002).
5. I. G. Alekseev *et al.*, "RHIC pC CNI Polarimeter: Status and Performance from the First Collider Run," in *Abstracts, Spin 2002*, Brookhaven National Laboratory, September 9-14, 2002.

Spin Asymmetry for Proton Deuteron Collisions at Forward Angles

N. H. Buttimore

School of Mathematics, University of Dublin, Trinity College, Dublin 2, Ireland

Abstract. The spin asymmetries for proton deuteron elastic scattering are studied at small angles in the context of understanding spin dynamics at high energy for the purposes of measuring the polarization of proton and deuteron beams. The effects of hadronic spin dependence, dispersive spin independent amplitudes, and the Coulomb phase are addressed in particular. Electromagnetic helicity amplitudes for proton deuteron collisions resulting from single photon exchange are presented, those prominent at low momentum transfer and high energy being highlighted. The character of the maximum analyzing power for colliding polarized protons, deuterons, and helium-3 is discussed, focusing on the dependence upon spin and phases that is important for polarimetry.

INTRODUCTION

In addition to the information on hadronic spin structure resulting from the availability of incident protons of known polarization there is a need to study other elememts of the isospin sector [1] using effective polarized neutrons at high energy. Polarized deuterons and helium-3 nuclei provide one source of such neutrons and an analysis of their electromagnetic and hadronic dynamics in collisions at small momentum transfers can assist the evaluation of their level of polarization.

The elastic scattering of spin polarized protons and deuterons is discussed here in the context of understanding its enriched spin dependence and of seeking a polarimeter for high energy deuterons. A comparison with the case of helium-3 of charge $Z = 2$ illuminates the discussion of the search for high energy neutrons of detectable spin polarization at squared momentum transfers in the interference region of [2]

$$-t_c = 8\pi Z\alpha/\sigma_{\text{tot}}. \tag{1}$$

Small angle proton carbon scattering has been used successfully [3] to evaluate high energy spin dependence [4] in the context of polarimetry. One of the significant results emerging from this study is that the hadronic single helicity flip amplitude appears to remain non-zero in the region of interference over a proton laboratory energy range extending to 100 GeV/c. There are hints of this in the proton proton case also [5].

With a deuteron spin of 1, the elastic scattering of protons on deuterons is another fermion boson process with rich spin properties. Elastic collisions at interference angles offer opportunities for probing a number of hadronic p d spin dependent amplitudes. Their rôle in deuteron relative polarimetry is highlighted.

CP675, *Spin 2002: 15th Int'l. Spin Physics Symposium and Workshop on Polarized Electron Sources and Polarimeters*, edited by Y. I. Makdisi, A. U. Luccio, and W. W. MacKay

SPIN 1/2 – 1 AMPLITUDES

The 36 helicity amplitudes of proton deuteron elastic scattering reduce to 12 independent ones under time reversal and parity invariance [6]. Of the hadronic amplitudes

$$H_i(\lambda'_p, \lambda'_d \mid \lambda_p, \lambda_d), \quad i \in \{1, 2, \ldots, 12\}, \quad \lambda_p \in \{+,-\}, \quad \lambda_d \in \{+,0,-\}$$

- four have imaginary parts relating to spin-dependent total cross sections

$$H_1(++\mid++), \quad H_2(++\mid-0), \quad H_3(+0\mid+0), \quad H_4(+-\mid+-)$$

- five amplitudes have the kinematic $\sqrt{-t}$ single helicity flip dependence as factor

$$H_5(++\mid+0), H_6(++\mid-+), H_7(++\mid--), H_8(+0\mid+-), H_9(+0\mid-0)$$

- two have a $(-t)$ factor and the twelfth amplitude a $-t\sqrt{-t}$ behavior near $t=0$

$$H_{10}(++\mid+-), \quad H_{11}(+0\mid-+); \qquad H_{12}(+-\mid-+).$$

In the reduction from 36 to 12 amplitudes, the multiplicity of the six amplitudes $H_1, H_3, H_4, H_7, H_9, H_{12}$ is two while that of the other six $H_2, H_5, H_6, H_8, H_{10}, H_{11}$ is four.

Coulomb amplitudes

One photon exchange helicity amplitudes have been calculated [7] in terms of the Dirac and Pauli proton form factors F_1, F_2, and deuteron form factors, F_1^{d}, F_2^{d}, G_1^{d}, which often appear in the linear combinations [8]

$$\begin{aligned} G_0^d &= F_1^d - \left[F_1^d + F_2^d\,(1 - t/4M^2) - G_1^d\right] t/6M^2 \\ G_2^d &= F_1^d + F_2^d\,(1 - t/4M^2) - G_1^d \end{aligned} \tag{2}$$

that have normalizations $G_0^d(0) = 1$, $G_1^d(0) = \mu_d$, and $G_2^d(0) = Q$, where the magnetic moment μ_d is in $e/2M$ units and the quadrupole moment Q is in units of e/M^2, the deuteron mass being $M = 1889.260$ MeV/c^2. In the case of electron deuteron scattering where $F_1 = 1$ and $F_2 = 0$ the amplitudes correctly reproduce the high energy unpolarized differential cross section for $ed \to ed$ collisions at center of mass scattering angle θ [9]

$$\frac{d\sigma}{dt} = 4\pi\left(\frac{\alpha}{t}\right)^2\left\{G_0^2 - \frac{t}{6M^2}\left[1 + 2\left(1 - \frac{t}{4M^2}\right)\tan^2\frac{\theta}{2}\right]G_1^2 + \frac{t^2}{18M^2}G_2^2\right\}. \tag{3}$$

Single photon exchange contributions to the $pd \to pd$ electromagnetic amplitudes have the following approximate form at asymptotic energies and low scattering angles

$$H_i^{\mathrm{em}}(+j\mid+j) = \frac{\alpha s}{t} F_1(t) F_1^{\mathrm{d}}(t), \quad i = 1, 3, 4 \tag{4}$$

in the helicity nonflip case for deuteron helicities $j \in \{+, 0, -\}$. The amplitude in which the proton of anomalous magnetic moment $\kappa_p = \mu_p - 1$, and mass m, flips its helicity is

$$H_6^{\text{em}}(++|-+) = \frac{\alpha s}{\sqrt{-t}} \frac{1}{2m} \kappa_p F_2(t) F_1^{\text{d}}(t) . \tag{5}$$

The two deuteron single helicity flip electromagnetic amplitudes are equal at large s

$$H_5^{\text{em}} = H_8^{\text{em}} = \frac{\alpha s}{\sqrt{-2t}} F_1(t) \frac{1}{2M} \left[G_1^{\text{d}}(t) - 2F_1^{\text{d}}(t) \right] \tag{6}$$

and the remaining photon exchange amplitudes, H_2^{em}, H_7^{em}, H_9^{em}, H_{10}^{em}, H_{11}^{em}, H_{12}^{em}, are negligible asymptotically at forward angles.

ASYMMETRY EXTREMA

The analyzing power for polarized protons scattering on deuterons at electromagnetic interference involves, in particular, the following significant amplitudes

$$A_N = \frac{3\,\text{Im}\left[H_6^{*}\left(H_1 + H_4\right) + \cdots\right]}{|H_1|^2 + 2\,|H_2|^2 + |H_3|^2 + |H_4|^2 + 2\,|H_5|^2 + 2\,|H_6|^2 + |H_7|^2 + 2\,|H_8|^2 + \cdots} \tag{7}$$

where each amplitude includes an electromagnetic contribution $H_i + e^{i\delta} H_i^{\text{em}}$ involving spin 1/2 and 1 currents, with $\delta \approx 0.02$ as a small Coulomb phase. A factor of 1/3 appears in the averaged differential cross section due to the spin one nature of the deuteron. The analysis for polarized proton deuteron collisions follows that of the proton proton case [5, 10].

The analyzing power for high energy polarized protons (or indeed other polarized light fermion nuclei with charge Z like helium-3) colliding with deuterons with hadronic slope b in the interference region $-t_c = 8\pi Z\alpha/\sigma_{\text{tot}}$ is given approximately by

$$\frac{mA_N}{\sqrt{-t}} \frac{16\pi}{\sigma_{\text{tot}}^2} \frac{d\sigma}{dt} e^{-bt} \quad = \quad (\kappa - 2\,\text{Im}\,r)\frac{t_c}{t} - 2\,\text{Re}\,r + 2\rho\,\text{Im}\,r \tag{8}$$

in which, neglecting the ratio of hadronic and electromagnetic form factors and other nuclear effects [11], the unpolarized differential cross section is

$$\frac{16\pi}{\sigma_{\text{tot}}^2} \frac{d\sigma}{dt} e^{-bt} \quad = \quad \left(\frac{t_c}{t}\right)^2 - 2(\rho + \delta)\frac{t_c}{t} + \left(1 + \rho^2\right)\left(1 + \beta^2\right) \tag{9}$$

with ratio $\rho = \text{Re}H_+/\text{Im}H_+$ where the hadronic H_+ is an average of H_1, H_3, and H_4. Spin dependent hadronic terms present in the forward direction are incorporated in $\beta^2 =$

$$\frac{|H_1 - H_3 + H_4|^2 + |H_1 + H_3 - H_4|^2 + |H_1 - H_3 - H_4|^2 - |H_1|^2 - |H_3|^2 - |H_4|^2 + 6|H_2|^2}{|H_1 + H_3 + H_4|^2}$$

and β here is zero in the case where H_1, H_3, and H_4 are all equal, and $H_2 = 0$. Given that the hadronic single helicity flip amplitude seems to persist at high energies in proton carbon scattering [3] it would be most interesting to determine if any of the proton deuteron amplitudes contributing to β, or any of the kinematically scaled hadronic helicity flip amplitudes defined by [12]

$$r_i = m(-t)^{-1/2} H_i / \mathrm{Im} H_+ , \quad i = 5, 6, 8 , \tag{10}$$

prevail in the asymptotic energy region also. The behavior of r_6 from $\mathrm{p}\!\uparrow \mathrm{d} \to \mathrm{p}\!\uparrow \mathrm{d}$, and of r_5 and r_8 from $\mathrm{d}\!\uparrow \mathrm{p} \to \mathrm{d}\!\uparrow \mathrm{p}$ at large s would be of considerable interest. The evaluation of such terms would facilitate the understanding of high energy proton deuteron spin dynamics. With its low magnetic moment of 0.85744 nuclear magnetons the extremal analyzing power for polarized deuteron scattering is very much less than that of the proton case and would require increased running time for comparable accuracy.

POLARIZED HELIUM-3

The study of another method for providing polarized neutons is instructive. Consider the elastic collision of a spin half particle of mass m' and charge Z' with a particle of charge Z and suppose that ζ is the sign of $Z'Z$. Magnetic moments μ provide $\kappa/m = \mu/m - Z'/m'$, when given in nuclear magneton units. In the absence of hadronic spin dependence, the single spin asymmetry for nuclear size effects $b < -1/t_c$ in the electromagnetic interference region has an extremum of

$$A_{\mathrm{N}}^{\mathrm{e}} = \frac{\kappa\zeta}{4m}\sqrt{-3t_{\mathrm{e}}} , \quad \text{at} \; -t_e = 8\pi\sqrt{3}\frac{|Z'Z|\alpha}{\sigma_{\mathrm{tot}}} \tag{11}$$

where for positive values of $\kappa\zeta$ the extremum is a maximum. In the proton case, for example, $\kappa = 1.79285$ nuclear magnetons, giving a maximum when scattering on a positive target. A helium-3 nucleus of charge $Z = 2$, mass $m' = 2808.392\ \mathrm{MeV}/c^2$, and $\kappa = -2.79569$ scattering on the same positive target would lead to a negative value of the analyzing power of about $-90\,\%$ times that of the proton case, assuming that the helium-3 total cross section is three times that of the proton total cross section at a corresponding energy. As in the proton case, the presence of high energy hadronic spin dependence suggests that only relative polarimetry is possible for helium-3 processes.

CONCLUSIONS

Polarimetry for high energy protons is being secured using the asymmetry in the elastic reaction $\mathrm{p\,C} \to \mathrm{p\,C}$. A study of the electromagnetic amplitudes for $\mathrm{p\,d} \to \mathrm{p\,d}$ reveals that such collisions may be used for either proton or deuteron polarization determination, but the small value of the magnetic moment of the deuteron hinders the evaluation of the spin polarization of the deuteron, in practice.

Another source of polarized neutron is provided by helium-3 nuclei. The minimum analyzing power for helium-3 elastic collisions at interference angles is expected to be about −90 % of that of the proton case at similar energies due to the relatively large and negative magnetic moment of helium-3. For scattering on a positively charged target the helium-3 analyzing power is negative in the interference region.

A study of the spin averaged differential cross section over a range of energies at low momentum transfers enables a test of causality by way of the analytic properties of spin independent amplitudes. The further study of asymmetries in the elastic process at forward angles suggests that the evaluation of dispersion integrals for spin dependent scattering amplitudes may eventually be within reach.

The analysis of the many helicity amplitudes related to the peripheral elastic processes of a number of particles with spin in the electromagnetic-hadronic interference region would assist the measurement of particle polarization and provide a deepening understanding of high energy spin dynamics.

REFERENCES

1. G. Bunce, N. Saito, J. Soffer and W. Vogelsang, Ann. Rev. Nucl. Part. Sci. **50**, 525 (2000) [arXiv:hep-ph/0007218].
2. B. Z. Kopeliovich and L. I. Lapidus, Sov. J. Nucl. Phys. **19**, 114 (1974); N. H. Buttimore, E. Gotsman and E. Leader, Phys. Rev. D **18**, 694 (1978).
3. J. Tojo *et al.*, Phys. Rev. Lett. **89**, 052302 (2002) [arXiv:hep-ex/0206057]; N. H. Buttimore, in *High Energy Spin Physics 1982*, edited by G. M. Bunce, AIP Conf. Proc. No. 95 (New York, 1983), p. 634.
4. Presentations by B. Z. Kopeliovich, K. Kurita, and O. Jinnouchi in these proceedings.
5. N. H. Buttimore, B. Z. Kopeliovich, E. Leader, J. Soffer and T. L. Trueman, Phys. Rev. D **59**, 114010 (1999) [arXiv:hep-ph/9901339].
6. C. Bourrely, J. Soffer and E. Leader, Phys. Rept. **59**, 95 (1980).
7. B. Corbett, M.Sc thesis, "Spin dependent proton deuteron scattering: electromagnetic and hadronic amplitudes", University of Dublin (1984).
8. S. Waldenstrøm and H. Olsen, Nuovo Cimento **3A**, 491 (1971); A. Pais, Nuovo Cimento **53**, 433 (1968).
9. M. Gourdin, Nuovo Cimento 28, 533 (1963); V. Glaser and B. Jak šić, Nuovo Cimento **5**, 1197 (1957).
10. N. Akchurin, N. H. Buttimore and A. Penzo, Phys. Rev. D **51**, 3944 (1995); A. T. Bates and N. H. Buttimore, Phys. Rev. D **65**, 014015 (2002) [arXiv:hep-ph/0010014].
11. B. Z. Kopeliovich and T. L. Trueman, Phys. Rev. D **64**, 034004 (2001) [arXiv:hep-ph/0012091].
12. N. H. Buttimore, E. Leader and T. L. Trueman, Phys. Rev. D **64**, 094021 (2001) [arXiv:hep-ph/0107013].

High Precision Electron Beam Polarization Measurement with Compton Polarimetry at Jefferson Laboratory

F. Marie, E. Burtin, C. Cavata, S. Escoffier, D. Lhuillier, D. Neyret, T. Pussieux* and P. Bertin†

*DSM/DAPNIA/SPhN - CEA/Saclay, 91191 Gif sur Yvette cedex, FRANCE
†CNRS/IN2P3 LPC - 63177 Aubiere cedex, FRANCE

Abstract. Since 1999, a Compton polarimeter based on a Fabry-Perot cavity to amplify the laser light is operational in the hall A of the Jefferson Laboratory. In 2000, the beam polarization has been continuously measured during $N-\Delta$ and G_e^p experiment providing a relative total uncertainty of 1.4% in 40 mn at 4.5 GeV. These unprecedented results have been obtained thanks to a scattered electron detector which has allowed to determine the response function of the photon calorimeter.

INTRODUCTION

The new generation of parity violation or spin sensible experiments require an accurate knowledge of the target or beam polarization. In this spirit, a Compton polarimeter has been constructed at Jefferson Laboratory in 1998, to measure and to monitor the polarization of the electron beam in the Hall A. The Compton polarimeter has been operated for the first time during the HAPPEX experiment[1, 2] and has provided an absolute monitoring of the electron beam polarization (100 μA, 3.3 GeV) during the entire run, with a total uncertainty of 3% in 1 hour[3]. In 2000, the Compton polarimeter has been running for $N-\Delta$[4] and G_e^P[5] experiments at $E_e = 4.5\ GeV$ and $I = 40\ \mu A$. The present contribution reports on the analysis method and the hardware improvements that have led to high precision Compton polarimetry measurements[6].

THE COMPTON POLARIMETER

When longitudinally polarized electrons scatter off circularly polarized photon, the experimental asymmetry of the counting rates of the scattered events A_{exp} is proportional to the polarization of the electron beam:

$$A_{exp} = \frac{N^+ - N^-}{N^+ + N^-} = P_e \times P_\gamma \times < A_{th} >, \tag{1}$$

where N^+ and N^- refer respectively to counting rates when the spin of the electron and the photon are parallel and anti-parallel, P_γ is the circular polarization of the photon

CP675, *Spin 2002: 15th Int'l. Spin Physics Symposium and Workshop on Polarized Electron Sources and Polarimeters*, edited by Y. I. Makdisi, A. U. Luccio, and W. W. MacKay

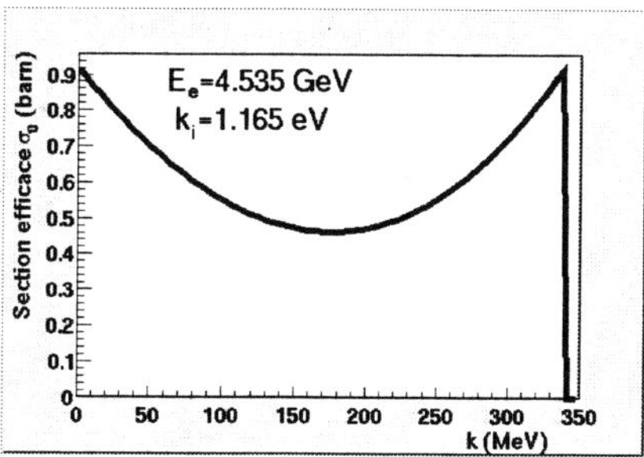

FIGURE 1. Compton cross section with respect to the energy of the scattered photon. The Compton edge is $K_{max} = 340\ MeV$.

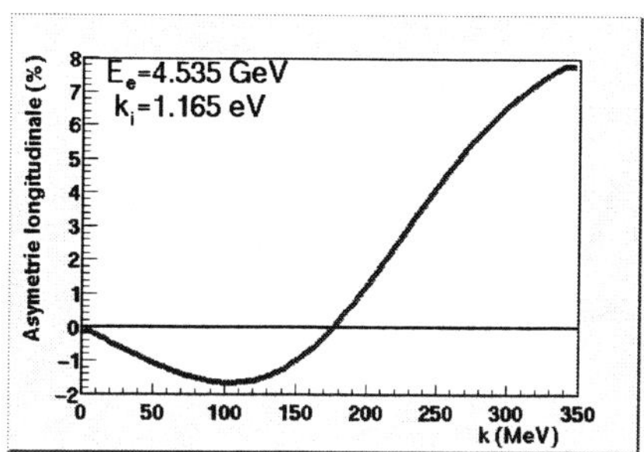

FIGURE 2. Theoretical asymmetry. The averaged asymmetry is 2%

beam and $< A_{th} >$ the analyzing power, entirely defined by QED and the experimental running conditions.

In Figures 1 and 2 are plotted the Compton cross section and the theoretical asymmetry for $E_e = 4.5\ GeV$ and $E_\gamma = 1.16\ eV$

The Compton polarimeter at Jlab hall A[8] consists in a magnetic chicane at the center of which the electron beam scatters off the photon beam. The main characteristic of this Compton polarimeter is a Fabry-Perot cavity[7] that amplifies the 250 mW IR laser light with a gain of 8000. The optical setup[8] provides a 1500 W laser light with a polarization of 99% at the Compton interaction point. The scattered photons are detected in the central crystal of a 25×25 $PbWO_4$ scintillators ($2 \times 2 \times 20\ cm^3$) matrix and the scattered electrons are detected in four planes made of 48 silicon strips (width=650 μm). The electron detector allows to determine the response function of the photon calorimeter and provides an complementary polarization measurement.

Scattered photons are detected in the central crystal and the deposited energy is measured with ADC's. Frequently, the laser light is turned off in order to measure the background due to Bremsstrahlung coming from the interaction of the beam with the residual atoms in the vacuum pipe and with cavity mirror holders. The sign of the photon polarization is also reversed alternatively to correct for false asymmetries. One of the major sources of false asymmetry is the variation of the luminosity related to electron beam position variations correlated to the helicity. Therefore, to lower this

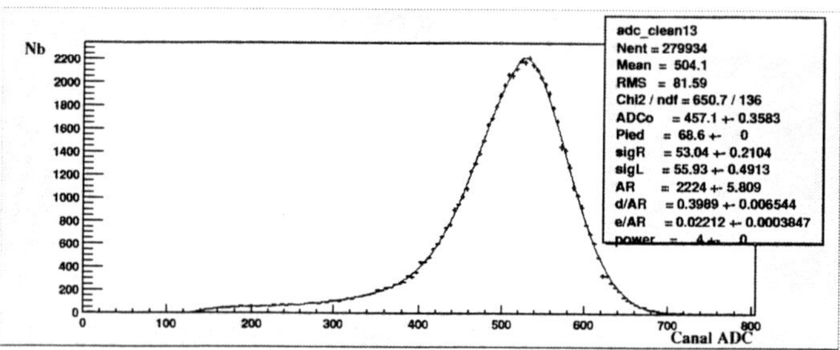

FIGURE 3. Fit of the response function of the central crystal for a 200 MeV incident photon.

effect, the beam in the magnetic chicane is locked to a nominal position corresponding to the maximum of the luminosity for the Compton interaction. Finally, the experimental asymmetry is determined from counting rates normalized to acquisition dead time and beam intensity. Runs are taken in both "coincidence" mode (to measure the response function) or in "only photon" mode (normal data acquisition).

ANALYZING METHOD

The analyzing method is called "*semi-integrated*" as the scattered photon counting rates are integrated from an energy threshold ($E_{th} = 200\ MeV$) to optimize the statistical precision and to minimize most of the systematics.

One of the major improvements of the Compton polarimeter data analysis since HAPPEX measurements[3] is the determination of the response function of the central crystal. For "coincidence" runs, this response function is determined using the electron detector where each strip defines a narrow energy band ($\sim 5\ MeV$). For a given scattered electron energy (i.e.d. scattered photon energy k) the central crystal ADC spectrum is fitted (see Figure 3) with an empirical function $g(ADC,k)$, of which the particularity is to reproduce the asymmetrical behavior of the deposited energy distribution and the low energy tail. For all the runs taken in "only photon" acquisition mode, the ADC spectrum is fitted (see Figure 4) with the theoretical Compton cross section convoluted with the response function:

$$\frac{d\sigma(ADC)}{dADC} = \int_0^{k_{max}} \frac{d\sigma_0(k)}{dk} g(ADC,k)dk \tag{2}$$

Possible run by run gain fluctuations λ can be thus taken into account. The response function allows thus to determine, the probability $p(k)$ for a photon with a given energy k to be detected in the calorimeter with an amplitude above ADC_{th} (corresponding to $E_{th} = 200\ MeV$):

$$p(k) = \frac{\int_{ADC_{th}/\lambda}^{\infty} g(ADC,k)dADC}{\int_0^{\infty} g(ADC,k)dADC} \tag{3}$$

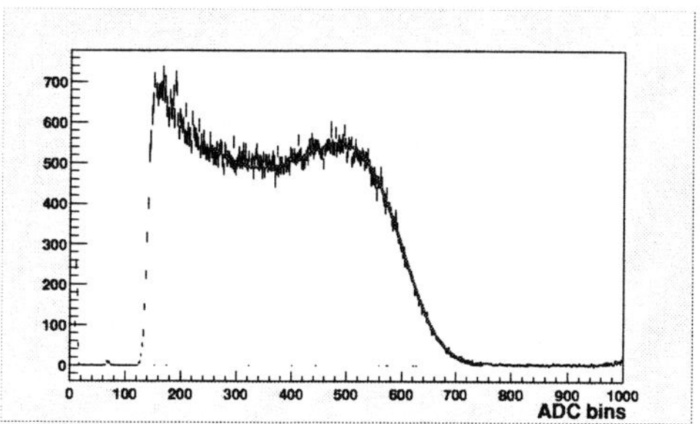

FIGURE 4. Fit of the ADC spectrum with the theoretical Compton cross section convoluted with the response function of the calorimeter.

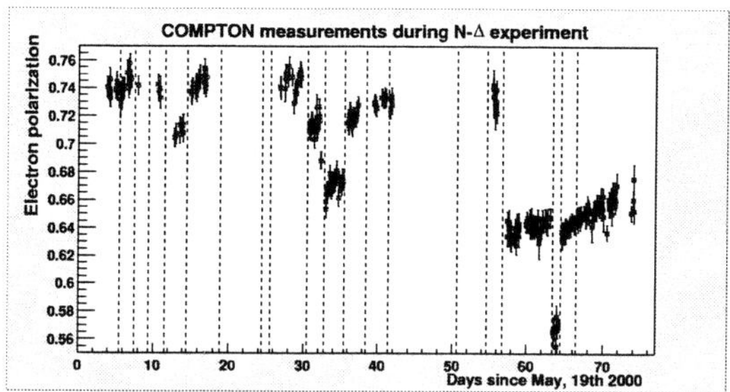

FIGURE 5. Errors are statistics plus run by run uncorrelated uncertainties. Dashed lines represent spot moves at the electron source .

and the analyzing power can be thus deduced from $p(k)$:

$$< A_{th} > = \frac{\int p(k) \frac{d\sigma_0(k)}{dk} A_{th}(k) dk}{\int p(k) \frac{d\sigma_0(k)}{dk} dk} \quad (4)$$

BEAM POLARIZATION RESULTS

330 polarization measurements have been performed during $N - \Delta$ experiment and 110 during G_e^P. The results are plotted in Figures 5 and 6. A 1% relative error, corresponding to common systematic uncertainty, must be added to all this points.

The typical relative uncertainty for a run of 40 mn is 0.8% statistics and 1.1% systematics, which gives a total uncertainty of 1.4%. The contribution of the different sources of errors are presented in table 1

TABLE 1. Error budget of Compton polarimeter measurements during $N-\Delta$ and G_e^P experiments

Error source	typical relative uncertainty (%)
Experimental asymmetry	
Statistical (40 mn)	0.80
Positions and angles	0.45
Events selection	0.10
Background asymmetry	0.05
Dead time	0.10
Laser beam	
Polarization P_γ	0.45
Analyzing power	
Modelization	0.45
Energy Calibration	0.60
Pile-up	0.45
Radiative corrections	0.25
total systematics	1.10
TOTAL	1.40

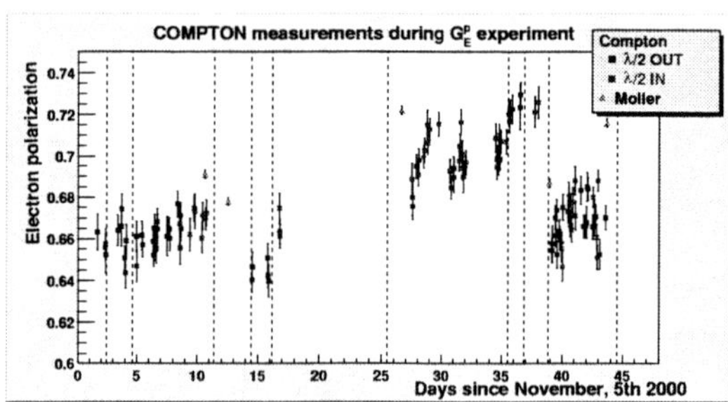

FIGURE 6. Errors are statistics plus run by run uncorrelated uncertainties. Dashed lines represent spot moves at the electron source. Preliminary Moller points (statistic only) have been added for comparison.

CONCLUSIONS

A Compton polarimeter operating a Fabry-Perot cavity is operational at JLAB since 1999. In 2000, during $N-\Delta$ and G_e^p experiments, a careful study of all sources of uncertainties and a new analysis approach[6] have allowed to measure continuously the beam polarization with a typical total relative uncertainty of 1.4% in 40 mn data taking, which are unprecedented results at these kinematical conditions (E=4.5 GeV, I=40 μA). For the first time, the electron detector has been used to determine the response function of the calorimeter. It will offer a complementary method to measure the beam polarization for the next generation of parity violation experiments above 3 GeV with a relative uncertainty better than 2%.

REFERENCES

1. K.A. Amiol & al. (HAPPEX collab.), Phys. Rev. Lett.**82** (1999) 1096-1100
2. K.A. Amiol & al. (HAPPEX collab.), Phys. Lett. **B 509** (2001) 211-216
3. M. Baylac & al, Phys. Lett. **B 539** (2002) 8-12
4. A. Sarty & al., "*Investigation of the* $N-\Delta$ *transition via polarization observables in hall A*", TJNAF Experiment Proposal E91-011 (1991)
5. O. Cadiou & al (TJNAF hall A collab.), Phys. Rev. Lett. **88** (2002)
6. S. Escoffier, "*Mesure précise de la polarisation du faisceau d'électrons à TJNAF par polarimetrie Compton, pour les expériences* G_e^p *et* $N-\Delta$", PhD Thesis Report, DAPNIA/SPhN-01-03-T (2001)
7. J.P. Jorda & al., Nucl. Instr. & Meth. **A 412** (1998) 1-18
8. N. Falletto & al, Nucl. Instr. & Meth. **A 459** (2001) 412-425

Polarized Ion Sources and Targets

2002 Spin

Summary of the PST 2001 Workshop

V.P.Derenchuk

Indiana University Cyclotron Facility, 2401 Milo B. Sampson Lane, Bloomington, IN 47405, USA
Laddie@IUCF.Indiana.edu

Abstract. "The Workshop on Polarized Sources and Targets (PST 2001) was held at the Brown County Inn in Nashville, Indiana from September 30th and October 4th, 2001. It was organized by the Indiana University Cyclotron Facility in Bloomington, Indiana. The Workshop is the most recent of a series held at about two year intervals, the last of which was held in Erlangen, Germany in 1999. About 80 scientists attended the Workshop. There were 12 invited talks, 30 contributed talks and 15 posters. E. Steffens (Erlangen) gave the summary talk at the conclusion of the Workshop. The subjects addressed in the Workshop included atomic beam polarized H and D targets, solid polarized targets, spin polarized HD, polarized electron sources, polarized ion sources, hadron polarimetry at intermediate to high energies, electron polarimetry, polarized neutrons and the use of polarized noble gases in medical imaging."[1]

PST 2001, NASHVILLE, INDIANA UNIVERSITY

Proceedings of the Ninth International Workshop on Polarized Sources and Targets, eds. Vladimir P. Derenchuk and Barbara von Przewoski, is now available. It is published with an ISBN number of 981-0204917-9 by World Scientific Publishing Co. Pte. Ltd., P.O. Box 128, Farrer Road, Singapore 912805. Copies may be purchased from the publisher or by contacting the author.

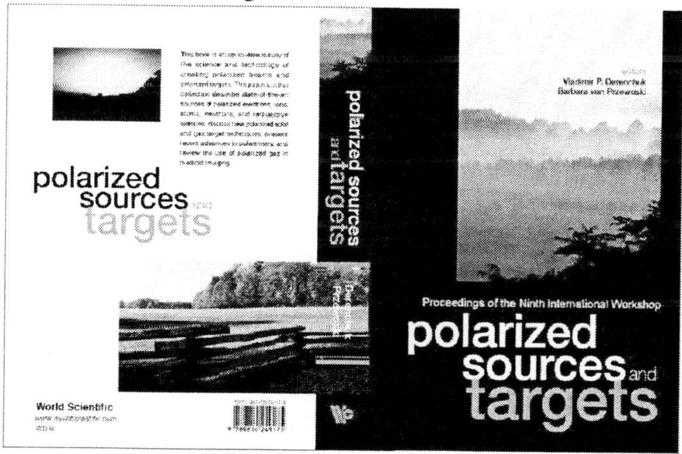

Figure 1. Front cover of the Proceedings.

[1] From the Preface to the Proceedings of the Ninth International Workshop on Polarized Sources and Targets, eds. Vladimir P. Derenchuk and Barbara von Przewoski, World Scientific, 2002.

CP675, *Spin 2002: 15th Int'l. Spin Physics Symposium and Workshop on Polarized Electron Sources and Polarimeters*, edited by Y. I. Makdisi, A. U. Luccio, and W. W. MacKay
© 2003 American Institute of Physics 0-7354-0136-5/03/$20.00

Workshop on Testing QCD Through Spin Observables in Nuclear Targets

D. G. Crabb

Physics Department. University of Virginia
Charlottesville, VA 22903, USA

Abstract. A summary of the Workshop is presented.

The Workshop was held at the University of Virginia, Charlottesville, Virginia from April 18th to April 20th, 2002. It was sponsored by the Institute of Nuclear and Particle Physics at the University of Virginia, Jefferson Laboratory and the International Spin Physics Committee. There were 60 registrants, 12 plenary talks and 27 parallel session contributions. Summary talks were given by Abhay Deshpande and Mark Strikman on the experimental and theoretical aspects, respectively, covered in the Workshop. The topics of the four parallel sessions were; 'New Initiatives', 'Polarized Beams and Targets', 'Inclusive, Exclusive and Semi-Inclusive Asymmetries' and 'Spin in Few Body Systems'. Other talks covered the programs at existing facilities, structure functions and form factors and transversity.

The Proceedings of the Workshop will be published by World Scientific Publishers.

CP675, *Spin 2002: 15th Int'l. Spin Physics Symposium and Workshop on Polarized Electron Sources and Polarimeters,* edited by Y. I. Makdisi, A. U. Luccio, and W. W. MacKay

Status of Frozen-spin Polarized HD Targets for Spin Experiments

Tsuneo Kageya*†, Christopher M. Bade**†, Anthony Caracappa†, Frank C. Lincoln†, Michael M. Lowry†, John C. Mahon**, Lino Miceli†, Andrew M. Sandorfi†, Craig E. Thorn†, Xiangdong Wei† and C. Steven Whisnant‡

*Virginia Polytechnic Institute and State University, Blacksburg, VA 24061, USA
†Brookhaven National Laboratory, Upton, NY 11973, USA
**Ohio University, Athens, OH 45701, USA
‡James Madison University, Harrisonburg, VA 22807, USA

Abstract. The first experiments have been carried out with polarized HD targets at the Laser-Electron-Gamma-Source (LEGS) facility. By holding targets at low temperature and high field (17 mK and 15 Tesla) in a Dilution Refrigerator (DF) for six weeks a frozen-spin state was reached, with equilibrium polarizations for protons and deuterons of 70% and 17%, respectively. Multiple measurements of the relaxation times and multiple transfers of the targets reduced these values so that experimental runs were carried out with polarizations of 30% and 6%, respectively. The relaxation times for protons and deuterons were observed to be 13 days and 36 days, respectively, in the beam line cryostat at 1.3 K and 0.7 Tesla magnetic field. For the future runs significantly higher D polarizations are possible by transfer of spin from the proton to the deuteron using an rf forbidden adiabatic fast passage. Higher polarizations and longer relaxation times are expected from ongoing development.

INTRODUCTION

Recently, experiments have started using circularly polarized photon beams with longitudinally polarized hydrogen and deuteron targets in Mainz [1], Bonn [2] and BNL [3][4]. Motivating these efforts have been tests of the Gerasimov-Drell-Hearn (GDH) sum rule first derived in 1966 [5] [6] and the forward nucleon spin-polarizability sum rule [7].

At LEGS, the first doubly polarized measurements were carried out in November 2001. The experiments used circularly and linearly polarized photon beams obtained by a laser induced backward scattering at the National Synchrotron Light Source at Brookhaven National Laboratory, using detectors for charged and neutral particles which cover almost the solid angle of four-π and a frozen-spin HD target. The solid HD target has a much smaller dilution factor than other solid target materials and background reactions with unpolarized nucleons are significantly reduced.

CP675, *Spin 2002: 15th Int'l. Spin Physics Symposium and Workshop on Polarized Electron Sources and Polarimeters*, edited by Y. I. Makdisi, A. U. Luccio, and W. W. MacKay

FROZEN-SPIN HD TARGET

The spins of two identical protons in an H_2 molecule at a low temperature are antiparallel in their lowest energy state (para); the para-H_2 can not be polarized. An ortho-H_2 molecule with a spin-parallel state for the two protons, is readily polarized at a high magnetic field and a low temperature. Three forth of all the H_2 molecules at room temperature are in the ortho state. At a low temperature, these decay into para-H_2 with a half life of about six days. An HD molecule consists of non-identical particles and there is no constraint as for the para-H_2.

A polarized ortho-H_2 transfers its polarization to HD due to a spin-spin coupling between an H in H_2 and an H in HD. This process happens on the order of hours. A small concentration of polarized ortho-H_2 (on order of 10^{-4}) is used to polarize HD and most of the ortho-H_2 decays into the para state in two months. There are no phonons to couple an S-wave HD to its crystal lattice; thus once polarized, HD has an extremely long relaxation time. A frozen spin state can be achieved by simply keeping the HD at high field and low temperature.

HD gas is isotopically purified by distillation, doped with a small amount of ortho-H_2 and frozen into a mesh of 2000 pure aluminum wires whose diameter is 51 microns and which conducts away the heat from the ortho to para conversion of the H_2 impurity. The target holder is made of Kel-F which does not contain hydrogen. The target contains one mole (3 grams) of HD and has a length of 5 cm with respect to the photon beam direction and a diameter of 2 cm. A schematic drawing of the HD ice, its target holder and aluminum wires in the nose of the InBeam Cryostat (IBC) is shown in Figure 1.

In September 2001, high field/low temperature equilibrium polarizations of 70% for H and 17% for D were obtained after aging for six weeks. In Figure 2 thermal equilibrium polarizations for hydrogen and deuterium are shown as a function of a temperature under a magnetic field of 15 Tesla. The frozen-spin targets were transfered to an IBC using a Transfer Cryostat (TC) which has a temperature of about 2 Kelvin and a minimum magnetic field of 0.016 Tesla. Multiple measurements of the relaxation times and multiple transfers of the targets reduced these values so that experimental runs in November 2001 were carried out with polarizations of 30% and 6%, respectively. The relaxation times in the IBC for protons and deuterons were observed to be 13 days and 36 days, respectively.

RECENT PROGRESSES

In a magnetic field of 15 Tesla, vibrations transfered along pumping lines induce eddie currents which heat the copper support tube to which three HD target holders are thermally connected. This heat results in increasing the temperatures of the targets and limiting the polarizations. A vibration isolator using gimbal mounts [8] in the pumping line has been recently installed and tested successfully [9]. This improvement is expected to increase the thermal equilibrium polarization to 80% and 22% for H and D, respectively.

Significantly higher D polarizations are possible by transfer of spin from the proton

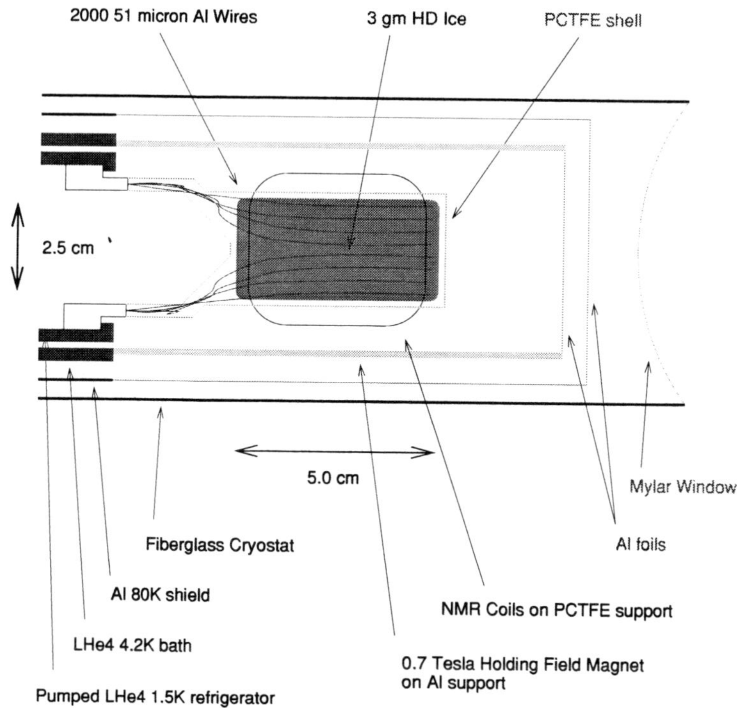

FIGURE 1. HD ice, its holder (Kel-F) and 2000 pure aluminum wires whose diameter is 51 microns to cool down the HD solid effectively.

to the deuteron using an rf forbidden adiabatic fast passage. In Figure 3, a transfer of proton polarization to deuterons by such a passage is shown on NMR signals. Coils for the passage have been installed in the DF. Our goal of the deuteron polarization is 50%.

A cross coil method has been used for the NMR system, leading to significant improvements [10].

FUTURE PLANS

Two new cryostats are under designing and fabrication. The present IBC was constructed by collaborators in Orsay, France. It operates as a ^{4}He evaporation cryostat in which 1.3 Kelvin under a magnetic field of 0.65 Tesla are obtained for holding the HD target as a frozen-spin target in the beam line. This will be replaced by a new IBC using a

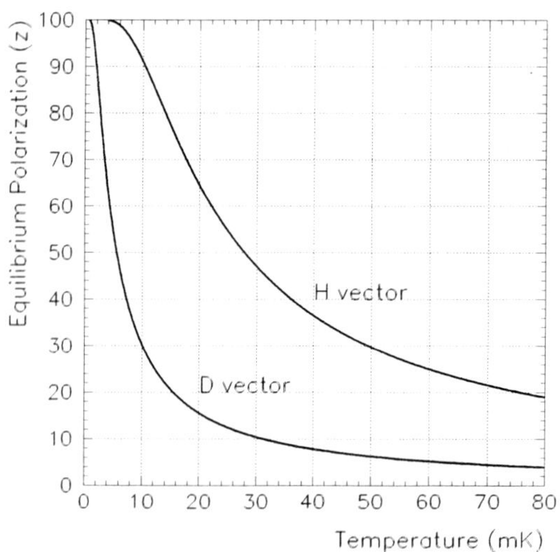

FIGURE 2. H and D thermal equilibrium polarization versus temperature under a magnetic field of 15 Tesla.

Forbidden Adiabatic Fast Passage

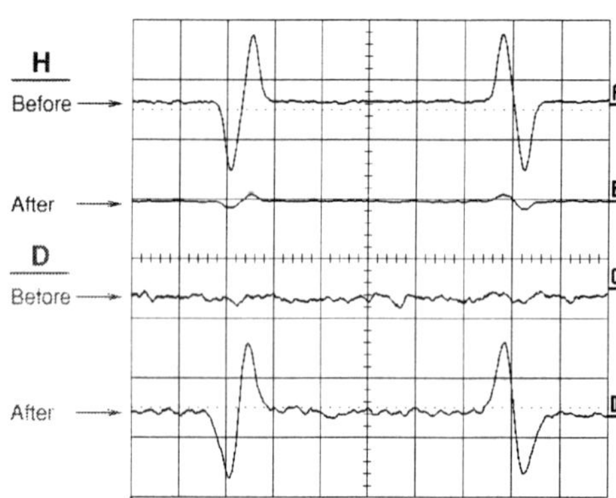

Efficiency of transfer = 67%

FIGURE 3. An example of the NMR signals for the rf forbidden adiabatic fast passage. The proton polarization before the passage (top) is transfered to deuteron (bottom). The second from the top shows an NMR signal for protons after the passage, while the second from the bottom shows one for deuterons before the passage. The NMR signal sizes are proportional to polarizations.

dilution refrigerator to achieve 0.2 Kelvin with a 1.0 Tesla field. The lower temperature and higher magnetic field are expected to increase the relaxation time for the hydrogen and deuterium at least a factor of 2. At some point, the production of free radicals by ionization from beam-generated $e^{\pm}$ pairs will become a limiting factor to the in-beam relaxation time, but this has yet to be determined.

The present TC, also fabricated by Orsay collaborators, keeps a target at about 2 Kelvin and has a minimum magnetic field of 0.016 Tesla. Relative polarization losses during the transfers from DF to IBC (lasting about 45 minutes) were 3% and 8% for protons and deuterons, respectively. To reduce these losses, a new TC has been designed by BNL, and will be fabricated and tested by Juelich in Germany. That will have a minimum magnetic field of 0.16 Tesla which is ten times higher than that of the present one.

Both cryostats are expected to be delivered to BNL in the middle of 2003.

SUMMARY

The first scattering experiments using a polarized HD target were performed at BNL in November 2001. Higher polarizations for the deuteron are expected for experiments in 2003 due to recent progress. Future improvements of the system will lead to longer relaxation times for proton and deuteron polarizations in beam and smaller polarization losses during target transfers.

ACKNOWLEDGMENTS

This work is supported by the U. S. Dept. of Energy under contract No. DEAC02-98CH10886, the U. S. National Science Foundation, and the Institute Nazionale di Fisica Nucleare - Italy.

REFERENCES

1. Ahrens, J. *et al..*, *Phys. Rev. Lett.*, **87**, 022003 (2001).
2. Helbing, K. *et al..*, *Nucl. Phys. Proc. Suppl.*, **105**, 113–116 (2002).
3. Hoblit, S. *et al..*, "First Beam-Target Double-Polarization Measurements Using Polarized HD at LEGS," in *this proceedings*, 2003.
4. Sandorfi, A. M., "First photo-pion dobule polarized experiments using polarized HD at LEGS," in *Proc. of 2nd International Symposium On The Gerasimov-Drell-Hearn Sum Rule And The Spin Structure Of The Nucleon*, 2003.
5. Gerasimov, S. B., *Sov. J. Nucl. Phys.*, **2**, 430 (1966).
6. Drell, S. D., and Hearn, A. C., *Phys. Rev. Lett.*, **16**, 906 (1966).
7. Blanpied, G. *et al..*, *Phys. Rev. C*, **64**, 025203 (2001).
8. Kirk, W. P., and Twerdochlib, M., *Rev. Sci. Instrum*, **49**, 765–769 (1978).
9. Bade, C. M. *et al..*, "Brute Force with Gentle Touch: Vibration Isolation Techniques used to Increase HD Target Polarization," in *this proceedings*, 2003.
10. Thorn, C. E. *et al..*, "Target Polarization Measurement With a Crossed Coil Polarimeter," in *this proceedings*, 2003.

Brute Force with a Gentle Touch: Vibration Isolation Techniques Used to Increase HD Target Polarization

Christopher M. Bade*†, Anthony Caracappa†, Tsuneo Kageya**†, Frank C. Lincoln†, Michael M. Lowry†, John C. Mahon*, Lino Miceli†, Andrew M. Sandorfi†, Craig E. Thorn†, Xiangdong Wei† and C. Steven Whisnant‡

**Ohio University, Athens, OH 45701, USA*
†Brookhaven National Laboratory, Upton, NY 11973, USA
***Virginia Polytechnic Institute and State University, Blacksburg, VA 24061, USA*
‡James Madison University, Harrisonburg, VA 22807, USA

Abstract. The performance of statically polarized high-field/low-temperature targets is a strong function of the base temperature during polarization. At the Laser-Electron Gamma Source (LEGS) facility, highly polarized Hydrogen Deuteride targets are created in a dilution refrigerator/15 tesla superconducting magnet system, and converted to a frozen spin state. This allows them to retain polarization when placed in a beam at a lower field (0.7 T) and higher temperature (1.3 K). An increase in temperature from the 0 T state to the 15 T state of the refrigerator suggested eddy currents were primarily responsible for heating of the cold finger. Vibration-isolation techniques have been developed to reduce the level of eddy currents due to vibration inside the polarizing field. These techniques reduced the amplitude of vibration due to the pumping system by two orders of magnitude and lowered the cold finger temperature with field energized from $\sim$ 17 mK to $\sim$ 12 mK. The potential gain in polarization is substantial.

INTRODUCTION

The Laser-Electron Gamma Source (LEGS) facility utilizes a polarized hydrogen deuteride target for double polarization photoproduction experiments. The highest possible level of nucleon polarization is desired for these experiments. The HD Target is prepared in a high magnetic field and low temperature condition. The target is then converted to a frozen-spin state and transfered via a transfer cryostat and placed in the LEGS beamline in a higher temperature, lower field condition.

The use of an HD target provides some advantages to a photonuclear experiment. The combination of a free proton with a bound proton and a bound neutron provides simplicity. The low background produced by aluminum cooling wires and Kel-F endcaps can be subtracted in a straightforward manner by pumping out the HD and running beam on the empty target cell. The frozen-spin feature allows for polarization of the target by brute-force in temperature and magnetic field conditions which are not practical to achieve in beam.

CP675, *Spin 2002: 15th Int'l. Spin Physics Symposium and Workshop on Polarized Electron Sources and Polarimeters*, edited by Y. I. Makdisi, A. U. Luccio, and W. W. MacKay

OPTIMIZING POLARIZATION

As the LEGS HD target is polarized by conventional brute-force, its maximum polarization is given by the equilibrium polarization value of H or D for the conditions in the dilution refrigerator. The equilibrium polarization value is given by the Brillouin function, Eq. (1), which is an increasing function in (B/T).

$$P_{TE}(x,I) = \frac{2I+1}{2I}\coth\left(\frac{2I+1}{2I}x\right) - \frac{1}{2I}\coth\left(\frac{x}{2I}\right)$$

$$x \equiv \frac{\mu B}{k_B T}, \qquad I = \frac{\mathbf{1}}{\mathbf{2}} \text{ for H and } \mathbf{1} \text{ for D} \tag{1}$$

This means to increase target polarization, one must either increase the magnetic field by buying a larger magnet, or lower the base temperature of the dilution refrigerator.

It was found that the temperature for this cryogenic system is about 9 mK without the magnet energized but about 17 mK when the magnetic field is at 15 T. A temperature decrease from 17 mK to 9 mK would result in a 23% increase in H polarization and a 83% increase in D polarization. This temperature disparity was attributed to vibrations from the main 10 cm pump line transfered down to the copper cold finger which lies within the primary solenoid magnet. As temperature decreases, effects other than eddy current heating will limit the optimum temperature as well.

Accelerometry was performed on the area where the pump line joins the refrigerator. These results are plotted on the predamping spectrum of Figure 1. The background spectrum is included to give a sense of the electronic noise in the accelerometer circuit. Examination of the relative power input due to vibrations showed that the contribution at 60 Hz was the dominant effect.

An established technique for the isolation of large pump lines was implemented in the dilution refrigerator. This vibration isolator consisted of cable-suspended bellows supported by double gimbals modeled after [1]. This design has the advantage of lowering the resonance frequency of the pump line to the point where vibration from most external sources is simply reflected. Figure 2 shows our implementation.

RESULTS

After the isolator was constructed, the accelerometer was again used to determine if the vibrations were reduced. These results are plotted as the after damping line in Figure 1. The spectrum after damping lies between the background and the spectrum before damping. The spectrum of power input after damping shows a decrease in 60 Hz vibration by three orders of magnitude.

The real effectiveness of this device is how it affects the temperature at polarizing conditions. Temperatures were measured before and after installation of the vibration isolator using a nuclear orientation thermometer with a ^{60}Co source [2]. Figure 3 shows the temperature of the cold finger before and after installation of the isolator. The reduction of the full-field temperature from 17 mK to 12 mK results in a quite significant increase in equilibrium polarization as shown in Figure 4.

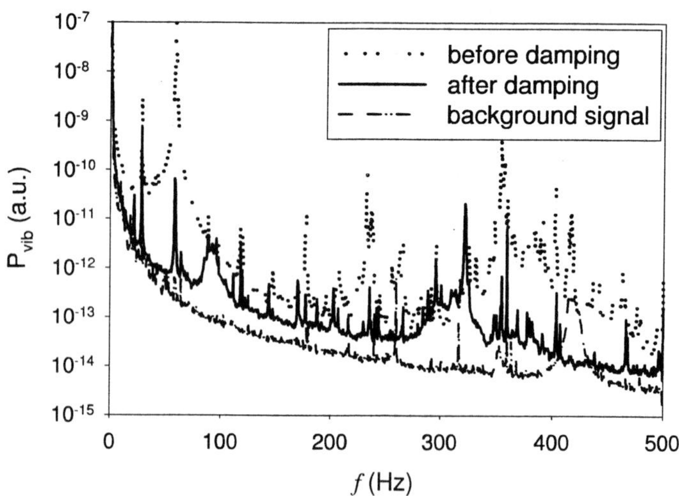

FIGURE 1. A power spectrum of relative heat input due to vibrations from pump line vibrations into the dilution refrigerator. The bottom line is from a measurement without any pumps activated. The top and middle are before and after installing the vibration isolator, respectively.

OUTLOOK

The construction of a pump line isolator was completed to reduce vibrations and eddy current heating. This device was successful in cutting down vibration input to the dilution refrigerator and lowering the polarizing temperature. This has led to a significant increase in equilibrium polarization.

ACKNOWLEDGMENTS

This work was supported by research grants from the U.S. Department of Energy under contract No. DE-AC02-98CH10886 and the U.S. National Science Foundation grant No. PHY-0072226.

REFERENCES

1. Kirk, W. P. and Twerdochlib, M. "Improved Method for Minimizing Vibrational Motion Transmitted by Pumping Lines" *Rev. Sci. Istrum.* **49** (6), pp. 765-769 (1978)
2. Berglund, P. M. et al. "The Design and Use of Nuclear Orientation Thermometers Employing ^{54}Mn and ^{60}Co Nuclei in Ferromagnetic Hosts" *J. Low Temp. Phys.* **6** (3/4), pp. 357-383 (1972)

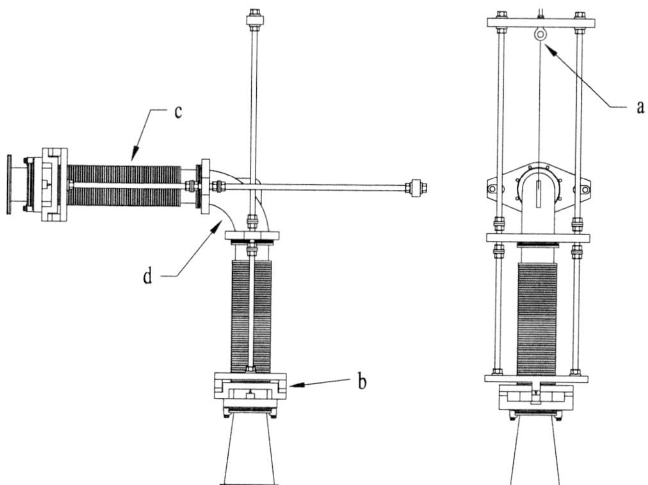

FIGURE 2. A drawing of the vibration isolator as constructed. The cables which suspend the elbow are suspended from U-joints (a) attached to double gimbals (b) resting on knife-edges. Of particular importance is the use of a low spring constant set of bellows (c) such as the thin-walled welded bellows in use here. The addition of 80 lbs. of lead to the elbow (d) also allowed for the lowering of the resonance frequency of the device.

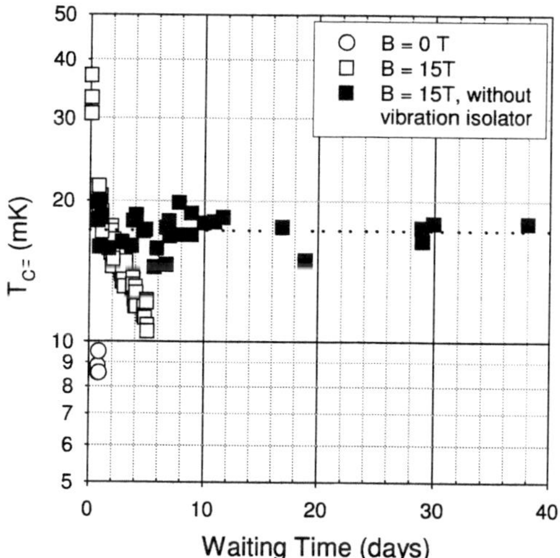

FIGURE 3. This plot shows temperature as a function of cooling time in the dilution refrigerator. The filled squares are measurements before the vibration isolator is installed. The open squares are measurements with the system in place. The isolator allows the refrigerator to reach temperatures below the previous limit of 17 mK.

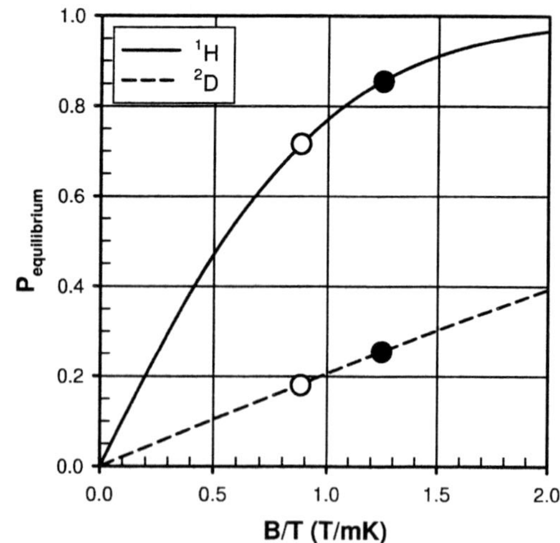

FIGURE 4. A plot of H and D equilibrium polarization as a function of B/T. These functions are given by Eq. (1). The open circles represent the vibration-limited temperature of 17 mK while the filled circles represent the new lower temperature condition.

Target Polarization Measurements with a Crossed-Coil NMR Polarimeter

Anthony Caracappa and Craig Thorn[†]

Brookhaven National Laboratory, Upton, NY 11973

Abstract. We have performed a complete electronic circuit analysis of the crossed coil NMR polarimeter (CC-meter), and from this analysis we have determined the optimum conditions for its operation. From this analysis, which is confirmed by NMR measurements on hydrogen, we conclude that the CC-meter can be operated to give a very high signal-to-noise ratio (SNR) for thermal equilibrium polarization while also producing a highly linear response for fully polarized targets. In general, a well-designed and properly constructed CC-meter can provide a larger SNR at similar non-linearity, compared to the more commonly used Q-meter.

INTRODUCTION

Accurate target polarimetry is essential to polarized target experiments. At the LEGS facility, our goal is to achieve 1% uncertainty in the measurement of highly polarized HD targets. To understand the conditions for achieving such high accuracy, we have performed a complete electronic circuit analysis of the crossed coil NMR polarimeter (CC-meter), and from this analysis we have determined the optimal conditions for its operation. These results have been compared to the very well studied and more commonly used Q-meter circuit [1-3]. The CC-meter replaces the single coil and resonant circuit of the Q-meter with a pair of coils arranged with orthogonal axes and a pair of associated resonant circuits, although a tuned input circuit is not essential. In any practical realization, both the CC-meter and Q-meter polarimeters suffer from a significantly non-linear response to susceptibility for highly polarized samples. One important purpose of the circuit analysis is to specify conditions under which these non-linearities become insignificant, or to allow the computation of a correction to the polarization deduced from the NMR signal in the presence of significant non-linearities.

CIRCUIT ANALYSIS

In order to understand and optimize the performance of the CC-polarimeter, we have constructed an analytic circuit model based on the equivalent circuit of Figure 1. The values of the circuit elements have been determined by direct measurement where possible (Rp, Rs, Lp, and Ls) and by fitting measurements of the voltage gain as a

[†] This work was supported by the U.S. Department of Energy under contract DE-AC02-98CH10886.

CP675, *Spin 2002: 15th Int'l. Spin Physics Symposium and Workshop on Polarized Electron Sources and Polarimeters,* edited by Y. I. Makdisi, A. U. Luccio, and W. W. MacKay

function of frequency (see Figure 2 for an example). The values determined for two slightly different coils are given in Table 1. The signal generator is a Rohde & Schwarz SMY01 and the PSD is a Stanford Research SR844 RF lock-in amplifier.

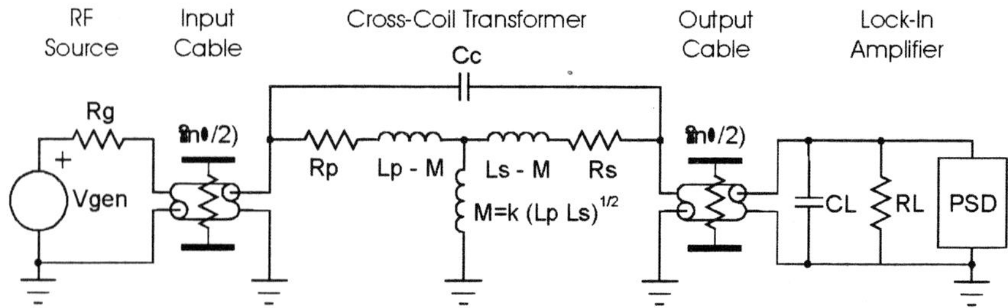

FIGURE 1. The equivalent circuit of the crossed-coil polarimeter.

TABLE 1. Measured Circuit Parameters

Coil	Lp (μH)	Ls (μH)	k	Cc (pF)	Rp (ohm)	Rs (ohm)
2	0.669	3.17	0.004	2.4	1.5	4.8
3	0.585	2.42	0.0001	0.8	0.8	3.4

Cables are 1.282 m of SR047FL inside the dewar and SR401 outside.

The sensitivity of the circuit to target polarization enters through the dependence of the inductance of a coil on the susceptibility (χ) of the enclosed sample:

$$L = L_0\,[1+\phi(\chi)] = L_0\,[1+\eta\chi]+O(\chi^2) \tag{1}$$

The change in inductance with susceptibility is not generally linear, but for the present analysis we have ignored the higher order terms. The constant relating susceptibility to change in inductance, η, is called the filling factor.

$$\eta \;=\; \frac{\int_{tgt} H_\perp dv}{2\int_\infty H_0 dv} + \frac{1}{\chi}\left[\frac{\int_\infty H dv}{\int_\infty H_0 dv} - 1\right] \tag{2}$$

Most importantly for the operation of the CC-polarimeter, the presence of a polarizable sample increases the small mutual inductance coupling the two coils:

$$M = \sqrt{L_p\,L_s}\,(k0+\eta\chi)+O(\chi^2) \tag{3}$$

For an NMR sample, the susceptibility is a complex valued resonance function (typically Lorentzian or Gaussian) with non-zero value only very near the Larmor frequency.

The voltage transfer function (gain) of the circuit is a rational polynomial in frequency with coefficients expressed in terms of the circuit elements. The response of the polarimeter to polarization can be determined by expanding the voltage gain in a power series in susceptibility.

$$G(\omega) = G_0(\omega)+G_1(\omega)\chi+G_2(\omega)\chi^2+O(\chi^3) \tag{4}$$

The first coefficient, $G_0(\omega)$, is the gain of the circuit in the absence of a polarized target. It is an undesirable *background gain* on which the signal due to the resonant susceptibility appears. The response to the susceptibility is determined by $G_1(\omega)$, the *transducer gain.* $G_2(\omega)$ is the lowest order *non-linearity* in the susceptibility response of the circuit, which ideally would be zero. An optimal circuit has G_0 and G_2 small and G_1 large.

Since the transducer gain is a complex function of the complex susceptibility, four gains can be defined: the derivatives of the real and imaginary parts of the circuit gain with respect to the real and imaginary parts of the susceptibility. However, the Cauchy-Riemann conditions reduce the matrix of four values to two gains in the form of a rotation matrix, so that a rotation in the complex plane produces a single transducer gain. The second order non-linearity $G_2(\omega)$ can similarly be reduced to a single number.

For fixed cable lengths, resonances appear in the gains G_0 and G_1 at frequencies slightly higher than that for which the length is a half integer multiple of the wavelength, λ. A cable nλ/2 long reproduces the terminating impedance at the receiving end. By adding tuning capacitor to the receiving end of the cable terminated by the NMR coil, the parallel resonance of the coil and capacitor can be made to match the frequency at which the cable length is nλ/2. The lowest frequency resonance corresponds to 0λ, that is, the parallel resonance of the coil inductance and the cable capacitance.

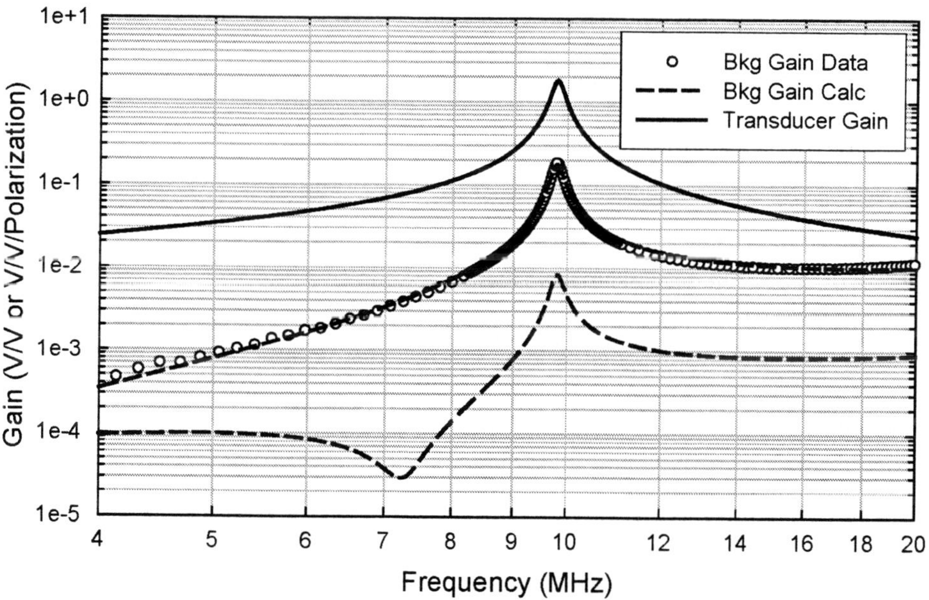

FIGURE 2. The computed background and transducer gains as a function of frequency for a typical cross-coil circuit at the coil-plus-cable capacitance (0 λ) resonance.

PERFORMANCE

We have made NMR measurements for both the 0λ and λ/2 resonances at frequencies between 8 and 12 MHz. Typical calculated resonance curves for the background and transducer gains are shown in Figure 2 along with the measured background gain for coil #2 in Table 1. Also shown is the background gain curve for a larger coil coupling capacitance, which moves the zero closer to the pole, reducing the background gain. We have not yet exploited this mode of operation.Noise is added to the NMR signal by four sources: amplitude noise in the RF generator, thermal noise from the real part of the impedance, amplifier noise from the lock-in and/or preamplifier, and flicker (1/|f|) noise from parametric fluctuations in the coil and cable properties induced by mechanical and thermal fluctuations. The noise of the RF generator, which is –90 dB, dominates the total error for RF levels above -40 dBm. Below that level, the total amplifier and resistor noise, which is less than $5\,nV/\sqrt{Hz}$, begins to contribute. For a single scan, the circuit adds no noise to that input to the circuit by the signal generator: the noise figure is 1.0, as shown in Figure 3. If multiple scans are averaged the RMS noise is reduced by the square root of the number of measurements, and the noise figure remains 1, as long as the noise is white. However, as the number of scans averaged increases, the measurement extends to

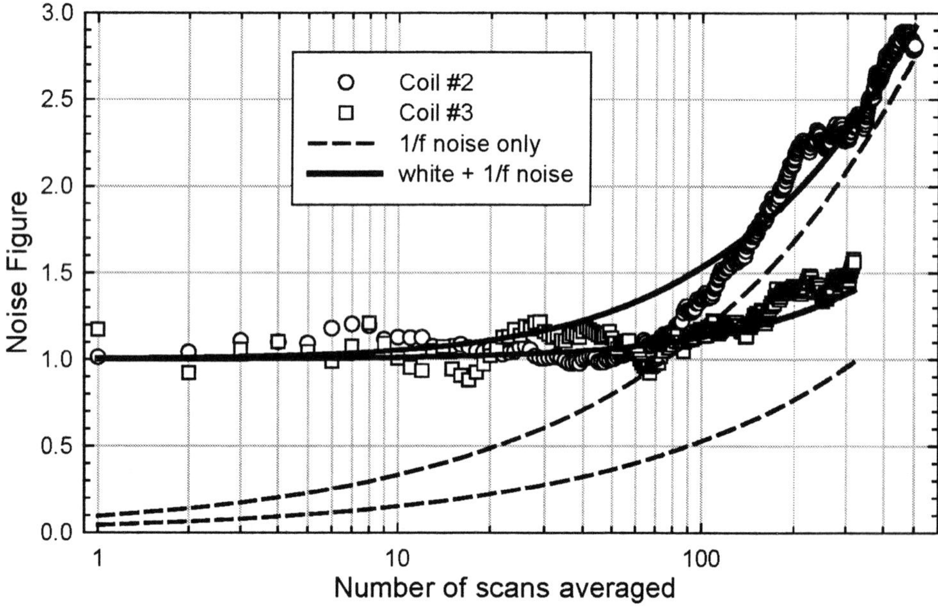

FIGURE 3. The measured noise performance of two NMR coils at –25 dBm. Noise figure is defined as (SNR out)/(SNR in), where the signal out is the transducer gain times the signal in. The dashed lines are the 1/f (flicker) noise that must be added to the white noise from the RF generator to fit the data [4]. As the bandwidth is extended to lower frequencies by averaging many scans, the 1/f noise dominates. Coil #3 has the windings more rigidly mounted and better aligned than Coil #2.

lower frequencies and 1/|f| noise dominates, causing the total noise to stop falling, or increase, with increasing measurement time [4]. The 1/|f| noise can be reduced, as was done for coil #3, by mounting the windings and the coil form more rigidly, and by reducing temperature fluctuations of the cables.

The second order non-linearity is proportional to the square of the susceptibility, and with the aid of the Cauchy-Riemann relations, can be reduced to the following form, as a fraction of the polarization.

$$\eta\chi \frac{G_2}{G_1} \times \int_{\omega 0-\Delta\omega}^{\omega 0+\Delta\omega} \left((\mathrm{Im}\,\chi)^2 - (\mathrm{Re}\,\chi)^2 \right) d\omega \Bigg/ \int_{-\omega 0-\Delta\omega}^{\omega 0+\Delta\omega} \mathrm{Im}\,\chi \, d\omega \tag{5}$$

A summary of the measured and calculated gains, noise, and non-linearities is given in Table 2.

TABLE 2. Measured and Calculated Gains and Noise

Resonance	Frequency MHz	Background Gain obs & calc	Transducer Gain obs (calc)	Transducer Gain Error (%) obs	SNR* obs (calc)	Non-linearity** calc, %
0 λ	9.257	0.175	1.66 (1.79)	3.2	7.0 (7.8)	9.7
½ λ	10.660	0.0168	0.265 (0.257)	1.3	12 (12)	2.4

* for 1 Hz bandwidth and protons in thermal equilibrium at 4.2 K

** for 80% proton polarization and integration over 10 linewidths

Under conditions of operation that produce similar, small, non-linearities for the two circuits, we have found that the CC-meter has the following significant advantages over the Q-meter. By proper design, the inductive and capacitive coupling between the two coils can be made very small (-60dB), so that the background signal under the NMR resonance signal is very small. This leads to small systematic errors in the background subtraction, which result from the inevitable drift of the background during long measurement times. Also, since most of the noise comes from the RF signal source, the low background results in greatly improved signal-to-noise ratios for the CC-meter. The inductances and geometries of the two coils of the CC-meter can be separately optimized to achieve specific requirements (such as RF field uniformity over the target) not possible with the typical embedded coils of the Q-meter. The principal disadvantage of the CC-meter is that two coils must be carefully constructed and rigidly mounted to avoid excess (flicker) noise. The meter is optimized by keeping the coil couplings small, the inductance of the receiver coil at about twice that of the transmitter coil, coil resistances low, using cables with high Q and high impedance, and using a low transmitter source impedance and a high receiver input impedance.

REFERENCES

1. G.R. Court, D.W. Gifford, P. Harrison, W.G. Heyes, and M.A. Houlden, *NIM* **A324,** 433 (1993).
2. T.O. Niinikoski, *NIM* **A356,** 62 (1995).
3. Y.K.Semertzidis, *NIM* **A356,** 83 (1995).
4. V. Radeka, *IEEE Trans. Nucl. Sci.* **NS-16**, 17 (1969).

Polarized Atomic Hydrogen Beam Tests in the Michigan Ultra-Cold Jet Target [1]

K. Yonehara*, Z.B. Etienne*, M.C. Kandes*, K.J. Klein*, A.D. Krisch*, M.A. Leonova*, V.G. Luppov*, V.S. Morozov*, C.C. Peters*, R.S. Raymond*, D.E. Saam*, D.L. Sisco*, D.R. Southworth*, N.S. Borisov†, V.V. Fimushkin† and A.F. Prudkoglyad**

*Spin Physics Center, University of Michigan, Ann Arbor, MI 48109-1120
†Joint Institute for Nuclear Research, RU-141980, Dubna, Russia
**Institute for High Energy Physics, RU-142284, Protvino, Russia

Abstract. Progress on the Michigan ultra-cold proton-spin-polarized atomic hydrogen Jet target is presented. We describe the present status of the Jet and some beam test results.

We are developing an ultra-cold high-density Jet target of proton-spin-polarized hydrogen atoms (Michigan Jet) to study spin effects in high energy collisions. The Jet uses a very high magnetic field and an ultra-cold separation cell coated with a superfluid ^{4}He-film to produce a slow monochromatic electron-spin-polarized atomic hydrogen beam. This beam is focused by a parabolic mirror coated with superfluid ^{4}He and a superconducting sextupole magnet. An rf transition unit will then convert this beam into a proton-spin-polarized beam [1].

The layout of the Michigan Jet is shown in Fig. 1. Atomic hydrogen is produced with a room-temperature rf dissociator and guided to the ultra-cold separation cell coated with superfluid ^{4}He to suppress the surface recombination of hydrogen atoms (see Fig. 2). The double walls of the cell form the mixing chamber of a dilution refrigerator. The cell's entrance and exit apertures are respectively located at about 95% and 50% of the superconducting solenoid's 12 Tesla magnetic field. After the hydrogen atoms are thermalized by collisions with the cell surface, the magnetic field gradient physically separates the atoms according to their electron-spin states. The "high-field-seeker" atoms in the two lowest hyperfine states ($|3\rangle$ and $|4\rangle$) are attracted up toward the high field region and escape from the cell. They quickly recombine on bare surfaces and are cryopumped. The "low-field-seeker" atoms in the two higher hyperfine states ($|1\rangle$ and $|2\rangle$) are pushed down toward the low field region and effuse from the exit aperture, forming a rather monochromatic electron-spin-polarized beam. To increase the Jet density, we use a gold-coated copper focusing mirror with a polished surface covered with a ^{4}He superfluid film similar to the prototype mirror [2]. After an rf transition unit, which changes state $|2\rangle$ atoms into state $|4\rangle$ atoms, the beam will pass through a supercon-

[1] Research supported by a Research Grants from the U.S. Department of Energy

CP675, *Spin 2002: 15th Int'l. Spin Physics Symposium and Workshop on Polarized Electron Sources and Polarimeters*, edited by Y. I. Makdisi, A. U. Luccio, and W. W. MacKay

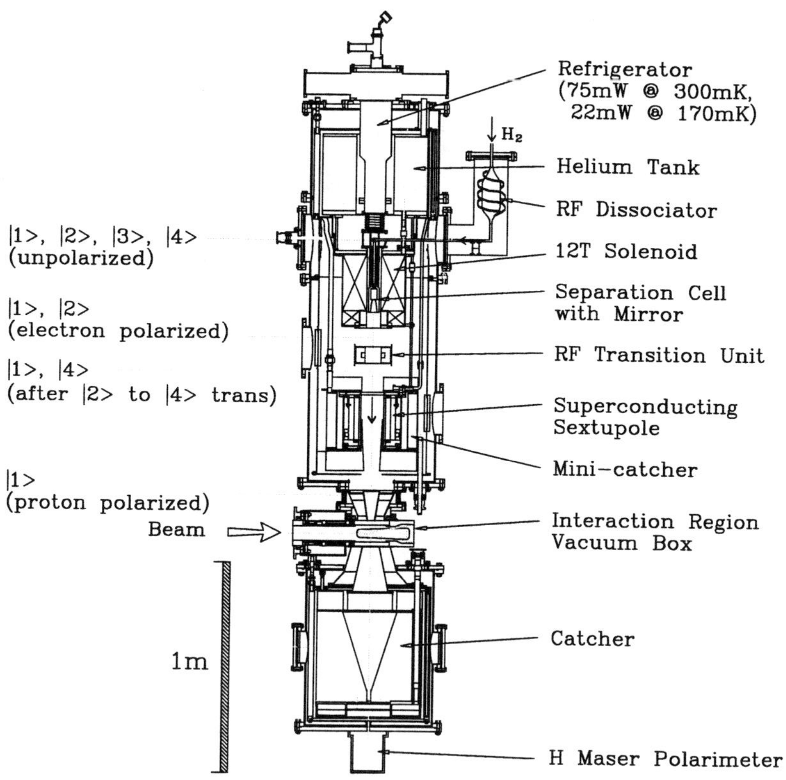

FIGURE 1. Layout of the Michigan ultra-cold Jet.

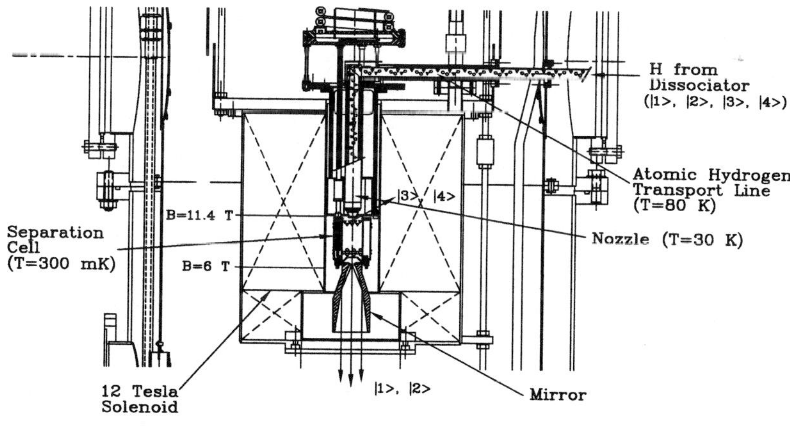

FIGURE 2. Details of the Michigan Jet's electron-spin separation region.

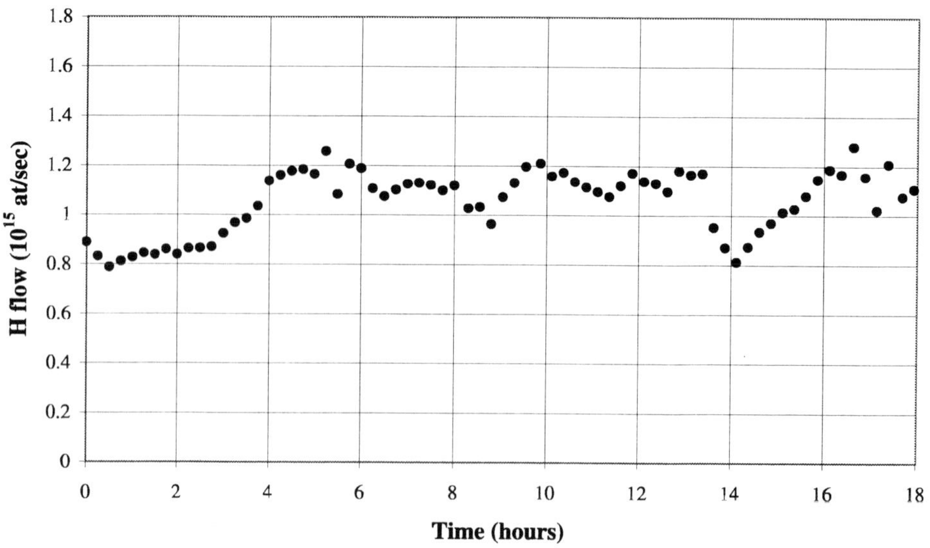

FIGURE 3. The observed hydrogen Jet intensity in an 18-hour run.

ducting sextupole magnet. The sextupole selects atoms in electron spin state +1/2, by focusing atoms in state $|1\rangle$ into the interaction region and defocusing atoms in state $|4\rangle$, which are then cryopumped away. The proton-spin-polarized beam then passes through the interaction region where it can collide with a proton beam in a high energy storage ring. The Jet beam is then captured below by a huge cryopumping catcher [3] to keep the interaction region and storage ring vacuum uncontaminated. A maser polarimeter below the catcher monitors the beam proton polarization.

Most of the Michigan Jet parts have been fabricated and successfully tested. This hardware includes a 12 Tesla superconducting solenoid with a very sharp gradient at the downstream end, a dilution refrigerator with a cooling power of about 22 mW at 170 mK, a 20 cm long superconducting sextupole magnet with a 0.31 T field at its iron poles and a 10.5 cm diameter bore, a cryocondensation catcher pump with a measured pumping speed of about 1.2×10^{7} liters s^{-1} (4.2×10^{26} atoms torr^{-1} sec^{-1}) [3], and a hydrogen maser polarimeter capable of monitoring the polarization with a precision of about $\pm2\%$ in a few minutes.

We studied a polarized beam of hydrogen atoms focused by the superconducting sextupole into a compression tube detector which measured the polarized atoms' intensity. By building a thick ^{4}He superfluid film, we were able to produce a high intensity spin-polarized hydrogen beam, which operated with good stability during an 18-hour run, until our liquid ^{4}He supply was depleted, as shown in Fig. 3. The average measured hydrogen intensity, into the 11 mm by 1.4 mm compression tube slot, was about 1.1×10^{15} H s^{-1}. This intensity corresponds to a hydrogen Jet thickness of 6×10^{11} H cm^{-2}. The maximum beam intensity fluctuation was about $\pm20\%$. It was not needed during this 18-hour run, but when necessary, we can heat the separation cell to about 40 K to remove the residual frozen H_2 molecules; this usually takes about 2 hours.

The Jet's highest measured spin-polarized atomic hydrogen Jet intensity was about 2.2×10^{15} H s^{-1}; this corresponds to a Jet thickness of about 1.1×10^{12} H cm^{-2}. The fully electron-spin polarized beam has a proton polarization of about 50%. We plan to convert the electron polarization into the proton polarization by an adiabatic passage of the beam through an rf transition unit, which has a novel dielectric-ring-resonator that accepts the 6 cm diameter beam. A room temperature prototype rf transition unit was built and successfully tested; its maximum measured transition efficiency was 97% at 235 mW rf power. A preliminary design and some tests of the cryogenic rf unit were also made [4]. The observed Q value of the cryogenic rf unit at 5 K was about 17,000, which is 2.5 times higher than that of the room temperature prototype rf unit; hence, less power should be needed for the cryogenic rf transition unit. We plan to soon fabricate and install this cryogenic rf transition unit, which should increase the proton polarization to over 90%.

ACKNOWLEDGMENTS

We thank D. Kleppner and T. Roser for their earlier help

REFERENCES

1. B.B. Blinov *et al.*, *"Michigan Ultra-Cold Polarized Atomic Hydrogen Jet"*, in *Proc. of 14th Internatioanl Spin Physics Symposium*, Osaka 2000, edited by K. Hatanaka *et al.*, American Institute of Physics, New York, 2001, AIP Conference Proceedings **570**, pp. 856-860
2. V.G. Luppov *et al.*, Phys. Rev. Lett. **71**, 2405 (1993)
3. J.D. Arnold *et al.*, Nucl. Instr. Meth. A**391**, 398 (1997)
4. R.S. Raymond, *"Development of a Large-bore Cryogenic 2-4 Transition Unit"*, in *Proc. of International Workshop on Polarized Source and Targets*, edited by A. Gute *et al.*, Erlangen, Germany (1999)

Summary Report of PESP-2002 Workshop

M. Farkhondeh

MIT-Bates Linear Accelerator Center
P.O.Box, 846, Middleton, MA 01949, USA

Abstract. A summary of the presentations for the 2002 Workshop on Polarized Electron Sources and Polarimeters (PESP-2002) will be given. PESP-2002 was held in the vicinity of MIT-Bates Linear Accelerator Center on September 4-6, 2002. It is a satellite workshop for the SPIN2002 symposium supported by the International and Local Organizing Committees of the Spin Physics Symposium, and MIT-Bates. The previous PESP workshop was held in Nagoya, Japan in October 2000. The workshop followed an interdisciplinary approach that included high-energy and low-energy physics, surface and material sciences, and the application of polarized sources in science and industry. In addition, PESP-2002 included contributions on polarized sources for parity-violating experiments that are underway at several medium- and high-energy accelerator centers worldwide. The summary report will include the operational experiences of medium and high-energy labs which are utilizing the latest developments in photocathodes to deliver highly polarized electron beams. Finally, a brief review of polarized sources in non-accelerator applications will be presented.

INTRODUCTION

The PESP2002 workshop was held at the Kings Grant Hotel in Danvers, Massachusetts. The workshop was near the site of the MIT-Bates Accelerator, which has over two decades of experience in producing polarized electron beams for medium- energy nuclear physics experiments. A total of 38 participants from six countries, thirteen universities, and two private U.S. companies attended the workshop. Unfortunately, several absentees from Russia were not granted US visas in time for the workshop. A mixture of high-energy and low-energy applications and physics of Polarized Electron Sources (PES) were discussed in this workshop. An excellent mix of participants from these areas, with (24) participants associated with accelerator centers and (14) from the low-energy and surface physics groups. One of the outcomes of this workshop has been the fruitful exchange between the accelerator-based groups with well-established high-polarization PES programs and the low-energy surface physics groups that are increasingly utilizing PES. This was demonstrated by the expressed interest of the surface physics groups in establishing collaborations with the major labs for utilizing high polarization strained photocathodes instead of "low" polarization bulk GaAs that they currently use. A total of 30 oral presentations and 5 poster presentations were divided into 8 sessions. Each speaker will contribute a 5-page manuscript for publication in these proceedings; each poster presentation is allocated a 3 page-manuscript also published in the same proceedings. The topic of each session is listed below:

CP675, *Spin 2002: 15th Int'l. Spin Physics Symposium and Workshop on Polarized Electron Sources and Polarimeters*, edited by Y. I. Makdisi, A. U. Luccio, and W. W. MacKay

Session 1: *NEA and Surface Physics*
Session 2: *Photocathode Spin Physics*
Session 3: E*lectron Beam Polarimeters*
Session 4: *Polarized Electron Sources at Accelerator Centers*
Session 5: *Development of PES and Lasers*
Session 6: *Applied Spin Physics and Technology*
Session 7: *Polarized Sources and Parity-Violation Experiments*
Session 8: *Developments of Laser Systems for Polarized Source*

It is not possible to discuss in this paper all topics and issues presented during the workshop. Instead, I will present a brief discussion of selected issues that are of great importance for this field. These issues include the basics of photoemission, surface charge limit effects, PES at accelerator centers, electron polarimetry, new lasers for accelerator based PES, and surface magnetization studies with spin polarized currents.

CURRENT ISSUES IN PES

In this section, I have chosen several topics that are of great importance in the field of polarized electron sources and polarimetry. Each is briefly discussed here.

Basics of photoemission

There are two fundamental principles in basic photoemission from GaAs-based photocathodes: a) the excitation of the electrons in the valence band to the conduction band with circularly polarized photons in III-V materials in the period table, and b) achieving the state of Negative Electron Affinity (NEA) by lowering the work function to increase the probability of the conduction band electrons to escape to vacuum. The photoemission process and the polarization mechanisms are shown schematically in Figure 1. Photoemission from bulk GaAs yields electron polarizations that are theoretically limited to 50% due to a degeneracy in $P_{3/2}$ state in the valence band (heavy hole and light hole). In strained GaAs (GaAsP, InGaAsP) or in InGaAs-AlGaAs superlattice, this degeneracy is removed by several tens of meV in the valence band. Individual energy levels are then preferentially pumped with laser radiation of specific wavelength providing electron beam polarizations that theoretically can approach 100%. In practice, high polarizations of 70-85% are now routinely achieved with polarization of 90% as an achievable goal.

Surface charge limit effect and remedies

A NEA prepared surface can suffer from a limitation on the maximum current that can be extracted from a photocathode; this is referred to as surface charge limit (SCL). A fraction of the excited electrons from the leading edge of the bunch are trapped on the surface creating a barrier for the subsequent electrons. In this process, the magnitude

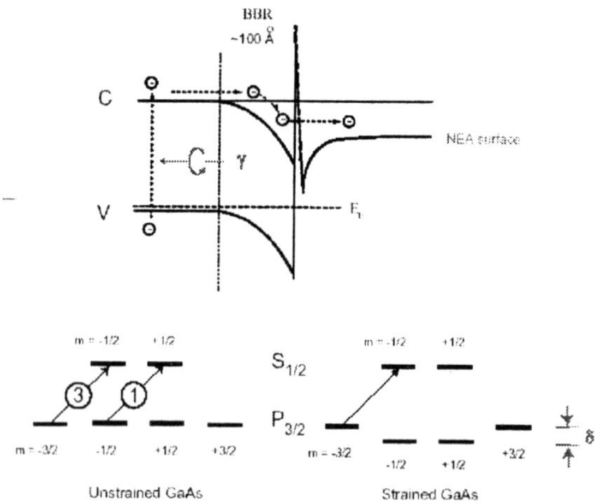

Figure 1. Polarization mechanism in photoemission process with circularly polarized optical pumping of GaAs based photocathode.

of the band bending (photovoltage effect) is reduced resulting in an increase in the electron affinity. The magnitude of the SCL effect depends strongly on the Quantum Efficiency (QE) and is more pronounced in strained photocathodes that are known to have low QE's. The trapped electrons dissipate eventually by recombining with holes. The SCL causes a reduction in bunch charge when closely spaced bunches (few ns apart) of high bunch charges are produced. It also manifests itself as a reduction in "apparent" QE where peak current or charge per bunch does not increase linearly with laser energy (see Figure 2.). To reduce the SCL effect one can either increase the recombination rate of trapped electrons with holes in the valence band, or reduce the trapped electrons in the band-bending region. The former is achieved by increasing the p-dopant rate on the surface and the latter is made possible by using the photocathode with a superlattice (SL) structure. The high dopant method has been pursued at SLAC both on unstrained GaAs and strained GaAsP [1]. The latest method is the development and use of samples grown by Bandwidth Semiconductor Inc. (BWSC) [2] with 10-nm thick GaAs final layer with high dopant rate of $5x10^{19}$ cm^{-3}. The use of the superlattice structure for reducing the SCL effect has been pursued by the Nagoya Group [3]. In these structures, the energy levels in the conduction bands are shifted apart causing an increase in the escape probability for the electrons and increase the recombination probability thus reducing the SCL effect.

PES at Accelerator Centers

Presentations by five accelerator centers (Bonn, J-Lab, Mainz, MIT-Bates and SLAC) discussed the status of their PES. All these labs are now routinely delivering polarized beams with polarizations exceeding ~75%. The polarized source at SLAC is successfully meeting the stringent requirements of the E158 parity-violating experiment [4] aimed at measuring a physics asymmetry as low as a few parts per billion (ppb). The polarized source at Jefferson Lab has been routinely delivering high polarization beams simultaneously to three experimental halls for the last several years. Recently, the source laser system has been modified to accommodate a new parity-violating experiment (the G_0 experiment) that has the challenging requirement

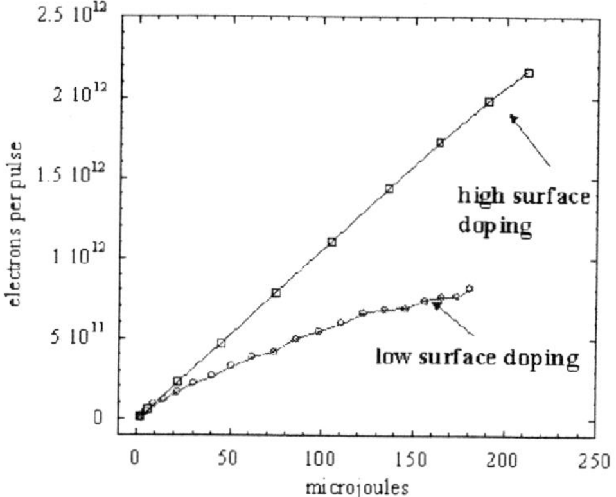

Figure 2. Plots of charge per bunch vs. laser energy for two photocathodes at SLAC [5]. The lower curve displaying the SCL effect is from a standard strained GaAs with low doping. The upper curve is for a high-gradient-doped strained GaAs [2] that does not show any sign of SCL effect.

experiment (the G_0 experiment) that has the challenging requirement of 40 μA high polarization (>70%) average current with micro-bunches 30 ns apart required for time-of-flight measurement [6]. At Mainz, the MAMI polarized source routinely delivers beams with polarization exceeding 75% at ~25 uA for parity violating and other medium energy nuclear physics experiments. Strained superlattice photocathodes are used in this source. The stable nature of the MAMI microtron accelerator reduced the needs for multiple feedback system often needed for taming helicity correlated effects in parity violating experiments. The polarized source at Bonn, feeding the ELSA pulsed stretcher ring is also routinely delivering high polarization beam using a flash lamp-pumped Ti:Sapphire laser and a superlattice photocathode. At MIT-Bates, the beam delivery to three SAMPLE experiments is now complete. A new high power diode laser system and high-gradient-doped strained GaAsP grown by BWSC [2] are now used for injecting and stacking over 100 mA highly polarized beams into the South Hall Ring for commissioning of the BLAST spectrometer.

Electron Beam Polarimetry

A total of five presentations in electron beam polarimetry included a new novel method of electron polarimetry using noble gas targets [7], spin filters as very high-performance spin polarimeters [8], and a high efficiency electron polarimeter based on exchange scattering from a magnetic target [9]. Data were also presented on the MIT-Bates laser back-scattering Compton polarimeter that pointed to the potential for this polarimeter and the South Hall Ring for studying the spin dynamics in storage rings [10]. Electron beam polarimetry with systemic errors of 3-5% is now routine at various high-energy laboratories. Work needs to be done to achieve routine electron polarimetry with 1% systematic errors.

Low Energy and Surface Physics

The low energy surface physics groups around the world have been using polarized electron sources for varieties of applications. These applications include manipulation of the surface magnetization by spin currents spin resolved inverse photoemission spectroscopy [11], and Spin-Polarized Low Energy Electron Microscopy (SPLEEM). These applications are utilized in studying thin magnetic film, non-magnetic films on ferromagnetic substrates, and imaging of magnetic domain structure of thin ferromagnetic film systems during growth. In SPLEEM process, as reported by A Shcmid of LBL [12], the spin-dependent reflectivity images of magnetic domain structures of thin ferromagnetic films are made by measuring the intensity asymmetry with polarized beams (I^{up}-I^{down}). The contrast of magnetic domain provided by polarized beam (SPLEEM) is far greater than the contrast with unpolarized beam (LEEM). A comparison of two images is shown in Figure 3 for a Co/Au (111) film system. Utilizing highly polarized PES from strained photocathodes instead of low polarization beams from bulk GaAs will greatly enhance the contrast and increase the signal to noise ratio in these measurements.

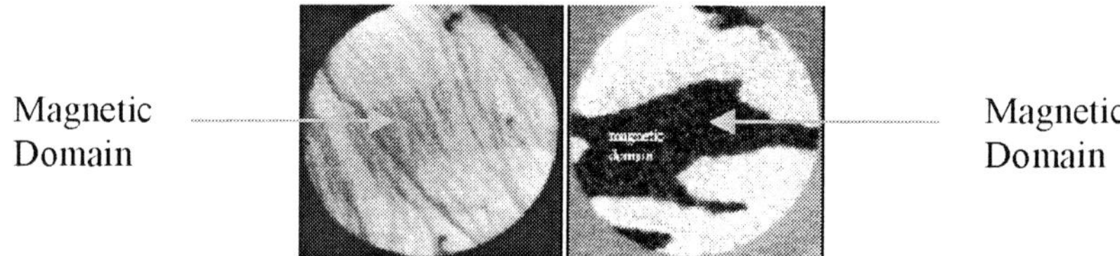

Figure 3. Comparison of images of magnetic domain structure for a thin ferromagnetic Co/Au (111) film obtained by two methods [11]. The left image is obtained by single spin-state beam and the right image is from a difference image using two spin states of the electron beam.

REFERENCES

1. T. Maruyama *et al*, Nucl. Instrum. Methods A492, 199 (2002).
2. Bandwidth Semiconductor Inc., Bedford, NH.
3. K. Wada, *et al., Surface Charge Limit effect of superlattice photocathode,* these proceedings.
4. P. Mastromarino, *et al.*, SLAC E-158: beam helicity-correlations in 2002 physics run, these proceedings.
5. J. Clendenin, *et al., The SLAC polarized electron source,* these proceedings.
6. J. Grames *et al., The polarized source and Parity experiments at J-LAB,* these proceedings.
7. T. Gay, *et al.*, J. Phys. B: At. Mol. Phy. 16(1983) and these proceedings.
8. N. Rougemaille, *et al., Spin filters as very high-performance spin polarimeters,* these proceedings.
9. L. Duò, *A high efficiency electron polarimeter based on exchange scattering from a magnetic target,* these proceedings.
10. W. Franklin, *et al., The MIT-Bates Compton Polarimeter,* these proceedings.
11. F. Ciccacci et *al.*, Spin *resolved inverse photoemission,* these proceedings.
12. A. Schmid, *et al., "In-situ imaging of magnetization reversal",* these proceedings.

Polarization Optimization Studies in the RHIC Optically-Pumped Polarized H⁻Ion Source

A.Zelenski[1], J.Alessi[1], B.Briscoe[1], A.Kponou[1], S.Kokhanovski[2], V.Klenov[2], V.LoDestro[1], D.Raparia[1], J.Ritter[1], V.Zubets[2].

1 - Brookhaven National Laboratory, USA, 2 - INR, Moscow, Russia

Abstact. The performance of the RHIC Optically-Pumped Polarized H⁻ Ion Source (OPPIS) in 2000-2002 runs in AGS and RHIC is reviewed. The OPPIS met the RHIC requirements for beam intensity with the reliable delivery of about 500 μA polarized H- ion current in 400 μs pulse duration (current can be increased to over 1.0 mA, if necessary). The beam intensity at 200 MeV was $(5\text{-}6)\cdot 10^{11}$ H⁻/pulse, which is sufficient to obtain the required $2\cdot 10^{11}$ polarized protons per bunch in RHIC. The polarization dilution by molecular ions, which are produced in the ECR primary proton source is discussed. The molecular component can be reduced to about 5% by further ECR source-operation optimization. The molecular component is suppressed by optimization of the extraction electrode optics and by the decelerating einzel lens in the 35 keV LEBT line. As a result, the proton polarization of the accelerated beam was increased to over 80%, as measured in the 200 MeV proton-deuterium polarimeter.

INTRODUCTION

A luminosity of $2\cdot 10^{32}$ $cm^{-2}\,sec^{-1}$ for polarized proton collisions in RHIC at up to $\sqrt{S}$ = 500 GeV energy will be produced by colliding 120 bunches in each ring at $2\cdot 10^{11}$ protons/bunch intensity[1]. A primary polarized H- ion beam is produced in the OPPIS. The RHIC OPPIS produces routinely 0.5-1.0 mA (maximum 1.5 mA) current in a 400 μs pulse duration. Polarized H^- ions are produced in the OPPIS at 35 keV beam energy. The beam is accelerated to 200 MeV with an RFQ and linac for charge-exchange strip-injection into the Booster . About 50% of the OPPIS beam intensity can be accelerated to 200 MeV. The 400 μs H^- ion pulse is captured in a single Booster bunch which contains about $4\cdot 10^{11}$ polarized protons The single bunch is accelerated in the Booster to 1.5 GeV kinetic energy and then transferred to the AGS, where it is accelerated to 25 GeV for injection to RHIC.

The OPPIS initial longitudinal polarization is converted to the transverse direction while the beam passes two bending magnets .The second 47.4 degree bending magnet switches linac injection between polarized and unpolarized high intensity (up to 100 mA) H^- ion beam. The magnet is pulsed and either beam can be accelerated pulse-to-pulse in the same RFQ. A pulsed focusing solenoid in front of the RFQ is tuned for optimal transmission for either beam. It rotates the polarization direction for about 420 degrees, but still keeps it in the transverse plane, and a final polarization alignment to the vertical direction can be adjusted by a spin-rotator solenoid in the 750 keV beam transport line before injection to the linac[2]. The AGS cycle for polarized beam

CP675, *Spin 2002: 15th Int'l. Spin Physics Symposium and Workshop on Polarized Electron Sources and Polarimeters,* edited by Y. I. Makdisi, A. U. Luccio, and W. W. MacKay

operation is 3 seconds. The OPPIS operates at 1 Hz repetition rate and additional source pulses were directed to the 200 MeV p-Carbon polarimeter for polarization monitoring by switching of another pulsed bending magnet in the high-energy beam transport line. The spin-rotator tuning is done using vertical polarimeter arms.

Recent source component upgrades have significantly improved the OPPIS performance and reliability. The OPPIS polarization technique is described elsewhere [3,4]. The schematic OPPIS layout is presented in Fig.1.

The ECR-primary proton source upgrade: The ECR operation in a pulsed mode was studied at 1-7 Hz repetition rate and 5-100 ms pulse duration. A significant reduction in the optimal hydrogen feeding flow was observed, which might be explained by gas adsorption on the quartz tube, then desorption at the beginning of the RF pulse. This produces additional gas contribution sufficient to maintain an optimal density for about 5-10 msec at the beginning of the pulse. Pulsed ECR operation reduces gas load to the vacuum system and heat load to the ECR and Rb cell. In dc operation an oxygen gas admixture is required to optimize the ECR-source proton production and proton to molecular H_2^+ ions dissociation ratio. In the pulsed operation sufficient amount of oxygen is supplied to the discharge from residual gas.

Advantages of the pulsed operation were not fully used in the RHIC run, because the heat load difference between pulsed and dc operation eventually caused leaks in quartz to copper cavity seal. In the new ECR cavity silicone O-rings (silicone has lower RF-power absorption than other rubber –like materials) are used instead of a teflon-indium seal. The silicone O-ring exposure to 29 GHz microwave power is reduced by the seal design. The quartz tube air-cooling was improved to prevent seal overheating through the contact with the hot quartz tube.

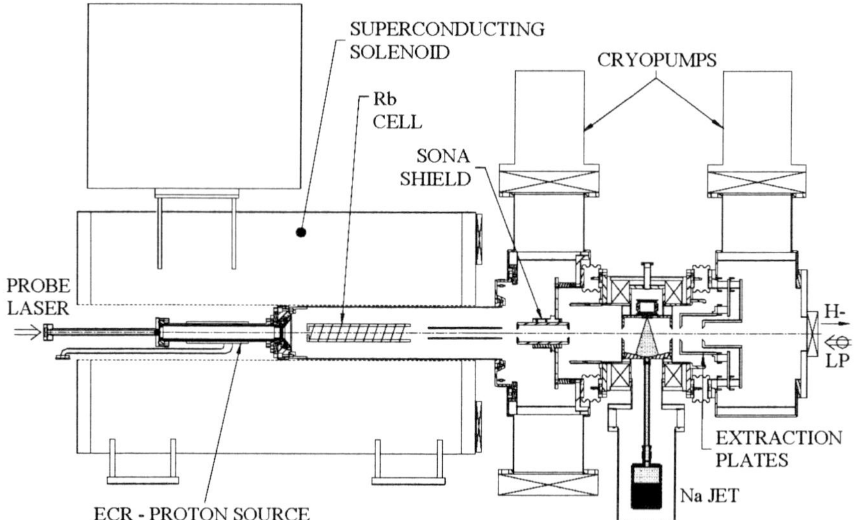

Figure1. The RHIC OPPIS layout.

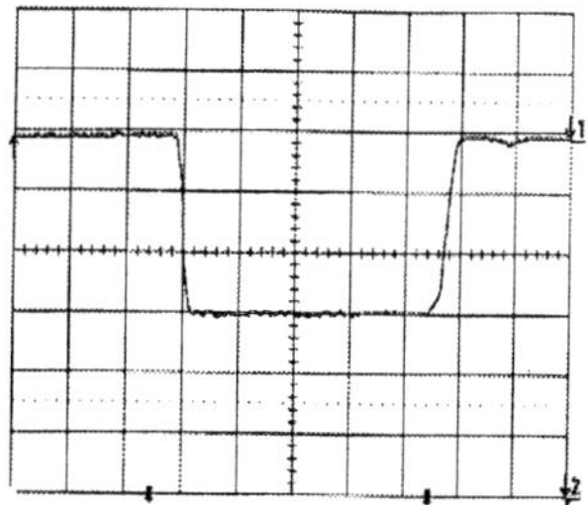

Figure 2. The polarized H- ion beam current pulse in a 35 keV LEBT Faradey cup. Vertical scale-500 μA/div, horizontal-100μsec/div.

The silicone O-rings provide the flexibility to compensate the difference in thermal expansion between quartz and copper, which allows frequent switching between dc and pulsed modes for optimal source operation. The new ECR-cavity length was also increased to match the magnetic field shape in the ECR-discharge region. As a result the maximum polarized H^- ion current was increased to 1.5 mA (see Fig. 2).

Sodium-jet ionizer upgrade: The sodium-jet ionizer was redesigned from horizontal [2] to vertical jet geometry (see Fig.3). A new ionizer vacuum chamber with the expansion between the solenoid coils provides sufficient room for the vertical jet with the nozzle at the top moved further away from the beam axis for better jet collimation. The larger collector cooling is provided by water circulation in attached stainless steel tube (1/8" in diameter). The thermal conductance was adjusted to maintain the collector at 120-140°C with the room temperature water. The return line was extended to the reservoir and immersed in the liquid sodium metal. It has significantly reduced the heat transfer to the trap through the heat-pipe (evaporation-condensation cycle) heat transfer process and reduced the reservoir heater power. Additional trap cooling is no longer required , which simplified the cell assembly. The sodium-jet operation and liquid sodium metal circulation stability is greatly improved. The sodium vapor is better confined within the ionizer-cell, and much less sodium is deposited at the acceleration electrodes. This ensured trouble-free 12 week OPPIS operation in the RHIC run.

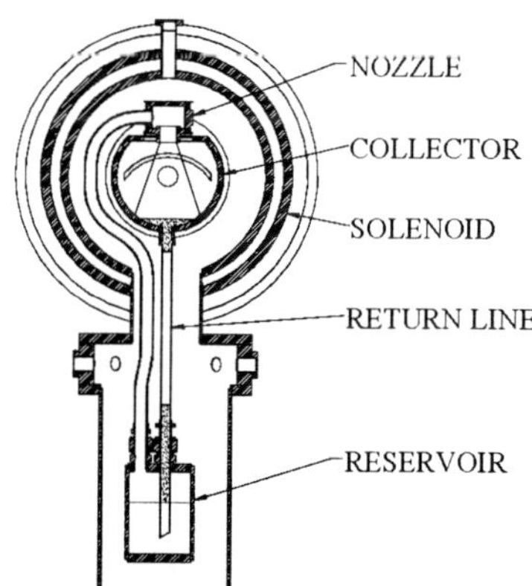

Figure 3. Sodium-jet ionizer cell with a vertical jet geometry.

Laser system upgrade: The laser pulse duration was extended to 450 us for optical pumping during 400 μs H^- current pulse. On-line rubidium polarization measurements were implemented, by probe laser linear polarization rotation measurements (Faraday rotation polarimeter). These measurements give a reliable polarization readout for confirmation of the spin sequence pattern, which is injected to RHIC. Recently, the laser was successfully tested at a 7 Hz OPPIS repetition rate.

POLARIZATION STUDIES

Over 85% polarization was obtained during the final RHIC OPPIS tests at TRIUMF where only electrostatic beam optics were used [4]. The strong spin precession in the LEBT line as described above, might produce some depolarization or longitudinal polarization component. A 70-72% polarization was measured in the first year 2000 run at 200 MeV in the p-carbon polarimeter, at a reduced intensity of less than 10 μA. (The polarimeter was designed for the low current atomic beam source and detectors were greatly overloaded at 200 μA current.) The polarimeter was upgraded for high current operation by extending target to detector distance from 70 to 250 cm and additional collimator installation to suppress particles scattered from the target holder. Polarization of about 75% was obtained at 180 μA beam intensity during spin-rotator tuning. The p-carbon polarimeter is inclusive and relied on the calibration measurements which were done 10 years ago under different conditions. The absolute polarization is an important reference point for polarization loss measurements in Booster and AGS and for evaluation of the OPPIS development status. E. Stephenson suggested an upgrade to a polarimeter using p-Deutron elastic scattering, where the analyzing power is precisely known at 200 MeV. Four additional detector arms for the proton-deutron coincidence measurements from CD_2 target (deuteriated polyethylene) were installed in the summer of 2001. During the November 2001-January 2002 run, p.-C and p.-D scattering data were accumulated and the old analyzing power for 12° degree inclusive pC scattering of a 0.62 was closely reproduced. Therefore the 70-75% polarization at 200 MeV seems to be real and lower than expected.

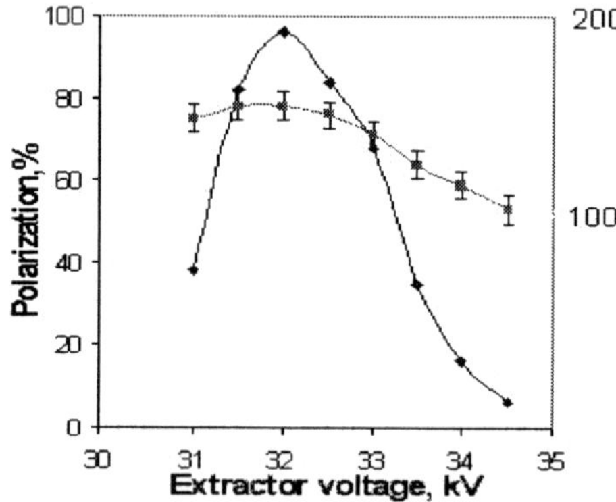

Figure 4. The H- ion current and polarization (squares) at 200 MeV dependence on the extractor voltage. Diamonds-current at 200MeV, right scale-100 uA/div.

At the same time about 80% polarization was measured in the Lamb-shift polarimeter at the source energy. It suggests the existence of significant polarization losses between the OPPIS and 200 MeV polarimeter. Polarization loss in LEBT was observed in the first 2000 run. One observed a strong polarization loss dependence on the beam energy in LEBT (see Fig.4).This dependence was too strong to be attributed to the difference in the spin rotation at different energies. Nevertheless, to check spin direction misalignment contribution a 5% Wien-Filter was built and installed in the LEBT. The polarization measurement results didn't show any significant polarization misalignment.

POLARIZATION DILUTION BY H_2^+ MOLECULAR IONS

There exists a molecular H_2^+ ion component in any plasma ion source. In the OPPIS molecular ions after dissociation will appear as H^- ions with the half of the primary beam energy. The polarization of this beam might be different from the main beam (measurement in Lamb-shift polarimeter gives about half the polarization for this molecular component). This component was observed at the TRIUMF OPPIS, but it was efficiently suppressed by electrostatic lenses in the 3 keV LEBT. In the RHIC OPPIS the H^- beam is accelerated for 32 keV immediately after ionization, producing 35 keV main beam and 33.5 keV beam from molecular ion admixture. These beams are not well separated in the LEBT, and the molecular component is responsible for polarization dilution. At lower acceleration energy these beams are separated, and the half energy component was directly observed (see Fig.5).

The value of molecular component of about 10-40 % was measured under different ECR conditions. Since every H_2^+ ion is dissociated to two half energy atoms, it means 5-20% molecular component out of the ECR-source. The molecular component is increased at higher ECR extraction voltage. At 4.0 keV its value is about 40%. The H^- yield drops at atomic beam energy above 3.0 keV, but for half energy beam of a 1.5 – 2.0 keV the yield is at maximum value. It explains an increase of molecular component up to 40% at 4.0 keV ECR extraction energy, and correspondent polarization decrease, which was also observed in polarization measurements.

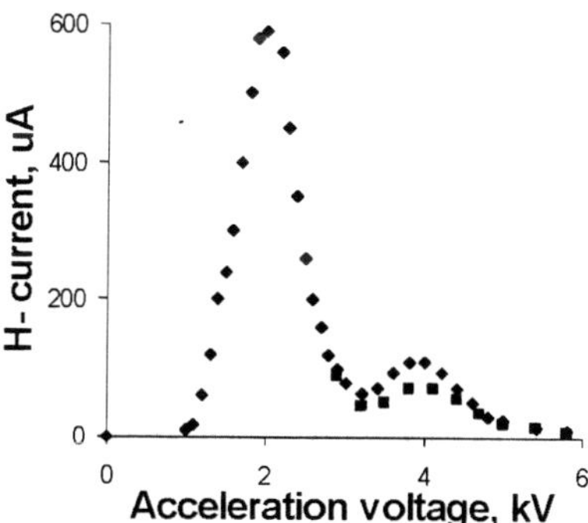

Figure 5. Molecular H2+ beam component is appeared as a second bump shifted to about a half primary ECR proton energy at a fixed bending magnet setting . Diamonds-dc operation; squares-pulsed

operation

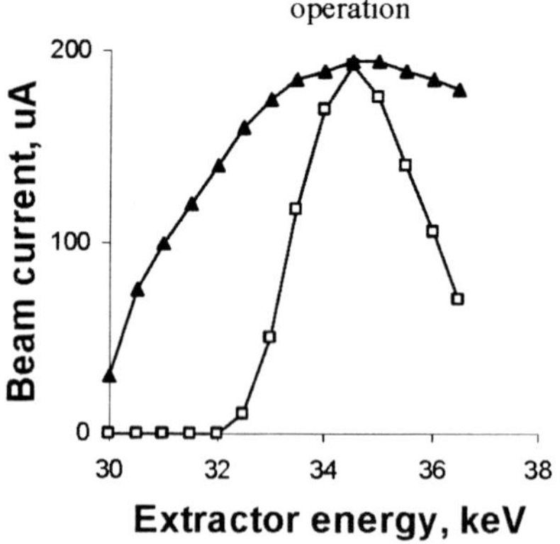

Figure 6. Linac transmission vs extractor voltage (acceleration voltage applied to the jet-ionizer cell). Triangles-LEBT optics with the magnetic quadrupole triplet; squares-quad triplets replaced with the decelerating Einzel-lens.

The increase of the ECR current was obtained earlier with an admixture of a few percent of gaseous oxygen to the hydrogen in the source [2]. The oxygen admixture also reduces the molecular H_2^+ ion production i.e. improves the dissociation ratio in the source. The magnetic field shape in the ECR region also affects the dissociation ratio, The optimization of the ECR source parameters gave rise to an increase of the main beam intensity and reduction of the half energy, lower polarization component to below 10%. The ECR operation in a pulsed mode was also studied. A significant molecular component suppression was also observed in a pulsed operation (see Fig.5), due to difference in a rise time for main and half energy beam components. In a pulsed operation molecular component is about 5%. As a result the polarization at 200 MeV was increased to 75% .The next step was a suppression of lower energy component at acceleration to 35 keV after ionizer and in the LEBT. The two-gap acceleration system was upgraded to three-gaps and the voltage in the first gap was tuned to suppress the half energy component. The voltages at the second and third gaps were adjusted to minimize the main component losses. The triplet of magnetic quadrupole lenses at the OPPIS exit was replaced by a single deceleration einzel lens.

The combined effect of these modifications is a significant (almost ten times) suppression of the beam transmission in LEBT for the lower energy molecular origin beam (see Fig.6). As a result of these upgrades, a polarization at 200 MeV of 80-82% was measured in both p-Carbon and p-Deutron polarimeters.

REFERENCES

1. G.Bunce et al., “Polarized protons at RHIC”, Particle World, 3, p.1, (1992).
2. A.Zelenski et al., Rev.Sci.Instr., **73**, p.888, (2002).
3. A.Zelenski et al., Nucl.Instr.Meth., **A402**, p.185, (1998).
4. A.Zelenski, in Proc. of SPIN-2000, AIP Conf. Proc. 570, p.179, (2001).

Polarized D⁻ Operation and Development of the IUCF Ion Source CIPIOS†

V.P.Derenchuk[1], A.S. Belov[2]

[1]*Indiana University Cyclotron Facility, Bloomington, Indiana*
[2]*Institute for Nuclear Research of the Russian Academy of Sciences, Troitsk, Russia*

Abstract. The Cooler Injector Polarized IOn Source (CIPIOS)[1] has most recently been used to provide polarized and unpolarized beams of negative deuterium ions for filling the injector synchrotron. More than 1.8 mA of up to 90% polarized D^- was available for injection into the RFQ pre-accelerator and several milliamperes of unpolarized beam was available. The addition of an electron blocker in a charge exchange ionizer with a two-stage converter[2] improved the source operation by reducing the total electron current extracted for the maximum 300 A arc discharge current available. A doubling of this discharge current is now possible and should result in a corresponding increase in polarized current.[3]

INTRODUCTION

At Indiana University (IUCF) an atomic beam source (ABS) with a resonant charge-exchange ionizer is used to provide D^- (or H^-) ions[1,7] for acceleration through an RFQ/DTL to an energy of 4 MeV (7 MeV for H^-). This D^- beam fills the Cooler Injector Synchrotron (CIS)[4] using the strip injection technique and after acceleration to 90 MeV is extracted. Between January and July of 2002, the source was used to provide polarized and unpolarized beams of D^- for injection into the Cooler synchrotron[5]. The IUCF experiment CE-82, a search for the isospin-forbidden $d+d\Rightarrow\alpha+\pi^0$ [6] and CE-64, a measurement of three body spin observables by the Pintex group were two of several that utilized this beam. At the completion of the CE-82 experiment on July 28th, 2002, the nuclear physics program with the Cooler synchrotron ended. The source is now used to provide unpolarized beam for acceleration by the RFQ/DTL and for CIS. A purchase agreement is being negotiated by the Forschungszentrum Jülich where the source will be used as the polarized ion source of a new Linac based injector for the COSY ring. The source is expected to be shipped in the late spring of 2003.

SOURCE DEVELOPMENT AND OPERATION

A new two stage converter assembly built and tested at INR [2] is installed in CIPIOS and together with a four electrode extraction system results in an increase of

† Work supported by grants from Indiana University and National Science Foundation PHY-9724216 and PHY-9314783.

CP675, *Spin 2002: 15th Int'l. Spin Physics Symposium and Workshop on Polarized Electron Sources and Polarimeters*, edited by Y. I. Makdisi, A. U. Luccio, and W. W. MacKay

the polarized D$^-$ beam intensity to 2 mA with up to 90% polarization. Source operation with the pulse length increased from 200 μs to 0.5 ms FWHM is also improved with an average polarized beam current during the pulse of about 1.5 mA. An electron blocker added to the ionizer converter assembly reduces the extracted electron current. The electron current is reduced from several hundred milliamperes to less than 100 mA during high intensity operation.

Ionizer Development

The design and development of a new lengthened plasma source and two stage converter with a redesigned extraction system is reported by Belov[2]. Similar modifications are installed on the CIPIOS ionizer. An electron blocker is added to the converter assembly as shown in Figure 1 and results in significantly reduced electron current extracted from the ionizer. The magnetic field generated by the plasma injector solenoid is opposite to the ionizer solenoid which results in a transition between negative to positive axial field in the vicinity of the neutralizer. Electrons and ions from the plasma source follow the magnetic field and collide with the inside surface of the neutralizer. Electrons streaming from the inside of the neutralizer surface follow the ionizer field lines into the blocker. The ions have a higher rigidity and are not focused as strongly into the blocker.

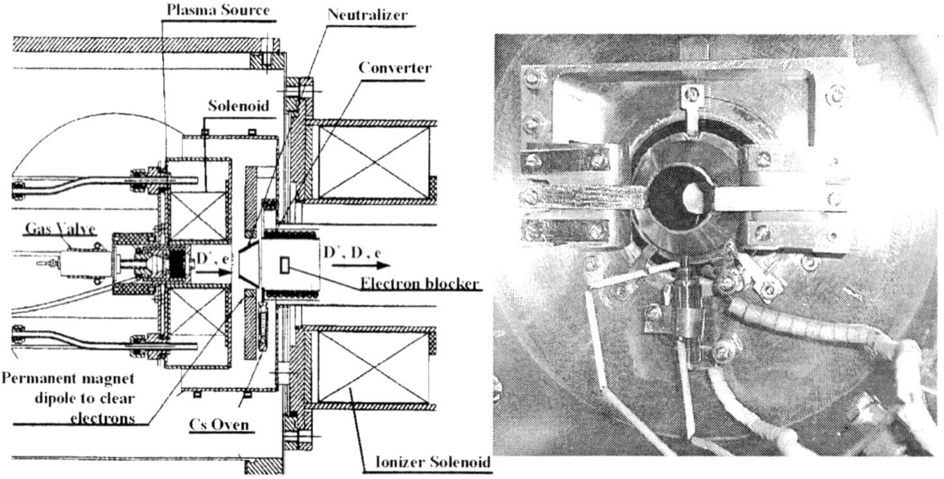

FIGURE 1. A schematic of the two stage converter at INR, Moscow and photograph of the implementation with electron blocker at IUCF. The photograph is taken from the vantage point of the plasma source solenoid. The electron blocker was added later at IUCF.

Without the blocker, the electron current exceeds 500 mA resulting in a reduction of polarized beam intensity, possibly due to the destruction of the polarized negative ions by plasma electrons inside the ionizer[8]. High electron currents also cause the extraction voltage to sag and damage the extraction grids on the high voltage electrodes. As a result, the beam brightness and pulse shape are degraded at the RFQ entrance.

The reduced electron current will allow a significant increase in the plasma injector arc current. The unpolarized H^- current extracted is 40 mA for an arc current of 300 A. With 450 A of arc current Belov reports[2] 90 mA of extracted unpolarized D^- ion current and 150 mA of unpolarized H- ion current with a corresponding improvement in plasma density. A comparable improvement in the polarized beam current is expected[8]. Higher arc currents have not yet been tested on CIPIOS.

The new four electrode extraction system built at INR for CIPIOS is designed to increase the space-charge neutralization in the vicinity of the source exit. The unpolarized negative ion current extracted from the ionizer together with the polarized ions total several tens of milliamperes before they are separated in the mass analysis magnet. In this region, positive ions will act to compensate the space-charge with the negative ion beam. The third electrode of the extraction system is biased slightly positive in order to block the positive ions from accelerating back into the ionizer.

Source Operation and Future

The nuclear physics research program based on the IUCF accelerators ended on the final day of the CE-82 experiment, July 28th, 2002. Since it was installed in 1999, CIPIOS was used exclusively to provide polarized and unpolarized beams to dozens of Cooler based nuclear physics and accelerator physics experiments. The total hours of operation, are tabulated in Table 1. The operation is very reliable, maintenance on the ABS cold nozzle and cryo-pumps is required once every two weeks and the plasma injector cathode replacement occurs on an interval of 4 to 6 weeks.

TABLE 1. CIPIOS Hours of Operation

Year	Total Hours of Operation	Unpolarized Operation
1999	3,100	672
2000	4,500	840
2001	4,620	1,200
2002 (to July 28th)	3,700	1,775

During the final months of operation in 2002, the polarized D^- peak beam intensity reached 2.2 mA, measured after mass analysis. The polarization for most states is between 85% to 90% (Table 2). The lower polarization of the –Vector state is due to

TABLE 2. D^- Polarization Results

State Name	Nominal Pz	Measured	Nominal Pzz	Measured
+ Vector	+1	0.909 (31)	+1	0.891 (13)
- Vector	-1	-0.684 (30)	+1	0.695 (14)
+ Tensor	0	0.003 (32)	+1	0.875 (13)
- Tensor	0	0.020 (33)	-2	-1.591 (13)

an inefficient weak field transition which was not repaired due to scheduling constraints.

CIPIOS will be disassembled and shipped to the Forschungszentrum Jülich in 2003 where it will be prepared for operation with a new Linac injector for the COSY ring. For this new purpose, it is required that the intensity of the polarized beam be optimized for a 0.5 ms pulse length with a low duty factor. The lengthened plasma injector and two stage converter resulted in an improvement of the polarized and unpolarized beam pulse shapes. With the source tuned for maximum average intensity during a 0.5 ms width pulse, an average beam current during the pulse of 1.5 mA is measured after mass analysis. The unpolarized beam current is 35 mA and the arc current is 350 A.

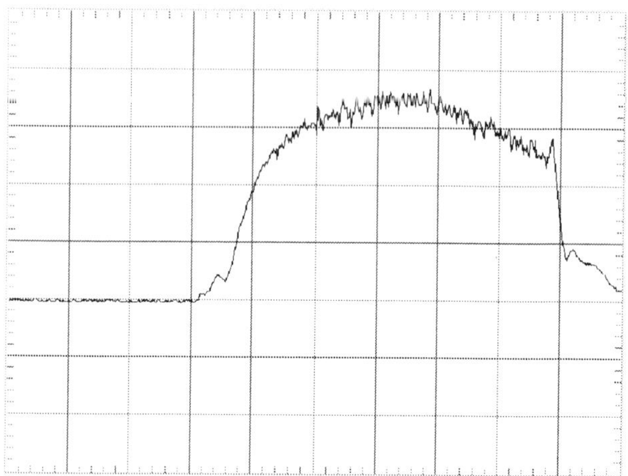

FIGURE 2. Polarized H^- beam tuned for wide pulse operation is measured after a mass analysis bending magnet. The horizontal scale is 0.1 ms/div and the vertical scale is 0.5 mA/div with zero current at the third division.

ACKNOWLEDGMENTS

The results reported would have been impossible without the help of Ron Kupper, Bill Lozowski and many other technical and professional staff at IUCF. The assistance and many important suggestions by Ralf Gebel during his visit to IUCF in November, 2001 is greatly appreciated.

This work is funded by NSF grants PHY-97-24216 and PHY-93-14783 and by the Indiana University.

REFERENCES

1. V.P.Derenchuk, et al, 2001 Particle Acc. Conf., eds. P. Lucas, S. Webber, IEEE 01CH37268, 2001, pp. 2093-2905.
2. A.S. Belov, et al, Proc. of Polarized Sources and Targets 2001, eds. V. P. Derenchuk and B. v. Przewoski, World Scientific, 2002, pp. 205-209.
3. A.S. Belov, et al, Nucl. Instr. Methods A333, 1993, pp. 256-259.
4. D.L. Friesel et al, Proc. 1997 Particle Acc. Conf., IEEE 97CH36167, 1997, p. 2811.
5. D.L. Friesel et al, Proc. 7th European Particle Acc. Conf., Vienna, Austria, eds. J.-.Laclare, W.Mitaroff, Ch.Petit-Jean-Genaz, J.Poole, M.Regler, Austrian Academy of Sciences Press, 2000, p. 539.
6. Browse to http://www.iucf.indiana.edu/Experiments/COOLCSB/.
7. V.P. Derenchuk, A.S. Belov, Proc. of Polarized Sources and Targets 2001, eds. V. P. Derenchuk and B. v. Przewoski, World Scientific, 2002, pp. 210-214.
8. A.S.Belov, et al, Proc. of 14th International Spin Physics Symposium, eds. K. Hatanaka et al, Osaka, Japan, 2000, *AIP Conf. Proc.*, **570** , 2001, pp. 835-840.

D- Charge Exchange Ionizer for the JINR Polarized Ion Source POLARIS

V.P. Ershov, V.V.Fimushkin, G.I.Gai, L.V.Kutuzova, Yu.K.Pilipenko, V.P.Vadeev, A.I.Valevich * and A.S. Belov †

*Joint Institute for Nuclear Research, Dubna, Russia
†Institute for Nuclear Research RAN, Troitsk, Russia

Abstract.
The spin physics program is an important part of the LHE JINR scientific program. An intensive study of polarization phenomena was carried out at the Dubna 4.5 Gev/nucl. synchrophasotron, using cryogenic source of polarized deuterons ($\uparrow D^+$) POLARIS. There is a proposal to make a polarized deuteron beam in the new accelerator Nuclotron to continue the spin physics experiments. The Nuclotron has a short one-turn injection (8 μs) of positive ions at present. To increase the intensity of the accelerated polarized beam up to 0.7-1*10^{10}↑d/pulse a multyturn charge exchange injection of negative ions (20-30 turns) should be applied. It is realized by injection into the accelerator of $\uparrow D^-$ ions and stripping them inside the ring. The source POLARIS is reconstructed into the $\uparrow D^-$ source now. The existing $\uparrow D^+$ cryogenic charge exchange ionizer modified into the $\uparrow D^-$ ionizer. The 10 mA H^- plasma beam have been measured at a test bench. The tests of the ionizer are in progress.

$\uparrow D^+$ SOURCE POLARIS

An intensive study of polarization phenomena in high energy spin physics was carried out at the Dubna 10 GeV synchrophasotron using the cryogenic polarized deuteron source POLARIS [1,2,3]. It is the cryogenic atomic beam source with two tapered sextupole superconducting magnets, SC solenoid, internal 4.2 K cryopanels for gas pumping. The source POLARIS consists of two LHe cryostats: a pulsed atomic beam stage and the Penning plasma ionizer (Fig.1). A new permanent sextupole magnet (inside bore - 30 mm, l- 60mm) is planned to installed behind the SC sextupoles for additional focussing of the atomic beam. The energy of the deuteron beam at the output of the source is about 3 keV,the current: 0.3-0.4 mA. The vector and tensor polarizations are: $P_z=\pm0.54$; $P_{zz}=\pm0.76$.

The commissioning of the new superconducting accelerator Nuclotron supposes to continue the spin physics program. Simulation shows that depolarizating resonances are absent under polarized deuteron acceleration almost at all energy range of that accelerator [4]. There is a project to produce a polarized deuteron beam. In this project it is supposed to realize the following:
- to upgrade the atomic stage of the source POLARIS,
- to modify the existing $\uparrow D^+$ charge exchange ionizer into the $\uparrow D^-$ ionizer,
- to realize multyturn charge exchange injection in the Nuclotron ring.

CP675, *Spin 2002: 15th Int'l. Spin Physics Symposium and Workshop on Polarized Electron Sources and Polarimeters*, edited by Y. I. Makdisi, A. U. Luccio, and W. W. MacKay

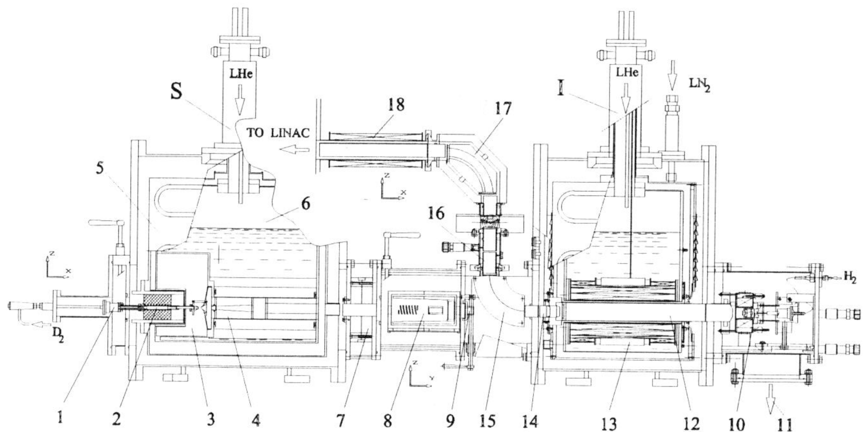

FIGURE 1. Schematic view of the D^- polarized source POLARIS. S - polarized atomic source, I - charge exchange ionizer. 1- pulsed D_2 valve, 2- dissociator, 3- nozzle chamber, 4- SC sextupole magnets, 5- nitrogen shield, 6- helium cryostat, 7- permanent sextupole, 8- RF cell, 9- vacuum gate, 10- H^- plasma generator, 11- two 500 l/s turbopumps, 12- HV charge exchange space, 13- SC solenoid, 14- extraction electrodes, 15- 90^0 bending magnet, 16- ion optics, 17- electrostatic mirror, 18- solenoid of spin-precessor.

CHARGE EXCHANGE IONIZER

The new machine Nuclotron has a short one turn injection (8 μs) of positive ions (factor 50 less compared to the old one). So to get a large intensity of the accelerator a new plasma charge exchange ionizer has been developed [5,6,7] (Fig.1). It has similar the Penning ionizer LHe cryostat with the 60 mm cold bore SC solenoid. A short pulse (300 μs) H^+ arc plasma generator is installed at the solenoid entrance. The potentials of the plasma source and the HV shield are +12 kV. A nuclear polarized deuterium atomic beam is injected into the solenoid space towards the plasma beam. The charge exchange reaction between polarized deuterium atoms and hydrogen plasma ions $\uparrow D^0 + H^+ = \uparrow D^+ + H^0$ takes place inside the HV shield.

Using POLARIS atomic beam stage, the 0.8 mA polarized D^+ beam, accompanying the 9 mA background H_2^+ plasma current have been measured behind a 90^0 bending magnet output of the charge exchange ionizer. An efficiency of the new ionizer was 3-5% instead of 1-2% for the our old Penning ionizer.

Vacuum in the ionizer is provided by cryopumping. Hydrogen vapor pressure at 4.2 K cryostat is 2-4*10^{-6} mbar. An adsorption charcoal panel, attached to the LHe cryostat improves vacuum to 6-7*10^{-7} mbar. Two 500 l/s turbopumps are installed at the plasma generator region to reduce hydrogen gas load to the cryostat.

To reach the accelerated polarized beam intensities up to 0.7-1*10^{10} ↑d/pulse a multy-turn charge exchange injection (20-30 turns) should be applied. It is realized by injection into the Nuclotron ring of $\uparrow D^-$ ions and stripping them inside the ring. A polarized $\uparrow D^-$ beam from the source is required.

The existing D^+ plasma charge exchange ionizer has been modified into $\uparrow D^-$ ionizer using an external converter-emitter [7,8,9]. At output of the H^+ plasma generator, a

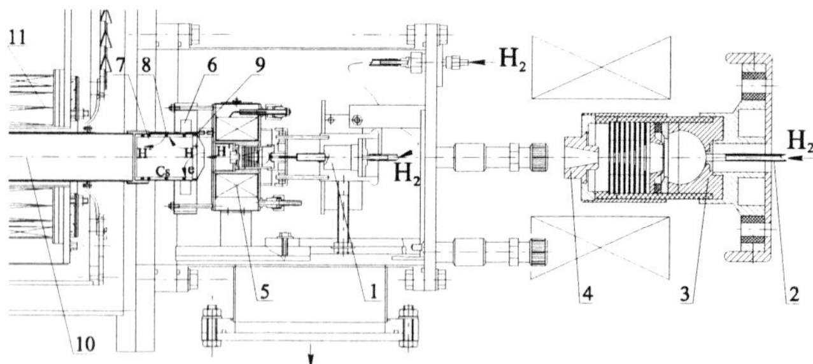

FIGURE 2. H^- plasma generator. 1- pulsed H_2 valve, 2- H_2 tube and ignition electrode, 3- Cu cathode, 4- anode, 5- magnetic coil, 6- traverse magnetic filter, 7- molybdenum converter, 8- heater, 9- Ti-Cs pellet volume, 10- HV charge exchange shield, 11- SC solenoid.

molybdenum surface converter is placed to produce H^- ions (Fig.2). Titanium-cesium chromate pellets are loaded into an O-ring groove of the converter and heated up to 300-500^{0}C. In time a plasma source pulse a cesiated molybdenum surfaces of the converter are exposed to an intense flux of superheated hydrogen atoms, positive ions and effectively generate H^- ions.

Fast plasma electrons, accompanying H^+ plasma beam, are removed by a 150 Gs transverse magnetic field of a permanent magnet, to avoid of H^- ion destruction. H^- ions, generated inside the converter, space charge compensated by residual H^+ ions, are entered axial region and fill up a charge exchange volume of the HV shield (-20 kV). The reaction $\uparrow D^0 + H^- = \uparrow D^- + H^0$ takes place. Polarized negative deuterium ions, confined in the radial direction by magnetic field of the solenoid, drift to the extracting grid and electrodes. The 90^0 bending magnet separates accelerated H^- plasma and polarized $\uparrow D^-$ ions.

Reconstruction of the ionizer cryostat is finished, cryogenic tests were started. A study of the surface conversion technology H^+ to H^- ions takes place simultaneously. Some configurations of the H^+ generator and molybdenum converter have been tested. A test bench ionizer with a sectioned conventional was used to exclude cryogenic and to have mobile runs. In time of the plasma generator tests were modified:

- converter diameter (32-45 mm), cone shape ($60^0, 180^0, 120^0$), length (50-90 mm),
- plasma source magnetic coil length (23-44 mm),
- arc discharge configuration space (cathode, divider plates, preanode volume),
- filter magnetic field (150-250 Gs),
- extraction electrode configuration,
- arc discharge feeding scheme (inductive transformer pulse or direct capacitor discharge).

About 0.1 cm^3 H_2 gas at pressure 0.5-1.2 bar is injected into a hollow cathode space by a fast thyristor triggered valve. High current arc discharge and H^+ plasma beam are started in the 0.9 ms due to ignition pulse. High voltage power supply sends acceleration pulse to the charge exchange shield, plasma generator, converter and H_2 valve simultaneously. The coils of the solenoid, bending magnet and plasma generator have been pulse charged beforehand.

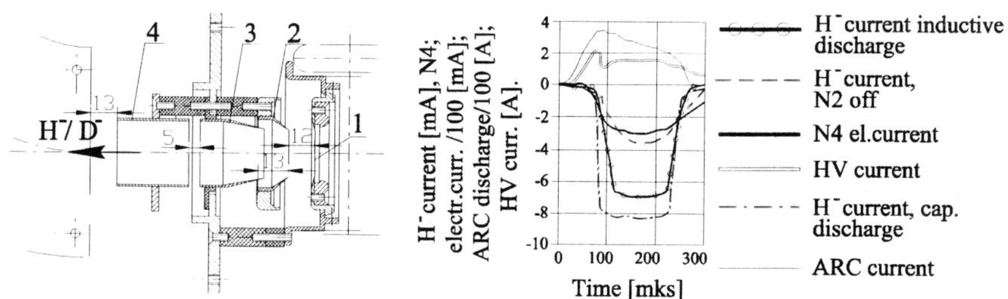

FIGURE 3. Scheme of test bench extraction electrodes and H^- current pulses behind bending magnet. 1- HV plasma wire mesh, 2,3,4- electrodes.

A scheme of the extraction electrodes is shown on Fig.3. Extraction of positive and negative ions from the charge exchange space are different for the same acceleration voltage (U_{acc}=12-17 kV). An extraction grid length of positive ions is short (two fine wire mesh, with gap 4-6 mm) and required strong electrical field gradient (3-4 kV/mm). Extraction electrodes (N2,N3) for negative ions have a cone shape. The first electrode gap is 12-15 mm and optimal gradient is only 0.5 kV/mm. A changing of the N2 electrode potential from 0 to -9 kV is heighten H^- current double from 4 to 8 mA. H^- current pulse is raised if two middle sections of the solenoid magnetic coils are tuned [10]. A capacitor discharge scheme of the arc feeding is preferable. Duration of the plasma current pulse is about 200 μs in that case. It is correspond to 25 turns of the injected beam. H^- currents at the output of 90^0 bending magnet, electron current of the N2 electrode and total current of the HV power supply are shown on Fig.3.

A conditioning of the converter surfaces and theirs covering by Cs are very important. There is small H^- current (0.2-0.5 mA) accompanying large quantity of plasma electrons first time after installation. The current is slowly raised with time. It is depend on arc voltage, pulse frequency, clean vacuum, converter temperature, Cs covering. Quantity of the injected gas strongly influences on the negative current value. Most of those plasma source parameters should be controlled remoutely. Surface activation process takes place in a pulse time while an adsorption of residual gas molecules occur between pulses. It is a disadvantage for the small pulse duty circle sources. The 10 mA H^- current was observed at the frequency 0.3 pulse/s. It is expected to get the larger current in real ionizer configuration, using cryostat with SC solenoid and cryopumping. Estimation shows, 0.3-0.5 mA $\uparrow D^-$ polarized beam should be got, using POLARIS atomic stage.

The plasma charge exchange ionizer of the polarized source has another important feature. It can be used as the intensive source of unpolarized proton or deuteron beams (10-20 mA). In that case the bending separation magnet is adjusted not for polarized but for the plasma beam.

Development of the ionizer will be continued.

ACKNOWLEDGMENTS

The authors are thankful to Dr. V.P.Derenchuk for many useful discussions. This work is supported by the Russian Fund of Fundamental Research, the Grant 01-02-16406.

REFERENCES

1. N.G. Anischenko *et al*, -In: The 5th Int. Symp. on High Energy Spin Physics, Brookhaven 1982, AIP Conf. Proc. N.Y. **N95**, 445 1983.
2. N.G. Anischenko *et al*, 6th Int. Symp. on High Energy Spin Physics, Marseille 1984, Jorn.De Phys., Colloquia C2, Supplement an n° 2, **46**, C2-703 1985.
3. V.P. Ershov *et al*, - In: Int. Workshop on Polarized Beams and Polarized Gas Targets, Cologne 1995, (World Scientific, Singapure) 193 1996.
4. A.M. Kondratenko *et al*, - In: The 6th Workshop on High Energy Spin Physics. Protvino 1995, **v.2**, 212 1996.
5. A.S. Belov *et al*, Nucl. Instr. and Meth., vol.**A255**, 3, 442 1987.
6. V.P. Ershov *et al*, -In: The 13th Int. Symp. on High Energy Spin Physics Protvino 1998, (World Scientific, Singapure) 615 1999.
7. V.P. Ershov *et al*, -In: Int. Workshop on Polarized Sources and Targets, Erlangen 1999, (Druckerei Lengenfelder, Erlangen) 456 1999.
8. A.S. Belov *et al*, Rev. Sci. Instr. vol.**67**, 3, 1293 1996.
9. V.P. Ershov *et al*, -In: Int. Workshop on Polarized Sources and Targets, Nashville 2001, (World Scientific, Singapure) 225 2002.
10. V. Derenchuk and A.S.Belov, -In: Int. Workshop on Polarized Sources and Targets, Nashville 2001, (World Scientific, Singapure) 210 2002.

A Precision Lamb-shift Polarimeter for the Polarized Gas Target at ANKE/COSY

R. Engels*, R. Emmerich†, J. Ley†, M. Mikirtytchiants***, H. Paetz gen. Schieck†, F. Rathmann*, H. Seyfarth* and A. Vassiliev**

*Institut für Kernphysik, Forschungszentrum Jülich, Leo-Brandt-Str., 52425 Jülich, Germany
†Institut für Kernphysik, Universität zu Köln, Zülpicher Str. 77, 50937 Köln, Germany
**High Energy Physics Dept., St. Petersburg Nucl. Phys. Inst., 188350 Gatchina, Russia

Abstract. The Lamb-shift polarimeter [1], described here, allows the polarization measurement of a beam of hydrogen (deuterium) atoms, or of a slow proton (deuteron) beam (500 - 2000 eV) with an absolute precision better than 1% within a few s. For a hydrogen beam of $3 \cdot 10^{16}$ atoms/s, $3 \cdot 10^{6}$ photons/s are registered in a photomultiplier. The signal-to-background ratio in the Lyman–α spectrum is excellent, thus beam intensities of one to two orders of magnitude less would still be sufficient to carry out a precise measurement. Therefore, it will be possible to extract a few percent of the atoms from a storage cell to measure the polarization in the cell. In order to obtain absolute nuclear polarization values the polarization derived from the Lyman-α spectrum has to be corrected by a number of correction factors calculated with high precision from the behaviour of different components of the polarimeter.

INTRODUCTION

To measure the polarization of a hydrogen or deuterium beam the atoms will be ionized in an electron-impact ionizer [3],[4] with an efficiency of about 10^{-3}. The ionization volume is located in a strong magnetic field ($\sim$ 160 mT) to decouple the nuclear and the electron spins. At the same time the field increases the collision probability of the accelerated electrons and the beam particles. The vertical ion beam is deflected into the horizontal beamline of the Lamb-shift polarimeter (LSP) by an electrostatic deflector. The necessary precession of the polarization back onto the beamline direction is archieved by a Wien filter. This Wien filter also acts as a mass filter and therefore only protons (deuterons) arrive in the cesium cell. By charge exchange with the cesium metastable atoms in the 2S state are produced. To define the polarization of the metastable hydrogen (deuterium) a strong magnetic field ($\sim$ 55 mT) in the cesium cell is required. The spin filter [5] transmits only metastable atoms in different single hyperfine states depending on the magnetic field. All other atoms are quenched into the ground state. In the vacuum chamber at the end of the polarimeter the residual metastable atoms are then quenched by an electric field (Stark effect). The emitted Lyman-α photons are registered by a photomultiplier sensitive to photons with a wavelength of 121 nm. The number of photons counted in one hyperfine state is proportional to the number of atoms with the same nuclear spin in the primary beam. From the ratio of the photons, which are produced at 53.5 mT and 60.5 mT in the spin filter, the polarization of the atomic beam or a slow ion beam of 500-2000 eV is calculated. With an overall efficiency of the polarimeter of

CP675, *Spin 2002: 15th Int'l. Spin Physics Symposium and Workshop on Polarized Electron Sources and Polarimeters,* edited by Y. I. Makdisi, A. U. Luccio, and W. W. MacKay

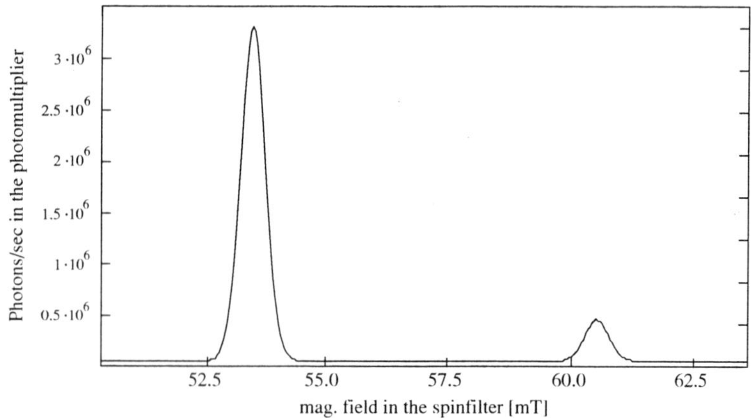

FIGURE 1. The Lyman-α spectrum of a polarized hydrogen beam in the hyperfine state 1. The polarization $P_{Ly} = 0.780 \pm 0.002$ was measured in less than 30 s.

$\sim 10^{-10}$ a beam of 10^{16} atoms/s will produce some 10^6 photons/s in the photomultiplier (Fig. 1). Thus a statistical error of $\Delta P_{Ly} \sim \pm 0.003$ is obtained within 30 s.

CORRECTION FACTORS

The polarization P_{Ly} resulting from the Lyman-α spectrum is lower than the real polarization p_z of the atomic beam. Some corrections are necessary. The knowledge of the correction factors determines the absolute accuracy which can be obtained with the LSP.

1. The ionizer can produce protons from residual gas molecules like H_2O, H_2, C_nH_n or H_2 after the recombination of atoms in the ionizer. With the atomic beam entering the ionizer an H_2^+ ion current of the same order of magnitude as for the H^+ ions will be produced. So up to 10% of the proton beam is made from unpolarized particles and has to be subtracted in the Lyman-α spectrum. This will increase the polarization value by up to 10%.
2. While the polarization of hydrogen atoms in hyperfine substates 2 and 4 depends on the magnetic field, it is constant for the pure substates 1 and 3 (Fig. 2). Therefore, it is necessary for the calculation of the polarization in arbitrary magnetic fields B to know the occupation numbers of the atoms in these substates for $B \to \infty$. By varying the magnetic field in the ionizer it is possible to determine the correction to the polarization. For example at 18 A in the solenoid coil, the magnetic field is 133 mT and only $93.5\% \pm 0.1\%$ of p_z will be measured. With this procedure it is even possible to identify the single hyperfine substates or some mixtures of these states in the beam. This resulting correction factor of about 7% is not needed for the pure hyperfine states 1 and 3.

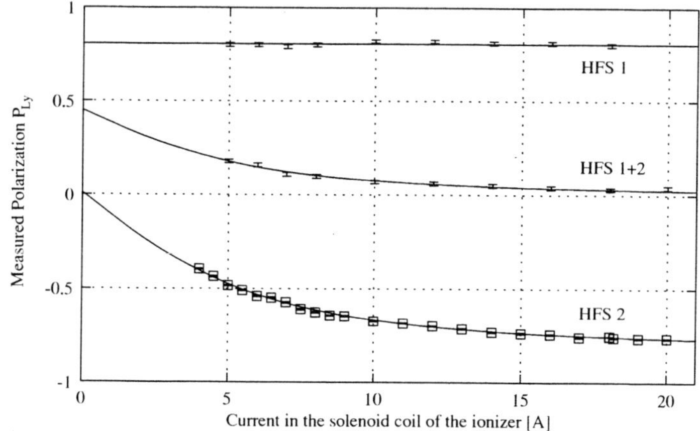

FIGURE 2. The measured polarization P_{Ly} of an atomic hydrogen beam in the hyperfine states 1, 2 or 1+2 as a function of the magnetic field in the ionizer.

3. With the Wien filter the polarization vector is rotated onto the horizontal beam line. During this process 0.8% of the polarization will be lost for protons. For deuterons this loss will increase up to $\sim$ 1%.
4. Like in the ionizer the magnetic field in the cesium cell influences the measured polarization because the production of the α_2 $(j = +1/2; I = -1/2)$ metastable atoms is a function of the magnetic field. Due to the small critical field (6.34 mT) of the metastable atoms, 99.4% of the polarization is conserved at 55 mT on the beam axis.
5. The transmission of the metastable atoms through the spin filter depends on the homogeneities of the magnetic field: at 53.5 mT the transmission is higher by almost 1% than at 60.5 mT. Therefore the measured polarization of the hyperfine states 1 and 4 will be lower by 0.2% and increased for the states 2 and 3.
6. The ionization efficiency of the ionizer is different for individual hyperfine states [7], when the electrons in the ionization volume are polarized [6]. Together with the unpolarized electrons from the hot filament electrons produced by ionization of polarized atoms are captured in the potential trap. The contribution of these electrons depends sensitively on the potentials and the magnetic field in the ionizer. Therefore the changes in the ionization efficiency can be between 0 and 7%. The resulting polarization-dependent correction factor varies usually between 0 and 0.7%.

SENSITIVITY

All measurements shown here have been made with a beam intensity up to the full intensity of the ANKE-Jülich atomic beam source [2] (about $7 \cdot 10^{16}$ particles/s in two spin states). The LSP is sensitive enough to measure the polarization of a proton beam

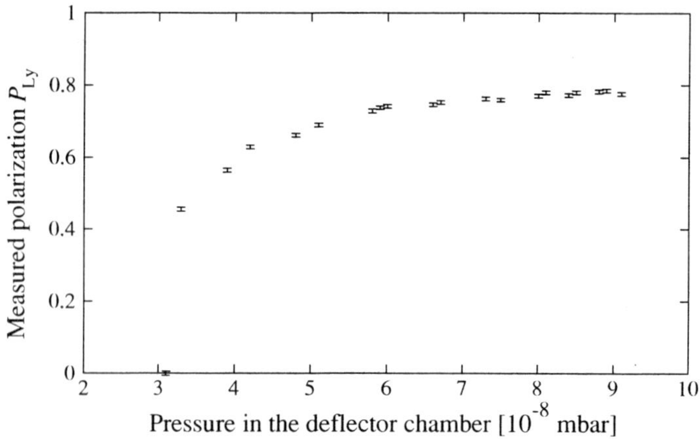

FIGURE 3. Measured polarization P_{Ly} for a beam with atoms in the hyperfine state 1 as a function of the pressure in the deflector chamber. The presuure decrease is due to the reduction of the incoming beam intensity.

with strongly reduced intensity in a reasonable time. When the efficiency of the ionizer is decreased by a factor of 100, the resulting 20 nA of polarized protons are sufficient to get a good Lyman-α spectrum.

These ion beam intensities correspond to atomic beams of $6 \cdot 10^{14}$ atoms/s with the ionizer working at an efficiency of 10^{-3}. However, it was observed that the measured polarization P_{Ly} decreases (Fig. 3), when the beam flux is reduced and the pressure in the deflector chamber drops. This is due to the constant beam-independent background in the Lyman-α spectrum, which causes a reduction of the ratio of the number of protons produced from the beam and from residual gas. The nuclear polarization p_z can still be determined by applying the proper correction factors.

DISCUSSION

The LSP ist an excellent *relative* polarimeter with a statistical error of 0.3% within 30 s. Therefore it can be used for online tuning of the transition units in an atomic beam source.

To derive the *absolute* polarization of the atomic beam with a high precision some known corrections have to be applied. The correction factors increase the measured polarization P_{Ly} of the hyperfine states 1 and 3 by about 11% and for the states 2 and 4 by about 19% with an error of 1%. These corrections are dominated by effects in the ionizer. A new ionizer with a stronger magnetic field and a getter pump around the ionization volume was designed and built to reduce these overall corrections to 5% at an error of 0.5%. This ionizer will be tested before the end of 2002.

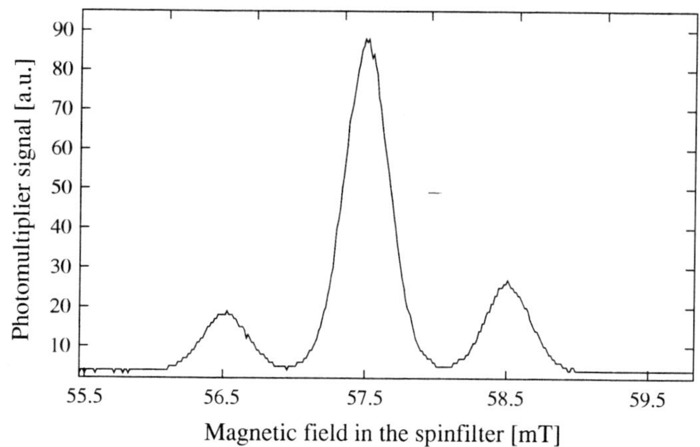

FIGURE 4. The Lyman-α spectrum of a polarized deuterium beam. ($p_{z_{Ly}} = -0.06$; $p_{zz_{Ly}} = -1.09$).

The LSP can be used for deuterium as well (Fig. 4). In this case the correction factors are smaller due to the smaller critical field and the lower background of the residual gas in the ionizer.

With the present LSP it will be possible to measure the polarization of the primary atomic beam with intensities down to 1% of the present beam. It is planned to measure the polarization of atoms in a storage cell at ANKE/COSY by extracting a small amount of the atoms. First tests are under preparation.

ACKNOWLEDGMENTS

This work has been supported by the German Ministry of Education and Research (BMBF) and FZ Jülich (FFE contract).

REFERENCES

1. A. J. Mendez, C. D. Roper, J. D. Dunham and T. B. Clegg, Rev. Sci. Instr. **67** (1996) 9.
2. R. Rathmann et al., contribution to these proceedings.
3. H. F. Glavish, Nucl. Instr. Meth. **65** (1968) 1.
4. R. Emmerich, diploma thesis, Universität zu Köln (2000).
5. J.L. McKibben et al., Phys. Lett. **28B** (1969) 594.
6. M. J. Alguard, V. W. Hughes, M. S. Lubell and P. F. Wainwright, Phys. Rev. Lett. **39** (1977) 334.
7. S. Jaccard, Conf. on Pol. Proton Ion Sources, ed. A. D. Krisch and A. T. M. Lin, Ann Arbor, 1981; AIP Conf. Proc. **80** (1982) 95.

Solid State Polarized Targets for the Study of Nucleon Spin Structure

G.R.Court

Physics Department, University of Liverpool
Liverpool L69 7ZE UK

Abstract. The dynamically polarized solid state nucleon target has played a very important role in the experiments to study the spin structure of the nucleon which have been carried out over the last two decades. The quality of the data obtainable is critically dependent the properties of the target materials used. The developments which have been made to improve both target materials and overall target performance during this period are reviewed. The effects that these developments have had on present and future experiments is discussed.

INTRODUCTION

A talk entitled "Polarized Targets – what is happening ?" [1] was given at the Spin Symposium held here at Brookhaven exactly twenty years ago. At that time, dynamic nuclear polarization (DNP) in solid hydrogenous materials was the technique being used to obtain polarized proton and deuteron targets for most high energy spin experiments. This technique, which enables high values of nucleon polarization to be obtained in a limited range of hydrogenous solids, requires that the solid material be maintained in a magnetic field of 2.5T or above and cooled to a temperature in the region of 1K. The technological and experimental problems which arise from these requirements were reviewed in some detail in the talk, and it was concluded that the technology of this type of target was well developed and that it would continue in use for some time to come. In this talk, the progress which has been made to improve the performance of the solid state polarized target, with particular emphasis on its use for the study of the spin structure of the nucleon, will be reviewed.

DYNAMIC NUCLEAR POLARISATION

The static nucleon polarization generated in solids, with practical values of applied magnetic field and temperature, is always much less than 1%. DNP works via free electron spins introduced into the solid matrix by chemical or other means. These spins are essentially 100% polarized under the same conditions and this high electron polarization can be transferred to the nucleons by microwave pumping. The target spin direction relative to the static magnetic field is then reversed by making small changes in the frequency. The electron spin concentration must be maintained within relatively close limits to obtain the highest values of nucleon polarization. DNP targets have a number of intrinsic limitations which need to be accounted for in the design of

CP675, *Spin 2002: 15th Int'l. Spin Physics Symposium and Workshop on Polarized Electron Sources and Polarimeters*, edited by Y. I. Makdisi, A. U. Luccio, and W. W. MacKay

experiments and their operation. Firstly, DNP will not work in solid hydrogen so the free nucleon to total nucleon ratio (nucleon dilution factor) in the target material is always less than unity. Secondly, the beam used in experiments causes radiation damage and hence generates free electron spins in the material which can interfere with the DNP process so reducing the polarisation. This can normally be restored to a value close to the initial one by annealing the target material, but only at the cost of lost data taking time. Finally, all nucleons with spins present in the material become polarized to some degree. The DNP process is described in detail in reference [2]

SPIN STRUCTURE FUNCTION MEASUREMENTS

For over twenty years polarized targets have been an essential tool in the experimental study of the spin structure of the nucleon via DIS. The first series of experiments was done at SLAC in the early 1980s with 20 GeV electrons on a proton target. The results demonstrated that in the kinematic region where the valence quarks carry the bulk of the nucleon's momentum, they account, as expected, for its spin. The target material used in the early stages was an alcohol (butanol), which has a dilution factor of 0.14, doped with a free radical to provide the free electron spins. Proton polarizations in the region of 70% were obtained at an operating field of 5T and a temperature of 1K. These high values of field and temperature were chosen to minimize the effect of beam heating of the target material. The radiation damage effects were severe, and limited the overall statistical accuracy attainable in the experiments.

During the same period a new target material was developed, which was solid ammonia (NH_3) with the electron spins introduced via radiation damage at a temperature in the region of 77 K. This material has a dilution factor of 0.17. The polarizations attainable are high, and the tolerance to beam induced radiation damage much better (~ x30) than for the alcohol materials. It was used in the later stages of the SLAC series of experiments.

The next experiment to study nucleon spin structure was the EMC experiment at CERN which used a 200 GeV polarized muon beam with an NH_3 target. This target was unique in two features[1]. Firstly, it had a much larger target material volume (x20) than any target built previously, and secondly it had two separate sections with opposite polarization directions which were mounted in tandem. This latter arrangement helped to minimize systematic uncertainties caused by beam intensity fluctuations and loss of data during polarization direction reversals. The proton polarization was in the region of 80% at a field of 2.5T and a temperature of less than 0.5K. Radiation damage was not an issue as the beam intensities were very low. It was necessary to make small corrections for the relatively small spectator nucleon (^{14}N) polarization which was measured separately. The data from this experiment revealed that the initial nucleon structure picture was overly simple, and that proton spin was attributable to a combination of a contribution from sea quarks and gluons and possibly angular momentum.

This unexpected result inevitably generated a great deal of theoretical interest, which stimulated programmes at both CERN (SMC) and SLAC to further explore the nucleon spin structure. The main motivation was to confirm and improve the precision of the EMC result for the proton and obtain data for the neutron. The latter is normally achieved by using polarized deuterons, which have spin one, and are effectively a polarized proton combined with a polarized neutron. The deuteron polarization attainable is always lower than the proton, with values of around 50% being attainable in deuterated butanol, and between 20 and 40% in deuterated ammonia. A third entirely new programme (HERMES) was started at DESY using gaseous state polarized targets of hydrogen, deuterium and ^3He in the HERA electron storage ring. This type of target has the advantage of dilution fraction at or close to unity, but it can only provide practical luminosities when used with stored beams. It is an important tool for nucleon spin structure studies but will not be discussed further in this talk.

The SMC programme started with a modified version of the EMC target which was later replaced with a new target of similar design. The target materials used for most of the data taking were ammonia for protons and deuterated butanol for deuterons. The latter was used because it has a better figure of merit in this situation., which was further enhanced when it was discovered that frequency modulation of the microwave source increased the polarisation from its normal value of around 30% to, on average, over 50%.

The SLAC programme started with a gaseous ^3He target for neutron measurements. This was later replaced by a solid state target using both NH_3 and ND_3 operating at 5T and 1K. Radiation damage was again a serious problem, with the consequence that with protons the mean value of the polarization was of the order 70% when the initial value was close to 100%. With ND_3 it was found that the deuteron polarization actually increased by a factor of up to three during the initial phase of beam irradiation, before returning to normal behavior at higher dose levels, so giving a higher than expected figure of merit. The materials actually used were $^{15}NH_3$ and $^{15}ND_3$. This obviates the need to make a correction for the effective neutron polarization in the ^{14}N as well as simplifying the measurement of spectator nuclear polarization.

In the later stages of the SLAC programme an entirely new material, lithium deuteride (^6LiD), was introduced as an alternative deuteron target material. The electron centres needed for DNP were again generated by radiation damage of the solid, but at a temperature of 185 K. The ^6Li and the deuteron both become polarized to a level of about 25% under normal conditions. If it is assumed that the ^6Li nucleus looks, as seen by the DIS process, like an ^4He nucleus (spin zero) plus a polarized deuteron, then the effective dilution factor, and hence figure of merit, is much larger than the equivalent figure for ND_3.

Spin structure functions for both the proton and the neutron have now been measured to high precision over a wide range of both Bjorken-x and Q^2. The HERMES data, obtained with a gaseous target, agrees very well with that obtained by both groups using different solid state targets. This gives confidence that the systematic errors in the target polarization measurement for both types of target are known and well understood.

POLARIZATION MEASUREMENT

The standard technique for monitoring and measuring the polarization in solid state targets involves measuring spin state populations by observing NMR with a linear RF Q-meter system. A high precision version of this system was originally developed for the EMC target and the same system is still being used with virtually all targets in current operation [3]. The overall systematic uncertainty attainable with this system is in the range 2 to 5% for both protons and deuterons, depending on the specific target operating conditions. In some recent experiments the uncertainty in the target polarization has been higher than this, and has dominated the overall uncertainty in the experiment. Development work has therefore been has carried out to improve both the linearity and intrinsic signal to noise ratio of the system [4]. Lower systematic uncertainties in target polarization measurement should now be attainable using this new system.

CURRENT AND FUTURE EXPERIMENTS

A precise interpretation of this world data set on spin structure functions suggests that the gluons carry a large proportion of the nucleon spin, so a number of experiments have been proposed to make a direct measurement of this contribution. HERMES has already attempted such a measurement, which was severely limited by statistical and systematic uncertainties. Another experiment of this type is under way at CERN (COMPASS) using an upgraded version of the SMC target with ^{6}LiD as target material. Polarizations of over 50% are being attained largely due to the low target operating temperature (< 500 mK). There is also an approved experiment at SLAC (E161) which will use ^{6}LiD with a polarized real photon beam, so measuring the gluon contribution via real photon interactions. The lower effective beam intensity will reduce the effect of radiation damage and allow operation at temperatures below 500 mK. It is also proposed to use a 6.5T static field which in combination with this low operating temperature should enable polarizations in the region of 70% to be achieved.

There are currently a number of experiments proposed or already under way, at SLAC and TJNAF and other laboratories, to study spin structure in a lower Q^2 regime. These generally need to operate with high intensity beams and therefore a target material with a high radiation damage tolerance is required . However, the assumption made about the nuclear properties of ^{6}Li which make it equivalent to a polarized deuteron do not apply in all situations and it is then necessary to use ND_3 .

CONCLUSIONS

A critical factor in the design of all nucleon spin structure experiments using fixed targets is the properties of the target material itself. The chemically doped target materials available in the early stages of the nucleon spin structure programme had a relatively small dilution fraction (~0.14) and a low tolerance to beam induced radiation damage. The proton polarization attainable was typically in the region of 85%, while the deuteron polarization in fully deuterated materials, was in the range 30 to 45%. A new material, solid ammonia with electron spin centre introduced via radiation damage, came into use at an early stage in the programme. This gives proton polarizations close to 100% with a somewhat improved dilution fraction (0.17) and greatly improved tolerance to radiation damage. Ammonia is still the best available material for proton targets. The deuteron polarization attainable in the deuterated material can, depending on target operating conditions, have values up to 50%.

The successful development of lithium deuteride (^{6}LiD) as a target material using radiation damage centres, and its first use in a nucleon spin structure experiment in 1995, were major breakthroughs. The dilution factor is 0.5 if it is assumed that a ^{6}Li nucleus acts like a pseudo deuteron plus an α-particle. The effective deuteron polarization attainable depends on both operating field and temperature with typical values of 25% at 5T and 1K and over 50% at 2.5T and < 500mK. It is likely that even higher polarization will be attained when operation at low enough temperature and higher magnetic field is possible. This material also has an even better radiation damage tolerance than ammonia. It is interesting to note that ^{6}LiD was first suggested as a target material in 1978 [5] and this perhaps highlights the magnitude of the problems which have to be overcome when developing new materials for DNP.

The dynamically polarized solid state target has been used in many nuclear and particle physics experiments during the last twenty years. It is a versatile tool which has played, and still is playing, a dominant role in the experimental study of the spin structure of the nucleons and in many other areas. It seems likely that it will continue to be used to make high precision measurements in many areas of spin physics for some time to come.

REFERENCES

The scope and nature of this talk makes a detailed list of references impractical., therefore two review papers [1],[2] are listed, plus some source references needed to clarify specific points.

1. Court, G. R., *Proc. High Energy Spin Physics Ed. G.Bunce AIP Conf. Proc. 95 (1982) p 464.*
2. Crabb, D.G., Meyer, W., *Annu. Rev. Nucl. Part. Sci 47 (1997) p 67.*
3. Court, G.R. et al, *Nucl. Instr. & Meth. A324 (1993) p 433.*
4. Court, G.R., Houlden M.A. and Crabb D.G. *9th Int.Workshop on Pol. Sources and Targets, World Scientific Press (2002) Eds V.Derenchuk and B.von Prezwoski p 111.*
5. Abragam, A., *Proc.High Energy Physics with Pol. Beams and Pol. Targets, Ed G.H.Thomas AIP Conf. Proc. 51 (1979) p 1.*

Thin Scintillating Polarized Targets for Spin Physics

B. van den Brandt [1], E.I. Bunyatova‡, P. Hautle and J.A. Konter

Paul Scherrer Institute, CH-5232 Villigen PSI, Switzerland
‡Joint Institute for Nuclear Research, Dubna, Head P.O. Box 79, 10100 Moscow, Russia

Abstract. At PSI polarized scintillating targets are available since 1996. Proton polarizations of more than 80%, and deuteron polarizations of 25% in polystyrene-based scintillators can be reached under optimum conditions in a vertical dilution refrigerator with optical access, suited for nuclear and particle physics experiments. New preparation procedures allow to provide very thin polarizable scintillating targets and widen the spectrum of conceivable experiments.

INTRODUCTION

Polarized scintillating targets are instruments, in which nuclei, present in scintillating organic substances, can be polarized and the light produced in the scintillator by scattering particles is forwarded to a photo-multiplier at room temperature. In the past years scintillator blocks of 5x18x18 mm were successfully polarized [1, 2, 3] and used in $\vec{n}\vec{p}$ experiments to measure the neutron-proton spin correlation parameter at forward angles at 68 MeV [4] and in $\pi\vec{p}$ scattering to measure the analyzing powers at 45-87 MeV [5]. Polarized scintillating targets offer the unique capability of coincident *in situ* detection of low energy recoil protons (or other nuclei) in the target itself, and the possibility to suppress background scattering by time-of-flight methods or by deposited energy level discrimination.

In this communication we report on the preparation of thin polarizable scintillating foils with an embedded NMR coil. A possible configuration is proposed for an instrument in which a scintillating thin foil, surrounded by an extremely small quantity of material, can be polarized.

SAMPLE PREPARATION

Samples were prepared by dissolving powder or small pieces of scintillating polymer (PMMA) in methyl-ethyl-ketone, warming the mixture to 120 °C during 1-2 hours. Subsequently the free radical (2,2,6,6-tetramethyl-4-acetooxypiperidine-1-oxyl: "aceto-TEMPO") was added, the mixture was stirred during 10 minutes, and filled into a syringe. A mould made of a glass substrate and a mask of PTFE plate (2 mm thickness)

---

[1] e-mail: `Ben.vandenBrandt@psi.ch`

CP675, *Spin 2002: 15th Int'l. Spin Physics Symposium and Workshop on Polarized Electron Sources and Polarimeters*, edited by Y. I. Makdisi, A. U. Luccio, and W. W. MacKay

was prepared (see figure 1). The PTFE plate was shaped to form little rectangular "bays", and in each of them a 0.3 mm diameter copper wire (the NMR coil) was placed. Using the syringe the solution was carefully injected to form a liquid layer of ca. 2 mm thickness, held in place by the PTFE mask on the glass substrate and by surface tension. After drying at room temperature during 2-4 hours and further drying in an oven at 60 °C, the solvent was completely evaporated, as could be concluded by weighing the samples. The concentration of the free radical in the remaining polymer was estimated by EPR measurements. Scintillating polymer films with an integrated NMR coil were produced with a thickness between 20 and 100 micrometers. Their scintillation properties were tested with an UV lamp.

In a similar way combinations of other scintillating polymers, solvents and free radicals

FIGURE 1. The mould (left) to produce scintillating foils. On the right a 70 micrometer foil of PMMA, doped with aceto-TEMPO.

were prepared and tested, as reported elsewhere [6].

RESULTS AND OUTLOOK

The foil shown in figure 1 was placed in the mixing chamber of our dilution refrigerator (lowest temperature 50 mK) and polarized at 2.5 T. A maximum proton polarization of +77% resp. -76% was obtained. It took 20 min to reach 60% and 40 min to reach 70% polarization. The relaxation time was 200 sec at 1.19 K and 2.5 T. The sensitivity of the NMR coil is demonstrated in figure 2, showing the T.E. signal of this 0.025 g sample at 2.17 K.

In scattering experiments background scattering from materials surrounding the polarized target should be in any case minimized, but in some experiments it is not permitted to have more than nothing around the target. In the past we have shown [7] that it is possible to polarize thin polymer films cooled by a 0.12 μm thin layer of superfluid ^{4}He. This suggests to build a target consisting of two parallel polarizable scintillating foils, held in place by the embedded NMR coil and glued together at the rims to form a rectangular box, covered on the inner sides with a layer of superfluid ^{4}He (see figure 3). The two foils should be glued together without adding glue that contains unwanted protons, e.g. using pure solvent. The helium layer should be in thermal contact with the

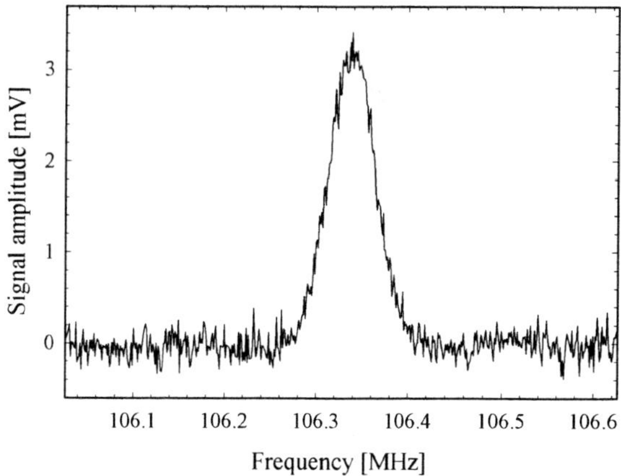

FIGURE 2. NMR signal of 0.025 g PMMA doped with 2,2,6,6-tetramethyl-4-acetooxypiperidine-1-oxyl free radical (Thermal equilibrium signal taken at $T = 2.17$ K and 2.5 T).

mixing chamber via a suitable heat exchanger, while the scintillating "sandwich" would be hanging free in the vacuum chamber of the polarized target cryostat, with close to nothing surrounding it. The scintillation light could be coupled to the lower end of the lightguide via an optical window of quartz in the bottom of the mixing chamber. A configuration consisting of a transparent disc of 3 mm thickness, glued with Stycast 2850 in a stainless steel 316L flange appeared to be reliable at low temperatures.

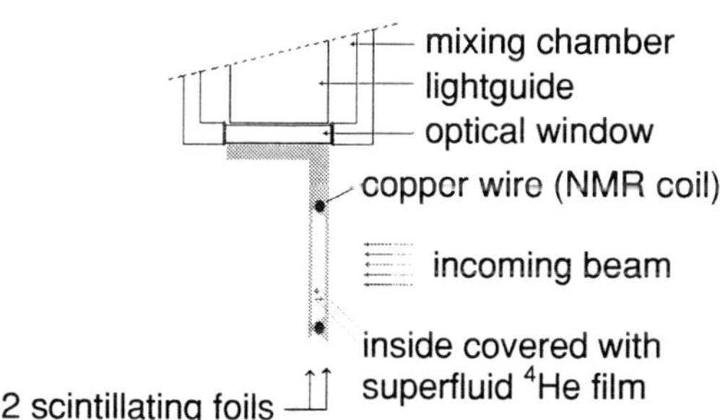

FIGURE 3. A possible configuration of a thin target consisting of 2 scintillating foils, cooled via a superfluid ^{4}He film, that is cooled in turn by a ^{3}He-^{4}He dilution refrigerator.

CONCLUSIONS

Thin scintillator foils with an embedded NMR coil have been produced and polarized. A new scintillating polarized target, in principle suited for experiments with heavy ions, radioactive beams etc. has been proposed. A further study of the specific technical requirements is necessary.

ACKNOWLEDGEMENTS

The continuous interest of Dr. S. Mango is gratefully acknowledged.

REFERENCES

1. B. van den Brandt, E.I. Bunyatova, P. Hautle, J.A. Konter, S. Mango, Proc. of SPIN96, 12th Internat. Symp. on High-Energy Spin Physics, Sept. 10-14, 1996, Amsterdam, (World Scientific, Singapore, 1997), p. 238
2. B. van den Brandt, E.I. Bunyatova, P. Hautle, J.A. Konter, S. Mango, Nucl. Instr. and Meth. **A446** (2000) 592-599 and references therein
3. B. van den Brandt, E.I. Bunyatova, P. Hautle, J.A. Konter, S. Mango and I. Nemchonok, Proc. SPIN2000, 14h Internat. Spin Phys. Symp., 16-21 oct 200, Osaka, AIP conf. Proc. **570** (2001) 866.
4. M. Hauger, Measurement of the neutron-proton spin correlation parameter A_{zx}, Inauguraldissertaion, Universität Basel, Basel 2002.
5. R. Bilger *et al.*, PSI Annual Rep. **1** (1997) 22
6. B. van den Brandt, E.I. Bunyatova, P. Hautle and J.A. Konter, Proc. of the GDH 2002 conference, Genova, Italy, July 3-6, 2002, World Scientific Publishing Co., to be published.
7. B. van den Brandt, P. Hautle, Yu. Kisselev, J.A. Konter and S. Mango, Nucl. Instr. and Methods **A 381** (1996) 219.

Polarized Solid Proton Target for RI Beam Experiments

T. Wakui*[1], M. Hatano†, H. Sakai†**, A. Tamii† and T. Uesaka**

*RIKEN, Wako, Saitama 351-0198, Japan
†Department of Physics, University of Tokyo, Tokyo 113-0033, Japan
**Center for Nuclear Study, University of Tokyo, Wako, Saitama 351-0198, Japan

Abstract. A polarized solid proton target system that can be used for radioisotope beam experiments has being developed. A high-power Ar-ion laser has been installed to improve proton polarization. With the laser, proton polarization of 36.8±4.2% has been achieved in 0.3 T at 77 K. The new target system has been constructed toward the first nuclear physics experiment scheduled in 2003.

INTRODUCTION

For the purpose to study nuclear structure of unstable nuclei with radioisotope (RI) beam, we are developing a polarized solid proton target that can be operated in a magnetic field lower than 0.3 T at a temperature higher than 77 K [1]. In the modest operating condition, low energy recoil protons can be detected in $\vec{p}$-RI elastic scattering experiments that carried out under the inverse kinematics condition. This allows us to descriminate true events from backgrounds and also to achieve high angular resolution.

As a target material, a crystal of aromatic molecules such as naphthalene doped with pentacene is used. Protons in the crystal can be polarized by means of a pulsed dynamic nuclear polarization [2]. In this method, pentacene molecules are excited to the lowest triplet state by laser irradiation [3]. A population difference appears spontaneously in the triplet state [4, 5]. Subsequently, the population difference is transferred to proton polarization by a cross polarization technique.

For the excitation of pentacene molecules, a flushlamp-pumped dye laser has been used as a light source. Iinuma *et al.* has successfully obtained proton polarization of 32% with the dye laser [6]. However, the dye laser is not suited for application in the nuclear physics experiments because of the rather short lifetime of the dye. It is necessary to change the dye once a day during the experiments to maintain the laser power. This causes that the proton polarizing process is interrupted by the changing of the dye for a few hours. Thus, we decided to use other type of laser, Ar-ion laser, which demands less maintenance.

---

[1] Present address : Center for Nuclear Study, University of Tokyo, Wako, Saitama 351-0198, Japan

CP675, *Spin 2002: 15th Int'l. Spin Physics Symposium and Workshop on Polarized Electron Sources and Polarimeters,* edited by Y. I. Makdisi, A. U. Luccio, and W. W. MacKay

In this paper we describe the recent progress in our polarized proton target project, including upgrade of the laser and construction of the new target system.

PROTON POLARIZATION WITH AR-ION LASER

A prototype of proton polarizing system has been constructed and used to study the effectiveness of Ar-ion laser as a light source for excitation of pentacene. By using this system, we obtained proton polarization of 18.4 ±3.9% in 0.3 T at 100 K [7]. In the experiment, a continuous wave Ar-ion laser having the maximum power of 4.2 W in the multiline operation was used for excitation of pentacene. Since the lifetime of the most populated sublevel, m_s=0, is shorter than that of other levels, m_s=±1, as shown in Fig. 1, the laser beam was mechanically pulsed by an optical chopper to obtain a large population difference. The pulse width and the repetition rate were 20 μs and 1 kHz, respectively. The resulting average power was 84 mW.

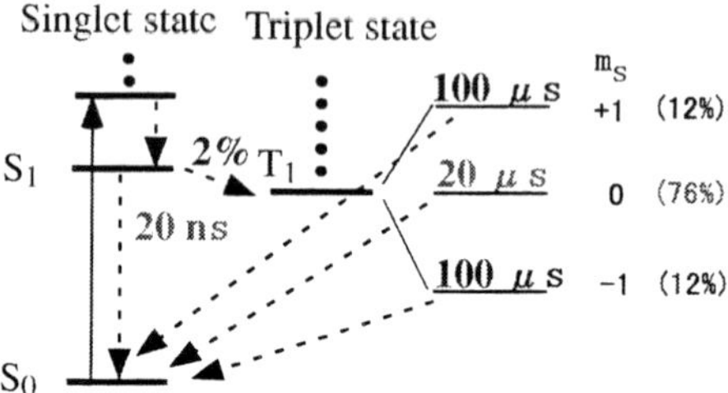

FIGURE 1. Energy levels of pentacene. The lifetime of m_s=0 state is 20 μs, while that of m_s=±1 states are 100 μs.

In order to improve proton polarization, we have installed a high-power Ar-ion laser. The laser has the maximum power of 25 W in multiline operation and has a standard power specification for the wavelengths ranging from 454.5 nm to 514.5 nm. Operation modes of the laser can easily be changed between multiline and single-line operation. For excitation of pentacene, the laser is used in the single-line operation at wavelength of 514.5 nm. This is because pentacene molecules in the lowest triplet state have absorption maxima at the wavelengths of 490.0 nm and 457.0 nm. The absorption will cause relaxation of proton polarization.

Figure 2 shows a result of the proton polarization as a function of time during the buildup process. We have succeeded in polarizing protons up to 36.8±4.2% in a magnetic field of 0.3 T at a temperature of 100 K. The laser power was 10 W, which corresponds to the average power of 200 mW. The crystal was naphthalene doped with 0.01% pentacene. The size of the crystal was $4 \times 4 \times 3$ mm^3. The error in the polarization is mainly dominated by the uncertainty in the calibration measurement. This proton polarization is comparable to that obtained with the dye laser, which shows that an ordinary Ar-ion laser can be a reasonable light source for the optical pumping.

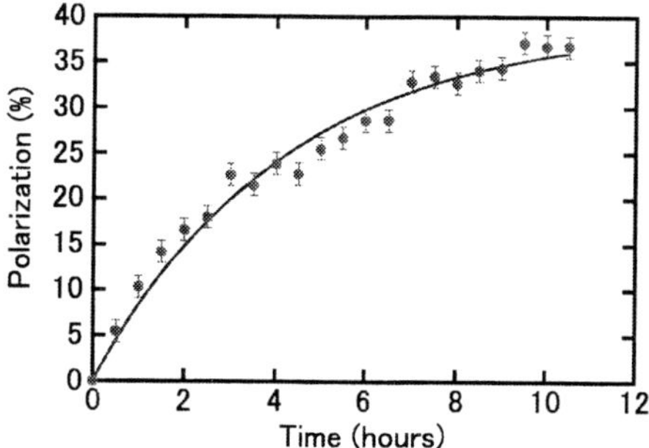

FIGURE 2. Proton polarization as a function of time in 0.3 T at 100 K. The obtained polarization is 36.8±4.3% and the extrapolated maximum proton polarization is 39.3±4.6%.

NEW TARGET SYSTEM

Figure 3 shows a schematic of the newly constructed target system. The magnetic field is produced by a C-type magnet, which can generate the maximum field of 0.7 T. The measured inhomogeneity of magnetic field is 1.1×10^{-3} in central 10 mmϕ sphere at 0.3 T. Between the pole gap, a scattering chamber is mounted. The scattering chamber has a double-layered structure to cool the target sample. A cooling chamber, in which the target sample is placed, is mounted in a vacuum chamber that is connected to a beam line. Both of the cooling and vacuum chambers have glass windows for laser irradiation and Kapton windows in the path of the RI beam and recoil protons.

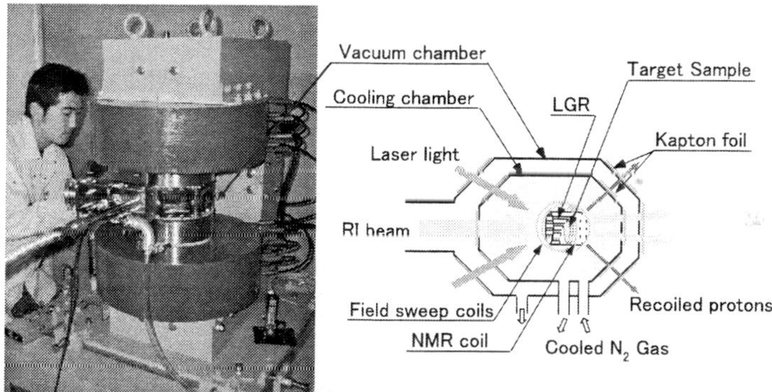

FIGURE 3. The new target system. The scattering chamber mounted between the pole gap consists of a vacuum chamber and a cooling chamber. The target sample is placed inside a copper-film loop-gap resonator.

In the prototype system, a cylindrical microwave cavity is used to produce oscillating magnetic fields. The cavity is enclosed by thick walls that prevent low energy recoil protons from reaching to detectors. Instead of the microwave cavity, a copper-film loop-gap resonator (LGR) [8] is introduced to the new system so that recoil protons can reach to detectors. Figure 4 shows a schematic view of the LGR. The LGR is made of 25-μm

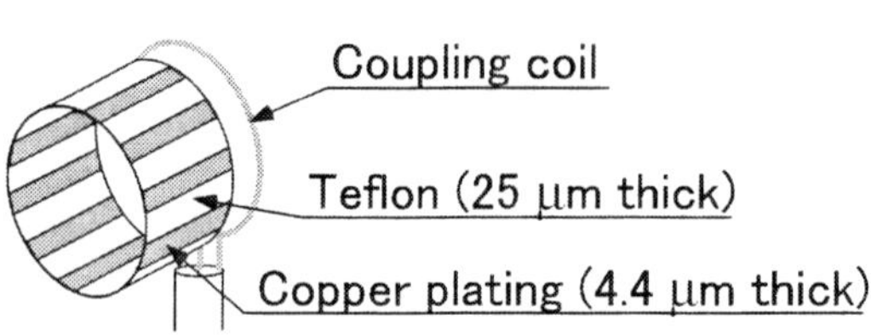

FIGURE 4. Schematic view of a copper-film loop-gap resonator. The resonator is used to generate oscillating magnetic fields.

thick Teflon sheet coated on both sides with 4.4-μm thick copper metal. The copper is etched to create capacitive gaps in the overlapping regions of strips. The etched sheet is formed into a cylindrical loop which act as an inductive element. The LGR and a microwave circuit are inductively coupled by a coupling coil. The radius of the LGR is 8 mm and the axial length is 10 mm. The resonance frequency of the LGR is 3.2 GHz, which is the ESR frequency in the magnetic field of 85 mT. An NMR coil is wound around the LGR. The NMR frequency is 3.6 MHz in 85 mT. A pair of coils are placed for magnetic field sweep and it generates the magnetic field of 6 mT by applying a current of 50 A.

Figure 5 shows a recent result of buildup of proton polarization with the new system. We have obtained an enhancement of proton polarization. To provide the absolute polarization caliblation factor, a measurement of proton polarization in the thermal equilibrium is required. For the measurement, we are planning to improve NMR sensitivity by installing a movable device on which the LGR is mounted. The LGR, which reduces NMR signal intensity, will be removed from the target position, when NMR measurements are carried out.

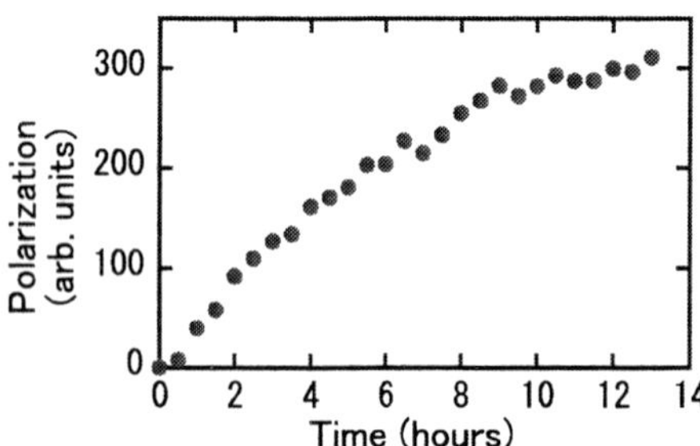

FIGURE 5. Enhancement of proton polarization in 85 mT at 90 K. The absolute value of polarization is not calibrated.

SUMMARY

A high-power Ar-ion laser has been installed to improve proton polarization. With the laser, proton polarization of 36.8±4.2% has been obtained in 0.3 T at 100K. The result shows that an Ar-ion laser can provide proton polarization comparable to that obtained with the dye laser.

We have also begun to construct the new target system that will be used in nuclear physics experiments. An enhancement of proton polarization has been observed with the LGR. An improvement of NMR sensitivity to measure the polarization in thermal equilibrium is under way. The first experiment, the measurement of a vector analyzing power in the elastic scattering of $\vec{p}$-^{6}He at 71 MeV/A, will be carried out with the new target system and a ^{6}He beam from the projectile fragment separator RIPS in 2003.

ACKNOWLEDGMENTS

The authors would like to thank Dr. I. Tanihata and Dr. T. Suda at RIKEN for their constant support of this study. One of us (T.W.) would like to acknowledge the Special Postdoctral Researchers Program.

REFERENCES

1. T. Wakui, M. Hatano, H. Sakai, A. Tamii, and T. Uesaka, 14th Int. Spin Physics Symposium, Osaka. 2000, AIP Conference Proceedings **570**, 861 (2001).
2. A. Henstra, P. Dirksen, and W. Th. Wenckebach, *Phys. Lett* **A 134**, 134 (1988).
3. W. H. Hesselink and D. A. Wiersma, *Phys. Rev. Lett.* **43**, 1991 (1979).
4. M. S. de Groot, I. A. M. Hesselmann, J. Schmidt, and J. H. van der Waals, *Mol. Phys.* **15**, 17 (1968).
5. W. S. Veeman and J. H. van der Waals, *Mol. Phys.* **18**, 63 (1970).
6. M. Iinuma, Y. Takahashi, I. Shaké, M. Oda, A. Masaike, and T. Yabuzaki, *Phys. Rev. Lett.* **84**, 171 (2000).
7. T. Wakui, M. Hatano, H. Sakai, A. Tamii, and T. Uesaka, Proceedings of the Ninth International Workshop on Polarized Sources and Targets, 2002, eds. V. P. Derenchuk and B. Przewoski, p. 133.
8. B. Ghim, G. A. Rinard, R. W. Quine, S. S. Eaton, G. R. Eaton, *J. Magn. Reson. A* **120**, 72 (1996).

The Status of the University of Michigan Polarized Proton Target*

R.S. Raymond[1], D.G. Crabb[2], V.V. Fimushkin[3], A.D. Krisch[1], A.M.T. Lin[1], V.G. Luppov[1], C.C. Peters[1], A.I. Mysnik[4], A.F. Prukoglyad[4], and K. Yonehara[1]

[1] *Randall Lab of Physics, University of Michigan, Ann Arbor MI 48109-1120, U.S.A.*
[2] *Department of Physics, University of Virginia, Charlottesville VA 22901, U.S.A.*
[3] *JINR, Dubna, Russia*
[4] *IHEP, Protvino, Russia*

Abstract The University of Michigan Solid Polarized Proton Target (PPT) was built in the late 1980's; it uses irradiated NH_3 at 5 T and 1 K. It was used in a 1990 experiment in a 24 GeV intense proton beam at the Brookhaven AGS, where its average polarization was about 85%. It was recently upgraded for the 70 GeV SPIN@U-70 experiment at IHEP-Protvino in Russia. Improvements were made to its superconducting magnet, its refrigerator, and its NMR and microwave systems.

The University of Michigan Solid Polarized Proton Target (PPT) [1], shown in Fig.1, is a 5 T and 1 K target using frozen ammonia (NH_3) as the target material. It was built in the late 1980's and was very successfully used in a p-p elastic scattering experiment [2] at the Brookhaven AGS in 1990. The target thickness is about 2 x 10^{23} polarized protons/cm^2. The extracted AGS beam intensity of 10^{11} protons/s on target was limited by quenches in the PPT magnet; hence, the polarized proton luminosity was about 2 x $10^{34}/cm^2s$, with frozen, pre-irradiated ammonia as the target material. The PPT's proton polarization averaged about 85% during the 3-month experiment. Many subsystems were upgraded for use in the SPIN@U-70 p-p elastic scattering experiment [3] at the 70 GeV U-70 accelerator at IHEP-Protvino in Russia.

The PPT magnet consists of a set of superconducting coils which produce a vertical 5 T field with a uniformity of better than 10^{-4} inside the 32 mm-long material holder. The magnet's original main power supply failed; thus we purchased a new American Magnetics uni-polar power supply with an energy absorber; this system was successfully tested and then used with the PPT magnet.

*Supported by a U.S. Department of Energy Research Grant.

CP675, *Spin 2002: 15th Int'l. Spin Physics Symposium and Workshop on Polarized Electron Sources and Polarimeters*, edited by Y. I. Makdisi, A. U. Luccio, and W. W. MacKay

The 213 MHz NMR system, used to measure the proton polarization, was upgraded with a new PC using LabView with National Instruments boards. A voltage ramp to an FM signal generator provides a frequency ramp into two Liverpool NMR Boxes. Signals from the Boxes are processed by a home-built manual amplifier/offset module and then fed back to the PC for analysis. Two new 8-mm-diameter NMR coils are both perpendicular to the beam axis, but are centered at different radii to monitor polarization non-uniformity caused by the high intensity beam. These coils are small and equal in size to avoid some problems encountered in earlier NMR polarization measurements at 213 MHz.

The PPT target is cooled by a ^{4}He evaporation refrigerator. A new calibrated RuO resistor from Scientific Instruments was installed and tested; it agrees with temperature measurements from ^{4}He vapor-pressure to within about 1% at 1.5 K. A new holder for the ammonia target material was made of Kel-F, with aluminum beam windows. The holder design allows fast replacement, with no soldering or unsoldering of the NMR coils. The 70 GeV extracted proton beam at the U-70 accelerator may be initially rastered only vertically; thus, a rectangular material holder, with inside dimensions of 5 mm wide by 20 mm high by 32 mm long, was also fabricated and successfully tested.

For the earlier AGS experiment, the helium pumping system consisted of three Roots blowers in series (6000 m^3/hr followed by 3000 m^3/hr followed by 350 m^3/hr), backed by a 60 m^3/hr mechanical pump. For SPIN@U-70, we would use two 5400 Roots blowers in parallel, backed by three 227 m^3/hr mechanical pumps in parallel. These blowers and mechanical pumps were purchased and successfully tested. A pump system stand was fabricated, and remotely-controlled valves were purchased.

A system of rails and linear bearings was designed to allow easy movement of the PPT, either to remove it from the beam path or adjust its position slightly, for the different $P_\perp^2$ settings of the experiment's 35-meter-long Recoil Proton Spectrometer. We plan to use a 300 l liquid helium buffer dewar inside the radiation shielding enclosure. We recently purchased a 630 l helium dewar to bring liquid helium from the IHEP liquifier; the helium would be transferred from the 630 l dewar to the buffer dewar using a long transfer line through the 2-meter-thick shielding roof.

To avoid air transportation difficulties, we may irradiate the frozen ammonia at an electron accelerator at Moscow State University. The 140 GHz microwaves would be produced with the Varian EIO tube used at the AGS, which has an output of about 22 W. A more modern Varian power supply for the tube was obtained to replace the 35-year old supply used in the AGS experiment.

The schedule for SPIN@U-70 is now uncertain, because after a brief test run, SPIN@U-70 as delayed due to Russian customs difficulties.

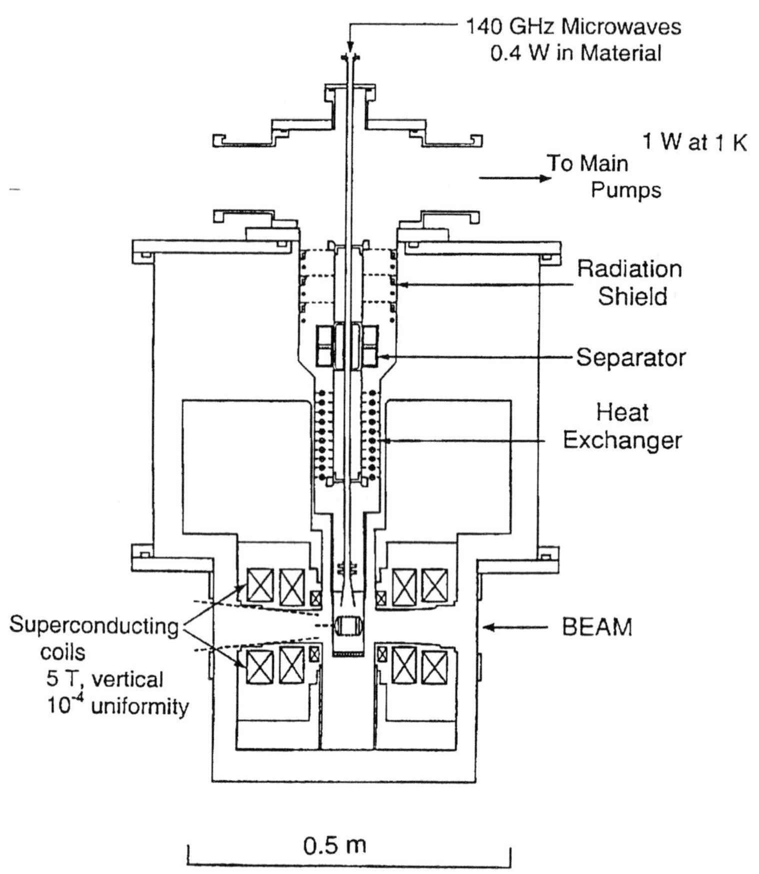

Fig 1. The Michigan Solid Polarized Proton Target

REFERENCES

1. D.G. Crabb, C.B. Higley, A.D. Krisch, R.S. Raymond, T. Roser, J.A. Stewart, G.R.Court, Observation of a 96% Proton Polarization in Irradiated Ammonia, Phys. Rev. Lett. **64**, 2627 (1990).
2. D.G. Crabb *et al.*, High-Precision Measurement of the Analyzing Power in Large-$P_\perp^2$ Spin-Polarized 24 GeV/c Proton-proton Elastic Scattering, Phys. Rev. Lett. **65**, 3241 (1990).
3. For SPIN@U-70 Proposal and details, refer to the Michigan Spin Physics Center website: *http://spinbud.physics.lsa.umich.edu.*

Radiation Damage Effects in Polarized Deuterated Ammonia

Paul M. McKee†

University of Virginia
for the E93-026 Collaboration

Abstract. Solid polarized targets utilizing deuterated ammonia, $^{15}ND_3$, offer an attractive combination of high polarization, high dilution factor and high resistance to polarization losses from radiation damage. Jefferson Laboratory Experiment E93-026 used $^{15}ND_3$ as a target material in a five-month form factor measurement, allowing a detailed study of it's performance. The dependence of the deuteron polarization on received dose by the ammonia and the effectiveness of annealing the material to recover performance lost to radiation damage will be discussed.

INTRODUCTION

A large number of high energy physics experiments of interest today require control of the spin degrees of freedom of the target and/or the beam. In such experiments the accuracy of the measurement (or, equivalently, the running time of the experiment) depends strongly on the degree of polarization attained, thus it is important to be able to characterize the polarization performance of the target used.

This article discusses the performance of a solid polarized target using deuterated ammonia, $^{15}ND_3$. This target was operated by the University of Virginia Polarized Target Group for experiment E93-026, a measurement of the electric form factor of the neutron, G_E^n, which ran in late 2001 in Hall C of Jefferson Laboratory, Newport News, Virginia. The focus will be on patterns observed in the polarization performance of the material as a function of deposited beam charge, and the techniques used to maximize this performance.

POLARIZED TARGET OVERVIEW

The target uses the technique of Dynamic Nuclear Polarization (DNP) to enhance the polarization of the spin species in the material. In DNP, the material is first doped to create paramagnetic centers which couple to the nuclei to be polarized. In the case of ammonia, this doping is accomplished by placing it in a beam of ionizing radiation.

The material is then placed in a high, uniform magnetic field, cooled to 1 K or less, and irradiated with microwave energy designed to drive transitions in the coupled-spin system. Using this technique, a system running at 5 T and 1 K can produce deuteron polarizations above 40%. The polarization typically takes a portion of an hour to reach its

CP675, *Spin 2002: 15th Int'l. Spin Physics Symposium and Workshop on Polarized Electron Sources and Polarimeters,* edited by Y. I. Makdisi, A. U. Luccio, and W. W. MacKay

maximum value, at which point the beam may be put on the target and the experimental measurement begun.

The beam has both a prompt and a cumulative effect on the target polarization. First, since the beam adds a small amount of heat to the material, it reduces the efficiency of the polarizing process, resulting in a polarization loss of 2 or 3% (absolute) over a period of about a minute. Over time, the beam also damages the material, producing additional paramagnetic centers that can allow some of the spins to relax, reducing the overall polarization of the target. In deuterated ammonia it is thought that the principle source of this relaxation is the production of atomic deuterium in the material. A side effect is that the microwave frequency for optimum polarization slowly changes with dose and must be tracked by the target operator. As the radiation damage of the material continues, the target polarization continues to decrease until a point at which it is no longer practical to run the experiment.

The performance of the damaged material may be largely restored by annealing the target. This involves shutting off the beam and the microwaves and warming the material for a length of time. The optimal temperature and duration depend on the history of the material, but range from 70 to 115 K and 10 to 60 minutes. After the anneal, the procedure is similar to that for new material: the target is cooled, the microwaves restored (at their original frequency) and the polarization builds to a maximum value, at which point beam may be reintroduced.

MATERIAL PREPARATION

As mentioned in the previous section, the deuterated ammonia is prepared by doping it with ionizing radiation. This may be done with the material in a special preparation dewar at 87 K ("warm irradiation") or in the actual polarized target at 1 K ("cold irradiation"). Irradiation produces the paramagnetic centers needed for the DNP process, and also changes the color of the icy material from a milky white to a deep purple.

Warm-irradiated material may undergo an additional step, known as tempering, in which the material is held above a liquid nitrogen bath until the purple color fades to a dirty white color. For experiment E93-026, two batches of cold-irradiated material and four batches of warm-irradiated, tempered material were used. A seventh batch consisted of a mixture of cold-irradiated and warm/tempered material.

PERFORMANCE

Experiment E93-026 was unique in that it put more beam on a deuterated ammonia target than any previous experiment. In all, 350 mC was deposited on the various batches of deuterated ammonia. This allowed the observation of long-term behavior difficult to isolate in shorter experiments, and further exploration of previously known effects.

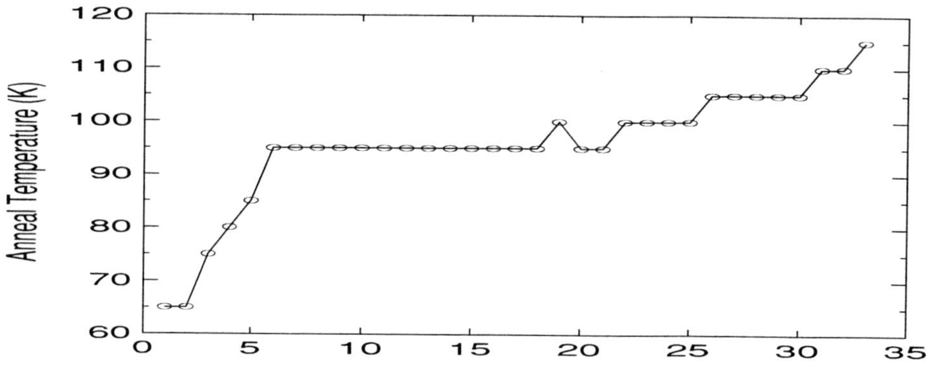

FIGURE 1. Temperatures used in anneal cycles on one load of material.

Anneal Temperature

The degree to which an anneal restores the performance of the material depends on both the anneal temperature and the length of time spent at that temperature. Many time/temperature combinations were tried during E93-026 and the degree of restoration noted for each.

The basic pattern that emerged is that for the first few anneal cycles, relatively cold and short anneals (70 to 80 K and 10 to 20 minutes) produce satisfactory restoration. After that, for perhaps the next 15 cycles, the best performance comes with 95 K/60 minute anneals. Eventually, increases in temperature up to as high as 115 K become necessary, although there is no evidence that anneals longer than 60 minutes are needed. Figure 1 shows the history of temperatures used for a warm-irradiated, tempered batch of material.

In judging whether a given anneal is successful it is important not to focus solely on the maximum polarization achieved after the anneal, because often deuterated ammonia will polarize higher after an anneal than in the previous cycle even if the anneal conditions were not optimal. This effect is most pronounced during the first several cycles of the material's life.

A more reliable method is to observe the microwave frequency that produces the maximum polarization. As mentioned above, the optimum frequency slowly drifts as radiation damage in the material accumulates (to lower frequencies for positive enhancement and to higher frequencies for negative enhancement). If the conditions of the anneal were well suited to the history of the material, the frequency will usually return to the value it had at the beginning of the previous anneal cycle.

Rate of Polarization Decay

In $^{15}NH_3$, a pattern emerges in which the rate of polarization decay increases with each anneal cycle. Although the maximum polarization may not decline much, the

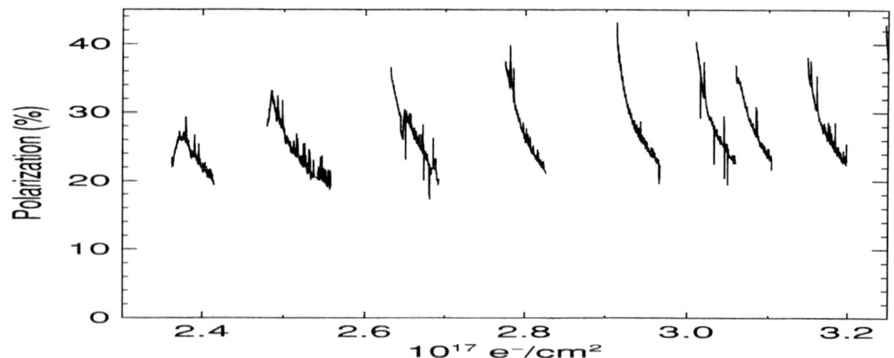

FIGURE 2. Evidence of slight increase in decay rate with dose.

decrease in running efficiency caused by the faster decays and more frequent anneals eventually requires the material to be replaced with a fresh sample.

In most experiments, this effect is not seen with deuterated ammonia. In E93-026 there is evidence of a somewhat faster decay rate, though nothing near the rate increase in NH_3, which can be more than a factor of two. An example of this effect in deuterated ammonia is shown in Figure 2.

Irradiation Temperature

The more common method of irradiating ammonia is to use a separate irradiation dewar, often at a different lab than the one at which the experiment will take place, filled with liquid Argon to produce an 87 K bath. We call this warm irradiation.

Cold-irradiation is done in the polarized target apparatus itself, with the refrigerator running at 1 to 1.5 K. It is often more convenient because beam time that does not require a polarized target, such as commissioning or calibration periods, can be used to prepare target material.

There is a noticeable difference in performance between warm- and cold-irradiated material. Warm-irradiated material polarizes more quickly, polarizes higher, and experiences more gradual radiation damage than cold-irradiated material. Figure 3 shows the performance of both types of material.

Material Tempering

The process of tempering the material before use in the target shows promise. The technique was first attempted during a test run of the target with old, poorly performing material that had been used in a past experiment. After tempering, it polarized to 50%.

For E93-026, all four warm-irradiated batches of material were tempered before use. Since the performance of tempered, cold-irradiated material is unknown, it would not

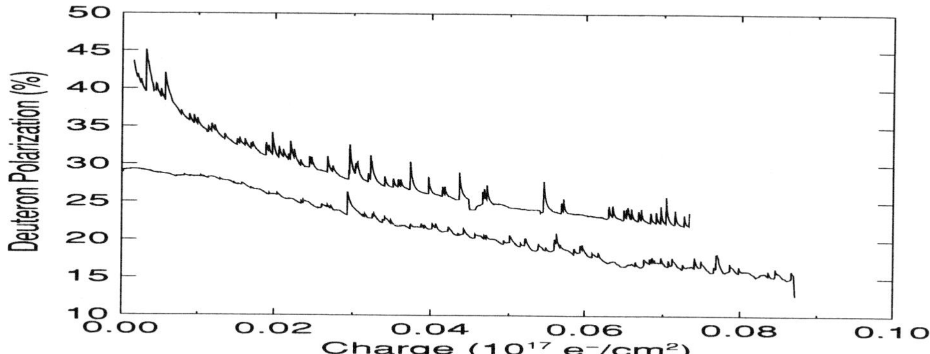

FIGURE 3. Characteristic performance of cold-irradiated (lower plot) and warm-irradiated (upper plot) material.

have been appropriate to use it for the first time during an experiment. Future tests will investigate the performance of this type of material.

CONCLUSION

Experiment E93-026 provided an opportunity to study the behavior of deuterated ammonia under accumulated doses larger than in any previous experiment. In analyzing the over 370,000 polarization measurements, several interesting aspects of the material's performance emerged.

First, the optimal annealing conditions appear to be 95 K for 60 minutes for all but the first few and last several cycles of a material's life. Second, it appears that the polarization decay rate does increase after several anneal cycles, but only slightly. Third, the performance of material irradiated at 87 K is better in several ways than material irradiated at 1 K. Finally, tempering a load of material by warming it slowly until the purple color disappears results in higher polarizations and faster polarization buildups.

Taken together, these new observations indicate that although deuterated ammonia has a long history of success as a polarized target material, further study provides information useful for improving its performance in future experiments.

ACKNOWLEDGMENTS

The author would like to thank D. G. Crabb for many helpful discussions. This work was supported by Department of Energy contract DE-FG05-86ER40261, and by the Institute of Nuclear and Particle Physics of the University of Virginia.

The Polarized Internal Gas Target of ANKE at COSY[1]

F. Rathmann*, R. Brüggemann†, R. Engels*, S. Geisler*, A. Gussen**, P. Jansen**, H. Kleines‡, F. Klehr**, P. Kravtsov§, S. Lemaître†, B. Lorentz*, S. Lorenz¶, M. Mikirtytchiants*§, M. Nekipelov*§, V. Nelyubin§, H. Paetz gen. Schieck†, U. Rindfleisch*, J. Sarkadi*, H. Seyfarth*, E. Steffens¶, H. Ströher*, V. Trofimov§, A. Vassiliev§ and K. Zwoll‡

*Institut für Kernphysik, Forschungszentrum Jülich, Germany
†Institut für Kernphysik, Universität zu Köln, Germany
**Zentralabteilung Technologie, Forschungszentrum Jülich, Germany
‡Zentrallabor für Elektronik, Forschungszentrum Jülich, Germany
§Petersburg Nuclear Physics Institute, Gatchina, Russia
¶Physikalisches Institut II, Universität Erlangen-Nürnberg, Germany

Abstract. For future few–nucleon interaction studies with polarized beams and targets at COSY–Jülich, a polarized internal–storage cell gas target is currently being developed and will be implemented in the near future at ANKE. The polarized atomic beam source, which will feed the target, provides beam intensities of $7.4 \cdot 10^{16}$ atoms/s in two hyperfine states of hydrogen. The implementation of the target at the internal spectrometer ANKE constitutes a major technological enterprise. The differential pumping system at ANKE has already been installed, as well as a new large target chamber to accomodate storage cells in the future, a new set of small–aperture horizontal and vertical beam position monitors, and a system of target–near detectors. In order to determine the nuclear polarization of the target, a Lamb–Shift polarimeter is currently set up. Tests with prototype storage cells aiming at the identification of cell dimensions suitable for the ANKE target are underway.

1. INTRODUCTION

At present two new polarized internal gas targets (PIT's) are being developed. One, intended for the physics programme at the BLAST facility at Bates [1], will utilize a refurbished source, formerly used at the PIT of the AmPS of NIKHEF [2]. With the closing of the Cooler operation at IUCF in 2002, COSY at Jülich remains the only ring capable to store polarized protons and deuterons on a worldwide scale. The polarized atomic beam source (ABS) described in this paper is intended to feed a PIT at COSY. One of the first experiments that will be carried out with the target deals with the proton–induced deuteron breakup[2] at the ANKE spectrometer. A presently developed Lamb–shift polarimeter [5] will be employed to measure the polarization of atoms extracted

[1] This work has been supported by the BMBF (contracts RUS 649-96, RUS 99/686, 06 ER 831, 06 ER 930, 06 OK 862, and WTZ 99/686), by DFG (contract 436 RUS 113/430), by the Forschungszentrum Jülich (FFE contracts 41149451, 41445283 (COSY–59)), and by the Russian Ministry of Sciences.

[2] Described in a separate contribution to these proceedings [4].

CP675, *Spin 2002: 15th Int'l. Spin Physics Symposium and Workshop on Polarized Electron Sources and Polarimeters*, edited by Y. I. Makdisi, A. U. Luccio, and W. W. MacKay

from the storage cell. The storage cell target at ANKE will be operated initially in a vertical guide field, provided by the stray field of the spectrometer magnet, which varies in magnitude along the axis of the storage cell. At a later stage also orientations other than vertical will be made available.

2. THE POLARIZED ATOMIC BEAM SOURCE

Details about the source developement have been reported elsewhere, e.g. ref. [3], therefore the description of the setup given here is only brief.

The spatial conditions at the magnetic spectrometer ANKE [6] require vertical mounting of the source. The atomic beam source has to move together with the target chamber, when the central spectrometer dipole magnet is set to a different beam deflection angle. For that reason, the atomic beam source is designed around a central plate (label 6 in Fig. 1), which serves as the main support for mounting and reference for alignment of internal elements, as well as with respect to the external environment. Other external support is not required, i.e. no optical bench like in conventional sources oriented horizontally. The layout of the vacuum vessel of the atomic beam source is shown in Fig. 1. Two cylindrical chambers are attached above and below a massive, 400×500 mm^2 steel plate of 50 mm thickness. The inner diameter of the upper chamber is 390 mm, it houses the first three stages of the differential pumping system, separated from each other by two baffles. Mounted on rods attached to the central plate are the first three magnets of the sextupole system [7], and the medium–field rf transition unit (MFT). The lower chamber has an inner diameter of 200 mm. It makes up stage IV of the differential pumping system, houses the second set of magnets and in a separate appendix chamber, provides space for the two transition units behind the magnet system. In front of the last two magnets, a beam chopper is installed (label 9 in Fig. 1), which consists of a cylindrical Al body with rectangular cutouts on opposite sides that rotates about an axis perpendicular to the beam. The lateral extension of the source, mostly defined by the large turbomolecular pumps, was minimized. Therefore, shutters on the cryopumps, commonly used in other sources, were omitted in the design. The horizontally oriented turbomolecular pumps and the cryopumps near the beam pipe have to be operated in the stray field of the central spectrometer magnet of a few hundred Gauss.

3. SOURCE PERFORMANCE

The source performance has been optimized by means of a calibrated compression tube device [8]. The pressure–to–flow dependence has been calibrated prior to the measurements and afterwards. The entrance tube of the compression tube (inner diameter of 10mm, length of 100 mm) correponds to that of the future cells at ANKE. The distance between the exit of the last magnet to the entrance of the compression tube amounts to 300 mm. The dependence of the beam intensity on the primary hydrogen flow into the dissociator is depicted in Fig. 2. The highest hydrogen beam intensity of $(7.4 \pm 0.3) \cdot 10^{16}$ atoms/s in two hyperfine states is found at a flow of 1.2 mbar·ℓ/s, a nozzle temperature of $T = 62$ K, and an admixture of O_2 of $1 \cdot 10^{-3}$mbar ·ℓ/s.

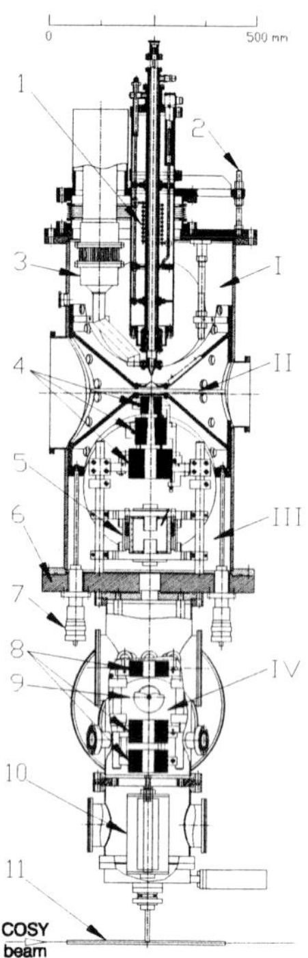

FIGURE 1. Cut through the mid plane of the atomic beam source. 1: dissociator, 2: adjustment screws to move the nozzle transversely and longitudinally, 3: coldhead with heat–bridge for cooling of the nozzle, 4: first set of sextupole magnets, 5: medium–field rf transition unit (MFT), 6: central 50 mm thick stainless steel support plate, 7: rotational feedthroughs for adjustment of lower baffle, 8: second set of sextupole magnets, 9: rotating beam chopper, 10: weak– and strong–field rf transition units (WFT and SFT), and 11: storage cell located on the COSY beam axis. Roman numbers, I–IV, denote the four stages of the differential pumping system.

The system of hyperfine transition units to provide polarized beams of hydrogen and deuterium atoms in the states listed in Table 1 has been completely assembled and successfully tested. The tests could be carried out efficiently with the Lamb–shift polarimeter. Results of these tests are described in more detail in ref. [5].

4. THE INTERNAL TARGET FOR COSY

The implementation of a PIT at a storage ring requires a powerful differential pumping system, in particular for a storage–cell target. In case of a polarized jet, a beam dump can be used. At the ANKE spectrometer space is severely limited, thus the design of such a system turns out to be quite complicated. With the implementation of the new target chamber at the end of the year 2002 at ANKE, a major fraction of the preparations for the installation of the polarized source have been completed (Fig. 3). The design of the support structure for the ABS between magnets D1 and D2 is underway. The

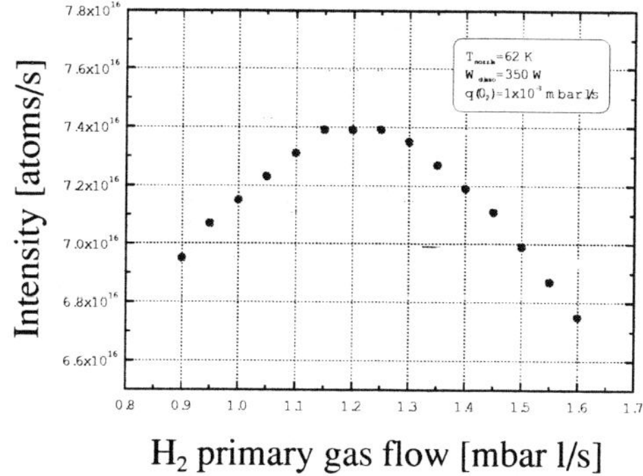

FIGURE 2. Atomic beam intensity as function of the primary hydrogen flow into the discharge tube. The parameters listed in the insert correspond to those for which the maximum intensity was achieved.

TABLE 1. System of hyperfine transitions employed to provide nuclear vector polarization for hydrogen (H) and nuclear vector and tensor polarization for deuterium atoms (D). The sextupole magnets are located before and behind the MFT.

Pol	H	H	D	D	D	D
P_z P_{zz}	$+1$ —	-1 —	$+1$ $+1$	-1 $+1$	0 $+1$	0 -2
Hyperfine Transition 1	MFT $(2 \leftrightarrow 3)$	MFT $(2 \leftrightarrow 3)$	MFT $(3 \leftrightarrow 4)$	MFT $(3 \leftrightarrow 4)$	MFT $(1 \leftrightarrow 4)$	MFT $(1 \leftrightarrow 4)$
Hyperfine Transition 2	—	WFT $(1 \leftrightarrow 3)$	—	WFT $\binom{1 \leftrightarrow 4}{2 \leftrightarrow 3}$	—	—
Hyperfine Transition 3	—	—	SFT $(2 \leftrightarrow 6)$	—	SFT $(2 \leftrightarrow 6)$	SFT $(3 \leftrightarrow 5)$
into cell	$H(\lvert 1\rangle)$	$H(\lvert 3\rangle)$	$D\binom{\lvert 1\rangle}{\lvert 6\rangle}$	$D\binom{\lvert 3\rangle}{\lvert 4\rangle}$	$D\binom{\lvert 3\rangle}{\lvert 6\rangle}$	$D\binom{\lvert 2\rangle}{\lvert 5\rangle}$

new target chamber also accomodates movable horizontal and vertical beam position monitors (BPM). Together with a set of similar BPM's in the section between the two magnets D2 and D3 (Fig. 3), these monitors will facilitate a determination of the beam position during acceleration of the beam. In addition, it should be possible to determine the beam angle at the target location from the measured positions in front and behind the target.

Among other aspects, the new chamber provides sufficient space to install the storage cells of the PIT. As a next step we will carry out tests at ANKE to identify dimensions suitable for those cells. The setup for these tests inside the new target chamber is depicted in Fig. 4. Details regarding these tests can be found in ref. [9].

FIGURE 3. 3D view of the setup at ANKE with the PIT. The ABS is located between the dipole magnet D1 and the central spectrometer magnet D2. The COSY beam enters from the left.

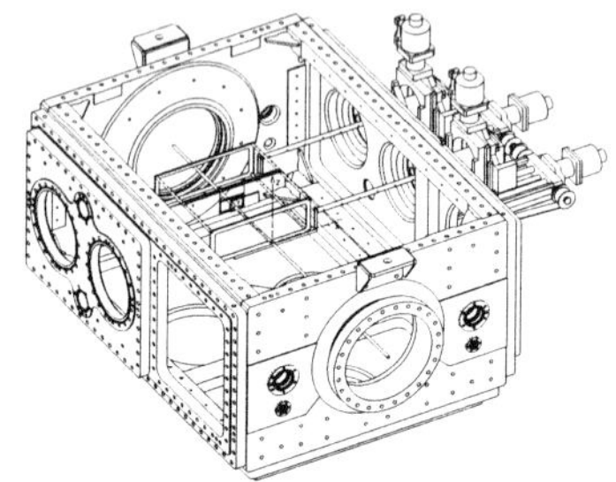

FIGURE 4. Target chamber with two *xy* manipulators to move a frame carrying the cells perpendicular to the COSY beam. The beam enters from the lower right. The movable BPM system is not shown. The dimensions of the chamber are $\ell = 800$ mm, $w = 600$ mm, and $h = 400$ mm.

REFERENCES

1. H. Kolster *et al.*, Proc. 9th Int. Workshop on Polarized Sources and Targets (PST01), Nashville, Indiana, USA, 2001. V. P. Derenchuk and B. von Przewoski (Eds.), World Scientific, p. 37 (2002).
2. L.D. van Buuren *et al.*, Nucl. Instr. Meth. **A 474**, 209 (2001).
3. M. Mikirtytchiants *et al.*, p. 47 of ref [1].
4. F. Rathmann *et al.*, *The Polarized Deuteron Break-up Experiment at COSY*, contribution to these proceedings.
5. R. Engels *et al.*, *A precision Lamb–shift polarimeter for the polarized gas target at ANKE*, contribution to these proceedings.
6. S. Barsov *et al.*, Nucl. Instr. Meth. **A 462**, 364 (2001).
7. A. Vassiliev *et al.*, Rev. Sci. Instrum. **71**, 3331 (2000).
8. M. Nekipelov, *Device for absolute beam intensity measurements at the ANKE atomic beam source*, Diploma thesis Saint–Petersburg State Technical University, 1999. (Available upon request.)
9. F. Rathmann, *Storage Cell Tests*, Proceedings of the 4th ANKE workshop on *Study of proton-deuteron Interactions*, Dubna, Russia 2002. A. Kacharava, V. Komarov, and F. Rathmann (Eds.), to be published as JÜL–Bericht **4012** (2003).

The HERMES Polarized Atomic Beam Source

A.Nass (for the HERMES target group)

University of Erlangen - Nürnberg, E. Rommel - Str. 1, 91058 Erlangen, Germany

Abstract. The atomic beam source (ABS) provides nuclear polarized hydrogen or deuterium atoms for the HERMES target at flow rates of about $6.5 \cdot 10^{16}\vec{\mathrm{H}}/s$ (hydrogen in two hyperfine substates) and $6.0 \cdot 10^{16}\vec{\mathrm{D}}/s$ (deuterium in three hyperfine substates). The degree of dissociation of 93% for H (95% for D) at the entrance of the storage cell and the nuclear polarization of around 0.97 (H) and 0.92 (D) have been found to be constant within a a couple of percent over the whole running period of the HERMES experiment. A new dissociator (MWD) based on a microwave discharge at 2.45 GHz has been developed and installed into the HERMES–ABS in 2000. Since the velocity distribution of the MWD differs from that of the RFD the intensity could be increased further with a modified sextupole magnet system. For this purpose the way for a new start generator for sextupole tracking calculations was opened. Monte-Carlo simulations were successfully used to describe the gas expansion between nozzle, skimmer and collimator. A new type of beam monitor was used to study the beam formation after the nozzle.

The HERMES experiment studies the spin structure of the nucleon by means of lepton deep inelastic scattering off an internal gaseous target. A polarized atomic beam is injected by an atomic beam source (ABS) into a storage cell [1]. A sample of the target gas is extracted from the centre of this storage cell in order to analyze the target polarization by means of a target gas analyzer (TGA) [2] and a Breit-Rabi polarimeter [3].

PRINCIPLE OF OPERATION AND SETUP

The setup of the HERMES ABS is shown in Figure 1. Molecular hydrogen or deuterium is dissociated via electron impact in a cold plasma provided by a radio frequency dissociator (1). The atomic gas expands through a cooled nozzle into the vacuum of chamber I supported by a powerful pumping system. A high brilliance beam is then formed using a skimmer (2) and a collimator. Based on the Stern-Gerlach principle, sextupole magnets (3) focus atoms with electron spin +1/2 (hyperfine states $|1\rangle$ and $|2\rangle$ for H, $|1\rangle$, $|2\rangle$ and $|3\rangle$ for D) and deflect atoms with electron spin -1/2 (states $|3\rangle$ and $|4\rangle$ for H, $|4\rangle$, $|5\rangle$ and $|6\rangle$ for D). Nuclear polarization is obtained by an interchange of the hyperfine state populations using high frequency transitions.

The plasma source of the radio frequency dissociator (RFD) [4] applied consists of a LC-circuit as a field applicator, tuned to a resonance frequency of 13.56 MHz, and a water-cooled pyrex discharge tube. A degree of dissociation of $\alpha = 80$ % (H) and 75 % (D) is achieved at throughputs of $Q = 0.9\ldots1.5$ mbarl/s and RF-power $P_{\mathrm{RF}} = 200\ldots350$ W . In 2000 a microwave dissociator (MWD) [5], based on a plasma source which

CP675, *Spin 2002: 15th Int'l. Spin Physics Symposium and Workshop on Polarized Electron Sources and Polarimeters,* edited by Y. I. Makdisi, A. U. Luccio, and W. W. MacKay

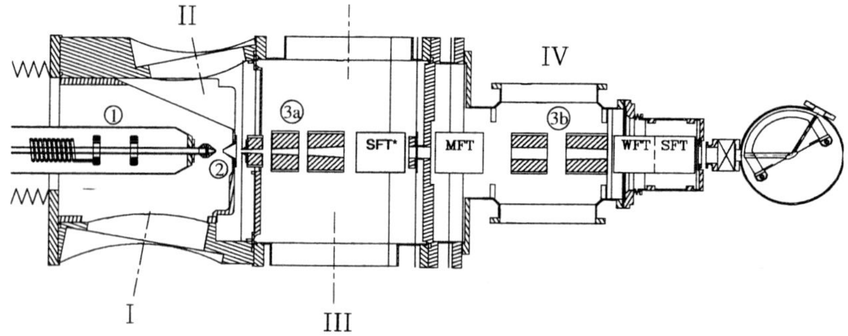

FIGURE 1. Schematic view on the HERMES ABS with the radio frequency dissociator (1) and the skimmer (2) for beam formation. Two sets of sextupole magnets (3a, 3b) are located along the beam axis in chamber III and IV as are the high frequency transitions (SFT*, MFT, WFT and SFT).

couples a 2.45 GHz surface wave to the discharge in an air-cooled pyrex glass tube, was installed. With typical throughputs of $Q = 1\ldots3$ mbarl/s and microwave power of about $P_{\mathrm{MW}} = 600$ W, α was in excess of 80 %. Further developments on a liquid cooled MWD are in progress in order to outperform the water-cooled RFD.

The sextupole magnets [6] are high gradient permanent magnets consisting of 24 segments made of Vacodym[1]. The maximum poletip field is 1.5 T. To prevent chemical destruction by hydrogen, the magnets are enclosed in vacuum tight stainless steel cans. In order to reduce the residual gas pressures due to the deflected atoms inside the magnets, the set of sextupole magnets (3a) in the first sextupole chamber (III) is split into 3 sections (figure 1). Each one is tapered to have the largest possible acceptance of the diverging atomic beam and to provide achromatic focussing. Two more magnets (3b) in the second sextupole chamber (IV) focus the atomic beam into the entrance tube of the target cell. The transmission probabilities of the sextupole system have been calculated with a sextupole tracking calculation.

Compact high frequency transitions are employed to nuclear polarize the atomic beam with high efficiency. They consist of coils for the static and gradient field, and a resonator cavity in the case of the strong field transition (SFT) or a high frequency coil in the case of the weak and medium field transition (WFT and MFT) [7, 8]. For hydrogen the SFT is tuned to a frequency of 1430 MHz, the WFT to 14 MHz and the MFT to 90 MHz. For deuterium the different hyperfine splitting energy requires a lower frequency of 370 MHz for the SFT, 7 MHz for the WFT and 25 MHz for the MFT.

The settings of the static and gradient fields are chosen using the BRP with its own high frequency transitions off. Due to the separation in the BRP sextupole system only the hyperfine states with electron spin +1/2 reach the quadrupole mass spectrometer (QMS). Figure 2 shows the dependence of the QMS signal on the current through the magnetic field coils of the respective transitions. The MFT can be operated exchanging

---

[1] Brand name of Vacuumschmelze GmbH, Postf. 2253, D63412 Hanau, Germany

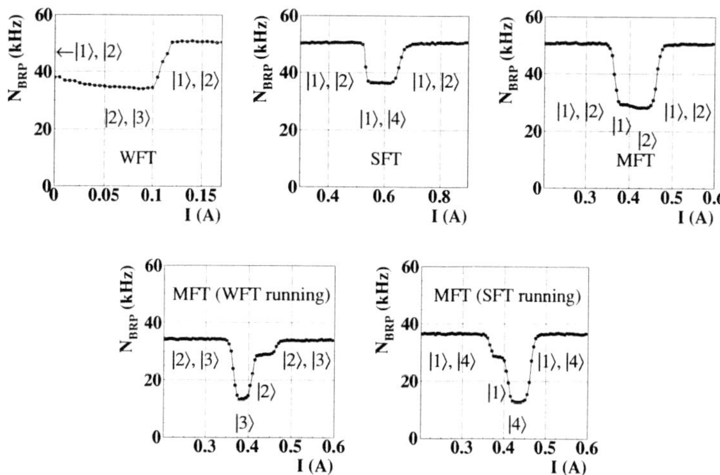

FIGURE 2. H_1 signal of the BRP-QMS as a function of the magnetic field B∼I. The injected states of hydrogen into the target cell are shown in every part of each graph.

states $|2\rangle$ and $|3\rangle$ or $|1\rangle$ and $|3\rangle$ at different static magnetic fields. The combination of the different high frequency transitions makes it is possible to inject every single state or even zero states. The latter is essential to determine the ballistic flux from the ABS. The efficiencies of the high frequency transitions could be obtained by spin relaxation measurements with the BRP ([9]).

OPTIMIZATION PROCEDURES

The output intensity can be optimized by changing the flux Q and the nozzle temperature T_{nozzle}. The maximum intensity values measured by a calibrated compression tube are shown in table 1 together with the parameters of the measurements for both dissociator types. Due to the higher degree of dissociation at the nozzle exit for higher throughputs a gain of 15 % in intensity of the ABS by using the MWD compared to the RFD has been detected. A QMS in front of the entrance of the compression tube has been used to

TABLE 1. The hydrogen and deuterium intensity of the ABS. P is the applied RF power, q_{CT} the intensity measured with the calibrated compression tube (not corrected for α) [4] and q_{BRP} the intensity determined with the BRP via spin exchange collisions [9].

	gas	Q (mbar·l/s)	P (W)	T_{nozzle} (K)	q_{CT} (atoms/s)	q_{BRP} (atoms/s)
RFD	H	1.5	290	115	$6.5 \cdot 10^{16}$	$6.6 \cdot 10^{16}$
	D	1.0	200	115	$5.2 \cdot 10^{16}$	$4.5 \cdot 10^{16}$
MWD	H	1.5	600	80	$6.2 \cdot 10^{16}$	–
	D	1.5	600	80	$6.0 \cdot 10^{16}$	$5.1 \cdot 10^{16}$

TABLE 2. Nuclear (P_z) and tensor (P_{zz}) polarization of the atomic beam injected into the storage cell [9, 10].

gas	injected HFS	P_z	P_{zz}
H	$\|1\rangle\ \|4\rangle$	$+0.973 \pm 0.010$	-
	$\|2\rangle\ \|3\rangle$	-0.974 ± 0.010	-
D	$\|1\rangle\ \|6\rangle$	$+0.924 \pm 0.010$	$+0.884 \pm 0.018$
	$\|3\rangle\ \|4\rangle$	-0.911 ± 0.015	$+0.941 \pm 0.022$
	$\|3\rangle\ \|6\rangle$	-0.015 ± 0.014	$+0.990 \pm 0.023$
	$\|2\rangle\ \|5\rangle$	-0.022 ± 0.013	-1.774 ± 0.020

measure the associated degree of dissociation α_{ABS}:

$$\alpha_{\mathrm{ABS}} = \frac{S_{\mathrm{a}}^{*}}{S_{\mathrm{a}}^{*} + 2\,\kappa_{\mathrm{ion}}\,\kappa_{\mathrm{det}}\,\kappa_{\mathrm{v}}\,S_{\mathrm{m}}} \tag{1}$$

where $S_{\mathrm{a}}^{*} = (S_{\mathrm{a}} - \delta^{\mathrm{di}} S_{\mathrm{m}})$ is the atomic signal corrected for dissociative ionization in the QMS, κ_{ion} the ratio of the ionization cross sections, κ_{det} the ratio of the detection probabilities and κ_{v} the ratio of the velocities of the atoms and molecules. At the working conditions degrees of dissociation of $92.8^{+3.3}_{-0.4}$ for hydrogen and $94.5^{+3.3}_{-0.6}$ for deuterium were determined using the MWD.

PERFORMANCE WITHIN THE HERMES EXPERIMENT

The HERMES ABS was continuously running from 1996 until 2002. The thickness of the HERMES target and therefore the ABS beam intensity has been determined via spin exchange collisions in the target cell using the BRP (table 1) confirming the expected MWD improvement of the deuterium intensity of the ABS by 15 %.

The nuclear polarization of the atoms depends on the efficiencies of the high frequency transitions and the transmission probabilities of the sextupole magnet system. The resulting polarization values of the injected atoms are listed in table 2.

DIRECT SIMULATION MONTE-CARLO OF THE EXPANSION

The understanding of the processes that occur in the expansion of the hydrogen gas into the vacuum and the formation of the atomic beam is essential for an improvement of the ABS. The use of continuous flow models is problematic because of their restricted validity in the transition region between laminar and molecular flow. Thus a direct simulation Monte-Carlo method was used [11] to describe the processes in the gas expansion. The simulated velocity and density distributions agree well with the values of the time-of-flight and beam profile measurements [12, 13]. As an example the measured and calculated resistances of a beam profile monitor [14] are shown in figure 3. The results of these Monte-Carlo simulations can be used to extract e.g. the input parameters for sextupole tracking calculation.

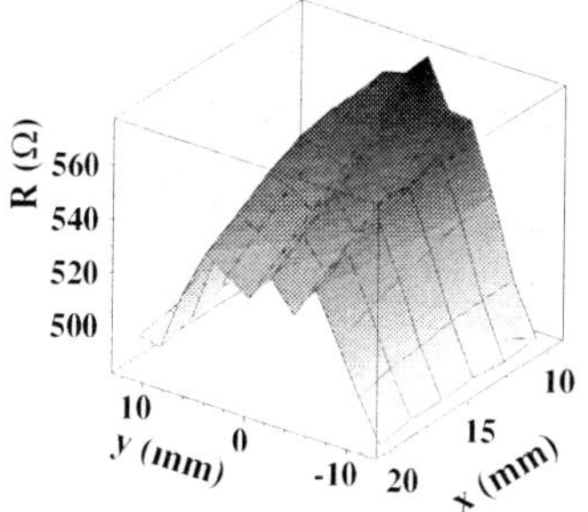

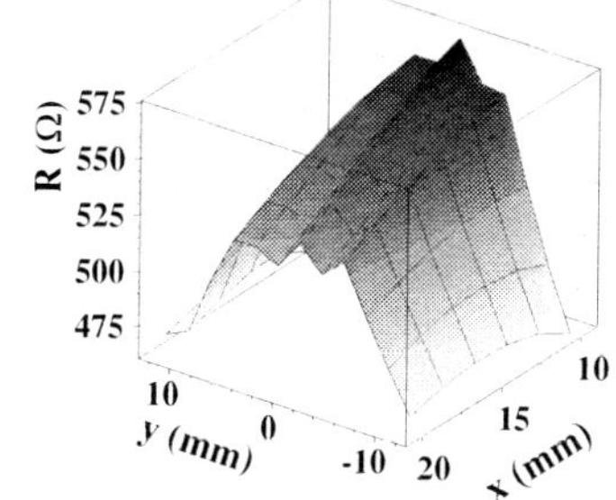

FIGURE 3. Measured (left) and calculated (right) resistances for an expansion of hydrogen (at $Q_{H_2} = 1\,\text{mbar}\cdot\text{l/s}, T_{\text{nozzle}} = 100\,\text{K}, \alpha = 80\%$). x – distance nozzle monitor and y – distance from the beam axis.

CONCLUSIONS

The ABS is found to be a very reliable source of polarized atoms for a storage cell target and other applications with hydrogen (deuterium) intensities up to $6.5 \cdot 10^{16}$ atoms/s ($6.0 \cdot 10^{16}$ atoms/s) in 2 (3) hyperfine substates. Nuclear polarization values of 0.97 (0.92) at a degree of dissociation of 93 % (95 %) for H (D) were reached. A smooth and stable operation within the HERMES experiment could be observed. The direct simulation Monte-Carlo is an excellent tool to describe the formation of atomic beams.

REFERENCES

1. Baumgarten, C. et al., *to appear in Nuclear Instruments and Methods A* (2003).
2. Henoch, M. et al., *to be submitted to Nuclear Instruments and Methods A* (2003).
3. Baumgarten, C. et al., *Nuclear Instruments and Methods A*, **482**, 606 (2002).
4. Stock, F., "The HERMES Target Source for Pol. H and D Atoms," in *Workshop on Pol. Beams and Pol. Gas Targets*, edited by H. Paetz and L. Sydow, World Scientific, Köln, Germany, 1996, p. 260.
5. Koch, N., and Steffens, E., *Review of Scientific Instruments*, **70**, 1631 (1999).
6. Schiemenz, P., Ross, A., and Graw, G., *Nuclear Instruments and Methods A*, **305**, 15 (1991).
7. Drewes, W., Jänsch, H., Koch, E., and Fick, D., *Physical Review Letters*, **50**, 1759 (1983).
8. Gaul, H. G., and Steffens, E., *Nuclear Instruments and Methods A*, **316**, 297 (1992).
9. Baumgarten, C., *Studies of Spin Relaxation and Recombination at the HERMES H/D Gas Target*, Ph.D. thesis, University of München (2000), also available as DESY-THESES-2000-038.
10. Henoch, M., *Absolute Calibration of a Polarized Deuterium Gas Target*, Ph.D. thesis, University of Erlangen-Nürnberg (2002), also available as DESY-THESES-2002-026.
11. Bird, G. A., *Molecular Gas Dynamics and the direct Simulation of Gas Flows*, Oxford, 1998.
12. Nass, A., *Low-Pressure Supersonic Gas Expansions*, Ph.D. thesis, University of Erlangen-Nürnberg (2002), also available as DESY-THESES-2002-012.
13. Nass, A. et al., "Studies on Beam Formation in the HERMES-ABS," in *Workshop on Pol. Sources and Targets*, edited by V. P. Derenchuk and B. von Przewoski, World Scientific, Nashville, Indiana, USA, 2001, p. 42.
14. Vassiliev, A. et al., "Investigation of the Atomic Hydrogen Beam with a Two-dimensional Multiwire Monitor," in *Workshop on Pol. Sources and Targets*, edited by A. Gute, S. Lorenz, and E. Steffens, FAU Erlangen-Nürnberg, Erlangen, Germany, 1999, p. 200.

Design of a Polarized Atomic H Source for a Jet Target at RHIC

T. Wise*, M. A. Chapman*, W. Haeberli*, H. Kolster†, P. A. Quin*

* *Department of Physics, University of Wisconsin, Madison. WI 53706, USA*
† *Laboratory for Nuclear Science, MIT,Cambridbe MA 02139, USA*

Abstract. As part of a project to calibrate the polarization of the RHIC proton beams we designed a polarized atomic hydrogen beam source for use as a jet target at RHIC. The model we developed agrees well with measured outputs from two working sources and for the system we designed predicts $9x10^{16}$ atoms/s into a 9 mm diameter aperture 285 mm from the last focussing magnet and a target thickness of $9x10^{11}$atoms/cm^2, 316 mm from the last magnet.

MODEL DESCRIPTION

We report on the design of a polarized hydrogen jet target to be inserted into RHIC at the 12:00 o'clock location. The jet will be used to calibrate polarimeters located elsewhere in the ring. Our model predicts a target thickness of $9x10^{11}$ atoms/cm^2 or a factor 2.5 larger than the recently designed EDDA target at COSY.

In the early stages of this study we compared outputs from different ray tracing codes. The comparisons revealed programming errors in the two codes we intended to use. The corrected versions more accurately predict the output of existing sources. It is interesting to note that both codes had previously been used as the basis for magnet purchases. To optimize the atomic beam density we added a gradient search routine to one of the codes and a random parameter generator to the other.

The intensity of an atomic beam source delivering a *single* hyperfine state of H (or D) into an aperture can be expressed as the product of terms in Eq. 1.

$$I(Q,T)_{\text{atoms/s}} = \frac{1}{4} Q\left(\frac{N_A}{22.4\cdot 1013}\right)\left(\frac{\Omega}{2\pi}\right)\cdot 2\cdot 2\cdot 1.15\cdot A(Q,T)\cdot \alpha(Q,T)\cdot t(Q,T,G) \tag{1}$$

Q is the flow of H_2 gas into the dissociator in mbar liter/s and is followed by the required conversion into atoms/s using Avagadro's number. Ω is the solid angle subtended by the aperture of the first six-pole element as seen from the nozzle of the dissociator and A(Q,T) represents attenuation of the beam due to scattering. In our model A(Q,T) was experimentally determined [1] by measurements with H_2 gas at various nozzle temperatures and gas flows. To allow for the significant variation in the degree of dissociation of H_2 gas as a function of gas flow and nozzle temperature, we

CP675, *Spin 2002: 15th Int'l. Spin Physics Symposium and Workshop on Polarized Electron Sources and Polarimeters*, edited by Y. I. Makdisi, A. U. Luccio, and W. W. MacKay

include the term α(Q,T). We factored α into flow and temperature dependent terms: α(Q,T) = α(Q)*α(T) with the Q and T dependence derived from experimental data in [1] and [2] respectively.

The last term, t(Q,T,G) represents the fractional transmission of the focussing N-Fe-B permanent magnet 6-poles [3,4] as calculated by the ray tracing code. The value of *t* depends strongly on the geometry, G, of the 6-pole magnets. The Q and T dependence of *t* results from the observation that the velocity distribution of atoms at the entrance to the focussing magnets depends on those variables. We relied on measurements of the atomic velocity distribution leaving a dissociator of similar design reported in [5]. In that reference the velocity distribution *f(v)* is expressed as

$$f(v) = v^2 e^{\left(\frac{-m}{2k_b T_{beam}}(v - v_{drift})^2\right)} \tag{2}$$

with v_{drift} and T_{beam} measured functions of the flow Q and nozzle temperature T.

The factor 1/4 in Eq.1 arises from the fact that we calculate trajectories in one of four hyperfine states. One of the factors 2 arises from the dissociation of H_2 into 2 atoms. The second factor 2 arises from the assumption of a simple cos(θ) angular distribution from the nozzle which has an on-axis intensity twice that of a flat distribution. Finally, we add the factor 1.15 based on the observation [1,2] that the angular distribution is slightly peaked on axis compared to a simple cos(θ) distribution.

To evaluate *t* for a particular magnet geometry we typically calculate $2\text{x}10^5$ individual atom trajectories. The magnitude of the atom velocities is randomly chosen with weighting factor *f(v)* from Eq. 2 and the initial direction is established by assuming a straight line trajectory between random positions on the 2 mm diameter dissociator nozzle and on a disc 50 mm downstream. The straight-line trajectory is extended an additional 10 mm to the first magnet element.

To compare the relative merit of one magnet geometry over another one must consider how the atomic beam will be used. Eq. 1 is the correct function to maximize for a beam entering a storage cell. In our application we need to instead optimize density of the jet along a line crossing perpendicular to the jet axis. In that case we weight each atom which passes through the target region by the factor 1/rv as indicated in Eq. 3. The 1/v weight accounts for the time atoms spend in the interaction region. The factor 1/r is related to the conversion from areal density of the beam to linear density along the RHIC beam whose cross-section is small compared to the diameter of the atomic beam.

$$\text{THICKNESS} \propto I(Q,T) * \sum_{\text{atoms}} 1/\text{rv} \tag{3}$$

In practice optimization with and without the 1/rv weighting gives nearly the same result.

MAGNET OPTIMIZATION

The process of optimization can be thought of as a search in n dimensional parameter space with the parameters being, for example, the length, diameter, taper, and spacing of the various magnet elements as well as the nozzle temperature and gas flow. To treat the problem adequately we estimate 17 parameters are needed but this results in a rather enormous computational problem. For example if only 5 values of each parameter are selected one has $7x10^{11}$ systems to compare. The problem is aggravated by the presence of numerous local maxima that render a simple gradient search routine ineffective. To begin, we reduced the problem to a more manageable 10 parameters. Systems were generated by randomly selecting parameters within a range we deemed acceptable. For example nozzle temperatures were selected between 30-200 K. Slopes of the magnet bores were constrained to be flat to diverging for the first group of magnet elements and flat to converging for the last.

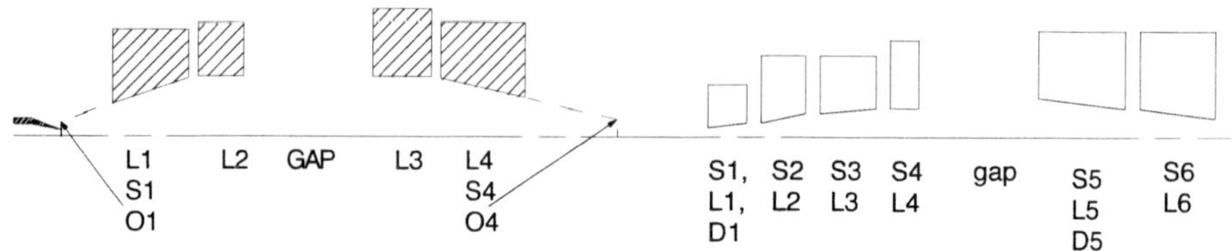

Figure 1. Simplified and complete parameterizations. The parameters S, L ,and D represent the slope, length, and diameter of magnet elements. In the left parameterization O is the offset after projecting the inner bore of sloped magnet elements to the nozzle and to the target aperture.

Computations continued until at least 10,000 separate magnet systems were calculated. We noted regions of parameter space unoccupied by the top 1% systems, constrained the range of some parameters accordingly, and calculated another 10,000 systems. The process was repeated until it became possible to apply the more general 17 parameter model. At this stage this method had already generated magnet geometries which were predicted to yield a maximum intensity of $1.0x10^{17}$ atoms/s into a 10 mm diameter target. Unfortunately these systems were massive and impractical to construct. The problem of massive magnets was resolved by adding an additional filter into the parameter selection process; only systems whose magnet volume was below a specified threshold were calculated. The best system out of 50,000 calculated this way was fine-tuned by applying the gradient search portion of the code. It is interesting to note that only an additional 1% output was found by the gradient search routine. This system is predicted to generate $9x10^{16}$ atoms/s into a 9 mm diameter aperture 285 mm from the last magnet element. It has a total volume of approximately 1230 cm^3 which is nearly 1/2 the volume of the more massive systems we rejected. The geometry of this system is shown in Fig. 2 followed by predictions of the beam profile at the interaction region and velocity distributions at three locations: nozzle, the jet interaction region, and at the location of the Breit-Rabi polarimeter detector (not shown).

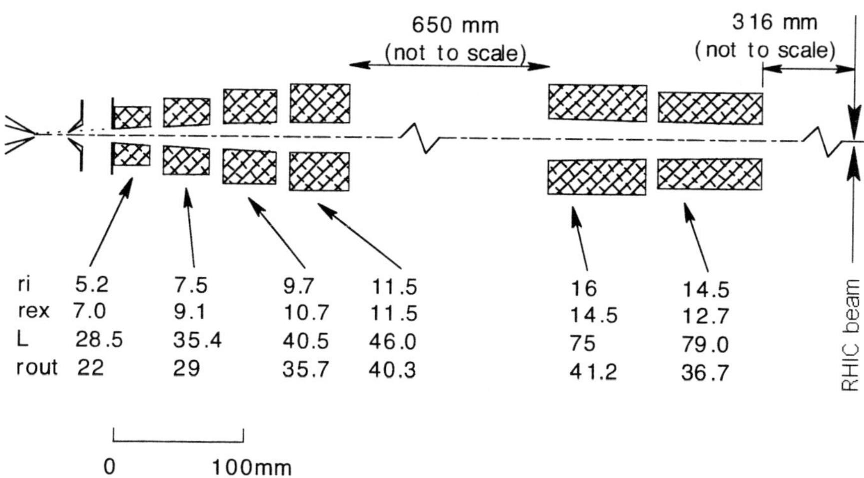

Figure 2. Final magnet geometry for the RHIC hydrogen jet polarimeter target. The entrance and exit radii, overall length, and outer radius of each magnet element are indicated. The BRP magnet elements to the right of the interaction region are not shown. All dimensions are in mm.

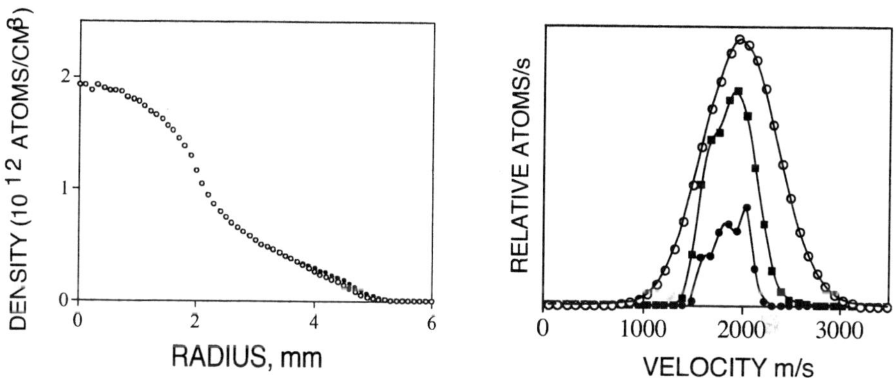

Figure 3. Left: calculated beam density profile at the RHIC-J ET interaction region (316 mm from the last focussing magnet) with and without a 9 mm diameter aperture 285 mm from the last focussing magnet. Right: calculated velocity distributions at the dissociator nozzle (open circles), at the interaction region (squares), and at the Breit-Rabi detector, (solid circles). All plots are calculated at the predicted optimum operating point of T_{nozzle} = 80K and Q_{H2} = 2.0 mbar-liter/s.

CONCLUSION

A major concern for any simulation is to what extent the model reflects the physical situation. To verify the reliability of our model, we entered the magnet geometry, operating parameters, and compression tube details for two atomic beam sources [1,6] whose outputs had previously been measured. The comparison between measurement and prediction is shown in Fig. 4. An improved operating point for the COSY-ANKE source was reported at this conference and is included in the left-hand plot of Fig. 4. Although the code appears to slightly overestimate the optimum gas flow and nozzle temperature it predicts the maximum output to better than 5%.

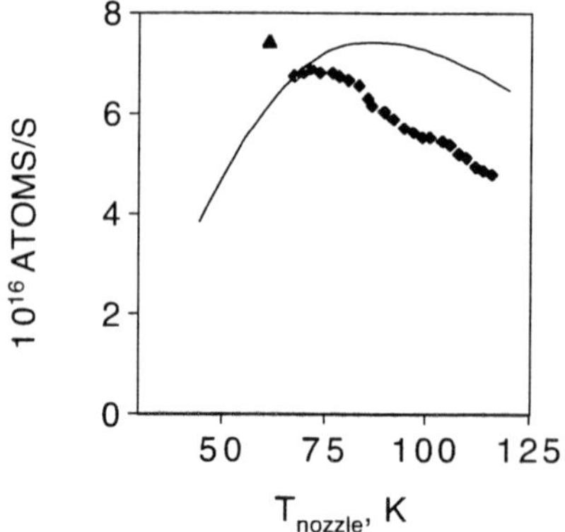

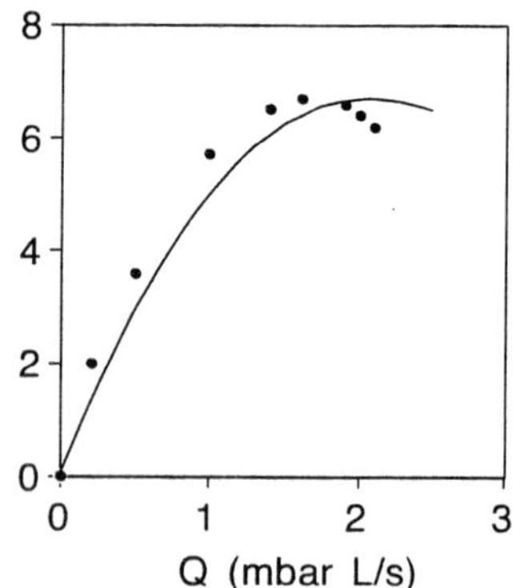

Figure 4. Left: Prediction of this model (solid line) compared to measurements on the COSY-ANKE atomic beam source as reported in [6]. The solid triangle is an improved operating point reported at this conference. Right: prediction of this model (solid line) compared to measurements of the Wisconsin ABS reported in [1].

ACKNOWLEDGEMENTS

This work was supported in part by United States Department of Energy under contract number DE-FG02-88ER40438.

REFERENCES

1. T. Wise, A. D. Roberts and W. Haeberli, Nucl. Inst. Meth. **A336**, 410-422 (1993).
2. N. Koch, PHD Thesis DESY-THESIS-1999-015, (May 1999). A Study on the Production of Intense Cold Atomic Beams for Polarized Hydrogen and Deuterium Targets
3. K. Halbach, Nucl. Inst. Meth. **169**, 1-10 (1980)
4. A. Vassiliev, et al., Rev. Sci. Inst. **71**, 3331-3341 (2000)
5. B. Lorentz, Diplomarbeit Max-Plank-Institut für Kernphysik, Heidelberg Germany (1993)
6. M. Mikirtychiants, et al., The Polarized Gas Target for the ANKE Spectrometer at COSY/Jülich in *Proceedings of the Ninth International Workshop on Polarized Sources and Targets* edited by V. P. Derenchuk and B. von Przewoski, World Scientific, River Edge, New Jersey 2002 pp. 47-51.

Nuclear Polarization of Molecular Hydrogen Recombined on Drifilm

P. Lenisa*† and U. Stösslein***

*on behalf of the HERMES Collaboration
†Universitá di Ferrara and INFN - Sez. di Ferrara, 44100 Ferrara, Italy
**Nuclear Physics Laboratory, University of Colorado, Boulder, Colorado 80309-0446
and DESY, Deutsches Elektronen Synchrotron, 22603 Hamburg, Germany

Abstract. The nuclear polarization of H_2 molecules formed by recombination of polarized H atoms on a Drifilm coated storage cell has been measured by using the longitudinal double spin asymmetry in deep inelastic positron-proton scattering. From the result of the measurement, a non-zero nuclear polarization for the atoms on the surface can be derived.

INTRODUCTION

During the past years, increased use has been made of polarized hydrogen and deuterium gas targets, which are placed in the circulating beams of storage rings. In order to increase the target thickness over that obtained by a jet of polarized H atoms, the beam from atomic beam sources is directed in an open, cooled cell (*storage cell*) in which the atoms make several hundred collisions before escaping. In order to inhibit recombination and depolarization processes, the cells are usually coated with teflon like materials. The HERMES experiment uses such a target to study deep inelastic scattering of the 27.6 GeV positrons of the HERA storage ring from polarized H nuclei. The polarization of the atoms is measured by a Breit-Rabi atomic polarimeter (BRP) which determines the populations of the four hyperfine states of H. However a fraction of the atoms recombines to form H_2, the amount of which is measured, but whose nuclear polarization is not known; this reflects itself in an increased systematic uncertainty in the target polarization.

The nuclear polarization of recombined H_2 molecules has been recently studied in a separate experiment [1]: a polarized atomic beam has made recombine using a copper surface and the nuclear polarization of the molecules has been measured by elastic proton-proton scattering. This result is not directly applicable to the HERMES storage cell which is coated with Drifilm [2], which has different surface characteristics. The only other existing measurement on this subject concerns recombination concerns tensor polarized D atoms, but it has also been performed on a copper surface [3]. The present work reports the first meaurement of the polarization of molecules which have formed via recombination of atoms on Drifilm.

CP675, *Spin 2002: 15th Int'l. Spin Physics Symposium and Workshop on Polarized Electron Sources and Polarimeters,* edited by Y. I. Makdisi, A. U. Luccio, and W. W. MacKay

THE HERMES EXPERIMENT

The Hermes experiment is installed in the HERA ring where the positron beam becomes transversely polarized to their momentum by emission of synchrotron radiation. A pair of spin rotators provide longitudinal polarization at Hermes interaction point. Positron identification in the momentum range 0.1 to 27.6 GeV is accomplished with an identification efficiency that exceeds 98 % and a negligible hadron contamination. The HERMES spectrometer is described in [4].

The Hydrogen Target

A beam of Hydrogen atoms is generated in a radio-frequency dissociator which forms part of the atomic beam source (ABS) [5]. The beam of nuclear polarized atoms is injected into the center of a thin-walled storage cell [6] via a side tube and the atoms then diffuse to the open ends of the cell where they are removed by a high speed pumping system. The storage cell is coated with Drifilm in order to minimize wall interaction effects. A magnetic holding field provides a quantization axis for the spins and inhibits nuclear spin relaxation by effectively decoupling nucleon and electron spins. The beam emerging from a second side tube is analysed by a Breit-Rabi polarimeter (BRP) [7] to measure its atom polarization and a target gas analyser (TGA) [8] to determine its atomic fraction [9]. During the atom diffusion process, relaxation by wall and spin exchange collisions and wall recombination [10] changes the polarization and the atomic fraction of the target gas. The atom polarization and atomic fraction values measured by the BRP and TGA must be corrected for these effects to obtain the average target polarization as seen by the positron beam [11], which is is described by the following expression:

$$P_T = \alpha_0 [\alpha_r + (1-\alpha_r)\beta] P_a, \tag{1}$$

where α_0 is the atomic fraction accounting for unpolarized molecules (i.e. not coming from recombination), α_r is the relative atomic fraction surviving recombination, $(1-\alpha_r)$ is the relative fraction of recombined atoms, P_a is the nuclear polarization of the atoms and $\beta = P_m/P_a$ the relative polarization of the recombined molecules respect to the atomic polarization ($0 \leq \beta \leq 1$), which is the quantity we extracted in the measurement presented below.

MEASUREMENT

The reported measurement is based on the 1997 data taking period using a Hydrogen target. The method adopted to extract the molecular polarization, exploits the double spin asymmetry in the deep inelastic scattering of a longitudinally polarized positron beam from a longitudinally polarized proton target. The asymmetry can be derived from the cross sections difference for the positron and the proton spin aligned antiparallel ($\overrightarrow{\Leftarrow}$)

and parallel ($\rightrightarrows$) respectively:

$$A_{||}^{meas} = \frac{\sigma^{\rightleftarrows} - \sigma^{\rightrightarrows}}{\sigma^{\rightleftarrows} + \sigma^{\rightrightarrows}} = \frac{1}{P_b P_T} \frac{(N/L)^{\rightleftarrows} - (N/L)^{\rightrightarrows}}{(N/L)^{\rightleftarrows} + (N/L)^{\rightrightarrows}}. \tag{2}$$

Here, N denotes the number of events per spin state corrected for the background arising from charge symmetric processes and L is the corresponding luminosity measured with Bhabha scattering. P_b (P_T) is the beam (target) polarization.

According to Eq. 1, the sensitivity to measure the relative molecular polarization β is best if the conditions are such that the recombination rate is high (low α_r). This could be achieved by increasing the temperature of the target cell from 100 K (normal running conditions) to 260 K. The temperature dependence of the recombination is described in [10].

Employing the fact that the cross section asymmetry is nearly independent from the experimental conditions, $A_{||}$ has to be the same for both target conditons. This leads to:

$$\frac{C_{||}^{100K}}{P_T^{100K}} = \frac{C_{||}^{260K}}{P_T^{260K}} \tag{3}$$

where $C_{||}$ indicates the quantity:

$$C_{||} = \frac{1}{P_b} \frac{(N/L)^{\rightleftarrows} - (N/L)^{\rightrightarrows}}{(N/L)^{\rightleftarrows} + (N/L)^{\rightrightarrows}} \tag{4}$$

and P_T^{100K} and P_T^{260K} are the target polarizations at 100 K and 260 K, which, by using Eq. (1), can be expressed by:

$$P_T^{100K} = \alpha_0^{100K}[\alpha_r^{100K} + (1-\alpha_r^{100K})\beta^{100K}]P_a^{100K} \tag{5}$$

$$P_T^{260K} = \alpha_0^{260K}[\alpha_r^{260K} + (1-\alpha_r^{260K})\beta^{260K}]P_a^{260K} \tag{6}$$

The values for $\alpha_0^{100K,260K}$, $\alpha_r^{100K,260K}$, $P_a^{100K,260K}$ are reported in Table 1. As the surface conditions at 100 K and 260 K are different, β^{100K} and β^{260K} has to be considered independent, so that two unknowns are present in equations (5) and (6) and enter in Eq. 3. For the determination of β^{260K} a minimization procedure has been adopted where the sum goes over the number of kinematic bins of the asymmetry measurement:

$$F(\beta) = \sum_{bins} \left[\frac{\frac{C_{||}^{260K}}{P_T^{260K}(\beta^{260K})} - \frac{C_{||}^{100K}}{P_T^{100K}(\beta^{100K})}}{\left((\delta_{C||}^{260K})^2 + (\delta_{C||}^{100K})^2\right)^{1/2}} \right]^2 = min. \tag{7}$$

where $\delta_{C||}$ is the statistical uncertainty of each bin. The minimization has been studied as a function of the parameter β^{100K} as reported in Fig. 1. The plot shows that the assumption taken for β^{100K} have low impact on the result for β^{260K}.
The following value has finally been extracted for β^{260K} (for $0 \leq \beta^{100K} \leq 1$):

TABLE 1. Atomic polarization and atomic fraction at 100 K and 260 K

T_{cell}	P_a	α_0	α_r
100 K	0.906 ± 0.01	0.96 ± 0.03	0.945 ± 0.035
260 K	0.939 ± 0.015	0.96 ± 0.03	0.26 ± 0.04

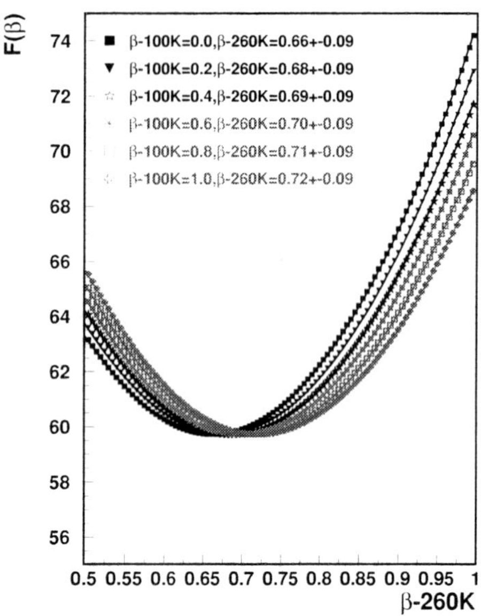

FIGURE 1. $F(\beta)$ as a function of β^{260K} for fixed values of β^{100K} in the allowed range $0 \leq \beta \leq 1$.

$$\beta^{260K} = 0.68 \pm 0.09_{stat} \pm 0.06_{syst} \tag{8}$$

The systematic uncertainty is mainly given by the uncertainty in the target polarization.

Under the experimental conditions in which the measurement has been performed, the main mechanism responsible for recombination active in the target cell is the Eley-Rideal mechanism [12] in which an atom coming from the volume hits a chemically bound atom on the surface with enough kinetic energy to overcome the activation barrier [10]. Assuming that the nucleon spins are not affected by the recombination process, the nuclear polarization of the molecule at its formation (P_m^0) can be evaluated by taking the average value of the polarization of the atom coming from the volume (P_a) and of the one sitting on the surface (P_s):

$$P_m^0 = \frac{P_a + P_s}{2} \tag{9}$$

The loss of polarization of the molecule after recombination has been well described in [1]. In free flight, the internal molecular fields B_c from the spin rotation interaction and the direct dipole-dipole interaction, cause the nuclei to rapidly precess around a direction which is skew to the external field by B_c/B. The orientation of B_c is randomized at each wall collision. Between successive wall collision, the component of the polarization along the external field decreases by an amount $(B_c/B)^2$ and after n wall bounces:

$$P_m = P_m^0 e^{-n(B_c/B)^2} \tag{10}$$

where B_c for H_2 is 6.1 mT. For the HERMES cell, we have the values: $n \approx 300$ [11], $B \approx 330$ mT so that Eq. 10 allows to conclude $P_m \approx 0.9P_m^0$. From the extracted value of β^{260K} and making use of Eq. 9 and of the value for P_a^{260K} (see Table 1), we are able to give an estimation for the residual polarization of the atoms on the surface:

$$P_s^{260K} = 0.46 \pm 0.22_{tot}. \tag{11}$$

CONCLUSIONS

The longitudinal double spin asymmetry in deep inelastic positron-proton scattering has been used to measure for the first time the nuclear polarization of the molecules produced by recombination of Hydrogen atoms on a Drifilm coated storage cell. The measurement indicates that the atoms on the surface show nonzero nuclear polarization and can represent an important point in the possible development of a polarized molecular target. The extension of the result to the normal working conditions of the HERMES target (100 K), will sensibly decrease the systematic uncertainty of the target polarization.

ACKNOWLEDGMENTS

We would like to thank C. Baumgarten, M. Contalbrigo and H. Kolster for the support and the fruitful discussions during the development of the presented analysis.

REFERENCES

1. Wise T. et al. *Phys. Rev. Lett.* **87**, 042701, (2001)
2. Thomas G. E. et. al.; *Nucl.Instrum. and Meth. A* **257**, 32 (1987)
3. van den Brand J.F.J et al. *Phys. Rev. Lett.* **78**, 1235 (1997)
4. Ackerstaff et al.; *Nucl. Instr. Meth. A* **482**, 606 (2002)
5. Nass A. et al. *The HERMES polarized Atomic Beam Source*, submitted
6. Baumgarten C. et al. *The polarized H/D internal target storage cell for the HERMES experiment at HERA*, to appear on NIM A
7. Baumgarten C. et al. *Nucl. Instrum. and Meth. A* **482**, 606 (2002)
8. Simani M. C. et al. *The gas analyzer of the target of the HERMES experiment*, submitted
9. a schematic rapresentation of the HERMES target is given in: P. Lenisa *The HERMES Polarized Internal Target*, Proceedings of PST2001, Nashville-Indiana, (2001).
10. Baumgarten C. et al. *Measurements of recombination at the HERMES H/D gas target* to appear on NIM A
11. Baumgarten C.et al. *Eur. Phys. J. D* **18**, 37 (2002)
12. Haeberli W.; *Ann. Rev. Nucl. Sci.* **37**, 297 (1967)

Beam induced depolarizing resonances in the HERMES hydrogen/deuterium target

D. Reggiani (for the HERMES Collaboration)

Istituto Nazionale di Fisica Nucleare, Laboratori Nazionali di Frascati, 00044 Frascati, Italy
Dipartimento di Fisica, Università di Ferrara, 44100 Ferrara, Italy

Abstract. Nuclear polarized hydrogen and deuterium gas targets employed in high-energy storage rings have become an important tool in the study of spin dependent processes in nuclear and particle physics. A severe problem in the use of this type of targets in bunched beams is the nucleon depolarization which can take place when the transient magnetic fields generated by the beam interact with the polarized nucleons and change their spin state. These depolarization process can be studied experimentally with a fully operational target installed in a storage ring. This is the case of the HERMES target (at HERA - DESY) where this effects have been extensively studied in the past with H longitudinally polarized with respect to beam axis. In the presentation, besides of the results related to the past longitudinal running, the new problematics related to the present running with transersally polarized H will be addressed.

INTRODUCTION

The HERMES gaseous polarized hydrogen/deuterium internal target is operational since 1996 in the HERA electron storage ring at DESY[1]. The polarized gas, produced by an atomic beam source (ABS), is injected into a storage cell[2] through which the 27.5 GeV HERA electron beam passes. The target atomic polarization is measured by a Breit-Rabi polarimeter[3]. A magnet sorrounding the storage cell provides a holding field defining the polarization axis and preventing spin relaxation by effectively decoupling the spin of electrons and nucleons. The orientation of the field, longitudinal to the electron beam axis from 1996 to 2000, was switched to transverse in 2001.
A potential source of target depolarization is related to the interaction of the transient magnetic fields generated by the bunched electron beam with the polarized nucleons of the target. This effect has been studied with both field orientations and beam induced depolarizing resonances have been clearly seen. The occurence of new densely spaced resonances makes this problem particularly critical in the transverse case, where it can be avoided only by designing a very uniform target magnetic field.

THE HERMES TARGET POLARIMETER

The atomic polarization is constantly monitored by a Breit-Rabi polarimeter (BRP) (fig. 1). A sample of target gas leaving the storage cell enters the BRP encountering first two hyperfine transition units and then a sextuple magnets system. A flux composed by the two (three) upper hyperfine states of atomic hydrogen (deuterium) is focused

CP675, *Spin 2002: 15th Int'l. Spin Physics Symposium and Workshop on Polarized Electron Sources and Polarimeters,* edited by Y. I. Makdisi, A. U. Luccio, and W. W. MacKay

into a quadrupole mass spectrometer (QMS) and detected by a Channeltron. A beam blocker placed inside the sextuple system ensures that no atoms in the lower states can reach the QMS. The background measurement is carried out by using a chopper which periodically shuts the flux in front of the QMS. A differential pumping system keeps the pressure in the detector chamber at $1 \cdot 10^{-10}$ mbar.

For any given ABS injection status, the BRP transition units are operated in at least four (six) different modes in the hydrogen (deuterium) case. In this way, a number of signals equal to or larger than the number of hyperfine levels can be collected. Knowing the efficiencies of the transitions units and the relative transmissions of the sextupole system for different hyperfine states, the four (six) hyperfine populations can be calculated. Finally, applying the knowledge of the target field intensity, the polarization of the sampled atomic beam is computed. The standard acquisition time for a polarization measurement lasts roughly 60 s for hydrogen and 90 s for deuterium.

The BRP calibration is carried out by operating the transition units in all possible ways. By doing so for different ABS injections modes, it is possible to collect a number of signals larger than the number of unknown (efficiencies, transmission ratios and hyperfine populations) and therefore determine the transition units efficiencies and the sextupoles transmission ratios.

The BRP sextupole system has been recently optimized by replacing its magnets. Due to this improvement, the current statistical uncertainty for 60 s polarization measurement is less than 0.5 %. The systematic error is in the order of 1 %.

Because of its capability of measuring the individual hyperfine populations, the Breit-Rabi polarimeter is particularly suited for identifying the different kinds of transitions caused by the beam induced time dependent fields.

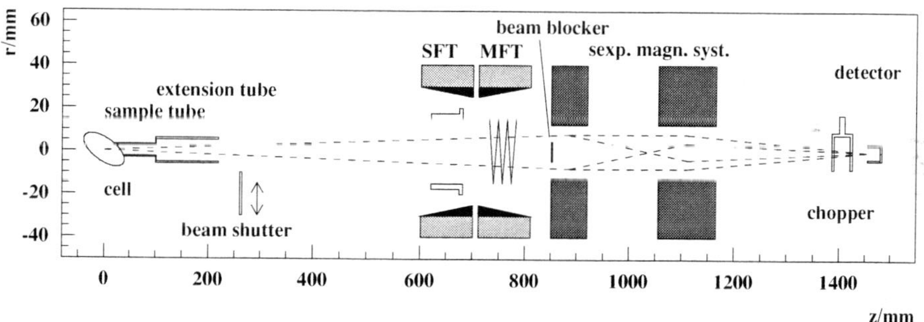

FIGURE 1. Schematic of the BRP. The light grey elements are the two transition units. As in the ABS case, the strong filed transition (SFT) needs to be replaced and the medium field transition (MFT) retuned when switching between hydrogen and deuterium. The dark grey elements are the sextupoles. The beam shutter is used to measure the hydrogen contribution coming from dissociative water.

TARGET DEPOLARIZATION BY BEAM INTERACTION

Bunch field induced resonant depolarization in the HERMES target may originate when the frequency of an rf-harmonic induced by the HERA e-beam matches the frequency difference between two different hyperfine states present in the storage cell. The probability of such an event is proportional to the square of the beam current. In order to determine for which values of the target field this mechanism can take place, one has to study the harmonic structure of the time dependent magnetic field induced by the 220 bunches of the electron beam. As the distance between two adjacent bunches is τ=96 ns, the frequency spacing between two harmonics is given by $\nu=\frac{1}{\tau}$=10.41 MHz. Since the width of the gaussian shaped bunch is very narrow (σ_t=37.7 ps), a huge number of harmonics with non-negligible amplitude (more than 400 within one sigma of the Fourier spectrum) can contribute to induce rf-fields.

Depending on the pair of hyperfine states involved, transitions are distinguished between π, occurring when the rf-field component is perpendicular to the static one, and σ, taking place when the two fields are parallel. Around the working point of the target magnet (300 to 340 mT), the π resonances are easily avoidable with a field uniformity of the percent level. This is unfortunately not the case for the σ resonances, whose spacing is only 0.37 mT. The resonance conditions for the hydrogen case are shown in figure 2. For deuterium the spacing between two σ_s of the same kind is again 0.37 mT, but the situation is complicated by the presence of two possible transitions ($|2\rangle \leftrightarrow |6\rangle$ and $|3\rangle \leftrightarrow |5\rangle$). On the other hand, the interval between two π_s is larger in this case. Due to the relative orientations of the beam induced magnetic field and the static holding

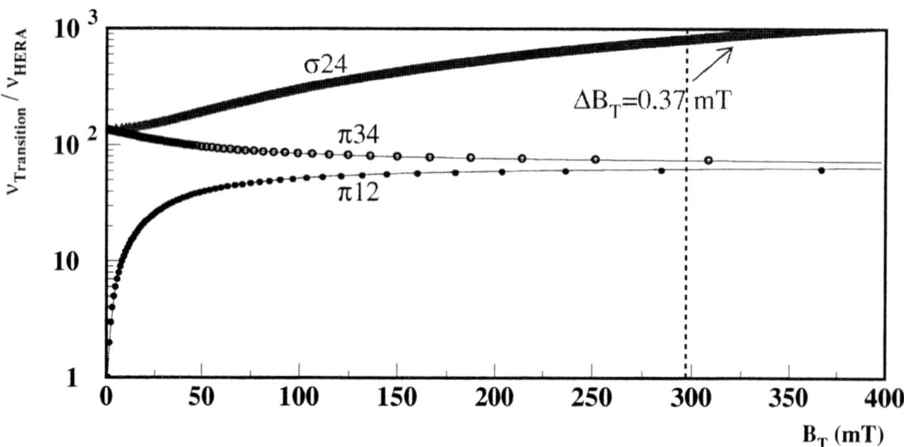

FIGURE 2. Nuclear depolarizing resonances in the hydrogen case. The frequency difference between pairs of hydrogen hyperfine states whose transitions would lead to a nuclear depolarization are plotted as function of the holding field. The frequency values are normalized for ν_{HERA} = 10.41 MHz. The marks, representing the resonance condition, are clearly distinguishable for the π transitions, while they overlap with each other in the σ case.

field, the σ transitions are present in the transverse case only.

THE MEASUREMENTS

Beam induced depolarizing resonances have been observed in the HERMES target during measurements taken in 1997 with the longitudinal hydrogen target[4], and in 1999 during a transverse target test run[5].

In the longitudinal case, the field was produced by a superconducting solenoid capable of an intensity up to 400 mT and a uniformity better than 2%. During normal condition the field strength was set to 335 mT, between two π resonances. For the observation of the depolarizing resonances two different techniques were applied. In both cases hyperfine states $|3\rangle$ and $|4\rangle$ produced by the ABS were injected into the storage cell. The first method, called flip-in, was designed to detect the two possible π transitions $|1\rangle \leftrightarrow |2\rangle$ and $|3\rangle \leftrightarrow |4\rangle$ in the target field range between 220 to 400 mT. During a slow field scan, the polarimeter was arranged to detect the presence of states $|2\rangle$ and $|3\rangle$ only. The position and shape of each resonance in that field range could be observed, as it is shown in the left plot of figure 3. In the second approach, the field was scanned within the range of the 62th harmonic, and the BRP arranged to perform a complete polarization measurement at each step. The right plot of figure 3 shows the loss of nuclear polarization of the target atomic sample during this measurement.

Similar measurements were taken during a special run in 1999 using a tranverse conventional dipole magnet capable of a field intensity up to 180 mT and a uniformity along the logitudinal direction z of 0.6 mT (at 150 mT). The π resonances were observed using the flip-in technique in a field range between 110 and 162 mT. Unfortunately, due to the non-uniformity of the field and to a failure of the QMS of the polarimeter, the indivudual σ resonances could not be seen and their effect could not be measured.

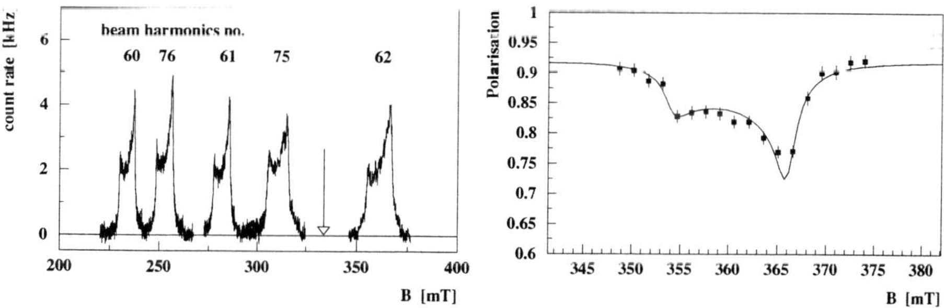

FIGURE 3. Beam induced depolarizing resonances observed in the HERMES longitudinal hydrogen target. The left plot shows the results of the flip-in measurement The signal peaks cause by the harmonics number 60 to 62 are caused by $\pi\ |1\rangle \leftrightarrow |2\rangle$ resonances, whereas the 75 and 76 are due to the $\pi\ |3\rangle \leftrightarrow |4\rangle$. The arrow shows the working point of the longitudinal magnet. In the right plot, the polarization measurement within the field range of the 62th harmonic is displayed. The fit represents the depolarization calculated taking into account the atomic density distribution inside the target cell, the expected resonance probability and the target field shape[6].

THE TRANSVERSE MAGNET DESIGN

In 1999, the HERMES collaboration decided to run with a tranverse polarized hydrogen target starting from 2001. For this pourpose, a new target dipole magnet was constructed. The most severe field requirement which had to be faced was related to the problem of the possible beam induced depolarization via σ transitions. In order to avoid any resonance inside the target cell, a field uniformity better than 0.14 mT was requested. On the other hand, in order to maintain the atomic polarization above 85 % and the total polarization relative uncertainty below 4.5 %, a field strength of around 300 mT was necessary. Moreover, a requirement on the maximum applicable field originated from the bending of the HERA electron beam passing through the vertical field and the resulting emission of synchrotron radiation towards the HERMES spectrometer. An estimation showed that the e-beam trajectory could have been compensated up to 340 mT target field by using two correction magnets already existing upstream and downstream of the target. The emitted synchrotron light would not have hit the HERMES detector. Due to geometrical constraints imposed by the HERMES setup, the design of a magnet fulfilling all the listed requirements was not possible. At a field intensity of B=297 mT, uniformities of ΔB_z=0.05 mT, ΔB_y=0.15 mT, ΔB_x=0.60 mT were achieved in the cell volume (z is the logitudinal direction, while y and x are the two transverse ones). The magnet was installed in the HERMES hydrogen target in July 2001. Due to the poor performance of the HERA machine after the startup in September 2001, a serious study of beam induced depolarization, with particular concern for the highest non-uniforminty along x, has not been possible up to now. A solution making use of two correction coils embedded inside the storage cell support structure is currently under test.

CONCLUSION

Beam induced resonant depolarization has been observed in the HERMES target. The problem turns to be particularly critical for the recently installed transverse polarized target, where it can be suppressed only with a holding field uniformity at the edge of the technical feasibility.

REFERENCES

1. P. Lenisa, "The HERMES polarized internal target", in *Polarized Sources and Target, Proceedings of the Ninth International Workshop*, edited by V. Derenchuk et al., World Scientific Publishing Co.
2. C. Baumgarten et al., The Storage Cell of the Polarized H/D Internal Target of the HERMES Experiment at HERA, to appear in *Nucl. Instrum. and Meth. A*
3. C. Baumgarten et al., *Nucl. Instrum. and Meth. A*, **482 (2002)**, 606
4. K. Ackerstaff et al., *Phys. Rev. Lett.*, **82 (1999)** 1164-1168
5. D. Reggiani, Diploma Thesis, Università degli Studi di Ferrara, December 1999
6. H. Kolster, Ph.D. Thesis, Ludwig-Maximilians Universitaet Muenchen, February 1998

Polarized Internal Target Experiments (PINTEX) at the Indiana Cooler

B. v.Przewoski

IUCF, Milo B. Sampson Lane, Bloomington, IN 47405, USA

Abstract. The PINTEX[1] facility at the IUCF Cooler consists of a polarized, internal target an an azimuthally symmetric detection system for charged particles. The polarized atomic beam source can be used to either produce a hydrogen target or a deuterium target. The target thickness obtained with a 12mm diameter and 25cm long storage cell is on the order of 10^{13} atoms/cm^2. The variable gradient of the two transition units facilitates changeover from hydrogen to deuterium. The ABS is capable of producing pure deuteron vector polarization with a theoretical maximum value of +2/3 and pure tensor polarization of theoretical maximal values of +/-1. The direction of the vector polarization can be reversed by reversing the direction of the holding field at the target. A hydrogen target was used for measurements of spin correlation coefficients in pp elastic sattering and pion production. The deuterium target served to measure spin correlation coefficients in pd elastic scattering and pd breakup.

THE ATOMIC BEAM SOURCE

The atomic beam source (ABS)[1] has been operational since 1993. In 2000 it was upgraded by installing two remotely controlled medium-field transition units to produce either polarized deuterium or hydrogen. Atoms from an 18 MHz dissociator emerge through an aluminum nozzle which is kept at liquid nitrogen temperature. The atoms then pass along the axis of a set of sextupole magnets where they are separated according to their electron polarization. In the following medium-field transition unit (MFT-1), transitions between hyperfine states are induced. The atoms pass along the axis of a second set of sextupole magnets whereby in the case of hydrogen an atomic beam in a pure spin state is prepared. At last, depending on which polarization state is desired, another transition between hyperfine states is induced in a second medium field transition unit (MFT-2). The atoms are then injected into the storage cell which is located in a weak holding field generated by a set of Helmholtz coils.

Medium Field Transitions

An MFT operates in magnetic fields of $B \sim 0.1B_c$ to $B \sim 0.2B_c$, where B_c is the hyperfine interaction field of 50.7 mT for hydrogen and 11.7 mT for deuterium. An appropriate field gradient along the beam direction is required to satisfy the condition

[1] Polarized INternal Target EXperimets

CP675, *Spin 2002: 15th Int'l. Spin Physics Symposium and Workshop on Polarized Electron Sources and Polarimeters,* edited by Y. I. Makdisi, A. U. Luccio, and W. W. MacKay

of adiabatic passage at a given, fixed RF frequency. Multiple transitions can be made by adjusting the static field so that the beam passes in sequence through field regions where the populations of different pairs of hyperfine states are interchanged.

Originally, the atomic beam source was equipped with a single, fixed-gradient MFT located after the first set of sextupole magnets. In order to facilitate switching between a proton or a deuteron target, two new transition units with variable gradient and variable static field were installed. The linearity of the gradient field over the transition region as well as the homogeneity of the static field were determined prior to installation of the units in the ABS. For deuterium the gradient field is set to +0.2 mT/cm. The RF coil of each MF unit consists of 12-turn solenoids of 1.6 mm diameter wire with a length of 70 mm and an I.D. of 34 mm. The units are water cooled and operated at 60.5 MHz for hydrogen and 30 MHz for deuterium. The currents used to drive the offset and gradient coils are controlled remotely . This makes it possible to quickly change, during the experiment, between vector-, positive tensor- and negative tensor polarization.

Since the operation of an ABS with hydrogen has been discussed extensively elsewhere[2], we limit the following discussion to deuterium. After the first set of sextupoles the atomic beam consists of states 1+2+3, where the states are labeled in order of decreasing energy in a non-zero magnetic field[3]. One or more transitions can be made sequentially in MFT-1.

Transitions are selected by changing the static field while the gradient field is kept constant. For small static fields no transitions are made. When the static field is increased, the atoms first undergo a $3 \leftrightarrow 4$ transition. When the field is further increased, the atoms undergo the $3 \leftrightarrow 4$ transition followed by the $2 \leftrightarrow 3$ transition. When the static field is increased even further, the atoms undergo the $3 \leftrightarrow 4$, $2 \leftrightarrow 3$ and $1 \leftrightarrow 2$ transitions sequentially. The second set of sextupoles eliminates state 4, so that one is left with states 1+2+3, 1+2, 1+3 and 2+3 depending on how many of the sequential transitions are made. The corresponding maximum polarizations of the atomic beam, before entering MFT-2, are $(P_z,P_{zz}) = (+1/3,-1/3)$, $(P_z,P_{zz}) = (+2/3,0)$, $(P_z,P_{zz}) = (+1/3,0)$ and $(P_z,P_{zz}) = (0,-1)$, where P_z,P_{zz} are vector and tensor polarization. MFT-2 is only needed to produce positive tensor polarization. After passing MFT-1 the beam contains states 1+3 with polarizations $(P_z,P_{zz}) = (+1/3,0)$. If the parameters of MFT-1 are set for the the $3 \leftrightarrow 4$ transition, the atomic beam contains states 1+4 with polarizations $(P_z,P_{zz}) = (0,+1)$ when it is injected into the target cell.

OPERATION

In order to minimize systematic errors of measured polarization observables, it is desirable to be able to alternate between target states frequently, preferably while beam is stored in the Cooler so that data in different target states are taken with the same stored beam. This was accomplished by remotely controlling the currents to the transition units. A sinusoidal function with exponentially decreasing amplitude was programmed to reproducably degauss each unit before every change to a new current. Fig. 1 shows the static field current through MFT-1 (trace a), the beam current (trace b) and the trigger rate (trace c) as a function of time. Trace (a) shows how MFT-1 is degaussed prior to

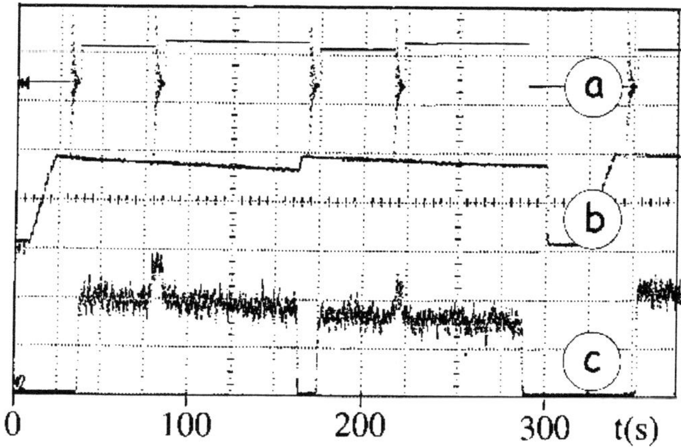

FIGURE 1. MF1 static current (a), stored beam current (b) and trigger rate (c) as function of time.

each new current setting. Trace (b) shows the Cooler being filled, a flattop at 135 MeV during which data are accumulated, a ramp to 200 MeV (which corresponds to an increase in beam current), another data taking flattop and the reset of all Cooler magnets during which the stored beam is discarded before the ring is filled with protons of the opposite spin state. Note that the trigger rate (trace (c) in Fig. 1 increases during degaussing, because the target density increases by 1/3. Events during degaussing are ignored in the final analysis.

SILICON DETECTORS IN AN ATOMIC HYDROGEN ENVIRONMENT

The target cell is surrounded by 18 position sensitive silicon detectors (see elsewhere in these proceedings). When we comissioned this setup we found that the ambient atomic deuterium or hydrogen gas apparently "poisons" silicon detectors. Even short exposures ($\sim$30 min) to ambient hydrogen in the region surrounding the target caused an increase in leakage current that rendered the detectors useless for data acquisition. Fortunately, we found that the detectors recovered once they were removed from atomic deuterium. Deuterium poisoning of the silicon detectors was eventually avoided by shielding them from ambient hydrogen using aluminized mylar foil. In addition, copper recombination baffles were placed around the feedtube and the ends of the storage cell. In this way, atomic deuterium was quickly recombined into harmless molecular deuterium. Although the problem was solved by isolating the detectors from the harmful atomic deuterium, we conducted an offline study in an attempt to understand the problem. Several 300μm thick detectors were tested. A detector was exposed to atomic hydrogen from the ABS, while another served as the reference detector under otherwise identical conditions. In this way changes in leakage current due to temperature change could be distinguished. Fig. 2 shows exposure and recovery curves for two different detectors. Although both

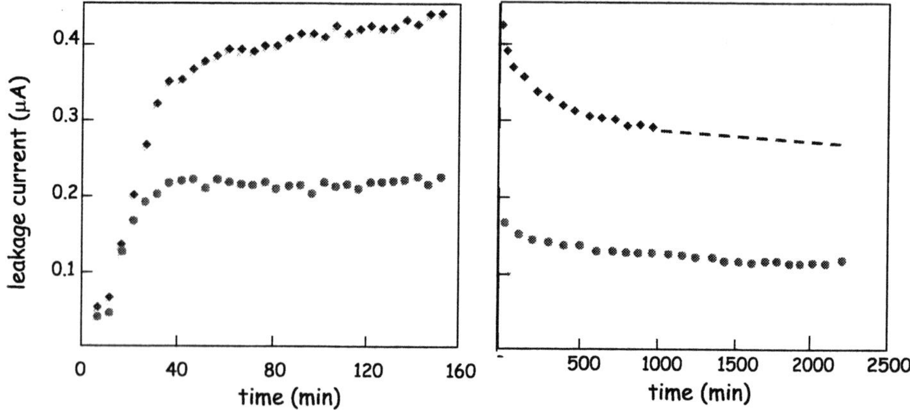

FIGURE 2. Silicon detector leakage current as a function of time. The left figure shows the increase in leakage current due to exposure to atomic hydrogen while the right figure shows the decrease in leakage current after the exposure. The two curves correspond to two apparently identical detectors.

detectors are seemingly identical and have the same initial leakage current of 0.05μA, their leakage currents after exposure to atomic hydrogen differ significantly. It thus appears that the susceptibility to atomic hydrogen poisoning also depends on some (unidentified) intrinsic property of the individual silicon wafer.

SPIN EXCHANGE

When we began using the deuterium target, we noted that the tensor polarization was significantly lower than expected. We investigated obvious causes for the deficiency, such as insufficient background subtraction, incomplete rejection of unwanted states by the sextupoles, wall depolarization and inefficiency of the transition units. None of these mechanisms quantitatively explained the low tensor polarization.

Spin exchange between deuterium atoms has a significant effect only on the tensor polarization. It also depends on the target density, in such a way that the polarization grows larger as the target density is decreased. We performed a series of measurements where we determined the tensor polarization as a function of target thickness by reducing the gas flow in the dissociator compared to its normal level. For these measurements we operated the ABS without medium-field transitions in order to be independent from inefficiencies of the transition units. In this case the target polarization is a mixture of vector and tensor polarization $(P_z, P_{zz}) = (+1/3, -1/3)$. The tensor component is shown in Fig. 3 as a function of relative target thickness. The data are normalized with a common, arbitrary factor. The curve shown is from Walker at al.[5] Note that the tensor polarization is negative and that polarizations of larger magnitude are indeed found at lower target densities. It can be seen from Fig. 3 that the calculation explains the observed relative change in target tensor polarization with target density.

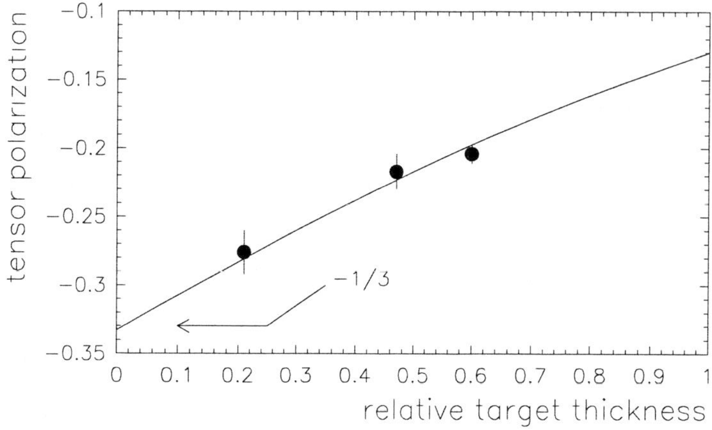

FIGURE 3. Target tensor polarization as a function of target thickness. The curve is the calculated effect from spin exchange.

CONCLUSIONS

We have commissioned a vector and tensor polarized deuterium target for the IUCF Cooler. To date, we have used this target to measure beam and target analyzing powers as well as spin correlation coefficients in pd elastic scattering at 135 and 200 MeV.

ACKNOWLEDGMENTS

This work was supported by NSF grants PHY-9602872, PHY-9722556, PHY-9901529 and DOE grant DOE-FG02-88ER40438.

REFERENCES

1. Wise, T. et al.*Nucl. Instr. Meth. A*, **336**, 410 (1993).
2. Haeberli, W. et al. *Phys. Rev. C*, **55**, 597 (1997).
3. Haeberli, W. *Ann. Rev. Nucl. Sci.*, **17**, 373 (1967).
4. Roberts, A.D. et al.*Nucl. Instr. Meth. A*, **322**, 6 (1992).
5. Walker, T. et al. *Nucl. Instr. Meth. A*, **334**, 313 (1993).

Polarized H⁻ Jet Polarimeter For Absolute Proton Polarization Measurements in RHIC

A.Zelenski[1], J.Alessi[1], A.Bravar[1], G.Bunce[1], M.A.Chapman[2], D.Graham[1], W.Haeberli[2], H.Hseuh[1], V.Klenov[4], H.Kolster[5], S.Kokhanovski[1], A.Kponou[1], V.Lodestro[1], W.MacKay[1], G.Mahler[1], Y.Makdisi[1], W.Meng[1], J.Ritter[1], T.Roser[1], E.Stephenson[3], T.Wise[2], V.Zoubets[4]

1-Brookhaven National Laboratory, 2-University.of Wisconsin, 3-IUCF, 4-INR, Moscow, 5-MIT, Bates

Abstract. Status of the H-jet polarimeter development is reviewed. A number of design issues are discussed including vacuum system, integration into the RHIC storage ring, scattering chamber, and uniform vertical holding field magnet design. The absolute proton polarization of the atomic hydrogen-jet target will be measured to 3% accuracy by a Breit- systematic error contribution to the jet-target polarization measurements is also discussed.

INTRODUCTION

An absolute polarimeter is an essential tool for polarized proton acceleration and polarization studies at high-energy colliders. Elastic scattering in the Coulomb-nuclear interference region was proposed for the absolute polarization measurements in RHIC [1]. Particle identity provides a unique opportunity for absolute polarization measurements in elastic proton-proton scattering of the polarized proton beam on a polarized proton target. Since analyzing powers for scattering of the polarized beam on an unpolarized target and unpolarized beam on the polarized target are equal, the beam polarization can be directly expressed in terms of the target polarization value, which can be precisely measured by Breit-Rabi polarimeter [2]. The polarimeter target is a free atomic beam jet, which crosses the RHIC beam in the vertical direction. With a state-of-art atomic polarized source, the maximum H-jet target thickness of about 0.9 10^{12} atoms/cm^2 can be achieved [3], which limits the polarimeter counting rate to less than 100 events/sec (a storage cell technique cannot be used, due to background considerations). Therefore, an integration time of about 10 hrs will be required for 2% statistical accuracy measurements [4]. These measurements will be used for calibration of a much faster p-Carbon CNI polarimeter, which is proven to be a very effective instrument for depolarization studies and polarization time-evolution monitoring during storage in RHIC [5].

CP675, *Spin 2002: 15th Int'l. Spin Physics Symposium and Workshop on Polarized Electron Sources and Polarimeters*, edited by Y. I. Makdisi, A. U. Luccio, and W. W. MacKay

POLARIZED H-JET POLARIMETER

Polarimeter location

The H-jet polarimeter will be installed at the RHIC beams intersection IP-12. This intersection is presently not occupied, is close to the p-Carbon CNI polarimeter, and infrastructure is available, so minimal construction work is required. The drawbacks are the very limited available space and difficult acccss. The decisive factor for the choice of IP-12 was the least interference with the other experiments and RHIC equipment. Due to the very small H-jet target thickness ($\sim 10^{12}$ atoms/cm^2), the polarimeter can be operated continuously, without any effect on the RHIC beams, and the remote location excludes any background generation for the other experiments.

Polarimeter vacuum system

The H-jet polarimeter includes three major parts: polarized Atomic Beam Source (ABS), scattering chamber, and Breit-Rabi polarimeter (see Fig.1). The polarimeter axis is vertical and the recoil protons are detected in the horizontal plane. The common vacuum system is assembled from nine identical vacuum chambers, which provide nine stages of differential pumping. The system building block is a cylindrical vacuum chamber 50 cm in diameter and 32 cm long with the four 20 cm (8.0") ID pumping ports.

Vacuum pumps

Each chamber will be pumped by two VT1000HT –Varian oil-free turbomolecular pumps (TMP) with ceramic bearings. A third pump can be installed if necessary on some stages, but even if one of two pumps fails the polarimeter still can be operated with somewhat reduced target density. The "Macrotorr" model TMP features high $\sim 10^6$ compression ratio for H_2 pumping and 900 l/sec pumping speed. The TMP cable length can be extended to 60 m, which allows the power supplies to be located outside the RHIC tunnel. Oil-free, piston-type EcoDry-15 pumps (Leybold) will be used for TMP backing.

Polarized atomic beam source

The ABS part of the polarimeter includes five vacuum chamber and five differential vacuum stages (see Fig.1). The ABS isolation valve is situated in the drift space between two groups of sextupole magnets, where space is available without any sacrifice in source performance. An isolation valve will be used for dissociator maintenance and it will be included in the interlock system to protect RHIC in case of a vacuum leak in the dissociator. With the hydrogen flow in the dissociator ~1-2 scc/sec, and 2·900 l/sec pumping speed, the vacuum is expected to be: 1-dissociator chamber $\sim 5\cdot 10^{-4}$ mbar; 2– skimmer chamber $\sim 10^{-4}$ mbar; 3-sextupole magnet chamber

$\sim 10^{-5}$ mbar; 4-drift space chamber $\sim 3 \cdot 10^{-6}$ mbar; 5- focusing sextupole, RF-transition chamber $\sim 5 \cdot 10^{-7}$ mbar, scattering chamber$\sim 5 \cdot 10^{-8}$. In conventional dissociator design the nozzle is cooled to 30-100°K to produce a "cold"- (low velocity) atomic beam. The acceptance of the sextupole separating magnets is higher at lower beam velocity, and the polarized beam density (for the same beam intensity) is therefore higher too. The high-intensity polarized proton (or H- ion) ABS operating with the 30°K nozzle temperature, would produce atomic beams of $\sim 10^{5}$ cm/sec velocity and about $5 \cdot 10^{11}$ atoms/cm^{3} atomic beam density in the ionizer region. A 30°K ABS is successfully implemented in the EDDA experiment at COSY with a polarized jet target [6].

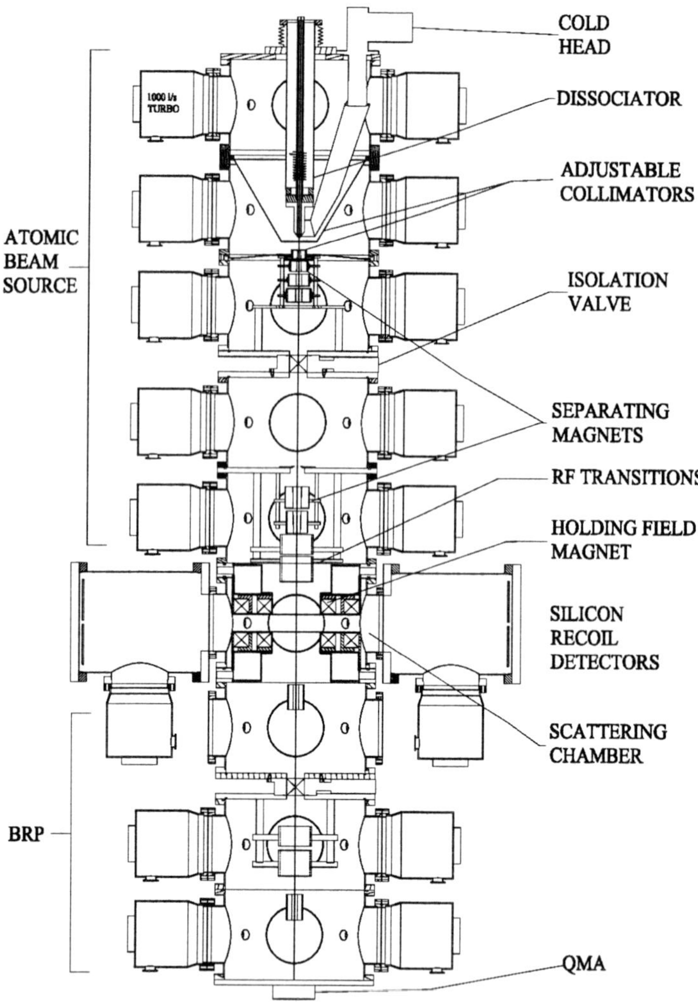

Figure 1: Polarized H-jet polarimeter assembly. BRP-Bret-Rabi polarimeter.

The $5 \cdot 10^{11}$ atoms/cm^3 atomic beam density was obtained with a lower beam flux and lower gas loading to the vacuum system than in the high flux ABS for storage cells feeding (HERMES, PINTEX, BLAST, ANKE), which operate at about 80°K. However, reliable and reproducible long term operation is easier to sustain at 80°K, and a somewhat higher atomic beam fraction (atomic/molecular beam ratio) was obtained at 80 °K (due to the recombination minimum at this temperature). Therefore, in spite of the fact that beam density, not flux, is the figure of merit for the jet-target, the RHIC polarimeter atomic beam energy was chosen at ~80°K. The sextupole separating magnet system was designed for an atomic beam velocity of about $2.0 \cdot 10^5$ cm/sec, which was measured for a dissociator nozzle temperature cooled to 80°K. The separating magnet system calculations and optimization are presented in the T.Wise et al., paper in this conference.

The dissociator design is conventional at a 13.56 MHz frequency. The tuning procedure developed by H. Kolster [7] will be used for optimization of the RF power coupling. A cryocooler will have sufficient power (~100 W at 80°K) to maintain the dissociator nozzle at the optimal temperature, which is expected to be in 60-80°K range. The dissociator and the first skimmer positions are designed to be adjustable without breaking vacuum. This will allow fine-tuning of the nozzle-skimmer and skimmer-sextupole magnet gaps, which are critical for the optimal ABS performance. The 32 cm drift space chamber #4 can be replaced with a shorter 20 cm chamber to study the effect of the drift space length on the ABS performance.

Scattering chamber

The CNI elastic p-p collision asymmetry is peaked at a momentum transfer of about $|t|$~0.001- 0.01 GeV^2, which corresponds to the recoil proton scattering angles 85-89 degrees for RHIC beam energies of 25-250 GeV/4/. The silicon strip recoil detectors are situated at 80cm distance from the jet-target in the recoil-arm cylindrical chambers, which are attached to the standard central chamber (see Fig.1). The pumping of the collision region is produced by the two 900 l/sec TMP's, backed by an additional TMP to ensure a high vacuum level with the mostly hydrogen residual gas composition. The scattering chamber will be bakeable with metal seals to keep the base pressure at 10^{-8} mbar level (without atomic beam, isolation valves are closed). The scattering angle is defined by the collision point within ~10 mm H-jet and the position sensitive silicon detector. The recoil proton energy is defined from silicon detectors and the time-of-flight is used to suppress the background. The scattering angle can be further constrained by remotely adjustable collimators located at ~4.0 cm distance from the target.

At the intersection point colliding bunches arrive at the same time and cannot be resolved by TOF. Therefore, the two beams will be separated spatially in the horizontal plane by about 10 mm, and the entire polarimeter apparatus can be scanned across the two beams in horizontal direction. In this way the scattering for one beam at a time can be selected in the polarimeter target, since the H-jet size is less than 10mm. To reduce the wake-field effect on the RHIC beams, the center part of the scattering chamber of 6 cm in diameter is enclosed in a metal grid screen. The grid screen has a sufficient vacuum conductance, and two 1.0cm openings for the recoil protons.

Holding field magnet

The vertical direction of the target polarization in the collision region is defined by a 1.0-1.5 kG vertical magnetic produced by the holding field magnet (see Fig.1). The coils and magnetic steel plates of this magnet are enclosed in stainless steel casings which are sealed in-between standard vacuum chambers, so that high-current vacuum feed throughs are not required. The magnet design has to satisfy a number of requirements: a) sufficiently high field is required to produce high proton polarization of the atomic beam (at 1.0 kG the maximum proton polarization is 95% and at 1.5 kG –97.5%); b) magnetic field homogeneity better than 10^{-3} within ~4.0 cm center region is required to tune the holding field magnitude in-between the depolarizing resonances and avoid depolarization by the bunch wake-field induced transitions in the atomic beam; c) holding magnet steering effect to the recoil protons must be minimized by compensation coils, whose fields are adjusted to keep the total field integral along the proton path near zero. The RF-transition cavities, which produce the proton polarization in the atomic beam, are closely positioned to the magnet and must be shielded from the holding field by additional magnetic screens.

TARGET POLARIZATION MEASUREMENTS

To achieve the polarimeter design accuracy of 5% for accelerated protons, the effective proton polarization of the target must be measured to at least 3% accuracy. According to the atomic beam transport calculations about 15% of the total H^{-}jet beam intensity will be within the BRP acceptance. The proton polarization of this beam part can be measured to ~1% accuracy by taking a set of atomic beam intensity measurements at different RF-transition settings. To derive the effective target proton polarization, as seen by the RHIC beams, from BRP measurements a number of systematic error contributions have to be determined with the high accuracy.

Beam induced depolarization

The effect of bunched-beam induced resonance depolarization can be directly measured in the BRP. This depolarization is not uniform across the target cross-section because the accelerated beam size of ~ 1.0 mm in diameter is much smaller than the jet-target. The BRP measures just a fraction of atomic beam, and the correction error might be large even if depolarization is just a few percent. To avoid this problem, the plan is to suppress the depolarization resonances by tuning the strength of the holding field in-between the adjacent resonances. The resonance position as a function of the holding field value will be measured using the BRP. The success of this plan depends on the width of the depolarizing resonances, which in turn is defined by the holding field uniformity and stability. To ensure depolarization suppression below the 1% level is a difficult task. It is easier to achieve at 1.5kG field strength due to larger resonance separation.

Target polarization dilution by molecular hydrogen and water vapor

The molecular hydrogen contamination of the atomic beam can be at the 5-10% level. The target polarization dilution by the molecular hydrogen (or water vapor) is the main systematic error contribution to the polarization measurements. Detailed studies of polarization dilution will be required to reduce the contamination to the lowest possible level and to develop techniques for precision correction measurements to meet the polarimeter accuracy design goal.

The contamination in the beam, which has passed the scattering chamber, can be measured in the first BRP chamber, before the sextupole magnets. The nondestructive beam diagnostics based on TOF measurements of the ion beams produced by electron beam ionizer can be used for on-line measurements. In off-line studies, a beam mass-analyzer can be installed in the scattering chamber and correlation with the dissociation ratio measurements (which will be monitored by another TOF mass-analyzer in the ABS) will be established, which will help to derive the polarization dilution correction factor.

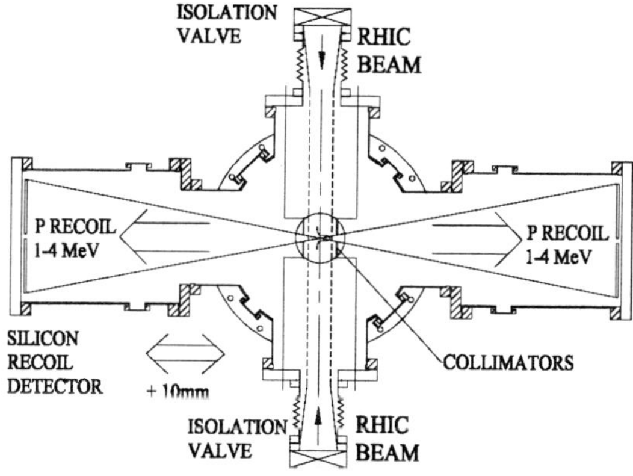

Figure 2. Scattering chamber.

H^- JET POLARIMETER INTEGRATION INTO THE RHIC

The polarimeter construction and tests will be done at the test facility in the linac injector complex. The polarimeter will be moved to the RHIC tunnel just before the polarized proton run and moved out for the further development to the test facility. The polarimeter is designed to be movable, i.e. it can be taken apart and reinstalled in either place in 2-3 days. The polarimeter will be removed from RHIC during the heavy-ion run, which relaxes the requirement to the polarimeter vacuum system (the polarimeter chambers cannot be baked because the permanent sextupole magnets do not tolerate heating above 50°C). At the test facility the polarimeter will be available for the precision polarization diagnostics development and systematic error studies of the jet-target polarization.

The polarimeter will be installed in RHIC between two isolation valves (see Fig.2). The isolation valves allow only a short part of the ring (about 1.0-1.5 m long) to be vented for polarimeter installation; therefore vacuum recovery time is reduced. These valves will also be included in the interlock system to protect the RHIC vacuum. In case of a leak in the polarimeter (most likely in dissociator) the vacuum sensors in the scattering chamber will trigger the close of the ABS and BRP isolation valves, if vacuum is improved RHIC operation can be continued without the polarimeter. Otherwise RHIC beams will be damped and the polarimeter isolation valves will be closed.

The first H-jet polarimeter operation in RHIC is planned for the 2003-2004 polarized run. The goals for the first run will be obtaining reliable operation at the designed intensity and high (over 80%) atomic beam polarization. The atomic beam polarization measurement accuracy might be at +/-5% level. It will require another one or two years to achieve the absolute accuracy of +/-2% for measurements of effective jet-target polarization.

REFERENCES

1. Akchurin N., et al., Polarimetry for high energy polarized proton colliders with a jet target, *Proc.12th Int. Symp.on high-energy Spin Physics 1996*, World Scientific, ed by C.W.De Jager , p.810, 1997.
2. Baumgarten C. et al., The Hermes Breit-Rabi polarimeter, *Nucl. Instr. andMeth.* **A482**, p.606, (2002).
3. Wise T., et al., Design of a polarized atomic H source for a jet target at RHIC, These Proceedings.
4. Bravar A., et al., The absolute polarimeter for RHIC, These Proceedings.
5. Eversheim D., et al., The EDDA H-jet polarimeter for COSY, These proceedings.
6. Kolster H., The BLAST polarized target, *Proc. "Polarized Sources and Targets*", PST 2001, p.37, World Scientific 2002.
7. Reggiani D., Beam induced depolarizing resonances in the HERMES target, These proceedings.

Future Facilities and Experiments

The Physics Programme of HERMES Run-II

Delia Hasch
(On behalf of the HERMES Collaboration)

INFN - Laboratori Nazionali di Frascati, 00044 Frascati, Italy

Abstract. The HERMES experiment has started its Run-II which is expected to run at least until 2006. The physics programme for this Run is based on two major axes of research: transversity and exclusive reactions. In the first phase the HERMES target will provide transversely polarized hydrogen atoms. The main aim of this part is to measure single-spin asymmetries in the semi-inclusive production of hadrons. From these asymmetries the transversity distribution $h_1(x)$ will be determined. This is the last twist-2 quark distribution function, which has not been measured at all up to now. Data from the first run of HERMES with a longitudinally polarized target suggest that this transversity distribution is accessible. During the last 2 years of the run a new recoil detector will be installed surrounding the HERMES target. The aim is to fully determine the kinematics for exclusive reactions by detecting also the slow recoil nucleon. This will enable HERMES to make virtually background-free measurements of exclusive reactions, in particular of Deeply Virtual Compton Scattering (DVCS).

INTRODUCTION

Historically, the investigation of the spin structure of the nucleon in electron scattering has been synonymous with inclusive measurements. In recent times, semi-inclusive measurements have substantially extended the understanding of the nucleon spin. From inclusive and semi-inclusive polarized deep inelastic scattering (DIS) the helicity distributions for u and d quarks are now known with reasonably good precision, while only a first glimpse has been obtained for the sea-quark and gluon helicity distributions. Moreover, the transversity distribution h_1 (also denoted as helicity-*flip* distribution δq) as well as the contribution from the orbital angular momentum of quarks and gluons are still unknown. While the experimental hunt for the former started very recently, a possible measurement of the latter is still under discussion on both theoretical and experimental grounds. The transversity distribution h_1 represents the transverse spin distribution of quarks in a nucleon polarized transversely to the virtual photon in the infinite momentum frame [1]. Transversity is a chiral-odd distribution which implies that it is not directly observable in an inclusive measurement as chirality is conserved in electromagnetic and strong interactions. Therefore, a second chiral-odd object has to be involved in the process. In semi-inclusive scattering this may be a chiral-odd fragmentation function, the so-called Collins function [2].

Nowadays exclusive processes, where all reaction products are detected, are becoming a promising and powerful experimental tool to access further information about the spin structure of the nucleon. The recently developed formalism of Generalised Parton Distributions (GPDs) [3] provides a unified description of inclusive and exclusive processes

CP675, *Spin 2002: 15th Int'l. Spin Physics Symposium and Workshop on Polarized Electron Sources and Polarimeters*, edited by Y. I. Makdisi, A. U. Luccio, and W. W. MacKay

which takes into account the dynamical correlations between partons of different momenta in the nucleon. The well-known parton distributions and form factors turn out to be the limiting cases and moments of GPDs. A particular feature in the context of spin physics is that the second moment of the sum of the two unpolarized GPDs give the total angular momentum carried by quarks [4].

THE HERMES EXPERIMENT

The HERMES experiment has been taken data at the HERA accelerator in Hamburg, Germany since 1995. HERMES records the scattering of the longitudianally polarized electron or positron beam of 27.6 GeV from polarized gas targets internal to the beam pipe. Pure atomic H, D and ^{3}He have been used as well as a variety of unpolarized nuclear targets. Featuring polarized beams and targets and an open-geometry spectrometer with good particle identification (PID), HERMES is well suited to a study of the spin-dependent azimuthal moments of the SIDIS cross-section. The PID capabilities of the experiment were significantly enhanced in 1998 when the threshold Čerenkov detector (used to identify pions above a momentum of 4 GeV) was upgraded to a Ring Imaging system (RICH). This new detector provides full separation between charged pions, kaons and protons over essentially the entire momentum range of the experiment. In September 2000 HERMES completed its first phase of data taking, using longitudinally-polarized targets. HERMES is now entering its second running phase which will continue until at least 2006. The main physics subject of the period 2002 to 2004 will be measurements from transversely-polarized hydrogen target with the specific goal of exploring the transversity structure function. In 2004 a recoil detector surrounding the target will be installed, allowing to fully determine the kinematics for exclusive reactions by detecting also the slow recoil nucleon. In conjunction with high density unpolarized nuclear targets, HERMES will be well suited for measurements of exclusive reactions with a recoiling proton in the final state in 2005 and 2006.

FUTURE TRANSVERSITY MEASUREMENTS

The Collins mechanism, proposed as a means to access to h_1, manifests itself in an azimuthal single-spin asymmetry which could be observed in semi-inclusive meson electroproduction. The moment of this asymmetry is related to the product of $h_1(x)$ and the Collins fragmentation function (FF) $\mathrm{H}_1^{\perp}(z)$, where the latter correlates the angular distribution of the produced hadrons with the transverse spin of the primary quark. HERMES has made the first measurements of single-spin azimuthal asymmetries (SSA) for semi-inclusive pion and kaon production in DIS, using an unpolarized beam and *longitudinally* polarized proton and deuteron targets [5, 6]. The observed $\sin\phi$ dependence of the asymmetry can be well described by model calculations based on the Collins mechanism. The results imply that the Collins FF could be sizeable, hence forseen measurement of h_1 with a transversely polarized target at HERMES is very promising. However, it cannot be excluded that at least part of the observed asymmetries

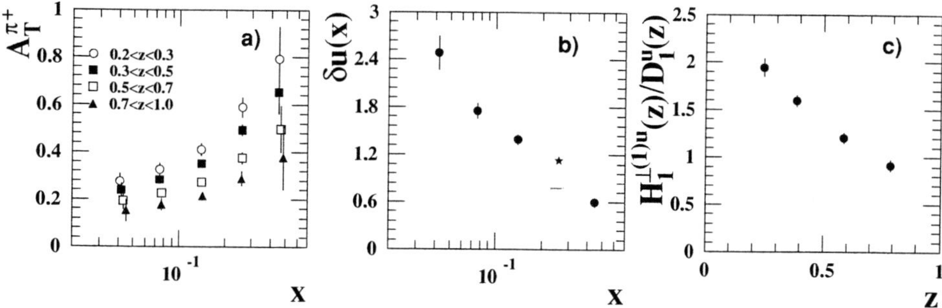

FIGURE 1. *a)* The weighted asymmetry $A_T(x)$ for different intervals of z; *b)* the transversity distribution $\delta u(x)$, and *c)* the ratio of the fragmentation functions $H_1^{\perp(1)u}(z)$ to $D_1^u(z)$ as they would be measured by HERMES. The asterisk in *b)* shows the normalization point.

is due to the interaction of the struck quark with the target remnant in the final state through the exchange of a single gluon [7]. This mechanism was shown [8] to be identical to the already known Sivers effect [9]. By scattering on a *transversely* polarized target, it will be possible to distinguish the Collins and Sivers mechanism through their different dependence on the angle ϕ_S between the transverse target polarization vector and the lepton scattering plane.

For the particular case of an unpolarized beam and a transversely (with respect to the incoming lepton momentum) polarized target (T) the following weighted asymmetry is sensitive at leading twist to the product of $h_1(x)$ and $H_1^{\perp}(z)$ [10]

$$A_T = \frac{\int d\phi \int d^2 P_\perp \frac{|P_\perp|}{M_h} \sin(\phi)\,(d\sigma^\uparrow - d\sigma^\downarrow)}{\int d\phi \int d^2 P_\perp\,(d\sigma^\uparrow + d\sigma^\downarrow)} = P_T \cdot D_{nn} \cdot \frac{h_1(x)\, z H_1^{\perp(1)}(z)}{f_1(x)\, D_1(z)}. \quad (1)$$

Here $\uparrow(\downarrow)$ denotes target up (down) transverse polarization, M_h is the mass of the produced hadron, P_T is the magnitude of the target polarization, D_{nn} is the transverse spin transfer coefficient and $D_1(z)$ is the familiar unpolarized FF. The angle ϕ is the azimuthal angle between the transverse target polarization vector and the transverse momentum $P_\perp$ of the produced hadron relative to the virtual photon direction. To evaluate the level of accuracy that could be obtained for a measurement of A_T, simulations [11] were done assuming a target polarization of $P_T = 75\%$ and 7.0 million reconstructed DIS events. Assuming u-quark dominance for π^+ electroproduction, the asymmetry is given by

$$A_T^{\pi+} = P_T \cdot D_{nn} \cdot \frac{\delta u(x)}{u(x)} \cdot \frac{H_1^{\perp(1)u}(z)}{D_1^u(z)}. \quad (2)$$

The expected asymmetry $A_T^{\pi+}(x)$ as it would be measured by HERMES is presented in Fig. 1*a)* for different intervals of z. The factorized form of the expression in Eq. 2 with respect to the variables x and z allows the reconstruction of the shape for both of the unknown functions δu and $H_1^{\perp(1)u}(z)/D_1^u(z)$, while the relative normalization cannot be fixed without further assumptions. The normalization ambiguity has been resolved by assuming $\delta u(x=0.25) = \Delta u(x=0.25)$, as indicated by the asterisk in Fig. 1*b)*. A very good statistical precision is expected for a first measurement of the x- and z-dependence of the transversity distribution $\delta u(x)$ and of the ratio of the fragmentation functions $H_1^{\perp(1)u}(z)/D_1^u(z)$.

FUTURE MEASUREMENTS OF EXCLUSIVE REACTIONS

Experimental information on GPDs can be obtained by measuring cross sections or asymmetries of exclusive reactions with different final states. Due to the experimental difficulties arising from the small cross sections involved and the high energy resolution required, first results in this field are appearing only now.

The theoretically cleanest way to access GPDs appears to be using DVCS process, i.e. the hard exclusive leptoproduction of real photons with the target nucleon remaining intact. Since the Bethe-Heitler (BH) process has an identical final state, the amplitudes of both processes add coherently, and the interference between them can be used to access the DVCS amplitude. The leading order and leading twist interference term [12]

$$I \propto e_l \left[\cos\phi \frac{1}{\sqrt{\varepsilon(\varepsilon-1)}} \mathrm{Re}\tilde{\mathcal{M}}^{1,1} - P_l \sin\phi \sqrt{\frac{1+\varepsilon}{\varepsilon}} \mathrm{Im}\tilde{\mathcal{M}}^{1,1} \right] \tag{3}$$

depends on the charge and the helicity of the incident lepton with polarization P_l. Here, ε is the polarization parameter of the virtual photon.

Experimental results of exclusive measurements at HERMES have been reported on this conference [13, 14]. However, the missing mass resolution of the HERMES spectrometer is not sufficient to exactly identify exclusive events individually and to separate them from non-exclusive events, like e.g. from those events where an intermediate Δ-resonance was created. Therefore exclusivity could be better established only on the level of a data sample by restrictive cuts and/or by background estimation and subtraction. The upgrade of the spectrometer with a recoil detector planned for 2004 will substantially improve this situation by establishing exclusivity on the event level for exclusive reactions with a recoiling proton in the final state. The HERMES recoil detector [15] will basically consist of a barrel of detectors around the target region. This will allow the detection of the slowly recoiling proton at large angle without interfering with the acceptance of the high momentum mesons and photons in the HERMES spectrometer. The recoil detector will consist of two layers of silicon detectors surrounding the target cell inside the beam vacuum and two layers of scintillating fibre detector in a longitudinal magnetic field of about 1 Tesla. The combination of both detectors allows for tracking in the momentum range of $0.1 < \mathrm{P} < 1.4$ GeV. Particle identification is provided via the energy deposition. Additional scitillators will be used for π^0 detection through their decay photons. The exclusive final state will be selected by applying co-planarity cuts, i.e., ensuring that the recoiling proton is in the same plane as the produced meson or photon. For exclusive production of photons used to study the DVCS-BH interference, it is estimated that the background due to Δ production (the primary contaminating process) can be reduced to 1%.

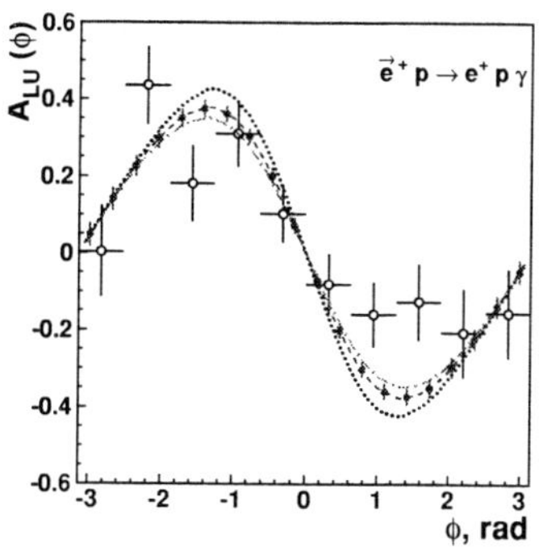

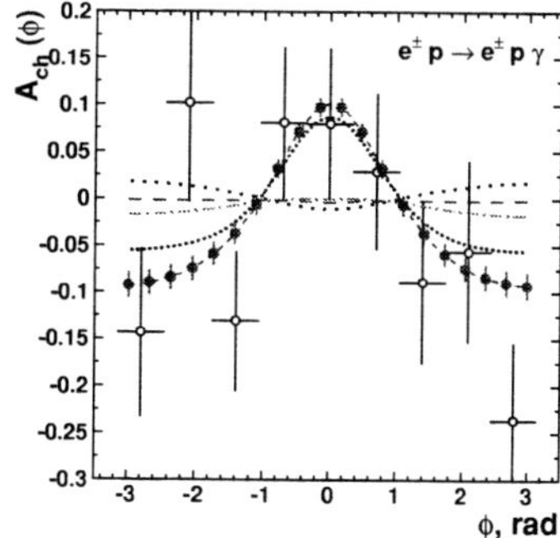

FIGURE 2. Projections of statistical accuracies (closed circles) for measurements of the beam-spin (left) and beam-charge (right) azimuthal asymmetries for hard electroproduction of photons as function of the azimuthal angle ϕ. The HERMES results from existing data are shown as open circles. The curves represent various GPD model predictions.

Fig. 2 shows the projected accuracy for measurements of beam-spin and beam-charge asymmetries associated with the DVCS-BH interference for one year of high-luminosity running (corresponding to 2 fb^{-1}) with the HERMES recoil detector. The calculations [16] were performed using different GPD parameterizations as shown by the different curves in Fig. 2. The open points shows the HERMES results [17, 13] from the 1996/97 data taking periods representing 0.13 fb^{-1} for the beam-spin asymmetry and from the 1998/2000 data taking periods for the beam-charge asymmetry. The remarkable improved accuracy in measuring the variable t with the recoil detector will allow one to study kinematic dependences of the asymmetries.

REFERENCES

1. For a recent review see V. Barone, A. Drago and P. G. Ratcliffe, *Phys. Rept.* **359**, 1-168 (2001).
2. J. Collins, *Nucl. Phys.* B **396**, 161 (1993).
3. For a recent review see K. Goeke, M.V. Polyakov and M. Vanderhaegen, *Prog. Part. Nucl. Phys.* **47**, 401-515 (2001).
4. X. Ji, *Phys. Rev. Lett.* **78**, 610 (1997).
5. HERMES collaboration, A. Airapetian et al., *Phys. Rev. Lett.* **84**, 4047 (2000), *Phys. Rev.* D **64**, 097101 (2001).
6. D. Hasch, Contribution to this Symposium.
7. S. Brodsky, D. Hwang and I. Schmidt, *Phys. Lett.* B **530**, 99 (2002).
8. J. Collins, hep-ph/0204004.
9. D.W. Sivers, *Phys. Rev.* D **41**, 83 (1990) and *Phys. Rev.* D **43**, 261 (1991).
10. A.M. Kotzinian and P.J. Mulders, *Phys. Lett.* B **406**, 373 (1997)
11. V.A. Korotkov, W.-D. Nowak and K. Oganessyan, Eur. Phys. J. C **18**, 639 (2001).
12. M. Diehl, PLB **411**, 193 (1997).
13. F. Ellinghaus, Contribution to this Symposium.
14. C. Schill, Contribution to this Symposium.
15. HERMES collaboration, Recoil Detector - Technical design report., DESY-PRC 02-01, 2002.
16. V.A. Korotkov and W.-D. Nowak, Eur. Phys. J. C **23**, 455 (2001).
17. HERMES collaboration, A. Airapetian et al., *Phys. Rev. Lett.* **87**, 182001 (2001).

Future of COMPASS experiment

Damien Neyret, on behalf of the COMPASS collaboration

DAPNIA/SPhN, CEA Saclay, F-91191 Gif sur Yvette, France

Abstract. The COMPASS physics program covers two domains, 1/ the study of nucleon spin structure using a muon beam, which includes the measurement of the gluon contribution to the nucleon spin and of the transversely polarized parton distributions; 2/ the hadron spectroscopy using hadron beams, in particular the study of charmed hadrons, hadron polarisabilities and exotic hadrons. The COMPASS spectrometer has been commissioned in 2001 and the physics data taking has started in 2002. We discuss here the plans for the future data taking and for the set-up improvements.

PHYSICS OF COMPASS

The COMPASS experiment uses the CERN high energy muon/hadron M2 beam line. Different targets will be used according to the program: polarized deuterons (^{6}LiD) or protons (NH_3), Carbon, Lead, Copper, and liquid H_2. The spectrometer has a wide angular acceptance for particle characterization and identification [1]. This allows to study exclusive reactions for several topics in hadron structure and hadron spectroscopy physics.

Nucleon spin structure

Several nucleon structure topics will be covered by COMPASS. In particular this experiment will have a large impact on two of them, the gluon contribution to the nucleon spin (ΔG) and the parton transverse polarization measurement.

Measurement of the gluon contribution to the nucleon spin

Previous experiments have shown that the spin contribution of the constituent quarks amount to only 20-30% to the total nucleon spin. Therefore other contributions like the gluon contribution ΔG may not be negligible. This value can be measured at COMPASS by probing polarized nucleons with a polarized muon beam, and selecting photon-gluon fusion events (γ-g $\rightarrow$ $q\bar{q}$). In particular, the open charm production of D^0 or D^* mesons is a good signature of γ-g fusion events [2, 3] (see Figure 1).

With the high current muon beam used at COMPASS (160 GeV, 2.10^8 μ/spill), the following statistical accuracy will be reached after two years and a half of data taking (i.e. around 225 days): for $D^0 \rightarrow K + \pi$ events we expect a statistical error on $\Delta G/G$ of 0.16, and for $D^* \rightarrow K + \pi + \pi_{soft}$ an error better that 0.15 [1]

CP675, *Spin 2002: 15th Int'l. Spin Physics Symposium and Workshop on Polarized Electron Sources and Polarimeters,* edited by Y. I. Makdisi, A. U. Luccio, and W. W. MacKay

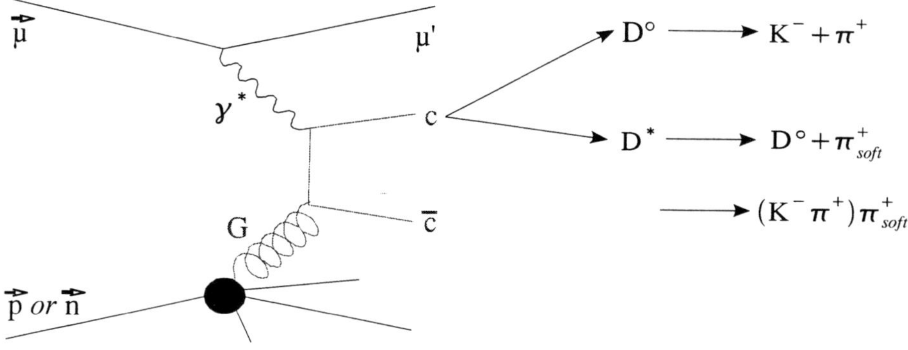

Figure 1: Photon-gluon fusion charm production

The ΔG measurement can also be done by using γ-g fusion events tagged with high transverse momentum hadrons [4]. This channel is affected by a higher systematical error due to the QCD Compton background.

Transversity measurement

The transverse spin distribution functions of the partons have never been measured. The COMPASS experiment can access these distribution functions by measuring the transverse structure function $h_1(x)=\frac{1}{2}\sum_q e_q^2[\Delta_T q(x)+\Delta_T \bar{q}(x)]$. This measurement is done through muon semi-inclusive deep inelastic scattering off a transversely polarized proton or deuteron target. The asymmetry of the Collins angle φ_C between the final hadron momentum and the final quark spin distribution is related to h_1 [5]. With 30 days of data taking for each target, using a 160 GeV muon beam of 2.10^8 μ/spill, one can reach a statistical accuracy better than 6 to 17 % on $x.h_1(x)$, depending of the x ranges.

Other physics topics can be studied with the muon beam, like g_1, g_2 or Λ baryons polarization measurements, but they will not been developed here.

Hadron spectroscopy physics

Several topics of hadron spectroscopy physics will be covered by the COMPASS experiment.

One important goal of COMPASS is to use Primakov scattering events to measure the polarisability of mesons [6] like pions, with an error level of 5%, or kaons or Σ for which polarisabilities have never been measured. Primakov scattering is an inverse Compton scattering: mesons are scattered off heavy nuclei target and the Coulomb field of the nucleus exposes the beam mesons to a high electromagnetic field which induces a polarization of the meson.

The COMPASS experiment has also a large program of studies on charmed hadrons: lifetime measurements, semi-leptonic decays, charm hadroproduction, etc.... There are also plans to search for double charmed baryon, which have never been seen up to know, and to measure their characteristics (lifetime, mass, decay modes).

Exotic hadrons can be also searched for in COMPASS experiment. Glueballs may be produced in central collisions of a proton beam on a proton target, via a double pomeron exchange process. They may also be observed in non-diffractive scattering or on Coulomb excitation, using a heavy nuclei target. Other exotics hadrons like charmed tetraquarks or pentaquarks will be also searched for.

COMPASS IN 2002

The COMPASS spectrometers

The COMPASS apparatus consists of two spectrometers for small and large angle scattered particles (Figure 2). High energy muon or hadron beams (up to 300 GeV for hadrons and 190 GeV for muons) can be scattered off different kind of targets. We presently use a long polarized ^{6}LiD target (two 60 cm long cells), inserted in the 2.5 Tesla superconducting solenoid magnet from the previous SMC experiment, since the foreseen large aperture COMPASS solenoid is not yet operating.

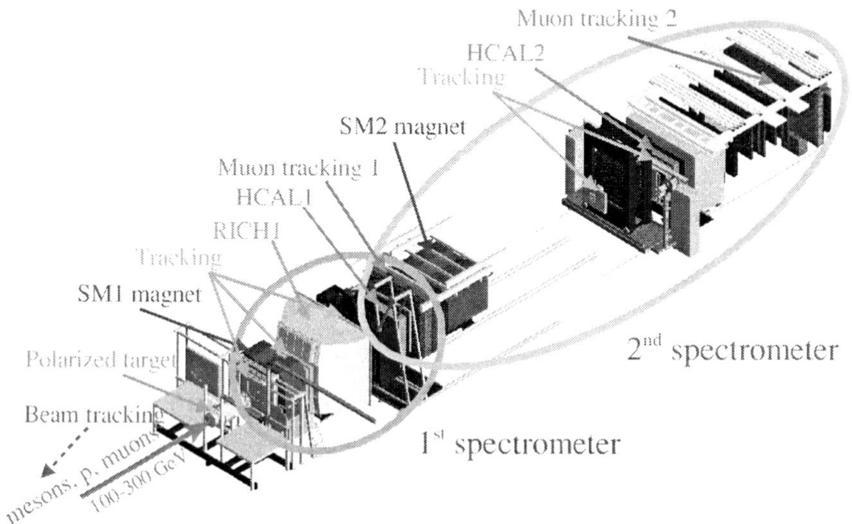

Figure 2: The COMPASS spectrometers

The first spectrometer, used for large angle particles, is composed of a vertical dipole magnet SM1 surrounded by a tracking system with scintillating fibers detectors, Micromegas microstrip chambers, drift chambers, GEM microstrip chambers, straw-tubes detectors and multi-wires proportional chambers. A RICH placed after the dipole magnet provides the particle identification of protons, pions and kaons for momentum between 3 and 30 GeV. It is followed by a hadronic calorimeter, and a muon wall system for muon tracking and identification.

The second spectrometer is dedicated to small angle particles, it uses the dipole magnet SM2 with the same tracking system as the first spectrometer, and uses MWPC,

GEMS and drift chambers after the magnet. An hadronic calorimeter and a muon wall system completes the second spectrometer but the RICH detector is not yet available.

COMPASS data taking

First physics data have been taken during the three-months run of 2002 (see also [7]). The 160 GeV polarized muon beam (2.10^8 μ per spill of 5 seconds each), was scattered off the ^{6}LiD target; the target polarization reached 50 %, and its direction was reversed every 8 hours. After one month of detector commissioning, 80 days were devoted to physics data (taking into account beam off days and hardware problems). During this period, 19 days have been dedicated to transversity studies, with the polarized target used in transverse mode. In total, $3.8.10^9$ events were taken in the longitudinal mode, and $1.2.10^9$ events in the transverse mode.

PLANS FOR COMPASS FUTURE

COMPASS plans for 2003 and 2004

Several improvements are foreseen on COMPASS detectors for the next years. Straw tubes detectors and large drift chambers will be added to improve the tracking efficiency at large angle. Electromagnetic calorimeters will be added and fully equipped in front of the hadronic calorimeters. The COMPASS solenoid magnet should be ready by 2004 or 2005; it will increase the angular acceptance from 70 to 180 mrad.

In 2003 and 2004, four months per year (about 100 days of full data taking) of muon beam are foreseen. The ^{6}LiD polarized target will be used in 2003; in 2004 it may be replaced by the NH_3 polarized target. 80% of these data will be taken with a longitudinal polarization, and the rest with a transverse one.

Tentative plans for 2006 and after

During the year 2005, the SPS accelerator will not run. In 2006 and beyond, a large part of the data taking will probably be devoted to the hadron beam physics program. In particular, Primakov reactions and charmed hadron physics may be studied with a set of data taken during one or two years. Glueballs and exotics programs request more data taking time.

If necessary, additional data with the polarized muon beam and the upgraded spectrometer can also been taken.

Several spectrometers upgrades are foreseen. The major one is the installation of a RICH detector in the second spectrometer in 2007.

Different targets (lead, copper, carbon, liquid H_2) and a silicon tracker to be placed just behind the target, will be prepared for the hadron beam program. In addition, upgrades in the trigger systems and the data acquisition will be necessary.

Deep Virtual Compton Scattering studies

Some studies [8] show that Deep Virtual Compton Scattering (DVCS) can be studied

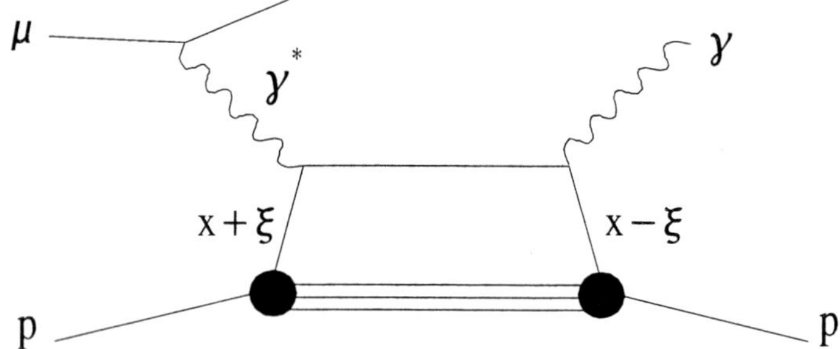

Figure 3: Deep Virtual Compton Scattering diagram

at COMPASS if some specific upgrades are made. DVCS (Figure 3) gives access to the generalized partons distributions in the nucleon [9, 10], which provide new insights on quark correlation in nucleon structure. A possible goal could be to measure the DVCS total cross section (which dominates the Bethe-Heitler process at energies of ~ 190 GeV) and the beam charge asymmetry between positive and negative muon (which is sensitive to the the DVCS - Bethe-Heitler interference at ~ 100 GeV).

The DVCS studies require the following modifications of the COMPASS set-up. Recoil protons must be detected with a time-of-flight and dE/dX detector for proton-pion discrimination at large angle (30 to 80°). The electromagnetic calorimetry acceptance must be increased to detect photons up to 20°. A liquid H_2 target (which may be common to the hadron program) is necessary. Finally a veto detector must be added to remove halo events.

CONCLUSION

The COMPASS experiment will cover a vast physics program in nucleon spin structure and hadron spectroscopy. Physics data taking has started in 2002, with a 160 GeV polarized muon beam scattered on a polarized deuteron target, for the measurement of ΔG and the transversity. This program will continue until 2004 and perhaps beyond. No beam is foreseen in 2005. Hadron spectroscopy physics is planed to be studied from 2006 on, and the opportunity to study Deep Virtual Compton Scattering is investigated.

REFERENCES

1. The COMPASS Collaboration, COMPASS Proposal, CERN/SPLC 96-14, SPSLC/P297, 1996
2. G. Altarelli and W.J. Stirling, Particle World Vol. 1, 40 (1989)
3. M. Glück and E. Reya, Z. Phys. C 39, 569 (1988)
4. A. Bravar, D. Von Harrach and A. Kotzinian, Phys. Lett. B 421, 349 (1998)
5. J. Collins, Nucl. Phys. B 396, 161 (1993)
6. M. Adamovitch et al., Phys. Lett. B305, 402 (1993)
7. F. Tessarotto for the COMPASS Collaboration, Status of the COMPASS experiment, these proceedings
8. E. Burtin et al., Outline for a Deeply Virtual Compton Scattering Experiment using COMPASS at CERN, to be published
9. X. Ji, Phys. Rev. Lett. 78, 610 (1997)
10. M. Vanderhaeghen, these proceedings

Future Spin Experiments at SLAC

Stephen Rock for the Real Photon Collaboration

University of Mass, Amherst MA 01003

Abstract. A series of three photo-production experiments using a new polarized coherent quasi-mono-energetic photon beam have been approved at SLAC. Experiment E159 will measure the high energy end of the GDH sum rule. E160 will measure the A dependence of J/ψ and ψ' to determine ordinary nuclear attenuation and compare with that observed in heavy ion collisions. E161 will measure the gluon polarization in the nucleon using open charm production.

THE PHOTON BEAM

The three approved experiments require a mono-energetic high intensity photon beam and two of them require that the beam be longitudinally polarized. This beam will be produced using collimated coherent bremsstrahlung of a polarized electron beam hitting an oriented diamond crystal. The electron beams will have energies from 9.9 to 48.5 GeV; an intensity of up to 5×10^{10} electrons per pulse; repetition rate of 120 Hz; pulse length of 500 nsec; and polarization of about 83%. Typical photon beam energy spectra are shown in Fig. 1. The coherent peak is produced by constructive interference of the photons produced at different planes of the diamond crystal. Different orientations of the crystal produce peaks at different energies and intensities. Each experiment has different criteria for optimum results. The beamline elements for E160 are shown in Fig. 2.

E159: THE GDH SUM RULE

The Gerasimov-Drell-Hearn (GDH) sum rule [1] is one of the most fundamental relations in hadronic physics, and its experimental test is one of the major challenges for photo-production experiments over the next decade. The GDH sum rule relates the difference in total hadronic photo-absorption cross sections for left- ($\sigma_A^{\gamma N}$) and right-handed ($\sigma_P^{\gamma N}$) circularly polarized photons interacting with longitudinally polarized nucleons to the square of the nucleon's anomalous magnetic moment κ,

$$\int_{k_\pi}^{\infty} \frac{dk}{k} \Delta\sigma^{\gamma N}(k) = \frac{2\pi^2 \alpha \kappa^2}{M^2} \tag{1}$$

where k is the photon energy, $\Delta\sigma^{\gamma N}(k) = \sigma_P^{\gamma N}(k) - \sigma_A^{\gamma N}(k)$, M is the nucleon mass, and k_π is the threshold energy needed to produce at least one pion. An alternate notation is $\Delta\sigma^{\gamma N} = \sigma_{3/2}^{\gamma N} - \sigma_{1/2}^{\gamma N}$, where $1/2$ $(3/2)$ refers to the spin of the nucleon-photon system. Experimental data is need up to high energies to measure the complete integral. SLAC

CP675, *Spin 2002: 15th Int'l. Spin Physics Symposium and Workshop on Polarized Electron Sources and Polarimeters*, edited by Y. I. Makdisi, A. U. Luccio, and W. W. MacKay

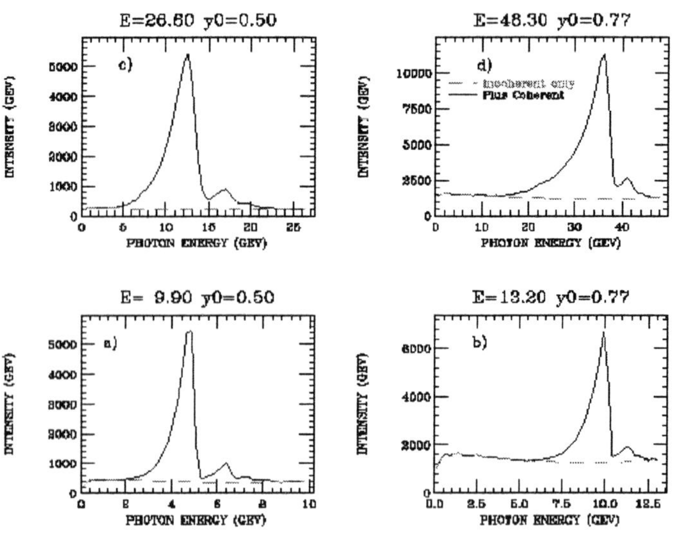

FIGURE 1. Calculated intensity (flux times energy) for collimated coherent bremsstrahlung at four settings. The dashed lines are incoherent radiation only, while the solid lines include coherent contributions.

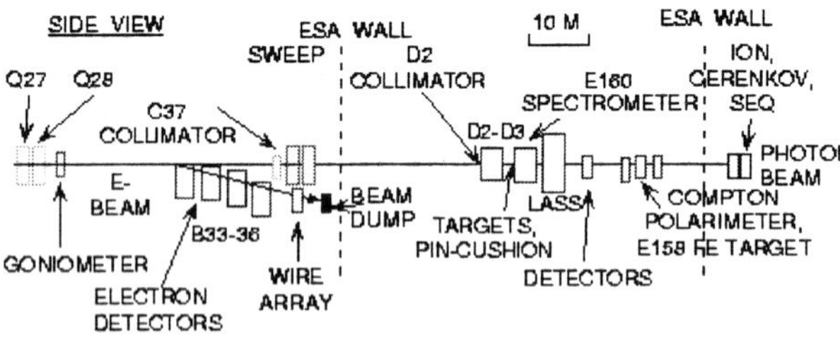

FIGURE 2. Overall view of the main components of the photon beam (horizontal scale is approximate, vertical scale is exaggerated)

is the only place where measurements above 6 GeV can presently be done. Numerically, the GDH sum rule prediction is 204 μb for the proton, 232 μb for the neutron, and 219 μb (-15 μb) for the average isoscalar (ISO-vector) combinations. The fundamental meaning of the sum rule is that any particle with a non-zero anomalous magnetic moment must have an excitation spectrum and internal structure. The energy scale at which the sum rule is saturated gives an indication of the energy scale beyond which nucleonic excitations become asymptotically spin-independent. Many authors [2] have analyzed the connection between the GDH sum rule, valid for real photons

($Q^2 = 0$), and the analogous sum rules for virtual photons, in particular the equally fundamental Bjorken Sum Rule [3] for $Q^2 \to \infty$, which has been the the subject of intense experimental study in the past decade. A generalized GDH Sum Rule can be formed that smoothly connects the $Q^2 = 0$ and large Q^2 limits. The goal of the present proposal is to augment the large body of data recently acquired at SLAC with virtual photons, with new data using real photons. This will provide valuable $Q^2 = 0$ anchor points for the study of spin-dependent photo-absorption cross sections in the 5 to 40 GeV energy range, thought to be responsible for much of the ISO-vector sum rule strength. Because there are no existing data in this energy region, it will be an exciting experimental venture, with surprising results certainly possible. Such was the case with the $Q^2 > 0$ measurements, which have concluded that the Ellis-Jaffe sum rules for the proton (p) and neutron (n) are strongly violated, although the more fundamental ISO-vector ($p - n$) Bjorken Sum Rule rule has been validated within experimental errors. Data from this proposal will compliment the vast body of spin-averaged data on $\sigma^{\gamma N}(k)$. Measurements of the spin degree of freedom will help to gain insight into underlying reaction mechanism for photo-absorption, such as the role of reggeon exchange and a possible pomeron cut contribution. A good understanding of soft Regge physics is essential for the interpretation of the Q^2-dependence of data taken with virtual photons. The energy region of 5 to 40 GeV accessible at SLAC extends the upper limit of integration by about an order of magnitude, which will likely be high enough if $\Delta\sigma^{\gamma N}$ is smoothly decreasing in strength with increasing energy. However, as in the resonance region, specific degrees of freedom for nucleon excitations can come in to play to cause oscillations in the magnitude of $\Delta\sigma^{\gamma N}$ which may well be significant. As an example, the SLAC energy range spans charm threshold. It is entirely possible that $\Delta\sigma^{\gamma N}$ is approximately constant in the SLAC energy range, which would imply that excitations at even higher energies play a crucial role in preventing the GDH integral from becoming divergent.

E160: PROPAGATION OF J/ψ AND ψ' IN NUCLEAR MATTER

The search for J/ψ suppression in heavy ion collisions is generally regarded as one of the most compelling signatures for the onset of the quark-gluon plasma [4](e.g. Fig.3 This search is one of the main goals of the newly constructed RHIC facility. However, modeling this suppression requires accurate knowledge of the J/ψ-nucleon and $\psi\prime$-nucleon cross section. The best way to measure this quantity is through the A-dependence of photo-production.

The cross section is dominated by quasi-elastic production. In this process, a photon undergoes a hard $\gamma N \to J/\psi N$ interaction with a nucleon in a nucleus with a kinematically dependent production time (or equivalently distance) scale. It takes a significantly longer time for the J/ψ to evolve from the point-like configuration of the hard interaction, to a full on-shell physical J/ψ particle. This formation time should be small compared to the nuclear radius R_A in order to use Glauber theory extract to the physical J/ψ-nucleon cross section. By varying both the production time. the formation time and the distance to be traveled in the nuclear medium (radius of nucleus) we can get

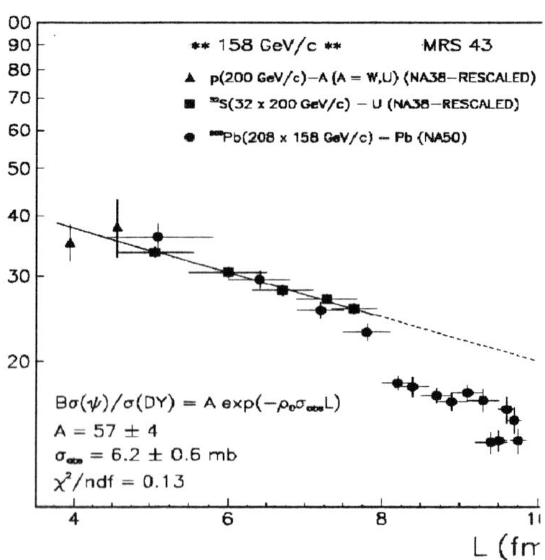

FIGURE 3. Suppression of J/ψ in heavy ion collision from NA50

a complete description of the propagation of the J/ψ in ordinary matter. We will use 4 different target materials and photon beam energies of 15, 25 and 35 GeV to span the relevant parameter space.

E161: THE GLUON POLARIZATION IN THE NUCLEON

To complete our understanding of the components of the nucleon spin it is necessary to make an accurate measurement of the contribution of the gluon. The unpolarized parton distributions have been measured over the last 30 years by many experiments. More recently the polarized up and down quark distributions have been measured in Deep Inelastic Scattering. The gluon spin density in the nucleon is poorly known since it comes mostly from the pQCD evolution of the DIS data. Current models differ widely. They indicate that the gluon spin makes very large contributions to the nucleon spin and cannot be ignored. Experiments at CERN, RHIC and SLAC will measure the gluon spin directly using a variety of theoretical and experimental techniques which complement one another in the x range covered and experimental and theoretical challenges.

At SLAC the gluon spin density within the nucleon will be determined by measuring the asymmetry of polarized photo-production of open charmed quarks from a polarized target. The hard scattering process is dominated by photon-gluon fusion and has been calculated to next to leading order in pQCD. Other mechanisms for producing charm are suppressed because of the large ($\sim$ 1.5 GeV) mass of the charm quark. Intrinsic charm contributions to the nucleon are expected to be very small at our kinematics. At photon energies above 40 GeV effects from associated production of Λ_c are estimated to be small. The highly polarized quasi-mono-chromatic photon beam will interact with

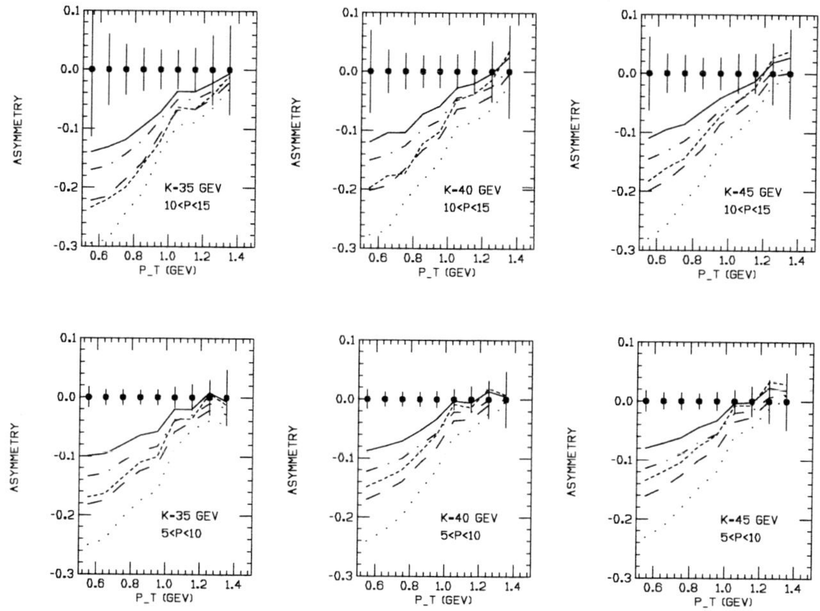

FIGURE 4. Estimated uncertainties for the asymmetry in open charm photo-production compared to various models of $\Delta g(x)$

longitudinally polarized nucleons in LiD. A polarization of greater than 60% will be attained using a temperature of 300 mK, and a 6.5 T magnetic field.

The open charm final state will be tagged by decays of D mesons into high transverse momentum muons. The parallel/anti-parallel asymmetry for producing open charm is closely related to the fundamental polarized gluon spin density $\Delta g(x)$. The asymmetry for single muons will be measured as a function of muon momentum, muon transverse momentum, and photon beam energies with sufficient precision to discriminate among models of $\Delta g(x)$ that differ from each other by as little as 10% in the range $0.1 < x < 0.2$. Significant constraints will be placed on both the shape and magnitude of $\Delta g(x)$ as illustrated in Fig. 4 The projected errors are significantly smaller than for other experiments that plan to make direct measurements of $\Delta g(x)$.

REFERENCES

1. S. D. Drell and A. C. Hearn, Phys. Rev. Lett 16, 908 (1966); S. B. Gerasimov, Yad. Fiz. 2, 598 (1966); S.J. Brodsky and J.R. Primack, Ann. Phys. 52 (1969) 315.
2. L. Tiator, D. Drechsel, and S.S. Kamalov, http://arXiv.org/abs/nucl-th/000561 and references therein; http://arXiv.org/abs/hep-ph/0008306.
3. J. D. Bjorken, Phys. Rev. 148 (1966) 1467; Phys. Rev. D 1 (1970) 1376.
4. T. Matsui and H. Satz, Phys. Lett. B178, 416 (1986).

Research Perspectives at Jefferson Lab: 12 GeV and Beyond

Kees de Jager

Jefferson Laboratory, Newport News, VA 23606, USA

Abstract. The plans for upgrading the CEBAF accelerator at Jefferson Lab to 12 GeV are presented. The research program supporting that upgrade are illustrated with a few selected examples. The instrumentation under design to carry out that research program is discussed. Finally, a conceptual design of a future upgrade which combines a 25 GeV fixed-target facility and an electron-ion collider facility at a luminosity of up to 10^{35} cm^{-2}s^{-1} and a CM energy of over 40 GeV.

1. INTRODUCTION

The design parameters of the Continuous Electron Beam Accelerator Facility (CEBAF) at the Thomas Jefferson National Accelerator Facility (JLab) were defined nearly two decades ago. In that period our understanding of the behaviour of strongly interacting matter has evolved significantly, providing important classes of experimental questions which can be optimally adressed by a CEBAF-type accelerator at higher energy. The original design of the facility, coupled to developments in superconducting RF technology, makes it feasible to triple the initial design value of CEBAF's beam energy to 12 GeV in a cost-effective manner (at about 15% of the cost of the initial facility).

The research program with the 12 GeV upgrade will provide breakthroughs in two key areas: (1) the experimental confirmation of the origin of quark confinement by QCD flux tubes and (2) the detailed mapping of the quark and gluon wavefunctions. In addition, the upgrade will provide important advances in areas already under study. A detailed overview of the upgrade research program is given in the recent White Paper[1].

2. ACCELERATOR

CEBAF was originally designed to accelerate electrons to 4 GeV by recirculating the beam four times through two superconducting linacs, each producing an energy gain of 400 MeV per pass. The beam can be split arbitrarily between three interleaved 499 MHz bunch trains. One such bunch train can be peeled off to any one of the Halls after each linac pass using RF separators and septa, while all Halls can simultaneously receive the maximum energy beam.

Each linac tunnel provides sufficient space to install five additional newly designed cryomodules. The new cryomodules will each provide over 100 MV (compared to the 28 MV from the existing ones), by increasing the gradient to 20 MV/m and the number

CP675, *Spin 2002: 15th Int'l. Spin Physics Symposium and Workshop on Polarized Electron Sources and Polarimeters,* edited by Y. I. Makdisi, A. U. Luccio, and W. W. MacKay

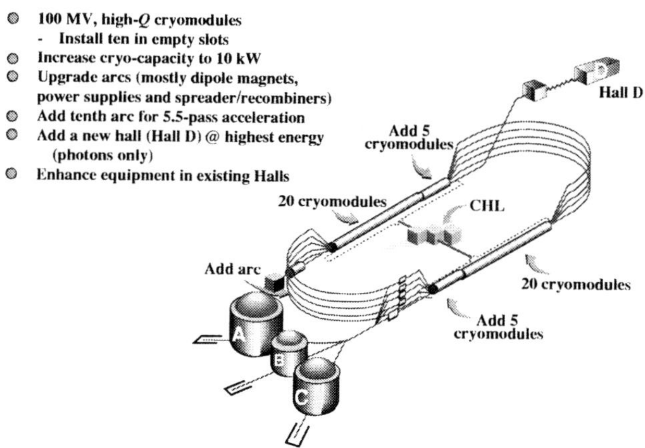

FIGURE 1. Overview of the accelerator upgrade to 12 GeV.

of cavity cells from five to seven. This will result in a maximum energy gain per pass of 2.2 GeV, providing a maximum beam energy to Halls A, B and C of 11 GeV. Hall D will be provided with the desired maximum energy of 12 GeV by adding a tenth arc and recirculating the beam a fifth time through one linac. A total of 90 μA of CW beam can be provided at the maximum beam energy. Further modifications required are changing the dipoles in the arcs from C-type to H-type magnets, replacing a large number of power supplies and doubling the central helium liquifier capacity to 10 kW. An overview of the upgrade of the accelerator is shown in Fig. 1.

3. HALL A

The present base instrumentation in Hall A has been used with great success for experiments which require high luminosity and high resolution in momentum and/or angle of at least one of the reaction products. The central elements are the two High Resolution Spectrometers (HRS). Both of these devices provide a momentum resolution of better than 2×10^{-4} and an angular resolution of better than 1 mrad with a design maximum central momentum of 4 GeV/c.

With the upgrade a large kinematic domain becomes available for studies of deep inelastic scattering. The combination of high luminosity and high polarization of beam and targets offers a unique opportunity to make significant contributions to the understanding of nucleon and nuclear structure and of the strong interaction in the high-x region. For example, the spin structure functions g_1 and A_1 of the neutron will be measured accurately by using a polarized ^{3}He target, establishing unambiguously the trend of A_1^n when x goes to 1, which will provide a benchmark test of pQCD and constituent quark models. Measurements of the g_2^n spin structure function and its moments will provide a clean measure of a higher twist effect (twist 3), which is related to quark-gluon correlations.

Two instrumentation upgrades are proposed to allow an optimal study of the intended experiments: a magnetic spectrometer, dubbed MAD (Medium Acceptance Detector), and an electro-magnetic calorimeter.

The proposed MAD device is a magnetic spectrometer built from two combined-function (quadrupole and dipole) superconducting magnets. The design provides a momentum acceptance (resolution) of ± 15 (0.2)% and an angular acceptance (resolution) of 30 msr (2 mrad). The maximum central momentum is 6 GeV/c at a total bend angle of 20°. The basic detector package for the MAD spectrometer, covering the full momentum and angular acceptance, includes: fast high-resolution tracking chambers, a hydrogen-gas Čerenkov counter, trigger scintillator counters and a lead-glass hadron rejector. For the detection of hadrons a variable-pressure gas Čerenkov counter, two diffuse-reflective aerogel counters, a Ring Imaging Čerenkov counter and a Focal Plane Polarimeter will also be available.

4. HALL B

The CEBAF Large Acceptance Spectrometer (CLAS) in Hall B is used for experiments that require the detection of several, loosely correlated particles in the hadronic final state at a limited luminosity. CLAS is a magnetic toroidal multi-gap spectrometer. Its magnetic field is generated by six superconducting coils. The detection system consists of drift chambers to determine the track of a charged particle, gas Čerenkov counters for particle identification, scintillation counters for the trigger and for measuring time-of-flight and electromagnetic calorimeters to detect showering particles like electrons and photons. CLAS presently operates at a luminosity of 10^{34} cm^{-2}s^{-1}.

With the 12 GeV upgrade the CLAS research program will focus on Generalized Parton Distributions (GPD) through the study of exclusive processes at large momentum transfer. The GPD's can be considered as overlap integrals between different components of the hadronic wave function[2], governed by the selection of the final state. Measurements of these GPD's will thus make it possible to map out quark and gluon wave functions. One way to access the quark orbital angular momentum contribution to the nucleon spin is through GPD's. Factorization is an essential ingredient in the extraction of GPD's. For Deeply Virtual Compton Scattering (DVCS) scaling will have been achieved at 11 GeV, but this has to be established experimentally for other processes.

The CLAS upgrade incorporates two major improvements: (1) increasing the luminosity by an order of magnitude to account for the lower cross section values, (2) provide more complete detection of the hadronic final state. The use of major components (torus magnet, scintillators, Cerenkov counters and EM calorimeters) will be retained. The tracking chambers will be replaced and a new central detector added.

5. HALL C

The Hall C facility has generally been used for experiments which require high luminosity at moderate resolution. The core spectrometers are the High Momentum Spectrometer HMS and the Short Orbit Spectrometer (SOS). These two devices have been used flexibly as either electron or hadron arms, at times in coincidence with each other, at times in coincidence with a third experiment-specific arm. The HMS has a maximum momentum of 7.6 GeV/c, the SOS a value limited to only 1.7 GeV/c.

The Hall C research program after the 12 GeV upgrade will be focused on electron-hadron coincidence experiments at large $z \equiv E_h/\nu$, where E_h and ν are the hadron energy and the energy loss, respectively. Examples of such experiments are measurements of the pion form factor at large Q^2 and a study of quark-hadron duality. This research program requires particle detection at a high luminosity, a small minimum scattering angle and a high maximum momentum. These conditions will be met by the so-called Super High Momentum Spectrometer (SHMS) which will replace the SOS spectrometer. The SHMS design consists of two superconducting quadrupoles and one combined-function magnet. It will have a maximum momentum of 11 GeV/c, a minimum scattering angle of 5.5°, a momentum acceptance (resolution) of ± 10 (0.2)% and an angular acceptance (resolution) of 2 msr (2 mrad). The basic configuration of the detector stack would consist of DC tracking chambers, trigger scintillator hodoscopes and a lead glass calorimeter.

6. HALL D

The Hall D research program will be focused on a definitive measurement of the spectrum of exotic hybrid mesons, which are expected in a mass range from 1 to 2.5 GeV/c^2. Lattice QCD calculations have convincingly illustrated[3] the linear quark-quark potential necessary for confinement. However, very little is still known about the direct excitation of the flux tube. The observation of such direct manifestations of gluonic degrees of freedom will provide understanding of confinement[4]. The quantum numbers of the flux tube, added to those of a $q\bar{q}$ meson, can produce exotic hybrids with unique J^{PC} quantum numbers. These excitations can be probed far more effectively with photons than with π- or K-mesons, because the quark spins are aligned in the virtual vector-

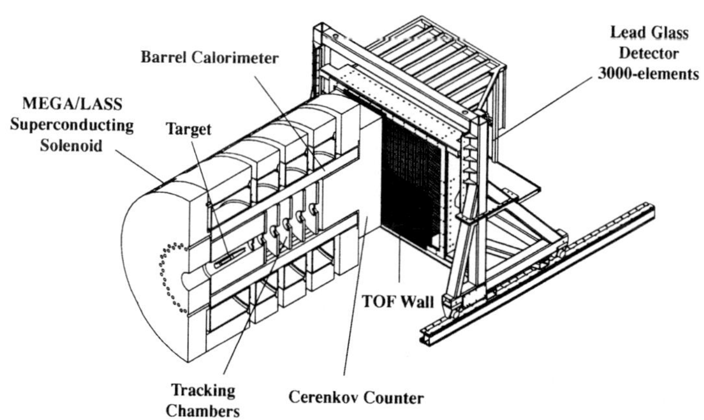

FIGURE 2. Schematic view of the detector in Hall D.

meson component of the photon. For a full partial-wave analysis of such excitations linearly polarized photons are a requisite.

The optimum photon energy for production of exotic hybrids in its expected mass range is between 8 and 9 GeV. Linearly polarized photons in this energy range are optimally produced by coherent bremsstrahlung with 12 GeV electrons. The Hall D detector provides a nearly hermetic acceptance for both charged and neutral particles and includes several particle identification systems. Figure 2 is a schematic representation of the proposed detector. Momentum analysis of charged particles is achieved with a superconducting solenoid and tracking chambers. The final planned photon flux is 10^8 photons/s. At this flux the experiment will acumulate in one year of running a factor of 100 more meson data than are presently available even from pion production.

7. ELECTRON-ION COLLIDER

A conceptual design has been initiated for a high-luminosity asymmetric (5 GeV on 250 GeV) electron-ion collider at Jefferson Lab. The ELIC (Electron Light-Ion Collider) proposal[5] is based on the following concepts. A "figure 8" booster and storage ring is added for final acceleration to a maximum of 250 GeV, spin preservation and flexible manipulation of ^{1}H, ^{2}H and ^{3}He ions. All of the remaining 20 MV cryo-modules in CEBAF are replaced by new 100 MV modules, so that electrons can be accelerated to 5 GeV in a single pass. After circulating for appr. 100 turns, the electrons are re-injected into CEBAF for deceleration and energy recovery. This design results in a maximum CM energy of over 40 GeV. A luminosity of 10^{35} cm^{-2}s^{-1} appears feasible through the use of electron cooling and crab crossings. This design will also provide a 25 GeV beam in fixed-target mode.

ACKNOWLEDGMENTS

This work was supported by DOE contract DE-AC05-84ER40150 under which the Southeastern Universities Research Association (SURA) operates the Thomas Jefferson National Accelerator Facility.

REFERENCES

1. The 12 GeV Upgrade of CEBAF, White Paper prepared for the NSAC Long Range Planning Exercise, 2000, L.S. Cardman *et al.*, editors.
2. X. Ji, *Phys. Rev. Lett.* **78**, 610 (1997); A. Radyushkin, *Phys. Lett.* B **380**, 417 (1996).
3. G.S. Bali *et al.*, Proceedings of Int. Conf. on Quark Confinement and the Hadron spectrum, World Scientific, 1995, p. 225.
4. N. Isgur, R. Kokoski and J. Paton, *Phys. Rev. Lett.* **54**, 869 (1985);
   S. Godfrey and J. Napolitano, *Rev. Mod. Phys.* **71**, 1411 (1999).
5. L. Merminga *et al.*, Proceedings of EPAC 2002, June 2002, Paris.

Perspectives for a Next-Generation Electron-Nucleon Scattering Facility in Europe

R.Kaiser

Department of Physics and Astronomy, University of Glasgow, Glasgow G12 8QQ, United Kingdom

Abstract. This paper discusses perspectives for a future fixed-target electron-nucleon scattering facility in Europe. Based on the intended measurements the requirements on accelerator, target and spectrometer are presented and their possible realisation is discussed.

INTRODUCTION

With the advent of Generalised Parton Distributions (GPDs)[1, 2, 3] a unified approach to the description of nucleon structure has become possible. Formerly separate aspects, such as form factors and (forward) parton distribution functions, can now be described in a common framework. An investigation of the currently approved experiments [4] shows that the study of the momentum and spin structure of the nucleon cannot be completed with the current generation of accelerators and spectrometers.

In a European context a next-generation facility is being discussed in the form of a fixed-target experiment, combining a linear accelerator with a large acceptance spectrometer. Two related proposals (ELFE@DESY [5], TESLA-N [6]) have been included in the appendix of the TESLA technical design report. The 'Declaration of Ferrara'[1], outlining the need for a next-generation facility, has already been signed by more than 160 European scientists and a European network is being formed. The alternative idea, a colliding beam facility, has also originally been discussed in Europe [7] and may now be realised in North-America in the form of the Electron Ion Collider (EIC) [8]. A collider and a fixed-target facility would complement each other in an ideal way, as the different kinematics would allow better access to gluon or quark distributions, respectively.

This paper outlines the main physics questions and the resulting requirements of a next-generation facility. Possible accelerator options are discussed and the basic design parameters for a fixed-target spectrometer are presented.

[1] `http://www.fe.infn.it/qcd-n02/`

CP675, *Spin 2002: 15th Int'l. Spin Physics Symposium and Workshop on Polarized Electron Sources and Polarimeters,* edited by Y. I. Makdisi, A. U. Luccio, and W. W. MacKay

PHYSICS AND RESULTING REQUIREMENTS

The complete set of GPDs consists of eight distributions for each parton species $f = u,d,s,g$. These distributions are classified according to whether they they flip *parton helicity* (subscript T) or not, if they conserve *nucleon helicity* (H) or not (E) and if they are unpolarised or polarised (marked with $\tilde{}$) [9, 10]. When all currently planned experiments will be completed and analysed, some information will be available on the unpolarised u-quark GPD H^u. All other GPDs will remain completely unmeasured [11].

The limiting cases of GPDs, parton distributions and form factors, have been studied for some time and will mostly be well known after the completion of the current or currently planned experiments. Notable exceptions are the transversity distributions and the polarized gluon distribution Δg - they will be measured, but still lack in precision. An overview of the experimental status of the relevant quantities is given in [4].

Generalised Parton Distributions can be accessed through measurements of exclusive reactions, e.g. Deeply Virtual Compton Scattering (DVCS), while transversity and Δg require semi-inclusive measurements. A common requirement for both types of measurement is the increase of the available luminosity by two orders of magnitude compared to present facilities to $10^{35}\mathrm{cm}^{-2}\mathrm{s}^{-1}$. The following sections will give a short overview of the requirements on the different parts of a future facility (accelerator, target, spectrometer) and review the available options.

ACCELERATOR OPTIONS

The present beam energy at HERMES can be seen as the lower limit for a future facility, while the energy available at TESLA-N has been shown to be sufficient for the semi-inclusive measurements. The technically feasible spectrometer resolution (see below) limits exclusive measurements to beam energies below about 50 GeV. The accelerator for a next-generation fixed-target facility should therefore ideally have a variable beam energy from 30-250 GeV.

As beam charge asymmetries are part of the measurement program for GPDs, both beam charges are required. The necessary beam intensity to achieve the desired luminosity of $10^{35}\mathrm{cm}^{-2}\mathrm{s}^{-1}$ can only be reached with electrons/positrons. The luminosity also leads to the requirement of a large duty factor $> 10\%$ to avoid pile-up. The condition of exclusivity implies that the energy spread of the beam should be of the order of only 1/3 of the pion mass or less.

Up to now three proposals for an accelerator have been made:

- ELFE@DESY [5]: use part of TESLA as injector and HERA as stretcher ring
- TESLA-N [6]: fill e^- into 'empty buckets' in the e^+-arm of TESLA
- EVELIN [12]: use TESLA cavities to put a 4.5 km linac into the HERA tunnel

All of these proposals fulfill some, but not all requirements: TESLA-N has a low duty factor of 0.5 % and delivers only electrons, the beam energy is limited to 27 GeV at ELFE@DESY and to 75 GeV at EVELIN. The duty factor of TESLA-N could perhaps be improved to 12.5% if two different RF sources could be used in TESLA.

TARGETS

Beam currents of the order of 1 A would be necessary to achieve a luminosity of $10^{35}cm^{-2}s^{-1}$ with gas targets. This does not appear to be feasible and therefore it is unavoidable to use solid polarised and solid or liquid unpolarised targets. A liquid H_2 'spaghetti' could be used as unpolarised target, while NH_3, ND_3 or ^6LiD would be suitable materials for polarised targets [5, 6, 13]. While there is no problem in principle, a series of technical problems must be solved, especially concerning the heat load from the beam and the design of a target geometry that allows for the combination with a recoil detector.

It is instructive to compare the luminosity of polarised experiments with solid state targets and of experiments at a collider or with a pure target. In such a comparison an effective polarised luminosity L_{eff} should be used that includes an appropriate correction factor. Assuming a NH_3 target with 80% polarisation and a purity of 0.176 and a proton beam/target with 70% polarisation and purity 1.0 this correction factor is about 25. Figure 1 shows the effective luminosity versus the center-of-mass energy for present and future high-energy eN-scattering facilities.

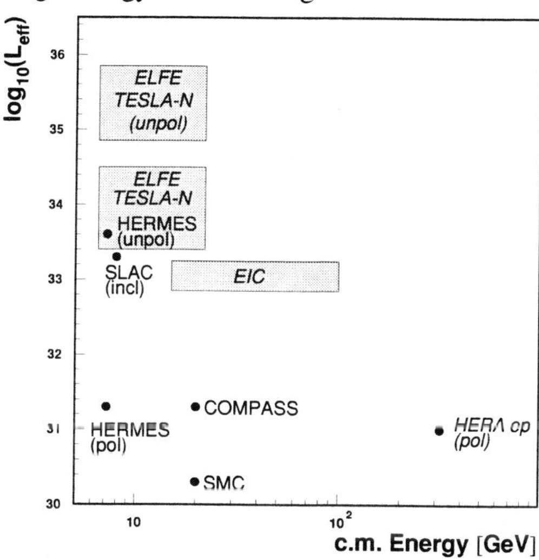

FIGURE 1. Effective Luminosity L_{eff} for present and future high-energy eN-scattering facilities. Shaded (yellow) areas mark the anticipated regions of ELFE/TESLA-N and EIC, future projects are labeled in tilted font. The point for HERMES (unpolarised) appears within the area of ELFE/TESLA-N (polarised).

SPECTROMETER

In the absence of perfect hermeticity and detector efficiency, exclusivity can only be guaranteed through a missing mass resolution that is small compared to the pion mass. An even stronger condition arises if also excited intermediate states should be detected

($\Delta M_{min} = M_\Sigma - M_\Lambda = 77$ MeV). The key issue of the spectrometer is therefore the momentum resolution $\frac{\sigma_p}{p}$, which can be approximated as [5]

$$\frac{\sigma_p}{p} \approx \frac{0.0138\sqrt{\delta/X_0}}{0.3 \cdot \int B\, dl} \frac{z_1}{z_m} \tag{1}$$

where $\int B\, dl$ is the integrated field strength, $\sqrt{\delta/X_0}$ is the thickness of material in front of the second tracking detector and $\frac{z_1}{z_m}$ the ratio of the position of the first tracking detector to that of the magnet center. For a sufficient missing mass resolution (≤ 40 MeV) a momentum resolution of $\leq 0.1\%$ is required at a beam energy of 50 GeV.

While factors like the maximisation of the integrated magnetic field over a short distance are important, the main technical challenge arises from the requirement to keep the material between target and the second tracking detector as thin as possible. This implies the necessity of a large vacuum vessel reaching from the target to the second tracking detector. Detector 1 is therefore located inside the vacuum (see fig. 2). One technical solution is the use of scintillating fibres: 5 layers of 500 μm fibres with a pitch of 340 nm offer 80 μm resolution, 99% efficiency and a thickness of only 0.6% of a radiation length. However, to measure both coordinates the thickness must be doubled and this is already crossing the limits of the thickness allowed for the required resolution. A possible alternative could be the use of a large area double sided silicon detector: 200 μm silicon plus Kapton foils with printed copper lines add up to only about 0.5% of a radiation length. Such a detector would have the problem of acceptance gaps while the size of this detector (about 7 m^2) would still be small compared to the largest silicon detectors (CMS silicon tracking detector, 206 m^2 [14]).

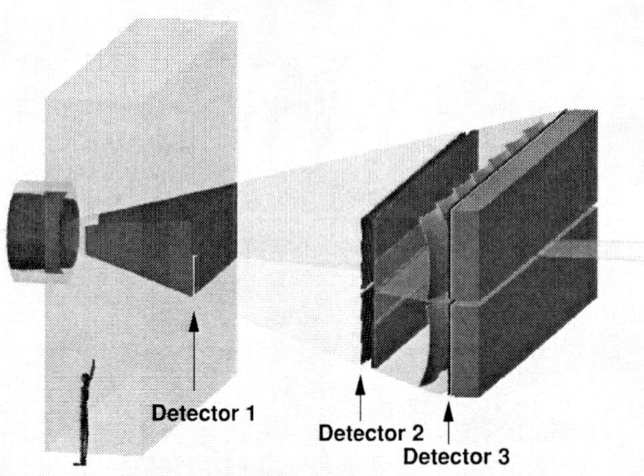

FIGURE 2. Schematic view of the ELFE spectrometer [5].

The conceptual layout of a forward spectrometer has already been studied for the ELFE proposal [5] at a beam energy of 25 GeV. It is shown in figure 2. This layout needs to be extended to allow for higher beam energies and to include flexibility towards a possible second stage. Besides the previously discussed tracking detectors the spectrometer should consist of a transition radiaton detector (TRD), ring imaging Cherenkov (RICH) detector and electromagnetic calorimeter for particle identification. The set-up is completed by a recoil detector. For each component detectors exist or are currently being developed that can serve as models: the HERMES TRD [15] or ALICE TRD [16], the LHCb RICH [17] or the COMPASS RICH2 [18], the CMS [19] or ALICE calorimeters [20] and the HERMES Recoil Detector [21]. A detailed discussion of some of these choices can be found in [22].

CONCLUSION

In principle, it appears to be feasible to realise a next-generation electron-nucleon scattering facility with luminosities of $\geq 10^{35}\mathrm{cm}^{-2}\mathrm{s}^{-1}$ as a (polarised) fixed-target experiment. However, further research into experimental technologies is necessary, especially concerning the accelerator design, the combination of targets and recoil detectors and the realisation of ultra-thin large area tracking detectors operating in vacuum.

ACKNOWLEDGMENTS

I would like to thank W.-D.Nowak, G.Rosner and D.v.Harrach for their support and for the careful reading of the manuscript.

REFERENCES

1. Müller, D. *et al.*, *Fortschr.Phys.* 42, 101 (1994).
2. Ji, X., Rev. Lett. **78**, 610 (1997); Phys. Rev. **D 55**, 7114 (1997).
3. Radyushkin, A.V., Phys. Lett. **B 380**, 417 (1996); Phys. Rev. **D 56** 5524 (1997).
4. Nowak, W.-D., in *Proc. of NATO Advanced Spin Physics Workshop, Yerevan*, Kluwer Academic Publishers, 2002, hep-ph/0210409
5. DeSanctis, E., Laget J.-M. and Rith, K. (edts.), *TESLA TDR Appendix 4*, ISBN 3-935702 06 X
6. Kaiser, R. and Nowak, W.-D. (edts.), *TESLA TDR Appendix 3*, ibid.
7. von Harrach, D., GSI Report 97-04 (1997)
8. Deshpande, A. *et al.*, *The Electron Ion Collider*, White Paper, February 2002, BNL-68933
9. Hoodbhoy, P. and Xi, J., Phys.Rev. **D 58**, 054006 (1998)
10. Diehl, M., Eur.Phys.J. **C 19**, 485 (2001)
11. Korotkov, V.A. and Nowak, W.-D., Nucl. Phys. **A 711**, 174 (2002)
12. von Harrach, D., Nucl. Phys. **A 711** (2002)
13. Meyer, W., private communication
14. Hartmann, F., Nucl.Instrum.Meth. **A478**,285-287 (2002)
15. HERMES Collab., Ackerstaff, K. *et al.*, Nucl.Instrum.Meth. **A417**, 230-265 (1998)
16. ALICE Collab., *ALICE TRD Technical Design Report*, CERN-LHCC-2001-021 (2001)
17. LHC-B Collab., *LHCB RICH Technical Design Report*,CERN-LHCC-2000-037 (2000)
18. Dalla Torre, S., private communication
19. Organtini, G., Nucl.Instrum.Meth. **A478**, 333-335 (2002)
20. Ippolitov, M. *et al.*, Nucl.Instrum.Meth. **A486**, 121-125 (2002)
21. HERMES Collab., Kaiser, R. (contact person), *The HERMES Recoil Detector*, DESY PRC 02-01
22. Chudakov, E., "High Rate Precision Experiments", in *Proc. of QCD-N 02, Ferrara*, 2002

The Electron-Ion Collider: Status and Plans

Richard G. Milner

Bates Linear Accelerator Center,
Laboratory for Nuclear Science
Massachusetts Institute of Technology
21 Manning Road,
Middleton, MA 01949, USA

Abstract. In the last several years, the realization of an Electron-Ion Collider (EIC) with luminosity greater than 10^{33} cm^{-2} s^{-1}, a center-of-mass energy in the range of 30 to 100 GeV and employing spin-polarized electron and nucleon beams as well as beams of low mass to heavy ions has developed into a leading initiative in hadronic physics worldwide. Using the precisely determined electroweak interaction to probe hadronic matter, EIC would open a new window on the fundamental quark and gluon structure of the nucleon and address completely new aspects of hadron structure like the origin of nuclear binding and the search for highly saturated gluonic matter. A promising realization of EIC utilizes the existing Relativistic Heavy Ion Collider (RHIC) at Brookhaven National Laboratory, New York, USA which has accelerated both heavy ions as well as polarized protons. At present, an effort is underway to develop a conceptual design for an electron-ion collider using RHIC within the next several years.

INTRODUCTION

The study of the fundamental structure of matter has been a central goal of physicists through the ages. The tremendous advances in the 20th century culminated in the 1970's in the development of the Standard Model. This is an elegant theoretical framework based on experiment which describes the structure of matter in terms of point-like particles interacting by the exchange of gauge bosons. In the Standard Model, Quantum Chromodynamics (QCD) describes strongly interacting particles in terms of a color interaction between quarks mediated by gluons. An essential goal of present research is to investigate and understand the strong interactions between quarks and gluons that underpin the structure and interactions of nucleons and nuclei. The Electron-Ion Collider (EIC) is proposed as the next essential step needed to understand the fundamental structure of matter.

EIC is motivated by the desire to use lepton scattering to precisely study nucleon and nuclear structure in a regime where the fundamental quark and gluon structure is directly probed using the precisely known QED interaction. The EIC design considerations are shaped by three decades of experimental work carried out with stationary or fixed targets at high energy physics facilities such as SLAC, CERN, DESY, and Fermilab. The inherent limitations of these facilities points to the need for a facility with the following characteristics:

- collider geometry where electron beams collide with beams of protons or nuclei

CP675, *Spin 2002: 15th Int'l. Spin Physics Symposium and Workshop on Polarized Electron Sources and Polarimeters,* edited by Y. I. Makdisi, A. U. Luccio, and W. W. MacKay

· wide range of collision energies (from E_{CM}/nucleon = 15 to 100 GeV)
· high luminosity L = 10^{33} nucleon $cm^{-2}s^{-1}$
· high polarization (> 70%) of electron and nucleon spins
· preferably, two interaction regions with dedicated detectors

The collider geometry offers two major advantages over fixed target electron-proton/ion studies. Firstly, the collider delivers vastly increased energy to the quark or gluon in the collision, providing a greater range for investigating partons with small momentum fraction (x) and their behaviour over a wide range of momentum transfers (Q^2). Secondly. the collider geometry is far superior to fixed-target experiments since it allows detection of complete final-states of the target. High luminosities of order $10^{33}cm^{-2}s^{-1}$ for electron-nucleon scattering are a necessary and crucial charcteristic of EIC. Figure 1. shows the center-of-mass energy vs. luminosity compared to a selection of other existing and planned facilities. EIC will have higher energies than any existing fixed-target machine and a higher luminosity than any existing collider.

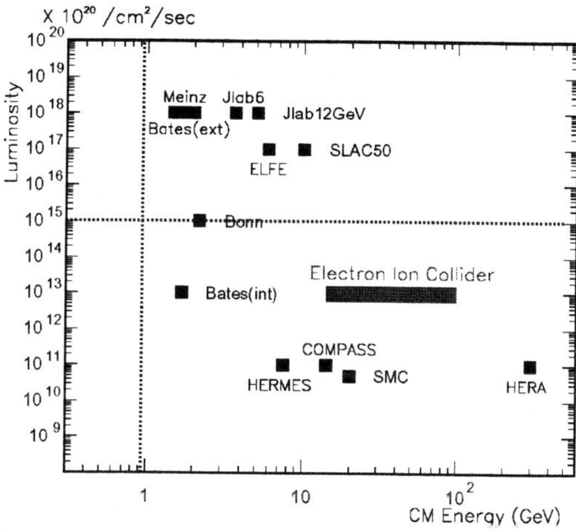

FIGURE 1. The Center-of-Mass Energy vs. Luminosity of EIC Relative to Other Facilities

SCIENTIFIC MOTIVATION

While a great deal has been learned about the quark and gluon structure of matter, some crucial questions about the structure of hadronic matter remain open:

· What is the structure of hadrons in terms of their quark and gluon constituents?
· How do quarks and gluons evolve into hadrons through the dynamics of confinement?
· How do quarks and gluons manifest themselves in the properties of atomic nuclei?
· Does partonic matter saturate in a universal high-density state?
· To what degree can QCD be demonstrated as an exact theory of the strong interaction?

The potential of EIC to open up new frontiers in the study of the fundamental structure of matter can be appreciated by consideration of the following highlights:

Quark and Gluon Distributions in the Nucleon

EIC offers a unique capability for measuring `flavor tagged' structure functions by providing access to a wide range of final states arising from the the fragmentation of the virtual photon. For example, with clean kaon identification both the momentum and spin distributions of the strange quarks can be determined with high precision down to $x \sim 10^{-3}$. The ability to tag the hadronic final-state will allow measurements of the neutron structure function at large x, so that a reliable and precise determination of the ratio of the quark distributions in neutrons and protons can be made in a regime where several competing theoretical predictions exist.

Spin Structure of the Nucleon

EIC operating at the highest CM energy can probe lower x than presently possible and so can search for the dramatic QCD prediction that the proton spin-dependent structure function turns negative at low x. In addition, EIC can provide a direct, clean probe of the gluon polarization by means of a number of measurements, namely charm production, and inclusive scattering.

The Role of Quarks and Gluons in Nuclei

Correlations between Partons

A complete charcterization of the partonic substructure of the nucleon must go beyond a pictureof collinear non-interacting partons. It must include a description of the correlations between the parton densities over impact parameters, and a comparison of the parton wave functions of different baryons. Progress in this direction can be realized by measuring hard, exclusive processes where, in the final state a photon, a meson, or several mesons are produced along the virtual photon direction and a baryon is produced in the nucleon fragmentation region. These processes are expressed, as a result of the new QCD factorization theorems, through a new class of parton distributions termed Generalized Parton Distributions (GPD's). The collider kinematics are optimal for detecting these processes.

The Role of Quarks and Gluons in Nuclei

Most hadronic matter exists in the form of nuclei. The ability of EIC to collide electrons with light and heavy nuclei opens horizons fundamental to nuclear physics. For example, the role of quarks and gluons in nuclei may be investigated by comparing the changes in parton distributions per nucleon as a function of the number of nucleons. Studies of parton modifications at $x \sim 0.1$ will be most sensitive to the underlying quark-gluon structure of the internucleon interactions that are usually described within low energy mesonic theories. It is particularly important to establish the quark distributions at small values of x where the presence of the other nucleons in the nucleus will alter (`shadow') the partonic distributions.

Hadronization in Nucleons and Nuclei

How do the colored quarks and gluons knocked out of nucleons in DIS evolve into the colorless hadrons that must eventually appear? This process is one of the clearest manifestations of confinement: the asymptotic physical states must be color-neutral. Hadronization is a complex process that involves both the structure of hadronic matter and the long range nonperturbative dynamics of confinement. A fundamental question related to hadronization is how and to what extent the spin of the quark is transferred to its hadronic daughters. The ability to `tag flavor' and a facility that creates detectable jets are crucial for these experiments. EIC makes it possible to strike quarks and observe the complete array of decay products from the nucleon or nucleus.

Partonic Matter under Extreme Conditions

Very high energy DIS on nuclear targets with electromagnetic probes offers new opportunities for studying partonic matter under extreme conditions. Particularly intriguing is the regime of very low $x<10^{-3}$ where gluons dominate. Measurements of the proton structure function at HERA showed that the gluon distribution grows rapidly at small x for Q^2 greater than a few $(GeV/c)^2$. When the density of gluons becomes large, they may saturate and give rise to a new form of partonic matter: a color glass condensate. It is a colored glass because the properties of the color-saturated gluons are analogous to that of a spin glass system in condensed matter physics. It is a condensate because the gluons have a large occupation number and are peaked in momentum about a typical scale of the saturation momentum Q_s. The gluon density per unit area is enhanced in nuclei relative to that in individual nucleons by a factor $A^{\frac{1}{3}}$. Therefore, high parton density effects will appear at much lower energies in nuclei than in protons. EIC, with its nuclear beams and e-A CM energies of at least 60 GeV, and its ability to study inclusive and semi-inclusive observables, willprobe this novel regime of QCD.

MACHINE CONSIDERATIONS

Realization of an electron-ion collider with a luminosity of order $10^{33}cm^{-2}s^{-1}$is a formidable task. While such a luminosity for e^+ -e^- colliders has been attained with the B-factories at SLAC and KEK, the requirements of EIC impose special considerations. Firstly, it is essential that the ion beam be cooled to reduce the transverse emittance and thus increase the luminosity. Secondly, the demand that the beams be spin polarized constrains the optics and interaction region. Thirdly, the high luminosity implies significant magnetic focussing near the interaction region and so the detector design is intimately connected with the machine design.

In March 2002 at the EIC workshops at Brookhaven National Laboratory, some important decisions were made concerning the realization of an electron-ion collider. Firstly, it was decided to produce a conceptual machine design within three years. This will require development of an excellent physics case based on a single, optimized machine design and utilizing a suite of carefully designed research equipment. To realize this goal it was decided at the meeting to identify the leading scenario for realization of the collider. To this end, it was decided

- to use the existing Relativistic Heavy Ion Collider (RHIC) for the ion beam.

RHIC can provide both polarized nucleon and heavy ion beams over the large energy range required by EIC. Further, it is clear that using RHIC can greatly reduce the cost of realization of EIC. In addition, the RHIC heavy ion and spin communities are strong participants in EIC and thus BNL provides a natural home for this facility.

- to choose a 10 GeV external electron-ring to RHIC.

This maximizes the freedom to design the interaction region and allows a progression of options to feed the electron-ring: a 2 GeV linac injector using self-polarization and ramping to deliver polarized electron and positron beams in the range of 5 to 10 GeV; a 10 GeV linac with polarized electron source to allow full energy injection of polarized electrons in the energy range from 2 to 10 GeV; a 10 GeV energy recovery linac and polarized electron source to allow full energy injection of polarized electrons in the energy range from 2 to 10 GeV.

FUTURE PLANS

Work on developing the EIC conceptual design is proceeding.

ACKNOWLEDGEMENTS

The work reported here is the fruit of a collaboration of many people. In particular, I would like to acknowledge discussions with A. Bruell, J. Cameron, A. Deshpande, G. Garvey, R. Holt, V. Hughes, R. Jaffe, T. Londergan, L. McLerran, P. Paul, J.-C. Peng, D. van Harrach, and R. Venugopalan on the scientific case for EIC. I have learned much from discussions with D. Barber, I. Ben Zvi, I. Koop, S. Peggs, T. Roser, Y. Shatunov, and C. Tschalaer on the EIC machine design. The author's work is supported by the United States Department of Energy under Cooperative Agreement.

REFERENCES

1. Slide-report of the Joint DESY/GSI/NuPecc Workshop on Electron-Nucleon/Nucleus Collisions, March 3-4 1997, Lufthansa-Zentrum Seeheim, Germany, GSI REPORT 97-04.
2. Proceedings of the Workshop on Physics with a High Luminosity Polarized Electron Ion Collider (EPIC 99), April 8-11 1999, Bloomington, Indiana, USA, Editors L.C. Bland, J.T. Londergan, and A.P. Szczepaniak, World Scientific.
3. Proceedings of the eRHIC Workshop, December 3-4, 1999, Brookhaven National Laboratory.
4. Proceedings of the Second eRHIC Workshop, April 6-8, 2000, Yale University, New Haven, Connecticut, USA, BNL Report 52592.
5. Proceedings of the Second Workshop on Physics with an Electron Polarized Light Ion Collider (EPIC 2000), September 14-16, 2000, MIT, Cambridge, USA, Editor R.G. Milner, AIP Conference Proceedings No. 588.

Status of the e-Ring Design for EIC

D. E. Berkaev, A. V. Otboev, Yu. M. Shatunov*, R. Milner, C. Tschalaer, F. Wang†, B. Parker, V. Ptitsyn** and D. P. Barber‡

*Budker Institute of Nuclear Physics, Novosibirsk, 630090, Russia
†MIT-Bates Linear Accelerator Center, Middleton, MA 01949, USA
**BNL, Upton, NY 19793, USA
‡DESY, Hamburg, Germany

Abstract. The layout and main parameters of the e-ring for EIC project are presented. Optics properties to fulfil so-called spin-transparency conditions to obtain sufficient polarization degree at IP are given. The possibility of using super-bend magnets for polarization time in a wide energy range to decrease is also discussed.

INTRODUCTION

In the Brookhaven National Laboratory (BNL) experiments at new collider RHIC have successfully started with both ion-ion and polarized proton-proton beams [1]. To enhance the experimental capability of the RHIC complex, different schemes of $e-p$ collisions arrangement are under discussion at few last years. High luminosity polarized $e-p$ scattering will open unique opportunity for physics beyond limits of today experiments in polarized DIS.

This paper presents a study of the ring-ring option of EIC. A project of the electron ring with the energy 5–10 GeV was developed in collaboration between BINP (Novosibirsk), BATES-MIT Laboratory and BNL. We suggest (see the Fig. 1) to construct mainly outside the RHIC tunnel the electron storage ring which will have the circumference $\frac{4}{15}$ of the RHIC orbit and an intersection with ions in the one of the existing RHIC experimental area (on 12 o'clock).

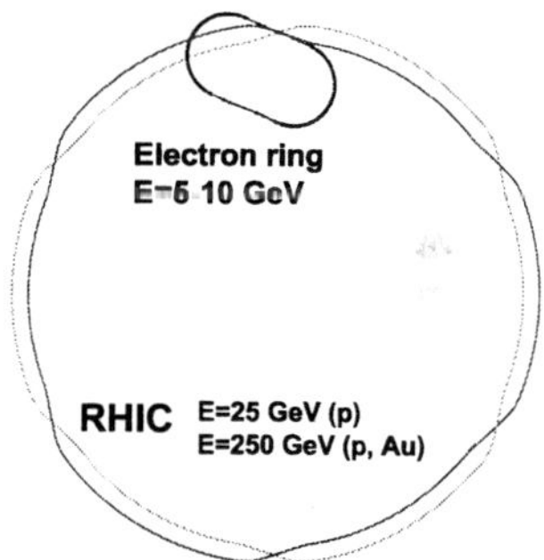

FIGURE 1. The general layout of the e-ring installed into the RHIC complex.

The radiative polarization of the electron beam and a combination of solenoids and bending magnets will provide high degree of the longitudinal polarization of the electron beam in the IP. To minimize the reconstruction of the RHIC rings while adding the new electron ring two possible schemes of the interaction region arrangement are proposed: so-called horizontal "dog-leg" scheme and vertical one. Spin-transparency conditions which are needed for obtaining sufficient polarization degree in electron beam have been found for both options of the IP layout.

CP675, *Spin 2002: 15th Int'l. Spin Physics Symposium and Workshop on Polarized Electron Sources and Polarimeters*, edited by Y. I. Makdisi, A. U. Luccio, and W. W. MacKay

THE LUMINOSITY CONSIDERATION

Achieving the high luminosity value of 1×10^{33} cm^{2}s^{-1} is a main challenge and needs a special consideration that has to take into account both a world wide experience of many machines, either electron's and proton's, and results of beam-beam interaction simulations. In particular, the simulations predict a number of advantages for round beam geometry by the collision due to a conservation of the angular momentum [2]. To satisfy the last requirement we should meet 2 conditions: equal beam sizes and equal tunes of betatron oscillations. Since origins of a forming of the beam emittances (ε_e; ε_i) are quite different, so as their dependences on the energy, lattices of the electron and ion rings have to provide some flexibility to control $\beta^\star$-functions in the IP. In favour of the round beams the HERA and SPS experiences witness a bad life time and high background for unmatched beam sizes even with moderate beam currents.

The round beam luminosity is given by the equation:

$$L = F_{coll} \left(\frac{4\pi\gamma_e\gamma_i}{r_e r_i} \right) \cdot \xi_e \cdot \xi_i \cdot \sqrt{\frac{\varepsilon_e}{\beta_e^\star} \cdot \frac{\varepsilon_i}{\beta_i^\star}} \,, \tag{1}$$

where F_{coll} is the collision repetition frequency, γ and r are the relativistic factors and classical radii for electrons and ions correspondingly. Assuming the matched beam sizes, the space charge parameters ξ_e and ξ_i for electrons and ions are determined by the expressions:

$$\xi_e = \frac{N_i r_e Z}{4\pi\gamma_e\varepsilon_e} \,; \qquad \xi_i = \frac{N_e r_i}{4\pi\gamma_i V_i \varepsilon_i Z} \,, \tag{2}$$

where N_e and N_i are electron and ion bunch populations; V_i is the ions velocity (c=1). The world wide experience shows that achievable values of the space charge parameters due to the beam-beam effects do not exceed 0.05 for electrons and 0.005 for protons.

Collision frequency F_{coll} is determined in our case practically by the ion bunch spacing in the RHIC. For realistic case of 360 ion bunches at RHIC, $F_{coll} = 28$ MHz. Single bunch populations N_e and N_i are limited except the beam-beam interaction by different kinds of instabilities. For electrons the most severe intensity threshold is set by the head-tail transverse mode-coupling instability, that limits the one bunch population. The modern accelerator experience (for instance, in both B-factories or LEP collider), tells us, that $N_e = 1 \cdot 10^{11}$ is more or less a safe number. The proton bunch population is admitted to $N_p = 1 \cdot 10^{11}$, which is based on BNL and FNAL experimental results. To achieve the luminosity of $L = 1 \cdot 10^{33}$ cm^{2}s^{-1} the beam size at the IP should be $\sigma_e^* = \sigma_p^* = 48\ \mu m$ together with the other fixed above parameters.

RADIATIVE POLARIZATION AND E-RING DESIGN

The radiative polarization have been observed at the many electron storage rings. According this experience the energy range 5–10 GeV is quite comfortable for the obtaining the polarization degree about 80 percents. If the equilibrium polarization direction (vector $\mathbf{n}$) is vertical in the arcs we can expect a relatively low polarization losses caused by spin manipulations around the IP.

A radiative polarization time strongly depends on the bending field ($\tau_p \sim B^{-3}$). On the other hand the high magnetic field increases the energy losses for the synchrotron radiation ($\Delta E_{turn} \sim B^2$). A possible compromise here may be a special design of the bending magnets. We propose to use so-called super-bends magnets with relatively high field in a short central part of each magnet. It allows us strongly decrease the polarization time at low energies and suppress spin resonances by the relatively minimal energy losses. The possible optimum is to use high field in the super-bends at low energy (so to keep the polarization time at the level of 15 minutes at 5 GeV) and the uniform field at 10 GeV. But a final strategy of using the super-bends can be found during practical work with the beam polarization.

We considered the e-ring which consists of two arcs with regular FODO structure and two straight sections: one for the beams collisions and other for a technical usage. To deliver spin longitudinal into the IP we need to install two spin rotators on both sides of the interaction area. At first, spin is rotated by a solenoidal field to horizontal plane and then by low field dipoles (including final focus quadrupole magnets) exactly to the longitudinal direction at the medium energy 7.5 GeV.

The $\pm 90°$ spin rotator consists of two super-conducting solenoids, each 3 m long, and with the field of about 6 T. Between solenoids a focusing structure is placed, which cancels the betatron coupling and minimizes negative effects from a spin perturbation $\mathbf{w}$ for momentum-off particles. The last requirement is so called spin transparency condition, which should be satisfied by the insertion optics, namely, the integral of the perturbation through the insertion azimuth θ should be made zero:

$$I = \int_{\theta_1}^{\theta_2} \mathbf{w}\eta d\theta = 0 \tag{3}$$

Here $\eta = \eta_1 - i\eta_2$ is a complex vector, which is composed from the unity vectors η_1 and η_2, which in turn are the two orthogonal to the vector $\mathbf{n}$ solutions of the spin motion equation for the equilibrium particle [3].

The spin perturbation components are[3, 4]:

$$w_x = \nu_0 z'' + K_x \frac{\Delta\gamma}{\gamma}; \quad w_z = -\nu_0 x'' + K_z \frac{\Delta\gamma}{\gamma}; \quad w_x = K_y \frac{\Delta\gamma}{\gamma}, \tag{4}$$

where $\nu_0 = \gamma a$ is the dimensionless spin tune, z'' and x'' are the second derivatives of the vertical or horizontal displacements over the azimuth θ ; $K_{x,y,z}$ are respectively the normalized horizontal, longitudinal or vertical magnetic fields.

We found a scheme of the focusing structure, that contains only regular quadrupoles inside the solenoid insertions and cancels the betatron coupling as well as creates the spin transparency. Transfer matrices of the whole insertion:

$$T_x = \begin{pmatrix} 0 & -2r \\ (2r)^{-1} & 0 \end{pmatrix} \quad T_z = \begin{pmatrix} 0 & 2r \\ -(2r)^{-1} & 0 \end{pmatrix} \tag{5}$$

Here r is a curvature radius in the solenoidal field B_y: $r = B\rho / B_y$

The solenoids are located in the drift spaces, where the velocity vector **v** has an angle $\pm 7.55°$ with respect to the collision axis. After the solenoids spin precess around the vertical magnetic field (in case of horizontal "dog-leg" scheme) or around the horizontal magnetic field (in case of vertical "dog-leg" scheme), becoming at the medium energy 7.5 GeV purely longitudinal at the IP. On the opposite side of the interaction straight, spin is restored to vertical direction by the mirror symmetrical spin rotator. As a result, due to this antisymmetry and the spin transparency of the solenoidal rotators, the spin phase advance along the whole interaction region is zero, spin is always restored to the vertical direction in the next arc at arbitrary energy and the polarization behavior is mainly the same as without the spin rotators.

TABLE 1. General parameters of the eRHIC

Parameter	e-ring	p-ring
Circumference, m	1022	3833
Energy, GeV	5–10	250
Number of bunches	96	360
Bunch population	$1 \cdot 10^{11}$	$1 \cdot 10^{11}$
Beam current, A	0.45	0.45
RMS emittance, mmμrad	45–63	9–13
Beta function at IP, cm	10	50
Beam size at IP, mm	0.07–0.08	0.07–0.08
Beam-beam parameter	≤ 0.05	≤ 0.005
Luminosity, cm^{-2} s^{-1}	$(0.3-0.5) \cdot 10^{33}$	

The main parameters of the eRHIC for the current variants of electron and proton ring lattices are listed in the Table 1.

The reoptimization of the electron ring lattice, in order to produce smaller electron beam emmitance, and the decrease of β^* value in the proton ring should provide the luminosity increase up to $1 \cdot 10^{33}$ level, as indicated by preliminary studies.

THE DETECTOR AREA LAYOUT

Besides the spin manipulations there are other issues for the IR design: beam separation to avoid the parasitic beam-beam interactions; focusing to the low beta; detector background; protection from the synchrotron radiation, etc. Both suggested schemes have transverse fields around the IP, that additionally to the spin rotation will separate the beams due to their big energy difference. The same fields could be used for a detector momentum analysis. In the case of longitudinal field in the detector compensating solenoids are needed to keep the zero spin rotation along the IR.

The first option supposes the lift up (about 1 m) one of the RHIC ion ring for the zero angle intersection with the flat electron ring. The Fig. 2 (left) shows schematically the interaction region (IR) and the spin vector behavior. Since the vector **n** is lying in the horizontal plane between the two solenoidal spin rotators, some depolarization comes from the bending magnets in this area. A choice of moderate field magnitudes (few KGauss) helps to avoid substantial polarization losses. Calculations with the ASPIRRIN code [5] give the equilibrium polarization degree about 90% and the polarization time about 500 s at 10 GeV (see the Fig. 3).

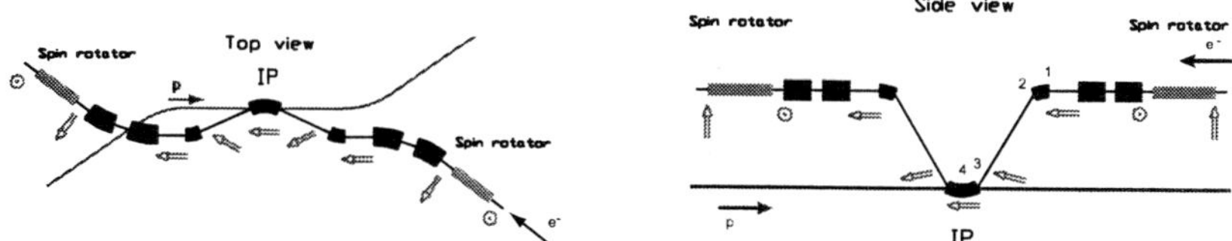

FIGURE 2. The layout of the e-ring interaction region.

One can see this scheme looks well for the electron polarization, but it might require serious rebuilding in the RHIC machine. That's why we considered other scheme with flat ion ring and a vertical orbit bump (about 0.5 m) in the electron ring, see the Fig. 2 (right).

In this variant the proton ring of the RHIC is almost unchanged except a new final focusing to get the low beta. The vertical profile of the e-ring have to be done also with the respect to the spin transparency, that in this case leads to demands on the betatron trajectory slopes at fixed points from 1 to 4:

$$z'_4 - z'_1 = 0; \qquad x'_3 - z'_2 = 0. \tag{6}$$

As the polarization calculations by ASPIRRIN shown, despite of the spin transparency, the vertical bend initiates some spin resonances even in the ring without any imperfections (the Fig. 3). The situation is dramatically changed due to random vertical fluctuations of the arc quadrupoles positioning. The polarization does not exceed 50 percents with the RMS shift 0.5 mm.

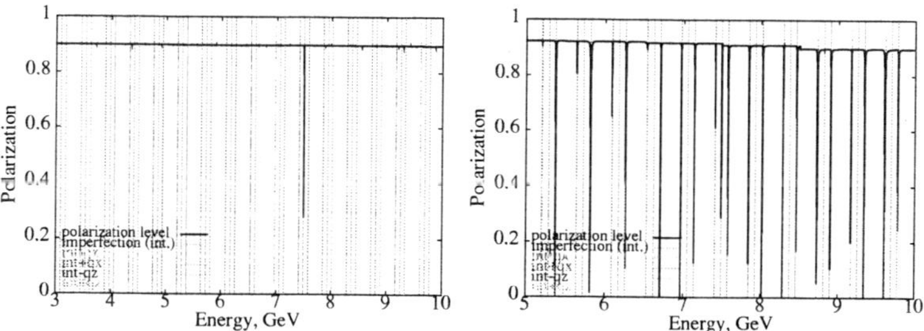

FIGURE 3. The equilibrium polarization vs. the energy in e-ring: left — horizontal dog-leg scheme, right — vertical dog-leg scheme. Intrinsic and imperefection resonances are also shown.

CONCLUSION

The present study shown that the ring-ring option of the electron-proton collider is able to provide the luminosity up to 1×10^{33} cm^{-2}s^{-1} in the SCM energy range 15-100 GeV. The project of the electron ring with the super bend magnets and the solenoidal spin rotators performs to obtain not less 70 percents of the longitudinal polarization in the IP.

Two possible layouts of the interaction region are considered. The scheme with flat electron ring (horizontal "dog leg") looks preferable for the electron polarization. A serious consideration of a new RHIC final focus design for the low beta is needed.

There are a number of topic which have not be mentioned in this paper: a flexible arcs lattice to control the electron beam emittance, chromaticity corrections, dynamical aperture, required cooling of ion beam, etc. That have to be subjects of further investigations.

REFERENCES

1. D.Trbojevic *et al.*, in: *Proc. of EPAC 2002*, Paris, (2002), p.380.
   T.Roser *et al.*, in: *Proc. of EPAC 2002*, Paris, (2002), p.290.
2. A.N. Filippov *et al.*, in: *Proc. 15th Int. Conf. High Energy Accelerators*, Hamburg (Germany), (1992), p.1145.
3. Ya. S. Derbenev, A. M. Kondratenko and A. N. Skrinsky, *Sov. Phys. JETP*, **33**, 658 (1971).
4. Ya. S. Derbenev, A. M. Kondratenko and A. N. Skrinsky, “Radiative polarization at ultra-high energies”, in *Particle Accelerators*, **9**, 247 (1979).
5. E. A. Perevedentsev, V. Ptitsyn and Yu. M. Shatunov, "Spin-Orbital Function Formalism and ASPIRRIN Code", *these proceedings*.

PART TWO

Workshop on Polarized Electron Sources and Polarimeters

Spin Filters as High-Performance Spin Polarimeters

N. Rougemaille[1], G. Lampel[1], J. Peretti[1], H.-J. Drouhin[1], Y. Lassailly[1], A. Filipe[1], T. Wirth[1] and A. Schuhl[2]

[1]*Laboratoire de Physique de la Matière Condensée, UMR 7643 - CNRS, Ecole Polytechnique, 91128 Palaiseau cedex, France.*

[2]*Laboratoire de Physique des Matériaux, UMR 7556 - CNRS, Université Henri Poincaré, 54506 Vandoeuvre-Les-Nancy, France.*

Abstract. A spin-dependent transport experiment in which hot electrons pass through a ferromagnetic metal / semiconductor Schottky diode has been performed. A spin-polarized free-electron beam, emitted in vacuum from a GaAs photocathode, is injected into the thin metal layer with an energy between 5 and 1000 eV above to the Fermi level. The transmitted current collected in the semiconductor substrate increases with injection energy because of secondary - electron multiplication. The spin-dependent part of the transmitted current is first constant up to about 100 eV and then increases by 4 orders of magnitude. As an immediate application, the solid-state hybrid structure studied here leads to a very efficient and compact device for spin polarization detection.

INTRODUCTION

Electron spin polarization detection yields additional information on the electronic structure of solids and spin-dependent interactions inside the matter. As spin measurements can be applied to electron microscopy and spectroscopy techniques, many efforts are still made to improve the efficiency and the convenience of the spin detectors [1]. But, the polarimeters in use up to now (Mott, SPLEED or absorbed / reflected current detectors) suffer from a low spin-discriminating power, a small collection efficiency and severe operating conditions (surface preparation, high voltages, large and complex equipment). The Mott polarimeter, based on the spin-orbit interaction of electrons with high atomic weight materials, remains the conventional spin detector for standard measurements but is not convenient enough for routine application of spin detection.

The spin filter effect in ferromagnetic thin films (the preferential transmission of a spin direction depending on the relative orientation of the electron spin and the layer

CP675, *Spin 2002: 15th Int'l. Spin Physics Symposium and Workshop on Polarized Electron Sources and Polarimeters,* edited by Y. I. Makdisi, A. U. Luccio, and W. W. MacKay

magnetization), which originates from exchange interaction, offers a new way for measuring the free electron spin polarization [2]. Direct transmission experiments of spin-polarized electrons through free-standing Au/Co/Au films have been already performed to characterize the spin-filtering efficiency of magnetic thin layers [3]. Such structures containing asymmetrical ferromagnetic cobalt bilayer have shown interesting properties to realize self-calibrated spin polarimeters [4]. When operated at very low injection energy (a few eV above the Fermi level), these spin filters exhibit a large spin-discriminating power (Sherman function) but are limited by a poor transmission efficiency. Their figure of merit is therefore at best comparable with the one of the Mott polarimeter. Moreover, it is usually admitted that the Sherman function of multilayer spin filters should decrease with increasing energy. Up to now, all the experiments performed at injection energy up to about 100 eV have indeed confirmed this decrease of the Sherman function and of the figure of merit.

Starting from a previous experiment of spin-dependent transmission through a thin ferromagnetic layer deposited on a semiconductor [5], we demonstrate here that, under operation at injection energy in the keV-range, a ferromagnetic metal / semiconductor junction constitutes a very efficient and convenient solid-state spin detector compatible with all standard techniques involving electrons.

EXPERIMENTAL CONDITIONS

The principle of the experiment presented here (Fig. 1) consists in measuring the spin-dependent transmission of a spin-polarized electron beam through a thin magnetic layer deposited on a semiconductor substrate.

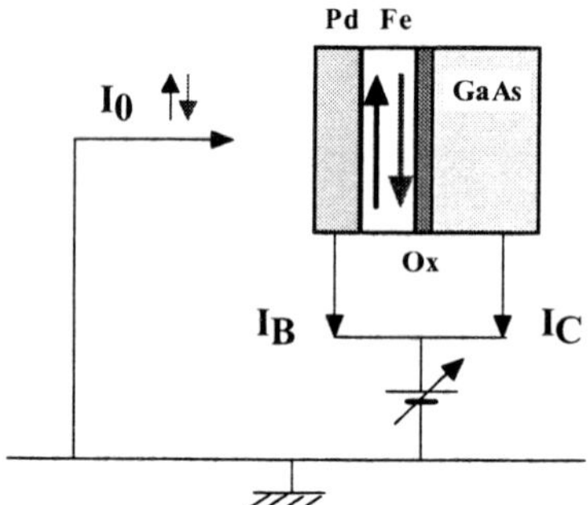

FIGURE 1. Schematic representation of the detection principle. The sample is in-plane magnetized and the spin polarization of the incident electron beam is modulated between -25% and +25%. The currents I_B and I_C can then be measured in four configurations depending on the relative orientation of the spin polarization and on the sample magnetization.

The sample is made of a 3.5 nm-thick iron layer grown onto a n-doped GaAs substrate. To avoid iron inter-diffusion inside the GaAs, the substrate is previously oxidized, leading to a typical 2 nm-thick oxide layer. A 4 nm-thick palladium cap layer is finally deposited to prevent iron from oxidation. The iron layer exhibits an in-plane easy-magnetization axis and a square hysteresis loop with a coercive field of about 20 Oe and a remanence in zero external magnetic field of 90%. Pulsed operation of magnetic coils allows to control in-situ the magnetization of the iron layer.

The electron source is a p^+-doped GaAs photocathode under optical pumping conditions. Before measurements the source is activated to Negative Electron Affinity by co-deposition of cesium and oxygen. Under excitation with a circularly polarized laser beam of wavelength 780 nm, the source yields an electron beam with a longitudinal spin-polarization P = 25%.

This beam is then focused onto the sample using electrostatic optics, after a 90° deflection which converts the longitudinal spin polarization into a transverse one, parallel to the Fe layer magnetization. A typical current I_0 of 200 nA is injected into the sample. The injection energy of the polarized electrons entering the sample is changed by varying the voltage between the source and the Schottky diode. Experiments at injection energies up to 1 keV (referred to the metal Fermi level) have been performed.

No bias voltage is applied to the Schottky diode. The electrons which have enough energy to overcome the Schottky barrier are collected in the GaAs substrate and yields the "transmitted" current I_C. The electrons which have an energy lower than the barrier height are detected in the front contact of the metallic base as a current I_B which is measured independently of I_C. We have checked that, in the whole injection energy range we used, the current balance verifies the relation $I_0 = I_B + I_C$ (no back-scattered electrons). The spin-dependent part of I_C, ΔI_C, is measured when the incident spin polarization is modulated between +P and -P.

RESULTS

The "transmission" $T=I_C/I_0$ and the spin-dependent transmission $\Delta T=\Delta I_C/I_0$ are plotted in Fig. 2 as a function of the injection energy. Let us remark that T and ΔT vary respectively over 6 and 4 orders of magnitude in the studied energy range.

When entering the palladium layer, the incident electrons suffer inelastic scattering, mainly by electron / electron interaction [6]. The energy lost in the collisions promotes secondary electrons. This secondary-electron production together with the increase of the electron mean-free path is responsible for the increase of the collected current I_C.

Due to this cascade process in the palladium layer, the spin polarization of the electrons before the spin filter is diluted by the unpolarized secondaries which are generated. Therefore, at low injection energy (below 80 eV), the spin-dependent transmission ΔT does not follow the increase of T and is energy independent as already observed in previous works [3,5]. But very surprisingly, at higher injection energies, ΔT is no longer energy independent and increases by 4 orders of magnitude. The reason of this ΔT increase is not yet fully understood and will be discussed elsewhere [7].

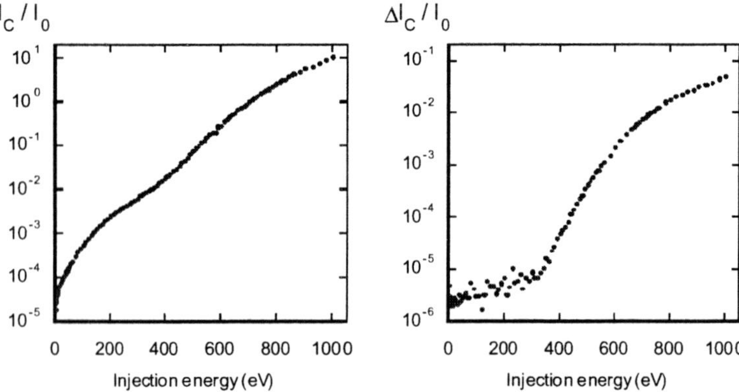

FIGURE 2. The left and right curves show the variations of the transmission T and the spin-dependent transmission ΔT respectively as a function of the injection energy in logarithmic scale for an injected current I_0=200 nA. ΔI_C is either obtained by modulating the incident spin polarization between +P and -P while keeping constant the sample magnetization or modulation the sample magnetization for a fixed spin polarization.

At the particular energy E_0 = 715 eV, the overall transmission reaches unity, meaning that the collected current I_C is exactly equal to the current I_0 injected from the vacuum and consequently $I_B = 0$. As ΔI_C verifies $\Delta I_C = -\Delta I_B$, the large magnetic signal ΔT observed at high injection energy can thus be conveniently measured in the metal contact with no background signal. Such a cancelled-background configuration has already been used in absorbed-current spin detectors [8]. Depending on the relative orientation of the sample magnetization and on the magnetic moment of the injected electrons, ΔI_B changes its sign.

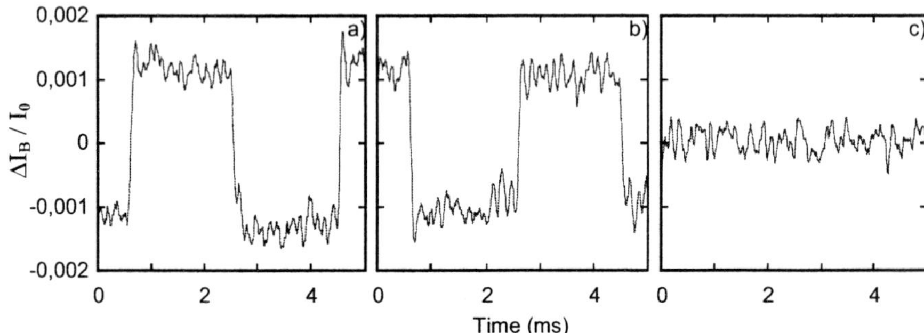

FIGURE 3. Magnetic signal measured at injection energy E_0 on the metallic base when modulating the spin polarization between +P and -P at 250 Hz. The signal is obtained in a single shot without any electronic treatment.

Fig. 3 shows the variations of ΔI_B measured when the incident spin polarization is reversed at a frequency 250 Hz between ±25% with a magnetization -M (curve a) and +M (curve b). Fig. 3c shows that $\Delta I_B/I_0$ remains zero when the light polarization is modulated between two orthogonal linear polarization directions (unpolarized electron beam), demonstrating that there is no experimental asymmetry. The measurements of ΔT presented here have been acquired with a band width of 100 kHz. This indicates that the P=25% spin-polarisation of the incident beam can be measured in a very short time with a signal-to-noise ratio of the order of 10.

CONCLUSION AND PERSPECTIVES

We have shown here that a ferromagnetic metal / semiconductor Schottky diode leads to a new kind of spin detector. A spin polarimeter is characterized by its figure of merit which yields the signal-to-noise ratio of the detection. The figure of merit is given by the quantity $F = S^2\varepsilon$ [9], where ε is the scattering efficiency and S is the Sherman function. In our case, $\varepsilon = T$ and $S = \Delta T/2TP$. At 1 keV injection energy, $F = 10^{-3}$. This value has to be compared with the figure of merit of the Mott polarimeter which is at best 10^{-4}. Moreover, our spin polarimeter present other advantages: a very small size (few cubic centimeters), a compatibility with ultra-high vacuum and high vacuum (no specific surface preparation and cleanness are required), a quite low voltage operation (about 1 kV to be compared with 100 kV for a self-calibrated Mott detector). Let us also mention that the figure of merit can again easily be improved. On the one hand, beyond 1 keV injection energy, F still increases. On the other hand, the noise is here limited by the Johnson noise of the Schottky junction which can be lowered by decreasing the dark current of the junction. Finally, this solid-state device is a very efficient spin polarimeter compatible with all electron spectroscopy and microscopy techniques.

REFERENCES

1. For a review on spin detectors for polarized electrons, see *Polarized electrons in surface physics*, edited by R. Feder, World Scientific, 1985.

2. Schönhense, G., and Siegmann, H.C., Ann. Physik **2**, 465-474 (1993).

3. Lassailly, Y., Drouhin, H.-J., van der Sluijs, A., Lampel, G., and Marliere, C., Phys. Rev. B **50**, 13054-13057 (1994); Drouhin, H.-J., van der Sluijs, A., Lassailly, Y., and Lampel, G., J. Appl. Physics **79**, 4734-4739 (1996); Oberli, D., Burgermeister, R., Riesen, S., Weber, W., and Siegmann, H.C., Phys. Rev. Letters **81**, 4228-4231 (1998).

4. Cacho, C., Lassailly, Y., Drouhin, H.-J., Lampel, G., and Peretti, J., Phys. Rev. Letters **88**, 066601-066604 (2002).

5. Filipe, A., Drouhin, H.-J., Lampel, G., Lassailly, Y., Nagle, J., Peretti, J., Safarov, V.I., and Schuhl, A., Phys. Rev. Letters **80**, 2425-2428 (1998).

6. Dekker, A.J., "Secondary Electron Emission", in *Solid State Physics,* vol. **6**, edited by F. Seitz and D. Turnbulln, Academic Press, 1958, pp. 251-311.

7. Rougemaille, N., Lampel, G., Peretti, J., Drouhin, H.-J., Lassailly, Y., and Schuhl, A., to be published.

8. Siegmann, H.C., Pierce, D.T., and Celotta, R.J., Phys. Rev. Letters **46**, 452-455 (1981).

9. Kessler, J., *Polarized electrons*, Springer-Verlag, Berlin 1985.

Strained Gaasp Photocathode With GaAs Quantum Well.

Yu.Yashin[1], Yu.Mamaev[1], A.Rochansky[1] and D.Vinokurov[2]

1- State Polytechnical University, Polytekhnicheskaya 29, 195251, St. Petersburg, Russia
2 - Ioffe Physico-Technical Institute, RAS, St. Petersburg, 194021, Russia

Abract. By varying of the phosphorous contents "x" and "y" at the $GaAs_{1-x}P_x/GaAs_{1-y}P_y$ cathodes they can be tuned to the wavelength, corresponding to maximum light power of the certain accelerator laser system. The parameters of strained GaAsP sample have been modified to enhance the quantum yield value at polarization maximum. The modification consisted of the incorporation of heavily doped thin GaAs quantum well layer at the top part of the structure. At the polarization maximum the yield enhancement of up to ten times has been achieved.

INTRODUCTION

Strained $GaAs_{1-x}P_x/GaAs_{1-y}P_y$ photocathodes, which we have developed [1,2], have been already applied in the source of polarized electrons attached to the MAMI accelerator in Mainz [3]. The measurement of polarization transfer from the electron to the neutron in the quasielastic reaction at 855 MeV has been performed successfully with such cathodes installed. In more than 1000 hours of beamtime the source produced a 20 μA electron beam with a polarization of 75%.

It turned out that the sources at different accelerators are equipped with various lasers. By varying of the phosphorous contents "x" at the strained overlayer $GaAs_{1-x}P_x$ and "y" at the $GaAs_{1-y}P_y$ buffer, such cathodes can be tuned to the wavelength, corresponding to maximum light power of the certain accelerator laser system (see fig.1). The values of the phosphorus fraction x and y were designed with help of computer "Band - Edge" program, developed by A. Subashiev, to have welcome energy gap E_g of the strained overlayer $GaAs_{1-x}P_x$, high coherent strain and, thus, sufficient energy splitting δ_{def} (about 60 meV) of the Heavy holes and Light holes bands. The samples were grown in the horizontal MOCVD reactor at the top of commercial GaAs (001) wafer. The growth of the epitaxial structures was carried out at the reactor pressure 50 Torr. Trimethylgallium (TMG) and trimethylindium (TMIn) were used as sources of III group elements; 30% PH_3-H_2 and 20% AsH_3-H_2 mixtures were used as the sources of V group elements. Growth temperature was varied from 600 to 750^0 C and carrier gas flow rate was varied from 3 to 11 l/min. TMGa and TMIn bubblers temperatures were kept constant at -10^0 C and +18^0 C, respectively. TMGa bubbler pressure was 1050 Torr. Low pressure in the growth reactor made it possible to use reduced pressure in TMIn bubbler and, hence, to obtain increased vapor concentration of TMIn, which has the low pressure of the saturation vapor (approx. 1.8 Torr at room temperature). We kept the pressure in TMIn bubbler at 2000 Torr level. The ratio of the III and V group elements in the vapor phase was varied from 50 to 200. Growth rate was typically 5 – 10 Å/sec.

CP675, *Spin 2002: 15th Int'l. Spin Physics Symposium and Workshop on Polarized Electron Sources and Polarimeters,* edited by Y. I. Makdisi, A. U. Luccio, and W. W. MacKay

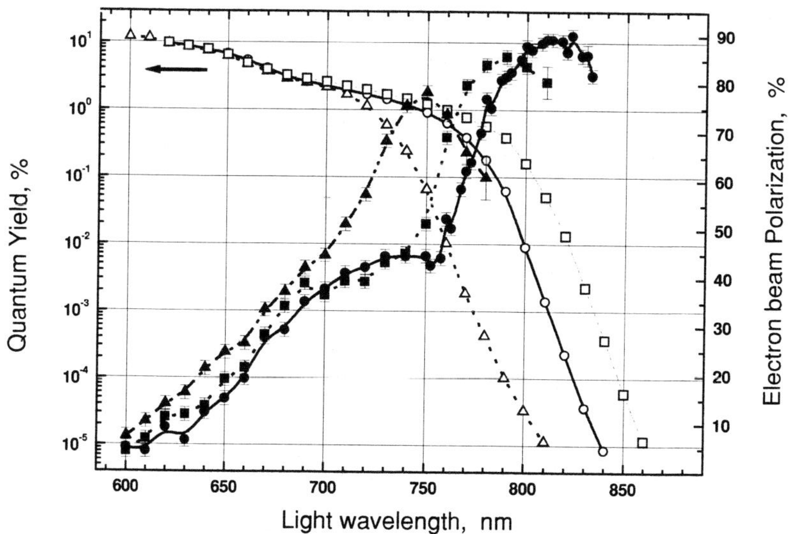

FIGURE 1. Electron spin polarization (solid symbols) and quantum yield (open symbols) as a function of excitation energy for the GaAsPx/GaAsPy strained samples with various phosphorous fractions "x" and "y". Room temperature. Sample 1 (x=0.08, y=0.38) - circles, sample 2 (x=0.12, y=0.36) - squares, sample 3 (x=0.18, y=0.37) – up triangles.

Layer	Thickness	Doping
GaAs quantum well	**20 nm**	**gradient doping of up to $5 \cdot 10^{19}$ cm^{-3} Mg**
$GaAs_{0.91}P_{0.09}$ strained overlayer	**140 nm**	**$5 \cdot 10^{18}$ cm^{-3} Mg**
$GaAs_{0.68}P_{0.32}$ buffer	**1.0 μm**	**Uniform Mg Doping $1 \cdot 10^{18}$ cm^{-3}**
SL 10 pairs $GaAs_{0.55}P_{0.45}$ $GaAs_{0.85}P_{0.15}$	**7 nm 7 nm**	
$GaAs_{0.68}P_{0.32}$	**500 nm**	
$GaAs_{0.8}P_{0.2}$	**500 nm**	
$GaAs_{0.9}P_{0.1}$	**500 nm**	
GaAs(100) – substrate	**0.5 mm**	**Intrinsic**

Table I. Structure of MOCVD grown GaAsP photocathode with GaAs quantum well.

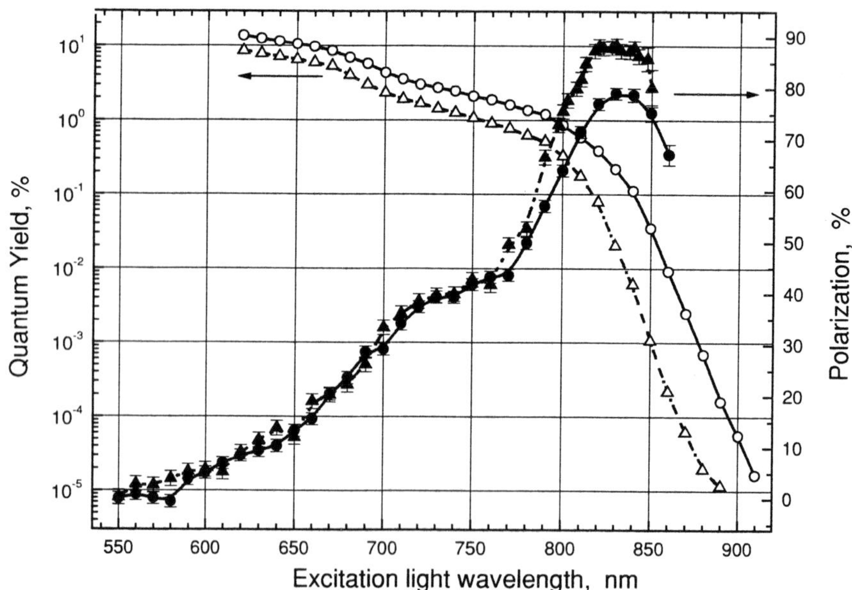

FIG. 2. Electron spin polarization (solid symbols) and quantum yield (open symbols) as a function of excitation energy for the GaAsP strained sample with (circles) and without (triangles) thin GaAs quantum well at room temperature.

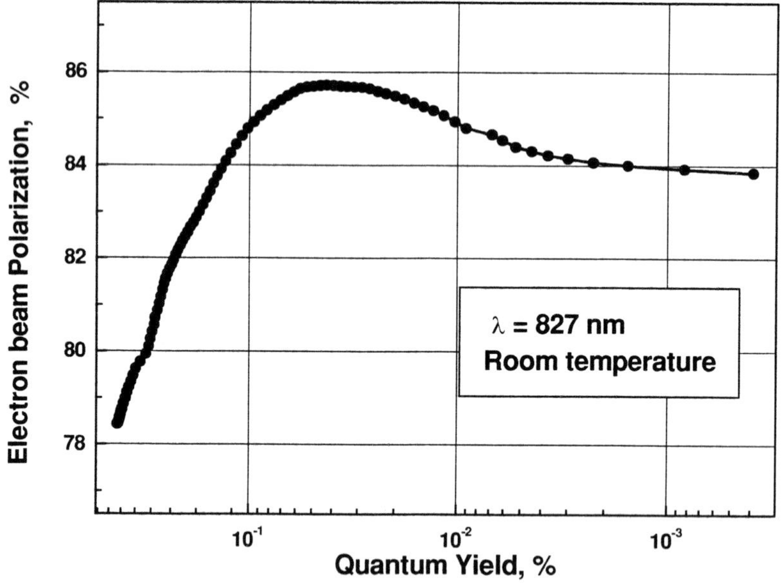

FIGURE 3. Polarization evolution upon the degradation of the GaAs/$GaAs_{0.91}P_{0.09}$/$GaAs_{0.68}P_{0.32}$ sample at room temperature. Excitation light wavelength 827 nm.

Strained GaAsP photocathodes allowed to achieve the polarization a little less than 90%. Nevertheless, rather low value of quantum yield at polarization maximum was still a disadvantage of such cathodes. To make the photocathodes features better the parameters of the samples once more have been improved on the base of X-ray, Raman and polarized photoluminescence studies. The modification consisted of the incorporation of heavily doped thin GaAs quantum well layer at the top part of the structure, based upon GaAsP strained overlayer, as it was discussed in [4]. The goal was to enhance electron emission from strained wide-gap low doped overlayer through heavily doped thin GaAs quantum well layer (one can call it as "field assistant emission"). Table I shows the composition of a photocathode under consideration. In the sample the intermediate layers with graded phosphorus fraction "z" serve to adjust the lattice parameters of the wafer and the buffer layer, and to withdraw unwelcome strain in the last one. The existence of the intermediate $GaAsP_{0.45}/GaAsP_{0.15}$ Superlattice is of crucial importance for the growing of the photocathodes with rather thick high quality strained overlayers. Thorough characterisation of MOCVD grown heterostructures with AlGaInAs strained overlayers by the Transmission Electron Microscopy and X-ray diffraction techniques has shown [5] that the intermediate SL significantly diminishes the initial density of structural defects, and, hence, influences on the process of elastic strain relaxation in lattice mismatched epitaxial structure. As a result the main part of thick lattice mismatched overlayer remains crystal lattice perfection.

The set of spectral and temperature experiments has been realised with the modified $GaAs/GaAs_{0.91}P_{0.09}/GaAs_{0.68}P_{0.32}$ and $GaAs_{0.91}P_{0.09}/GaAs_{0.68}P_{0.32}$ strained MOCVD grown heterostructures both at room and 130K temperatures. All measurements have been performed at the experimental set-up [6], which includes the polarized electron source and the spin detector and is based upon the Russian commercial UHV system USU-4. It has an additional cryogenic pump and two chambers separated by a valve. The smaller one has an activation system, Auger analyser and the load-lock system and equipped with a manipulator with cooling and temperature control systems. The bigger chamber serves as a chamber for mini-Mott polarimetry. All experiments are performed under computer control. Fig. 2 shows **P**(λ) and quantum yield **Y**(λ) curves both for $GaAs/GaAs_{0.91}P_{0.09}/GaAs_{0.68}P_{0.32}$ and $GaAs_{0.91}P_{0.09}/GaAs_{0.68}P_{0.32}$ cathodes at room temperature.

One can see that the polarization maximum value for $GaAs_{0.91}P_{0.09}/GaAs_{0.68}P_{0.32}$ cathode is higher, than for $GaAs/GaAs_{0.91}P_{0.09}/GaAs_{0.68}P_{0.32}$ one. But the value of quantum yield at the polarization maximum for $GaAs/GaAs_{0.91}P_{0.09}/GaAs_{0.68}P_{0.32}$ cathode is about ten times higher, than in the case of $GaAs_{0.91}P_{0.09}/GaAs_{0.68}P_{0.32}$ cathode, which means that thin GaAs quantum well is in fact very effectively improves the electrons escape conditions. The reason of it is the absence of a potential barrier, even at extremely high doping of the GaAs quantum well. At the same time highly doped GaAs quantum well helps to achieve really Negative Electron Affinity surface and, hence, high quantum yield. Fig 3 illustrates the polarization and quantum yield evolution upon the degradation of the $GaAs/GaAs_{0.91}P_{0.09}/GaAs_{0.68}P_{0.32}$ sample at room temperature, excitation light wavelength being equal to 827 nm. A considerable increase of a polarization can be explained by high polarization of the "hot" electrons, which have high probability to be escaped prior the spin relaxation.

The depolarization of such electrons in GaAs QW is less, than for slow electrons. It is seen that upon the quantum yield decreasing of about ten times, the polarization value increases of up to 86%, which means that the initial electron polarization is really very high. Further perfection and optimisation of the GaAsP strained structures is underway.

CONCLUSIONS

The tuning of the GaAsP photocathodes has been experimentally demonstrated. The value of quantum yield of strained GaAsP sample has been improved by incorporating of a thin heavily doped GaAs layer. At the polarization maximum the yield enhancement of up to ten times has been achieved.

ACKNOWLEDGMENTS

This work was supported by INTAS under grant 99-00125, CRDF under grant RP1-2345-ST-02 and Russian Fond for Basic Research under grant 00-02-16775.

REFERENCES

1. Mamaev, Yu. A. et al., Phys.Low-Dim.Structures, **7**,127 (1994).
2. Subashiev, A.V., Mamaev, Yu.A., Yashin, Yu.P. and Clendenin, J.E., Phys. Low-Dim. Structures, **1/2**, 1 (1999).
3. Drescher, P. et al., Appl. Phys. **A 63**, 203 (1996).
4. Maruyama T. et. al, "Investigation of the charge limit phenomenon in GaAs photocathodes", SPIN 2000 proceedings, AIP, Melville, New York, 976 (2001).
5. Mamaev, Yu. et al., "Photocathodes for Spin - Polarized Electron Source with Strained AlGaInAs Layers ", Proc. of Int. Workshop on Polarized beams and Polarized Gas Targets, June 1995, Cologne, Germany, World Scientific, 303 (1996).
6. Yashin, Yu.P., Ambrajei, A.N. and Mamaev, Yu.A., Instruments for Experimental Techniques, **43, #2**, 245 (2000).

Transmission Polarimetry at MIT Bates

T. Zwart*, E.C. Booth°, M. Farkhondeh*, W.A.Franklin*, E.Ihloff*, J.L.Matthews*, E.Tsentalovich*, W.Turchinetz*

**MIT-Bates Linear Accelerator Center, 21 Manning Rd. Middleton MA 01949*
°Boston Universit, Boston MA 02215

Abstract. The polarization dependence of Compton scattering in magnetized iron can be used to determine the polarization of an incident photon beam. This can in turn be related to the polarization of the electron beam which radiated the photons. It is difficult to calculate the analyzing power of these devices absolutely, however they are of great utility for rapid, relative measurements of electron beam polarization. These devices have been used at Bates as relative electron polarization monitors at 20 and 200 MeV. Efforts are now being made to use the device at 850 MeV as an online measure of the beam polarization in the South Hall Ring. A technique to calibrate these devices and build an affordable, absolute polarimeter is also being explored.

INTRODUCTION

Transmission polarimeters have been in use at the Bates Linear Accelerator Center over the last several years. The initial design was very similar to a transmission polarimeter used at the Mainz Microtron for a parity violating experiment on Beryllium (1), but subsequent devices at Bates have been modified for optimal performance at various locations throughout the facility. The transmission polarimeter is a simple, inexpensive device with a relatively small analyzing power, $0.1\% < A < 5\%$, but high counting rates. The devices can deliver rapid, precise measurements of the longitudinal electron beam polarization, but to date have relied on calibration against polarimeters with better known analyzing powers, principly Moller (2) and Compton (3).

TRANSMISSION POLARIMETRY TECHNIQUE

The technique of transmission polarimetry is illustrated in Fig. 1. A high energy electron beam is incident on a thin target where high energy photons are radiated. A portion of the electron polarization is transferred to the photon beam. At the photon endpoint, where the photon receives the full momentum of the incident electron, helicity conservation requires full transfer of electron polarization to the photon. This polarized photon beam is subsequently attenuated in a magnetized iron bar. Due to the polarization dependence of the Compton cross section, a helicity dependent asymmetry develops in the downstream photon yield. Although helicity independent processes dominate in the iron absorber, the cumulative effect of the spin dependent

CP675, *Spin 2002: 15th Int'l. Spin Physics Symposium and Workshop on Polarized Electron Sources and Polarimeters,* edited by Y. I. Makdisi, A. U. Luccio, and W. W. MacKay

scattering over tens of cm results in asymmetries between 0.1-1% depending on the beam energy and the precise details of the polarimeter.

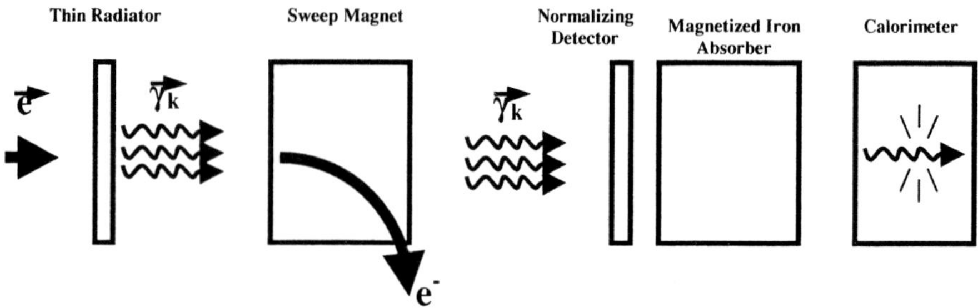

FIGURE 1. The technique of transmission polarimetry. See text for details.

Equations 1-4 below describe the details of the asymmetry generating processes. The total cross section, σ_T, in the magnetized iron depends on photon wavenumber, k, and includes a helicity independent term, σ_0, and a helicity dependent term, σ_p,

$$\sigma_T(k) = \sigma_0(k) + P_e P_\gamma \sigma_p(k) \quad . \tag{1}$$

The absorption coefficient, C(k), is then a simple exponential which depends on the cross section, the electron number density in the iron, n, and the absorber length, L,

$$C(k) = e^{-\sigma_0(k)nL} e^{-P_e P_\gamma \sigma_p(k)nL} \quad . \tag{2}$$

An asymmetry, A(k), is formed by varying the sign of the magnet polarization (or incident electron polarization) and the spin independent term cancels giving,

$$A(k) = \tanh\left(-P_e P_\gamma \sigma_p(k)nL\right) \quad . \tag{3}$$

For small values of the argument tanh is linear and A(k) can be approximated,

$$A(k) \approx -P_e P_\gamma \sigma_p(k)nL \quad , \tag{4}$$

depending linearly on the length. A figure of merit proportional to NA^2 can be defined, where N is the counting rate. Optimization of this quantity gives an absorber length of ~ 6 cm. However, if the counting rate is large, it is sensible to increase the length to maximize the asymmetry and limit the influence of systematic effects.

TRANSMISSION POLARIMETERS AT BATES

The first transmission polarimeter used at Bates was installed at 200 MeV in 1996. The geometry was extremely favorable. A large dipole magnet swept away the low energy shower downstream of the radiator, in this case a thin stainless steel beampipe exit window. The absorber magnet was a 7.5 cm dia. soft iron cylinder 20 cm in length with adequate return so that a 2200 turn coil at 3 A gave a field of 2 T in the bar corresponding to an electron polarization of ~8%. An 8x8x12" lead glass cerenkov detector was used to measure the photon yield 5 m downstream of the absorber. Due to the large peak intensity of the Bates Linac's 600 Hz pulse structure it was necessary to integrate each pulse and single photons could not be resolved. The measured analyzing power of the device was 0.8 %

Following this result a smaller 15 cm long device was constructed for use at 20 MeV at the beginning of the accelerator. Space constraints here lead to a less favorable geometry so only a 2.5 cm dia. iron cylinder could be used. The radiator was a 0.5 mm BeO viewing screen. The initial use of a 2" dia. x 2" long NaI crystal for the photon detector caused extreme saturation in the first stages of the photomultiplier so a Lucite calorimeter of the same size was used instead. In this location it was necessary to place the photon detector within 10 cm of the absorber magnet so it was not possible to make use of a sweeping magnet. The measured analyzing power of this device was 1.7%.

The 20 MeV transmission polarimeter was used routinely each day during the second and third SAMPLE experiments (4). Each measurement took about ½ hour and the bulk of that time was devoted to switching the beam into and out of a small chicane in the injector. The actual data taking for a 2% absolute polarization error (5% relative) took 5 minutes at 4 uA electron current. Much of this error is still attributed to instrumentation noise and improvements in the signal amplification are planned in the next year.

The 20 MeV polarimeter has also proven useful for rapid calibration of a 60 keV Wien Filter, used to rotate the electron spin, and for verification of the polarization of newly installed photocathodes on the injector.

A transmission polarimeter for use at higher energies, 400 – 1000 MeV, is being installed downstream of the internal gas target in the South Hall Ring. This device will have lower analyzing power ~0.2% due to the higher electron beam energy, but will be useful for optimizing the orientation of the injected polarization into the South Hall Ring. The transmission polarimeter could also provide a redundant online monitor of the stored electron polarization (a laser back-scattering polarimeter has recently been commissioned) using the internal gas target as a radiator.

CALIBRATION OF A TRANSMISSION POLARIMETER THROUGH LASER COMPTON SCATTERING

In conjunction with the laser backscattering polarimeter, a 20 cm long transmission polarimeter has been used to analyze the polarization of the backscattered photons. This is useful for two reasons. First, since the backscattered photon polarization is kinematically determined and follows a cosine like distribution as a function of backscattered photon energy, the measured zero crossing in the asymmetry provides an energy calibration point for the laser backscattering calorimeter. This energy calibration of the calorimeter is one of the largest systematic errors in the Compton polarimeter. The zero crossing in the asymmetry will only be accurate when the energy response of the calorimeter is correctly modeled. Second, since the backscattered photon polarization is determined by the incident laser polarization, it is very high since P_{Laser}>99%. At the photon endpoint (and at the lowest backscatter photon energy) the backscattered photons have the full polarization of the laser. This provides an accurate calibration of the analyzing power of the transmission polarimeter on polarized photons without any particular knowledge of the processes in the iron.

Figure 2 shows the analyzing power of all transmission polarimeters at Bates scaled to a 15 cm long absorber magnet. The solid curve with the larger analyzing power corresponds to the analyzing power at the photon endpoint. The solid curve with the smaller analyzing power is a weighted with a bremsstrahlung spectrum of the radiated photons with appropriate polarization transfer included. All the measured points fall between these two bounding curves indicating that the scale of the analyzing power can easily be predicted.

The solid line labeled Laser Backscattering Calibration reflects measured data for a laser backscattering run at 850 MeV electron energy and 532 nm laser wavelength. This data should exactly lie on the endpoint analyzing curve, but we find disagreement of order 20%. Future measurements with the laser backscattering polarimeter will investigate this discrepancy in the coming year.

The maximum endpoint analyzing power of this device is predicted to lie at 3 MeV electron energy. The laser backscattering calibration has determined an analyzing power of 5% at 5 MeV photon energy. As indicated by the upright star in Figure 2, Jefferson Laboratory has a low energy beamline where a transmission polarimeter could be installed with an analyzing power between 5-10%. A longer absorber magnet is possible considering both available space (a few m) and available average current (~10 uA). This device could be absolutely calibrated against laser backscattered photons at the Bates South Hall Ring.

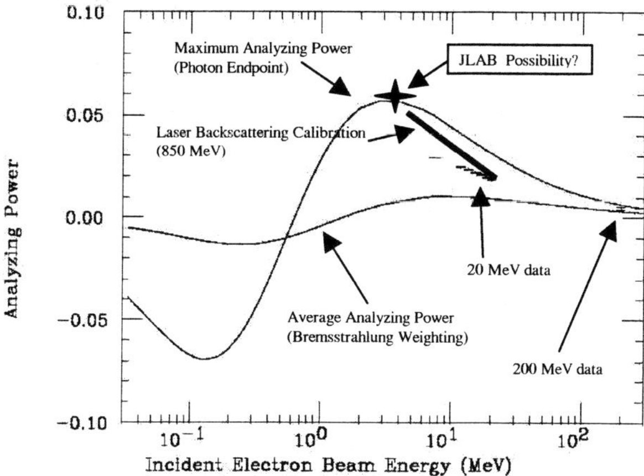

FIGURE 2. Predicted and measured transmission polarimeter analyzing powers. See text for details.

CONCLUSIONS

Over the last several years Transmission Polarimeters have proven very useful at the Bates facility. They provide an affordable, rapid relative meaure of the electron polarization. The polarimeters have an optimal operating energy of ~5 MeV, but are being used at higher energies elsewhere in the lab. In the coming year we hope to calibrate the device absolutely, simplify the design and commission an on-line polarimeter for the internal taget physics program in the South Hall Ring.

REFERENCES

1. Bellanca, J., and Wilson, R., in *Parity Violation in Electron Scattering*, edited by E.J. Beise and R.D. McKeown (World Scientific, New Jersey, 1990), p. 111.
2. Arrington, J. *et al.*, *Nucl. Inst. And Methods A* **311,** 39 (1992).
3. Franklin, W. *et al.*, "The MIT-Bates Compton Polarimeter" These Proceedings.
4. Hasty, R., *et. al., Science* **290**, 2117 (2000).

A Novel Imaging Spectrometer for Energy-Distribution Measurements of Photoelectrons from GaAs Cathodes

C. D. Schröter, A. Rudenko, A. Dorn, R. Moshammer and J. Ullrich

Max-Planck-Institut für Kernphysik, 69029 Heidelberg, Germany

Abstract. The investigation of the photoelectron-escape mechanism from GaAs cathodes with negative electron affinity requires the detection of very low energy electrons. We have built a novel, UHV-compatible, spectrometer where the photoelectrons are imaged by a homogeneous electric field onto a position-sensitive detector. The time-of-flight of each single emitted electron and its position on the detector is measured. From these informations energy-distribution curves are extracted. The spectrometer has run successfully and preliminary energy-distribution curves have been measured. The system is now under improvement. With the optimized spectrometer, an excellent energy resolution (a few meV) can be achieved.

INTRODUCTION

To investigate the photoelectron-escape mechanism from a photocathode with negative electron affinity, a spectrometer is required that allows the simultaneous measurement of the energy and angular distributions of very low energy electrons. In the past, measurements of the longitudinal energy distributions have been performed by several groups. Quite recently, complete energy distributions have been studied as a function of longitudinal and transverse energies. The employed method is based on selecting photoelectrons of a fixed longitudinal energy using a retarding field analyzer, and subsequently measuring the associated differential transverse energy distribution by applying an adiabatic magnetic compression technique [1].

Recently we have built up a novel imaging spectrometer which allows the simultaneous measurement of the longitudinal and transverse momenta for each individual photoelectron emitted from a GaAs surface [2]. Our spectrometer has run successfully and preliminary energy-distribution curves (EDC's) have been extracted from this information. However, the first tests have been performed under non-ideal conditions with high extraction fields and, hence, the thus far measured EDC's have lower resolution than what is ultimately achievable. Design changes have been made to be able to measure EDC's of photoelectrons with an improved energy resolution in the near future.

EXPERIMENTAL SET-UP

A schematic diagram of the spectrometer set-up is shown in figure 1. The photocathode is illuminated by a short-pulsed laser diode (pulse width $\sim$ 100 ps). The laser focus spot size and its position on the cathode is controlled by a CCD camera. For the mea-

CP675, *Spin 2002: 15th Int'l. Spin Physics Symposium and Workshop on Polarized Electron Sources and Polarimeters*, edited by Y. I. Makdisi, A. U. Luccio, and W. W. MacKay

surements, highly doped p-type reflection mode GaAs/AlGaAs heterostructure crystals (6×10^{18} Zn/cm^3) are used with an active GaAs layer of 0.9 μm. For such a thin layer, the long tail of the electron pulse, generated by a δ-pulse light excitation, extends out to less than 200 ps [3]. Hence, the photoelectron pulse width will be less than 300 ps and therefore short enough to not limit the final resolution of the electrons' time-of-flight (TOF) measurement. Details of the preparation technique of the photocathodes are described elsewhere [2, 4].

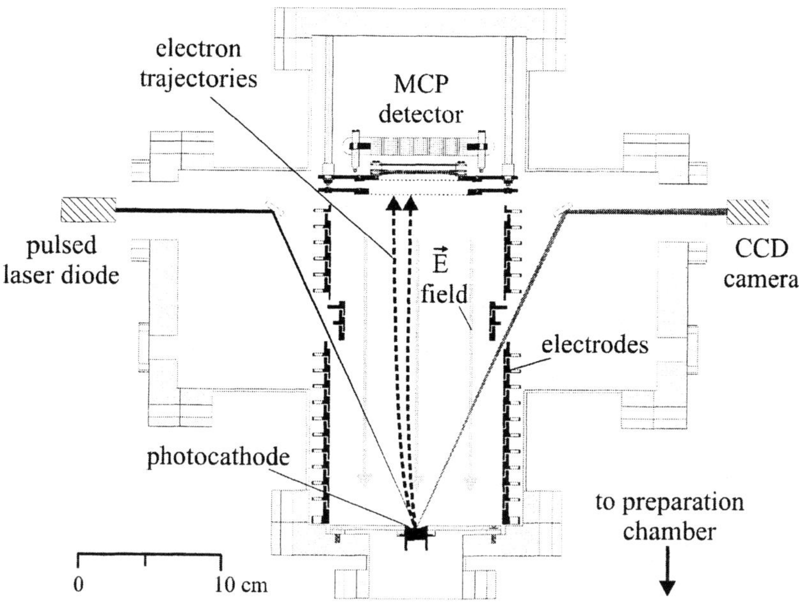

FIGURE 1. UHV chamber with spectrometer and MCP detector.

Emitted photoelectrons are projected by a homogeneous electric field onto a position-sensitive micro-channel plate (MCP) detector. The homogeneous field is produced by means of 20 cylindrical electrodes on which equidistant potentials are applied. The TOF of each single emitted photoelectron and its position on the detector are measured. From the TOF, the longitudinal momentum of the electron can be determined and from the position on the MCP detector, its transverse momentum can be extracted.

Computer simulations show that with the spectrometer an excellent energy resolution can be achieved (see figure 2). But in order to reach the highest resolution a very homogeneous electric field is required to guide the electrons to the detector. In the transverse direction, high resolution can be obtained even at high extraction fields; however, in the longitudinal direction, high resolution is achieved only for the smallest electric-field strengths. Then, even small electric and magnetic stray fields will influence the parabolic trajectories of the photoelectrons. Thus, for the whole apparatus, only UHV-compatible materials with very low permeability have been used. External fields are compensated by three pairs of Helmholtz coils installed pairwise, perpendicular to each other. The magnetic stray fields originating from ion getter pumps are shielded by μ-metal housings.

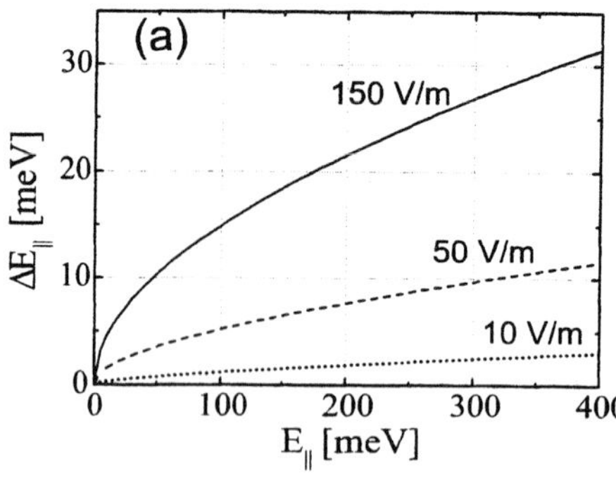

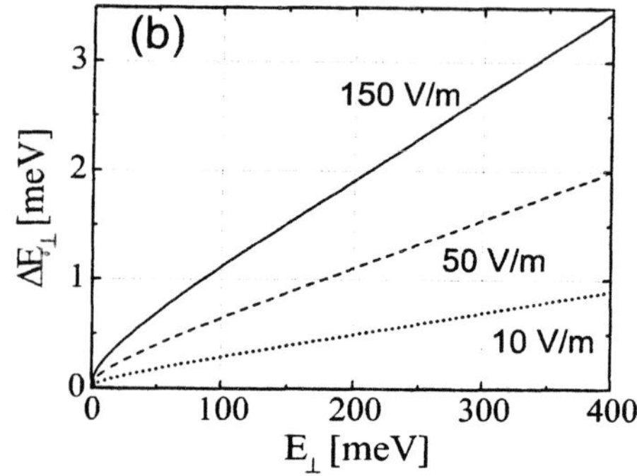

FIGURE 2. (a) Longitudinal and (b) transverse energy resolution of the spectrometer for electric field strengths of 150 V/m, 50 V/m and 10 V/m, respectively. The curves in (b) are plotted for $E_{\parallel} = 0$ meV.

Great efforts have been undertaken to guarantee that the work functions of the different photocathode mounting components are homogeneous. Here, the path of a photoelectron is most sensitive to stray fields because the electron is not yet accelerated by the electric field. The potential of the photocathode mounts can be tuned relative to the potential of the first electrode of the spectrometer. This is important to minimize contact-potential differences in the photocathode mounting region. Moreover, the electrodes of the spectrometer are gold plated.

Using low-outgasing materials for all of the mounts, and pumping with ion getter pumps and volume getter strips should enable us to reach a vacuum base pressure in the $\sim 10^{-12}$ mbar range. This will ensure a long lifetime of the photocathodes and will lead to EDC measurements on undegraded cathodes in the near future.

ACKNOWLEDGMENTS

This work was partially supported by the Deutsche Forschungsgemeinschaft within the Leibniz-program. The heterostructure material was kindly put at our disposal by A. S. Terekhov. We are grateful to A. S. Terekhov as well as to D. A. Orlov and A. Wolf for their continuous support.

REFERENCES

1. Orlov D. A., Hoppe M., Weigel U.,Schwalm D., Terekhov A. S., and Wolf A., *Appl. Phys. Lett.* **78**, 2721-2723 (2001).
2. Schröter C. D., Dorn A., Deipenwisch J., Höhr C., Moshammer R. and Ullrich J., International Workshop on Polarized Electron Source and Polarimeters, Nagoya 2000, in *SPIN 2000, AIP Conference Proceedings* **570**, 996-999 (2001).
3. Hartmann P., Bermuth J., v. Harrach D., Hoffmann J., and Köbis S., Reichert E., Aulenbacher K., and Schuler J., Steigerwald M., *J. Appl. Phys.* **86**, 2245-2249 (1999).
4. Schröter C. D., Dorn A., Moshammer R., Höhr C. and Ullrich J., International Workshop on Polarized Sources and Targets, Nashville 2001, *Conference Proceedings*, ed. by Derenchuk V. P. and von Przewoski B., (World Scientific, Singapore 2002), p. 166-169.

High Power Diode Laser System For SHR

D.Cheever, M.Farkhondeh, W.Franklin, E. Tsentalovich, T.Zwart

MIT-Bates Linear Accelerator Center, Middleton, MA, USA

Abstract. Experiments with a polarized electron beam stored in the South Hall Ring (SHR) at MIT-Bates Linear Accelerator Center will begin in 2003. Currently, the commissioning of BLAST detector is under way. The polarized injector uses for the first time high power diode array laser for photoemission. The laser operates at a wavelength of 808 nm and produces peak power up to 150 W at a duty factor of 1.5%. Higher power is available at lower duty factor. The laser is coupled to a fiber; laser beam emitting from the fiber has an emittance of 200 mm·mrad and a set of lenses is used to deliver the beam through polarizing optics to a strained GaAs crystal inside the electron gun. The photocathode has a diameter of 11 mm and the laser illuminates the entire area. The gun optics has been specifically designed for such a large beam spot size. The diode laser provides an excellent stability and convenience of operation. At the same time, large divergence of the laser beam requires special attention to the transport system in general, and to polarizing optics in particular.

INTRODUCTION

The MIT-Bates accelerator complex operates in three different modes. The electron beam accelerated in the linac can be immediately used for the experiments ("pulse" mode). In the second mode the beam is injected into the SHR, and then slowly extracted into the experimental area, thus increasing the duty factor from 1% to almost 100% ("stretcher" mode). Finally, beam is stored and stacked to high average currents in the SHR for internal target experiments ("storage" mode). The requirements for the polarized electron injector are presented in the following table.

	I (Peak), mA	ΔT, μsec	Rep.Rate, Hz	I(Aver.), μA	P(laser) (QE~1%)	P(laser) (QE~.05%)
Pulse mode	~ 10	~ 25	600 Hz	~ 120	~ 3 W	~ 60 W
Stretcher mode	~ 10	~ 4	600 Hz	~ 20	~ 3 W	~ 60 W
Storage mode	~ 10	~ 4	< 1 Hz	<< 1	~ 3 W	~ 60 W

The existing cw Ti:Sa laser is quite adequate for the operation with bulk GaAs photocathodes. It is tunable, has reasonable stability, and any temporal structure can be tailored out of the cw beam using Pockels cell based shutter. But the maximum peak power available from this laser doesn't exceed 4 W, and it is insufficient for the operations with high polarization (strained or superlattice) photocathodes.

Flash-lamp pumped Ti:Sa laser can be used for the "storage" mode. It provides hundreds of Watts of laser power and is tunable. However, the stability of this laser is

CP675, *Spin 2002: 15th Int'l. Spin Physics Symposium and Workshop on Polarized Electron Sources and Polarimeters,* edited by Y. I. Makdisi, A. U. Luccio, and W. W. MacKay very low, and the maximum repetition rate is only about 10 Hz. Therefore, this laser doesn't meet the requirements for the "pulse" and "stretcher" modes.

FIGURE 1. Diode array laser system. Two diodes are installed on the cooling plate to maintain a constant temperature. The two armored fibers conducting the light from the lasers merge together into a single fiber.

In 2001 we acquired a multimode fiber-coupled diode array laser system [1] that meets the requirements for all modes listed.

LASER PARAMETERS AND PERFORMANCE

The diode laser can be operated in both cw and pulse modes. In cw mode the power can be as high as 60 W. In the pulsed mode the maximum peak power is at least 150-200 W. Higher power can be achieved at low duty factor, but it may shorten the effective life time of the laser. The rise time for the pulses is limited by the inductance of the leads in the current design to about 0.1 μsec, which is quite adequate for our application.

The laser provides excellent stability, extremely convenient in operation and virtually requires no maintenance.

The drawbacks of the laser are the fixed wavelength (λ = 808 nm) and a very high emittance ε = 200 mm·mrad (compare with $\varepsilon \approx$ 1mm·mrad of the diffraction-limited beam from the Ti:Sa laser).

The advances in the technology of photocathodes production allowed designing the crystals matching the fixed wavelength of the laser. The first photocathodes for our injector with a maximum of polarization obtained at about 800-810 nm have been produced in the laboratory of Spin-Polarized Spectroscopy, State Technical University, St. Petersburg (Russia). Fig.2 demonstrates QE and polarization wavelength dependence measured at Bates.

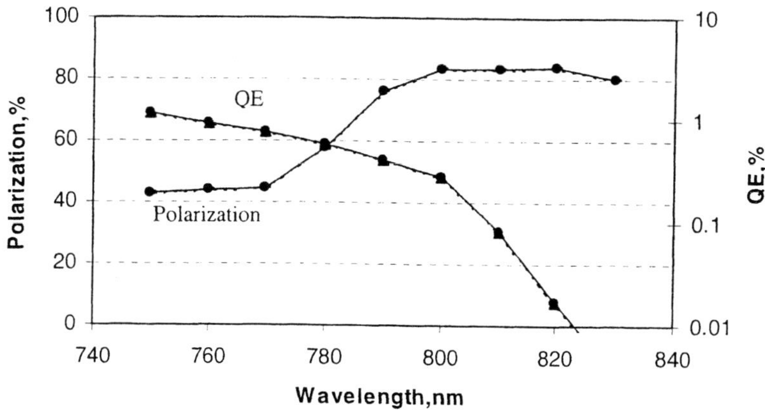

FIGURE 2. Polarization and Quantum Efficiency (QE) for the St. Petersburg photocathode.

Most recently we are using high gradient doped strained GaAsP photocathodes grown by Bandwidth Semiconductor Inc. [2] with the SLAC specification [3]. The phosphate fraction in these crystals (≈5%) is adjusted to obtain the maximum polarization at our wavelength. The top 10 nm layer is heavily doped to minimize surface charge effects. We measured a polarization of more than 80% and QE of 0.2-0.3 % at our wavelength of $\lambda = 808$ nm.

Very high divergence of the diode laser beam precluded use of the existing 20 m long transport line for the Ti:Sa laser. A new wide aperture transport line was designed. All optical elements are installed on an optical breadboard located next to the gun, and the laser beam is transported straight to the cathode without any mirror bounces. For the "storage" mode operation we are using only waveplates to produce and reverse circular polarization. A $\lambda/4$ plate following a linear polarizer transforms linear polarization into circular. For slow reversal of the helicity a remotely controlled $\lambda/2$ plate mounted on a pneumo-driven actuator is inserted into or removed from the laser path. The maximum rate of slow helicity reversal is about 1 Hz.

"Stretcher" and "pulse" modes might require much faster rate of helicity reversal. For these modes, a Helicity Pockels Cell (HPC) must be used. Pockels cells performance deteriorates with large beam divergence. To minimize the divergence of the beam with a given emittance passing through the HPC, the beam size should be maximized. We obtained a very large Pockels cell with a clear aperture of 75 mm [4]. A set of lenses is used to expand the beam to a diameter of about 50 mm. A 2" linear polarizer (which is also sensitive to the beam divergence) and the HPC are located in this section. An extinction ratio of about 140 was achieved in this set up. Two $\lambda/2$ plates are installed after the HPC: one for slow helicity reverse and second for the alignment of the direction of residual linear polarization with the photocathode axis in order to minimize helicity-correlated effects. Since wave plates are less sensitive to beam divergence we could use 0.5" plates. A waist was designed in the transport line for these plates.

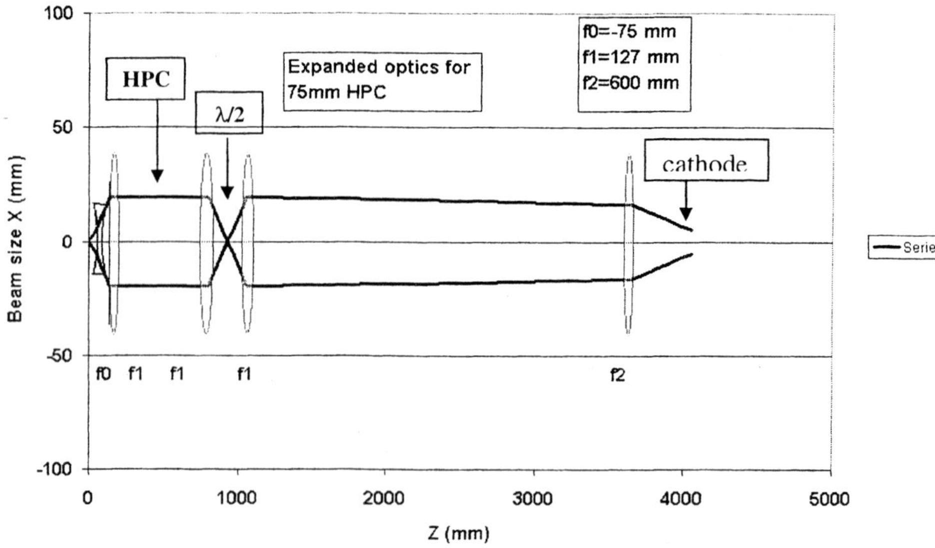

FIGURE 3. Diode laser transport line.

The final lens focuses the beam unto the photocathode. The beam size is adjusted by varying the lens location. Usually the laser beam illuminates the whole photocathode (11 mm in diameter). Large beam spot size allows reducing surface charge effect.

DRIVER FOR THE LASER

The laser requires a current pulse with sharp leading edge and amplitude of up to 200 A. We developed a driver with very low inductance. It is designed for long (up to 20-30 μsec) pulses and sharp (about 70 nsec) leading edge. This driver was specifically designed for parity-violating experiments and allows controlling the intensity asymmetry for different helicity states with an accuracy of better than 1 ppm.

RESULTS

Currently the polarized electron source with strained GaAs crystal and diode laser is used for the commissioning of BLAST detector. It operates in the storage mode and provides highly polarized (>80%) electron beam to the SHR. Excellent quality and stability of the beam have been achieved. With very low duty factor in this mode, the lifetime of the photocathode is very long. We perform cesiation approximately once a week, and full activation once in 1-2 months.

The preliminary measurements indicate that we will be able to control helicity-correlated effects with this system to a level required for parity-violating experiments, but further developments are needed. Also, further tests are required to prove our ability to deliver high average current with high polarization photocathodes over long periods of time.

REFERENCES

1. Spectra-Physics, Opto Power diode laser model OPC-DO60-FC.
2. Bandwidth Semiconductor Inc., Bedford, NH.
3. T. Maruyama et al, NIM A492, 199 (2002).
4. Electro-Optical Prod. Co., model QC-70I, max extinction ratio 280:1.

Helicity-Correlated Effects For SAMPLE Experiment

M.Farkhondeh, W.Franklin, E. Tsentalovich, T.Zwart

MIT-Bates Linear Accelerator Center, Middleton, MA, USA

Abstract. In 1998-2001 three series of SAMPLE experiment [1-3] were conducted at the MIT-Bates Linear Accelerator Center. SAMPLE measures parity-violating effects in electron scattering from protons and deuterons. The measured asymmetry associated with electron helicity is very small (about 1 ppm). In order to reduce systematic errors, the properties of the beam (intensity, position, size and energy) must remain unchanged when the helicity of polarized electrons is reversed. In this paper we analyze the sources of the helicity-correlated effects in the electron beam and our approach to minimize them.

INTRODUCTION

The electron gun at MIT-Bates [4] is mounted on the top of an accelerator column and surrounded by a Faraday cage maintained at 300 keV relative to ground. A Ti:Sa tunable cw laser provides the light source for photoemission. The laser beam travels over 20 m between the optical table and the GaAs crystal in the gun. Although the output power of Ti:Sa laser (about 7 W) is sufficient to run the experiments with bulk GaAs photocathodes, the Quantum Efficiency (QE) of high-polarization (strained or superlattice) crystals is too low to be used with this laser system. Recently, we installed a multimode diode array laser on the injector. This laser produces up to 200 W of laser power at fixed wavelength λ=810 nm, has excellent stability and is very convenient to operate. The drawback of the laser is the high emittance of the beam ε=200mm·mrad, in comparison with $\varepsilon \approx$ 1mm·mrad of the diffraction-limited beam from the Ti:Sa laser. The existing laser beam transport system was inadequate for the new laser, and a new wide-aperture 4-m long transport system was designed and installed (Fig.1). With the new system, the laser beam strikes the photocathode with an angle of ~37°.

POLARIZATION REVERSAL

Circularly polarized light is required to produce polarized electrons. There are two different ways to transform the linear polarization of light into circular polarization and reverse the helicity as needed. The first way is to pass the laser beam through a $\lambda/4$ wave plate, producing circular polarization. The helicity can be reversed by inserting a $\lambda/2$ wave plate. This method is very simple and provides excellent results even if the beam divergence is high, but the rate of helicity reversal is limited by the

CP675, *Spin 2002: 15th Int'l. Spin Physics Symposium and Workshop on Polarized Electron Sources and Polarimeters*, edited by Y. I. Makdisi, A. U. Luccio, and W. W. MacKay

necessity to insert/retrieve the λ/2 wave plate and it is very difficult to exceed 1 Hz. The SAMPLE experiment requires helicity reversal with a rate of 600 Hz; therefore this approach is inadequate for this experiment.

The second approach involves a Helicity Pockels Cell (HPC). The positive (negative) helicity is achieved by applying positive (negative) λ/4 voltage to HPC. Slow helicity reversal by inserting a λ/2 wave plate changes the sense of helicity, providing a powerful tool to suppress systematic errors. The performance of Pockels cells crucially depends on the quality of the alignment, and this performance deteriorates with large beam divergence. A correlation between angle and position within a laser beam leads to a gradient of the polarization within a beam profile.

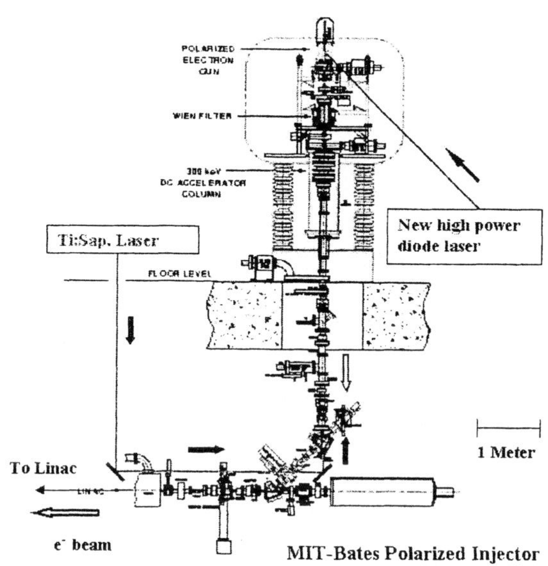

FIGURE 1. A schematic view of the MIT-Bates polarized injector with the laser beam paths.

ORIGINATION OF HELICITY-CORRELATED EFFECTS

The main source of all helicity-correlated effects in the beam is Polarized Induced Transport Asymmetry (PITA), produced by different reflectivity for S and P waves. Since some fraction of residual linear polarization always exists in the circularly polarized beam, and the direction and amplitude of this linear polarization might be different for positive and negative helicity states, the helicity-correlated asymmetries appear. Let us define the analyzing power ε of the transport system as an asymmetry in the transition of light linearly polarized in orthogonal directions. For an air-glass interface (n_1=1.0, n_2=1.5) ε varies from 0 to 5 % at incident angles between 0° and 45°. For GaAs crystal (n_2=4.5) ε ≈25 % at 45°. The analyzing power of strained GaAs crystals is governed by different mechanism, but it is high (ε ≈5-15 %) even at normal incidence.

Modern mirrors with antireflective coating have relatively low analyzing power (0.2 – 0.5 %). Still, a transport line containing several mirrors might produce significant helicity-correlated effects. It is important to pair the mirrors in such a way that the plane of reflection rotates by 90° in the pair in order to balance P- and S-reflections. That allows canceling in the first order PITA effect.

When the HPC for the Ti:Sa laser beam line was located at the beginning of the transport line, and circularly polarized beam traveled through 4 mirrors, the imperfections in the mirrors balancing dominated and the resulting analyzing power of the transport system was of the order of 10^{-3}-10^{-4}. When we moved the HPC to the end of the transport line, between the last mirror and the input vacuum window, the only analyzing elements left after the HPC were the vacuum window and the photocathode itself, with almost normal incident beam. With a bulk GaAs crystal the total analyzing power dropped greatly, but with a strained crystal the anisotropy of the cathode becomes dominant and the analyzing power grows to 5-15%.

The beam from the diode array laser strikes the crystal at an angle of 37°. The analyzing power is very large (about 20 %) for both bulk and strained crystals.

Since the direction and amplitude of the linear component of polarization is non-uniform across the laser beam, traveling through the transport system with some analyzing power results in a positional asymmetry. Another important source of positional asymmetry for strained crystals is the non-uniformity of the analyzing power across the surface of the crystal. Pockels cells produce very small steering effect on the beam (angles of the order of 10^{-7} - 10^{-8}), and this effect may result in the positional asymmetry as well if the transport line is sufficiently long. This effect can be suppressed by imaging Pockels cell location onto the crystal.

If the electron beam in the accelerator has an intensity asymmetry, loading effects in the accelerating structure lead to an energy asymmetry.

All three asymmetries (intensity, position, energy) are mutually dependent. If the electron beam with an energy asymmetry travels through a section with a large dispersion, differential scraping (losing electrons on one side of the beam for positive helicity, and on another side for negative helicity) may occur, resulting in a positional asymmetry. The beam with a positional asymmetry traveling through a narrow aperture may develop an intensity asymmetry.

There could be other, more exotic asymmetries, which we considered negligible for the SAMPLE experiment, but they could be significant for more demanding experiments ([5] for instance): beam size asymmetry, beam shape asymmetry, asymmetries with a temporal profile (for instance, an intensity asymmetry can be positive at the beginning of the pulses, and negative at the end. Averaged over the pulse, the asymmetry becomes zero, but non-linear effects could affect the systematic errors.)

SUPPRESSION OF HELICITY-CORRELATED EFFECTS

The most general recipe in dealing with the helicity-correlated effects is to minimize the asymmetries using miscellaneous alignment techniques, and to control the residual asymmetries with active feedback loops.

Since the effects are usually proportional to the degree of linear polarization in the laser beam, the first step is to maximize the circular polarization. This is a rather challenging task for the diode laser beam due to the very large emittance. In order to minimize the angular spread of the beam traveling through the HPC we have to maximize the size of the beam (we didn't want to lose a significant fraction of laser power using collimators). In order to achieve an extinction ratio of about 150 we had to increase the beam size to 50 mm. The Pockels cell used had a clear aperture of 75 mm.

Balancing P- and S-reflection, careful alignment of Pockels cells and polarizers are the obvious steps to minimize helicity-correlated effects. When strained crystals with a high anisotropy are used, the direction of the residual linear polarization must be lined up along the axis of the photocathode. If the $\lambda/2$ wave plate is placed after the HPC, the intensity asymmetry depends on the rotation angle θ of the wave plate as

$$\alpha = \varepsilon[\Delta \cdot \sin(4\theta - a) + \gamma \cdot \sin(2\theta - b) + \beta]$$

where ε is the analyzing power, Δ term corresponds to asymmetric retardation errors for the HPC, γ terms corresponds to the imperfection of the $\lambda/2$ plate itself and β term corresponds to the birefringent components in the transport line [6]. Since the positional asymmetries usually have the same origin as intensity asymmetry, they generally exhibit the same behavior. Fig.2 demonstrates the results of intensity and positional asymmetry measurements with a strained crystal, and the laser beam striking the cathode at 37° angle.

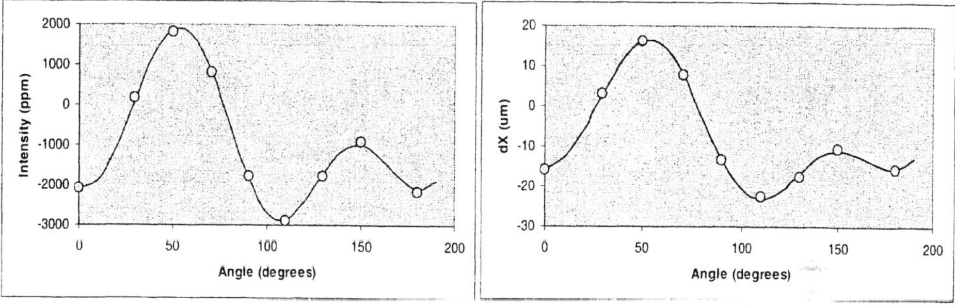

FIGURE 2. The intensity and positional asymmetries as a function of wave plate angle. The solid lines represent the fit.

The helicity-correlated effects are large: thousands of ppm in intensity, tens of microns in position. Our feedback systems were designed to control the initial asymmetries of the order of hundreds ppm and hundreds nanometers. Therefore, several conditions must be satisfied: the lines on Fig.2 must cross zero, the lines for intensity and both positional asymmetries must cross zero at about the same rotational angle of the $\lambda/2$ plate, and the asymmetries must be stable with time. Our preliminary measurements didn't provide a clear answer to this question. It was concluded that further tests were needed, and due to scheduling constraints, the production runs for SAMPLE were carried out with a bulk GaAs cathodes and Ti:Sa laser.

Three feedback systems were used in SAMPLE experiments: two positional (x and y) and one intensity feedbacks. We found that it is essential to have these three systems separate and orthogonal.

Our first approach of adjusting the HPC voltage to correct for the intensity asymmetries failed to adhere to this rule. Adjustment of the HPC voltage essentially changes the fraction of linear polarization in the beam, and it affects both intensity and positional asymmetries. In order to separate the intensity and positional feedbacks, we introduced an additional Correction Pockels cell (CPC), installed between two linear polarizers before the HPC. CPC is biased to a voltage of V0~500 V, and small (>10V) helicity-correlated corrections on the bias control the intensity asymmetry without affecting the positional asymmetries. The sensitivity can be adjusted by altering the V0 voltage, and is usually set to about 50 ppm/V.

Since some Pockels cells deteriorate with a constant bias applied, we developed another scheme, where CPC bias is set to zero, but a $\lambda/10$ plate is inserted in front of CPC. The plate introduces some elliptical polarization in the laser beam and in this scheme the sensitivity is set by changing the $\lambda/10$ plate rotational angle.

A piezo-electric system [7] for positional feedback was developed by the SAMPLE collaboration. A thin optical flat was inserted in a laser beam path. The angle of the plate could be adjusted with a frequency exceeding 1 kHz, thus altering the beam position by as much as several hundreds nanometers. The flat was installed in the 3-point suspension holder, allowing to control x and y components independently.

RESULTS AND CONCLUSIONS

During the production SAMPLE runs the intensity feedback functioned automatically, updating the measurements every 3 minutes. The statistical error in each measurement was of the order of 10 ppm. Averaged over a full day of running, the error reduced to less than 1 ppm.

Positional asymmetries were very stable, and required only occasional adjustments. When the piezo-electric system was used for the first time, it allowed reducing positional asymmetries from several hundreds to about 30 nm. In the last SAMPLE production run the positional differences averaged over several days were less than 10 nm.

As a result, all three SAMPLE experiments were successfully completed using a bulk GaAs photocathodes and Ti:Sa laser. More development is required to conduct parity-violating experiments with strained photocathodes and high power diode laser.

REFERENCES

1. B. Mueller et al, Phys. Lett. 78, 3824 (1997).
2. D.T.Spayde et al, Phys. Lett. 84, 1106 (2000).
3. R.Hasty et al, Science 290, 2117 (2000).
4. G.D.Cates et al, NIM A278, 293 (1989).
5. SLAC Proposal E-158, 1997.
6. B.Humensky, Princeton University, to be published.
7. T.Averett et al, NIM A438, 246 (1999)

Recent Polarized Photocathode R&D at SLAC

D.-A. Luh[a], A. Brachmann[a], J. E. Clendenin[a], T. Desikan[a], E. L. Garwin[a], S. Harvey[a], R. E. Kirby[a], T. Maruyama[a], C. Y. Prescott[a], and R. Prepost[b]

[a]*Stanford Linear Accelerator Center, Menlo Park, CA 94025*
[b]*Department of Physics, University of Wisconsin, Madison, WI 53706*

Abstract. The SLAC high-gradient-doped MOCVD-grown GaAs cathode presently in use consists of a strained GaAs low-doped layer (with a small admixture of P) capped by a few nanometers of highly Zn-doped GaAs, which is heat-cleaned at relatively high temperature and then activated by Cs/NF_3 co-deposition. The high-gradient-doped structure solves the problem of the surface charge limit that the previously-used SLAC cathodes had, and this preparation procedure has produced satisfactory results. However, the preparation procedure has a few weaknesses that prevent cathodes from achieving the ultimate desired performance. The peak polarization is limited to 80% due to strain relaxation in the relatively thick strained layers. Also dopant loss causes the surface charge limit effect to reappear after multiple high-temperature heat-cleanings. In this paper, we will discuss recent progress made at SLAC that addresses these limitations, including using the MBE growth technique with Be doping and using the superlattice structure. In addition, to reduce the heat-cleaning temperature, an atomic hydrogen cleaning technique is explored.

CURRENT PROCEDURES

The cathode used in the current accelerator operation at SLAC is a high-gradient-doped strained GaAs/GaAsP [1]. The MOCVD-grown cathode consists of a strained GaAs low-doped layer (with a small admixture of P) capped by a few nanometers of highly Zn-doped GaAs. When the 2-inch cathode wafer is received, it is anodized at 2.5V to form a 3-nm oxide protecting layer, and then waxed on glass and cut to the required circular shape. When the cathode is ready for installation, the glass is removed, and the cathode is degreased in boiling trichloroethane. After degreasing, the surface oxide of the cathode is stripped by ammonium hydroxide, and the cathode is immediately transferred into a loadlock. The cathode is heat-cleaned at 600°C for one hour and activated by Cs/NF_3 co-deposition. In our standard procedure, heat cleaning and activation are done twice before the cathode is transferred into the polarized electron gun.

WEAKNESSES

Although our standard procedures for cathode preparation have yielded satisfactory results, it has a few weaknesses that have prevented cathodes achieving the ultimate desired performance.

CP675, *Spin 2002: 15th Int'l. Spin Physics Symposium and Workshop on Polarized Electron Sources and Polarimeters,* edited by Y. I. Makdisi, A. U. Luccio, and W. W. MacKay

MOCVD: The base pressure in MOCVD growth chambers is usually in high-vacuum range. Compared with ultra-high-vacuum techniques, the growth environment of MOCVD may not be as clean. Furthermore, because chemical reactions are needed on wafer surfaces during growth, MOCVD requires higher growth temperature, and its growth mechanism is complicated. Due to the nature of the MOCVD process, it may be difficult to grow thin films of the highest quality.

Zn dopant: The diffusion coefficient of Zn in GaAs is high at the heat-cleaning temperature in our preparation procedure (600°C). Because of the high diffusion coefficient, the heat-cleaning capability of Zn-doped cathodes is very limited. In high-gradient-doped cathodes, the highly doped surface layer is responsible for removing the surface-charge-limit effect in high photocurrent operations [1], and dopant loss is very undesirable.

The SIMS (Secondary Ion Mass Spectroscopy) analysis shown in Fig. 1 confirms the Zn-dopant loss after cathodes receive heat-cleaning treatments. Before the cathode is heat-cleaned, the Zn-dopant concentration is $5\times10^{19}cm^{-3}$ as specified [1]. After 5 hours of heat-cleaning at 600°C, the Zn-doping level at surface drops significantly. A test on a Zn-doped high-gradient-doped cathode indicates that the cathode starts to show the surface-charge-limit effect after three hours of heat-cleaning at 600°C.

Strain relaxation in single strained layer: The high-gradient-doped cathode has a 90nm thick strained GaAsP as its active layer. The strain in the active layer is responsible for the polarization of photoelectrons, and strain relaxation in the active layer will lower the polarization.

A strained layer starts relaxing when its thickness exceeds the critical thickness. It relaxes completely when its thickness exceeds the practical thickness. It is partially relaxed when its thickness is between the critical thickness and the practical thickness. In the case of GaAs or GaAsP, the critical thickness is about 10nm, and the practical thickness is about 100nm. Table 1 shows the comparison of the performance between two cathodes. Both cathodes are high-gradient-doped strained GaAs/GaAsP. The only difference between the two cathodes is the thickness of their active layers. The polarization of the cathode MO5-6007 is considerably lower due to the strain relaxation in its active layer.

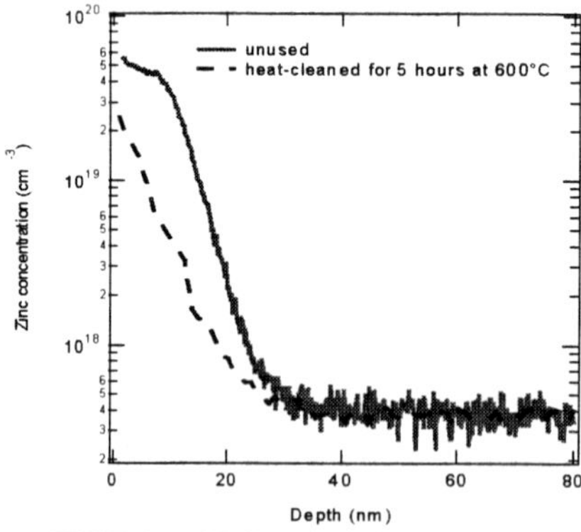

FIGURE 1. SIMS analysis for the cathode before and after heat cleanings.

TABLE 1.

Cathode No.	Active Layer Thickness (nm)	Polarization (%)
MO5-5868	90	82
MO5-6007	170	70

IMPROVEMENTS AND PROGRESS

To address the weaknesses in the standard cathode growth and preparation procedure, a few improvements can be made.

MBE and Be/C doping: Our cathode growth technique has been switched from MOCVD with Zn doping to MBE with Be doping. The ultra-high vacuum of MBE growth ensures a clean growth environment. MBE growth usually requires a lower growth temperature and has a simpler growth mechanism. All these advantages make the growth of high-quality layers easier. Be and C have much lower diffusion coefficients than Zn, and Be-doped and C-doped cathodes will have better heat-cleaning capability.

Fig. 2 shows the performance of the two cathodes from wafers SVT-3982 (MBE-grown Be-doped) and MO5-5868 (MOCVD-grown Zn-doped), tested in Cathode Test Laboratory (CTL) at SLAC. Both cathodes are high-gradient-doped strained GaAsP with the same structure. The result shows that the performance of MBE-grown SVT-3982 is better than MOCVD-grown SVT-5868. The heat-cleaning performance of the Be-doped SVT-3982 is yet to be tested.

As-capped cathodes: MBE allows As capping at the end of the wafer growth. A thin As cap layer is sufficient to protect cathode surface from exposure to air, and can be removed by heat-cleaning at lower temperature.

Atomic hydrogen cleaning: Another method to lower the heat-cleaning temperature is to use atomic hydrogen cleaning. The idea is to use atomic hydrogen generated by a RF dissociator to react with and remove oxide and carbon contamination from cathode surface at a lower temperature. Fig. 3 shows the preliminary results from the atomic hydrogen cleaning system in CTL [2]. The reference cathode is GaAs, stripped of its oxide layer by NH_4OH, heat-cleaned, and activated by Cs/NF_3 co-deposition. The result shows that the QE increases with increasing heat-cleaning temperature, which indicates that the thin oxide layer on the cathode surface from the short air exposure is gradually removed as the temperature increases. The GaAs test cathode is treated as indicated in the figure and then activated by Cs/NF_3 co-deposition. Because the test cathode is not stripped by NH_4OH, the cathode surface starts with a thick native oxide layer. As the data indicates, repetitive

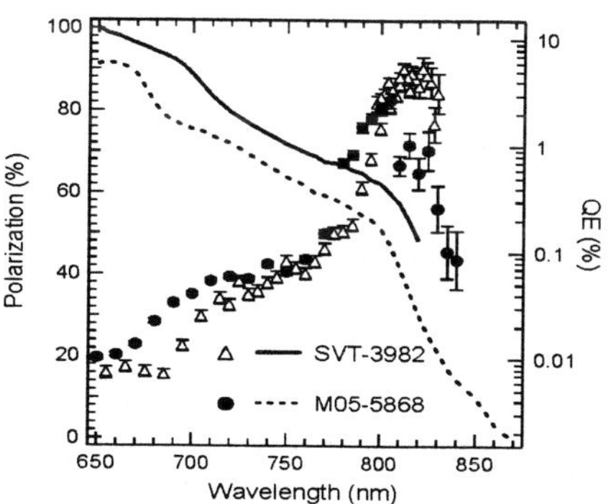

FIGURE 2. The performance comparison between MBE-grown Be-doped (SVT-3982) and MOCVD-grown Zn-doped (MO5-5868) cathodes.

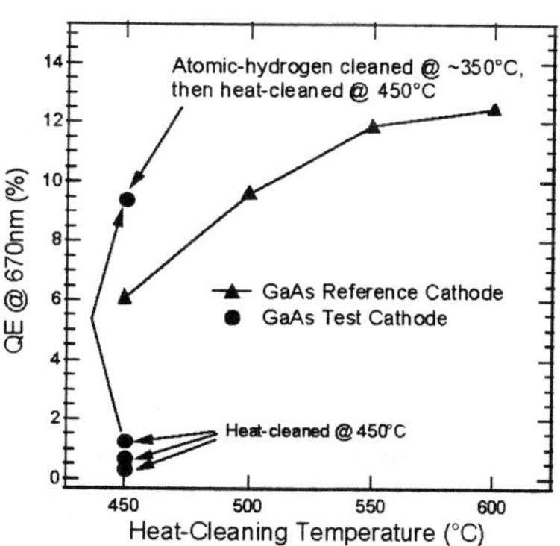

FIGURE 3. Preliminary results from atomic-hydrogen cleaning system.

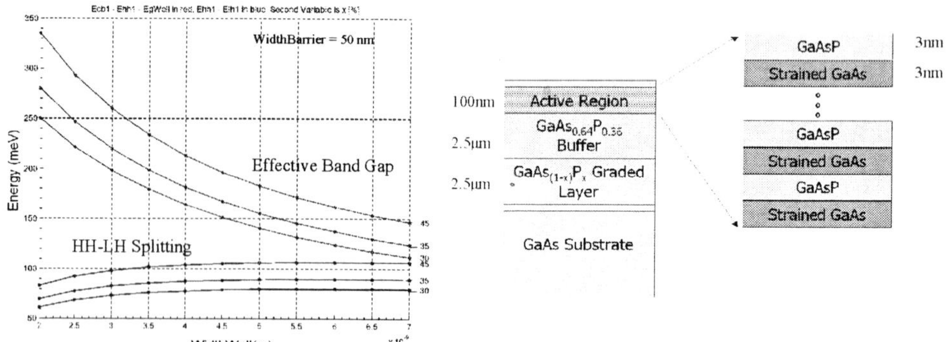

FIGURE 4. Typical results from band structure calculation.

FIGURE 5. The structure of high-gradient-doped superlattice cathodes (SVT-3682 and SVT-3984).

heat-cleanings at 450°C cannot remove the oxide layer effectively, and thus the QE is very low. After one hour of atomic hydrogen cleaning, the QE is greatly improved and is similar to that from the reference cathode heat-cleaned at 500°C. Later studies have demonstrated that it is possible to produce cathode with a QE higher than 14% from unstripped GaAs cathode by atomic-hydrogen cleaning (not shown in the figure). However, these studies also indicate that excessive atomic hydrogen cleaning yields GaAs cathode surfaces with low QE. Work continues to optimize the conditions for atomic-hydrogen cleaning.

Superlattice photocathodes: The superlattice structure is employed to preserve strain in the active layer of photocathodes. The idea is to grow strained layers sandwiched between unstrained layers, where the thickness of each strained layer is less than the critical thickness.

1. A band structure calculation is performed to determine the proper structure parameters to grow superlattice cathodes. The transfer matrix method [3] is used for the calculation. Fig. 4 shows typical results of effective band gap and heavy-hole-light-hole (HH-LH) splitting from the calculation. In this figure, the barrier width is fixed at 50 nm, while the well width and the phosphorus fraction are allowed to change. Because the cathode QE is related to the band gap, and the polarization is related to HH-LH splitting, the result from calculation can be used to predict roughly how the cathode QE and the polarization will change when the structure parameters are varied.

2. Photoluminescence measurements are performed on cathode wafers to check the cathode band structure. This provides a way to check the accuracy of the band structure calculation. Furthermore, it also checks the uniformity of cathode wafers if the photoluminescence measurements are performed on entire wafers.

3. X-ray diffraction is done to characterize the cathode structure when cathode wafers are received from vendors. By studying x-ray diffraction patterns, structure parameters can be determined, including layer thickness, composition, and strain.

Fig. 5 shows the structure of the first superlattice cathode we studied. The structure parameters of the cathode are similar to the ones reported in Ref. [4]. In this superlattice cathode, strained GaAs layers are sandwiched by GaAsP layers. Both the well and barrier widths are 3nm. The phosphorus fraction in the GaAsP is 0.36. A 5nm

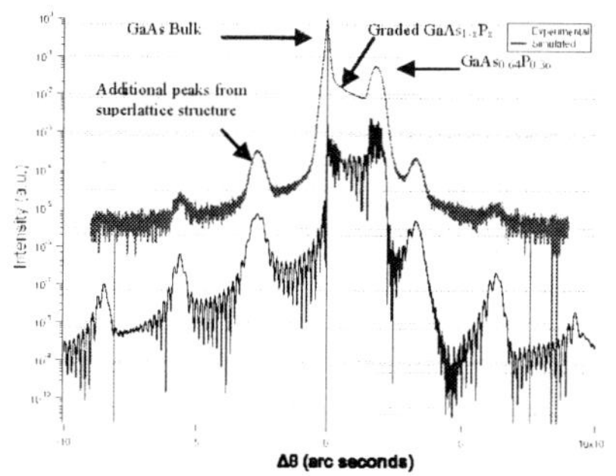

FIGURE 6. Results of (004) scan on a superlattice cathode from experiments (top) and simulations (bottom).

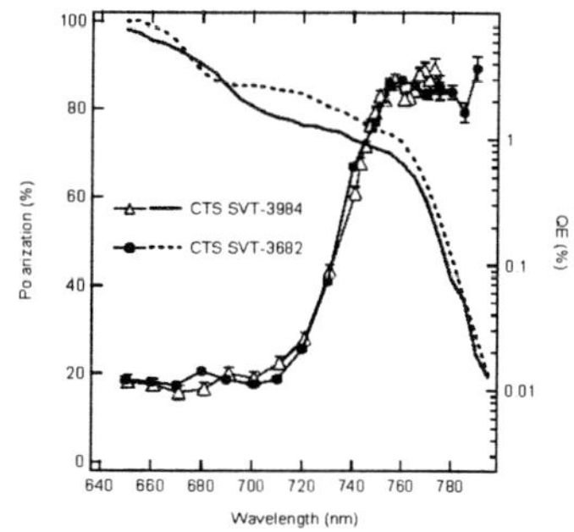

FIGURE 7. Performance of high-gradient-doped superlattice cathodes SVT-3682 and SVT-3984.

highly-doped GaAs layer is grown on the surface of the superlattice cathode to address the surface-charge-limit problem.

Fig. 6 shows the result from a (004) scan in an x-ray diffraction measurement. The experimental results from show all the familiar features. The Bragg peaks from GaAs bulk, graded GaAsP, and GaAsP are clearly identified. Additional peaks from the superlattice structure can also be seen. The simulation results show that the barrier width and the well width are about 32Å with the phosphorus fraction about 0.36. These numbers are very close to the structure design.

The performance of two superlattice cathodes is shown in Fig. 7. Both cathodes show good QE, and their peak polarizations are over 85%. One cathode from the wafer SVT-3984 has been tested in Gun Test Laboratory at SLAC, and there is no surface charge limit observed with available laser energy.

CONCLUSIONS

Although the standard cathode growth and preparation procedure at SLAC has yielded satisfactory results, there is still room for improvement. Dopant loss during high-temperature heat-cleaning and strain relaxation in single strained layer have prevented cathodes from achieving ultimate performance. MBE-grown Be-doped cathodes are expected to be of higher quality and higher heat-cleaning capability than MOCVD-grown Zn-doped cathodes. To reduce the heat-cleaning temperature, an atomic hydrogen cleaning system has been set up. Preliminary tests show promising results although the operational conditions are yet to be optimized. The superlattice structure is employed to preserve strain. The first superlattice cathodes have shown both good QE and good polarization.

REFERENCES

1. T. Maruyama, A. Brachmann, J. E. Clendenin, T. Desikan, E. L. Garwin, R. E. Kirby, D.-A. Luh, J. Turner, and R. Prepost, *Nucl. Instrum. Meth.* **A492**, 199-211(2002).
2. Many thanks to Matt Poelker of Jefferson Lab for his help during system design and installation.
3. Shun-Lien Chuang, *Physics of Photoelectronic Devices*, New York, Wiley, 1995.
4. T. Nishitani, *et al.*, *Proceedings of the 14th International Spin Physics Symposium*, Osaka, Japan, 2000, AIP Conference Proceedings 570, 2001, pp. 1021-1023.

A Zero-Degree Inline Optical Electron Polarimeter

A.S. Green, M.A. Rosenberry, and T.J. Gay

Behlen Laboratory of Physics, University of Nebraska, Lincoln, NE 68588-0111, USA

Abstract. We have used a new configuration for a noble gas optical electron polarimeter. This polarimeter is part of an experiment involving dichroic scattering of longitudinally polarized electrons from chiral molecules. Our polarimeter sits along the electron beam axis at the end of the apparatus and measures the polarization of noble gas fluorescence emitted at 0°. The polarimeter is maximally sensitive to longitudinal electron polarization, and it maintains the axial symmetry of our experiment.

INTRODUCTION

In a noble gas electron polarimeter, electrons collisionally excite a ground-state noble gas through exchange scattering. Because of spin-orbit coupling, the excited state spin angular momentum is partially transferred to orbital angular momentum. When the atom decays, it emits circularly polarized light represented by the Stokes parameter P_3. This parameter can be related to electron polarization P_e through the use of integrated-state multipoles [1]. For fluorescence emitted along the electron beam axis (0°), P_3 is given for the heavy noble gases by

$$P_3 = -\frac{\begin{Bmatrix} 1 & 1 & 1 \\ J & J & J_f \end{Bmatrix} \sqrt{2} \left\langle t(J)^{+}_{10} \right\rangle}{\frac{2(-1)^{J+J_f}}{3\sqrt{2J+1}} - \begin{Bmatrix} 1 & 1 & 2 \\ J & J & J_f \end{Bmatrix} 2\sqrt{\frac{1}{6}} \left\langle t(J)^{+}_{20} \right\rangle} \tag{1}$$

The symbols $\{...\}$ are $6j$ coefficients, and J and J_f are the total angular momenta of the excited and final states, respectively. The relative integrated-state multipole $\left\langle t(J)^{+}_{10} \right\rangle$ is the atomic orientation parameter which is proportional to P_e; and $\left\langle t(J)^{+}_{20} \right\rangle$ depends on the linear polarization fraction P_1, and is a measure of the alignment of the atomic charge cloud. With this information, we can relate electron polarization to light polarization by introducing the *analyzing power*, A, of the target gas: $P_e = P_3 / A$. We calculate A to be

CP675, *Spin 2002: 15th Int'l. Spin Physics Symposium and Workshop on Polarized Electron Sources and Polarimeters*, edited by Y. I. Makdisi, A. U. Luccio, and W. W. MacKay

$$A \equiv \frac{2}{3}\left[\frac{1-0.254P_1}{1+P_1}\right]. \tag{2}$$

In this expression, P_1 is measured at 90° to the beam axis. Since our apparatus does not permit such a measurement, we must use data from another experiment [1, 2] performed at 90° to determine P_1.

PROCEDURE AND RESULTS

The electron lens portion of our optical electron polarimeter is shown in Fig. 1. As a preliminary test, we used argon as the target gas because argon has the largest analyzing power of the noble gases and is inexpensive [3]. The transition of interest in argon is $3p^5 4p\ ^3D_3 \rightarrow 3p^5 4s\ ^3P_2$ (8115 Å), with a threshold energy of 13.07 eV.

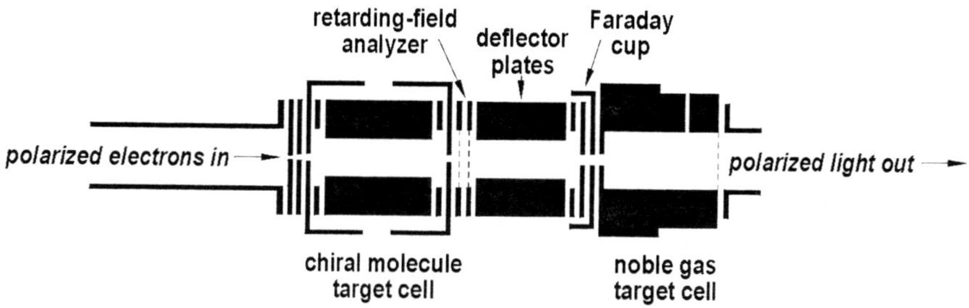

FIGURE 1. Electron lens configuration. When chiral molecule studies are being performed within the cell on the left, deflector plates send the transmitted electron beam onto the Faraday cup. When electron polarization measurements are necessary, electrons pass through the Faraday cup into the noble gas cell on the right. Noble gas fluorescence is collected outside the vacuum chamber by a photomultiplier tube with an appropriate interference filter. Polarization (P_3) is measured by rotating a quarter-wave plate that sits in front of a fixed linear polarizer.

We focus a circularly-polarized 785 nm diode laser onto a wafer of strained GaAs to produce an electron beam. After attenuating the beam by 1/e with argon, we ramp the electron energy and measure excitation functions (near threshold) for orthogonal settings of a quarter wave plate in our optical polarimeter (Fig. 2a). From these results, we calculate P_3 (Fig. 2c). To determine the electron polarization, we first fit the argon data of Wijayaratna [1, 2] for P_1 at 90° (Fig. 2b). Then we calculate the analyzing power for argon as a function of energy (Fig. 2d) and finally determine electron polarization (Fig. 2e).

Our first test shows that $P_e \approx 0.26$ for this strained GaAs crystal. This result is low when compared with a polarization of 0.55 for 785nm light reported by Maruyama *et al.* [4]. Unfortunately, our measurements were made with a GaAs wafer just before it needed to be replaced. The surface had fogged over significantly from repeated heat cleaning and activation sessions. More polarization tests will be made soon with a fresh wafer of the strained GaAs.

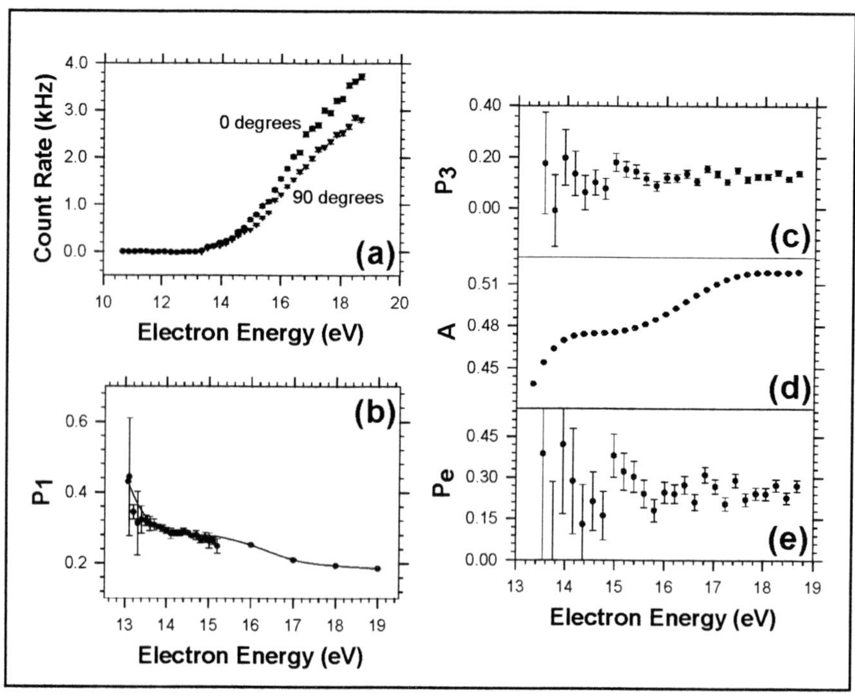

FIGURE 2. (a) Argon excitation functions near threshold for quarter-wave plate at 0° and 90°. (b) Fit to Wijayaratna's P_1 data for argon. (c-e) Polarimetry results (P_3, A, and P_e) for argon near threshold.

ACKNOWLEDGMENTS

We thank P.D. Burrow and K. Aflatooni of the University of Nebraska-Lincoln for advice regarding the design of our electron optics. We also thank M. Poelker of Jefferson Lab for providing us with pieces of strained GaAs. This project was funded by NSF Grant No. PHY0099363.

REFERENCES

1. J.E. Furst, W.M.K.P. Wijayaratna, D.H. Madison, and T.J. Gay, *Phys. Rev. A* **47**, 3775 (1993).
2. W.M.K.P. Wijayaratna, Doctoral Dissertation, University of Missouri-Rolla, (unpublished).
3. T.J. Gay, J.E. Furst, K.W. Trantham, and W.M.K.P. Wijayaratna, *Phys. Rev. A* **53**, 1623 (1996).
4. T. Maruyama, E.L. Garwin, R. Prepost, and G.H. Zaplac, *Phys. Rev. B* **46**, 4261 (1992).

Polarized Electrons Using the PWT RF Gun*

J. E. Clendenin[†], R. Kirby[†], Y. Luo[¶], D. Newsham[¶], D. Yu[¶]

[†]*Stanford Linear Accelerator Center, 2575 Sand Hill Rd., Menlo Park, CA 94025, USA*
[¶]*DULY Research Inc., Rancho Palos Verdes, CA 90275, USA*

Abstract. Future colliders that require low-emittance highly-polarized electron beams are the main motivation for developing a polarized rf gun. However there are both technical and physics issues in generating highly polarized electron beams using rf guns that remain to be resolved. The PWT design offers promising features that may facilitate solutions to technical problems such as field emission and poor vacuum. Physics issues such as emission time now seem to be satisfactorily resolved. Other issues, such as the effect of magnetic fields at the cathode—both those associated with the rf field and those imposed by schemes to produce flat beams—are still open questions. Potential solution of remaining problems will be discussed in the context of the PWT design.

INTRODUCTION

Colliders presently being designed require ~ 1 nC of charge per microbunch with an emittance of 10^{-8} m in the vertical plane for both the electron and positron beams. The electron beam is required to be polarized. To date all polarized electron beams for accelerators have been produced by dc-biased photocathode guns. For a bias of 100 kV, the typical transverse emittance for a solid-state photocathode gun is on the order of 10^{-5} m for a 1-nC bunch. Primarily because of the longitudinal space-charge force associated with high charge density, a relatively long pulse is generated at the dc gun, then an rf buncher following the gun is used to reduce the microbunch length sufficiently to allow injection into one cycle of the accelerating rf, where additional compression can take place. The rf buncher increases the emittance to the 10^{-4}-m level. RF guns operate with much higher fields at the cathode, which permits generation of short pulses without significant blow up of the longitudinal emittance, which in turn makes an rf buncher unnecessary. A transverse emittance of 1.2×10^{-6} m has been demonstrated for an S-band rf gun for a charge of 1-nC bunch and square-shaped bunch length of 9 ps FWHM [1]. In addition, the potential to reduce the transverse emittance in one plane to the 10^{-8}-m level using an optical transformation [2] is being actively investigated [3].

The proposal here is to adapt an rf gun to the production of polarized electron beams by using a highly p-doped GaAs crystal for the cathode and exciting with

*Work supported by Department of Energy contract DE–AC03–76SF00515.

CP675, *Spin 2002: 15th Int'l. Spin Physics Symposium and Workshop on Polarized Electron Sources and Polarimeters*, edited by Y. I. Makdisi, A. U. Luccio, and W. W. MacKay

circularly polarized laser light of the required wavelength (in the range of 750-850 nm). The plane-wave-transformer (PWT) gun is chosen as a design that best matches the requirements for continuously operating an NEA GaAs crystal in a high rf field for long time periods. These requirements will next be outlined.

REQUIREMENTS FOR A POLARIZED SOURCE

There is much experience operating GaAs crystals in dc guns biased at about 100 kV [4]. An atomically clean GaAs crystal must be activated with Cs and an oxidizer to produce a negative electron affinity (NEA) surface. This is best done in a UHV chamber isolated from the gun by a gate valve, after which the crystal must be inserted into the gun without breaking vacuum. For an rf gun, the crystal diameter must be kept small, typically about 1-cm, and the holder designed to eliminate rf breakdown. To produce a low emittance beam, the laser spot on the cathode should have a radius of about 1 mm and a pulse length that is $\leq 20°$ rf phase, which for S-band is ≤ 20 ps. Since this results in extremely high current densities at the cathode, care must be taken to choose a cathode design that is unaffected by the surface charge limit [5]. For high polarization, the active layer of the GaAs crystal will be only about 100 nm thick, which ensures that the low-charge extraction time is only a few picoseconds [6].

Successful dc guns operate with a vacuum of $<10^{-11}$ Torr. The best pressures achieved with presently operating rf guns are on the order of 10^{-10} Torr. Some gas species, such as H_2O and $C0_2$, affect the quantum efficiency (QE) of the cathode more than others, such as H_2 or CO. The dynamics of contamination and surface damage of the GaAs by gas molecules and ions formed in the rf cavity have yet to be explored experimentally.

HV breakdowns in dc guns have a devastating effect on the QE. One can assume the same will happen with rf breakdowns. Thus the gun rf cavities must be designed and processed to eliminate this phenomenon. The rf processing must be done with a dummy cathode. Based on experience with dc guns, the resulting dark current must be <<1 nC per μs of rf. The large dark current observed in the BINP experiment with an activated GaAs cathode in a prototype S-band rf gun was attributed to secondary electron emission [7], which can be high for NEA GaAs. Reducing electron bombardment as well as keeping the emission time short may prove to be crucial.

Magnetic fields at the cathode can potentially decrease the beam polarization. Since the initial spin vector is axial, transverse magnetic fields are the principal concern. Azimuthal rf magnetic fields, $B_\theta(r)$, exist on the cathode during extraction that are of zero magnitude at $r=0$ and increase as ~ $\frac{r}{R_0} E_0\sqrt{\varepsilon\mu}$ up to $R_0/2$, where E_0 is the electric field on the cathode at the time of extraction and R_0 is the radius of the cavity. In this proposal these fields are negligible at $r=1$ mm $\approx 0.025R_0$.

THE INTEGRATED PWT PHOTOINJECTOR

An example of an S-band integrated PWT photojector is shown in Fig. 1. It consists of a standing wave structure of ½+10+½ cells operating in the π-mode. The disks, which are suspended from 4 water-carrying rods, are mounted inside a large cylindrical tank. The rf power couples first into the annular region of the tank in a TEM-like mode, then couples into the accelerating cells in a TM-like mode on axis—thus the name.

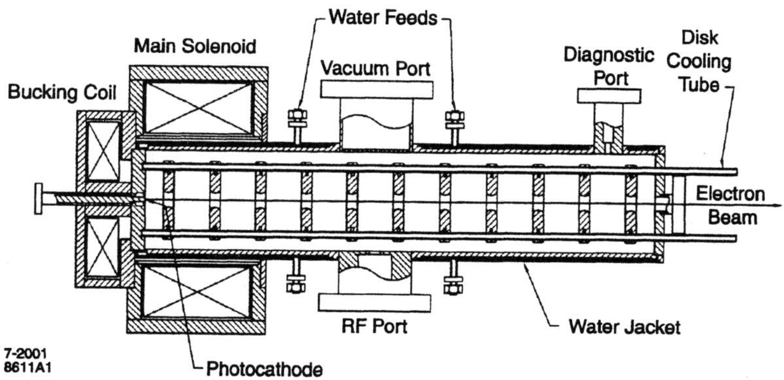

FIGURE 1. The integrated PWT photoinjector.

There are several reasons for choosing the PWT design over the 1.6-cell gun plus booster. Parmela simulations indicate that a peak field of 120-140 MV/m at the cathode is required to achieve a transverse emittance of 10^{-6} m for a 1-nC bunch, 10 ps pulse. However, the growth in emittance as the peak field is lowered to 60 MV/m is much less for the PWT injector. Thus for collider applications the PWT injector can be operated at lower fields, which generally will reduce dark current and avoids any field emission from the GaAs due to an inversion layer.[1] In addition, the separation of the tank from the disks improves the conductance for vacuum pumping and also allows a wide range of Q values and thus filling times. Long filling times may reduce rf breakdown.

For polarized beam applications, the PWT gun shown in Fig. 1 can be improved as follows. To reduce dark current and rf breakdown, Class 1 OFHC Cu forged using the hot isostatic pressure (HIP) method and single-point diamond machining to a roughness of 0.5 μm or better can be used. The final assembly should undergo a simple rinsing in ultra-pure water [8]. Vacuum pumping can be improved by increasing the tank diameter and coating its inner surface with a thin film getter

[1] When the field into the surface of a p-doped semiconductor is strong enough that the conduction band at the surface is bent closer to the Fermi level than the valence band, then the surface becomes n-type with electrons occupying states in the conduction band near the surface. See ref. [9] for a fuller discussion.

material such as TiZr or TiZrV [10]. The increased tank diameter will also increase the stored energy, which is desirable for pulse trains. It is possible to increase the water cooling to handle 25 MW peak rf power at 180 Hz. The design parameters for a prototype PWTS-band polarized photoinjector are given in Table 1.

Table 1. S-Band PWT Design Parameters

Parameter	Value
Frequency	2856 MHz
Energy	20 MeV
Charge per bunch	1 nC
Normalized emittance	**1 μm**
Energy spread	<0.1%
Bunch length	2 ps rms
Rep. rate	5 Hz
Peak current	100 A
Linac length	58 cm
Beam radius	<1 mm
Peak B field	1.8 kG
Peak gradient	**60 MV/m**
Peak brightness	2×10^{14} A/(m-rad)^2

GaAs crystals must be prepared to have an NEA surface in a separate UHV chamber, and then inserted into the gun without venting. Such systems are in routine use for dc-biased guns, but will require some careful development for an rf gun since an rf seal must be made while the crystal itself, which is quite fragile, must not undergo significant stress.

FUTURE PLANS AND CONCLUSIONS

An SBIR-I has recently been awarded to DULY Research, Inc. to begin the engineering design of an S-band PWT integrated photoinjector with load lock for polarized electron beams. If this is followed by a Phase II award, the photoinjector will be tested at the Gun Test Facility at SLAC. The test will be for QE in a high rf field of an activated GaAs cathode at 20 MeV. The QE during the test will be monitored using a green laser (532 nm) of 2-10 ps pulse length. Polarization and emittance measurements will be performed in subsequent testing.

A PWT integrated rf photoinjector built specifically for polarized electrons has many features that will be important for successful operation of polarized electron beams. Such a gun can be built and tested in the near future.

REFERENCES

1. J. Wang et al., "Experimental Studies of Photocathode RF Gun with Laser Pulse Shaping," *Proc. of European Part. Acc. Conf.* (2002), p. 1828, http://accelconf.web.cern.ch/AccelConf/e02/default.htm .
2. R. Brinkmann et al., "A Low Emittance, Flat-Beam Electron Source for Linear Colliders," *Proc. of European Part. Acc. Conf.* (2000), p. 453, http://accelconf.web.cern.ch/accelconf/e00/ .
3. D. Edwards et al., "Status of Flat Electron Beam Production," *Proc. of Part. Acc. Conf.* (2001), p. 73, http://accelconf.web.cern.ch/AccelConf/p01/INDEX.HTM .
4. R. Alley et al., "The Stanford linear accelerator polarized electron source," *Nucl. Instrum. and Meth. A* 365 (1995) 1.
5. T. Maruyama et al., "A very high charge, high polarization gradient-doped strained GaAs photocathode," *Nucl. Instrum. and Meth. A* 492 (2002) 199.
6. J. Schuler et al., "Latest Results from Time Resolved Intensity and Polarization Measurements at MAMI," in *Spin 2000, 14th Int. Spin Phys. Symp.*, Osaka, 2000, ed. by K. Hatanaka et al., AIP, Melville, NY, 2001, p. 926.
7. N.S. Dikansky et al., "Experimental Study of GaAs Photocathode Performance in RF Gun," *Proc. of European Part. Acc. Conf. (2000)*, p. 1645, http://accelconf.web.cern.ch/accelconf/e00/ .
8. C. Suzuki et al., "Fabrication of ultra-clean copper surface to minimize field emission dark currents," *Nucl. Instrum. and Meth A* 462 (2001) 337.
9. S.M. Sze, *Physics of Semiconductor Devices*, John Wiley, New York, 1981, ch. 7.
10. C. Benvenuti et al., "Nonevaporable getter films for ultrahigh vacuum applications," *J. Vac. Sci. Technol. A* 16 (1998) 148.

The SLAC Polarized Electron Source*

J. E. Clendenin, A. Brachmann, T. Galetto, D.-A. Luh, T. Maruyama, J. Sodja, and J. L. Turner

Stanford Linear Accelerator Center, 2575 Sand Hill Rd., Menlo Park, CA 94025

Abstract. The SLAC PES, developed in the early 1990s for the SLC, has been in continuous use since 1992, during which time it has undergone numerous upgrades. The upgrades include improved cathodes with their matching laser systems, modified activation techniques and better diagnostics. The source itself and its performance with these upgrades will be described with special attention given to recent high-intensity long-pulse operation for the E-158 fixed-target parity-violating experiment.

1 INTRODUCTION

The polarized electron source (PES) [1] was initially commissioned for the SLC in early 1992. The present post-SLC configuration is shown in Fig. 1. The all solid-state Nd:YLF-pumped Ti:sapphire short-pulse laser is used for filling PEP-II and for other experiments not requiring polarized electrons. The flash lamp-pumped long-pulse Ti:sapphire laser[1] has been recently improved for a fixed-target experiment[2] that requires a high-charge polarized beam. The characteristics of the PES are listed in Table 1. Which of the two laser beams is delivered to the cathode is chosen on a pulse-to-pulse basis. Improvements to the source are discussed in Section 2 and its present performance in Section 3.

2 IMPROVEMENTS

Given the uncertainties of operating a major accelerator with a polarized electron source, four guns were built for the following purposes: one operating, one on standby, one being repaired, and one for experiments in the off-line laboratory consisting of a duplicate of the first few meters of the 3-km linac. However, the addition of a load-lock[3] in 1993 improved the reliability of the PES to such an extent

*Work supported by Department of Energy contract DE–AC03–76SF00515.
[1]A. Brachmann *et al.*, this Workshop.
[2]P. Mastromarino *et al.*, this Workshop.
[3]Before biasing the cathode, most of the load-lock apparatus must be removed and a corona shield installed around the high-voltage (HV) electrode. Because of space constraints in the linac housing, the HV insulator and corona shield is surrounded by dry gas contained within a sealed shroud.

CP675, *Spin 2002: 15th Int'l. Spin Physics Symposium and Workshop on Polarized Electron Sources and Polarimeters,* edited by Y. I. Makdisi, A. U. Luccio, and W. W. MacKay

that certainly no more than 3 guns are now needed. All activations[4] are done in the load-lock with the gun isolated. Re-application of Cs to the cathode is done periodically in the gun chamber under computer control (operator initiated) with the bias voltage reduced to 1 kV. Cesium channels from SAES are used exclusively. Additional NF_3 is not added inbetween activations. To monitor the cathode quantum efficiency (QE), a low-power cw diode laser is mounted on a side window of the gun with a view of the cathode. The laser is modulated at a few hertz to minimize the cw beam in the linac. A fiber-optically coupled nanoammeter installed at the gun HV electrode support flange supplies a modulated current signal to a lock-in amplifier as shown in Fig. 2. The control chassis also blanks off the diode laser during the linac rf accelerating pulse to prevent any acceleration of photoelectrons generated by this source.[5]

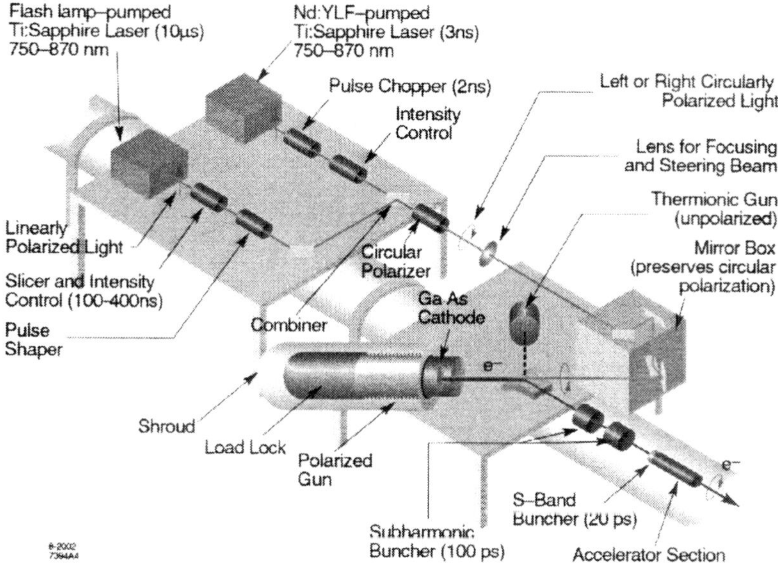

FIGURE 1. The SLAC polarized electron source in its present configuration.

The addition of the load-lock required the cathode to be mounted on a detachable puck that is supported by a 2.5-cm diameter emitter tube made from a 8-cm long Mo cup brazed to a Kovar tube, which is welded to a 304L stainless-steel (SST) tube. During operation, cold N_2 is circulated inside the tube to lower the cathode temperature to about 0 °C. When activating, a resistive heater is installed. To bring the GaAs crystal to the desired 600 °C, even with a heat shield on the vacuum side of

[4] Activations consist of heat cleaning for 1-hour at 600 °C followed by application of Cs and florine in the form of NF_3.
[5] The low intensity of this laser precludes a radiation issue.

the crystal, the puck end of the emitter tube reaches about 800°C, which formerly resulted in frequent vacuum failures of the Mo-Kovar brazed joint. Ni-plating the Kovar prior to brazing helped.[6] Eventually an ANSYS 5.1 study of the temperature response was used to determine the optimum length of the Mo cup to keep the temperature of the brazed joint below 250 °C [2]. A new Mo-length of 31 cm was chosen. Computer controlled temperature-cycling tests resulted in no failures after 70 cycles. To reduce hazards, the gas mixture in the shroud can now be readily switched from SF_6 to dry air or to a mixture of both. Generally SF_6 is used only in the first couple days of operation following closure of the shroud.

TABLE 1. Source Characteristics

Component	Value
Cathode diameter	20 mm (emitting area)
Cathode holder	Mo puck with Ta clip
Electrodes	317L SST with low carbon content, low inclusion density
Bias voltage alumina insulator	39.5-cm long by 19.0-cm ID, OD fluted
Design/nominal bias voltage	-150/-120 kV dc
Dark current at nominal voltage	30 nA typical
Vacuum pumping	120 lps DI and 200 lps NEG pumps
Vacuum	10^{-11} Torr dominated by H_2
Ti:sapphire laser systems:	
Nd:YLF pumped	2-ns pulse for PEP II, FFTB
Flashlamp pumped	Polarized 100-400 ns pulse for ESA

The major improvement is in the cathode itself. The progress toward higher polarization is illustrated in Table 2. There has been only limited success in improving the cathode preparation technique. A vacuum transfer vessel allows the cathode to be activated in the laboratory and then transferred to the load-lock system at the gun under vacuum. The vessel has a small battery-operated IP, but nonetheless a final re-activation in the load-lock is generally necessary. A glove-box for performing the initial cathode cleaning in an N_2 atmosphere has also been constructed, but so far not used. Presently H* cleaning at 400 °C is being studied.[7]

3 PERFORMANCE

Since 1992 the polarized electron source has provided ~80% of the electron beams for the SLAC linac, accumulating nearly 40,000 h of operation with availability >>95%. Except for 1992 (when there was no load-lock) the pattern has been as follows, independent of the type of cathode used (listed in Table 2). After initial activation in the load-lock, the cathode is inserted into the gun. Neither the cathode bias HV nor the presence of the electron beam are observed to have any

[6]Suggested by G. Collet and R. Kirby (SLAC).
[7]D.-A. Luh *et al.*, this Workshop.

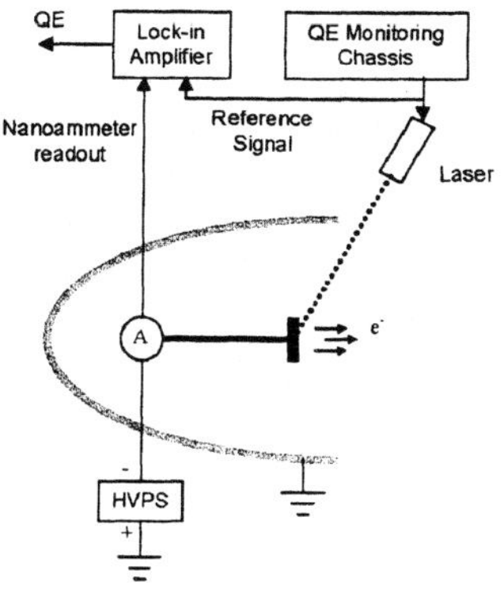

FIGURE 2. Use of nanoammeter (A) cathode bias for continuous monitoring of yield.

TABLE 2. Cathode History

Cathode	P_e (%)	Experiment
Bulk GaAs	25	1992 SLC
Bulk AlGaAs	35	1992 E142
300-nm strained GaAs	65	1993 SLC
100-nm strained GaAs	80	1993 143
		1994-98 SLC
		1995 E154
		1996 E155
		1999 E155X
Gradient-doped strained GaAs	80	2001-present

effect on the QE. When the QE decreases to the point where the required current can no longer be obtained—because of limited laser energy and/or because of the surface charge limit—the cathode is re-cesiated in the gun with the cathode bias reduced to – 1 kV. This typically occurs every 3-5 days and takes ~20 min., the time dominated by cycling of the bias voltage. About 0.5 ML Cs is estimated to be added per re-cesiation. No NF_3 is added. The re-cesiation completely restores the original QE. The rate of decrease of the QE (QE lifetime) changes over long periods (generally to shorter lifetimes), but not between re-cesiations. Re-cesiations are repeated up to 50 times over the period of a year, limited only by operational considerations to preemptively change the cathode during the accelerator scheduled major mainte-

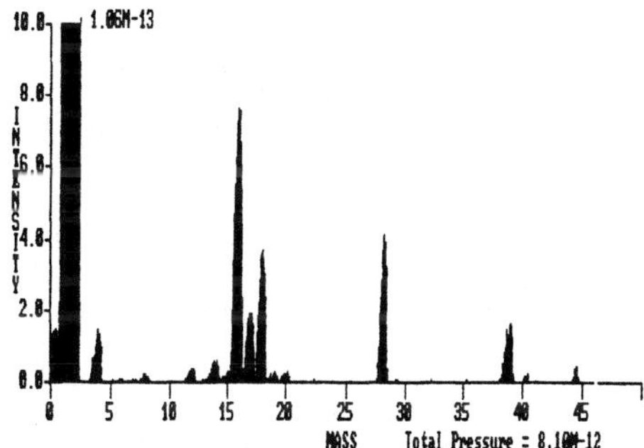

FIGURE 3. Recent gun-chamber RGA mass spectrum. The intensity is in units of 10^{-13} Torr. The m=2 (offscale) and m=4 peaks are $\sim10^{-11}$ and $\sim10^{-13}$ Torr respectively.

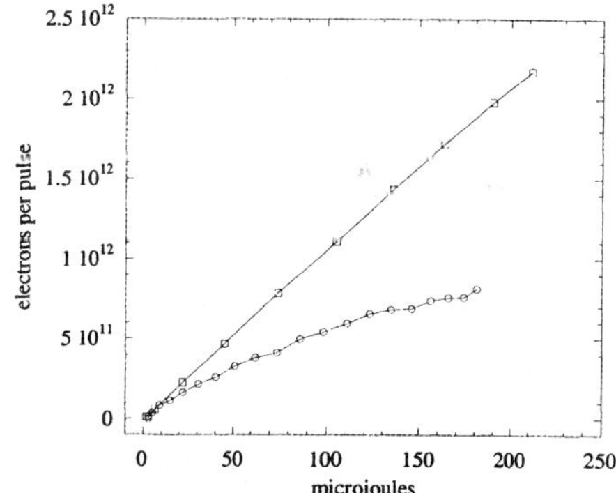

FIGURE 4. Comparison of emission from gradient-doped cathode (top curve, squares) using 100-ns laser pulse with medium-doped (SLC type) cathode (bottom curve, circles) and 300-ns laser pulse.

nance periods. An important element in the cathode performance is the gun vacuum. An RGA scan for the present gun, which has not been back-filled for several years, is shown in Fig. 3. The RGA is operated continuously.

The high charge that can be generated with the new gradient-doped strained-layer GaAsP-GaAs cathodes [3] is illustrated in Fig. 4. Note the absence of saturation effects in the upper curve, which is generated using a 100-ns laser pulse. Due to the longer pulse length of 300-ns, the current density in the lower curve is much less than the upper one for the same laser energy. For both curves, the laser illuminates a 20-mm diameter spot on the cathode. The parameters achieved for the currently running E158 using this new cathode are compared with those required for NLC in Table 3.

Table 3. Comparison of Parameter Achieved in E158 with NLC Requirements

Source Parameter	E158	NLC
Electrons per macropulse	8×10^{11} e^-	27×10^{11} e^-
Repetition rate	120 Hz	120 Hz
Macropulse length	270 ns	266 ns
Micropulse spacing	DC	1.4 ns
Polarization	~80%	≥80%
Intensity jitter	<0.5%	0.5%
Transverse jitter	5% of spot size	22% σ_x, 50% σ_y

4 CONCLUSIONS

The SLAC polarized electron source has proven to be an outstanding performer since its initial commissioning in 1992. The source produces a charge for the presently running E158 that is close to that required for an NLC macropulse. Even higher charge has been demonstrated. Work continues at SLAC to develop a reliable low-temperature activation technique to provide more flexibility for the gradient-doped cathodes. Higher polarization is being explored, principally using superlattice semiconductor structures.

5 REFERENCES

1. R. Alley *et al.*, "The Stanford linear accelerator polarized electron source," Nucl. Instrum. and Meth. A 365 (1995) 1.
2. G. Mulhollan and G. Kraft (SLAC), private communication.
3. T. Maruyama *et al.*, at this Workshop and "A very high charge, high polarization gradient-doped strained GaAsP photocathode," Nucl. Instrum. and Meth. A 492 (2002) 199.

Status of the Jefferson Lab Polarized Beam Physics Program and Preparations for Upcoming Parity Experiments

J. Grames, P. Adderley, M. Baylac, J. Clark, A. Day, J. Hansknecht, M. Poelker, M. Stutzman

Thomas Jefferson National Accelerator Facility, 12000 Jefferson Avenue, Newport News, VA 23606, USA

Abstract. An ambitious nuclear physics research program continues at Jefferson Lab with Users at three experiment halls receiving reliable, highly polarized electrons at currents to 100 μA. The polarized photoguns and drive lasers that contribute to Jefferson Lab's success will be described as well as significant events since PES2000. Typical of conditions at accelerators worldwide, success brings new challenges. Beam quality specifications continue to become more demanding as Users conduct more challenging experiments. In the months that follow this workshop, two parity violation experiments will begin at Jefferson Lab, G0 and HAPPEx2. The photogun requirements for these experiments will be discussed as well as our plans to eliminate/minimize systematic errors. Recent efforts to construct high power Ti-Sapphire drive lasers for these experiments also will be discussed.

POLARIZED PHOTOINJECTOR

The Jefferson Lab photoinjector delivers highly polarized electrons to three Users simultaneously at beam currents spanning six decades (100 pA to 100 μA). Two identical -100 kV DC photoemission electron guns (see Figure 1) provide all of the electrons for the nuclear physics program at the laboratory. One is used for production beam delivery to Users and the second serves as a spare. We use strained layer GaAs photocathodes from Bandwidth Semiconductor (formerly Spire Corporation). These GaAs-on-GaAsP photocathodes are grown to the SLAC specification and the best samples provide beam polarization 70 to 80%. Cathode preparation entails brief (15 minute) atomic hydrogen cleaning in a dedicated chamber after which the cathode is transferred under nitrogen purge and bath to the gun chamber which is then baked. Subsequently, the cathode is heated and then activated with cesium and nitrogen tri-fluoride to establish negative electron affinity (NEA). The guns and past performance are further described in ref. [1,2].

Two laser tables straddle the beamline and provide a direct optical path to either gun. The drive lasers emit rf-pulsed light with picosecond pulsewidths, synchronous to the CEBAF accelerating frequency (1.497 GHz). A variety of lasers and rf pulse repetition rates can be chosen to meet the demands of the physics program which varies on a monthly basis. Diode laser systems [3] have been used to satisfy most of

CP675, *Spin 2002: 15th Int'l. Spin Physics Symposium and Workshop on Polarized Electron Sources and Polarimeters,* edited by Y. I. Makdisi, A. U. Luccio, and W. W. MacKay

the beam current needs of the physics Users. Diode lasers are reliable and relatively maintenance free and can be easily configured to produce different repetition rates; 499 MHz for beam to a single experiment hall and 1497 MHz for beam to all three halls. The demand for both high current and high polarization has spurred our development of high power Ti-Sapphire drive lasers with high repetition rates [4]. A new parity violation experiment at Hall C (G0) has extended the need for drive laser flexibility by requiring beam with a pulse repetition rate of 31.1875 MHz. To accommodate this experiment we have developed a variety of home made modelocked lasers and we also have purchased a commercial laser. The optical pulse width of all of the CEBAF drive lasers is typically 70 ps which is compatible with the pre-accelerator phase acceptance of the photoinjector. The laser beams are focused at the photocathode to a spot diameter of 500 to 700 microns.

FIGURE 1. The two CEBAF photoemission guns (the gun at left is covered with a "shroud" to prevent accidental contact with high voltage). The guns are identical and positioned symmetrically on either side of a 15 degree bend magnet. The polarity of the bend magnet is flipped to switch between guns. A temperature controlled, plastic-walled laser hut is shown in the background.

SOURCE PERFORMANCE

Over the course of years, we have refined our photocathode preparation procedures to obtain consistent results; namely, high quantum efficiency and high polarization [M. Baylac et al., these proceedings]. Typical values of quantum efficiency and polarization at two standard operating wavelengths are ~ 0.15% QE at 840 nm for polarization > 70% and ~ 1.0% QE at 770 nm for polarization ~ 30%.

The operating lifetime of the CEBAF polarized photoguns is very good. At high current (250 μA extracted gun current), the 1/e photocathode lifetime is approximately 300 C. At lower current (< 100 μA), lifetime increases to approximately 600 C. The endstations receive roughly 90% of the extracted gun current; the rest is dumped on apertures that limit the transverse and longitudinal acceptance of the injector.

We attribute our excellent operating lifetime to low vacuum pressure within the guns. Pressure in the 10^{-12} Torr range is obtained using an array of non-evaporable getter pumps (NEG) surrounding the cathode/anode gap. The beamline downstream of the gun activation chamber is also coated with NEG material. Vacuum quality is maintained during beam delivery by eliminating unintentional photoemission from portions of the photocathode that are not directly illuminated with light from the drive lasers. Electrons that originate from the edge of the photocathode can have extreme trajectories, eventually hitting the vacuum chamber walls, liberating gas and degrading vacuum pressure. The QE of the edge of the photocathode is killed using the process of anodization. These features and techniques are described in past Polarized Source Workshop proceedings.

Gun maintenance and associated "downtime" varies according to the nuclear physics program and User demands for current. At high current (~ 250 μA extracted gun current), we can deliver uninterrupted beam for 2 to 3 weeks before we are forced to move the laser to a fresh photocathode location. We can move the laser spot approximately 5 times before the photocathode must be heated and reactivated to restore QE. The process of heating and reactivating takes 8 hours and QE restores to initial values. At lower gun currents, QE degrades more slowly and beam interruptions occur less frequently. An example of a particularly long period of uninterrupted beam delivery occurred during the summer months of 2001, when we delivered beam (60 μA gun current) to three experiments for three months from one photocathode location.

Unfortunately, limited experience suggests catastrophic field emission develops within each gun after approximately four activations. Field emission necessitates cathode electrode re-polishing and the installation of a new photocathode sample, a time consuming, labor intensive process. A load lock gun is being developed which eliminates the need to apply cesium on the photocathode within the high voltage chamber [M. Stutzman, these proceedings]. By keeping cesium from the cathode electrode, we hope to eliminate field emission from our guns.

DRIVE LASER

The drive laser system is the most dynamic component of our polarized photoinjector. Three Users make for a dynamic physics program that requires the laser table be easily configured to achieve different combinations of beam intensity (laser power), electron polarization (laser wavelength), RF synchronization (1.497 GHz or sub-harmonic) and high availability (little maintenance). Other important laser features include cost and availability (sometime diode laser vendors take a product off the market), intensity noise at the helicity reversal rate, and the extent that each laser turns off between optical pulses (so that one laser does not inadvertently provide beam to all three halls).

RF-synchronous gain-switched diode lasers are easy to use, low maintenance, and reliable. The intensity noise at the helicity reversal rate is very good, <0.1%. However, the diode laser power is low (<100 mW) and the laser wavelength is fixed for either high current (770 nm) or high polarization (840 nm). Additionally, amplified spontaneous emission produces DC beam that leaks into the RF buckets for other Users. For high current Users this is not problematic, but DC leakage to low current (nanoamp) Users can cause polarization dilution. Finally, the vendor Spectra Diode

Labs no longer sells the diode optical amplifier used to boost laser power from the gain switched seed laser. We are testing another vendor (Toptica) although the lifetime performance appears questionable.

An alternative to gain switched diode lasers is the Ti-Sapphire laser that can provide both high current and high polarization. The laser power exceeds 300 mW and the wavelength is tunable from 770 to 860 nm, which means the laser can be used for different photocathodes with peak polarization occurring at different wavelengths. These desirable features (i.e., high power and tunable wavelength) come with the cost of added complexity and more frequent need for maintenance (i.e., mirrors must be cleaned and realigned). The Ti-Sapphire laser can also provide the low pulse repetition rate for the G0 experiment (31.1875 MHz), something not achievable with diode lasers and the method of gain switching.

HELICITY CORRELATED SYSTEMATIC CONTROL

Any parameter of the beam (e.g., intensity, position, or energy) that may change when the laser helicity is reversed is called a helicity correlated (HC) effect. To satisfy beam requirements for the parity violation experiments G0 and HAPPEx2, the accelerator must provide beam with integrated HC intensity variation below one part per million and HC position variation at the target less than one nanometer.

To first order, the sources of helicity correlated effects originate at the Pockels cell and the strained layer photocathode. Ref. [5] describes in great detail the various factors that contribute to different helicity correlated effects. In this paper, only the feedback controls employed at JLab are described.

JLab multi-User operation imposes constraints on the laser configuration and choice of optical elements used to control HC systematics. The most stringent are those imposed on the optical elements common to all lasers. These elements include transport mirrors, a Pockels cell for creating/reversing the circular polarization, an insertable half-wave plate for systematic reversal, and a rotatable half-wave plate used to null the effect of QE anisotropy of the high polarization photocthodes. During the parity violation HAPPEx experiment the Pockels cell was used to control intensity asymmetry using the PITA effect. The rotating half-wave plate was used to minimize HC position sensitivity. Active HC position feedback was not performed. However, these common optical elements affect all Users.

The demands for the upcoming Hall C (G0) and Hall A (HAPPEx2) experiments require the capability to independently control HC intensity and position for each parity User. This meant adding additional parity controls specific to each laser system. Independent intensity control was achieved with an Intensity Attenuator (IA), a low voltage Pockels cell inserted between two similarly oriented optical polarizers after the laser, but prior to the combining optic. The cell has a low insertion loss (~10%) and is very compact, necessary for installing into the region prior to the laser combination optic. The cell voltage is remotely controlled by a DAC and is modulated with the helicity reversal sequence. The transmitted intensity is sinusoidal with applied voltage and has maximum transmission at zero voltage. To shift the sinusoid achieving linear, bipolar intensity modulation, and to operate at a non-biased cell voltage, a tenth-order waveplate is inserted upstream of the cell. The resultant sensitivity of the IA has been measured to be ~300 ppm/volt.

To achieve independent position control we retrofit some of our remotely controlled picomotor mirror mounts upstream of the combination optic with kinematic mounts containing piezoelectric (PZT) stacks. A PZT stack voltage is set remotely by a DAC and is modulated with the helicity reversal sequence providing deflection of the laser beam toward the cathode. The result is an electron beam modulated in position from the photocathode. An IA and PZT have been added to the optical paths for both parity lasers. The G0 experiment demonstrated benchmark capability in September 2002 with these controls in place.

Finally, a software lock has been implemented to manage the interface between measurement of helicity correlated beam parameters at the experimental targets and actuation of the parity controls on the laser table. It is worth noting that the injector and accelerator diagnostics detect all three beams and therefore we use only diagnostics located near the experimental targets for actuating the parity controls.

TI-SAPPHIRE LASER DEVELOPMENT

The time-of-flight proton resolution of the G0 detector requires that the experiment operate at the 48th sub-harmonic (31.1875 MHz) of the accelerator fundamental. As noted, this repetition rate is not suited for diode lasers. Instead, a home built Ti-Sapphire laser was built at JLab using a 4-mirror folded cavity design and a solid state green pump laser (Coherent Verdi-10). Despite an overall cavity length of 4.84 m, the laser fits on a 1.2 m x 0.25 m optical breadboard. An acousto-optic modulator at 15.59375 MHz is used to actively mode-lock the laser. The laser power is >300 mW at 825 nm with <0.2% intensity noise at the helicity reversal rate. However, the optical pulse length of 180 ps FWHM presented difficulties for pre-accelerator transmission.

In parallel, a commercial Ti-Sapphire laser was purchased from Time-Bandwidth. This laser uses a passive SESAM mode-locking technique. The performance of the laser has been turn-key with little maintenance. The output power is >300 mW at 840 nm and is tunable from 770 to 860 nm. The measured pulse length is 70 ps FWHM and provides adequate injector transmission. Importantly, the phase noise is ~1 ps and the HC intensity noise at 30 Hz has been measured between 0.1 and 0.02%. This laser is presently installed in the tunnel and is being used for the G0 experiment.

Finally, the HAPPEx2 experiment, to begin in early 2003, requires a high polarization, high current (>80μA) electron beam at 499 MHz. A second laser operating at this repetition rate has been ordered from Time-Bandwidth.

ACKNOWLEDGMENTS

We thank members of the G0 and Hall A Parity collaborations for continued effort developing the laboratory's parity controls. Brian Bevins contributed substantially by writing the parity control software locks. This work was supported by the USDOE under contract DE-AC05-84ER40150.

REFERENCES

1. Poelker, M., et al., "Polarized Source Performance and Developments at Jefferson Lab" in *14th International Spin Physics Symposium,* edited by K. Hatanaka et al., AIP Conference Proceedings 570, New York: American Institute of Physics, 2001, pp. 943-948.
2. See contribution by M. Stutzman, same proceedings.
3. Poelker, M., *Appl. Phys. Lett.* **67**, 2762 (1995).
4. Poelker, M., and Hansknecht, J., "A High Power and High Repetition Rate Modelocked Ti-Sapphire Laser for Photoinjectors" in *Proceedings of the 2001 Particle Accelerator Conference,* edited by P. Lucas et al., IEEE Catalog Number 01CH37268C, 2001, pp. 95-97.
5. Humensky, T.B., et al., "SLAC's Polarized Electron Source Laser and Optics Systems and Minimization of Helicity Correlations for the E-158 Parity Violation Experiment", Submitted to *Nucl. Instrum. Meth. A* (2002) and SLAC-PUB-9381 (2002).

The Polarized Electron Source at ELSA

Wolther von Drachenfels, Frank Frommberger, Michael Gowin,
Wolfgang Hillert, Markus Hoffmann, Bernhold Neff

University of Bonn, Dept. of Physics, Nussallee 12, 53115 Bonn Germany

Abstract. At the electron stretcher accelerator ELSA in Bonn a pulsed 50 kV inverted gun of polarized electrons has been in operation since February 2000. A strained-layer superlattice crystal is used to deliver a beam with a polarization of about 80 %. A flashlamp-pumped Ti-Sapphire laser with a pulse repetition rate of 50 Hz serves as source of light. The gun is operated in space charge limitation. The current can be chosen by varying the distance between cathode and anode. With 1 μs pulses of 100 mA the source was particularly used together with a polarized target for a GDH sum rule experiment. The high photocathode lifetime allows continuous operation at 100 mA typically for periods of about two weeks without maintenance. So far no change of the crystal was necessary.

INTRODUCTION

Polarization is of increasing importance in hadron physics and therefore efforts have been undertaken at the electron stretcher accelerator ELSA (see Fig. 1) to provide polarized electron beams for external fixed target experiments.

For this purpose in 1997 a pulsed source of polarized electrons came into operation at ELSA to study depolarizing resonances which occur during acceleration of the beam in the circular machines. These resonances might depolarize severely the beam and therefore have to be corrected for. To produce the polarized electrons the source used a GaAs superlattice crystal illuminated by a pulsed Ti:Sapphire laser. From the measurements of the polarization [1] schemes were developed to compensate the influences of the depolarizing resonances. But the reliability of the source was not sufficient for experiments over several weeks. Therefore a second source was developed which started operation in February 2000. In the following construction and operation experiences will be described.

SETUP OF THE POLARIZED SOURCE

In comparison with the old source the new source has several improvements in respect of reliability, polarization, intensity and life time (see also [2]).

An important change is the lower high voltage of -50 kV compared with -120 kV of the old source. The new source runs together with a second linac which meanwhile came into operation. This linac operates with pulses of 1 μs and a repetition rate of 50 Hz but needs only 50 keV injection energy.

CP675, *Spin 2002: 15th Int'l. Spin Physics Symposium and Workshop on Polarized Electron Sources and Polarimeters*, edited by Y. I. Makdisi, A. U. Luccio, and W. W. MacKay

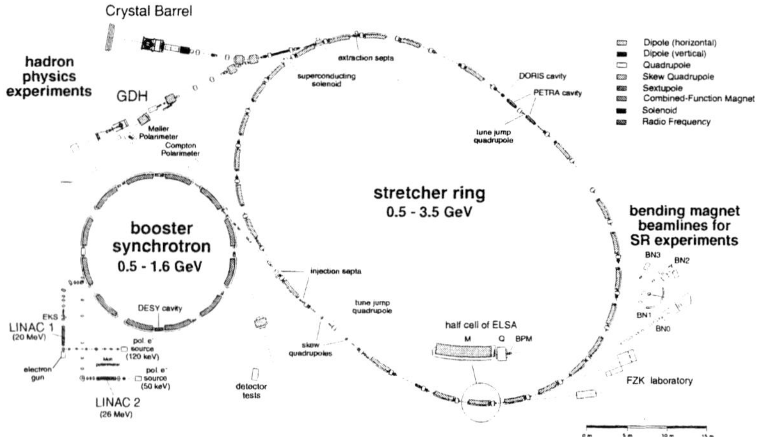

FIGURE 1. Site plan of the electron stretcher accelerator facility ELSA.

In addition the source was designed as an inverted type gun, that means only the cathode is on −50 kV, all other components are on ground potential. A careful HV-design (surface fields below 4 MV/m) led to dark currents below 1 nA @ 60 kV and to a very save operation of the source.

The polarization was increased by using a Be-InGaAs/Be-AlGaAs strained superlattice crystal available from Nagoya University [3]. With a flashlamp-pumped Ti-Sapphire laser at $\lambda = 830$ nm an electron polarization of about 80 % is obtained. From the 10 μs pulses of the Ti-Sapphire laser a pulse slicer cuts 1 μs pulses which are fed via a 85 m long optical fiber to the source. The cathode anode arrangement allows to operate the source in space charge limitation. The perveance of the gun is adjustable. The pulse intensity can be chosen by varying the distance between cathode and anode between 45 and 70 mm. A distance of 61 mm yields 100 mA. The strong so called "spiking behavior" of the Ti-Sapphire does not influence the shape of the electron pulse. Rectangular pulses of 100 mA are delivered even when the quantum efficiency (QE) decreases down to about 0.1 % during longer operation. The very slow decrease in QE means excellent life times of up to 5000 hours for the polarized source. Therefore only seldom a new activation of the photocathode is necessary (see Fig. 2).

The setup of the source is shown in Fig. 3. The source mainly consists of a gun chamber and a small preparation chamber which has to be vented for crystal exchange. The materials (low carbon stainless steel, molybdenum and marcor ceramic) for the gun chamber and the preparation chamber have been chosen carefully and vacuum fired at 1050 °C or 800 °C respectively before assembling. Silver plated copper sealings with an additional blocking layer of nickel have been used. With combinations of ion getter pumps and non-evaporable getter (NEG) pumps a vacuum of 1×10^{-11} mbar in the gun chamber is obtained during operation. With a manipulator the photocathode can be pulled back into the preparation chamber for activation. Here heat cleaning is performed with a filament heater at 19 W (equivalent to 450 °C). The temperature measurement was calibrated in the temperature range of 500 – 600 °C with an infrared thermometer

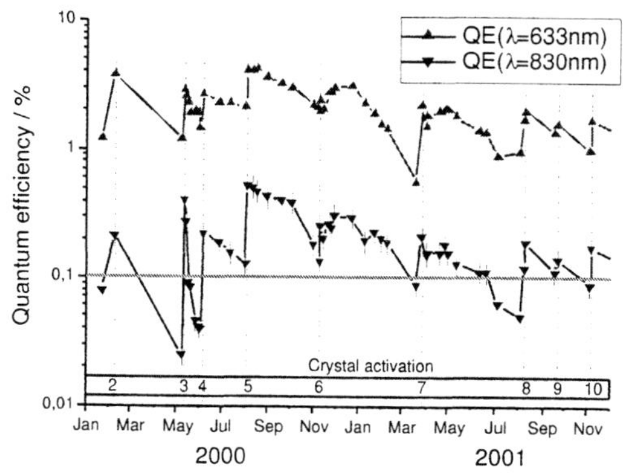

FIGURE 2. The quantum efficiency measured at two different wavelengths.

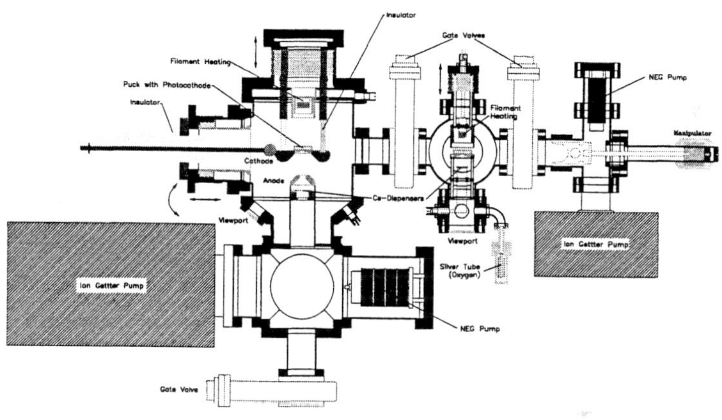

FIGURE 3. Setup of the 50 keV source: Inverted gun and preparation chamber.

using a bulk GaAs crystal and then extrapolated towards lower temperatures . After heat cleaning of about 1 hour and cooling down of about 6–7 hours a typical activation lasts about $1^1/_2$ hour. Since beginning of 2000 only one crystal was in use, no change was necessary.

To connect the source with the linac a 6.2 m long transfer line was installed (see Fig. 4). The vacuum tube has a diameter of 35 mm. Two α-magnets and an electrical bending by 90° together with several ion getter pump stations serve for differential pumping from 10^{-7} mbar at the linac (vacuum dominated by water vapor) to partial pressures below 10^{-13} mbar at the gun for poisoning gases like H_2O and CO_2. Using a large number of

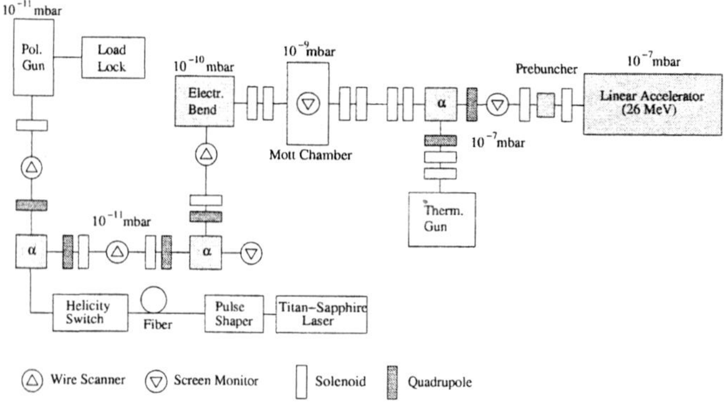

FIGURE 4. The transfer line from the polarized source to the linac.

steerers, solenoids and additional quadrupoles a transfer efficiency to the linac of close to 100 % is obtained. The use of wire scanners and screens is essential to align the beam. All components of the source and the transfer line are under computer control and integrated in the ELSA control system. Parameter sets can be stored and reloaded easily. An optimized set was derived in February 2000 and was used for more than two years without changes.

In table 1 the main parameters of the source are listed.

TABLE 1. Parameters of the polarized electron source.

crystal	Be-InGaAs/AlGaAs strained-superlattice
light source	flashlamp-pumped Ti:Sapphire laser
laser power at the photocathode	$\approx$ 5 kW
laser spot size on the cathode	8 mm Ø
electron polarization	about 80 % at 830 nm
electron current, pulsed	100 mA
pulse length	1 μs
repetition rate	50 Hz
time of operation	typically two weeks without maintenance

OPERATING THE SOURCE WITH ELSA

The studies of depolarizing resonances have been continued with the new source. The depolarizing influence of imperfection resonances could be compensated by special vertical orbit corrections. In addition two fast pulsed quadrupoles have been installed in the main ELSA ring to minimize the influence of intrinsic resonances. The measurements of the beam polarization were done by Møller scattering on the extracted beam. When accelerating from 1.2 GeV at injection into the main ring up to 3.2 GeV nine resonances have to be crossed. The results are shown in Fig. 5. It shows that the source together with ELSA can provide beam polarization of more than 55 % (even above 3 GeV) for the experiments.

In the last two years the source delivered polarized electrons over more than 2500 hours for the Gerassimov-Drell-Hearn (GDH) sum rule experiment [4]. Single runs last up to three weeks. During a run no maintenance at the source was necessary. A source availability of close to 100 % has been obtained.

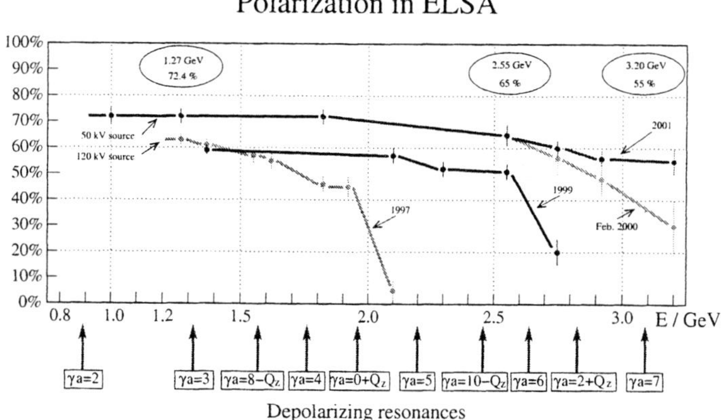

FIGURE 5. Polarization of the electron beam in ELSA.

ACKNOWLEDGMENTS

We would like to thank T. Nakanishi and his group from Nagoya University for providing us photocathodes and for the fruitful collaboration.

This work was supported as part of the SPP 1034 by the Deutsche Forschungsgemeinschaft.

REFERENCES

1. Nakamura S. et al., *Nucl. Instr. and Meth.* **A 411**, 93-106 (1998)
2. Hillert W. et al., *SPIN 2000 Proc. of the 14th International Spin Physics Symposium*, 961-64 (2001)
3. Nakanishi T. et al., *Proc. Low Energy Polarized Electron Workshop, St. Petersburg*, 118-124 (1998)
4. Helbing K. et al., *see this proceedings*, and Speckner T., thesis, Univ. Erlangen (2002) Zeitler G., thesis, Univ. Erlangen (2002)

The MIT-Bates Compton Polarimeter [1]

W.A. Franklin, T. Akdogan, D. Dutta, M. Farkhondeh, M. Hurwitz, J.L. Matthews, E. Tsentalovich, W. Turchinetz, T. Zwart[*†] and E. Booth[**]

[*] *MIT-Bates Linear Accelerator Center, Middleton, MA 01949, USA*
[†] *Massachusetts Institute of Technology, Cambridge, MA 02139, USA*
[**] *Boston University, Boston, MA 02215, USA*

Abstract.
The MIT-Bates Compton Polarimeter, a laser back-scattering device, is used to measure the polarization of electron beams circulating in the South Hall Ring for a range of energies between 300 and 1000 MeV. The apparatus is described in detail and compared to other polarimeters operating within a similar energy region. Preliminary polarization results at 850 MeV are presented.

Compton polarimeters based on the laser back-scattering technique have been successfully employed at a number of labs. Because the analyzing power for Compton scattering rises with electron energy, and because the measurement technique does not destroy the electron beam, these types of polarimeters have been implemented at several high-energy storage rings and colliders, including HERA [1, 2], LEP [3], and SLAC [4]. As shown in Fig. 1, the analyzing powers for polarimeters at such multi-GeV machines are typically 0.5 or higher, but at electron energies below 1 GeV, the analyzing power for Compton scattering is at most a few percent.

Low analyzing powers present many challenges for the application of the laser back-scattering technique. The minimization of systematic false asymmetries is crucial. Furthermore, the relatively low energy of back-scattered gamma rays necessitates accurate modeling of detector response. Polarimeters which have successfully confronted such challenges include the Jefferson Lab Hall A Polarimeter [5, 6] and the NIKHEF Compton Polarimeter [7]. The latter, the first laser backscattering polarimeter to operate below 1 GeV, was used in some very precise measurements at the AmPS Ring [8], but suffered from an inability to operate at high stored beam currents. These experiences were considered carefully in the design of the MIT-Bates Compton Polarimeter.

The MIT-Bates Compton Polarimeter is designed to make accurate absolute measurements of stored electrons in the South Hall Ring (SHR) during nuclear physics experiments with internal targets. Polarized beams are generated by a photoemission-based source [9] and accelerated to energies up to 1 GeV using the Bates Linac and Recirculator [10]. Although the linac duty cycle is less than one percent, the beam in the SHR is effectively continuous wave. The polarimeter has been designed to measure for SHR energies in the range of 300 MeV $< E_e <$ 1000 MeV range. Currents above 100 mA have

[1] This research has been supported in part by a grant from the United States Department of Energy under Cooperative Agreement BEFC294ER40818.

CP675, *Spin 2002: 15th Int'l. Spin Physics Symposium and Workshop on Polarized Electron Sources and Polarimeters*, edited by Y. I. Makdisi, A. U. Luccio, and W. W. MacKay

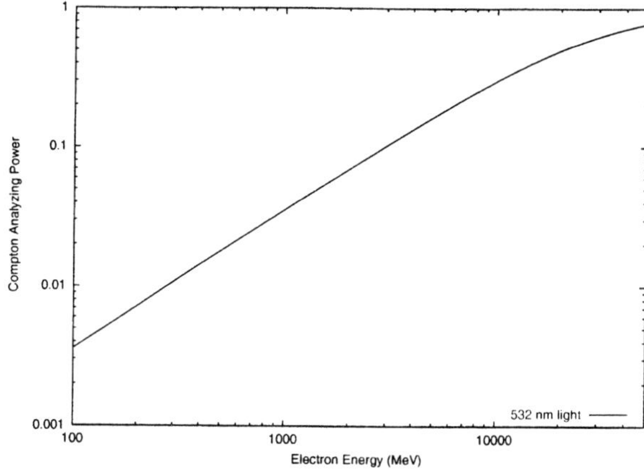

FIGURE 1. Calculated analyzing power as a function of electron beam energy for 532 nm photons undergoing Compton scattering at 180 degrees.

been achieved in the SHR, and the polarimeter should be able to provide continuous data under such conditions for experiments with the newly commissioned Bates Large Acceptance Spectrometer Toroid (BLAST) [11].

The MIT-Bates Compton Polarimeter is shown schematically in Fig. 2. Photons are produced by a Coherent Verdi[2], a solid-state laser, and transported over a flight path of twenty meters to the Interaction Region (IR). The laser produces up to 5 W of continuous-wave output at 532 nm, which is mechanically chopped by a rotating wheel at a frequency of 10 Hz. Background measurements are made during intervals when the laser is blocked. Optical elements in the transport path include seven mirrors, two lenses, a polarizer, a vacuum window, and a helicity Pockels cell (HPC). The HPC, which permits rapid reversal of the laser helicity, is operated at quarter-wave voltage to impart circular polarization to the laser. Two of the mirror mounts have remotely controllable actuators, which allow for optimization of the scattering rate. Two of the SHR dipole chambers have been modified to allow the laser to enter and exit the vacuum system through transparent windows which are not directly exposed to the backscattered flux.

The interaction between the laser and electron beam takes place in a four-meter-long straight section of beam line between the two SHR dipoles immediately preceding the BLAST internal target. Geometric constraints require the vertical crossing angle between the beams to be less than 1 mrad. In setting up for polarization measurements, the electron beam must be carefully tuned to minimize steering in the IR quadrupoles, as the alignment of the scattered photon flux with the polarimeter detector is very sensitive to the electron trajectory. The SHR optics dictate that the beam polarization must be longitudinal at the BLAST target. A correction is made to account for precession of the spin between the polarimeter and the internal target due to the 22.5° bend.

[2] Coherent Laser Group, Auburn, CA 95602

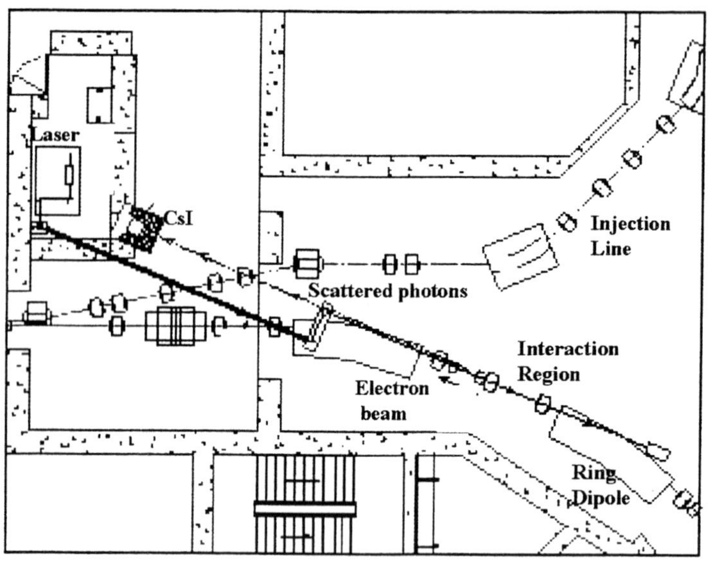

FIGURE 2. Schematic view of MIT-Bates Compton Polarimeter

Back-scattered photons are detected by a CsI detector located at zero degrees with respect to the electron beam, 16 m from the center of the IR. The detector's acceptance is defined by an adjustable lead collimator, 2.5 cm in diameter. For operation at high electron currents, it is necessary to attenuate the photon flux to avoid overwhelming the CsI detector. To keep the signal-to-background ratio high, the laser has been run at its full power and absorbers have been introduced into the backscattered gamma ray line. Stainless steel gives relatively uniform attenuation of the photon flux as a function of gamma ray energy. Energy spectra taken with absorber thicknesses ranging up to 7.5 cm have exhibited little change in shape, and the analyzing power of the polarimeter is not affected. Following the absorber, a permanent magnet is used to sweep away charged particles. The calorimeter consists of a single CsI crystal of dimensions $10x10x25$ cm located inside a shielded hut and preceded by a thin plastic veto scintillator. The efficiency of detection and the effects of adding absorbers to the backscatter line have been modelled by a GEANT simulation.

Data are analyzed by converting histograms into cumulative energy spectra sorted by spin state. The spectrum of scattered photons exhibits a steep edge at an energy determined by the electron energy. The energy calibration of the detector is established and monitored by a combination of radioactive sources, the location of the Compton edge, and an LED pulser embedded in the CsI light guide. The asymmetry is calculated

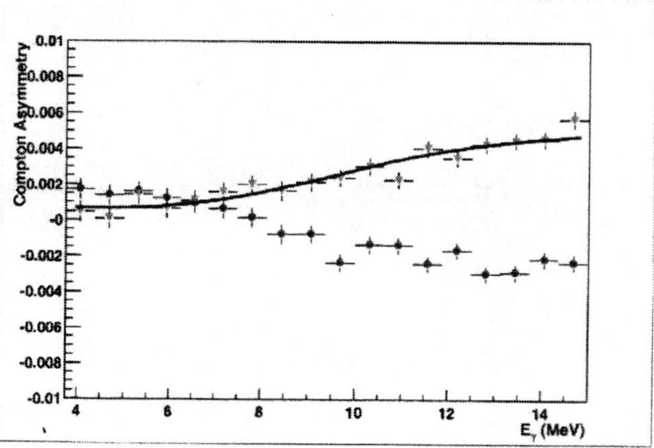

FIGURE 3. Compton polarimeter asymmetry data as a function of gamma ray energy for 669 MeV electrons. The two data sets represent postive and negative polarization from the polarized source.

as a function of gamma ray energy based on the helicity-sorted gamma ray spectra. The appropriate normalization factor for the background can be determined either from scalers characterizing the polarimeter's luminosity as a function of spin state or from the bremsstrahlung background level at energies above the Compton edge. At currents below 10 mA, the signal-to-background ratio generally exceeds 10:1. The background increases nonlinearly with beam current, thereby worsening this ratio for higher currents. The Compton scattering asymmetries are typically of the order of one percent. Unpolarized electrons have been used to verify that false asymmetries are of the order of 10^{-3} or smaller.

The beam polarization is determined by fitting the asymmetry data with a function, determined through a detailed simulation, characterizing the polarimeter's effective analyzing power. The accuracy of polarization extraction is greatly enhanced by averaging over multiple SHR fills with opposite polarization directions. Because an exact spin flip of the electron at the polarized source and a helicity reversal of the laser should leave the asymmetry unchanged, the cross-ratio method [12] can be used in the calculation of the mean Compton polarimeter asymmetry, which eliminates many sources of systematic error associated purely with laser properties or electron beam properties.

Stored polarized beams were first observed in 2001 in the SHR with 669 MeV electrons. Asymmetries exhibiting the expected dependence on back-scattered gamma ray energy were observed, as shown in Fig. 3. The injection orientation for the beam polarization was optimized by varying the voltage on a Wien filter in the polarized injector. The measured beam polarization with the Compton polarimeter exhibited the expected sinusoidal dependence on the Wien filter voltage. Systematics were poorly understood at the time of the run, so a detailed comparison with the polarization at the source is of limited utility.

Since the initial tests, upgrades to the data acquisition system have improved the efficiency and reproducibility of polarimeter measurements. Recently, data were taken at 850 MeV with a strained GaAs crystal installed in the polarized injector. The beam polarization near the source was measured to be 0.75 ± 0.07 by a 20 MeV transmission polarimeter [10]. For a series of measurements with the Compton polarimeter carried out with 7 mA stored in the SHR, an average polarization of 0.73 ± 0.01 (statistical uncertainty only) was measured. While the polarization results for the Ring remain preliminary, their reproducibility and consistency with the transmission polarimeter measurements is encouraging.

The results of polarization measurements with higher stored beam currents have been less straightforward to interpret. The first polarization measurements carried out with large currents in the SHR yielded asymmetries only about half as large as expected. The possibility of instrumental error was ruled out by comparing polarization results for two different runs taken with a beam current of 20 mA. In the first case, the SHR was filled to 20 mA, while in the second case, 70 mA were injected into SHR, but the beam current was allowed to decay to 20 mA. This comparison showed that the polarization was clearly reduced when SHR was filled at the higher current. Furthermore, it was found that full polarization of the stored beam could be restored by changing the betatron tune of the SHR in a certain manner. Further tests explored the sensitivity of the beam polarization to SHR tune in more detail. During these tests, it became clear that tune spreading occurred for large injected currents. This spreading of the tune often produced some degree of depolarization in the stored beam. Furthermore, it was found that for large injected currents, tune changes occurred as the beam circulated in SHR with deleterious effects on the beam polarization. Such effects were not observed for low beam currents where the betatron tune remains much more stable.

The measurements with the Bates Compton Polarimeter qualitatively resemble results obtained at the AmPS Ring [7, 8]. Polarization measurements with the NIKHEF Compton Polarimeter were usually lower than measurements of the beam polarization at the source, and acute polarization loss was seen for large injected currents. However, rate limitations prevented these effects from being mapped out in detail. The Bates polarimeter allows an opportunity to study spin dynamics for intense electron currents in a previously unexplored regime. Such studies will be important for BLAST experiments, and could also prove valuable for other devices which seek intense highly polarized electron beams, such as a proposed electron-ion collider.

REFERENCES

1. Barber, D., et al., *Nucl. Inst. and Meth. A*, **329**, 79 (1993).
2. Barber, D., et al., *Phys. Lett. B*, **343**, 436 (1995).
3. Knudsen, L., et al., *Phys. Lett. B*, **270**, 97 (1991).
4. Woods, M., *SLAC-PUB-7319* (1996).
5. Baylac, M., et al., *Phys. Lett. B*, **539**, 8 (2002).
6. Falletto, N., et al., *Nucl. Inst. and Meth. A*, **459**, 412 (2001).
7. Passchier, I., et al., *Nucl. Inst. and Meth. A*, **414**, 446 (1998).
8. Passchier, I., et al., *Phys. Rev. Lett.*, **88**, 102302 (2002).
9. Farkhondeh, M., et al., *AIP Conf. Proc.*, **570**, 955 (2001).
10. Zwart, T., et al., *AIP Conf. Proc.*, **588**, 343 (2001).
11. Alarcon, R., et al., *Nucl. Phys. A*, **663-664**, 1111c (2000).
12. Wells, S., et al., *Nucl. Inst. and Meth. A*, **325**, 205 (1993).

200 keV Polarized Electron Source at Nagoya University

K. Wada, M. Yamamoto, T. Nakanishi, S. Okumi, T. Gotoh, C. Suzuki, F. Furuta, T. Nishitani, M. Miyamoto, M. Kuwahara, T. Hirose, R. Mizuno, N. Yamamoto, H. Matsumoto [a)] and M. Yoshioka [a)]

Nagoya University, Department of Physics, Nagoya 464-8602, Japan
a) KEK High Energy Accelerator Research Organization, Tsukuba 305-0801, Japan

Abstract. 200 keV polarized electron source with load-lock system has been constructed to produce a beam with high peak current and low emittance that are required by a future linear collider. GaAs photocathode was cleaned by atomic hydrogen and dark currents between the electrodes of the gun that degrade an NEA (Negative Electron Affinity) surface of photocathode could be reduced to less than 1 nA at 200 kV. Recent data on photocathode preparation and dark current measurement are reported in this paper.

1. INTRODUCTION

Japan linear collider requires a highly spin-polarized electron beam that has multi-bunch structure with high intensity and low emittance [1]. For C-band linac scheme, each micro-bunch must contain 2×10^{10} electrons in 700 ps bunch that corresponds the peak current of more than 3 A. The normalized emittance required at exit of the gun should be less than 10π mm·mrad (r.m.s.). A GaAs-type photocathode with NEA (Negative Electron Affinity) surface is expected to produce such a beam.

70 kV DC-gun at Nagoya University have already produced a 2 bunch polarized electron beam separated by 2.8 ns without the NEA surface charge-limit phenomenon by use of a GaAs-GaAsP strained superlattice photocathode with a heavy surface doping [2][3]. It has been confirmed that this type photocathode has high spin-polarization (~ 90 %) and high QE (Quantum Efficiency, ~ 0.5 %) simultaneously [4]. However, the peak current was limited to be 1.6 A by space charge effect of the gun.

In order to lift this limit, a 200 kV DC polarized electron gun system with a load-lock mechanism has been constructed. It is also expected that this gun can produce a beam with lower emittance. Lifetime of an NEA photocathode is so important in practical operation that the reduction of the field emission dark current is considered as a key technique for the construction of this gun. The dark currents may cause the stimulated desorption of gases from electrodes, and a part of these gases are ionized by the collisions with electron beam. In addition to adsorption of the neutral gases, the back-bombardment of these ions should induce the degradation of the good NEA surface state. It is well known that the dark current level must be kept below 10 nA to preserve the good NEA surface state. An outline of the design for this gun,

CP675, *Spin 2002: 15th Int'l. Spin Physics Symposium and Workshop on Polarized Electron Sources and Polarimeters,* edited by Y. I. Makdisi, A. U. Luccio, and W. W. MacKay

photocathode preparation and the data of dark current measurement are reported in this paper.

2. 200 KEV POLARIZED ELECTRON GUN

A schematic view of this system is shown in Figure 1. Without breaking the ultra-high vacuum (UHV), a photocathode can be transferred between three chambers (gun, activation and loading chamber) by using two transporters. In order to operate the load-lock system at grand level, the high voltage (HV) is supplied to a central flange of a pair of ceramic insulators and the support tube for the cathode electrode is fixed to this flange. Each ceramic insulator and HV power supply are divided into some segments with shield rings and each segment is connected with 500 MΩ divider resistor to assure the uniform HV distribution. These HV components are installed in an insulation tank. Dry nitrogen gas is flowed into this tank and pressurized up to 3.6 atm. As a result, 200 kV can be supplied without leakage current along the ceramic surface and corona discharge. The insulation tank and the HV power supply are electrically isolated from ground level respectively, and the dark current from the cathode electrode can be distinguished from the discharge current between the gun and the insulation tank.

The shape of accelerating electrodes was determined by using the simulation code of EGUN and POISSON [5]. The field gradient on the photocathode was estimated to be 3.0 MV/m for the diameter of 18 mm. The peak current limited by space charge effect with full laser illumination to the cathode surface was estimated to be 30 A. To reduce the dark current from the cathode electrode, the size and the shape were carefully designed so that the surface field gradient was 7.8 MV/m at maximum.

As non-metallic impurities on the electrode surface would become the emitting sites of the dark current, the super-clean stainless steel made by remelting method was used as the material of electrodes. Surface polishing by electro-chemical buffing method and the subsequent rinsing with ultra-pure water were also employed to

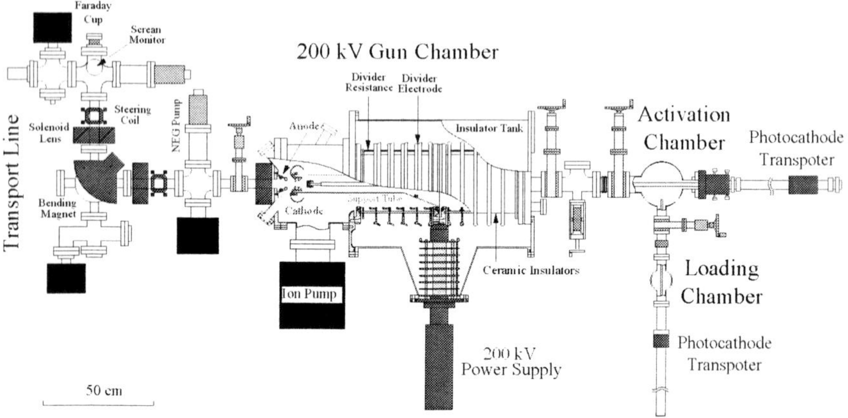

FIGURE 1. Schematic view of the 200 keV polarized electron gun system.

remove the contamination or dust on the electrode surface. For the test sample with an area of about 7 mm^2, the dark current could be about 90 pA under a high field gradient of 34 MV/m [6].

UHV is also indispensable to maintain a good NEA surface state. The 360 l/s ion pump (PST400AX2, ULVAC) and two NEG pump (GP500 and GP100 ST707, SAES Getters) are used as main pumps. The vacuum conditions are monitored by an extractor gage and a quadrupole mass spectrometer. After the careful bakeout at 200°C for 100 hours, the basic pressure reached about 1×10^{-11} torr.

3. PREPARATION OF PFOTOCATHODE

A GaAs photocathode is placed on a Mo pack and hold down by Ta cap. The pack with GaAs wafer is installed from atmosphere to the loading chamber and cleaned by atomic hydrogen cleaning method (AHC). Heat cleaning (HC) by RF induction heating method and NEA activation by deposition of cesium and oxygen are done at the activation chamber. Finally the photocathode is transferred to the gun chamber. In order to avoid the increase of dark current, the cesium is never deposited in the gun chamber.

3.1. Laser-cutting for GaAs photocathode

The cathode electrode that has a hole of 20 mmϕ in its center must hide the edges of cleaved GaAs or Ta cap from high field gradient as shown in the left of Figure 2, otherwise these edges can become the emitting sites of dark current. Therefore the shape of photocathode should be round and set completely into a dent of the cathode pack (23 mmϕ). The laser-cutting method was considered to be useful to make such a round-shaped GaAs wafer without the mechanical contact.

A pulsed Nd:YAG laser at Nagoya University with a repetition rate of 100 Hz and a pulse width of 0.3 ms was used for the laser cutting. Laser power was 0.3 J/pulse and the spot size on the wafer was 0.1 mmϕ. Circular cathodes with 23 mmϕ were cut out from a GaAs wafer covered with paraffin to protect the surface. The cutting edge was blown by dry nitrogen gas to prevent the oxidation of cathode. The paraffin was removed by diethyl ether after the laser cutting. The wafer was attached on the cathode pack by Ta cap with a hole of 20 mmϕ as shown in the photograph of Figure 2. Using

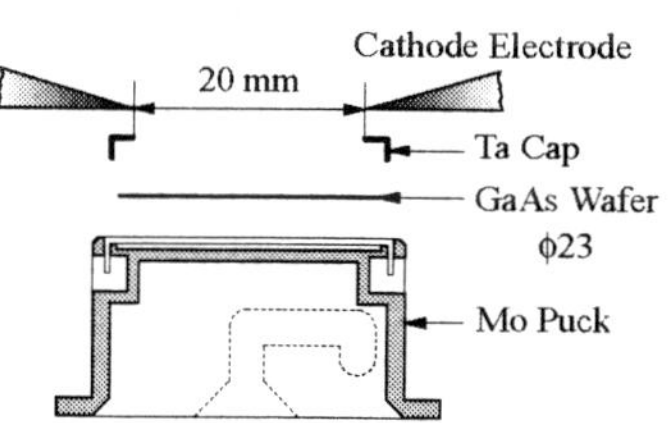

FIGURE 2. Cross section of cathode pack assembly (Left) and photograph of a cathode pack with attached GaAs photocathode (Right).

this round-shaped GaAs wafer cleaned by chemical etching as substrate layer, it is scheduled that the active layer of GaAs-GaAsP superlattice structure is grown on its wafer by a MOCVD apparatus at Nagoya University for high polarization experiments [7].

3.2. Hydrogen cleaning method and NEA activation

Figure 3 shows a schematic view of preparation chambers. Atomic hydrogen was produced by the RF dissociation [8]. The pressure of hydrogen gas in the dissociation tube was 2×10^{-2} torr, and RF frequency was tuned to ~100 MHz. Atomic hydrogen in the dissociation tube travels toward the cathode surface through a hole of 0.8 mmϕ. During AHC, the vacuum pressure in the loading chamber was about 10^{-5} torr and the photocathode was heated at 400°C for 30 min. It is reported that the surface layer of 7.5 nm was removed from GaAs cathode for one hour heating cycle at 600°C [9]. Therefore, typical HC temperature of about 600°C is considered too high to protect the highly p-doped surface-layer of the superlattice photocathode. In this point, AHC has a great advantage in removing oxides and carbides from the surface by heating at relatively low temperature. The cleaned bulk-GaAs photocathode was transferred to the activation chamber and the NEA surface was made after HC at temperature less than 450°C. Low QE of 4 % at excitation wavelength of 633 nm obtained without AHC treatment was greatly improved to 13 % by AHC effect [10]. This result suggests that AHC at 400°C could make the GaAs surface to be atomically clean one.

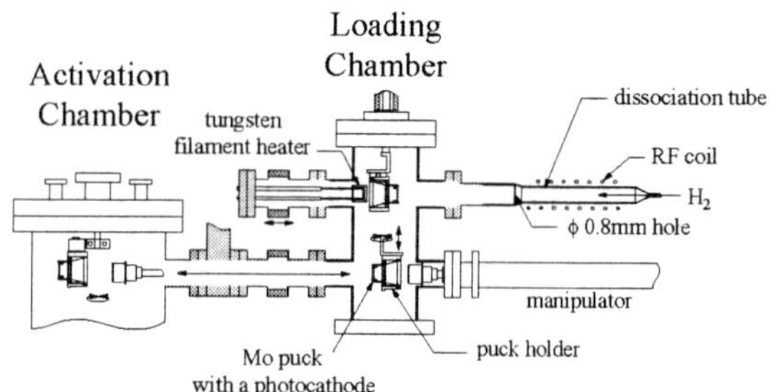

FIGURE 3. Cross-section of photocathode preparation chamber.

4. REDUCTION OF DARK CURRENT AND EXTRACTION OF 200 KEV BEAM

A bulk-GaAs crystal was installed to the center of cathode electrode in the gun chamber for dark current measurement. As shown in Figure 4, dark current began to rise at 120 kV and was expected to grow up much greater than 10 nA at 200 kV. In order to improve this situation, the HV processing method was employed by introducing 99.9999 % pure nitrogen gas into the gun chamber until the pressure became to 1×10^{-6} torr [9]. The HV was kept at constant value until the dark current was decreased and then it was raised up step by step. As a result, dark current could be reduced to less than 1 nA at 200 kV for the NEA activated GaAs cathode.

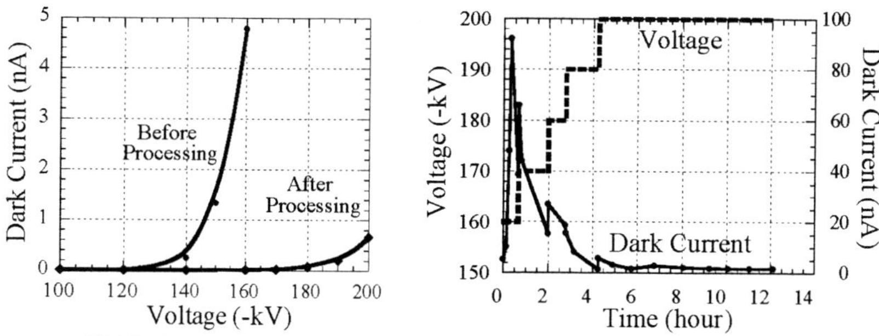

FIGURE 4. Reduction of dark current by HV processing using nitrogen gas.

The beam extraction from the gun was already done. The installation of the GaAs photocathode from the activation chamber to the gun chamber took less than 2 minutes and the QE achieved in the activation chamber was preserved during the cathode transportation. The NEA-GaAs cathode was illuminated by a He-Ne laser with a spot size of 1.2 mmϕ on the cathode. The 200 keV beam with 1 μA-level current was extracted and transported through a 90° bending magnet, three solenoid lenses and two pair of steering coils to a Faraday cup mounted on the end of transport line, as shown in Figure 1. Transport efficiency from cathode to Faraday cup was maximized by tuning the transport parameters and it reached up to 98 %.

5. SCHEDULED PLAN

This 200 keV polarized electron source is considered to be useful to make two kinds of R/D works for establishing the technologies that can satisfy the source-requirements of future LC (Linear Collider) and ERL (Energy Recovery Linac) projects [11]. Concerning the LC, the experiments to produce the sub-nanosecond multi-bunch beam with peak current more than 3 A are scheduled using the GaAs-GaAsP photocathode. Concerning the ERL, the high field gradient DC-gun such as our 200 keV gun must solve three tough-problems to reach the goal, these are, highest average current (~100 mA), lowest emittance (≤ 1.0π mm·mrad) and enough long lifetime. Many efforts seem necessary to find the complete solutions in these subjects.

REFERENCES

1. T. Nakanishi et al., "Polarized Electron Source" in *JLC Design Study*, KEK Report **97-01**, 36-48 (1997)
2. K. Togawa et al., *Nucl. Inst. and Meth. in Phys. Res. A* **455**, 118-122 (2000)
3. K. Togawa et al., AIP Conf. Proc. **570** 982-987 (2001)
4. T. Nishitani et al., AIP Conf. Proc. **570** 1021-1023 (2001)
5. K. Wada et al., HEACC2001, http://conference.kek.jp/heacc2001/, P1lc13 (2001)
6. C. Suzuki et al., AIP Conf. Proc. **570** 1009-1011 (2001)
7. O. Watanabe et al., AIP Conf. Proc. **570** 1024-1026 (2001)
8. C. Sinclair, AIP Conf. Proc. **421** 218-228 (1998)
9. R. Alley et al., *Nucl. Inst. and Meth. in Phys. Res. A* **365**, 1-27 (1995)
10. M. Yamamoto et al., LINAC2002, http://linac2002.postech.ac.kr/db/proceeding/TH435.PDF (2002)
11. T. Nakanishi, LINAC2002, http://linac2002.postech.ac.kr/db/proceeding/FR103.PDF (2002)

Basic R&D Studies for Lower Emittance Polarized Electron Guns

C. Suzuki, T. Nakanishi, S. Okumi, F. Furuta, K. Wada, T. Nishitani, M. Yamamoto, T. Hirose, M. Kuwahara, R. Mizuno, N. Yamamoto, H. Matsumoto [a)], M. Yoshioka [a)], H. Horinaka [b)], K. Wada [b)], T. Matsuyama [b)] and H. Kobayakawa [c)]

Nagoya University, Department of Physics, Nagoya 464-8602, Japan
a) KEK High Energy Accelerator Research Organization, Tsukuba 305-0801, Japan
b) Osaka Prefecture University, Department of Physics and Electronics, Osaka 599-8531, Japan
c) Nagoya University, Department of Materials Processing Engineering, Nagoya 464-8603, Japan

Abstract. In order to produce the lower emittance electron beam, the higher field gradient must be used for the gun. From this point of view, we try to develop both of a polarized DC-gun and RF-gun. In case of a 200 keV polarized DC-gun at Nagoya University, the accelerating gradient at the photocathode surface was designed to be 3 MV/m, and the dark current emitted from the SUS electrodes was suppressed below 1 nA. To increase the gradient, we tested the property of pure Titanium as a new electrode material. The tested electrode showed small dark current (~1 nA) even at field gradient of 88 MV/m, which is as twice as higher than that of SUS electrode. Concerning the feasibility of a polarized RF-gun, it seems difficult for the NEA surface to survive in high field gradient of 100 MV/m. Therefore, we proposed a new type of polarized electron source using two-photon excitation method, for which it is not necessary to use the NEA surface to extract electrons into vacuum. For this method, the polarization higher than 90 % was already demonstrated by the photoluminescence measurement using the bulk GaAs crystal.

1. INTRODUCTION

Recently, an electron gun using an NEA (Negative Electron Affinity) GaAs photocathode has been noticed not only as a polarized electron source but also as a good candidate of an electron source for the FEL (Free Electron Laser) machine. It has the possibility to produce a beam with emittance less than 1.0π mm·mrad/mm (emitter radius) [1]. In order to keep such a low emittance against the space charge effect at the gun, the accelerating gradient higher than 10 MV/m should be required, which corresponds to the gun bias-voltage of 300~500 kV.

As well known, the NEA performance is extremely sensitive to the surface contamination and is easily degraded by back-bombardment of ionized residual gas produced by the collisions with electron beam. In addition of UHV level of 10^{-11} torr, the field emission dark currents between HV electrodes must be kept below 10 nA, since the dark currents stimulate the desorption of gases from the anode [2]. There are two ways for approaching the lower emittance by the higher field-gradient DC-gun or by RF-gun. However, the technical difficulties are much greater for these guns, as the

CP675, *Spin 2002: 15th Int'l. Spin Physics Symposium and Workshop on Polarized Electron Sources and Polarimeters,* edited by Y. I. Makdisi, A. U. Luccio, and W. W. MacKay

field gradient at the cathode surface is assumed to be one or two orders of magnitude higher (10~100 MV/m) than those of existing DC-guns. At moment, both approaches have the common NEA lifetime problem and many R/D studies are required to overcome this problem. At first, the effort to reduce the dark current for high gradient DC-gun is described, and next the idea of a new PES using two-photon excitation method is described and discussed.

2. DARK CURRENT STUDY OF TITANIUM ELECTRODE

The future project to build the FEL machine, which is based on ERL (Energy Recovery Linac) as the fourth generation of synchrotron light source, has been discussed at Cornell and KEK intensively. The key technology is considered to develop a 500 keV DC-gun that can produce the lowest level of emittance (≤1.0π mm·mrad/mm) for the highest level of average current (~100 mA) from the NEA-GaAs photocathode. In this gun, the accelerating gradient at NEA surface is required to be more than 10 MV/m, and the dark current problem becomes more serious. In a series of our dark current studies, the property of Titanium as an electrode material has been just recently tested.

2.1 Experimental Apparatus

A study to reduce the field emission dark current has been continued by our group for several years. The same test apparatus was used to supply a high DC field gradient (≤100 MV/m) and to measure the dark current from a pair of sample electrodes under UHV (~10^{-11} torr) condition.

A schematic view of this test stand and the geometrical shapes of the cathode and anode are shown in Fig. 1. The sample electrode can be replaced to compare the performance of various kinds of electrodes. The field gradients can be changed by control both of the gap separation of the electrodes (0.5~20 mm) and of the bias voltage (0~150 kV). The dark current emitted from the cathode is collected at the anode that is electrically isolated from ground level and measured by a pico-ampere meter. The dark current study for SUS and copper electrodes were already done by this test stand [3].

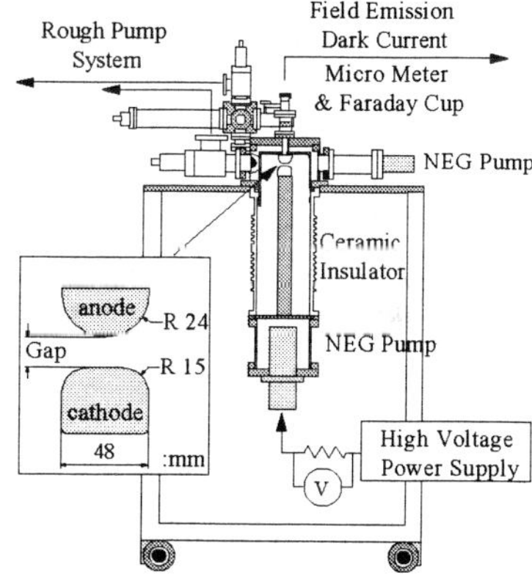

FIGURE 1. A schematic view of test stand.

2.2 Experiment

A new titanium electrode was made of JIS grade-2, which contains 99.4 % of Ti. After the machining, the electrode was finished to mirror like surface by buff polishing.

The surface was cleaned by the 5 minutes high pressure rinsing (80 kg/cm^2) with final ultra-pure-water rinsing. The electrode was installed to the chamber in class-100 clean room immediately after this rinsing. Then, the apparatus was pumped down to 10^{-11} torr after the baking at 250°C for 1 week. The dark current was measured after careful current conditioning for 2 weeks, which means that no break down was occurred during the high voltage conditioning up to 100 kV. Then, the gap separation was set to 1.0 mm for dark current measurement. Fig.2 shows the results for Ti electrode together with SUS and Cu data for comparison. The preparation procedures for each electrode are summarized in Table 1.

Obviously, Ti electrode achieved the highest field gradient of 88 MV/m for the small dark current level of 1 nA. This value is as twice as higher than that of carefully fabricated SUS electrode. As well known, the attainable field gradient depends on the electrode material. Ti has higher melting point, lower secondary electron emission rate, and lower ion-sputtering rate than SUS and Cu. However, it is not yet understood completely which property of metal contributes to achieve high field gradient. Further studies should be needed from this point of view.

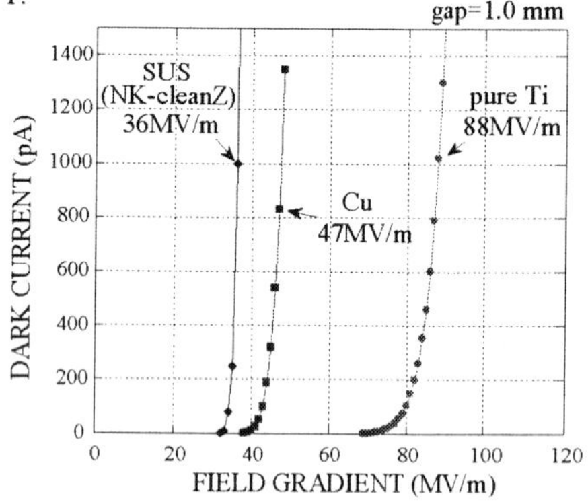

FIGURE 2. Results of dark current measurement.

TABLE 1. Preparation procedures for Ti, SUS, and Cu electrodes.

Electrode	Material	Surface finish	Rinsing
Ti	JIS grade-2	Buff polishing	High pressure rinsing
SUS	Clean-Z[1]	Electro-chemical buffing	Hot ultra pure water
Cu	Class-1 OFHC with HIP[2]	Diamond turning with ethanol	Ultra pure water

[1] Clean-Z is a new type of SUS material, which contains much fewer impurities than normal SUS 316L.
[2] Hot Iso-static Pressing.
The details of preparation procedures for SUS and Cu electrodes are described in Reference [3].

2.3 Future Plan of Dark Current Study

Now we are planning to fabricate Ti electrodes for our 200 keV polarized electron gun in order to obtain higher accelerating gradient to minimize the emittance growth. Basic studies for the dark current have been also continued with the aims of revealing its origin and mechanism. New electrodes made of Mo and Nb will be tested following the study of pure and alloy of Ti material.

3. A NEW PES USING TWO PHOTON EXCITATION

RF-guns are now used at several laboratories as the injector system of accelerator, because they can produce an electron beam with high peak current and short bunch length of ps-range. However, the feasibility of a polarized RF-gun is not yet established. It was reported by Novosibirsk group that the lifetime of GaAs with NEA surface is really too short in the RF-cavity [4]. From this result, we expected that a mechanism with something new for production of polarized electron beam is needed to solve the lifetime problem under 100 MV/m gradient, and proposed a new possibility to use the two-photon excitation mechanism.

3.1 Principle

The principle of two-photon excitation method is rather simple. Fig. 3 (b) shows two-photon adsorption mechanism. If we use right-circular polarized photon, only one type of transition is allowed, that is, from the valence band with m_j=-3/2 to the conduction band with m_j=+1/2. The expected advantages of this two-photon excitation are summarized as follows, 1) even a bulk-GaAs can give the highest polarization of 100 % in principle so that the super-lattice or strained layer is not necessary. 2) The electrons excited by two-photons have enough energy to escape into vacuum through the small PEA surface. For example, if the laser wavelength of 1500 nm is used, the obtained energy by two-photon adsorption is 1.65 eV that is larger by 0.23 eV than the band gap energy of GaAs (~1.42 eV). It means the complete NEA surface is not required to extract electrons. This method seems applicable for a polarized RF-gun for which the ultra-short (0.1 ps) laser pulse with highest peak power can be used.

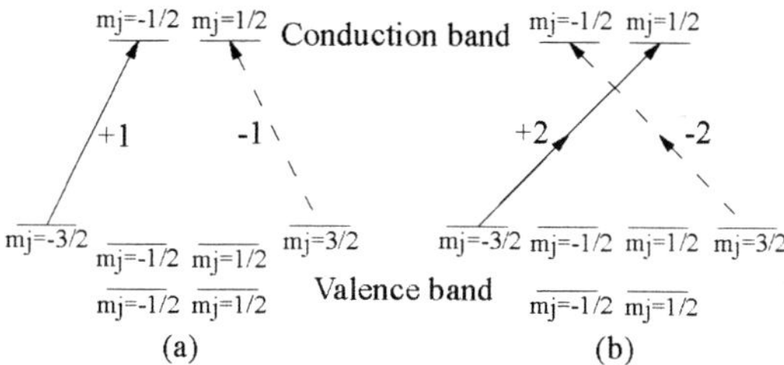

FIGURE 3. Optical transitions in (a) superlattice or strained GaAs layer by circular polarized single-photon excitation, and in (b) bulk-GaAs by circular polarized two-photon excitation.

3.2 Measurement of Photoluminescence Polarization

Photoluminescence spectra were measured to estimate the spin polarization of conduction band electrons excited by two-photon absorption. Fig. 4 shows the experimental set up, where a bulk-GaAs wafer was mounted in a cryostat and cooled down to 90 K. An optical parametric oscillator was used as a light source that has wavelength of 1500 nm, pulse width of 120 fs, repetition rate of 82 MHz, and average power of 80 mW. The luminescence from the sample through a monochrometer was detected by photomultiplier tube. The results are summarized in Fig. 5, where I^+ (I^-) indicates the right (left) circularly polarized component. From this result, the average photoluminescence polarization (P_l) was evaluated to be 29 %, which corresponds to 58 % of the electron polarization. The spin polarization of electrons in conduction band is relaxed during drifting toward the surface. The initial spin polarization (P_i) at the instant of excitation is estimated from the equation as

$$P_l = 0.5\frac{T_s}{T_s + \tau}P_i \qquad (1)$$

where T_s is the spin relaxation time and τ is the lifetime of conduction band electron. The measurement of time-resolved-photoluminescence was carried out at the same time by using a Ti:Sapphire laser and a streak scope. T_s of 78 ps and τ of 50 ps were obtained by analyzing of the decay curves of the photoluminescence polarization and the photoluminescence intensity [5]. Substituting P_l, T_s and τ values into the equation (1), the initial spin polarization was estimated to be as high as 92 %. Therefore, it was really confirmed that the highest polarization could be obtained by this new method.

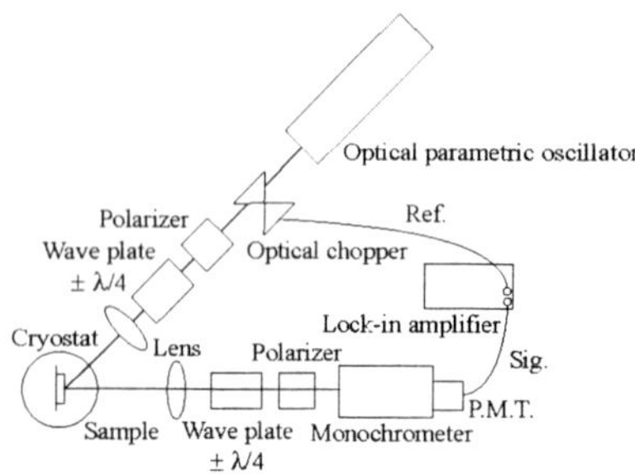

FIGURE 4. Experimental set up to measure the photoluminescence polarization.

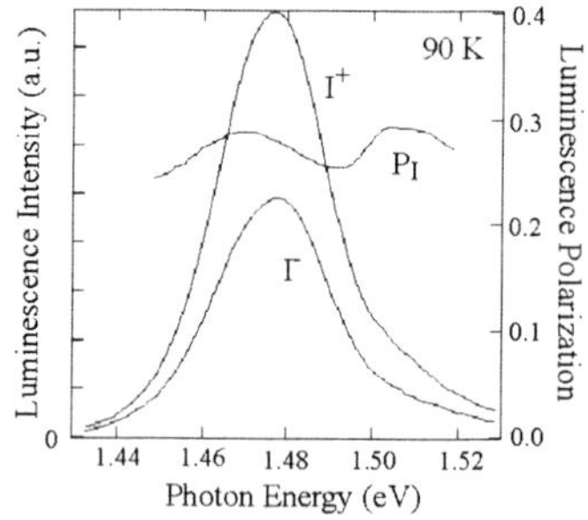

FIGURE 5. Spectra of right- and left-circular polarized lights of photoluminescence.

3.3 Further plan of Two-photon excitation study

Following the photoluminescence measurement, the first attempt to extract polarized electrons excited by two-photon absorption was tested by using a 70 kV DC-gun and a Mott polarimeter. In this experiment, a bulk GaAs was illuminated by Nd:YAG laser which has the shorter wavelength of 1064 nm and the much longer pulse width of 20 ns than those of the photoluminescence measurement. As a result, the extracted beam polarization was rather small, although the polarization was really increased from 0 % to 13 % when the laser power was increased up to 55 μJ/pulse. It is obvious that more shorter pulse width ($\leq$ 1.0 ps) laser must be used to extract high polarization and high QE beam. Otherwise, the highest polarization by two-photon absorption is diluted by the large background contributions induced by single photon absorption process. Concerning the RF-gun, we have started collaboration with KEK to put the Cs_2Te photocathode for practical use of ATF damping ring. There, we are planning to continue the two-photon experiment by using this laser system which has 7.2 ps pulse width, ~10 μJ/pulse energy, and 357 MHz repetition rate.

REFERENCES

1. T. Nakanishi, "Polarized Electron Source using NEA-GaAs Photocathode" in Proceedings of Linac 2002, Kyongju in Korea (http://linac2002.postech.ac.kr/db/proceeding/FR103.PDF)
2. K. Togawa et al., *Nucl. Instr. and Meth. A* **414** 431-455 (1998).
3. C. Suzuki et al., *Nucl. Instr. and Meth. A* **462** 337-348 (2001) and AIP Conf. Proc. **570**, 1009-1011 (2000).
4. A. V. Aleksandrov et al., Proceedings of EPAC98 1450- (1998), http://accelconf.web.cern.ch/AccelConf/e98/PAPERS/TUP02J.PDF
5. T. Matsuyama et al., *Jpn. J. Appl. Phys.* **40** L555 (2001)

Effect of Atomic Hydrogen Exposure on Electron Beam Polarization from Strained GaAs photocathodes

M. Baylac, P. Adderley, J. Clark, T. Day, J. Grames, J. Hansknecht, M. Poelker, P. Rutt, C. Sinclair, M. Stutzman

Thomas Jefferson National Accelerator Facility
12000 Jefferson Avenue, Newport News, VA 23606

Abstract. Strained-layer GaAs photocathodes are used at Jefferson Lab to obtain highly polarized electrons. Exposure to atomic hydrogen (or deuterium) is used to clean the wafer surface before the activation with cesium and nitrogen trifluoride to consistently produce high quantum yield photocathodes. The hydrogen-cleaning method is easy, reliable and inexpensive. However, recent tests indicate that exposure to atomic hydrogen may affect the polarization of the electron beam. This paper presents preliminary results of a series of tests conducted to study the effect of atomic H exposure on the polarized electron beam from a strained–layer GaAs sample. The experimental setup is described and the first measurements of the beam polarization as a function of exposure dose to atomic hydrogen are presented.

INTRODUCTION

Jefferson Lab is a nuclear physics research facility where highly polarized electrons can be delivered to three experimental endstations. The beam is produced by photoemission from strained layer GaAs photocathodes, optically pumped by circularly polarized laser light. The photocathode is formed, or activated, with Cs and NF3 to build a NEA surface. Preparation of an atomically clean surface is an essential step in the fabrication of an NEA surface.

Wet chemical etching techniques were initially used at Jlab but results often varied, something we attribute to variations in chemical quality and purity. The method of hydrogen cleaning, which is widely used in the semiconductor industry [1], was implemented at JLab and provided a drastic improvement in our ability to create photocathodes with consistently high QE. The method is particularly well suited for cleaning thin strained layer photocathode samples because, in principle, only contaminants are removed from the surface. Other laboratories recognize the benefits of hydrogen cleaning; we have helped implement this cleaning method at MAMI, Nagoya, Bates and SLAC.

Despite obtaining high QE from strained layer photocathodes, our initial experience with high polarization photocathodes was frustrating. Polarization from some samples was often very low (~ 50 to 60%) and polarization could vary from a single sample as the laser beam was moved across the active area of the photocathode (few mm's).

CP675, *Spin 2002: 15th Int'l. Spin Physics Symposium and Workshop on Polarized Electron Sources and Polarimeters,* edited by Y. I. Makdisi, A. U. Luccio, and W. W. MacKay

Over time, at least some of this behavior was attributed to excessive cleaning with atomic hydrogen. This paper presents results from a dedicated experiment to quantify the effects of hydrogen exposure on beam polarization from a single photocathode sample.

EXPERIMENTAL SETUP

The photocathode sample used for this test was obtained from Bandwidth Semiconductor (formerly Spire Corporation) and was grown to SLAC specifications [2]. The sample was cleaved to proper dimensions (15.5 mm x 15.5 mm) and indium soldered to a standard JLab stalk. The sample was not treated with wet chemicals at any time throughout the experiment (no acid/base etching, degreasing or anodization).

Beam polarization and QE measurements were performed on a test stand that includes a -100 kV photogun and Mott polarimeter. Hydrogen cleaning was performed on a separate vacuum chamber that is used for cleaning all of the photocathode samples used at the CEBAF photoinjector. When not in vacuum, great care was taken to ensure the sample was exposed only to clean, inert nitrogen. Venting of the vacuum chambers was conducted using nitrogen filled glove bags and the sample was transferred between chambers within a nitrogen filled transport vessel.

The hydrogen cleaning vacuum chamber consists of a 6 way cross with 4.5" flanges [3]. There are two vacuum pumps; a 20 L/s Perkin Elmer ion pump and a Balzers 50 L/s turbo pump which serves as the dominant pump during hydrogen cleaning. Molecular hydrogen (or deuterium) flows through a leak valve into a Pyrex glass dissociator. The molecular hydrogen is dissociated with an RF inductive discharge created by coil, which is part of a tuned LC circuit. The LC circuit resonates around 100 MHz and 20 W of RF power is absorbed by the hydrogen when the dissociator-region pressure is 15 mTorr. Atomic H exits the dissociator through a 1 mm diameter hole to reach the photocathode approximately 15 cm away, at a temperature ~ 300° C.

On the beam teststand, a high power DC Nd:YVO4 laser pumps a Titanium:Sapphire crystal inside a four-fold optical cavity. The Ti:Sap laser light is wavelength tunable between 740 and 860 nm. It is circularly polarized by a Pockels cell and the helicity of the beam is reversed pseudo-randomly at 10 Hz. Laser light enters the gun chamber through a vacuum window to reach the photocathode. The vertical electron gun consists of a sample loading section, a preparation chamber and a high voltage section [5]. An isolated load lock chamber enables loading of a stalk while keeping the gun under vacuum. This small chamber can be quickly pumped and baked. The preparation chamber, or main gun chamber, accommodates the different ports necessary to perform the NEA activation of the semiconductor surface. An alumina ceramic is used to hold off the 100 kV acceleration voltage. The stalk is placed in contact with the cathode seating at high voltage. Electron beam exits the gun through the center of a donut-shaped anode. It travels through a simple beam line consisting of a bend magnet, correctors and lenses, beam viewers and an electrostatic bend used as a spin rotator. At the end of the beam line, electrons impinge on a gold foil in the Mott chamber. The Mott polarimeter contains two identical silicon detectors in the horizontal plane tracking electrons backscattered ($\theta=\pm 120°$)[6]. Measuring the experimental counting rate asymmetry between the 2 helicity states and/or between the 2 detectors yields the electron beam polarization, when one knows the analyzing power of the polarimeter (~33% for Mott scattering off a 300 A gold foil at 100 keV).

EXPERIMENT

The experiment consisted of first characterizing the sample prior to exposure to atomic hydrogen. Following this, the sample was repeatedly exposed to atomic hydrogen in ~15 minute time intervals for a total cumulative dose of 100 minutes. The experiment was time consuming, requiring successful implementation of many steps including; a) sample loading into the –100 kV gun, b) vacuum chamber bakeout, c) sample heating, d) sample activation, e) beam measurement of polarization and QE versus wavelength, f) vacuum chamber venting and sample transport to the hydrogen cleaning chamber, g) hydrogen cleaning and then finally reinstallation of the sample into the –100 kV gun where the process repeats. Every effort was made to ensure that the steps were carried out under identical conditions following successive hydrogen exposures. Some specific details on individual steps of the test are given below.

After the sample is installed in the load lock chamber of the test gun, this chamber is evacuated and baked at a temperature around 250° C for 12 hours. After cool down to room temperature (RT), the load lock chamber is open to the main chamber of the electron gun and the stalk holding the photocathode inserted into the gun. The sample is then heated again to ~ 570° C for 2 hours. We activate the wafer after it cooled down to RT in the gun chamber. Approximately one monolayer of cesium is applied to the GaAs surface and oxidized by nitrogen trifluoride (NF3) while photocurrent yield is measured during illumination with white light.

Once the characterization is complete (QE, polarization), the sample is retracted from the gun chamber into the load lock chamber. The load lock chamber is vented to atmospheric pressure while the gun and the beamline remain under vacuum at all times. The stalk is transported and loaded into the chamber equipped with the hydrogen dissociator; the chamber is pumped down. Once the pressure reaches ~ 10-8 Torr, the sample is heated to 500° C for a few minutes and quickly cooled to 300° C. Atomic deuterium is released into the chamber at a pressure of 15 mTorr and the dissociator parameters are adjusted quickly while a paddle on top of the glassware protects the sample surface from the gas flow. Once the RF power absorbed by deuterium is maximized (~40 W), the paddle is removed and the ion counter is engaged. The RF power is switched off and the sample is cooled down as the ion count reaches the desired value corresponding to a 15 minute cleaning time. The sample is transferred back into the gun for the next series of measurements.

RESULTS

Polarization and QE results for the unexposed sample are shown in Fig. 1. It is interesting to note that QE is quite good (0.35% at 840 nm of incident laser light), even without hydrogen dose, indicating that our methods for handling samples may have improved at Jlab (photocathode cleaning, in one form or another, was essential to obtain high QE years ago). The QE of the photocathode surface is scanned and is measured to be uniform within ~10%. Polarization measurements were made at 5 cathode locations. Conditions were kept constant for all measurements (foil thickness, beam steering, counting rates) to minimize systematic discrepancies between different runs. The systematic uncertainty of the absolute polarization with this Mott polarimeter is estimated ~10%, the relative comparison of one measurement to another is smaller. A maximum polarization ~80% is measured at 840 nm.

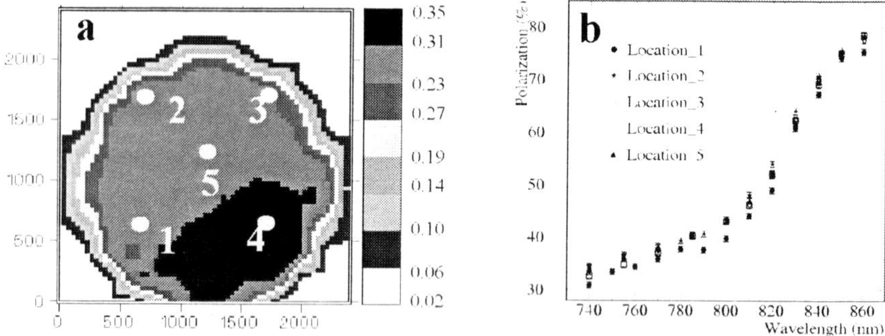

FIGURE 1. (a) Quantum efficiency profile of the photocathode at 840 nm and 5 locations selected for polarization measurements; (b) Beam polarization vs. wavelength for the 5 locations (statistical errors only).

Polarization as a function of hydrogen dose is shown in Fig. 2 for locations 1 and 3. We observe a strong depolarization as the H dose is increased (60 minutes and more). This effect varies significantly with wavelength: the relative polarization variation $\Delta P(t)=[P(t)-P(0)]/P(0)$ where t is the cleaning time, is maximum at high wavelengths whereas ΔP exhibits very little dependence below the band-gap (< 800 nm). However, a minimal H dose may seem to increase slightly the polarization, $0< \Delta P(t) <+10$ %.

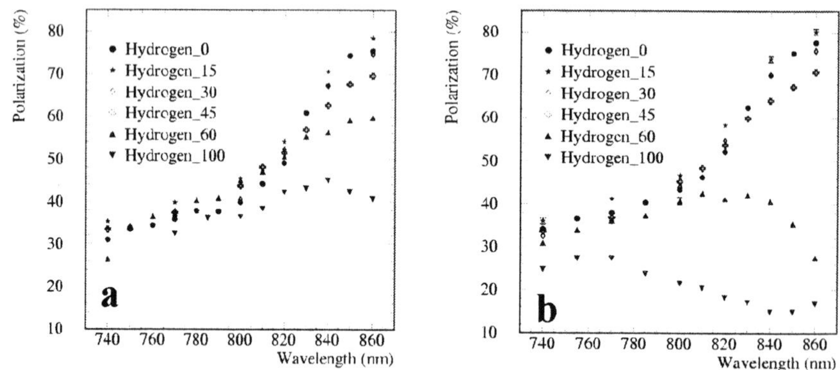

FIGURE 2. Beam polarization vs. wavelength for 6 total H exposure (0, 15, 30, 45, 60 and 100 minutes) for locations 1 (a) and 3 (b). Errors are statistical only.

Moreover, the depolarization varies strongly across the surface: $-80\%<\Delta P(t)<-40\%$ for the 5 measured locations whereas the initial polarization profile was uniform.

The quantum yield is measured to be decreasing as the H dose increases: from 0.035% at 840 nm with a bare surface down to 0.002% after the photocathode has been exposed to the maximum dose. This trend was confirmed by a similar test performed in a different chamber equipped with a dissociator, which remained under vacuum throughout all measurements.

The original sample used for this test was removed from the gun and sent for Atomic Force Microscope imaging. The surface analysis indicates some differences when compared to a bare uncleaned strained GaAs sample but did not provide any conclusive results regarding structure changes of the semiconductor surface.

CONCLUSIONS & OUTLOOKS

This experiment carried out the first characterization of the polarization of an electron beam photoemitted from a strained layer GaAs semiconductor as a function of atomic hydrogen exposure of the surface. The preliminary results show that the hydrogen cleaning process of our photocathodes needs to be revisited. Indeed, exposure to a heavy dose of hydrogen can depolarize the sample in a dramatic manner. A strong dependence on the wavelength of the incident light is also observed, but not yet understood, as the depolarization increases significantly with wavelength. The past benefit of this cleaning procedure is called into question, as wafers now provide high QE directly from manufacturers. This experiment brings some new light on intriguing past measurements (low polarization, non-uniformity) as poor performances can now be attributed to overexposure to hydrogen. However, one may also notice indications of a slight increase in polarization with a small H dose, which can create some new interest in the H exposure.

Since this test was done, a hydrogen dissociator has been installed on the preparation chamber of the test gun to continue the experiment. We will repeat this test with a strained layer GaAs sample to confirm the observed behavior. A similar series of measurements is in progress using a bulk GaAs wafer. This aims to pinpoint the origin of the depolarization mechanism, as it will help differentiate between effects associated with the surface or with the strain of the semiconductor. Study is also underway regarding the characteristics of the H source, as the energy and angle of incidence onto the wafer surface of the atomic hydrogen may play a significant role. Optimizing the H cleaning source parameters may help reaching atomically flat surfaces. Whereas the use of atomic H for cleaning semiconductors seems now irrelevant, it may become a tool to increase the beam polarization.

REFERENCES

1. E.J. Petit *et al*, J.V.S.T. A 10, 2172 (1992); E.J. Petit & F. Houzay, J.V.S.T. A 12, 547 (1994)
2. C.K. Sinclair *et al*, proc. 1997 PAC
3. M. Poelker *et al*, proc. 1997 JLab Contamination workshop: measurement and control in vacuum sys.
4. W.W. Mc Alpine, R.O. Schildknecht, proceedings of the IRE, 2099 (1959)
5. B. Dunham, Ph. D thesis, University of Urbana-Champaign, Illinois (1993)

Status of Jefferson Lab's Load Locked Polarized Electron Gun

M. L. Stutzman*, P. Adderley, M. Baylac, J. Clark, A. Day, J. Grames, J. Hansknecht, and M. Poelker

Thomas Jefferson National Accelerator Facility
12000 Jefferson Ave., Newport News, VA 23606

Abstract. A new 100 kV load locked polarized electron gun has been built at Jefferson Lab. The gun is installed in a test stand on a beam line that resembles the first few meters of the CEBAF nuclear physics photoinjector. With this gun, a GaAs photocathode can be loaded from atmosphere, hydrogen cleaned, activated and taken to high voltage in less than 8 hours. The gun is a three chamber design, with all of the moving parts remaining at ground potential during gun operation. Studies of gun performance, photocathode life times, transverse emittance at high bunch charge, helicity correlated effects and beam polarizations from new photocathode samples will all be greatly facilitated by the use of this load locked gun.

INTRODUCTION

The JLab photoinjector must be extremely reliable to meet the demands of the three hall nuclear physics program. The present two-gun design[1], one production gun and one spare, was chosen as an alternative to load lock gun-based designs used at labs such as SLAC[2], MAMI[3], ELSA[4], and NIKHEF[5]. The two-gun design has been successful because each gun provides exceptional photocathode lifetime. Lifetimes, measured in Coulombs extracted before the QE becomes unacceptably low, are approximately 300 C when operating at high current (250μA from gun) and 600 C at low current (65μA from gun). A gun can typically provide uninterrupted beam for two to three weeks at high current (250 μA) before the laser spot must be moved to a higher quantum efficiency (QE) location on the photocathode. The laser spot is 0.5 mm in diameter and a 5 mm active area is defined on the photocathode by anodization[6]. There are approximately five locations that can provide beam before the cathode must be heated and reactivated. A cathode can often be used for three months at typical production currents before it needs to be heated and re-activated. Cathode replacement and gun swaps have been occurring approximately yearly.

Despite the good performance of ~~our~~ the dual horizontal guns, there are drawbacks to a non-load-locked polarized electron gun. Field emission begins to develops after about the fourth successive activation, as suggested by our limited data. Severe field emission may require time consuming repolishing of the cathode electrode.

Installing a new photocathode within a gun necessitates venting the gun to atmospheric pressure and a subsequent gun bake to 250 C, which takes 3 days to complete. If extensive vacuum work is necessary, such as opening the gun to repolish the electrode, a sacrificial bulk GaAs has to be installed to check the quality of the

* marcy@jlab.org

CP675, *Spin 2002: 15th Int'l. Spin Physics Symposium and Workshop on Polarized Electron Sources and Polarimeters,* edited by Y. I. Makdisi, A. U. Luccio, and W. W. MacKay

work before a more expensive strained layer GaAs photocathode is installed. This prolongs the effective downtime of the gun.

LOAD-LOCK GUN DESIGN

A load-lock polarized electron gun has been designed at Jefferson Lab to improve the performance of the Jlab photoinjector[7]. The load-lock has been designed to enable replacement of the GaAs photocathode within an eight hour time period. In addition, the gun design incorporates the best features of the existing photoguns including extreme high vacuum, excellent pumping conductance, and electrostatic optic elements without short focal lengths. Finally, to keep the system as simple as possible, the entire loading assembly is kept at ground potential and no moving parts are taken to high voltage.

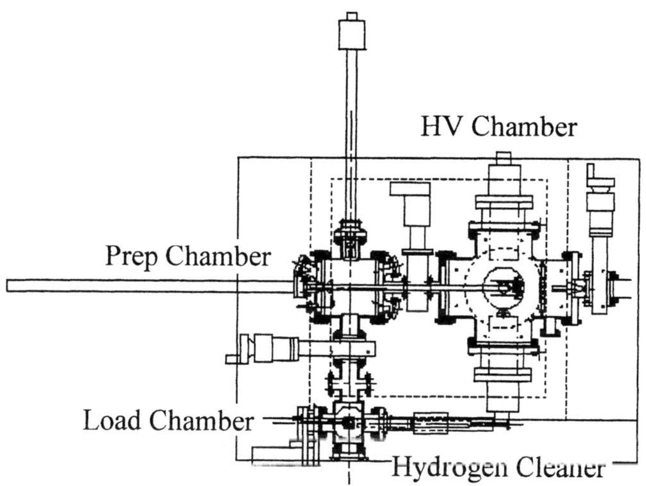

FIGURE 1. Overhead view of Load-Lock Gun. Load, hydrogen clean and heat chamber is shown at the bottom, preparation chamber with manipulators attached is at the top left, and the high voltage chamber is on the right.

Load, Hydrogen and Heating Chamber

The first chamber in the load-lock gun is the load, hydrogen clean and heat chamber. It is a small chamber which is initially baked but then vented with a dry nitrogen purge and only opened into a glove bag over pressurized with nitrogen to minimize water vapor contamination. The system is pumped down with a combination of a 60 l/s turbo pump and an 11 l/s ion pump. The cathode is mounted on a machined molybdenum puck and secured with indium solder and a tantalum cup. The puck is loaded in this chamber on stainless steel fingers attached to a 4.5 inch flange. The back of the puck is hollow, and a ceramic heater can be inserted into the hollow in the puck for heating. The temperature of the puck is measured by a thermocouple that can be moved in against the side of the puck on a bellows. Cross calibration of the temperature read by the thermocouple to the actual temperature of the cathode has been made by using both an optical pyrometer and by heating the wafer until the non-congruent evaporation of As from the surface leaves the surface frosted at ~620°C. A hydrogen cleaner mounted to this chamber allows the surface of the GaAs to be cleaned by RF dissociated hydrogen or deuterium. The puck and cathode can be cooled more quickly by the insertion of a copper cooling finger, which is on a second bellows coincident in the port with the thermocouple. This chamber also has a convectron gauge to measure the pressure of the hydrogen during cleaning and a Residual Gas Analyzer (RGA) to detect leaks and to determine the contaminant species in the chamber. Base pressure in this chamber is ~1e-8 Torr.

Preparation Chamber

The second chamber in this load-lock gun is the preparation chamber. The puck is moved from the loading to the preparation chamber on the transverse manipulator, which is a Surface Interface (now Transfer Engineering) magnetic manipulator. This manipulator has translational motion only, and grabs the puck with fingers spaced somewhat wider than the dock fingers in the load chamber and holds the puck with springs. The puck is then transferred to the longitudinal manipulator, which has both rotational and translational mobility. This longitudinal manipulator fits into the hollow back of the puck and can hold the puck firmly with ears that fit into internal grooves in the recess of the puck.

Bias for activation in the preparation chamber is applied by means of a ring anode biased positively just in front of the puck. Cesium (Cs) is deposited using of two SAES[8] alkali metal dispensers on a bellows and the oxidant used is NF_3 which is introduced through a Balzers dosing valve. The preparation chamber has 5 optical ports for either white light or laser activation and to assist in the handoff between the two manipulators. Pumping in this chamber is accomplished through a combination of a 60 l/s ion pump and a GP100 NEG cartridge pump, with a pumping speed of 600 l/s for H2 and 300 l/s for CO. Base pressure in this chamber is ~1e-10 Torr.

High Voltage Chamber

The high voltage chamber has been designed to have excellent pumping conductance and minimal material at high voltage. The HV chamber is a 6 way, 10 inch cross, made of electropolished 316L stainless steel. The system is pumped by 3 GP500 SAES NEG pumps, with a pumping speed of 1900 l/s for H2 and 650 l/s for CO. Base pressure as measured by an extractor gauge is 9e-12 Torr. Only the top of the ceramic, the cathode support, the cathode electrode and the puck with the photocathode are taken to high voltage. The puck snaps into place in the cathode support by means of sapphire rollers mounted on stainless steel springs which fit against the beveled back edge of the puck, holding the puck and cathode securely in place. The longitudinal manipulator can then be retracted, leaving no moving parts at high voltage. The cathode electrode is made out of titanium, which should provide less secondary electron emission.

The electrostatic optics in this HV chamber are identical to those in the production gun, with a 25 degree angle between the surface of the cathode and the cathode electrode. A bias of –100 kV is applied to the cathode, and the anode is 6 cm away, giving a maximum field of 4MΩ/m at the highest part of the cathode electrode and ~1MΩ/m at the surface of the cathode.

The 2.5 inch beamline leading away from this HV chamber is coated with a NEG material to maintain the lowest possible vacuum in the HV chamber. This same NEG coated beamline is used in the production nuclear physics machine.

INSTRUMENTED BEAMLINE

The beamline connected to the load-lock polarized electron gun in the Injector Test Stand consists of an 5 meter long beamline similar to the production beamline. Downstream of the gun is a NEG coated beamline which is then followed by a 15° bend and a Y chamber lined with additional NEG pumps to limit conductance from the beamline back into the gun. The beamline has 5 viewer screens, 1 harp scanner and 2 Beam position monitors (BPM). When the incident laser light has RF structure, the 100 kV beam can be seen with the BPMs and real time information about the current and position of the beam can be obtained. This will allow remote monitoring of beam during lifetime studies and allow us to monitor changes in beam intensity or position with changing helicity which is of critical importance for parity violation experiments.

At the moment, the beamline has no Wien filter or Mott polarimeter, though spaces have been designated for these pieces of equipment. These polarimetry devices will be installed in the near future.

EXPERIMENTAL RESULTS

Initial tests of this load-lock gun have been done by loading bulk GaAs[9] from air, hydrogen cleaning, heating to liberate the hydrogen and activating with Cs/NF_3. Following this loading, cleaning and activation, the QE of the photocathode was measured at two diode laser wavelengths. QE at 770 nm was measured to be 9.2% and QE at 860 nm was measured at 6.6%. These values are close to the best QE measurements that we have made in a baked, NEG pumped test chamber. Whether heating in the unbaked loading chamber is detrimental to the cathode performance will be verified in the near future.

The HV chamber has been taken to -110kV for processing then returned to -100kV with minimal leakage current. A current of 100 nA can be measured at the power supply, but these losses are primarily explained by losses in the cables and the corona. A test of the dark current in this gun will be performed and reported on at a later date.

COMMISSIONING EXPERIMENTS

The initial experiments planned for this load-lock gun are a series of lifetime measurements. Jefferson Lab has been anodizing all but the central portion of the GaAs cathodes to limit the emission of unwanted electrons from the edges of the cathode. A systematic study of the active area size, the anodization technique and the radial position of the anodized area will be made. In addition, the effect of different wavelengths and contaminant species on cathode lifetime will be studied. Due to the quick loading time, this gun is an ideal platform for studying different photocathode materials. Following the lifetime studies, a series of experiments will be made to gain an understanding of the helicity correlated effects caused by cathode anisotropy along with beam based and laser based feedback systems that can be used to control these effects for parity violation experiments.

If the performance of this load-lock gun proves superior to those in the production machine, the design is such that it can replace one of the conventional guns with minimal reconfiguration.

ACKNOWLEDGMENTS

The authors thank Bruce Dunham, Peter Hartmann, Reza Kazimi, Danny Machie, Ganapati Rao Myneni, Scott Price, Paul Rutt, Bill Schneider, Charlie Sinclair, and Michael Steigerwald for valuable contributions. This work was supported by US DOE Contract No. DE-ACO5-84-ER40150.

REFERENCES

[1] Sinclair, C.K., "Recent advances in polarized electron sources," *Contributed to IEEE Particle Accelerator Conference (PAC 99), New York, 29 Mar-2Apr 1999.*

[2] Clendenin, J.E, *et al.* "The SLAC polarized electron source," SLAC-PUB-9509 *Presented at this conference*; Alley, R. *et al., Nucl. Instrum. Meth. A,* **365**, 1 (1995).

[3] Aulenbacher, K. *et al.*, "Status of the Polarized Source at MAMI," *Presented at this conference;* Aulenbacher, K. *et al., Nucl. Instrum. Meth. A,* **391**, 498 (1996).

[4] von Drachenfels, W. *et al.*, "Polarized Electron Source at ELSA," *Presented at this conference;* Nakamura, S. *et al., Nucl. Instrum. Meth. A*, **411**, 93 (1998).

[5] Koop, I. *et al.* "Polarized Electrons In Amps," *Nucl. Instrum. Meth. A* **427**, 36 (1999).; Bolkhovityanov, Y.B., *et al.*

[6] Rutt, P.M. and A.R.Day, "Active Area Definition of GaAs Photocathodes via Anodization," JLAB-TN-01-030; Poelker, M., *et al.*, "Polarized source performance and developments at Jefferson Lab," *Presented at 14th International Spin Physics Symposium (SPIN 2000), Osaka, Japan, 16-21 Oct 2000.*

[7] Schneider, W.J., *et al.* "A Load-Locked Gun for the Jefferson Lab Polarized Injector" *Contributed to IEEE Particle Accelerator Conference (PAC99), New York, 29 Mar - 2 Apr 1999.*

[8] SAES Getters USA Inc., 1122 East Cheyenne Mountain Blvd., Colorado Springs, CO; GP100 and GP500 pumps, and Cs alkali metal dispensers.

[9] American Xtal Technology, 4311 Solar Way, Fremont, CA; Zn doped GaAs at 1.5e19, (100) surface, Etch Pit Density <5000/cm^2, 600±25 um thick.

Suppression of the Surface Charge Limit in Strained GaAs Photocathodes*

T. Maruyama[a], A. Brachmann[a], J.E. Clendenin[a], T. Desikan[a], E.L. Garwin[a], R.E. Kirby[a], D.-A. Luh[a], C. Y. Prescott[a], J. Turner[a], and R. Prepost[b]

[a]*Stanford Linear Accelerator Center, Stanford, CA 94309*
[b]*Department of Physics, University of Wisconsin, Madison, WI 53706*

Abstract. Single strained, medium-doped (5×10^{18} /cm^3) GaAs photocathodes show the surface charge limit (SCL). The SCL poses a serious problem for operation of polarized electron sources at future linear colliders such as the NLC/JLC. A high-gradient-doping technique has been applied to address this problem. A 5 -7.5 nm *p*-type surface layer doped to 5×10^{19}/cm^3 is found sufficient to overcome the SCL, while maintaining high beam polarization. This technique can be employed to meet the charge requirements of the NLC with a polarization approaching 80%.

INTRODUCTION

Strained, medium-doped (5×10^{18} cm^{-3}) GaAs photocathodes were introduced for the SLAC polarized electron source in 1993. These photocathodes are capable of producing up to 1×10^{11} electrons in a 2 ns pulse with 75-85% polarization. However, they are susceptible to the surface charge limit (SCL). The SCL was attributed to a photovoltage effect in the band bending region, which momentarily increases the surface work function and thus suppresses emission. The SCL poses a serious problem for operation of polarized electron sources at future linear colliders such as the Next Linear Collider (NLC) and the Japan Linear Collider (JLC). The NLC is presently designed to operate with a train of 190 micro-bunches, spaced 1.4 ns apart. Each micro-bunch is required to have as much as 1.4×10^{10} e$^-$ per pulse in 0.5 ns at the gun, totaling 2.7×10^{12} e$^-$ per train. The standard SLAC strained photocathodes saturate at a level of 8×10^{11} e$^-$ for a 300-ns pulse, well below the space charge limit of the gun. The SCL problem can be reduced significantly by increasing the *p*-type doping concentration to at least 2×10^{19} cm^{-3}. However, increasing the doping level leads to some depolarization of the electron spin. To increase the doping level without spin depolarization, a high-gradient-doping technique has been explored. While a standard SLAC photocathode is uniformly doped at 5×10^{18} cm^{-3}, we have reduced the active layer dopant density to 5×10^{17} cm^{-3} while increasing the density of a 10-nm surface layer to 5×10^{19} cm^{-3}. The lower-than-standard doping level in the active layer avoids depolarization, while the high surface doping addresses the charge limit problem [1].

* This work was supported in part by Department of Energy Contract No. DE-AC03-76SF00515

CP675, *Spin 2002: 15th Int'l. Spin Physics Symposium and Workshop on Polarized Electron Sources and Polarimeters*, edited by Y. I. Makdisi, A. U. Luccio, and W. W. MacKay

DOPANT CONCENTRATION DEPENDENCIES

A series of measurements were performed using unstrained 100 nm thick GaAs to study the SCL dependence on the doping concentration [2]. The p-type doping concentrations were 5×10^{18} cm^{-3}, 1×10^{19} cm^{-3}, 2×10^{19} cm^{-3}, and 5×10^{19} cm^{-3} for a set of four samples. The experiments were performed using the 121 kV diode gun in the SLAC Gun Test Laboratory (GTL). Figure 1(a) and 1(b) show the temporal profiles of the emission current pulses measured for a number of light pulse energies for the 5×10^{18} cm^{-3} sample and for the 2×10^{19} cm^{-3} sample, respectively. As seen in Figure 1(a), the emission current follows the rectangular laser shape when the laser energy is low. As the laser energy is increased, the emission peak at the start of the light pulse reflects retardation in the build-up of the photovoltage due to the finite time of the charging of the surface states. As the photovoltage builds up, the emission current decreases exponentially. The suppression of the later portion of the emission pulse manifests the decrease of the surface escape probability with the growth of the photovoltage and its relaxation to the steady state. As the laser energy is increased, the photovoltage builds up more quickly and the emission current suppression is more pronounced. For the 1×10^{19} cm^{-3} sample (not shown), the charge limit behavior is much reduced. As seen in Figure 1(b), the 2×10^{19} cm^{-3} sample does not show the charge limit effect.

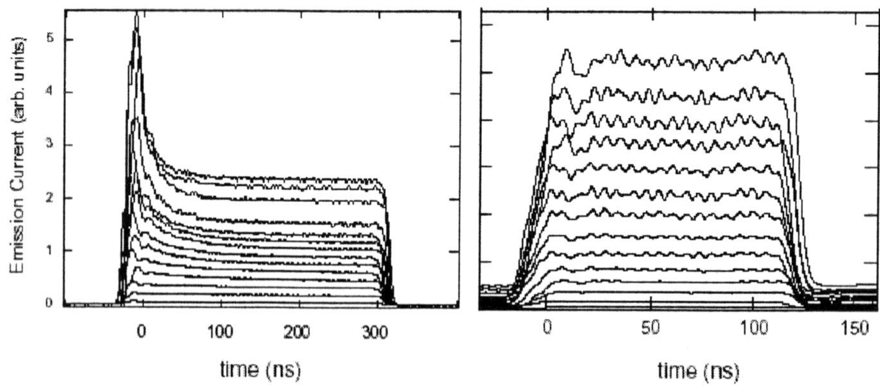

FIGURE 1. The temporal profiles of the emission pulses for a) the 5×10^{18} cm^{-3} sample and b) the 2×10^{19} cm^{-3} sample.

HIGH GRADIENT DOPED STRAINED GaAs

While high doping concentration is necessary to suppress the SCL, electron polarization will be significantly reduced as the doping level is increased. One way to achieve the high doping level without suffering from polarization loss is to dope only ~10 nm of the surface layer at $>2\times10^{19}$ cm^{-3}. To test this technique, polarization measurements were performed at the SLAC Cathode Test System (CTS) using high gradient doped GaAs. While the active layer is 100 nm thick GaAs, the surface layer is doped at 4×10^{19} cm^{-3} and the rest at 3×10^{17} cm^{-3}. Two samples with a highly doped layer thickness of 25 nm and 40 nm were used. The thickness of the highly doped surface layer was reduced by anodization/stripping, and the polarization and quantum efficiency (QE) measurements were repeated. Figure 2 shows the peak polarization as a function of the highly-doped layer thickness. The polarization increased significantly as the highly-doped layer thickness was reduced, reaching 80% with about 7.5 nm of the highly-doped layer.

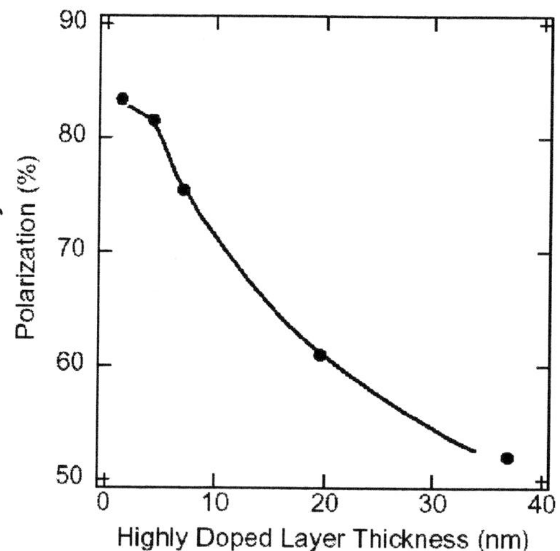

FIGURE 2. The peak polarization as a function of the highly-doped layer thickness.

The next question is whether 7.5 nm of highly-doped layer is thick enough to overcome the SCL problem. The GTL was used to study the charge performance. Figure 3 shows the temporal profiles of the emission current pulses measured using varying laser energies. As the laser energy was increased, the temporal profile simply scaled without developing the leading edge spike typically observed in the surface-charge-limited photoemission. Figure 4 shows the charge output as a function of the laser energy. The observed charge output increases linearly with the laser energy, reaching 8×10^{11} e^{-}/pulse.

Although the sample did not show any charge limit behavior, the QE was lower than expected. The low QE was thought to be due to the energy barrier in the conduction band.

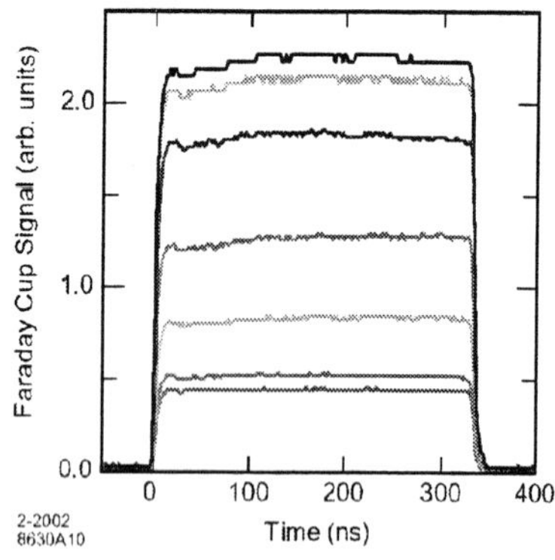

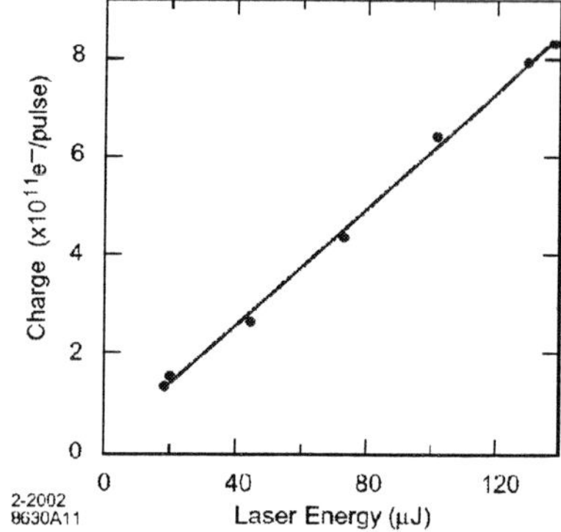

FIGURE 3. The temporal profiles of the electron emission current.

FIGURE 4 The photoemission charge per pulse as a function of the laser energy

HIGH GRADIENT DOPED STRAINED GaAsP

To remove the energy barrier in the conduction band, the active layer was changed to $GaAs_{0.95}P_{0.05}$. The 5% phosphorus raised the conduction band by 58 meV compensating the energy difference due to the high gradient doping. A QE of 0.2% and peak polarization of 80% were measured at room temperature at 805 nm. The charge output was measured using two laser systems, a flashlamp-pumped Ti:Sapphire laser (Flash-Ti) with 270 ns pulse and a Nd:YAG pumped Ti:Sapphire laser (YAG-Ti) with 4 ns pulse. The laser spot size of both lasers was set to 14 mm. Figure 5 shows the charge output as a function of the Flash-Ti laser energy with and without the YAG-Ti laser. The YAG-Ti laser by itself produced $2.3\times10^{11}e^-$/pulse, equivalent to 9.2 A peak current, twice the NLC requirement. The charge output was linear up to the maximum laser energy, producing a maximum charge of 2.2×10^{12} e^- in 270 ns. If one assumes a linear scaling with spot size, this charge is equivalent to 4.5×10^{12} e^- (2.6 A) from a 20 mm diameter cathode.

A SLAC high energy experiment, E158, to measure parity violation in Möller scattering ($e^-e^- \rightarrow e^-e^-$) will use a 48 GeV polarized electron beam scattering off unpolarized electrons in a liquid hydrogen target. The experiment requires a beam intensity of 8×10^{11} e^- in a 370 ns pulse at the gun. Since the beam intensity requirement is difficult to meet using the standard SLAC photocathodes, it was proposed to use the newly developed high-gradient-doped GaAsP photocathode described above. A sample was installed in the SLAC polarized electron source and heat-cleaned and activated twice. The initial QE was 0.4% at 805 nm. Figure 6 shows the charge output as a function of the laser energy for a 100-ns long pulse. The charge output was linear up to the maximum laser energy, producing a maximum charge of 2.3×10^{12} e^- in 100 ns (3.7 A), which is nearly equal to the peak current required for the NLC. Finally the charge output was observed to scale with the laser pulse length, reaching more than 8×10^{12} e^- for 370 ns. The charge output corresponds to ten times the E158 requirement and more than twice the NLC-train charge.

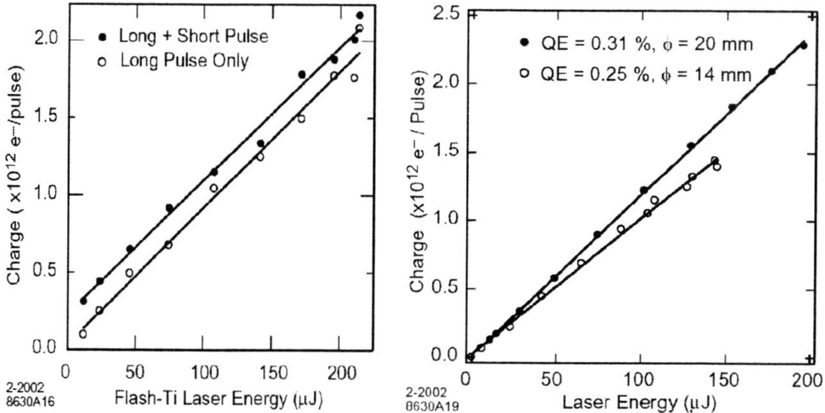

FIGURE 5. The photoemission charge per pulse as a function of the laser energy for Flash-Ti only (open circles) and Flash-Ti and YAG-Ti overlaid (solid circles).

FIGURE 6. The photoemission charge per pulse as a function of the laser energy measured at the SLAC injector.

CONCLUSIONS

The SCL effect has a strong doping concentration dependence, decreasing with increased doping and diminishing to zero at a doping level of 5×10^{19} cm^{-3}. The charge capabilities of high-gradient-doped strained photocathodes have been determined. A charge as large as 2.2×10^{12} e$^-$ per pulse was produced from a 14-mm diameter laser spot in 270 ns. By overlaying a short pulse laser, the peak current capability was also determined. A peak current as high as 9.2 A was extracted. There were no indications of a surface charge limit even when the QE decreased, and the charge output was limited by the laser energy. A cathode of this type is being used in a current high energy polarization experiment at SLAC, for which the charge performance has been consistent with, if not superior to, the GTL results. This is the first demonstration that a NLC compatible beam with polarization approaching 80% is achievable.

REFERENCES

1. T. Maruyama, A. Brachmann, J.E. Clendenin, T. Desikan, E.L. Garwin, R.E. Kirby, D.-A. Luh, J. Turner, and R. Prepost, *Nucl. Instrum. Meth. A* 492, 199 (2002).
2. G.A. Mulhollan, A.V. Subashiev, J.E. Clendenin, E.L. Garwin, R.E. Kirby, T. Maruyama, R. Prepost, *Phys. Letters* A 282, 309 (2001).

Status of the polarized source at MAMI

Kurt Aulenbacher, Valeri Tioukine, Markus Wiessner, Konrad Winkler

Institut für Kernphysik der Universität Mainz, 55118 Mainz, Germany

Abstract. This talks addresses the operation of the polarized source at the Mainz Microtron MAMI. The source is operating with selected photocathodes of modulation doped, uniaxially strained layer photocathodes, which results in an average spin polarization of 80% and a quantum efficiency of typically $2\,\mu A/mW$. The operative lifetime has been improved by employing a novel activation technique which reduces transmission losses in the vicinity of the cathode. In addition a considerable simplification of the laser system has become possible by improving the power output of laser diode seed lasers so that it is not necessary to employ power amplifier units. It was shown that the potential for increasing the laser power is limited in our setup because of the thermal resistance between cathode and the surrounding electrode.

PHOTOCATHODE PROPERTIES AND LIFETIME IMPROVEMENT

This talk deals with some results and observations concerning the attempt to improve the operative stability for operation with optimized polarization in our setup operating at the MAinz MIcrotron MAMI.

The photocathode that is employed at MAMI is a 150 nm thick strained layer cathode [1], with an active region that consists of uniaxially strained $GaAs_{0.95}P_{0.05}$. In recent years the fabrication at Joffe institute in St. Petersburg/Russia has been optimized, so that ever higher polarizations and quantum efficiencies are available for production runs at MAMI. The latest samples achieve typical quantum efficiencies (q.e.'s) of $2\,\mu A/mW$ in the maximum of the polarization spectrum. At this maximum the exciting photon energy is 1.50 eV corresponding to 826 nm wavelength. The available average polarization is about 80%. An increase of polarization is frequently observed during the quantum efficiency decay, a drop of a factor two may increase the polarization by 6% (absolute). This was also observed by Mamaev et al. with the same type of cathodes [2].

The decrease of q.e. is believed to be caused by contamination of the Cesium/Oxygen (Cs/O) activation layer and is undesirable because of the need for stable operating conditions. It has been long guessed that very small transmission losses in the vicinity of the anode of the source lead to this contamination, and that the losses are generated by photoemission from the edge area of the cathode. It is not possible to suppress edge emission by mechanical masks because of unwanted focussing effects.

The attempts to create a non mechanical mask have lead to the anodization technique [3] which was successfully applied at JLAB, where the cathode is oxidized outside a small area around the main laser spot, so that the quantum efficiency is close to zero on the outer parts of the cathode. Even though we have demonstrated that this method indeed leads to reduced transmission losses and to an increased photocathode lifetime [4]

CP675, *Spin 2002: 15th Int'l. Spin Physics Symposium and Workshop on Polarized Electron Sources and Polarimeters*, edited by Y. I. Makdisi, A. U. Luccio, and W. W. MacKay

we decided to make a further modification to the process. We have applied the Cs/O evaporation through an aperture of 3 mm ('mask'). This method has the advantage that no chemical treatment of the sensitive cathodes is necessary, and that it is possible to do a comparative measurement of lifetimes with the same cathode. The mask activation suppresses all photoemission outside the 3 mm diameter activation area. Even after several weeks of observation we have not been able to detect photoemission outside this area, what indicates that the activation layer does not spread laterally on the semiconductor surface. In order to implement this technique we make use of the possibility to move the cathode holder in vacuum by manipulators, so that the cathode is activated in the mask and then removed from the mask and transferred to its operational position in the source. In figure 1 a comparison is presented, it shows the behavior of the same cathode operated at 200 Mikroamperes, once activated 'nude' and then masked.

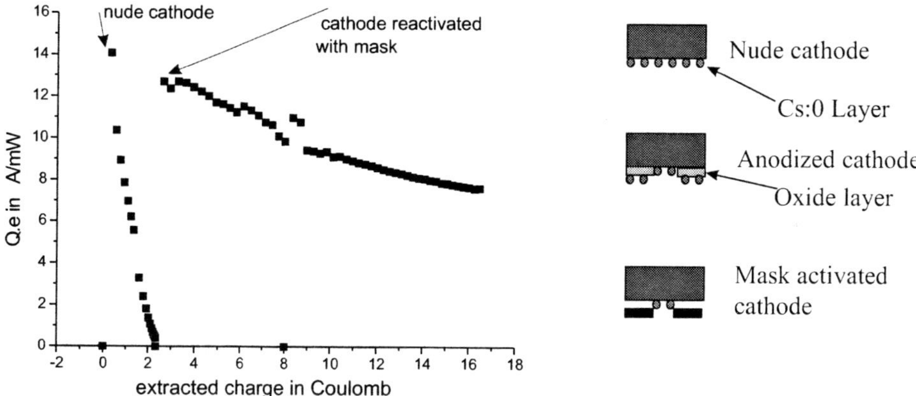

FIGURE 1. Right: Illustration of the Cs/O coverage of 'nude', 'anodized' and 'mask activated' photocathodes. Left: Comparison q.e.-decay during beam current production.

It is evident that the decrease of q.e. is much smaller in the mask activated case. The mask activation allows therefore to extract more charge. A measure for the available charge in practical operation is given by $Q = I \cdot \tau$ where τ (the 'lifetime') is defined as the time for a 1/e decrease of q.e. from its initial value.

We observe a local q.e.-decrease which is restricted to the vicinity of the emission spot: The local extractable charge from one beamspot of the laser (of the area A_l) is restricted to $Q/A_l = 3 \cdot 10^4 C/cm^2$ or about 20 Coulomb per beamspot. This has negligible effect on availability, since the beam spot position can be changed remotely during operation with a minimum impact on the injection parameters. For the total charge that can be extracted from the cathode (for one activation) we observe differences if operated on long time scale (moderate average current of 10-30 μA, typical for operation with strained layer cathodes) and on short time scale (test experiments with bulk GaAs cathodes of high q.e. and high average currents of 200 μA). In the first case we found Q=40 and in the second case Q>100 Coulomb.

Even in the first regime it was possible to operate all experiments in the last year continuously with high polarization over their scheduled runtime.

LASER STUDIES FOR 2.5 GHZ PULSED OPERATION

The Master oscillator power amplifier (MOPA) system was first described with 0.5 GHz operation for accelerator injection by Poelker [5] and transferred to 2.5 GHz for our purposes later on [6]. It was considered as a mature system which allows to produce sufficient optical power at a high stability. However an unforeseen problem occurred when the vendor of the power amplifier elements stopped production in 1999, whereas the master oscillator laser diodes remain available.

Our last amplifier burnt out in Spring 2002 after about 4000 hours of operation. Even though power amplifiers are available from a new manufacturer[1], we found favorable circumstances to omit the power amplifier and hence to obtain a considerable simplification of the system. Presently our laser system consists out of the master oscillator laser alone. The reason for this possibility is found in the ever higher quantum efficiencies (and good lifetimes) that can be achieved with strained layer cathodes. The master oscillator is a high power single mode laser diode which is capable of producing 200 mW of average power in c.w. operation. With the typical initial efficiency of $2\mu A/mW$ the diode can in principle deliver enough power to drive the accelerator to the limit of its capabilities (100 μA) and still have a considerable reserve for optical transmission losses and decreasing q.e..

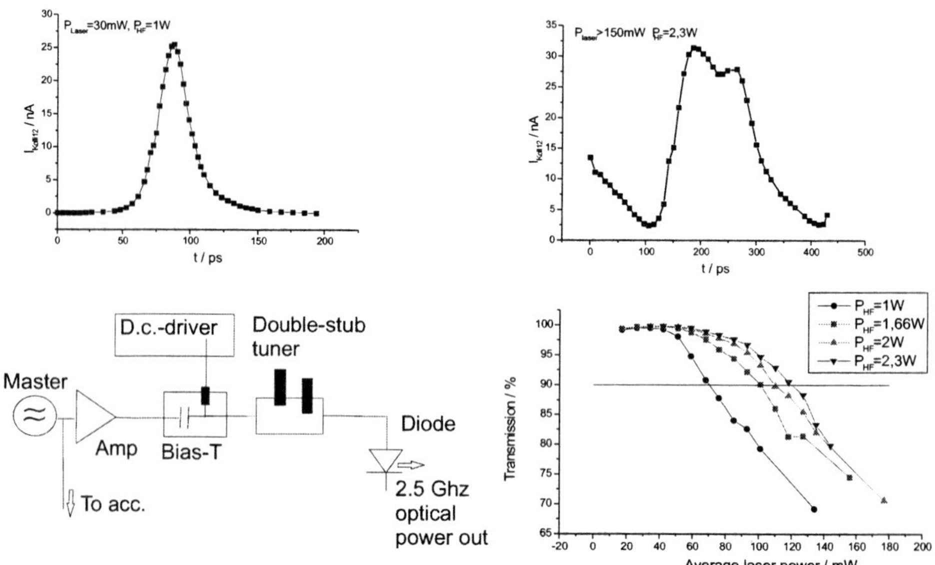

FIGURE 2. Upper part: Pulse distortion at high average powers. Lower left: schematic of diode r.f.-drive. Lower right: transmission as a function of average output power for different r.f. power levels.

The remaining problem for high power output is the requirement for 2.5 GHz pulse operation which is essential to achieve good transmission through the accelerator. The

[1] Toptica AG, Munich, Germany, model Toptica TA100

typical operation consists of driving the diode with a d.c.-current and superimposing r.f.-power by a bias-t (see figure 2). Improved r.f.-coupling to the diode is achieved with the help of a double stub tuner.

We found that increasing the average power of the diode resulted in severe pulse distortion and pulse lengthening (see upper part of figure 2). Consequently the accelerator transmission decreases with increasing power. However we found that this can be compensated to some extent by increasing the r.f.-power at the diode. The lower right part of figure 2 shows the dependence on the r.f. power level. We find that more than 100 mW of output power can be realized at > 90% transmission. Since the total charge from the source is limited a high transmission leads to an optimized continuous operation of the cathode.

Presently all running experiments at MAMI can be supplied by this simple system. However, in order to be prepared for future eventualities (e.g. cathodes with smaller q.e.'s) we will install a MOPA system in parallel to the existing system. An additional problem that must be solved before the high power capabilities of the MOPA can be employed, is discussed in the next section.

TEMPERATURE EFFECTS IN CATHODES

We investigated the temperature increase that is generated by the operation at average laser powers which would be typical for operation of MAMI at the limits of its capabilities ($100\mu A$). In order to do this under realistic conditions we have employed a non invasive technique. The luminescence radiation generated by the laser absorption in the semiconductor was observed and spectrally resolved (see figure 3). For a bulk-GaAs cathode - which is assumed to behave thermally identical to a strained layer cathode - the position of the luminescence spectrum is related to the temperature by the Varshni equation [7].

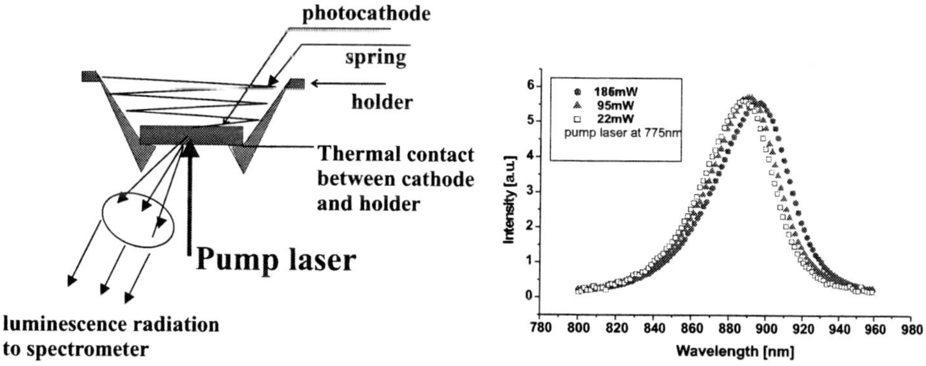

FIGURE 3. Left part: schematic situation for the laser-heating experiment, Right side: Observed spectral shift with average power

The thermal conditions in this experiment where chosen in a way that they are similar to the operation in the polarized source: We used the same photocathode holder, and the holder was coupled to room temperature by a good thermal conductance. The limiting thermal resistance should therefore be the thermal contact between cathode and holder. Figure 3 shows the spectral shift that is observed when the optical power is increased. From the spectral shift we conclude that the temperature increase is about 40 degree per 100 mW.

Under this circumstances it seems not appropriate to use higher average powers than 100 mW because it is believed that the lifetime of the Cs/O surface layer will be reduced drastically at temperatures higher than 50°C. From this we conclude that the operation of strained GaAs at average currents of 100 μA will possibly suffer from the increased temperature. We will therefore start work to decrease the thermal resistance of the cathode/holder interface in the near future.

ACKNOWLEDGMENTS

This work was supported by the Deutsche Forschungsgemeinschaft (DFG) through the SFB 443 and by the European Union through INTAS grant 9900125.

Further we want to thank Dima Orlov and Uli Waigel from MPI Heidelberg for productive discussions concerning photoluminescence spectroscopy.

REFERENCES

1. Drescher, P., Andresen, H. G., Aulenbacher, K., Bermuth, J., Dombo, T., Euteneuer, H., Faleev, N. N., Fischer, H., Galaktionov, M. S., v. Harrach, D., Hartmann, P., Hoffmann, J., Jennewein, P., Kaiser, K.-H., Köbis, S., Kovalenkov, O. V., Kreidel, H. J., Langbein, J., Mamaev, Y. A., Nachtigall, C., Petri, M., Plützer, S., Reichert, E., Schemies, M., Steffens, K.-H., Steigerwald, M., Subashiev, A. V., Trautner, H., Vinokurov, D. A., Yashin, Y. P., and Yavich, B. S., *Appl. Phys. A*, **63**, 203–206 (1996).
2. Mamaev, Y. A., Yashin, Y. P., Subashev, A. V., Ambrajei, A. N., and Rochansky, A. V., "Temperature dependence of electron spin dynamics," in *Spin 2000, 14th international spin physics symposium, AIP proceedings, Vol. 570*, edited by K. Hatanaka, T. Nakano, K. Imai, and H. Ejiri, AIP-publishing, Melville, New York, 2001, pp. 920–925.
3. Sinclair., C., "Performance of the Jefferson Laboratory polarized electron source at high average current," in *International workshop on polarized sources and targets, (PST99)*, edited by A. Gute, S. Lorenz, and E. Steffens, University of Erlangen, D-91058 Erlangen, 1999, pp. 222–230.
4. Aulenbacher, K., Euteneuer, H., Jennewein, P., Kaiser, K.-H., Kreidel, H. J., v. Harrach, D., Reichert, E., Schuler, J., Tioukine, V., Wiessner, M., and Winkler, K., "New results from the Mainz polarized electron facilities," in *Spin 2000, 14th international spin physics symposium, AIP proceedings, Vol. 570*, edited by K. Hatanaka, T. Nakano, K. Imai, and H. Ejiri, AIP-publishing, Melville, New York, 2001, pp. 949–954.
5. Poelker, M., *Appl. Phys. Lett.*, **67**, 2762–2764 (1995).
6. Aulenbacher, K., Euteneuer, H., v. Harrach, D., Hartmann, P., Hoffmann, J., Jennewein, P., Kaiser, K.-H., Kreidel, H. J., Leberig, M., Nachtigall, C., Reichert, E., Schemies, M., Schuler, J., Steigerwald, M., and Zalto, C., "High capture efficiency for the polarized beam at MAMI by r.f. synchronized photoemission," in *Proceedings sixth European accelerator conference (EPAC98)*, edited by S. Myers, L. Lilijeby, C. Petit-Jean-Genaz, J. Poole, and K.-G. Rensfeldt, Institute of physics publishing, Bristol and Philadalphia, 1998, pp. 1388–1390.
7. Varshni, Y. P., *Physica (Utrecht)*, **34**, 149–154 (1967).

Ultra-stable flashlamp-pumped laser*

A. Brachmann, J. Clendenin, T.Galetto, T. Maruyama, J.Sodja, J. Turner, M. Woods

Stanford Linear Accelerator Center, 2575 Sand Hill Rd., Menlo Park, CA 94025

Abstract. We present the design and experimental results for the flashlamp-pumped Ti:Sapphire laser system used at the Stanford Linear Accelerator Center (SLAC). This laser system is used in conjunction with the Polarized Electron Source to generate polarized electron beams for fixed target experiments (e.g. the E-158 experiment). The unique capabilities such as high pulse-to-pulse stability, long pulse length and high repetition rate is discussed. Emphasis is placed on recent modifications of the laser system, which allow ultra-stable operation with 0.5% rms intensity jitter.

1 INTRODUCTION

The flashlamp-pumped Ti:sapphire laser system initially was installed at the Stanford Linear Accelerator Center's Polarized Electron Source in 1993 and has been described previously [1, 2]. Since then, the laser system has had significant upgrades [3, 4]. This paper documents recent modifications and the performance of the laser system mainly during the past year.

2 LASER SYSTEM

The scheme of the laser system is depicted in figure 1. The Ti:sapphire laser cavity consists of a 2 meter concave high reflector and a flat output coupler with a reflectivity of 85 % and a spacing of 1m. Both mirrors are coated for a ~ 50 nm bandwidth. A Ti:sapphire laser rod (0.1 % Ti doping) of 6 inch length and 4 mm diameter is pumped by 2 flashlamps. Rhodium coated, double elliptical reflector surfaces focus the pump light into the center of the laser rod. Within the cavity, a birefringent tuner (BRT) oriented at Brewster's angle allows for wavelength tuning with a typical bandwidth of ~ 0.7 nm FWHM.

The extra-ordinary optical axis of the Ti:sapphire crystal is oriented parallel to the plane of the rod's optical surfaces. If no precautions are taken to align the

* Work supported by Department of Energy contract DE-AC03-76SF00515

CP675, *Spin 2002: 15th Int'l. Spin Physics Symposium and Workshop on Polarized Electron Sources and Polarimeters,* edited by Y. I. Makdisi, A. U. Luccio, and W. W. MacKay

crystallographic axis of the Ti:sapphire rod with respect to the Brewster angle of the BRT, a rotatable half wave plate in between laser rod and BRT is necessary to maximize the amount of p-polarized light transmitted.

Cooling of the laser head is provided by a closed loop of ultra-pure water flow (conductivity 18 MΩ) at ~ 2.5 GPM. The closed loop water temperature is maintained by a 3 or 5 ton chiller.

The flashlamps are pulsed by a SLAC built modulator / power supply. The modulator provides the high voltage pulse needed to fire the flashlamps. A 1.2 μF capacitor is charged by a 10 kV, 8 kJ/s power supply. Upon ignition of a thyratron, the capacitor is discharged through the flashlamps in series. The pulse has a peak current of 1 kA and a duration of 22 μs. Between pulses, a current through the flashlamps is maintained by a simmer power supply in parallel. The simmer current reduces the high voltage needed for conduction in the lamps and thereby extends their lifetime. The modulator and the laser cavity were designed to operate at a maximum repetition rate of 120 Hz.

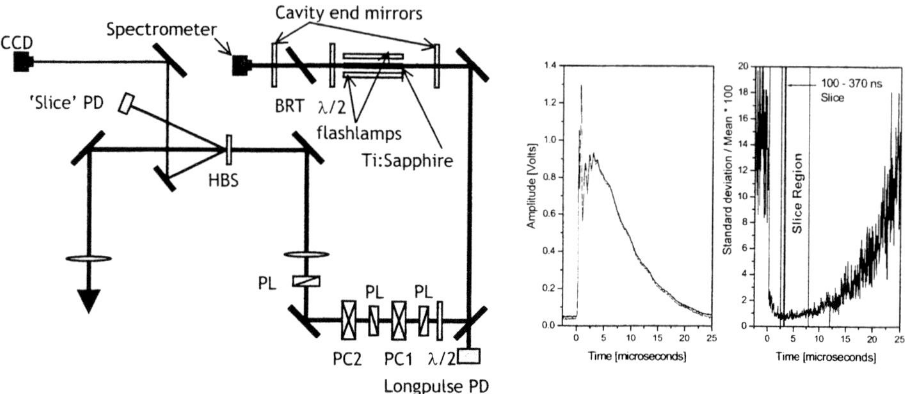

FIGURE 1: Scheme of laser system setup including pulse shaping components and diagnostics.

FIGURE 2: 100 shot envelope of cavity pulse (left); shot to shot intensity stability across pulse, region of lowest jitter and 'Slice' timing (right).

Downstream of the cavity, polarizer / Pockels cell combinations allow duration and intensity modifications of the laser pulse. The 'Slice' pockels cell is used to separate a 50 to 370 ns 'slice' out of the ~ 20 μs long pulse delivered by the cavity. Slicing is applied within the region of lowest intensity jitter of the total pulse length. Figure 2 gives an example of the pulse shape generated by the cavity. Also, the stability of the laser pulse as a function of time and the corresponding region of slicing is depicted. A fast pockels cell allows temporal modification of the sliced laser pulse (<u>T</u>op <u>H</u>at

Pulse Shaper – TOPS). A trapezoidal pulse shape is needed to achieve a flat energy profile in the electron beam due to beam loading effects. The high voltage pulse shape of the pockels cell driver is set by a function generator. A 25 ns time resolution of the applied function was found to be sufficient.

Further downstream, a pair of Pockels cells are used to generate circularly polarized light of either helicity at the photocathode. An optical transport system (OTS), ~ 20 meter in length, delivers the laser beam to the photocathode. The OTS preserves the laser helicity and images the polarization Pockels cell onto the photocathode. A detailed description of helicity control and methods to minimize helicity correlated beam assymetries has been published by Humensky et al. [4].

3 DIAGNOSTICS

The laser performance is monitored by photodiodes, a CCD camera and a spectrometer (see figure 1). Leakage light through 45° folding mirrors or sampled beams provide signals for routine intensity, beam profile, intensity jitter and wavelength measurements. A photodiode installed upstream of the pulse-shaping optics monitors the cavity output (Longpulse PD). Downstream of the pulse-shaping optics, two one-percent samples of the laser beam are separated by a holographic beam sampler. One sample is used to monitor the intensity of the sliced pulse (Slice PD). The second sample is focused onto a CCD camera and provides an image of the beam cross section. The laser wavelength is measured by a fiberoptic spectrometer, using the leakage light of the cavity end mirror.

4 CAVITY MODELING

A significant improvement compared to previous laser performance has been achieved by measurement of the thermal lensing of the cavity and inclusion of the results into the cavity design. The focal length of the thermal lens under typical operating conditions has been derived by measurement of the laser spot size as a function of distance from the cavity center until a minimum of the beam waist was observed. With a 1.1 to 1.2 m focal length, the thermal lens dominates the cavity optics. As a result of high voltage pulse instabilities, fluctuations of the thermal lens may occur. One goal of cavity design was to minimize thermal lens induced spot size changes within the active medium. Modeling of the beam waist within the laser rod was conducted for a 5 meter concave end mirror (used historically) and a 2 meter concave end mirror. Also, the separation of the end mirrors was included in our calculations. We have found a minimum sensitivity to thermal lensing using a 2 meter concave end mirror and a separation of ~ 60 cm from the cavity center. A

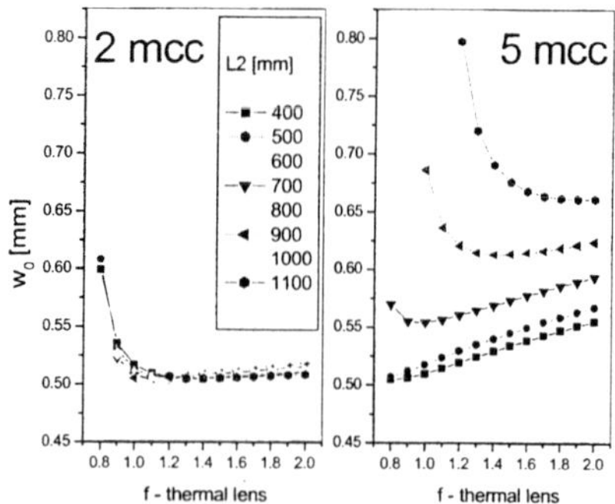

second set of calculations was performed to study matching and stability of wave front curvature at the end mirror curvature as a function of thermal lensing and mirror separation. The calculations also suggest improved performance for a 2 meter concave mirror. The results are plotted in figure 3.

FIGURE 3: Results of cavity modeling: Spotsize at the cavity center as function of thermal lens focal length and end mirror location (L2) for a 2 and 5 meter concave mirror curvature; spot size resulting from a 2 mcc mirror shows less sensitivity to thermal lensing.

5 CAVITY HALF WAVE PLATE AND LASER ROD CRYSTALLOGRAPHY

As described above, the half wave plate installed in the cavity insures the proper orientation of the plane of p-polarized light incident on the BRT (installed at Brewster's angle). High fluences within the cavity and multimodal operation leads to 'hotspots' within the beam profile resulting in a high probability for optical damage of cavity components. Indications of optical damage typically increase laser jitter and decrease cavity output power. During extended periods of operation, the cavity half wave plate frequently needed replacement due to damage of coatings and bulk material.

To alleviate this problem, the laser rod was installed with controlled crystallographic orientation. With the Ti:sapphire rod mounted in the laser head assembled to a degree where rod manipulation is still possible, the assembly was placed between a polarizer and analyzer. A collimated diode laser beam (830 nm) was aligned through polarizer, laser rod and analyzer. The polarizer ensures linearly polarized light incident to the laser rod (with the plane of p-polarization parallel to the plane required by the Brewster angle of the BRT). The laser rod acts as a thick, multiple order quarter wave plate and transforms linearly polarized light into an elliptical state. The degree of ellipticity is a function of the angle between the polarization plane of incident light and orientation of the Ti:sapphire's extraordinary optical axis. The amount of ellipticity caused by laser rod's optical orientation is detected by the measurement of the light transmitted through laser rod and analyzer. A high degree of ellipticity maximizes the light transmitted through the crossed polarizers. By rotation of the laser rod around its geometrical axis, generation of elliptically polarized light can be minimized, which is the case if one of its optical axis is parallel to the plane of incident linearly polarized light. Under this condition, the light passing through polarizer, laser rod and analyzer shows maximum extinction. Using this procedure, the laser rod can be mounted in an oriented fashion and the need for the cavity half wave plate is eliminated. As a result, the peak to peak intensity stability as well as the total laser power increases.

6 LASER PERFORMANCE

The performance of the laser system is summarized in Table 1. The E-158 2002 spring and fall experimental runs have shown ultra-stable operation for several months without interruption except scheduled flashlamp changes. Flashlamps were changed after ~ 1.45×10^8 pulses (2 weeks at 120 Hz). The maximum laser power of the cavity is ~ 7 W at 120 Hz (or 58mJ). In our development laboratory we achieved a 0.3% rms laser intensity jitter, while at the injector 0.5% was typical. After slicing and pulse shaping we obtained up to 500 µJ in a 370 ns long (sliced) pulse. All downstream optics account for an additional attenuation of 10 – 15%.

TABLE 1. Laser operating parameters

Mode structure	Multimodal
Wavelength	Tunable (805nm,850 nm)
Bandwidth	0.7nm
Repetition rate	120 Hz
Peak power	58 mJ
'Sliced power' (370 ns)	500 µJ
Stability	0.5 % rms

7 CONCLUSIONS

Our results show that the careful setup of a flashlamp pump laser system can achieve a level of performance required for generation of highly stable polarized e^-beams. Despite advances in diode laser technology, flashlamp pumping remains a viable option if tunable and high power lasers are required for the generation of polarized e^- beams for high energy physics experiments.

8 REFERENCES

1. K.H. Witte. *A Reliable Low Maintenance Flashlamp Pumped Ti:Sapphire Laser Operating at 120-PPS. Presented at Lasers '93, Lake Tahoe, NV, Dec. 6-9, 1993;* SLAC-PUB-6443.

2. R. Alley et al., *The Stanford Linear Accelerator Polarized Electron Source, Nucl. Instrum. Meth., A365:1-27, 1995;* SLAC-PUB-95-6489.

3. A. Brachmann et al., *SLAC's polarized electron source laser system for the E-158 parity violation experiment, Proc. SPIE Vol. 4632, 211-222, 2002;* SLAC-PUB-9145.

4. T. B. Humensky et al., *SLAC's Polarized Electron Source Laser System and Minimization of Electron Beam Helicity Correlations for the E-158 Parity Violation Experiment, Nucl. Instrum. Meth., submitted.*

MIT-Bates Polarized Source†

M. Farkhondeh, W. Franklin, E. Tsentalovich, T. Zwart and E. Ihloff

MIT-Bates Linear Accelerator Center
P.O.Box, 846, Middleton, MA 01949, USA

Abstract. During the fall of 2001, the polarized source at MIT delivered over 140 Coulombs of high quality polarized beams to the SAMPLE-III 125 MeV parity violating experiment. Prior to the experiment, the source was reconfigured to deliver highly polarized beam using the new high power diode array laser system with large aperture beam optics, and a strained layer GaAsP photocathode from St. Petersburg. The results of these tests will be presented. The production run for SAMPLE-III was then carried out with a bulk GaAs and the Ar-Ti:Sapphire laser system. Since April of this year, the source has been delivering high polarization beams to the South Hall Ring for commissioning of the BLAST spectrometer. The stored current in the ring exceeds 100 mA. This is accomplished using the high power diode laser system and high-gradient-doped strained GaAsP photocathodes from Bandwidth Semiconductor Inc. tuned for 810 nm. The operational lifetime of the photocathode is excellent. A status report of the Bates polarized source and the operational experience of delivering high polarization beam to SHR will be presented.

INTRODUCTION

The MIT-Bates accelerator complex is capable of operation in three distinct modes. These are the long pulsed mode used for the SAMPLE parity violating experiment [1] in the North Hall, the storage mode, and the extraction mode using the South Hall Ring (SHR). A schematic diagram of the MIT-Bates accelerator complex is shown in Figure 1. The pulsed mode consists of pulses up to 35 μs long, 10-mA peak at 600 Hz providing 40 μA average currents for the SAMPLE experiment. For the storage mode, multiple 1.3 μs long pulses at low repetition rates are injected into the SHR and stacked to circulating currents exceeding 80 mA for internal target experiments. The extraction mode employs 1.3-μs long ~20 mA pulses from the source at 600 Hz for injection into and cw extraction from the SHR to a fixed target beam line. The requirements for injecting high polarization beams for the storage mode are relatively modest, but very demanding in peak current and repetition rate for the extraction mode. In this paper the polarized source for the SAMPLE experiments and for the commissioning of the BLAST with SHR will be described.

POLARIZED BEAMS FOR SAMPLE

The 200 MeV SAMPLE experiments on hydrogen and deuterium [2,3] were successfully completed in 1998-1999 with beams of unprecedented quality and

† Work supported in part by the Department of Energy under a Cooperative agreement # BEFC294ER40818.

CP675, *Spin 2002: 15th Int'l. Spin Physics Symposium and Workshop on Polarized Electron Sources and Polarimeters*, edited by Y. I. Makdisi, A. U. Luccio, and W. W. MacKay

stability using bulk GaAs photocathodes and an Ar-Ti:Sapphire laser. The polarized injector and the laser trajectories are shown schematically in Figure 2. Prior to the start of the 125 MeV SAMPLE-III experiment, a period of 1.5 months was dedicated

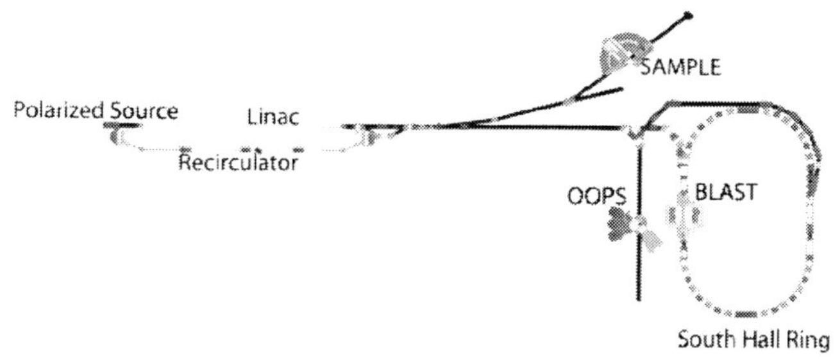

Figure 1. A schematic view of the MIT-Bates Linear Accelerator Center showing the linac, recirculator, the South Hall Ring, and the three experimental areas.

to the evaluation of the feasibility of utilizing a newly available high polarization beam for SAMPLE. The results of these studies will be presented in the next section. The SAMPLE-III production run was completed using 140 Coulombs from a bulk GaAs photocathode and the existing Ti:Sapphire laser system.

HIGH POLARIZATION BEAM DEVELOPMENT

To meet the Bates requirements for highly polarized beams, a 60 keV test beam setup with a Mott polarimeter was completed in 2000 and is now in routine use. This setup allows research and development with high power diode lasers and high polarization photocathodes independent of the main accelerator. In fall of 2001, a multimode fiber-coupled diode array laser system [4] capable of producing peak power up to 150 W unpolarized radiation at λ=810 ±3 nm was installed on the main injector. The parameters of this laser system are listed in Table 1. The large 200 mm-mrad emittance of this laser precluded use of the existing 20-m transport line for the Ti:Sapphire laser. A new wide-aperture 4-m long laser transport system matching the emittance of the diode laser system was designed and installed. The transport system consists of an optical board installed on a precision six-strut alignment system, 2-inch aperture focusing lenses and polarizers, and a 70-mm aperture helicity Pockels Cell [5] (HPC).

TABLE 1. The parameters for the high power diode array laser systems.

	Value	Unit
Wavelength	810±3	nm
Emittance	200	mm.mr
Peak power	150	W
Repetition rate	1-cw	Hz
Pulse-to-pulse jitter	≤ 0.1	%

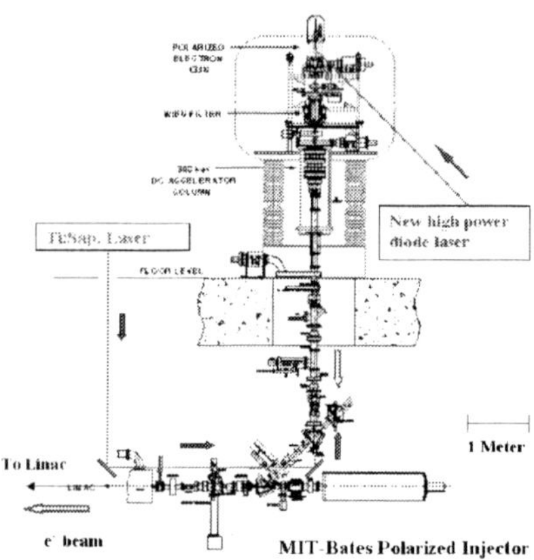

Figure 2. A schematic view of the MIT-Bates polarized injector showing the gun, the injector beam line and the laser paths for both the Ti:Sapphire and the new diode lasers.

The laser beam spot size can be adjusted to illuminate half to full 11-mm diameter of the photocathode by adjusting the longitudinal location of the last converging lens near the end of the transport line. The trajectory of this laser is also shown in Figure 2. The laser system and the associated hardware are installed at ground potential outside the Faraday cage and the acceleration column both at 300 keV potential. The laser beam enters the gun chamber directly through a vacuum port and strikes the photocathode at an incident angle of 37° with respect to the normal. Because the index of refraction of GaAs based material is 4.5, the refracted ray in the GaAs is close to the normal ($\theta_r \approx 8°$), and there is no significant reduction in the electron beam polarization. However, the ~37° incident angle has influence on the helicity-correlated effects due to differences between the reflectivities of the residual linear polarization of the S and P waves.

High Polarization Beam Tests for SAMPLE-III

A gun was prepared with an As-capped strained GaAs photocathode from the St. Petersburg group [6] and installed for high polarization beam tests for SAMPLE-III experiment. Measurement with the SAMPLE Moller polarimeter at 125 MeV yielded a beam polarization of only ~55% at the diode wavelength of 810 nm. It was then discovered that this sample was not optimized for our laser and was designed to produce maximum polarization of ~75% at 850 nm. However, the lifetime and the pulse-to-pulse stability were excellent and exceeded the SAMPLE requirements. The experiment placed stringent limits on helicity-correlated beam position (<100 nm) and intensity (< 10 ppm) differences at the SAMPLE target (averaged over 1/2 hour). Due to birefringing effects in the optical elements and the vacuum window, an asymmetry is generated in the transport of the laser light for the two helicities due to residual linear polarization in the light. In addition, strained photocathodes are known to have a preferred optical axis causing an analyzing power of the order 5-10 %. The non-zero incident angle of laser adds to these asymmetries. These analyzing powers create helicity correlated asymmetries in the position and intensity of the beam. A halfwave plate was inserted on the laser table between the HPC and the photocathode and rotated to minimize these differences. A piezo-electric system [7] was also used for slow feedback to reduce the position differences. More details on these tests are found in the presentation of E. Tsentalovich in this workshop [8].

High Polarization Beam for SHR

A gun with a high-gradient-doped strained GaAsP photocathode grown by Bandwidth Semiconductor Inc. [9] with the SLAC specification [10] was prepared and installed on the injector in 2002. With 5% phosphate concentration, the peak of polarization is lowered from 850 nm to ~810 nm matching the wavelength of our high power diode laser. The top 10 nm layer of this sample is GaAs and is heavily Zinc doped ($5x10^{19}$) to minimize the surface charge limit effect often present in strained samples at high laser power densities. However, diffusion of Zinc during heat cleaning at ~600 °C will cause a depletion of the dopant concentration in the top layer. This will lead to a gradual appearance of surface charge limit after several one-hour long heat cleanings. The degradation is shown in Figure 4 for the first high gradient doped photocathode that was in use on the main injector for a five months period between April and September of this year.

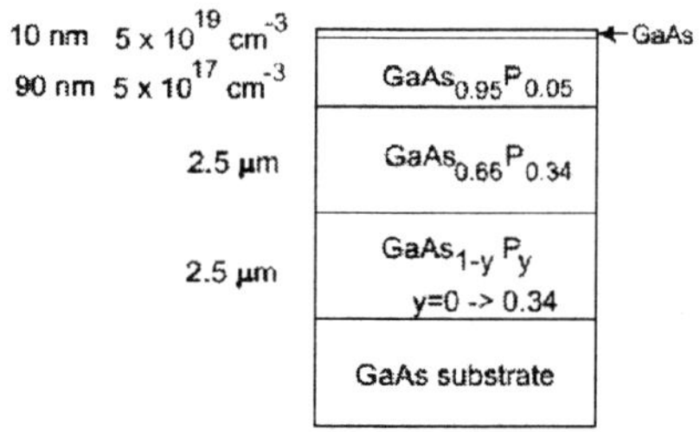

Figure 3. A schematic diagram of a high-gradient-doped strained GaAsP photocathode grown by Bandwidth Semiconductor Inc. [9] with the SLAC specification [10].

This type of photocathode and the high power diode array laser system [11] have been in routine use injecting beams with polarizations exceeding ~70% for the BLAST commissioning. The second high-gradient-doped strained photocathode has been in use since September and has displayed a superior performance compared to the first sample. One reason is that for this second sample, we are limiting the heat cleaning duration and temperature to ~10 minutes and ~575 °C respectively. This should reduce the delusion rate of the Zinc dopant out of the heavily doped layer, thus reducing the magnitude of the charge limit effect. The beam polarization in the SHR is measured using a laser back-scattering Compton polarimeter [12] in the last arc upstream of the internal target and BLAST area. This polarimeter provides noninvasive continuous monitoring of the beam polarization in the ring with roughly 5% statistical accuracy in a half-hour run. The Compton polarimeter and the SHR form a good combination for studying the spin dynamics in storage rings. A transmission polarimeter [13] is periodically used at the 20 MeV region of the linac for fast relative measurements of the beam polarization near the injector.

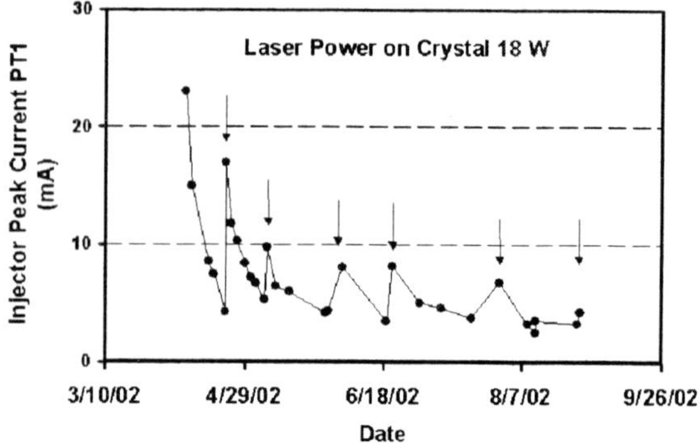

Figure 4. A graph of peak current in mA vs. calendar time for the 5 months period that the first high high-gradient-doped strained GaAsP sample from Bandwidth Semiconductor Inc. [9] was in use for the BLAST commissioning. To show the degradation of the QE over this period, the peak current is plotted for a fixed laser power of 18 W. The location of each activation is shown by a vertical arrow.

CONCLUSIONS

Three SAMPLE experiments were successfully completed using over 450 Coulombs of polarized beams of high quality originated from unstrained bulk GaAs photocathodes. A test of high polarization beams using strained GaAsP and a high power diode laser system showed photocathode lifetimes and intensity stabilities exceeding SAMPLE-III requirement, but helicity correlated beam effects require further development. The polarized injector is now in continuous operation, injecting highly polarized beams into the SHR for the BLAST commissioning runs.

REFERENCES

1. B. Mueller, et al., Phys. Lett. **78**, 3824 (1997).
2. D. T. Spayde, et al., Phys. Lett. **84**, 1106 (2000).
3. R. Hasty, *et al.*, Science **290** 2117 (2000).
4. Spectra-Physics, Opto Power diode laser model OPC-DO60-mmm-FC.
5. Electro-Optical Prod. Co., model QC-70I, max. extinction ratio 280:1.
6. Lab. of Spin-Polarized Electron Spectroscopy, State Technical University, St. Petersburg, Russia.
7. T. Averett, *et al.*, NIM A 438 (1999) 246.
8. E. Tsentalovich, *et al.*, "Helicity-Correlated Effects for SAMPLE Experiment", these proceedings.
9. Bandwidth Semiconductor Inc., Bedford, NH.
10. T. Maruyama *et al*, Nucl. Instrum. Methods A492, 199 (2002).
11. E. Tsentalovich, *et al.*, "High power diode laser system for SHR", these proceedings.
12. W. Franklin, *et al.*, "The MIT-Bates Compton Polarimeter", these proceedings.
13. T. Zwart, *et al.*, "Transmission Polarimetry for electron beams", these proceedings.

Appendix A

PHENIX Collaboration

MEMBERS

S.S. Adler,[5] S. Afanasiev,[17] C. Aidala,[5] N.N. Ajitanand,[43] Y. Akiba,[20] J. Alexander,[43] R. Amirikas,[12] L. Aphecetche,[45] S.H. Aronson,[5] R. Averbeck,[44] T.C. Awes,[34] R. Azmoun,[44] V. Babintsev,[15] A. Baldisseri,[39] K.N. Barish,[6] P.D. Barnes,[27] B. Bassalleck,[32] S. Bathe,[29] S. Batsouli,[10] V. Baublis,[36] A. Bazilevsky,[15,38] S. Belikov,[15,16] M.J. Bennett,[27] Y. Berdnikov,[40] S. Bhagavatula,[16] J.G. Boissevain,[27] H. Borel,[39] S. Borenstein,[25] S. Botelho,[41] M.L. Brooks,[27] D.S. Brown,[33] N. Bruner,[32] D. Bucher,[29] H. Buesching,[29] V. Bumazhnov,[15] G. Bunce,[5,38] J.M. Burward-Hoy,[26,44] S. Butsyk,[36,44] X. Camard,[45] J.-S. Chai,[18] P. Chand,[4] J. Chang,[6] W.C. Chang,[2] S. Chernichenko,[15] C.Y. Chi,[10] J. Chiba,[20] M. Chiu,[10] I.J. Choi,[53] J. Choi,[19] R.K. Choudhury,[4] T. Chujo,[5,49] V. Cianciolo,[34] Y. Cobigo,[39] B.A. Cole,[10] P. Constantin,[16] D.G. D'Enterria,[45] G. David,[5] H. Delagrange,[45] A. Denisov,[15] A. Deshpande,[38] E.J. Desmond,[5] A. Devismes,[44] O. Dietzsch,[41] O. Drapier,[25] A. Drees,[44] R. du Rietz,[28] A. Durum,[15] D. Dutta,[4] Y.V. Efremenko,[34] K. El Chenawi,[50] A. Enokizono,[14] H. En'yo,[24,37,38] S. Esumi,[49] L. Ewell,[5] T. Ferdousi,[6] D.E. Fields,[32] F. Fleuret,[25] S.L. Fokin,[23] B.D. Fox,[38] Z. Fraenkel,[52] J.E. Frantz,[10] A. Franz,[5] A.D. Frawley,[12] S.-Y. Fung,[6] S. Garpman,[28,*] T.K. Ghosh,[50] A. Glenn,[46] A.L. Godoi,[41] G. Gogiberidze,[46] M. Gonin,[25] J. Gosset,[39] Y. Goto,[38] R. Granier de Cassagnac,[25] N. Grau,[16] S.V. Greene,[50] M. Grosse Perdekamp,[38] W. Guryn,[5] H.-Å. Gustafsson,[28] T. Hachiya,[14] J.S. Haggerty,[5] H. Hamagaki,[9] A.G. Hansen,[27] E.P. Hartouni,[26] M. Harvey,[5] R. Hayano,[9] N. Hayashi,[37] X. He,[13] M. Heffner,[26] T.K. Hemmick,[44] J.M. Heuser,[44] M. Hibino,[51] J.C. Hill,[16] D.S. Ho,[53] W. Holzmann,[43] K. Homma,[14] B. Hong,[22] A. Hoover,[33] T. Ichihara,[37,38] V.V. Ikonnikov,[23] K. Imai,[24,37] L.D. Isenhower,[1] M. Ishihara,[37,38] M. Issah,[43] A. Isupov,[17] B.V. Jacak,[38,44] W.Y. Jang,[22] Y. Jeong,[19] J. Jia,[44] O. Jinnouchi,[37] B.M. Johnson,[5] S.C. Johnson,[26,44] K.S. Joo,[30] D. Jouan,[35] S. Kametani,[9,51] N. Kamihara,[47,37] J.H. Kang,[53] M. Kann,[36] S.S. Kapoor,[4] K. Katou,[51] S. Kelly,[10] B. Khachaturov,[52] A. Khanzadeev,[36] J. Kikuchi,[51] D.H. Kim,[30] D.J. Kim,[53] D.W. Kim,[19] E. Kim,[42] G.-B. Kim,[25] H.J. Kim,[53] Y.G. Kim,[53] W.W. Kinnison,[27] E. Kistenev,[5] A. Kiyomichi,[49] K. Kiyoyama,[31] C. Klein-Boesing,[29] H. Kobayashi,[37,38] L. Kochenda,[36] V. Kochetkov,[15] D. Koehler,[32] T. Kohama,[14] M. Kopytine,[44] D. Kotchetkov,[6] A. Kozlov,[52] P.J. Kroon,[5] C.H. Kuberg,[1,27] K. Kurita,[37,38] Y. Kuroki,[49] M.J. Kweon,[22] Y. Kwon,[53] G.S. Kyle,[33] R. Lacey,[43] V. Ladygin,[17] J.G. Lajoie,[16] J. Lauret,[43] A. Lebedev,[16,23] S. Leckey,[44] D.M. Lee,[27] S. Lee,[19] M.J. Leitch,[27] X.H. Li,[6] Z. Li,[7,37] D.J. Lim,[53] H. Lim,[42] A. Litvinenko,[17] M.X. Liu,[27] Y. Liu,[35] C.F. Maguire,[50] Y.I. Makdisi,[5] A. Malakhov,[17] V.I. Manko,[23] Y. Mao,[7,37] G. Martinez,[45] M.D. Marx,[44]

H. Masui,[49] F. Matathias,[44] T. Matsumoto,[9,51] P.L. McGaughey,[27] E. Melnikov,[15] F. Messer,[44] Y. Miake,[49] J. Milan,[43] T.E. Miller,[50] A. Milov,[44,52] S. Mioduszewski,[5,46] R.E. Mischke,[27] G.C. Mishra,[13] J.T. Mitchell,[5] A.K. Mohanty,[4] D.P. Morrison,[5] J.M. Moss,[27] F. Mühlbacher,[44] D. Mukhopadhyay,[52] M. Muniruzzaman,[6] J. Murata,[37,38] S. Nagamiya,[20] Y. Nagasaka,[31] J.L. Nagle,[10] T. Nakamura,[14] B.K. Nandi,[6] M. Nara,[49] J. Newby,[46] P. Nilsson,[28] S. Nishimura,[9] A.S. Nyanin,[23] J. Nystrand,[28] E. O'Brien,[5] C.A. Ogilvie,[16] H. Ohnishi,[5,14] I.D. Ojha,[3,50] K. Okada,[37] M. Ono,[49] V. Onuchin,[15] A. Oskarsson,[28] I. Otterlund,[28] K. Oyama,[9,48] K. Ozawa,[9] D. Pal,[52] A.P.T. Palounek,[27] V.S. Pantuev,[44] V. Papavassiliou,[33] J. Park,[42] A. Parmar,[32] S.F. Pate,[33] T. Peitzmann,[29] J.-C. Peng,[27] V. Peresedov,[17] A.N. Petridis,[16] C. Pinkenburg,[5,43] R.P. Pisani,[5] F. Plasil,[34] K. Pope,[46] M.L. Purschke,[5] A. Purwar,[44] J. Rak,[16] I. Ravinovich,[52] K.F. Read,[34,46] M. Reuter,[44] K. Reygers,[29] V. Riabov,[36,40] Y. Riabov,[36] G. Roche,[8] A. Romana,[25] M. Rosati,[16] A.A. Rose,[50] P. Rosnet,[8] S.S. Ryu,[53] M.E. Sadler,[1] N. Saito,[37,38] T. Sakaguchi,[9,51] M. Sakai,[31] S. Sakai,[49] V. Samsonov,[36] L. Sanfratello,[32] R. Santo,[29] H.D. Sato,[24,37] S. Sato,[5,49] S. Sawada,[20] Y. Schutz,[45] V. Semenov,[15] R. Seto,[6] M.R. Shaw,[1,27] T.K. Shea,[5] T.-A. Shibata,[37,47] K. Shigaki,[20] T. Shiina,[27] Y.H. Shin,[53] C.L. Silva,[41] D. Silvermyr,[27,28] K.S. Sim,[22] J. Simon-Gillo,[27] C.P. Singh,[3] V. Singh,[3] M. Sivertz,[5] A. Soldatov,[15] R.A. Soltz,[26] W.E. Sondheim,[27] S. Sorensen,[34,46] I.V. Sourikova,[5] F. Staley,[39] P.W. Stankus,[34] E. Stenlund,[28] M. Stepanov,[33] A. Ster,[21] S.P. Stoll,[5] M. Sugioka,[37,47] T. Sugitate,[14] J.P. Sullivan,[27] M. Suzuki,[49] E.M. Takagui,[41] A. Taketani,[37,38] M. Tamai,[51] K.H. Tanaka,[20] Y. Tanaka,[31] K. Tanida,[37] M.J. Tannenbaum,[5] P. Tarján,[11] J.D. Tepe,[1,27] T.L. Thomas,[32] J. Tojo,[24,37] H. Torii,[24,37] R.S. Towell,[1,27] I. Tserruya,[52] H. Tsuruoka,[49] S.K. Tuli,[3] H. Tydesjö,[28] N. Tyurin,[15] T. Ushiroda,[31] H.W. van Hecke,[27] C. Velissaris,[33] J. Velkovska,[5,44] M. Velkovsky,[44] V. Veszprémi,[11] L. Villatte,[46] A.A. Vinogradov,[23] M.A. Volkov,[23] E. Vznuzdaev,[36] X.R. Wang,[13] Y. Watanabe,[37,38] S.N. White,[5] F.K. Wohn,[16] C.L. Woody,[5] W. Xie,[6,52] Y. Yang,[7] A. Yanovich,[15] S. Yokkaichi,[37,38] G.R. Young,[34] I.E. Yushmanov,[23] W.A. Zajc,[10] C. Zhang,[10] Z. Zhang,[44] S. Zhou,[7,52] and L. Zolin,[17]

PHENIX COLLABORATION AFFILIATIONS

1 Abilene Christian University, Abilene, TX 79699, USA

2 Institute of Physics, Academia Sinica, Taipei 11529, Taiwan

3 Department of Physics, Banaras Hindu University, Varanasi 221005, India

4 Bhabha Atomic Research Centre, Bombay 400 085, India

5 Brookhaven National Laboratory, Upton, NY 11973-5000, USA

6 University of California - Riverside, Riverside, CA 92521, USA

7 China Institute of Atomic Energy (CIAE), Beijing, People's Republic of China

8 Laboratoire de Physique Corpusculaire (LPC), Univsersite de Clermont-Ferrand, 63170 Aubiere, Clermont-Ferrand, France

9 Center for Nuclear Study, Graduate School of Science, University of Tokyo, 7-3-1 Hongo, Bunkyo, Tokyo 113-0033, Japan

10 Columbia University, New York, NY 10027 and Nevis Laboratories, Irvington, NY 10533, USA
11 Debrecen University, H-4010 Debrecen, Egyetem tér 1, Hungary
12 Florida State University, Tallahassee, FL 32306, USA
13 Georgia State University, Atlanta, GA 30303, USA
14 Hiroshima University, Kagamiyama, Higashi-Hiroshima 739-8526, Japan
15 Institute for High Energy Physics (IHEP), Protvino, Russia
16 Iowa State University, Ames, IA 50011, USA
17 Joint Institute for Nuclear Research, 141980 Dubna, Moscow Region, Russia
18 Cyclotron Application Laboratory, KAERI, Seoul, South Korea
19 Kangnung National University, Kangnung 210-702, South Korea
20 KEK, High Energy Accelerator Research Organization, Tsukuba-shi, Ibaraki-ken 305-0801, Japan
21 KFKI Research Institute for Particle and Nuclear Physics (RMKI), H-1525 Budapest 114, POBox 49, Hungary
22 Korea University, Seoul, 136-701, Korea
23 Russian Research Center "Kurchatov Institute", Moscow, Russia
24 Kyoto University, Kyoto 606, Japan
25 Laboratoire Leprince-Ringuet, Ecole Polytechnique, CNRS-IN2P3, Route de Saclay, F-91128, Palaiseau, France
26 Lawrence Livermore National Laboratory, Livermore, CA 94550, USA
27 Los Alamos National Laboratory, Los Alamos, NM 87545, USA
28 Department of Physics, Lund University, Box 118, SE-221 00 Lund, Sweden
29 Institut fuer Kernphysik, University of Muenster, D-48149 Muenster, Germany
30 Myongji University, Yongin, Kyonggido 449-728, Korea
31 Nagasaki Institute of Applied Science, Nagasaki-shi, Nagasaki 851-0193, Japan
32 University of New Mexico, Albuquerque, NM, USA
33 New Mexico State University, Las Cruces, NM 88003, USA
34 Oak Ridge National Laboratory, Oak Ridge, TN 37831, USA
35 IPN-Orsay, Universite Paris Sud, CNRS-IN2P3, BP1, F-91406, Orsay, France
36 PNPI, Petersburg Nuclear Physics Institute, Gatchina, Russia
37 RIKEN (The Institute of Physical and Chemical Research), Wako, Saitama 351-0198, JAPAN
38 RIKEN BNL Research Center, Brookhaven National Laboratory, Upton, NY 11973-5000, USA
39 Dapnia, CEA Saclay, Bat. 703, F-91191, Gif-sur-Yvette, France
40 St. Petersburg State Technical University, St. Petersburg, Russia
41 Universidade de São Paulo, Instituto de Física, Caixa Postal 66318, São Paulo CEP05315-970, Brazil
42 System Electronics Laboratory, Seoul National University, Seoul, South Korea

43 Chemistry Department, Stony Brook University, SUNY, Stony Brook, NY 11794-3400, USA

44 Department of Physics and Astronomy, Stony Brook University, SUNY, Stony Brook, NY 11794, USA

45 SUBATECH (Ecole des Mines de Nantes, IN2P3/CNRS, Universite de Nantes) BP 20722 - 44307, Nantes, France

46 University of Tennessee, Knoxville, TN 37996, USA

47 Department of Physics, Tokyo Institute of Technology, Tokyo, 152-8551, Japan

48 University of Tokyo, Tokyo, Japan

49 Institute of Physics, University of Tsukuba, Tsukuba, Ibaraki 305, Japan

50 Vanderbilt University, Nashville, TN 37235, USA

51 Waseda University, Advanced Research Institute for Science and Engineering, 17 Kikui-cho, Shinjuku-ku, Tokyo 162-0044, Japan

52 Weizmann Institute, Rehovot 76100, Israel

53 Yonsei University, IPAP, Seoul 120-749, Korea

* Deceased.

ACKNOWLEDGEMENTS

We thank the staff of the RHIC Project, Collider-Accelerator, and Physics Departments at Brookhaven National Laboratory and the staff of the other PHENIX participating institutions for their vital contributions. We acknowledge support from the Department of Energy, National Science Foundation, Research Foundation of SUNY, and Dean of the College of Arts and Sciences, Vanderbilt University (U.S.A), Ministry of Education, Culture, Sports, Science, and Technology and the Japan Society for the Promotion of Science (Japan), Russian Academy of Science, Ministry of Atomic Energy of Russian Federation, Ministry of Industry, Science, and Technologies of Russian Federation (Russia), Bundesministerium fuer Bildung und Forschung, Deutscher Akademischer Austausch Dienst, and Alexander von Humboldt Stiftung (Germany), VR and the Wallenberg Foundation (Sweden), Conselho Nacional de Desenvolvimento Científico e Tecnológico and Fundação de Amparo à Pesquisa do Estado de São Paulo (Brazil), Natural Science Foundation of China (People's Republic of China), IN2P3/CNRS (France), Hungarian National Science Fund, OTKA (Hungary), Department of Atomic Energy and Department of Science and Technology (India), Israel Science Foundation (Israel), Korea Research Foundation and Center for High Energy Physics (Korea), the U.S. Civilian Research and Development Foundation for the Independent States of the Former Soviet Union, the US-Hungarian NSF-OTKA-MTA, and the US-Israel Binational Science Foundation.

MEMBERS

J. Adams[3], C. Adler[11], Z. Ahammed[23], C. Allgower[12], J. Amonett[14], B.D. Anderson[14], M. Anderson[5], G.S. Averichev[9], J. Balewski[12], O. Barannikova[9,23], L.S. Barnby[14], J. Baudot[13], S. Bekele[20], V.V. Belaga[9], R. Bellwied[31], J. Berger[11], H. Bichsel[30], A. Billmeier[31], L.C. Bland[2], C.O. Blyth[3], B.E. Bonner[24], A. Boucham[26], A. Brandin[18], A. Bravar[2], R.V. Cadman[1], H. Caines[33], M. Calderón de la Barca Sánchez[2], A. Cardenas[23], J. Carroll[15], J. Castillo[15], M. Castro[31], D. Cebra[5], P. Chaloupka[20], S. Chattopadhyay[31], Y. Chen[6], S.P. Chernenko[9], M. Cherney[8], A. Chikanian[33], B. Choi[28], W. Christie[2], J.P. Coffin[13], T.M. Cormier[31], M.M. Corral[16], J.G. Cramer[30], H.J. Crawford[4], A.A. Derevschikov[22], L. Didenko[2], T. Dietel[11], J.E. Draper[5], V.B. Dunin[9], J.C. Dunlop[33], V. Eckardt[16], L.G. Efimov[9], V. Emelianov[18], J. Engelage[4], G. Eppley[24], B. Erazmus[26], P. Fachini[2], V. Faine[2], J. Faivre[13], R. Fatemi[12], K. Filimonov[15], E. Finch[33], Y. Fisyak[2], D. Flierl[11], K.J. Foley[2], J. Fu[15,32], C.A. Gagliardi[27], N. Gagunashvili[9], J. Gans[33], L. Gaudichet[26], M. Germain[13], F. Geurts[24], V. Ghazikhanian[6], O. Grachov[31], M. Guedon[13], S.M. Guertin[6], E. Gushin[18], T.J. Hallman[2], D. Hardtke[15], J.W. Harris[33], M. Heinz[33], T.W. Henry[27], S. Heppelmann[21], T. Herston[23], B. Hippolyte[13], A. Hirsch[23], E. Hjort[15], G.W. Hoffmann[28], M. Horsley[33], H.Z. Huang[6], T.J. Humanic[20], G. Igo[6], A. Ishihara[28], Yu.I. Ivanshin[10], P. Jacobs[15], W.W. Jacobs[12], M. Janik[29], I. Johnson[15], P.G. Jones[3], E.G. Judd[4], M. Kaneta[15], M. Kaplan[7], D. Keane[14], J. Kiryluk[6], A. Kisiel[29], J. Klay[15], S.R. Klein[15], A. Klyachko[12], T. Kollegger[11], A.S. Konstantinov[22], M. Kopytine[14], L. Kotchenda[18], A.D. Kovalenko[9], M. Kramer[19], P. Kravtsov[18], K. Krueger[1], C. Kuhn[13], A.I. Kulikov[9], G.J. Kunde[33], C.L. Kunz[7], R.Kh. Kutuev[10], A.A. Kuznetsov[9], M.A.C. Lamont[3], J.M. Landgraf[2], S. Lange[11], C.P. Lansdell[28], B. Lasiuk[33], F. Laue[2], J. Lauret[2], A. Lebedev[2], R. Lednický[9], V.M. Leontiev[22], M.J. LeVine[2], Q. Li[31], S.J. Lindenbaum[19], M.A. Lisa[20], F. Liu[32], L. Liu[32], Z. Liu[32], Q.J. Liu[30], T. Ljubicic[2], W.J. Llope[24], H. Long[6], R.S. Longacre[2], M. Lopez-Noriega[20], W.A. Love[2], T. Ludlam[2], D. Lynn[2], J. Ma[6], D. Magestro[20], R. Majka[33], S. Margetis[14], C. Markert[33], L. Martin[26], J. Marx[15], H.S. Matis[15], Yu.A. Matulenko[22], T.S. McShane[8], F. Meissner[15], Yu. Melnick[22], A. Meschanin[22], M. Messer[2], M.L. Miller[33], Z. Milosevich[7], N.G. Minaev[22], J. Mitchell[24], C.F. Moore[28], V. Morozov[15], M.M. de Moura[31], M.G. Munhoz[25], J.M. Nelson[3], P. Nevski[2], V.A. Nikitin[10], L.V. Nogach[22], B. Norman[14], S.B. Nurushev[22], G. Odyniec[15], A. Ogawa[2], V. Okorokov[18], M. Oldenburg[16], D. Olson[15], G. Paic[20], S.U. Pandey[31], Y. Panebratsev[9], S.Y. Panitkin[2], A.I. Pavlinov[31], T. Pawlak[29], V. Perevoztchikov[2], W. Peryt[29], V.A Petrov[10], M. Planinic[12], J. Pluta[29],

N. Porile[23], J. Porter[2], A.M. Poskanzer[15], E. Potrebenikova[9], D. Prindle[30], C. Pruneau[31], J. Putschke[16], G. Rai[15], G. Rakness[12], O. Ravel[26], R.L. Ray[28], S.V. Razin[9,12], D. Reichhold[23], J.G. Reid[30], G. Renault[26], F. Retiere[15], A. Ridiger[18], H.G. Ritter[15], J.B. Roberts[24], O.V. Rogachevski[9], J.L. Romero[5], A. Rose[31], C. Roy[26], V. Rykov[31], I. Sakrejda[15], S. Salur[33], J. Sandweiss[33], I. Savin[10], J. Schambach[28], R.P. Scharenberg[23], N. Schmitz[16], L.S. Schroeder[15], A. Schüttauf[16], K. Schweda[15], J. Seger[8], P. Seyboth[16], E. Shahaliev[9], K.E. Shestermanov[22], S.S. Shimanskii[9], F. Simon[16], G. Skoro[9], N. Smirnov[33], R. Snellings[15], P. Sorensen[6], J. Sowinski[12], H.M. Spinka[1], B. Srivastava[23], E.J. Stephenson[12], R. Stock[11], A. Stolpovsky[31], M. Strikhanov[18], B. Stringfellow[23], C. Struck[11], A.A.P. Suaide[31], E. Sugarbaker[20], C. Suire[2], M. Šumbera[20], B. Surrow[2], T.J.M. Symons[15], A. Szanto de Toledo[25], P. Szarwas[29], A. Tai[6], J. Takahashi[25], A.H. Tang[15], D. Thein[6], J.H. Thomas[15], M. Thompson[3], S. Timoshenko[18], M. Tokarev[9], M.B. Tonjes[17], T.A. Trainor[30], S. Trentalange[6], R.E. Tribble[27], V. Trofimov[18], O. Tsai[6], T. Ullrich[2], D.G. Underwood[1], G. Van Buren[2], A.M. Vander Molen[17], I.M. Vasilevski[10], A.N. Vasiliev[22], S.E. Vigdor[12], S.A. Voloshin[31], M. Vznuzdaev[18], F. Wang[23], H. Ward[28], J.W. Watson[14], R. Wells[20], G.D. Westfall[17], C. Whitten Jr. [6], H. Wieman[15], R. Willson[20], S.W. Wissink[12], R. Witt[33], J. Wood[6], N. Xu[15], Z. Xu[2], A.E. Yakutin[22], E. Yamamoto[15], J. Yang[6], P. Yepes[24], V.I. Yurevich[9], Y.V. Zanevski[9], I. Zborovský[9], H. Zhang[33], W.M. Zhang[14], R. Zoulkarneev[10], A.N. Zubarev[9]

STAR COLLABORATION AFFILIATIONS

1 Argonne National Laboratory, Argonne, Illinois 60439
2 Brookhaven National Laboratory, Upton, New York 11973
3 University of Birmingham, Birmingham, United Kingdom
4 University of California, Berkeley, California 94720
5 University of California, Davis, California 95616
6 University of California, Los Angeles, California 90095
7 Carnegie Mellon University, Pittsburgh, Pennsylvania 15213
8 Creighton University, Omaha, Nebraska 68178
9 Laboratory for High Energy (JINR), Dubna, Russia
10 Particle Physics Laboratory (JINR), Dubna, Russia
11 University of Frankfurt, Frankfurt, Germany
12 Indiana University, Bloomington, Indiana 47408
13 Institut de Recherches Subatomiques, Strasbourg, France
14 Kent State University, Kent, Ohio 44242
15 Lawrence Berkeley National Laboratory, Berkeley, California 94720
16 Max-Planck-Institut fuer Physik, Munich, Germany
17 Michigan State University, East Lansing, Michigan 48824
18 Moscow Engineering Physics Institute, Moscow Russia
19 City College of New York, New York City, New York 10031

20 Ohio State University, Columbus, Ohio 43210
21 Pennsylvania State University, University Park, Pennsylvania 16802
22 Institute of High Energy Physics, Protvino, Russia
23 Purdue University, West Lafayette, Indiana 47907
24 Rice University, Houston, Texas 77251
25 Universidade de Sao Paulo, Sao Paulo, Brazil
26 SUBATECH, Nantes, France
27 Texas A&M University, College Station, Texas 77843
28 University of Texas, Austin, Texas 78712
29 Warsaw University of Technology, Warsaw, Poland
30 University of Washington, Seattle, Washington 98195
31 Wayne State University, Detroit, Michigan 48201
32 Institute of Particle Physics, CCNU (HZNU), Wuhan, 430079 China
33 Yale University, New Haven, Connecticut 06520

Muon *g*-2 Collaboration

MEMBERS

E. P. Sichtermann[11], G. W. Bennett[2], B. Bousquet[9], H. N. Brown[2], G. Bunce[2], R. M. Carey[1], P. Cushman[9], G. T. Danby[2], P. T. Debevec[7], M. Deile[11], H. Deng[11], W. Deninger[7], S. K. Dhawan[11], V. P. Druzhinin[3], L. Duong[9], E. Efstathiadis[1], F. J. M. Farley[11], G. V. Fedotovich[3], S. Giron[9], F. E. Gray[7], D. Grigoriev[3], M. Grosse-Perdekamp[11], A. Grossmann[6], M. F. Hare[1], D. W. Hertzog[7], X. Huang[1], V. W. Hughes[11], M. Iwasaki[10], K. Jungmann[5], D. Kawall[11], B. I. Khazin[3], J. Kindem[9], F. Krienen[1], I. Kronkvist[9], A. Lam[1], R. Larsen[2], Y. Y. Lee[2], I. Logashenko[1,3], R. McNabb[9], W. Meng[2], J. Mi[2], J. P. Miller[1], W. M. Morse[2], D. Nikas[2], C. J. G. Onderwater[7], Y. Orlov[4], C. S. Özben[2], J. M. Paley[1], Q. Peng[1], C. C. Polly[7], J. Pretz[11], R. Prigl[2], G. zu Putlitz[6], T. Qian[9], S. I. Redin[3,11], O. Rind[1], B. L. Roberts[1], N. Ryskulov[3], P. Shagin[9], Y. K. Semertzidis[2], Y. M. Shatunov[3], E. Solodov[3], M. Sossong[7], A. Steinmetz[11], L. R. Sulak[1], A. Trofimov[1], D. Urner[7], P. von Walter[6], D. Warburton[2], and A. Yamamoto[8]

MUON *g*-2 COLLABORATION AFFILIATIONS

1 Department of Physics, Boston University, Boston, Massachusetts 02215, USA
2 Brookhaven National Laboratory, Upton, New York 11973, USA
3 Budker Institute of Nuclear Physics, Novosibirsk, Russia
4 Newman Laboratory, Cornell University, Ithaca, New York 14853, USA
5 Kernfysisch Versneller Instituut, Rijksuniversiteit Groningen, 9747 AA Groningen, The Netherlands
6 Physikalisches Institut der Universität Heidelberg, 69120 Heidelberg, Germany
7 Department of Physics, University of Illinois at Urbana-Champaign, Illinois 61801, USA
8 KEK, High Energy Accelerator Research Organization, Tsukuba, Ibaraki 305-0801, Japan
9 Department of Physics, University of Minnesota, Minneapolis, Minnesota 55455, USA
10 Tokyo Institute of Technology, Tokyo, Japan
11 Department of Physics, Yale University, New Haven, Connecticut 06520, USA

Appendix B

Spin 2002 Symposium Agenda

Sunday September 8, 2002

Special Student Tutorial and Workshop

09:30 – 11:00 Basics of QCD spin physics, D. Boer
11:00 – 12:00 20 years of polarized lepton-nucleon scattering, E. Hughes,
12:00 - 13:00 Lunch
13:00 - 14:00 Heavy-Ion physics at RHIC, D. Morrison
14:00 - 15:00 Spin physics at RHIC, M. Grosse-Perdekamp
Coffee Break
15:30 – 16:30 Basics of polarized-proton acceleration, W. MacKay
16:30 – 17:30 Techniques for polarimetry in pp and ep physics, H. Spinka

18:00 – 21:00 Symposium Registration and Reception

The Symposium

Monday September 9, 2002

8:00	Registration continues	
9:00	Opening Session,	Chair T. Kirk, BNL
9:00 – 9:10	Welcoming Remarks, P. Paul (BNL Directorate)	
9:10 – 9:15	Welcoming remarks, T. Roser, BNL (International Committee Chair)	
9:15 – 9:20	Welcome & House Keeping, Y. Makdisi, BNL (Symposium Chair)	
9:20 – 10:15	Opening presentation	
	A Beautiful Spin, Xiang-Dong Ji, U. Maryland	
10:15 – 10:30	Coffee Break	
10:30 – 13:00	Plenary Session,	Chair V. Hughes, Yale.
	New Results from the Muon g-2 Experiment, E. Sichtermann, Yale	
	Semi Inclusive Deep-Inelastic-Scattering, A. Miller, TRIUMF	
	Experimental Verification of the GDH Sum Rule, K. Helbing, Erlangen	
13:00 – 14:00	Lunch	
14:00 – 16:30	Parallel Sessions	
	Quark and Gluon Polarization (I) Chair: A. Bruell, MIT	
	Spin & Hadron Dynamics (I), Chair: L. Trueman, BNL	
	Pionic 0^- State and Nuclear Structure, Chair: E. Stepehenson, IUCF	
	Polarized Sources and Targets, Chair: G. Court, Liverpool	
16:30 – 16:45	Coffee Break	

16:45 – 18:30 Parallel sessions
Quark and Gluon Polarization (II), Chair: W. Vogelsang, BNL/RBRC
Spin and Hadron Dynamics (II), Chair: B. von Przewoski, IUCF
Nucleon Structure: GDH Experiments, Chair: K. de Jager, Jlab
Acceleration & Storage of Polarized Beams (I), Chair: A. Luccio, BNL

Tuesday September 10, 2002

Solovianov Memorial Session
9:00 – 10:45 Plenary Session Chair: N. Tyurin, ITEP
Inclusive Asymmetries, M. N. Ukhanov, IHEP
Samll Angle Scattering of Polarized Protons, B. Kopeliovich, Max Planck Inst.
10:45 – 11:00 Coffee Break
11:00 – 12:40 Plenary Session Chair: V. Soergel, Heidelberg
Spin and The Three-Nucleon Force, B. von Przewoski, IUCF
Nucleon Electromagnetic Form Factors and Densities, J. Kelly, U. Maryland
12:40 – 14:00 Lunch
14:00 – 16:30 Parallel Sessions
Quark and Gluon Polarization (III), Chair: M. Gross-Perdekamp, RBRC/Illinois
Nucleon EM Form Factors, Chair: J. Kelly, Maryland
Few Body Physics & Nuclear Properties, Chair: H. Sakai, Tokyo
Polarized Ion Sources, Chair: A. Zelenski, BNL
Future experiments, Chair: D-W Nowak, DESY Zeuthen
16:30 – 16:45 Coffee Break
16:45 – 18:30 Parallel Sessions
Nucleon Structure: Single Spin Effects (I), Chair: N. Saito, Kyoto
Spin and Hadron Dynamics (III), Chair: W. Guryn, BNL
Photon Experiments etc. Chair: K. de Jager, JLab
Polarimetry proton and electron beams (I), Chair: Ed Stephenson, IUCF

Wednesday September 11, 2002

Nucleon Spin Structure and RHIC
9:00 – 10:45 Plenary Session Chair: K. Imai, Kyoto
Nucleon Spin Structure functions (g^1, g^2) From Polarized Inclusive Scattering, T. Averett, William & Mary
Spin Physics at RHIC, L. Bland, BNL
10:45 – 11:00 Coffee Break
11:00 – 12:40 Plenary Session Chair: Y. Shatunov, Novosibirsk
Exploring Nucleon Spin Structure in Longitudinally Polarized Collisions, M. Stratmann, Regensburg U.
Acceleration of Polarized Protons at RHIC, H. Huang, BNL
12:40 – 14:00 Lunch

14:00 – 16:30 Parallel Sessions
Spin and Symmetry (I), Chair: W. van Oers, TRIUMF/ Manitoba
Nucleon Structure: Perspectives on Current Spin Experiments, Chair: Richard Milner, MIT-Bates
Acceleration and Storage of Polarized Beams (II), Chair: E. Courant, BNL
Polarized Solid Targets Chair: D. Crabb, Virginia
16:30 – 18:30 RHIC tours

Thursday September 12, 2002

9:00 – 10:45 Plenary Session Chair: C. Prescott, SLAC
Results on DIS Exclusives and Generalized Parton Distributions, M. Vanderhaeghen, Mainz
Parity Violating Electron Scattering, K. Kumar, UMass
10:45 – 11:00 Coffee Break
11:00 – 12:40 Plenary Session Chair: D. Barber, DESY
Low Energy Photon Beam Experiments, A. Sandorfi, BNL
Acceleration of Polarized Protons and Deuterons at COSY, A. Lehrach, Julich.
12:40 – 14:00 Lunch
13:00 – 19:00 Excursion to the Rose Center and Planetarium NYC
20:00 – 21:00 buses return to Hotels / BNL

Friday September 13, 2002

9:00 – 10:45 Plenary Session Chair: R. Jaffe, MIT
Spin On The Lattice, K. Orginos, RBRC
Single Spin Asymmetries and Transversity, P. Ratcliff, dell'Insubria
10:45 – 11:00 Coffee Break
11:00 – 12:40 Plenary Session Chair: W. Haeberli, Wisconsin
Polarized Jets and Storage Cell Targets for Storage Rings, E. Steffens, Erlangen
Parity Violation in pp and pn Experiments, W. D. Ramsay, Manitoba /TRIUMF
12:40 – 14:00 Lunch
14:00 – 16:30 Parallel sessions
Spin and Symmetry (II), Chair: P. Souder, Syracuse
Nucleon Structure: Single Spin effects (II), Chair: A.Bruell, MIT-Bates
Acceleration & Storage of Polarized Beams (III), Chair: A. Lehrach, Julich
Polarized ABS Internal Targets (I), Chair: T. Wise, Wisconsin
16:30 – 16:45 Coffee Break

16:45 – 18:30	Parallel Sessions
	Nucleon Structure: Single Spin Effects (III) Chair: W. Vogelsang, BNL/RBRC
	Polarized ABS Internal Targets (II), Chair: F. Rathmann, Julich
	Polarimery Protons and electron beams (II), Chair: K. Kurita, Rykkio
	Future Facilities, Chair: A. Desphande, RBRC
20:00	Symposium Banquet

Saturday September 14, 2002

9:00 – 10:30	Plenary Session Chair: S. Ozaki, BNL
	A Future Linear Collider with Polarized Beams; Searches for New Physics, G. Moortgat-Pick, DESY
	Laser-Polarized Noble Gases for Magnetic Resonance Imaging, G. Cates, Virginia
10:30 – 10:45	Coffee Break
10:45 – 11:45	Plenary Session Chair: A. Krisch, Michigan
	Summary of DUBNA-SPIN 01 Workshop, A. Efromov, Dubna
	Highlights from the Praha Workshop, M. Finger, Charles Univ.
11:45 – 12:45	Concluding talk
	Looking into the Future of Spin and QCD, J. Soffer, Marseille
12:45 - 13:00	Remarks by the Chair of the International Committee, T. Roser, BNL
13:00	Symposium ends

SPIN 2002 List of Participants

Leif Ahrens
Brookhaven National Laboratory
Building 911B
P O Box 5000
Upton, NY 11973
ahrens@bnl.gov

Igor G Alekseev
ITEP
B. Cheremushkinskaya 25
Moscow 117259
Russian Federation
Igor.Alekseev@itep.ru

Mauro Anselmino
Department of Theoretical Physics
Via Giuria 1
Torino 10125
Italy
anselmino@to.infn.it

David S Armstrong
College of William & Mary
P.O. Box 8795
Physics Dept.
Williamsburg, VA 23187-8795
armd@physics.wm.edu

Samuel H Aronson
Brookhaven National Laboratory
Physics Dept - Bldg 510A
P O Box 5000
Upton, NY 11973
aronsons@bnl.gov

Harut Avakian
Jefferson Lab
12000 Jefferson Ave.
MS 12H
Newport News, VA 23606
avakian@jlab.org

Todd D Averett
College of William and Mary
Dept. of Physics
P.O. Box 8795
Williamsburg, VA 23185
averett@physics.wm.edu

Mei Bai
Brookhaven National Laboratory
Building 911B
P O Box 5000
Upton, NY 11973
mbai@bnl.gov

Christopher Bade
Ohio University
Physics Dept. Clippinger Laboratories
Athens, OH 45701
bade@helios.phy.ohiou.edu

Jan T Balewski
Indiana University
2401 Milo B Sampson Ln
Bloomington, Indiana 47408
balewski@iucf.indiana.edu

Desmond P Barber
Deutsches Elektronen-Synchrotron
Notkestrasse 85
Hamburg 22607
Germany
mpybar@mail.desy.de

Kenneth N Barish
University of California at Riverside
Department of Physics
UC Riverside
Riverside, CA 92521
Kenneth.Barish@ucr.edu

Peter D. Barnes
Los Alamos National Laboratory
MS H846 P-25
Los Alamos, NM 87545
pdbarnes@lanl.gov

Frank Bauer
UC Riverside, Department of Physics
University of California
Riverside, CA 92521
fbauer@rcf.rhic.bnl.gov

Les C Bland
Brookhaven National Laboratory
Building 510A
P O Box 5000
Upton, NY 11973
bland@bnl.gov

Charles O Blyth
University of Birmingham
Edgbaston Birmingham B13 9SU
United Kingdom
c.o.blyth@bham.ac.uk

Daniel Boer
Free University Amsterdam
De Boelelaan 1081
Amsterdam NL-1081 HV
Netherlands
dboer@nat.vu.nl

Ingo Bojak
CSSM
The University of Adelaide
Adelaide SA 5005
Australia
ibojak@physics.adelaide.edu.au

Franco Bradamante
University of Trieste
via A., Valerio 2
Trieste, 34127
Italy
franco.bradamante@ts.infn.it

Alessandro Bravar
Brookhaven National Laboratory
Bldg. 510A
P O Box 5000
Upton, NY 11793
bravar@bnl.gov

David S Brown
New Mexico State University
c/o Brookhaven National Laboratory
P. O. Box 754
Upton, NY 11973
brownds@rcf2.rhic.bnl.gov

Antje Bruell
MIT
77 Massachusetts Avenue
Cambridge, MA 02139
abr@mitlns.mit.edu

Nichelle L Bruner
University of New Mexico
800 Yale Blvd NE
Albuquerque, NM 87131
bruner@dot.phys.unm.edu

Stephen L Bueltmann
Brookhaven National Laboratory
Building 510D
P O Box 5000
Upton, NY 11973
bueltmann@bnl.gov

Gerry Bunce
Brookhaven National Laboratory
Building 510A
P O Box 5000
Upton, NY 11973
bunce@bnl.gov

Nigel H Buttimore
University of Dublin
School of Mathematics
Trinity College, Dublin 2
Ireland
nhb@maths.tcd.ie

Robert V, Cadman
Argonne National Laboratory
Building 362/HEP
9700 South Cass Avenue
Argonne, IL 60439
rvc@anl.gov

Peter Cameron
Brookhaven National Laboratory
Building 817
P O Box 5000
Upton, NY 11973
cameron@bnl.gov

Anthony Caracappa
Brookhaven National Laboratory
Physics Dept - Building 510A
P O Box 5000
Upton, NY 11973
caracappa@legs.BNL.gov

Roger Carlini
Jefferson Lab
12000 Jefferson Avenue
Newport News, VA 23606
carlini@jlab.org

Gordon D. Cates Jr.
Department of Physics and Radiology
University of Virginia
Charlottesville, VA 22094
cates@virginia.edu

H. Cynthia Chiang
Univ. of Illinois at Urbana-Champaign
Loomis Laboratory of Physics
1110 W. Green St.
Urbana IL 61801
hchiang@uiuc.edu

William B Christie
Brookhaven National Laboratory
Building 510A
P O Box 5000
Upton, NY 11973
christie@bnl.gov

Eugene Chudakov
Jefferson Lab
12000 Jefferson Avenue
Newport News, VA 23606
gen@jlab.org

Marco Contalbrigo
INFN – Ferrara University
Via Paradiso 12
Ferrara 44100
Italy
contalbrigo@fe.infn.it

Mario Conte
INFN and Genoa University
Via Dodecaneso 33
Genova 16146
Italy
Mario.Conte@ge.infn.it

Ernest D Courant
Brookhaven National Laboratory
Building 911B
P O Box 5000
Upton, NY 11973
courant@bnl.gov

Geoffrey R Court
University of Liverpool
Physics Department Liverpool
None L69 7ZE United Kingdom
grc@hep.ph.liv.ac.uk

Donald G Crabb
University of Virginia, Physics Dept.
382 McCormick Rd.
Charlottesville, VA 22903
dgc3q@virginia.edu

Umberto D'Alesio
Dipartimento di Fisica
University of, Cagliari
Cittadella Universitaria di Monserrato
C.P. 170
Monserrato (CA) 09042
umberto.dalesio@ca.infn.it

Kees de Jager
Jefferson Lab
12000 Jefferson Avenue
Newport News, VA 23606
kees@jlab.org

Igor Degtyarev
Institute for High Energy Physics
Pobeda str. 1
Protvino Moscow 142281
Russian Federation
degtyarev@mx.ihep.su

Vladimir P Derenchuk
Indiana University Cyclotron Facility
2401 Milo B. Sampson Lane
Bloomington IN 47408-1398
Laddie@iucf.indiana.edu

Abhay L Deshpande
Riken BNL Research Center
Physics Dept.
P.O. Box 5000
Upton, NY 11973
abhay@bnl.gov

K.Vipuli Dharmawardane
Old Dominion University
4600 Elkhorn Ave.
Department of Physics
Norfolk, VA 23529
vipuli@physics.odu.edu

Elvio Di Salvo
Universita" di Genova –
Dipartimento di Fisica
Via dodecaneso 33
Genova GE I-16146
Italy
disalvo@ge.infn.it

Gail E Dodge
Old Dominion Univ. Physics Dept.
4600 Elkhorn Ave.
Norfolk, VA 23529
dodge@physics.odu.edu

Anatoli V Efremov
Joint Institute for Nuclear Research
Lab. Theor. Phys.
JINR Dubna Moscow reg. 141980
Russian Federation
efremov@thsun1.jinr.ru

Frank Ellinghaus
DESY Zeuthen
Platanenalle 6
Zeuthen 15738
Germany
Frank.Ellinghaus@desy.de

Ralf Engels
Forschungszentrum Jülich
Institut für Kernphysik
Leo-Brandt Str.
Jülich 52425
Germany
r.w.engels@fz-juelich.de

Geary Eppley
Rice University
Rice University MS315
Houston, TX 77005
eppley@physics.rice.edu

Vladimir Ershov
Joint Institute for Nuclear Research
141980 Dubna Moscow Region
Russian Federation
ershov@sunhe.jinr.ru

Dieter Eversheim
Helmholtz Institut fuer Strahlen- und
Kernphysik
University Bonn
Nussallee 14-16, Bonn D-53115
Germany
evershei@iskp.uni-bonn.de

Riccardo Fabbri
University of Ferrara
Department of Physics
Via Paradiso 12
Ferrara 44100
Italy
rfabbri@fe.infn.it

Manouchehr Farkhondeh
MIT-Bates
MIT Bates Accelerator Center
26, Manning Road
Middleton,, MA 01949
manouch@mit.edu

Douglas E Fields
University of New Mexico/RBRC
800 Yale NE
Albuquerque, NM 87131
fields@unm.edu

Victor Fimushkin
Joint Institute for Nuclear Research
Laboratory of High Energy
Dubna 141980
Moscow Region
Russian Federation
fimush@sunhe.jinr.ru

Miroslav Finger
Charles University
Faculty of, Mathematics and Physics
V Holesovickach 2
Praha 8 CZ-180 00
Czech Republic
finger@mbox.troja.mff.cuni.cz

Brendan D Fox
RIKEN-BNL Research Center
Brookhaven National Laboratory
Building 510A – Physics Dept
P O Box 5000
Upton, NY 11973-5000
bfox@bnl.gov

Yoshinori Fukao
Kyoto University
Kyoto city
Kyoto Prefecture 606-8502
Japan
fukao@nh.scphys.kyoto-u.ac.jp

Leonard Gamberg
Penn State-Berks
Tulpehocken Road
PO BOX 7009
Reading, PA 19610
lpg10@psu.edu

Ron Gill
Brookhaven National Laboratory
Building 510D
P O Box 5000
Upton, NY 11973
rongill@bnl.gov

Serguey V Goloskokov
Bogoliubov Lab. of Theor.Phys.
Joint Institute for Nuclear Research
Joliot Curie 6
Dubna Moscow Region 141980
Russian Federation
goloskkv@thsun1.jinr.ru

Mehrdad Goshtasbpour
Dept. of Physics
Shahid Beheshti University
Evin Tehran 19834
Iran
goshtasb@alborz.sbu.ac.ir

Yuji Goto
RIKEN / RBRC
Brookhaven National Lab
Building 510A
P O Box 5000
Upton, NY 11973
goto@bnl.gov

Oleg A Grachov
Wayne State University, Physics Dept.
666 West Hancock
Detroit, Mi 48201
grachov@physics.wayne.edu

Matthias Grosse Perdekamp
Brookhaven National Laboratory
Building 510
P O Box 5000
Upton, NY 11973
matthias@bnl.gov

Paolo Guazzoni
Milan University, Physics Department
via Celoria 16
Milano I 20133
Italy
paolo.guazzoni@mi.infn.it

Wlodek Guryn
Brookhaven National Laboratory
Building 510D
P O Box 5000
Upton, NY 11973
guryn@bnl.gov

Willy Haeberli
University of Wisconsin
1150 University Ave
Madison, WI 53706
haeberli@wisc.edu

Mark C Harvey
Brookhaven National Laboratory
Physics - Building 510C
P O Box 5000
Upton, NY 11973
mharvey@bnl.gov

Delia Hasch
INFN-LNF
via E. Fermi 40
Frascati, Rome 00044
Italy
delia.hasch@lnf.infn.it

Douglas K Hasell
M.I.T. 26-415
77 Massachusetts Ave
Cambridge, MA 02139
hasell@mit.edu

Richard D Hasty
Univ. of Illinois at Urbana-Champaign
505 W. Green St.
Urbana Il 61801
rhasty@uiuc.edu

Kazumi Hasuko
RIKEN/RBRC
2-1 Hirosawa
Wako Saitama 351-0198
Japan
hasuko@riken.go.jp

Kichiji Hatanaka
Research Center for Nuclear Physics
Osaka University
10-1 Mihogaoka
Ibaraki Osaka 567-0047
Japan
hatanaka@rcnp.osaka-u.ac.jp

Arata Hayashigaki
University of Tokyo
7-3-1 Hongo Bunkyou-ku
Tokyo 113-0033
Japan
arata@nt.phys.s.u-tokyo.ac.jp

Klaus Helbing
University of Erlangen
Physikalisches Institut
Erwin-Rommel-Str. 1
Erlangen Bavaria 91058
Germany
klaus.helbing@physik.uni-erlangen.de

Thomas W Henry
Texas A&M University, Physics Dept.
College Station, TX 77843-3366
thenry@physics.tamu.edu

Frank Hinterberger ISKP
Univ. Bonn
Nussallee 14-1
Bonn D-53115
Germany
fh@iskp.uni-bonn.de

Masanori Hirai
RIKEN
Hirosawa 2-1
Wako Saitama 351-0198
Japan
mhirai@rarfaxp.riken.go.jp

Ishaq H Hleiqawi
Ohio University, Dept. of Physics
Clippinger 251B
Athens, OH 45701
hleiqawi@helios.phy.ohiou.edu

Sam Hoblit
Brookhaven National Laboratory
Bldg 510A- LEGS Group
P O Box 5000
Upton, NY 11973
hoblit@bnl.gov

Haixin Huang
Brookhaven National Laboratory
C-AD - Building 911B
P O Box 5000
Upton, NY 11973
huanghai@bnl.gov

Emlyn W Hughes
California Institute of Technology
Kellogg Radiation Laboratory 304-38
Pasadena, CA 91125
emlyn@its.caltech.edu

Vernon W Hughes
Yale University, Physics Dept.
Rm. 465JWG
P. O. Box 208121
New Haven CT 06520 U.S.A
Vernon.Hughes@yale.edu

George J. Igo
Dept of Physics & Astronomy/UCLA
405 Hilgard Ave.
Los Angeles, CA 90095-1547
igo@physics.ucla.edu

Kenichi Imai
Kyoto University
Department of Physics
Kyoto 606-8224
Japan
imai@nh.scphys.kyoto-u.ac.jp

Donald Isenhower
Abilene Christian University
ACU Box 27963
Abilene, TX 79699
isenhowe@acu.edu

William W Jacobs
Indiana University Cyclotron Facility
2401 Milo B. Sampson Lane
Bloomington, Indiana 47408
jacobs@iucf.indiana.edu

Robert L Jaffe
MIT, 6-311
77 Mass Ave
Cambridge, MA 02139
jaffe@mit.edu

Barbara Jaeger
University of Regensburg
Institute for Theoretical Physics
D-93040 Regensburg 93040
Germany
barbara.jaeger@physik.uni-regensburg.de

Xiangdong Ji
University of, Maryland
Department of Physics
College Park MD 20742
xji@physics.umd.edu

Xiaodong Jiang
Rutgers University
MS 12 H12000 Jefferson Ave
Newport News, VA 23606
jiang@jlab.org

Osamu Jinnouchi
RIKEN 2-1 Hirosawa
Wako Saitama 351-0198
Japan
josamu@bnl.gov

Brant M Johnson
Brookhaven National Laboratory
Building 510C
P O Box 5000
Upton, NY 11973-5000
brant@bnl.gov

Tsuneo Kageya
Brookhaven National Lab
Building 510A, LEGS Group
Upton, NY 11973-5000
kageya@legsux5.phy.bnl.gov

Ralf Kaiser
University of Glasgow
University Avenue
Glasgow Scotland G12 8QQ
United Kingdom
r.kaiser@physics.gla.ac.uk

Nobuyuki Kamihara
Tokyo Institute of Technology
2-12-1ohokayama
Tokyo Institute of Technology H1-58
meguro-ku Tokyo 152-8551
Japan
kamihara@bnl.gov

Juniciro Kamiya
RCNP Osaka University
Mihogaoka10-1 Ibaraki-shi
Osaka-fu 567-0047
Japan
kamiya@rcnp.osaka-u.ac.jp

Takahiro Kawabata
Brookhaven National Laboratory
RIKEN BNL Research Center
Building 510A Physics
P O Box 5000
Upton, NY 11973
kawabata@bnl.gov

James J Kelly
University of Maryland
Dept. of Physics
College Park, MD 20742
jjkelly@physics.umd.edu

Cynthia E Keppel
Hampton University/Jefferson Lab
Physics Department
Olin 102
Hampton, VA 23668
keppel@jlab.org

Edward R Kinney
University of Colorado
390 UCB
Boulder CO 80309-0390
Edward.Kinney@colorado.edu

Thomas Kirk
Brookhaven National Laboratory
Building 510F
Upton, NY 11973
tkirk@bnl.gov

Joanna Kiryluk
University of California
UCLA Physics & Astronomy
405 Hilgard Ave.
Los Angeles, CA 90095-1547
joanna@physics.ucla.edu

Hideyuki Kobayashi
RIKEN BNL Research Center
Building 510A
20 Pennsylvania
Upton, NY 11973
hyuki@bnl.gov

Kai U Koehler
ETH Zurich
Paul Scherrer Institut Villigen
WLGA
E 27 Villigen-PSI AG 5232
Switzerland kai.koehler@psi.ch

Yuji Koike
Dept. Phys. Niigata Univ.
Ikarashi Niigata
Niigata 950-2181
Japan
koike@nt.sc.niigata-u.ac.jp

Hauke Kolster
Massachusetts Institute of Technology
77 Massachusetts Ave.
Cambridge, MA 02139
kolster@mit.edu

Boris Z Kopeliovich
Max-Planck-Institut fuer Kernphysik
Heidelberg
Postfach 103980
Heidelberg 69026
Germany
Boris.Kopeliovich@mpi-hd.mpg.de

Kevin M Kramer
The College of William and Mary
12000 Jefferson Ave.
Newport News, VA 23606
kramerk@jlab.org

A D Krisch
Spin Physics Center
University of Michigan
Ann Arbor, MI 48109-1120
krisch@umich.edu

Krishna S Kumar
University of Massachusetts
Department of Physics
Amherst, MA 01002
kkumar@physics.umass.edu

Kazuyoshi Kurita
Rikkyo University
3-34-1 Nishi-Ikebukuro
Toshima Tokyo 171-8501
Japan kurita@ne.rikkyo.ac.jp

Larry Lee
U. Manitoba/TRIUMF
4004 Wesbrook, MAll
Vancouver BC V6T2A3
Canada
lrylee@triumf.ca

Andreas Lehrach
Fz Jülich
Leo Brandt Street
Jülich NRW 52428
Germany
a.lehrach@fz-juelich.de

Paolo Lenisa
Universita and INFN –
Sez. Ferrara -Via Paradiso 12
Ferrara 44100
Italy
lenisa@hermes.desy.de

Cheng-Pang Liu
TRIUMF Research Facility
4004 Wesbrook, MAll
Vancouver BC V6T 2A3
Canada
cpliu@triumf.ca

Ming X Liu
Los Alamos National Lab
P-25 MS H846
Los Alamos, NM 87545
mliu@lanl.gov

Nilanga K Liyanage
University of Virginia
383 McCormick Rd.
Charlottesville, VA 23693
nilanga@virginia.edu

Wolfgang B Lorenzon
University of Michigan
2477 Randall Lab
Ann Arbor, MI 48109-1120
lorenzon@umich.edu

Derek I Lowenstein
Brookhaven National Laboratory
C-A Dept - Building 911B
P O Box 5000
Upton, NY 11973-5000
lowenstein@bnl.gov

Michael Lowry
Brookhaven National Laboratory
Physics Dept Bldg 510A
P O Box 5000
Upton, NY 11973
Lowry@bnl.gov

Lanchun Lu
University of Virginia, Physics Dept.
382 McCormick Rd.
P.0. Box 400714
Charlottesville, VA 22904
lulc@fnal.gov

Alfredo U. Luccio
Brookhaven National Laboratory
Building 911B
P O Box 5000
Upton, NY 11973
luccio@bnl.gov

Vladimir G Luppov
University of Michigan, Physics Dept.
500 East University
Ann Arbor, MI 48109-1120
vluppov@umich.edu

William W. MacKay
Brookhaven National Laboratory
Building 911B
P O Box 5000
Upton, NY 11973-5000
waldo@bnl.gov

Yousef I, Makdisi
Brookhaven National Laboratory
Building 911B
P O Box 5000
Upton, NY 11973
makdisi@bnl.gov

Sateesh R, Mane
Convergent Computing Inc.
P.O. Box 561
Shoreham, NY 11786
srmane@optonline.net

Frederic M. Marie
CEA/Saclay
DAPNIA/SPhN
Gif sur Yvette 91191
France
fmarie@cea.fr

Paul M McKee
University of Virginia
1575 Cool Spring Rd.
Charlottesville, VA 22901-1386
pmm3w@virginia.edu

Melvin Mckenzie
Colonel US Army, Retired
8 Alexandra Drive
Middle Island, NY 11953
melnsel@msn.com

Falk Meissner
Lawrence Berkeley National Laboratory
Once Cyclotron Rd.
MS 70R0319
Berkeley, CA 94704
FMeissner@lbl.gov

Federica Messer
RIKEN-BNL Research Center
Building 510
P O Box 5000
Upton, NY 11973
federica@bnl.gov

Lino Miceli
Brookhaven National Laboratory
Bldg 510A
P.O. Box 5000
Upton, NY 11973
miceli@bnl.gov

Andy Miller
TRIUMF
4004 Wesbrook, MAll
UBC, Campus
Vancouver BC V6T 2A3
Canada
miller@triumf.ca

Richard G Milner
MIT-Bates 26-411
Cambridge, MA 02139
milner@mitlns.mit.edu

Richard E Mischke
Los Alamos National Laboratory
MS H846
Los Alamos, NM 87545
mischke@lanl.gov

Rory A Miskimen
University of Massachusetts, Physics
Lederle GRT 417C
Amherst, MA 01003
miskimen@physics.umass.edu

Christoph H Montag
Brookhaven National Laboratory
P.O. Box 5000
Upton, NY 11973
montag@bnl.gov

Gudrid A Moortgat-Pick
University of Hamburg
Institute for Theoretical Physics and DESY
Deut Luruper Chaussee 149
D-22761 Hamburg
Germany
gudrid@mail.desy.de

Vassili S Morozov
University of Michigan
1239 Kipke Dr. Rm. 2341
Ann Arbor, Mchigan 48109
morozov@umich.edu

Genis Musulmanbekov
Joint Institute for Nuclear Research
Dubna Moscow region, 141980
Russian Federation
genis@jinr.ru

Pavel M Nadolsky
Southern Methodist University
Department of Physics
Fondren Science Building
Dallas, TX 75275-0175
nadolsky@mail.physics.smu.edu

Alexander Nass
University of Erlangen
C/O Brookhaven National Laboratory
C-AD Building 930
Upton, NY 11973
alexander.nass@desy.de

Damien Neyret
CEA Saclay
Bat 703 – Orme
Gif sur Yvette F91190
France
dneyret@cea.fr

Wolf-Dieter Nowak
DESY Zeuthen
Platanenallee 6
Zeuthen D-15738
Germany
Wolf-Dieter.Nowak@ifh.de

Karo Oganessyan
INFN - LNF
via Enrico Fermi 40, C.P. Box 13
Frascati, Rome 00044
Italy
kogan@lnf.infn.it

Akio Ogawa
Brookhaven National Laboratory
Physics 510A
P O Box 5000
Upton, NY 11973
akio@bnl.gov

Kazumasa, Ohkuma
Yokohama National University
79-5 Tokiwadai
Hodogaya-ku Yokohama
Kanagawa 240-8501
Japan
ohkuma@phys.ynu.ac.jp

Shigemi, Ohta
The High-Energy Accelerator Research Organization (KEK)
Tsukuba Ibaraki 305-0801
Japan
shigemi.ohta@kek.jp

Kensuke Okada
RIKEN
Hirosawa 2-1 Wako
Saitama 351-0198
Japan
okada@bnl.gov

Hiroyuki Okamura
Saitama University
255 Shimo-Ohkubo Saitama
Saitama 338-8570
Japan
okamura@phy.saitama-u.ac.jp

Konstantinos Orginos
Brookhaven National Laboratory
RIKEN/BNL Research Center
Building 510A
P O Box 5000
Upton, NY 11973-5000
kostas@bnl.gov

Satoshi Ozaki
Brookhaven National Laboratory
Building 510F
P O Box 5000
Upton, NY 11973
ozaki@bnl.gov

Hans G Paetz gen. Schieck
Universität zu Köln
Zülpicher Strasse 77
D-50765 Köln none
Germany
schieck@ikp.uni-koeln.de

Barbara Pasquini
Strada delle Tabarelle
286 Villazzano
Trento 38050
Italy
pasquini@ect.it

Charles F Perdrisat
College of William and Mary
Physics Department
Williamsburg, VA 23185
perdrisa@jlab.org

Philip Pile
Brookhaven National Laboratory
Building 911B
P O Box 5000
Upton, NY 11973
pile@bnl.gov

Nikolai Piskunov
Joint Institute for Nuclear Research
6 Jolliot-Curie str.
Dubna Moscow region, 141980
Russian Federation
piskunov@sunhe.jinr.ru

Bradley R Plaster
Massachusetts Institute of Technology
77 Massachusetts Avenue #26-650B
Cambridge, MA 02139
plaster@jlab.org

Barry M Preedom
University of South Carolina
Physics Dept.
Columbia, SC 29208
preedom@sc.edu

Richard Prepost
University of Wisconsin
1150 University Ave
Madison WI 53706
prepost@hep.wisc.edu

Charles Y Prescott
Stanford Linear Accelerator Center
SLAC MS 78
PO Box 20450
Stanford, CA 94309
prescott@slac.stanford.edu

Yelena Prok
University of Virginia
382 McCormick Road
Charlottesville, VA 22904-4714
yprok@jlab.org

Vadim Ptitsyn
Brookhaven National Laboratory
Building 911B
P O Box 5000
Upton, NY 11973
vadimp@bnl.gov

Greg Rakness
Indiana University Cyclotron Facility
2401 Milo B. Sampson Ln.
Bloomington, Indiana 47408
rakness@iucf.indiana.edu

Desmond Ramsay
U of, Manitoba/TRIUMF
4004 Wesbrook Mall
Vancouver BC V6T 2A3
Canada
rams@triumf.ca

Gordon P Ramsey
Loyola University Chicago
Physics Dept.
6525 N. Sheridan
Chicago IL 60626
gpr@hep.anl.gov

Vahid H Ranjbar
Fermilab
P.O. Box 500
MS-341
Batavia, IL 60510
ranjbar@fnal.gov

Philip G Ratcliffe
Dip.to di Scienze CC.FF.MM.
Via, VAlleggio 11
Como I-22100
Italy
philip.ratcliffe@uninsubria.it

Frank Rathmann
Forschungszentrum Juelich
Institut fuer Kernphysik
Leo-Brandt Str. 1
Juelich 52425
Germany
f.rathmann@fz-juelich.de

Richard S Raymond
University of Michigan
Randall Lab of Physics
University of Michigan
Ann Arbor, MI 48109
rraymond@umich.edu

Davide Reggiani
University of Ferrara
via del Paradiso
12 Ferrara 44100
Italy
reggiani@mail.desy.de

Bodo Reitz
Jefferson Lab
12000 Jefferson Avenue
Newport News, VA 23606
reitz@jlab.org

David R Relyea
Princeton University
3350 Park Blvd
Palo Alto, CA 94306
relyea@slac.stanford.edu

Hans Georg Ritter
LBNL
433 Boynton
Berkeley, CA 94707
HGRitter@lbl.gov

Stephen E Rock
SLAC
MS 44
2575 Sand Hill Road
Menlo Park, CA 94025
ser@slac.stanford.edu

Heiko Rohdjess
ISKP Universitaet Bonn
Nussallee 14-16
Bonn 53115
Germany
rohdjess@iskp.uni-bonn.de

Thomas Roser
Brookhaven National Laboratory
Collider-Accelerator Department
Building 911B
P O Box 5000
Upton, NY 11973-5000
roser@bnl.gov

Vladimir Rykov,
RIKEN, Radiation Laboratory, 2-1
Hirosawa Room 254
Wako Saitama 351-0198
Japan
rykov@bnl.gov

Naohito Saito
Kyoto University
Physics Department
Kitashirakawa-Oiwake-cho
Sakyo-ku Kyoto 606-8502
Japan
saito@bnl.gov

Takaaki M Saito
University of Tokyo
Hongo 7-3-1
Bunkyo Tokyo 113-0033
Japan
saito@nucl.phys.s.u-tokyo.ac.jp

Hide Sakai
University of Tokyo
Department of Physics
Hongo 7-3-1
Bunkyo Tokyo 113-0033
Japan
sakai@phys.s.u-tokyo.ac.jp

Andrzej M Sandacz
Soltan Institute for Nuclear Studies
Warsaw Soltan Institute for Nuclear
Studies
ul. Hoza 69
Warsaw PL 00681
Poland
sandacz@fuw.edu.pl

Andrew M Sandorfi
Brookhaven National Laboratory
Physics Dept Bldg 510A
P O Box 5000
Upton, NY 11973
sandorfi@bnl.gov

Hiroki Sato
RIKEN BNL Research Center
Building 510A
P O Box 5000
Upton New York 11973
satohiro@bnl.gov

Yoshiteru Satou
Department of Physics
Tokyo Institute of Technology
1-12-1 O-Okayama Meguro
Tokyo 152-8551
Japan
satou@ap.titech.ac.jp

Nikolai A Savvinov
University of Maryland
082 Regents Dr.
College Park MD 20742
nsavvinv@physics.umd.edu

Christian Schill
LNF-INFN Frascati
Via Enrico Fermi 40
Frascati (Rome) I-00044
Italy
Christian.Schill@desy.de

Michael Seimetz
Universität, Mainz
Institut für Kernphysik
Becherweg 45
Mainz 55099
Germany
seimetz@kph.uni-mainz.de

Kimiko Sekiguchi
RIKEN
Hirosawa 2-1
Wako Saitama 351-0198
Japan
kimiko@rarfaxp.riken.go.jp

Oleg Selyugin
BLTBH,
Joint Institute for Nuclear Research
Joliot-Curie 6
Dubna Moscow region, 141980
Russian Federation
selyugin@jinr.ru

Yuri M. Shatunov
Budker Institute of Nuclear Physics
11 Lavrentiev str.
Novosibirsk 630090
Russian Federation
shatunov@inp.nsk.su

Vassili Sharov
Joint Institute for Nuclear Research
Joliot-Curie 6
Dubna Moscow Region
141980 Russian Federation
sharov@sunhe.jinr.ru

Ernst Sichtermann
Yale University
Physics Department
Sloane Physics Laboratory
217 Prospect Street
New Haven, CT 06520-8120
sichtermann@bnl.gov

Viktor Siegle
RBRC
12 Rhododendron Dr
Center Moriches, NY 11934
siegle@bnl.gov

Sergey M. Sikach
Sergey Sikach
Institute of Physics
National Academy of Sciences
F.Skorina Av. 68
Minsk, 220072, Belarus
Sikach@dragon.bas-net.by

Dennis Sivers
Portland Physics Institute
4730 SW Macadam
Portland, Oregon 97201
densivers@sivers.com

Volker Soergel
Max-Planck-Institute for Physics
Munich Physics Institute
University of Heidelberg
Philosophenweg 12
Heidelberg 69120
Germany
sparenb@physi.uni-heidelberg.de

Jacques F Soffer
Centre de Physique Theorique - CNRS
Luminy, CAse 907
Marseille Cedex 09 13288
France
soffer@cpt.univ-mrs.fr

Paul A Souder
Syracuse University
1722 West Lake Road
Skaneateles, NY 13152
souder@phy.syr.edu

Harold M Spinka
Argonne National Laboratory
Building 362 - HEP
9700 South Cass Ave.
Argonne, Illinois 60439
hms@anl.gov

Michelle Stancari
INFN Ferrara
Via Paradiso 12
Ferrara FE I-44100
Italy
michelle@fe.infn.it

Erhard Steffens
Univ. of Erlangen-Nürnberg
Erwin-Rommel-Str.1
Erlangen D-91058
Germany
steffens@physik.uni-erlangen.de

Edward J Stephenson
Indiana University Cyclotron Facility
2401 Milo B. Sampson Lane
Bloomington, IN 47408
stephens@iucf.indiana.edu

Marco Stratmann
University of Regensburg
Institute for Theoretical Physics
D-93040 Regensburg
Germany
marco.stratmann@physik.uni-regensburg.de

Kazutaka Sudoh
RIKEN
2-1 Hirosawa Wako
Saitama 351-0198
Japan
sudou@rarfaxp.riken.go.jp

Riad S Suleiman
Massachusetts Institute of Technology
Jefferson Lab MS 12H
12000 Jefferson Ave
Newport News, VA 23606
suleiman@jlab.org

Bernd Surrow
Brookhaven National Laboratory
P O Box 5000
Upton, NY 11973-5000
surrow@bnl.gov

Dmitry N Svirida
ITEP 25 B.Cheremushkinskaya
Moscow 117259
Russian Federation
Dmitry.Svirida@itep.ru

Hiroyuki Takeda
Kyoto University, Physics Department
Kitashirakawa-Oiwakecho
Sakyo Kyoto 606-8502
Japan
takeda@nh.scphys.kyoto-u.ac.jp

Atsushi Taketani
RBRC Brookhaven National Laboratory
RIKEN Building 510A
P O Box 5000
Upton, NY 11973
taketani@bnl.gov

Michael J Tannenbaum
Brookhaven National Laboratory
Physics Dept - Building 510C
P O Box 5000
Upton, NY 11973-5000
mjt@bnl.gov

Fulvio Tessarotto
CERN and INFN Trieste INFN
Area di Ricerca
PADRICIANO 99
Trieste TS 34012
Italy
fulvio.tessarotto@cern.ch

Dylan T Thein
UCLA
440 Kelton
Los Angeles, CA 90024
dthein@physics.ucla.edu

Craig Thorn
Brookhaven National Laboratory
Physics Dept - Building 510A
P O Box 5000
Upton, NY 11973
THORN@bnl.gov

Feodor F Tikhonin
Institute for High Energy Physics
Pobeda str. 1
Protvino Moscow region 142284
Russian Federation
tikhonin@mx.ihep.su

Leonid Tkatchev
Joint Institute for Nuclear Research (JINR)
Laboratory of Nuclear Problems
Juliot Curie 6
Dubna Moscow region 141980
Russian Federation
tkatchev@nusun.jinr.ru

Manabu Togawa
Kyoto University
20436-1 Dounomaetyo
Kitashirakawa Kyoto 606-8277
Japan
togawa@nh.scphys.kyoto-u.ac.jp

Larry Trueman
Brookhaven National Laboratory
Bldg 510A
PO Box 5000
Upton, NY 11973
trueman@bnl.gov

Nikolai Tyurin
IHEP Protvino Russia
Podeby 1 Protvino Moscow Region
142281 Russian Federation
tyurin@mx.ihep.su

Mikhail N Ukhanov
IHEP Protvino Russia
Pobeda Street 1 Protvino
Serpukhov dst Moscow Region
142281 Russian Federation
ukhanov@mx.ihep.su

Kay Horst
Ulbrich University of Bonn
Nussallee 14-16
Bonn 53225
Germany
ulbrich@iskp.uni-bonn.de

Ben Van den Brandt
Paul Scherrer Institute
WLGA-D23
Villigen CH-5232
Switzerland
Ben.vandenBrandt@psi.ch

Brandon I. Van der Ventel
University of Stellenbosch
Private Bag X1
Matieland Stellenbosch 7602
South Africa
ventel@physics.sun.ac.za

Willem T.H. Van Oers
University of Manitoba
Department of Physics and Astronomy
Winnipeg MB R3T 2N2
Canada
vanoers@triumf.ca

Marc Vanderhaeghen
University, Mainz
Institut fuer Kernphysik
J.J. Becher Weg 45, Mainz 55099
Germany
marcvdh@kph.uni-mainz.de

Steven E Vigdor
Indiana University,Physics Department
Swain Hall West
Bloomington, IN 47405
vigdor@iucf.indiana.edu

Werner Vogelsang
Brookhaven National Laboratory
Building 510A
PO Box 5000
Upton, NY 11973
wvogelsang@bnl.gov

Barbara von Przewoski
IUCF
2401 Milo B. Sampson Lane
Bloomington, IN 47408
przewoski@iucf.indiana.edu

Masashi Wakamatsu
Osaka University
Machikaneyama-cho 1-1 Toyonaka
Osaka 560-0043
Japan
wakamatu@miho.rcnp.osaka-u.ac.jp

Tomotsugu Wakasa
RCNP Osaka University
Mihogaoka 10-1 Ibaraki
Osaka 567-0047
Japan
wakasa@rcnp.osaka-u.ac.jp

Takashi Wakui
RIKEN
2-1 Hirosawa Wako
Saitama 351-0198
Japan
wakui@rarfaxp.riken.go.jp

Roger P Ward
Keele University
School of Chemistry and Physics
Keele Staffordshire ST5 5BG
United Kingdom
r.p.ward@phys.keele.ac.uk

Xiangdong Wei
Brookhaven National Laboratory
Building 510A
PO Box 5000
Upton, NY 11973
xwei@bnl.gov

Frank R Wesselmann
University of Virginia
1005 Woodrow Ave #3F
Norfolk, VA 23507
frw@jlab.org

C. Steven Whisnant
James Madison University
Physics Department
MSC7702, Miller Hall
Harrisonburg, VA 22807
whisnacs@jmu.edu

Charles, A Whitten
UCLA Department of Physics
405 Hilgard Avenue
Los Angeles, CA 90095-1547
whitten@physics.ucla.edu

Roland H Windmolders
CERN-EP (blg.892)
CH-1211 Geneva 23
Switzerland
roland.windmolders@cern.ch

Thomas S Wise
University of Wisconsin
1150 University Ave.
Madison, WI 53706
wise@uwnuc0.physics.wisc.edu

Jeff Wood
Brookhaven National Laboratory
Building 902B
PO Box 5000
Upton, NY 11973
wood@physics.ucla.edu

Craig Woody
Brookhaven National Laboratory
Physics Dept - Building 510C
PO Box 5000
Upton, NY 11973
woody@bnl.gov

Kentaro Yako
Center for Nuclear Study (CNS)
University of Tokyo
RIKEN 2-1 Hirosawa
Wako Saitama 351-0198
Japan
yakou@cns.s.u-tokyo.ac.jp

Katsuya Yonehara
University of Michigan
1239 Kipke Dr.
2341 CSSB
Ann Arbor, MI 48109
yonehara@umich.edu

Hidetomo P Yoshida
Research Center for Nuclear Physics
Osaka University
10-1 Mihogaoka Ibaraki
Osaka 567-0047
Japan
hidetomo@rcnp.osaka-u.ac.jp

Feng Yuan
Department of Physics-TQHN
University of Maryland
College Park, MD 20742
fyuan@physics.umd.edu

Anatoli Zelenski
Brookhaven National Laboratory
Building 930
PO Box 5000
Upton, NY 11973
zelenski@bnl.gov

Luisa Zetta
Milan University –
Department of Physics
via Celoria 16
Milano I 20133
Italy
luisa.zetta@mi.infn.it

Xiaochao Zheng
M. I. T.
12000 Jefferson Avenue
MS 16B
Newport News, VA 23606
xiaochao@jlab.org

Vitaliy Ziskin
MIT MIT-Bates
21 Manning Rd
Middleton, MA 01949
vziskin@mit.edu

Piotr A Zolnierczuk
IUCF
2401 Milo B Sampson Lane
Bloomington IN 47408
zolnie@iucf.indiana.edu

No Food
Beverages
in Berkner Hall

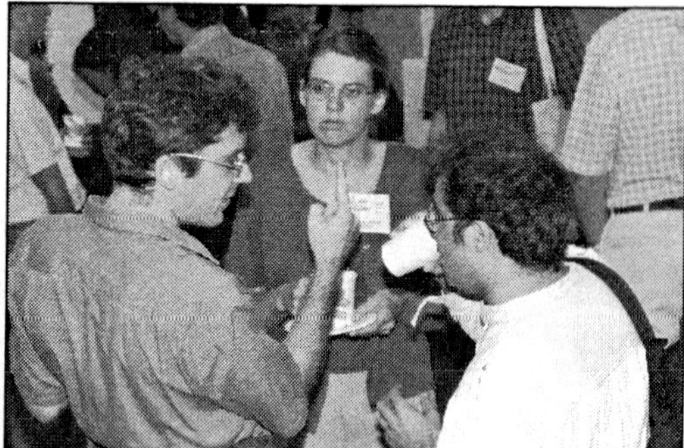

BERKNER HALL

Thursday 5, 2002

Session 4:	**Polarized Electron Sources Chair:** E. Tsentalovich
8:30-9:00	Status of MAMI polarized source K. Aulenbacker, Mainz
9:00-9:25	Commissioning of the JLAB load-lock gun M. Stutzman, JLAB
9:30-10:00	The SLAC polarized electron source J. Clendenin, SLAC
10:00-10:30	The MIT-Bates Polarized electron Source M. Farkhondeh, MIT-Bates
10:30-10:55	Coffee Break
10:55-11:20	Polarized Source at ELSA W. von Drachenfels, Bonn
Session 5:	**Development of polarized Electron Sources and Lasers** Chair: J. Clendenin
11:20-11:45	A 200 keV PES at Nagoya K. Wada, Nagoya
11:45-12:10	Recent Polarized Photocathode R&D at SLAC D.-A. Luh, SLAC
12:10-12:35	Basic R&D studies for lower emittance polarized electron guns C. Suzuki, Nagoya
12:40-14:00	Lunch
Session 6:	**Applied spin physics and technology** Chair: Y. Mamaev
14:00-14:30	Manipulation of the magnetization by spin currents H. Siegmann, ETH-Zurich
14:30-15:00	Spin resolved inverse photoemission F. Ciccacci, Polit. Di Milano
15:00-15:30	Prospects of electron RF photoinjector using the plane-wave transformer J. Clendenin, SLAC
15:30-15:50	Coffee break
15:50-16:20	Spin-polarized low energy electron microscopy A. Schmid, LBL

Excursion: Departure for Boston harbor cruise and dinner

Friday, September 6

Session 7:	**Polarized Sources and parity-violating experiments** Chair: M. Farkhondeh
8:30-9:00	Helicity-correlated effects for SAMPLE experiment E. Tsentalovich, MIT-Bates
9:00-9:30	The polarized source and Parity experiments at JLAB J. Grames, JLAB
9:30-10:00	SLAC E-158: beam helicity-correlations in 2002 physics run P. Mastromarino, SLAC

10:00-10:25	Helicity correlated effects for the Parity violation experiment at Mainz K. Aulenbacker, Mainz
10:25-10:50	Coffee break
Session 8:	**Developments of Laser systems for polarized source** Chair: T. Zwart
10:50-11:20	Ultra-stable flashlamp-pumped laser A. Brachmann, SLAC
11:20-11:45	High power diode laser system for SHR E. Tsentalovich, MIT-Bates
12:30-13:00	Bus ride to MIT-Bates
13:00-14:30	Lunch at MIT-Bates
Session 9:	**Poster Session at MIT-Bates**
14:45-16:30	poster session
16:30-17:30	Tour of Bates
17:30	Workshop ends

PESP List of Participants

Kurt Aulenbacher
Institute fur Kernphysik
Universitat Mainz
J.J. Becher Weg 45
55099 Mainz
Germany
aulenbac@kph.uni-mainz.de

Juergen Baehr
DESY Zeuthen
Platanenallee 6
D-15738 Zeuthen
Germany
baehr@ifh.de

Maud Baylac
Jefferson Laboratory
Injector Group
MS-5A, 12000 Avenue
Newport News, VA 23606

Rudy Benz
ITT Industries Night Vision
7635 Plantation Road
Roanoke, VA 24019
Rudy.benz@itt.com

Axel Brachmann
SLAC
Mail Stop 18
2574 Sand Hill Road
Menlo Park, CA 94025
brachman@slac.stanford.edu

Franco Ciccacci
Dipartimento di Fisica, Politecnico di Milano
Piazza Leonardo da Vinci 32
Milano I-20133 Italy
Franco.ciccacci@fisi.polimi.it

James Clendenin
SLAC
MS 18
2575 Sand Hill Road
Menlo Park, CA 94025
clen@slac.stanford.edu

Lamberto Duo
Dipartimento di Fisica - Politecnico di Milano
Piazza L. da Vinci 32
Milano 20133
Italy
Lamberto.duo@fisi.polimi.it

Manouchehr Farkhondeh
MIT-Bates Laboratory
P. O. Box 846
Middleton, MA 01949
manouch@mit.edu

Wilbur Franklin
MIT-Bates Laboratory
P. O. Box 846
Middleton, MA 01949
wfrankl@mit.edu

Timothy Gay
University Nebraska
Behlen Laboratory of Physics
Lincoln, NE 68588-0111
Tgay1@unl.edu

Peter Goodwin
MIT-Bates Labortory
P. O. Box 846
Middleton, MA 01949
pgoodwin@batespop.mit.edu

Joseph Grames
Jefferson Laboratory
MS 5A
12000 Jefferson Avenue
Newport News, VA 23606
grames@jlab.org

Adam Green
University of Nebraska - Lincoln
Behlen Laboratory of Physics
Lincoln, NE 68588-0111
agreen@unlserve.unl.edu

Ernie Ihloff
MIT-Bates Laboratory
P. O. Box 846
Middleton, MA 01949
zwart@bates.mit.edu

Georges Lampel
Ecole Polytechnique
Palaiseau
91128
France
Georges.lampel@polytechnique.fr

Dah-An Luh
Stanford Linear Accelerator Center
2575 Sandhill Road, MS 18
Menlo Park, CA 94025
luh@slac.stanford.edu

Yuri Mamaev
Department of Experimental Physics
State Polytechnical University
Polytekhnicheskaya 29
St. Petersburg, 195251
Russia
mamaev@spes.stu.neva.ru

Takashi Maruyama
SLAC
MS 78
2575 Sand Hill Road
Menlo Park, CA 94025
tvm@slac.stanford.edu

Peter Mastromarino
SLAC
Mail Stop 78
2574 Sand Hill Road
Menlo Park, CA 94025
maestro@caltech.edu

Richard Milner
MIT-Bates Laboratory
P. O. Box 846
Middleton, MA 01949
milner@mitlns.edu

Aaron Moy
SVT Associates
7620 Executive Drive
Eden Prairie, MN 55344
moy@exetek.com

Dmitry Orlov
Max-Planck-Institut fuer Kernphysik
Saupfercheckweg 1
Heidelberg 69117
Baden-Wurten
Germany
Dmitry.Orlov@mpi-hd.mpg.de

Keith Passmore
ITT Industries Night Vision
7635 Plantation Road
Roanoke, VA 24019
Keith.passmore@itt.com

Matt Poelker
Jefferson Laboratory
MS 5
12000 Jefferson Avenue
Newport News, VA 23606
poelker@jlab.org

Charles Prescott
SLAC
MS 78
P. O. Box 20450
Stanford, CA 94309
Prescott@slac.stanford.edu

Bob Redwine
Massachusetts Institute of Technology
4-110
77 Massachusetts Avenue
Cambridge, MA 02139
redwine@mit.edu

Nicolas Rougemaille
Ecole Polytechnique
Palaiseau 91128
France
Nicolas.rougemaille@polytechnique.fr

Andreas Schmid
Lawrence Berkeley Lab
One Cyclotron Rd.
Berkeley, CA 94720
akschmid@lbl.gov

Claus Dieter Schroter
Max-Planck-Institut fur Kernphysik
Postfach 10 39 80
69029 Heidelberg
Germany
Claus.schroeter@mpi-hd.mpg.de

Hans C. Siegmann
Stanford Linear Accelerator
365 Texas Street
San Francisco, CA 94107
siegmann@slac.stanford.edu

Marcy Stutzman
Jefferson Laboratory
12000 Jefferson Avenue
Newport News, VA 23606
marcy@jlab.org

Chihiro Suzuki
Nagoya University
Department of Physics
Furo-cho, Chikusa-ku
Nagoya 464-8602
Japan
chihiro@spin.phys.nagoya-u.ac.jp

Alexander Terekhov
Institute of Semiconductor Physics
Zolotodolinskaya str. 33, 5
Novosibirsk 630090
Siberia
Russia
terek@thermo.isp.nsc.ru

Evgeni Tsentalovich
MIT - Bates Laboratory
21 Manning Avenue
Middleton, MA 01949
evgeni@mit.edu

Wolther von Drachenfels
University of Bonn
Dept. of Physics
Nussallee 12
Bonn 53115
Germany
drachen@physik.uni-bonn.de

Kouji Wada
Nagoya University
Department of Physics
Nagoya
Aichi 464-8602
Japan
wada@spin.phys.nagoya-u.ac.jp

Steven Wang
Shanghai China
Shanghai 200331
China
Stianfo_wang@octasoft.com

Yuri Yashin
St. Petersburg State Technical Univ.
Lab Spin-Polarized Electron
Spectroscopy
Div. of Experimental Physics
Polytechnicheskaya str., 29
St. Petersburg, 195251
Russia
yashin@spes.stu.neva.ru

Abbi Zolfaghari
MIT-Bates Laboratory
21 Manning Avenue
P. O. Box 846
Middleton, MA 01949
abbi@mit.edu

Townsend Zwart
MIT Bates Laboratory
21 Manning Avenue
Middleton, MA 01949
zwart@bates.mit.edu

PESP-2002 Scientific Program (Revised)

Wednesday September 4, 2002

8:15-9:00	**Registration**		
Opening Remarks:			
9:00-9:05	M. Farkhondeh	MIT-Bates	Opening Remarks
9:05-9:10	Richard Milner	MIT-Bates	Welcoming remarks from the Bates Director
Session 1:	**NEA and surface physics:/ Chair: C. Prescott**		
9:15-9:50	G. Lampel	Polytechnique	*A tutorial on photoemission from GaAs based photocathodes*
9:50-10:20	T. Maruyama	SLAC	*Suppression of the surface charge limit in strained GaAs photocathodes*
10:20:10:45	Y. Mamaev	St. Petersburg	*Tuning of strained photocathodes for matching the Wavelength of the excitation light*
10:45-11:15	**Coffee Break**		
11:15-11:45	K. Wada	Nagoya	*Surface Charge Limit effect of superlattice photocathode*
11:45-12:10	A. Moy	SVT Associates	*Growth and quality control of production of strained layer/SL photocathodes*
Session 2:	**Photocathode-Spin Physics / Chair: T. Maruyama**		
12:10-12:40	M. Baylac	JLAB	*Effect of atomic hydrogen exposure on electron beam polarization from strained GaAs photocathodes*
12:40-14:00	**Lunch**		
14:0-14:30	Y. Mamaev	St. Petersburg	*Polarized Emission from strain-compensated superlattices*
Session 3 :	**Polarimeters / Chair: K. Aulenbacker**		
14:35-15:05	T. Gay	U. Nebraska	*Electron polarimetry using noble gas targets*
15:05-15:35	B. Franklin	MIT-Bates	*The MIT-Bates Compton Polarimeter*
15:35-16:10	**Coffee Break**		
16:10-16:40	N. Rougemaille	Polytechnique	*Spin filters as very high-performance spin polarimeters*
16:40-17:05	T. Zwart	MIT-Bates	*Transmission Polarimetry for electron beams*
17:05-17:35	L. Duò	Polit. Di Milano	*A high efficiency electron polarimeter based on exchange scattering from a magnetic target*

Thursday 5, 2002

Session 4: Polarized Electron Sources /Chair: E. Tsentalovich			
8:30-9:00	K. Aulenbacker	Mainz	*Status of MAMI polarized source*
9:00-9:25	M. Stutzman	JLAB	*Commissioning of the JLAB load-lock gun*
9:30-10:00	J. Clendenin	SLAC	*The SLAC polarized electron source*
10:00-10:30	M. Farkhondeh	MIT-Bates	*The MIT-Bates Polarized electron Source*
10:30-10:55	**Coffee Break**		
10:55-11:20	W. von Drachenfels	Bonn	*Polarized Source at ELSA*
Session 5:	**Development of polarized Electron Sources and Lasers /Chair : J. Clendenin**		
11:20-11:45	K. Wada	Nagoya	*A 200 keV PES at Nagoya*
11:45-12:10	D.-A. Luh	SLAC	*Recent Polarized Photocathode R&D at SLAC*

12:10-12:35	C. Suzuki	Nagoya	*Basic R&D studies for lower emittance polarized electron guns*
12:40-14:00	**Lunch**		
Session 6 :	**Applied spin physics and technology / Chair: Y. Mamaev**		
14:00-14:30	H. Siegmann	ETH-Zurich	*Manipulation of the magnetization by spin currents*
14:30-15:00	F. Ciccacci	Polit. Di Milano	*Spin resolved inverse photoemission*
15:00-15:30	J. Clendenin	SLAC	*Prospects of electron RF photoinjector using the plane-wave transformer*
15:30-15:50	**Coffee break**		
15:50-16:20	A. Schmid	LBL	*Spin-polarized low energy electron microscopy*
16:20-16:55			
Excursion:	**Departure for Boston harbor cruise and dinner**		

Friday, September 6

Session 7	**Polarized Sources and parity-violating experiments /Chair: M. Farkhondeh**		
8:30-9:00	E. Tsentalovich	MIT-Bates	*Helicity-correlated effects for SAMPLE experiment*
9:00-9:30	J. Grames	JLAB	*The polarized source and Parity experiments at JLAB*
9:30-10:00	P. Mastromarino	SLAC	*SLAC E-158: beam helicity-correlations in 2002 physics run*
10:00-10:25	K. Aulenbacker	Mainz	*Helicity correlated effects for the Parity violation experiment at Mainz*
10:25-10:50	**Coffee break**		
Session 8	**Developments of Laser systems for polarized source/ Chair: T. Zwart**		
10:50-11:20	A. Brachmann	SLAC	*Ultra-stable flashlamp-pumped laser*
11:20-11:45	E. Tsentalovich	MIT-Bates	*High power diode laser system for SHR*
11:45-12:30			
12:30-13:00	**Bus ride to MIT-Bates**		
13:00-14:30	**Lunch at MIT-Bates**		
Session 9	**Poster Session at MIT-Bates**		
14:45-16:30	**poster session**		
16:30-17:30	**Tour of Bates**		
17:30	**Workshop ends**		

Author Index

A

Adachi, T., 691
Adderley, P., 1047, 1073, 1078
Ado, Y. M., 538
Afonin, A. G., 538
Aghalaryan, A., 625
Ahmidouch, A., 625
Ahrens, L., 751, 794
Akdogan, T., 1058
Akimune, H., 681, 720
Akiyoshi, H., 705
Alekseev, I. G., 558, 812, 817
Alessi, J., 881, 954
Alexeeva, L. V., 538
Amalsky, G. M., 730
Anderson, B. D., 625
Andreev, V. A., 730
Andronenko, M. N., 730
Anferov, V. A., 538, 766, 776
Anischenko, N. G., 563
Anselmino, M., 474
Antonenko, V. G., 563
Ardashev, K., 651
Arefiev, V. A., 826
Arenhövel, H., 625
Armstrong, D. S., 267
Asaturyan, R., 625
Asilyan, L. S., 439
Atanasov, I., 826
Aulenbacher, K., 1088
Avakian, H., 434
Averett, T. D., 88
Averichev, S. A., 563
Azhgirey, L. S., 563, 826

B

Bade, C. M., 651, 857, 862
Bai, M., 751, 771, 794
Baker, O., 625
Baldwin, A. R., 625
Balewski, J., 418
Barber, D. P., 741, 993
Barish, K. N., 323
Barsov, S., 553
Bartenev, V. D., 563
Basilev, S. N., 826
Baylac, M., 1047, 1073, 1078
Bayman, B., 686
Bazhanov, N. A., 558, 563
Bazilevsky, A., 584
Bechstedt, U., 153
Beloglazov, Y. A., 558
Belostotski, S. L., 730
Belousov, V. I., 538
Belov, A. S., 887, 892
Belyaev, A. A., 563
Berg, G. P. A., 676, 681
Berkaev, D. E., 993
Bertin, P., 846
Bland, L. C., 98, 584, 817
Blecher, M., 651
Blinov, B. B., 776
Blinov, N. A., 563
Bodek, K., 241
Boer, D., 479
Bogdanov, A., 584
Bojak, I., 338
Booth, E. C., 1011, 1058
Borisov, N. S., 538, 563, 872
Borzakov, S. B., 563
Borzunov, Y. T., 563
Brachmann, A., 1029, 1042, 1083, 1093
Bravar, A., 812, 817, 830, 954
Breuer, H., 625
Briscoe, B., 881
Brown, B. A., 681
Brown, K. A., 751, 794
Bruell, A., 343, 548
Brüggemann, R., 924
Bruner, N., 375
Budkovksy, P. E., 558
Budzanowski, A., 241
Bültmann, S., 528
Bunce, G., 584, 812, 817, 954
Bunyatova, E. I., 558, 907
Burnstein, R. A., 251
Burtin, E., 846
Bushuev, Y. P., 563, 826
Buttimore, N. H., 841

C

Cacace, C., 651
Cadman, R., 812, 817
Cameron, P., 781, 786
Cantalbrigo, M., 284
Caracappa, A., 651, 857, 862, 867
Carlini, R., 625
Cates, Jr., G. D., 217
Cavata, C., 846
Chakravorty, A., 251
Chan, A., 251
Chapman, M. A., 934, 954
Cheever, D., 1019
Chen, Y. C., 251
Chernenko, L. P., 563
Chernykh, E. V., 563
Chiang, H. C., 543
Choong, W. S., 251
Christy, E., 625
Chujko, B. V., 538
Chumakov, V. F., 563
Churchwell, S., 625
Cichocki, A., 651
Clark, J., 1047, 1073, 1078
Clark, K., 251
Clarke, N. M., 725
Clendenin, J. E., 1029, 1037, 1042, 1083, 1093
Cole, L., 625
Commeaux, C., 651
Conte, M., 781, 786
Courant, E. D., 538, 799
Court, G. R., 902
Covello, A., 686
Crabb, D. G., 538, 856, 916
Crawford, M. Q., 766
Crouse, E., 625

D

D'Alesio, U., 469, 474
Danagoulian, S., 625
d'Angelo, A., 651
Danneberg, N., 241
Davidenko, A. N., 538
Davinson, T., 725
Davis, N. J., 725
Day, A., 1047, 1078
Day, D., 625
Day, T., 1073
Deininger, R., 651
de Jager, K., 978
de Lesquen, A., 563
Delheij, P. P. J., 538
Derbenev, Y. S., 538
Derenchuk, V. P., 855, 887
De Sanctis, E., 439
Deshpande, A., 584, 812, 817
Desikan, T., 1029, 1083
Dharmawardane, K. V., 601
Dhawan, S., 812, 817
Didelez, J.-P., 651
Dietrich, J., 153
di Nezza, P., 313
Di Salvo, E., 494
di Salvo, R., 651
Dodge, G. E., 596
Dolgii, S. A., 563
Domchenkov, O. A., 730
Dorn, A., 1016
Drechsel, D., 646
Drouhin, H.-J., 1001
Dukes, E. C., 251
Durandet, C., 251
Dutta, D., 1058
Dymov, S., 553

E

Efremov, A. V., 235, 459
Eisermann, Y., 686
Elaasar, M., 625
Ellinghaus, F., 303
Emmerich, R., 897
Engels, R., 897, 924
Ent, R., 625
En'yo, H., 584
Ermisch, K., 711
Ershov, V. P., 892
Escoffier, S., 846
Etienne, Z. B., 766, 776, 872

F

Fabbri, R., 313
Farkhondeh, M., 625, 876, 1011, 1019, 1024, 1058, 1098
Fedorov, A. N., 563
Fedorov, O. Y., 730
Felix, J., 251
Fenker, H., 625
Fetscher, W., 241
Fidecaro, G., 538
Fidecaro, M., 538
Fields, D. E., 812, 817
Filipe, A., 1001
Fimushkin, V. V., 538, 563, 872, 892, 916
Finger, M., 563
Finger Jr., M., 563
Finn, J. M., 625
Fox, B., 412, 584
Franklin, W. A., 1011, 1019, 1024, 1058, 1098
Frommberger, F., 1053
Fu, Y., 251
Fujimura, H., 676, 681, 720
Fujita, H., 681
Fujita, K., 676, 691
Fujita, Y., 681
Fujiwara, M., 681
Fukao, Y., 584
Furuta, F., 1063, 1068

G

Gai, G. I., 892
Galetto, T., 1042, 1093
Galyaev, N. A., 538
Gamberg, L., 489
Gan, L., 625
Gargano, A., 686
Garkusha, V. I., 538
Garrow, K., 625
Garwin, E. L., 1029, 1083
Gay, T. J., 1034
Gebel, R., 153
Geisler, S., 924
Gibson, C., 651
Gidal, G., 251
Glagolev, V. V., 826
Glenn, J. W., 794
Glenn, W., 751
Goeke, K., 459
Goldstein, G. R., 489
Goloskokov, S. V., 464
Golovanov, L. B., 563
Gómez-Camacho, J., 725
Gorchtein, M., 646
Goshtasbpour, M., 299
Goto, Y., 499, 584
Gotoh, T., 1063
Gowin, M., 1053
Graham, D., 954
Grames, J., 1047, 1073, 1078
Graw, G., 686
Green, A. S., 1034
Greenfield, M. B., 700, 734
Griffith, J. A. R., 725
Grishin, V. N., 538
Grosse Perdekamp, M., 454
Gu, P., 251
Guazzoni, P., 686
Gueye, P., 625
Gurevich, G. M., 563
Gussen, A., 924
Gustafson, H. R., 251

H

Haeberli, W., 934, 954
Haggerty, J., 584
Hall, S. J., 725
Hansknecht, J., 1047, 1073, 1078
Hara, K., 681, 691
Harvey, S., 1029
Hasch, D., 429, 963
Hasuko, K., 454
Hatanaka, K., 676, 681, 691, 700, 705, 730, 734
Hatano, M., 696, 700, 711, 715, 734, 911
Hautle, P., 907
Hayashigaki, A., 569
Helbing, K., 33
Hertenberger, R., 686
Hicks, K., 651
Hilbes, C., 241
Hillert, W., 1053
Hirai, M., 365
Hirose, T., 1063, 1068

Ho, C., 251
Hoblit, S., 651
Hoffmann, M., 1053
Holmstrom, T., 251
Honig, A., 651
Horinaka, H., 1068
Hosono, K., 681
Howell, C., 625
Hseuh, H., 954
Hu, B., 625
Huang, H., 122, 751, 794, 812, 817
Huang, M., 251
Hughes, V., 812, 817
Hurwitz, M., 1058

I

Ichihara, T., 696
Ichimura, M., 734
Igo, G., 812, 817, 836
Ihloff, E., 1011, 1098
Imai, K., 584, 812, 817
Ireland, D. G., 725
Ishida, S., 696
Ishida, T., 705
Ishikawa, T., 681, 720
Isupov, A. Y., 715
Itoh, M., 676, 681, 720
Izotov, A. A., 730

J

Jaganathan, R., 741
Jäger, B., 353
James, C., 251
Janata, A., 563
Jansen, P., 924
Jarczyk, L., 241
Jaskola, M., 686
Jenkins, M., 251
Jgoun, A. A., 730
Ji, X., 3, 484
Jiang, X., 656
Jinnouchi, O., 812, 817
Jones, M. K., 625, 826
Jones, T., 251

K

Kachanov, V. A., 538
Kacharava, A., 553
Kageya, T., 651, 857, 862
Kaiser, R., 983
Kalantar-Nayestanaki, N., 711
Kamiya, J., 676, 681, 691, 700, 705, 730, 734
Kanavets, V. P., 558, 812, 817
Kandes, M. C., 766, 872
Kaplan, D. M., 251
Karban, O., 725
Kato, H., 696, 711, 715, 734
Kawabata, T., 676, 681, 691, 720
Kawahigashi, K., 734
Kelly, J. J., 78, 625
Keppel, C., 625
Keppel, C. E., 294
Khandaker, M., 625, 651
Kharlov, Y. V., 538
Khiari, F. Z., 538
Khoukaz, A., 553
Kim, W.-Y., 625
Kirby, R. E., 1029, 1037, 1083
Kirch, K., 241
Kirillov, A. D., 563
Kirillov, D. A., 826
Kiryluk, J., 424, 812, 817
Kisselev, A. Y., 730
Kistner, O., 651
Kistryn, S., 241
Kitamura, Y., 676, 691, 700
Klehr, F., 924
Klein, K. J., 872
Kleines, H., 924
Klement, J., 241
Klenov, V., 881, 954
Kobayakawa, H., 1068
Kobayashi, H., 390
Köhler, K.-U., 241
Koike, Y., 449, 574
Kokhanovski, S., 881, 954
Kolomiets, V. G., 563
Kolster, H., 934, 954
Komarov, V., 553
Komogorov, E. V., 563
Konter, J. A., 907
Kopeliovich, B. Z., 58
Kopytin, M. A., 730

Koroleva, L. I., 558
Korovin, P. P., 826
Kotov, V. I., 538
Koutin, S. V., 538
Kovalenko, A. D., 563
Kovalev, A. I., 558, 563
Kowalski, S., 625
Kozela, A., 241
Kponou, A., 881, 954
Kramer, K. M., 615
Krasnov, V. A., 563
Kravtsov, P., 924
Krisch, A. D., 46, 538, 766, 776, 872, 916
Krstonoshich, P., 563
Kruglov, S. P., 558
Kulikov, A., 553
Kumar, K. S., 142
Kumasaka, H., 671, 715
Kumbartzky, G. F., 826
Kurbatov, A., 553
Kurita, K., 812, 817
Kusnetsov, A. V., 538
Kutuzova, L. V., 892
Kuwahara, M., 1063, 1068
Kuzmin, E. S., 563

L

Ladygin, V. P., 563, 705, 715
Ladygina, N. B., 715
Lampel, G., 1001
Lang, J., 241
Lang, N., 553
Lassailly, Y., 1001
Lazarev, A. B., 563
Lederman, L. M., 251
Lee, S. Y., 751
Lee, G, L., 272
Lehar, F., 563
Lehmann, A., 651
Lehmann, I., 553
Lehrach, A., 153
Lemaître, S., 924
Lenisa, P., 279, 939
Lenz, W., 584
Leonova, M. A., 538, 766, 776, 872
Leros, N., 251
Ley, J., 897
Lhuillier, D., 846
Li, Z., 812, 817
Liburg, M. Y., 563
Lin, A. M. T., 538, 776, 916
Lincoln, F. C., 651, 857, 862
Liu, C.-P., 262
Liu, M., 584
Livanov, A. N., 563
Livingston, K., 725
Llosá Llácer, G., 241
LoDestro, V., 881, 954
Longo, M. J., 251
Lopez, F., 251
Lorentz, B., 153, 553, 924
Lorenz, S., 924
Lorenzon, W., 538, 776
Lowry, M. M., 651, 857, 862
Lozowski, W., 812, 817
Lu, L. C., 251
Lu, L.-C., 251
Lucas, M., 651
Luccio, A. U., 746, 751, 771, 781, 786, 794
Luebke, W., 251
Luh, D.-A., 1029, 1042, 1083
Luk, K. B., 251
Lukhanin, A. A., 563
Lung, A., 625
Luo, Y., 1037
Luppov, V. G., 538, 872, 916

M

Macdonald, E. W., 725
Macharashvili, G., 553
Mack, D., 625
MacKay, W. W., 751, 771, 781, 786, 794, 812, 817, 954
MacLachlan, G., 625
Madey, R., 625
Maeda, Y., 696, 700, 705, 711, 715, 734
Mahler, G., 954
Mahon, J. C., 651, 857, 862
Maier, R., 153
Makdisi, Y., 584, 812, 817, 954
Malakhov, A. I., 715
Mamaev, Y., 1006
Mane, S. R., 756
Maniakov, P. K., 563

Manley, D. M., 625
Manyakov, P. K., 826
Marie, F., 846
Markowitz, P., 625
Martel-Bravo, I., 725
Maruyama, T., 1029, 1042, 1083, 1093
Matafonov, V. N., 563
Matsumoto, H., 1063, 1068
Matsuyama, T., 1068
Matthews, J. L., 1011, 1058
Matyushevsky, E. A., 563
McKee, P. M., 919
Medvedev, V. A., 538
Melnik, Y. M., 538
Meng, W., 954
Metz, A., 646
Meyer, H., 651
Meyer, H. O., 766
Miceli, L., 651, 857, 862
Mikirtytchiants, M., 897, 924
Miklukho, O. V., 730
Miller, C. A., 23
Milner, R. G., 988, 993
Minami, S., 705
Mischke, R. E., 246
Mitchell, J., 625
Miyamoto, M., 1063
Mizuno, R., 1063, 1068
Mkrtchyan, H., 625
Mochalov, V. V., 538
Montag, C., 794
Moortgat-Pick, G., 206
Morii, T., 333
Morizzianni, D., 651
Moroz, V. D., 563
Morozov, A. A., 563
Morozov, B. V., 558
Morozov, V. S., 538, 766, 776, 872
Morris, C. L., 700
Moshammer, R., 1016
Muccifora, V., 439
Murgia, F., 469, 474
Mušinský, J., 826
Mussgiller, A., 553
Musulmanbekov, G., 358
Muto, R., 584
Mysnik, A. I., 538, 916

N

Nadolsky, P. M., 380
Nakamura, M., 681
Nakanishi, T., 1063, 1068
Nakaoka, Y., 734
Nass, A., 929
Neff, B., 1053
Neganov, A. B., 563
Nekipelov, M., 924
Nelson, J. M., 725
Nelson, K. S., 251
Nelyubin, V., 924
Nesterov, V. M., 558
Newsham, D., 1037
Neyret, D., 846, 968
Nikitin, V. A., 538
Nikolaevsky, G. P., 563
Nishikawa, J., 711, 715
Nishitani, T., 1063, 1068
Nomofilov, A. A., 563
Nomokonov, P. V., 538
Noro, T., 681, 691, 705, 720, 730
Norum, B., 651
Novinsky, D. V., 558
Nowak, W.-D., 591
Nurushev, S. B., 538, 584

O

Obayashi, E., 676, 681, 691, 720, 730
Oganessyan, K. A., 439, 445, 489
Ogawa, A., 407, 454, 504, 817
Ohkuma, K., 333
Ohnishi, T., 696, 711, 715, 734
Okada, K., 395
Okamura, H., 671, 696, 700, 711, 715, 734
Okumi, S., 1063, 1068
Opper, A. K., 625
Orginos, K., 166
Ormand, W. E., 686
Otboev, A. V., 993
Otsu, H., 734

P

Paetz gen. Schieck, H., 553, 897, 924
Page, R. D., 725
Palazzi, M., 781, 786

Panteleev, T., 563
Park, H. K., 251
Parker, B., 993
Pascuzzi, E., 584
Pasquini, B., 646
Peaslee, D. C., 538
Penchev, L., 826
Perdrisat, C. F., 625, 639, 826
Peretti, J., 1001
Perevedentsev, E. A., 761
Perroud, J.-P., 251
Peters, C. C., 538, 776, 872, 916
Pilipenko, Y. K., 538, 563, 892
Pisarev, I. L., 563
Piskunov, N. M., 826
Plaster, B., 625
Plis, Y. A., 563
Poelker, M., 1047, 1073, 1078
Polunin, Y. P., 563
Prasuhn, D., 153
Preedom, B. M., 651
Prepost, R., 1029, 1083
Prescott, C. Y., 1029, 1083
Prokofiev, A. N., 563, 730
Prokofiev, D. A., 730
Prudkoglyad, A. F., 538, 872, 916
Prytkov, V. Y., 563
Ptitsyn, V., 746, 751, 761, 794, 993
Punjabi, V., 625, 639, 826
Purschke, M. L., 584
Pussieux, T., 846
Pusterla, M., 781, 786

Q

Quin, P. A., 934

R

Rajaram, D., 251
Rakness, G., 400
Ramsay, W. D., 196
Ramsey, G. P., 348
Ranjbar, V. H., 746, 751, 771, 794
Rapaport, J., 700
Raparia, D., 881
Ratcliffe, P. G., 176
Rathmann, F., 553, 897, 924
Raue, B., 625
Raymond, R. S., 538, 872, 916
Reggiani, D., 944
Reichelt, T., 625
Reinhold, J., 625
Reitz, B., 634
Rescia, S., 812, 817
Reznikov, S. G., 715
Rindfleisch, U., 924
Ritter, J., 881, 954
Rochansky, A., 1006
Roche, J., 625
Rock, S., 289, 516, 973
Rohdjeß, H., 523
Rosenberry, M. A., 1034
Roser, T., 751, 771, 794, 812, 817, 954
Rougemaille, N., 1001
Rubin, H. A., 251
Rudenko, A., 1016
Rufanov, I. A., 538
Rukoyatkin, P. A., 563
Rusek, K., 725
Rutt, P., 1073
Ryltsov, V. V., 558

S

Saam, D. E., 872
Sagara, K., 705
Saito, N., 584, 812, 817
Saito, T., 700, 705, 711, 715
Saitoh, T., 651
Sakaguchi, H., 676, 681, 691, 720, 730
Sakai, H., 671, 696, 700, 705, 711, 715, 734, 911
Sakamoto, N., 671, 676, 691, 696, 711, 715
Sakemi, Y., 676, 691, 700, 705
Sakoda, S., 711, 715, 734
Sakuma, F., 584
Sandorfi, A. M., 651, 857, 862
Sarkadi, J., 924
Sato, H., 776
Sato, Y., 625
Sato, H. D., 328
Satou, Y., 696, 711
Savvinov, N., 630
Schaerf, C., 651
Schedrov, V. A., 563

Schevelev, O. N., 563
Schill, C., 308
Schleichert, R., 553
Schnase, A., 153
Schneider, H., 153
Schröter, C. D., 1016
Schuhl, A., 1001
Schwandt, P., 766, 776
Schweitzer, P., 459
Schweizer, T., 241
Seimetz, M., 620
Sekiguchi, K., 696, 700, 705, 711, 715, 734
Sellin, P. J., 725
Selyugin, O. V., 533
Semenov, A. Y., 625
Semenov, P. A., 538
Semenova, I. A., 625
Seo, W., 625
Seyfarth, H., 553, 897, 924
Shafi'i, A., 299
Sharov, V. I., 563
Shatunov, Y. M., 761, 993
Shchedrov, V. A., 558
Shepherd-Themistocleous, C. H., 725
Shilov, S. N., 563
Shimbara, Y., 681, 691
Shimizu, Y., 676, 691, 700, 705
Shindin, R. A., 563
Shishov, Y. A., 563
Shotter, A. C., 725
Shutov, V. B., 563
Shvedchikov, A. V., 730
Siegle, V., 454
Sikach, S. M., 661, 666
Simicevic, N., 625
Sinclair, C., 1073
Sisco, D. L., 872
Sitink, I. M., 826
Sivers, D. W., 538, 766, 776, 791
Slepnev, I. V., 826
Slepnev, V. M., 826
Slunečká, M., 563
Slunečková, V., 563
Smith, G., 625
Smyrski, J., 241
Sodja, J., 1042, 1093
Soeren Lange, J., 454
Soffer, J., 225
Solomey, N., 251
Solovianov, V. L., 538
Southworth, D. R., 872
Spinka, H., 794, 807, 812, 817
Sromicki, J., 241
Starikov, A. Y., 563
Stassen, R., 153
Steffens, E., 186, 553, 924
Stepanov, V. P., 538
Stepanyan, S., 625
Stephan, E., 241
Stephenson, E. J., 766, 776, 954
Stewart, J. A., 538
Stockhorst, H., 153
Stoletov, G. D., 563
Stoll, S., 584
Stösslein, U., 939
Stratmann, M., 112, 353
Ströher, H., 553, 924
Strunov, L. N., 563
Strzałkowski, A., 241
Stutzman, M. L., 1047, 1073, 1078
Suda, K., 671, 696, 700, 711, 715, 734
Sudoh, K., 370
Suleiman, R., 256
Sulimov, A. D., 558
Sulimov, V. V., 730
Sumachev, V. V., 558
Surrow, B., 318, 812, 817
Suzuki, C., 1063, 1068
Suzuki, K., 569
Suzuki, R., 671, 715
Svetov, A. L., 563
Svirida, D. N., 558, 812, 817

T

Tadevosyan, V., 625
Tajima, S., 625
Takeda, H., 676, 681, 720, 730
Taketani, A., 390
Taki, T., 681, 720
Tamii, A., 671, 681, 696, 700, 705, 711, 715, 720, 734, 911
Tanaka, K., 569
Tang, L., 625
Tanida, K., 584
Tanihata, I., 836
Taylor, S., 625

Teng, P. K., 251
Teplyakov, V. A., 538
Terashima, S., 720
Tessarotto, F., 509
Thorn, C. E., 651, 857, 862, 867
Tioukine, V., 1088
Tireman, W., 625
Togawa, M., 584
Tojo, J., 584, 812, 817
Tölle, R., 153
Tomasi-Gustafsson, E., 826
Torun, Y., 251
Toyokawa, H., 681
Trautman, V. Y., 558
Trofimov, V., 924
Troshin, S., 579
Troshin, S. M., 538
Trush, S. I., 730
Tschalaer, C., 993
Tsentalovich, E., 1011, 1019, 1024, 1058, 1098
Tsoupas, N., 751, 794
Tsukahara, N., 720
Tsuruta, K., 705
Tungate, G., 725
Turchinetz, W., 1011, 1058
Turko, B., 251
Turner, J. L., 1042, 1083, 1093
Tyurin, N., 47, 579

U

Uchida, M., 676, 681, 691, 720
Uchigashima, N., 700, 711, 715
Ueno, H., 681
Uesaka, T., 671, 696, 711, 715, 911
Ufimtsev, A. G., 538
Ukhanov, M. N., 48, 538
Ullrich, J., 1016
Ulmer, P., 625
Underwood, D. G., 794, 812, 817
Usov, Y. A., 563
Uzikov, Y., 553

V

v.Przewoski, B., 949
Vadeev, V. P., 892
Valevich, A. I., 892
van den Brandt, B., 907
Vanderhaeghen, M., 132, 646
van Oers, W. T. H., 538
Varzar, S. M., 538
Vasiliev, T., 563
Vassiliev, A., 897, 924
Vikhrov, V. V., 730
Vinokurov, D., 1006
Vogelsang, W., 353
Vogt, M., 741
Volk, J., 251
Volkov, V. I., 563
von Drachenfels, W., 1053
von Lintig, D., 584
von Przewoski, B., 69, 766, 776
Vorobiev, E. I., 563
Vulcan, W., 625

W

Wada, K., 1063, 1068, 1068
Wakamatsu, M., 385
Wakasa, T., 676, 681, 691, 696, 700, 705, 711, 730, 734
Wakui, T., 911
Wang, F., 993
Wang, K., 651
Ward, R. P., 725
Watanabe, Y., 584
Watson, J. W., 625
Wei, X., 651, 857, 862
Wells, S., 625
Wesselmann, F., 625
Wesselmann, F. R., 606
Whisnant, C. S., 651, 857, 862
White, C. G., 251
White, S. L., 251
Wiessner, M., 1088
Winkler, K., 1088
Wirth, H.-F., 686
Wirth, T., 1001
Wise, T., 934, 954
Wong, V. K., 538, 766, 776
Wood, J., 812, 817

Wood, S., 625
Woods, M., 1093
Woods, P. J., 725
Woody, C., 584

Y

Yagita, T., 705, 734
Yako, K., 696, 700, 705, 715, 734
Yakou, K., 711
Yakutin, A. E., 538
Yamamoto, M., 1063, 1068
Yamamoto, N., 1063, 1068
Yamasaki, K., 681
Yan, C., 625, 625
Yang, S., 625
Yaschenko, S., 553
Yashin, Y., 1006
Yasuda, Y., 676, 681, 691, 720, 730
Yochida, H. P., 681
Yonehara, K., 538, 766, 776, 872, 916
Yoshida, H. P., 676, 691, 705, 720, 730
Yoshimura, M., 705, 720
Yoshioka, M., 1063, 1068
Yosoi, M., 676, 681, 691, 720
Yu, D., 1037
Yuan, C.-P., 380
Yuan, F., 484
Yuan, L., 625
Yudin, I. P., 563

Z

Zaitsev, I. V., 563
Zalikhanov, B., 553
Zapolsky, V. N., 538
Zarucheisky, V. G., 538
Zejma, J., 241
Zelenski, A. N., 538, 881, 954
Zeno, K., 794
Zetta, L., 686
Zhang, W.-M., 625
Zhdanov, A. A., 563, 730
Zheng, X., 610
Zhmyrov, V. N., 563
Zhu, H., 625
Zhu, X., 625
Zhurkin, V. V., 558
Zoubets, V., 954
Zubets, V., 881
Zwart, T., 1011, 1019, 1024, 1058, 1098
Zwoll, K., 924
Zyla, P., 251